LDA	lithium diisopropylamide
LDMAN	lithium 1-(dimethylamino)naphthalenide
LHMDS	= LiHMDS
LICA	lithium isopropylcyclohexylamide
LiHMDS	lithium hexamethyldisilazide
LiTMP	lithium 2,2,6,6-tetramethylpiperidide
LTMP	= LiTMP
LTA	lead tetraacetate
lut	lutidine
m-CPBA	*m*-chloroperbenzoic acid
MA	maleic anhydride
MAD	methylaluminum bis(2,6-di-*t*-butyl-4-methylphenoxide)
MAT	methylaluminum bis(2,4,6-tri-*t*-butylphenoxide)
Me	methyl
MEK	methyl ethyl ketone
MEM	(2-methoxyethoxy)methyl
MIC	methyl isocyanate
MMPP	magnesium monoperoxyphthalate
MOM	methoxymethyl
MoOPH	oxodiperoxomolybdenum(pyridine)-(hexamethylphosphoric triamide)
mp	melting point
MPM	= PMB
Ms	mesyl (methanesulfonyl)
MS	mass spectrometry; molecular sieves
MTBE	methyl *t*-butyl ether
MTM	methylthiomethyl
MVK	methyl vinyl ketone
n	refractive index
NaHDMS	sodium hexamethyldisilazide
Naph	naphthyl
NBA	*N*-bromoacetamide
nbd	norbornadiene (bicyclo[2.2.1]hepta-2,5-diene)
NBS	*N*-bromosuccinimide
NCS	*N*-chlorosuccinimide
NIS	*N*-iodosuccinimide
NMO	*N*-methylmorpholine *N*-oxide
NMP	*N*-methyl-2-pyrrolidinone
NMR	nuclear magnetic resonance
NORPHOS	bis(diphenylphosphino)bicyclo[2.2.1]-hept-5-ene
Np	= Naph
PCC	pyridinium chlorochromate
PDC	pyridinium dichromate
Pent	*n*-pentyl
Ph	phenyl
phen	1,10-phenanthroline
Phth	phthaloyl
Piv	pivaloyl
PMB	*p*-methoxybenzyl
PMDTA	*N,N,N′,N″,N″*-pentamethyldiethylenetriamine
PPA	polyphosphoric acid
PPE	polyphosphate ester
PPTS	pyridinium *p*-toluenesulfonate
Pr	*n*-propyl
PTC	phase transfer catalyst/catalysis
PTSA	*p*-toluenesulfonic acid
py	pyridine
RAMP	(R)-1-amino-2-(methoxymethyl)pyrrolidine
rt	room temperature
salen	bis(salicylidene)ethylenediamine
SAMP	(*S*)-1-amino-2-(methoxymethyl)pyrrolidine
SET	single electron transfer
Sia	siamyl (3-methyl-2-butyl)
TASF	tris(diethylamino)sulfonium difluorotrimethylsilicate
TBAB	tetrabutylammonium bromide
TBAF	tetrabutylammonium fluoride
TBAD	= DBAD
TBAI	tetrabutylammonium iodide
TBAP	tetrabutylammonium perruthenate
TBDMS	*t*-butyldimethylsilyl
TBDPS	*t*-butyldiphenylsilyl
TBHP	*t*-butyl hydroperoxide
TBS	= TBDMS
TCNE	tetracyanoethylene
TCNQ	7,7,8,8-tetracyanoquinodimethane
TEA	triethylamine
TEBA	triethylbenzylammonium chloride
TEBAC	= TEBA
TEMPO	2,2,6,6-tetramethylpiperidinoxyl
TES	triethylsilyl
Tf	triflyl (trifluoromethanesulfonyl)
TFA	trifluoroacetic acid
TFAA	trifluoroacetic anhydride
THF	tetrahydrofuran
THP	tetrahydropyran; tetrahydropyranyl
Thx	thexyl (2,3-dimethyl-2-butyl)
TIPS	triisopropylsilyl
TMANO	trimethylamine *N*-oxide
TMEDA	*N,N,N′,N′*-tetramethylethylenediamine
TMG	1,1,3,3-tetramethylguanidine
TMS	trimethylsilyl
Tol	*p*-tolyl
TPAP	tetrapropylammonium perruthenate
TBHP	*t*-butyl hydroperoxide
TPP	tetraphenylporphyrin
Tr	trityl (triphenylmethyl)
Ts	tosyl (*p*-toluenesulfonyl)
TTN	thallium(III) nitrate
UHP	urea–hydrogen peroxide complex
Z	= Cbz

Handbook of Reagents for Organic Synthesis

Chiral Reagents for Asymmetric Synthesis

OTHER TITLES IN THIS COLLECTION

Oxidizing and Reducing Agents
Edited by Steven D. Burke and Rick L. Danheiser
ISBN 0 471 97926 0

Acidic and Basic Reagents
Edited by Hans J. Reich and James H. Rigby
ISBN 0 471 97925 2

Activating Agents and Protecting Groups
Edited by Anthony J. Pearson and William R. Roush
ISBN 0 471 97927 9

Reagents, Auxiliaries, and Catalysts for C-C Bond Formation
Edited by Robert M. Coates and Scott E. Denmark
ISBN 0 471 97924 4

Handbook of Reagents for Organic Synthesis

Chiral Reagents for Asymmetric Synthesis

Edited by

Leo A. Paquette
The Ohio State University, Columbus, OH, USA

WILEY

Telephone (+44) 1243 779777

E-mail (for orders and customer service enquiries): cs-books@wiley.co.uk
Visit our Home Page on www.wileyeurope.com or www.wiley.com

Other Wiley Editorial Offices

John Wiley & Sons Inc., 111 River Street,
Hoboken, NJ 07030, USA

Jossey-Bass, 989 Market Street,
San Francisco, CA 94103-1741, USA

Wiley-VCH Verlag GmbH, Boschstr. 12,
D-69469 Weinheim, Germany

John Wiley & Sons Australia Ltd, 33 Park Road,
Milton, Queensland 4064, Australia

John Wiley & Sons (Asia) Pte Ltd, 2 Clementi Loop #02-01,
Jin Xing Distripark, Singapore 129809

John Wiley & Sons Canada Ltd, 22 Worcester Road,
Etobicoke, Ontario, Canada M9W 1L1

Wiley also publishes its books in a variety of electronic formats. Some content that appears in print may not be available in electronic books.

Library of Congress Cataloguing-in-Publication Data

Handbook of reagents for organic synthesis.
p. cm.
Includes bibliographical references.
Contents: [1] Reagents, auxiliaries, and catalysts for C–C bond formation / edited by Robert M. Coates and Scott E. Denmark [2] Oxidising and reducing agents / edited by Steven D. Burke and Riek L. Danheiser [3] Acidic and basic reagents / edited by Hans J. Reich and James H. Rigby [4] Activating agents and protecting groups / edited by Anthony J. Pearson and William R. Roush [5] Chiral Reagents for Asymmetric Synthesis / edited by Leo A. Paquette
ISBN 0-471-97924-4 (v. 1). ISBN 0-471-97926-0 (v. 2)
ISBN 0-471-97925-2 (v. 3) ISBN 0-471-97927-9 (v. 4)
ISBN 0-470-85625-4 (v. 5)
1. Chemical tests and reagents. 2. Organic compounds–Synthesis.
QD77.H37 1999
547′.2 dc 21 98-53088
CIP

British Library Cataloguing in Publication Data

A catalogue record for this book is available from the British Library

ISBN 0 470 85625 4

Typeset in $9\frac{1}{2}/11\frac{1}{2}$ pt Times Roman by Thomson Press (India) Ltd., New Delhi
Printed and bound in Great Britain by Antony Rowe, Chippenham, Wiltshire
This book is printed on acid-free paper responsibly manufactured from sustainable forestry in which at least two trees are planted for each one used for paper production.

e-EROS Editorial Board

EROS 1st Print Edition Editorial Board

Contents

Preface

As stated in its Preface, the major motivation for our undertaking publication of the *Encyclopedia of Reagents for Organic Synthesis* was "to incorporate into a single work a genuinely authoritative and systematic description of the utility of all reagents used in organic chemistry." By all accounts, this reference compendium succeeded admirably in approaching this objective. Experts from around the globe contributed many relevant facts that define the various uses characteristic of each reagent. The choice of a masthead format for providing relevant information about each entry, the highlighting of key transformations with illustrative equations, and the incorporation of detailed indexes serve in tandem to facilitate the retrieval of desired information.

Notwithstanding these accomplishments, the editors came to recognize that the large size of this eight-volume work and its cost of purchase often deterred the placement of copies of the *Encyclopedia* in or near laboratories where the need for this type of information is most critical. In an effort to meet this demand in a cost-effective manner, the decision was made to cull from the major work that information having the highest probability for repeated consultation and to incorporate the same into a set of handbooks. The latter would also be purchasable on a single unit basis.

The ultimate result of these deliberations was the publication of the *Handbook of Reagents for Organic Synthesis*, the first four volumes of which appeared in 1999:

Reagents, Auxiliaries, and Catalysts for C-C Bond Formation
edited by Robert M. Coates and Scott E. Denmark

Oxidizing and Reducing Agents
edited by Steven D. Burke and Rick L. Danheiser

Acidic and Basic Reagents
edited by Hans J. Reich and James H. Rigby

Activating Agents and Protecting Groups
edited by Anthony J. Pearson and William R. Roush

Each of the volumes contains a selected compilation of those entries from the original *Encyclopedia* that bear on the specific topic. Ample listings can be found to functionally related reagents contained in the original work. For the sake of current awareness, references to recent reviews and monographs have been included, as have relevant new procedures from *Organic Syntheses*.

The present volume entitled *Chiral Reagents for Asymmetric Synthesis* constitutes the fifth entry into a continuing series of utilitarian reference works. As with its predecessors, this handbook is intended to be an affordable, enlightening compilation that will hopefully find its way into the laboratories of all practicing synthetic chemists. Every attempt has been made to be of the broadest possible relevance and the expectation is that my colleagues will share in this opinion.

Leo A. Paquette
Department of Chemistry
The Ohio State University
Columbus, OH, USA

Introduction

All of us are aware of the sharp increase in demand for enantiomerically pure reagents and products that has transpired over the past twenty-five years or so. To some extent, the move in this direction has been brought on by the quest by synthetic organic chemists for optically pure natural product targets and for effective asymmetric catalysts. More significantly, this activity has been spurred on throughout the world by governmental oversight agencies whose responsibility it is to guarantee the availability of pure drugs for human consumption. As a consequence, the international medicinal chemistry community continues to upgrade its search for economic ways to develop chiral technology. The need for chiral, nonracemic raw materials, intermediates, and bioactive end products continues to grow at a rapid rate. In the light of these developments, this seemed an appropriate time for assembly into a single volume of a compilation listing many of the optically active reagents and catalysts in use at the present time.

The selection covered in this volume comes from two sources. The first is the *Encyclopedia of Reagents for Organic Synthesis (EROS)* which was published in 1995. In the intervening time, new entries have been written by many experts in the field for incorporation into the ever-expanding electronic version of the same work (*e-EROS*). As to be expected, the compilation includes both well recognized and lesser known reagents and ligands. In order to assist the researcher searching for relevant information, this *Introduction* is followed by a listing of *Recent Reviews and Monographs* on subjects related to this general theme. Following that, there is a section that illustrates those procedures appearing in volumes 68–78 of *Organic Syntheses* that feature the detailed preparation of enantiomerically enriched end-products. The overall intent is to assemble in manageable format as much indispensable information dealing with the subject of *Chiral Reagents for Asymmetric Synthesis* as possible. To this end, the entries are grouped into the following categories: alcohols, aldehydes, amides and lactams, amino compounds, carbohydrate derivatives, diols, esters and lactones, heterocycles, ketones, sulfur compounds, phosphines, and miscellaneous. In the majority of cases, asymmetric reactions are involved. Enantioselective applications of transition metal catalysts can be found throughout the volume. In the body of the text, no attempt has been made to group the reagents in other than alphabetical order. The benefit derived from scanning its pages is thereby maximized.

Finally, we hope that the reader will find this volume to constitute the useful and convenient handbook it was designed to be. This goal will have been reached if the present compilation develops into a valued adjunct to mainstream synthetic practice.

Leo A. Paquette
Department of Chemistry
The Ohio State University
Columbus, OH, USA

Recent Review Articles and Monographs

General Considerations and Theory

Reetz, M. T. Molecular Recognition and Stereotopic Group Recognition, *Pure Appl. Chem.* **1996**, *68*, 1279–1284.

Gawley, R. E.; Aube, J., Eds. *Principles of Asymmetric Synthesis*. Elsevier:Amsterdam, The Netherlands, 1996.

Ager, D. J.; East, M. B., Eds. *Asymmetric Synthetic Methodology*. CRC:Boca Raton, FL, 1996.

Stecher, H.; Faber, K. Biocatalytic Deracemization Techniques. Dynamic Resolutions and Stereoinversions, *Synthesis* **1997**, 1–16.

Caddick, S.; Jenkins, K. Dynamic Resolutions in Asymmetric Synthesis, *Chem. Soc. Rev.* **1996**, *25*, 447–457.

Seebach, D.; Sting, A. R.; Hoffmann, M. Self-Regeneration of Stereocenters (SRS)-Applications, Limitations and Abandonment of a Synthetic Principle, *Angew. Chem., Int. Ed. Engl.* **1997**, *35*, 2708–2748.

Ebbers, E. J.; Ariaans, G. J. A.; Houbiers, J. P. M.; Bruggink, A.; Zwanenburg. B. Controlled Racemization of Optically Active Compounds: Prospects for Asymmetric Transformation, *Tetrahedron* **1997**, *53*, 9417–9476.

Avalos, M.; Babiano, R.; Cintas, P.; Jimenez, J. L.; Palacios, J. C. Nonlinear Stereochemical Effects in Asymmetric Reactions, *Tetrahedron:Asymmetry* **1997**, *8*, 2997–3017.

Somfai, P. Nonenzymic Kinetic Resolution of Secondary Alcohols, *Angew. Chem., Int. Ed. Engl.* **1998**, *36*, 2731–2733.

Avalos, M.; Babiano, R.; Cintas, P.; Jimenez, J. L.; Palacios, J. C.; Barron, L. D. Absolute Asymmetric Synthesis Under Physical Fields: Facts and Fictions, *Chem. Rev.* **1998**, *98*, 2391–2404.

Girard, C.; Kagan, H. B. Nonlinear Effects in Asymmetric Synthesis and Stereoselective Reactions: Ten Years of Investigation, *Angew. Chem., Int. Ed. Engl.* **1998**, *37*, 2923–2959.

Fenwick, D. R.; Kagan, H. B. Asymmetric Amplification, *Topics Stereochem.* **1999**, *22*, 257–296.

Mislow, K. Molecular Chirality, *Topics Stereochem.* **1999**, *22*, 1–82.

Adcock, W.; Trout, N. A. Nature of the Electronic Factor Governing Diastereofacial Selectivity in Some Reactions of Rigid Saturated Model Substrates, *Chem. Rev.* **1999**, *99*, 1415–1435.

Kaselj, M.; Chung, W.-S.; Le Noble, W. J. Face Selection in Addition and Elimination in Sterically Unbiased Systems, *Chem. Rev.* **1999**, *99*, 1387–1413.

Gung, B. W. Structure Distortions in Heteroatom Substituted Cyclohexanones, Adamantanones, and Adamantanes, *Chem. Rev.* **1999**, *99*, 1377–1386.

Cieplak, A. S. Inductive and Resonance Effects of Substituents on π-Face Selection, *Chem. Rev.* **1999**, *99*, 1265–1336.

Mengel, A.; Reiser, O. Around and Beyond Cram's Rule, *Chem. Rev.* **1999**, *99*, 1191–1223.

Reetz, M. T. Synthesis and Diastereoselective Reactions of *N,N*-Dibenzylamino Aldehydes and Related Compounds, *Chem. Rev.* **1999**, *99*, 1121–1162.

Mahrwald, R. Diastereoselection in Lewis-Acid-Mediated Aldol Reactions, *Chem. Rev.* **1999**, *99*, 1095–1120.

Lippmann, D. Z. Possible Mechanism for Spontaneous Production of Enantiomeric Excess, *Adv. Biochirality*, Palyi, G.; Zucchi, C.; Caglioti, L., Eds.; Elsevier: Amsterdam, The Netherlands, 1999.

Eames, J. Parallel Kinetic Resolutions, *Angew. Chem. Int. Ed.* **2000**, *39*, 885–888.

Nicolaou, K. C.; Vourloumis, D.; Winssinger, N.; Baran, P. S. The Art and Science of Total Synthesis at the Dawn of the Twenty-First Century, *Angew. Chem. Int. Ed.* **2000**, *39*, 44–122.

Feringa, B. L.; Van Delden, R. A. Absolute Asymmetric Synthesis: The Origin, Control, and Amplification of Chirality, *Angew. Chem. Int. Ed.* **1999**, *38*, 3419–3428.

Blackmond, D. G. Kinetic Aspects of Nonlinear Effects in Asymmetric Catalysis, *Acc. Chem. Res.* **2000** *33*, 402–411.

Soai, K.; Shibata, T.; Sato, I. Enantioselective Automultiplication of Chiral Molecules by Asymmetric Autocatalysis, *Acc. Chem. Res.* **2000**, *33*, 382–390.

Seoane, G. Enzymatic C-C Bond-Forming Reactions in Organic Synthesis, *Curr. Org. Chem.* **2000**, *4*, 283.

Cook, G. R. Transition Metal-Mediated Kinetic Resolution, *Curr. Org. Chem.* **2000**, *4*, 869–885.

Keith, J. M.; Larrow, J. F.; Jacobsen E. N. Practical Considerations in Kinetic Resolution Reactions, *Adv. Synth. Catal.* **2001**, *34*, 5–26.

Kagan, H. B. Practical Consequences of Non-Linear Effects in Asymmetric Synthesis, *Adv. Synth. Catal.* **2001**, *343*, 227–233.

Kagan, H. B. Nonlinear Effects in Asymmetric Catalysis: A Personal Account, *Synlett* **2001**, 888–899.

Kolb, H. C.; Finn, M. G.; Sharpless, K. B. Click Chemistry: Diverse Chemical Function from a Few Good Reactions, *Angew. Chem. Int. Ed.* **2001**, *40*, 2004–2021.

Stutz, A. E., Ed. Glycoscience: Epimerization, Isomerization, and Rearrangement Reactions of Carbohydrates, *Top. Curr. Chem.* **2001**, 215.

Asymmetric Synthesis

Stephenson, G. R., Ed., *Advanced Asymmetric Synthesis*, Chapman & Hall: London, U.K., 1996.

Procter, G. *Asymmetric Synthesis*, Oxford University Press: Oxford, U.K. 1995.

De Lucchi, O. High Symmetry Chiral Auxiliaries Containing Heteroatoms, *Pure Appl. Chem.* **1996**, *68*, 945–650.

Ruano, J. L. G.; Carretero, J. C.; Carreno, M. C.; Cabrejas, L. M. M.; Urbano, A. The Sulfinyl Group as a Chiral Inductor in Asymmetric Diels-Alder Reactions, *Pure Appl. Chem.* **1996**, *68*, 925–930.

Trost, B. M. Asymmetric Introduction of Heteroatoms, *Pure Appl. Chem.* **1996**, *68*, 779–784.

Cardillo, G.; Tomasini, C. Asymmetric Synthesis of β-Amino Acids and α-Substituted β-Amino Acids, *Chem. Soc. Rev.* **1996**, *25*, 117–128.

Keinan, E.; Sinha, S. C.; Shabat, D.; Itzhaky, H.; Raymond, J.-L. Asymmetric Organic Synthesis with Catalytic Antibodies, *Acta Chem. Scand.* **1996**, *50*, 679–687.

Gothelf, K. V.; Jorgensen, K. A. Metal-Catalyzed Asymmetric 1,3-Dipolar Cycloaddition Reactions, *Acta Chem. Scand.* **1996**, *50*, 652–660.

Studer, A. Amino Acids and Their Derivatives as Stoichiometric Auxiliaries in Asymmetric Synthesis, *Synthesis* **1996**, 793–815.

Pete, J.-P. Asymmetric Photoreactions of Conjugated Enones and Esters, *Adv. Photochem.* **1996**, *21*, 135–216.

Frederickson, M. Optically Active Isoxazolidines via Asymmetric Cycloaddition Reactions of Nitrones with Alkenes-Applications in Organic Synthesis, *Tetrahedron*, **1997**, *53*, 403–425.

Enders, D.; Klatt, M. Asymmetric Synthesis with (*S*)-2-Methoxymethyl Pyrrolidine (SMP)-A Pioneer Auxiliary, *Synthesis* **1996**, 1403–1418.

Nakai, T.; Tomooka, K. Asymmetric [2,3]-Wittig Rearrangement as a General Tool for Asymmetric Synthesis, *Pure Appl. Chem.* **1997**, *69*, 595–600.

Ojima, I. Asymmetric Syntheses by Means of the Lactam Synthon Method, *Adv. Asymm. Synth., Vol. 1*, Hassner, A., Ed., JAI:Greenwich, CT, 1995.

Williams, R. M. Asymmetric Synthesis of α-Amino Acids, *Adv. Asymm. Synth., Vol. 1*, Hassner, A., Ed., JAI:Greenwich, CT, 1995.

Mikami, K. Supramolecular Chemistry in Asymmetric Carbonyl-Ene Reactions, *Adv. Asymm. Synth., Vol. 1*, Hassner, A., Ed. JAI: Greenwich, CT, 1995.

Aversa, M. C.; Barattucci, A.; Bonaccorsi, P.; Giannetto, P. Chiral Sulfinyl-1, 3-dienes. Synthesis and Use in Asymmetric Reactions, *Tetrahedron:Asymmetry* **1997**, *8*, 1339–1367.

Regan, A. C. Asymmetric Processes *Contemp. Org. Synth.* **1997**, *4*, 1–21.

Ager, D. J.; Prakash, I.; Schaad, D. R. Chiral Oxazolidinones in Asymmetric Synthesis, *Aldrichim. Acta* **1997**, *30*, 3–12.

Hudlicky, T., Ed., Asymmetric Synthesis, *Curr. Org. Chem.* **1997**, *1*, 1–107.

Enders, D.; Reinhold, U. Asymmetric Synthesis of Amines by Nucleophilic 1,2-Addition of Organometallic Reagents to the CN-Double Bond, *Tetrahedron:Asymmetry* **1997**, *8*, 1895–1946.

Osborn, H. M. I.; Sweeney, J. The Asymmetric Synthesis of Aziridines, *Tetrahedron:Asymmetry* **1997**, *8*, 1693–1715.

Shibasaki, M.; Boden, C. D. J.; Kojima, A. The Asymmetric Heck Reaction, *Tetrahedron* **1997**, *53*, 7371–7393.

Aube, J. Oxaziridine Rearrangements in Asymmetric Synthesis, *Chem. Soc. Rev.* **1997**, *26*, 269–278.

Li, A.-H.; Dai, L.-X.; Aggarwal, V. K. Asymmetric Ylide Reactions: Epoxidation, Cyclopropanation, Aziridination, Olefination, and Rearrangement, *Chem. Rev.* **1997**, *97*, 2341–2372.

Mikolajczk, M.; Drabowicz, J.; Kielbasinski, P. *Chiral Sulfur Reagents: Applications in Asymmetric and Stereoselective Synthesis*, CRC: Boca Raton, FL, 1997.

Kagan, H. B.; Riant, O. Preparation of Chiral Ferrocenes by Asymmetric Synthesis or by Kinetic Resolution, *Adv. Asym. Synth.*, Vol. 2, **1997**, Hassner, A. Ed. JAI:Greenwich, CT.

Guingant, A. Asymmetric Syntheses of Nonracemic Amines, *Adv. Asym. Synth.*, Vol. 2, **1997**, Hassner, A., Ed. JAI:Greenwich, CT.

Shimizu, T.; Kamigata, N. Optically Active Selenium and Tellurium Compounds. Synthesis and Application for Asymmetric Synthesis. A Review, *Org. Prep. Proc. Int.* **1997**, *29*, 605–629.

Bennani, Y. L.; Hanessian, S. *trans*-1,2-Diaminocyclohexane Derivatives as Chiral Reagents, Scaffolds, and Ligands for Catalysis: Applications in Asymmetric Synthesis and Molecular Recognition, *Chem. Rev.* **1997**, *97*, 3161–3195.

Davis, F. A.; Zhou, P.; Chen, B.-C. Asymmetric Synthesis of Amino Acid Using Sulfilimines (Thiooxime *S*-Oxides), *Chem. Soc. Rev.* **1998**, *27*, 13–18.

Takayama, S.; McGarvey, G. J.; Wong, C.-H. Enzymes in Organic Synthesis: Recent Developments in Aldol Reactions and Glycosylations, *Chem. Soc. Rev.* **1997**, *26*, 407–415.

Gothelf, K. V.; Jorgensen, K. A. Asymmetric 1,3-Dipolar Cycloaddition Reactions, *Chem. Rev.* **1998**, *98*, 863–909.

Ghosh, A. K.; Fidanse, S.; Senanayake, C. H. *cis*-1-Aminoindan-2-ol in Asymmetric Synthesis, *Synthesis* **1998**, 937–961.

Agami, C.; Couty, F.; Puchot-Kadouri, C. Asymmetric Synthesis of α-Amino Acids from a Chiral Masked Form of Glyoxal, *Synlett* **1998**, 449–456.

Allin, S. M.; Page, P. C. B. The Development and Application of 1,3-Dithiane 1-Oxide Derivatives as Chiral Auxiliaries and Asymmetric Building Blocks for Organic Synthesis. A Reveiw, *Org. Prep. Proc. Int.* **1998**, *30*, 145–172.

Ingate, S. T.; Marco-Contelles, J. The Asymmetric Pauson-Khand Reaction, *Org. Prep. Proc. Intl.* **1998**, *30*, 121–143.

Senanayake, C. H. Applications of *cis*-1-Amino-2-indanol in Asymmetric Synthesis, *Aldrichimica Acta* **1998**, *31*, 3–15.

Cowden, C. J.; Paterson, I. Asymmetric Aldol Reactions Using Boron Enolates, *Org. React.* **1997**, *51*, 1–200.

Hudlicky, T., Ed. Asymmetric Synthesis, *Curr. Org. Chem.*, **1998**, No. 2.

Ito, H.; Taguchi, T. Asymmetric Claisen Rearrangement, *Chem. Soc. Rev.* **1999**, *28*, 43–50.

Casiraghi, G.; Rassu, G.; Zanardi, F.; Battistini, L. Asymmetric Access to Functional, Structurally Diverse Molecules Exploiting Five-Membered Heterocyclic Silyloxy Dienes, *Adv. Asym. Synth.*, Vol. 3, Hassner, A., Ed., JAI Press, Stamford, CT, 1998.

Yamamoto, Y.; Asao, N.; Tsukada, N. Asymmetric Synthesis of β-Amino Acids and β-Lactam Derivatives via Conjugate Addition of Metal Amides, *Adv. Asym. Synth.*, Vol. 3, Hassner, A., Ed. JAI Press, Stamford, CT, 1998.

O'Brien, P.; Sharpless, K. B. Asymmetric Aminohydroxylation: Scope, Limitations, and Use in Synthesis, *Angew. Chem. Int. Ed.* **1999**, *38*, 326–329.

Najera, C.; Yus, M. Pyroglutamic Acid: A Versatile Building Block in Asymmetric Synthesis, *Tetrahedron:Asymmetry* **1999**, *10*, 2245–2303.

Zhou, W.-S.; Lu, Z.-H.; Xu, Y.-H.; Liao, L.-X.; Wang, Z.-M. Syntheses of Optically Active α-Furfurylamine Derivatives and Application to the Asymmetric Syntheses, *Tetrahedron* **1999**, *55*, 11959–11983.

Palomo, C.; Aizpurua, J. M.; Ganboa, I.; Oiarbide, M. Asymmetric Synthesis of β-Lactams by Staudinger Ketene-Imine Cycloaddition Reaction, *Eur. J. Org. Chem.* **1999**, 3223–3235.

Ruano, J. L. G.; De la Plata, B. C. Asymmetric [4+2] Cycloadditions Mediated by Sulfoxides, *Topics Curr. Chem.* **1999**, *204*, 1–126.

Arya, P.; Qin, H. Advances in Asymmetric Enolate Methodology, *Tetrahedron* **2000**, *56*, 917–947.

Arai, Y. Remote Asymmetric Induction Using Chiral (*p*-Tolylsulfinyl)-Furyl, -Thienyl, and -Pyrrolyl Carbonyl Compounds, *Rev. Heteroat. Chem.* **1999**, *21*, 65–91.

Comins, D. L. Asymmetric Synthesis and Synthetic Utility of 2,3-Dihydro-4-pryridones, *J. Heterocycl. Chem.* **1999**, *36*, 1491–1500.

Langer, P. New Strategies for the Development of an Asymmetric Version of the Baylis-Hillman Reaction, *Angew. Chem. Int. Ed.* **2000**, *39*, 3049–3052.

Groaning, M. D.; Meyers, A. I. Chiral Non-Racemic Bicyclic Lactams. Auxiliary-Based, Asymmetric Reactions, *Tetrahedron* **2000**, *56*, 9843–9873.

Bandini, M.; Cozzi, P. G.; Umani-Ronchi, A. Asymmetric Synthesis with "Priviledged" Ligands, *Pure Appl. Chem.* **2001**, *73*, 325–329.

Kim, Y. H. Dual Enantioselective Control in Asymmetric Synthesis, *Acc. Chem. Res.* **2001**, *34*, 955–962.

Asymmetric Catalysis

Hiroi, K. Transition Metal or Lewis Acid-Catalyzed Asymmetric Reactions with Chiral Organosulfur Functionality, *Rev. Heteroatom Chem.* **1996**, *14*, 21–58.

Noyori, T.; Hashiguchi, S.; Iwasawa, Y. Asymmetric Transfer Hydrogenation Catalyzed by Chiral Ruthenium Complexes, *Acc. Chem. Res.* **1997**, *30*, 97–102.

Tokunaga, M.; Larrow, J. F.; Kakiuchi, F.; Jacobsen, E. N. Asymmetric Catalysis with Water: Efficient Kinetic Resolution of Terminal Epoxides by Means of Catalytic Hydrolysis, *Science* **1997**, *276*, 936–938.

Shibasaki, M.; Sasai, H.; Arai, T. Asymmetric Catalysis with Heterobimetallic Compounds, *Angew. Chem., Int. Ed. Engl.* **1997**, *36*, 1237–1256.

Ghosh, A. K.; Mathivanan, P.; Cappiello, J. C_2 Symmetric Chiral Bis(Oxazoline)-Metal Complexes in Catalytic Asymmetric Synthesis, *Tetrahedron:Asymmetry* **1998**, *9*, 1–45.

Doyle, M. P.; Forbes, D. C. Recent Advances in Asymmetric Catalytic Metal Carbene Transformations, *Chem. Rev.* **1998**, *98*, 911–935.

Davies, H. M. L. Asymmetric Synthesis Using Rhodium-Stabilized Vinylcarbenoid Intermediates, *Aldrichimica. Acta* **1997**, *30*, 107–114.

Nelson, S. G. Catalyzed Enantioselective Aldol Additions of Latent Enolate Equivalents, *Tetrahedron:Asymmetry* **1998**, *9*, 357–389.

Aggarwall, V. K. Catalytic Asymmetric Epoxidation and Aziridination Mediated by Sulfur Ylides. Evolution of a Project, *Synlett* **1998**, 329–336.

Corey, E. J.; Guzman Perez, A. The Catalytic Enantioselective Construction of Molecules with Quanternary Carbon Centers, *Angew. Chem., Int. Ed. Engl.* **1998**, *37*, 388–401.

Moberg, C. C_3 Symmetry in Asymmetric Catalysis and Chiral Recognition, *Angew. Chem., Int. Ed. Engl.* **1998**, *37*, 249–268.

Doyle, M. P.; Protopopova, M. N. New Aspects of Catalytic Asymmetric Cyclopropanation, *Tetrahedron* **1998**, *54*, 7919–7946.

Doyle, M. P. New Catalysts and Methods for Highly Enantioselective Metal Carbene Reactions, *Pure Appl. Chem.* **1998**, *70*, 1123–1128.

Nugent, W. A.; Licini, G.; Bonchio, M.; Bortolini, O.; Finn, M. G.; McCleland, B. W. Homogeneous Catalysis as a Tool for Organic Synthesis, *Pure Appl. Chem.* **1998**, *70*, 1041–1046.

Iseki, K. Catalytic Asymmetric Synthesis of Chiral Fluoroorganic Compounds, *Tetrahedron* **1998**, *54*, 13887–13914.

Bolm, C.; Muniz, K. Planar Chiral Arene Chromium (0) Complexes: Potential Ligands for Asymmetric Catalysis, *Chem. Soc. Rev.* **1999**, *28*, 51–59.

Shibasaki, M.; Sasai, H. Asymmetric Catalysis Using Heterobimetallic Compounds, *Adv. Asym. Synth.*, Vol. 3, Hassner, A., Ed., JAI Press, Stamford, CT, 1998.

Strauss, U. T.; Feller, U.; Faber, K. Biocatalytic Transformations of Racemates into Chiral Building Blocks in 100% Chemical Yield and 100% Enantiomer Excess, *Tetrahedron:Asymmetry* **1999**, *10*, 107–117.

Togni, A.; Dorta, R.; Kollner, C.; Pioda, G. Some New Aspects of Asymmetric Catalysis with Chiral Ferrocenyl Ligands, *Pure Appl. Chem.* **1998**, *70*, 1477–1485.

Shibasaki, M.; Sasai, H. Asymmetric Catalysis with Chiral Lanthanoid Complexes, *Topics Stereochem.* **1999**, *22*, 201–225.

Hilvert, D. Stereoselective Reactions with Catalytic Antibiodies, *Topics Stereochem.* **1999**, *22*, 83–135.

Pfaltz, A. From Corrin Chemistry to Asymmetric Catalysis. A Personal Account, *Synlett* **1999**, 835–842.

Soai, K. Asymmetric Autocatalysis and Biomolecular Chirality, *Adv. Biochirality*, Palyi, G.; Zucchi, C.; Caglioti, L., Eds.; Elsevier: Amsterdam, The Netherlands, 1999.

Bäckvall, J.-E. Synthesis of Natural Products via Stereocontrolled Palladium-Catalyzed Reaction, *Pure, Appl. Chem.* **1999**, *71*, 1065–1070.

Tenaglia, A.; Heumann, A. Palladium-Catalyzed Enantioselective Organic Transformations, *Angew. Chem. Int. Ed.* **1999**, *38*, 2180–2184.

Pagenkopf, B. L.; Carreira, E. M. Transition Metal Fluoride Complexes in Asymmetric Catalysis, *Chem. Eur. J.* **1999**, *5*, 3437–3442.

Jacobsen, E. N.; Pfaltz, A.; Yamamoto, H., Eds. *Comprehensive Asymmetric Catalysis*; Vol. 1–3. Springer:Berlin, Germany, 1999.

Denmark, S. E.; Stavenger, R. A. Asymmetric Catalysis of Epoxide Ring-Opening Reactions, *Acc. Chem. Res.* **2000**, 33, 432–440.

Fu, G. C. Enantioselective Nucleophilic Catalysis with "Planar-Chiral" Heterocycles, *Acc. Chem. Res.* **2000**, *33*, 412–420.

Jacobsen, E. N. Asymmetric Catalysis of Epoxide Ring-Opening Reactions, *Acc. Chem. Res.* **2000**, *33*, 421–431.

Burk, M. J. Modular Phospholane Ligands in Asymmetric Catalysis, *Acc. Chem. Res.* **2000**, *33*, 363–372.

Hayashi, T. Chiral Monodentate Phosphine Ligand MOP for Transition Metal-Catalyzed Asymmetric Reactions, *Acc. Chem. Res.* **2000**, *33*, 354–362.

Feringa, B. L. Phosphoramidites: Marvellous Ligands for Asymmetric Catalysis, *Acc. Chem. Res.* **2000**, *33*, 346–358.

Dias, L. C. Chiral Lewis Acid-Catalyzed Ene Reactions, *Curr. Org. Chem.* **2000**, *4*, 283–304.

Fache, F.; Schulz, E.; Tommasino, M. L.; Lemaire, M. Nitrogen-Containing Ligands for Asymmetric Homogeneous and Heterogeneous Catalysis, *Chem. Rev.* **2000**, *100*, 2091–2157.

Ratovelomanana-Vidal, V.; Genet, J.-P. Synthetic Applications of the Ruthenium-Catalyzed Hydrogenation via Dynamic Kinetic Resolution, *Can. J. Chem.* **2000**, *78*, 846–851.

Machajewski, T. D.; Wong, C.-H; Lerner, R. A. The Catalytic Asymmetric Aldol Reaction, *Angew. Chem. Int. Ed.* **2000**, *39*, 1352–1374.

Mikami, K.; Terada, M.; Korenaga, T.; Matsumoto, Y.; Matsukawa, S. Enantiomer-Selective Activation of Racemic Catalysts, *Acc. Chem. Res.* **2000**, *33*, 391–401.

Ojima, I., Ed. *Catalytic Asymmetric Synthesis*, 2nd Ed. Wiley: New York, 2000.

Krause, N.: Hoffmann-Roder, A. Recent Advances in Catalytic Enantioselective Michael Reactions, *Synthesis* **2001**, 171–196.

Frohn, M.; Shi, Y. Chiral Ketone-Catalyzed Asymmetric Epoxidation of Olefins, *Synthesis* **2000**, 1979–2000.

Guiry, P. J.; McCarthy, M.; Lacey, P. M.; Saunders, C. P.; Kelly, S.; Connolly, D. J. Axially Chiral Phosphinamine Ligands in Asymmetric Catalysis, *Curr. Org. Chem.* **2000**, *4*, 821–836.

Noyori, R.; Okhuma, T. Asymmetric Catalysis by Architectural and Functional Molecular Engineering. Practical Chemo- and Stereoselective Hydrogenation of Ketones, *Angew. Chem. Int. Ed.* **2001**, *40*, 40–73.

Mikami, K.; Shimizu, M.; Zhang, H. C.; Maryanoff, B. E. Acyclic Stereocontrol Between Remote Atom Centers via Intramolecular and Intermolecular Stereo-Communication, *Tetrahedron* **2001**, *57*, 2917–2951.

Hoveyda, A. H.; Schrock, R. R. Catalytic Asymmetric Olefin Synthesis, *Chem. Eur. J.* **2001**, *7*, 945–950.

Pu, L.; Yu, H.-B. Catalytic Asymmetric Organozinc Additions to Carbonyl Compounds, *Chem. Rev.* **2001**, *101*, 757–824.

Yet, L. Recent Developments in Catalytic Asymmetric Strecker-Type Reactions, *Angew. Chem. Int. Ed.* **2001**, *40*, 875–877.

Groger, H.; Wilken, J. The Application of L-Proline as an Enzyme Mimic and Further New Asymmetric Syntheses using Small Organic Molecules as Chiral Catalysts, *Angew. Chem. Int. Ed.* **2001**, *40*, 529–532.

McCarthy, M.; Guiry, P. J. Axially Chiral Bidentate Ligands in Asymmetric Catalysis, *Tetrahedron* **2001**, *57*, 3809–3844.

Imamoto, T. New P-Chirogenic Diphosphines and Their Use in Catalytic Asymmetric Reactions, *Pure Appl. Chem.* **2001**, *73*, 373–376.

Fu, G. C. Asymmetric Catalysis with "Planar-Chiral" Heterocycles, *Pure. Appl. Chem.* **2001**, *73*, 347–349.

Adams, N. J.; Bargon, J.; Brown, J. M.; Farrington, E. J.; Galardon, E.; Giernoth, R.; Heinrich, H.; John, B. D.; Maeda, K. Interplay of Synthesis and Mechanism in Asymmetric Homogeneous Catalysis, *Pure Appl. Chem.* **2001**, *73*, 343–346.

Genet, J. P.; Marinetti, A.; Ratovelomanana-Vidal, V. Recent Advances in Asymmetric Catalysis. Synthetic Applications to Biologically Active Compounds, *Pure Appl. Chem.* **2001**, *73*, 299–303.

Mikami, K.; Korenaga, T.; Matsumoto, Y.; Ueki, M.; Terada, M.; Matsukawa, S. Racemic Catalysis through Asymmetric Activation, *Pure Appl. Chem.* **2001**, *73*, 255–259.

Negishi, E.-I. Some Newer Aspects of Organozirconium Chemistry of Relevance to Organic Synthesis. Zr-Catalyzed Enantioselective Carbometalation, *Pure Appl. Chem.* **2001**, *73*, 239–242.

Noyori, R.; Koizumi, M.; Ishii, D.; Ohkuma, T. Asymmetric Hydrogenation via Architectural and Functional Molecular Engineering, *Pure Appl. Chem.* **2001**, *73*, 227–232.

Pavlov, V. A Mechanism of Asymmetric Induction in Catalytic Hydrogenation, Hydrosilylation, and Cross-Coupling Reactions on Metal Complexes, *Russ. Chem. Rev.* **2002**, *71*, 39–56.

Spivey, A. C.; Andrews, B. I. Catalysis of the Asymmetric Desymmetrization of Cyclic Anhydrides by Nucleophilic Ring-Opening with Alcohols, *Angew. Chem. Int. Ed.* **2001**, *40*, 3131–3134.

Groger, H. Enzymatic Routes to Enantiomerically Pure Aromatic α-Hydroxy Carboxylic Acids:A Further Example for the Diversity of Biocatalysis, *Adv. Synth. Catal.* **2001**, *343*, 547–558.

Enantioselective Transformations

Rahman, A.-u, Ed., Stereoselective Synthesis, *Stud. Nat. Prod. Chem.* **1996.** *Volume 18*, Elsevier: Lausanne, Switzerland, 1996.

Hoveyda, A. H.; Morken, J. P. Enantioselective C-C and C-H Bond Formation Mediated by Chiral Ebthi [ethylene bis(tetrahydroindenyl)] Complexes of Titanium and Zirconium, *Angew. Chem., Int. Ed. Engl.* **1996**, *35*, 1262–1284.

Sardina, F. J.; Rapoport, H. Enantiospecific Synthesis of Heterocycles from α-Amino Acids, *Chem. Rev.* **1996**, *96*, 1825–1872.

Bäckvall, J. E. Enantiocontrol in Some Palladium- and Copper-Catalyzed Reactions, *Acta Chem. Scand.* **1996**, *50*, 661–665.

Remuzon, P. *trans*-4-Hydroxy-L-proline, a Useful and Versatile Chiral Building Block, *Tetrahedron* **1996**, *52*, 13803–13836.

Fehr, C. Enatioselective Protonation of Enolates and Enols, *Angew. Chem., Int. Ed. Engl.* **1996**, *35*, 2566–2587.

Beak, P.; Basu, A.; Gallagher, D. J.; Park, Y. S.; Thayumanavan, S. Regioselective Diastereoselective, and Enantioselective Lithiation-Substitution Sequences-Reaction Pathways and Synthetic Applications, *Acc. Chem. Res.* **1996**, *29*, 552–560.

Hudlicky, T.; Reed, J. W. An Evolutionary Perspective of Microbial Oxidations of Aromatic Compounds in Enantioselective Synthesis: History, Current Status, and Perspectives,

Adv. Asymm. Synth., Vol. 1, Hassner, A., Ed., JAI: Greenwich, CT, 1995.

Mori, K. Enantioselective Synthesis of Bioactive Natural Products: Examples in the Field of Insect Chemistry, *Adv. Asym. Synth.*, Vol. 1, Hassner, A., Ed., JAI:Greenwich, CT, 1995.

Brown, H. C.; Ramachandra, P. V. Synthesis via Chiral Organoboranes Based on α-Pinene, *Adv. Asym. Synth.,* Vol. 1, Hassner, A., Ed., JAI:Greenwich, CT 1995.

Lundt, I. Aldonolactones as Chiral Synthons, *Top. Curr. Chem.* **1997**, *187*, 117.

Negishi, E.-I.; Kotora, M. Regio- and Stereoselective Synthesis of γ-Alkylidenebutenolides and Related Compounds, *Tetrahedron* **1997**, *53*, 6707–6738.

Juaristi, E., Ed., Enantioselective Synthesis of β-Amino Acids, Wiley-VCH:New York, 1997.

Coppola, G. M.; Schuster, H. F. *Chiral α-Hydroxy Acids in Enantioselective Synthesis,* Wiley-VCH:Weinheim, Germany, 1997.

Pridgen, L. N. Synthesis of Nonracemic Amines, *Adv. Asym. Synth,* Vol. 2, **1997**, Hassner, A., Ed. JAI:Greenwich, CT.

Hultin, P. G.; Earle, M. A.; Sudharshan, M. Synthetic Studies with Carbohydrate-Derived Chiral Auxiliaries, *Tetrahedron* **1997**, *53*, 14823–14870.

Cozzi, P. G.; Tagliavini, E.; Umani-Ronchi, A. Enantioselective Addition of Allylic Silanes and Stannanes to Aldehydes Mediated by Chiral Lewis Acids, *Gazz. Chim. Ital.* **1997**, *127*, 247–254.

Hoppe, D.; Hense, T. Enantioselective Synthesis of Lithium/(−)-Sparteine Carbanion Pairs, *Angew. Chem., Int. Ed. Engl.* **1997**, *36*, 2282–2316.

Enders, D.; Jandeleit, B.; von Berg, S. Synthesis of Highly Enantioenriched Compounds via Iron Mediated Allylic Substitutions, *Synlett* **1997**, 421–431.

Juaristi, E.; Escalante, J.; Leon-Romo, J. L.; Reyes, A. Recent Applications of α-Phenylethylamine in the Preparation of Enantiopure Compounds, *Tetrahedron:Asymmetry* **1998**, *9*, 1279–1332.

Du, Y.; Linhardt, R. J.; Vlahov, I. R. Recent Advances in Stereoselective *C*-Glycoside Synthesis, *Tetrahedron* **1998**, *54*, 9913–9959.

Crimmins, M. T. New Developments in the Enantioselective Synthesis of Cyclopentyl Carbocyclic Nucleosides, *Tetrahedron* **1998**, *54*, 9229–9272.

Corey, E. J; Helal, C. J. Reduction of Carbonyl Compounds with Chiral Oxazaborolidine Catalysts: A New Paradigm for Enantioselective Catalysis and a Powerful New Synthetic Method, *Angew. Chem., Int. Ed. Engl.* **1998**, *37*, 1987–2012.

Sesay, S. J.; Williams, J. M. J. Palladium-Catalyzed Enantioselective Allylic Substitution Reactions, *Adv. Asym. Synth.*, Vol. 3, Hassner, A., Ed., JAI Press, Stamford, CT, 1998.

Gawley, R. E. Stereoselective Addition of Chiral α-Aminoorganometallics to Aldehydes, *Adv. Asym. Synth.*, Vol. 3, Hassner, A., Ed. JAI Press, Stamford, CT 1998.

Majewski, M. Enantioselective Deprotonation of Cyclic Ketones, *Adv. Asym. Synth.*, Vol. 3, Hassner, A., Ed. JAI Press, Stamford, CT, 1998.

Cativiella, C.; Diaz-De-Villegas, M. D. Stereoselective Synthesis of Quaternary α-Amino Acids. Part 1. Acyclic Compounds, *Tetrahedron: Asymmetry* **1998,** *9*, 3517–3599.

Sibi, M. P.; Porter, N. A. Enantioselective Free Radical Reactions, *Acc. Chem. Res.* **1999**, *32*, 163–171.

Denmark, S. E.; Wu, Z. The Development of Chiral, Nonracemic Dioxiranes for the Catalytic Enantioselective Epoxidation of Alkenes, *Synlett* **1999**, 847–859.

Mehta, G.; Chandrasekhar, J. Electronic Control of Facial Selection in Additions to Sterically Unbiased Ketones and Olefins, *Chem. Rev.* **1999**, *99*, 1437–1467.

Kobayashi, S.; Ishitani, H. Catalytic Enantioselective Addition to Imines, *Chem. Rev.* **1999**, *99*, 1069–1094.

Putala, M. Synthetic Approaches to Axially Chiral C_2-Symmetrical Nonracemic Binaphthyl Derivatives, *Enantiomer* **1999**, *4*, 243.

Ho, T.-L. *Stereoselectivity in Synthesis.* Wiley-Interscience: New York, 1999.

Solladie, G. Applications of Chiral Sulfoxides in Enantioselective Synthesis of Diols and Total Synthesis of Natural Products, *Enantiomer* **1999**, *4*, 183–193.

Arend, M. Asymmetric Catalytic Aminoalkylations:New Powerful Methods for the Enantioselective Synthesis of Amino Acid Derivatives, Mannich Bases, and Homoallylic Amines, *Angew. Chem. Int. Ed.* **1999**, *38*, 2873–2874.

Hudlicky, T.; Gonzalez, D.; Gibson, D. T. Enzymatic Dihydroxylation of Aromatics in Enantioselective Synthesis. Expanding Asymmetric Methodology, *Aldrichim. Acta* **1999**, *32*, 35–62.

Soloshonok, V. A., Ed. *Enantiocontrolled Synthesis of Fluoro-Organic Compounds: Stereochemical Challenges and Biomedical Targets.* Wiley. Chichester, U.K., 1999.

Gawronski, J.; Gawronska, K. *Tartaric and Malic Acids in Synthesis:A Source Book of Building Blocks, Ligands, Auxiliaries, and Resolving Agents.* Wiley: New York, 1998.

Cativiela, C.; Diaz-De-Villegas, M. D. Stereoselective Synthesis of Quaternary α-Amino Acids. Part 2: Cyclic Compounds, *Tetrahedron: Asymmetry* **2000**, *11*, 645–732.

Breit, B. Controlling Stereoselectivity with the Aid of a Reagent-Directing Group. Hydroformylation, Cuprate Addition, and Domino Reaction Sequences, *Chem. Eur. J.* **2000**, 1519–1524.

Faul, M. M.; Huff, B. E. Strategy and Methodology Development for the Total Synthesis of Polyether Ionophores, *Chem. Rev.* **2000**, *100*, 2407–2473.

Franz, A. K.; Woerpel, K. A. Development of Reactions of Silacyclopropanes as New Methods for Stereoselective Organic Synthesis, *Acc. Chem. Res.* **2000**, *33*, 813–820.

Taber, D. F.; Campbell, C. L.; Louey, J. P.; Wang, Y.; Zhang, W. Predicting the Diastereoselectivity of Intramolecular Diene Cyclozirconation: Applications to Natural Product Synthesis, *Curr. Org. Chem.* **2000**, *4*, 809–819.

Hanessian, S.; Lou, B. Stereocontrolled Glycosyl Transfer Reactions with Unprotected Glycosyl Donors, *Chem. Rev.* **2000**, *100*, 4443–4463.

Hollingsworth, R. I.; Wang, G. Toward a Carbohydrate-Based Chemistry: Progress in the Development of General-Purpose Chiral Synthons from Carbohydrates, *Chem. Rev.* **2000**, *100*, 4267–4282.

Wirth, T. Organoselenium Chemistry in Stereoselective Reactions, *Angew. Chem. Int. Ed.* **2000**, *39*, 3740–3749.

Bringmann, G.; Hinricks, J.; Pabst, T.; Henschel, P.; Peters, K.; Peters, E.-M. From Dynamic to Non-Dynamic Kinetic Reslution of Lactone-Bridged Biaryls: Synthesis of Mastigophorene B. *Synthesis* **2001**, 155–167.

Laschat, S.; Dicker, T. Stereoselective Synthesis of Piperidines, *Synthesis* **2000**, 1781–1813.

Dirat, O.; Kouklovsky, C.; Mauduit, M.; Langlois, Y. Oxazoline-*N*-oxide Mediated Asymmetric Cycloadditions. Recent Progress in the Stereoselective Syntheses of β-Lactone and β-Lactams, *Pure Appl. Chem.* **2000**, *72*, 1721–1737.

Hassner, A.; Ghera, E.; Yechezkel, T.; Kleiman, V.; Balasubramanian, T.; Ostercamp, D.; Stereoselective and Enantioselective Synthesis of Five-Membered Rings via Conjugate Additions to Allylsulfone Carbanions, *Pure Appl. Chem.* **2000**, *72*, 1671–1683.

Hodgson, D. M.; Pierard, F. Y. T. M.; Stupple, P. A. Catalytic Enantioselective Rearrangements and Cycloadditions Involving Ylides from Diazo-Compounds, *Chem. Soc. Rev.* **2001**, *30*, 50–61.

Seebach, D.; Beck, A. K.; Heckel, A. TADDOLS, Their Derivatives, and TADDOL Analogs: Versatile Chiral Auxiliaries, *Angew. Chem. Int. Ed.* **2001**, *40*, 92–138.

Eames, J.; Weerasooriya, N. Recent Avances into the Enantioselective Protonation of Prostereogenic Enol Derivatives, *Tetrahedron:Asymmetry* **2001**, *12*, 1–24.

Karlsson, S.; Hogberg, H.-E. Asymmetric 1,3-Dipolar Cycloadditions for the Construction of Enantiomerically Pure Heterocycles, *Org. Prep. Proced. Intern.* **2001**, *33*, 103–172.

Hartung, J. Stereoselective Contruction of the Tetrahydrofuran Nucleus by Alkoxyl Radical Cyclization, *Eur. J. Org. Chem.* **2001**, 619–632.

Nicolaou, K. C.; Mitchell, H. J. Adventures in Carbohydrate Chemistry. New Synthetic Methodologies, Chemical Synthesis, Molecular Design, and Chemical Biology, *Angew. Chem. Int. Ed.* **2001**, *40*, 1576–1624.

Komarov, I. V.; Borner, A. Highly Enantioselective or Not? Chiral Monodentate Monophosphorus Ligands in the Asymmetric Hydrogenation, *Angew. Che. Int. Ed.* **2001**, *40*, 1197–1200.

O'Donnell, M. J. The Preparation of Optically Active α-Amino Acids from the Benzophenone Imines of Glycine Derivatives, *Aldrichim. Acta* **2001**, *34*, 3–15.

Hayashi, T. Rhodium-Catalyzed Asymmetric 1,4-Addition of Organoboronic Acids and their Derivatives to Electron-Deficient Olefins, *Synlett* **2001**, 879–887.

Enders, D. Efficient Stereoselective Syntheses of Piperidine, Pyrrolidine, and Indolizidine Alkaloids, *Pure Appl. Chem.* **2001**, *73*, 573–578.

Lu, X.; Zhang, Q. Effect of Ligands on the Divalent Palladium-Catalyzed Carbon-Carbon Coupling Reactions. Highly Enantioselective Synthesis of Optically Active γ-Butyrolactones, *Pure Appl. Chem.* **2001**, *73*, 247–250.

Sibi, M. P.; Liu, M. Reversal of Stereochemistry in Enantioselective Transformations. Can They be Planned or are They Just Accidental? *Curr. Org. Chem.* **2001**, *5*, 719–755.

Katsuki, T. The Catalytic Enantioselective Synthesis of Optically Active Epoxides and Tetrahydrofurans. Asymmetric Epoxidation, the Desymmetrization of *meso*-Tetrahydrofurans, and Enantiospecific Ring-Enlargement, *Curr. Org. Chem.* **2001**, *5*, 663–678.

Jotter, N.; Vogel, P. Recent Progress in the Synthesis of Zaragozic Acids and Analogs, *Curr. Org. Chem.* **2001**, *5*, 637–661.

Kumar, R.; Chandra, R. Stereocontrolled Additions to Dihydropyridines and Tetrahydropyridines: Access to *N*-Heterocyclic Compounds Related to Natural Products. *Adv. Heterocycl. Chem.* **2001**, *78*, 269–313.

Bringmann, G.; Menche, D. Stereoselective Total Synthesis of Axially Chiral Natural Products via Biaryl Lactones, *Acc. Chem. Res.* **2001**, *34*, 615–624.

Liang, X.; Anderrsch, J.; Bols, M. Garner's Aldehyde, *J. Chem. Soc., Perkin Trans. 1* **2001**, 3136–3157.

Organic Syntheses Procedures Featuring Chiral, Non-Racemic Reagent Preparation, Volumes 68–78

ALCOHOLS

(1*R*)-1-Methyl-2-ethynyl-*endo*-3,3-dimethyl-2-norbornanol

Midland, M.M.; McLoughlin, J.I.; Werley, R.T., Jr. *Org. Synth.* **1990**, *68*, 14.

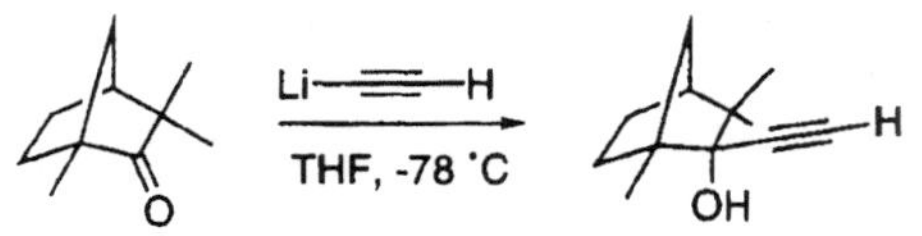

(2*S*,3*S*)-3-Hydroxy-3-phenyl-2-methylpropanoic Acid

Gage, J.R.; Evans, D.A. *Org. Synth.* **1990**, *68*, 83.

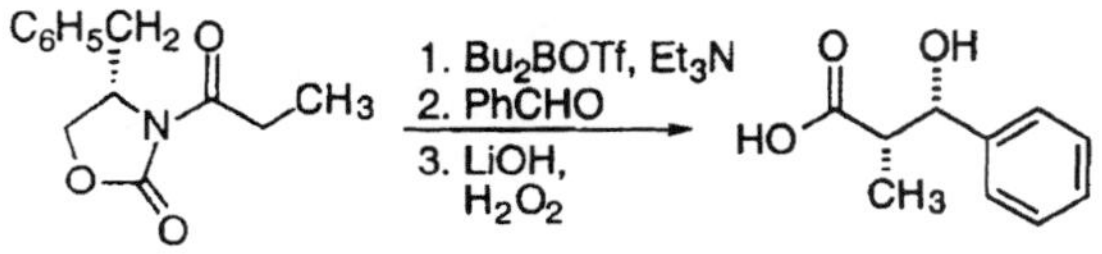

(−)-(1*R*,2*S*)- and (+)-(1*S*,2*R*)-*trans*-2-Phenylcyclohexanol

Schwartz, A.; Madan, P.; Whitesell, J.K.; Lawrence, R.M. *Org. Synth.* **1990**, *69*, 1.

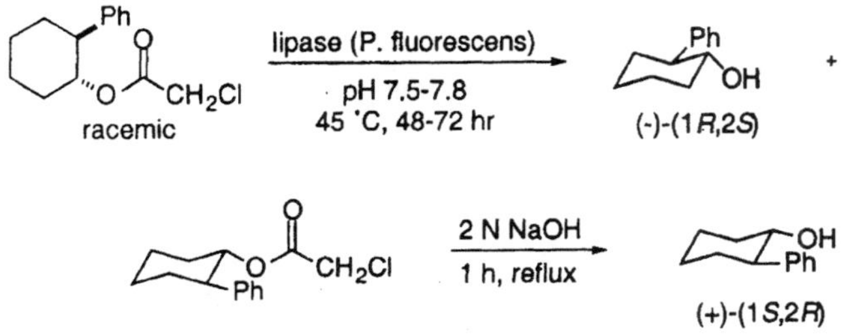

(1*S*,2*S*,3*R*)-3-Hydroxy-2-nitrocyclohexyl Acetate

Eberle, M.; Missbach, M.; Seebach, D. *Org. Synth.* **1990**, *69*, 19.

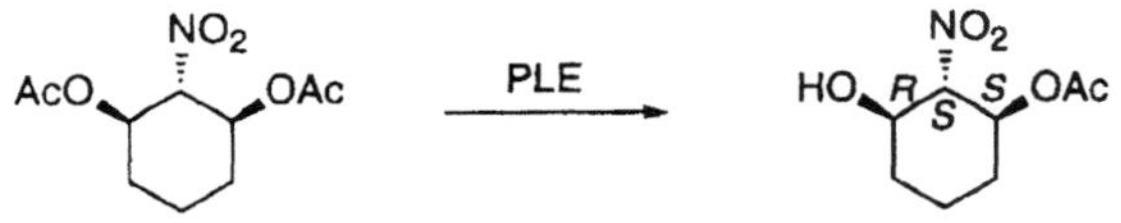

(*R*)-3-Hydroxybutanoic Acid

Seebach, D.; Beck, A.K.; Breitschuh, R.; Job, K. *Org. Synth.* **1993**, *71*, 39.

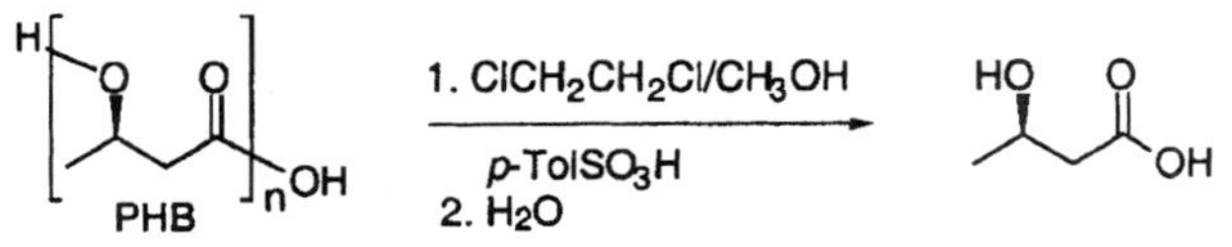

(*R*)-(+)-2-Hydroxy-1,2,2-triphenylethyl Acetate

Braun, M.; Gräf, S.; Herzog, S. *Org. Synth.* **1995**, *72*, 32.

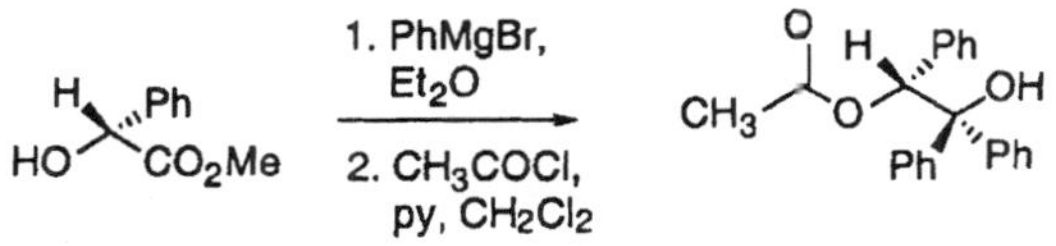

(*R*)-3-Hydroxy-4-methylpentanoic Acid

Braun, M.; Gräf, S. *Org. Synth.* **1995**, *72*, 38.

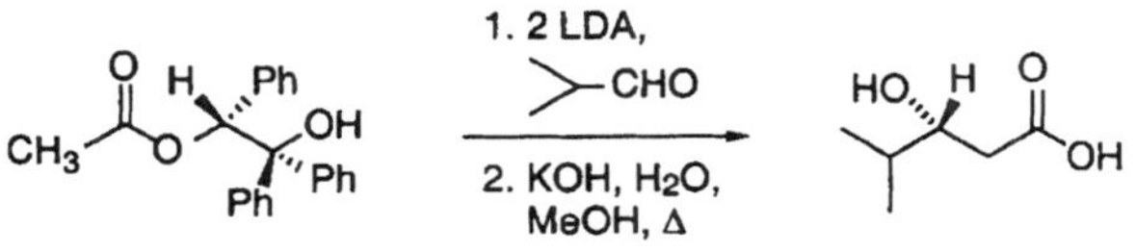

(*S*)-(−)-Citronellol

Takaya, H.; Ohta, T.; Inoue, S.-i.; Tokunaga, M.; Kitamura, M.; Noyori, R. *Org. Synth.* **1995**, *72*, 74.

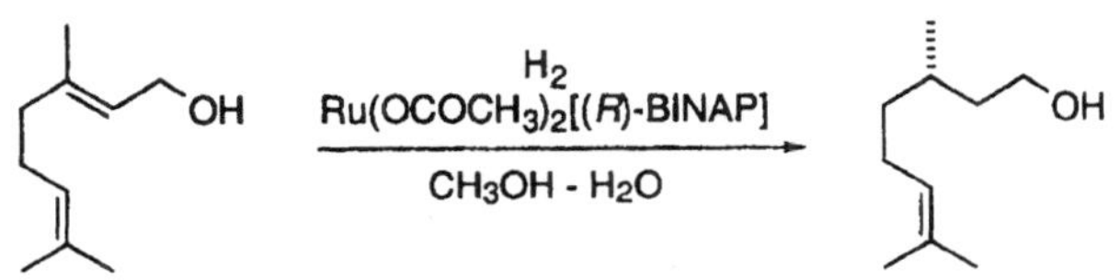

(1*R*,4*S*)-(+)-4-Hydroxy-2-cyclopentenyl Acetate

Deardorff, D.R.; Windham, C.Q.; Craney, C.L. *Org. Synth.* **1996**, *73*, 25.

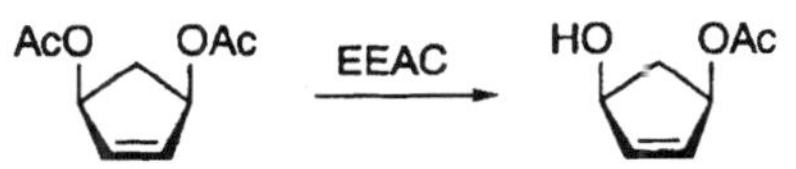

(+)-Isopinocampheol

Kabalka, G.W.; Maddox, J.T.; Shoup, T.; Bowers, K.R. *Org. Synth.* **1996**, *73*, 116.

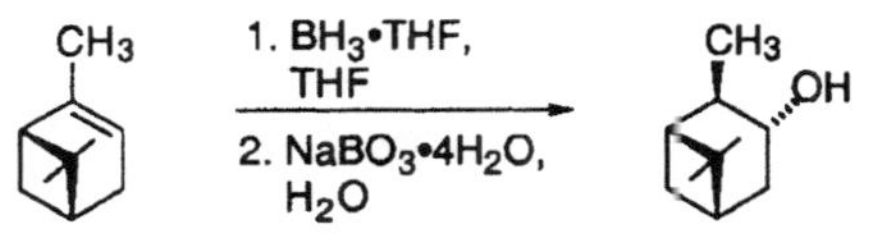

Ethyl (*R*,*E*)-4-*O*-Benzyl-4,5-dihydroxy-2-pentenoate

Steuer, B.; Wehner, V.; Lieberknecht, A.; Jäger, V. *Org. Synth.* **1997**, *74*, 1.

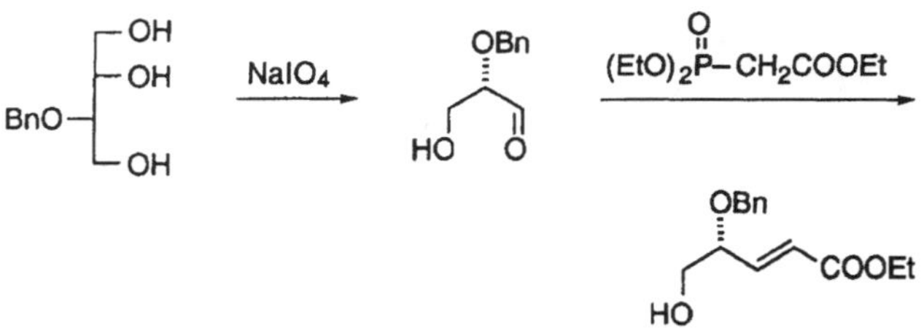

(*R*)-(−)-2,2-Diphenylcyclopentanol

Denmark, S.E.; Marcin, L.R.; Schnute, M.E.; Thorarensen, A. *Org. Synth.* **1997**, *74*, 33.

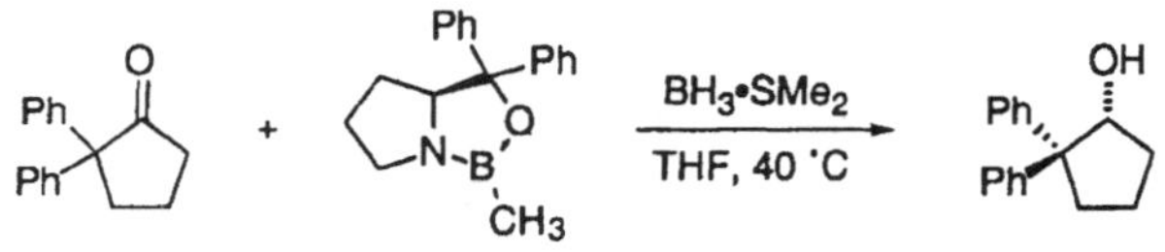

(*R*)-2,3-Dihydro-1*H*-inden-1-ol

Xavier, L.C.; Mohan, J.J.; Mathre, D.J.; Thompson, A.S.; Carroll, J.D.; Corley, E.G.; Desmond, J. *Org. Synth.* **1997**, *74*, 50.

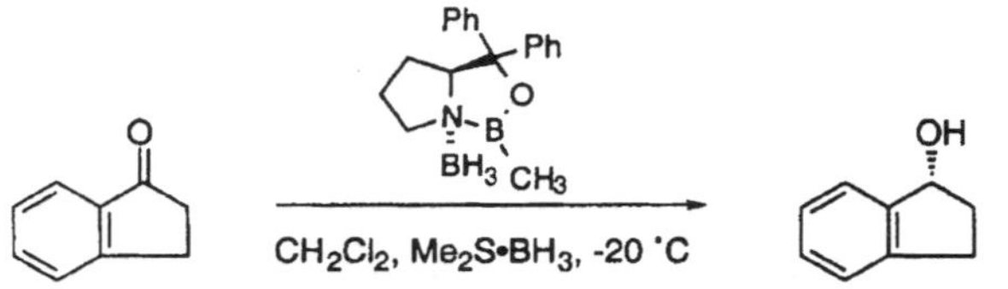

(*S*)-1-(Phenylmethoxy)-4-penten-2-ol

Keck, G.E.; Krishnamurthy, D. *Org. Synth.* **1998**, *75*, 12.

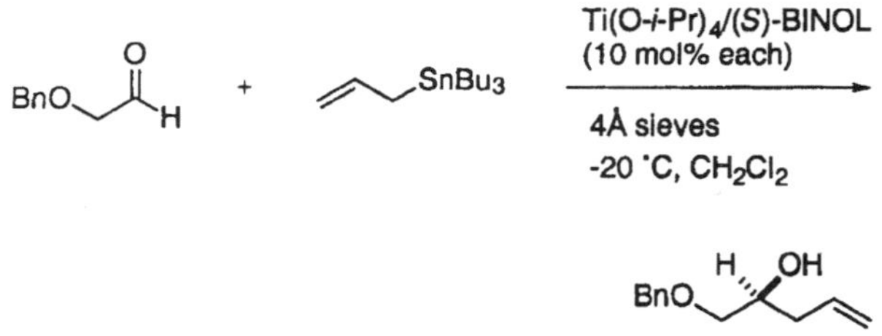

(2*S*,3*S*)-(+)-(3-Phenylcyclopropyl)methanol

Charette, A.B.; Lebel, H. *Org. Synth.* **1999**, *76*, 86.

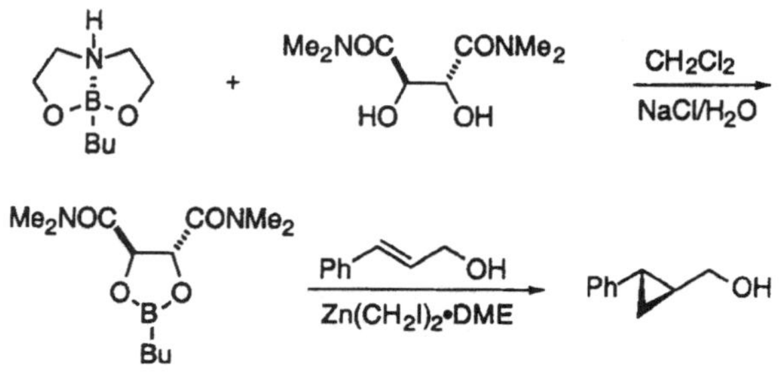

Vitamin D_2

Okabe, M. *Org. Synth.* **1999**, *76*, 275.

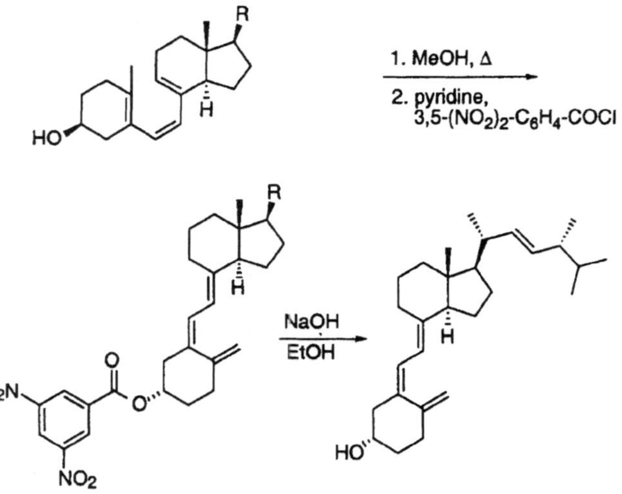

[*R*-(*R,*S**)]-β-Methyl-α-phenyl-1-pyrrolidineethanol**

Zhao, D.; Chen, C.-y.; Xu, F.; Tan, L.; Tillyer, R.; Pierce, M.E.; Moore, J.R. *Org. Synth.* **2000**, *77*, 12.

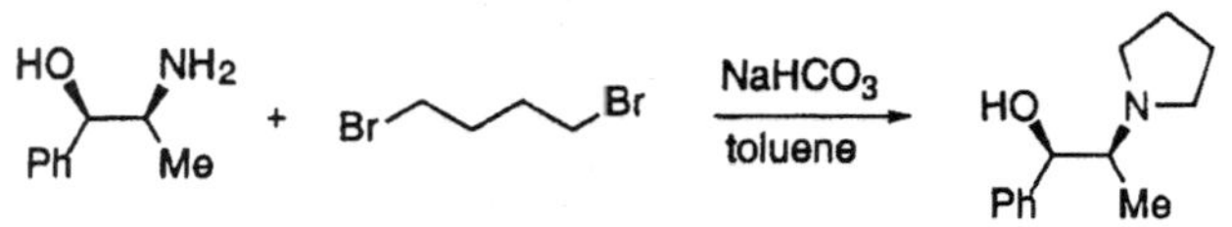

(*R*)-β-Methylbenzenepropanol

Myers, A.G.; Yang, B.H.; Chen, H. *Org. Synth.* **2000**, *77*, 29.

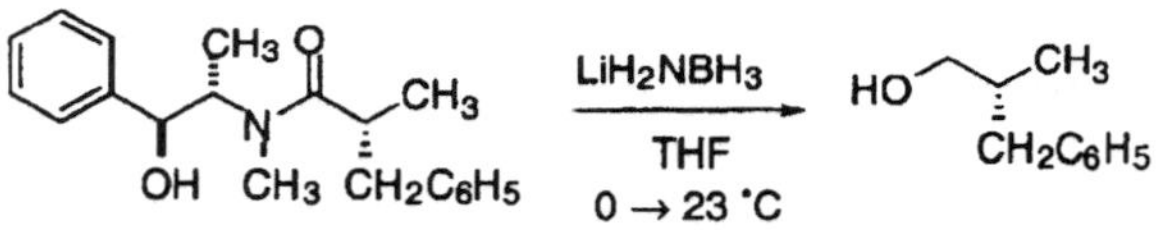

(*R*)-(+)-2-Hydroxy-1,2,2-triphenylethyl Acetate

Macor, J.; Sampognaro, A.J.; Verhoest, P.R.; Mack, R.A. *Org. Synth.* **2000**, *77*, 45.

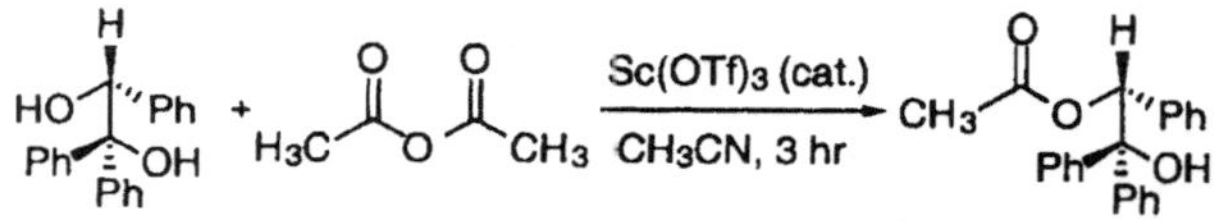

ALDEHYDES

(*S*)-(+)-2-Methylbutanal

Anelli, P.L.; Montanari, F.; Quici, S. *Org. Synth.* **1990**, *69*, 212.

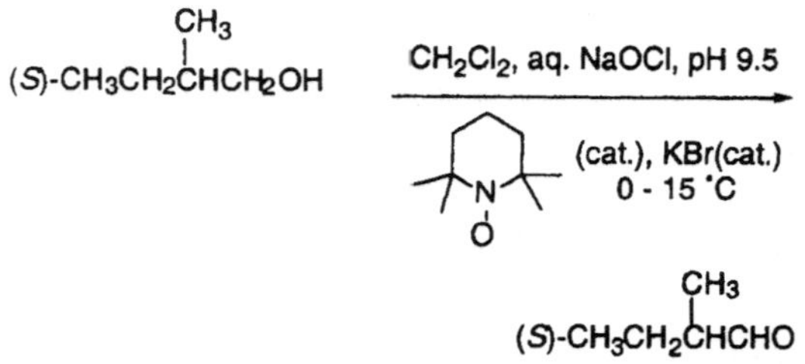

(*R*)-4-Formyl-2,2-dimethyl-3-oxazolidinecarboxylate

Garner, P.; Park, J.M. *Org. Synth.* **1992**, *70*, 18.

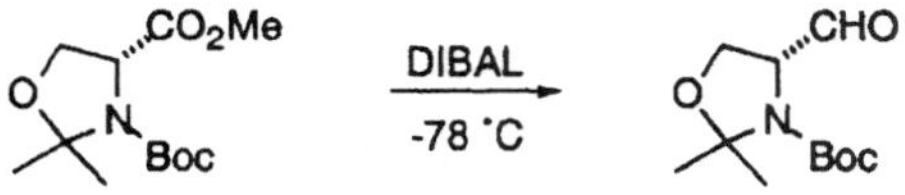

L-(*S*)-Glyceraldehyde Acetonide

Hubschwerlen, C.; Specklin, J.-L.; Higelin, J. *Org. Synth.* **1995**, *72*, 1.

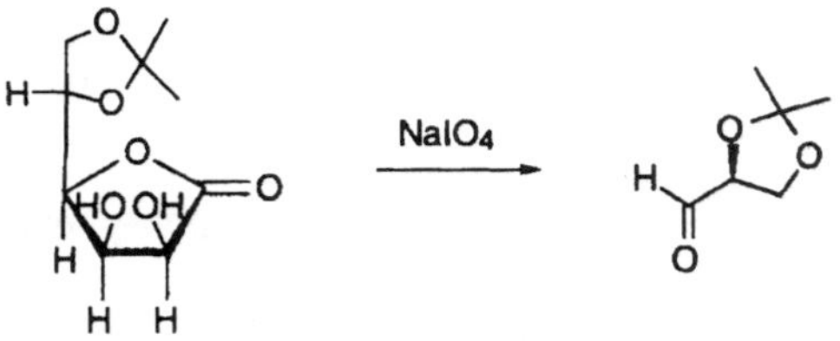

2-*O*-Benzyl-3,4-isopropylidene-D-erythrose

Dondoni, A.; Merino, P., *Org. Synth.* **1995**, *72*, 21.

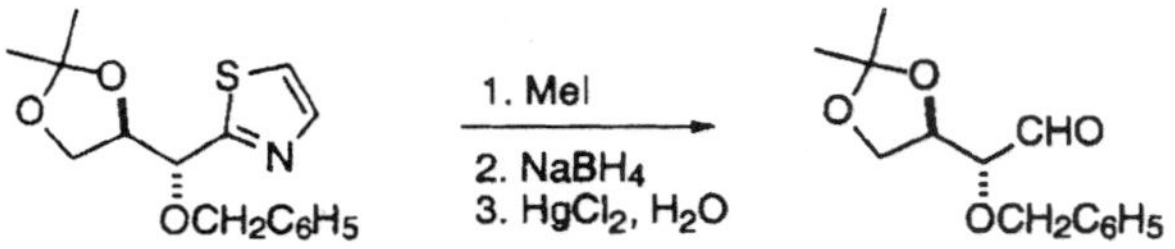

D-(*R*)-Glyceraldehyde Acetonide

Schmid, C.R.; Bryant, J.D. *Org. Synth.* **1995**, *72*, 6.

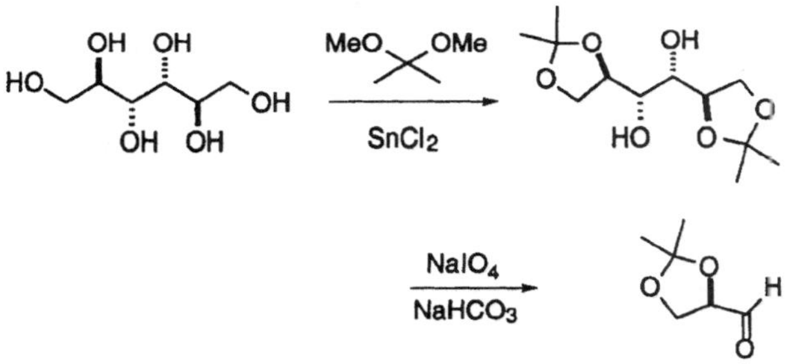

(1*R*)-1,3,4-Trimethyl-3-cyclohexene-1-carboxaldehyde

Furuta, K.; Gao, Q.-z.; Yamamoto, H. *Org. Synth.* **1995**, *72*, 86.

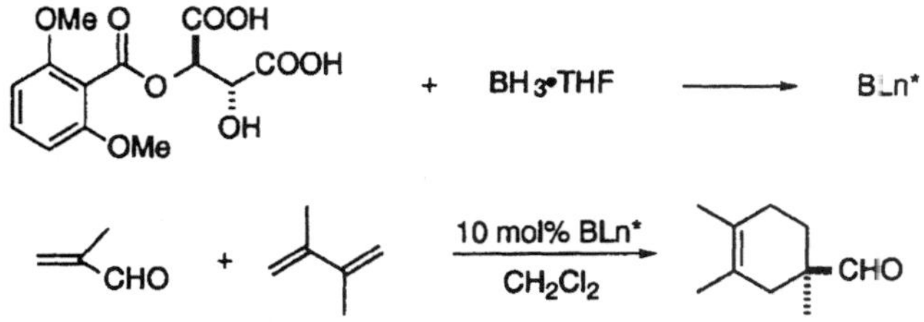

Methyl (*S*)-2-Phthalimido-4-oxobutanoate

Meffre, P.; Durand, P.; Le Goffic, F. *Org. Synth.* **1999**, *76*, 123.

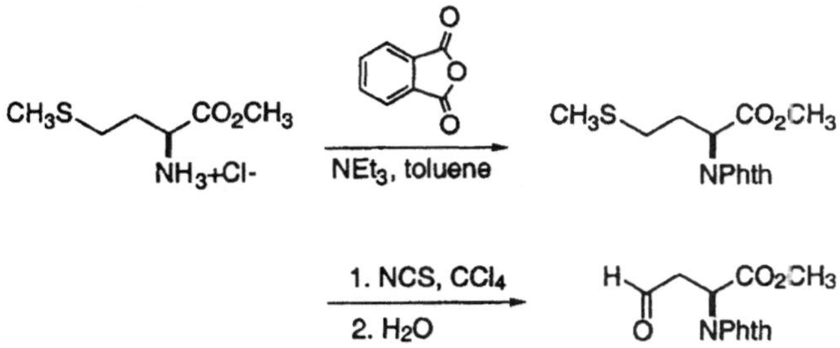

(*S*)-2-(*N*,*N*-Dibenzylamino)-3-phenylpropanal

Reetz, M.T.; Drewes, M.W.; Schwickardi, R. *Org. Synth.* **1999**, *76*, 110.

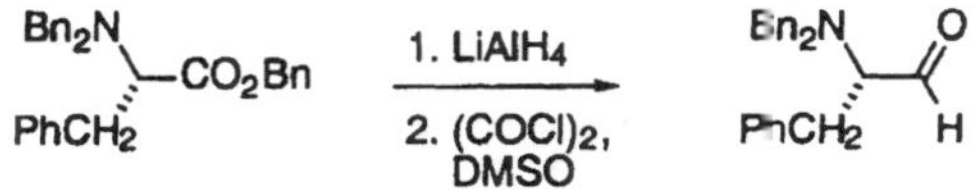

(*R*)-α-Methylbenzenepropanal

Myers, A.G.; Yang, B.H.; Chen, H. *Org. Synth.* **2000**, *77*, 29.

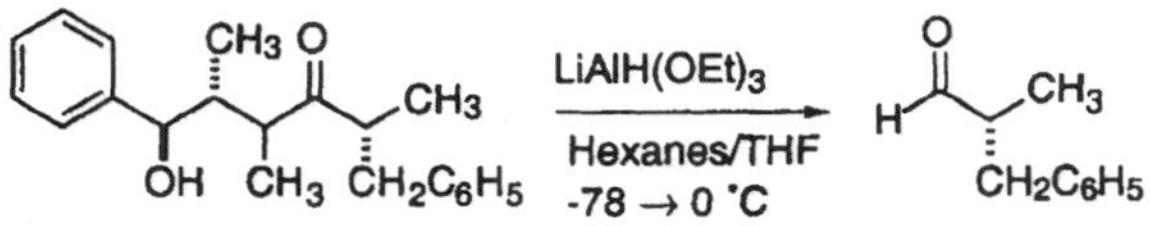

1,1-Dimethylethyl (*S*)-4-Formyl-2,2-dimethyl-3-oxazolidinecarboxylate

Dondoni, A.; Perrone, D. *Org. Synth.* **2000**, *77*, 64.

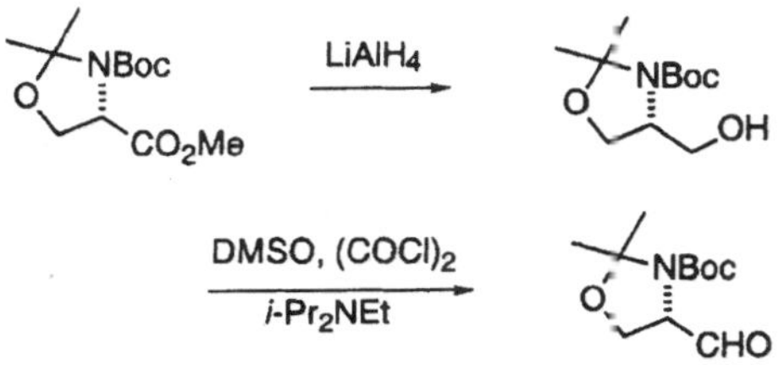

(*S*)-2-[(4*S*)-*N-tert*-Butoxycarbonyl-2,2-dimethyl-1,3-oxazolid-inyl]-2-*tert*-butyldimethylsiloxyethanal

Dondoni, A.; Perrone, D. *Org. Synth.* **2000**, *77*, 78.

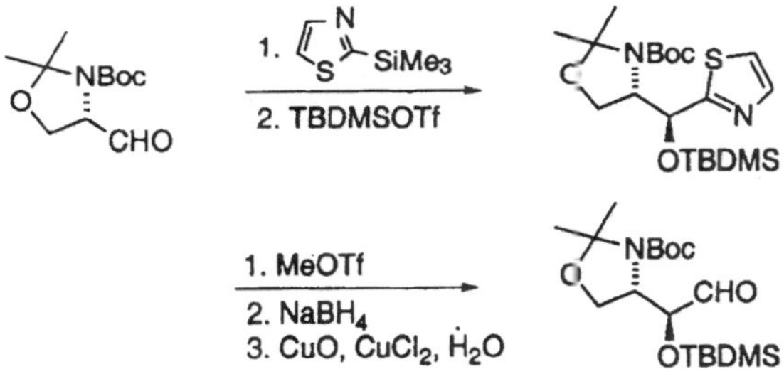

AMIDES and LACTAMS

(3*S*,4*S*)-3-Amino-1-(3,4-dimethoxybenzyl)-4-[(*R*)-2,2-dimethyl-1,3-dioxolan-4-yl]-2-azetidinone

Hubschwerlen, C.; Specklin, J.-L. *Org. Synth.* **1995**, *72*, 14.

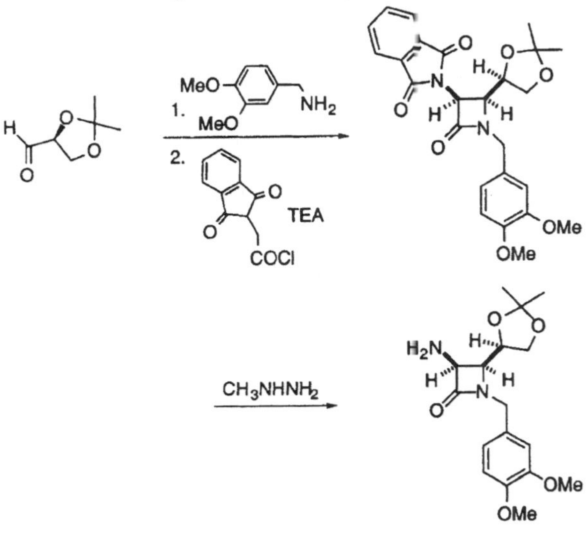

(*S*)-(−)-Heptyl-2-pyrrolidinone

Tschantz, M.A.; Burgess, L.E.; Meyers, A.I. *Org. Synth.* **1996**, *73*, 221.

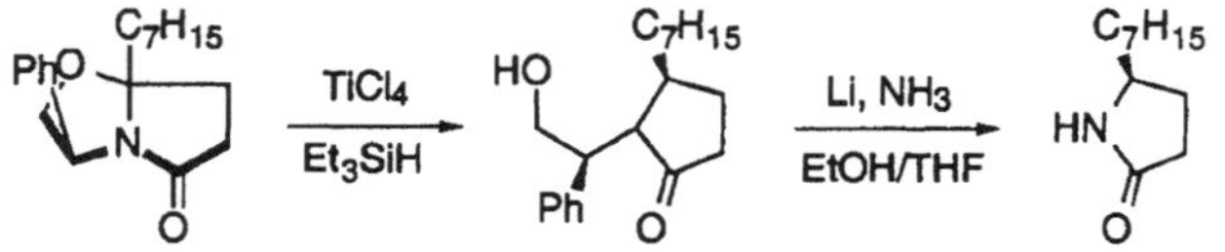

(1*S*,2*S*)-Pseudoephedrine-(*R*)-2-methylhydrocinnamamide

Myers, A.G.; Yang, B.H. *Org. Synth.* **2000**, *77*, 22.

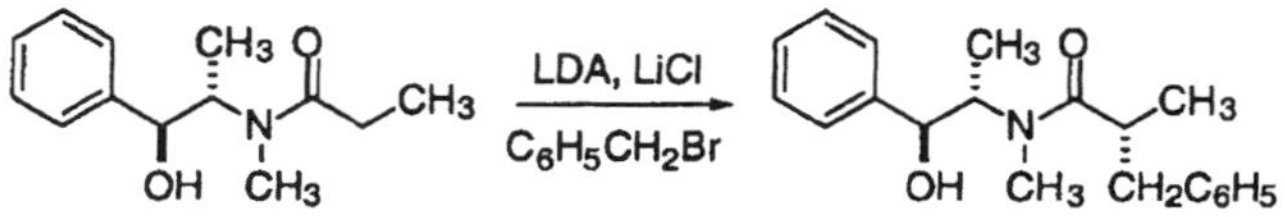

AMINO COMPOUNDS

(1*R*,2*R*)-(+)- and (1*S*,2*S*)-(−)-1,2-Diphenyl-1,2-ethylenediamine

Pikul, S.; Corey, E.J. *Org. Synth.* **1993**, *71*, 22.

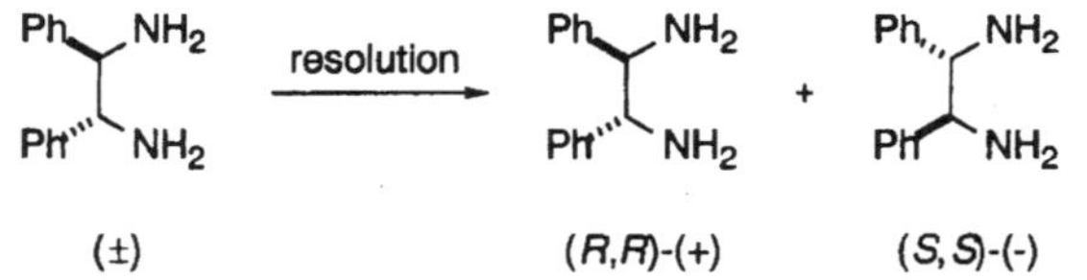

(*S*)-2-Methylproline

Beck, A.K.; Blank, S.; Job, K.; Seebach, D.; Sommerfeld, Th. *Org. Synth.* **1995**, *72*, 62.

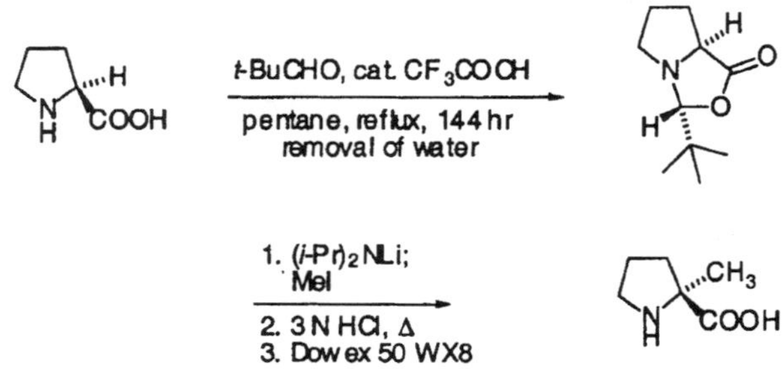

(*R*)-3-Amino-3-(*p*-methoxyphenyl)propionic Acid

Lakner, F.J.; Chu, K.S.; Negrete, G.R.; Konopelski, J.P. *Org. Synth.* **1996**, *73*, 201.

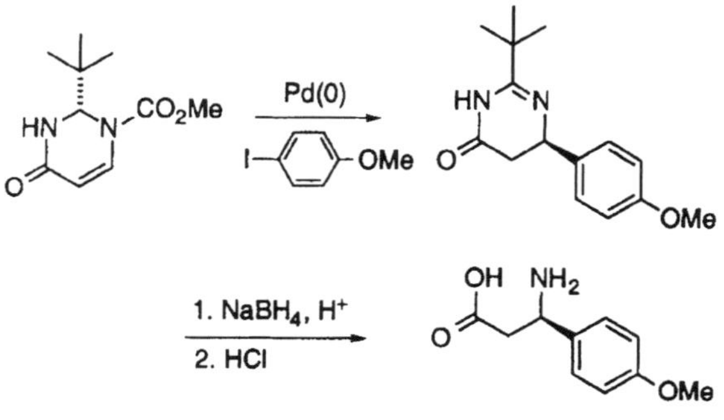

(*R*)-(+)-2-(Diphenylhydroxymethyl)pyrrolidine

Nikolic, N.A.; Beak, P. *Org. Synth.* **1997**, *74*, 23.

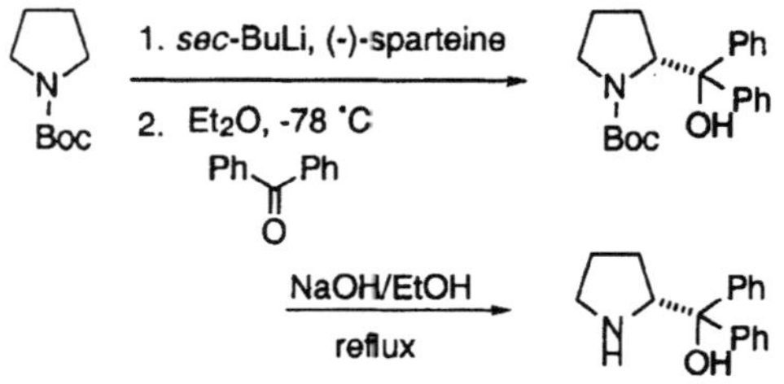

Diethyl (*R*)-(−)-(1-Amino-3-methylbutyl)phosphonate

Smith, A.B. III.; Yager, K.M.; Phillips, B.W.; Taylor, C.M. *Org. Synth.* **1998**, *75*, 19.

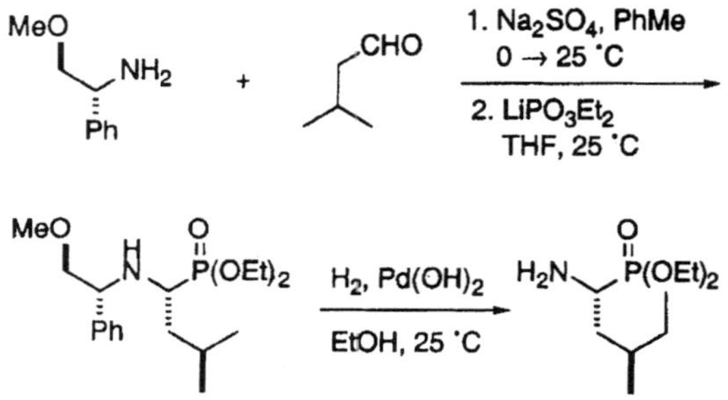

(*R*,*R*)- and (*S*,*S*)-*N*,*N*′-Dimethyl-1,2-diphenylethylene-1,2-diamine

Alexakis, A.; Aujard, I.; Kanger, T.; Mangeney, P. *Org. Synth.* **1999**, *76*, 23.

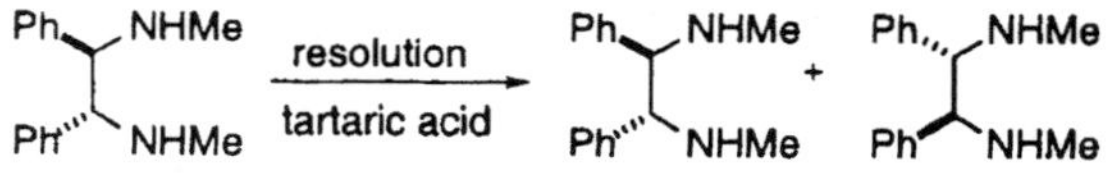

(1*S*,2*R*)-1-Aminoindan-2-ol

Larrow, J.F.; Roberts, E.; Verhoeven, T.R.; Ryan, K.M.; Senanayake, C.H.; Reider, P.J.; Jacobsen, E.N. *Org. Synth.* **1999**, *76*, 46.

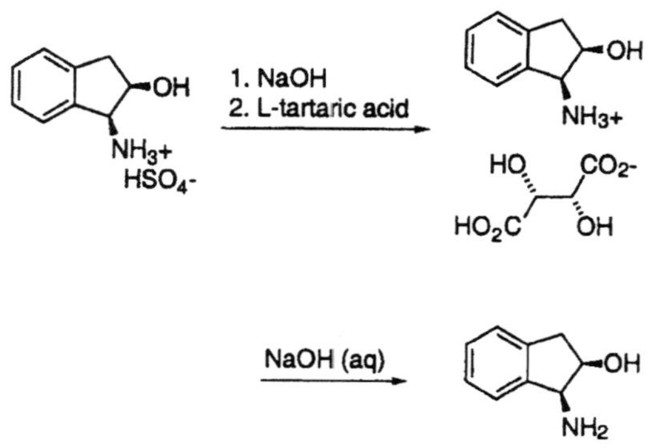

L-Allylglycine

Myers, A.G.; Gleason, J.L. *Org. Synth.* **1999**, *76*, 57.

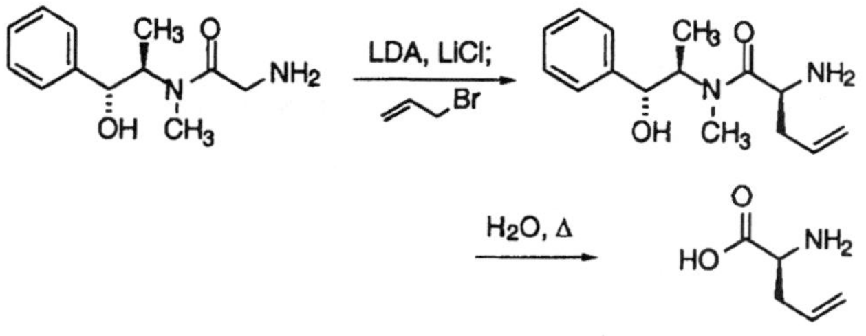

CARBOHYDRATE DERIVATIVES

1,3,4,6-Tetra-*O*-acetyl-2-deoxy-α-D-glucopyranose

Giese, B.; Gröninger, K.S. *Org. Synth.* **1990**, *69*, 66.

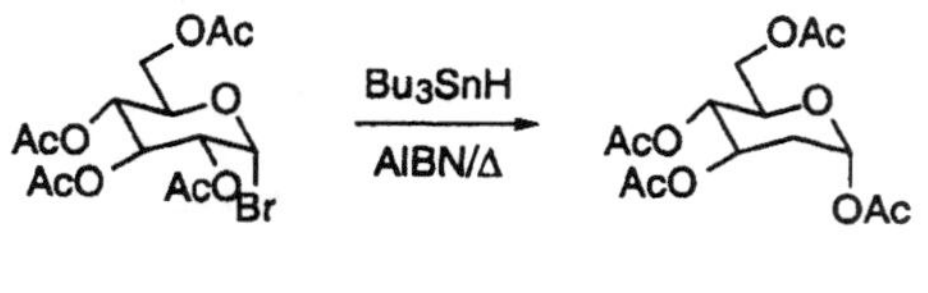

(1′*S*,2′*S*)-Methyl-3*O*,4*O*-(1′,2′-dimethoxy-cyclohexane-1′,2′-diyl)-α-D-mannopyranoside

Ley, S.V.; Osborn, H.M.I.; Priepke, H.W.M.; Warriner, S.L.; *Org. Synth.* **1998**, *75*, 170.

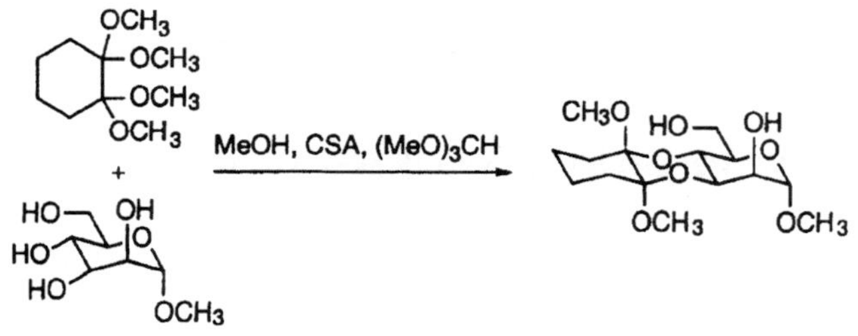

***O*⁴,*O*⁵-Isopropylidene-1,2:3,6-dianhydro-D-glucitol**

Ejjiyar, S.; Saluzzo, C.; Amouroux, R. *Org. Synth.* **2000**, *77*, 91.

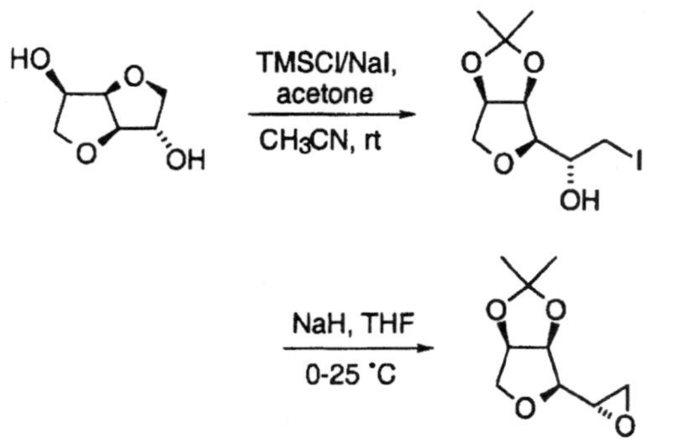

Methyl 2,3-*O*-(6,6′-Octahydro-6,6′-bi-2*H*-pyran-2,2′-diyl)-α-D-galactopyranoside

Ley, S.V.; Osborn, H.M.I. *Org. Synth.* **2000**, *77*, 212.

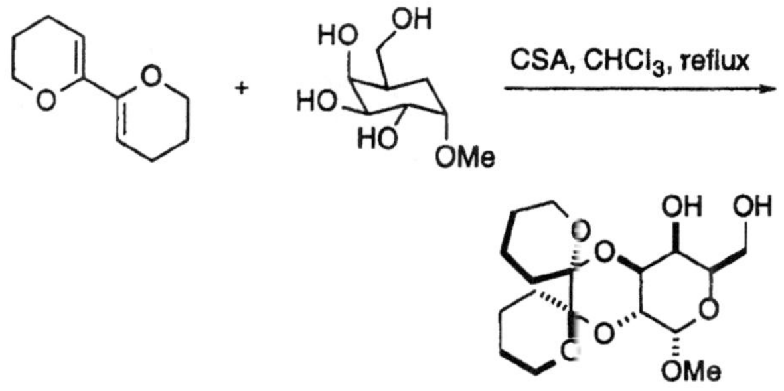

3-Deoxy-1,2:5,6-bis-*O*-(1-methylethylidene)-α-D-ribohexofuranose

Tormo, J.; Fu, G. C. *Org. Synth.* **2001**, *78*, 239.

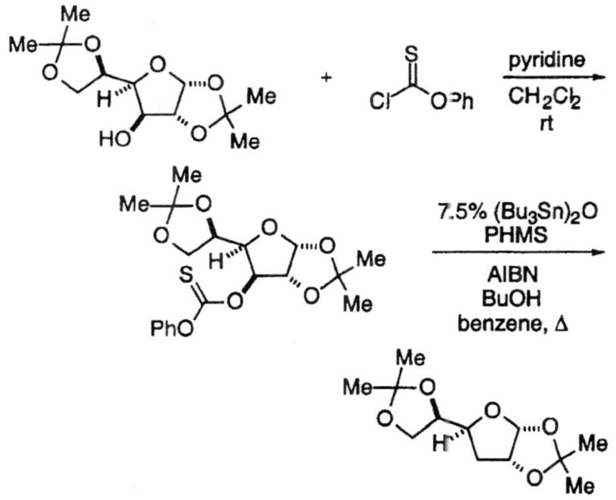

DIOLS

1,4-Di-*O*-benzyl-L-threitol

Mash, E.A.; Nelson, K.A.; Van Deusen, S.; Hemperly, S.B. *Org. Synth.* **1990**, *68*, 92.

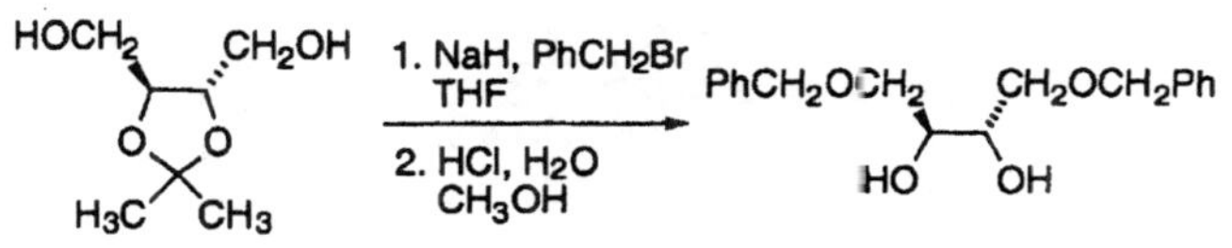

(*R*,*R*)-1,2-Diphenyl-1,2-ethanediol

McKee, B.H.; Gilheany, D.G.; Sharpless, K.B. *Org. Synth.* **1992**, *70*, 47.

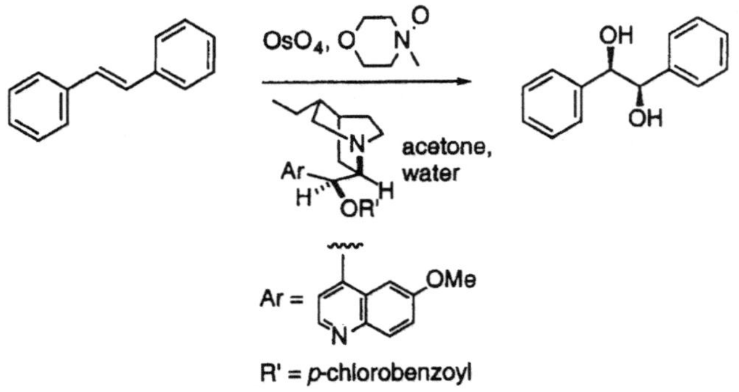

(*S*)-(−)- and (*R*)-(+)-1,1′-Bi-2-naphthol

Kazlauskas, R.J. *Org. Synth.* **1992**, *70*, 60.

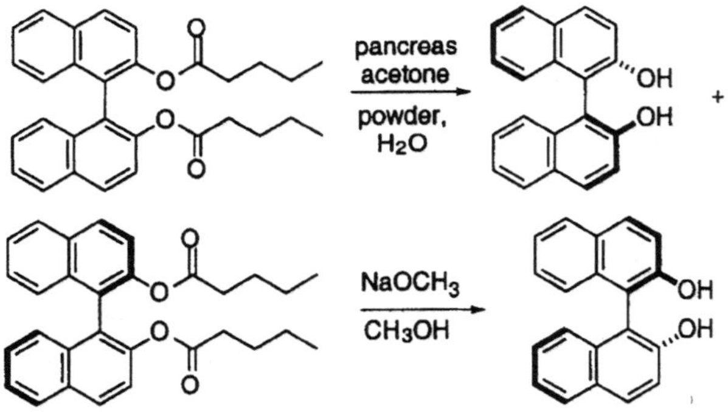

3-[(1S)-1,2-Dihydroxyethyl]-1,5-dihydro-3*H*-2,4-benzodioxepine

Oi, R.; Sharpless, K.B. *Org. Synth.* **1996**, *73*, 1.

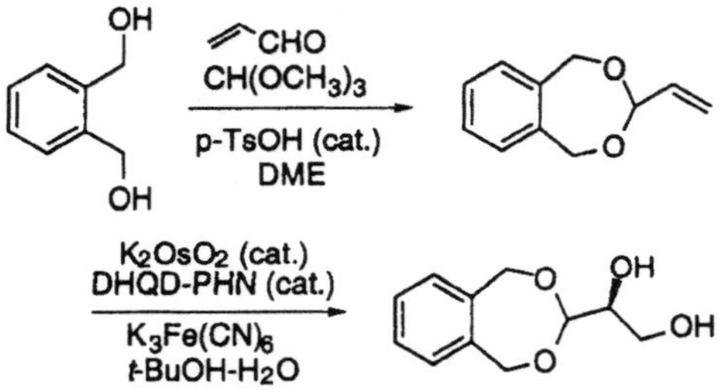

16α-Methylcortexolone

Horiguchi, Y.; Nakamura, E.; Kuwajima, I. *Org. Synth.* **1996**, *73*, 123.

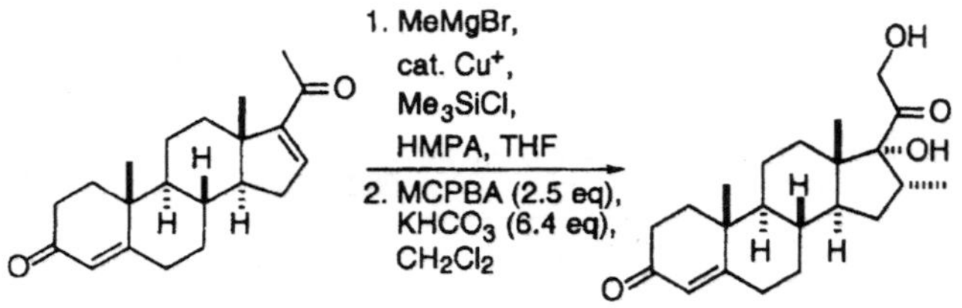

(*R*)-(+)- and (*S*)-(−)-1,1′-Bi-2-naphthol

Cai, D.; Hughes, D.L.; Verhoeven, T.R.; Reider, P.J. *Org. Synth.* **1999**, *76*, 1.

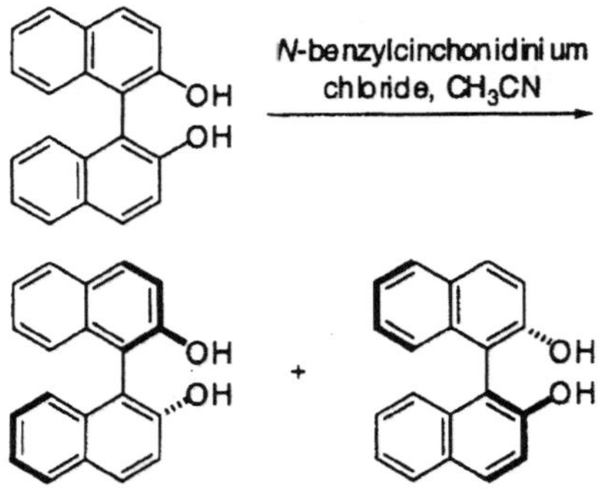

(*R*)-(+) (*S*)-(-)

(4*R*,5*R*)-2,2-Dimethyl-α,α,α′,α′-tetra(naphth-2-yl)-1,3-dioxolane-4,5-dimethanol

Beck, A.K.; Gysi, P.; La Vecchia, L.; Seebach, D. *Org. Synth.* **1999**, *76*, 12.

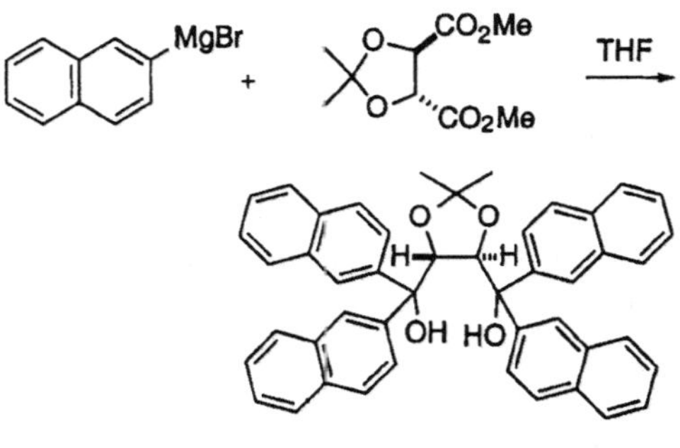

1-Chloro-(2*S*,3*S*)-dihydroxycyclohexa-4,6-diene

Hudlicky, T.; Stabile, M.R.; Gibson, D.T.; Whited, G.M.; *Org. Synth.* **1999**, *76*, 77.

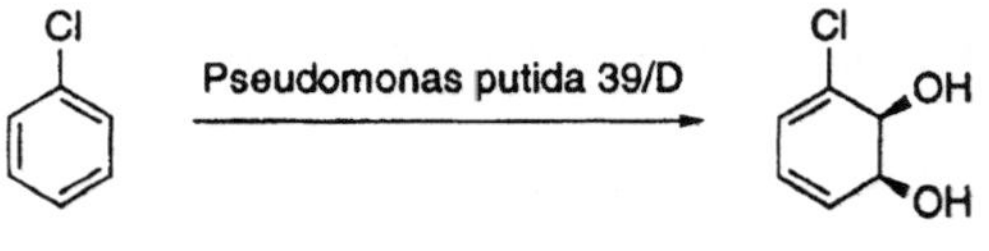

(2*S*,3*S*)-Dihydroxy-1,4-diphenylbutane

Robbins, M.A.; Devine, P.N.; Oh, T. *Org. Synth.* **1999**, *76*, 101.

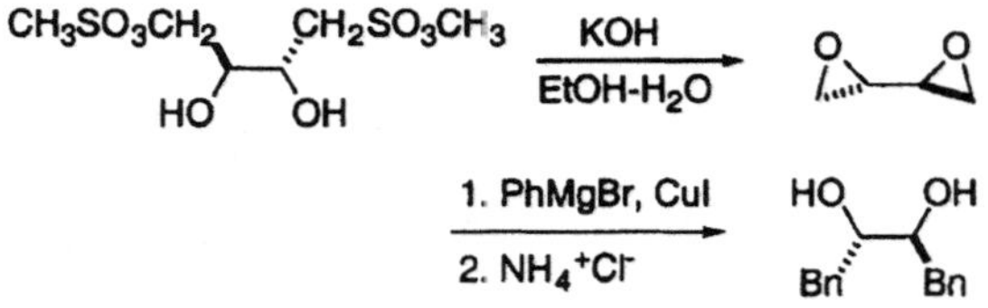

(2*S*,3*S*)-2-Nitro-5-phenyl-1,3-pentanediol

Sasai, H.; Watanabe, S.; Suzuki, T.; Shibasaki, M. *Org. Synth.* **2001**, *78*, 14.

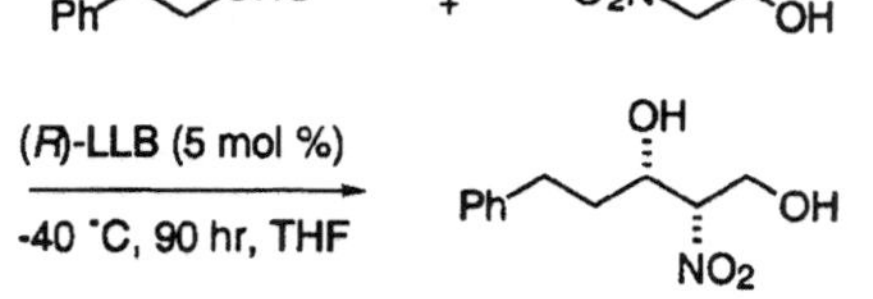

ESTERS and LACTONES

(−)-Menthyl Cinnamate

Meth-Cohn, O. *Org. Synth.* **1990**, *68*, 155.

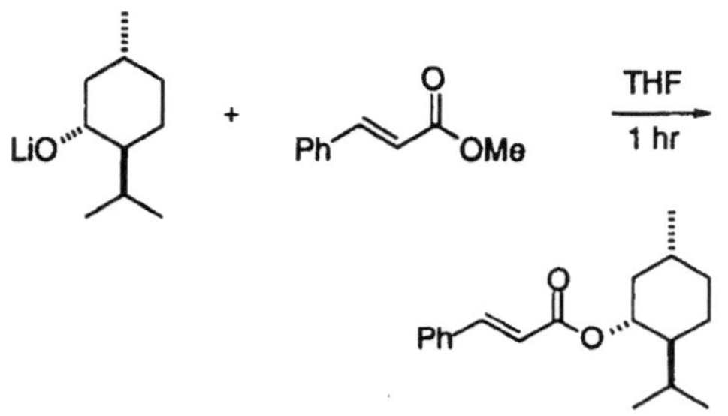

Ethyl (*R*)- and (*S*)-2-Fluorohexanoate

Kalaritis, P.; Regenye, R.W. *Org. Synth.* **1990**, *69*, 10.

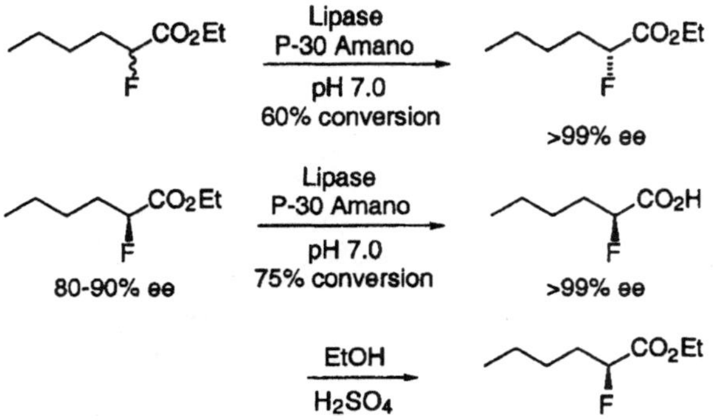

(4*S*, 4a*S*, 6*S*, 8a*S*)-4-Methoxycarbonyl-1, 1, 6-trimethyl-1, 4, 4a, 5, 6, 7, 8, 8a-octahydro-2,3-benzopyrone

Tietze, L.F.; Kiedrowski, G.V.; Fahlbusch, K.-G.; Voss, E. *Org. Synth.* **1990**, *69*, 31.

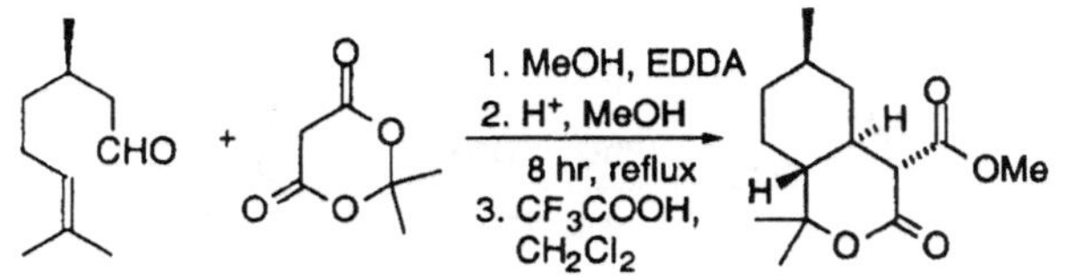

(*S*)-3-Amino-2-oxetanone *p*-Toluenesulfonate Salt

Pansare, S.V.; Arnold, L.D.; Vederas, J.C. *Org. Synth.* **1992**, *70*, 10.

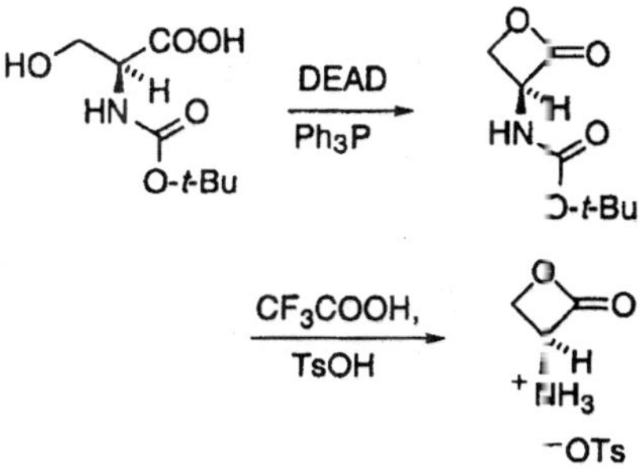

***N*-(Benzyloxycarbonyl)-L-vinylglycine Methyl Ester**

Carrasco, M.; Jones, R.J.; Kamel, S.; Rapoport, H.; Truong, T. *Org. Synth.* **1992**, *70*, 29.

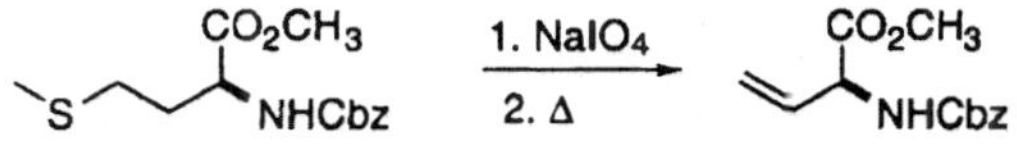

(*R*)-(−)-Methyl 3-Hydroxybutanoate

Kitamura, M.; Tokunaga, M.; Ohkuma, T.; Noyori, R. *Org. Synth.* **1993**, *71*, 1.

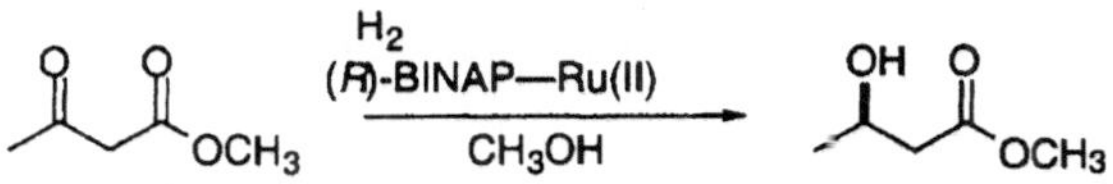

Methyl (2*R*)-2-Hydroxy-4-phenyl-4-pentenoate

Mikami, K.; Terada, M.; Narisawa, S.; Nakai, T. *Org. Synth.* **1993**, *71*, 14.

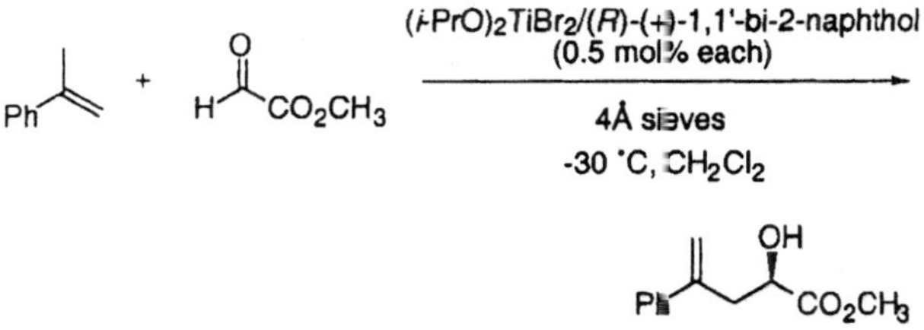

(−)-(1*S*, 4*R*)-Camphanoyl Chloride

Gerlach, H.; Kappes, D.; Boeckman, R.K. Jr.; Maw, G.N. *Org. Synth.* **1993**, *71*, 48.

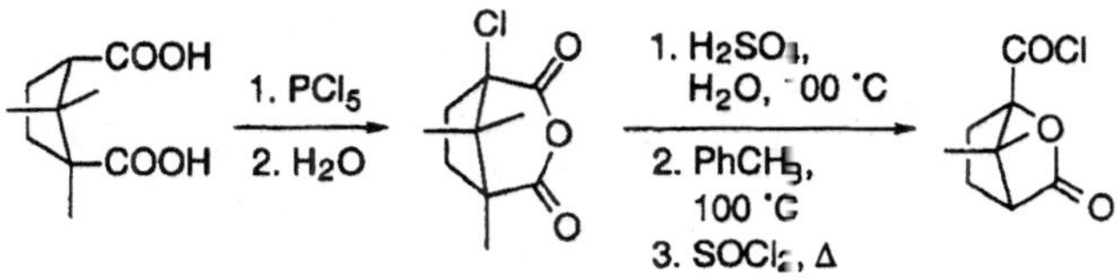

Dimethyl (1′*R*, 2′*R*, 5′*R*)-2-(2′-Isopropenyl-5′-methylcyclohex-1′-yl)propane-1, 3-dioate

Tietze, L.F.; Beifuss, U. *Org. Synth.* **1993**, *71*, 167.

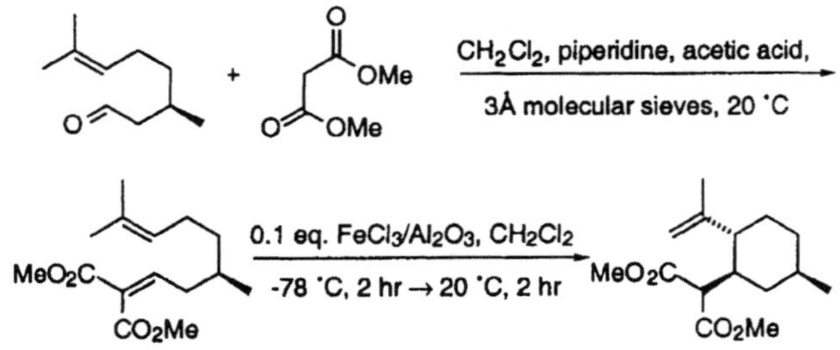

(3a*S*, 7a*R*)-Hexahydro-(3*S*, 6*R*)-dimethyl-2(3*H*)-benzofuranone

Jefford, C.W.; Li, Y.; Wang, Y. *Org. Synth.* **1993**, *71*, 207.

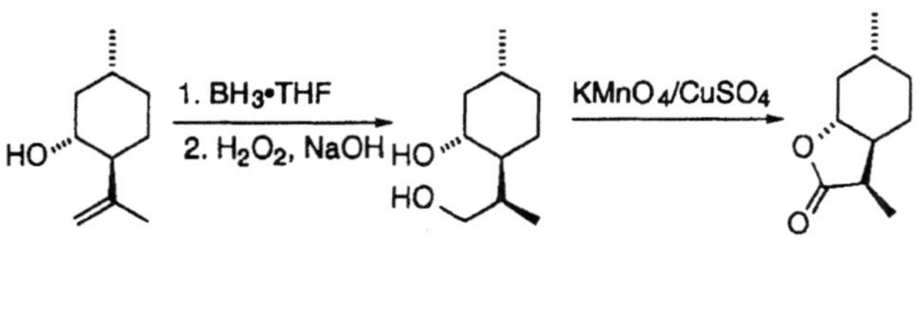

(*S*)-Dimethyl *N*-(9-Phenylfluoren-9-yl)aspartate

Jamison, T.F.; Rapoport, H. *Org. Synth.* **1993**, *71*, 226.

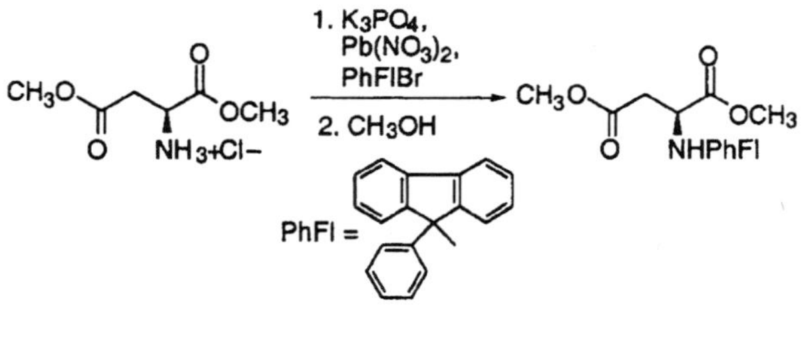

(2*S*, 4*S*)-2, 4, 5-Trihydroxypentanoic Acid 4, 5-Acetonide Methyl Ester

Sun, R.C.; Okabe, M. *Org. Synth.* **1995**, *72*, 48.

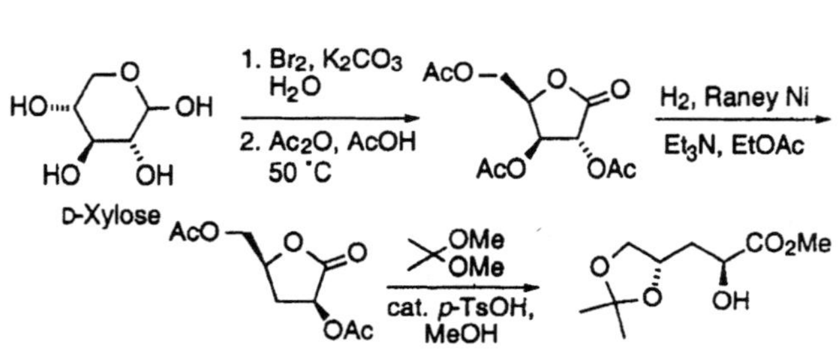

(1*R*, 5*S*)-(−)-6,6-Dimethyl-3-oxabicyclo[3.1.0]-hexan-2-one

Doyle, M.P.; Winchester, W.R.; Protopopova, M.N.; Kazala, A.P.; Westrum, L.J. *Org. Synth.* **1996**, *73*, 13.

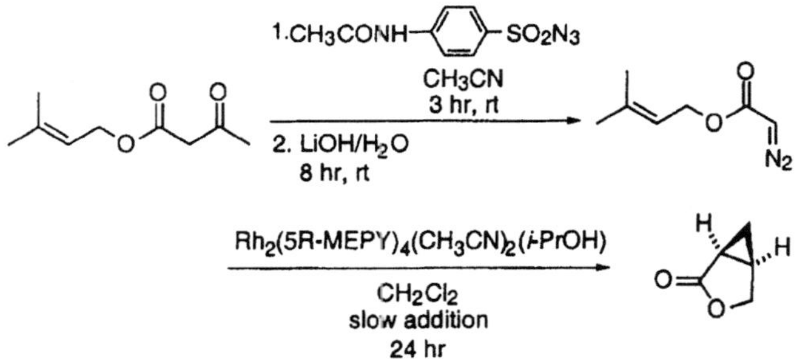

(1*S*, 2*S*, 5*R*)-5-Methyl-2-(1-methylethyl)cyclohexyl 4-Nitrobenzoate

Dodge, J.A.; Nissen, J.S.; Presnell, M. *Org. Synth.* **1996**, *73*, 110.

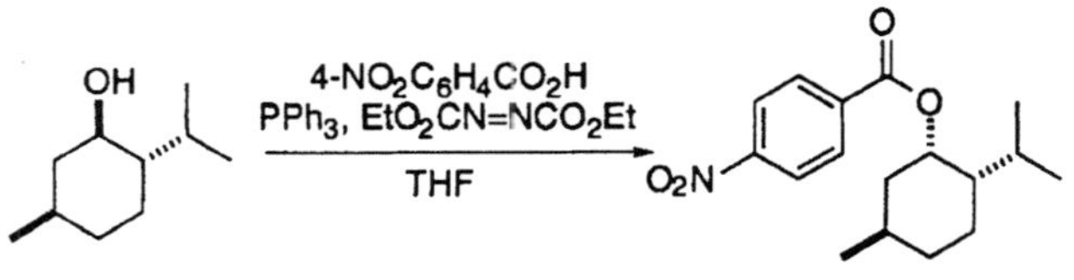

Diethyl (2*S*, 3*R*)-2-(*N*-*tert*-Butoxycarbonyl)amino-3-hydroxysuccinate

Saito, S.; Komada, K.; Moriwake, T. *Org. Synth.* **1996**, *73*, 184.

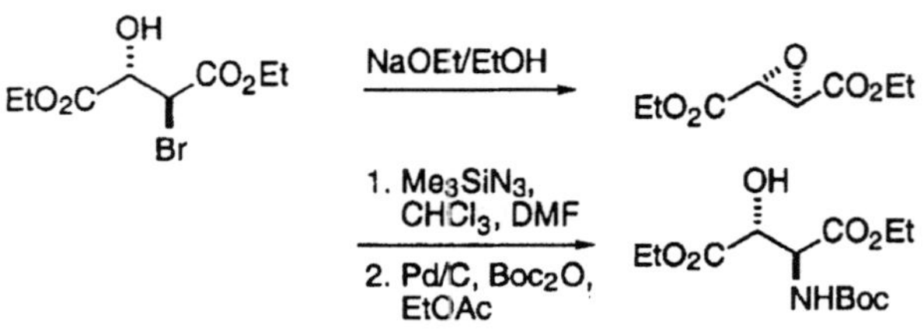

Ethyl (*R*)-2-Azidopropionate

Thompson, A.S.; Hartner, F.W. Jr.; Grabowski, E.J.J. *Org. Synth.* **1998**, *75*, 31.

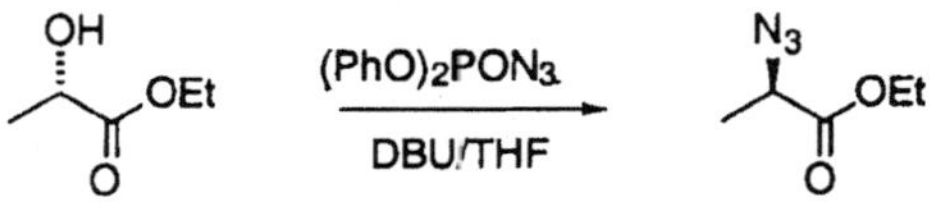

(4*R*, 5*S*)-4-Hydroxymethyl-(5,*O*-*tert*-butyldimethylsiloxy-methyl)furan-2(5*H*)-one

Mann, J.; Weymouth-Wilson, A.C. *Org. Synth.* **1998**, *75*, 139.

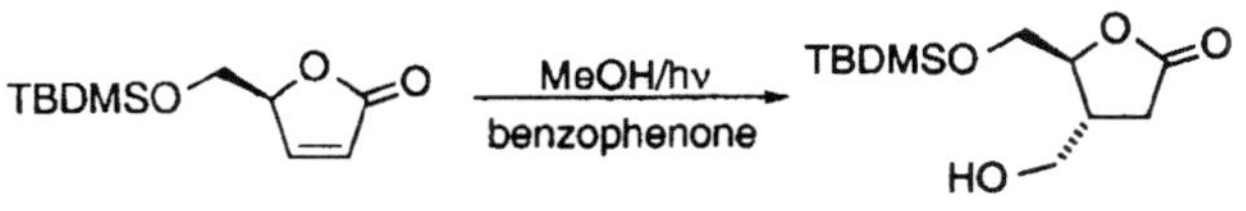

[3*R*- and 3*S*]-(4*E*)-Methyl 3-(Dimethylphenylsilyl)-4-hexenoate

Beresis, R.T.; Solomon, J.S.; Yang, M.G.; Jain, N.F.; Penek, J.S. *Org. Synth.* **1998**, *75*, 78.

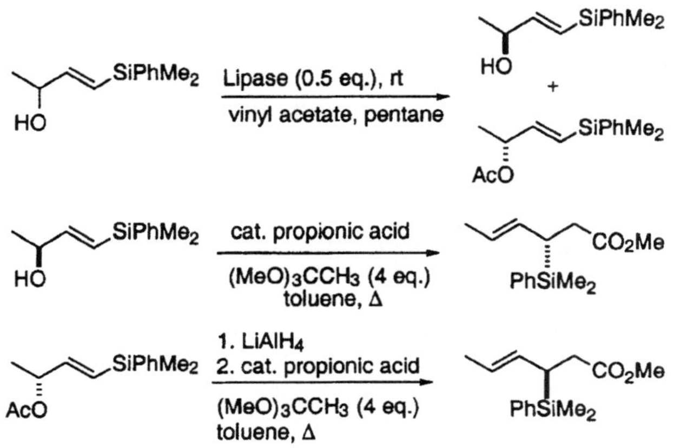

Methyl (*R*)-(+)-β-Phenylalanate

Fanelli, D.L.; Szewczyk, J.M.; Zhang, Y.; Reddy, G.V.; Burns, D.M.; Davis, F. A. *Org. Synth.* **2000**, *77*, 50.

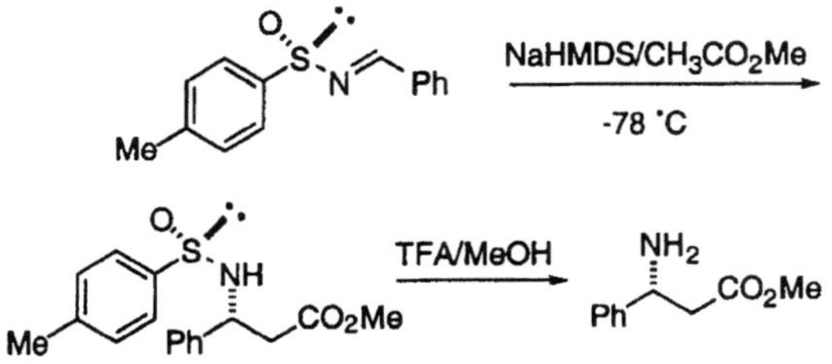

Methyl (*S*)-2-Isocyanato-3-phenylpropanoate

Tsai, J.H.; Takaoka, L.R.; Powell, N.A.; Nowick, J.S. *Org. Synth.* **2001**, *78*, 220.

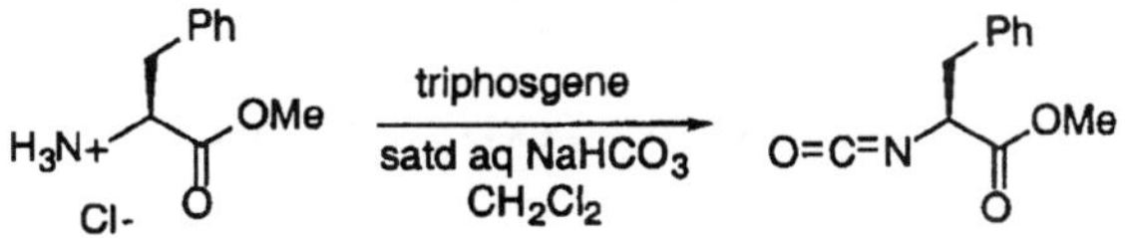

Ethyl (*E*)-(−)-4,6-*O*-Ethylidene-(4*S*, 5*R*, 1′*R*)-4,5,6-Trihydroxy-2-hexenoate

Fengler-Veith, M.; Schwardt, O.; Kautz, U.; Krämer, B.; Jäger, V. *Org. Synth.* **2001**, *78*, 123.

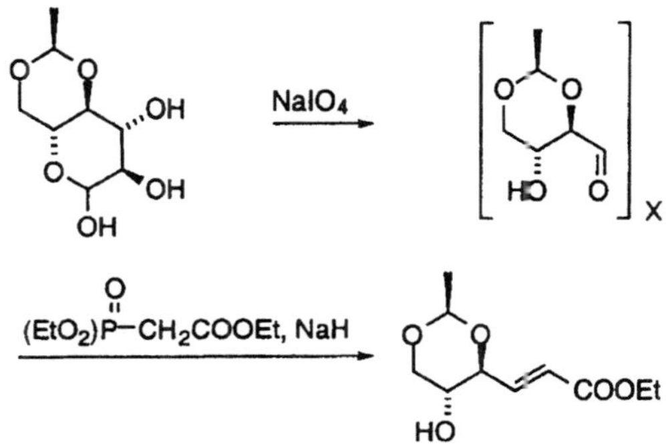

HETEROCYCLES

(*S*)-4-(Phenylmethyl)-2-oxazolidinone

Gage, J.R.; Evans, D.A. *Org. Synth.* **1990**, *68*, 77.

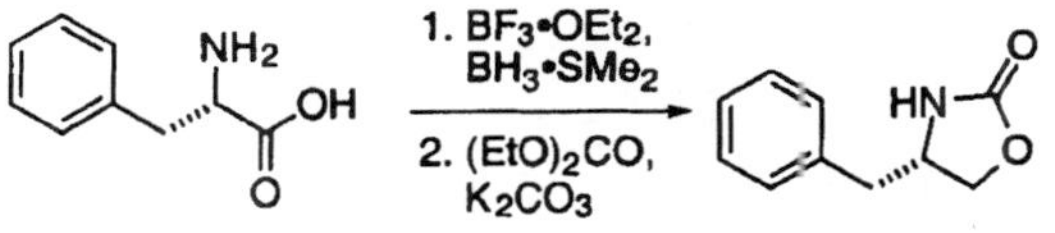

(−)-Menthyl Nicotinate

Meth-Cohn, O. *Org. Synth.* **1990**, *68*, 155.

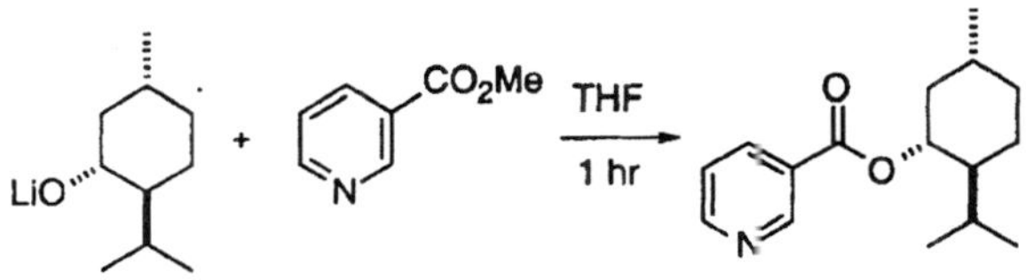

***N*$^{\alpha}$-(Benzyloxycarbonyl)-β-(pyrazol-1-yl)-L-alanine**

Pansare, S.V.; Huyer, G.; Arnold, L.D.; Vederas, J.C. *Org. Synth.* **1992**, *70*, 1.

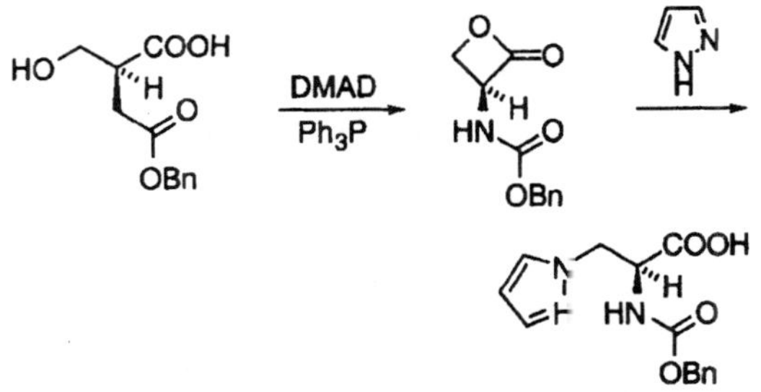

2-Cyano-6-phenyloxazolopiperidine

Bonin, M.; Grierson, D.S.; Royer, J.; Husson, H.-P. *Org. Synth.* **1992**, *70*, 54.

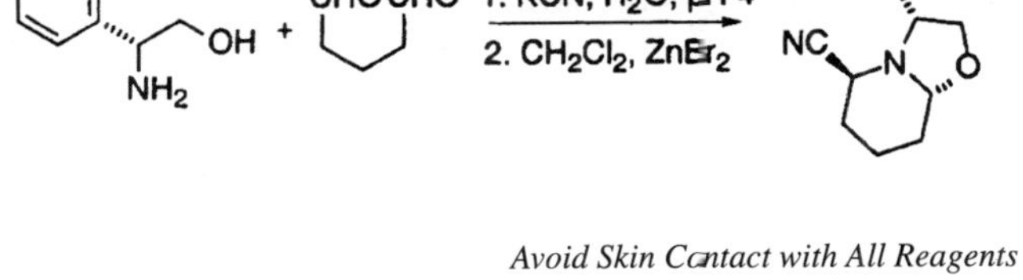

(1*S*-*endo*)-3-(Bicyclo[2.2.1]hept-5-en-2-ylcarbonyl)-2-oxazolidinone

Pikul, S.; Corey, E.J. *Org. Synth.* **1993**, *71*, 30.

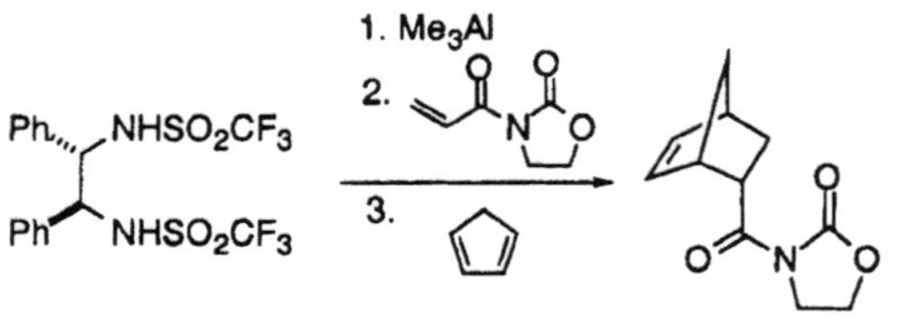

(2*S*, 3*S*)-3-Acetyl-8-carboethoxy-2,3-dimethyl-1-oxa-8-azaspiro[4.5]decane

Overman, L.E.; Rishton, G.M. *Org. Synth.* **1993**, *71*, 63.

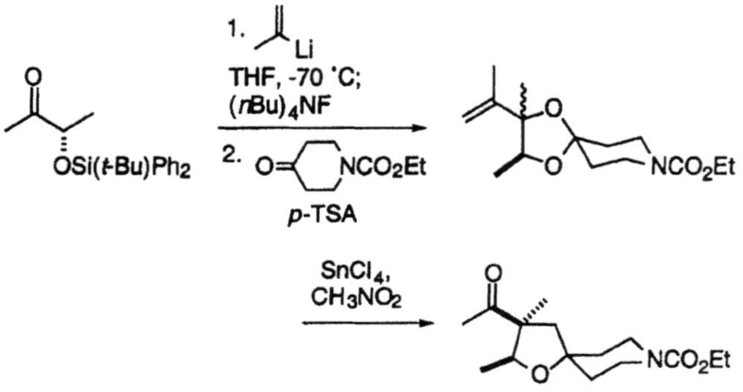

Ethyl (*R*)-(+)-2,3-Epoxypropanoate

Petit, Y.; Larchevêque, M. *Org. Synth.* **1998**, *75*, 37.

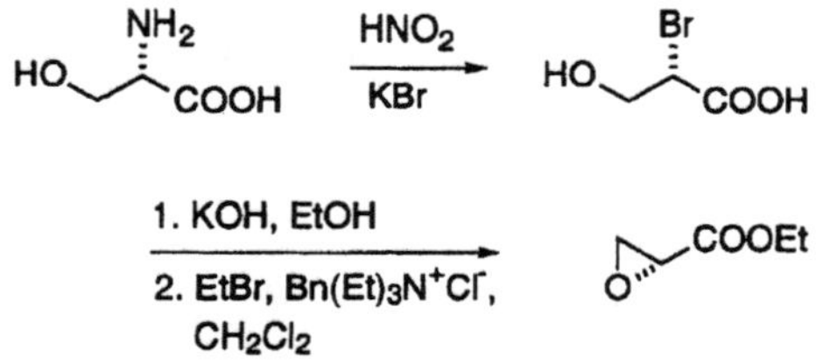

(4*R*, 5*S*)-4,5-Diphenyl-3-vinyl-2-oxazolidinone

Akiba, T.; Tamura, O.; Terashima, S. *Org. Synth.* **1998**, *75*, 45.

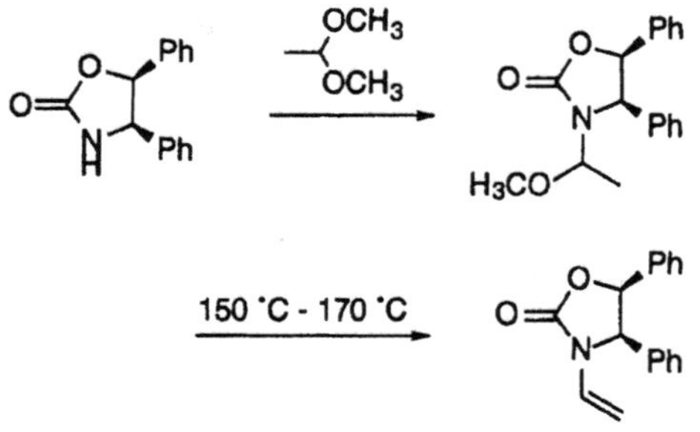

(*R*,*R*)-1,2:4,5-Diepoxypentane

Rychnovsky, S.D.; Griesgraber, G.; Powers, J.P. *Org. Synth.* **2000**, *77*, 1.

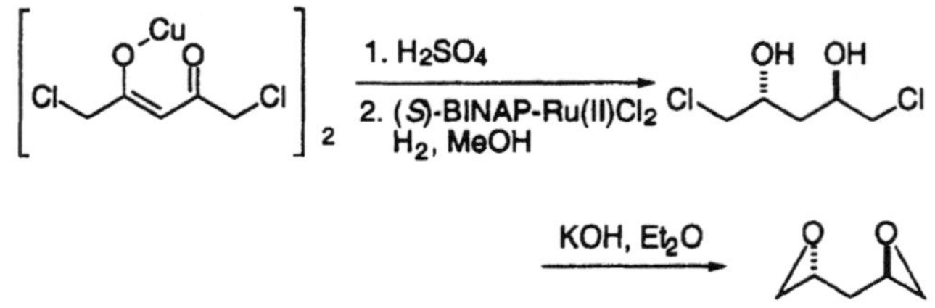

KETONES

(*S*)-(+)-3-Hydroxy-2,2-dimethylcyclohexanone

Mori, K.; Mori, H. *Org. Synth.* **1990**, *68*, 56.

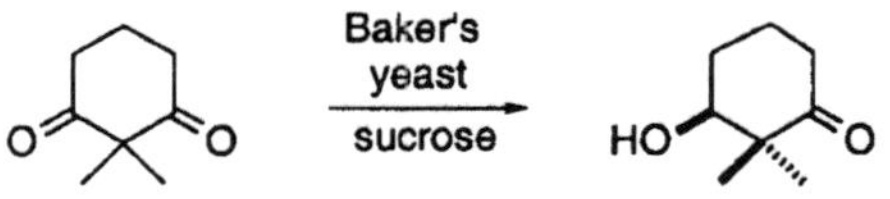

(*R*)-4-Ethyl-4-allyl-2-cyclohexen-1-one

Meyers, A.I.; Berney, D. *Org. Synth.* **1990**, *69*, 55.

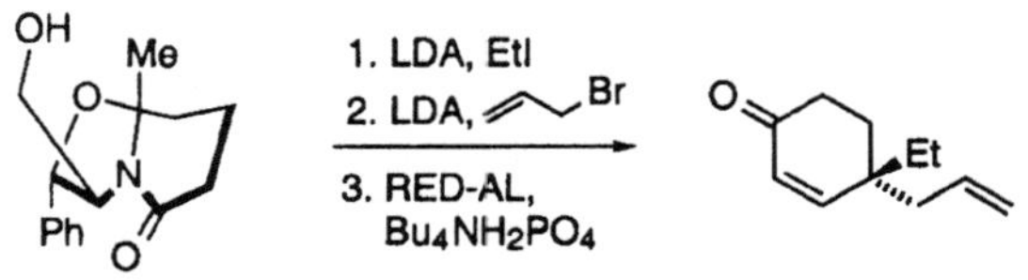

16α-Fluoro-3-methoxy-1,3-5(10)-estratrien-17-one

Umemoto, T.; Tomita, K.; Kawada, K. *Org. Synth.* **1990**, *69*, 129.

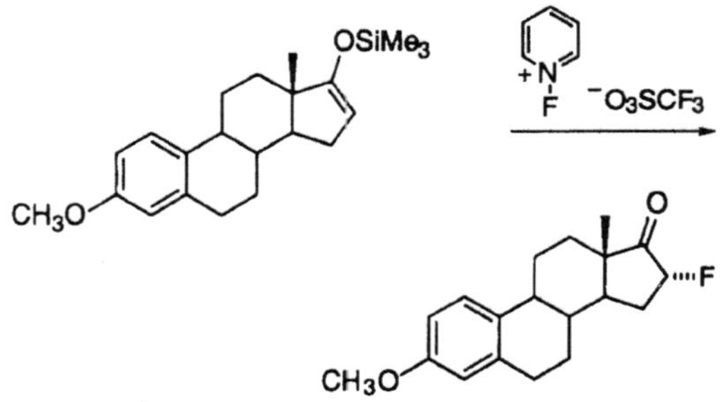

(*R*)-(−)-10-Methyl-1(9)-octal-2-one

Revial, G.; Pfau, M. *Org. Synth.* **1992**, *70*, 35.

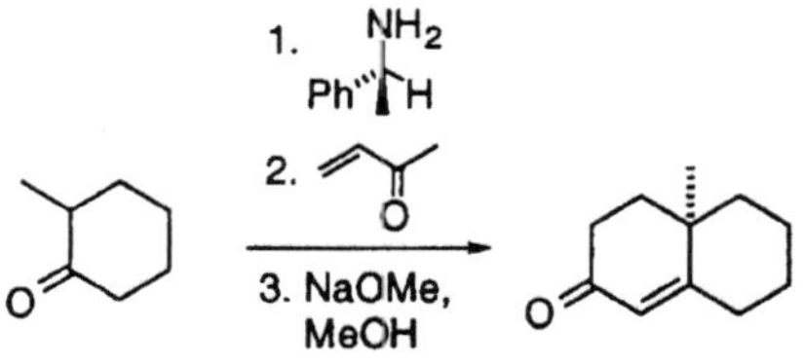

3-(*S*)-[(*tert*-Butyldiphenylsilyl)oxy]-2-butanone

Overman, L.E.; Rishton, G.M. *Org. Synth.* **1993**, *71*, 56.

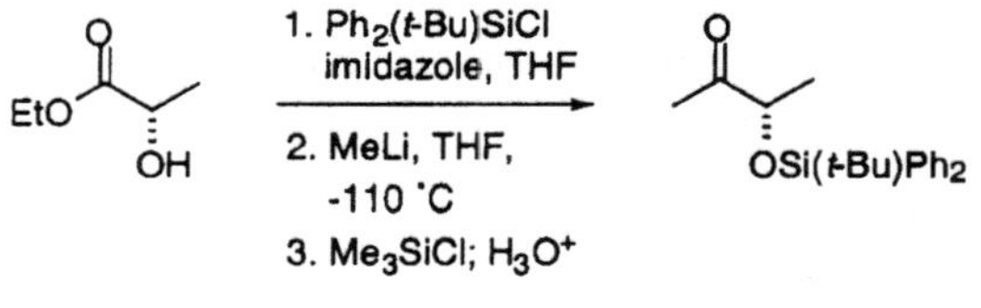

(1*R*, 5*R*)-(+)-Verbenone

Sivik, M.R.; Stanton, K.J.; Paquette, L.A. *Org. Synth.* **1995**, *72*, 57.

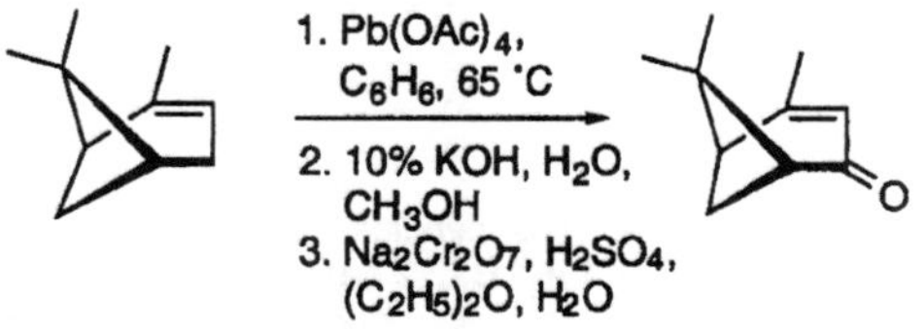

(4*R*)-(+)-tert-Butyldimethylsiloxy-2-cyclopenten-1-one

Paquette, L.A.; Earle, M.J.; Smith, G.F. *Org. Synth.* **1996**, *73*, 36.

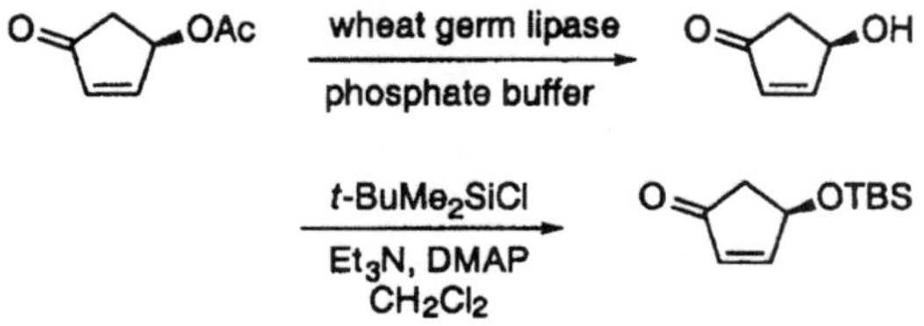

(4*S*)-(−)-tert-Butyldimethylsiloxy-2-cyclopenten-1-one

Paquette, L.A.; Heidelbaugh, T.M. *Org. Synth.* **1996**, *73*, 44.

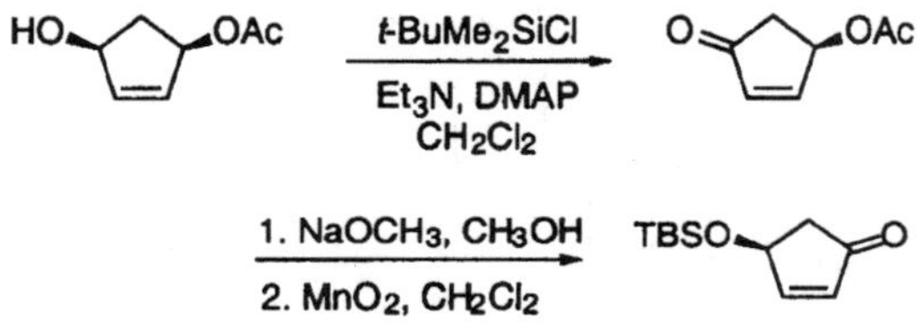

4a(*S*), 8a(*R*)-2-Benzoyloctahydro-6(2*H*)-isoquinolinone

Hutchinson, D.R.; Khau, V.V.; Martinelli, M.J.; Nayyar, N.K.; Peterson, B.C.; Sullivan, K.A. *Org. Synth.* **1998**, *75*, 223.

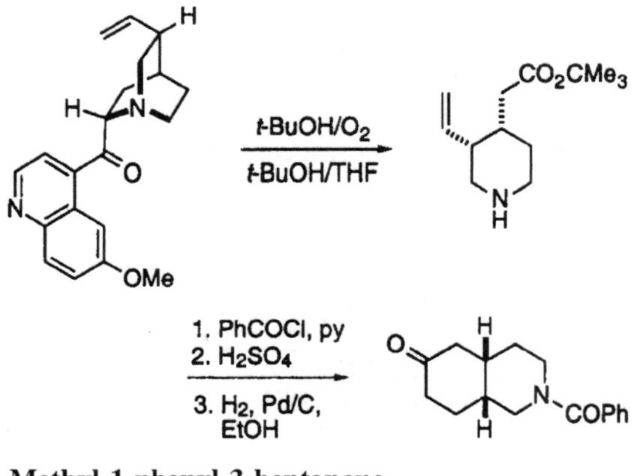

(*R*)-2-Methyl-1-phenyl-3-heptanone

Myers, A.G.; Yang, B.H.; Chen, H. *Org. Synth.* **2000**, *77*, 29.

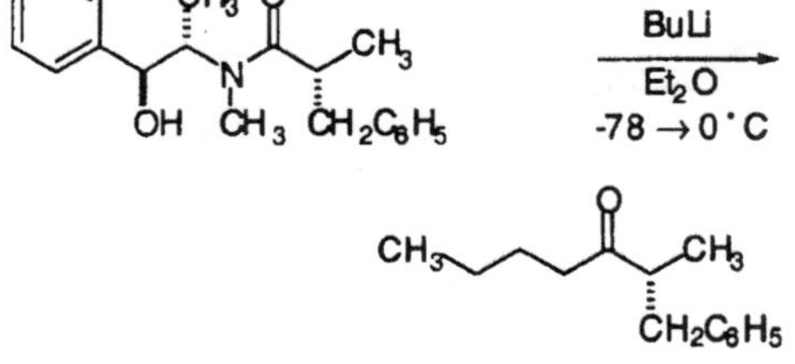

SULFUR COMPOUNDS

(*S*)-(−)-Methyl p-Tolyl Sulfoxide

Zhao, S.H.; Samuel, O.; Kagan, H.B. *Org. Synth.* **1990**, *68*, 49.

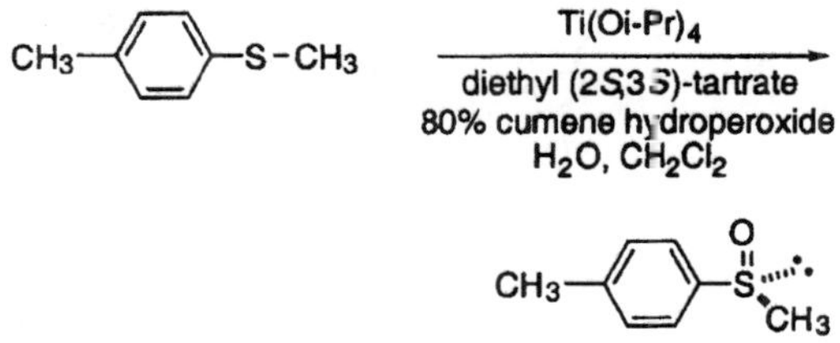

(−)-D-2, 10-Camphorsultam

Weismiller, M.C.; Towson, J.C.; Davis, F.A. *Org. Synth.* **1990**, *69*, 154.

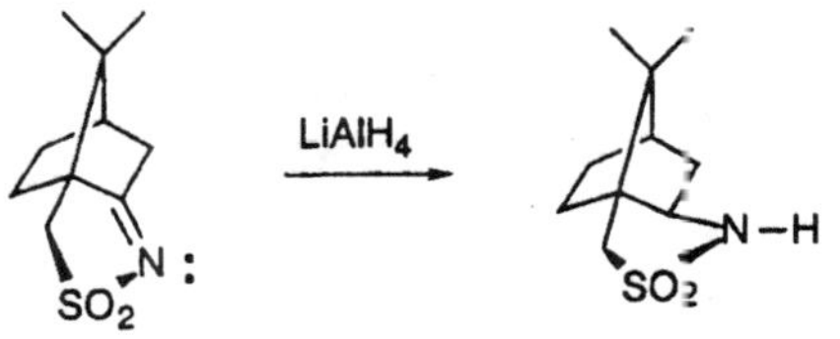

(+)-(2*R*, 8a*R)-[(8,8-Dimethyoxycamphoryl)-sulfonyl]oxaziri-dine**

Chen, B.-C.; Murphy, C.K.; Kumar, A.; Reddy, R.T.; Clark, C.; Zhou, P.; Lewis, B.M.; Gala, D.; Mergelsberg, I.; Scherer, D.; Buckley, J.; DiBenedetto, D.; Davis, F.A. *Org. Synth.* **1996**, *73*, 159.

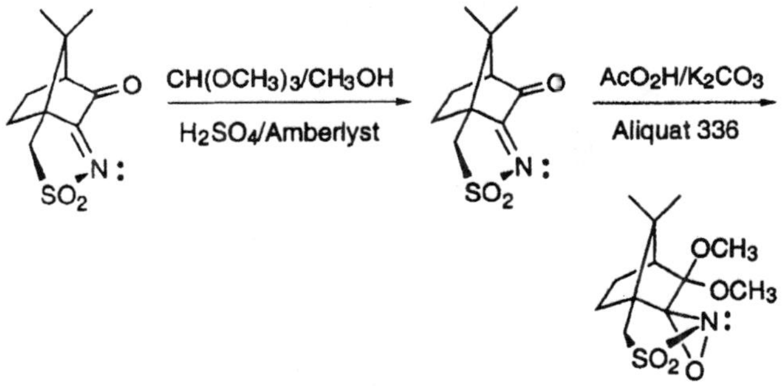

(+)-(2*R*, 8a*S*)-10-(Camphorylsulfonyl)oxaziridine

Towson, J.C.; Weismiller, M.C.; Lal, G.S.; Sheppard, A.C.; Davis, F.A. *Org. Synth.* **1990**, *69*, 158.

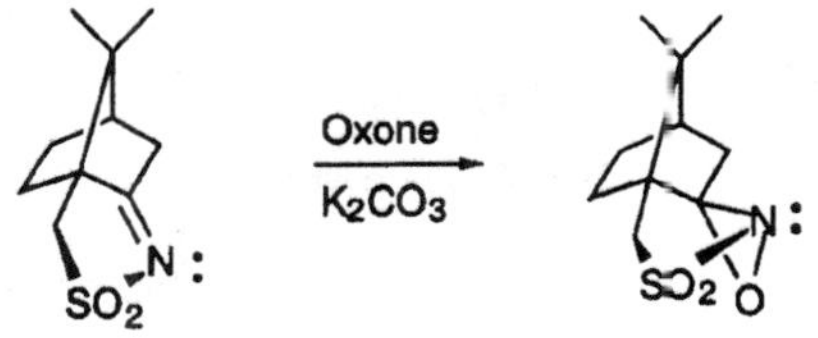

(+)-(2*R*, 8a*R)-[(8,8-Dichlorocamphoryl)sulfonyl]oxaziridine**

Chen, B.-C.; Murphy, C.K.; Kumar, A.; Reddy, R.T.; Clark, C.; Zhou, P.; Lewis, B.M.; Gala, D.; Mergelsberg, I.; Scherer, D.;

Buckley, J.; DiBenedetto, D.; Davis, F.A. *Org. Synth.* **1996**, *73*, 159.

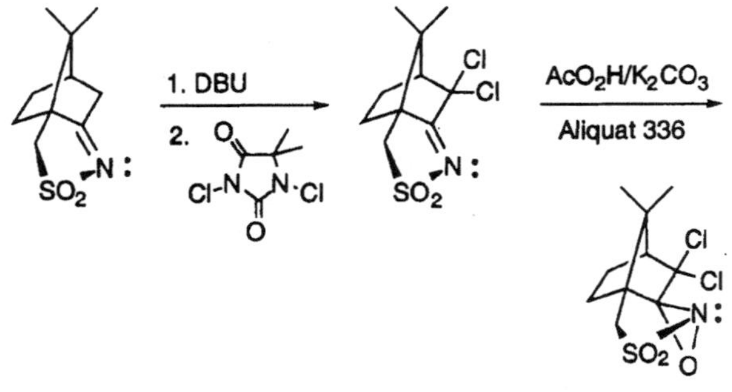

(1*S*)-(−)-1,3-Dithiane 1-Oxide

Bulman Page, P.C.; Heer, J.P.; Bethell, D.; Collington, E.W.; Andrews, D.M. *Org. Synth.* **1999**, *76*, 37.

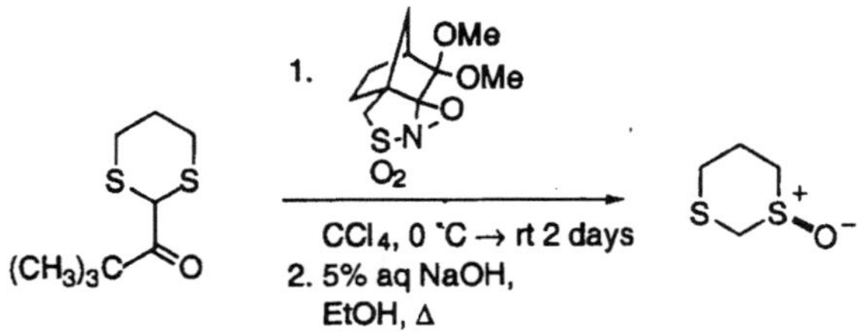

(−)-(*E,S*)-3-(Benzyloxy)-1-butenyl Phenyl Sulfone

Enders, D.; von Berg, S.; Jandeleit, B. *Org. Synth.* **2001**, *78*, 177.

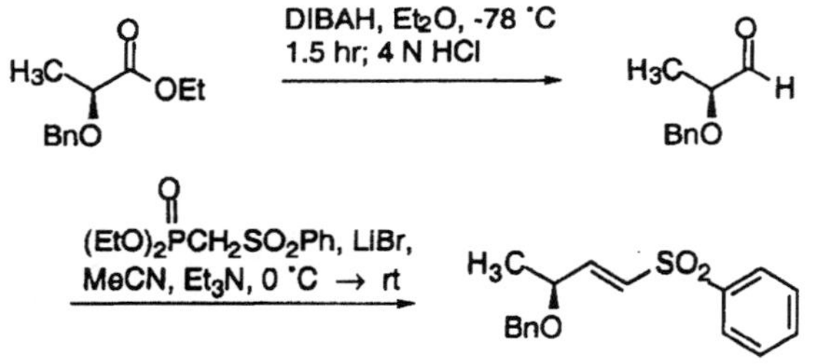

PHOSPHINES

(*R*)-2-Diphenylphosphino-2′-methoxy-1,1′-binaphthyl

Uozumi, Y.; Kawatsura, M.; Hayashi, T. *Org. Synth.* **2001**, *78*, 1.

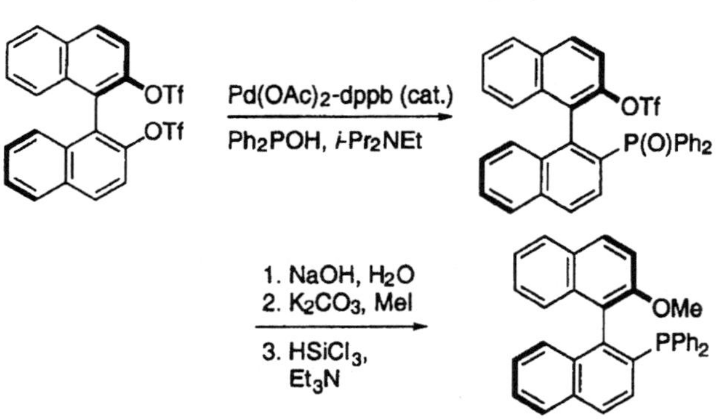

(*R*)-(+)- and (*S*)-(−)-2,2′-Bis(diphenyl-phosphino)-1,1′-binaphthyl (BINAP)

Cai, D.; Payack, J.F.; Bender, D.R.; Hughes, D.L.; Verhoeven, T.R.; Reider, P.J. *Org. Synth.* **1999**, *76*, 6.

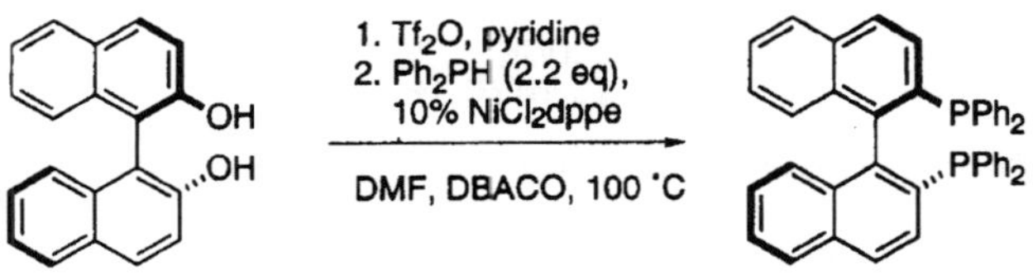

MISCELLANEOUS

Cholesta-3,5-diene

Cacchi, S.; Morera, E.; Ortar, G. *Org. Synth.* **1990**, *68*, 138.

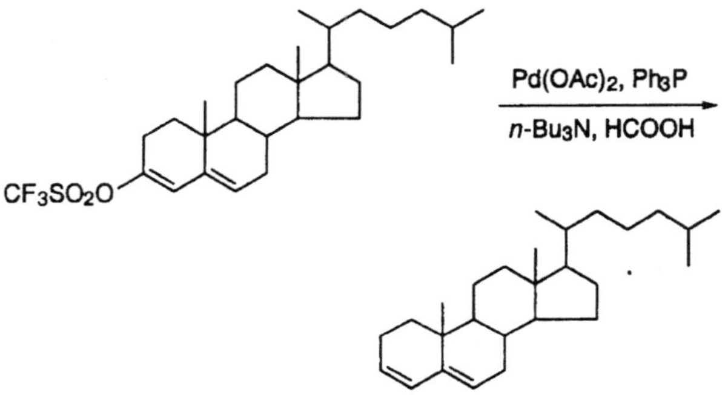

(*Z*)- and (*E*)-1-Menthoxy-1-butene

Kann, N.; Bernardes, V.; Greene, A.E. *Org. Synth.* **1997**, *74*, 13.

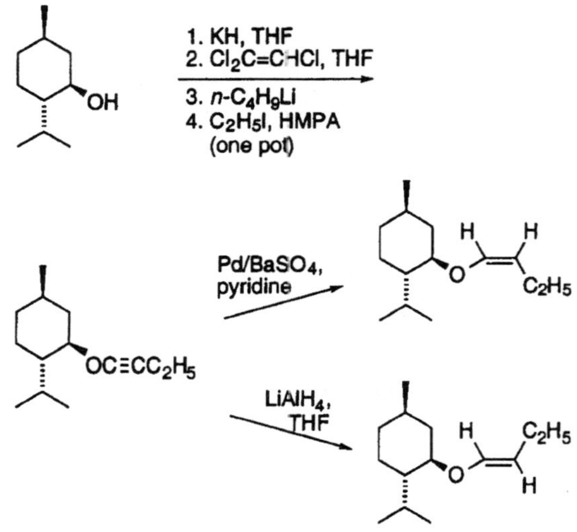

A list of General Abbreviations appears on the front Endpapers

(1*R*)-9,9-Dimethyltricyclo[6.1.1.0^{2,6}]deca-2,5-diene

Paquette, L.A.; McLaughlin, M.L. *Org. Synth.* **1990**, *68*, 220.

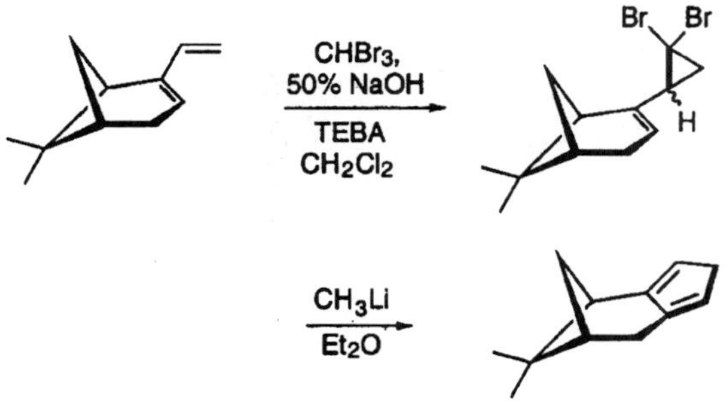

(*S,E*)-1-(Methoxymethoxy)-1-tributylstannyl-2-butene

Marshall, J.A.; Garafalo, A.W.; Hinkle, K.W. *Org. Synth.* **2000**, *77*, 98.

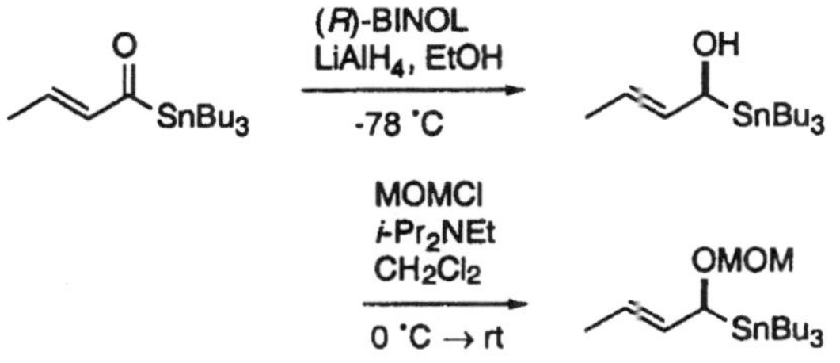

A

(*S*)-Aceto(carbonyl)(cyclopentadienyl)-(triphenylphosphine)iron

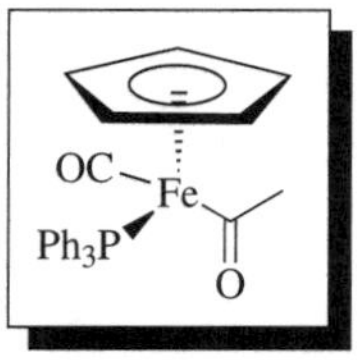

[*36548-60-4*] $C_{26}H_{23}FeO_2P$ (MW 454.29)

(chiral acetate enolate equivalent which can be deprotonated with BuLi or LDA in THF at −78 °C; the enolate reacts stereoselectively with a variety of achiral, prochiral, and chiral electrophiles to generate functionalized organoiron compounds from which the iron can subsequently be removed[1])

Physical Data: mp 142 °C; $[\alpha]^{22}$ +288° (*c* 0.004, C_6H_6), $[\alpha]^{20}$ +160° (*c* 0.04, C_6H_6).
Solubility: insol H_2O; sol THF, $CHCl_3$, CH_2Cl_2, acetone.
Form Supplied in: orange solid.
Handling, Storage, and Precautions: can be stored in air for days with little decomposition. The solid is best stored under nitrogen for long periods of time. More air sensitive when in solution, especially chlorinated hydrocarbons. Like all metal carbonyls, it is best handled in a fume hood.

Preparation and Determination of Absolute Stereochemistry. The preparation of racemic aceto(carbonyl)(cyclopentadienyl)(triphenylphosphine)iron (**2**) was first reported in 1966.[2] Brunner subsequently recognized that the iron atom in the complex was a chiral center and first reported the preparation of the (+)- and (−)-enantiomers in 1972.[3] Two groups[4,5] recognized in the early 1980s that the enolate of this compound might serve as a chiral acetate enolate equivalent. The absolute configuration of the (+)-enantiomer was reported in 1986[6a] and confirmed in 1988.[7] Alternative preparations of (*S*)-(+)-(**2**) involving kinetic reductions were published in late 1993.[6b,c] The first reported preparation[3a] of (*S*)-(+)-(**2**) begins with treatment of $[CpFe(CO)_2(PPh_3)]^+PF_6^-$ with sodium mentholate to produce the menthol ester diastereomers which were separated by fractional crystallization.[3b] The (−)-menthol ester (**1**) was then treated with MeLi (1.5 M in Et_2O) (slow addition of the MeLi produces acetyl (**2**) of the highest optical purity)[3c] in THF at −30 °C (eq 1). Aqueous workup and alumina chromatography produced the (*S*)-(+)-acetyl (**2**). A resolution procedure for the separation of the related trimethyl- and triethylphosphine substituted acetyls has also been reported.[8]

Enolate Generation and Subsequent Alkylation or Oxidation. The (*S*)-(+)-acetyl complex (**2**) is cleanly deprotonated to the configurationally stable enolate anion (**3**) upon treatment with *n-Butyllithium* or *Lithium Diisopropylamide* in THF at −78 °C.[1] The enolate (**3**) is alkylated in an almost quantitative yield when treated with a variety of alkyl iodides (eq 2). Alkylation of (**3**) with (*R*)-chloromenthyl ether produced (**5**) which was used to establish the absolute stereochemistry of (**2**) (eq 3).[6,1f] Subsequent deprotonation of these alkylated products (**4**) is believed to produce (*E*)-enolates (**6**) which can be alkylated cleanly from the enolate face away from the triphenylphosphine ligand (eq 4).[1]

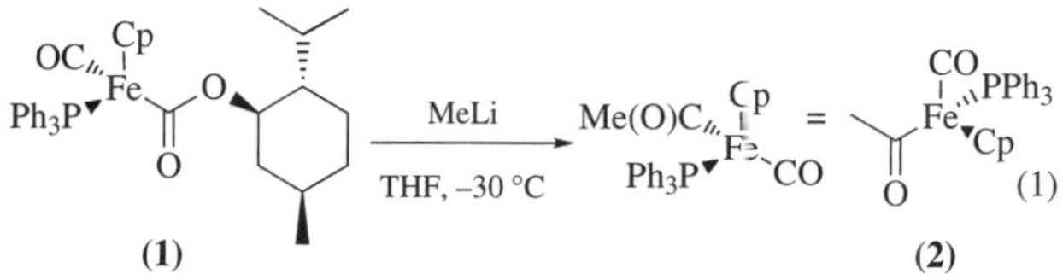

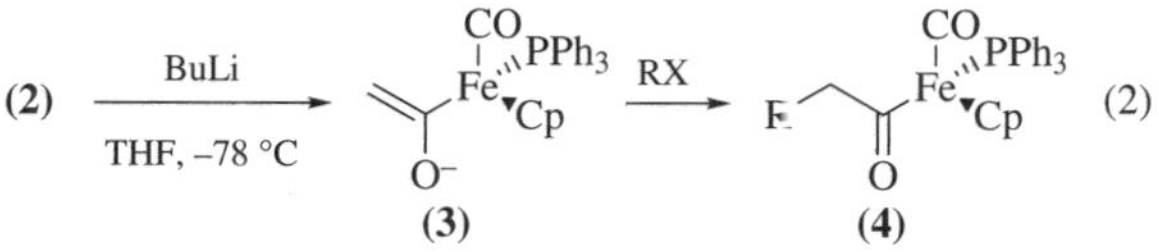

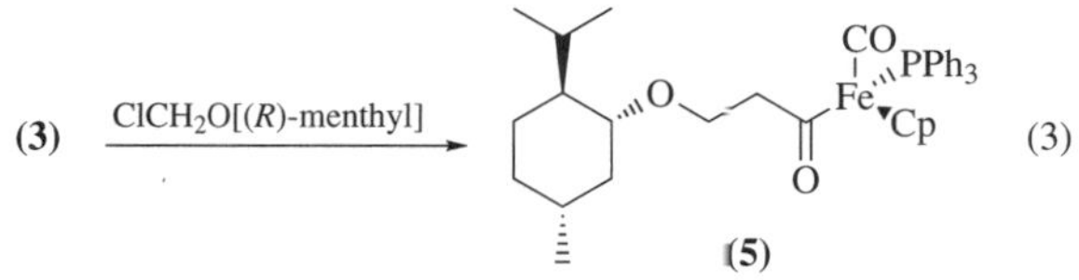

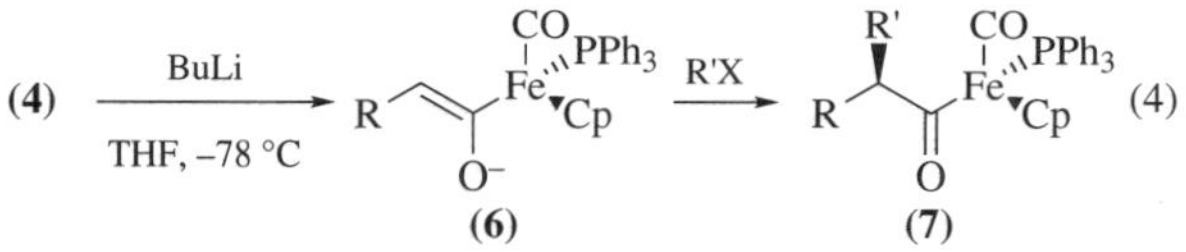

This sequential alkylation chemistry has been used to synthesize (−)-epicaptopril (**9**) (eq 5)[1b,9] as well as (*S*)-(+)-2-methylhept-4-ynoate (**10**), a key intermediate for the side chain of prostacyclin ZK96480 (eq 6).[10]

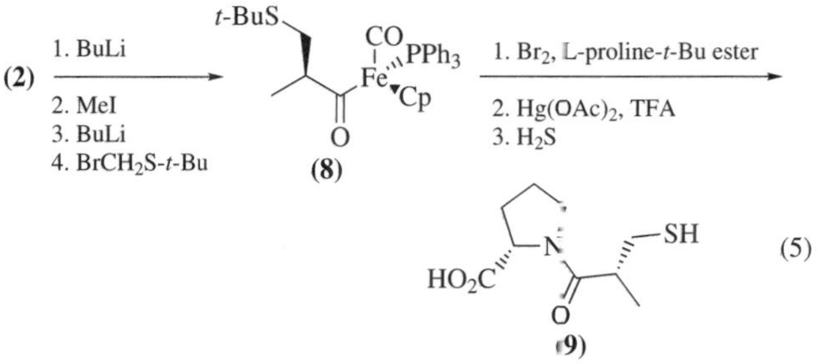

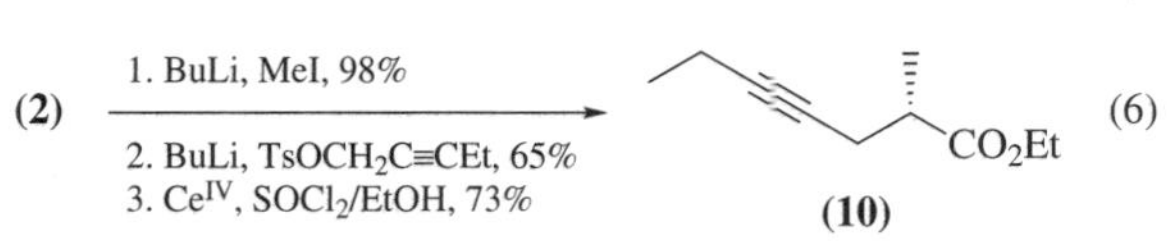

Related alkylation chemistry of (**2**) has provided a route to chiral succinic acid derivatives (**11**) (eq 7).[11] In this chemistry, acetyl complex (**2**) was used to produce a succinoyl complex which was subsequently deprotonated α to the ester and alkylated with primary and secondary alkyl halides. This sequence of reactions has been used to produce the succinate fragment of actinonin,[11a] as well as a variety of other succinic acid and amide derivatives.[11b,c]

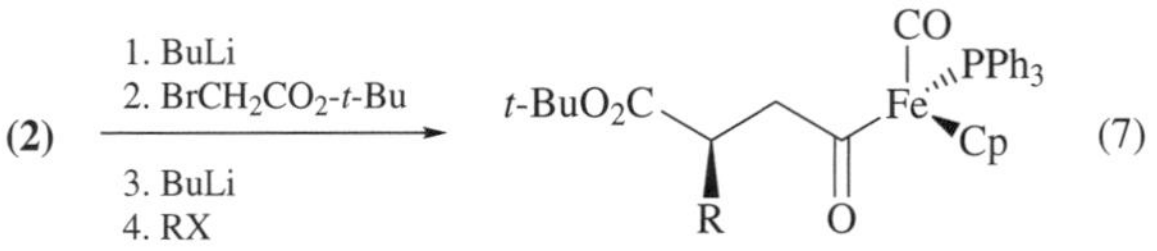

(11)

α-Oxidation of the enolate (**3**) has also been reported. Deprotonation of (**4**) (R = Me) followed by treatment with *Diphenyl Disulfide* generated a thioether complex which was oxidized to the sulfoxide and used in an asymmetric synthesis of sulfoxides.[12] Enolate (**3**) has also been trapped with MoOPH (*Oxodiperoxymolybdenum(pyridine)(hexamethylphosphoric triamide)*) and benzylated to produce the α-benzyloxyacetyl complex.[13] Enolate generation from the α-benzyloxyacetyl followed by alkylation with two equivalents of racemic 1-phenylethyl bromide provided almost exclusively the alkylation product derived from reaction with (*S*)-1-phenylethyl bromide. This sequence demonstrated the power of the $CpFe(CO)(PPh_3)R$ fragment as a chiral recognition element. Recently, treatment of (**3**) with *Chlorodiphenylphosphine* has been shown to produce chiral phosphine (**12**) which was subsequently used as a ligand in the preparation of several palladium complexes (eq 8).[14]

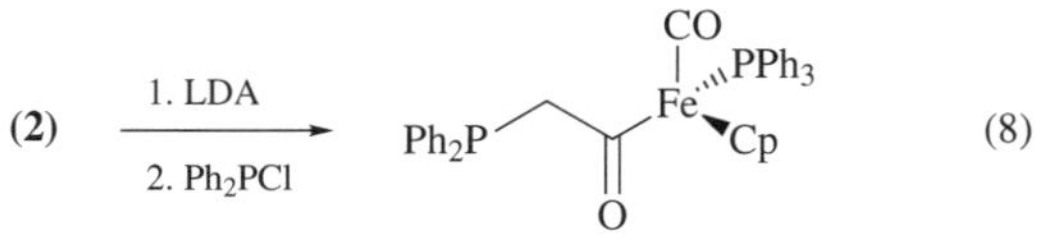

(12)

Aldol and Related Condensations. The fact that the enolate of the racemic iron acetyl will participate in diastereoselective aldol condensations and that enolate counterions have a large effect on the diastereoselectivity of those reactions has been known for some time.[1] More recently, an alkylation/aldol sequence involving (**2**) was used to prepare complex (**13**) of known absolute configuration. Complex (**13**) was used to assign, by chemical correlation, the absolute configuration to a series of marine cyclic epoxides (eq 9).[15] Similar aldol chemistry followed by iron–carbon bond cleavage using *Bromine* has been used to prepare a series of optically active β-lactones (**15**)[16] including tetrahydrolipstatin (eq 10).[17]

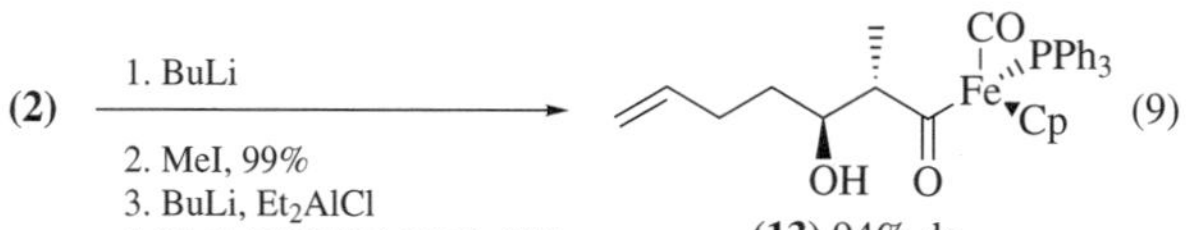

(13) 94% de

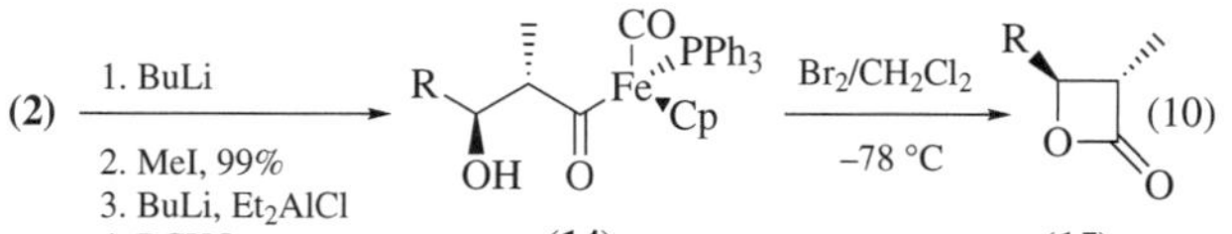

(14) **(15)**

Examples of aldol condensations involving chiral aldehydes have also been reported. Condensation of the aluminum enolate derived from (**2**) with Boc-L-prolinal has been shown to proceed in a highly stereoselective manner and the iron chirality overpowered the latent stereoselectivity inherent in Boc-L-prolinal.[18] Subsequent deprotection and demetalation of (**16**) produced (−)-(1*R*,8*S*)-1-hydroxypyrrolizidin-3-one (**17**) (eq 11). Aldol condensation between the tin(II) and aluminum enolates derived from (**2**) and 2,3-isopropylidene-D-glyceraldehyde also proceeded with high stereoselectivity and the iron once again had the overwhelming directing effect on the stereochemical outcome of these condensations (eq 12).[19]

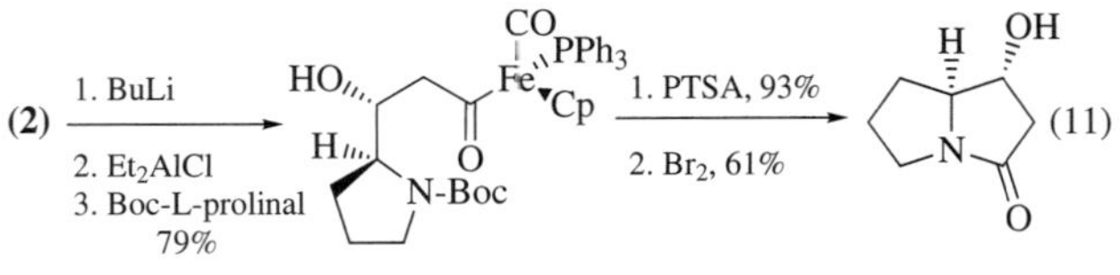

(16) 300:1 de **(17)**

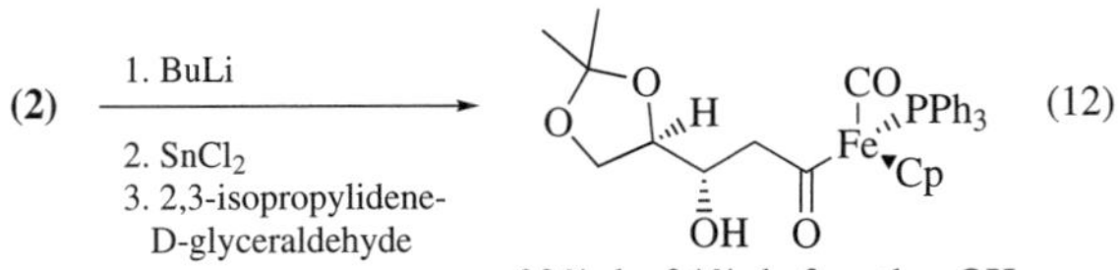

92% de, 94% de for other OH stereochem with Al enolate

Synthesis and Reaction Chemistry of α,β-Unsaturated Acyl Complexes Derived from (2). Two methods for the preparation of optically active (*E*)- and (*Z*)-α,β-unsaturated iron acyls from (**2**) have been reported.[1f] One method involves aldol condensation of (**2**) with aldehydes followed by *O*-methylation to produce diastereomeric acyls (**18**). This mixture (**18**) is then treated with *Sodium Hydride* to produce predominantly (*E*)-α,β-unsaturated acyl complexes (**19**) (eq 13).[20] Alternatively, (**2**) can be deprotonated and treated with *Chlorotrimethylsilane* to produce the *C*-silylated complex which is subsequently deprotonated and treated with an aldehyde.[1f,21] This Peterson alkenation produced mixtures of the isomers (*E*)-(**19**) and (*Z*)-(**20**) which could be separated via chromatography (eq 14). The (*Z*) isomers (**20**) with γ-protons are deprotonated when treated with strong bases and selectively alkylated α to the carbonyl.[1f] The (*E*) isomers and (*Z*) isomers without γ protons participate in stereoselective Michael additions and Michael addition/alkylation sequences.[1f,20–22] Most often this chemistry has been used to prepare optically active β-lactams and amides (**21**) and (**22**) (eq 15 and eq 16).

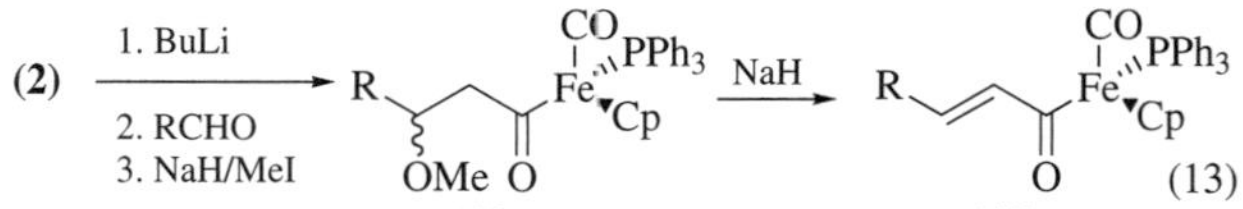

(18) **(19)**

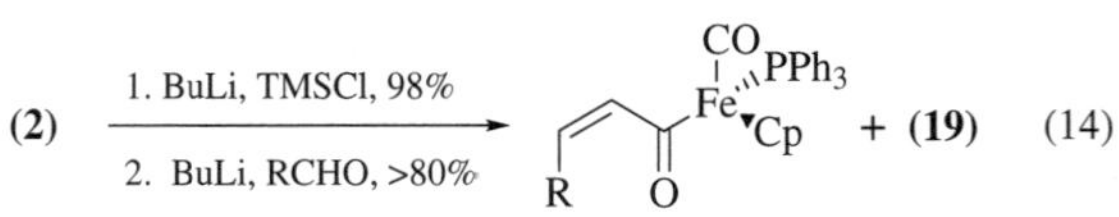

(20)

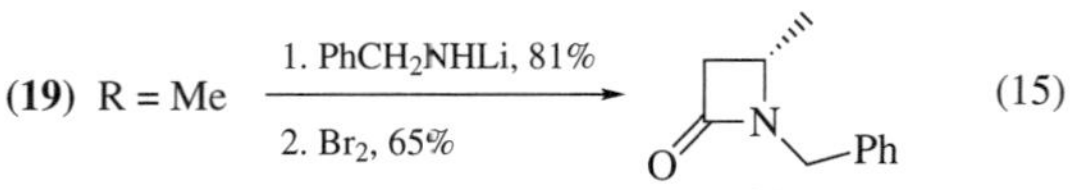

(21)

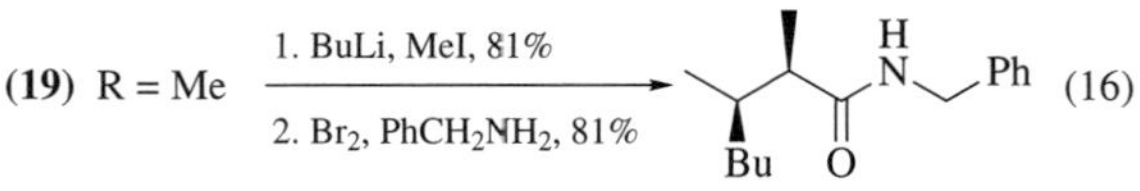

(22)

The (*E*)- and (*Z*)-α,β-unsaturated acyls (**19**) and (**20**) have also been methylenated as part of an asymmetric route to cyclopropanecarboxylic acid derivatives.[23] The acryloyl complex, which has been used in asymmetric Diels–Alder reactions[24] as well as a verapamil precursor synthesis,[25] has not been prepared in optically active form from (**2**), but can be prepared via an elimination reaction from (**5**).[24,25]

1. (a) Fatiadi, A. J. *J. Res. Nat. Inst. Stand. Technol.* **1991**, *96*, 1. (b) Davies, S. G. *Aldrichim. Acta* **1990**, *23*, 31. (c) Blackburn, B. K.; Davies, S. G.; Whittaker, M. *Stereochemistry of Organometallic and Inorganic Compounds*; Bernal, I., Ed.; Elsevier: Amsterdam, 1989; Vol. 3, pp 141–223. (d) Davies, S. G. *Pure Appl. Chem.* **1988**, *60*, 13. (e) Davies, S. G.; Bashiardes, G.; Beckett, R. P.; Coote, S. J.; Dordor-Hedgecock, I. M.; Goodfellow, C. L.; Gravatt, G. L.; McNally, J. P.; Whittaker, M. *Philos. Trans. R. Soc. London, Ser. A* **1988**, *326*, 619. (f) Davies, S. G.; Dordor-Hedgecock, I. M.; Easton, R. J. C.; Preston, S. C.; Sutton, K. H.; Walker, J. C. *Bull. Soc. Chem. Fr.* **1987**, 608.
2. Bibler, J. P.; Wojcicki, A. *Inorg. Chem.* **1966**, *5*, 889.
3. (a) Brunner, H.; Schmidt, E. *J. Organomet. Chem.* **1972**, *36*, C18. (b) Brunner, H.; Schmidt, E. *J. Organomet. Chem.* **1973**, *50*, 219. (c) Brunner, H.; Strutz, J. *Z. Naturforsch., Teil B* **1974**, *29*, 446.
4. Aktogu, N.; Davies, S. G.; Felkin, H. *Chem. Commun.* **1982**, 1303.
5. Liebeskind, L. S.; Welker, M. E. *J. Organomet. Chem.* **1983**, *2*, 194.
6. (a) Davies, S. G.; Dordor-Hedgecock, I. M.; Sutton, K. H.; Walker, J. C.; Bourne, C.; Jones, R. H.; Prout, K. *Chem. Commun.* **1986**, 607. (b) Baker, R. W.; Davies, S. G.; *Tetrahedron: Asymmetry* 1993, *4*, 1479. (c) Case-Green, S. C.; Costello, J. F.; Davies, S. G.; Heaton, N.; Hedgecock, C. J. R.; Prime, J. C. *Chem. Commun.* **1993**, 1621.
7. Bernal, I.; Brunner, H.; Muschiol, M. *Inorg. Chim. Acta* **1988**, *142*, 235.
8. Brookhart, M.; Liu, Y.; Goldman, E. M.; Timmers, D. A.; Williams, G. D. *J. Am. Chem. Soc.* **1991**, *113*, 927.
9. Bashiardes, G.; Davies, S. G. *Tetrahedron Lett.* **1987**, *28*, 5563.
10. Bodwell, G. J.; Davies, S. G. *Tetrahedron: Asymmetry* **1991**, *2*, 1075.
11. (a) Bashiardes, G.; Davies, S. G. *Tetrahedron Lett.* **1988**, *29*, 6509. (b) Bashiardes, G.; Collingwood, S. P.; Davies, S. G.; Preston, S. C. *J. Chem. Soc., Perkin Trans. 1* **1989**, 1162. (c) Bashiardes, G.; Collingwood, S. P.; Davies, S. G.; Preston, S. C. *J. Organomet. Chem.* **1989**, *364*, C29.
12. Davies, S. G.; Gravatt, G. L. *Chem. Commun.* **1988**, 780.
13. Davies, S. G.; Middlemiss, D.; Naylor, A.; Wills, M. *Chem. Commun.* **1990**, 797.
14. Douce, L.; Matt, D. *CR(2)* **1990**, *310*, 721.
15. Capon, R. J.; MacLeod, J. K.; Coote, S. J.; Davies, S. G.; Gravatt, G. L.; Dordor-Hedgecock, I. M.; Whittaker, M. *Tetrahedron* **1988**, *44*, 1637.
16. Case-Green, S. C.; Davies, S. G.; Hedgecock, C. J. R. *Synlett* **1991**, 779.
17. Case-Green, S. C.; Davies, S. G.; Hedgecock, C. J. R. *Synlett* **1991**, 781.
18. (a) Beckett, R. P.; Davies, S. G. *Chem. Commun.* **1988**, 160. (b) Beckett, R. P.; Davies, S. G.; Mortlock, A. A. *Tetrahedron: Asymmetry* **1992**, *3*, 123.
19. Bodwell, G. J.; Davies, S. G.; Mortlock, A. A. *Tetrahedron* **1991**, *47*, 10077.
20. Davies, S. G.; Dordor-Hedgecock, I. M.; Sutton, K. H.; Walker, J. C.; Jones, R. H.; Prout, K. *Tetrahedron* **1986**, *42*, 5123.
21. Davies, S. G.; Dupont, J.; Easton, R. J. C. *Tetrahedron: Asymmetry* **1990**, *1*, 279.
22. Davies, S. G.; Dordor-Hedgecock, I. M.; Sutton, K. H.; Walker, J. C. *Tetrahedron Lett.* **1986**, *27*, 3787.
23. (a) Ambler, P. W.; Davies, S. G. *Tetrahedron Lett.* **1988**, *29*, 6979. (b) Ambler, P. W.; Davies, S. G. *Tetrahedron Lett.* **1988**, *29*, 6983.
24. Davies, S. G.; Walker, J. C. *Chem. Commun.* **1986**, 609.
25. Brunner, H.; Forster, S.; Nuber, B. *J. Organomet. Chem.* **1993**, *12*, 3819.

Mark E. Welker
Wake Forest University, Winston-Salem, NC, USA

Allylcyclopentadienyl[(4*R*, *trans*)- and (4 *S*,*trans*)-α,α,α′, α′-tetraphenyl-1,3-dioxolane-4,5-dimethanolato-*O*,*O*′] titanium [Cp(*R*,*R*)-Ti[All] and Cp(*S*,*S*)-Ti[All]][1]

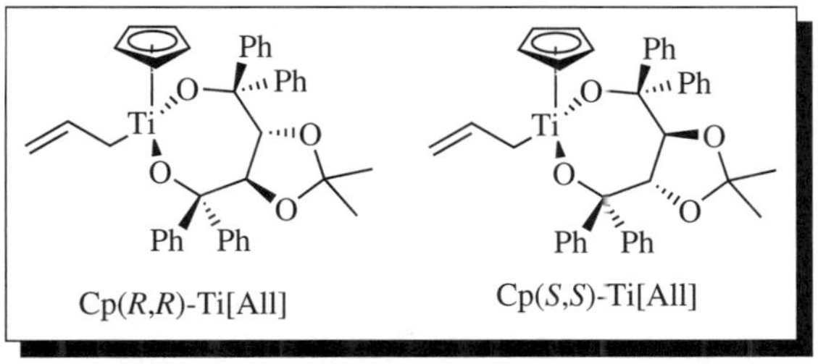

[139 354-61-3]	$C_{39}H_{38}O_4Ti$	(MW 618.23)
[139 354-59-4]	$C_{39}H_{38}O_4Ti$	(MW 618.23)

(reagent for the asymmetric allyltitanation of aldehydes to produce homoallylic alcohols)[2]

Solubility: used as the crude preparation in Et_2O.
Analysis of Reagent Purity: ^{1}H NMR and ^{13}C NMR (CD_2Cl_2).[2]
Preparative Methods: it can be prepared in two steps from either (4*R*,*trans*)-2,2-dimethyl-α,α,α′,α′-tetraphenyl-1,3-dioxolane-4,5-dimethanol (*R*,*R*-Taddol)[3] or (4*S*,*trans*)-2,2-dimethyl-α,α,α′,α′-tetraphenyl-1,3-dioxolane-4,5-dimethanol (*S*,*S*-Taddol)[3] (eq 1).[2]

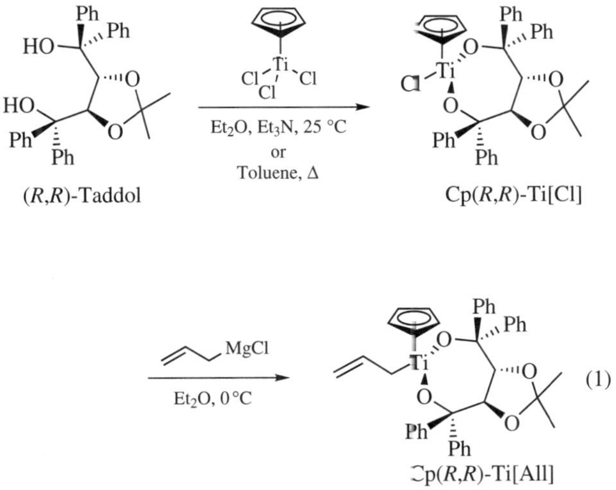

Handling, Storage, and Precautions: best handled as stock solution in Et_2O (ca. 0.82 M) which must be protected from moisture. Should be used just after preparation. Reactions should be carried out in dry equipment and with absolute solvents under Ar or N_2.

Addition to Aldehydes. Condensation of Cp(*R*,*R*)-Ti[All] or Cp(*S*,*S*)-Ti[All] with aldehydes occurs in good yield (50–94%)

and with excellent enantioselectivity (93–98%) to provide secondary homoallylic alcohols (eqs 2 and 3).[2–5]

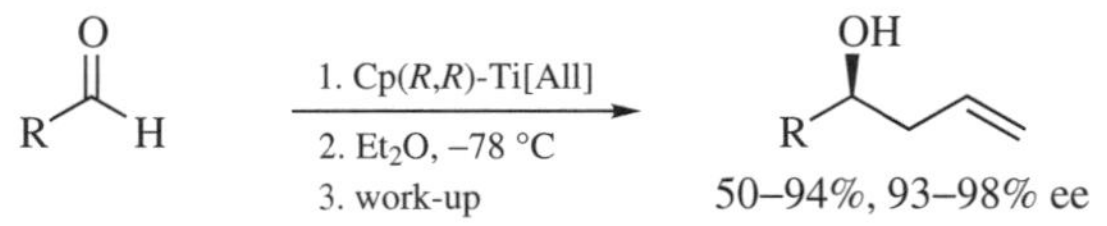

(2)

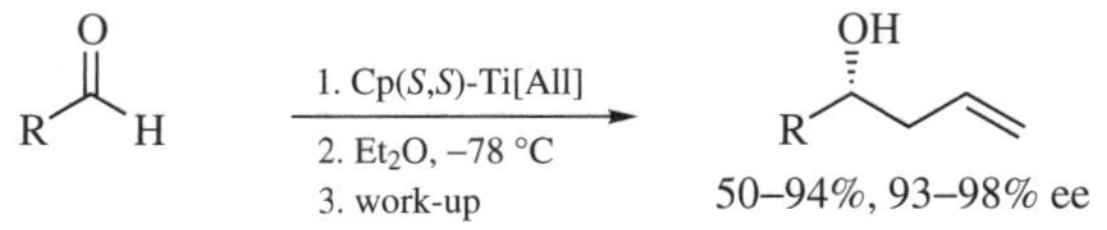

(3)

R = alkyl, alkenyl, alkynyl, Ar

One method has been employed in the reaction work-up. The reaction mixture is treated with aqueous 45% NH_4F solution (or with water when silyl groups are present), stirred for 12 h at rt, filtered through Celite, and extracted twice with ether. The combined organic phases were washed with brine, dried over $MgSO_4$, and concentrated. The solid residue was stirred with pentane. Subsequent filtration furnished the crystalline (*R*,*R*)-Taddol or (*S*,*S*)-Taddol (ligand) which can be recycled after crystallization. The filtrate was evaporated and the residue was purified by chromatography to afford the homoallylic alcohol.

The Cp(*R*,*R*)-Ti[All] and Cp(*S*,*S*)-Ti[All] reagents have been condensed with a variety of aldehydes always with good enantioselectivities (eqs 4–11). The degree of enantioface discrimination of these allyltitanium reagents is very high. The *Si* face attack is preferred for the Cp(*R*,*R*)-Ti[All] reagent and the *Re* face attack is preferred for the Cp(*S*,*S*)-Ti[All] reagent.

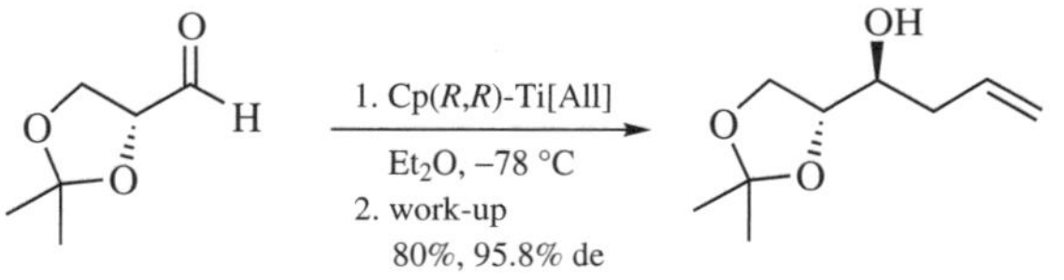

(4)

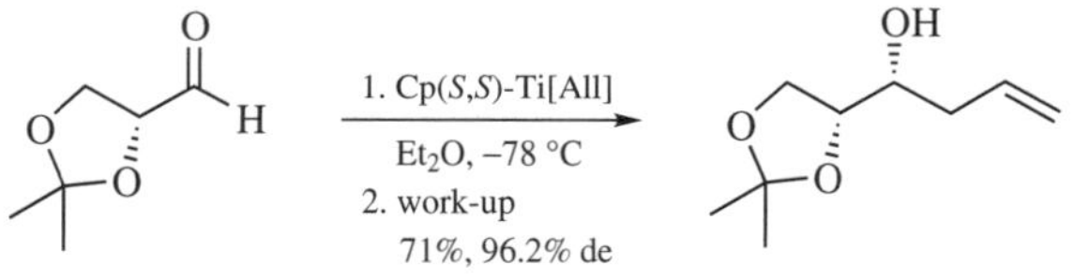

(5)

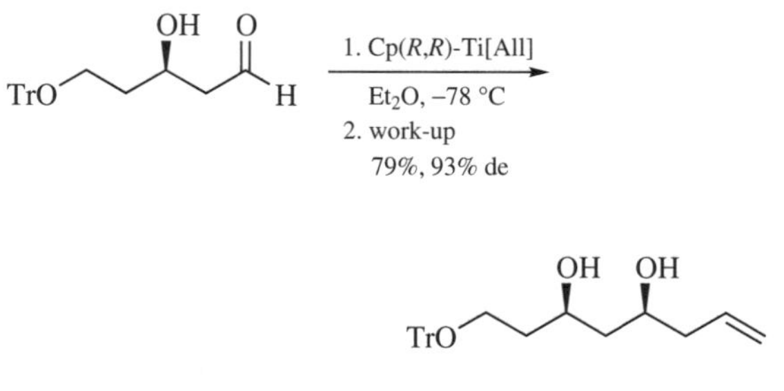

(6)

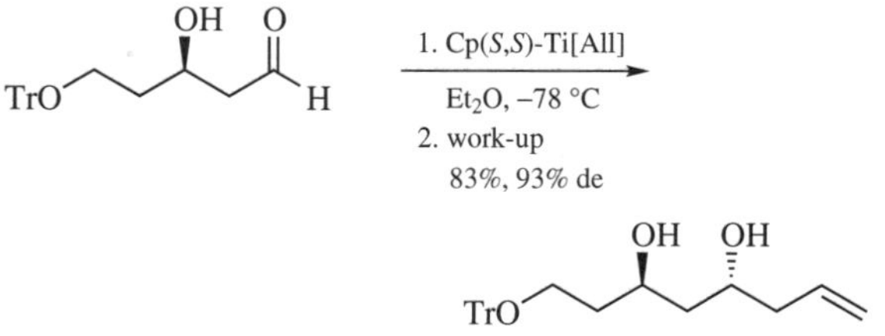

(7)

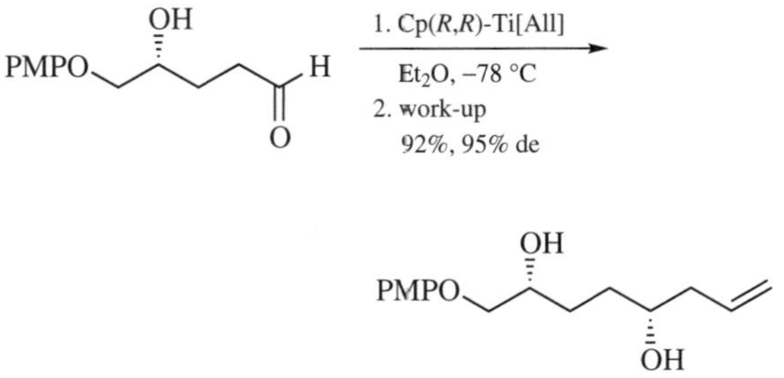

(8)

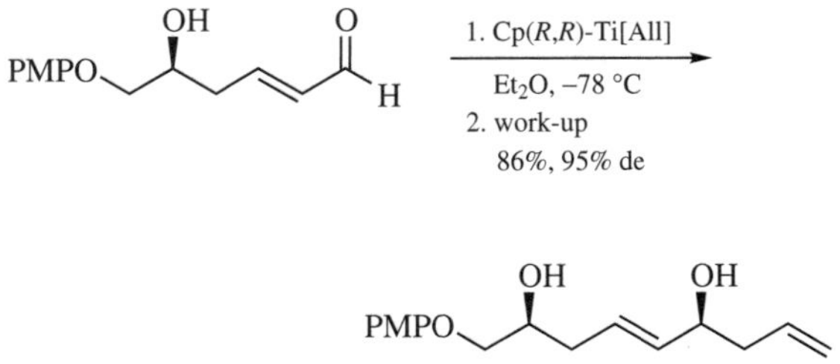

(9)

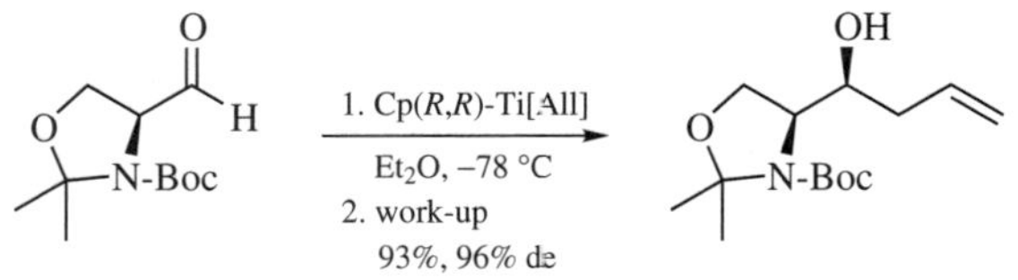

(10)

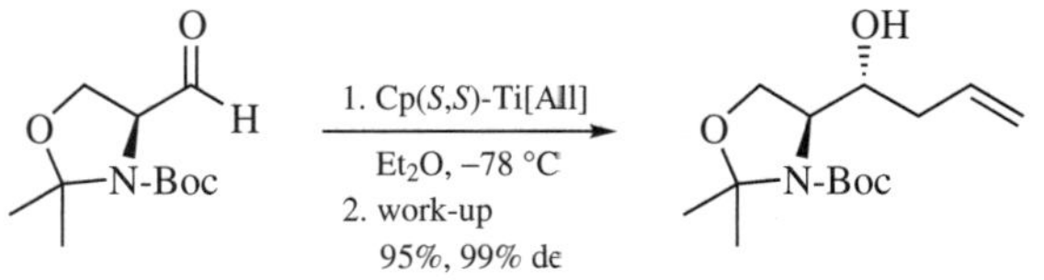

(11)

The Cp(*R*,*R*)Ti[All] and the Cp(*S*,*S*)Ti[All] reagents have been used to prepare precursors of 1,2-diols from α-oxygenated aldehydes (eqs 4 and 5),[6] 1,3-diols from 3-hydroxy aldehydes (eqs 6 and 7),[7] 1,4-diols from 4-hydroxy aldehydes (eq 8),[8] 1,5-diols from 5-hydroxy aldehydes (eq 9),[5] and 1,2-amino alcohols from 2-amino aldehydes (eqs 10 and 11)[2] with high diastereoselectivities and high enantioselectivities. Whatever the substituent and the position of the substituent present in the aldehyde, the diastereofacial selectivity has still been retained. Furthermore, no complexation of these reagents was observed with polar substituents such as protected or non-protected hydroxy groups.[7]

These reagents have been used for the conversion of C_s symmetrical chains into chiral non-racemic products. Condensation of *meso*-dialdehydes with Cp(*R*,*R*)-Ti[All] or Cp(*S*,*S*)-Ti[All] reagents led (after reduction with $NaBH_4$) to optically active polyketides (eq 12).[9]

Cp(*R*,*R*)-Ti[All] and Cp(*S*,*S*)-Ti[All] reagents discriminate, respectively, the pro-(*S*) and the pro-(*R*) face of the *meso*-dialdehydes. Furthermore, the allyltitanation reactions have been shown to closely follow a Felkin-Anh attack.

Cp(*R*,*R*)-Ti[All] and Cp(*S*,*S*)-Ti[All] complexes are much better in yield and enantioselectivity than alternate reagents of allyltitanation such as allyl(cyclopentadienyl)bis[3-*O*-1,2:5,6-di-*O*-isopropylidene-α-D-glucofuranosyl)]titanium[10] which is only available as one enantiomer favoring, as the Cp(*R*,*R*)-Ti[All] complex, the *Si* face attack of aldehydes. Chiral allyltitanocenes are less selective than the Cp(*R*,*R*)-Ti[All] and Cp(*S*,*S*)-Ti[All] reagents.[11]

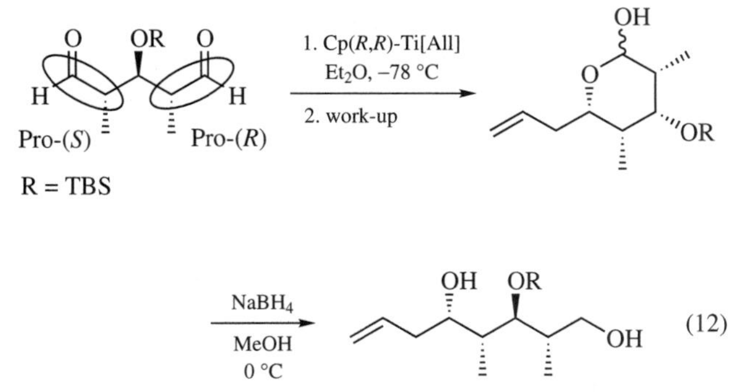

49% (for the two steps), > 95% de (12)

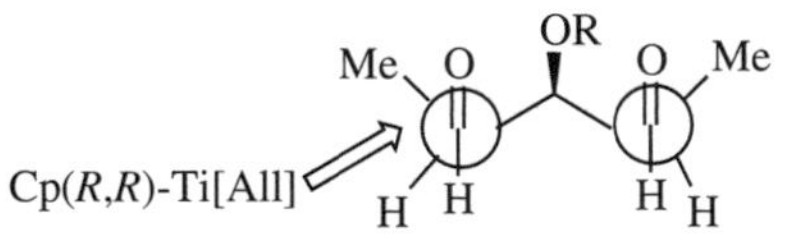

Felkin-Anh, favored attack

Addition to Ketones. Addition of Cp(*R*,*R*)-Ti[All] and Cp(*S*,*S*)-Ti[All] reagents to arylketones proceeds with good enantioselectivity (ee = 80%) (eq 13).[10,12]

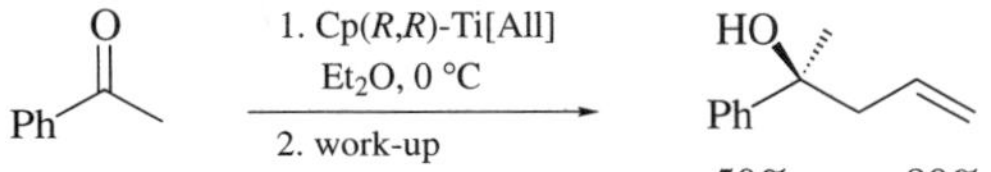

50%, ee = 80% (13)

Related Reagents.

E-But-2-enylcyclopentadienyl[(4R,trans)- and (4S,trans)-α,α,α′,α′-tetraphenyl-1,3-dioxolane-4,5-dimethanolato-O-O′] titanium[2]

The derivatives of these reagents, containing different substituted allyl groups, have been used to synthesize functionalized homoallylic alcohols. In general, these reagents are prepared in a manner analogous to Cp(*R*,*R*)-Ti[All] and Cp(*S*,*S*)-Ti[All] from Cp(*R*,*R*)-Ti[Cl] and Cp(*S*,*S*)-Ti[Cl] complexes by chloride substitution with allyl Grignard reagents or direct allylic metallation of the corresponding allylic chloride with *s*-BuLi, *n*-BuLi/K(O*t*-Bu),[13] or lithium tetramethylpiperidide/K(O*t*-Bu).[14]

The reagents containing alkyl substituents, such as but-2-enylcyclopentadienyl[(4*R*,*trans*)- and (4*S*,*trans*)-α,α, α′,α′-tetraphenyl-1,3-dioxolane-4,5-dimethanolato-*O*,*O*′]titanium [Cp(*R*,*R*) Ti[Crotyl] and Cp(*S*,*S*)Ti[Crotyl]] condense with aldehydes to give the corresponding substituted homoallylic alcohols in *anti* form with excellent diastereoselectivities and enantioselectivities (eq 14).[1] These reagents have been used to synthesize stereopentads with excellent diastereoselectivities and enantioselectivities by desymmetrization of *meso*-dialdehydes (eqs 15 and 16).[15]

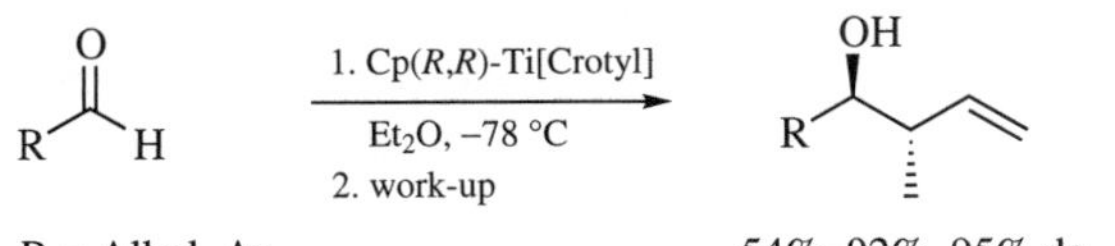

R = Alkyl, Ar

54%–92%, 95% de, (14)

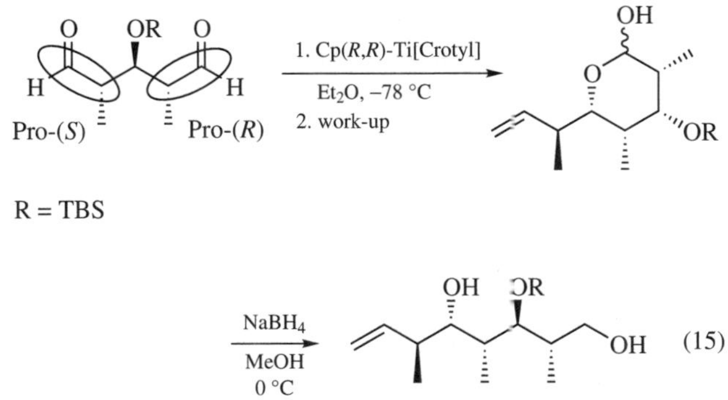

54% (for the two steps)
> 95% de (15)

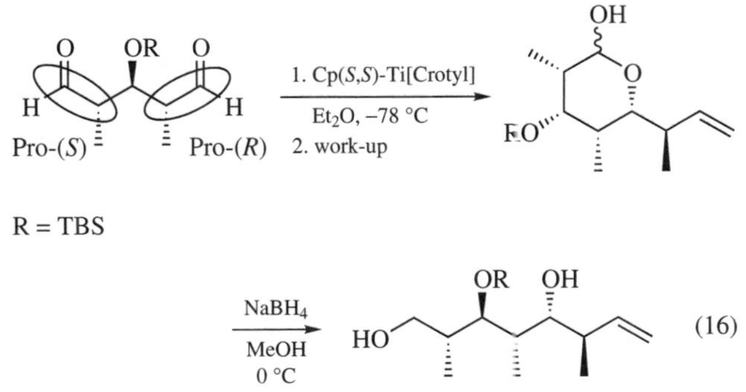

56% (for the two steps)
> 95% de (16)

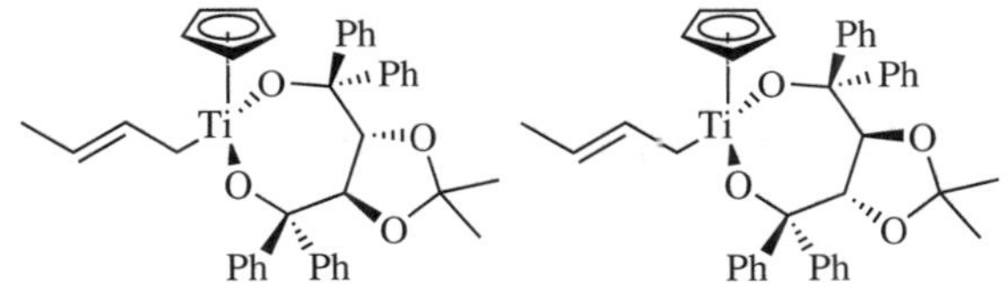

Cp(*R*,*R*)-Ti[Crotyl] Cp(*S*,*S*)-Ti[Crotyl]

Derivatives containing ether and silyl substituents, respectively, allow for the synthesis of *anti*-allylic monoprotected 1,2-diols (eq 17) and *anti* α-hydroxy allylsilanes (eq 18).[2]

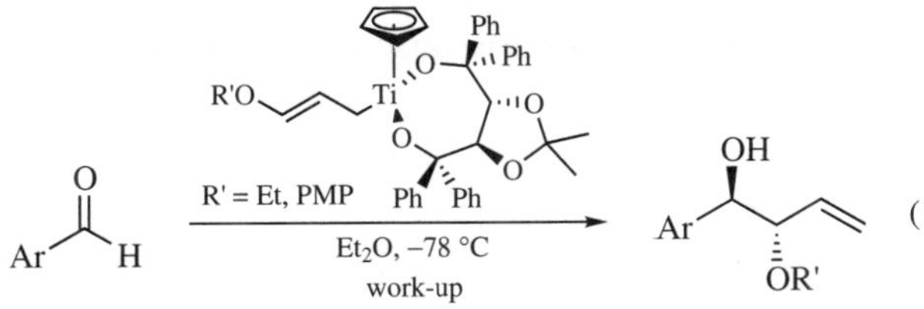

70–93 %,
70%–99% de, 95% ee (17)

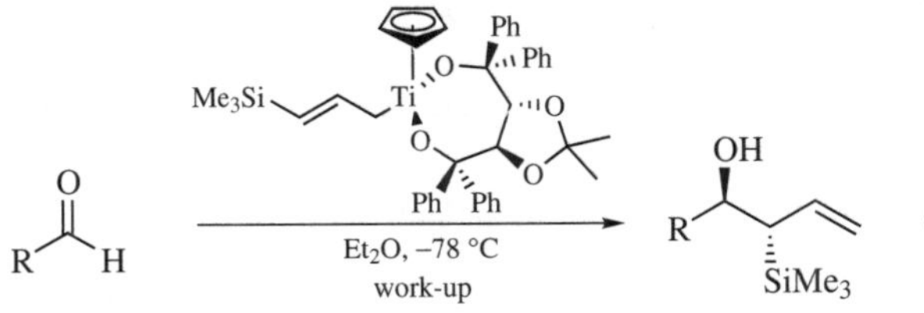

R = Alkyl, Ar

69%,
> 98% de, > 95% ee (18)

The Cp(*R*,*R*)-Ti[Crotyl] reagent is *Si* face selective and the Cp(*S*,*S*)Ti[Crotyl] reagent is *Re* face selective. A clear drawback of these reagents is the formation of the *anti* isomers as only the (*E*)-crotyltitanium reagents can be obtained.

1. Duthaler, R. O.; Hafner, A. *Chem. Rev.* **1992**, *92*, 807.
2. Hafner, A.; Duthaler, R. O.; Marti, R.; Rihs, G.; Rothe-Streit, P.; Schwarzenbach, F. *J. Am. Chem. Soc.* **1992**, *114*, 2321.
3. Seebach, D.; Beck, A. K.; Imwinvelried, R.; Roggo, S.; Wonnacott, A. *Helv. Chim. Acta* **1987**, *70*, 954.
4. BouzBouz, S.; Pradaux, F.; Cossy, J.; Ferroud, C.; Falguières, A. *Tetrahedron Lett.* **2000**, *41*, 887.
5. BouzBouz, S.; Cossy, J. *Org. Lett.* **2001**, *3*, 1451.
6. Cossy, J.; BouzBouz, S.; Caille, J.-C. *Tetrahedron: Asymmetry* **1999**, *10*, 3859.
7. BouzBouz, S.; Cossy, J. *Org. Lett.* **2000**, *2*, 501
8. BouzBouz, S.; Cossy, J. *Org. Lett.* **2000**, *2*, 3975.
9. BouzBouz, S.; Popkin, M.-E.; Cossy, J. *Org. Lett.* **2000**, *3*, 3449.
10. Duthaler, R.-O.; Riediker, M. *Angew. Chem., Int. Ed. Engl.* **1989**, *28*, 494.
11. Reetz, M.-T.; Kyung, S.-H.; Westermann, J. *Organometallics* **1984**, *3*, 1716.
12. Cossy, J.; BouzBouz, S., unpublished results.
13. Schlosser, M. *J. Organomet. Chem.* **1967**, *8*, 9.
14. Brandsma, L.; Verkruijsse, H.-D., In *Preparative Polar Organometallic Chemistry*; Springer-Verlag: Berlin, 1987, Vol. 1.
15. BouzBouz, S.; Cossy, J. *Org. Lett.* **2001**, *3*, 3995.

Janine Cossy & Samir Bouzbouz
ESPCI,
75231 Paris Cedex 05, France

B-Allyldiisocaranylborane[1]

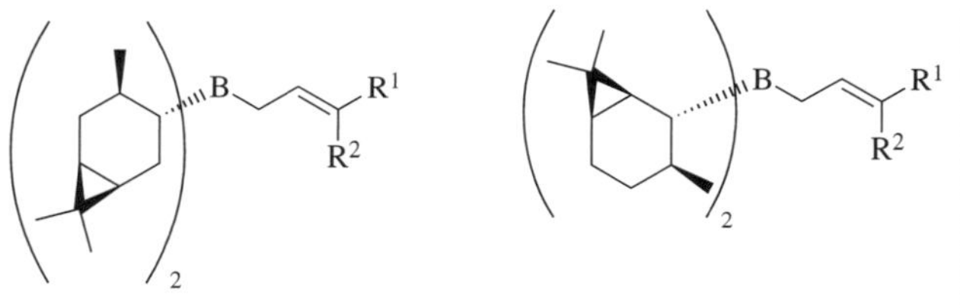

R^1, R^2 = (H, H), (Me, H), (H, Me)

(4-^{d}Icr$_2$BAll)		
[92055-65-7]	$C_{23}H_{39}B$	(MW 326.38)
(2-^{d}Icr$_2$BAll)		
[124821-92-7]	$C_{23}H_{39}B$	(MW 326.38)
(4-^{d}Icr$_2$BCrt^Z)		
[103818-03-7]	$C_{24}H_{41}B$	(MW 340.41)
(4-^{d}Icr$_2$BCrt^E)		
[103882-37-7]	$C_{24}H_{41}B$	(MW 340.41)
(2-^{d}Icr$_2$BCrt^Z)		
[130540-34-0]	$C_{24}H_{41}B$	(MW 340.41)
(2-^{d}Icr$_2$BCrt^E)		
[130609-25-5]	$C_{24}H_{41}B$	(MW 340.41)

(reagents for the asymmetric allyl- and crotylboration of aldehydes to produce secondary homoallylic alcohols[2] and β-methylhomoallylic alcohols[3])

Solubility: these reagents are prepared and used in situ at −78 °C in Et_2O (Icr$_2$BAll) or THF (Icr$_2$BCrt).

Preparative Methods: both reagents are derived from the corresponding *B*-methoxydiisocaranylborane which is prepared in two steps from (+)-3- or (+)-2-carene (eq 1 and eq 2). Note: only the (+) isomer of 3-carene occurs naturally; (+)-2-carene is obtained by base-catalyzed isomerization, and both are commercially available.

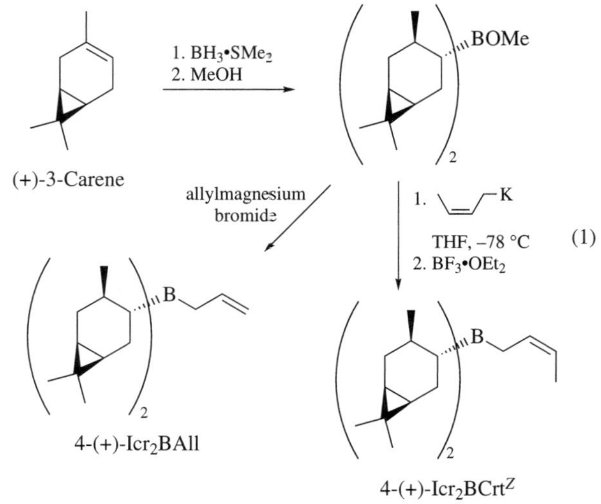

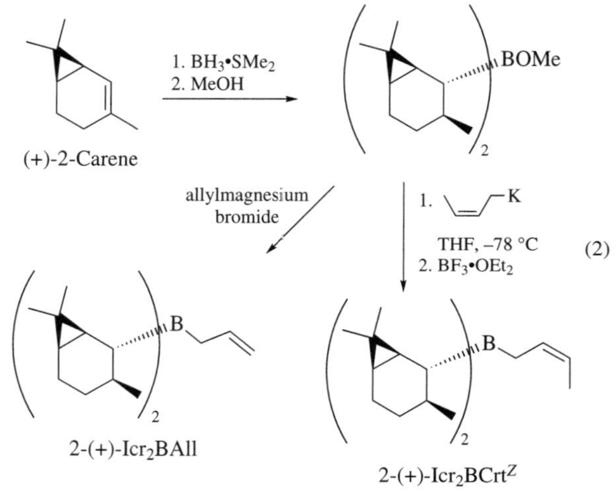

Allylboration of Aldehydes. The *B*-allyldiisocaranylboranes (Icr$_2$BAll) condense with aldehydes with exceptional levels of enantioselectivity (98–>99% ee) to form, upon workup, secondary homoallylic alcohols.[2] While both regioisomers of this reagent (2-^{d}Icr$_2$BAll and 4-^{d}Ipc$_2$BAll) are derived from the (+)-terpene, they exhibit complementary reactivity to form enantiomeric products (eq 3 and eq 4). The degree of enantioselection achieved in the condensation of Icr$_2$BAll with aldehydes is superior to other means of allylboration[1] (see also *B-Allyldiisopinocampheylborane*; *Diisopropyl 2-Allyl-1,3,2-dioxaborolane-4,5-dicarboxylate*; *(R,R)-2,5-Dimethylborolane*).

Crotylboration of Aldehydes. The crotyl analogs of this reagent (Icr$_2$BCrt) likewise provide very high levels of enantio- and diastereoselectivity (>99% de) when condensed with aldehydes.[3] By varying the geometry of the crotyl group (*Z* or *E*) and the (+)-carene isomer used in the reagent preparation (2- or 3-carene), all four of the possible β-methyl homoallylic alcohol products can be obtained (eq 5 and eq 6). As with allylborations, the condensation of *B*-crotyldiisocaranylboranes with aldehydes provide higher levels of enantioselection than the other crotylboration reagents[3] (see also *B-Crotyldiisopinocampheylborane*; (*Z*) and (*E*)*-Diisopropyl 2-Allyl-1,3,2-dioxaborolane-4,5-dicarboxylate*; *(R,R)-2,5-Dimethylborolane*).

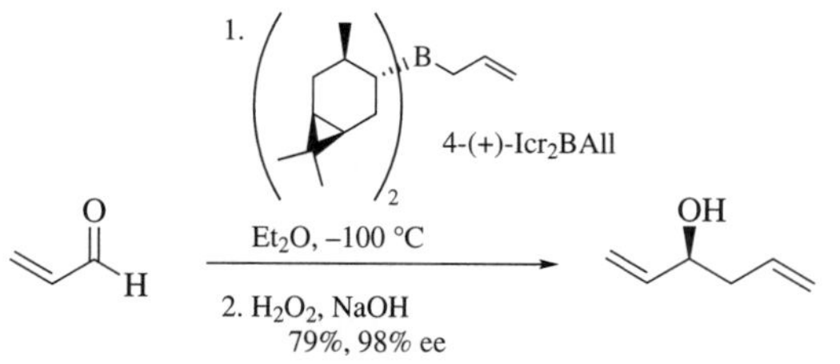

(3)

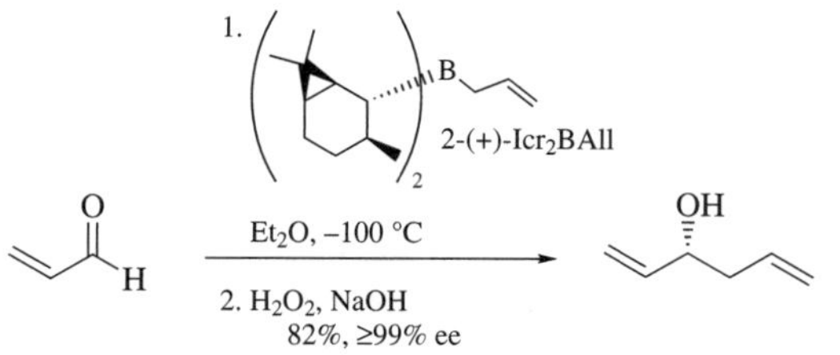

(4)

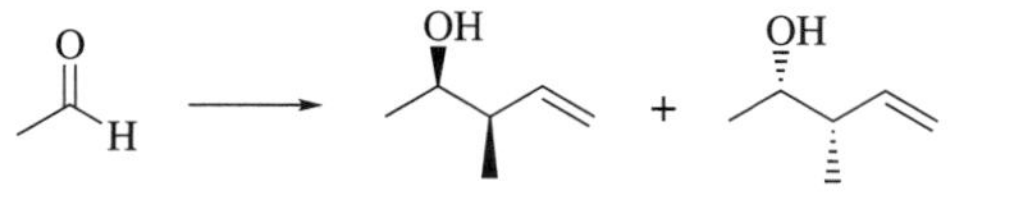

(5)

2-(+)-IcrBCrt^Z 3:97
4-(+)-IcrBCrt^Z 97:3

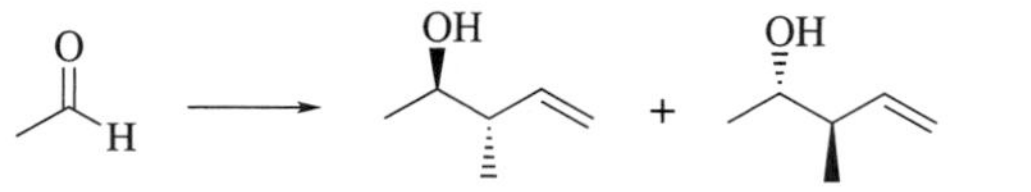

(6)

2-(+)-IcrBCrt^E 2:98
4-(+)-IcrBCrt^E 97:3

1. Brown, H. C.; Ramachandran, P. V. *Pure Appl. Chem.* **1991**, *63*, 307.
2. (a) Brown, H. C.; Jadhav, P. K. *J. Org. Chem.* **1984**, *49*, 4089. (b) Brown, H. C.; Randad, R. S.; Bhat, K. S.; Zaidlewicz, M.; Racherla, U. S. *J. Am. Chem. Soc.* **1990**, *112*, 2389. (c) Racherla, U. S.; Brown, H. C. *J. Org. Chem.* **1991**, *56*, 401.
3. (a) Brown, H. C.; Bhat, K. S. *J. Am. Chem. Soc.* **1986**, *108*, 5919. (b) Brown, H. C.; Randad, R. S. *Tetrahedron* **1990**, *46*, 4457.

Mark T. Goulet
Merck Research Laboratories, Rahway, NJ, USA

(1*R*,2*S*)-1-Amino-2,3-dihydro-1*H*-inden-2-ol[1]

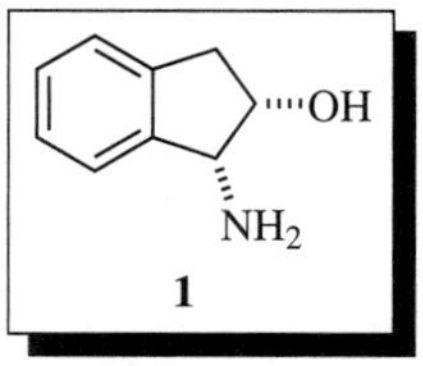

[136030-00-7] $C_9H_{11}NO$ (MW 149.19)

(synthetic building block used in pharmaceutical compounds, and as an asymmetric control element in chiral auxiliaries and asymmetric catalysts)

Physical Data: mp 122–124 °C.
Solubility: soluble in ethanol, isopropanol, dichloromethane, and toluene (hot).
Form Supplied in: colorless crystalline solid; commercially available in either enantiomeric form.
Analysis of Reagent Purity: NMR/CHN analysis. Chiral HPLC to confirm ee.
Preparative Methods: several routes from indene. See main text for details.
Purification: recrystallization from toluene.[2d]
Handling, Storage, and Precautions: relatively air- and moisture-stable, colorless, and odorless crystalline solid, irritating to skin, eyes, and respiratory system.

Preparation. A number of methods have been reported for both the racemic and asymmetric preparations of 1-amino-2,3-dihydro-1*H*-inden-2-ol (**1**), most commonly starting from inexpensive and readily available indene. These methods have been described in detail in recent reviews.[1] The valuable properties of **1** as both a component of a medicinally active compound and as a chirality control element, derive primarily from its rigid and well-defined stereochemical structure. As a result, the compound is most desirable in enantiomerically pure form. One of the most efficient asymmetric syntheses of **1**, which may be employed for the synthesis of either enantiomer of the target molecule, involves an asymmetric epoxidation (89% yield, 88% ee) of indene to give epoxide **2** using the well-established Jacobsen catalyst. This is followed by a Ritter reaction using oleum in acetonitrile resulting in conversion to the oxazoline (**3**) which is subsequently hydrolysed to the amino alcohol. Fractional crystallization with a homochiral diacid permits purification to >99% ee (eq 1).[2]

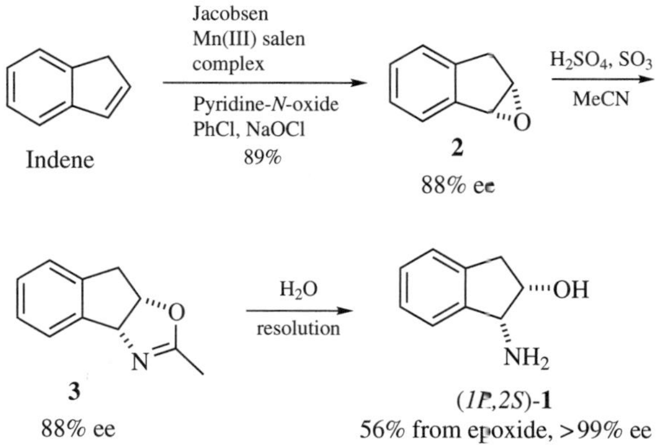

(1)

The enantioselective synthesis of **1** has also been achieved by a number of methods including enzymatic resolution of a keto ester precursor to the racemate followed by conversion of the ester to an amino group,[3a] enzymatic resolution of an amino azide precursor followed by reduction,[4] enzymatic resolution through *O*-acylation of a racemic *N*-benzylcarbamate derivative of **1**,[3b] and the resolution via the formation of an amide with a homochiral amino acid.[5] Bioconversion of indene to *trans-2S,1S*-bromoindanol furnishes a key intermediate towards the synthesis of **1**.[6a,6b] Desymmetrization of 2-TBS-protected indanol through an enantioselective oxidation provides access to a ketone precursor of **1** in up to 70% ee.[6c]

Application as a Synthetic Building Block in Pharmaceutical Compounds. The best-known application of the (1*S*,2*R*)-enantiomer of *cis*-aminoindanol is as a component of Indinavir (**4**), the primary component of a Crixivan® combination therapy (with other reverse transcriptase inhibitors) for AIDS.[7] An excellent account of the synthetic approach to Indinavir, as well as the use of **1** in other drugs, can be found in a recent review.[1a]

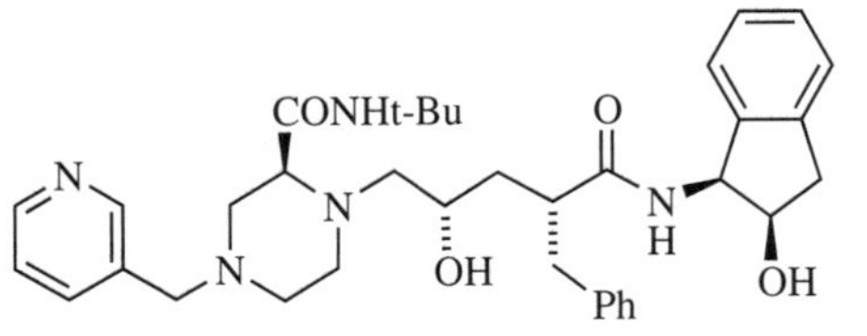

4

Application as a Component of a Chiral Auxiliary. The rigid structure and well-defined conformational rigidity of **1** makes it an ideal building block for a chiral auxiliary. Three different types have been described in some detail. The 'Evans-auxiliary'-type oxazolidinone derivative **5** has given excellent results in aldol reactions with aldehydes (eq 2).[8] The reaction illustrated, proceeding via a boron enolate, is selective for the syn diastereoisomer of product, i.e. **6**, often with >99% de. Following the reaction, the aldol product can be removed from the auxiliary using lithium hydroxide in a water/THF mixture.

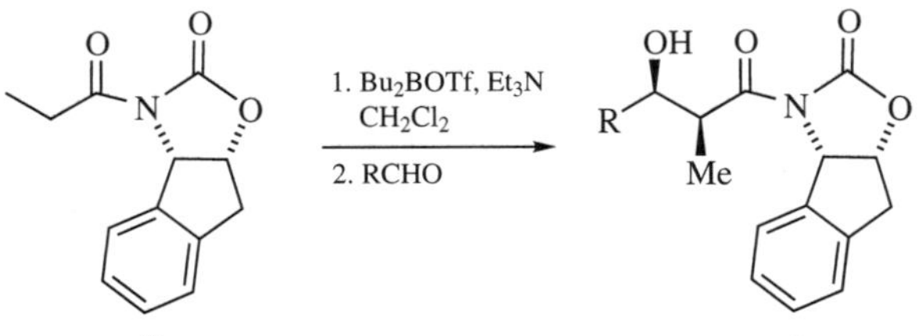

5 **6** (2)

The aldol reaction illustrated in eq 2 has been applied to the targeted synthesis of a number of complex molecules including Tylosin,[8] Hapalosin,[9] the antibiotic Sinefungin,[10] and the HIV protease Saquinavir® inhibitor.[11] Oxazolidinone-type chiral auxiliaries derived from **1** have also been employed for the control of Diels–Alder reactions of attached acryloyl or crotonyl groups.[12]

Asymmetric aldol reactions may also be controlled with high diasteroselectivity, but this time for the anti isomer, in reactions of *N*-tosyl derivatives of esters derived from **7** (eq 3).[13] Diastereoselectivities of up to 99:1 were achieved in the illustrated titanium(IV)-mediated reaction, which has been employed for the synthesis of dipeptide isosteres for incorporation into pharmaceutical building blocks.[14] The selectivity reverses when α- or β-alkoxy aldehydes are employed as electrophiles.

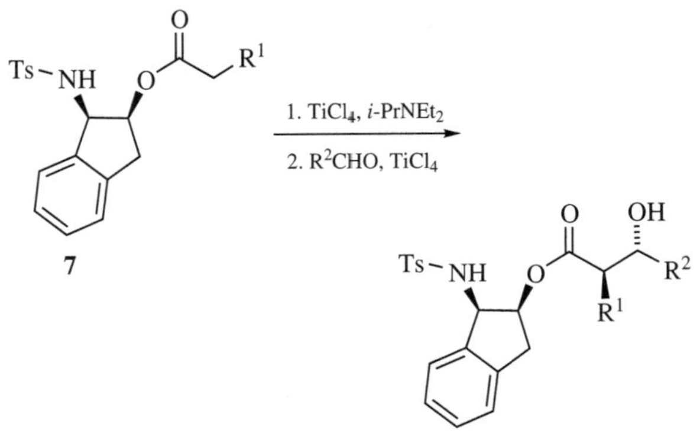

7 (3)

The *N*-tosyl class of auxiliaries derived from **1** have also been successfully applied to the diastereocontrol of Diels–Alder reactions[15,7] and the selective reduction of attached α-keto esters to furnish α-hydroxy ester products.[16]

A third class of chiral auxiliary derived from **1** contains a bridging isopropylidene group between the oxygen and nitrogen atoms, the removable group being appended to nitrogen. This class of auxiliary has been employed in homoaldol reactions via Zn(II) species,[17] to the stereocontrol (in several cases >99% de) of [2,3]-sigmatropic rearrangements[18] and, in the example illustrated in eq 4, the asymmetric synthesis of amino acids through electrophilic amination of attached copper(I) enolates.[19] α-Amino acids may also be prepared through the diastereoselective alkylation of glycine derivatives of the same auxiliary.[20] Addition of organometallic reagents to α-keto amides derived from the same auxiliary provides a means for the asymmetric synthesis of α,α-disubstituted-α-hydroxy acids with excellent enantioselectivity.[21]

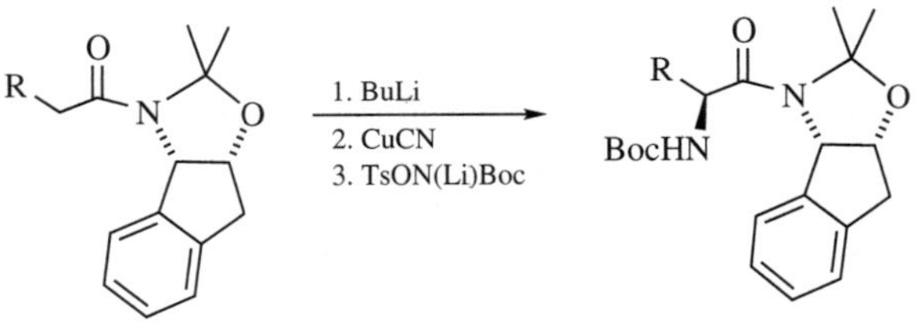

(4)

In a recent application, *cis*-aminoindanol has been employed as a rigid diastereocontrol element in the alkylation of bicyclic lactams and thiolactams of which they are a component.[22] The resulting products form the basis of an enantioselective synthesis of alkaloids.

Application as a Component of an Asymmetric Catalyst. Amino alcohol (**1**) has proven to be a highly versatile ligand for use in asymmetric catalysts for a series of reactions.[1] One of the most comprehensively studied uses is as an oxazaborolidine derivative such as **8** for the asymmetric control of the reduction of ketones by borane. Although its use was first described with stoichiometric levels of **1** being employed for the reduction of both ketones and oximes,[3] development of the system has delivered a catalytic method requiring only 5–10 mol % catalyst.[23] Enantiomeric excesses of over 85% and as high as 96% have been achieved for a range of ketone substrates. α-Chloro and α-bromo ketones are particularly excellent substrates; the reaction in eq 5 is the key step in a highly efficient asymmetric synthesis of the asthma drug (*R,R*)-formoterol[24b,c] and the histamine receptor antagonist Fexofenadine.[24d]

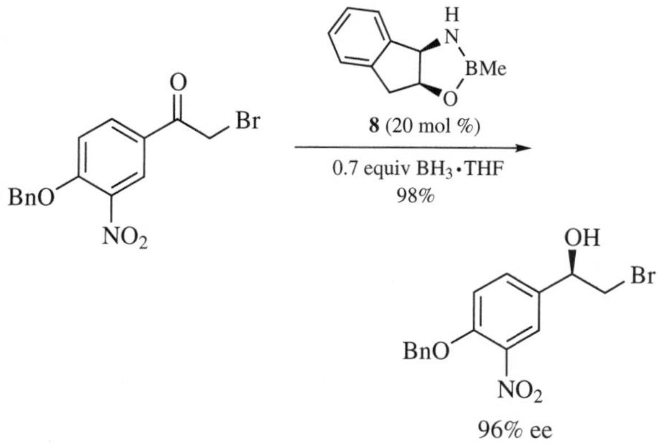

(5)

N,N-Dialkyl derivatives of **1** have been successfully applied to the asymmetric addition of dialkylzinc reagents to aldehydes, giving products of moderate enantiomeric excess.[23] In addition, ruthenium(II) complexes of **1** have been demonstrated to be excellent catalysts for the control of the enantioselective transfer hydrogenation of ketones to alcohols at catalyst loadings as low as 1 mol %.[25] The ruthenium/**1** complex has been applied to a range of ketone substrates, including cyclic enones and α-amino and alkoxy substituted derivatives.

Metal complexes of bis-oxazoline derivatives (**9**) of **1** have been employed for the asymmetric catalysis of the Diels–Alder reactions of acryloyl-*N*-oxazolines with dienes. Detailed studies have been carried out into the effect of the bite angle of the ligand and the nature of the bridging group, on the efficiency of the reaction.[26] In the reaction shown in eq 6, the copper(II) complex of the six-membered chelate ligand catalyzes the addition reaction to give a product in 94% yield and 98% ee at a loading of only 8 mol %. Use of the same ligand with magnesium(II) in place of copper(II) resulted in a reversal of the enantioselectivity, an effect which has been rationalized by a change in coordination at the metal from square planar to tetrahedral. Hetero Diels–Alder reactions have also been achieved using metal complexes of bis-oxazolines derived from **1**.[27] In addition, magnesium(II) complexes of the bis-oxazolines act as effective asymmetric control elements for the asymmetric conjugate [1,4] addition of free-radicals to oxazolidinone-bound cinnamates.[28] Copper(I) complexes of **9** have been employed for the control of carbenoid insertions into silicon–hydrogen bonds.[29]

Bis-oxazoline ligands bearing a bridging 2,6-pyridine group (often referred to as 'pybox' ligands) have been employed for the asymmetric catalysis of alkene cyclopropanations, giving products in good enantioselectivity and diastereoselectivity at very low (0.2 mol % metal) catalyst loadings.[26] However, ligands derived from certain other 1,2-amino alcohols gave superior results.

In a detailed study, bis-oxazoline (**9**) has been employed for the enantiocontrol of a palladium-catalyzed annulation of allenes with aryl and vinylic iodides. This procedure provided an efficient means for the asymmetric synthesis of several classes of heterocyclic target structures including indoles, cyclic ethers, and lactones.[30] Bis-oxazoline (**9**), and certain derivatives bearing different bridging groups, have been employed in the copper-catalyzed allylic acyloxylation reaction of cyclic alkenes. Enantiomeric excesses of up to 78% were achieved using the most efficient catalyst.[31]

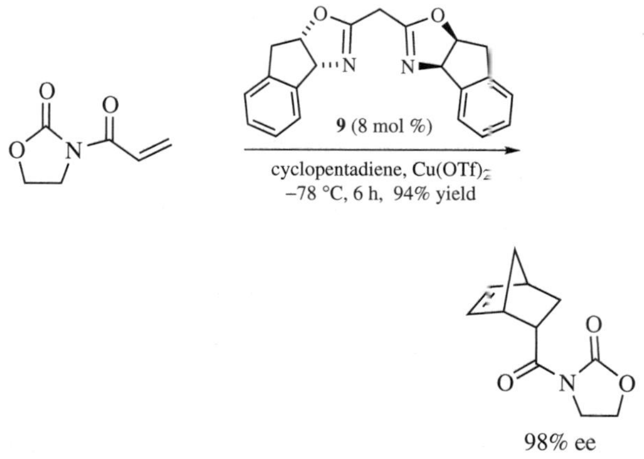

(6)

Tridentate salen ligands (**10**) derived from **1** have given excellent results in the enantiocontrol of the hetero Diels–Alder addition reaction of dienes with aldehydes (eq 7)[32] and in the asymmetric additions of TMS-azide to *meso*-epoxide[33] and trimethylsilyl cyanide to benzaldehyde (up to 85% ee).[34] Phosphino-oxazolines derived from **1** have been employed for the asymmetric control of palladium-catalyzed allylic substitution reactions; products of 70–90% ee were obtained.[35] Photolysis of crystalline adducts of enantiomerically pure **1** with prochiral alcohols results in asymmetric inductions of up to 79% in a rare example of a solid-state enantioselective reaction.[36]

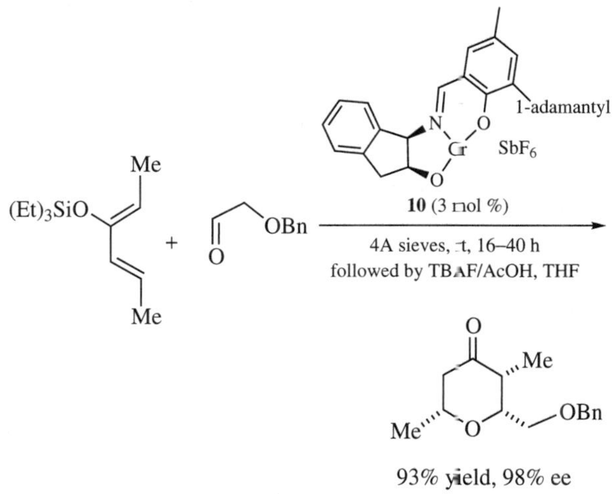

(7)

Related Reagents. (*1S,2R*) enantiomer [*26456-43-7*].

1. (a) Ghosh, A. K.; Fidanze, S.; Senanayake, C. H. *Synthesis* **1998**, 937. (b) Senanayake, C. H. *Aldrichimica Acta.* **1998**, *31*, 3.

2. (a) Senanayake, C. H.; Roberts, F. E.; DiMichele, L. M.; Ryan, K. M.; Liu, J.; Fredenburgh, L. E.; Foster, B. S.; Douglas, A. W.; Larsen, R. D.; Verhoeven, T. R.; Reider, P. J. *Tetrahedron Lett.* **1995**, *36*, 3993.

(b) Senanayake, C. H.; Smith, G. B.; Ryan, K. M.; Fredenburgh, L. E.; Liu, J.; Roberts, F. E.; Hughes, D. L.; Larsen, R. D.; Verhoeven, T. R.; Reider, P. J. *Tetrahedron Lett.* **1996**, *37*, 3271. (c) Hughes, D. L.; Smith, G. B.; Liu, J.; Dezeny, G. C.; Senanayake, C. H.; Larsen, R. D.; Verhoeven, T. R.; Reider, P. J. *J. Org. Chem.* **1997**, *62*, 2222. (d) Larrow, J. F.; Roberts, E.; Verhoeven, T. R.; Ryan, K. M.; Senanayake, C. H.; Reider, P. J.; Jacobsen, E. N.; Lodise, S. A.; Smith, A. B. Jr *Org. Synth.* **1999**, *76*, 46.

3. (a) Didier, E.; Loubinoux, B.; Ramos Tombo, G. M.; Rihs, G., *Tetrahedron* **1991**, *47*, 4941. (b) Luna, A.; Maestro, A.; Astorga, C.; Gotor, V. *Tetrahedron: Asymmetry* **1999**, *10*, 1969.
4. (a) Ogasawara, K.; Takahashi, M. *Synthesis* **1996**, 954. (b) Ghosh, A. K.; Kincaid, J. F.; Haske, M. G. *Synthesis* **1997**, 541. (c) Ghosh, A. K.; Cesti, P.; Battistel, E. *J. Org. Chem.* **1988**, *53*, 5531.
5. Thompson, W. J.; Fitzgerald, P. M. D.; Holloway, M. K.; Emini, E. A.; Darke, P. L.; McKeever, B. M.; Schleif, W. A.; Quintero, J. C., Zugay, J. A.; Tucker, T. J.; Schwering, J. E.; Homnick, C. F.; Nunberg, J.; Springer, J. P.; Huff, J. R. *J. Med. Chem.* **1992**, *35*, 1685.
6. (a) Zhang, J.; Roberge, C.; Reddy, J.; Conners, N.; Chartrain, M.; Buckland, B.; Greasham, R. *Enzyme and Microbial Technology* **1999**, *24*, 86. (b) Igarashi, Y.; Otsutomo, S.; Harada, M.; Nakano, S. *Tetrahedron: Asymmetry* **1997**, *8*, 2833. (c) Komiya, N.; Noji, S.; Murahashi, S.-I. *Tetrahedron Lett.* **1998**, *39*, 7921.
7. Vacca, J. P.; Dorsey, B. D.; Schleif, W. A.; Levin, R. B.; McDaniel, S. L.; Darke, P. L.; Zugay, J.; Quintero, J. C.; Blahy, D. M.; Roth, E.; Sardana, V. V.; Schlabach, A. J.; Graham, P. I.; Condra, J. H.; Gotlib, L.; Holloway, M. K.; Lin, J.; Chen, I.-W.; Vastag, K.; Ostovic, D.; Anderson, P. S.; Emini, E. A.; Huff, J. R. *Proc. Natl. Acad. Sci. USA* **1994**, *91*, 4096.
8. Ghosh, A. K.; Duong, T. T.; McKee, S. P. *Chem. Commun.* **1992**, 1673.
9. Ghosh, A. K.; Liu, W.; Xu, Y.; Chen, Z. *Angew. Chem.* **1996**, *108*, 73; *Angew. Chem., Int. Edn.* **1996**, *35*, 74.
10. Ghosh, A. K.; Liu, W.; *J. Org. Chem.* **1996**, *61*, 6175.
11. Ghosh A. K.; Hussain, K. A.; Fidanze, S. *J. Org. Chem.* **1997**, *62*, 6080.
12. Davies, I. W.; Senanayake, C. H.; Castonguay, L.; Larsen, R. D.; Verhoeven, T. R.; Reider, P. J. *Tetrahedron Lett.* **1995**, *36*, 7619.
13. (a) Ghosh, A. K.; Onishi, M. *J. Am. Chem.* **1996**, *118*, 2527. (b) Ghosh, A. K.; Fidanze, S.; Onishi, M.; Huissain, K. A. *Tetrahedron Lett.* **1997**, *38*, 7171.
14. (a) Ghosh, A. K.; Fidenze, S. *J. Org. Chem.* **1998**, *63*, 6146. (b) Ghosh, A. K.; Bischoff, A. *Org. Lett.* **2000**, *2*, 1573.
15. Ghosh, A. K.; Mathivanan, P. *Tetrahedron: Asymmetry* **1996**, *7*, 375.
16. Ghosh A. K.; Chen, Y. *Tetrahedron Lett.* **1995**, *36*, 6811.
17. (a) Armstrong, J. D., Jr; Hartner, F. W., Jr; DeCamp, A. E.; Volante, R. P.; Shinkai, I. *Tetrahedron Lett.* **1992**, *33*, 6599. (b) McWilliams, J. C.; Armstrong, J. D., Jr; Zheng, N.; Bhupathy, M.; Volante, R. P.; Reider, P. J. *J. Am. Chem. Soc.* **1996**, *118*, 11970.
18. Kress, M. H.; Yang, C.; Yasuda, N.; Grabowski, E. J. J. *Tetrahedron Lett.* **1997**, *38*, 2633.
19. Zheng N.; Armstrong, J. D., Jr; McWilliams, C.; Volante, R. P., *Tetrahedron Lett.* **1997**, *38*, 2817.
20. Lee, J.; Choi, W.-B.; Lynch, J. E.; Volante, R. P.; Reider, P. J. *Tetrahedron Lett.* **1998**, *39*, 3679.
21. Senanayake, C. H.; Fang, K.; Grover, P.; Bekale, R. P.; Vandenbossche, C. P.; Wald, S. A. *Tetrahedron Lett.* **1999**, *40*, 819.
22. (a) Watson, D. J.; Lawrence, C. M.; Meyers, A. I. *Tetrahedron Lett.* **2000**, *41*, 815. (b) Mechelke, M. F.; Meyers, A. I. *Tetrahedron Lett.* **2000**, *41*, 4339.
23. (a) Hong, Y.; Gao, Y.; Nie, X.; Zepp, C. M. *Tetrahedron Lett.* **1994**, *35*, 6631. (b) DiSimone, B.; Savoia, D.; Tagliavini, E.; Umani-Ronchi, A. *Tetrahedron: Asymmetry* **1995**, *6*, 301. (c) Jones, S.; Atherton, J. C. C. *Tetrahedron: Asymmetry* **2000**, *11*, 4543.
24. (a) Hett, R.; Fang, Q. K.; Gao, Y.; Hong, Y.; Butler, H. T.; Nie, X.; Wald, S. A. *Tetrahedron Lett.* **1997**, *38*, 1125. (b) Hett, R.; Senanayake, C. H.; Wald, S. A. *Tetrahedron Lett.* **1999**, *39*, 1705. (c) Hett, R.; Fang, Q. K.; Gao, Y.; Wald, S. A.; Senanayake, C. H. *Org. Proc. Res and Dev.* **1998**, *2*, 96. (d) Fang, Q. K.; Senanayake, C. H.; Wilkinson, H. S.; Wald, S. A.; Li, H. *Tetrahedron Lett.* **1998**, *39*, 2701.
25. (a) Palmer, M.; Walsgrove, T.; Wills, M. *J. Org. Chem.* **1997**, *62*, 5226. (b) Kenny, J. A.; Palmer, M. J.; Smith, A. R. C.; Walsgrove, T.; Wills, M. *Synlett* **1999**; 1615. (c) Wills, M.; Gamble, M.; Palmer, M.; Smith, A.; Studley, J.; Kenny, J. *J. Mol. Catal A: Chemical* **1999**, *146*, 139. (d) Kawamoto, A.; Wills, M. *Tetrahedron: Asymmetry* **2000**, *11*, 3257. (e) Kenny, J. A.; Versluis, K.; Heck, A. J. R.; Walsgrove, T.; Wills, M. *Chem. Commun.* **2000**; 99. (f) Hennig, M.; Puntener, K.; Scalone, M.; *Tetrahedron: Asymmetry* **2000**, *11*, 1849.
26. (a) Davies, I. W.; Senanayake, C. H.; Larsen, R. D.; Verhoeven, T. R.; Reider, P. J. *Tetrahedron Lett.* **1996**, *37*, 1725. (b) Kanemasa, S.; Oderaotoshi, Y.; Yamamoto, H.; Tanaka, J.; Wada, E.; Curran, D. P. *J. Org. Chem.* **1997**, *62*, 6454. (c) Ghosh, A. K.; Mathivnan, P.; Cappiello, J. *Tetrahedron Lett.* **1996**, *37*, 3815. (d) Davies, I. W.; Garena, L.; Castonguay, L.; Senanayake, C. H.; Larsen, R. D.; Verhoeven, T. R.; Reder, P. J. *Chem. Commun.* **1996**, 1753. (e) Davies, I. W.; Garena, L.; Cai, D.; Larsen, R. D.; Verhoeven, T. R.; Reider, P. J. *Tetrahedron Lett.* **1997**, *38*, 1145. (f) Davies, I. W.; Senenayake, C. H.; Larsen, R. D.; Verhoeven, T. R.; Reider, P. J. *Tetrahedron Lett.* **1996**, *37*, 813.
27. (a) Ghosh, A. K.; Mathivanan, P.; Cappiello, J.; Krishnan, K. *Tetrahedron: Asymmetry* **1996**, *7*, 2165. (b) Ghosh, A. K.; Mathivanan, P.; Cappiello, J. *Tetrahedron Lett.* **1997**, *38*, 2427.
28. (a) Wu, J. H.; Radinov, R.; Porter, N. A. *J. Am. Chem. Soc.* **1995**, *117*, 11029. (b) Sibi, M. P.; Ji, J.; Wu, J.-H.; Gurtler, S.; Porter, N. A. *J. Am. Chem. Soc.* **1996**, *118*, 9200. (c) Sibi, M. P.; Ji, J. *J. Org. Chem.* **1997**, *62*, 3800. (d) Sibi, M. P.; Shay, J. J.; Ji, J. *Tetrahedron Lett.* **1997**, *38*, 5955.
29. Dakin, L. A.; Schaus, S. E.; Jacobsen, E. N.; Panek, J. S. *Tetrahedron Lett.* **1998**, *39*, 8947.
30. Zenner, J. M.; Larock, R. C. *J. Org. Chem.* **1999**, *64*, 7312.
31. Clark, J. S.; Tolhurst, K. F.; Taylor, M.; Swallow, S. *J. Chem. Soc., Perkin Trans 1* **1998**; 1167.
32. Dossetter, A. G.; Jamison, T. F.; Jacobsen, E. N. *Angew. Chem., Int. Edn.* **1999**, *38*, 2398.
33. Li, Z.; Fernandez, M.; Jacobsen, E. N. *Org. Lett.* **1999**, *1*, 1611.
34. Flores-Lopez, L. Z.; Parra-Hake, M.; Somanathan, R.; Walsh, P. J., *Organometallics* **2000**, *19*, 2153.
35. Weise, B.; Helmchen, G. *Tetrahedron Lett.* **1998**, *32*, 5727.
36. Rademacher, K.; Scheffer, J. R.; Trotter, J. *Tetrahedron Lett.* **2000**, *56*, 6739.

Martin Wills and Richard Eaves
Warwick University, UK

(*S*)-1-Amino-2-hydroxymethylindoline

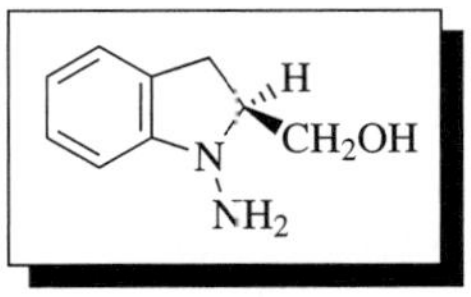

[27719-98-8] $C_9H_{12}N_2O$ (MW 164.208)

(stereospecific synthesis of α-substituted α-amino acids from α-keto acids[1])

Physical Data: mp 81.5–82.7 °C (racemate).
Solubility: readily sol methanol, ethanol; insol hexane, ether.

Preparative Methods: 2-hydroxymethylindoline is resolved with (*S*)-(+)-*Mandelic Acid* to give (*S*)-(+)-2-hydroxymethylindoline, which is then nitrosated and reduced to produce (*S*)-(**1**) (eq 1).[1] (*R*)-(**1**) is also available by a similar procedure.

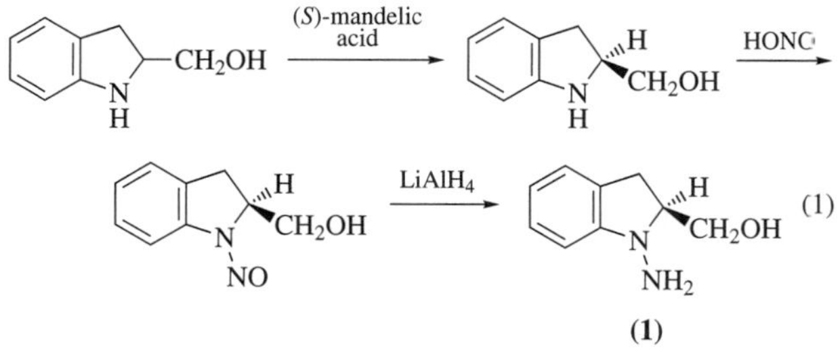

Purification: recrystallization from methanol/ether.

Handling, Storage, and Precautions: the reagent is very air sensitive. Store under nitrogen or argon at $-20\,^{\circ}$C. Use in a fume hood.

The enantiospecific synthesis of natural and unnatural α-amino acids has been reviewed.[2] Some of the most successful approaches involve the stereoselective hydrogenation of chiral dehydroamino acid derivatives. Many of these transformations are equivalent to the stereoselective reductive amination of α-keto acids (eq 2).[3] For example, catalytic reduction of the imines of α-keto acids with chiral α-methylbenzylamine gives α-substituted α-amino acids with 12–80% ee (eq 3).[4]

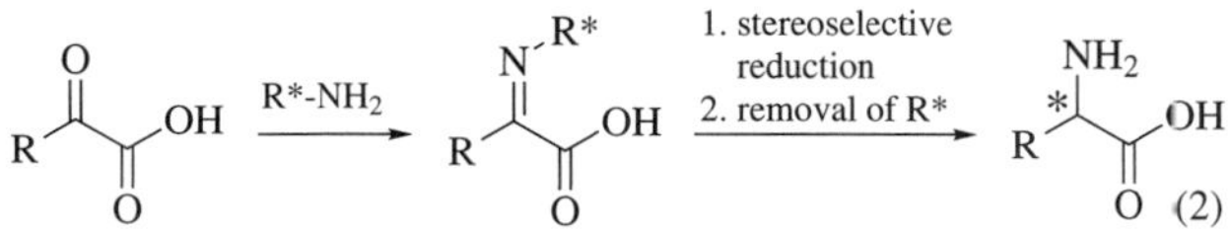

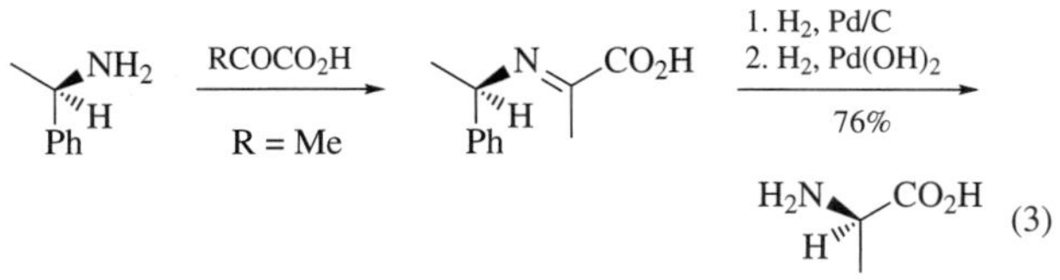

Compound (**1**) and its enantiomer provide a variation on the same theme of stereospecific reductive amination. In this case, reduction of a chiral cyclic hydrazone (derived from an α-keto acid and (**1**)) with *Aluminum Amalgam* in wet DME proceeds with high stereoselectivity. Reductive cleavage of the N–N bond and ester hydrolysis complete the procedure, which produces α-amino acids with high optical purity (eq 4).[1] The source of chirality is recovered by conversion of the resulting indoline-2-methanol back into (**1**).[1]

Related reagents that have been used successfully for these purposes include (**2**) and (**3**).[1,5] In general, (**3**) appears to generate product with higher enantiomeric purity than either (**1**) or (**2**).

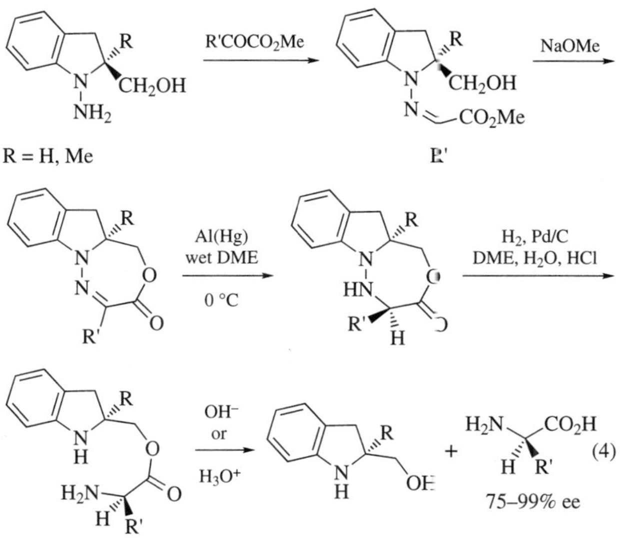

(**2**) (**3**)

Other chiral amine reagents that have been used to effect similar stereospecific reductive aminations include 1,2-diamines[6] and 1,2-amino alcohols (eq 5).[7,8]

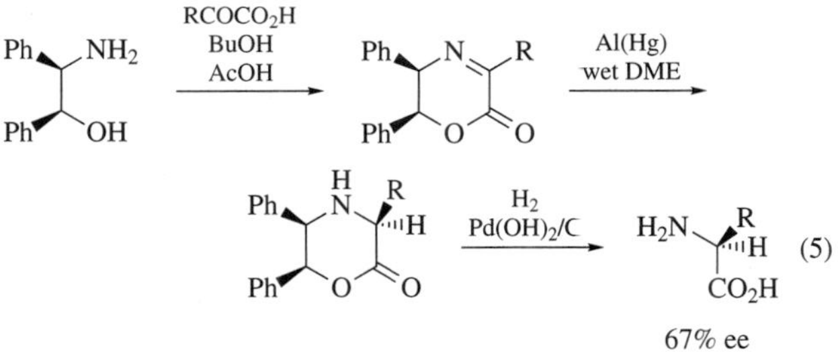

Stereospecific reduction of chiral diketopiperazine derivatives from proline and α-keto acids also provide a versatile route to α-amino acids (eq 6).[9,10] The selectivity of the reduction is highly dependent on the nature of the R′ group on the nitrogen atom.

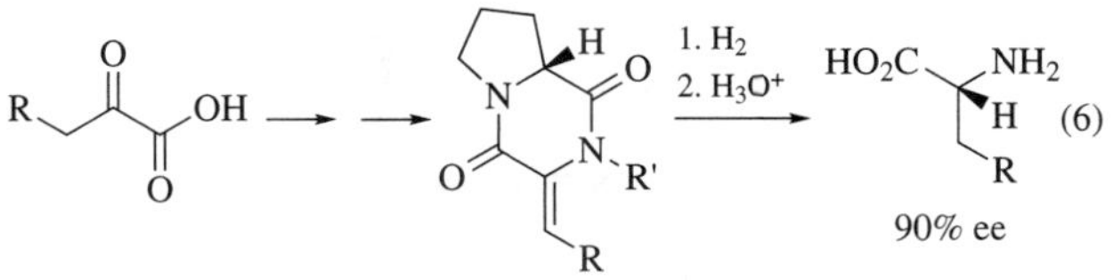

A related synthetic approach involves the biomimetic transamination of α-keto acids with chiral pyridoxamine analogs (eq 7)[11] or achiral pyridoxamine analogs in the presence of a chiral ligand.[12,13]

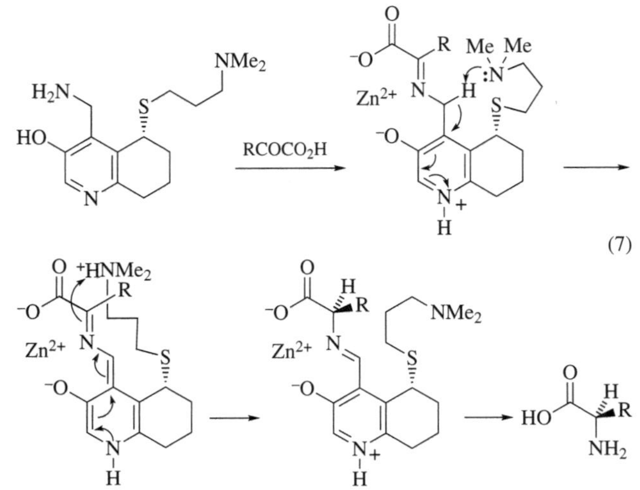

(7)

Related Reagents. (*S*)-1-Amino-2-methoxymethylpyrrolidine.

1. Corey, E. J.; McCaully, R. J.; Sachdev, H. S. *J. Am. Chem. Soc.* **1970**, *92*, 2476.
2. (a) Haemers, A.; Mishra, L.; Van Assche, I.; Bollaert, W. *Pharmazie* **1989**, *44*, 97. (b) *Asymmetric Synthesis, Chiral Catalysis*; Morrison, J. D. Ed.; Academic: Orlando, 1985; Vol. 5.
3. Babievskii, K. K.; Latov, V. K. *Russ. Chem. Rev.* **1969**, *38*, 456.
4. (a) Harada, K.; Matsumoto, K. *J. Org. Chem.* **1967**, *32*, 1794. (b) Hiskey, R. G.; Northrop, R. C. *J. Am. Chem. Soc.* **1961**, *83*, 4798.
5. Corey, E. J.; Sachdev, H. S.; Gougoutas, J. Z.; Saenger, W. *J. Am. Chem. Soc.* **1970**, *92*, 2488.
6. Meric, R.; Vigneron, J. P. *Tetrahedron Lett.* **1974**, *15*, 2059.
7. Jiao, X. Y.; Chen, W. Y.; Hu, B. F. *Synth., Coll.* **1992**, *22*, 1179.
8. Vigneron, J. P.; Kagan, H.; Horeau, A. *Tetrahedron Lett.* **1986**, *9*, 5681.
9. Bycroft, B. W.; Lee, G. R. *Chem. Commun.* **1975**, 988.
10. Poisel, H.; Schmidt, U. *Ber. Dtsch. Chem. Ges.* **1973**, *106*, 3408.
11. Zimmerman, S. C.; Breslow, R. *J. Am. Chem. Soc.* **1984**, *106*, 1490.
12. Deschenaux, R.; Bernauer, K. *Helv. Chim. Acta* **1984**, *67*, 373.
13. Bernauer, K.; Deschenaux, R.; Taura, T. *Helv. Chim. Acta* **1983**, *66*, 2049.

Juan C. Jaen
Parke-Davis Pharmaceutical Research, Ann Arbor, MI, USA

(*S*)-1-Amino-2-methoxymethylpyrrolidine[1]

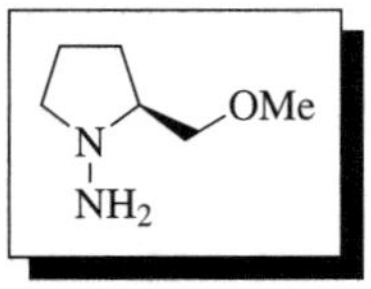

[59983-39-0] $C_6H_{14}N_2O$ (MW 130.19)

(chiral auxiliary; Enders' reagent; diastereo- and/or enantioselective alkylations,[2,3] aldol reactions,[4] Michael additions[5] and reductive or alkylative aminations,[6] resolutions,[7] ee determinations[8])

Alternate Name: SAMP.
Physical Data: bp 186–187 °C; *d* 0.977 g cm^{-3}; n_D^{20} 1.4650; α_D^{20} −80 to −82° (neat).
Solubility: sol H_2O, ether, dichloromethane.
Form Supplied in: colorless liquid or as crystalline colorless oxalate.
Handling, Storage, and Precautions: storage at 0–4 °C under argon atmosphere.

Since the pioneering times of the mid-1970s, (*S*)-1-amino-2-methoxymethylpyrrolidine (SAMP) and its enantiomer RAMP have been among the most powerful chiral auxiliaries in asymmetric synthesis, with a very broad range of applications. As a proline derivative it generally shows high stereoselectivities due to the rigidity of the five-membered ring and the ability to coordinate metal fragments[9] [see also *(S)-2-Methoxymethylpyrrolidine*, SMP].

The procedure involves the transformation of carbonyl compounds to the corresponding SAMP or RAMP hydrazones, metalation, trapping of the intermediate azaenolates with various electrophiles, and either hydrazone cleavage (carbonyl compounds) or hydrazone reduction/N–N bond cleavage (amines).

The synthetic utility of the SAMP/RAMP hydrazone method is demonstrated in particular in the stereoselective alkylation of aldehyde[2] and ketone[3] SAMP/RAMP hydrazones. A great number of natural products have been synthesized using this method, like the principal alarm pheromone of the leaf cutting ant *Atta texana* (eq 1),[3a] the C(1)–C(15) segment of FK 506 (eq 2),[2b] the amino acid MeBMT (eq 3),[2c] and (−)-methyl kolavenate (eq 4).[3b]

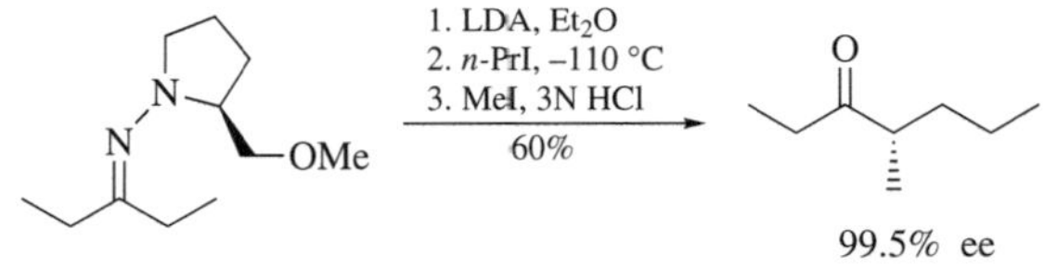

(1)

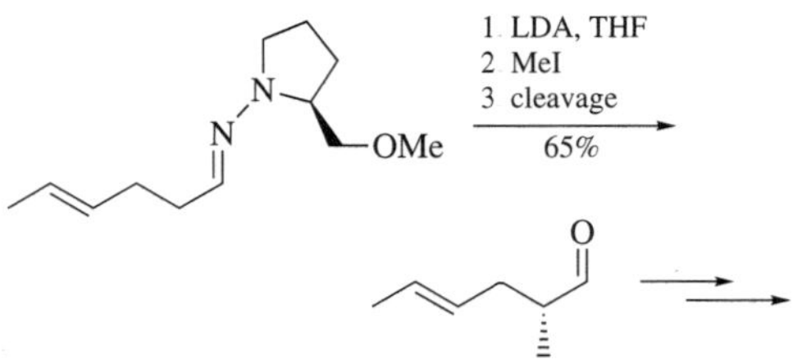

MeBMT (2)

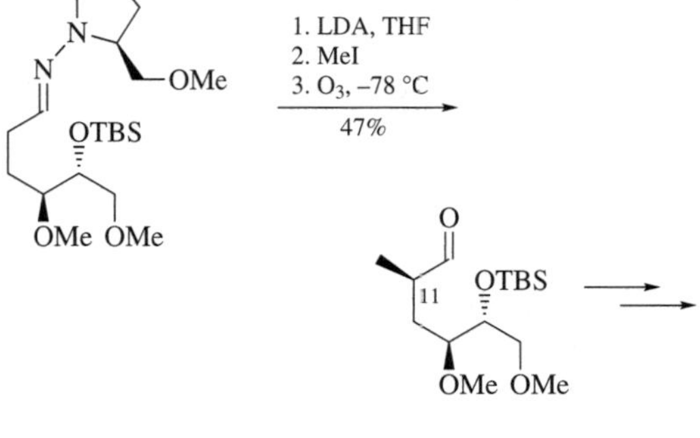

FK 506 (3)

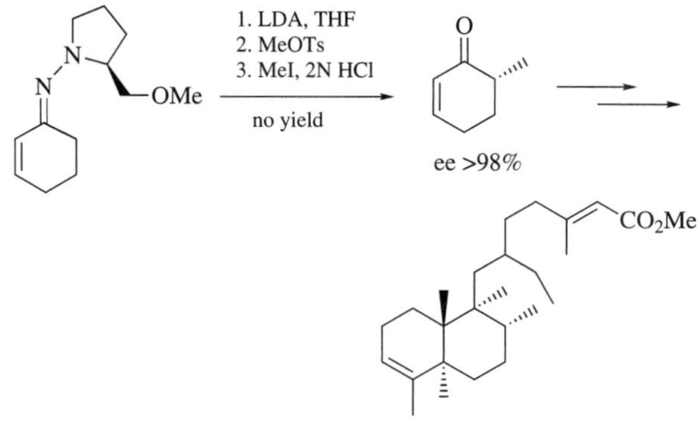

(4)

(−)-Methyl kolavenate

2,2-Dimethyl-1,3-dioxan-5-one SAMP/RAMP hydrazones[3f–j] were used as dihydroxyacetonephosphate equivalents in the synthesis of C_2 symmetric ketones (eq 5),[3g] aza sugars with novel substitution patterns,[3h] or C_5 to C_9 deoxy sugars.[3i] SAMP hydrazones of 2-oxo esters represent novel phosphoenolpyruvate (PEP) equivalents.[3k,l] α,α-Disubstituted spiroacetals are accessible via the alkylation of ketone SAMP/RAMP hydrazones.[3m]

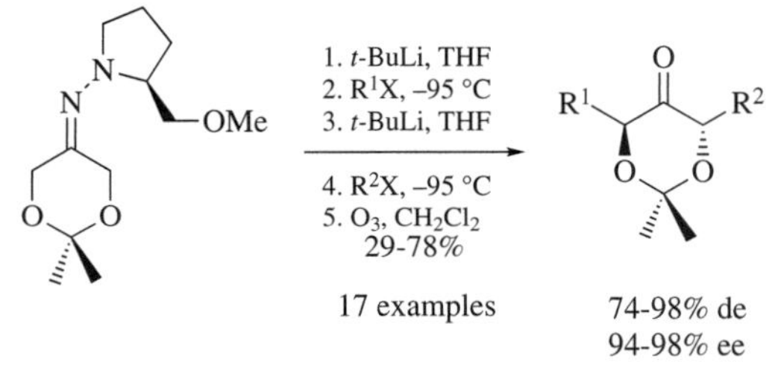

(5)

The aggregation pheromone of *Drosophila mulleri*, (*S*)-2-tridecanol acetate, is obtainable by alkylation of propiophenone SAMP hydrazone followed by a Baeyer–Villiger reaction of the ketone (eq 6).[3n]

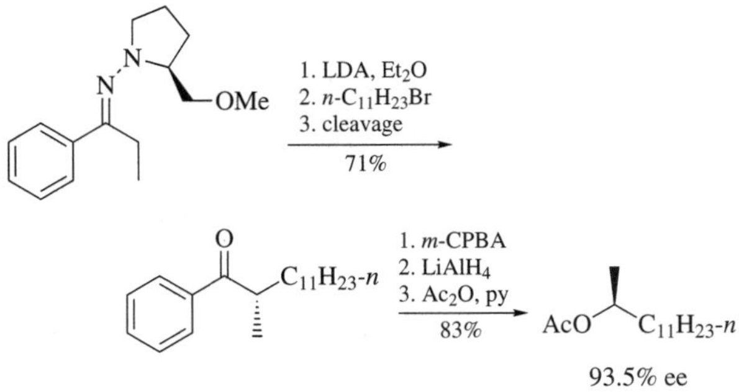

(6)

The relative and absolute configuration of Stigmatellin A, one of the most potent inhibitors of the electron transport chain, was determined via alkylation of diethyl ketone SAMP hydrazone.[3o]

The aldol reaction is the preferred method for the stereoselective synthesis of 1,3-dioxygenated building blocks. In 1978, Enders et al.[4a,b] reported the first enantioselective aldol reaction in the difficult case of α-unsubstituted β-ketols using SAMP and RAMP. Diastereo- and enantiomerically pure *syn*-β-ketols are available by aldol reaction of SAMP/RAMP hydrazones (eq 7).[4c]

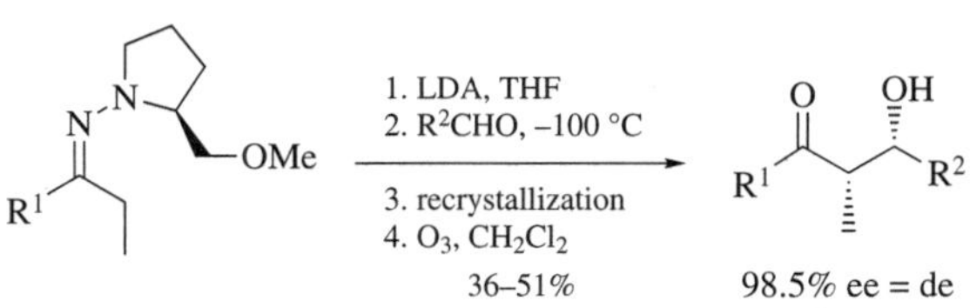

(7)

The aggregation pheromone of the rice and maize weevil was synthesized by aldol reaction of an enantiomerically pure α-silyl ketone, obtained by the SAMP/RAMP hydrazone method,[10a–d] with various aldehydes (eq 8).[4d]

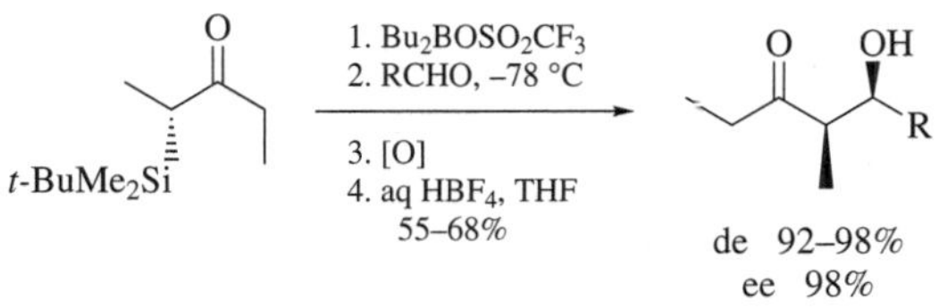

(8)

The utility of the SAMP/RAMP hydrazone method in diastereo- and enantioselective Michael additions was demonstrated in the synthesis of 5-oxo esters[5a–e] (eq 9),[5b] δ-lactones (eq 10),[5c,e,f] oxo diesters and dinitriles,[5g] heterocyclic compounds (eq 11),[5h] MIRC (Michael initiated ring closure) reactions,[5i] and 2-substituted 4-oxo sulfones.[5i]

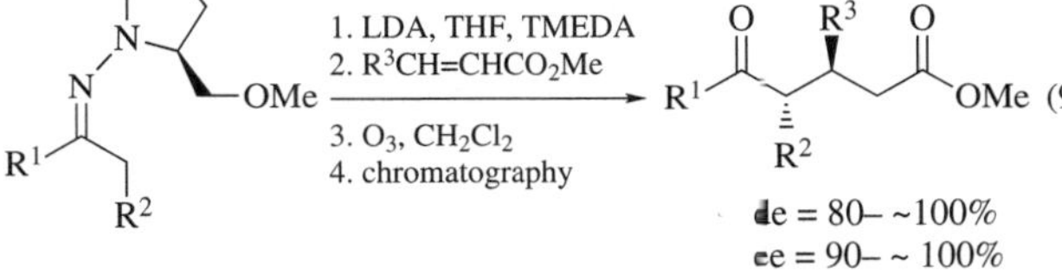

(9)

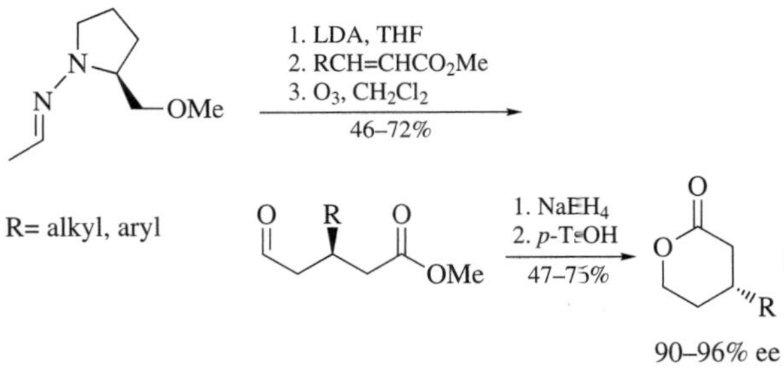

(10)

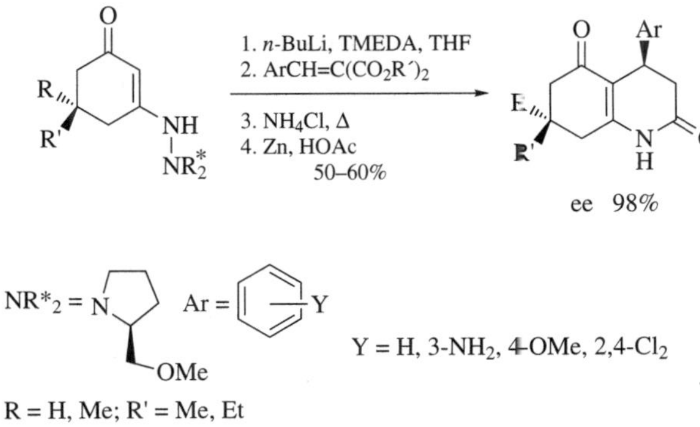

(11)

Organotin reagents can be added to cyclic α,β-unsaturated SAMP/RAMP hydrazones (eq 12).[5k]

Lithiated *N*-protected SAMP can be used as an ammonia equivalent in Michael and tandem Michael additions to α,β-unsaturated esters (eq 13).[5l] Furthermore, SAMP and RAMP

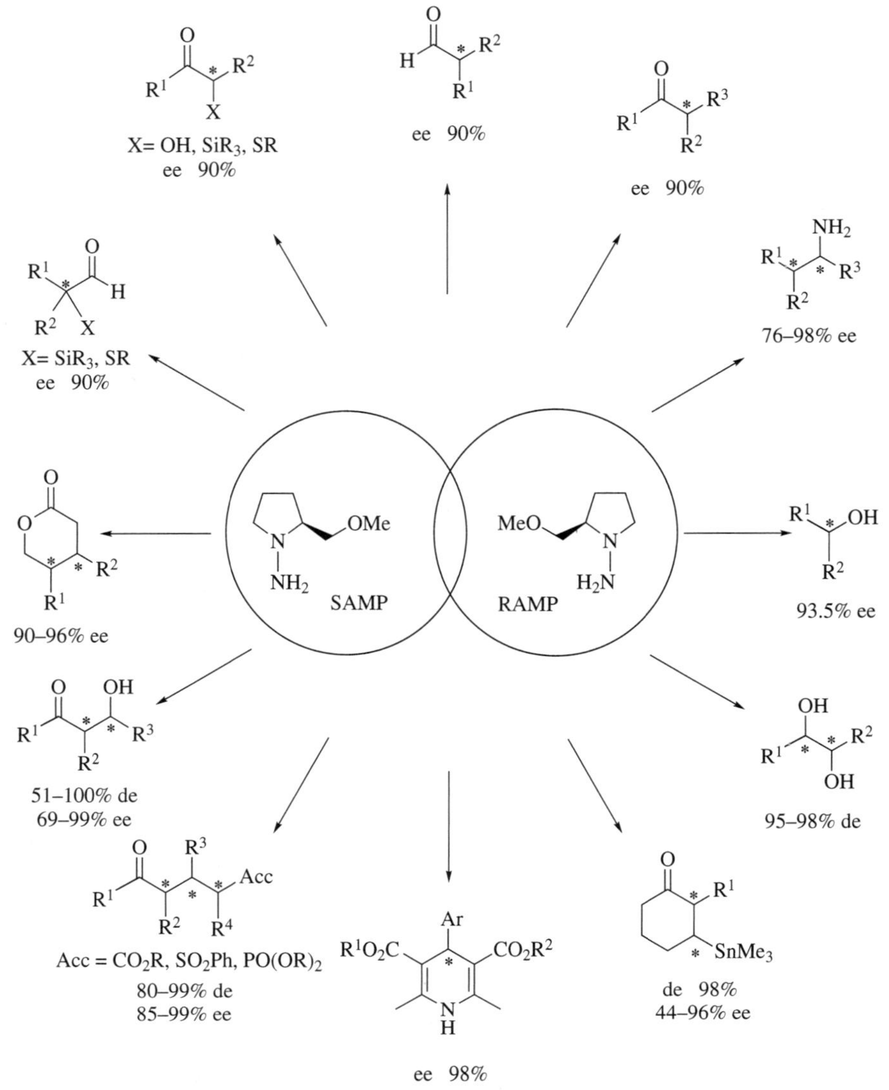

Scheme 1

can be employed for the asymmetric synthesis of α- and/or β-substituted primary amines with high regio-, diastereo-, and enantioselectivities.[6] This variant involves hydrazone reduction with a subsequent N–N bond cleavage and can be combined with a prior α-alkylation, as described above.

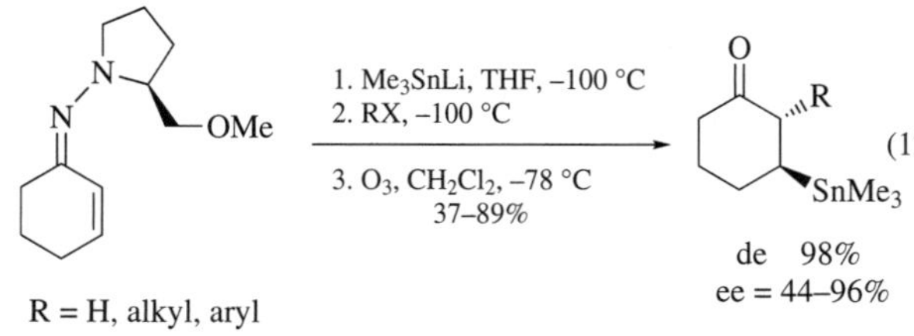

(12)

This method was recently used in the synthesis of different natural products, like the ladybug defence alkaloid harmonine,[6d] α- and β-amino acetals and acids (eq 14),[6e,f] and both enantiomers of the hemlock alkaloid coniine,[6g] utilizing the nucleophilic 1,2-addition of organolithium and -lanthanoid reagents to SAMP/RAMP hydrazones.

The alkylation of SAMP/RAMP hydrazones with heteroelectrophiles leads to enantiomerically pure α-silyl aldehydes and ketones (eq 15),[10a–d] α-sulfenyl aldehydes and ketones (eq 16),[10e] and α-hydroxy aldehydes and ketones (eq 17).[10f]

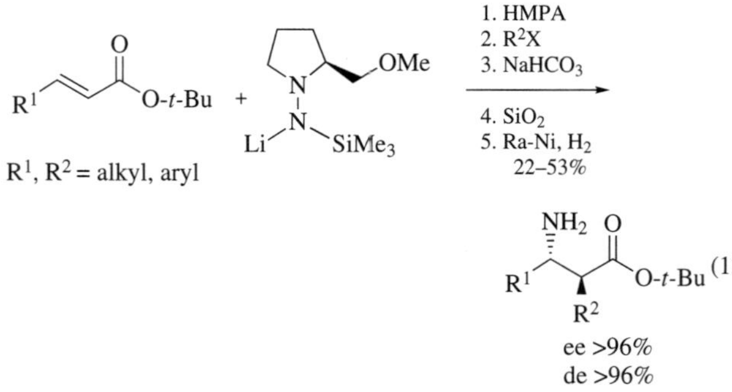

(13)

These very interesting chiral building blocks are employed in aldol reactions,[4d] and in the synthesis of enantiomerically pure vicinal diols (eq 18)[10g,h] and 3-oxo esters and acids bearing quarternary stereogenic centers (eq 19).[10i]

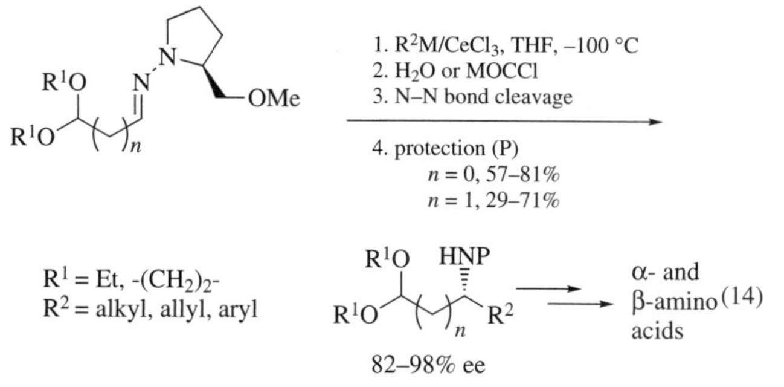

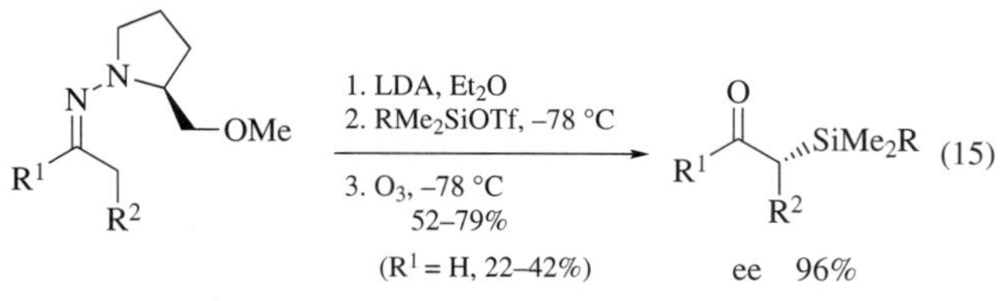

R[1], R[2] = H, alkyl, aryl; R= *t*-Bu, Me_2CHCMe_2

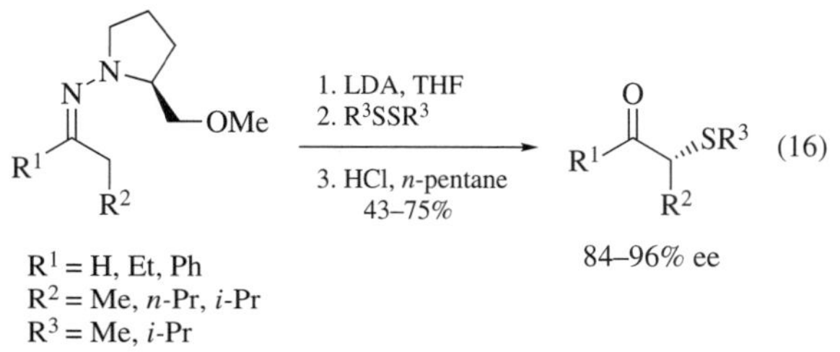

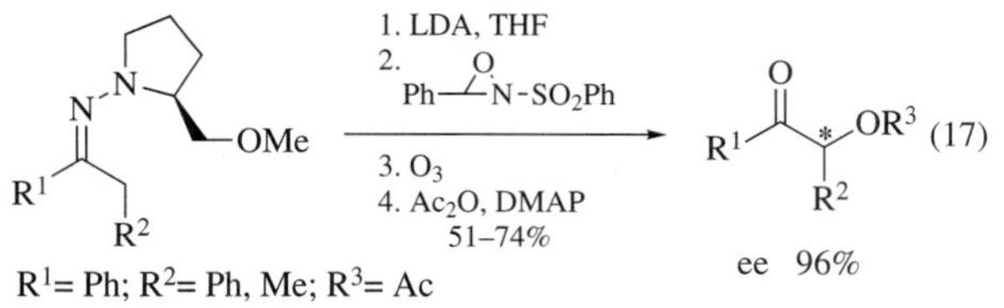

Further applications can be mentioned briefly. SAMP was used in the resolution of 4-demethoxy-7-deoxydaunomycinone,[7] in ee determinations (Scheme 1),[8] as a chelate for tetracarbonylmolybdenum complexes,[11] in intramolecular Heck reactions,[12] as polysilylated hydrazine,[13] in the enantioselective synthesis of isoquinuclidines,[14] and in the conversion of hydrazones to aldehydes[15] and nitriles.[16] The structure of a chiral lithium SAMP hydrazone azaenolate has been determined.[17] In cases where SAMP did not lead to satisfactory inductions, a modified auxiliary, (*S*)-1-amino-2-dimethylmethoxymethylpyrrolidine (SADP),[18] enhanced the stereochemical control.

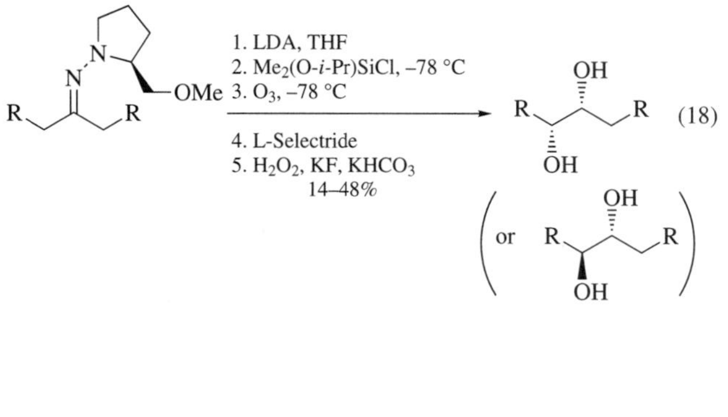

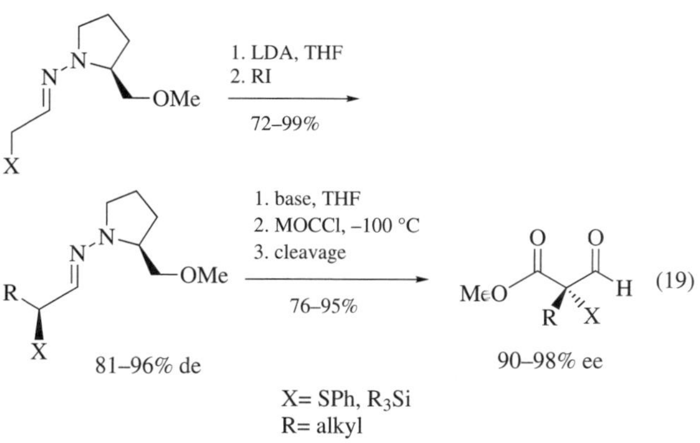

Related Reagents. (*S*)-2-Methoxymethylpyrrolidine.

1. (a) Enders, D., In *Asymmetric Synthesis*; Morrison, J. D., Ed.; Academic Press: New York, 1984, Vol. 3, p 275. (b) Enders D.; Fey, P.; Kipphardt, H. *Org. Synth.* **1987**, *65*, 173, 183.
2. (a) Nicolaou, K. C.; Papahatjis, D. P.; Claremon, D. A.; Dolle, R. E. *J. Am. Chem. Soc.* **1981**, *103*, 6967. (b) Kocienski, P.; Stocks, M.; Donald, D.; Cooper, M.; Manners, A. *Tetrahedron Lett.* **1988**, *29*, 4481. (c) Beulshausen, T.; Groth, U. M.; Schöllkopf, U. *J. Am. Chem. Soc.* **1994**, in press. (d) Enders, D.; Dyker, H. *Liebigs Ann. Chem.* **1990**, 1107. (e) Schmidt, U.; Siegel, W.; Mundinger, K. *Tetrahedron Lett.* **1988**, *29*, 1269. (f) Hauck, R. S.; Wegner, C.; Blumtritt, P.; Fuhrhop, J. H.; Nau, H. *Life Sci.* **1990**, *46*, 513. (g) Kündig, P.; Liu, R.; Ripa, A. *Helv. Chim. Acta* **1992**, *75*, 2657.
3. (a) Enders, D.; Eichenauer, H. *Angew. Chem.* **1979**, *91*, 425; *Angew. Chem., Int. Ed. Engl.* **1979**, *18*, 397. (b) Hideo, I.; Mitsugu, M.; Kimikazu, O.; Tokoroyama, T. *Chem. Commun.* **1987**, 358. (c) Pennanen, S. I. *Acta. Chem. Scand.* **1981**, *B35*, 555. (d) Mori, K.; Nomi, H.; Chuman, T.; Kohno, M.; Kato, K.; Noguchi, M. *Tetrahedron* **1982**, *38*, 3705. (e) Fischer, J.; Kilpert, C.; Klein, U.; Steglich, W. *Tetrahedron* **1986**, *42*, 2063. (f) Enders, D.; Bockstiegel, B. *Synthesis* **1989**, 493. (g) Enders, D.; Gatzweiler, W.; Jegelka, U. *Synthesis* **1991**, 1137. (h) Enders, D.; Jegelka, U. *Synlett* **1992**, 999. (i) Enders, D.; Jegelka, U.; Dücker, B. *Angew. Chem.* **1993**, *105*, 423; *Angew. Chem., Int. Ed. Engl.* **1993**, *32*, 423. (j) Enders, D.; Jegelka, U. *Tetrahedron Lett.* **1993**, *34*, 2453. (k) Enders, D.; Dyker, H.; Raabe, G. *Angew. Chem.* **1992**, *104*, 649; *Angew. Chem., Int. Ed. Engl.* **1992**, *31*, 618. (l) Enders, D.; Dyker, H.; Raabe, G., *Synlett* **1992**, 901. (m) Enders, D.; Gatzweiler, W.; Dederichs, E. *Tetrahedron* **1990**, *46*, 4757. (n) Enders, D.; Plant, A. *Liebigs Ann.* **1991**, 1241. (o) Enders, D.; Osborne, S. *Chem. Commun.* **1993**, 424. (p) Sainsbury, M.; Williams, C. S; Naylor, A.; Scopes, D. I. C. *Tetrahedron Lett.* **1990**, *31*, 2763. (q) Sainsbury, M.; Mahon, M. F.; Williams, C. S.; Naylor, A.; Scopes, D. I. C. *Tetrahedron* **1991**, *47*, 4195. (r) Ziegler, F. E.; Becker, M. R., *J. Org. Chem.* **1990**, *55*, 2800. (s) Warshawsky, A. M.; Meyers, A. I. *J. Am. Chem. Soc.* **1990**, *112*, 8090. (t) Andersen, M. W.; Hildebrandt, B.; Hoffmann, R. W. *Angew. Chem.* **1991**, *103*, 90; *Angew. Chem., Int. Ed. Engl.* **1991**, *30*, 90. (u) Clark, J. S.; Holmes, A. B. *Tetrahedron Lett.* **1988**, *29*, 4333. (v) Carling, R. W.; Curtis, N. R.; Holmes, A. B. *Tetrahedron Lett.* **1989**, *30*, 6081. (w) Curtis, N. R.; Holmes, A. B.; Looney, M. G.; Pearson, N. D.; Slim, G. C. *Tetrahedron Lett.* **1991**, *32*, 537. (x) Hart, T. W.; Guillochon, D.; Perrier, G.; Sharp, B. W.; Toft, M. P.; Vacher, B.; Walsh, R. J. A. *Tetrahedron Lett.* **1992**, *33*, 7211.
4. (a) Enders, D.; Friedrich, E.; Lutz, W.; Pieter, R. *Angew. Chem.* **1978**, *90*, 219; *Angew. Chem., Int. Ed. Engl.* **1978**, *17*, 206. (b) Enders, D.; Eichenauer, H.; Pieter, R. *Chem. Ber.* **1979**, *112*, 3703. (c) Enders, D. *Chem. Scr.* **1985**, *25*, 139. (d) Enders, D.; Lohray, B. B. *Angew. Chem.* **1988**, *100*, 594; *Angew. Chem., Int. Ed. Engl.* **1988**, *27*, 581. (e) Enders, D.; Dyker, H.; Raabe, G. *Angew. Chem.* **1993**, *105*, 420; *Angew. Chem., Int. Ed. Engl.* **1993**, *32*, 421.

5. (a) Enders, D.; Papadopoulos, K. *Tetrahedron Lett.* **1983**, *24*, 4967. (b) Enders, D.; Papadopoulous, K.; Rendenbach, B. E. M.; Appel, R.; Knoch, F. *Tetrahedron Lett.* **1986**, *27*, 3491. (c) Enders, D.; Rendenbach, B. E. M. *Pestic. Sci. Biotechnol., Proc. Int. Congr. Pestic. Chem., 6th* **1986**, 17. (d) Enders, D.; Rendenbach, B. E. M. *Tetrahedron* **1986**, *42*, 2235. (e) Enders, D.; Rendenbach, B. E. M. *Chem. Ber.* **1987**, *120*, 1223. (f) Tietze, L. F.; Schneider, C. *J. Org. Chem.* **1991**, *56*, 2476. (g) Enders, D.; Demir, A. S.; Rendenbach, B. E. M. *Chem. Ber.* **1987**, 1731. (h) Enders, D.; Demir, A. S.; Puff, H.; Franken, S. *Tetrahedron Lett.* **1987**, *28*, 3795. (i) Enders, D.; Scherer, H. J.; Raabe, G. *Angew. Chem.* **1991**, *103*, 1676; *Angew. Chem., Int. Ed. Engl.* **1991**, *30*, 1664. (j) Enders, D., Papadopoulos, K.; Herdtweck, E. *Tetrahedron* **1993**, *49*, 1821. (k) Enders, D.; Heider, K. *Angew. Chem.* **1993**, *105*, 592; *Angew. Chem., Int. Ed. Engl.* **1993**, *32*, 598. (l) Enders, D.; Wahl, H.; Bettray, W. *Angew. Chem.* **1995**, *107*, 527; *Angew. Chem., Int. Ed. Engl.* **1995**, *34*, 455.
6. (a) Enders, D.; Schubert, H.; Nübling, C. *Angew. Chem.* **1986**, *98*, 1118; *Angew. Chem., Int. Ed. Engl.* **1986**, *25*, 1109. (b) Denmark, S. E.; Weber, T.; Piotrowski, D. W. *J. Am. Chem. Soc.* **1987**, *109*, 2224. (c) Weber, T.; Edwards, J. P.; Denmark, S. E. *Synlett* **1989**, 20. (d) Enders, D.; Bartzen, D. *Liebigs Ann. Chem.* **1991**, 569. (e) Enders, D.; Funk, R.; Klatt, M.; Raabe, G.; Hovestreydt, E. R. *Angew. Chem.* **1993**, *105*, 418; *Angew. Chem., Int. Ed. Engl.* **1993**, *32*, 418. (f) Enders, D.; Klatt, M.; Funk, R. *Synlett* **1993**, 226. (g) Enders, D.; Tiebes, J., *Liebigs Ann. Chem.* **1993**, 173. (h) Denmark, S. E.; Nicaise, O., *Synlett* **1993**, 359.
7. Dominguez, D.; Ardecky, R. J.; Cava, M. P. *J. Am. Chem. Soc.* **1983**, *105*, 1608.
8. (a) Günther, K.; Martens, J.; Messerschmidt, M. *J. Chromatogr.* **1984**, *288*, 203. (b) Effenberger, F.; Hopf, M.; Ziegler, T.; Hudelmeyer, J. *Chem. Ber.* **1991**, *124*, 1651. (c) Harden, R. C.; Rackham, D. M. *J. High Resolut. Chromatogr.* **1992**, *15*, 407.
9. (a) Enders, D.; Eichenauer, H. *Angew. Chem.* **1976**, *93*, 579. (b) Enders, D.; Fey, P.; Kipphardt, H. *Org. Prep. Proced. Int.* **1985**, *17*, 1.
10. (a) Enders, D.; Bhushan, B. B. *Angew. Chem.* **1987**, *99*, 359; *Angew. Chem., Int. Ed. Engl.* **1987**, *26*, 351. (b) Enders, D.; Bhushan, B. B. *Angew. Chem.* **1988**, *100*, 594; *Angew. Chem., Int. Ed. Engl.* **1988**, *27*, 581. (c) Bhushan, B. B.; Enders, D. *Helv. Chim. Acta* **1989**, *72*, 980. (d) Bhushan, B. B.; Zimbinski, R. *Tetrahedron Lett.* **1990**, *31*, 7273. (e) Enders, D.; Schäfer, T. publication in preparation. (f) Enders, D.; Bhushan, V. *Tetrahedron Lett.* **1988**, *29*, 2437. (g) Enders, D.; Nakai, S. *Helv. Chim. Acta* **1990**, *73*, 1833. (h) Enders, D.; Nakai, S. *Chem. Ber.* **1991**, *124*, 219. (i) Enders, D.; Zamponi, A.; Raabe, G. *Synlett* **1992**, 897.
11. Ehlers, J.; Tom Dieck, H. *Z. Anorg. Allg. Chem.* **1988**, *560*, 80.
12. Grigg, R.; Dorrity, M. J. R.; Malone, J. F.; Mongkolaussavaratana, T.; Norbert, W. D. J. A.; Sridharan, V. *Tetrahedron Lett.* **1990**, *31*, 3075.
13. Hwu, J. R.; Wang, N. *Tetrahedron* **1988**, *44*, 4181.
14. (a) Mehmandoust, M.; Marazano, C.; Singh, R.; Cesario, M.; Fourrey, J. L.; Das, B. C. *Tetrahedron Lett.* **1988**, *29*, 4423. (b) Genisson, Y.; Marazano, C.; Mehmandoust, M.; Gnecco, D.; Das, B. C., *Synlett* **1992**, 431.
15. (a) Enders, D.; Bhushan V. *Z. Naturforsch., Teil B* **1987**, *42B*, 1595. (b) Enders, D.; Plant, A. *Synlett* **1990**, 725.
16. (a) Moore, J. S.; Stupp, S. I. *J. Org. Chem.* **1990**, *55*, 3374. (b) Enders, D.; Plant, A. *Synlett* **1994**, 1054.
17. Enders, D.; Bachstädter, G.; Kremer, K. A. M.; Marsch, M.; Harms, K.; Boche, G. *Angew. Chem.* **1988**, *100*, 1580; *Angew. Chem., Int. Ed. Engl.* **1988**, *27*, 1522.
18. (a) Enders, D.; Kipphardt, H.; Gerdes, P.; Breña-Valle, J.; Bushan, V. *Bull. Soc. Chim. Belg.* **1988**, *97*, 691. (b) Enders, D.; Müller, S.; Demir, A. S. *Tetrahedron Lett.* **1988**, *29*, 6437. (c) Enders, D.; Dyker, H.; Raabe, G. *Angew. Chem.* **1993**, *105*, 420; *Angew. Chem., Int. Ed. Engl.* **1993**, *32*, 421. (d) Enders, D.; Bhushan, V. *Tetrahedron Lett.* **1988**, *29*, 2437.

Dieter Enders & Martin Klatt
RWTH Aachen, Germany

2-Amino-3-methyl-1,1-diphenyl-1-butanol[1]

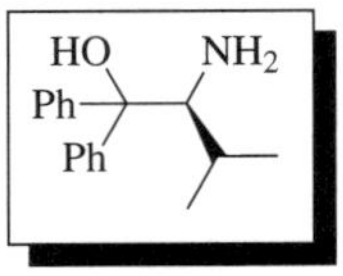

(S)
[78603-95-9] $C_{17}H_{21}NO$ (MW 255.36)
(·HCl)
[130432-39-2]
(R)
[86695-06-9]
(·HCl)
[56755-20-5]

(precursor to Itsuno's reagent, a chiral oxazaborolidine catalyst[1] used for the enantioselective reduction of prochiral ketones,[2] oxime *O*-ethers,[2c,d,3] and imines[4])

Alternate Name: 1,1-diphenylvalinol.
Physical Data: mp 94–95 °C; $[\alpha]_{589}$ −127.7° (*c* 0.693, $CHCl_3$) for the (*S*) enantiomer.
Solubility: very sol THF, CH_2Cl_2, MeOH; not very sol water.
Preparative Methods: from valine methyl ester hydrochloride and excess *Phenylmagnesium Bromide* in 56% yield.[2]
Purification: recrystallization from ethanol–water (10:1).
Handling, Storage, and Precautions: no special information available. In general, however, it is advisable that all reactions with this reagent be conducted in a well ventilated fume hood. Care should be exercised to avoid contact of this reagent and the derived oxazaborolidine catalyst with the eyes and skin.

Stoichiometric Enantioselective Ketone Reduction. The oxazaborolidine–borane complex (**3**) prepared in situ from diphenylvalinol (**1**) and *Borane–Tetrahydrofuran* (2 mol equiv) (eq 1) will enantioselectively reduce a variety of prochiral ketones (eq 2, Table 1).[2] Free oxazaborolidine (**2**) is ineffective stoichiometrically as an enantioselective reducing agent. The asymmetric catalyst (**3**) works best for the reduction of aryl alkyl ketones, providing very high levels of enantioselectivity (94–100% ee). In the case of dialkyl ketones, the best enantioselectivity (78% ee) is obtained for the reduction of *t*-butyl methyl ketone. Reduction of acylsilanes affords the corresponding carbinols in moderate to high enantioselectivity (50–94% ee).[5]

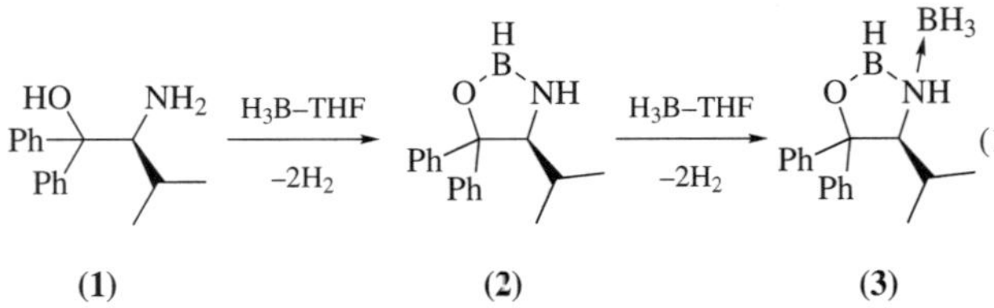

(**1**) (**2**) (**3**)

Reduction of α-chloroacetophenone using the catalyst prepared from the related (*S*)-diphenylisoleucinol (**4**) and borane gives (*S*)-chlorohydrin (**5**), which is readily transformed to (*S*)-

styrene oxide (eq 3).[3a] The reduction affords the (*S*) enantiomer of (**5**) due to chlorine having a higher priority in nomenclature, not a change in the enantiofacial selectivity of the asymmetric catalyst.

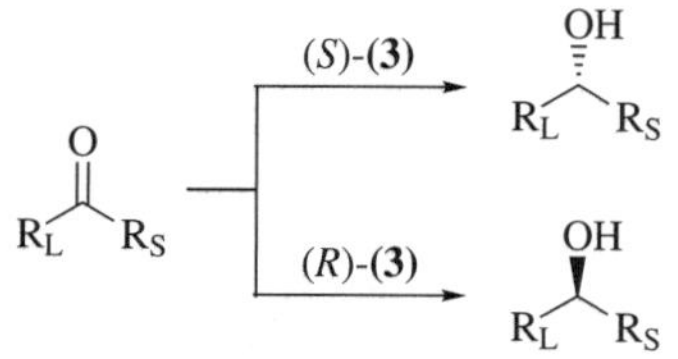

(2)

Table 1 Stoichiometric Enantioselective Reduction of Prochiral Ketones

R_L	R_S	Enantiomeric purity[a] (%)	Absolute configuration[b]
Ph	Me	94	*R*
Ph	Et	94	*R*
Ph	*n*-Pr	96	*R*
Ph	*n*-Pr	6[c]	*R*
Ph	*n*-Bu	100	*R*
n-Pr	Me	55	*R*
i-Pr	Me	60	*R*
t-Bu	Me	78	*R*
Ph	CH_2Cl	96[d]	*S*

[a] Enantiomeric purity based on the optical rotation of the product vs. the maximum value reported. [b] Absolute configuration of the product obtained using the catalyst derived from (*S*)-diphenylvalinol (**1**). [c] Reaction using free oxazaborolidine (**2**). [d] Catalyst derived from (*S*)-diphenylleucinol.

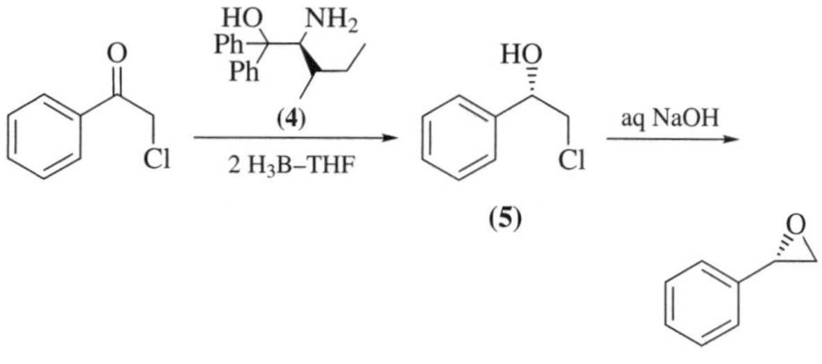

(3)

96% ee

Catalytic Enantioselective Ketone Reduction. Although free oxazaborolidine (**2**) is ineffective by itself as an enantioselective reducing agent, Corey demonstrated that it can be used catalytically (0.025–0.5 equiv) with excess borane (0.6 mol equiv) for the enantioselective reduction of acetophenone (eq 4).[6] A mechanism was proposed to explain the behavior of the catalyst. Initial coordination between the Lewis acidic ring boron and the ketonic oxygen activates the ketone towards reduction. Intramolecular hydride transfer from the BH_3 coordinated to the ring nitrogen then occurs via a six-membered ring cyclic transition state. In addition, oxazaborolidine (**6**) prepared from α,α-diphenyl-2-pyrrolidinemethanol (eq 5) was reported to be 'an even better catalyst for the reduction of ketones'. For a more detailed discussion, see the entries for *α,α-Diphenyl-2-pyrrolidinemethanol* and *Tetrahydro-1-methyl-3,3-diphenyl-1H,3H-pyrrolo[1,2-c][1,3,2]oxazaborole*.

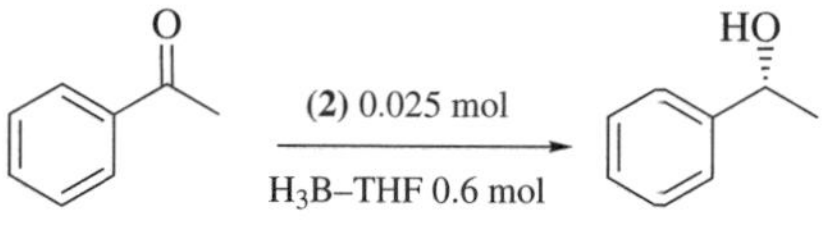

(4)

94.7% ee

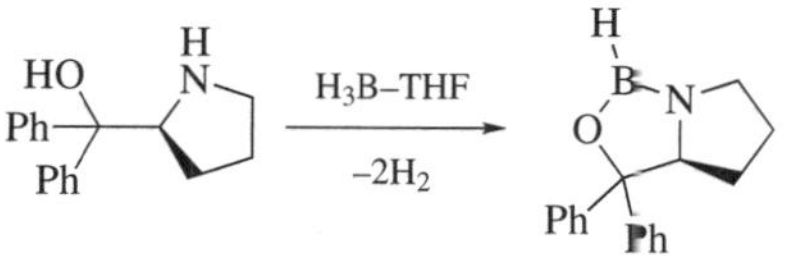

(**6**)

(5)

Recently, Katz employed oxazaborolidine (**2**) to catalyze the enantioselective reduction of ketone (**7**) (eq 6). The resultant carbinol was used for the synthesis of optically active helical metallocene oligimers.[7]

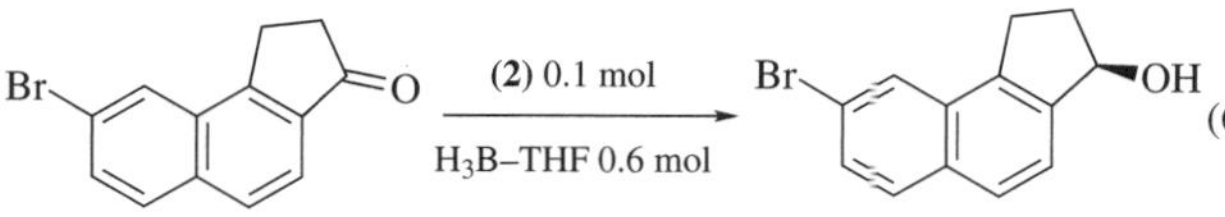

(**7**)

(6)

Enantioselective Reduction of Oxime *O*-Ethers. In addition to the reduction of prochiral ketones, oxazaborolidine (**3**) has been used (both stoichiometrically and catalytically with borane–THF) to catalyze the enantioselective reduction of prochiral ketoxime *O*-ethers to the corresponding amine (eq 7).[2c–d,3] Unlike the ketone reduction described above, the (*S*)-oxazaborolidine catalyst gives (*S*)-amines. The best enantioselectivity is obtained for the case where R = Me (Table 2). Addition of a Lewis acid, such as *Aluminum Chloride*, to the oxime increases the rate of reduction (complete reaction in 3 h vs. 24 h).

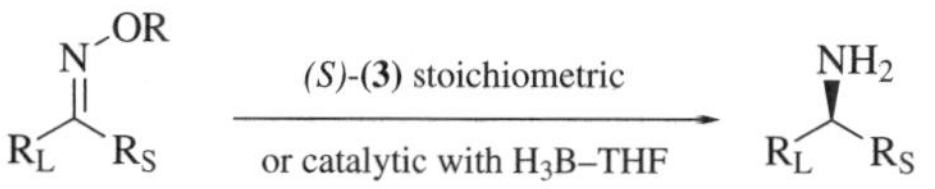

(7)

Enantioselective Reduction of Imines. Oxazaborolidine (**3**) also enantioselectively reduces *N*-substituted ketimines to the corresponding *N*-substituted amine in low to moderate ee (eq 8, Table 3).[4a] In this case the enantioselectivity is the same as the reduction of ketones; thus the (*S*)-oxazaborolidine catalyst gives (*R*)-amines. Oxazaborolidine (**3**) is reported to provide higher enantioselectivity than oxazaborolidine (**6**). An interesting application of this reaction is the preparation of a (a*RS*,*S*) diastereomerically enriched (63% de) sample of the more active atropisomers of the herbicide Metalochlor (eq 9).[4b]

Table 2 Enantioselective Reduction of Prochiral Ketoxime *O*-Ethers

R_L	R_S	R	Enantiomeric purity[a] (%)	Absolute configuration[b] (%)
Ph	Me	H	0.6	*S*
Ph	Me	Me	99	*S*
Ph	Me	Et	81	*S*
Ph	Me	Bn	95	*S*
Ph	Me	Bn	90[c]	*S*
Ph	Me	Bn	94[d]	*S*
Ph	Me	Bn	89[d,e]	*S*
Ph	Me	TMS	62	*S*
Ph	Me	Ac	8.7	*R*

[a] Enantiomeric purity based on the optical rotation of the product vs. the maximum value reported. [b] Absolute configuration of the product obtained using a stoichiometric amount of the (*S*)-catalyst (**3**). [c] Reduction using 0.25 mol equiv of the catalyst with excess borane–THF. [d] Reduction using the catalyst derived from (*S*)-diphenyl-*O*-benzyltyrosenol. [e] $AlCl_3$ added to the ketoxime *O*-ether prior to reduction.

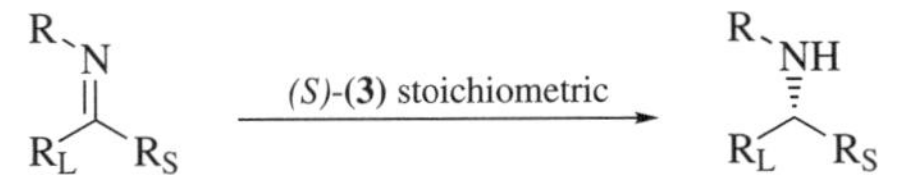

(8)

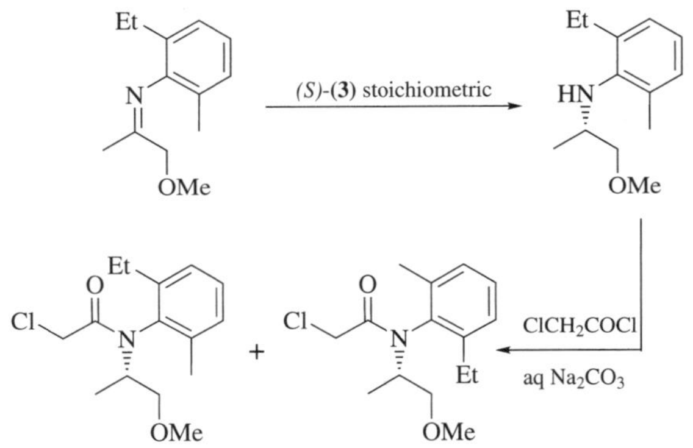

(9)

Table 3 Enantioselective Reduction of Prochiral Imines

R_L	R_S	R	Enantiomeric purity[a] (%)	Absolute configuration[b]
Ph	Me	Ph	73	*R*
Ph	Et	Ph	87	*R*
Ph	*n*-Pr	Ph	88	*R*
Ph	Me	Bn	46	*R*
Et	Me	Ph	9[c]	*R*

[a] Enantiomeric purity based on the optical rotation of the product vs. the maximum value reported. [b] Absolute configuration of the product obtained using a stoichiometric amount of the catalyst derived from (*S*)-1,1-diphenylvalinol.

Related Reagents. α,α-Diphenyl-2-pyrrolidinemethanol Ephedrine-borane Norephedrine–Borane Tetrahydro-1-methyl-3,3-diphenyl-1*H*,3*H*-pyrrolo[1,2-*c*][1,3,2]oxazaborole.

1. (a) Wallbaum S.; Martens, J. *Tetrahedron: Asymmetry* **1992**, *3*, 1475. (b) Singh, V. K. *Acta Chem. Scand.* **1992**, 605.
2. (a) Itsuno, S.; Ito, K.; Hirao, A.; Nakahama, S. *Chem. Commun.* **1983**, 469. (b) Itsuno, S.; Hirao, A.; Nakahama, S.; Yamazaki, N., *J. Chem. Soc., Perkin Trans. 1* **1983**, 1673. (c) Itsuno, S.; Ito, K.; Hirao, A.; Nakahama, S. *J. Org. Chem.* **1984**, *49*, 555. (d) Itsuno, S.; Nakano, M.; Miyazaki, K.; Masuda, H.; Ito, K.; Hirao, A.; Nakahama, S. *J. Chem. Soc., Perkin Trans. 1* **1985**, 2039. (e) Itsuno, S.; Nakano, M.; Ito, K.; Hirao, A.; Owa, M.; Kanda, N.; Nakahama, S. *J. Chem. Soc., Perkin Trans. 1* **1985**, 2615.
3. (a) Itsuno, S.; Sakurai, Y.; Ito, K ; Hirao, A.; Nakahama, S., *Bull. Chem. Soc. Jpn.* **1987**, *60*, 395. (b) Itsuno, S.; Sakurai, Y.; Shimizu, K.; Ito, K. *J. Chem. Soc., Perkin Trans. 1* **1989**, 1548. (c) Itsuno, S.; Sakurai, Y.; Shimizu, K.; Ito, K. *J. Chem. Soc., Perkin Trans. 1* **1990**, 1859.
4. (a) Cho, B. T.; Chun, Y. S. *J. Chem. Soc., Perkin Trans. 1* **1990**, 3200. (b) Cho, B. T.; Chun, Y. S. *Tetrahedron: Asymmetry* **1992**, *3*, 337.
5. Buynak, J. D.; Strickland, J. B.; Hurd, T.; Phan, A. *Chem. Commun.* **1989**, 89.
6. Corey, E. J.; Bakshi, R. K.; Shibata, S. *J. Am. Chem. Soc.* **1987**, *109*, 5551.
7. Katz, T. J.; Sudhakar, A.; Teasley, M. F.; Gilbert, A. M.; Geiger, W. E.; Robben, M. P.; Wuensch, M.; Ward, M. D. *J. Am. Chem. Soc.* **1993**, *115*, 3182.

David J. Mathre & Ichiro Shinkai
Merck Research Laboratories, Rahway, NJ, USA

3-Amino-2,6,6-trimethylbicyclo[3.1.1]heptan-2-ol

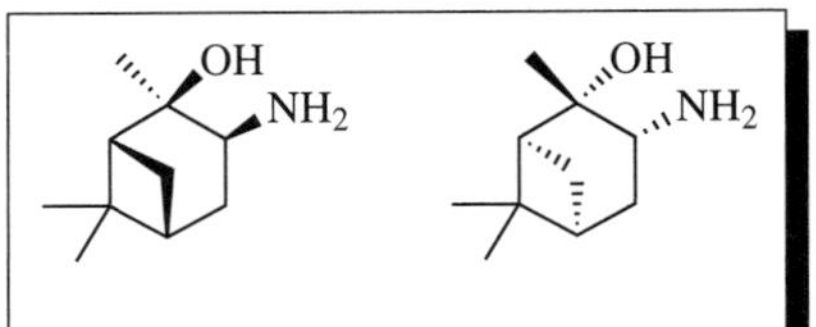

(1*R*,2*R*,3*S*,5*R*) (1*S*,2*S*,3*R*,5*S*)

(1*R*,2*R*,3*S*,5*R*)
[168286-10-0];
(1*S*,2*S*,3*R*,5*S*)
[69363-09-3] $C_{10}H_{19}NO$ (MW 169.26)

(reagent used as a chiral source in stereoselective reactions)

Alternate Name: ATBH, 3-amino-2-hydroxypinane
Physical Data: (1*R*,2*R*,3*S*,5*R*) mp 48–49.5 °C, $[\alpha]_D^{22}$ +13.1 (*c* 1.0, $CHCl_3$);[1] (1*S*,2*S*,3*R*,5*S*) mp 45–46.5 °C, $[\alpha]_D^{22}$ −14.3 (*c* 1.0, $CHCl_3$).[1]
Solubility: soluble in most organic solvents; e.g. THF, CH_2Cl_2, $CHCl_3$, EtOAc.
Form Supplied in: colorless crystals; not commercially available.
Analysis of Reagent Purity: ^{1}H NMR, IR, elemental analysis.
Preparative Methods: optically pure (1*R*,2*R*,3*S*,5*R*)-3-amino-2,6,6-trimethylbicyclo[3.1.1]heptan-2-ol (ATBH) can be prepared from optically impure (1*R*,5*S*)-α-pinene (eq 1).[1] Oxidation of (1*R*,5*S*)-α-pinene (81% ee) with potassium permanganate and subsequent reaction with hydroxylamine affords the α-hydroxy oxime, which is re-

crystallized from ethyl ether-hexane to give the enantiomerically enriched product. Reduction of the α-hydroxy oxime with lithium aluminum hydride furnishes (1*R*,2*R*,3*S*,5*R*)-ATBH in optically pure form. (1*S*,2*S*,3*R*,5*S*)-ATBH can also be prepared in the same way from (1*S*,5*R*)-α-pinene (91% ee).[1]

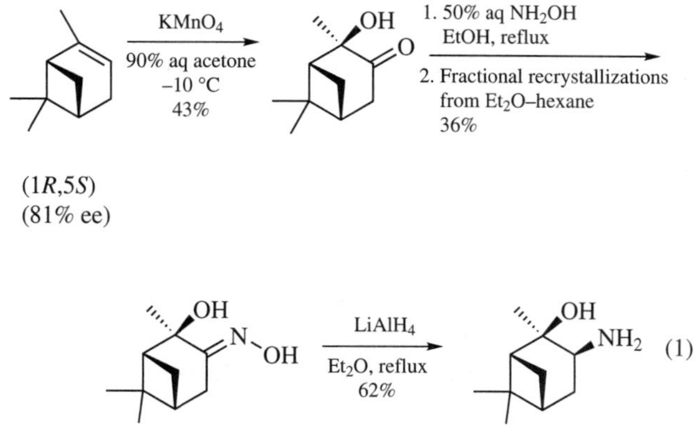

(1*R*,5*S*) (81% ee)

(>99.9% ee) (1*R*,2*R*,3*S*,5*R*)

Purification: recrystallization from ethyl acetate-hexane.

Handling, Storage, and Precautions: hygroscopic crystals. Use in a fume hood.

Asymmetric Borane Reduction. The reaction of ATBH with trimethylboroxine by refluxing in toluene affords the chiral *B*-methyl oxazaborolidine in high yield (eq 2).[1] This oxazaborolidine can serve as an efficient catalyst for the asymmetric borane reduction of prochiral ketones (eq 3).[2] The corresponding chiral secondary alcohols are obtained in high yields with good enantioselectivities.

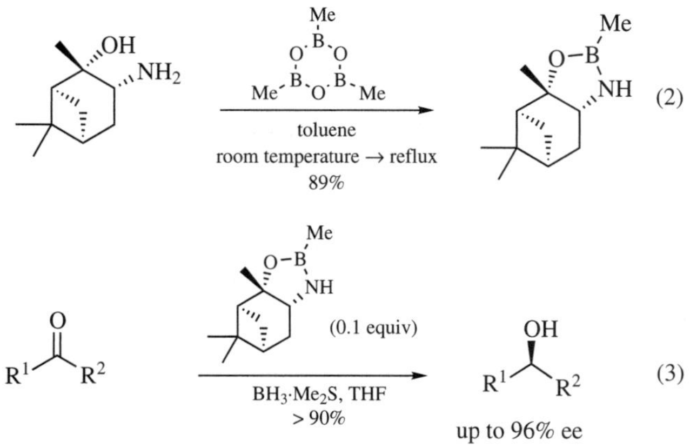

R^1 = Ph, *t*-Bu

R^1 = Me, Et, Bn, CH_2Cl

Reaction of ATBH with trimethyl borate in THF presumably affords the *B*-methoxy oxazaborolidine, which effectively catalyzes asymmetric borane reduction of prochiral ketones. Thus the borane reduction of acetophenone with the reagent prepared in situ from 0.1 equiv of ATBH and 0.12 equiv of trimethyl borate provides (*S*)-2-phenethyl alcohol in 93% yield with 95.5% ee (eq 4).[3] This method offers some advantages in its inexpensiveness and simplicity of the procedure.

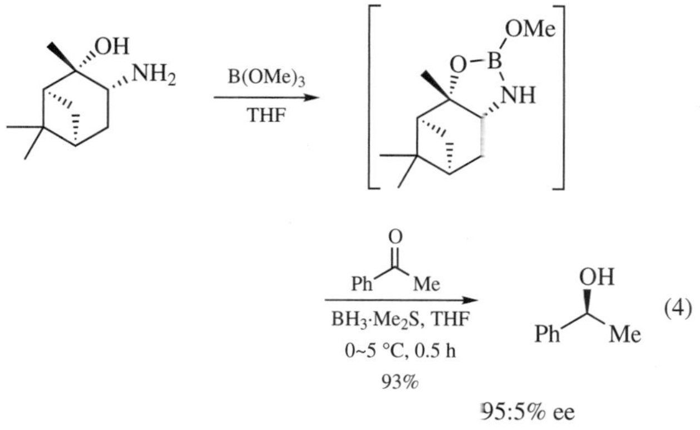

Stereoselective Reduction of α-Oxoketoxime Ethers. The in situ generated catalyst from ATBH and trimethyl borate has also been used in the stereoselective reduction of α-oxoketoxime ethers to prepare the corresponding chiral 1,2-amino alcohols.[4] Thus the asymmetric borane reduction of buta-2,3-dione monoxime ether followed by acidic work-up and subsequent reaction with benzyloxycarbonyl chloride affords a 90% yield of *N*-(*Z*)-3-aminobutan-2-ol with excellent enantioselectivities (eq 5). A trityl group in the oxime ether is required for high enantioselectivity. This method has been successively applied to both cyclic and acyclic α-oxoketoxime ethers.

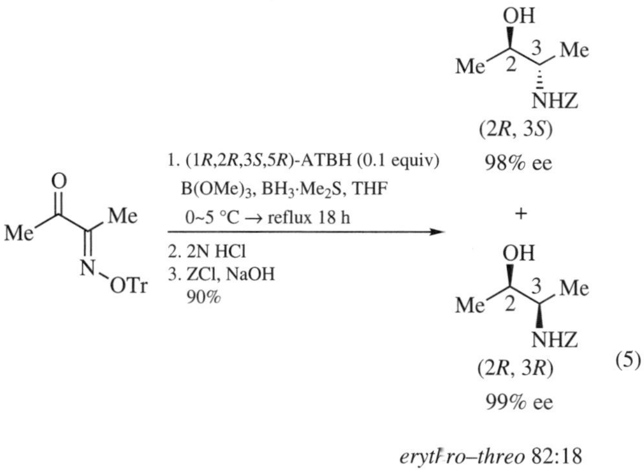

The modified procedure for asymmetric borane reduction is applicable to the stereoselective synthesis of *N*-benzoylsphinganine.[5] Reduction of α-oxoketoxime trityl ethers **1** and **2** using catalyst prepared in situ from (1*R*,2*R*,3*S*,5*R*)-ATBH and trimethyl borate proceeds in high yields with high enantioselectivities (eq 6). Satisfactory results are obtained by employing the borane-*N*,*N*-diethylaniline complex as a reducing agent. In the reduction of substrate **1**, the predominant diastereomer is *threo*. On the other hand, the reduction of **2** proceeds with excellent *erythro* selectivity.

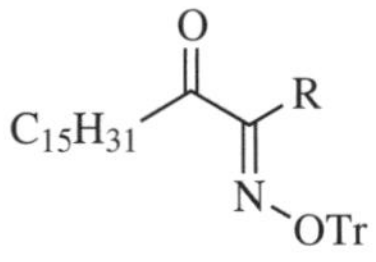

R = CO_2Me (**1**)
R = CH_2OTBS (**2**)

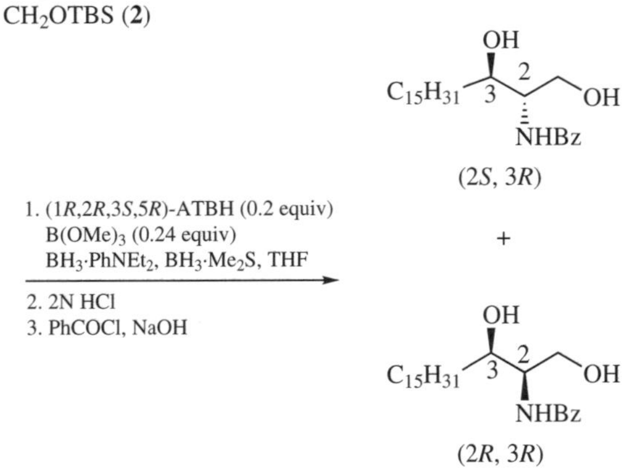

(6)

from **1** 90% yield 93% ee 95% ee
erythro–threo = 20:80

from **2** 94% yield 97% ee 91% ee
erythro–threo = 98:2

Stereoselective Alkylation. Chiral tricyclic lactams can be prepared from (1*R*,2*R*,3*S*,5*R*)-ATBH and γ-keto acids by heating in toluene with a catalytic amount of *p*-toluenesulfonic acid (eq 7).[6] Enolization of the resulting lactams with *sec*-butyllithium, followed by trapping with methyl iodide, furnishes the methylated products in high diastereoselectivity. Subsequent enolization and alkylation with benzyl bromide affords a single diastereomer in 82% yield. Further acidic hydrolysis in butanol provides the desired ester with a quaternary asymmetric center (eq 7).[6]

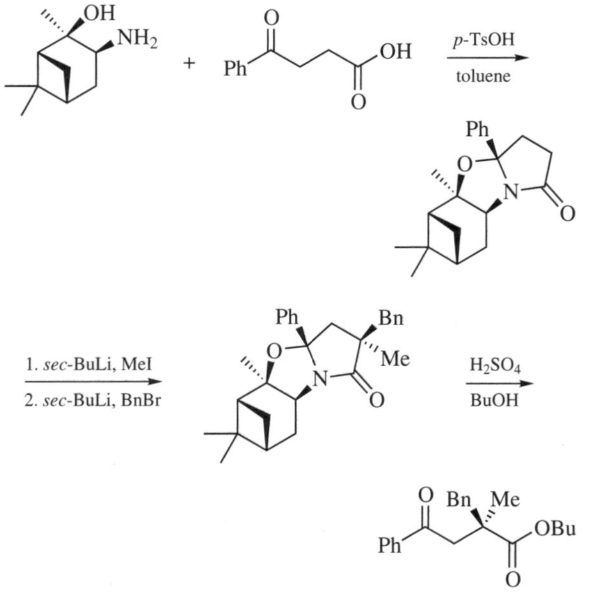

(7)

Related Reagents. α,α-Diphenyl-2-pyrrolidinemethanol.

1. Masui, M.; Shioiri, T. *Tetrahedron* **1995**, *51*, 8363.
2. Masui, M.; Shioiri, T. *Synlett* **1996**, 49.
3. Masui, M.; Shioiri, T. *Synlett* **1997**, 273.
4. Masui, M.; Shioiri, T. *Tetrahedron Lett.* **1998**, *39*, 5195.
5. Masui, M.; Shioiri, T. *Tetrahedron Lett.* **1998**, *39*, 5199.
6. Roth, G. P.; Leonard, S. F.; Tong, L. *J. Org. Chem.* **1996**, *61*, 5710.

Moriyasu Masui
Aburahi Laboratories, Shionogi & Co., Ltd., Japan

(*S*)-4-Anilino-3-methylamino-1-butanol[1]

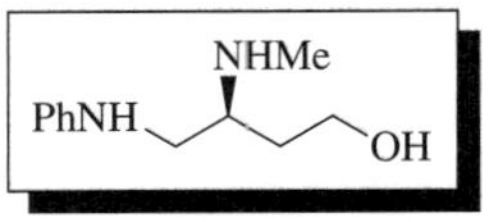

[88733-39-5] $C_{11}H_{18}N_2O$ (MW 194.27)

(tridentate chiral ligand to modify $LiAlH_4$ for the enantioselective reduction of alkyl phenyl ketones[2] and α,β-unsaturated ketones[3])

Physical Data: bp 139–140 °C; $[\alpha]_D^{23}$ −13.7° (*c* 1.0, $CHCl_3$).
Solubility: sol THF.
Preparative Methods: β-benzyl *N*-benzyloxycarbonyl-aspartate[4] is treated with ethyl chloroformate, *N*-methymorpholine, and aniline. Subsequent reduction of the corresponding anilide with $LiAlH_4$ gives the title reagent in overall 80% yield.[2]

Chiral Ligand of $LiAlH_4$ for the Enantioselective Reduction of Alkyl Phenyl Ketones. Optically active alcohols are important synthetic intermediates. There are two major chemical methods for synthesizing optically active alcohols from carbonyl compounds. One is asymmetric (enantioselective) reduction of ketones.[1] The other is asymmetric (enantioselective) alkylation of aldehydes.[5] Extensive attempts have been reported to modify *Lithium Aluminum Hydride* with chiral ligands in order to achieve enantioselective reduction of ketones.[1] However, most of the chiral ligands used for the modification of $LiAlH_4$ are unidentate or bidentate, such as alcohol, phenol, amino alcohol, or amine derivatives.

Unlike many other chiral ligands, (*S*)-4-amilino-3-methylamino-1-butanol (**1**) is designed as a tridentate chiral ligand anticipating a more rigid complex formation with $LiAlH_4$. The chiral reducing reagent is prepared in situ by mixing $LiAlH_4$ and (**1**) in THF. To this chiral reducing reagent is added alkyl phenyl ketone at −100 °C. Optically active (*S*)-*s*-alcohols with 51–88% ee's are obtained in 84–93% yields (eq 1). The results are summarized in Table 1.

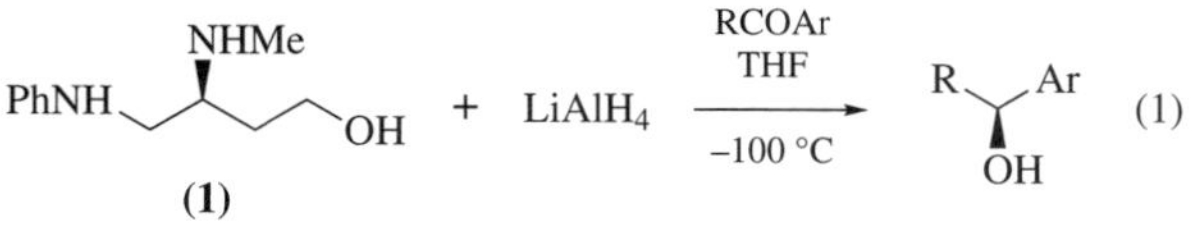

Table 1 Enantioselective Reduction of Alkyl Phenyl Ketones

Ketone	Yield (%)	ee (%)
PhCOMe	87	51
PhCOEt	93	68
PhCOPr-*i*	93	77
PhCOBu-*t*	84	86
α-Tetralone	89	88

The ee's of the obtained alcohols increase according to the increase in steric bulkiness of the alkyl substituents of prochiral ketones. Thus the reduction of *t*-butyl phenyl ketone occurs with 86% ee whereas reduction of acetophenone gives 51% ee. The enantioselective reduction of *t*-butyl phenyl ketone and α-tetralone (86 and 88% ee, respectively) are among the most selective of those reported.[6]

After quenching the reaction, the amino alcohol (**1**) is recovered in a yield of over 85% without any racemization.

Chiral Ligand of $LiAlH_4$ for the Enantioselective Reduction of α,β-Unsaturated Ketones. Enantioselective reductions of α,β-unsaturated ketones afford optically active allylic alcohols which are useful intermediates in natural product synthesis.[7] Enantioselective reduction of α,β-unsaturated ketones with $LiAlH_4$ modified with chiral amino alcohol (**1**) affords optically active (*S*)-allylic alcohols with high ee's. When 2-cyclohexen-1-one is employed, (*S*)-2-cyclohexen-1-ol with 100% ee is obtained in 95% yield (eq 2). This is comparable with the results obtained using $LiAlH_4$–chiral binaphthol[8] and chiral 1,3,2-oxazaborolidine.[9]

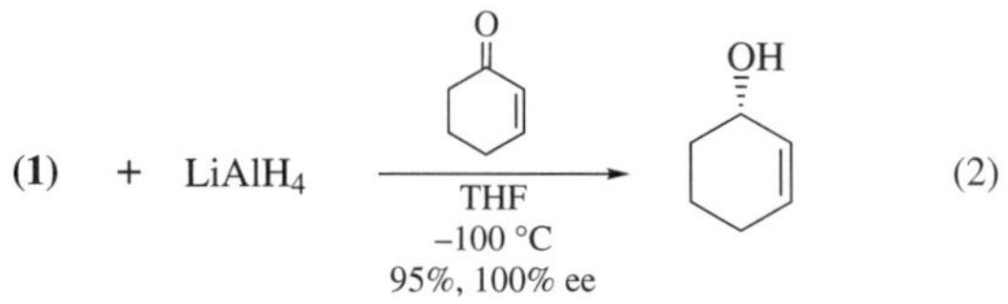

When (*S*)-4-(2,6-xylidino)-3-methylamino-1-butanol (**2**) is used instead of (**1**), (*R*)-2-cyclohexen-1-ol is obtained with less enantioselectivity (13% ee) (eq 3).

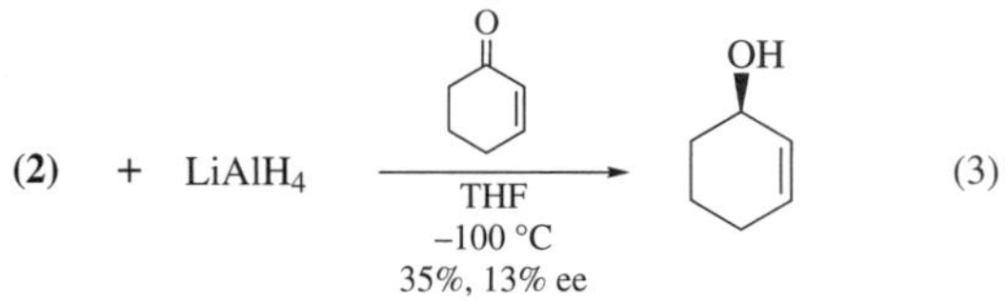

1. Grandbois, E. R.; Howard, S. I.; Morrison, J. D., In *Asymmetric Synthesis*, Morrison, J. D., Ed.; Academic: New York, 1983; Vol. 2, Chapter 3.
2. Sato, T.; Goto, Y.; Fujisawa, T. *Tetrahedron Lett.* **1982**, *23*, 4111.
3. Sato, T.; Goto, Y.; Wakabayashi, Y.; Fujisawa, T. *Tetrahedron Lett.* **1983**, *24*, 4123.
4. Benoiton, L. *Can. J. Chem.* **1962**, *40*, 570.
5. Soai, K.; Niwa, S. *Chem. Rev.* **1992**, *92*, 833.
6. (a) Landor, S. R.; Miller, B. J.; Tatchell, A. R. *J. Chem. Soc. (C)* **1967**, 197. (b) Noyori, R.; Tomino, I.; Tanimoto, Y. *J. Am. Chem. Soc.* **1979**, *101*, 3129. (c) Terashima, S.; Tanno, N.; Koga, K. *Chem. Lett.* **1980**, 981. (d) Yamaguchi, S.; Mosher, H. S. *J. Org. Chem.*, **1973**, *38*, 1870. (e) Asami, M.; Mukaiyama, T., *Heterocycles* **1979**, *12*, 499.
7. (a) Terashima, S.; Tanno, N.; Koga, K. *Tetrahedron Lett.*, **1980**, *21*, 2753. (b) Suzuki, M.; Sugiura, S.; Noyori, R. *Tetrahedron Lett.*, **1982**, *23*, 4817.
8. Noyori, R.; Tomino, I.; Nishizawa, M. *J. Am. Chem. Soc.* **1979**, *101*, 5843.
9. Corey, E. J.; Bakshi, R. K. *Tetrahedron Lett.* **1990**, *31*, 611.

Kenso Soai
Science University of Tokyo, Japan

(*S*)-2-(Anilinomethyl)pyrrolidine[1]

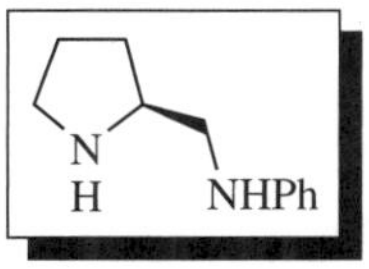

[64030-44-0] $C_{11}H_{16}N_2$ (MW 176.25)

(chiral ligand of $LiAlH_4$ for enantioselective reduction of ketones;[2] can form chiral aminals for diastereoselective alkylation,[3,4] 1,2-,[5] and 1,4-additions[6])

Physical Data: bp 111–112 °C/0.55 mmHg; $[\alpha]_D^{24}$ +19.7° (*c* 1.04, EtOH).
Solubility: sol ether, THF.
Form Supplied in: colorless oil.
Preparative Methods: reaction of (*S*)-*N*-(benzyloxycarbonyl)proline with (1) ethyl chloroformate and *N*-methylmorpholine, (2) aniline, (3) hydrogenolysis with Pd/C, and (4) reduction with $LiAlH_4$ affords the title compound in 59% overall yield.[2b]

Enantioselective Reduction of Ketones. (*S*)-2-(Anilinomethyl)pyrrolidine (**1**) is a chiral ligand which, in combination with *Lithium Aluminum Hydride*, generates a chiral reagent for the enantioselective reduction of ketones. (*S*)-1-Phenylethanol with 92% ee is obtained from the enantioselective reduction of acetophenone with $LiAlH_4$–(**1**) (eq 1).[2a,3] When (*S*)-2-(2,6-xylidinomethyl)pyrrolidine is used instead of (**1**), the ee of the alcohol obtained reaches 95%.[2c] The ee's of aromatic alcohols obtained using $LiAlH_4$–(**1**) are comparable with other highly enantioselective reductions[1] and with highly enantioselective alkylation of aldehydes.[7]

Diastereoselective Alkylation of Chiral Keto- and Formylaminals. Diamine (**1**) forms a chiral ketoaminal by condensation with phenylglyoxal monohydrate. Diastereoselective addition of a Grignard reagent to the ketoaminal and subsequent hydrolysis affords optically active *t*-α-hydroxyaldehydes with >94% ee (eq 2).[3a] Various α-hydroxyaldehydes are synthesized by this method.[3b] Diastereoselectivities are comparable with those of the reactions of chiral 1,3-oxathiane.[8]

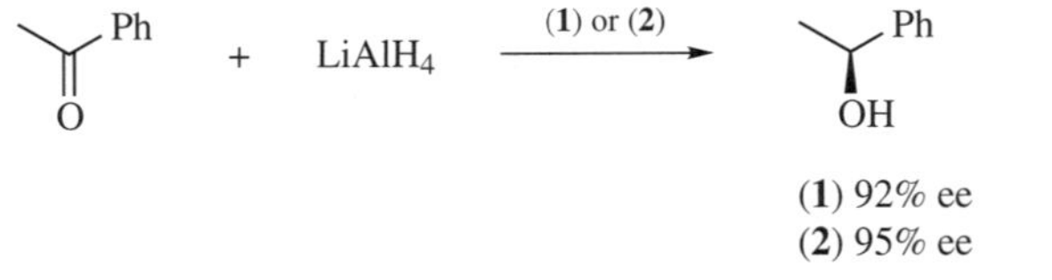

(1)

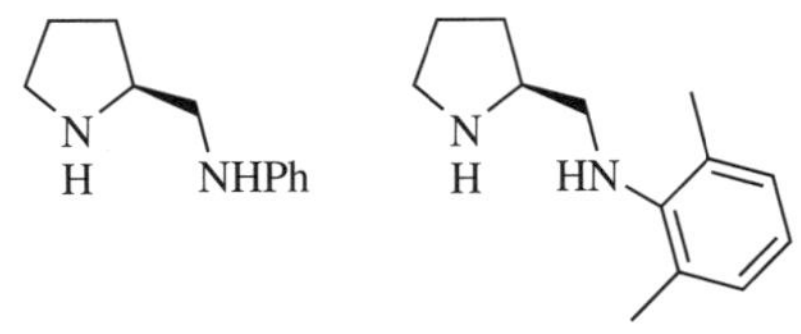

(1) (2)

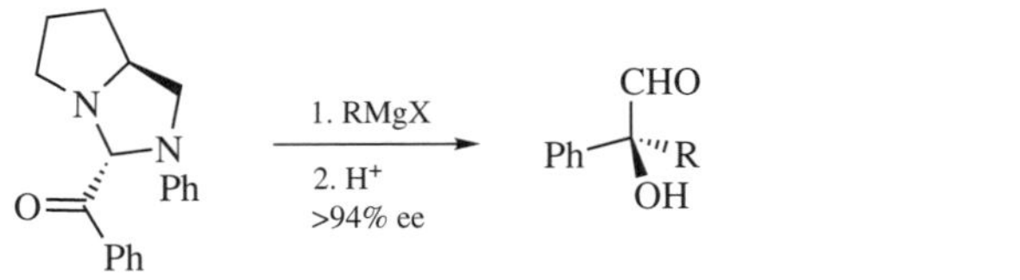

(2)

The method has been applied to the diastereoselective synthesis of naturally occurring compounds such as frontalin (84–100% ee)[3c] and (−)-malyngolide (95% ee).[3d] On the other hand, diastereoselective alkylation of chiral formylaminal with Grignard reagents and the subsequent hydrolysis afford optically active *S*-α-hydroxyaldehydes with moderate stereoselectivity (60% ee).[4]

Diastereoselective 1,2- and 1,4-Additions of Chiral Aminals. The organolithium reagent derived from a chiral bromoaminal and BuLi adds to pentanal to afford an optically active lactol. Subsequent oxidation affords the optically active lactone with 88% ee (eq 3).[5]

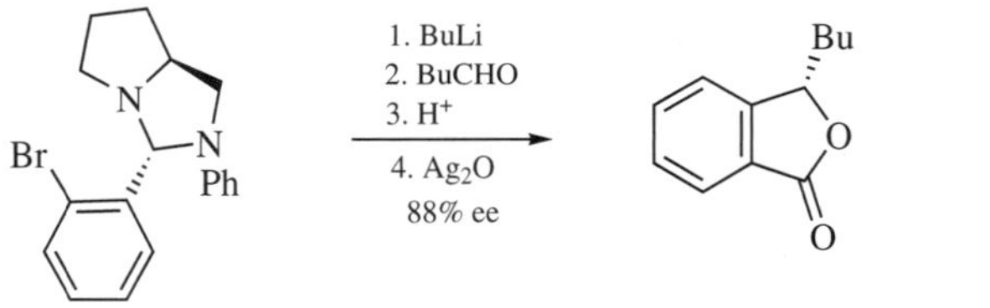

(3)

Diastereoselective 1,4- (conjugate) additions of Grignard reagents to a chiral α,β-unsaturated aminals afford optically active 3-substituted succinaldehydic acid methyl esters with 85–93% ee (eq 4).[6]

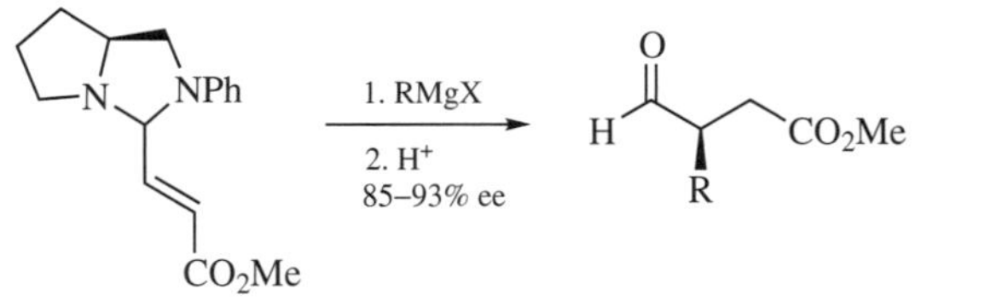

(4)

Related Reagents. (2*S*,4*S*)-2-(Anilinomethyl)-1-ethyl-4-hydroxypyrrolidine (*R*,*R*)-1,2-Diphenyl-1,2-diaminoethane *N*,*N*′-Bis[3,5-bis(trifluoromethyl)benzenesulfonamide] (2*S*,2′*S*)-2-Hydroxymethyl-1-[(1-methylpyrrolidin-2-yl)methyl]pyrrolidine

1. Grandbois, E. R.; Howard, S. I.; Morrison, J. D., In *Asymmetric Synthesis*, Morrison, J. D., Ed.; Academic: New York, 1983; Vol. 2, Chapter 3.
2. (a) Mukaiyama, T,; Asami, M.; Hanna, J.; Kobayashi, S. *Chem. Lett.* **1977**, 783. (b) Asami, M.; Ohno, H.; Kobayashi, S.; Mukaiyama, T. *Bull. Chem. Soc. Jpn.* **1978**, *51*, 1869. (c) Asami, M.; Mukaiyama, T. *Heterocycles* **1979**, *12*, 499.
3. (a) Mukaiyama, T.; Sakito, Y.; Asami, M. *Chem. Lett.* **1978**, 1253. (b) Mukaiyama, T.; Sakito, Y.; Asami, M. *Chem. Lett.* **1979**, 705. (c) Sakito, Y.; Mukaiyama, T. *Chem. Lett.* **1979**, 1027. (d) Sakito, Y.; Tanaka, S.; Asami, M.; Mukaiyama, T. *Chem. Lett.* **1980**, 1223.
4. Asami, M.; Mukaiyama, T. *Chem. Lett.* **1983**, 93.
5. Asami, M.; Mukaiyama, T. *Chem. Lett.* **1980**, 17.
6. Asami, M.; Mukaiyama, T. *Chem. Lett.* **1979**, 569.
7. Soai, K.; Niwa, S. *Chem. Rev.* **1992**, *92*, 833.
8. Eliel, E. L.; Koskimies, J. K.; Lohri, B. *J. Am. Chem. Soc.* **1978**, *100*, 1614.

Kenso Soai
Science University of Tokyo, Japan

L-Aspartic Acid[1]

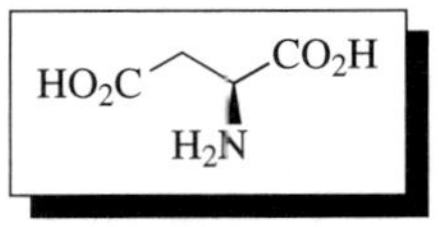

(L)		
[56-84-8]	$C_4H_7NO_4$	(MW 133.10)
(D)		
[1783-96-6]		
(DL)		
[617-45-8]		

(chiral reagent used in diastereoselective alkylations[2] and as a ligand for $LiAlH_4$ in asymmetric reductions[3] of enones)

Physical Data: (*S*)-(+),[4a] $[\alpha]_D^{20}$ +25°, mp >300 °C (dec.); (*R*)-(−),[4a] $[\alpha]_D^{20}$ −24°, mp >300 °C; (±) mp 325–348 °C (dec.).
Solubility: sol in acid and alkali; sol water (1 g/222.2 mL at 20 °C), forming supersaturated solutions easily; insol alcohol.
Form Supplied in: commercially available as a white solid in racemic and optically pure forms.[4]

Diastereoselective Alkylations. Esters derived from L-aspartic acid have been alkylated at both the α- and β-positions (eq 1).[1,2] β-Alkylations have been more widely used. The amino acid moiety is responsible for the diastereoselection in the β-alkylation process.

Alkylation of cyclic derivatives of L-aspartic acid (**1**) occurs exclusively at the β-position with good to excellent diastereoselection. One application is the synthesis of chiral β-dicarbonyl equivalents.[5] Equivalents of either enantiomer can be prepared depending on whether the alkylation is performed on a lactone (eq 2) or an oxazoline (eq 3).

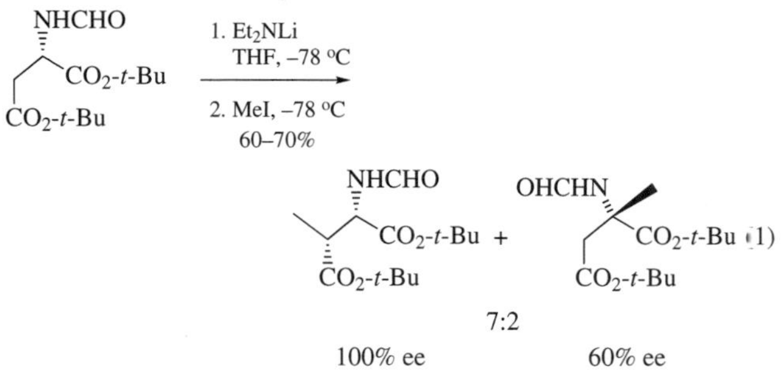

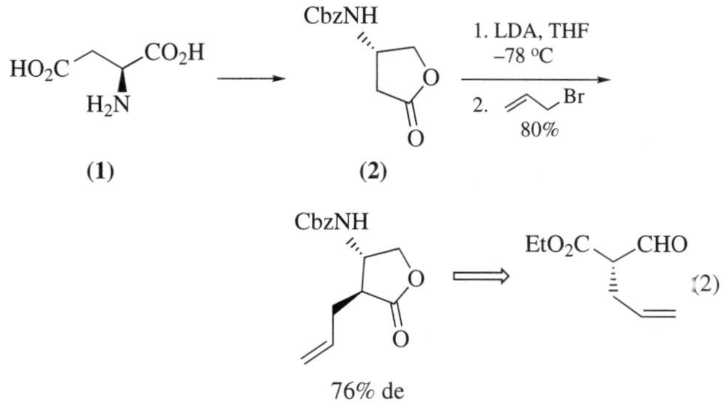

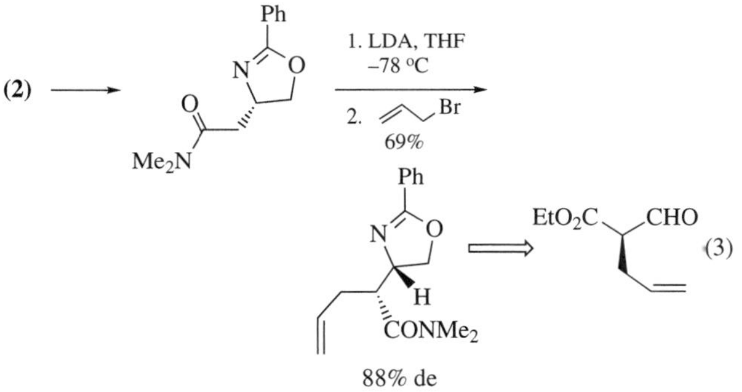

The most commonly used cyclic derivatives of L-aspartic acid are β-lactams (eq 4).[6] For example, excellent regioselectivity and diastereoselectivity are observed in the alkylation of the dianion of (**3**). Other compounds related to (**3**) have been prepared from L-aspartic acid[7] and used in highly diastereoselective alkylations en route to a variety of natural and nonnatural products[8] including β-lactams,[9] γ-lactams,[10] and dihydroisocoumarin derivatives.[11]

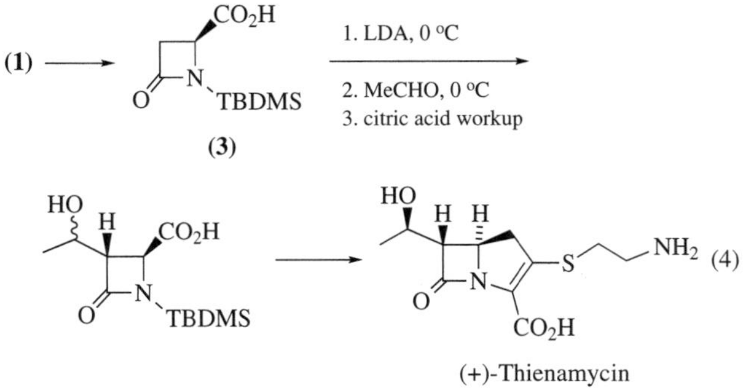

Asymmetric Reductions. Asymmetric reductions of prochiral ketones to optically active secondary alcohols have been extensively studied.[3] The most common method involves the use of chiral unidentate or bidentate ligands in conjunction with *Lithium Aluminum Hydride*. However, an (*S*)-aspartic acid derived tridentate ligand has been shown to be very effective in certain cases, presumably due to the rigidity of aluminum complex (**4**) (eq 5–7).[12]

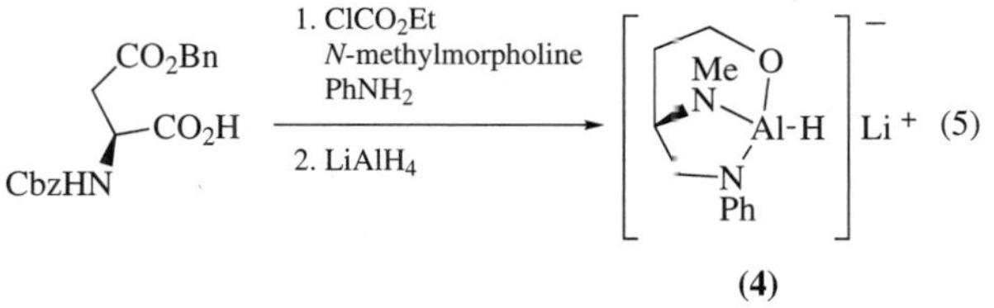

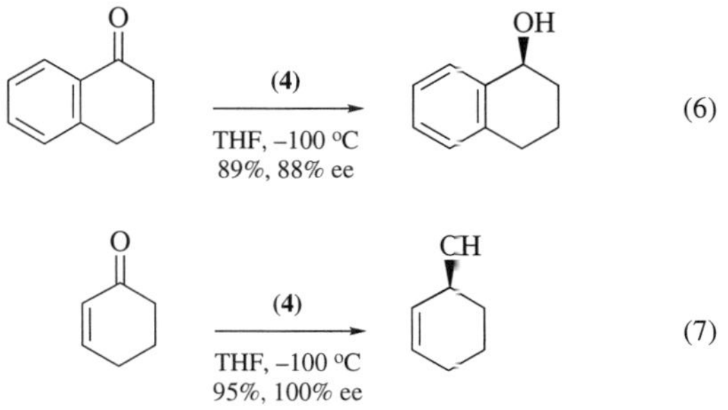

Unfortunately, the complete enantiofacial differentiation of cyclohexenone (eq 7) appears to be an isolated case, as reaction with 3-methylcyclohexenone afforded the corresponding (*S*)-cyclohexanol in only 28% ee. In the absence of a general trend for the outcome of these reductions, the scope of this method seems limited at this point, as opposed to (*S*)-BINAL-H mediated reductions (see *Lithium Aluminum Hydride* and subsequent articles).

---

1. (a) Coppola, G. M.; Schuster, H. F. *Asymmetric Synthesis*; Wiley: New York, 1987; Chapter 7. (b) Greenstein, W. *Chemistry of the Amino Acids*; Wiley: New York, 1961; Vol. 3, Chapter 23.
2. Seebach, D.; Wasmuth, D. *Angew. Chem., Int. Ed. Engl.* **1981**, *20*, 971.
3. Nishizawa, M.; Noyori, R. *Comprehensive Organic Synthesis* **1991**, *8*, Chapter 1.7.
4. (a) Harada, K. *Bull. Chem. Soc. Jpn.* **1964**, *37*, 1383. (b) For syntheses of DL-aspartic acid, see Dunn, M. S.; Smart, B. W. *OCS* **1963**, *4*, 55.
5. McGarvey, G. J.; Hiner, R. N.; Matsubara, Y.; Oh, T. *Tetrahedron Lett.* **1983**, *24*, 2733.
6. Reider, P. J.; Grabowski, E. J. J. *Tetrahedron Lett.* **1982**, *23*, 2293.
7. Labia, R.; Morin, C. *Chem. Lett.* **1984**, 1007.
8. (a) Nordlander, J. E.; Payne, M. J.; Njoroge, F. G.; Vishwanath, V. M.; Han, G. R.; Laikos, G. D.; Balk, M. A. *J. Org. Chem.* **1985**, *50*, 3619. (b) Baldwin, J. E.; North, M.; Flinn, A. *Tetrahedron Lett.* **1987**, *28*, 3167.
9. (a) Salzmann, T. N.; Ratcliffe, R. W.; Christensen, B. G.; Bouffard, F. A. *J. Am. Chem. Soc.* **1980**, *102*, 6163. (b) Salzmann, T. N.; Ratcliffe, R. W.; Christensen, B. G. *Tetrahedron Lett.* **1980**, *21*, 1193.
10. Baldwin, J. E.; Adlington, R. M.; Gollins, D. W.; Schofield, C. J., *Tetrahedron* **1990**, *46*, 4733.
11. Broady, S. D.; Rexhausen, J. E.; Thomas, E. J. *Chem. Commun.* **1991**, 708.

12. (a) Sato, T.; Gotoh, Y.; Fujisawa, T. *Tetrahedron Lett.* **1982**, *23*, 4111. (b) Sato, T.; Gotoh, Y.; Wakabayashi, Y.; Fujisawa. T. *Tetrahedron Lett.* **1983**, *24*, 4123.

Alyx-Caroline Guével
The Ohio State University, Columbus, OH, USA

2-Azabicyclo[2.2.1]hept-5-en-3-one

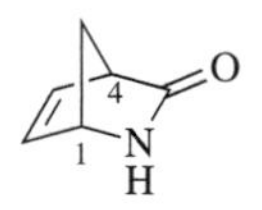

(1*R*,4*S*)-(−)-(**1**)

[49805-30-3] C_6H_7NO (MW 109.13)
(±)
[61865-48-3]
(1*R*)
[79200-56-9]
(1*S*)
[130931-83-8]

(building block for the synthesis of carbocyclic nucleosides, GABA inhibitors, etc.)

Physical Data: mp 55–57 °C; bp 102–106 °C/0.25 mmHg.
Solubility: sol H_2O, MeOH, CH_2Cl_2.
Form Supplied in: off-white solid.
Preparative Methods: from reaction of *Cyclopentadiene* with either sulfonyl cyanides[1,2] or *Chlorosulfonyl Isocyanate*.[3] Both enantiomers can be obtained via enzymic kinetic resolution of the racemate.[4]

2-Azabicyclo[2.2.1]hept-5-en-3-one (**1**) has been used as a building block for the synthesis of an array of compounds with potential pharmacological applications. Either enantiomer of *cis*-3-aminocyclopentanecarboxylic acid (**2**) is accessible from a single enantiomer of (**1**) (eq 1). Compound (**2**) has been shown to act as an agonist at the γ-aminobutyric acid (GABA) receptor.[5]

The (1*R*,4*S*) enantiomer ((−)-**1**) has been used for the preparation of the carbocyclic nucleoside (−)-carbovir (**3**) (eq 2),[6] which has been shown to have similar activity against HIV (RF strain) as AZT (Zidovudine).

Treatment of (**1**) with *Potassium Monoperoxysulfate* (oxone) gave an 80% yield of the *exo*-epoxide (**4**) (eq 3), which is of potential use for the preparation of carbocyclic analogs of 2′-or 3′-deoxyribofuranosylamines.

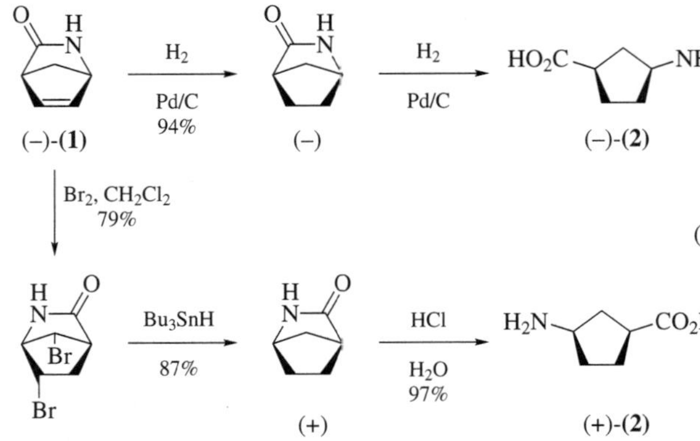

(1)

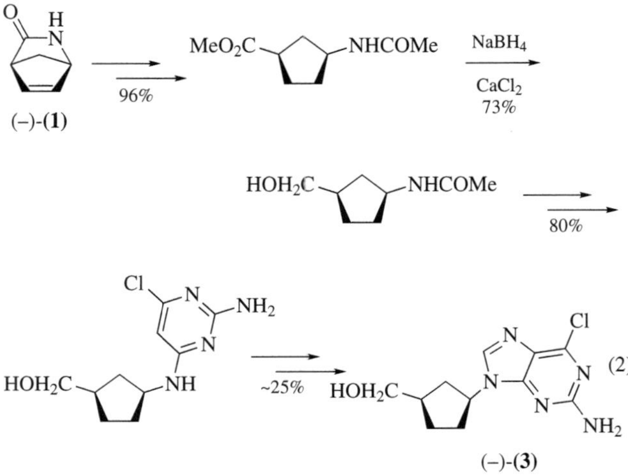

(2)

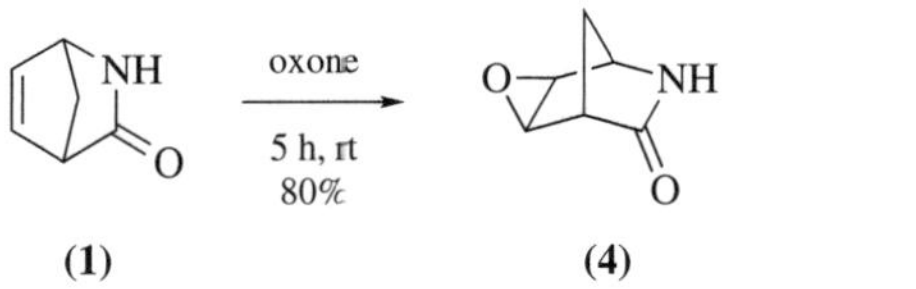

(3)

---

1. Daluge, S; Vince, R. *J. Org. Chem.* **1978**, *43*, 2311.
2. Griffiths, G.; Previdoli, F. Eur. Patent 508 352 (*Chem. Abstr.* **1993**, *118*, 59 591).
3. Malpass, J. R.; Tweddle, N. J. *J. Chem. Soc., Perkin Trans. 1* **1977**, 874.
4. Taylor, S. J. C.; Sutherland, A. G.; Lee, C.; Wisdom, R.; Thomas, S.; Roberts, S. M.; Evans, C. *Chem. Commun.* **1990**, 1120.
5. Evans, C.; McCague, R.; Roberts, S. M.; Sutherland, A. G. *J. Chem. Soc., Perkin Trans. 1* **1991**, 656.
6. Evans, C. T.; Roberts, S. M.; Shoberu, K. A.; Sutherland, A. G. *J. Chem. Soc., Perkin Trans. 1*, **1992**, 589.
7. Legraverend, M.; Bisagni, E.; Huel, C. *J. Heterocycl. Chem.*, **1989**, *26*, 1881.

Gareth J. Griffiths
Lonza, Visp, Switzerland

B

Baker's Yeast

(microorganism used as biocatalyst for the reduction of carbonyl groups and double bonds,[1] either under fermenting conditions, immobilized, or ultrasonically stimulated)

Solubility: insol cold and warm H_2O; used as a slurry.
Form Supplied in: yellowish pressed cakes, commercially available as cubes from bakeries or supermarkets, usually produced by brewery companies.
Handling, Storage, and Precautions: the wet cake must be stored in the refrigerator (0–4 °C) and used within the date indicated by the manufacturer.

Baker's Yeast-Mediated Biotransformations. Baker's yeast (BY, *Saccharomyces cerevisiae*) is readily available and inexpensive, and its use does not require any special training in microbiology. For these reasons, this biocatalyst has enjoyed a wide popularity among organic chemists, so that it can be considered as a microbial reagent for organic synthesis.[2] BY is generally used as whole cells, in spite of the problems connected with rates of penetration and diffusion of the substrates into, and the product from, the cells. However, the crude system is an inexpensive reservoir of cofactor-dependent enzymes such as oxidoreductases.[3] These benefits overcome the complication caused by undesired enzymatic reactions which lead to the formation of byproducts. Well-defined experimental procedures for BY-mediated bioreductions can be found in *Organic Syntheses*.[4] In the usual applications the biotransformations lead to optically active compounds with variable, but generally high, enantioselectivity.[5] The reaction is easily carried out in a heterogeneous medium containing a slurry of the yeast in tap water, in aerobic and fermenting conditions. Typically, the experimental conditions require a variable yeast:substrate ratio (1–40 g $mmol^{-1}$). The yeast is suspended in an aqueous solution of glucose or sucrose (0.1–0.3 M) to start the fermentation and to the fermenting yeast the substrate is added neat or in a suitable solvent, and therefore dispersed into the heterogeneous medium. The reaction is kept at 25–30 °C and, if necessary, additional fermenting BY can be added. At the end, the yeast is filtered off through Celite and the product extracted with organic solvents.

Carbonyl Group Reductions. Early applications of BY date back to the end of the 19th century and the first examples are reductions of carbonyl compounds.[1c] The widespread applications of this biotransformation are based on some systematic investigations on various ketones[6] and the stereochemical outcome of the reaction is generally described by the so-called Prelog's rule[7] which successfully applies to a great number of structures (eq 1).

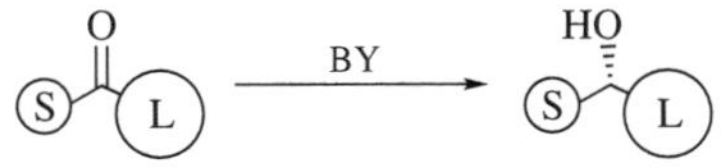
(1)

The structural variety of carbonyl compounds appears to be almost unlimited since aliphatic, aromatic, and cyclic ketones are good substrates for the bioreduction.[1,5] Also, organometallic carbonyl compounds such as $Cr(CO)_3$-complexed aromatic aldehydes (eq 2)[8] or ketones (eq 3)[9] are enantioselectively reduced by BY.

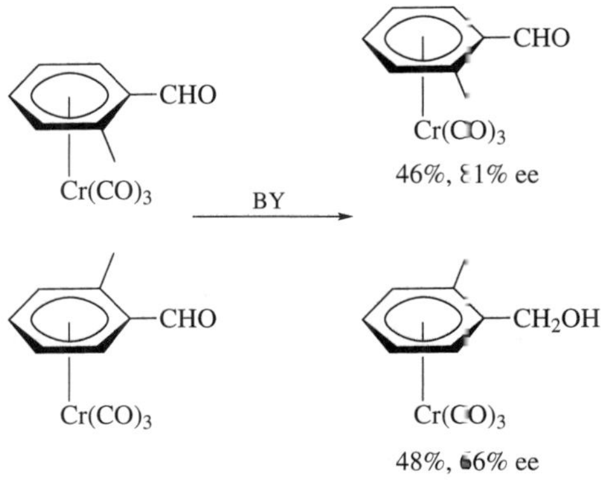

(2)

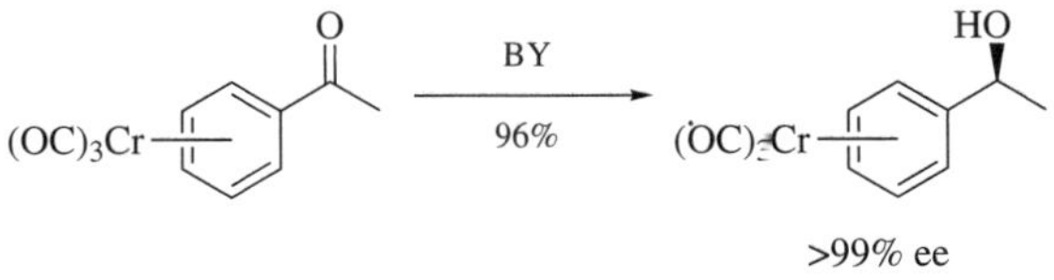

(3)

In general, the enantiomeric excess and the configuration of the optically active alcohols are strongly dependent on the structure of the starting carbonyl compound; many examples of diastereoselective reduction have also been reported.[10] The reduction of an epoxy ketone is accompanied by a stereocontrolled epoxide hydrolytic opening to afford a racemic triol, diastereomerically pure (eq 4).[11]

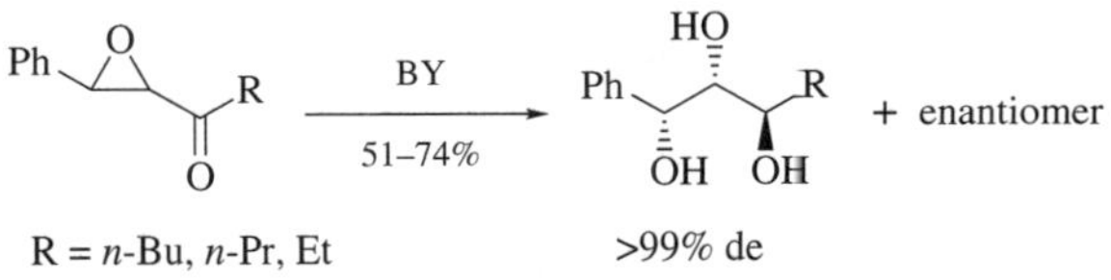

(4)

Many experimental procedures have been developed in order to influence the enantioselectivity and the stereochemistry of the products: use of organic media,[12] the addition of various compounds to the incubation mixture,[13] or enclosure in a dialysis tube[14] can be helpful. Immobilized BY can be used in water or in organic solvents for the same purpose.[15] Slight modifications of the substrate can obtain the same result and many examples are available.[16] Several other groups can be present in the carbonyl-containing substrate.[5] For instance, the asymmetric reduction of keto groups in compounds containing a cyclopropyl moiety has been achieved (eq 5).[17]

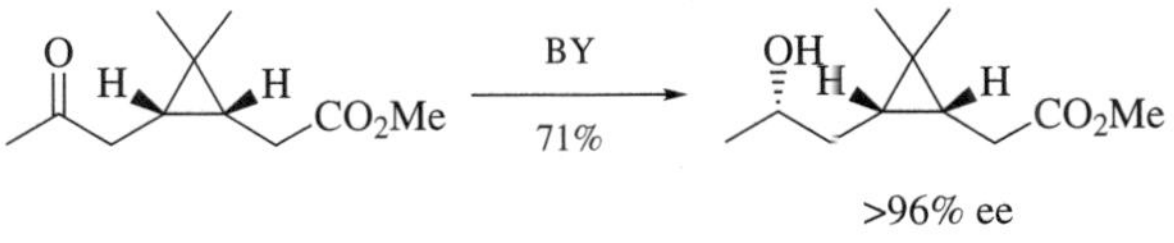

(5)

β-Keto esters are reduced to the corresponding hydroxy esters but, since more oxidoreductases are present in the yeast,[18] occasionally different stereochemistry or lowered enantioselectivity are observed. This is well illustrated by the stereochemical outcome of the reduction of a β-keto ester such as ethyl

4-chloroacetoacetate,[16b] when compared to ethyl acetoacetate (eq 6).[4a]

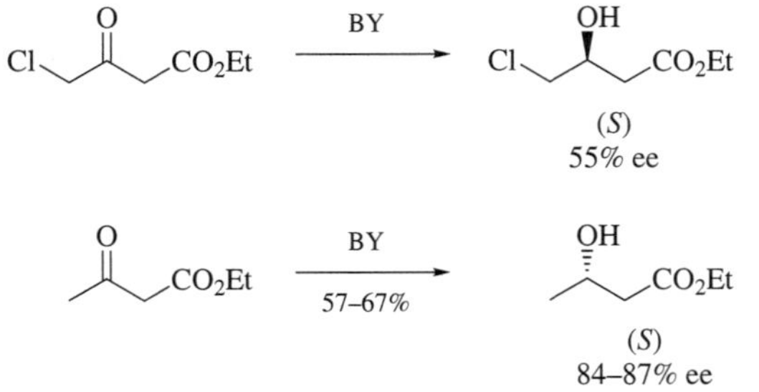

(6)

Both γ- and δ-keto acids are reduced to hydroxy acids, which directly cyclize to the corresponding lactones in the incubation media.[19] The pheromone (*R*)-(+)-hexadecanolide has been prepared in this way by reduction of the corresponding δ-keto acid (eq 7).[20]

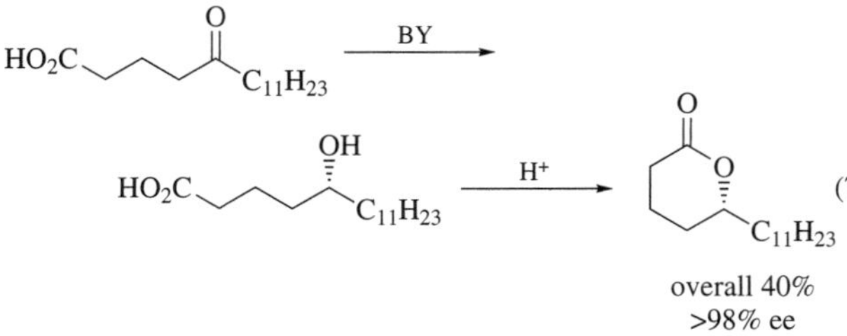

(7)

α-Hydroxy ketones are good substrates for the bioreduction and several optically active 1,2-diols have been prepared.[21] The monobenzoate of dihydroxyacetone is reduced to the corresponding optically pure glycerol derivative (eq 8).[22] In many instances, simple protection of the α-hydroxy group may afford the opposite enantiomer.[23]

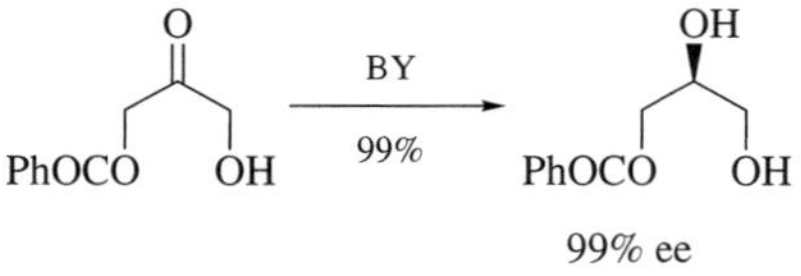

(8)

Activated Double-Bond Hydrogenation. Fermenting BY is able to carry out the hydrogenation of double bonds which bear certain functional groups. A compound containing an unsaturated acetal and an ester function is directly transformed enantioselectively in a hydroxy acid, later chemically cyclized to the corresponding lactone (eq 9).[24] Other α,β-unsaturated alcohols and aldehydes are efficiently and enantioselectively converted to the corresponding saturated alcohols.[25] 2-Chloro-2-alkenoates (eq 10)[26] or nitroalkenes (eq 11)[27] are enantioselectively hydrogenated, the stereochemistry of the reaction depending on the double bond configuration.

(9)

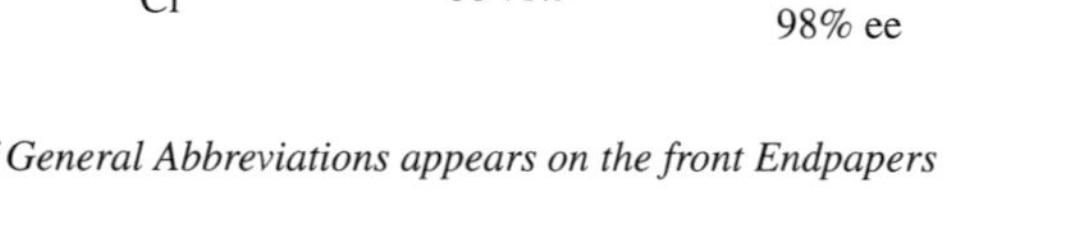

(10)

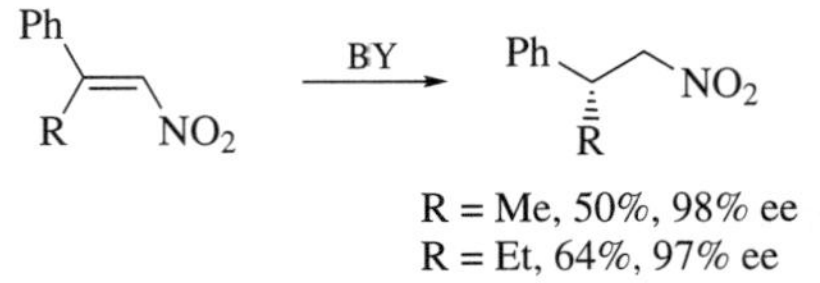

(11)

Acyloin Condensations. The condensation between furfural or benzaldehyde and a two-carbon unit to afford a hydroxy ketone has long been known.[1b] The extension of the reaction to α,β-unsaturated aldehydes has provided access to optically active functionalized diols (eq 12),[28] which are used as chiral intermediates for the synthesis of natural products.[1b]

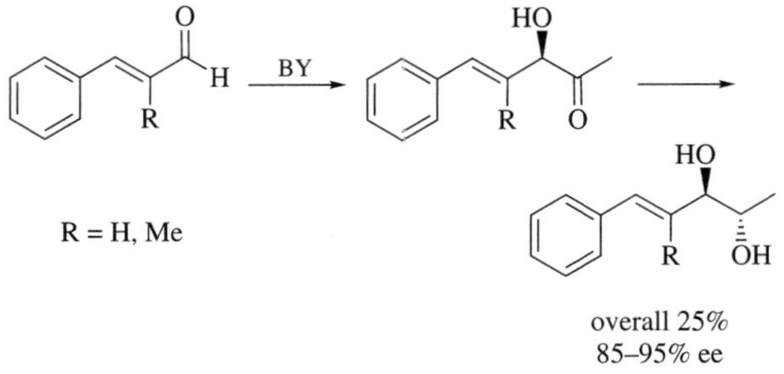

(12)

Cyclization of Squalene-like Substrates. Ultrasonically stimulated BY is a source of sterol cyclase, which catalyzes the cyclization of squalene oxide and squalenoid compounds to lanosterol derivatives (eq 13).[29]

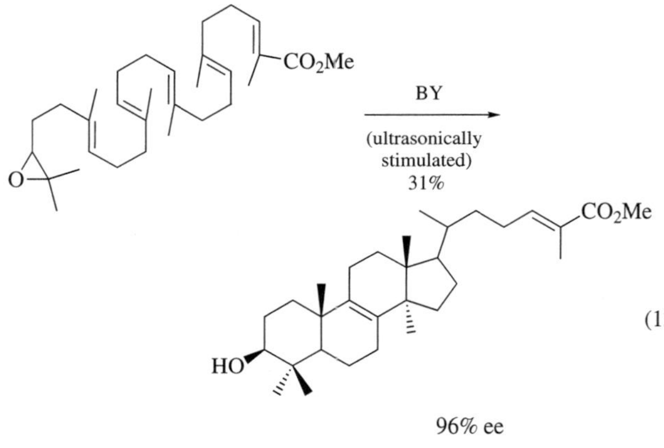

(13)

Hydrolyses. The presence of hydrolytic enzymes in BY is well documented.[30] However, the use of the yeast for biocatalytic ester hydrolysis suffers because of the availability of commercially available purified hydrolases. Nonetheless, the hydrolytic ability of fermenting BY has been proposed for the resolution of various amino acid esters (eq 14).[31] The BY-mediated enantioselective hydrolysis has also been applied to the resolution of acetates of hydroxyalkynes[32] and a hydroxybutanolide.[33]

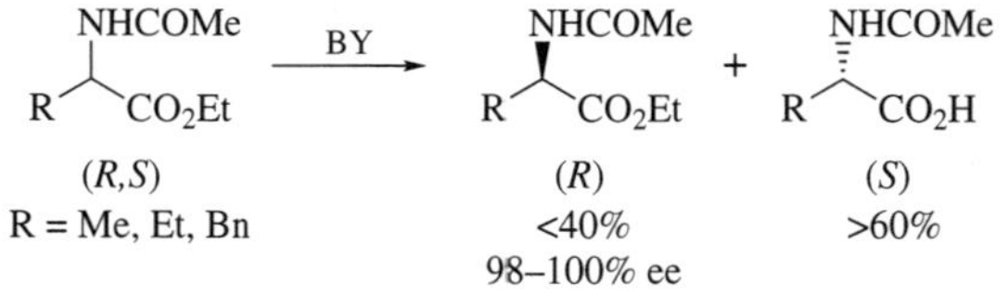

(14)

Oxidations. Reductions are by far the most exploited reactions carried out in the presence of BY. However, a few interesting examples of oxidations are available,[34] such as the dehydrogenation of thiastearates.[35] The regeneration of protected functional groups is possible with BY, which can effect the deprotection of hydrazones[36] and, if ultrasonically stimulated, may release from the oximes the corresponding carbonyl compounds, without further reduction to the alcohols (eq 15).[37]

 (15)

R^1 = Ph, R^2 = H, 96%
R^1 = C_5H_{11}, R^2 = H, 95%
R^1 = Et, R^2 = Me, 93%

An attractive reaction has been reported for BY, which is able in ethanol to catalyze the oxidative coupling of various thiols to disulfides (eq 16).[38]

$$\text{PhSH} \xrightarrow[97\%]{\text{BY}} \text{PhSSPh} \quad (16)$$

Miscellaneous Reactions. Many other reactions can be realized in the presence of BY. An interesting hydrolysis–reduction process transformed a derivative of secologanin into two different cyclic compounds, depending on the pH of the incubation.[39] Here, a glucosidase activity afforded the intermediate aldehyde, which could be reduced or rearranged to different products (eq 17).

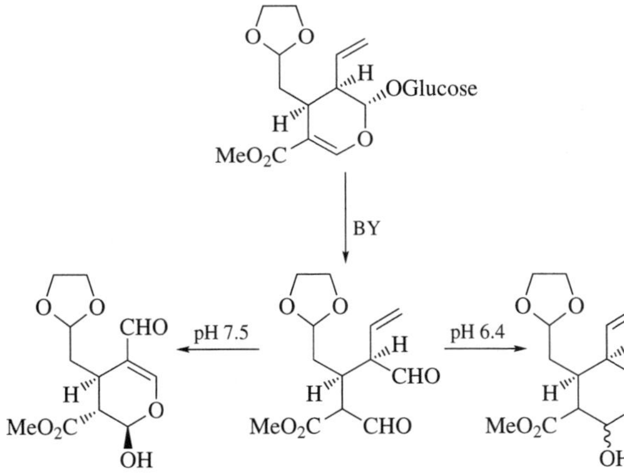 (17)

With BY, the reduction of α,β-unsaturated aldehydes can act together with a hydration process, affording optically active diols in acceptable yields (eq 18).[40]

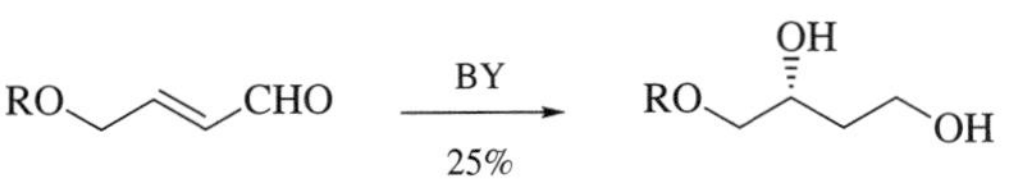 (18)

R = $PhCH_2$, PhCO 90%

Some cycloaddition reactions have also been carried out in the presence of BY.[41] The asymmetric 1,3-dipolar cycloaddition of benzonitrile *N*-oxides to various dipolarophiles led to optically active 2-oxazolines (eq 19).[42]

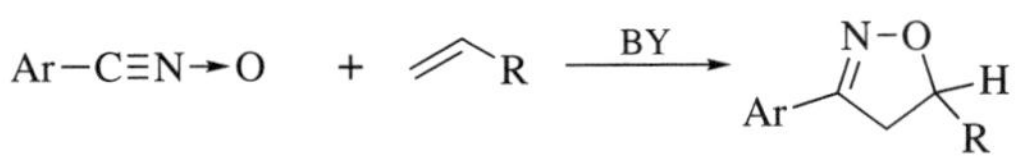 (19)

R = pyridine, 85%, 64% ee
R = carbazole, 78%, 51% ee

The regio- and enantioselectivity of the reactions depend on the structure of dipolarophiles and the addition of β-cyclodextrin. BY is also able to carry out a Diels–Alder condensation,[41] and a few Michael-type additions are enantioselectively performed in the presence of BY.[43] For example, the addition of amines to α,β-unsaturated esters affords optically active β-amino acid esters (eq 20).[44]

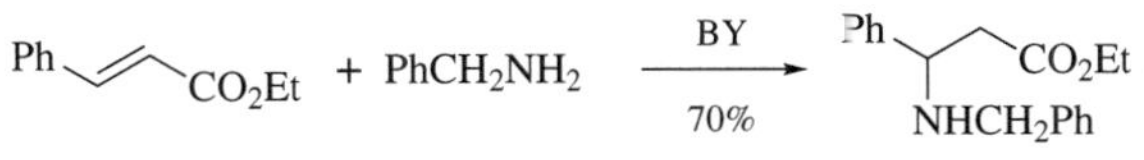 (20)

72.5% ee

1. (a) Sih, C. J.; Chen, C.-S. *Angew Chem., Int. Ed. Engl.* **1984**, *23*, 570. (b) Servi, S. *Synthesis* **1990**, 1. (c) Csuk, R.; Glänzer, B. I. *Chem. Rev.* **1991**, *91*, 49.
2. *Microbial Reagents in Organic Synthesis*; Servi, S., Ed.; Kluwer: Dordrecht, 1992.
3. Ward, O. P.; Young, C. S. *Enz. Microb. Technol.* **1990**, *12*, 482.
4. (a) Seebach, D.; Sutter, M. A.; Weber, R. H.; Züger, M. F. *Org. Synth.* **1985**, *63*, 1. (b) Mori, K.; Mori, H. *Org. Synth.* **1990**, *68*, 56.
5. Santaniello, E.; Ferraboschi, P.; Grisenti, P.; Manzocchi, A. *Chem. Rev.* **1992**, *92*, 1071.
6. (a) MacLeod, R.; Prosser, H.; Fikentscher, L.; Lanyi, J.; Mosher, H. S. *Biochemistry* **1964**, *3*, 838. (b) Červinka, O.; Hub, L. *Collect. Czech. Chem. Commun.* **1966**, *31*, 2615.
7. Prelog, V. *Pure Appl. Chem.* **1964**, *9*, 119.
8. (a) Top, S.; Jaouen, G.; Gillois, J.; Baldoli, C.; Maiorana, S. *Chem. Commun.* **1988**, 1284. (b) Top, S.; Jaouen, G.; Baldoli, C.; Del Buttero, P.; Maiorana, S. *J. Organomet. Chem.* **1991**, *413*, 125.
9. Gillois, J.; Jaouen, G.; Buisson, D.; Azerad, R. *J. Organomet. Chem.* **1989**, *367*, 85.
10. (a) Ticozzi, C.; Zanarotti, A. *Liebigs Ann. Chem.* **1989**, 1257. (b) Itoh, T.; Fukuda, T.; Fujisawa, T. *Bull. Chem. Soc. Jpn.* **1989**, *62*, 3851. (c) Fujisawa, T.; Yamanaka, K.; Mobele, B. I.; Shimizu, M. *Tetrahedron Lett.* **1991**, *32*, 399.
11. Fouché, G.; Horak, R. M.; Meth-Cohn, O. *Molecular Mechanisms in Bioorganic Processes*; Bleasdale, C.; Golding, B. T., Eds.; Royal Society of Chemistry: London, 1990, p 350.
12. (a) Haag, T.; Arslan, T.; Seebach, D. *Chimia* **1989**, *43*, 351. (b) Nakamura, K.; Kondo, S.; Kawai, Y.; Ohno, A. *Tetrahedron Lett.* **1991**, *32*, 7075.
13. (a) Nakamura, K.; Kawai, Y.; Miyai, T.; Ohno, A. *Tetrahedron Lett.* **1990**, *31*, 3631. (b) Nakamura, K.; Kawai, Y.; Ohno, A. *Tetrahedron Lett.* **1990**, *31*, 267. (c) Nakamura, K.; Kawai, Y.; Oka, S.; Ohno, A. *Bull. Chem. Soc., Jpn.* **1989**, *62*, 875. (d) Ushio, K.; Ebara, K.; Yamashita, T. *Enz. Microb. Technol.* **1991**, *13*, 834.
14. Spiliotis, V.; Papahatjis, D.; Ragoussis, N. *Tetrahedron Lett.* **1990**, *31*, 1615.
15. (a) Nakamura, K.; Inoue, K.; Ushio, K.; Oka, S.; Ohno, A. *J. Org. Chem.* **1988**, *53*, 2589. (b) Nakamura, K.; Kawai, Y.; Oka, S.; Ohno, A. *Tetrahedron Lett.* **1989**, *30*, 2245. (c) Naoshima, Y.; Maeda, J.; Munakata, Y. *J. Chem. Soc., Perkin Trans. 1* **1992**, 659.
16. (a) Nakamura, K.; Ushio, K.; Oka, S.; Ohno, A.; Yasui, S. *Tetrahedron Lett.* **1984**, *25*, 3979. (b) Zhou, B.-N.; Gopalan, A. S.; VanMiddlesworth, F.; Shieh, W.-R.; Sih, C. J. *J. Am. Chem. Soc.* **1983**, *105*, 5925.
17. Tkachev, A. V.; Rukavishnikov, A. V.; Gatilov, Y. V.; Bagrjanskaja, I. Yu. *Tetrahedron: Asymmetry* **1992**, *3*, 1165.
18. Shieh, W.-R.; Gopalan, A. S.; Sih, C. J. *J. Am. Chem. Soc.* **1985**, *107*, 2993.

19. (a) Muys, G. T.; Van der Ven, B.; de Jonge, A. P. *Nature* **1962**, *194*, 995. (b) Gessner, M.; Günther, C.; Mosandl, A. *Z. Naturforsch. Teil C* **1987**, *42c*, 1159. (c) Utaka, M.; Watabu, H.; Takeda, A. *J. Org. Chem.* **1987**, *52*, 4363. (d) Aquino, M.; Cardani, S.; Fronza, G.; Fuganti, C.; Pulido-Fernandez, R.; Tagliani, A. *Tetrahedron* **1991**, *47*, 7887.
20. Utaka, M.; Watabu, H.; Takeda, A. *Chem. Lett.* **1985**, 1475.
21. (a) Levene, P. A.; Walti, A. *Org. Synth., Coll. Vol.* **1943**, *2*, 545. (b) Guetté, J.-P.; Spassky, N. *Bull. Soc. Chim. Fr., Part 2* **1972**, 4217. (c) Barry, J.; Kagan, H. B. *Synthesis* **1981**, 453. (d) Kodama, M.; Minami, H.; Mima, Y.; Fukuyama, Y. *Tetrahedron Lett.* **1990**, *31*, 4025. (e) Ramaswamy, S.; Oehlschlager, A. C. *Tetrahedron* **1991**, *47*, 1145.
22. Aragozzini, F.; Maconi, E.; Potenza, D.; Scolastico, C. *Synthesis* **1989**, 225.
23. (a) Manzocchi, A.; Fiecchi, A.; Santaniello, E. *J. Org. Chem.* **1988**, *53*, 4405. (b) Ferraboschi, P.; Grisenti, P.; Manzocchi, A.; Santaniello, E. *J. Chem. Soc., Perkin Trans. 1* **1990**, 2469.
24. Leuenberger, H. G. W.; Boguth, W.; Barner, R.; Schmid, M.; Zell, R. *Helv. Chim. Acta* **1979**, *62*, 455.
25. (a) Gramatica, P.; Manitto, P.; Poli, L. *J. Org. Chem.* **1985**, *50*, 4625. (b) Gramatica, P.; Manitto, P.; Monti, D.; Speranza, G. *Tetrahedron* **1986**, *42*, 6687. (c) Gramatica, P.; Manitto, P.; Monti, D.; Speranza, G. *Tetrahedron* **1988**, *44*, 1299. (d) Fuganti, C.; Grasselli, P.; Servi, S.; Högberg, H.-E. *J. Chem. Soc., Perkin Trans. 1* **1988**, 3061. (e) Högberg, H.-E.; Hedenström, E.; Fägerhag, J.; Servi, S. *J. Org. Chem.* **1992**, *57*, 2052.
26. Utaka, M.; Konishi, S.; Mizuoka, A.; Ohkubo, T.; Sakai, T.; Tsuboi, S.; Takeda, A. *J. Org. Chem.* **1989**, *54*, 4989.
27. Ohta, H.; Kobayashi, N.; Ozaki, K. *J. Org. Chem.* **1989**, *54*, 1802.
28. Fuganti, C.; Grasselli, P. *Chem. Ind. (London)* **1977**, 983.
29. (a) Bujons, J.; Guajardo, R.; Kyler, K. S. *J. Am. Chem. Soc.* **1988**, *110*, 604. (b) Medina, J. C.; Kyler, K. S. *J. Am. Chem. Soc.* **1988**, *110*, 4818. (c) Medina, J. C.; Guajardo, R.; Kyler, K. S. *J. Am. Chem. Soc.* **1989**, *111*, 2310. (d) Xiao, X.-Y.; Prestwich, G. D. *Tetrahedron Lett.* **1991**, *32*, 6843.
30. (a) Rose, A. H. *The Yeast*; Harrison, J. S., Ed.; Academic Press: London, 1969, Vol. I; 1971, Vol III. (b) Glänzer, B. I.; Faber, K.; Griengl. H.; Roehr, M.; Wöhrer, W. *Enz. Microb. Technol.* **1988**, *10*, 744.
31. (a) Glänzer, B. I.; Faber, K.; Griengl, H. *Tetrahedron Lett.* **1986**, *27*, 4293. (b) Glänzer, B. I.; Faber, K.; Griengl, H. *Tetrahedron* **1987**, *43*, 771.
32. Glänzer, B. I.; Faber, K.; Griengl, H. *Tetrahedron* **1987**, *43*, 5791.
33. Glänzer, B. I.; Faber, K.; Griengl, H. *Enz. Microb. Technol.* **1988**, *10*, 689.
34. Sato, T.; Hanayama, K.; Fujisawa, T. *Tetrahedron Lett.* **1988**, *29*, 2197.
35. (a) Buist, P. H.; Dallmann, H. G.; Rymerson, R. T.; Seigel, P. M. *Tetrahedron Lett.* **1987**, *28*, 857. (b) Buist, P. H.; Dallmann, H. G. *Tetrahedron Lett.* **1988**, *29*, 285. (c) Buist, P. H.; Dallmann, H. G.; Rymerson, R. T.; Seigel, P. M.; Skala, P. *Tetrahedron Lett.* **1988**, *29*, 435.
36. Kamal, A.; Rao, M. V.; Meshram, H. M. *Tetrahedron Lett.* **1991**, *32*, 2657.
37. Kamal, A.; Rao, M. V.; Meshram, H. M. *J. Chem. Soc., Perkin Trans. 1* **1991**, 2056.
38. Rao, K. R.; Sampath Kumar, H. M. *Bioorg. Med. Chem. Lett.* **1991**, *1*, 507.
39. Brown, R. T.; Dauda, B. E. N.; Santos, C. A. M. *Chem. Commun.* **1991**, 825.
40. Fronza, G.; Fuganti, C.; Grasselli, P.; Poli, G.; Servi, S. *J. Org. Chem.* **1988**, *53*, 6153.
41. Rao, K. R.; Srinivasan, T. N.; Bhanumathi, N. *Tetrahedron Lett.* **1990**, *31*, 5959.
42. (a) Rao, K. R.; Bhanumathi, N.; Sattur, P. B. *Tetrahedron Lett.* **1990**, *31*, 3201. (b) Rao, K. R.; Bhanumathi, N.; Srinivasan, T. N.; Sattur, P. B. *Tetrahedron Lett.* **1990**, *31*, 899. (c) Rao, K. R.; Nageswar, Y. V. D.; Sampathkumar, H. M. *J. Chem. Soc., Perkin Trans. 1* **1990**, 3199.
43. Kitazume, T.; Ishikawa, N. *Chem. Lett.* **1984**, 1815.
44. Rao, K. R.; Nageswar, Y. V. D.; Sampath Kumar, H. M. *Tetrahedron Lett.* **1991**, *32*, 6611.

Enzo Santaniello, Patrizia Ferraboschi & Paride Grisenti
Università di Milano, Italy

(1*R*,5*R*)-2*H*-1,5-Benzodithiepin-3(4*H*)-one 1,5-Dioxide

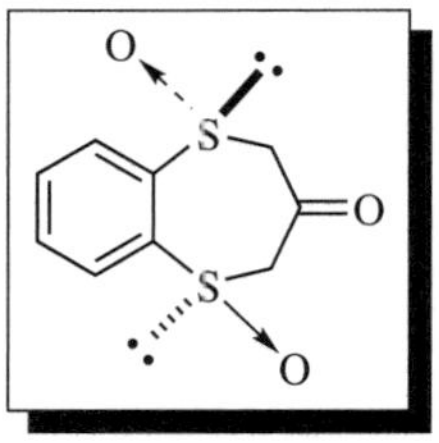

[183595-53-1] $C_9H_8O_3S_2$ (MW 228.29)

(chiral auxiliary for asymmetric desymmetrization of cyclic *meso*-1,2-diols)

Alternate Name: 2*H*-1,5-benzodithiepin-3(4*H*)-one, 1,5-dioxide, (1*R*-*trans*)-; (1*R*,5*R*)-1,5-benzodithiepan-3-one 1,5-dioxide.

Physical Data: colorless prisms, mp 195.0–196.0 °C (decomposes) (from hexane/EtOAc), $[\alpha]_D^{25}$ −100.3 (*c* 0.29, $CHCl_3$).

Solubility: soluble in MeOH, acetone, EtOAc, THF, CH_2Cl_2, and $CHCl_3$.

Form Supplied in: colorless powder; not commercially available.

Analysis of Reagent Purity: ^1H and ^{13}C NMR; elemental analysis.

Preparative Methods: the title reagent can be prepared from commercially available (1,2-benzenedithiol[1] and 1,3-dichloroacetone. After condensation of these reagents in the presence of DMAP, the resulting 1,5-benzodithepan-3-one is enantioselectively oxidized to the (*R*)-monosulfoxide by modified Sharpless oxidation [cumene hydroperoxide, Ti(O-*i*-Pr)$_4$] in the presence of (+)-diethyl tartrate as a chiral ligand.[2,3] Subsequent dry ozonation[4] of the (*R*)-monosulfoxide affords (1*R*,5*R*)-bis-sulfoxide **1**, having >98% optical purity. Alternative use of (−)-diethyl tartrate in the modified Sharpless oxidation makes possible convenient access to enantiomeric (1*S*,5*S*)-**1**.[5,6]

Purification: purification is performed by column chromatography. Since unpurified (1*R*,5*R*)-bis-sulfoxide **1** is only slightly soluble in the eluent, the following procedure is convenient. The crude material is dissolved in EtOAc and mixed with silica gel (ca. 5 g silica gel per 1 g of crude reagent). After solvent evaporation, the silica gel residue containing **1** is added to the top of the column and eluted with hexane–EtOAc (2:1).

Handling, Storage, and Precautions: the reagent can be stored for at least 1 month at room temperature without loss of its chemical and optical purities.

Introduction. (1*R*, 5*R*)-2*H*-1,5-Benzodithiepin-3(4*H*)-one 1,5-dioxide (C_2-symmetric bis-sulfoxide **1**) has been used as a chiral auxiliary for asymmetric desymmetrization of cyclic *meso*-

1,2-diols via diastereoselective acetal cleavage reaction. The procedure consists of three steps (eq 1), that is, acetalization (step 1), acetal cleavage reaction followed by benzylation (step 2), and hydrolysis of the vinyl ether (step 3). Due to the C_2-symmetry of **1**, the chiral auxiliary gives only one product in step 1. In addition, no regio- or geometric isomers of the enol ether are formed in step 2. This reagent can be recovered by acid-promoted hydrolysis and reused.

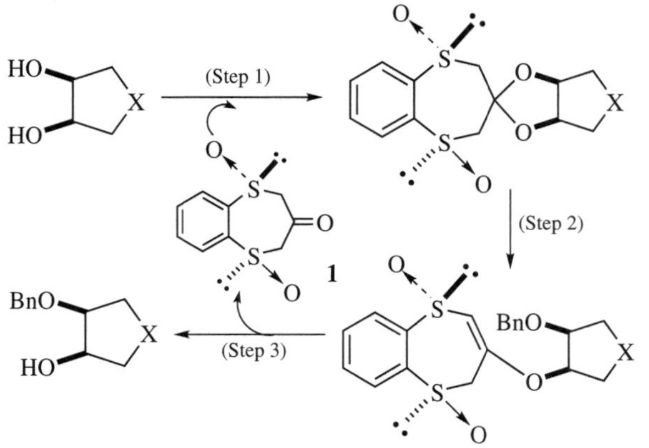

(1)

Acetal Formation Involving C_2-Symmetric bis-Sulfoxide and *meso*-1,2-Diols (Step 1). Acetalization of *meso*-1,2-diols with this reagent should be conducted with TMSOTf and 2,6-lutidine in dichloromethane below 4 °C.[7] Higher temperatures and prolonged reaction times cause undesirable racemization and decomposition of the reagent. When the reactivity of *meso*-1,2-diols with the chiral auxiliary is low, acetalization using the mono-TMS ether of *meso*-diols and TMSOTf is recommended.[8]

Diastereoselective Acetal Fission Followed by Benzylation (Step 2). Upon treatment with KHMDS and 18-crown-6 in THF at −78 °C, the acetal from the (*R*,*R*)-bis-sulfoxide is rapidly converted into the alkoxide having the (1*S*,2*R*) configuration. The counter cation of the base is very important for high selectivity. Diastereoselectivity was seen to increase in the order LiHMDS (8% de) <NaHMDS (90% de) <KHMDS (>96% de).

Hydrolysis of the Vinyl Ether and Reagent Recovery (Step 3). The resulting vinyl ether can be hydrolyzed with 10% HCl in acetone at room temperature. The chiral auxiliary is recovered without loss of optical purity and is reusable.

Desymmetrization of Functionalized *meso*-1,2-diols. Using this methodology, various cyclic *meso*-1,2-diols can be desymmetrized with very high (>96% ee) and predictable selectivity. The enantiomers are obtained through use of an appropriate chiral auxiliary. Bis-sulfoxide **1** has been applied to the desymmetrization of a poly-oxygenated *meso*-diol containing five stereogenic centers (eq 2).

However, the same chiral auxiliary is not suited to the desymmetrization of acyclic *meso*-diols since the acetalization step is sluggish. Nonetheless, an acyclic *meso*-diol such as erythritol can be desymmetrized by prior protection of the terminal primary hydroxyl groups as an *o*-xylyl ether (eq 3).

Desymmetrization by means of this methodology was successfully applied to a synthesis of key intermediates for mosin B,[9] aspicilin,[10] *gala*-quercitol,[11] and allosamizoline.[8]

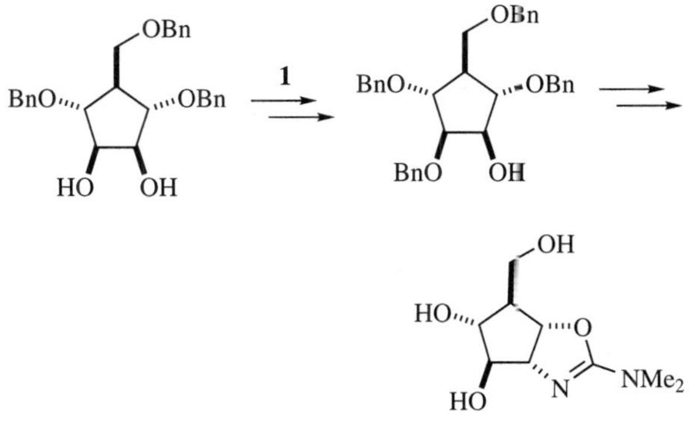

(2)

Allosamizoline

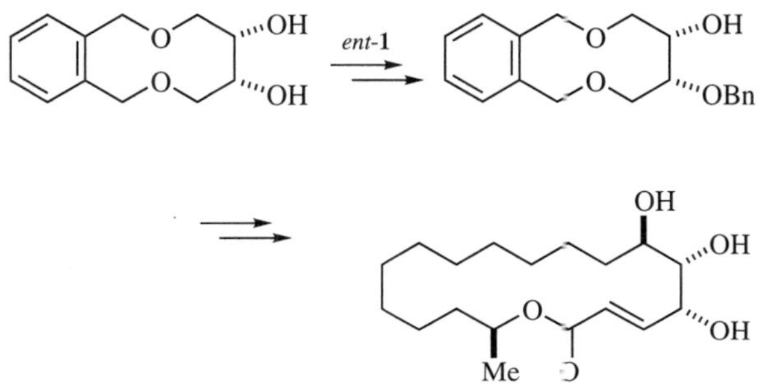

(3)

Aspicilin

Related Reagents. (1*S*,5*S*)-2*H*-1,5-benzodithiepin-3(4*H*)-one 1,5-dioxide.

1. Giolando, D. M.; Kirschbaum, K. *Synthesis* **1992**, 451.
2. Di Furia, F.; Modena, G.; Seraglia, R. *Synthesis* **1984**, 325.
3. Zhao, S. H.; Samuel, O.; Kagan, H. B. *Tetrahedron* **1987**, *43*, 5135.
4. Cohen, Z.; Keinan, E.; Mazur, Y.; Varkony, T. H. *J. Org. Chem.* **1975**, *40*, 2141.
5. Maezaki, N.; Sakamoto, A.; Nagahashi, N.; Soejima, M.; Li, Y. X.; Imamura, T.; Kojima, N.; Ohishi, H.; Sakaguchi, K.; Iwata, C.; Tanaka, T. *J. Org. Chem.* **2000**, *65*, 3284.
6. Maezaki, N.; Sakamoto, A.; Soejima, M.; Sakamoto, I.; Li, Y. X.; Tanaka, T.; Ohishi, H.; Sakaguchi, K.; Iwata, C. *Tetrahedron: Asymmetry* **1996**, *7*, 2787.
7. Matsuda, F.; Terashima, S. *Tetrahedron* **1988**, *44*, 4721.
8. Maezaki, N.; Sakamoto, A.; Tanaka, T.; Iwata, C. *Tetrahedron: Asymmetry* **1998**, *9*, 179.
9. Maezaki, N.; Kojima, N.; Sakamoto, A.; Iwata, C.; Tanaka, T. *Org. Lett.* **2001**, *3*, 429.
10. Maezaki, N.; Li, Y. X.; Ohkubo, K.; Goda, S; Iwata, C.; Tanaka, T. *Tetrahedron* **2000**, *56*, 4405.
11. Maezaki, N.; Nagahashi, N.; Yoshigami, R.; Iwata, C.; Tanaka, T. *Tetrahedron Lett.* **1999**, *40*, 3781.

Naoyoshi Maezaki & Tetsuaki Tanaka
Osaka University, Suita, Japan

1-Benzoyl-2-*t*-butyl-3,5-dimethyl-4-imidazolidinone[1]

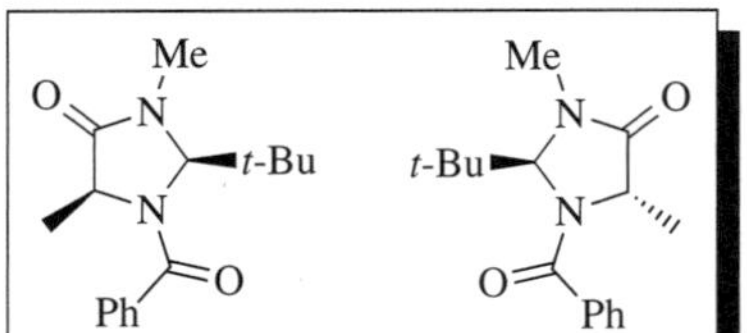

(2*R*,5*S*)-(**1**) (2*S*,5*S*)-(**1**)

(2*R*,5*S*)
[97443-91-9] $C_{16}H_{22}N_2O_2$ (MW 274.36)
(2*S*,5*S*)
[97443-88-4]

(imidazolidinones for generating the enantiomeric enolates derived from alanine;[2,3] reagents for the preparation of α-methylated amino acids through enolate alkylation, benzylation, and nitroalkene addition[3–5])

Physical Data: (2*R*,5*S*)-(**1**): mp 125 °C; $[\alpha]_D^{rt} = -47.7°$ ($c = 1.04$, $CHCl_3$); (2*S*,5*S*)-(**1**): mp 175 °C; $[\alpha]_D^{rt} = +44.5°$ ($c = 1.0$, $CHCl_3$).

Solubility: sol THF.

Preparative Methods: the imine from (*S*)-alanine *N*-methylamide and pivalaldehyde is cyclized by heating with benzoic anhydride to give mainly *cis*-(**1**) [(2*R*,5*S*)-(**1**)] in modest yields; alternatively, the imine cyclizes to the *trans*-substituted heterocycle by treatment with HCl in MeOH, and subsequent benzoylation produces *trans*-(**1**) [(2*S*,5*S*)-(**1**)] in high yield (eq 1).[2]

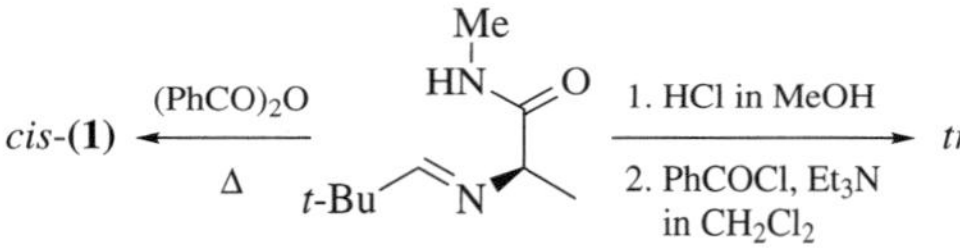

(1)

Handling, Storage, and Precautions: both diastereoisomers are readily crystallizable compounds which are stable at rt for years.

Alkylations of the Lithium Enolates. Treatment of the reagents with *Lithium Diisopropylamide* (LDA) generates the enolates (**2**) or *ent*-(**2**) (crystal structure of *rac*-TBDMS-(**2**)[6]) which can be alkylated[3,5] to give, for instance, (*R*)-α-methyl-dopa (**3**) or triacetyl (*S*)-α-methyl-dopa.[3]

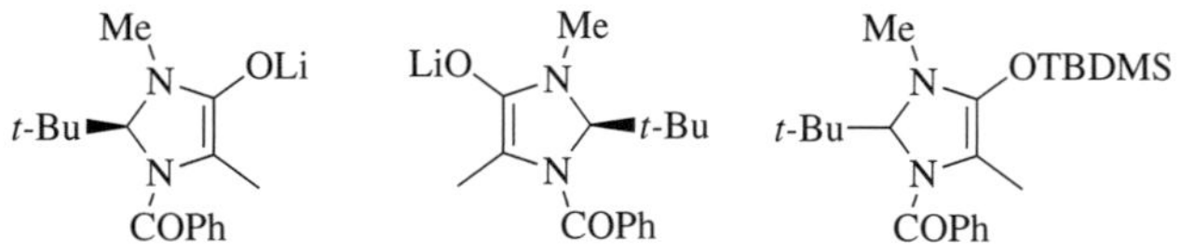

(**2**) *ent*-(**2**) *rac*-TBDMS-(**2**)

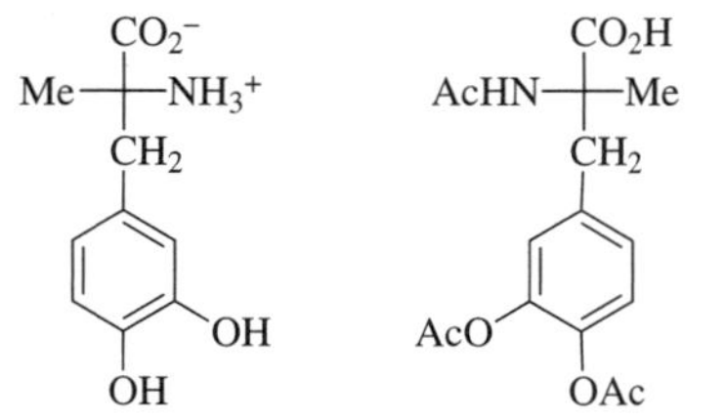

(**3**) *ent*-$(Ac)_3$-(**3**)

Benzoylimidazolidinones of Other Amino Acids. In a similar way, other proteinogenic and nonproteinogenic amino acids have been converted to imidazolidinones (**4**) and alkylated through enolates to give derivatives of α-branched α-amino acids. Examples of the R group in (**4**) are as follows: *i*-Pr,[2–4] Bn,[2] $(CH_2)_2SMe$,[2,7,8] $CH{=}CH_2$,[8] Ph,[2] $(CH_2)_3NHCO_2Bn$,[5] $(CH_2)_4NHCO_2Bn$,[5] CH_2CO_2H,[9] $(CH_2)_2CO_2H$.[9] The limitation of the method is given by the fact that certain amino acid *N*-methylamides, the intermediates on the way from the 3-methylimidazolidinones to free amino acids, are *very* difficult to hydrolyze.[10] Except for the procedure described here and for those methods involving enantioselective catalysis, all other syntheses of amino acids rest upon the use of a covalently attached chiral auxiliary which has to be discarded, recovered, or destroyed after use.[1c]

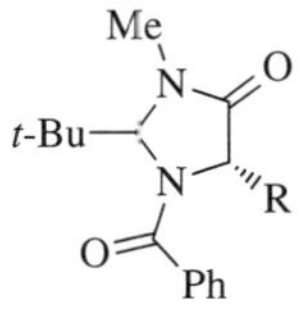

(**4**)

The Principle of Self-Regeneration of Stereogenic Centers.[1,11,12] In the absence of additional chirality the generation of an enolate from a simple amino acid will lead to racemization.[13] There are two ways around this: (i) attachment of a chiral auxiliary, and (ii) diastereoselective generation of an additional stereogenic center which makes sure that the subsequently generated enolate is still chiral. In the case of alanine, described here, the acetal chirality center serves this purpose. The most general case is described in eq 2. Thus an α- or β-amino, -hydroxy-, and -mercaptocarboxylic acid may be converted to one of two diastereoisomeric acetals. The original stereogenic center can now be eliminated without forming an achiral species; subsequent reactions at the newly formed trigonal center should be diastereoselective, so that the product of acetal hydrolysis is nonracemic. The trigonal center at the site of the original stereogenic center may be part of an electrophilic or a nucleophilic double bond system, or may be a radical center. In the overall process, a substituent (mostly a hydrogen) at the one and only chirality center of the starting material is replaced by a new substituent stereoselectively. Since no chiral auxiliary is employed, this has been termed the principle of self-regeneration of the stereogenic center (SRSC). The auxiliary is actually the aldehyde, used for generating the second stereogenic center, and it is removed in the final hydrolysis step.

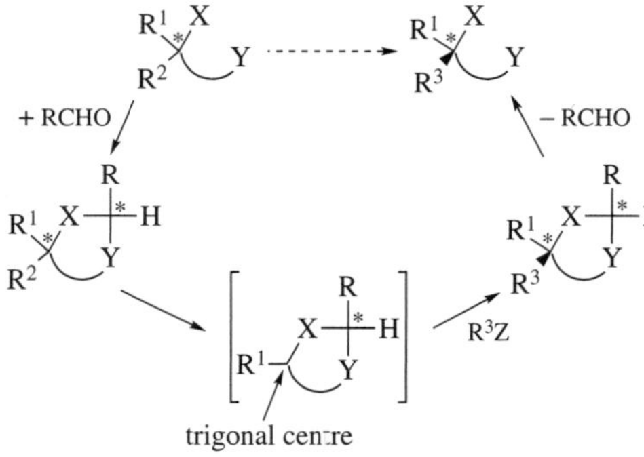

(2)

Oxazoline, oxazolidine, dihydropyrimidine, and 1,3-dioxine derivatives can also be used in this way.

Related Reagents. (2*S*,4*S*)-3-Benzoyl-2-*t*-butyl-4-methyl-1,3-oxazolidin-5-one; *t*-Butyl 2-*t*-Butyl-3-methyl-4-oxo-1-imidazolidinecarboxylate; (*R*)-2-*t*-Butyl-6-methyl-4*H*-1,3-dioxin-4-one; (*R*,*R*)-2-*t*-Butyl-5-methyl-1,3-dioxolan-4-one; Diethyl Acetamidomalonate; *N*-(Diphenylmethylene)aminoacetonitrile; Ethyl *N*-(Diphenylmethylene)glycinate; Ethyl Isocyanoacetate; Methyl *N*-Benzylidenealaninate; (*R*)-Methyl 2-*t*-Butyl-3(2*H*)-oxazolecarboxylate.

---

1. (a) Seebach, D.; Imwinkelried, R.; Weber, T. In *Modern Synthetic Methods*; Springer: New York, 1986; Vol. 4, pp 125–259. (b) Seebach, D.; Roggo, S.; Zimmermann, J. In *Workshop Conferences Hoechst*; Verlag Chemie: Weinheim, 1987; Vol. 17, pp 85–126. (c) Williams, R. M. *Synthesis of Optically Active α-Amino Acids*; Pergamon: Oxford, 1989. (d) Duthaler, R. O. *Tetrahedron* **1994**, *50*, 1539.
2. Naef, R.; Seebach, D. *Helv. Chim. Acta* **1985**, *68*, 135 (*Chem. Abstr.* **1985**, *103*, 71 633q).
3. Seebach, D.; Aebi, J. D.; Naef, R.; Weber, T. *Helv. Chim. Acta* **1985**, *68*, 144.
4. Calderari, G.; Seebach, D. *Helv. Chim. Acta* **1985**, *68*, 1592 (*Chem. Abstr.* **1986**, *105*, 133 326u).
5. Gander-Coquoz, M.; Seebach, D. *Helv. Chim. Acta* **1988**, *71*, 224 (*Chem. Abstr.* **1988**, *109*, 110 880p).
6. Seebach, D.; Maetzke, T.; Petter, W.; Klötzer, B.; Plattner, D. A. *J. Am. Chem. Soc.* **1991**, *113*, 1781.
7. Weber, T.; Aeschimann, R.; Maetzke, T.; Seebach, D. *Helv. Chim. Acta* **1986**, *69*, 1365 (*Chem. Abstr.* **1987**, *107*, 97 075s).
8. Seebach, D.; Juaristi, E.; Miller, D. D.; Schickli, C.; Weber, T. *Helv. Chim. Acta* **1987**, *70*, 237.
9. Aebi, J. D.; Seebach, D. *Helv. Chim. Acta* **1985**, *68*, 1507 (*Chem. Abstr.* **1986**, *105*, 97 883n).
10. Seebach, D.; Gees, T.; Schuler, F. *Justus Liebigs Ann. Chem./Liebigs Ann. Chem.* **1993**, 785.
11. Seebach, D.; Naef, R.; Calderari, G. *Tetrahedron* **1984**, *40*, 1313.
12. Strijtveen, B.; Kellogg, R. M. *Tetrahedron* **1987**, *43*, 5039.
13. For three exceptions see the alkylations of an aspartic acid,[13a] of an aziridine carboxylic acid,[13b] and of a cysteine derivative.[13c] (a) Seebach, D.; Wasmuth, R. *AC(E)* **1981**, *20*, 971. (b) Häner, R.; Olano, B.; Seebach, D. *Helv. Chim. Acta* **1987**, *70*, 1676. (c) Beagley, B.; Betts, M. J.; Pritchard, R. G.; Schofield, A.; Stoodley, R. J.; Vohra, S. *Chem. Commun.* **1991**, 924.

Armido Studer & Dieter Seebach
Eidgenössische Technische Hochschule, Zürich, Switzerland

(2*S*,4*S*)-3-Benzoyl-2-*t*-butyl-4-methyl-1,3-oxazolidin-5-one[1]

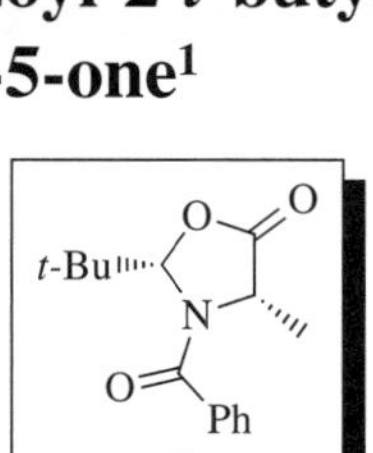

(1)

[104057-64-9] $C_{15}H_{19}NO_3$ (MW 261.32)

(cyclic acetal of *N*-benzoylalanine; reagent for the preparation of enantiopure α-methyl-α-aminocarboxylic acids;[2] precursor to the 4-methylidene derivative;[3a] a radicalophile;[3b] a Michael acceptor;[4] and an ene component for Diels–Alder additions,[5] with formation of more complex α-amino acids)

Physical Data: mp 94.2–94.5 °C; $[\alpha]_D^{rt} = -29.3°$ ($c = 1.0$, $CHCl_3$).
Solubility: sol THF, poorly sol hexane.
Preparative Methods: the sodium salt of the pivalaldehyde imine of (*S*)-alanine is cyclized to the oxazolidinone by reaction with PhCOCl in CH_2Cl_2 at low temperature; the *cis*/*trans* ratio in the crude product depends upon the exact reaction conditions, it is usually 4:1; purification by a combination of crystallization and chromatography gives the pure *cis* derivative in 60–70% yield.[2]
Handling, Storage, and Precautions: the solid material is stable for years when stored in a bottle.

Generation of the Enolate and Reactions with Electrophiles. The enolate of the title reagent (**1**) is generated with lithium amide bases and reacts with electrophiles such as benzylic bromides or tricarbonyl(fluorobenzene)chromium to give the corresponding 4,4-disubstituted derivatives of type (**2**), which are hydrolyzed under acidic conditions[2,6–8] to, for instance, the amino acids[2,7] (**3**) and (**4**).

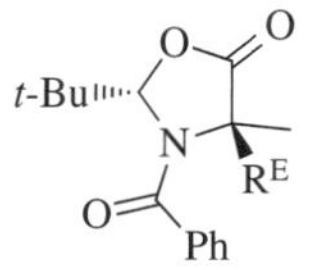

(**2**)

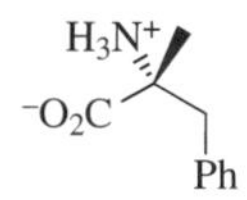

(**3**)

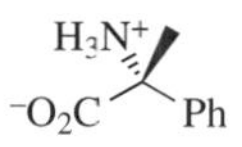

(**4**)

In the overall process the α-hydrogen of the alanine has been replaced by benzyl or phenyl stereoselectively (an application of the principle of self-regeneration of the stereogenic center). The analogous 2-*t*-butyl-3-benzoyloxazolidinones of valine,[2] phenylalanine,[2] methionine,[2,9] and lysine[6] have also been prepared and alkylated to give α-branched amino acids.

Other Oxazolidine as well as Thiazolidine Derivatives for Branching Amino Acids. The cyclic derivative of alanine and other amino acids employed most frequently for α-alkylation is not (**1**) but rather the benzaldehyde acetal (**5**), either with a benzoyl[5,10] or with a Cbz[11] group on nitrogen. These compounds were used for the preparation of 2-methyl-2-aminobutanoic acid, α-methylphenylalanine, α-methyllysine, 2-methylaspartic acid, and 2-methylglutamic acid. Bicyclic compounds containing oxazolidinone rings such as (**6**) (from alanine, leucine, and phenylalanine)[12] and (**7**) (from azetidinecarboxylic acid,[13] proline,[14] hydroxyproline,[15] and cysteine[16]) have also been applied to the synthesis of branched amino acids.

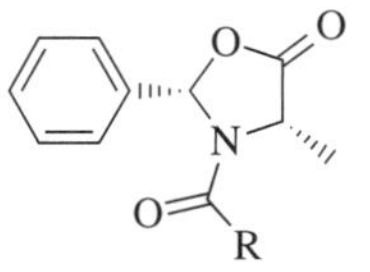

(**5**)
R = Ph, OBn

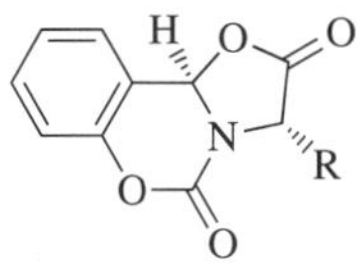

(**6**)

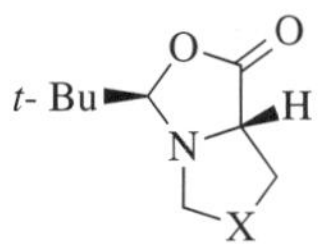

(**7**)
X = $(CH_2)_{0,1}$
X = CHOR
X = S

Finally, the enolates (**8**), (**9**), and (**10**) of oxazolidine and thiazolidine carboxylates have been used for the synthesis of enantiopure α-substituted serine,[17] cysteine,[18] threonine,[17] *allo*-threonine,[17] and β-hydroxyleucine.[19]

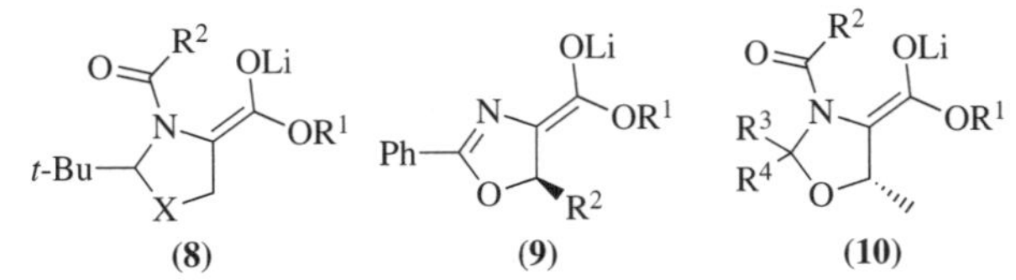

α-Bromination–dehydrobromination of oxazolidinone (**1**) (NBS, then DBU) or an alternative preparation starting from cysteine gives the methylene derivative (**11**) which combines with cyclopentadiene and cyclohexadiene in [4 + 2] cycloadditions; these reactions were used for the synthesis of (*R*) and (*S*)-2-aminobicyclo[2.2.1]heptane-2-carboxylic acids as shown in eq (1).[3–5]

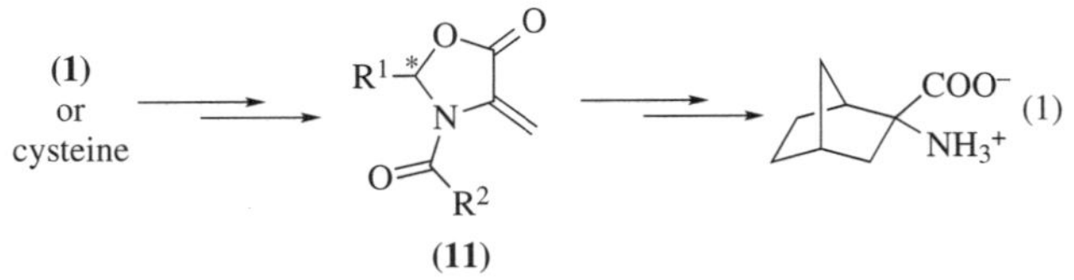

(*R*)- and (*S*)-2-*t*-Butyl-1,3-oxazolidin-5-ones: Chiral Glycine Derivatives. Since the oxazolidinone enolates are good diastereoselective nucleophiles, and since the products are much more readily hydrolyzed to the free amino acids than those derived from imidazolidinones, it was also desirable to make available the corresponding glycine derivative (**12**). This was achieved by preparative HPLC resolution on Chiraspher, Chiracel OD, or Pirkle columns (up to 10 g per injection, separation factors α up to 2.35).[8,20] The enolate of the 2-*t*-butyl-substituted N-Cbz oxazolidinone was especially useful for the synthesis of a large variety of (*R*)- and (*S*)-threonines; see, for instance, the cyclosporin component MeBmt (**13**)[21] and the *p*-nitrophenylserine (**14**)[22] in eq (2).

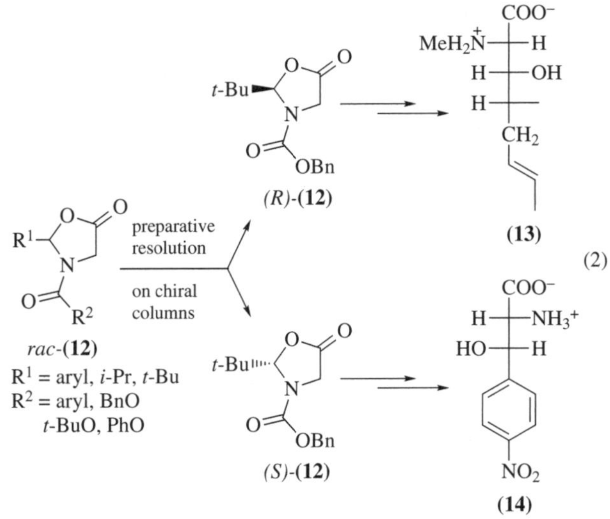

Related Reagents. 1-Benzoyl-2-*t*-butyl-3,5-dimethyl-4-imidazolidinone; *t*-Butyl 2-*t*-Butyl-3-methyl-4-oxo-1-imidazolidinecarboxylate; (*R*,*R*)-2-*t*-Butyl-5-methyl-1,3-dioxolan-4-one; Diethyl Acetamidomalonate *N*-(Diphenylmethylene)aminoacetonitrile; Ethyl *N*-(Diphenylmethylene)glycinate; Methyl *N*-Benzylidenealaninate.

1. Seebach, D.; Imwinkelried, R.; Weber, T. In *Modern Synthetic Methods*; Springer: Berlin, 1986; Vol. 4, pp 125–259.
2. Seebach, D.; Fadel, A. *Helv. Chim. Acta* **1985**, *68*, 1243.
3. (a) Zimmermann, J.; Seebach, D. *Helv. Chim. Acta* **1987**, *70*, 1104. (b) Beckwith, A. L. J.; Chai, C. L. L. *Chem. Commun.* **1990**, 1087.
4. Crossley, M. J.; Tansey, C. W. *Aust. J. Chem.* **1992**, *45*, 479.
5. Pyne, S. G.; Dikic, B.; Gordon, P. A.; Skelton, B. W.; White, A. H. *Aust. J. Chem.* **1993**, *46*, 73.
6. Gander-Coquoz, M.; Seebach, D. *Helv. Chim. Acta* **1988**, *71*, 224 (*Chem. Abstr.* **1988**, *109*, 110 880p).
7. Chaari, M.; Jenhi, A.; Lavergne, J.-P.; Viallefont, Ph. *J. Organomet. Chem.* **1991**, *401*, C10 (*Chem. Abstr.* **1991**, *114*, 164 736t).
8. Seebach, D.; Gees, T.; Schuler, F. *Justus Liebigs Ann. Chem.* **1993**, 785.
9. Beck, A. K.; Seebach, D. *Agric. Biol. Chem.* **1988**, *42*, 142.
10. Nebel, K.; Mutter, M. *Tetrahedron* **1988**, *44*, 4793; Fadel, A.; Salaün, J. *Tetrahedron Lett.* **1987**, *28*, 2243.
11. Karady, S.; Amato, J. S.; Weinstock, L. M. *Tetrahedron Lett.* **1984**, *25*, 4337; Abell, A. D.; Taylor, J. M. *J. Org. Chem.* **1993**, *58*, 14; Altmann, E.; Nebel, K.; Mutter, M. *Helv. Chim. Acta* **1991**, *74*, 800.
12. Zydowsky, T. M.; de Lara, E.; Spanton, S. G. *J. Org. Chem.* **1990**, *55*, 5437.
13. Seebach, D.; Vettiger, T.; Müller, H. M.; Plattner, D. A.; Petter, W. *Justus Liebigs Ann. Chem.* **1990**, 687.
14. Seebach, D.; Naef, R. *Helv. Chim. Acta* **1981**, *64*, 2704; Seebach, D.; Boes, M.; Naef, R.; Schweizer, B. *J. Am. Chem. Soc.* **1983**, *105*, 5390; Calderari, G.; Seebach, D. *Helv. Chim. Acta* **1985**, *68*, 1592 (*Chem. Abstr.* **1986**, *105*, 133 326u); Thaisrivongs, S.; Pals, D. P.; Lawson, J. A.; Turner, S. R.; Harris, D. W. *J. Med. Chem.* **1987**, *30*, 536; Williams, R. M.; Glinka, T.; Kwast, E. *J. Am. Chem. Soc.* **1988**, *110*, 5927; Orsini, F.; Pelizzoni, F.; Forte, M.; Sisti, M.; Bombieri, G.; Benetollo, F. *J. Heterocycl. Chem.* **1989**, *26*, 837; Beck. A. K.; Blank, S.; Job, K.; Seebach, D.; Sommerfeld, Th. *Comprehensive Organic Synthesis* **1994**, *72*, in press.
15. Weber, T.; Seebach, D. *Helv. Chim. Acta* **1985**, *68*, 155 (*Chem. Abstr.* **1985**, *103*, 88 182q).
16. Seebach, D.; Weber, T. *Tetrahedron Lett.* **1983**, *24*, 3315; Seebach, D.; Weber, T. *Helv. Chim. Acta* **1984**, *67*, 1650 (*Chem. Abstr.* **1985**, *102*, 185 449u).
17. Seebach, D.; Aebi, J. D.; Gander-Coquoz, M.; Naef, R. *Helv. Chim. Acta* **1987**, *70*, 1194 (*Chem. Abstr.* **1988**, *108* 187 232r) and earlier work cited therein.
18. Mulqueen, G. C.; Pattenden, G.; Whiting, D. A. *Tetrahedron* **1993**, *49*, 5359.
19. Sunazuka, T.; Nagamitsu, T.; Matsuzaki, K.; Tanaka, H.; Ōmura, S. *J. Am. Chem. Soc.* **1993**, *115*, 5302.
20. Seebach, D.; Müller, S. G.; Gysel, U.; Zimmermann, J. *Helv. Chim. Acta* **1988**, *71*, 1303 (*Chem. Abstr.* **1989**, *110*, 114 764x); Kinkel, J. N.; Gysel, U.; Blaser, D.; Seebach, D. *Helv. Chim. Acta* **1991**, *74*, 1622.
21. Blaser, D.; Ko, S. Y.; Seebach, D. *J. Org. Chem.* **1991**, *56*, 6230.
22. Blaser, D.; Seebach, D. *Justus Liebigs Ann. Chem.* **1991**, 1067.

Andrea Rolf Sting & Dieter Seebach
Eidgenössische Technische Hochschule, Zürich, Switzerland

1-Benzoyl-2(*S*)-*tert*-butyl-3-methylperhydropyrimidin-4-one

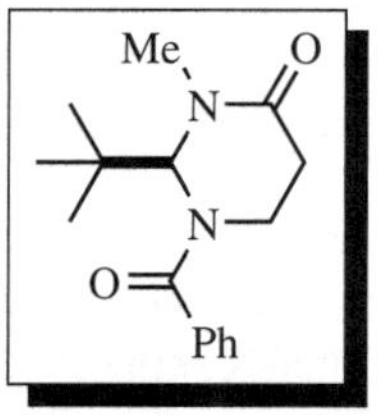

[139119-52-1] $C_{16}H_{22}N_2O_2$ (MW 274.36)

(reagent used for the enantioselective synthesis of β-amino acids; in particular, α- and α,α-substituted β-amino acids)

Physical Data: mp 99–100 °C; $[\alpha]_D^{29}$ +51.2 (*c* 1, $CHCl_3$).
Solubility: soluble in THF and most organic solvents.
Preparative Methods: [1,2,3]

Step 1:

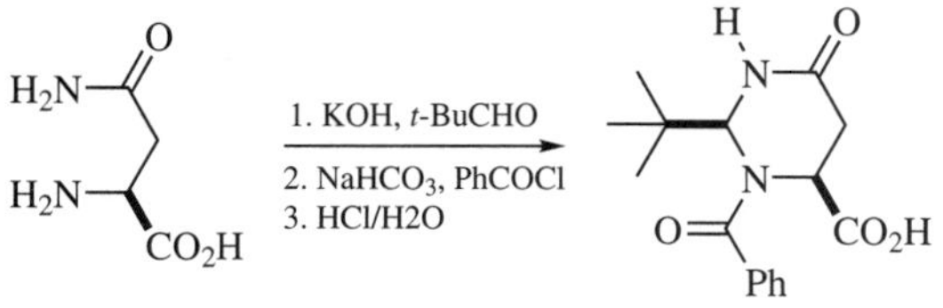

(1)

(*S*)-Asparagine cis-**1**

According to the procedure described by Lakner et al.,[1] with benzoyl chloride instead of methyl chloroformate, 13.2 g (0.2 mol) of KOH and 300 mL of water are placed in a 1 L round-bottomed flask before the addition of 30.0 g (0.2 mol) of L-(*S*)-asparagine with vigorous stirring. The resulting mixture is cooled to 0 °C and treated with 25.0 mL (19.8 g, 0.23 mol) of pivalaldehyde. Stirring is continued for 1 h at 0 °C and for 5 h at ambient temperature. The reaction mixture is cooled to 0 °C before the addition of 16.8 g (0.2 mol) of $NaHCO_3$ and 23.2 mL (28.1 g, 0.2 mol) of benzoyl chloride, and stirring is continued for 2 h at ambient temperature prior to quenching with 73 mL of 10% aqueous HCl. The desired product, which precipitates from solution, is filtered, washed with cold water, and dried under vacuum to afford 54.0 g (88%) of 1-benzoyl-2(*S*)-*tert*-butyl-6(*S*)-carboxy-perhydropyrimidin-4-one [(2*S*,6*S*)-**1**], mp 202-203 °C; $[\alpha]_D{}^{29}$ -107.0 (*c* 1, EtOH).

Step 2:

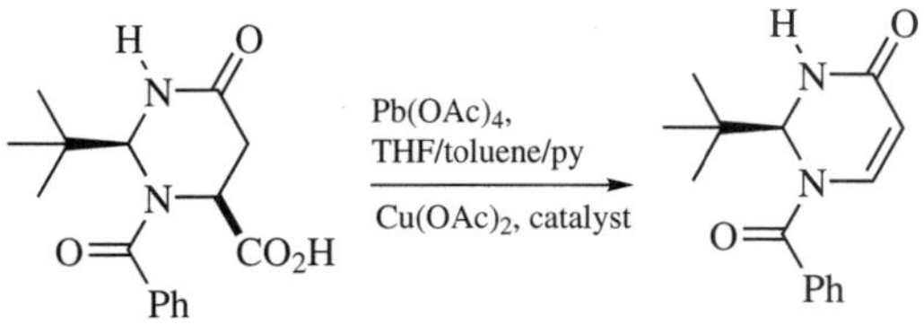

(2)

cis-**1** (*S*)-**2**

In a 3 L round-bottomed flask provided with a magnetic stirrer, 28.0 g (92 mmol) of (2*S*,6*S*)-**1** is dissolved in 900 mL of dry THF, 700 mL of toluene, and 11.2 mL (11.0 g, 138.4 mmol) of pyridine. The resulting solution is treated with 3.7 g (18.4 mmol) of copper diacetate monohydrate, and the resulting suspension is stirred at ambient temperature for 2 h. The reaction flask is then submerged in an ice-water bath before the addition of 61.6 g (138.8 mmol) of lead tetraacetate. The cooling bath is removed and the reaction mixture is heated to 80–90 °C for 12 h. The precipitate is removed by filtration and washed several times with EtOAc until the extracts come out free of product (TLC). The organic extracts are combined with the original filtrate, dried over anhydrous Na_2SO_4, and concentrated. The crude product is purified by flash chromatography (eluent hexane-EtOAc, 80:20 → 50:50) to give 16.0 g (73%) of 1-benzoyl-2(*S*)-*tert*-butyl-2,3-dihydro-4(*H*)-pyrimidin-4-one [(*S*(-**2**], mp 209-210 °C; $[\alpha]_D^{29}$ +556.4 (*c* 1, $CHCl_3$).

Step 3:

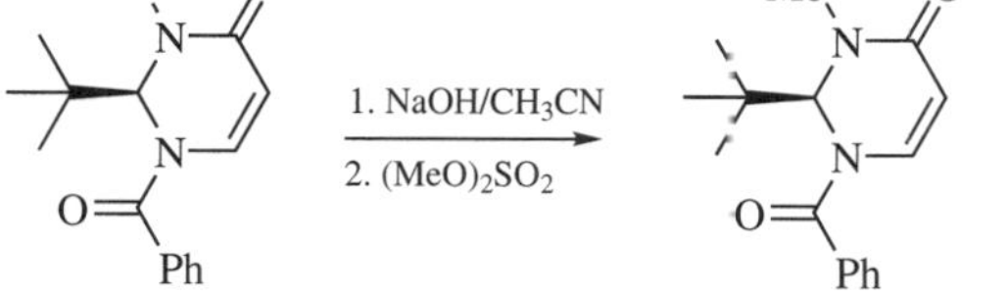

(3)

(*S*)-**2** (*S*)-**3**

N-Methylation of (*S*)-**2** is best accomplished following the procedure of Juaristi et al.[3] In a 50 mL round-bottomed flask provided with magnetic stirrer is placed 3.0 g (11.6 mmol) of (*S*)-**2** and 15 mL of acetonitrile. The resulting solution is treated with 1.1 mL (11.6 mmol) of dimethyl sulfate (slow addition) and 0.5 g (11.6 mmol) of NaOH (slow addition). The reaction mixture is stirred for 3 h at 45–50 °C (mineral oil bath) and the acetonitrile is removed at reduced pressure. The residue is suspended in 50 mL of water and extracted with three 50 mL portions of EtOAc. The combined organic extracts are dried over anhydrous Na_2SO_4 and concentrated to give the crude product that is purified by flash chromatography (hexane-EtOAc, 80:20 → 50:50) to afford 2.7 g (84%) of 1-benzoyl-2(*S*)-*tert*-butyl-3-methyl-2,3-dihydro-4(*H*)-pyrimidin-4-one, mp 141-142 °C; $[\alpha]_D^{29}$ +560.2 (*c* 1, $CHCl_3$).

Step 4:

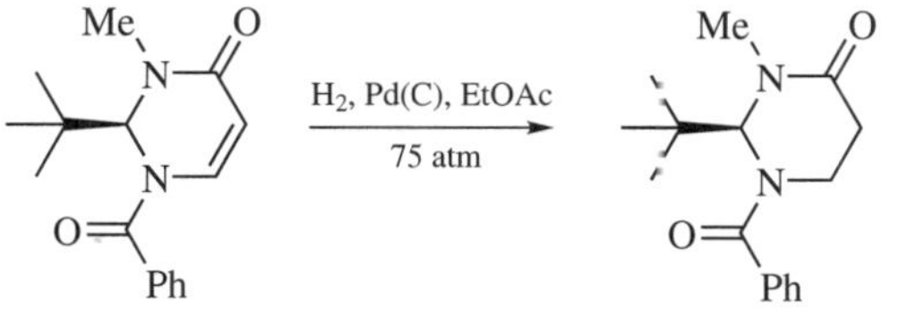

(4)

(*S*)-**3** (*S*)-**4**

According to the procedure of Juaristi et al.,[2] heterocycle (*S*)-**3** (5.0 g, 18.4 mmol), 40 mL of EtOAc, 0.5 g of 10% Pd(C), and 0.4 mL of acetic acid is placed in a hydrogenation flask. The reaction mixture is pressurized to 75 atm of hydrogen, heated to 45 °C, and stirred for 24 h. The catalyst is removed by filtration over Celite, the filtrate is washed with aqueous 10% $NaHCO_3$, dried over anhydrous Na_2SO_4, filtered, and evaporated at reduced pressure to give 4.5 g (90%) of 1-benzoyl-2(*S*)-*tert*-butyl-3-methylperhydropyrimidin-4-one [(*S*(-**4**], mp

99–100 °C; $[\alpha]_D^{29}$ +51.2 (*c* 1, $CHCl_3$). The overall yield in the preparation of (*S*)-**4** from (*S*)-asparagine (four steps) is 49%.

Preparation of Enantiopure α-Substituted β-Amino Acids.[2,4] In preliminary studies, racemic 2-*tert*-butylperhydropyrimidinone, *rac*-**4**, was alkylated with high diastereoselectivity via its corresponding enolate (eq 5).[5] The high stereoselectivity encountered in the reaction of *rac*-**4**-Li with various electrophiles was ascribed to steric hindrance generated by the axial disposition of the *tert*-butyl group at C(2),[5,6] which directs approach to the electrophile from the enolate face opposite to this group.

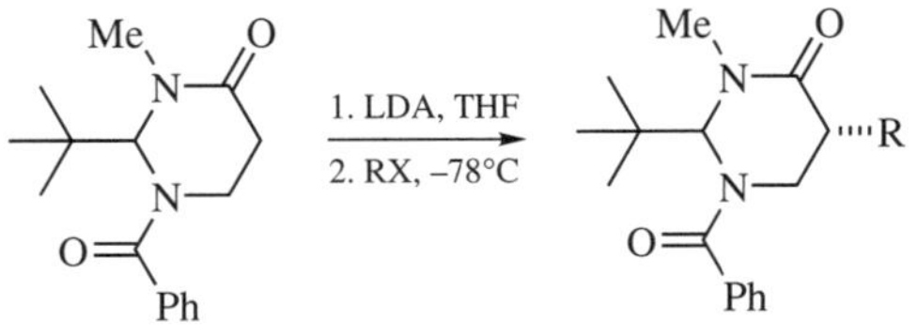

(5)

rac-**4** *trans*

RX = CH_3I, $PhCH_2Br$, *n*-BuI, *n*-$C_6H_{13}I$, CH_2=$CHCH_2Cl$

These observations paved the road for the development of a new method for the asymmetric synthesis of α-substituted β-amino acids. Thus, an efficient protocol for the preparation of enantiopure pyrimidinone (*S*)–**4** was developed (vide supra, eq 1–4).

Enolate (*S*)–**4**-Li was generated upon treatment of the heterocycle with lithium diisopropylamide (LDA) in THF solution and under a nitrogen atmosphere. The electrophile was then added at −78 °C to afford the *trans*-alkylated products with high diastereoselectivity and in good yields (eq 6, Table 1).

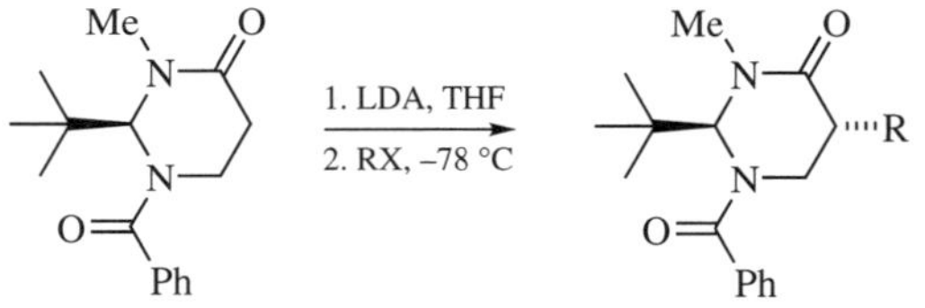

(6)

(*S*)-**4** (2*S*,5*R*)-**5**-**8**

Table 1 Diastereoselectivity of enolate (*S*)-**4**-Li alkylation

Product	RX	ds (%)	mp (°C)	$[\alpha]_D^{29}$	Yield (%)
5	MeI	>96	121–2	+39.5	77
6	*n*-BuI	95	80–1	+26.7	75
7	*n*-$C_6H_{13}I$	96	70–1	+31.2	80
8	$PhCH_2Br$	>96	173–4	−64.0	80

The final step of the overall conversion of (*S*)-asparagine to 2-alkyl-3-aminopropanoic acid, the hydrolysis of heterocycles **5**–**8**, was achieved by heating with 6 N HCl in a sealed tube at 90–100 °C. The free amino acids **9**–**12** were purified by chromatography on an ion-exchange column (eq 7, Table 2).

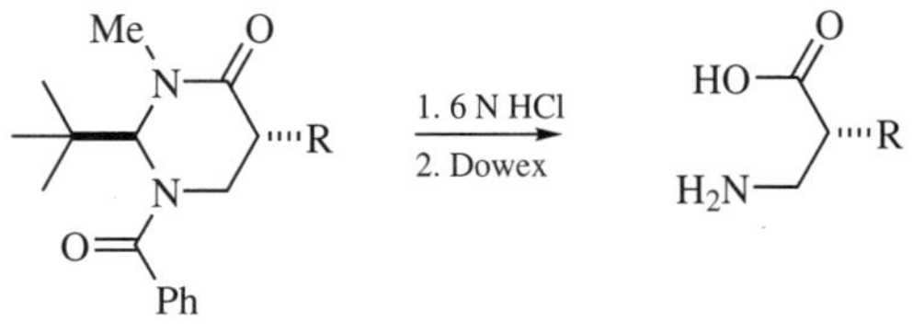

(7)

(2*S*,5*R*)-**5**–**8** (*R*)-**9**–**12**

Table 2 Hydrolysis of products **5**–**8**

R	mp (°C)	$[\alpha]_D^{29}$	Yield (%)
Me	185–6	−11.8	80
n-Bu	170–1	+5.3	80
n-C_6H_{13}	219–20	+6.6	80
$PhCH_2$	225–6	+11.3	85

Epimerization of Adducts 5-8, and Hydrolysis to give the Enantiomeric α-Alkylated β-Aminopropionic Acids

In principle, α-substituted β-amino acids of opposite configuration, (*S*)-**9**–**12**, can be obtained when enantiomeric pyrimidinone (*R*)-**4** [from (*R*)-asparagine] is used as the starting material, following the reaction sequence described above. Nevertheless, a practical alternative consisted in the epimerization of *trans*-adducts **5**–**8** to afford the *cis*-diastereoisomers **13**–**16** (eq 8). Hydrolysis of *cis*-**13**–**16** provided the desired (*S*)-α-substituted β-amino acids, (*S*)-**17**–**20** (eq 9).[2]

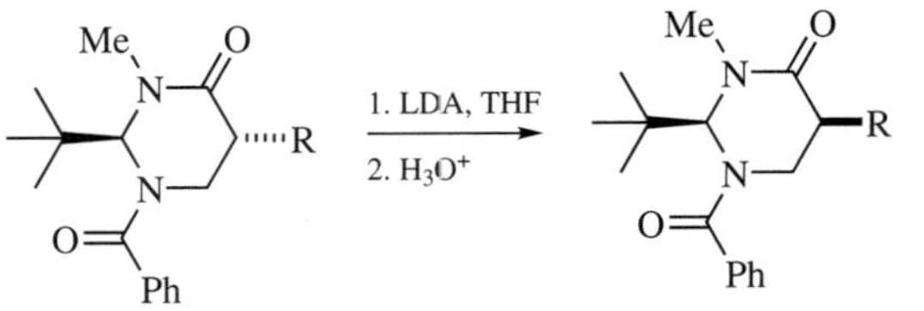

(8)

(2*S*,5*R*)-**5**–**8** (2*S*,5*S*)-**13**–**16**

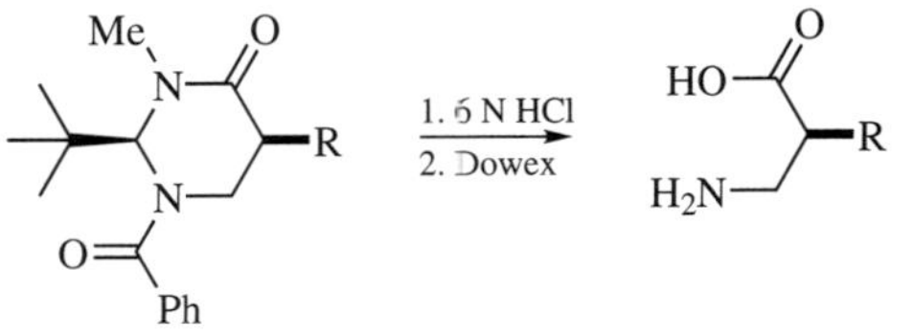

(9)

(2*S*,5*S*)-**13**–**16** (*S*)-**17**–**20**

Clearly, protonation (aqueous NH_4Cl) of the enolates generated from **5**–**8** takes place on the face opposite to the *tert*-butyl group, and this reaction is also highly stereoselective.

Table 3 Diastereoselectivity of enolate (2*S*)-**5**–**8**-Li alkylations

R^1	R^2	ds (%)	mp (°C)	$[\alpha]_D^{29}$	Yield (%)
Me	*n*-Bu	>97	184–5	+30.3	81
n-Bu	Me	>97	147–8	+12.6	88
Me	CH_2Ph	>97	209–10	−48.8	96
CH_2Ph	Me	>97	129–30	−38.7	92
CH_2Ph	Et	>97	106–7	−35.3	76
CH_2Ph	*n*-Bu	>97	[a]	−11.3	63

[a] Viscous oil.

Preparation of Enantiopure α, α-Disubstituted β-Amino Acids[7,8]

(2*S*,5*R*)-**5**–**8** were alkylated via their corresponding enolates to provide suitable precursors of enantiopure α,α-dialkylated β-amino acids. Thus, enolates (2*S*)-**5**–**8**-Li were generated upon treatment of the appropriate heterocycle with LDA in THF solution and under nitrogen atmosphere. The electrophile (methyl iodide, ethyl iodide, *n*-butyl bromide, *n*-hexyl iodide, or benzyl bromide) in solvent *N*,*N*′-dimethylpropyleneurea (DMPU) was then added at −78 °C to afford the dialkylated products with high diastereoselectivity and in excellent yields (Table 3). The use of

DMPU as cosolvent was necessary to effect the dialkylation in high yield (eq 10).[9]

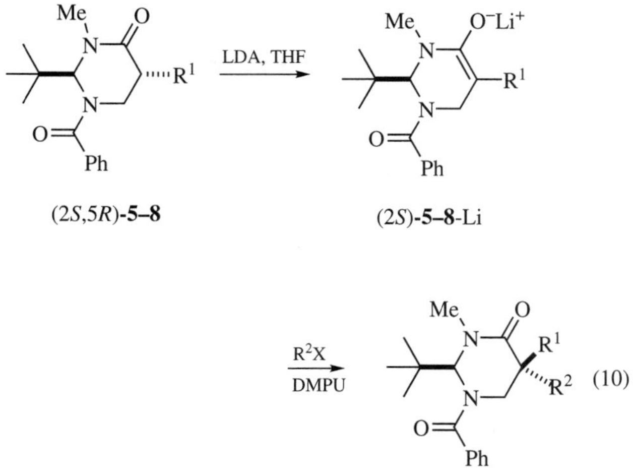

(2*S*,5*R*)-**5–8** (2*S*)-**5–8**-Li **21–26** (10)

Hydrolysis of geminal disubstituted perhydropyrimidinones **21–26** necessitated drastic conditions: 8 N HCl at 100–140 °C in a sealed tube. While these harsh conditions may not be tolerated by sensitive amino acids,[10] they proved harmless to the α,α-disubstituted β-amino acids **27–32**. Nevertheless, milder conditions could be employed when *p*-dioxane was used as cosolvent, since improved solubility of the substrate in the aqueous medium resulted in much faster hydrolysis (eq 11).

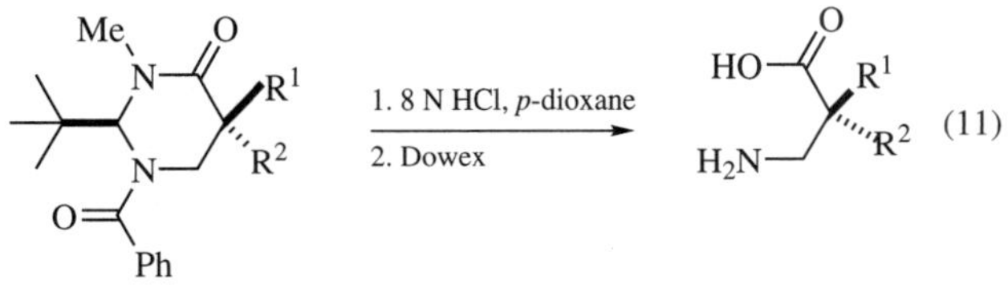

21–26 **27–32** (11)

Related Reagents. Owing to the high price of pivalaldehyde, we have substituted this aldehyde with isobutyraldehyde in the synthesis of imino ester (2*S*,6*S*)-**33**, which proved to be a convenient substrate for the enantioselective synthesis of α-substituted aspartic acids (eq 12).[11,12]

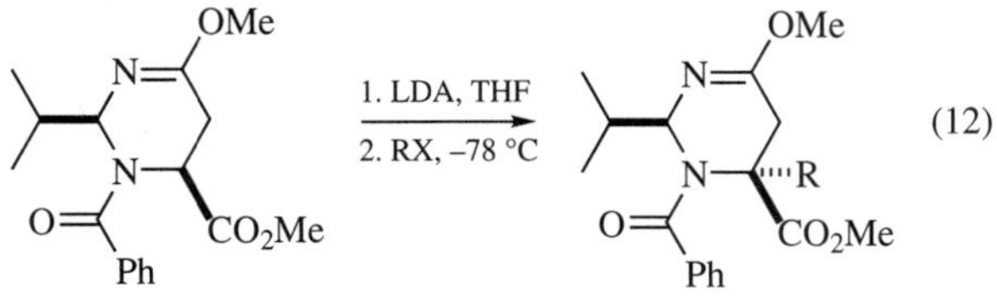

(2*S*,6*S*)-**33** (2*S*,6*S*)-**34-37** (12)

Hydrolysis of the alkylated products (2*S*,6*S*)-**34-37** and isolation of the desired amino acids was facilitated by the presence of the labile imino group.[13] Indeed, hydrolysis was achieved by heating with 17% HCl in a sealed tube at 95 °C. The free amino acids **38–41** were purified by chromatography on an ion-exchange column (eq 13).[8,11]

On the other hand, Beaulieu et al.[14] have reported the synthesis of unusually functionalized optically active β-substituted β-amino acids via the highly diastereoselective ((>95:5) transformation of enantiopure *N*-(*o*-iodobenzoyl)-2-*tert*-butylperhydropyrimidinone [(*S*)-**42**] (eq 14).

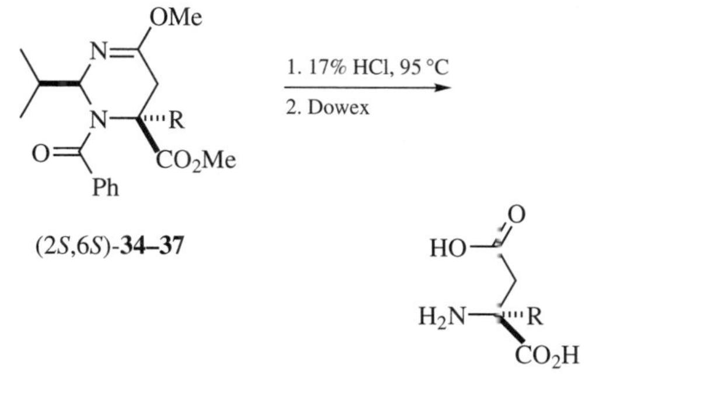

(2*S*,6*S*)-**34–37** (*S*)-**38–41**

R= Me, Et, *n*-Bu, CH_2Ph (13)

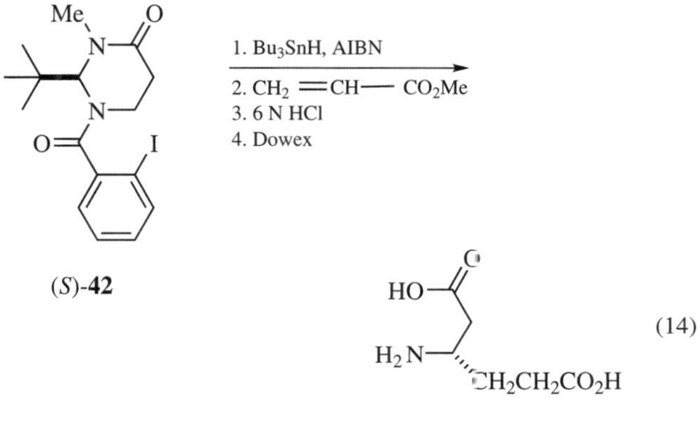

(*S*)-**42** (*S*)-**43** (14)

In this context, dihydropyrimidinone (*R*)-**44** has been exploited by Chu et al.[15] and Konopelski et al.[16] in the enantioselective synthesis of β-alkyl β-amino acids. Furthermore, *N*-phenethylperhydropyrimidinone (*S*)-**45**,[17,18] as well as the 6-substituted analogs **46–48** (configuration not indicated)[19–22] are useful substrates for the asymmetric synthesis of α,β-disubstituted β-amino acids.

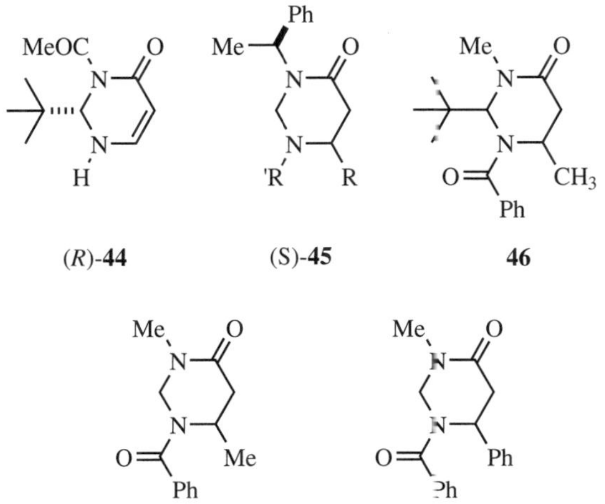

(*R*)-**44** (S)-**45** **46** **47** **48**

1. Lakner, F. J.; Ch, K. S.; Negrete, G. R.; Konopelski, J. P., *Org. Synth.* **1995**, *73*, 201.
2. Juaristi, E.; Quintana, D.; Balderas, M.; García-Pérez, E. *Tetrahedron: Asymmetry* **1996**, *7*, 2233.
3. Juaristi, E.; Rizo, B.; Natal, V.; Escalante, J.; Regla, I. *Tetrahedron: Asymmetry* **1991**, *2*, 821.
4. Juaristi, E.; Quintana, D. *Tetrahedron: Asymmetry* **1992**, *3*, 723.
5. Juaristi, E.; Quintana, D.; Lamatsch, B.; Seebach, D. *J. Org. Chem.* **1991**, *56*, 2553.
6. Seebach, D.; Lamatsch, B.; Amstutz, R.; Beck, A. K.; Dobler, M.; Egli, M.; Fitzi, R.; Gautschi, M.; Herradón, B.; Hidber, P. C.; Irwin, J. J.; Locher, R.; Maestro, M.; Maetzke, T.; Mouriño, A.; Pfammatter, E.; Plattner, D. A.; Schickli, C.; Schweizer, W. B.; Seiler, P.; Stucky, G.; Petter, W.; Escalante, J.; Juaristi, E.; Quintana, D.; Miravitlles, C.; Molins, E. *Helv. Chim. Acta* **1992**, *72*, 913.
7. Juaristi, E.; Balderas, M.; Ramírez-Quirós, Y. *Tetrahedron: Asymmetry* **1998**, *9*, 3881.
8. Juaristi, E.; Balderas, M.; López-Ruiz, H.; Jímenez-Pérez, V. M.; Kaiser-Carril, M. L.; Ramírez-Quirós, Y. *Tetrahedron: Asymmetry* **1999**, *10*, 3493.
9. DMPU has been recommended as solvent in various alkylation reactions: Juaristi, E.; Murer, P.; Seebach, D. *Synthesis* **1993**, 1243, and references cited therein.
10. Seebach, D.; Juaristi, E.; Miller, D. D.; Schickli, C.; Weber, T. *Helv. Chim. Acta* **1987**, *70*, 237.
11. Juaristi, E.; López-Ruiz, H.; Madrigal, D.; Ramírez-Quirós, Y.; Escalante, J. *J. Org. Chem.* **1998**, *63*, 4706.
12. See also: Seebach, D.; Boog, A.; Schweizer, W. B. *Eur. J. Org. Chem.* **1999**, 335.
13. Compare mild conditions employed to hydrolyze bis-lactimethers: Schöllkopf, U.; Tiller, T.; Bardenhagen, J. *Tetrahedron* **1998**, *44*, 5293. Compare mild conditions used to hydrolyze dihydroimidazoles: Blank, S.; Seebach, D. *Angew. Chem., Int. Ed. Engl.* **1993**, *32*, 1765.
14. Beaulieu, F.; Arora, J.; Vieth, U.; Taylor, N. J.; Chapell, B. J.; Snieckus, V. *J. Am. Chem. Soc.* **1996**, *118*, 8727.
15. (a) Chu, K. S.; Negrete, G. R.; Konopelski, J. P. *J. Org. Chem.* **1991**, *56*, 5196. (b) Chu, K. S.; Konopelski, J. P. *Tetrahedron* **1993**, *49*, 9183.
16. Konopelski, J. P. In *Enantioselective Synthesis of β-Amino Acids*, Juaristi, E., Ed. Wiley: New York, 1997, pp 249–259.
17. (a) Amoroso, R.; Cardillo, G.; Tomasini, C.; Tortoreto, P. *J. Org. Chem.* **1992**, *57*, 1082. (b) Cardillo, G.; Tolomelli, A.; Tomasini, C. *Tetrahedron* **1995**, *51*, 11831.
18. Cardillo, G.; Tomasini, C. In *Enantioselective Synthesis of β-Amino Acids*, Juaristi, E., Ed. Wiley: New York, 1997, pp 211–248.
19. Juaristi, E.; Escalante, J.; Lamatsch, B.; Seebach, D. *J. Org. Chem.* **1992**, *57*, 2396.
20. Juaristi, E.; Escalante, J. *J. Org. Chem.* **1993**, *58*, 2282.
21. Escalante, J.; Juaristi, E. *Tetrahedron Lett.* **1995**, *36*, 4397.
22. Juaristi, E.; Seebach, D. In *Enantioselective Synthesis of β-Amino Acids*, Juaristi, E., Ed. Wiley: New York, 1997, pp 261–277.

Eusebio Juaristi
Centro de Investigación y de Estudios Avanzados del Instituto Politécnico Nacional, México, D.F., México

Benzyl(methoxymethyl)methylamine

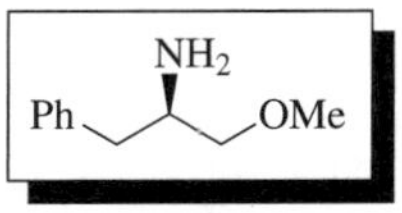

(*S*)
[64715-80-6] $C_{10}H_{15}NO$ (MW 165.23)
(*R*)
[59919-07-2]

(chiral auxiliary for the enantioselective alkylation of ketones[1] and aldehydes;[2] can form chiral cuprate reagents[3])

Physical Data: (*S*) bp 55–59 °C/0.1 mmHg; mp HCl salt 151–152 °C; $[\alpha]_D^{25}$ −14.4° (*c* 5.7, benzene); HCl salt $[\alpha]_D^{25}$ +19.7° (*c* 2.5, EtOH).

Preparative Methods: these chiral methoxy amines are readily prepared from (*S*)- or (*R*)-phenylalanine via reduction followed by methylation.[2]

Handling, Storage, and Precautions: conversion of the freshly distilled amine to its hydrochloride salt is a convenient way to store and handle the compound. The free amine reacts with atmospheric carbon dioxide to produce the respective carbonate. The free amine should be stored tightly sealed under argon or nitrogen immediately after distillation to avoid CO_2 adsorption.

Enantioselective Alkylation. Both antipodes of this chiral amine have been used in the enantioselective alkylation of ketones and aldehydes via their respective chiral, nonracemic lithioenamines (eq 1). The enantioselectivity in alkylation results from the induced rigidity of the lithioenamine upon chelation with the methoxy group, providing the bias necessary to influence the direction and rate of entry of the electrophile.

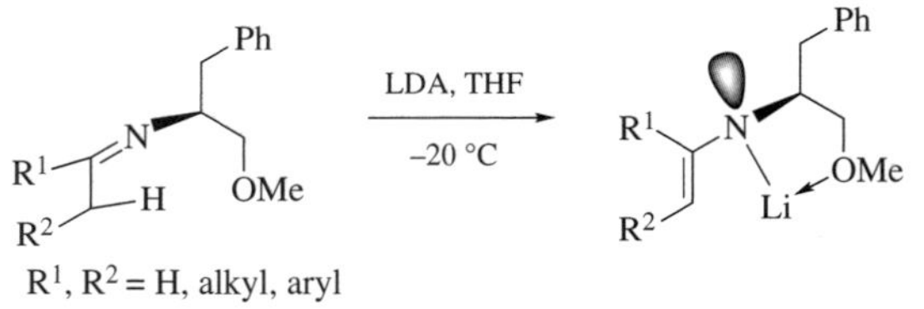

(1)

Medium-sized cyclic ketones have been enantioselectively alkylated via their chiral lithioenamines to yield 2-alkylcycloalkanones in 80–100% ee.[4] This procedure has also furnished α,α′-dialkyl cyclohexanones in good enantiomeric excess (eq 2).[4] Based on this protocol, regiospecific deuteration of 3-methylcyclohexanones has been achieved with good enantioselectivity.[5]

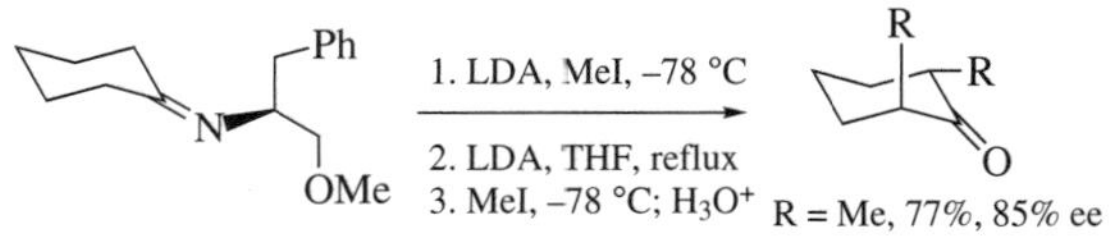

(2)

In contrast with medium-sized cyclic ketones, alkylation of macrocyclic ketones can afford either optical antipode depending on whether the lithioenamine is formed via kinetic (*E*-) or thermodynamic conditions (*Z*-enamine) (eq 3).[6] Optically active α-alkyl macrocyclic ketones have been formed in 30–82% enantiomeric excess, with chemical yields of 62–90%.

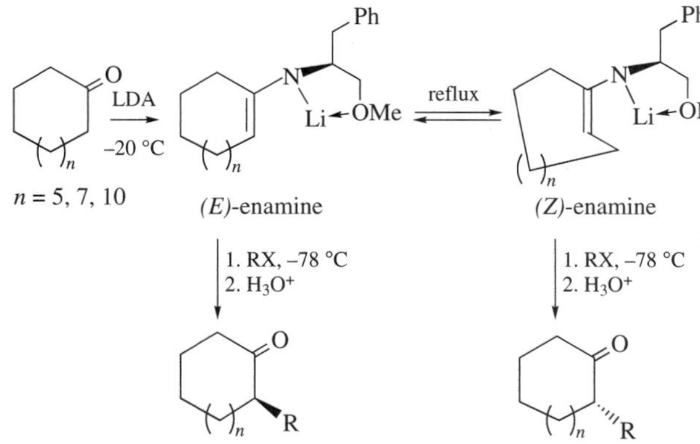

(3)

In a similar manner, aldehydes can also be enantioselectively alkylated by this procedure. However, the enantiomeric excess obtained is much lower (47%).[2] A special application of this method is the enantioselective alkylation of aldehydes for the construction of quaternary stereogenic centers. An example is the formation of the chiral quaternary carbon in 4-methyl-4-phenylcyclohex-2-en-1-one in high enantiomeric excess using this methodology (eq 4).[7]

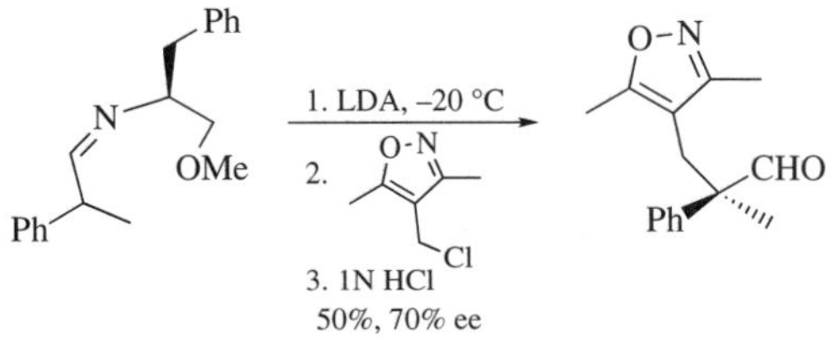

(4)

Chiral Cuprate Reagents. This chiral amine has also found application in asymmetric conjugate addition of copper azaenolates to cyclic enones. Lithium azaenolates of optically active acetone imines have been used in the preparation of chiral cuprate reagents. However, the asymmetric induction is low (17–28% ee) when this amine is employed (eq 5).[3]

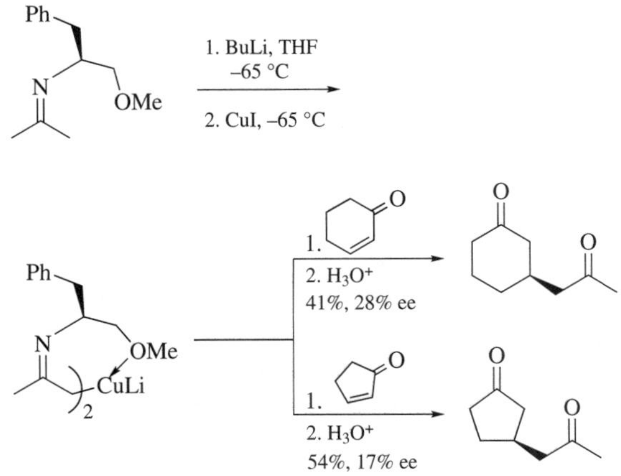

(5)

Enantioselective alkylation of aldehydes and ketones can also be accomplished using Enders' reagents SAMP and RAMP.[8] In contrast with Meyer's chiral auxiliary, the synthesis of Enders' reagents is lengthy and the recovery is inconvenient because cleavage of the auxiliary does not afford back the reagent. It also generates nitrosoamines (via ozonolysis) which are considered carcinogenic compounds. Therefore, the ease of preparation, availability of the starting material, and efficient cleavage and recovery of this chiral amine make it a convenient chiral auxiliary.

---

1. Meyers, A. I.; Williams, D. R.; Druelinger, M. *J. Am. Chem. Soc.* **1976**, *98*, 3032.
2. Meyers, A. I.; Poindexter, G. S.; Brich, Z. *J. Org. Chem.* **1978**, *43*, 892.
3. Yamamoto, K.; Iijima, M.; Ogimura, Y. *Tetrahedron Lett.* **1982**, *23*, 3711.
4. Meyers, A. I.; Williams, D. R.; Erickson, G. W.; White, S.; Druelinger, M. *J. Am. Chem. Soc.* **1981**, *103*, 3081.
5. Kallmerten, J.; Knopp, M. A.; Durham, L. L.; Holak, I. *J. Label. Compound Radiopharm.* **1986**, *23*, 329.
6. Meyers, A. I.; Williams, D. R.; White, S.; Erickson, G. W. *J. Am. Chem. Soc.* **1981**, *103*, 3088.
7. Marron, B. E.; Schlicksupp, L.; Natale, N. R. *J. Heterocycl. Chem.* **1988**, *25*, 1067.
8. Enders, D.; Eichenauer, H. *Angew. Chem.* **1976**, *88*, 579; *Angew. Chem., Int. Ed. Engl.* **1976**, *15*, 549.

Eduardo A. Véliz & Joseph P. Konopelski
University of California, Santa Cruz, CA, USA

(*S*)-4-Benzyl-2-oxazolidinone

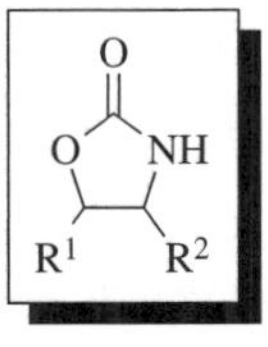

(**1**; R^1 = H, R^2 = ◀ Bn) (*S*)
[90719-32-7] $C_{10}H_{11}NO_2$ (MW 177.20)
(Li salt)
[123731-35-1]
(**2**; R^1 = H, R^2 = ⋯ Bn) (*R*)
[102029-44-7] $C_{10}H_{11}NO_2$ (MW 177.20)
(Li salt)
[128677-61-2]
(**3**; R^1 = H, R^2 = ◀ *i*-Pr) (*S*)
[17016-83-0] $C_6H_{11}NO_2$ (MW 129.16)
(Li salt)
[96021-69-1]
(**4**; R^1 = ⋯ Ph, R^2 = ⋯ Me) (4*R*,5*S*)
[77943-39-6] $C_{10}H_{11}NO_2$ (MW 177.20)
(Li salt)
[92061-65-7]
(**5**; R^1 = ◀ Ph, R^2 = ◀ Me) (4*S*,5*R*)
[16251-45-9] $C_{10}H_{11}NO_2$ (MW 177.20)
(Li salt)
[127882-97-7]
(**6**; R^1 = H, R^2 = ◀ Ph) (*S*)
[99395-88-7] $C_9H_9NO_2$ (MW 163.18)
(**7**; R^1 = H, R^2 = ⋯ Ph) (*R*)
[90319-52-1] $C_9H_9NO_2$ (MW 163.18)
(**8**; R^1 = H, R^2 = ◀ *t*-Bu) (*S*)
[54705-42-9] $C_7H_{13}NO_2$ (MW 143.19)

(chiral auxiliaries used in asymmetric alkylations,[1] acylations,[2] halogenations,[3] aminations,[4] hydroxylations,[5] aldol reactions,[6] conjugate additions,[7,8] Diels–Alder reactions,[9] acyl transfer,[10] and sulfinyl transfer[11])

Physical Data: (**1**) mp 87–89 °C; (**2**) mp 85–87 °C; (**3**) mp 71–72 °C; (**4**) mp 118–121 °C; (**5**) mp 118–121 °C; (**6**) mp 130–132 °C; (**7**) mp 130–132 °C; (**8**) mp 118–120 °C.
Solubility: sol most polar organic solvents.
Form Supplied in: white crystalline solid; commercially available.
Analysis of Reagent Purity: 99% purity attainable by GLC.
Handling, Storage, and Precautions: no special handling or storage precautions are necessary. There is no known toxicity. It may be harmful by inhalation, ingestion, or skin absorption and may cause skin or eye irritation.

Synthesis of the Chiral Oxazolidinone Auxiliaries. (*S*)-4-Benzyl- (**1**), (*R*)-4-benzyl- (**2**), (*S*)-4-*i*-propyl- (**3**), (4*R*,5*S*)-4-methyl-5-phenyl- (**4**), (*S*)-4-*t*-butyl- (**8**), and (*S*)-4-phenyl-2-oxazolidinones (**6**) are commercially available. Typical procedures to form these chiral auxiliaries involve the reduction of α-amino acids to the corresponding amino alcohols or the purchase of amino alcohols, followed by formation of the cyclic carbamate (eq 1). A number of high-yielding methods of reduction have been employed for this transformation, including *Boron Trifluoride Etherate/Borane–Dimethyl Sulfide*,[12] *Lithium Aluminum Hydride*,[1,6,13,14] *Sodium Borohydride/Iodine*,[15] and *Lithium Borohydride/Chlorotrimethylsilane*.[8] Selection among these methods is largely based upon cost of reagents and ease of performance. Reagents for effecting the second transformation include *Diethyl Carbonate/Potassium Carbonate*[12] or *Phosgene*,[16–18] with the former being preferable for large-scale production. Ureas,[19,20] dioxolanones,[21] chloroformates,[22] trichloroacetate esters,[22,23] *N,N′-Carbonyldiimidazole*,[24] and *Carbon Monoxide* with catalytic elemental *Sulfur*[25] or *Selenium*[26,27] provide alternatives for the transformation of amino alcohols to the derived oxazolidinones.

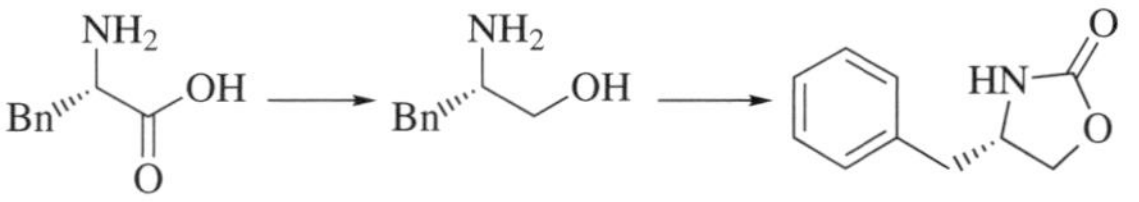

(1)

Conversion of the appropriate α-amino acids to oxazolidinones may also be performed as a one-pot procedure, obviating the need to isolate the intermediate amino alcohols (eq 2 and 3).[28,29] Overall isolated yields for these procedures are 70–80%.

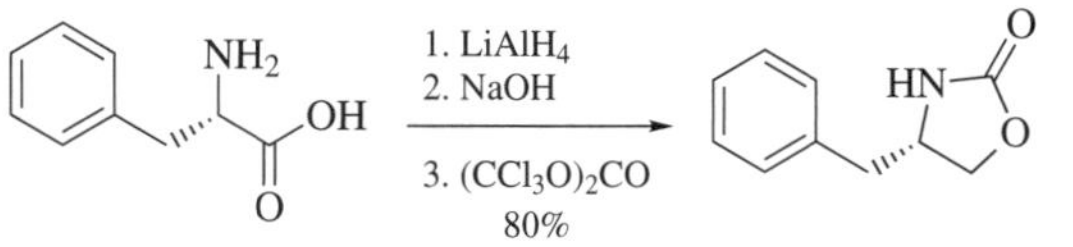

(2)

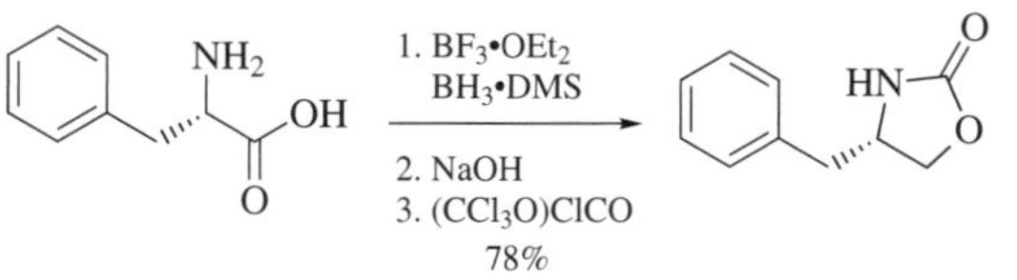

(3)

Carbamate-protected amino alcohols also yield oxazolidinones upon treatment with base (eq 4)[30,31] or *p-Toluenesulfonyl Chloride* (eq 5).[32] The latter reaction requires the protection of the amino group as the *N*-methylated carbamate for selective inversion of the hydroxyl-bearing center.

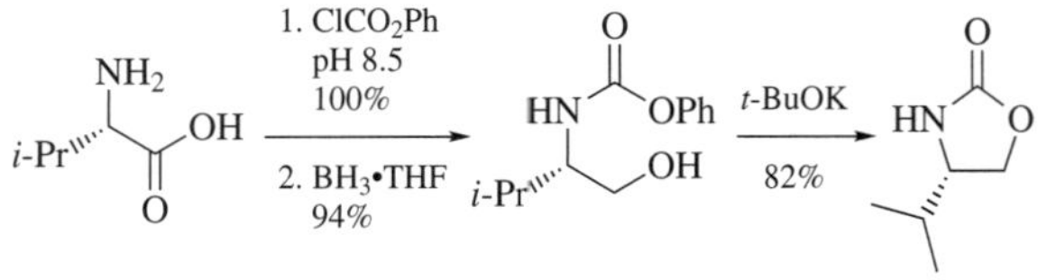

(4)

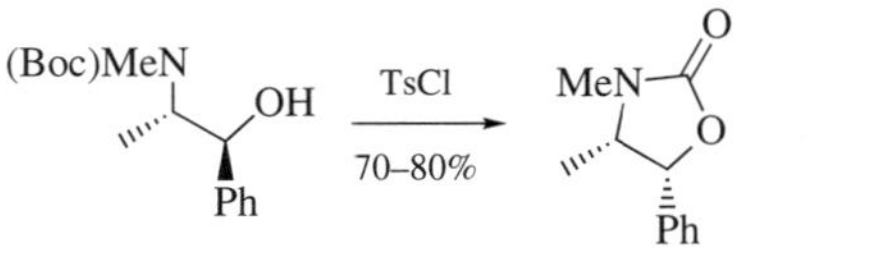

(5)

Resolution of racemic oxazolidinones affords either enantiomer of the auxiliary and provides a versatile route to unusually substituted derivatives (eq 6).[33]

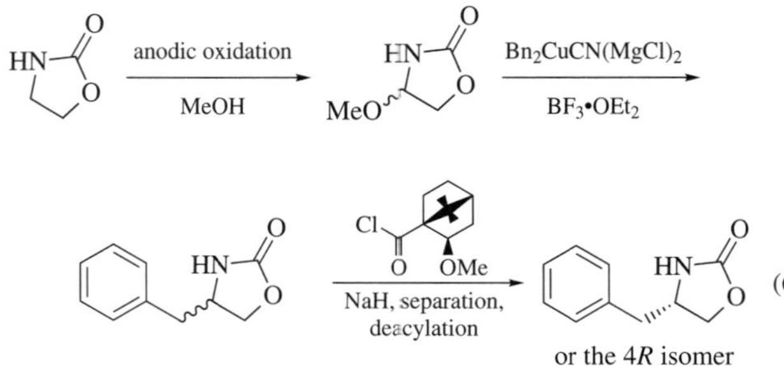

(6)

Methods of *N*-Acylation. Lithiated oxazolidinones add to acid chlorides (eq 7)[6,34] and mixed anhydrides (eq 8)[35,36] in high yields to form the derived *N*-acyl imides. In the latter case the anhydride may be formed in situ with *Trimethylacetyl Chloride*, and then condensed with the lithiated oxazolidinone selectively at the less hindered carbonyl moiety.

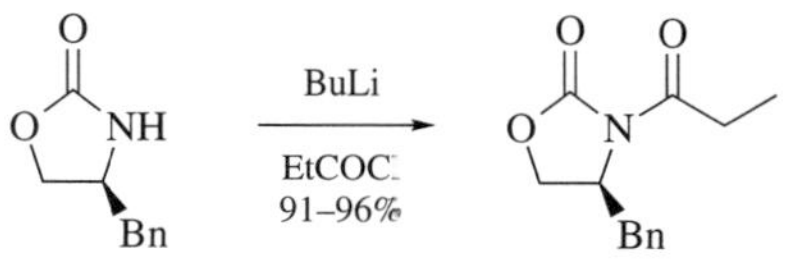

(7)

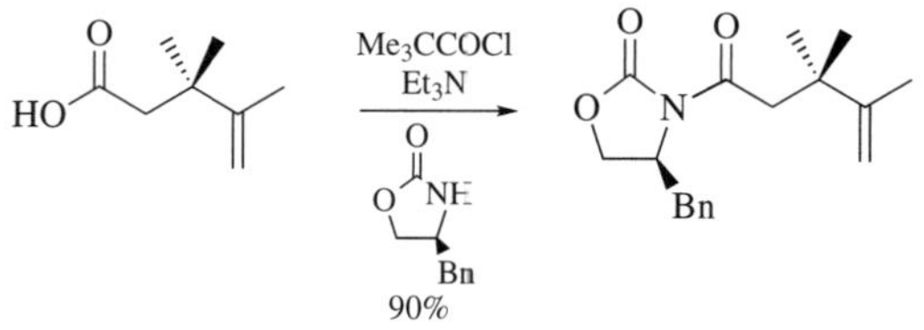

(8)

Acryloyl adducts cannot be formed through traditional acylation techniques due to their tendency to polymerize. These adducts may be obtained through reaction of acryloyl chloride with the bromomagnesium salt of the oxazolidinone auxiliary[9,37] or the *N*-trimethylsilyl derivative in the presence of *Copper(II) Chloride* and *Copper* powder.[38] These methods yield products in the range of 50–70%.

Methods of *N*-Alkylation. In analogy with acylation techniques, metalated oxazolidinones add to alkyl halides to afford the *N*-alkylated products in high yields.[39–42]

Enolization of *N*-Acyloxazolidinones. Various methods have been developed to effect the enolization of chiral *N*-

acyloxazolidinones. In alkylation reactions, both *Lithium Diisopropylamide* and *Sodium Hexamethyldisilazide* deprotonate these imides to provide the (*Z*)-enolates in >100:1 selectivity.[1]

Di-n-butylboryl Trifluoromethanesulfonate with a tertiary amine also provides the (*Z*)-enolates of chiral acyl oxazolidinones in >100:1 selectivity for use in subsequent aldol additions.[6,14] With *Triethylamine*, *Diisopropylethylamine* (Hünig's base), or *2,6-Lutidine* the order of addition is of no consequence to enolization.[43] Triethylamine has traditionally seen the greatest utilization in these reactions based upon cost considerations; however, with certain sensitive aldehyde substrates, lutidine provides milder reaction conditions.[44]

The (*Z*)-enolate is also accessed exclusively using titanium enolization procedures.[45,47] Irreversible complexation of *Titanium(IV) Chloride* with tertiary amine bases demands complexation of the substrate with the Lewis acid prior to treatment with either triethylamine or Hünig's base. Reactions using Hünig's base occasionally display higher diastereoselectivities, particularly in Michael additions.[7,45] Of the alkoxy titanium species employed in imide enolization, only $TiCl_3$(O-*i*-Pr) is capable of quantitative enolate formation. In these reactions, order of addition of reagents is not significant. These enolates demonstrate enhanced nucleophilicity, albeit with somewhat diminished diastereoselectivity.

Other Lewis acids have been demonstrated to provide moderate levels of enolization, including *Aluminum Chloride*, *Magnesium Bromide*, and *Tin(II) Trifluoromethanesulfonate*.[45,46] However, $SnCl_4$, Me_2AlCl, and $ZrCl_4$ failed to provide detectable enolization.[45]

Enolate Alkylation. Alkylation of chiral *N*-acyloxazolidinones by simple alkyl and allylic halides occurs through the chelated lithiated (*Z*)-imide enolates to afford products in greater than 93:7 diastereoselectivities (eqs 9 and 10).[1] For small electrophiles such as *Iodomethane* and *Ethyl Iodide*, NaHMDS proved to be the enolization base of choice. On selected substrates, alkyl triflates also demonstrate promise as alkylating agents.[34]

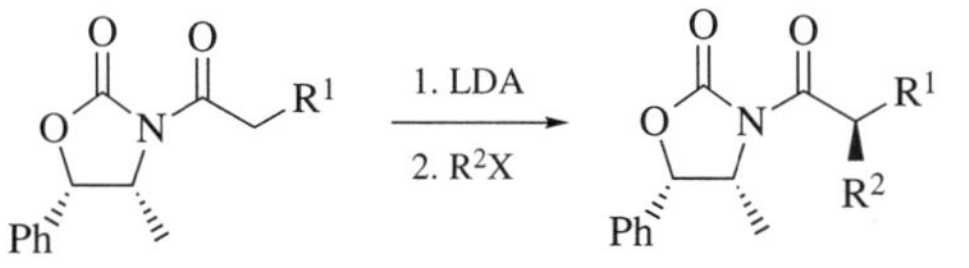

(9)

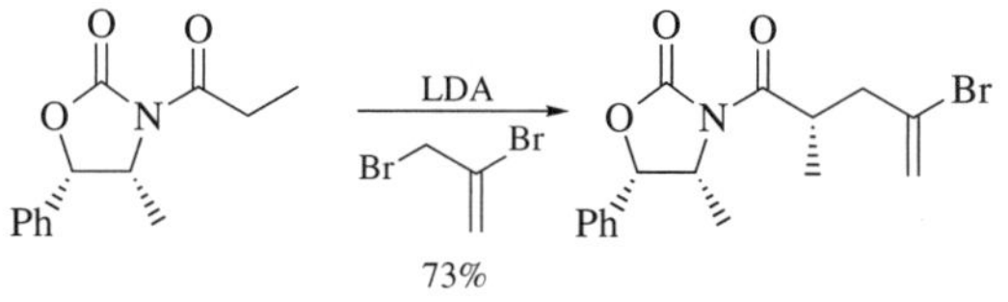

(10)

For benzyloxymethyl electrophiles, titanium enolates are superior to the corresponding lithium enolates in both yield and alkylation diastereoselectivity (eq 11). Unfortunately, the analogous *p*-methoxybenzyl-protected β-hydroxy adducts cannot be obtained by this method. In other cases the titanium methodology complements the corresponding reactions of the lithium and sodium enolates for S_N1-like electrophiles.[47] It is noteworthy that imides may be selectively enolized under all of the preceding conditions in the presence of esters (eq 12).

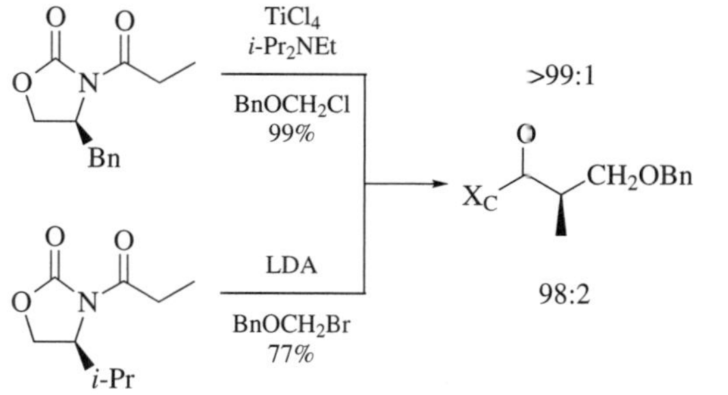

(11)

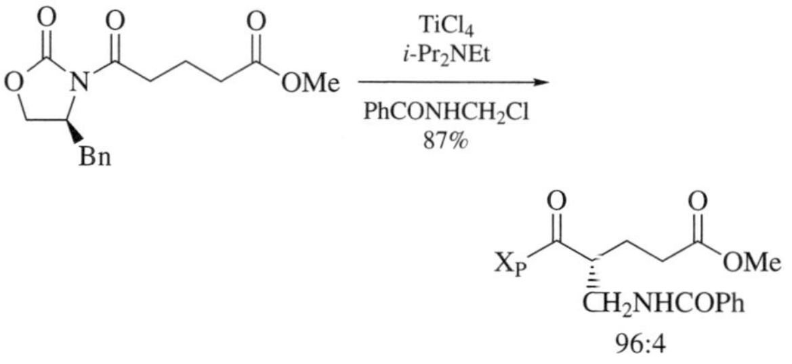

(12)

Treatment of the silyl enol ethers of *N*-acyloxazolidinones with selected electrophiles that do not require Lewis acid activation similarly results in high induction of the same enolate face (eq 13).[48] The facial bias of this conformationally mobile system improves with the steric bulk of the silyl group.

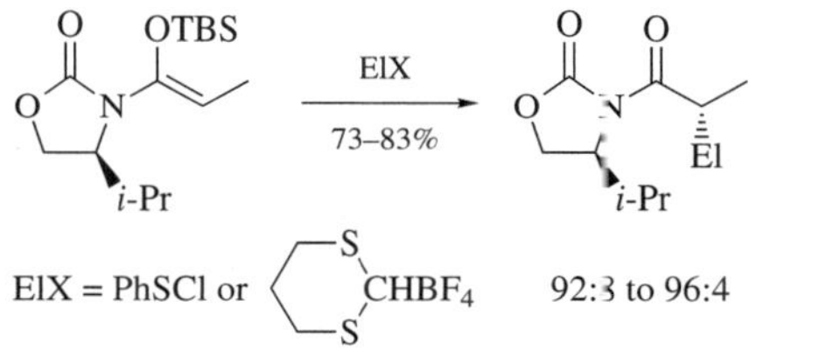

(13)

Chiral oxazolidinones have also been used to induce chirality in $TiCl_4$-mediated allylsilane addition reactions to α-keto imides (eq 14).[49]

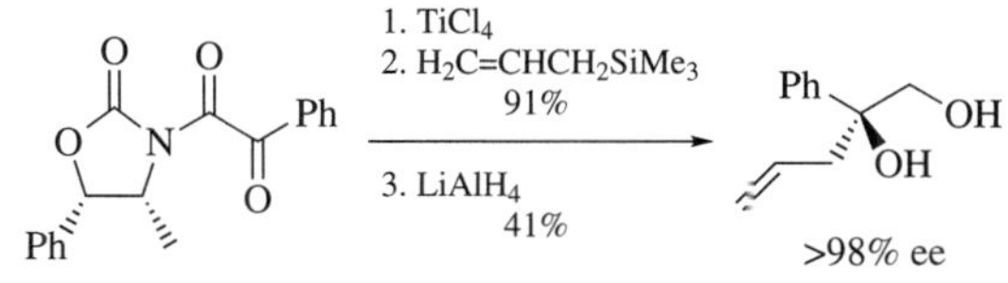

(14)

Enolate Alkylations with Transition Metal Coordinated Electrophiles. Coordination of various transition metals to dienes and aromatic compounds sufficiently activates these compounds to nucleophilic addition, resulting in high asymmetric induction at the α-center. However, the manganese complexes of various benzene derivatives couple with lithium enolates in low selectivity at the nascent stereogenic center on the ring (eq 15).[50]

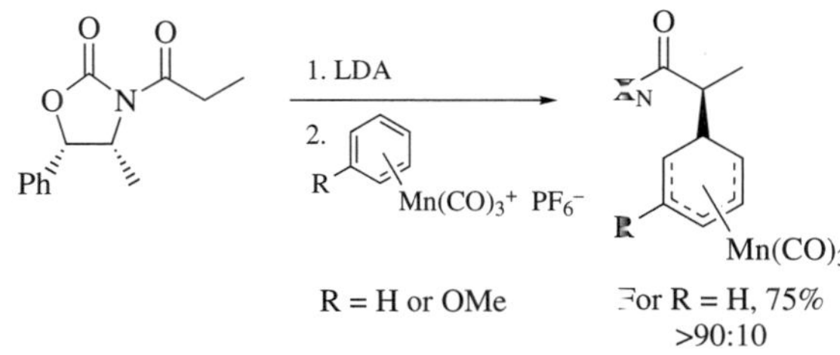

(15)

In contrast, molybdenum and iron diene complexes undergo the same type of reaction with chiral lithium imide enolates, with moderate to good induction at the β-position (eq 16).[51–53]

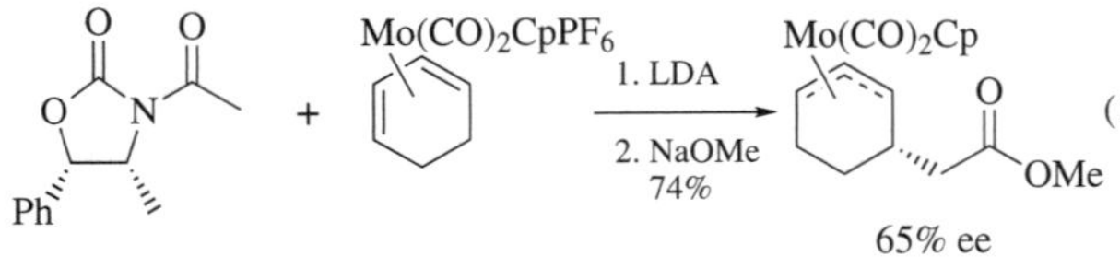

(16)

Dicobalt hexacarbonyl-coordinated propargyl ethers also combine with imide boron enolates through a kinetic resolution of the rapidly interconverting propargylic cation isomers to afford a 92:8 mixture of isomers at the β-center in 80% yield (eq 17). Stereocontrol of the α-center is 97:3.[54]

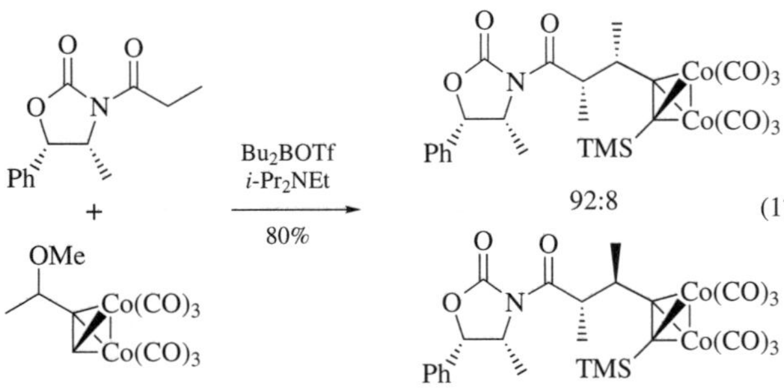

(17)

Enolate Acylation. Acylation of these enolates provides a direct route to β-dicarbonyl systems. Acylations generally proceed with >95% diastereoselection in 83–95% yields, with the valine-derived auxiliary providing slightly higher selectivity (eq 18).[2] The sense of induction is consistent with reaction through the chelated lithium (*Z*)-enolate, and the newly generated stereocenter is retained through routine manipulations.

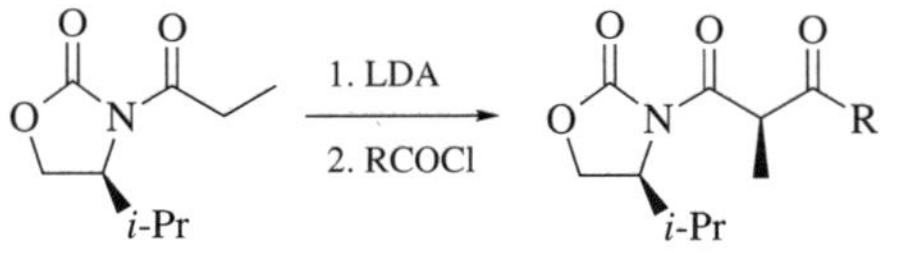

(18)

An alternate approach to these useful 1,3-dicarbonyl substrates may be achieved through enolate orthoester acylation. Titanium enolates have been employed to effect this transformation (eq 19).[45,47] Similarly, treatment of the titanium enolate of β-ketoimide with dioxolane orthoesters results in the formation of a masked tricarbonyl compound (eq 20). Trimethyl orthoacetate and *Triethyl Orthoacetate* are not appropriate partners in these coupling reactions.[45,47]

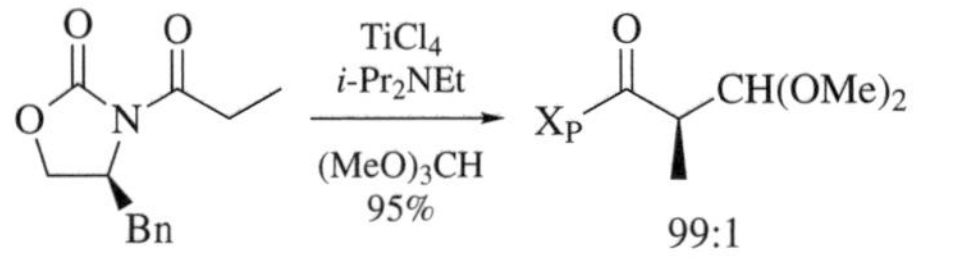

(19)

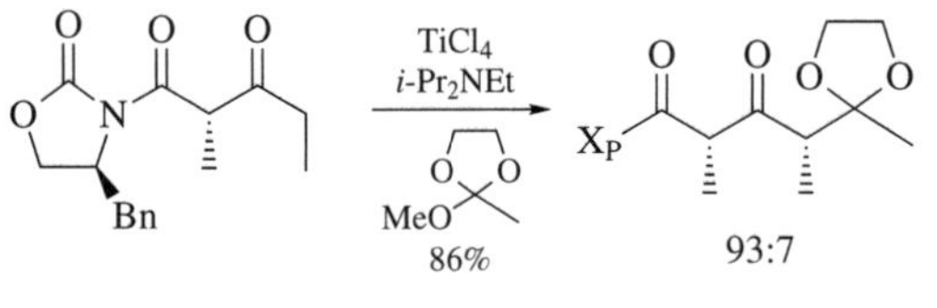

(20)

Michael Addition. Titanium imide enolates are excellent nucleophiles in Michael reactions. Michael acceptors such as ethyl vinyl ketone, *Methyl Acrylate*, *Acrylonitrile*, and *t*-butyl acrylate react with excellent diastereoselection (eq 21).[7,45] Enolate chirality transfer is predicted by inspection of the chelated (*Z*)-enolate. For the less reactive unsaturated esters and nitriles, enolates generated from $TiCl_3(O\text{-}i\text{-}Pr)$ afford superior yields, albeit with slightly lower selectivities. The scope of the reaction fails to encompass β-substituted, α,β-unsaturated ketones which demonstrate essentially no induction at the prochiral center. Furthermore, substituted unsaturated esters do not act as competent Michael acceptors at all under these conditions.

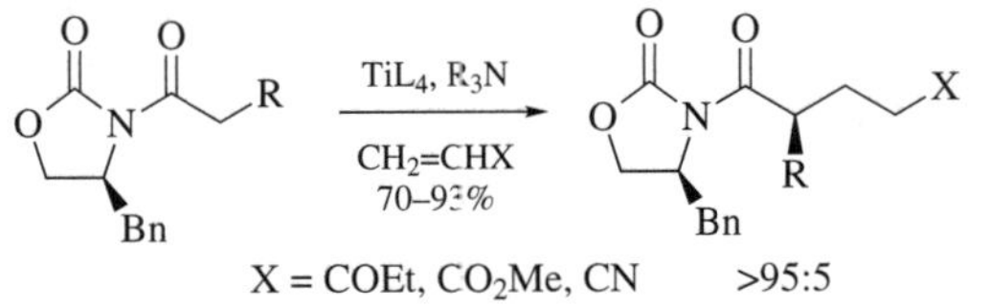

(21)

Various chelated lithium imide enolates have also served as nucleophiles in Michael additions to 3-trifluoromethyl acrylate, favoring the *anti* isomer (eq 22).[55]

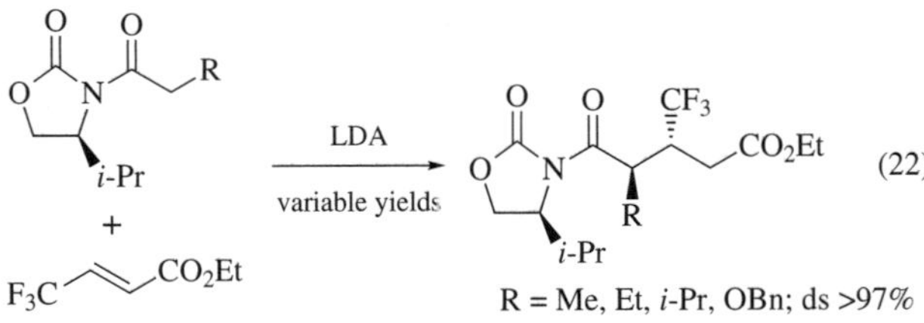

(22)

Enolate Hydroxylation. Treatment of the sodium enolates with the Davis oxaziridine reagent affords the hydroxylated products with the same sense of induction as the alkylation products (eq 23).[5,35] Although high diastereoselectivity may be achieved with *Oxodiperoxymolybdenum(pyridine)(hexamethylphosphoric triamide)* (MoOPH), such reactions proceed in lower yields.

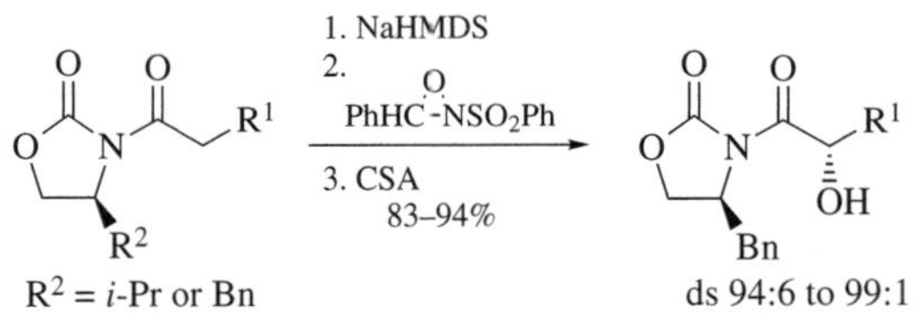

(23)

Enolate Amination. Amination likewise can be effected using *Di-t-butyl Azodicarboxylate* (DBAD).[4,56] Despite the excellent yields and diastereoselectivity obtained using this methodology (eq 24), the harsh conditions required for further transformation of the resultant hydrazide adducts (*Trifluoroacetic Acid* and hydrogenation at 500 psi over *Raney Nickel* catalyst) limit its synthetic utility.

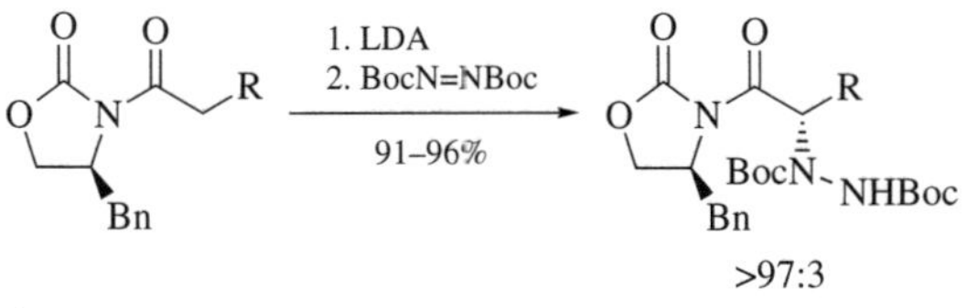

(24)

As a method for the synthesis of α-amino acids, the hydrazide methodology has now largely been supplanted by direct enolate azidation (eq 25).[4,57] These adducts are susceptible to mild chemical modification to afford *N*-protected α-amino acid derivatives. Under optimal conditions, yields range from 74–91% and selectivities from 91:9 to >99:1. Imide enolization can be carried out selectively in the presence of an enolizable *t*-butyl ester and suitably protected amino groups.

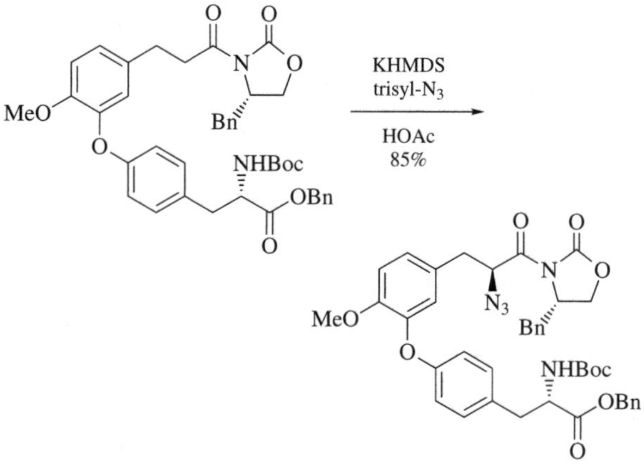

(25)

Hydrogenation of the azide moiety readily provides the amine using *Palladium on Carbon* and H_2 or *Tin(II) Chloride*. This methodology has been extended to the synthesis of arylglycines (eq 26).[58]

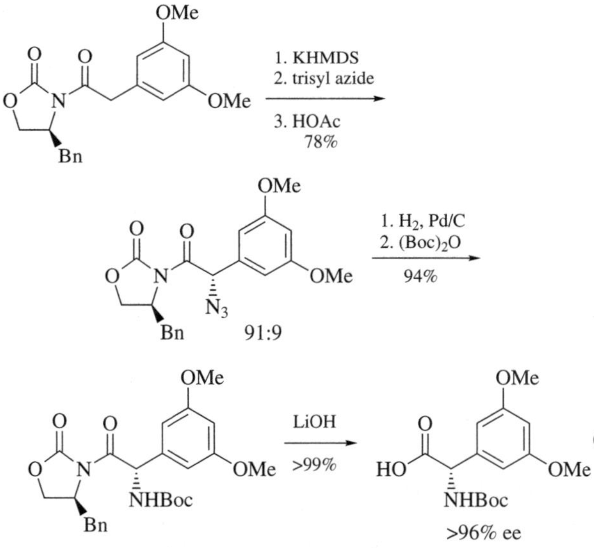

(26)

Failure to use *2,4,6-Triisopropylbenzenesulfonyl Azide* results in substantial diazo imide formation. However, optimization for the formation of the α-diazo imide compounds can be achieved with NaHMDS and *p*-nitrobenzenesulfonyl azide, followed by a neutral quench (eq 27).[4] These diazo compounds, however, have failed to demonstrate utility in asymmetric carbenoid chemistry.[59]

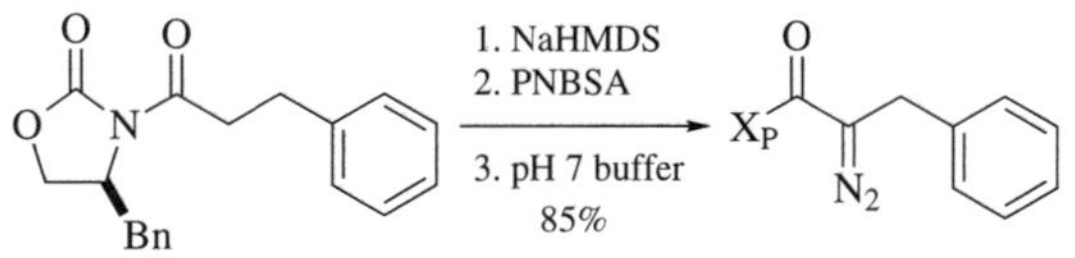

(27)

Enolate Halogenation. Enolate halogenation is achieved by reaction of the boryl enolate with *N-Bromosuccinimide*, affording configurationally stable α-bromo imides in >94:6 diastereoselectivity in 80–98% yield (eq 28).[3,4] The sense of induction suggests halogenation of the chelated (*Z*)-enolate. Introduction of an α-fluoro substituent can be effected by the treatment of imide enolates or α,β-unsaturated enolates with *N*-fluoro-*o*-benzenedisulfonimide (eq 29).[60]

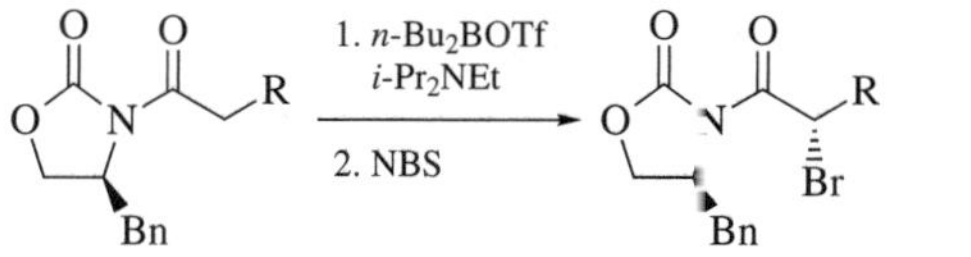

(28)

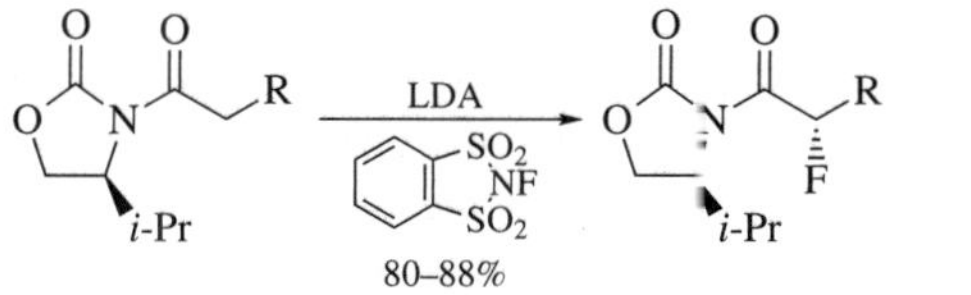

(29)

R = *n*-Bu, *t*-Bu, Bn, Ph

86–97% de

Displacement of a halide at the α-position with tetramethylguanidinium azide (TMGA) introduces nitrogen functionality with inversion of the original halide configuration and <1% epimerization (eq 30).[3,4]

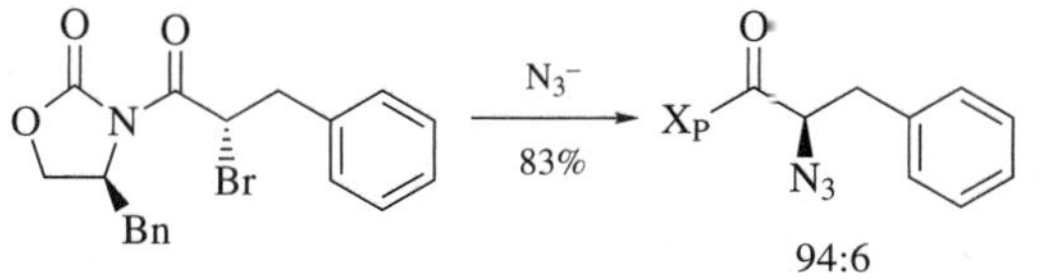

(30)

In those transformations where other stereogenic centers reside proximal to the prochiral center, auxiliary control is dominant in most cases (eq 31).[61]

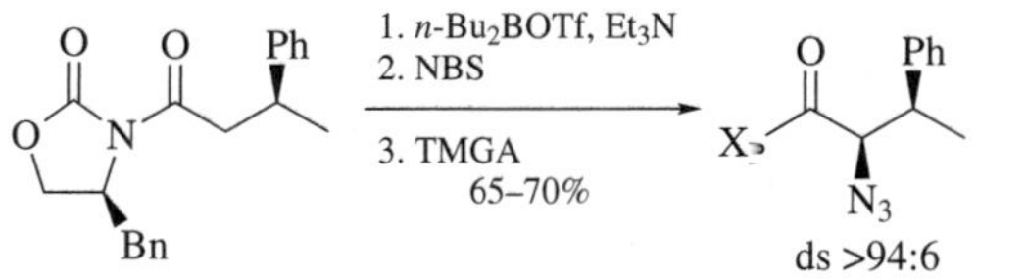

(31)

Aldol Reactions. The dibutyl boryl enolates of chiral acyloxazolidinones react to afford the *syn*-aldol adducts with virtually complete stereocontrol (eq 32).[6,13,14,43,61–64] Notably, the sense of induction in these reactions is opposite to that predicted from the analogous alkylation reactions. This reaction is general for a wide range of aldehydes and imide enolates.[3c,65–69] Enolate control overrides induction inherent to the aldehyde reaction partner.

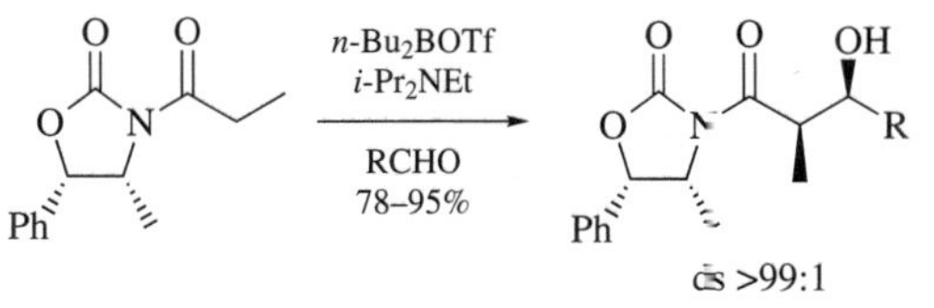

(32)

Titanium enolates of propionyloxazolidinones also undergo aldol reactions with the same sense of induction as the boryl counterparts, but require two or more equivalents of amine base to afford adducts in marginally higher yields but diminished selectivity (eq 33).[45]

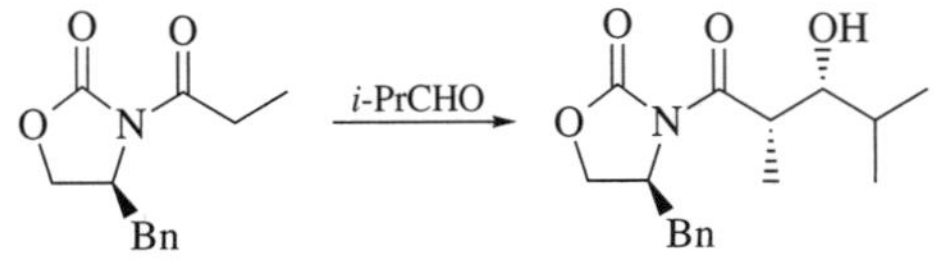

(33)

Enolization	Yield	Stereoselection
$TiCl_4$, *i*-Pr_2NEt	87%	94:6
$TiCl_4$, TMEDA	84%	98:2
n-Bu_2BOTf, Et_3N	83%	>99:1

A second entry to dicarbonyl substrates utilizes the aldol reaction to establish the α-methyl center prior to oxidation of the β-hydroxyl moiety. Commonly, this oxidation is performed using the *Sulfur Trioxide–Pyridine* complex, which results in <1% epimerization of the methyl-bearing center (eq 34).[2] Interestingly, this procedure procures the opposite methyl stereochemistry from that obtained through enolate acylation of the same enantiomer of oxazolidinone.

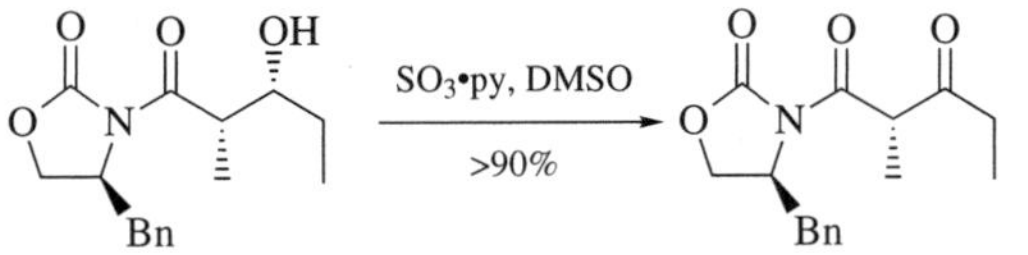

(34)

Non-Evans Aldol Reactions. Either the *syn*- or *anti*-aldol adducts may be obtained from this family of imide-derived enolates, depending upon the specific conditions employed for the reaction. Although the illustrated boron enolate affords the illustrated *syn*-aldol adduct in high diastereoselectivity, the addition reactions between this enolate and Lewis acid-coordinated aldehydes afford different stereochemical outcomes depending on the Lewis acid employed (eq 35).[70] Open transition states have been proposed for the *Diethylaluminum Chloride* mediated, *anti*-selective reaction. These *anti*-aldol reactions have been used in kinetic resolutions of 2-phenylthio aldehydes.[71]

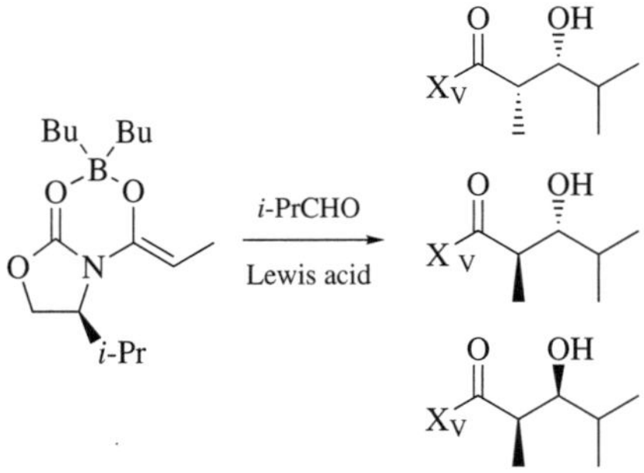

(A) (B) (C) (35)

Lewis acid	Yield	A:B:C
$TiCl_4$ (2 equiv)	83%	0:16:84
$SnCl_4$ (2 equiv)	60%	0:13:87
Et_2AlCl (2 equiv)	63%	0:95:5

Enolates derived from α-haloimides also exhibit metal-dependent *syn/anti*-aldol diastereoselection. The derived Li, Sn^{IV}, and Zn enolates afford the *anti* isomer in reactions with aromatic aldehydes, while the corresponding B and Sn^{II} enolates lead to the conventional *syn* products.[72,73] The 'non-Evans' *syn* adducts have also been observed in reactions organized by *Chlorotitanium Triisopropoxide*.[74,75]

Crotonyl Enolate Aldol Reactions. Boron enolates of the *N*-crotonyloxazolidinones have been shown to afford the expected *syn*-aldol adducts (eq 36).[76,77] The propensity for self-condensation during the enolization process is minimized by the use of triethylamine over less kinetically basic amines.

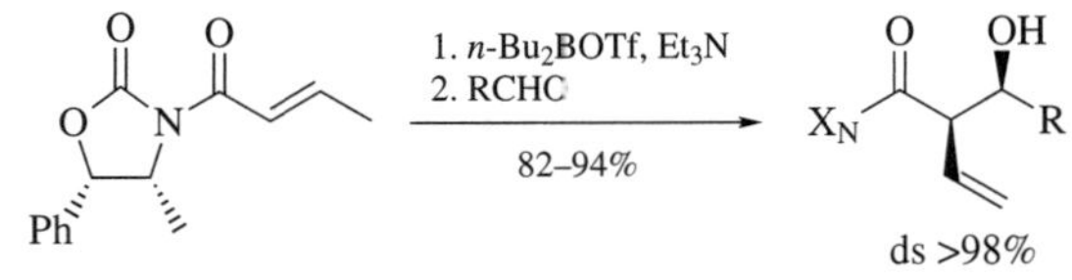

(36)

α-Alkoxyacetate Aldol Reactions. The enolates derived from *N*-α-alkoxyacetyloxazolidinones also provide good yields of aldol adducts. Proper choice of reaction conditions leads to either the *syn* (eq 37)[78] or *anti* (eq 38)[46] adducts. In an application of this aldol reaction in the synthesis of cytovaricin, a complex chiral aldehyde was found to turnover the expected *syn* diastereoselectivity of the boron enolate.[66]

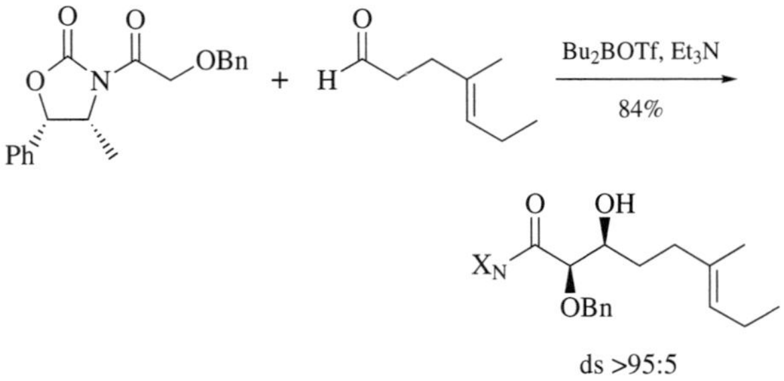

(37)

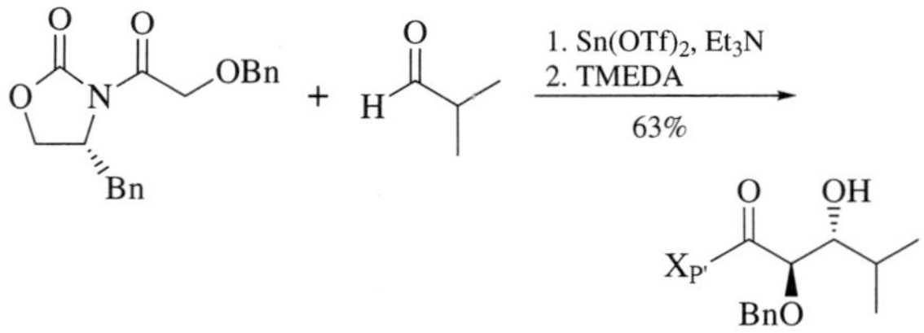

(38)

ds of desired to the total amount of other isomers = 77:23

***N*-Isothiocyanoacetyl Aldol Reactions.** Auxiliary-controlled masked glycine enolate aldol reactions afford the chiral oxazolidine-2-thiones which can be cleaved to provide the *syn*-aldol adducts regardless of aldehyde stereochemistry (eq 39).[79]

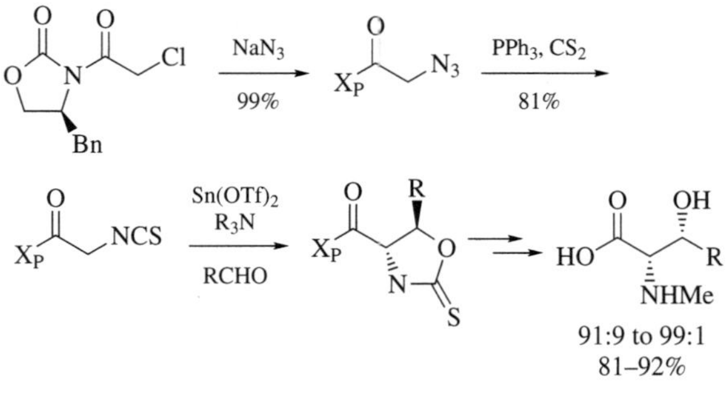

(39)

***N*-Haloacetyl Aldol Reactions.** *N*-Haloacetyloxazolidinones form suitable enolate partners in aldol reactions, although complete aldehyde conversion requires the use of a slight excess of imide (eq 40). The products can be chromatographed to diastereomeric purity.[4,80] Nucleophilic azide displacement of α-halo-β-hydroxy *syn* aldol adducts affords the corresponding *anti* α-amino-β-hydroxy compounds (eq 41).[4,80] Intramolecular displacement of the halogen to form the α-amino product is also possible (eq 42).[80]

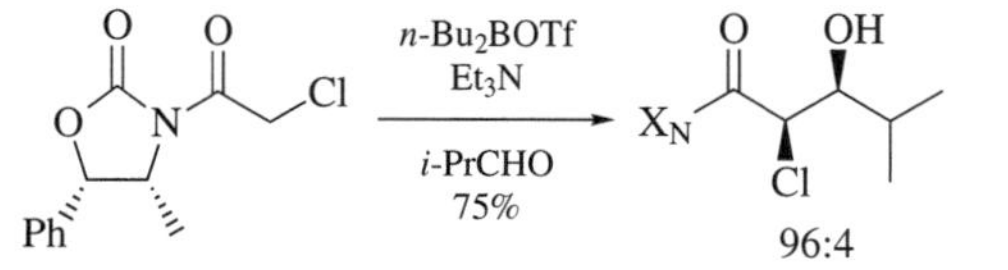

(40)

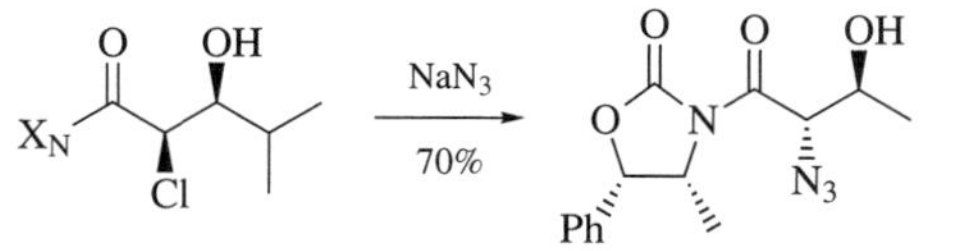

(41)

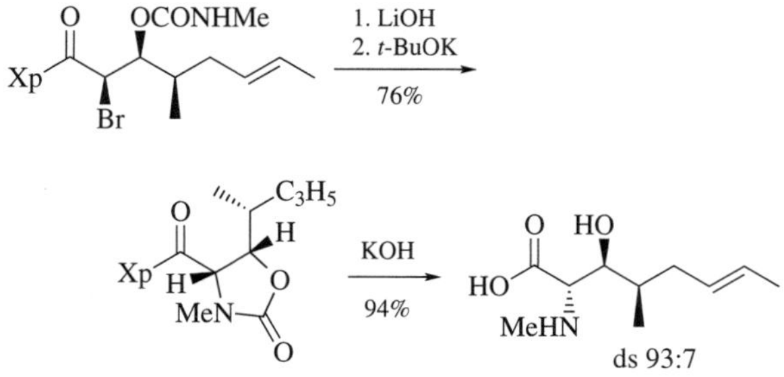

(42)

Acetate Aldol Equivalents. In contrast to the reliably excellent selectivities of α-substituted dibutylboryl imide enolates, boron enolates derived from *N*-acetyloxazolidinones lead to a statistical mixture of aldol adducts under the same reaction conditions. Acetate enolate equivalents may be obtained from these enolates bearing a removable α-substituent. To this end, thiomethyl- or thioethylacetyloxazolidinones (eq 43)[13] as well as haloacetyloxazolidinones can be submitted to highly selective boron-mediated aldol reactions. Products can be transformed to the acetate aldol products via desulfurization with either Raney Ni[81] or *Tri-n-butyltin Hydride* and *Azobisisobutyronitrile*,[82] or via dehalogenation with *Zinc–Acetic Acid* (eq 44).[81] This latter procedure provides several advantages over the sulfur methodology, including ease of imide preparation and improved overall yields.

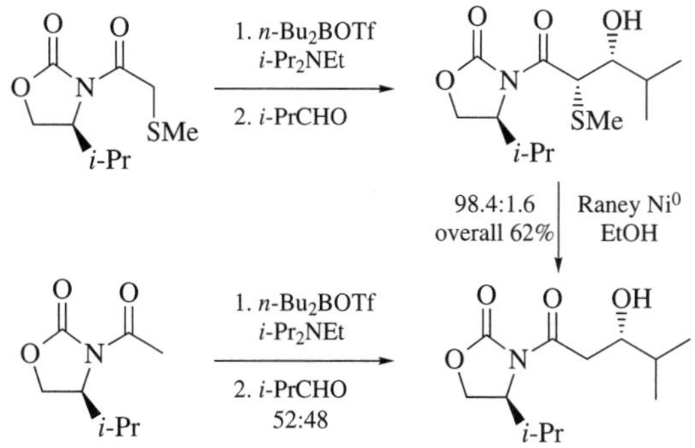

(43)

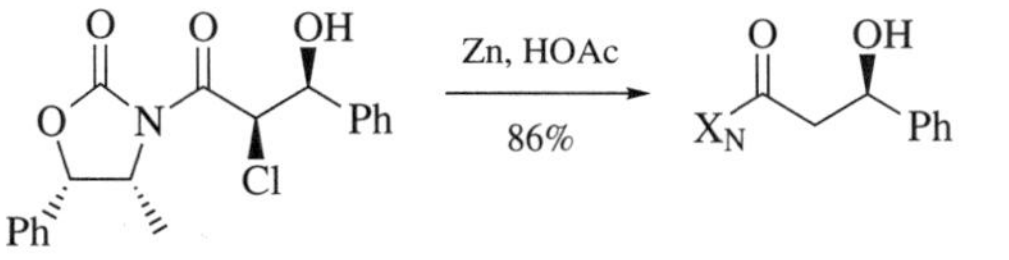

(44)

β-Ketoimide Aldol Reactions. As has been demonstrated, chiral oxazolidinones provide a gateway into asymmetric β-ketoimides via either an aldol–oxidation sequence, or enolate acylation. These substrates can then undergo an iterative aldol reaction, where chirality is induced by the methyl-bearing α-center. To date, three of the four diastereomeric aldol adducts may be selectively obtained with a variety of aldehydes (eq 45).[36,83–86]

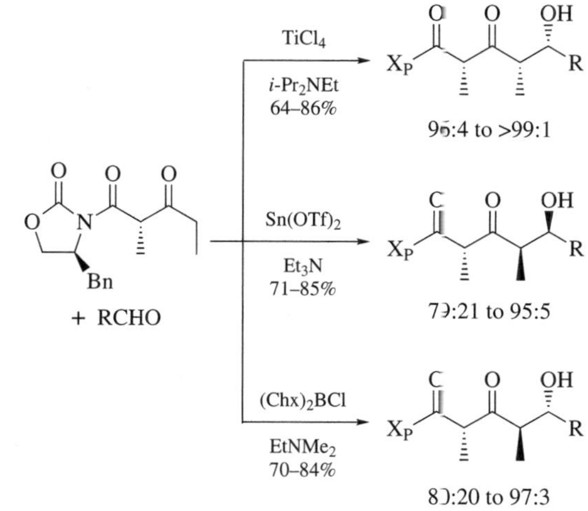

(45)

Reformatsky Reactions. The Reformatsky reaction of α-halooxazolidinones provides an alternative to the more conventional aldol reaction. Although the traditional zinc-mediated Reformatsky using valine-derived compounds proceeds nonselectively,[87,88] the Sn^{II} modification with 2-bromo-2-methylpropionyloxazolidinone proceeds well (eq 46).[89,90] In this particular case, however, the geminal dialkyl substituents favor the endocyclic carbonyl acyl transfer of the auxiliary by the aldolate oxygen.

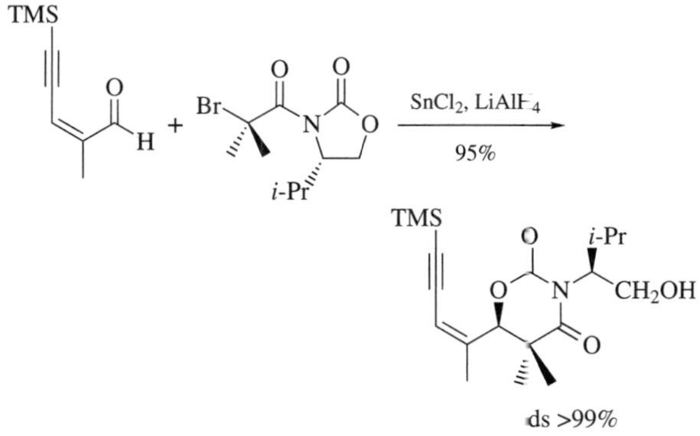

(46)

Acyl Transfer Reactions. (*S*)-*N*-benzoyloxazolidinones have been used as acyl transfer reagents to effect the kinetic resolution of racemic alcohols.[10] The bromomagnesium alkoxides formed from phenyl *n*-alkyl alcohols selectively attack the exocyclic benzoyl moiety to afford recovered auxiliary and the derived (*R*)-benzoates in >90% ee and >90% yield (eq 47). The scope of this reaction seems to be limited to this class of substrates as selectivity drops with increasing the steric bulk of the alkyl group.

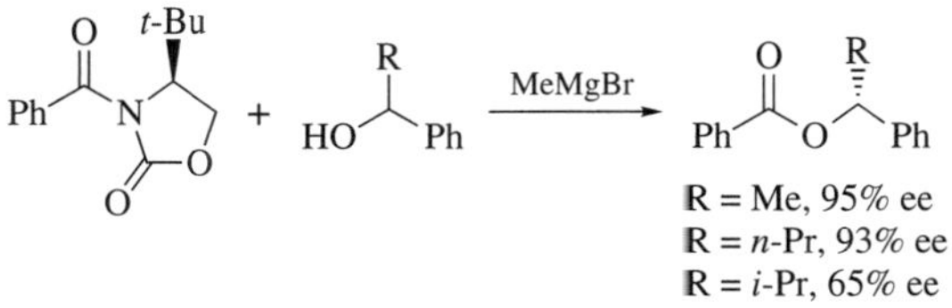

(47)

Sulfinyl Transfer Reactions. Grignard reagents add to diastereomerically pure *N*-arylsulfinyloxazolidinones with inversion of configuration at sulfur to afford enantiopure dialkyl or aryl alkyl sulfoxides in excellent yields (eq 48).[11] Although broader in synthetic utility than the menthyl sulfinate esters,[91,92] this methodology is comparable to Kagan's chiral sulfite substrates as a strategy for constructing chiral sulfoxides.[93,94]

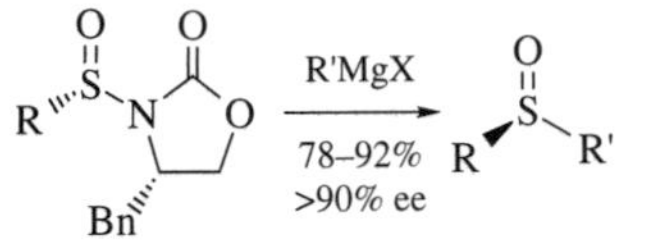

(48)

R = Me, *t*-Bu, *p*-Tol
R' = alkyl, aryl

The *N*-arylsulfinyloxazolidinone methodology is readily extended to the formation of sulfinylacetates, sulfinates, and sulfinamides with >95% ee and high yields (eq 49).

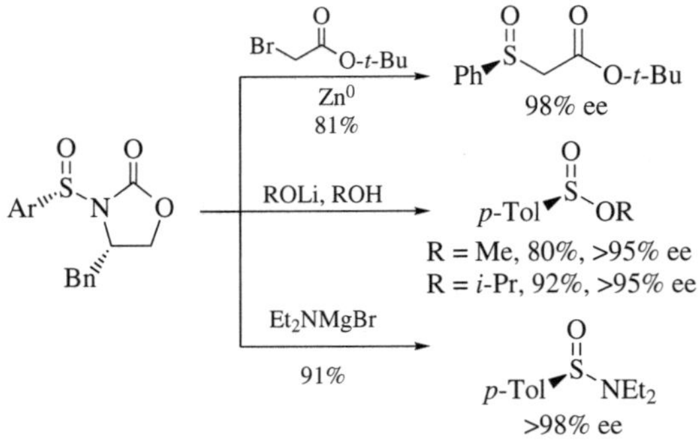

(49)

Diels–Alder Reactions. Chiral α,β-unsaturated imides participate in Lewis acid-promoted Diels–Alder cycloaddition reactions to afford products in uniformly excellent *endo/exo* and *endo* diastereoselectivities (eq 50 and 51).[9,37,95,96] Unfortunately, this reaction does not extend to certain dienophiles, including methacryloyl imides, β,β-dimethylacryloyl imides, or alkynic imides. Cycloadditions also occur with less reactive acyclic dienes with high diastereoselectivity (eq 52). Of the auxiliaries surveyed, the phenylalanine-derived oxazolidinones provided the highest diastereoselectivities. This methodology has been recently extended to complex intramolecular processes (eq 53).[68,95,97] In this case, use of the unsubstituted achiral oxazolidinone favored the undesired diastereomer.

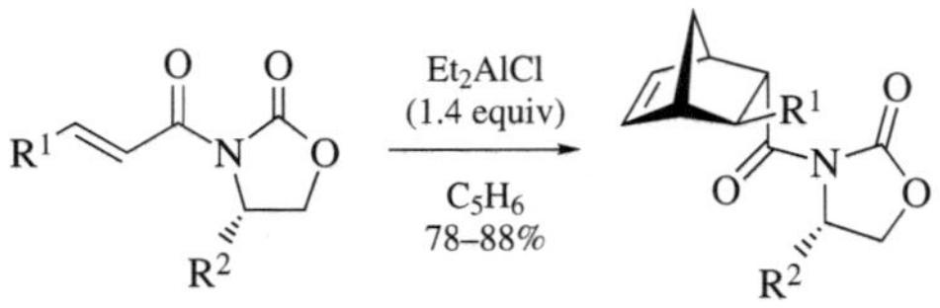

(50)

R^1 = H or Me; R^2 = *i*-Pr or Bn

endo:exo >98:2
endo ds >95:5

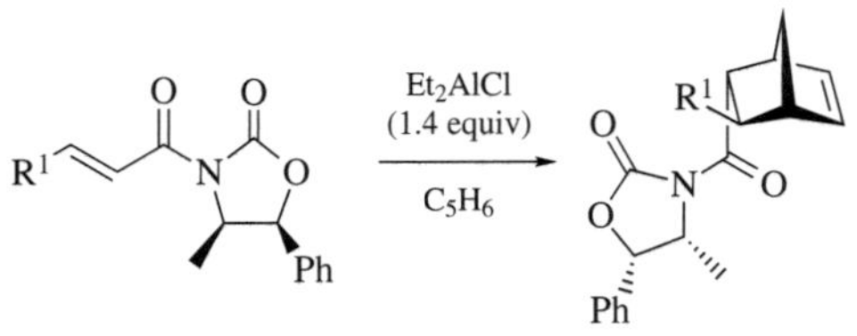

(51)

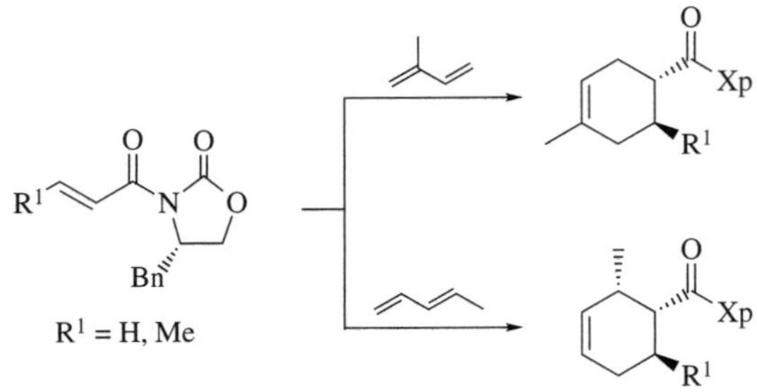

(52)

ds from 94:6 to >99:1
77–85%

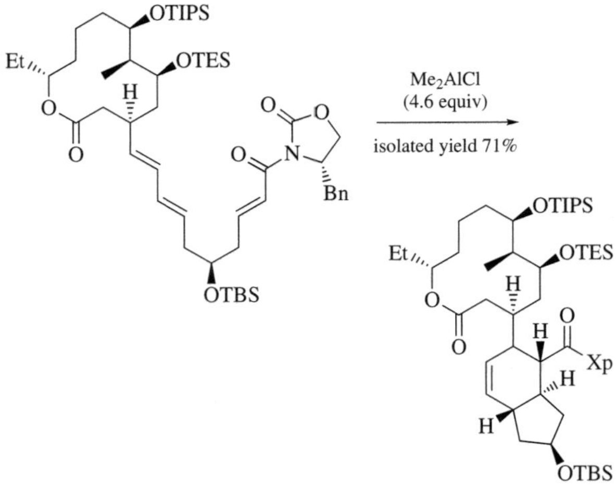

(53)

ds 91:9

Staudinger Reactions. Chiral oxazolidinones have been employed as the chiral control element in the Staudinger reaction as well as the ultimate source of the α-amino group in the formation of β-lactams.[41] Cycloaddition of ketene derived from 4-(*S*)-phenyloxazolidylacetyl chloride with conjugated imines affords the corresponding β-lactams in 80–90% yields with excellent diastereoselectivity (eq 54). The auxiliary can then be reduced under Birch conditions to reveal the α-amino group.

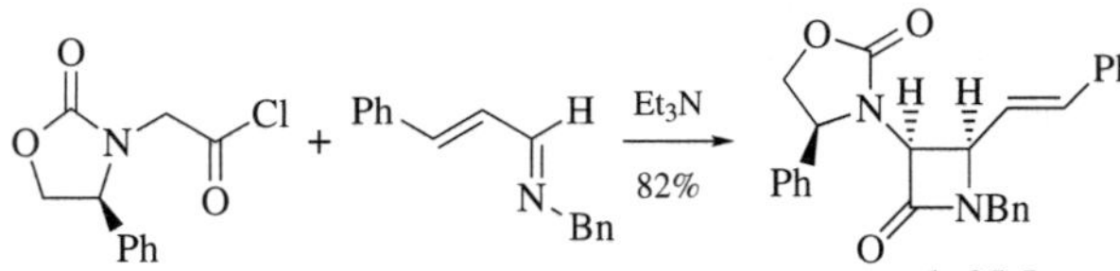

(54)

ds 95:5

Conjugate Addition Reactions. α,β-Unsaturated *N*-acyloxazolidinones have been implemented as Michael acceptors, inducing chirality in the same sense as in enolate alkylation reactions. Chiral α,β-unsaturated imides undergo 1,4-addition when treated with diethylaluminum chloride (eq 55). Photochemical initiation is required for the analogous reaction with *Dimethylaluminum Chloride*.[96]

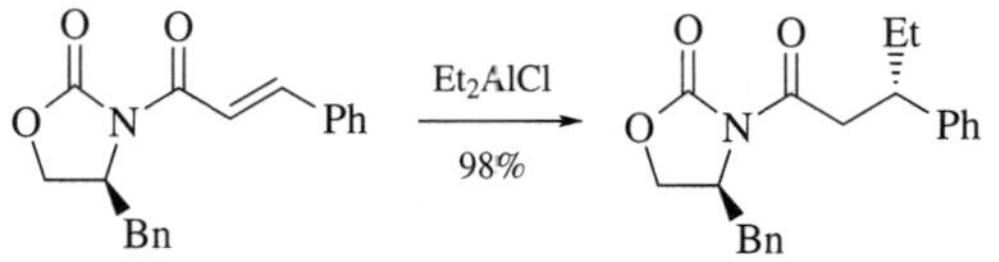

(55)

ds 93:7

Organocuprates also undergo conjugate addition with chiral α,β-unsaturated imides.[98] Treatment of the imides derived from

4-phenyl-2-oxazolidinone with methyl- or arylmagnesium halides and CuBr affords conjugate addition products in yields over 80% with few exceptions (eq 56).[8] Reaction diastereoselectivity appears to be contingent upon the use of the 4-phenyloxazolidinone auxiliary. The preceding methodology has been applied to the synthesis of β-methyltryptophan (eq 57).[99]

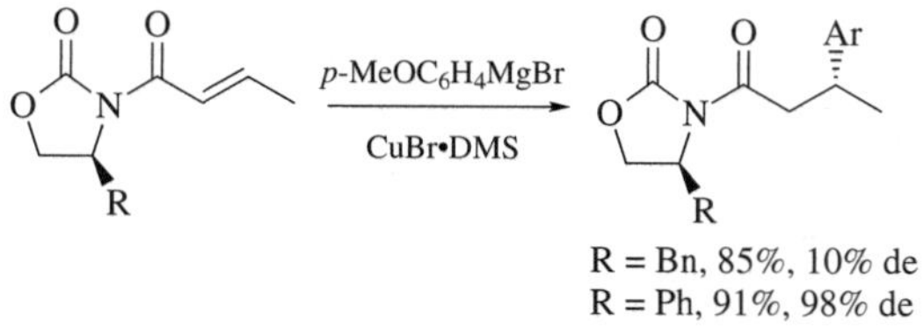

(56)

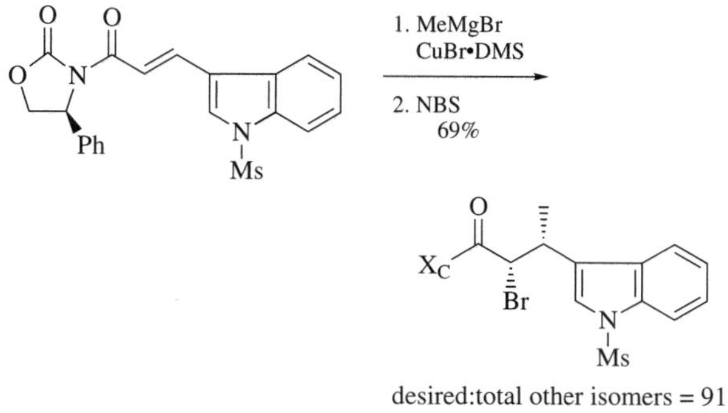

(57)

Similar chemistry using chiral sultam auxiliaries demonstrates superior yields and selectivities for specific cases of cuprate conjugate additions, but have not yet been extended to the more complex multistep transformation series illustrated above.[100,101] Moderate selectivities have been obtained in alkyl cuprate additions to γ-aminocrotonate equivalents where the nitrogen is derived from the oxazolidinone.[102]

The 4-phenyl-2-oxazolidinone auxiliary has also been employed in the $TiCl_4$-mediated conjugate additions of allylsilanes (eq 58).[103] Analogous reactions using the phenylalanine-derived auxiliary with dimethylaluminum chloride afforded lower selectivities.[104] In these reactions the oxazolidinones perform better than the sultams.

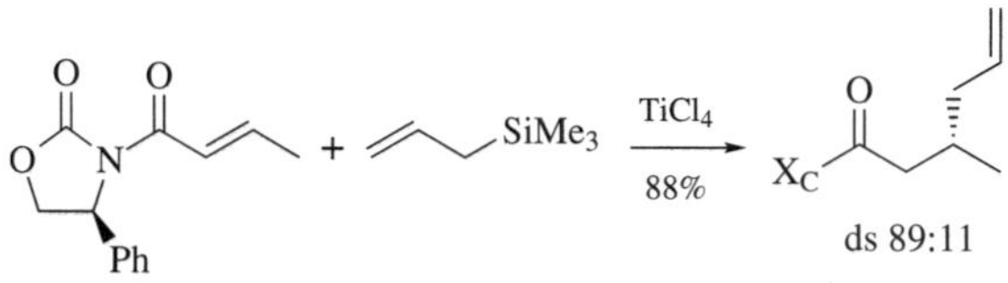

(58)

Nucleophilic addition of thiophenol to chiral tiglic acid-derived imides proceeds in excellent yields and diastereoselectivities (eq 59).[105] Complete turnover of both the α- and β-centers results from the use of the (*Z*) rather than (*E*) isomer. Poor β-induction was found with the imides derived from cinnamic acid.

Dimethylaluminum chloride also catalyzes the ene reactions of chiral α,β-unsaturated imides with 1,1-disubstituted alkenes in moderate yields and selectivities.[104]

Oxazolidinone-Substituted Carbanions. Oxazolidinone-substituted organostannanes readily undergo transmetalation with alkyllithium reagents to the organolithium derivatives which then can undergo nucleophilic addition reactions. *N*-Substituted oxazolidinones can act in this capacity as both a nitrogen source and source of chirality (eq 60). Although the α-stereoselection in these reactions is excellent, a greater variety of reactant alkylstannanes are available using chiral imidazolidinones in place of oxazolidinones.[39,40,42]

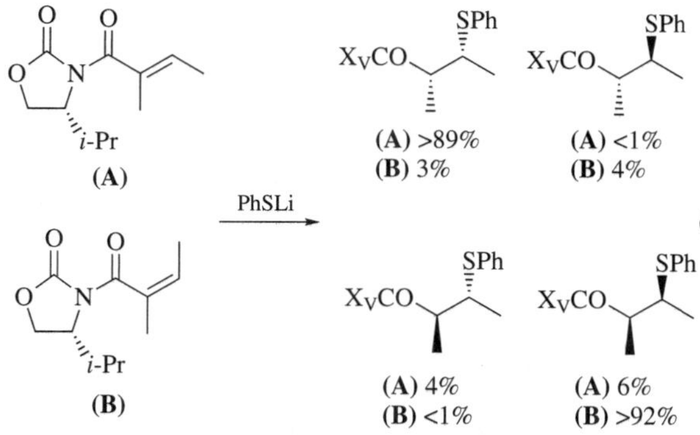

(59)

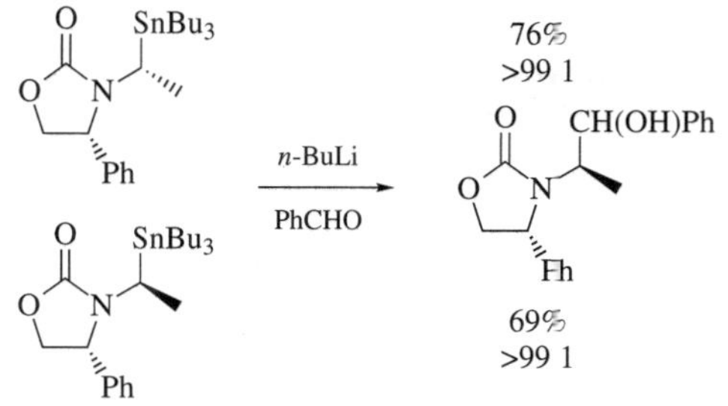

(60)

Synthesis of Cyclopropanes. Chiral imide enolates which contain γ-halide substituents undergo intramolecular displacement to form cyclopropanes.[106] Halogenation of γ,δ-unsaturated acyl imides occurs at the γ-position in 85% yield with modest stereoinduction. The (*Z*) sodium enolates of these compounds then cyclize through an intramolecular double stereodifferentiating reaction (eq 61).

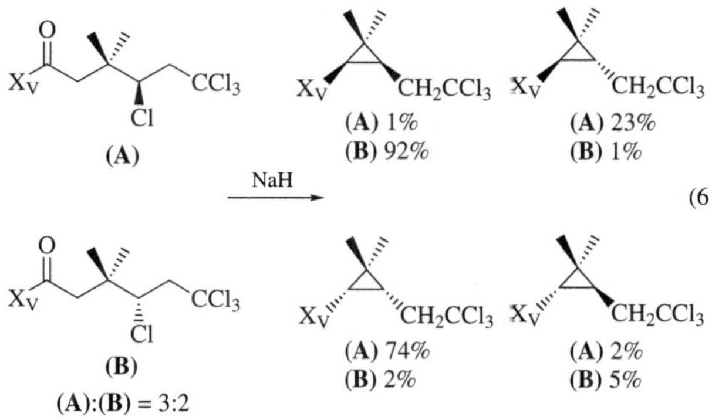

(61)

Stereoselective Cyclizations. Sultams have been demonstrated to be superior sources of chirality in selected cases of iodolactonizations,[107] oxidative 1,5-diene cyclizations,[108] and Claisen-type rearrangements of β-acetoxyl substrates.[109]

Chiral Ligands. Bidentate chelation of dirhodium(II) compounds by chiral oxazolidinones creates asymmetric sites on the metal, leading to induction in cyclopropanations and carbon–hydrogen insertion reactions. The oxazolidinones are less effective in this capacity than are the pyrrolidinones.[110]

Removal of the Chiral Auxiliary. In each of the following transformations, the oxazolidinone auxiliary is recovered in high yields (eq 62).

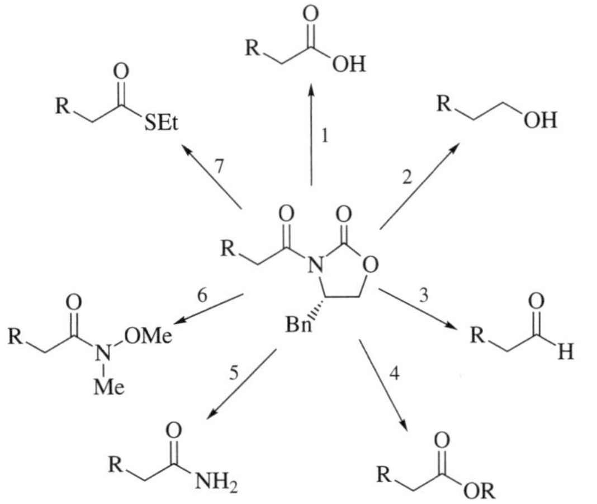

(62)

1a) KOH, MeOH; 1b) LiOH, MeOH or THF; 1c) LiOH, H_2O_2.
2a) $LiAlH_4$; 2b) $LiBH_4$; 2c) $LiBH_4$, H_2O; 2d) $LiBH_4$, MeOH;
2e) $LiAlH_4$, H_2, Lindlar cat., TFA; 2f) Bu_3B, HOAc, $LiBH_4$.
3a) i) $LiBH_4$, H_2O; ii) DMSO, $(COCl)_2$, Et_3N;
3b) i) $Me_2AlN(OMe)Me$; ii) DIBAL;
3c) i) LiSEt; ii) Et_3SiH, Pd/C.
4a) LiOBn; 4b) $Ti(OBn)_4$; 4c) ROMgBr; 4d) NaOMe;
4e) $Ti(OEt)_4$.
5) i) N_2H_4; ii) isopentyl nitrite, NH_4Cl.
6) $Me_2AlN(OMe)Me$.
7) LiSEt.

Conversion to the Acid. Hydroxide[6,111] and peroxide[112] agents saponify acyl imides in excellent yields; however, with sterically hindered acyl groups endocyclic cleavage may predominate upon treatment with *Lithium Hydroxide*. *Lithium Hydroperoxide*, however, is highly selective for the exocyclic carbonyl moiety.

Conversion to the Alcohol. Reduction of acyl imides to their corresponding alcohols is effected by a number of reagents, including *Lithium Aluminum Hydride*,[1] *Lithium Borohydride*,[1] $LiAlH_4/H_2$/Lindlar's cat./TFA,[113] $LiBH_4/H_2O/Et_2O$,[114] $LiBH_4$/MeOH/THF,[36] and Bu_3B/HOAc/$LiBH_4$.[77] Although the sole use of $LiAlH_4$ or $LiBH_4$ affords product often in low yields, the addition of an equivalent of H_2O or MeOH greatly enhances reaction efficiency. The MeOH/THF modification occasionally produces more consistent results. The last of the methods outlined above is effective in preventing retro-aldol cleavage in sensitive substrates such as crotyl or α-fluoro aldol adducts (eq 63).

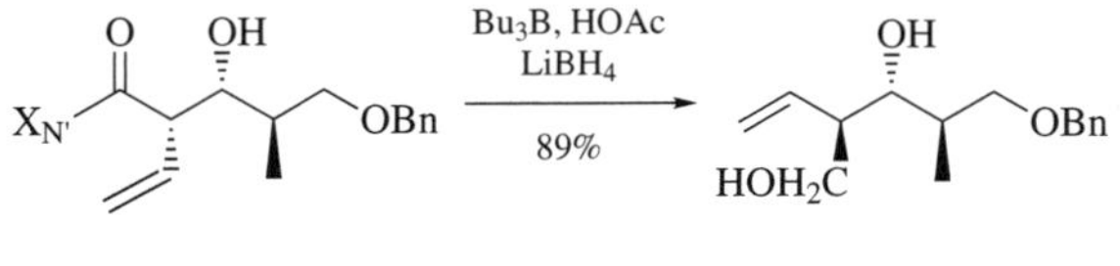

(63)

Conversion to the Aldehyde. This transformation is accomplished through a two-step procedure. One such variant requires reduction to the alcohol (e.g. $LiAlH_4$, H_2O) and subsequent oxidation (e.g. Swern conditions).[36,85] Alternatively, Weinreb transamination[78,115–117] followed by *Diisobutylaluminum Hydride*,[78] or conversion to the thioester (see below) and subsequent *Triethylsilane* reduction,[86] afford the desired aldehyde in excellent yields. Weinreb transamination proceeds with minimal endocyclic cleavage when there is a β-hydroxy moiety free for internal direction of the aluminum species.

Conversion to Esters. Ester formation is readily achieved by conventional alcoholysis with alkoxides such as LiOBn,[1] NaOMe,[6] or ROMgBr (eq 64).[76,118] In hindered cases, endocyclic cleavage becomes competitive. Various titanium(IV) alkoxides have also been employed to effect this transformation.[4,119]

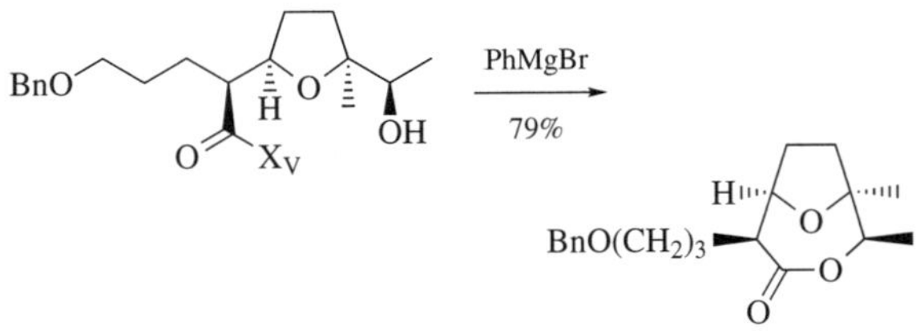

(64)

Conversion to Amides. *N*-Acyloxazolidinones may be converted to the primary amide via the corresponding hydrazide.[120] Alternatively, trimethylaluminum/amine adducts form active transamination reagents,[36,117] providing amides of β-hydroxy acyloxazolidinones through intramolecular amine addition of the aminoaluminum species (eq 65).[105]

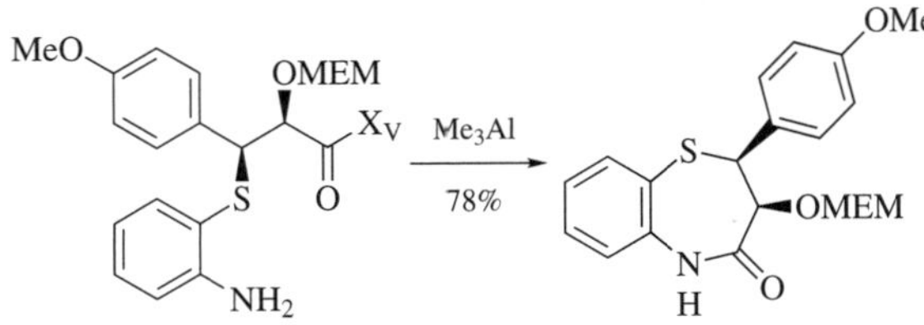

(65)

Conversion to Thioesters. The transformation of *N*-acyl imides into thioesters with lithium thiolate reagents proceeds with exceptional selectivity for the *exo* carbonyl moiety even in exceptionally hindered cases.[121] A recent application of this reaction in a complex setting has been reported (eq 66).[68,97] This transformation is significant in that the normally reliable peroxide hydrolysis procedure proved to be nonselective. The recently reported high yield reduction of thioesters to aldehydes[86] enhances the utility of these thioester intermediates.

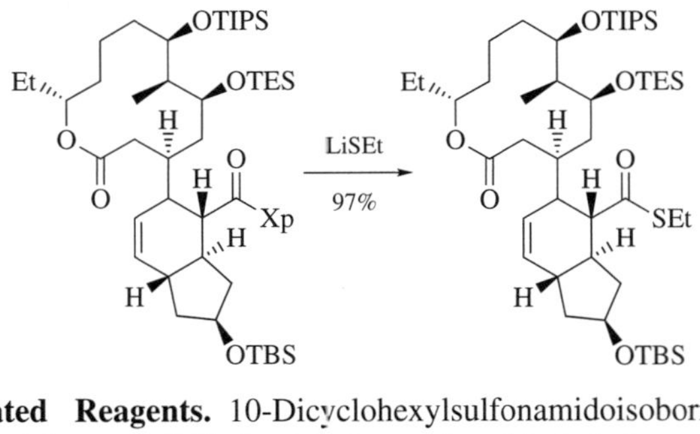

(66)

Related Reagents. 10-Dicyclohexylsulfonamidoisoborneol; (S)-Ethyl Lactate; 3-Hydroxyisoborneol; α-Methyltoluene-2,α-sultam.

1. Evans, D. A.; Ennis, M. D.; Mathre, D. J. *J. Am. Chem. Soc.* **1982**, *104*, 1737.
2. Evans, D. A.; Ennis, M. D.; Le, T.; Mandel, N.; Mandel, G. *J. Am. Chem. Soc.* **1984**, *106*, 1154.
3. Evans, D. A.; Ellman, J. A.; Dorow, R. L. *Tetrahedron Lett.* **1987**, *28*, 1123.
4. Evans, D. A.; Britton, T. C.; Ellman, J. A.; Dorow, R. L. *J. Am. Chem. Soc.* **1990**, *112*, 4011.
5. Evans, D. A.; Morrissey, M. M ; Dorow, R. L. *J. Am. Chem. Soc.* **1985**, *107*, 4346.
6. Evans, D. A.; Bartroli, J.; Shih, T. L. *J. Am. Chem. Soc.* **1981**, *103*, 2127.
7. Evans, D. A.; Bilodeau, M. T.; Somers, T. C.; Clardy, J.; Cherry, D.; Kato, Y. *J. Org. Chem.* **1991**, *56*, 5750.

8. Nicolás, E.; Russell, K. C.; Hruby, V. J. *J. Org. Chem.* **1993**, *58*, 766.
9. Evans, D. A.; Chapman, K. T.; Bisaha, J. *J. Am. Chem. Soc.* **1988**, *110*, 1238.
10. Evans, D. A.; Anderson, J. C.; Taylor, M. K. *Tetrahedron Lett.* **1993**, *34*, 5563.
11. Evans, D. A.; Faul, M. M.; Colombo, L.; Bisaha, J. J.; Clardy, J.; Cherry, D. *J. Am. Chem. Soc.* **1992**, *114*, 5977.
12. (a) Gage, J. R.; Evans, D. A. *Org. Synth.* **1990**, *68*, 77. (b) Gage, J. R.; Evans, D. A. *Org. Synth., Coll. Vol.* **1993**, *8*, 528.
13. Evans, D. A.; Takacs, J. M.; McGee, L. R.; Ennis, M. D.; Mathre, D. J.; Bartroli, J. *Pure Appl. Chem.* **1981**, *53*, 1109.
14. Evans, D. A.; Nelson, J. V.; Taber, T. R. *Top. Stereochem.* **1982**, *13*, 1.
15. McKennon, M. J.; Meyers, A. I.; Drauz, K.; Schwarm, M. *J. Org. Chem.* **1993**, *58*, 3568.
16. Crowther, H. L.; McCombie, R. *J. Chem. Soc.* **1913**, 27.
17. Newman, M. S.; Kutner, A. *J. Am. Chem. Soc.* **1951**, *73*, 4199.
18. Hyne, J. B. *J. Am. Chem. Soc.* **1959**, *81*, 6058.
19. Stratton, J. M.; Wilson, F. J. *J. Chem. Soc.* **1932**, 1133.
20. Close, W. J. *J. Org. Chem.* **1950**, *15*, 1131.
21. Lynn, J. W. U.S. Patent 2 975 187 (*Chem. Abstr.* **1955**, *49*, 16 568d).
22. Lesher, G. Y.; Surrey, A. R. *J. Am. Chem. Soc.* **1955**, *77*, 632.
23. Caccia, G.; Gladiali, S.; Vitali, R.; Gardi, R. *J. Org. Chem.* **1973**, *38*, 2264.
24. Saund, A. K.; Prashad, B.; Koul, A. K.; Bachhawat, J. M.; Mathur, N. K. *Int. J. Peptide Protein Res.* **1973**, *5*, 7.
25. Applegath, F. U.S. Patent 2 857 392 (*Chem. Abstr.* **1953**, *47*, 5286d).
26. Koch, P.; Perrotti, E. *Tetrahedron Lett.* **1974**, 2899.
27. Sonoda, N.; Yamamoto, G.; Natsukawa, K.; Kondo, K.; Murai, S. *Tetrahedron Lett.* **1975**, 1969.
28. Correa, A.; Denis, J.-N.; Greene, A. E. *Synth. Commun.* **1991**, *21*, 1.
29. Pridgen, L. N.; Prol, J., Jr.; Alexander, B.; Gillyard, L. *J. Org. Chem.* **1989**, *54*, 3231.
30. Kano, S.; Yokomatsu, T.; Iwasawa, H.; Shibuya, S. *Tetrahedron Lett.* **1987**, *28*, 6331.
31. Wuts, P. G. M.; Pruitt, L. E. *Synthesis* **1989**, 622.
32. Agami, C.; Couty, F.; Hamon, L.; Venier, O. *Tetrahedron Lett.* **1993**, *34*, 4509.
33. Ishizuka, T.; Kimura, K.; Ishibuchi, S.; Kunieda, T. *Chem. Lett.* **1992**, 991.
34. Koch, S. S. C.; Chamberlin, A. R. *J. Org. Chem.* **1993**, *58*, 2725.
35. Evans, D. A.; Gage, J. R. *J. Org. Chem.* **1992**, *57*, 1958.
36. Evans, D. A.; Gage, J. R.; Leighton, J. L. *J. Am. Chem. Soc.* **1992**, *114*, 9434.
37. Evans, D. A.; Chapman, K. T.; Bisaha, J. *J. Am. Chem. Soc.* **1984**, *106*, 4261.
38. Thom, C.; Kociénski, P. *Synthesis* **1992**, 582.
39. Pearson, W. H.; Lindbeck, A. C. *J. Am. Chem. Soc.* **1991**, *113*, 8546.
40. Pearson, W. H.; Lindbeck, A. C.; Kampf, J. W. *J. Am. Chem. Soc.* **1993**, *115*, 2622.
41. Evans, D. A.; Sjogren, E. B. *Tetrahedron Lett.* **1985**, *26*, 3783.
42. Pearson, W. H.; Lindbeck, A. C. *J. Org. Chem.* **1989**, *54*, 5651.
43. Evans, D. A.; Nelson, J. V.; Vogel, E.; Taber, T. R. *J. Am. Chem. Soc.* **1981**, *103*, 3099.
44. Carreira, E. M. Ph.D. Thesis, Harvard University, 1990.
45. Evans, D. A.; Bilodeau, M. Unpublished results.
46. Evans, D. A.; Gage, J. R.; Leighton, J. L.; Kim, A. S. *J. Org. Chem.* **1992**, *57*, 1961.
47. Evans, D. A.; Urpí, F.; Somers, T. C.; Clark, J. S.; Bilodeau, M. T. *J. Am. Chem. Soc.* **1990**, *112*, 8215.
48. Alexander, R. P.; Paterson, I. *Tetrahedron Lett.* **1985**, *26*, 5339.
49. Soai, K.; Ishizaki, M.; Yokoyama, S. *Chem. Lett.* **1987**, 341.
50. Miles, W. H.; Smiley, P. M.; Brinkman, H. R. *Chem. Commun.* **1989**, 1897.
51. Green, M.; Greenfield, S., Kersting, M. *Chem. Commun.* **1985**, 18.
52. Pearson, A. J.; Khetani, V. D.; Roden, B. A. *J. Org. Chem.* **1989**, *54*, 5141.
53. Pearson, A. J.; Zhu, P. Y.; Youngs, W. J.; Bradshaw, J. D.; McConville, D. B. *J. Am. Chem. Soc.* **1993**, *115*, 10376.
54. Schreiber, S. L.; Klimas, M. T.; Sammakia, T. *J. Am. Chem. Soc.* **1987**, *109*, 5749.
55. Yamazaki, T.; Haga, J.; Kitazume, T. *Chem. Lett.* **1991**, 2175.
56. Evans, D. A.; Britton, T. C.; Dorow, R. L.; Dellaria, J. F. *J. Am. Chem. Soc.* **1986**, *108*, 6395.
57. Evans, D. A.; Britton, T. C. *J. Am. Chem. Soc.* **1987**, *109*, 6881.
58. Evans, D. A.; Evrard, D. A.; Rychnovsky, S. D.; Früh, T.; Whittingham, W. G.; DeVries, K. M. *Tetrahedron Lett.* **1992**, *33*, 1189.
59. Doyle, M. P.; Dorow, R. L.; Terpstra, J. W.; Rodenhouse, R. A. *J. Org. Chem.* **1985**, *50*, 1663.
60. Davis, F. A.; Han, W. *Tetrahedron Lett.* **1992**, *33*, 1153.
61. Dharanipragada, R.; Nicolas, E.; Toth, G.; Hruby, V. J. *Tetrahedron Lett.* **1989**, *30*, 6841.
62. Evans, D. A.; Taber, T. R. *Tetrahedron Lett.* **1980**, *21*, 4675.
63. Evans, D. A. *Aldrichim. Acta* **1982**, *15*, 23.
64. Hamada, Y.; Hayashi, K.; Shioiri, T. *Tetrahedron Lett.* **1991**, *32*, 931.
65. Evans, D. A.; Bender, S. L. *Tetrahedron Lett.* **1986**, *27*, 799.
66. Evans, D. A.; Kaldor, S. W.; Jones, T. K.; Clardy, J.; Stout, T. J. *J. Am. Chem. Soc.* **1990**, *112*, 7001.
67. Evans, D. A.; Miller, S. J.; Ennis, M. D.; Ornstein, P. L. *J. Org. Chem.* **1992**, *57*, 1067.
68. Evans, D. A.; Black, W. C. *J. Am. Chem. Soc.* **1993**, *115*, 4497.
69. Evans, D. A.; Dow, R. L. *Tetrahedron Lett.* **1986**, *27*, 1007.
70. Walker, M. A.; Heathcock, C. H. *J. Org. Chem.* **1991**, *56*, 5747.
71. Chibale, K.; Warren, S. *Tetrahedron Lett.* **1992**, *33*, 4369.
72. Abdel-Magid, A.; Pridgen, L. N.; Eggleston, D. S.; Lantos, I. *J. Am. Chem. Soc.* **1986**, *108*, 4595.
73. Pridgen, L. N.; Abdel-Magid, A.; Lantos, I. *Tetrahedron Lett.* **1989**, *30*, 5539.
74. Nerz-Stormes, M.; Thornton, E. R. *Tetrahedron Lett.* **1986**, *27*, 897.
75. Nerz-Stormes, M.; Thornton, E. R. *J. Org. Chem.* **1991**, *56*, 2489.
76. Evans, D. A.; Dow, R. L.; Shih, T. L.; Takacs, J. M.; Zahler, R. *J. Am. Chem. Soc.* **1990**, *112*, 5290.
77. Evans, D. A.; Sjogren, E. B.; Bartroli, J.; Dow, R. L. *Tetrahedron Lett.* **1986**, *27*, 4957.
78. Evans, D. A.; Bender, S. L.; Morris, J. *J. Am. Chem. Soc.* **1988**, *110*, 2506.
79. Evans, D. A.; Weber, A. E. *J. Am. Chem. Soc.* **1986**, *108*, 6757.
80. Evans, D. A.; Sjogren, E. B.; Weber, A. E.; Conn, R. E. *Tetrahedron Lett.* **1987**, *28*, 39.
81. Sjogren, E. B. Ph.D. Thesis, Harvard University, 1986.
82. Evans, D. A.; Shumsky, J. Unpublished results.
83. Evans, D. A.; Clark, J. S.; Metternich, R.; Novack, V. J.; Sheppard, G. S. *J. Am. Chem. Soc.* **1990**, *112*, 866.
84. Evans, D. A.; Ng, H. P.; Clark, J. S.; Rieger, D. L. *Tetrahedron* **1992**, *48*, 2127.
85. Evans, D. A.; Sheppard, G. S. *J. Org. Chem.* **1990**, *55*, 5192.
86. Evans, D. A.; Ng, H. P. *Tetrahedron Lett.* **1993**, *34*, 2229.
87. Ito, Y.; Sasaki, A.; Tamoto, K.; Sunagawa, M.; Terashima, S. *Tetrahedron* **1991**, *47*, 2801.
88. Ito, Y.; Terashima, S. *Tetrahedron* **1991**, *47*, 282[illegible].
89. Kende, A. S.; Kawamura, K.; Orwat, M. J. *Tetrahedron Lett.* **1989**, *30*, 5821.

90. Kende, A. S.; Kawamura, K.; DeVita, R. J. *J. Am. Chem. Soc.* **1990**, *112*, 4070.
91. Andersen, K. K.; Gaffield, W.; Papnikolaou, N. E.; Foley, J. W.; Perkins, R. I. *J. Am. Chem. Soc.* **1964**, *86*, 5637.
92. Andersen, K. K. *Tetrahedron Lett.* **1962**, 93.
93. Rebiere, F.; Samuel, O.; Ricard, L.; Kagan, H. B. *J. Org. Chem.* **1991**, *56*, 5991.
94. Rebiere, F.; Kagan, H. B. *Tetrahedron Lett.* **1989**, *30*, 3659.
95. Evans, D. A.; Chapman, K. T.; Bisaha, J. *Tetrahedron Lett.* **1984**, *25*, 4071.
96. Rück, K.; Kunz, H. *Angew. Chem., Int. Ed. Engl.* **1991**, *30*, 694.
97. Evans, D. A.; Black, W. C. *J. Am. Chem. Soc.* **1992**, *114*, 2260.
98. Pourcelot, G.; Aubouet, J.; Caspar, A.; Cresson, P. *J. Organomet. Chem.* **1987**, *328*, C43.
99. Boteju, L. W.; Wegner, K.; Hruby, V. J. *Tetrahedron Lett.* **1992**, *33*, 7491.
100. Oppolzer, W.; Moretti, R.; Bernardinelli, G. *Tetrahedron Lett.* **1986**, *27*, 4713.
101. Oppolzer, W.; Poli, G. *Tetrahedron Lett.* **1986**, *27*, 4717.
102. Le Coz, S.; Mann, A. *Synth. Commun.* **1993**, *23*, 165.
103. Wu, M.-J.; Wu, C.-C.; Lee, P.-C. *Tetrahedron Lett.* **1992**, *33*, 2547.
104. Snider, B.; Zhang, Q. *J. Org. Chem.* **1991**, *56*, 4908.
105. Miyata, O.; Shinada, T.; Ninomiya, I.; Naito, T. *Tetrahedron Lett.* **1991**, *32*, 3519.
106. Kleschick, W. A.; Reed, M. W.; Bordner, J. *J. Org. Chem.* **1987**, *52*, 3168.
107. Yokomatsu, T.; Iwasawa, H.; Shibuya, S. *Chem. Commun.* **1992**, 728.
108. Walba, D. M.; Przybyla, C. A.; Walker, C. B., Jr. *J. Am. Chem. Soc.* **1990**, *112*, 5624.
109. Brandänge, S.; Leijonmarck, H. *Tetrahedron Lett.* **1992**, *33*, 3025.
110. Doyle, M. P.; Winchester, W. R.; Hoorn, J. A. A.; Lynch, V.; Simonsen, S. H.; Ghosh, R. *J. Am. Chem. Soc.* **1993**, *115*, 9968.
111. Šavrda, J.; Descoins, C. *Synth. Commun.* **1987**, *17*, 1901.
112. Evans, D. A.; Britton, T. C.; Ellman, J. A. *Tetrahedron Lett.* **1987**, *28*, 6141.
113. Tietze, L. F.; Schneider, C. *J. Org. Chem.* **1991**, *56*, 2476.
114. Penning, T. D.; Djuric, S. W.; Haack, R. A.; Kalish, V. J.; Miyashiro, J. M.; Rowell, B. W.; Yu, S. S. *Synth. Commun.* **1990**, *20*, 307.
115. Levin, J. L.; Turos, E.; Weinreb, S. M. *Synth. Commun.* **1982**, *12*, 989.
116. Basha, A.; Lipton, M.; Weinreb, S. M. *Tetrahedron Lett.* **1977**, 4171.
117. Evans, D. A.; Sjogren, E. B. *Tetrahedron Lett.* **1986**, *27*, 3119.
118. Evans, D. A.; Weber, A. E. *J. Am. Chem. Soc.* **1987**, *109*, 7151.
119. Harre, M.; Trabandt, J.; Westermann, J. *Liebigs Ann.* **1989**, 1081.
120. Bock, M. G.; DiPardo, R. M.; Evans, B. E.; Rittle, K. E.; Boger, J. S.; Freidinger, R. M.; Veber, D. F. *Chem. Commun.* **1985**, 109.
121. Damon, R. E.; Coppola, G. M. *Tetrahedron Lett.* **1990**, *31*, 2849.

David A. Evans & Annette S. Kim
Harvard University, Cambridge, MA, USA

N-Benzyloxycarbonyl-L-serine β-Lactone

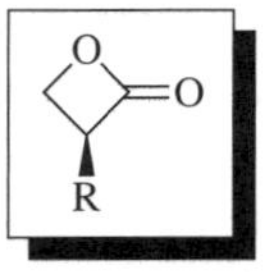

(R = NH-Cbz) (**1**)
[26054-60-4] $C_{11}H_{11}NO_4$ (MW 221.21)
(R = NH-Boc) (**2**)
[98541-64-1] $C_8H_{13}NO_4$ (MW 187.20)
(R = NH_3^+ OTs^-) (**3**)
[112839-95-9] $C_{10}H_{13}NO_5S$ (MW 259.28)

(reagent for the synthesis of β-substituted alanines[1,2])

Alternate Name: (*S*)-3-(*N*-benzyloxycarbonyl)aminooxetan-2-one.

Physical Data: (**1**): mp 133–134 °C, $[\alpha]_D^{25} = -26.8°$ ($c = 1$, MeCN); (**2**): mp 119.5–120.5 °C, $[\alpha]_D^{25} = -26.2°$ ($c = 1$, MeCN, 24 °C); (**3**): mp 133–135 °C (darkening), dec. 173 °C, $[\alpha]_D^{25} = -15.9°$ ($c = 2.2$, DMF).

Solubility: (**1**) and (**2**) sol most organic solvents; (**3**) sol polar aprotic organic solvents and H_2O.

Form Supplied in: white solids.

Preparative Methods: (**1**) is easily made by cyclization of commercially available *N*-Cbz-L-serine (**4**) under modified Mitsunobu conditions, using a preformed complex of dimethyl azodicarboxylate (DMAD) and *Triphenylphosphine* (eq 1).[3] The reaction proceeds via hydroxy group activation, and labeling studies show that the 3-hydroxy group is lost in a 4-*exo-tet* cyclization mechanism.[5] The β-lactone must be separated quickly from the reaction mixture,[3] and a slight excess of DMAD improves the yield because unreacted triphenylphosphine can cause polymerization.[6] The Boc (*t*-butoxycarbonyl) analog (**2**) is prepared similarly, and the *p*-toluenesulfonate (tosylate) salt (**3**) is synthesized from (**2**) by acidic cleavage.[4]

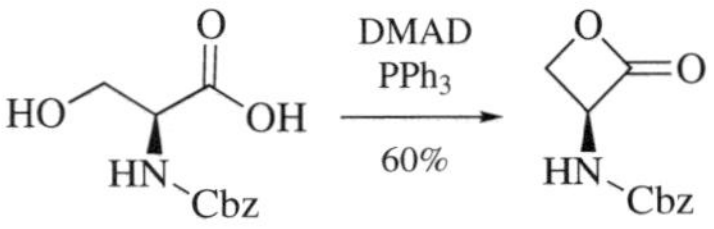

(**4**) (**1**) (1)

Purification: see Pansare et al.[3,4]

Handling, Storage, and Precautions: the β-lactones (**1**)–(**3**) are stable for many months at 4 °C in dry form. Neutral or slightly acidic solutions of (**1**) and (**2**) are stable for at least 1 day; (**3**) must be used in situ; basic and strongly acidic solutions rapidly decompose these β-lactones. They should be handled in a fume hood.

Ring-Opening Reactions. *N*-Cbz-β-lactone (**1**) is a very useful tool for the synthesis of optically active α-amino acids. Unlike other well established procedures,[2] this method does not generate the chiral center at the α-carbon but rather homologates optically pure serine derivatives at the β-carbon. Although this review is limited to the use of *N*-Cbz-L-serine β-lactone to generate L-amino acids in most applications, the corresponding D-serine β-lactones are available analogously from inexpensive D-serine derivatives. Ring-opening of β-lactones can occur in two different modes. 'Soft' nucleophiles (e.g. carboxylate, thiolate) usually attack the β-carbon, whereas 'hard' nucleophiles (e.g. hydroxide, methoxide, organolithium compounds) tend to target the carbonyl group. In certain cases, altering the conditions (e.g. N-substituent, solvent) directs the mode of addition. For example, ammonia in THF at 0 °C attacks the β-position of (**1**) to give the protected α,β-diaminopropanoic acid (**6**), whereas the same nucleophile in acetonitrile reacts with acyl oxygen cleavage to produce serine amide (**5**) (eq 2).[7] However, acetonitrile enhances β-attack by ammonia in the case of the *N*-Boc lactone (**2**).[8]

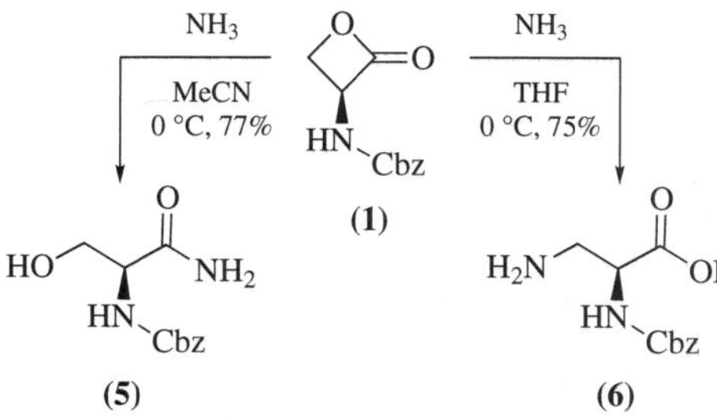

(2)

A variety of heteroatom nucleophiles have proven suitable for the desired ring-opening at the β-position (eq 3), such as $Nu^- = NH_3$, NMe_3, OAc^-, SBn^-, Cl^-, Br^-, pyrazole, and thiourea (S-attack).[7] Sulfur nucleophiles appear to require protic solvent for good results in this process.

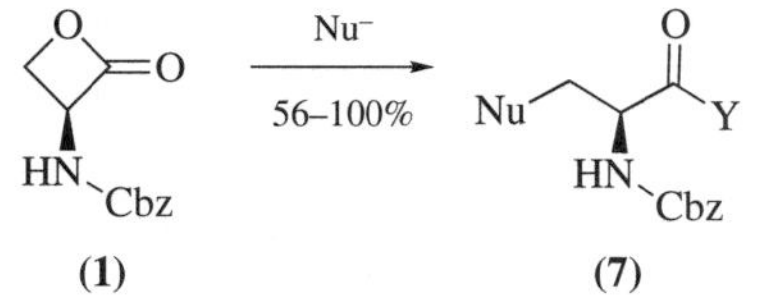

(3)

Nu^- = heteroatom nucleophile
Y = OH, O^-

Carbon–carbon bond formation can be achieved with copper-catalyzed Grignard reagents (eq 4), or less cleanly with organolithium-derived cuprate reagents (R_2CuLi or $R_2Cu(CN)Li_2$).[9] Thus Cbz-serine β-lactones behave like 'chiral enone equivalents', with '1,4-attack' of the carbanion at the β-position leading to amino acids and '1,2-attack' leading to ketones or even alcohols. 'Hard' organometallics, such as copper-free organolithium or Grignard reagents, react primarily at the carbonyl group, but the organocuprates add at the β-carbon to produce the desired amino acids (**8**) in good to moderate yields. The magnesium plays a key role in this reaction in terms of regioselectivity, optical purity (>99.4 % ee), and yield. The lithium cuprate reagents show less regioselectivity and in certain cases (e.g. R = Ph) can lead to some epimerization at the α-carbon. Temperature control (−23 °C is often ideal) is critical in reactions of (**1**) with organometallic reagents.

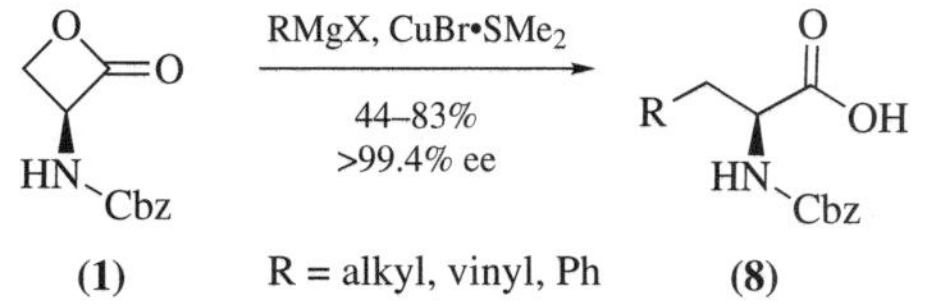

(4)

Acyl-oxygen cleavage of the serine β-lactone, although not desired in most cases, has been employed for the synthesis of (*S*)-2-methyl-4-benzyloxycarbonylaminopyrazolidine (**10**). Methylhydrazine in dichloromethane adds regioselectively to the carbonyl group of (**1**) to afford the hydrazide (**9**), which after several steps affords (**10**) (eq 5).[10]

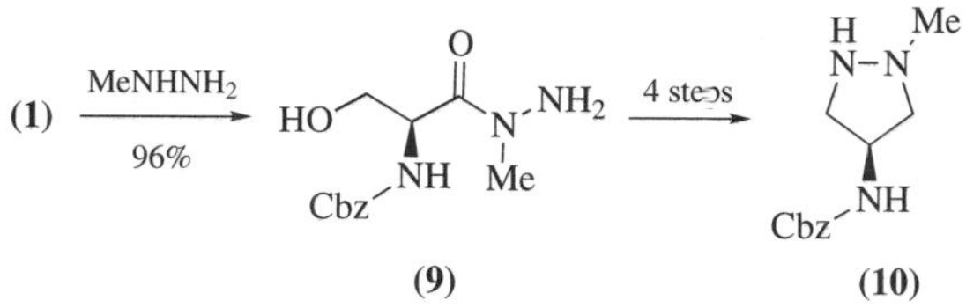

(5)

The anionic polymerization of *N*-benzyloxycarbonyl-L-serine β-lactone leads to poly(*N*-acyl-L-serine ester) (M_w = ca. 40 000), from which poly(L-serine ester hydrochloride) can be obtained by hydrogenation.[11]

Other 3-Aminooxetanones. Various other α-amino-β-lactones have been prepared and used in amino acid syntheses. The Boc derivative (**2**),[6–9,12–14] the tosylate salt (**3**),[4,15] and the recently published *N*-Fmoc-L-serine β-lactone[16] generally exhibit similar reactivity towards nucleophiles (e.g. phosphites[14,16]), and the choice of β-lactone can often be determined by requirements of subsequent synthetic steps. However, some differences have been observed. In addition to the ammonia reaction described above,[7,8] the condensation of β-mercaptoethylamine (**11**) with (**1**) results in nucleophilic attack by nitrogen leading to (**12**), whereas the same nucleophile attacks the tosylate salt (**3**) with reverse chemoselectivity giving the corresponding amino acid (**13**) (eq 6).[7,15] The salt (**3**) has the advantage of producing free amino acids in cases where deprotection of nitrogen may affect the β-substituent.

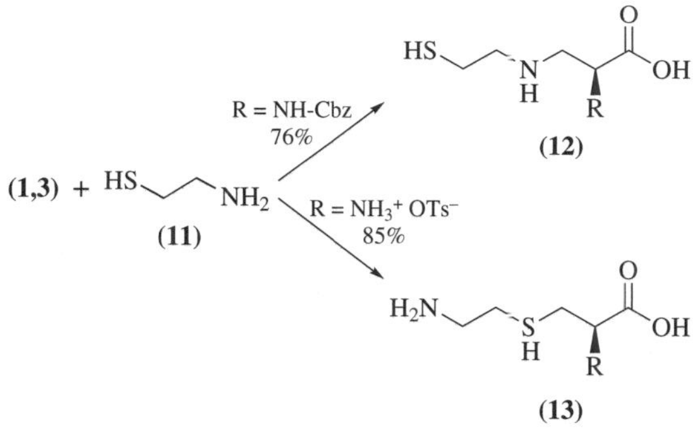

(6)

The L-threonine β-lactone (**14**), in contrast to the serine β-lactones, has been synthesized by carboxyl group activation (4-*exo-trig*).[17] Initial experiments have shown that ring opening of such β-substituted lactones tends to proceed by attack at the carbonyl except with certain nucleophiles (e.g. thiourea, halides). However, correct choice of N-protecting group allows the synthesis of a range of other β-substituted α-amino-β-lactones,[17b,c] such as the antiobiotic obafluorin (**15**).[18]

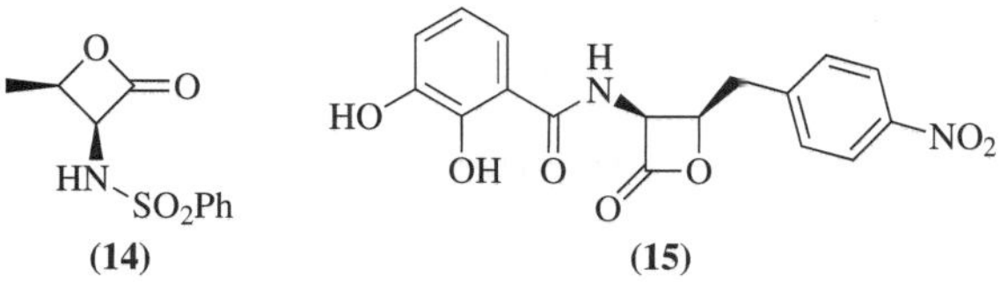

In summary, ring opening of serine-β-lactones is an attractive method for generating optically pure β-substituted alanines; the synthesis usually occurs with little or no epimerization.

Related Reagents. Diketene; β-Ethynyl-β-propiolactone; β-Methyl-β-propiolactone; β-Propiolactone.

1. For a review on β-lactone chemistry, see: Pommier, A.; Pons, J.-M. *Acta Chem. Scand.* **1993**, 441.
2. For a review on amino acid synthesis, see: Williams, R. M. *Synthesis of Optically Active α-Amino Acids*; Pergamon: Oxford, 1989.
3. Pansare, S. V.; Huyer, G.; Arnold, L. D.; Vederas, J. C. *Comprehensive Organic Synthesis* **1991**, *70*, 1.
4. Pansare, S. V.; Arnold, L. D.; Vederas, J. C. *Comprehensive Organic Synthesis* **1991**, *70*, 10.
5. Ramer, S. E.; Moore, R. N.; Vederas, J. C. *Can. J. Chem.* **1986**, *64*, 706.
6. Lodwig, S. N.; Unkefer, C. J. *J. Labelled Compds. Radiopharm.* **1991**, *31*, 95.
7. Arnold, L. D.; Kalantar, T. H.; Vederas, J. C. *J. Am. Chem. Soc.* **1985**, *107*, 7105. Ratemi, E. S.; Vederas, J. C. *Tetrahedron Lett.* **1994**, *35*, 7605.
8. Kucharczyk, N.; Badet, B.; Le Goffic, F. *Org. Synth., Coll. Vol.* **1989**, *19*, 1603.
9. Arnold, L. D.; Drover, J. C. G.; Vederas, J. C. *J. Am. Chem. Soc.* **1987**, *109*, 4649.
10. Kim, K. S.; Ryan, P. C. *Heterocycles* **1990**, *31*, 79.
11. (a) Zhou, Q.-X.; Kohn, J. *Macromol.* **1990**, *23*, 3399. (b) Gelbin, M. E.; Kohn, J. *J. Am. Chem. Soc.* **1992**, *114*, 3962.
12. Soucy, F.; Wernic, D.; Beaulieu, P. *J. Chem. Soc., Perkin Trans. 1* **1991**, 2885.
13. Rosenberg, S. H.; Spina, K. P.; Woods, K. W.; Polakowski, J.; Martin, D. L.; Yao, Z.; Stein, H. H.; Cohen, J.; Barlow, J. L.; Egan, D. A.; Tricarico, K. A.; Baker, W. R.; Kleinert, H. D. *J. Med. Chem.* **1993**, *36*, 449.
14. Smith, E. C. R.; McQuaid, L. A.; Paschal, J. W.; DeHoniesto, J. *J. Org. Chem.* **1990**, *55*, 4472.
15. Arnold, L. D.; May, R. G.; Vederas, J. C. *J. Am. Chem. Soc.* **1988**, *110*, 2237.
16. Hutchinson, J. P. E.; Parkes, K. E. B. *Tetrahedron Lett.* **1992**, *33*, 7065.
17. (a) Pansare, S. V.; Vederas, J. C. *J. Org. Chem.* **1989**, *54*, 2311; (b) Pu, Y.; Martin, F. M.; Vederas, J. C. *J. Org. Chem.* **1991**, *56*, 1280. (c) Rao, M. N.; Holkar, A. G.; Ayyangar, N. R. *Chem. Commun.* **1991**, 1007.
18. Lowe, C.; Pu, Y.; Vederas, J. C. *J. Org. Chem.* **1992**, *57*, 10. Pu, Y.; Lowe, C.; Sailer, M.; Vederas, J. C. *J. Org. Chem.* **1994**, *59*, 3642.

Michael Klinge & John C. Vederas
University of Alberta, Edmonton, AB, Canada

2-[2-[(Benzyloxy)ethyl]-6,6-dimethylbicyclo[3.3.1]-3-nonyl]-9-borabicyclo-[3.3.1]nonane

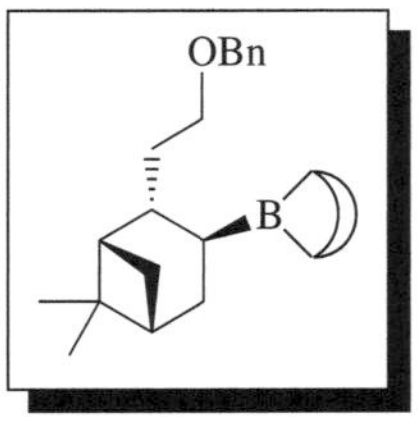

[81971-15-5] $C_{26}H_{39}B$ (MW 378.41)

(asymmetric reducing agent[1])

Alternate Name: NB-Enantrane®.
Solubility: sol most organic solvents.
Form Supplied in: 0.5M solution in THF; commercially available.
Preparative Methods: NB-Enantrane[2] may be prepared by hydroboration of nopol benzyl ether (commercially available, or from nopol and benzyl bromide[1a]) with *9-Borabicyclo[3.3.1]nonane* (9-BBN) (eq 1).

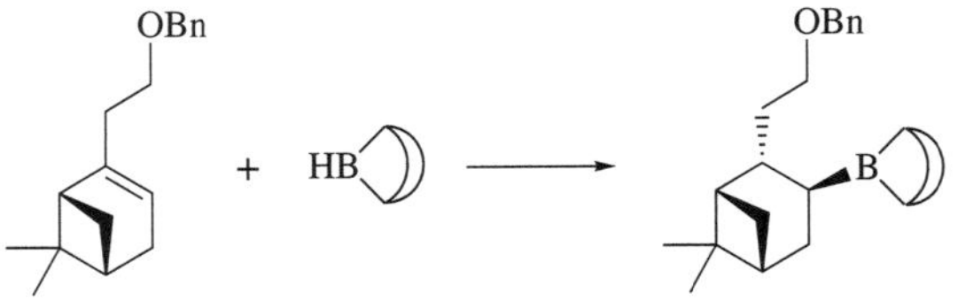

(1)

Nopol benzyl ether 9-BBN NB-Enantrane

Handling, Storage, and Precautions: NB-Enantrane is an air-sensitive reagent and must be handled under an inert atmosphere. Use in a fume hood.

Asymmetric Reductions. NB-Enantrane is an analog of *B-3-Pinanyl-9-borabicyclo[3.3.1]nonane* (Alpine-Borane[1b]). Since the absolute configuration of nopol is the opposite of (+)-α-pinene, the reagent provides the opposite mode of asymmetric induction. The reagent provides higher asymmetric induction than Alpine-Borane for cases where the two groups flanking the ketone are relatively small, such as methyl or ethyl alkynyl ketones (eq 2).

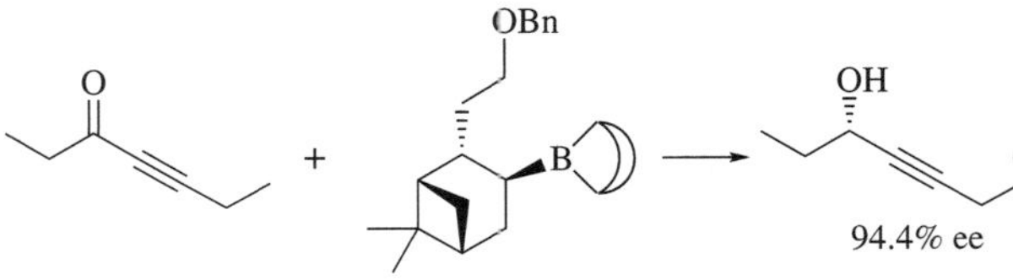

(2)

94.4% ee

Reduction occurs via transfer of the tertiary hydride, which is β to the boron. Since this is the hydride added during the hydroboration process, labeled material may be obtained by using labeled 9-BBN in the initial preparation. The absolute configuration of the product may be predicted by using a simple six-membered ring model (**1a**) for the transition state. The larger group is placed away from the reagent in this arrangement.

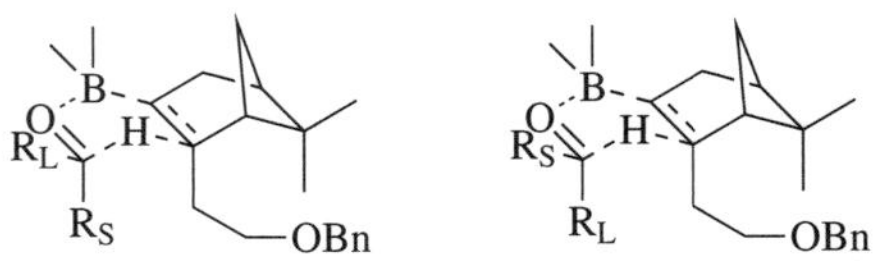

(**1a**) favored (**1b**) disfavored

The reagent presumably leads to higher asymmetric induction because of the increased size of the nopol side chain. Brown has shown that replacing the vinylic methyl group of α-pinene with larger groups also leads to higher asymmetric induction (eq 3).[3]

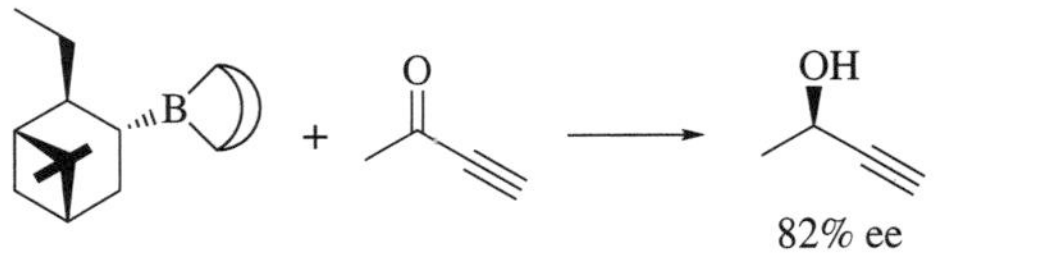

(3)

82% ee

Borohydride Reagent. Treatment of NB-enantrane with *t-Butyllithium* provides the lithium trialkylborohydride NB-Enantride® (eq 4).[4] This reagent is an effective asymmetric reducing agent for acetophenone and alkyl methyl ketones such as 2-octanone (eq 5). Few reagents show selectivity for such alkyl ketones.

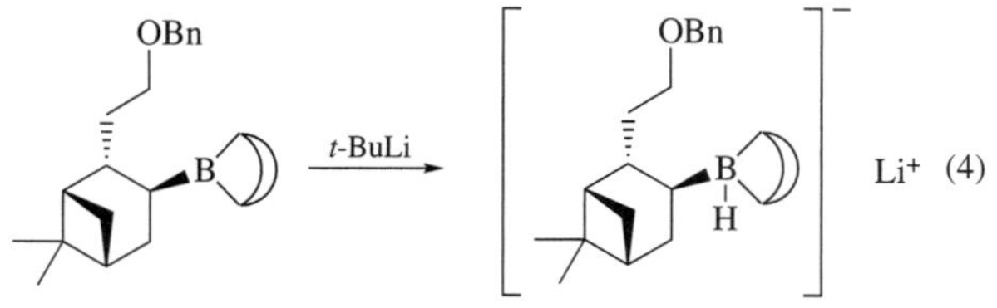

NB-Enantride

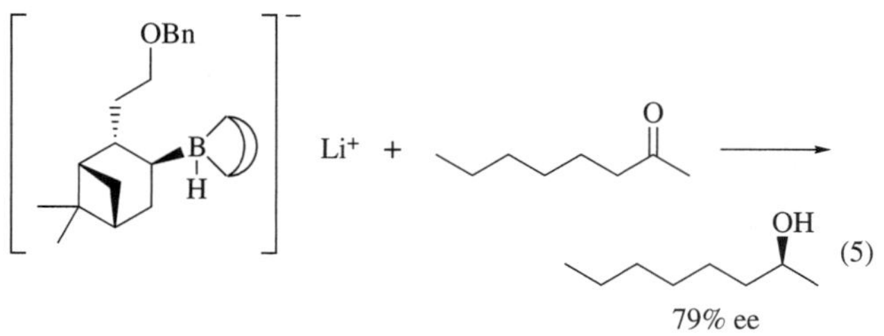

Related Reagents. (+)-B-Chlorodiisopinocampheylborane B-3-Pinanyl-9-borabicyclo[3.3.1]nonane

1. (a) Midland, M. M.; Kazubski, A. *J. Org. Chem.* **1982**, *47*, 2814. (b) Midland, M. M. *Chem. Rev.* **1989**, *89*, 1553.
2. NB-Enantrane is a trademark of Aldrich Chemical Company.
3. Brown, H. C.; Ramachandran, P. V.; Weissman, S. A.; Swaminathan, S. J. *J. Org. Chem.* **1990**, *55*, 6328.
4. Midland, M. M.; Kazubski, A.; Woodling, R. E. *J. Org. Chem.* **1991**, *56*, 1068.

M. Mark Midland
University of California, Riverside, CA, USA

(*R*,*R*)-1-(2′-Benzyloxymethylphenyl)-2,5-dimethylphospholane[1]

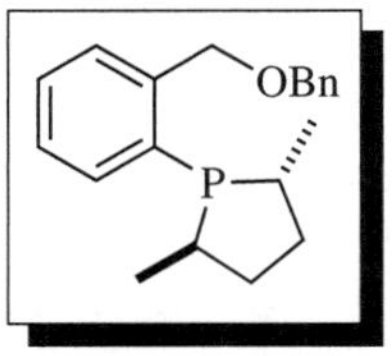

[252554-72-6] $C_{20}H_{25}OP$ (MW 312.39)

(monophosphine ligand for asymmetric catalysis;[1] nickel(II) complexes are effective catalyst precursors for enantioselective hydrovinylation with non-coordinating counterions)

Solubility: soluble in THF, diethyl ether, hydrocarbons, and chlorinated solvents.
Analysis of Reagent Purity: ^{1}H NMR, ^{13}C NMR, ^{31}P NMR, and GLC analysis.
Preparative Methods: prepared by the reaction of potassium dianion of 2′-(benzyloxymethyl)phenylphosphane[1] and (2*S*,5*S*)-hexane-2,5-diol cyclic sulfate. The preparation of the latter reagent is described in the literature.[2]

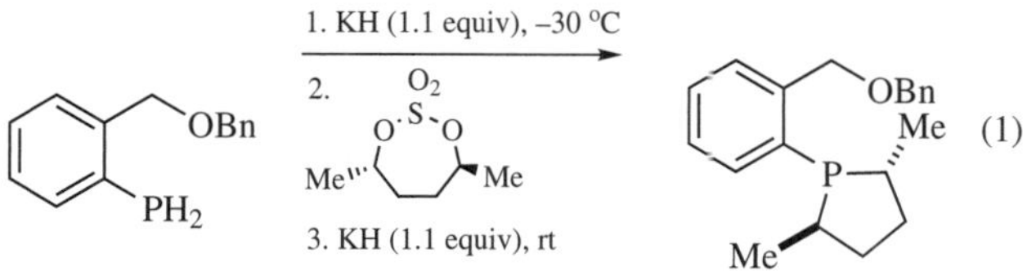

Purification: by column chromatography under an inert atmosphere.
Handling, Storage, and Precautions: readily oxidizes to the phospholane oxide upon exposure to atmospheric oxygen. The corresponding (allyl)Ni(BARF) complex is stable upon storage at room temperature for several days under a nitrogen atmosphere.

Asymmetric Hydrovinylation. Heterodimerization of olefins is a reaction of enormous synthetic potential, since it has been demonstrated that excellent yields and selectivities can be achieved under exceptionally mild conditions in many cases (eq 2, Table 1).[3,4] In eq 2, the catalytically active species is thought to be $[Ni–H]^+$ generated via β-hydride elimination from a nickel precursor. In this reaction, the steric environment provided by ligand as well as the appropriate positioning of possible hemilabile coordinating group (OR) in the ligand are critical factors in yield, regioselectivity, and enantioselectivity.

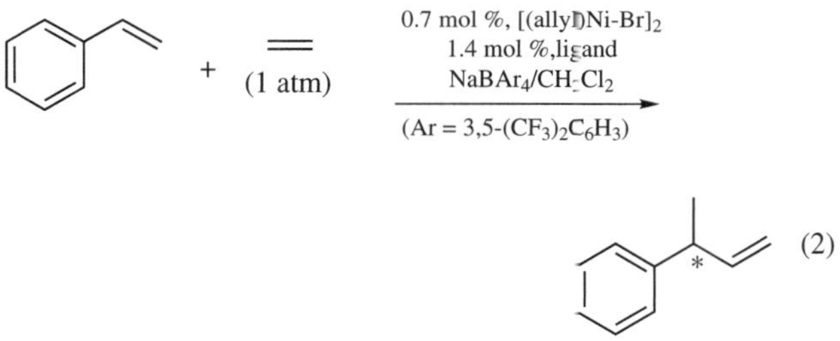

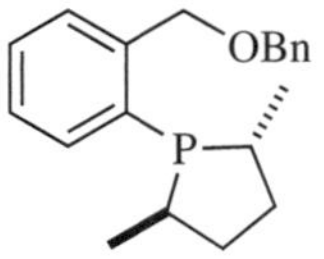

Table 1 Heterodimerization of olefins

Ligand	Conditions	Yield (%)	% ee
1a	−40°C/2.5 h	97	50 (*S*)
1b	−30°C/2 h	>99	63 (*R*)

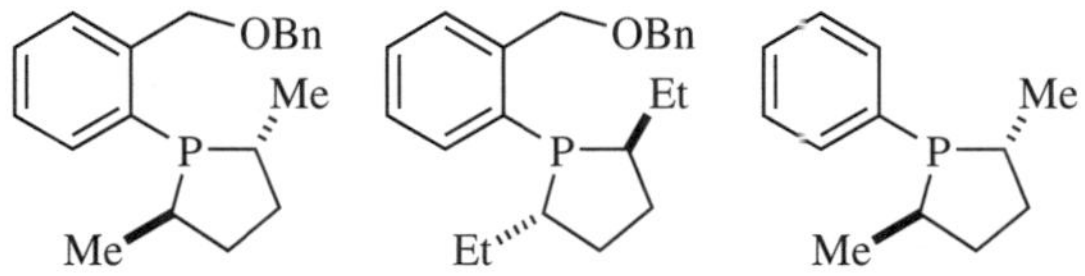

1a **1b** **2**

Effect of Counterions on the Hydrovinylation. In the hydrovinylation of styrene using a ligand with a hemilabile atom (e.g. OBn in **1a**, **1b**), the choice of appropriate counter anion is very important. For example, when **2** and a weakly coordinating anion such as AgOTf is used in the reaction of styrene, the yield of hydrovinylation product is 94%. However, with phospholane ligands having a hemilabile coordinating group (-OR), the externally added coordinating anion (AgOTf) suppresses the reaction, while a non-coordinating anion ($NaBAr_4$, Ar = 3,5-$(CF_3)_2C_6H_3$) completely restores the activity and selectivity (Table 2).[1]

Table 2 Effect of counterions on the hydrovinylation of styrene using 'hemilabile' ligand

	Yield of product (% ee)	
Additive	**1a**	**2**
AgOTf	<4	94 (37, *S*)
$AgClO_4$	<2	95
$AgNTf_2$	48 (47, *S*)	<2
$AgSbF_6$	94 (48, *S*)	<2
$NaBAr_4$	97 (50, *S*)	<2

1. Nandi, M.; Jin, J.; RajanBabu, T. V. *JACS* **1999**, *121*, 9899.
2. Burk, M. J. *JACS* **1991**, *113*, 8518.
3. (a) Jolly, P. W.; Wilke, G. In *Applied Homogeneous Catalysis with Organometallic Compounds*, Cornils, B. and Herrmann, W. A., Eds.; VCH: New York, 1996; p. 1024-1048. (b) RajanBabu, T. V.; Nomura, N.; Jin, J.; Radetich, B.; Park, H.; Nandi, M. *Chem. Eu. J.* **1999**, *5*, 1963.
4. Nomura, N.; Jin, J.; Park, H.; RajanBabu, T. V. *JACS* **1998**, *120*, 459.

Seunghoon Shin & T. V. RajanBabu
The Ohio State University, Columbus, Ohio, USA

N-Benzylquininium Chloride

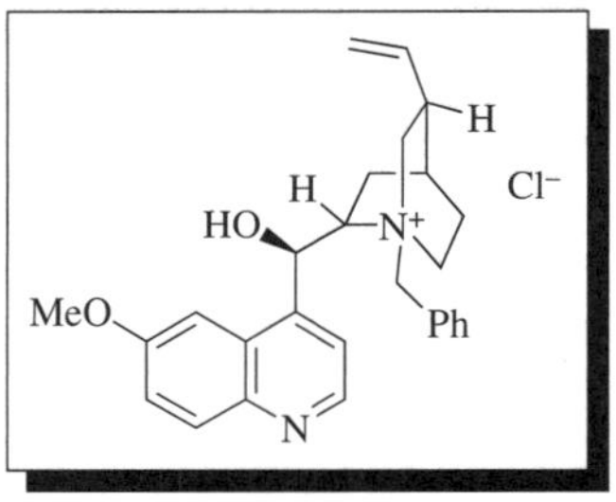

[67174-25-8] $C_{27}H_{31}ClN_2O_2$ (MW 451.01)

(chiral, nonracemic quaternary ammonium salt; catalyst for a variety of phase transfer reactions under basic conditions[1])

Alternate Name: Quibec; BQC.
Physical Data: mp 180–181 °C, $[\alpha]_D^{19}$ −235.5° (*c* = 1.5, H_2O);[2] monohydrate: mp 169–172 °C (dec), $[\alpha]_D$ −212.5° (*c* = 0.5, EtOH).[3]
Solubility: freely sol H_2O, alcohols, acetone; slightly sol EtOAc; sparingly sol $CHCl_3$.[4]
Form Supplied in: solid.
Purification: recrystallized from absolute EtOH.[3]

The combined use of catalytic amounts of *N*-benzylquininium chloride (**1**) with hydroxide bases (NaOH or KOH) has been explored for a variety of phase transfer reactions, including epoxidations, alkylations, and Michael reactions. The degree of stereoselectivity in product formation induced by the reagent can vary widely.[5]

BQC is derived from quinine, which is a member of the *cinchona* family of alkaloids. Ammonium salts derived from quinidine, a diastereomer of (**1**) at the hydroxyl substituent, have been used less frequently in catalysis than BQC. Quinidinium salts often give rise to products with enantioselectivity opposite to that from (**1**). Other related compounds, such as those derived from cinchonine and cinchonidine (which lack the methoxy substituent on the quinoline nucleus), have found application in organic synthesis. The *cinchona* alkaloids, as well as salt derivatives in which the benzyl group bears various substituents, have also been studied.[6] Results from polymer-bound catalysts have not been promising.[7]

Benzylquininium chloride has shown good to excellent selectivity in the epoxidation of α,β-unsaturated ketones.[8] Oxidation of quinone (**2**) in the presence of (**1**) with aqueous *t-Butyl Hydroperoxide* and *Sodium Hydroxide* in toluene gave rise to a 95% chemical yield of epoxide (**3**) in 78% ee (eq 1).[2] Recrystallization improved the ee to 100% with 63% mass recovery. Aqueous *Hydrogen Peroxide* decreased both the yield (89%) and enantioselectivity (50% ee).

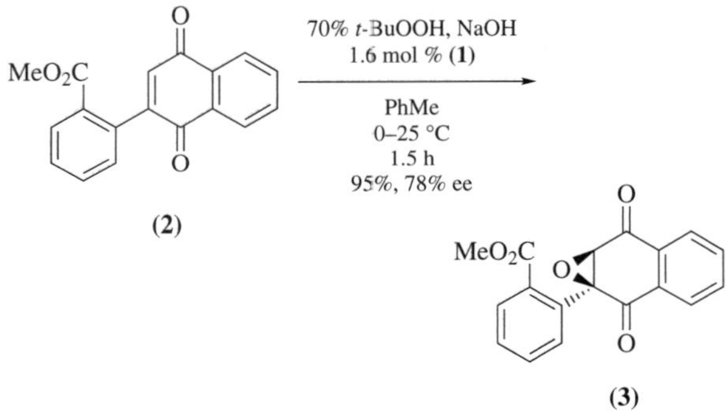

(1)

Enantiomerically enriched epoxides have also been generated using (**1**) in the Darzens reaction of α-chloro ketones with aldehydes, as well as through ring closure of racemic halohydrins.[9] The extent of enantioselectivity for both reactions is similar (optical purity 6–8% ±1), suggesting a moderate degree of kinetic resolution occurring in each case.

Benzylquininium chloride has been studied as a catalyst for the asymmetric Michael reaction. Reaction of amidomalonate (**5**) and chalcone (**4**) with catalytic base and a variety of chiral, nonracemic ammonium salts in the absence of solvent produced (**6**) in yields of 41–68% and 20–68% ee (eq 2).[10] The quinine-derived salt (**1**) was of intermediate effectiveness (38% ee, 47% yield) when compared to ephedrine-based catalysts. Although (**1**) was not specifically tested with regard to solvation effects, it is suggested that increased aggregation of reactive species under solid–liquid PTC conditions leads to enhanced organization and selectivity. Low levels of induction have been observed in other systems.[11] A study comparing various chiral alkaloid salts, bases, and reaction conditions on enantioselectivities in the conjugate addition of thiols and nitroalkanes to enones has been reported.[12] Remarkable results have also been obtained in the cinchonium/cinchonidinium-catalyzed reaction of an indanone with methyl vinyl ketone.[13] A direct comparison with a more efficient lanthanum–binaphthol complex has been reported.[14]

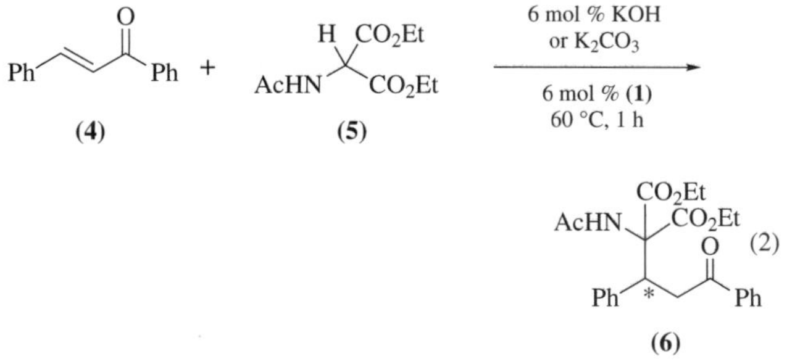

Phase-transfer alkylations have been studied using *cinchona* alkaloid derivatives; however, results have been more promising with the cinchonium/cinchonidinium series than with the quininium/quinidinium group.[15] For example, catalytic asymmetric alkylation in high yield and selectivity has been achieved with *N*-(*p*-trifluoromethylphenyl)cinchonium bromide (*p*-CF_3BCNB) (eq 3).[16] A systematic investigation of reaction variables (catalyst type, solvent, concentration, temperature, stirring rate, leaving group, etc.) has produced a general method to prepare α-amino acids in 44–62% ee.[17]

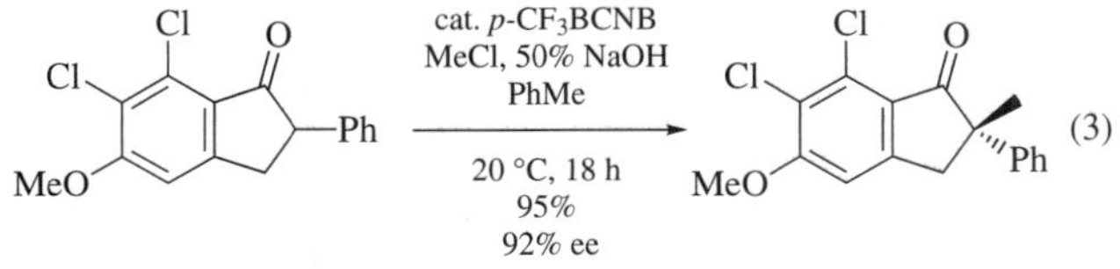

1. For reviews of phase transfer reactions, see for example (a) Keller, W. E. *Phase Transfer Reactions. Fluka Compendium*; Thieme: Stuttgart, 1986; Vols. 1, 2. (b) Dehmlow, E. V. *Phase Transfer Catalysis*; Verlag Chemie: Deerfield Beach, 1980. (c) Starks, C. M.; Liotta, C. *Phase Transfer Catalysis, Principles and Techniques*; Academic Press: New York, 1978. (d) Dockx, J. *Acta Chem. Scand.* **1973**, 441. (e) Dehmlow, E. V. *Angew. Chem., Int. Ed. Engl.* **1974**, *13*, 170. For a mechanistic review of hydroxide mediated reactions under PTC conditions, see: (f) Rabinovitz, M.; Cohen, Y.; Halpern, M. *Angew. Chem., Int. Ed. Engl.* **1986**, *25*, 960.
2. Harigaya, Y.; Yamaguchi, H.; Onda, M. *Chem. Pharm. Bull.* **1981**, *29*, 1321.
3. Colonna, S.; Fornasier, R. *J. Chem. Soc., Perkin Trans. 1* **1978**, 371.
4. Jacobs, W. A.; Heidelberger, M. *J. Am. Chem. Soc.* **1919**, *41*, 2090.
5. Some doubts have been expressed with regard to the accuracy of reported results: Dehmlow, E. V.; Singh, P.; Heider, J. *J. Chem. Res. (S)* **1981**, 292.
6. For example (a) Sera, A.; Takagi, K.; Katayama, H.; Yamada, H.; Matsumoto, K. *J. Org. Chem.* **1988**, *53*, 1157. (b) Wynberg, H.; Helder, R. *Tetrahedron Lett.* **1975**, 4057. (c) Helder, R.; Arends, R.; Bolt, W.; Hiemstra, H.; Wynberg, H. *Tetrahedron Lett.* **1977**, 2181.
7. Kelly, J.; Sherrington, D. C. *Polymer* **1984**, *25*, 1499.
8. (a) Helder, R.; Hummelen, J. C.; Laane, R. W. P. M.; Wiering, J. S.; Wynberg, H. *Tetrahedron Lett.* **1976**, 1831. (b) Wynberg, H.; Marsman, B. *J. Org. Chem.* **1980**, *45*, 158.
9. Hummelen, J. C.; Wynberg, H. *Tetrahedron Lett.* **1978**, 1089.
10. Loupy, A.; Sansoulet, J.; Zaparucha, A.; Merienne, C. *Tetrahedron Lett.* **1989**, *30*, 333.
11. For example: (a) Jianguo, C.; Lingchong, Y. *Org. Synth., Coll. Vol.* **1990**, *20*, 2895. (b) Brunner, H.; Zintl, H. *J. Organomet. Chem.* **1991**, *122*, 841. (c) Annunziata, R.; Cinquini, M.; Colonna, S. *J. Chem. Soc., Perkin Trans. 1* **1980**, 2422.
12. Colonna, S; Re, A.; Wynberg, H. *J. Chem. Soc. Perkin Trans. 1* **1981**, 547.
13. Conn, R. S. E.; Lovell, A. V.; Karady, S.; Weinstock, L. M. *J. Org. Chem.* **1986**, *51*, 4710.
14. Sasai, H.; Itoh, N.; Suzuki, T.; Shibasaki, M. *Tetrahedron Lett.* **1993**, *34*, 855.
15. (a) Julia, S.; Ginebreda, A.; Guixer, J.; Tomas, A. *Tetrahedron Lett.* **1980**, *21*, 3709. (b) Julia, S.; Ginebreda, A.; Guixer, J.; Masana, J.; Tomas, A.; Colonna, S. *J. Chem. Soc., Perkin Trans. 1* **1981**, 574.
16. (a) Hughes, D. L.; Dolling, U. H.; Ryan, K. M.; Schoenewaldt, E. F.; Grabowski, E. J. J. *J. Org. Chem.* **1987**, *52*, 4745. (b) Bhattacharya, A.; Dolling, U. H.; Grabowski, E. J. J.; Karady, S.; Ryan, K. M.; Weinstock, L. M. *Angew. Chem., Int. Ed. Engl.* **1986**, *25*, 476. (c) Dolling, U. H.; Davis, P.; Grabowski, E. J. J. *J. Am. Chem. Soc.* **1984**, *106*, 446.
17. The ee's can often be improved by recrystallization. O'Donnell, M. J.; Bennett, W. D.; Wu, S. *J. Am. Chem. Soc.* **1989** *111*, 2353.

Mary Ellen Bos
R. W. Johnson Pharmaceutical Research Institute, Raritan, NJ, USA

(*S*)-4-Benzyl-2,2,5,5-tetramethyl-oxazolidine

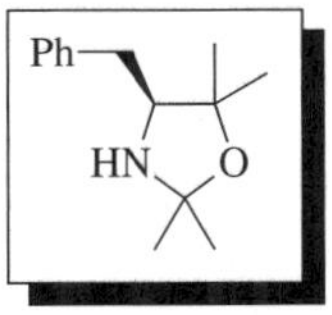

[144899-42-3] $C_{14}H_{21}NO$ (MW 219.33)

(chirality-controlling auxiliary[1])

Physical Data: colorless liquid; IR,[2] ^{1}H NMR,[1] and ^{13}C NMR[2] spectra are available.

Solubility: very sol in most organic solvents; decomposition occurs in protic solvents.

Handling, Storage, and Precautions: decomposes on silica gel column chromatography; recommended not to be stored for a long time.

Introduction. (*S*)-4-Benzyl-2,2,5,5-tetramethyloxazolidine is used as a chirality-controlling auxiliary. Its amide derivatives preferentially occupy the *syn* conformation[1] so that one of the π-facial reaction sites of the amide moiety becomes sterically less hindered. This chiral auxiliary will be especially useful for the asymmetric reactions which have to be performed in the absence of metallic additives.

Preparation and Stability. Since *N*-unsubstituted oxazolidines are labile to hydrolysis, they should be transformed immediately after preparation to amide derivatives, which show much higher stability. Thus reaction of (*S*)-3-amino-2-methyl-4-phenyl-2-butanol[3] with acetone in the presence of a catalytic amount of *p-Toluenesulfonic Acid* produces (*S*)-4-benzyl-2,2,5,5-tetramethyloxazolidine. *N*-Acylation using acryloyl, cinnamoyl, and propanoyl chloride (Et_3N, −78 °C) gives the corresponding amides. The *N*-crotonoyl derivative is better obtained by the crotonoylation of (*S*)-3-amino-2-methyl-4-phenyl-2-butanol followed by acetalization with $(Me)_2C(OMe)_2$.

Unsaturated Amides. Cycloaddition of (*S*)-3-acryloyl-4-benzyl-2,2,5,5-tetramethyloxazolidine with *Benzonitrile Oxide* proceeds smoothly at 0 °C to provide the corresponding oxazolidine in a 93:7 diastereomer ratio (eq 1),[4] which is quantitatively reduced with *Lithium Triethylborohydride* to give the isoxazoline-5-methanol derivative without loss of enantiomeric purity. A single diastereomer of isoxazoline can be obtained when 3-acryloyl-2,2-dimethyl-4-diphenylmethyloxazolidine is employed.[4]

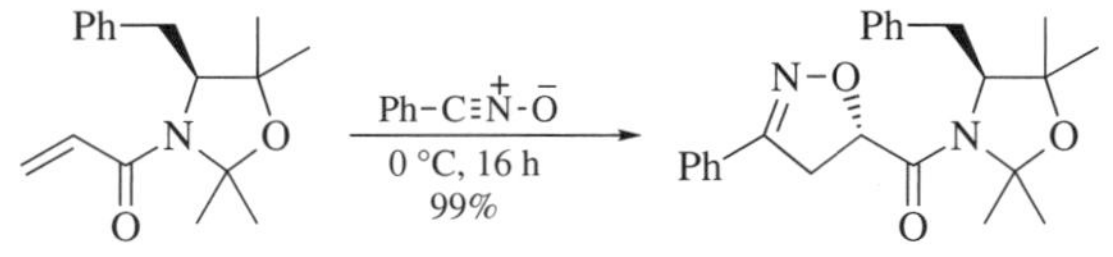

(1)

93:7

Conjugate additions of organocuprates to (*S*)-4-benzyl-3-crotonoyl-2,2,5,5-tetramethyloxazolidine in the presence of *Chlorotrimethylsilane* (1.2 equiv) also proceed in a highly diastereoselective manner (eq 2) to give, after the acid-catalyzed hydrolytic removal of the chiral auxiliary, optically pure carboxylic acids with β-chirality.[5]

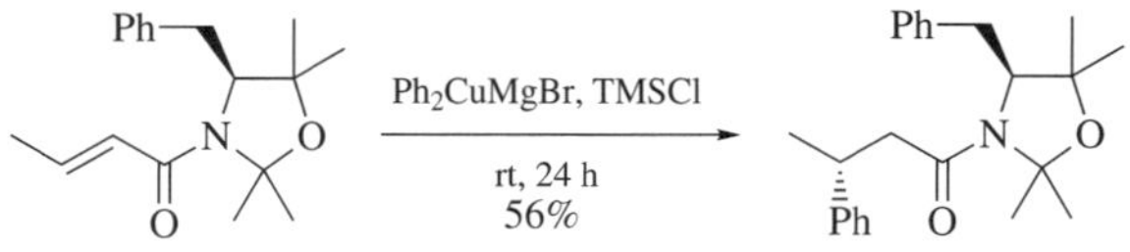

(2)

>99:1

Amide Enolates. The lithium (*Z*)-enolate can be generated from (*S*)-4-benzyl-3-propanoyl-2,2,5,5-tetramethyloxazolidine and *Lithium Diisopropylamide* in THF at −78 °C. Its alkylations[6] take place smoothly in the presence of *Hexamethylphosphoric Triamide* with high diastereoselectivity (eq 3), and its Michael additions[7] to α,β-unsaturated carbonyl compounds are also exclusively diastereoselective (eq 4). Synthetic applications have been made in the aldol reactions of the titanium (*Z*)-enolates of α-(alkylideneamino) esters.[8]

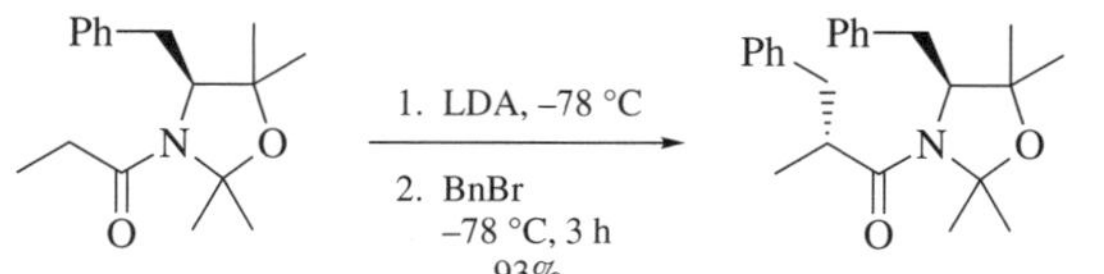

(3)

97:3

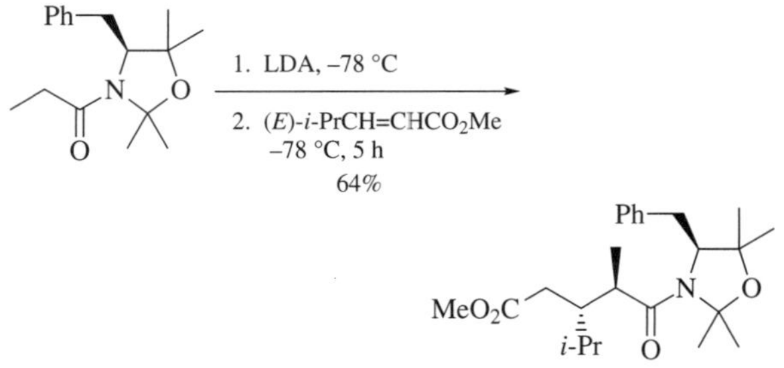

(4)

>99:1

1. Kanemasa, S.; Onimura, K. *Tetrahedron* **1992**, *48*, 8631.
2. IR (neat) 3350, 2950, 1410, 1360, 1100, and 800 cm^{-1}; ^{13}C NMR ($CDCl_3$) δ = 23.21, 27.37, 28.52, 29.60, 35.95, 67.04, 80.66, 92.75, 128.48, 128.62, 128.71, and 139.00.
3. (*S*)-3-Amino-2-methyl-4-phenyl-2-butanol is available from (*S*)-phenylalanine and MeMgI (4 equiv): diethyl ether, rt, 5 h, 58%.
4. Kanemasa, S.; Onimura, K. *Tetrahedron* **1992**, *48*, 8645.
5. Kanemasa, S.; Suenaga, H.; Onimura, K. *J. Org. Chem.* **1994**, in press.
6. Kanemasa, S.; Ueno, K.; Kikkawa, T.; Onimura, K. Unpublished results.
7. Nomura, M.; Kanemasa, S.; Yoshinaga, S. Unpublished results.
8. Kanemasa, S.; Mori, T.; Tatsukawa, A. *Tetrahedron Lett.* **1993**, *34*, 8293.

Shuji Kanemasa
Kyushu University, Kasuga, Japan

(Bicyclo[2.2.1]hepta-2,5-diene)[(2*S*,3*S*)-bis(diphenylphosphino)butane]rhodium Perchlorate[1]

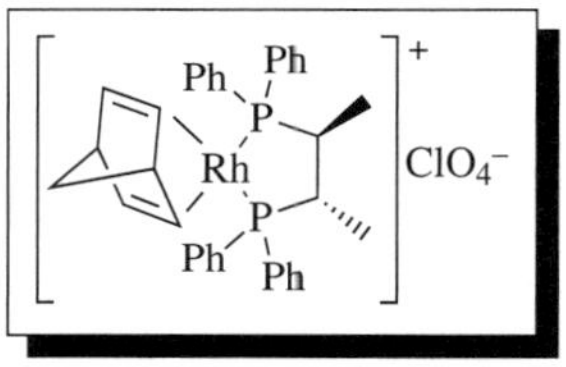

(ClO_4^-)
[65012-74-0] $C_{35}H_{36}ClO_4P_2Rh$ (MW 720.97)
(BF_4^-)
[79790-89-9] $C_{35}H_{36}BF_4P_2Rh$ (MW 708.33)
(PF_6^-)
[99340-29-1] $C_{35}H_{36}F_6P_3Rh$ (MW 766.49)

(catalyst precursor used in the asymmetric hydrogenation of amino acid precursors, as well as in transfer hydrogenation and hydrosilation reactions[2a,b])

Alternate Name: (bicyclo[2.2.1]hepta-2,5-diene)(chiraphos)-rhodium perchlorate; (norbornadiene)(chiraphos)rhodium perchlorate.

Physical Data: $^{31}P\{^1H\}$ NMR (CD_3OD) δ 58.4 ppm, J_{Rh-P} = 154 Hz.

Solubility: sol CH_2Cl_2, MeOH, THF/MeOH, and H_2O/MeOH mixtures; sparingly sol THF; insol Et_2O.

Form Supplied in: cationic rhodium–chiraphos complexes are usually prepared in situ or from the procedure given below. In reactions employing the various rhodium–chiraphos complexes, the diene ligand is frequently hydrogenated prior to catalysis to afford the active solvated catalyst precursors $[RhS_2(chiraphos)]^+$.

Preparative Methods: Bis(bicyclo[2.2.1]hepta-2,5-diene)-rhodium Perchlorate (0.290 g) and (*S,S*)-chiraphos[1,3] (0.308 g) (see *2,3-Bis(diphenylphosphino)butane*) are dissolved in methylene chloride (5 mL) and THF (5 mL) under nitrogen. Hexane (6 mL) is then added and, after the mixture is allowed to stand at 25 °C for 1 h and then for 2 h at 5 °C, orange-red crystals of [Rh(nbd)(*S,S*-chiraphos)]ClO_4 are collected (0.43 g).

Handling, Storage, and Precautions: cationic rhodium–chiraphos complexes are air-sensitive and should be handled and stored under an inert atmosphere.

Hydrogenation. Rhodium(I) complexes containing chiral phosphine ligands are extremely useful in the preparation of enantiometrically pure amino acids.[2,4] Asymmetric hydrogenation of α-(*N*-acylamino)acrylic acids proceeds in high chemical yields (95–100%) with essentially complete optical purity (eq 1). For example, DOPA, alanine, and tyrosine are obtained from asymmetric hydrogenations catalyzed by rhodium(I)–chiraphos cations with optical yields of 83, 91, and 92% respectively.[2,4,5] Catalyzed hydrogenations of the analogous aliphatic compounds are somewhat less enantioselective.[6] Indeed, hydrogenation of the (*E*) isomers gives products with even lower optical yields (eq 2).

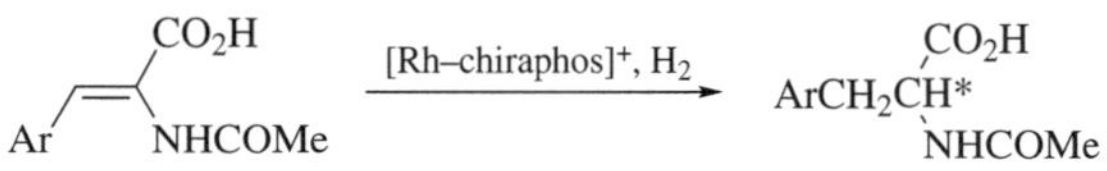

(1)

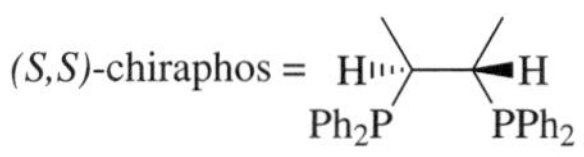

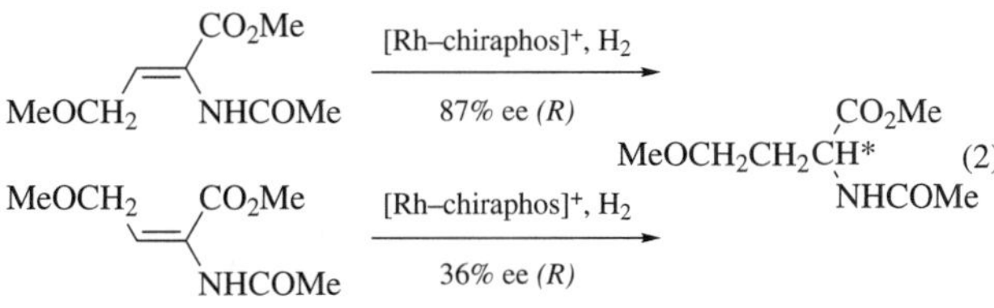

(2)

The active herbicidal agent L-phosphinothricin[7] is obtained with enantiomeric excesses up to 91% from the asymmetric hydrogenation of α-acylamidoacrylate precursors (eq 3). While yields of the primary hydrogenation products are quantitative, enantiomeric excesses increase slightly in reactions run in a H_2O/MeOH mixture. Adverse effects are observed in reactions carried out in THF/MeOH.

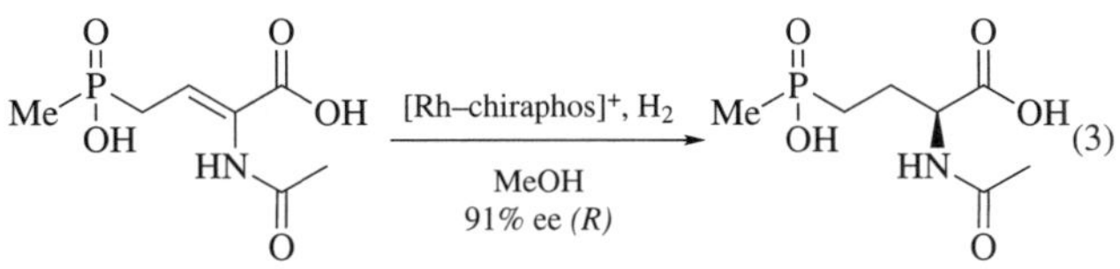

(3)

Reactions with 1,1,1-trifluoro-2-(acetyloxy)-2-propene provide the first example of a vinyl acetate with a fully saturated substituent geminal to the heteroatom to be hydrogenated asymmetrically with high efficiency (eq 4).[8]

Rhodium–chiraphos cations also hydrogenate ketone[9] and epoxide[10] functionalities, albeit with low optical yields, and are, therefore, not synthetically useful. While this rhodium system seems somewhat limited to the preparation of amino acids, other rhodium[11] and ruthenium[12] catalyst precursors are currently available which show enhanced activity and selectivity for a much broader group of hydrogenation substrates.

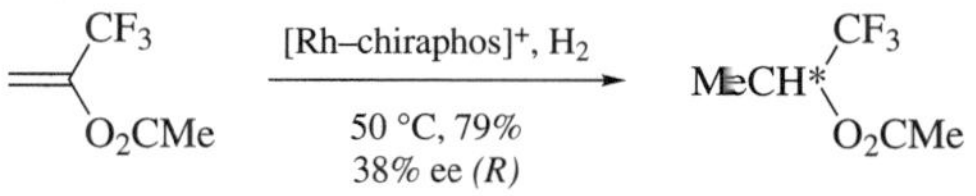

(4)

Transfer Hydrogenation. Transfer hydrogenation invokes the use of alcohols as a source of hydrogen for the reduction of organic functionalities. While rhodium–chiraphos cations catalyze the asymmetric transfer hydrogenation of acetophenone with high conversion (76%), optical yields are low (8.3%).[13] Slightly higher enantiomeric excesses are obtained for the asymmetric reduction of ethyl phenyl ketone to give the corresponding alcohol (eq 5).[14]

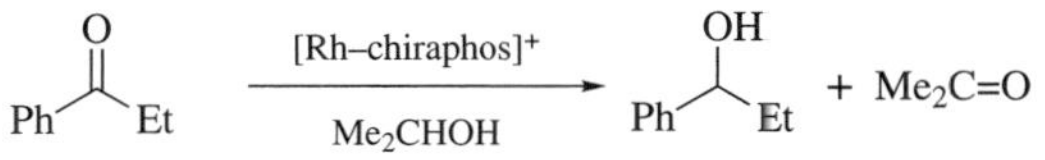

(5)

Miscellaneous. Intramolecular hydrosilation of internally substituted alkenes proceeds rapidly ($t = 6$ min) in the presence of rhodium–chiraphos cations to give cyclic siloxanes in high yields (eq 6).[15] Basic workup proceeds with retention of configuration converting allylic alcohol derivatives to chiral 1,3-diols. Higher optical yields are obtained for analogous aryl alkenes. Hydroboration of vinylarenes with catecholborane employing cationic rhodium–chiraphos complexes affords secondary alcohols, upon oxidative workup, in high yields but with low optical purity. Much higher enantiomeric excesses are obtained in reactions using analogous rhodium–BINAP catalyst precursors.[16]

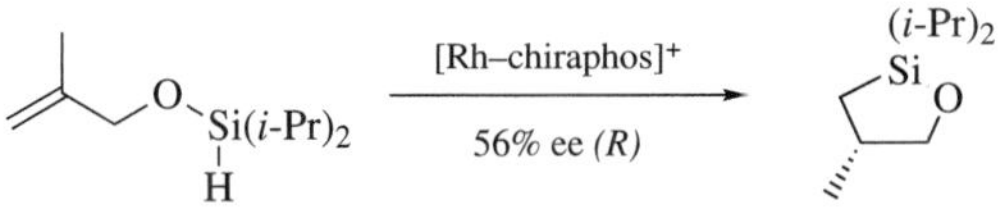

(6)

---

1. Fryzuk, M. D.; Bosnich, B. *J. Am. Chem. Soc.* **1977**, *99*, 6262.
2. (a) Ojima, I.; Clos, N.; Bastos, C. *Tetrahedron* **1989**, *45*, 6901. (b) Morrison, J. D., *Asymmetric Synthesis*; Academic: New York, 1983; Vol. 5. (c) Köttner, J.; Greber, G. *Ber. Dtsch. Chem. Ges.* **1980**, *113*, 2323.
3. Alcock, N. W.; Brown, J. M.; Maddox, P. J. *Chem. Commun.* **1986**, 1532. Reaction between resolved iridium enamide complexes and racemic chiraphos mixture is highly enantioselective and permits in situ resolution for use in asymmetric catalysis.
4. Nugent, W. A.; RajanBabu, T. V.; Burk, M. J. *Science* **1993**, *259*, 479.
5. Scott, J. W.; Keith, D. D.; Nix, Jr., G.; Parrish, D. R.; Remington, S.; Roth, G. P.; Townsend, J. M.; Valentine, Jr., D.; Yang, R. *J. Org. Chem.* **1981**, *46*, 5086.
6. Weissermel, K.; Kleiner, H. J.; Finke, M.; Felch, U. H., *Angew. Chem., Int. Ed. Engl.* **1981**, *20*, 223.
7. Zeiss, H-J. *J. Org. Chem.* **1991**, *56*, 1783.
8. Koenig, K. E.; Bachman, G. L.; Vineyard, B. D. *J. Org. Chem.* **1980**, *45*, 2362.

9. Genet, J. P.; Pinel, C.; Mallart, S.; Juge, S.; Thorimbert, S.; Laffitte, J. A. *Tetrahedron: Asymmetry* **1991**, *2*, 555.
10. Chan, A. S. C.; Landis, C. R. *J. Mol. Catal.* **1989**, *49*, 165.
11. Burk, M. J.; Feaster, J. E.; Nugent, W. A.; Harlow, R. L. *J. Am. Chem. Soc.* **1993**, *115*, 10125.
12. Noyori, R. *CHEMTECH* **1992**, *22*, 360.
13. Spogliarich, R.; Kaspar, J.; Graziani, M.; Morandini, F.; Piccolo, O. *J. Catal.* **1985**, *94*, 292.
14. Spogliarich, R.; Kaspar, J.; Graziani, M.; Morandini, F. *J. Organomet. Chem.* **1986**, *306*, 407.
15. Bergens, S. H.; Noheda, P.; Whelan, J.; Bosnich, B. *J. Am. Chem. Soc.* **1992**, *114*, 2121, 2129.
16. Hayashi, T.; Matsumoto, Y.; Ito, Y. *Tetrahedron: Asymmetry* **1991**, *2*, 601.

Stephen A. Westcott
University of North Carolina, Chapel Hill, NC, USA

(Bicyclo[2.2.1]hepta-2,5-diene)[1,4-bis(diphenylphosphino)butane]rhodium(I) Tetrafluoroborate

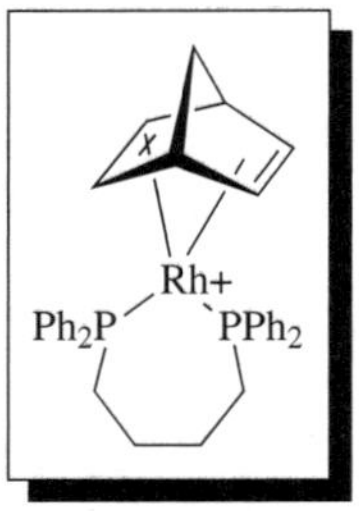

[82499-43-2] $C_{35}H_{36}BF_4P_2Rh$ (MW 708.4)

(reagent for catalytic hydrogenation;[9–11] hydrosilylation,[26] hydroboration,[27,28] and aldol condensation[32])

Physical Data: mp 211–212 °C (dec).
Solubility: insol Et_2O, pentane; sol CH_2Cl_2, MeOH, etc.
Form Supplied in: orange crystalline powder.
Drying: used as supplied in anhydrous solvent.
Analysis of Reagent Purity: ^{31}P NMR indicates presence of free phosphine ligand and phosphine oxides; the characteristic ^{103}Rh coupled doublet (J_{RhP} = 155 Hz) demonstrates pure material.
Preparative Methods: a solution of [Rh(nbd)acac] in THF is treated with fluoroboric acid, followed by addition of 1,4-bis(diphenylphosphino)butane (dppb). The solution becomes deep red and ether is added to precipitate the catalyst, which is then isolated by filtration.[3a] The unpurified product is generally adequately pure for most applications. Recrystallization from methanol can be performed to obtain orange needles of [Rh(nbd)(dppb)]BF_4. Although air-sensitive in solution, the crystalline complex is indefinitely stable when prepared pure and stored under N_2 or Ar below 0 °C. A convenient preparation of the corresponding trifluoromethanesulfonate has been described.[3b] The cyclooctadiene (cod) analog has been characterized by X-ray crystallography,[4] and the dynamic solution structure of related systems has been studied by multinuclear NMR techniques.[5]
Handling, Storage, and Precautions: store under argon in freezer; stable as solid; solutions are air sensitive. Literature describes the use of perchlorate salts but their use cannot be recommended due to risk of detonation. Use in a fume hood.

Introduction. Homogeneous catalytic hydrogenation with cationic rhodium catalysts has been extensively explored by Schrock and Osborn.[1] Use of these complexes in stereoselective organic synthesis has been a topic of more recent interest, and has been recently reviewed.[2] The reagent of choice for many of these directed hydrogenations has continued to be [Rh(nbd)(dppb)]BF_4 (**1**).

Directed Hydrogenation. By far the most significant application of (**1**) has been in diastereoselective hydrogenation reactions. The ability of (**1**) to retain a significant level of Lewis acidity under conventional hydrogenation conditions has facilitated the development of directed hydrogenations in which (**1**) engages in prior coordination to heteroatoms situated proximal to the alkene functionality. Scheme 1 shows a general mechanistic scheme by which (**1**) may catalyze hydrogenation.[6] This scheme is derived by analogy to mechanistic steps elucidated in the context of the hydrogenation of *N*-acyl dehydroamino acids, which have been demonstrated to coordinate to (**1**) in a bidentate fashion.

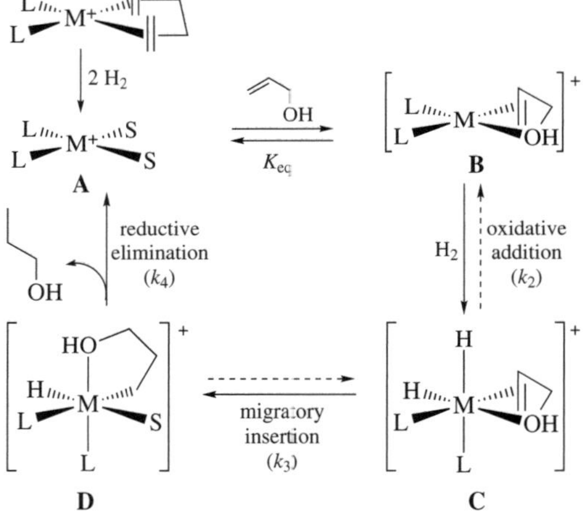

Scheme 1

Catalyst (**1**) readily absorbs 2 moles of H_2 to form complex A which may coordinate two solvent molecules.[7] The reversible binding of the substrate then occurs in a bidentate fashion to give complex B. Oxidative addition of dihydrogen then occurs to form the dihydride species C, which then undergoes migratory insertion to form the rhodium alkyl D. Although a primary Rh alkyl is shown in Scheme 1, a secondary Rh alkyl is possible as well. Finally, reductive elimination occurs to give the reduced product with concomitant regeneration of the active catalyst.

At low pressures the addition of hydrogen to a rhodium complex (schematically shown as B) is probably rate-determining, but at higher pressure, pre-equilibria (A to B) can contribute to the rate law. The effect of pressure on mechanism has impor-

Table 1 Hydroxyl-Directed Reductions with Catalyst (**1**)

Entry	Substrate	Major product	Mol % (**1**)	H_2 pressure (psi)	Selectivity
1	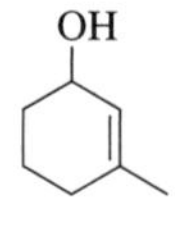	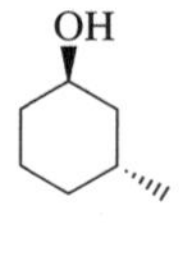	3.5	375	>200:1
2, 3	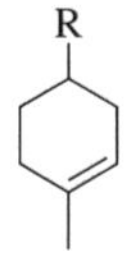		10	1000	R = OH 64:1 R = CH_2OH 19:1
4	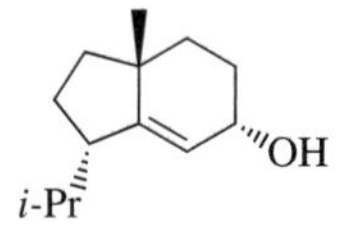	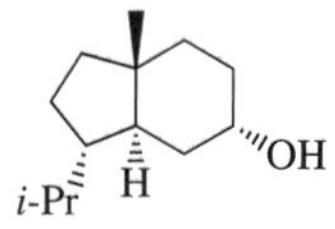	10	800	70:1
5	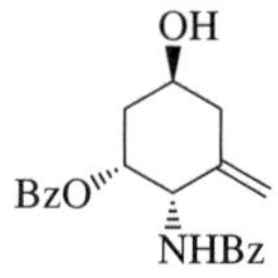	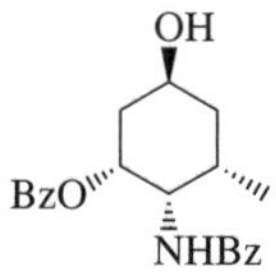	35	15	>19:1
6	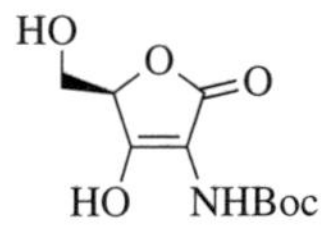	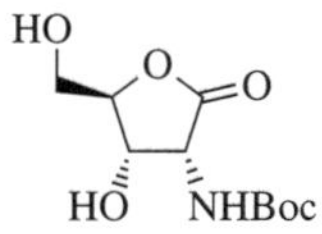	10	1400	91:9
7	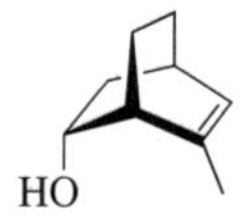	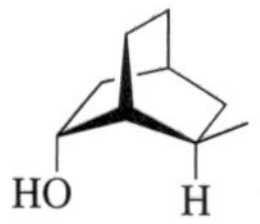	20	20	>99:1

tant implications in the directed hydrogenation of substituted alkenes which can undergo double bond isomerization. This issue has been addressed in synthetic and mechanistic studies employing (**1**) (see below).

An early observation of directed hydrogenation was made by Thompson[8a] using $ClRh(PPh_3)_3$ with a cyclic homoallylic alcohol as reactant; the stereoselectivity of hydrogenation could be enhanced by base so that delivery of hydrogen to the alkene from an alkoxide-coordinated rhodium was postulated. Acyclic diastereoselection was observed in hydrogenation of allylic alcohols with complex (**1**)[8b] and a high level of stereochemical control observed for hydrogenations catalyzed by *(1,5-Cyclooctadiene)(tricyclohexylphosphine)(pyridine)iridium(I) Hexafluorophosphate* (**2**).[8c,8d] The ensuing discussion will concentrate on catalysis by complex (**1**).

Cyclic Alkenes. The directing effect of alcohol substituents is perhaps most dramatically demonstrated with cyclic alkenes in which products are formed via hydrogenation from the more hindered face (Table 1). Entries 1–3 serve to illustrate that the directing alcohol may reside in either the allylic, homoallylic, or bis(homoallylic) position relative to the alkene undergoing hydrogenation, while synthetically useful levels of selectivity are retained.[9] Entry 4 is included to illustrate the dramatic steric congestion which can be overcome in directed hydrogenation reactions employing (**1**). Entries 5[10] and 6[11] illustrate the compatibility of (**1**) with protected amine functionality. In addition, when the alkenes in entry 5 or 6 are reduced under heterogeneous, nondirecting conditions, the other face of each alkene is reduced with high selectivity. Finally, entry 7 demonstrates that hydroxyl directivity in the [2.2.2] bicyclic system is also possible.[12]

Other Oxygen Directing Groups. The Lewis acidity of (**1**) is manifested in coordination to other heteroatom-based functional groups which can direct hydrogenation. Table 2 demonstrates that (**1**) has a sufficient affinity for ethers[11] and esters[13] so that directed hydrogenations can be achieved when these functional groups are proximal to alkenes, although the rates and selectivities are somewhat lower than with the corresponding alcohols. While ketones, acetals, carboxylic acids, and amides have been demonstrated to direct catalytic hydrogenation with the corresponding Ir^+ catalyst (**2**),[14] these functional groups do not provide good results when (**1**) is employed with cyclic alkenes. The lack of directivity from amides with (**1**) is particularly intriguing given the dramatic coordination properties of this moiety in the Rh^I catalyzed hydrogenation of *N*-acyl dehydroamino acids.[15]

Acyclic Alkenes. Acyclic alkenes also undergo stereoselective hydrogenation with catalysis by (**1**) (Table 3). It is sig-

Table 2 Oxygen Heteroatom-Directed Reductions with Catalyst (**1**)

Entry	Substrate	Major product	Mol %	Selectivity
1	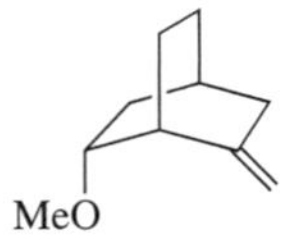	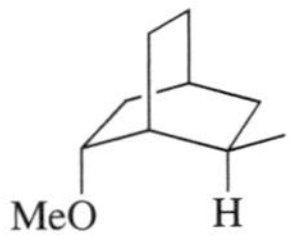	2	7:1
2	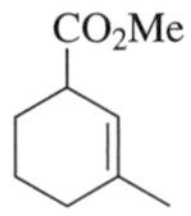	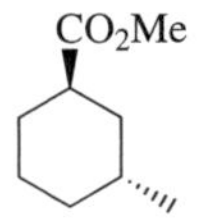	2	32:1
3	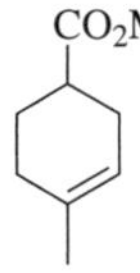	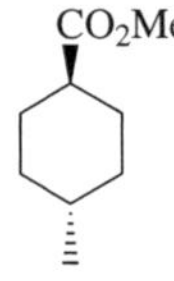	2	7:1
4	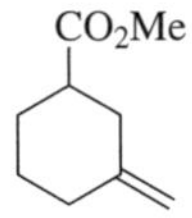	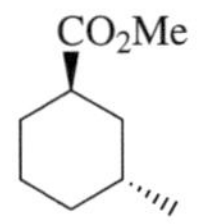	2	4:1

nificant to note that, in entry 1,[8b] although high selectivity is observed for the *anti* isomer, about 20% of the product is obtained as the corresponding methyl ketone, indicating that alkene isomerization is a competitive process. Entries 2 and 3 illustrate that either the *syn* or *anti* isomer can be obtained depending on the substitution pattern on the starting alkene.[8] In addition, this study established that alkene isomerization can be effectively suppressed at higher pressures. Entries 4–6 establish the efficiency of (**1**) for the reduction of unsaturated hydroxy esters,[16,17] which are not effectively reduced by (**2**). A model rationalizing the stereochemical outcome of these reductions has been proposed.[18] In this model, simultaneous coordination of the hydroxyl group and the alkene differentiates the diastereotopic faces. This analysis also explains the fact that alkene isomerization at low H_2 pressures contributes to diminished diastereoselectivities. This explanation has been confirmed by independent mechanistic experiments in which entries 2 and 3 were studied employing (**1**) and D_2.[19]

Catalyst (**1**) is effective in the directed reduction of homoallylic alcohols as well (Table 4). Entries 1–3 indicate that 1,1-disubstituted alkenes exhibit reasonable levels of diastereoselectivity in directed hydrogenations, although this level depends greatly on substitution pattern.[15,20] Entries 4–8 illustrate homoallylic trisubstituted alkenes undergo directed hydrogenation under catalysis by (**1**) with very high levels of selectivity, and that the configuration of the allylic substituent plays a significant role in modulating the level of selectivity. This study includes a rationale to explain the observed diastereoselectivities based on the principles of allylic strain.[1a,21]

As with cyclic alkenes, nonhydroxylic directing groups can be used in the directed hydrogenation of acyclic alkenes. The selective reductions of 3-substituted itaconate esters illustrate the directing capacity of esters (Table 5). It appears that the presence of coordinating allylic substituents can effect the level of selectivity.[22]

Applications in Total Synthesis. Two recent examples of directed hydrogenations employing (**1**) in the total synthesis of complex molecules are illustrated. In eq 1 a simultaneous diastereoselective reduction of the trisubstituted alkene and the α,β-unsaturated ester afforded the illustrated advanced intermediate in the asymmetric total synthesis of ionomycin.[23] In addition, a two-directional application has been utilized in an asymmetric synthesis FK-506 (eq 2).[24]

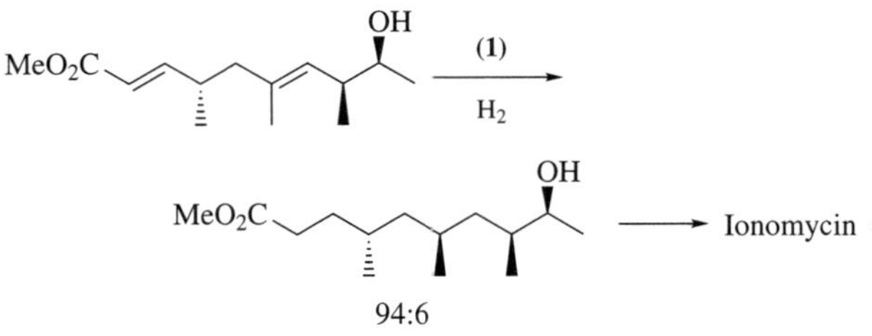

(1)

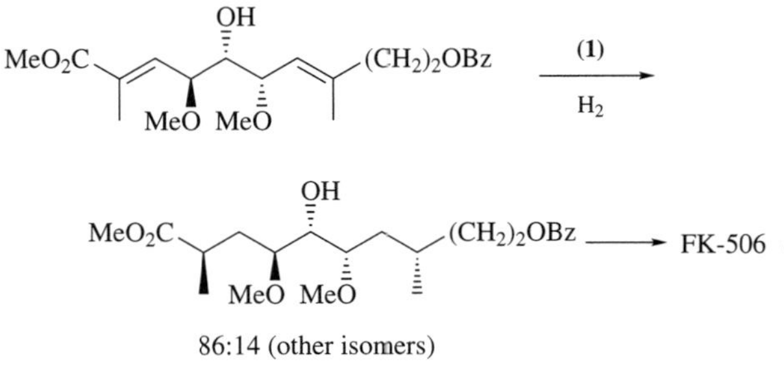

(2)

Miscellaneous Applications. The hydroxyl directed hydrogenation of vinylstannanes and -silanes has been demonstrated to proceed efficiently.[25] The authors present a transition state model which rationalizes the observed results (eq 3 and eq 4).

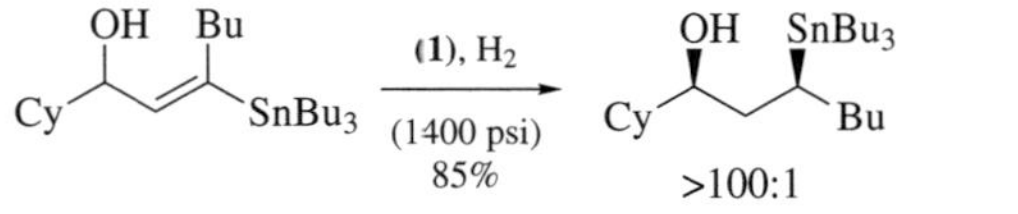

(3)

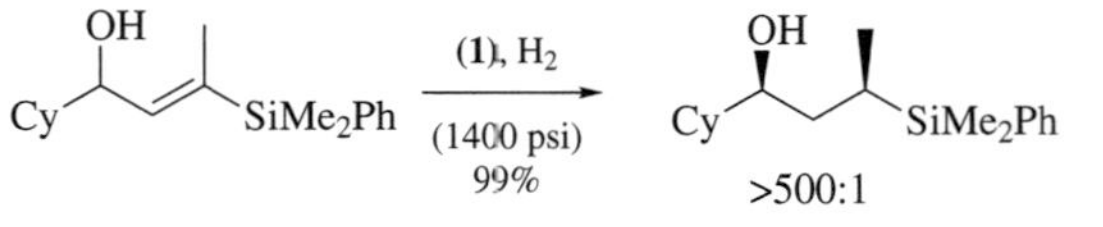

(4)

Hydrosilylation. Rhodium(I) complexes catalyze the asymmetric hydrosilylation of prochiral ketones (eq 5), in the presence of (−)-sparteine.[26] Secondary alcohols are obtained in up to 30% optical yield by this method.

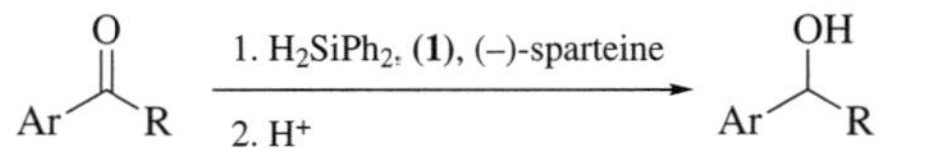

(5)

Ar = phenyl, naphthyl, furyl, thienyl; R = Me, *i*-Pr, *t*-Bu

Hydroboration. [Rh(cod)(dppb)]BF_4 is an efficient catalyst for the hydroboration of a range of vinylarenes.[27,28] Addition of *Catecholborane* to various styrene derivatives in the

Table 3 Reduction of Acyclic Allylic Alcohols with Catalyst **(1)**

Entry	Substrate	Major product	Mol % **(1)**	H_2 pressure (psi)	Selectivity
1	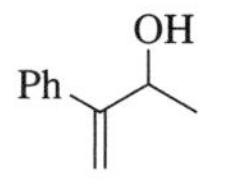	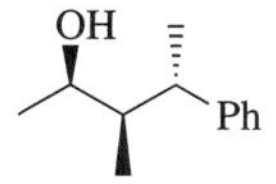	2	15	32:1
2	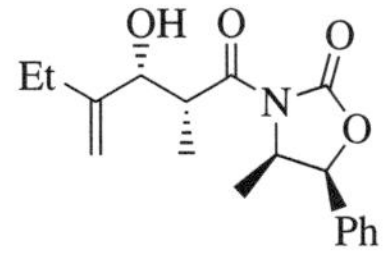	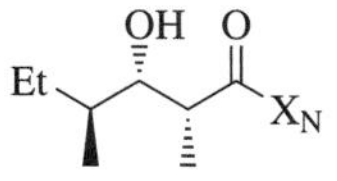	17.5	640	13:1
3	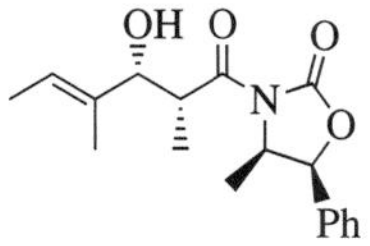	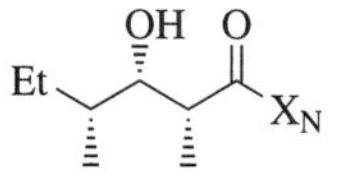	17.5	640	10:1
4	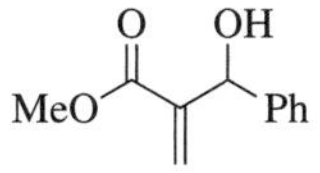	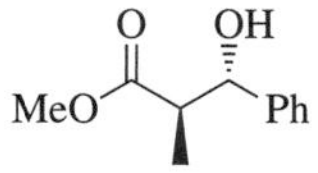	0.5	15	99:1
5	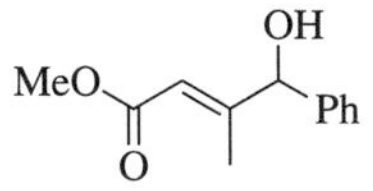	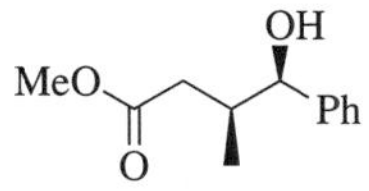	10	15	32:1
6	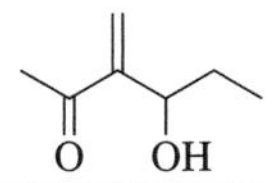	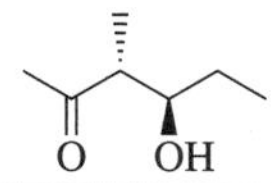	5	15	9:1

presence of 2 mol % catalyst gives, after oxidative workup, >99% secondary alcohol; in contrast, the corresponding uncatalyzed reaction gives the primary alcohol as the major product (eq 6).

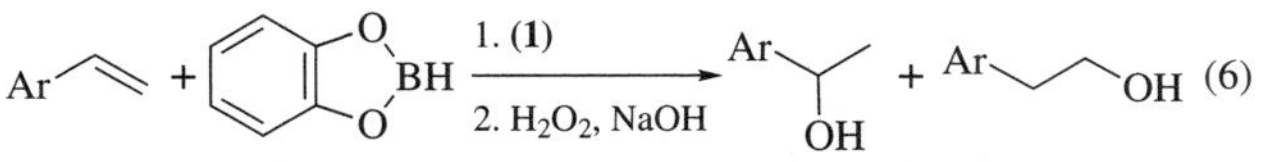

Ar = phenyl, 4-F, 4-Cl, 4-Me, 4-MeO, 3-F, 3-Cl 2-Cl, 2-MeOC$_6$H$_4$, naphthyl

This high regioselectivity is restricted to monosubstituted alkenes, indene, and (*E*)-1-phenylpropene. Analogous reactions with terminal aliphatic alkenes generally lead to primary alcohols as the major product, although alkene isomerization and BH_3-derived products arising from catalyst degradation can be problematic. 1,1-Disubstituted alkenes require longer reaction times, 1,2-disubstituted alkenes are still less reactive, and trisubstituted alkenes are essentially unreactive. This allows preferential hydroboration at the less hindered double bond in limonene (eq 7).[29]

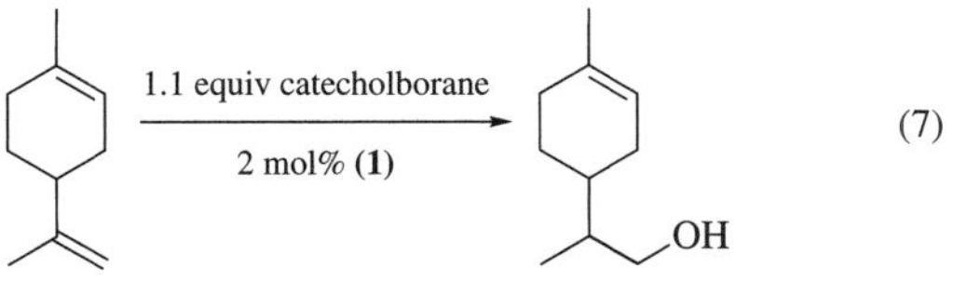

Asymmetric hydroboration of styrenes with the boranes derived from ephedrine and pseudoephedrine catalyzed by [Rh(nbd)(dppb)]OTf gives excellent regioselectivity, but poor enantioselectivity (eq 8). Better optical yields are obtained using rhodium complexes with more rigid [ferrocenyl]diphosphines.[30]

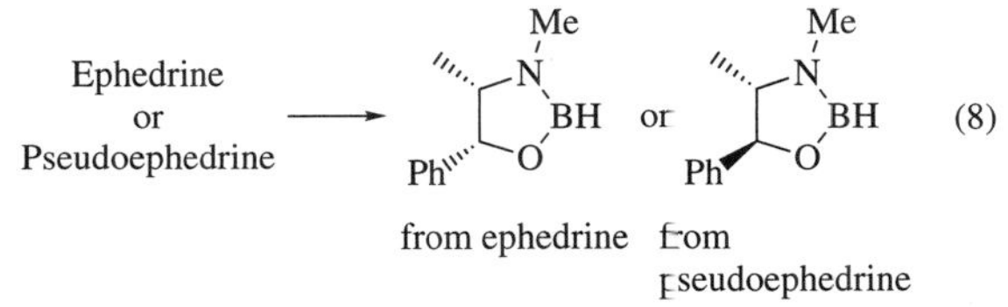

Diastereoselective rhodium-catalyzed hydroborations of allylic alcohol derivatives give results complementary to those observed in the uncatalyzed reaction with *9-Borabicyclo[3.3.1]nonane*. The *syn* selectivity of the catalyzed reaction increases as the bulk of the R group increases (*syn*:*anti* = 79:21 for R = TBDPS) (eq 9).[29]

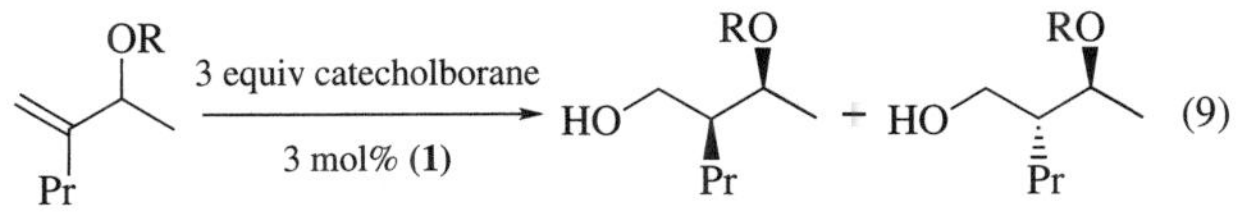

R = H, Bn, TBDMS, TBDPS

Exocyclic 1,1-disubstituted alkenes also give complementary selectivity (*syn*:*anti* = 93:7 for R = TBDMS) (eq 10) to that observed with 9-BBN.[29]

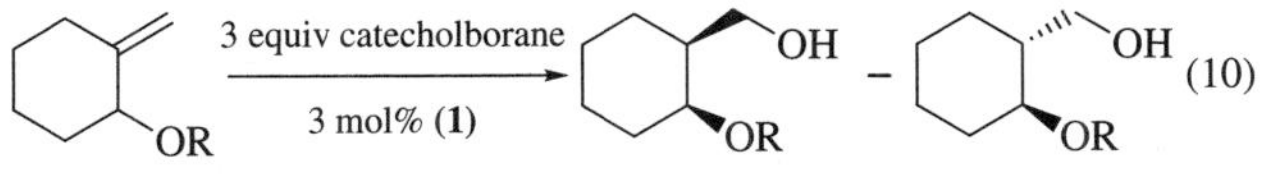

R = H, TBDMS

Table 4 Reduction of Acyclic Homoallylic Alcohols with Catalyst (**1**)

Entry	Substrate	Major product	Mol % (**1**)	H_2 pressure (psi)	Selectivity
1	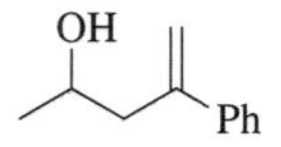	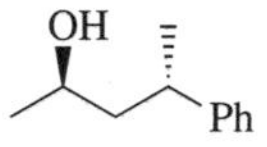	2	15	7:1
2	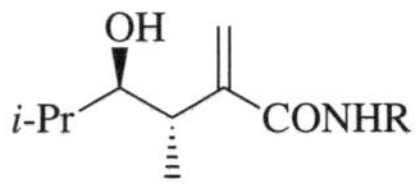	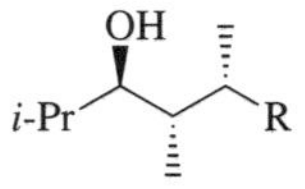	5	15	10:1
3	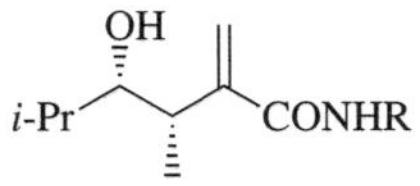	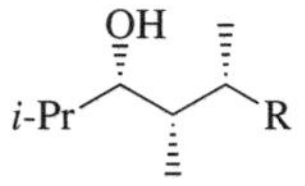	5	15	2:1
4	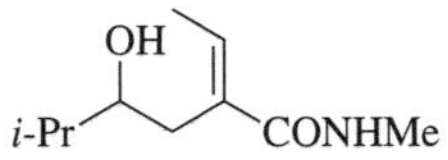	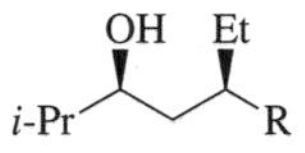	5	15	8:1
5	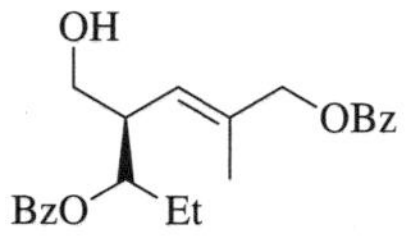	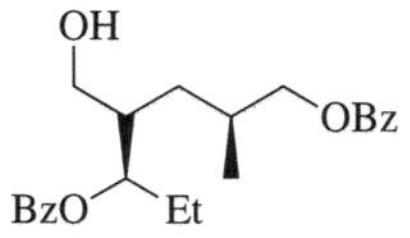	20	15	32:1
6	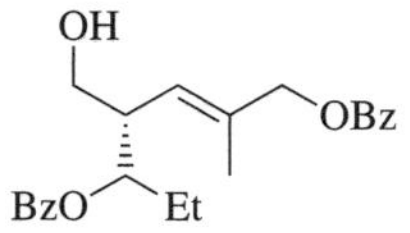	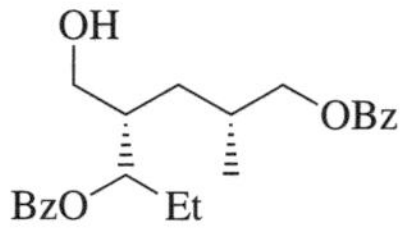	20	15	99:1
7	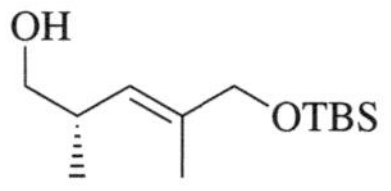	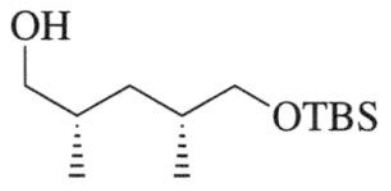	3	15	19:1
8	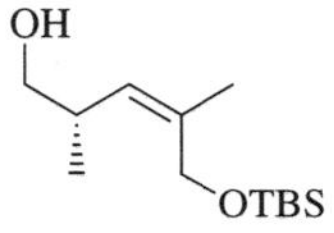	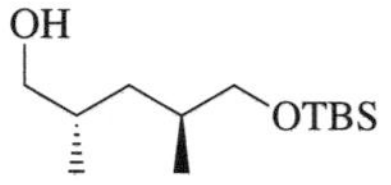	3	15	10:1

Table 5 Directed Reduction Of Unsaturated Esters

Entry	Substrate	Major product	Mol %	Selectivity
1	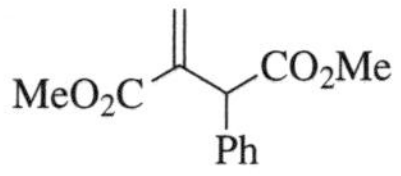	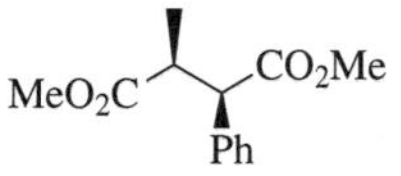	2	200:1
2	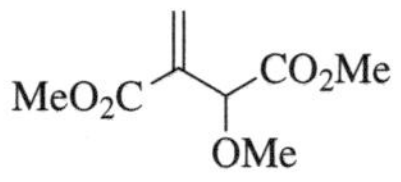	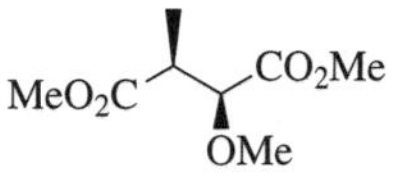	2	49:1
3	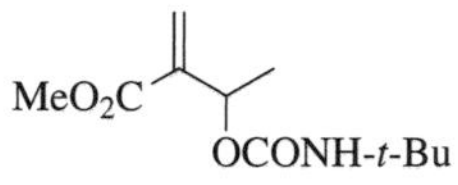		2	17:1
4	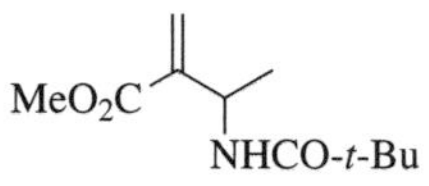	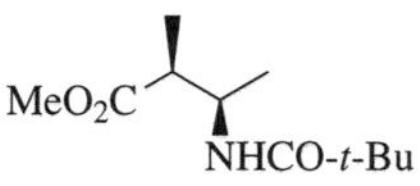	5	99:1
5	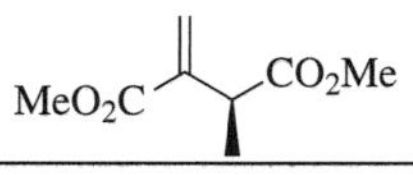	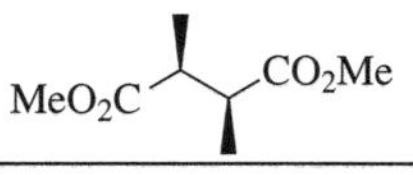	5	250:1

Aldol Condensations. The rhodium complex has been utilized as a catalyst in aldol condensation of silyl enol ethers and aldehydes[31] or aldehyde equivalents (eq 11 and eq 12).[32]

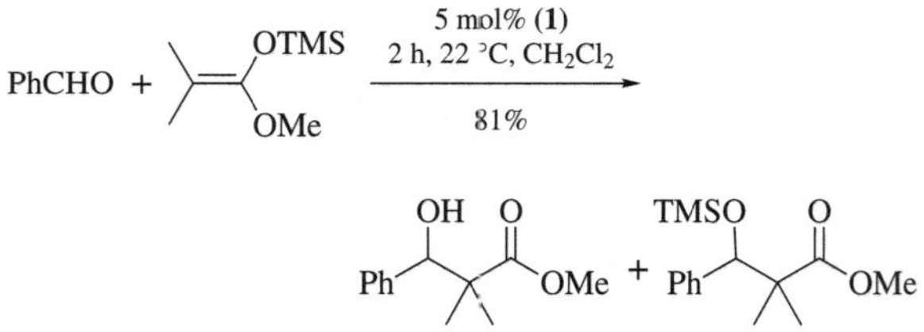

(11)

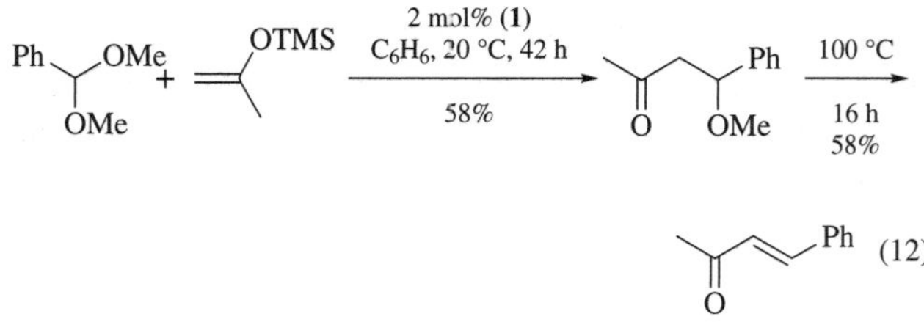

(12)

Although this rhodium complex has been studied in hydrocarbonylation and alkene isomerization, other rhodium catalysts give much higher yields and/or offer greater selectivity.

Finally, the ability of (**1**) to induce the isomerization of alkenes has been exploited in synthesis.[33] In the absence of H_2, the dominant product of the isomerization of the illustrated homoallylic silane is the allylic silane shown (eq 13). The reaction is presumably driven by the β-silicon effect.[34]

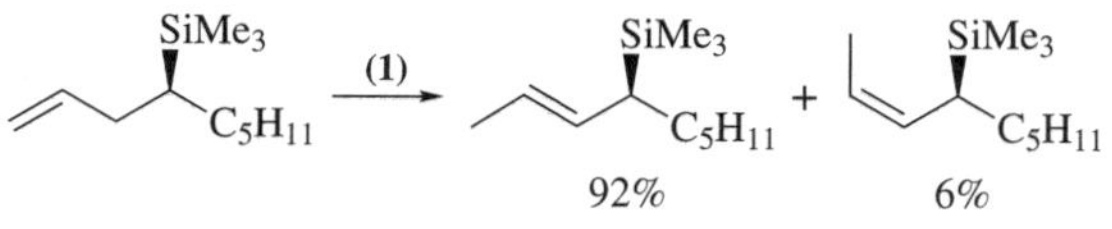

(13)

1. (a) Schrock, R. R.; Osborn, J. A. *J. Am. Chem. Soc.* **1976**, *98*, 2134. (b) Schrock, R. R.; Osborn, J. A. *J. Am. Chem. Soc.* **1976**, *98*, 2143. (c) Schrock, R. R.; Osborn, J. A. *J. Am. Chem. Soc.* **1976**, *98*, 4450.
2. (a) Hoveyda, A. H.; Evans, D. A.; Fu, G. C. *Chem. Rev.* **1993**, *93*, 1307. (b) Brown, J. M. *Angew. Chem., Int. Ed. Engl.* **1987**, *26*, 190.
3. (a) Brown, J. M.; Chaloner, P. A. *J. Am. Chem. Soc.* **1980**, *102*, 3040. (b) Brown, J. M.; Evans, P. L.; James, A. P. *Comprehensive Organic Synthesis* **1989**, *68*, 64.
4. Anderson, M. P.; Pignolet, L. H. *Inorg. Chem.* **1981**, *20*, 4101.
5. Chaloner, P. *J. Organomet. Chem.* **1984**, *266*, 191.
6. Halpern, J. *Science* **1982**, *217*, 401, and references cited therein.
7. Halpern, J.; Riley, D. P.; Chan, A. S. C.; Pluth, J. J. *J. Am. Chem. Soc.* **1977**, *99*, 8055.
8. (a) Thompson, H. W.; McPherson, E. *J. Am. Chem. Soc.* **1974**, *96*, 6232. (b) Brown, J. M.; Naik, R. G. *Chem. Commun.* **1982**, 348. (c) Crabtree, R. H.; Davis, M. W. *J. Organomet. Chem.* **1983**, *2*, 681. (d) Stork, G.; Kahne, D. E. *J. Am. Chem. Soc.* **1983**, *105*, 1072.
9. Evans, D. A.; Morrissey, M. M. *J. Am. Chem. Soc.* **1984**, *106*, 3866.
10. Machado, A. S.; Olesker, A.; Castillon, S.; Lukacs, G. *Chem. Commun.* **1985**, 330.
11. Hamada, Y.; Kawai, A.; Matsui, T.; Hara, O.; Shioiri, T. *Tetrahedron* **1990**, *46*, 4823.
12. Brown, J. M.; Hall, S. A. *Tetrahedron* **1985**, *41*, 4639.
13. Brown, J. M.; Hall, S. A. *J. Organomet. Chem.* **1985**, *285*, 333.
14. See for example (a) Schultz, A. G.; McCloskey, P. J. *J. Org. Chem.* **1985**, *50*, 5905. (b) Takagi, M.; Yamamoto, K. *Tetrahedron* **1991**, *47*, 8869.
15. Landis, C. R.; Halpern, J. *J. Am. Chem. Soc.* **1987**, *109*, 1746.
16. Brown, J. M.; Cutting, I. J. *Chem. Commun.* **1985**, 578.
17. Sato, S.; Matsuda, I.; Shibata, M. *J. Organomet. Chem.* **1989**, *377*, 347.
18. (a) Hoffman, R. W. *Chem. Rev.* **1989**, *89*, 1841. (b) Johnson, F. *Chem. Rev.* **1968**, *68*, 375.
19. (a) Morrissey, M. M. Ph.D. Thesis, Harvard University, 1986 (*Diss. Abstr. Int. B* **1987**, *48*, 444). (b) Hoveyda, A. H.; Evans, D. A.; Fu, G. C. *Chem. Rev.* **1993**, *93*, 1336.
20. Birtwistle, D. H.; Brown, J. M.; Herbert, R. H.; James, A. P.; Lee, K.-F.; Taylor, R. J. *Chem. Commun.* **1989**, 194.
21. Evans, D. A.; Morrissey, M. M.; Dow, R. L. *Tetrahedron Lett.* **1985**, *26*, 6005.
22. (a) Brown, J. M.; James, A. P. *Chem. Commun.* **1987**, 181. (b) Brown, J. M.; Cutting, I.; James, A. P. *Bull. Soc. Chem. Fr. Part 2* **1988**, 211.
23. Evans, D. A.; Dow, R. L.; Shih, T. L.; Takasc, J. M.; Zahler, R. *J. Am. Chem. Soc.* **1990**, *112*, 5290.
24. Villalobos, A.; Danishefsky, S. J. *J. Org. Chem.* **1990**, *55*, 2776.
25. Lautens, M.; Zhang, C.; Crudden, C. M. *Angew. Chem., Int. Ed. Engl.* **1992**, *31*, 232.
26. Goldberg, Y.; Alper, H. *Tetrahedron: Asymmetry* **1992**, *3*, 1055.
27. Westcott, S. A.; Blom, H. P.; Marder, T. B.; Baker, R. T. *J. Am. Chem. Soc.* **1992**, *114*, 8863.
28. Hayashi, T.; Matsumoto, Y.; Ito, Y. *Tetrahedron: Asymmetry* **1991**, *2*, 601.
29. Evans, D. A.; Fu, G. C.; Hoveyda, A. H. *J. Am. Chem. Soc.* **1992**, *114*, 6671.
30. Brown, J. M.; Lloyd-Jones, G. C. *Tetrahedron: Asymmetry* **1990**, *1*, 869.
31. (a) Sato, S.; Matsuda, I.; Izumi, Y. *Tetrahedron Lett.* **1987**, *28*, 6657. (b) Sato, S.; Matsuda, I.; Izumi, Y. *Tetrahedron Lett.* **1986**, *27*, 5517.
32. Reetz, M. T.; Vougioukas, A. E. *Tetrahedron Lett.* **1987**, *28*, 793.
33. Matsuda, I.; Kato, T.; Sato, S.; Izumi, Y. *Tetrahedron Lett.* **1986**, *27*, 5747.
34. Lambert, J. B.; Emblidge, R. W.; Malany, S. *J. Am. Chem. Soc.* **1993**, *115*, 1317, and references therein.

David A. Evans & Scott J. Miller
Harvard University, Cambridge, MA, USA
John M. Brown, Timothy P. Layzell, & James A. Ramsden
University of Oxford, UK

(1*R*,1′*R*,2*R*,2′*R*)-[1,1′-Bicyclopentyl-2,2′-diylbisdiphenylphosphine]

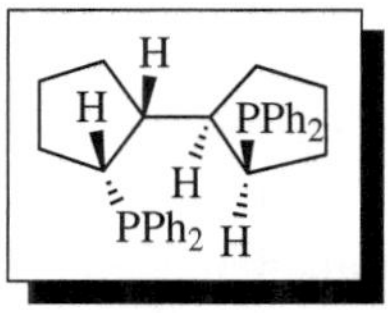

$C_{34}H_{36}P_2$ (MW 505.8)

(chiral, nonracemic phosphine ligand for asymmetric transition metal-catalyzed reactions)

Alternate Name: (2*R*,2′*R*)-bis(diphenylphosphino)-(1*R*,1′*R*)-dicyclopentane [(*R*,*R*)-BICP].

Solubility: soluble in common organic solvents (i.e., dichloroethane, THF, ethanol, ethyl acetate, and methylene chloride).

Analysis of Reagent Purity: ^{1}H-NMR.

Preparative Methods: prepared in three steps from 1,1-dicyclopentene by hydroboration with (+)-monoisopinocamphenylborane followed by oxidation with hydrogen peroxide, formation of the dimesylate and displacement with Li_2PPh.

Purification: purification was accomplished by chromatography of the corresponding borane complex. Decomplexation using $HBF_4{\cdot}O(C_2H_5)_2$ afforded the pure bisphosphine.

Handling, Storage, and Precautions: sensitive to atmospheric oxidation. Should be stored and handled under an inert atmosphere.

Introduction. This chiral, nonracemic phosphine ligand belongs to a group of chiral bisphosphines such as DIPAMP,[1] DIOP,[2] Chiraphos,[3] BINAP,[4] among many others,[5] that are capable of inducing high levels of asymmetry in metal-catalyzed processes. To expand the repertoire of reactions amenable to enantioselective catalysis, bicyclopentyl-2,2′-diylbisdiphenylphosphine [(*R*,*R*)-BICP] was designed to be conformationally restricted by the rigid bicyclopentane backbone.[6] Molecular modeling (MM2 force field) of the corresponding transition metal complexes suggests that the lowest energy conformation maintains a highly twisted seven-membered chelate (Figure 1). In this conformation, the phenyl groups occupy both axial and equatorial positions similar to other effective bidentate ligands.

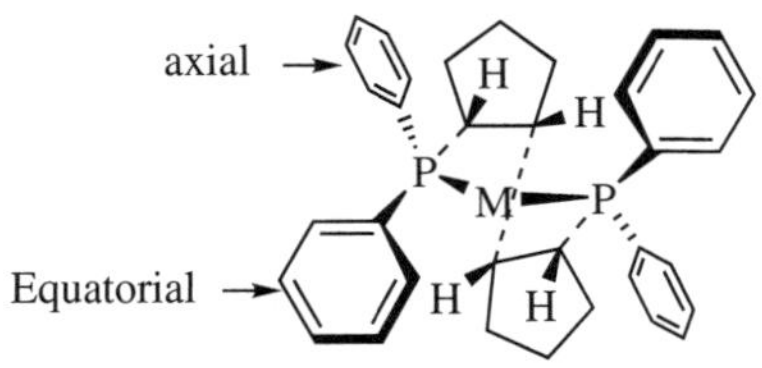

Asymmetric Hydrogenation. Enantioselective hydrogenation of α-acetamidocinnamic acid using a catalyst formed in situ from $[Rh(COD)_2]BF_4$ and (*R*,*R*)-BICP afforded α-amino acids with high enantioselectivities (eq 1, Table 1).[6] The optimized reaction conditions employing substoichiometric quantities of triethylamine afforded highest selectivities. The increase in enantioselectivity observed in the presence of a base may be due to the increased structural rigidity imparted to the metal complex that occurs following deprotonation and subsequent coordination of the carboxylic acid of the substrate to the metal center. The enantioselectivities were highest when a cationic rather than neutral Rh-complex was used as catalyst precursor.

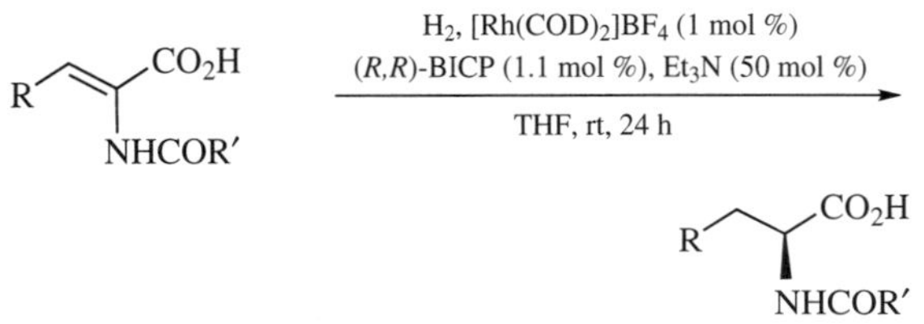

(1)

Table 1

Substrate	ee (%)
R = H, R' = CH_3	97.5
R= Ph, R' = CH_3	96.8
R= Ph, R' = CH_3	83.6[a]
R= Ph, R' = Ph	99.0
R= *p*-OAc-*m*-OMePh, R' = CH_3	98.2
R= *i*-Pr, R' = CH_3	92.6

[a] $[Rh(COD)Cl]_2$ used as catalyst precursor in EtOH.

The enantioselectivity of the hydrogenation was slightly lower for simple enamides (eq 2, Table 2).[7] The highest enantioselectivities were obtained in nonpolar solvents and at higher pressures of hydrogen. However, the effect of solvent polarity and hydrogen pressure on the selectivity was quite small. Substitution at the β-position of the enamide generally increased the enantioselectivity which was also relatively insensitive to the β-substituent or the geometry of the olefin (eq 3). If the β-substituent was a MOM-protected alcohol, hydrogenation produced enantiomerically pure β-amino alcohols after deprotection.[8]

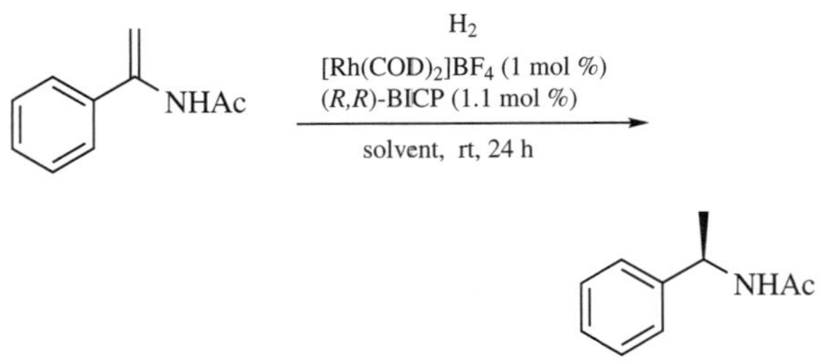

(2)

Table 2

Solvent	Pressure (psi)	ee (%)
Toluene	40	86.3
DMF	40	77.4
Toluene	14.7	80.2

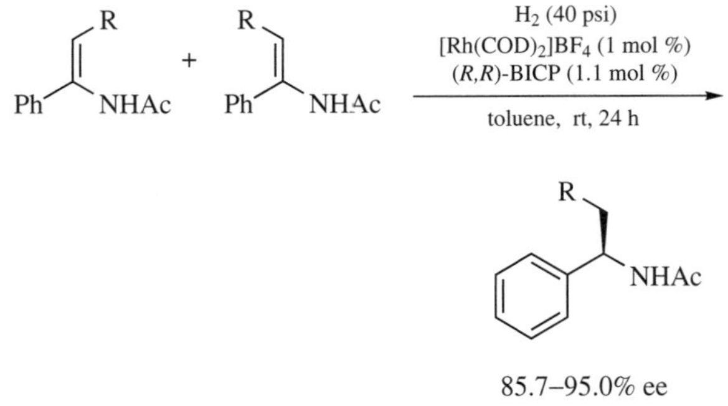

(3)

Asymmetic hydrogenation of 3-acylaminoacrylate derivatives affords enantiomerically enriched β-amino acids (eq 4).[9] In contrast to Ru-BINAP catalysts that hydrogenate the *Z* isomers most rapidly,[10] the Rh-BICP catalyst hydrogenates the *E* isomer fastest. Generally, the *E* isomers afford higher ee's than the *Z* isomers for Rh-BICP and Rh-Me-Duphos (Table 3). However, the Rh-BICP catalyst exhibits greater selectivity in hydrogenation of *Z* isomers than the Rh-Me-Duphos catalyst. Although the Rh-Me-Duphos is a more enantioselective catalyst for the hydrogenation of *E* isomers than Rh-BICP, the Rh-BICP catalyst hydrogenates *E*/*Z* mixtures more selectively.

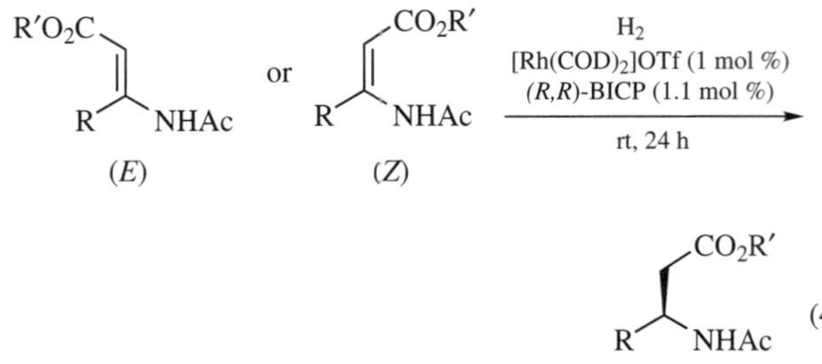

(4)

A ruthenium hydrogenation catalyst prepared by the reaction of $RuCl_2$[(*R*,*R*)-BICP](TMEDA) with (*R*,*R*)-1,2-diphenylethylenediamine has been shown to hydrogenate aromatic ketones in the presence of potassium hydroxide with high enantioselectivity (eq 5).[11] The catalyst provides enantioselectivities that range from 41–93% ee which are generally 10–20% lower than the selectivities previously reported for the Ru-BINAP-chiral diamine-KOH catalyst developed by Noyori.[12]

Table 3

Ligand	Substrate	R	R'	Pressure (psi)	ee (%)
(*R,R*)-BICP	(*E*)	Me	Me	40	95.1
Ru-BINAP	(*E*)	Me	Me	40	69.0
(*R,R*)-Me-DuPhos	(*E*)	Me	Me	40	99.0
(*R,R*)-Me-DuPhos	(*Z*)	Me	Me	294	63.7
(*R,R*)-BICP	(*Z*)	Me	Me	294	88.6
(*R,R*)-BICP	(*E*)	*i*-Bu	Me	40	90.9
(*R,R*)-BICP	(*Z*)	*i*-Bu	Me	294	92.9

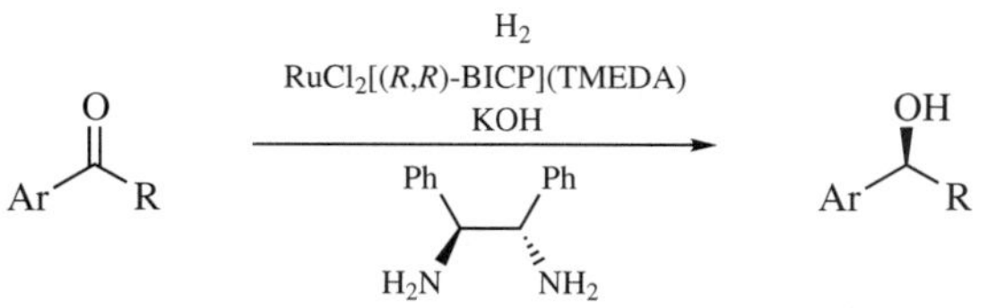

(5)

Other Metal-Catalyzed Reactions. Coordination of the BICP ligand to a palladium precursor permits the cyclocarbonylation of allylic alcohols to occur enantioselectively (eq 6).[13] It is noteworthy that this ligand provides significantly higher enantioselectivities than other ligands such as BINAP or DIOP.[14] In contrast to catalysts derived from BINAP or DIOP, the Pd-BICP catalyst promotes the cyclocarbonylation of β,γ-substituted allylic alcohols with high enantioselectivity (eq 7). The efficiency of the Pd-BICP catalyst has been rationalized by the greater flexibility of the seven-membered ligand-metal chelate which increases the rate of CO insertion.

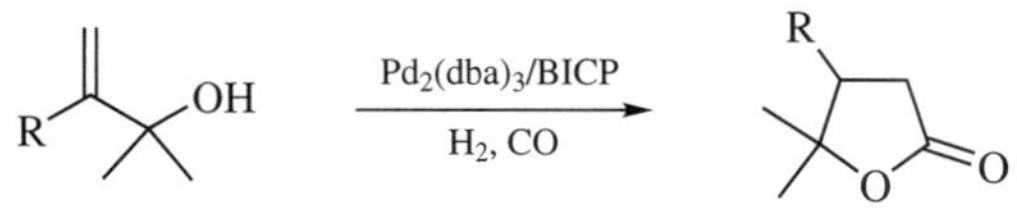

(6)

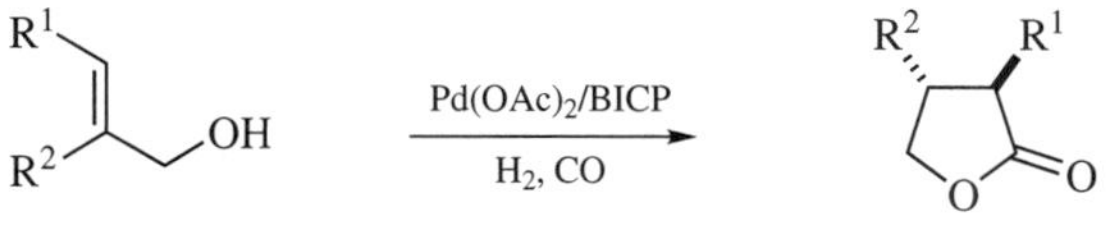

(7)

The cycloisomerization of 1,6-enynes can also be catalyzed enantioselectively using a Rh-BICP catalyst (eq 8).[15] However, Rh-Me-DuPhos affords a significantly higher enantioselectivity than Rh-BICP for this process (Table 4).

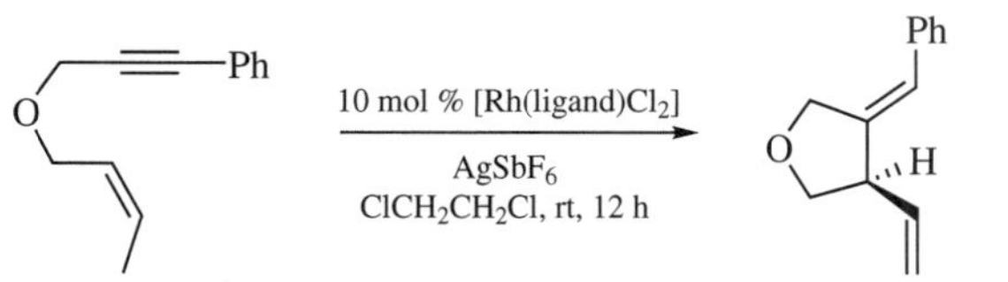

(8)

Table 4

Ligand	ee (%)
Me-DuPhos	95
BICP	74

1. (a) Knowles, W. S.; Sabacky, M. J.; Vineyard, B. D. *J. Chem. Soc., Chem. Commun.* **1972**, *1*, 10. (b) Knowles, W. S. *Acc. Chem. Res.* **1983**, *16*, 106. (c) Vineyard, B. D.; Knowles, W. S.; Sabacky, M. J.; Bachman, G. L.; Weinkauff, D. J. *J. Am. Chem. Soc.* **1977**, *99*, 5946.
2. Kagan, H. B.; Dang-Tuan-Phat, *J. Amer. Chem. Soc.* **1972**, *94*, 6429.
3. Fryzuk, M. D.; Bosnich, B. *J. Am. Chem. Soc.* **1977**, *99*, 6262.
4. (a) Noyori, R. *Acta Chem. Scand.* **1996**, *50*, 380. (b) Noyori, R.; Takaya, H. *Acc. Chem. Res.* **1990**, *23*, 345.
5. (a) Lee, S.; Hartwig, J. F. *J. Org. Chem.* **2001**, *66*, 3402. (b) Zhang, X. *Enantiomer* **1999**, *4*, 541. (c) Noyori, R. *Asymmetric Catalysis in Organic Synthesis*; Wiley: New York, 1994.
6. Zhu, G.; Cao, P.; Jiang, Q.; Zhang, X. *J. Am. Chem. Soc.* **1997**, *119*, 1799.
7. Zhu, G.; Zhang, X. *J. Org. Chem.* **1998**, *63*, 9590.
8. Zhu, G.; Casalnuovo, A. L.; Zhang, X. *J. Org. Chem.* **1998**, *63*, 8100.
9. Zhu, G.; Chen, Z.; Zhang, X. *J. Org. Chem.* **1999**, *64*, 6907.
10. Lubell, W. D.; Kitamura, M.; Noyori, R. *Tetrahedron: Asymmetry* **1991**, *2*, 543.
11. Cao, P.; Zhang, X. *J. Org. Chem.* **1999**, *64*, 2127.
12. (a) Hashiguchi, S.; Fujii, A.; Takehara, J.; Ikariya, T.; Noyori, R. *J. Am. Chem. Soc.* **1995**, *117*, 7562. (b) Ohkuma, T.; Ooka, H.; Hashiguchi, S.; Ikariya, T.; Noyori, R. *J. Am. Chem. Soc.* **1995**, *117*, 2675.
13. Cao, P.; Zhang, X. *J. Am. Chem. Soc.* **1999**, *121*, 7708.
14. Yu, W.-Y.; Bensimon, C.; Alper, H. *Chem. Eur. J.* **1997**, *3*, 417.
15. Cao, P.; Zhang, X. *Angew. Chem., Int. Ed.* **2000**, *39*, 4104.

Jon R. Parquette
The Ohio State University, OH, USA

1,1′-Binaphthalene-2,2′-dithiol[1]

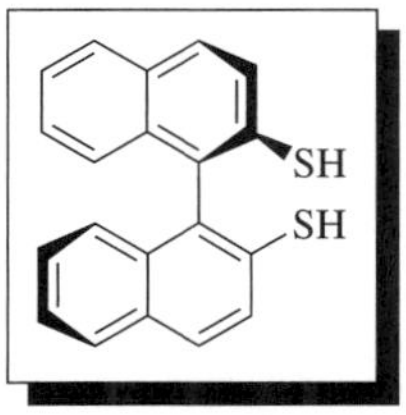

[102555-71-5] $C_{20}H_{14}S_2$ (MW 318.45)
(−)
[124414-36-4]
(±)
[55441-99-1]

(reagent for the preparation of chiral, atropisomeric organosulfur reagents of C_2 symmetry;[1] used as chiral ligand and in the preparation of chiral crown ethers)

Physical Data: mp 150–151 °C (benzene); (*R*)-(−): $[\alpha]_D^{22}$ −85.9°; $[\alpha]_{546}^{22}$ −103.8° ($c = 1$, $CHCl_3$).
Analysis of Reagent Purity: the presence of the disulfide can be checked by TLC on silica gel, eluting with dichloromethane.

Preparative Methods: the original procedure used Ullman coupling of 1-bromo-2-naphthalenesulfonic acid.[2a] The intermediate 1,1′-binaphthalene-2,2′-sulfonic acid can be resolved with strychnine.[2b] Lithiation of 2,2′-dibromo-1,1′-binaphthalene with *t*-butyllithium, quenching with sulfur,[2c] and reduction of the resulting disulfide is an alternative preparation of the racemic dithiol. More practical procedures entail Newman–Kwart rearrangement of the thioester derived from binaphthol and dimethylthiocarbamoyl chloride, followed by hydrolysis.[3–5] Use of enantiomerically pure binaphthol as starting material gives the enantiomerically pure reagent.[4] Another resolution procedure involves enantioselective oxidation of sulfides which can be further transformed into the dithiol.[6]

Handling, Storage, and Precautions: the reagent oxidizes easily to the disulfide and should be stored under inert atmosphere. Use in a fume hood.

Binaphthalene-2,2′-dithiol is the starting material for the preparation of a number of sulfur-containing heterocycles of synthetic utility. The basic principle lies in the generation of C_2 symmetric chiral variants of reagents that contain two sulfur atoms. For example, the achiral bis(phenylthio)methane becomes the chiral 1,3-dithiepine (eq 1), still maintaining similar structural features to the acyclic reagent.[7] This dithiepine belongs to the class of reagents that function as formyl anion synthons. The C_2 symmetry is also shared by the bisoxide (it forms stereoselectively as a single pseudoequatorial isomer) and the bis-sulfone. Reaction of the anion of the dithiepine with benzaldehyde gives an 8:2 mixture of diastereoisomers (eq 2).[7]

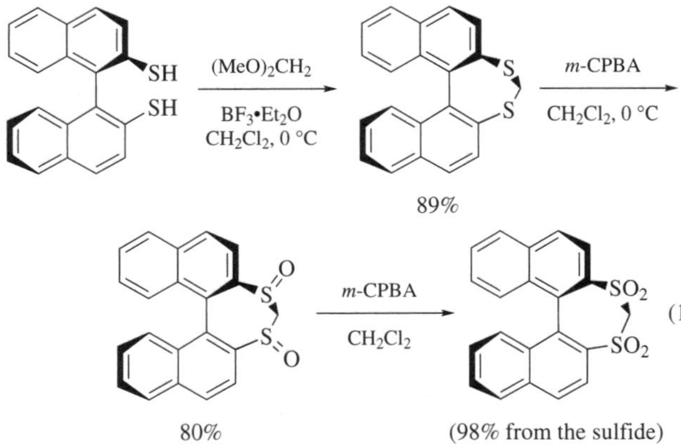

(1)

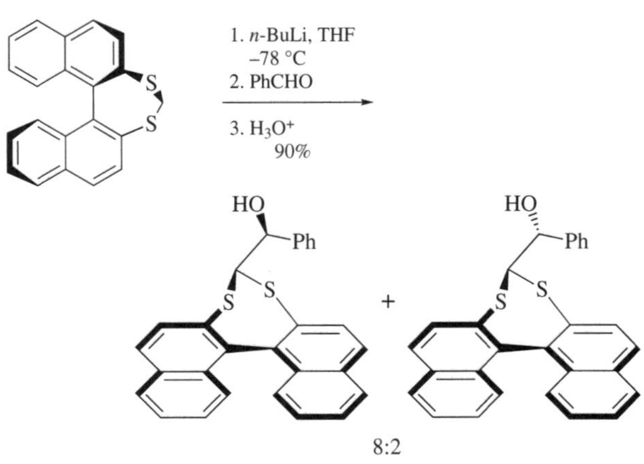

(2)

A single diastereoisomer is obtained in the reduction of the ketone with *Lithium Aluminum Hydride* as illustrated in eq 3.[8] The addition of other nucleophiles such as methylmagnesium iodide also gives single adducts.

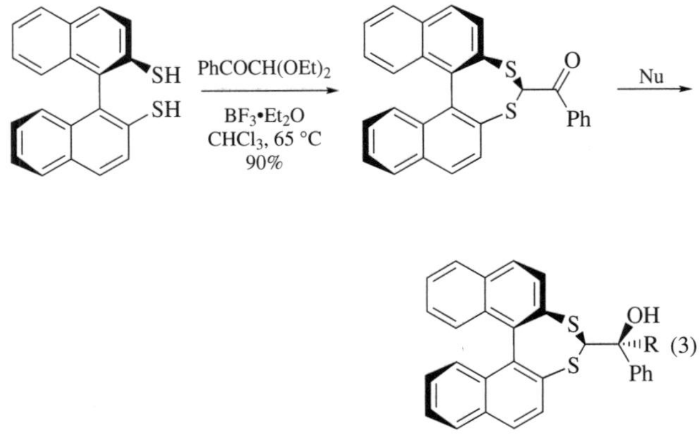

(3)

Nucleophile (Nu)	R	Reaction conditions	Diastereomeric ratio	Yield (%)
$LiAlH_4$	H	Et_2O, −78 °C, 45 min	100:0	98
$NaBH_4$	H	EtOH, H_2O, Et_3N, 20 °C, 4 h	86:14	99
MeMgI	Me	THF, Et_2O, 0 °C, 40 min	100:0	98

Monoxidation of the dithiepine gives single diastereomeric sulfoxides with pseudoequatorial configuration (eq 4).[7] In general, the oxides exhibit increased diastereoselectivity with respect to the unoxidized substrates, as in the examples of eq 5.[7]

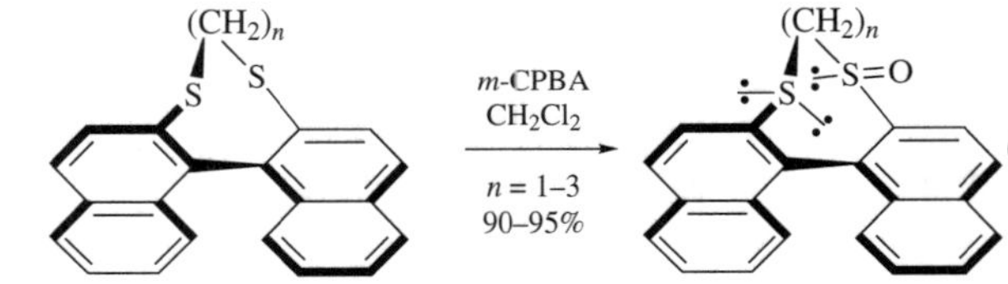

(4)

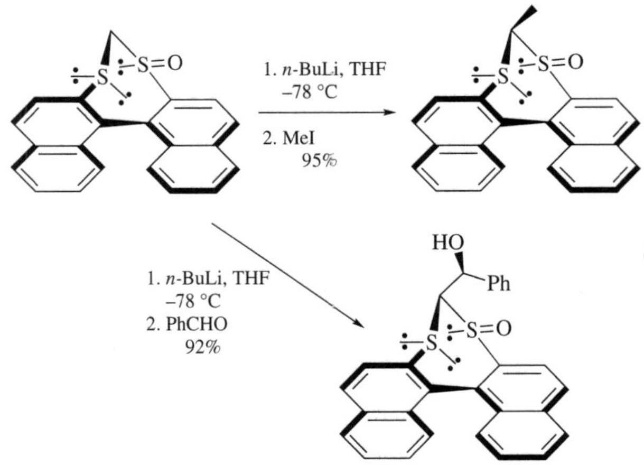

(5)

C_2 symmetric, chiral ketene dithioacetals containing the binaphthyl moiety can be prepared by Peterson alkenation of the title reagent (eq 6).[9] The corresponding bis-sulfone affords one *exo* and one *endo* adduct with cyclopentadiene (eq 7) which, once separated and desulfonylated, give the corresponding norbornenes (see *1,1-Bis(phenylsulfonyl)ethylene*).[10]

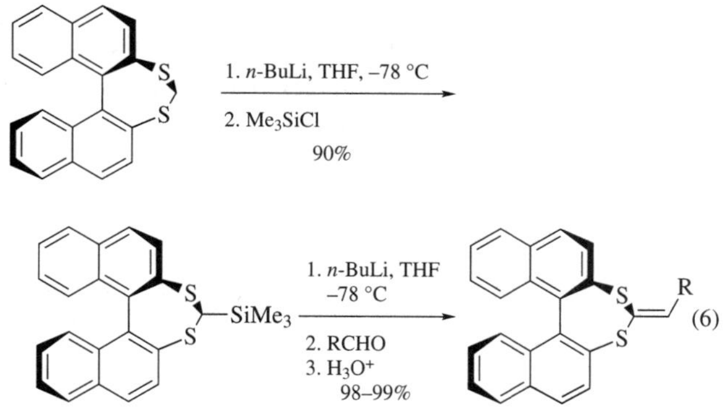

R = H, Me, Ph

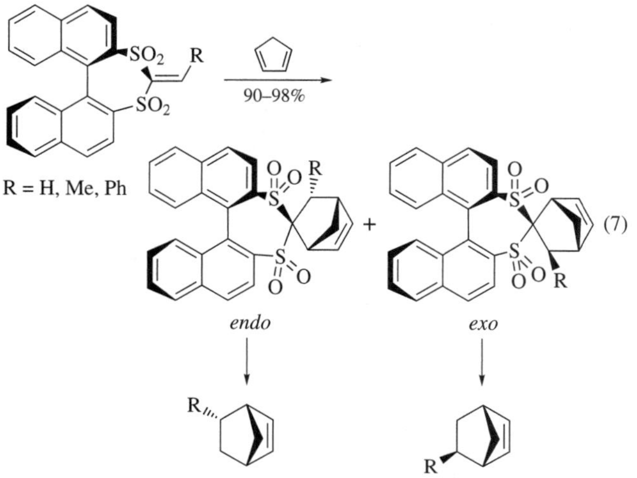

The dithiocine tetraoxide derived from cyclocondensation of binaphthodithiol with dichloroethylene and oxidation (eq 8) is a chiral version of the bis(phenylsulfonyl)ethylenes.[11] These compounds are useful acetylene equivalents in cycloaddition reactions (see *1,2-Bis(phenylsulfonyl)ethylene*).[9] Indeed, a chiral acetylene equivalent allows the preparation of optically active hydrocarbons which would be difficult to prepare by classical methods. The dithiocine tetroxide reacts with nonsymmetric dienes to give a single crystalline diastereomeric adduct in most cases. Adducts (**1**) and (**2**) were obtained from acyclic and cyclic dienes.

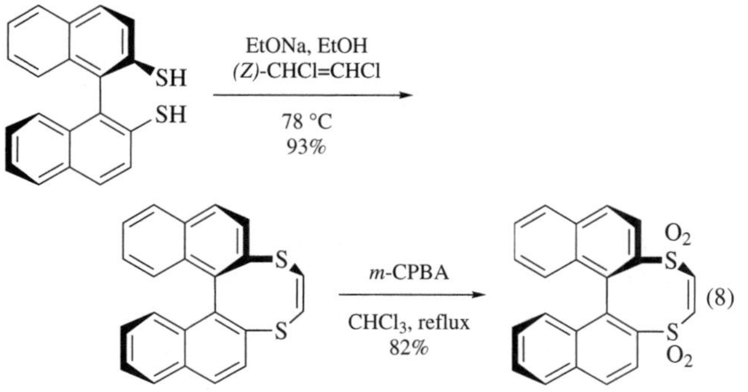

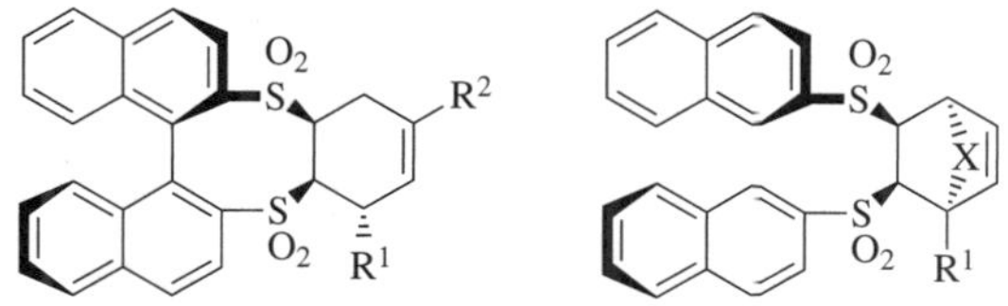

(**1**)
R^1 = OMe, R^2 = OTMS
R^1 = OMe, R^2 = H
R^1 = OTMS, R^2 = H
R^1 = Me, R^2 = H
R^1 = H, R^2 = OTMS
R^1 = H, R^2 = Me (8:2 mixture)

(**2**)
R^1 = OMe, X = O
R^1 = OMe, X = $-(CH_2)_2-$

The addition provides only one stereoisomer out of the four possible ones. *Sodium Amalgam* reduction in buffered methanol removes the binaphthyl residue to afford the hydrocarbon and recovered starting dithiol.[11]

Finally, it is notable that the title reagent has been used to prepare even larger ring systems such as chiral crown ethers,[3] and the use of 1,1′-binaphthalene-2,2′-dithiol as ligand for rhodium(I) in the asymmetric hydroformylation of styrene has been described.[12]

1. (a) De Lucchi, O. *J. Pharm. Sci.* **1993**, *74*, 195. (b) Cossu, S.; De Lucchi, O.; Fabbri, D.; Licini, G.; Pasquato, L. *Org. Prep. Proced. Int.* **1991**, *23*, 571. (c) Cossu, S.; De Lucchi, O.; Fabbri, D.; Fois M. P.; Maglioli, P. In *Heteroatom Chemistry: ICHAC-2*, Block, E., Ed.; VCH: New York, 1990; Chapter 8, pp 143–163.
2. (a) Barber, H. J.; Smiles, S. *J. Chem. Soc.* **1928**, 1141. (b) Armarego W. L. F.; Turner, E. E. *J. Chem. Soc.* **1957**, 13. (c) Murata, S.; Suzuki, T.; Yanagisawa, A.; Suga, S. *J. Heterocycl. Chem.* **1991**, *28*, 433.
3. Cram, D. M.; Helgeson, R. C.; Koga, K.; Kyba, E. P.; Madan, K.; Sousa, L. R.; Siegel, M. G.; Moreau, P.; Gokel, G. W.; Timko, J. M.; Sogah, G. D. Y. *J. Org. Chem.* **1978**, *43*, 2758.
4. Fabbri, D.; Delogu, G.; De Lucchi, O. *J. Org. Chem.* **1993**, *58*, 1748.
5. Cossu, S.; Delogu, G.; De Lucchi, O.; Fabbri, D.; Fois, M. P. *Org. Synth., Coll. Vol.* **1989**, *19*, 3431.
6. Di Furia, F.; Licini, G.; Modena, G.; De Lucchi, O. *Tetrahedron Lett.* **1989**, *30*, 2575.
7. (a) Delogu, G.; De Lucchi, O.; Licini, G. *Chem. Commun.* **1989**, 411. (b) Delogu, G.; De Lucchi, O.; Maglioli, P.; Valle, G. *J. Org. Chem.* **1991**, *56*, 4467.
8. Delogu, G.; De Lucchi, O.; Maglioli, P. *Synlett* **1989**, 28.
9. De Lucchi, O.; Fabbri, D.; Lucchini, V. *Synlett* **1991**, 565.
10. De Lucchi, O.; Fabbri, D.; Lucchini, V. *Tetrahedron* **1992**, *48*, 1485.
11. (a) Cossu, S.; Delogu, G.; De Lucchi, O.; Fabbri, D.; Licini, G. *Angew. Chem., Int. Ed. Engl.* **1989**, *28*, 766. (b) De Lucchi, O.; Fabbri, D.; Cossu, S.; Valle, G. *J. Org. Chem.* **1991**, *56*, 1858. (c) Pindur, U.; Lutz, G.; Fischer, G.; Schollmeyer, D.; Massa, W.; Schröder, L. *Tetrahedron* **1993**, *49*, 2863.
12. Claver, C.; Castillon, S.; Ruiz, N.; Delogu, G.; Fabbri, D.; Gladiali, S. *Chem. Commun.* **1993**, 1833.

Ottorino De Lucchi
Università di Venezia, Italy
Giulia Licini
Università di Padova, Italy

(*R*)-1,1′-Bi-2,2′-naphthol[1]

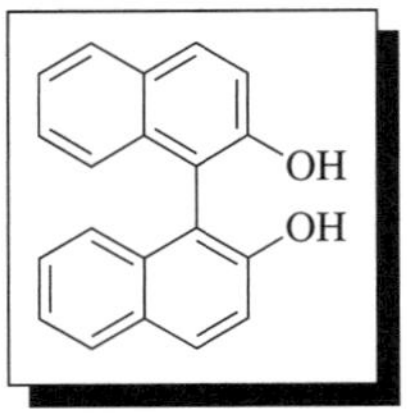

[18531-94-7] $C_{20}H_{14}O_2$ (MW 286.33)

(chiral ligand and auxiliary[1])

Alternate Name: BINOL.
Physical Data: mp 208–210 °C; $[\alpha]^{21}$ +34° ($c = 1$, THF).
Solubility: sol toluene, CH_2Cl_2, $EtNO_2$.
Form Supplied in: white solid; widely available.
Preparative Methods: racemic 1,1′-bi-2,2′-naphthol (BINOL) is most conveniently prepared by the oxidative coupling reaction of 2-naphthol in the presence of transition metal complexes (eq 1).[2] The resolution of racemic BINOL with cinchonine may be performed via the cyclic phosphate (eq 2).[3] An alternative procedure to provide directly optically active BINOL is the oxidative coupling of 2-naphthol catalyzed by Cu^{II} salt in the presence of chiral amines (eq 3).[4] The best procedure uses (+)-amphetamine as the chiral ligand and provides BINOL in 98% yield and 96% ee. Above 25 °C the Cu^{II}/(+)-amphetamine/(*S*)-BINOL complex precipitates, while the more soluble Cu^{II}/(+)-amphetamine/(*R*)-BINOL complex is slowly transformed into the former complex. 9,9′-Biphenanthrene-10,10′-diol has also been prepared in 86% yield and with 98% ee by a similar asymmetric oxidative coupling of 9-phenanthrol in the presence of (*R*)-1,2-diphenylethylamine.[5]

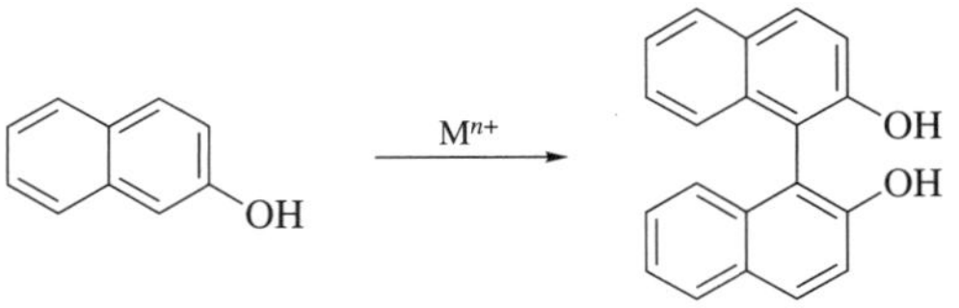

(1)

Handling, Storage, and Precautions: keep tightly closed, store in a cool dark place; on heating in butanol at 118 °C for 24 h, BINOL lost ~1% of its optical rotation; at 100 °C for 24 h in dioxane–1.2 N HCl, BINOL lost 56% of its rotation; after 24 h at 118 °C in butanol–0.7 N KOH, BINOL lost 20% of its rotation.

Hydrocarboxylation. The cyclic phosphate resolved according to eq 2 can be used as the chiral ligand in the palladium(II) catalyzed asymmetric hydrocarboxylation of arylethylenes.[6] The 1-arylpropanoic acid is obtained regiospecifically with high enantioselectivity (91% ee) (eq 4).

Crown Ethers. BINOL-derived crown ethers have been reported.[7] Crown ethers containing 3,3′-disubstituted BINOL derivatives are particularly effective for asymmetric synthesis. Thus complexes of these crown ethers (e.g. *18-Crown-6*) with *Potassium Amide* or *Potassium t-Butoxide* catalyze asymmetric Michael additions. The reaction of methyl 1-oxo-2-indancarboxylate with methyl vinyl ketone with the 3,3′-dimethyl-BINOL–crown ether/KO-*t*-Bu complex gives the Michael product in 48% yield and with 99% ee (eq 5).[8]

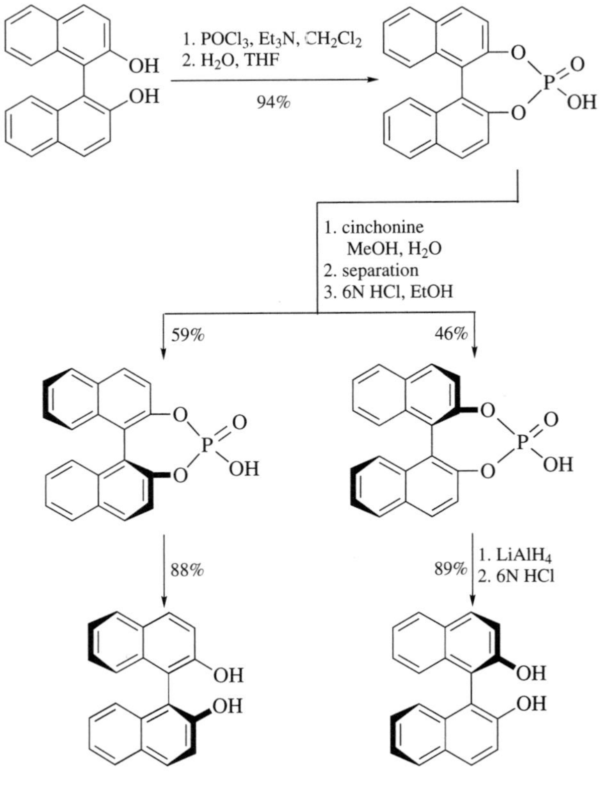

(2)

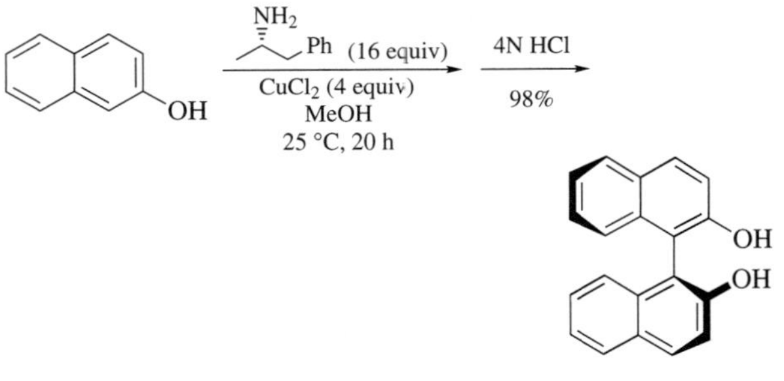

(3)

96% ee

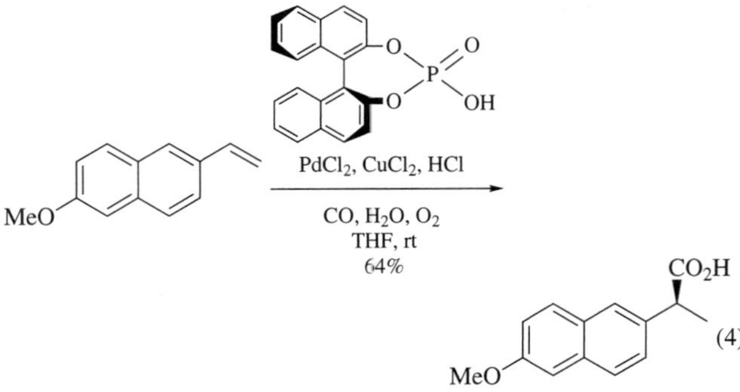

(4)

91% ee

Polymerization. Complexes of BINOL-derived crown ethers with KO-*t*-Bu or BuLi have been used as initiators in the asymmetric polymerization of methacrylates.[9] Thus optically active polymers are obtained with 80–90% isotacticity. Complexes of BINOL with *Diethylzinc* or $CdMe_2$ also initiate the asymmetric polymerization of heterocyclic monomers.[10] The chiral initiators selectively polymerize one enantiomer to give an optically active polymer. The unreacted monomer is recovered with 92% ee at 67% conversion in the polymerization of methylthiirane with (*S*)-BINOL/Et_2Zn.

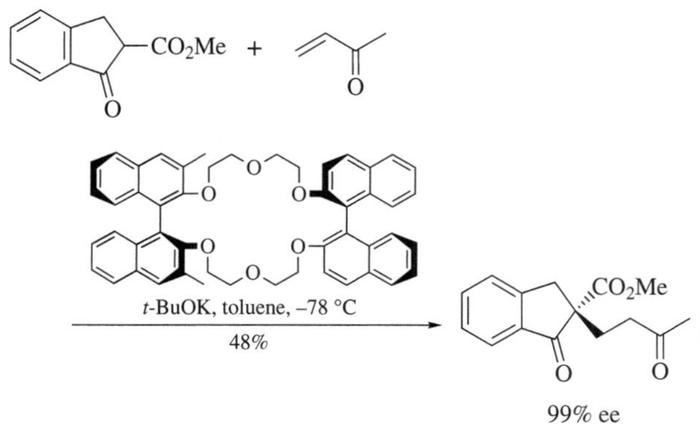
(5)

Ullmann Coupling Reaction. Axially dissymmetric biaryls have been synthesized via an intramolecular Ullmann coupling reaction of BINOL-derived aryl diesters (eq 6).[11] In the example shown, the functionalized binaphthyl is obtained with high ee after hydrolysis of the intermediate 12-membered cyclic diester.

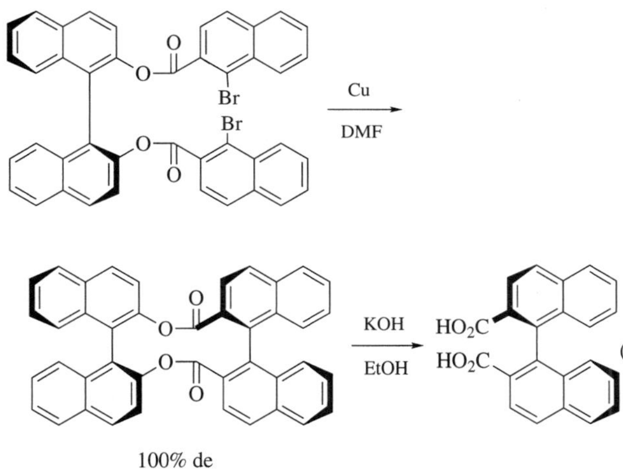
(6)

Reduction of Prochiral Ketones. BINOL has been used as the chiral ligand of the reagent BINAL-H (see *Lithium Aluminum Hydride–2,2′-Dihydroxy-1,1′-binaphthyl*, Vol. B) for asymmetric reduction.[12] The reagent reduces prochiral unsaturated ketones to the corresponding secondary alcohols in up to 90% yield and >90% ee (eq 7); (*R*)-BINAL-H leads to the (*R*)-alcohols while (*S*)-BINAL-H gives the (*S*)-alcohols.

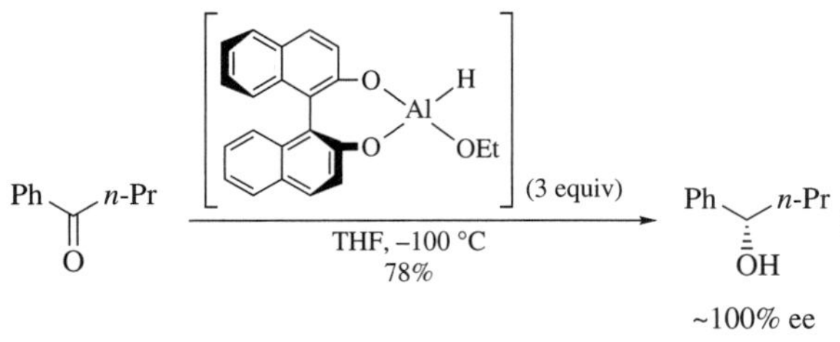
(7)

Addition Reactions of Chiral Titanium Reagents to Aldehydes. The preparation and use of the BINOL-derived titanium complexes in the enantioselective synthesis of some benzhydrols (>90% ee) have been reported (eq 8).[13]

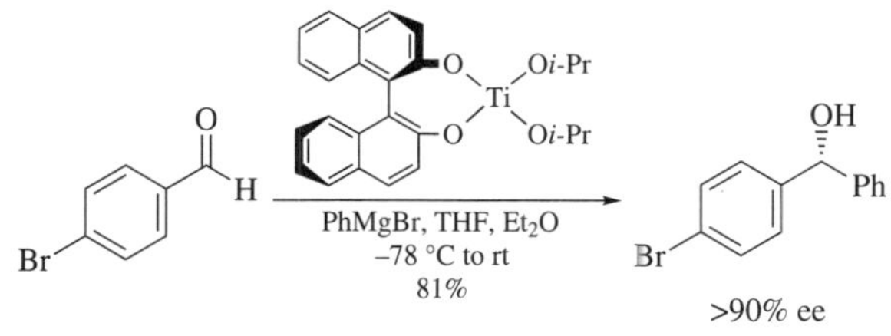
(8)

Cyanosilylation. The chiral titanium reagent, prepared from the lithium salt of BINOL with $TiCl_4$, has been used as a catalyst for the asymmetric addition of cyanotrimethylsilane to aldehydes.[14] In the example shown, the cyanohydrin is obtained with ≤82% ee (eq 9).

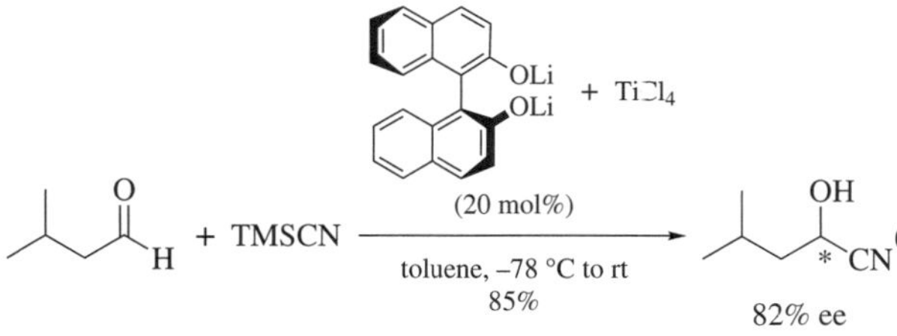
(9)

Diels–Alder Reactions. BINOL and its derivatives are used as the chiral ligand of chiral Lewis acid complexes for enantioselective Diels–Alder cycloadditions. BINOL-$TiCl_2$, prepared from the lithium salt of BINOL with *Titanium(IV) Chloride*, also catalyzes the enantioselective Diels–Alder reaction of cyclopentadiene with methacrolein (eq 10)[14,15] The *exo* adduct is obtained as the major product (56% yield), but with low enantioselectivity (16% ee). More recently, BINOL-TiX_2 (X = Br or Cl) have been prepared in situ from diisopropoxytitanium dihalides ($(i\text{-PrO})_2TiX_2$, X = Br[16] or Cl[17]) with BINOL in the presence of molecular sieves (MS 4A).[16] The Diels–Alder reaction of methacrolein with 1,3-dienol derivatives can be catalyzed by BINOL-TiX_2. The *endo* adducts are obtained in high enantioselectivity (eq 11).[18] Asymmetric catalytic Diels–Alder reaction of naphthoquinone derivatives as the dienophile (eq 12)[18] can in principle provide an efficient entry to the asymmetric synthesis of anthracyclinone aglycones. The reaction of the 5-hydroxynaphthoquinone with 1-acetoxy-1,3-diene in the presence of MS-free BINOL-$TiCl_2$ (10 mol %) provides the corresponding Diels–Alder product in high chemical yield and with high enantioselectivity (76–96% ee).[18b] The Diels–Alder product is also obtained by the use of 1 equiv of 3,3′-diphenyl-BINOL/borane complex (eq 13); the structure of the intermediate has been proposed.[19]

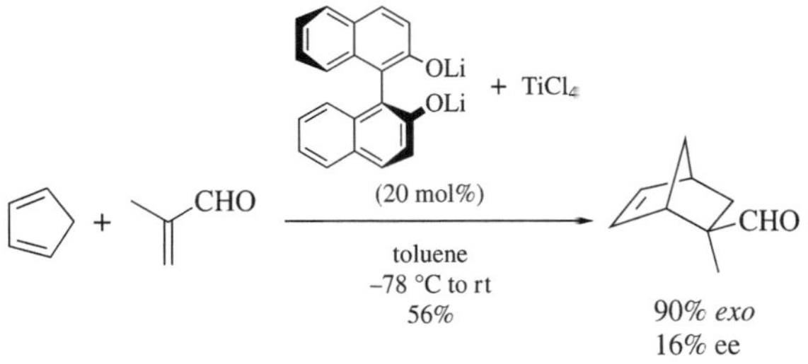
(10)

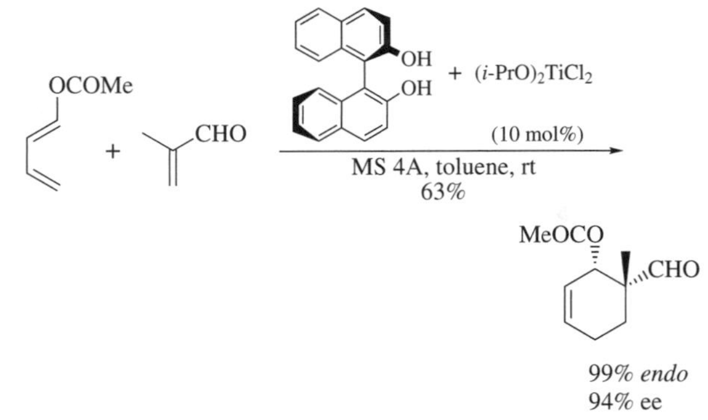
(11)

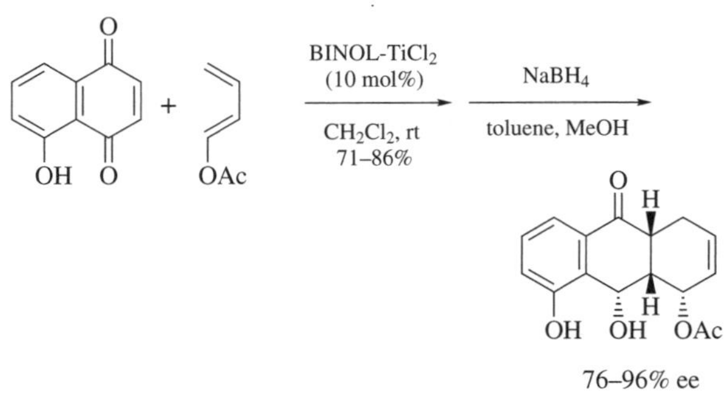
(12)

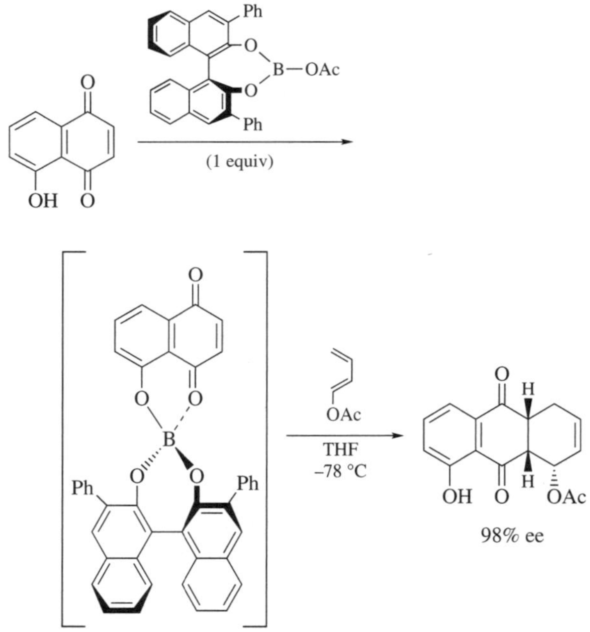
(13)

3,3′-Diphenyl-BINOL-derived chiral aluminum reagents are prepared in situ by addition of *Ethylaluminum Dichloride* or *Diethylaluminum Chloride* to 3,3′-diphenyl-BINOL. These chiral aluminum reagents promote the enantioselective Diels–Alder reaction of cyclopentadiene with the oxazolidone dienophile (eq 14).[20] *Endo* products are obtained with a high level of asymmetric induction (>90% ee); however, a stoichiometric amount of the Lewis acid is required. The preparation and use of a C_3 symmetric BINOL-derived boronate has been reported (eq 15).[21] BINOL-$B(OAr)_3$ complexes have recently been developed for the asymmetric Diels–Alder reaction with imines (eq 16).[22]

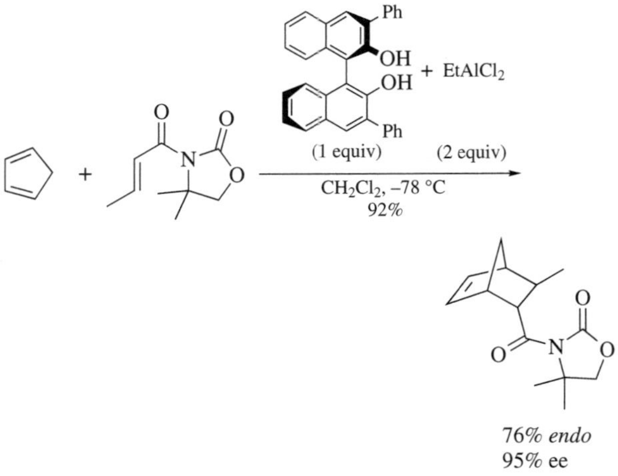
(14)

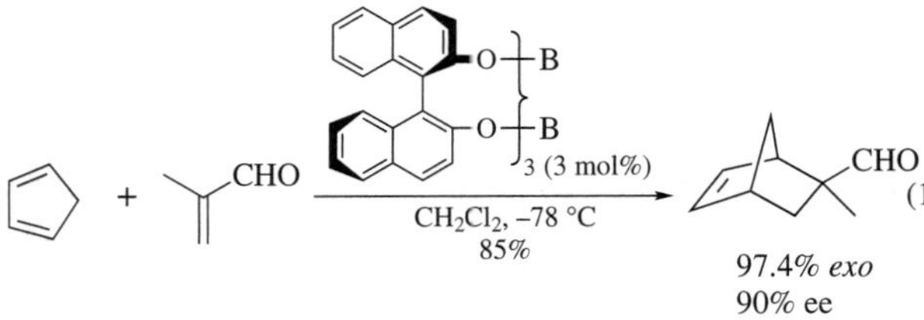
(15)

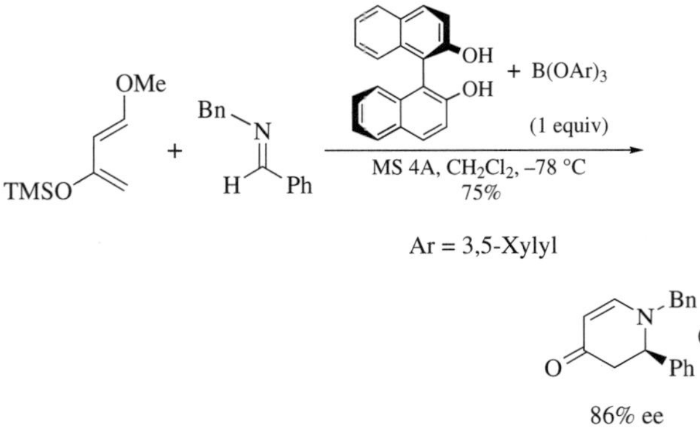
(16)

Hetero Diels–Alder Reaction. Modified BINOL-derived organoaluminum reagents have been used in the asymmetric hetero Diels–Alder reaction of aldehydes (eq 17). The dihydropyrones are obtained with high *cis* diastereoselectivity and enantioselectivity.[23] The hetero Diels–Alder reaction of glyoxylates proceeds smoothly with methoxydienes using BINOL-$TiCl_2$ as a catalyst to give the *cis* product in high enantiomeric excess (eq 18).[18b,24] The hetero Diels–Alder product thus obtained can be readily converted to the lactone portion of HMG-CoA inhibitors such as mevinolin or compactin.

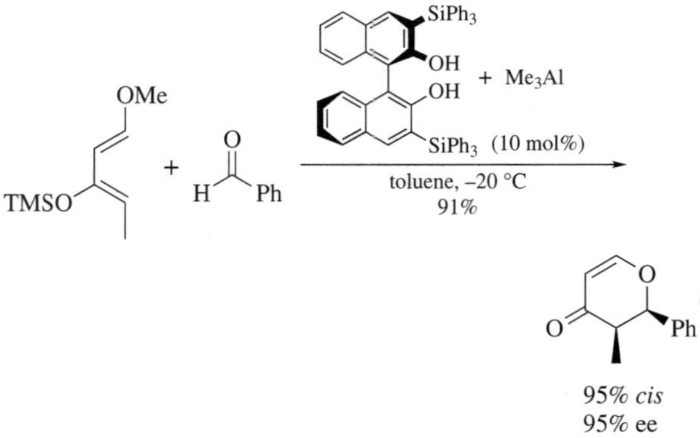
(17)

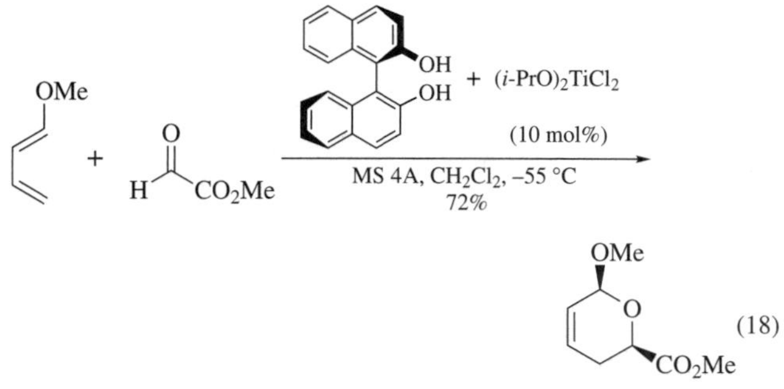

(18)

87% *cis*
96% ee

Carbonyl–Ene Reaction. BINOL-TiX_2 reagent exhibits a remarkable level of asymmetric catalysis in the carbonyl–ene reaction of prochiral glyoxylates, thereby providing practical access to α-hydroxy esters.[16,25] These reactions exhibit a remarkable positive nonlinear effect (asymmetric amplification) that is of practical and mechanistic importance (eq 19).[26] The desymmetrization of prochiral ene substrates with planar symmetry by the enantiofacial selective carbonyl–ene reaction provides an efficient solution to remote internal asymmetric induction (eq 20).[27] The kinetic resolution of a racemic allylic ether by the glyoxylate–ene reaction also provides efficient access to remote but relative asymmetric induction (eq 21).[27] Both the dibromide and dichloride catalysts provide the (2*R*,5*S*)-*syn* product with 97% diastereoselectivity and >95% ee.

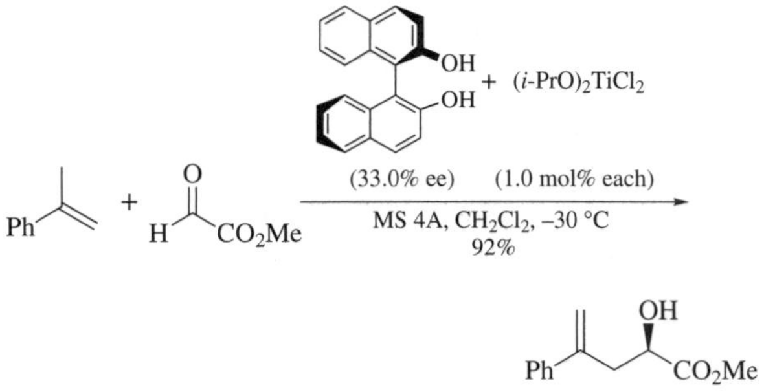

(19)

91.4% ee

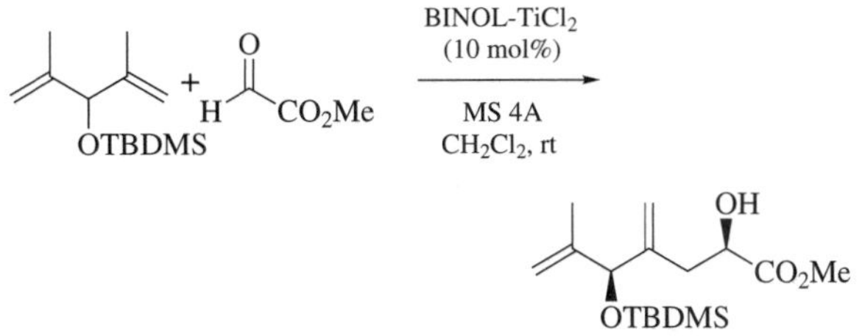

(20)

>99% *syn*
97% ee

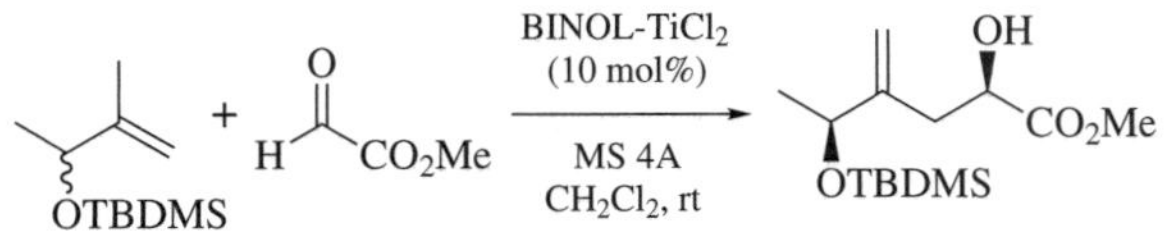

(21)

97% *syn*
>95% ee

Ene Cyclization. An intramolecular (3,4)-ene reaction of unsaturated aldehydes has been accomplished with the BINOL-derived zinc reagent.[28] Cyclization of 3-methylcitronellal with at least 3 equiv of BINOL-Zn reagent afforded the *trans*-cyclohexanol in 86% yield and 88% ee (eq 22). Asymmetric ene cyclizations of type (2,4) are also catalyzed by the BINOL-derived titanium complexes ((*R*)-BINOL-TiX_2, X = ClO_4 or OTf), modified by the perchlorate or trifluoromethanesulfonate ligand. The 7-membered cyclization of type 7-(2,4) gives the oxepane in high ee (eq 23).

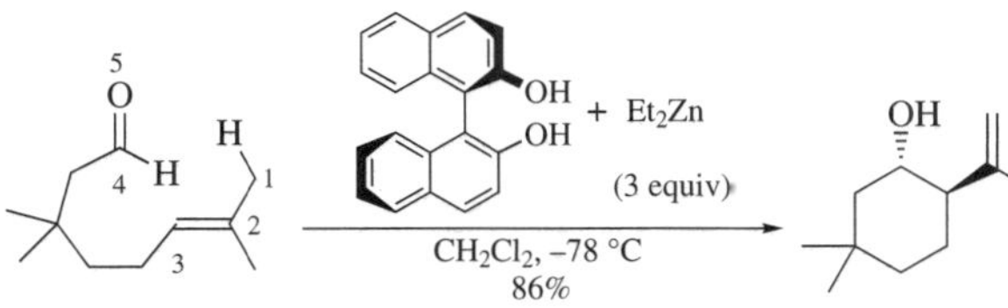

(22)

88% ee

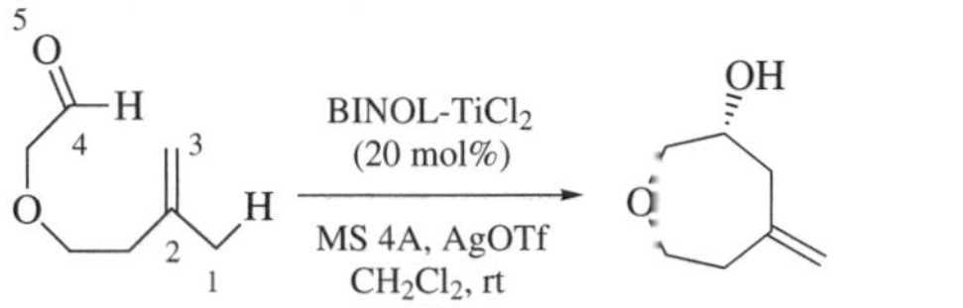

(23)

92% ee

Cationic Cyclization. A cationic cyclization of BINOL-derived neryl ether has been accomplished with an organoaluminum triflate catalyst.[29] Limonene is obtained in 54% yield and 77% ee (eq 24).

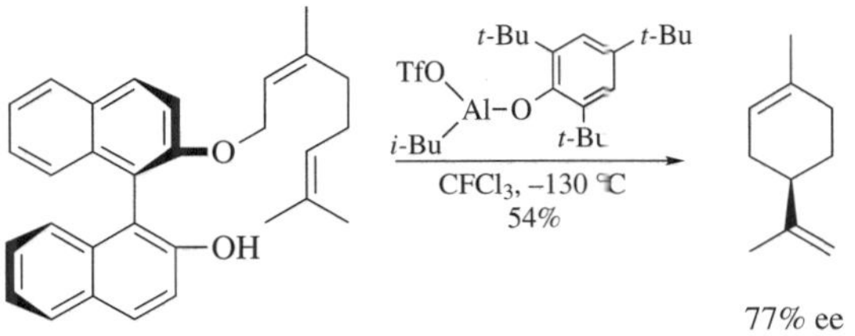

(24)

77% ee

Mukaiyama Aldol Condensation. The BINOL-derived titanium complex BINOL-$TiCl_2$ is an efficient catalyst for the Mukaiyama-type aldol reaction. Not only ketone silyl enol ether (eq 25),[30] but also ketene silyl acetals (eq 26)[31] can be used to give the aldol-type products with control of absolute and relative stereochemistry.

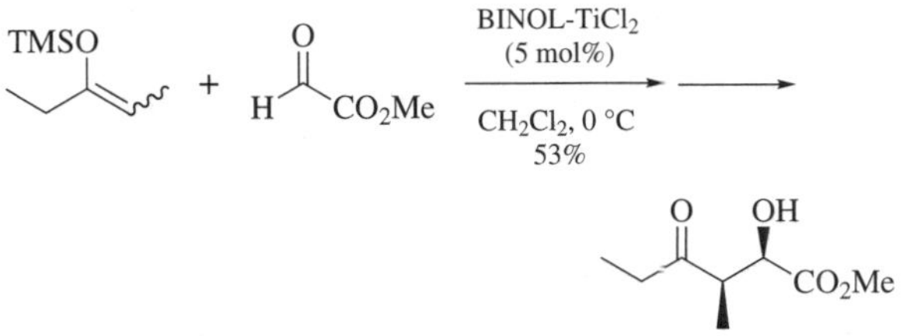

(25)

98% *syn*
99% ee

Nitro-Aldol Condensation. A BINOL-derived lanthanide complex has been used as an efficient catalyst for the nitro-aldol reaction (eq 27).[32] Interestingly enough, the presence of water and LiCl in the reaction mixture is essential to obtain the high level of asymmetric induction and chemical yield.

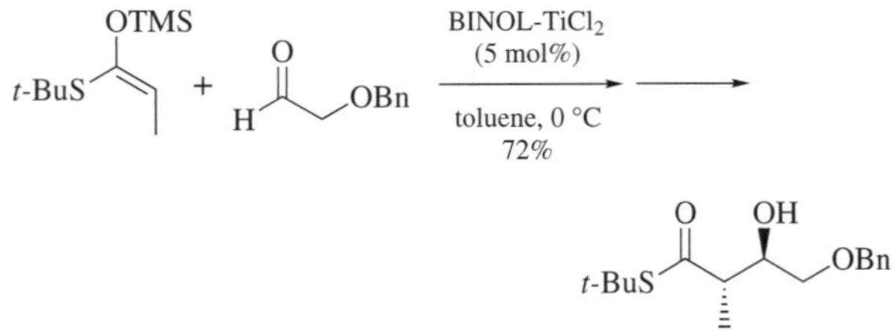

(26)

92% *anti*
90% ee

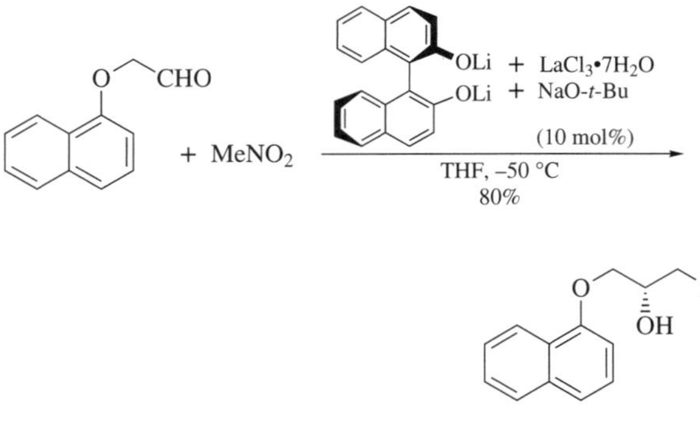

(27)

92% ee

Carbonyl Addition of Allylic Silanes and Stannanes. BINOL-$TiCl_2$ reagent also catalyzes the asymmetric carbonyl addition reaction of allylic silanes and stannanes.[33] Thus the addition reaction of glyoxylate with (*E*)-2-butenylsilane and -stannane proceeds smoothly to give the *syn* product in high enantiomeric excess (eq 28). The reaction of aliphatic and aromatic aldehydes with allylstannane is also catalyzed by BINOL-$TiCl_2$ or BINOL-Ti(O-*i*-Pr)$_2$ to give remarkably high enantioselectivity.[34]

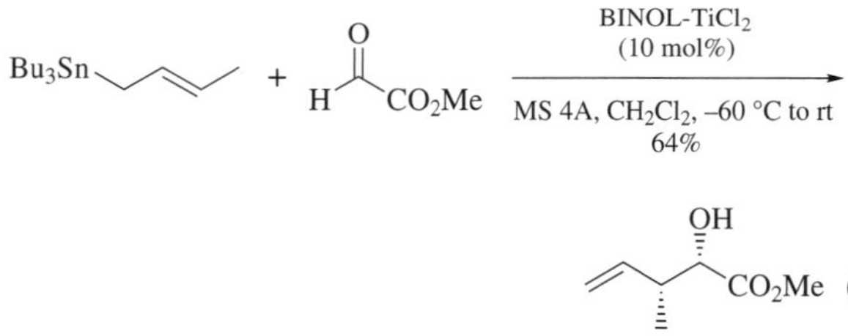

(28)

84% *syn*
86% ee

Claisen Rearrangements. A modified BINOL-derived aluminum reagent is an effective chiral catalyst for asymmetric Claisen rearrangement of allylic vinyl ethers (eq 29).[35] The use of vinyl ethers with sterically demanding C-3 substituents is necessary for the high level of asymmetric induction.

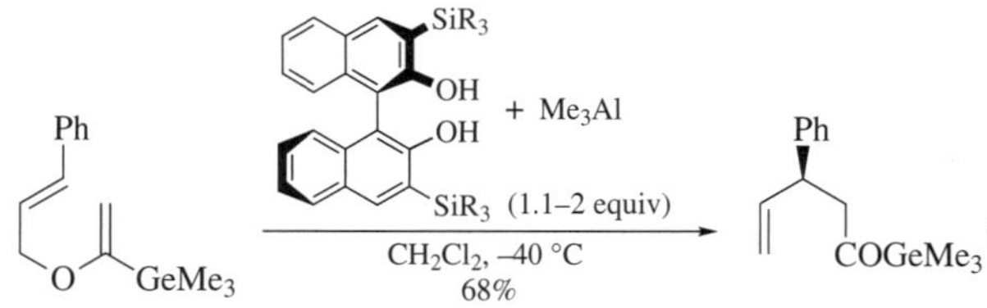

(29)

93% ee

R_3 = *t*-BuPh$_2$

Alkylation of BINOL-Derived Ester Enolates. The diastereoselective alkylation of BINOL-derived arylacetates affords the optically active 2-arylalkanoic acids (eq 30).[36]

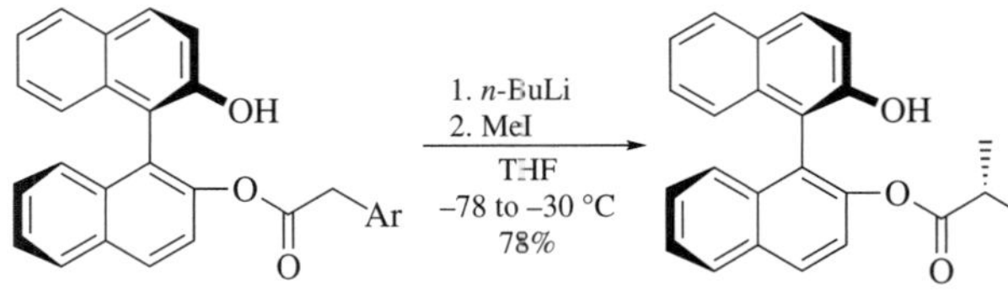

(30)

Ar = 3-pyridyl

90% de

Related Reagents. (*R*)-1,1′-Bi-2,2′-naphtholate; (*R*)-1,1′-Bi-2,2′-naphthotitanium Dichloride; (*R*)-1,1′-Bi-2,2′-naphthotitanium Diisopropoxide; (*R,R*)-[Ethylene-1,2-bis(η^5-4,5,6,7-tetrahydro-1-indenyl)]titanium (*R*)-1,1′-Bi-2,2′-naphtholate; (−)-[Ethylene-1,2-bis(η^5-4, 5, 6, 7-tetrahydro-1-indenyl)]zirconium (*R*)-1,1′-Bi-2,2′-naphtholate; Lithium Aluminum Hydride-2,2′-Dihydroxy-1,1′-binaphthyl.

1. (a) Rosini, C.; Franzini, L.; Raffaelli, A.; Salvadori, P. *Synthesis* **1992**, 503. (b) Miyano, S.; Hashimoto, H. *J. Synth. Org. Chem. Jpn.* **1986**, *44*, 713.
2. (a) Toda, F.; Tanaka, K.; Iwata, S. *J. Org. Chem.* **1989**, *54*, 3007. (b) Pirkle, W. H.; Schreiner, J. L. *J. Org. Chem.* **1981**, *46*, 4988. (c) McKillop, A.; Turrell, A. G.; Young, D. W.; Taylor, E. C. *J. Am. Chem. Soc.* **1980**, *102*, 6504. (d) Carrick, W. L.; Karapinka, G. L.; Kwiatkowski, G. T. *J. Org. Chem.* **1969**, *34*, 2388. (e) Dewar, M. J. S.; Nakaya, T. *J. Am. Chem. Soc.* **1968**, *90*, 7134. (f) Pummerer, R.; Prell, E.; Rieche, A. *Chem. Ber.* **1926**, *59*, 2159.
3. (a) Jacques, J.; Fouquey, C. *Org. Synth.* **1988**, *67*, 1. (b) Jacques, J.; Fouquey, C.; Viterbo, R. *Tetrahedron Lett.* **1971**, 4617. (c) Gong, B.; Chen, W.; Hu, B. *J. Org. Chem.* **1991**, *56*, 423.
4. (a) Brussee, J.; Groenendijk, J. L. G.; Koppele, J. M.; Jansen, A. C. A. *Tetrahedron* **1985**, *41*, 3313. (b) Smrcina, M.; Polakova, J.; Vyskocil, S.; Kocovsky, P. *J. Org. Chem.* **1993**, *58*, 4534.
5. Yamamoto, K.; Fukushima, H.; Nakazaki, M. *Chem. Commun.* **1984**, 1490.
6. Alper, H.; Hamel, N. *J. Am. Chem. Soc.* **1990**, *112*, 2803.
7. (a) Cram, D. J.; Cram, J. M. *Acc. Chem. Res.* **1978**, *11*, 8. (b) Helgeson, R. C.; Timko, J. M.; Moreau, P.; Peacock, S. C.; Mayer, J. M.; Cram, D. J. *J. Am. Chem. Soc.* **1974**, *96*, 6762.
8. Cram, D. J.; Sogah, G. D. Y. *Chem. Commun.* **1981**, 625.
9. Cram, D. J.; Sogah, G. D. Y. *J. Am. Chem. Soc.* **1985**, *107*, 8301.
10. Sepulchre, M.; Spassky, N. *Makromol. Chem. Rapid Commun.* **1981**, *2*, 261.
11. (a) Miyano, S.; Tobita, M.; Hashimoto, H. *Bull. Chem. Soc. Jpn.* **1981**, *54*, 3522. (b) Miyano, S.; Fukushima, H.; Handa, S.; Ito, H.; Hashimoto, H. *Bull. Chem. Soc. Jpn.* **1988**, *61*, 3249.
12. (a) Noyori, R. *Chem. Soc. Rev.* **1989**, *18*, 187. (b) Noyori, R.; Tomino, I.; Yamada, M.; Nishizawa, M. *J. Am. Chem. Soc.* **1984**, *106*, 6717. (c) Noyori, R.; Tomino, I.; Tanimoto, Y.; Nishizawa, M. *J. Am. Chem. Soc.* **1984**, *106*, 6709. (d) Noyori, R.; Tomino, I.; Tanimoto, Y. *J. Am. Chem. Soc.* **1979**, *101*, 3129.
13. (a) Olivero, A. G.; Weidmann, B.; Seebach, D. *Helv. Chim. Acta* **1981**, *64*, 2485. (b) Seebach, D.; Beck, A. K.; Roggo, S.; Wonnacott, A. *Chem. Ber.* **1985**, *118*, 3673. (c) Wang, J. T.; Fan, X.; Feng, X.; Qian, Y. M. *Synthesis* **1989**, 291.
14. Reetz, M. T.; Kyung, S.-H.; Bolm, C.; Zierke, T. *Chem. Ind. (London)* **1986**, 824.
15. Seebach, D.; Beck, A. K.; Imwinkelried, R.; Roggo, S.; Wonnacott, A. *Helv. Chim. Acta* **1987**, *70*, 954.
16. (a) Mikami, K.; Terada, M.; Narisawa, S.; Nakai, T. *Org. Synth.* **1992**, *71*, 14. (b) Mikami, K.; Terada, M.; Nakai, T. *J. Am. Chem. Soc.* **1990**, *112*, 3949.
17. Dijkgraff, C.; Rousseau, J. P. G. *Spectrochim. Acta* **1968**, *2*, 1213.

18. (a) Mikami, K.; Terada, M.; Motoyama, Y.; Nakai, T. *Tetrahedron: Asymmetry* **1991**, *2*, 643. (b) Mikami, K.; Motoyama, Y.; Terada, M. *J. Am. Chem. Soc.* **1994**, *116*, 2812. (c) For the use of a modified BINOL-derived titanium complex, see: Maruoka, K.; Murase, N.; Yamamoto, H. *J. Org. Chem.* **1993**, *58*, 2938.
19. Kelly, T. R.; Whiting, A.; Chandrakumar, N. S. *J. Am. Chem. Soc.* **1986**, *108*, 3510.
20. Chapuis, C.; Jurczak, J. *Helv. Chim. Acta* **1987**, *70*, 436.
21. Kaufmann, D.; Boese, R. *Angew. Chem., Int. Ed. Engl.* **1990**, *29*, 545.
22. (a) Hattori, K.; Yamamoto, H. *Tetrahedron* **1993**, *49*, 1749. (b) For the use in imine-aldol reactions, see: Hattori, K.; Miyata, M.; Yamamoto, H. *J. Am. Chem. Soc.* **1993**, *115*, 1151.
23. (a) Maruoka, K.; Itoh, T.; Shirasaka, T.; Yamamoto, H. *J. Am. Chem. Soc.* **1988**, *110*, 310. (b) For the use D.–A. reaction, see: Maruoka, K.; Concepcion, A. B.; Yamamoto, H. *Bull. Chem., Soc. Jpn.* **1992**, *65*, 3501.
24. Terada, M.; Mikami, K.; Nakai, T. *Tetrahedron Lett.* **1991**, *32*, 935.
25. (a) Glyoxylate-ene reaction with vinylic sulfides and selenides: Terada, M.; Matsukawa, S.; Mikami, K. *Chem. Commun.* **1993**, 327. (b) For the ene reaction of chloral, see: Maruoka, K.; Hoshino, Y.; Shirasaka, T.; Yamamoto, H. *Tetrahedron Lett.* **1988**, *29*, 3967.
26. (a) Mikami, K.; Terada, M. *Tetrahedron* **1992**, *48*, 5671. (b) Terada, M.; Mikami, K.; Nakai, T. *Chem. Commun.* **1990**, 1623.
27. Mikami, K.; Narisawa, S.; Shimizu, M.; Terada, M. *J. Am. Chem. Soc.* **1992**, *114*, 6566.
28. Sakane, S.; Maruoka, K.; Yamamoto, H. *Tetrahedron* **1986**, *42*, 2203.
29. Sakane, S.; Fujiwara, J.; Maruoka, K.; Yamamoto, H. *J. Am. Chem. Soc.* **1983**, *105*, 6154.
30. Mikami, K.; Matsukawa, S. *J. Am. Chem. Soc.* **1993**, *115*, 7039.
31. Mikami, K.; Matsukawa, S. *J. Am. Chem. Soc.* **1994**, *116*, 4077.
32. (a) Sasai, H.; Suzuki, T.; Itoh, N.; Shibasaki, M. *Tetrahedron Lett.* **1993**, *34*, 851. (b) Sasai, H.; Suzuki, T.; Arai, S.; Arai, T.; Shibasaki, M. *J. Am. Chem. Soc.* **1992**, *114*, 4418.
33. Aoki, S.; Mikami, K.; Terada, M.; Nakai, T. *Tetrahedron* **1993**, *49*, 1783.
34. (a) Costa, A. L.; Piazza, M. G.; Tagliavini, E.; Trombini, C.; Umani-Ronchi, A. *J. Am. Chem. Soc.* **1993**, *115*, 7001. (b) Keck, G. E.; Tarbet, K. H.; Geraci, L. S. *J. Am. Chem. Soc.* **1993**, *115*, 8467. (c) Keck, G. E.; Krishnamurthy, D.; Grier, M. C. *J. Org. Chem.* **1993**, *58*, 6543.
35. Maruoka, K.; Banno, H.; Yamamoto, H. *J. Am. Chem. Soc.* **1990**, *112*, 7791.
36. (a) Fuji, K.; Node, M.; Tanaka, F. *Tetrahedron Lett.* **1990**, *31*, 6553. (b) Fuji, K.; Node, M.; Tanaka, F.; Hosoi, S. *Tetrahedron Lett.* **1989**, *30*, 2825.

Koichi Mikami & Yukihiro Motoyama
Tokyo Institute of Technology, Japan

(*R*)-1,1′-Bi-2,2′-naphthotitanium Dichloride[1]

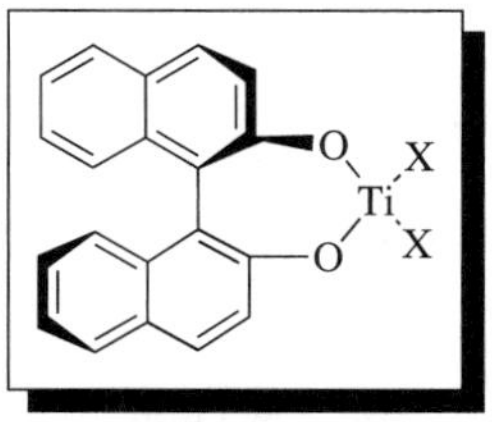

(X = Cl)		
[116051-73-1]	$C_{20}H_{12}Cl_2O_2Ti$	(MW 403.10)
(X = Br)		
[128030-80-8]	$C_{20}H_{12}Br_2O_2Ti$	(MW 492.00)
($X = ClO_4$)		
[138645-47-3]	$C_{20}H_{12}Cl_2O_{10}Ti$	(MW 531.09)
($X = OSO_2CF_3$)		
[139327-61-0]	$C_{22}H_{12}F_6O_8S_2Ti$	(MW 630.32)

(chiral Lewis acid for ene reactions,[2] Mukaiyama aldol reactions,[16] Diels–Alder reactions,[24] and cyanosilylations[27])

Alternate Name: BINOL-TiX_2.
Solubility: insol propionitrile; sol toluene, dichloromethane, and nitroethane.
Handling, Storage, and Precautions: titanium is reputed to be of low toxicity.

Introduction. The (*R*)-1,1′-bi-2,2′-naphthotitanium dihalides (BINOL-TiX_2; X = Br or Cl) are most conveniently prepared in situ from the reaction of diisopropoxytitanium dihalides ($(i\text{-PrO})_2TiX_2$; X = Br[2] or Cl[3]) with *(R)-1,1′-Bi-2,2′-naphthol* (BINOL) in the presence of molecular sieves (MS 4A) (eq 1).[2] When BINOL is mixed with *Dichlorotitanium Diisopropoxide* in the absence of MS 4A, almost no change is observed on the hydroxy-carbon signal of BINOL in the ^{13}C NMR spectrum. However, the addition of MS 4A to the solution of BINOL and $(i\text{-PrO})_2TiCl_2$ leads to a downfield shift of the hydroxy-carbon signal, indicating the formation of the BINOL-derived chiral catalyst. MS (zeolite) serves as an acid/base catalyst[4] and significantly facilitates the alkoxy ligand exchange in the in situ preparation of the chiral catalyst BINOL-TiX_2. A 1:1 mixture of $(i\text{-PrO})_2TiX_2$ and (*R*)-BINOL in the presence of MS 4A in dichloromethane provides a red-brown solution. The molecularity of BINOL-TiX_2 in dichloromethane is ca. 2.0, depending on the concentration, particularly of homochiral (*R*)(*R*)- or (*S*)(*S*)-dimer which tends to dissociate to the monomer in lower concentration.[5]

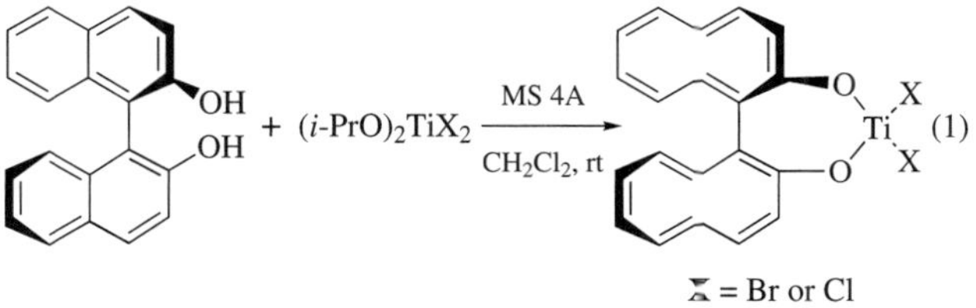

X = Br or Cl

The chiral titanium complexes modified by the perchlorate or trifluoromethanesulfonate ligand such as (*R*)-1,1′-bi-2,2′-naphthotitanium diperchlorate (BINOL-$Ti(ClO_4)_2$) or (*R*)-1,1′-bi-2,2′-naphthotitanium ditriflate ((*R*)-BINOL-$Ti(OTf)_2$) can easily be prepared by the addition of *Silver(I) Perchlorate* or *Silver(I) Trifluoromethanesulfonate* (2 equiv) to BINOL-$TiCl_2$ (eq 2).[6]

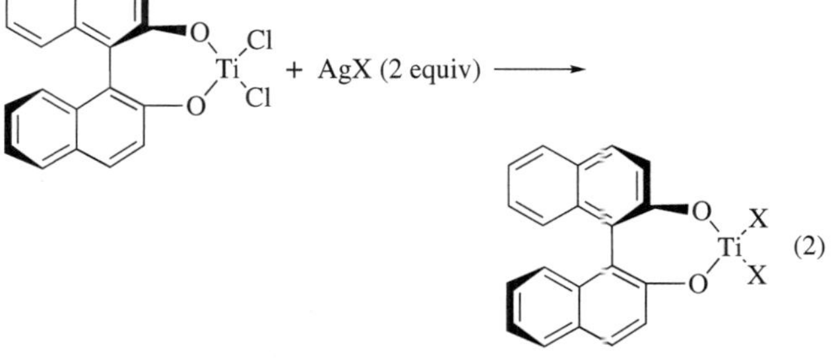

$X = ClO_4$ or OTf

Asymmetric Catalysis of Carbonyl–Ene Reaction. (*R*)-1,1′-Bi-2,2′-naphthotitanium dihalides exhibit a remarkable level of asymmetric induction in the carbonyl–ene reaction of prochiral glyoxylate to provide practical access to α-hydroxy esters, a class of compounds of biological and synthetic importance[7] (eq 3).[2] The catalyst derived from (*R*)-BINOL leads consistently to the (*R*)-alcohol product, whereas the catalyst derived from (*S*)-BINOL affords the (*S*)-enantiomer. Generally speaking, the dibromide is superior to the dichloride in both reactivity and enantioselectivity for the reactions involving a methylene hydrogen shift in particular. On the other hand, the dichloride is lower in reactivity but superior in enantioselectivity for certain reactions involving methyl hydrogen shift. The present asymmetric catalysis is applicable to a variety of 1,1-disubstituted alkenes to provide the ene products in extremely high enantiomeric excess by judicious choice of the dibromo or dichloro catalyst. The reactions of mono- and 1,2-disubstituted alkenes afford no ene product. However, vinylic sulfides and selenides serve as alternatives to mono- and 1,2-disubstituted alkenes, giving the ene products with virtually complete enantioselectivity along with high diastereoselectivity (eq 4).[8] The synthetic advantage of vinylic sulfides and selenides is exemplified by the synthesis of enantiomerically pure (*R*)-(−)-ipsdienol, an insect aggregation pheromone.

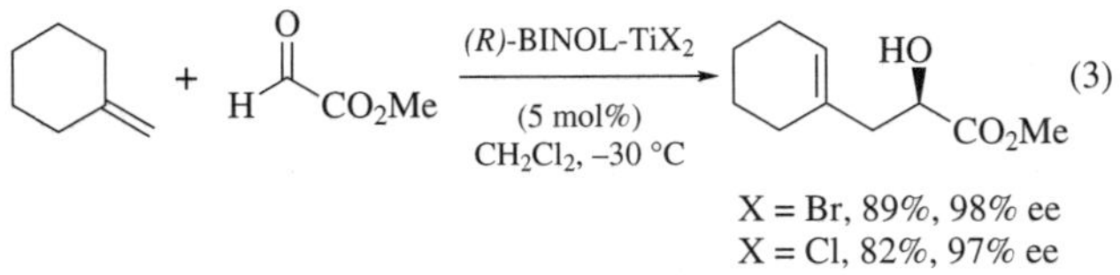

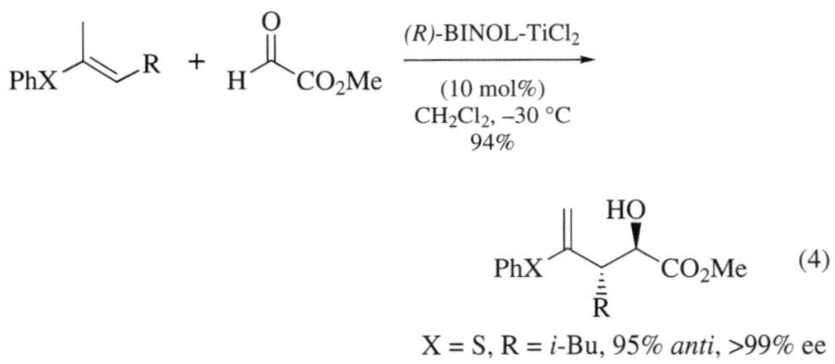

Positive Nonlinear Effect[5] (Asymmetric Amplification[9]). A nonclassical phenomenon of asymmetric catalysis by the chiral BINOL-derived titanium complex is the remarkable positive nonlinear effect observed, which is of practical and mechanistic importance.[5] Convex deviation is observed from the usually assumed linear relationship between the enantiomeric purity of the BINOL ligand and the optical yield of the product. The glyoxylate–ene reaction catalyzed by the chiral titanium complex derived from a partially-resolved BINOL of 33.0% ee, for instance, provides the ene product with 91.4% ee in 92% chemical yield (eq 5). The optical yield thus obtained with a partially resolved BINOL ligand is not only much higher than the % ee of BINOL employed but is also very close to the value of 94.6% ee obtained using the enantiomerically pure BINOL. Thus the use of 35–40% ee of BINOL is sufficient to provide the equally high (>90% ee) level obtained with enantiomerically pure BINOL.

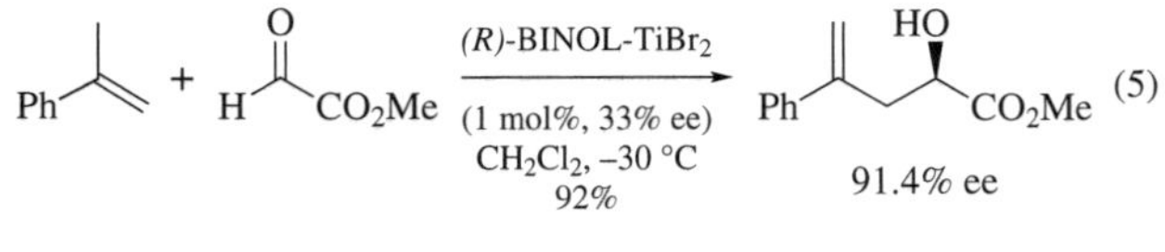

Asymmetric Desymmetrization.[10] Desymmetrization of an achiral, symmetrical molecule is a potentially powerful but relatively unexplored concept for the asymmetric catalysis of carbon–carbon bond formation. While the ability of enzymes to differentiate between enantiotopic functional groups is well known,[11] little is known about the similar ability of nonenzymatic catalysts to effect carbon–carbon bond formation. The desymmetrization by the enantiofacial selective carbonyl–ene reaction of prochiral ene substrates with planar symmetry provides an efficient access to remote internal[12] asymmetric induction which is otherwise difficult to attain (eq 6).[10] The (2*R*,5*S*)-*syn* product is obtained in >99% ee along with more than 99% diastereoselectivity. The desymmetrized product thus obtained can be transformed stereoselectively by a more classical diastereoselective reaction (e.g., hydroboration).

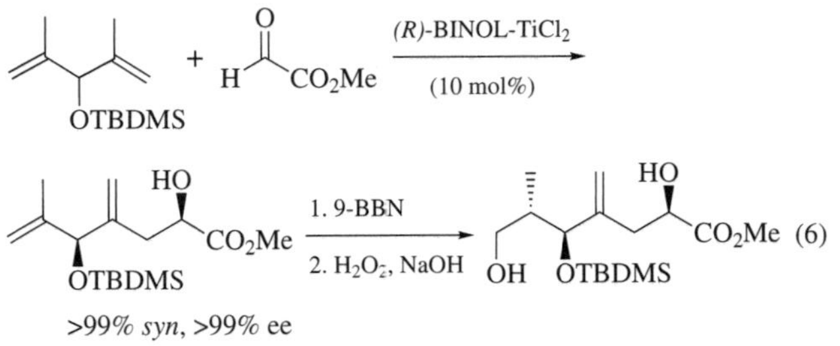

Kinetic Resolution.[13] On the basis of the desymmetrization concept, the kinetic resolution of a racemic substrate might be recognized as an intermolecular desymmetrization.[10] The kinetic resolution of a racemic allylic ether by the glyoxylate–ene reaction also provides an efficient access to remote relative[12] asymmetric induction. Both the dibromide and dichloride catalysts provide the (2*R*,5*S*)-*syn* product with >99% diastereoselectivity along with more than 95% ee (eq 7). The high diastereoselectivity, coupled with the high % ee, strongly suggests that the catalyst/glyoxylate complex efficiently discriminates between the two enantiomeric substrates to accomplish effective kinetic resolution. In fact, the relative rates with racemic ethers are quite large, ca. 60 and 700, respectively. As expected, the reaction of (*S*)-ene using the catalyst (*R*)-BINOL-$TiCl_2$ ('matched' catalytic system) provides complete (>99%) 1,4-*syn* diastereoselectivity in high chemical yield, whereas the reaction of (*R*)-ene using (*R*)-BINOL-$TiCl_2$ ('mismatched' catalytic system) affords a diastereomeric mixture in quite low yield (eq 8).

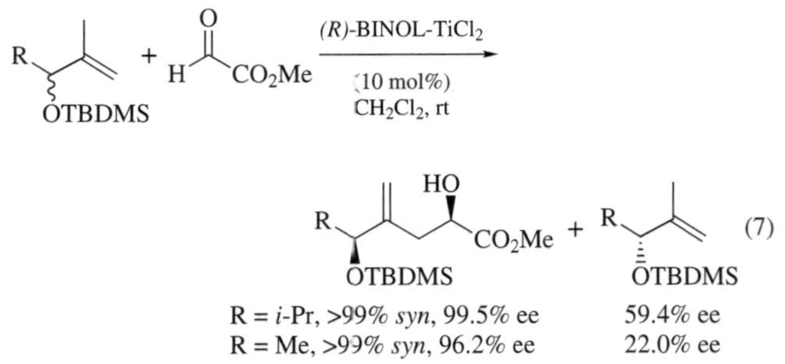

Ene Cyclization.[14] The asymmetric catalysis of the intramolecular carbonyl–ene reaction not only of type (3,4) but also (2,4) employs the BINOL-derived titanium complexes [(*R*)-BINOL-TiX_2; X = ClO_4 or OTf], modified by the perchlorate and trifluoromethanesulfonate ligands.[6] The *trans*-

tetrahydropyran is thus preferentially obtained in 84% ee (eq 9). The seven-membered cyclization of type 7-(2,4) gives the oxepane in high ee, where the *gem*-dimethyl groups are unnecessary (eq 10).

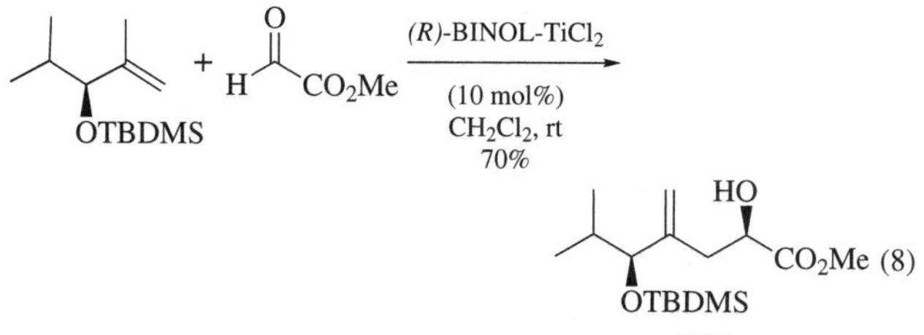

(8)

>99% *syn*

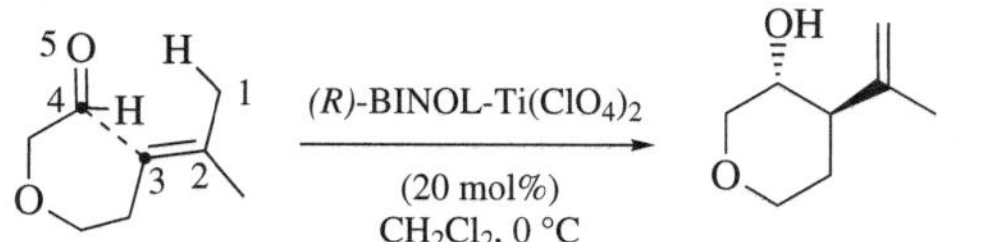

(9)

80% *trans*, 84% ee

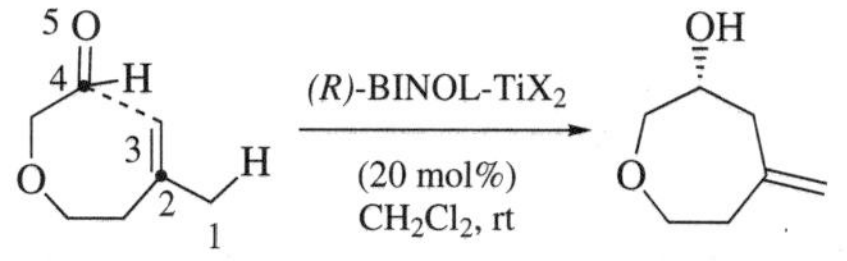

(10)

X = ClO_4, 91% ee
X = OTf, 92% ee

Mukaiyama Aldol Condensation. As expected, the chiral titanium complex is also effective for a variety of carbon–carbon bond forming processes such as the aldol and the Diels–Alder reactions. The aldol process constitutes one of the most fundamental bond constructions in organic synthesis.[15] Therefore the development of chiral catalysts that promote asymmetic aldol reactions in a highly stereocontrolled and truly catalytic fashion has attracted much attention, for which the silyl enol ethers of ketones or esters have been used as a storable enolate component (Mukaiyama aldol condensation). The BINOL-derived titanium complex BINOL-$TiCl_2$ can be used as an efficient catalyst for the Mukaiyama-type aldol reaction of not only ketone silyl enol ethers but also ester silyl enol ethers with control of absolute and relative stereochemistry (eq 11).[16]

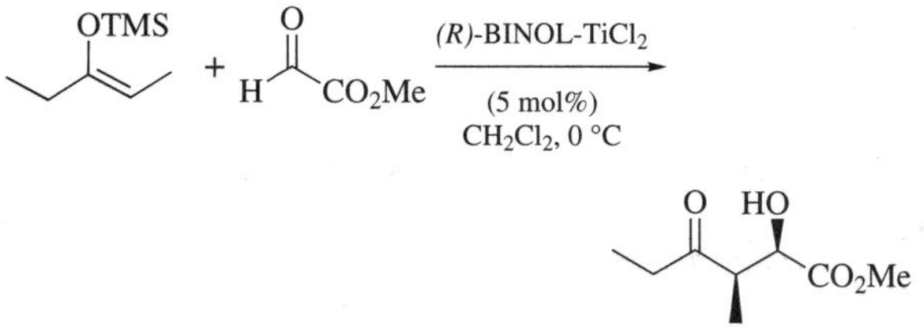

(11)

98% *syn*, 99% ee

Carbonyl Addition of Allylic Silanes and Stannanes.[17] The chiral titanium complex BINOL-$TiCl_2$ also catalyzes the asymmetric carbonyl addition reaction of allylic silanes and stannanes.[18] Thus the addition reaction of glyoxylate with (*E*)-2-butenylsilane and -stannane proceeds smoothly to give the *syn* product in high enantiomeric excess (eq 12). The *syn* product thus obtained can be readily converted to the lactone portion of verrucaline A. The reaction of aliphatic and aromatic aldehydes with allylstannane is also catalyzed by BINOL-$TiCl_2$ to give remarkably high enantioselectivity (eq 13).[19]

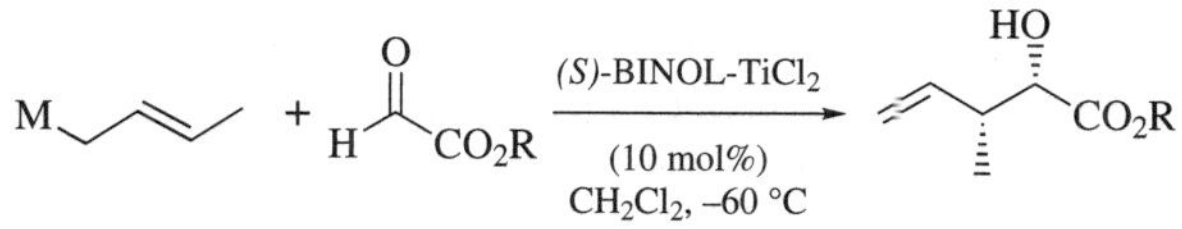

(12)

M = $SnBu_3$, R = *n*-Bu, 84% *syn*, 86% ee
M = $SiMe_3$, R = Me, 83% *syn*, 80% ee

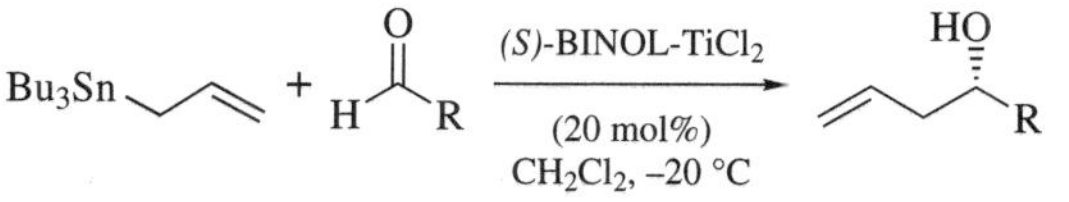

(13)

R = C_5H_{11}, 98.4% ee
R = PhCH=CH, 94% ee

Hetero Diels–Alder Reaction.[20] The hetero-Diels–Alder reaction involving glyoxylate as the dienophile provides an efficient access to the asymmetric synthesis of monosaccharides.[21] The hetero Diels–Alder reaction with methoxydienes proceeds smoothly with catalysis by BINOL-$TiCl_2$ to give the *cis* product in high enantiomeric excess (eq 14).[22] The dibromide affords a higher *cis* selectivity, however, with a lower enantioselectivity, particularly in the *trans* adduct. The product thus obtained can be readily converted to the lactone portion of HMG-CoA inhibitors such as mevinolin or compactin.[23]

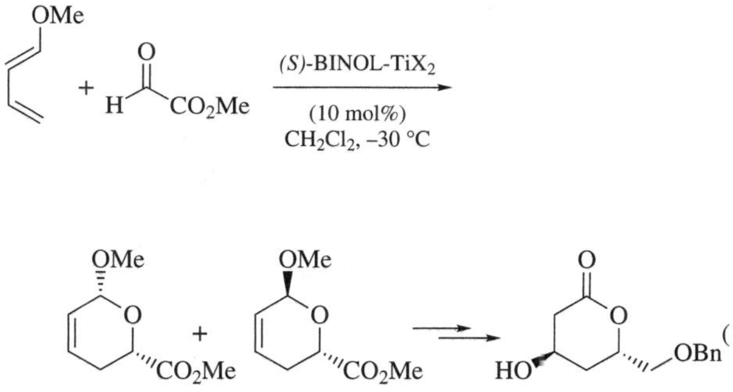

(14)

X = Cl ,78 (94% ee):22 (>90% ee)
X = Br, 84 (92% ee):16 (50% ee)

Diels–Alder Reaction.[24] The Diels–Alder reaction of methacrolein with 1,3-dienol derivatives can also be catalyzed by the chiral BINOL-derived titanium complex BINOL-$TiCl_2$. The *endo* adduct was obtained in high enantioselectivity (eq 15).[22a,25] The sense of asymmetric induction is exactly the same as observed for the asymmetric catalytic reactions shown above. Asymmetric catalytic Diels–Alder reactions with naphthoquinone derivatives as a dienophile provide an efficient entry to the asymmetric synthesis of anthracyclinone aglycones (eq 16).[26]

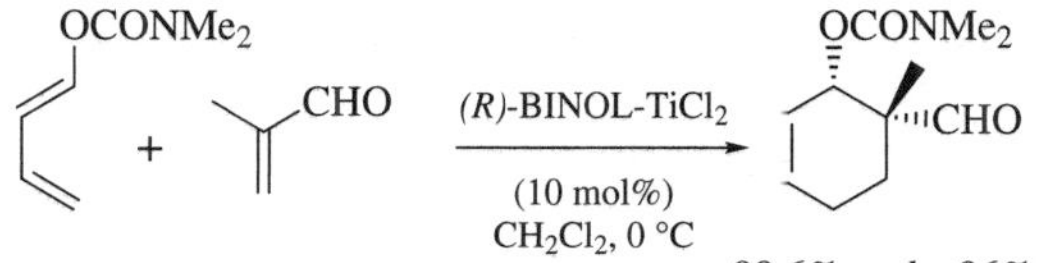

(15)

99.6% *endo*, 86% ee

Cyanosilylation.[27] Another preparative procedure of BINOL-$TiCl_2$ and the use thereof was reported in the asymmetric catalysis of the addition reaction of cyanotrimethylsilane to aldehydes.[28] The dilithium salt of BINOL in ether was treated with

Titanium(IV) Chloride, the red-brown mixture was warmed to room temperature, and the ether removed in vacuo. Dry benzene was added and the nondissolved solid was separated via filtration under nitrogen. Removal of the solvent delivered 50% of a sensitive red-brown solid which showed a single set of ^{13}C NMR signals (eq 17). The BINOL-$TiCl_2$ thus obtained was utilized to prepare the cyanohydrin of 3-methylbutanal in <82% ee (eq 18).

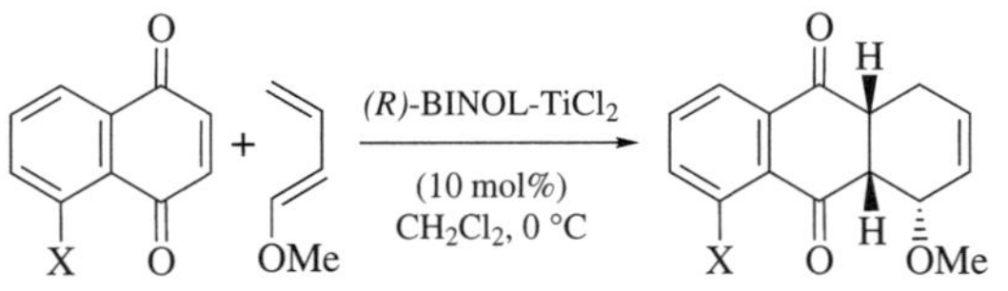

(16)

X = H, 85% ee

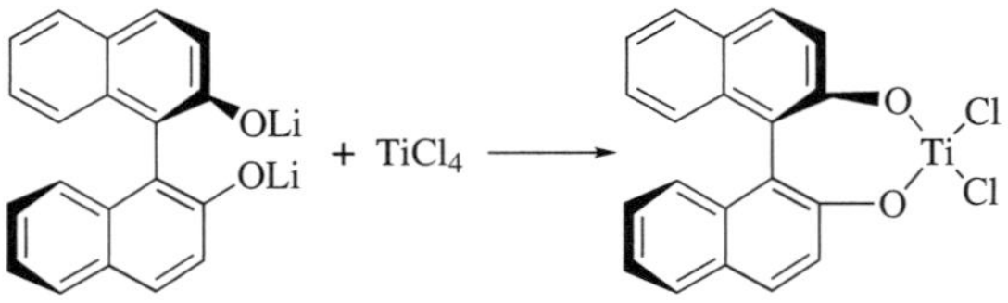

(17)

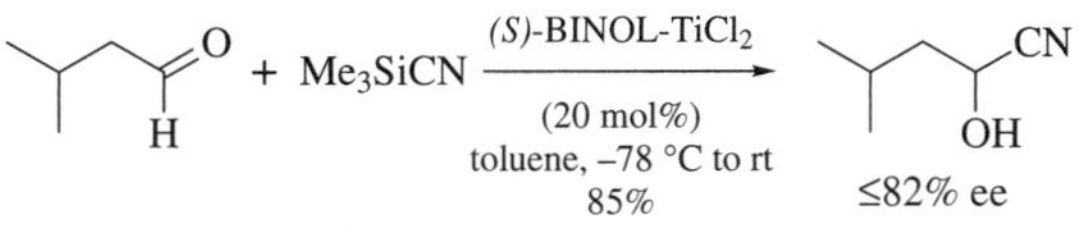

(18)

Related Reagents. (*R*)-1,1′-Bi-2,2′-naphthol; (*R*)-1,1′-Bi-2,2′-naphthotitanium Diisopropoxide; Titanium(IV) Chloride.

1. (a) Mikami, K.; Shimizu, M. *Chem. Rev.* **1992**, *92*, 1021. (b) Mikami, K.; Terada, M.; Narisawa, S.; Nakai, T. *Synlett* **1992**, 255.
2. (a) Mikami, K.; Terada, M.; Narisawa, S.; Nakai, T. *Org. Synth.* **1992**, *71*, 14. (b) Mikami, K.; Terada, M.; Nakai, T. *J. Am. Chem. Soc.* **1990**, *112*, 3949. (c) Terada, M.; Nakai, T. *J. Am. Chem. Soc.* **1989**, *111*, 1940.
3. Dijkgaaf, C.; Rousseau, J. P. G. *Spectrochim. Acta* **1968**, *24A*, 1213.
4. (a) Thomas, J. M.; Theocaris, C. R. *Modern Synthetic Methods*; Springer: Berlin, 1989. (b) Onaka, M.; Izumi, Y. *Yuki Gosai Kagaku Kyokaishi* **1989**, *47*, 233. (c) Dyer, A. *An Introduction to Zeolite Molecular Sieves*; Wiley: Chichester, 1988.
5. (a) Mikami, K.; Terada, M. *Tetrahedron* **1992**, *48*, 5671. (b) Terada, M.; Mikami, K.; Nakai, T. *Chem. Commun.* **1990**, 1623.
6. (a) Mikami, K.; Sawa, E.; Terada, M. *Tetrahedron: Asymmetry* **1991**, *2*, 1403. (b) Mikami, K.; Terada, M.; Sawa, E.; Nakai, T. *Tetrahedron Lett.* **1991**, *32*, 6571.
7. (a) Omura, S. *J. Synth. Org. Chem., Jpn.* **1986**, *44*, 127; (b) Hanessian, S. *Total Synthesis of Natural Products: The 'Chiron' Approach*; Pergamon: Oxford, 1983. (c) Seebach, D.; Hungerbuhler, E. *Modern Synthetic Methods*; Otto Salle: Frankfurt am Main, 1980.
8. Terada, M.; Matsukawa, S.; Mikami, K. *Chem. Commun.* **1993**, 327.
9. (a) Noyori, R.; Kitamura, M. *Angew. Chem., Int. Ed. Engl.* **1991**, *30*, 49. (b) Wynberg, H. *Chimia* **1989**, *43*, 150. (c) Puchot, C.; Samuel, O.; Duñach, E.; Zhao, S.; Agami, C.; Kagan, H. B. *J. Am. Chem. Soc.* **1986**, *108*, 2353.
10. Mikami, K.; Narisawa, S.; Shimizu, M.; Terada, M. *J. Am. Chem. Soc.* **1992**, *114*, 6566.
11. Ward, R. S. *Chem. Soc. Rev.* **1990**, *19*, 1.
12. Bartlett, P. A. *Tetrahedron* **1980**, *36*, 3.
13. (a) Kagan, H. B.; Fiaud, J. C. *Top. Stereochem.* **1988**, *18*, 249. (b) Brown, J. M. *Chem. Ind. (London)* **1988**, 612.
14. (a) Oppolzer, W.; Snieckus, V. *Angew. Chem., Int. Ed. Engl.* **1978**, *17*, 476. (b) Taber, D. F. *Intramolecular Diels–Alder and Alder Ene Reactions*; Springer: Berlin, 1984.
15. (a) Masamune, S.; Choy, W.; Peterson, J. S.; Sita, L. R. *Angew. Chem., Int. Ed. Engl.* **1985**, *24*, 1. (b) Heathcock, C. H. In *Asymmetric Synthesis*; Morrison, J. D., Ed.; Academic: New York, 1984. (c) Evans, D. A.; Nelson, J. V.; Taber, T. R. *Top. Stereochem.* **1982**, *13*, 1. (d) Mukaiyama, T. *Org. React.* **1982**, *28*, 203.
16. Mikami, K.; Matsukawa, S. *J. Am. Chem. Soc.* **1993**, *115*, 7039; **1994**, *116*, 4077.
17. (a) Sakurai, H. *Synlett* **1989**, 1. (b) Hosomi, A. *Acc. Chem. Res.* **1988**, *21*, 200. (c) Yamamoto, Y. *Acc. Chem. Res.* **1987**, *20*, 243. (d) Hoffmann, R. W. *Angew. Chem., Int. Ed. Engl.* **1982**, *21*, 555.
18. (a) Aoki, S.; Mikami, K.; Terada, M.; Nakai, T. *Tetrahedron* **1993**, *49*, 1783. (b) Mikami, K.; Matsukawa, S. *Tetrahedron Lett.* **1994**, *35*, 3133.
19. Costa, A. L.; Piazza, M. G.; Tagliavini, E.; Trombini, C.; Umani-Ronchi, A. *J. Am. Chem. Soc.* **1993**, *115*, 7001.
20. (a) Bednarski, M. D.; Lyssikatos, J. P. *Comprehensive Organic Synthesis*, **1991**, *2*, Chapter 2.5. (b) Boger, D. L.; Weinreb, S. M. *Hetero-Diels–Alder Methodology in Organic Synthesis*; Academic: New York, 1987. (c) Konowal, A.; Jurczak, J.; Zamojski, A. *Tetrahedron* **1976**, *32*, 2957.
21. (a) Konowal, A.; Jurczak, J.; Zamojski, A. *Tetrahedron* **1976**, *32*, 2957. (b) Danishefsky, S. J.; DeNinno, M. P. *Angew. Chem., Int. Ed. Engl.* **1987**, *26*, 15.
22. (a) Mikami, K.; Motoyama, Y.; Terada, M. *J. Am. Chem. Soc.* **1994**, *116*, 2812. (b) Terada, M.; Mikami, K.; Nakai, T, *Tetrahedron Lett.* **1991**, *32*, 935.
23. Rosen, T.; Heathcock, C. H. *Tetrahedron* **1986**, *42*, 4909.
24. (a) Kagan, H. B.; Riant, O. *Chem. Rev.* **1992**, *92*, 1007. (b) Oppolzer, W. *Comprehensive Organic Synthesis*, **1991**, *5*, Chapter 1.2. (c) Fringuelli, F.; Taticchi, A. *Dienes in the Diels–Alder Reaction*; Wiley: New York, 1990. (d) Taschner, M. J. *Org. Synth. Theory Appl.* **1989**, *1*, 1. (e) Paquette, L. A. In *Asymmetric Synthesis*; Morrison, J. D., Ed.; Academic: New York, 1984; Vol. 3B, Chapter 7.
25. Mikami, K.; Terada, M.; Motoyama, Y.; Nakai, T. *Tetrahedron: Asymmetry* **1991**, *2*, 643.
26. (a) Krohn, K. *Tetrahedron* **1990**, *46*, 291. (b) Krohn, K. *Angew. Chem., Int. Ed. Engl.* **1986**, *25*, 790. (c) Broadhurst, M. J.; Hassall, C. H.; Thomas, G. J. *Chem. Ind. (London)* **1985**, 106. (d) Arcamone, F. *Med. Res. Rev.* **1984**, *4*, 153.
27. Rasmussen, J. K.; Heilmann, S. M.; Krepski, L. R. *Adv. Silicon Chem.* **1991**, *1*, 65.
28. Reetz, M. T.; Kyung, S.-H.; Bolm, C.; Zierke, T. *Chem. Ind. (London)* **1986**, 824.

Koichi Mikami
Tokyo Institute of Technology, Japan

(*R*)-1,1′-Bi-2,2′-naphthotitanium Diisopropoxide[1]

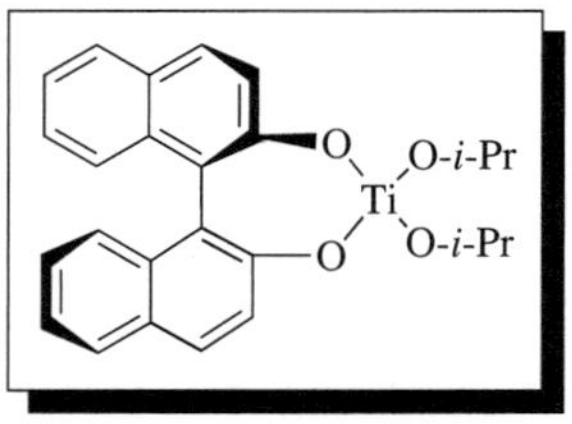

[123436-17-9] $C_{26}H_{26}O_4Ti$ (MW 450.37)

(mild Lewis acid catalyst for asymmetric oxidations[1,2] and allylations[7])

Physical Data: mp 127 °C (dec) (pentane/ether).
Solubility: sol dichloromethane, toluene, and ether.
Handling, Storage, and Precautions: titanium is reputed to be of low toxicity.

(*R*)-1,1′-Bi-2,2′-naphthotitanium diisopropoxide (BINOL-Ti(O-*i*-Pr)$_2$) (**1**), most conveniently prepared by the reaction of *Titanium Tetraisopropoxide* with *(R)-1,1′-Bi-2,2′-naphthol* through the azeotropic removal of isopropanol, is oxophilic (eq 1). A 1:1 mixture of Ti(O-*i*-Pr)$_4$ and (*R*)-BINOL in dichloromethane provides an orange-yellow solution and on removal of solvent gives a pale yellow solid. The molecularity of the 1:1 titanium–binaphthol species has been determined to be 2.3. However, the X-ray crystal structure is trimeric, containing a C_2 axis of symmetry.[3] This 1:1 complex provides, however, only low enantiomeric excess (ee) in the asymmetric epoxidation of allylic alcohols (eq 2).[1,3] However, the oxidation of sulfides to sulfoxides by *t-Butyl Hydroperoxide* proceeds catalytically with (**1**) to afford higher enantioselectivity than Kagan's catalytic method using diethyl tartrate as a chiral ligand (eq 3).[4] As Kagan has already reported, the amount of water added exhibits a significant effect upon ee value. A high ee is obtained when 0.5–3.0 equiv of H_2O was added to the sulfide, while a decrease of ee is observed when less than 0.5 equiv or more than 3.0 equiv of H_2O is used. In the absence of H_2O, only a low ee is obtained. Addition of water to (**1**) provides binaphthol-titanium oxide (BINOL–Ti=O; **2**) (eq 4),[5] which has been reported to serve as an asymmetric catalyst for the Mukaiyama aldol reaction (eq 5).[6]

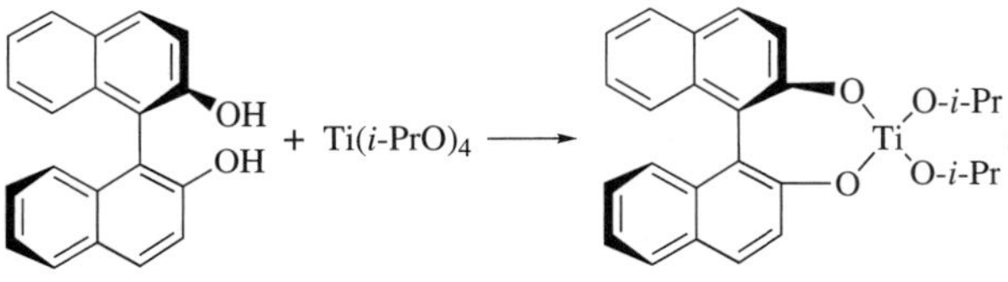

(1)

(**1**)

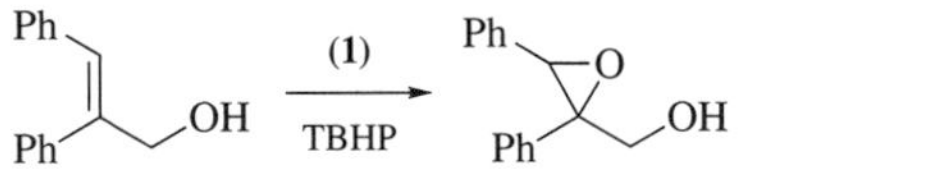

(2)

9% ee

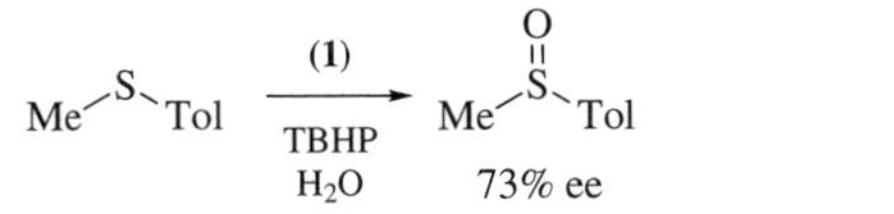

(3)

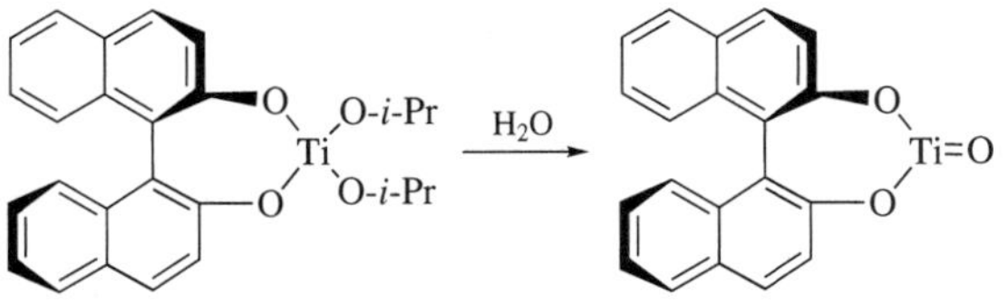

(4)

(**2**)

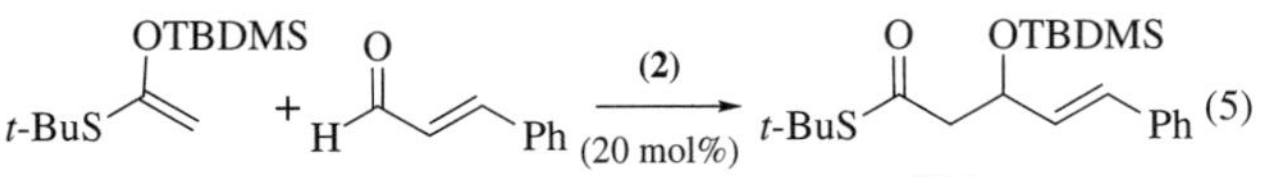

(5)

85% ee

Related Reagents. (*R*)-1,1′-Bi-2,2′-naphthol; (*R*)-1,1′-Bi-2,2′-naphthotitanium Dichloride Titanium(IV) Chloride.

---

1. Finn, M. G.; Sharpless, K. B. In *Asymmetric Synthesis*, Morrison, J. D., Ed.; Academic: New York, 1985; Vol. 5, pp 247–308.
2. Komatsu, N.; Nishibayashi, Y.; Sugita, T.; Uemura, S. *Tetrahedron Lett.* **1992**, *33*, 5391. Komatsu, N.; Hashizume, M.; Sugita, T.; Uemura, S. *J. Org. Chem.* **1993**, *58*, 4529.
3. Martin, C. A. Ph.D. Thesis, Massachusetts Institute of Technology, 1988.
4. Kagan, H. B.; Rebiere, F. *Synlett* **1990**, 643.
5. Bradley, D. C.; Gaze, R.; Wardlaw, W. *J. Chem. Soc.* **1955**, 721.
6. Mukaiyama, T.; Inubushi, A.; Suda, S.; Hara, R. Kobayashi, S. *Chem. Lett.* **1990**, 1015.
7. Keck, G. E., Tarbet, K. H.; Geraci, L. S. *J. Am. Chem. Soc.* **1993**, *115*, 8467. Keck, G. E.; Krishnamurthy, O.; Grier, M. C. *J. Org. Chem.* **1993**, *58*, 6543.

Koichi Mikami
Tokyo Institute of Technology, Japan

(*S*)-2,2′Binaphthoyl(*R*,*R*)-di(1-phenylethyl)aminoylphosphine[1]

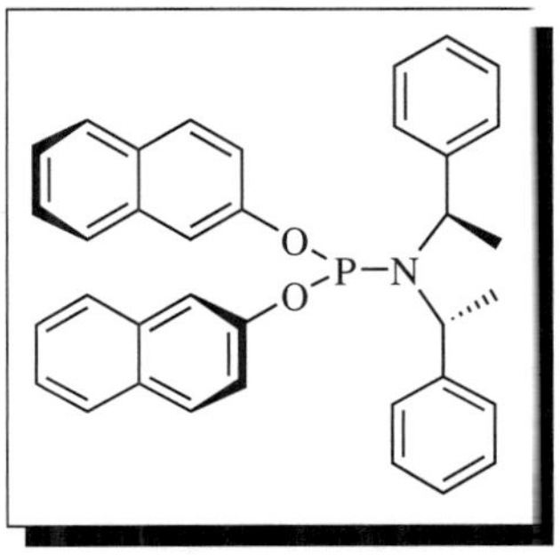

[201732-49-2] $C_{36}H_{30}NO_2P$ (MW 539.61)

(chiral phosphoramidite ligand for copper-catalyzed asymmetric conjugate addition of dialkylzinc reagents to α,β-unsaturated acyclic and cyclic ketones,[2] kinetic resolution of diene epoxides,[3] and ring annulation[4])

Alternate Name: *O,O*′-(*S*)-(1,1′-dinaphthyl-2,2′-diyl)-*N,N*′-di-(*R,R*)-1-phenylethylphosphoramidite.
Physical Data: $[\alpha]_D = +456.0$ (c 0.79, $CHCl_3$); ^1H NMR: δ 7.98–8.08 (m, 4H), 7.17–7.74 (m, 18H), 4.63 (q, $J = 7.2$ Hz, 2H), 1.85 (d, $J = 7.2$ Hz, 6H); ^{31}P NMR; δ 145.3.[5]
Solubility: soluble in chloroform, tetrahydrofuran, and diethyl ether.
Form Supplied in: crystalline solid.
Analysis of Reagent Purity: NMR, MS.
Preparative Methods: the phosphoramidite ligand can be prepared by the nucleophilic substitution of phosphoryl chloride (formed from the reaction of PCl_3 and (*S*)-2,2′-binaphthol in presence of triethylamine) with (*R,R*)-bis(1-phenylethyl)amine.[5]
Purification: recrystallization from diethyl ether/dichloromethane.
Handling, Storage, and Precautions: air and moisture stable solid; no special handling and storage precautions are indicated.

Tandem Asymmetric Conjugate Addition. Enantioselective conjugate addition of an organometallic reagent to a prochiral

Michael acceptor is a powerful tool for carbon-carbon bond formation with simultaneous introduction of a new stereogenic center.[6] The binaphthyl derived phosphoramidite {(*S,R,R*)-**1** ligand} showed remarkable stereoselectivities in copper-catalyzed 1,4-addition of alkylzinc reagents to α,β-unsaturated carbonyl compounds.[1]

Cyclic Enones

High levels of enantioselectivity (94–98% ee) and good chemical yield (72–95%) were observed in the catalytic conjugate addition of dialkylzinc reagents to numerous cyclic enones (eq 1) using a catalyst prepared in situ from $Cu(OTf)_2$ and this chiral phosphoramidite ligand. Here the steric properties of the substrate and the reagent appear to be unimportant.

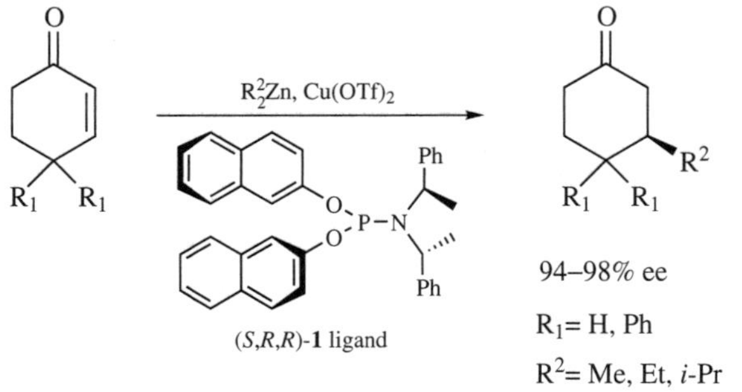
(1)

Cyclohexadienones

Enantioselective copper phosphoramidite-catalyzed conjugate addition of dialkylzinc reagents to several 4,4-disubstituted cyclohexadienones is achieved with diastereomeric ratios ranging from 1/1 to 99/1 with 85% to 99% ee. When the two substituents are equal (eq 2), selective *Re* versus *Si* face-selective addition of the zinc reagent affords a single isomer.[7] Sequential catalytic 1,4-addition to the prochiral dienones gave *cis* or *trans* bis-adducts with high enantio and diastereoselectivity.[8]

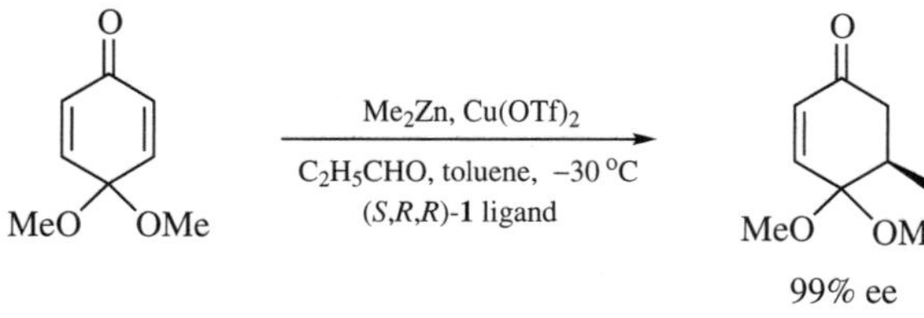
(2)

Asymmetric allylation of cinnamyl halides,[9] alkylation of alkynyl epoxides,[10] and 1,4-addition of nitro olefins[11] are also successfully demonstrated by combination of an organozinc reagent and a chiral copper phosphoramidite.

1,4-Addition—Aldol Reaction. The zinc enolates resulting in situ from conjugate addition are trapped by an appropriate electrophile (aldehyde) in a subsequent aldol reaction to achieve the regio and enantioselective catalytic three-component coupling (eq 3).[12] The ligand-accelerated 1,4-addition using chiral copper phosphoramidite catalyst is developed for highly enantioselective annulation methodology (eq 4) for cyclohexanones, cycloheptanones, and cyclooctanones.[4]

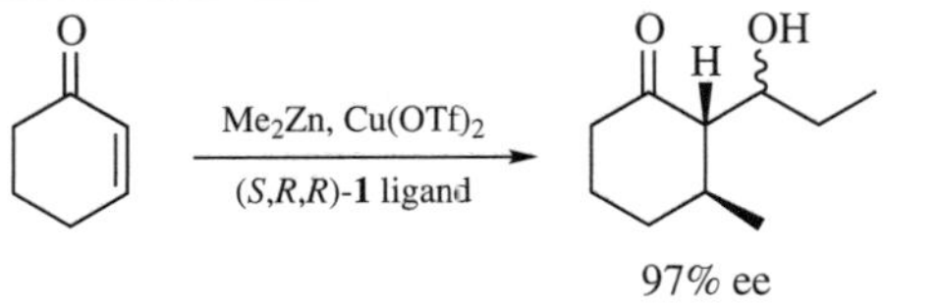
(3)

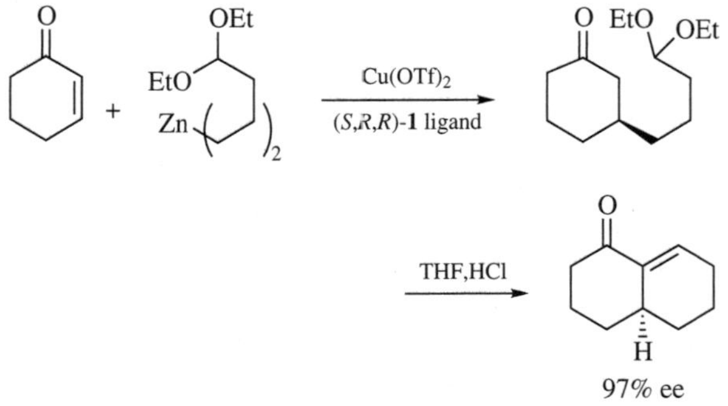
(4)

Kinetic Resolution and Desymmetrization. A variety of substituted 2-cyclohexenones are obtained in enantiomerically pure form employing chiral copper phosphoramidite catalyst for kinetic resolution (eq 5) (>99% ee at 52% conversion, selectivity $S > 200$).[13] The enantioselective desymmetrization of methylidene epoxycycloalkanes is also reported.[3]

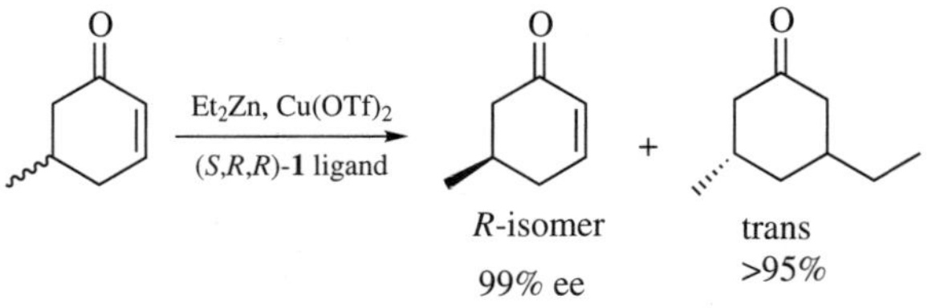
(5)

Related Reagents. (*S*)-2-(2,6-dimethoxyphenyl)oxazole; hexamethylphosphoric triamide; (*S*)-BINAP; (+)-DIOP; trialkyl phosphines; triaryl phosphines etc.

1. (a) Feringa, B. L. *Acc. Chem. Res.* **2000**, *33*, 346–353. (b) Krause, N. *Angew. Chem., Int. Ed. Engl.* **1998**, *37*, 283–285. (c) Feringa, B. L.; deVries, A. H. M. In *Advances in Catalytic Processes*, Doyle, M. P., Ed.; JAI Press Inc; Greenwich, Connecticut, 1995, pp 151–192.
2. deVries, A. H. M.; Meetsma, A.; Feringa, B. L. *Angew. Chem., Int. Ed. Engl.* **1996**, *35*, 2374–2376.
3. Bertozzi, F.; Crotti, P.; Macchia, F.; Pineschi, M.; Arnold, A.; Feringa, B. L. *Org. Lett.* **2000**, *2*, 933–936.
4. Naasz, R.; Arnold, L. A.; Pineschi, M.; Keller, E.; Feringa, B. L. *J. Am. Chem. Soc.* **1999**, *121*, 1104–1105.
5. Arnold, L. A.; Imbos, R.; Mandoli, A.; deVries, A. H. M.; Naasz, R.; Feringa, B. L. *Tetrahedron* **2000**, *56*, 2865–2878.
6. (a) Rossiter, B. E.; Swingle, N. M. *Chem. Rev.* **1992**, *92*, 771–806. (b) Krause, N.; Gerold, A. *Angew. Chem., Int. Ed. Engl.* **1997**, *36*, 186–204.
7. Imbos, R.; Brilman, M. H. G.; Pineschi, M.; Feringa, B. L. *Org. Lett.* **1999**, *1*, 623–626.
8. Imbos, R.; Minnaard, A. J.; Feringa, B. L. *Tetrahedron* **2001**, *57*, 2485–2489.
9. Malda, H.; van Zijl, A. W.; Arnold, L. A.; Feringa, B. L. *Org. Lett.* **2001**, *3*, 1169–1171.
10. Bertozzi, F.; Crotti, P.; Macchia, F.; Pineschi, M.; Arnold, A.; Feringa, B. L. *Tetrahedron Lett.* **1999**, *40*, 4893–4896.
11. (a) Sewald, N.; Wendisch, V. *Tetrahedron: Asymmetry* **1998**, *9*, 1341. (b) Versleijen, J. P. G.; van Leusen, A. M.; Feringa, B. L. *Tetrahedron Lett.* **1999**, *40*, 5803–5806.

12. (a) Feringa, B. L.; Pineschi, M.; Arnold, L. A.; Imbos, R.; deVries, A. H. M. *Angew. Chem., Int. Ed. Engl.* **1997**, *36*, 2720–2623. (b) Alexakis, A.; Trevitt, G. P.; Bernardinelli, G. *J. Am. Chem. Soc.* **2001**, *123*, 4358–4359.
13. (a) Bertozzi, F.; Crotti, P.; Macchia, F.; Pineschi, M.; Feringa, B. L. *Angew. Chem., Int. Ed. Engl.* **2001**, *40*, 930–932. (b) Naasz, R.; Arnold, L. A.; Minnaard, A. J.; Feringa, B. L. *Angew. Chem., Int. Ed. Engl.* **2001**, *40*, 927–929.

Robert S. Coleman & Srinivas Reddy Gurrala
The Ohio State University, Columbus, OH, USA

1,1′-Binaphthyl-2,2′-diyl Hydrogen Phosphate[1]

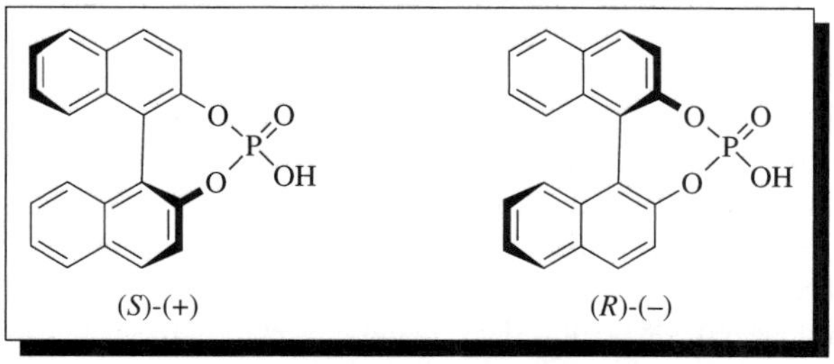

(*S*)-(+)
[35193-64-7] $C_{20}H_{13}O_4P$ (MW 348.30)
(*R*)-(−)
[39648-67-4]
(±)
[50574-52-2]

(reagent for optical resolution of a variety of organic bases[2] and helicenes;[3] NMR shift reagent for determining the enantiomeric purity of secondary and tertiary amines;[4] chiral ligand for homogeneous asymmetric catalysts[5,6])

Alternate Name: BNPPA.
Physical Data: (*S*)-(+)-BNPPA: $[\alpha]_D^{20}$ +605° (*c* 1.35, MeOH). (*R*)-(−)-BNPPA: $[\alpha]_D^{20}$ −605° (*c* 1.35, MeOH). Decomposes without melting at ca. 300 °C.
Solubility: racemate: sol ethanol (10 g/100 mL), sol methanol (3 g/100 mL), slightly sol water and other organic solvents. Enantiomer: sol ethanol (6 g/100 mL), sol methanol (2 g/100 mL), slightly sol water and other organic solvents.
Form Supplied in: white solid; racemic and optically active BNPPA are commercially available.
Preparative Methods: racemic BNPPA is obtained by condensation of (±)-binaphthol (see *(R)-1,1′-Bi-2,2′-naphthol*) and *Phosphorus Oxychloride* followed by hydrolysis (eq 1).[1] Racemic BNPPA can be resolved by recrystallization of its cinchonine salt.[1]

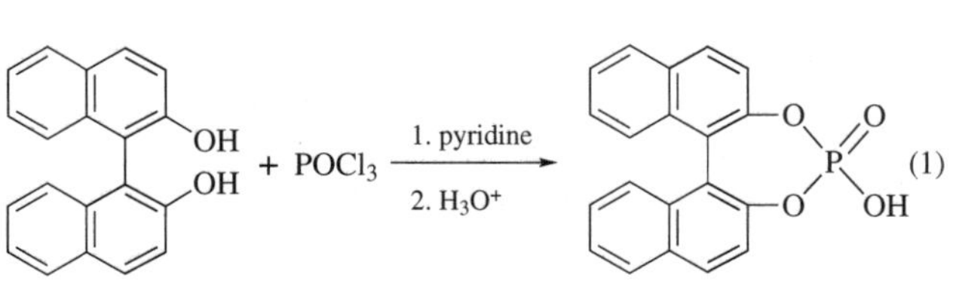

Chiral Ligand for Asymmetric Catalysts. (*S*)-(+)- and (*R*)-(−)-BNPPA are efficient chiral ligands for the Pd-catalyzed hydrocarboxylation of alkenes.[5] Naproxen can be obtained regioselectively in 91% ee (eq 2).

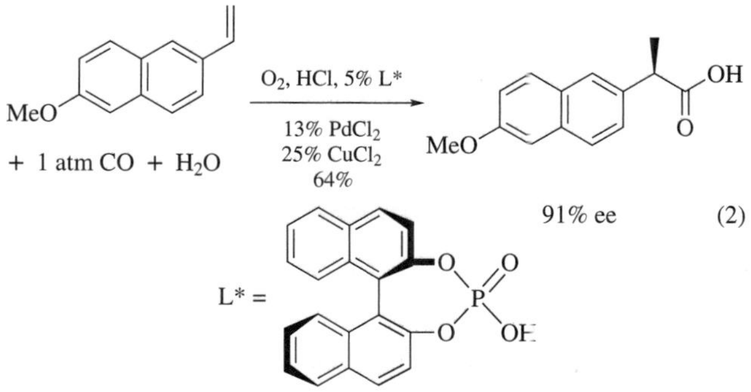

Dinuclear Rh complexes modified with optically active BNPPA catalyze asymmetric carbene reactions with moderate enantioselectivity (eq 3 and eq 4).[6]

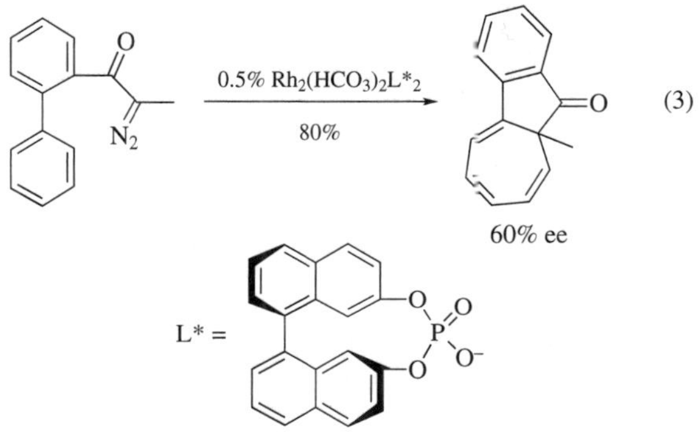

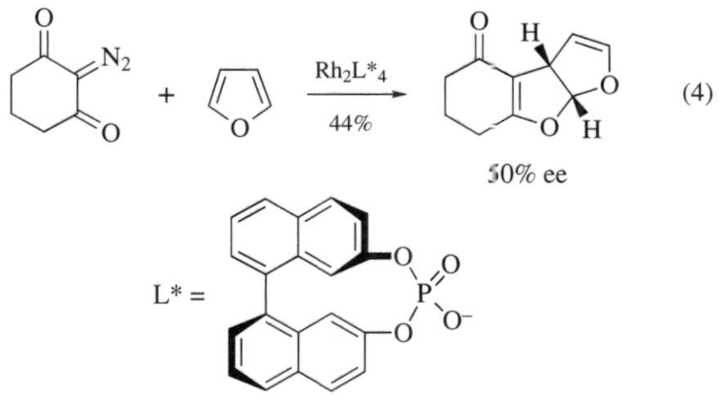

1. Jacques, J.; Fouquey, C. *Comprehensive Organic Synthesis* **1989**, *67*, 1.
2. (a) Jacques, J.; Fouquey, C.; Viterbo, R. *Tetrahedron Lett.* **1971**, 4617. (b) Arnold, W.; Daly, J. J.; Imhof, R.; Kyburz, E. *Tetrahedron Lett.* **1983**, *24*, 343.
3. Mikes, F.; Boshart, G. *Chem. Commun.* **1978**, 173.
4. Shapiro, M. J.; Archinal, A. E.; Jarema, M. A. *J. Org. Chem.* **1989**, *54*, 5826.
5. Alper, H.; Hamel, N. *J. Am. Chem. Soc.* **1990**, *112*, 2803.
6. (a) McCarthy, N.; McKervey, M. A.; Ye, T.; McCann, M.; Murphy, E.; Doyle, M. P. *Tetrahedron Lett.* **1992**, *33*, 5983. (b) Pirrung, M. C.; Zhang, J. *Tetrahedron Lett.* **1992**, *33*, 5987.

Ryoji Noyori & Kazuhiko Sato
Nagoya University, Japan

Bis(α-camphorquinone dioximato)cobalt

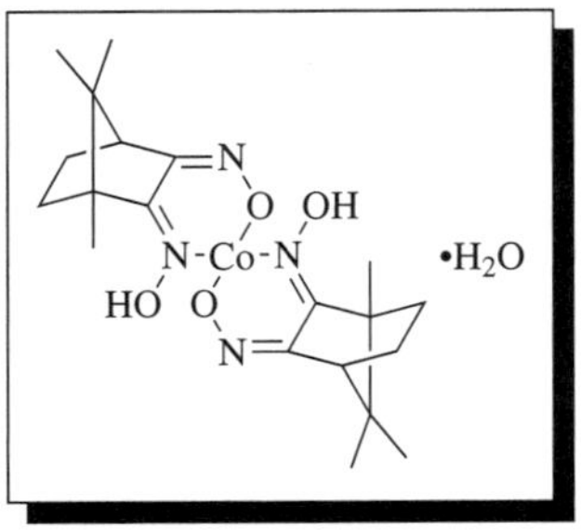

[67770-21-2] $C_{20}H_{32}CoN_4O_5$ (MW 467.43)

(catalyst for asymmetric cyclopropanation, especially for styrene, butadiene, and conjugated alkenes such as acrylates)

Physical Data: mp 240 °C (under nitrogen).
Solubility: sol hydrocarbons or similar organic solvents.
Form Supplied in: ligand may be available but the catalyst is sensitive to air and must be prepared just before use.
Analysis of Reagent Purity: a change in color from brown to pale brown or yellow indicates decomposition. Only the soluble part should be used.
Preparative Methods: Co(α-cqd)$_2$ is prepared from *Cobalt(II) Chloride* hexahydrate and the corresponding isomer of camphorquinone dioxime (cqd) (from optically active natural camphor) in ethanol with addition of an aqueous solution of NaOH under nitrogen or argon.[1] This complex is best when freshly prepared before use under nitrogen or preferably under argon. The starting material, L-camphor, is easily obtained in optically pure form.
Handling, Storage, and Precautions: the dry solid must be kept under nitrogen or argon, preferably in the cold. The solution of the catalyst in organic solvents such as benzene, acetone, ethyl acetate, hexanes, or acetophenone deteriorates on standing even under nitrogen. Therefore use just after preparation is recommended.

Although many chiral cyclopropanation catalysts are known, this class of complexes is superior for the alkenes containing vinyl, phenyl, or alkoxycarbonyl groups. Some relevant examples are shown in eq 1–5. In eq 5, the enantiomeric excess of the product is not known due to the absence of enantiomerically pure isomer. The absolute configuration is not known.

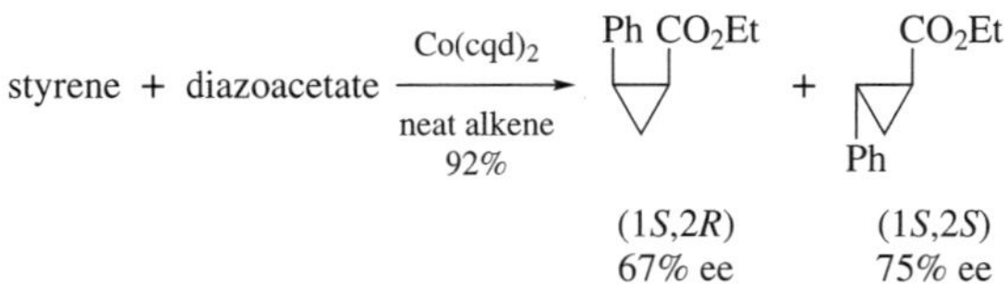

(1)

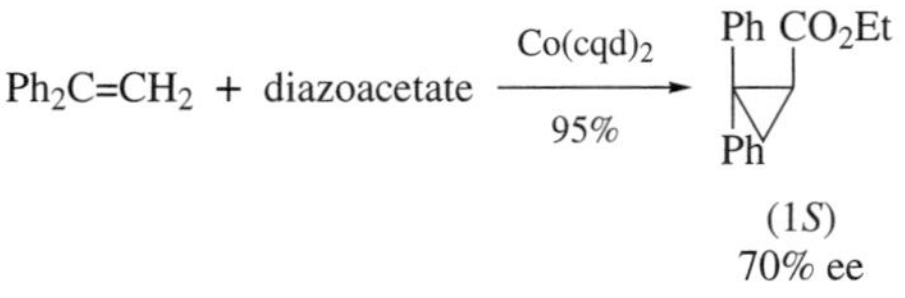

(2)

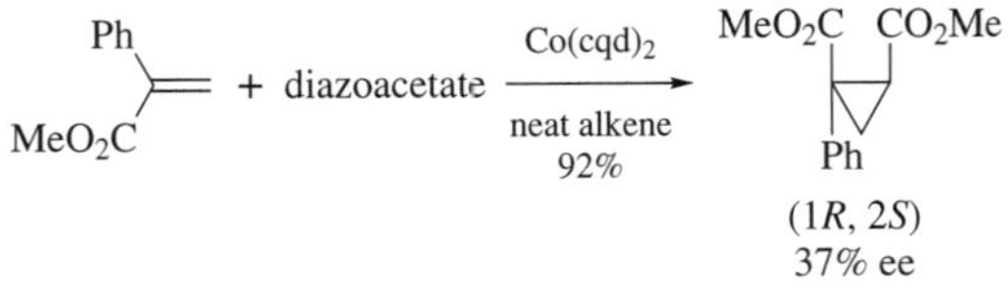

(3)

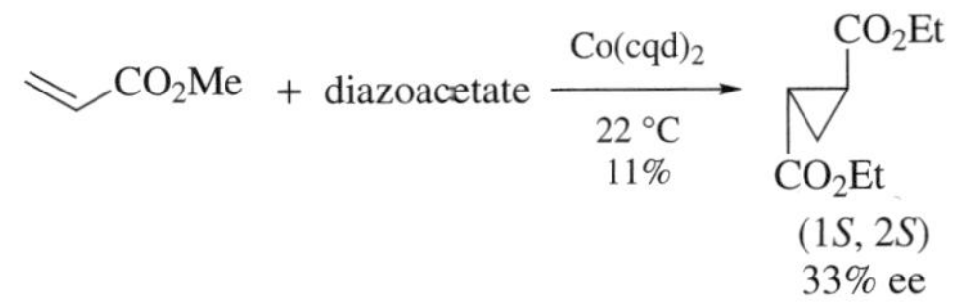

(4)

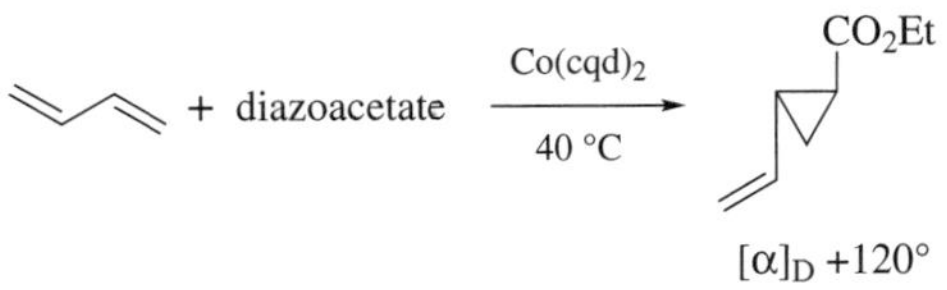

(5)

The related vicinal dioximatocobalt(II) complexes such as Co(dmg)$_2$ and Co(nqd)$_2$ (dmg = dimethylglyoximato, nqd = nopinoquinone dioximato), are also catalytically active but the enantioselectivity varies with the structure of the alkenes.

The steric bulk of the ester alkyl group generally enhances the ee values (eq 6–8). Thus, the neopentyl ester of diazoacetate gives the highest ee value (88%) for the reaction with styrene (eq 6).

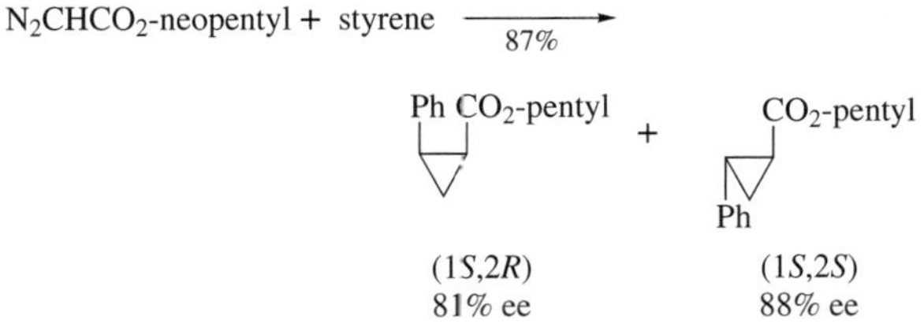

(6)

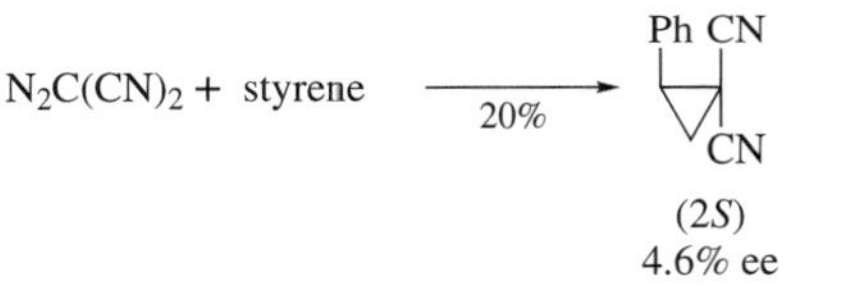

(7)

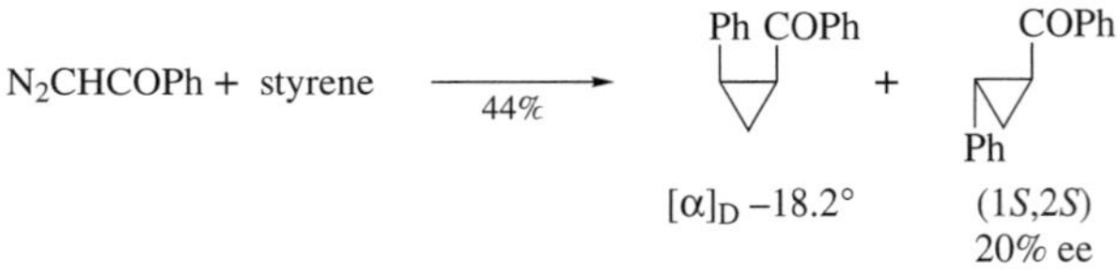

(8)

Other than diazoacetates, diazoacetophenone and diazodicyanomethane may also be used for cyclopropanation. The ee values are, however, lower than those obtained for diazoacetates.

Various organic solvents can be used, e.g. benzene, toluene, hexanes, acetone, acetophenone, diethyl ether. However, the substrate must be in large excess to the diazo compounds. The effect of additives has been examined. Pyridine or similar donor molecules retard the catalytic rates and decrease the optical yields.[3]

The enantioselectivity increases upon decreasing temperature. Thus the reaction between 0 and about −30 °C gives the best enantioselectivity in neat alkene.

Related chiral dioximato ligands have also been prepared. Isomeric nopinoquinone dioximato ligands (β- and δ-nqd) are prepared from 1-pinene.[4] The δ-isomer has been found to work to give an enantiomerically opposite isomer relative to the isomer obtainable with cqd by the cyclopropanation.

The enantiomeric purity of the chiral cyclopropanes may be enhanced by recrystallization of the acid obtained after mild alkaline hydrolysis of the chiral ester when the ee values are over 60%.

Related Reagents. (*S,S*)-2,2′-(Dimethylmethylene)bis-(4-*t*-butyl-2-oxazoline); Bis(dimethylglyoximato)(methyl)(pyridine)-cobalt(III); Copper(II) Trifluoromethanesulfonate; (1*S*,9*S*)-1,9-Bis{[(*t*-butyl)dimethylsilyloxy]methyl}-5-cyanosemicorrin; (*S*)-2-[2-(Diphenylphosphino)phenyl]-4-phenyloxazoline; Ethyl Diazoacetate.

1. Nakamura, A.; Konishi, A.; Tatsuno, Y.; Otsuka, S. *J. Am. Chem. Soc.* **1978**, *100*, 3443.
2. Nakamura, A.; Konishi, A.; Tsujitani, R.; Kudo, M.; Otsuka, S. *J. Am. Chem. Soc.* **1978**, *100*, 3449.
3. Nakamura, A. *Pure Appl. Chem.* **1978**, *50*, 37.
4. Tatsuno, Y.; Konishi, A.; Nakamura, A.; Otsuka, S. *Chem. Commun.* **1974**, 588.
5. Nakamura, A.; Konishi, A.; Otsuka, S. *J. Chem. Soc., Dalton Trans.* **1979**, 488.

Akira Nakamura
Osaka University, Japan

(*R,R*)-1,2-Bis(aminocarbonylphenyl-2′-diphenylphosphino)cyclohexane

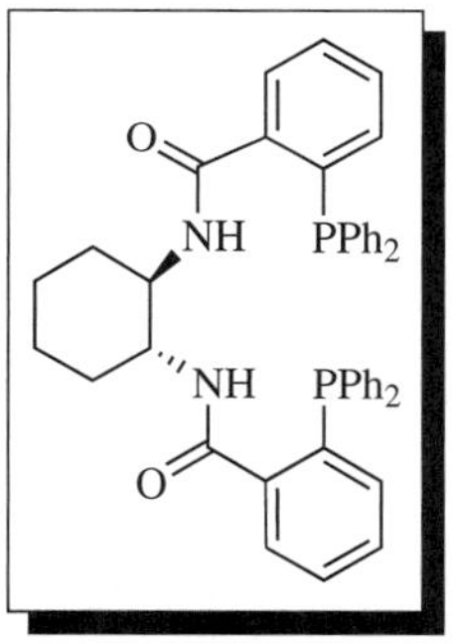

[138517-61-0] $C_{44}H_{40}N_2O_2P_2$ (MW 690.76)

(chiral phosphine ligand used in asymmetric Pd^0-catalyzed allylic substitution reactions)[1]

Physical Data: mp 134–136 °C.[2]

Solubility: soluble in chlorinated solvents, ethers, alcohols, toluene and most organic solvents. Partially soluble in acetonitrile.

Form Supplied in: white to off-white crystalline solid. Major impurity is the corresponding monophosphine oxide (<1%).

Analysis of Reagent Purity: NMR and IR details are available.[3,4] Optical rotation $[\alpha]_D^{25}$ +46.7 (*c* 2.4, CH_2Cl_2).[3] $[\alpha]_D^{25}$ +88 (*c* 7.0, CH_2Cl_2);[2] crystalline material. Chiral HPLC: Chiracel OD-R, UV 210 nm, 0.8 mL min^{-1}, 100% MeOH. Retention time (*R,R*)=6.3 min, retention time (*S,S*)=9.4 min. Achiral HPLC: Hypersil BDS C8, UV 254 nm, 1.0 mL min^{-1}, 85% MeOH, 15% H_2O. Retention time ligand=9.5 min, retention time monophosphine oxide=5.2 min, retention time bisphosphine oxide=3.6 min.

Preparative Methods: commercially available. The ligand can be prepared by the coupling of (1*R*,2*R*)-(−)-1,2-diaminocyclohexane [20439-47-8] with 2-(diphenylphosphino)benzoic acid [17261-28-8],[5] using reagents such as DCC.[3] An alternative procedure has been developed whereby (1*R*,2*R*)-(+)-1,2-diaminocyclohexane L-tartrate salt [39961-95-0][6] is coupled to a mixed anhydride of 2-(diphenylphosphino)benzoic acid and diphenylchlorophosphate.[2] The procedure is reproduced below 2-(Diphenylphosphino)benzoic acid (20 g, 65.3 mmol, 2 equiv) is suspended in dichloromethane (150 mL) and cooled in an ice-water bath to 0 °C (internal temperature). Triethylamine (10.1 mL, 71.8 mmol, 2.2 equiv) is added dropwise and a clear solution is obtained. This process is exothermic and a rise in temperature to 5 °C is observed. The solution is re-cooled to 0 °C and diphenylchlorophosphate (13.4 mL, 64.7 mmol, 1.98 equiv) is added slowly, maintaining the internal temperature between 0–5 °C. The yellow solution is stirred for 1 h at 0 °C. (1*R*,2*R*)-(+)-1,2-Diaminocyclohexane-L-tartrate salt (8.63 g, 32.65 mmol, 1 equiv) is suspended in water (50 mL, 5.8 vol) and potassium carbonate (15 g, 107.8 mmol, 3.3 equiv) is added. This process is exothermic and a clear solution is obtained after approximately 10 min. After 30 min, the clear aqueous solution of diamine is added to the mixed anhydride solution at 0 °C, and the resulting yellow two-phase mixture is stirred for 2 h at 0 °C, then allowed to warm to room temperature. After 14 h, the mixture is poured into a separating funnel and 200 mL of dichloromethane and 100 mL of water are added. The organic phase is separated, washed with 2 N HCl (100 mL) and saturated aqueous $NaHCO_3$ solution (100 mL), then dried over magnesium sulfate. The dried organic phase is filtered through a silica pad and the pad is washed with dichloromethane (50 mL). The combined filtrates are evaporated to dryness under reduced pressure, producing a yellow foam (22.3 g, 99% crude). The foam is crystallized from boiling acetonitrile (390 mL, 17.5 vol) to afford a white crystalline solid. The solid is dried under vacuum to provide the phosphine ligand (15 g, 67%).

Purification: column chromatography on silica gel, eluting with 15–30% ethyl acetate/hexanes.[3,4] Recrystallization from hot acetonitrile.[2,7]

Handling, Storage, and Precautions: store under nitrogen at room temperature. Oxidation to phosphine oxides may occur upon prolonged exposure to air. No known toxicology data.

Chiral Diphosphine Ligands. In 1992, Professor Barry Trost introduced a family of chiral diphosphine ligands for palladium(0)-catalyzed asymmetric allylic substitution reactions. The first ligands were based on 2-(diphenylphosphino)benzoic acid with a variety of chiral backbones.[3] The most useful of these backbones are *trans*-1,2-diaminocyclohexane, *trans*-1,2-diphenylethanediamine, and *trans*-11,12-diamino-9,10-dihydro-9,10-ethanoanthracene. The ligand based on the diaminocyclohexane backbone has proved to be the most generically useful ligand, with the most reported applications. Throughout this text the term ‘ligand’ refers to (*R,R*)-1,2-bis(aminocarbonylphenyl-2′-diphenylphosphino)cyclohexane.

Catalyst Preparation Two palladium sources are generally used to form the active precatalysts (**1**) and (**2**) in situ, the allylpalladium chloride dimer [12012095-2] and the tris(dibenzylideneacetone)dipalladium(0)–chloroform adduct

[52522-40-4]. For the palladium dibenzylideneacetone complex (**1**), NMR data to support the proposition that the bis-phosphine acts as a bidentate ligand has been reported.[8] A triflate salt of the π-allyl palladium complex has been isolated and is stable in the solid state. However, no crystals suitable for X-ray analysis were obtained.[7] An X-ray crystal structure of the ligand and a bis-palladium complex has been reported.[7] The palladium complexes are generated just before use under an inert atmosphere; exposure to air affords a catalytically inactive tetra-coordinated palladium(II) species.[8]

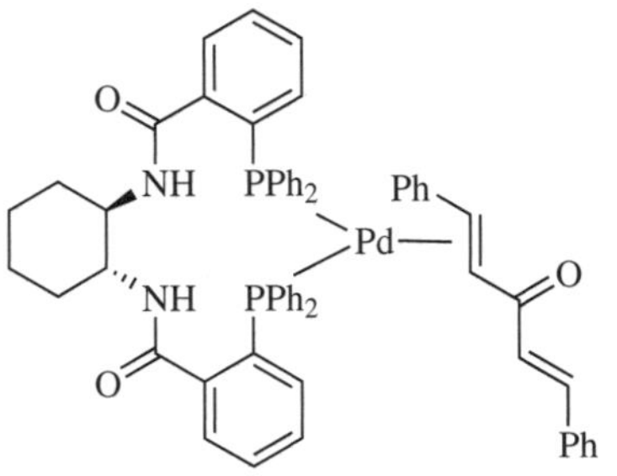

1

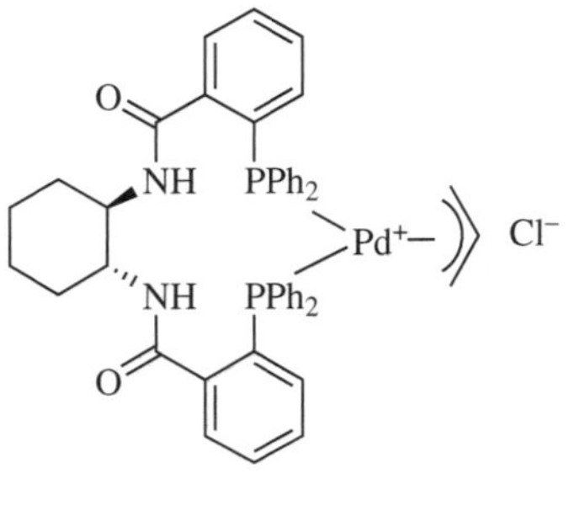

2

Reactions with Carbon Nucleophiles. A wide range of carbon nucleophiles have been used in asymmetric allylic alkylation reactions (AAA). The first reported reactions involved the use of the sodium salt of malonates as the nucleophile. Five-, six-, and seven-membered ring allylic acetates and carbonates (in general, the carbonates are more reactive substrates) are ionized by the catalyst, prepared from the ligand and a palladium source, to provide a single palladium(0) intermediate (eq 1). Reaction of the intermediate with the sodium malonate in the presence of tetra-*n*-hexylammonium bromide gives the malonate product in high yield (*n*=1, 81%) and enantiomeric excess (*n*=1, 98%).[9] The use of microwave radiation has been reported to accelerate the rate of this reaction with carbon, oxygen and nitrogen nucleophiles.[10]

In addition to cyclic allylic substrates, malonate nucleophiles have been used in reactions with both symmetrical[11] and unsymmetrical[12] acyclic systems, and with geminal dicarboxylates.[13] Malonates and Meldrum's acid have also been used as nucleophiles in the desymmetrization of *meso* diesters.[14]

Azlactones have been used as nucleophiles to provide access to a variety of α-alkylated amino acid derivatives. This has been demonstrated with 3-acetoxycyclohexene and with geminal dicarboxylates (eq 2).[15] The enantiomeric and diastereoisomeric excess of the products increase with more bulky R groups.

This method has also been applied to unsymmetrical acyclic allyl esters.[16] The reaction of an azlactone with a geminal diacetate substrate gave access to an advanced intermediate for the synthesis of sphingofungin F.[17]

This methodology has been used to provide efficient protocols for the asymmetric allylic alkylation of β-keto esters,[18] ketone enolates,[19] barbituric acid derivatives,[20] and nitroalkanes.[21]

Several natural products and analogs have been accessed using asymmetric desymmetrization of substrates with carbon nucleophiles. The palladium-catalyzed reaction of a dibenzoate with a sulfonylsuccinimide gave an advanced intermediate in the synthesis of L-showdomycin (eq 3).[22]

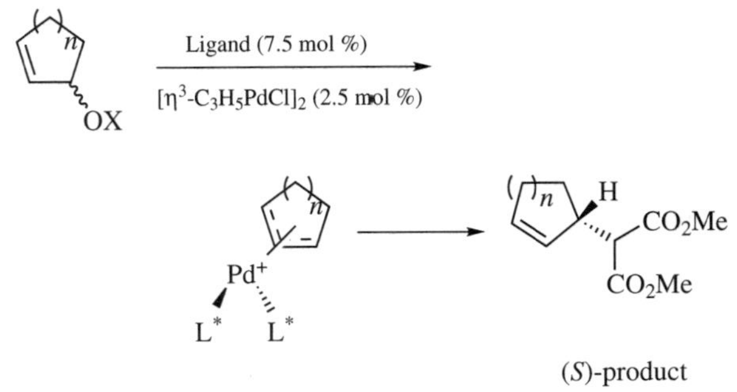

(1)

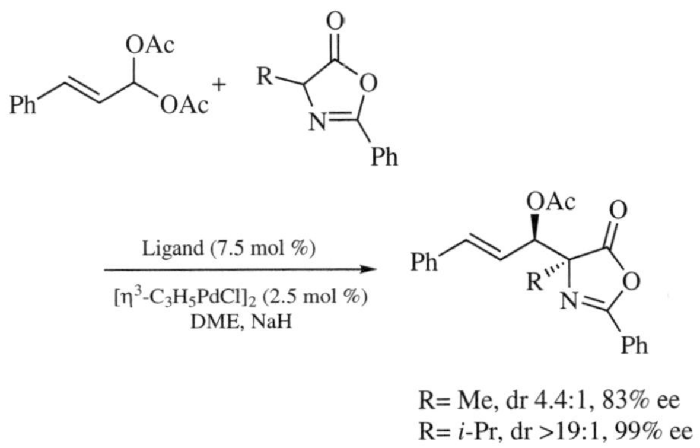

(2)

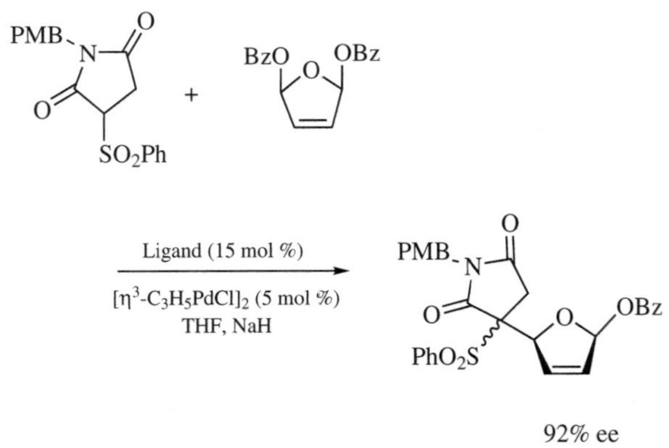

(3)

The alkylation of a dibenzoate with (phenylsulfonyl)nitromethane gave an intermediate for the synthesis of (+)-valienamine.[23]

The reaction of azlactones or a Meldrum's acid derivative with 2-phenylbut-3-ene-2-yl acetate, in the presence of the racemic ligand and a palladium source has provided a new method for controlling alkene geometry. By varying the reaction conditions excellent selectivities for either *E* or *Z* geometry could be obtained.[24]

Reactions with Nitrogen Nucleophiles. The palladium(0)-catalyzed asymmetric desymmetrization of *cis*-3,5-dibenzoyloxy-1-cyclopentene, with 6-chloropurine and 2-amino-6-chloropurine as nucleophiles, has been utilized in the synthesis of (−)-carbovir[25] and (−)-neplanocin.[26] In these examples, the diphenylethanediamine[3] and the anthracenyldiamine[3] based ligands were found to be superior to the standard ligand.

Phthalimide has been used as a nucleophile with cyclic (as depicted for carbon nucleophiles in eq 1)[9] and acyclic allylic carbonates.[27] In addition, phthalimide has been used for the amination of 3,4-epoxybut-1-ene and, in this case, the 1,2-bis(aminocarbonyl-1′-naphthyl-2′-diphenylphosphino)cyclohexane ligand was found to provide the catalyst of choice.[28]

Azide has been used as a nucleophile in the desymmetrization of a dicarbonate derivative (eq 4).[29] In this example, a key intermediate in the synthesis of (+)-pancratistatin was produced.

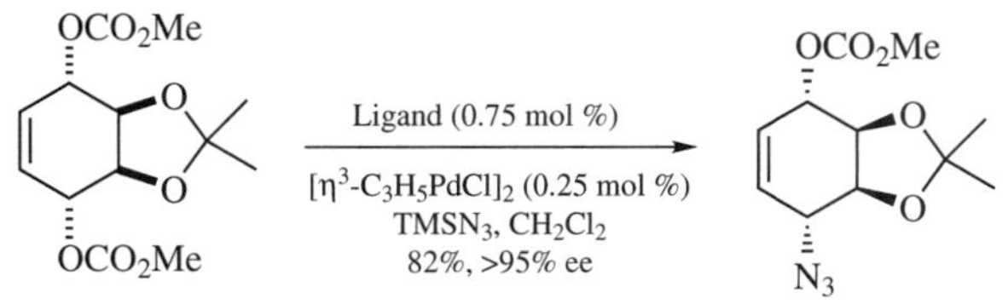

(4)

Basic hydrolysis of the allylic azide affords the rearranged 1,2-isomer, which was an intermediate in the synthesis of (+)-conduramine E.[30] Following a similar strategy, but starting with *cis*-3,6-dibenzoyloxycyclohex-1-ene, a total synthesis of the non-opioid analgesic (−)-epibatidine was developed.[31]

Trost has reported enhanced enantioselectivity in the desymmetrization of *meso*-biscarbamates in the presence of triethylamine.[32] Under these conditions, high yields (>80%) and enantiomeric excesses (93-99% ee) are obtained. This methodology has been applied to the synthesis of (−)-swainsonine.[33]

α-Amino esters have been used as nucleophiles in the reaction with acyclic allylic esters and isoprene monoepoxide, providing access to diastereoselective *N*-alkylated α-amino esters.[34] By employing the feature ligand, asymmetric palladium(0)-catalyzed cyclization of 2-(tosylamino)phenol with (*Z*)-1,4-bis[(methoxycarbonyl)oxy]but-2-ene provides 2-vinylbenzomorpholine in 79% ee.[35] A number of alternative diphosphine ligands were studied and found to be inferior.

The asymmetric synthesis of indolizidine alkaloids is described utilizing a palladium-catalyzed amination process. Ionization of an allylic carbonate provides a symmetrical π-allyl palladium complex, subsequent reaction with a protected homoallylamine gave the product in 93% yield and >99.5% ee (eq 5).[36]

The product of the allylic amination process is set up for a ring-closing-ring-opening metathesis process, and subsequent elaboration to alkaloid derivatives.

Reactions with Oxygen Nucleophiles. The first report of the reaction of oxygen nucleophiles was for the deracemization of cyclic allylic ethers, for example, the palladium(0)-catalyzed reaction of 2-cyclohexenyl-1-methyl carbonate with sodium pivalate afforded the pivalate ester in 94% yield and 92% ee.[37] This reaction was extended to other cyclic allylic carbonates.

Racemic conduritol B acetates and carbonates provide very versatile substrates for asymmetric allylic substitution reactions. Reaction of conduritol B tetraacetate with sodium pivalate in the presence of a palladium catalyst, generated from the ligand and allylpalladium chloride dimer, resulted in a kinetic resolution to give the monosubstituted product in 44% yield (>99% ee) and the recovered tetraacetate in 50% yield (83% ee) (eq 6). This method provided a key intermediate for the synthesis of (+)-cyclophellitol.[38]

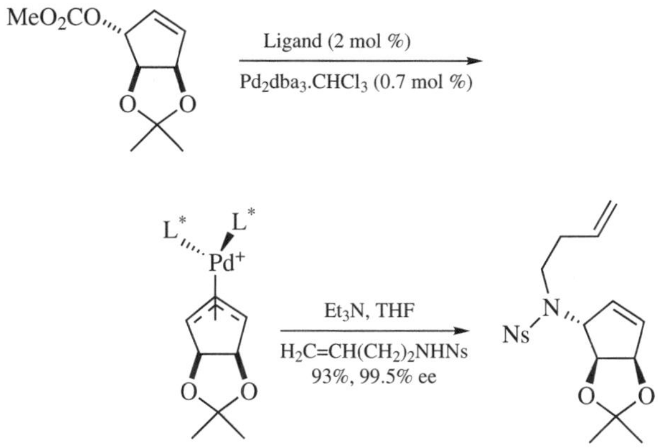

(5)

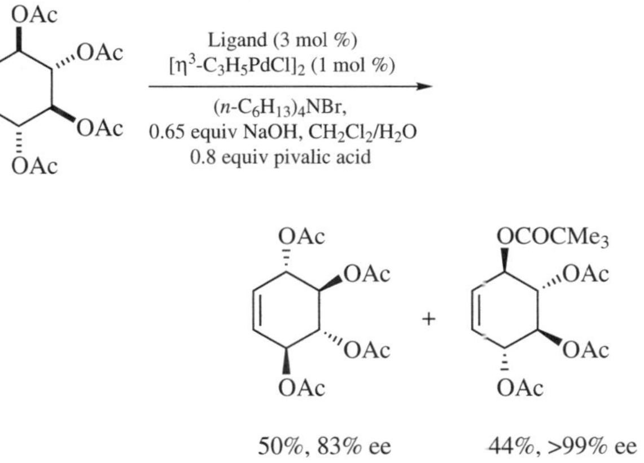

(6)

Later work has shown that a dynamic kinetic asymmetric transformation could be obtained if the acetates were converted into carbonate groups. With the tetra(2,2,2-trichloroethyl) carbonate derivative, reactions with carbon and nitrogen nucleophiles gave exclusively the monosubstituted products in high yield (61–95%) and excellent enantiomeric excesses (95–99%).[39] However, carboxylate nucleophiles afforded the disubstituted products in high yield and enantiomeric excess (eq 7).[39] This allowed an efficient synthesis of D-myo-inositol-1,4,5-trisphosphate to be devised.

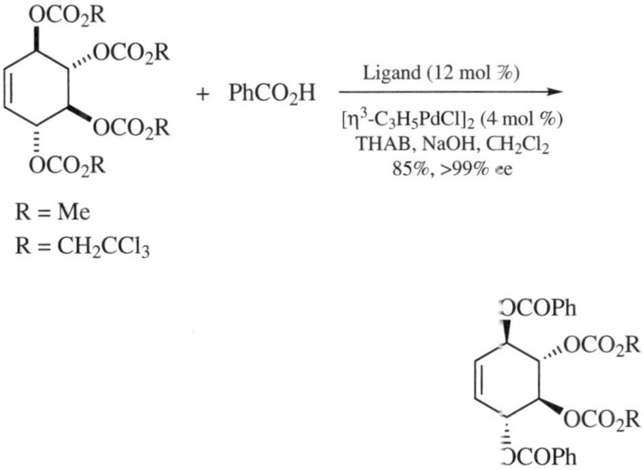

(7)

The reaction of isoprene monoxide with a range of alcohol pronucleophiles in the presence of the ligand (3 mol%), $Pd_2dba_3.CHCl_3$ (1 mol%) and triethylboron (1 mol%) gave the glycol monoethers in excellent yield and enantiomeric excess.[40] The use of *p*-methoxybenzyl alcohol and 3-nonyl-3,4-epoxybut-1-ene afforded an intermediate that was converted into (−)-malyngolide (eq 8).[41]

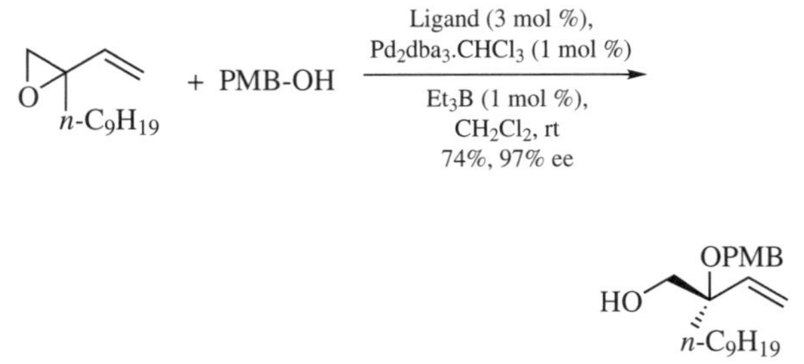

(8)

Extending this methodology to 3,4-epoxybut-1-ene was not successful with the featured ligand and the more sterically encumbered 1,2-bis(aminocarbonyl-1′-naphthyl-2′-diphenylphosphino)cyclohex ane ligand was required.[40] The use of inorganic carbonates for the asymmetric synthesis of vinylglycidols has also been reported.[42] Reaction of isoprene monoxide with sodium bicarbonate, or sodium carbonate in the presence of the ligand, $Pd_2dba_3.CHCl_3$and triethylboron afforded the diol in 91% yield and 97% ee. In the absence of triethylboron a cyclic carbonate was formed. Again, the 2-naphthyl ligand was required to provide optimum selectivity with 3,4-epoxybut-1-ene.[42]

The palladium(0)-catalyzed asymmetric *O*-allylation of phenols has been described using five-, six- and seven-membered ring allylic carbonates and acyclic allylic carbonates (eq 9).[43] The products from these reactions were subjected to a Claisen rearrangement to provide *C*-alkylated phenols. A study of various ligands for the reaction of phenol with 2-cyclohexenyl-1-methyl carbonate clearly showed that the Trost ligand is superior.[44]

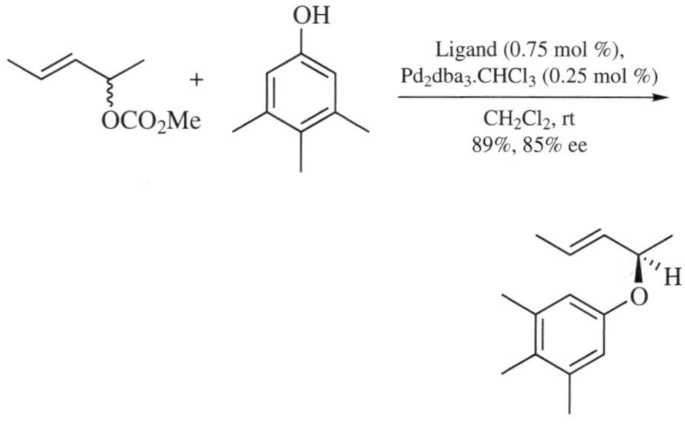

(9)

This methodology has been expanded to geranyl methyl carbonate for the synthesis of the vitamin E nucleus, and to tiglyl methyl carbonate for the synthesis of (−)-calanolide A and B.[45] In the latter example, the anthracenyldiamine[3]-based ligand was required for optimum selectivity. The synthesis of (−)-aflatoxin B lactone utilizes a dynamic kinetic asymmetric transformation, whereby a suitably functionalized phenol reacts with a racemic 5-acyloxy-2-(5*H*)-furanone to provide a single product in 89% yield.[46] One final example of phenol as a nucleophile is for the deracemization of Baylis-Hillman adducts.[47]

Cyclic 1,2-diketones, such as 3-methylcyclopentane-1,2-dione, act as oxygen nucleophiles in palladium(0)-catalyzed reactions with a range of cyclic and acyclic allylic esters.[48] The products of these reactions were subjected to a lanthanide-catalyzed Claisen rearrangement to access the *C*-alkylated products.

Reactions with Sulfur Nucleophiles. The use of sulfur nucleophiles in palladium-catalyzed allylic substitution reactions is less well documented than that of carbon, nitrogen and oxygen nucleophiles. The asymmetric synthesis of allylic sulfones utilizing a catalytic phase transfer system has been used to produce (3*S*)-(phenylsulfonyl)cyclohex-1-ene on a 45 g scale (eq 10).[49] In many cases, it has been reported that allylic carbonates are more reactive than allylic acetates in asymmetric allylic substitution reactions.[49,50]

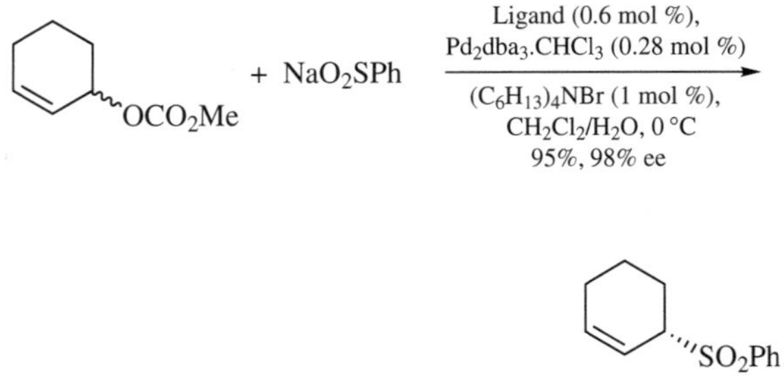

(10)

A range of cyclic allylic carbonates was found to be useful in this process and a myriad of useful functionalized building blocks were accessed via dihydroxylation and epoxidation reactions.[49] The reaction of lithium *tert*-butylsulfinate with acyclic allylic acetates in the presence of the ligand, $Pd_2(dba)_3.CHCl_3$ and tetrahexylammonium bromide under phase-transfer conditions (CH_2Cl_2/H_2O) led to a kinetic resolution whereby the starting material was isolated in 96% ee and the *tert*-butyl sulfone in 95% ee.[50] With cyclic allylic carbonates, a single *tert*-butyl sulfone is obtained in 76–92% yield and 89–93% ee.[50] However, stopping the reaction at 54% conversion gave the sulfone (49% yield, 98% ee) and the carbonate (34% yield, >99% ee), this kinetic resolution protocol was later extended to thiols with cyclic and acyclic allylic carbonates.[51] In general, the synthesis of allylic sulfides requires higher catalyst loading and was found to be unsuccessful for *tert*-butyl thiol and thiophenol.[52] However, cyclic and acyclic allylic *S*-*p*-chlorophenyl, *S*-2-pyridyl and *S*-2-pyrimidyl sulfides could be obtained in high yield and enantiomeric excess, in the presence of the ligand and $Pd_2(dba)_3.CHCl_3$ in organic solvent.[52]

A more efficient method to access single enantiomer thiols and sulfides has been developed using a palladium(0)-catalyzed rearrangement of *O*-allylic thiocarbamates (eq 11).[53]

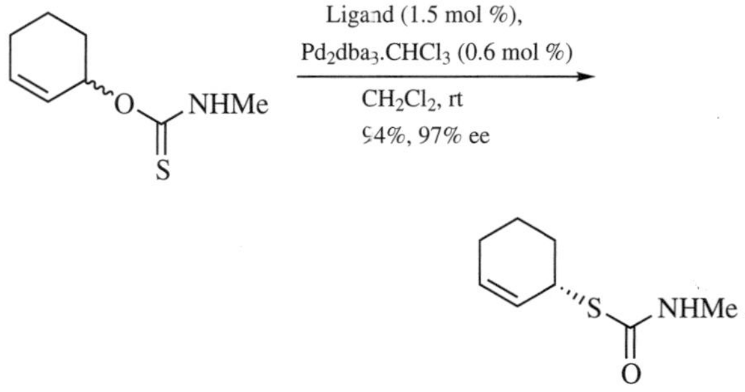

(11)

This reaction was carried out on cyclic and acyclic allylic carbonates. The *S*-allylic thiocarbamate products were hydrolyzed to the corresponding thiol or reacted with 2-chloropyrimidine in the presence of potassium hydroxide to provide the sulfide without any loss in stereochemical purity for either example.[53]

α-Acetoxysulfones can be regarded as acid-stable, but base-labile, chiral aldehyde equivalents. These can be accessed through the palladium(0)-catalyzed reaction of geminal esters with sodium benzenesulfinate under phase-transfer conditions (eq 12).[54]

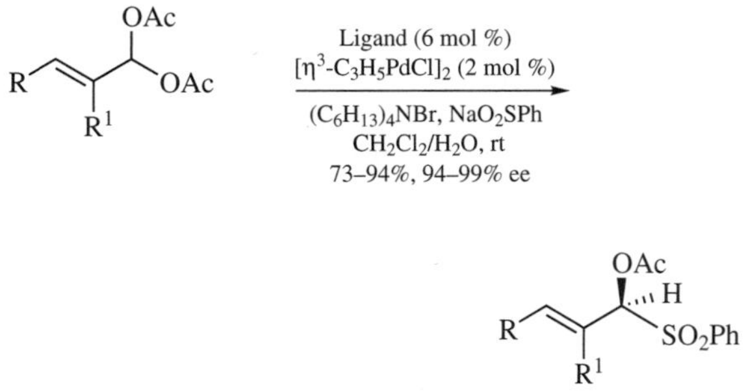

(12)

Osmium tetroxide-catalyzed dihydroxylation of the chiral α-acetoxysulfones and acetonide formation affords versatile chemical intermediates. Reduction with DIBAL-H provides primary alcohols, and addition of Grignard reagents provides secondary alcohols with excellent stereochemical control of the newly formed chiral center.[54]

---

1. (a) Trost, B. M.; Van Vranken, D. L. *Chem. Rev.* **1996**, *96*, 395. (b) Trost, B. M.; Lee, C. In *Catalytic Asymmetric Synthesis*, 2nd ed.; Ojima, I., Ed.; Wiley-VCH, Inc.: New York, 2000, p 593.
2. Lennon, I. C.; Berens, U. WO 99/51614 (October 1999).
3. Trost, B. M.; Van Vranken, D. L.; Bingel, C. *J. Am. Chem. Soc.* **1992**, *114*, 9327.
4. Trost, B. M.; Van Vranken, D. L.; Bunt, R. C. US Patent 5,739, 396 (April 1998).
5. Hoots, J. E.; Rauchfuss, T. B.; Wrobleski, D. A. *Inorganic Syntheses* **1982**, *21*, 175.
6. Larrow, J. F.; Jacobsen, E. N.; Gao, Y.; Hong, Y.; Nie, X.; Zepp, C. M. *J. Org. Chem.* **1994**, *59*, 1939.
7. Butts, C. P.; Crosby, J.; Lloyd-Jones, G. C.; Stephen, S. C. *Chem. Commun.* **1999**, 1707.
8. Trost, B. M.; Breit, B.; Organ, M. G. *Tetrahedron Lett.* **1994**, *35*, 5817.
9. Trost, B. M.; Bunt, R. C. *J. Am. Chem. Soc.* **1994**, *116*, 4089.
10. Bremberg, U.; Lutsenko, S.; Kaiser, N.-F.; Larhed, M.; Hallberg, A.; Moberg, C. *Synthesis* **2000**, 1004.
11. Trost, B. M.; Breit, B.; Peukert, S.; Zambrano, J.; Ziller, J. W. *Angew. Chem., Int. Ed. Engl.* **1995**, *34*, 2386.
12. Trost, B. M.; Toste, F. D. *J. Am. Chem. Soc.* **1999**, *121*, 4545.
13. (a) Trost, B. M.; Lee, C. B.; Weiss, J. M. *J. Am. Chem. Soc.* **1995**, *117*, 7247. (b) Trost, B. M.; Lee, C. B.; Weiss, J. M. *J. Am. Chem. Soc.* **2001**, *123*, 3671. (c) Trost, B. M.; Lee, C. B.; Weiss, J. M. *J. Am. Chem. Soc.* **2001**, *123*, 3687.
14. Trost, B. M.; Tanimori, S.; Dunn, P. T. *J. Am. Chem. Soc.* **1997**, *119*, 2735.
15. Trost, B. M.; Ariza, X. *Angew. Chem., Int. Ed. Engl.* **1997**, *36*, 2635.
16. Trost, B. M.; Ariza, X. *J. Am. Chem. Soc.* **1999**, *121*, 10727.
17. Trost, B. M.; Lee, C. B. *J. Am. Chem. Soc.* **1998**, *120*, 6818.
18. Trost, B. M.; Radinov, R.; Grenzer, E. M. *J. Am. Chem. Soc.* **1997**, *119*, 7879.
19. Trost, B. M.; Schroeder, G. M. *J. Am. Chem. Soc.* **1999**, *121*, 6759.
20. Trost, B. M.; Schroeder, G. M. *J. Org. Chem.* **2000**, *65*, 1569.
21. (a) Trost, B. M.; Surivet, J.-P. *Angew. Chem., Int. Ed. Engl.* **2000**, *39*, 3122. (b) Trost, B. M.; Surivet, J.-P. *J. Am. Chem. Soc.* **2000**, *122*, 6291.
22. Trost, B. M.; Kallander, L. S. *J. Org. Chem.* **1999**, *64*, 5427.
23. Trost, B. M.; Chupak, L. S.; Lübbers, T. *J. Am. Chem. Soc.* **1998**, *120*, 1732.
24. Trost, B. M.; Heinemann, C.; Ariza, X.; Weigand, S. *J. Am. Chem. Soc.* **1999**, *121*, 8667.
25. Trost, B. M.; Madsen, R.; Guile, S. G.; Elia, A. E. H. *Angew. Chem., Int. Ed. Engl.* **1996**, *35*, 1569.
26. Trost, B. M.; Madsen, R.; Guile, S. G. *Tetrahedron Lett.* **1997**, *38*, 1707.
27. Trost, B. M.; Krueger, A. C.; Bunt, R. C.; Zambrano, J. *J. Am. Chem. Soc.* **1996**, *118*, 6520.
28. (a) Trost, B. M.; Bunt, R. C. *Angew. Chem., Int. Ed. Engl.* **1996**, *35*, 99. (b) Trost, B. M.; Bunt, R. C.; Lemoine, R. C.; Calkins, T. L. *J. Am. Chem. Soc.* **2000**, *122*, 5968. (c) Harris, M. C. J. Jackson, M.; Lennon, I. C.; Ramsden, J. A.; Samuel, H. *Tetrahedron Lett.* **2000**, *41*, 3187.
29. Trost, B. M.; Pulley, S. R. *J. Am. Chem. Soc.* **1995**, *117*, 10143.
30. Trost, B. M.; Pulley, S. R. *Tetrahedron Lett.* **1995**, *36*, 8737.
31. Trost, B. M.; Cook, G. R. *Tetrahedron Lett.* **1996**, *37*, 7485.
32. Trost, B. M.; Patterson, D. E. *J. Org. Chem.* **1998**, *63*, 1339.
33. Trost, B. M.; Patterson, D. E. *Chem. Eur. J.* **1999**, *5*, 3279.
34. Trost, B. M.; Calkins, T. L.; Oertelt, C.; Zambrano, J. *Tetrahedron Lett.* **1998**, *39*, 1713.
35. Lhoste, P.; Massacret, M.; Sinou D. *Bull. Soc. Chim. Fr.* **1997**, *134*, 343.
36. Ovaa, H.; Stragies, R.; van der Marel, G. A.; van Boom, J. H.; Blechert, S. *Chem. Commun.* **2000**, 1501.
37. Trost, B. M.; Organ, M. G. *J. Am. Chem. Soc.* **1994**, *116*, 10320.
38. Trost, B. M.; Hembre E. J. *Tetrahedron Lett.* **1999**, 40, 219.
39. Trost, B. M.; Patterson, D. E.; Hembre, E. J. *J. Am. Chem. Soc.* **1999**, *121*, 10834.
40. Trost, B. M.; McEachern, E. J.; Toste, F. D. *J. Am. Chem. Soc.* **1998**, 120, 12702.
41. Trost, B. M.; Tang, W.; Schulte, J. L. *Organic Lett.* **2000**, *2*, 4013.
42. Trost, B. M.; McEachern, E. J. *J. Am. Chem. Soc.* **1999**, *121*, 8649.
43. Trost, B. M.; Toste, F. D. *J. Am. Chem. Soc.* **1998**, *120*, 815.
44. Iourtchenko, A.; Sinou, D. *J. Mol. Cat. A: Chem.* **1997**, *122*, 91.
45. (a) Trost, B. M.; Toste, F. D. *J. Am. Chem. Soc.* **1998**, *120*, 9074. (b) Trost, B. M.; Asakawa, N. *Synthesis* **1999**, 1491.
46. Trost, B. M.; Toste, F. D. *J. Am. Chem. Soc.* **1999**, *121*, 3543.
47. Trost, B. M.; Tsui, H.-C.; Toste, F. D. *J. Am. Chem. Soc.* **2000**, *122*, 3534.
48. Trost, B. M.; Schroeder, G. M. *J. Am. Chem. Soc.* **2000**, *122*, 3785.
49. Trost, B. M.; Organ, M. G.; O'Doherty, G. A. *J. Am. Chem. Soc.* **1995**, *117*, 9662.
50. Gais H.-J, Eichelmann, H.; Spalthoff, N.; Gerhards, F.; Frank, M.; Raabe, G. *Tetrahedron: Asymmetry* **1998**, *9*, 235.
51. Gais, H.-J.; Spalthoff, N.; Jagusch, T.; Frank, M.; Raabe, G. *Tetrahedron Lett.* **2000**, *41*, 3809.
52. Frank, M.; Gais, H.-J. *Tetrahedron: Asymmetry* **1998**, *9*, 3353.
53. Böhme, A.; Gais, H.-J. *Tetrahedron: Asymmetry* **1999**, *10*, 2511.
54. Trost, B. M.; Crawley, M. L.; Lee, C. B. *J. Am. Chem. Soc.* **2000**, *122*, 6120.

Ian C. Lennon
Chirotech Technology Limited, Cambridge, UK

Bis(bicyclo[2.2.1]hepta-2,5-diene)-rhodium Perchlorate–(*R*)-1-(*S*)-1′,2-Bis-(diphenylphosphino)ferrocenylethanol[1]

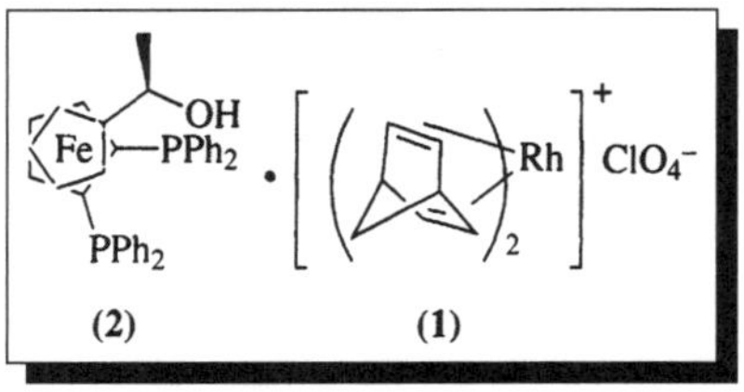

(**1**)
[60576-58-1] $C_{14}H_{16}ClO_4Rh$ (MW 386.64)
(**2**)
[71049-99-5] $C_{36}H_{32}FeOP_2$ (MW 598.44)

(catalyst for asymmetric hydrogenation of functionalized carbonyl compounds[2,3] and enol phosphinates[4])

Alternate Name: (**1**): $[Rh(nbd)_2]ClO_4$; (**2**): (*R*)-(*S*)-BPPFOH.
Physical Data: (*R*)-(*S*)-BPPFOH: $[\alpha]_D^{25}$ −285° (*c* 0.5, $CHCl_3$); mp 154–155 °C.
Solubility: $[Rh(nbd)_2]ClO_4$: sol CH_2Cl_2; insol THF, hexane.
Form Supplied in: $[Rh(nbd)_2]ClO_4$: rust-brown crystals containing 1 mol THF solvent of recrystallization.
Preparative Methods: (*R*)-(*S*)-BPPFOH is prepared from optically resolved (*R*)-*N,N*-dimethyl-1-ferrocenylethylamine (**3**) as shown in eq 1.[5] Dilithiation of (*R*)-(**3**) with *n-Butyllithium* followed by treatment with *Chlorodiphenylphosphine* affords (*R*)-(*S*)-BPPFA stereoselectively. It is treated with an excess of *Acetic Anhydride* at 100 °C to give (*R*)-(*S*)-BPPFOAc with retention of configuration. Finally, treatment with BuLi followed by hydrolysis affords (*R*)-(*S*)-BPPFOH in enantiomerically pure form.

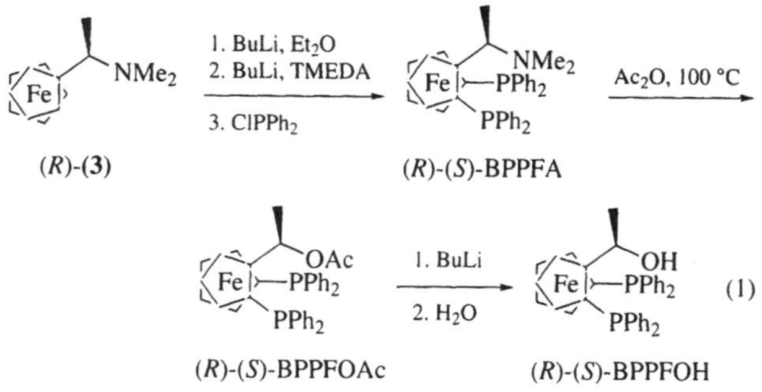

(1)

Purification: (*R*)-(*S*)-BPPFOH: chromatography on alumina (EtOAc) followed by recrystallization from ethanol.
Handling, Storage, and Precautions: $[Rh(nbd)_2]ClO_4$: *explosive* when heated.

Asymmetric Hydrogenation of Functionalized Carbonyl Compounds. (*R*)-(*S*)-BPPFOH is designed to have a secondary interaction with a carbonyl substrate through the hydroxy group on the side chain of the ferrocene ring. This additional interaction enables a substrate to coordinate to the metal center with preferential recognition of one prochiral face. In addition to the title combination reagent, $[Rh(cod)_2]ClO_4$ and (*R*)-(*S*)-BPPFOH and the isolated complex, [Rh{(*R*)-(*S*)-BPPFOH}(diene)]ClO_4, will be considered together below.

A high optical yield is attained with [Rh{(*R*)-(*S*)-BPPFOH}-(cod)]ClO_4 in the hydrogenation of pyruvic acid to afford (*R*)-2-hydroxypropionic acid (eq 2).[2] Addition of 1 equiv of *Triethylamine* is necessary to obtain optimal enantioselectivity. The beneficial effect of the hydroxy group on the side chain of the ferrocene ring is demonstrated by the fact that use of (*R*)-1-(*S*)-1′,2-bis(diphenylphosphino) ferrocenylethyldimethylamine ((*R*)-(*S*)-BPPFA), which is analogous to (*R*)-(*S*)-BPPFOH but lacks the hydroxy group, gives a much inferior result. In the case of a simple ketone, e.g. acetophenone or methyl *t*-bytyl ketone, the enantioselectivity is around 40–50% ee.

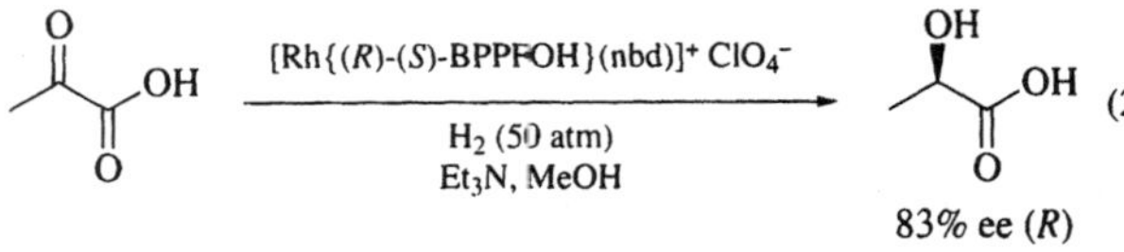

(2)

Hydrogenation of aminomethyl aryl ketone hydrocholorides is also catalyzed by [Rh{(*R*)-(*S*)-BPPFOH}(nbd)]ClO_4 to give (*R*)-2-amino-1-arylethanol hydrochlorides in high ee in the presence of triethylamine (eq 3).[3] 3,4-Disubstitution on the aromatic ring by hydroxy or alkoxy groups affords higher selectivity. The principal sympathomimetic hormone, epinephrine, especially is produced in enantiometrically pure form from (*N*-methylamino)methyl 3,4-dihydroxyphenyl ketone hydrochloride.

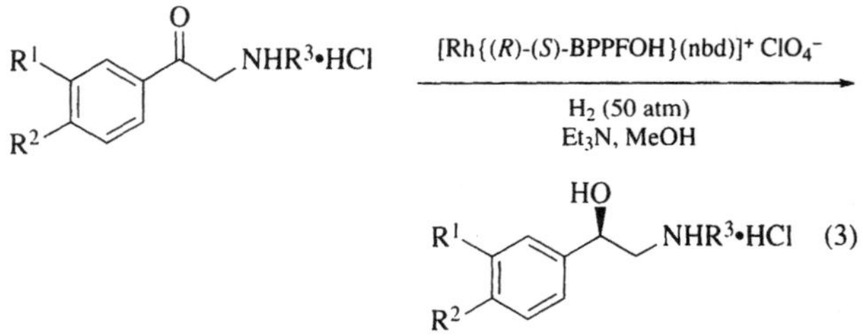

(3)

$R^1 = R^2 = OMe$, $R^3 = H$	86% ee
$R^1 = R^2 = OH$, $R^3 = Me$	95% ee
$R^1 = R^2 = R^3 = H$	52% ee
$R^1 = H$, $R^2 = OH$, $R^3 = Me$	69% ee

Asymmetric Hydrogenation of Enol Phosphinates. Catalytic asymmetric synthesis of secondary alkyl alcohols in up to 78% ee is accomplished by asymmetric hydrogenation of enol diphenylphosphinates followed by hydrolysis (eq 4).[4] The highest enantioselectivity is obtained in the hydrogenation of 1-phenylvinyldiphenylphosphinate, though in the case of phosphinates derived from dialkyl ketones, selectivities are rather low. Substitution of the diphenylphosphinyl group for other phosphorus-containing functional groups lowers the stereoselectivity. Since enol phosphinates are easily prepared from prochiral ketones, this sequence provides an alternative method for the asymmetric hydrogenation of prochiral ketones.

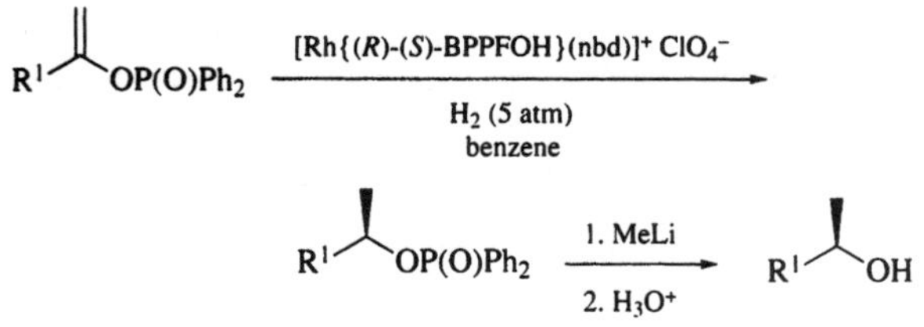

(4)

R^1 = Ph, 78% ee (*R*); R^1 = *i*-Pr, 60% ee (*R*); R^1 = *t*-Bu, 39% ee (*S*)

Related Reagents. Bis(bicyclo[2.2.1]hepta-2,5-diene)-rhodium Perchlorate.

1. Harada, K.; Munegumi, T. *Comprehensive Organic Synthesis* **1991**, *8*, Chapter 1.6.
2. Hayashi, T.; Mise, T.; Kumada, M. *Tetrahedron Lett.* **1976**, 4351
3. Hayashi, T.; Katsumura, A.; Konishi, M.; Kumada, M. *Tetrahedron Lett.* **1979**, 425.
4. Hayashi, T.; Kanehira, K.; Kumada, M. *Tetrahedron Lett.* **1981**, *22*, 4417.
5. Hayashi, T.; Mise, T.; Fukushima, M.; Kagotani, M.; Nagashima, N.; Hamada, Y.; Matsumoto, A.; Kawakami, S.; Konishi, M.; Yamamoto, K.; Kumada, M. *Bull. Chem. Soc. Jpn.* **1980**, *53*, 1138.

Yoshihiko Ito & Michinori Suginome
Kyoto University, Japan

(1*S*,9*S*)-1,9-Bis{[(*t*-butyl)dimethyl-silyloxy]methyl}-5-cyanosemicorrin[1]

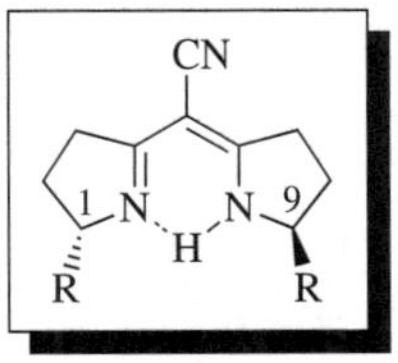

(**1**; R = CH_2OSiMe_2-*t*-Bu)
[105251-52-3] $C_{24}H_{45}N_3O_2Si_2$ (MW 463.81)
(**2**; R = CMe_2OH)
[105251-53-4] $C_{16}H_{25}N_3O_2$ (MW 291.39)
(**3**; R = CO_2Me)
[105251-49-8] $C_{14}H_{17}N_3O_4$ (MW 291.31)

(chiral ligands for enantiocontrol of metal-catalyzed reactions such as cobalt-catalyzed conjugate reduction of α,β-unsaturated carboxylic esters and amides or copper-catalyzed cyclopropanation of alkenes)[1]

Physical Data: (**1**) mp 75–76 °C, $[\alpha]_D$ −64.7° (*c* 1.0, $CHCl_3$ at rt); (**2**) mp 162 °C, $[\alpha]_D$ −82.0°; (**3**) mp 78–79 °C, $[\alpha]_D$ −145°.

Solubility: insol H_2O; (**1**) sol in all common organic solvents, including *n*-hexane; (**2**) and (**3**) sol CH_2Cl_2, alcohol, THF, and EtOAc, insol hexane, slightly sol diethyl ether.

Form Supplied in: white crystalline solid; (**1**) and (**3**) are commercially available.

Handling, Storage, and Precautions: as crystalline solids, semicorrins of this type are stable at ambient temperature; for longer periods, storage at −20 °C is recommended.

Preparation of Semicorrin Ligands and Metal Complexes. The crystalline diesters (*S,S*)-(−)-(**3**) and (*R,R*)-(+)-(**3**) are readily synthesized in enantiomerically pure form starting from L-pyroglutamic acid (−)-(**4**) or its enantiomer (eq 1).[2] By selective transformation of the ester groups, a wide range of semicorrin derivatives with different substituents at the stereogenic centers is accessible.[2,3] Among the various derivatives that have been prepared, semicorrins (**1**) and (**2**) proved to be the most versatile ligands for the stereocontrol of metal-catalyzed reactions.

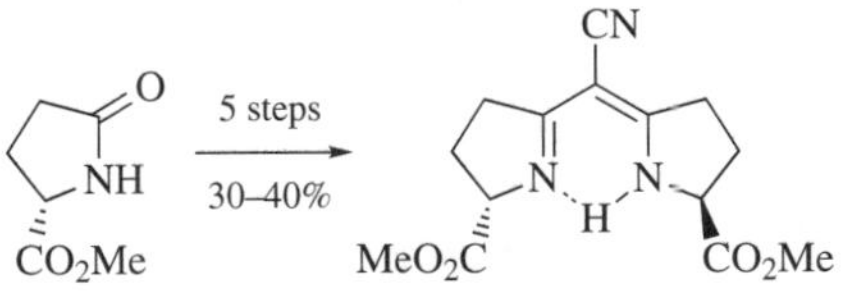

(**4**) (**3**) (1)

Semicorrins form stable chelate complexes with a variety of metal ions such as Co^{II}, Rh^{I}, Pd^{II}, or Cu^{II}. Depending on the metal ion, the ligand structure, and the reaction conditions, mono- or bis(semicorrinato) complexes are obtained.[2,3]

Enantioselective Conjugate Reduction of α,β-Unsaturated Carboxylic Esters and Amides. Cobalt semicorrin complexes are highly efficient catalysts for the reduction of electrophilic C=C bonds, using *Sodium Borohydride* as reducing agent.[1] In the presence of 0.1–1 mol % of catalyst, formed in situ from *Cobalt(II) Chloride* and ligand (**1**), esters of β-disubstituted α,β-unsaturated carboxylic acids are cleanly reduced to the corresponding saturated esters in essentially quantitative yield and with high enantioselectivity.[1,3b,4] The best results are obtained in a mixture of ethanol and a polar aprotic solvent such as DMF or diglyme under careful exclusion of oxygen. The reduction of ethyl geranate (eq 2) and the corresponding (*Z*) isomer (eq 3) are typical examples. Both reactions lead to ethyl citronellate with 94% ee. Depending on the double bond geometry, either the (*R*) or (*S*) enantiomer is obtained. The isolated double bond is inert under these conditions. During aqueous workup, the chiral ligand (**3**) forms a catalytically inactive bis(semicorrinato)cobalt(II) complex, and can be recovered by decomplexation with acetic acid.

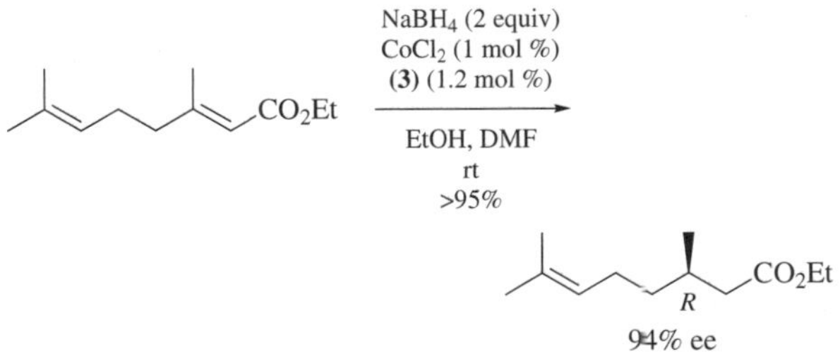

(2)

94% ee

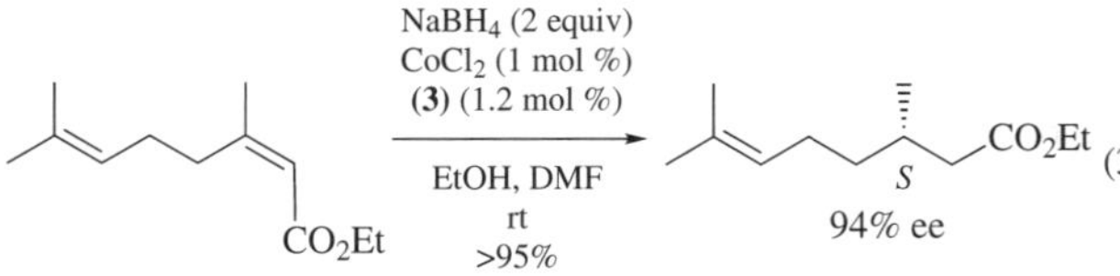

(3)

Even higher selectivities approaching 99% ee have been obtained with primary and secondary carboxamides (eq 4 and eq 5).[5] With substrates of this type, the catalyst system can undergo more than 5000 turnovers without significant loss of selectivity. An analogous diene-carboxamide was found to react with high regio- and enantioselectivity to give the corresponding γ,δ-unsaturated amide with a preference of >95:5 over the α,β-unsaturated isomer (eq 5). Tertiary carboxamides react rather sluggishly and with distinctly lower selectivity. The method cannot be applied to α,β-unsaturated ketones because the uncatalyzed nonstereoselective reaction with $NaBH_4$ proceeds at a similar rate as the cobalt-catalyzed process.

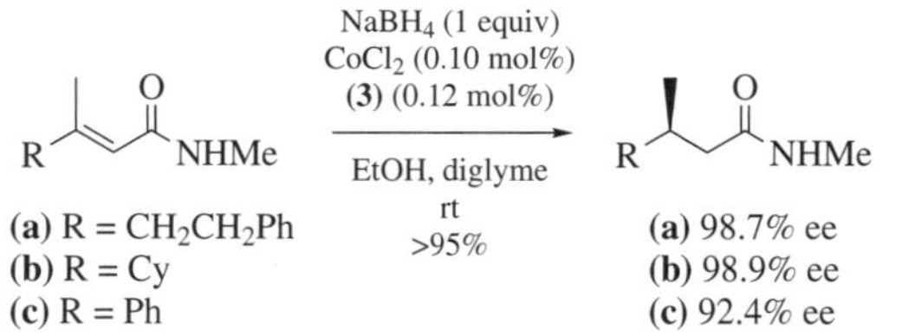

(4)

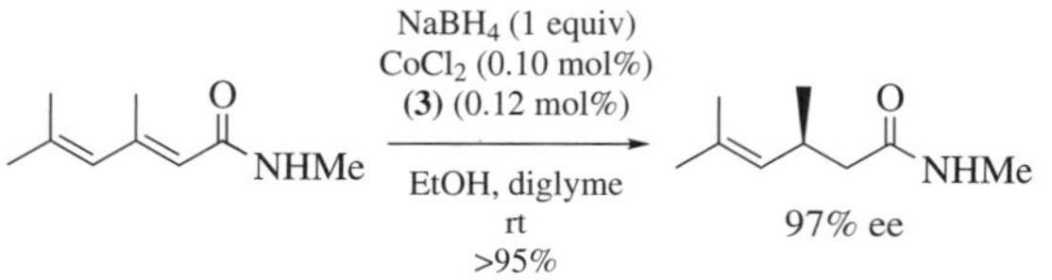

(5)

Deuteration experiments showed that the β-H atom in the product stems from borohydride whereas the α-H atom is introduced by proton transfer from ethanol.[3b] Formation of the α-(C–H) bond is nonstereoselective; accordingly, the reduction of analogous substrates with an α- instead of a β-disubstituted double bond leads to racemic products (a mechanistic model rationalizing the stereoselectivity of (semicorrinato)cobalt catalysts is available[1]).

Enantioselective Cyclopropanation of Alkenes. Semicorrin copper complexes catalyze the reaction of diazo compounds with alkenes leading to optically active cyclopropanes.[1,6] The highest enantiomeric excesses have been obtained with the bulky ligand (**2**). The stable crystalline bis(semicorrinato)copper(II) complex (**5**) serves as a convenient catalyst precursor. The actual catalyst, which is presumed to be a mono(semicorrinato)copper(I) complex, is generated in situ by heating in the presence of the diazo compound or by reduction with *Phenylhydrazine* at rt. Alternatively, the catalyst can be prepared from the free ligand and *Copper(I) t-Butoxide*. Reactions are usually carried out at rt in an apolar solvent such as dichloroethane using 1 mol % of catalyst. The best results are obtained with terminal alkenes or dienes and certain 1,2-disubstituted alkenes which react with alkyl diazoacetates to give the corresponding cyclopropanecarboxylates with high enantioselectivity (eq 6–8). The relatively poor *trans*/*cis* selectivity is a general problem which is also encountered with other catalysts.[7] Recently, even higher enantiomeric excesses have been achieved with substrates of this type, using cationic Cu^I complexes of C_2-symmetric bis(oxazolines) (**6**) (see *(S,S)-2,2′-(Dimethylmethylene)bis(4-t-butul-2-oxazoline)* or 5-aza-semicorrins (**7**).[1,8]

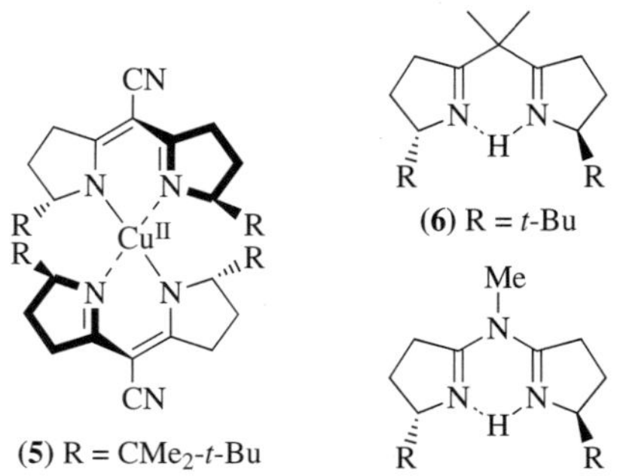

(**5**) R = CMe_2-*t*-Bu

(**6**) R = *t*-Bu

(**7**) R = $CMe_2OSiR'_3$

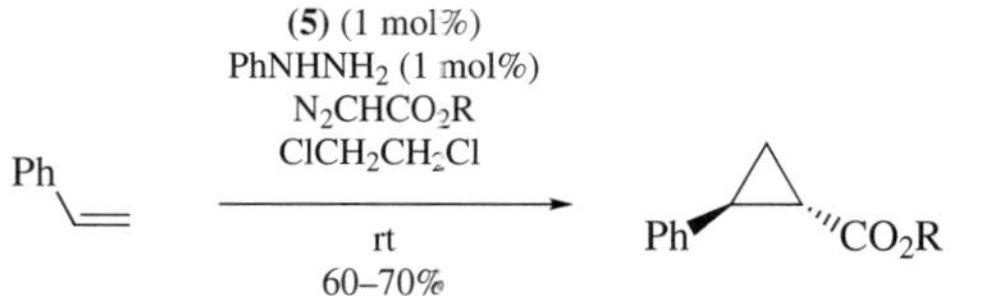

(6)

R = Et, 92% ee; *trans*:*cis* = 73:27
R = (+)-menthyl, 97% ee; *trans*:*cis* = 82:18

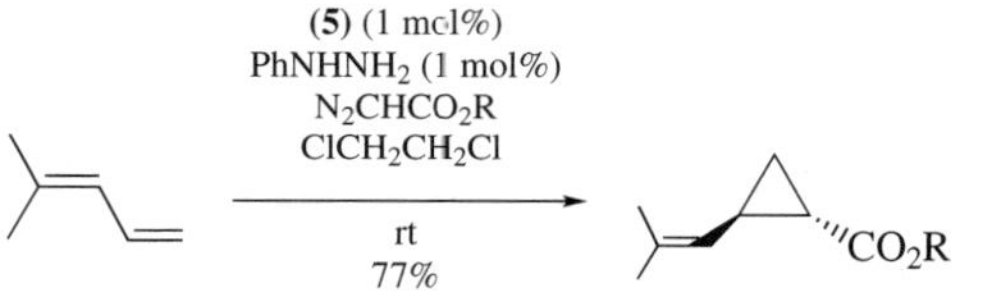

(7)

R = (+)-menthyl, 97% ee
trans:*cis* = 63:37

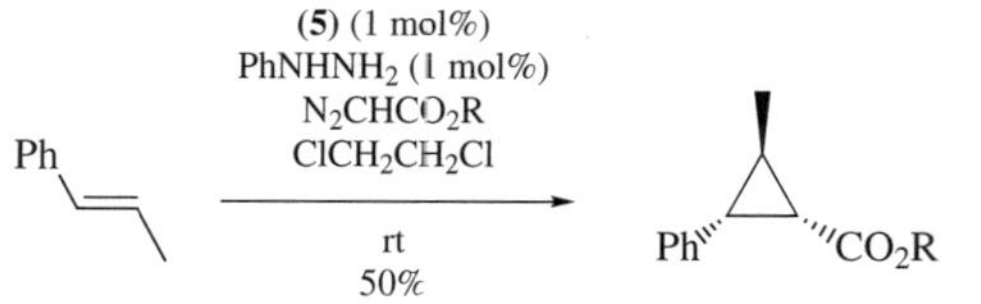

(8)

R = Et, 97% ee
trans:*cis* = 20:80

(Semicorrinato)copper catalysts have also been used for intramolecular cyclopropanation reactions of alkenyl diazo ketones (eq 9 and eq 10).[9] In this case the (semicorrinato)copper catalyst derived from complex (**5**) proved to be superior to related methylene-bis(oxazoline)copper complexes. Interestingly, analogous allyl diazoacetates react with markedly lower enantioselectivity under these conditions, in contrast to the results obtained with chiral Rh^{II} complexes which are excellent catalysts for intramolecular cyclopropanations of allyl diazoacetates but give poor enantioselectivities with alkenyl diazo ketones (see *Dirhodium(II) Tetrakis(methyl 2-pyrrolidone-5(S)-carboxylate)*).[7] Moderate enantioselectivities in the reactions shown in eq 9 and eq 10 have been reported for (salicylaldiminato)copper catalysts (77% and 34% ee, respectively).[10]

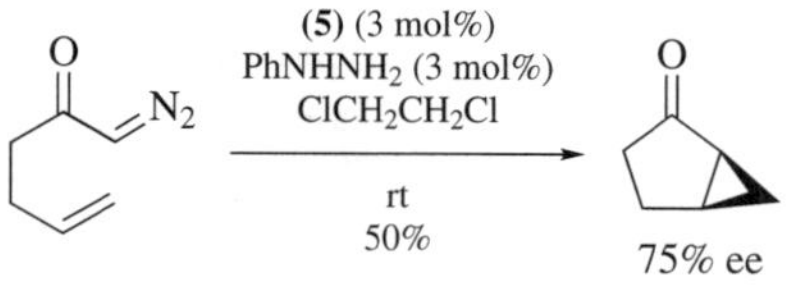

(9)

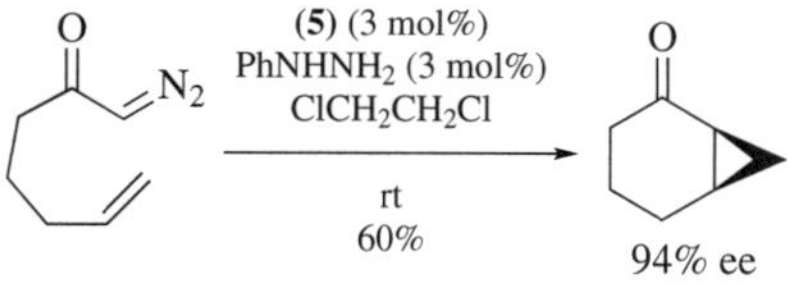

(10)

1. (a) Pfaltz, A. *Acc. Chem. Res.* **1993**, *26*, 339. (b) Pfaltz, A. In *Modern Synthetic Methods*; Scheffold, R., Ed.; Springer: Berlin, 1989; pp 199–248.
2. Fritschi, H.; Leutenegger, U.; Siegmann, K.; Pfaltz, A.; Keller, W.; Kratky, Ch. *Helv. Chim. Acta* **1988**, *71*, 1541.
3. (a) Fritschi, H. Dissertation, ETH-Zürich, No. 8951, 1989. (b) Leutenegger, U. Dissertation, ETH-Zürich No. 9091, 1990.
4. Leutenegger, U.; Madin, A.; Pfaltz, A. *Angew. Chem.* **1989**, *101*, 61; *Angew. Chem., Int. Ed. Engl.* **1989**, *28*, 60.
5. (a) von Matt, P.; Pfaltz, A. *Tetrahedron: Asymmetry* **1991**, *2*, 691. (b) von Matt, P. Dissertation, University of Basel, 1993.
6. Fritschi, H.; Leutenegger, U.; Pfaltz, A. *Helv. Chim. Acta* **1988**, *71*, 1553.
7. Doyle, M. P. *Recl. Trav. Chim. Pays-Bas* **1991**, *110*, 305.
8. Leutenegger, U.; Umbricht, G.; Fahrni, Ch.; von Matt, P.; Pfaltz, A. *Tetrahedron* **1992**, *48*, 2143.
9. C. Piqué, Dissertation, University of Basel, 1993.
10. Dauben, W. G.; Hendricks, R. T.; Luzzio, M. J.; Ng, H. P. *Tetrahedron Lett.* **1990**, *31*, 6969.

Andreas Pfaltz
University of Basel, Switzerland

(*R,R*)-Bis(*tert*-butylmethylphosphino)-methane[1]

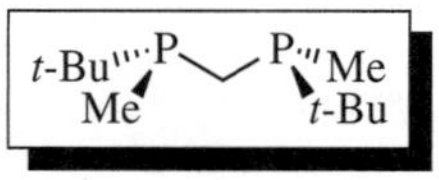

[224618-29-5] $C_{11}H_{26}P_2$ (MW 220.27)

(a chiral ligand for transition metal-catalyzed asymmetric reactions)

Alternate Name: MiniPHOS.
Physical Data: colorless oil.
Solubility: soluble most organic solvents.
Preparative Methods: phosphorus trichloride is allowed to react sequentially with alkylmagnesium chloride, methylmagnesium bromide, and BH_3–THF complex in THF to give alkyldimethylphosphine-borane. The phosphine-borane so obtained is enantioselectively deprotonated by *s*-BuLi in the presence of (−)-sparteine in Et_2O at −78 to −50°C, followed by treatment with alkyldichlorophosphine, methylmagnesium bromide, and BH_3-THF complex to give a MiniPHOS-borane complex as a diastereomixture. After removal of the *meso*-isomer by silica gel chromatography, the enantiomerically pure product is obtained by recrystallization from methanol or ethanol. The boranato group is removed by reaction with trifluoromethanesulfonic acid in toluene, followed by treatment with aqueous KOH to give the desired diphosphine **1** (eq 1).

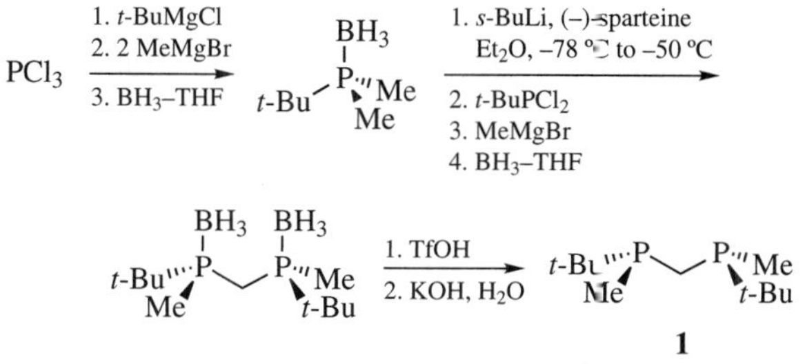

(1)

Purification: filtered through Al_2O_3 with Et_2O elution under N_2 or Ar atmosphere.
Handling, Storage, and Precautions: stench. Undergoes oxidation to the phosphine oxide on standing in air. Usually prepared before use by deboranation of the air stable phosphine-borane.

Rhodium-Catalyzed Asymmetric Hydrogenation of Olefins. MiniPHOS (**1**) can be used in rhodium-catalyzed asymmetric hydrogenation of olefinic compounds.[1] The complexation with rhodium is carried out by treatment of **1** with $[Rh(nbd)_2]BF_4$in THF (eq 2). The hydrogenation of α-(acylamino)acrylic derivatives proceeds at room temperature and an initial H_2 pressure of 1 or 6 atm in the presence of the 0.2 mol% MiniPHOS-Rh complex **2**. The reactions are complete within 24–48 h to afford almost enantiomerically pure α-amino acids (eq 3). Itaconic acids,[2] enamides,[3] and dehydro-β-amino acids[4] can also be hydrogenated with excellent enantioselectivity (eq 4–6).

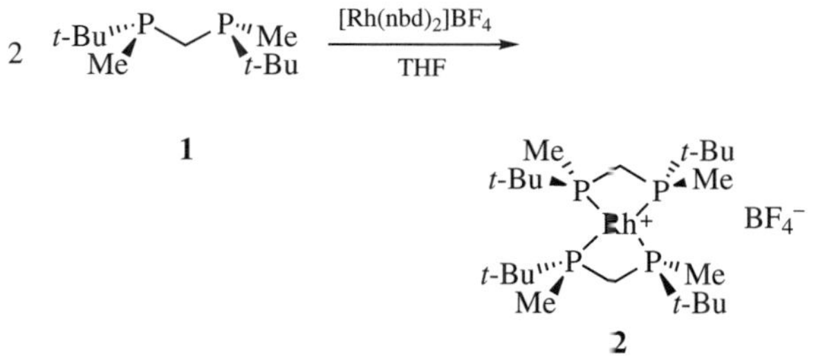

(2)

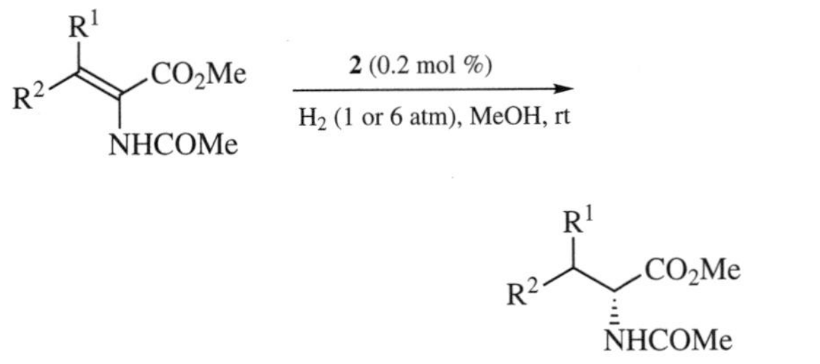

(3)

$R^1 = H$; $R^2 = Ph$, 98% ee
$R^1 = H$; $R^2 = H$, > 99% ee
R^1, $R^2 = –(CH_2)_5–$, 97% ee

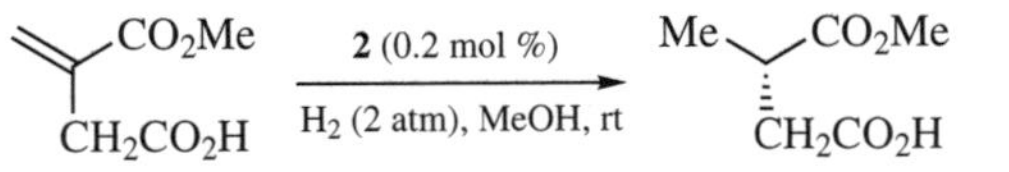

(4)

> 99% ee

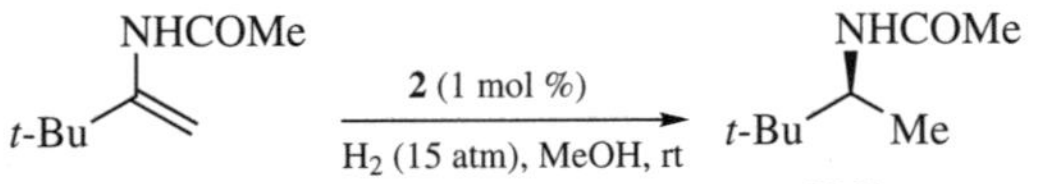

(5)

99% ee

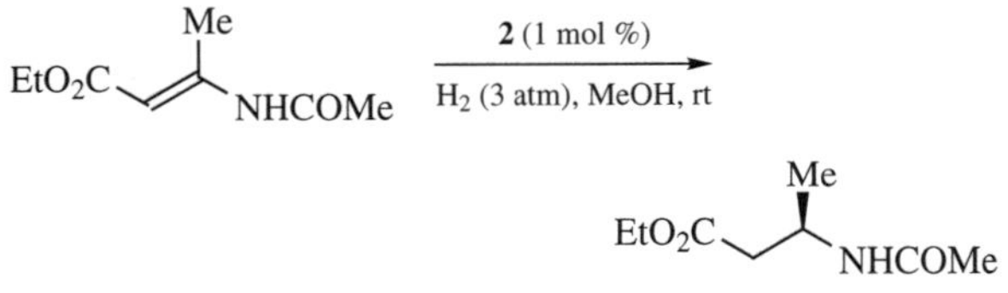

(6)

> 99% ee

Rhodium-Catalyzed Asymmetric Hydrosilylation of Ketones. Complex **2** is a good catalyst for catalytic asymmetric hydrosilylation of ketones (eq 7).[1] The reactions are carried out by using 1-naphthylphenylsilane at −40 °C in THF in the presence of **2** (1 mol%) for 3–4 days. Several types of ketones are hydrosilylated to afford optically active alcohols after acidic work-up.

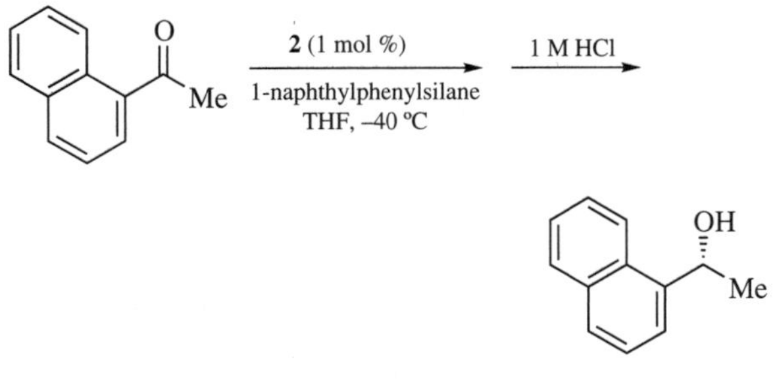

(7)

90% yield, 97% ee

Related Reagents. BisP*;[5] DIOP;[6] BINAP;[7] CHIRAPHOS;[8] DIPAMP;[9] DuPHOS;[10] BPE;[10] TRAP;[11] PHANEPHOS.[12]

1. Yamanoi, Y.; Imamoto, T. *J. Org. Chem.* **1999**, *64*, 2988.
2. Gridnev, I. D.; Yamanoi, Y.; Higashi, N.; Tsuruta, H.; Yasutake, M.; Imamoto, T. *Adv. Synth. Catal.* **2000**, *1*, 343.
3. Gridnev, I. D.; Yasutake, M.; Higashi, N.; Imamoto, T. *J. Am. Chem. Soc.* **2001**, *123*, 5268.
4. Yasutake, M.; Gridnev, I. D.; Higashi, N.; Imamoto, T. *Org. Lett.* **2001**, *3*, 1701.
5. Imamoto, T.; Watanabe, J.; Wada, Y.; Masuda, H.; Yamada, H.; Tsuruta, H.; Matsukawa, S.; Yamaguchi, K. *J. Am. Chem. Soc.* **1998**, *120*, 1635.
6. Kagan, H. B.; Dang, T. P. *J. Am. Chem. Soc.* **1972**, *94*, 6429.
7. Noyori, R.; Takaya, H. *Acc. Chem. Res.* **1990**, *23*, 345.
8. Fryzuk, M. D.; Bosnich, B. *J. Am. Chem. Soc.* **1978**, *100*, 5491.
9. (a) Scott, J. W.; Keith, D. D.; Nix Jr, G.; Parrish, D. R.; Remington, S.; Roth, G. P.; Townsend, J. M.; Valentine Jr, D.; Yang, R. *J. Org. Chem.* **1981**, *46*, 5086. (b) Knowles, W. S. *Acc. Chem. Res.* **1983**, *16*, 106.
10. Burk, M. J.; Gross, M. F.; Martinez, J. P. *J. Am. Chem. Soc.* **1995**, *117*, 9375.
11. Sawamura, M.; Kuwano, R.; Ito, Y. *J. Am. Chem. Soc.* **1995**, *117*, 9602.
12. Pye, P. J.; Rossen, K.; Reamer, R. A.; Tsou, N. N.; Volante, R. P.; Reider, P. J. *J. Am. Chem. Soc.* **1997**, *119*, 6207.

Hiroshi Danjo & Tsuneo Imamoto
Chiba University, Chiba, Japan

2,2-Bis{[2-[4(*S*)-*tert*-butyl-1,3-oxazolinyl]}propane[1]

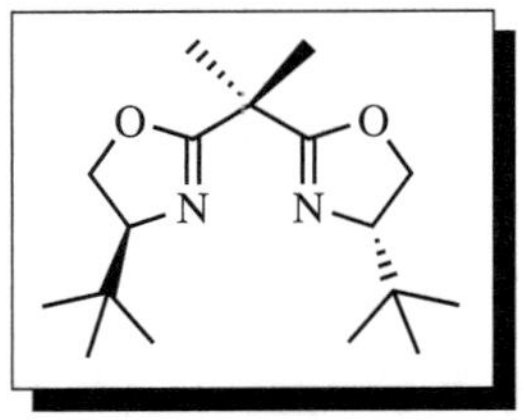

[131833-93-7] $C_{17}H_{30}N_2O_2$ (MW 294.44)

(reagent used as C_2-symmetric ligand for enantioselective catalysis[2])

Alternate Name: (*S*,*S*)-*t*-Bu-box.
Physical Data: mp 89–91 °C; $[\alpha]^{20}$ −120 ($c = 5$, $CHCl_3$).
Form Supplied in: white powder.
Preparative Methods: several methods for the synthesis of this ligand have been reported.[3] The preparation usually starts with the reduction of commercially available (*S*)-*tert*-leucine to the corresponding amino alcohol, followed by acylation with 0.5 equiv of dimethylmalonyl dichloride. The resulting dihydroxy malonodiamide is cyclized via the bis(alkyl chloride) or via the bis(tosylate) as described in an improved procedure (eq 1).[4]

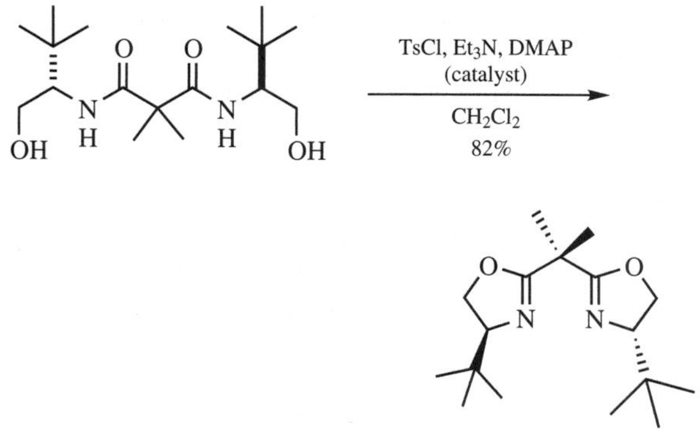

(1)

Handling, Storage, and Precautions: (*S*,*S*)-*t*-Bu-box is irritating to eyes, respiratory system and skin. In case of contact with eyes, rinse immediately with plenty of water and seek medical advice. For the purification of (*S*,*S*)-*t*-Bu-box, crystallization from pentane can be used.

Cyclopropanation. The cationic Cu(I) complex, which is readily prepared from (*S*,*S*)-*t*-Bu-box and CuOTf, is the most efficient catalyst available today for the cyclopropanation of mono- and 1,1-disubstituted olefins with diazoacetates. For example, in the reaction of ethyl diazoacetate with 2-methylpropene, >99% ee and high yields can be obtained with this catalyst using substrate to catalyst ratios as high as 1000:1.

The reaction is carried out at ambient temperature and nearly complete enantioselectivity (>99%) is observed for mono- and 1,1-disubstituted olefins with diazoacetates.[5] With all copper catalysts, the *trans*/*cis* selectivities in the cyclopropanation of monosubstituted olefins are only moderate. The *trans*/*cis* ratio depends, in this case, mainly on the structure of the diazo ester rather than the chiral ligand (eq 2). It increases with the steric bulk of the ester group of the diazo compound. With the BHT ester, the more stable *trans* isomer is formed with selectivities up to >10:1. The steric hindrance usually prevents ester hydrolysis, but the BHT group can be removed by reduction with $LiAlH_4$. The *trans* isomer is even enriched by the reduction procedure because the *cis* isomer reacts more slowly.

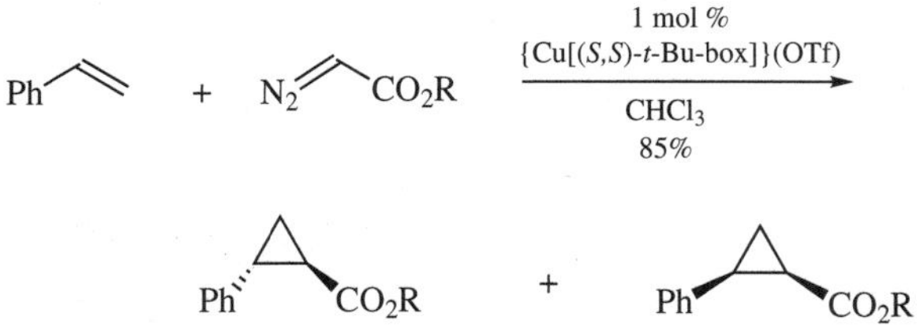

(2)

R = Et	99% ee	73:27	97% ee
R = BHT	99% ee	94:6	

On the other hand, with 1,2-disubstituted or certain trisubstituted olefins, the chiral ligand also influences the *trans*/*cis* selectivity. For example, treatment of a glucose-derived enol ether with diazomethyl acetate in the presence of {Cu[(*S*,*S*)-*t*-Bu-box]}(OTf) complex affords the cyclopropanation product with an excellent *trans*/*cis* ratio but only moderate *trans*-enantioselectivity (eq 3).[6]

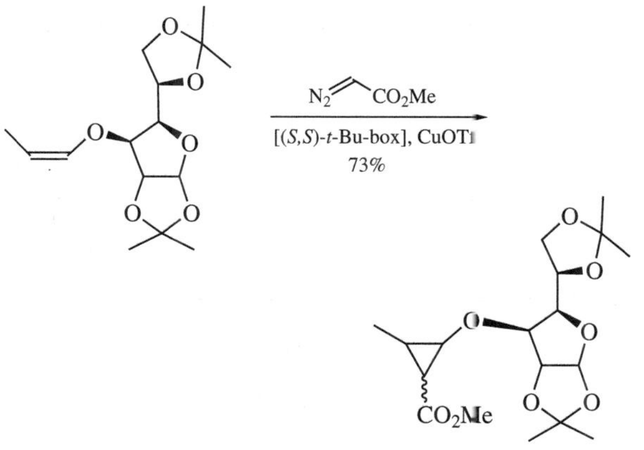

(3)

trans/*cis* >97:3
trans de 60%

Intramolecular reactions using (*S*,*S*)-*t*-Bu-box and $[Cu(MeCN)_4]PF_6$ complexes as catalysts have emerged as remarkably effective for the synthesis of macrocycles from ω-alkenyl diazoacetates. Also 10- and 15-membered ring lactones can be obtained in high enantiomeric purity with high efficiency.[7] If two double bonds are present in the molecule, the unique preference of copper-bisoxazoline catalysts to promote the formation of the larger ring is demonstrated (eq 4).[8]

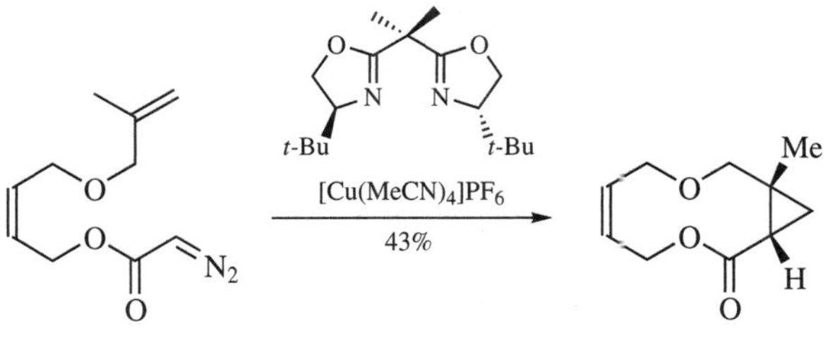

(4)

87% ee

Diels–Alder Reactions. It has been demonstrated that the ligand–metal complexes derived from (*S*,*S*)-*t*-Bu-box and a mild Lewis acid such as $Cu(OTf)_2$ are very efficient chiral catalysts for the Diels–Alder reaction with cyclopentadiene and substituted acylimide derivatives. Among various ligands examined, the (*S*,*S*)-*t*-Bu-box ligand consistently provided a very high level of *endo*/*exo* selectivity as well as *endo* enantioselectivity (90–98% ee with 5–10 mol % catalyst) and yield (82–92%) with a number of substituted dienophiles.

The counterion in these complexes plays a significant role for both catalyst activity and reaction enantioselectivity (eq 5). The hexafluoroantimonate-derived complex is 20 times more reactive in the Diels–Alder reaction than its triflate counterpart. This discovery resulted in a significantly broader scope (e.g. 1,3-cyclohexadiene, furan, isoprene and many other dienes can also be used successfully) of the reaction.[9] The crystalline aquo complexes {Cu[(*S,S*)-*t*-Bu-box](H_2O)$_2$}(OTf)$_2$ and {Cu[(*S,S*)-*t*-Bu-box](H_2O)$_2$}(SbF_6)$_2$ have also be evaluated as Lewis acid catalysts. The results indicate that hydration of the triflate complex effectively terminates catalysis. In contrast, hydration of the SbF_6 complex leads to a catalyst which is nearly as effective as its anhydrous counterpart.[10]

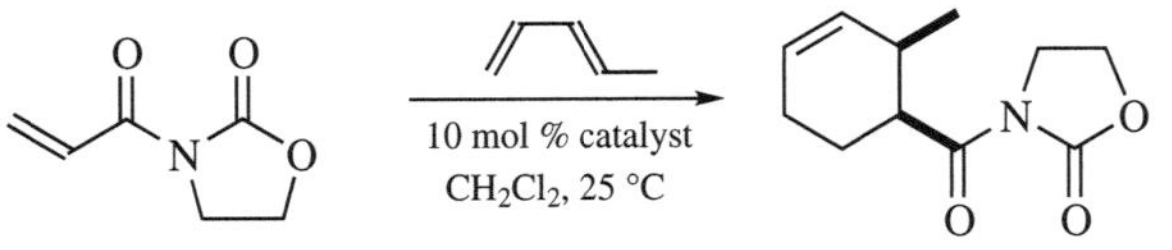
(5)

{Cu[(*S,S*)-*t*-Bu-box]}(OTf)$_2$: 15 h, 94% conversion, 84% ee

{Cu[(*S,S*)-*t*-Bu-box]}(SbF_6)$_2$: 50 min, 100% conversion, 95% ee

{Cu[(*S,S*)-*t*-Bu-box](H_2O)$_2$}(SbF_6)$_2$: 70 min, 100% conversion, 94% ee

For the enantioselective intramolecular Diels–Alder cycloaddition process, complex {Cu[(*S,S*)-*t*-Bu-box]}(SbF_6)$_2$ has also shown to be a very effective catalyst. In comparison, the complex prepared from Cu(OTf)$_2$ displays a very slow reaction, together with poor yields and selectivities. For example, the reaction of the substituted trienimide with 5 mol % of the hexafluoroantimonate complex provided the cycloaddition product as a single diastereomer within 5 h at 25 °C in good yield and 96% ee. The cycloadduct can afterwards be converted into (−)-isopulo'upone in a number of synthetic steps (eq 6).[11]

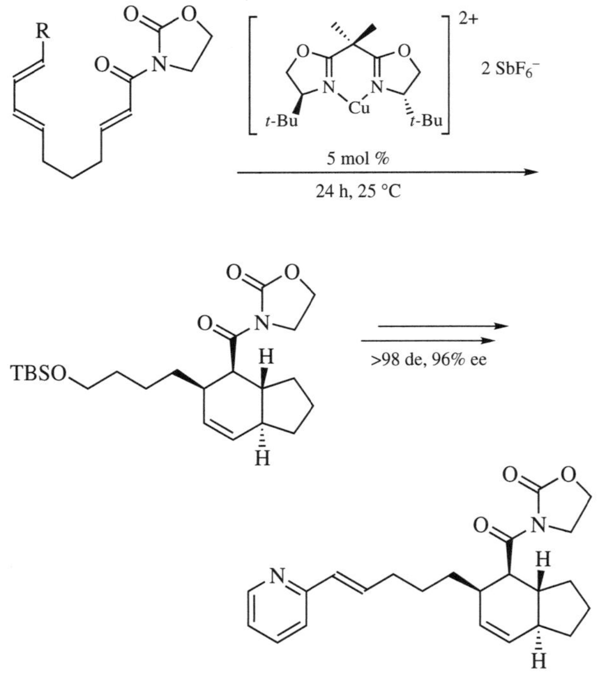
(6)

(–)-isopulo´upone

Hetero-Diels–Alder Reactions of Aldehydes. Cyclic conjugated dienes, such as 1,3-cyclohexadiene, are excellent substrates for the hetero-Diels–Alder reaction with ethyl glyoxylate catalyzed by {Cu[(*S,S*)-*t*-Bu-box]}(OTf)$_2$ (eq 7). The rate of this reaction is dependent on the counterion and the solvent. To obtain a highly diastereo- and enantioselective transformation, it is necessary to use Cu[(*S,S*)-*t*-Bu-box](OTf)$_2$ as a catalyst and $MeNO_2$ as a solvent, giving exclusively the *endo* adduct in more than 90% isolated yield with enantiomeric excess >97% ee.[12]

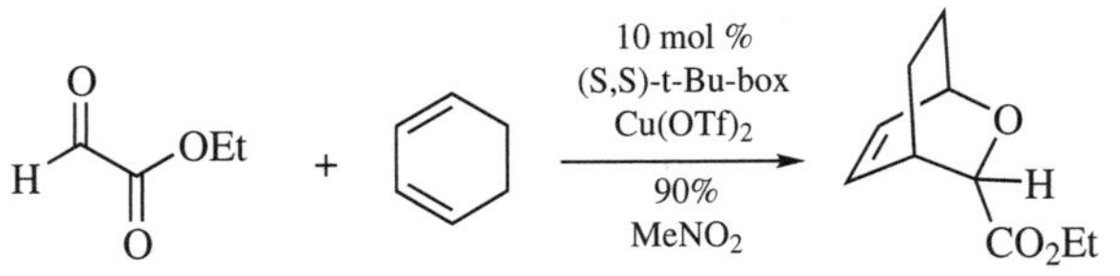
(7)

95% *endo*, > 97% ee

The product formed in this hetero-Diels–Alder reaction of ethyl glyoxylate with a cyclic diene catalyzed by (*S,S*)-*t*-Bu-box in combination with a copper(II) salt was used in the simple synthetic approach to enantiopure synthons for a class of natural products. Saponification of the bicyclic adduct followed by acidification with aqueous HCl provides the enantiopure (>99% ee) rearrangement product (eq 8).[13]

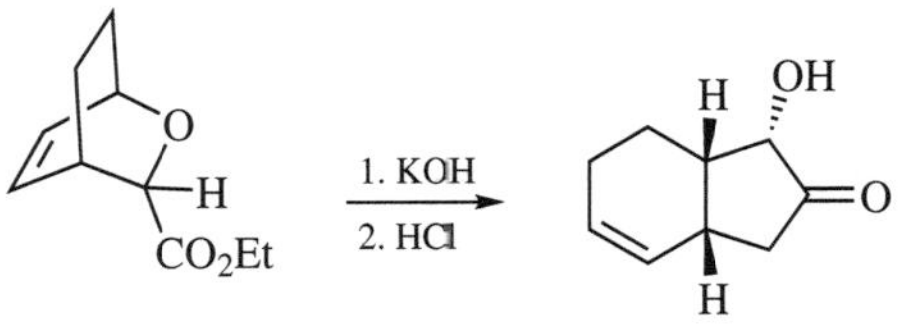
(8)

> 99% ee

Hetero-Diels–Alder Reactions of Ketones. Ketonic substrates such as ethyl pyruvate (eq 9, R^1 = Me, R^2 = OEt) do not react with simple dienes such as cyclopentadiene or 1,3-cyclohexadiene in the presence of (*S,S*)-*t*-Bu-box and a metal salt as catalyst. However, using activated dienes such as *trans*-1-methoxy-3-[(trimethylsilyl)oxyl]-1,3-butadiene (Danishefsky's diene), a hetero-Diels–Alder reaction with ethyl pyruvate and similar substrates catalyzed by 10 mol% of {Cu[(*S,S*)-*t*-Bu-box]}(OTf)$_2$ takes place in good yields and enantioselectivities (eq 9).[14] Surprisingly, it was even possible to reduce the catalyst loading to only 0.5 mol % without affecting the yield of the product, and in some cases the enantiomeric excess was even improved.

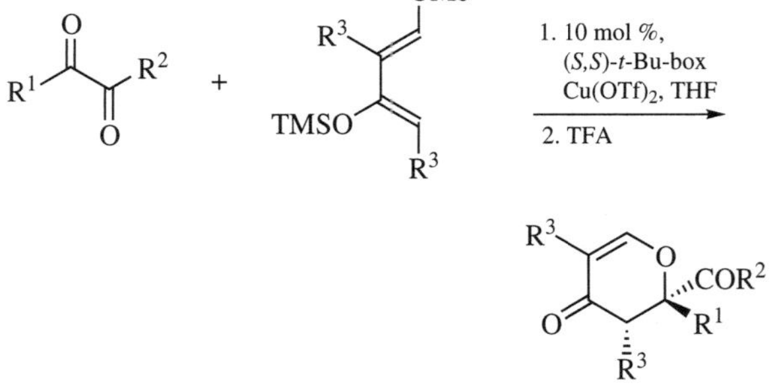
(9)

R^1 = R^2 = alkyl; yields up to 90%, up to 88% ee

R^1 = alkyl, aryl; R^2 = OEt; yields up to 96%, up to 99% ee

Inverse electron demand hetero-Diels–Alder reactions of acyl phosphonates[15] or α-keto ester heterodienes and enol ethers are also catalyzed by (*S,S*)-*t*-Bu-box complexes. High levels of enantioselectivity are obtained with γ-alkyl-, -aryl-, -alkoxy-

and -thioalkyl-substituted β,γ-unsaturated α-keto esters using 2 mol % of the aquo complex {Cu[(*S,S*)-*t*-Bu-box](H_2O)$_2$}(OTf)$_2$ (eq 10).[16] This reaction shows a number of practical advantages. First, the aquo complex, a bench-stable pale blue powder that can be stored indefinitely without special precaution, provides not only uniformly high levels of enantioselection but also excellent control of regioselectivity. A second feature is the possibility of reusing the catalyst following a simple recycling protocol involving hexane, a solvent in which the catalyst is apparently insoluble.

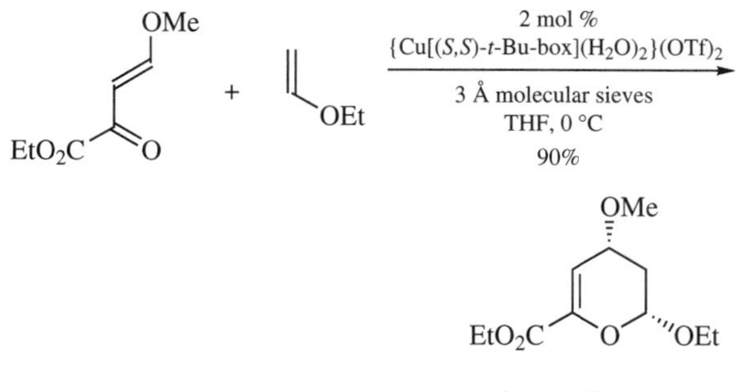

(10)

endo/exo 59:1, 98% ee

Ene Reaction. The dicarbonyl moiety of ethyl glyoxylate was found to react with a broad range of unactivated olefins to afford γ,δ-unsaturated α-hydroxy esters in high enantioselectivity and high yields (eq 11).

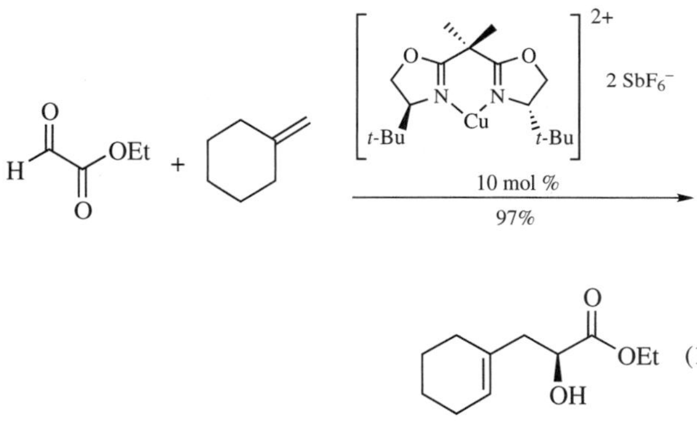

(11)

97% ee

In this reaction, several attractive features can be noted. The bench-stable aquo {Cu[(*S,S*)-*t*-Bu-box](H_2O)$_2$}(SbF$_6$)$_2$ complex was as effective as the analogous anhydrous {Cu[(*S,S*)-*t*-Bu-box]}(SbF$_6$)$_2$ complex, even with catalyst loadings as low as 0.1 mol %, without significant loss of yield and enantioselectivity. A testament for the Lewis acidity of the copper(II) (*S,S*)-*t*-Bu-box complexes is that weakly nucleophilic olefins such as hex-1-ene and cyclohexene had not been previously employed in catalytic asymmetric ene reactions.[17]

Besides the symmetrical 1,1-disubstituted alkenes, unsymmetrical 1,1-disubstituted, 1,2-disubstituted, and monosubstituted alkenes also react in a highly enantioselective manner in the presence of the copper(II) (*S,S*)-*t*-Bu-box catalyst.

A short and efficient asymmetric total synthesis of (−)-α-kainic acid, which is an important neurotransmitter, has been achieved by means of a metal-promoted, enantioselective ene reaction. This approach provides entry into the kainic acid ring system from a very simple precursor (eq 12).[18] One of the key steps involved (*S,S*)-*t*-Bu-box-promoted magnesium(II) catalysis. In this case, cyclization favored strongly the desired *cis*-diastereomer, which can be converted to the desired acid in a number of synthetic steps.

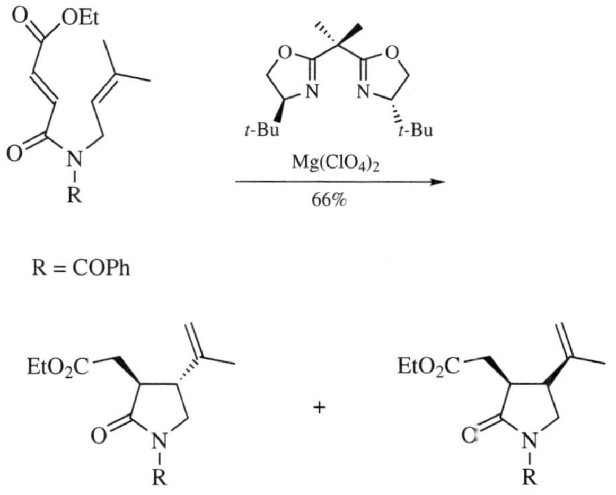

(12)

1:20

Enol Amination. The {Cu[(*S,S*)-*t*-Bu-box]}(OTf)$_2$ complex was found to be optimal for promoting the enantioselective conjugated addition of enolsilanes to azodicarboxylate derivatives (eq 13). This methodology provides an enantioselective catalytic route to differentially protected α-hydrazino carbonyl compounds. Isomerically pure enolsilanes of aryl ketones, acylpyrroles, and thioesters add to the azo-imide in greater than 95% ee. The use of an alcohol additive was critical to achieve catalyst turnover. Amination of cyclic enolsilanes was also possible. For example, the enolsilane of 2-methylindanone provides the adduct containing a tetrasubstituted stereogenic center in 96% ee and high yield. Acyclic (*Z*)-enolsilanes react in the presence of a protic additive with enantioselection up to 99%.[19]

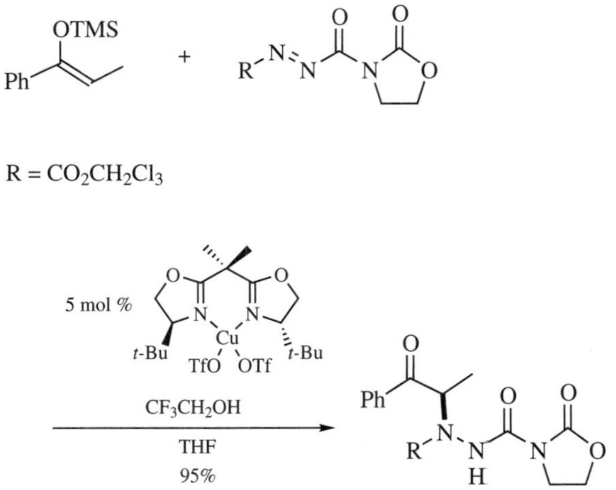

(13)

99% ee

Aldol Addition. A catalyst generated upon treatment of Cu(OTf)$_2$ with the (*S,S*)-*t*-Bu-box ligand has been shown to be an effective Lewis acid for the enantioselective Mukaiyama aldol reaction.[20] The addition of substituted and unsubstituted enolsilanes at −78 °C in the presence of 5 mol % catalyst was reported to be very general for various nucleophiles, including silyl dienolates and enol silanes prepared from butyrolactone as well as acetate and propionate esters.

Mukaiyama aldol reactions of silylketene acetal and pyruvate ester (eq 14) in the presence of 10 mol % {Cu[(*S*,*S*)-*t*-Bu-box]}$(OTf)_2$ catalyst furnish the corresponding aldol product in excellent enantiomeric excess (98%). Furthermore, the addition reactions of ketene acetals derived from *t*-butyl thioacetate and benzyloxyacetaldehyde with only 5 mol % catalyst afford the aldol product in 91% ee (eq 15).[21] It is also noteworthy that the addition of both propionate-derived (*Z*)- and (*E*)-silylketene acetals stereoselectively forms the *syn*-adduct in 97% and 85% ee, respectively.

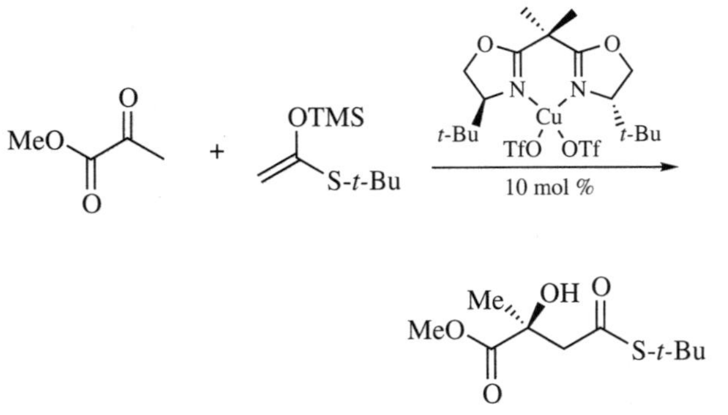

(14)

98% ee

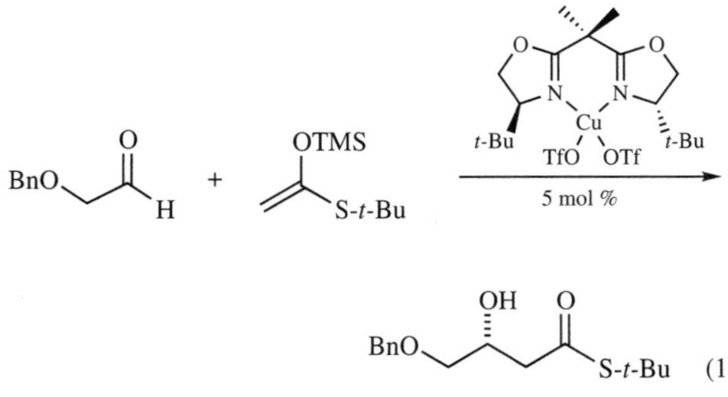

(15)

91% ee

Michael Additions. The {Cu[(*S*,*S*)-*t*-Bu-box]}$(SbF_6)_2$ complex catalyzes the enantioselective addition of enolsilanes to fumaroyl imides with enantioselectivities of up to 99% ee and in good yields (up to 91%).[22] Here, the diastereoselectivity correlates with the geometry of the nucleophile; (*E*)-silylketene acetals preferentially deliver *anti* adducts (eq 16), while (*Z*)-silylketene acetals afford *syn* products (eq 17).

Alkylidene malonates also react with silylketene thioacetals under catalysis by the {Cu[(*S*,*S*)-*t*-Bu-box]}$(SbF_6)_2$ complex. The reaction adducts are obtained with good efficiency (up to 91% yield) and high levels of enantiocontrol (up to 93% ee),[23] especially for alkylidene malonates bearing sterically demanding substituents in the β-position.

Oxidations. A widely used method for allylic oxidation is the Kharash–Sosnovsky reaction using a peroxide and a copper(I) salt system. Enantioselective allylic oxidations of cycloalkenes such as cyclopentene, cyclohexene and cycloheptene with *tert*-butyl perbenzoate were investigated with a variety of catalysts derived from bis(oxazoline) ligands and copper(I) triflate complexes (eq 18). The ligand–copper(I) complexes from the *t*-Bu-box, Ph-box and *i*-Pr-box have shown comparable results.[24] In the presence of 5 mol % {Cu[(*S*,*S*)-*t*-Bu-box]}(OTf), a remarkable 84% ee (61% yield at 68% conversion) was achieved in the transformation of cyclopentene to 2-cyclopentenyl benzoate. Acetonitrile was the solvent of choice. The reactions were typically run at −20 °C for 5 days. At these temperatures acyclic olefins exhibited only very low or no optical activity. However, at 55 °C for 2 days, allylbenzene and oct-1-ene afforded 36% ee and 30% ee, respectively.[25]

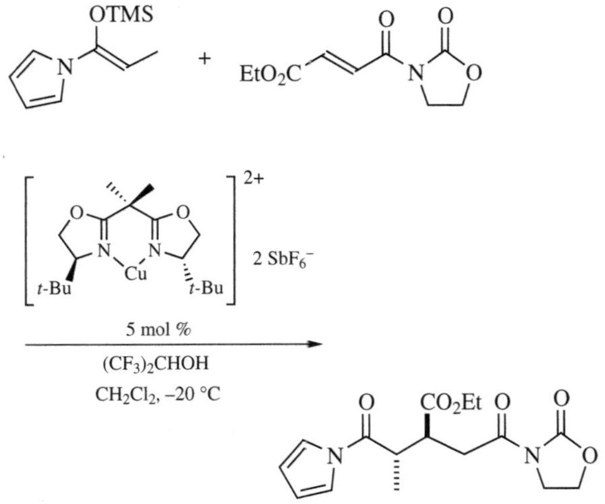

(16)

94% ee *anti*/*syn* 97:3

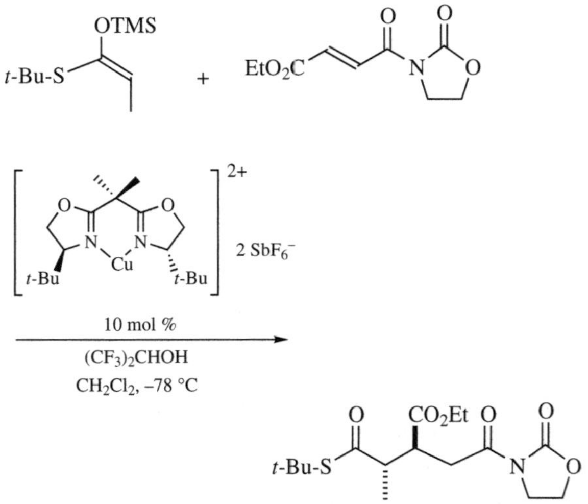

(17)

99% ee *anti*/*syn* 7:93

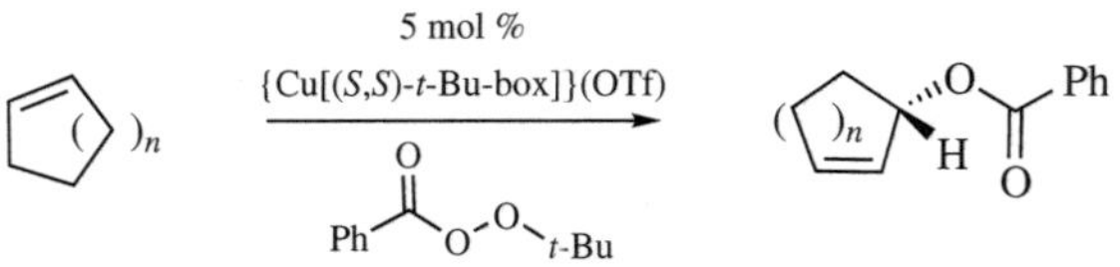

(18)

n = 1; 61%, 78% ee
n = 2; 44%, 79% ee

Aziridination Reactions. CuOTf–bis(oxazoline) complexes are efficient catalysts for the aziridination of olefins. Olefins with aryl substituents have proven to be the most efficient substrates for this reaction. For styrene, the corresponding *N*-tosylaziridine was obtained in good yield (89%), but only moderate enantiomeric excess (66% ee) (eq 19).[26] Catalysts derived from other bis(oxazoline) ligands, like for example Ph-box, have exhibited superior results over the sterically demanding (*S*,*S*)-*t*-Bu-box giving rise to enantioselectivities of up to 97% ee.[27]

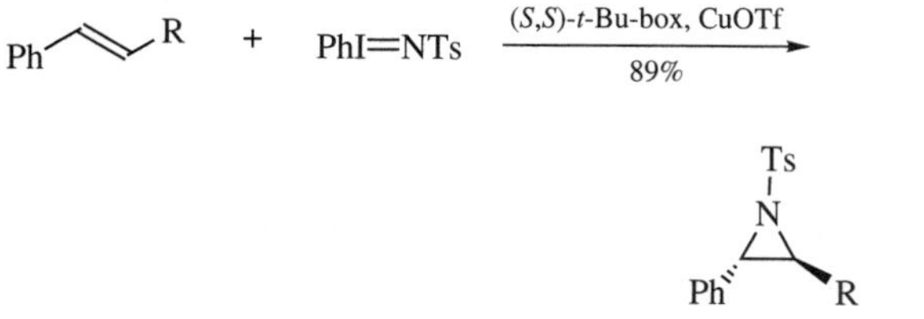

(19)

R = H, 63% ee
R = Me, 70% ee

Radical Reactions. For radical additions, chiral Lewis acids can be complexed to radical traps which undergo enantioselective attack at the β-centers. One solution to the problem of acyclic diastereoselection in β-radical additions has been the use of bis(oxazolines) in conjunction with Lewis acid additives. (*S*,*S*)-*t*-Bu-box-derived, Lewis acid-promoted free radical conjugate additions to β-substituted, α,β-unsaturated *N*-oxazolidinone derivatives, with stoichiometric amounts of Lewis acid and ligand, proceed in excellent chemical yields and high enantioselectivities (eq 20). From the variety of tested Lewis acids for this reaction, usually magnesium or zinc salts, $MgBr_2$ gave the best results with the (*S*,*S*)-*t*-Bu-box ligand.[28]

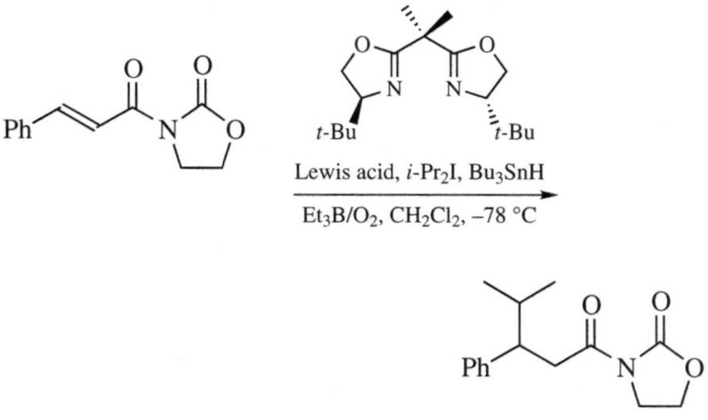

(20)

Lewis acid
$Zn(OTf)_2$, 37% ee
$MgBr_2$, 77% ee

Radical allylation of bromides derived from β-substituted, α,β-unsaturated *N*-oxazolidinones with several allylsilanes have been carried out with 1 equiv of Lewis acid and ligand in dichloromethane initiated by Et_3B at −78 °C.[29] The use of the (*S*,*S*)-*t*-Bu-box ligand in combination with MgI_2 proceeds with good selectivity and yield (eq 21).

Polymerization Reactions. The enantioselective co-polymerization of styrenes and carbon monoxide has been achieved by the use of a palladium catalyst based on the (*S*,*S*)-*t*-Bu-box ligand. Co-polymerization of *p*-*tert*-butylstyrene (TBS) and carbon monoxide in the presence of 0.1 mol % chiral catalyst afforded the alternating co-polymer with a highly isotactic microstructure and excellent optical purity (eq 22). The stereoregularity of the polymer is >98% and the polymer exhibits high molar rotation.[30]

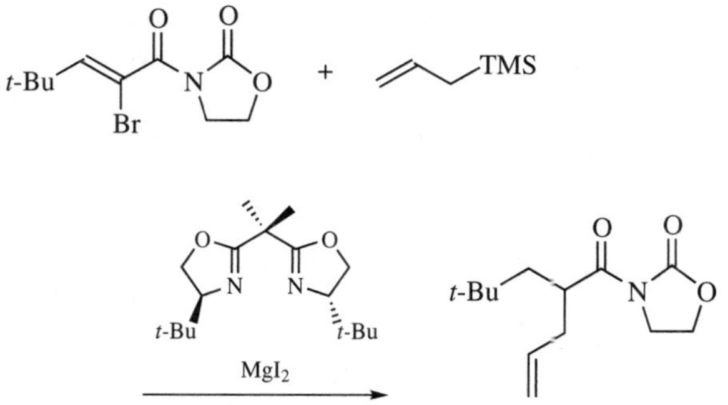

(21)

R/S 98:1

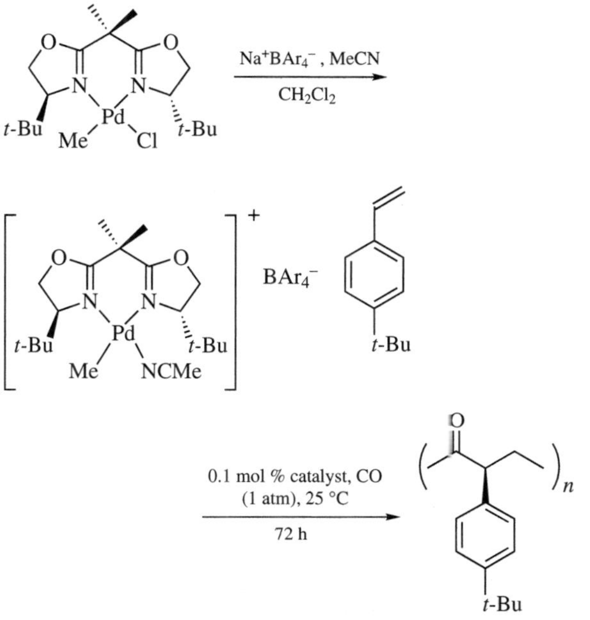

(22)

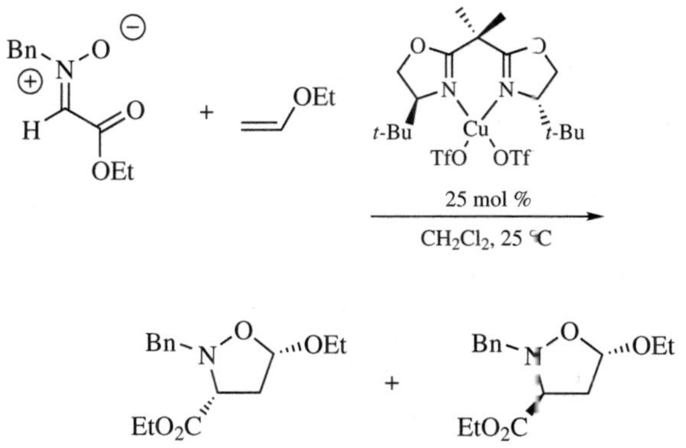

(23)

84:16
89:35 % ee

1,3-Dipolar Cycloadditions. 1,3-Dipolar cycloadditions provide a powerful method for the synthesis of five-membered heterocyclic rings. The use of (*S*,*S*)-*t*-Bu-box in combination with $Cu(OTf)_2$ as catalyst for the reaction of a nitrone with ethyl vinyl ether leads to the products in 93% yield (eq 23). The diastereoselectivity is *exo*-selective, as the product was obtained in an *endo*/*exo* ratio of 83:16. A change of the counterion in the catalyst from triflate to antimonate leads to a nonselective reaction. The use of a catalyst prepared from (*S,S*)-*t*-Bu-box and $Zn(OTf)_2$ displays weaker Lewis acidity than the corresponding copper(II) catalyst, which results in lower conversion (73%) and selectivity (*endo*/*exo* 66:34).[31]

Enantioselective Friedel–Crafts Reactions. The copper(II) complex of the (*S,S*)-*t*-Bu-box ligand has been used as catalyst for the reaction of *N,N*-dimethylaniline with ethyl glyoxylate and it has been found that a highly regio- and enantioselective Friedel–Crafts reaction takes place.[32] This reaction proceeds with the exclusive formation of the *para*-substituted isomer in up to 91% yield and 94% ee (eq 24).

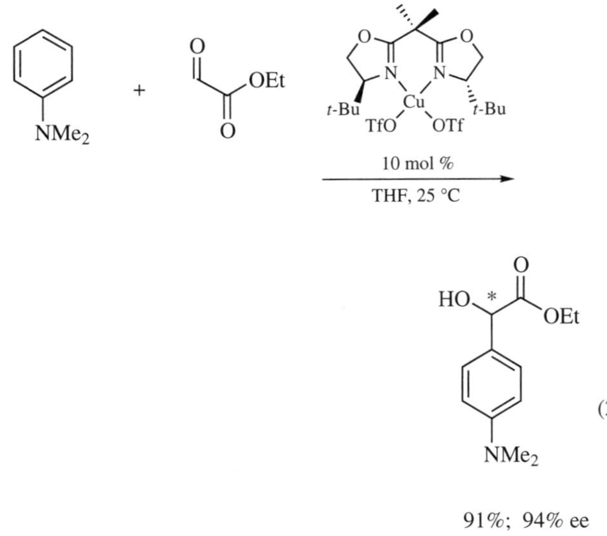

(24)

The reaction has been investigated for *N,N*-dimethylaniline under different reaction conditions and has been developed into a highly enantioselective catalytic reaction for *meta*-substituted *N,N*-dimethylanilines containing either electron-withdrawing or electron-donating substituents. The reaction also proceeds well for catalytic aromatic amines such as *N*-methylindoline, *N*-methyltetrahydroquinoline, and julolidine, where up to 91% yield and 93% ee are obtained. For polyaromatic amines, high yields but only moderate ee values for the Friedel–Crafts products are obtained. To enhance the potential of the reaction, the *N,N*-dimethyl- and *N*-methyl substituents, respectively, can be removed, successfully leading to the mono *N*-methyl product or the free amine, which allows the introduction of a variety of other substituents. Moreover, the catalytic enantioselective reaction also proceeds for heteroaromatic compounds such as 2-substituted furans, which react with ethyl glyoxylate as well as trifluropyruvates, giving up to 89% ee of the Friedel–Crafts products.

Poly(ethylene glycol)-supported (*S,S*)-*t*-Bu-box Ligands. (*S,S*)-*t*-Bu-box-supported on a modified poly(ethylene glycol) (PEG) has been prepared by a reaction sequence that involves formation of a suitably functionalized ligand and its attachment to the polymer matrix by means of a spacer and a linker. The solubility properties of PEG allowed the successful use of the supported ligand in the enantioselective cyclopropanation carried out under homogeneous conditions, and allow recovery of the ligand as if bound to an insoluble support.

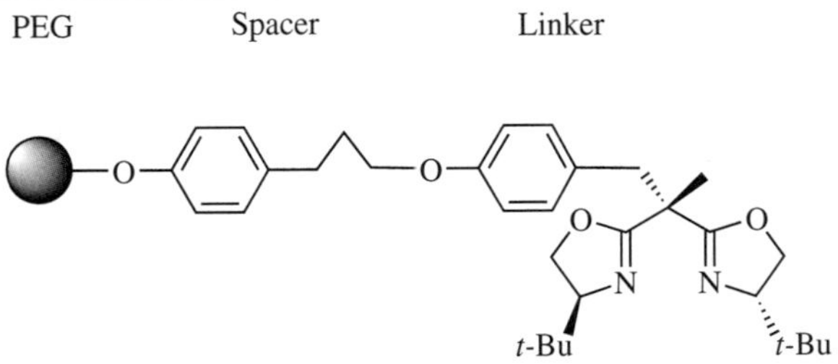

The cyclopropanation of styrene carried out with ethyl diazoacetate in the presence of 10 mol % supported ligand and 10 mol % CuOTf gave a 77:23 mixture of the *trans*/*cis* cyclopropane adducts in 63% yield and 91% ee for the major isomer (eq 25).[33] These results were comparable to those obtained with the free (*S,S*)-*t*-Bu-box ligand.[5a]

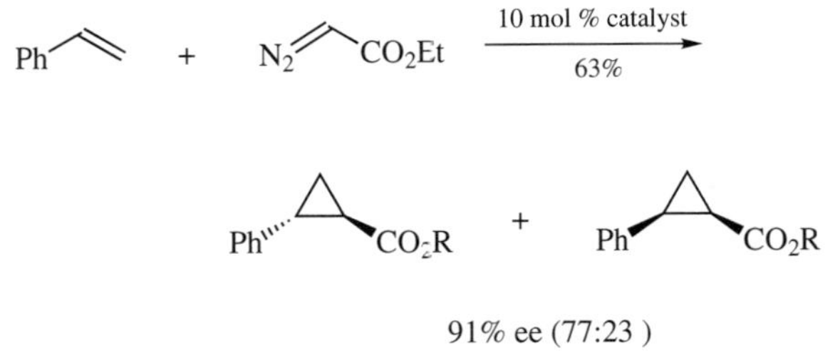

(25)

1. (a) Gosh, A. K.; Mathivanan P.; Cappiello J. *Tetrahedron Asymm.* **1998**, *9*, 1. (b) Jørgensen, K. A.; Johannsen, M.; Yao, S.; Audrain, H.; Thorhauge, J. *Acc. Chem. Res.* **1999**, *32*, 605. (c) Pfaltz, A. *Acc. Chem. Res.* **1993**, *26*, 339. (d) Bolm, C. *Angew. Chem., Int. Ed. Engl.* **1991**, *30*, 542.
2. Johnson, J. S.; Evans, D. A. *Acc. Chem. Res.* **2000**, *33*, 325.
3. (a) Evans, D. A.; Woerpel, K. A.; Hinman, M. M.; Faul, M. M. *J. Am. Chem. Soc.* **1991**, *113*, 726. (b) Desimoni, G.; Faita, G.; Mella, M. *Tetrahedron* **1996**, *52*, 13649. (c) Denmark, S. E.; Nakajima, N.; Nicaise, O. J.-C.; Faucher, A.-M.; Edwards, J. P. *J. Org. Chem.* **1995**, *60*, 4884. (d) Davies, I. W.; Gerena, L.; Lu, N.; Larsen, R. D.; Reider, P. J. *J. Org. Chem.* **1996**, *61*, 9629. (e) Bedekar, A. V.; Koroleva, E. B.; Andersson, P. G. *J. Org. Chem.* **1997**, *62*, 2518.
4. Evans, D. A.; Peterson, G. S.; Johnson, J. S.; Barnes, D. M.; Campos, K. R.; Woerpel, K. A. *J. Org. Chem.* **1998**, *63*, 4541.
5. (a) Evans, D. A.; Woerpel, K. A.; Hinman, M. M.; Faul, M. M. *J. Am. Chem. Soc.* **1991**, *113*, 726. (b) Evans, D. A.; Woerpel, K. A.; Scott, M. J. *Angew. Chem., Int. Ed. Engl.* **1992**, *31*, 430.
6. Schumacher, R.; Reissig, H.-U. *Synlett* **1996**, 1121.
7. Doyle, M. P.; Hu, W. *J. Org. Chem.* **2000**, *65*, 8839.
8. (a) Doyle, M. P.; Protopopova, M. N. *Tetrahedron* **1998**, *54*, 7919. (b) Doyle, M. P.; Peterson, C. S.; Zhou, Q.-L.; Nishiyama, H. *J. Chem. Soc., Chem. Commun.* **1997**, 211.
9. (a) Evans, D. A.; Murry, J.; von Matt, P.; Norcross, R. D.; Miller, S. J. *Angew. Chem., Int. Ed. Engl.* **1995**, *34*, 798. (b) Evans, D. A.; Barnes, D. M.; Johnson, J. S.; Lectka, T.; von Matt, P.; Miller, S. J.; Murry, J. A.; Norcross, R. D. *J. Am. Chem. Soc.* **1999**, *121*, 7582.
10. Evans, D. A.; Miller, S. J.; Lectka, T.; von Matt, P. *J. Am. Chem. Soc.* **1999**, *121*, 7559.
11. Evans, D. A.; Johnson, J. S. *J. Org. Chem.* **1997**, *62*, 786.
12. (a) Johannsen, M.; Jørgensen, K. A. *Tetrahedron* **1996**, *52*, 7321. (b) Johannsen, M.; Yao, S.; Graven, A.; Jørgensen, K. A. *Pure Appl. Chem.* **1998**, *70*, 1117.
13. (a) Johannsen, M.; Jørgensen, K. A. *J. Org. Chem.* **1995**, *60*, 5757. (b) Johannsen, M.; Jørgensen, K. A. *J. Chem. Soc., Perkin Trans. 2* **1997**, 1183.
14. Yao, S. L.; Johannsen, M.; Audrain, H.; Hazell, R. G.; Jørgensen, K. A. *J. Am. Chem. Soc.* **1998**, *120*, 8599.

15. Schuster, T.; Evens, D. A. *Phosphorus Sulfur Silicon Relat. Elem.* **1995**, *103*, 259.
16. (a) Evans, D. A.; Olhava, E. J.; Johnson, J. S.; Janey, J. M. *Angew. Chem., Int. Ed.* **1998**, *37*, 3372. (b) Thorauge, J.; Johannsen, M.; Jørgensen, K. A. *Angew. Chem., Int. Ed. Engl.* **1998**, *37*, 2404.
17. Evans, D. A.; Burgey, C. S.; Paras, N. A.; Vojkovsky, T.; Tregay, S. *J. Am. Chem. Soc.* **1998**, *120*, 5824.
18. Xia, Q.; Ganem, B. *Org. Lett.* **2001**, *3*, 485.
19. Evans, D. A.; Johnson, D. S. *Org. Lett.* **1999**, *1*, 595.
20. Evans, D. A.; Kozlowski, M. C.; Brugey, C. S.; MacMillan, D. W. C. *J. Am. Chem. Soc.* **1997**, *119*, 7893.
21. Evans, D. A.; Murry, J. A.; Kozlowski, M. C. *J. Am. Chem. Soc.* **1996**, *118*, 5814.
22. Evans, D. A.; Willis, M. C.; Johnston, J. N. *Org. Lett.* **1999**, *1*, 865.
23. Evans, D. A.; Rovis, T.; Kozlowsky, M. C.; Tedrow, J. S. *J. Am. Chem. Soc.* **1999**, *121*, 1994.
24. Gokhale, A. S.; Minidis, A. B. E.; Pfalz, A. *Tetrahedron Lett.* **1995**, *36*, 1831
25. Andrus, M. B.; Argade, A. B.; Pamment, M. G. *Tetrahedron Lett.* **1995**, *36*, 2945.
26. Llewellyn, D. B.; Adamson, D.; Arndtsen, B. A. *Org. Lett.* **2000**, *2*, 4165.
27. (a) Evans, D. A.; Faul, M. M.; Bilodeau, M. T.; Anderson, B. A.; Barnes, D. M. *J. Am. Chem. Soc.* **1993**, *115*, 5328. (b) Hansen, K. B.; Finney, N. S.; Jacobsen, E. N. *Angew. Chem., Int. Ed.* **1995**, *34*, 676. (c) Juhl K.; Hazell RG.; Jørgensen, K. A. *J. Chem. Soc., Perkin Trans. 1* **1999**, 2293.
28. Sibi, M. P.; Ji, J. *J. Am. Chem. Soc.* **1996**, *118*, 9200.
29. Porter, N. A.; Wu, J.; Zhang, G.; Reed, A. D. *J. Org. Chem.* **1997**, *62*, 6702.
30. Brookhart, M.; Wagner, M. I. *J. Am. Chem. Soc.* **1994**, *116*, 3641.
31. (a) Jensen, K. B.; Hazell, R. G.; Jørgensen, K. A. *J. Org. Chem.* **1999**, *64*, 2353. (b) Gothelf, K. V.; Hazell, R. G.; Jørgensen, K. A. *J. Org. Chem.* **1996**, *61*, 346.
32. Gatherdood, N.; Zhuang, W.; Jørgensen, K. A. *J. Am. Chem. Soc.* **2000**, *122*, 12517.
33. Annunziata, R.; Benaglia, M.; Cinquini, M.; Cozzi, F.; Pitillo, M. *J. Org. Chem.* **2001**, *66*, 3160.

Angelika S. Magnus & Pher G. Andersson
Uppsala University, Sweden

Bis(cyclohexyl isocyanide)gold(I) Tetrafluoroborate–(*R*)-*N*-[2-(*N*,*N*-Dimethylamino)ethyl]-*N*-methyl-1-[(*S*)-1′,2-bis(diphenylphosphino)-ferrocenyl]ethylamine[1]

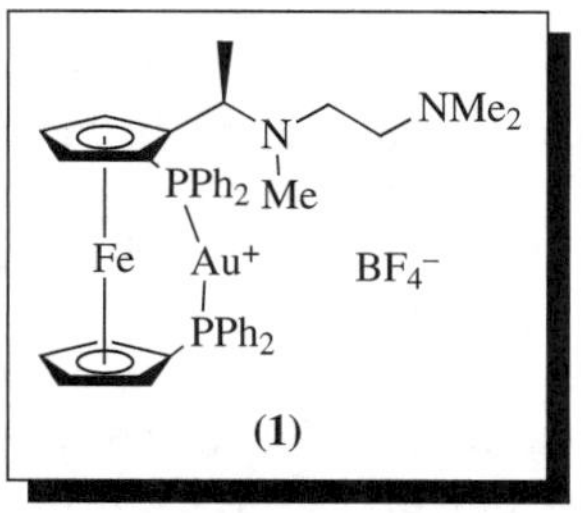

(1)

[–] (gold component)	$C_{41}H_{44}AuBF_4FeN_2P_2$	(MW 966.38)
[43067-36-3]	$C_{14}H_{22}AuBF_4N_2$	(MW 502.14)
(ferrocene component) *[119477-31-5]*	$C_{41}H_{44}FeN_2P_2$	(MW 682.62)

(chiral catalyst for asymmetric aldol reactions giving high diastereo- and enantioselectivity;[2] enantioselective synthesis of β-hydroxy-α-aminophosphonates;[3] asymmetric allylation[4])

Solubility: sol dichloromethane, 1,2-dichloroethane, and diethylene glycol dimethyl ether; insol diethyl ether and pentane.

Preparative Methods: the complex is prepared in situ by the reaction of bis(cyclohexyl isocyanide)gold(I) tetrafluoroborate[5] with *(R)-N-[2-(N,N-Dimethylamino)ethyl]-N-methyl-1-[(S)-1′,2-bis (diphenylphosphino)ferrocenyl]ethylamine*,[6] typically in dichloromethane.[7]

Handling, Storage, and Precautions: prepare under anhydrous conditions and use under a dry inert atmosphere of nitrogen.

Enantioselective Aldol Reactions. The development of synthetic methodology for the diastereoselective and enantioselective formation of C–C bonds through the use of catalytic quantities of chiral transition-metal catalysts is a topic of fundamental importance. In 1986, Ito and Hayashi reported an elegant asymmetric synthesis of dihydrooxazolines by the gold(I)-catalyzed aldol reaction (more correctly a Knoevenagel reaction) of an aldehyde with an isocyanoacetate ester in the presence of a chiral ferrocenylamine ligand.[7] The chiral catalyst (*R*)-(*S*)-(**1**) is conveniently prepared in situ as described above.[5,6] The dihydrooxazolines obtained provide a convenient precursor to enantiomerically pure β-hydroxy-α-amino acid derivatives. The *trans* (4*S*,5*R*)-oxazoline in high ee is obtained predominantly in the reaction of aldehydes with α-isocyanoacetate esters catalyzed by (*R*)-(*S*)-(**1**) (eqs 1–3).[7–9] High stereoselectivity is retained with alkyl-substituted α-isocyanoacetate esters (eq 4), although reduced diastereoselectivity and enantioselectivity is often obtained with large α-substituents.[10]

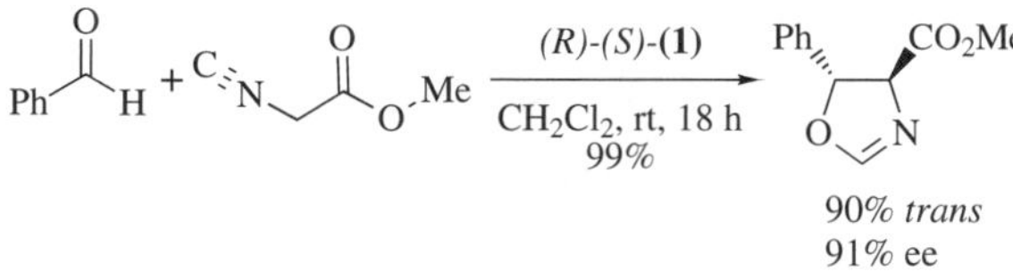

(1)

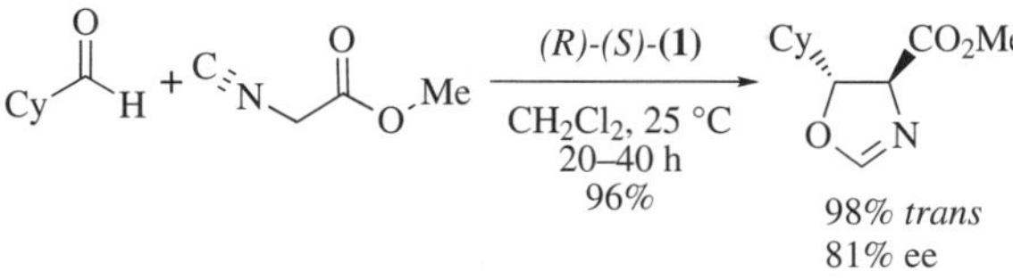

(2)

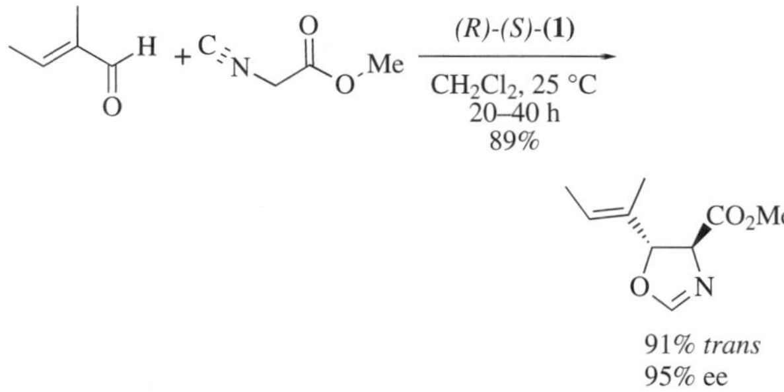

(3)

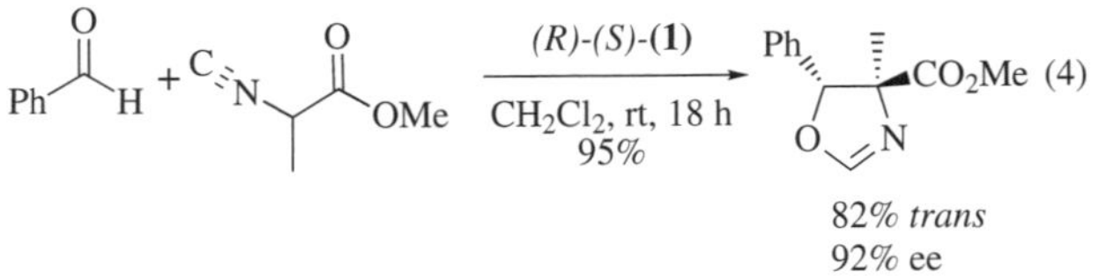

A wide variety of ester functionalities are tolerated (eq 5).[11] A single stereocenter is formed in high ee when formaldeyde is utilized as a reaction component, which leads to an efficient asymmetric synthesis of α-alkylserines (eq 6).[12] The utilization of α-keto esters provides a facile route to β-alkyl-β-hydroxyaspartic acid. Higher diastereo- and enantioselectivity are obtained by the reaction of the α-keto ester with the corresponding *N,N*-dimethyl-α-isocyanoacetamide (eq 7).[13] In certain cases, the use of α-isocyanoacetamides is advantageous for improving stereoselectivity in the corresponding reaction with aldehydes.[14] The presence of α-heteroatoms or certain electronegative groups in the aldehyde component of the reaction may lead to dramatic changes in diastereo- and enantioselectivity.[8,11] Opposite product chirality can be obtained in the gold(I)-catalyzed aldol reaction by using the (*S*)-(*R*) enantiomer of (**1**).

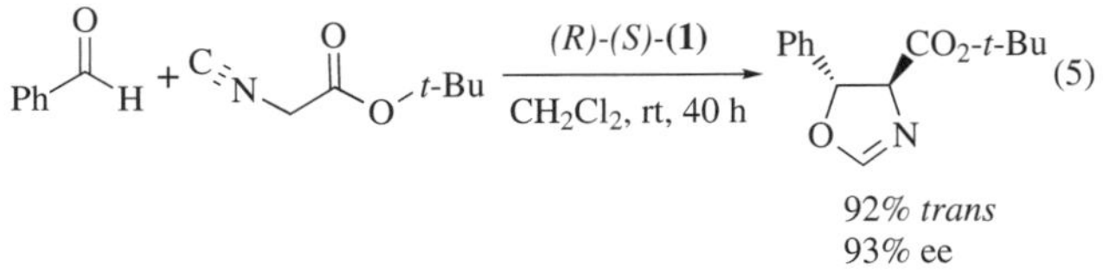

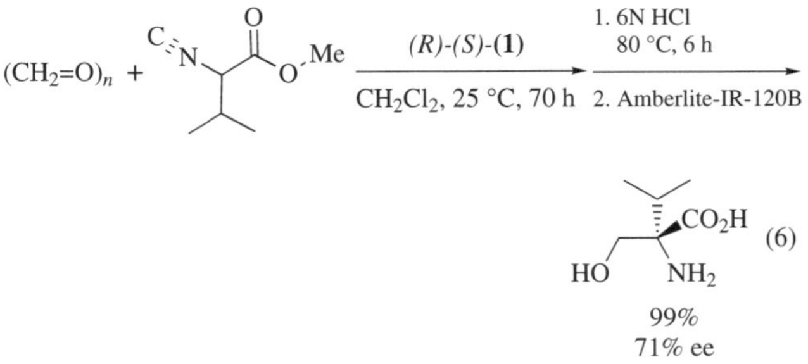

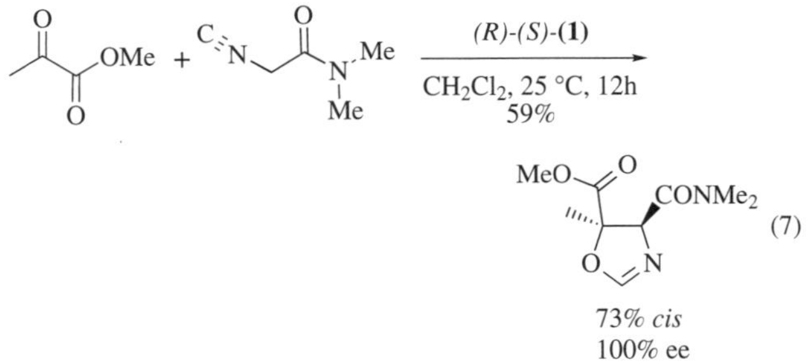

Double stereodifferentiation (double asymmetric induction) between a chiral substrate and the chiral ferrocenylamine ligand has been demonstrated (eq 8).[11] The stereocenter (central chirality) as well as the stereoaxis (axial chirality) affects both product diastereoselectivity and enantioselectivity.[15] Chiral cooperativity (or internal cooperativity) refers to individual chirotopic segments of the ligand molecule acting in a cooperative manner to promote a particular diastereo- and enantioselectivity in the product.[8,15–18] The effects of distant stereocenters in ligands analogous to (*R*)-(*S*)-(**1**) upon the stereoselectivity of the gold(I)-catalyzed aldol reaction has been studied.[19,20] Improvements in stereoselectivity can be obtained in certain cases by modification of the terminal *N,N*-dimethylamino substituent in the side chain of (*R*)-(*S*)-(**1**),[2,9,10,12,21] (compare eq 4 and eq 9) as well as the aryl substituents on phosphorus.[22] Several recent studies have appeared dealing with the elucidation of the stereoselective transition-state geometry.[8,23–25] An elegant study of aldol stereochemistry has important implications on the stereoselective transition-state geometry for the gold(I)-catalyzed aldol reaction.[26] A report has appeared describing the use of a neutral gold(I) ferrocenylamine catalyst for asymmetric aldol reactions, albeit in lower diastereo- and enantioselectivity.[27]

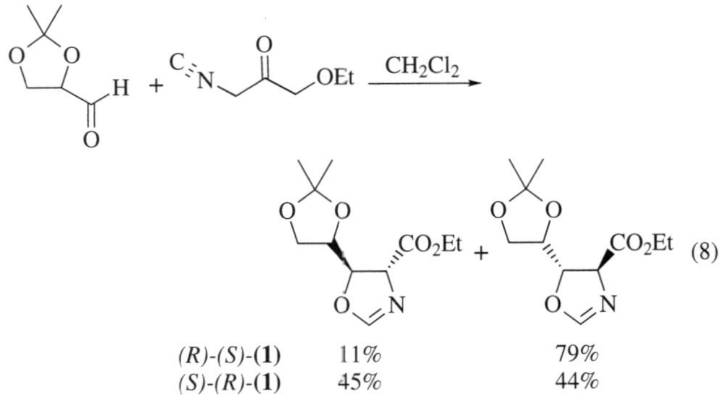

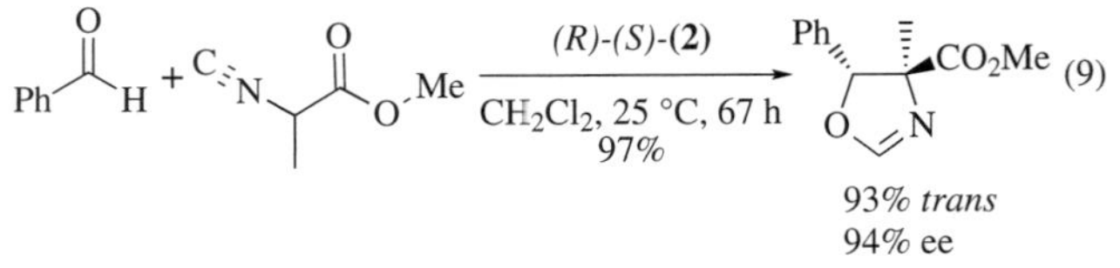

(*R*)-(*S*)-(**2**) =

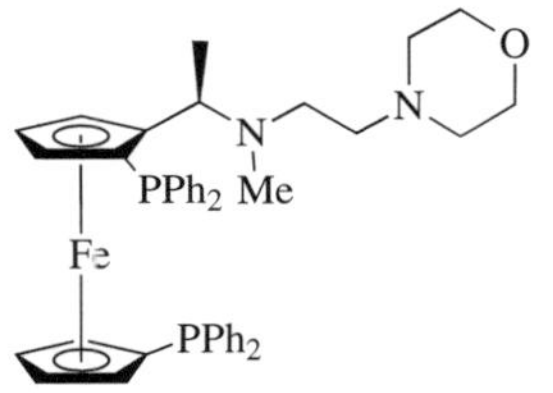

The gold(I)-catalyzed aldol reaction has been applied to the synthesis of several natural products including cyclosporin's unusual amino acid MeBmt, in which two of the three product stereocenters were generated by the reaction of 2-(*R*)-methyl-4-hexenal with ethyl α-isocyanoacetate (eq 10).[28] Although either enantiomer of the *trans*-dihydrooxazole can be obtained by using (*R*)-(*S*)-(**1**) or (*S*)-(*R*)-(**1**), a modest effect due to matching and mismatching of substrate and ligand chirality is apparent (double stereodifferentiation).[29] Methylation of the *trans*-dihydrooxazole with *Trimethyloxonium Tetrafluoroborate* followed by aqueous sodium bicarbonate hydrolysis gives the formamido ester enantiomerically pure after crystallization. The absolute configuration of the two stereocenters formed was confirmed by X-ray crystallography. Careful hydrolysis of the formamido ester to avoid skeletal rearrangement yields MeBmt. In this kilogram scale synthesis, the catalyst can be recovered by precipitation with either diethyl ether or pentane, and can be recycled a number of times without any loss of activity or selectivity.

The gold(I)-catalyzed asymmetric aldol reaction with the chiral ligand (*R*)-(*S*)-(**2**) has been applied in the synthesis of D-*erythro*- and *threo*-sphingosines (eq 11).[30] The D-*erythro*-sphingosine can be prepared from the *threo* isomer by inversion of the C-3 hydroxyl group.

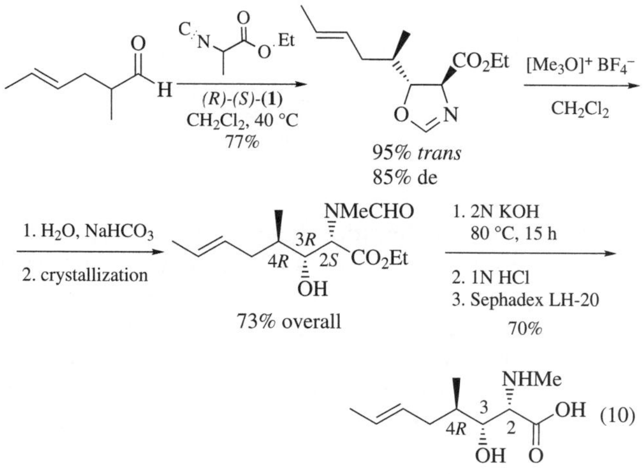

MeBmt

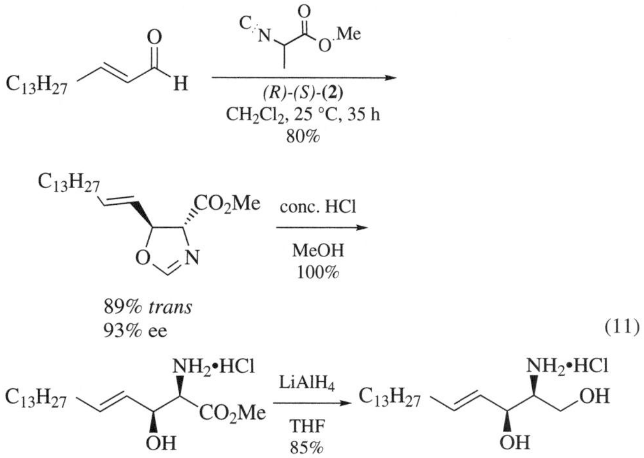

Chiral β-Hydroxy-α-Aminophosphonic Acids. An enantioselective synthesis of substituted dihydrooxazolin-4-yl phosphonates was reported by the reaction of an aldehyde with α-isocyanomethylphosphonate ester catalyzed by (*R*)-(*S*)-(**1**) (eq 12).[31] The enantiomeric purity of the product was determined by $^{31}P\{^1H\}$ NMR spectroscopy using the chiral solvating reagent (*S*)-(+)-2,2,2-trifluoro-1-(9-anthryl)ethanol. Independently, an asymmetric synthesis of α-aminophosphonic acids was reported using the chiral ferrocenylamine catalyst (*R*)-(*S*)-(**3**) (eq 13).[32]

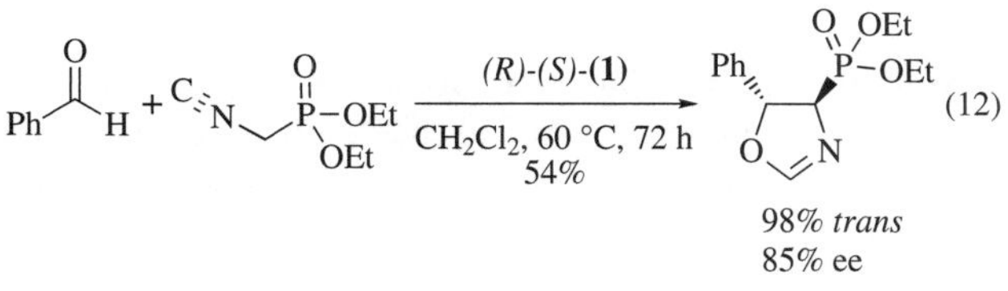

Asymmetric Allylation. Asymmetric allylation of β-diketones using the palladium analog of (**1**) has been described.[4] Higher enantioselectivity can be achieved in this case using ferrocenylamines with a modified alkyl side chain.[4] For synthetically useful ferrocenylamine complexes of other metals, see *(R)-N-[2-(N,N-Dimethylamino)ethyl]-N-methyl-1-[(S)-1′,2-bis-(diphenylphosphino)ferrocenyl]ethylamine*.

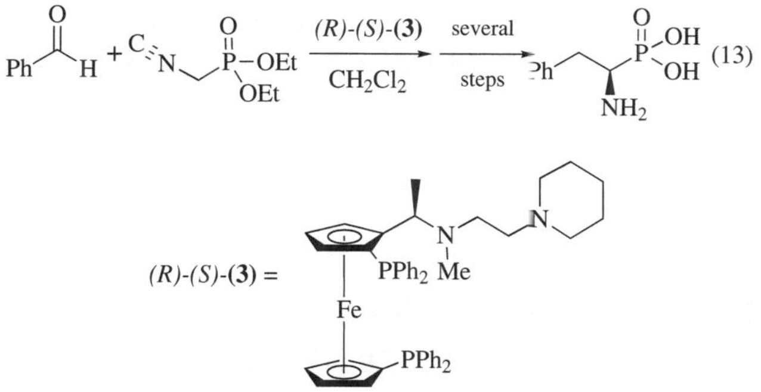

1. Sawamura, M.; Ito, Y. *Chem. Rev.* **1992**, *92*, 857.
2. Hayashi, T.; Sawamura, M.; Ito, Y. *Tetrahedron* **1992**, *48*, 1999.
3. Mastalerz, P. In *Handbook of Organophosphorus Chemistry*; Engel, R., Ed.; Dekker: New York, 1992; pp 336–339.
4. Sawamura, M.; Nagata, H.; Sakamoto, H.; Ito, Y. *J. Am. Chem. Soc.* **1992**, *114*, 2586.
5. Bonati, F.; Minghetti, G. *Angew. Chem.* **1973**, *103*, 373.
6. Hayashi, T.; Mise, T.; Fukushima, M.; Kagotani, M.; Nagashima, N.; Hamada, Y.; Matsumoto, A.; Kawakami, S.; Konishi, M.; Yamamoto, K.; Kumada, M. *Bull. Chem. Soc. Jpn.* **1980**, *53*, 1138.
7. Ito, Y.; Sawamura, M.; Hayashi, T. *J. Am. Chem. Soc.* **1986**, *108*, 6405.
8. Togni, A.; Pastor, S. D. *J. Org. Chem.* **1990**, *55*, 1649.
9. Ito, Y.; Sawamura, M.; Hayashi, T. *Tetrahedron Lett.* **1987**, *28*, 6215.
10. Ito, Y.; Sawamura, M.; Shirakawa, E.; Hayashizaki, K.; Hayashi, T. *Tetrahedron* **1988**, *44*, 5253.
11. Togni, A.; Pastor, S. D. *Helv. Chim. Acta* **1989**, *72*, 1038.
12. Ito, Y.; Sawamura, M.; Shirakawa, E.; Hayashizaki, K.; Hayashi, T. *Tetrahedron Lett.* **1988**, *29*, 235.
13. Ito, Y.; Sawamura, M.; Hitoshi, H.; Emura, T.; Hayashi, T. *Tetrahedron Lett.* **1989**, *30*, 4681.
14. Ito, Y.; Sawamura, M.; Kobayashi, M.; Hayashi, T. *Tetrahedron Lett.* **1988**, *29*, 6321.
15. Pastor, S. D.; Togni, A. *J. Am. Chem. Soc.* **1989**, *111*, 2333.
16. Togni, A.; Pastor, S. D. *Chirality* **1991**, *3*, 331.
17. (a) Togni, A.; Häusel, R. *Synlett* **1990**, 633. (b) Togni, A.; Rihs, G.; Blumer, R. E. *J. Organomet. Chem.* **1992**, *11*, 613.
18. (a) Nagel, U.; Rieger, B. *J. Organomet. Chem.* **1989**, *8*, 1534. (b) Burgess, K.; Ohlmeyer, M. J.; Whitmire, K. H. *J. Organomet. Chem.* **1992**, *11*, 3588 and references therein.
19. Pastor, S. D.; Togni, A. *Tetrahedron Lett.* **1990**, *31*, 839.
20. Pastor, S. D.; Togni, A. *Helv. Chim. Acta* **1991**, *74*, 905.
21. Hayashi, T. *Pure Appl. Chem.* **1988**, *60*, 7.
22. Hayashi, T.; Yamazaki, A. *J. Organomet. Chem.* **1991**, *413*, 295.
23. Sawamura, M.; Ito, Y.; Hayashi, T. *Tetrahedron Lett.* **1990**, *31*, 2723.
24. Togni, A.; Blumer, R. E.; Pregosin, P. S. *Helv. Chim. Acta* **1991**, *74*, 1533.
25. Pastor, S. D.; Kesselring, R.; Togni, A. *J. Organomet. Chem.* **1992**, *429*, 415.
26. Denmark, S. E.; Henke, B. R. *J. Am. Chem. Soc.* **1991**, *113*, 2177.
27. Togni, A.; Pastor, S. D.; Rihs, G. *J. Organomet. Chem.* **1990**, *381*, C21.
28. Togni, A.; Pastor, S. D.; Rihs, G. *Helv. Chim. Acta* **1989**, *72*, 1471.
29. (a) Masamune, S.; Choy, W.; Petersen, J. S.; Sita, L. R. *Angew. Chem., Int. Ed. Engl.* **1985**, *24*, 1. For earlier reports, see (b) Heathcock, C. H.; White, C. T. *J. Am. Chem. Soc.* **1979**, *101*, 7076. (c) Horeau, A.; Kagan, H.-B.; Vigneron, J.-P. *Bull. Soc. Chem. Fr.* **1968**, 3795.
30. Ito, Y.; Sawamura, M.; Hayashi, T. *Tetrahedron Lett.* **1988**, *29*, 239.

31. Togni, A.; Pastor, S. D. *Tetrahedron Lett.* **1989**, *30*, 1071.
32. Sawamura, M.; Ito, Y.; Hayashi, T. *Tetrahedron Lett.* **1989**, *30*, 2247.

Stephen D. Pastor
Ciba-Geigy Corporation, Ardsley, NY, USA

Bis(1,5-cyclooctadiene)rhodium Tetrafluoroborate–(*R*)-2,2′Bis(diphenylphosphino)-1,1′-binaphthyl[1]

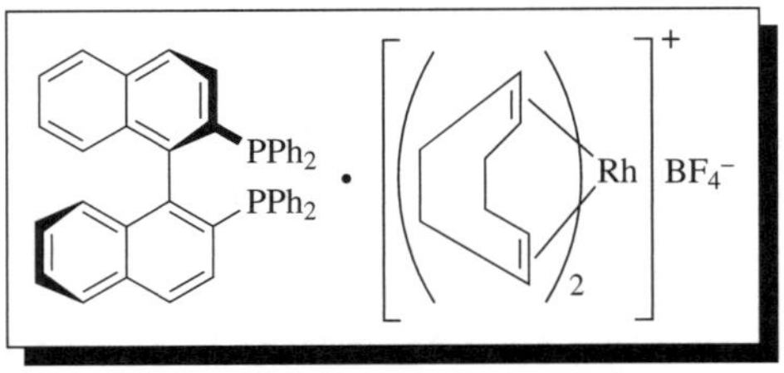

($[Rh(cod)_2]BF_4$)		
[35138-22-8]	$C_{16}H_{24}BF_4Rh$	(MW 406.07)
((*R*)-BINAP)		
[76189-55-4]	$C_{44}H_{32}P_2$	(MW 622.68)

(catalyst for asymmetric hydrogenation,[2] isomerization,[3] hydroboration,[4] and intramolecular hydrosilation[5] of alkenes)

Physical Data: $[Rh(cod)_2]BF_4$: mp 206–8 °C; (*R*)-BINAP: mp 240–241 °C; $[\alpha]_D^{25}$ +229° (c = 0.32, benzene).
Form Supplied in: $[Rh(cod)_2]BF_4$: orange-red crystals; (*R*)-BINAP: colorless crystals.
Analysis of Reagent Purity: (*R*)-BINAP: ^{31}P NMR (4:1 C_6D_6–CD_3OD): δ −12.8 (s); mp and optical rotation shown above are also useful for analysis of the purity.
Purification: $[Rh(cod)_2]BF_4$: recrystallization from CH_2Cl_2 and ether; (*R*)-BINAP: recrystallization from a mixture of toluene and EtOH.
Handling, Storage, and Precautions: $[Rh(cod)_2]BF_4$: hygroscopic; corrosive.

Asymmetric Hydrogenation. The diene-free cationic rhodium complex of (*R*)-BINAP catalyzes the enantioselective hydrogenation of dehydroamino acids. α-(Benzoylamino)acrylic acid is hydrogenated at rt to afford (*S*)-*N*-benzoylphenylalanine in 100% ee (eq 1).[2] To obtain maximal stereoselectivity the reaction should be carried out under a low concentration of substrate (100% in 0.013 M vs. 62% in 0.15 M) and low initial hydrogen pressure (100% at 1 atm, but 71% at 50 atm).

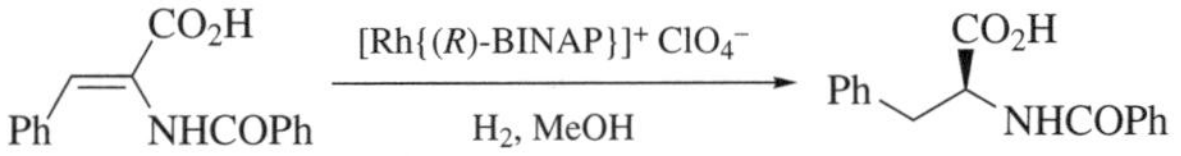

(1)

100% ee (*S*)

Optically active homoallylic alcohols are hydrogenated with differentiation of the diastereofaces (eq 2).[6] Use of the matched ligand, i.e. (*R*)-BINAP, gives a product of 96% de, while the mismatched (*S*)-ligand affords low selectivity.

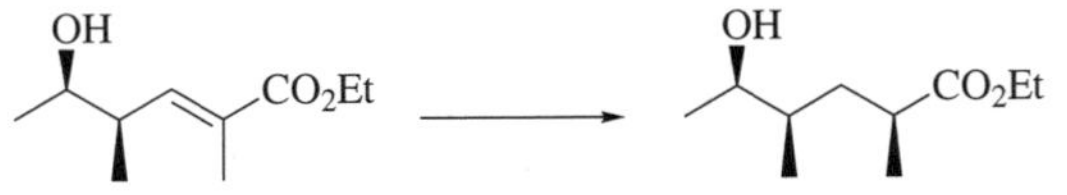

(2)

$[Rh\{(R)\text{-BINAP}\}]^+ ClO_4^-$	98:2
$[Rh\{(S)\text{-BINAP}\}]^+ ClO_4^-$	67:33

Allylic Hydrogen Migration. Cationic Rh^I diphosphine complexes are very active catalysts for allylic hydrogen migration of tertiary and secondary allylamines to give the corresponding (*E*)-enamines and imines, respectively. Highly enantioselective isomerization is accomplished by use of (*R*)-BINAP as a diphosphine ligand.[3] Diethylnerylamine, which has (*Z*) geometry, gives (*R*)-(*E*)-diethylcitronellenamine in 95% ee in the presence of 1 mol % of $[Rh(R)\text{-BINAP}(cod)]ClO_4$, while the isomeric diethylgeranylamine gives (*S*)-(*E*)-diethylcitronellenamine in 96% ee (eq 3). Thus the method presented above offers a desired enantiomer by proper choice of alkene geometry and chirality of BINAP.

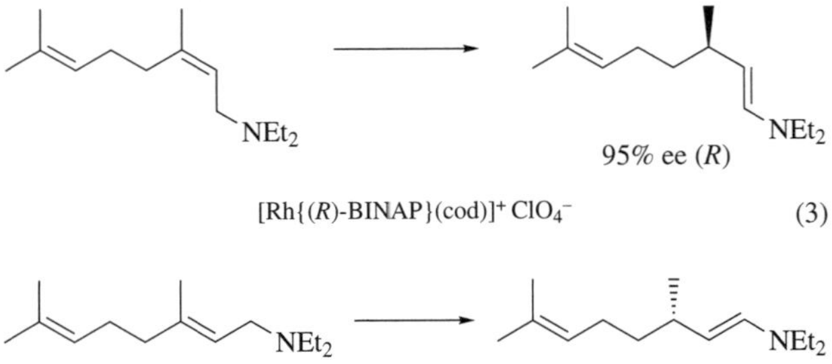

(3)

Cationic Rh–(*R*)-BINAP complexes also catalyze the allylic hydrogen migration of racemic 4-hydroxy-2-cyclopentenone to 1,3-cyclopentanedione with 5:1 enantiomeric discrimination. The racemate is kinetically resolved to (*R*)-4-hydroxy-2-cyclopentenone of 91% ee at 72% conversion at 0 °C (eq 4).[7]

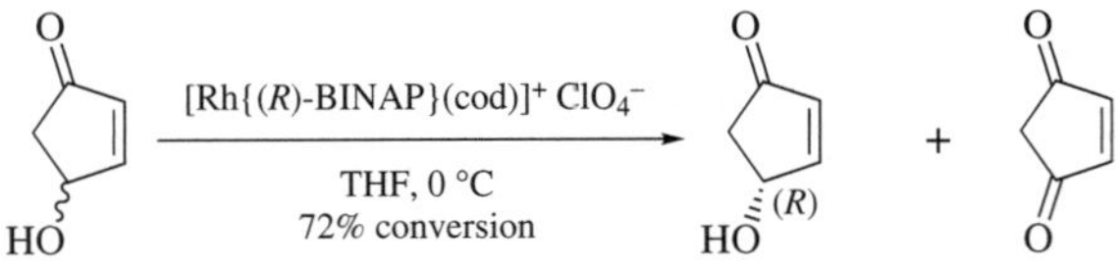

(4)

91% ee (27% yield)

Asymmetric Hydroboration. Rhodium complexes are known to catalyze hydroboration of alkenes with unreactive borane derivatives, e.g. catecholborane and oxaborolidine.[8] This reaction proceeds enantioselectively by use of BINAP as a ligand for neutral[9–11] or cationic[4,12] rhodium complexes. Reaction of styrene with catecholborane followed by oxidation affords (*R*)-1-phenylethanol in 96% ee in the presence of (*R*)-BINAP and $[Rh(cod)_2]BF_4$ (eq 5).[4]

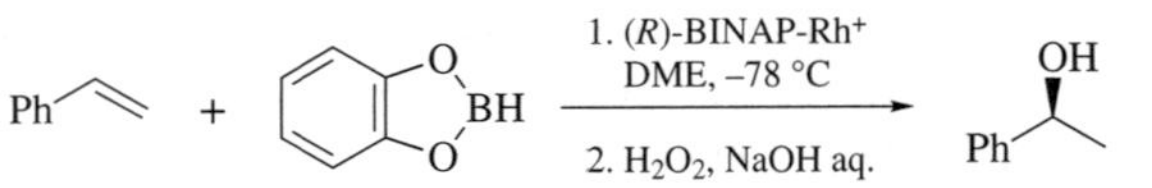

(5)

96% ee

Asymmetric Intramolecular Hydrosilation. Intramolecular hydrosilation of allylic alcohols followed by oxidation is a convenient method for the stereoselective preparation of 1,3-diols.[13] An enantioselective version is achieved by use of diene-free BINAP–Rh$^+$ (eq 6).[5] Both silyl ethers derived from cinnamyl alcohol and its *cis* isomer give (*R*)-1-phenylpropane-1,3-diol in high ee regardless of alkene geometry.

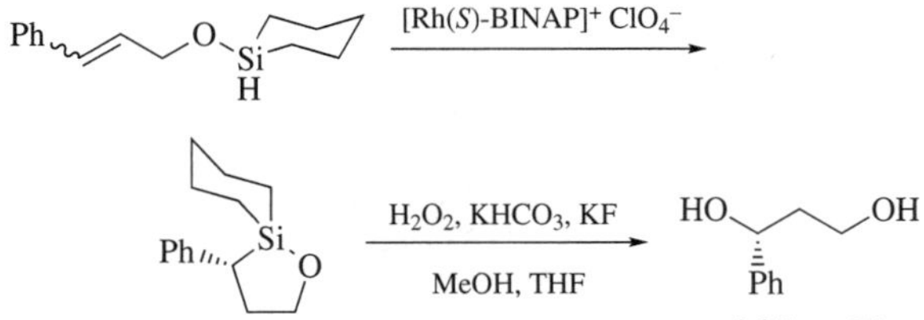

(6)

96% ee (*R*)

Related Reagents. Bis(bicyclo[2.2.1]hepta-2,5-diene)rhodium Perchlorate; Bis(bicyclo[2.2.1]hepta-2,5-diene)rhodium Perchlorate–(*R*)-1-(*S*)-1′, 2-Bis(diphenylphosphino)ferrocenylethanol; 2,2′-Bis(diphenylphosphino)-1,1′-binaphthyl; Chlorotris(triphenylphosphine)rhodium(I).

1. (a) Takaya, H.; Noyori, R. *Comprehensive Organic Synthesis* **1991**, *8*, Chapter 3.2. (b) Smith, K.; Pelter, A. *Comprehensive Organic Synthesis* **1991**, *8*, Chapter 3.10. (c) Hiyama, T.; Kusumoto, T. *Comprehensive Organic Synthesis* **1991**, *8*, Chapter 3.12. (d) Noyori, R.; Kitamura, M. In *Modern Synthetic Methods*, Sheffold, R., Ed.; Springer: Berlin, 1989; Vol. 5, p. 115. (e) *Comprehensive Organometallic Chemistry*; Wilkinson, G., Ed.; Pergamon: Oxford, 1982; Vol. 8.
2. (a) Miyashita, A.; Yasuda, A.; Takaya, H.; Toriumi, K.; Ito, T.; Souchi, T.; Noyori, R. *J. Am. Chem. Soc.* **1980**, *102*, 7932. (b) Miyashita, A.; Takaya, H.; Souchi, T.; Noyori, R. *Tetrahedron* **1984**, *40*, 1245.
3. Tani, K.; Yamagata, T.; Akutagawa., S.; Kumobayashi, H.; Taketomi, T.; Takaya, H.; Miyashita, A.; Noyori, R.; Otsuka, S. *J. Am. Chem. Soc.* **1984**, *106*, 5208.
4. (a) Hayashi, T.; Matsumoto, Y.; Ito, Y. *J. Am. Chem. Soc.* **1989**, *111*, 3426. (b) Hayashi, T.; Matsumoto, Y.; Ito, Y. *Tetrahedron: Asymmetry* **1991**, *2*, 601.
5. Bergens, S. H.; Noheda, P.; Whelan, J.; Bosnich, B. *J. Am. Chem. Soc.* **1992**, *114*, 2121.
6. Evans, D. A.; Morrissey, M. M.; Dow, R. L. *Tetrahedron Lett.* **1985**, *26*, 6005.
7. Kitamura, M.; Manabe, K.; Noyori, R.; Takaya, H. *Tetrahedron Lett.* **1987**, *28*, 4719.
8. Männig, D.; Nöth, H. *Angew. Chem., Int. Ed. Engl.* **1985**, *24*, 878.
9. Burgess, K.; Ohlmeyer, M. J. *J. Org. Chem.* **1988**, *53*, 5178.
10. Sato, M.; Miyaura, N.; Suzuki, A. *Tetrahedron Lett.* **1990**, *31*, 231.
11. Zhang, J.; Lou, B.; Guo, G.; Dai, L. *J. Org. Chem.* **1991**, *56*, 1670.
12. Brown, J. M.; Lloyd-Jones, G. C. *Tetrahedron: Asymmetry* **1990**, *1*, 869.
13. (a) Tamao, K.; Nakajima, T.; Sumiya, R.; Arai, H.; Higuchi, N.; Ito, Y. *J. Am. Chem. Soc.* **1986**, *108*, 6090. (b) Tamao, K.; Tohma, T.; Inui, N.; Nakayama, O.; Ito, Y. *Tetrahedron Lett.* **1990**, *31*, 7333.

Yoshihiko Ito & Michinori Suginome
Kyoto University, Japan

(*S*,*S*)-1,2-Bis(2,5-diethylphospholano)-benzene

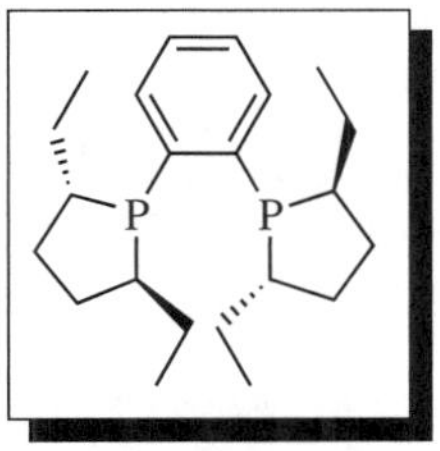

[136779-28-7] $C_{22}H_{36}P_2$ (MW 362.47)

(ligand for asymmetric catalysis; rhodium(I) complexes are efficient catalysts in highly enantioselective hydrogenation of various unsaturated substrates [enol acylates,[1,2] (*N*-acylamino)acrylates,[1,3,4] and *N*-acylhydrazones[5]])

Alternate Name: (*S*,*S*)-ethyl-DuPHOS.
Physical Data: bp 138–145 °C/0.04 mmHg; $[\alpha]_D$ +265 (*c* 1, hexane).
Solubility: soluble in most organic solvents.
Form Supplied in: colorless oil; commercially available.
Analysis of Reagent Purity: optical rotation; NMR spectroscopy.
Preparative Methods: the preparation of (*S*,*S*)-ethyl-DuPHOS is based on (3*R*,6*R*)-octane-3,6-diol as an enantiomerically pure starting compound.[1,3] The latter is synthesized by a three-step procedure[6,7] starting from methyl 3-oxopentanoate, which is transformed to methyl (*R*)-3-hydroxypentanoate (99% ee) by enantioselective hydrogenation with a Ru-(*R*)-BINAP catalyst,[8] followed by hydrolysis to the hydroxy acid. The subsequent electrochemical Kolbe coupling reaction leads to (3*R*,6*R*)-octane-3,6-diol in a protocol that can be scaled up to multigram quantities (eq 1).[3,6]

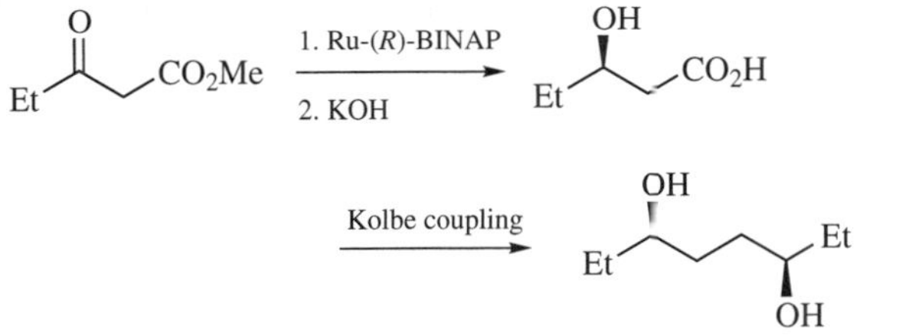

(1)

(3*R*,6*R*)-octane-3,6-diol
(36% overall yield)

The chiral octanediol in turn is converted into the corresponding cyclic sulfate by reaction with **thionyl chloride** and subsequent oxidation with **sodium periodate** and a catalytic amount of **ruthenium(III) chloride** (0.1 mol%) (eq 2).[3] In the final step, 1,2-diphosphinobenzene[9] is lithiated by treatment with ***n*-butyllithium** (*n*-BuLi; 2 equiv, 1.6 mol% in hexane) followed by the addition of the (3*R*,6*R*)-octane-3,6-diol cyclic sulfate (2 equiv) and a further addition of 2.2 equiv of *n*-BuLi. (*S*,*S*)-Ethyl-DuPHOS is obtained in a yield of over 70% [78% yield was described for the (*R*,*R*)-enantiomer by an analogous method[3]].[1] In addition to (*S*,*S*)-ethyl-DuPHOS, a variety of related bisphospholanes either linked by an ethylene bridge, or bearing other 2,5-alkyl substituents, or with opposite configuration have been prepared by this methodology.[1,3]

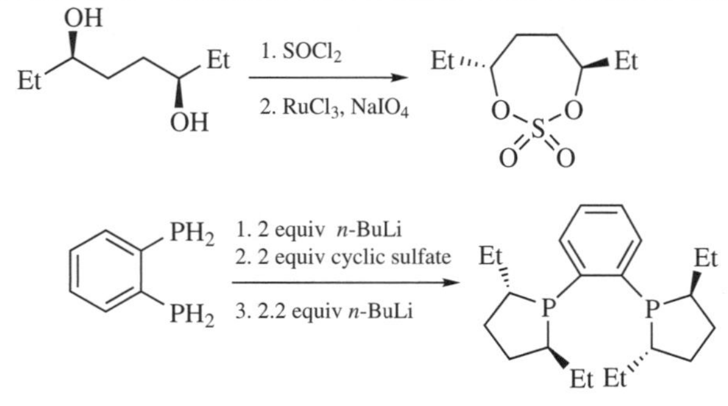

(2)

(*S,S*)-Ethyl-DuPHOS

Handling, Storage, and Precautions: the reagent is sensitive to air and should be handled and stored under argon or nitrogen.

Catalyst Precursors

Rhodium Complexes

Cationic rhodium(I) complexes such as {Rh[(*S,S*)-ethyl-DuPHOS](cod)}$^+$X$^-$ (X=OTf, PF_6, BF_4, SbF_6) are usually employed as precatalysts for enantioselective hydrogenation[1–4] or hydrosilylation[10] reactions. The precatalysts can be prepared from the chiral ligand and $[Rh(cod)_2]^+X^-$-complexes by a standard method.[3,11] The corresponding {Rh[(*S,S*)-ethyl-DuPHOS](nbd)} complex can be accessed equally well by the method of Schrock and Osborne[11] or by exchange of cod in the relevant rhodium complex by norbornadiene (nbd).[12]

Enantioselective Hydrogenations

C=C Double Bond

The most commonly used reactions that employ {Rh[(*S,S*)-Et-DuPHOS](cod)}$^+$X$^-$ complexes involve the enantioselective hydrogenation of α-(*N*-acyl)enamide carboxylates. α-Amino acids are obtained in quantitative yield with high optical purity (95–99% ee) (eq 3).[1,3,4]

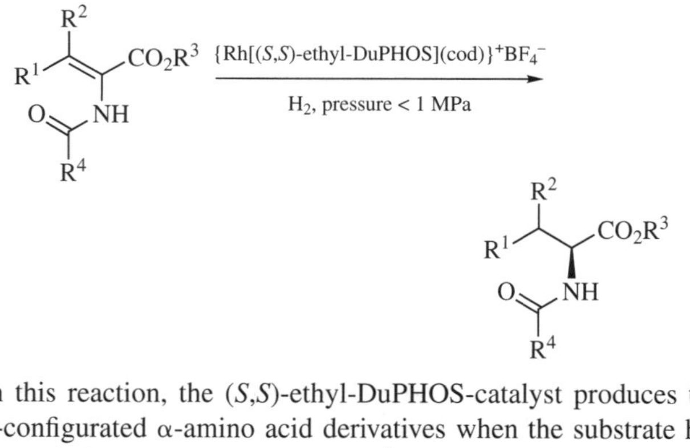

(3)

In this reaction, the (*S,S*)-ethyl-DuPHOS-catalyst produces the *S*-configurated α-amino acid derivatives when the substrate has one substituent in the β-position (R^2=H). The catalyst tolerates a range of substituents R^1 in the unsaturated substrate. The reaction conditions are mild (25 °C, MeOH, low hydrogen pressure) and the reaction proceeds rapidly (turnover frequencies >5000 h^{-1}). High substrate-to-catalyst ratios can be employed (up to 50 000). In general, high enatiomeric excesses are observed, independent of the geometry of the enamide (*E*- or *Z*-isomer) used. This feature is advantageous because frequently the unsaturated α-enamide carboxylic acids derivatives are synthesized as a mixture of both stereoisomers. Due to the high stereodiscriminating ability of the catalyst, separation prior to hydrogenation is not required. Access to a great number of natural, unnatural and nonproteogenic amino acids is possible in this manner (Table 1).

Table 1 Enantioselective hydrogenation of substituted α-(*N*-acetylamido)acrylates (with R^2=H, R^3=R^4=Me) with Rh-ethyl-DuPHOS catalyst (MeOH, 25 °C, 0.2 MPa H_2)[3]

R^1	% ee	R^1	% ee
H	99.4	2-Naphthyl	>99.0
Me	>99.0	2-Thienyl	>99.0
Et	>99.0	Ferrocenyl	>99.0
n-Pr	> 99.0	2-Quinolinyl	94.0[13]
i-Pr	99.0	2-(6-Methyl)pyridyl	97.0[13]
Ph	99.0	2-Bromophenyl	99.0[14]
1-Naphthyl	>99.0	3-Ac-2,2-Me-cyclobutylmethyl	96.0[a, 15]

[a] % de.

The highly enantioselective hydrogenation of the corresponding dehydroamino acids (R^3=H) and the synthesis of *N*-Cbz-protected α-amino acids (R^4=OBn) are likewise possible.[3,16] Enantioselectivities of >99% can be achieved after 20–40 hours. Amino acid esters can be used directly for the synthesis of peptides. Deprotection of the amino groups can be carried out under mild conditions, thus avoiding racemization reactions.

Evidence has been given that the use of the Rh(P_2)(nbd) precatalyst is favored over the application of the corresponding cod-precatalyst. Börner and Heller[17] found that hydrogenation of the cod of the precatalyst takes place in parallel to the enantioselective hydrogenation of methyl (*Z*)-*N*-acetylaminocinnamate.[17,18] About 50% of the Rh-precatalyst remained unchanged after complete hydrogenation of the prochiral substrate. Therefore, precious ligand and Rh complex are wasted. This can be avoided by the application of {Rh[(*S,S*)-ethyl-DuPHOS](nbd)}$^+$X$^-$ as precatalyst. The hydrogenation of nbd proceeds much faster than that of cyclooctadiene. As an alternative, prehydrogenation of the cod-precatalyst in MeOH is possible for generating the catalytically active species.

Excellent enantio- and regioselectivities were also observed when *N*-acetylamino acrylates bearing additional functional groups, e.g. alkenes, were applied as substrates (eq 4).[3,4]

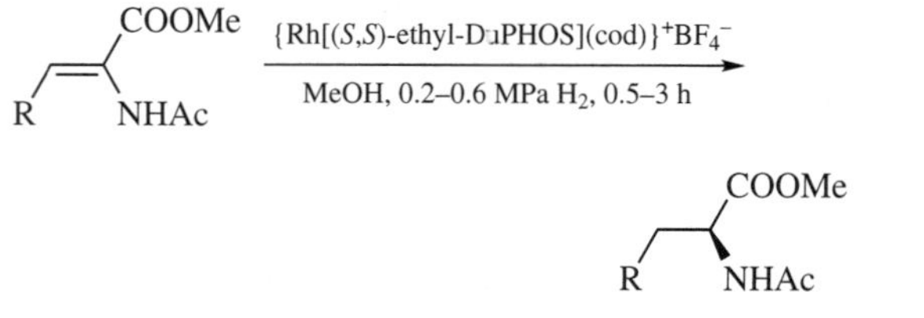

(4)

regioselectivity: > 98 %
enantioselectivity: > 99 %

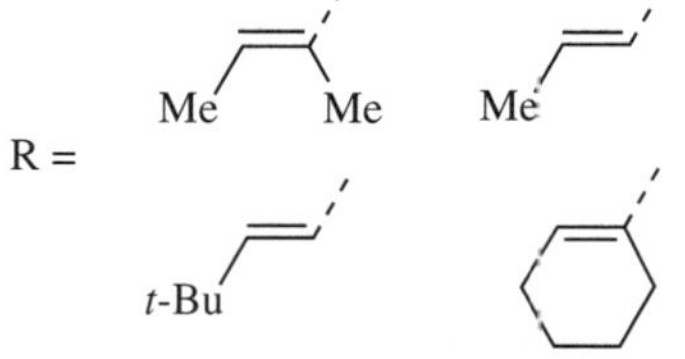

In these hydrogenations, less than 2% of the γ,δ-double bond was reduced. This feature indicates the ethyl-DuPHOS ligand to be superior in comparison to related DuPHOS/BPE-ligands or other chelating diphosphines. Other functional groups that are generally sensitive to reduction such as carbonyl groups, nitro groups, and halides also survive under the mild conditions applied to hydrogenation of the double bond adjacent to the acylamino group.

Diastereoselective hydrogenation of a bis(dehydroamino acid) derivative, recognized to be important for the syntheses of isotyrosine, in the presence of $\{Rh[(S,S)\text{-ethyl-DuPHOS}](cod)\}^+OTf^-$ as catalyst yielded excellent results (eq 5).[19]

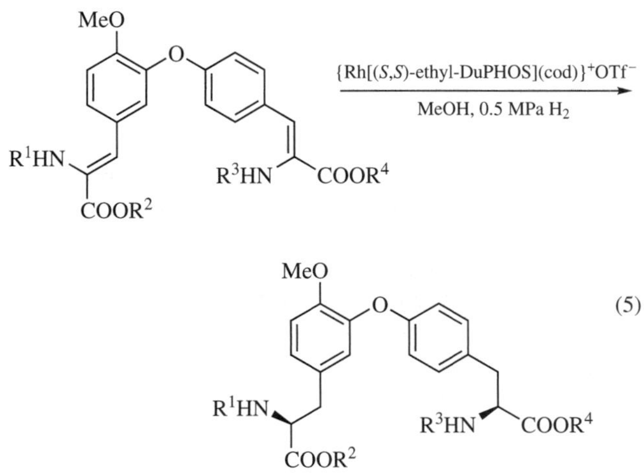

(5)

a–c: $R^1 = R^3$ = Cbz, Boc, Ac, $R^2 = R^4$ = Me

d: R^1 = Cbz, R^2 = TMSE, R^3 = Boc, R^4 = Me

The possibility of preparing an isotyrosine derivative with four orthogonal protecting groups gives access to a highly versatile building block for several biologically active natural compounds. Enantioselectivities in excess of 98% ee for the *S,S*-enantiomer and diastereoselectivities above 84% have been observed. In general, the yields exceeded 90%.

The formation of stereogenic C–N bonds by hydrogenation of the enamine structure is not only limited to amino acids. Likewise, chiral 1,2-aminoalcohols or 1,2-diamines can be produced by the enantioselective hydrogenation of dehydro-β-amino alcohols (or their esters) and of dehydro-α-amino aldoximes, respectively (eq 6 and eq 7, Table 2).[20] Esters and aldoximes thus obtained can be converted into the corresponding alcohols or diamines by standard methods. By this means, simple amines with one aryl group attached to the double bond can also be hydrogenated with high enantioselectivity.[21]

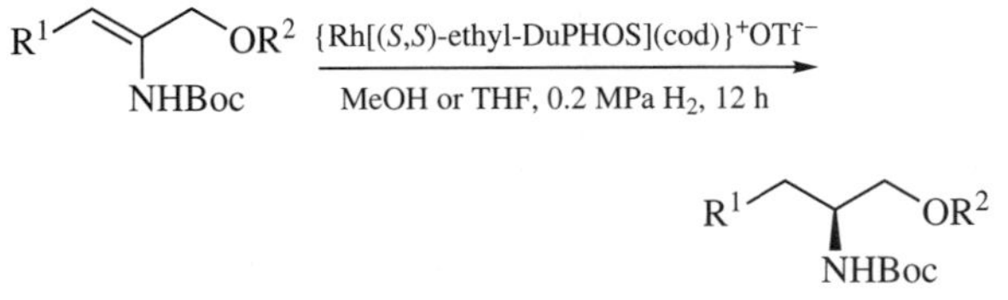

(6)

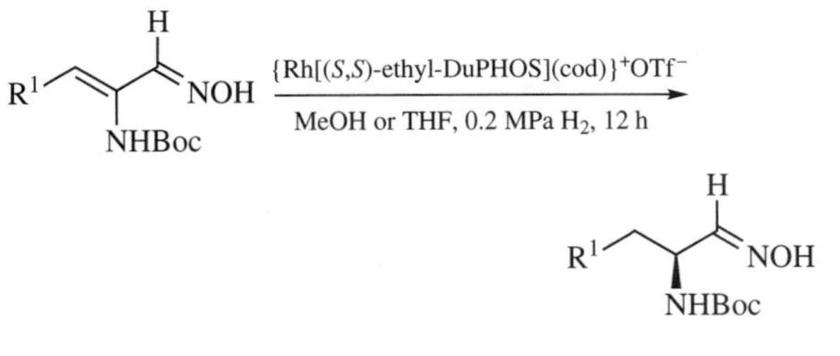

(7)

Table 2 Enantioselective hydrogenation of dehydroamino alcohols and dehydro-α-amino aldoximes with $\{Rh[(S,S)\text{-ethyl-DuPHOS}](cod)\}^+OTf^-$ (MeOH or THF, 25 °C, 0.2 MPa H_2, 12 h, S/C = 1000:1)

R^1	R^2	% ee amino alcohol	% ee amino aldoxime
Me	H	95	94
Ph	H	93	90
Cy	H	96	99
2-MeOPh	H	93	88
Bn	H	>99	98
2-Naphthyl	Ac	98	81
2-Furanyl	Ac	98	–
2-Thienyl	Ac	96	–

Enol acetates and corresponding derivatives constitute another class of unsaturated compounds that can advantageously be hydrogenated with high enantiomeric excess. This reaction is related to the enantioselective reduction of ketones. Acylated enol carboxylates (as an equivalent of α-keto carboxylic acid) can likewise be successfully reduced with rhodium(I) catalysts based on (*S,S*)-ethyl-DuPHOS (eq 8).[2] Subsequent deprotection of the hydroxyl group or reduction of the carboxylic acid derivatives so obtained deliver chiral α-hydroxy carboxylates and 1,2-diols, respectively.

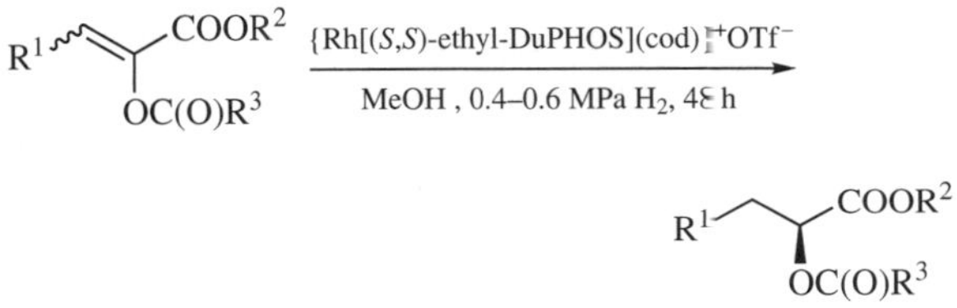

(8)

Burk et al.[2] showed that the Rh-(*S,S*)-ethyl-DuPHOS complex is able to reduce acylated α-hydroxy carboxylates with high enantiomeric excess independently of the *E*:*Z* ratio of the alkene substrate (Table 3). However, the reaction failed when substrates branched in β-position were tested.

Table 3 Enantioselective hydrogenation of various enol esters with $\{Rh[(S,S)\text{-ethyl-DuPHOS}](cod)\}^+OTf^-$ complex (MeOH, 25 °C, 0.4–0.6 MPa H_2, 48 h, S/C = 500:1)

R^1	R^2	R^3	E/Z ratio	% ee[a]
H	Et	Ac	–	>99
Me	Et	Bz	3	96.0
n-Pr	Me	Bz	3	98.0
i-Pr	Et	Ac	6	96.1
i-Pr	Et	Bz	6	96.9
Ph	Et	Ac	9	95.6
Ph	Et	Bz	10	98.0
α-Naphthyl	Et	Bz	3	93.2

[a] *S*-configurated products were obtained.

Finally, several examples of the enantioselective hydrogenation of unsaturated substrates without any heteroatom attached to the olefinic double bond are noteworthy. Of particular relevance to the production of pharmaceutics, agrochemicals, flavors and aroma stuffs is the formation of the chiral 2-substituted succinates based on relevant itaconic acid derivatives. Burk et al.[22] demonstrated that a rhodium(I) catalyst derived from (*S,S*)-ethyl-DuPHOS is able to hydrogenate aryl- or alkyl-substituted itaconic acid derivatives with excellent enantioselectivity (eq 9, Table 4).

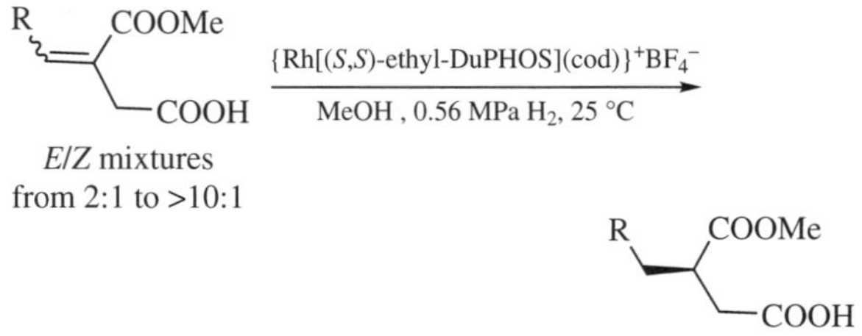

(9)

Table 4 Enantioselective hydrogenation of itaconic acid derivatives with $\{Rh[(S,S)\text{-ethyl-DuPHOS}](cod)\}^+BF_4^-$ (MeOH, 25 °C, 0.56 MPa H_2)

R	S/C	Time (h)	% ee[a]
Et	1000	1	99
i-Pr	3000	2	99
n-Bu	1500	2	97
CH_2CH_2Ph	2000	2	99
Cyclohexyl	1500	3	98
Ph	3000	12	97
1-Naphthyl	3000	12	98

[a] *R*-configuration was obtained.

Prior separation of the *E*:*Z* mixtures of the itaconic acid substrates (usually representing mixtures of 2:1 to more than 10:1) and also by-products of the synthesis do not affect the results.

A catalyst based on ethyl-DuPHOS also showed its high potential in the enantioselective hydrogenation of a sodium glutarate as the pivotal intermediate in the multi-step synthesis of candoxatril (eq 10).[23]

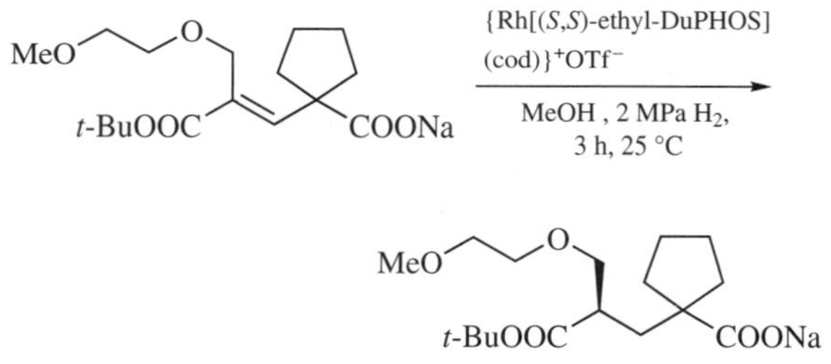

(10)

98% ee (*R*)

By application of the relevant Rh-(*S*,*S*)-ethyl-DuPHOS catalyst, no isomerization of the starting material to the enol ether occurred by migration of the double bond. This side reaction operates in the presence of the corresponding ruthenium catalysts. When (*S*,*S*)-ethyl-DuPHOS was applied as ligand the *R*-enantiomer was formed instead of the desired *S*-enantiomer, necessary for the synthesis of candoxatril.

Another example of the synthesis of a compound with pharmaceutical relevance is the chemical transformation of *rac*-warfarin into enantiomerically pure (*R*)- or (*S*)-warfarin.[24] In the first step, *rac*-warfarin is oxidized to the corresponding α,β-unsaturated ketone. The latter can be easily hydrogenated to the desired enantiomer by application of the appropriate DuPHOS-catalyst (eq 11). Prior transformation of dehydrowarfarin into the sodium salt or its methyl ether improved the yield and suppressed side reactions. Simultaneously, the enantioselectivity of the hydrogenation product was enhanced. The (*S*,*S*)-ethyl-DuPHOS-complex leads to *R*-configured warfarin.

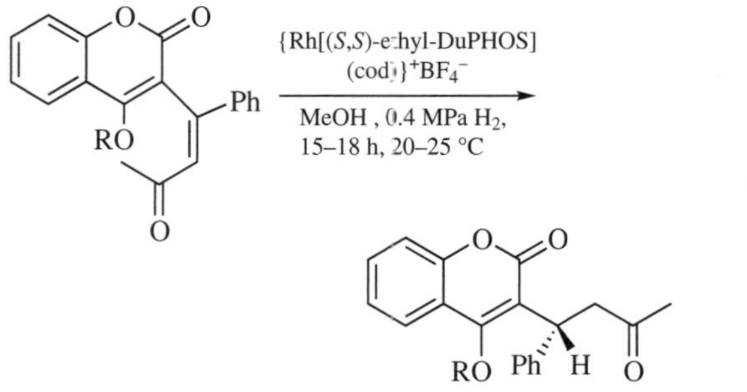

(11)

R = Me: 95% yield, 89% ee (*R*)
R = Na: >98% yield, 86% ee (*R*)

C=N Double Bond

The enantioselective reduction of a C=N double bond is an interesting alternative for the production of chiral amines by hydrogenation of enamides. Required imines or oximes can be prepared by reaction of ketones with amines or hydroxylamines. However, to date, trials to reduce these substrates with ethyl-DuPHOS catalysts gave no satisfying results. Therefore, transformation of ketones or α-keto acids into acylhydrazones and subsequent enantioselective hydrogenation has proven advantageous (eq 12, Table 5).[25,26]

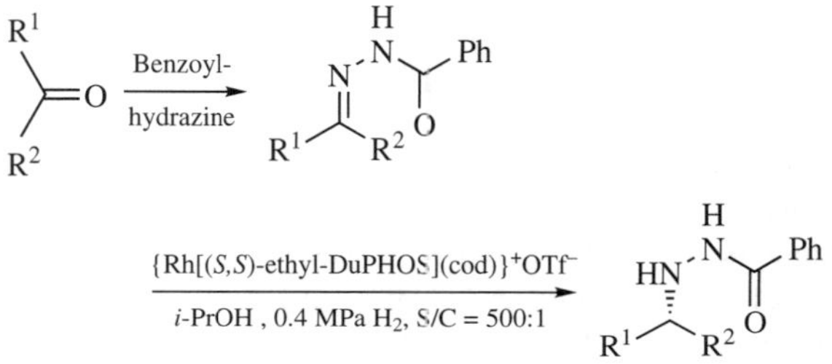

(12)

Table 5 Enantioselective hydrogenation of benzoylhydrazones with $\{Rh[(S,S)\text{-ethyl-DuPHOS}](cod)\}^+OTf^-$ catalyst (*i*-PrOH, 0.4 MPa H_2, S/C = 500:1)[a]

R^1	R^2	Temperature (°C)	Time (h)	% ee
Ph	Me	−10	24	95
p-MeO-Ph	Me	0	12	88
p-EtO_2C-Ph	Me	0	12	96
Ph	Bn	−10	24	84
CO_2Me	Et	0	36	91
CO_2Me	Pr	0	36	90
CO_2Me	Ph	0	36	91
i-Pr	Me	−10	36	73
Et	Me	−10	36	43

[a] The reactions were preferentially carried out by employment of (*R*,*R*)-ethyl-DuPHOS giving rise to the *S*-configurated product. However, as Burk et al.[26] pointed out, the antipodal (*S*,*S*)-configurated ligand gave same yields and enatiomeric excess, but products with *R*-configuration.

Chiral hydrazines can be transformed to α-amino acids and amines by cleavage of the N–N bond. Conversion to α-hydrazino acids by hydrolysis of the esters or into hydrazines by deacylation is likewise possible.[26]

(*S*,*S*)-Ethyl-DuPHOS has been employed mainly for enantioselective hydrogenation. Several types of reaction can be run very successfully. In other enantioselective reactions, this ligand has been used only rarely, one example being the enantioselective allylation of benzaldehyde by application of the corresponding silver(I) catalyst. However, in this example, the reaction failed.[27]

It is likely that several results obtained with the homologous ligands of R-DuPHOS (R=Me, Pr) or their opposite enantiomers can be related also to (*S,S*)-ethyl-DuPHOS. Recently, Burk has published a review about the application of phospholane ligands in asymmetric catalysis, which gives a good survey of the use of DuPHOS- and BPE-ligands.[28]

Related Reagents. The homologous derivatives of DuPHOS- and BPE-ligands. RoPHOS;[29–33] PennPHOS;[34] BASPHOS;[35] CnrPHOS.[36–39]

1. Burk, M. J. *J. Am. Chem. Soc.* **1991**, *113*, 8515.
2. Burk, M. J.; Kahlberg, C. S.; Pizzano, A. *J. Am. Chem. Soc.* **1998**, *120*, 4345.
3. Burk, M. J.; Feaster, J. E.; Nugent, W. A.; Harlow, R. L. *J. Am. Chem. Soc.* **1993**, *115*, 10125.
4. Burk, M. J.; Allen, J. G.; Kiesman, W. F. *J. Am. Chem. Soc.* **1998**, *120*, 657.
5. Burk, M. J.; Feaster, J. E. *J. Am. Chem. Soc.* **1992**, *114*, 6266.
6. Burk, M. J.; Feaster, J. E.; Harlow, R. L. *Tetrahedron: Asymmetry* **1991**, *2*, 569.
7. Burk, M. J.; Feaster, J. E.; Harlow, R. L. *Organometallics* **1990**, *9*, 2653.
8. Noyori, R.; Ohkuma, T.; Kitamura, M.; Takaya, H.; Sayo, N.; Kumobayashi, H.; Ohkuma, T.; Inoue, S. *J. Am. Chem. Soc.* **1987**, *109*, 5856.
9. Kyba, E. P.; Liu, S. T.; Harris, R. L. *Organometallics* **1983**, *2*, 1877.
10. Burk, M. J.; Feaster, J. E. *Tetrahedron Lett.* **1992**, *33*, 2099.
11. Schrock, R. R.; Osborne, J. A. *J. Am. Chem. Soc.* **1971**, *93*, 2397.
12. Heller, D.; Borns, S.; Baumann, W.; Selke, R. *Chem. Ber.* **1996**, *129*: 85.
13. Jones, S. W.; Palmer, C. F.; Paul, J. M.; Tiffin, P. D. *Tetrahedron Lett.* **1999**, *40*: 1211.
14. Wagaw, S.; Rennels, R. A.; Buchwald, S. L. *J. Am. Chem. Soc.* **1997**, *119*, 8451.
15. Aguado, G. P.; Alvarez-Larena, A.; Illa, O.; Moglioni, A. G.; Ortuno, R. M. *Tetrahedron: Asymmetry* **2001**, *12*, 25.
16. Stammers, T. A.; Burk, M. J. *Tetrahedron Lett.* **1999**, *40*, 3325.
17. Börner, A.; Heller, D. *Tetrahedron Lett.* **2001**, *42*, 223.
18. Drexler, H. J.; Baumann, W.; Spannenberg, A.; Fischer, C.; Heller, D. *J. Organomet. Chem.* **2001**, *621*, 89.
19. Jørgensen, K. B.; Gautun, O. R. *Tetrahedron* **1999**, *55*: 10527.
20. Burk, M. J.; Johnson, N. B.; Lee, J. R. *Tetrahedron Lett.* **1999**, *40*: 6685.
21. Burk, M. J.; Wang, Y. M.; Lee, J. R. *J. Org. Chem.* **1996**, *118*: 5142.
22. Burk, M. J.; Bienewald, F.; Harris, M.; Zanotti-Gerosa, A. *Angew. Chem., Int. Ed. Engl.* **1998**, *37*, 1931.
23. Burk, M. J.; Bienewald, F.; Challenger, S.; Derrick, A.; Ramsden, J. A. *J. Org. Chem.* **1999**, *64*, 3290.
24. Robinson, A.; Li, H. Y.; Feaster, J. *Tetrahedron Lett.* **1996**, *37*: 8321.
25. Burk, M. J.; Feaster, J. E. *J. Am. Chem. Soc.* **1992**, *114*, 6266.
26. Burk, M. J.; Martinez, J. P.; Feaster, J. E.; Cosford, N. *Tetrahedron* **1994**, *50*: 4399.
27. Yanagisawa, A.; Nakashima, H.; Ishiba, A.; Yamamoto, H. *J. Am. Chem. Soc.* **1996**, *118*, 4723.
28. Burk, M. J. *Acc. Chem. Res.* **2000**, *33*, 363.
29. Holz, J.; Quirmbach, M.; Schmidt, U.; Heller, D.; Stürmer, R.; Börner, A. *J. Org. Chem.* **1998**, *63*, 8031.
30. Li, W.; Zhang, Z.; Xiao, D.; Zhang, X. *Tetrahedron Lett.* **1999**, *40*: 6701.
31. Yan, Y.-Y.; RajanBabu, T. V. *J. Org. Chem.* **2000**, *65*, 900.
32. Li, W.; Zhang, Z.; Xiao, D.; Zhang, X. *J. Org. Chem.* **2000**, *65*, 3489.
33. Yan, Y.-Y.; RajanBabu, T. V. *Org. Lett.* **2000**, *2*, 199.
34. Jiang, Q.; Jiang, Y.; Xiao, D.; Cao, P.; Zhang, X. *Angew. Chem., Int. Ed. Engl.* **1998**, *37*, 1100.
35. Holz, J.; Heller, D.; Stürmer, R.; Börner, A. *Tetrahedron Lett.* **1999**, *40*, 7059.
36. Marinetti, A.; Genêt, J. P.; Jus, S.; Blanc, D.; Ratovelomanana-Vidal, V. *Chem. Eur. J.* **1999**, *5*, 1160.
37. Marinetti, A.; Labrue, F.; Genêt, J. P. *Synlett* **1999**, *12*, 1975.
38. Marinetti, A.; Jus, S.; Genêt, J. P. *Tetrahedron Lett.* **1999**, *40*, 8365.
39. Marinetti, A.; Jus, S.; Genêt, J. P.; Ricard, L. *Tetrahedron* **2000**, *56*, 95.

Armin Börner & Jens Holz
Institut für Organische Katalyseforschung, Rostock, Germany

1,2-Bis((2*S*,5*S*)-2,5-dimethylphospholano)benzene (*S*,*S*)-Me-DuPhos, 1,2-Bis((2*R*,5*R*)2,5-dimethylphospholano)benzene (*R*,*R*)-Me-DuPhos

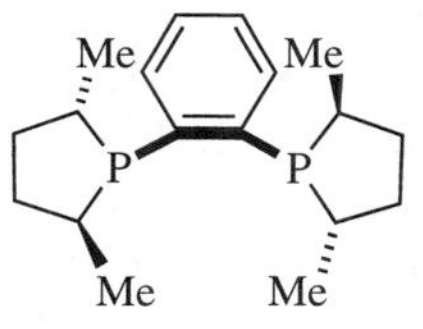

(*S,S*)-Me-DuPhos

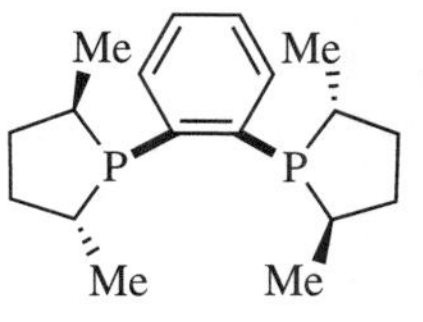

(*R,R*)-Me-DuPhos

[147253-67-6], $C_{18}H_{28}P_2$,
[136735-95-0] $C_{18}H_{28}P_2$

Physical Data: mp 79–81 °C, $[\alpha]_D^{25}$ + 476 +/- 5 (*c* 1, hexanes) for the *S,S*-enantiomer.

Form Supplied in: colorless crystals, Strem chemicals.

Preparative Methods: the DuPhos ligands are readily synthesized from the corresponding chiral 1,4-diols via the derived cyclic sulfate (eq 1).[1] The intermediate cyclic sulfate is isolable as a crystalline solid, and can be recrystallized from hexane/diethyl ether. The ligands are then obtained by treatment with lithiated 1,2-phenylene bisphosphine. After nucleophilic ring opening, treatment with two additional equivalents of *n*-butyllithium gives facile ring closure to generate the five-membered phospholane ligands. The analogous four-membered phospholane has been prepared in the same manner.[2]

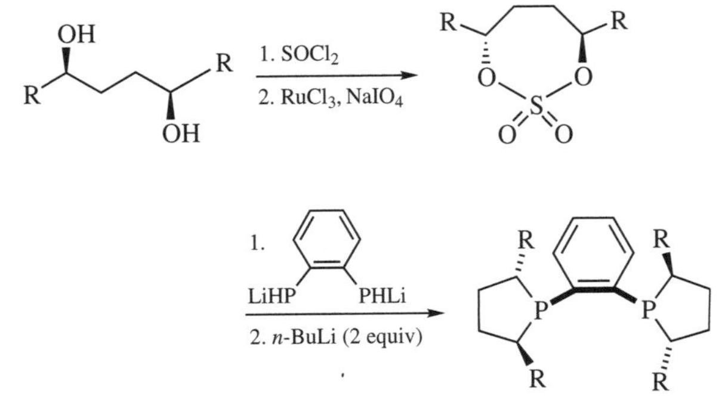

(1)

Purification: recrystallized from methanol at −10 °C.

Handling, Storage, and Precautions: crystalline Me-DuPhos is stable to air oxidation for over 10 days. However, it is generally prudent to store the DuPhos ligands under an inert atmosphere. In benzene solution, the DuPhos ligands are prone to oxidation with ca. 65% conversion to phosphine oxide after 3 weeks. Toxicity data are not available.

Bisphospholanes as Ligands in Asymmetric Catalysis. Bisphospholanes such as **1–5**, first reported by Burk,[3] have found use as ligands in transition metal-catalyzed asymmetric transformations. While metal complexes derived from the bis(phospolano)ethane (BPE) ligands (**5**) exhibit dynamic behavior, those of the more rigid DuPhos ligands (**1–4**) do not. In contrast to chiral triaryl phosphines, the DuPhos and BPE ligands are more basic and provide more electron-rich metal centers which often lead to differences in reactivity and selectivity. The modular ligand synthesis also allows access to a rich array of steric environments and allows for a significant degree of steric tuning between the ligand and substrate.

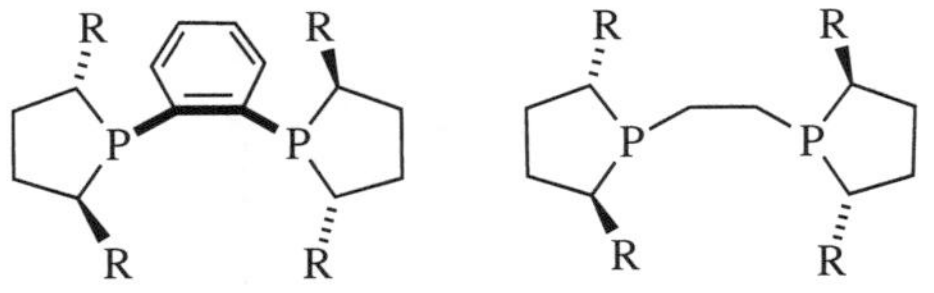

1, R = Me
2, R = Et
3, R = n-Pr
4, R = *i*-Pr

5

Rhodium-Catalyzed Asymmetric Hydrogenation. Cationic rhodium catalysts derived from the DuPhos ligands are highly effective in catalytic enantioselective alkene hydrogenation. α-Amino acids are produced in a predictable fashion by reduction of the corresponding enamide esters. Coordination of the enamide group to the metal center is a prerequisite for reduction and, as a result, regioselective hydrogenations are possible (eq 2).[4] When the β-position of the substrate is not prochiral, both alkene stereoisomers provide the same enantiomer of amino acid, such that readily available *E*/*Z* mixtures of the enamide ester may be used (eq 3). In addition to enamide coordinating groups, hydrogenation may also be directed by benzoates which provides a route to chiral α-hydroxy esters[5] and α-hydroxy phosponates (eq 4).[6] Catalytic hydrogenation of more elaborate substrates has also been employed for the synthesis of candoxatril and C-linked glycopeptides.[7]

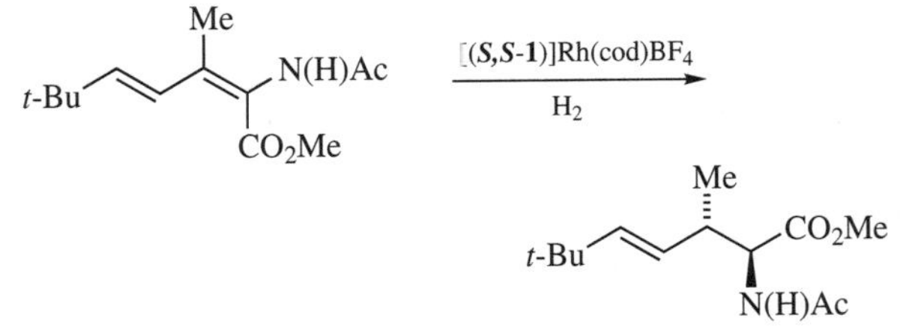

(2)

92.4% ee

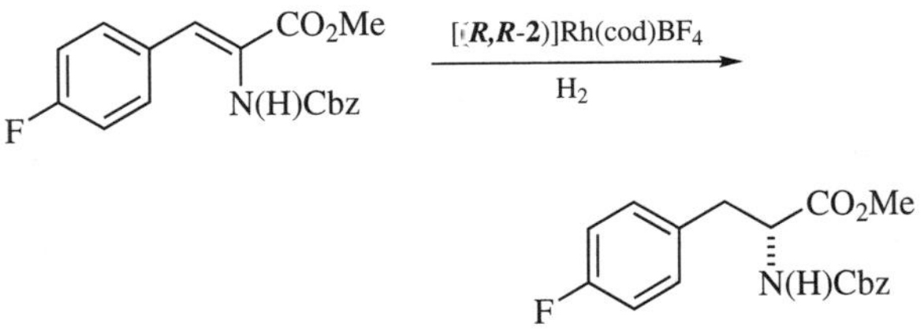

(3)

>99% ee

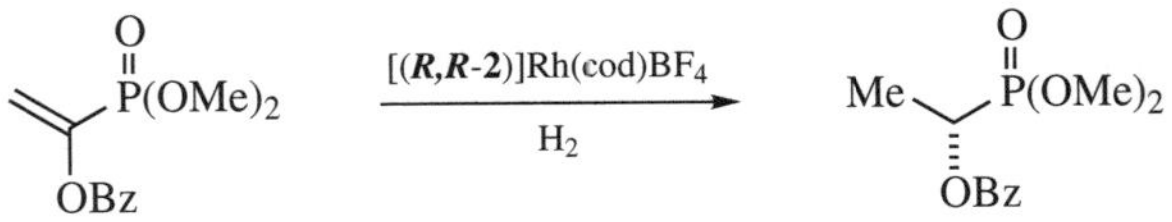

(4)

96% ee

Ruthenium-Catalyzed Hydrogenations. Ru-DuPhos complexes are commonly prepared by reacting $[RuCl_2(cod)]_n$ with methallylmagnesium chloride to generate $[Ru(cod)(methallyl)_2]$ which when treated with DuPhos and HX forms the catalytically active complex (DuPhos)RuX_2. The procedure can be performed in a single pot or in stepwise fashion.[8] Ru-DuPhos complexes effectively reduce a variety of substrates to provide chiral materials.

Both aromatic and aliphatic β-ketoesters are hydrogenated yielding the corresponding β-hydroxy esters in good yields and enantioselectivities (eq 5).[9] The ruthenium-DuPhos catalyst generated in situ from (cod)$RuBr_2$, methallylmagnesium bromide, and (*R*,*R*)-**1** efficiently reduces phenylpyruvate in quantitative yield and high enantioselectivity (eq 6). Symmetric 1,3-diketones are effectively hydrogenated to *anti*-1,3-diols by the same Ru-DuPhos complexes (eq 7). In these examples Me-DuPhos affords products in slightly higher selectivity than Et-DuPhos (93 versus 85% ee).[10]

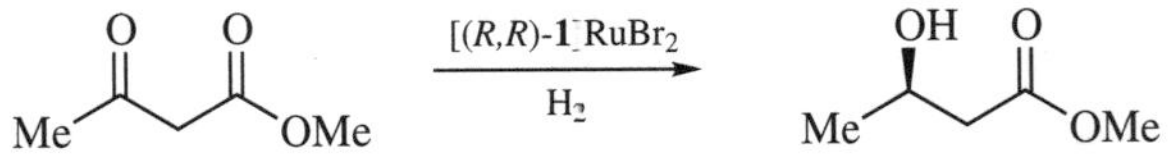

(5)

87% ee

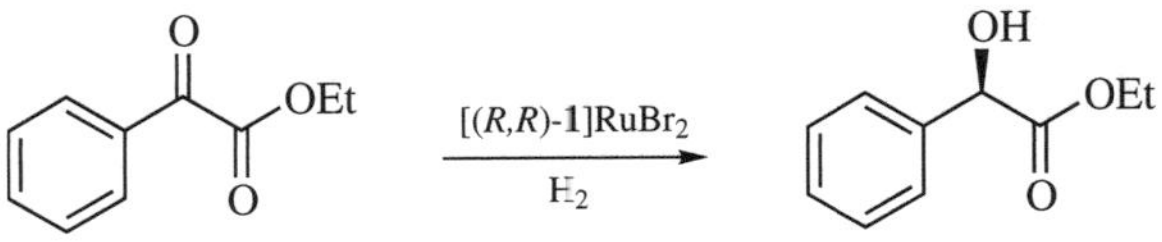

(6)

80% ee

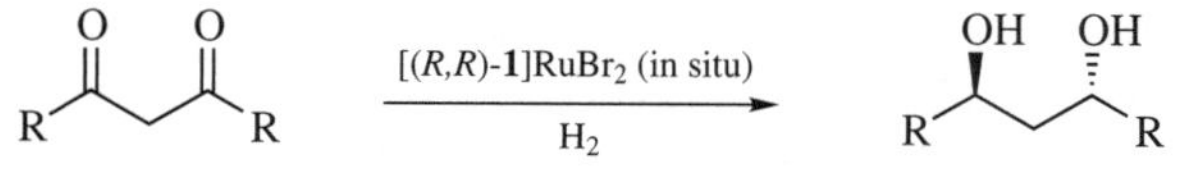

(7)

R = Me, Et, *i*-Pr, Cy, *n*-C_5H_{11}

90–97% de
85–97% ee

Simple α-substituted styrenes are reduced in the presence of $RuCl_2$(DuPhos)$(DMF)_n$. The reactivity of the ruthenium catalyst is enhanced by the addition of potassium *tert*-butoxide, which may facilitate generation of a ruthenium hydride. The products are obtained under low hydrogen pressures and selectivities obtained are up to 89% ee (eq 8).[11] Neutral Rh-DuPhos complexes catalyze the hydrogenation of α,β-unsaturated acids such as tiglic acid (eq 9). The product is obtained in quantitative yield and good enantioselectivity.[9]

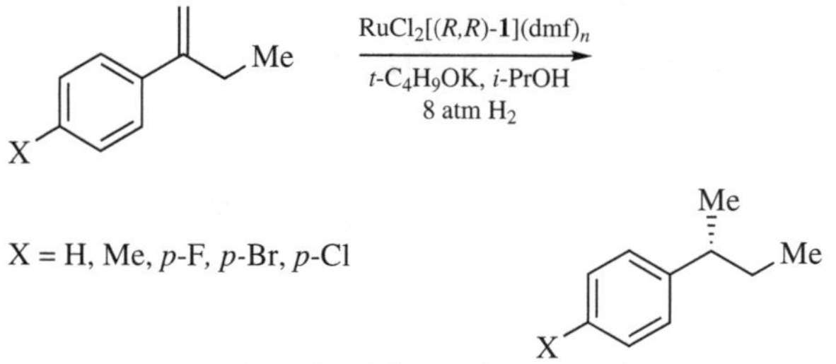

X = H, Me, *p*-F, *p*-Br, *p*-Cl

(8)

69–89% ee

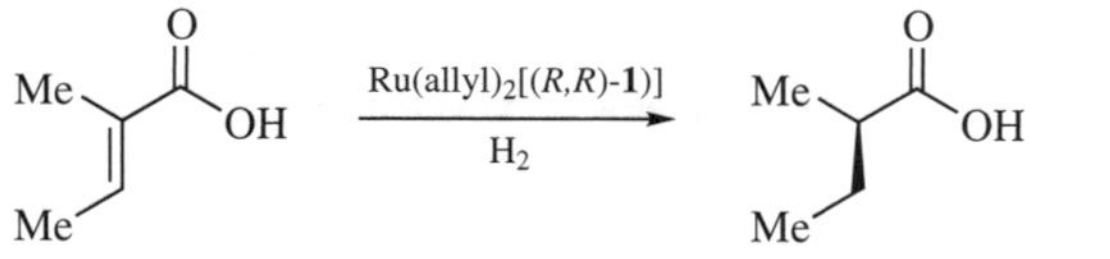

(9)

80% ee

Miscellaneous Reactions. The cationic species [(cod) Rh(*R,R*)-**4**]OTf efficiently performs the intramolecular hydrosilation of α-hydroxy esters (eq 10). In these transformations, increased steric congestion at the prochiral center correlates with increased enantioselection. It is also noteworthy that, in some cases, the absolute configuration of the product inverts when (*R,R*)-Me DuPhos is employed as opposed to (*R,R*)-*i*-PrDuPhos. In these examples, reactions in the presence of DuPhos were found to proceed in higher selectivity than either BINAP or chiraPhos.[12] Platinum DuPhos catalysts have also been applied to the hydrophosphination of various acrylonitriles and acrylates. High regioselectivity is obtained in the hydrophosphination reaction with phosphine attachment beta to the cyano or ester moiety (eq 11).[13] Platinum salts complexed with DuPhos also effectively desymmetrize *meso*-cyclohexanones (eq 12).[14] The reactions are performed under Baeyer–Villager oxidation conditions utilizing hydrogen peroxide as the terminal oxidant. Enantioselection is dependent on the substitution pattern in the starting *meso* ketone. For 2,6-dimethylcyclohexanone, Pt-DuPhos oxidation yields the lactone product with enantioselectivity comparable to Pt-BINAP, both of which are superior to Pt-DIOP and Pt-norPhos.

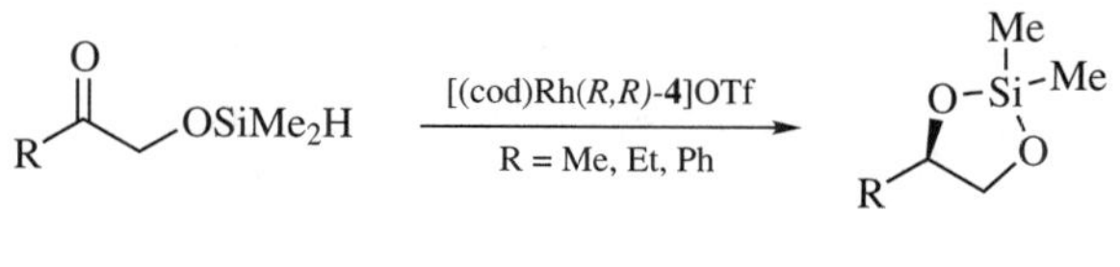

(10)

61–93% ee

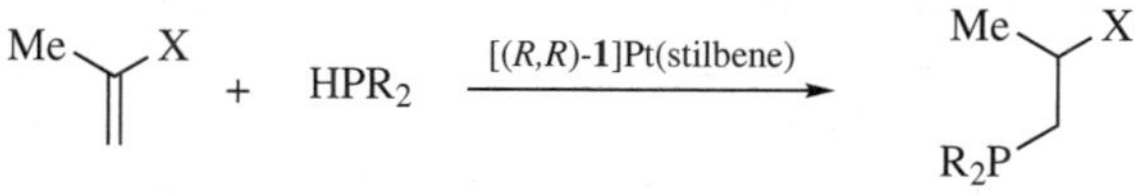

(11)

X = CN, CO_2*t*-Bu

0–27% ee

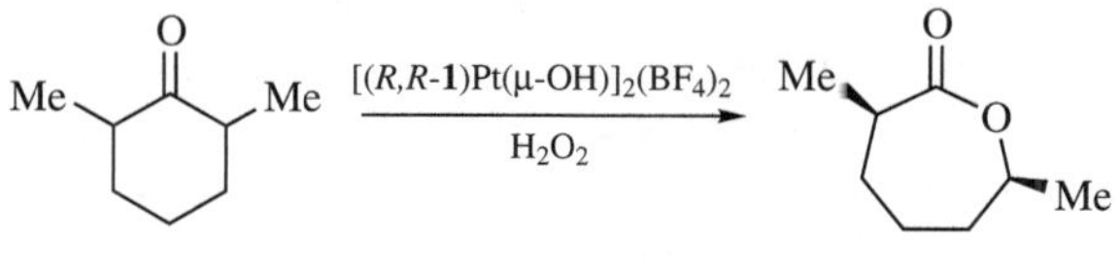

(12)

80% ee

Hydroacylation of 4-substituted pentenals using a cationic Rh-DuPhos catalyst provides chiral cyclopentanones ranging in enantioselectivity from 91 to 94% ee (eq 13).[15] In these examples, the nature of the R substituent on the olefin has little effect on the selectivity of the reaction. Aldol products are obtained from the reductive coupling between aldehydes methyl acrylate and dichloromethylsilane when employing $[(cod)RhCl]_2$ and Me-DuPhos as the catalyst system. The β-hydroxy esters are obtained in excellent diastereoselectivity for a range of aldehyde-acrylate combinations, although yields are inferior with aliphatic aldehydes (eq 14).[16] Frauenrath and co-workers have reported that chiral dihalogeno nickel complexes are efficient catalyst precursors for the asymmetric isomerization of cyclic allylic acetals (eq 15).[17] It was previously shown that the ring size of the acetal as well as the ring size of the metal–ligand chelate have a bearing on the enantioselectivity of the reaction. Me-DuPhos was thus found to be a suitable ligand for these types of reactions. It was later discovered that yields and enantioselectivities are also dependent on the counter ion. When iodide was used as the counter ion and upon activation with $LiBHEt_3$, the reaction proceeded with high enantioselectivity at low temperatures.

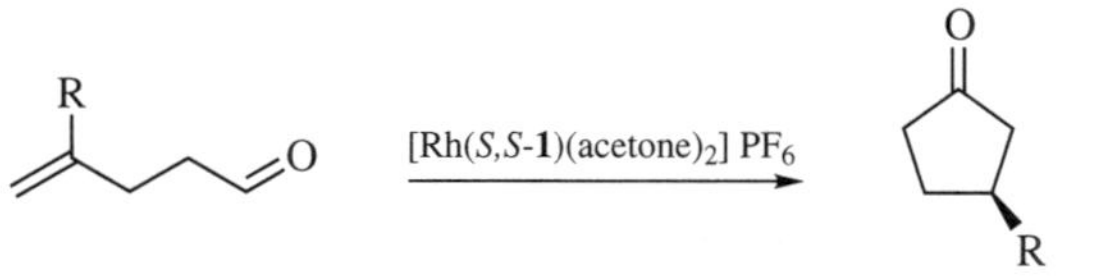

(13)

R = Me, Et, Bn, C_5H_{11}

93–98% ee

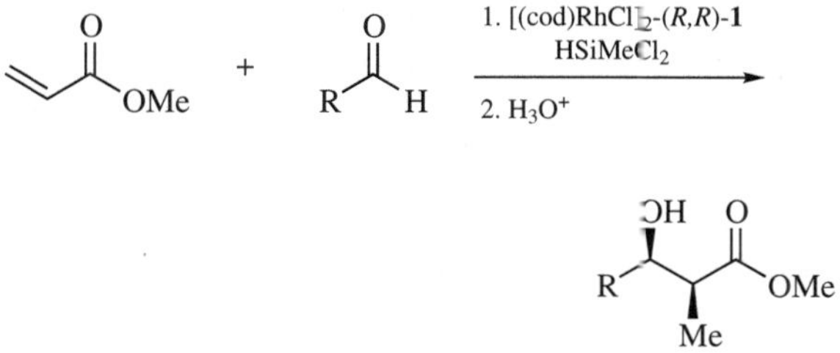

(14)

racemic
>6:1 dr

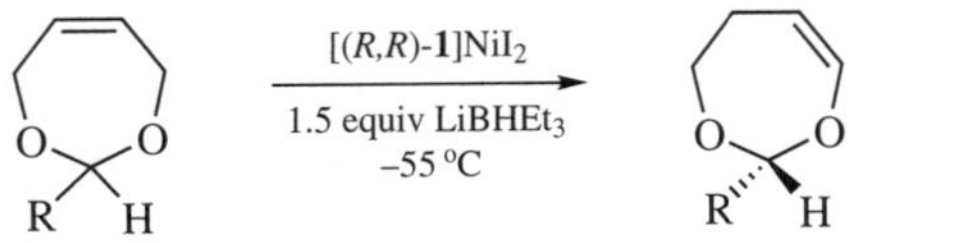

(15)

90–98% ee

Polymerization Reactions. Lee and Alper have shown that the use of a Pd(II)-Me-DuPhos catalyst produces highly functionalized alternating polyketones derived from CO and α-olefins (eq 16).[18] Notably, the polymers are exclusively head-to-tail selective, isotactic, high molecular weight, and, when prepared with the Pd-DuPhos catalyst, can contain functionality in the polymer side chain. Sen et al. have demonstrated that Pd(II)-Me-DuPhos complexes are excellent catalysts for alternating copolymerizations of ethene, propene, and cyclopentene with SO_2 (eq 17).[19] The copolymers that were produced were 1:1 alternating and atactic with exclusive head-to-tail enchainment as shown for propene. The ability of the Pd(II) catalyst to promote the polymerization depended strongly on the nature of the ligand bound to palladium. It was discovered that monodentate phosphines as well as bidentate nitrogen ligands were ineffective as ligands for catalysis. Only bidentate phosphines acted as ligands to generate the active catalyst.

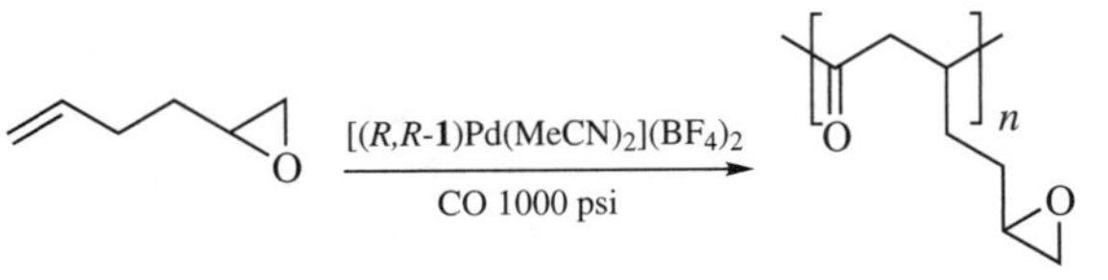

(16)

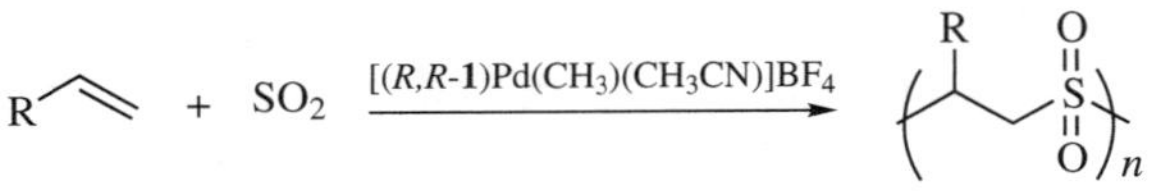

(17)

1. Burk, M. J.; Feaster, J. E.; Nugent, W. A.; Harlow, R. L. *J. Am. Chem. Soc.* **1993**, *115*, 10125
2. Marinetti, A.; Kruger, V.; Fancois-Xavier, B. *Tetrahedron Lett.* **1997**, *38*, 2947.
3. Burk, M. J. *J. Am. Chem. Soc.* **1991**, *113*, 8518.
4. Burk, M. J.; Allen, J. G.; Kiesman, W. F. *J. Am. Chem. Soc.* **1998**, *120*, 657.
5. Burk, M. J.; Kalberg, C. S.; Pizzano, A. *J. Am. Chem. Soc.* **1998**, *120*, 4345.
6. Burk, M. J.; Stammer, T. A.; Straub, J. A. *Organic Lett.* **1999**, *1*, 387.
7. (a) Burk, M. J.; Bienewald, F.; Challenger, S.; Derrick, A.; Ramsden, J. A. *J. Org. Chem.* **1999**, *64*, 3290. (b) Debenham, S. D.; Debenham, J. S.; Burk, M. J.; Toone, E. J. *J. Am. Chem. Soc.* **1997**, *119*, 9897.
8. Guerreiro, P.; Cano de Andrade, M.; Henry, J.; Tranchier, J.; Phansavath, P.; Ratovelomana-Vidal, V.; Genet, J.; Homri, T.; Touati, A. R.; Ben Hassine, B. *C. R. Acad. Sci. Paris* **1999**, 175.
9. Genet, J. P.; Pinel, C.; Ratovelomana-Vidal, V.; Mallart, S.; Pfister, X.; Bischoff, L.; Cano de Andrade, M.; Darses, S.; Galopin, C.; Lafitte, J. A. *Tetrahedron: Asymm.* **1994**, *5*, 675.
10. Blanc, D.; Ratovelomana-Vidal, V.; Marinetti, A.; Genet, J. P. *Synlett* **1999**, 480.
11. Forman, G. S.; Ohkuma, T.; Hems, W. P.; Noyori, R. *Tetrahedron Lett.* **2000**, *41*, 9471.
12. Burk, M. J.; Feaster, J. E. *Tetrahedron Lett.* **1992**, *33*, 2099.
13. Kovacik, I.; Wicht, D. K.; Grewal, N. S.; Glueck, D. S. *Organometallics* **2000**, *19*, 950.
14. Paneghetti, C.; Gavagnin, R.; Pinna, F.; Strukul, G. *Organometallics* **1999**, *18*, 5057.
15. Barnhart, R. W.; McMorran, D. A.; Bosnich, B. *Chem Commun.* **1997**, 589.
16. Taylor, S. J.; Morken, J. P. *J. Am. Chem. Soc.* **1999**, *121*, 12202.
17. Frauenrath, H.; Brethauer, D.; Reim, S.; Maurer, M.; Raabe, G. *Angew. Chem. Int. Ed. Engl.* **2001**, *40*, 177.
18. Lee, J. T.; Alper, H. *Chem Commun.* **2000**, 2189.
19. Wojcinski, L. M.; Boyer, M. T.; Sen, A. *Inorg. Chem. Acta* **1998**, *270*, 8.

James P. Morken, Albert E. Russell & Steven J. Taylor
UNC Chapel Hill, North Carolina, USA

[Bis(4*R*,5*S*)-4,5-diphenyl-1,3-oxazolin-2-yl]methane [Bis(4*S*,5*R*)-4,5-diphenyl-1,3-oxazolin-2-yl]methane

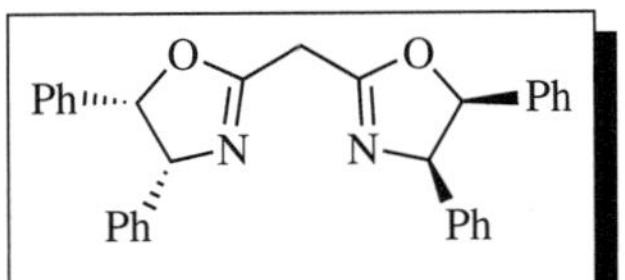

(4*S*,5*R*)-**1**

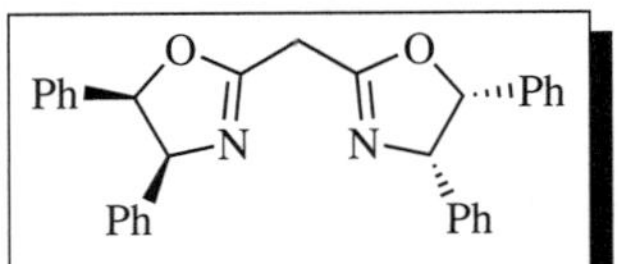

(4*R*,5*S*)-**1**

[157904-66-0]	$C_{31}H_{26}N_2O_2$	(MW 458.55)
[139021-82-8]	$C_{31}H_{26}N_2O_2$	(MW 458.55)

(asymmetric catalyst for copper-catalyzed cyclopropanation,[1] direct asymmetric α-amination of 2-keto esters,[2] aza-Henry reactions of nitronates,[3] reduction of aromatic ketones[4])

Alternate Name: *cis*-DiPh-Box.
Physical Data: (4*S*,5*R*)-**1** mp 205–208 °C, $[\alpha]_D^{24}$ +160 (*c* 0.5, $CHCl_3$).
Solubility: soluble in $CHCl_3$, CH_2Cl_2, CH_3CN, Et_2O, toluene, and THF.
Form Supplied in: (4*S*,5*R*)-**1** commercially available as a white, crystalline powder.
Analysis of Reagent Purity: by NMR and specific rotation determination.
Preparative Methods: both enantiomers are readily prepared in one step from diethyl malonate or malono-bis-imidate and the corresponding amino alcohols.[1,5]
Purification: chromatograph on neutral alumina.[5]

Asymmetric Copper-Catalyzed Cyclopropanation.[6] Since the pioneering work of Nozaki and his co-workers, several chiral ligands have been designed to achieve high enantio and diastereoselectivity in copper-catalyzed asymmetric cyclopropanation of olefins. Masamune introduced C_2-symmetric bisoxazoline ligands,[1,5,7] which are readily prepared in one step from diethyl malonate or malono-bis-imidate and the corresponding amino alcohols, for highly diastereo (trans/cis ratio) and enantioselective cyclopropanation of olefins. The *cis*-DiPh-Box ligand[1] proved to be highly effective in the cyclopropanation of trisubstituted and *cis*-1,2-disubstituted olefins (eq 1). Diastereoselectivity depends on the alkoxy moiety of diazoacetates and the trans/cis ratio can be improved to 99:1 by the use of dicyclohexylmethyl diazoacetate. Although this ligand showed considerable generality in Masamune's report, it failed to give a satisfactory result in the cyclopropanation of 3-methyl-2-butenyl acetate.[8]

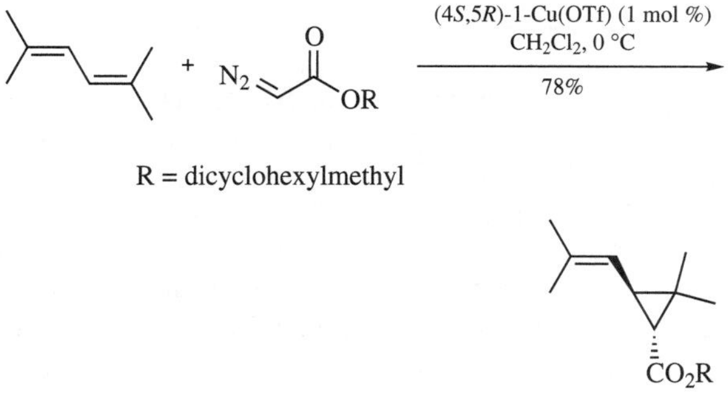

(1)

90% de, 94% ee

Direct Asymmetric α-Amination Reaction of 2-Keto Esters.[2] The *cis*-DiPh-Box copper complex catalyzes highly enantioselective direct α-amination reaction of 2-keto esters with dialkyl azodicarboxylates and thus provides convenient access to optically active *syn*-β-amino-α-hydroxy esters (eq 2). This enantioselective, direct α-amination is applicable to a range of 2-keto esters when dibenzyl azodicarboxylate is used as the nitrogen source. The immediate product of the amination reaction is prone to racemization. Stereoselective reduction of the keto functionality by L-selectride enables further synthetic operations to be carried out without loss of enantiopurity.

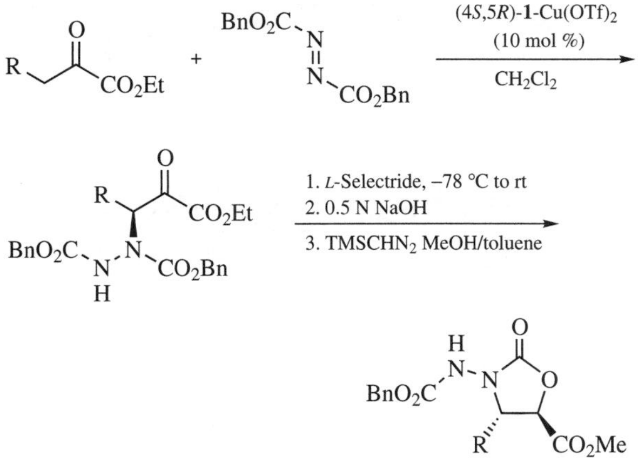

(2)

R	% Yield	% ee
benzyl	55	77
methyl	45	90
pentyl	63	93
allyl	62	93
isobutyl	53	96
isopropyl	78	95
c-hexylmethyl	54	96

Asymmetric Aza-Henry Reactions of Nitronates with Imines.[3] Although the Henry reaction and its aza-analogs are powerful C–C bond-forming reactions, there are few reports of catalytic asymmetric versions of these reactions. The *cis*-DiPh-Box copper complexes are excellent catalysts for highly diastereo and enantioselective aza-Henry reactions of a variety of trimethylsilylnitronates with *N*-(*p*-methoxyphenyl)-α-imino-esters (eq 3). The use of an *N*-(*p*-methoxyphenyl) group for protection prevents undesirable side reactions and can be easily removed. The aza-Henry reaction products can be further derivatized to the corresponding α,β-diamino acids whose syntheses have rarely been reported.

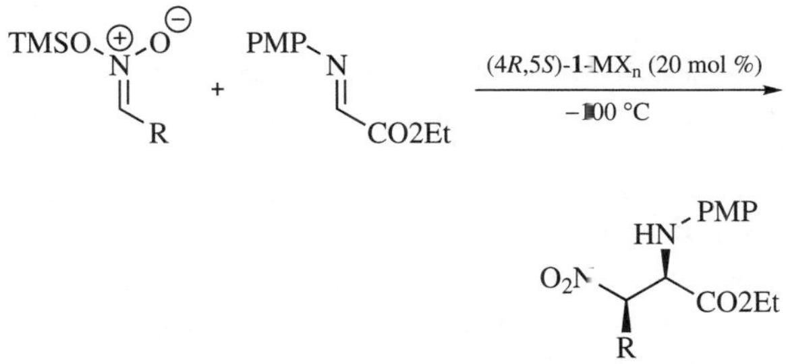

(3)

R	MX$_n$	% Yield	*erythro*: *threo*	% ee (*erythro*)	Solvent
ethyl	Cu(I)PF$_6$	94	25:1	95	CH_2Cl_2
pentyl	Cu(II)(OTf)$_2$	87	39:1	83	CH_2Cl_2
benzyl	Cu(II)(SbF$_6$)$_2$	93	32:1	88	THF

Enantioselective Reduction of Aromatic Ketones.[4] Aromatic substituted ketones and α-halo ketones are reduced by $(EtO)_3SiH$ with good enantioselectivity in the presence of bis-oxazoline titanium complex [Ti(*cis*-DiPh-Box)$_2$F$_2$], prepared from chiral bis-oxazoline, BuLi, and TiF_4 (eq 4).

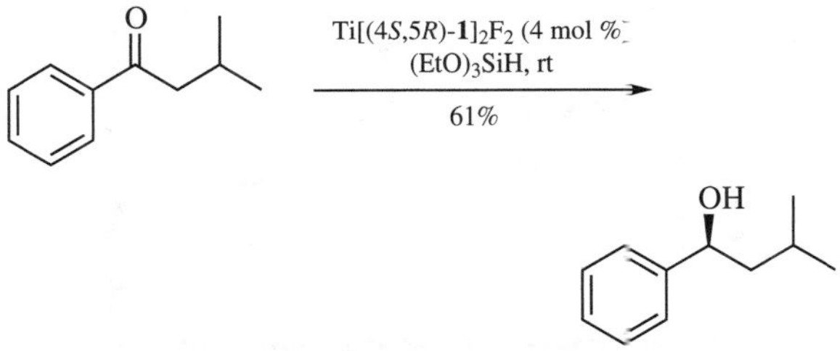

(4)

85% ee

Miscellaneous. There are several other reports on the application of this ligand to catalytic asymmetric reactions, although enantioselectivities are modest. Those reports include the Mukaiyama–Michael reaction,[9] allylation of aldehydes,[10] asymmetric Diels–Alder reaction,[11] Mukaiyama–Aldol reaction of ketomalonate,[12] aziridination reaction of α-imino esters,[13] and asymmetric hetero-Diels–Alder reaction.[14]

Related Reagents. 2,2′-Methylenebis[(4*S*)-4-*tert*-butyl-2-oxazoline]; 2,2′-bis{2-[(4*S*)-*tert*-butyl-1,3-oxazolinyl]}propane; (1*S*,9*S*)-1,9-bis(1-hydroxy-1-methylethyl)semicorrin-5-carbonitrile; (1*S*,9*S*)-1,9-bis[(*tert*-butyl)diemthylsiloxy]methylsemicorrin-5-carbonitrile.

---

1. Lowenthal, R.; Masamune, S. *Tetrahedron Lett.* **1991**, *32*, 7373.
2. Juhl, K.; Jørgensen, K. A. *J. Am. Chem. Soc.* **2002**, *124*, 2420.
3. Knudsen, K. R.; Risgaard, T.; Nishikawa, N.; Gothelf, K. V.; Jørgensen, K. A. *J. Am. Chem. Soc.* **2001**, *123*, 5843.
4. Bandini, M.; Cozzi, P. G.; Negro, L.; Umani-Ronchi, A. *Chem. Commun.* **1999**, 39.
5. Lowenthal, R.; Abiko, A.; Masamune, S. *Tetrahedron Lett.* **1990**, *31*, 6005.
6. (a) Doyle, M. P. *Catalytic Asymmetric Synthesis*; Ojima, I., Ed.; Wiley: New York, 2000; pp 191–228. (b) Pfaltz, A., *Comprehensive Asymmetric Catalysis*; Jacobsen, E. N.; Pfaltz, A.; Yamanoto, H., Eds; Springer: Berlin, 1999; Vol. 2, pp 513–538.
7. Ghosh, A. K.; Mathivanan, P.; Cappiello, J. *Tetrahedron: Asymmetry* **1998**, *9*, 1.
8. Østergaard, N.; Jensen, J.; Tanner, D. *Tetrahedron* **2001**, *57*, 6083.
9. Bernardi, A.; Colombo, G.; Scolastico, C. *Tetrahedron Lett.* **1996**, *37*, 8921.
10. Cozzi, P. G.; Orioli, P.; Tagliavini, E.; Umani-Ronchi, A. *Tetrahedron Lett.* **1997**, *38*, 145.
11. Brimble, M. A.; McEwan, J. F. *Tetrahedron: Asymmetry* **1997**, *8*, 4069.
12. Reichel, F.; Fang, X.; Yao, S.; Ricci, M.; Jørgensen, K. A. *Chem. Commun.* **1999**, 1505.
13. Juhl, K.; Hazell, R. G.; Jørgensen, K. A. *J. Chem. Soc., Perkin Trans. 1* **1999**, *37*, 2293.
14. Yao, S.; Roberson, M.; Reichel, F.; Hazell, R. G.; Jørgensen, K. A. *J. Org. Chem.* **1999**, *64*, 6677.

Erik J. Sorensen, Hirofumi Seike & Jason M. Rohde
The Scripps Research Institute, La Jolla, CA, USA

(*R*)- & (*S*)-2,2′-Bis(diphenylphosphino)-1,1′-binaphthyl[1]

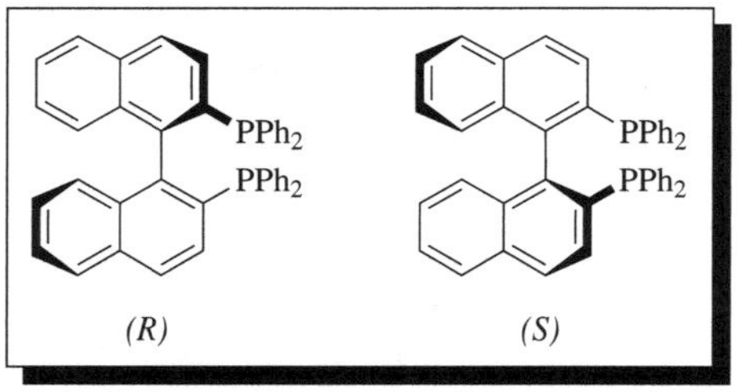

[76189-55-4] $C_{44}H_{32}P_2$ (MW 622.70)

(chiral diphosphine ligand for transition metals;[2] the complexes show high enantioselectivity and reactivity in a variety of organic reactions)

Alternate Name: BINAP.
Physical Data: mp 241–242 °C; $[\alpha]_D^{25}$ −229° (c = 0.312, benzene) for (*S*)-BINAP.[3]
Solubility: sol THF, benzene, dichloromethane; modestly sol ether, methanol, ethanol; insol water.
Form Supplied in: colorless solid.
Analysis of Reagent Purity: GLC analysis (OV-101, capillary column, 5 m, 200–280 °C) and TLC analysis (E. Merck Kieselgel 60 PF_{254}, 1:19 methanol–chloroform); R_f 0.42 (BINAPO, dioxide of BINAP), 0.67 (monoxide of BINAP), and 0.83 (BINAP). The optical purity of BINAP is analyzed after oxidizing to BINAPO by HPLC using a Pirkle column (Baker bond II) and a hexane/ethanol mixture as eluent.[3]
Preparative Methods: enantiomerically pure BINAP is obtained by resolution of the racemic dioxide, BINAPO, with camphorsulfonic acid or 2,3-di-*O*-benzoyltartaric acid followed by deoxygenation with *Trichlorosilane* in the presence of *Triethylamine*.[3]
Handling, Storage, and Precautions: solid BINAP is substantially stable to air, but bottles of BINAP should be flushed with N_2 or Ar and kept tightly closed for prolonged storage. BINAP is slowly air oxidized to the monoxide in solution.

BINAP–Ru^{II} Catalyzed Asymmetric Reactions. Halogen-containing BINAP–Ru complexes are most simply prepared by reaction of $[RuCl_2(cod)]_n$ or $[RuX_2(arene)]_2$ (X = Cl, Br, or I) with BINAP.[4] Sequential treatment of $[RuCl_2(benzene)]_2$ with BINAP and sodium carboxylates affords $Ru(carboxylate)_2(BINAP)$ complexes. The dicarboxylate complexes, upon treatment with strong acid HX,[5] can be converted to a series of Ru complexes empirically formulated as $RuX_2(BINAP)$. These Ru^{II} complexes act as catalysts for asymmetric hydrogenation of various achiral and chiral unsaturated compounds.

α,β-Unsaturated carboxylic acids are hydrogenated in the presence of a small amount of $Ru(OAc)_2(BINAP)$ to give the corresponding optically active saturated products in quantitative yields.[6] The reaction is carried out in methanol at ambient temperature with a substrate:catalyst (S:C) ratio of 100–600:1. The sense and degree of the enantioface differentiation are profoundly affected by hydrogen pressure and the substitution pattern of the substrates. Tiglic acid is hydrogenated quantitatively with a high enantioselectivity under a low hydrogen pressure (eq 1), whereas naproxen, a commercial anti-inflammatory agent, is obtained in 97% ee under high pressure (eq 2).[6a]

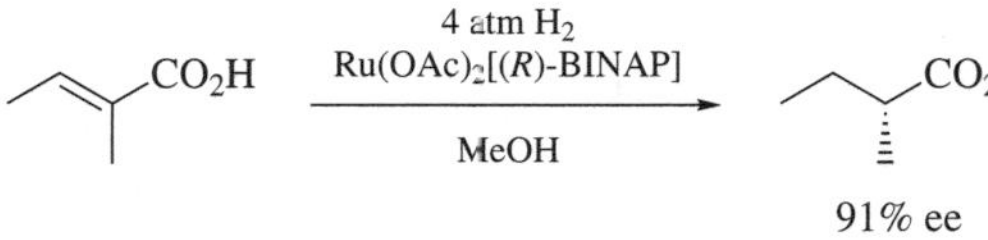

(1)

91% ee

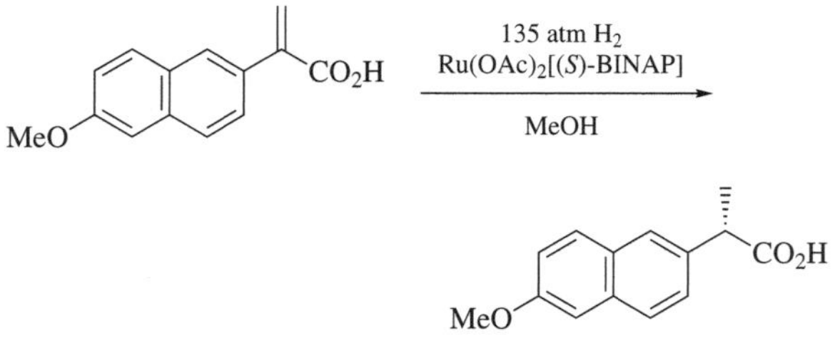

(2)

97% ee

Enantioselective hydrogenation of certain α- and β-(acylamino)acrylic acids or esters in alcohols under 1–4 atm H_2 affords the protected α- and β-amino acids, respectively (eqs eq 3 and eq 4).[2a,7] Reaction of *N*-acylated 1-alkylidene-1,2,3,4-tetrahydroisoquinolines provides the 1*R*- or 1*S*-alkylated products. This method allows a general asymmetric synthesis of isoquinoline alkaloids (eq 5).[8]

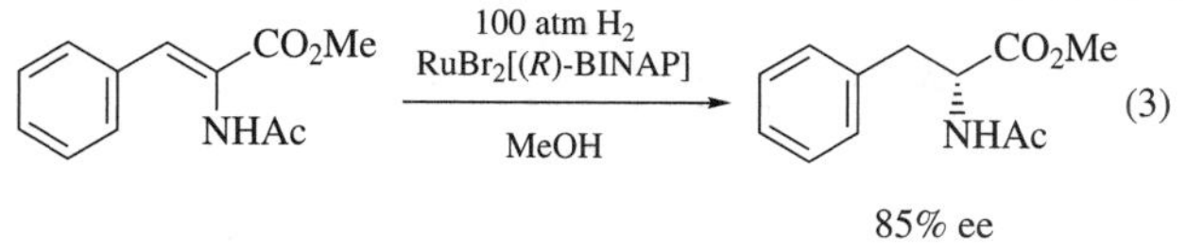

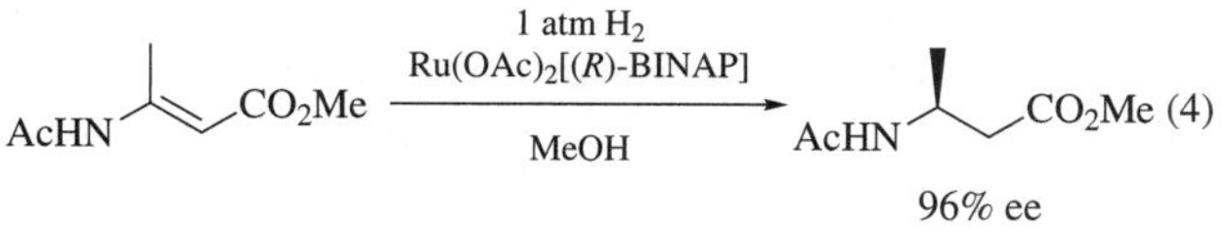

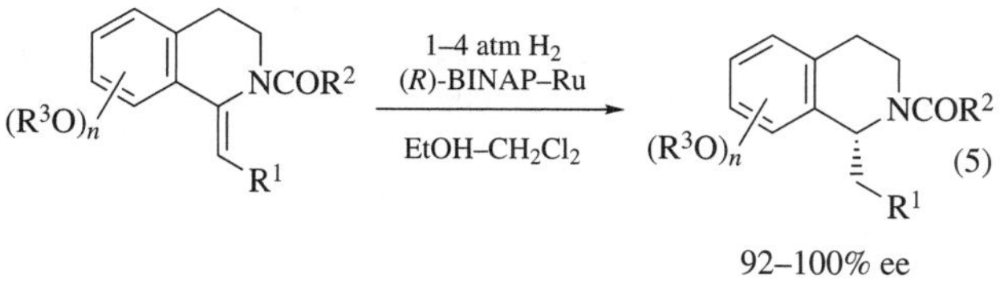

Geraniol or nerol can be converted to citronellol in 96–99% ee in quantitative yield without saturation of the C(6)–C(7) double bond (eq 6).[9] The S:C ratio approaches 50 000. The use of alcoholic solvents such as methanol or ethanol and initial H_2 pressure greater than 30 atm is required to obtain high enantioselectivity. Diastereoselective hydrogenation of the enantiomerically pure allylic alcohol with an azetidinone skeleton proceeds at atmospheric pressure in the presence of an (*R*)-BINAP–Ru complex to afford the β-methyl product, a precursor of 1β-methylcarbapenem antibiotics (eq 7).[10] Racemic allylic alcohols such as 3-methyl-2-cyclohexenol and 4-hydroxy-2-cyclopentenone can be effectively resolved by the BINAP–Ru-catalyzed hydrogenation (eq 8).[11]

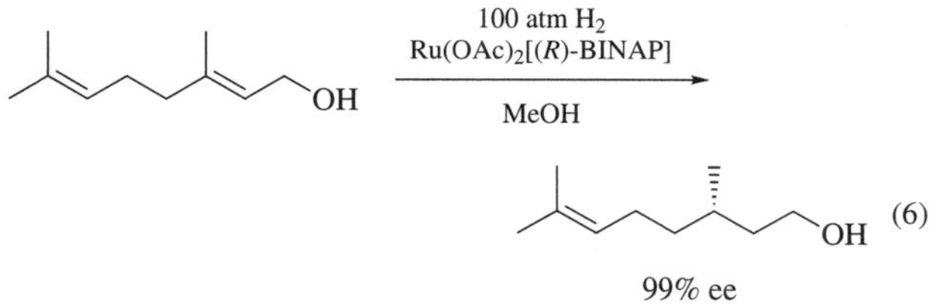

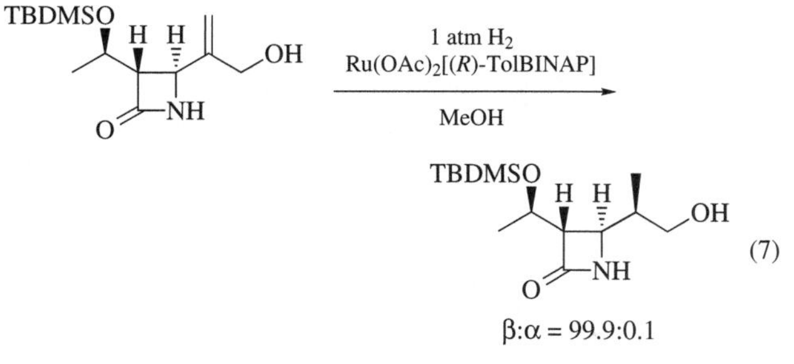

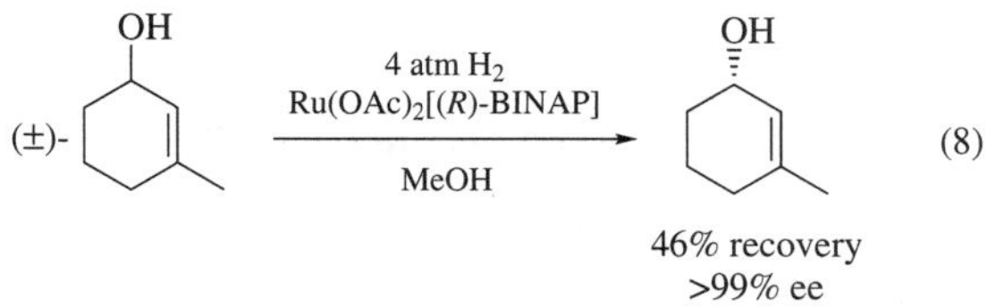

Diketene is quantitatively hydrogenated to 3-methyl-3-propanolide in 92% ee (eq 9). Certain 4-methylene- and 2-alkylidene-4-butanolides as well as 2-alkylidenecyclopentanone are also hydrogenated with high enantioselectivity.[12]

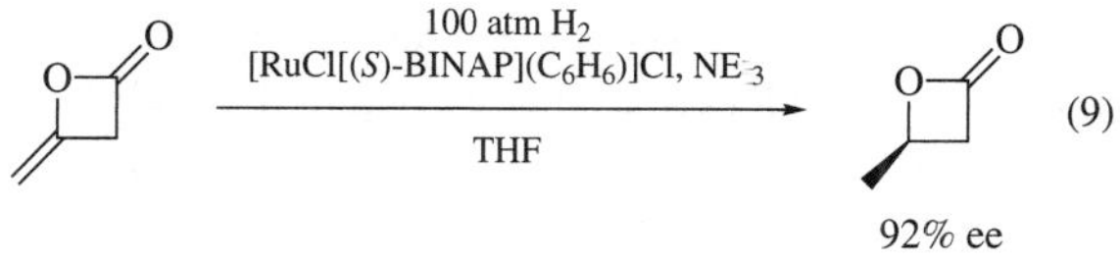

Hydrogenation with halogen-containing BINAP–Ru complexes can convert a wide range of functionalized prochiral ketones to stereo-defined secondary alcohols with high enantiomeric purity (eq 10).[13] 3-Oxocarboxylates are among the most appropriate substrates.[13a,4d] For example, the enantioselective hydrogenation of methyl 3-oxobutanoate proceeds quantitatively in methanol with an S:C ratio of 1000–10 000 to give the hydroxy ester product in nearly 100% ee (eq 11). Halogen-containing complexes RuX_2(BINAP) (X = Cl, Br, or I; polymeric form) or $[RuCl_2(BINAP)]_2NEt_3$ are used as the catalysts. Alcohols are the solvents of choice, but aprotic solvents such as dichloromethane can also be used. At room temperature the reaction requires an initial H_2 pressure of 20–100 atm, but at 80–100 °C the reaction proceeds smoothly at 4 atm H_2.[4c,4d]

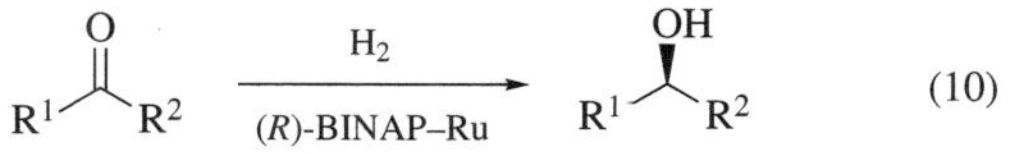

R^1 = alkyl, aryl; R^2 = CH_2OH, CH_2NMe_2, CH_2CH_2OH, CH_2Ac, CH_2CO_2R, CH_2COSR, CH_2CONR_2, $CH_2CH_2CO_2R$, etc.

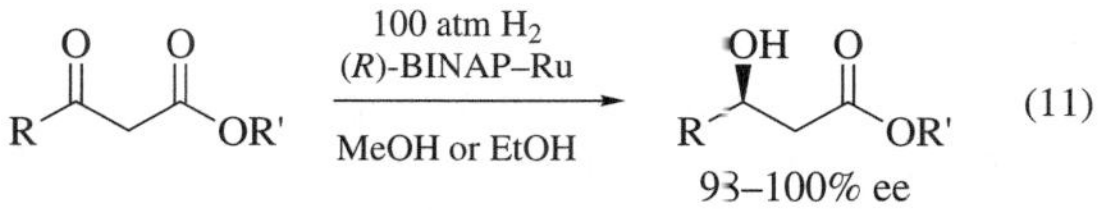

R = Me, Et, Bu, *i*-Pr; R' = Me, Et, *i*-Pr, *t*-Bu

3-Oxocarboxylates possessing an additional functional group can also be hydrogenated with high enantioselectivity by choosing appropriate reaction conditions or by suitable functional group modification (eq 12).[13b,13c]

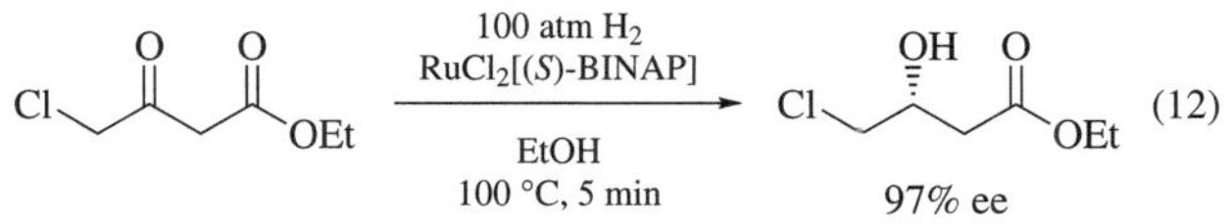

The pre-existing stereogenic center in the chiral substrates profoundly affects the stereoselectivity. The (*R*)-BINAP–Ru-catalyzed reaction of (*S*)-4-(alkoxycarbonylamino)-3-oxocarboxylates give the statine series with (3*S*,4*S*) configuration almost exclusively (eq 13).[14]

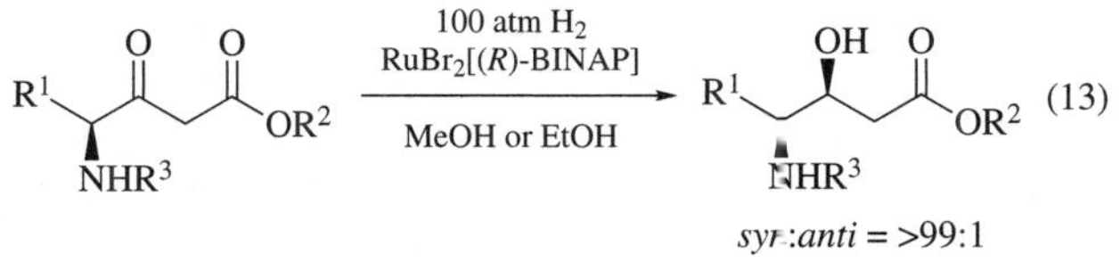

Hydrogenation of certain racemic 2-substituted 3-oxocarboxylates occurs with high diastereo- and enantioselectivity via dynamic kinetic resolution involving in situ racemization of the substrates.[15] The (*R*)-BINAP–Ru-catalyzed reaction of 2-acylamino-3-oxocarboxylates in dichloromethane allows preparation of threonine and DOPS (anti-Parkinsonian agent)

(eq 14).[16] In addition, a common intermediate for the synthesis of carbapenem antibiotics is prepared stereoselectively on an industrial scale from a 3-oxobutyric ester (**1**) with an acylaminomethyl substituent at the C(2) position.[16a] The second-order stereoselective hydrogenation of 2-ethoxycarbonylcycloalkanones gives predominantly the *trans* hydroxy esters (**2**) in high ee, whereas 2-acetyl-4-butanolide is hydrogenated to give the *syn* diastereomer (**3**).[17]

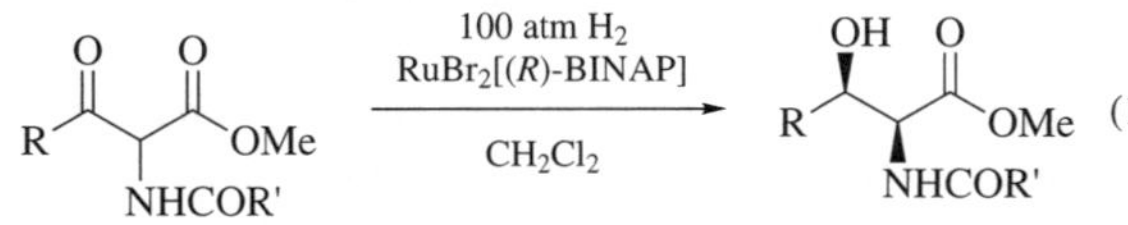

(14)

syn:*anti* = 99:1
92–98% ee

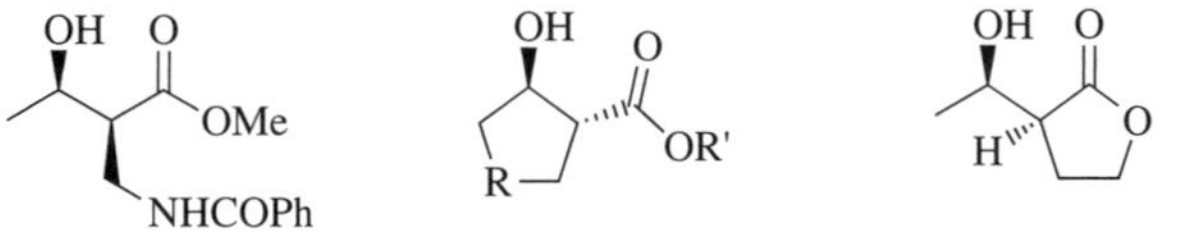

(**1**)
syn:*anti* = 94:6
98% ee

(**2**)
R = CH_2, $(CH_2)_2$, $(CH_2)_3$
trans:*cis* = 93:7–99:1
90–93% ee

(**3**)
syn:*anti* = 98:2
94% ee

Certain 1,2- and 1,3-diketones are doubly hydrogenated to give stereoisomeric diols. 2,4-Pentanedione, for instance, affords (*R,R*)- or (*S,S*)-2,4-pentanediol in nearly 100% ee accompanied by 1% of the *meso* diol.[13b]

A BINAP–Ru complex can hydrogenate a C=N double bond in a special cyclic sulfonimide to the sultam with >99% ee.[18]

The asymmetric transfer hydrogenation of the unsaturated carboxylic acids using formic acid or alcohols as the hydrogen source is catalyzed by Ru(acac-F_6)(η^3-C_3H_5)(BINAP) or $[RuH(BINAP)_2]PF_6$ to produce the saturated acids in up to 97% ee (eq 15).[19]

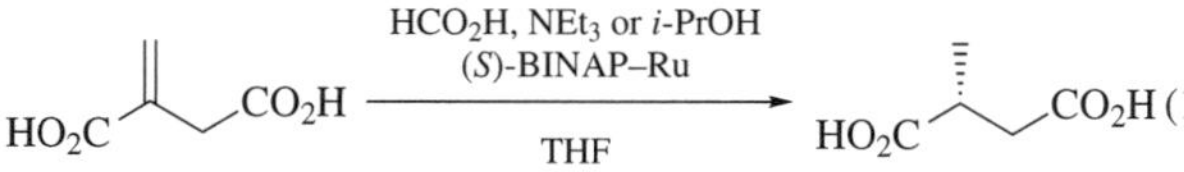

(15)

93–97% ee

BINAP–Ru complexes promote addition of arenesulfonyl chlorides to alkenes in 25–40% optical yield.[20]

BINAP–Rh^I^ Catalyzed Asymmetric Reactions. The rhodium(I) complexes $[Rh(BINAP)(cod)]ClO_4$, $[Rh(BINAP)(nbd)]ClO_4$, and $[Rh(BINAP)_2]ClO_4$, are prepared from $[RhCl(cod)]_2$ or *Bis(bicyclo[2.2.1]hepta-2,5-diene)dichlorodirhodium* and BINAP in the presence of $AgClO_4$.[21] $[Rh(BINAP)S_2]ClO_4$ is prepared by reaction of [Rh(BINAP)(cod or nbd)]ClO_4 with atmospheric pressure of hydrogen in an appropriate solvent, S.[21a] BINAP–Rh complexes catalyze a variety of asymmetric reactions.[2]

Prochiral α-(acylamino)acrylic acids or esters are hydrogenated under an initial hydrogen pressure of 3–4 atm to give the protected amino acids in up to 100% ee (eq 16).[21a] The BINAP–Rh catalyst was used for highly diastereoselective hydrogenation of a chiral homoallylic alcohol to give a fragment of the ionophore ionomycin.[22]

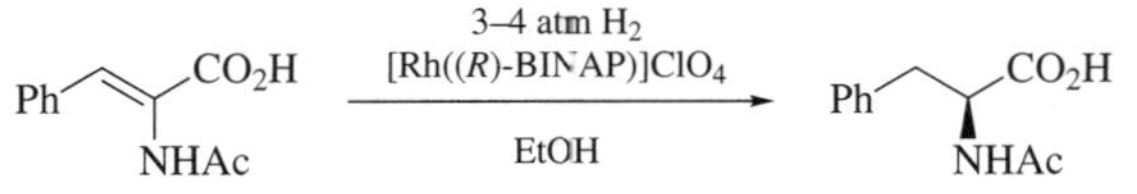

(16)

100% ee

The cationic BINAP–Rh complexes catalyze asymmetric 1,3-hydrogen shifts of certain alkenes. Diethylgeranylamine can be quantitatively isomerized in THF or acetone to citronellal diethylenamine in 96–99% ee (eq 17).[23] This process is the key step in the industrial production of (−)-menthol. In the presence of a cationic (*R*)-BINAP–Rh complex, (*S*)-4-hydroxy-2-cyclopentenone is isomerized five times faster than the (*R*) enantiomer, giving a chiral intermediate of prostaglandin synthesis.[24]

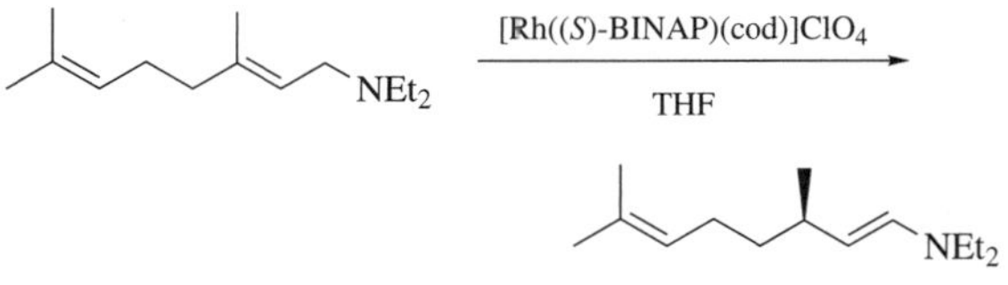

(17)

99% ee

Enantioselective cyclization of 4-substituted 4-pentenals to 3-substituted cyclopentanones in >99% ee is achieved with a cationic BINAP–Rh complex (eq 18).[25]

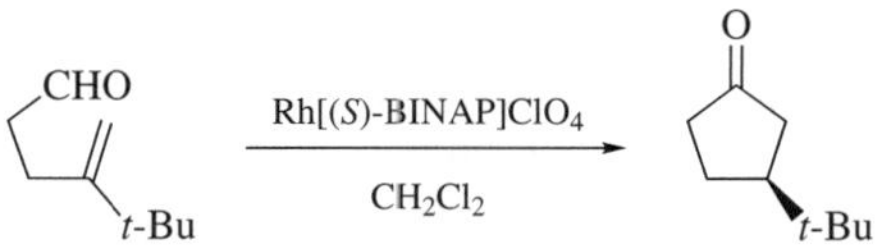

(18)

>99% ee

Reaction of styrene and catecholborane in the presence of a BINAP–Rh complex at low temperature forms, after oxidative workup, 1-phenylethyl alcohol in 96% ee (eq 19).[26]

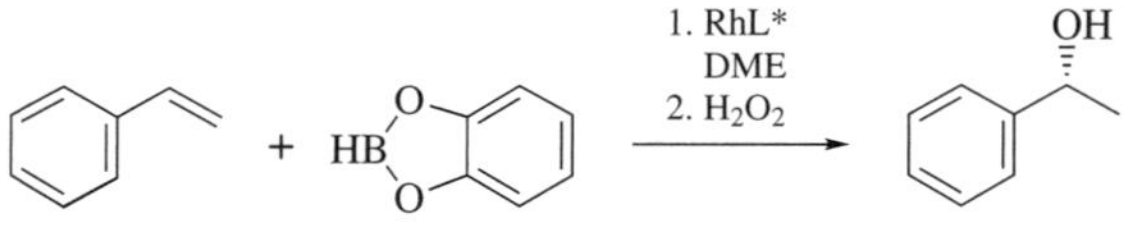

(19)

RhL* = $[Rh(cod)_2]BF_4$ + (*R*)-BINAP

96% ee

Neutral BINAP–Rh complexes catalyze intramolecular hydrosilylation of alkenes. Subsequent *Hydrogen Peroxide* oxidation produces the optically active 1,3-diol in up to 97% ee (eq 20).[27]

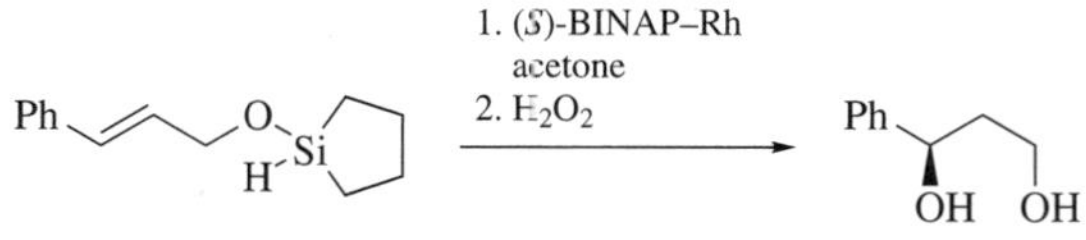

(20)

97% ee

BINAP–Pd Catalyzed Asymmetric Reactions. BINAP–Pd^0^ complexes are prepared in situ from *Bis(dibenzylideneacetone)palladium(0)* or $Pd_2(dba)_3 \cdot CHCl_3$ and BINAP.[28] BINAP–Pd^II^ complexes are formed from *Bis(allyl)di-μ-chlorodipalladium*, *Palladium(II) Acetate*, or $PdCl_2(MeCN)_2$ and BINAP.[29–31]

A BINAP–Pd complex brings about enantioselective 1,4-disilylation of α,β-unsaturated ketones with chlorinated disilanes, giving enol silyl ethers in 74–92% ee (eq 21).[29]

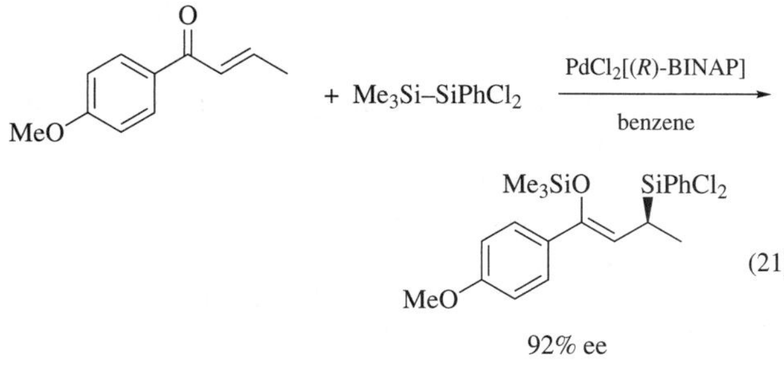

(21)

A BINAP–Pd^{II} complex catalyzes a highly enantioselective C–C bond formation between an aryl triflate and 2,3-dihydrofuran (eq 22).[30] The intramolecular version of the reaction using an alkenyl iodide in the presence of $PdCl_2$[(*R*)-BINAP] and *Silver(I) Phosphate* allows enantioselective formation of a bicyclic ring system (eq 23).[31]

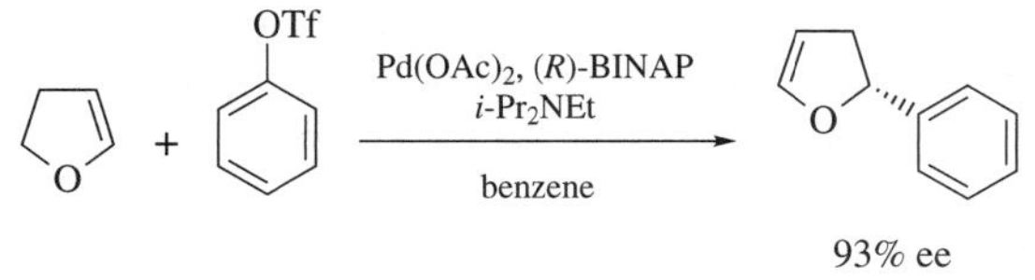

(22)

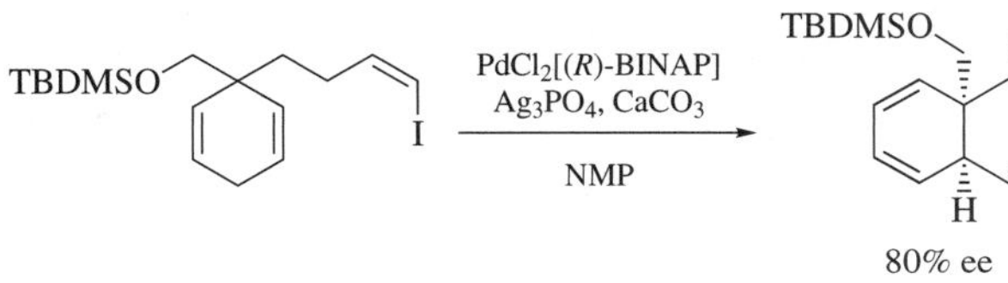

(23)

Enantioselective electrophilic allylation of 2-acetamidomalonate esters is effected by a BINAP–Pd^0 complex (eq 24).[32]

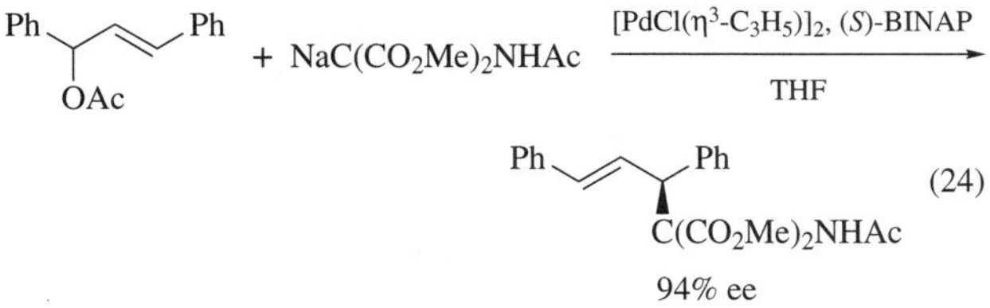

(24)

A BINAP–Pd^0 complex catalyzes hydrocyanation of norbornene to the *exo* nitrile with up to 40% ee.[28]

BINAP–Ir^I Catalyzed Asymmetric Reactions. [Ir(BINAP)(cod)]BF_4 is prepared from [Ir(cod)$(MeCN)_2$]BF_4 and BINAP in THF.[33]

A combined system of the BINAP–Ir complex and bis(*o*-dimethylaminophenyl)phenylphosphine or (*o*-dimethylaminophenyl)diphenylphosphine catalyzes hydrogenation of benzylideneacetone[33a] and cyclic aromatic ketones[33b] with modest to high enantioselectivities (eq 25).

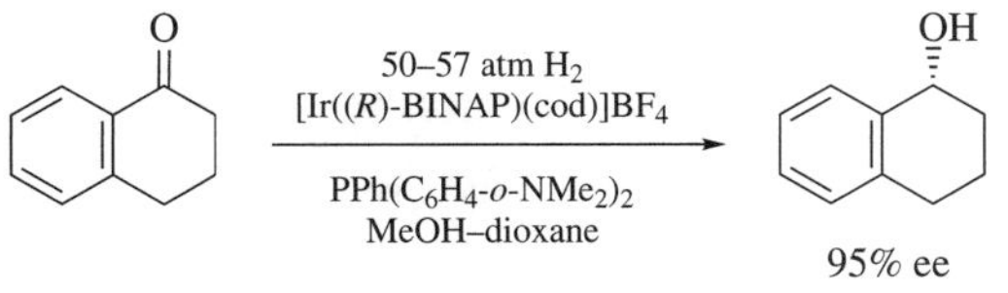

(25)

1. (a) Miyashita, A.; Yasuda, A.; Takaya, H.; Toriumi, K.; Ito, T.; Souchi, T.; Noyori, R. *J. Am. Chem. Soc.* **1980**, *102*, 7932. (b) Noyori, R.; Takaya, H. *Acta Chem. Scand.* **1985**, *25*, 83.
2. (a) Noyori, R.; Kitamura, M. In *Modern Synthetic Methods*; Scheffold, R., Ed.; Springer: Berlin, 1989; p 115. (b) Noyori, R. *Science* **1990**, *248*, 1194. (c) Noyori, R.; Takaya, H. *Acc. Chem. Res.* **1990**, *23*, 345. (d) Noyori, R. *Chemtech* **1992**, *22*, 360.
3. Takaya, H.; Akutagawa, S.; Noyori, R. *Comprehensive Organic Synthesis* **1988**, *67*, 20.
4. (a) Ikariya, T.; Ishii, Y.; Kawano, H.; Arai, T.; Saburi, M.; Yoshikawa, S.; Akutagawa, S. *Chem. Commun./J. Chem. Soc., Chem. Commun.* **1985**, 922. (b) Ohta, T.; Takaya, H.; Noyori, R. *Inorg. Chem.* **1988**, *27*, 566. (c) Kitamura, M.; Tokunaga, M.; Ohkuma, T.; Noyori, R. *Tetrahedron Lett.* **1991**, *32*, 4163. (d) Kitamura, M.; Tokunaga, M.; Ohkuma, T.; Noyori, R. *Comprehensive Organic Synthesis* **1992**, *71*, 1.
5. Kitamura, M.; Tokunaga, M.; Noyori, R. *J. Org. Chem.* **1992**, *57*, 4053.
6. (a) Ohta, T.; Takaya, H.; Kitamura, M.; Nagai, K.; Noyori, R. *J. Org. Chem.* **1987**, *52*, 3174. (b) Saburi, M.; Takeuchi, H.; Ogasawara, M.; Tsukahara, T.; Ishii, Y.; Ikariya, T.; Takahashi, T.; Uchida, Y. *J. Organomet. Chem.* **1992**, *428*, 155.
7. Lubell, W. D.; Kitamura, M.; Noyori, R. *Tetrahedron: Asymmetry* **1991**, *2*, 543.
8. (a) Noyori, R.; Ohta, M.; Hsiao, Y.; Kitamura, M.; Ohta, T.; Takaya, H. *J. Am. Chem. Soc.* **1986**, *108*, 7117. (b) Kitamura, M.; Hsiao, Y.; Noyori, R.; Takaya, H. *Tetrahedron Lett.* **1987**, *28*, 4829.
9. (a) Takaya, H.; Ohta, T.; Sayo, N.; Kumobayashi, H.; Akutagawa, S.; Inoue, S.; Kasahara, I.; Noyori, R. *J. Am. Chem. Soc.* **1987**, *109*, 1596, 4129. (b) Takaya, H.; Ohta, T.; Inoue, S.; Tokunaga, M.; Kitamura, M.; Noyori, R. *Comprehensive Organic Synthesis* **1994**, *72*, 74.
10. Kitamura, M.; Nagai, K.; Hsiao, Y.; Noyori, R. *Tetrahedron Lett.* **1990**, *31*, 549.
11. Kitamura, M.; Kasahara, I.; Manabe, K.; Noyori, R.; Takaya, H. *J. Org. Chem.* **1988**, *53*, 708.
12. Ohta, T.; Miyake, T.; Seido, N.; Kumobayashi, H.; Akutagawa, S.; Takaya, H. *Tetrahedron Lett.* **1992**, *33*, 635.
13. (a) Noyori, R.; Ohkuma, T.; Kitamura, M.; Takaya, H.; Sayo, N.; Kumobayashi, H.; Akutagawa, S. *J. Am. Chem. Soc.* **1987**, *109*, 5856. (b) Kitamura, M.; Ohkuma, T.; Inoue, S.; Sayo, N.; Kumobayashi, H.; Akutagawa, S.; Ohta, T.; Takaya, H.; Noyori, R. *J. Am. Chem. Soc.* **1988**, *110*, 629. (c) Kitamura, M.; Ohkuma, T.; Takaya, H.; Noyori, R. *Tetrahedron Lett.* **1988**, *29*, 1555. (d) Kawano, H.; Ishii, Y.; Saburi, M.; Uchida, Y. *Chem. Commun./J. Chem. Soc., Chem. Commun.* **1988**, 87. (e) Ohkuma, T.; Kitamura, M.; Noyori, R. *Tetrahedron Lett.* **1990**, *31*, 5509.
14. Nishi, T.; Kitamura, M.; Ohkuma, T.; Noyori, R. *Tetrahedron Lett.* **1988**, *29*, 6327.
15. (a) Kitamura, M.; Tokunaga, M.; Noyori, R. *J. Am. Chem. Soc.* **1993**, *115*, 144. (b) Kitamura, M.; Tokunaga, M.; Noyori, R. *Tetrahedron* **1993**, *49*, 1853.
16. (a) Noyori, R.; Ikeda, T.; Ohkuma, T.; Widhalm, M.; Kitamura, M.; Takaya, H.; Akutagawa, S.; Sayo, N.; Saito, T.; Taketomi, T.; Kumobayashi, H. *J. Am. Chem. Soc.* **1989**, *111*, 9134. (b) Genet, J. P.; Pinel, C.; Mallart, S.; Juge, S.; Thorimbert, S.; Laffitte, J. A. *Tetrahedron: Asymmetry* **1991**, *2*, 555. (c) Mashima, K.; Matsumura, Y.; Kusano, K.; Kumobayashi, H.; Sayo, N.; Hori, Y.; Ishizaki, T.; Akutagawa, S.; Takaya, H. *Chem. Commun./J. Chem. Soc., Chem. Commun.* **1991**, 609.
17. Kitamura, M.; Ohkuma, T.; Tokunaga, M.; Noyori, R. *Tetrahedron: Asymmetry* **1990**, *1*, 1.
18. Oppolzer, W.; Wills, M.; Starkemann, C.; Bernardinelli, G. *Tetrahedron Lett.* **1990**, *31*, 4117.
19. (a) Brown, J. M.; Brunner, H.; Leitner, W.; Rose, M. *Tetrahedron: Asymmetry* **1991**, *2*, 331. (b) Saburi, M.; Ohnuki, M.; Ogasawara, M.; Takahashi, T.; Uchida, Y. *Tetrahedron Lett.* **1992**, *33*, 5783.
20. Kameyama, M.; Kamigata, N.; Kobayashi, M. *J. Org. Chem.* **1987**, *52*, 3312.

21. (a) Miyashita, A.; Takaya, H.; Souchi, T.; Noyori, R. *Tetrahedron* **1984**, *40*, 1245. (b) Toriumi, K.; Ito, T.; Takaya, H.; Souchi, T.; Noyori, R. *Acta Crystallogr.* **1982**, *B38*, 807.
22. Evans, D. A.; Morrissey, M. M. *Tetrahedron Lett.* **1984**, *25*, 4637.
23. (a) Tani, K.; Yamagata, T.; Otsuka, S.; Akutagawa, S.; Kumobayashi, H.; Taketomi, T.; Takaya, H.; Miyashita, A.; Noyori, R. *Chem. Commun./J. Chem. Soc., Chem. Commun.* **1982**, 600. (b) Inoue, S.; Takaya, H.; Tani, K.; Otsuka, S.; Sato, T.; Noyori, R. *J. Am. Chem. Soc.* **1990**, *112*, 4897. (c) Yamakawa, M.; Noyori, R. *J. Organomet. Chem.* **1992**, *11*, 3167. (d) Tani, K.; Yamagata, T.; Tatsuno, Y.; Yamagata, Y.; Tomita, K.; Akutagawa, S.; Kumobayashi, H.; Otsuka, S. *Angew. Chem., Int. Ed. Engl.* **1985**, *24*, 217. (e) Otsuka, S.; Tani, K. *Acta Chem. Scand.* **1991**, 665.
24. Kitamura, M.; Manabe, K.; Noyori, R.; Takaya, H. *Tetrahedron Lett.* **1987**, *28*, 4719.
25. Wu, X.-M.; Funakoshi, K.; Sakai, K. *Tetrahedron Lett.* **1992**, *33*, 6331.
26. (a) Hayashi, T.; Matsumoto, Y.; Ito, Y. *J. Am. Chem. Soc.* **1989**, *111*, 3426. (b) Sato, M.; Miyaura, N.; Suzuki, A. *Tetrahedron Lett.* **1990**, *31*, 231. (c) Zhang, J.; Lou, B.; Guo, G.; Dai, L. *J. Org. Chem.* **1991**, *56*, 1670.
27. Tamao, K.; Tohma, T.; Inui, N.; Nakayama, O.; Ito, Y. *Tetrahedron Lett.* **1990**, *31*, 7333.
28. Hodgson, M.; Parker, D. *J. Organomet. Chem.* **1987**, *325*, C27.
29. Hayashi, T.; Matsumoto, Y.; Ito, Y. *J. Am. Chem. Soc.* **1988**, *110*, 5579.
30. Ozawa, F.; Hayashi, T. *J. Organomet. Chem.* **1992**, *428*, 267.
31. (a) Sato, Y.; Sodeoka, M.; Shibasaki, M. *Chem. Lett.* **1990**, 1953; Sato, Y.; Sodeoka, M.; Shibasaki, M. *J. Org. Chem.* **1989**, *54*, 4738. (b) Ashimori, A.; Overman, L. E. *J. Org. Chem.* **1992**, *57*, 4571.
32. Yamaguchi, M.; Shima, T.; Yamagishi, T.; Hida, M. *Tetrahedron Lett.* **1990**, *31*, 5049.
33. (a) Mashima, K.; Akutagawa, T.; Zhang, X.; Takaya, H.; Taketomi, T.; Kumobayashi, H.; Akutagawa, S. *J. Organomet. Chem.* **1992**, *428*, 213. (b) Zhang, X.; Taketomi, T.; Yoshizumi, T.; Kumobayashi, H.; Akutagawa, S.; Mashima, K.; Takaya, H. *J. Am. Chem. Soc.* **1993**, *115*, 3318.

Masato Kitamura & Ryoji Noyori
Nagoya University, Japan

(2*R*,3*R*)-2,3-Bis(diphenylphosphino)-butane[1]

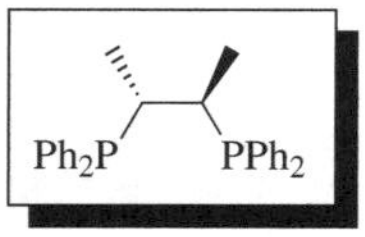

(2*R*,3*R*)
[74839-84-2] $C_{28}H_{28}P_2$ (MW 426.48)
(2*S*,3*S*)
[64896-28-2]

(ligand for asymmetric hydrogenation of alkenes[2] and β-keto esters;[3] allylic alkylation;[4] hydroarylation[5])

Alternate Name: (*R*,*R*)-CHIRAPHOS.
Physical Data: mp 107–109 °C; $[\alpha]_D^{20}$ +195° (c = 1.5, $CHCl_3$).
Form Supplied in: white solid; widely available.
Handling, Storage, and Precautions: is indefinitely stable in air in the solid state. Solutions of CHIRAPHOS are readily oxidized to the phosphine oxide and should be handled under N_2 or Ar.[2]

Asymmetric Hydrogenation. CHIRAPHOS has been employed in the enantioselective hydrogenation of a variety of unsaturated functional groups. The asymmetric hydrogenation of cinnamic acid derivatives has been extensively studied due to its relevance to the commercial synthesis of amino acids, such as L-Dopa.[1] Hydrogenation of (*Z*)-α-benzoylaminocinnamic acid is catalyzed in quantitative yield and 98% ee by a cationic rhodium complex prepared from $[Rh(cod)_2](ClO_4)$ and (*S,S*)-CHIRAPHOS (eq 1).[2] Asymmetric hydrogenations have also been performed using racemic CHIRAPHOS which has been resolved in situ by a substrate-induced kinetic resolution.[6] When racemic CHIRAPHOS (2 equiv) was reacted with complex (**1**) (R = (−)-menthyl), (*S,S*)-CHIRAPHOS selectively coordinated to Ir. This resulting (*S,S*)-CHIRAPHOS–Ir complex is catalytically inactive for alkene hydrogenation under typical conditions. In the presence of this Ir complex the remaining uncoordinated (*R,R*)-CHIRAPHOS is then utilized for rhodium-catalyzed hydrogenation of methyl (*Z*)-α-acetylaminocinnamate (eq 2). The enantiomeric excess using this route is identical to that observed with authentic (*R,R*)-CHIRAPHOS.

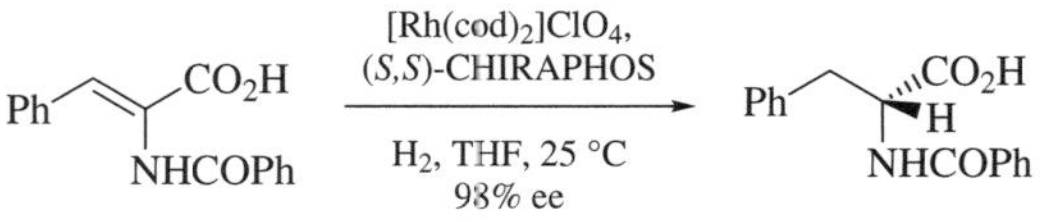

(1)

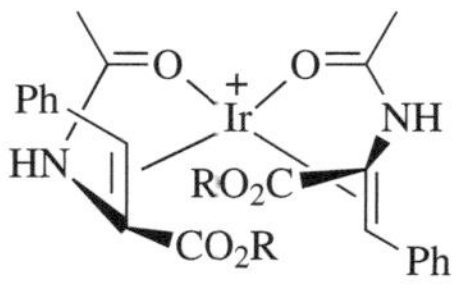

(**1**)

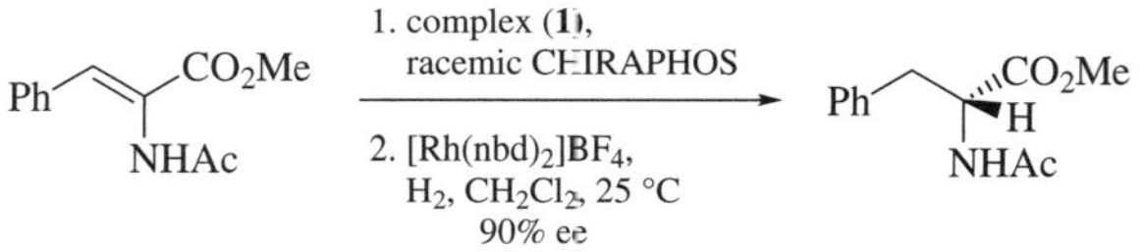

(2)

Rhodium and ruthenium complexes of CHIRAPHOS are also useful for the asymmetric hydrogenation of β-keto esters. Dynamic kinetic resolution of racemic 2-acylamino-3-oxobutyrates was performed by hydrogenation using ((*S,S*)-CHIRAPHOS)$RuBr_2$ (eq 3).[3] The product yields and enantiomeric excesses were dependent upon solvent, ligand, and the ratio of substrate to catalyst. Under optimum conditions a 97:3 mixture of *syn* and *anti* β-hydroxy esters was formed, which was converted to D-threonine (85% ee) and D-allothreonine (99% ee) by hydrolysis and reaction with propylene oxide.

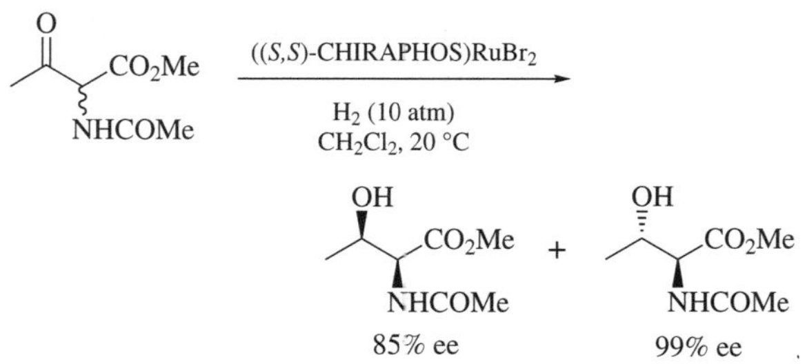

(3)

Allylic Alkylation. The palladium-catalyzed asymmetric alkylation of 1,3-diphenyl-2-propenyl acetate with dimethyl sodiomalonate produces (**2**) in 86% yield and 90% ee (eq 4).[4] CHIRAPHOS was found to give higher enantioselectivity than both *(R)- & (S)-2,2′-Bis(diphenylphosphino)-1,1′-binaphthyl* (BINAP) and *(+)-trans-(2S,3S)-Bis(diphenylphosphino)bicyclo[2.2.1]hept-5-ene* (NORPHOS). Alkylation using other malonic acid derivatives gave similar optical yields. The product enantiomeric excess was reported to be greatly dependent upon the method of catalyst preparation.

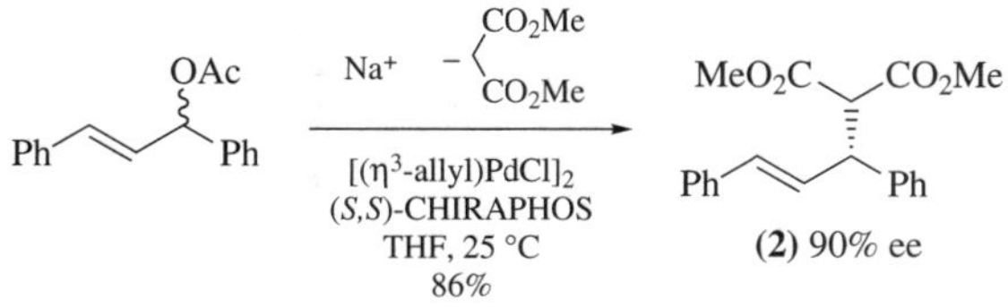

(4)

Alkene Hydroarylation. The enantioselective addition of aryl iodides to norbornene has been reported using a palladium(II) complex of (*S,S*)-CHIRAPHOS. The reaction of norbornadiene with 4-methoxyiodobenzene proceeded with 30% ee (eq 5).[5] Enantioselectivities were dependent upon phosphine structure (see *(+)-trans-(2S,3S)-Bis(diphenylphosphino)bicyclo[2.2.1]hept-5-ene*).

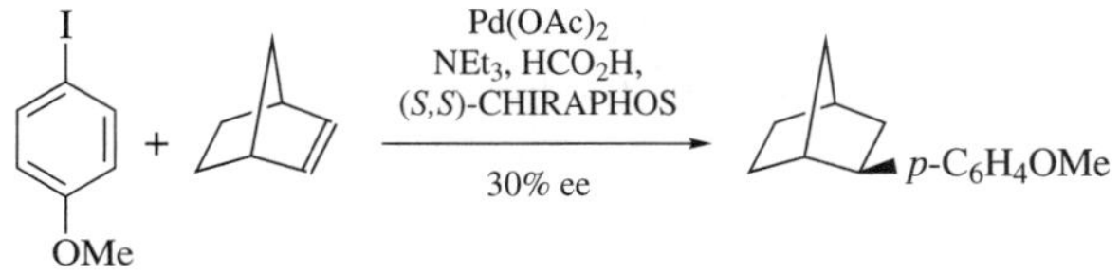

(5)

1. Ojima, I.; Clos, N.; Bastos, C. *Tetrahedron* **1989**, *45*, 6901.
2. Fryzuk, M. D.; Bosnich, B. *J. Am. Chem. Soc.* **1977**, *99*, 6262.
3. Genet, J. P.; Pinel, C.; Mallart, S.; Juge, S.; Thorimbert, S.; Laffitte, J. A. *Tetrahedron: Asymmetry* **1991**, *2*, 555.
4. Yamaguchi, M.; Shima, T.; Yamagishi, T.; Hida, M. *Tetrahedron: Asymmetry* **1991**, *2*, 663.
5. Brunner, H.; Kramler, K. *Acta Chem. Scand.* **1991**, *12*, 1121.
6. Alcock, N. W.; Brown, J. M.; Maddox, P. J. *Chem. Commun./J. Chem. Soc., Chem. Commun.* **1986**, 1532.

Gregory T. Whiteker
Union Carbide Corporation, South Charleston, WV, USA

(*R*)-7,7′-Bis(diphenylphosphinomethyl)-2,2′-dimethoxy-1,1′-binaphthyl

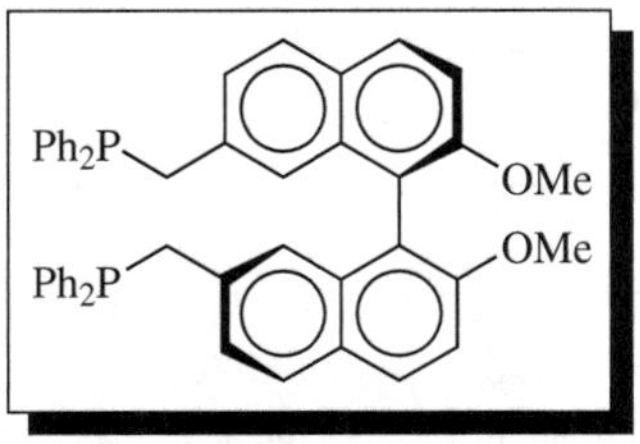

[188563-54-4] $C_{48}H_{40}O_2P_2$ (MW 710.78)

(chiral and C_2 symmetric bisphosphine ligand having a large natural bite angle)

Physical Data: mp 146–148 °C; $[\alpha]_D^{26} = -205.2$ (c 1.0, $CHCl_3$); ^{31}P NMR δ −10.63.

Solubility: soluble in organic solvents such as benzene, toluene, and methanol.

Form Supplied in: not commercially available.

Preparative Methods: [1] the title reagent **1** is prepared from (*R*)-7,7′-bis(triflyloxy)-2,2′-dimethoxy-1,1′-binaphthyl (**2**) in four steps. Treatment of **2** with carbon monoxide (4 atm) and methanol in the presence of palladium(II) acetate gives a 7,7′-bis(methoxycarbonyl) derivative, which is reduced to diol, converted to dibromide, and then reacted with lithium diphenylphosphide to finally give **1** (eq 1).

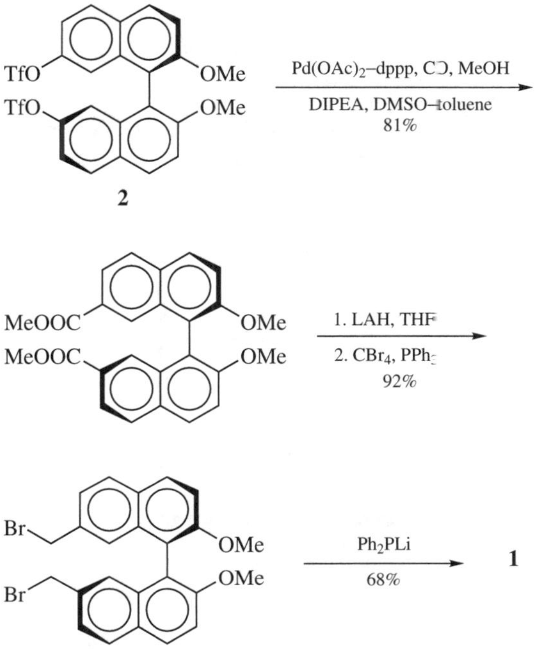

(1)

Asymmetric Michael Addition of α-Cyano Esters.[1] Due to a large natural bite angle, **1** works as a trans-chelating bisphosphine ligand (TRAP) similar to 2,2″-bis[1-(diphenylphosphino)ethyl]-1,1″-biferrocene.[2] The rhodium catalyst generated by mixing **1** and $Rh(acac)(CO)_2$ promotes Michael addition of methyl 2-cyanopropanoate to methyl vinyl ketone (eq 2). With 1 mol % of the catalyst, the reaction at 0 °C is completed in 13 h to give an (*R*)-adduct of 73% ee in 86% yield. The substrates amenable to this reaction are rather limited. When the nucleophile is methyl 2-cyanobutanoate, the product ee is decreased to 14%. With methyl acrylate as the electrophile, addition does not take place at all. In contrast to the reaction with the ferrocene derivative, use of isopropyl 2-cyanopropanoate does not improve the product ee.

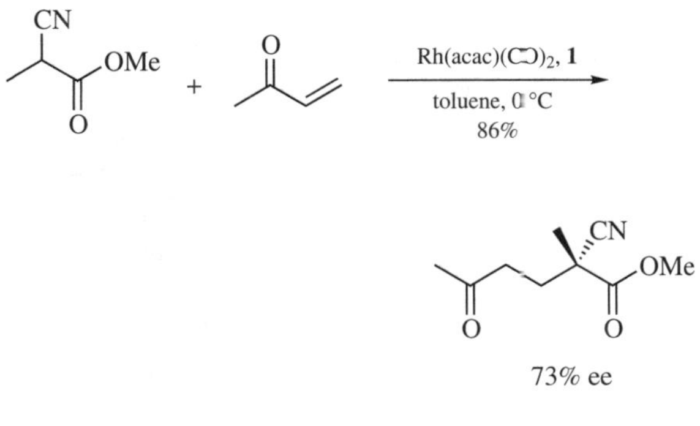

(2)

Other Asymmetric Reactions. Asymmetric synthesis using the new ligand **1** is still limited. When **1** is used for Pd-clay catalyzed hydroesterification of styrene with carbon monoxide and methanol, a chiral methyl 2-phenylpropanoate is obtained in 12% ee at low conversion.[3]

1. Inagaki, K.; Nozaki, K.; Takaya, H. *Synlett* **1997**, 199.
2. (a) Sawamura, M.; Hamashima, H.; Ito, Y. *J. Am. Chem. Soc.* **1992**, *114*, 8295. (b) Sawamura, M.; Hamashima, H.; Ito, Y. *Tetrahedron* **1994**, *50*, 4439. (c) Sawamura, M.; Hamashima, H.; Shinoto, H.; Ito, Y. *Tetrahedron Lett.* **1995**, *36*, 6479.
3. Nozaki, K.; Kantam, M. L.; Horiuchi, T.; Takaya, H. *J. Mol. Catal. A: Chem.* **1997**, *118*, 247.

Takashi Sugimura
Himeji Institute of Technology, Hyogo, Japan

(η^5,η^5-1*S*,2*R*,4*S*,5*R*-1,4-Bis(indenyl)-2,5-diisopropylcyclohexane)titanium Dichloride

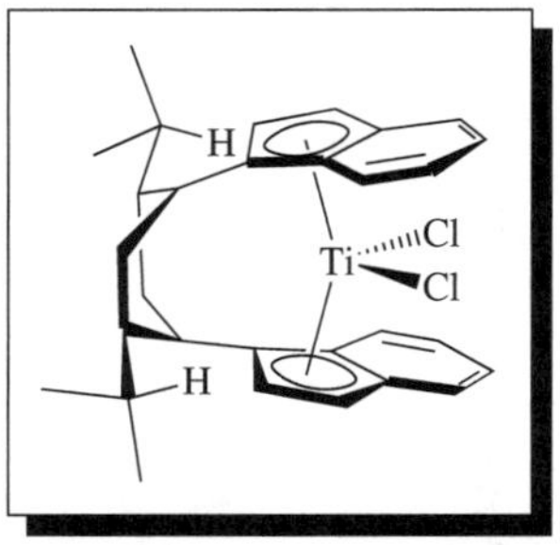

[139561-09-4] $C_{30}H_{34}Cl_2Ti$ (MW 513.38)

(alkene isomerization; titanium;[4] *ansa*-bis(indenyl) ligand; chiral ligand; C_2 symmetric ligand; asymmetric catalysis[8–10])

Physical Data: mp 264–265 °C (dec 240 °C); *d* 1.339 g cm^{-3}. Dark green crystals from hexane–CH_2Cl_2. $[\alpha]_D^{23} = +2900°$ (*c* 0.0202, CH_2Cl_2).

Solubility: sol $CHCl_3$ and CH_2Cl_2; insol hexane.

Preparative Methods: [1] the enantiomerically pure, chiral ligand (**2**) is prepared in two steps from the known (1*S*,2*R*,4*S*,5*R*)-2,5-diisopropylcyclohexane-1,4-diol (**3**) (eq 1).[2,3] It is obtained in 60% yield as a mixture of alkene regioisomers. The chiral titanocene reagent (**1**) is prepared by treatment of (**2**) with 2 equiv of *n-Butyllithium* followed by reaction with *Titanium(III) Chloride* and subsequent oxidation (HCl/air in $CHCl_3$) of the product (eq 2). It is obtained in 80% yield as a single stereoisomer.

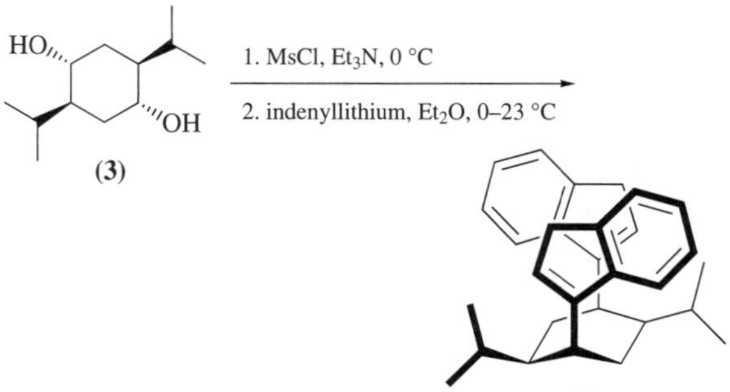

(**3**) (**2**) (1)

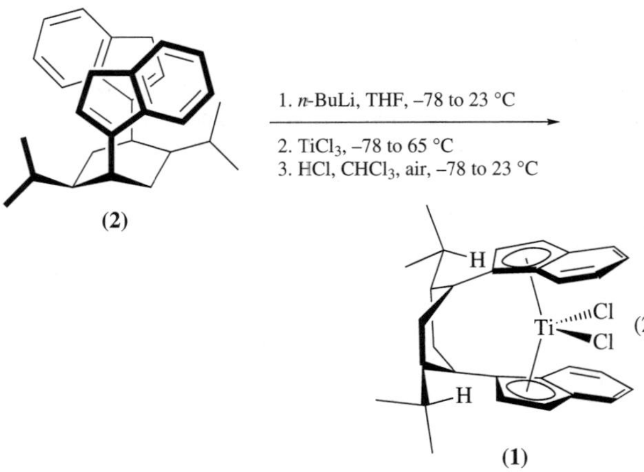

(**2**) (**1**) (2)

Asymmetric Alkene Isomerization.[1] The chiral titanocene reagent (**1**) serves as precatalyst for the isomerization of alkene (**4**) (eq 3). Active isomerization catalyst is obtained by in situ reduction of (**1**) with *Lithium Aluminum Hydride* (164 °C, 30 min). Treatment of the achiral substrate (**4**) with 2 mol % catalyst produced axially dissymmetric product (*S*)-(**5**) in 44–76% ee (100% yield). The reaction is slow at room temperature (120 h required for complete reaction); faster rates are obtained at higher temperatures, but at the expense of lower product enantiomeric purity.

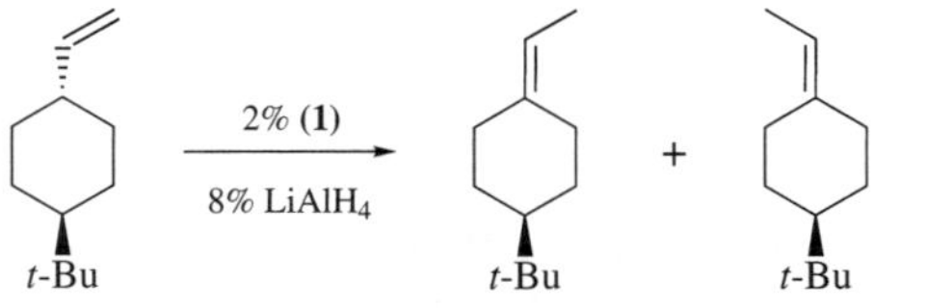

(**4**) (*S*)-(**5**) major (*R*)-(**5**) minor (3)

Other Chiral Cyclopentadienylmetal Complexes. The chemistry of chiral cyclopentadienylmetal complexes is covered in a review by Halterman.[4] Brintzinger's 1,2-ethylenebis(1-indenyl) ligand is the one most commonly used in the preparation of chiral early metal metallocene catalysts.[5,6] Compared to Brintzinger's ligand, (**2**) has the advantage of producing chiral metallocene complexes as single stereoisomers. A related chiral bis(1-indenyl) ligand developed by Burk and Halterman, possessing the same advantages as (**2**), incorporates a binaphthyl unit as the chiral, enantiomerically pure bridging group.[7]

Other Reactions of Chiral Titanocene Derivatives. Buchwald has recently reported the catalytic asymmetric hydrogenation of imines[8,9] and unfunctionalized alkenes[10] using chiral titanocene catalysts.

1. Chen, Z. L.; Halterman, R. L. *J. Am. Chem. Soc.* **1992**, *114*, 2276.
2. Chen, Z.; Halterman, R. L. *Synlett* **1990**, 103.
3. Chen, Z.; Halterman, R. L. *J. Organomet. Chem.* **1991**, *10*, 3449.
4. Halterman, R. L. *Chem. Rev.* **1992**, *92*, 965.
5. Wild, F. R. W. P.; Wasiucionek, M.; Brintzinger, H. H. *J. Organomet. Chem.* **1982**, *288*, 63.
6. Wild, F. R. W. P.; Zsolnai, L.; Huttner, G.; Brintzinger, H. H. *J. Organomet. Chem.* **1982**, *232*, 233.

7. Burk, M. J.; Colletti, S. L.; Halterman, R. L. *J. Organomet. Chem.* **1991**, *10*, 2998.
8. Willoughby, C. A.; Buchwald, S. L. *J. Am. Chem. Soc.* **1992**, *114*, 7562.
9. Willoughby, C. A.; Buchwald, S. L. *J. Org. Chem.* **1993**, *58*, 7627.
10. Broene, R. D.; Buchwald, S. L. *J. Am. Chem. Soc.* **1993**, *115*, 12569.

William E. Crowe
Emory University, Atlanta, GA, USA

2,6-Bis[(4*S*)-4-isopropyloxazolin-2-yl]pyridine

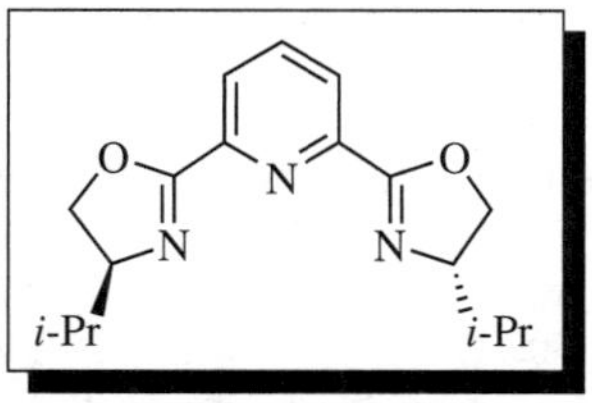

[118949-61-4] $C_{17}H_{23}N_3O_2$ (MW 301.38)

(reagent used as a ligand for various asymmetric metal-catalyzed reactions)

Physical Data: crystalline solid, mp 152–153 °C; $[\alpha]_D^{26}$ −116.8 (c 1.01, CH_2Cl_2).
Solubility: soluble in most organic solvents.
Form Supplied in: solid; commercially available.
Handling, Storage, and Precautions: toxicity unknown; as with all organic chemicals, use only in a well-ventilated fume hood is recommended.

Iso-propyl-substituted pyridinyl bisoxazoline (**1**) has been used exclusively as a ligand for metals, primarily late transition metals and lanthanides. The resulting complexes are effective in a variety of enantioselective transformations.

Reductive Transformations. The utility of **1** was first demonstrated in the enantioselective hydrosilylation of ketones. Uniformly high enantioselectivity, yield, and turnover were observed for aromatic (and some aliphatic) ketones when using the complex derived from $RhCl_3$ (eq 1).[1] Lower enantioselection is observed with *t*-Bu-pybox or *i*-Pr-pybox·cobalt(I).[2] The derived **1**·$Sn(OTf)_2$ complex gives alcohol products with up to 58% ee using methanolic polymethylhydrosiloxane.[3] A cationic ruthenium(III) catalyst diverts the usual reduction pathway to enolsilane formation, particularly when the nature of the silane is modified (eq 2).[4]

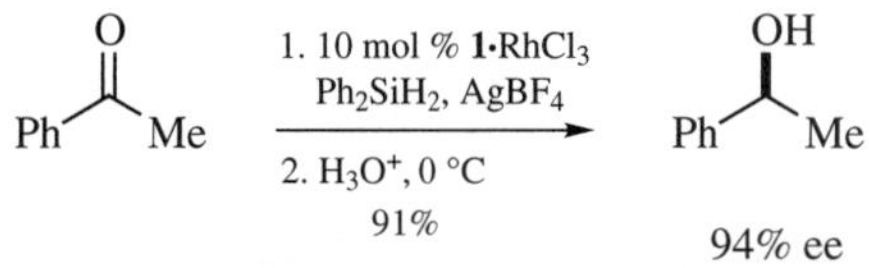

(1)

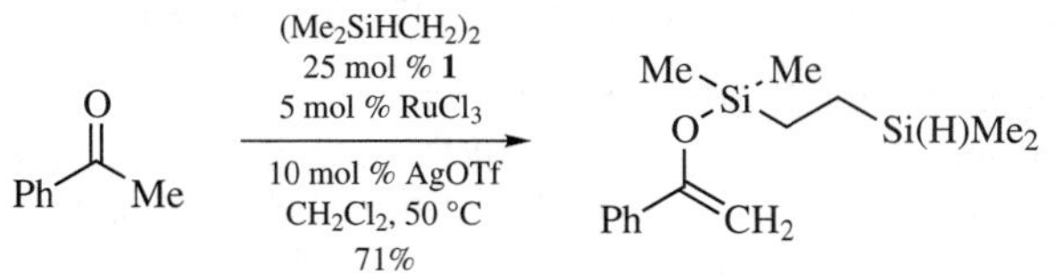

(2)

Oxidative Transformations. The use of **1** in enantioselective oxidations remains limited at the present time. Among the promising developments, allylic perester oxidation proceeds with significant enantioselection.[5] The copper(I)-catalyzed oxidation of cyclohexene furnished the protected cyclic allylic alcohol with modest enantioselection (eq 3).[6]

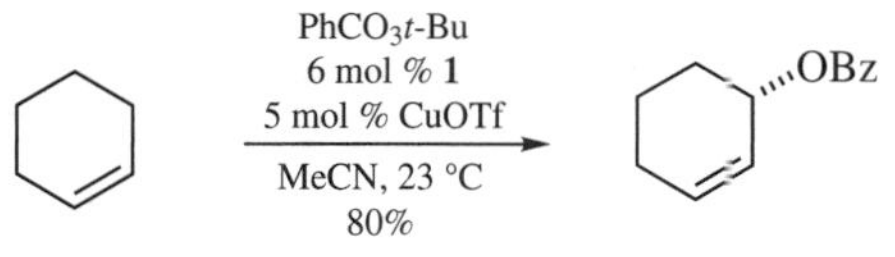

(3)

7: % ee

Epoxidation of simple olefins can be effected using a ruthenium catalyst employing a mixed ligand system (eq 4).[7] Using this method, epoxystilbene was generated in 74% ee. Bis(acetoxy)iodobenzene is used as the oxidant in the enantioselective epoxidation of *trans*-stilbene. Both homogeneous and heterogeneous aziridination proceed with low levels of enantioselection using **1**.[8]

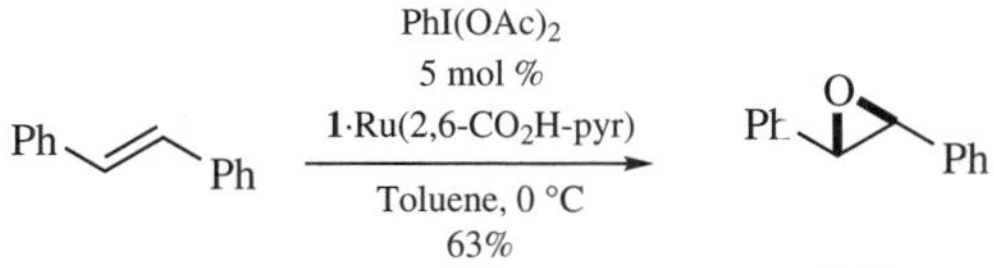

(4)

74% ee

Carbon–Carbon Bond Forming Reactions. The effectiveness of **1** in asymmetric transformations is most pronounced in those resulting in the formation of carbon–carbon bonds. Highly enantioselective aldol addition of enolsilanes to benzyloxyacetaldehyde and 1,2-diketones are possible (eq 5).[9] Use of less nucleophilic olefins is less effective as evidenced by low levels of enantioselection in the glyoxylate ene reaction.[10] Organometal addition to aldehydes using diethylzinc[11] and allylindium reagents are moderately effective, with the latter providing homollylic alcohols in 92% ee with stoichiometric cerium(III) triflate hydrate.[12] 1,2-Addition of phenyllithium or phenylmagnesium bromide to a discrete *i*-Pr-pybox ruthenium(III)–acrolein complex furnished the allylic alcohol in 63–87% ee.[13] Cyanohydrin synthesis with metal complexes of **1** can be effective. Although the aluminum(III) complex of **1** provides the silylated cyanohydrin in moderate ee, recent studies using lanthanides offer a slight improvement (to 89% ee).[14]

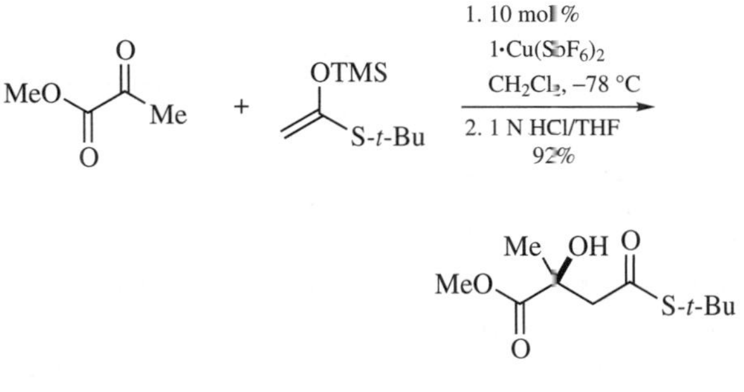

(5)

95% ee

Alkylation reactions of ketones and esters using **1** have been reported with good enantioselection. Free radical-mediated alkylation of a γ-lactam proceeded with good enantioselection (eq 6).[15] Malonate alkylation provides the 1,3-diphenyl allylation product with 86% ee (45% yield) through the intermediacy of the **1**·Pd(0) complex.[16]

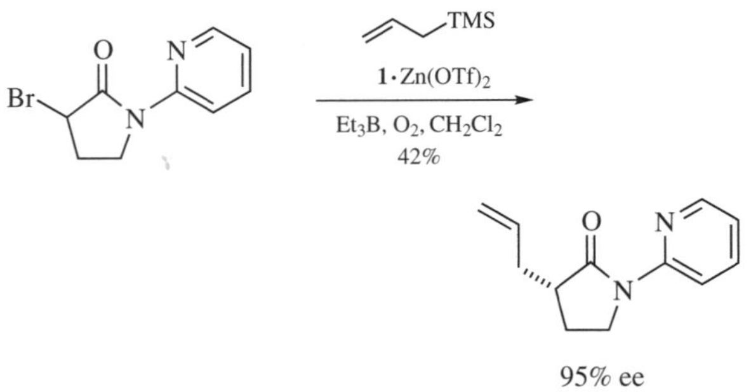

(6)

95% ee

Numerous highly enantioselective ring-forming reactions have also been discovered with the assistance of **1**. Cyclopropanation with the rhodium complex of **1** furnishes *trans*-cyclopropanes selectively (eq 7).[17] A discrete ruthenium vinyl carbene was similarly successful in the stoichiometric cyclopropanation,[18] whereas enantioselection in the copper(I)-catalyzed variant was nonselective.[19]

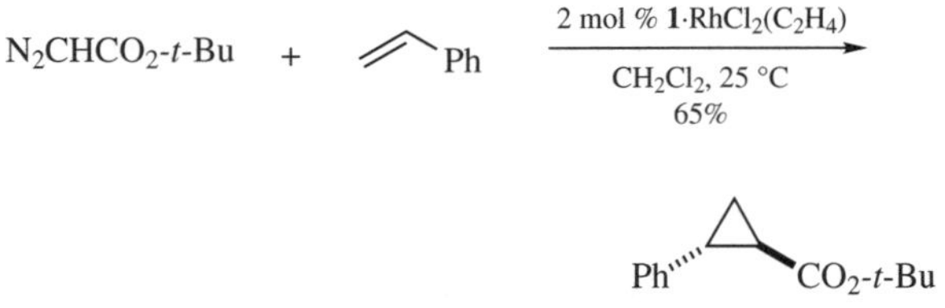

(7)

97:3 dr
94% ee

Both Diels–Alder and hetero-Diels–Alder reactions can be rendered stereoselective using **1**·copper(II) salts, but inferior levels of stereoselection were observed relative to other pybox derivatives.[20] Lanthanide-catalyzed 1,3-dipolar cycloaddition also exhibited moderate (61%) enantioselection.[21]

Lanthanide catalysis was again effective in ring-opening reactions of cyclic epoxides (eq 8). Finally, MAO-activated **1**·$RuCl_3$ provides block copolymers of ethylene and hex-1-ene.[22]

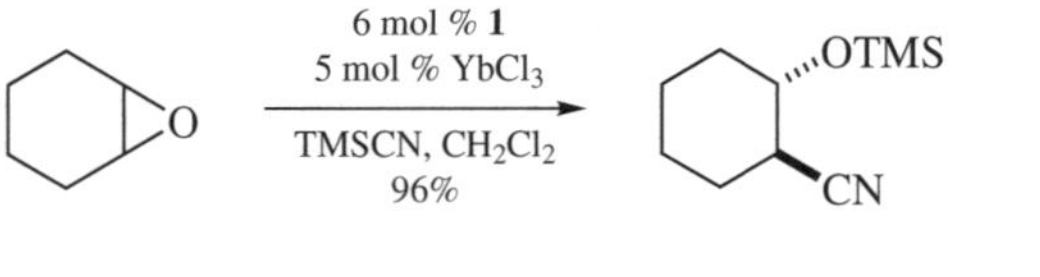

(8)

47% ee

1. (a) Nishiyama, H.; Sakaguchi, H.; Nakamura, T.; Horihata, M.; Kondo, M.; Itoh, K. *Organometallics* **1989**, *8*, 846. (b) Nishiyama, H.; Kondo, M.; Nakamura, T.; Itoh, K. *Organometallics* **1991**, *10*, 500.
2. Brunner, H.; Amberger, K. *J. Organometallic Chem.* **1991**, *417*, C63.
3. Lawrence, N. J.; Bushell, S. M. *Tetrahedron Lett.* **2000**, *41*, 4507.
4. Nagashima, H.; Ueda, T.; Nishiyama, H.; Itoh, K. *Chem. Lett.* **1993**, 347.
5. Schulz, M.; Kluge, R.; Gelalcha, F. G. *Tetrahedron: Asymm.* **1998**, *9*, 4341.
6. Gokhale, A. S.; Minidis, A. B. E.; Pfaltz, A. *Tetrahedron Lett.* **1995**, *36*, 1831.
7. Nishiyama, H.; Shimada, T.; Itoh, H.; Sugiyama, H.; Motoyama, Y., *Chem. Commun.* **1997**, 1863.
8. (a) Rasmussen, K. G.; Jorgensen, K. A. *J. Chem. Soc., Perkin Trans. I* **1997**, 1287. (b) Langham, C.; Piaggio, P.; Bethell, D.; Lee, D. F.; McMorn, P.; Bulman Page, P. C.; Witlock, D. J.; Sly, C.; Hancock, F. E.; King, F.; Hutchings, G. *J. Chem. Commun.* **1998**, 1601.
9. (a) Evans, D. A.; Kozlowski, M. C.; Murry, J. A.; Burgey, C. S.; Campos, K. R.; Connell, B. T.; Staples, R. J. *J. Am. Chem. Soc.* **1999**, *121*, 669. (b) Evans, D. A.; Burgey, C. S.; Kozlowski, M. C.; Tregay, S. W. *J. Am. Chem. Soc.* **1999**, *121*, 686.
10. Qian, C.; Wang, L. *Tetrahedron: Asymm.* **2000**, *10*, 2347.
11. Ding, K.; Ishii, A.; Mikami, K. *Angew. Chem., Int. Ed. Engl.* **1999**, *38*, 497.
12. Loh, T. P.; Zhou, J. R. *Tetrahedron Lett.* **1999**, *40*, 9115.
13. Motoyama, Y.; Kurihara, O.; Murata, K.; Aoki, K.; Nishiyama, H. *Organometallics* **2000**, *19*, 1025.
14. (a) Iovel, I.; Popelis, Y.; Fleisher, M.; Lukevics, E. *Tetrahedron: Asymm.* **1997**, *8*, 1279. (b) Aspinall, H. C.; Greeves, N.; Smith, P. M. *Tetrahedron Lett.* **1999**, *40*, 1763.
15. Porter, N. A.; Feng, H.; Kavrakova, I. K. *Tetrahedron Lett.* **1999**, *40*, 6713.
16. Chelucci, G.; Deriu, S.; Pinna, G. A.; Saba, A.; Valenti, R. *Tetrahedron Lett.* **1999**, *10*, 3803.
17. (a) Nishiyama, H.; Itoh, Y.; Matsumoto, H.; Park, S. B.; Itoh, K. *J. Am. Chem. Soc.* **1994**, *116*, 2223. (b) Nishiyama, H.; Itoh, Y.; Sugawara, Y.; Matsumoto, H.; Aoki, K.; Itoh, K. *Bull. Chem. Soc. Jpn.* **1995**, *68*, 1247.
18. Nishiyama, H.; Park, S. B.; Itoh, K. *Chem. Lett.* **1995**, 599.
19. Muller, P.; Bolea, C. *Synlett* **2000**, *6*, 826.
20. (a) Evans, D. A.; Barnes, D. M.; Johnson, J. S.; Lectka, T.; von Matt, P.; Miller, S. J.; Murry, J. A.; Norcross, R. D.; Shaughnessy, E. A.; Campos, K. R. *J. Am. Chem. Soc.* **1999**, *121*, 7582. (b) Qian, C.; Wang, L. *Tetrahedron Lett.* **2000**, *41*, 2203.
21. Sanchez-Blanko, A. I.; Gothelf, K. V.; Jorgensen, K. A. *Tetrahedron Lett.* **1997**, *38*, 7923.
22. Nomura, K.; Sikokmai, W.; Imanishi, Y. *Bull. Chem. Soc. Jpn.* **2000**, *73*, 599.

Jeffrey N. Johnston
Indiana University, Bloomington, IN, USA

2,6-Bis[(*S*)-4′-isopropyloxazolin-2′-yl]-(pyridine)rhodium Trichloride[1]

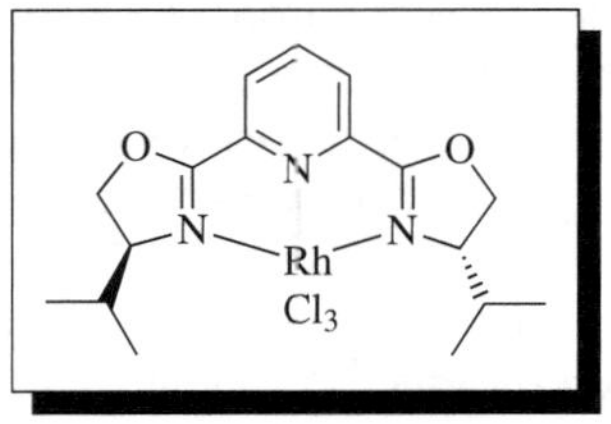

([Pybox-(*S,S*)-ip]$RhCl_3$)		
[119038-51-6]	$C_{17}H_{23}Cl_3N_3O_2Rh$	(MW 510.65)
(Pybox-(*S,S*)-ip)		
[118949-61-4]	$C_{17}H_{23}N_3O_2$	(MW 301.39)

(highly enantioselective catalyst for hydrosilylative reduction of methyl ketones in combination with Ph_2SiH_2 and $AgBF_4$[5,6])

Alternate Name: [Pybox-(*S,S*)-ip]$RhCl_3$.

Physical Data: [Pybox-(*S,S*)-ip]$RhCl_3$: mp 279 °C (dec); $[\alpha]_D^{20} = +543°$ (*c* 1.09, CH_2Cl_2); Pybox-(*S,S*)-ip: white solid, mp 152–153 °C, $[\alpha]_D^{26} = -116.8°$ (*c* 1.0, CH_2Cl_2).[6]

Solubility: sol CH_2Cl_2, $CHCl_3$, and alcohols; slightly sol THF and ethyl acetate; insol diethyl ether, benzene, and H_2O.

Form Supplied in: orange solid; synthesized with $RhCl_3(H_2O)_3$ and Pybox-(*S,S*)-ip in ethanol at reflux for 3 h. The crude solid was purified by silica gel column chromatography with ethyl acetate and methanol as eluents; TLC $R_f = 0.45$ (ethyl acetate/methanol = 5:1, Merck Art 5715).

Drying: although the rhodium compound is nonhygroscopic, it contains some quantity of water or solvents after the purification by chromatography. It should be dried under vacuum (< ca. 1 Torr) for 1 day at 20–60 °C before use.

Handling, Storage, and Precautions: stable in air and moisture.

Enantioselective Reduction of Ketones. Hydrosilylation of ketones with chiral metal catalysts and hydrosilanes followed by hydrolysis provides optically active secondary alcohols from ketones. Most of the enantioselective hydrosilylations of ketones have been carried out with rhodium catalysts and chiral phosphine ligands. These systems give a middle range of enantioselectivities, especially with diphenylsilane and 1-naphthylsilane.[2] In the 1980s, many splendid results of more than 90% ee were reported with rhodium catalysts of chiral nitrogen-containing ligands such as pyridinethiazolidine[3] and pyridineoxazoline-*t*-Bu,[4,5] which are easily accessible from readily available, optically active amino acids and amino alcohols.

The 2,6-bis(oxazolinyl)pyridine ligand [Pybox-(*S,S*)-ip] was also developed as a chiral adjuvant for the asymmetric hydrosilylation of ketones. The Pybox ligand can be synthesized in large scale as a white crystalline solid by condensation of (*S*)-valinol and pyridine-2,6-dicarboxylic acid.[6] Heating of the ethanol solution of Pybox-(*S,S*)-ip and *Rhodium(III) Chloride* trihydrate gave the stable complex [Pybox-(*S,S*)-ip]$RhCl_3$ in 70% yield. Diphenylsilane and the catalytic amount of the Pybox–rhodium(III) complex with the aid of *Silver(I) Tetrafluoroborate* reduced aromatic and aliphatic methyl ketones in THF solution at −5 to 25 °C. After hydrolysis of the product silyl ethers in acidic methanol at 0 °C, optically active secondary alcohols were obtained in high yields and high enantioselectivities, e.g. from acetophenone to 1-phenylethanol in 91% yield and 94% ee (*S*) (eq 1; Table 1).[6]

Table 1 Hydrosilylative Reduction of Methyl Ketones with [Pybox-(*S,S*)-ip]$RhCl_3$ and Diphenylsilane

Ketones	Product (85–93% yield)	ee (%)
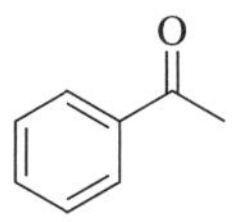	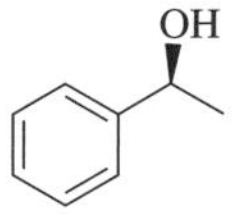	94
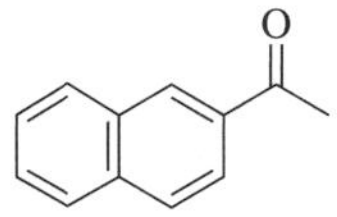	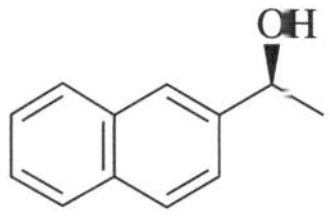	93
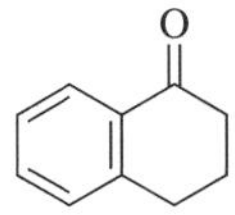	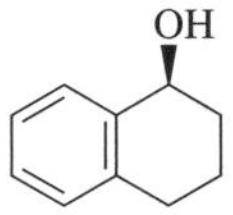	99
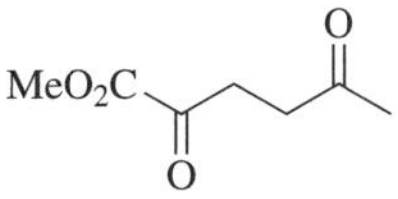	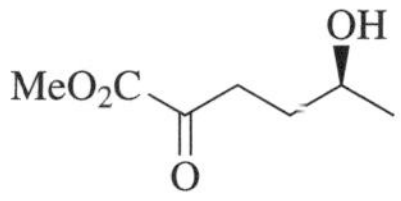	95
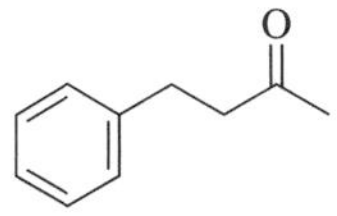	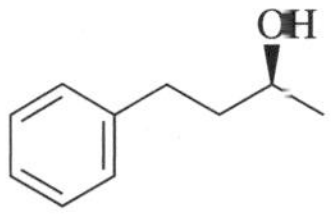	66
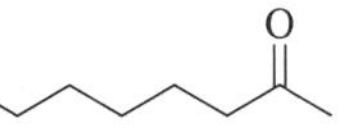	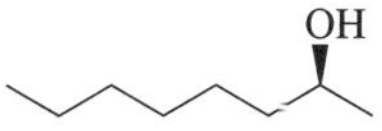	63

The stereoselectivity of the reduction of 4-*t*-butylcyclohexanone has been shown to give high proportions (>90%) of the *cis* (equatorial) alcohol in hydrogenations with heterogeneous rhodium catalysts or in transfer hydrogenations with homogeneous rhodium or iridium catalysts.[7] However, the Pybox–rhodium catalyst gave a ratio of 67:33 of the *trans/cis* alcohol,[8] similar to that obtained with the Wilkinson catalyst (eq 2).[9] Despite the low axial/equatorial selectivity, the enantioselectivities of the reduction of 2-phenyl- and 2-methoxycarbonylmethylcyclohexanone were extremely high (eq 3 and eq 4).[8] Chalcone was also reduced to the allylic alcohol in 71% ee (eq 5).[6]

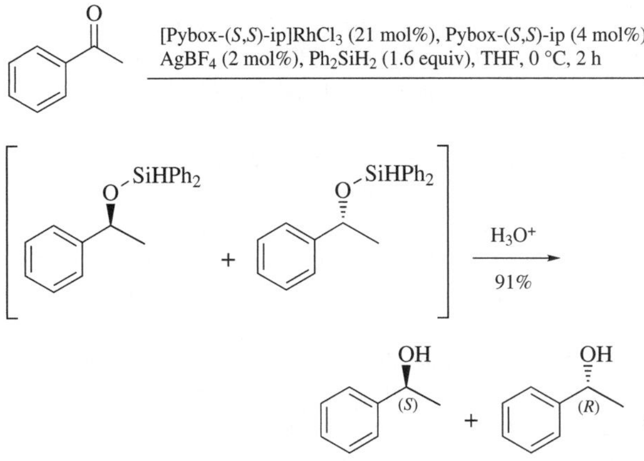

(1)

94% ee, *S*

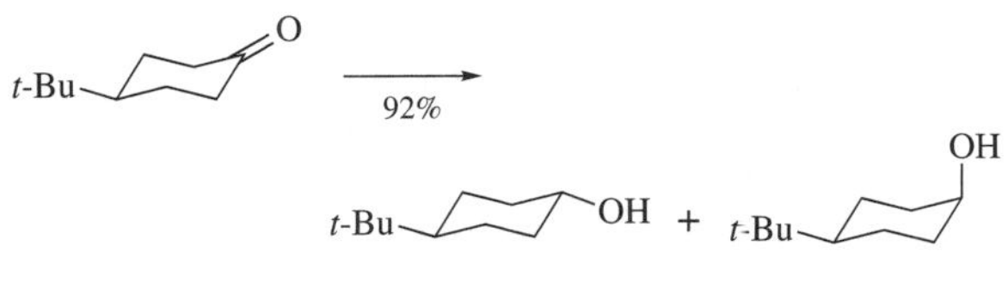

(2)

67:33

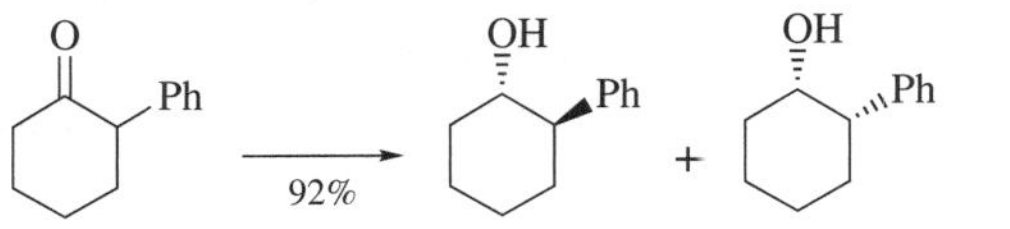

(3)

51:49

99% ee 96% ee

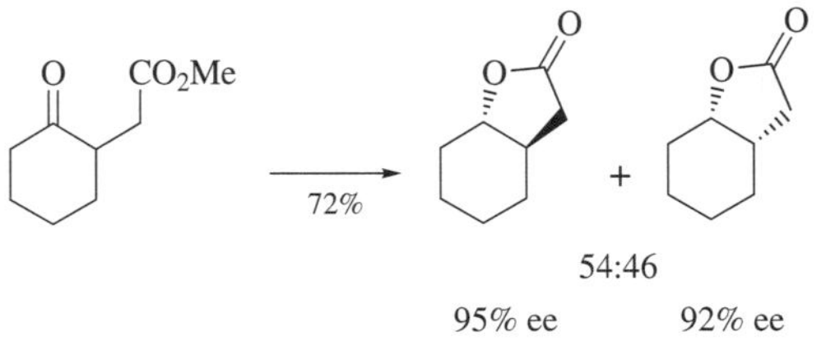

(4)

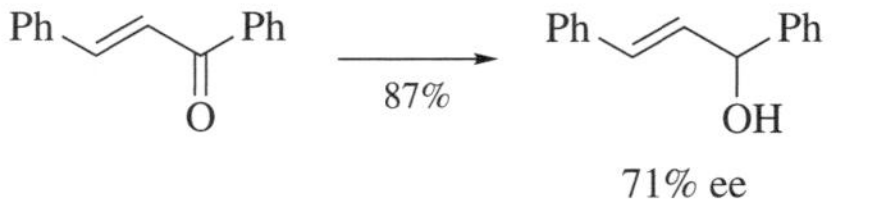

(5)

In place of diphenylsilane, 1,2-bis(dimethylsilyl)ethane was applied to the reduction of several ketones by the combination of the Pybox–rhodium complex and *Silver(I) Trifluoromethanesulfonate* to give the corresponding silyl enol ether exclusively (eq 6).[10]

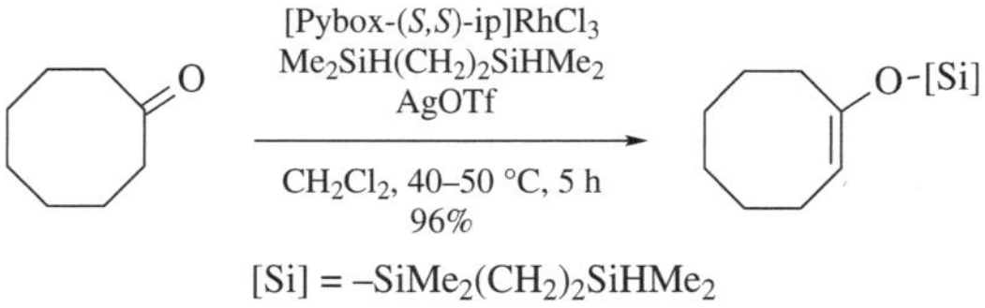

(6)

Chiral 4-substituted Pybox derivatives (**1**) were synthesized as chiral adjuvants to study remote electronic effects of the substituents in the asymmetric hydrosilylation.[11] The 4-Cl-Pybox-(*S,S*)-ip–Rh catalyst afforded the highest result (80% ee (*S*)) for the reduction of 2-octanone to 2-octanol in 88% yield.

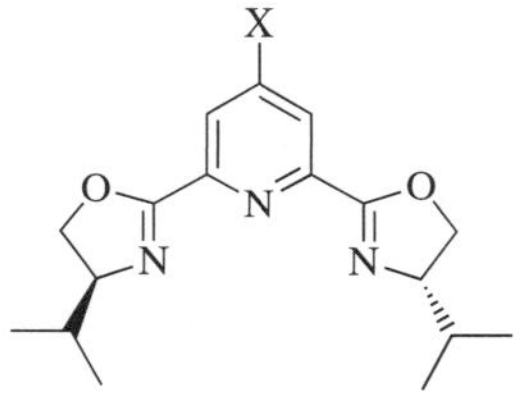

(**1**) 4-X-Pybox-(*S,S*)-ip

1. (a) Brunner, H.; Nishiyama, H.; Itoh, K. *Catalytic Asymmetric Synthesis*; Ojima, I., Ed.; VCH: New York, 1993, Chapter 6. (b) Bolm, C. *Angew. Chem., Int. Ed. Engl.* **1991**, *30*, 542.
2. (a) Ojima, I.; Clos, N.; Bastos, C. *Tetrahedron* **1989**, *45*, 6901. (b) Ojima, I., *The Chemistry of Organic Silicon Compounds, Part 2*; Patai, S.; Rappoport, Z., Ed.; Wiley: New York, 1989; pp 1479–1526. (c) Brunner, H. *Acta Chem. Scand.* **1988**, 645. (d) Ojima, I.; Hirai, K., *Asymmetric Synthesis*; Morrison, J. D., Ed.; Academic: Orlando, Fl, 1985; Vol. 5, Chapter 4, pp 103–146.
3. (a) Brunner, H.; Riepl, G.; Weitzer, H. *Angew. Chem., Int. Ed. Engl.* **1983**, *22*, 331. (b) Brunner, H.; Becker, R.; Riepl, G. *J. Organomet. Chem.* **1984**, *3*, 1354. (c) Brunner, H.; Kürzinger, A. *J. Organomet. Chem.* **1988**, *346*, 413.
4. Brunner, H.; Obermann, U. *Ber. Dtsch. Chem. Ges.* **1989**, *122*, 499.
5. Nishiyama, H.; Sakaguchi, H.; Nakamura, T.; Horihata, M.; Kondo, M.; Itoh, K. *J. Organomet. Chem.* **1989**, *8*, 846.
6. Nishiyama, H.; Kondo, M.; Nakamura, T.; Itoh, K. *J. Organomet. Chem.* **1991**, *10*, 500.
7. (a) Henbest, H. B.; Mitchell, T. R. B. *J. Chem. Soc. (C)* **1970**, 785. (b) Kaspar, J.; Spogliarich, R.; Graziani, M. *J. Organomet. Chem.* **1982**, *231*, 71. (c) Felföldi, K.; Kapocsi, I.; Bartók, M. *J. Organomet. Chem.* **1984**, *277*, 439. (d) Bennett, M. A.; Mitchell, T. R. B. *J. Organomet. Chem.* **1985**, *295*, 223. (e) Smith, T. A.; Maitlis, P. M. *J. Organomet. Chem.* **1985**, *289*, 385.
8. Nishiyama, H.; Park, S.-B.; Itoh, K. *Tetrahedron: Asymmetry* **1992**, *3*, 1029.
9. (a) Ishiyama, J.; Senda, Y.; Shinoda, I.; Imaizumi, S. *Bull. Chem. Soc. Jpn.* **1979**, *52*, 2353. (b) Semmelhack, M. F.; Misra, R. N. *J. Org. Chem.* **1982**, *47*, 2469.
10. Nagashima, H.; Ueda, T.; Nishiyama, H.; Itoh, K. *Chem. Lett.* **1993**, 347.
11. Nishiyama, H.; Yamaguchi, S.; Kondo, M.; Itoh, K. *J. Org. Chem.* **1992**, *57*, 4306.

Hisao Nishiyama
Toyohashi University of Technology, Japan

trans-2,5-Bis(methoxymethyl)pyrrolidine

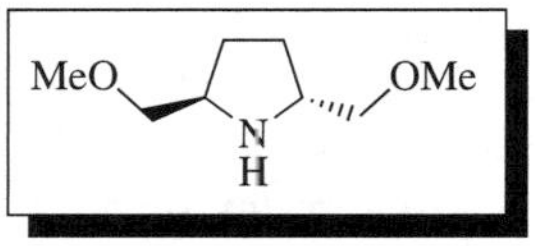

(±)
[144993-81-7] $C_8H_{17}NO_2$ (MW 159.23)
(*S,S*)-(+)
[93621-94-4]
(*R,R*)-(−)
[90290-05-4]

(chiral auxiliary with C_2 symmetry;[1] used as a chiral auxiliary for alkylations, acylations, reductions, cycloadditions, and radical additions)

Physical Data: colorless oil; bp 110–115 °C/15 mmHg; for *S,S*-form, $[\alpha]_D = +7.8°$ ($c = 3.0$, EtOH).
Solubility: sol methylene chloride, chloroform, ethanol.
Form Supplied in: not commercially available.
Analysis of Reagent Purity: ^{13}C NMR (δ 27.7, 56.7, 58.8, 76.1 ppm).
Handling, Storage, and Precautions: irritant; use only in a fume hood.

Chiral Amines with C_2 Symmetry. *trans*-2,5-Dimethylpyrrolidine (**1**)[2] was the first chiral amine possessing C_2 symmetry used as a chiral auxiliary in asymmetric synthesis.[1,2a] Since that time a number of related systems have been developed including the title compound (**2**) and (**4**).[1] These amines were developed as C_2-symmetric analogs to the commercially available prolinol derivative (**5**). While proline-derived chiral auxiliaries have been widely used in asymmetric synthesis,[3] the C_2-symmetric chiral auxiliaries often give enhanced stereoselectivity when compared directly to the prolinol derivatives. Unfortunately the preparation of the C_2-symmetric compounds is more tedious and, at the time of writing, none are commercially available. For example, the standard route to chiral pyrrolidines (**2**) and (**3**) involves the resolution of *trans-N*-benzylpyrrolidine-2,5-dicarboxylic acid,[4] although other preparations have been reported.[5] In general, pyrrolidines (**2**) and (**3**) have been used interchangeably and will be the primary focus of this entry.

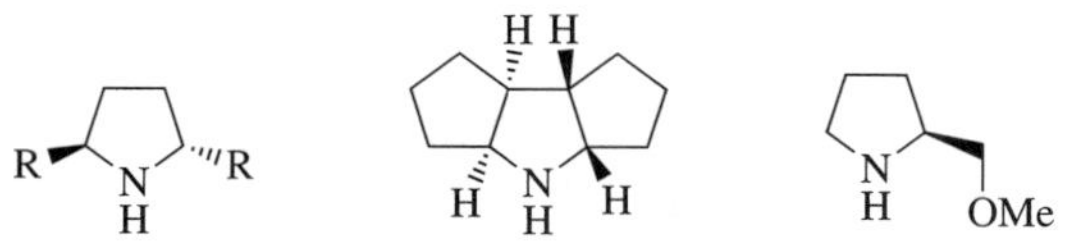

(**1**) R = Me
(**2**) R = CH_2OMe
(**3**) R = $[CH_2O]_2Me$

(**4**) (**5**)

Alkylation of Amides Derived from C_2-Symmetric Pyrrolidines. Alkylation of amides derived from either pyrrolidine (**2**) or (**3**) are highly stereoselective (eq 1).[4a] The reaction is successful for a large variety of amide derivatives[4a,6] and alkylating agents.[7] Upon hydrolysis of the amide, chiral acids are produced (eq 2). This method has been used to prepare α-amino[6d] and α-hydroxy acids.[6a] The diastereoselectivity observed is greater than for amides derived from prolinol (**5**).[3] While the degree of the stereoselectivity is almost identical for amides derived from either (**2**) or (**3**), the method for removal of the chiral auxiliary does differ (eq 2). In keeping with earlier results from the prolinol-derived amides,[3] the hydrolysis is best effected via the hydroxymethyl derivative (**10**). For derivative (**8**) this means that the methyl ether is first cleaved with either *Boron Trichloride*[4a] or *Boron Tribromide*.[8] In the case of the methoxymethyl derivative (**9**), refluxing in aqueous acid simultaneously effects both cleavage of the methoxymethyl ether and hydrolysis to the chiral acid.

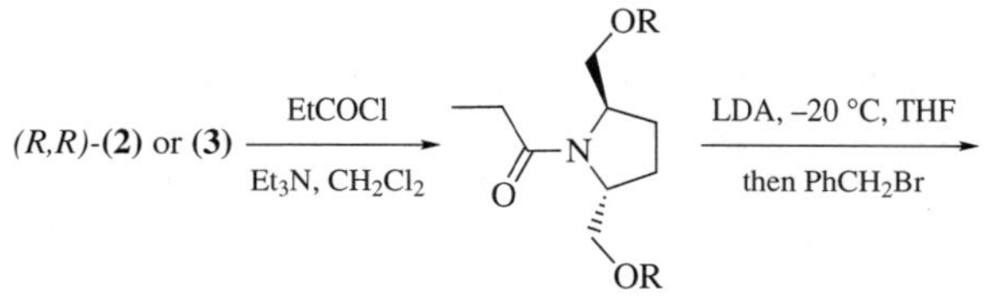

(**6**) R = Me
(**7**) R = CH_2OMe

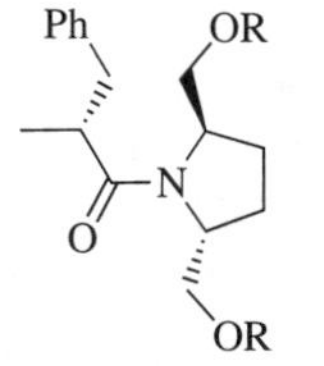

(1)

(**8**) R = Me, >95% de
(**9**) R = CH_2OMe, >95% de

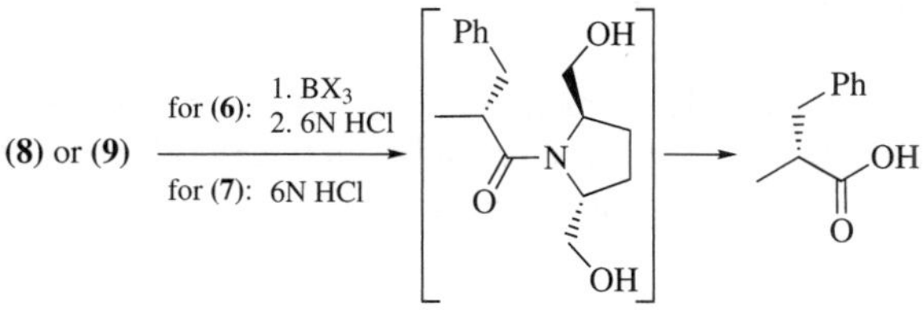

(2)

(**10**)

Reduction of α- and β-Ketoamides Derived from C_2-Symmetric Pyrrolidines. By analogy to the alkylation reaction discussed above, acylation of the enolate derived from amide (**8**) produces the β-ketoamide (**11**) as a single diastereomer (eq 3).[9] Subsequent reduction of the ketone produces either the *syn*[9] or *anti*[10] β-hydroxyamides with high diastereoselectivity (eqs 4 and 5). Pyrrolidine-derived α-ketoamides have also been shown to react stereospecifically with reducing agents,[11] as well as with other organometallic reagents.[12]

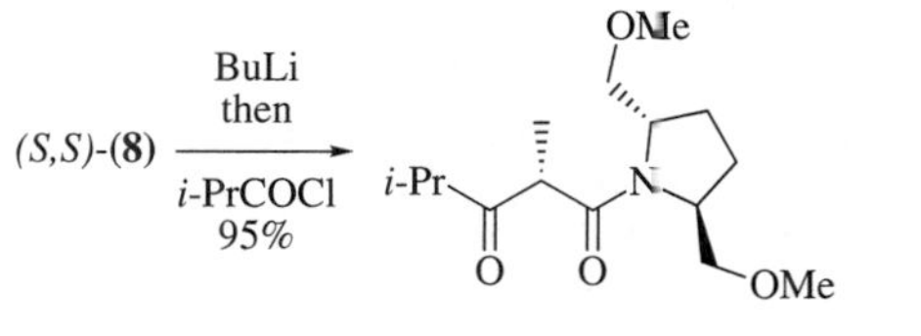

(3)

(**11**)

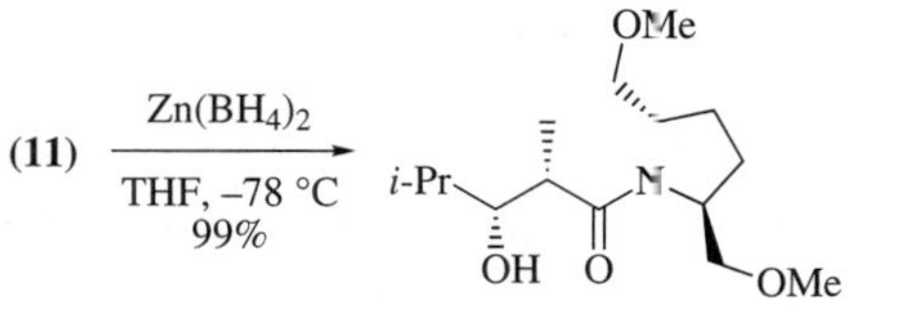

(4)

syn:anti = >99:1

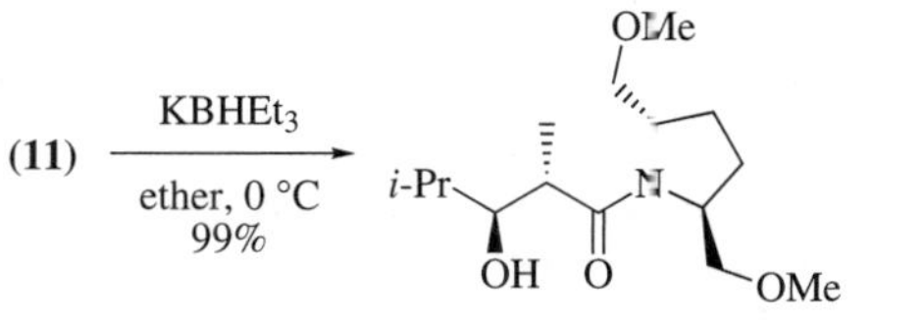

(5)

anti:syn =>99:1

Cycloaddition Reactions. Chiral acrylamides derived from pyrrolidines (**2**) or (**3**) undergo stereoselective [4 + 2] cycloaddition reactions with a variety of cyclic dienes.[8a,13] Similarly, nitroso compounds derivatized with pyrrolidine (**2**) and generated in situ give cycloadducts with a high degree of stereoselectivity (eq 6).[8a,14] Intramolecular [2 + 2] cycloadditions involving pyrrolidine-derived keteniminium salts have been shown to produce chiral cyclobutanones.[4b]

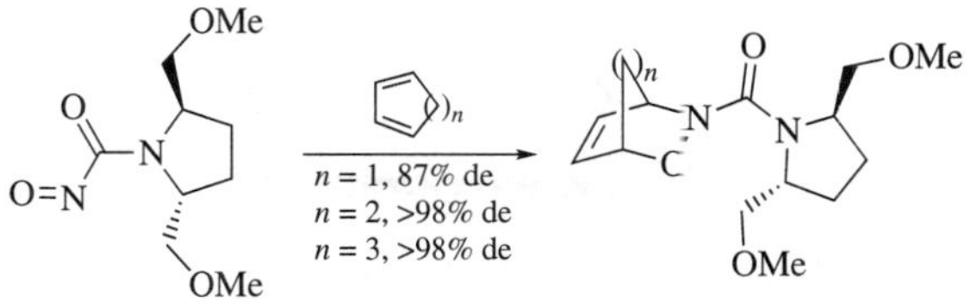

(6)

Other Reactions. Other reactions in which pyrrolidines (**2**) and (**3**) have been used as chiral auxiliaries include radical additions,[8b] electrocyclizations,[15] and Wittig rearrangements.[7b]

1. Whitesell, J. K. *Chem. Rev.* **1989**, *89*, 1581.
2. (a) Whitesell, J. K.; Felman, S. W. *J. Org. Chem.* **1977**, *42*, 1663. (b) Schlessinger, R. H.; Iwanowicz, E. J. *Tetrahedron Lett.* **1987**, *28*, 2083. (c) Short, R. P.; Kennedy, R. M.; Masamune, S. *J. Org. Chem.* **1989**, *54*, 1755.
3. Evans, D. A.; Takacs, J. M.; McGee, L. R.; Ennis, M. D.; Mathre, D. J.; Bartroli, J. *Pure Appl. Chem.* **1981**, *53*, 1109.

4. (a) Kawanami, Y.; Ito, Y.; Kitagawa, T.; Taniguchi, Y.; Katsuki, T.; Yamaguchi, M. *Tetrahedron Lett.* **1984**, *25*, 857. (b) Chen, L-y.; Ghosez, L. *Tetrahedron Lett.* **1990**, *31*, 4467.
5. (a) Takano, S.; Moriya, M.; Iwabuchi, Y.; Ogasawara, K. *Tetrahedron Lett.* **1989**, *30*, 3805. (b) Marzi, M.; Minetti, P.; Misiti, D. *Tetrahedron* **1992**, *48*, 10 127. (c) Yamamoto, Y.; Hoshino, J.; Fujimoto, Y.; Ohmoto, J.; Sawada, S. *Acta Chem. Scand.* **1993**, 298.
6. (a) Enomoto, M.; Ito, Y.; Katsuki, T.; Yamaguchi, M. *Tetrahedron Lett.* **1985**, *26*, 1343. (b) Hanamoto, T.; Katsuki, T.; Yamaguchi, M. *Tetrahedron Lett.* **1986**, *27*, 2463. (c) Katsuki, T.; Yamaguchi, M. *Tetrahedron Lett.* **1987**, *28*, 651. (d) Ikegami, S.; Uchiyama, H.; Hayama, T.; Katsuki, T.; Yamaguchi, M. *Tetrahedron* **1988**, *44*, 5333.
7. (a) Katsuki, T.; Yamaguchi, M. *Tetrahedron Lett.* **1985**, *26*, 5807. (b) Uchikawa, M.; Hanamoto, T.; Katsuki, T.; Yamaguchi, M. *Tetrahedron Lett.* **1986**, *27*, 4577.
8. (a) Lamy-Schelkens, H.; Ghosez, L. *Tetrahedron Lett.* **1989**, *30*, 5891. (b) Veit, A.; Lenz, R.; Seiler, M. E.; Neuburger, M.; Zehnder, M.; Giese, B. *Helv. Chim. Acta* **1993**, *76*, 441.
9. Ito, Y.; Katsuki, T.; Yamaguchi, M. *Tetrahedron Lett.* **1984**, *25*, 6015.
10. Ito, Y.; Katsuki, T.; Yamaguchi, M. *Tetrahedron Lett.* **1985**, *26*, 4643.
11. Kawanami, Y.; Fujita, I.; Asahara, S.; Katsuki, T.; Yamaguchi, M. *Bull. Chem. Soc. Jpn.* **1989**, *62*, 3598.
12. Kawanami, Y.; Fujita, I.; Ogawa, S.; Katsuki, T. *Chem. Lett.* **1989**, 2063.
13. Kawanami, Y.; Katsuki, T.; Yamaguchi, M. *Bull. Chem. Soc. Jpn.* **1987**, *60*, 4190.
14. (a) Gouverneur, V.; Ghosez, L. *Tetrahedron Lett.* **1990**, *1*, 363. (b) Gouverneur, V.; Ghosez, L. *Tetrahedron Lett.* **1991**, *32*, 5349.
15. Fuji, K.; Node, M.; Naniwa, Y.; Kawabata, T. *Tetrahedron Lett.* **1990**, *31*, 3175.

Patrick G. McDougal
Reed College, Portland, OR, USA

Bis[(4*S*)-(1-methylethyl)oxazolin-2-yl]-methane

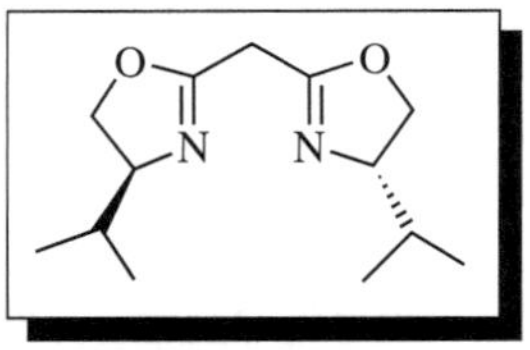

[131833-90-4] $C_{13}H_{22}N_2O_2$ (MW 238.33)

(chiral ligand for enantiocontrol of metal-catalyzed reactions)

Physical Data: $[\alpha]_D^{20}$ −113 (*c* 1.08, CH_2Cl_2).
Solubility: soluble in most organic solvents.
Form Supplied in: clear, oily, low-melting solid.
Preparative Methods: Bis[(4*S*)-(1-methylethyl)oxazolin-2-yl]methane is prepared by acylation of L-valinol followed by cyclization (eq 1).[1,2,3] Thus, transamination of dimethyl malonate with 2 equiv of L-valinol afforded the corresponding amide in 72% yield. Chlorination with $SOCl_2$ followed by cyclization with LiOMe in refluxing MeOH afforded bis[(4*S*)-(1-methylethyl)oxazolin-2-yl]methane in 79% yield.

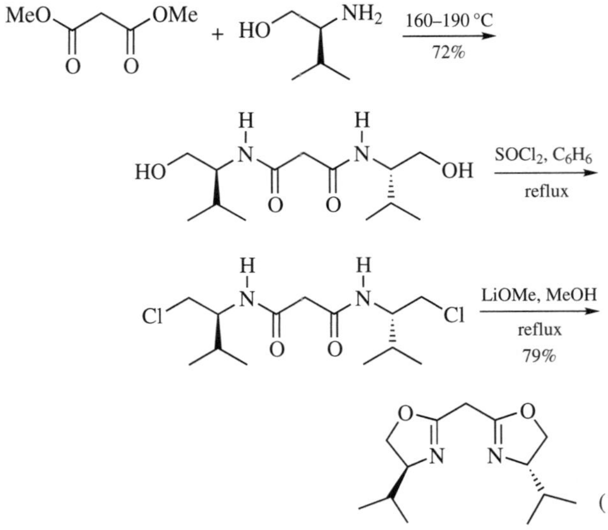

(1)

Handling, Storage, and Precautions: stable at ambient temperature.

Bis(oxazoline)–Metal Complexes. Metal complexes of bis[(4*S*)-(1-methylethyl)oxazolin-2-yl]methane are efficient catalysts in numerous asymmetric reactions. The ligand-metal complex is prepared in situ by mixing the metal salt and ligand. The formation of a monomeric or dimeric complex depends upon the reaction conditions and the reactivity of the metal ion. In the asymmetric reaction, the C_2-symmetric axis in the ligand minimizes the number of possible transition states in a reaction.[4] The metal chelate is conformationally constrained and the chiral centers are located in close proximity to the donor ligands, thereby imposing a strong directing effect on the catalytic site. The metal complexes of bis[(4*S*)-(1-methylethyl)oxazolin-2-yl]methane with Ti(IV),[5] zinc(II)[6] and magnesium(II)[7] have been reported.

The Ti(IV) complexes were prepared by treatment of bis[(4*S*)-(1-methylethyl)oxazolin-2-yl]methane with TiX_4 (X=Cl, NEt_2, O-*i*-Pr) in toluene. In the infrared spectra the absence of absorption for an (NH) vibration in the region 3500–3200 cm^{-1}, suggests that the ligand is deprotonated. The presence of absorption due to the (C=N) and (C=C) vibrations in the region 1602–1540 cm^{-1} indicate a bidentate ligand pattern.[8] The far-infrared region contains contributions from (Ti–O) at 472 cm^{-1}, (Ti–N)[9] at 360 cm^{-1} and (Ti–Cl)[10] at 280 cm^{-1} supporting a monomeric trigonal bipyramidal structure where the ligand is coordinated to Ti(IV) in a bidentate fashion and the nitrogen atoms of the ligand occupy the equatorial sites. The structure of the TiX_3L complex (X = Cl, O-*i*-Pr_2, NEt_2) is shown (reprinted from reference 5, pg 157 and 160, by courtesy of Marcel Decker. Inc.).

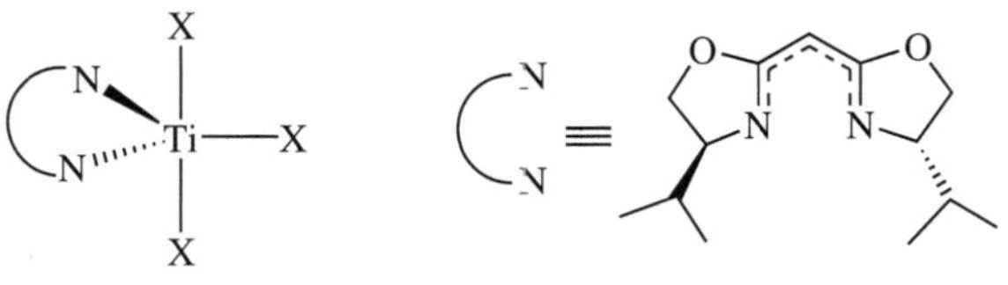

The zinc(II) complex ClZnL was prepared by treatment of EtZnCl with bis[(4*S*)-(1-methylethyl)oxazolin-2-yl]methane. Treatment of ClZnL with 1.0 or 2.0 equiv of PhSH yielded (PhS)ZnL and $(PhS)_2ZnLH$, respectively. The infrared spectra of ClZnL and (PhS)ZnL reveal that the ligand acts as a bidentate donor to zinc(II). However, the infrared spectra of $(PhS)_2ZnLH$ show a band at $3200\,cm^{-1}$ due to coordinated (NH) and bands at $1660–1550\,cm^{-1}$ due to (C=N) and (C=C) vibrations. Contributions from (Zn–N)[11] at $485–470\,cm^{-1}$, (Zn–S)[12] at $388–355\,cm^{-1}$ and (Zn–Cl)[13] at $280–260\,cm^{-1}$ are also evident, consistent with a dimeric structure for ClZnL and (PhS)ZnL and a monomeric structure for $(PhS)_2ZnLH$.

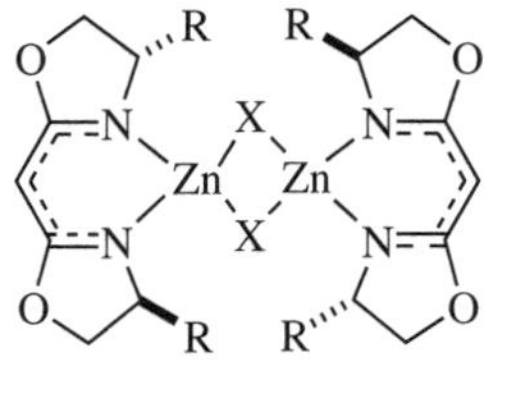

or

X = Cl, SPh

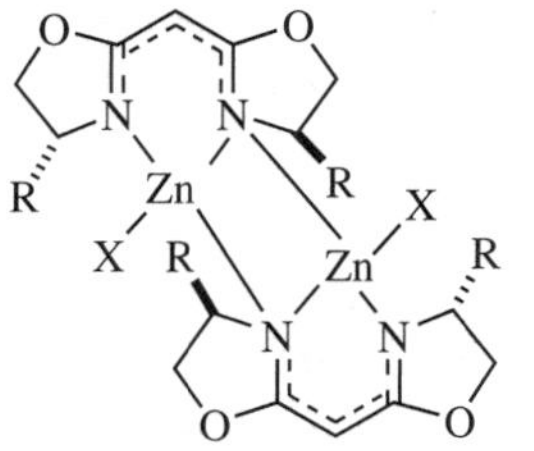

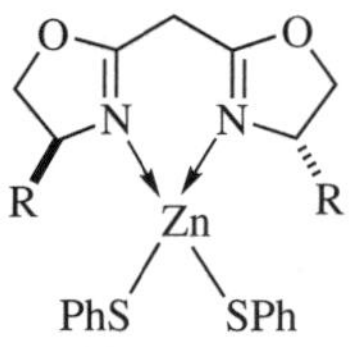

The magnesium(II) complex ClMgL was prepared by treatment of EtMgCl with bis[(4*S*)-(1-methylethyl)oxazolin-2-yl]methane. Treatment of Et_2Mg with 1.0 or 2.0 equiv bis[(4*S*)-(1-methylethyl)oxazolin-2-yl]methane afforded EtMgL and MgL_2, respectively. As observed with the other complexes, the infrared spectra reveal that the ligand acts as a bidentate donor to magnesium(II).[14] The far infrared region contains contributions from (Mg–C)[15] at $820\,cm^{-1}$, (Mg–N)[16] at $355\,cm^{-1}$ and (Mg–Cl)[17] at $280\,cm^{-1}$, consistent with a dimeric structure for ClMgL and EtMgL and a monomeric structure for MgL_2) (reprinted from reference 7, p 1716, with permission from Elsevier Science).

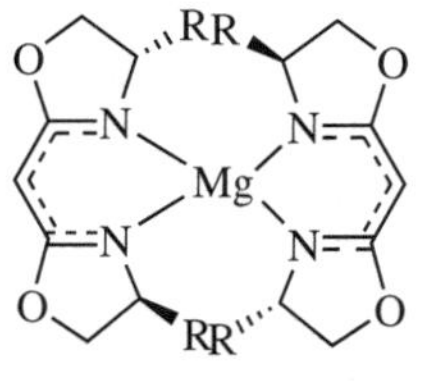

or

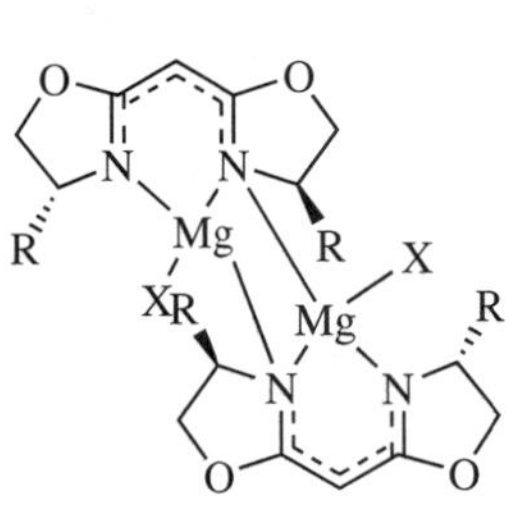

MgL_2 MXL

X = Et, Cl; L = bisoxazoline

The ^1H and ^{13}C NMR data for all complexes are similar. The most significant difference in ^1H NMR spectra of the complexes relative to the free ligand is the shift in the C–CH_2–C of the ligand from δ 3.3 ppm to δ 4.7 ppm for C–CH–C in the complex upon deprotonation (eq 2). In the ^{13}C NMR spectra, the CH_2 carbon of C–CH_2–C appeared at δ 28.4 ppm in the free ligand but was shifted to δ55 ppm upon complexation. In addition, the carbon resonance due to C=N which appeared in the ligand at δ161 ppm shifted to δ 172 ppm. It is interesting to note that ^{13}C NMR spectra of $(PhS)_2ZnLH$ showed little shifting due to the coordination of the ligand in the protonated form.

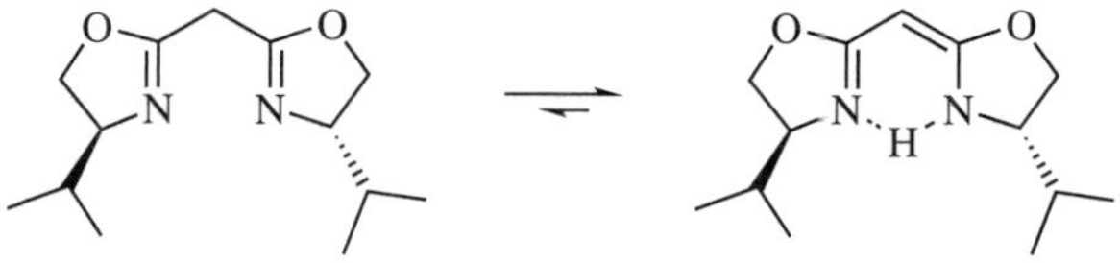
(2)

A carboxylate-bridged triiron(II) complex of bis[(4*S*)-(1-methylethyl)oxazolin-2-yl]methane has also been prepared and its antiferromagnetic properties examined.[18]

Asymmetric Reactions[19]

Cyclopropanation

A number of reports on the use of bis[(4*S*)-(1-methylethyl)-oxazolin-2-yl]methane in the asymmetric cyclopropanation of styrene have been reported (eq 3, Table 1).[20] Although the yields of the cyclopropanes are good, the enantioselectivities are not as high as those observed with other bis(oxazoline) ligands.[2,20]

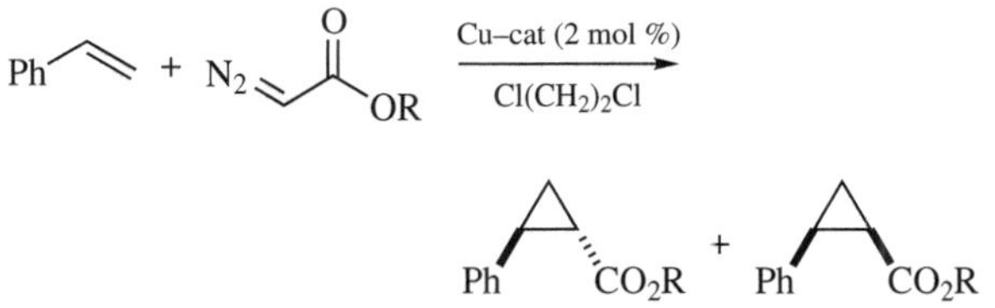
(3)

Epoxidation

Epoxidation of styrene or stilbene with the ruthenium $[RuCl_2(cod)L]$ complex of bis[(4*S*)-(1-methylethyl)oxazolin-2-yl]methane afforded only racemic epoxide, suggesting that the reaction is not metal centered.[21] In fact, mechanistic studies of this reaction indicate that the metal acts as a promoter for the production of *i*-PrO_3H and that it is this species that carries out the epoxidation, either directly or by the formation of an intermediate oxo-ruthenium species.

Allylation and Addition Reactions

The enantioselective allylzincation of cyclopropyl acetals catalyzed by bis[(4*S*)-(1-methylethyl)oxazolin-2-yl]methane has been reported (eq 4, Table 2).[22] The allylzinc complex, prepared by reaction of deprotonated ligand with allylzinc bromide, reacted readily with cyclopropenone acetals **1** and **2** at room temperature to provide the optically active cyclopropanone acetals in good yield and high enantioselectivity (Table 2, entries 1–6). The ethyl-substituted cyclopropenone acetal **3** afforded the optically active cyclopropanone acetal possessing a quaternary chiral center (Table 2, entry 7).

Table 1 Cyclopropanation of styrene with several diazoacetates using bis[(4*S*)-(1-methylethyl)oxazolin-2-yl]methane

Entry	Yield (%)	de (*trans*/*cis*)	ee (*trans*/*cis*)	ConÞguration	R
1	72	71:29	46:31	(1*R*,2*R*)/(1*S*,2*R*)	Et[19b]
2	60Ð80	84:16	13:5	(1*R*,2*R*)/(1*R*,2*S*)	[(1*R*,3*R*,4*S*)-menthyl][2]

Table 2 Ligand-induced enantioselective allylzincation of cyclopropyl acetals

Entry	CPA	R^1	R^2	Yield (%)[a]	de[b]	ee[c]
1	1	H	H	89	Ð	98%
2	1	Me	Me	90	Ð	93%
3	1	H	Ph	86	73:27	56
4	1	H	C_6H_{11}	94	83:17	62
5	2	H	H	76	Ð	0
6	2	H	Ph	94	86:14	1:99
7	3	H	H	48	Ð	74

CPA, cyclopropyl acetal.
[a]Isolated yield based upon CPA.
[b]C(3)ÐC(4)diastereoselectivity.
[c]The enantioselectivity was determined for the major C(3)ÐC(4)diastereomer; the ratio refers to C(4)ÐHdown versus up.

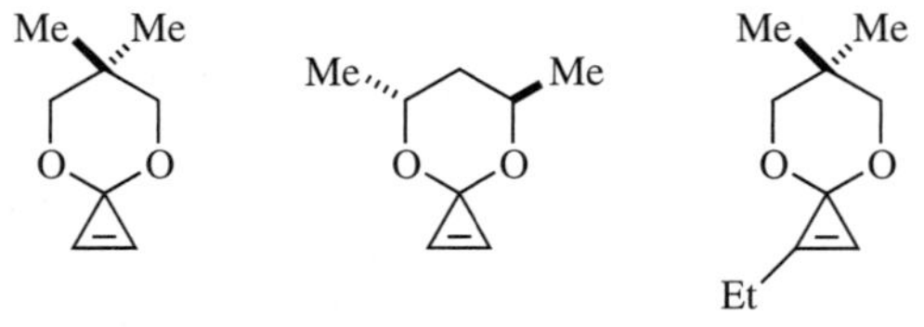

1 **2** **3**

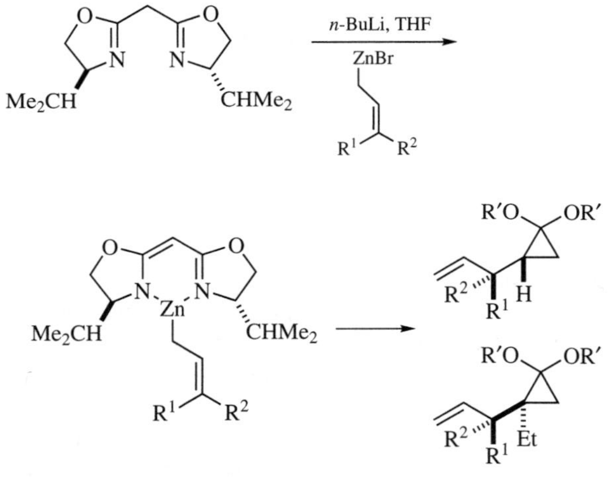

(4)

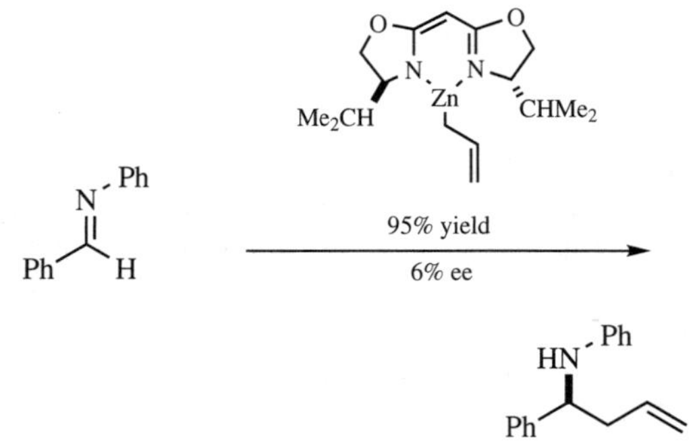

(5)

The chiral allylzinc complex also reacts with chiral aldimines to afford allylated secondary amines in high enantioselectivity.[23] For the acyclic (*E*)-benzaldehyde *N*-phenylimine, the amine was obtained in high yield (95%), although the enantioselectivity was low (6%) (eq 5). However cyclic imines afforded good yields and enantioselectivities (eq 6 and 7).

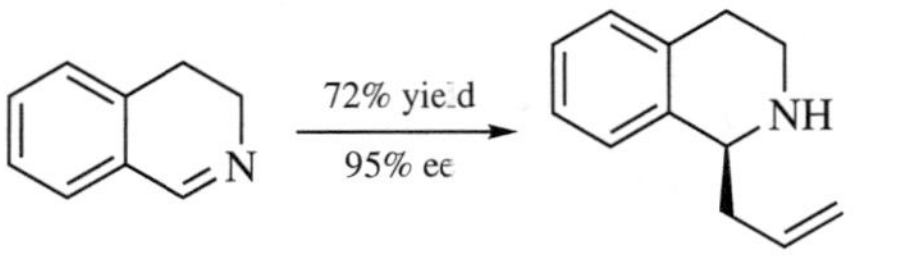

(6)

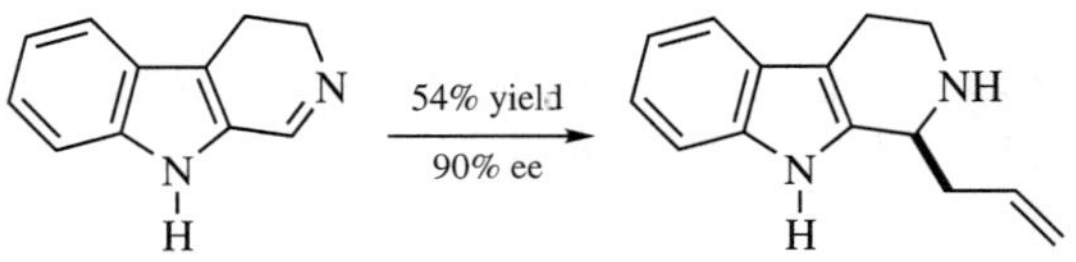

(7)

The reaction is thought to proceed through a chair-like transition state in which the steric interactions between the imine substituents and the C4-substitutent of the ligand are minimized. This model is consistent with the observed selectivities.

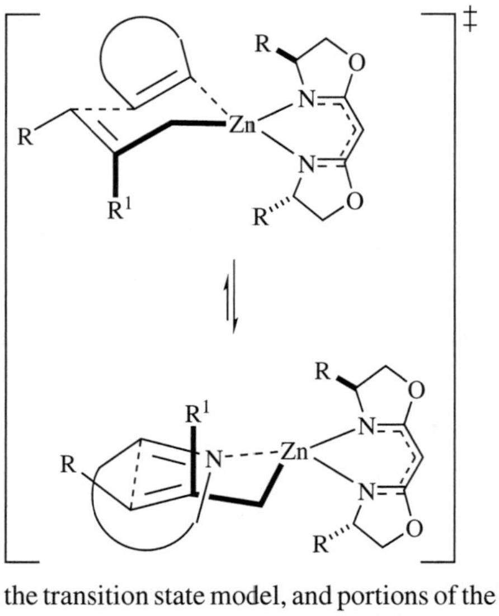

(Eq 5–7, the transition state model, and portions of the text relating to allylzincation reactions of bis[(4*S*)-(1-methylethyl)oxazolin-2-yl]methane were reproduced from reference 21 with permission from Elsevier Science.)

Hydrosilation

Enantioselective hydrosilylation of acetophenone using either bis[(4*S*)-(1-methylethyl)oxazolin-2-yl]methane or its rhodium(I) complex has been reported, although the enantioselectivity was only 12% (eq 8).[24]

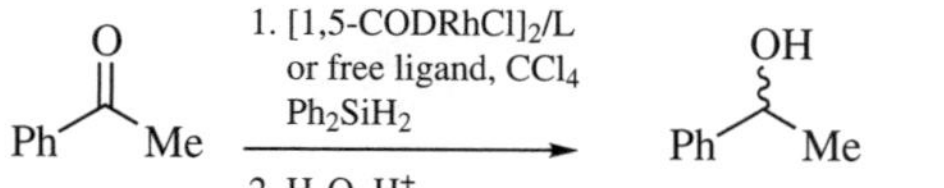

No yield reported
ee: 12%

(8)

Reduction of α-Alkoxy Ketones

Enantioselective reduction of α-alkoxy ketones with catecholborane and the $Zn(OTf)_2$-ligand complex afforded the diol in 70% yield albeit with low enantioselectivity (15%) (eq 9).[25]

Yield 70%; ee 15%

(9)

Radical Cyclizations

Cyclization of *N*-trichloroacetamides using copper(I)–bis[(4*S*)-(1-methylethyl)oxazolin-2-yl]methane complex afforded the corresponding γ-lactams in high yield (85–95%) but low diastereoselectivity (6%) (eq 10).[26]

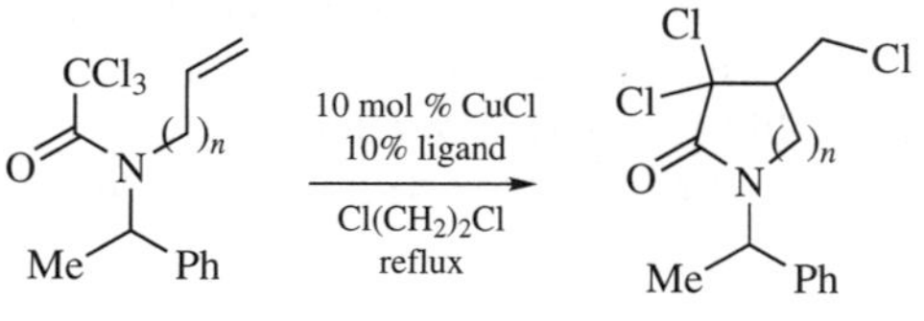

n = 1, 2

yield 85–95%; de 6%

(10)

Incorporation of substituents in the alkenyl side chain resulted in the formation of the *trans*-γ- and δ-lactams in high conversion and 60% diastereoselectivity (eq 11 and eq 12). In all experiments, the diastereoselectivity was similar to that previously reported by Nagashima et al.[27] using a 2,2′-bipyridine complex.

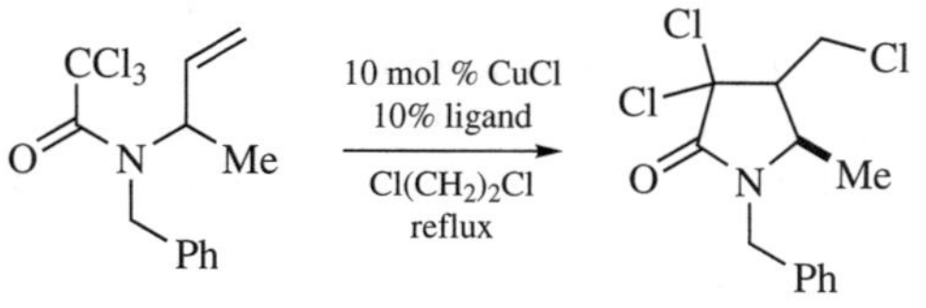

yield 92%; de 60%

(11)

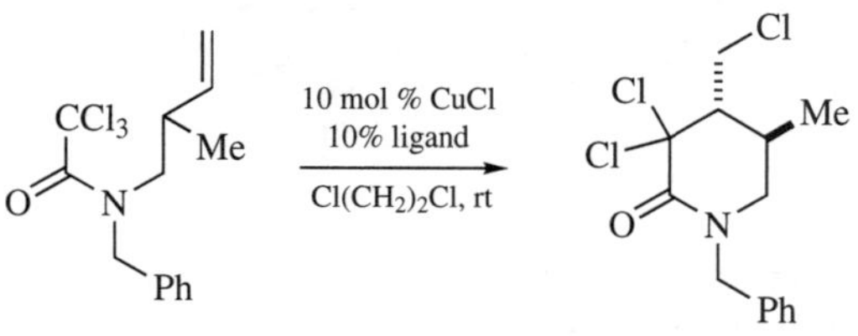

yield 100%; de 60%

(12)

Synthesis of Optically Active Polyguanidines

Polyguanidines have been prepared from achiral carbodimides using the copper(II)-bis[(4*S*)-(1-methylethyl)oxazolin-2-yl]-methane as a catalyst, although the yield and enantioselectivity was low (eq 13).[28]

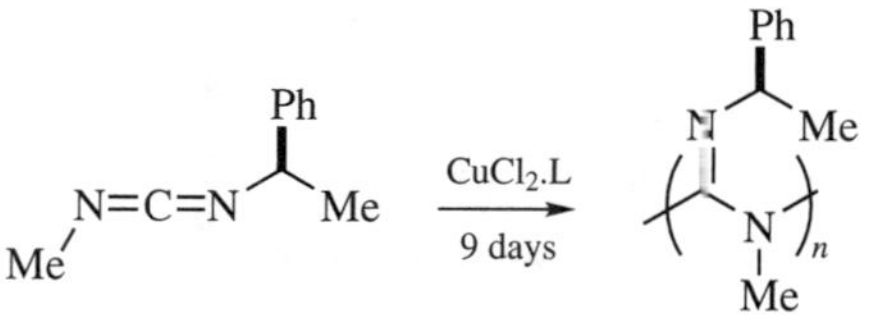

5% yield

(13)

1. Butula, I.; Karlovic, G. *Leibigs. Ann. Chem.* **1976**, *7–8*, 1455.
2. Müller, D.; Umbricht, G.; Weber, B.; Pfaltz, A. *Helv. Chim. Acta.* **1991**, *74*, 232
3. Helmchen, G.; Krotz, A.; Ganz, K. T.; Hansen, D. *Synlett* **1991**, *4*, 257.
4. Whitesell, J. K. *Chem. Rev.* **1989**, *89*, 1581.
5. Singh, R. P. *Synth. React. Inorg. Met. Org. Chem.* **1997**, *27*, 155.
6. Singh, R. P. *Bull. Soc. Chim. Fr.* **1997**, *134*, 765
7. Singh, R. P. *Spectrochimica Acta A* **1997**, *53*, 1713.
8. Brookhart, M.; Wagner, M. I.; Balavoine, G. G. A.; Haddou, H. A. *J. Am. Chem. Soc.* **1994**, *116*, 3641.
9. Ohkahu, N.; Nakamoto, K. *Inorg. Chem.* **1973**, *12*, 2440.
10. Alvarez-Boo, P.; Martinez, E. G.; Casas, J. S.; Sorodo, J. *Synth. React. Inorg. Met. Org. Chem.* **1995**, *25*, 115.
11. Nakamoto, K. *Infrared and Raman spectra of Inorganic and Coordination Compounds*; Wiley Interscience: New York, 1978, p 213.
12. Mishra, L.; Pandey, A. K.; Singh, R. P. *Indian J. Chem.* **1992**, *31A*, 1995.
13. Coates, G. E.; Ridley, D. J. *J. Chem. Soc.* **1964**, *(Jan)*, 166.
14. Ashby, E. C.; Nackashi, J.; Paris, G. E. *J. Am. Chem. Soc.* **1975**, *97*, 3162.
15. Fujiwara, M.; Matsushita, T.; Sono, T. *Polyhedron* **1984**, *3*, 3162.
16. Einarsrud, M. A.; Justnes, H.; Tytter, E.; Oyb, E. A. *Polyhedron* **1987**, *6*, 975.
17. Goldberg, D. P.; Telser, J.; Bastos, C. M.; Lippard, S. J. *Inorg. Chem.* **1995**, *34*, 3011.
18. Ghosh, A. K.; Mathivanan, P.; Cappiello, J. *Tetrahedron Asymmetry* **1998**, *9*, 1.
19. Lowenthal, R. E.; Abiko, A.; Masamune, S. *Tetrahedron Lett.* **1990**, *31*, 6005.
20. Bennett, S.; Brown, S. M.; Conole, G.; Kessler, M.; Rowling, S.; Sinn, E.; Woodward, S. *J. Chem. Soc. Dalton. Trans.* **1995**, *3*, 367.
21. (a) Nakamura, M.; Arai, M.; Nakamura, E. *J. Am. Chem. Soc.* **1995**, *117*, 1179. (b) Nakamura, M.; Inoue, T.; Sato, A.; Nakamura, E. *Org. Lett.* **2000**, *2*, 2193.
22. Nakamura, M.; Hirai, A.; Nakamura, E. *J. Am. Chem. Soc.* **1996**, *118*, 8489.
23. Bandini, M.; Cozzi, P. G.; de Angelis, M.; Umani-Ronchi, A., *Tetrahedron Lett.* **2000**, *41*, 1601.
24. Clark, A. J.; De Campo, F.; Deeth, R. J.; Filik, R. P.; Gatard, S.; Hunt, N. A.; Lastécouères, D.; Thomas, G. H.; Verlhac, J.-B.; Wongtap, H. *J. Chem. Soc. Perkin Trans. 1.* **2000**, *5*, 671.
25. Nagashima, H.; Ozaki, N.; Ishii, M.; Seki, K.; Washiyama, M.; Itoh, K. *J. Org. Chem.* **1993**, *58*, 464.
26. Heintz, A. M.; Novak, B. M. *Polym. Prepr. (Am. Chem. Soc., Div. Polym. Chem)* **1998**, *39*, 429.

Margaret M. Faul
Eli Lilly and Company, Indianapolis, IN, USA

(*R*)-3,3′-Bis(triphenylsilyl)-binaphthomethylaluminum

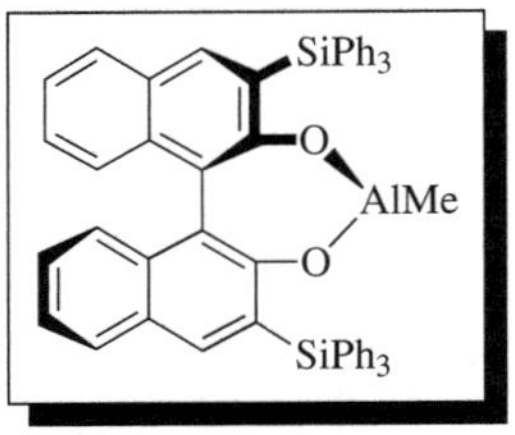

[118724-91-7] $C_{57}H_{43}AlO_2Si_2$ (MW 843.16)

(chiral Lewis acid for asymmetric hetero-Diels–Alder,[2,3] Diels–Alder,[6] and ene reactions,[7] Claisen rearrangement,[8,9] and polymerization[10])

Solubility: sol toluene and CH_2Cl_2; slightly sol hexane.
Form Supplied in: prepared and used in situ.
Preparative Methods: can be prepared by treatment of (*R*)-3,3′-bis(triphenylsilyl)binaphthol[1] in CH_2Cl_2 or toluene with a 1 M hexane solution of *Trimethylaluminum* room temperature for 1–2 h.[2] This reagent can be also generated in situ by discrimination of its racemates with optically active (−)-bromocamphor.[3]
Handling, Storage, and Precautions: the dry solid and solutions are highly flammable and must be handled in the absence of oxygen and moisture. The solution should be used as prepared for best results. Use in a fume hood.

Asymmetric Hetero-Diels–Alder Reaction. This chiral organoaluminum reagent has been developed for effecting the hetero-Diels–Alder reaction of aldehydes and siloxydienes with high enantioselectivity.[2,4] Thus reaction of benzaldehyde with 1-methoxy-2-methyl-3-trimethylsiloxy-1,3-pentadiene under the influence of 10 mol % of the chiral aluminum reagent affords *cis*-dihydropyrone as a major product in 95% ee (eq 1). The enantioselectivity is highly dependent on the bulk of the triarylsilyl moiety of the aluminum reagent. Thus replacement of the triphenylsilyl group by the tris(3,5-xylyl)silyl group enhances the enantio as well as *cis* selectivity, but replacement by trimethylsilyl groups lowers the enantioselectivity. This chiral aluminum reagent is generated in situ by discrimination of racemic organoaluminum reagents with (−)-bromocamphor by diastereoselective complexation, and utilized as a chiral Lewis acid for the asymmetric hetero-Diels–Alder reaction.[3]

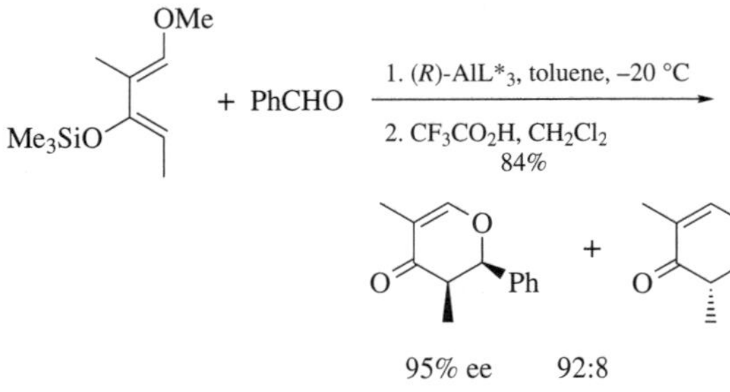

(1)

Asymmetric Diels–Alder Reaction. Although the asymmetric Diels–Alder reaction of cyclopentadiene with methacrolein is quite disappointing (17–23% ee) with the chiral aluminum reagent,[5] the use of α,β-unsaturated esters as dienophiles gives good enantioselectivity (eq 2).[6]

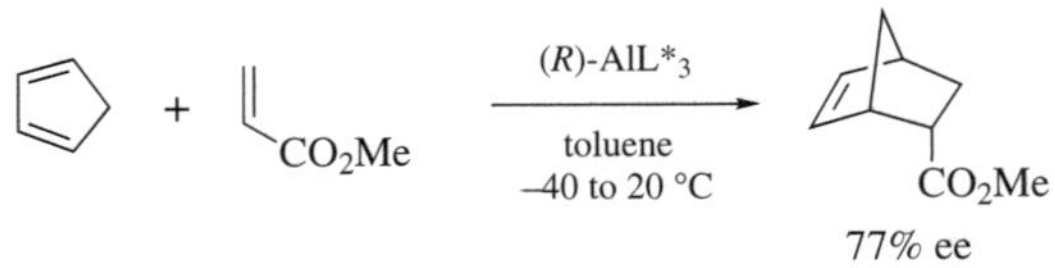

(2)

Asymmetric Ene Reaction. The enantioselective activation of carbonyl groups with the chiral aluminum reagent also enabled the asymmetric ene reaction of electron-deficient aldehydes with various alkenes (eq 3).[7] In the presence of powdered 4 Å molecular sieves, the chiral aluminum reagent can be utilized as a catalyst without loss of enantioselectivity.

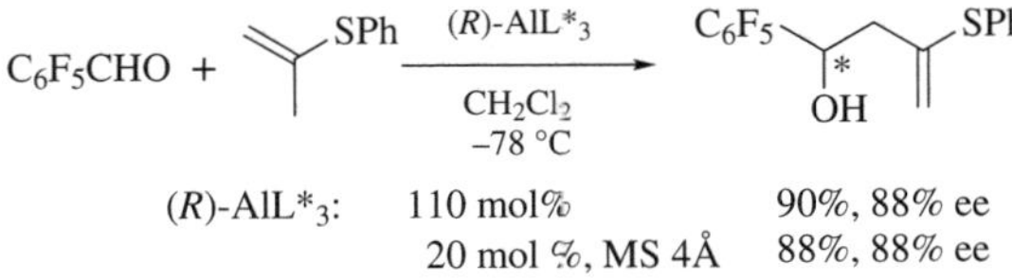

(3)

Asymmetric Claisen Rearrangement. The enantioselective activation of an ether oxygen with the chiral organoaluminum reagent allows for the first example of the asymmetric Claisen rearrangement of allylic vinyl ethers.[8] This method provides a facile asymmetric synthesis of various acylsilanes and acylgermanes from allylic α-(trimethylsilyl)vinyl ethers and allylic α-(trimethylgermyl)vinyl ethers, respectively (eq 4). Among various trialkylsilyl substituents of the chiral aluminum reagent, use of the bulkier *t*-butyldiphenylsilyl group results in the highest enantioselectivity.

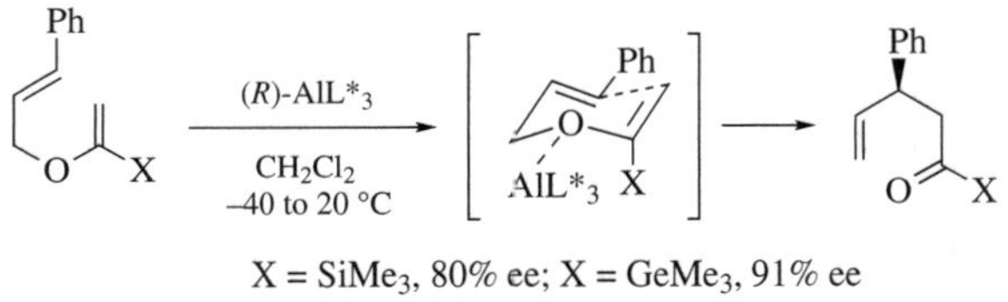

(4)

Notably, the asymmetric Claisen rearrangement of *cis*-allylic α-(trimethylsilyl)vinyl ethers with the chiral aluminum reagent produced optically active acylsilanes with the same absolute configuration as those from *trans*-allylic α-(trimethylsilyl)vinyl ethers (eq 5).[9]

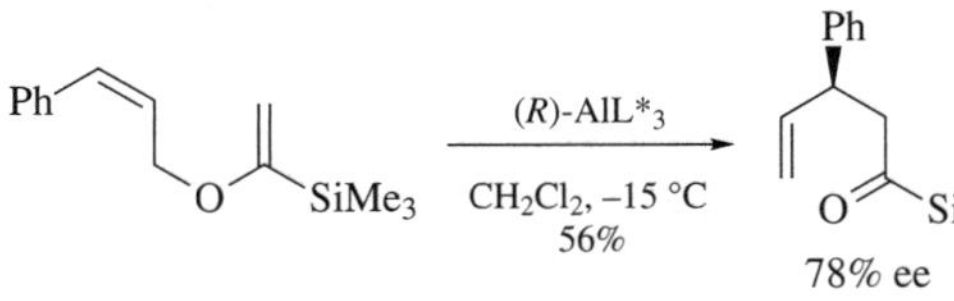

(5)

Asymmetric Polymerization. The chiral organoaluminum catalyst is utilized for asymmetric polymerization of racemic α-methyl and β-methyl β-lactones.[10] Optically active polymers possessing negative optical rotation values are produced, suggesting that (*S*) enantiomers of racemic β-lactones are preferentially activated by the aluminum catalyst.

1. Maruoka, K.; Itoh, T.; Araki, Y.; Shirasaka, T.; Yamamoto, H. *Bull. Chem. Soc. Jpn.* **1988**, *61*, 2975.
2. Maruoka, K.; Itoh, T.; Shirasaka, T.; Yamamoto, H. *J. Am. Chem. Soc.* **1988**, *110*, 310.
3. Maruoka, K.; Yamamoto, H. *J. Am. Chem. Soc.* **1989**, *111*, 789.
4. Burns, C.; Sharpless, K. B. *Chemtracts: Org. Chem.* **1988**, *1*, 123.
5. Bao, J.; Wulff, W. D. *J. Am. Chem. Soc.* **1993**, *115*, 3814.
6. Maruoka, K.; Concepcion, A. B.; Yamamoto, H. *Bull. Chem. Soc. Jpn.* **1992**, *65*, 3501.
7. Maruoka, K.; Hoshino, Y.; Shirasaka, T.; Yamamoto, H. *Tetrahedron Lett.* **1988**, *29*, 3967.
8. (a) Maruoka, K.; Banno, H.; Yamamoto, H. *J. Am. Chem. Soc.* **1990**, *112*, 7791. (b) Maruoka, K.; Banno, H.; Yamamoto, H. *Tetrahedron: Asymmetry* **1991**, *2*, 647.
9. Maruoka, K.; Yamamoto, H. *Synlett* **1991**, 793.
10. Sato, R.; Miyaki, N.; Takeishu, M. *Polymer Reprints, Jpn.* **1992**, *41*, 331.

Keiji Maruoka & Hisashi Yamamoto
Nagoya University, Japan

(–)-*endo*-Bornyltriazolinedione

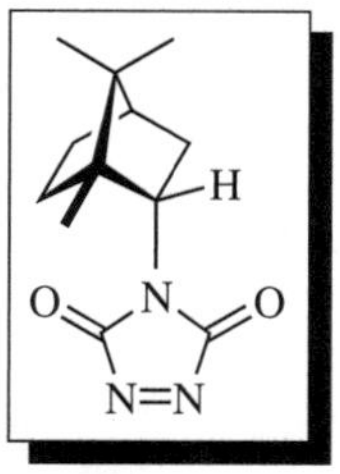

[73462-83-6] $C_{12}H_{17}N_3O_2$ (MW 235.29)

(reagent used in cycloaddition reactions[1] as a dienophile of high reactivity for the trapping of unstable intermediates,[2,3] and 1,3-dienes,[4] the preparation of optically active axially symmetric molecules,[5] and the resolution of hydrocarbons and chiral dienes[6,7])

Alternate Name: 4-(1,7,7-trimethylbicyclo[2.2.1]hept-2-yl)-[1,2,4]triazole-3,5-dione.

Physical Data: mp 152–154 °C; sublimation point 75–85 °C at 0.1 Torr; $[\alpha]_D^{20}$ −77 (*c* 5.7, CH_2Cl_2)

Form Supplied in: red solid prepared from optically pure *d*-camphor oxime[8] in a five-step sequence (eq 1): reduction with sodium in *n*-amyl alcohol followed by fractional recrystallization of the resulting hydrochloride salts of bornylamines[9,10] gives the *endo*-isomer in enantiomerically pure form; treatment with phosgene and direct condensation of the isocyanate with (ethoxycarbonyl)hydrazine gives a compound which cyclizes upon treatment with base; subsequent nitrogen dioxide oxidation furnishes (–)-*endo*-bornyltriazolinedione as a red crystalline solid.[11–13]

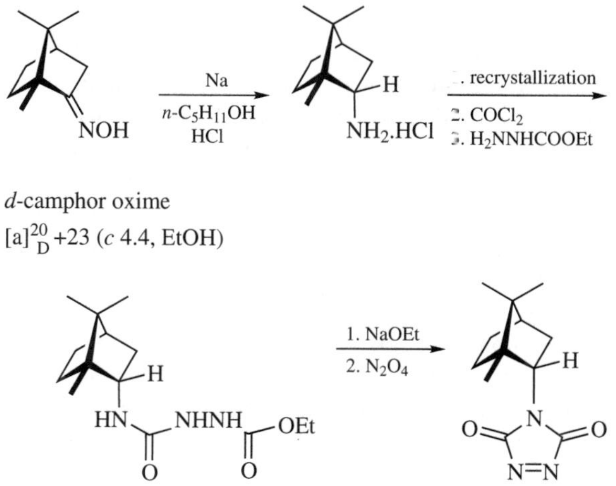

(1)

Handling, Storage, and Precautions: stable for prolonged periods when stored cold in the absence of light; easy to handle; solid.

The main interest in (–)-*endo*-bornyltriazolinedione resides with its high dienophilic character in cycloaddition processes to yield separable mixtures of diastereoisomeric urazoles. The non-destructive resolution of cyclooctatetraenes, which allows direct access to optically pure derivatives, is a typical illustration and has been amply demonstrated.[14–19] Typically, (–)-*endo*-bornyltriazolinedione is heated with racemic 1,2,3-trimethylcyclooctatetraene in ethyl acetate to afford a mixture of diastereoisomeric adducts, which can be separated by fractional recrystallization from ethyl acetate and hexane. HPLC is an alternative separation technique leading to both enantiomerically pure antipodes. The chiral auxiliary is subsequently removed by basic hydrolysis-manganese dioxide oxidation to afford the optically pure cyclooctatetraenes (eq 2).

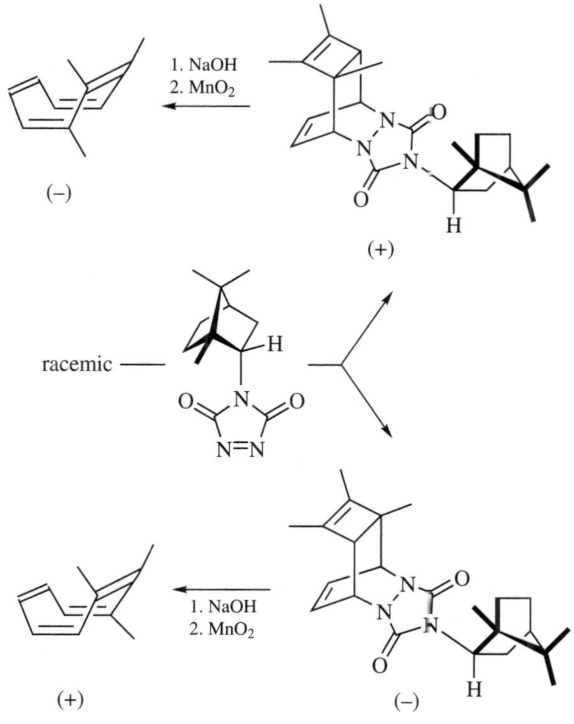

(2)

This method of resolution of polyolefins has been extensively studied for cyclooctatetraene systems where excellent enantiomeric excesses are normally observed. Lanthanide-induced shifting can be used to determine the diastereoisomeric composition of the urazoles.[14] Alternate means for the resolution of polyenes based on kinetic resolution using (+)-tetra-2-pinanylborane have been described,[20,21] but this reagent consumes valuable substrate. Chiral platinum complexes[22] can also be used but at prohibitive cost on a large scale and with poor regioselectivity when several coordination sites are present.

The utilization of an optically active triazolinedione for asymmetric transfer has also led to the preparation of enantiomerically pure polycyclic hydrocarbons. The method provides a straightforward means for introducing optical activity into chiral propellanes that possess a conjugated diene unit. Racemic propellane[5] reacts with (–)-*endo*-bornyltriazolinedione in ethyl acetate at $-78\,^{\circ}$C to give, after HPLC separation, the two optically pure urazoles. Subsequent reduction with lithium aluminum hydride affords the two propellanes in enantiomerically pure form (eq 3).

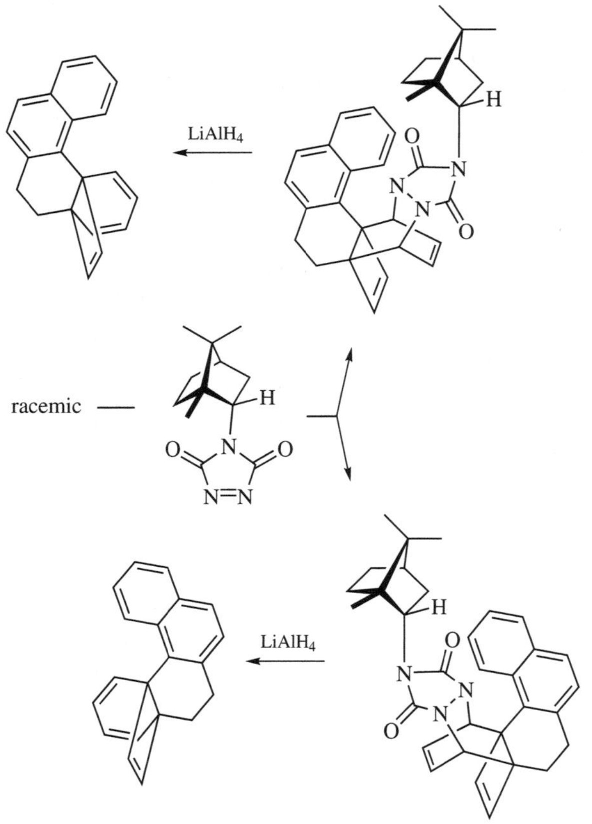

(3)

From a purely synthetic viewpoint, triazolinedione adducts have served as substrates for gaining access to numerous target molecules such as prismane,[23] semibullvalene,[24] bridged semibullvalenes,[25] elassovalene,[26] caged compounds,[27] and azoalkanes. Indeed, the title reagent can be used not only as a chiral source, but also as an azo donor. In the synthesis of 4,5-diazatwis-4-ene,[28] for example, (–)-*endo*-bornyltriazolinedione was a pivotal reactant that allowed incorporation of the azo unit in addition to providing a means for resolution. The first step involved cycloaddition to cyclooctatetraene dibromide with formation of a separable mixture of diastereoisomers. Eventual removal of the chiral moiety in a modified hydrolysis-oxidation sequence yielded the desired azo compound (eq 4).

The high crystallinity of the urazole obtained by cycloaddition has also made (–)-*endo*-bornyltriazolinedione a useful reagent for obtaining crystalline derivatives. Diels-Alder cycloaddition of (–)-*endo*-bornyltriazolinedione to a diene of unknown configuration resulted in the formation of a single cycloadduct[4] whose structure was confirmed by X-ray diffraction analysis of the urazole (eq 5).

Applications of (–)-*endo*-bornyltriazolinedione to asymmetric synthesis in cycloaddition reactions have shown low levels of induction. In the examples studied, (–)-*endo*-bornyltriazolinedione reacted almost instantaneously with various dienes even at low temperature ($<96\,^{\circ}$C) resulting in low asymmetric induction ($<10\%$).[29] The high reactivity of triazolinedione in [4+2] π-cycloadditions due to its high dienophilicity minimized differentiation between the transition states and has to date impeded its use in asymmetric synthesis.

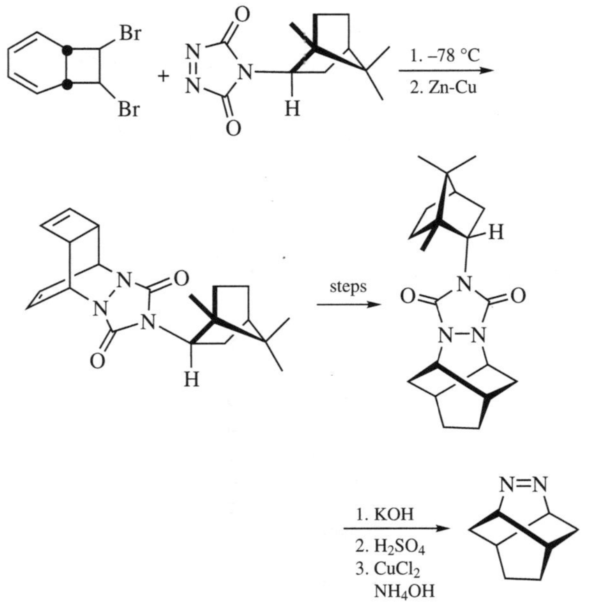

(4)

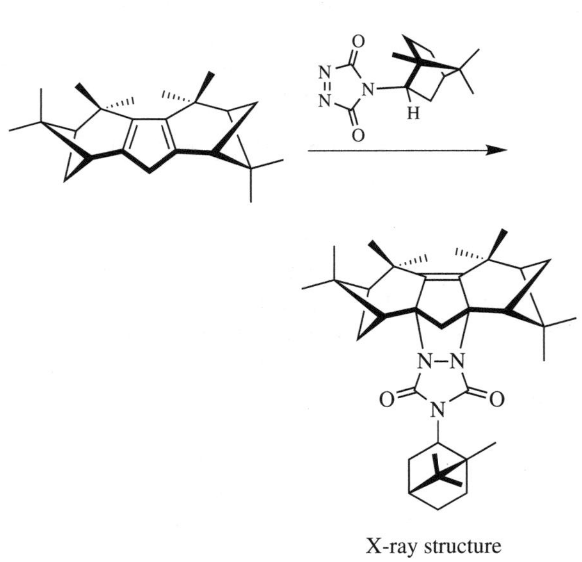
(5)

X-ray structure

Related Reagents. 4-phenyl-1,2,4-triazoline-3,5-dione; (−)-(α)-(methylbenzyl) triazolinedione; (dehydroabiethyl) triazolinedione; (+)-tetra-2-pinanylborane; chiral platinum complexes.

1. Diels, O. *Chem. Ber.* **1914** *47*, 2183.
2. Paquette, L. A.; Wang, T. Z. *J. Am. Chem. Soc.* **1988** *110*, 3663.
3. Horn, K. A.; Browne, A. R.; Paquette, L. A. *J. Org. Chem.* **1980** *45*, 5381.
4. Paquette, L. A.; Bzowej, E. I.; Kreuzholz, R. *Organometallics* **1996** *15*, 4857.
5. Klobucar, W. D.; Paquette, L. A.; Blount, J. F. *J. Org. Chem.* **1981** *46*, 4021.
6. Gardlik, J. M.; Paquette, L. A. *Tetrahedron Lett.* **1979** *20*, 3597.
7. Paquette, L. A.; Doehner, R. F.; Jenkins, J. A.; Blount, J. F. *J. Am. Chem. Soc.* **1980** *102*, 1188.
8. von Auwers, K. *Chem. Ber.* **1889** *22*, 605.
9. Forster, M. O. *J. Chem. Soc.* **1898**, 386.
10. Hückel, W.; Rieckmann, P. *Justus Liebigs Ann. Chem.* **1959** *625*, 1.
11. Cookson, R. C.; Gilani, S. S. H.; Stevens, I. D. R. *Tetrahedron Lett.* **1962**, *3*, 615.
12. Cookson, R. C.; Gilani, S. S. H.; Stevens, I. D. R. *J. Chem. Soc. (C)* **1967**, 1905.
13. Goering, H. L.; Eikenberry, J. N.; Koermer, G. S. *J. Am. Chem. Soc.* **1971**, *93*, 5913.
14. Paquette, L. A.; Trova, M. P.; Luo, J.; Clough, A. E.; Anderson, L. B. *J. Am. Chem. Soc.* **1990**, *112*, 228.
15. Paquette, L. A.; Trova, M. P. *Tetrahedron Lett.* **1986**, *27*, 1895.
16. Paquette, L. A.; Hanzawa, Y.; Hefferon, G. J.; Blount, J. F. *J. Org. Chem.* **1982**, *47*, 265.
17. Klobucar, W. D.; Burson, R. L.; Paquette, L. A. *J. Org. Chem.* **1981**, *46*, 2680.
18. Paquette, L. A.; Hanzawa, Y.; McCullough; K. J.; Tagle, B.; Swenson, W.; Clardy, J. *J. Am. Chem. Soc.* **1981**, *103*, 2262.
19. Paquette, L. A.; Gardlik, J. M.; Johnson, L. K.; McCullough, K. J. *J. Am. Chem. Soc.* **1980**, *102*, 5026.
20. Brown, H. C.; Ayyangar, N. R.; Zweifel, G. *J. Am. Chem. Soc.* **1964**, *86*, 397.
21. Brown, H. C.; Ayyangar, N. R.; Zweifel, G. *J. Am. Chem. Soc.* **1964**, *86*, 1071.
22. Cope, A. C.; Moore, W. R.; Bach, R. D.; Winkler, J. S. *J. Am. Chem. Soc.* **1970**, *92*, 1243.
23. Katz, T. J.; Acton, N. *J. Am. Chem. Soc.* **1973**, *95*, 2738 and references therein.
24. Paquette, L. A. *J. Am. Chem. Soc.* **1970**, *92*, 5765 and references therein.
25. Burson, R. L.; Paquette, L. A. *Tetrahedron* **1978**, *34*, 1307 and references therein.
26. Paquette, L. A.; Wallis, T. G.; Kempe, T.; Christoph, G. G.; Springer, J.; Clardy, J. *J. Am. Chem. Soc.* **1977**, *99*, 6946 and references therein.
27. Paquette, L. A.; James, D. R.; Birnberg, G. H. *J. Am. Chem. Soc.* **1974**, *96*, 7454 and references therein.
28. Jenkins, J. A.; Doehner, R. F.; Paquette, L. A. *J. Am. Chem. Soc.* **1980**, *102*, 2131.
29. Paquette, L. A.; Doehner, R. F. *J. Org. Chem.* **1980**, *45*, 5105.

Fabrice Gallou
The Ohio State University, Columbus, OH, USA

(4*R*,5*R*)-2-Bromo-1,3-bis[(4-methylphenyl)sulfonyl]-4,5-diphenyl-1,3,2-diazaborolidine

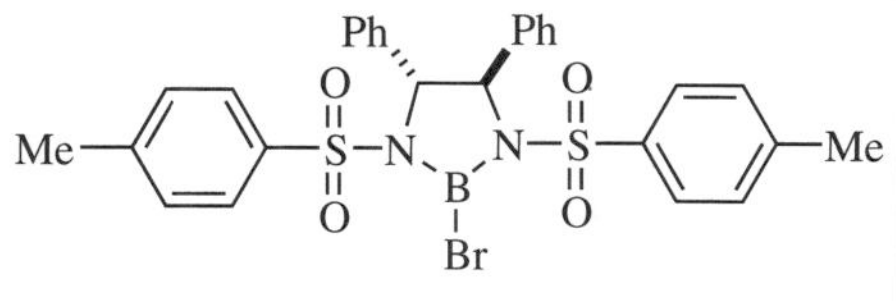

(*R*,*R*)-**1**

[121758-17-6] $C_{28}H_{26}BBrN_2O_4S_2$ (MW 609.36)

(chiral controller group for asymmetric carbonyl allylations,[1–8] allenations,[9] and propargylations,[9] enantioselective Claisen rearrangements,[10,11] and enantioselective enolborinations[12])

Alternate Name: 2-bromo-4,5-diphenyl-1,3-bis-(toluene-4-sulfonyl)-[1,3,2]diazaborolidine.
Solubility: soluble in dichloromethane.
Form Supplied in: not commercially available.
Preparative Methods: prepared from the corresponding bissulfonamide, (*R*,*R*)-1,2-diphenyl-1,2-diaminoethane-*N*,*N*′-bis(4-methylbenzenesulfonamide)[13–16] and BBr_3 in dichloromethane. After drying under high vacuum (0.1 mmHg) overnight (8–16 h) at 80–100 °C in a Schlenk flask, the sulfonamide (1.4 equiv) is dissolved in dichloromethane (0.1 M), and cooled to 0 °C. Care should be taken during drying, as temperatures above 100–110 °C produce a brownish-colored material which is insoluble in dichloromethane at 0.1 M and ineffective for chemical transformations. BBr_3 (1.0 M in dichloromethane; 1.4 equiv) is added, the mixture is stirred at 0 °C for 10 min, warmed to room temperature, and stirred for 1 h. The solvent and HBr are then carefully removed under high vacuum, kept at room temperature under high vacuum for 15 min after all solvent has been removed, and the white to tan residue is then redissolved in dichloromethane (0.1 M). This evaporation–redissolution procedure is repeated two additional times, giving a 0.1 M solution of **1** in CH_2Cl_2 dichloromethane.[7]
Handling, Storage, and Precautions: highly moisture-sensitive; should be prepared immediately prior to use under inert atmosphere, preferably using standard Schlenk techniques (use of a glove box not required). Best results are obtained when fresh solutions of BBr_3 are used.

Carbonyl Allylations. Reagent **1** was originally introduced by Corey in 1989 for the asymmetric allylation of aldehydes with tri-n-butylallylstannane,[2] as previously reviewed.[1] Subsequent studies have demonstrated the utility of this reagent for the stereocontrolled generation of complex homoallylic alcohols via the convergent coupling of various functionalized, C_2-symmetric allylstannanes and substituted aldehydes.[4–8] The absolute stereochemistry of the newly formed alcohol stereocenter is predictable using a Zimmerman–Traxler model, and product formation generally is governed by the absolute stereochemistry of **1** (eq 1).

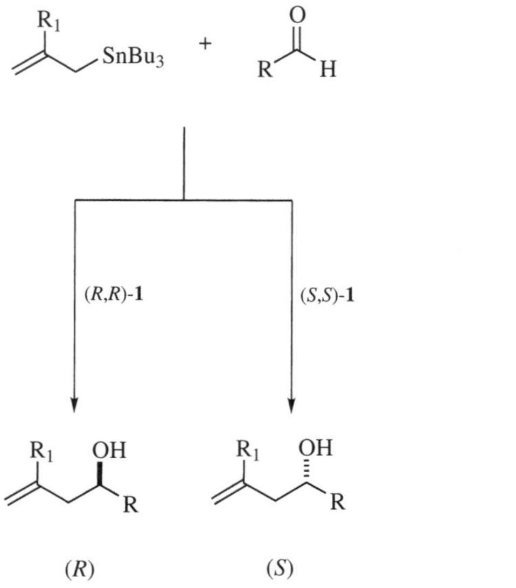

(1)

In situ transmetallation of the starting allylstannane to an intermediate allylic borane is rationalized via a 1,3-transposition pathway. In reactions with chiral aldehydes, matched and mismatched diastereotopic pathways are possible based upon the asymmetry of **1**, and the intrinsic face selectivity exhibited for the carbonyl addition process. Yields are generally high (85–99%) with good to excellent stereoselectivity. Numerous functional groups are tolerated in the starting allylstannane, including esters, silyl and benzyl or *para*-methoxybenzyl ethers, dithioketals, and vinylstannanes. Lewis acid sensitive functionalities (acetals, ketals, tetrahydropyranyl ethers) are not compatible. The aldehyde component may contain a wide variety of common protecting groups and additional functionality, including basic heteroaromatic systems such as pyridines and oxazoles.

Reactions of achiral aldehydes and homochiral stannanes exhibit stereoselectivity which is predominantly dictated by the chiral auxiliary **1** if the pre-existing asymmetry of the stannane is located at least two carbons or more (β) from the reactive allyl unit (eqs 2–4).[4,5]

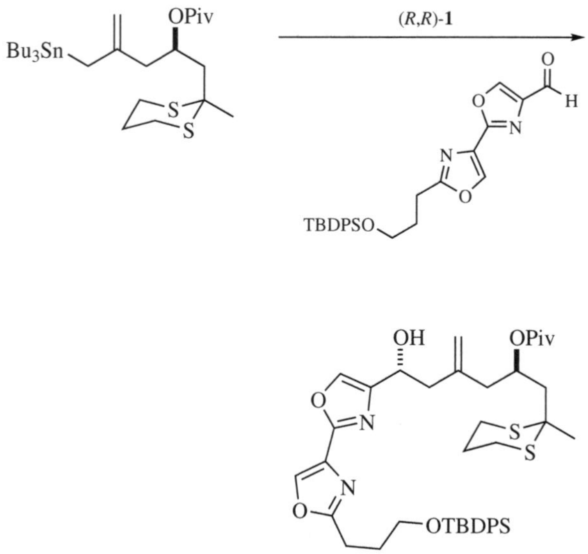

(2)

95%, 10.5:1 dr

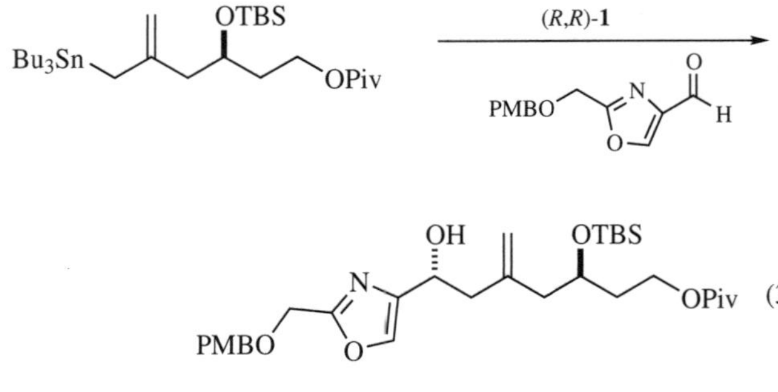

(3)

92%, ≥20:1 dr

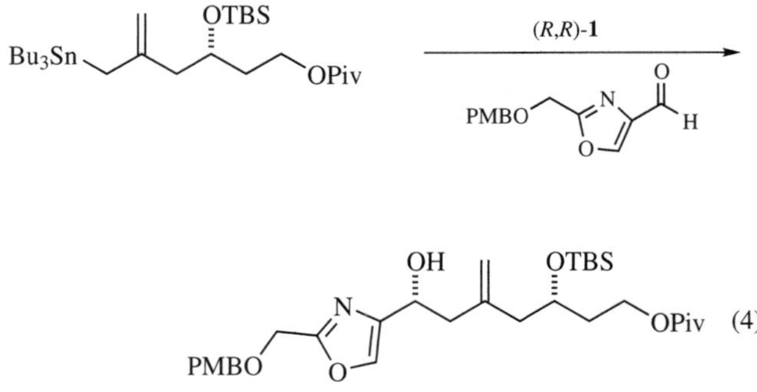

(4)

98%, ≥20:1 dr

Achiral stannanes undergo reactions with aldehydes bearing α-asymmetry, and provide examples of matched diastereoselectivity with respect to **1** (eq 5),[7] as well as cases of mismatched diastereoselection of these controlling factors (eq 6).[4]

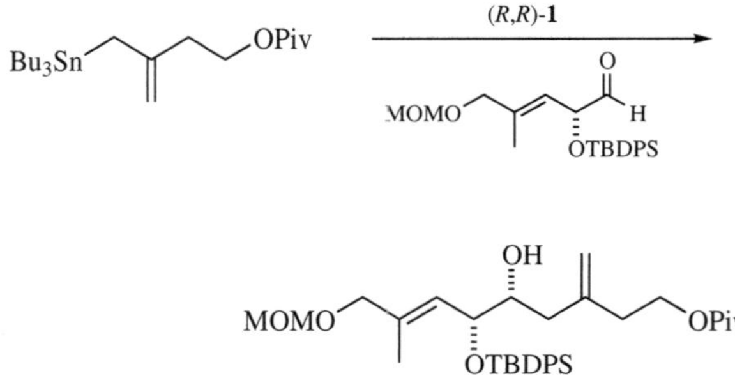

(5)

98%, 10:1 dr

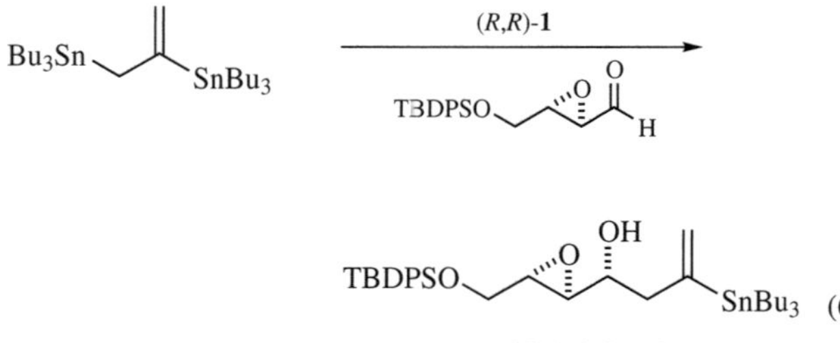

(6)

85%, 3.2:1 dr

In a similar fashion, asymmetric allylations with **1** and chiral aldehydes bearing β-substitution also display the expected behavior of diastereotopic transition states (eqs 7 and 8).[4]

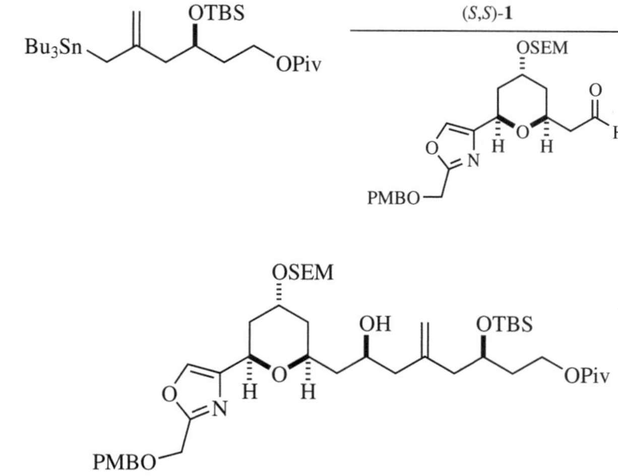
(7)

99%, 11.4:1 dr

The presence of α-asymmetry in the stannane component can have a dramatic impact on diastereoselection (eq 9).[4] The minimization of $A^{1,3}$ strain in the allylic component is a factor that influences the face selectivity enforced by the auxiliary **1**.

In complex examples, high levels of stereodifferentiation require the consideration of the conjoined influences of α-asymmetry in the allylstannane, and chirality of the starting aldehyde, in addition to the choice of auxiliary **1** (eq 10).[4,8]

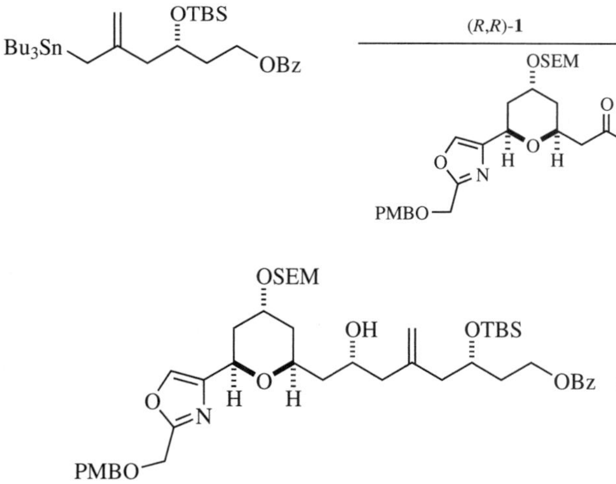
(8)

88%, 4:1 dr

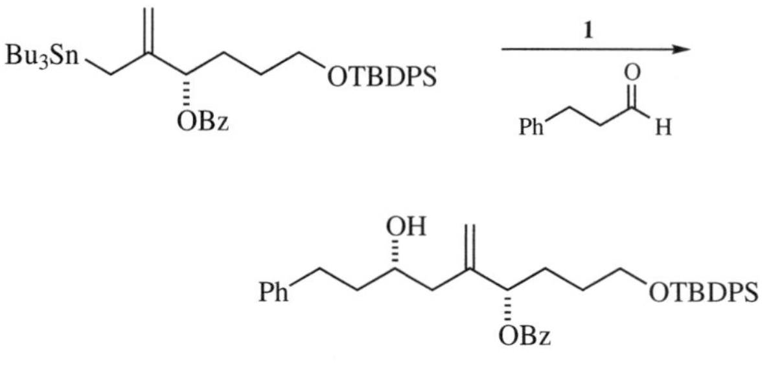
(9)

(*R*,*R*)-**1**: 96%, ≥20:1 dr
(*S*,*S*)-**1**: 94%, 1:1 dr

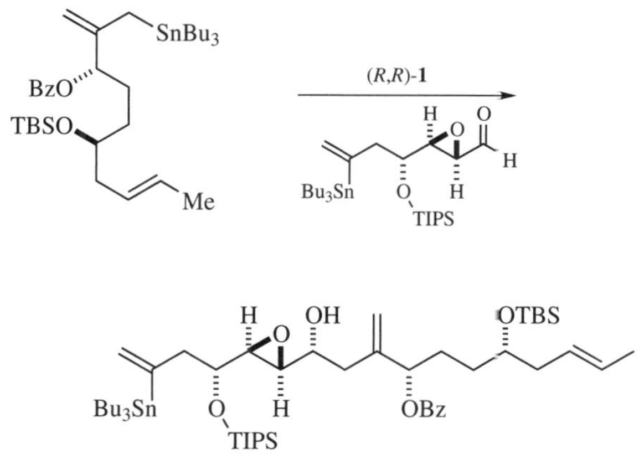
(10)

72%, 17:1 dr

Claisen Rearrangements. Claisen rearrangements of catechol allylic ethers, which avoid production of the 'abnormal' Claisen product, have been achieved using 1.5 equiv **1** and 1.5 equiv Et_3N at low temperature in dichloromethane with excellent (80–97%) yields and high (86–95%) enantioselectivities (eqs 11 and 12). The absolute configuration of the newly created benzylic stereocenter is dependent upon both the olefin geometry and the configuration of the controller. Lewis acid catalysis with (*S*,*S*)-**1** and *E*-olefins led to vinylic substituents bearing the *S* configuration (eq 11), whereas (*S*,*S*)-**1** and *Z*-olefins yielded products with *R* stereochemistry (eq 12).[10]

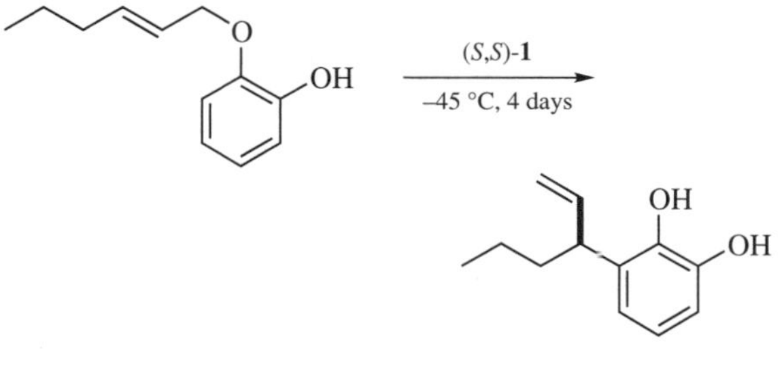
(11)

89%, 94% ee

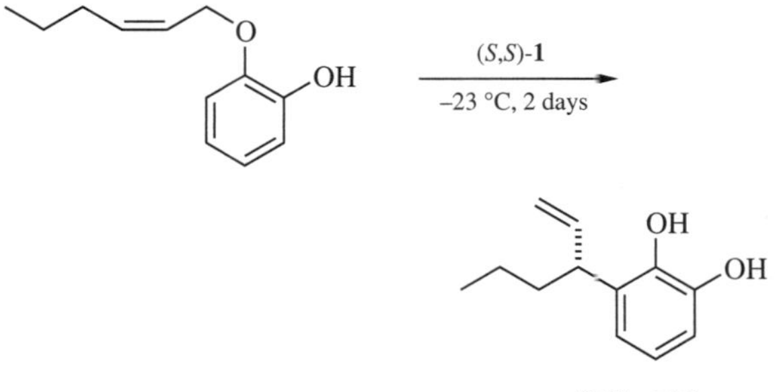
(12)

92%, 95% ee

Similarly, Claisen rearrangements of difluorovinyl allyl ethers occurred with moderate to excellent yields (39–90%) and moderate enantioselectivities (eq 13). Simple alkyl-substituted olefins rearrange at −15 °C with modest stereocontrol (41–56% ee) whereas vinylsilanes rearrange at −78 °C with good (85% ee) selectivity. The absolute configuration of the newly formed benzylic stereocenter appears to depend upon both the geometry (*E* or *Z*) of the starting olefin as well as the configuration of **1**, although the absolute stereochemistry of the product was proven only in the case cited below.[11]

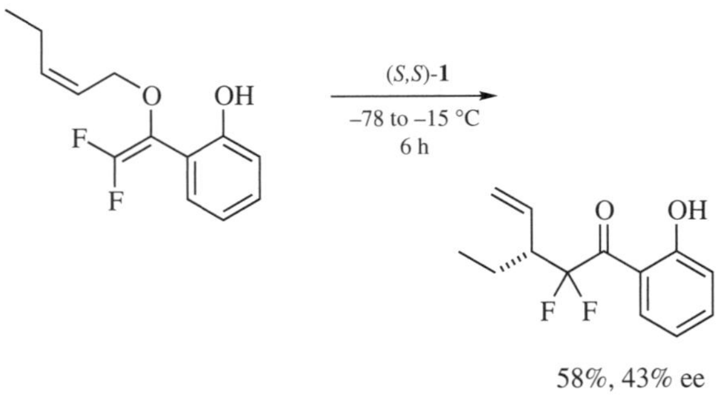

(13)

Other Uses. Reagent **1** has been used for enantioselective enolborination, albeit with poor (1.1:1) selectivity.[12] Similar bis-sulfonamide-derived boron Lewis acids have been used for aldol additions,[17–23] ester-Mannich reactions,[24] Diels–Alder reactions,[13,25,26] Ireland–Claisen reactions,[27,28] and [2,3]-Wittig rearrangements.[29,30] Similar bis-sulfonamide-derived aluminum Lewis acids have been used for aldol additions,[13] Diels–Alder reactions,[13,31–34] [2 + 2] ketene–aldehyde cycloadditions,[35,36] cyclopropanation of allylic alcohols,[37–39] and polymerization.[40,41]

Related Reagents. Boron-bissulfonamide Lewis acids: (*R*,*R*)-1,3-bis{[3,5-bis(trifluoromethyl)phenyl]sulfonyl}-2-bromo-4,5-diphenyl-1,3,2-diazaborolidine; (*R*,*R*)-1,3-bis(methylsulfonyl)-2-bromo-4,5-diphenyl-1,3,2-diazaborolidine; (*R*,*R*)-1,3-bis[(trifluoromethyl)sulfonyl]-2-bromo-4,5-diphenyl-1,3,2-diazaborolidine; (*R*,*R*)-1,3-bis(phenylsulfonyl)-2-bromo-4,5-diphenyl-1,3,2-diazaborolidine; (*R*,*R*)-1,3-bis[(4-fluorophenyl)sulfonyl]-2-bromo-4,5-diphenyl-1,3,2-diazaborolidine; (*R*,*R*)-1,3-bis[(4-nitrophenyl)sulfonyl]-2-bromo-4,5-diphenyl-1,3,2-diazaborolidine; (*R*,*R*)-1,3- bis-(2-naphthalenylsulfonyl)-2-bromo-4,5-diphenyl-1,3,2-diazaborolidine; (*R*,*R*)-1,3-bis(phenylsulfonyl)-2-bromooctahydro-1*H*-1,3,2-benzodiazaborole; (*R*,*R*)-1,3-bis[(4-methylphenyl)sulfonyl]-2-bromooctahydro-1*H*-1,3,2-benzodiazaborole.
Aluminum-bissulfonamide Lewis acids: (*R*,*R*)-[*N*,*N*′-(1,2-diphenyl-1,2-ethanediyl)bis(1,1,1-trifluoromethanesulfonamidato)](2-)- *N*,*N*′methylaluminum; (*R*,*R*)-{[*N*,*N*′-(1,2-diphenyl-1,2-ethanediyl)bis(1,1,1-trifluoromethanesulfonamidato)](2-)-*N*,*N*′}(2-methylpropyl)aluminum; (*R*,*R*)-{[*N*,*N*′-(1,2-diphenyl-1,2-ethanediyl)bis(4-methylbenzenesulfonamidato)](2-)-*N*,*N*′}chloroaluminum; (*R*,*R*)-{[*N*,*N*′-(1,2-diphenyl-1,2-ethanediyl)bis[3,5-bis(trifluoromethyl)benzenesulfonamidato]](2-)-*N*,*N*′}ethylaluminum; (*S*,*S*)-[*N*,*N*′-(1,2-diphenyl-1,2-ethanediyl)bis[2,4,6-trimethylbenzenesulfonamidato](2-)-*N*,*N*′]methylaluminum; (*S*,*S*)-[*N*,*N*′-(1,2-diphenyl-1,2-ethanediyl)(2,4,6-trimethylbenzenesulfonamidato)-2,4,6-tris(1-methylethyl)benzenesulfonamidato(2-)-*N*,*N*′]methylaluminum; (*S*,*S*)-{[*N*,*N*′-(1,2-diphenyl-1,2-ethanediyl)bis[4-(1,1-dimethylethyl)-2,6-dimethylbenzenesulfonamidato]](2-)-*N*,*N*′}methylaluminum; (*S*,*S*)-{*N*,*N*′-(1,2-diphenyl-1,2-ethanediyl)[4-(1,1-dimethylethyl)-2,6-dimethylbenzenesulfonamidato]-2,4,6-tris(1-methylethyl)benzenesulfonamidato(2-)-*N*,*N*′}trimethylaluminum; (*R*,*R*)-{(*N*,*N*′-(1,2-diphenyl-1,2-ethanediyl)-bis[2,4,6-tris(1-methylethyl)benzenesulfonamidato])(2-)-*N*,*N*′}-ethylaluminum; (*S*,*S*)-{[*N*,*N*′-[1,2-bis(3,5-dimethylphenyl)-1,2-ethanediyl]bis(1,1,1-trifluoromethanesulfonamidato)](2-)-*N*,*N*′}methylaluminum; (*R*,*R*)-{[*N*,*N*′-1,2-cyclohexanediylbis(1,1,1-trifluoromethanesulfonamidato)](2-)-*N*,*N*′}methylaluminum; (*R*,*R*){[*N*,*N*′-1,2-cyclohexanediylbis(benzenesulfonamidato)]-(2-)-*N*,*N*′}(2-methylpropyl)aluminum; (*R*,*R*)-{[*N*,*N*′-1,2-cyclohexanediylbis(4-nitrobenzenesulfonamidato)](2-)-*N*,*N*′}methylaluminum; (*R*,*R*)-{[*N*,*N*′-1,2-cyclohexanediylbis(4-nitrobenzenesulfonamidato)](2-)-*N*,*N*′}ethylaluminum; (*R*,*R*)-{[*N*,*N*′-1,2-cyclohexanediylbis(4-nitrobenzenesulfonamidato)](2-)-*N*,*N*′}(2-methylpropyl)aluminum; (*R*,*R*)-{(*N*,*N*′-1,2-cyclohexanediylbis[4-(trifluoromethyl)benzenesulfonamidato])(2-)-*N*,*N*′}(2-methylpropyl)aluminum; (*R*,*R*)-{(*N*,*N*′-1,2-cyclohexanediylbis[3,5-bis(trifluoromethyl)benzenesulfonamidato])(2-)-*N*,*N*′}(2-methylpropyl)aluminum.
Other chiral controllers for allylation: (*R*)-[(1,1′-binaphthalene)-2,2′-diolato(2-)-κ*O*, κ*O*′]dichlorotitanium; (*R*)-[(1,1′-binaphthalene)-2,2′-diolato(2-)-κ*O*, κ*O*′]bis(2-propanolato)titanium; (*R*)-[(1,1′-binaphthalene)-2,2′-diolato(2-)-κ*O*, κ*O*′]bis(2-propanolato)zirconium; (*R*)-[(1,1′-binaphthalene)-2,2′-diylbis(diphenylphosphine-κ*P*)]trifluoromethanesulfonato-κ*O*-silver; chloro(η^5-cyclopentadienyl)[(4*R*, *trans*)-2,2-dimethyl-α,α,α′,α′-tetraphenyl-1,3-dioxolane-4,5-dimethanolato(2-)-*O*α,*O*α′]titanium; 2,2-dimethyl-α,α,α′,α″-tetraphenyl-1,3-dioxolane-4,5-dimethanolatotitanium diisopropoxide; chloro(cyclopentadienyl)bis[3-*O*-(1,2:5,6-di-*O*-isopropylidene-α-D-glucofuranosyl)]titanium; {2,2′-methylenebis[(4*S*,5*R*)-4,5-dihydro-4,5-diphenyloxazole-κ*N*3]}bis(trifluoromethanesulfonato)-κ*O*-zinc; aqua{2,6-bis[(4*S*)-4,5-dihydro-4-(1-methylethyl)-2-oxazolyl-κ*N*3]phenyl-κ*C*}dichlororhodium; (*S*,*S*)-[2,6-bis(1-methylethoxy)benzoyl]-oxy-5-oxo-3,2-dioxaborolane-4-acetic acid; *B*-methoxydiisopinocampheylborane; 1,3,2-benzodioxastannol-2-ylidene complex with diisopropyl tartrate; 2,2,2-trifluoro-*N*-[(1*R*, 2*R*)-1-methyl-2-phenyl-2-(trimethylsilyl)oxy]ethylacetamide; (*R*,*R*)-octahydro-1,3-dimethyl-2-(1-piperidinyl)-1*H*-1,3,2-benzodiazaphosphole-2-oxide.

1. Gage, J. R., In *Encyclopedia of Reagents for Organic Synthesis*; Paquette, L. A., Ed; Wiley: Chichester, 1995, Vol. 4, p 2207.
2. Corey, E. J.; Yu, C.-M.; Kim, S. S. *J. Am. Chem. Soc.* **1989**, *111*, 5495.
3. Corey, E. J.; Huang, H.-C. *Tetrahedron Lett.* **1989**, *30*, 5235.
4. Williams, D. R.; Brooks, D. A.; Meyer, K. G.; Clark, M. P. *Tetrahedron Lett.* **1998**, *39*, 7251.
5. Williams, D. R.; Brooks, D. A.; Berliner, M. A. *J. Am. Chem. Soc.* **1999**, *121*, 4924.
6. Williams, D. R.; Clark, M. P.; Berliner, M. A. *Tetrahedron Lett.* **1999**, *40*, 2287.
7. Williams, D. R.; Clark, M. P.; Emde, U.; Berliner, M. A. *Org. Lett.* **2000**, *2*, 3023.
8. Williams, D. R.; Meyer, K. G. *J. Am. Chem. Soc.* **2001**, *123*, 765.
9. Corey, E. J.; Yu, C.-M.; Lee, D.-H. *J. Am. Chem. Soc.* **1990**, *112*, 879.
10. Ito, H.; Sato, A.; Taguchi, T. *Tetrahedron Lett.* **1997**, *38*, 4815.
11. Ito, H.; Sato, A.; Kobayashi, T.; Taguchi, T. *Chem. Commun.* **1998**, 2441.
12. Ward, D. E.; Lu, W.-L. *J. Am. Chem. Soc.* **1998**, *120*, 1098.
13. Corey, E. J.; Imwinkelried, R.; Pikul, S.; Xiang, Y. B. *J. Am. Chem. Soc.* **1989**, *111*, 5493.
14. Pikul, S.; Corey, E. J. *Org. Synth.* **1993**, *71*, 22.
15. Wang, S.-M.; Sharpless, B. A. *J. Org. Chem.* **1994**, *59*, 8302.
16. Pini, D.; Iuliano, A.; Rosini, C.; Salvadori, P. *Synthesis* **1990**, 1023.

17. Corey, E. J.; Kim, S. S. *J. Am. Chem. Soc.* **1990**, *112*, 4976.
18. Corey, E. J.; Kim, S. S. *Tetrahedron Lett.* **1990**, *31*, 3715.
19. Corey, E. J.; Soongyu, C. *Tetrahedron Lett.* **1991**, *32*, 2857.
20. Corey, E. J.; Lee, D.-H.; Soongyu, C. *Tetrahedron Lett.* **1992**, *33*, 6735.
21. Corey, E. J.; Lee, D.-H. *Tetrahedron Lett.* **1993**, *34*, 1737.
22. Corey, E. J.; Soongyu, C. *Tetrahedron Lett.* **2000**, *41*, 2765.
23. Corey, E. J.; Soongyu, C. *Tetrahedron Lett.* **2000**, *41*, 2769.
24. Corey, E. J.; Decicco, C. P.; Newbold, R. C. *Tetrahedron Lett.* **1991**, *32*, 5287.
25. Bienayme, H.; Longeau, A. *Tetrahedron* **1997**, *53*, 9637.
26. Richardson, B. M.; Day, C. S.; Welker, M. E. *J. Organomet. Chem.* **1999**, *577*, 120.
27. Corey, E. J.; Lee, D.-H. *J. Am. Chem. Soc.* **1991**, *113*, 4026.
28. Corey, E. J.; Roberts, B. E.; Dixon, B. R. *J. Am. Chem. Soc.* **1995**, *117*, 193.
29. Fujimoto, K.; Nakai, T. *Tetrahedron Lett.* **1994**, *35*, 5019.
30. Fujimoto, K.; Matsuhashi, C.; Nakai, T. *Heterocycles* **1996**, *42*, 423.
31. Corey, E. J.; Imai, N.; Pikul, S. *Tetrahedron Lett.* **1991**, *32*, 7517.
32. Corey, E. J.; Sarchar, S.; Bordner, J. *J. Am. Chem. Soc.* **1992**, *114*, 7938.
33. Pikul, S.; Corey, E. *J. Org. Synth.* **1993**, *71*, 30.
34. Corey, E. J.; Sarshar, S.; Lee, D.-H. *J. Am. Chem. Soc.* **1994**, *116*, 12089.
35. Dymock, B. W.; Kocienski, P. J.; Pons, J.-M. *Chem. Commun.* **1996**, 1053.
36. Miyano,S.; Hattori, T.; Uesugi, S.; Tamai, Y.; Sayo, N., Jpn. Kokai Tokkyo Koho (CAN 131:228639) 1999.
37. Imai, N.; Takahashi, H.; Kobayashi, S. *Chem. Lett.* **1994**, 177.
38. Kobayashi, S.; Imai, N.; Takahashi, H., Jpn. Kokai Tokkyo Koho (CAN 123:169276) 1995.
39. Kobayashi, S.; Imai, N.; Takahashi, H., Jpn. Kokai Tokkyo Koho (CAN 124:146482) 1995.
40. Itsuno, S.; Tada, S.; Ito, K. *Chem. Commun.* **1997**, 933.
41. Kamahori, K.; Tada, S.; Ito, K.; Itsuno, S. *Macromolecules* **1999**, *32*, 541.

David R. Williams & David C. Kammler
Indiana University, Bloomington, Indiana, USA

3-Bromocamphor-8-sulfonic Acid[1]

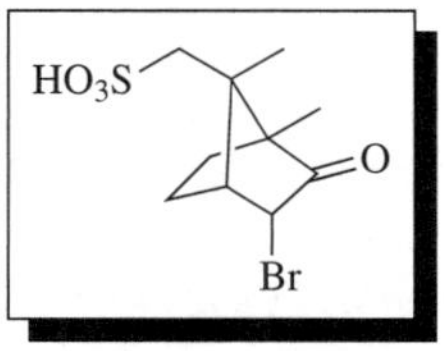

(1*S*)		
[46472-20-2]	$C_{10}H_{15}BrO_4S$	(MW 311.19)
(1*R*)		
[5344-58-1]		
(1*S*) (NH_4^+ salt)		
[55870-50-3]	$C_{10}H_{18}BrNO_4S$	(MW 328.22)
(1*R*) (NH_4^+ salt)		
[14575-84-9]		
(±) (NH_4^+ salt)		
[122519-23-7]		

(chemical resolutions;[1] starting material for the preparation of chiral reagents[2])

Alternate Name: α-bromocamphor-π-sulfonic acid; 3-bromocamphor-9-sulfonic acid.

Physical Data: free acid: mp 195–196 °C; (1*R*): $[\alpha]_D$ 88.3° (*c* 2.6, H_2O). NH_4^+ salt: mp 284 °C (dec); (1*R*): $[\alpha]_D$ 84.8° (*c* 4, H_2O)

Solubility: the ammonium salt is sol in water, slightly sol in EtOH, but essentially insol in acetone and Et_2O. The free acid is sol in EtOAc, MeCN, and 5% aq. NaOH.

Form Supplied in: both enantiomers are commercially available as ammonium salts.

Preparative Methods: by sulfonation of bromocamphor with *Chlorosulfonic Acid* in $CHCl_3$.[3] Alternatively, fuming *Sulfuric Acid* can be used as both the solvent and sulfonating agent.[4] Recently, an improved preparation with an easier isolation procedure was reported (34% yield).[5] The acid can be prepared by passing a solution of the NH_4^+ salt in H_2O through a Dowex resin (H^+) column.[6] Alternatively, it can be obtained by adding *Acetyl Chloride* to a suspension of the ammonium salt in a 2:1 mixture of $CHCl_3$ and absolute EtOH.[7] The corresponding sulfonyl chloride is readily prepared from the acid or the NH_4^+ salt upon treatment with *Phosphorus(V) Chloride* and *Phosphorus Oxychloride*.[8,9]

Purification: crystallized from H_2O.

Handling, Storage, and Precautions: a 0.5 M solution of the free acid in dry MeCN is stable for at least 30 days at 5 °C in a closed vessel under N_2.[6]

Chemical Resolution of Compounds Containing Basic Groups. 3-Bromocamphor-8-sulfonic acid has been widely used as a resolving agent for compounds containing basic groups. A number of primary (**1**),[10] secondary (**2**),[11] and tertiary (**3**) amines[12] as well as oxazolines (**4**)[6] have been resolved by the formation of diastereomeric salts derived from 3-bromo-8-camphorsulfonic acid.

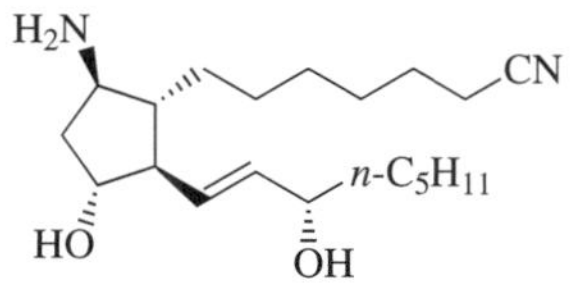

(**1**)

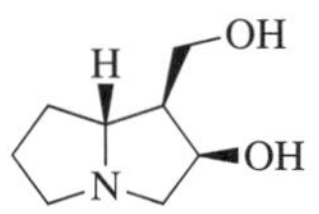

(**3**)

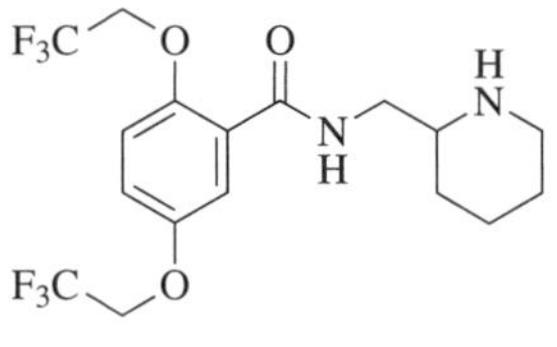

(**2**) Flecainide

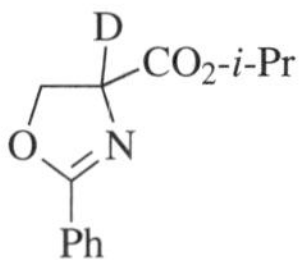

(**4**)

The optical resolution of racemic *p*-hydroxyphenylglycine with 3-bromocamphor-8-sulfonic acid has also been achieved.[13] This resolving agent has also been widely used in the preparation of optically pure chromium[14] and cobalt complexes.[15]

Preparation of Chiral Reagents. 3-Bromocamphor-8-sulfonic acid has been used as a starting material for the synthesis of chiral reagents.[16] Although the oxidation of sulfides to sulfoxides can be accomplished with the oxaziridine (**5**) or (**6**), other camphor-derived oxaziridines (**7** and **8**) are the reagents of choice to accomplish this transformation (eq 1).[2,9,17]

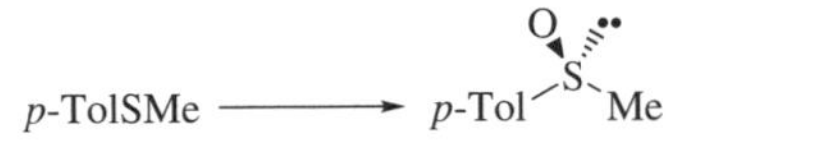

(1)

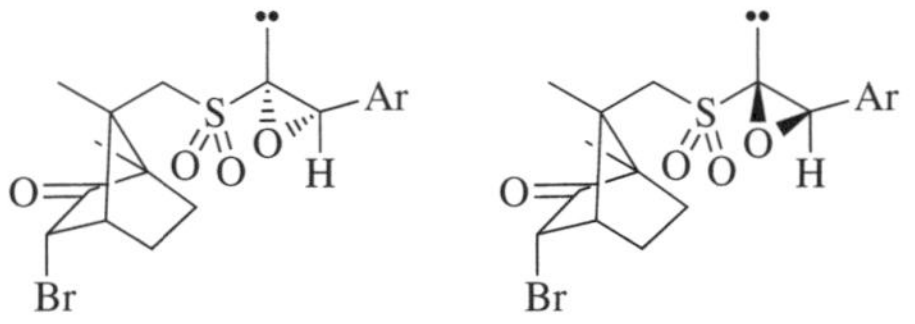

(5) 21% ee (*S*), 75–80% yield **(6)** 26% ee (*R*), 75–80% yield

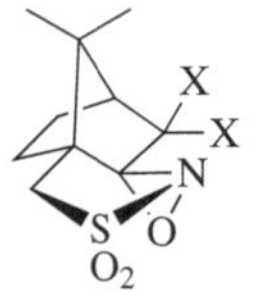

(7) X = H, 28% ee, (*S*), 80% yield
(8) X = Cl, 95% ee, (*S*), 95% yield

1. (a) Newman, P. *Optical Resolution Procedures for Chemical Compounds*; Optical Resolution Information Center, Manhattan College: Riverdale, NY, 1978–1993; Vol. I–IV. (b) Wilen, S. H. *Top. Stereochem.* **1971**, *6*, 107.
2. Davis, F. A.; Chen, B.-C. *Chem. Rev.* **1992**, *92*, 919.
3. Kipping, F. S.; Pope, W. J. *J. Chem. Soc.* **1895**, *67*, 354.
4. Kauffman, G. B. *J. Prakt. Chem.* **1966**, *33*, 295.
5. Hammershøi, A.; Hansson, E.; Springborg, J. *Inorg. Synth.* **1989**, *26*, 24.
6. Reider, P. J.; Conn, R. S. E.; Davis, P.; Grenda, V. J.; Zambito, A. J.; Grabowski, E. J. J. *J. Org. Chem.* **1987**, *52*, 3326.
7. Schowen, K. B.; Smissman, E. E.; Stephen, W. F., Jr. *J. Med. Chem.* **1975**, *18*, 292.
8. Cremlyn, R.; Bartlett, M.; Wu, L. *J. Pharm. Sci.* **1988**, *39*, 173.
9. Davis, F. A.; Jenkins, R. H., Jr.; Awad, S. B.; Stringer, O. D.; Watson, W. H.; Galloy, J. *J. Am. Chem. Soc.* **1982**, *104*, 5412.
10. Corey, E. J.; Vlattas, I.; Harding, K. *J. Am. Chem. Soc.* **1969**, *91*, 535.
11. Banitt, E. H.; Schmid, J. R.; Newmark, R. A. *J. Med. Chem.* **1986**, *29*, 299.
12. Aasen, A. J.; Culvenor, C. C. J. *J. Org. Chem.* **1969**, *34*, 4143.
13. Yamada, S.; Hongo, C.; Yoshioka, R.; Chibata, I. *Agric. Biol. Chem.* **1979**, *43*, 395.
14. Sakabe, Y. *Inorg. Chim. Acta* **1990**, *168*, 237.
15. (a) Kauffman, G. B.; Lindley, E. V., Jr. *Inorg. Synth.* **1976**, *16*, 93. (b) Kauffman, G. B.; Lindley, E. V., Jr. *J. Chem. Educ.* **1974**, *51*, 424.
16. Dauphin, G.; Kergomard, A.; Scarset, A. *Bull. Soc. Chem. Fr. Part 2* **1976**, 862.
17. (a) Davis, F. A.; Towson, J. C.; Weismiller, M. C.; Lal, S.; Carroll, P. J. *J. Am. Chem. Soc.* **1988**, *110*, 8477. (b) Davis, F. A.; Reddy, R. T.; Weismiller, M. C. *J. Am. Chem. Soc.* **1989**, *111*, 5964.

André B. Charette
Université de Montréal, QC, Canada

3-Bromo-5,6-diphenyl-2,3,5,6-tetrahydro-4*H*-oxazin-2-one[1]

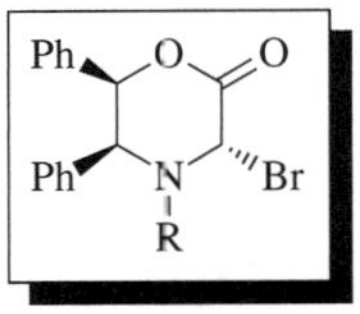

(R = *t*-Boc) (3*S*,5*S*,6*R*)
[112741-51-2] $C_{21}H_{22}BrNO_4$ (MW 432.31)
(R = *t*-Boc) (3*R*,5*R*,6*S*)
[127420-01-3]
(R = Cbz) (3*S*,5*S*,6*R*)
[111934-06-6] $C_{24}H_{20}BrNO_4$ (MW 466.33)
(R = Cbz) (3*R*,5*R*,6*S*)
[117527-28-3]

(electrophilic glycine equivalent useful for the preparation of α-substituted-α-amino acids in high enantiomeric excess[2])

Alternate Name: 3-bromo-5,6-diphenylmorpholin-2-one.
Physical Data: white solid, decomposes upon heating.
Solubility: sol THF, CH_2Cl_2.
Preparative Methods: N-t-Boc- and *N*-Cbz-3-bromo-5,6-diphenyl-2,3,5,6-tetrahydrooxazin-2-ones are not commercially available. They are prepared by addition of 1 equiv of *N-Bromosuccinimide* to a solution of the parent oxazinone (commercially available[3] as the individual enantiomers or as racemates) in CCl_4 at reflux. Upon cooling of the reaction mixture to 0 °C and filtering off the succinimide, the CCl_4 is removed under reduced pressure and the bromooxazinone is obtained in essentially quantitative yield as a white solid and is used without further purification.[2]
Handling, Storage, and Precautions: generally prepared immediately prior to use. Chromatography on silica gel results in decomposition.

General Reactivity. The *N*-protected 3-bromo-5,6-diphenyl-2,3,5,6-tetrahydrooxazin-2-ones serve as chiral electrophilic glycine equivalents. They are prepared as discussed above to yield the *anti*-diastereomer exclusively (eq 1).[2] The bromide is subject to displacement by a variety of reagents under a range of conditions to afford the substituted oxazinone generally with the newly introduced substituent oriented *anti* to the C(5) and C(6) phenyl groups. Deprotection of the heterocyclic amino acid precursor is accomplished by scission of the benzylic carbon–heteroatom bonds via reductive or oxidative cleavage. The deprotection routes afford the amino acid zwitterion or *N-t*-Boc amino acid directly but also result in destruction of the chiral auxiliary. Hydrogenolysis of the bromooxazinone with deuterium[4c] or tritium[4a,b] using *Palladium(II) Chloride* as catalyst occurs with net retention of configuration to afford the chiral isotopically labeled glycine (eq 2). Ease of preparation and introduction of the isotope in the final step make this a valuable synthesis of chiral glycines.

Coupling with Allylsilanes. Allyltrimethylsilanes react with the bromooxazinone in the presence of *Zinc Chloride* in THF to afford the allylated heterocycles with high selectivity (eq 3).[2] The coupling is presumed to take place by an S_N1 mechanism in which the Lewis acid promotes expulsion of bromide resulting in iminium ion formation. The heterocyclic iminium ion then undergoes attack by the nucleophile on the least hindered face, giving the *anti* diastereomer. Hydrogenolysis of the Cbz protected oxazinone (20–50 psi) affords the amino acid zwitterion in good yield and high chemical purity (eq 4).

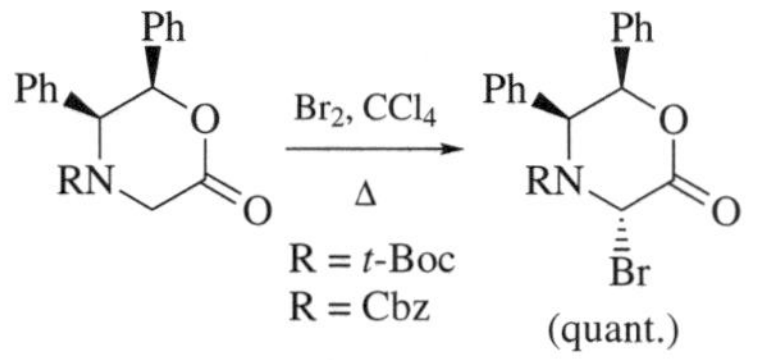
(1)

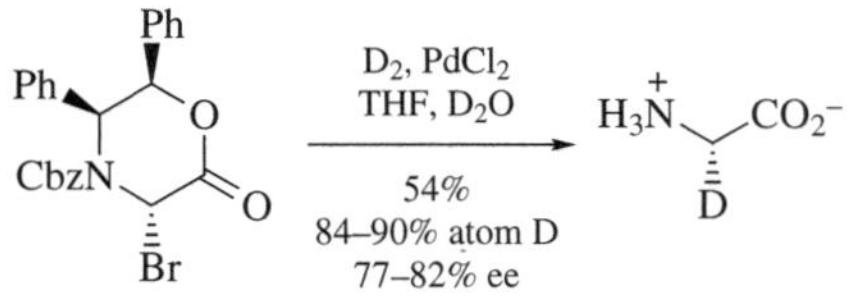
(2)

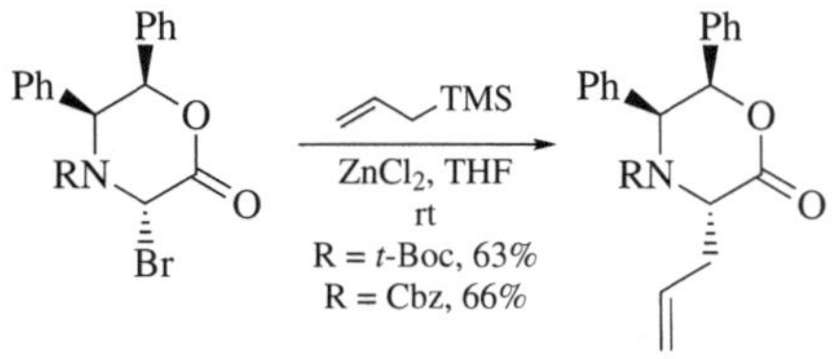
(3)

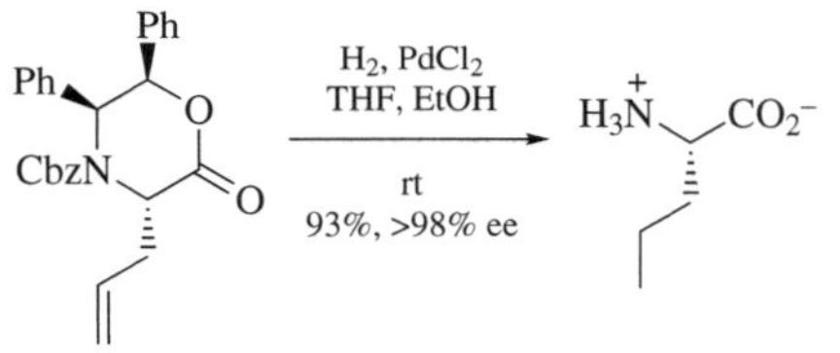
(4)

The amino acids can also be liberated by dissolving metal reduction.[2] Treatment of the oxazinone with *Lithium* or *Sodium* metal in liquid ammonia at −33 °C effects deprotection (eq 5). Ion exchange chromatography yields the zwitterionic amino acid free of inorganic salts. This procedure has the advantage of permitting the synthesis of amino acids possessing unsaturated side chains. When the dissolving metal reduction is carried out on the *t*-Boc protected oxazinone, the *N*-*t*-Boc amino acid is obtained directly (eq 6).

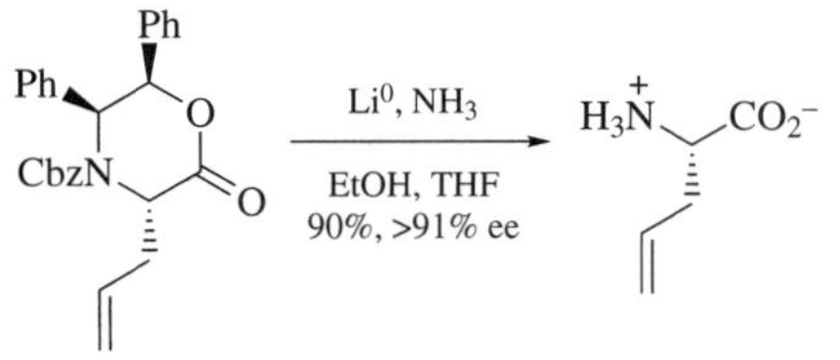
(5)

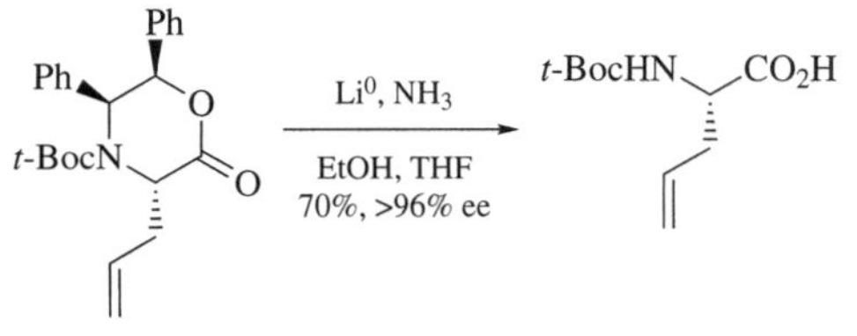
(6)

Coupling with Tin Acetylides. Trialkyltin acetylides react with the bromooxazinone in the presence of $ZnCl_2$ to furnish the alkyne-substituted heterocycle (eq 7).[5] Hydrogenation of the Cbz-protected acetylide adducts yields the aliphatic amino acids in good yield and high enantiomeric excess (eq 8).[5b] Dissolving metal reduction affords the (*E*)-vinylglycines, though some racemization is observed. The use of sodium metal in the deprotection results in higher chemical yields (71–80%) and lower enantiomeric excess (56–68%) while the use of lithium metal gives better enantiomeric excess (65–98%) but much lower chemical yields (16–20%) (eq 9).[5a]

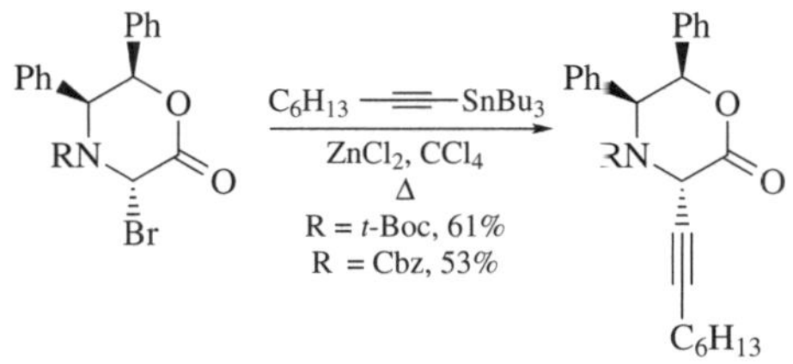
(7)

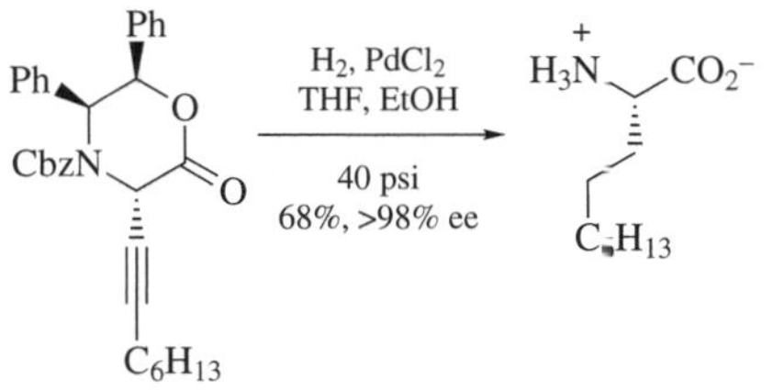
(8)

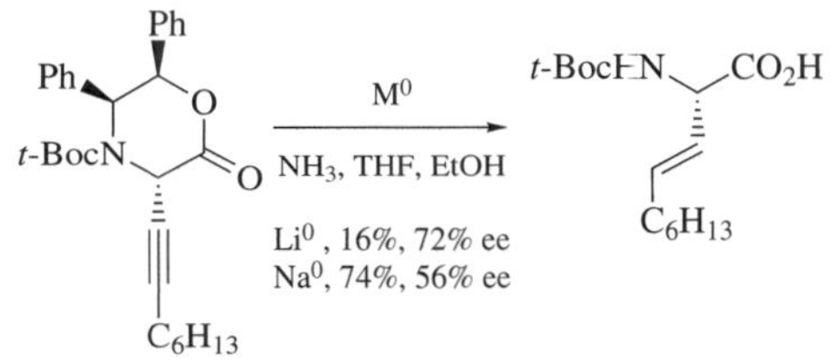
(9)

Coupling with Electron-Rich Arenes. Electron-rich aromatics such as trimethoxybenzene, furan, and 2-methylfuran also couple the bromooxazinone in the presence of $ZnCl_2$ to afford the 3-aryloxazinones stereoselectively (eq 10).[2a,6] This process introduces a third benzylic carbon–heteroatom bond into the molecule and thereby precludes the reductive deprotections described. An alternative oxidative deprotection was developed.[6] Removal of the *t*-Boc protecting group followed by acid catalyzed opening of the heterocycle and subsequent oxidative cleavage with *Sodium Periodate* affords the arylglycines in moderate yield (eq 11).

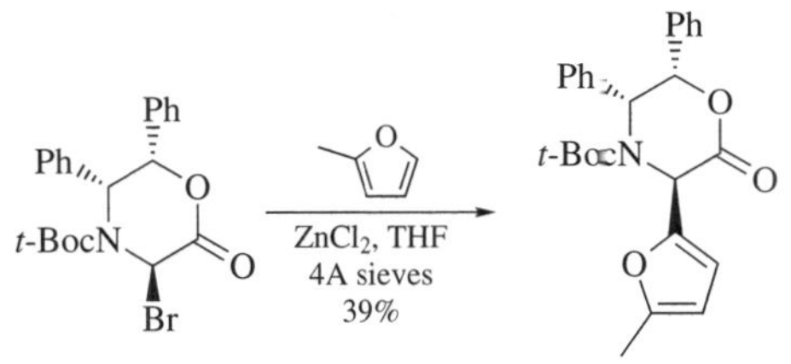
(10)

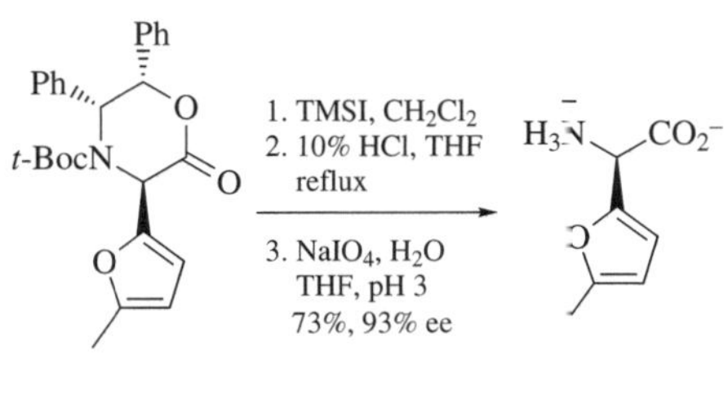
(11)

Coupling with Silyl Enol Ethers and Silyl Ketene Acetals. Silyl enol ethers can couple to the bromooxazinone to give both the *syn* and *anti* diastereomers.[2,7] The reaction can proceed via the S_N1 mechanism discussed above or by a Lewis acid assisted S_N2 displacement of the bromide. The reaction conditions can be manipulated to favor the S_N1 (stronger Lewis acids, more polar solvents) or S_N2 path (weaker Lewis acids, less polar solvents) (eq 12 and eq 13).[2a]

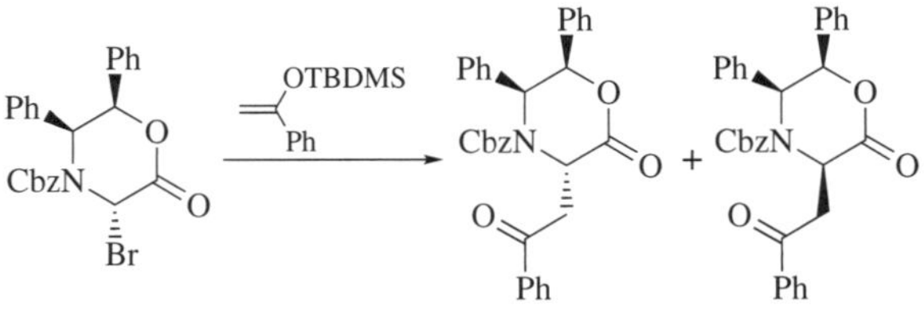

(12)

AgOTf, THF	24.5	1
$ZnCl_2$, MeCN	14.5	1
$ZnCl_2$, THF	7	1
$ZnCl_2$, CH_2Cl_2	1	34

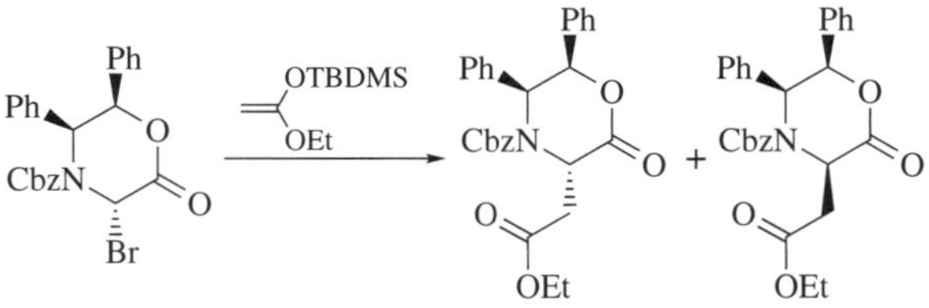

(13)

AgOTf, THF	1	2
$ZnCl_2$, THF	1	14–45
$ZnCl_2$, CH_2Cl_2	1	45

Coupling with Organozincs and Organocuprates. Alkylzinc chlorides and alkyl- and arylcuprates couple with the bromooxazinones with a high degree of diastereoselection but in lower yields.[2,6] Reduction of the bromide to the parent oxazinone is a significant side reaction and is attributed to the reaction taking place via an electron-transfer, radical–radical coupling.[2a] Substituted phenyl and naphthyl glycines have been prepared by coupling of the bromide with the corresponding organocuprate and employing the oxidative deprotection described above (eq 14 and eq 15).[6]

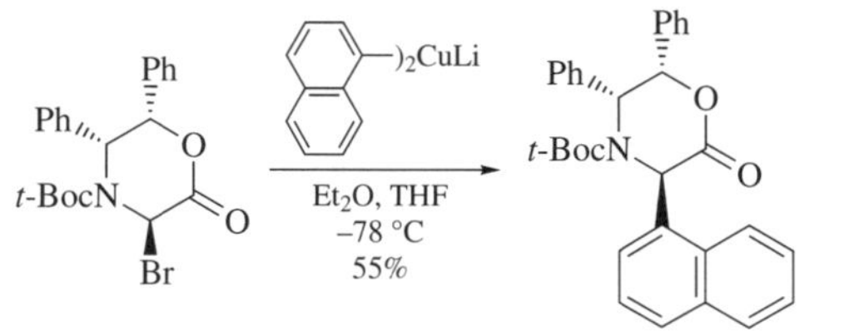

(14)

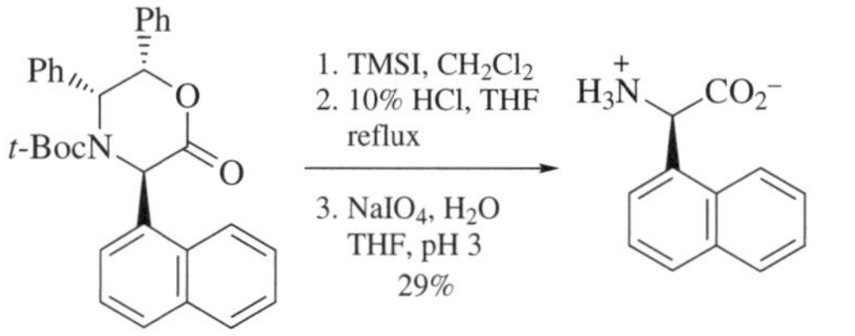

(15)

1-Aminocyclopropane-1-carboxylic Acids. These amino acids are prepared by a multistep procedure involving treatment of the bromooxazinone with *Trimethyl Phosphite* to give the corresponding phosphonate at the 3-position. Ylide formation and condensation with an aldehyde produces the α,β-dehydrooxazinone adduct possessing the (*E*) configuration (eq 16). Cyclopropanation with either *Diazomethane* or *Dimethylsulfoxonium Methylide* occurs with little diastereoselectivity. In contrast, cyclopropanation with (diethylamino)phenylsulfoxonium methylide is highly selective (eq 17).[8] Unexpectedly, delivery of the methylene occurs on the face of the heterocycle *syn* to the phenyl rings. The reason for this selectivity has not yet been determined. Deprotection with Li^0/NH_3 yields the *N*-*t*-Boc-1-aminocyclopropane-1-carboxylic acids (eq 18).

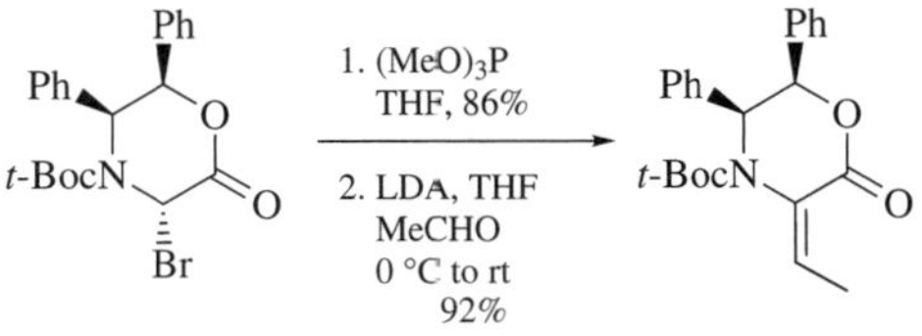

(16)

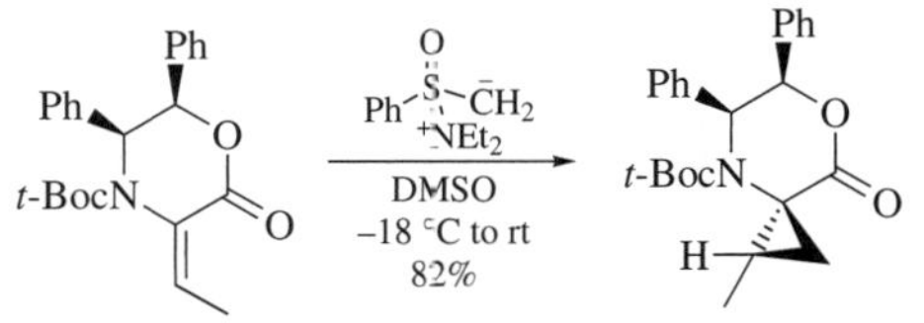

(17)

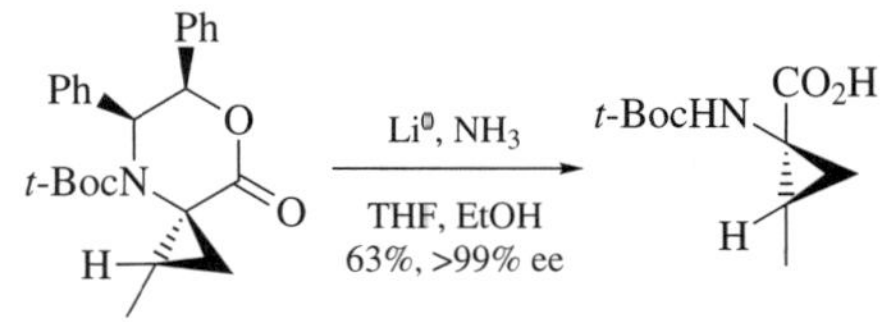

(18)

The α,β-dehydrooxazinone adducts can also undergo 1,3-dipolar cycloadditions as demonstrated in the synthesis of *S*-(−)-cucurbitine (eq 19 and eq 20).[9]

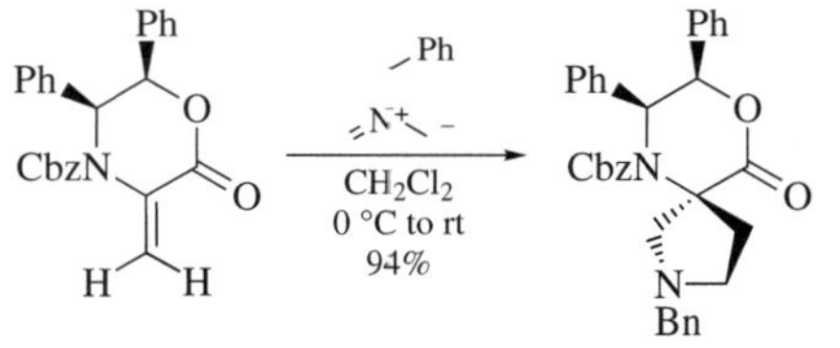

(19)

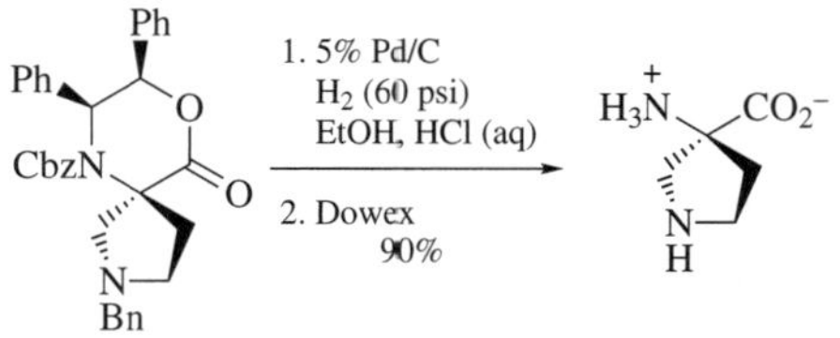

(20)

For the complementary synthesis of α-substituted-α-amino acids via a chiral glycine enolate equivalent see *4-t-Butoxycarbonyl-5,6-diphenyl-2,3,5,6-tetrahydro-4H-oxazin-2-one*.

1. (a) Williams, R. M. *Aldrichim. Acta* **1992**, *25*, 11. (b) Williams, R. M. *Synthesis of Optically Active α-Amino Acids*; Pergamon: Oxford, 1989.
2. (a) Williams, R. M.; Sinclair, P. J.; Zhai, D.; Chen, D. *J. Am. Chem. Soc.* **1988**, *110*, 1547. (b) Sinclair, P. J.; Zhai, D.; Reibenspies, J.; Williams, R. M. *J. Am. Chem. Soc.* **1986**, *108*, 1103.

3. Listed in the Aldrich Catalog as *t*-butyl 6-oxo-2,3-diphenyl-4-morpholinecarboxylate and benzyl 6-oxo-2,3-diphenyl-4-morpholinecarboxylate.
4. (a) Ramer, S. E.; Cheng, H.; Vederas, J. C. *Pure Appl. Chem.* **1989**, *61*, 489. (b) Ramer, S. E.; Cheng, H.; Palcic, M. M.; Vederas, J. C. *J. Am. Chem. Soc.* **1988**, *110*, 8526. (c) Williams, R. M.; Zhai, D.; Sinclair, P. J. *J. Org. Chem.* **1986**, *51*, 5021.
5. (a) Williams, R. M.; Zhai, W. *Tetrahedron* **1988**, *44*, 5425. (b) Zhai, D.; Zhai, W.; Williams, R. M. *J. Am. Chem. Soc.* **1988**, *110*, 2501.
6. (a) Williams, R. M.; Hendrix, J. A. *Chem. Rev.* **1992**, *92*, 889. (b) Williams, R. M.; Hendrix, J. A. *J. Org. Chem.* **1990**, *55*, 3723.
7. Williams, R. M.; Sinclair, P. J.; Zhai, W. *J. Am. Chem. Soc.* **1988**, *110*, 482.
8. Williams, R. M.; Fegley, G. J. *J. Am. Chem. Soc.* **1991**, *113*, 8796.
9. Williams, R. M.; Fegley, G. J. *Tetrahedron Lett.* **1992**, *33*, 6755.

Peter J. Sinclair
Merck Research Laboratories, Rahway, NJ, USA

Brucine[1]

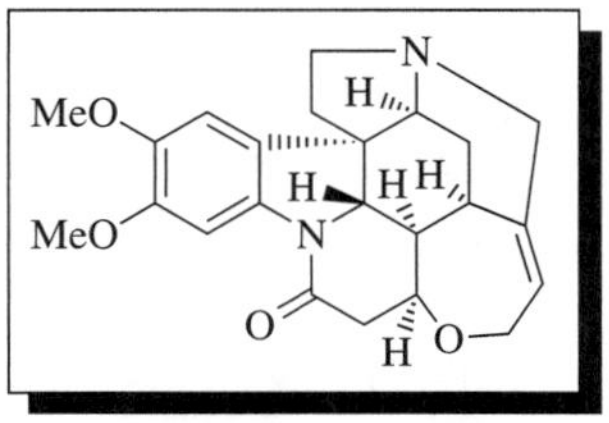

[357-57-3] $C_{23}H_{26}N_2O_4$ (MW 394.47)

(reagent for the resolution of acids, alcohols, and other neutral compounds[1])

Physical Data: colorless needles (acetone/water) mp 178 °C; $[\alpha]_D$ −79.3° (*c* 1.3, EtOH).

Solubility: very sol methanol, ethanol, and chloroform; mod sol ethyl acetate or benzene.

Form Supplied in: colorless needles or plates. The free base, which is available from multiple commercial sources, is usually hydrated. Dihydrated and tetrahydrated forms have been characterized. Anhydrous brucine can be obtained by heating at 100–120 °C in vacuo for 24 h. The hydrated forms can be used for most applications.

Purification: the commercial reagent is often used without further purification. However, the reagent can be purified by recrystallization from ethanol/water (1:1).[2] Recovered reagent[3] should be purified before reuse.

Handling, Storage, and Precautions: EXTREMELY POISONOUS. Oral LD_{50} in rats is 1 mg kg^{-1}. Handle in well-ventilated hood only.

Introduction. The alkaloid brucine has been a key resolving agent for over a century, in spite of its highly toxic nature. The group of chiral bases represented primarily by brucine, its homolog strychnine, and the cinchona alkaloids quinine, quinidine, cinchonidine, and cinchonine, has been extremely useful for the resolution of all types of acids.[1] No empirical rules have emerged from all of this work to help in predicting the optimal resolving agent for a given type of acid. Acid resolution is still primarily an empirical process that requires the evaluation of several diastereomeric salts. An inherent limitation to the use of alkaloids as resolving agents for acids is the availability of only one antipode, which sometimes allows the practical isolation of only one of the acid enantiomers in a pure form. Nevertheless, there are reports of resolutions with brucine that are so efficient that the less crystalline enantiomer can be isolated directly from the mother liquors (see below for examples). In other cases, pairs of pseudoenantiomeric cinchona alkaloids (i.e., quinine and quinidine, cinchonine and cinchonidine), or brucine and another alkaloid, display opposite selectivities for the enantiomers of a racemic acid (see below).[1a]

Resolution of Acids. The number of acids resolved with brucine is too large to attempt to list even a small portion of them in this synopsis. An excellent tabulation of all published resolutions with brucine up to 1972 is available.[1a] Only a few representative examples will be described here (eqs 1–4).[4–9] In all these cases, the resolved acids were obtained in high yield and with almost absolute enantiomeric purity. The solvents most frequently used for brucine resolutions are acetone and alcohol solvents. However, water, hexane, and others have also been used as cosolvents.

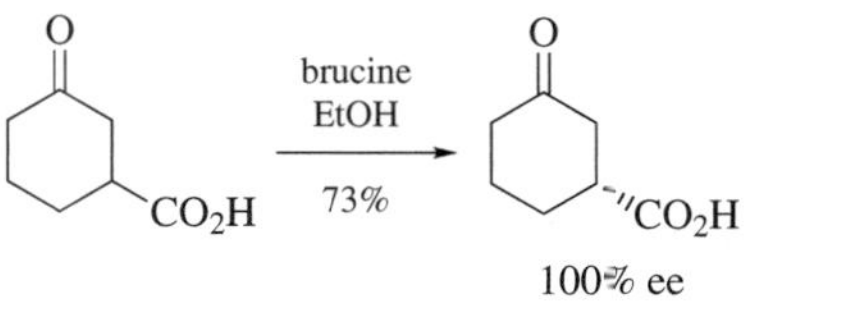

(1)

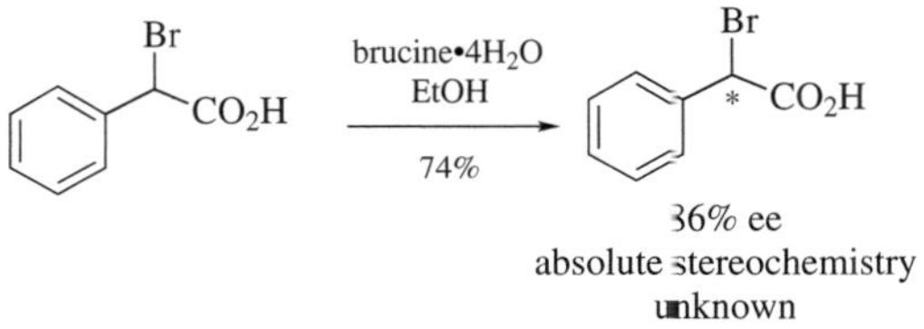

(2)

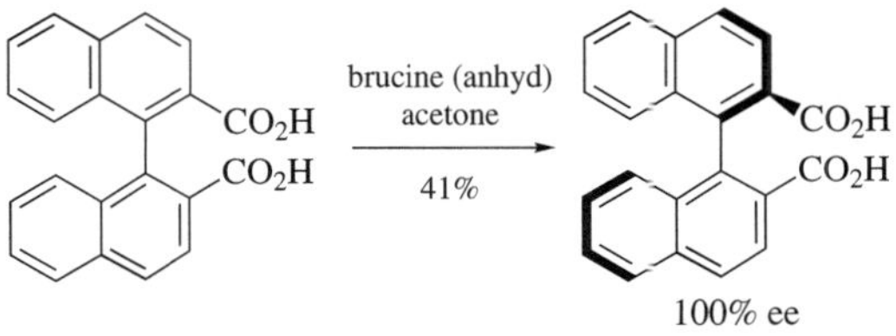

(3)

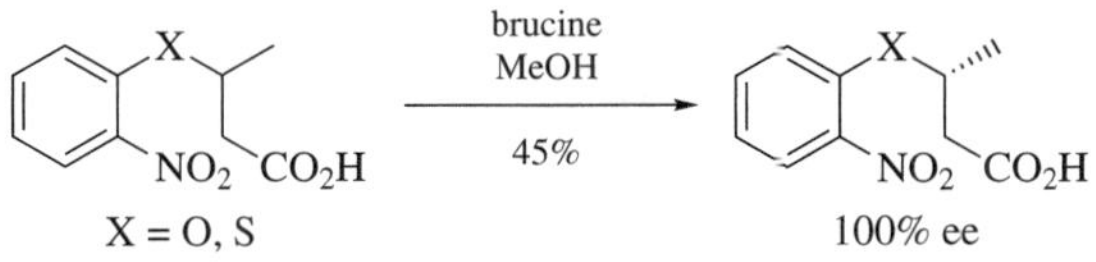

(4)

Additional types of carboxylic acids that have been successfully resolved with brucine are represented by structures (**1**)–(**5**).[10–14]

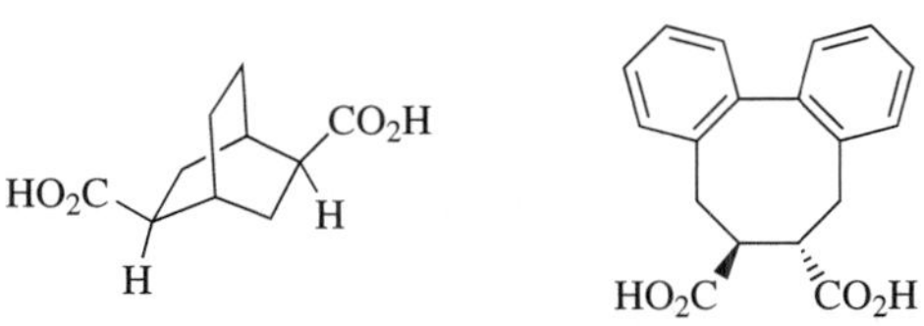

(**1**) (**2**)

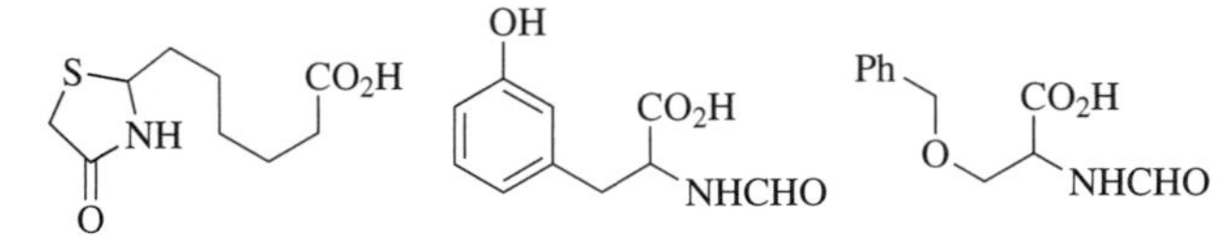

(3) (4) (5)

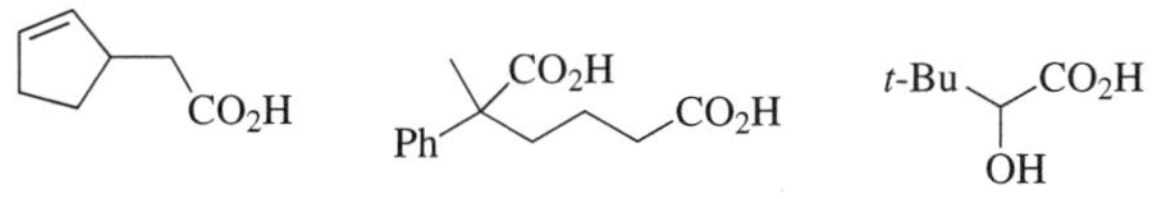

(6) (7) (8)

As mentioned above, one of the limitations of using naturally occurring resolving agents is that only one enantiomer of the compound being resolved may be readily accessible by resolution. However, many examples have been described where brucine and some other alkaloid favor crystallization with opposite enantiomers of a given acid. For example, resolution of acid (**6**) with brucine yields the (+)-enantiomer, while cinchonidine provides material that is enriched in the (−)-enantiomer of the acid.[15] Similarly, diacid (**7**) is resolved into its (−)-enantiomer by brucine and into its (+)-enantiomer by strychnine.[16] The (+)-enantiomer of acid (**8**) can be obtained with brucine, while the (−)-enantiomer crystallizes with cinchonidine.[17] Additional examples of the same phenomenon can be found in the literature.[1a]

Resolution of Alcohols. Although not a well exploited use of brucine, a variety of secondary benzylic alcohols have been resolved by complexation and crystallization with brucine (eq 5).[18] About a dozen alcohols were obtained in close to enantiomeric purity by this procedure.[18] Also resolved by crystallization of their brucine inclusion complexes were a series of tertiary propargylic alcohols (eq 6).[19] In this case, the enantiomer that does not crystallize with brucine can be obtained in almost complete optical purity from the mother liquors.

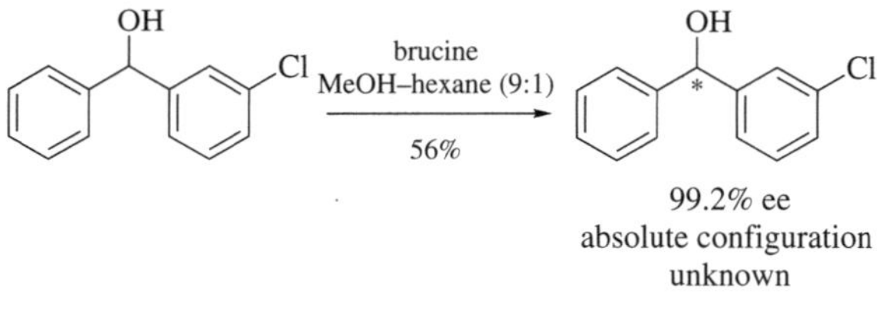

(5)

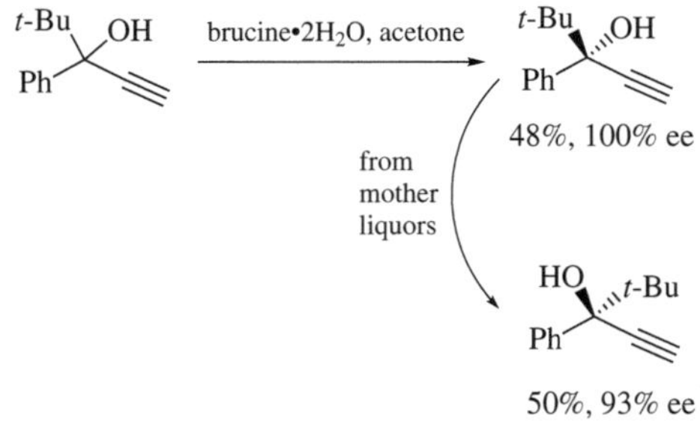

(6)

A more traditional and general approach to the resolution of alcohols is the formation of the corresponding hemiphthalate or hemisuccinate esters, followed by resolution of these acidic derivatives with brucine or some other chiral base (eqs 7–9).[20–23] The resolved alcohols are liberated by alkaline hydrolysis of the esters. High enantiomeric purity is frequently achieved by this procedure, which has been applied successfully to primary, secondary, and tertiary alcohols.

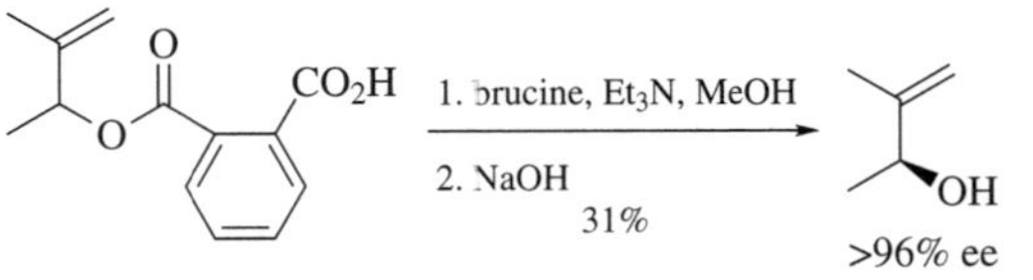

(7)

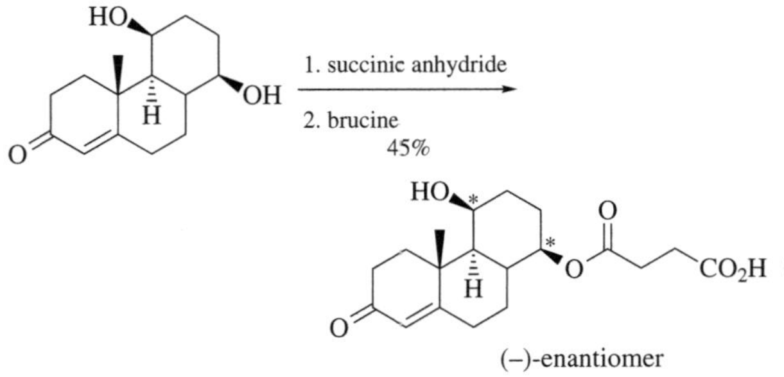

(8)

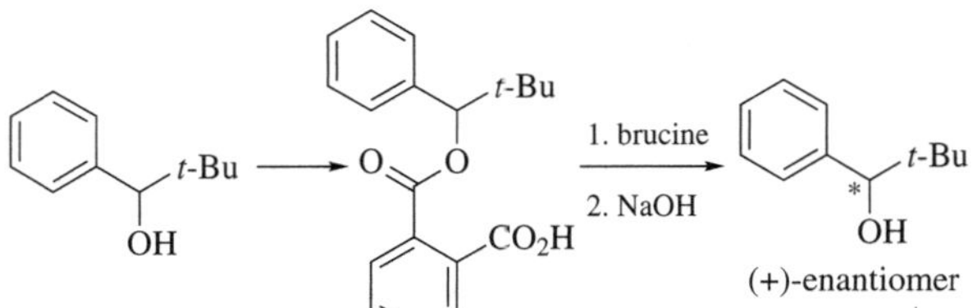

(9)

Resolution of Ketones. Brucine has not been used very extensively for the resolution of neutral compounds. However, in some cases, ketones or ketone derivatives may form diastereomeric inclusion complexes with brucine, providing an opportunity for their resolution. For example, the cyanohydrin of a bicyclic ketone has been resolved by this procedure (eq 10).[24] Following resolution of the cyanohydrin, the ketone was regenerated and determined to be of 94% ee.

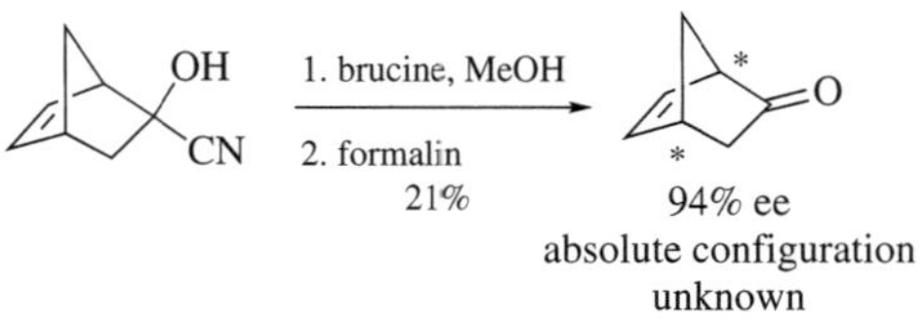

(10)

Resolution of Sulfoxides. Although it can be considered as the resolution of an unique type of carboxylic acid, some racemic sulfoxides containing carboxylic acids have been resolved via diastereomeric crystalline complexes with brucine (eq 11).[25]

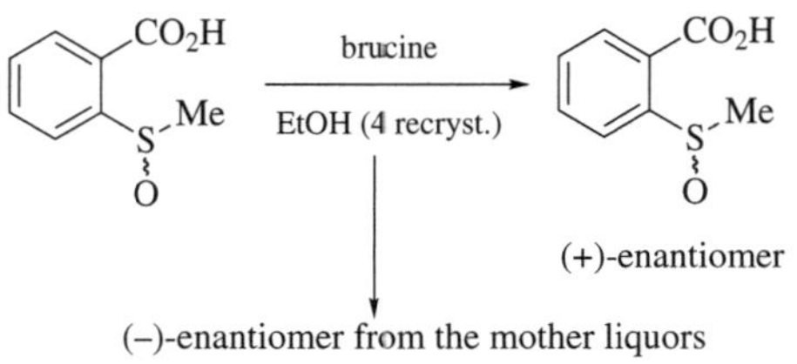

(11)

Chiral Catalysis. Brucine has been utilized as chiral catalyst in a variety of reactions. For example, its incorporation into a polymer support provides a chiral catalyst for performing enantioselective benzoin condensations.[26] It has also been used as a chiral catalyst in the asymmetric synthesis of (*R*)-malic acid via the corresponding β-lactone, which results from the asymmetric cycloaddition of chloral and ketene (eq 12).[27] Though brucine yields malic acid with 68% ee, quinidine was found to be a more selective catalyst (98% ee).

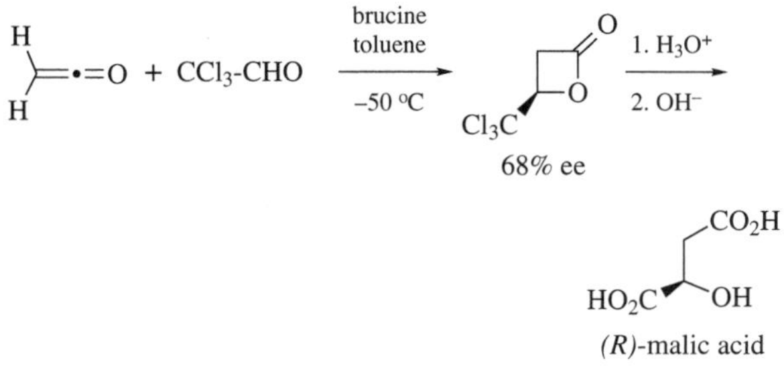

(12)

Brucine has been used as an enantioselective catalyst in the kinetic resolution of alcohols. For example, an azirinylmethanol was reacted with 0.5 equiv of *Acetic Anhydride* in the presence of 25 mol % brucine. The resulting acetate was found to possess 24% ee (eq 13).[28]

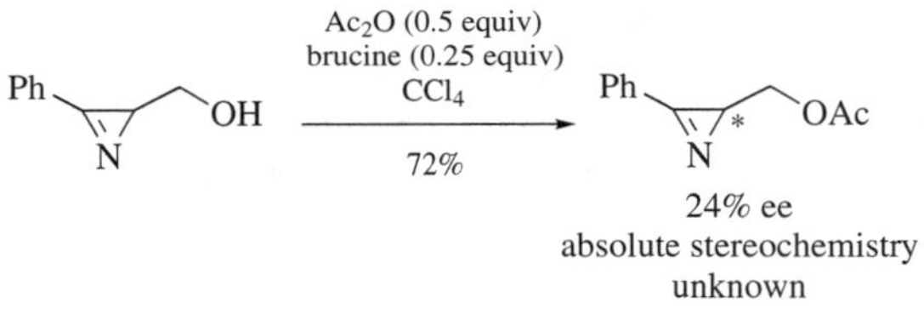

(13)

Brucine has been used to produce enantiomerically enriched compounds by selective reaction with or destruction of one of the enantiomers. The optical purity of the resulting compound is usually modest, although some exceptions have been described. For example, dibromo compound (**9**) was obtained [enriched in the (−)-enantiomer] by selective destruction of the (+)-enantiomer with brucine in chloroform.[29] The resolution of (±)-2,3-dibromobutane may have also been a case of enantioselective destruction,[30] although more recent reports suggest that it is more likely a case of enantioselective entrapment in the brucine crystals (eq 14).[31]

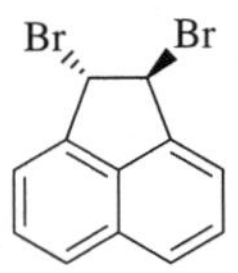

(**9**)

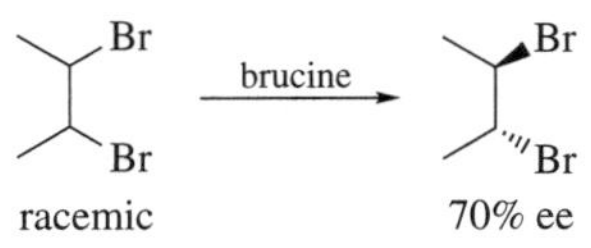

(14)

Miscellaneous. Brucine greatly accelerates the decarboxylation of certain β-oxo carboxylic acids at rt (eq 15),[32] as well as the decarbalkoxylation of β-oxo esters.[33] In some cases the products of these reactions possess some (modest) enantiomeric excess.[34]

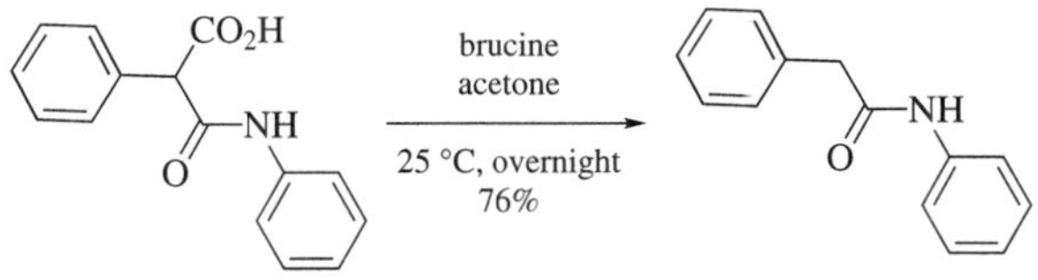

(15)

Related Reagents. (1*R*,2*S*)-Ephedrine; (2*S*,2′*S*)-2-Hydroxymethyl-1-[(1-methylpyrrolidin-2-yl)methyl]pyrrolidine; (*S*)-α-Methylbenzylamine; 1-(1-Naphthyl)ethylamine; Quinine.

1. (a) Wilen, S. H. In *Tables of Resolving Agents and Optical Resolutions*; Eliel, E. L., Ed.; University of Notre Dame Press: Notre Dame, 1972. (b) Jacques, J.; Collet, A. In *Enantiomers, Racemates and Resolutions*; Wilen, S. H., Ed.; Wiley: New York, 1981.
2. DePuy, C. H.; Breitbeil, F. W.; DeBruin, K. R. *J. Am. Chem. Soc.* **1966**, *88*, 3347.
3. Vogel, A. I. *Practical Organic Chemistry*; Longmans: London, 1957, p 507.
4. Allan, R. D.; Johnston, G. A. R.; Twitchin, B. *Aust. J. Chem.* **1981**, *34*, 2231.
5. Kaifez, F.; Kovac, T.; Mihalic, M.; Belin, B.; Šunjic, V. *J. Heterocycl. Chem.* **1976**, *13*, 561.
6. Kanoh, S.; Hongoh, Y.; Motoi, M.; Suda, H. *Bull. Chem. Soc. Jpn.* **1988**, *61*, 1032.
7. Hasaka, N.; Okigawa, M.; Kouno, I.; Kawano, N. *Bull. Chem. Soc. Jpn.* **1982**, *55*, 3828.
8. Lévai, A.; Ott, J.; Snatzke, G. *Monatsh. Chem.* **1992**, *123*, 919.
9. Puzicha, G.; Lévai, A.; Szilágyi, L. *Monatsh. Chem.* **1988**, *119*, 933.
10. Tichy, M.; Sicher, J. *Tetrahedron Lett.* **1969**, *52*, 4609.
11. Dvorken, L. V.; Smyth, R. B.; Mislow, K. *J. Am. Chem. Soc.* **1958**, *80*, 486.
12. McLamore, W. M.; Celmer, W. D.; Bogert, V. V.; Pennington, F. C.; Sobin, B. A.; Solomons, I. A. *J. Am. Chem. Soc.* **1953**, *75*, 105.
13. Sealock, R. R.; Speeter, M. E.; Schweet, R. S. *J. Am. Chem. Soc.* **1951**, *73*, 5386.
14. Dutta, A. S.; Morley, J. S. *Chem. Commun.* **1971**, 883.
15. Mislow, K.; Strinberg, I. V. *J. Am. Chem. Soc.* **1955**, *77*, 3807.
16. Hoffman, T. D.; Cram, D. J. *J. Am. Chem. Soc.* **1969**, *91*, 1000.
17. Tanabe, T.; Yajima, S.; Imaida, M. *Bull. Chem. Soc. Jpn.* **1968**, *41*, 2178.
18. Toda, F.; Tanaka, K.; Koshiro, K. *Tetrahedron: Asymmetry* **1991**, *2*, 873.
19. Toda, F.; Tanaka, K. *Tetrahedron Lett.* **1981**, *22*, 4669.
20. Crout, D. H. G.; Morrey, S. M. *J. Chem. Soc., Perkin Trans. 1* **1983**, 2435.
21. Lukes, R. M.; Sarett, L. H. *J. Am. Chem. Soc.* **1954**, *76*, 1178.
22. MacLeod, R.; Welch, F. J.; Mosher, H. S. *J. Am. Chem. Soc.* **1960**, *82*, 876.
23. Eliel, E. L.; Kofron, J. T. *J. Am. Chem. Soc.* **1953**, *75*, 4585.
24. Black, K. A.; Vogel, P. *Helv. Chim. Acta* **1984** *67*, 1612.
25. Barbieri, G.; Davoli, V.; Moretti, I.; Montanari, F.; Torre, G. *J. Chem. Soc. (C)* **1969**, 731.
26. Castells, J.; Duñach, E. *Chem. Lett.* **1984**, 1859.
27. Wynberg, H.; Staring, E. G. J. *J. Am. Chem. Soc.* **1982**, *104*, 166.

28. Stegmann, W.; Uebelhart, P.; Heimgartner, H.; Schmid, H. *Tetrahedron Lett.* **1978**, *34*, 3091.
29. Greene, F. D.; Remers, W. A.; Wilson, J. W. *J. Am. Chem. Soc.* **1957**, *79*, 1416.
30. Tanner, D. D.; Blackburn, E. V.; Kosugi, Y.; Ruo, T. C. S. *J. Am. Chem. Soc.* **1977**, *99*, 2714.
31. Pavlis, R. R.; Skell, P. S. *J. Org. Chem.* **1983**, *48*, 1901.
32. Hargreaves, M.; Khan, M. *Monatsh. Chem.* **1978**, *109*, 799.
33. Miles, D. H.; Stagg, D. D. *J. Org. Chem.* **1981**, *46*, 5376.
34. Toussaint, O.; Capdevielle, P.; Maumy, M. *Tetrahedron Lett.* **1987**, *28*, 539.

Juan C. Jaen
Parke-Davis Pharmaceutical Research, Ann Arbor, MI, USA

4-*t*-Butoxycarbonyl-5,6-diphenyl-2,3,5,6-tetrahydro-4*H*-oxazin-2-one[1]

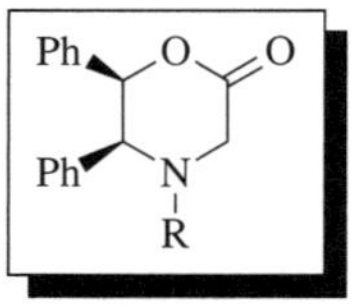

(5*R*,6*S*)-(R = *t*-Boc)
[112741-50-1] $C_{21}H_{23}NO_4$ (MW 353.45)
(5*S*,6*R*)-(R = *t*-Boc)
[112741-49-8]
(5*R*,6*S*)-(R = Cbz)
[105228-46-4] $C_{24}H_{21}NO_4$ (MW 387.46)
(5*S*,6*R*)-(R = Cbz)
[100516-54-9]

(chiral glycine enolate equivalent useful for the preparation of α-substituted α-amino acids and α,α-disubstituted α-amino acids in high enantiomeric excess[2,3])

Alternate Name: t-butyl 6-oxo-2,3-diphenyl-4-morpholinecarboxylate.
Physical Data: R = *t*-Boc: mp 206 °C. R = Cbz: mp 205 °C.
Solubility: sol THF, CH_2Cl_2.
Form Supplied in: commercially available as the individual enantiomers or as racemates.
Preparative Methods: via a three-step procedure from *erythro*-2-amino-1,2-diphenylethanol.[4]
Handling, Storage, and Precautions: no special handling required.

α-Substituted α-Amino Acids. The *N*-protected 5,6-diphenyl-2,3,5,6-tetrahydrooxazin-2-ones serve as chiral nucleophilic glycine equivalents by deprotonation at C-3 to give the corresponding enolates.[2,3] The enolates are formed using *Lithium Hexamethyldisilazide* or *Sodium Hexamethyldisilazide* in THF at low temperature.[2] Alkylation is best achieved by addition of the base to a solution of the oxazinone and electrophile in THF–HMPA (10:1) at −78 °C (eq 1)[2a] or by formation of the sodium enolate in the presence of *15-Crown-5* at −78 °C and subsequent addition of the electrophile (eq 2).[3] Alkylation takes place with very high stereoselectivity to afford the C-3 modified oxazinone with the new substituent oriented *anti* to the C-5 and C-6 phenyl groups.[2]

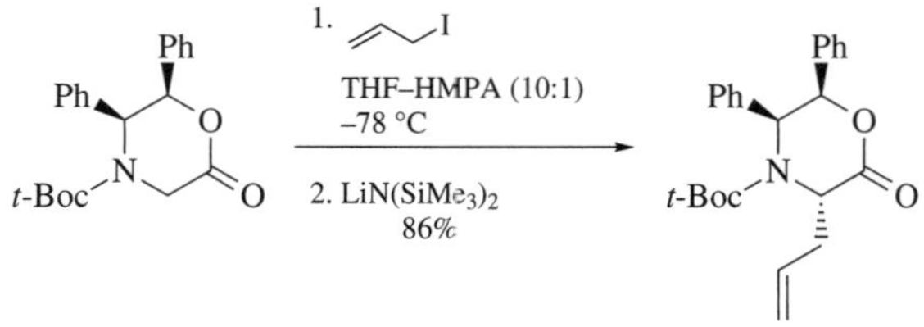
(1)

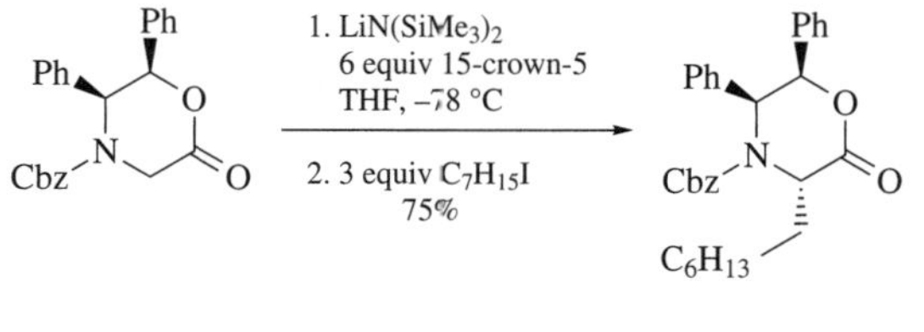
(2)

The amino acid is released by scission of the benzylic carbon–heteroatom bonds, generally via a reductive process.[2–5] Hydrogenolysis and dissolving metal reduction of the Cbz-protected oxazinone gives the amino acid zwitterion directly (eq 3). The *t*-Boc protected oxazinone undergoes dissolving metal reduction to give the *t*-Boc amino acid (eq 4). Alternatively, the *t*-Boc group can be removed from the oxazinone by treatment with *Iodotrimethylsilane* or *Trifluoroacetic Acid* and the resulting compound can then be hydrogenated to afford the amino acid zwitterion. Amino acids possessing alkyl, allyl, (ethoxycarbonyl)methyl, hydroxyalkyl, ω-aminoalkyl, and methyl-β-D-ribofuranose functionality have been prepared via this enolate chemistry.[2,3,6,7]

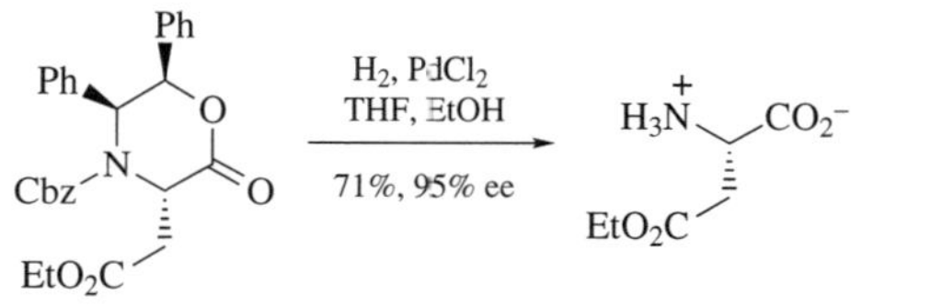
(3)

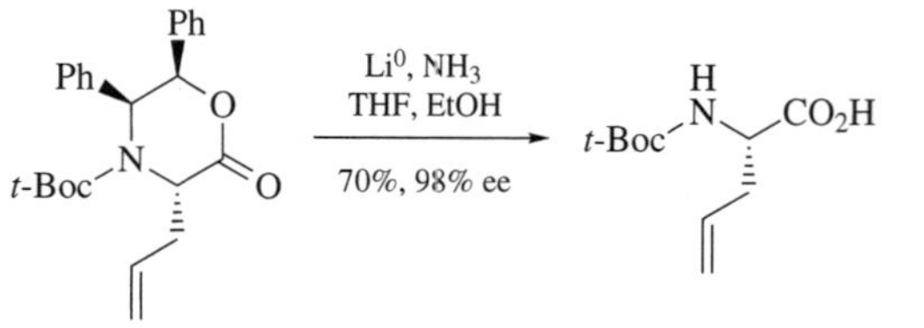
(4)

α,α-Disubstituted α-Amino Acids. The 3-alkyloxazinones also undergo enolization and subsequent alkylation to afford, upon deprotection, α,α-dialkyl α-amino acids.[2a,3,8,9a] Again, the newly introduced alkyl group is oriented *anti* to the phenyl groups of the oxazinone. Alkylation by addition of the base to a solution of the oxazinone and electrophile works only with allylic and benzylic halides and can require as much as 2–5 equiv of *Potassium Hexamethyldisilazide* (eq 5).[2a,8] Alternatively, formation of the sodium enolate in the presence of 15-crown-5 followed by addition of the electrophile permits coupling of the enolate even to simple alkyl halides (eq 6).[3,9a] The disubstituted amino acids are liberated as described above (eq 7).

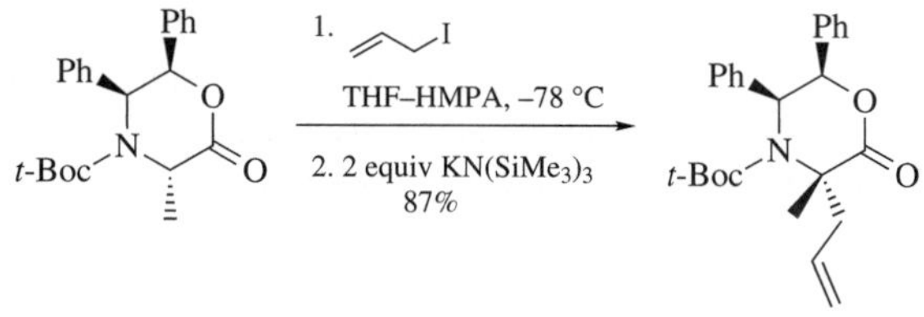
(5)

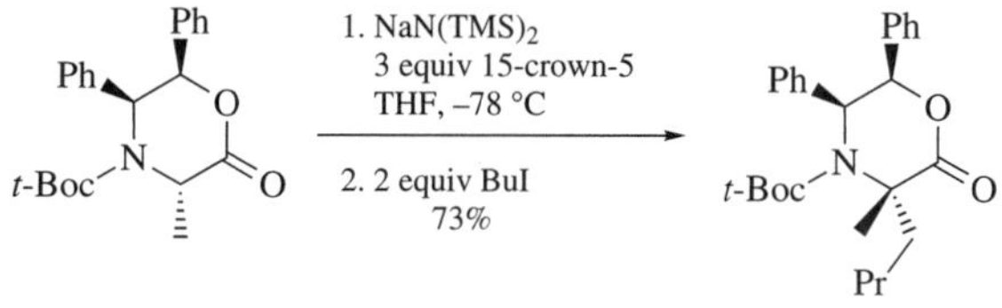
(6)

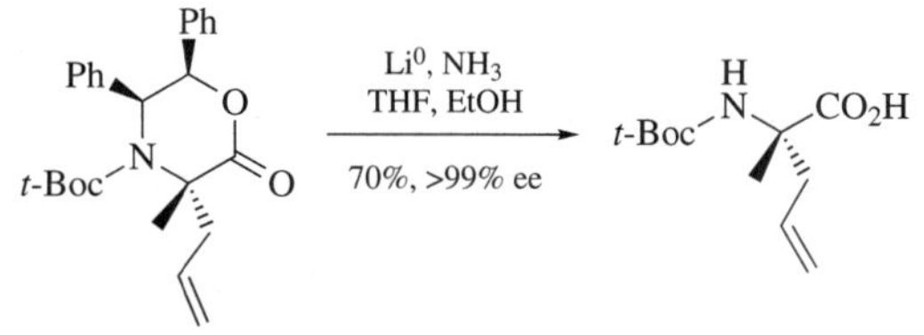
(7)

Boron Enolates. The oxazinones can be converted to their corresponding boron enolates by treatment with *Di-n-butylboryl Trifluoromethanesulfonate* and *Triethylamine* in CH_2Cl_2.[9] The boron enolates react with aldehydes at −78 °C to give β-(hydroxy)alkyl-substituted oxazinones. Condensation of the boron enolate with acetaldehyde followed by recrystallization of the major product and then deprotection affords allothreonine (eqs 8 and 9).[9c] This approach has been used in the asymmetric synthesis of diaminopimelic acid and derivatives thereof.[9a,b]

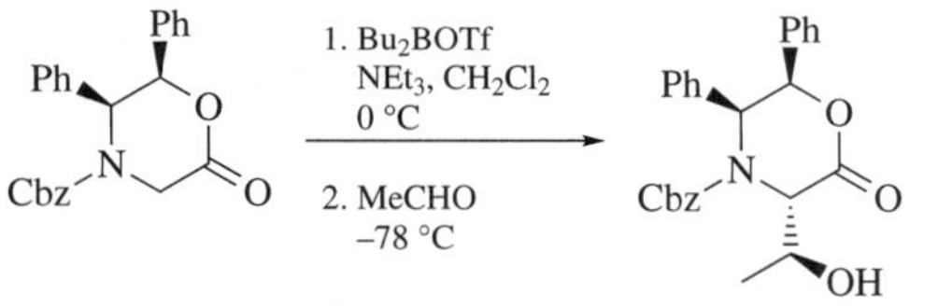
(8)

major product, 57%

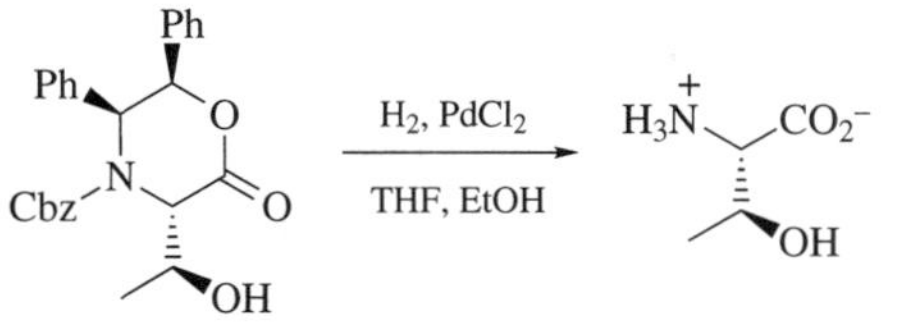
(9)

[3 + 2] Dipolar Cycloadditions. Highly substituted proline derivatives can be prepared by removal of the *t*-Boc protecting group from the oxazinone followed by condensation of the heterocycle with an aldehyde in the presence of *p-Toluenesulfonic Acid* in benzene. Under these conditions, Schiff base formation and ylide generation occur. Subsequent [3 + 2] cycloaddition with a dipolarophile affords the bicyclic heterocycle, which is then deprotected to yield the desired proline derivative (eqs 10 and 11).[10]

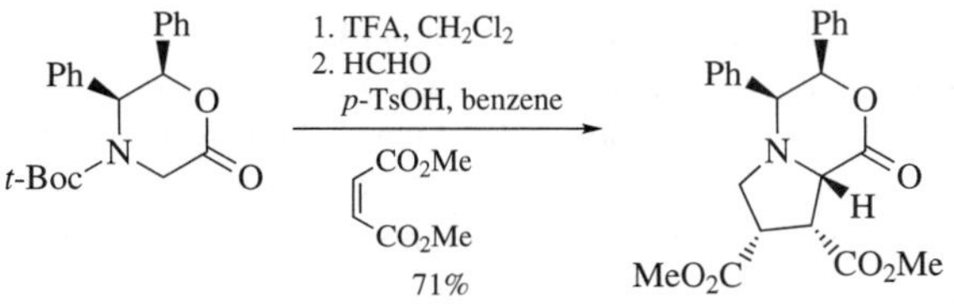
(10)

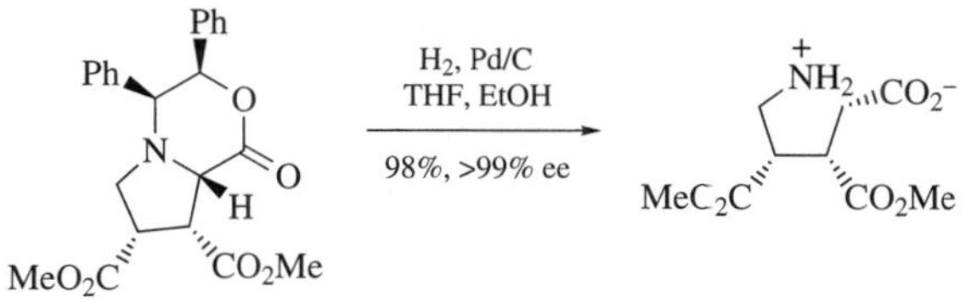
(11)

Related Reagent. For the complementary synthesis of α-substituted α-amino acids via a chiral electrophilic glycine equivalent, see *3-Bromo-5,6-diphenyl-2,3,5,6-tetrahydro-4H-oxazin-2-one*.

1. (a) Williams, R. M. *Aldrichim. Acta* **1992**, *25*, 11. (b) Williams, R. M. *Synthesis of Optically Active α-Amino Acids*; Pergamon: Oxford, 1989.
2. (a) Williams, R. M.; Im, M.-N. *J. Am. Chem. Soc.* **1991**, *113*, 9276. (c) Williams, R. M.; Im, M.-N. *Tetrahedron Lett.* **1988**, *29*, 6075.
3. Baldwin, J. E.; Lee, V.; Schofield, C. J. *Synlett* **1992**, 249.
4. (a) Williams, R. M.; Sinclair, P. J.; Zhai, D.; Chen, D. *J. Am. Chem. Soc.* **1988**, *110*, 1547. (b) Sinclair, P. J.; Zhai, D.; Reibenspies, J.; Williams, R. M. *J. Am. Chem. Soc.* **1986**, *108*, 1103.
5. For an oxidative deprotection see: Williams, R. M.; Hendrix, J. A. *J. Org. Chem.* **1990**, *55*, 3723.
6. Dong, Z. *Tetrahedron Lett.* **1992**, *33*, 7725.
7. Dudycz, L. W. *Nucleosides Nucleotides* **1991**, *10*, 329.
8. Baldwin, J. E.; Lee, V.; Schofield, C. J. *Heterocycles* **1992**, *34*, 903.
9. (a) Williams, R. M.; Yuan, C. *J. Org. Chem.* **1992**, *57*, 6519. (b) Williams, R. M.; Im, M.-N.; Cao, J. *J. Am. Chem. Soc.* **1991**, *113*, 6976. (c) Reno, D. S.; Lotz, B. T.; Miller, M. J. *Tetrahedron Lett.* **1990**, *31*, 827.
10. Williams, R. M.; Zhai, W.; Aldous, D. J.; Aldous, S. C. *J. Org. Chem.* **1992**, *57*, 6527.

Peter J. Sinclair
Merck Research Laboratories, Rahway, NJ, USA

(*R,R*)-2-Butyl-4,5-bis(dimethylaminocarbonyl)-1,3-dioxaborolane[1]

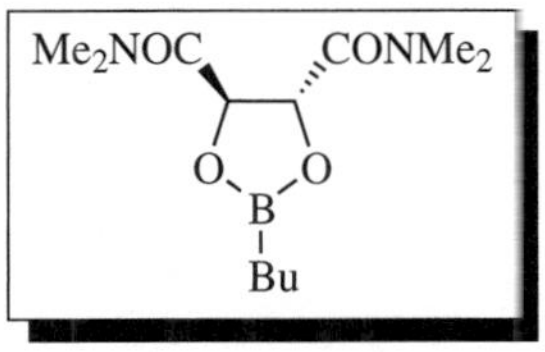

[161344-85-0] $C_{12}H_{23}BN_2O_4$ (MW 270.13)

(enantioselective cyclopropanation[1])

Solubility: soluble in CH_2Cl_2, $ClCH_2CH_2Cl$, toluene, benzene and most organic solvents.
Form Supplied in: colorless oil, not commercially available.
Analysis of Reagent Purity: NMR (^1H, ^{11}B)
Preparative Methods: the reagent is easily prepared from commercially available butylboronic acid (or its more stable diethanolamine complex) and (*R,R*)-(+)-*N,N,N′,N′*-tetramethyltartaric acid diamide.[2] The other enantiomer is also

readily available from (*S,S*)-(+)-*N,N,N′,N′*-tetramethyltartaric acid diamide.

Purification: not easily purified since the reagent hydrolyzes slowly in the presence of moisture and oxidizes slowly in the presence of oxygen. The formation of crystals over time is an indication of decomposition.

Handling, Storage, and Precautions: the reagent is stable indefinitely when stored under an inert atmosphere.

Enantioselective Cyclopropanation of Allylic Alcohols. (*R, R*)-2-Butyl-4, 5-bis(dimethylaminocarbonyl)-1,3-dioxaborolane (**1**) is one of the most effective chiral additives for the enantioselective cyclopropanation of allylic alcohols.[3] The synthesis of a wide range of substituted cyclopropylmethanols proceeds with excellent enantiocontrol (85–93% ee) when a solution of the alcohol and the dioxaborolane ligand is added to bis(iodomethyl)zinc. The use of the DME complex of bis(iodomethyl)zinc is preferable on large scale (eq 1).[4]

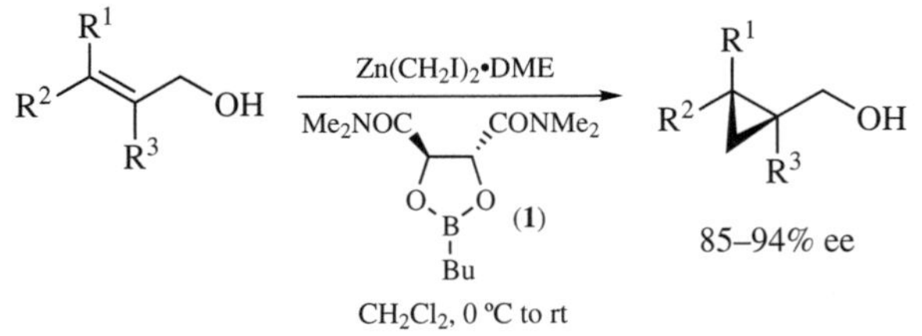

(1)

The reaction can also be used in bidirectional chain synthesis to generate bis(cyclopropyl) derivatives simultaneously (eq 2 and 3).[5] This reaction was used as the key step for elaboration of the polycyclopropane natural products FR-900848 and U-106305.

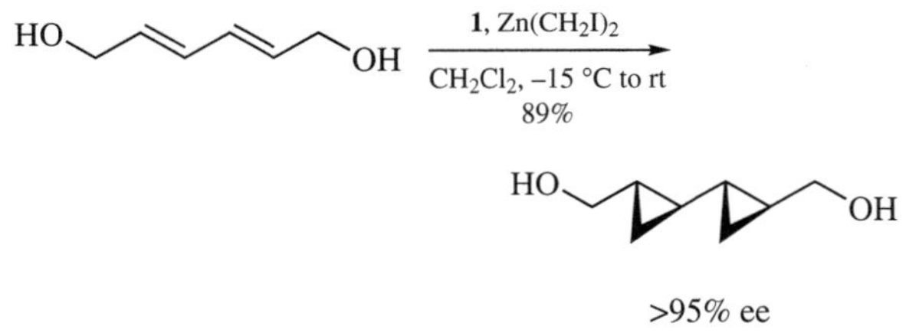

(2)

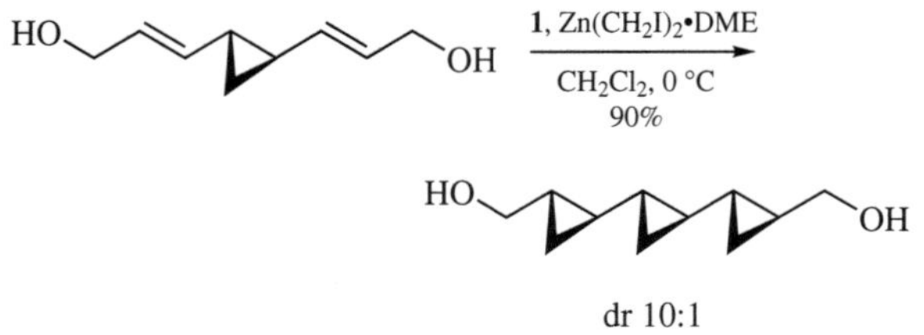

(3)

Extremely high diastereoselectivities are also observed when the antipode of the starting material is used, providing efficient access to the other diastereomer (eq 4).

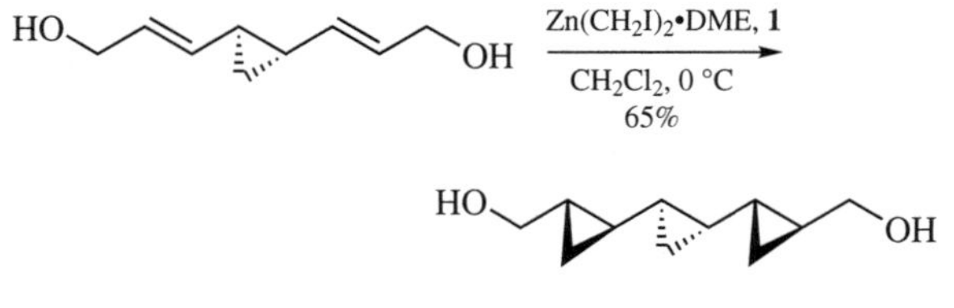

(4)

The reaction has also been extended to the enantioselective cyclopropanation of 3-tributylstannylprop-2-en-1-ol,[5a] to 3-iodo-[3a] and 3-chloroprop-2-en-1-ol.[6] The first two are useful precursors in palladium-catalyzed cross-coupling reactions[7] while the last was used in the total synthesis of callipeltoside A.

The enantioselective cyclopropanation reaction is quite general and practical. For example, the cyclopropanation reaction has been used to synthesize 3-methylcyclopropylmethanol, a precursor to curacin A.[8] Tri- and tetrasubstituted allylic alcohols are also converted into their corresponding cyclopropanes with high enantiocontrol (eq 5).

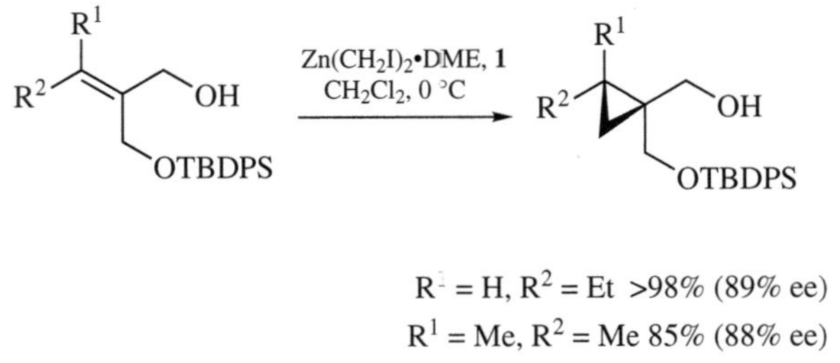

(5)

Polyenes can also be cyclopropanated at the allylic alcohol position with high chemo- and enantioselectivities due to the strong directing ability of the chiral ligand. This reaction has been used to generate key precursors of bicyclohumulenone (eq 6)[9] and noranthoplone (eq 7).[3a]

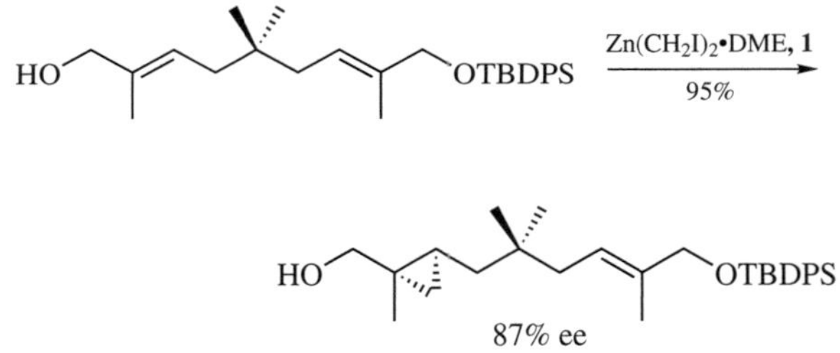

(6)

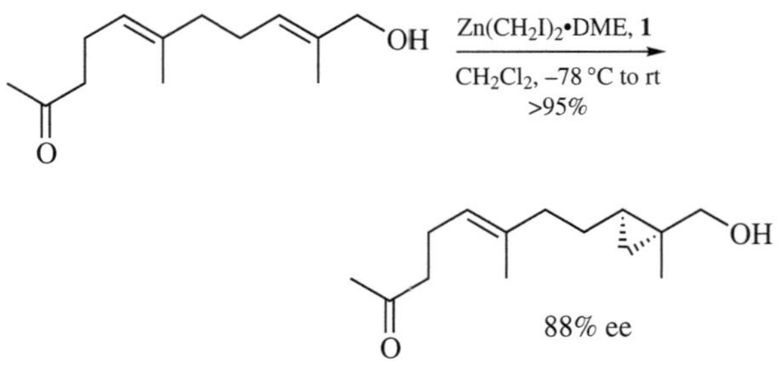

(7)

Unprecedented high *anti*-selectivities are obtained when *E*-substituted chiral allylic alcohols are treated with bis(iodomethyl)zinc and the dioxaborolane ligand (eq 8).[10] In contrast, the *syn*-isomer is obtained if the substrate is treated with the zinc reagent in the absence of the chiral ligand.[11] The method complements that involving the direct reduction of cyclopropylketones with $LiAlH_4$ or DIBAL-H.[12]

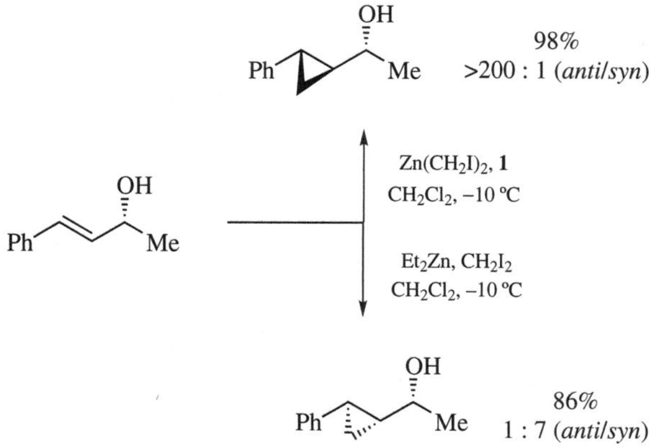

(8)

Enantioselective Synthesis of 1,2,3-trisubstituted Cyclopropanes. The chiral dioxaborolane ligand can also be used to generate 1,2,3-substituted cyclopropyl units when the appropriate 1,1-diiodoalkane is used in the preparation of the zinc reagent (eq 9).[13] The reaction affords 1,2,3-trisubstituted cyclopropanes with excellent enantio- and diastereocontrol, including those obtained from functionalized zinc reagents (eq 10).

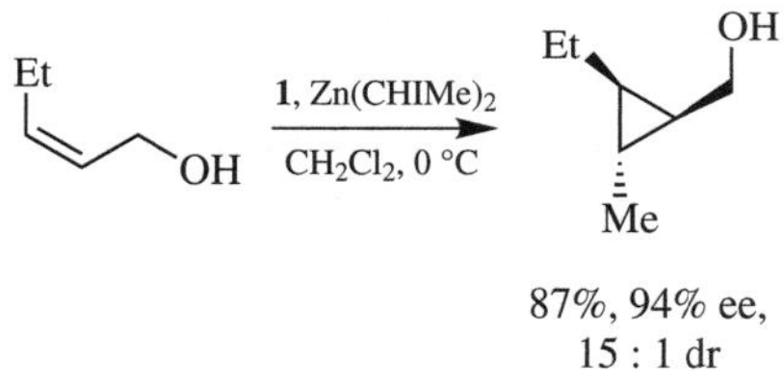

(9)

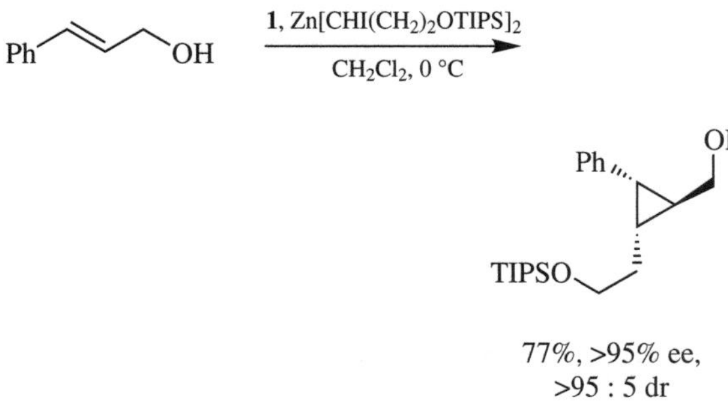

(10)

Use as a Chiral Auxiliary: Synthesis of Cyclopropylboronic Acids. The chiral dioxaborolane unit can also be used as an effective chiral auxiliary in the synthesis of enantiomerically enriched cyclopropylboronic acids. For example, 1-alkenylboronic esters bearing the tetramethyltartramide group undergo diastereoselective cyclopropanations to afford the cyclopropylboronic acid (eq 11).[14] These products can be used for *in situ* Suzuki coupling reactions[15] or can be oxidized to produce 2-substituted cyclopropanols.

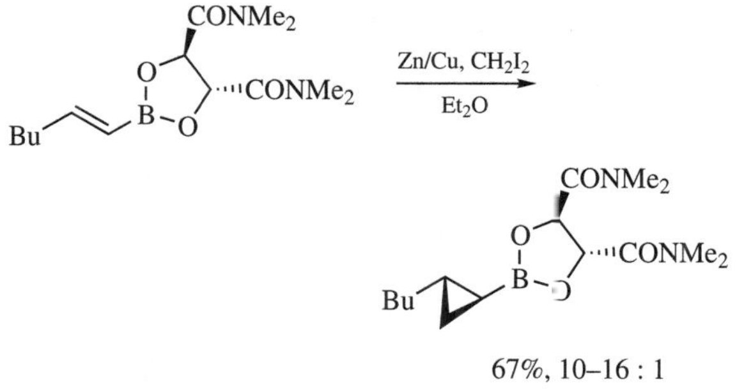

(11)

1. (a) Charette, A. B.; Marcoux, J.-F. *Synlett* **1995**, 1197. (b) Charette, A. B. In *Organozinc reagents. A practical approach*; Knochel, P.; Jones, P., Eds.; Oxford University Press: Oxford, 1999, pp 263-283. (c) Charette, A. B.; Lebel, H. In *Comprehensive Asymmetric Catalysis*; Jacobsen, E. N.; Pflatz, A.; Yamamoto, H., Eds.; Springer-Verlag: Berlin, 1999, Vol. 2, pp 581-606. (d) Charette, A. B.; Molinaro, C. In *Organoboranes for Syntheses*; Ramachandran, P. V.; Brown, H. C., Eds.; ACS: Washington DC, 2001, pp 136–147.
2. Charette, A. B.; Lebel, H. *Org. Syn.* **1998**, *76*, 86.
3. (a) Charette, A. B.; Juteau, H.; Lebel, H.; Molinaro, C. *J. Am. Chem. Soc.* **1998**, *120*, 11943. (b) Charette, A. B.; Juteau, H. *J. Am. Chem. Soc.* **1994**, *116*, 2651.
4. Charette, A. B.; Prescott, S.; Brochu, C. *J. Org. Chem.* **1995**, *60*, 1081.
5. (a) Falck, J. R.; Mekonnen, B.; Yu, J. R.; Lai, J. Y. *J. Am. Chem. Soc.* **1996**, *118*, 6096. (b) Barrett, A. G. M.; Kasdorf, K. *Chem. Commun.* **1996**, 325. (c) Barrett, A. G. M.; Hamprecht, D.; White, A. J. P.; Williams, D. J. *J. Am. Chem. Soc.* **1996**, *118*, 7863. (d) Charette, A. B.; Lebel, H. *J. Am. Chem. Soc.* **1996**, *118*, 10327.
6. Paterson, I.; Davies, R. D. M.; Marquez, R. *Angew. Chem., Int. Ed. Engl.* **2001**, *40*, 603.
7. (a) Charette, A. B.; Giroux, A. *J. Org. Chem.* **1996**, *61*, 8718. (b) Charette, A. B.; De Freitas-Gil, R. P. *Tetrahedron Lett.* **1997**, *38*, 2809. (c) Piers, E.; Coish, P. D. *Synthesis* **1995**, 47.
8. White, J. D.; Kim, T. S.; Nambu, M. *J. Am. Chem. Soc.* **1995**, *117*, 5612.
9. Charette, A. B.; Juteau, H. *Tetrahedron* **1997**, *53*, 16277.
10. Charette, A. B.; Lebel, H.; Gagnon, A. *Tetrahedron* **1999**, *55*, 8845.
11. Charette, A. B.; Lebel, H. *J. Org. Chem.* **1995**, *60*, 2966.
12. Lautens, M.; Delanghe, P. H. M. *Tetrahedron Lett.* **1994**, 9513.
13. Charette, A. B.; Lemay, J. *Angew. Chem., Int. Ed. Engl.* **1997**, *36*, 1090.
14. Sugimura, T.; Yoshikawa, M.; Futagawa, T.; Tai, A. *Tetrahedron* **1990**, *46*, 5955.
15. (a) Fontani, P.; Carboni, B.; Vaultier, M.; Mass, G. *Synthesis* **1991**, 605. (b) Wang, X. Z.; Deng, M. Z. *J. Chem. Soc., Perkin Trans. 1.* **1996**, *21*, 2663. (c) Pietruszka, J.; Widenmeyer, M. *Synlett.* **1997**, 977. (d) Zhou, S. M.; Deng, M. Z.; Xia, L. J.; Tang, M. H. *Angew. Chem., Int. Ed. Engl.* 1998, **37**, 2845. (e) Zhou, S. M.; Yan, Y. L.; Deng, M. Z. *Synlett* **1998**, 2. (f) Luithle, J. E. A.; Pietruszka, J. *J. Org. Chem.* **1999**, *64*, 8287. (g) Zhou, S. M.; Deng, M. Z. *Tetrahedron Lett.* **2000**, *41*, 3951. (h) Yao, M. L.; Deng, M. Z. *Synthesis* **2000**, 1095. (i) Chen, H.; Deng, M. Z. *Org. Lett.* **2000**, *2*, 1649. (j) Hildebrand, J. P.; Marsden, S. P. *Synlett* **1996**, 893

André B. Charette
Université de Montréal, Montréal, QC, Canada

t-Butyl 2-*t*-Butyl-3-methyl-4-oxo-1-imidazolidinecarboxylate

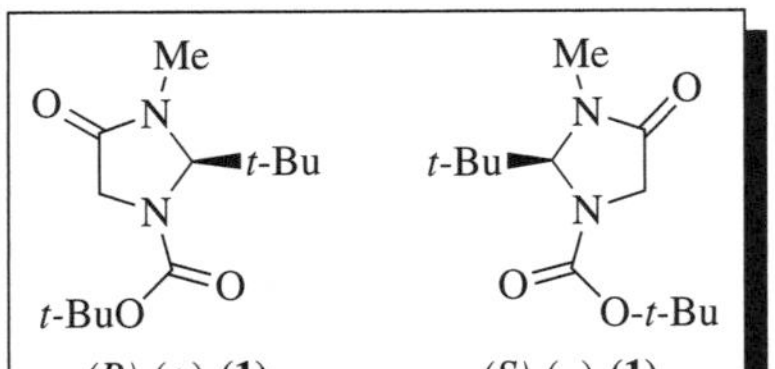

(*R*)-(+)-(**1**)   (*S*)-(−)-(**1**)

(*R*)-(+)-(**1**)
[119838-44-7] $C_{13}H_{24}N_2O_3$ (MW 256.34)
(*S*)-(−)-(**1**)
[119838-38-9]

(chiral glycine derivatives[1] for the synthesis of amino acids)

Alternate Name: *N-t*-butoxycarbonyl-2-*t*-butyl-3-methylimidazolidin-4-one; Boc-BMI.
Physical Data: mp 68–70 °C; $[\alpha]_D^{rt} = 14.6°$ ($c = 1.18$, $CHCl_3$).
Solubility: sol all common organic solvents.
Form Supplied in: colorless crystalline solid.
Preparative Methods: the commercial glycine *N*-methylamide hydrochloride is converted to the racemic imidazolidinone (**2**) by imine formation with *Pivalaldehyde* and cyclization under acidic conditions (eq 1).[1,2] The mandelate salt of like configuration is less soluble and is used for highly efficient resolution; subsequent treatment with Boc anhydride (*Di-t-butyl Dicarbonate*) gives the enantiomeric Boc-BMI (**1**) (eq 2).

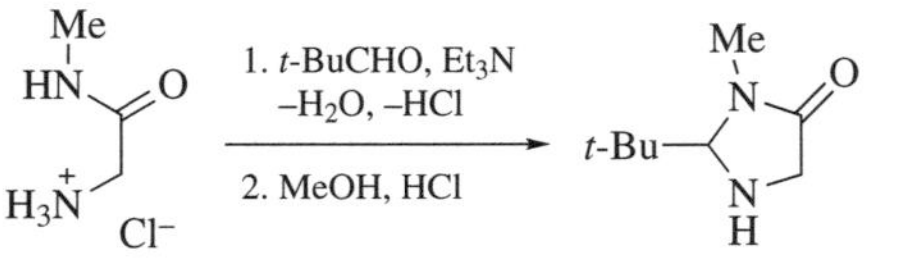

(1)

rac-(**2**)

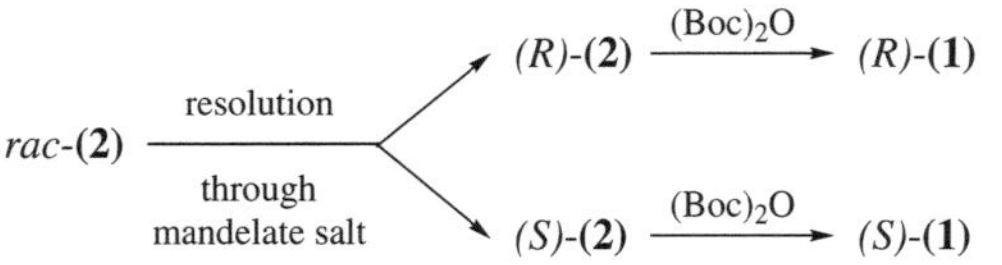

(2)

Handling, Storage, and Precautions: stable in a bottle at rt for years.

Reactions of Boc-BMI with Electrophiles. The enolate of Boc-BMI is generated with *Lithium Diisopropylamide* in THF at −75 °C; the resulting solutions of this highly nucleophilic reagent are stable up to 0 °C. All reactions occur from the face of the enolate *trans* to the *t*-Bu group at C(2). Alkylations[1,3–6] even with secondary alkyl halides are so efficient, to give (**3**), that one-pot double alkylations, which yield (**4**), are possible; the sequence in which two different alkyl halides are employed determines the absolute configuration of the α-branched α-amino acids eventually obtained.

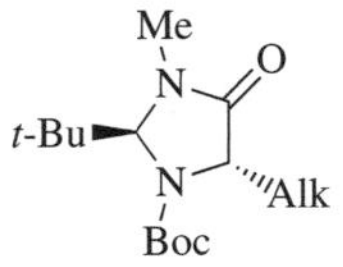

(**3**)

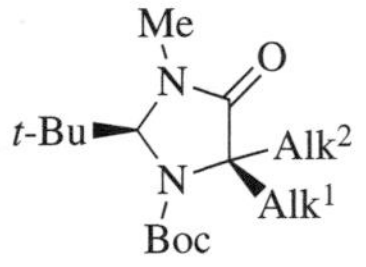

(**4**)

The method has been used to prepare isotopically labelled amino acids.[3,7] While Boc-BMI enolate adds to aldehydes with only moderate diastereoselectivity, reduction of the acylation products (**5**) gives allothreonine derivatives (**6**).[8] Michael additions to α,β-unsaturated esters,[9] ketones,[10] and nitro compounds[1] lead to products of type (**7**) and (**8**) (for a general discussion see Suzuki and Seebach[9]).

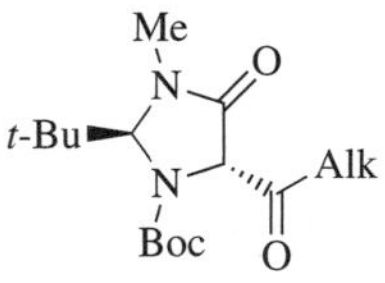

(**5**)

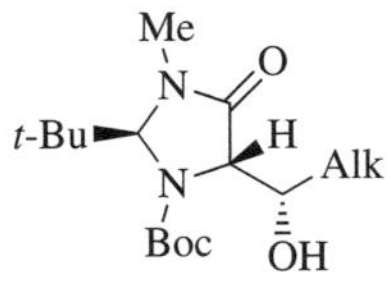

(**6**)

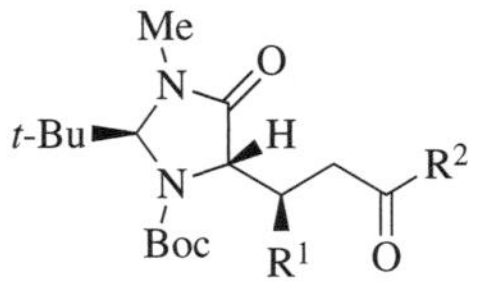

(**7**)

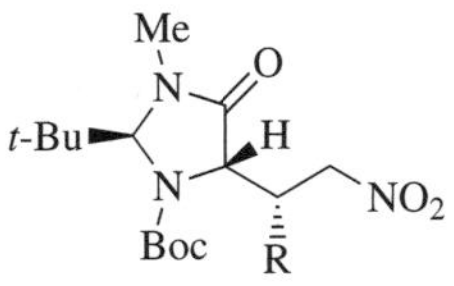

(**8**)

Preparation of and Michael Addition to 5-Alkylidene Boc-BMI. Radical bromination to give (**9**) (eq 3),[11] Arbuzov reaction, and alkenation lead to (*E*)-5-alkylidene-Boc-BMI, to which cuprates add highly diastereoselectively with formation of the imidazolidinones (**10**) containing two new stereogenic centers (eq 4).[12]

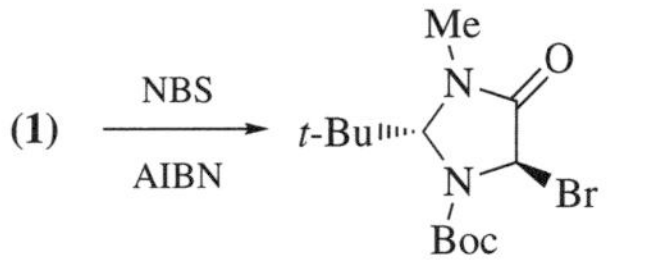

(3)

(**9**)

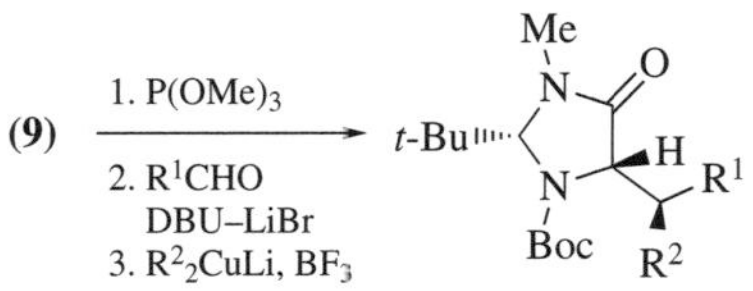

(4)

(**10**)

Hydrolysis to Nonproteinogenic Amino Acids. Numerous amino acids, including α-branched ones, have been prepared from Boc-BMI. Only a few examples can be alluded to here: (**11**) from (*S*)-(**1**), EtI, and $Br(CH_2)_4Cl$;[3] (**12**) from (*S*)-(**1**), 3-methylbutanoyl chloride, and *Lithium Triethylborohydride*;[8] (**13**) from (*R*)-(**1**), butanal, and dibutylcuprate;[12] (**14**) from (*S*)-(**1**) and 4-phenylbut-2-enoate.[9]

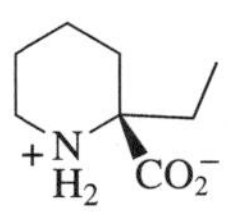

(11)

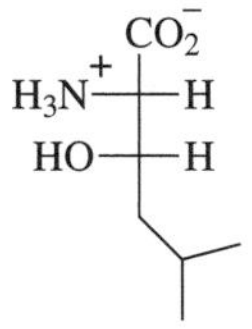

(12)

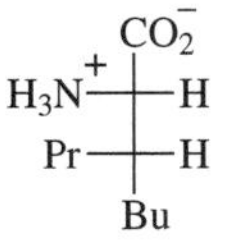

(13)

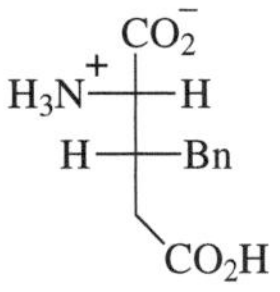

(14)

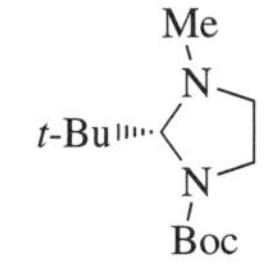

(16)

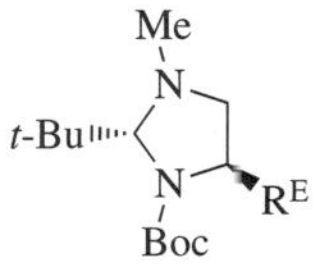

(17)

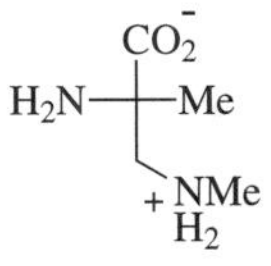

(18)

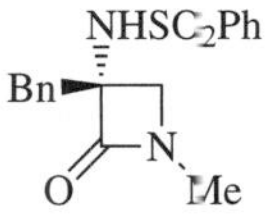

(19)

The method is applicable to the synthesis of amino acids with extremely bulky substituents in the α-position.[4b]

Chiral Imidazolidinones with Other Substitution Patterns. The intermediate imidazolidinone (**2**) can also be *N*-acylated by *Benzoyl Chloride* or *Benzyl Chloroformate*, and other substituents on the acetal center and/or on the N(3) nitrogen can be present, depending upon the aldehyde used for the imine formation and upon the glycine amide employed at the beginning of the synthesis. Chiral substituents, such as the 1-phenylethyl group may be placed on N(3), and the imidazolidinone may be derived from a dipeptide. Some examples are collected for (**15**), with references, in Table 1.[1,5,13–17] These different derivatives have advantages of their own, depending on the particular synthetic application. An access to Boc-BMI by kinetic resolution has recently been described.[18]

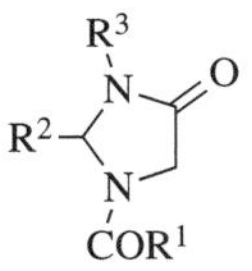

(15)

Other Synthetic Building Blocks from Boc-BMI. Deoxygenation of Boc-BMI leads to the imidazolidine (**16**) which can be lithiated on the methylene group next to N(1), and thus converted to compounds (**17**); from the corresponding ester ($R^E = CO_2Me$), 2,3-diaminopropionic acid derivatives (**18**) and (**19**) are available.[19]

Table 1 Chiral Imidazolidinones (**15**) with Different Substitution Patterns

R^1	R^2	R^3
t-BuO[5]	*i*-Pr	Me
t-BuO[1]	*t*-Bu	Bn
BnO[1]	*t*-Bu	Me
BnO[1]	*t*-Bu	Bn
BnO[13]	*t*-Bu	CH_2CO_2R
BnO[13]	*t*-Bu	(*S*)-$CHMeCO_2Me$
BnO[14]	H	(*S*)-CHMePh
Ph[15,16]	*t*-Bu	Me
Ph[1]	*t*-Bu	Bn
Ph[17]	*t*-Bu	(*S*)- or (*R*)-CHMePh

Related Reagents. 1-Benzoyl-2-*t*-butyl-3,5-dimethyl-4-imidazolidinone (2S,4S)-3-Benzoyl-2-*t*-butyl-4-methyl-1,3-oxazolidin-5-one N-Benzyloxycarbonyl-L-serine β-Lactone N-*t*-Butoxycarbonyl-N-methylaminomethyllithium (R)-2-*t*-Butyl-6-methyl-4H-1, 3-dioxin-4-one N, N-Diethylaminoacetonitrile Ethyl N-(Diphenylmethylene)glycinate Ethyl Isocyanoacetate Methyl α-Phenylglycinate

1. Fitzi, R.; Seebach, D. *Tetrahedron* **1988**, 5277.
2. Seebach, D.; Fitzi, R. Ger. Patent 3 604 591, 1986 (*Chem. Abstr.* **1988**, *108*, 94 944j).
3. Seebach, D.; Dziadulewicz, E.; Behrendt, L.; Cantoreggi, S.; Fitzi, R. *Justus Liebigs Ann. Chem./Liebigs Ann. Chem.* **1989**, 1215.
4. (a) Seebach, D.; Gees, T.; Schuler, F. *Justus Liebigs Ann. Chem./Liebigs Ann. Chem.* **1993**, 785. (b) Studer, A.; Seebach, D. *Justus Liebigs Ann. Chem./Liebigs Ann. Chem.* **1995**, 217.
5. Müller, W.; Lowe, D. A.; Neijt, H.; Urwyler, S.; Herrling, P. L.; Blaser, D.; Seebach, D. *Helv. Chim. Acta* **1992**, *75*, 855.
6. Morton, M. E.; Leanna, M. R. *Tetrahedron Lett.* **1993**, *34*, 4481; Hawthorne, M. F. *Angew. Chem.* **1993**, *105*, 997
7. Lemaire, C.; Plenevaux, A.; Cantineau, R.; Christiaens, L.; Guillaume, M.; Comar, D. *Appl. Radiat. Isot.* **1993**, *44*, 737.
8. Blank, S.; Seebach, D. *Justus Liebigs Ann. Chem./Liebigs Ann. Chem.* **1993**, 889.
9. Suzuki, K.; Seebach, D. *Justus Liebigs Ann. Chem./Liebigs Ann. Chem.* **1992**, 51.
10. Seebach, D.; Pfammatter, E.; Gramlich, V.; Bremi, T.; Kühnle, F.; Portmann, S.; Tironi, I. *Justus Liebigs Ann. Chem./Liebigs Ann. Chem.* **1992**, 1145.
11. Zimmermann, J.; Seebach, D. *Helv. Chim. Acta* **1987**, *70*, 1104.
12. Schickli, C. P.; Seebach, D. *Justus Liebigs Ann. Chem./Liebigs Ann. Chem.* **1991**, 655 (*Chem. Abstr.* **1991**, *115*, 72 163w). Seebach, D.; Bürger, H. M.; Schickli, C. P. *Justus Liebigs Ann. Chem./Liebigs Ann. Chem.* **1991**, 669 (*Chem. Abstr.* **1991**, *115*, 72 164x).
13. Polt, R.; Seebach, D. *J. Am. Chem. Soc.* **1989**, *111*, 2622.
14. Amoroso, R.; Cardillo, G.; Tomasini, C. *Tetrahedron Lett.* **1990**, *31*, 6413.
15. Seebach, D.; Miller, D. D.; Müller, S.; Weber, T. *Helv. Chim. Acta* **1985**, *68*, 949. Seebach, D.; Juaristi, E.; Miller, D. D.; Schickli, C. P.; Weber, T. *Helv. Chim. Acta* **1987**, *70*, 237.
16. Lowe, C.; Pu, Y.; Vederas, J. C. *J. Org. Chem.* **1992**, *57*, 10.
17. Juaristi, E.; Rizo, B.; Natal, V.; Escalante, J.; Regla, I. *Tetrahedron: Asymmetry* **1991**, *2*, 821.
18. Coggins, P.; Simpkins, N. S. *Synlett* **1991**, 515.

19. Pfammatter, E.; Seebach, D. *Justus Liebigs Ann. Chem./Liebigs Ann. Chem.* **1991**, 1323.

Armido Studer & Dieter Seebach
Eidgenössische Technische Hochschule, Zürich, Switzerland

(*R*)-2-*t*-Butyl-6-methyl-4*H*-1,3-dioxin-4-one[1]

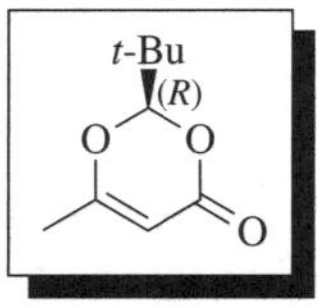

(*R*)
[107289-20-3] $C_9H_{14}O_3$ (MW 170.21)
(*S*)
[139973-88-9]

(enantiopure derivative of acetoacetic acid,[2–4] highly reactive Michael acceptor for Cu^I-doped Grignard and for Gilman reagents;[2,5] component for [2 + 2] photocycloadditions;[2] catalytic hydrogenation leads to the *cis*-disubstituted dioxanone;[2,5] the dienolate generated from the reagent can be used for chain elongations at the C(6)-Me carbon[6])

Physical Data: mp 59.8–60.2 °C; $[\alpha]_D^{rt} = -215°$ ($c = 1$, $CHCl_3$).
Solubility: sol most common organic solvents; poorly sol pentane at low temperature.
Preparative Methods: acid-catalyzed acetalization of *Pivalaldehyde* with (*R*)-3-hydroxybutanoic acid[8] gives the *cis*-1,3-dioxan-4-one (**2**) in 40% yield after recrystallization from ether/pentane. (Up to 60% yield can be obtained by using freshly loaded acidic ion-exchange resin and following the procedure of Seebach et al.[7b] Runs with up to 120 g hydroxy butanoic acid were performed in this way). Bromination with *N-Bromosuccinimide* leads to a mixture of brominated dioxinones which are debrominated hydrogenolytically to (**1**) (eq 1). The yield (**2**) → (**1**) is ~45% after recrystallization from pentane/ether (50:3) at −20 °C.[4,7] The enantiomer *ent*-(**1**) is of course equally readily available from (*S*)-3-hydroxybutanoic acid.[9] Both enantiomers of 3-hydroxybutanoic acid are commercially available.

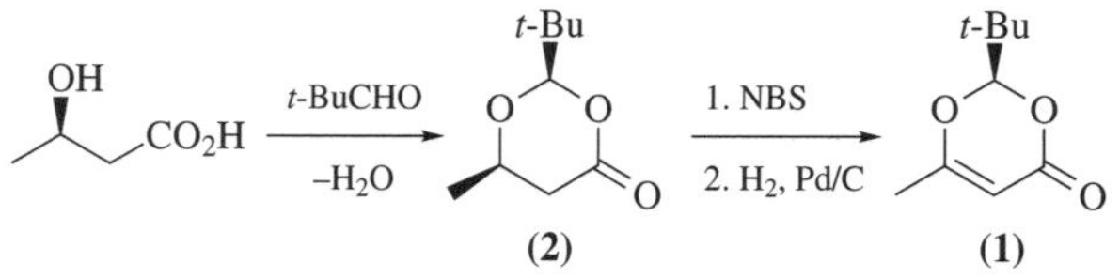
(1)

Handling, Storage, and Precautions: dioxinone (**1**) is commercially available and is indefinitely stable as a crystalline solid stored in a dark bottle at rt.

Reactions of 4*H*-1,3-Dioxin-4-one (1). The cuprate additions to (**1**) occur preferentially from the face *trans* to the *t*-Bu group. An example is the preparation of and correlation with mevalonolactone (**4**) in eq 2 by Michael addition of *Lithium Diallylcuprate* to give (**3**) and ozonolysis for degradation by one carbon.[5] Two examples of the use of the dienolate derived from dioxinone (**1**) are shown in eqs 3 and 4. The dienolate adds to aromatic aldehydes in a 1,2-fashion with reasonable diastereoselectivities at the exocyclic carbon atom. Oxidative degradation of the major diastereoisomer (**5**), obtained with benzaldehyde, leads to the β-hydroxy acid (**6**) of (*S*) configuration (eq 3).[6] With α,β-unsaturated aldehydes the exocyclic dienolate carbon reacts in a Michael addition. Thus the adduct (**7**) is isolated (53%) in a diastereomer ratio of 20:1 (eq 4).[6] Activation of the exocyclic methyl group in (**1**) is also realized by *N-Bromosuccinimide* bromination.[3,4,10] The resulting 6-bromomethyldioxinone has been employed in a vineomycinone B2 synthesis: see the intermediate (**8**) in eq 5.[11]

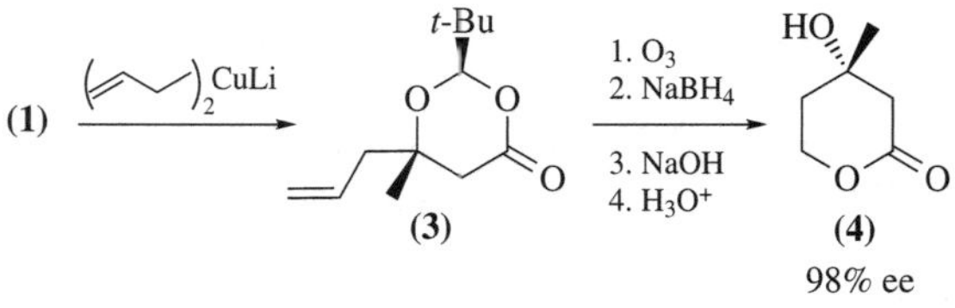
(2)

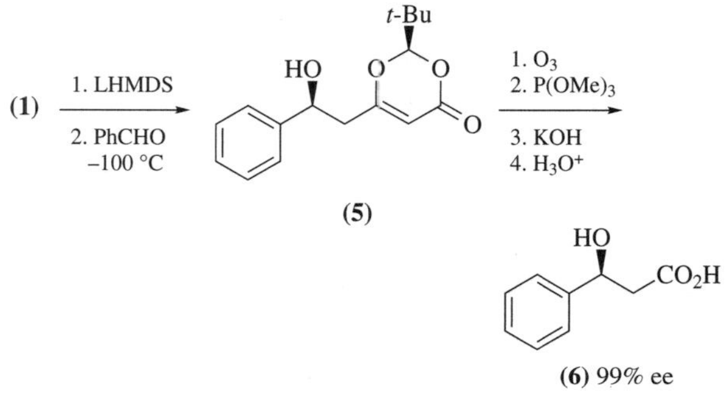
(3)

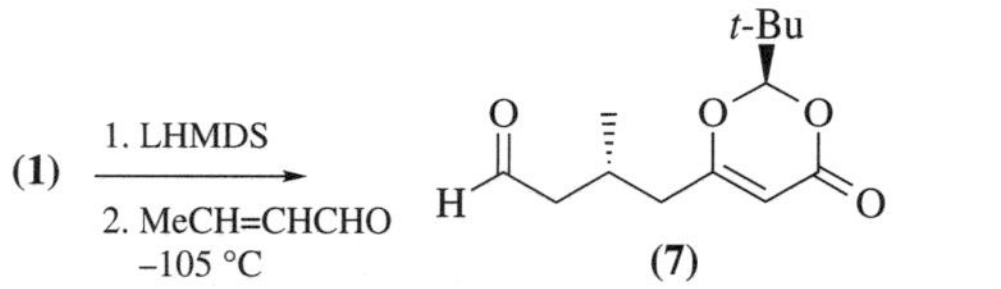
(4)

Other Enantiopure Dioxinones for Self-regeneration of the Stereogenic Center. The principle of preparing dioxinones from enantiopure β-hydroxy acids was also applied to 3-hydroxypentanoic acid,[3,6] 4,4,4-trifluoro-3-hydroxybutanoic acid,[12] 4,4,4-trichloro-3-hydroxybutanoic acid,[13] and (*S*)-serine;[3,14] aldehydes other than pivaldehyde and ketones[15] may be used for dioxinone preparation as well. Furthermore, numerous other dioxinones have been prepared from the parent compound (**1**) (eqs 3, 4, 6 and 7). The dibromide intermediate (**9**) in the preparation of (**1**) can be converted to an aldehyde, which after undergoing Wittig alkenation, followed by catalytic hydrogenation, leads to the 3-hydroxyadipic acid derivative (**10**), shown in eq 6, in an overall yield of 55%.[10] Aldol condensations of dioxanone (**2**) with aldehydes and shift of the double bond from the exo- to the endocyclic position produce 2,5,6-trisubstituted dioxinones such as (**11**), which can be used for the preparation of 2,3-disubstituted β-hydroxycarboxylic acids: see (**12**) in eq 7.[16,17] Such compounds are not accessible by current enantioselective aldol addition methodology. An example of the preparation of a CF_3-branched 3-hydroxycarboxylic acid derivative is shown in eq 8; trifluoromethyldioxinone (**13**) and *Lithium Di-n-butylcuprate* give a dioxanone which is solvolyzed in methanol to the hydroxy ester (**14**).[12b]

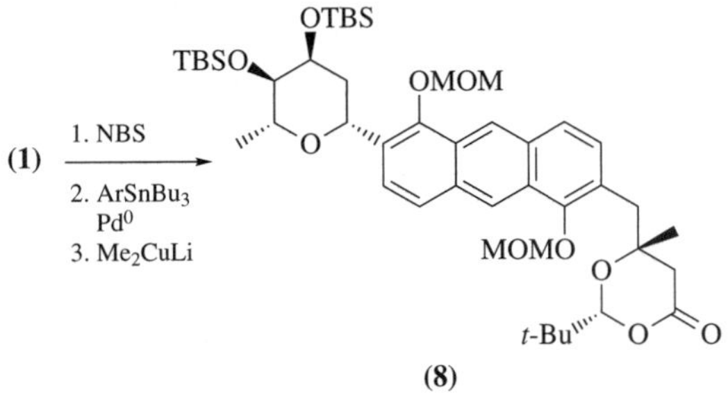
(5)

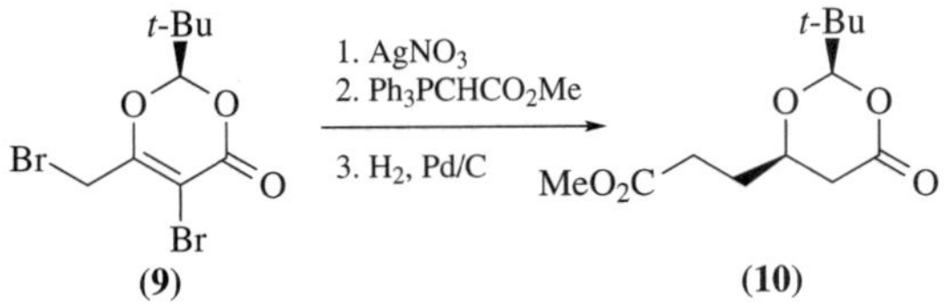
(6)

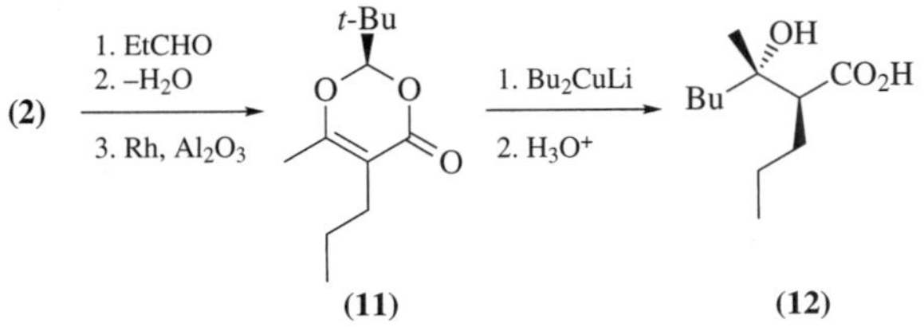
(7)

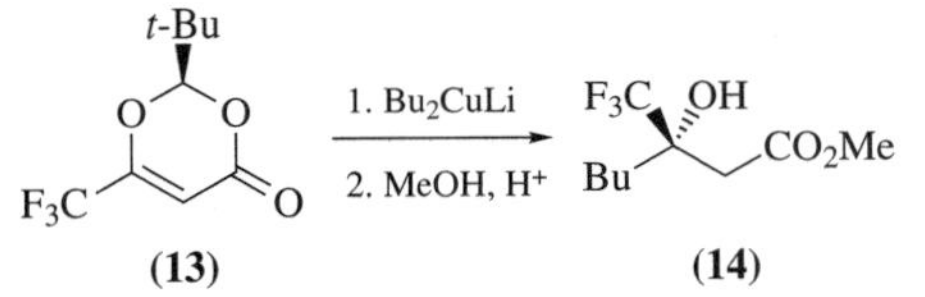
(8)

Dioxinones Obtained by Resolution or Prepared with a Chiral Auxiliary. 2-Phenyl-4*H*-1,3-dioxin-4-ones (**15**) derived from formylacetate or acetoacetate can be readily prepared in enantiopure form by preparative resolution[14,18] on cellulose triacetate.[18] These have been used for Michael additions and hydrolysis to long-chain β-hydroxycarboxylic acids, for example the tridecanoic acid (**16**) from (*R*)-(**15a**).[18] The cuprate adducts formed with the methylphenyldioxinone (*S*)-(**15b**) can be hydrogenolytically cleaved directly to β-branched β-hydroxy acids with benzyl protection of the hydroxy functional group; see (**17**) in eq 9.[18]

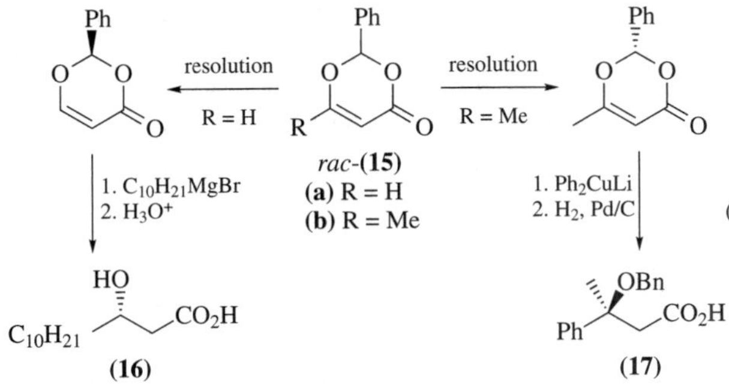
(9)

The chiral auxiliary approach involving dioxinones has been chosen by Demuth et al.[19] and, most extensively, by Kaneko and his collaborators.[20–22] They have used menthol esters (**18**) and (**19**) for typical diastereoselective reactions of dioxinones, with subsequent hydrolysis, for the preparation of various enantiopure products. For a review, also referring to the work of Winkler about photoreactions of *rac* or achiral dioxinones, see the articles by Kaneko.[23,24] For a table with enantiopure dioxinones as of mid-1991, see Kinkel et al.[14]

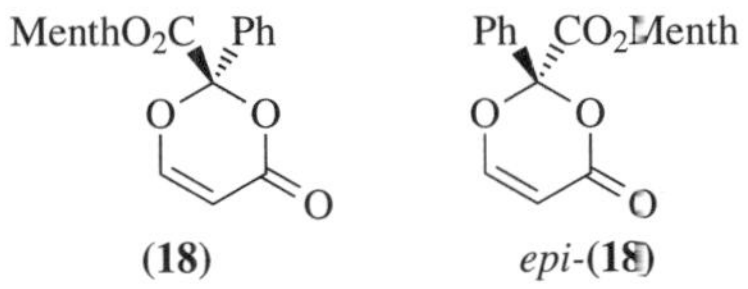

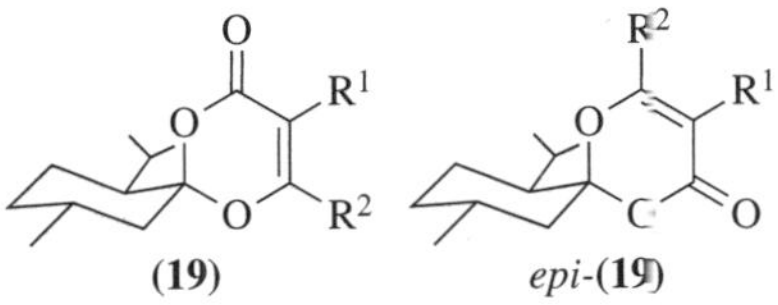

Related Reagents. (2*S*,4*S*)-3-Benzoyl-2-*t*-butyl-4-methyl-1,3-oxazolidin-5-one; (*S*)-4-Benzyl-2-oxazolidinone; (*R*,*R*)-2-*t*-Butyl-5-methyl-1,3-dioxolan-4-one; 10,2-Camphorsultam; 10-Dicyclohexylsulfonamidoisoborneol; (*R*,*R*)-2,5-Dimethylborolane; Ethyl 3-Hydroxybutanoate; 2-Hydroxy-1,2,2-triphenylethyl Acetate; (*S*)-4-Benzyl-2-oxazolidinone; (*R*,*R*)-1,2-Diphenyl-1,2-diaminoethane *N*,*N*′-Bis[3,5-bis(trifluoromethyl)-benzenesulfonamide]; 2,2,6-Trimethyl-4*H*-1,3-dioxin-4-one.

---

1. Seebach, D.; Roggo, S.; Zimmermann, J. *Stereochemistry of Organic and Bioorganic Transformations*; Proceedings of the Seventeenth Workshop Conferences Hoechst; Bartmann, W.; Sharpless, K. B., Eds; VCH: Weinheim, 1987, Vol. 17, pp 85–126.
2. Seebach, D.; Zimmermann, J. *Helv. Chim. Acta* **1986**, *69*, 1147.
3. Zimmermann, J.; Seebach, D. *Helv. Chim. Acta* **1987**, *70*, 1104.
4. Seebach, D.; Gysel, U.; Job, K.; Beck, A. K. *Synthesis* **1992**, 39.
5. Seebach, D.; Zimmermann, J.; Gysel, U.; Ziegler, R.; Ha, T.-K. *J. Am. Chem. Soc.* **1988**, *110*, 4763.
6. (a) Seebach, D.; Misslitz, U.; Uhlmann, P. *Angew. Chem., Int. Ed. Engl.* **1989**, *28*, 472. (b) Seebach, D.; Misslitz, U.; Uhlmann, P. *Chem. Ber.* **1991**, *124*, 1845 (*Chem. Abstr.* **1991**, *115*, 92 177g).
7. (a) Seebach, D.; Imwinkelried, R.; Stucky, G. *Angew. Chem., Int. Ed. Engl.* **1986**, *25*, 178. (b) Seebach, D.; Imwinkelried, R.; Stucky, G. *Helv. Chim. Acta* **1987**, *70*, 448 (*Chem. Abstr.* **1988**, *108*, 55 448f).
8. (a) Seebach, D.; Beck, A. K.; Breitschuh, R.; Job, K. *Org. Synth.* **1992**, *71*, 39. (b) Kitamura, M.; Tokunaga, M.; Ohkuma, T.; Noyori, R. *Org. Synth.* **1992**, *71*, 1.
9. (a) Seebach, D.; Sutter, M. A.; Weber, R. H.; Züger, M. F. *Org. Synth.* **1985**, *63*, 1; *Org. Synth., Coll. Vol.* **1990**, *7*, 215. (b) Ehrler, J.; Giovannini, F.; Lamatsch, B.; Seebach, D. *Chimia* **1986**, *40*, 172.
10. Noda, Y.; Seebach, D. *Helv. Chim. Acta* **1987**, *70*, 2137.
11. Tius, M. A.; Gomez-Galeno, J.; Gu, X.-q.; Zaidi, J. H. *J. Am. Chem. Soc.* **1991**, *113*, 5775.
12. (a) Acs, M.; von dem Bussche, C.; Seebach, D. *Chimia* **1990**, *44*, 90. Beck, A. K.; Gautschi, M.; Seebach, D. *Chimia* **1990**, *44*, 291 (*Chem. Abstr.* **1991**, *114*, 101 862k). (b) Gautschi, M.; Seebach, D. *Ann. Chim.(E)* **1992**, *31*, 1083. Gautschi, M.; Schweizer, W. B. Seebach, D. *Chem. Ber.* **1994**, *127*, 565 (*Chem. Abstr.* **1994**, *121*, 107 565g).
13. Beck, A. K.; Brunner, A.; Montanari, V.; Seebach, D. *Chimia* **1991**, *45*, 379 (*Chem. Abstr.* **1992**, *116*, 174 083h).

14. Kinkel, J. N.; Gysel, U.; Blaser, D.; Seebach, D. *Helv. Chim. Acta* **1991**, *74*, 1622.
15. (a) Lange, G. L.; Organ, M. G. *Tetrahedron Lett.* **1993**, *34*, 1425. (b) Organ, M. G.; Froese, R. D. J.; Goddard, J. D.; Taylor, N. J.; Lange, G. L. *J. Am. Chem. Soc.* **1994**, *116*, 3312.
16. Amberg, W.; Seebach, D. *Chem. Ber.* **1990**, *123*, 2429 (*Chem. Abstr.* **1991**, *114*, 23 106a).
17. Pietzonka, T.; Seebach, D. *Chem. Ber.* **1991**, *124*, 1837.
18. Seebach, D.; Gysel, U.; Kinkel, J. N. *Chimia* **1991**, *45*, 114 (*Chem. Abstr.* **1991**, *115*, 136015j).
19. Demuth, M.; Palomer, A.; Sluma, H.-D.; Dey, A. K.; Krüger, C.; Tsay, Y.-H. *Angew. Chem., Int. Ed. Engl.*, **1986**, *25*, 1117. Demuth, M.; Mikhail, G. *Synthesis* **1989**, 145.
20. Kaneko, C. *Organic Synthesis in Japan. Past, Present, and Future*, Noyori, R., Ed.; Tokyo Kagaku Dozin: Tokyo, 1992, pp 175–183.
21. Sato, M.; Murakami, M.; Kaneko, C.; Furuya, T. *Tetrahedron* **1993**, *49*, 8529.
22. See also: Jansen, U.; Runsink, J.; Mattay, J. *Liebigs Ann. Chem.* **1991**, 283.
23. Kaneko, C.; Sato, M.; Sakaki, J.-i.; Abe, Y. *J. Heterocycl. Chem.* **1990**, *27*, 25.
24. (a) Cf. also: Takeshita, H.; Cui, Y.-S.; Kato, N.; Mori, A.; Nagano, Y. *Bull. Chem. Soc. Jpn.* **1992**, *65*, 2940. (b) Winkler, J. D.; Shao, B. *Tetrahedron Lett.* **1993**, *34*, 3355.

Albert K. Beck & Dieter Seebach
Eidgenössische Technische Hochschule, Zürich, Switzerland

(R,R)-2-t-Butyl-5-methyl-1,3-dioxolan-4-one[1]

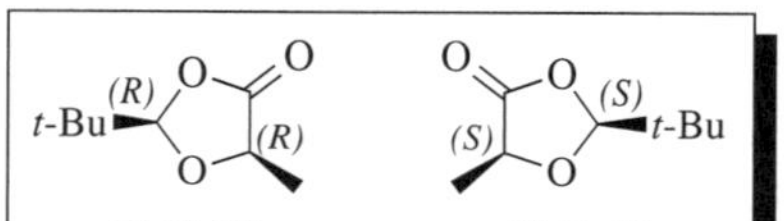

(R,R)-(1) (S,S)-(1)

(R,R) *[104194-02-7]* $C_8H_{14}O_3$ (MW 158.20)
(S,S) *[81037-06-1]*

(cyclic acetals from (*R*)- and (*S*)-lactic acid and pivaldehyde;[1b,2–6] reagents for the preparation of enantiopure α-hydroxy-α-methyl carboxylic acids by alkylation of the corresponding lithium enolate with alkyl,[1b,2–4] allyl,[1b,2,7,8] and benzyl[1b] halides, by hydroxyalkylation with aldehydes and ketones,[1b,3,9,10] and by Michael addition to nitroalkenes;[9,11] precursor to the 5-bromo derivative used for radical reactions;[12,13] precursor to 2-*t*-butyl-5-methylene-1,3-dioxolan-4-one;[13–15] an acceptor for radical additions;[15,16] and an ene component for Diels–Alder reactions leading to cyclic, heterocyclic, and bicyclic α-hydroxy carboxylic acids[14,17–19])

Physical Data: mp ca. 5 °C; bp 80 °C/20 mmHg; $[\alpha]^{rt}_D = +44.8°$ ($c = 1.83$, $CHCl_3$) for (*S,S*)-(**1**) containing 4% (2*R*,5*S*) epimer after two recrystallizations from ether/pentane at −75 °C.
Solubility: good to excellent in all common organic solvents.
Preparative Methods: on a 0.5 mol scale reagent (*S,S*)-(**1**) is prepared by condensation of *Pivalaldehyde* and (*S*)-lactic acid under acid catalysis in pentane, with azeotropic removal of the water formed. The crude product is distilled in vacuo to give 93% of a 4:1 *cis/trans* mixture. Two recrystallizations from pentane/ether at −75 °C furnish 60% (*S,S*)-(**1**) (*cis/trans* = 96:4).
Handling, Storage, and Precautions: stable for many months under an inert atmosphere in a refrigerator.

Reactions of the Enolate of (1) with Electrophiles. Addition of the dioxolanones (**1**) to solutions of *Lithium Diisopropylamide* or *Lithium Hexamethyldisilazide* in THF at dry-ice temperature generates the corresponding enolates which react with alkyl halides,[1b,2–4,7,8] carbonyl compounds,[1b,3,9,10] and nitroalkenes[9,11] almost exclusively from the face remote from the *t*-Bu group to give products of type (**2**). These can be hydrolyzed to simple α-hydroxy-α-methyl carboxylic acids or further elaborated. Four examples are shown in (**3**)–(**6**) in which the part of the molecule originating from lactic acid is indicated in bold.

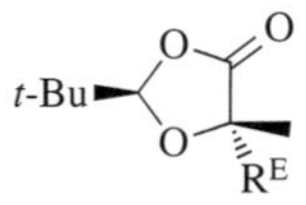

(**2**)

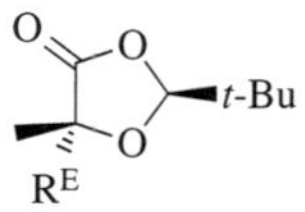

ent-(**2**)

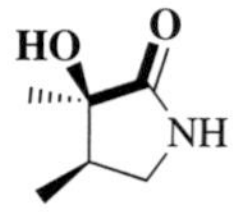

(**3**) γ-lactam from (*S*)-(+)-lactic acid

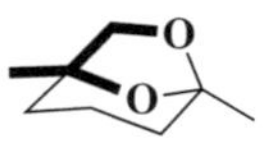

(**4**) (*S*)-(−)-Frontalin from (*R*)-(−)-lactic acid

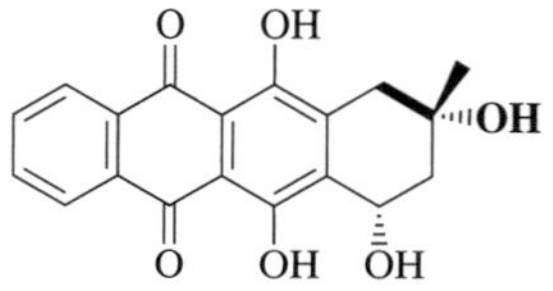

(**5**) 4-Demethoxyfeudomycinone C from (*S*)-(+)-lactic acid

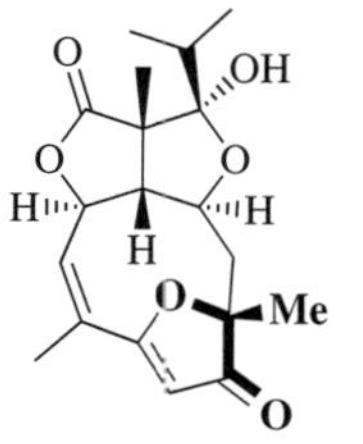

(**6**) (+)-Eremantholid A from (*S*)-(+)-lactic acid

Analogous Transformations with other α-Hydroxy Carboxylic Acids. The conversion of lactic acid to products (**2**)–(**6**) is an example of the principle of self-regeneration of the stereogenic centers (SRSC) which is also applicable to β-hydroxy-, α- and β-amino acids (see Related Reagents). Many α-hydroxy carboxylic acids occur naturally, and most α-amino acids can be converted to hydroxy acids by diazotization, with retention of configuration,[20] producing a host of readily available starting materials for this kind of conversion, for example in the synthesis of the antitumor alkaloid (**7**).[21] Table 1 lists various dioxolanones (**8**) made from the corresponding hydroxy acids and aldehydes or ketones, together with information on reactions carried out with them. This method is also applicable to α-mercapto carboxylic acids.[1,22]

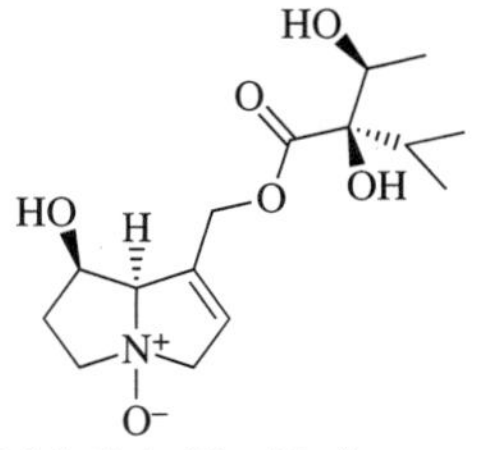

(**7**) (+)-Indicin *N*-oxide from (*S*)-2-hydroxy-3-methylbutanoic acid (from valine)

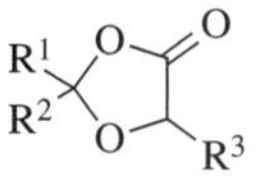

(**8**)

Table 1 Various Dioxolanones (**8**) Derived from the Corresponding α-Hydroxy Acids

R^1	R^2	R^3	Reported reaction
Me	H	Me	–[2]
Ph	H	Me	–[6b]
Cy	H	Me	–[6b]
i-Pr	H	Me	–[6b]
Me	Bu	Me	–[23]
Me	Ph	Me	–[6a,23]
t-Bu	Me	Me	–[6a]
t-Bu	Ph	Me	–[6a]
t-Bu	Ph	Me	Aldol addition[24]
t-Bu	H	Et	Alkylation[7]
t-Bu	H	*i*-Pr	Aldol addition[21]
t-Bu	H	Bu	–[6b]
t-Bu	H	*s*-Bu	–[6b]
t-Bu	H	CH_2CO_2H	–[5]
t-Bu	H	CH_2CO_2H	Alkylation[25]
t-Bu	H	CH_2CO_2H	Barton reaction[15]
t-Bu	H	CH_2SPh	Oxidation/elimination[17a]
Cy	H	CH_2SPh	Oxidation/elimination[17a]
t-Bu	H	Bn	–[6b]
Me	H	Ph	–[6b]
i-Pr	H	Ph	Alkylation[2]
t-Bu	H	Ph	–[5,6b]
t-Bu	H	Ph	Alkylation[2]
Me	Me	Ph	–[23]
C(Me)=CH_2	Me	Ph	–[23]
Bu	Me	Ph	–[23]
Ph	H	Ph	–[23]

(*R*)- and (*S*)-*t*-Butyl-5-methylene-1,3-dioxolan-4-one, a Chiral α-Alkoxy Acrylate. It is also possible to introduce an exocyclic double bond onto the dioxolanone ring, as in compounds (**9**)–(**11**), derived from lactic[13,14,17–19,26] and malic[12,27] acids. These α,β-unsaturated carbonyl derivatives are acceptors for radical additions[12,15,16] and undergo cycloadditions with dienes[14,17–19,26,27] and heterodienes.[18] The Diels–Alder adduct (**12**) of *ent*-(**9**) with cyclopentadiene is formed[14,17a,19] with *exo* selectivity (96:4) and serves as a precursor to norbornenone (**13**).[14,19] Cycloadduct (**14**), obtained from methylenedioxolanone (**9**) and an open-chain triene, is also the result of an *exo* addition and is used in tetronolide synthesis.[17b]

Menthone-Derived Dioxolanones: Chiral Glycolic Acid Derivatives.[28] Acetalization of menthone and phenylmenthone with glycolic acid leads to chromatographically separable mixtures of diastereoisomeric dioxolanones (**15**) and (**16**); they are precursors for chiral enolate derivatives of glycolic acid. Alkylations occur highly selectively, and the products can be solvolyzed with ethanol to give ethyl α-hydroxy carboxylates of either (*R*) or (*S*) configuration. Thus spiro compound (**16b**) gives the allylation product (**17**) (84%), from which pure ethyl (*R*)-2-hydroxypent-4-enoate is obtained.

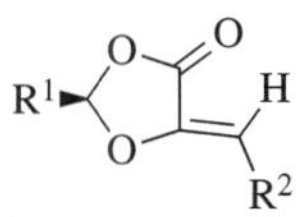

(**9**) R^1 = *t*-Bu, R^2 = H
(**10**) R^1 = C_6H_{11}, R^2 = H
(**11**) R^1 = *t*-Bu, R^2 = CO_2Me or CO_2Et

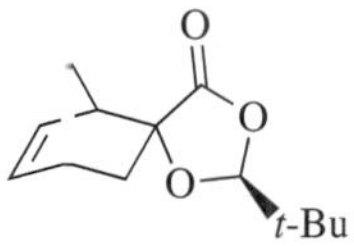

(**12**)

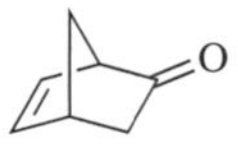

(**13**)

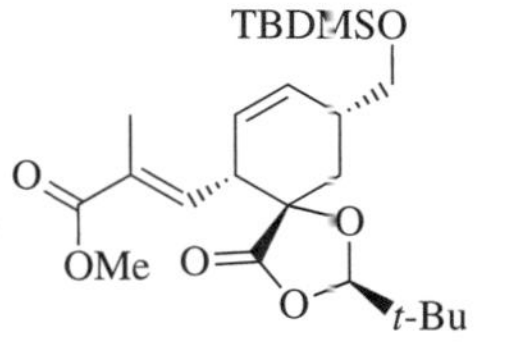

(**14**)

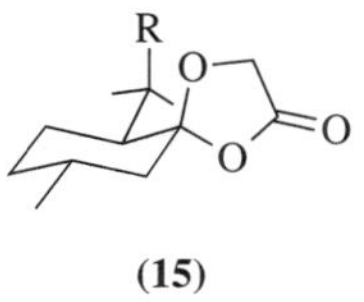

(**15**)
(**a**) R = H
(**b**) R = Ph

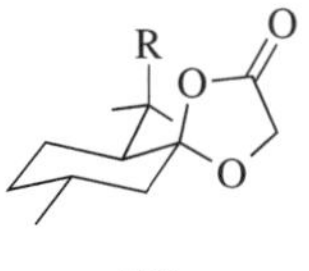

(**16**)
(**a**) R = H
(**b**) R = Ph

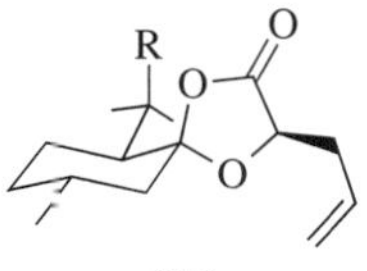

(**17**)
123:1 from (**16b**)

Related Reagents. 1-Benzoyl-2-*t*-butyl-3,5-dimethyl-4-imidazolidinone; (2*S*,4*S*)-3-Benzoyl-2-*t*-butyl-4-methyl-1,3-oxazolidin-5-one; *t*-Butyl 2-*t*-Butyl-3-methyl-4-oxo-1-imidazolidinecarboxylate; (*R*)-2-*t*-Butyl-6-methyl-4*H*-1,3-dioxin-4-one; Ethyl 3-Hydroxybutanoate; Ethyl Mandelate; (*R*)-Methyl 2-*t*-Butyl-3(2*H*)-oxazolecarboxylate; Methyl *O*-Methyllactate; Phenoxyacetic Acid; (−)-8-Phenylmenthol.

1. (a) Seebach, D.; Imwinkelried, R.; Weber, T. In *Modern Synthetic Methods*; Springer: New York, 1986, Vol. 4, pp 125–259. (b) Seebach, D.; Naef, R.; Calderari, G. *Tetrahedron* **1984**, *40*, 1313.
2. Fráter, G.; Müller, U.; Günther, W. *Tetrahedron Lett.* **1981**, *22*, 4221.
3. Seebach, D.; Naef, R. *Helv. Chim. Acta* **1981**, *64* 2704.
4. Naef, R.; Seebach, D. *Liebigs Ann. Chem.* **1983**, 1930.
5. Hoye, T. R.; Peterson, B. H.; Miller, J. D. *J. Org. Chem.* **1987**, *52*, 1351.
6. Greiner, A.; Ortholand, J.-Y. *Tetrahedron Lett.* **1990**, *31*, 2135.
7. Krohn, K. *Tetrahedron* **1990**, *46*, 291.
8. Boeckman, R. K.. Jr.; Yoon, S. K.; Heckendorn, D. K. *J. Am. Chem. Soc.* **1991**, *113*, 9682.
9. Suzuki, K.; Seebach, D. *Liebigs Ann. Chem.* **1992**, 51.
10. Naef, R. Dissertation ETH Nr. 7442, **1983**.
11. Calderari, G.; Seebach, D. *Helv. Chim. Acta* **1985**, *68*, 1592 (*Chem. Abstr.* **1986**, *105*, 133 326u).
12. Kneer, G.; Mattay, J. *Tetrahedron Lett.* **1992**, *33*, 8051.
13. Zimmermann, J.; Seebach, D. *Helv. Chim. Acta* **1987**, *70*, 1104.

14. Mattay, J.; Mertes, J.; Maas, G. *Chem. Ber.* **1989**, *122*, 327.
15. Beckwith, A. L. J.; Chai, C. L. L. *Chem. Commun.* **1990**, 1087.
16. Beckwith, A. L. J. *Chem. Soc. Rev.* **1993**, *22*, 143.
17. (a) Roush, W. R.; Essenfeld, A. P.; Warmus, J. S.; Brown, B. B. *Tetrahedron Lett.* **1989**, *30*, 7305. (b) Roush, W. R.; Koyama, K. *Tetrahedron Lett.* **1992**, *33*, 6227.
18. Mattay, J.; Kneer, G.; Mertes, J. *Synlett* **1990**, 145.
19. Roush, W. R.; Brown, B. B. *J. Org. Chem.* **1992**, *57*, 3380.
20. Brewster, P.; Hiron, F.; Hughes, E. D.; Ingold, C. K.; Rao, P. A. D. S. *Nature* **1950**, *166*, 179.
21. Ogawa, T.; Niwa, H.; Yamada, K. *Tetrahedron* **1993**, *49*, 1571.
22. Strijtveen, B.; Kellogg, R. M. *Tetrahedron* **1987**, *43*, 5039.
23. Neveux, M.; Seiller, B.; Hagedorn, F.; Bruneau, C.; Dixneuf, P. H. *J. Organomet. Chem.* **1993**, *451*, 133.
24. Greiner, A.; Ortholand, J.-Y. *Tetrahedron Lett.* **1992**, *33*, 1897.
25. Krohn, K.; Rieger, H. *Liebigs Ann. Chem.* **1987**, 515 (*Chem. Abstr.* **1987**, *107*, 58 713d).
26. Roush, W. R.; Brown, B. B. *Tetrahedron Lett.* **1989**, *30*, 7309.
27. Kneer, G.; Mattay, J.; Raabe, G.; Krüger, C.; Lauterwein, J. *Synthesis* **1990**, 599.
28. (a) Pearson, W. H.; Cheng, M.-C. *J. Org. Chem.* **1986**, *51*, 3746; **1987**, *52*, 1353, 3176. (b) Pearson, W. H.; Hines, J. V. *J. Org. Chem.* **1989**, *54*, 4235.

Andrea Rolf Sting & Dieter Seebach
Eidgenössische Technische Hochschule, Zürich, Switzerland

(*R*)-(+)-*t*-Butyl 2-(*p*-Tolylsulfinyl)acetate

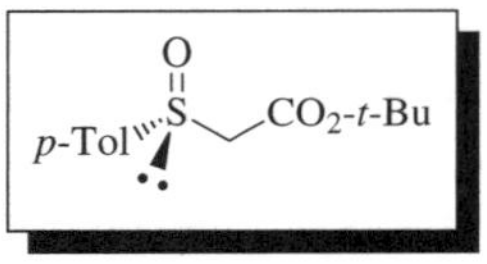

[58059-08-8] $C_{13}H_{18}O_3S$ (MW 254.34)

(reagent for asymmetric aldol-type condensation;[1] used for the synthesis of sulfinyl dienophiles[13])

Physical Data: $[\alpha]_D^{20} = +149°$ (EtOH, $c = 2.25$).

Preparative Methods: conveniently prepared[2,3] by reaction of the magnesium enolate of *t*-butyl acetate (readily made with *Bromomagnesium Diisopropylamide*) with *(−)-(1R,2S,5R)-Menthyl (S)-p-Toluenesulfinate* (eq 1). It was also made in 91% yield by reacting a solution of *Lithium Diisopropylamide* with (*R*)-(+)-methyl *p*-tolyl sulfoxide and *t*-butyl carbonate (eq 2).[4] It should be noted that asymmetric oxidation of *t*-butyl 2-(*p*-tolylsulfinyl)acetate with a modified Sharpless reagent gave a poor ee.[5]

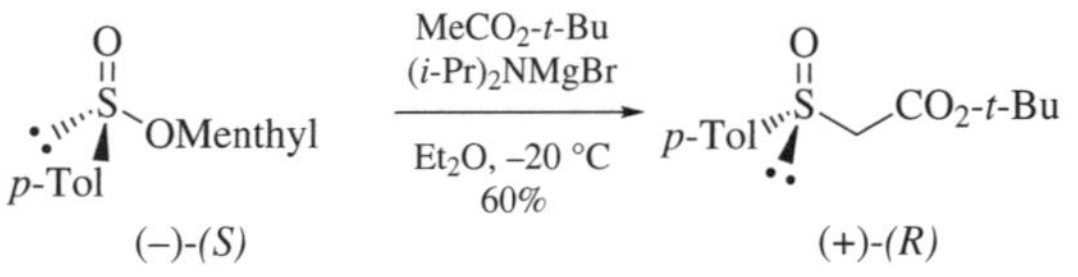
(1)

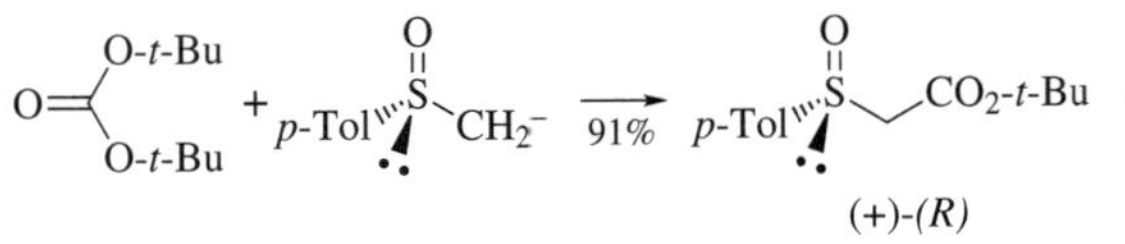
(2)

Aldol-Type Addition. Aldol-type addition of the magnesium enolate of (*R*)-(+)-*t*-butyl 2-(*p*-tolylsulfinyl)acetate, prepared with *t*-butylmagnesium bromide, with aldehydes and ketones afforded, after desulfurization with *Aluminum Amalgam*, β-hydroxy esters in very high diastereoselectivity (eq 3).[3,6,7] Two chiral centers are created in the first step with very high diastereoselectivity (mainly one diastereomer is formed). A model M based on the structure of the sulfinyl ester enolate (determined by ^{13}C NMR)[8] and on electrophilic assistance of magnesium to the carbonyl approach, was proposed to explain and predict the absolute configuration of the two created chiral centers.[3]

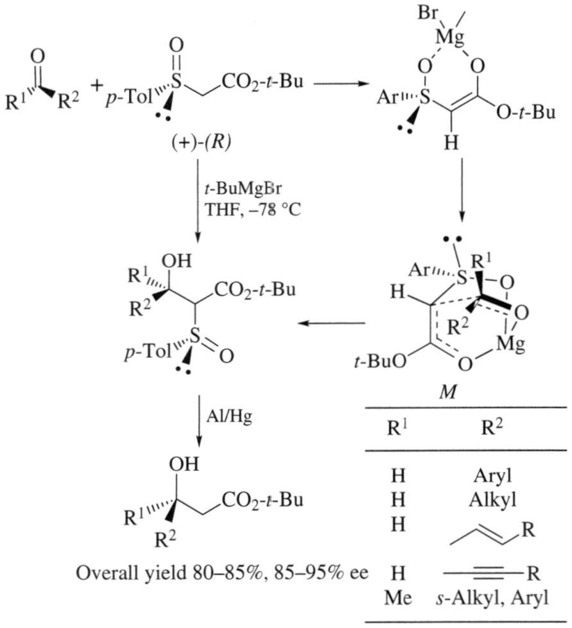
(3)

The first application of this aldol-type asymmetric synthesis was reported by Corey during the later stages of the total synthesis of maytansine.[9] This result (eq 4) showed that the *t*-butyl ester could be replaced by a phenyl ester as long as the same base, *t*-BuMgBr, is used for the condensation. The reaction of the α,β-unsaturated aldehyde gave, after desulfurization, the corresponding β-hydroxy ester in 80% yield and 86% de.

Optically active five- and six-membered lactones were also prepared by this aldol-type addition (eq 5).[10]

Propargylic aldehyde was also used to prepare, by condensation with (+)-(*R*)-*t*-butyl *p*-tolylsulfinylacetate, a precursor of the C-3–C-8 fragment of leukotriene B_4 (eq 6).[11]

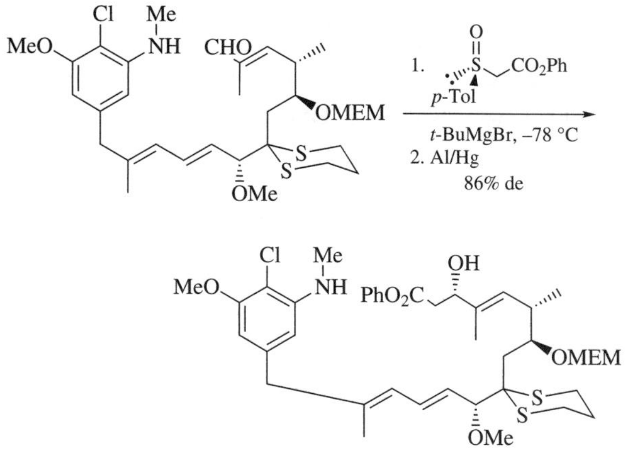
(4)

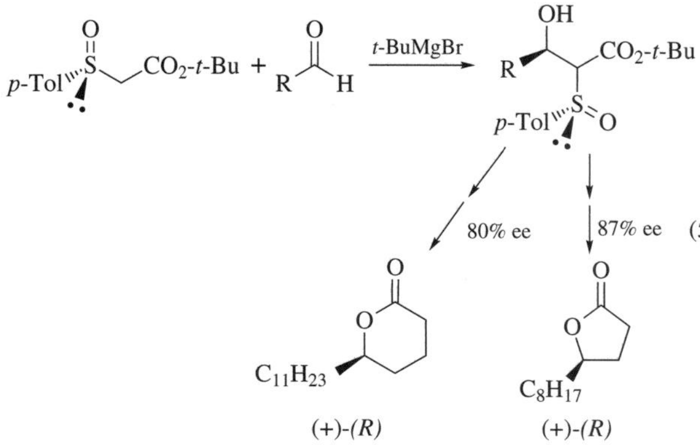
(5)

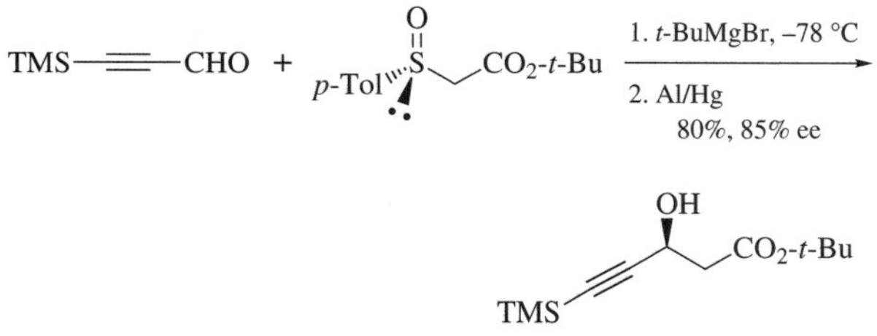
(6)

It should be noted that a poor ee was observed during the Michael addition of (+)-(*R*)-*t*-butyl *p*-tolylsulfinylacetate to an α,β-unsaturated ester.[12]

Preparation of Sulfinyl Dienophiles. This sulfinyl ester was also used to prepare optically active sulfinyl dienophiles by a Knoevenagel-type condensation of *Glyoxylic Acid* (eq 7).[13,14]

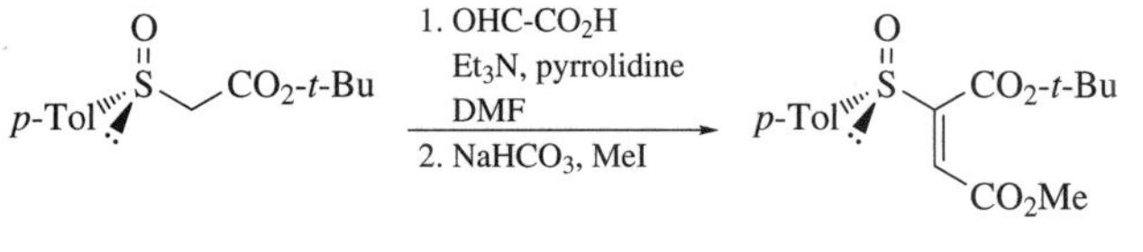
(7)

Related Reagents. (*R*)-(+)-*t*-Butyl 2-(*p*-Tolylsulfinyl)-propionate; (*R*)-(+)-Methyl *p*-Tolyl Sulfoxide; (*R*)-(+)-Phenyl (*p*-Toluenesulfinyl)acetate.

1. Solladié, G. *Synthesis* **1981**, 185.
2. Solladié, G. In *Asymmetric Synthesis*; Morrison, J. D., Ed.; Academic: New York, 1983, Vol. 2, pp 157–198.
3. (a) Mioskowski, C.; Solladié, G. *Tetrahedron Lett.* **1975**, 3341. (b) Mioskowski, C.; Solladié, G. *Tetrahedron* **1980**, *36*, 227.
4. Abushanab, E.; Reed, D.; Suzuki, F.; Sih, C. J. *Tetrahedron Lett.* **1978**, 3415.
5. Duñach, E.; Kagan, H. B. *Nouv. J. Chim.* **1985**, *9*, 1.
6. Mioskowski, C.; Solladié, G. *Chem. Commun.* **1977**, 162.
7. Solladié, G.; Fréchou, C.; Demailly, G. *Nouv. J. Chim.* **1985**, *9*, 21.
8. Solladié-Cavallo, A.; Mioskowski, C. *Org. Magn. Reson.* **1981**, *16*, 273.
9. Corey, E. J.; Weigel, L. O.; Chamberlin, A. R. Cho, H.; Hua, D. H. *J. Am. Chem. Soc.* **1980**, *102*, 6613.
10. Solladié, G.; Matloubi-Moghadam, F. *J. Org. Chem.* **1982**, *47*, 91.
11. Solladié, G.; Hamdouchi C. *Synthesis* **1991**, 979.
12. Matloubi, F.; Solladié, G. *Tetrahedron Lett.* **1979**, 2141.
13. Alonso, I.; Carretero, J. C.; García Ruano, J. L. *Tetrahedron Lett.* **1991**, *32*, 947.
14. Alonso, I.; Cid, M. B.; Carretero, J. C.; García Ruano, J. L.; Hoyos, M. A. *Tetrahedron: Asymmetry* **1991**, *2*, 1193.

Guy Solladié & Françoise Colobert
University Louis Pasteur, Strasbourg, France

(*R*)-(+)-*t*-Butyl 2-(*p*-Tolylsulfinyl)-propionate

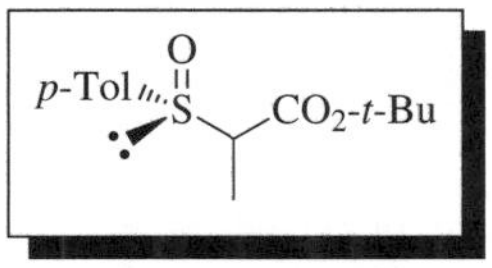

(*R,R*)
[*83909-72-2*] $C_{14}H_{20}O_3S$ (MW 268.37)
(*R,S*)
[*83909-83-5*]

(reagent for asymmetric aldol-type condensation)

Physical Data: $[\alpha]_D = +148°$ (EtOH, $c = 0.3$) for the 1:1 mixture of diastereomers.

Preparative Methods: the synthesis of this compound was first reported by methylation of *(R)-(+)-t-Butyl 2-(p-Tolylsulfinyl)acetate* via enolate generation with lithium bases such as *n-Butyllithium* or *t-Butyllithium* at 0 °C and with only *Iodomethane* as the alkylating agent (eq 1).[1] The diastereomeric ratio was shown by ^{1}H NMR to be 50:50 with BuLi and 42:58 with *t*-BuLi. The title compound was also prepared from *(−)-(1R,2S,5R)-Menthyl (S)-p-Toluenesulfinate* and the magnesium enolate of *t*-butyl propionate in 68% yield as a 1:1 ratio of the two possible diastereomers (eq 2).[1]

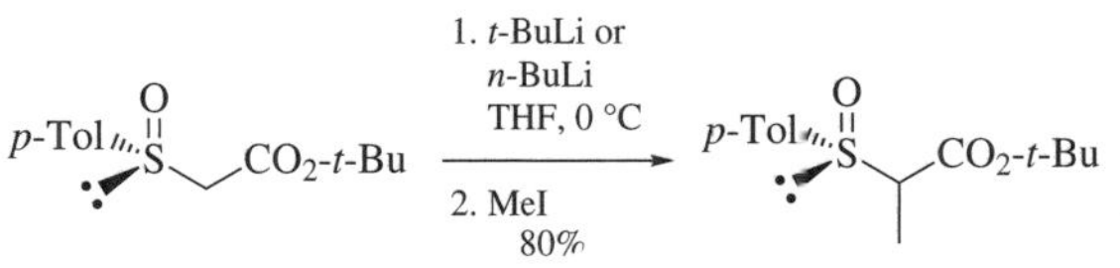
(1)

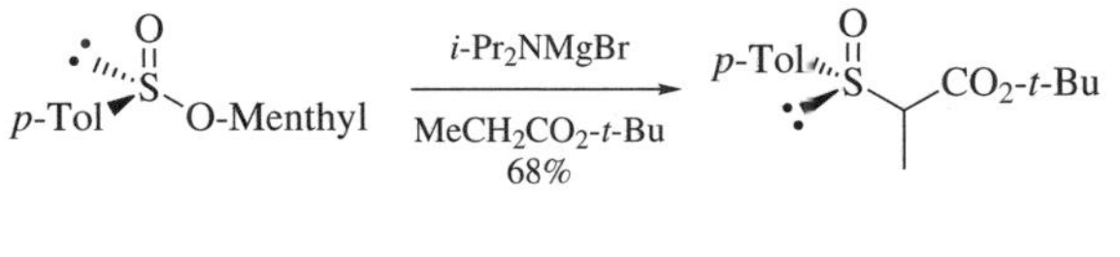
(2)

Aldol-Type Condensation. The aldol-type condensation of the enolate anion of sulfinylpropionate, which was prepared as usual with the base *t*-butylmagnesium bromide, with aldehydes afforded after desulfurization with *Aluminum Amalgam* the corresponding β-hydroxy esters in high yield (90%) and, with aliphatic aldehydes, high diastereoselectivity (eq 3).[2,3] The amount of asymmetric induction was determined by transformation of the β-hydroxy esters to the corresponding isopropyl-substituted alcohols.

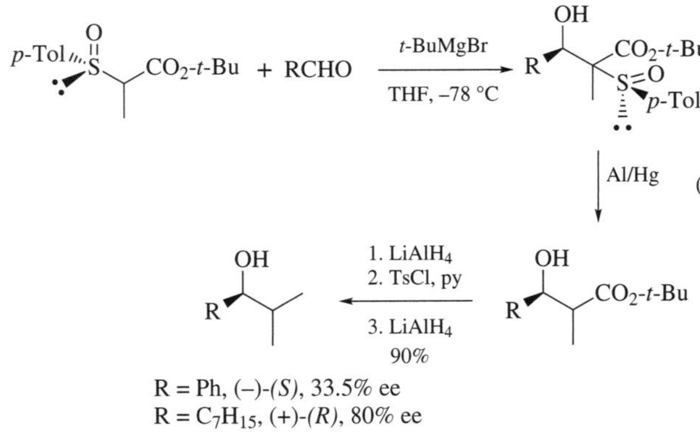

(3)

1. Solladié, G.; Matloubi-Moghadam, F.; Luttmann, C.; Mioskowski, C. *Helv. Chim. Acta* **1982**, *65*, 1602.
2. Solladié, G.; Mioskowski, C. *Tetrahedron* **1980**, *36*, 227.
3. Solladié, G. *Acta Chem. Scand.* **1981**, 185.

Guy Solladié & Françoise Colobert
University Louis Pasteur, Strasbourg, France

C

(−)-(1*S*,4*R*)-Camphanic Acid[1]

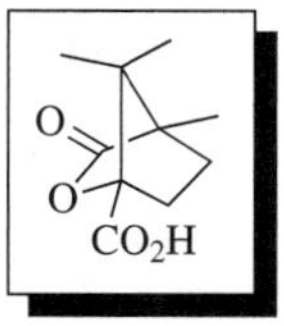

(1*S*)		
[13429-83-9]	$C_{10}H_{14}O_4$	(MW 198.22)
(1*R*)		
[67111-66-4]		
(±)		
[465-48-5]		

(enantiomeric purity determination;[2] chemical resolution;[3] chiral auxiliaries[4])

Physical Data: mp 201–204 °C; (1*S*): $[\alpha]_D$ −20.4° (*c* 1.71, dioxane); $[\alpha]_D$ −18° (*c* 1, dioxane).
Solubility: sol EtOH, ether, boiling H_2O, and AcOH.
Form Supplied in: white solid.
Preparative Methods: commercially available. Alternatively, the acid can be prepared in two steps from camphoric acid (1. PCl_5; 2. H_2SO_4; 65% overall yield). The acid can be converted to the corresponding acid chloride upon treatment with *Thionyl Chloride* (99% yield).[1,5]
Purification: crystallized from hot toluene.
Handling, Storage, and Precautions: stable; no special precautions.

Analysis of the Enantiomeric Purity of Alcohols and Amines. It has been shown that camphanic acid is an efficient chiral derivatizing agent for the determination of the enantiomeric purity of alcohols and amines.[2] A typical procedure involves mixing a solution of the amine or the alcohol (in CH_2Cl_2, py, or benzene) with camphanoyl chloride in the presence of a base (Et_3N, py, DMAP, or $NaHCO_3$). Alternatively, the substrate can be coupled directly with camphanic acid in the presence of DCC/DMAP. These conditions, however, can potentially lead to significant kinetic resolution.[6] Camphanic acid was initially developed for the analysis of the enantiomeric purity of α-deuterated primary alcohols[7] and amines.[8] Distinct signals by ^1H NMR for the two diastereomers can usually be observed upon addition of a chiral shift reagent or when using C_6D_6 as the solvent.[9] Since then, this chiral derivatizing agent has been widely used for measuring the enantiomeric excess of several classes of compounds such as α-monodeuterated glycine derivatives (**1**),[10] α- and β-amino acids (**2, 3**),[11,12] α,α-disubstituted α-amino acids (**4**),[13] secondary alcohols (**5**),[14] 1,2-amino alcohols (**6**),[15] and sulfoximines.[16]

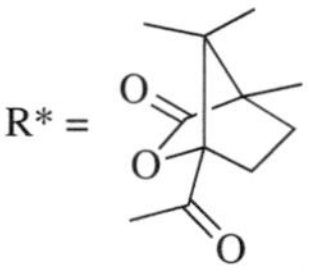

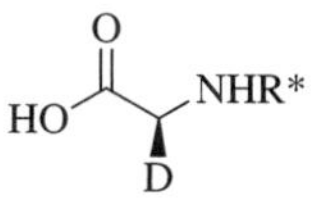

(1)

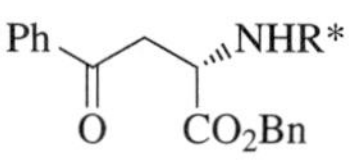

(2)

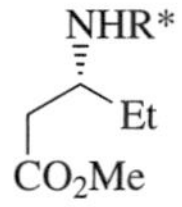

(3)

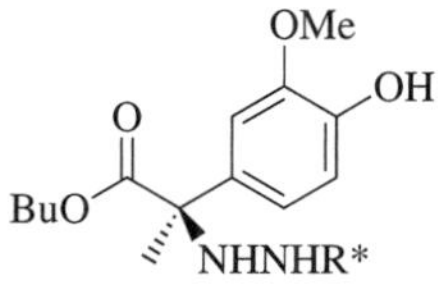

(4)

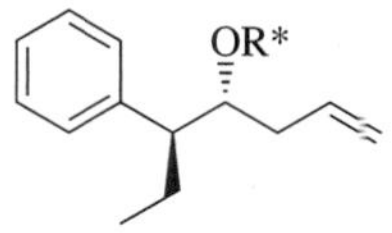

(5)

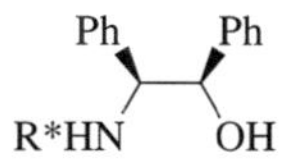

(6)

Resolution of Alcohols. In addition to generally providing highly crystalline derivatives that are usually suitable for X-ray crystallographic studies,[17] diastereomeric esters derived from camphanic acid have been widely used in organic synthesis for the resolution of racemic alcohols by fractional crystallization or chromatography.[18] This is one of the methods of choice to resolve inositol derivatives.[19] Selected examples are shown in (**7**)–(**10**).[20]

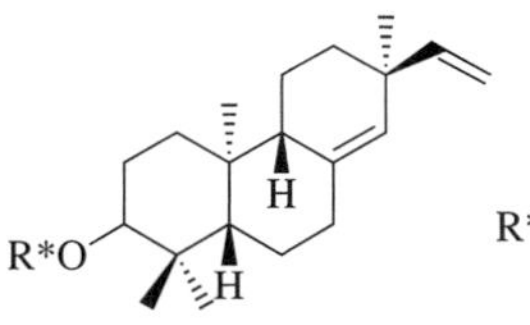

(7)

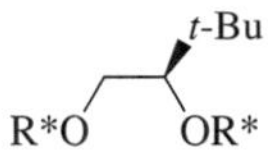

(8)

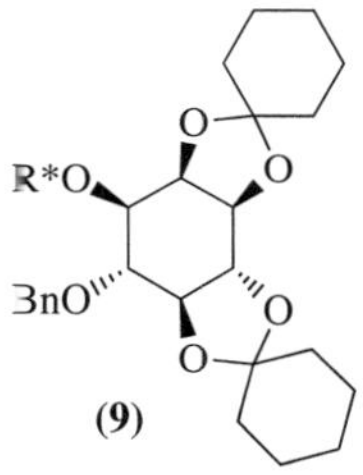

(9)

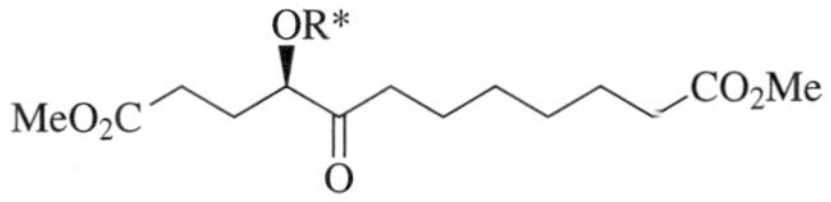

(10)

Chiral Auxiliary for Cycloaddition Reactions. Camphanate ester (**11**) has been used as a chiral dienophile in cycloaddition reactions with substituted furans to produce 7-oxabicyclo[2.2.1]heptene derivatives (eq 1).[4,21]

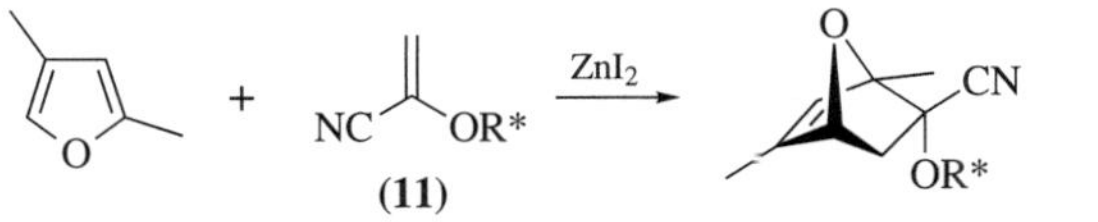

(1)

(11)

54%
(after 2 recrystallizations)

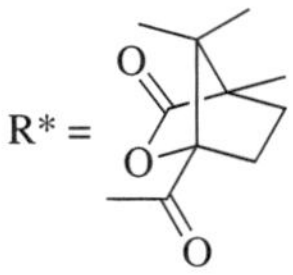

---

1. Gerlach, H.; Kappes, D.; Boeckman, Jr., R. K.; Maw, G. N. *Org. Synth.* **1993**, *71*, 48.
2. Parker, D. *Chem. Rev.* **1991**, *91*, 1441.
3. Wilen, S. H. *Top. Stereochem.* **1971**, *6*, 107.
4. Vieira, E.; Vogel, P. *Helv. Chim. Acta* **1983**, *66*, 1865.
5. Kappes, D.; Gerlach, H. *Synth., Coll.* **1990**, *20*, 581.
6. Chinchilla, R.; Najera, C.; Yus, M.; Heumann, A. *Tetrahedron: Asymmetry* **1990**, *1*, 851.
7. Gerlach, H.; Zagalak, B. *Chem. Commun./J. Chem. Soc., Chem. Commun.* **1973**, 274.
8. (a) Parker, D. *J. Chem. Soc., Perkin Trans. 2* **1983**, 83. (b) Parker, D.; Taylor, R. J.; Ferguson, G.; Tonge, A. *Tetrahedron* **1986**, *42*, 617.
9. (a) Schwab, J. M.; Ray, T.; Ho, C.-K. *J. Am. Chem. Soc.* **1989**, *111*, 1057. (b) Prabhakaran, P. C.; Gould, S. J.; Orr, G. R.; Coward, J. K. *J. Am. Chem. Soc.* **1988**, *110*, 5779. (c) Schwab, J. M.; Li, W.; Thomas, L. P. *J. Am. Chem. Soc.* **1983**, *105*, 4800.
10. (a) Armarego, W. L. F.; Milloy, B. A.; Pendergast, W. *J. Chem. Soc., Perkin Trans. 1* **1976**, 2229. (b) Hamon, D. P. G.; Massy-Westropp, R. A.; Razzino, P. *Tetrahedron* **1993**, *49*, 6419. (c) Hegedus, L. S.; Lastra, E.; Narukawa, Y.; Snustad, D. C. *J. Am. Chem. Soc.* **1992**, *114*, 2991. (d) Williams, R. M.; Zhai, D.; Sinclair, P. J. *J. Org. Chem.* **1986**, *51*, 5021.
11. (a) Jackson, R. F. W.; Wishart, N.; Wood, A.; James, K.; Wythes, M. J. *J. Org. Chem.* **1992**, *57*, 3397. (b) Lowe, C.; Pu, Y.; Vederas, J. C. *J. Org. Chem.* **1992**, *57*, 10. (c) Arnold, L. D.; Drover, J. C. G.; Vederas, J. C. *J. Am. Chem. Soc.* **1987**, *109*, 4649. (d) Trimble, L. A.; Vederas, J. C. *J. Am. Chem. Soc.* **1986**, *108*, 6397.
12. Jefford, C. W.; Wang, J. *Tetrahedron Lett.* **1993**, *34*, 1111.
13. Yee, C.; Blythe, T. A.; McNabb, T. J.; Walts, A. E. *J. Org. Chem.* **1992**, *57*, 3525.
14. Hafner, A.; Duthaler, R. O.; Marti, R.; Rihs, G.; Rothe-Streit, P.; Schwarzenbach, F. *J. Am. Chem. Soc.* **1992**, *114*, 2321.
15. Williams, R. M.; Sinclair, P. J.; Zhai, D.; Chen, D. *J. Am. Chem. Soc.* **1988**, *110*, 1547.
16. Shiner, C. S.; Berks, A. H. *J. Org. Chem.* **1988**, *53*, 5542.
17. See, for examples: (a) Oppenländer, T.; Schönholzer, P. *Helv. Chim. Acta* **1989**, *72*, 1792. (b) Eberle, M.; Egli, M.; Seebach, D. *Helv. Chim. Acta* **1988**, *71*, 1. (c) Estermann, H.; Prasad, K.; Shapiro, M. J.; Repic, O.; Hardtmann, G. E.; Bolsterli, J. J.; Walkinshaw, M. D. *Tetrahedron Lett.* **1990**, *31*, 445.
18. Gerlach, H. *Helv. Chim. Acta* **1968**, *51*, 1587.
19. Billington, D. C.; Baker, R.; Kulagowski, J. J.; Mawer, I. M. *Chem. Commun./J. Chem. Soc., Chem. Commun.* **1987**, 314.
20. (a) (**7**): Mori, K.; Waku, M. *Tetrahedron* **1985**, *41*, 5653. (b) (**8**): Naemura, K.; Ueno, M. *Bull. Chem. Soc. Jpn.* **1990**, *63*, 3695. (c) (**9**): Vacca, J. P.; DeSolms, S. J.; Huff, J. R. *J. Am. Chem. Soc.* **1987**, *109*, 3478. (d) (**10**): Tochtermann, W.; Scholz, G.; Bunte, G.; Wolff, C.; Peters, E.-M.; Peters, K.; Von Schnering, H. G. *Justus Liebigs Ann. Chem./Liebigs Ann. Chem.* **1992**, 1069.
21. (a) Wagner, J.; Vogel, P. *Tetrahedron Lett.* **1991**, *32*, 3169. (b) Kernen, P.; Vogel, P. *Tetrahedron Lett.* **1993**, *34*, 2473.

André B. Charette
Université de Montréal, QC, Canada

10-Camphorsulfonic Acid

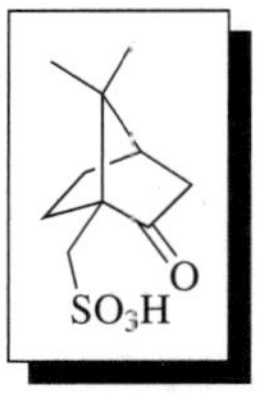

(1*S*)-(+)
[3144-16-9] $C_{10}H_{16}O_4S$ (MW 232.30)
(1*R*)-(−)
[35963-20-3]
(±)
[5872-08-2]

(acid catalyst,[1–8] resolving agent,[9,10] chiral auxiliary[11–20])

Physical Data: mp 203–206 °C (dec).
Solubility: sol dichloromethane, methanol, benzene; insol ether.
Form Supplied in: white crystals, racemic (±).
Analysis of Reagent Purity: melting point, NMR.
Preparative Methods: commercially available from several sources; can be prepared by sulfonation of camphor with acetic–sulfuric anhydride.[21]
Purification: recrystallize from ethyl acetate.
Handling, Storage, and Precautions: hygroscopic; corrosive.

Acid Catalyst. Camphorsulfonic acid (CSA) has been used extensively in synthetic organic chemistry as an acid catalyst. It has particularly been used in protecting group chemistry. For example, hydroxyl groups can be protected as tetrahydropyranyl (THP) ethers using dihydropyran and a catalytic amount of CSA (eq 1).[1] Both 1,2- and 1,3-diols can be selectively protected by reaction with orthoesters in the presence of camphorsulfonic acid to form the corresponding cyclic orthoester (eq 2).[2] This method of protection is particularly useful in that reduction of the orthoester with *Diisobutylaluminum Hydride* forms the monoacetal, which allows for preferential protection of a secondary alcohol in the presence of a primary alcohol. Ketones have also been protected using catalytic CSA (eq 3).[3]

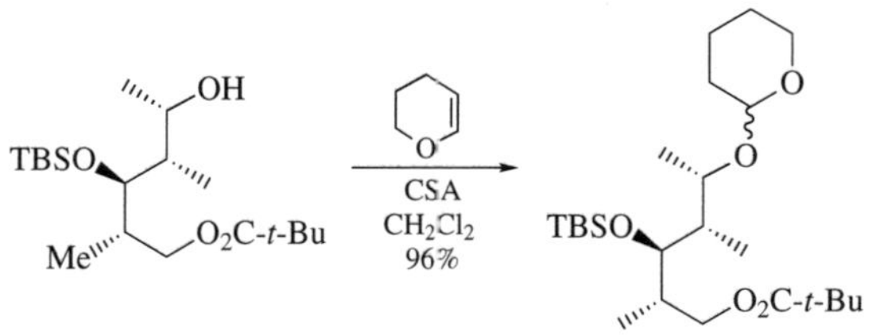

(1)

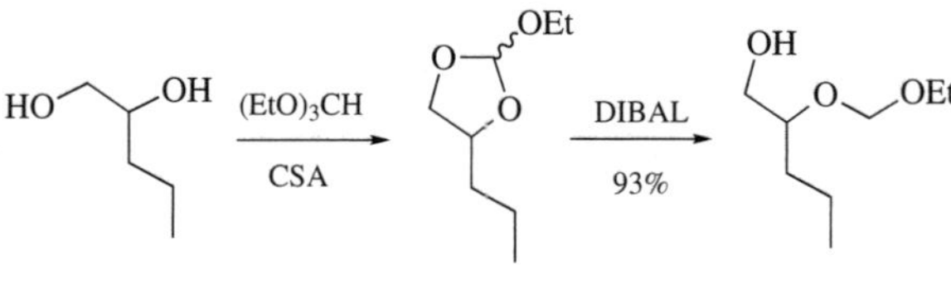

(2)

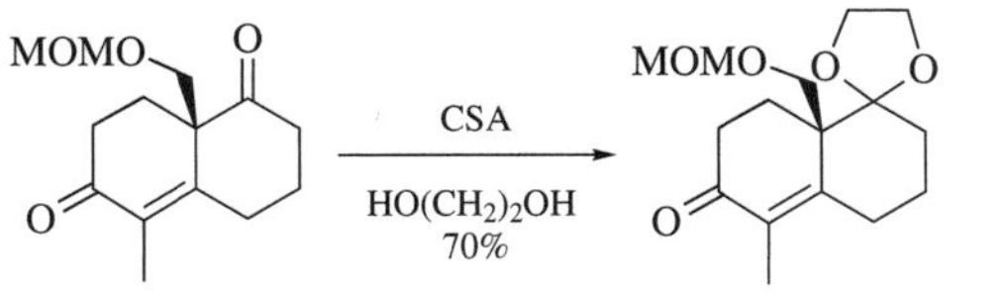
(3)

Overman has shown that camphorsulfonic acid can also be used in nucleophile-promoted alkyne–iminium cyclizations.[22] Alkylamines can react with formaldehyde and sodium iodide to yield piperidines in good yield. This methodology has been applied in the total synthesis of pumiliotoxin A (eq 4).[4]

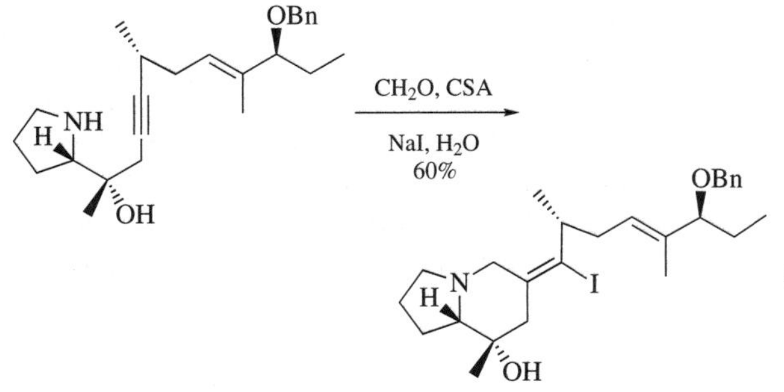
(4)

The most efficient catalyst for intramolecular opening of epoxides is CSA.[5,6] The formation of tetrahydrofurans or tetrahydropyrans is highly dependent on the structure of the hydroxy epoxide. The presence of a saturated chain at the secondary epoxide position leads to the formation of tetrahydrofurans (eq 5)[5] via 5-*exo* ring closure, whereas an electron-rich double bond at this position gives tetrahydropyrans (eq 6)[6] via 6-*endo* ring closure. This methodology has also been extended to the synthesis of oxepanes (eq 7).[6]

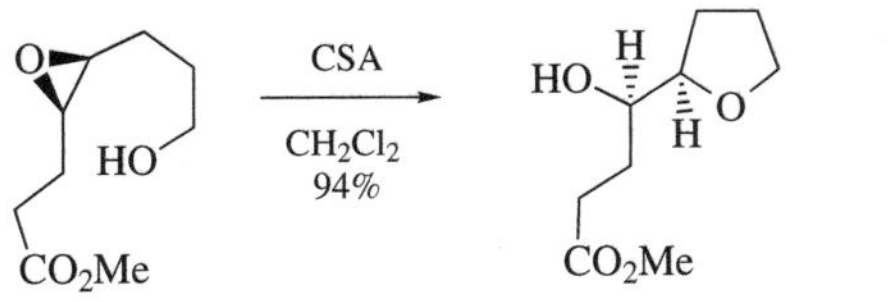
(5)

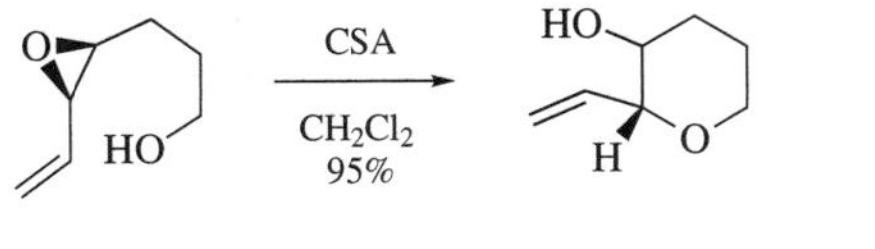
(6)

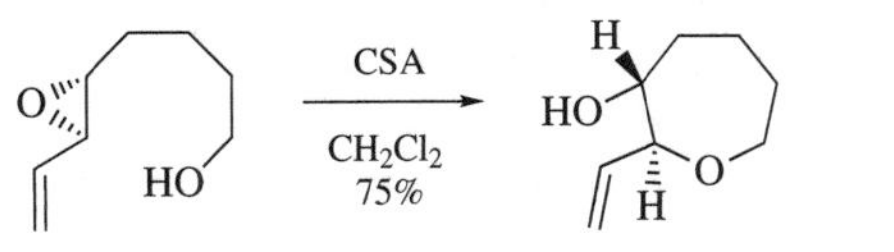
(7)

CSA is also the acid of choice for use in phenylselenation reactions.[7] It has been used as an acid catalyst in hydroxyselenation reactions of alkenes (eq 8)[7] and organoselenium-induced cyclizations (eq 9) using *N-Phenylselenophthalimide* (NPSP).[7]

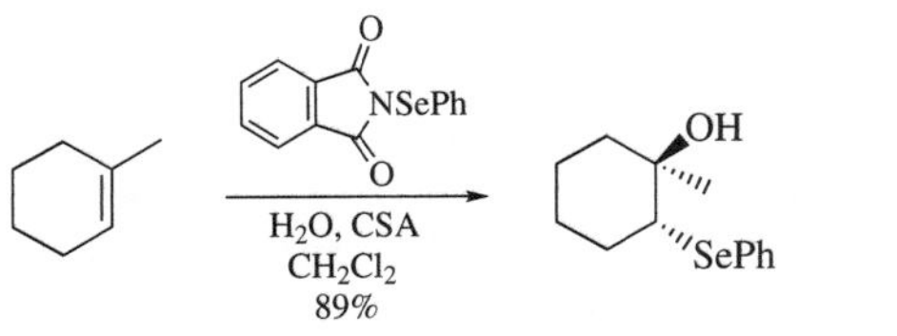
(8)

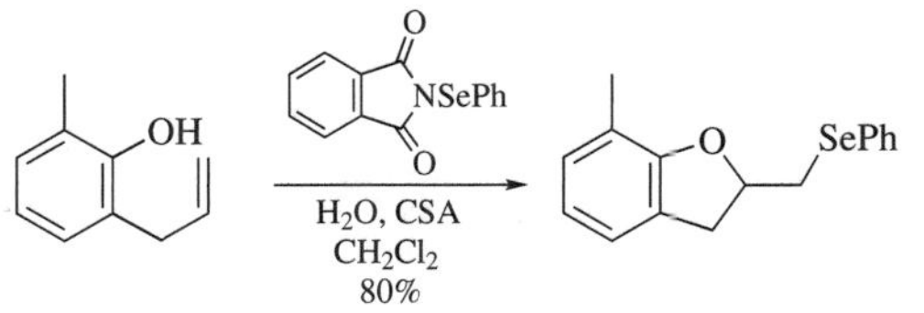
(9)

CSA has also been used to catalyze spiroacetalizations.[8,22] In Schreiber's approach to the talaromycins, he utilized a CSA-catalyzed spiroacetalization (eq 10) and found that the use of different solvents led to varying percentages of isomeric products.[8] Other approaches to the talaromycins also utilize CSA for the required spiroacetalization.[23]

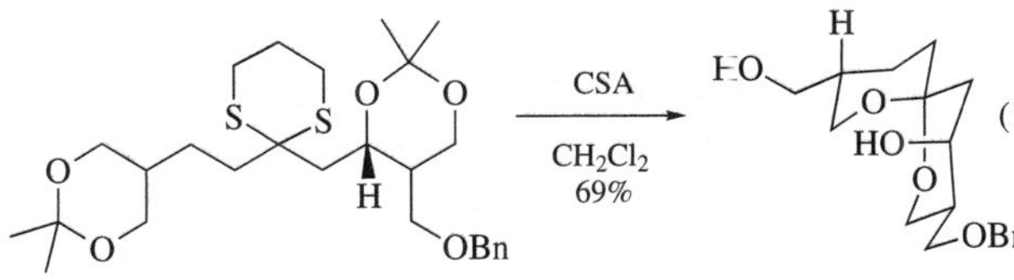
(10)

Resolving Agent. Scalemic CSA has been used to resolve amines by forming diastereomeric salts which can be separated by fractional crystallization (eq 11).[9] In this instance, after obtaining the desired crystalline diastereomeric salt, the undesired diastereomer was completely transformed into the desired one by a resolution–racemization procedure (eq 12).[9] Additionally, racemic ketones can be resolved by forming enantiomeric iminium salts (eq 13).[10] Two different procedures have been devised depending on the ease of enamine formation.

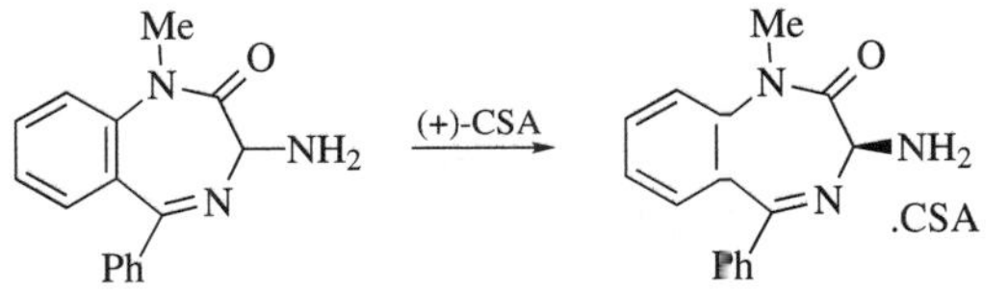
(11)

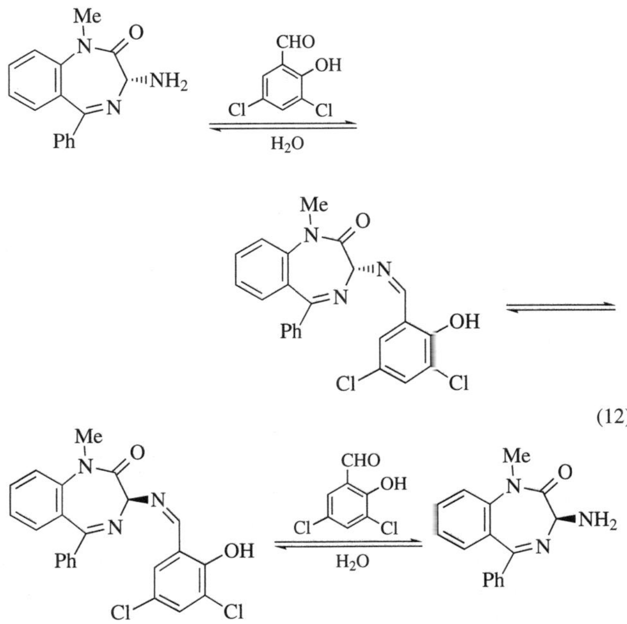
(12)

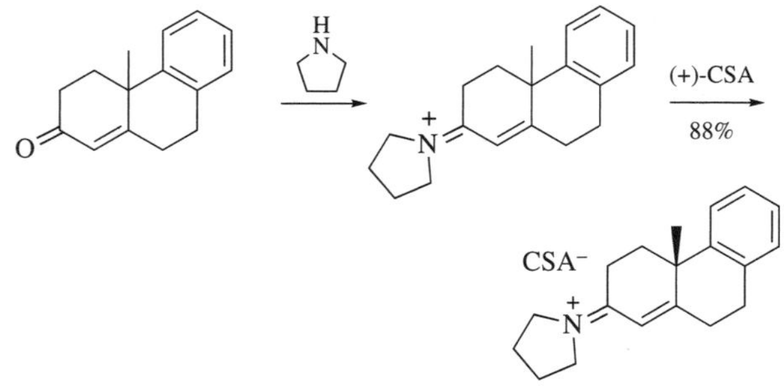

(13)

Chiral Auxiliaries

Asymmetric Diels–Alder Reactions. The commercial availability of either enantiomer of camphorsulfonic acid has made it quite useful in asymmetric Diels–Alder reactions. Reaction of the sultone (generated from CSA) with *Lithium Diisopropylamide* followed by esterification and β-elimination yields the crystalline acrylate (eq 14).[11] The Lewis acid-catalyzed [4 + 2] cycloaddition of 1,3-dienes with this acrylate affords the corresponding scalemic adduct which can be reduced with *Lithium Aluminum Hydride* to yield an enantiomerically pure alcohol (eq 15).[12]

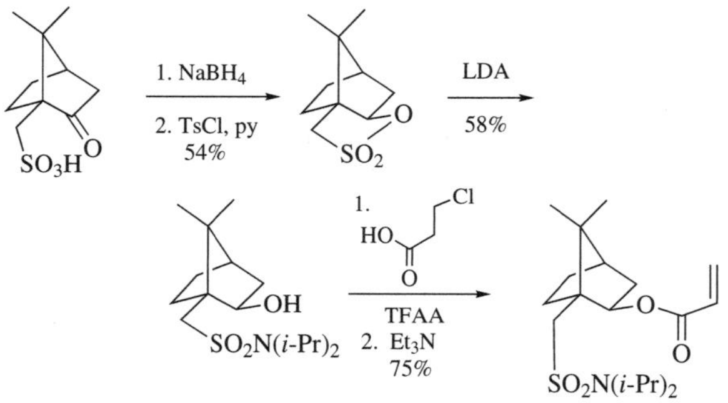

(14)

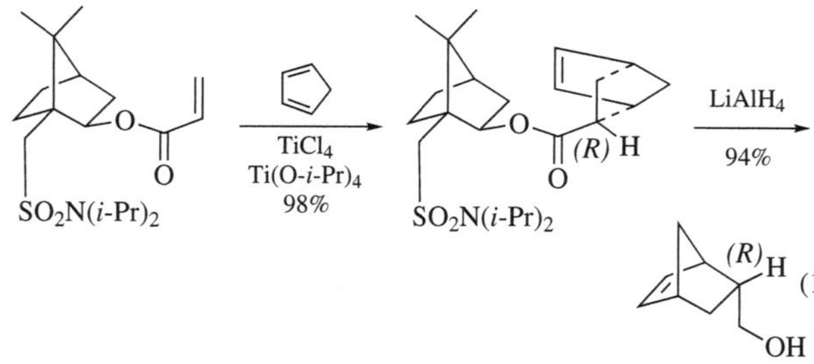

(15)

A different approach to the asymmetric Diels–Alder reaction involves the use of the sultam derived from CSA. Lewis acid-promoted reaction with dienes followed by reductive removal of the chiral auxiliary is analogous to that previously discussed for the sultone. Smith has successfully utilized this approach to synthesize the chiral acid used in the synthesis of the immunosuppressant FK-506 (eq 16).[13]

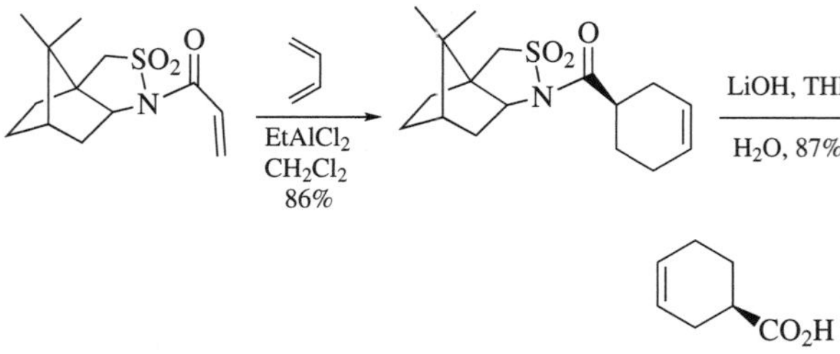

(16)

Oxaziridines. Davis has developed the use of chiral 2-sulfonyloxaziridines derived from camphorsulfonic acid as chiral auxiliaries in the asymmetric oxidation reactions.[24] Although other oxaziridines may be preferable, the camphor-derived oxaziridines can be used for the oxidation of sulfides and disulfides to sulfoxides and thiosulfinates as well as for the epoxidation of alkenes.[24] On the other hand, the camphoryloxaziridines are the preferred reagents for hydroxylation of lithium enolates of esters, amides, and ketones, as utilized in the synthesis of kjellmanianone (eq 17).[14]

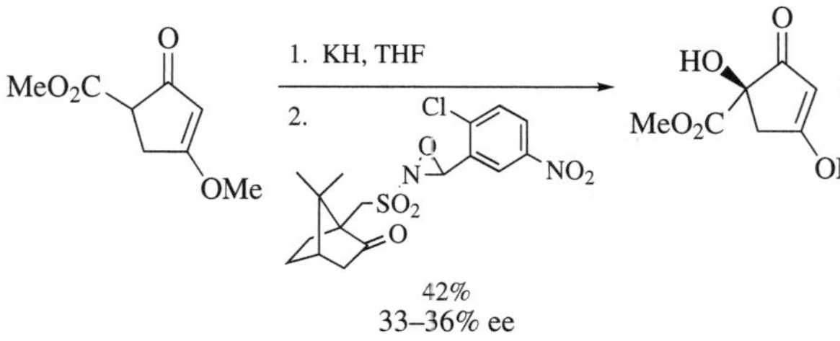

(17)

Chiral Sulfides. Optically active sulfides prepared from (+)-CSA can be used to prepare optically active 1,2-diaryloxiranes (eq 18).[15]

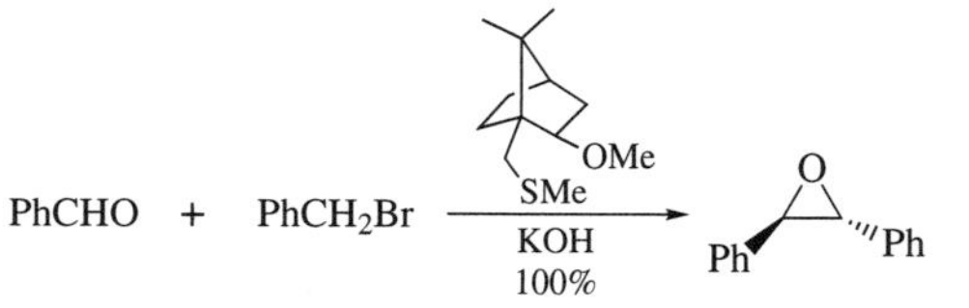

(18)

Grignard Addition to Enones. The sultam generated from camphorsulfonic acid can also be used as a chiral auxiliary in the conjugate addition of Grignard reagents to enones. Simple alkylmagnesium chlorides add in a 1,4-fashion to afford imides (eq 19).[16]

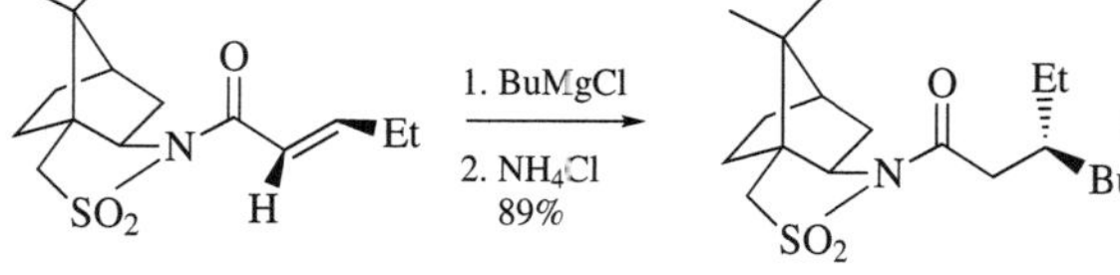

(19)

Asymmetric Hydrogenation of Camphor-Derived Sultamides. The sultamide of CSA can be used as a chiral auxiliary for synthesis of β-substituted carboxylic acids (eq 20).[17]

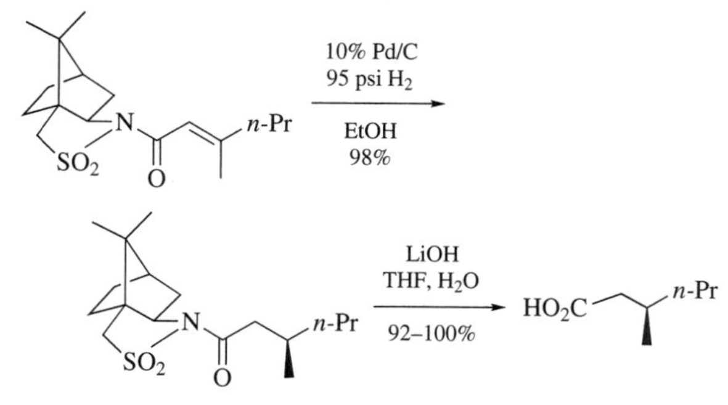

(20)

Asymmetric Acetoxylation of Esters. The silyl enol ether derived from CSA reacts with *Lead(IV) Acetate* to yield the α-acetoxy ester with good diastereoselectivity. Hydrolysis of the chiral auxiliary gives the α-hydroxy acid, whereas reduction affords the terminal α-glycol (eq 21).[18]

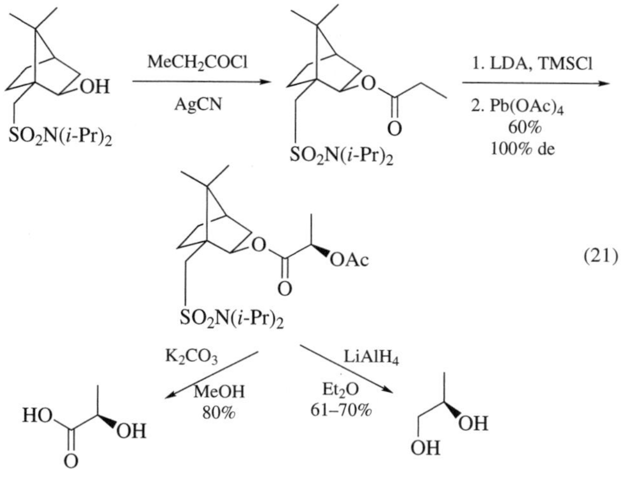

(21)

Allylation of Aldehydes. Synthesis of enantiomerically pure allyl alcohols can be accomplished by catalytic asymmetric addition of divinylzinc to aldehydes using a camphorsulfonic acid-derived catalyst (eq 22).[19]

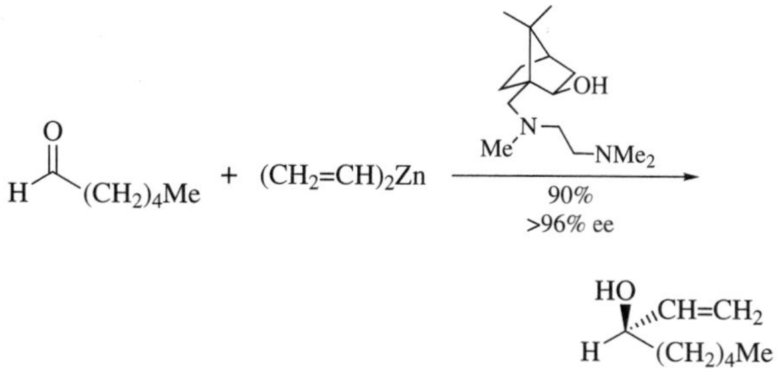

(22)

Synthesis of Epoxides from Chiral Chlorohydrins. Asymmetric halogenation of CSA-derived esters allows for the formation of enantiomerically pure halohydrins and terminal epoxides (eq 23).[20]

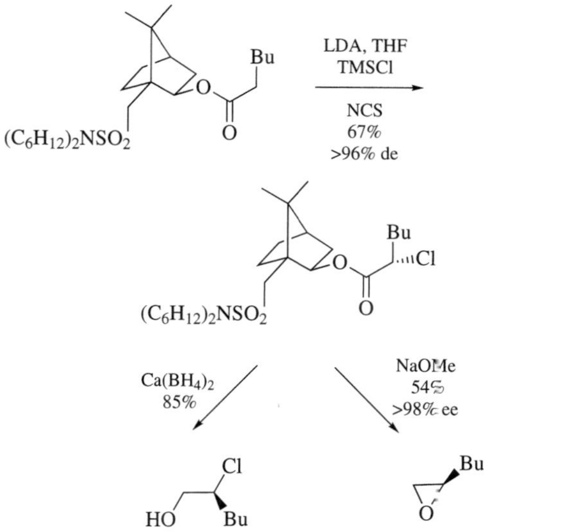

(23)

Related Reagents. 3-Bromocamphor-3-sulfonic Acid; (−)-(1*S*,4*R*)-Camphanic Acid; 2,4,6-Collidinium *p*-Toluenesulfonate; Pyridinium *p*-Toluenesulfonate; *p*-Toluenesulfonic Acid; Trifluoromethanesulfonic Acid.

1. Nicolaou, K. C.; Chakraborty, T. K.; Daines, R. A.; Simpkins, N. S. *Chem. Commun.* **1986**, 413.
2. Takasu, M.; Naruse, Y.; Yamamoto, H. *Tetrahedron Lett.* **1988**, *29*, 1947.
3. Tamai, Y.; Hagiwara, H.; Uda, H. *J. Chem. Soc Perkin Trans. 1* **1986**, 1311.
4. Overman, L. E.; Sharp, M. J. *Tetrahedron Lett.* **1988**, *29*, 901.
5. Nicolaou, K. C.; Prasad, C. V. C.; Somers, P. K.; Hwang, C.-K., *J. Am. Chem. Soc.* **1989**, *111*, 5330.
6. Nicolaou, K. C.; Prasad, C. V. C.; Somers, P. K.; Hwang, C.-K. *J. Am. Chem. Soc.* **1989**, *111*, 5335.
7. Nicolaou, K. C.; Petasis, N. A.; Claremon, D. A. *Tetrahedron* **1985**, *41*, 4835.
8. Schreiber, S. L.; Sommer, T. J.; Satake, K. *Tetrahedron Lett.* **1985**, *26*, 17.
9. Reider, P. J.; Davis, P.; Hughes, D. L.; Grabowski, E. J. J. *J. Org. Chem.* **1987**, *52*, 955.
10. Adams, W. R.; Chapman, O. L.; Sieja, J. B.; Welstead, W. J., Jr. *J. Am. Chem. Soc.* **1966**, *88*, 162.
11. Oppolzer, W.; Chapuis, C.; Kelly, M. J. *Helv. Chim. Acta* **1983**, *66*, 2358.
12. Oppolzer, W.; Chapuis, C.; Bernardinelli, G. *Tetrahedron Lett.* **1984**, *25*, 5885.
13. Smith, A. B., III; Hale, K. J.; Laakso, L. M.; Chen, K.; Riéra, A. *Tetrahedron Lett.* **1989**, *30*, 6963.
14. Boschelli, D.; Smith, A. B., III; Stringer, O. D.; Jenkins, R. H., Jr. Davis, F. A. *Tetrahedron Lett.* **1981**, *22*, 4385.
15. Furukawa, N.; Sugihara, Y.; Fujihara, H. *J. Org. Chem.* **1989**, *54*, 4222.
16. Oppolzer, W.; Poli, G.; Kingma, A. J.; Starkemann, C.; Bernardinelli, G. *Helv. Chim. Acta* **1987**, *70*, 2201.
17. Oppolzer, W.; Mills, R. J.; Réglier, M. *Tetrahedron Lett.* **1986**, *27*, 183.
18. Oppolzer, W.; Dudfield, P. *Helv. Chim. Acta* **1985**, *68*, 216.
19. Oppolzer, W.; Radinov, R. N. *Tetrahedron Lett.* **1988**, *29*, 5645.
20. Oppolzer, W.; Dudfield, P. *Tetrahedron Lett.* **1985**, *26*, 5037.
21. Bartlett, P. D.; Knox, L. H. *Org. Synth., Coll. Vol.* **1973**, *5*, 194.
22. Overman, L. E.; Sharp, M. J. *J. Am. Chem. Soc.* **1988**, *110*, 612.
23. Baker, R.; Boyes, A. L.; Swain, C. J. *Tetrahedron Lett.* **1989**, *30*, 985.

24. Davis, F. A.; Jenkins, R. H., Jr; Awad, S. B.; Stringer, O. D.; Watson, W. H.; Galloy, J. *J. Am. Chem. Soc.* **1982**, *104*, 5412.

Ellen M. Leahy
Affymax Research Institute, Palo Alto, CA, USA

10-Camphorsulfonyl Chloride[1]

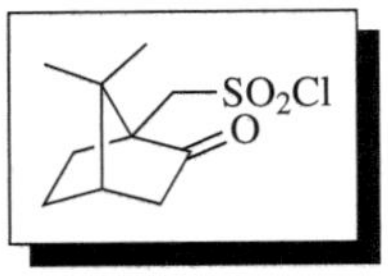

(1*R*)
[21286-54-4] $C_{10}H_{15}ClO_3S$ (MW 250.75)
(1*S*)
[39262-22-1]
(±)
[6994-93-0]

(enantiomeric excess determination;[2] chemical resolution;[3] synthesis of chiral auxiliaries;[4] chiral precursor for natural product synthesis;[1] synthesis of chiral reagents[5])

Physical Data: mp 65–67 °C; (1*S*)-(+): $[\alpha]_D$ +32.1° (*c* 1, $CHCl_3$).
Solubility: sol CH_2Cl_2; slightly sol ether; insol H_2O.
Form Supplied in: both enantiomers and the racemic sulfonyl chloride are commercially available.
Preparative Methods: can be prepared from *10-Camphorsulfonic Acid* upon treatment with *Phosphorus(V) Chloride* or *Thionyl Chloride*.[6]
Purification: crystallized from hexane or from MeOH.
Handling, Storage, and Precautions: corrosive and moisture-sensitive. This reagent should be handled in a fume hood.

Reagent for Determination of Enantiomeric Excesses and for Chemical Resolution of Alcohols and Amines. 10-Camphorsulfonyl chloride has been widely used as a chiral derivatizing agent for the assay of enantiomeric purity of alcohols and amines by NMR techniques.[2] A typical procedure for the preparation of the sulfonate ester or sulfonamide involves mixing a solution of the alcohol or amine in CH_2Cl_2 with camphorsulfonyl chloride in the presence of an amine base (Et_3N, py, or DMAP). This reagent has been particularly valuable for determining the enantiomeric purity of secondary alcohols (**1, 2**) and β-hydroxy esters (**3**).[7]

R* = 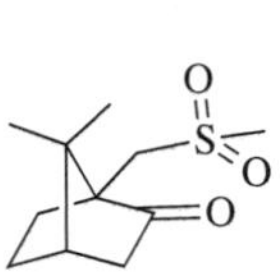

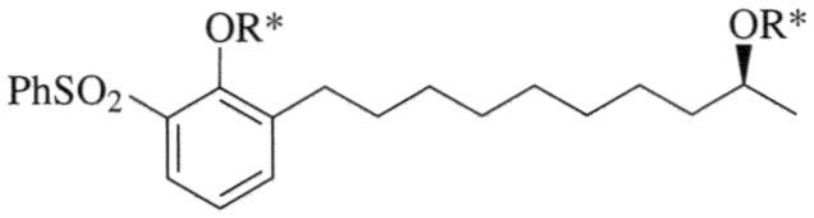

(1)

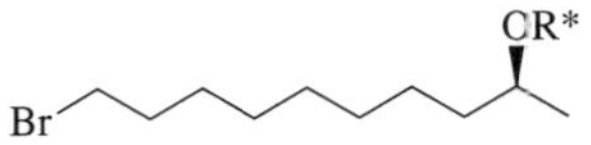

(2)

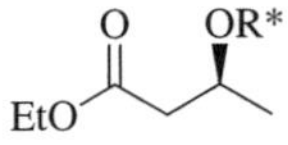

(3)

In some cases (**3**), the addition of a chiral shift reagent ($Eu(hfc)_3$) is necessary to obtain baseline separation of the signals corresponding to the β-proton of both diastereomers by ^1H NMR. Diastereomeric mixtures derived from secondary alcohols have also been analyzed by HPLC.[8] The resolution of a secondary alcohol (**4**) could be achieved by a selective crystallization of one of the two diastereomeric camphorsulfonate esters.[3]

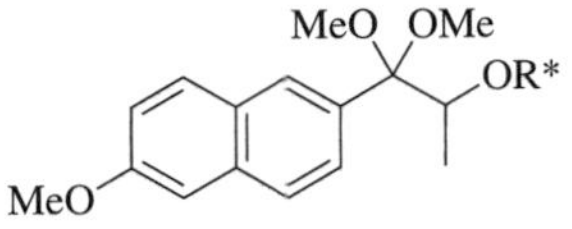

(4)

The enantiomeric purities of primary and secondary amines have also been established by ^1H NMR spectroscopy by their conversion into the corresponding sulfonamide. These derivatives often produce crystalline compounds that are suitable for X-ray crystallographic studies. For example, the enantiomeric purities of amines (**5**),[9] (**6**),[10] and (**7**)[11] were determined by ^1H NMR spectroscopy and the absolute stereochemistry of (**7**) was unequivocally established by X-ray crystallography.

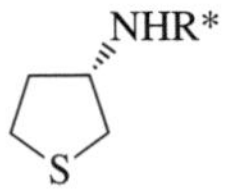

(5)

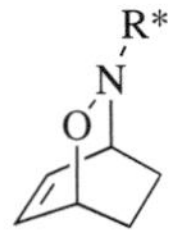

(7)

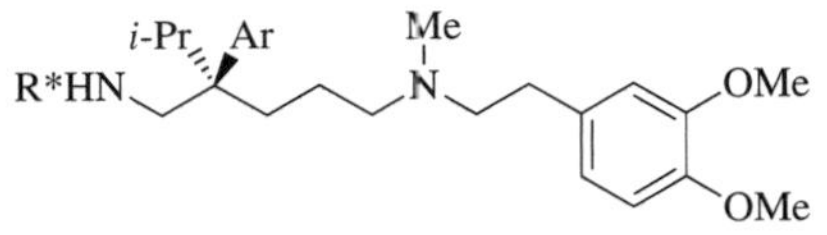

(6) Ar = 3,4-$(OMe)_2C_6H_3$

A general protocol for the HPLC separation of diastereomeric camphorsulfonamides[12] derived from racemic α-amino acids has been developed (eq 1).[13] More complex amino acids, such as (**8**), were successfully analyzed by this procedure.[14]

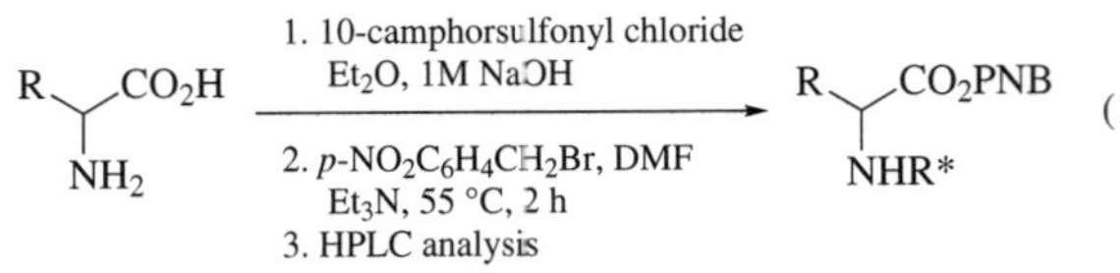

(1)

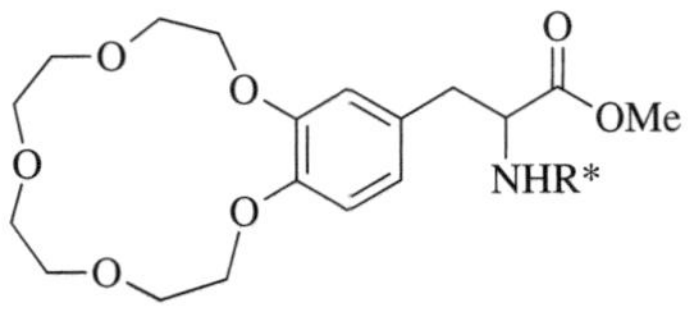

(8)

Synthesis of Chiral Auxiliaries. Their availability and crystalline nature has made camphor derivatives the precursors of choice for the design and synthesis of chiral auxiliaries.[4] 10-Camphorsulfonyl chloride is the starting material for the synthesis of chiral auxiliaries (**9**)–(**12**) (eq 2). Sulfonamides (**9**) and (**10**)[15] have been used as chiral auxiliaries in a number of reactions, e.g. the Lewis acid-catalyzed Diels–Alder reaction, the [3 + 2] cycloaddition of a nitrile oxide to an acrylate, and the stereoselective conjugate addition reaction of organocopper reagents to α,β-unsaturated esters.[4]

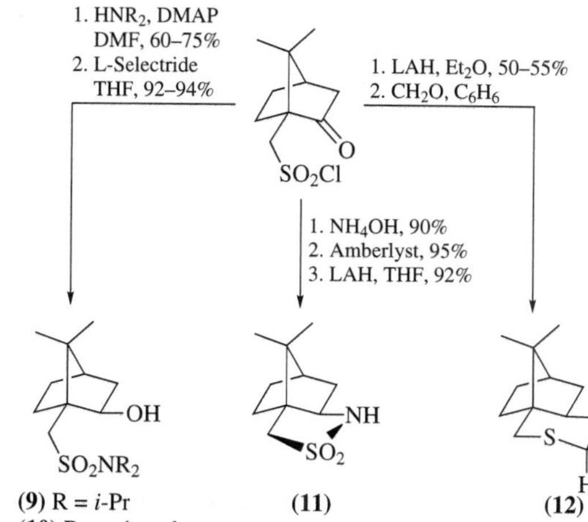

(2)

(**9**) R = *i*-Pr
(**10**) R = *c*-hexyl
(**11**)
(**12**)

In addition to being an efficient chiral controller in a number of stereoselective transformations of chiral acrylates, (i.e. the Diels–Alder reaction,[4] the conjugate reduction,[16] the asymmetric dihydroxylation,[17] and the nitrile oxide cycloaddition[18]) the bornanesultam (**11**)[19] has been shown to be an exceptionally efficient chiral auxiliary for stereoselective aldol condensations (eqs eq 3 and eq 4). Depending upon the reaction conditions, *N*-propionylsultam can produce either the *syn* or *anti* aldol product with an excellent diastereoselectivity.[20] Furthermore, good diastereoselectivities are also observed for the corresponding acetate aldol reaction (eq 5).[21]

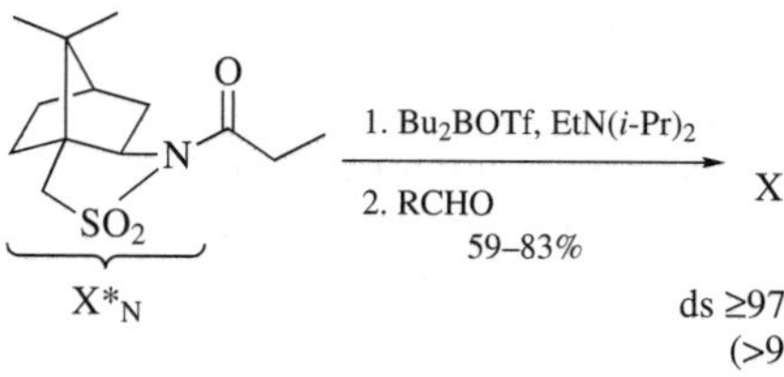

(3)

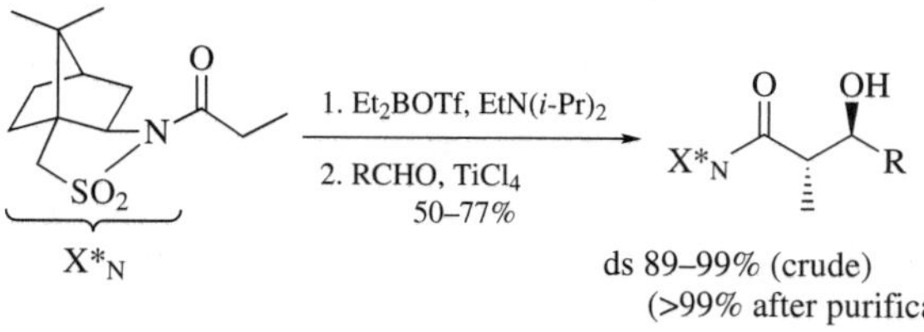

(4)

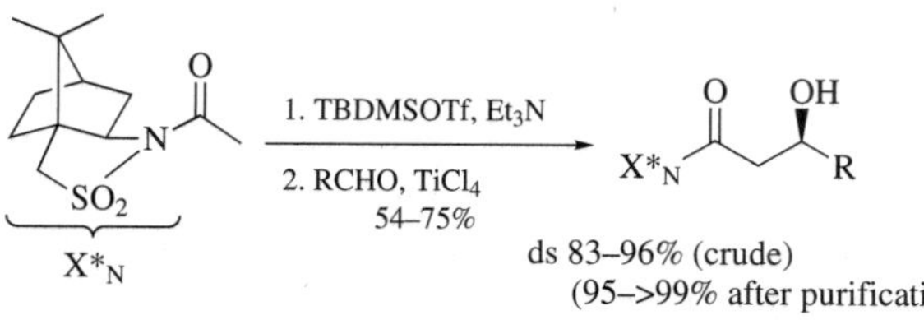

(5)

Oxathiane (**12**) has been shown to be an efficient chiral auxiliary in the nucleophilic addition to carbonyl compounds.[22]

10-Camphorsulfonyl chloride has also been widely used as a useful precursor to chiral dienophiles in hetero-Diels–Alder reactions.[23]

An elegant use of the chirality and the leaving group ability of the camphorsulfonate ester has been reported in the synthesis of a chiral C_2 symmetric cyclopentadienyl ligand (eq 6).[24]

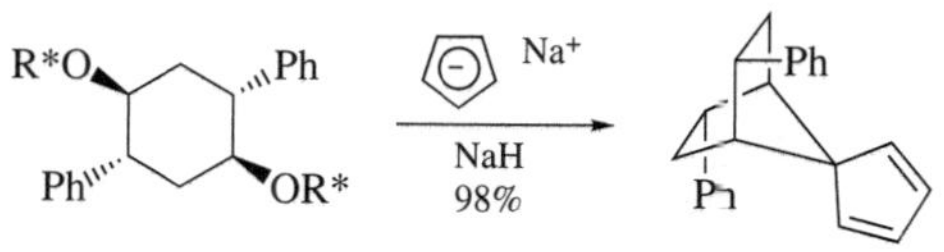

(6)

Synthesis of Chiral Reagents. An efficient chiral α-chloro-α-nitroso reagent derived from 10-camphorsulfonyl chloride (1. Cy_2NH; 2. NH_2OH; 3. *t*-BuOCl; 70–78%) has been developed for the asymmetric α-amination of ketone enolates (eq 7).[25] The resulting β-keto *N*-hydroxylamine can be converted to the *anti*-1,2-hydroxyamine under reducing conditions ($NaBH_4$; Zn, HCl, AcOH).

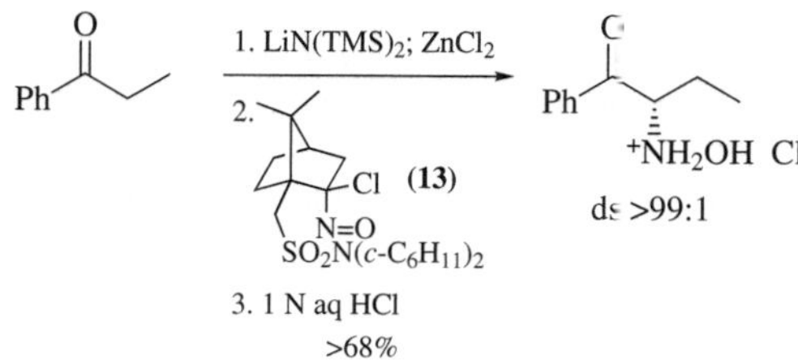

(7)

Several oxaziridines related to (**14**)[5] (eq 3) have been used, most notably in the enantioselective oxidation of sulfides to sulfoxides,[26] of selenides to selenoxides,[27] and of alkenes to oxiranes.[28] It is also the reagent of choice for the hydroxylation of lithium and Grignard reagents[29] and for the asymmetric oxidation of enolates to give α-hydroxy carbonyl compounds.[5,30] A similar chiral fluorinating reagent has also been developed.[31]

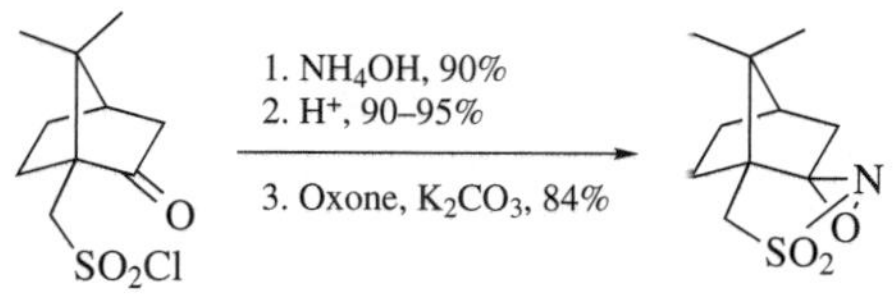

(8)

(**14**)

Chiral Precursor for Natural Product Synthesis. 10-Camphorsulfonyl chloride has been used as a chiral starting material for the synthesis of a number of products[1] such as ketopinic acid[32] (eq 9), which has been used to resolve alcohols[33] and hemiacetals.[34]

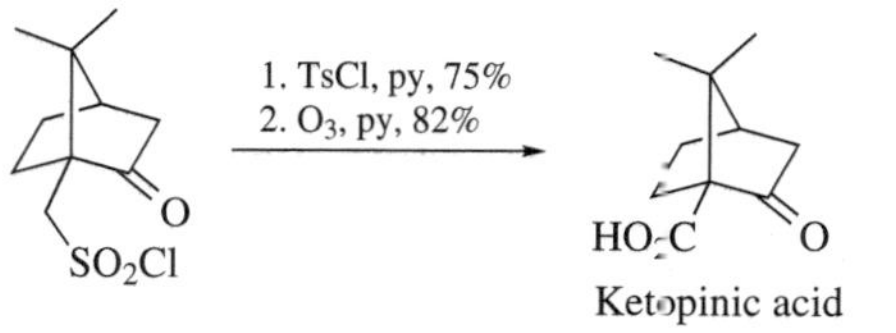

(9)

Ketopinic acid

1. Money, T. *Nat. Prod. Rep.* **1985**, *2*, 253.
2. (a) Parker, D. *Chem. Rev.* **1991**, *91*, 1441. (b) Weisman, G. R. *Asymm. Synth.* **1983**, *1*, 153.
3. Tsuchihashi, G.-I.; Mitamura, S.; Kitajima, K.; Kobayashi, K. *Tetrahedron Lett.* **1982**, *23*, 5427.
4. Oppolzer, W. *Tetrahedron* **1987**, *43*, 1969.
5. Davis, F. A.; Chen, B.-C. *Chem. Rev.* **1992**, *92*, 919.
6. Bartlett, P. D.; Knox, L. H. *Org. Synth., Coll. Vol.* **1973**, *5*, 196
7. (a) Quinkert, G.; Küber, F.; Knauf, W.; Wacker, M.; Koch, U.; Becker, H.; Nestler, H. P.; Dürner, G.; Zimmermann, G.; Bats, J. W.; Egert, E. *Helv. Chim. Acta* **1991**, *74*, 1853. (b) Quinkert, G.; Döller, U.; Eichhorn, M.; Küber, F.; Nestler, H. P.; Becker, H.; Bats, J. W.; Zimmermann, G.; Dürner, G. *Helv. Chim. Acta* **1990**, *73*, 1999. (c) Quinkert, G.; Fernholz, E.; Eckes, P.; Neumann, D.; Dürner, G. *Helv. Chim. Acta* **1989**, *72*, 1753.
8. Mori, K.; Kisida, H. *Justus Liebigs Ann. Chem./Liebigs Ann. Chem.* **1989**, 35.
9. Dehmlow, E. V.; Westerheide, R. *Synthesis* **1992**, 947.
10. Theodore, L. J.; Nelson, W. L. *J. Org. Chem.* **1987**, *52*, 1309.
11. Braun, H.; Felber, H.; Kresze, G.; Schmidtchen, F. P.; Prewo, R.; Vasella, A. *Justus Liebigs Ann. Chem./Liebigs Ann. Chem.* **1993**, 261.
12. Burke, T. R., Jr.; Nelson, W. L.; Mangion, M.; Hite, G. J.; Mokler, C. M.; Ruenitz, P. C. *J. Med. Chem.* **1980**, *23*, 1044.
13. (a) Furukawa, H.; Mori, Y.; Takeuchi, Y.; Ito, K. *J. Chromatogr.* **1977**, *136*, 428. (b) Furukawa, H.; Sakakibara, E.; Kamei, A.; Ito, K. *Chem. Pharm. Bull.* **1975**, *23*, 1625.
14. Berthet, M.; Sonveaux, E. *Chem. Commun./J. Chem. Soc., Chem. Commun.* **1983**, 10.
15. Oppolzer, W.; Chapuis, C.; Bernardinelli, G. *Tetrahedron Lett.* **1984**, *25*, 5885.
16. Oppolzer, W.; Poli, G.; Starkemann, C.; Bernardinelli, G. *Tetrahedron Lett.* **1988**, *29*, 3559.
17. Oppolzer, W.; Barras, J.-P. *Helv. Chim. Acta* **1987**, *70*, 1666.
18. Curran, D. P.; Kim, B. H.; Daugherty, J.; Heffner, T. A. *Tetrahedron Lett.* **1988**, *29*, 3555.
19. (a) Davis, F. A.; Towson, J. C.; Weismiller, M. C.; Lal, S.; Carroll, P. J. *J. Am. Chem. Soc.* **1988**, *110*, 8477. (b) Weismiller, M. C.; Towson, J. C.; Davis, F. A. *Org. Synth.* **1990**, *69*, 154. (c) Towson, J. C.; Weismiller, M. C.; Lal, G. S.; Sheppard, A. C.; Davis, F. A. *Org. Synth.* **1990**, *69*, 158.
20. *Syn* aldol: (a) Oppolzer, W.; Blagg, J.; Rodriguez, I.; Walther, E. *J. Am. Chem. Soc.* **1990**, *112*, 2767. *Anti* aldol: (b) Oppolzer, W.; Lienard, P. *Tetrahedron Lett.* **1993**, *34*, 4321. (c) Oppolzer, W.; Starkemann, C.; Rodriguez, I.; Bernardinelli, G. *Tetrahedron Lett.* **1991**, *32*, 61.
21. Oppolzer, W.; Starkemann, C. *Tetrahedron Lett.* **1992**, *33*, 2439.
22. Eliel, E. L.; Frazee, W. J. *J. Org. Chem.* **1979**, *44*, 3598.
23. (a) Braun, H.; Felber, H.; Kresze, G.; Schmidtchen, F. P.; Prewo, R.; Vasella, A. *Justus Liebigs Ann. Chem./Liebigs Ann. Chem.* **1993**, 261. (b) Blanco, J. M.; Caamaño, O.; Eirin, A.; Fernandez, F.; Medina, L. *Bull. Soc. Chim. Belg.* **1989**, *98*, 923. (c) Caamaño, O.; Eirín, A.; Fernandez, F.; Gómez, G.; Uriarte, E. *Heterocycles* **1988**, *27*, 2839. (d) De Lucchi, O.; Lucchini, V.; Marchioro, C.; Valle, G.; Modena, G. *J. Org. Chem.* **1986**, *51*, 1457.
24. Halterman, R. L.; Vollhardt, K. P. C.; Welker, M. E.; Bläser, D.; Boese, R. *J. Am. Chem. Soc.* **1987**, *109*, 8105.
25. Oppolzer, W.; Tamura, O.; Sundarababu, G.; Signer, M. *J. Am. Chem. Soc.* **1992**, *114*, 5900.
26. Davis, F. A.; McCauley, Jr., J. P.; Chattopadhyay, S.; Harakal, M. E.; Towson, J. C.; Watson, W. H.; Tavanaiepour, I. *J. Am. Chem. Soc.* **1987**, *109*, 3370.
27. Davis, F. A.; Reddy, R. T. *J. Org. Chem.* **1992**, *57*, 2599.
28. Davis, F. A.; Chattopadhyay, S. *Tetrahedron Lett.* **1986**, *27*, 5079.
29. (a) Davis, F. A.; Wei, J.; Sheppard, A. C.; Gubernick, S. *Tetrahedron Lett.* **1987**, *28*, 5115. (b) Davis, F. A.; Lal, G. S.; Wei, J. *Tetrahedron Lett.* **1988**, *29*, 4269.
30. Davis, F. A.; Kumar, A. *J. Org. Chem.* **1992**, *57*, 3337.
31. Differding, E.; Lang, R. W. *Tetrahedron Lett.* **1988**, *29*, 6087.
32. Haslanger, M. F.; Heikes, J. *Synthesis* **1981**, 801.
33. Paulsen, H.; Brauer, O. *Ber. Dtsch. Chem. Ges./Chem. Ber.* **1977**, *110*, 331.
34. Woodward, R. B.; Gosteli, J.; Ernest, I.; Friary, R. J.; Nestler, G.; Raman, H.; Sitrin, R.; Suter, C.; Whitesell, J. K. *J. Am. Chem. Soc.* **1973**, *95*, 6853.

André B. Charette
Université de Montréal, QC, Canada

10,2-Camphorsultam[1]

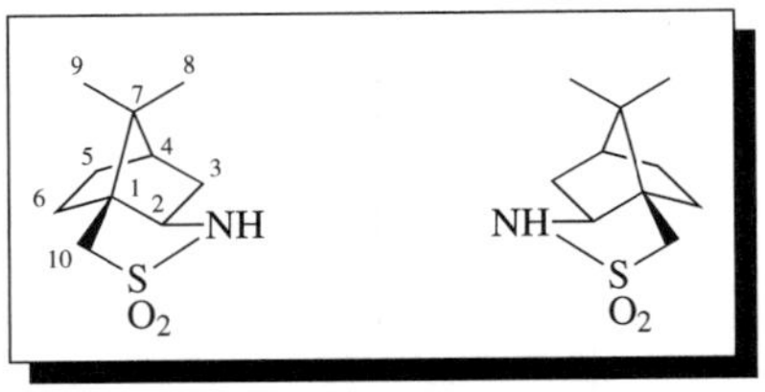

(−)-D-(2*R*)
[94594-90-8] $C_{10}H_{17}NO_2S$ (MW 215.31)
(+)-L-(2*S*)
[108448-77-7]

(versatile chiral auxiliary: *N*-enoyl derivatives undergo highly stereoselective [2+4] Diels–Alder[2] and [2+3][3] cycloadditions, cyclopropanations,[4] aziridinations,[5] dihydroxylations,[6] hydrogenations,[7] azido-iodinations[8] and conjugate hydride,[9] Grignard,[10] cuprate,[11] allylsilane[12] and thiolate[13] additions; radical additions[14] and S_N2' reactions[15] at the α-position also occur stereoselectively; enolates of *N*-acyl derivatives participate in highly stereoselective aldolizations,[16] alkylations,[17] halogenations, and 'aminations', the latter three types of reactivity being useful for α-amino acid preparation;[18] free radicals generated at the α-position of *N*-acyl derivatives participate in stereoselective intra- and intermolecular addition reactions;[19] the *N*-fluoro derivative functions as an enantioselective, electrophilic fluorinating reagent[20])

Alternate Name: bornane-10,2-sultam.

Physical Data: mp 183–185 °C (EtOH). (−)-D-(2*R*) enantiomer: $[\alpha]_D^{20}$ −31±1° ($CHCl_3$, *c* 2.3). (+)-L-(2*S*) enantiomer: $[\alpha]_D^{20}$ +34±1° (EtOH, *c* 1.00).

Form Supplied in: white crystalline solid; both enantiomers are commercially available (~same price) or may be readily prepared (3 steps, >70% overall yield) from *10-Camphorsulfonic Acid*.[21]

Handling, Storage, and Precautions: stable indefinitely at ambient temperature in a sealed container; mild irritant.

Introduction. Exploitation of chiral auxiliary controlled face discrimination in the reaction of a reactant with a prochiral molecule or functional group is a powerful strategy in asymmetric synthesis.[22] Clearly the choice of auxiliary for a desired chemical transformation is crucial for optimal synthetic efficiency. Hence, the ease with which the auxiliary can be introduced, the extent of stereoselection it imparts to the desired transformation, and the ease of its nondestructive removal are of critical importance. The 10,2-camphorsultam not only meets these criteria for a range of transformations, but also generally imparts crystallinity to all derived intermediates, thereby facilitating purification and isolation of enantiomerically pure products. Indeed, 10,2-camphorsultam derivatization alone allows for facile crystallographic determination of absolute configuration.[23]

Preparation of Derivatives. *N*-Acyl- and *N*-enoylsultam derivatives are routinely prepared in good yields using either sodium hydride–acid chloride[16a] or trimethylaluminum–methyl ester[18g] single-step protocols. A variant of the former method employing in situ stabilization of labile enoyl chlorides with CuCl/Cu has also been reported.[3k] A two-step procedure via the *N*-TMS derivative (**1**) is useful when a nonaqueous work-up is desirable and for synthesis of the *N*-acryloyl derivative.[24] *N*-Enoyl derivatives may also be prepared via the phosphonate derivative (**2**) by means of an Horner–Wadsworth–Emmons reaction (eq 1).[2c,2d]

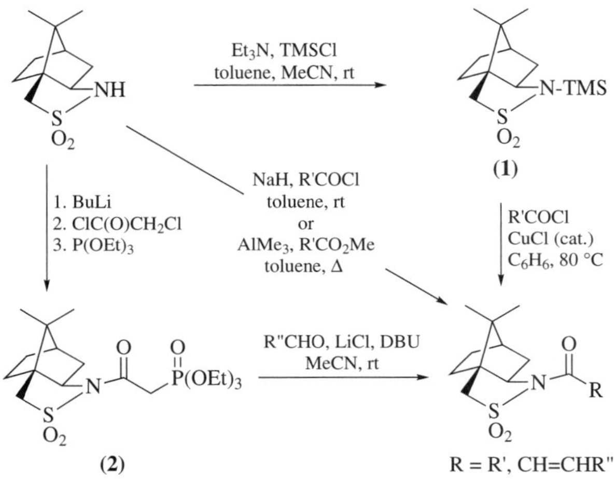

(1)

An *N*-acyl-β-keto derivative has been prepared by reaction with a diketene equivalent[17b] and the *trans*-N-cinnamoyl derivative by a Heck type coupling reaction.[4] The *N*-fluoro derivative (**3**) is prepared by direct fluorination (eq 2).[20]

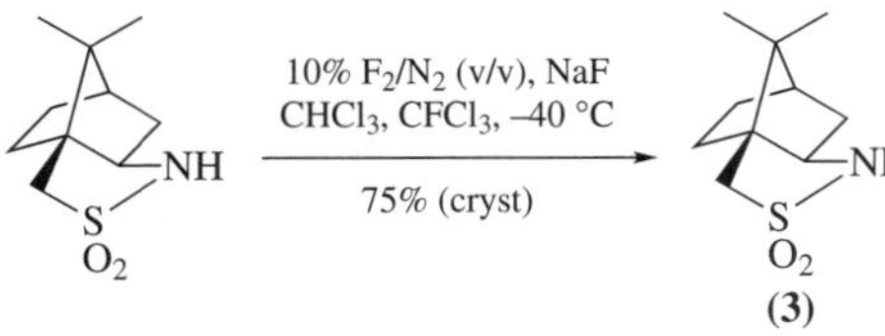

(2)

Reactions of *N*-Enoyl Derivatives.

[4+2] Diels–Alder Cycloadditions (Alkene → Six-Membered Cycloadduct)[2] *N*-Enoylsultam derivatives were originally devised as 'activated chiral dienophiles' for stereoselective Diels–Alder reactions.[1,2a]

Thermal reactions of *N*-enoylsultams generally show only moderate *endo* and π-face selectivity, e.g. *N*-acryloyl- and *N*-crotonoyl-10,2-camphorsultams (**4**) and (**6**) with cyclopentadiene (eq 3, Table 1).[2g] The thermal hetero-Diels–Alder reaction of *N*-glyoxaloyl-10,2-camphorsultam with 1-methoxybuta-1,3-diene also proceeds with moderate *exo* and π-face selectivity (57% *exo*, 46% de).[2h] Thermal hetero-Diels–Alder reactions of *N*-acylnitroso-10,2-camphorsultam with cyclopentadiene and 1,3-cyclohexadiene, however, proceed with excellent selectivity (>98% ee, π-face selectivity not established).[2i]

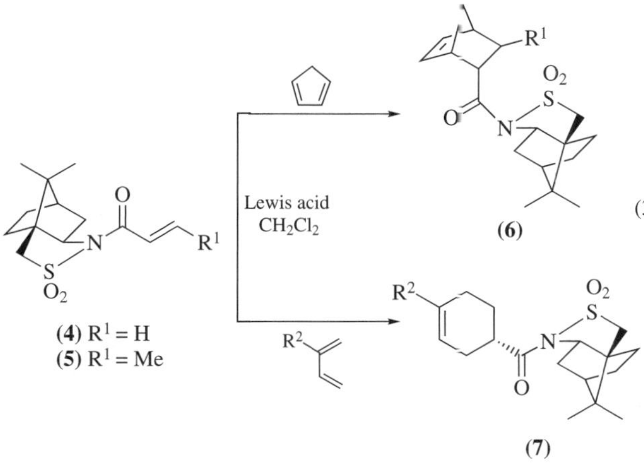

(3)

Lewis acid-mediated reactions of *N*-enoylsultams, on the other hand, occur under very mild conditions and with high levels of *endo* and π-face selectivity (eq 3, Table 1).[2b,2g] Dicoordinate $TiCl_4$, $EtAlCl_2$, and Me_2AlCl are particularly effective and their role in the stereodifferentiating process, which results in almost exclusive C(α)-*re* face dienophile attack, has been rationalized.[2g] Both inter- and intramolecular reactions proceed well even on a preparative scale (e.g. >100 g), often requiring just a single recrystallization to furnish isomerically pure products, valuable as synthetic intermediates (eq 4, the key step in a synthesis of (−)-pulo'upone).[2d] The hetero-Diels–Alder reaction of *N*-glyoxaloyl-10,2-camphorsultam with 1-methoxybuta-1,3-diene also proceeds efficiently and with high *endo* and π-face selectivity in the presence of 2% Eu(fod)$_3$ (90% *endo*, 88% de).[2h] These levels of asymmetric induction compare very favorably with those obtained using alternative auxiliaries (see Related Reagents below) for most substrates.

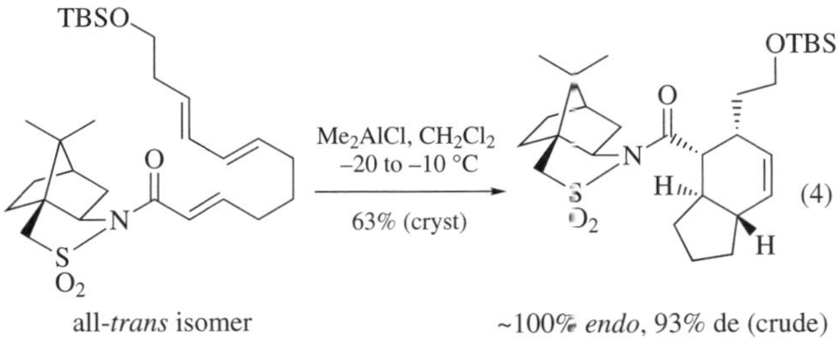

(4)

[3+2] Cycloadditions (Alkene → Five-Membered Cycloadduct)[3] The levels of selectivity found for 1,3-dipolar cycloaddition reactions are not as high as those obtained for Lewis acid-catalyzed Diels–Alder reactions However, the 10,2-camphorsultam auxiliary can achieve synthetically useful levels of induction in these reactions, and this has been attributed to efficient enoyl conformational control by the sultam moiety leading to preferred C(α)-*re* face attack even in the absence of metal complexation.[1d]

The reactions of *N*-enoyl-10,2-camphorsultams with various nitrile oxides to give isoxazolines have been well studied.[3a–c]

Table 1 Intermolecular Diels–Alder Reactions of *N*-Enoylsultams (**4**)/(**5**) → (**6**) and (**4**) → (**7**)

Dienophile	Diene	Lewis acid[a]	Temp (°C)/time (h)	Adduct	Yield crude (cryst)[b] (%)	de crude (cryst) (%)
(**4**)	Cyclopentadiene	None	21 (72)	(**6**) R^1 = H	80[c]	66
(**5**)	Cyclopentadiene	None	21 (96)	(**6**) R^1 = Me	51[d]	52
(**4**)	Cyclopentadiene	$EtAlCl_2$	−130 (6)[e]	(**6**) R^1 = H	96 (83)	95 (99)
(**5**)	Cyclopentadiene	$TiCl_4$	−78 (1)	(**6**) R^1 = Me	98 (83)	93 (99)
(**4**)	1,3-Butadiene	$EtAlCl_2$	−78 (18)	(**7**) R^2 = H	93 (81)	94 (99)
(**4**)	Isoprene	$EtAlCl_2$	−94 (18)	(**7**) R^2 = Me	88 (68)	94 (99)

[a] $EtAlCl_2$ (1.5 equiv), $TiCl_4$ (0.5 equiv). [b] >98% *endo*. [c] 89% *endo*. [d] 79% *endo*. [e] EtCl as solvent.

Indeed, the high regioselectivity and high π-face selectivity (62–90% de)[3a] observed in reactions with the *N*-acryloyl compound (**4**) have been exploited in synthesis (eq 5, the key step in a synthesis of (+)-hepialone[3b]), although related toluene-2,α-sultam auxiliaries provide still higher selectivity (see *α-Methyltoluene-2,α-sultam*). Isoxazolines may also be obtained by regioselective and similarly π-face selective cycloadditions of silyl nitronates followed by acid catalyzed elimination of TMS alcohol[3d–f] (eq 6, the key step in a synthesis of (+)-methylnonactate).[3f] A cyclic, photochemically generated azomethine ylide also participates in *exo* and π-face selective 1,3-dipolar cycloaddition with (*ent*-**4**), a reaction for which alternative auxiliaries were significantly less effective (eq 7, the key step in a synthesis of (−)-quinocarcin).[3g–i]

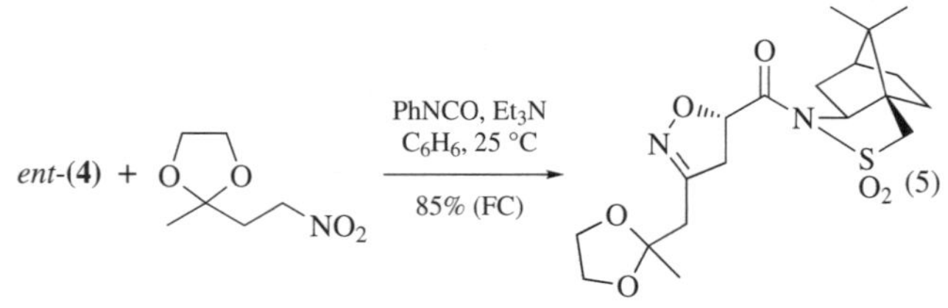

76% de (crude)

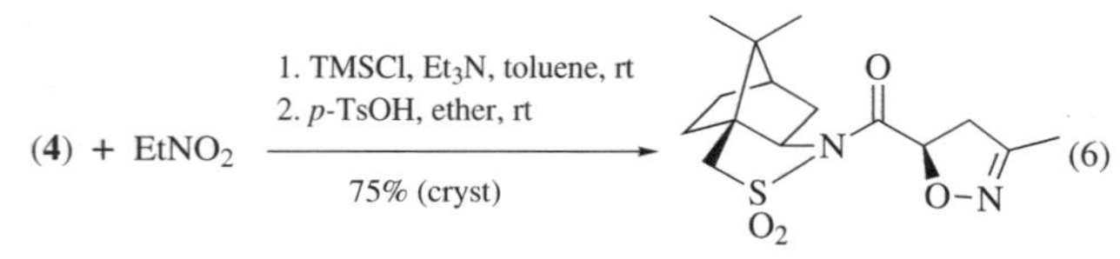

86% de (crude)

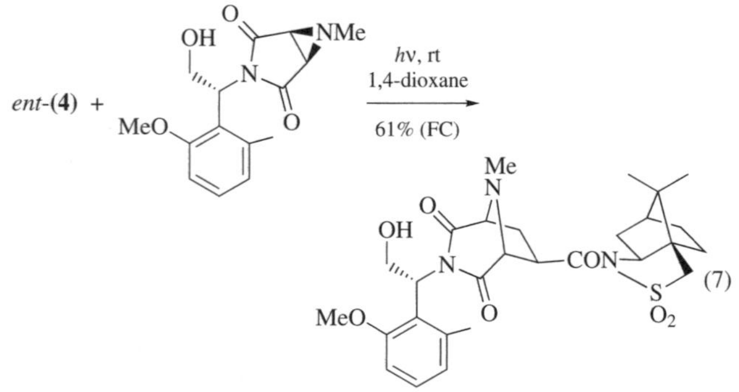

~100% *exo*, 92% de (crude)

Nickel catalyzed [3 + 2] cycloadditions of methylenecyclopropane and 2,2-dimethylmethylenecyclopropane with (**4**) afford 3-methylenecyclopentane derivatives with extremely high π-face selectivities (91% and 98% de respectively); five alternative auxiliaries were found to be less effective.[3j,3k] Palladium catalyzed [3 + 2] cycloaddition of 2-(TMS-methyl)-3-acetoxy-1-propene with an *N*-enoylsultam, however, proceeds with disappointing selectivity (4–26% de).[3l] A norephedrine derived auxiliary ((4*R*,5*S*)-4-methyl-5-phenyl-2-oxazolidinone) was similarly ineffective in this instance.[3l]

Cyclopropanation and Aziridination (Alkene → Three-Membered Cycloadduct)[4,5] Cyclopropanation of various *trans*-*N*-enoyl derivatives using diazomethane with $Pd(OAc)_2$ as catalyst affords cyclopropyl products with good C(α)-*re* π-facial control (eq 8).[4] Similarly, aziridination with *N*-aminophthalimide–lead tetraacetate affords *N*-phthalimidoaziridines with variable but generally good π-face selectivity (33–95% de).[5]

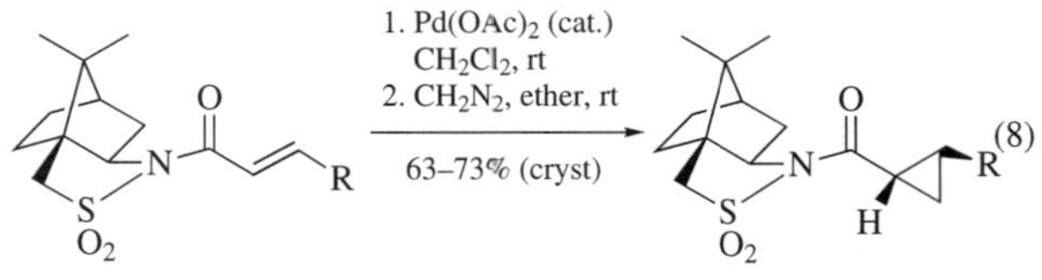

R = Me, Ph, *o*-$MeOC_6H_4$, *m*-$MeOC_6H_4$, 2-furyl
72.3–92.1% de (crude)

Dihydroxylation, Azido-Iodination and Hydrogenation (Alkene → α,β-Addition Product)[6–8] *syn*-Dihydroxylation of β-substituted *N*-enoylsultams using *N*-methylmorpholine with a catalytic amount of OsO_4 affords vicinal diol products with good C(α)-*re* π-facial selectivity (80–90% de) (eq 9, the key step in a synthesis of (+)-LLP 880β).[6a] Similar levels of selectivity but lower chemical yields are obtained using $KMnO_4$ and *N*-dienoylsultams.[6b] Regioselective but poorly stereoselective *trans* addition of iodine azide to *N*-crotonoyl- and *N*-cinnamoyl-10,2-camphorsultams has also been reported (34% and 47% de, respectively). The sense of addition corresponds to iodonium ion formation from the C(α)-*re* face followed by S_N2 attack of azide at the β-position.[8] Heterogeneous *syn* hydrogenation of β,β-disubstituted enoylsultams over Pd/C using gaseous hydrogen (100 psi) affords reduced products, again with excellent C(α)-*re* topicity (90–96% de).[7]

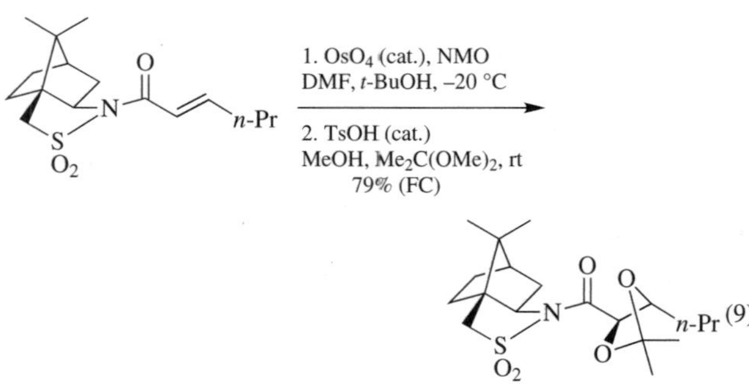

83% de (crude)

1,4-Hydride, Grignard, Cuprate, Allylsilane, and Thiolate Addition (Alkene → β- or α,β-Functionalized Product)[9–13] β,β-Disubstituted enoylsultams undergo efficient reduction with L-Selectride.[9a] The *syn* hydrogenated products obtained result from conjugate hydride delivery (and protonation) on the opposite π-face [i.e. C(α)-*si*] to that from hydrogenation (90–94% de) (eq 10).[9b] Similarly, simple alkylmagnesium chlorides also undergo 1,4-addition–protonation with *trans*-β-substituted enoylsultams from this face (72–89% de).[10] Use of α-substituted *N*-enoyl substrates,[9a] or trapping of the intermediate aluminum or magnesium enolates with other electrophiles, allows creation of two asymmetric centers in one synthetic operation.[9a,10] The observed topicity is that of *syn* addition from the C(α)-*si* face. As PBu_3 stabilized alkylcopper reagents,[11a,11b] Grignard reagents (in the presence of copper salts),[11c] and cuprates (Gilman reagents)[11d] participate in analogous reactions but show reversed π-face selectivity, an appropriate 1,4-addition–trapping protocol can be devised to generate products with any desired configuration at both the α- and β-positions (eq 11).[11c] This complementarity has been rationalized.[11] Phosphine stabilized alkyl- and alkenylcopper reagents also add to *N*-(β-silylenoyl)sultams (giving aldols after C–Si oxidative bond cleavage). In this case, either π-face selectivity can be achieved, depending on the promoting Lewis acid employed.[11b] Similar Lewis acid dependent selectivity is observed for addition of allyltrimethylsilane to *N*-enoylsultams.[12]

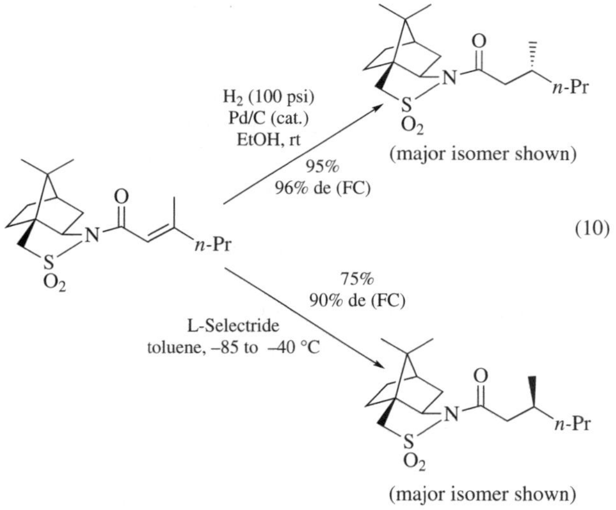

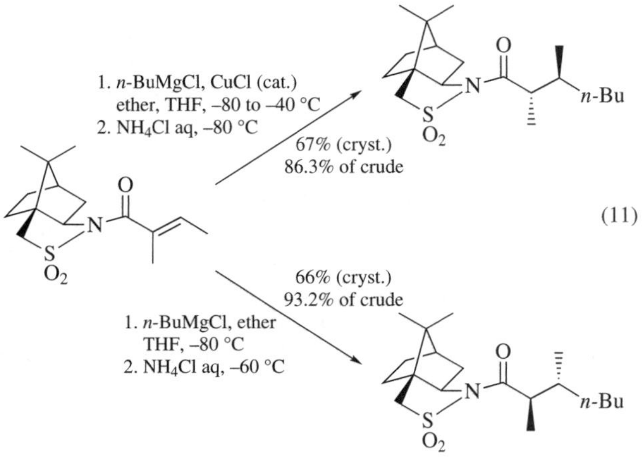

Stereoselective *anti* addition of thiophenol to *N*-[β-(*n*-butyl)methacryloyl]-10,2-camphorsultam [the key step in a synthesis of (+)-*trans* whiskey lactone] has been explained by a sulfur-induced, stereoelectronically directed protonation following C(β)-*re* face conjugate addition.[13]

Radical Addition and S_N2' Displacement (Alkene → α-Functionalized Product)[14,15] Stereoselective radical additions to *N*-enoylsultams occur at the α-position, while additions to the β-position are essentially nonselective.[1d,14] The S_N2' displacement of γ-bromo-*N*-enoylsultams with higher order cyanocuprates occurs with good π-face selectivity (90–96% de).[15]

Reactions of *N*-Acyl Derivatives.

Aldolization (Acyl Species → β-Hydroxyacyl Product)[16] Chiral oxazolidin-2-ones and 10,2-camphorsultams presently represent 'state of the art' aldol reaction mediators. Both auxiliaries have similarly high π-facial preferences (totally overwhelming any modest facial preference of most chiral aldehydes), allowing the predictable formation of essentially one (of four possible) diastereomeric aldol type products by judicious choice of auxiliary antipode and reaction conditions.[16a] Although sultam mediated aldolizations generally require a 2–3 fold excess of aldehyde to go to completion (cf. 1–2 equiv when oxazolidin-2-one mediated), which is clearly wasteful when employing a valuable aldehyde, the superior crystallinity and cleavage properties of the sultam adducts makes the choice of auxiliary for a given aldolization dependent on the specific substrate.

syn-Aldols with (*R*) configuration at the α-position are obtained from boryl enolates of (−)-10,2-camphorsultam derivatives (**8**) on condensation with aldehydes.[16b] The observed topicity is consistent with C(α)-*si*/C=O-*re* interaction of the 'nonchelated' (*Z*)-enolate and the aldehyde.[16b] *syn*-Aldols with (*S*) configuration at the α-position are obtained from lithium (BuLi–THF) or better tin(IV) enolates of the same derivatives (**8**), and this outcome is consistent with C(α)-*re*/C=O-*si* interaction of the 'chelated' (*Z*)-enolate and the aldehyde.[16b] *anti*-Aldols with (*S*) configuration at the α-position are obtained from in situ prepared *O*-silyl-*N,O*-ketene acetals of sultams (**8**) on condensation with aldehydes in the presence of $TiCl_4$[16c] (Mukaiyama aldolization). This topicity arises from C(α)-*re*/C=O-*re* interaction of the *(Z)-N,O*-ketene acetal and the Lewis acid coordinated aldehyde (eq 12).[16c] These same *anti*-aldols can also be obtained from sultams (**8**) with similarly excellent stereocontrol using boryl enolates in the presence of $TiCl_4$, and this unique procedure is the method of choice when using crotonaldehyde or methacrolein.[16d] *anti*-Aldols with (*R*) configuration at the α-position should be obtained from sultams (*ent*-**8**) using the above Mukaiyama conditions. Enantiocontrolled synthesis of α-unsubstituted β-hydroxy carbonyl compounds from the *N*-acetyl derivative is best accomplished using the Mukaiyama conditions (58–93% de).[16e] The synthesis of beetle sex pheromone (−)-serricorole serves to highlight the power of the above methods.[16f,16g]

Alkylation (Acyl Species → α-Alkylated Acyl Product)[17] An efficient procedure for the C(α)-*re* alkylation of lithium and sodium enolates of *N*-acylsultams with various (even nonactivated) primary halides in the presence of HMPA has been developed (88.7–99% de).[17a] 'Alkylation' with $ClCH_2NMeCO_2Bn$ enables a two-step β-lactam synthesis.[1c,17a] Michael-type alkylation of a β-keto derivative with arylidenemalononitriles in toluene containing piperidine has been reported to give 4*H*-pyrans (60–70% de).[17b]

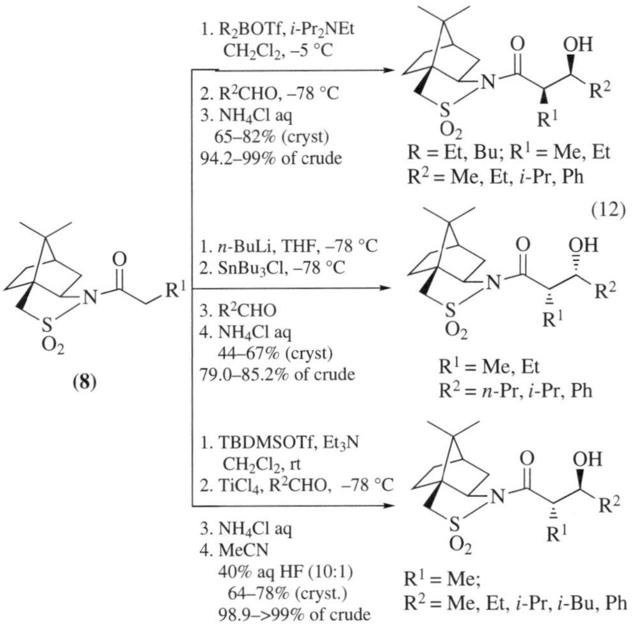

α-Amino Acid Preparation[18] Three distinct strategies for the asymmetric preparation of α-amino acids using the 10,2-camphorsultam auxiliary have been developed. The first is a glycine anion strategy[18a] centered on alkylation, with excellent C(α)-*si* π-face stereocontrol, of lithium enolates of sultam derivative (**9a**) [mp 107–109 °C (EtOH)][18b,18c] to give adducts (**10**) (eq 13). α-Amino acids are obtained simply by Schiff base hydrolysis (0.5N HCl, rt) and auxiliary cleavage (LiOH, aq THF). Compound (**9b**) has also been reported to participate in analogous chemistry,[18d] but it is not crystalline and its derivatives require more vigorous hydrolysis. Commercially available (**9a**) is thus the preferred reagent, comparing favorably with other 'glycine anion' synthetic equivalents (see *4-t-Butoxycarbonyl-5,6-diphenyl-2,3,5,6-tetrahydro-4H-oxazin-2 -one*). Promising preliminary results of deprotonation–alkylation of (**9a**) under phase transfer catalysis have also been disclosed.[1c,18c,18j]

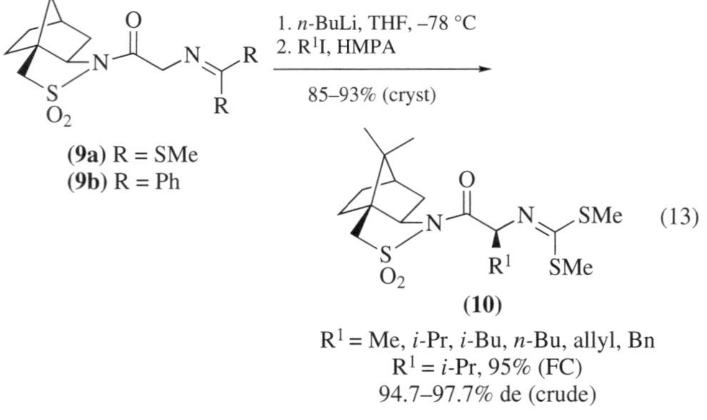

The second strategy involves a bromination–azide displacement–hydrogenolysis protocol. Treatment of boryl enolates of *N*-acylsultams (**8**) with NBS provides the key α-bromo derivatives (**11**) with good C(α)-*re* topicity (eq 14).[1c,18b] Stereospecific substitution with tetramethylguanidinium azide [$(Me_2N)_2C{=}NH_2{}^+N_3{}^-$], hydrogenolysis ($H_2$–Pd/C), and auxiliary cleavage provides α-amino acids in good overall yield.[1c,18b] As with the previous strategy, given that an appropriate derivative is crystallized to enantiomeric homogeneity, the enantiomeric purity of the product will reflect the extent of racemization during auxiliary hydrolysis (e.g. phenylglycine: 90.3% ee, isoleucine: >99% ee).[1c] This problem can be circumvented by the use of 'nonbasic' Ti(O-*i*-Pr)$_4$ assisted 'transesterification' with allyl alcohol then rhodium-catalyzed 'deprotection'. This allows for the preparation of either 'free' or *N*-Fmoc α-amino acids of excellent enantiomeric purity.[18e]

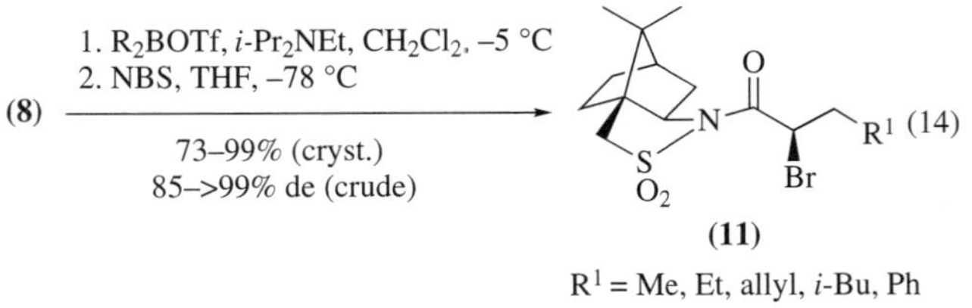

The third strategy involves electrophilic 'amination' of sodium enolates of *N*-acylsultams (**8**) using *1-Chloro-1-nitrosocyclohexane* as an [$NH_2{}^+$] equivalent.[18f–i] The reaction proceeds via nitrone intermediates which are routinely hydrolyzed without isolation to give the key α-*N*-hydroxyamino derivatives (**13**) with outstanding C(α)-*re* π-facial control (eq 15). Nitrogen–oxygen bond hydrogenolysis (Zn, aq HCl, AcOH), then auxiliary cleavage, affords α-amino acids.[18f,18g] Omission of the hydrogenolysis step allows access to *N*-hydroxy-α-amino acids, which are extremely difficult to prepare by alternative means.[18f] The scope of the reaction has been extended to encompass the use of 1-chloro-1-nitrosocyclohexane as an electrophilic partner in conjugate addition–trapping reactions [allowing an expedient preparation of (2*S*,3*S*)-isoleucine],[18f,18g] *N*-alkyl-α-amino acid preparation,[18h] and enantiomerically pure α-substituted cyclic nitrone formation [giving a concise preparation of the piperidine alkaloid (−)-pinidine].[18i]

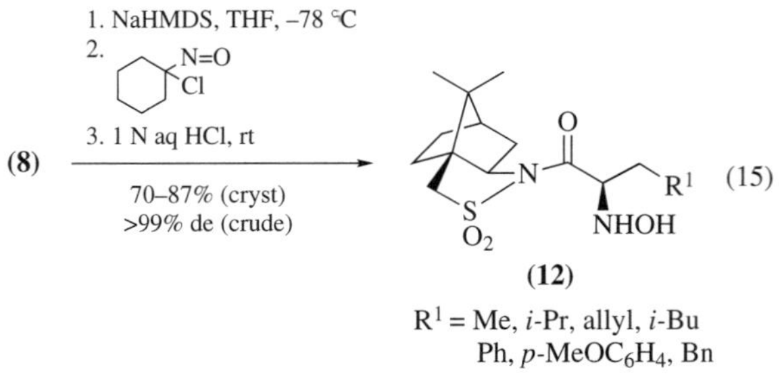

α-Radical Addition (Acyl Species → α-Functionalized Acyl Product)[19] Radicals derived from α-iodo-*N*-acylsultams give high levels of asymmetric induction in intramolecular addition reactions with allyltributylstannanes (85–>94% de)[19a] and 5-*exo*-dig type cyclizations and annulations (eq 16).[19a] In addition, 'zipper' type manganese(III) promoted oxidative radical cyclization of *N*-(*trans*-4-methyl-4,9-nonadienoyl)-10,2-camphorsultam gives a *cis*-fused hydrindane derivative with modest (50% de) selectivity at the α-center.[19b] All these reactions proceed at or above room temperature, making the levels of induction remarkable. Furthermore, effective alternative auxiliaries are scarce.[1d,14]

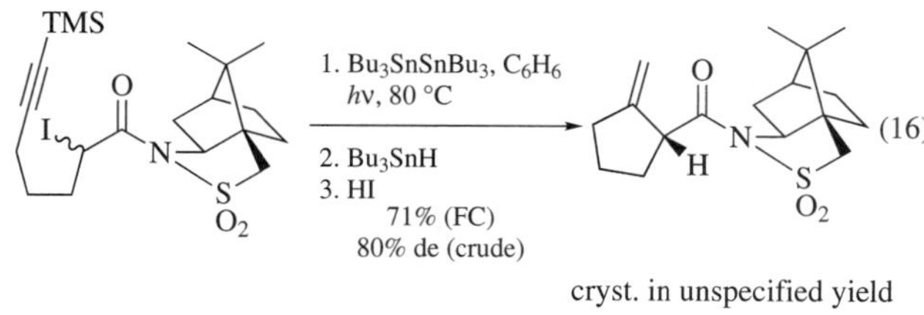

Nondestructive Auxiliary Cleavage. One feature which makes the sultam chiral auxiliary, and to an even greater extent the related toluene-2,α-sultam auxiliaries (see *α-Methyltoluene-2,α-sultam*), so versatile is the ease with which *N*-acyl bond fission occurs in derivatives. A great variety of extremely mild, bimolecular and intramolecular nondestructive cleavage protocols have been developed which tolerate a wide array of molecular functionality, simple extraction and crystallization usually providing almost quantitative auxiliary recovery without loss of enantiomeric purity.

Saponification with LiOH[6a] or H_2O_2–LiOH[16b] in aqueous THF is routinely employed for conversion of *N*-acylsultams to enantiomerically pure carboxylic acids. A variant conducted in aprotic media with phase transfer catalysis has also been reported.[18d] If base sensitive functionality is present, then the corresponding esters can be prepared by 'nonbasic' titanium mediated 'alcoholysis'. This can be accomplished with ethyl,[11b] benzyl,[4] or allyl[18e] alcohols, and in the latter two instances the carboxylic acids can be subsequently liberated by 'neutral' hydrogenolysis or $RhCl(PPh_3)_3$ catalyzed hydrolysis,[18e] respectively. Lactones and esters can also be formed by intra-[18j] and intermolecular[2d] sultam cleavage with lithium alkoxides and bromomagnesium alkoxides.[11d] β-Lactams can be prepared by intramolecular ring closure of metallated β-aminomethyl derivatives[1b,17a] and an aluminum 'thiobenzyloxy ate' complex has been used to obtain thioester derivatives.[13] Reductive cleavage of *N*-acylsultams using lithium aluminum hydride[2f] or L-Selectride®[3b,3f] in THF gives rise to the corresponding primary alcohols.

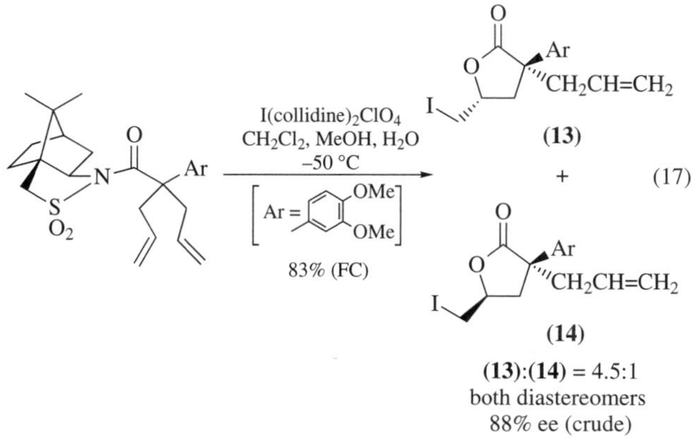

Auxiliary cleavage with concomitant carbon–carbon bond formation is a particularly attractive option, which has been demonstrated in a bimolecular sense using the dianion of methyl sulfone (giving a methyl ketone),[16f] and in an intramolecular sense using a Claisen-type condensation of a β-acetoxy enolate (giving a δ-lactone).[25] An interesting 'halolactonization' procedure has also been devised; for certain α-aryl-bis-(γ-unsaturated)-*N*-acyl derivatives this allows for highly efficient auxiliary cleavage and asymmetric formation of two stereocenters, one of which is quaternary (eq 17), the key step in a synthesis of (−)-mesembrine).[26]

Enantioselective, Electrophilic Fluorination. (−)-*N*-Fluoro-10,2-camphorsultam (**3**) [mp 112–114 °C (CH_2Cl_2–pentane)] is an enantioselective, electrophilic fluorinating agent.[20] Fluorination of stabilized enolates occurs with highly variable yield (5–63%) and stereoselectivity (10–70% de).

Related Reagents. 10-Dicyclohexylsulfonamidoisoborneol; 2-Hydroxy-1,2,2-triphenylethyl Acetate; (4*S*,5*S*)-4-Methoxymethyl-2-methyl-5-phenyl-2-oxazoline; α-Methyltoluene-2,α-sultam; (*S*)-4-Benzyl-2-oxazolidinone.

---

1. (a) Oppolzer, W. *Tetrahedron* **1987**, *43*, 1969. (b) Oppolzer, W. *Pure Appl. Chem.* **1988**, *60*, 39. (c) Oppolzer, W. *Pure Appl. Chem.* **1990**, *62*, 1241. (d) Kim, B. H.; Curran, D. P. *Tetrahedron* **1993**, *49*, 293.
2. (a) Oppolzer, W. *Angew. Chem., Int. Ed. Engl.* **1984**, *23*, 876. (b) Oppolzer, W.; Chapuis, C.; Bernardinelli, G. *Helv. Chim. Acta* **1984**, *67*, 1397. (c) Oppolzer, W.; Dupuis, D. *Tetrahedron Lett.* **1985**, *26*, 5437. (d) Oppolzer, W.; Dupuis, D.; Poli, G.; Raynham, T. M.; Bernardinelli, G. *Tetrahedron Lett.* **1988**, *29*, 5885. (e) Smith III, A. B.; Hale, J. K.; Laahso, L. M.; Chen, K.; Riera, A. *Tetrahedron Lett.* **1989**, *30*, 6963. (f) Vandewalle, M.; Van der Eycken, J.; Oppolzer, W.; Vullioud, C. *Tetrahedron* **1986**, *42*, 4035. (g) Oppolzer, W.; Rodriguez, I.; Blagg, J.; Bernardinelli, G. *Helv. Chim. Acta* **1989**, *72*, 123. (h) Bauer, T.; Chapuis, C.; Kozac, J.; Jurczak, J. *Helv. Chim. Acta* **1989**, *72*, 482. (i) Gouverneur, V.; Dive, G.; Ghosez, L. *Tetrahedron: Asymmetry* **1991**, *2*, 1173.
3. (a) Curran, D. P.; Kim, B. H.; Daugherty, H.; Heffner, T. A. *Tetrahedron Lett.* **1988**, *29*, 3555. (b) Curran, D. P.; Heffner, T. A. *J. Org. Chem.* **1990**, *55*, 4585. (c) Kim, K. S.; Kim, B. H.; Park, W. M.; Cho, S. J.; Mhin, B. J. *J. Am. Chem. Soc.* **1993**, *115*, 7472. (d) Kim, B. H.; Lee, J. Y.; Kim, K.; Whang, D. *Tetrahedron: Asymmetry* **1991**, *2*, 27. (e) Kim, B. H.; Lee, J. Y. *Tetrahedron: Asymmetry* **1991**, *2*, 1359. (f) Kim, B. H.; Lee, J. Y. *Tetrahedron Lett.* **1992**, *33*, 2557. (g) Garner, P.; Ho, W. B. *J. Org. Chem.* **1990**, *55*, 3973. (h) Garner, P.; Ho, W. B.; Grandhee, S. K.; Youngs, W. J.; Kennedy, V. O. *J. Org. Chem.* **1991**, *56*, 5893. (i) Garner, P.; Ho, W. B.; Shin, H. *J. Am. Chem. Soc.* **1992**, *114*, 2767. (j) Binger, P.; Schafer, B. *Tetrahedron Lett.* **1988**, *29*, 529. (k) Binger, P.; Brinkmann, A.; Roefke, P.; Schafer, B. *Liebigs Ann.* **1989**, 739. (l) Trost, B. M.; Yang, B.; Miller, M. L. *J. Am. Chem. Soc.* **1989**, *111*, 6482.
4. Vallgarda, J.; Hacksell, U. *Tetrahedron Lett.* **1991**, *32*, 5625, and corrigendum *ibid. Tetrahedron Lett.* **1991**, *32*, 7136.
5. Kapron, J. T.; Santarsiero, B. D.; Vederas, J. C. *Chem. Commun.* **1993**, 1074.
6. (a) Oppolzer, W.; Barras, J.-P. *Helv. Chim. Acta* **1987**, *70*, 1666. (b) Walba, D. M.; Przybyla, C. A.; Walker, Jr, C. B. *J. Am. Chem. Soc.* **1990**, *112*, 5624.
7. Oppolzer, W.; Mills, R. J.; Reglier M. *Tetrahedron Lett.* **1986**, *27*, 183.
8. Lee, P.-C.; Wu, C.-C.; Cheng. M.-C.; Wang, Y.; Wu, M.-J. *J. Chinese Chem. Soc.* **1992**, *39*, 87.
9. (a) Oppolzer, W.; Poli, G. *Tetrahedron Lett.* **1986**, *27*, 4717. (b) Oppolzer, W.; Poli, G.; Starkemann, C.; Bernardinelli, G. *Tetrahedron Lett.* **1988**, *29*, 3559.
10. Oppolzer, W.; Poli, G.; Kingma, A. J.; Starkemann, C.; Bernardinelli, G. *Helv. Chim. Acta* **1987**, *70*, 2201.
11. (a) Oppolzer, W.; Mills, R. J.; Pachinger, W.; Stevenson, T. *Helv. Chim. Acta* **1986**, *69*, 1542. (b) Oppolzer, W.; Schneider, P. *Helv. Chim. Acta* **1986**, *69*, 1817. (c) Oppolzer, W.; Kingma, A. J. *Helv. Chim. Acta* 1989, *72*, 1337. (d) Oppolzer, W.; Kingma, A. J.; Poli, G. *Tetrahedron* **1989**, *45*, 479.
12. Wu, M.-J.; Wu, C.-C.; Lee, P.-C. *Tetrahedron Lett.* **1992**, *33*, 2547.

13. Miyata, O.; Shinada, T.; Kawakami, N.; Taji, K.; Ninomiya, I.; Naito, T.; Date, T.; Okamura, K. *Chem. Pharm. Bull.* **1992**, *40*, 2579.
14. Porter, N. A.; Giese, B.; Curran, D. P. *Acc. Chem. Res.* **1991**, *24*, 296.
15. Girard, C.; Mandville, G.; Bloch, R. *Tetrahedron: Asymmetry* **1993**, *4*, 613.
16. (a) Heathcock, C. H. In *Modern Synthetic Methods*, VCH-VHCA: Basel, 1982, 1. (b) Oppolzer, W.; Blagg, J.; Rodriguez, I.; Walther, E. *J. Am. Chem. Soc.* **1990**, *112*, 2767. (c) Oppolzer, W.; Starkemann, C.; Rodriguez, I.; Bernardinelli, G. *Tetrahedron Lett.* **1991**, *32*, 61. (d) Oppolzer, W.; Lienard, P. *Tetrahedron Lett.* **1993**, *34*, 4321. (e) Oppolzer, W.; Starkemann, C. *Tetrahedron Lett.* **1992**, *33*, 2439. (f) Oppolzer, W.; Rodriguez, I. *Helv. Chim. Acta* **1993**, *76*, 1275. (g) Oppolzer, W.; Rodriguez, I. *Helv. Chim. Acta* **1993**, *76*, 1282.
17. (a) Oppolzer, W.; Moretti, R.; Thomi, S. *Tetrahedron Lett.* **1989**, *30*, 5603. (b) Martin, N.; Martinez-Grau, A.; Seoane, C.; Marco, J. L. *Tetrahedron Lett.* **1993**, *34*, 5627.
18. (a) Williams, R. M. *Synthesis of Optically Active α-Amino Acids*, Pergamon: Oxford, 1989. (b) Oppolzer, W. *Arch. Pharm. (Weinheim, Ger.)* **1990**, 190. (c) Oppolzer, W.; Moretti, R.; Thomi, S. *Tetrahedron Lett.* **1989**, *30*, 6009. (d) Josien, H.; Martin, A.; Chassaing, G. *Tetrahedron Lett.* **1991**, *32*, 6547. (e) Oppolzer, W.; Lienard, P. *Helv. Chim. Acta* **1992**, *75*, 2572. (f) Oppolzer, W.; Tamura, O. *Tetrahedron Lett.* **1990**, *31*, 991. (g) Oppolzer, W.; Tamura, O.; Deerberg, J. *Helv. Chim. Acta* **1992**, *75*, 1965. (h) Oppolzer, W.; Cintas-Moreno, P.; Tamura, O. *Helv. Chim. Acta* **1993**, *76*, 187. (i) Oppolzer, W.; Merifield, E. *Helv. Chim. Acta* **1993**, *76*, 957. (j) Oppolzer, W.; Bienayme, H.; Genevois-Borella, A. *J. Am. Chem. Soc.* **1991**, *113*, 9660.
19. (a) Curran, D. P.; Shen, W.; Zhang, Z.; Heffner, T. A. *J. Am. Chem. Soc.* **1990**, *112*, 6738. (b) Zoretic, P. A.; Weng, X.; Biggers, C. K.; Biggers, M. S.; Caspar, M. L. *Tetrahedron Lett.* **1992**, *33*, 2637.
20. Differding, E.; Lang, R. W. *Tetrahedron Lett.* **1988**, *29*, 6087.
21. Weismiller, M. C.; Towson, J. C.; Davis, F. A. *Org. Synth.* **1990**, *69*, 154.
22. Davies, S. G. *Chem. Br.* **1989**, *25*, 268.
23. Harada, N.; Soutome, T.; Nehira, T.; Uda, H. *J. Am. Chem. Soc.* **1993**, *115*, 7547.
24. Thom, C.; Kocienski, P. *Synthesis* **1992**, 582.
25. Brandange, S.; Leijonmarck, H. *Tetrahedron Lett.* **1992**, *33*, 3025.
26. (a) Yokomatsu, T.; Iwasawa, H.; Shibuya, S. *Chem. Commun.* **1992**, 728. (b) Yokomatsu, T.; Iwasawa, H.; Shibuya, S. *Tetrahedron Lett.* **1992**, *33*, 6999.

Alan C. Spivey
University of Cambridge, UK

(Camphorylsulfonyl)oxaziridine[1]

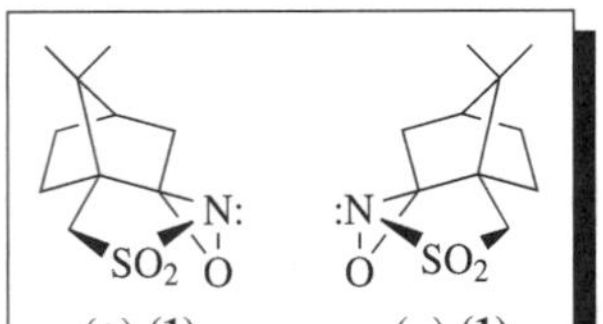

(+)-(1) (−)-(1)

(+)-(**1**)
[104322-63-6] $C_{10}H_{15}NO_3S$ (MW 229.30)
(−)-(**1**)
[104372-31-8]

(neutral, aprotic, electrophilic, and asymmetric oxidizing agents for the chemoselective oxidation of many nucleophilic substrates such as sulfides, enamines, enol esters, carbanions, and enolates[1])

Physical Data: (+)-(**1**): mp 165–167 °C, $[\alpha]_D$ +44.6° ($CHCl_3$, *c* 2.2); (−)-(**1**): mp 166–167 °C, $[\alpha]_D$ −43.6° ($CHCl_3$, *c* 2.2).
Solubility: sol THF, CH_2Cl_2, $CHCl_3$; slightly sol isopropanol, ethanol; insol hexane, pentane, water.
Form Supplied in: commercially available as a white solid.
Analysis of Reagent Purity: by mp and specific rotation determination.
Preparative Methods: the enantiopure (+)- and (−)-(camphorylsulfonyl)oxaziridines (**1**) and [(8,8-dichlorocamphor)sulfonyl]oxaziridines (**2**) are commercially available. They can also be prepared on a large scale via the oxidation of corresponding camphorsulfonimines with buffered *Potassium Monoperoxysulfate* (Oxone)[2] or buffered peracetic acid.[3] Since oxidation takes place from the *endo* face of the C=N double bond, only a single oxaziridine isomer is obtained. The precursor camphorsulfonimines can be prepared in 3 steps (>80% yield) from inexpensive (+)- and (−)-*10-Camphorsulfonic Acids*. A variety of (camphorylsulfonyl)oxaziridine derivatives such as (**2**)–(**4**) are also readily available via the functionalization of the camphorsulfonimines followed by oxidation.[1,2–6]

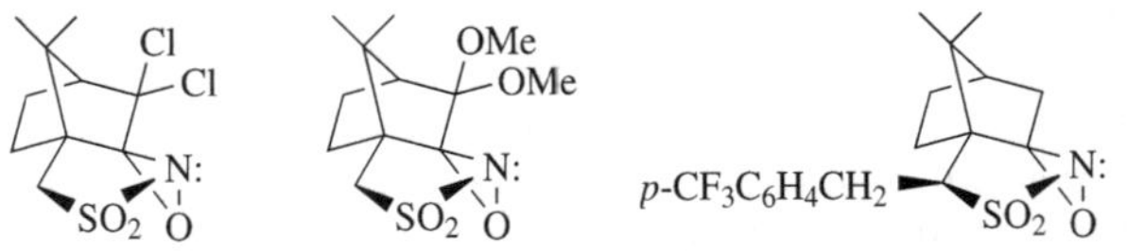

(+)-(**2**) (+)-(**3**) (+)-(**4**)

Purification: by recrystallization.
Handling, Storage, and Precautions: indefinitely stable to storage at room temperature and to exposure to air.

Asymmetric Oxidation of Sulfides. Prochiral sulfides are oxidized by (camphorylsulfonyl)oxaziridine (**1**) to optically active sulfoxides. Over-oxidation to sulfones is not observed (eq 1).[7] However, the best chiral *N*-sulfonyloxaziridines for the asymmetric oxidation of sulfides to sulfoxides are the (+)- and (−)-*N-(Phenylsulfonyl)(3,3-dichlorocamphoryl)oxaziridines*.[8]

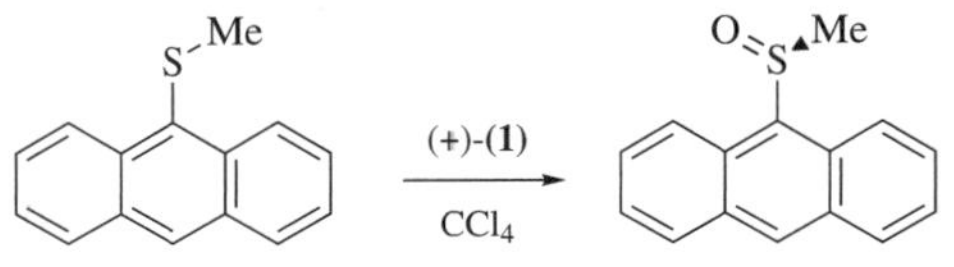

(1)

(*S*) 73% ee

Oxidation of Enamines. Enamines are rapidly oxidized by (+)-(camphorylsulfonyl)oxaziridine (**1**). Disubstituted enamines give rise to racemic α-amino ketones, while trisubstituted enamines afford, after hydrolysis, α-hydroxy ketones (eq 2).[9] A mechanism involving initial oxidation of the enamine to an α-amino epoxide is suggested to account for these products.

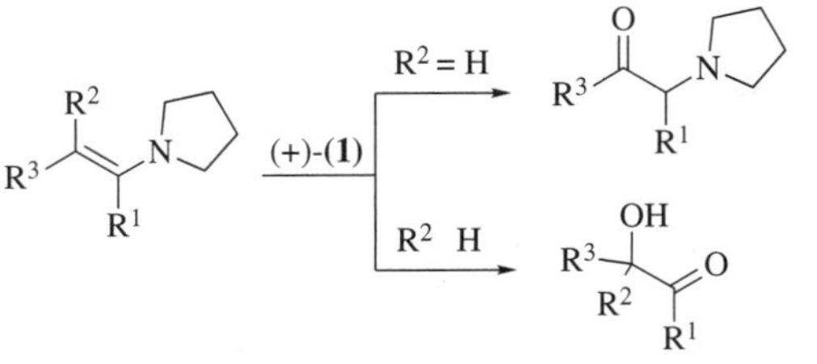

(2)

Oxidation of Oxaphospholenes. Reaction of oxaphospholene (**5**) with (+)-[(8,8-dichlorocamphoryl)sulfonyl] oxaziridine (**2**) affords β-hydroxy-γ-keto-phosphonate in 49% ee with undetermined absolute configuration (eq 3).[10] Higher temperatures accelerate the reaction but lower the stereoselectivity.

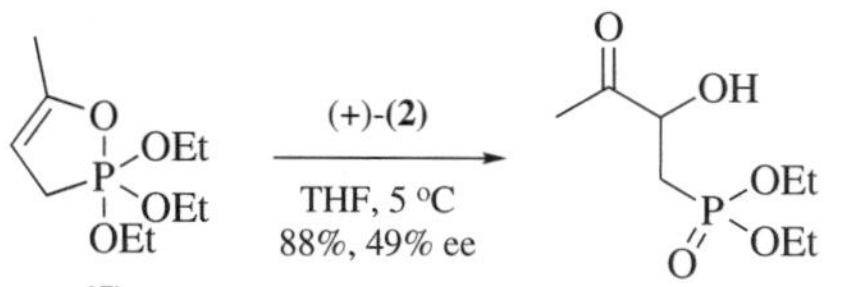

(5) (3)

Oxidation of Organolithium and Organomagnesium Compounds. Oxidation of phenylmagnesium bromide and phenyllithium with *(±)-trans-2-(Phenylsulfonyl)-3-phenyloxaziridine* or (camphorylsulfonyl)oxaziridine (**1**) gives phenol (eq 4).[11] Products are cleaner with the latter reagent because addition of the organometallic reagent to the C=N double bond of the imine is not observed. Oxidation of (*E*)- and (*Z*)-vinyllithium reagents with (+)-(**1**) affords enolates. The reaction is fast and represents useful methodology for the stereo- and regioselective formation of enolates.[12] While the enolates can be trapped with *Chlorotrimethylsilane* to give silyl enol ethers, better yields and higher stereoselectivity are obtained with *Bis(trimethylsilyl) Peroxide* (eq 5).[12]

$$\text{PhMgBr (or PhLi)} + (+)\text{-}(\mathbf{1}) \xrightarrow[-78\ ^\circ\text{C}]{\text{THF}} \text{PhOH} \qquad (4)$$

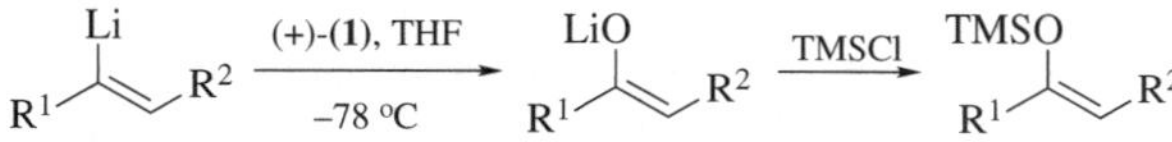

(5)

Oxidation of Phosphoranes. Monosubstituted phosphoranes (ylides) are rapidly oxidized to *trans*-alkenes by (+)-(**1**), while disubstituted phosphoranes give ketones (eq 6). A mechanism involving initial attack of the carbanion of phosphorane to the electrophilic oxaziridine oxygen atom of (+)-(**1**) is proposed.[13]

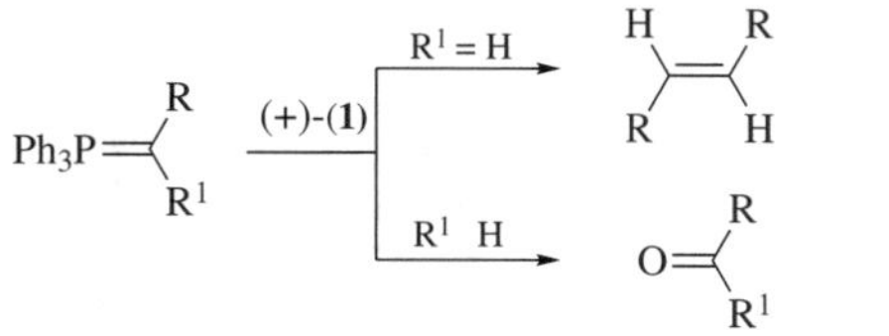

(6)

Asymmetric α-Hydroxylation of Enolates. α-Hydroxylation of enolates represents one of the simplest and most direct methods for the synthesis of α-hydroxy carbonyl compounds, a key structural unit found in many natural products.[1b] Enolate oxidations using (+)- and (−)-(**1**) and their derivatives generally effect this transformation in good to excellent yields with a minimum of side reactions (e.g. over-oxidation). Furthermore, these reagents are the only aprotic oxidants developed to date for the direct asymmetric hydroxylation of prochiral enolates to optically active α-hydroxy carbonyl compounds.

By choice of the appropriate reaction conditions and (camphorylsulfonyl)oxaziridine derivative, acyclic α-hydroxy ketones of high enantiomeric purity have been prepared.[1] An example is the oxidation of the sodium enolate of deoxybenzoin with (+)-(**1**). The reaction proceeds very fast at −78 °C, affording (+)-(*S*)-benzoin in 95% ee. Both benzoin enantiomers are readily available by choice of (+)- or (−)-(**1**), because the configuration of the oxaziridine controls the absolute stereochemistry of the product (eq 7).[14] Detailed studies have indicated that the generation of a single enolate regioisomer is a pre-condition for high enantioselectivity, although this does not necessarily always translate into high ee's. Hydroxylation of tertiary substituted acyclic ketone enolates usually gives lower stereoselectivities due to the formation of (*E*/*Z*) enolate mixtures (eq 8).[14] In addition to enolate geometry, the molecular recognition depends on the structure of the oxidant, the type of enolate, and the reaction conditions.[1b] Generally the stereoselectivity can be predicted by assuming that the oxaziridine approaches the enolate from the least sterically hindered direction.

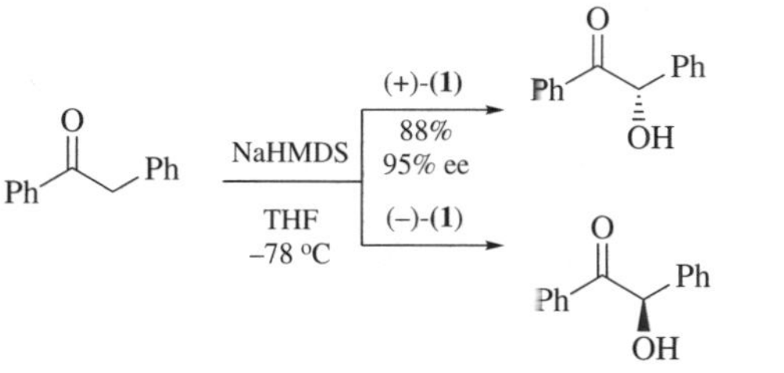

(7)

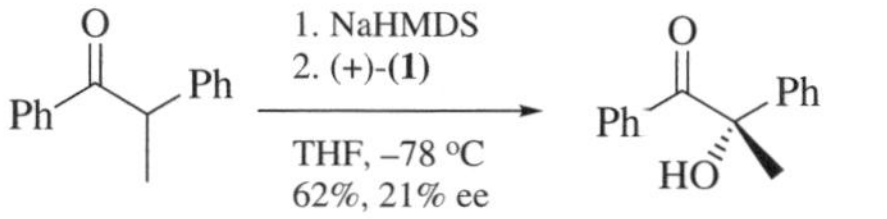

(8)

The asymmetric hydroxylation of cyclic ketone enolates, particularly the tetralone and 4-chromanone systems, has been studied in detail because the corresponding α-hydroxy carbonyl compounds are found in many natural products.[1b] Some general trends have been observed. 2-Substituted 1-tetralones having a variety of groups at C-2 (Me, Et, Bn) are best oxidized by chlorooxaziridine (**2**) in >90% ee (eq 9).[2c,15,16] However, substitution of a methoxy group into the 8-position lowers the stereoselectivity. For the 8-methoxytetralones, (8,8-dimethoxycamphorylsulfonyl)oxaziridine (**3**) is the reagent of choice. Similar trends have also been observed in 4-chromanones.[1b] Oxidation of the lithium enolate of (**6**) with (8,8-methoxycamphorsulfonyl)oxaziridine (**3**) affords 5,7-dimethyleucomol (**7**) in >96% ee (eq 10).[17] Hydroxylation of the enolate of 1-methyl-2-tetralone (**8**) to (**9**) gives poor to moderate stereoselectivities. The optimum result, 76% ee, is obtained using the sodium enolate and oxaziridine (+)-(**1**) (eq 11).[15]

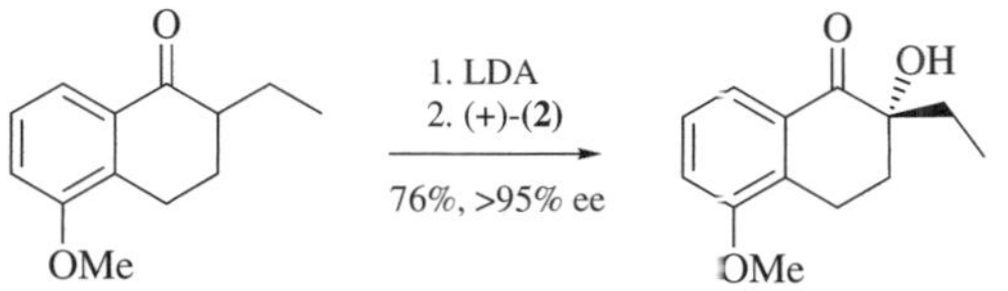

(9)

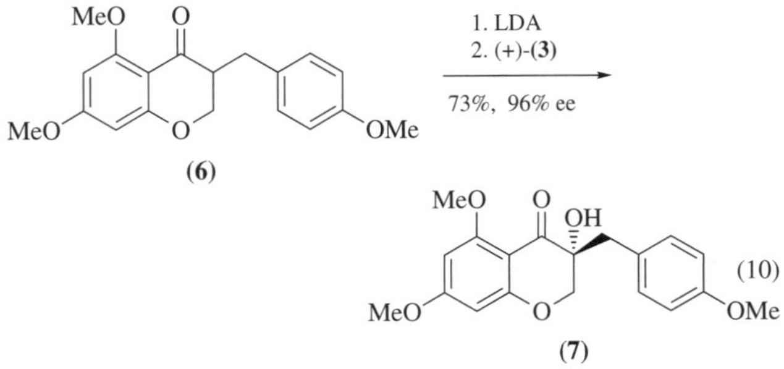

(6) (7) (10)

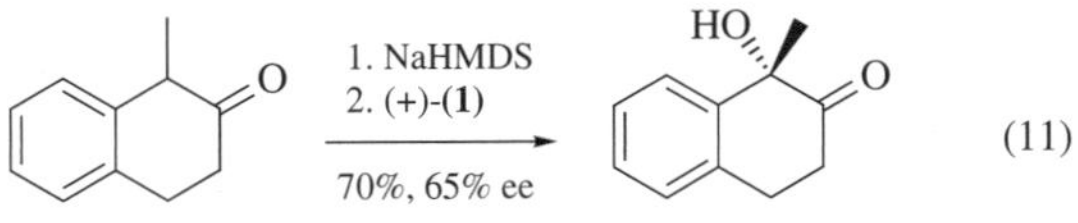

(8) (9) (11)

It should be pointed out that enolates are oxidized by the (camphorylsulfonyl)oxaziridine at a much faster rate than sulfides. An example is the preparation of α-hydroxy ketone sulfide (**9**), an intermediate for the total synthesis of (±)-breynolide (eq 12).[18]

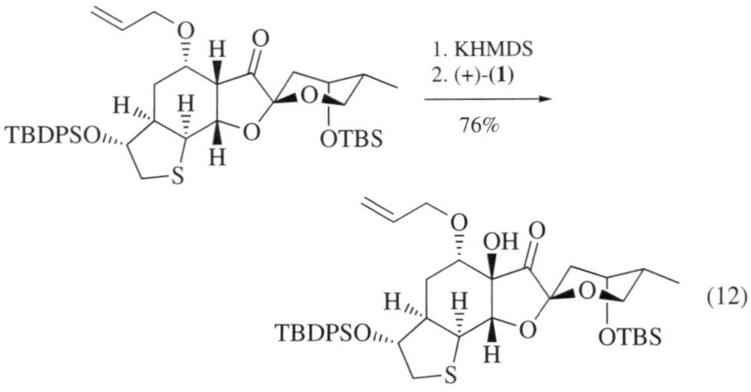

(12)

(9)

The asymmetric hydroxylation of ester enolates with *N*-sulfonyloxaziridines has been less fully studied.[1b] Stereoselectivities are generally modest and less is known about the factors influencing the molecular recognition. For example, (*R*)-methyl 2-hydroxy-3-phenylpropionate (**10**) is prepared in 85.5% ee by oxidizing the lithium enolate of methyl 3-phenylpropionate with (+)-(**1**) in the presence of HMPA (eq 13).[19] Like esters, the hydroxylation of prochiral amide enolates with *N*-sulfonyloxaziridines affords the corresponding enantiomerically enriched α-hydroxy amides. Thus treatment of amide (**11**) with LDA followed by addition of (+)-(**1**) produces α-hydroxy amide (**12**) in 60% ee (eq 14).[19] Improved stereoselectivities were achieved using double stereodifferentiation, e.g., the asymmetric oxidation of a chiral enolate. For example, oxidation of the lithium enolate of (**13**) with (−)-(**1**) (the matched pair) affords the α-hydroxy amide in 88–91% de (eq 15).[20] (+)-(Camphorsulfonyl)oxaziridine (**1**) mediated hydroxylation of the enolate dianion of (*R*)-(**14**) at −100 to −78 °C in the presence of 1.6 equiv of LiCl gave an 86:14 mixture of *syn*/*anti*-(**15**) (eq 16).[21] The *syn* product is an intermediate for the C-13 side chain of taxol.

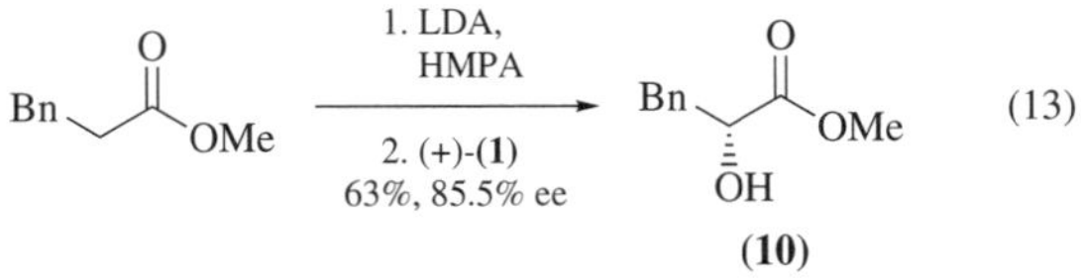

(13)

(10)

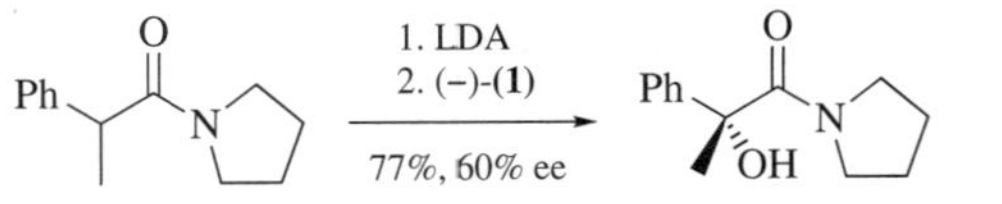

(11) (12) (14)

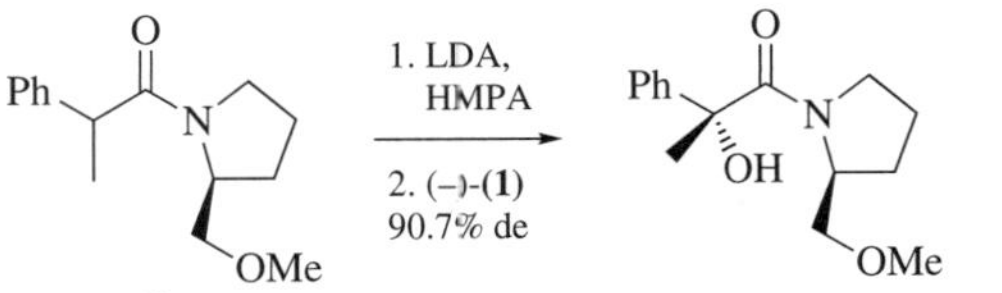

(13) (15)

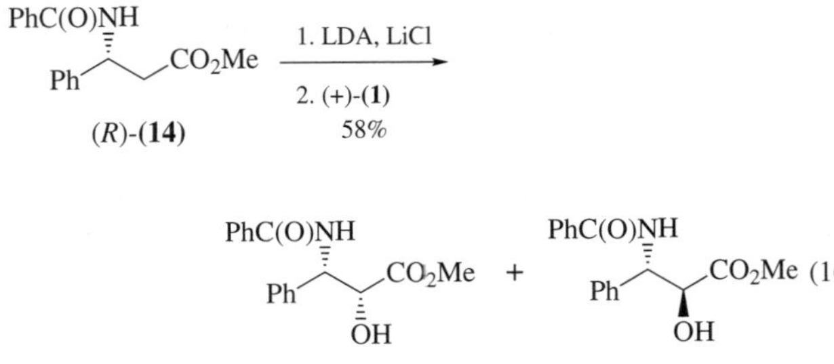

(*R*)-(**14**) *syn*-(**15**) *anti*-(**15**) (16)

Hydroxylation of the sodium enolate of lactone (**16**) with (+)-(**1**) gives α-hydroxy lactone in 77% ee (eq 17).[15] Kinetic resolution and asymmetric hydroxylation with (camphorsulfonyl)oxaziridines has been applied to the synthesis of enantiomerically enriched α-hydroxy carbonyl compounds having multiple stereocenters, which may otherwise be difficult to prepare.[22] Thus hydroxylation of the enolate of racemic 3-methylvalerolactone with substoichiometric amounts of (−)-(**1**) affords (2*S*,3*R*)-verrucarinolactone in 60% ee (eq 18) which on recrystallization is obtained enantiomerically pure.[22]

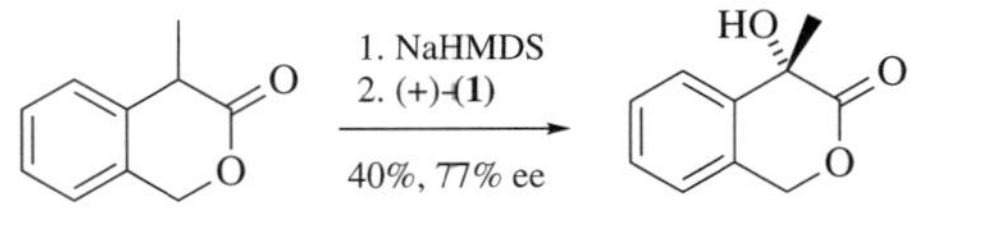

(16) (17)

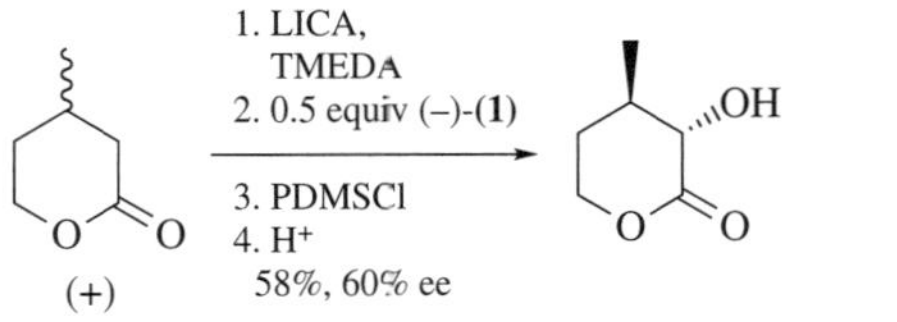

(+) (18)

Oxidation of the dienolate of (**17**) with (+)-(**1**) affords α-hydroxy ester (**18**), a key intermediate in the enantioselective synthesis of the antibiotic echinosporin (eq 19);[23] whereas oxidation of enolates derived from 1,3-dioxin vinylogous ester (**19**) gives rise to both α′- and γ-hydroxylation depending on the reaction conditions (eq 20).[24] With (+)-(**1**) the lithium enolate of (**19**) gives primarily the α′-hydroxylation product (**20**), while the sodium enolate gives γ-hydroxylation product (**21**). Only low levels of asymmetric induction (ca. 16% ee) are found in these oxidations. Birch reduction products are also asymmetrically hydroxylated in situ by (+)-(**1**) (eq 21).[25]

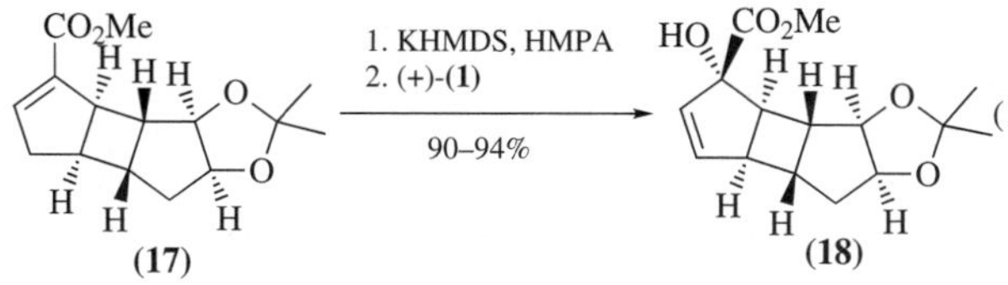
(17) (18) (19)

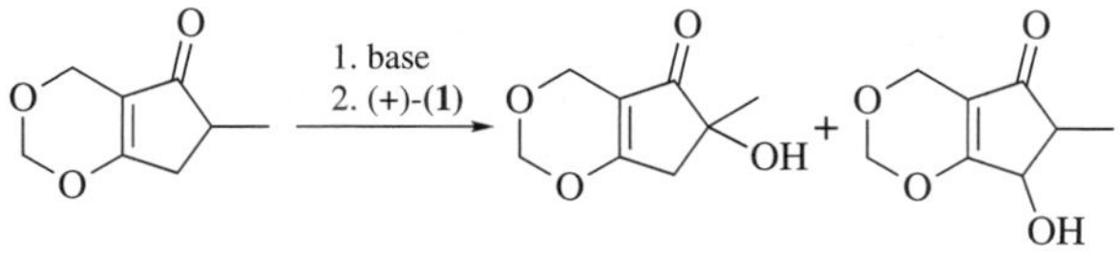
(19) (20) (21) (20)

	(20)	(21)
LDA	68%	4%
NaHMDS	3%	28%

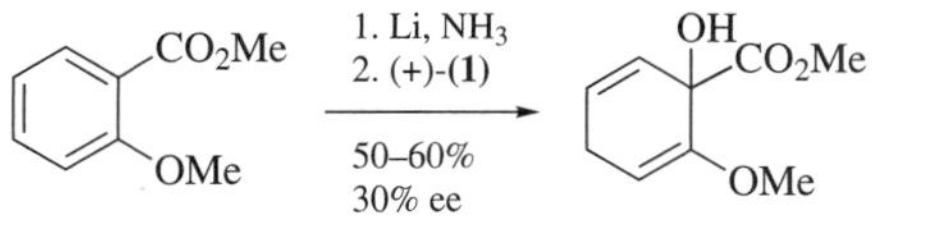
(21)

Few reagents are available for the hydroxylation of stabilized enolates such as β-keto esters, e.g., Vedejs' MoOPH reagent *(Oxodiperoxymolybdenum(pyridine)(hexamethylphosphoric triamide))* fails.[26] On the other hand, oxaziridines hydroxylate such enolates in good yield with good to excellent stereoselectivities.[1b] For example, enantioselective hydroxylation of the potassium enolate of the β-keto ester (**22**) with methoxyoxaziridine (−)-(**3**) affords (*R*)-(+)-2-acetyl-5,8-dimethoxy-1,2,3,4-tetrahydro-2-naphthol (**23**), a key intermediate in the asymmetric synthesis of the anthracycline antitumor agents demethoxyadriamycin and 4-demethoxydaunomycin (eq 22).[27] Hydroxylation of the sodium enolate of enone ester (**24**) furnishes kjellmanianone (**25**), an antibacterial agent isolated from marine algae (eq 23).[5] With (+)-(**1**) the ee's are modest (ca 40%), but improved to 69% ee with benzyloxaziridine (**4**).

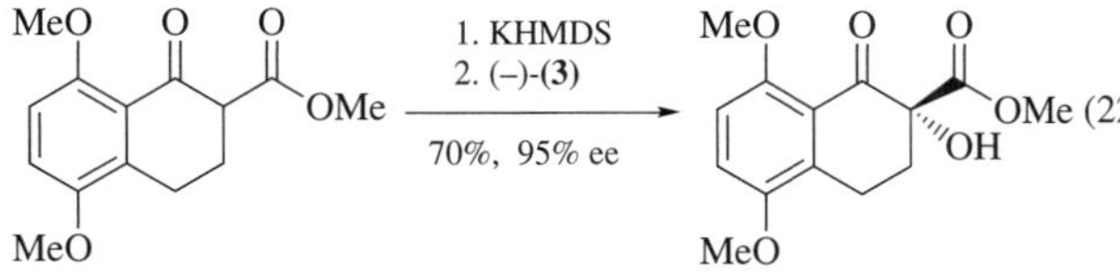
(22) (23) (22)

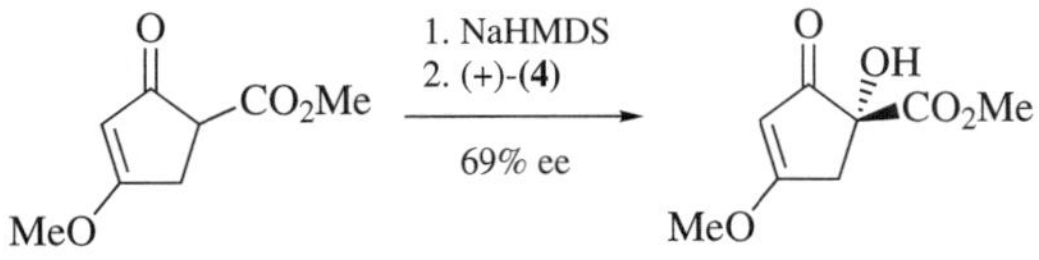
(24) (25) (23)

1. (a) Davis, F. A.; Sheppard, A. C. *Tetrahedron* **1989**, *45*, 5703. (b) Davis, F. A.; Chen, B.-C. *Chem. Rev.* **1992**, *92*, 919.
2. (a) Towson, J. C.; Weismiller, M. C.; Lal, G. S.; Sheppard, A. C.; Davis, F. A. *Org. Synth.* **1990**, *69*, 158. (b) Towson, J. C.; Weismiller M. C.; Lal, G. S.; Sheppard, A. C.; Kumar, A.; Davis, F. A. *Org. Synth.* **1993**, *72*, 104. (c) Davis, F. A.; Weismiller, M. C.; Murphy, C. K.; Thimma Reddy, R.; Chen, B.-C. *J. Org. Chem.* **1992**, *57*, 7274.
3. Mergelsberg, I.; Gala, D.; Scherer, D.; DiBenedetto, D.; Tanner *Tetrahedron Lett.* **1992**, *33*, 161.
4. Davis, F. A.; Kumar, A.; Chen, B.-C. *J. Org. Chem.* **1991**, *56*, 1143.
5. Chen, B.-C.; Weismiller, M. C.; Davis, F. A.; Boschelli, D.; Empfield, J. R.; Smith, III, A. B. *Tetrahedron* **1991**, *47*, 173.
6. (a) Glahsl, G.; Herrmann, R. *J. Chem. Soc. Perkin Trans. 1* **1988**, 1753. (b) Meladinis, V.; Herrmann, R.; Steigelmann, O.; Muller, G. *Z. Naturforsch. Teil B* **1989**, *44b*, 1453.
7. Davis, F. A.; Towson, J. C.; Weismiller, M. C.; Lal, G. S.; Carroll, P. J. *J. Am. Chem. Soc.* **1988**, *110*, 8477.
8. (a) Davis, F. A.; Thimma Reddy, R.; Weismiller, M. C. *J. Am. Chem. Soc.* **1989**, *111*, 5964. (b) Davis, F. A.; Thimma Reddy, R.; Han, W.; Carroll, P. J. *J. Am. Chem. Soc.* **1992**, *113*, 1428.
9. Davis, F.; Sheppard, A. C. *Tetrahedron Lett.* **1988**, *29*, 4365.
10. McClure, C. K.; Grote, C. W. *Tetrahedron Lett.* **1991**, *32*, 5313.
11. Davis, F. A.; Wei, J.; Sheppard, A. C.; Gubernick, S. *Tetrahedron Lett.* **1987**, *28*, 5115.
12. Davis, F. A.; Lal, G. S.; Wei, J. *Tetrahedron Lett.* **1988**, *29*, 4269.
13. Davis, F. A.; Chen, B.-C. *J. Org. Chem.* **1990**, *55*, 360.
14. Davis, F. A.; Sheppard, A. C.; Chen, B.-C.; Haque, M. S. *J. Am. Chem. Soc.* **1990**, *112*, 6679.
15. Davis, F. A.; Weismiller, M. C. *J. Org. Chem.* **1990**, *55*, 3715.
16. Davis, F. A.; Kumar, A. *Tetrahedron Lett.* **1991**, *32*, 7671.
17. Davis, F. A.; Chen, B.-C. *Tetrahedron Lett.* **1990**, *31*, 6823.
18. Smith, III, A. B.; Empfield, J. R.; Rivero, R. A.; Vaccaro, H. A. *J. Am. Chem. Soc.* **1991**, *113*, 4037.
19. Davis, F. A; Haque, M. S.; Ulatowski, T. G.; Towson, J. C. *J. Org. Chem.* **1986**, *51*, 2402.
20. Davis, F. A.; Ulatowski, T. G.; Haque, M. S. *J. Org. Chem.* **1987**, *52*, 5288.
21. Davis, F. A.; Thimma Reddy, R.; Reddy, R. E. *J. Org. Chem.* **1992**, *57*, 6387.
22. Davis, F. A.; Kumar, A. *J. Org. Chem.* **1992**, *57*, 3337.
23. Smith, III, A. B.; Sulikowski, G. A.; Fujimoto, K. *J. Am. Chem. Soc.* **1989**, *111*, 8039.
24. Smith, III, A. B.; Dorsey, B. D.; Ohba, M.; Lupo, Jr, A. T.; Malamas, M. S. *J. Org. Chem.* **1988**, *53*, 4314.
25. Schultz, A. G.; Harrington, R. E.; Holoboski, M. A. *J. Org. Chem.* **1992**, *57*, 2973.
26. (a) Vedejs, E.; Larsen, S. *Org. Synth.* **1985**, *64*, 127. (b) Vedejs, E.; Engler, D. A.; Telschow, J. E. *J. Org. Chem.* **1978**, *43*, 188. (c) Vedejs, E. *J. Am. Chem. Soc.* **1974**, *96*, 5944.
27. Davis, F. A.; Kumar, A.; Chen, B.-C. *Tetrahedron Lett.* **1991**, *32*, 867. Davis, F. A.; Clark, C.; Kumar, A.; Chen, B.-C. *J. Org. Chem.* **1994**, *59*, 1184.

Bang-Chi Chen
Discovery Chemistry, Bristol-Myers Squibb Pharmaceutical Research Institute, Princeton, NJ, USA
Franklin A. Davis
Temple University, Philadelphia, USA

(*R,S*)-CAMPHOS

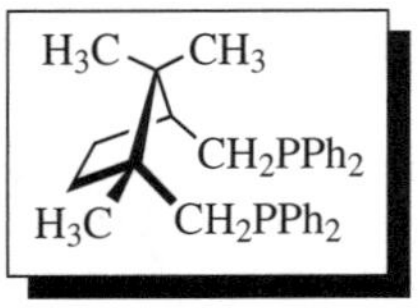

[60989-76-6] $C_{34}H_{38}P_2$ (MW 508.62)

(optically active reagent used as a ligand in homogeneous transition metal-chiral phosphine catalyst systems for asymmetric homogeneous reactions)

Alternate Name: (+)-1,2,2-trimethyl(*1R,3S*)-1,3-bis[(diphenylphosphino)methyl]cyclopentane.

Physical Data: viscous oil; $[\alpha]_D^{20}$ +79.15°(*c* 2.792, benzene);[1] $[\alpha]_D^{22}$ +99.0°(*c* 2.02, CH_2Cl_2).[2c]

Solubility: soluble in benzene, methylene chloride, and other common organic solvents.

Form Supplied in: CAMPHOS is not commercially available.

Analysis of Reagent Purity: [2c]IR (neat) no bands at 1175 or 1100 cm^{-1} (P=O or P=S); ^1H NMR ($CDCl_3$) δ 0.80 (d, 9 H), 1.0–2.04 (m, 7 H), 2.16 (d, $J = 4$ Hz, 2 H), 7.00–7.34 (m, 20 H).

Preparative Methods: CAMPHOS can be prepared from (+)-camphoric acid in three synthetic steps (eq 1).

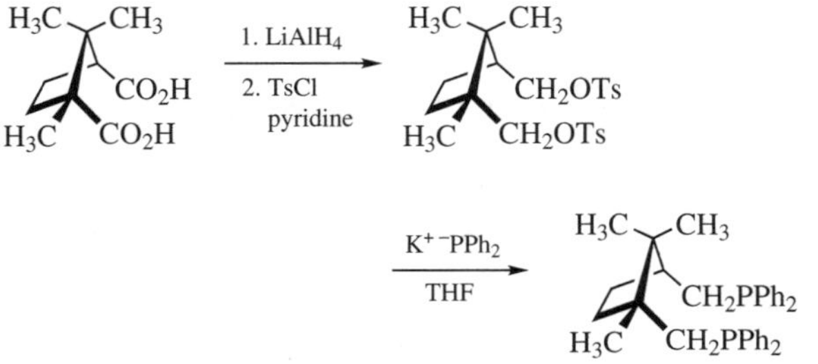

(1)

Purification: (*R,S*)-CAMPHOS cannot be purified by distillation or crystallization; purification is best carried out via column chromatography on silica gel or alumina by eluting with benzene.[2]

Handling, Storage, and Precautions: no specific information is available for CAMPHOS. In general, alkyldiarylphosphines are air-sensitive materials; all operations, including handling and storage, should be carried out under an inert atmosphere. In addition, compounds of this type are irritants; skin contact should be avoided, and vapors should not be inhaled.

Enantioselective Hydrogenation. (*R,S*)-CAMPHOS has been employed in combination with rhodium(I) to reduce alkene carbon-carbon double bonds. Thus, the Rh(I) complex formed from (*R,S*)-CAMPHOS and $[Rh(cyclooctene)_2Cl]_2$ in toluene–EtOH–Et_3N solution catalyzes the hydrogenation (1 atm H_2, 20 °C) of atropic acid and of α-acetamidocinnamic acid. The products, (*S*)-hydratropic acid and *N*-acetyl-(*S*)-phenylalanine, respectively, are formed in low optical yield (i.e., 4% ee and 7% ee, respectively).[1] When hydrogenation of atropic acid was performed with the Rh(I)-(*R,S*)-CAMPHOS catalyst under more vigorous conditions [i.e., 300 psi (gauge) H_2, 60 °C, 24 h], (*S*)-hydroatropic acid was formed in 69% synthetic yield, 6.05% ee.[2a] However, significantly higher optical yields of (*S*)-hydratropic acid were obtained when hydrogenation of atropic acid was performed with other Rh(I)-phosphine ligands [i.e., Rh(I)-L, L = **(*R,R*)-(−)-DIOP(1)**:[3] synthetic yield 70%, 43.9% ee; **(*S*)-(+)-NMDPP(2)**:[4] synthetic yield 49.2%, 29.6% ee].[2a] The optical yield of *N*-acetyl-(*S*)-phenylalanine formed via hydrogenation of α-acetamidocinnamic acid with Rh(I)-(*R,S*)-CAMPHOS catalyst was increased to 17.0% ee when the reaction was performed at 1000 psi (gauge) H_2 at 100 °C for 48 h.[5] Under these same conditions, hydrogenation of methyl α-acetamido-cinnamate afforded *N*-acetyl-(*S*)-phenylalanine methyl ester (22.4% ee).[5]

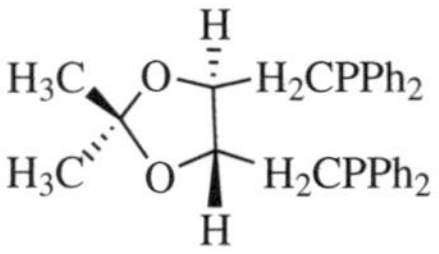

1

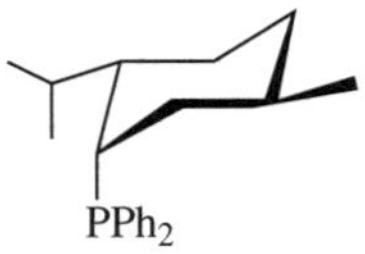

2

Enantioselective Hydroformylation. Hydroformylation of α-methylstyrene, when performed with equimolar quantities of CO and H_2 (1 atm) in benzene at 80 °C in the presence of Rh(I)-(*R,S*)-CAMPHOS catalyst, afforded a 30% chemical yield of optically active 3-phenylbutanal with low optical rotation (eq 2).[1]

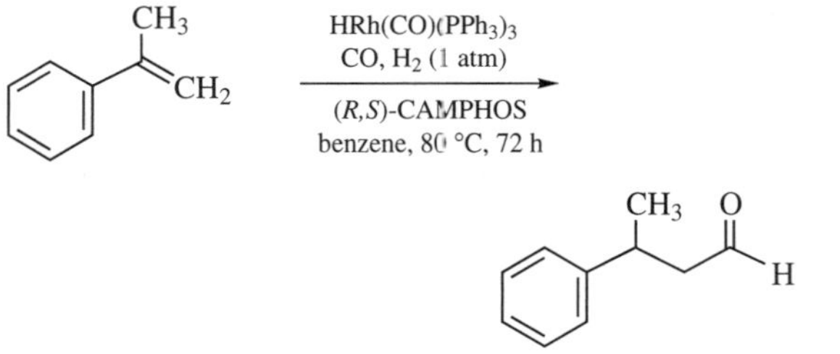

(2)

30% chemical yield

$[\alpha]_D^{20}$ −0.22° (*c* 28, Et_2O)

Enantioselective Allylic Alkylation. Reaction of 1-(α-acetoxyethyl)cyclopentene with the sodium salt of dimethyl malonate in refluxing 1,2-dimethoxyethane, when performed in the presence of tetrakis(triphenylphosphine)palladium and (*R,S*)-CAMPHOS catalyst, afforded the corresponding α-alkylated malonic ester in 99% chemical yield (37% ee)(eq 3).[6] Comparable results were obtained when this reaction was performed in THF solution by using $Pd(PPh_3)_4$-**(*S,S*)-(+)-DIOP** as catalyst.[6]

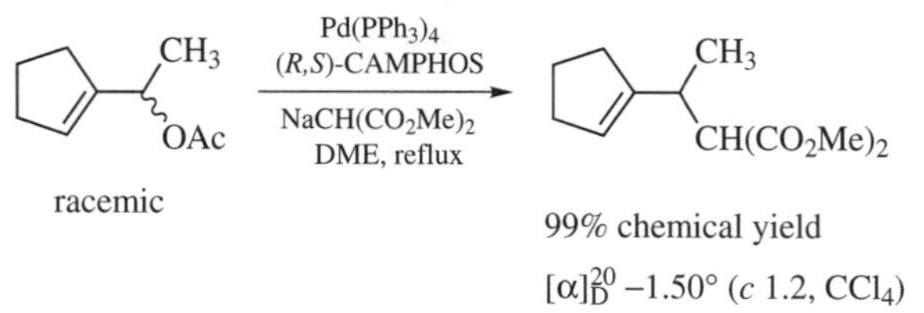

(3)

Enantioselective Intramolecular Wittig Reaction. Asymmetric induction has been reported to accompany intramolecular Wittig cyclization of a phosphonium ylide performed in the presence of (*R,S*)-CAMPHOS catalyst. However, this reaction proceeds to afford the corresponding cyclization product with only 10% ee (eq 4).[7]

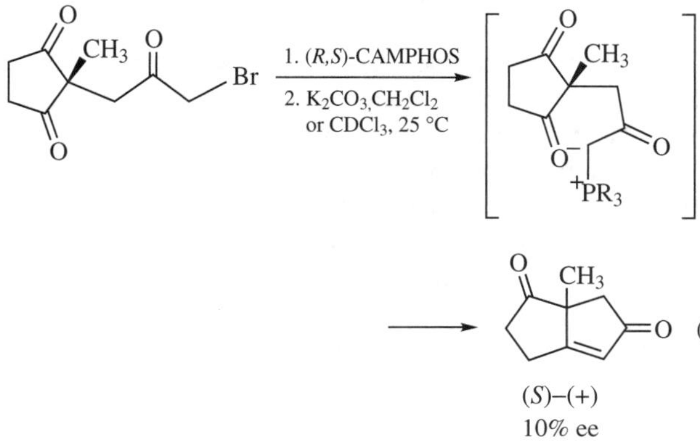

(4)

Related Reagents. (*S,S*)-(+)-DIOP;[4] (*R,R*)-(−)-DIOP;[4] (*S*)-(+)-NMDPP;[5] $Pd(PPh_3)_4$; $[Rh(cyclooctene)_2Cl]_2$; $HRh(CO)(PPh_3)_3$.

1. Beck, W.; Menzel, H. *J. Organometal. Chem.* **1977**, *133*, 307–310.
2. (a) Morrison, J. D.; Masler, W. F.; Neuberg, M. K. *Adv. Catal.* **1976**, *25*, 81–124. (b) Morrison, J. D.; Masler, W. F.; Hathaway, S. *Catalysis in Organic Synthesis–1976*; Academic Press: New York, 1976, pp 203–233. (c) Masler, W. F., III, Ph.D. Dissertation, University of New Hampshire, 1974; *Diss. Abstr. Int. B* **1974**, *35*, 2660.
3. Kagan, H. B.; Dang, T.-P. *J. Am. Chem. Soc.* **1972**, *94*, 6429–6433.
4. (a) Morrison, J. D.; Burnett, R. E.; Aguiar, A. M.; Morrow, C. J.; Phillips, C. *J. Am. Chem. Soc.* **1971**, *93*, 1301–1303. (b) Morrison, J. D.; Masler, W. F. *J. Org. Chem.* **1974**, *39*, 270–272.
5. Johnson, T. H.; Pretzer, D. K.; Thomen, S.; Chaffin, V. J. K.; Rangarajan, G. *J. Org. Chem.* **1979**, *44*, 1878–1879.
6. Trost, B. M.; Strege, P. E. *J. Am. Chem. Soc.* **1977**, *99*, 1649–1651.
7. Trost, B. M.; Curran, D. P. *Tetrahedron Lett.* **1981**, *22*, 4929–4932.

Alan P. Marchand
University of North Texas, Denton, TX, USA
Jaroslaw Romanski
University of Lodz, Poland
T. Pavan Kumar
University of North Texas, Denton, TX, USA

Chloro(cyclopentadienyl)bis[3-*O*-(1,2:5,6-di-*O*-isopropylidene-α-D-glucofuranosyl)]titanium

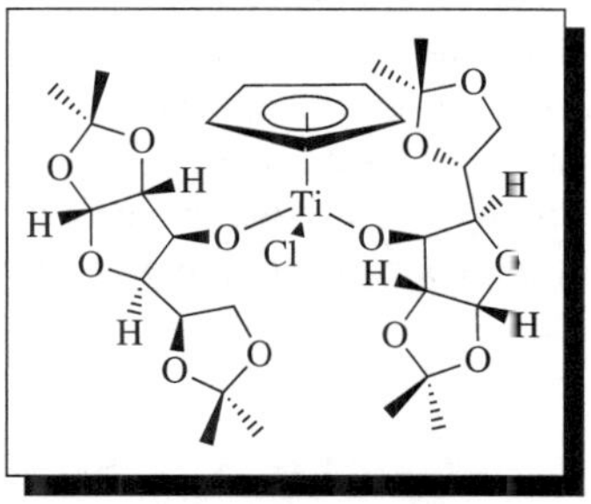

[119528-80-2] $C_{29}H_{43}ClO_{12}Ti$ (MW 667.05)

(highly enantio- and diastereoselective aldol reactions of acetic acid,[1] propionic acid,[2] and glycine[3] ester enolates with various aldehydes; stereoselective addition of allyl groups to aldehydes[4])

Physical Data: crystal structure; ^1H and ^{13}C NMR.[5]
Solubility: toluene (not determined, 0.155 M possible); Et_2O (not determined, 0.09 M possible).[4]
Analysis of Reagent Purity: ^1H NMR; test reaction.
Preparative Methods: see *Trichloro(cyclopentadienyl)titanium.*
Handling, Storage, and Precautions: best handled as stock solution either in Et_2O (ca. 0.1 M) or toluene (ca. 1.5 M), which must be protected from moisture and UV light. If handled under an inert atmosphere (argon), such solutions can be stored in a refrigerator (8 °C) for several months (possibly much longer) without deterioration. Reactions should be carried out in dry equipment and with absolute solvents under Ar or N_2.

Aldol Reactions. The titanium enolate (**2**) is obtained by addition of ca. 1.3 equiv of the title reagent (**1**) as a 0.1–0.15 M solution in toluene to the Li enolate of *t*-butyl acetate (**3**) generated at −78 °C with lithium dicyclohexylamide in Et_2O. This transmetalation takes about 24 h at −78 °C but is completed within 1 h at −30 °C (eq 1).[1a,b] The medium might also be important, as it has recently been reported that *12-Crown-4* has to be added for reproducible results in THF–Et_2O.[1c] The solution of (**2**) usually is recooled to −78 °C for the reaction with aldehydes, affording β-hydroxy esters (**4**) of high optical purity (90–96% ee) upon hydrolytic workup. Byproducts are insoluble cyclopentadienyltitanium oxides (**5**) and the ligand diacetone-glucose (DAGOH, **6**). The oxides (**5**) can be separated by filtration and may be recycled to $CpTiCl_3$. Ligand (**6**) and product are either separated by conventional methods (crystallization, distillation, chromatography), or glucose is extracted into the aqueous phase after acetonide cleavage in 0.1 N HCl (1.5 h at rt).[1a,b] In the case of isovaleraldehyde (R = *i*-Bu) it could be shown that the enantioselectivity (92–96% ee) is retained up to rt (27 °C).[1a,b]

A clear drawback of this reagent is the availability of only one enantiomer, the one favoring the *re* attack to the aldehyde carbonyl, as only D-glucose is readily available. *si* Attack is observed with the analogous enolate prepared from *Chloro(η^5-cyclopentadienyl)[(4R,trans)-2,2-dimethyl-α,α,α′,α′-tetraphenyl-1,3-dioxolane-4,5-dimethanolato(2−)-O^{α},$O^{\alpha\alpha}$]ti-*

tanium, but only with moderate enantioselectivity (78% ee).[1c,6] The reagent (**2**) is probably the most versatile chirally modified acetate enolate. Good results have also been obtained with the Mg enolate of 2-acetoxy-1,1,2-triphenylethanol[7] and with boron enolates derived from 2,4-dialkylborolanes.[8] Chiral Fe–acetyl complexes, which can be considered as acetate equivalents, give impressive stereocontrol upon enolization and aldol reaction.[9] Except for unsaturated residues R, β-hydroxy esters (**4**) of excellent optical purity can also be obtained by enantioselective hydrogenation of the corresponding β-keto esters catalyzed by $RuCl_2$(BINAP).[10]

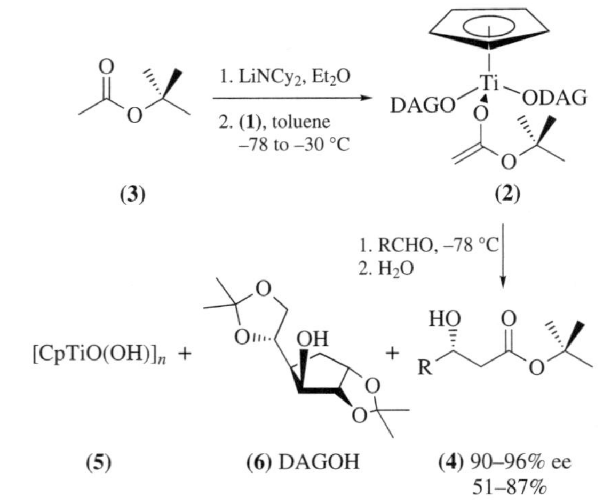

(1)

For the propionate aldol reaction the Li enolate (**7**), generated by deprotonation of 2,6-dimethylphenyl propionate with *Lithium Diisopropylamide* in Et_2O,[11] was chosen.[2] Transmetalation with 1.25 equiv of an ethereal solution of (**1**) takes 24 h at −78 °C. The completion of this step is evident by the disappearance of racemic *anti*-aldol (**9**) in favor of optically active *syn*-isomer (**10**) (91–98% ee) upon reaction with an aldehyde (RCHO) and aqueous workup. At this point, 3–11% of *anti*-aldol (**9**) remaining in the reaction mixture is optically active as well (eq 2). This *anti*-isomer (**9**) (94–98% ee) becomes the major product if the reaction mixture, containing the putative (*E*)-titanium enolate derived from (**7**), is warmed for 4–5 h to −30 °C before reaction with an aldehyde (RCHO) again at −78 °C. Isomerization to the (*Z*)-titanium enolate is a possible explanation of this behavior. Some substrates, aromatic and unsaturated aldehydes, behave exceptionally, as a high proportion of *syn*-isomer (**10**) (19–77%) of lower optical purity (47–66% ee) is formed in addition to (**9**) (94–98% ee). After hydrolysis of the acetonide (**6**) the products (**9/10**) are isolated and separated by chromatography in 50–87% yield. The reactions of pivalaldehyde (R = *t*-Bu) are sluggish at −78 °C and have therefore been carried out at −50 to −30 °C.

As above (eq 1), a major drawback of this reagent is the lack of a readily available enantiomer. There are many alternative methods for the enantioselective propionate aldol reaction. The most versatile chirally modified propionate enolates or equivalents are *N*-propionyl-2-oxazolidinones,[12] α-siloxy ketones,[13] boron enolates with chiral ligands,[14] as well as tin enolates.[15] Especially rewarding are new chiral Lewis acids for the asymmetric Mukaiyama reaction of *O*-silyl ketene acetals.[16] Most of these reactions afford *syn*-aldols; good methods for the *anti*-isomers have only become available recently.[8,17]

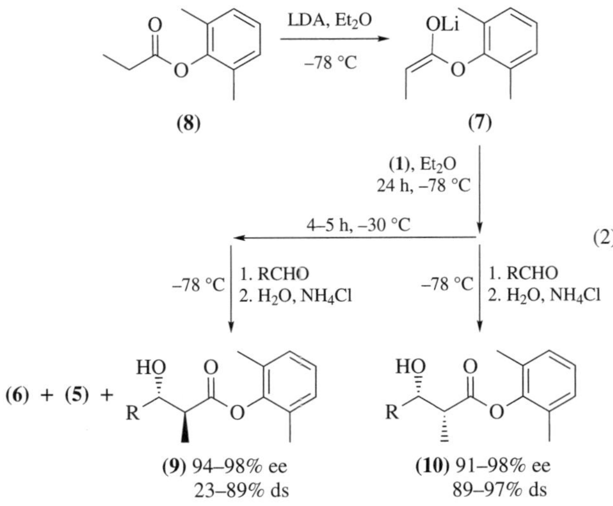

(2)

Transmetalation of the (*E*)-*O*-Li-enolate derived from the 'stabase'-protected glycine ethyl ester (**11**) with 1.1 equiv of (**1**) affords the chiral Ti enolate (**12**), which adds with high *re* selectivity to various aldehydes.[3,18] By mild acidic cleavage of the silyl protecting group, the primary product (**13**) can be transformed to various *N*-derivatives (**14**) of D-*threo*-α-amino-β-hydroxy acids in 45–66% yield and with excellent enantio- and *syn*(*threo*) selectivity (97–99%) (eq 3). An exception with lower enantioselectivity is glyoxylic ester (ethyl ester 78% ee; *t*-butyl ester 87% ee).

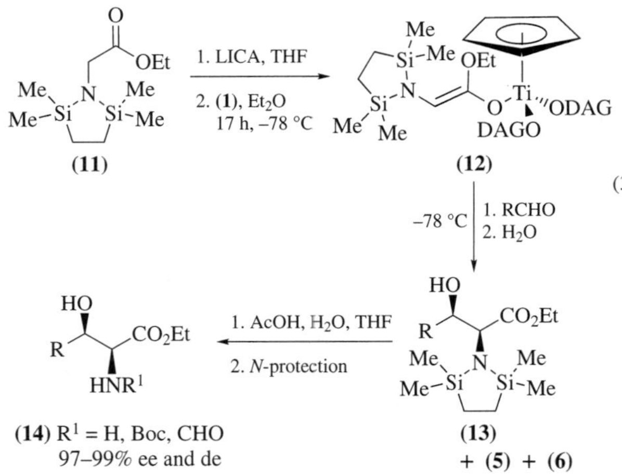

(3)

In this case the enantiomers are available by the analogous conversion of glycine *t*-butyl ester using *Chloro(η*[5]*-cyclopentadienyl)[(4R,trans)-2, 2-dimethyl-α, α, α′, α′-tetraphenyl-1,3-dioxolane-4,5-dimethanolato(2−)-O*[α]*,O*[α′]*]titanium*. An elegant alternative is the enantioselective addition of isocyanoacetate to aldehydes under the catalysis of a chiral Au[I] complex.[19] Further methods, also for the *anti*(*erythro*) epimers, can be found in recent reviews of enantioselective α-amino acid synthesis.[20]

Allyltitanation of Aldehydes. The allyltitanium complex (**15**) is obtained by reaction of chloride (**1**) (1.1 equiv) with allylmagnesium chloride in Et_2O for 1 h at 0 °C.[4] The compound (**15**) has been characterized by ^{13}C NMR.[5] Reaction with various aldehydes (RCHO) at −78 °C and hydrolysis affords the homoallyl alcohols (**16**) (55–88%) of high optical purity (85–94% ee) (eq 4).[4] The isolation of the product is analogous to the aldol reactions (cf. eq 1).

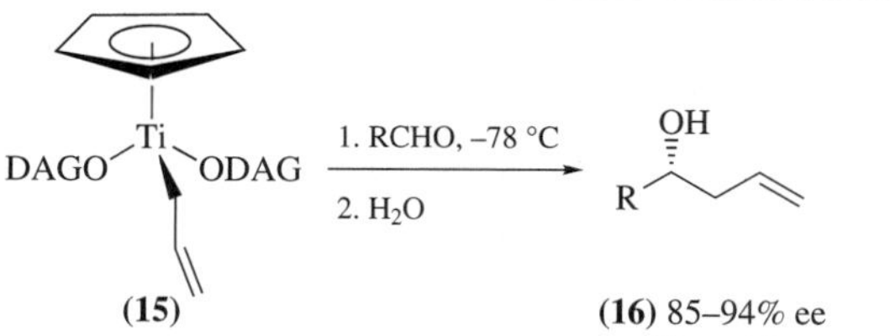

(4)

(15) (16) 85–94% ee + (5) + (6)

The enantiomers of (**16**) are obtained analogously by using *Chloro(η^5-cyclopentadienyl)[(4R,trans)-2, 2-dimethyl-α, α, α′, α′-tetraphenyl-1,3-dioxolane-4,5-dimethanolato(2−)-O^{α}, $O^{\alpha'}$]titanium.*[21] The stereoselectivity of this cyclic Ti complex in allyltitanations is better than the diacetone–glucose system (**1**). It is therefore advisable to use the (4*S,trans*) enantiomer instead of (**1**) for controlling the *re* addition to problematic substrates. For further examples of this method and for analogous reagents see the discussion provided in *Chloro(η^5-cyclopentadienyl)[(4R,trans)-2,2-dimethyl-α, α, α′, α′-tetraphenyl-1,3-dioxolane-4,5-dimethanolato(2−)-O^{α}, $O^{\alpha'}$]titanium.*

1. (a) Duthaler, R. O.; Herold, P.; Lottenbach, W.; Oertle, K.; Riediker, M. *Angew. Chem., Int. Ed. Engl.*, **1989**, *28*, 495. (b) Oertle, K.; Beyeler, H.; Duthaler, R. O.; Lottenbach, W.; Riediker, M.; Steiner, E. *Helv. Chim. Acta* **1990**, *73*, 353. (c) Cambie, R. C.; Coddington, J. M.; Milbank, J. B. J.; Paulser, M. G.; Rustenhoven, J. J.; Rutledge, P. S.; Shaw, G. L.; Sinkovich, P. I. *Aust. J. Chem.* **1993**, *46*, 583.
2. Duthaler, R. O.; Herold, P.; Wyler-Helfer, S.; Riediker, M. *Helv. Chim. Acta* **1990**, *73*, 659.
3. Bold, G.; Duthaler, R. O.; Riediker, M. *Angew. Chem., Int. Ed. Engl.* **1989**, *28*, 497.
4. Riediker, M.; Duthaler, R. O. *Angew. Chem., Int. Ed. Engl.* **1989**, *28*, 494.
5. Riediker, M.; Hafner, A.; Piantini, U.; Rihs, G.; Togni, A. *Angew. Chem., Int. Ed. Engl.* **1989**, *28*, 499.
6. Duthaler, R. O.; Hafner, A.; Riediker, M. *Pure Appl. Chem.* **1990**, *62*, 631.
7. Braun, M. *Angew. Chem., Int. Ed. Engl.* **1987**, *26*, 24.
8. (a) Masamune, S.; Sato, T.; Kim, B. M.; Wollmann, T. A. *J. Am. Chem. Soc.* **1986**, *108*, 8279. (b) Reetz, M. T.; Rivadeneira, E.; Niemeyer, C. *Tetrahedron Lett.* **1990**, *31*, 3863.
9. (a) Liebeskind, L. S.; Welker, M. E. *Tetrahedron Lett.* **1984**, *25*, 4341. (b) Davies, S. G.; Dordor, I. M.; Warner, P. *Chem. Commun./J. Chem. Soc., Chem. Commun.* **1984**, 956. (c) Brunner, H. *Angew. Chem.* **1991**, *103*, A310.
10. Noyori, R.; Ohkuma, T.; Kitamura, M.; Takaya, H.; Sayo, N.; Kumobayashi, H.; Akutagawa, S. *J. Am. Chem. Soc.* **1987**, *109*, 5856.
11. Montgomery, S. H.; Pirrung, M. C.; Heathcock, C. H. *Org. Synth.* **1985**, *63*, 99.
12. Evans, D. A.; Nelson, J. V.; Taber, T. R. *Top. Stereochem.* **1982**, *13*, 1.
13. (a) Masamune, S.; Choy, W.; Petersen, J. S.; Sita, L. R. *Angew. Chem., Int. Ed. Engl.* **1985**, *24*, 1. (b) Heathcock, C. H. *Aldrichim. Acta* **1990**, *23*, 99.
14. (a) Corey, E. J.; Imwinkelried, R.; Pikul, S.; Xiang, Y. B. *J. Am. Chem. Soc.* **1989**, *111*, 5493. (b) Corey, E. J.; Kim, S. S. *J. Am. Chem. Soc.* **1990**, *112*, 4976.
15. Mukaiyama, T.; Kobayashi, S.; Sano, T. *Tetrahedron* **1990**, *46*, 4653.
16. (a) Kobayashi, S.; Uchiro, H.; Fujishita, Y.; Shiina, I.; Mukaiyama, T. *J. Am. Chem. Soc.* **1991**, *113*, 4247. (b) Kobayashi, S.; Uchiro, H.; Shiina, I.; Mukaiyama, T. *Tetrahedron* **1993**, *49*, 1761.
17. (a) Helmchen, G.; Leikauf, U.; Taufer-Knöpfel, I. *Angew. Chem., Int. Ed. Engl.* **1985**, *24*, 874. (b) Gennari, C.; Schimperna, G.; Venturini, I. *Tetrahedron* **1988**, *44*, 4221. (c) Oppolzer, W.; Starkemann, C.; Rodriguez, I.; Bernardinelli, G. *Tetrahedron Lett.* **1991**, *32*, 61. (d) Van Draanen, N. A.; Arseniyadis, S.; Crimmins, M. T.; Heathcock, C. H. *J. Org. Chem.* **1991**, *56*, 2499. (e) Myers, A. G.; Widdowson, K. L.; Kukkola, P. J. *J. Am. Chem. Soc.* **1992**, *114*, 2765. (f) Cardani, S.; De Toma, C.; Gennari, C.; Scolastico, C. *Tetrahedron* **1992**, *48*, 5557. (g) Oppolzer, W.; Lienard, P. *Tetrahedron Lett.* **1993**, *34*, 4321.
18. Bold, G.; Allmendinger, T.; Herold, P.; Moesch, L.; Schär, H.-P.; Duthaler, R. O. *Helv. Chim. Acta* **1992**, *75*, 865.
19. Ito, Y.; Sawamura, M.; Hayashi, T. *J. Am. Chem. Soc.* **1986**, *108*, 6405.
20. (a) Williams, R. M. *Synthesis of Optically Active α-Amino Acids*; Pergamon: Oxford, 1989. (b) Duthaler, R. O. *Tetrahedron* **1994**, *50*, 1539.
21. Hafner, A.; Duthaler, R. O.; Marti, R.; Rihs, G.; Rothe-Streit, P.; Schwarzenbach, F. *J. Am. Chem. Soc.* **1992**, *114*, 2321.

Andreas Hafner
Ciba-Geigy, Marly, Switzerland
Rudolf O. Duthaler
Ciba-Geigy, Basel, Switzerland

Chloro (η^5-cyclopentadienyl) [(4*R*, *trans*)-2,2-dimethyl-α,α,α′,α′-tetraphenyl-1,3-dioxolane-4,5-dimethanolato(2−)-O^{α},$O^{\alpha'}$]titanium

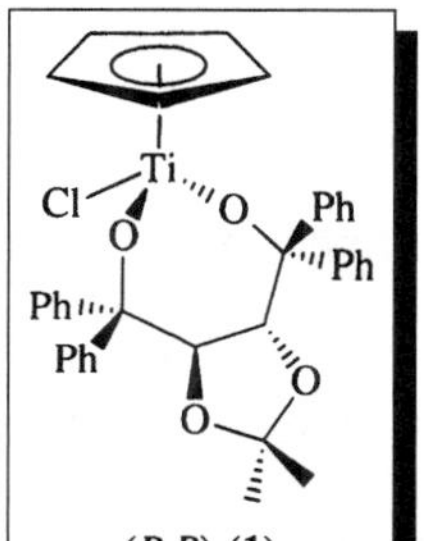

(*R*,*R*)-(**1**)

(4*R,trans*)
[132068-98-5] $C_{36}H_{33}O_4ClTi$ (MW 613.0)

(highly enantio- and diastereoselective addition of allyl and terminally monosubstituted allyl groups to achiral and chiral aldehydes;[1] can also be used for enantioselective aldol reactions,[2] especially of glycine ester enolates[2a])

Physical Data: mp 214 °C (Mettler DSC); $[\alpha]_D = -243.4°$ ($c = 1$, $CHCl_3$); X-ray, ^1H, ^{13}C NMR.[1a]
Solubility: sol toluene (ca. 370 mg mL^{-1}), THF (ca. 470 mg mL^{-1}), diethyl ether (ca. 150 mg mL^{-1}).
Form Supplied in: pale yellow powder or crystals.
Analysis of Reagent Purity: ^1H NMR ($CDCl_3$);[1] may contain 1–5% of (4*R*, *trans*)-2,2-dimethyl-α,α,α′,α′-tetraphenyl-1,3-dioxolane-4,5-dimethanol (but does not affect its efficiency).
Preparative Methods: see *Trichloro(cyclopentadienyl)titanium.*
Handling, Storage, and Precautions: the dry solid must be stored under exclusion of moisture and UV at rt (brown, tightly sealed bottle). It can, however, be handled quickly in the open, e.g. for weighing. Reactions should be carried out in dry equipment and with absolute solvents under argon or N_2.

Allyltitanation of Aldehydes. The two-stage, one-pot procedure involves first the generation of the allyltitanium reagents (*R*)-(**2a–f**) (eq 1) by transmetalation of allyl–Grignard or allyl–Li compounds with a slight excess (1.2 equiv) of (*R,R*)-(**1**). It is advisable to analyze the content of the allylmetal precursor solution. Optimal conditions (time and temperature) of these transmetalations are preferably determined by test reactions with a simple aldehyde, assessing for maximal diastereo- and enantioselectivity. For stable allylmetal compounds, 1–3 h at 0° C is usually sufficient. The allyltitanates (**2**) can be analyzed by ^{1}H and ^{13}C NMR.[1a] Fast allylic rearrangements are responsible for the formation of the thermodynamically most stable allyltitanium isomer with terminal substitution and *trans* double bond, irrespective of the nature of the organometallic precursor.

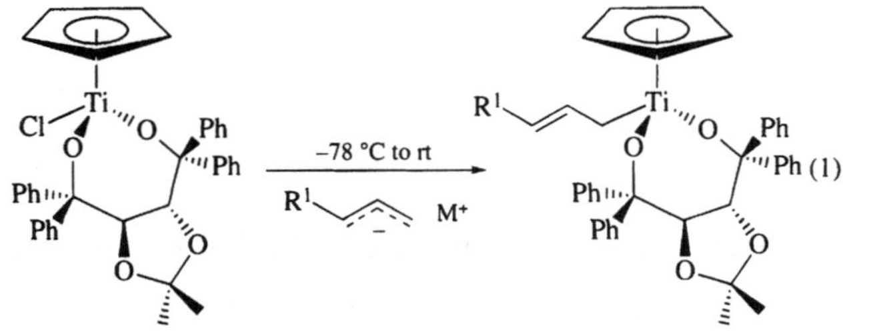

(*R,R*)-(**1**)

M =	
MgCl	(*R*)-(**2a**), R^1 = H
MgCl	(*R*)-(**2b**), R^1 = Me
Li	(*R*)-(**2c**), R^1 = CH=CH_2
Li	(*R*)-(**2d**), R^1 = Ph
Li	(*R*)-(**2e**), R^1 = $SiMe_3$
Li	(*R*)-(**2f**), R^1 = 4-$MeOC_6H_4O$

The second step, performed in situ, is the addition of ca. 0.75 equiv (based on **1**) of an aldehyde at −78 °C (eq 2). Most reactions are completed within 3 h and the resulting cyclopentadienyltitanium trialkoxides are hydrolyzed by stirring overnight with aqueous NH_4F (45%) at rt. The precipitated cyclopentadienyltitanium oxide (**3**) is removed by filtration and can be recycled to $CpTiCl_3$. The chiral ligand (**4**) and the product (**5**) are separated by crystallization or precipitation of (**4**) followed by distillation and/or chromatography.

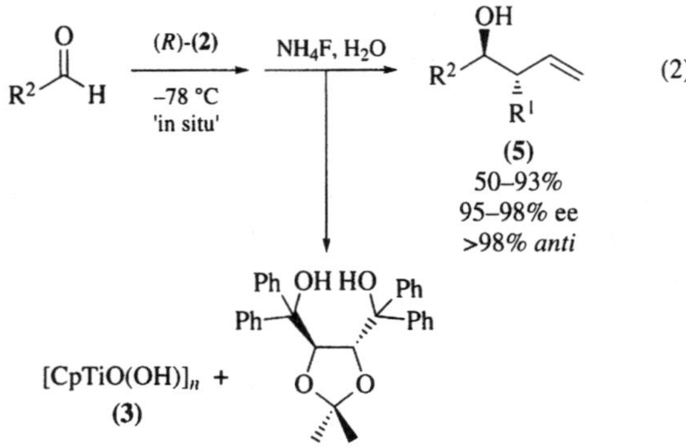

(**5**)
50–93%
95–98% ee
>98% *anti*

(**3**)

(**4**)

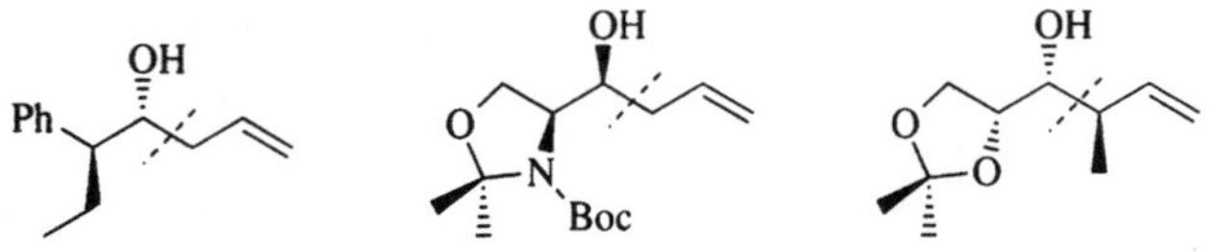

(**6**) (*S*)-(**2a**) 95% ds

(**7**) (*R*)-(**2a**) 98% ds

(**8**) (*S*)-(**2b**) 98% ds

The homoallylic alcohols (**5**) are isolated in fair to good yields. Their optical purity often exceeds 95% and the *anti* diastereomers are produced almost exclusively (>98% ds). Exceptions with up to 33% *syn* epimer have, however, been obtained for (**2f**).[1b] The reagent controlled stereoselectivity of (*R*)-(**2**) and the enantiomers (*S*)-(**2**) is best documented by conversions with chiral aldehydes, examples being (**6**),[1a] (**7**),[1a] and (**8**).[1b]

The allylation of (2*S*)-2-phenylbutyraldehyde to give (**6**) allows a comparison with other similar reagents. Whereas the transformation to (**6**) with (*S*)-(**2a**) (mismatched pair) proceeds with 95% ds, lower selectivity is observed for the allyltitanium reagent prepared analogously from *Chloro (cyclopentadienyl)-bis[3-O-(1,2:5,6-di-O-isopropylidene-α-D-glucofuranosyl)]titanium* (79% ds) or from *B-Allyldiisopinocampheylborane* (74% ds).[3a] Due to the fast *E/Z* isomerization of (**2**), the *syn* isomers of (**5**) cannot be obtained as the major product by using allyltitanium compounds, e.g. (*R*)-(**2b**), derived from chloride (**1**). *Syn*-crotyl adducts are efficiently obtained with chiral allylboron reagents,[3] e.g. diisopinocampheyl-(*Z*)-crotylborane[3a,b] or diisopropyl tartrate (*Z*)-crotylboronate.[3c] Inferior stereoselectivity is also exhibited by 2-substituted and 1,3-disubstituted allyltitanium reagents derived from (**1**).[1,2b] The use of similar chiral titanium reagents[4] and stereoselective additions with other allylmetal compounds[5] have recently been reviewed. While all these methods rely on stoichiometric amounts of a chiral reagent, catalytic amounts of a chiral acyloxyborane complex mediate the enantioselective addition of allylsilanes[6a] and allylstannanes[6b] to aldehydes. Very recently, the addition of allyltributyltin was also found to be catalyzed by a complex formed from 1,1′-bi-2-naphthol and $TiCl_4$ (see *(R)-1,1′-Bi-2,2′-naphthotitanium Dichloride*).[7]

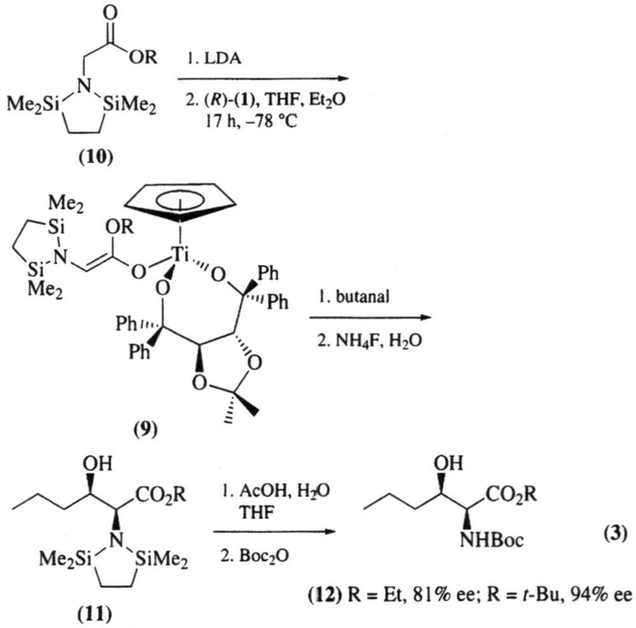

(**10**)

(**9**)

(**11**)
+ (**3**) + (**4**)

(**12**) R = Et, 81% ee; R = *t*-Bu, 94% ee

Aldol Reaction. In addition to the allyl derivatives (**2**) (eq 1), titanium ester enolates derived from chloride (**1**) react with aldehydes, affording aldol products after hydrolysis. Compared to the analogous reagents prepared from *Chloro(cyclopentadienyl)bis[3-O-(1,2:5,6-di-O-isopropylidene-*α-D-glucofuranosyl)]-titanium the enantioselectivity of these enolates is only moderate, 78% ee for the enolate of *t*-butyl acetate[2] and 26–78% ee in the case of 2,6-dimethylphenyl propionate.[4] Better selectivity (81–94% ee) was, however, obtained for Ti enolates (**9**) derived from 'stabase'-protected glycine esters (**10**) (eq 3).[2a] The primary *N*-bis-silyl-protected adduct (**11**) can easily be transformed to other *N*-derivatives, e.g. the *t*-butyl carbamate (**12**). This method thus gives access to L-*threo*-α-amino-β-hydroxy acids. Further details and references to other methods are provided in the entry for *Chloro(cyclopentadienyl)bis[3-O-(1,2:5,6-di-O-isopropylidene-α-D-glucofuranosyl)]titanium*, an analogous reagent affording the enantiomer of (**12**).

---

1. (a) Hafner, A.; Duthaler, R. O.; Marti, R.; Rihs, G.; Rothe-Streit, P.; Schwarzenbach, F. *J. Am. Chem. Soc.* **1992**, *114*. 2321. (b) Duthaler, R. O.; Hafner, A.; Alsters, P. L.; Rothe-Streit, P.; Rihs, G. *Pure Appl. Chem.* **1992**, *64*, 1897.
2. (a) Duthaler, R. O.; Hafner, A.; Riediker, M. *Pure Appl. Chem.* **1990**, *62*, 631. (b) Cambie, R. C.; Coddington, J. M.; Milbank, J. B. J.; Pausler, M. G.; Rustenhoven, J. J.; Rutledge, P. S.; Shaw, G. L.; Sinkovich, P. I. *Aust. J. Chem.* **1993**, *46*, 583.
3. (a) Brown, H. C.; Bhat, K. S.; Randad, R. S. *J. Org. Chem.* **1989**, *54*, 1570. (b) Racherla, U. S.; Brown, H. C. *J. Org. Chem.* **1991**, *56*, 401. (c) Roush, W. R.; Ando, K.; Powers, D. B.; Palkowitz, A. D.; Halterman, R. L. *J. Am. Chem. Soc.* **1990**, *112*, 6339. (d) Short, R. P.; Masamune, S. *J. Am. Chem. Soc.* **1989**, *111*, 1892. (e) Corey, E. J.; Yu, Ch.-M.; Kim, S. S. *J. Am. Chem. Soc.* **1989**, *111*, 5495. (f) Stürmer, R. *Angew. Chem. Int. Ed. Engl.* **1990**, *29*, 59.
4. Duthaler, R. O.; Hafner, A. *Chem. Rev.* **1992**, *92*, 807.
5. Yamamoto, Y.; Asao. N. *Chem. Rev.* **1993**, *93*, 2207.
6. (a) Furuta, K.; Mouri, M.; Yamamoto, H. *Synlett* **1991**, 561. (b) Marshall, J. A.; Tang, Y. *Synlett* **1992**, 653.
7. (a) Costa, A. L.; Piazza, M. G.; Tagliavini, E.; Trombini, C.; Umani-Ronchi, A. *J. Am. Chem. Soc.* **1993**, *115*, 7001. (b) Keck, G. E.; Tarbet, K. H.; Geraci, L. S. *J. Am. Chem. Soc.* **1993**, *115*, 8467.

Andreas Hafner & Rudolf O. Duthaler
Ciba-Geigy, Marly, Switzerland

(+)-*B*-Chlorodiisopinocampheylborane

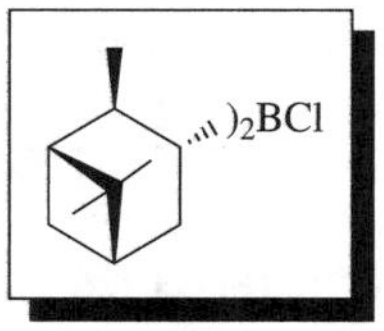

(+)
[112246-73-8] $C_{20}H_{34}BCl$ (MW 320.80)
[85116-37-6]

(chiral reducing agent for various prochiral ketones;[1] reagent for asymmetric aldol condensation of methyl alkyl ketones;[2] reacts with *meso*-epoxides to give nonracemic chlorohydrins[3])

Alternate Name: (+)- and (−)-DIP-Chloride™.
Physical Data: mp 52–56 °C; (−)-DIP-Chloride $[\alpha]_D$ −67.07° (c = 13.5, CH_2Cl_2); ^{11}B NMR (diethyl ether) singlet at δ = 74 ppm (with reference to $BF_3 \cdot OEt_2$).
Solubility: sol in both polar and nonpolar aprotic solvents like diethyl ether, THF, methylene chloride, pentane, hexane, etc.
Form Supplied in: white crystalline solid, available commercially.
Handling, Storage, and Precautions: the dry solid and its solutions are moisture and air sensitive. The reagent reacts instantaneously with water and protic solvents to liberate HCl. Containers of DIP-Chloride should be stored in the absence of moisture. Cans or bottles of DIP-Chloride should be flushed with N_2 and kept tightly sealed to avoid contact of oxygen and moisture. The solid reagent should be crushed, transferred, or weighed only in glove bag or dry box under N_2 atmosphere. The reagent can be stored for several years under N_2 atmosphere below 25 °C. Use in a fume hood.

Asymmetric Reduction of Ketones. The reagent, developed by Brown and co-workers, is used primarily for the asymmetric reduction of a variety of prochiral ketones to form secondary alcohols with high enantiomeric purity.[1] It has been demonstrated that reduction of aryl alkyl ketones with DIP-Chloride proceeds with extraordinary consistency and predictable stereochemistry (eq 1).[4,5] Reduction of a variety of substituted aryl alkyl ketones demonstrates that representative aromatic substituents do not affect the stereochemical outcome.[1a] The reagent shows poor enantioselectivity with unhindered aliphatic ketones; however, hindered aliphatic ketones like 3,3-dimethyl-2-butanone provide the corresponding alcohol in 95% ee.[5,6] Bicyclic ketones like α-tetralone, 1-indanone, and 2′-acetonaphthone are reduced by the reagent in 87%, 97%, and 98% ee, respectively.[5] Substrates with heteroaromatic groups show some decrease in the enantioselectivity (e.g. 2′-acetylthiophene and 3-acetylpyridine are reduced in 91% and 92% ee, respectively).[5] The reagent has been recently applied to the reduction of fluoro ketones.[7] It has been shown that good enantioselectivity can be obtained if the reduction of fluoro ketones are carried out neat (without solvent) (eq 2). The reagent shows poor selectivity in the reduction of unhindered alkynic ketones. However, hindered alkynic ketones are reduced in relatively good optical purity (eq 3).[8] A variety of hindered alkynic ketones have been recently synthesized and converted by the reagent to the corresponding propargylic alcohol in high yields and in essentially optically pure form.[8] It has also been demonstrated that DIP-Chloride is a remarkably effective reagent for the asymmetric reduction of acylsilanes to form corresponding α-silyl alcohols in 96–98% ee (eq 4).[9]

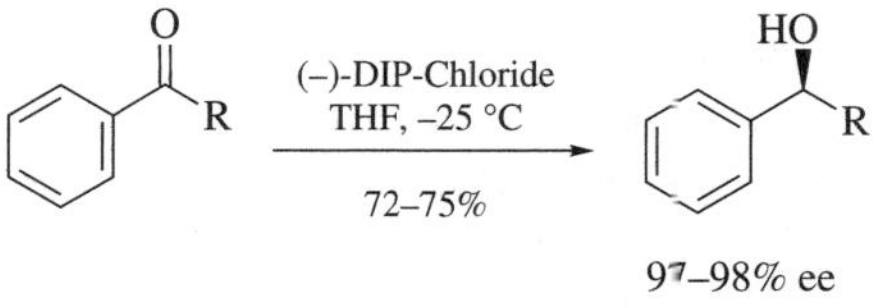

(1)

R = Me, Et, *n*-Pr, *n*-nonyl

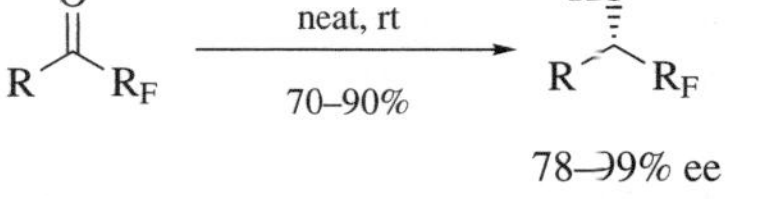

(2)

R = aryl or alkyl; R_F = CF_3, C_2F_5, C_3F_7

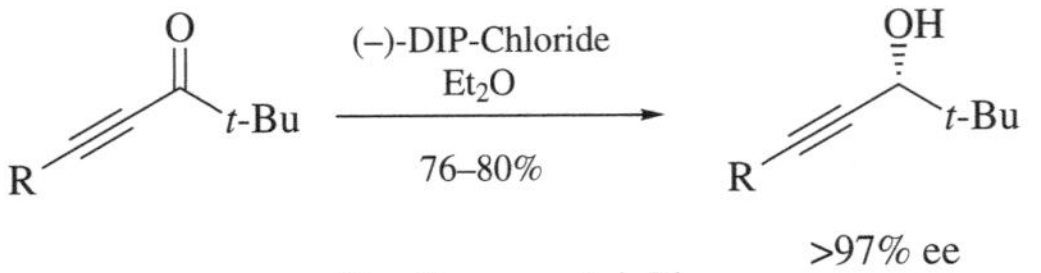

(3)

R = Cyp, *n*-octyl, Ph

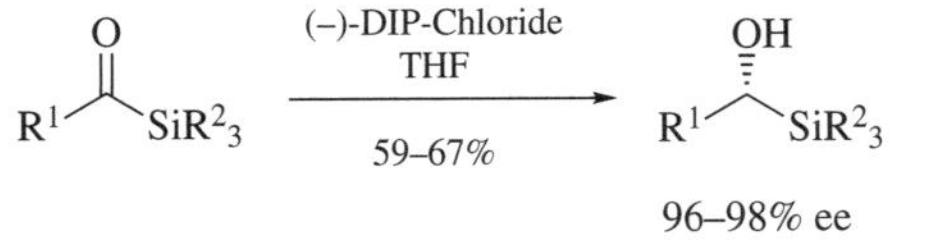

(4)

R^1 = Me, *n*-Bu, *i*-Pr; R^2 = Me, Et, *i*-Pr

Chiral secondary alcohols are potentially of great importance in biological and medicinal science. For example, 3-chloro-1-phenyl-1-propanol, which is obtained in >99% ee via DIP-Chloride reduction of the corresponding ketone, provides access to a highly enantioselective synthesis of antidepressant agents such as Tomoxetine, Fluoxetine, and Nisoxetine.[10] Another representative application of DIP-Chloride is found in the synthesis of (1*R*,3*S*)-1-aminomethyl-3,4-dihydro-5,6-dihydroxy-3-phenyl-1*H*-2-benzopyran, a potent and selective D1 agonist.[11] The reagent has also been applied to the synthesis of a prostaglandin intermediate[1a] and for the synthesis of a dolaphenine intermediate which is the C-terminal unit of dolastatin (a promising anticancer agent).[12]

Enolboration of Ketones and Opening of *meso*-Epoxides. Methyl alkyl ketones have been successfully enolized by Ipc_2BX (X = OTf or Cl) in the presence of a tertiary amine. The corresponding enolborinates have been used in asymmetric aldol condensations (eq 5).[2] The reagent has also been applied to the enantioselective opening of *meso*-epoxides to form the corresponding nonracemic chlorohydrins (eq 6).[3]

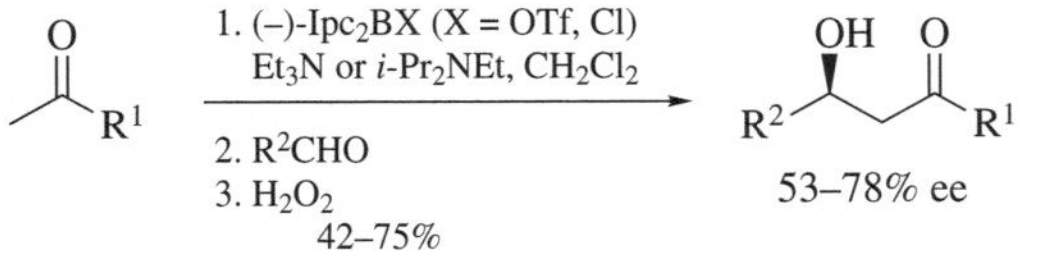

(5)

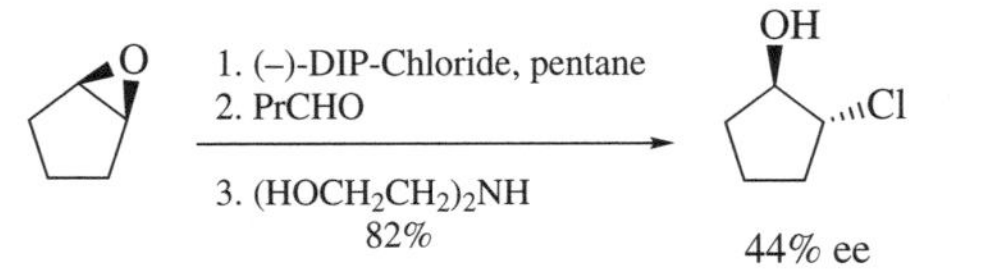

(6)

Related Reagents. Diisopinocampheylboron Trifluoromethanesulfonate.

1. (a) Brown, H. C.; Ramachandran, P. V. *Acc. Chem. Res.* **1992**, *25*, 16. (b) Singh, V. K. *Synthesis* **1992**, *7*, 605. (c) Midland, M. M. *Chem. Rev.* **1989**, *89*, 1553. (d) Brown, H. C.; Park, W. S.; Cho, B. T.; Ramachandran, P. V. *J. Org. Chem.* **1987**, *52*, 5406. (d) Dhar, R. K. *Aldrichim. Acta* **1994**, *27*, 43.
2. (a) Paterson, I.; Osborne, S. *Tetrahedron Lett.* **1990**, *31*, 2213. (b) Paterson, I.; Goodman, J. M.; Lister, M. A.; Schumann, R. C.; McClure, C. K.; Norcross, R. D. *Tetrahedron* **1990**, *46*, 4663.
3. (a) Joshi, N. N.; Srebnik, M.; Brown, H. C. *J. Am. Chem. Soc.* **1988**, *110*, 6246. (b) Srebnik, M.; Joshi, N. N.; Brown, H. C. *Isr. J. Chem.* **1989**, *29*, 229.
4. Chandrasekharan, J.; Ramachandran, P. V.; Brown, H. C. *J. Org. Chem.* **1985**, *50*, 5446.
5. Brown, H. C.; Chandrasekharan, J.; Ramachandran, P. V. *J. Am. Chem. Soc.* **1988**, *110*, 1539.
6. Brown, H. C.; Chandrasekharan, J.; Ramachandran, P. V. *J. Org. Chem.* **1986**, *51*, 3394.
7. Ramachandran, P. V.; Teodorovic, A. V.; Brown, H. C. *Tetrahedron* **1993**, *49*, 1725.
8. Ramachandran, P. V.; Teodorovic, A. V.; Rangaishenvi, M. V.; Brown, H. C. *J. Org. Chem.* **1992**, *57*, 2379.
9. (a) Cirillo, P. F.; Panek, J. S. *Org. Prep. Proced. Int.* **1992**, *24*, 555. (b) Soderquist, J. A.; Anderson, C. L.; Miranda, I. R.; Rivera, I.; Kabalba, G. W. *Tetrahedron Lett.* **1990**, *31*, 4677. (c) Buynak, J. D.; Strickland, J. B.; Hurd, T.; Phan, A. *Chem. Commun./J. Chem. Soc., Chem. Commun.* **1989**, 89.
10. Srebnik, M.; Ramachandran, P. V.; Brown, H. C. *J. Org. Chem.* **1988**, *53*, 2916.
11. DeNinno, M. P.; Schoenleber, R.; Asin, K. E.; Mackenzie, R.; Kebabian, J. W. *J. Med. Chem.* **1990**, *33*, 2948.
12. (a) Shioiri, T.; Hayashi, K.; Hamada, Y. *Tetrahedron* **1993**, *49*, 1913. (b) Irako, N.; Hamada, Y.; Shioiri, T. *Tetrahedron* **1992**, *48*, 7251.

Raj K. Dhar
Aldrich Chemical Company, Sheboygan Falls, WI, USA

N,N′-(1*R*,2*R*)-1,2-Cyclohexanediylbis-2-pyridinecarboxamide

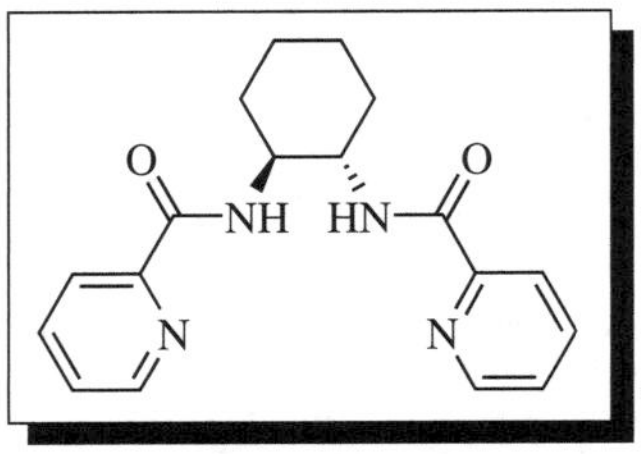

[201551-23-7]	$C_{18}H_{20}N_4O_2$	(MW 324.38)

(chiral bis-picolinic amide ligand for metal complexes used in a variety of asymmetric processes, particularly in regio and enantioselective allylic alkylations)

Alternate Name: $bpchH_2$.
Physical Data: [1] mp 201–202 °C.
Solubility: soluble in ethanol, EtOAc, pyridine; sparingly soluble in chloroform; insoluble in hexanes.
Form Supplied in: colorless or slightly brown crystals.
Analysis of Reagent Purity: IR: 3300, 3050, 2940, 2850, 1655, 1535; ^{1}H NMR: 8.6 (m), 8.4 (br), 8.1 (m, 8H), 7.8 (m, 2H), 7.4 (m, 2H), 4.1 (br, 2H), 2.3 (br), 1.6 (br, 6H); MS.
Preparative Methods: the ligand can be obtained through reaction of 2 equiv of 2-pyridinecarboxylic acid with (1*R*,2*R*)-1,2-diaminocyclohexane using Mukaiyama's reagent[2] or via the acid chloride.[3]
Purification: recrystallization from chloroform.
Handling, Storage, and Precautions: relatively safe reagent; no special instructions for its storage and handling are mentioned in the literature. Proper caution should be used as with all picoline amide reagents. Stable at room temperature.

Metal Complexes. Complexes of $\mathbf{bpchH_2}$, both protonated and deprotonated, with a variety of transition state metal ions

(Cu^{II}, Ni^{II}, Pd^{II}, Pt^{II}, Zn^{II}, Co^{II}, Fe^{II}, Ag^{I}) have been obtained, isolated and well characterized based on their physical, spectroscopical properties[4,5] and by X-ray diffraction.[6] The structure of each complex depends on the nature of the metal; the four nitrogen atoms (N_4) coordinate the metal in a square planar, square pyramidal, octahedral or tetrahedral geometry; N_2O_2 coordination is also possible.[5]

Catalytic Applications

Lewis Acid Catalyzed Ring Opening of Epoxides Ti^{IV} and Zr^{IV} complexes of bispicolinic amides in general and of *N,N'*-(1*R*,2*R*)-1,2-cyclohexanediylbis-2-pyridinecarboxamide, in particular, catalyze the ring opening of cyclohexene oxide (**1**) with trimethylsilyl azide ($TMSN_3$) as nucleophile (eq 1). The product, (1*R*,2*R*)-1-azido-2-trimethylsilyloxycyclohexane (**2**), is obtained in 42% yield and 36% ee when using Ti(OPr-*i*)$_4$; 33% yield and 35% ee when using Zr(OBu-*t*)$_4$.[7]

Asymmetric Allylic Alkylations **Molybdenum** (or **tungsten**) complexes of *N,N'*-(1*R*,2*R*)-1,2-cyclohexanediylbis-2-pyridinecarboxamide catalyze thermal regio and enantioselective allylic alkylation of cinnamyl-like[8] and polyenyl[9] systems with nucleophiles such as sodium alkyl malonates. While Pd-catalyzed reactions normally provide products from attack at the less substituted terminus, Mo and W catalysts favor attack at the more substituted position. Reaction of cinnamyl carbonate (**3**) with dimethyl sodiomalonate (**4**) in THF (0.1 M) at reflux in the presence of 10 mol % of $(C_2H_5CN)_3Mo(CO)_3$ and 15 mol % of **bpchH$_2$** as ligand affords a mixture of **5** and **6** in 55% yield in a 97:3 ratio in favor of **5**, with the latter having an ee of 99% (eq 2). Lowering the temperature to room temperature improves regioselectivity while maintaining a high level of enantiomeric excess.[8]

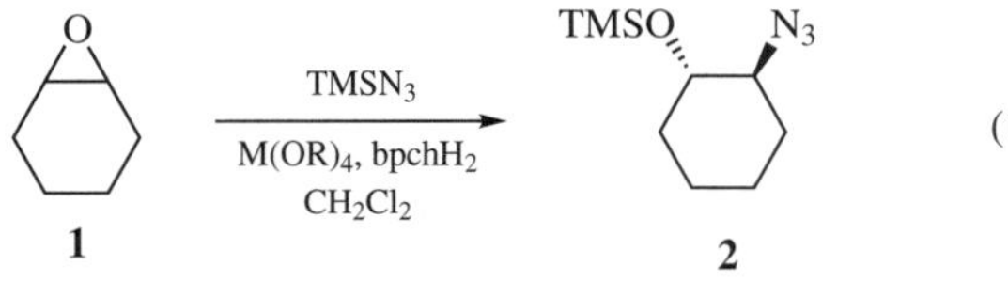

(1)

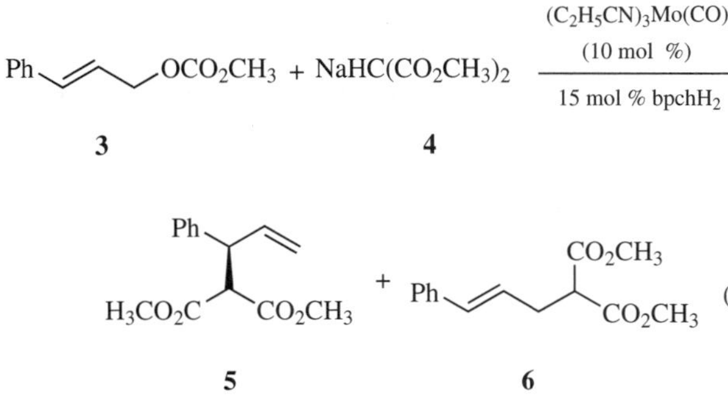

(2)

This type of allylic alkylation has been applied in the asymmetric total synthesis of (−)-anatoxin-a (**7**) using a *N*-nucleophile and Pd-based **bpchH$_2$** complexes in an intramolecular reaction to form the 9-azabicyclo[4.2.1]non-2-ene system (eq 3).[10]

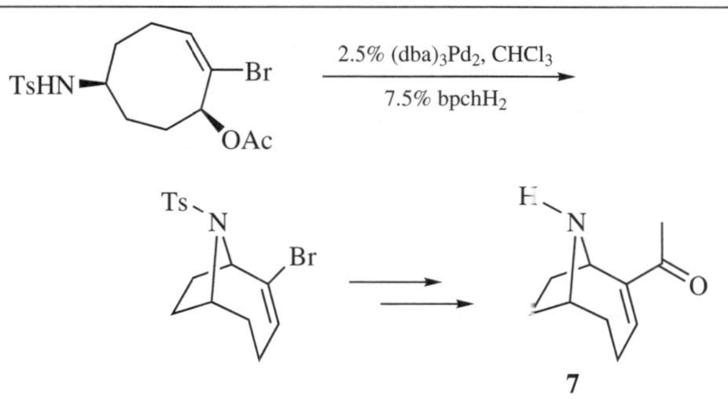

(3)

In addition, polyene substrates such as 5-phenylpentadienyl methyl carbonate (**8**) (eq 4) react with **4** using the same Mo-complex as in eq 2 furnishing the corresponding products **9** and **10** in 95% yield and 6.1:1 ratio. None of the product derived from attack at the benzylic position was observed.[9]

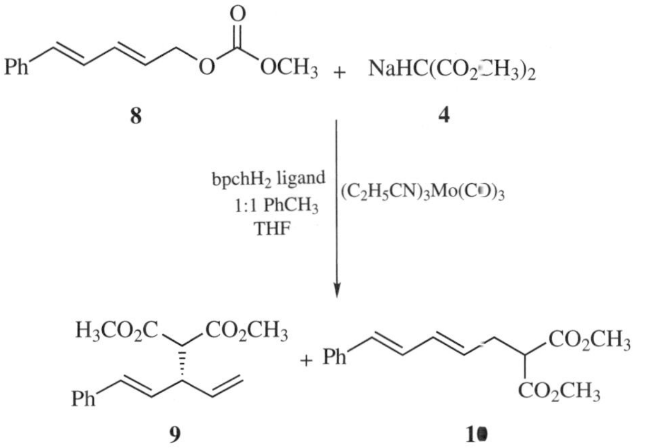

(4)

The molybdenum-catalyzed asymmetric allylic alkylation can also be carried out under non-inert conditions with microwave activation (eq 5).[11] In comparison to the thermally promoted reaction, reaction times are shorter (5 min vs 4 h). Nevertheless, high yields (87%) and ee (98%) are obtained, the regioselectivity is lower in the microwave reactions (19:1). This can be improved (41:1) by introducing an electron-donating substituent (4-methoxypyridine derivative) into the title ligand.[12]

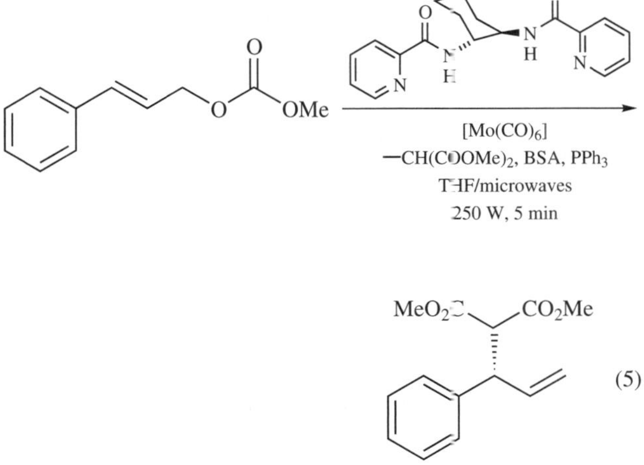

(5)

Related Reagents. *N,N'*-1,2-ethanediylbis-2-pyridinecarboxamide; *N,N'*-1,2-propanediylbis-2-pyridinecarboxamide; *N,N'*-1,

2-benzenediylbis-2-pyridinecarboxamide; *N,N'*-piperazinediylbis-2-pyridinecarboxamide; *N,N'*-1,2-cyclohexanediylbis-2-(4-substituted)pyridinecarboxamide; *N,N'*-2,2'-(α-binaphthyl)-bis-2-pyridinecarboxamide; *N,N'*-1,2-(1',2'-diphenyl)ethanediylbis-2-pyridinecarboxamide.

1. Barnes, D. J.; Chapman, R. L.; Vagg, R. S.; Watton, E. C. *J. Chem. Eng. Data* **1978**, *23*, 349–350.
2. Armstrong, A., In *Encyclopedia of Reagents for Organic Synthesis*; Paquette, L. A., Ed.; Wiley: New York, 1995, pp 1174–1175.
3. Belda, O.; Kaiser, N.-F.; Bremberg, U.; Larhed, M.; Hallberg, A.; Moberg, C. *J. Org. Chem.* **2000**, *65*, 5868–5870.
4. Mulqi, M.; Stephens, F. S.; Vagg, R. S. *Inorg. Chim. Acta* **1981**, *53*, L91–L93.
5. Moberg, C.; Adolfosson, H.; Warnmark, K. *Acta Chem. Scan.* **1996**, *50*, 195–202.
6. Mulqi, M.; Stephens, F. S.; Vagg, R. S. *Inorg. Chim. Acta* **1982**, *52*, 221.
7. Adolfsson, H.; Moberg, C. *Tetrahedron: Asymmetry* **1995**, *6*, 2023–2031.
8. Trost, B. M.; Hachiya, I. *J. Am. Chem. Soc.* **1998**, *120*, 1104–1105.
9. Trost, B. M.; Hildebrand, S.; Dogra, K. *J. Am. Chem. Soc.* **1999**, *121*, 10416–10417.
10. Trost, B. M.; Oslob, J. D. *J. Am. Chem. Soc.* **1999**, *121*, 3057–3064.
11. Kaiser, N.-F.; Bremberg, U.; Larhed, M.; Moberg, C.; Hallberg, A. *Angew. Chem. Int. Ed.* **2000**, *39*, 3595–3598.
12. Belda, O.; Kaiser, N.-F.; Bremberg, U.; Larhed, M.; Hallberg, A.; Moberg, C. *J. Org. Chem.* **2000**, *65*, 5869–5870.

Robert S. Coleman & Antonio Navarro
The Ohio State University, Columbus, OH, USA

(*R*)-(+)-Cyclohexyl(2-anisyl)methylphosphine[1]

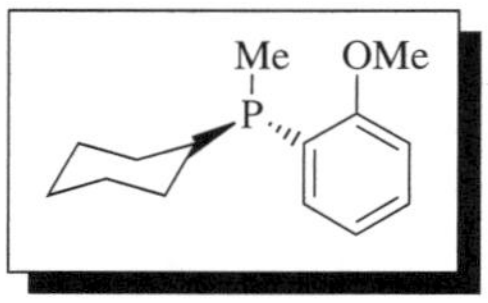

(*R*)
[52885-02-6] $C_{14}H_{21}OP$ (MW 236.29)
(*S*)
[35144-03-7]
(±)
[36293-81-9]

(ligand for transition metal catalyzed asymmetric hydrogenation of alkenes;[2] reagent for asymmetric Wittig[3] and Baylis–Hillman reactions[4])

Alternate Name: (+)-CAMP.
Preparative Methods: by hydrogenation (5% Rh/C) of optically pure (2-anisyl)phenylmethylphosphine oxide[5] followed by reduction of the phosphine oxide with *Trichlorosilane*.[2]
Handling, Storage, and Precautions: readily oxidized to the phosphine oxide and should be handled under N_2 or Ar. This reagent should be handled in a fume hood.

Asymmetric Hydrogenation. The synthesis of L-Dopa via asymmetric rhodium-catalyzed hydrogenation was reported using a catalyst prepared from rhodium(I) complexes and (+)-CAMP.[2,6] Activation of the catalyst by pretreatment with 1 atm H_2 was necessary. Hydrogenation at lower pressures (10 psi H_2) gave the dihydrocinnamic acid in 90% yield and 88% ee (eq 1). Similar yields and enantioselectivities have been reported with related cinnamic acid derivatives which are capable of bifunctional binding to the rhodium catalyst.[2] Bidentate diphosphine ligands generally lead to higher enantioselectivities in such reactions and are, therefore, employed in the commercial synthesis of L-Dopa.[1]

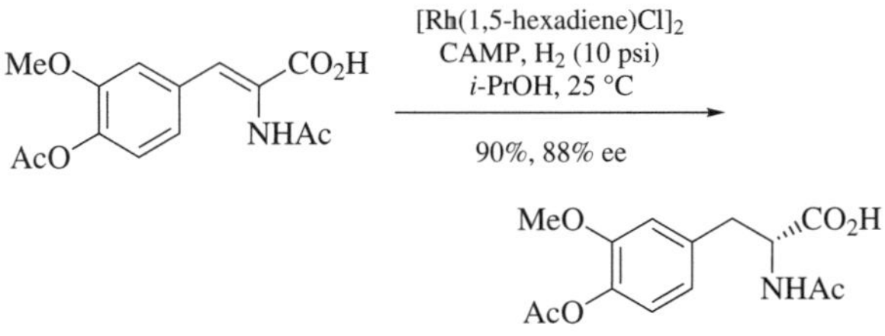

(1)

Wittig Reaction. The intramolecular Wittig reaction of an ylide obtained from (+)-CAMP occurred with 77% ee to give the (*S*)-bicyclic diketone.[3] Although a stoichiometric amount of the optically pure phosphine was required, the phosphine oxide (+)-CAMPO could be recycled. Enantiomeric excesses using CAMP were much higher than similar reactions which employed other chiral phosphines.

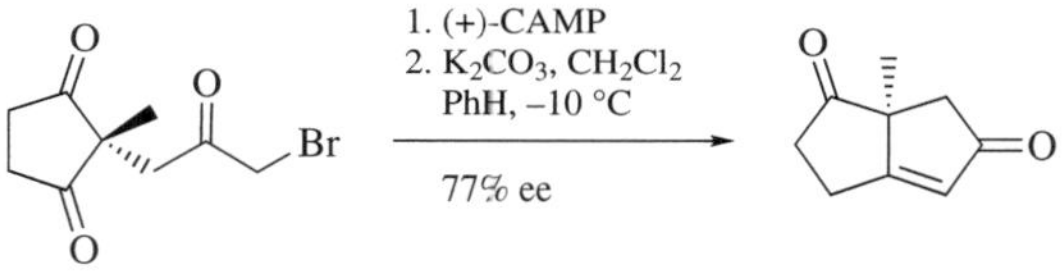

(2)

Baylis–Hillman Reaction. Intramolecular cyclization of $MeCO(CH_2)_2CH{=}CHCO_2Et$ using (+)-CAMP produced the cyclopentene in 40% isolated yield.[4] A 3:1 equilibrium mixture which favored the product cyclopentene was formed after 10 days at 25 °C. CAMP was found to be superior to other phosphines, such as PBu_3. DABCO and other nitrogen bases were ineffective for the cyclization reaction. However, the enantioselectivity of the product using CAMP was only 14%.

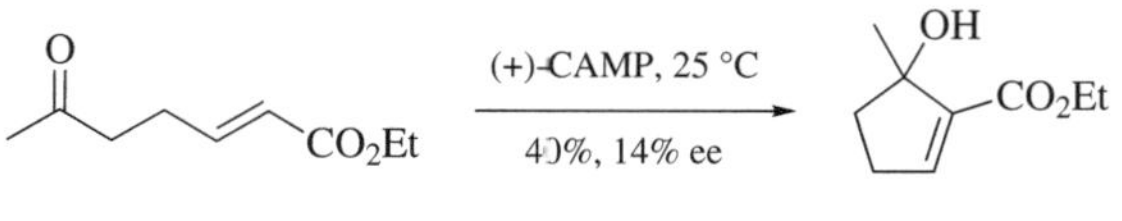

(3)

1. Knowles, W. S. *Acc. Chem. Res.* **1983**, *16*, 106.
2. Knowles, W. S.; Sabacky, M. J.; Vineyard, B. D. *Chem. Commun. J. Chem. Soc., Chem. Commun.* **1972**, 10.
3. Trost, B. M.; Curran, D. P. *Tetrahedron Lett.* **1981**, *22*, 4929.
4. Roth, F.; Gygax, P.; Frater, G. *Tetrahedron Lett.* **1992**, *33*, 1045.

5. Nauman, K.; Zon, G.; Mislow, K. *J. Am. Chem. Soc.* **1969**, *91*, 7012.
6. (a) Knowles, W. S.; Sabacky, M. J.; Vineyard, B. D. *Adv. Chem. Ser.* **1974**, *132*, 274. (b) Knowles, W. S.; Sabacky, M. J.; Vineyard, B. D. U.S. Patent 4 005 127, 1977.

Gregory T. Whiteker
Union Carbide Corporation, South Charleston, WV, USA

(1,5-Cyclooctadiene)[1,4-bis-(diphenylphosphino)butane]iridium(I) Tetrafluoroborate

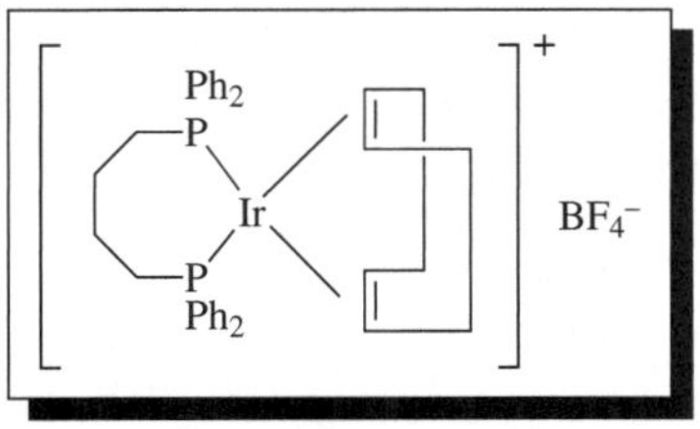

[78036-20-1] $C_{36}H_{40}BF_4IrP_2$ (MW 813.685)

(reagent for hydrogenation;[1] isomerization of butenyl- to allylsilanes[2])

Physical Data: dec. on attempted melting.
Solubility: insol Et_2O, pentane; sol CH_2Cl_2, MeOH, etc.
Form Supplied in: orange powder.
Drying: used as supplied in anhydrous solvent.
Analysis of Reagent Purity: ^{31}P NMR indicates presence of free phosphine ligand and phosphine oxides.
Handling, Storage, and Precautions: store under argon in freezer; stable as solid; solutions are air sensitive. This compound should be handled in a fume hood.

The complex has enjoyed relatively little use in organic synthesis. For iridium-catalyzed homogeneous hydrogenation of alkenes, Crabtree's iridium complex (*(1,5-Cyclooctadiene)(tricyclohexylphosphine)(pyridine)iridium(I) Hexafluorophosphate*) is generally preferred, although this readily prepared Ir complex is active.[1] It is more reactive than its rhodium counterpart in the catalytic isomerization of butenyl- to allylsilanes.[2]

---

1. Brown, J. M.; Derome, A. E.; Hall, S. A. *Tetrahedron* **1985**, *41*, 4647.
2. Matsuda, I.; Kato, T.; Sato, S.; Izumi, Y. *Tetrahedron Lett.* **1986**, *27*, 5747.

John M. Brown & James A. Ramsden
University of Oxford, UK

(1,5-Cyclooctadiene)[(3*R*,4*R*)-3,4-bis(diphenylphosphino)-1-methylpyrrolidine]rhodium Tetrafluoroborate[1]

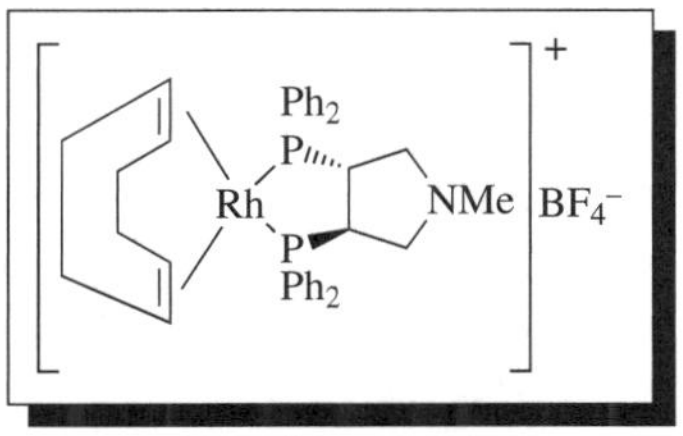

[100366-06-1] $C_{37}H_{41}BF_4NP_2Rh$ (MW 737.39)

(catalyst for asymmetric hydrogenation of α-acylaminoacrylic acid derivatives[2])

Solubility: insol ether; sol CH_2Cl_2, MeOH
Analysis of Reagent Purity: ^{31}P NMR (CD_2Cl_2): δ 33.17 (d, $^2J_{Rh,P}$ = 148.2 Hz).
Preparative Methods: the ligand (3*R*,4*R*)-3,4-bis(diphenylphosphino)-1-methylpyrrolidine (**1a**) is prepared from L-tartaric acid via (3*R*,4*R*)-3,4-bis(diphenylphosphino)-1-benzylpyrrolidine (**1b**) as shown in eq 1. The title cationic rhodium complex is made by mixing bis(1,5-cyclopentadiene)rhodium tetrafluoroborate and (**1a**) in CH_2Cl_2 under nitrogen.

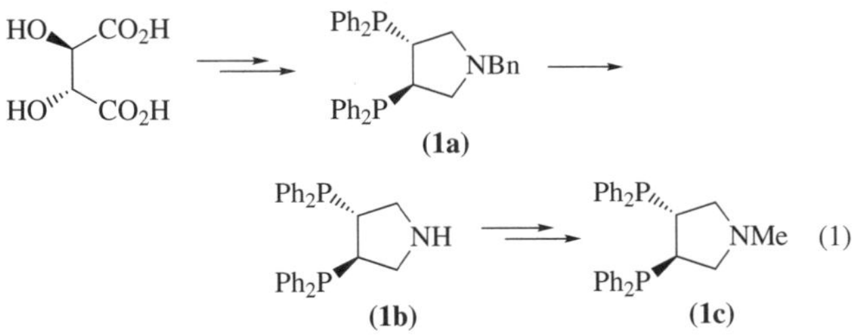

Asymmetric Hydrogenation Catalyst. Since catalytic asymmetric hydrogenation of α-acylaminoacrylic acid derivatives is an important process to afford α-amino acid derivatives, many rhodium complexes possessing optically active phosphine ligands have been developed in last two decades.[3–7] [Rh{(*R*)-(**1a**)}(cod)]BF_4 (**2a**) is one of the most efficient catalysts for this purpose with respect to activity and enantioselectivity. Besides (**2a**), several derivatives, which have various substituents on the nitrogen atom of the pyrrolidine ring, will be mentioned below.

α-(Acetylamino)cinnamic acid derivatives (**3**) are hydrogenated in the presence of (**2**), which are derived from natural L-tartaric acid, to afford natural (*S*)-*N*-acetylphenylalanine derivatives with high enantioselectivity (eq 1).[8] Some examples are shown in Table 1. Less than 0.01 mol % of catalyst is needed to complete the reaction under mild conditions with high enantioselectivity. The substituents on the nitrogen atom of (**2**) have no marked influence on the catalytic activity or the selectivity. This reaction is also insensitive to the variation of substituents on the phenyl ring of (**3**).

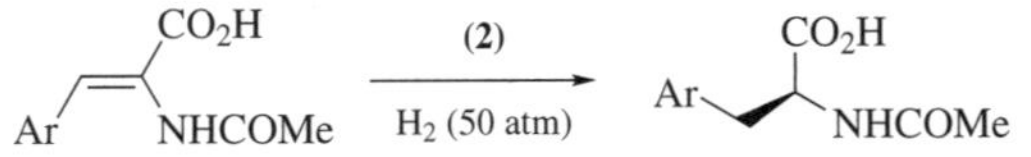

(2)

(**3a**) Ar = Ph
(**3b**) Ar = 4-MeOC_6H_4
(**3c**) Ar = 3-MeO-4-HOC_6H_3
(**3d**) Ar = 3,4-(OCH_2O)C_6H_3
(**3e**) Ar = 3-indolyl

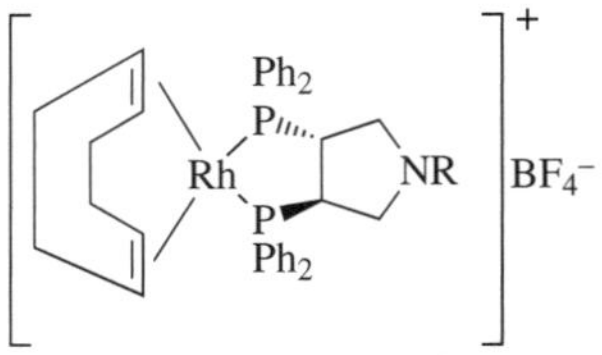

(**2a**) R = Me
(**2b**) R = Bn
(**2c**) R = H
(**2d**) R = COPh

Water-soluble (1,5-cyclooctadiene)[(3*R*,4*R*)-3,4-bis(diphenylphosphino)-1,1-dimethylpyrrolidinium]rhodium bis(tetrafluoroborate) (**4**) catalyzes the hydrogenation of the sodium salt of α-(acetylamino)cinnamic acid to afford the (*S*)-acetylphenylalanine sodium salt in 90% ee at 22 °C.[9]

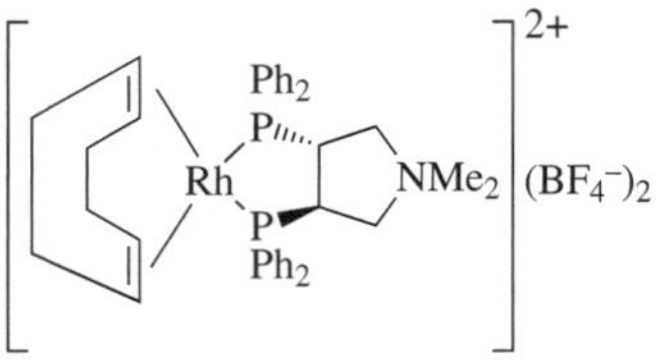

(**4**)

Table 1 Asymmetric Hydrogenation of Dehydro Amino Acids

Catalyst (**2**)	Dehydro amino acid (**3**)	Substrate catalyst	Temp. (°C)	ee (%)
a	**a**	880	22	93
b	**a**	8000	20	98
c	**a**	8000	50	96
d	**a**	15500	22–60	96.5
d	**b**	850	22	98
d	**c**	850	22	100
d	**d**	800	22	93
d	**e**	1500	22	82.5

A silica-supported rhodium complex of (3*R*,4*R*)-3,4-bis(diphenylphosphino)pyrrolidine (**5**) is also an effective catalyst.[10] This heterogeneous catalyst mediates asymmetric hydrogenation of (**3a**) as well as its methyl ester in comparable selectivities to the homogeneous one shown above.

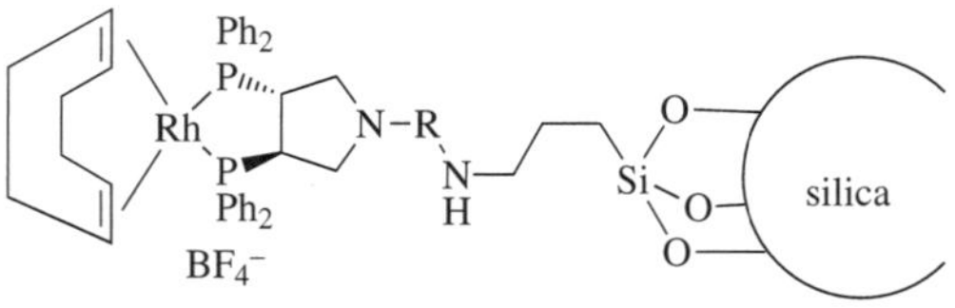

(**5**)

Related Reagents. Bis(bicyclo[2.2.1]hepta-2,5-diene)rhodium Perchlorate; Bis(bicyclo[2.2.1]hepta-2,5-diene)rhodium Perchlorate–(*R*)-1-(*S*)-1′,2-Bis(diphenylphosphino)ferrocenylethanol; Bis(1,5-cyclooctadiene)rhodium Tetrafluoroborate–(*R*)-2,2′-Bis(diphenylphosphino)-1,1′-binaphthyl.

1. (a) Takaya, H.; Noyori, R. *Comprehensive Organic Synthesis* **1991**, *8*, Chapter 3.2. (b) Noyori, R.; Kitamura, M., In *Modern Synthetic Methods*; Sheffold, R. Ed.; Springer: Berlin, 1989; Vol. 5, p 115.
2. Nagel, U.; Kinzel, E. *Ber. Dtsch. Chem. Ges./Chem. Ber.* **1986**, *119*, 1731.
3. Knowles, W. S.; Sabacky, M. J. *Chem. Commun./J. Chem. Soc., Chem. Commun.* **1968**, 1445.
4. Horner, L.; Siegel, H.; Büthe, H. *Angew. Chem., Int. Ed. Engl.* **1968**, *7*, 942.
5. Kagan, H. B.; Dang, T. P. *J. Am. Chem. Soc.* **1972**, *94*, 6429.
6. Miyashita, A.; Takaya, H.; Souchi, T.; Noyori, R. *Tetrahedron* **1984**, *40*, 1245.
7. Burk, M. J. *J. Am. Chem. Soc.* **1991**, *113*, 8518.
8. Nagel, U. *Angew. Chem., Int. Ed. Engl.* **1984**, *23*, 435; *Angew. Chem.* **1984**, *96*, 425.
9. Nagel, U.; Kinzel, E.; Andrade, J.; Prescher, G. *Ber. Dtsch. Chem. Ges./Chem. Ber.* **1986**, *119*, 3326.
10. Nagel, U.; Kinzel, E. *Chem. Commun./J. Chem. Soc., Chem. Commun.* **1986**, 1098.

Yoshihiko Ito & Michinori Suginome
Kyoto University, Japan

Cyclopentadienyl(3,5-dimethoxybenzyl)(nitrosyl)(triphenylphosphine)rhenium[1]

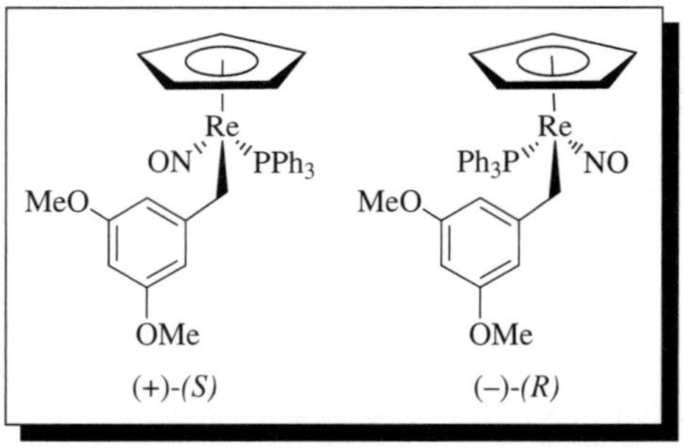

(+)-(*S*)
[109283-17-2] $C_{32}H_{31}NO_3PRe$ (MW 694.78)
(−)-(*R*)
[109362-39-2]

(reagents for enantioselective synthesis of organic compounds with chiral methyl groups[1])

Physical Data: mp 204–205 °C; $[\alpha]^{21}_{589}$ ±116° (CH_2Cl_2, *c* 0.3–0.6 mg cm^{-3}).
Solubility: sol CH_2Cl_2, benzene, ether, and THF; slightly sol hexane.
Form Supplied in: bright orange crystals.
Analysis of Reagent Purity: IR, and ^{1}H, ^{13}C, and ^{31}P NMR spectroscopies, microanalysis, and polarimetry.[1]

Preparative Methods: reaction of the methyl ester (+)-(*S*)-(Cp)Re(NO)(PPh_3)(CO_2Me) (*S*-**1**)[2] with 3,5-dimethoxyphenylmagnesium iodide in toluene at −24 °C gives 3,5-dimethoxybenzoyl complex (+)-(*S*)-(Cp)Re(NO)(PPh_3)-(CO(3,5-$(MeO)_2C_6H_3$)) (*S*-**2**; 85%). Refluxing (+)-(*S*)-(**2**) with $BH_3 \cdot THF$ or $BD_3 \cdot THF$ in THF gives the title reagents (+)-(*S*)-(**3**) or (+)-(*S*)-(**3**)-α-d_2 (95–84%). The opposite enantiomers (−)-(*R*)-(**3**) or (−)-(*R*)-(**3**)-α-d_2 can be similarly made from (−)-(*R*)-(**1**).[1]

Purification: crystallization from benzene/hexane.

Handling, Storage, and Precautions: the solid reagent is stable for days in air. However, it should be prepared, stored, and reacted under a dry nitrogen atmosphere.

Compounds with 'chiral methyl groups' (RCHDT) play important roles in the elucidation of biological and abiological reaction mechanisms.[3] The title compound can be utilized to prepare chiral 3,5-dimethoxytoluene, which can in turn be degraded to $CHDTCO_2H$. The enantiomeric purity of the latter can be assayed enzymatically.[1] Analogs (Cp)Re(NO)(PPh_3)(CH_2R), which can be similarly synthesized, can usually be converted to the corresponding RCHDT compounds.

In practice, (+)-(*S*)-(**3**)-α-d_2 and $Ph_3C^+PF_6^-$ are allowed to react to give the alkylidene complex (+)-(*S*)-[(Cp)Re(NO)-(PPh_3)(=CD(3,5-$(MeO)_2C_6H_3$)]$^+PF_6^-$ ((+)-(*S*)-(**4t**)-α-d_1; 91%) (eq 1). Addition of $NaBT_4$ gives (+)-(*S,S*)-(**3**)-α-d_1t_1 (87%). Subsequent reaction with HBr gives (+)-(*R*)-(Cp)Re(NO)(PPh_3)(Br) ((+)-(*R*)-(**5**); 93%) and (*R*)-dimethoxytoluene-α-d_1t_1 ((*R*)-(**6**); 85%) with retention of configuration at carbon and rhenium. The latter is treated with O_3 to give, after addition of NaOH, the chiral acetate salt (*S*)-$CHDTCO_2^-Na^+$ in 93% ee. The opposite enantiomer, (*R*)-$CHDTCO_2^-Na^+$, is made from (−)-(*R*)-(**3**)-α-d_2 in 86% ee.[1]

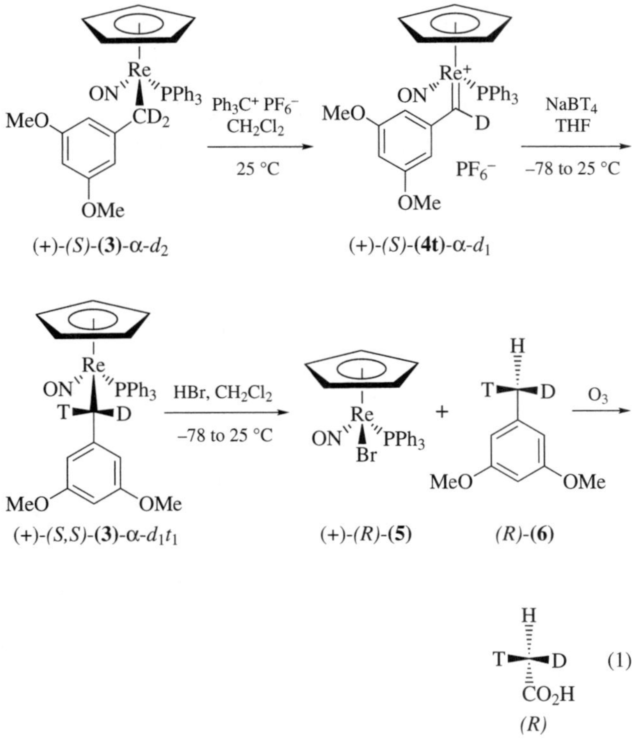

Thus the chiral rhenium auxiliary allows the highly stereoselective introduction of all hydrogen isotopes. The generalization of this methodology to other substrates is shown in eq 2. No complications are encountered for cases where R = *n*-alkyl or aryl. The method fails when R is a secondary alkyl group, as the reaction of (Cp)Re(NO)(PPh_3)($CH_2CHR^1R^2$) and $Ph_3C^+PF_6^-$ gives an alkene complex (β-hydride abstraction).[4]

There are several elements of synthetic flexibility. In most cases, both enantiomers of the target can be generated from the *same* enantiomer of the precursor alkyl complex. For example, the hydrogen isotopes can be introduced in different orders. Alternatively, depending upon reaction temperature, either of the two alkylidene complex Re=C geometric isomers (**7k**) and (**7t**) (eq 2) can be generated in >90% isomeric purity. The hydrogen isotope nucleophile attacks from a direction *anti* to the bulky PPh_3 ligand in each case, giving different diastereomers.[3] These in turn give different product enantiomers. Finally, if a benzylic rhenium complex is treated with a deuterated or tritiated acid in the rhenium–carbon bond-cleavage step, some aryl C–H bonds are also labeled. The optically active bromide complex (+)-(*R*)-(**5**) can be recycled to the methyl complex (+)-(*S*)-(Cp)Re(NO)(PPh_3)(Me) without racemization.[5] The latter is easily converted to the methyl ester (+)-(*S*)-(**1**),[2] or directly to alkyl complexes.[4]

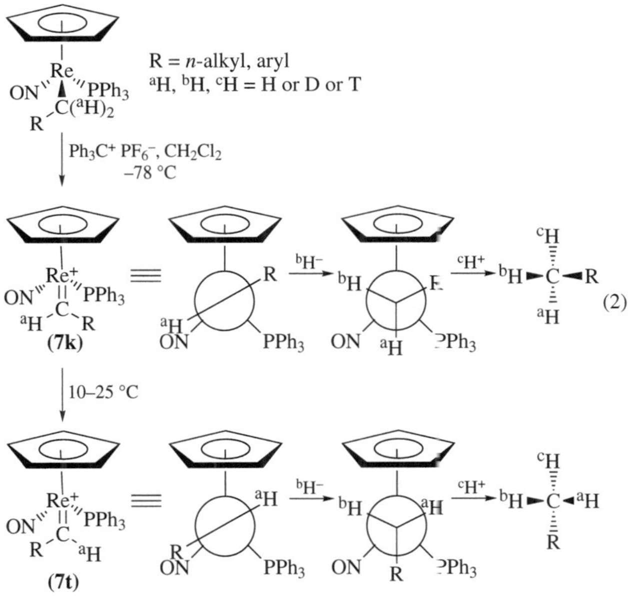

---

1. O'Connor, E. J.; Kobayashi, M.; Floss, H. G.; Gladysz, J. A. *J. Am. Chem. Soc.* **1987**, *109*, 4837.
2. Agbossou, F.; O'Connor, E. J.; Garner, C. M.; Quirós Méndez, N.; Fernández, J. M.; Patton, A. T.; Ramsden, J. A.; Gladysz, J. A. *Inorg. Synth.* **1992**, *29*, 211.
3. Floss, H. G., In *Mechanisms of Enzymatic Reactions: Stereochemistry*; Frey, P. A., Ed.; Elsevier: New York, 1986; pp 71–88.
4. Kiel, W. A.; Lin, G.-Y.; Bodner, G. S.; Gladysz, J. A. *J. Am. Chem. Soc.* **1983**, *105*, 4958.
5. Ramsden, J. A.; Peng, T.-S.; Gladysz, J. A. *Bull. Soc. Chim. Fr.* **1992**, *129*, 625.

Tang-Sheng Peng & J. A. Gladysz
University of Utah, Salt Lake City, UT, USA

(2*R*,3*R*)-(*Z*)-*cyclo*-Phenylalanine

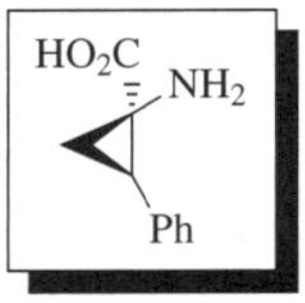

(·HCl)

[110716-95-5] $C_{10}H_{12}ClNO_2$ (MW 213.68)

(reagent for syntheses of conformationally constrained peptidomimetics)

Physical Data: (HCl salt) mp 201 °C (dec); $[\alpha]_D^{25}$ +105° (*c* 0.69, H_2O).

Solubility: the compounds of this series closely resemble the parent amino acids; they are sol water, moderately sol lower alcohols, and insol apolar solvents.

Form Supplied in: colorless solid; not currently commercially available on a routine basis.

Analysis of Reagent Purity: chemical purities are accessed by mp and NMR; optical purities are deduced by forming diastereomeric derivatives, or by derivatization and analysis in the presence of chiral NMR shift reagents.

Preparative Methods: substituted 2,3-methanoamino acids are difficult to prepare. Unfortunately, most of the reported syntheses give racemic materials whereas stereochemically pure compounds are required for studies of cyclopropane-based peptidomimetics. The only 2,3-methanologs of protein amino acids prepared in optically active form are (*E*)- and (*Z*)-*cyclo*-Phe[1–4] and -Tyr,[5] all four stereoisomers of *cyclo*-Met,[6] (*Z*)-*cyclo*-Arg[7] and (2*S*,3*S*)-(*Z*)-*cyclo*-Trp,[8] although several routes to enantio-enriched 2,3-methanologs of simple nonproteogenic amino acids have been reported.[9–12] The most practical synthesis of the title compound is that based on a diastereoselective, rhodium-catalyzed cyclopropanation reaction.[3]

Handling, Storage, and Precautions: some compounds in this series are slightly hygroscopic, but they are otherwise quite stable, and indefinitely so under an inert atmosphere in a freezer. For good results in peptide syntheses these compounds must be used with the same precautions taken for any common amino acid derivative.

Background[13]. One of the least drastic perturbations of amino acid structure is to link the α- and β-carbons of the side chain with a methylene group, giving 2,3-methanoamino acid analogs (or 'methanologs', e.g. the cyclopropyl analogs of phenylalanine, *cyclo*-Phe, and of methionine, *cyclo*-Met). Incorporation of a cyclopropane ring locks the side chain substituent *cis* or *trans* to the amino functionality. Designation of the absolute configurations of the two chiral centers completely defines the stereochemistry, whereas (*Z*) and (*E*) nomenclature specifies the diastereomeric, but not the enantiomeric, form. Both systems are shown on the examples in Figure 1, even though it is redundant to specify (*Z*) or (*E*) stereochemistry if the absolute configuration is marked.

These amino acid surrogates have side chains locked *cis* or *trans* to the amino functionality and the cyclopropane ring also restricts rotations about the N–C_α and C_α–CO bonds (i.e. Φ and ψ, respectively).[14–16] A Ramachandran plot[17] for a related compound indicates these conformational restrictions are severe.[14–16,18,19] Consequently, systematic variations of stereoisomeric 2,3-methanolog substitutions facilitate controlled restrictions on the conformational freedom of peptidomimetics,[20,21] and it is possible to constrain them into molecular orientations that resemble bioactive conformations of the parent peptide. Substitution of a protein amino acid with its 2,3-methanolog therefore can enforce or preclude structural shapes required for various bioactivities. They can also impart considerably enhanced resistance to proteolytic degradation.[22–27] These are the main applications of this class of compounds.

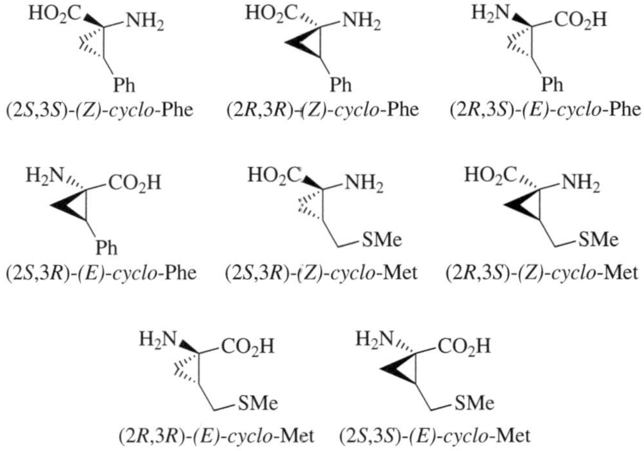

Figure 1

At the present time, 2,3-methanoamino acid analogs can only be viewed as *reagents* in the context of syntheses of peptidomimetics. Consequently, this entry describes only that chemistry related to incorporation of protein amino acid methanologs into peptide sequences.

Incorporation into Peptidomimetics. Solution phase methods have been used to incorporate *cyclo*-Phe stereoisomers into enkephalin analogs. The mixed anhydride method (*i*-BuOCOCl, *Isobutyl Chloroformate*) was used as illustrated in Figure 2, which depicts the deprotection/coupling sequence used in the preparation of Tyr–D-Ala–Gly–(*cyclo*-Phe)–Leu–OH. Thus the first step was *N*-Cbz protected *cyclo*-Phe being coupled with Leu–OMe via the mixed anhydride method; the dipeptide so formed was coupled with Cbz–Tyr–D-Ala–Gly–OH after *Trifluoroacetic Acid* deprotection.[22] Methyl ester protecting groups were used at the *C*-terminus, and benzyloxycarbonyl (Cbz) protection was employed at the *N*-termini. The Cbz groups were removed by acid deprotection and not hydrogenolysis, since the latter conditions cause extremely facile ring-opening of methanologs with aromatic substituents connected to the strained ring.[2]

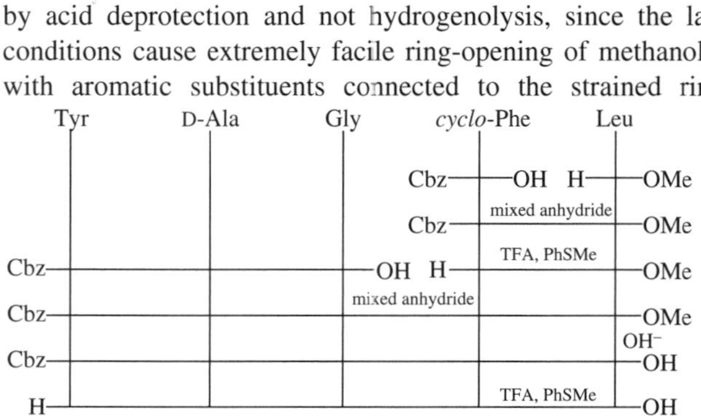

Figure 2

Peptide syntheses have undergone extensive improvements in the decade since the preparations described above were performed.[28,29] A more contemporary approach to syntheses of peptidomimetics using the solid phase Fmoc (9-fluorenylmethoxycarbonyl) approach is illustrated in Figure 3.[23,30] The couplings were performed using Castro's reagent (BOP, *Benzotriazol-1-yloxytris(dimethylamino)phosphonium Hexafluorophosphate*) in the presence of *1-Hydroxybenzotriazole* (HOBt).[31]

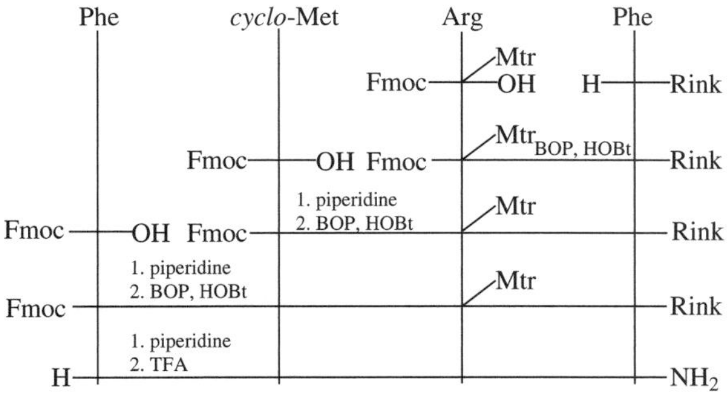

Figure 3

Other peptides in the F*cyclo*-MRF–NH_2 series have been prepared using a more acid stable resin (MBHA, methoxybenzhydrylamine) and Boc (*t*-butoxycarbonyl) protecting groups, but the Fmoc approach above gives superior yields.[23,30]

---

1. King, S. W.; Riordan, J. M.; Holt, E. M.; Stammer, C. H. *J. Org. Chem.* **1982**, *47*, 3270.
2. Kimura, H.; Stammer, C. H. *J. Org. Chem.* **1983**, *48*, 2440.
3. Davies, H. M. L.; Cantrell, W. R., Jr. *Tetrahedron Lett.* **1991**, *32*, 6509.
4. Fernández, M. D.; Frutos, M. P. D.; Marco, J. L.; Fernández-Alverez, E.; Bernabé, M. *Tetrahedron Lett.* **1989**, *30*, 3101.
5. Ahmad, S.; Phillips, R. S.; Stammer, C. H. *J. Med. Chem.* **1992**, *35*, 1410.
6. Burgess, K.; Ho, K.-K. *J. Org. Chem.* **1992**, *57*, 5931.
7. Burgess, K.; Ho, K.-K. *Tetrahedron Lett.* **1992**, *33*, 5677.
8. Bruncko, M.; Crich, D. *Tetrahedron Lett.* **1992**, *33*, 6251.
9. Baldwin, J. E.; Adlington, R. M.; Rawlings, B. J.; Jones, R. H. *Tetrahedron Lett.* **1985**, *26*, 485.
10. Pirrung, M. C.; Dunlap, S. E.; Trinks, U. P. *Helv. Chim. Acta* **1989**, *72*, 1301.
11. Alami, A.; Calmes, M.; Daunis, J.; Escale, F.; Jacquier, R.; Roumestant, M-L.; Viallefont, P. *Tetrahedron: Asymmetry* **1991**, *2*, 175.
12. Williams, R. M.; Fegley, G. J. *J. Am. Chem. Soc.* **1991**, *113*, 8796.
13. Stammer, C. H. *Tetrahedron* **1990**, *46*, 2231.
14. Varughese, K. I.; Srinivasan, A. R.; Stammer, C H. *Int. J. Pept. Protein Res.* **1985**, *26*, 242.
15. Varughese, K. I.; Wang, C. H.; Kimura, H.; Stammer, C. H. *Int. J. Pept. Protein Res.* **1988**, *31*, 299.
16. Taylor, E. W.; Wilson, S.; Stammer, C. H. *ACS Symp. Ser.* **1991**, *450*, 162.
17. Ramachandran, G. N.; Sasisekharan, V. *Adv. Protein. Chem.* **1968**, *23*, 283.
18. Barone, V.; Fraternali, F.; Cristinziano, P. L.; Lelj, F.; Rosa, A. *Biopolymers* **1988**, *27*, 1673.
19. Nitz, T. J.; Shimohigashi, Y.; Costa, T.; Chen, H. C.; Stammer, C. H. *Int. J. Pept. Protein Res.* **1986**, *27*, 522.
20. Mapelli, C.; Elrod, L. F.; Switzer, F. L.; Stammer, C. H.; Holt, E. M. *Biopolymers* **1989**, *28*, 123.
21. Mapelli, C.; Van Halbeck, H.; Stammer, C. H. *Biopolymers* **1990**, *29*, 407.
22. Kimura, H.; Stammer, C. H.; Shimohigashi, Y. Cui, R. L.; Stewart, J. *Biochem. Biophys. Res. Commun.* **1983**, *115*, 112.
23. Malin, D. H.; Lake, J. R.; Ho, K.-K.; Corriere, L. S.; Garber, T. M.; Waller, M.; Benson, T.; Smith, D. A.; Luu, T.-A.; Burgess, K. *Peptides* **1993**, in the press.
24. Ogawa, T.; Shimohigashi, Y.; Yoshitomi, H.; Sakamoto, H.; Kodama, H.; Waki, M.; Stammer, C. H. *Pept. Chem.* **1988**, *26*, 25.
25. Ogawa, T.; Shimohigashi, Y.; Shiota, M.; Waki, M.; Stammer, C. H.; Ohno, M. *Pept. Chem.* **1989**, *27*, 379.
26. Ogawa, T.; Yoshitomi, H.; Kodama, H.; Waki, M.; Stammer, C. H.; Shimohigashi, Y. *FEBS Lett.* **1989**, *250*, 227.
27. Breckenridge, R. J.; Suckling, C. J. *Tetrahedron* **1986**, *42*, 5665.
28. Atherton, E.; Sheppard, R. C., In *Solid Phase Peptide Synthesis: A Practical Approach*; IRL: Oxford, 1989.
29. Fields, G. B.; Noble, R. L. *Int. J. Pept. Protein Res.* **1990**, *35*, 161.
30. Malin, D. H.; Payza, K.; Lake, J. R.; Corriere, L. S.; Benson, T. M.; Smith, D. A.; Kelley, R. S.; Ho, K. K.; Burgess, K. *Peptides* **1993**, *14*, 47.
31. Knorr, R.; Trzeciak, A.; Bannwarth, W.; Gillessen, D. *Tetrahedron Lett.* **1989**, *30*, 1927.

Kevin Burgess & Kwok-Kan Ho
Texas A & M University, College Station, TX, USA

D

(1*S*,2*S*)-1,2-Diaminocyclohexane[1]

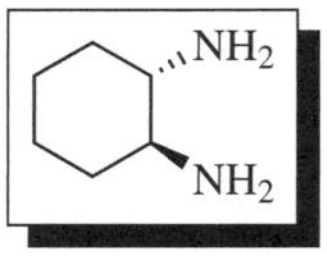

[21436-03-3] $C_6H_{14}N_2$ (MW 114.2)

(chiral compound, chiral auxiliary, chiral ligand)

Physical Data: mp 42–45 °C; bp 104–114 °C 40 mm Hg; $[\alpha]_D^{20}$ + 25 (*c* 5, 1 N HCl).
Solubility: soluble in aqueous acidic solution, alcohols, and most organic solvents.
Form Supplied in: colorless liquid; both enantiomers are commercially available.
Preparative Methods: cheap and readily available racemic *trans*-1,2-diaminocyclohexane can be resolved with D-(−)-tartaric acid, giving (1*S*,2*S*)-diaminocyclohexane with >98% enantiomeric excess. Detailed procedures for the resolution have been published.[2,3] Determination of enantiomeric excess is made by HPLC analysis of the *N,N′*-bis(*m*-toluyl) derivative on a Pirkle L-Leucine-DNB column. Direct separation of enantiomers by preparative HPLC on a chiral column has also been described.[4]
Purification: bulb-to-bulb vacuum distillation.
Handling, Storage, and Precautions: air- and CO_2-sensitive; bottles should be stored under an inert gas atmosphere, preferably away from direct light. 1,2-Diaminocyclohexane is harmful by inhalation and contact with skin. May be fatal if swallowed. Incompatible with strong acids and strong oxidizing reagents.

Chiral C_2-symmetric vicinal diamines have emerged as powerful tools for the synthesis of enantiomerically pure compounds and are now commonly used as chiral auxiliaries or ligands for a wide array of asymmetric chemical transformations,[5] with efficiencies comparable to those obtained with the closely related 1,2-diols. (1*S*,2*S*)-1,2-Diaminocyclohexane (also named (1*S*,2*S*)-1,2-cyclohexanediamine), together with its (1*R*, 2*R*) enantiomer, allows excellent levels of asymmetric induction in many reactions, for which it has become the ligand of choice. Its applications in asymmetric synthesis and catalysis often involves the preparation of various *N,N′*-substituted derivatives.

Resolving Reagent. Although (1*S*,2*S*)-1,2-diaminocyclohexane is not a commonly used resolving reagent, it has been used for the resolution of 1,2-diols, the most important application being the resolution of 1,1-binaphthol (BINOL) and other atropoisomeric alcohols.[6]

Bicyclic phosphoramines derived from *N,N′*-dimethyl-diaminocyclohexane have been used for the determination of enantiomeric purities of chiral alcohols, amines, and thiols using spectroscopic and chromatographic techniques.[7] The thiophosphoramide derived from **1** allows the determination of enantiomeric purity of alcohols by ^{31}P NMR analysis via the hydroxy adduct **2** with excellent resolution of diastereomers (eq 1).

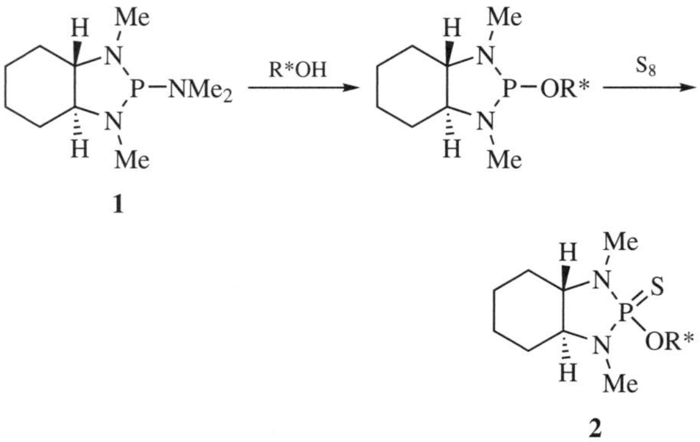

(1)

Chiral Auxiliary. Chiral 1,2-diamines have often been used as chiral auxiliaries in various carbon-carbon bond-forming reactions. The reaction of a diamine with an aldehyde gives a chiral aminal which can undergo stereoselective reactions. This was applied in the synthesis of enantiomerically pure α-hydrazino aldehydes by stereoselective addition of carbon nucleophiles onto the aminal of glyoxal monohydrazone (eqs 2 and 3).[8] In this reaction, the use of 1,2-diaminocyclohexane gave lower diastereomeric excesses than with the related 1,2-diphenyl ethylenediamine.

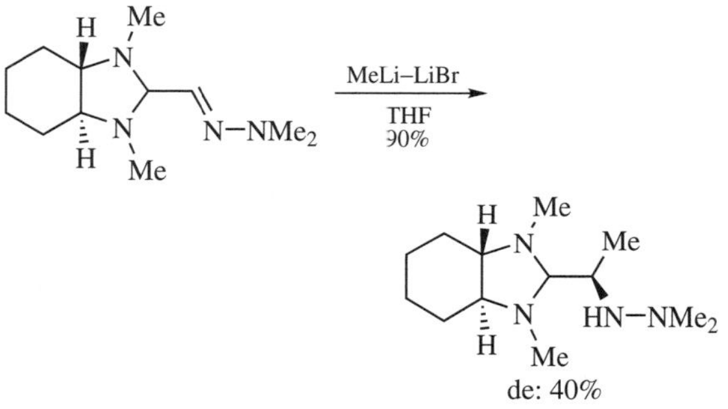

(2)

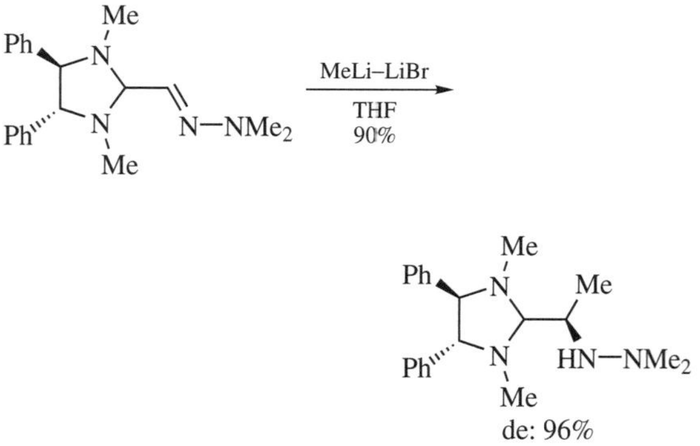

(3)

The same chiral auxiliary has also been used for the stereoselective synthesis of arene-chromium complexes:[9] treatment of an aromatic aminal with chromium hexacarbonyl gives the corresponding complex with high diastereomeric excess. This protocol was recently applied in a total synthesis of (−)-lasubine (eq 4).[10]

A successful application of 1,2-diaminocyclohexane (as its 1*R*,2*R* enantiomer) as a chiral auxiliary is illustrated by the diastereoselective alkylation of the potassium enolate of bis-amide (**3**) with electrophiles such as benzyl bromide to give bis-alkylated products with excellent diastereoselectivity (eq 5).[11] Lower levels of induction were obtained using related 1,2-diphenyl ethylenediamine.

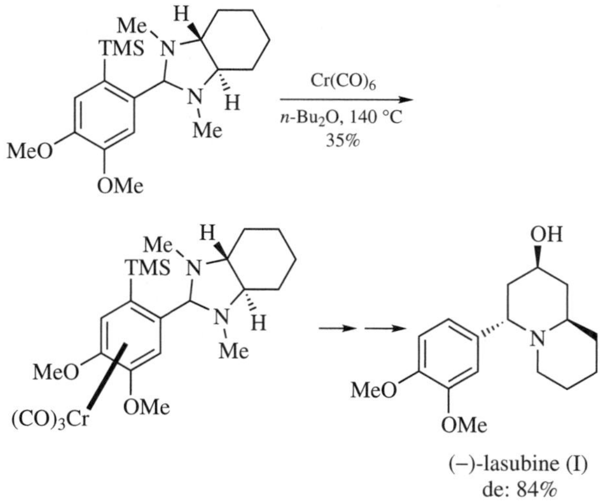
(4)

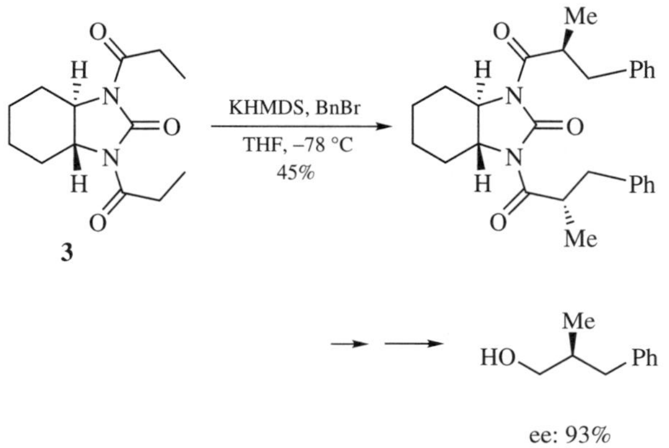
(5)

Chiral imidazolines such as **4**, obtained by condensation of iminoether hydrochlorides with (1*S*,2*S*)-1,2-diaminocyclohexane, may be metalated and alkylated with high stereoselectivity.[12] This process is highly efficient for the stereoselective synthesis of quaternary benzylic stereogenic centers, and has been applied to a total synthesis of mesembrine (eq 6).[13] (1*S*,2*S*)-1,2-Diaminocyclohexane here again gives higher diastereomeric excesses than 1,2-diphenyl ethylenediamine in this reaction.

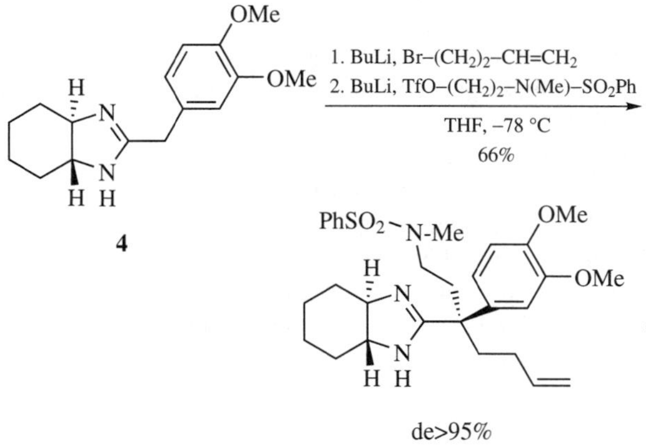
(6)

The asymmetric synthesis of chiral phosphonic acids has been accomplished starting from alkyl phosphonamides **5** derived from *N,N′*-dimethyl-diaminocyclohexane, which are easily prepared by condensation with alkyl phosphonic dichlorides (eq 7).[14] Upon metalation with a strong base, the corresponding anion reacts with a great variety of electrophiles with high stereoselectivity. This has been applied to conjugate addition,[15] cyclopropanation,[16] α-amination,[17] and enantioselective Wittig reactions (eq 8).[14]

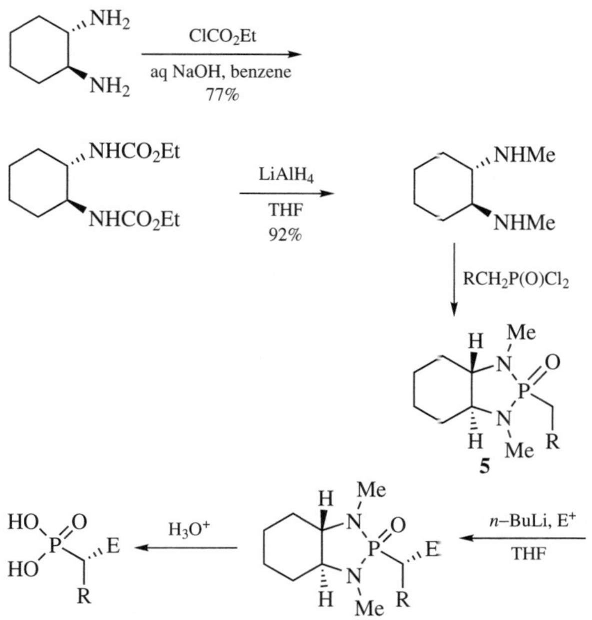
(7)

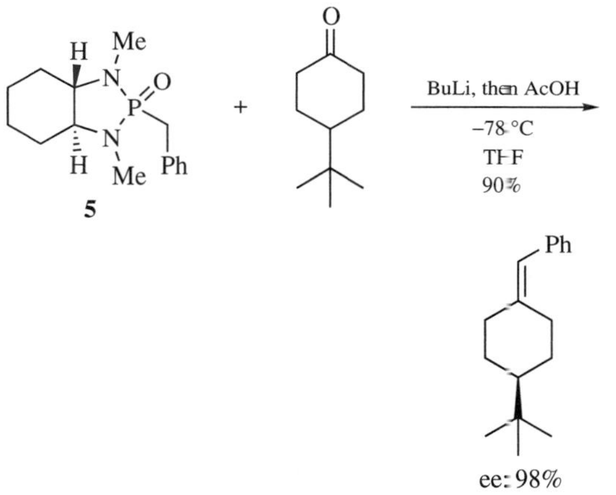
(8)

Chiral Reagent. The diamino phenyl borane (**6**) derived from (1*S*,2*S*)-1,2-diaminocyclohexane has been used as a chiral proton source for the enantioselective protonation of prochiral cyclic lithium enolates, with ee's up to 93% (eq 9).[18] (1*S*,2*S*)-1,2-Diaminocyclohexane proved to be highly superior to 1,2-diphenyl ethylenediamine or bis-naphthylamine.

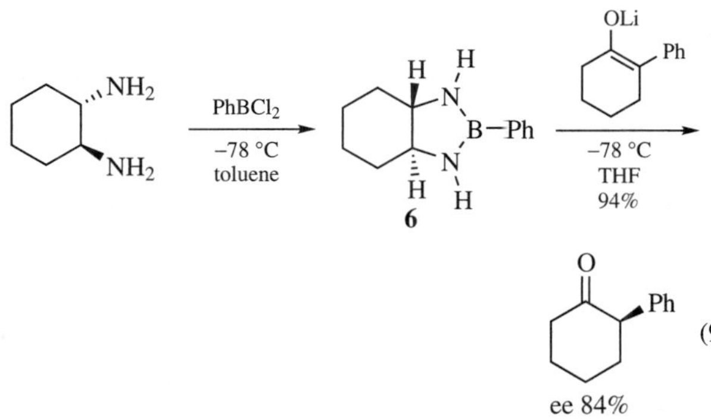
(9)

Chiral Ligand for Asymmetric Catalysis. (1*S*,2*S*)-1,2-Diaminocyclohexane and its derivatives are among the most frequently used chiral ligands for a variety of catalytic asymmetric transformations. *N,N′*-disubstitution allows fine tuning of steric and electronic properties of each ligand, using either electron-withdrawing or electron-donating substituents. The main derivatives are bis-alkyl, bis-acyl, bis-sulfonyl, and bis-imino compounds.

(1*S*,2*S*)-1,2-Diaminocyclohexane. (1*S*,2*S*)-1,2-Diaminocyclohexane along with other chiral 1,2-diamines has been used as ligand for the ruthenium-catalyzed hydrogenation of ketones. Chiral bisphosphine-ruthenium(II) diamine complexes have shown high efficiency in the catalytic hydrogen transfer from isopropanol to aromatic and conjugated ketones (eq 10).[19,20] Complexes including (1*S*,2*S*) 1,2-diaminocyclohexane gave slightly lower ee's than those with 1,2-diphenyl ethylenediamine or 1,1-dianisyl-2-isosopropyl-1,2-ethylenediamine (DAIPEN). The enantioselective reduction of ketones may also be performed with polyhydroxysilane in the presence of a zinc diamine complex.[21] Using (1*S*,2*S*)-1,2-diaminocyclohexane or its alkylated derivatives, ee's were lower than with other chiral diamines.

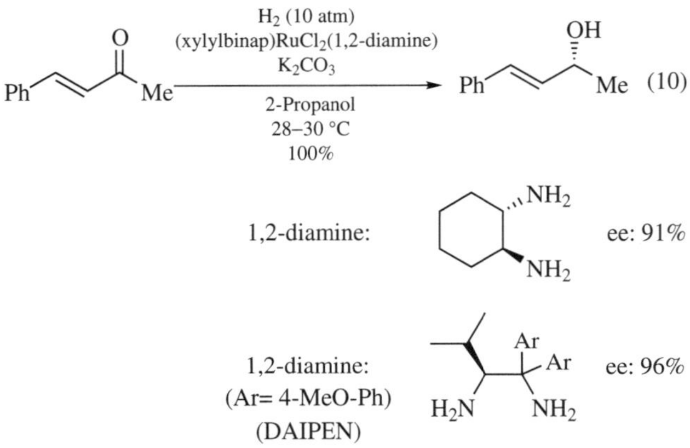

Another application of the free diamine as a ligand for asymmetric catalysis is the Michael reaction of cyclic β-keto esters with methyl vinyl ketone, which has been accomplished with a nickel(II)-(1*S*,2*S*)-1,2-diaminocyclohexane complex, with ee's up to 91% (eq 11).[22]

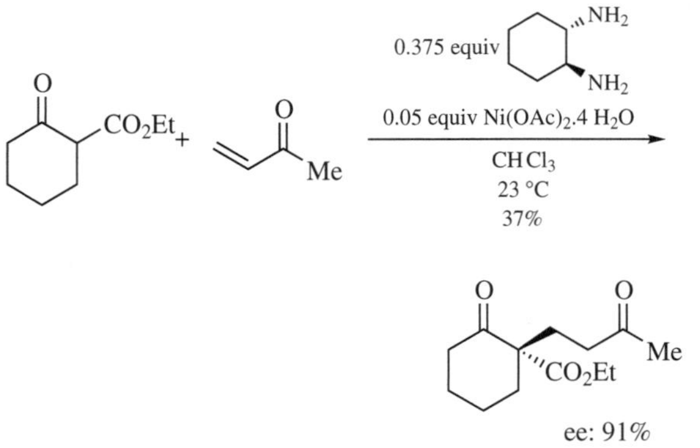

Alkyl and Aryl derivatives. Except for *N,N′*-dimethylaminocyclohexane, which is commonly prepared in two steps by treatment of 1,2-diaminocyclohexane with ethyl chloroformate,[14] followed by reduction, *N,N′*-alkyl derivatives are better prepared by the reductive amination of aldehydes with 1,2-diaminocyclohexane. The most important application of these alkyl derivatives in asymmetric catalysis involves the osmium-catalyzed dihydroxylation of alkenes: a great variety of substrates have been transformed into diols with high yields and enantiomeric excesses using *N,N′*-bis-neohexyl diaminocyclohexane (eq 12).[23] The *N,N,N′*, *N′*-tetramethyl derivative has also been used for this reaction.[24]

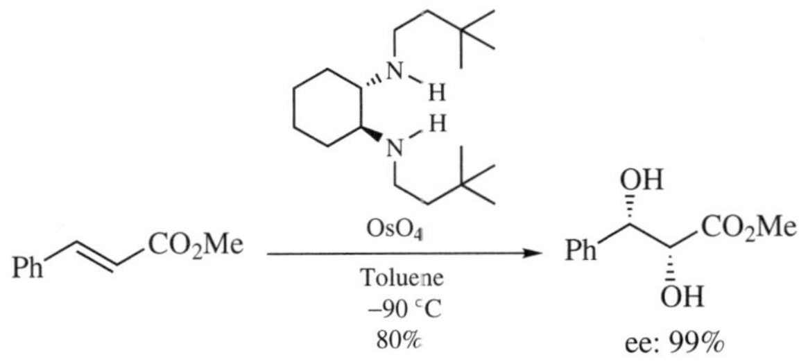

(12)

A method for the monoarylation of (1*S*,2*S*)-1,2-diaminocyclohexane by means of palladium-catalyzed aromatic amination has been recently described (eq 13).[25] The resulting new ligands were tested in the catalytic asymmetric transfer hydrogenation of acetophenone.

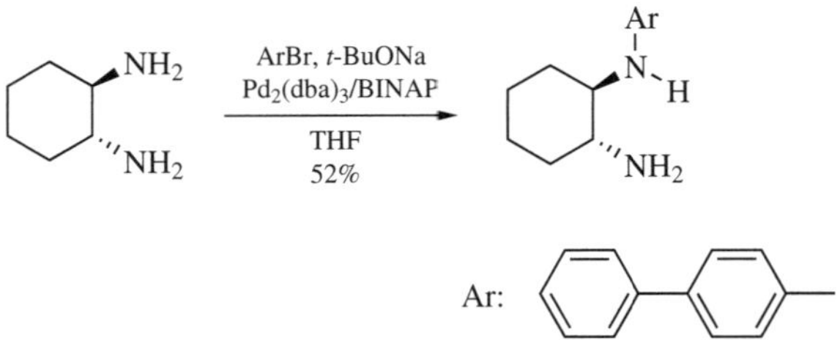

(13)

***N,N′*-Bis-Acyl and *N,N′*-Bis-Sulfonyl derivatives.** The preparation of *N,N′*-bis-acyl derivatives of 1,2-diaminocyclohexane involves the coupling of the diamine with 2 equiv of a carboxylic acid in the presence of a coupling reagent such as DCC,[26] whereas *N,N′*-bis-sulfonyl derivatives are better prepared by treatment with alkyl (or aryl) sulfonyl chlorides in the presence of Hünig's base.[27] These electron-poor compounds have been widely used as chiral ligands for palladium, zinc or titanium-catalyzed reactions such as allylic substitution, cyclopropanation, and 1,2-addition to carbonyl compounds. A great variety of acyl and sulfonyl groups may be introduced on the chiral diamine for ligand tuning, including chiral substituents.[28] The development of new ligands and the screening of their efficiency by parallel synthesis has been described.[29]

The palladium-catalyzed allylic substitution of allylic acetates with soft nucleophiles has attracted increasing attention in recent years,[30,31] and many different chiral ligands have been used for this reaction, including those derived from 1,2-diaminocyclohexane. The tetradentate ligand *N,N′*-bis(2-diphenylphosphino-benzoyl)-(1*S*,2*S*)-1,2-diaminocyclohexane (**7**) has shown very high levels of chirality transfer for acyclic and cyclic substrates with low amounts of catalyst loading (eq 14).[32] An application of this reaction to the synthesis of alkylated α-amino acids is shown in (eq 15).[33] Other diamines were also tested for this reaction, with slightly lower ee's resulting. Modification catalyst structure has been proposed, with naphthyl,[32] ferrocenyl[34] groups replacing the benzene ring. More recently, a molybdenum complex with a bis(2-pyridyl) ligand has been reported to give excellent regio and enantioselectivities in the allylic substitution of linear acetates.[35] In an analogous study, a series of C_2 symmetrical bis-oxazolines of type **8** with a 1,2-diaminocyclohexane backbone were synthesized and tested for the molybdenum-catalyzed allylic substitution with high levels of enantiomeric excess and excellent branched/linear ratio of products (eq 16).[36]

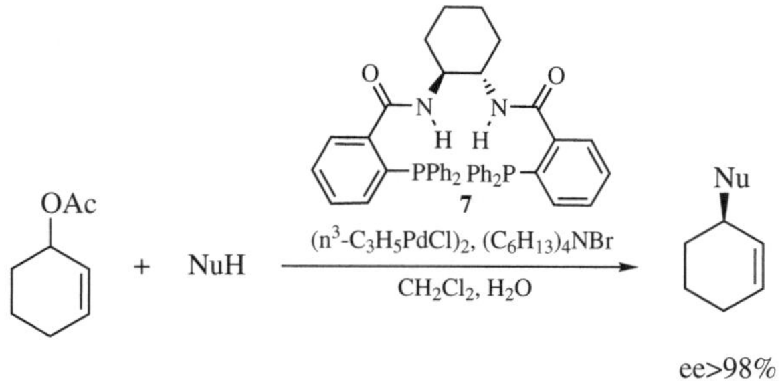
(14)

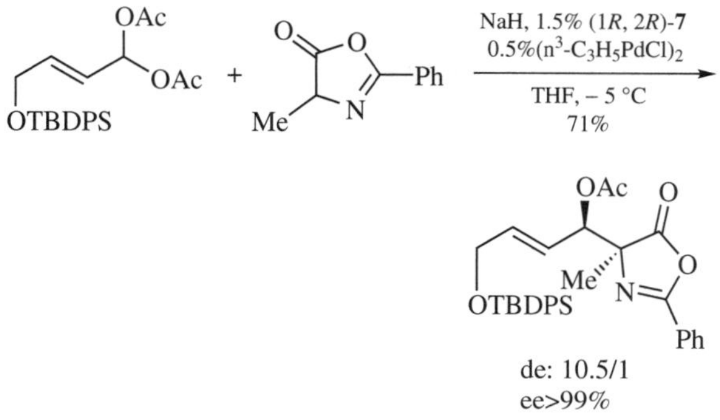
(15)

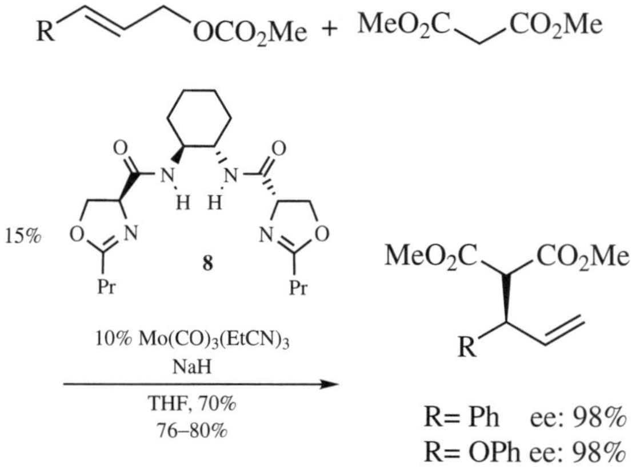
(16)

The first application of bis-sulfonyl derivatives of (1*S*,2*S*)-1,2-diaminocyclohexane as chiral ligands was for the titanium-catalyzed addition of diethylzinc to aldehydes,[37] which occurred with high enantioselectivity when a titanium (IV) salt was added.[27] Various sulfonyl derivatives were screened, the best results being obtained with the bis-triflamide **9**, which gave ee's up to 99% (eq 17). The scope of this reaction has been extended to aliphatic substrates including functionalized ones, and to other zinc organometallic reagents (eq 18).[38] Other ligands have been recently introduced for this reaction, the most selective being a tetradentate ligand, which gives very high ee's with α,β-unsaturated and aromatic aldehydes.[39] Although the use of other chiral diamines does not seem to have been thoroughly investigated, asymmetric addition of organozinc reagents to aldehydes may also be performed using the chiral 1,2-diol TADDOL with comparable selectivities.

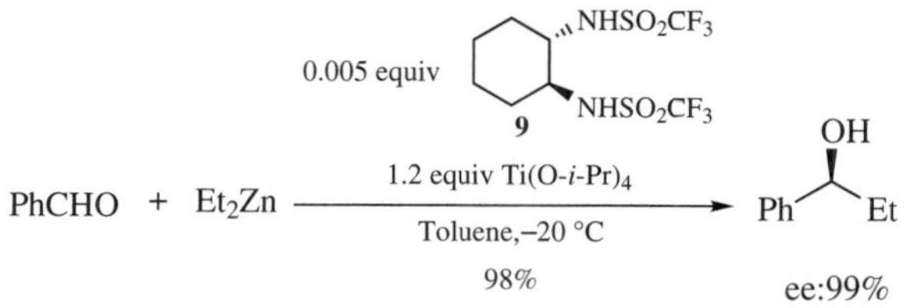
(17)

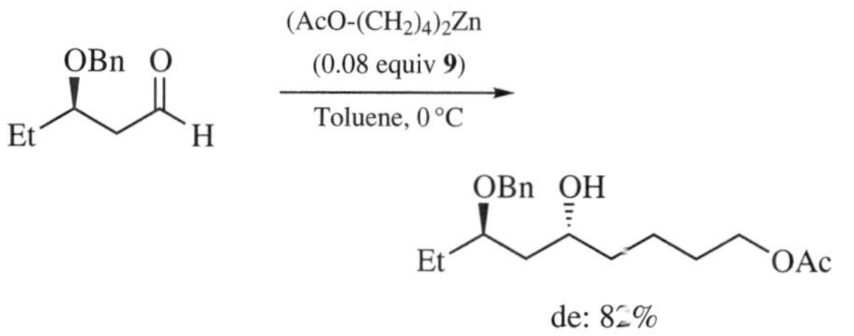
(18)

Bis-sulfonamides of (1*S*,2*S*)-1,2-diaminocyclohexane are also good ligands for the asymmetric cyclopropanation of allylic alcohols using the Simmon–Smith reagent.[40] Early reports recommended the use of the *N,N′*-bis-*p*-nitrophenylsulfonyl derivative, with ee's in the range of 60–80% with *trans*-allylic alcohols.[41] Further studies of reaction conditions and ligand screening have resulted in an enhancement of enantiomeric excess, best results being obtained with *N,N′*-bis-methanesulfonyl (1*S*,2*S*)-1,2-diaminocyclohexane (**10**), giving an ee of 80% for cinnamyl alcohol (eq 19).[42] More recently, a mixed sulfonyl-imine derivative of (1*S*,2*S*) 1,2-diaminocyclohexane has been reported to give good selectivities in the cyclopropanation of *cis*- and *trans*-allylic alcohols,[43] although increased catalyst loading (up to 50%) was necessary for achieving good enantioselectivity.

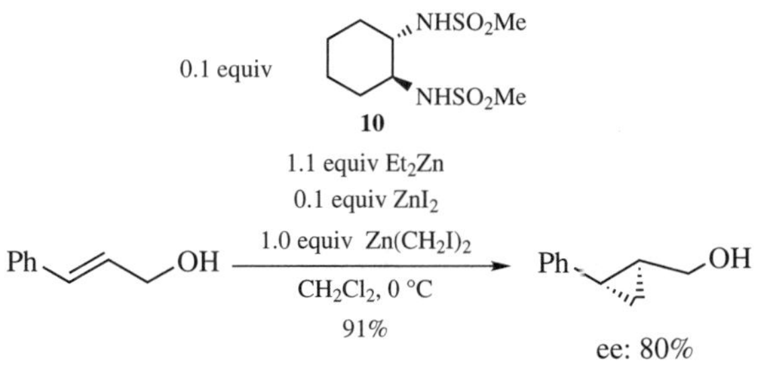
(19)

Bis-Imine Derivatives. Bis-imine derivatives of (1*S*,2*S*)-1,2-diaminocyclohexane are prepared in high yields and on a large scale by condensation of 2 equiv of an aromatic aldehyde with (1*S*,2*S*)-1,2-diaminocyclohexane. These compounds are very important and efficient chiral ligands for various transition metal-catalyzed reactions. Bis-imine derivatives of 1,2-diphenyl ethylenediamine have also been prepared but have been less often used.

Copper complexes derived from bis(-2,6-dichlorophenyledene)-(1*S*,2*S*)-1,2-diaminocyclohexane (**11**) catalyze various reactions such as Diels–Alder reaction,[44] aziridination (eq 20),[45] cyclopropanation,[46] and silyl enol ether addition to pyruvate esters.[47] Although the scope of these reactions may be sometimes limited, enantioselectivities are generally high. The same complex (with copper(I) salts) catalyzes the asymmetric insertion of silicon– hydrogen bond into carbenoids.[48]

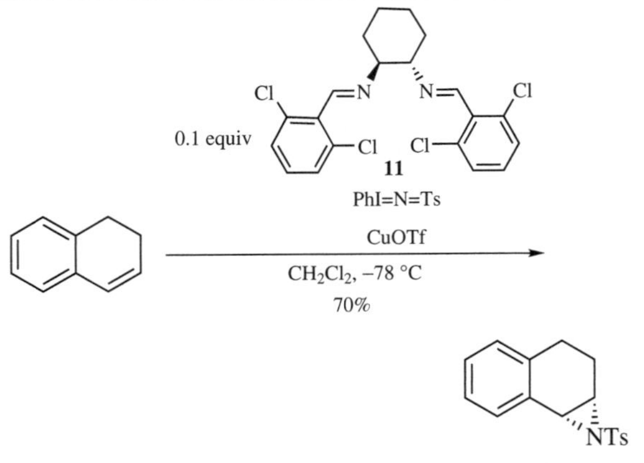

(20)

ee: 87%

Bis-salicylidene (or bis-salen) derivatives of 1,2-diaminocyclohexane are prepared by treatment of the diamine (or its tartrate salt) with an *o*-hydroxybenzaldehyde derivative and are used for asymmetric manganese-, cobalt- or chromium-catalyzed reactions. The most important ligand of this type is *N,N'*-bis(3,5-di-*tert*-butylsalicylidene)-diaminocyclohexane (**12**), the detailed synthesis of which has been published.[2] Another important ligand is the atropoisomeric derivative (**13**) (eq 21).[49]

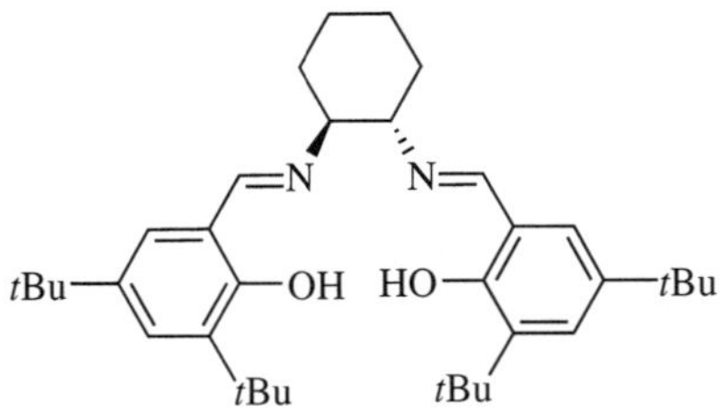

12

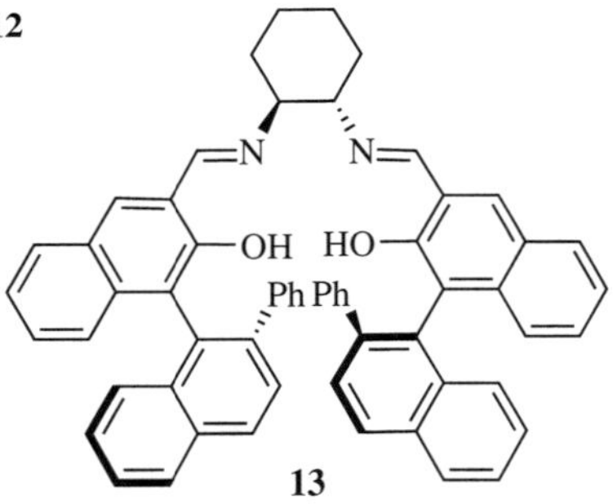

(21)

13

The most important application of these ligands in asymmetric catalysis is the manganese-catalyzed epoxidation of *cis*-alkenes using simple oxidants such as bleach or iodosobenzene (eq 22).[50] The manganese complexes of **12** and **13** give good to excellent ee's in the asymmetric oxygen-transfer to unfunctionalized alkenes. The manganese complex of **13** also oxidizes *meso*-cyclic ethers to give the corresponding lactols (eq 23).[49] Another important reaction is the ring-opening of achiral epoxides with nucleophiles such as trimethylsilyl azide or carboxylic acids or alcohols to give chiral bifunctional compounds (eq 24).[51] The reaction is catalyzed by chromium or cobalt complexes of salen ligand **12** and shows high levels of facial discrimination. Kinetic resolution of chiral racemic epoxides may also be accomplished with high efficiency (eq 25).[52] Chromium complex of salen ligand **12** has also been used in an asymmetric allylation of aromatic and aliphatic aldehydes. The resulting allylic alcohols were obtained in 65% to 89% enantiomeric excesses.[53] These complexes may be supported on polymers and used for combinatorial or parallel synthesis.[54] A fluorinated equivalent of this complex has been prepared for epoxidation reactions in perfluorinated solvents.[55]

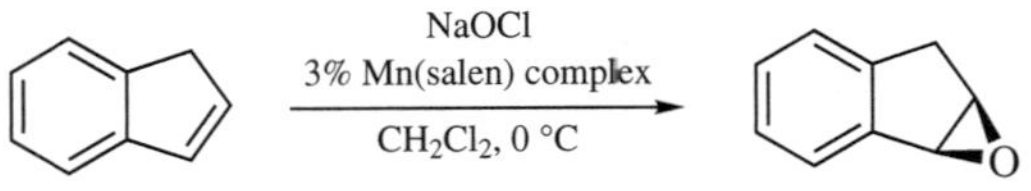

(22)

complex with **12**: ee: 86%
complex with **13**: ee: 98%

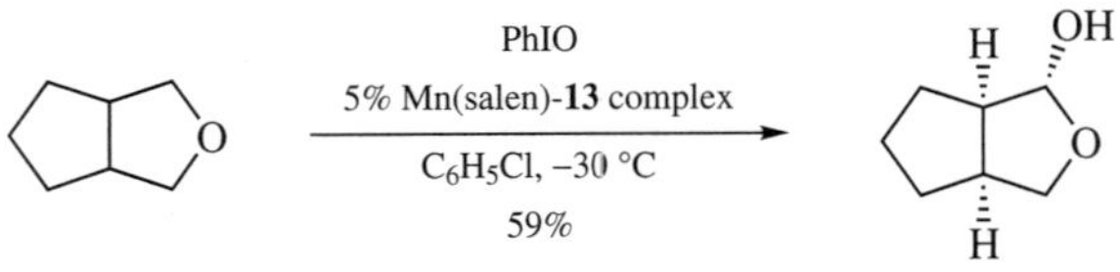

(23)

ee: 82%

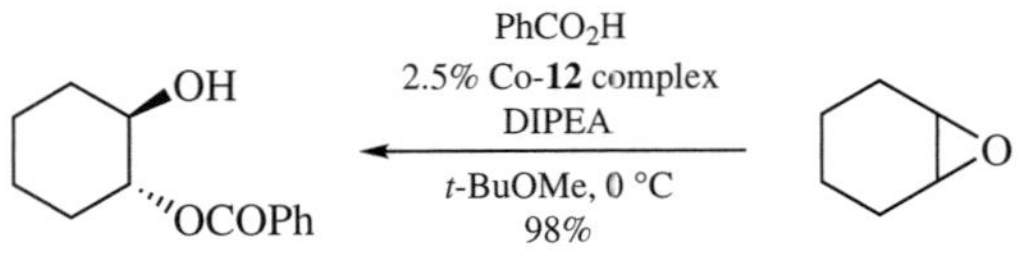

ee: 77%
ee: 98% after recrystallisation

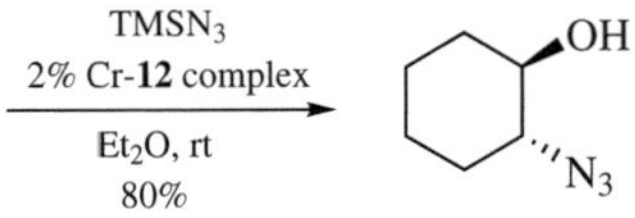

(24)

ee: 88%

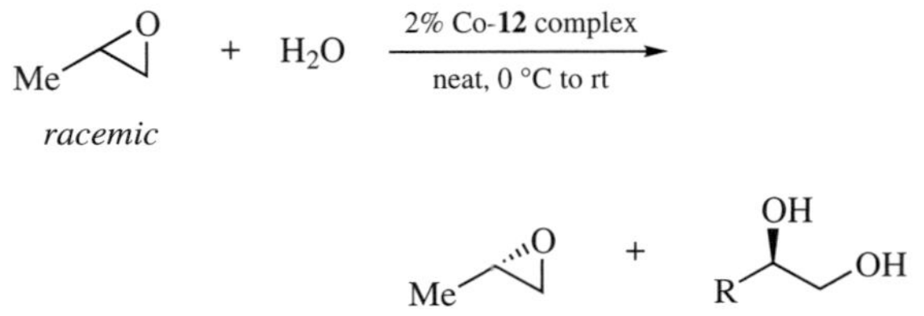

(25)

yield: 92%
ee>99%

yield: 89%
ee: 99%

Molecular Recognition. 1,2-Diaminocyclohexane has been often used as a scaffold for the syntheses of chiral host molecules and artificial receptors. Most of the examples relevant to this field may be found in reference 1.

C_2- and D_2-symmetric dioxatetraaza 18-membered macrocycles such as **14** and **15** have been prepared by a chemoenzymatic method involving (±)-*trans*-diaminocyclohexane as starting material (eq 26).[56] Good enantiomeric discrimination was observed with tetraprotonated species (*R,R*)-**14** and the D-enantiomer of *N*-acetyl aspartate.

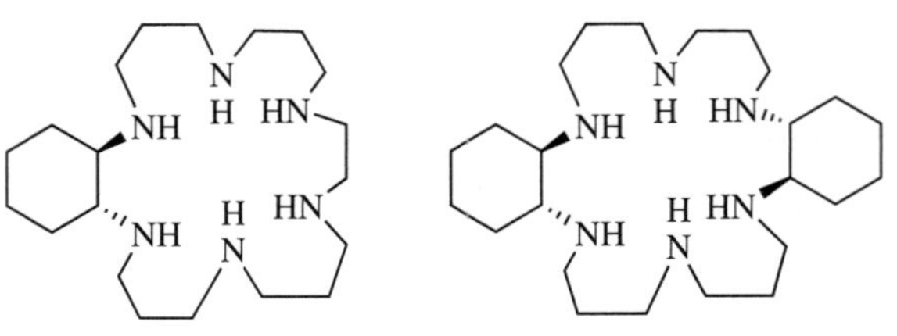

14 **15**

(26)

A complete study concerning a new class of supramolecular structures (supraminols) has been recently published.[57] These supramolecular structures are formed by an enantiodifferentiating self-assembly between (1*R*, 2*R*)-diaminocyclohexane and various trans-1,2-cyclohexane diols. An example is shown in eq 27. When racemic diols were used, a homochiral cristalline adduct is formed with an efficient kinetic resolution.

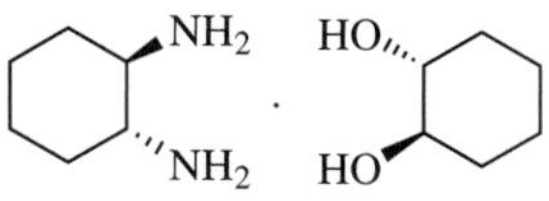

(27)

Related Reagents. 1,2-Diphenyl ethylenediamine; 2-aminomethyl pyrrolidine; 1,2-aminoalcohols; and 1,2-diols.

1. Bennani, Y. L.; Hanessian, S. *Chem. Rev.* **1997**, *97*, 3161.
2. Larrow, J. F.; Jacobsen, E. N. *Org. Synth.* **1998**, *75*, 1.
3. Walsh, P. J.; Smith, D. K.; Castello, C. *J. Chem. Ed.* **1998**, *75*, 1459.
4. Ôi, N.; Kitahara, H.; Aoki, F. *J. Chromatogr. A.* **1995**, *707*, 380.
5. Lucet, D.; Mioskowski, C.; Le Gall, T. *Angew. Chem. Int. Ed. Engl.* **1998**, *37*, 2580.
6. Kawashima, M.; Hirayama, A. *Chem. Lett.* **1990**, 2299.
7. Alexakis, A.; Mutti, S.; Mangeney, P. *J. Org. Chem.*, **1992**, *57*, 1224.
8. Alexakis, A.; Tranchier, J.-P.; Mangeney, P.; Feneau-Dupont, J. *Synthesis* **1995**, 1038.
9. Alexakis, A.; Mangeney, P.; Marek, I.; Rose-Munch, F.; Rose, E.; Semra, A., Robert, F. *J. Am. Chem. Soc.* **1992**, *114*, 8288.
10. Ratni, H.; Kündig, P. *Org. Lett.* **1999**, *1*, 1997.
11. Davies, S. G.; Mortlock, A. A. *Tetrahedron Lett.* **1992**, *33*, 117.
12. Dalko, P. I.; Langlois, Y. *J. Org. Chem.*, **1998**, *63*, 8107.
13. Dalko, P. I.; Brun, V.; Langlois, Y. *Tetrahedron Lett.* **1998**, *39*, 8979.
14. Hanessian, S.; Delorme, D.; Beaudoin, S.; Leblanc, Y. *J. Am. Chem. Soc.* **1984**, *106*, 5754.
15. Hanessian, S.; Gomtsyan, A.; Payne, A.; Hervé, Y.; Beaudoin, S. *J. Org. Chem.* **1993**, *58*, 5032.
16. Hanessian, S.; Andreotti, D.; Gomtsyan, A. *J. Am. Chem. Soc.* **1995**, *117*, 10393.
17. Hanessian, S.; Bennani, Y. L. *Synthesis* **1994**, 1272.
18. Yanagisawa, A.; Inanami, H.; Yamamoto, H. *Chem. Commun.* **1998**, 1573.
19. Okhuma, T.; Koizumi, M.; Doucet, H.; Pham, T.; Kozawa, M.; Murata, K.; Katayama, A.; Yokozawa, T.; Ikariya, T.; Noyori, R. *J. Am. Chem. Soc.* **1998**, *120*, 13529.
20. Burk, M. J.; Hems, W.; Herzberg, D.; Malan, C.; Zanotti-Gerosa, A. *Org. Lett.* **2000**, *2*, 4173.
21. Mimoun, H.; de Saint Laumer, J.-Y.; Giannini, L.; Scopelliti, R.; Floriani, C. *J. Am. Chem. Soc.* **1999**, *121*, 6158.
22. Christoffers, J.; Rössler, U.; Werner, T. *Eur. J. Org. Chem.* **2000**, 701.
23. Hanessian, S.; Meffre, P.; Girard, M.; Beaudoin, S.; Sancéau, J.-Y.; Bennani, Y. *J. Org. Chem.*, **1993**, *58*, 1991.
24. Tokles, M.; Snyder, J. K. *Tetrahedron Lett.* **1986**, *27*, 3951.
25. Frost, C. G.; Mendonça, P. *Tetrahedron: Asymmetry* **1999**, *10*, 1831.
26. Trost, B. M.; Van Vranken, D. L.; Bingel, C. *J. Am. Chem. Soc.* **1992**, *114*, 9327.
27. Takahashi, H.; Kawakita, T.; Ohno, M.; Yoshioka, M.; Kobayashi, S. *Tetrahedron* **1992**, *27*, 5691.
28. Flores-Lopez, L. Z.; Parra-Hake, M.; Somanathan, R.; Ortega, F.; Aguirre, G. *Synth. Commun.* **2000**, *30*, 147.
29. Gennari, C.; Ceccarelli, S.; Piarulli, U.; Montalbetti, C. A. G. N.; Jackson, R. F. W. *J. Org. Chem.* **1998**, *63*, 5312.
30. Trost, B. M.; Van Vranken, D. L. *Chem. Rev.* **1996**, *96*, 395.
31. Pfaltz, A.; Lautens, M., In *Comprehensive Asymmetric Catalysis*; Jacobsen, E. N.; Pfaltz, A.; Yamamoto, H., Eds. Springer: Berlin, 1999, Vol. 2, p 833.
32. Trost, B. M. *Acc. Chem. Res.* **1996**, *29*, 355.
33. Trost, B. M.; Lee, C. B. *J. Am. Chem. Soc.* **1998**, *120*, 6818.
34. Longmire, J.; Wang, B.; Zhang, X. *Tetrahedron Lett.* **2000**, *41*, 5435.
35. Trost, B. M.; Hachiya, I. *J. Am. Chem. Soc.* **1998**, *120*, 1104.
36. Glorius, F.; Pfaltz, A. *Org. Lett.* **1999**, *1*, 141.
37. Soai, K.; Shibata, T., In *Comprehensive Asymmetric Catalysis*; Jacobsen, E. N.; Pfaltz, A.; Yamamoto, H., Eds.; Springer Berlin, **1999**, Vol. 2, p 911.
38. Knochel, P.; Singer, R. D. *Chem. Rev.* **1993**, *93*, 2117.
39. Qiu, J.; Guo, C.; Zhang, X. *J. Org. Chem.* **1997**, *62*, 2665.
40. Charette, A.; Lebel, H., In *Comprehensive Asymmetric Catalysis*; Jacobsen, E. N.; Pfaltz, A.; Yamamoto, H., Eds.; Springer: Berlin, 1999, Vol 2, p 581.
41. Takahashi, H.; Yoshioka, M.; Ohno, M.; Kobayashi, S. *Tetrahedron Lett.* **1992**, *33*, 2575.
42. Denmark, S. E.; O'Connor, S. P. *J. Org. Chem.* **1997**, *62*, 5284.
43. Balsells, J.; Walsh, P. J. *J. Org. Chem.*, **2000**, *65* 5005.
44. Evans, D. A.; Lectka, T.; Miller, S. J. *Tetrahedron Lett.* **1993**, *34*, 7027.
45. Jacobsen, E. N., In *Comprehensive Asymmetric Catalysis*; Jacobsen, E. N.; Pfaltz, A.; Yamamoto, H., Eds.; Springer: Berlin, 1999, Vol. 2, p 607.
46. Pfaltz, A., In *Comprehensive Asymmetric Catalysis*; Jacobsen, E. N.; Pfaltz, A.; Yamamoto, H., Eds.; Springer: Berlin, 1999, Vol 2, p 513.
47. Evans, D. A.; Burgey, C. S.; Kozlowski, M. C. Tregay, S. W. *J. Am. Chem. Soc.* **1999**, *121*, 686.
48. Dakin, L. A.; Schaus, S. E.; Jacobsen, E. N.; Panek, J. S. *Tetrahedron Lett.* **1998**, *39*, 8947.
49. Miyafuji, A.; Katsuki, T. *Tetrahedron* **1998**, *54*, 10339.
50. Jacobsen, E. N.; Wu, M. H., In *Comprehensive Asymmetric Catalysis*; Jacobsen, E. N.; Pfaltz, A.; Yamamoto, H., Eds.; Springer: Berlin, 1999, Vol. 2, p 649.
51. Jacobsen, E. N.; Wu, M. H., In *Comprehensive Asymmetric Catalysis*, Jacobsen, E. N.; Pfaltz, A.; Yamamoto, H., Eds.; Springer: Berlin, 1999, Vol. 3, p 1309.
52. Ready, J. M.; Jacobsen, E. N. *J. Am. Chem. Soc.* **1999**, *121*, 6086.
53. Bandini, M.; Cozzi, P. G.; Melchiorre, P.; Umani-Ronchi, A. *Angew. Chem. Int. Ed.* **1999**, *38*, 3357.
54. Peukert, S.; Jacobsen, E. N. *Org. Lett.* **1999**, *1*, 1245.
55. Pozzi, G.; Cinato, F.; Montanari, F.; Quici, S. *Chem. Commun.* **1998**, 877.
56. Alfonso, I.; Rebodello, F.; Gotor, V. *Chem Eur. J.* **2000**, *6*, 3331.
57. Hanessian, S.; Saladino, R.; Margarita, R.; Simard, M. *Chem Eur. J.* **1999**, *5*, 2169[YL1].

Cyrille Kouklovsky & Yves Langlois
Université de Paris-Sud, Orsay, France

(*R,R*)-1,2-Diamino-1,2-di-*tert*-butylethane

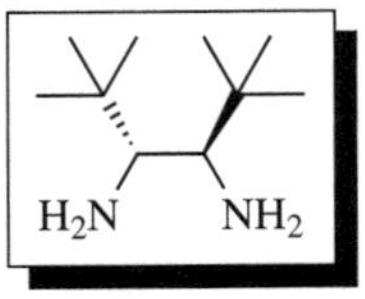

[171357-23-6] $C_{10}H_{24}N_2$ (MW 172.31)

(vicinal diamine as a source of chirality; precursor for the synthesis of bidentate ligands)

Physical Data: bp 240 °C; $[\alpha]_D^{25}$ −15 (*c* 0.145, CH_2Cl_2).
Solubility: most organic solvents.
Form Supplied in: colorless liquid; not commercially available.
Purification: shake with powdered K_2CO_3 in diethyl ether, filter, concentrate, and distill under reduced pressure.
Analysis of Reagent Purity: ^{1}H and ^{13}C NMR.
Handling, Storage, and Precautions: store under argon or nitrogen atmosphere; may be corrosive, like other vicinal diamines.
Preparative Methods: although syntheses of enantiomerically pure diamines have been developed mainly through resolution of racemic diamines, few methods have been reported for the diastereoselective synthesis of vicinal diamines. In particular, the addition of Grignard or zinc reagents to the carbon-nitrogen double bonds of chiral 1,2-bisimines[1] derived from glyoxal and (*S*)- or (*R*)-phenethylamine, followed by removal of the phenethyl group, has been shown to be an attractive alternative method. Following a similar procedure, 1,2-diamino-1,2-di-*tert*-butylethane **3** can be synthesized in an optically pure form in three steps, as the *R,R*- or *S,S*-enantiomer, starting, respectively, from glyoxal and (*S*)- or (*R*)-phenethylamine.[2] Indeed, addition of the chiral (*S,S*)-1,2-bis-imine **1** to a suspension of *tert*-butylmagnesium chloride in hexane at 50 °C leads cleanly and in good yield to a single diastereomer of the (*R,R*)-diamine (**2**) (eq 1).

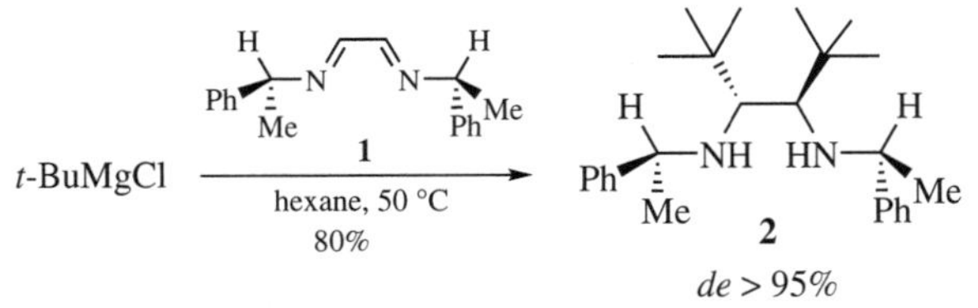

(1)

Deprotection of the chiral auxiliary groups is performed using ammonium formate, acetic acid, and palladium hydroxide in refluxing ethanol (eq 2). This sequence gives, after purification by distillation, the free optically pure (*R,R*)-diamine **3**.[3]

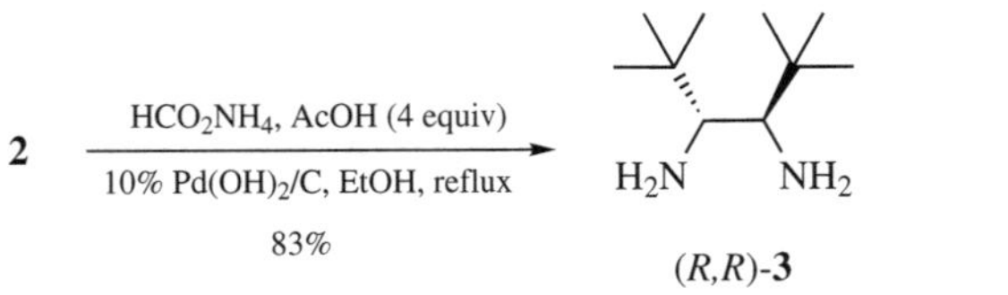

(2)

The *N,N'*-dimethyl derivative **4** is obtained in three steps by formation of the corresponding imidazolidine with acetone, alkylation with iodomethane, and hydrolysis (eq 3).

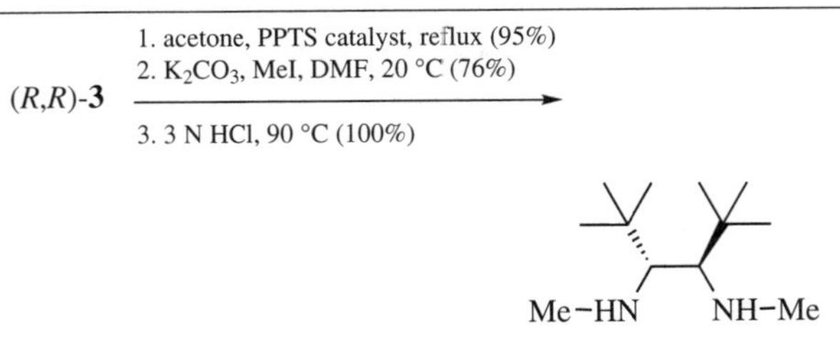

(3)

4

The *N*-tosyl derivative **5**, a possible precursor for the synthesis of ruthenium hydrogen transfer catalysts of type **6**,[4] is obtained in good yield by slow addition of TsCl in dichloromethane at 0 °C (eq 4).

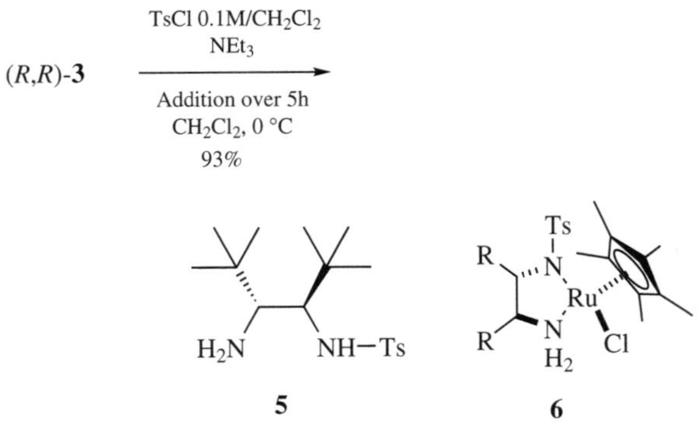

(4)

5 **6**

Utility. Many asymmetric syntheses have been developed using vicinal diamines as the source of chirality. The major interest lies in their use as precursors for the synthesis of a broad family of bidentate ligands.[5] Many reactions have also been described using the *N*-alkyl derivatives of these diamines as chiral auxiliaries and protecting groups of aldehydes.[6] Most of these applications generally use the framework of 1,2-diphenyl-1,2-diaminoethane (**7**) or 1,2-diaminocyclohexane (**8**), whose preparations have been fully described.[7]

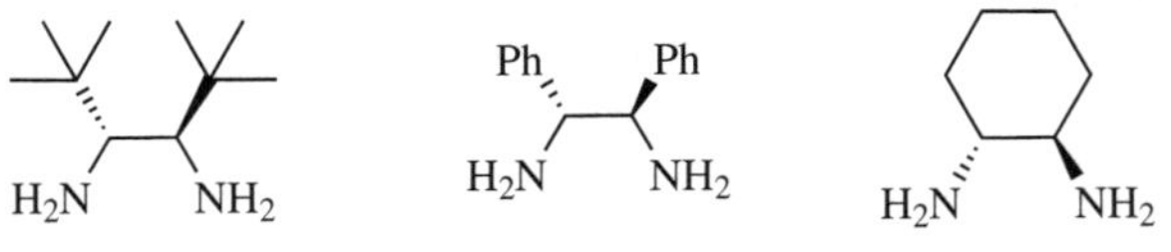

3 **7** **8**

However, 1,2-diamino-1,2-di-*tert*-butylethane (**3**) holds particular interest because of its increased steric bulk and the absence of benzylic protons. Its recent ready availability should render it as attractive as the frequently used vicinal diamines **7** and **8**. To our knowledge, only one application of this diamine has been previously described in the literature (eq 5),[3b] where the regio- and enantioselective epoxidation of conjugated aliphatic dienes were studied using the chiral manganese salen complex (**9**).

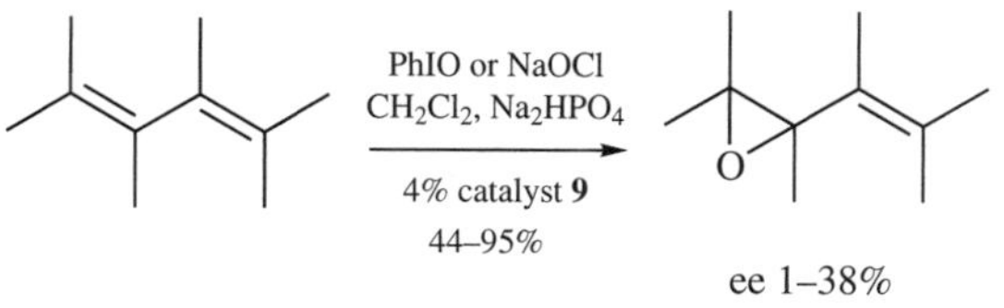

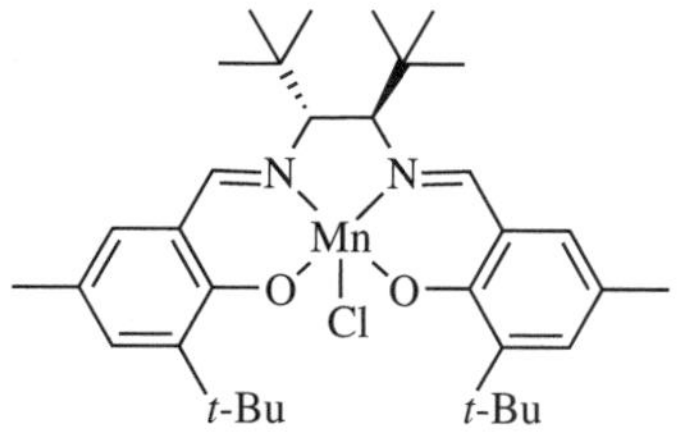

9 (5)

Related Reagents. 1,2-diphenyl-1,2-diaminoethane; 1,2-diaminocyclohexane.

1. (a) Tom Dieck, H.; Dietrich, J. *Chem. Ber.* **1984**, *117*, 694. (b) Neumann, W. L.; Rogic, M. M.; Dunn, T. J. *Tetrahedron Lett.* **1991**, *32*, 5865. (c) Alvaro, J.; Grepioni, F.; Savoia, D. *J. Org. Chem.* **1997**, *62*, 4180. (d) Alexakis, A.; Tomassini, A.; Chouillet, C.; Roland, S.; Mangeney, P.; Bernardinelli, G. *Angew. Chem., Int. Ed. Engl.* **2000**, *39*, 4093.
2. (a) Roland, S.; Mangeney, P.; Alexakis, A. *Synthesis* **1999**, 228. (b) Roland, S.; Mangeney, P. *Eur. J. Org. Chem.* **2000**, 611.
3. To our knowledge, the only method previously described for the synthesis of **3** involved the coupling of a nitrile or an *N*-(trimethylsilyl)imine, promoted by $NbCl_4$(THF). In this procedure the (+)-diamine was obtained pure in 18% yield by resolution with (−)-mandelic acid: (a) Roskamp, E. J.; Pedersen, S. F. *J. Am. Chem. Soc.* **1987**, *109*, 3152. (b) Rasmussen, K. G.; Thomsen, D. S.; Jørgensen, K. A. *J. Chem. Soc., Perkin Trans 1* **1995**, 2009.
4. (a) Murata, K.; Ikariya, T. *J. Org. Chem.* **1999**, *64*, 2186. (b) Murata, K.; Okano, K.; Miyagi, M.; Iwane, H.; Noyori, R.; Ikariya, T. *Org. Lett.* **1999**, *7*, 1119. (c) Uematsu, N.; Fujii, A.; Hashiguchi, S.; Ikariya, T.; Noyori, R. *J. Am. Chem. Soc.* **1996**, *118*, 4916.
5. (a) Tomioka, K. *Synthesis* **1990**, 541. (b) Corey, E. J.; Sarshar, S.; Bordner, J. *J. Am. Chem. Soc.* **1992**, *114*, 7938. (c) Corey, E. J.; Kim, S. S. *J. Am. Chem. Soc.* **1990**, *112*, 4976. (d) Bennani, Y. L.; Hanessian, S. *Chem. Rev.* **1997**, *97*, 3161. (e) Jacobsen, E. J. In *Catalytic Asymmetric Synthesis*; Ojima, I., Ed.; VCH, 1993, p 159-179. (f) Katsuki, T. *J. Mol. Cat.* **1996**, *113*, 87. (g) Mukaiyama, T. *Aldrichimica Acta* **1996**, *29*, 59. (h) Mukaiyama, T.; Yamada, T. *Bull. Chem. Soc. Jpn* **1995**, *68*, 17 and 1455. (i) Trost, B. M.; Van Vrancken, D. L. *Chem. Rev.* **1996**, *96*, 395. (j) Lucet, D.; Le Gall, T.; Mioskowski, C. *Angew. Chem., Int. Ed. Engl.* **1998**, *37*, 2580.
6. (a) Alexakis, A.; Mangeney, P. In *Advanced Asymmetric Synthesis*; Stephenson, G. R., Ed.; Chapman & Hall, London, UK, 1996; p 93. (b) Barettm A, G, M.; Doubledaym W. W.; Tustin, G. J.; White, A. J. P.; Williams, D. J. *J. Chem. Soc., Chem. Commun.* **1994**, 2739.
7. For the diamine **4**, see: (a) Pini, D.; Iuliano, A.; Rosini, C.; Salvadori, P. *Synthesis* **1990**, 1023. (b) Lohray, B. B.; Ahuja, J. R. *J. Chem. Soc., Chem. Commun.* **1991**, 95. (c) Oi, R.; Sharpless, K. B. *Tetrahedron Lett.* **1991**, *32*, 999. (d) Corey, E. J.; Imwinkelried, R.; Pikul, S.; Xiang, Y. B. *J. Am. Chem. Soc.* **1989**, *111*, 4486. (e) Shimizu, M.; Kamei, M.; Fujisawa, T. *Tetrahedron Lett.* **1995**, *36*, 8607. For the diamine **5**, see: (f) Wieland, A.; Schlichtung, O.; Langsdorf, W. V. Z. *Phys. Chem.* **1926**, *161*, 74. (g) Swift, G.; Swern, D. *J. Org. Chem.* **1967**, *32*, 511. (h) Whitney, T. A. *J. Org. Chem.* **1980**, *45*, 4214.

Sylvain Roland & Pierre Mangeney
Université Pierre et Marie Curie, Paris, France

Dibornacyclopentadienyltrichlorozirconium[1]

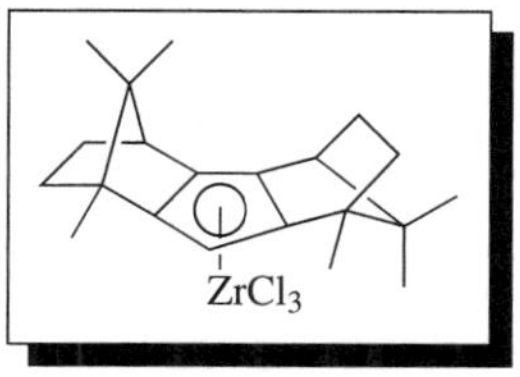

[126035-93-6] $C_{21}H_{29}Cl_3Zr$ (MW 479.07)

(chiral Lewis acid for enantioselective C–C coupling of pyruvic esters with active arenes[1])

Physical Data: ^{1}H NMR ($CDCl_3$, 200 MHz): δ 2.73, 3.05 (4,4′-H), 0.8–1.0, 1.8–2.2, 2.68 (5,5′,6,6′-H), 6.11 (8-H), 0.26, 0.92, 0.93, 0.95, 1.19, 1.46 (1,1′,7,7′-CH_3). ^{13}C NMR ($CDCl_3$, 50 MHz, ^1J(CH) in Hz): δ 11.6, 12.5 (1,1′-CH_3), 19.8, 20.6 (double intensity), 29.8 (7,7′-CH_3), 25.8 (122) and 00.0 (125; 5,5′), 32.1 (131) and 38.5 (137; 6,6′), 50.5 (149) and 51.6 (147; 4,4′), 55.8 (triple intensity), 68.6 (1,1′,7,7′), 106.4 (173; 8), 144.9, 146.9, 152.7, 159.7, 159.2 (2,2′,3,3′).

Solubility: sol dichloromethane.

Preparative Methods: reaction of dibornacyclopentadienyllithium with $ZrCl_4$ (40% yield).[1]

Preparation. Both the title zirconium catalyst (**1**), and the corresponding hafnium derivative (**2**) have been synthesized according to Scheme 1. Addition of 2 equiv of 2-bornen-2-yllithium (**4**) (generated via the Shapiro reaction[2] from (+)-camphor (**3**)) to ethyl formate gives bis(2-bornen-2-yl)methanol (**5**). Cyclization of alcohol (**5**) is carried out with *Potassium Hydrogen Sulfate* at 100–150 °C to afford dibornacyclopentadiene (**6**). Deprotonation of (**6**) with *n-Butyllithium* followed by reaction with *Zirconium(IV) Chloride* affords (**1**). Reaction with $HfCl_4$ affords (**2**).

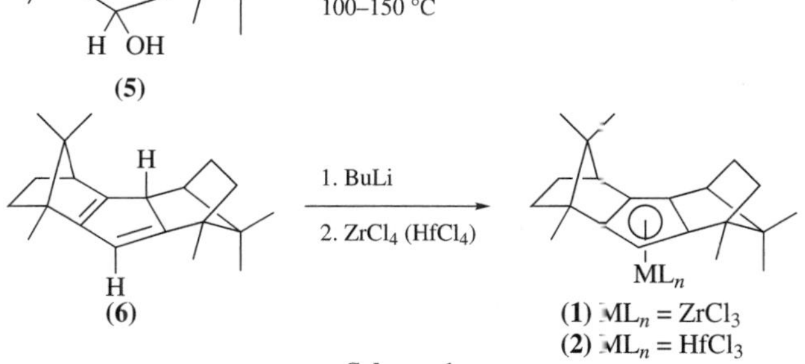

Scheme 1

Arylation of α-Keto Esters. Zirconium complex (**1**) has been used as a catalyst for the synthesis of optically active 2-(2-hydroxyaryl)lactic acid ethyl ester (**7**) (eq 1).[1]

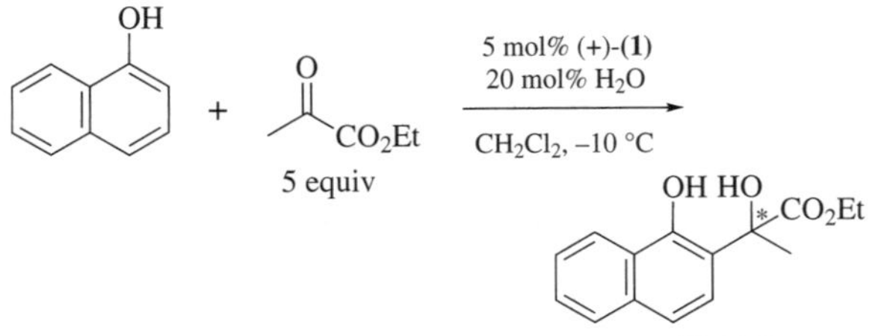

(+)-(**7**) 89% ee (1)

It was found that lowering the reaction temperature from rt to −10 °C resulted in an increased enantiomeric excess (from 27% ee to 54% ee). Surprisingly, the addition of small amounts of water (ca. 20 mol %) also increased the chiral induction observed (up to 89% ee).

1. Erker, G.; van der Zeijden, A. A. H. *Angew. Chem., Int. Ed. Engl.* **1990**, *29*, 512.
2. Chamberlin, A. R.; Stemke, J. E.; Bond, F. T. *J. Org. Chem.* **1978**, *43*, 147.

Eduardo Peña-Cabrera
Emory University, Atlanta, GA, USA

(*R*)-2,10-Dichloro-5*H*-dinaphtho[2,1-g:1,2-i] [1,5]dioxacycloundecin-3,6,9(7*H*)-trione[1]

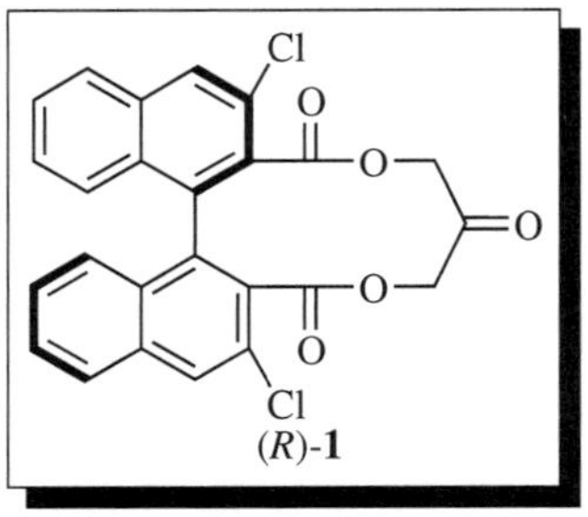

(*R*)-**1**

[184034-09-1] $C_{25}H_{14}Cl_2O_5$ (MW 465.28)

(a chiral ketone reagent used for enantioselective olefin epoxidations, kinetic resolution of acyclic secondary allylic silyl ethers)

Physical Data: a white solid.
Solubility: soluble in most organic solvents, such as CH_2Cl_2, $CHCl_3$, CH_3CN, DME, and ethyl acetate.
Analysis of Reagent Purity: IR, ^1H NMR, ^{13}C NMR, HRMS, and chiral HPLC.
Preparative Methods: Direct Synthesis Approach[2] (eq 1):

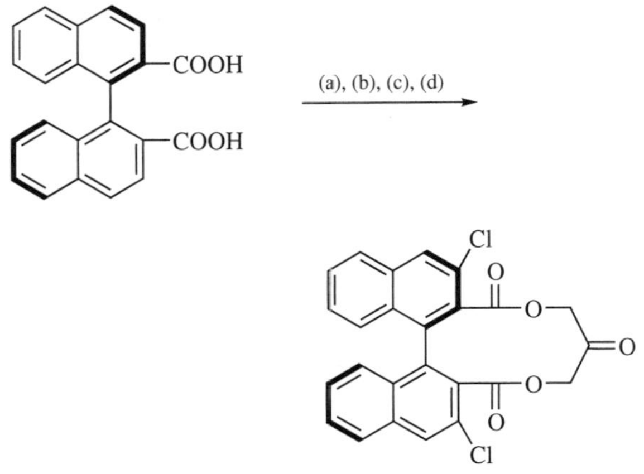

(*R*)-**1** (1)

Conditions:
(a) *sec*-BuLi, TMEDA, THF, −90 °C, 1.5 h.
(b) Cl_3CCCl_3, −78 °C.
(c) 3-chloro-2-chloromethyl-1-propene, Cs_2CO_3, DMF, 100 °C.
(d) $RuCl_3$, $NaIO_4$, CCl_4, CH_3CN, H_2O, rt.

Preparation of (*R*)-**1** involves the chlorination of (*R*)-1,1-binaphthyl-2,2-dicarboxylic acid with *sec*-BuLi and hexachloroethane. Condensation of the resulting dichorinated product with 3-chloro-2-chloromethyl-1-propene, followed by oxidative cleavage of the olefin moiety with ruthenium trichlorde[3] affords (*R*)-**1** in modest yields.
Enzymatic Resolution Approach[4] (eq 2): Ketone (*R*)-**1** can also be obtained by enzymatic resolution of racemic acetate (±)-**2**, which is prepared by reduction and acetylation of racemic (±)-**1** (prepared according to the direct synthesis approach). Oxidation of the resulting (*R*)-alcohol affords (*R*)-**1** in high enantioselectivity (> 99% ee). This method has been employed for large-scale synthesis of (*R*)-**1**.
Purification: silica gel flash column chromatography (30% ethyl acetate in *n*-hexane).
Handling, Storage, and Precautions: very stable at room temperature in air.

Table 1 Enantioselective epoxidation of *trans*-olefins using (*R*)-**1** as catalyst

Substrate	Product	Yield (%)	ee (%)
3a (R = H)	**4a**	95	76
3b (R = Me)	**4b**	>90	80
3c (R = Et)	**4c**	>90	85
3d (R = *i*-Pr)	**4d**	>90	85
3e (R = *t*-Bu)	**4e**	>90	91
3f	**4f**	96	76
3g	**4g**	75	65

Catalytic Enantioselective Epoxidation of Unfunctionalized *trans*-Olefins and Trisubstituted Olefins. (*R*)-**1** is an efficient catalyst for enantioselective epoxidation of unfunctionalized *trans*-olefins and trisubstituted olefins (eq 3).[2,5] In a homogenous CH_3CN/H_2O solvent system, (*R*)-**1** reacts with oxone to generate a chiral dioxirane in situ,[6] which can epoxidize *trans*-stilbenes (**3a**–**3e**) with high yields (> 90%) and high enantioselectivity (76–91% ee). The enantioselectivity of epoxidation is generally

Table 2 Kinetic resolution of racemic silyl ethers catalyzed by *R*-**1**

		(*S*)-**7a–7e**		**8a–8e**			
Substrate	conversion (%)	yield (%)	ee (%)	*erythro/threo*	yield (%)	ee (%)	S^a
7a (R = H)	55	84	96	>49:1	85	77	30
7b (R = Me)	45	86	74	>49:1	90	89	39
7c (R = Et)	56	86	99	>49:1	76	77	37
7d (R = *i*-Pr)	48	94	87	>49:1	88	93	72
7e (R = *t*-Bu)	50	87	94	>49:1	82	93	100

^{a}The selectivity (S) is calculated by the equation $S = \ln[(1 - C)(1 - \text{ee})]/\ln[(1 - C)(1 + \text{ee})]$, where C is the conversion and ee is the percentage enantiomeric excess of the recovered substrate.

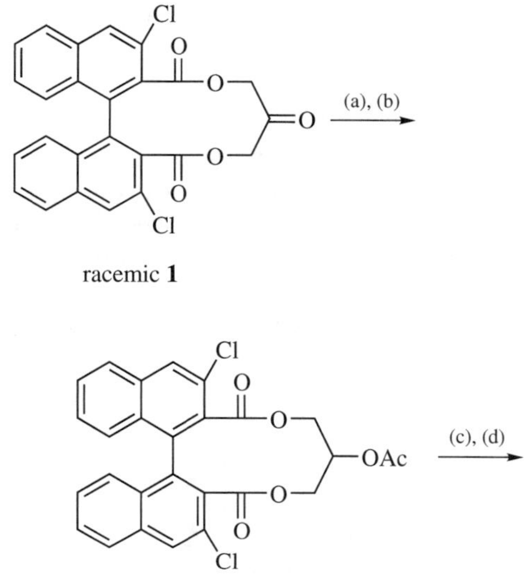

racemic **1**

racemic **2**

(2)

Conditions:
(a) $NaBH_4$, methanol.
(b) Ac_2O, DMAP.
(c) Lipoprotein lipase, toluene, 0.1 M tris-HCl buffer (pH 7.5), 30 °C, 48 h.
(d) MnO_2.

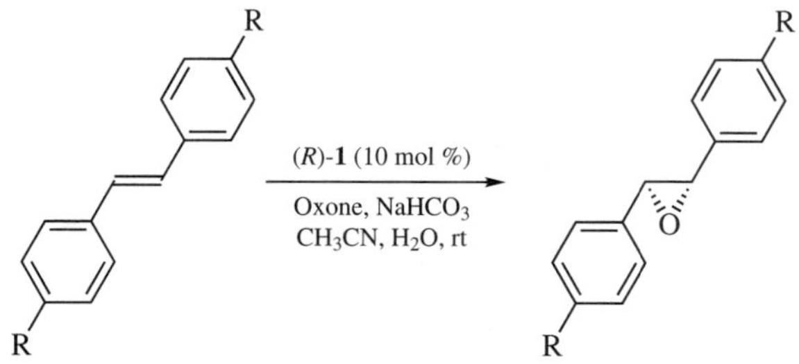

3a–3e **4a–4e** (3)

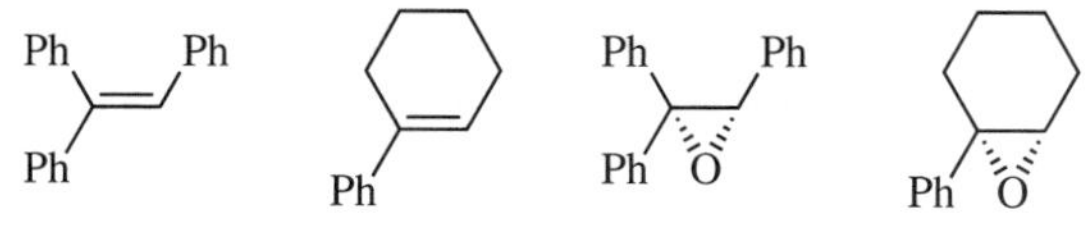

3f **3g** **4f** **4g**

higher when the para substituents of the *trans*-stilbenes become larger. Although the enantioselectivities for the trisubstituted olefins are generally lower than the *trans*-stilbenes, the ee value for olefin **3g** can be improved by replacing the chlorine atoms of (*R*)-**1** with more extended steric sensors (Table 1).[2] This epoxidation system requires only 10 mol % of the chiral ketone catalyst, and the epoxidation reactions are usually complete in 2–3 h at room temperature. Under these conditions, Baeyer–Villiger oxidation of the catalyst is not observed and (*R*)-**1** can be recovered with over 80% efficiency without the loss of catalytic activity and chiral induction.

Asymmetric Epoxidation of Electron-deficient *trans*-Olefins. (*R*)-**1** can also catalyze epoxidation of electron-deficient *trans*-olefins, especially (*E*)-cinnamate derivatives (eq 4).[7] With 5 mol % of (*R*)-**1**, epoxidation of acrylate (**5**) is completed in 27 h with 74% yield and 85% ee. The crude product can be purified using a continuous dissolution and crystallization process to afford enantiomerically pure product and recover the ketone catalyst simultaneously. A similar practical method[7] has been employed for large-scale synthesis of a key intermediate for diltiazem hydrochloride[8] (a potent calcium antagonist for treatment of cardiovascular disease).

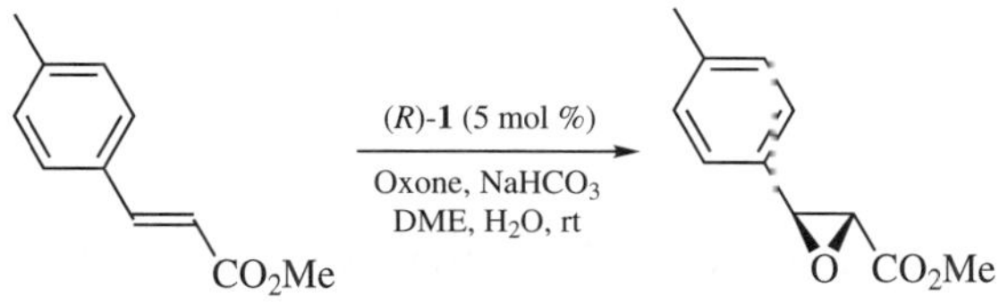

5 **6**: 74%, 85% ee (4)

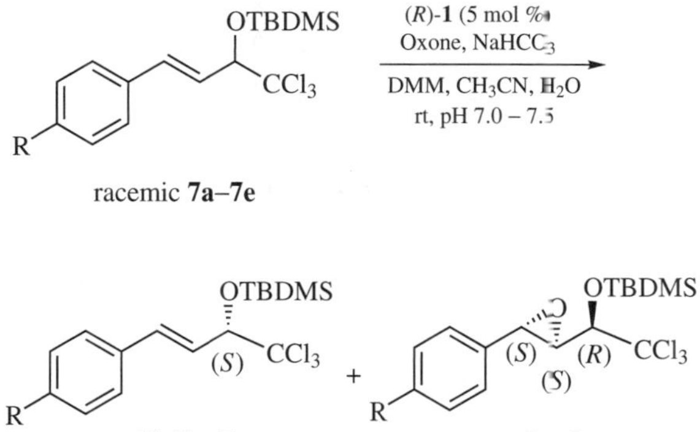

racemic **7a–7e**

(*S*)-**7a–7e** **8a–8e** (5)

Kinetic Resolution of Acyclic Secondary Allylic Silyl Ethers. (*R*)-**1** can catalyze kinetic resolution of acyclic secondary allylic silyl ethers (eq 5).[9] When racemic silyl ethers (**7a–7e**) are submitted to the optimized epoxidation conditions, the recovered starting materials are found to be enriched in the (*S*)-enantiomers, and the resulting epoxides (**8a–8e**) are single diastereomers (*ery-*

thro/threo ratio > 49:1) with (*R*)-configuration at the C-2 position. High selectivities (*S* up to 100) have been obtained for these α-trichloromethyl allylic silyl ether substrates (Table 2). This kinetic resolution approach provides both the recovered substrates and the resulting epoxides with high enantiomeric excess. These chiral compounds are useful building blocks for natural product synthesis, and there is no direct and obvious synthetic method for this class of compounds.

Related Reagents. (*R*)-5*H*-Dinaphtho[2,1-g:1,2-i][1,5]dioxacycloundecin-3,6,9(7*H*)-trione; oxone; *trans*-stilbenes; (*E*)-cinnamates; α-trichloromethyl allylic silyl ethers.

---

1. (a) Denmark, S. E.; Wu, Z. *Synlett* **1999**, 847. (b) Frohn, M.; Shi, Y. *Synthesis* **2000**, 1979.
2. Yang, D.; Wong, M. K.; Yip, Y. C.; Wang, X. C.; Tang, M. W.; Zheng, J. H.; Cheung, K. K. *J. Am. Chem. Soc.* **1998**, *120*, 5943.
3. Yang, D.; Zhang, C. *J. Org. Chem.* **2001**, *66*, 4814.
4. Seki, M.; Furutani, T.; Hatsuda, M.; Imashiro, R. *Tetrahedron Lett.* **2000**, *41*, 2149.
5. Yang, D.; Wang, X. C.; Wong, M. K.; Yip, Y. C.; Tang, M. W. *J. Am. Chem. Soc.* **1996**, *118*, 11311.
6. Yang, D.; Yip, Y. C.; Tang, M. W.; Wong, M. K.; Zheng, J. H.; Cheung, K. K. *J. Am. Chem. Soc.* **1996**, *118*, 491.
7. Seki, M.; Furutani, T.; Imashiro, R.; Kuroda, T.; Yamanaka, T.; Harada, N.; Arakawa, H.; Kusama, M.; Hashiyama, T. *Tetrahedron Lett.* **2001**, *42*, 8201.
8. Nagao, T.; Sato, M.; Nakajima, H.; Kiyomoto, A. *Chem. Pharm. Bull.* **1973**, *21*, 92.
9. Yang, D.; Jiao, G. S.; Yip, Y. C.; Lai, T. H.; Wong, M. K. *J. Org. Chem.* **2001**, *66*, 4619.

Dan Yang & Chi Sing Lee
The University of Hong Kong, Hong Kong

(−)-Dichloro(ethylene)(α-methylbenzylamine)platinum(II)

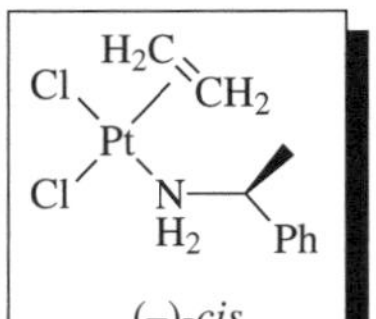

(−)-*cis*

(1) (−)-*cis*
[12084-44-5] $C_{10}H_{15}Cl_2NPt$ (MW 415.22)
(2) (−)-*trans*
[12084-42-3]

(determination of % ee of alkenes and allenes via ^{195}Pt NMR;[1] resolution of alkenes;[2] asymmetric epoxidation of alkenes[3])

Physical Data: trans isomer: mp 71 °C, $[\alpha]_D^{21}$ +15.5–16.9° (*c* 1.0, CH_2Cl_2), usually obtained as a viscous oil but can be crystallized upon further purification;[2g] *cis* isomer: mp 164 °C, $[\alpha]_D^{25}$ −54.5° (*c* 1.3, acetone), obtained as pale green-yellow needles.[4]

Solubility: both the *cis* and *trans* isomers are sol acetone and dichloromethane.

Analysis of Reagent Purity: trans isomer: ^{1}H NMR (CD_2Cl_2) δ 7.39 (s, 5H), 4.62 (t, *J* = 30.5 Hz, 4H), 1.78 (d, *J* = 6.5 Hz, 3H);[2g] for ^{13}C and ^{195}Pt NMR data see Pregosin et al.[5]

Preparative Methods: trans isomer: synthesized from Zeise's salt [K_2PtCl_3(ethylene)] by adding (+)- or (−)-α-methylbenzylamine under acidic conditions;[2g] *cis* isomer: synthesized by adding 2 equiv of (+)- or (−)-α-methylbenzylamine to an aqueous solution of K_2PtCl_4 and then exposing this material to ethylene under pressure.[4]

Purification: trans isomer: chromatography over silica with CH_2Cl_2 as eluent;[2g] *cis* isomer: recrystallization from 1:1 acetone/*n*-heptane or toluene.[4]

Handling, Storage, and Precautions: both isomers are air and water stable.

Determination of % ee via ^{195}Pt NMR. The *cis* chiral complex has been used to determine enantiomeric purity of asymmetric allenes[1a] and allylic alcohols and ethers[1b] via complexation and ^{195}Pt NMR spectroscopy. The complexes (**3**) and (**4**) are generated by displacement of ethylene from (**1**) by the alkene; recovery of the alkene and (**1**) is effected by the reverse sequence employing excess ethylene.[1b]

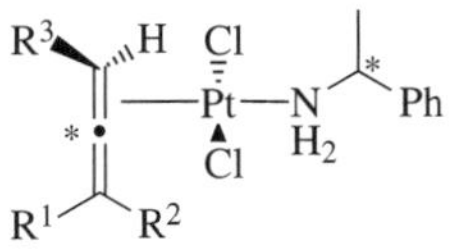

(**3**)

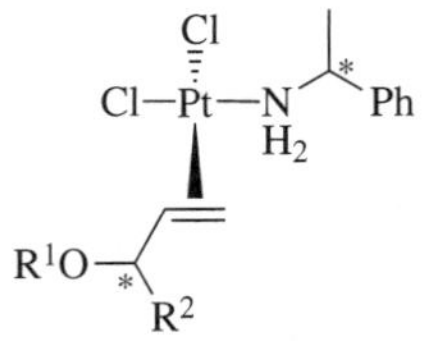

(**4**)

Resolving Agent. A variety of chiral alkenes have been resolved via complexation and crystallization including alkenes (**5**)–(**7**) (via **1**),[2a] *cis,trans*-1,5-cyclooctadiene (**8**) (via **2**),[2b] 2-vinyltetrahydropyran (**9**) (via **1**),[2c,d] 1,2-cyclononadiene (**10**) (via **2**),[2e] *trans*-cyclooctene (**11**) (via **2**),[2f] and spiro[3.3]hepta-1,5-diene (**12**) (via **2**).[2g]

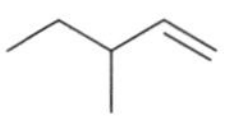

(**5**) 24% ee

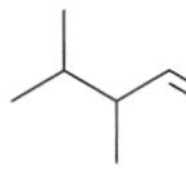

(**6**) 93% ee

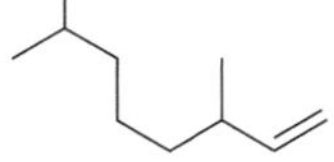

(**7**) 42% ee

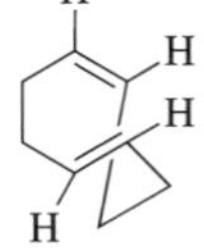

(**8**) 80% ee

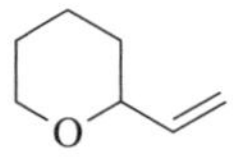

(**9**) 92% ee

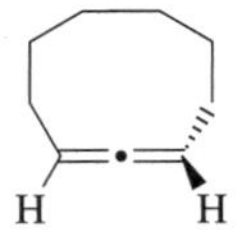

(**10**) 44% ee

(**11**) >99% ee

(**12**)

Asymmetric Epoxidation. McKervey and co-workers have effected the asymmetric epoxidation of humulene. Complexation of humulene with (**2**) in a 1:2 ratio followed by peracid oxidation gave (−)-humulene 1,2-epoxide in 37% ee (eq 1).[3]

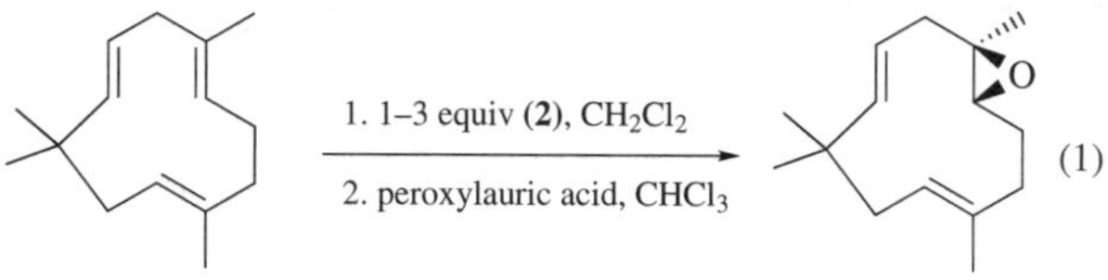

(1)

1. (a) Salvadori, P.; Uccello-Barretta, G.; Lazzaroni, R.; Caporusso, A. M. *Chem. Commun./J. Chem. Soc., Chem. Commun.* **1990**, 1121. (b) Salvadori, P.; Uccello-Barretta, G.; Bertozzi, S.; Settambolo, R.; Lazzaroni, R. *J. Org. Chem.* **1988**, *53*, 5768.
2. (a) Lazzaroni, R.; Salvadori, P.; Pino, P. *Tetrahedron Lett.* **1968**, *9*, 2507. (b) Cope, A. C.; Hecht, J. K.; Johnson, H. W., Jr.; Keller, H.; Winkler, H. J. S. *J. Am. Chem. Soc.* **1966**, *88*, 761. (c) Lazzaroni, R.; Uccello-Barretta, G.; Pini, D; Pucci, S.; Salvadori, P. *J. Chem. Res. (S)* **1983**, 286. (d) Lazzaroni, R.; Uccello-Barretta, G.; Bertozzi, S.; Bertucci, C.; Marchetti, F. *J. Chem. Res. (S)* **1984**, 286. (e) Cope, A. C.; Moore, W. R.; Bach, R. D.; Winkler, H. J. S. *J. Am. Chem. Soc.* **1970**, *92*, 1243. (f) Cope, A. C.; Ganellin, C. R.; Johnson, H. W., Jr.; Van Auken, T. V.; Winkler, H. J. S. *J. Am. Chem. Soc.* **1963**, *85*, 3276. (g) Hulshof, L. A.; McKervey, M. A.; Wynberg, H. *J. Am. Chem. Soc.* **1974**, *96*, 3906.
3. Chamberlain, T. R.; McKervey, M. A. *Chem. Commun./J. Chem. Soc., Chem. Commun.* **1969**, 366.
4. Paiaro, G.; Panunzi, A. *Tetrahedron Lett.* **1965**, *6*, 441.
5. Pregosin, P. S.; Sze, S. N.; Salvadori, P.; Lazzaroni, R. *Helv. Chim. Acta* **1977**, *60*, 2514.

Steven D. Paget
The Ohio State University, Columbus, OH, USA

Dichloro[2,3-*O*-isopropylidene-2,3-dihydroxy-1,4-bis(diphenylphosphino)-butane]palladium(II)

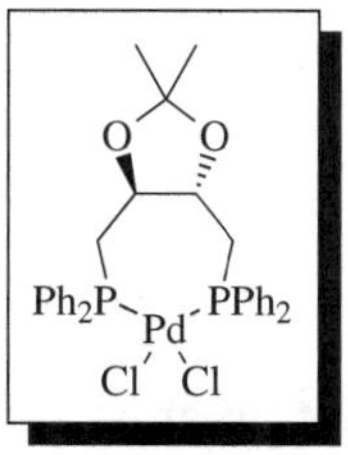

(2*R*,3*R*) or (−)
[63598-08-3] $C_{31}H_{32}Cl_2O_2P_2Pd$ (MW 675.87)
(2*S*,3*S*) or (+)
[59634-23-0]

(chiral catalyst used in asymmetric hydrocarboxylation[1] or hydroalkoxycarbonylation reactions,[2] allylation of β-diketones or β-keto esters,[3] double carbonylation,[4] cross-coupling reactions,[5] preparation of optically active Pd^0 derivatives and their subsequent reactions[6])

Alternate Name: (DIOP)$PdCl_2$.
Physical Data: crystalline pale yellow solid; ((−)-(2*R*,3*R*)-DIOP)$PdCl_2$, $[\alpha]_D$ −7.9° ($CHCl_3$, *c* 1.0);[6b] mp 278–279 °C (dec).[7b]
Solubility: sol dichloromethane; insol heptane.
Analysis of Reagent Purity: for ((−)-(2*R*,3*R*)-DIOP)$PdCl_2$: crystal structure,[7a] ^1H NMR,[6b,7b] ^{13}C NMR,[6b] ^{31}P NMR.[6b]
Preparative Methods: prepared from commercially available *Palladium(II) Chloride* and (−)-(2*R*,3*R*)-*(2,3-O-Isopropylidene)-2,3-dihydroxy-1,4-bis(diphenylphosphino)butane* in ether,[7a] or by treating a stoichiometric amount of (−)-(2*R*,3*R*)-DIOP in ethanol with Li_2PdCl_4 in water,[5] or from (+)-(2*S*,3*S*)-DIOP and $Pd(PhCN)_2Cl_2$ in benzene[6b] or acetone.[3c]
Handling, Storage, and Precautions: relatively stable to air oxidation.

Asymmetric Hydrocarboxylation. The title reagent was used in the first example of an asymmetric hydrocarboxylation (eq 1).[1a,1c] With the α-methylstyrene, the straight chain isomer was formed. The regiospecificity was much less pronounced, however, for other alkenic substrates.[1b] The influence of some reaction variables on the reaction shown in eq 1 was studied. For example, the presence of a solvent such as THF or benzene, the alcohol source, the effect of CO pressure, the effect of substitution on the phenyl ring, the $PdCl_2$/DIOP molar ratio, or the presence of PPh_3 along with DIOP, were varied to improve the optical yield.[1c–e]

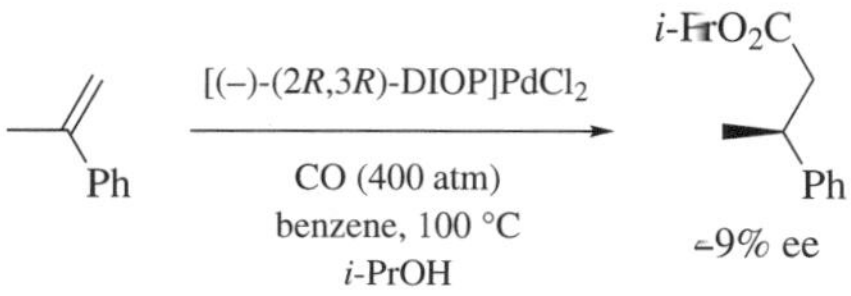

(1)

Methyl methacrylate has been hydromethoxycarbonylated with ((−)-(2*R*,3*R*)-DIOP)$PdCl_2$ as a catalyst to afford (*S*)-dimethyl succinate with excellent regioselectivity, but only modest enantioselectivity (eq 2).[2a,2b]

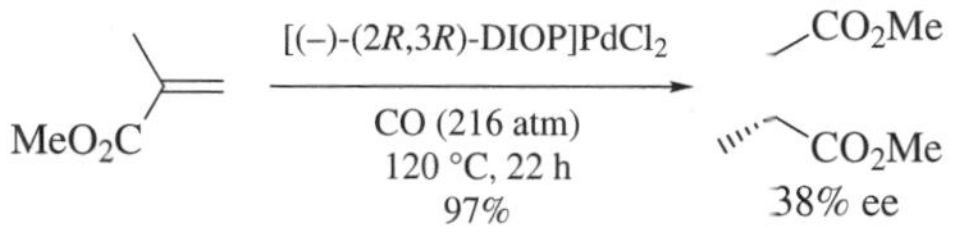

(2)

Asymmetric Allylic Alkylation. In early investigations of Pd-mediated allylations of alkenes and allylic acetates, Trost and his co-workers used (+)-(2*S*,3*S*)-DIOP along with $PdCl_2$ or $Pd(PPh_3)_4$ and obtained optical yields in the range of 12–46%.[3a,3b] In the allylation of β-diketones or β-keto esters using allyl phenyl ether or allylic esters, use of the title reagent as a chiral catalyst afforded allylated compounds in good yields, but with low enantioselectivity (eq 3).[3c]

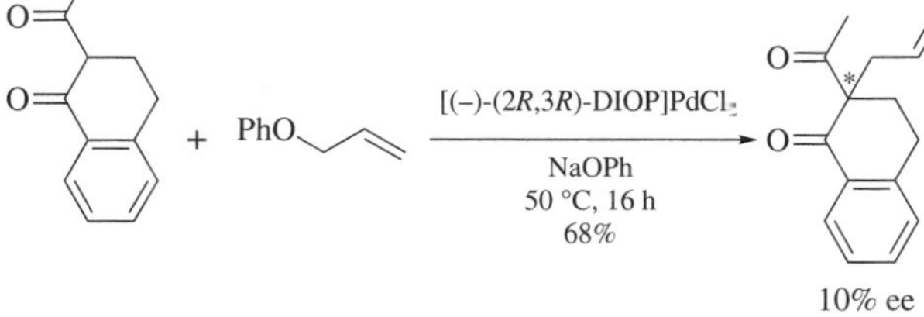

(3)

Double Carbonylation of Aryl Halides. (DIOP)$PdCl_2$ catalyzed the double carbonylation of phenyl iodide in the presence of diethylamine to afford the α-ketoamide in a very good yield along with a minor amount of the benzamide derivative (eq 4).[4]

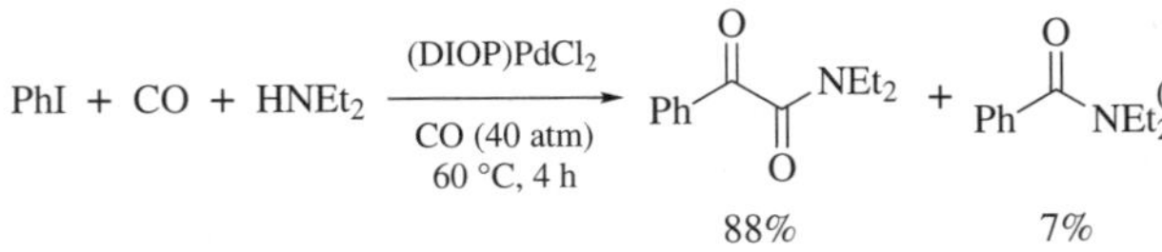

(4)

Cross-Coupling of Halides. 4,4-Dimethyl-1-phenylpenta-1,2-diene has been prepared using a ((−)-(2*R*,3*R*)-DIOP)$PdCl_2$-catalyzed cross-coupling reaction, although low enantioselectivity was observed (eq 5).[5]

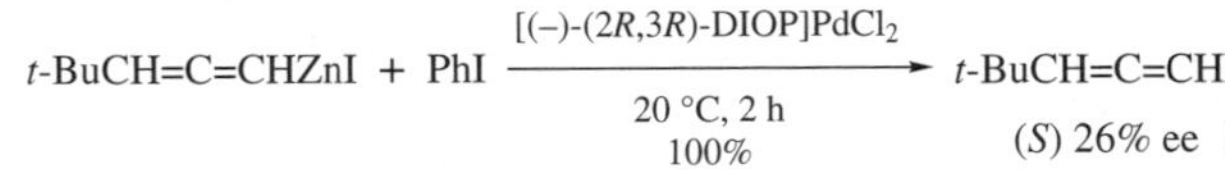

(5)

Other Pd^0 Derivatives and Related Reactions. Other chiral palladium complexes, such as $(DIOP)_2Pd^0$ or (DIOP)(alkene)Pd^0, can be prepared from (DIOP)$PdCl_2$.[6a,6d] These catalysts have afforded low levels of asymmetric induction (10% ee) in the hydrocyanation of norbornene derivatives.[6b,6c]

Oxirane Formation. The reaction of acetonyltributyltin with α-bromoacetophenone to yield 2-acetonyl-2-phenyloxirane has been investigated using ((+)-(2*S*,3*S*)-DIOP)$PdCl_2$ as a catalyst, but no asymmetric induction was observed.[8]

1. (a) Botteghi, C.; Consiglio, G.; Pino, P. *Chimia* **1973**, *27*, 477. (b) Consiglio, G.; Marchetti, M. *Chimia* **1976**, *30*, 26. (c) Consiglio, G.; Pino, P. *Chimia* **1976**, *30*, 193. (d) Consiglio, G. *J. Organomet. Chem.* **1977**, *132*, C26. (e) Consiglio, G.; Roncetti, L. *Chirality* **1991**, *3*, 341.
2. (a) Consiglio, G.; Kollár, L.; Kölliker, R. *J. Organomet. Chem.* **1990**, *396*, 375. (b) Consiglio, G.; Nefkens, S.; Pisano, C.; Wenzinger, F. *Helv. Chim. Acta* **1991**, *74*, 323.
3. (a) Trost, B. M.; Dietsche, T. J. *J. Am. Chem. Soc.* **1973**, *95*, 8200. (b) Trost, B. M.; Strege, P. E. *J. Am. Chem. Soc.* **1977**, *99*, 1649. (c) Fiaud, J. C.; Gournay, A. H.; Larcheveque, M.; Kagan, H. B. *J. Organomet. Chem.* **1978**, *154*, 175.
4. Kobayashi, T.; Tanaka, M. *J. Organomet. Chem.* **1982**, *233*, C64.
5. De Graaf, W.; Boersma, J.; Van Koten, G.; Elsevier, C. J. *J. Organomet. Chem.* **1989**, *378*, 115.
6. (a) Brown, K.; Chaloner, P. A. *J. Organomet. Chem.* **1981**, *217*, C25. (b) Elmes, P. S.; Jackson, W. R. *Aust. J. Chem.* **1982**, *35*, 2041. (c) Hodgson, M.; Parker, D.; Taylor, R. J.; Ferguson, G. *Organometallics* **1988**, *7*, 1761. (d) Hodgson, M.; Parker, D.; Taylor, R. J.; Ferguson, G. *Chem. Commun./J. Chem. Soc., Chem. Commun.* **1987**, 1309.
7. (a) Gramlich, V.; Consiglio, G. *Helv. Chim. Acta* **1979**, *62*, 1016. (b) Chaloner, P. A. *J. Organomet. Chem.* **1984**, *266*, 191.
8. Pri-Bar, I.; Pearlman, P. S.; Stille, J. K. *J. Org. Chem.* **1983**, *48*, 4629.

Ronan Guével

The Ohio State University, Columbus, OH, USA

10-Dicyclohexylsulfonamidoisoborneol[1]

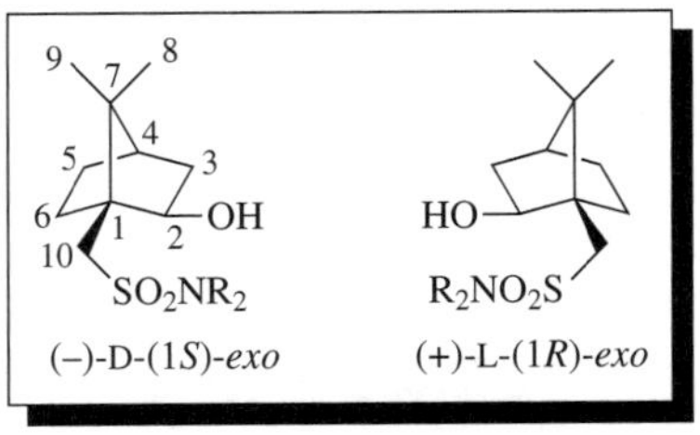

(1*S*)-*exo*; R = cyclohexyl		
[96303-88-7]	$C_{22}H_{39}NO_3S$	(MW 397.69)
(1*S*)-*exo*; R = isopropyl		
[89156-11-6]	$C_{16}H_{31}NO_3S$	(MW 317.55)

(chiral auxiliary: enoate derivatives undergo stereoselective Diels–Alder[2,3] and 1,3-dipolar[3] cycloadditions and 1,4-cuprate additions;[4] enol ether derivatives undergo stereoselective [2 + 2] cycloadditions with dichloroketene;[5] ester enolate derivatives participate in stereoselective imine condensation,[6] alkylation,[4a] aldolization,[7] acetoxylation,[8] halogenation,[9] and 'amination'[10] reactions)

Physical Data: R = cyclohexyl: mp (from hexane) 163–164 °C; $[\alpha]_D^{21}$ −25.7 (*c* = 0.76, EtOH). R = isopropyl: mp (from hexane) 102–103 °C; $[\alpha]_D^{21}$ −34.4 (*c* = 4.74, EtOH).

Form Supplied in: white crystalline solids.

Preparative Methods: crystalline, enantiomerically pure 10-diisopropyl- and 10-dicyclohexylsulfonamidoisoborneol auxiliaries are readily prepared from the appropriate enantiomer of *10-Camphorsulfonyl Chloride* by successive amidation and *exo*-selective reduction (eq 1).[2]

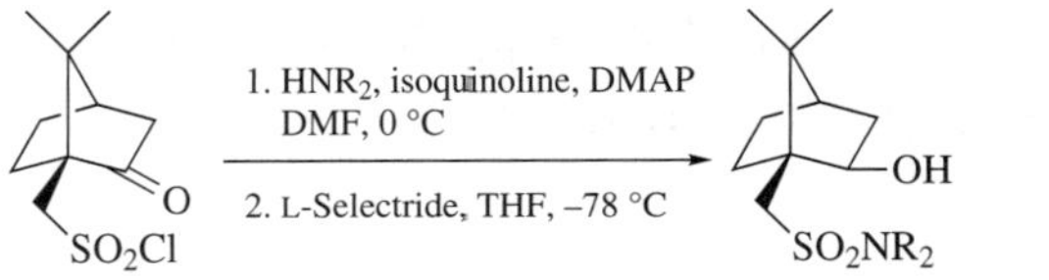

(1)

(**1a**) R = Cy, 55% (cryst)
(**1b**) R = *i*-Pr, 67% (cryst)

Simple acyl derivatives are prepared in good yields from carboxylic acids using Mukaiyama's *2-Chloro-1-methylpyridinium Iodide* coupling reagent[2] or from carboxylic acid chlorides using *Silver(I) Cyanide*.[9a] The former method is also suitable for the preparation of enoyl derivatives, although a Horner–Wadsworth–Emmons reaction has also been employed for this purpose.[4b] The *cis*-propenyl enol ether derivative of 10-diisopropylsulfonamidoisoborneol was prepared by base-promoted isomerization of the corresponding allyl ether (the preparation of which was not described).[5]

Handling, Storage, and Precautions: these reagents are stable indefinitely at ambient temperature in sealed containers.

Introduction. The 10-dialkylsulfonamidoisoborneol auxiliaries exert a powerful topological bias over the π-facial reactivity of enoate, enol ether, and ester enolate derivatives in a wide range of asymmetric transformations. However, the subsequently developed *10,2-Camphorsultam* chiral auxiliary outperforms these auxiliaries both in terms of stereoinduction and ease of nondestructive cleavage for most applications. Consequently, only transformations for which the 10-dialkylsulfonamidoisoborneol auxiliaries are particularly advantageous, or for which the analogous transformations of the 10,2-camphorsultam have not been reported, are described here. It should be noted, however, that the origin of the stereoinduction provided by these two camphor-derived auxiliaries is fundamentally different;[1] hence key references for all transformations are provided above.

Reactions of Enoate, Enol Ether, and Acyl Derivatives.

1,4-Organocopper Addition (Alkene to β-Functionalized Product).[4] Tri-*n*-butylphosphine-stabilized organocopper reagents add in a conjugate fashion to *trans*-enoate derivatives of the 10-dicyclohexylsulfonamidoisoborneol auxiliary from the less hindered C(α)-*si* π-face with excellent selectivity (eq 2) (Table 1). This type of reaction has formed the basis of several natural product syntheses.[4]

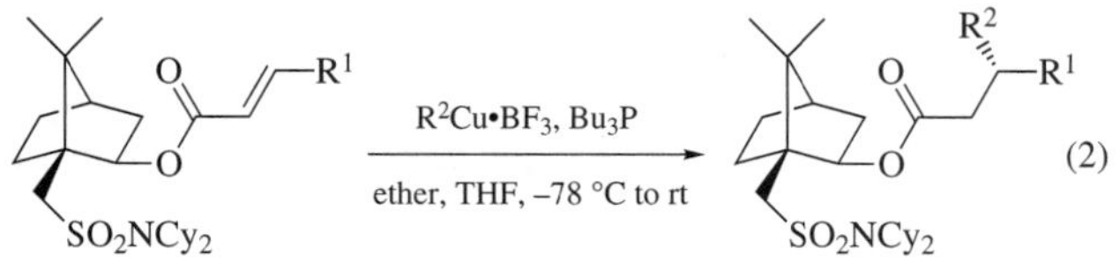

(2) (3)

Table 1 Conjugate Addition of Organocopper Reagents (2) → (3)

R^1	R^2	Yield (%)	de (%)
Bu	Me	93	97
Me	Bu	89	97
Pr	Me	89	94
Me	Pr	98	95
Me	CH_2=CH	80	98
Me	CH_2=CMe	84	94

[2+2] Dichloroketene Addition (Enol Ether to β-Alkoxy-α-dichlorocyclobutanone).[5] Of six different chiral auxiliaries screened for their ability to control stereochemistry in the reaction of dichloroketene with derived *cis*-propenyl enol ethers, the 10-diisopropylsulfonamidoisoborneol auxiliary was the best. Thus, following ring expansion of the initially formed cyclobutanone (**4**) with *Diazomethane–Chromium(II) Perchlorate*, α-chloro-γ-methylcyclopentenone was isolated in ~60% yield and 80% ee [C(α)-*si* face attack of the ketene] (eq 3). The auxiliary was also recovered in unspecified yield.

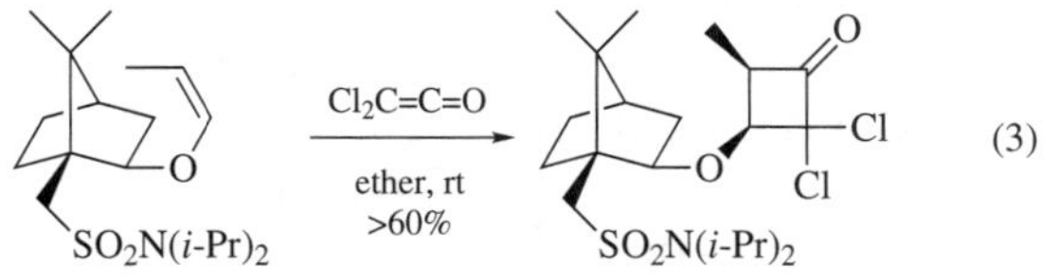

(4)
80% de
(major isomer shown)

Imine Condensation (Acyl Species to β-Lactam).[6] Lithium enolates of acyl 10-diisopropylsulfonamidoisoborneols condense with *N*-aryl aldimines to give *cis*-disubstituted β-lactams with 56–92% ee, accompanied by 2.5–9% of their *trans* isomers (in undetermined ee) (eq 4) (the key step in a synthesis of the carbapenem antibiotic (+)-PS-5). Menthol was found to be a less efficient auxiliary for this application.[6]

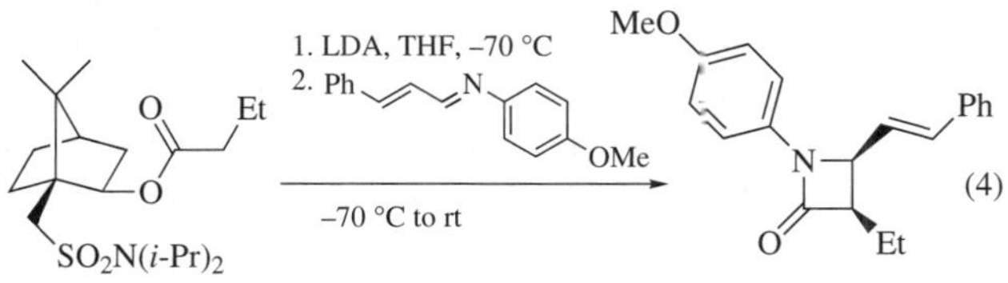

79% (FC), 91% ee
(+ recovered auxiliary 95%)

α-Acetoxylation and α-Halogenation (Acyl Species to α-Acetoxy or α-Halo Acyl Product).[8,9] α-Acetoxylations of *O*-silyl enol ether derivatives of acyl 10-dicyclohexylsulfonamidoisoborneols with *Lead(IV) Acetate* proceed in high yield with excellent π-facial stereocontrol (95–100% de, with C(α)-*re* topicity).[8] Mechanistically related α-halogenations with *N*-halosuccinimides also proceed smoothly to afford α-halo acyl products in 76–96% de, but with C(α)-*si* topicity.[9] The observed topicities are consistent with initial attack of the electrophilic species from the less hindered C(α)-*si* face to give transient plumbonium/bromonium/chloronium ions. The plumbonium intermediates undergo S_N2-type attack by acetate at the β-position, whereas the bromonium/chloronium intermediates fragment with retention at C(β).[9a] The stereofacial influence of the auxiliary overrides any preexisting β-stereocenter. Hence, consecutive alkylcopper conjugate addition, then α-acetoxylation or α-bromination, allows the concise and stereocontrolled formation of two contiguous stereocenters. α-Acetoxy ester derivative (**6**) formed in this way is a precursor to a key intermediate for the synthesis of the elm bark beetle pheromone (eq 5), and α-bromo ester derivative (**7**) was converted via azide displacement, transesterification, and hydrogenolysis into L-*allo*-isoleucine (eq 6).[9b] α-Halo esters are also useful precursors of enantiomerically pure epoxides.[9a]

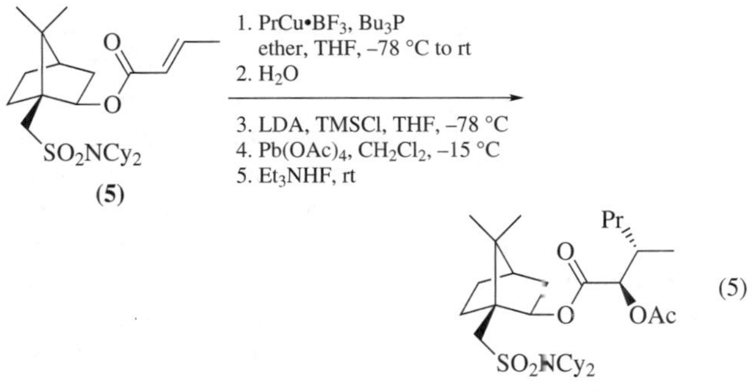

(6)
66% (cryst)
97% de C(α), 92% de C(β) (crude)

α-'Amination' (Acyl Species to α-Amino Acyl Product).[10] Although asymmetric bromination and stereospecific azide displacement of *O*-silyl enol ethers of acyl 10-dicyclohexylsulfonamidoisoborneols (as described above) is a generally applicable route to optically active α-amino acids,[9b] a complementary and more direct approach to this important class of compounds is via electrophilic 'amination' of these same compounds using *Di-t-butyl Azodicarboxylate* (DBAD). The initially formed α-(di-*N*-Boc-hydrazido)amino acid derivatives (**8**) (eq 7) may be efficiently converted to the corresponding α-amino acid hydrochlorides by successive deacylation, hydrogenolysis, transesterification, and hydrolysis. This reaction sequence has been shown to be efficient for the preparation of a wide range of α-amino acids in excellent enantiomeric purity[10] and compares favorably with closely related methods using alternative auxiliaries.[11]

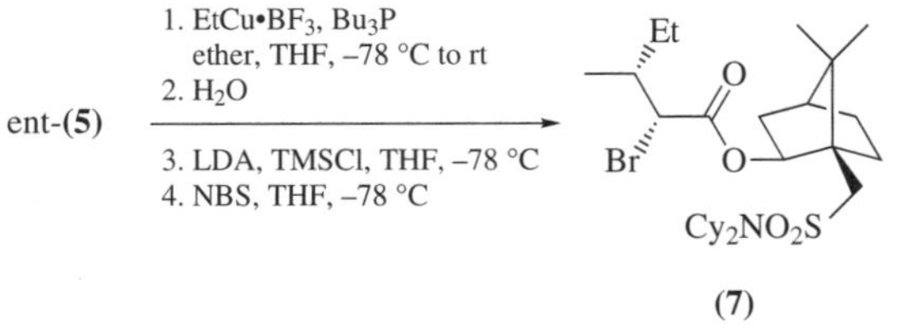

(6)

(**7**)
60% (cryst)
99.3% de C(α), 97.8% de C(β)

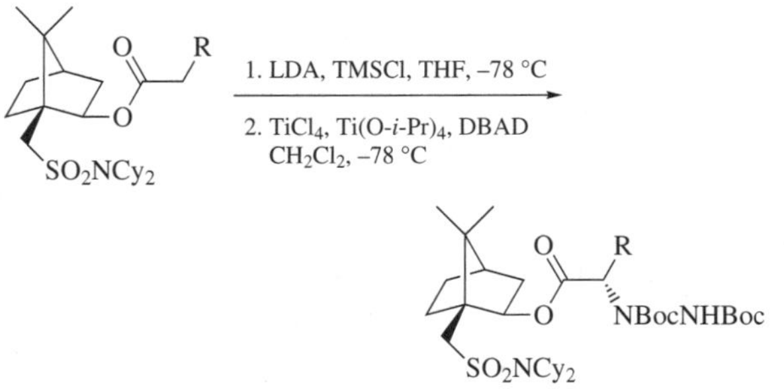

(7)

(**8**)
R = Me, Et, Pr, *i*-Pr, Bu, *i*-Bu, Bn, Cy
69–84% (FC) [92.6–96.4% de (crude)]

Nondestructive Auxiliary Cleavage. The hindered ester linkage present in acyl derivatives of 10-dialkylsulfonamidoisoborneols is less readily cleaved than the corresponding sulfonamidic linkage of *N*-acyl-10,2-camphorsultam derivatives. However, it can be hydrolyzed and the auxiliary recovered intact under basic conditions using *Potassium Hydroxide*[7] or *Potassium Carbonate*[8] in MeOH, *Sodium Hydroxide* in aq EtOH,[4a] or *Lithium Hydroxide* in aq THF. Elevated temperatures are required to achieve acceptable reaction rates for all but the latter procedure which, although sluggish at ambient temperature, was employed for unmasking sensitive aldol products.[7] 'Nonbasic' transesterification using $Ti(OBn)_4$/BnOH affords benzyl esters which may be subject to hydrogenolysis to give the corresponding carboxylic acids.[9b] Alternatively, transesterification with $Ti(OEt)_4$/EtOH[10b] may be followed by hydrolysis under acidic conditions.[10]

Primary alcohols can be obtained by hydride reduction using either *Lithium Aluminum Hydride* in ether[2,8] or $Ca(BH_4)_2$ in THF,[9a] and this latter reagent is compatible with halogen functionality. A dimethyl tertiary alcohol was obtained by addition of 2 equiv of methyllithium in ether.[4b]

Related Reagents. α-Methyltoluene-2,α-sultam.

1. Oppolzer, W. *Tetrahedron* **1987**, *43*, 1969.
2. Oppolzer, W.; Chapuis, C.; Bernardinelli, G. *Tetrahedron Lett.* **1984**, *25*, 5885.
3. Curran, D. P.; Kim, B. H.; Piyasena, H. P.; Loncharich, R. J.; Houk, K. N. *J. Org. Chem.* **1987**, *52*, 2137.
4. (a) Oppolzer, W.; Dudfield, P.; Stevenson, T.; Godel, T. *Helv. Chim. Acta* **1985**, *68*, 212. (b) Oppolzer, W.; Moretti, R.; Bernardinelli, G. *Tetrahedron Lett.* **1986**, *27*, 4713.
5. Greene, A. E.; Charbonnier, F. *Tetrahedron Lett.* **1985**, *26*, 5525.
6. Hart, D. J.; Lee, C.-S.; Pirkle, W. H.; Hyon, M. H.; Tsipouras, A. *J. Am. Chem. Soc.* **1986**, *108*, 6054.
7. Oppolzer, W.; Marco-Contelles, J. *Helv. Chim. Acta* **1986**, *69*, 1699.
8. Oppolzer, W.; Dudfield, P. *Helv. Chim. Acta* **1985**, *68*, 216.
9. (a) Oppolzer, W.; Dudfield, P. *Tetrahedron Lett.* **1985**, *26*, 5037. (b) Oppolzer, W.; Pedrosa, R.; Moretti, R. *Tetrahedron Lett.* **1986**, *27*, 831.
10. (a) Oppolzer, W. In *Chirality in Drug Design and Synthesis*; Academic: New York, 1990. (b) Oppolzer, W.; Moretti, R. *Helv. Chim. Acta* **1986**, *69*, 1923. (c) Oppolzer, W.; Moretti, R. *Tetrahedron* **1988**, *44*, 5541.
11. (a) Gennari, C.; Colombo, L.; Bertolini, G. *J. Am. Chem. Soc.* **1986**, *108*, 6394. (b) Evans, D. A.; Britton, T. C.; Dorow, R. L.; Dellaria, J. F. *J. Am. Chem. Soc.* **1986**, *108*, 6395. (c) Evans, D. A.; Britton, T. C.; Dorow, R. L.; Dellaria, J. F. *Tetrahedron* **1988**, *44*, 5525. (d) Trimble, L. A.; Vederas, J. C. *J. Am. Chem. Soc.* **1986**, *108*, 6397.

Alan C. Spivey
University of Cambridge, UK

Dihydro-5-(hydroxymethyl)-2(3*H*)-furanone[1]

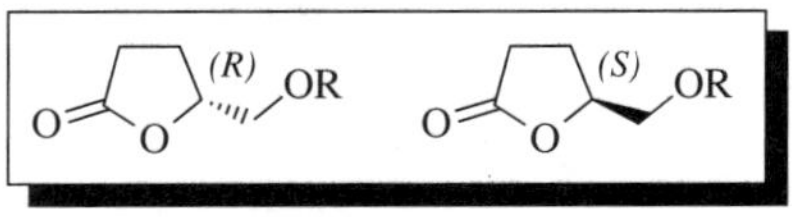

(*R*)-(−)-(**1**; R = H)		
[52813-63-5]	$C_5H_8O_3$	(MW 116.12)
(*S*)-(+)-(**2**; R = H)		
[32780-06-6]	$C_5H_8O_3$	(MW 116.12)
(*R*)-(−)-(**3**; R = Bn)		
[77697-15-5]	$C_{12}H_{14}O_3$	(MW 206.24)
(*S*)-(+)-(**4**; R = Bn)		
[32780-08-8]	$C_{12}H_{14}O_3$	(MW 206.24)
(*R*)-(−)-(**5**; R = Tr)		
[78158-90-4]	$C_{24}H_{22}O_3$	(MW 358.42)
(*S*)-(+)-(**6**; R = Tr)		
[73968-62-4]	$C_{24}H_{22}O_3$	(MW 358.42)
(*R*)-(−)-(**7**; R = TBDMS)		
[130767-09-8]	$C_{11}H_{22}O_3Si$	(MW 230.38)
(*S*)-(+)-(**8**; R = TBDMS)		
[62396-80-9]	$C_{11}H_{22}O_3Si$	(MW 230.38)
(*R*)-(−)-(**9**; R = TBDPS)		
[128075-94-5]	$C_{21}H_{26}O_3Si$	(MW 354.52)
(*S*)-(+)-(**10**; R = TBDPS)		
[102717-29-3]	$C_{21}H_{26}O_3Si$	(MW 354.52)
(*R*)-(−)-(**11**; R = Ts)		
[58879-33-7]	$C_{12}H_{14}O_5S$	(MW 270.30)
(*S*)-(+)-(**12**; R = Ts)		
[58879-34-8]	$C_{12}H_{14}O_5S$	(MW 270.30)

(versatile chiral building blocks used in the synthesis of a wide variety of natural products[1a–d] and other biologically important molecules,[2] including anti-HIV dideoxynucleosides;[1e,f] also useful for preparing optically active ligands[3])

Alternate Name: (*R*)- and (*S*)-γ-hydroxymethyl-γ-butyrolactones; 5-oxotetrahydrofuran-2-methanol; 5-hydroxymethylpentanolide.

Physical Data: (**1**) bp 101–102 °C/0.048 mmHg;[4,5] (**2**) bp 122–130 °C/0.6 mmHg;[5,6] (**3**) bp 160–170 °C/0.4–0.6 mmHg;[7] (**4**) bp 160–164 °C/0.02 mmHg; 152–160 °C/0.04 mmHg;[8,9] (**5**) mp 153–154 °C;[10] (**6**) mp 153–154 °C;[10] (**7**)–;[11] (**8**) bp 88–92 °C/0.05 mmHg;[12] (**9**) mp 72–73 °C;[13] (**10**) mp 75–76 °C;[14] (**11**) mp 84.5–86 °C;[15,16] (**12**) mp 85–87 °C.[15,17]

Preparative Methods: both enantiomers of dihydro-5-(hydroxymethyl)-2(3*H*)-furanone and their trityl derivatives are commercially available but expensive. The simplest and by far most popular method for preparing (*S*)-dihydro-5-(hydroxymethyl)-2(3*H*)-furanone (**2**)[1e,f] consists of enantiospecific deamination of L-glutamic acid[5] and subsequent selective reduction[15] of the resulting carboxylic acid (**13**) (eq 1). Purification of the intermediate acid (**13**) by crystallization[5] and not by distillation[18] is recommended in order to secure an excellent optical yield (>98% ee). Likewise, (*R*)-dihydro-5-(hydroxymethyl)-2(3*H*)-furanone (**1**) (>98% ee) can be obtained from D-glutamic acid.[5] As the latter is considerably more expensive than its natural antipode, an appealing option is to convert the (*S*)-lactone into its enantiomer (eq 2).[17] Also available and equally useful is an inversion route to (*R*)-dihydro-5-(trityloxymethyl)-2(3*H*)-furanone (**5**) by way of the Mitsunobu reaction (eq 3).[10]

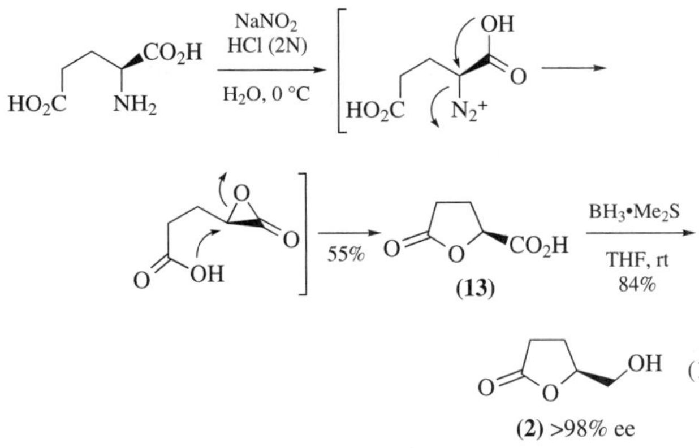
(1)

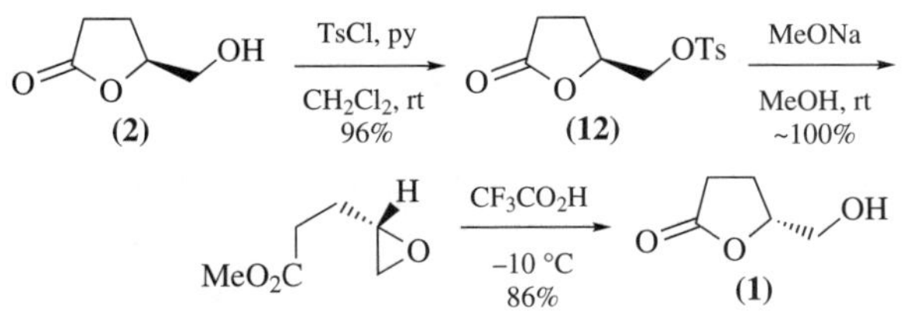
(2)

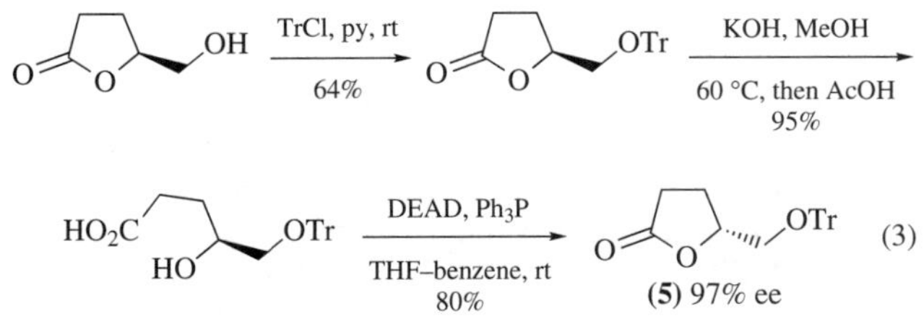
(3)

Synthetic Applications. (*S*)-Dihydro-5-(hydroxymethyl)-2(3*H*)-furanone (**2**) was first described in 1971 as an intermediate in the synthesis of D-ribose from L-glutamic acid.[19] Since then, this lactone and its (*R*)-enantiomer[4] have found widespread use as chirons for constructing a rich variety of natural products ranging from simple pheromones[16,20] to complex macrocycles[1b,21] and ionophore antibiotics.[22] The chemical manipulation of these chirons often involves lactone cleavage at an early stage,[3,21,23] as illustrated by the preparation of suitable intermediates for the synthesis of the *Vespa orientalis* pheromone (*R*)-δ-*n*-hexadecanolactone (**14**) (eq 4),[6] the antiviral fungal metabolite brefeldin A (**15**) (eq 5),[24] and (7aα)-*epi*-hemibrevetoxin B (**16**) (eq 6).[9]

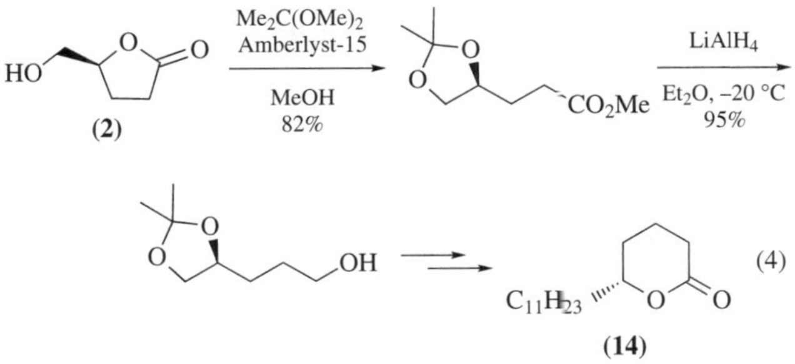
(4)

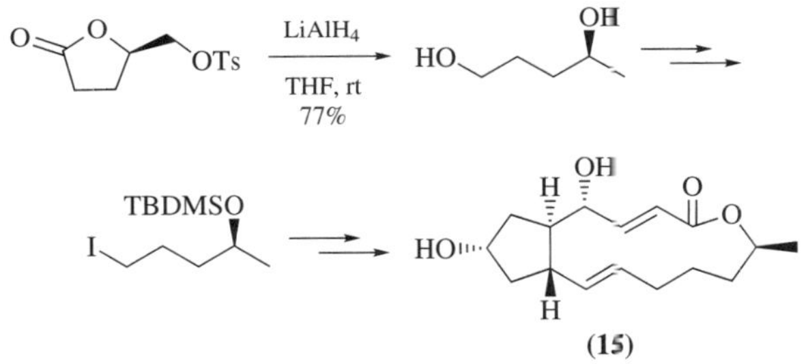
(5)

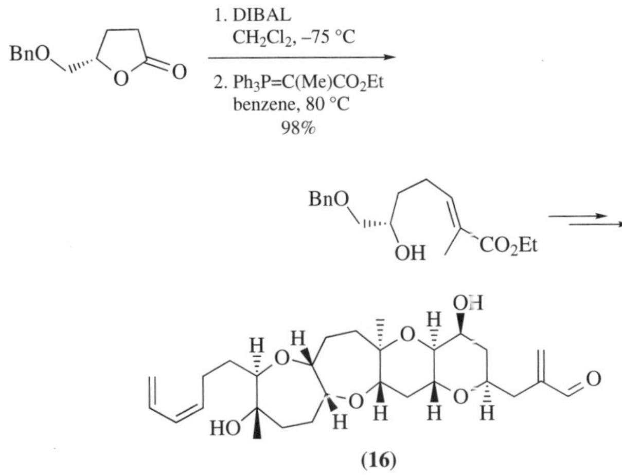
(6)

Optically active γ-alkyl-γ-butyrolactones are readily available from the tosylates through a one-pot reductive procedure[25a] or by alkylative side-chain elongation with the appropriate organocuprate (eq 7).[15]

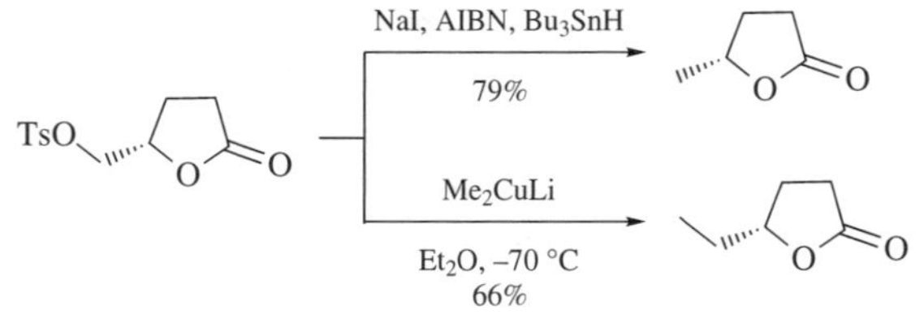

(7)

Protection of the alcohol moiety with a bulky group, such as trityl or *t*-butyldiphenylsilyl (TBDPS), shields the *syn* face of the lactone, thereby forcing incoming reagents to attack *anti* to the side chain.[26,2c] This scenario has been heavily exploited for the stereocontrolled introduction of one or two ring substituents through alkylation[2c,26–29] or aldolization.[30] For instance, sequential dialkylation of (*S*)-dihydro-5-(trityloxymethyl)-2(3*H*)-furanone provides products of high diastereomeric purity which can be transformed into β,β-disubstituted γ-butyrolactones (eq 8).[26] Various optically active lactones with predetermined substitution patterns, available in a similar way, have been utilized as key intermediates in the synthesis of indole alkaloids,[27] lignans,[1d,9] β-lactams,[31] and other natural products.[1a,32]

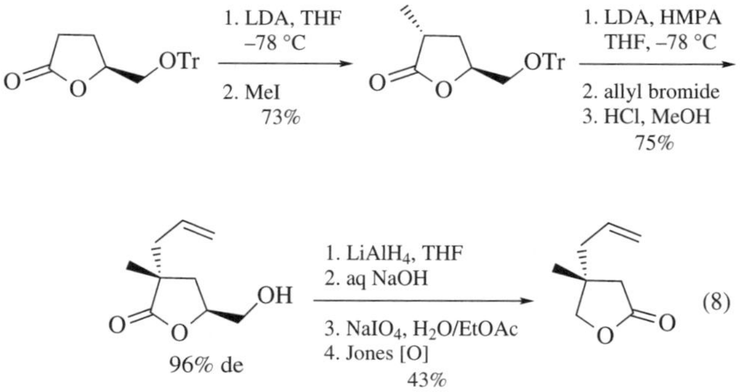

(8)

The initial *trans*-alkylated lactones can be easily epimerized in situ by enolization and protonation from the less hindered face[29] using *t*-butyl bromide as the proton source (eq 9).[2b] Further handling of the so-formed *cis* isomers allows for the total synthesis of dihydromevinolin (**17**) and several analogs of import as cholesterol lowering agents.[2b]

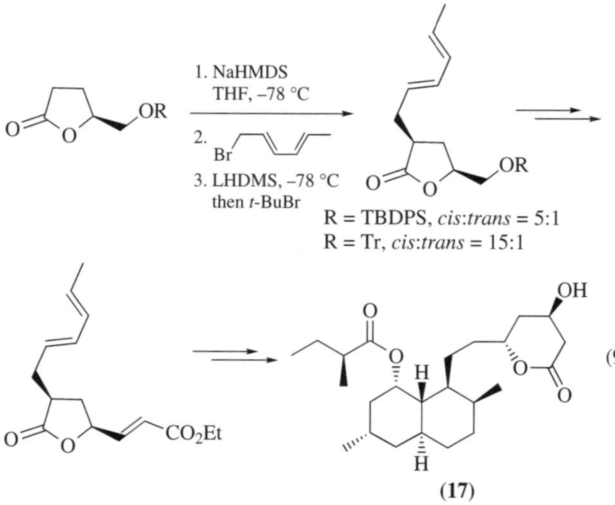

(9)

(17)

Stereoselective *trans* α-hydroxylation of (*S*)-dihydro-5-(*t*-butyldiphenylsiloxymethyl)-2(3*H*)-furanone can be realized in good yield by enolization and reaction with the *Oxodiperoxymolybdenum(pyridine)(hexamethylphosphoric triamide)* complex (MoOPH) (eq 10).[33] Appropriate manipulation of the resulting *trans*-hydroxylactone provides 1,3-polyols[1b] typified by (**18**),[33a] as well as tetrahydropyran (**19**) which is a key intermediate in mevinic acid syntheses.[33b]

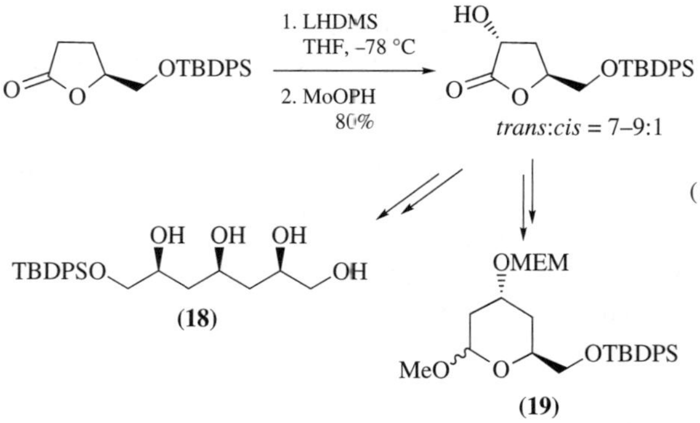

(10)

Owing to the urgent need of new drugs for the treatment of AIDS, many methods for converting the title lactones into antiviral dideoxynucleosides have been devised within the past few years.[1e,f,12,34] A viable synthesis of the potent anti-HIV agent 3′-deoxy-2′,3′-dehydrothymidine (d4T, **21**) relies on *trans* selective sulfenylation of the lactone-derived silyl ketene acetal (**20**) with *N*-(phenylthio)-ϵ-caprolactam (eq 11).[35] In comparable fashion, (**21**) and related nucleosides have been prepared through selenenylation of (**20**).[36]

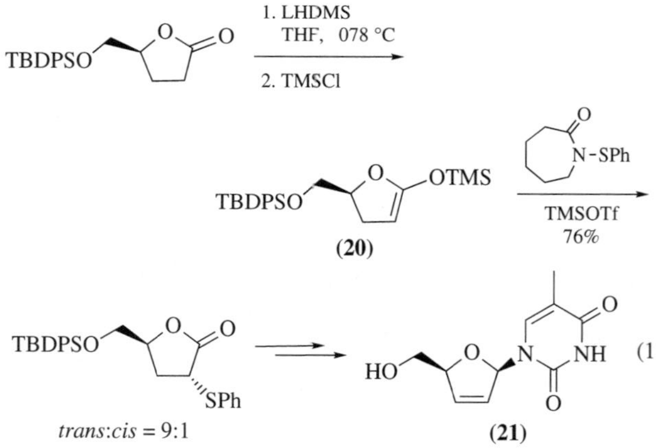

(11)

Related Chirons. The commercially available (*R*)-(−)- and (*S*)-(+)-enantiomers of 5-hydroxymethyl-2(5*H*)-furanone (**22** and *ent*-**22**) and their various protected derivatives have also been extensively used in the synthesis of natural products,[1b–d,37,38] nucleosides,[1e,f] and other bioactive substances.[39,40] A cost-effective, versatile route to these chirons[14,37,40b] is illustrated by the preparation of (**22**) (eq 12).[38]

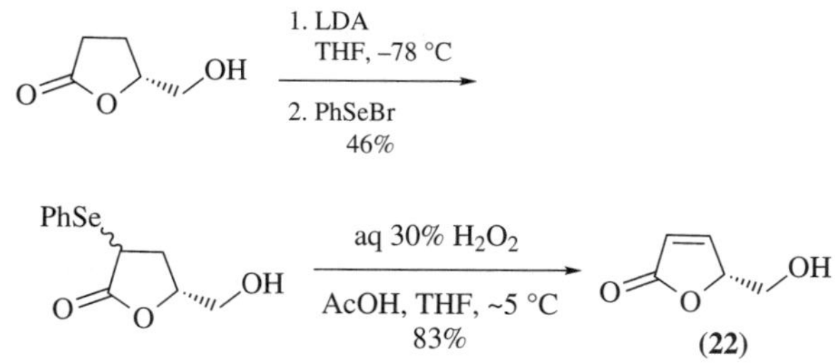

(22) (12)

Related Reagents. α,β-Butenolide; γ-Butyrolactone; (*R*)-Pantolactone; β-Propiolactone; 2-Trimethylsilyloxyfuran; β-Vinyl-α,β-buten olide.

---

1. (a) Coppola, G. M.; Schuster, H. F. *Asymmetric Synthesis, Construction of Chiral Molecules Using Amino Acids*; Wiley: New York, 1987; pp 237–256. (b) Hanessian, S. *Aldrichim. Acta* **1989**, *22*, 3. (c) Hanessian, S. *Pure Appl. Chem.* **1993**, *65*, 1189. (d) Ward, R. S. *Tetrahedron* **1990**, *46*, 5029. (e) Dueholm, K. L.; Pedersen, E. B. *Synthesis* **1992**, 1. (f) Huryn, D. M.; Okabe, M. *Chem. Rev.* **1992**, *92*, 1745.
2. (a) Mattes, H.; Benezra, C. *J. Org. Chem.* **1988**, *53*, 2732. (b) Blackwell, C. M.; Davidson, A. H.; Launchbury, S. B.; Lewis, C. N.; Morrice, E. M.; Reeve, M. M.; Roffey, J. A. R.; Tipping, A. S.; Todd, R. S. *J. Org. Chem.* **1992**, *57*, 5596. (c) Herdeis, C.; Lütsch, K. *Tetrahedron: Asymmetry* **1993**, *4*, 121. (d) Koert, U.; Stein, M.; Harms, K. *Tetrahedron Lett.* **1993**, *34*, 2299.
3. Brunner, H.; Lautenschlager, H.-J. *Synthesis* **1989**, 706.
4. Eguchi, C.; Kakuta, A. *Bull. Chem. Soc. Jpn.* **1974**, *47*, 1704.
5. Herdeis, C. *Synthesis* **1986**, 232. See also note 9 of ref. 2c.
6. Larchevêque, M.; Lalande, J. *Tetrahedron* **1984**, *40*, 1061.
7. Takano, S.; Goto, E.; Hirama, M.; Ogasawara, K. *Heterocycles* **1981**, *16*, 381.
8. (a) Taniguchi, M.; Koga, K.; Yamada, S. *Tetrahedron* **1974**, *30*, 3547. (b) Tomioka, K.; Mizuguchi, H.; Koga, K. *Chem. Pharm. Bull.* **1982**, *30*, 4304.
9. Nicolaou, K. C.; Reddy, K. R.; Skokotas, G.; Sato, F.; Xiao, X.-Y.; Hwang, C.-K. *J. Am. Chem. Soc.* **1993**, *115*, 3558.
10. Takano, S.; Yonaga, M.; Ogasawara, K. *Synthesis* **1981**, 265.
11. Takle, A.; Kocieński, P. *Tetrahedron* **1990**, *46*, 4503.
12. Okabe, M.; Sun, R.-C.; Tam, S. Y.-K.; Todaro, L. J.; Coffen, D. L. *J. Org. Chem.* **1988**, *53*, 4780.
13. Sato, M.; Ohuchi, H.; Abe, Y.; Kaneko, C. *Tetrahedron: Asymmetry* **1992**, *3*, 313.
14. Hanessian, S.; Murray, P. J. *Tetrahedron* **1987**, *43*, 5055.
15. Ravid, U.; Silverstein, R. M.; Smith, L. R. *Tetrahedron* **1978**, *34*, 1449.
16. Mori, K. *Tetrahedron* **1975**, *31*, 3011.
17. Ho, P.-T.; Davies, N. *Synthesis* **1983**, 462.
18. Gringore, O. H.; Rouessac, F. P. *Org. Synth., Coll. Vol.* **1990**, *7*, 99.
19. Koga, K.; Taniguchi, M.; Yamada, S. *Tetrahedron Lett.* **1971**, 263. See also ref. 8a.
20. Further examples: Vigneron, J. P.; Méric, R.; Larchevêque, M.; Debal, A.; Lallemand, J. Y.; Kunesch, G.; Zagatti, P.; Gallois, M. *Tetrahedron* **1984**, *40*, 3521. Chattopadhyay, S.; Mamdapur, V. R.; Chadha, M. S. *Synth. Commun.* **1990**, *20*, 1299.
21. Smith III, A. B.; Rano, T. A.; Chida, N.; Sulikowski, G. A.; Wood, J. L. *J. Am. Chem. Soc.* **1992**, *114*, 8008.
22. Hanessian, S.; Cooke, N. G.; DeHoff, B.; Sakito, Y. *J. Am. Chem. Soc.* **1990**, *112*, 5276.
23. Hirama, M.; Uei, M. *J. Am. Chem. Soc.* **1982**, *104*, 4251.
24. Kitahara, T.; Mori, K. *Tetrahedron* **1984**, *40*, 2935.
25. (a) Harmange, J.-C.; Figadère, B.; Hocquemiller, R. *Tetrahedron: Asymmetry* **1991**, *2*, 347. See also: (b) Ortuño, R. M.; Alonso, D.; Cardellach, J.; Font, J. *Tetrahedron* **1987**, *43*, 2191.
26. Tomioka, K.; Cho, Y.-S.; Sato, F.; Koga, K. *J Org. Chem.* **1988**, *53*, 4094.
27. (a) Takano, S.; Yonaga, M.; Ogasawara, K. *Chem. Commun./J. Chem. Soc., Chem. Commun.* **1981**, 1153. (b) Takano, S.; Tamura, N.; Ogasawara, K. *Chem. Commun./J. Chem. Soc., Chem. Commun.* **1981**, 1155. (c) Takano, S.; Yonaga, M.; Morimoto, M.; Ogasawara, K. *J. Chem. Soc., Perkin Trans. 1* **1985**, 305.
28. Hanessian, S.; Murray, P. J.; Sahoo, S. P. *Tetrahedron Lett.* **1985**, *26*, 5623.
29. Davidson, A. H.; Jones, A. J.; Floyd, C. D.; Lewis, C.; Myers, P. L. *Chem. Commun./J. Chem. Soc., Chem. Commun.* **1987**, 1786.
30. Pathirana, C.; Dwight, R.; Jensen, P. R.; Fenical, W.; Delgado, A.; Brinen, L. S.; Clardy, J. *Tetrahedron Lett.* **1991**, *32*, 7005.
31. Takano, S.; Kasahara, C.; Ogasawara, K. *Chem. Lett.* **1982**, 631.
32. Recent examples: (a) Hanessian, S.; Roy, P. J.; Petrini, M.; Hogdes, P. J.; Di Fabio, R.; Carganico, G. *J. Org. Chem.* **1990**, *55*, 5766. (b) Ezquerra, J.; He, W.; Paquette, L. A. *Tetrahedron Lett.* **1990**, *31*, 6979. (c) Maier, M. E.; Schöffling, B. *Tetrahedron Lett.* **1991**, *32*, 53.
33. (a) Hanessian, S.; Sahoo, S. P.; Murray, P. J. *Tetrahedron Lett.* **1985**, *26*, 5631. (b) Davidson, A. H.; Floyd, C. D.; Lewis, C. N.; Myers, P. L. *Chem. Commun./J. Chem. Soc., Chem. Commun.* **1988**, 1417.
34. Examples: (a) Kim, C. U.; Misco, P. F. *Tetrahedron Lett.* **1992**, *33*, 5733. (b) Secrist III, J. A.; Riggs, R. M.; Tiwari, K. N.; Motgomery, J. A. *J. Med. Chem.* **1992**, *35*, 533. (c) Zhang, H.-C.; Daves, Jr., G. D. *J. Org. Chem.* **1993**, *58*, 2557.
35. Wilson, L. J.; Liotta, D. C. *J. Org. Chem.* **1992**, *57*, 1948. See also: Wilson, L. J.; Liotta, D. C. *Tetrahedron Lett.* **1990**, *31*, 1815.
36. Beach, J. W.; Kim, H. O.; Jeong, L. S.; Nampalli, S.; Islam, Q.; Ahn, S. K.; Babu, J. R.; Chu, C. K. *J. Org. Chem.* **1992**, *57*, 3887.
37. Tomioka, K.; Sato, F.; Koga, K. *Heterocycles* **1982**, *17*, 311.
38. Ortuño, R. M.; Bigorra, J.; Font, J. *Tetrahedron* **1987**, *43*, 2199.
39. Additional examples: Boeckman, Jr.; R. K.; Heckendorn, D. K.; Chinn, R. L. *Tetrahedron Lett.* **1987**, *28*, 3551. Caine, D.; Venkataramu, S. D.; Kois, A. *J. Org. Chem.* **1992**, *57*, 2960. De Alvarenga, E. S.; Mann, J. *J. Chem. Soc., Perkin Trans. 1* **1993**, 2141.
40. (a) Mann, J.; Thomas, A. *Tetrahedron Lett.* **1986**, *27*, 3533. (b) Mattes, H.; Hamada, K.; Benezra, C. *J. Med. Chem.* **1987**, *30*, 1948. (c) Hanafi, N.; Ortuño, R. M. *Tetrahedron: Asymmetry* **1994**, *5*, 1657.

John Boukouvalas
Université Laval, Québec, Canada

(2*S*)-(+)-2,5-Dihydro-2-isopropyl-3,6-dimethoxypyrazine[1]

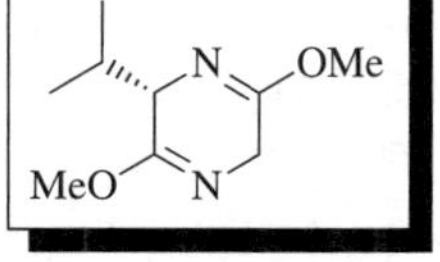

[78342-42-4] $C_9H_{16}N_2O_2$ (MW 184.24)

(Schöllkopf–Hartwig bislactim ether reagent for asymmetric synthesis of amino acids by reaction of the metalated reagent with alkyl halides,[2] aldehydes,[3] ketones,[4] epoxides,[5] and enones,[6] and subsequent hydrolysis of the resulting bislactim ether adduct)

Physical Data: bp 103–104 °C/15 mmHg; *d* 1.03 g mL^{-1}; $[\alpha]_D^{20}$ −109 (*c* = 1, EtOH).
Solubility: sol ether, THF, *n*-hexane.
Form Supplied in: colorless liquid.
Analysis of Reagent Purity: NMR.[2]
Purification: distillation.
Handling, Storage, and Precautions: store refrigerated.

Bislactim Ether Method. The commercially available bislactim ethers of cyclo(L- and D-Val-Gly) and cyclo(L- and D-Val-Ala)[7] are very versatile reagents for the preparation of nonproteinogenic amino acids in high yields and with excellent enantioselectivities (typically >95%).[1]

Reactions of the Lithiated Bislactim Ether. The procedure involves metalation with *n-Butyllithium* in THF at −70 °C and reaction of the resulting azaenolate with alkyl halides (eq 1).[2] The latter enters with high diasteroselectivity *trans* to the isopropyl group. The alkylation products are hydrolyzed under mild acidic conditions to give the desired amino acid ester and the chiral auxiliary L-Val-OMe, which can be separated by distillation or chromatography (eq 1). As well as reacting with primary and secondary alkyl halides and sulfonates,[8] the lithiated bislactim ether reacts in good yields and with high *trans* diastereoselectivities (in general >95%) with a variety of other electrophiles such as ketones,[9] acyl chlorides,[10] thioketones,[11] epoxides,[5] α,β-unsaturated esters,[12] and arene–manganese tricarbonyl complexes.[13]

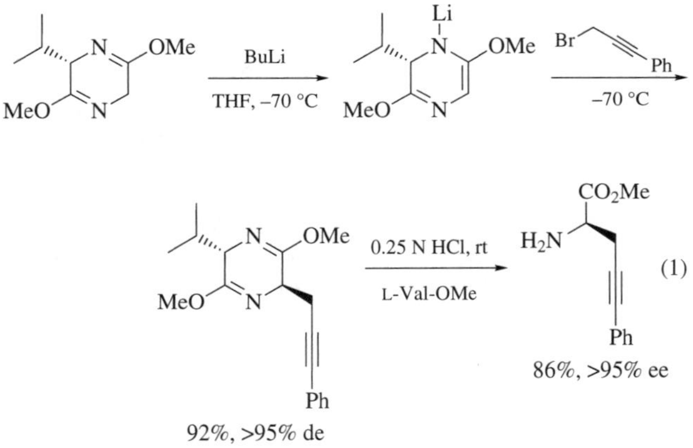

Reaction of the Titanated Bislactim Ether. The titanium derivative of the bislactim ether of cyclo(L-Val-Gly) reacts with alkyl aldehydes,[3] aryl aldehydes,[14] and α,β-unsaturated aldehydes[15] highly diastereoselectively to give almost exclusively the *syn* addition products (eq 2). Hydrolysis with dilute *Trifluoroacetic Acid*[3c] affords (2*R*, 3*S*)-β-hydroxy-α-amino acid methyl esters. α-Amino-γ-nitro amino acids can be obtained by 1,4-addition of the titanated bislactim ether to nitroalkenes and subsequent hydrolysis of the adduct.[16]

Reactions of the Bislactim Ether Cuprate. The lithiated bislactim ether can be converted to an azaenolate cuprate by treatment with CuBr · SMe_2 (see *Copper(I) Bromide*).[6] Conjugate addition of the cuprate to enones (eq 3)[6] and dienones,[17] or alkylation[6] with base labile electrophiles like ethyl 3-bromopropionate, proceeds with high *trans* diastereoselectivity. Hydrolysis of the Michael adducts and subsequent protection afford (2*R*,3*R*)-*N*-Boc-δ-oxo-α-amino acid methyl esters (eq 3).

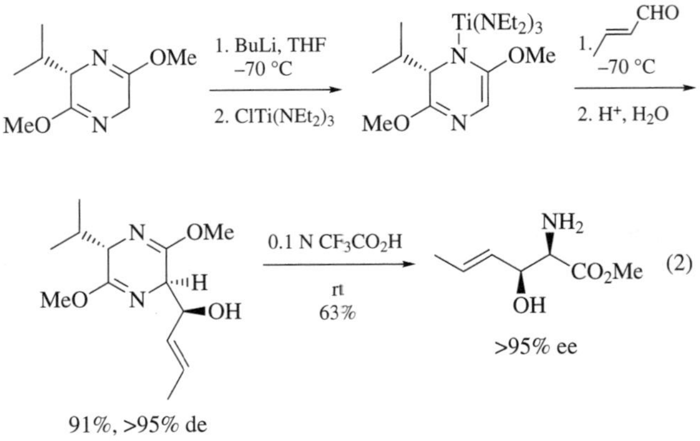

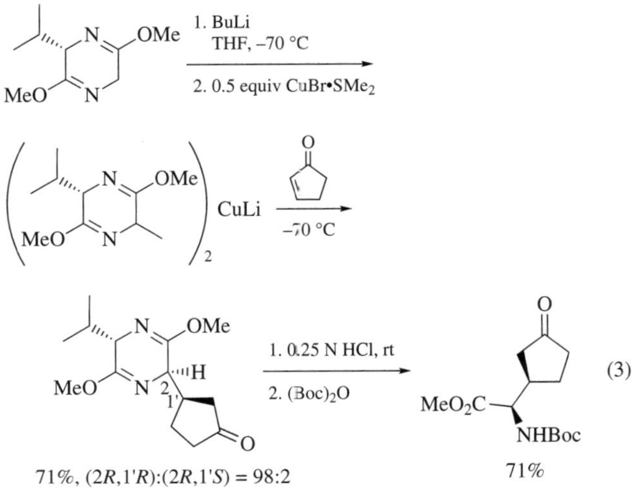

Reactions of the Bislactim Ether Carbene. Diazotization of the lithiated bislactim ether generates an electrophilic carbene species, which reacts in good yields and with high diastereomeric excess (<95%) with alkenes[18] and aryl alkynes[19] (eq 4) to give spirocyclopropanes and spirocyclopropenes, respectively. Hydrolysis of the latter affords the novel (*R*)-1-amino-2-arylcyclopropene-1-carboxylic acids.[19]

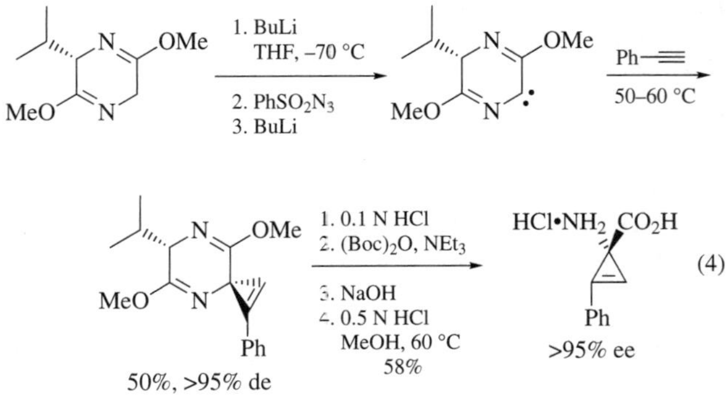

Related Reagents. 1-Benzoyl-2-*t*-butyl-3,5-dimethyl-4-imidazolidinone; (2*S*,4*S*)-3-Benzoyl-2-*t*-butyl-4-methyl-1,3-oxazolidin-5-one.

1. (a) Hartwig, W.; Schöllkopf, U. Ger. Patent 2 934 252, 1981. (b) Schöllkopf, U. In *Organic Synthesis: An Interdisciplinary Challenge*, Streith, H.; Prinzbach, G.; Schill, G., Eds.; Blackwell: Oxford, 1985; p 101. (c) Schöllkopf, U. *Pure Appl. Chem.* **1983**, *55*, 1799. (d) Schöllkopf, U. *Chem. Scr.* **1985**, *25*, 105. (e) Williams, R. M. *Synthesis of Optically Active α-Amino Acids*, Pergamon: Oxford, 1989, (f) Schöllkopf, U. *Top. Curr. Chem.* **1983**, *109*, 65.
2. (a) Schöllkopf, U.; Groth, U.; Deng, C. *Angew. Chem., Int. Ed. Engl.* **1981**, *20*, 798.
3. (a) Schöllkopf, U.; Nozulak, J.; Grauert, M. *Synthesis* **1985**, 55. (b) Grauert, M.; Schöllkopf, U. *Justus Liebigs Ann. Chem./Liebigs Ann. Chem.* **1985**, 1817. (c) Beulshausen, T.; Groth, U.; Schöllkopf, U. *Justus Liebigs Ann. Chem./Liebigs Ann. Chem.* **1991**, 1207.
4. (a) Schöllkopf, U.; Groth, U.; Gull, M.-R.; Nozulak, J. *Justus Liebigs Ann. Chem./Liebigs Ann. Chem.* **1983**, 1133. (b) Neubauer, H.-J.; Balza, J.; Freer, J.; Schöllkopf, U. *Justus Liebigs Ann. Chem./Liebigs Ann. Chem.* **1985**, 1508.
5. Gull, R.; Schöllkopf, U. *Synthesis* **1985**, 1052.
6. Schöllkopf, U.; Pettig, D.; Schulze, E.; Klinge, M.; Egert, E.; Benecke, B.; Noltemeyer, M. *Angew. Chem., Int. Ed. Engl.* **1988**, *27*, 1194.
7. Merck Suchardt D-6100 Hohenbrunn, Germany.
8. Baldwin, J. E.; Adlington, R. M.; Bebbington, D.; Russel, A. T. *Chem. Commun./J. Chem. Soc., Chem. Commun.* **1992**, 1249.
9. Schöllkopf, U.; Groth, U. *Angew. Chem., Int. Ed. Engl.* **1981**, *20*, 977.
10. Schöllkopf, U.; Westphalen, K.-O.; Schröder, J.; Horn, K. *Justus Liebigs Ann. Chem./Liebigs Ann. Chem.* **1988**, 781.
11. Schöllkopf, U.; Nozulak, J.; Groth, U. *Tetrahedron* **1984**, *40*, 1409.
12. (a) Hartwig, W.; Born, L. *J. Org. Chem.* **1987**, *52*, 4352. (b) Schöllkopf, U.; Pettig, D.; Busse, U. *Synthesis.* **1986**, 737.
13. (a) Pearson, A. J.; Bruhn, P. R.; Gouzoules, F.; Lee, S.-K. *Chem. Commun./J. Chem. Soc., Chem. Commun.* **1989**, *10*, 659. (b) Pearson, A. J.; Bruhn, P. R. *J. Org. Chem.* **1991**, *56*, 7092.
14. Schöllkopf, U.; Beulshausen, T. *Justus Liebigs Ann. Chem./Liebigs Ann. Chem.* **1989**, 223.
15. Schöllkopf, U.; Bendenhaben, J. *Justus Liebigs Ann. Chem./Liebigs Ann. Chem.* **1987**, 393.
16. (a) Schöllkopf, U.; Kühnle, W.; Egert, E.; Dyrbusch, M. *Angew. Chem., Int. Ed. Engl.* **1987**, *26*, 480. (b) Busch, K.; Groth, U.; Kühnle, W.; Schöllkopf, U. *Tetrahedron* **1992**, *27*, 5607.
17. Wild, H.; Born, L. *Angew. Chem., Int. Ed. Engl.* **1991**, *30*, 1685.
18. Schöllkopf, U.; Hauptreif, M.; Dippel, J.; Nieger, M.; Egert, E. *Angew. Chem., Int. Ed. Engl.* **1986**, *25*, 192.
19. Schöllkopf, U.; Hupfeld, B.; Küper, S.; Egert, E.; Dyrbusch, M. *Angew. Chem., Int. Ed. Engl.* **1988**, *27*, 433.

W. Hartwig & J. Mittendorf
Bayer, Wuppertal, Germany

Dihydroquinidine Acetate[1]

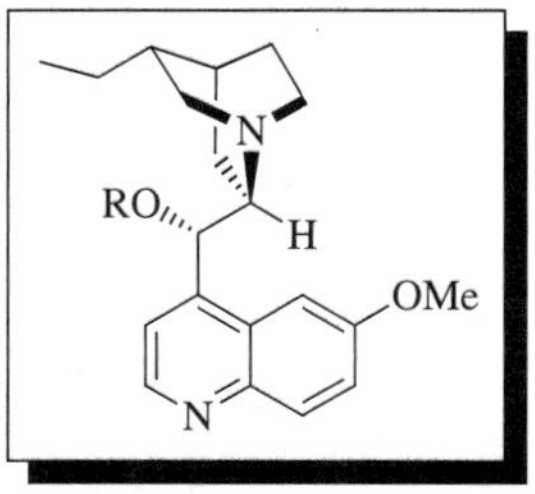

(R = Ac)
[72989-10-7] $C_{22}H_{28}N_2O_3$ (MW 368.47)
(R = p-$ClC_6H_4C(O)$–)
[113162-02-0] $C_{27}H_{29}ClN_2O_3$ (MW 464.99)
(R = H)
[1435-55-8] $C_{20}H_{26}N_2O_2$ (MW 326.44)

(asymmetric dihydroxylation;[2] conjugate additions;[3] carbonyl additions[3])

Alternate Name: DHQD-Ac.
Physical Data: *p*-$ClC_6H_4C(O)$–: mp 102–105 °C; $[\alpha]_D$ −73° (*c* = 1, EtOH).
Solubility: *p*-$ClC_6H_4C(O)$–: sol CH_2Cl_2, Et_2O, EtOH, EtOAc.
Form Supplied in: the *p*-chlorobenzoate is available as a white foam.
Preparative Methods: the acetate is prepared from dihydroquinidine[4] and the *p*-chlorobenzoate is commercially available. The phthalazine-derived bis(dihydroquinidine) ligand is commercially available.[5] A formulation of the standard reactants for the asymmetric dihydroxylation (AD-mix-β) on the small scale has been developed and is commercially available.[6] AD-mix-β (1 kg) consists of potassium osmate (0.52 g), the phthalazine-derived ligand (5.52 g), $K_3Fe(CN)_6$ (700 g), and powdered K_2CO_3 (294 g).
Purification: the recovery of dihydroquinidine *p*-chlorobenzoate after a dihydroxylation reaction is accomplished in the following manner:[7] the crude dihydroquinidine *p*-chlorobenzoate (DHQD-CLB) is dissolved in ether, cooled to 0 °C, and HCl gas is bubbled into the solution until a pH of 1–2 is obtained using wet pH paper. The pale yellow precipitate of the hydrochloride salt is collected by filtration and dried under high vacuum (0.01 mmHg). The free base is liberated by suspending the salt in EtOAc, cooling the heterogeneous mixture to 0 °C, and adding 28% NH_4OH (or 15% NaOH) until a pH = 11 is obtained. After separation, the aqueous layer is extracted with portions of EtOAc, the combined organic layers are dried over Na_2SO_4, and the solvent removed in vacuo to give the pure DHQD-CLB as a white foam.
Handling, Storage, and Precautions: toxic; use in a fume hood.

Chiral Ligand for the Asymmetric Dihydroxylation of Alkenes. Dihydroquinidine acetate (DHQD-Ac) was found to be one of the first efficient cinchona-derived chiral ligands for the asymmetric dihydroxylation reaction of substituted alkenes.[8] For example, styrene could be dihydroxylated in 61% ee (62% yield) using a mixture of 1.1 equiv of DHQD-Ac and 1.1 equiv of *Osmium Tetroxide* in toluene. An osmium-catalyzed asymmetric process, in which the co-oxidant is *N-Methylmorpholine N-Oxide* (NMO) and the chiral ligand is DHQD-CLB, was described later.[9] The other enantiomer of the diol could also be obtained by using the analogous dihydroquinine ester (*Dihydroquinine Acetate*), which acts as a pseudoenantiomer of the dihydroquinidine ester. Significant increases in the level of asymmetric induction of the dihydroxylation were later observed if the re-oxidant NMO was replaced by potassium hexacyanoferrate(III) ($K_3Fe(CN)_6$) (eq 1).[10]

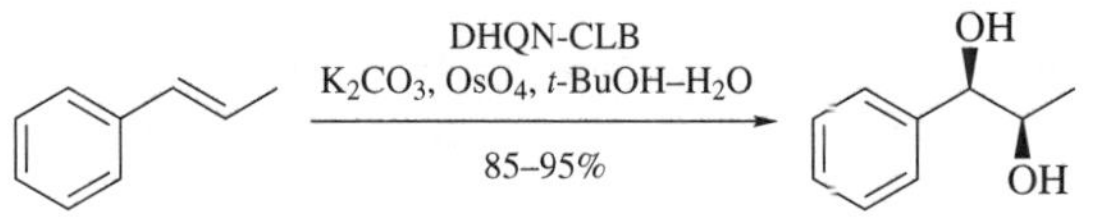

(1)

NMO (slow addition)	86% ee
NMO (no slow addition)	65% ee
$K_3Fe(CN)_6$	91% ee

The nature of the group attached to the 9-*O* position of dihydroquinidine was found to have a profound impact on the level of stereochemical induction in these reactions and a variety of new ligands have been developed (see **1–5**).[11]

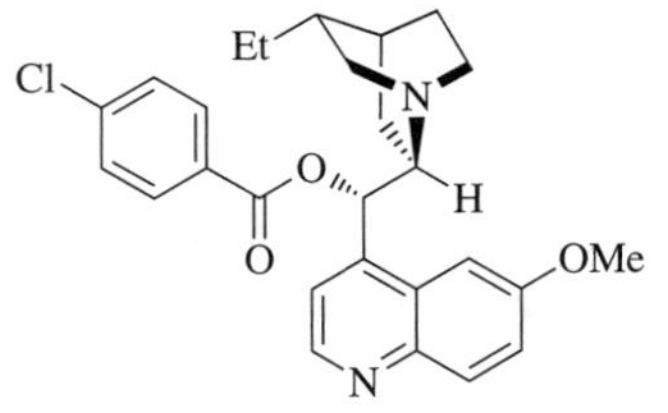

(1)
DHQD-CLB

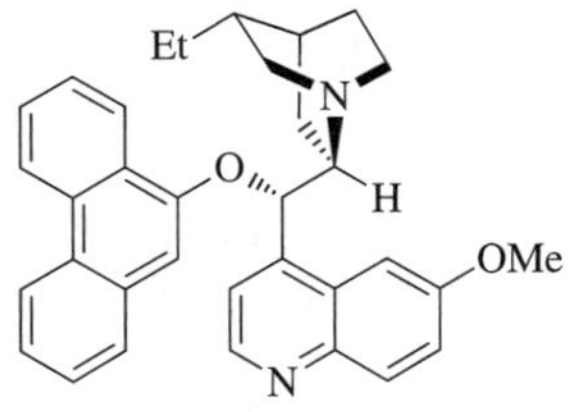

(2)
DHQD-PHN

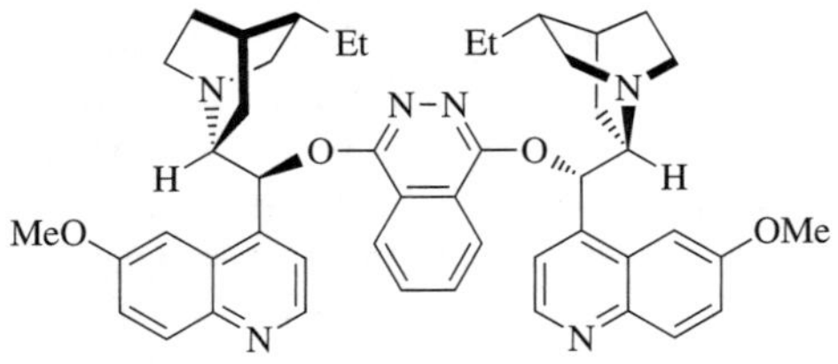

(3)
$(DHQD)_2$-PHAL
ligand used in AD-mix-β

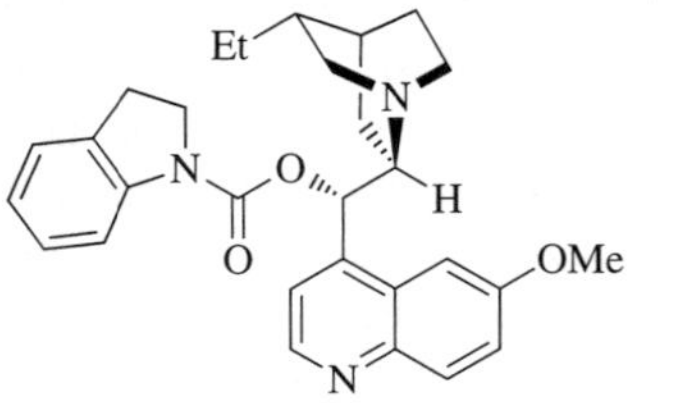

(4)
DHQD-IND

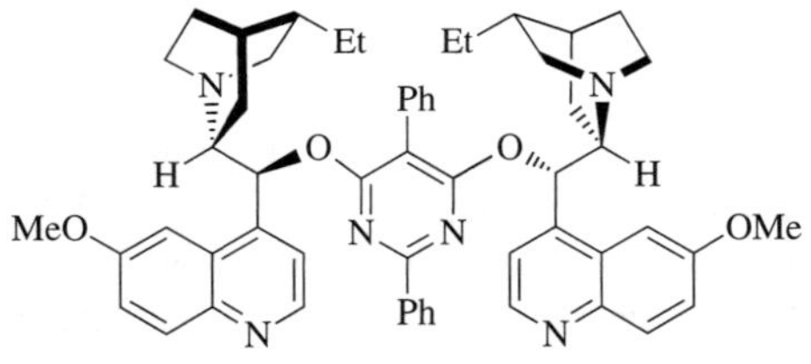

(5)
$(DHQD)_2$-PYR

The phthalazide bis(cinchona) derivatives [$(DHQD)_2$-PHAL][5,6,12] are the best ligands for the asymmetric dihydroxylation of *trans*, 1,1-disubstituted,[13] and trisubstituted alkenes, enol ethers,[14] α,β-unsaturated ketones,[15] and α,β- and β,γ-unsaturated esters,[16] whereas the DHQD-IND ligand[17] turns out to be superior for *cis*-alkenes (Table 1). The bis(cinchona) alkaloid-substituted pyrimidine ligand was found to be the best for monosubstituted terminal alkenes.[18] The addition of *Methanesulfonamide* to enhance the rate of osmate(VI) ester hydrolysis is recommended for all nonterminal alkenes.

Asymmetric dihydroxylation of substituted aryl allyl ethers also proceeds with high enantioselectivities (89–95% ee) providing that there are no *ortho* substituents on the aryl group.[19] Dienes,[20] polyenes,[21] and enynes[22] can also be regioselectively dihydroxylated (eq 2–5). In some cases, such as in the asymmetric dihydroxylation of α,β- and β,γ-unsaturated amides,[23] the amount of ligand and potassium osmate in the AD-mix content has to be increased fivefold to achieve good catalytic turnover rates.

Table 1 Alkenes Dihydroxylated using DHQD Ligands

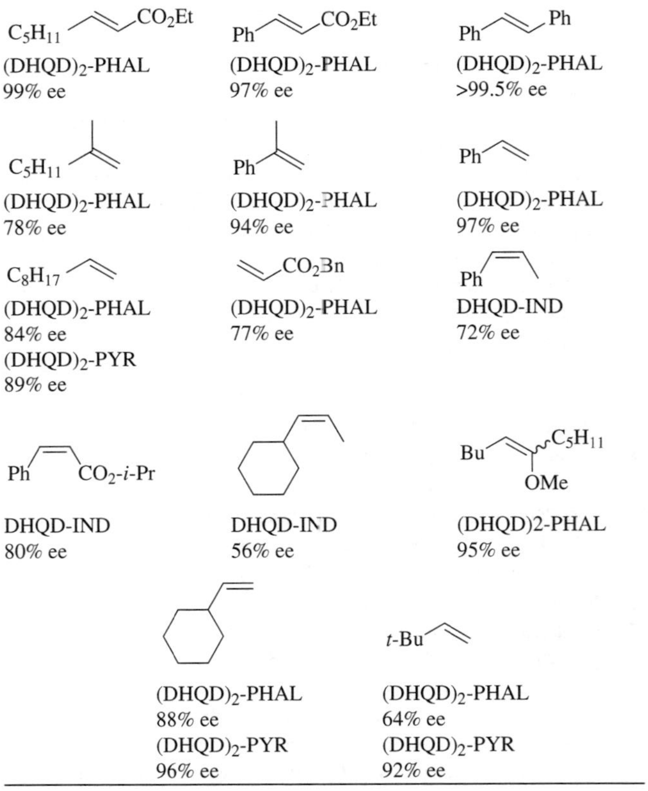

$(DHQD)_2$-PHAL 99% ee	$(DHQD)_2$-PHAL 97% ee	$(DHQD)_2$-PHAL >99.5% ee
$(DHQD)_2$-PHAL 78% ee	$(DHQD)_2$-PHAL 94% ee	$(DHQD)_2$-PHAL 97% ee
$(DHQD)_2$-PHAL 84% ee; $(DHQD)_2$-PYR 89% ee	$(DHQD)_2$-PHAL 77% ee	DHQD-IND 72% ee
DHQD-IND 80% ee	DHQD-IND 56% ee	(DHQD)2-PHAL 95% ee
$(DHQD)_2$-PHAL 88% ee; $(DHQD)_2$-PYR 96% ee	$(DHQD)_2$-PHAL 64% ee; $(DHQD)_2$-PYR 92% ee	

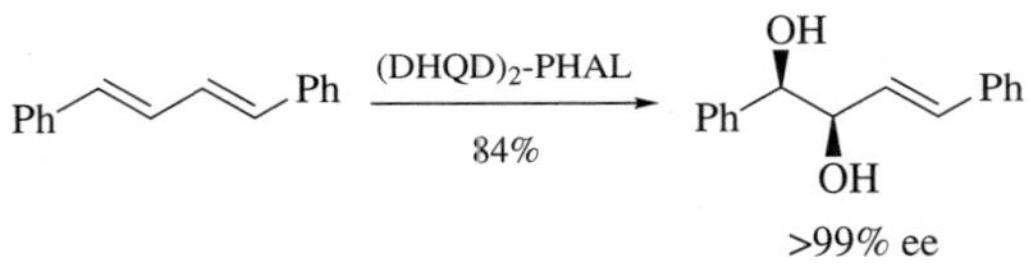

(2)

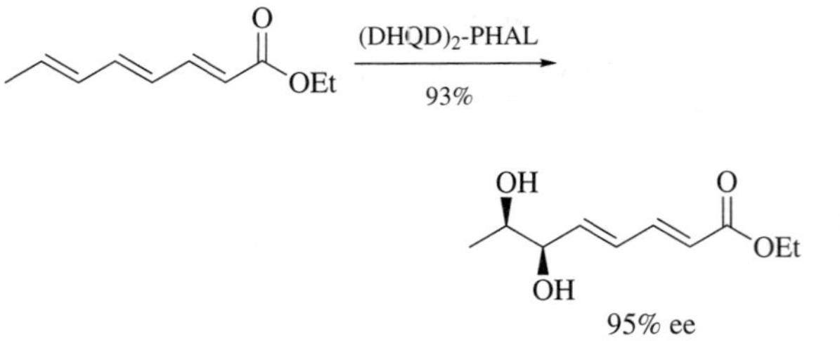

(3)

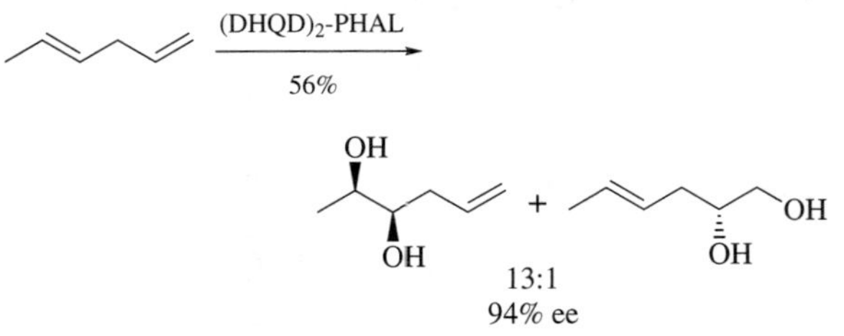

(4)

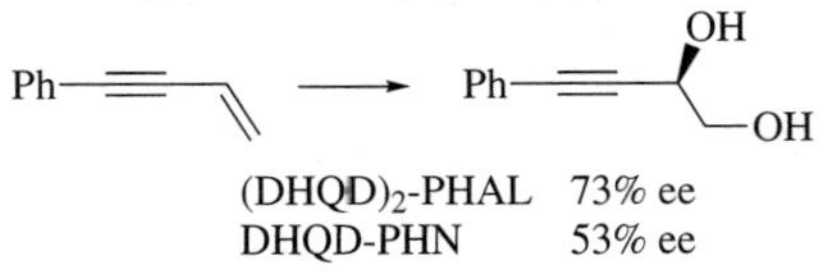

(5)

The prediction of the absolute stereochemistry of the predominant enantiomer obtained is provided by the model shown in Scheme 1.

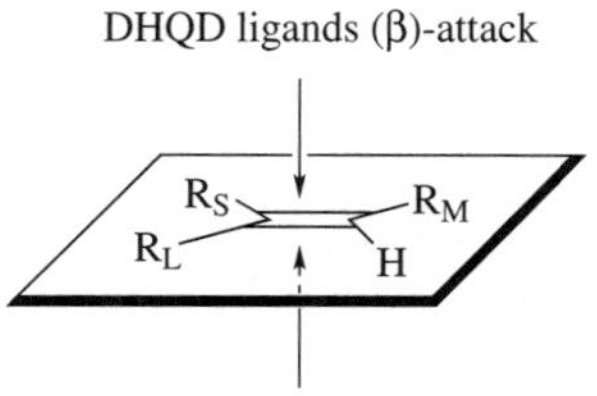

Scheme 1

By employing polymer-bound alkaloid derivatives, heterogeneous catalytic asymmetric dihydroxylation has been achieved with good to excellent enantioselectivities in the dihydroxylation of *trans*-stilbene.[24] These polymers can be recovered and reused while both the yields and the optical purities of diols were maintained.

Double Diastereoselection in the Dihydroxylation Reaction. The dihydroxylation reaction of chiral nonracemic substrates using the cinchona-derived ligand leads to a matched and mismatched pair (eq 6).[25] Kinetic resolution of several racemic secondary alcohols has also been examined.[26]

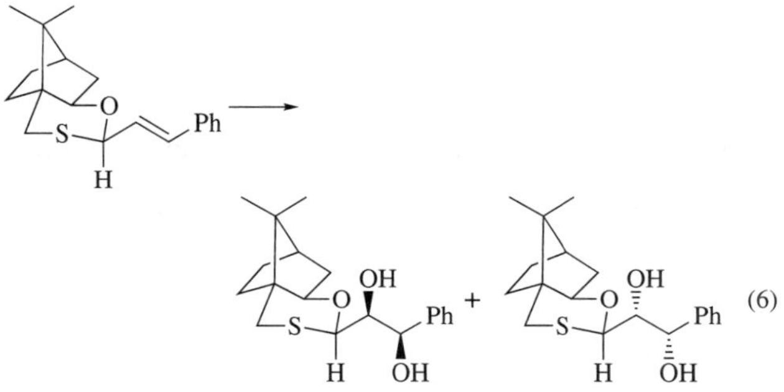

DHQD-OAc	40:1
DHQ-OAc	1:16

Additional examples of asymmetric dihydroxylation are provided under the entry for *Dihydroquinine Acetate*.

Chiral Ligand for other Stereoselective Reactions[1]. The effect of the addition of dihydroquinidine-derived alkaloids on the product enantioselectivity has also been investigated in the addition reaction of *Diethylzinc* to aldehydes,[27] in the addition of aromatic thiols to conjugated cycloalkenones,[28] and in the heterogeneous hydrohalogenation of α,α-dichlorobenzazepinone-2.[29] In these cases, the dihydroquinidine derivatives were not the optimal ligands.

---

1. Wynberg, H. *Top. Stereochem.* **1986**, *16*, 87.
2. (a) Lohray, B. B. *Tetrahedron: Asymmetry* **1992**, *3*, 1317. (b) Kolb, H. C.; VanNieuwenhze, M. S.; Sharpless, K. B. *Chem. Rev.* **1994**, *94*, 2483.
3. Blaser, H.-U. *Chem. Rev.* **1992**, *92*, 935.
4. (a) Rabe, P.; Huntenburg, W.; Schultze, A.; Volger, G. *Ber. Dtsch. Chem. Ges./Chem. Ber.* **1931**, *64*, 2487. (b) Hesse, O. *Justus Liebigs Ann. Chem./Liebigs Ann. Chem.* **1882**, *214*, 1. (c) Hesse, O. *Justus Liebigs Ann. Chem./Liebigs Ann. Chem.* **1887**, *241*, 255.
5. Amberg, W.; Bennani, Y. L.; Chadha, R. K.; Crispino, G. A.; Davis, W. D.; Hartung, J.; Jeong, K.-S.; Ogino, Y.; Shibata, T.; Sharpless, K. B. *J. Org. Chem.* **1993**, *58*, 844.
6. Sharpless, K. B.; Amberg, W.; Bennani, Y. L.; Crispino, G. A.; Hartung, J.; Jeong, K.-S.; Kwong, H.-L.; Morikawa, K.; Wang, Z.-M.; Xu, D.; Zhang, X.-L. *J. Org. Chem.* **1992**, *57*, 2768.
7. Jacobsen, E. N.; Markõ, I.; Mungall, W. S.; Schröder, G.; Sharpless, K. B. *J. Am. Chem. Soc.* **1988**, *110*, 1968.
8. Hentges, S. G.; Sharpless, K. B. *J. Am. Chem. Soc.* **1980**, *102*, 4263.
9. (a) Wai, J. S. M.; Markõ, I.; Svendsen, J. S.; Finn, M. G.; Jacobsen, E. N.; Sharpless, K. B. *J. Am. Chem. Soc.* **1989**, *111*, 1123. (b) Lohray, B. B.; Kalantar, T. H.; Kim, B. M.; Park, C. Y.; Shibata, T.; Wai, J. S. M.; Sharpless, K. B. *Tetrahedron Lett.* **1989**, *30*, 2041.
10. Kwong, H.-L.; Sorato, C.; Ogino, Y.; Chen, H.; Sharpless, K. B. *Tetrahedron Lett.* **1990**, *31*, 2999.
11. For 9-*O*-aryl dihydroquinidine ligands, see: (a) Shibata, T.; Gilheany, D. G.; Blackburn, B. K.; Sharpless, K. B. *Tetrahedron Lett.* **1990**, *31*, 3817. (b) Sharpless, K. B.; Amberg, W.; Beller, M.; Chen, H.; Hartung, J.; Kawanami, Y.; Lübben, D.; Manoury, E.; Ogino, Y.; Shibata, T.; Ukita, T. *J. Org. Chem.* **1991**, *56*, 4585. (c) Ogino, Y.; Chen, H.; Manoury, E.; Shibata, T.; Beller, M.; Lübben, D.; Sharpless, K. B. *Tetrahedron Lett.* **1991**, *32*, 5761.
12. For a similar C_2-symmetric bisether, see: Lohray, B. B.; Bhushan, V. *Tetrahedron Lett.* **1992**, *33*, 5113.
13. Wang, Z.-M.; Sharpless, K. B. *Synlett* **1993**, 603
14. Hashiyama, T.; Morikawa, K.; Sharpless, K. B. *J. Org. Chem.* **1992**, *57*, 5067.
15. Walsh, P. J.; Sharpless, K. B. *Synlett* **1993**, 605.
16. (a) Wang, Z.-M.; Zhang, X.-L.; Sharpless, K. B ; Sinha, S. C.; Sinha-Bagchi, A.; Keinan, E. *Tetrahedron Lett.* **1992**, *33*, 6407. (b) Keinan, E.; Sinha, S. C.; Sinha-Bagchi, A.; Wang, Z.-M.; Zhang, X.-L.; Sharpless, K. B. *Tetrahedron Lett.* **1992**, *33*, 6411.
17. Wang, L.; Sharpless, K. B. *J. Am. Chem. Soc.* **1992**, *114*, 7568.
18. Crispino, G. A.; Jeong, K.-S.; Kolb, H. C.; Wang, Z.-M.; Xu, D.; Sharpless, K. B. *J. Org. Chem.* **1993**, *58*, 3785.
19. Wang, Z.-M.; Zhang, X.-L.; Sharpless, K. B. *Tetrahedron Lett.* **1993**, *34*, 2267.
20. Xu, D.; Crispino, G. A.; Sharpless, K. B. *J. Am. Chem. Soc.* **1992**, *114*, 7570.
21. Crispino, G. A.; Sharpless, K. B. *Tetrahedron Lett.* **1992**, *33*, 4273.
22. Jeong, K.-S.; Sjö, P.; Sharpless, K. B. *Tetrahedron Lett.* **1992**, *33*, 3833.
23. Bennani, Y. L.; Sharpless, K. B. *Tetrahedron Lett.* **1993**, *34*, 2079.
24. (a) Kim, B. M.; Sharpless, K. B. *Tetrahedron Lett.* **1990**, *31*, 3003. (b) Lohray, B. B.; Thomas, A.; Chittari, P.; Ahuja, J. R.; Dhal, P. K. *Tetrahedron Lett.* **1992**, *33*, 5453.
25. (a) Annunziata, R.; Cinquini, M.; Cozzi, F.; Raimondi, L.; Stefanelli, S. *Tetrahedron Lett.* **1987**, *28*, 3139. (b) Annunziata, R.; Cinquini, M.; Cozzi, F.; Raimondi, L. *Tetrahedron* **1988**, *44*, 6897. (c) Gurjar, M. K.; Mainkar, A. S. *Tetrahedron: Asymmetry* **1992**, *3*, 21.
26. Lohray, B. B.; Bhushan, V. *Tetrahedron Lett.* **1993**, *34*, 3911.
27. Smaardijk, A. A.; Wynberg, H. *J. Org. Chem.* **1987**, *52*, 135.
28. (a) Hiemstra, H.; Wynberg, H. *J. Am. Chem. Soc.* **1981**, *103*, 417. (b) Kobayashi, N.; Iwai, K. *Tetrahedron Lett.* **1980**, *21*, 2167.
29. Blaser, H.-U.; Boyer, S. K.; Pittelkow, U. *Tetrahedron: Asymmetry* **1991**, *2*, 721.

André B. Charette
Université de Montréal, QC, Canada

Dihydroquinine Acetate[1]

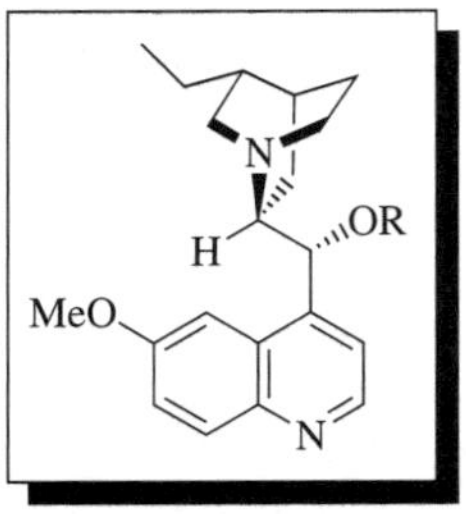

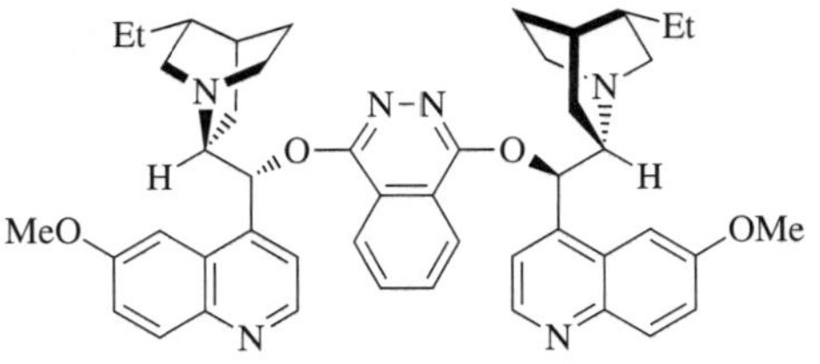

(1)
$(DHQ)_2$-PHAL
(ligand used in AD-mix-α)

(R = Ac)		
[75917-54-3]	$C_{22}H_{28}N_2O_3$	(MW 368.47)
(R = *p*-$ClC_6H_4C(O)$–)		
[113162-88-9]	$C_{27}H_{29}ClN_2O_3$	(MW 464.99)
(R = H)		
[522-66-7]	$C_{20}H_{26}N_2O_2$	(MW 326.44)

(asymmetric dihydroxylation;[2] conjugate additions;[3] carbonyl additions[3])

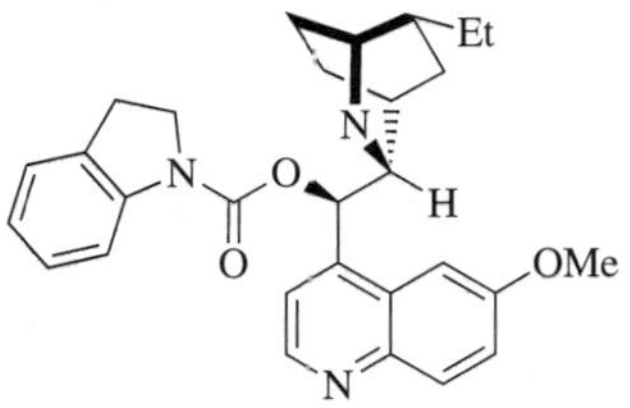

(2)
DHQ-IND

Alternate Name: DHQ-Ac.
Physical Data: *p*-$ClC_6H_4C(O)$–: mp 130–133 °C; $[\alpha]_D$ +150° (c = 1, EtOH).
Solubility: *p*-$ClC_6H_4C(O)$–: sol CH_2Cl_2, Et_2O, EtOH, EtOAc.
Form Supplied in: the *p*-chlorobenzoate is available as a white foam.
Preparative Methods: the acetate is prepared from dihydroquinine[4] and the *p*-chlorobenzoate is commercially available. The phthalazine-derived bis(dihydroquinine) ligand is commercially available.[5] A formulation of the standard reactants for the asymmetric dihydroxylation (AD-mix-α) on the small scale has been developed and is commercially available.[6] AD-mix-α (1 kg) consists of potassium osmate (0.52 g), the phthalazine-derived ligand (5.52 g), $K_3Fe(CN)_6$ (700 g), and powdered K_2CO_3 (294 g).
Purification: dihydroquinine *p*-chlorobenzoate is recovered after a dihydroxylation reaction using the same method as that described for *Dihydroquinidine Acetate*.
Handling, Storage, and Precautions: toxic; use in a fume hood.

Table 1 Alkenes Dihydroxylated using DHQ Ligands

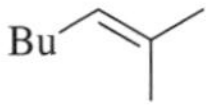	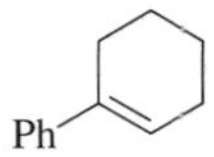	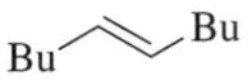
$(DHQD)_2$-PHAL 98% ee $(DHQ)_2$-PHAL 95% ee	$(DHQD)_2$-PHAL 99% ee $(DHQ)_2$-PHAL 97% ee	$(DHQD)_2$-PHAL 97% ee $(DHQ)_2$-PHAL 93% ee
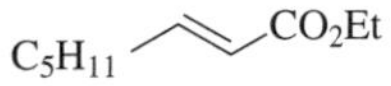	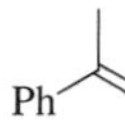	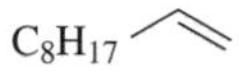
$(DHQD)_2$-PHAL 99% ee $(DHQ)_2$-PHAL 96% ee	$(DHQD)_2$-PHAL 94% ee $(DHQ)_2$-PHAL 93% ee	$(DHQD)_2$-PHAL 84% ee $(DHQ)_2$-PHAL 80% ee
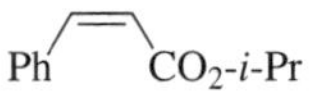	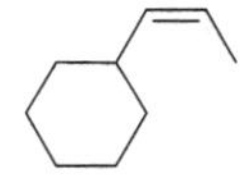	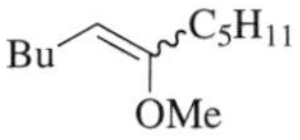
DHQD-IND 80% ee DHQ-IND 72% ee	DHQD-IND 56% ee DHQ-IND 44% ee	$(DHQD)_2$-PHAL 95% ee $(DHQ)_2$-PHAL 96% ee

Chiral Ligand for the Asymmetric Dihydroxylation of Alkenes. Dihydroquinine-derived chiral ligands have been used as pseudoenantiomers of the dihydroquinidine analog in the catalytic asymmetric dihydroxylation of alkenes. In general, the enantioselectivities with these ligands are as good as or slightly lower than those obtained with the dihydroquinidine ligand. For example, styrene could be dihydroxylated in 62% ee using a mixture of dihydroquinidine *p*-chlorobenzoate (DHQD-CLB, 0.13 equiv), *Osmium Tetroxide* (0.13 equiv), and *N-Methylmorpholine N-Oxide*, whereas the analogous reaction with dihydroquinine *p*-chlorobenzoate produced the diol of opposite absolute stereochemistry in 54% ee.[7]

As in the dihydroquinidine series, the phthalazide cinchona derivative [$(DHQ)_2$-PHAL] (**1**)[6] is the best ligand for the asymmetric dihydroxylation of terminal, *trans*, 1,1-disubstituted, and trisubstituted alkenes, and enol ether,[8] whereas the DHQ-IND ligand (**2**)[9] turns out to be superior for *cis*-alkenes (Table 1). The addition of *Methanesulfonamide* to enhance the rate of osmate(VI) ester hydrolysis is recommended for all nonterminal alkenes.

For additional examples of regioselective asymmetric dihydroxylations, see *Dihydroquinidine Acetate*.

Double Diastereoselection in the Dihydroxylation Reaction. The dihydroxylation reaction of chiral nonracemic substrates using the cinchona-derived ligand leads to a matched and mismatched pair.[10] The dihydroquinine-derived ligand was found to be superior to its pseudoenantiomer in the dihydroxylation of carbohydrate derivatives (eq 1).[11]

For additional examples and an extensive discussion on the use of these ligands in asymmetric dihydroxylation reactions, see *Dihydroquinidine Acetate*.

Chiral Ligand for Other Stereoselective Reactions. The effect of the addition of dihydroquinine-derived alkaloids on the product enantioselectivity has also been investigated in the addition reaction of *Diethylzinc* to aldehydes,[12] in the addition of aromatic thiols to conjugated cycloalkenones,[13] and in the heterogeneous hydrohalogenation of α,α-dichlorobenzazepinone-2.[14]

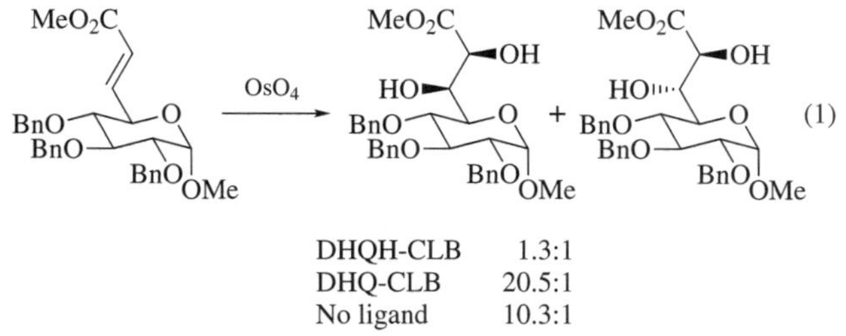

1. Wynberg, H. *Top. Stereochem.* **1986**, *16*, 87.
2. (a) Lohray, B. B. *Tetrahedron: Asymmetry* **1992**, *3*, 1317. (b) Kolb, H. C.; VanNieuwenhze, M. S.; Sharpless, K. B. *Chem. Rev.* **1994**, *94*, 2483.
3. Blaser, H.-U. *Chem. Rev.* **1992**, *92*, 935.
4. (a) Rabe, P.; Huntenburg, W.; Schultze, A.; Volger, G. *Ber. Dtsch. Chem. Ges./Chem. Ber.* **1931**, *64B*, 2487. (b) Hesse, O. *Justus Liebigs Ann. Chem./Liebigs Ann. Chem.* **1882**, *214*, 1. (c) Hesse, O. *Justus Liebigs Ann. Chem./Liebigs Ann. Chem.* **1887**, *241*, 255.
5. Amberg, W.; Bennani, Y. L.; Chadha, R. K.; Crispino, G. A.; Davis, W. D.; Hartung, J.; Jeong, K.-S.; Ogino, Y.; Shibata, T.; Sharpless, K. B. *J. Org. Chem.* **1993**, *58*, 844.
6. Sharpless, K. B.; Amberg, W.; Bennani, Y. L.; Crispino, G. A.; Hartung, J.; Jeong, K.-S.; Kwong, H.-L.; Morikawa, K.; Wang, Z.-M.; Xu, D.; Zhang, X.-L. *J. Org. Chem.* **1992**, *57*, 2768.
7. Jacobsen, E.; Markõ, I.; Mungall, W. S.; Schröder, G.; Sharpless, K. B. *J. Am. Chem. Soc.* **1988**, *110*, 1968.
8. Hashiyama, T.; Morikawa, K.; Sharpless, K. B. *J. Org. Chem.* **1992**, *57*, 5067.
9. Wang, L.; Sharpless, K. B. *J. Am. Chem. Soc.* **1992**, *114*, 7568.
10. (a) Annunziata, R.; Cinquini, M.; Cozzi, F.; Raimondi, L.; Stefanelli, S. *Tetrahedron Lett.* **1987**, *28*, 3139. (b) Annunziata, R.; Cinquini, M.; Cozzi, F.; Raimondi, L. *Tetrahedron* **1988**, *44*, 6897. (c) Gurjar, M. K.; Mainkar, A. S. *Tetrahedron: Asymmetry* **1992**, *3*, 21.
11. Brimacombe, J. S.; McDonald, G.; Rahman, M. A. *Carbohydr. Res.* **1990**, *205*, 422.
12. Smaardijk, A. A.; Wynberg, H. *J. Org. Chem.* **1987**, *52*, 135.
13. (a) Hiemstra, H.; Wynberg, H. *J. Am. Chem. Soc.* **1981**, *103*, 417. (b) Kobayashi, N.; Iwai, K. *Tetrahedron Lett.* **1980**, *21*, 2167.
14. Blaser, H.-U.; Boyer, S. K.; Pittelkow, U. *Tetrahedron: Asymmetry* **1991**, *2*, 721.

André B. Charette
Université de Montréal, QC, Canada

Diisopinocampheylborane

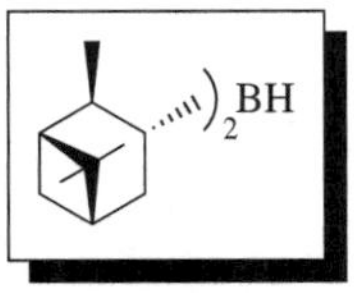

(+)
[21947-87-5] $C_{20}H_{35}B$ (MW 286.31)
(–)
[21932-54-7]

(chiral hydroborating reagent for asymmetric hydroboration of *cis*-alkenes to provide access to optically active secondary alcohols;[1] precursor for the preparation of a large number of chiral reagents for asymmetric synthesis.[1])

Alternate Name: Ipc_2BH.
Physical Data: white crystalline dimer.
Solubility: sparingly sol THF.
Analysis of Reagent Purity: active hydride is determined by hydrolysis of an aliquot and measuring the hydrogen evolved according to the standard procedure;[2] enantiomeric purity is determined by measuring the rotation of the α-pinene liberated in its reaction with 0.5 equiv of *N,N,N′,N′-Tetramethylethylenediamine* (TMEDA)[3] or reaction with aldehydes.[4]
Preparative Methods: (+)-diisopinocampheylborane is prepared in high enantiomeric purity and good yield (Table 1) by hydroboration of commercially available (−)-α-pinene (of low enantiomeric purity) with *Borane–Dimethyl Sulfide* (BMS) complex, carried out by mixing the two reagents to make a solution of known molarity in THF at 0 °C or rt (eq 1); the mixture is left without stirring at 0 °C for ~12 h for the development of crystals (the slow crystallization facilitates the incorporation of the major diastereomer in the crystalline product, leaving the undesired isomer in solution); the supernatant solution is decanted using a double-ended needle; the crystalline lumps are broken and washed with diethyl ether and dried under vacuum (~12 mmHg) at rt.[3,4]

Table 1 Synthesis of Diisopinocampheylborane (Ipc_2BH) of High Optical Purity via Selective Single Crystallization in THF (Optimized Conditions)

(+)-α-Pinene % ee	Molar ratio[a]	Molarity M (in borane)	Temp (°C)	Isolated (% yield)	(−)-Ipc_2BH % ee[b]
92.0	2.3:1	1	0	70–75	>99
91	2.5:1	1.25	20–25[c]	>90	>99
84	2:1	1	0	50–60	98.3

[a] Molar ratio of α-pinene to BMS. [b] Based on measuring the rotation of the (+)-α-pinene obtained from (−)-Ipc_2BH. [c] At times, Ipc_2BH starts precipitating immediately; in such cases the reaction mixture should be redissolved at 50–55 °C, followed by slow recrystallization.[4]

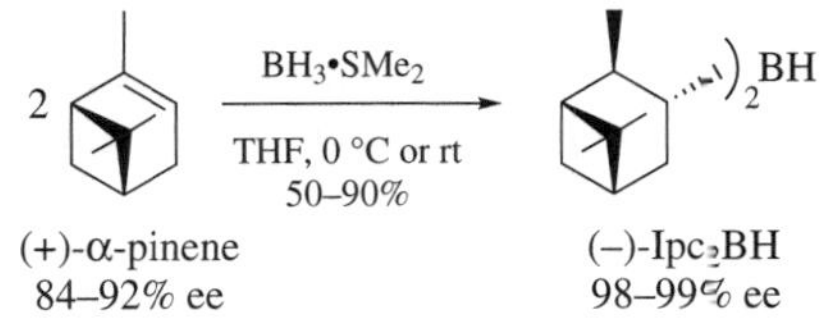

(1)

Handling, Storage, and Precautions: air sensitive, reacting instantaneously with protic solvents to liberate hydrogen; must be handled under an inert atmosphere (N_2 or Ar); can be stored at 0 °C under inert atmosphere for several months without loss of hydride activity.[4]

Asymmetric Hydroboration. Brown and Zweifel originally carried out the hydroboration of α-pinene to study the sensitivity of the α-pinene structure towards rearrangement. Surprisingly, the hydroboration reaction proceeded without rearrangement and stopped at the dialkylborane (R_2BH) stage.[5] This important reaction (reported in 1961) thus gave birth to a unique reagent, diisopinocampheylborane (Ipc_2BH). The failure of this reagent to hydroborate a third molecule of α-pinene suggested the possibility of its application in asymmetric hydroboration of less sterically hindered alkenes.

The first substrate which was asymmetrically hydroborated using Ipc_2BH was *cis*-2-butene, and the enantiomeric purity of the product 2-butanol (87% ee) obtained in this preliminary experiment was spectacular (eq 2), since Ipc_2BH was made from α-pinene of low optical purity.[5] This reaction represents the first nonenzymatic asymmetric synthesis for achieving high enantioselectivity. Its discovery marked the beginning of a new era of practical asymmetric synthesis obtained via reagent control.[1,5]

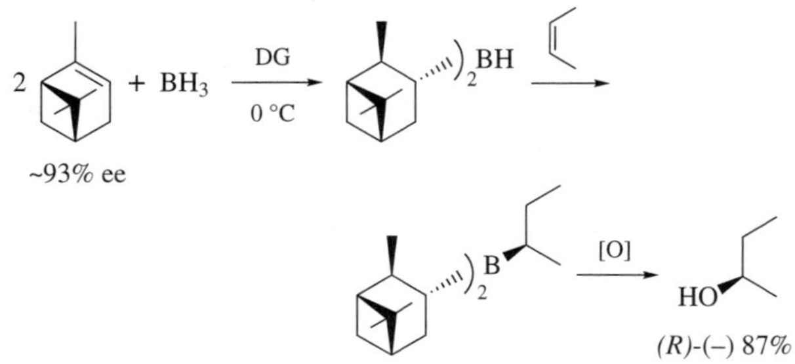

(2)

Later, Brown and co-workers developed the method described above for the preparation of enantiomerically pure Ipc_2BH (>99% ee)[3,4] and applied the reagent in the asymmetric hydroboration of prochiral alkenes. Oxidation of the trialkylboranes provided optically active alcohols. In the case of *cis*-alkenes, secondary alcohols were obtained in excellent enantiomeric purity (Figure 1). The reaction is general for most types of *cis*-alkene, e.g. *cis*-2-butene forms (*R*)-2-butanol in 98.4% ee, and *cis*-3-hexene is converted to (*R*)-3-hexanol in 93% ee. However, the reagent is somewhat limited in reactions with unsymmetrical alkenes; e.g. *cis*-4-methyl-2-pentene yields 4-methyl-2-pentanol with 96% regioselectivity but only 76% ee (Figure 1).[6]

Asymmetric hydroborations of heterocyclic alkenes are highly regio- and enantioselective. For example, hydroboration of 2,3-dihydrofuran with Ipc_2BH followed by oxidation provides 3-hydroxyfuran in 83% ee, which can be upgraded to essentially the enantiomerically pure form (>99% ee) (Figure 2).[7]

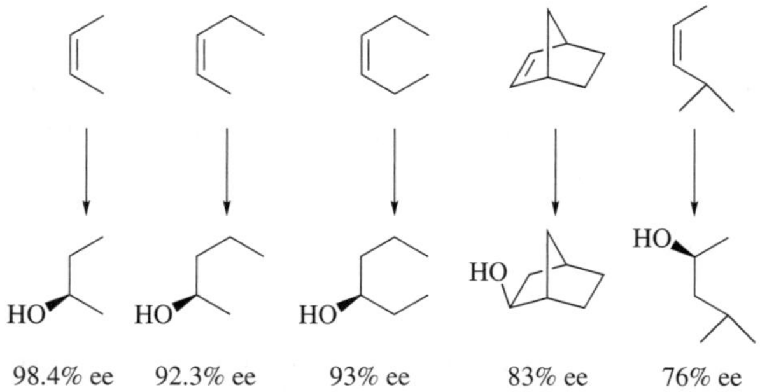

Figure 1 Asymmetric hydroboration–oxidation of *cis*-alkenes with Ipc_2BH

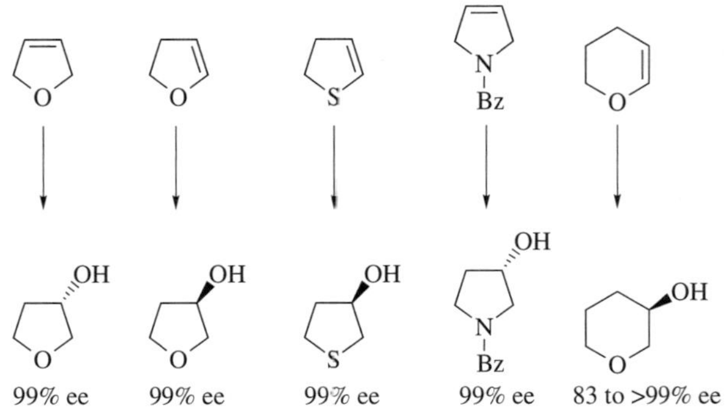

Figure 2 Asymmetric hydroboration–oxidation of some heterocyclic alkenes with Ipc_2BH

Applications. The ability of Ipc_2BH to hydroborate *cis*-alkenes has been elegantly applied to the preparation of key intermediates which have been utilized in syntheses of valuable target molecules.[1a] For example, asymmetric hydroboration–oxidation of 5-methylcyclopentadiene to the corresponding optically active alcohol has been applied in the synthesis of loganin (eq 3).[8a] In another example, a prostaglandin precursor was obtained by the asymmetric hydroboration–oxidation reaction of methyl cyclopentadiene-5-acetate (eq 4).[8b] Ipc_2BH has also been used in the preparation of $PGF_{2\alpha}$.[9]

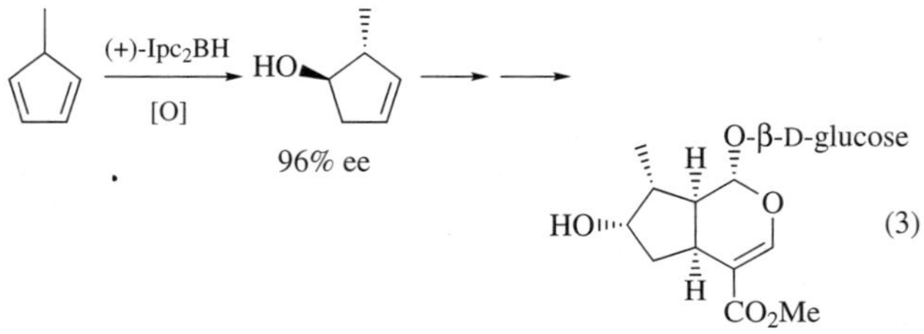

(3)

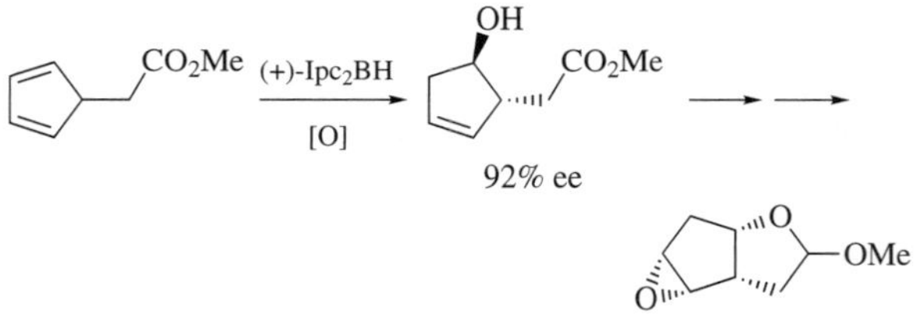

(4)

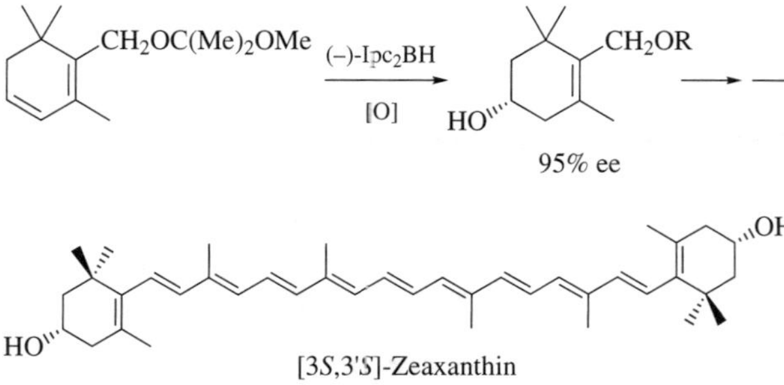

(5)

Both the enantiomers of Ipc_2BH have been elegantly applied in the asymmetric hydroboration of safranol isoprenyl methyl ether for the synthesis of carotenoids (3*R*,3′*R*)-, (3*S*,3′*S*)-, and (3R,3′S;*meso*)-zeaxanthins (eq 5).[10] (3*S*,5*R*,3′*S*,5′*R*)-Capsorubin, a carotenoid found in the red paprika *Capsicum annuum*, was synthesized via a key step involving asymmetric hydroboration of the unsaturated acetal followed by an aldol condensation (eq 6).[11]

Asymmetric hydroboration using Ipc_2BH was also applied in the stereocontrolled synthesis of a linearly fused triquinane, (+)-hirsutic acid (eq 7).[12]

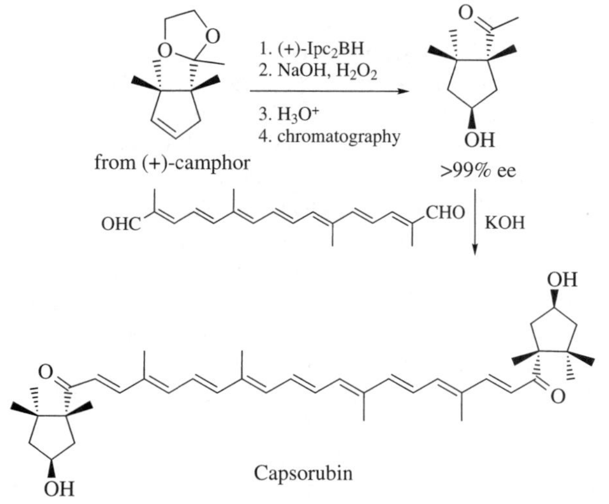
(6)

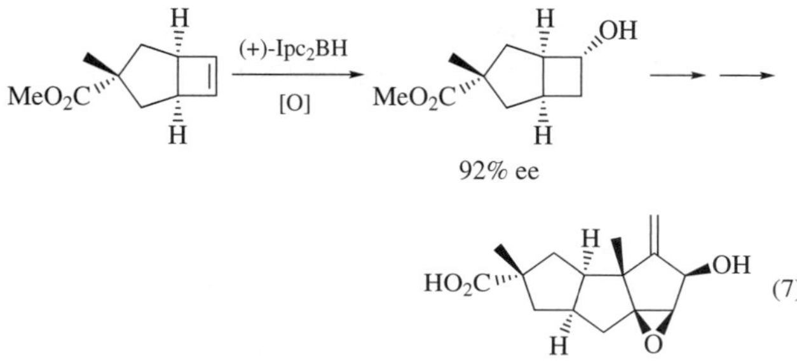
(7)

(+)-Hirsutic acid

Diisopinocampheylborane is not an effective asymmetric hydroborating agent for 2-substituted 1-alkenes. High selectivities have, however, been achieved where one of the substituents is very bulky. This aspect has been elegantly demonstrated by the synthesis of both enantiomers of a precursor of tylonolide, the aglycone of tylosin, which is one of the members of the polyoxomacrolide antibiotics. In both cases the isomeric ratio was at least 50:1 (eqs eq 8 and eq 9).[13]

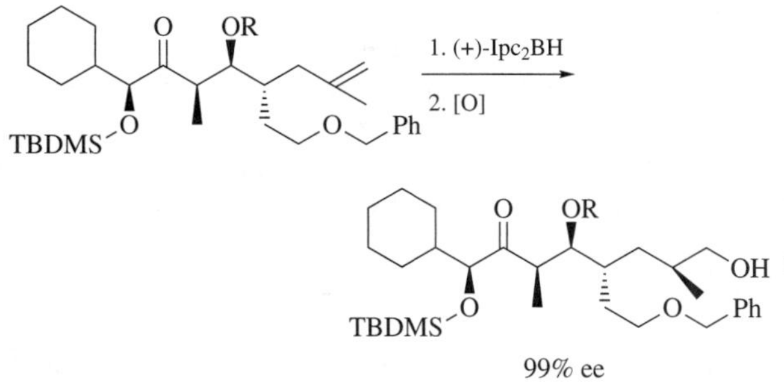
(8)

Application of Various Chiral Reagents Derived from Ipc_2BH. Diisopinocampheylborane does not normally yield satisfactory ee's in hydroboration reactions of 1,1-disubstituted alkenes, *trans*-alkenes, or trisubstituted alkenes. This problem has been partially solved by the introduction of *Monoisopinocampheylborane*, $IpcBH_2$, which is derived from Ipc_2BH. $IpcBH_2$ handles *trans*-alkenes and trisubstituted alkenes effectively, since it is of lower steric requirement than Ipc_2BH (Table 2). Moreover, $IpcBH_2$ and Ipc_2BH provide an entry into the synthesis of a large variety of optically active borinate and boronate esters. These esters have been successfully converted into α-chiral aldehydes, acids, amines, α-chiral *cis*- and *trans*-alkenes, α-chiral alkynes, β-chiral esters, ketones,[1] etc.

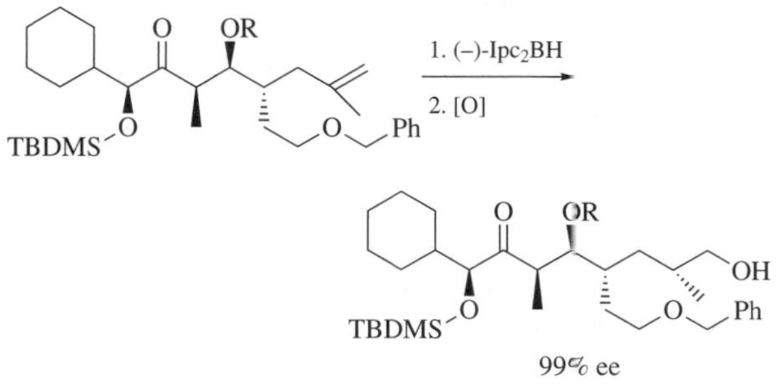
(9)

Table 2 Asymmetric Hydroboration of Alkenes with Ipc_2BH and $IpcBH_2$

Class[a]	Alkene	% ee of alcohol Ipc_2BH	% ee of alcohol $IpcBH_2$
I	2-Methyl-1-alkenes	~20	~1
II	*cis*-Alkenes	≥99	~25
III	*trans*-Alkenes	~20	70–90[b]
IV	Trisubstituted alkenes	~20	60–≥99[b]

[a] Steric requirement increases from class I to class IV. [b] The ee of initial product can be upgraded to 99% ee via crystallization.

Other reagents which have been derived from Ipc_2BH include *Diisopinocampheylboron Trifluoromethanesulfonate* (Ipc_2BOTf),[14] *B-Methoxydiisopinocampheylborane* (Ipc_2BOMe), and *(+)-B-Chlorodiisopinocampheylborane* and its bromo- and iodo analogs (Scheme 1).[1] Ipc_2BOTf and Ipc_2BOMe reagents are used in stereoselective C–C bond forming reactions (aldol condensation and allylboration); Ipc_2BCl (DIP-chloride) is used for asymmetric reduction of prochiral ketones, and Ipc_2BX (X = Br or I) for enantioselective opening of *meso*-epoxides to nonracemic halohydrins. Numerous applications of all these reagents have been reviewed in detail.[1]

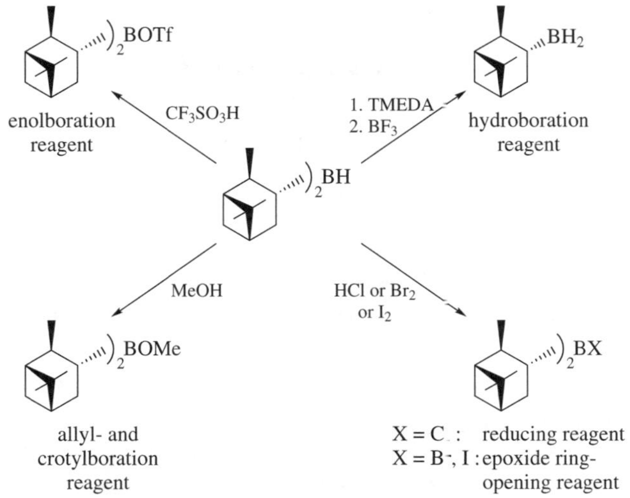

Scheme 1

Related Reagents. (+)-*B*-Chlorodiisopinocampheylborane; Diisopinocampheylboron Trifluoromethanesulfonate; Dilongifolylborane; (*R*,*R*)-2,5-Dimethylborolane; *B*-Methoxydiisopinocampheylborane; Monoisopinocampheylborane.

1. For some excellent reviews on synthetic applications of diisopinocampheylborane and related reagents, see: (a) Brown, H. C.; Ramachandran, P. V. *Advances in Asymmetric Synthesis*; Hassner, A., Ed.; JAI Press: Greenwich, CT, 1994; Vol. 2, in press. (b) Brown, H. C.; Ramachandran, P. V. *Pure Appl. Chem.* **1991**, *63*, 307. (c) Brown, H. C.; Singaram, B. *Acc. Chem. Res.* **1988**, *21*, 287. (d) Srebnik, M.; Ramachandran, P. V. *Aldrichim. Acta* **1987**, *20*, 9. (e) Matteson, D. S. *Synthesis* **1986**, 973.
2. Brown, H. C.; Kramer, G. W.; Levy, A. B.; Midland, M. M. *Organic Synthesis via Boranes*; Wiley: New York, 1975; p 239.
3. Brown, H. C.; Singaram, B. *J. Org. Chem.* **1984**, *49*, 945.
4. Brown, H. C.; Joshi, N. N. *J. Org. Chem.* **1988**, *53*, 4059.
5. Brown, H. C.; Zweifel, G. *J. Am. Chem. Soc.* **1961**, *83*, 486.
6. (a) Brown, H. C.; Desai, M. C.; Jadhav, P. K. *J. Org. Chem.* **1982**, *47*, 5065. (b) Brown, H. C.; Ayyangar, N. R.; Zweifel, G. *J. Am. Chem. Soc.* **1964**, *86*, 397.
7. Brown, H. C.; Prasad, J. V. N. V. *J. Am. Chem. Soc.* **1986**, *108*, 2049.
8. (a) Partridge, J. J.; Chadha, N. K.; Uskokovic, M. R. *J. Am. Chem. Soc.* **1973**, *95*, 532. (b) *J. Am. Chem. Soc.* **1973**, *95*, 7171.
9. Corey, E. J.; Noyori, R. *Tetrahedron Lett.* **1970**, 311.
10. Ruttimann, A.; Mayer, H. *Helv. Chim. Acta* **1980**, *63*, 1456.
11. Ruttimann, A.; Englert, G.; Mayer, H.; Moss, G. P.; Weedon, B. C. L. *Helv. Chim. Acta* **1983**, *66*, 1939.
12. Greene, A. E.; Luche, M.-J.; Serra, A. A. *J. Org. Chem.* **1985**, *50*, 3957.
13. Masamune, S.; Lu, L. D.-L.; Jackson, W. P.; Kaiho, T.; Toyoda, T. *J. Am. Chem. Soc.* **1982**, *104*, 5523.
14. Paterson, I.; Goodman, J. M.; Lister, M. A.; Schumann, R. C.; McClure, C. K.; Norcross, R. D. *Tetrahedron* **1990**, *46*, 4663.

Raj K. Dhar
Aldrich Chemical Company, Sheboygan Falls, WI, USA

Diisopinocampheylboron Trifluoromethanesulfonate

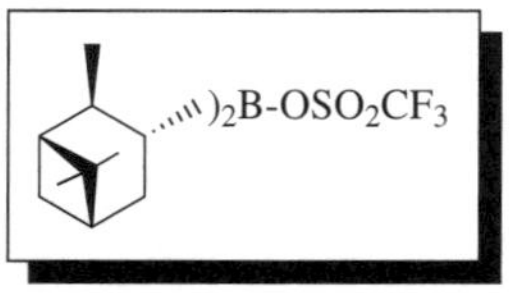

(+)
[108266-89-3] $C_{21}H_{34}BF_3O_3S$ (MW 434.36)
(−)
[108161-70-2]

(enolboration reagent for enantio- and diastereoselective aldol condensation of oxazolines[1] and ketones;[2] also used for Ireland–Claisen rearrangement[3])

Alternate Name: diisopinocampheylboron triflate; Ipc_2BOTf.
Physical Data: colorless, viscous oil; bp ≤150 °C/0.01 mmHg ('bulb-to-bulb distillation'); (−)Ipc_2BOTf $[\alpha]_D$ −43.5° (c = 30.4, hexane); ^{11}B NMR (hexane) broad singlet at $\delta = 60$ ppm (with reference to $BF_3 \cdot OEt_2$).
Solubility: highly sol in both polar and nonpolar aprotic solvents, e.g. diethyl ether, THF, dichloromethane, pentane, hexane, etc.
Preparative Methods: the (+) enantiomer of the reagent (first reported in 1981)[1] is prepared from commercially available (−)-α-pinene (∼87% ee) by hydroboration with *Borane–Dimethyl Sulfide* in THF at 0 °C, which generates *Diisopinocampheylborane*, (+)-Ipc_2BH, in more than 99% ee.[4] The crystalline Ipc_2BH is isolated and treated with *Trifluoromethanesulfonic Acid* at 0 °C either in dichloromethane[5a] or hexane.[5b] The reagent develops color in dichloromethane. However, it has been prepared in hexane as a clear and colorless solution which separates from an immiscible colored lower layer. In calculating the molarity of the reagent solution, a 60–70% conversion to triflate is assumed.[2] The reagent is usually prepared in situ from Ipc_2BH, and then its aldol reaction is carried out in the same flask by sequential addition of the required reagents[5,11] (procedure A).[6] Alternatively, the enantiomerically pure reagent can be conveniently prepared by treatment of commercially available (−)- or *(+)-B-Chlorodiisopinocampheylborane* (DIP-Cl) with triflic acid at 0 °C in hexane (Scheme 1).[7] This method (procedure B) generates Ipc_2BOTf instantaneously in almost quantitative yield. The reagent generated by procedure B can be utilized for aldol reaction in the same manner as described for procedure A.

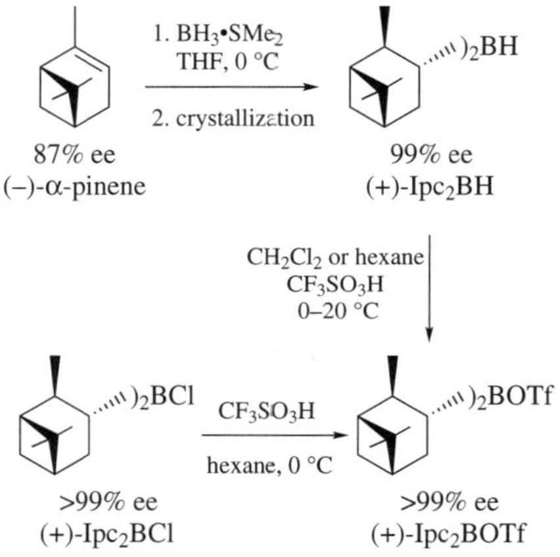

Scheme 1

Handling, Storage, and Precautions: air sensitive; reacts instantaneously with protic solvents to liberate triflic acid; should be freshly prepared prior to use; the freshly prepared reagent turns from pale yellow to clear red upon standing. All transformations involving this reagent should be carried out under N_2 using standard techniques for air sensitive reagents; use in a fume hood.

Boron Azaenolates from Oxazolines. The reagent is useful for asymmetric aldol condensations of achiral oxazolines. Treatment of 2-ethyl-4-dimethyl-2-oxazoline with Ipc_2BOTf in the presence of a tertiary amine furnishes a boron azaenolate. Without isolation, treatment with an aldehyde in ether at −78 °C provides an alkylated oxazoline, which is hydrolyzed and converted to β-hydroxy ester via treatment with *Diazomethane*. Although the yields for the four-step sequence are only moderate, the *anti* selectivities of the hydroxy acids are excellent with enantioselectivities of 77–85% ee (eq 1).[1]

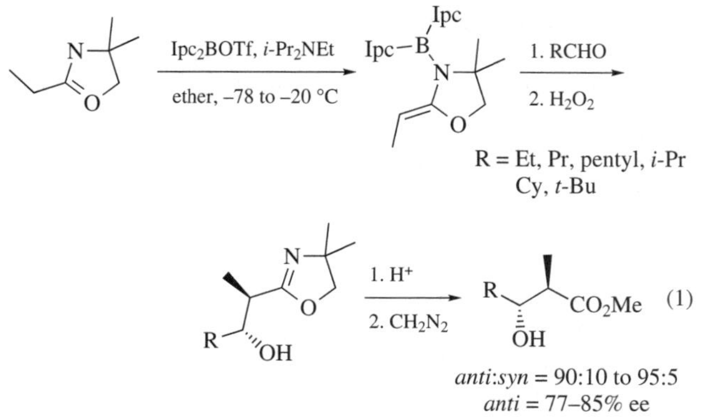

Boron Enolates from Ketones. Boron enolates are highly versatile intermediates in organic synthesis.[8] Their high reactivity and stereoselectivity are often utilized for aldol condensation reactions.[9,10] The reagent has been elegantly applied for regio- and stereoselective enolboration of ketones and subsequent enantio- and diastereoselective aldol reactions with aldehydes.[11] For example, the aldol reaction between ethyl ketones and aldehydes using the (+) or (−) reagent in the presence of a tertiary amine in dichloromethane gives (via the desired (*Z*)-enolborinate) *syn*-α-methyl-β-hydroxy ketones in good enantiomeric excess (66–93% ee) and with high diastereoselectivity (>95%) (eq 2). In contrast, the *anti* selectivity of the aldol product derived from diethyl ketone via formation of the (*E*)-enolate, derived from Ipc_2BCl (DIP-Cl) with *Methacrolein*, proceeds with negligible enantioselectivity.[11a] However, use of both the triflate and the chloride reagents in the aldol reaction of methyl ketones with aldehydes have been reported to give β-hydroxy ketones in moderate enantiomeric excess (53–78% ee) with a reversal in the enantioface selectivity of the aldehyde compared to the corresponding ethyl ketone *syn*-aldol. This variable selectivity is interpreted as evidence for the participation of competing chair and boat transition states.[11]

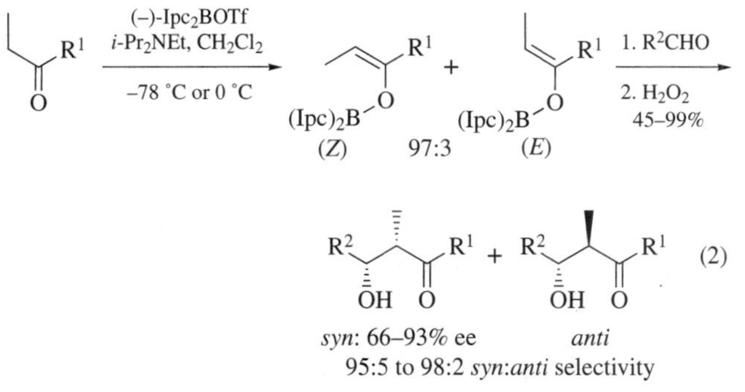

The aldol methodology mediated by Ipc_2BOTf was successfully applied to a macrolide antibiotic synthesis. Paterson reported a convenient asymmetric synthesis of a C_{19}–C_{27} segment of rifamycin S used in the Kishi synthesis, based on ethyl ketone aldol reactions mediated by optically pure reagent.[12a] He also reported the novel aldol approach to the synthesis of an enantiomerically pure C_7–C_{15} segment of tirandamycin A.[12b] This was prepared via enolboration of the (*R*)-ethyl ketone by (−)-Ipc_2BOTf in the presence of a tertiary amine. Addition of aldehyde to the corresponding enolborinate, followed by oxidative workup and chromatographic purification, led to the two separated *syn*-aldol isomers (8:1 ratio with 63% combined yield) with no *anti*-aldol product detected by HPLC. The major 1,2-*syn*-3,4-*syn* diastereomer is reported to be enantiomerically pure. Moreover, it was observed that enantiomeric excess of the major aldol isomer is significantly enhanced relative to the starting ketone. The corresponding aldol product is reduced to the 1,3-diol (eq 3) and subsequently converted to the enantiomerically pure C_7–C_{15} segment of tirandamycin A via pyranone synthesis.[12b]

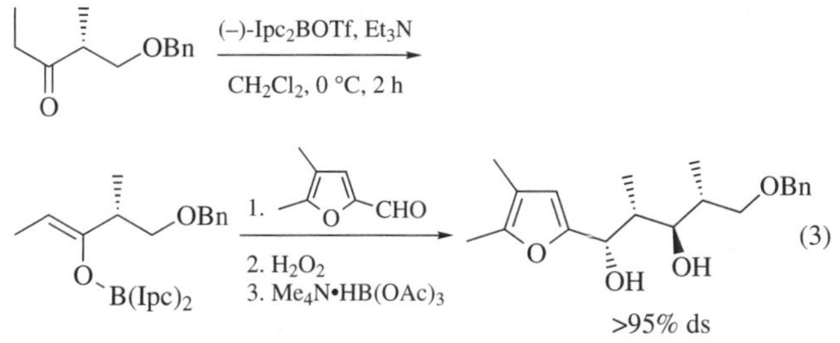

Dihydropyrones are valuable intermediates for the synthesis of a variety of substituted tetrahydropyran rings. Recently, stereoselective aldol reactions of β-chlorovinyl ketones using the dienol boronate derivative derived from chiral Ipc_2BOTf was utilized for enantioselective formation of dihydropyrones. No detectable racemization was reported on the cyclization step (eq 4).[12c]

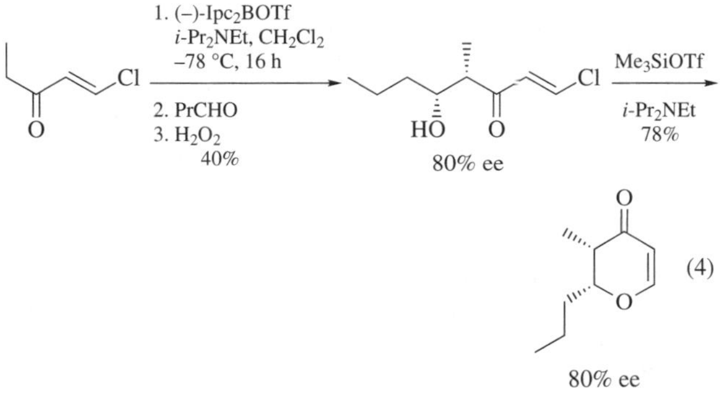

Ireland–Claisen Rearrangement. Oh et al recently reported the Ireland–Claisen rearrangement of a variety of *O*-protected 2-butenyl glycolates via chelated boron and tin triflates to give, after esterification, methyl 2-methoxy/benzyloxy-3-methyl-4-pentenoates.[3] In reactions using Ipc_2BOTf, diastereoselection as high as 99.5% was reported. The diastereoselection obtained in reactions using *Tin(II) Trifluoromethanesulfonate*, *Zinc Trifluoromethanesulfonate*, and *Di-n-butylboryl Trifluoromethanesulfonate* was far lower than in reactions with Ipc_2BOTf. Moreover, the rate of rearrangement with boron enolates was found to be higher than the rates of rearrangement of the silyl ketene acetals or lithium enolate. As anticipated the *cis*-alkene gives better diastereoselectivity than the *trans* isomer (eq 5). With this high diastereoselection obtained using Ipc_2BOTf, it is surprising that the enantioselectivity of this reaction is only 0–10% ee.[3]

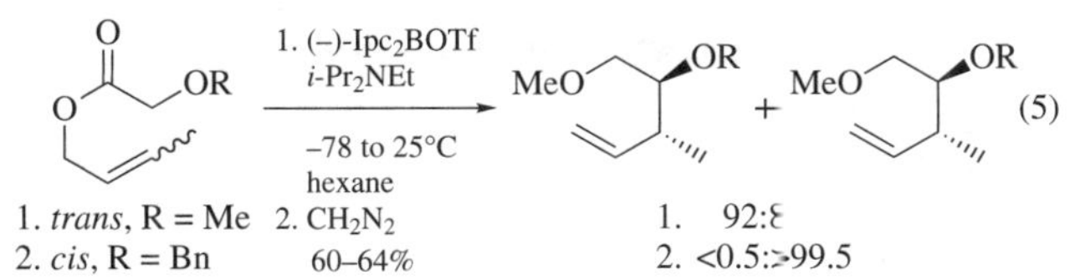

Related Reagents. (+)-*B*-Chlorodiisopinocampheylborane; Diisopinocampheylborane; (*R*,*R*)-2,5-Dimethylborolane.

1. (a) Meyers, A. I.; Yamamoto, Y. *Tetrahedron* **1984**, *40*, 2309. (b) *J. Am. Chem. Soc.* **1981**, *103*, 4278.
2. Paterson, I.; Goodman, J. M.; Lister, M. A.; Schumann, R. C.; McClure, C. K.; Norcross, R. D. *Tetrahedron* **1990**, *46*, 4663.
3. Oh, T.; Wrobel, Z.; Devine, P. N. *Synlett* **1992**, 81.
4. (a) Brown, H. C.; Singaram, B. *J. Org. Chem.* **1984**, *49*, 945. (b) Brown, H. C.; Joshi, N. N. *J. Org. Chem.* **1988**, *53*, 4059.
5. (a) Paterson, I.; Lister, M. A.; McClure, C. K. *Tetrahedron Lett.* **1986**, *27*, 4787. (b) Paterson, I.; Lister, M. A. *Tetrahedron Lett.* **1988**, *29*, 585.
6. Purification of Ipc_2BOTf by distillation is unnecessary and probably inadvisable as optimum results are obtained with freshly prepared undistilled reagent.[2]
7. Dhar, R. K.; Brown, H. C.; unpublished results.
8. (a) Kim, B. M.; Williams, S. F.; Masamune, S. *Comprehensive Organic Synthesis* **1991**, *2*, Chapter 5. (b) Evans, D. A. In *Asymmetric Synthesis*; Morrison, J. D.; Ed.; Academic: New York, 1984; Vol. 3, Chapter 1. (c) Heathcock, C. H. In *Asymmetric Synthesis*; Morrison, J. D.; Ed.; Academic: New York, 1984; Vol. 3, Chapter 2. (d) Evans, D. A.; Nelson, J. V.; Taber, T. R. *Top. Stereochem.* **1982**, *13*, 1.
9. For an examination of the effect of leaving group (X) on the stereoselective enolboration of ketones with various R_2BX reagents (X = OTf, OMs, Cl, Br, I), see: (a) Brown, H. C.; Ganesan, K.; Dhar, R. K. *J. Org. Chem.* **1993**, *58*, 147. (b) Brown, H. C.; Dhar, R. K.; Bakshi, R. K.; Pandiarajan, P. K.; Singaram, B. *J. Am. Chem. Soc.* **1989**, *111*, 3441. (c) Goodman, J. M.; Paterson, I. *Tetrahedron Lett.* **1992**, 7223.
10. Selective *trans* deprotonation of the ketone–L_2BOTf complex leads to formation of (*Z*)-enolborinate; see: Evans, D. A.; Nelson, J. V.; Vogel, E.; Taber, T. R. *J. Am. Chem. Soc.* **1981**, *103*, 3099. An alternative explanation for L_2BCl to (*E*)-enol borinate and L_2BOTf to (*Z*)-enol borinate has been proposed by Corey and Kim; see: Corey, E. J.; Kim, S. S. *J. Am. Chem. Soc.* **1990**, *112*, 4976.
11. (a) Paterson, I.; Goodman, J. M. *Tetrahedron Lett.* **1989**, *30*, 997. (b) Paterson, I.; Goodman, J. M.; Isaka, M. *Tetrahedron Lett.* **1989**, *30*, 7121. (c) Paterson, I.; McClure, C. K. *Tetrahedron Lett.* **1987**, *28*, 1229.
12. (a) Paterson, I.; McClure, C. K.; Schumann, R. C. *Tetrahedron Lett.* **1989**, *30*, 1293. (b) Paterson, I.; Lister, M. A.; Ryan, G. R. *Tetrahedron Lett.* **1991**, *32*, 1749. (c) Paterson, I.; Osborne, S. *Tetrahedron Lett.* **1990**, *31*, 2213.

Raj K. Dhar
Aldrich Chemical Company, Sheboygan Falls, WI, USA

(*R**,*R**)-α-(2,6-Diisopropoxybenzoyloxy)-5-oxo-1,3,2-dioxaborolane-4-acetic Acid[1]

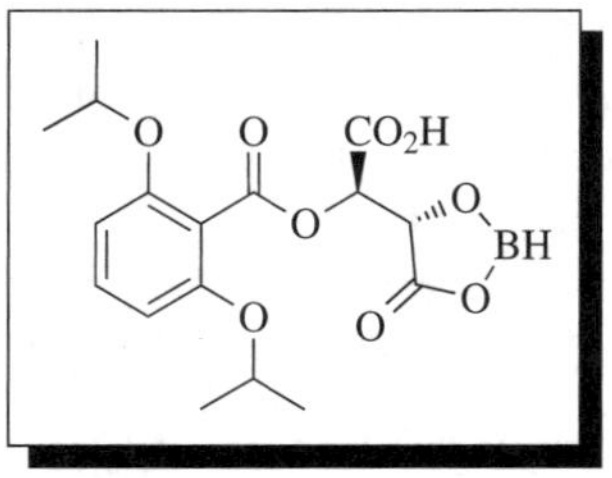

(*R*,*R*)
[*131703-55-4*] $C_{17}H_{21}BO_9$ (MW 380.16)
(*S*,*S*)
[*131703-56-5*]

(chiral Lewis acid catalyst for Diels–Alder,[1,2] aldol-type,[3] allylation,[4] and hetero Diels–Alder[5] reactions)

Solubility: sol dichloromethane, propionitrile, THF.

Form Supplied in: the acyloxyborane·THF complex is available as a 0.1–0.2 M solution in dichloromethane or propionitrile.

Analysis of Reagent Purity: ^1H NMR (CD_2Cl_2, −95 °C, 500 MHz) δ 1.07–1.13 (m, 6H, $2CH_3$), 1.24 (br, 6H, $2CH_3$), 4.50 (br, 2H, $2(CH_3)_2CH$), 4.70–4.92 (m, 1H, $CHCO_2B$), 5.45–5.72 (m, 1H, $CHCO_2H$), 6.48 (br, 2H, 2*m*-H), 7.21 (br, 1H, *p*-H).

Preparative Methods: to a solution of (*R*,*R*)- or (*S*,*S*)-mono-(2,6-diisopropoxybenzoyl)tartaric acid (74 mg, 0.2 mmol) in dry dichloromethane or propionitrile (1 mL) is added BH_3·THF (0.189 mL of 1.06 M solution in THF, 0.2 mmol) at 0 °C under an argon atmosphere. The reaction mixture is stirred for 1 h at 0 °C to produce the chiral acyloxyborane. Only 2 equiv of hydrogen gas should evolve under these reaction conditions (0 °C). See also Furuta.[1]

Handling, Storage, and Precautions: the acyloxyborane solution should be flushed with Ar and stored tightly sealed (to preclude contact with oxygen and moisture) below 0 °C. Use in a fume hood.

Acyloxyborane as an Activating Device for Carboxylic Acids[2a]. The reduction of carboxylic acids by borane is an important procedure in organic synthesis. The remarkable reactivity of borane towards carboxylic acids over esters is characteristic of this reagent. Such selectivity is rarely seen with other hydride reagents.

The rapid reaction between carboxylic acids and borane is related to the electrophilicity of the latter. The carbonyl group of the initially formed acyloxyborane intermediate, which is essentially a mixed anhydride, is activated by the Lewis acidity of the trivalent boron atom. Addition of 1/3 equiv of the *Borane–Tetrahydrofuran* complex to acrylic acid in dichloromethane followed by addition of a diene at low temperature results in the formation of Diels–Alder adducts in good yield (eq 1). Further, the reaction is successful even with a catalytic amount of borane.

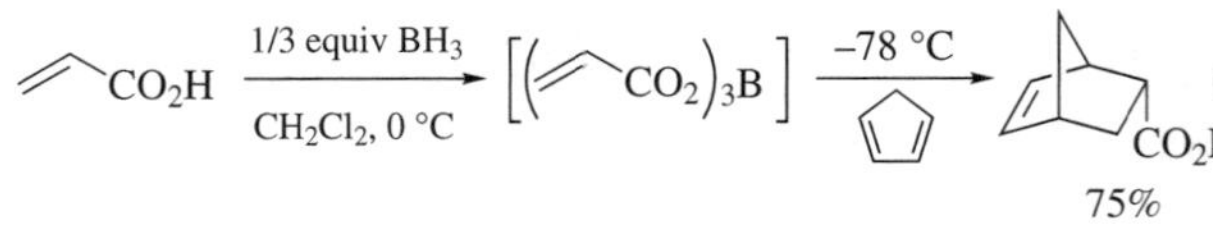

(1)

75%

Asymmetric Diels–Alder Reaction of Unsaturated Carboxylic Acids[2a]. A chiral acyloxyborane (CAB) complex (**1**) prepared from mono(2,6-dimethoxybenzoyl)tartaric acid and 1 equiv of borane is an excellent catalyst for the Diels–Alder reaction of α,β-unsaturated carboxylic acids and dienes. In the CAB-catalyzed Diels–Alder reaction, adducts are formed in a highly diastereo- and enantioselective manner under mild reaction conditions (eq 2). The reaction is catalytic: 10 mol % of catalyst is sufficient for efficient conversion, and the chiral auxiliary can be recovered and reused.

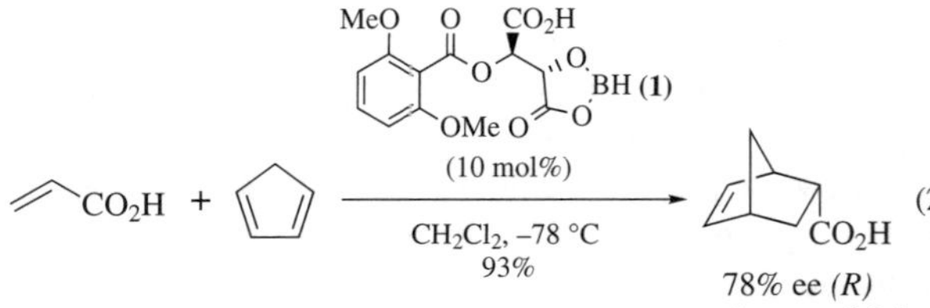

(2)

78% ee (*R*)
endo:*exo* = 96:4

Asymmetric Diels–Alder Reaction of Unsaturated Aldehydes[1,2b–e]. The boron atom of acyloxyborane is activated by the electron-withdrawing acyloxy groups, and consequently acyloxyborane derivatives are sufficiently Lewis acidic to catalyze certain reactions. Thus, asymmetric Diels–Alder reactions of α,β-enals with dienes using (**1**) as a Lewis acid catalyst have been developed. For example, the reaction of cyclopentadiene and methacrolein gives the adduct in 85% yield (*endo*:*exo* = 11:89) and 96% ee (major *exo* isomer) (eq 3). Some additional examples are listed in Figure 1. The α-substituent on the dienophile increases the enantioselectivity, while β-substitution dramatically decreases the selectivity. In the case of a substrate having substituents in both α- and β-positions, high enantioselectivity is observed; thus the α-substituent effect overcomes that of the β-substituent.

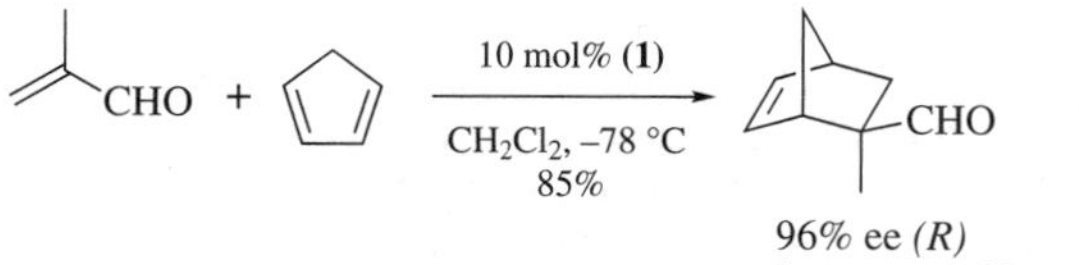

(3)

96% ee (*R*)
endo:*exo* = 11:89

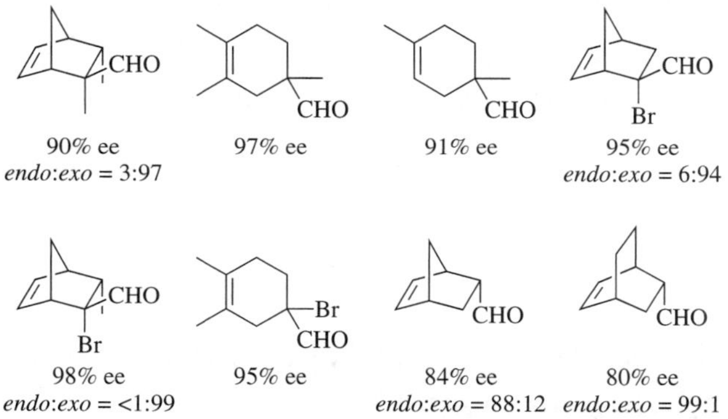

Figure 1 Products of asymmetric Diels-Alder reactions

The intramolecular Diels–Alder reaction of 2-methyl-(*E*,*E*)-2,7,9-decatrienal with CAB catalysis proceeds with high diastereo- and enantioselectivities.[2c]

Asymmetric Aldol-Type Reaction[3]. CAB complex (**2**) is an excellent catalyst for the Mukaiyama condensation of simple achiral enol silyl ethers of ketones with various aldehydes. The CAB-catalyzed aldol process allows the formation of adducts in a highly diastereo- and enantioselective manner (up to 96% ee) under mild reaction conditions (eqs 4 and 5). The reactions are catalytic: 20 mol % of catalyst is sufficient for efficient conversion, and the chiral auxiliary can be recovered and reused.

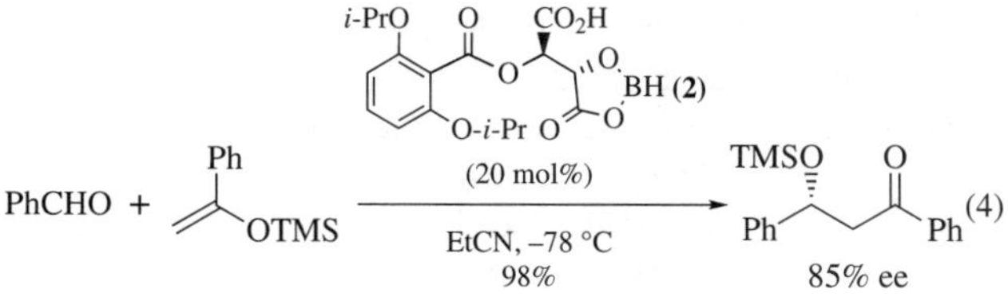

(4)

85% ee

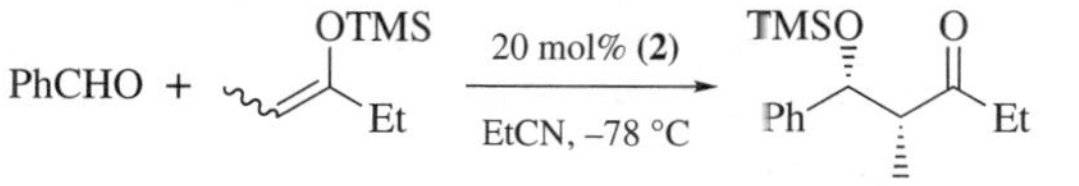

(5)

(*E*):(*Z*) = 4:1 96%, 96% ee, *syn*:*anti* = 94:6
(*E*):(*Z*) = 1:49 97%, 96% ee, *syn*:*anti* = 93:7

Almost perfect asymmetric induction is achieved in the *syn* adducts, reaching 96% ee, although a slight reduction in both the enantio- and diastereoselectivities is observed in the reactions with saturated aldehydes. It is noteworthy that, regardless of the stereochemistry of the starting enol silyl ethers, the CAB-catalyzed reaction is highly selective for *syn* adducts. The high *syn* selectivity and the independence of selectivity on the stereochemistry of silyl ethers in the CAB-catalyzed reactions are fully consistent with Noyori's *Trimethylsilyl Trifluoromethanesulfonate*-catalyzed aldol reactions of acetals,[6] and thus may reflect the acyclic extended transition state mechanism postulated in the latter reactions (Figure 2). Judging from the product configurations, the CAB catalyst (from natural tartaric acid) should effectively cover the *si*-face of carbonyl following its coordination and the selective approach of nucleophiles from the *re*-face should result.

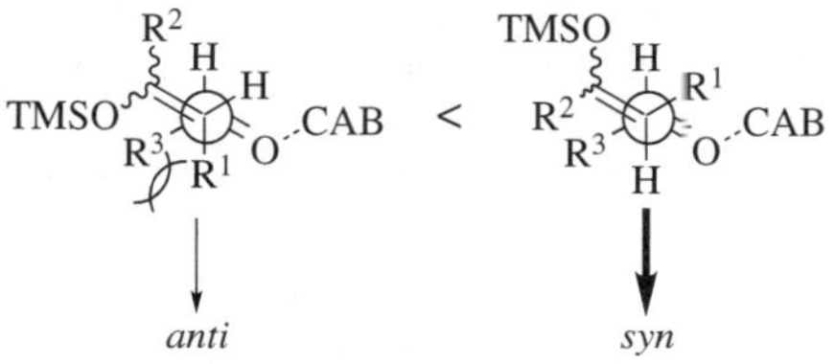

Figure 2 Extended transition state model

A catalytic asymmetric aldol-type reaction of ketene silyl acetals with achiral aldehydes also proceeds with the CAB catalyst (**2**), which can furnish *syn*-β-hydroxy esters with high enantioselectivity (eq 6).

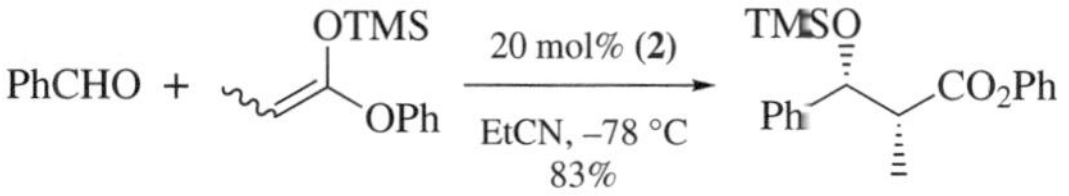

(6)

92% ee
syn:*anti* = 79:21

This reaction is sensitive to the substituents of the starting ketene acetals. The use of ketene silyl acetals from phenyl esters leads to good diastereo- and enantioselectivities with excellent chemical yields.

Analogous with the previous results of enol silyl ethers of ketones, nonsubstituted ketene silyl acetals are found to exhibit lower levels of stereoregulation, while the propionate-derived ketene silyl acetals display a high level of asymmetric induction. The reactions with aliphatic aldehydes, however, resulted in a slight reduction in optical and chemical yields. With phenyl ester-derived ketene silyl acetals, *syn* adducts predominate, but the selectivities are moderate in most cases in comparison with the reactions of ketone-derived silyl enol ethers. Exceptions are α,β-unsaturated aldehydes, which revealed excellent diastereo- and enantioselectivities. The observed *syn* selectivity and *re*-face attack of nucleophiles on the carbonyl carbon of aldehydes are consistent with the aforementioned aldol reactions of ketone-derived enol silyl ethers.

Asymmetric Allylation (Sakurai–Hosomi Allylation)[4]. Condensation of achiral aldehydes with allylsilanes promoted by CAB catalyst (**2**) (20 mol %) at −78 °C in propionitrile produces homoallylic alcohols with excellent enantioselectivity (eq 7).

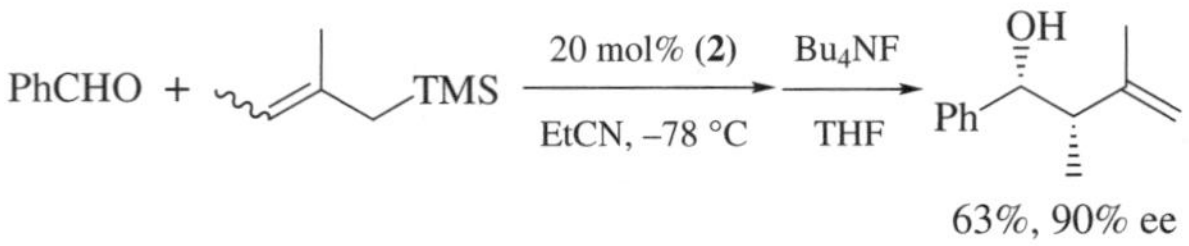

(7)

63%, 90% ee
syn:*anti* = 96:4

Alkyl substitution at the alkene of the allylsilanes increases the reactivity, permitting lower reaction temperature and improved asymmetric induction. γ-Alkylated allylsilanes exhibit excellent diastereo- and enantioselectivities, affording *syn* homoallylic alcohols with high enantiomeric purity. The *syn* selectivity of these reactions is independent of the allylsilane stereochemistry. Thus regardless of the geometry of the starting allylsilane, the predominant isomer in this reaction has *syn* configuration. The observed preference for relative and absolute configurations for the adduct alcohols derived from reaction catalyzed by the (2*R*,3*R*)-ligand–borane reagent can be rationalized on the basis of an extended transition state model similar to that for the CAB-catalyzed aldol reaction (see Figure Figure 2).

Allystannanes are more nucleophilic than allylsilanes. Addition of achiral allylstannanes to achiral aldehydes in the presence of (**1**) (20 mol %) and *Trifluoroacetic Anhydride* (40 mol %) also affords homoallylic alcohols with high diastereo- and enantioselectivities (eq 8).

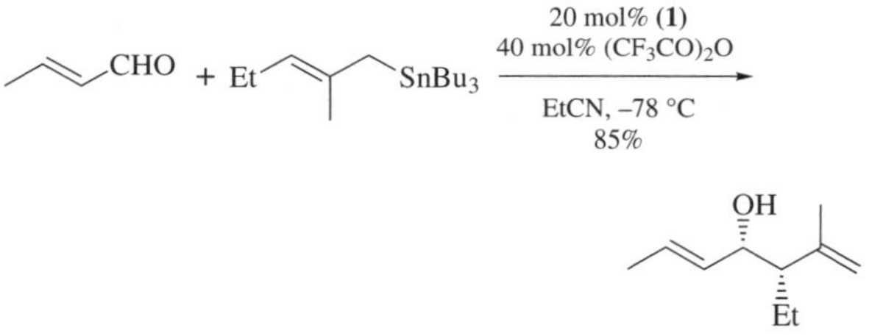

(8)

95% ee
syn:*anti* = 92:8

Asymmetric Hetero Diels–Alder Reaction[5]. In contrast to the CAB catalyst (**2**; R = H) which is stable and both air and moisture sensitive, the *B*-alkylated CAB catalyst (**3**; R = Ph or alkyl) is stable and can be stored in a closed container at rt. A solution of the CAB (**3**; R = Ph) catalyzes Diels–Alder, aldol, and Sakurai–Hosomi reactions. Although the asymmetric inductions achieved by these complexes are slightly less efficient than that of the corresponding hydride-type catalyst, the CAB catalyst (**3**; R = Ph) is shown to be an excellent system for hetero Diels–Alder reactions.

The *B*-alkylated CAB catalyst (**3**) is easily prepared in situ by mixing a 1:1 molar ratio of tartaric acid derivative and phenylboronic acid in dry propionitrile at room temperature for 0.5 h. The hetero Diels–Alder reaction of aldehydes with Danishefsky dienes is promoted by 20 mol % of this catalyst solution at −78 °C for several hours to produce dihydropyranone derivatives of high optical purity (eq 9).

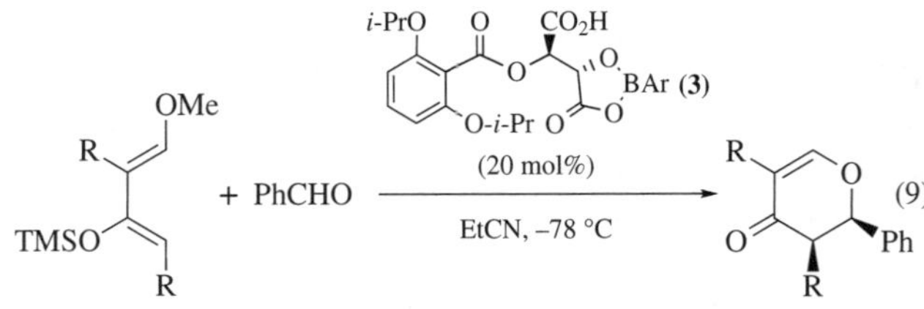

(9)

R = H	Ar = 2,4,6-*i*-$Pr_3C_6H_2$	95% ee (*R*)
R = Me	Ar = *o*-$MeOC_6H_4$	97% ee (2*R*, 3*R*)

1. Furuta, K.; Gao, Q.; Yamamoto, H. *Org. Synth.* **1995**, *72*, 86.
2. (a) Furuta, K.; Miwa, Y.; Iwanaga, K.; Yamamoto, H. *J. Am. Chem. Soc.* **1988**, *110*, 6254. (b) Furuta, K.; Shimizu, S.; Miwa, Y.; Yamamoto, H. *J. Org. Chem.* **1989**, *54*, 1481. (c) Furuta, K.; Kanematsu, A.; Yamamoto, H. *Tetrahedron Lett.* **1989**, *30*, 7231. (d) Ishihara, K.; Gao, Q.; Yamamoto, H. *J. Org. Chem.* **1993**, *58*, 6917. (e) Ishihara, K.; Gao, Q.; Yamamoto, H. *J. Am. Chem. Soc.* **1993**, *115*, 10412.
3. (a) Furuta, K.; Maruyama, T.; Yamamoto, H. *J. Am. Chem. Soc.* **1991**, *113*, 1041. (b) Furuta, K.; Maruyama, T.; Yamamoto, H. *Synlett* **1991**, 439. (c) Ishihara, K.; Maruyama, T.; Mouri, M.; Gao, Q.; Furuta, K.; Yamamoto, H. *Bull. Chem. Soc. Jpn.* **1993**, *66*, 3483.
4. (a) Furuta, K.; Mouri, M.; Yamamoto, H. *Synlett* **1991**, 561. (b) Marshall, J. A.; Tang, Y. *Synlett* **1992**, 653. (c) Ishihara, K.; Mouri, M.; Gao, Q.; Maruyama, T.; Furuta, K.; Yamamoto, H. *J. Am. Chem. Soc.* **1993**, *115*, 11490.
5. (a) Gao, Q.; Maruyama, T.; Mouri, M.; Yamamoto, H. *J. Org. Chem.* **1992**, *57*, 1951. (b) Gao, Q.; Ishihara, K.; Maruyama, T.; Mouri, M.; Yamamoto, H. *Tetrahedron* **1994**, *50*, 979.
6. Noyori, R.; Murata, S.; Suzuki, M. *Tetrahedron* **1981**, *37*, 3899.

Kazuaki Ishihara & Hisashi Yamamoto
Nagoya University, Japan

Diisopropyl 2-Allyl-1,3,2-dioxaborolane-4,5-dicarboxylate[1,2]

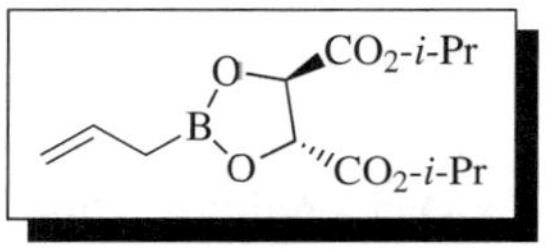

L-(*R,R*)
[99417-55-7] $C_{13}H_{21}BO_6$ (MW 284.12)
D-(*S,S*)
[99493-25-1]

(reagent for the asymmetric allylboration of aldehydes to produce homoallylic alcohols[2])

Alternate Name: tartrate allylboronate.
Physical Data: bp 88–90 °C/0.03 mmHg.
Solubility: sol toluene, THF, ether, CH_2Cl_2.
Analysis of Reagent Purity: ^{11}B NMR (δ 35, $CDCl_3$); capillary GC;[2c] solutions are easily standardized via reaction of an aliquot with cyclohexanecarbaldehyde.[2c]
Preparative Methods: prepared by the reaction of *Triisopropyl Borate* and *Allylmagnesium Bromide* in Et_2O followed by an acidic extractive workup and direct esterification with diisopropyl tartrate (DIPT) (eq 1).[2c]

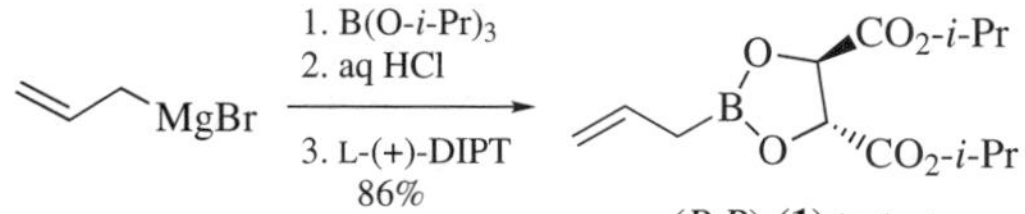

(1)

(*R,R*)-(**1**) tartrate allylboronate

Handling, Storage, and Precautions: allylboronates are typically handled as a solution in toluene (0.5–1.0 M) and transferred by syringe under an inert atmosphere. The reagent, stored neat or as a solution in toluene over 4Å molecular sieves under an argon atmosphere in a refrigerator (−20 °C), is stable for several months. In the presence of water, (**1**) rapidly hydrolyzes to DIPT and the achiral allylboronic acid.

Reactions with Achiral Aldehydes. The reaction of tartrate allylboronates with achiral aldehydes proceeds with moderate to excellent enantioselectivity (60–92% ee) and high yield (80–90%). Simple aliphatic aldehydes give good enantioselectivities (decanal 86% ee, CyCHO 87% ee, eq 2),[2] while β-alkoxy and conjugated aldehydes give diminished selectivities (60–80% ee) (eq 3).[3] The enantioselectivity is highly temperature and solvent dependent. Best results for reactions with the vast majority of aldehydes are obtained in toluene at −78 °C.[2c] 4Å molecular sieves are included to ensure that the reaction is anhydrous. Other tartrate esters (e.g. diethyl tartrate) may also be used without loss of enantioselectivity.

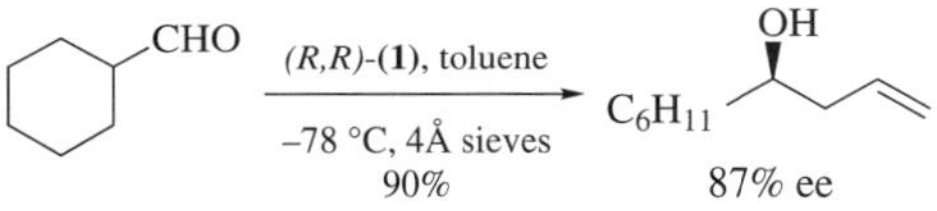

(2)

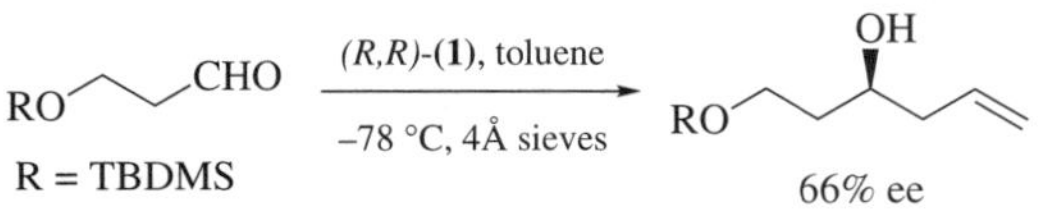

(3)

An allylboronate reagent with a conformationally rigid tartramide auxiliary was designed to improve the enantioselectivity of the reactions with achiral aldehydes.[4] The *N,N′*-dibenzyl-*N,N′*-ethylenetartramide modified allylboronate (**2**) ($R = CH_2Ph$) is considerably more enantioselective than (**1**) (CyCHO, 97% ee) but has very poor solubility in toluene at −78 °C. Consequently, reactions of (**2**) often require up to 48 h. *N,N′*-Bistrifluoroethyl-*N,N′*-ethylenetartramide modified allylboronate (**3**) ($R = CH_2CF_3$) is much more soluble at −78 °C than the dibenzyl derivative and therefore reacts with aldehydes much more efficiently (CyCHO, −78 °C, THF, 5 h, 91%, 94% ee).[5]

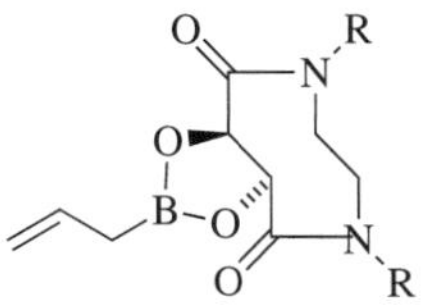

(**2**) $R = CH_2Ph$
(**3**) $R = CH_2CF_3$

The poor results obtained with various unsaturated aldehydes have been overcome by conversion of these substrates to the corresponding metal carbonyl complexes.[6,7] For example, allylboration of 2-decynal proceeds with 72% ee, while that of its cobalt complex proceeds with excellent enantioselection (eqs 4 and 5).[6]

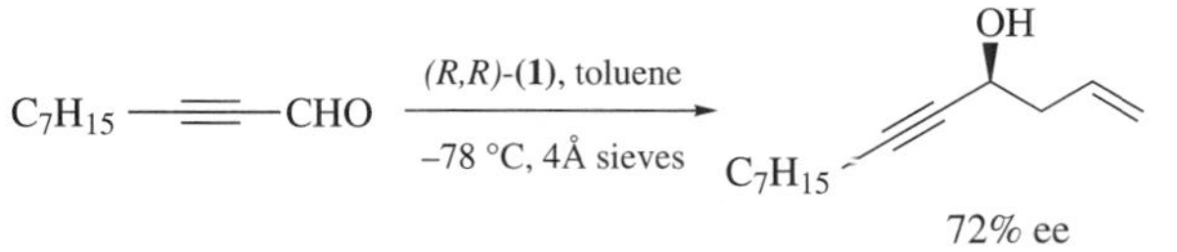

(4)

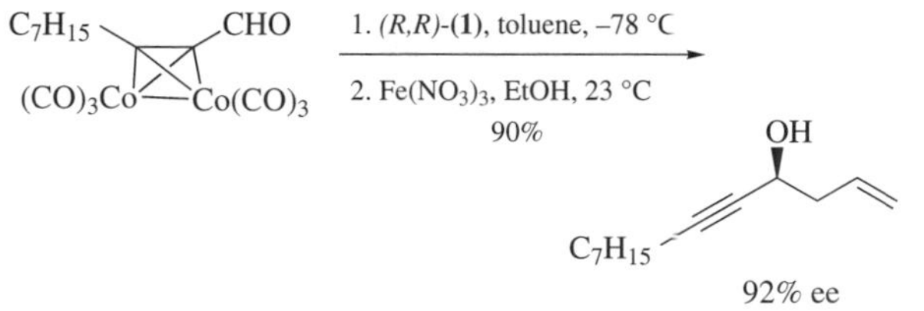

(5)

A second example of the allylboration of a metal carbonyl containing substrate is a highly group- and face-selective allylboration of a *meso* iron–diene dialdehyde complex (eq 6).[7] Efficient kinetic resolutions of racemic diene aldehyde–$Fe(CO)_3$ complexes have also been demonstrated.[7]

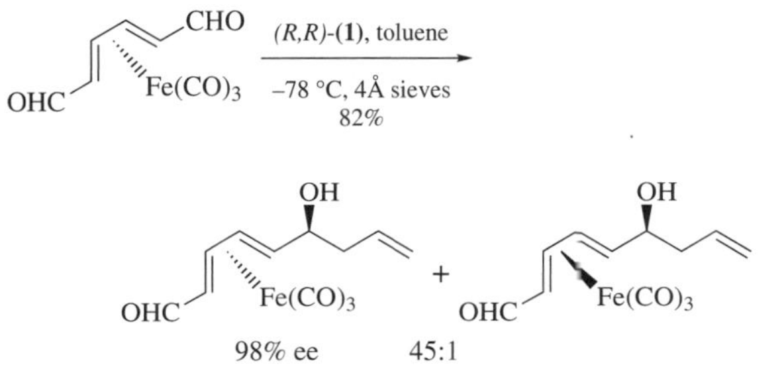

(6)

Reactions with Chiral Aldehydes[1,8]. The tartrate allylboronates have been shown to serve as highly useful chiral acetate enolate equivalents in the reactions with α-chiral aldehydes. The diastereoselectivities obtained are good to excellent, depending on whether the reaction is a matched or mismatched case (eqs 7 and 8).[3] These reagents have been applied to several complex problems in natural product synthesis.[8,9] As shown in eq 8, the diastereoselection is significantly improved by using the rigid tartramide reagent (**3**).[5]

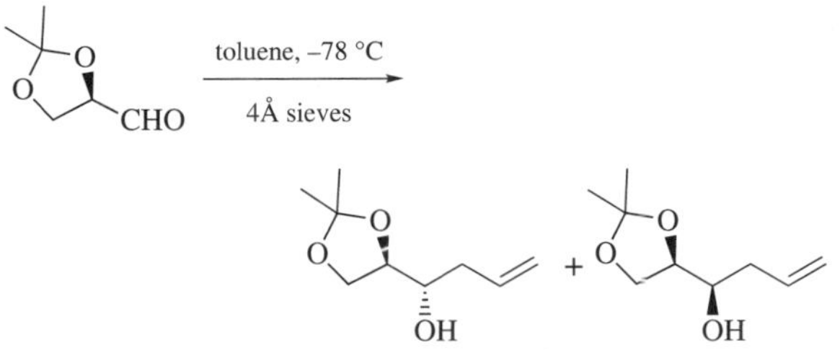

(7)

(*R,R*)-(**1**) matched case 98:2 (94%)
(*S,S*)-(**1**) mismatched case 7:93

The tartrate-derived allylboronate reagents in the best cases compare favorably with other allylboration reagents in their reactions with both achiral and chiral aldehydes (e.g. *B-Allyldi-*

isopinocampheylborane; β-allyl-2-(trimethylsilyl)borolane; 2,5-dimethyl-β-allylborolane; 1,2-diamino-1,2-diphenylethane modified allylboranes). The advantage of the tartrate-modified allylboronate reagent rests with its ease of preparation and its capability of prolonged storage without noticeable deterioration.

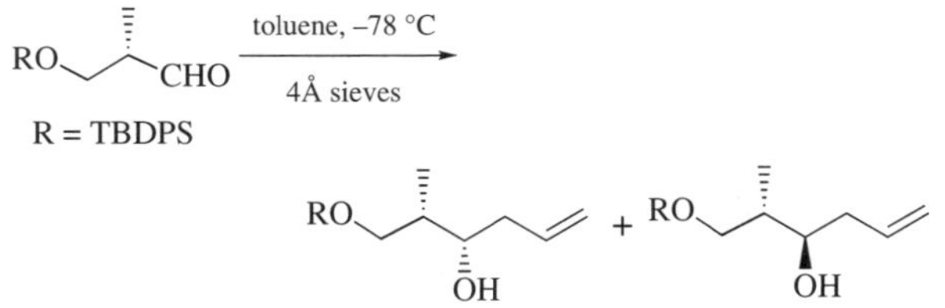

(8)

(*R,R*)-(**1**)	matched case	79:21 (72%)
(*S,S*)-(**1**)	mismatched case	13:87
(*R,R*)-(**3**)		8:92 (82%)
(*S,S*)-(**3**)		3:97

Related Allylboronate Reagents. A stereoselective synthesis of *anti* 1,2-diols has been achieved by using a DIPT-modified (*E*)-γ-[(cyclohexyloxy)dimethylsilyl]allylboronate reagent.[10] This reagent is best applied in double asymmetric reactions with chiral aldehydes such as D-glyceraldehyde acetonide (eq 9).

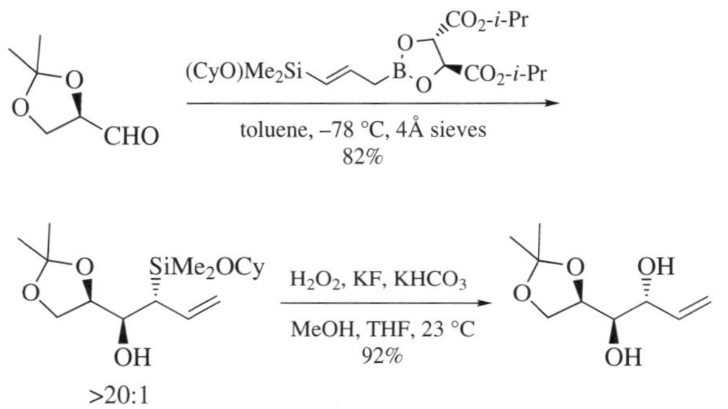

(9)

A chiral allylic alcohol β-carbanion equivalent has also been developed which utilizes a DIPT-modified (*E*)-γ-(dimethylphenylsilyl)allylboronate reagent.[10] This method involves treating the product homoallylic alcohol with *Dimethyldioxirane* and subjecting the derived epoxide to an acid-catalyzed Peterson elimination. This sequence has been applied in the synthesis of the trioxadecalin ring system of the mycalamides (eq 10).[11]

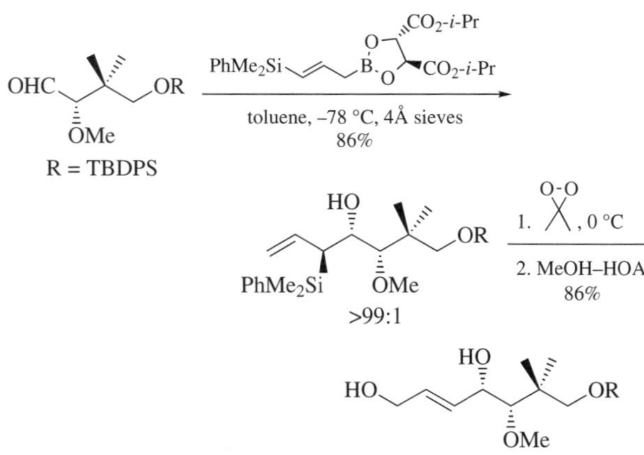

(10)

Highly stereoselective introduction of a β,β-dimethylhomoallylic alcohol subunit was also accomplished in this synthesis by using a DIPT-modified prenylboronate (eq 11).[11]

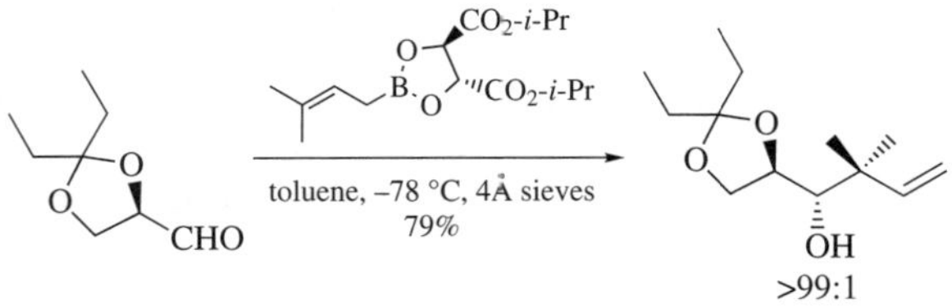

(11)

All the reagents discussed above are readily prepared using techniques described for the preparation of the tartrate-modified crotylboronates, and can be handled in a similar manner.

Related Reagents. *B*-Allyldiisopinocampheylborane; (*E*)-1-(*N,N*-Diisopropylcarbamoyloxy)crotyllithium.

1. Roush, W. R. *Comprehensive Organic Synthesis* **1991**, *2*, 1.
2. (a) Roush, W. R.; Walts, A. E.; Hoong, L. K. *J. Am. Chem. Soc.* **1985**, *107*, 8186. (b) Roush, W. R; Banfi, L.; Park, J. C.; Hoong, L. K. *Tetrahedron Lett.* **1989**, *30*, 6457. (c) Roush, W. R.; Hoong, L. K.; Palmer, M. A. J.; Park, J. C. *J. Org. Chem.* **1990**, *55*, 4109.
3. Roush, W. R.; Hoong, L. K.; Palmer, M. A. J.; Straub, J. A.; Palkowitz, A. D. *J. Org. Chem.* **1990**, *55*, 4117.
4. Roush, W. R.; Banfi, L. *J. Am. Chem. Soc.* **1988**, *110*, 3979.
5. Roush, W. R.; Grover, P. T. Unpublished results.
6. Roush, W. R.; Park, J. C. *J. Org. Chem.* **1990**, *55*, 1143.
7. Roush, W. R.; Park, J. C. *Tetrahedron Lett.* **1990**, *31*, 4707.
8. (a) Roush, W. R.; Kageyama, M. *Tetrahedron Lett.* **1985**, *26*, 4327. (b) Roush, W. R.; Palkowitz, A. D.; Ando, K. *J. Am. Chem. Soc.* **1990**, *112*, 6348. (c) Goulet, M. T.; Boger, J. *Tetrahedron Lett.* **1990**, *31*, 4845.
9. (a) Roush, W. R.; Straub, J. A.; VanNieuwenhze, M. S. *J. Org. Chem.* **1991**, *56*, 1636. (b) Roush, W. R.; Lin, X.; Straub, J. A. *J. Org. Chem.* **1991**, *56*, 1649.
10. Roush, W. R.; Grover, P. T. *Tetrahedron* **1992**, *48*, 1981.
11. Roush, W. R.; Marron, T. G. *Tetrahedron Lett.* **1993**, *34*, 5421.

David J. Madar
Indiana University, Bloomington, IN, USA

Diisopropyl 2-Crotyl-1,3,2-dioxaborolane-4,5-dicarboxylate[1,2]

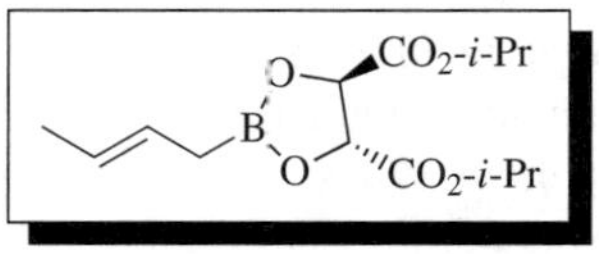

(**1**) L-(*R,R*)-tartrate (*E*)-crotyl
[99745-86-5] $C_{14}H_{23}BO_6$ (MW 298.14)
D-(*S,S*), (*Z*)
[99687-40-8]
L-(*R,R*), (*Z*)
[106357-20-4]
(**2**) D-(*S,S*), (*E*)
[106357-33-9]

(reagents for the asymmetric crotylboration of aldehydes to produce either *syn* or *anti* β-methylhomoallylic alcohols)[2]

Physical Data: bp 80 °C/0.1 mmHg.

Solubility: sol toluene, THF, ether, or CH_2Cl_2.

Analysis of Reagent Purity: ^{11}B NMR, data for (*E*)-crotyl: (δ 34.8, C_6D_6);[2b] the purity of the reagent is best determined by capillary GC;[2b] solutions of the reagent can be standardized using cyclohexanecarbaldehyde.[2b]

Preparative Methods: prepared by treatment of (*E*)- or (*Z*)-crotylpotassium with *Triisopropyl Borate* followed by acidic extractive workup and direct esterification with diisopropyl tartrate (DIPT) (eqs 1 and 2).[2]

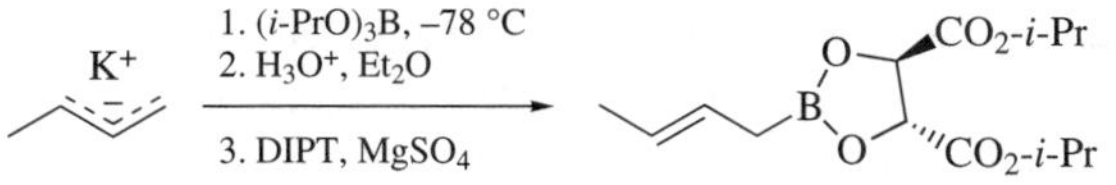

(1)

(*R,R*)-(**1**) tartrate
(*E*)-crotylboronate

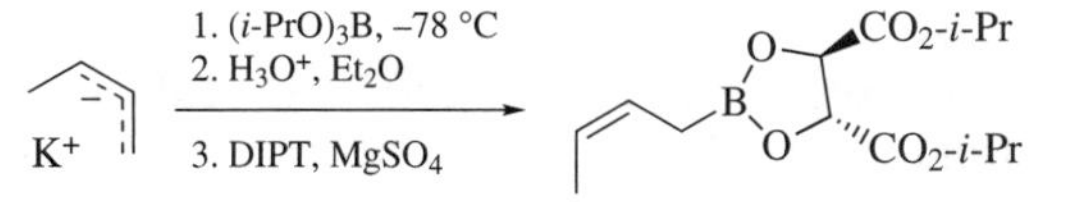

(2)

(*S,S*)-(**2**) tartrate
(*E*)-crotylboronate

Handling, Storage, and Precautions: typically the reagents are handled as solutions in toluene (0.5–1M) and transferred by syringe under an inert atmosphere; stored neat or as a solution in toluene over 4Å molecular sieves under an argon atmosphere in a refrigerator (−20 °C), the reagent is stable for many months. In the presence of water, (**1**) rapidly hydrolyzes to achiral crotylboronic acid, the presence of which leads to reduced enantioselectivity in reactions with aldehydes.

Reactions with Achiral Aldehydes. The tartrate ester modified (*E*)- and (*Z*)-crotylboronates undergo rapid additions to aldehydes at −78 °C. The enantioselectivities obtained for aliphatic linear or α-monobranched aldehydes range from 72 to 91% ee.[2] When cyclohexanecarbaldehyde is treated with the (*E*)-crotylboronate reagent at −95 °C in toluene, the homoallylic alcohol is obtained in 98% yield and 91% ee (eq 3). The (*Z*)-crotylboronate reagent gives slightly lower selectivity (83% ee, eq 4). The *anti*:*syn*/*syn*:*anti* ratios obtained are also excellent for this reagent (typically greater than 98:2 and 2:98 for (**1**) and (**2**), respectively).

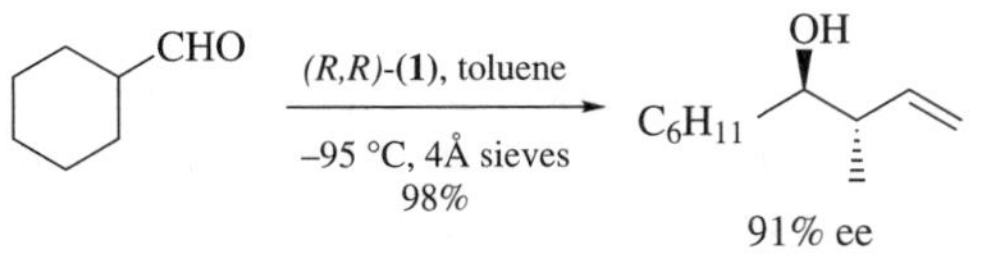

(3)

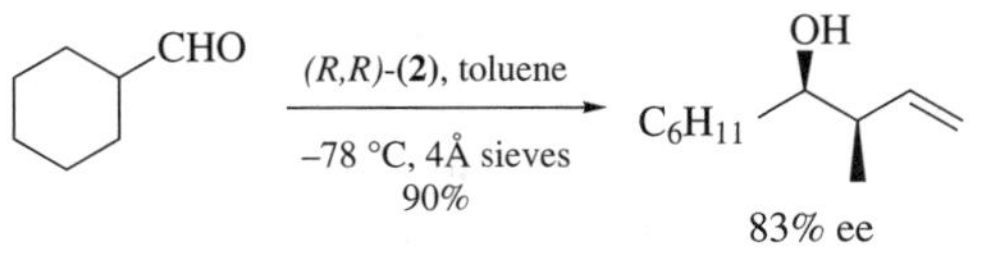

(4)

As with the corresponding allylboronate, the enantioselectivity of reactions with β-alkoxy and conjugated aldehydes are lower (55–74% ee). In the case of benzaldehyde (91%, 66% ee), selectivity can be improved by the use of the derived chromium tricarbonyl complex. The homoallylic alcohol is obtained after oxidative decomplexation in high yield and 92% enantiomeric purity (eq 5).[3]

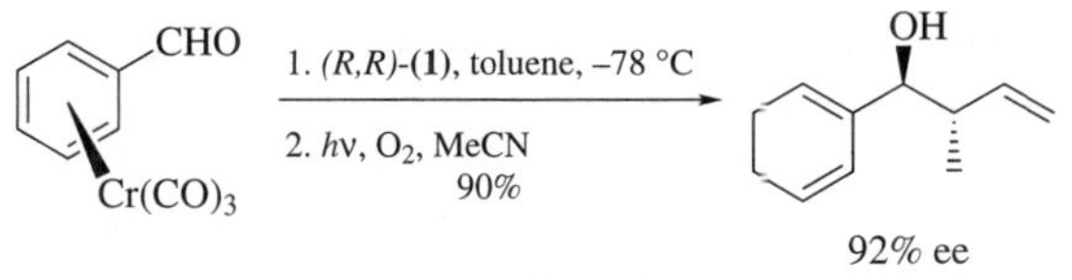

(5)

Reactions with Chiral Aldehydes. Addition of the (*E*)- or (*Z*)-crotylboronate reagent to optically active β-alkoxy-α-methylpropionaldehydes gives the corresponding polypropionate structures with good to excellent diastereoselection (eqs 6 and 7).[4] Three of the four stereochemical triads can be prepared in high yield with useful levels of selectivity. The all-*syn* stereoisomer of eq 7 is best prepared using other methods, such as the crotyltin methodology developed by Keck and co-workers.[5] The polypropionate structures with 1,3-*anti* relationships between branching methyl groups are prepared with excellent diastereoselection (via matched double asymmetric reactions). Those with a 1,3-*syn* relationship are more difficult to prepare. The relative diastereoselectivity of the reaction of α-methyl chiral aldehydes with (*E*)- and (*Z*)-crotylboronates can be predicted by use of the *gauche* pentane model.[6]

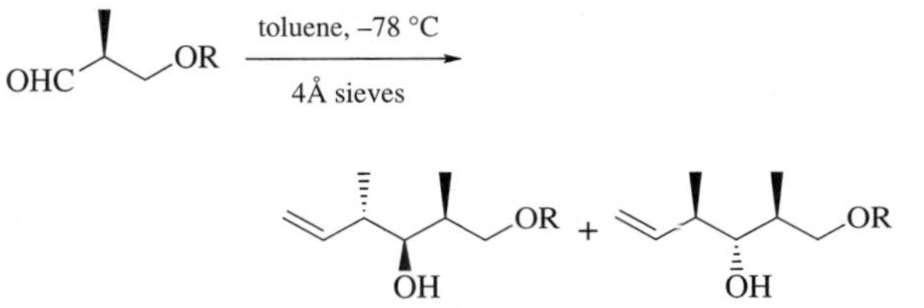

(6)

R = TBDMS, (*R,R*)-(**1**) 97:3 80%
R = TBDPS, (*S,S*)-(**1**) 10:90 77%

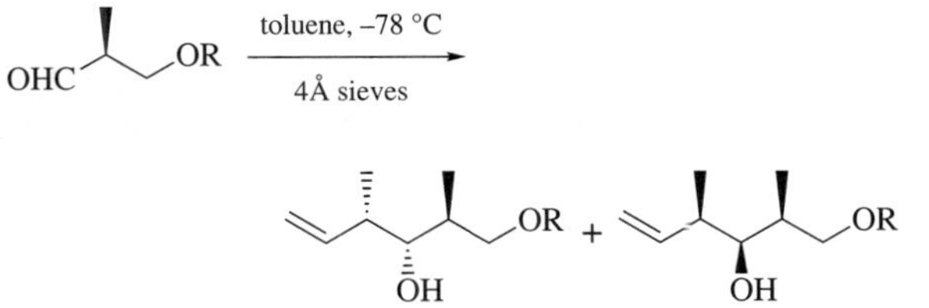

(7)

R = TBDMS, (*R,R*)-(**2**) 98:2 71%
R = TBDPS, (*S,S*)-(**2**) 27:73

Both (*E*)- and (*Z*)-crotylboronates have been used in several applications in natural product synthesis.[4b,7] One application of both the allylboronate and (*E*)-crotylboronate reagents is found in the synthesis of the C(19)–C(29) segment of rifamycin S. The desired stereochemistry at C(25)–C(26) of the rifamycin ansa chain is set with excellent stereocontrol (>95:5) and high yield (87%) (eq 8).[4b,7a]

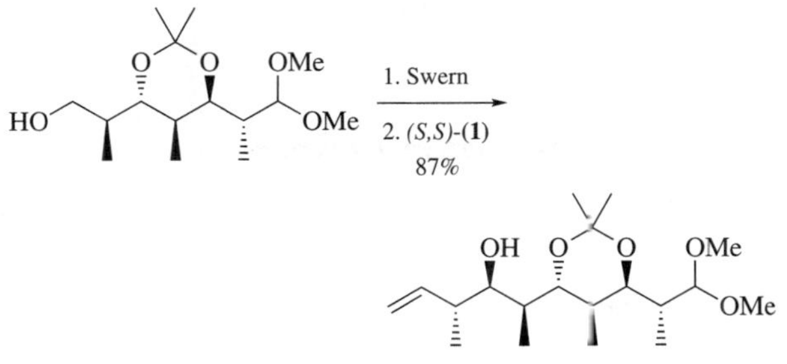

(8)

The (*E*)- and (*Z*)-crotylboronates provide selectivity in the best cases comparable to that obtained with other crotylboration procedures. Combining ease of preparation, stability, and selectivity the tartrate-modified (*E*)- and (*Z*)-crotylboronates are highly useful propionate enolate equivalents.

Related Reagents. *B*-Allyl-9-borabicyclo[3.3.1-nonane; *B*-Allyldiisocaranylborane; *B*-Allyldiisopinocampheylborane; *B*-Crotyldiisopinocampheylborane; (*R*,*R*)-2,5-Dimethylborolane.

---

1. Roush, W. R. *Comprehensive Organic Synthesis* **1991**, *2*, 1.
2. (a) Roush, W. R.; Halterman, R. L. *J. Am. Chem. Soc.* **1986**, *108*, 294. (b) Roush, W. R.; Ando, K.; Powers, D. B.; Palkowitz, A. D.; Halterman, R. L. *J. Am. Chem. Soc.* **1990**, *112*, 6339; *J. Am. Chem. Soc.* **1991**, *114*, 5133.
3. Roush, W. R.; Park, J. C. *J. Org. Chem.* **1990**, *55*, 1143.
4. (a) Roush, W. R.; Palkowitz, A. D.; Palmer, M. A. J. *J. Org. Chem.* **1987**, *52*, 316. (b) Roush, W. R.; Palkowitz, A. D.; Ando, K. *J. Am. Chem. Soc.* **1990**, *112*, 6348.
5. Keck, G. E.; Abbott, D. E. *Tetrahedron Lett.* **1984**, *25*, 1883.
6. (a) Evans, D. A.; Nelson, J. V.; Taber, T. R. *Top. Stereochem.* **1982**, *13*, 1. (b) Roush, W. R. *J. Org. Chem.* **1991**, *56*, 4151.
7. (a) Roush, W. R.; Palkowitz, A. D. *J. Am. Chem. Soc.* **1987**, *109*, 953. (b) Danishefsky, S. J.; Armistead, D. M.; Wincott, F. E.; Selnick, H. G.; Hungate, R. *J. Am. Chem. Soc.* **1987**, *109*, 8117. (c) Coe, J. W.; Roush, W. R. *J. Org. Chem.* **1989**, *54*, 915. (d) Roush, W. R.; Palkowitz, A. D. *J. Org. Chem.* **1989**, *54*, 3009. (e) Tatsuta, K.; Ishiyama, T.; Tajima, S.; Koguchi, Y.; Gunji, H. *Tetrahedron Lett.* **1990**, *31*, 709. (f) Akita, H.; Yamada, H.; Matsukura, H.; Nakata, T.; Oishi, T. *Tetrahedron Lett.* **1990**, *31*, 1735. (g) White, J. D.; Johnson, A. T. *J. Org. Chem.* **1990**, *55*, 5938. (h) Fisher, M. J.; Myers, C. D.; Joglar, J.; Chen, S.-H.; Danishefsky, S. J. *J. Org. Chem.* **1991**, *56*, 5826. (i) Roush, W. R.; Bannister, T. D. *Tetrahedron Lett.* **1992**, *33*, 3587. (j) Roush, W. R.; Brown, B. B. *J. Am. Chem. Soc.* **1993**, *115*, 2268. (k) White, J. D.; Porter, W. J.; Tiller, T. *Synlett* **1993**, 535.

David J. Madar
Indiana University, Bloomington, IN, USA

9-*O*-(1,2;5,6-Di-*O*-isopropylidene-α-D-glucofuranosyl)-9-boratabicyclo[3.3.1]-nonane, Potassium Salt

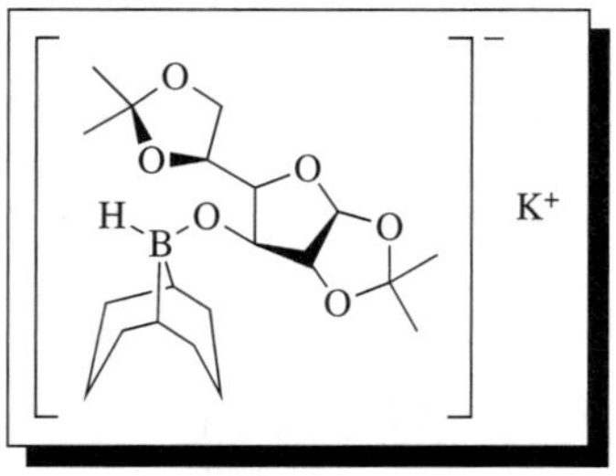

[101696-41-7] $C_{20}H_{34}BKO_6$ (MW 420.40)

(chiral borohydride reagent for enantioselective reduction of ketones[1])

Alternate Name: K-glucoride; K 9-*O*-DIPGF-9-BBNH.
Solubility: usually prepared and stored in THF solution.[1]
Analysis of Reagent Purity: ^{11}B NMR (δ 1.33, br s) and strong absorption at 2038 cm^{-1} in the IR spectrum (B–H str); stoichiometric ratio of K:B:H as 1:1:1 by analysis.[1]
Preparative Methods: 9-BBN, by reaction with the chiral alcohol 1,2;5,6-di-*O*-isopropylidene-α-D-glucofuranose (DPGF) (both commercially available), is transformed into the borinic ester 9-*O*-DIPGF-9-BBN which, by treatment with a modest excess (1.1–1.5 equiv) of potassium hydride, is completely transformed within 2 h into K-glucoride.[1]
Handling, Storage, and Precautions: K-glucoride is relatively stable toward disproportionation at rt, especially when the THF solution is stored over excess potassium hydride under a positive pressure of nitrogen.[1]

Enantioselective and Diastereoselective Reduction of Carbonyl Compounds. K-Glucoride is the first example of a well-defined chiral borohydride reagent containing a monosaccharide as chiral auxiliary;[2] moreover, it has only one hydride per reagent molecule.[1a] K-Glucoride allows the reduction of various ketones[1a–c] to the corresponding alcohols in THF, even at −78 °C; it was first used for hindered alkyl phenyl ketones (like pivalophenone)[1a] and it gave considerably higher enantioselectivity than Noyori's BINAL-H reagent (see *Lithium Aluminum Hydride–2,2′-Dihydroxy-1,1′-binaphthyl*).[1a] Reduction of unhindered aliphatic ketones (like 2-butanone) with the same reagent gave only very low optical yields; for such compounds the lithium hydrido-9-BBN-nopol benzyl ether adduct (NB-Enantride™)[1a] is much more favorable. On the other hand, the reduction of pinacolone, which is relatively hindered, gave a reasonable enantioselectivity (70%), while NB-Enantride gave only 2%. A significant effect of reaction temperature on optical induction was demonstrated for propiophenone,[1c] with ee's varying from 92% at −78 °C to 76% at 0 °C. It is noteworthy that the alcohols obtained were always enriched in the (*R*) enantiomer.

Prochiral ketones bearing various functionalities near the carbonyl group can be reduced by K-glucoride; for example very good optical yields can be obtained in the reduction of α-keto esters[3a] to the corresponding α-hydroxy esters. The ee's obtained are always close to 100%, even with relatively hindered derivatives. Moreover, all of the α-hydroxy esters obtained were enriched in the (*S*) enantiomer. In contrast, *B*-isopinocampheyl-9-borabicyclo[3.3.1]nonane (Alpine-borane®)[3a] usually gave lower optical yields and very slow reactions with relatively hindered compounds; in this case the absolute configuration of product depends on the starting α-keto ester. While α-hydroxy esters can be obtained with high ee by reduction with K-glucoride, the same procedure gave only poor results in the asymmetric reduction of β-keto esters.[3a]

Secondary or tertiary β-amino alcohols can be obtained by reduction of α-amino ketones with K-glucoride;[3b] best results were obtained starting from aromatic α-amino ketones (44–73% ee), while aliphatic amino ketones gave only low enantioselectivity (9–33% ee). Interestingly, the amino alcohols obtained are enriched in the (*S*) enantiomer and the enantioselectivity increases with the bulkiness of the substituents on the amino group.

α,β-Alkynic ketones can be reduced to the corresponding (*R*)-alkanols with K-glucoride;[3c] ee's are good with compounds bearing an internal triple bond (61–87%), while they drop with terminal alkynes (in this case the (*S*) configuration is preferred). K-glucoride can also be used for the diastereoselective reduction of chiral racemic cyclic and bicyclic ketones to give the less stable alcohol with excellent diastereoselectivity.[1c]

Other Reductions. There is an example of a reduction (resolution) of racemic 1,2-epoxyalkanes to give the corresponding (*R*)-2-alkanols with moderate ee (up to 43.3%).[4]

The enantioselective synthesis of optically active secondary amines via asymmetric reduction of prochiral ketimines was studied by screening various chiral hydrides.[5a,b] In this case, K-glucoride gave only disappointing results and was inferior to other reagents. Better results were obtained in the asymmetric reduction of prochiral *N*-diphenylphosphinylimines to chiral *N*-(diphenylphosphinyl)amines (eq 1),[5c] which can then be readily converted into optically active primary amines. For this reaction the stereochemical course depends dramatically on the relative bulkiness of the groups R^1 and R^2. The reaction conditions for reduction of C=N double bonds are the same as used for ketones, but the high reactivity of diphenylphosphinylimines dramatically reduces the reaction time.

(1)

Finally, K-glucoride can also be used for the enantioselective reduction with moderate ee (52%) of 1-substituted 2-methylisoquinolinium salts, which are employed in the preparation of 1-substituted 2-methyltetrahydroisoquinoline alkaloids.[5d]

1. (a) Brown, H. C.; Park, W. S.; Cho, B. T. *J. Org. Chem.* **1986**, *51*, 1934. (b) Brown, H. C.; Park, W. S.; Cho, B. T.; Ramachandran, P. V. *J. Org. Chem.* **1987**, *52*, 5406. (c) Brown, H. C.; Cho, B. T.; Park, W. S. *J. Org. Chem.* **1988**, *53*, 1231.
2. Kunz, H.; Rück, K. *Angew. Chem., Int. Ed. Engl.* **1993**, *32*, 336.
3. (a) Brown, H. C.; Cho, B. T.; Park, W. S. *J. Org. Chem.* **1986**, *51*, 3396. (b) Cho, B. T.; Chun, Y. S. *Tetrahedron: Asymmetry* **1992**, *3*, 341; (c) Cho, B. T.; Park, W. S. *Bull. Korean Chem. Soc.* **1987**, *8*, 257.
4. Cha, J. S.; Lee, K. W.; Yoon, M. S.; Lee, J. C.; Yoon, N. M. *Heterocycles* **1988**, *27*, 1713.
5. (a) Cho, B. T.; Chun, Y. S. *Tetrahedron: Asymmetry* **1992**, *3*, 1583. (b) Cho, B. T.; Chun, Y. S. *Tetrahedron: Asymmetry* **1992**, *3*, 337. (c) Hutchins, R. O.; Abdel-Magid, A.; Stercho, Y. P.; Wambsgans, A. *J. Org. Chem.* **1987**, *52*, 702. (d) Cho, B. T.; Han, C. K. *Bull. Korean Chem. Soc.* **1991**, *12*, 565.

Luca Banfi, Enrica Narisano, & Renata Riva
Università di Genova, Italy

Dilongifolylborane[1]

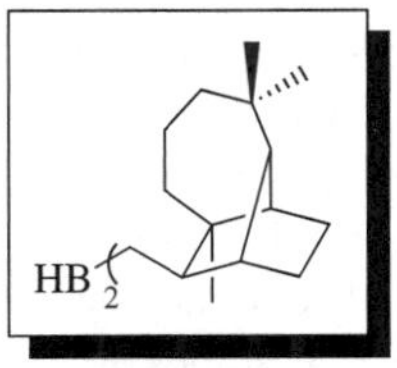

[77882-24-7] $C_{30}H_{51}B$ (MW 422.55)

(reagent for asymmetric hydroboration of prochiral alkenes[1,2])

Alternate Name: Lgf_2BH.
Physical Data: mp 160–161 °C (sealed evacuated capillary).
Solubility: sparingly sol common organic solvents, i.e. THF, CCl_4, CH_2Cl_2, $CHCl_3$. The suspended material is active.
Form Supplied in: white solid, mp 161–163 °C.
Preparative Methods: although now commercially available, the reagent is readily prepared by hydroboration of (+)-longifolene with *Borane–Dimethyl Sulfide* ($BH_3{\cdot}SMe_2$) in a 2:1 ratio (eq 1).

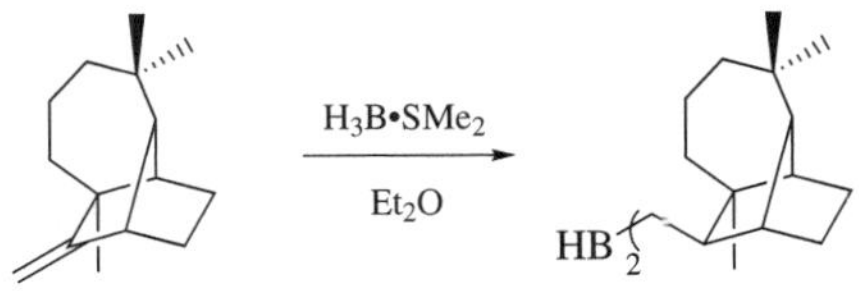

(1)

Handling, Storage, and Precautions: guidelines for handling air- and moisture-sensitive reagents should be followed.[1c] The pure reagent is described as an irritant.

Asymmetric Hydroboration[2]. Dilongifolylborane (Lgf_2BH) is a chiral dialkylborane intermediate in steric requirement between the two widely investigated chiral organoboranes derived from α-pinene: *Monoisopinocampheylborane* ($IpcBH_2$)[3] and *Diisopinocampheylborane* (Ipc_2BH).[4]

Due to the hindered *exo* face environment of the longifolene double bond, the addition of borane proceeds exclusively from the *endo* face. Even in the presence of excess longifolene, hydroboration stops cleanly at the dialkylborane stage.

The reported levels of asymmetric induction achieved with this reagent in the hydroboration–oxidation of representative alkenes are in the range of 59–78% ee for *cis* and 45–75% ee for trisubstituted alkenes. The highest levels of asymmetric induction have been recorded for *cis*-2-butene (eq 2) and 2-methyl-2-pentene (eq 3).[2]

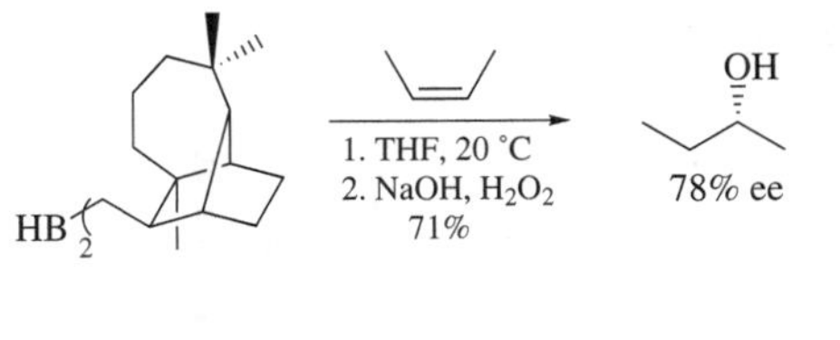

(2)

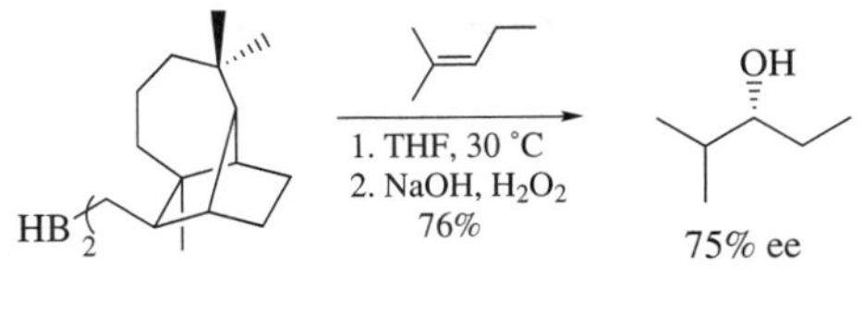

(3)

Trisubstituted alkenes such as 2,4,4-trimethyl-2-pentene and 1-phenylcyclopentene, which apparently exceed the steric requirements of Lgf_2BH, fail to react after 4 days at 35 °C.

Although the ready availability of only the (+)-longifolene enantiomer potentially limits the use of Lgf_2BH as a reagent for natural product synthesis, this form proved appropriate for the preparation of an optically enriched intermediate in a synthesis of the naturally occurring enantiomer of verrucarinic acid (eq 4).[5]

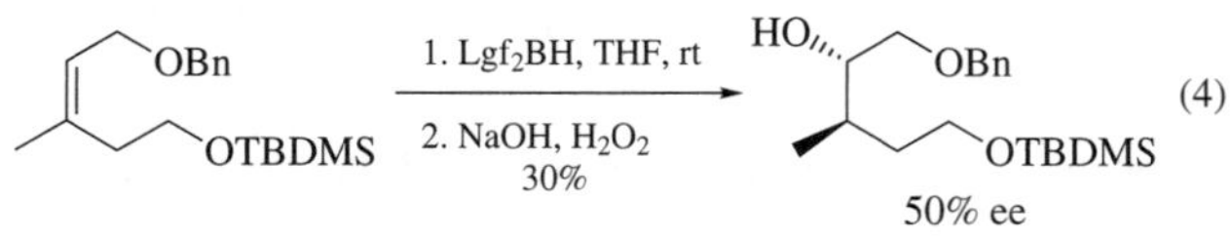

(+)-Longifolene of >99% ee can be liberated from Lgf_2BH which has been diastereomerically enriched by recrystallization from THF at 0 °C (eq 5).[6]

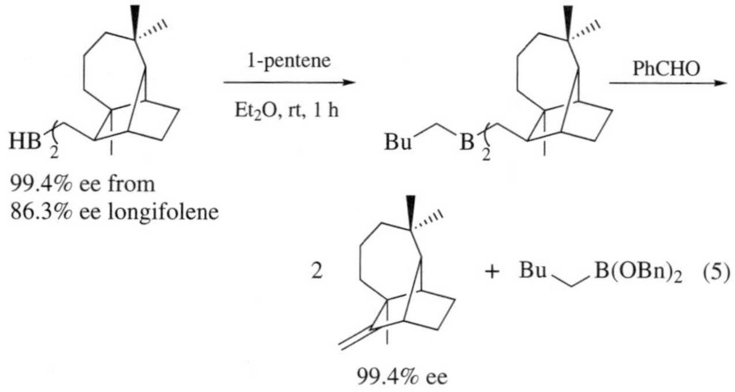

1. (a) Brown, H. C.; Jadhav, P. K.; Mandal, A. K. *Tetrahedron* **1981**, *37*, 3547. (b) Brown, H. C.; Jadhav, P. K. In *Asymmetric Synthesis*; Morrison, J. D., Ed.; Academic: New York, 1983; Vol. 2, Chapter 1. (c) Brown, H. C.; Kramer, G. W.; Levy, A. B.; Midland, M. M. In *Organic Synthesis via Boranes*; Wiley: New York, 1975. (d) Smith, K.; Pelter, A. *Comprehensive Organic Synthesis* **1991**, *8*, Chapter 3.10.
2. Jadhav, P. K.; Brown, H. C. *J. Org. Chem.* **1981**, *46*, 2988.
3. Brown, H. C.; Yoon, N. M. *J. Am. Chem. Soc.* **1977**, *99*, 5514.
4. Brown, H. C.; Zweifel, G. *J. Am. Chem. Soc.* **1961**, *83*, 486.
5. Herold, P.; Mohr, P.; Tamm, C. *Helv. Chim. Acta* **1983**, *66*, 744.
6. Jadhav, P. K.; Vara Prasad, J. V. N.; Brown, H. C. *J. Org. Chem.* **1985**, *50*, 3203.

Robert P. Short
Polaroid Corporation, Cambridge, MA, USA

(*R*)-2-[1-(Dimethylamino)ethyl]-benzenethiol

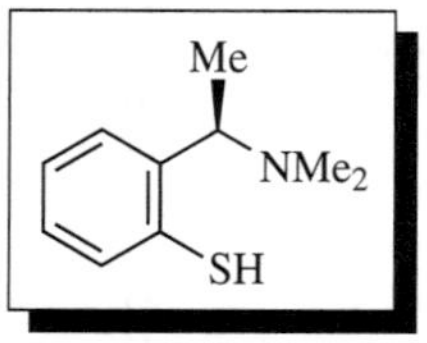

[135190-26-0] $C_{10}H_{15}NS$ (MW 181.30)

(catalyst precursor for enantioselective C–C bond forming reactions)

Physical Data: mp 133 °C.
Solubility: soluble in common organic solvents.
Form Supplied in: white solid; not commercially available.
Analysis of Reagent Purity: ^{1}H NMR, elemental analysis.
Preparative Methods: the title compound can be prepared by reaction of (*R*)-2-[1-(dimethylamino)ethyl]phenyllithium with elemental sulfur (eq 1).[1] A solution of pure (*R*)-2-[1-(dimethylamino)ethyl]phenyllithium[1] in THF is slowly added at −50 °C to a suspension of a stoichiometric amount of freshly sublimed sulfur. The solution is warmed to room temperature and quenched with an equimolar amount of a 10 M aqueous HCl solution. All volatiles are evaporated at reduced pressure and the residue is sublimed at 120 °C in vacuo (0.1 mmHg). The nitrogen-functionalized derivatives (*R*)-2-[1-(1-pyrrolidinyl) ethyl]benzenethiol[2] and (*R*)-2-[1-(1-piperidinyl)ethyl]benzenethiol[2] may be prepared in a similar way. It should be noted that reaction with Me_3SiCl instead of HCl after the sulfur insertion reaction affords the corresponding trimethylsilyl thio ether, which also is a valuable catalyst precursor.[1,2]

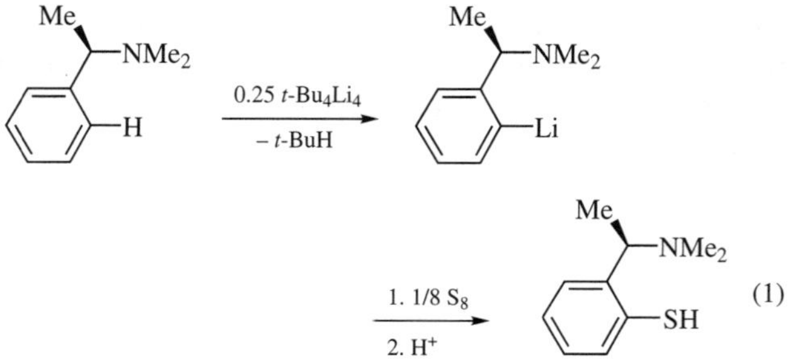

Purification: sublimation in vacuo (0.1 mmHg).

Application as Catalyst Precursor in the Enantioselective 1,2-Addition of Diorganozinc Compounds to Aldehydes. The enantioselective synthesis of secondary alcohols via a zinc-mediated 1,2 addition to aldehydes in the presence of a chiral catalyst, discovered by Mukaiyama[3] and by Oguni,[4] initiated a search for the ultimate catalyst system that has made this reaction one of the most studied.[5] The best catalytic systems possess a β-amino alkoxide skeleton, containing two chiral carbon atoms, since these have the capability of forming a five-membered chelate ring when bonded to a metal center.[6] A major disadvantage of this approach is the requirement of expensive enantiopure starting materials. The application of (*R*)-2-[1-(dimethylamino)ethyl]benzenethiol (**1**) as the catalyst precursor[7] overcomes this disadvantage because the enantiopure starting material is relatively cheap and, moreover, available in both enantiomeric forms. It was shown that these thiolate catalysts are at least as selective and active as the β-amino alkoxide catalyst. In particular, the pyrrolidinyl- **2** and piperidinyl- **3** analogs exhibit enhanced selectivity and reactivity in the 1,2-addition reaction.[2]

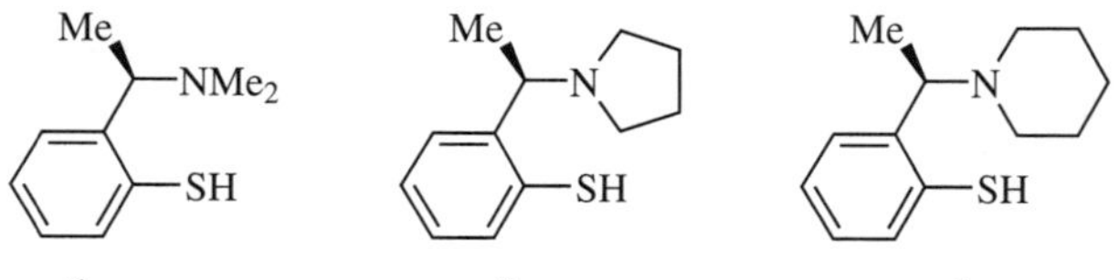

1 **2** **3**

Mechanistic studies have shown that the EtZn-thiolates derived from **1–3** are the actual catalysts. An X-ray crystal structure determination of the MeZn derivative of precursor **1** revealed a dimeric structure with bridging thiolate ligands, as shown below. In separate experiments, it was shown that reaction of either the thiol **1** or the corresponding trimethylsilyl thioether with Me_2Zn affords this dimeric methylzinc thiolate.[2]

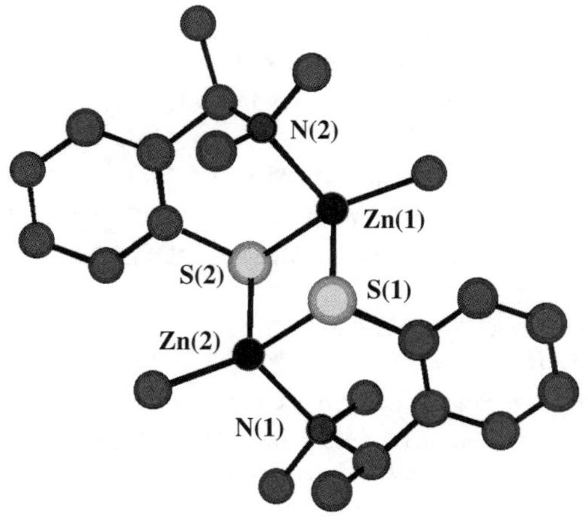

A mechanism has been put forward in which the rate-determining step in the 1,2-addition reaction is cleavage of the dimeric zinc thiolate into a transient species in which both the aldehyde substrate and the reagent, i.e. dialkylzinc, are present.[2]

Some representative results of the application of the catalyst precursors **1–3** in the enantioselective zinc-mediated 1,2-addition to aldehydes are compiled in Table 1 (eq 2).

These data show that application of one of the catalyst precursors **1–3** in the enantioselective 1,2-addition combines excellent chemical yield with high enantioselectivity.

Table 1 Enantioselective 1,2-addition of dialkyl zinc to aldehydes

Entry	Catalyst	R	R′	Yield (%)	ee (%)
1	**1**	Ph	Et	99	94 (*S*)
2	**2**	Ph	Et	93	98 (*S*)
3	**3**	Ph	Et	99	96 (*S*)
4	**1**	4-MeOC$_6$H$_4$	Et	94	95 (*S*)
5	**1**	(*E*)-C$_6$H$_5$CH CH	Et	95	75 (*S*)
6	**3**	C$_6$H$_5$CH$_2$CH$_2$	Et	99	82 (*S*)
7	**3**	Ph	Me	97	93 (*S*)
8	**3**	Ph	*i*-Pr	68	91 (*S*)

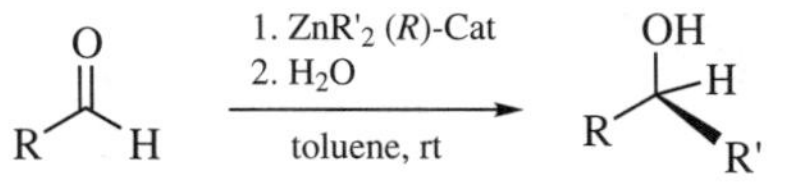

(2)

(*S*)-product

To overcome a main problem associated with homogeneous catalysis, i.e. the recovery of the catalyst, the latest development in this field comprises the functionalization of catalyst precursor **1–3** with perfluoroalkyl chains to enable catalysis to be carried out in an organic/perfluorinated two-phase solvent system. It has been demonstrated that this approach allows reuse of the catalyst several times.[8] Recently, an elegant synthetic route towards the enantioselective synthesis of chiral allylic alcohols in which catalyst precursor **1** is used has been reported (eq 3).[9] This reaction sequence involves the hydrozirconation of an acetylenic compound followed by a transmetallation reaction with Me_2Zn. The alkenyl group of the resulting heteroleptic alkenylzinc compound is selectively transferred to the aldehyde in the presence of catalyst precursor **1**, giving the chiral allylic alcohol with reported enantiomeric excesses of up to 90%.

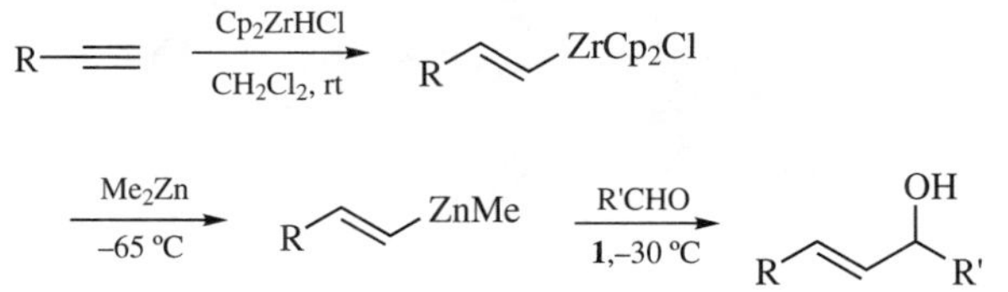

(3)

Application as Catalyst Precursor in Copper-Mediated Enantioselective C–C Bond Formation Reactions. At the present time, organocopper reagents are frequently used in synthetic organic chemistry. The discovery of the Gilman cuprate Me_2CuLi,[10] and the demonstration of its synthetic potential by House[11,12] and Corey[13] caused a major breakthrough in the applicability of organocopper compounds. A disadvantage, especially from a standpoint of 'atom economy,' in the application of stoichiometric cuprate reagents is the fact that one equiv of the potentially available organic groups is not used in the reaction and ends up as chemical waste. The idea of using a well-chosen non-transferable group has been applied in the enantioselective 1,4-addition of Grignard reagents to α,β-unsaturated enones, in the presence of catalytic amounts of a copper-arenethiolate derived from catalyst precursor **1**. The arenethiolate acts as a non-transferable group and induces enantioselectivity (eq 4).[14,15]

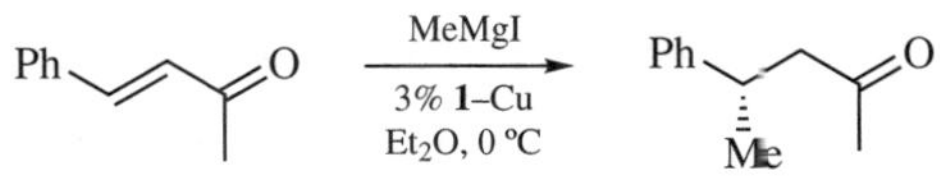

(4)

Copper arenethiolate **1**–Cu can be prepared starting from (*R*)-2-[1-(dimethylamino)ethyl]benzenethiol and Cu_2O,[1] or from the corresponding trimethylsilyl thio ether and CuCl (eq 5).[16]

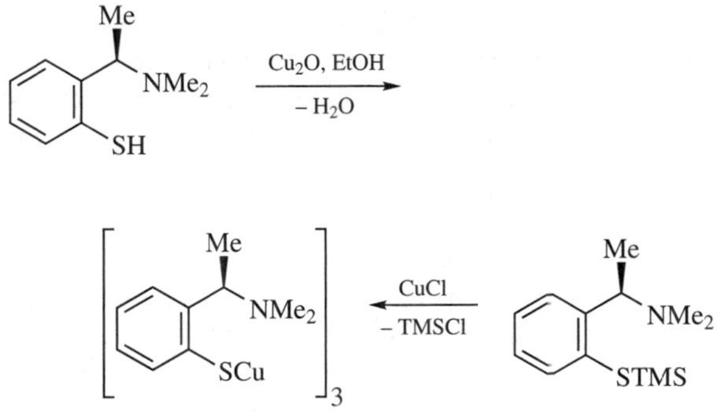

(5)

1 – Cu

The trimeric nature of catalyst **1**–Cu in the solid state was unambiguously proven by an X-ray crystal structure determination (see below).[1,17] This aggregate is retained in solution as shown by cryoscopic molecular weight determinations.[1]

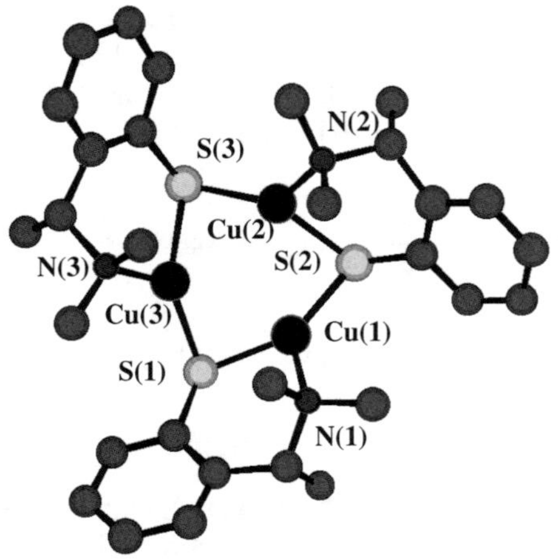

An interesting feature of the crystals of **1**–Cu is that they show triboluminescent behaviour.[18] Compound **1**–Cu has been applied successfully as a catalyst in the enantioselective Michael addition reaction involving a variety of substrates and Grignard reagents (eq 6).[19] The addition reaction proceeds with excellent chemical yields, and enantiomeric excesses of up to 70% have been reported. It was shown that the (*R*)-catalyst gives rise to the formation of (*S*)-products.[19]

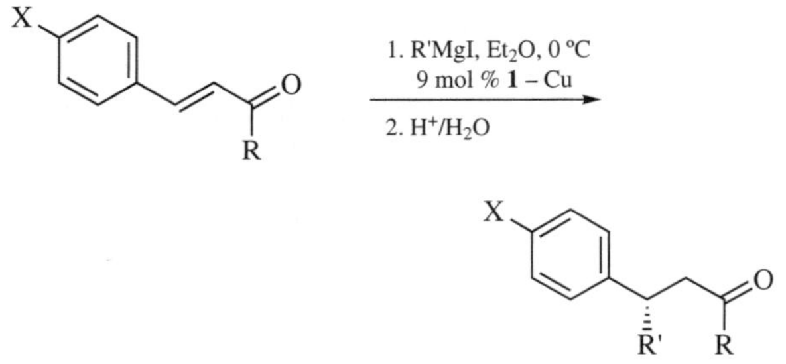

(6)

X = H, Cl,CN, Me or OMe
R = Me, *i*-Pr, *t*-Bu or Ph
R' = Me, *n*-Bu or *i*-Pr

Furthermore, **1**–Cu has been applied as a catalyst in the asymmetric substitution reaction of Grignard reagents with allylic substrates (eq 7).[20,21] Under optimized experimental conditions, the γ-product is obtained selectively in quantitative yield. However, the enantioselective induction is low to moderate (up to 40%).

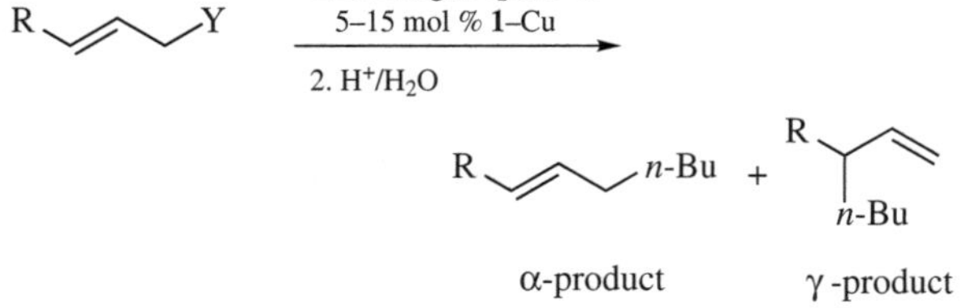

(7)

Y = $MeCO_2$, *t*-$BuCO_2$, CF_3CO_2 or $(EtO)_2P(O)O$

1. Knotter, D. M.; van Maanen, H. L.; Grove, D. M.; Spek, A. L.; van Koten, G. *Inorg. Chem.* **1991**, *30*, 3309.
2. Rijnberg, E.; Hovestad, N. J.; Kleij, A. W.; Jastrzebski, J. T. B. H.; Boersma, J.; Janssen, M. D.; Spek, A. L.; van Koten, G. *Organometallics* **1997**, *16*, 2847.
3. Mukaiyama, T.; Soai, K.; Kobayashi, S. *Chem. Lett.* **1978**, 219.
4. Oguni, N.; Omi, T. *Tetrahedron Lett.* **1984**, *25*, 2823.
5. Noyori, R.; Kitamura, M. *Angew. Chem., Int. Ed. Engl.* **1991**, *30*, 49.
6. Soai, K.; Niwa, S. *Chem. Rev.* **1992**, *92*, 833.
7. Rijnberg, E.; Jastrzebski, J. T. B H.; Janssen, M. D.; Boersma, J.; van Koten, G. *Tetrahedron Lett.* **1994**, *35*, 6521.
8. Kleijn, H.; Rijnberg, E.; Jastrzebski, J. T. B. H.; van Koten, G. *Org. Lett.* **1999**, *1*, 853.
9. Wipf, P.; Ribe, S. *J. Org. Chem.* **1998**, *63*, 6454.
10. Gilman, H.; Jones, R. G.; Woods, L. A. *J. Org. Chem.* **1952**, *17*, 1630.
11. House, H. O.; Respess, W. L.; Whitesides, G. M. *J. Org. Chem.* **1966**, *31*, 3128.
12. Whitesides, G. M.; Fisher, W. F., Jr; San Fulippo, J., Jr; Bashe, R. W.; House, H. O. *J. Am. Chem. Soc.* **1969**, *91*, 4871.
13. Corey, E. J.; Posner, G. H. *J. Am. Chem. Soc.* **1967**, *89*, 3911.
14. Lambert, F.; Knotter, D. M.; Janssen, M. D.; van Klaveren, M.; Boersma, J.; van Koten, G. *Tetrahedron Asymm.* **1991**, *2*, 1097.
15. van Koten, G. *Pure Appl. Chem.* **1994**, *66*, 1455.
16. Knotter, D. M.; Janssen, M. D.; Grove, D. M.; Smeets, W. J. J.; Horn, E.; Spek, A. L.; van Koten, G. *Inorg. Chem.* **1991**, *30*, 4361.
17. Knotter, D. M.; van Koten, G.; van Maanen, H. L.; Grove, D. M.; Spek, A. L. *Angew. Chem., Int. Ed. Engl.* **1989**, *28*, 341.
18. Knotter, D. M.; Blasse, G.; van Vliet, J. P. M.; van Koten, G. *Inorg. Chem.* **1992**, *31*, 2196.
19. van Klaveren, M.; Lambert, F.; Eijkelkamp, D. J. F. M.; Grove, D. M.; van Koten, G. *Tetrahedron Lett.* **1994**, *35*, 6135.
20. van Klaveren, M.; Persson, E. S. M.; del Villar, A.; Grove, D. M.; Bäckvall, J.-E.; van Koten, G. *Tetrahedron Lett.* **1995**, *36*, 3059.
21. Meuzelaar, G. J.; Karlström, A. S. E.; van Klaveren, M.; Persson, E. S. M.; del Villar, A.; van Koten, G ; Bäckvall, J.-E. *Tetrahedron* **2000**, *56*, 2895.

Johann T. B. H. Jastrzebski & Gerard van Koten
Debye Institute, Utrecht University, The Netherlands

(*R*)-*N*-[2-(*N,N*-Dimethylamino)ethyl]-*N*-methyl-1-[(*S*)-1′,2-bis(diphenylphosphino)ferrocenyl]ethylamine

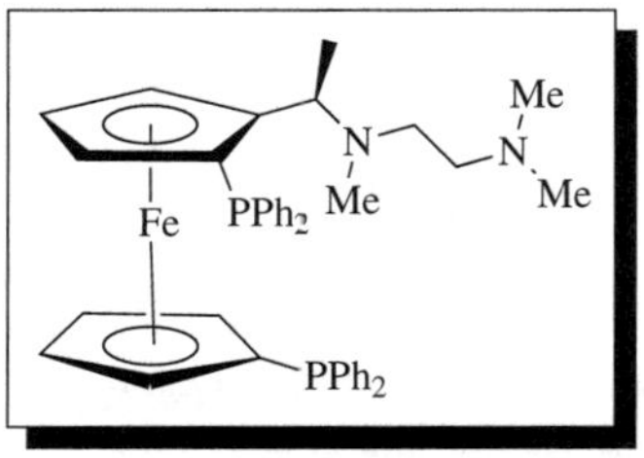

[119477-31-5] $C_{41}H_{43}FeN_2P_2$ (MW 682.62)

(chiral ligand for asymmetric synthesis;[1] gold(I)-catalyzed asymmetric aldol reaction;[2] silver(I)-catalyzed asymmetric aldol reaction;[3] enantioselective synthesis of β-hydroxy-α-aminophosphonates;[4] asymmetric allylic alkylation;[5] asymmetric allylic aminations;[6] asymmetric hydrogenations;[7] asymmetric [3 + 2] cycloaddition reactions[8])

Physical Data: viscous liquid, $[\alpha]_D^{25}$ +313° (c = 0.3, $CHCl_3$).

Solubility: sol dichloromethane, 1,2-dichloroethane, toluene, diethyl ether.

Preparative Methods: can be prepared[9] in two steps from commercially available (−)-(*R*)-*N*,*N*-dimethyl-1-[(*S*)-1′,2-bis(diphenylphosphino)ferrocenyl]ethylamine.

Handling, Storage, and Precautions: although air stable at rt, storage under anhydrous conditions under an inert atmosphere is recommended both to prevent the slow air oxidation of the phosphorus(III) ligating groups and absorption of atmospheric moisture.

Chiral Ferrocenylamine Ligands. Chiral ferrocenylamine ligands typified by the title reagent (**1**) have played a key role in the development of both methodology and ligand design[10] for asymmetric synthesis, particularly for the enantioselective formation of C–C bonds using catalytic quantities of chiral transition-metal catalysts. In the following discussion, both (**1**) and close analogs will be discussed and compared because small structural modifications of the alkyl side-chain of (**1**) can lead to significant increases in stereoselectivity.[11]

Gold(I)-Catalyzed Aldol Reaction. In 1986 an elegant enantioselective and diastereoselective synthesis of dihydrooxazolines was reported, using the aldol reaction of an aldehyde with an α-isocyanoacetate ester (formally a Knoevenagel reaction) using a cationic gold(I) complex of (**1**) (eq 1).[2]

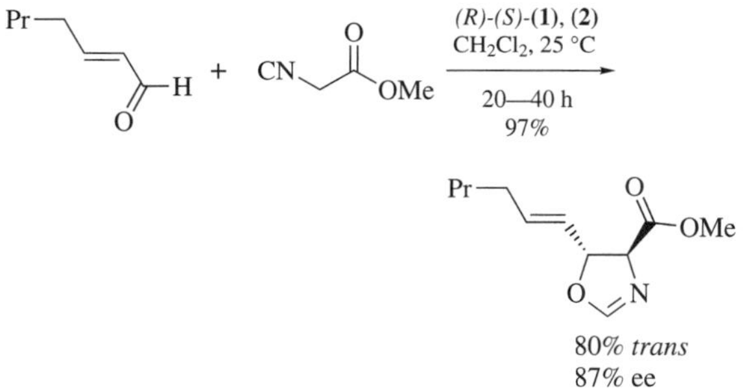

(1)

The gold(I) complex is prepared in situ by the reaction of (**1**) with bis(cyclohexyl isocyanide)gold(I) tetrafluoroborate (**2**),[12] typically in anhydrous dichloromethane. The dihydrooxazolines obtained provide a ready access to enantiomerically pure β-hydroxy-α-amino acid derivatives. High diastereo- and enantioselectivity are generally maintained with a wide variety of substituted aldehydes,[2,13,14] and α-isocyanoacetate esters.[15–17] *N*,*N*-Dimethyl-α-isocyanoacetamides[18] and α-keto esters[19] have been substituted for the α-isocyanoacetate ester and aldehyde component, respectively, sometimes with improved stereoselectivity. The effect of both the central and planar chirality of (**1**) on the diastereo- and enantioselectivity of the gold(I)-catalyzed aldol reaction has been studied.[20] The modification of the terminal dialkylamino group of (**1**) can lead to improvements in the stereoselectivity of the reaction, which in certain cases can be dramatic (eq 2).[11,21] The utility of the gold(I)-catalyzed aldol reaction in the synthesis of natural products has been demonstrated.[22,23] The gold(I)-catalyzed reaction of an α-isocyanomethylphosphonate ester with an aldehyde provides an enantioselective synthesis of β-hydroxy-α-aminophosphonic acid derivatives.[24–26]

Silver(I)-Catalyzed Aldol Reaction. In 1991 the silver(I)-catalyzed aldol reaction of an aldehyde with an α-isocyanoacetate ester was reported, analogous to the above mentioned gold(I)-catalyzed reaction.[3] The catalyst was prepared in situ from (**2**) and *Silver(I) Perchlorate*. The stereoselectivity of the silver(I)-catalyzed reaction was shown to be temperature dependent, which was attributed to the variation of the degree of metal coordination with temperature. Slow addition of the α-isocyanoacetate ester to a mixture of the aldehyde and catalyst, which favored the preferred tricoordinate Ag^I, gave high diastereo- and enantioselectivity (eq 3).

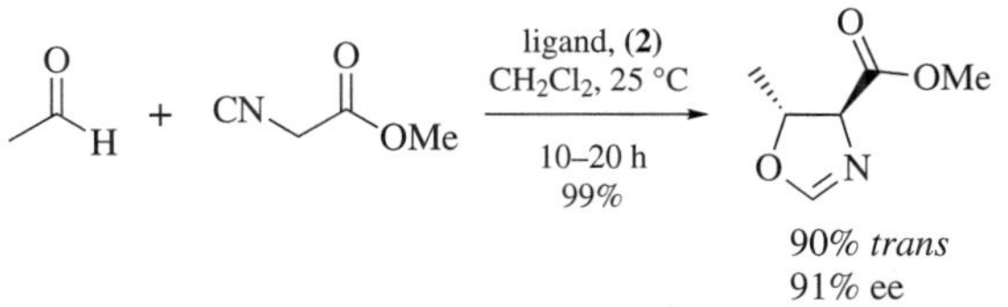

(2)

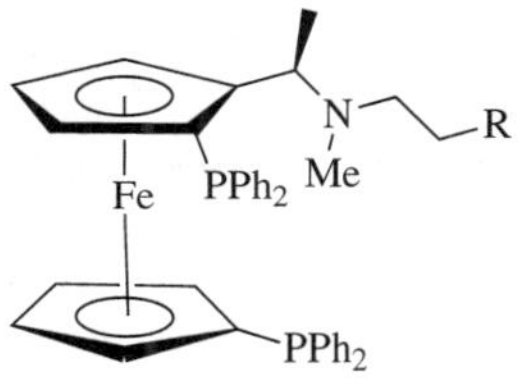

Ligand	R	% *trans*	% ee	% Yield
(*R*)-(*S*)-(**1**)	NMe_2	78	37	94
(*R*)-(*S*)-(**3**)	NEt_2	84	72	100
(*R*)-(*S*)-(**4**)	N 	85	85	100
(*R*)-(*S*)-(**5**)	N 	89	89	99

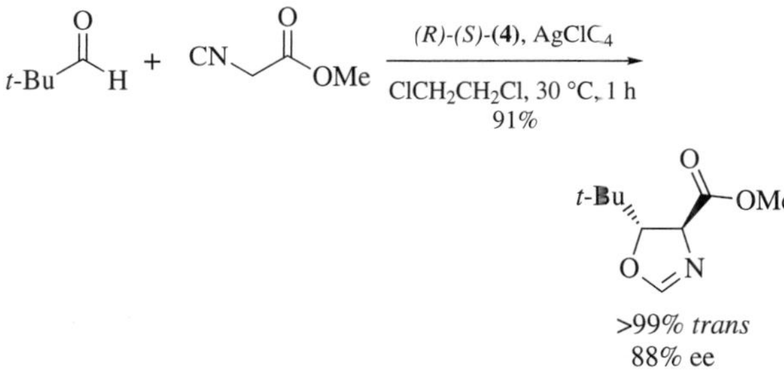

(3)

Earlier workers reported the silver(I)-catalyzed reaction of an aldehyde with *p-Tolylsulfonylmethyl Isocyanide* (eq 4).[27] The (*R*)-(*R*)-dihydrooxazolines formed can be reduced with *Lithium Aluminum Hydride* to provide a facile route to α-alkyl-β-(*N*-methylamino)ethanols in good to excellent yield.

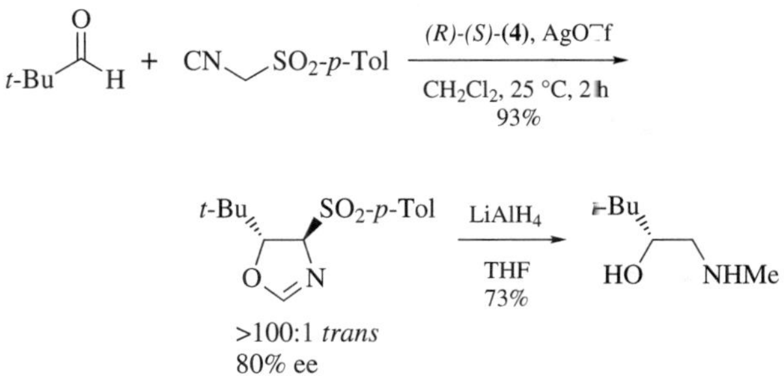

(4)

Asymmetric Allylations. The asymmetric allylation of β-diketones with π-allyl Pd^{II} complexes using the chiral ligand (**1**) was reported to proceed with low stereoselectivity.[5] Modification of the alkyl side-chain of (**1**) led to significant improvements in enantioselectivity (eq 5).[5,28–31]

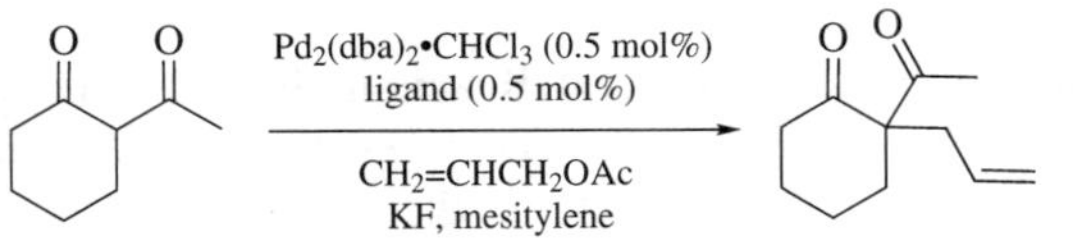

(5)

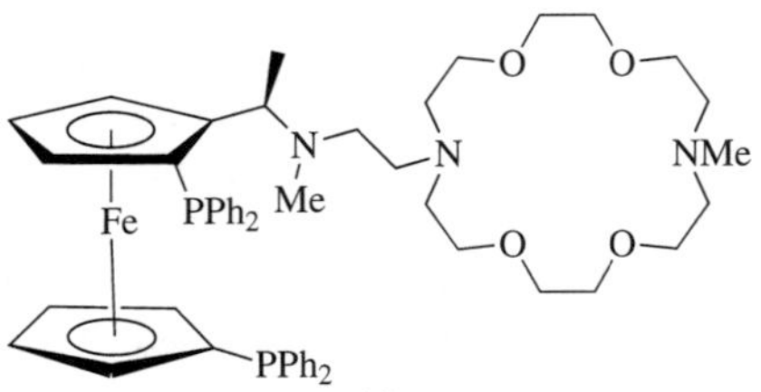

(**6**)

Ligand	Yield (%)	ee (*R*) (%)
(**1**)	57	22
(**6**)	92	75

The in situ formed Pd^{II} catalyst system prepared with the hydroxyalkyl-substituted ferrocenylamine (**7**) led to the opposite absolute configuration at the carbon stereocenter (eq 6).[28] A similar inversion of stereochemistry is observed with ferrocenylamine ligands containing a free hydroxyl substituent in the gold(I)-catalyzed aldol reaction.[21b] Although asymmetric allylic aminations can be achieved using the chiral ligand (**7**), significantly improved enantioselectivity is obtained with the bis(hydroxyalkyl)-substituted ligand (**8**) (eq 7).[6]

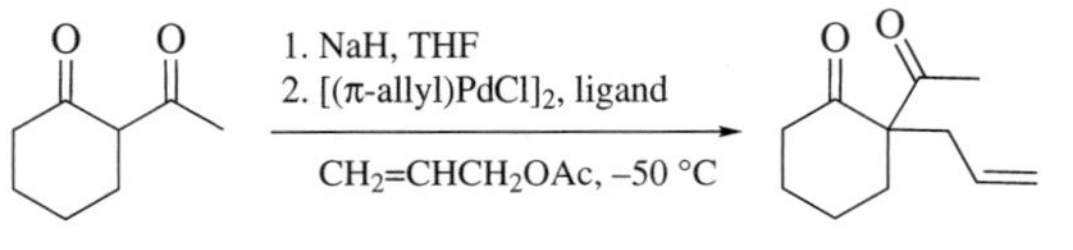

(6)

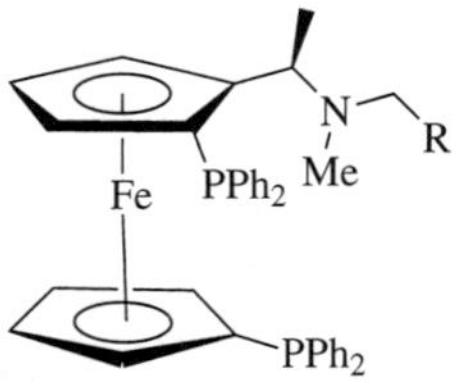

(**7**) R = CH_2CH_2OH
(**8**) R = $CH(CH_2OH)_2$

Ligand	Yield (%)	ee (*S*) (%)
(**7**)	100	73
(**8**)	55	49

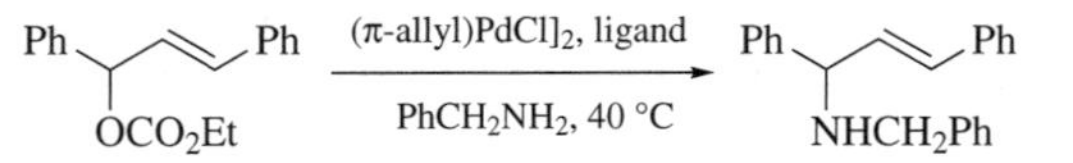

(7)

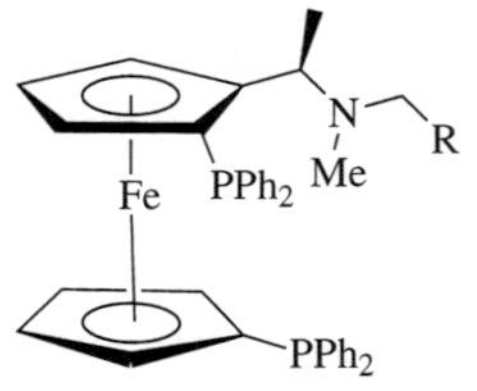

(**7**) R = CH_2CH_2OH
(**8**) R = $CH(CH_2OH)_2$

Ligand	Yield (%)	ee (*R*) (%)
(**7**)	80	79.3
(**8**)	93	97

An interesting intramolecular variation of this reaction provides oxazolidones, which may be hydrolyzed to synthetically useful optically active 2-amino-3-butenols (eq 8).[32] The absolute stereochemistry of the stereocenter formed is dependent upon the geometry about the double bond of the 2-butenylene dicarbamate substrate. A related Pd^{II}-promoted [3 + 2] cycloaddition of an activated alkene with a 2-(sulfonylmethyl)-2-propenyl carbonate, using the bis(hydroxyalkyl)-substituted ligand (**8**), gave methylenecyclopentane derivatives with high asymmetric induction.[8]

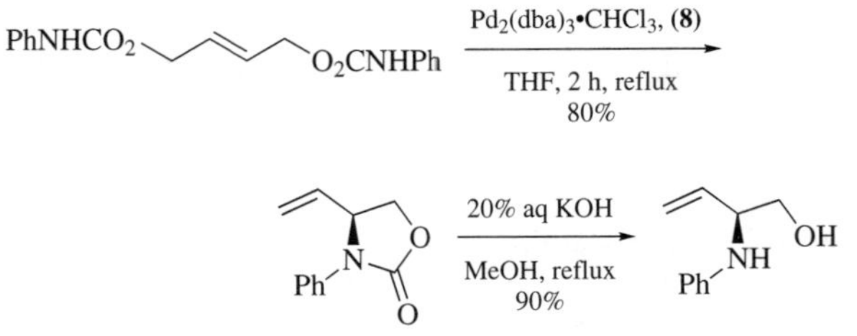

(8)

Asymmetric Hydrogenations. Catalytic asymmetric hydrogenations of β-disubstituted-α-phenylacrylic acids have been achieved using the Rh^{I} complex of (**4**) (eq 9).[7,33] Asymmetric hydrogenation of unsymmetrically substituted trisubstituted acrylic acids leads to the formation of two stereocenters in high ee.[7] The variation of the terminal dialkylamino substituents has little effect on enantioselectivity.[33] A study of a Ru^{II} complex of (**1**) was reported as a model for understanding the stereoselective transition state of asymmetric hydrogenations.[34]

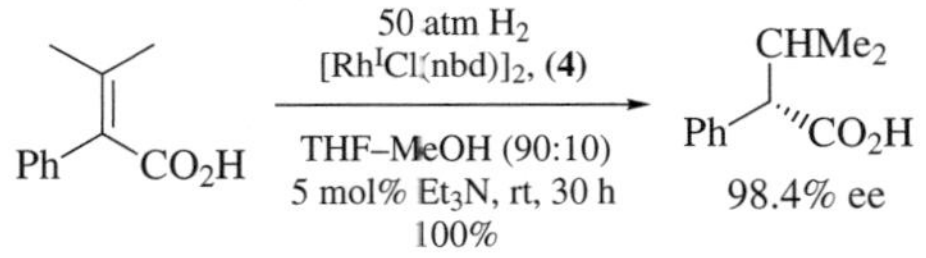

(9)

1. Hayashi, T. *Pure Appl. Chem.* **1988**, *60*, 7.
2. Ito, Y.; Sawamura, M.; Hayashi, T. *J. Am. Chem. Soc.* **1986**, *108*, 6405
3. Hayashi, T.; Uozumi, Y.; Yamazaki, A. *Tetrahedron Lett.* **1991**, *32*, 2799.
4. Mastalerz, P. In *Handbook of Organophosphorus Chemistry*; Engel, R., Ed.; Dekker: New York, 1992; pp 277–375.
5. Sawamura, M.; Nagata, H.; Sakamoto, H.; Ito, Y. *J. Am. Chem. Soc.* **1992**, *114*, 2586.
6. Hayashi, T.; Yamamoto, A.; Ito, Y.; Nishioka, E.; Miura, H.; Yanagi, K. *J. Am. Chem. Soc.* **1989**, *111*, 6301.
7. Hayashi, T.; Kawamura, N.; Ito, Y. *Tetrahedron Lett.* **1988**, *29*, 5969.
8. Yamamoto, A.; Ito, Y.; Hayashi, T. *Tetrahedron Lett.* **1989**, *30*, 375.
9. Hayashi, T.; Mise, T.; Fukushima, M.; Kagotani, M.; Nagashima, N.; Hamada, Y.; Matsumoto, A.; Kawakami, S.; Konishi, M.; Yamamoto, K.; Kumada, M. *Bull. Chem. Soc. Jpn.* **1980**, *53*, 1138.
10. Sawamura, M.; Ito, Y. *Chem. Rev.* **1992**, *92*, 857.

11. Hayashi, T.; Sawamura, M.; Ito, Y. *Tetrahedron* **1992**, *48*, 1999.
12. Bonati, F.; Minghetti, G. *Gazz. Chim. Ital.* **1973**, *103*, 373.
13. Togni, A.; Pastor, S. D. *J. Org. Chem.* **1990**, *55*, 1649.
14. Ito, Y.; Sawamura, M.; Hayashi, T. *Tetrahedron Lett.* **1987**, *28*, 6215.
15. Ito, Y.; Sawamura, M.; Shirakawa, E.; Hayashizaki, K.; Hayashi, T. *Tetrahedron* **1988**, *44*, 5253.
16. Togni, A.; Pastor, S. D. *Helv. Chim. Acta* **1989** *72*, 1038.
17. Ito, Y.; Sawamura, M.; Shirakawa, E.; Hayashizaki, K.; Hayashi, T. *Tetrahedron Lett.* **1988**, *29*, 235.
18. Ito, Y.; Sawamura, M.; Kobayashi, M.; Hayashi, T. *Tetrahedron Lett.* **1988**, *29*, 6321.
19. Ito, Y.; Sawamura, M.; Hamashima, H.; Emura, T.; Hayashi, T *Tetrahedron Lett.* **1989**, *30*, 4681.
20. Pastor, S. D.; Togni, A. *J. Am. Chem. Soc.* **1989**, *111*, 2333.
21. (a) Hayashi, T.; Yamazaki *J. Organomet. Chem.* **1991**, *413*, 295. (b) Pastor, S. D.; Togni, A. *Helv. Chim. Acta* **1991**, *74*, 905.
22. Togni, A.; Pastor, S. D.; Rihs, G. *Helv. Chim. Acta* **1989** *72*, 1471.
23. Ito, Y.; Sawamura, M.; Hayashi, T *Tetrahedron Lett.* **1988**, *29*, 239.
24. Togni, A.; Pastor, S. D. *Tetrahedron Lett.* **1989**, *30*, 1071.
25. Sawamura, M.; Ito, Y.; Hayashi, T. *Tetrahedron Lett.* **1989**, *30*, 2247.
26. For a more detailed discussion, see *Bis(cyclohexyl isocyanide)gold(I) Tetrafluoroborate–(R)-N-[2-(N,N-Dimethylamino)ethyl]-N-methyl-1-[(S)-1′,2-bis(diphenylphosphino)ferrocenyl]ethylamine*.
27. Sawamura, M.; Hamashima, H.; Ito, Y. *J. Org. Chem.* **1990**, *55*, 5935.
28. Hayashi, T.; Kanehira, K.; Hagihara, T.; Kumada, M. *J. Org. Chem.* **1988**, *53*, 113.
29. Hayashi, T.; Yamamoto, A.; Hagihara, T.; Ito, Y. *Tetrahedron Lett.* **1986**, *27*, 191.
30. Hayashi, T.; Yamamoto, A.; Ito, Y. *Chem. Commun./J. Chem. Soc., Chem. Commun.* **1986** 1090.
31. Ito, Y.; Sawamura, M.; Matsuoka, M.; Matsumoto, Y.; Hayashi, T. *Tetrahedron Lett.* **1987**, *28*, 4849.
32. Hayashi, T.; Yamamoto, A.; Ito, Y. *Tetrahedron Lett.* **1988**, *29*, 99.
33. Hayashi, T.; Kawamura, N.; Ito, Y. *J. Am. Chem. Soc.* **1987**, *109*, 7876.
34. Alcock, N. W.; Brown, J. M.; Rose, M.; Wienand, A. *Tetrahedron: Asymmetry* **1991**, *2*, 47.

Stephen D. Pastor
Ciba-Geigy Corporation, Ardsley, NY, USA

(1*R*,2*S*,3*R*,4*S*)-3-Dimethylamino-1,7,7-trimethylbicyclo[2.2.1]heptan-2-ol-[(−)DAIB]

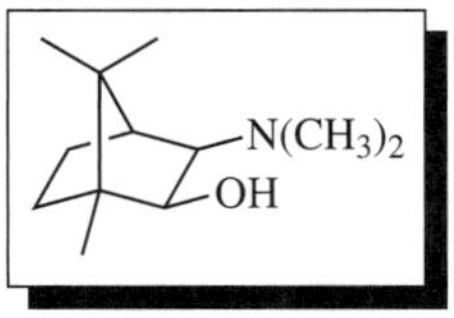

[103729-96-0] $C_{12}H_{23}NO$ (MW 197.18)

(chiral catalyst for the enantioselective addition of dialkylzinc reagents to aldehydes and ketones)

Alternate Name: (−)DAIB, *cis-exo-N,N*-dimethyl-3-aminoisoborneol, (−)-3-*exo*-(dimethylamino)isoborneol, (2*S*)-DAIB.
Physical Data: bp 120 °C (18 mm Hg),[1] 75 °C (0.05 mm Hg),[5] 70 °C (0.1 mm Hg).[2] $[\alpha]^{28}_D$ −14.7° (*c* 4.58, ethanol),[1] $[\alpha]^{20}_D$ −14.5° (*c* 0.93, ethanol),[5] $[\alpha]^{14}_D$ −9.40° (*c* 4.31, ethanol),[4] $[\alpha]_D$ −8.0° (*c* 4.3, ethanol).[2]
Solubility: soluble in most organic solvents, e.g., ethanol, toluene, and CH_2Cl_2.
Form Supplied in: yellow to colorless oil.
Analysis of Reagent Purity: ^{1}H NMR,[1,2] HPLC,[1,3] polarimetry,[1,2,4,5] ^{31}P NMR after derivatization.[5]
Preparative Methods: conversion of (1*R*)-camphor into vicinal amino alcohol,[6–8] followed by *N*-methylation using either aqueous formaldehyde and formic acid[6] or methyl iodide.[5,7]
Purification: bulb-to-bulb distillation,[1,5] column chromatography.[5,6]

Introduction. (1*R*,2*S*,3*R*,4*S*)-3-Dimethylamino-1,7,7-trimethylbicyclo[2.2.1]heptan-2-ol, herein referred to as (−)DAIB, has been shown to be an effective catalyst for enantioselective carbon-carbon bond formation utilizing dialkylzinc reagents with aldehydes and ketones. Attempts to utilize (−)DAIB for asymmetric inductions with other organometallic reagents such as aluminum alkyls, alkyl Grignards, and alkyl lithiums, have been unsuccessful.[9]

1,2-Additions of Dialkylzinc Reagents to Aromatic Aldehydes. The addition of dialkylzinc reagents to aldehydes for carbon-carbon bond formation was rarely utilized in organic synthesis because such reactions were sluggish and reduction by-products were typically observed. (−)DAIB has been shown to be an effective catalyst for highly enantioselective 1,2-additions of dialkylzinc reagents to aromatic aldehydes, with high yields (eq 1).[1,4,9] In nearly all examples, the resulting alcohols were identified as having the *S* configuration.

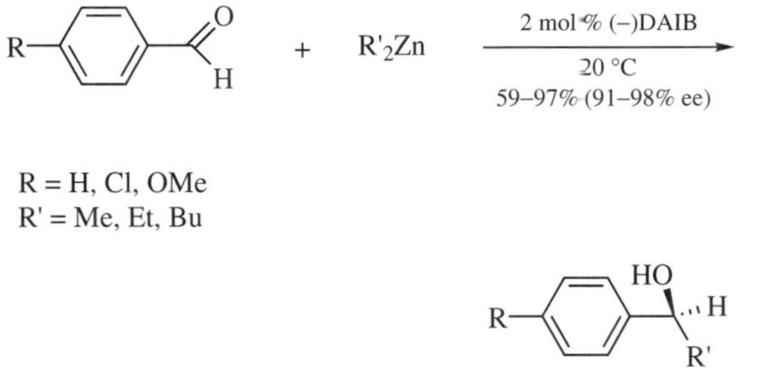

(1)

A variety of factors have been identified, which affect yield and enantioselectivity. The stoichiometry of the aldehyde, dialkylzinc reagent, and (−)DAIB has significant effects on the reaction rate and course.[4,9] Alkylation of benzaldehyde occurs only when the stoichiometry of dialkylzinc reagent/(−)DAIB is greater than 2; the optimized stoichiometry for yield and ee was found to be 50.[9,10] Non-polar solvents such as toluene and hexane are preferred; the use of THF retards alkylation and lowers product ee.[1] Use of halide-free dialkylzinc reagents is crucial for obtaining high enantioselectivity.[1] An investigation found that the presence of electron-withdrawing substituents in the para position tends to increase the enantioselectivity, ranging from 86% ee (*p*-OCH_3) to 96% ee (*p*-CF_3); rates also improved slightly with electron-withdrawing substituents.[11]

The effect of the enantiometric purity of (−)DAIB has also been well documented. The enantioselectivities obtained using optically pure (−)DAIB are higher than those obtained from (−)DAIB having lower ee's. However, the relationship between the enantiometric purity of (−)DAIB and that of the resulting product is nonlinear, and varies as a function of reactants, reactant stoichiometry, and reaction conditions.[10,12,13] For example, when benzaldehyde and diethylzinc were reacted in a 1:1 molar ratio in the presence of 8 mol % (−)DAIB, the enantioselectivity of the resulting product, (*S*)-1-phenyl-1-propanol, was 98% ee when pure (−)DAIB was used. The use of 15% ee (−)DAIB resulted in a product enantioselectivity of 95% ee (92% yield).[10] The mechanism for this dramatic effect has been thoroughly studied.[10–12,14–17]

Polystyrene-attached analogs of (−)DAIB have been synthesized, which show both high yields and high enantioselectivity similar to that observed for the homogeneous alkylations (eq 2).[2,18] All product alcohols which had their configurations determined possessed the *S* configuration.

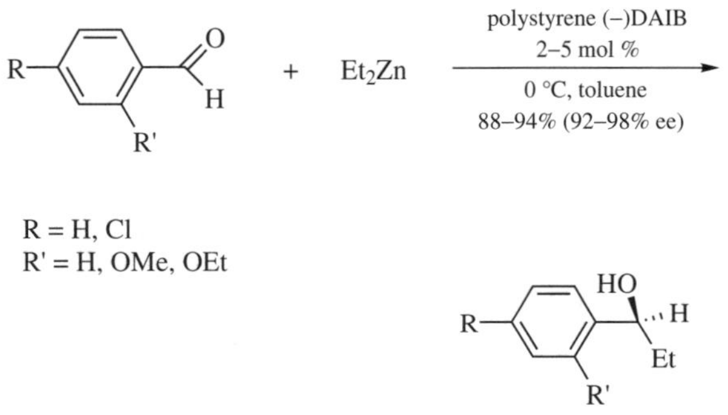

R = H, Cl
R' = H, OMe, OEt

(2)

1,2-Additions of Dialkylzinc Reagents to Conjugated Aldehydes. (−)DAIB has been evaluated as a catalyst for enantioselective 1,2-additions of diethylzinc and dipentylzinc to furanals.[1,19] Unfortunately, the limited data show that enantioselectivities vary widely according to furanal substitution and the dialkylzinc reagent. Although most enantioselectivites were 90% ee, reaction of 5-[2-(1,3-dioxylanyl)]-2-furanal with diethylzinc produced the corresponding alcohol in 56% yield with only 46% ee (eq 3).[19] Comparison of (−)DAIB with other chiral aminoalcohols showed (*S*)-(+)-diphenyl-(1-methylpyrrolidin-2-yl)methanol (DPMPM) to yield significantly higher enantioselectivities in this specific application.[19] Other examples of (−)DAIB-catalyzed 1,2-additions to conjugated aldehydes are shown below (eq 4); all reported examples have been shown to produce alcohols with *S* configuration.[1]

1,2-Additions of Dialkylzinc Reagents to Aliphatic Aldehydes. Few examples of (−)DAIB-catalyzed 1,2-additions to aliphatic aldehydes by diakylzinc reagents have been reported.[1] Although alkylation yields were high, enantioselectivities ranged from 90% to 0% ee (eq 5). All examples produced alcohols with *S* configuration.[1]

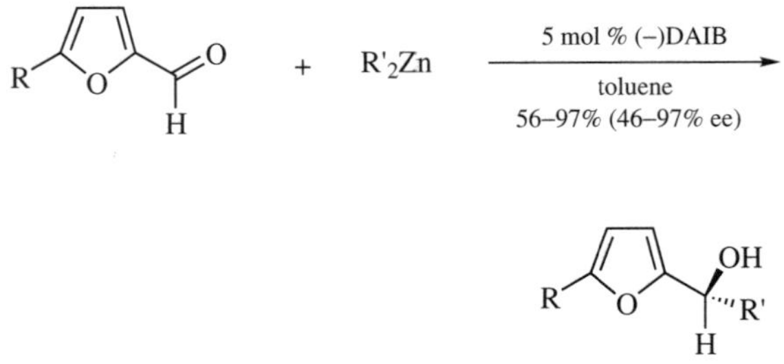

(3)

R = H, Me, $SiMe_3$, 2-(1,3)-dioxalanyl
R' = Et, *n*-Pen

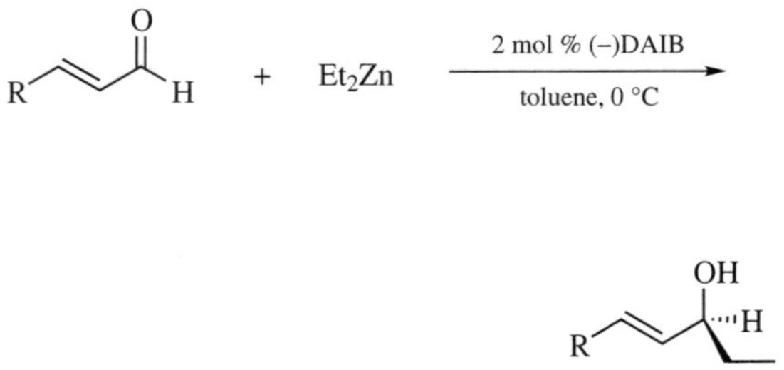

(4)

R = Ph, 81% (96% ee)
R = $SnBu_3$, 84% (85% ee)

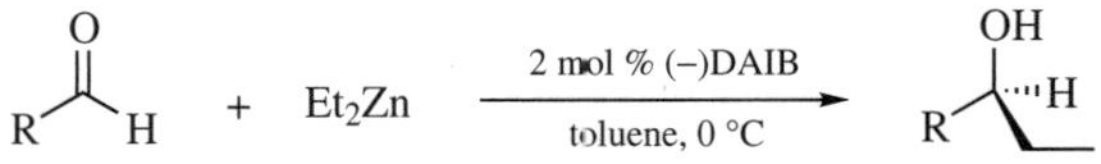

(5)

R = $PhCH_2CH_2$, 80% (90% ee)
R = $PhCH_2OCH_2$, 80% (0% ee)
R = *N*-Hexyl, 81% (61% ee)

1,2-Additions of (1-Alkenyl)alkylzinc Reagents to Aldehydes. (−)DAIB has been reported to be a superior catalyst for the enantioselective addition of (1-alkenyl)alkylzinc reagents to aromatic and aliphatic aldehydes (eq 6).[20] The [(*E*)-1-alkenyl]-alkylzinc reagents are generated in situ by the transmetallation reaction of dialkylzinc reagents with [(*E*)-1-alkenyl]boranes. Stereochemical retention in the transmetallation and subsequent reaction steps is evidenced by the absence of (*Z*)-allyl alcohol formation. The greatest enantioselectivities are realized for benzaldehyde with ee's ranging from 92% to 98%, while enantioselectivities for aliphatic aldehydes are more moderate, ranging from 73% to 91% ee. Absolute configurations of the resulting alcohols were not determined.

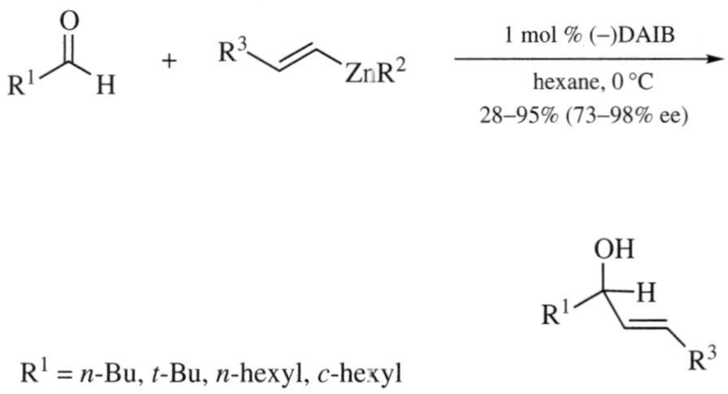

(6)

R^1 = *n*-Bu, *t*-Bu, *n*-hexyl, *c*-hexyl
R^2 = Me, Et
R^3 = Et, *n*-Bu, *t*-Bu, *c*-hexyl, Ph

The source of the (1-alkenyl)alkylzinc reagent may significantly affect the outcome of the addition. In the presence of (−)DAIB, (1-hexenyl)methylzinc alkylates benzaldehyde with 96% ee (87% yield).[20] However, negligible enantioselectivity (3% ee) was observed when the reaction was repeated using (1-hexenyl)methylzinc generated in situ from dimethylzinc and (1-hexenyl)zirconcene.[21]

1,4-Additions of Dialkylzinc Reagents to Conjugated Ketones. High regioselective and enantioselective diethylzinc additions to chalcones have been reported using a Ni^{II} complex and (−)DAIB as catalysts (eq 7).[5,22] A variety of factors has been identified, which affect enantioselectivity and the absolute configuration of the product. Increased (−)DAIB to Ni^{II} ratios, lower reaction temperatures, and shorter reaction times—all resulted in increased ee's.[5] The use of acetonitrile, propionitrile, or butyronitrile solvents was critical for high enantioselectivity. Chiral amplification similar to that observed for 1,2-additions of dialkylzinc reagents to aromatic aldehydes (vide supra) was reported, and decreasing the amount of Ni^{II} catalyst increased the amplification factor. Although it was originally reported that the addition of achiral amines improved enantioselectivity,[22] this effect disappeared when pure (−)DAIB was employed.[5]

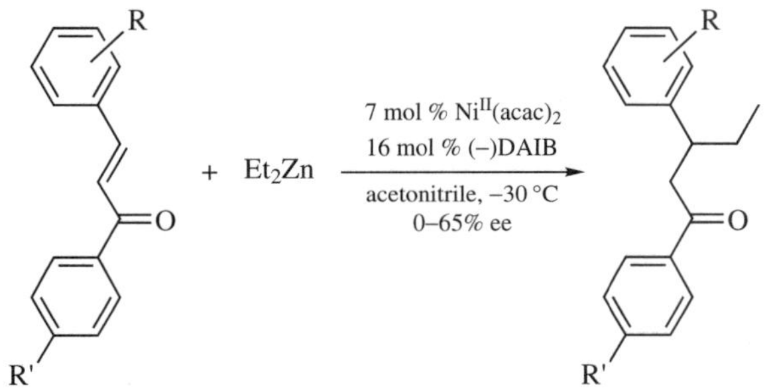

(7)

R = H, 4-(MeO), 4-Cl, 3-(NO_2)
R' = H, MeO, Cl

Similar yield and enantioselectivity were obtained with chalcones when $Co(acac)_2$ was substituted for $Ni(acac)_2$ catalysts under otherwise identical reaction conditions.[23] However, the cobalt-catalyzed reaction was significantly slower and produced a significant amount of reduced by-product (5%) compared to reactions catalyzed by nickel. For both the cobalt- and nickel-catalyzed reactions, both (−)DAIB and (+)DAIB were shown to be superior to several chiral aminoalcohols for enantioselectivity.

Related Reagents. Chiral aminoalcohols (prolinol, cinchonidine, quinidine) and polystyrene-attached analogs of (−)DAIB.

1. Noyori, R.; Suga, S.; Kawai, K.; Okada, S.; Kitamura, M.; Oguni, N.; Hayashi, M.; Kaneko, T.; Matsuda, Y. *J. Organomet. Chem.* **1990**, *382*, 19.
2. Sung, D. W. L.; Hodge, P.; Stratford, P. W. *J. Chem. Soc., Perkin Trans. 1* **1999**, *11*, 1463.
3. Kitamura, M.; Suga, S.; Niwa, M.; Noyori, R. *J. Am. Chem. Soc.* **1995**, *117*, 4832.
4. Kitamura, M.; Suga, S.; Kawai, K.; Noyori, R. *J. Am. Chem. Soc.* **1986**, *108*, 6071.
5. De Vries, A. H. M.; Jansen, J. F. G. A.; Feringa, B. L. *Tetrahedron* **1994**, *50*, 4479.
6. Chittenden, R. A.; Cooper, G. H. *J. Chem. Soc. (C)* **1970**, 49
7. Davies, S. R.; Mitchell, M. C.; Cain, C. P.; Devitt, P. G.; Taylor, R. J.; Kee, T. P. *J. Organomet. Chem.* **1998**, *550*, 29.
8. Daniel, A.; Pavia, A. A. *Bull. Soc. Chim. Fr.* **1971** 1060.
9. Noyori, R.; Suga, S.; Kawai, K.; Okada, S.; Kitamura, M. *Pure & Appl. Chem.* **1988**, *60*, 1597.
10. Kitamura, M.; Okada, S.; Suga, S.; Noyori, R. *J. Am. Chem. Soc.* **1989**, *111*, 4028.
11. Kitamura, M.; Oka, H.; Noyori, R. *Tetrahedron* **1999**, *55*, 3605.
12. Kitamura, M.; Suga, S.; Oka, H.; Noyori, R. *J. Am. Chem. Soc.* **1998**, *120*, 9800.
13. Goldfuss, B.; Houk, K. N. *J. Org. Chem.* **1998**, *63*, 8998.
14. Noyori, R.; Suga, S.; Oka, H.; Kitamura, M. *Chemical Record* **2001**, *1*, 85.
15. Noyori, R.; Kitamura, M. *Angew. Chem. Int. Ed. Engl.* **1991**, *30*, 49.
16. Kitamura, M.; Suga, S.; Niwa, M.; Noyori, R.; Zhai, Z.-X.; Suga, H. *J. Phys. Chem.* **1994**, *98*, 12776.
17. Yamakawa, M.; Noyori, R. *Organometallics* **1999** *18*, 128.
18. Itsuno, S.; Frechet, J. M. J. *J. Org. Chem.* **1987**, *52*, 4140.
19. Van Oeveren, A.; Menge, W.; Feringa, B. L. *Tetrahedron Lett.* **1989**, *30*, 6427.
20. Oppolzer, W.; Radinov, R. N. *Helv. Chim. Acta* **1992**, *75*, 170.
21. Wipf, P.; Ribe, S. *J. Org. Chem.* **1998**, *63*, 6454.
22. Jansen, J. F. G. A.; Feringa, B. L. *Tetrahedron: Asymmetry* **1992**, *3*, 581.
23. De Vries, A. H. M.; Feringa, B. L. *Tetrahedron: Asymmetry* **1977**, *8*, 1377

Grant R. Krow & Kevin C. Cannon
Temple University, Philadelphia, Pennsylvania, USA

(4*R*,5*R*)-2,2-Dimethyl-4,5-bis(hydroxydiphenylmethyl)-1,3-dioxolane–Titanium(IV) Chloride

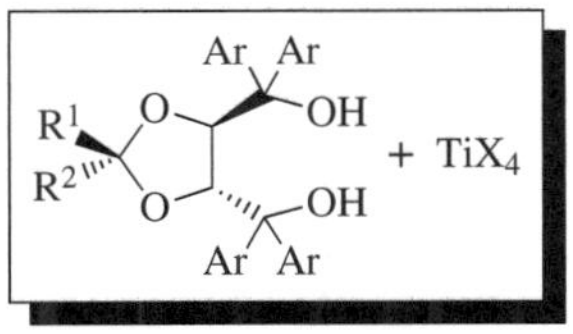

(R^1 = Me, R^2 = Me, Ar = Ph)
[93379-49-8] $C_{31}H_{30}O_4$ (MW 466.61)
(R^1 = mesityl, R^2 = H, Ar = Ph)
[114026-75-4] $C_{38}H_{36}O_4$ (MW 556.74)
(R^1 = Me, R^2 = Me, Ar = 1-naphthyl)
[137536-94-8] $C_{47}H_{38}O_4$ (MW 666.85)
(R^1 = *t*-Bu, R^2 = H, Ar = Ph)
[114026-72-1] $C_{33}H_{34}O_4$ (MW 494.67)
(R^1 = Ph, R^2 = Me, Ar = Ph)
[109306-21-0] $C_{36}H_{32}O_4$ (MW 528.65)
(R^1 = Et, R^2 = Et, Ar = 3,5-$Me_2C_6H_3$)
[138710-29-9] $C_{41}H_{50}O_4$ (MW 606.91)

(chiral alkyltitanium reagent for asymmetric alkylation reaction;[1] chiral Lewis acid for asymmetric Diels–Alder reaction, [2 + 2] cycloaddition reaction, and intramolecular ene reaction;[2] reagent for asymmetric hydrocyanation of aldehydes and for kinetic resolution of α-aryl carboxylic acid derivatives[2])

Preparative Methods: chiral titanates are usually prepared by mixing dichlorodiisopropoxytitanium and a chiral 1,4-diol in toluene. Other solvents such as ether and dichloromethane can also be employed. The alcohol exchange reaction takes place immediately at rt. Wherever necessary, liberated isopropyl alcohol is removed by azeotropic removal with toluene.[3] The chiral 1,4-diols are prepared from dimethyl (or diethyl) tartrate by a two-step procedure comprising acetalization followed by the addition of an aryl Grignard reagent.[4,5]

Handling, Storage, and Precautions: chiral titanates are usually prepared just before use under argon atmosphere. Care should be taken to avoid moisture, especially when a catalytic amount of the reagent is used.

Chiral Alkylating Reagent. The chiral methyltitanium reagent prepared from (**1**) and *Methyllithium* or methyl Grignard reagent adds to various aldehydes with moderate to good enantioselectivity (eq 1).[6] Furthermore, the ate complex prepared from the chiral tetraalkoxytitanium (**2**) and methyllithium adds to aromatic aldehydes with more than 90% ee (eq 2).[7] The chiral allyltitanium reagent (**3**) having a cyclopentadienyl group on titanium adds to various aldehydes to give the corresponding allylated products with high optical purity (eq 3).[8]

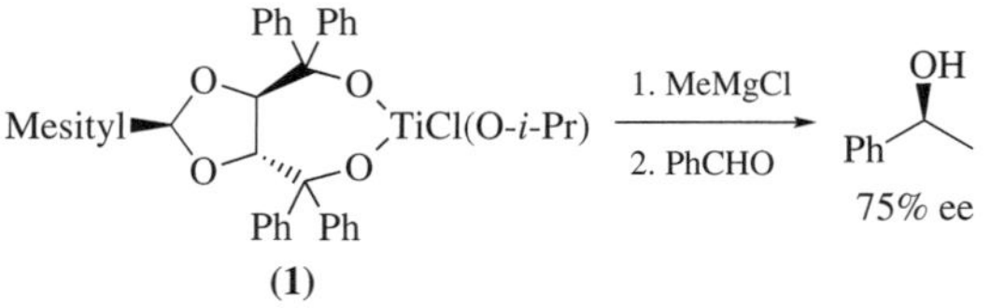

(1)

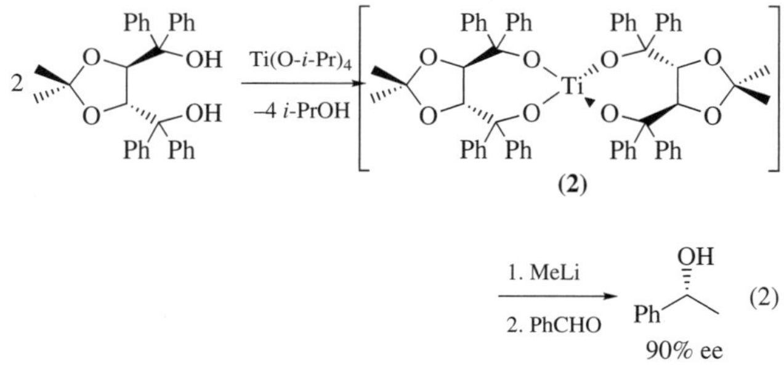

(2)

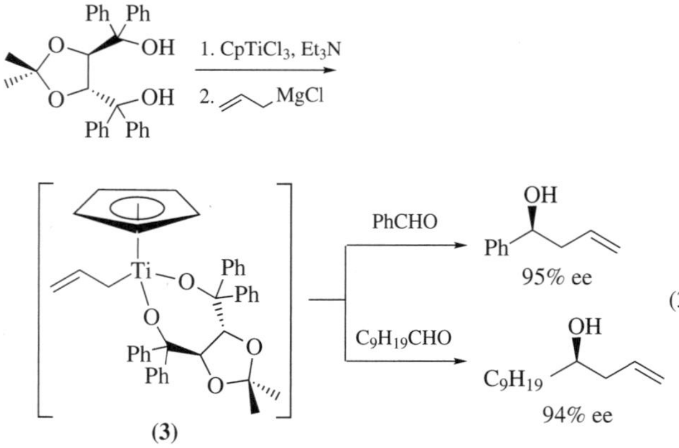

(3)

Chiral titanates can be employed as catalysts for the alkylation of aldehydes using dialkylzinc reagents. For example, by the use of a catalytic amount of the chiral titanium reagent (**4**), addition of *Diethylzinc* to various aldehydes occurs with high enantioselectivity in the presence of *Titanium Tetraisopropoxide* (eq 4).[9] Furthermore, by using the chiral tetraalkoxytitanium (**2**), the alkylation reaction can be carried out in ether, which enables the use of various dialkylzinc reagents prepared in situ from the corresponding Grignard reagents and *Zinc Chloride* (eq 5).[10]

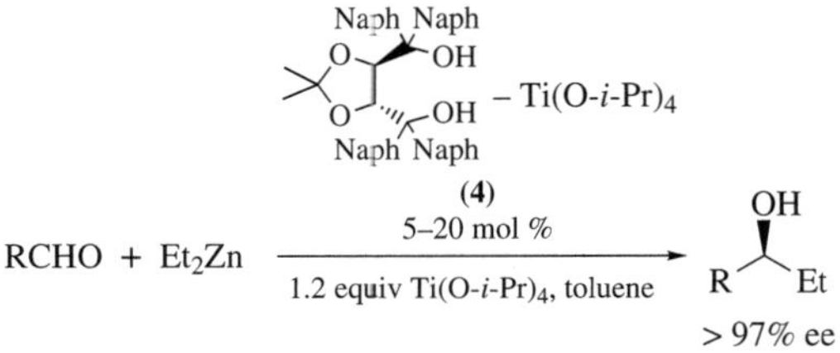

(4)

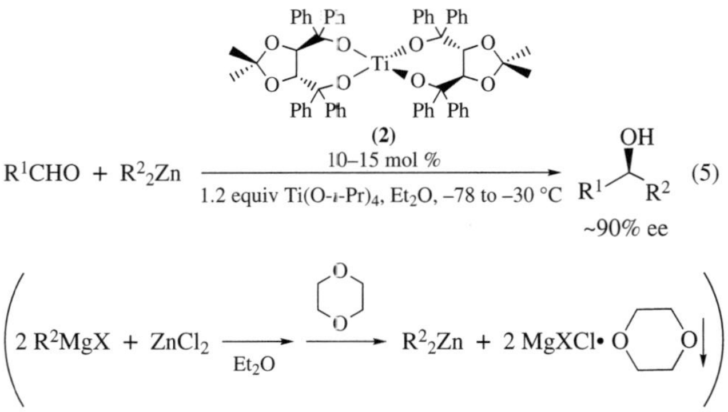

(5)

Chiral Lewis Acid. These chiral titanium reagents are widely used as chiral Lewis acid catalysts. The Diels–Alder reaction of methyl acrylate and cyclopentadiene affords the *endo* adduct in moderate enantioselectivity when a stoichiometric amount of the chiral titanium reagent (**5**) is employed (eq 6).[6] Use of 3-(2-alkenoyl)-1,3-oxazolidin-2-ones as dienophiles greatly improves the optical purity of the cycloadduct when the 2-phenyl-2-methyl-1,3-dioxolane derivative (**6**) is used as a chiral ligand. Most importantly, the reaction proceeds with the same high enantioselectivity for the combination of various dienophiles and dienes even when 5–10 mol % of the chiral titanium reagent is employed in the presence of molecular sieves 4A (eqs 7 and 8).[11]

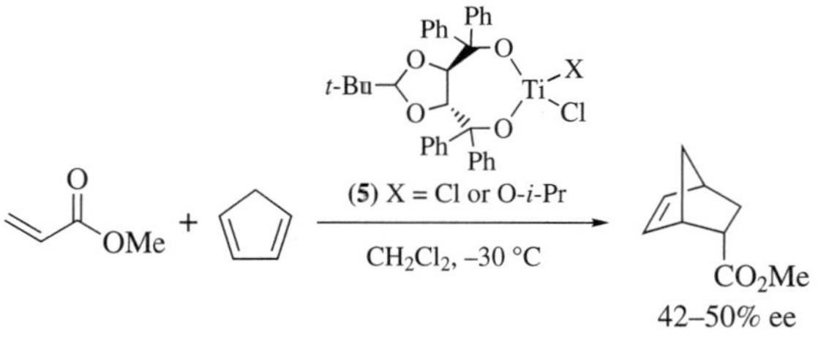

(6)

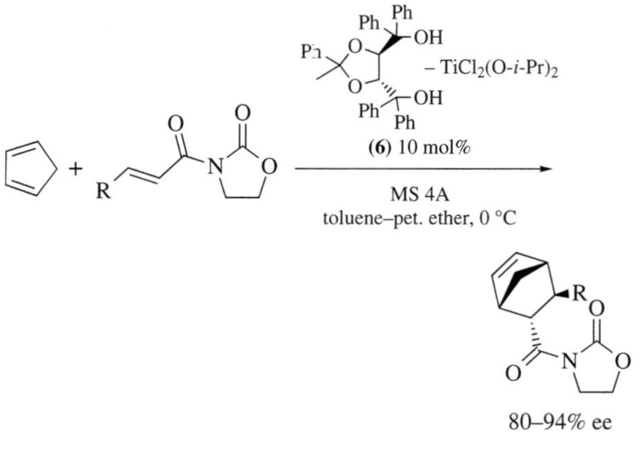

(7)

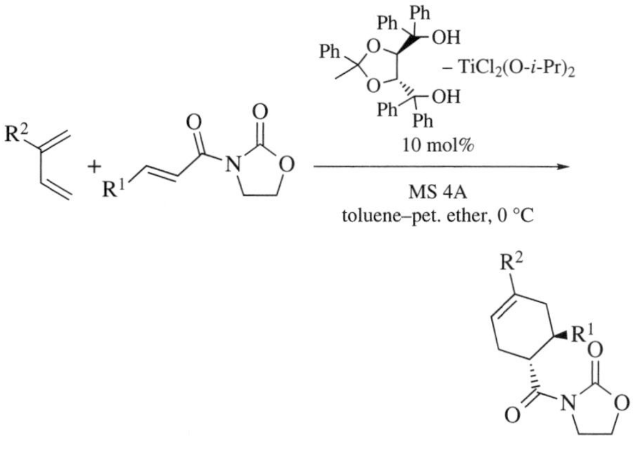

(8)

91–96% ee

The origin of the high enantioselectivity is found to lie in an attractive π–π interaction between the aryl group of the diol moiety and the dienophiles. Replacement of the phenyl group by a 3,5-dimethylphenyl group (as shown in **7**), which has higher π-basicity than a phenyl group, affords improved enantioselectivities in some cases (eq 9).[12]

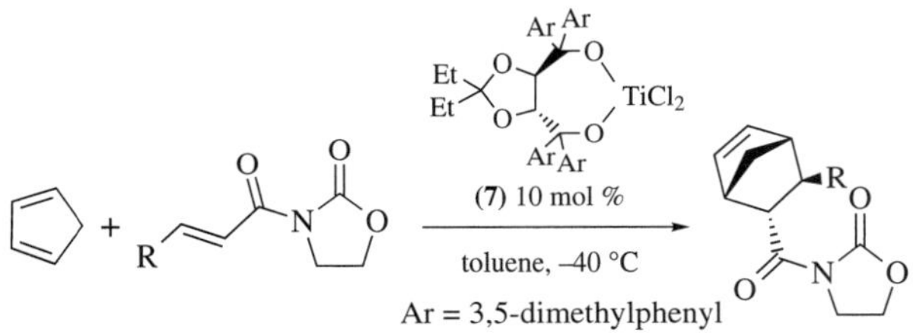

(9)

91–95% ee

The chiral titanium reagent (**6**) also catalyzes the [2+2] cycloaddition reaction of 1,3-oxazolidin-2-one derivatives of α,β-unsaturated carboxylic acids and ketene dithioacetals in the presence of MS 4A to give cyclobutanone dithioacetal derivatives with high optical purity (eq 10).[13] Vinyl sulfides, alkynyl sulfides, and 1,2-propadienyl sulfides can also be employed in this reaction to give the corresponding cyclobutanes, cyclobutenes and methylenecyclobutane derivatives with high optical purity (eqs 11 and 12).[13–15]

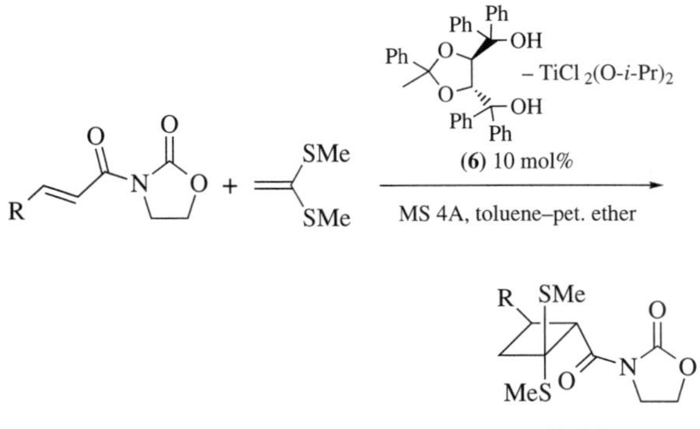

(10)

88–98% ee

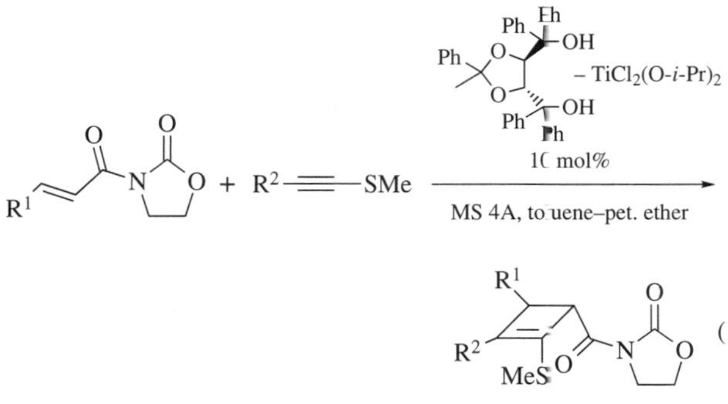

(11)

>98% ee

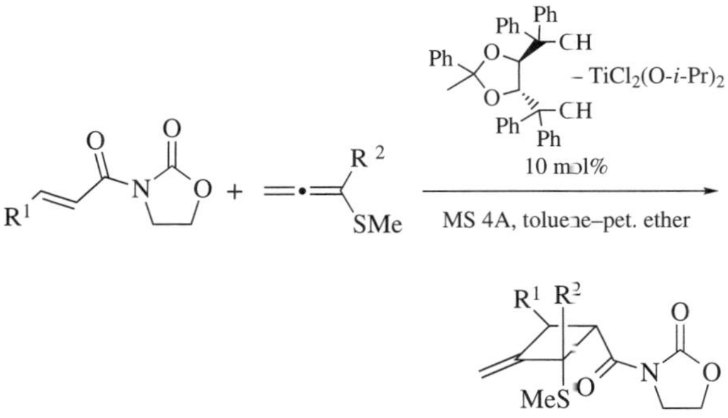

(12)

94–98% ee

By using a stoichiometric amount of the chiral titanium reagent prepared by mixing chiral diol, *Titanium(IV) Chloride*, and titanium tetraisopropoxide, the asymmetric [2+2] cycloaddition reaction of 1,4-benzoquinones and styrenes gives the corresponding cyclobutane derivatives with high optical purity. These rearrange to 2,3-dihydrobenzofuran derivatives on mild acid treatment (eq 13).[16]

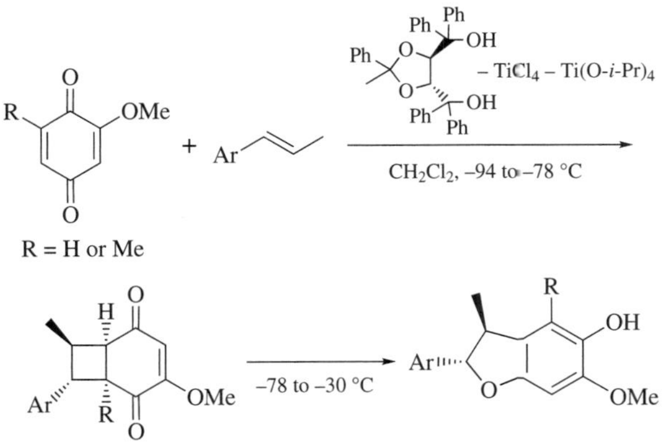

(13)

~90% ee

The asymmetric intramolecular ene reaction of 1,3-oxazolidin-2-one derivatives of a diene carboxylic acid is also promoted by a stoichiometric amount of the chiral titanium reagent (**6**) to give cyclopentane or cyclohexane derivatives with high optical purity (eq 14).[17]

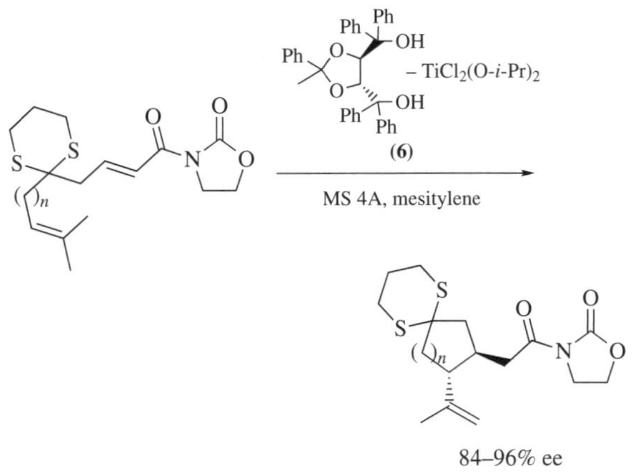

(14)

Other Reactions. Chiral titanates can be employed in several other asymmetric reactions. For example, the chiral titanate (**6**) promotes hydrocyanation of aryl aldehydes by *Cyanotrimethylsilane* at low temperature (−65 °C) to give the corresponding cyanohydrins with high optical purity (eq 15).[18] Alkyl aldehydes are also converted into their cyanohydrins in high optical purity by employing the chiral titanium dicyanide species prepared in situ from the chiral titanate (**6**) and TMSCN at rt (eq 16).[18]

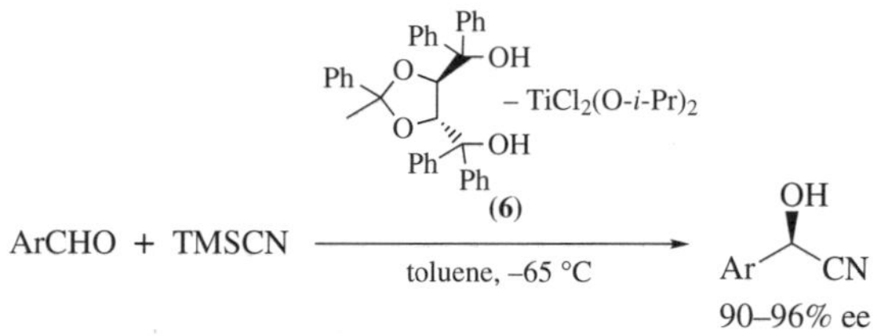

(15)

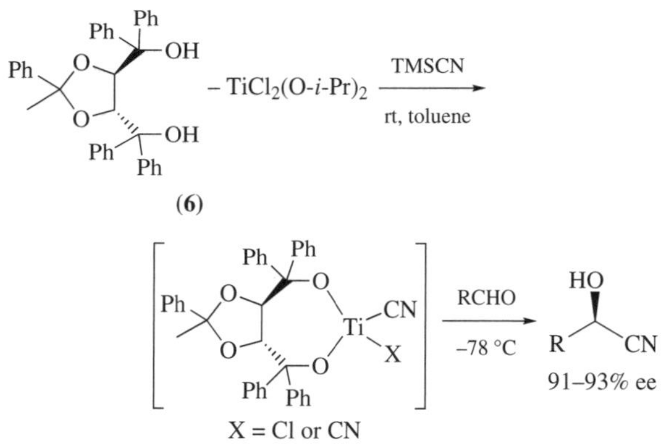

(16)

In the presence of a catalytic amount of the chiral titanium reagent (**8**) prepared from titanium tetraisopropoxide and the (*R*)-1,4-diol, kinetic resolution of *S*-(2-pyridyl) thioesters of α-aryl carboxylic acids is achieved with high relative rate of both the enantiomers to give the (*R*)-isopropyl esters with high optical purity (eq 17).[19]

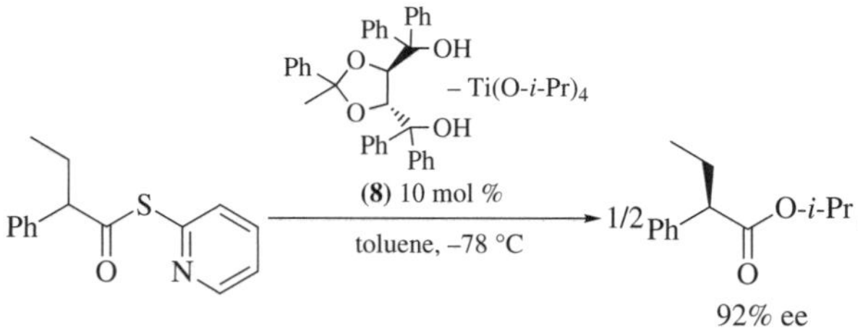

(17)

The enantioselective iodolactonization of α-hydroxy carboxylic acid derivatives is achieved by using a stoichiometric amount of the chiral titanium reagent (**8**) (eq 18).[20]

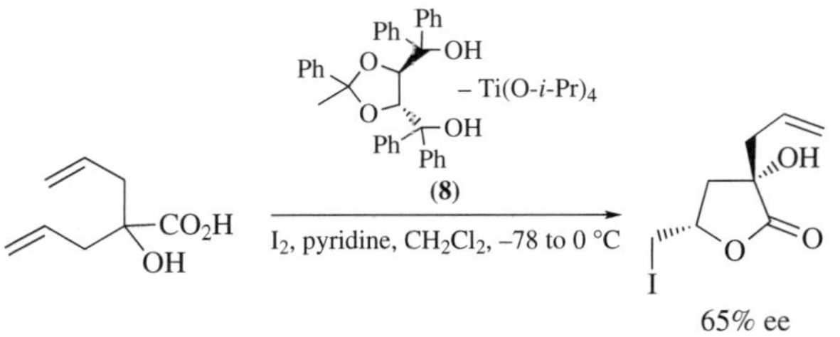

(18)

Related Reagents. Chloro(η^5-cyclopentadienyl)[(4*R*,*trans*)-2,2-dimethyl-α,α,α′,α′-tetraphenyl-1,3-dioxolane-4,5-dimethanolato(2−)-O^{α}, $O^{\alpha'}$]titanium; Dichlorotitanium Diisopropoxide; 2,2-Dimethyl-α,α, α′,α′-tetraphenyl-1,3-dioxolane-4,5-dimethanolatotitanium Diisopropoxide.

1. Duthaler, R. O.; Hafner, A. *Chem. Rev.* **1992**, *92*, 807.
2. Narasaka, K. *Synthesis* **1991**, 1.
3. Seebach, D.; Weidmann, B.; Widler, L. In *Modern Synthetic Methods*; Scheffold, R.; Ed.; Salle: Frankfurt, 1983; Vol. 3, pp 217–353.
4. Beck, A. K.; Bastani, B.; Plattner, D. A.; Petter, W.; Seebach, D.; Braunschweiger, H.; Gysi, P.; La Vecchia, L. *Chimia* **1991**, *45*, 238 (*Chem. Abstr.* **1991**, *115*, 279 866y).
5. Narasaka, K.; Iwasawa, N.; Inoue, M.; Yamada, T.; Nakashima, M.; Sugimori, J. *J. Am. Chem. Soc.* **1989**, *111*, 5340.
6. Seebach, D.; Beck, A. K.; Imwinkelried, R.; Roggo, S.; Wonnacott, A. *Helv. Chim. Acta* **1987**, *70*, 954.
7. Schmidt, B.; Seebach, D. *Angew. Chem., Int. Ed. Engl.* **1991**, *30*, 99.
8. (a) Riediker, M.; Duthaler, R. O. *Angew. Chem., Int. Ed. Engl.* **1989**, *28*, 494. (b) Hafner, A.; Duthaler, R. O.; Marti, R.; Rihs, G.; Rothe-Streit, P.; Schwarzenbach, F. *J. Am. Chem. Soc.* **1992**, *114*, 2321.
9. (a) Schmidt, B.; Seebach, D. *Angew. Chem., Int. Ed. Engl.* **1991**, *30*, 99. (b) Schmidt, B.; Seebach, D. *Angew. Chem., Int. Ed. Engl.* **1991**, *30*, 1321. (c) Seebach, D.; Plattner, D. A.; Beck, A. K.; Wang, Y. M.; Hunziker, D.; Petter, W. *Helv. Chim. Acta* **1992**, *75*, 2171.
10. (a) Seebach, D.; Behrendt, L.; Felix, D. *Angew. Chem., Int. Ed. Engl.* **1991**, *30*, 1008. (b) Bussche-Hunnefeld, J. L.; Seebach, D. *Tetrahedron* **1992**, *48*, 5719.
11. (a) Narasaka, K.; Inoue, M.; Okada, N. *Chem. Lett.* **1986**, 1109. (b) Narasaka, K.; Inoue, M.; Yamada, T. *Chem. Lett.* **1986**, 1967. (c) Narasaka, K.; Inoue, M.; Yamada, T.; Sugimori, J.; Iwasawa, N. *Chem. Lett.* **1987**, 2409. (d) Iwasawa, N.; Sugimori, J.; Kawase, Y.; Narasaka, K. *Chem. Lett.* **1989**, 1947. (e) Narasaka, K.; Tanaka, H.; Kanai, F. *Bull. Chem. Soc. Jpn.* **1991**, *64*, 387. (f) Narasaka, K.; Yamamoto, I. *Tetrahedron* **1992**, *48*, 5743.
12. Corey, E. J.; Matsumura, Y. *Tetrahedron Lett.* **1991**, *32*, 6289.
13. (a) Hayashi, Y.; Narasaka, K. *Chem. Lett.* **1989**, 793. (b) Narasaka, K.; Hayashi, Y.; Shimadzu, H.; Niihata, S. *J. Am. Chem. Soc.* **1992**, *114*, 8869.
14. Hayashi, Y.; Narasaka, K. *Chem. Lett.* **1990**, 1295.
15. Hayashi, Y.; Niihata, S.; Narasaka, K. *Chem. Lett.* **1990**, 2091.
16. Engler, T. A.; Letavic, M. A.; Reddy, J. P. *J. Am. Chem. Soc.* **1991**, *113*, 5068.
17. (a) Narasaka, K.; Hayashi, Y.; Shimada, S. *Chem. Lett.* **1988**, 1609. (b) Narasaka, K.; Hayashi, Y.; Shimada, S.; Yamada, J. *Isr. J. Chem.* **1991**, *31*, 261.
18. (a) Narasaka, K.; Yamada, T.; Minamikawa, H. *Chem. Lett.* **1987**, 2073. (b) Minamikawa, H.; Hayakawa, S.; Yamada, T.; Iwasawa, N.; Narasaka, K. *Bull. Chem. Soc. Jpn.* **1988**, *61*, 4379.

19. Narasaka, K.; Kanai, F.; Okudo, M.; Miyoshi, N. *Chem. Lett.* **1989**, 1187.
20. Kitagawa, O.; Hanano, T.; Tanabe, K.; Shiro, M.; Taguchi, T. *Chem. Commun./J. Chem. Soc., Chem. Commun.* **1992**, 1005.

K. Narasaka & N. Iwasawa
The University of Tokyo, Japan

(*R,R*)-2,5-Dimethylborolane[1]

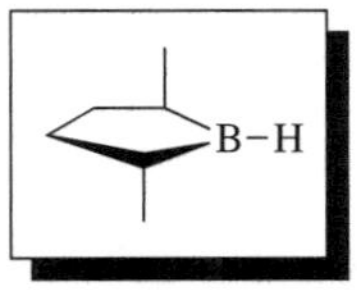

[97011-90-0] $C_6H_{13}B$ (MW 96.00)

(reagent for asymmetric hydroboration;[2] also used as an auxiliary for asymmetric ketone reduction,[3] aldol,[4] and crotylboration[5] reactions; derived reagents are used in double asymmetric syntheses[6])

Physical Data: the most stable precursor is a complex with (*S*)-(+)-prolinol: mp 225–226 °C; $[\alpha]_D^{21}$ +23.2° (*c* 1.28, $CHCl_3$).
Solubility: sol pentane, Et_2O, THF.
Analysis of Reagent Purity: optical purity is assayed by oxidation (NaOMe, H_2O_2) to give 2,5-hexanediol, which has a maximum published rotation $[\alpha]^{23}_D$ −35.6° (*c* 8.29, $CHCl_3$). Derivitization of the diol as the bis-MTPA ester followed by HPLC analysis gives the relative amounts of (*R,R*)-, (*R,S*)-, and (*S,S*)-diols present.
Preparative Methods: a ca. 1:1 mixture of *cis*- and *trans*-*B*-methoxy-2,5-dimethylborolanes is first obtained by reaction of the Grignard reagent derived from 2,5-dibromohexane with diethylaminodichloroborane, followed by acidic methanolysis (eq 1).[2]

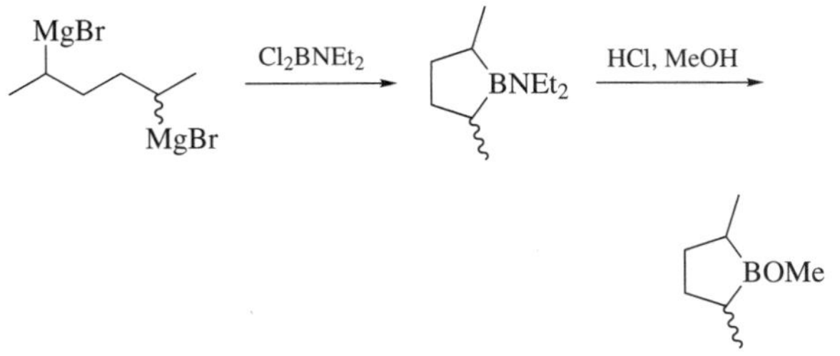

(1)

The *cis* isomer is then selectively complexed with *N,N*-dimethylaminoethanol, and the resolvable *trans* isomer is isolated by vacuum transfer (eq 2).

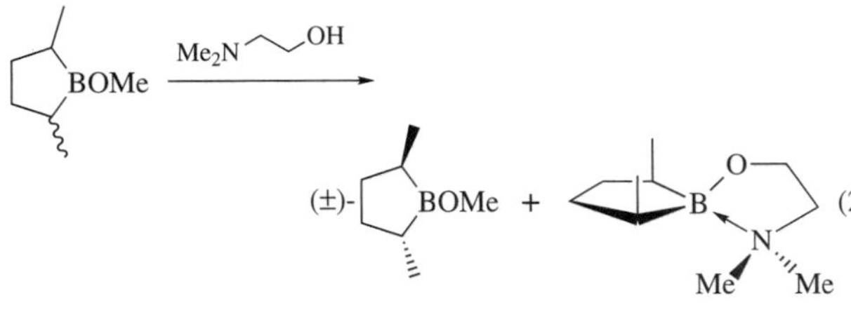

(2)

Reagent (**1**) is obtained by treatment of the racemic *trans* isomer with 0.45 equiv of (*S*)-prolinol, followed by vacuum transfer of the volatile fraction containing mostly (*S,S*)-2,5-dimethyl-*B*-methoxyborolane (eq 3). Through similar manipulation, (**2**) is obtained (eq 4).

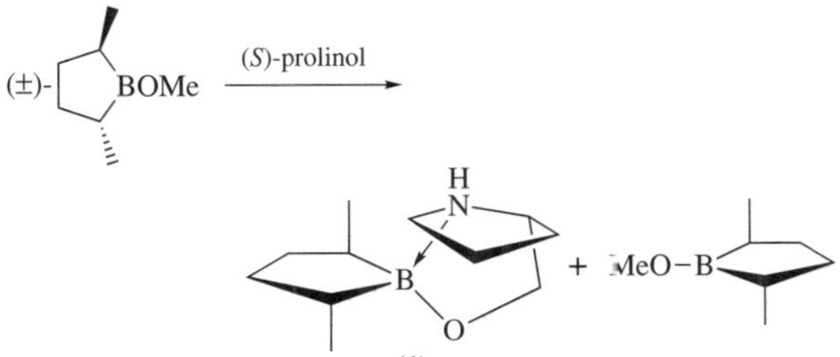

(**1**) (3)

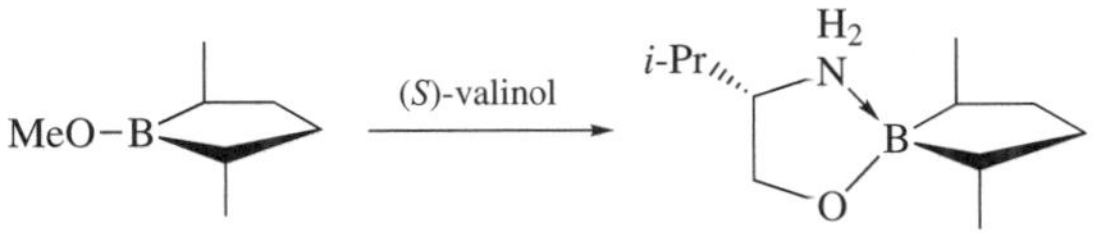

(**2**) (4)

Handling, Storage, and Precautions: (*R,R*)-2,5-dimethylborolane is stored as an air- and moisture-stable complex (**1**) with (*S*)-(+)-prolinol. The (*S,S*)-isomer is stored as a stable complex with either (*R*)-(−)-prolinol or as a complex (**2**) with (*S*)-(+)-valinol, which is more readily available. The reagent and many of its derivatives are extremely air- and moisture-sensitive, and may ignite when exposed to air. Precautions for the handling of such materials should be rigorously followed.[7]

(**1**) (**2**)

Asymmetric Hydroboration[2]. For reaction with a prochiral alkene, (*R,R*)-2,5-dimethyl-*B*-methoxyborolane is liberated from (**1**) and a standard solution of the corresponding lithium dihydridoborate in ether is prepared (eq 5). Hydroboration is effected by addition of *Iodomethane* to the solution of dihydridoborate and alkene (eq 6). After oxidation, chiral secondary alcohols of high enantiomeric purity and predictable configuration are obtained from *cis, trans*, and trisubstituted alkenes. As is the case with other known asymmetric hydroborating agents,[8] 2-methyl-1-alkenes react with low asymmetric induction.

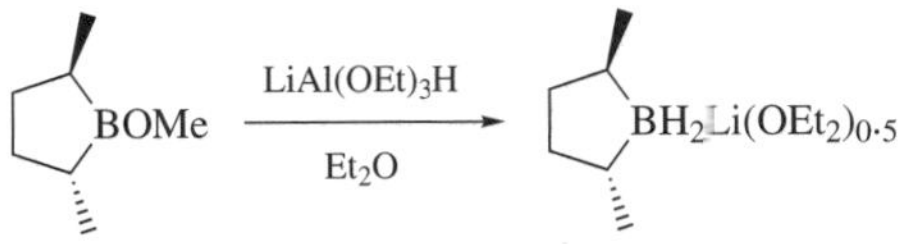

(5)

Asymmetric Reduction of Ketones.[3a,b] A reagent system consisting of (*R,R*)-2,5-dimethylborolane (1.0 equiv) and the corresponding borolanyl mesylate (0.2 equiv) reduces a variety of prochiral ketones with asymmetric induction in the range of 80–100% ee. The reagent system is prepared in situ by addition of 1.4 equiv of *Methanesulfonic Acid* to a solution of the lithium dihydridoborate, prepared as in eq 5 above (eq 7).

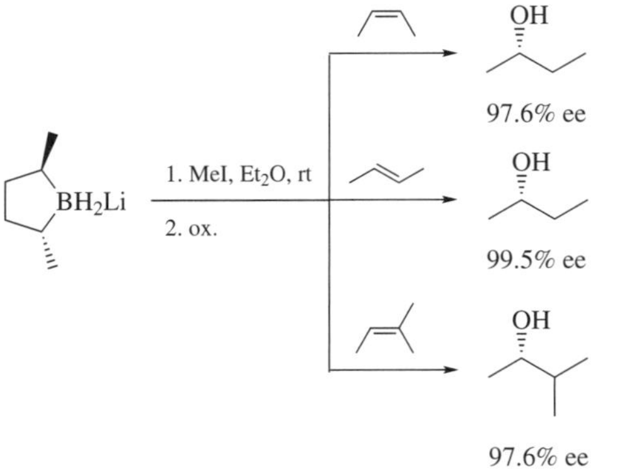

(6)

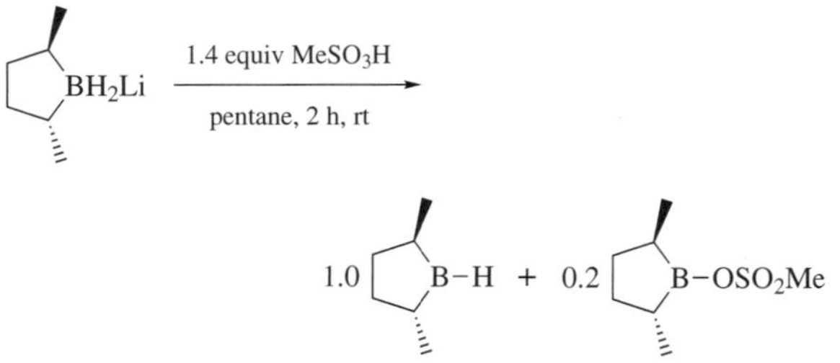

(7)

Kinetic and molecular modeling studies support the view that asymmetric ketone reduction proceeds through reaction of the borolane with a complex formed by coordination of the borolanyl mesylate *syn* to the smaller alkyl group (R^1) of the ketone (eq 8).[3b] After reaction is complete, the chiral borolane moiety is recovered as a crystalline complex with 2-amino-2-methyl-1-propanol (eq 9).

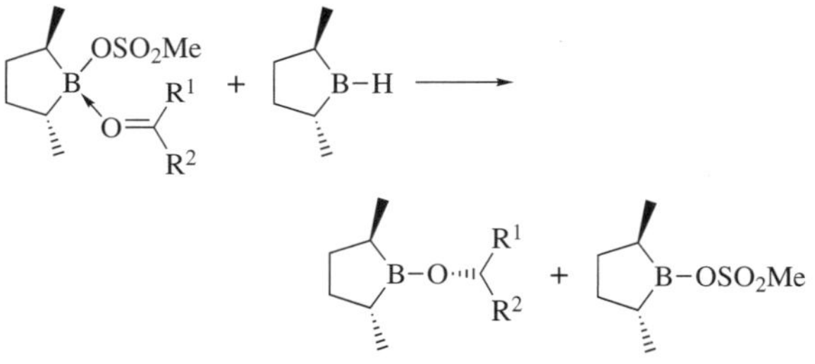
(8)

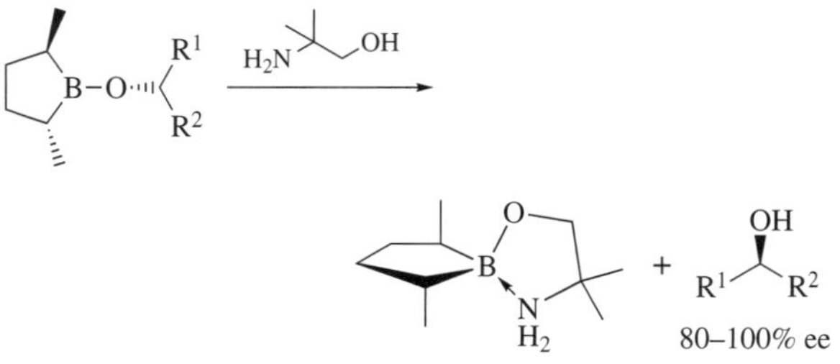

(9)

Asymmetric Aldol Reactions[4]**.** For use in asymmetric aldol reactions, the dihydridoborate (eq 5) is converted to the borolanyl triflate (eq 10). The derived boron enolates of 1,1-diethylpropyl propanethioate (R = Me) and ethanethioate (R = H) react with representative aldehydes to give β-hydroxythioates with good to excellent enantioselectivity (eq 11). In the propanethioate series (R = Me), the observed 2,3-*anti* selectivity (*anti*:*syn* ≥ 30:1) is related to the preponderance of *E*(O)-geometry in the enolate.[9] The 1,1-diethylpropyl group of the thioate was selected to maximize the *E*(O):*Z*(O) ratio. Due to their intrinsically high enantioselectivity, the above enolates undergo highly diastereoselective aldol reactions with chiral aldehydes.[10] These 'double asymmetric'[6] reactions have been employed in natural product syntheses.[11]

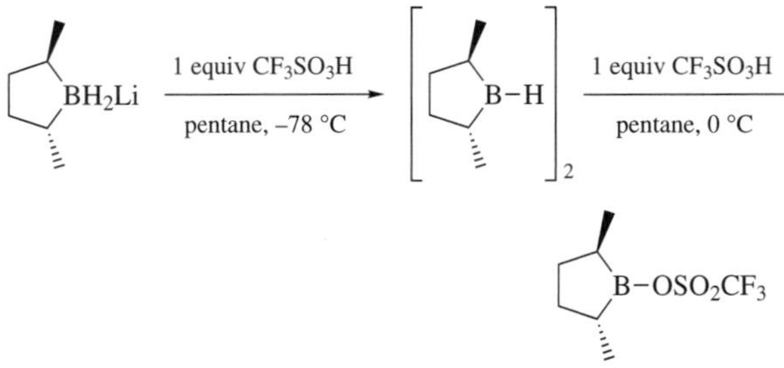

(10)

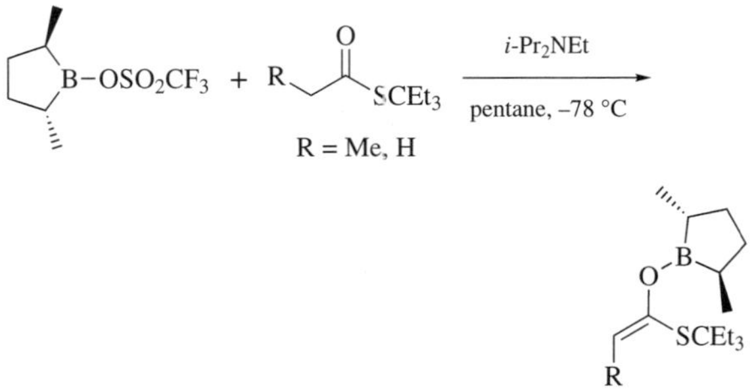

(11)

The borolanyl triflate (eq 10) has also been employed to form the chiral boron enolates of methyl ketones which have additional chiral centers present in their carbon framework. Reaction of these enolates with chiral aldehydes constitutes a 'triple asymmetric synthesis', in which the approximate multiplicativity of the three diastereofacial selectivities appears to be valid.[12]

Asymmetric Crotylboration[5]**.** Reagents for crotylboration are prepared from 2,5-dimethyl-*B*-methoxyborolane (eq 5) by addition of (*Z*)- or (*E*)-crotylpotassium under standard conditions. Reactions with representative achiral aldehydes are 93–96% diastereoselective and 86–97% enantioselective for the major diastereomer (eqs eq 12 and eq 13). Results with chiral aldehydes conform to the rule of double asymmetric synthesis.[6]

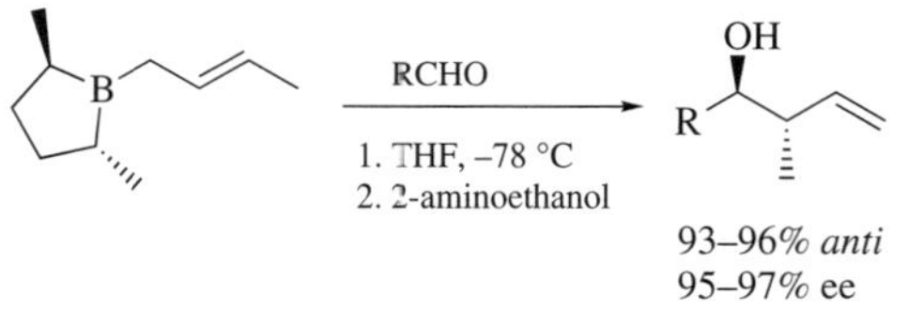

(12)

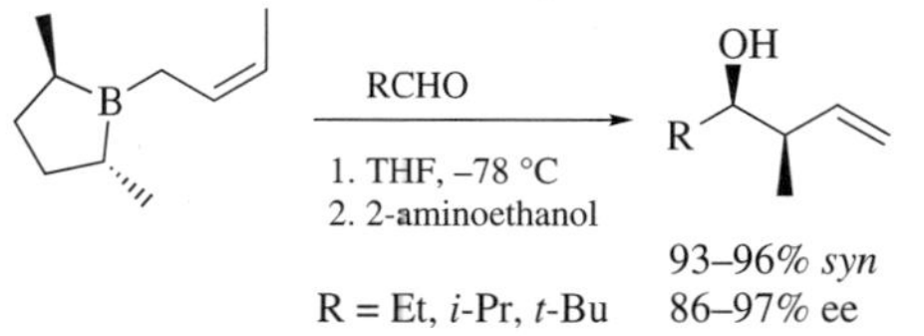

(13)

Related Reagents. Diisopinocampheylborane; Dilongifolylborane; Monoisopinocampheylborane.

---

1. (a) Roush, W. R. *Comprehensive Organic Synthesis* **1991**, *2*, Chapter 1.1 (b) Kim, B. M.; Williams, S. F.; Masamune, S. *Comprehensive Organic Synthesis* **1991**, *2*, Chapter 1.7. (c) Smith, K.; Pelter, A. *Comprehensive Organic Synthesis* **1991**, *8*, Chapter 3.10. (d) Nishizawa, M.; Noyori, R. *Comprehensive Organic Synthesis* **1991**, *8*, Chapter 1.7

2. Masamune, S.; Kim, B. M.; Petersen, J. S.; Sato, T.; Veenstra, S. J.; Imai, T. *J. Am. Chem. Soc.* **1985**, *107*, 4549.

3. (a) Imai, T.; Tamura, T.; Yamamuro, A.; Sato, T.; Wollmann, T. A.; Kennedy, R. M.; Masamune, S. *J. Am. Chem. Soc.* **1986**, *108*, 7402. (b) Masamune, S.; Kennedy, R. M.; Petersen, J. S. *J. Am. Chem. Soc.* **1986**, *108*, 7404.

4. Masamune, S.; Sato, T.; Kim, B. M.; Wollmann, T. A. *J. Am. Chem. Soc.* **1986**, *108*, 8279.

5. Garcia, J.; Kim, B. M.; Masamune, S. *J. Org. Chem.* **1987**, *52*, 4831.

6. Masamune, S.; Choy, W.; Petersen, J. S.; Sita, L. R. *Angew. Chem., Int. Ed. Engl.* **1985**, *24*, 1.

7. Brown, H. C.; Kramer, G. W.; Levy, A. B.; Midland, M. M. In *Organic Synthesis via Boranes*; Wiley: New York, 1975.

8. Brown, H. C.; Jadhav, P. K. In *Asymmetric Synthesis*; Morrison, J. D.; Ed.; Academic: New York, 1983; Vol. 2, Chapter 1.

9. Masamune, S. *Heterocycles* **1984**, *21*, 107.

10. Short, R. P.; Masamune, S. *Tetrahedron Lett.* **1987**, *28*, 2841.

11. (a) Blanchette, M. A.; Malamas, M. S.; Nantz, M. H.; Roberts, J. C.; Somfai, P.; Whritenour, D. C.; Masamune, S. *J. Org. Chem.* **1989**, *54*, 2817. (b) Kageyama, M.; Tamura, T.; Nantz, M. H.; Roberts, J. C.; Somfai, P.; Whritenour, D. C.; Masamune, S. *J. Am. Chem. Soc.* **1990**, *112*, 7407.

12. Duplantier, A. J.; Nantz, M. H.; Roberts, J. C.; Short, R. P.; Somfai, P.; Masamune, S. *Tetrahedron Lett.* **1989**, *30*, 7357.

Robert P. Short
Polaroid Corporation, Cambridge, MA, USA

(*S*)-*N,N*-Dimethyl-*N′*-(1-*t*-butoxy-3-methyl-2-butyl)formamidine[1,2]

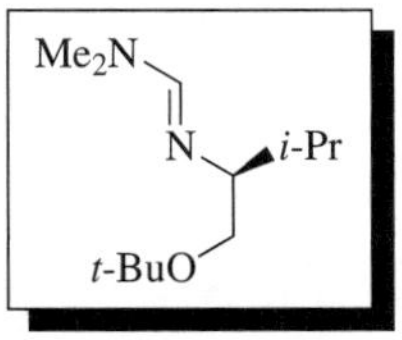

(*E*)-(*S*)		
[114318-94-4]	$C_{12}H_{26}N_2O$	(MW 214.40)
(*S*)		
[66919-83-3]		
(*R*)		
[90482-06-7]		

(chiral auxiliary for derivatization, directed lithiation, and asymmetric alkylation adjacent to nitrogen of benzylic or allylic secondary amines by formamidine exchange, metalation, alkylation, and hydrolysis[1,2])

Physical Data: bp 55–65 °C/0.05 mmHg; $[\alpha]_D$ −59.6° (*c* 3.5, EtOH).[2]

Preparative Methods: easily prepared from (*S*)-valinol in a high yielding four-step procedure (eq 1).[2] Other chiral formamidines can be prepared and used successfully in the methodology outlined here (eq 2).[2,3]

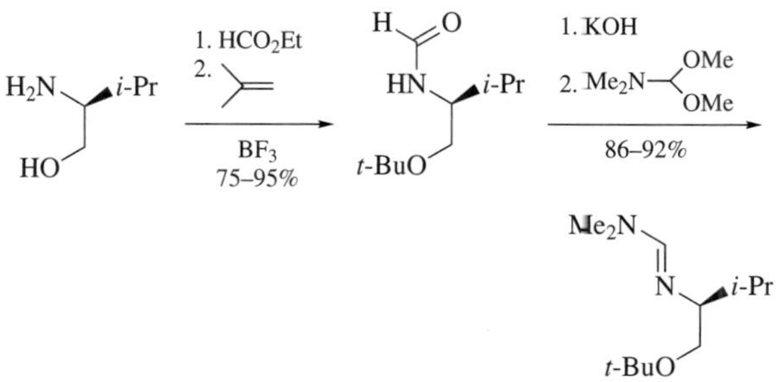

(1)

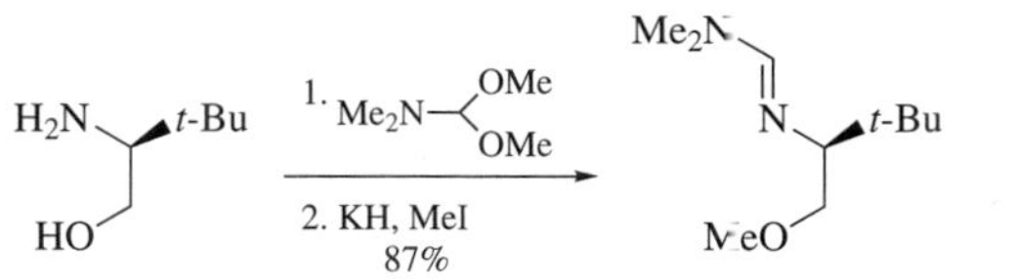

(2)

Handling, Storage, and Precautions: store under argon at rt; use in a fume hood.

Introduction. In the equations, 'VBE' will be used to depict the valinol *t*-butyl ether portion of the formamidine, while the *t*-leucinol methyl ether portion will be abbreviated 'LME.'

Upon heating the title formamidine with a secondary amine, dimethylamine is extruded, affording a chiral formamidine derivative of the original amine (eq 3).[4a]

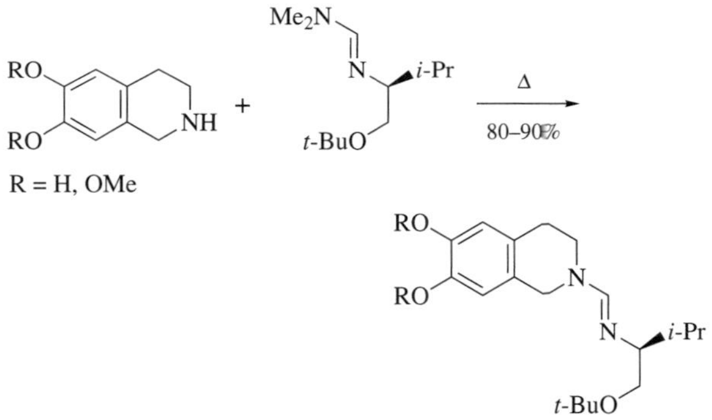

(3)

Deprotonation and alkylation followed by formamidine removal allows entry to a host of isoquinoline alkaloids (eq 4).[3b,4]

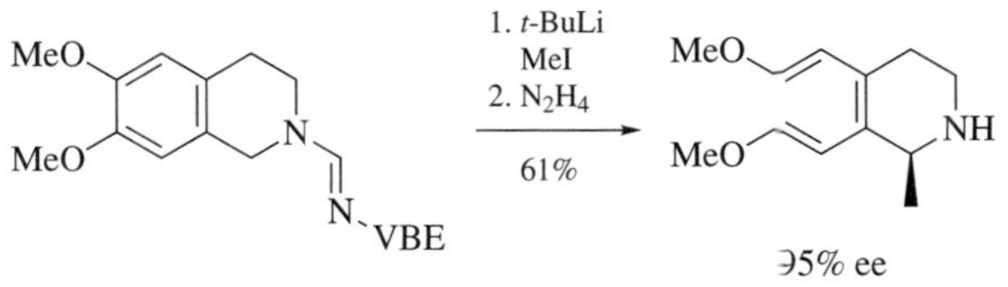

(4)

This protocol has also been an avenue to a variety of indole alkaloids (eq 5).[5]

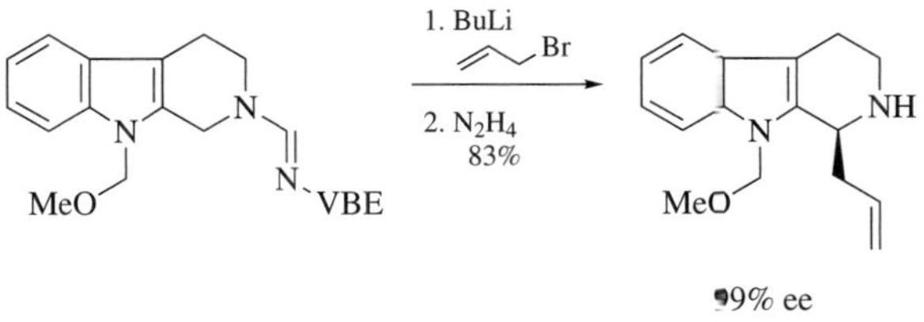

(5)

Chiral 1-alkyl-2-benzazepines can be formed by utilization of the same method (eq 6).[3a]

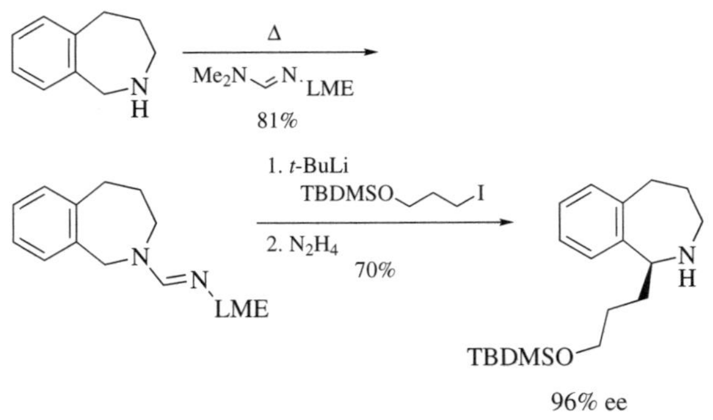

(6)

This strategy also works well for the asymmetric alkylation of 3-pyrrolines (eq 7)[6] and tetrahydropyridines (eq 8).[7]

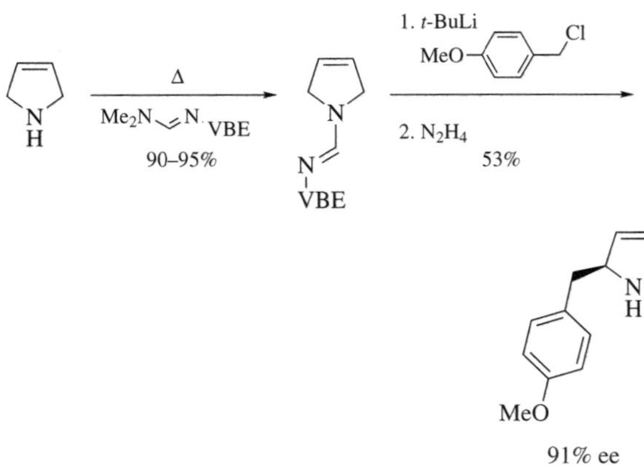

(7)

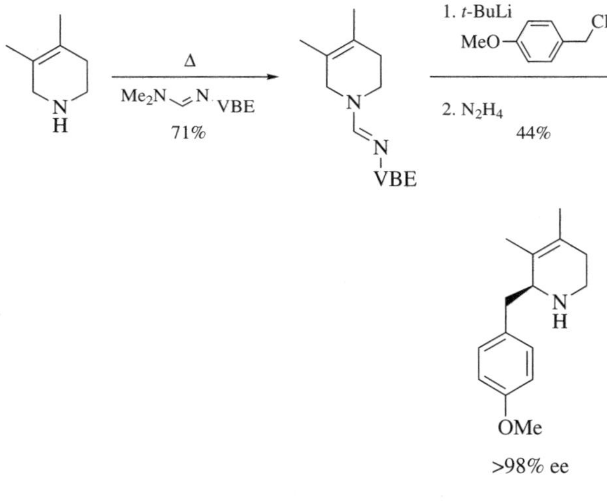

(8)

The mechanistic pathway for these asymmetric alkylations and the configurational stability of the chiral lithioformamidines have been investigated.[8] A limitation of this strategy is that saturated, cyclic, secondary amines (e.g., pyrrolidines and piperidines) cannot be successfully alkylated in an asymmetric fashion.

Related Reagents. *N*-*t*-Butoxycarbonyl-*N*-methylaminomethyllithium; N′-*t*-Butyl-*N*,*N*-dimethylformamidine; *N*′-*t*-Butyl-*N*-methyl-*N*-trimethylsilylmethylformamidine; *N*,*N*-Dimethylformamide Diethyl Acetal; (*R*)-Methyl 2-*t*-Butyl-3(2*H*)-oxazolecarboxylate; *N*-Nitrosodimethylamine.

1. (a) Meyers, A. I. *Tetrahedron* **1992**, *48*, 2589. (b) Meyers, A. I.; Highsmith, T. K. In *Advances in Heterocyclic Natural Product Synthesis*; Pearson, W. H., Ed.; JAI Press: Greenwich, CT, 1990. (c) Meyers, A. I. *Aldrichim. Acta* **1985**, *18*, 59. (d) Meyers, A. I. *Heterocycles* **1984**, *21*, 360.
2. (a) Dickman, D. A.; Boes, M.; Meyers, A. I. *Org. Synth., Coll. Vol.* **1993**, *8*, 204. (b) Meyers, A. I.; Boes, M.; Dickman, D. A. *Org. Synth., Coll. Vol.* **1993**, *8*, 573.
3. (a) Meyers, A. I.; Hutchings, R. H. *Tetrahedron* **1993**, *49*, 1807. (b) Meyers, A. I.; Elworthy, T. R. *J. Org. Chem.* **1992**, *57*, 4732.
4. (a) Meyers, A. I.; Dickman, D. A.; Boes, M. *Tetrahedron* **1987**, *43*, 5095. (b) Meyers, A. I.; Sielecki, T. M.; Crans, D. C.; Marshman, R. W.; Nguyen, T. H. *J. Am. Chem. Soc.* **1992**, *114*, 8483. (c) Sielecki, T. M.; Meyers, A. I. *J. Org. Chem.* **1992**, *57*, 3673. (d) Guiles, J. W.; Meyers, A. I. *J. Org. Chem.* **1991**, *56*, 6873. (e) Meyers, A. I.; Sielecki, T. M., *J. Am. Chem. Soc.* **1991**, *113*, 2789. (f) Gottlieb, L.; Meyers, A. I. *J. Org. Chem.* **1990**, *55*, 5559. (g) Meyers, A. I.; Guiles, J. *Heterocycles* **1989**, *28*, 295.
5. (a) Meyers, A. I.; Highsmith, T. K.; Buonora, P. T. *J. Org. Chem.* **1991**, *56*, 2960. (b) Beard, R. L.; Meyers, A. I. *J. Org. Chem.* **1991**, *56*, 2091.
6. (a) Meyers, A. I.; Dupre, B. *Heterocycles* **1987**, *25*, 113. (b) Warmus, J. S.; Dilley, G. J.; Meyers, A. I. *J. Org. Chem.* **1993**, *58*, 270.
7. Meyers, A. I.; Dickman, D. A.; Bailey, T. R. *J. Am. Chem. Soc.* **1985**, *107*, 7974.
8. (a) Castonguay, L. A.; Guiles, J. W.; Rappé, A. K.; Meyers, A. I. *J. Org. Chem.* **1992**, *57*, 3819. (b) Meyers, A. I.; Warmus, J. S.; Gonzalez, M. A.; Guiles, J.; Akahane, A. *Tetrahedron Lett.* **1991**, *32*, 5509. (c) Meyers, A. I.; Guiles, J.; Warmus, J. S.; Gonzalez, M. A. *Tetrahedron Lett.* **1991**, *32*, 5505. (d) Meyers, A. I.; Gonzalez, M. A.; Struzka, V.; Akahane, A.; Guiles, J.; Warmus, J. S. *Tetrahedron Lett.* **1991**, *32*, 5501.

Todd D. Nelson & Albert I. Meyers
Colorado State University, Fort Collins, CO, USA

(−)-(*S*,*S*)-α, α′-Dimethyldibenzylamine

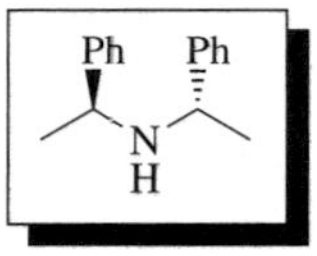

[56210-72-1] $C_{16}H_{19}N$ (MW 225.36)

(starting material for the formation of chiral amide reagents; useful in the stereospecific deprotonation of prochiral ketones, and as a chirality transfer agent in the reactions of prochiral enolates; stereoselective conjugate addition of organometallic reagents to unsaturated carbonyl systems[1])

Physical Data: (free base): bp 103–105 °C/0.5 mmHg;[2] $[\alpha]_D$ −157° (*c* 2.4, EtOH);[1a] −197.3° (*c* 3.65, benzene);[1b] −187.9° (*c* 6.87, benzene);[3] −171.6° (*c* 6.71, chloroform).[4] (HCl salt): mp >300 °C; $[\alpha]_D$ −84.1° (*c* 3, EtOH);[1a] −72.1° (*c* 2.94, EtOH).[1b]

Solubility: readily sol common organic solvents (ether, THF, chloroform, etc.); insol H_2O.

Form Supplied in: available commercially.

Analysis of Reagent Purity: diastereomeric purity can be assessed by the ^{1}H NMR chemical shift of the methyl groups,[1a] and by GC analysis.[5] Optical purity can be assessed by derivatization with (*R*)-(−)-1-(1-naphthyl)ethyl isocyanate and ^{1}H NMR analysis of the product.[4]

Preparative Methods: minor improvements to the original catalytic hydrogenation procedure[6] have been described (eq 1).[1a]

This method provides (*S,S*)-(−)-(**1**) with an optical purity of only 70%. Enantiomerically pure (*S,S*)-(−)-(**1**) can be obtained by recrystallization of the hydrochloride salt of this enriched material from water[1b] or the benzoate salt from isopropanol.[3] A chemical reduction procedure has also been described that yields optically active (*S,S*)-(**1**) with 74% enantiomeric excess (eq 2).[2] A significant improvement to the former procedures is the diastereoselective hydrogenation of imines catalyzed by rhodium/chiral diphosphines, which yields (*S,S*)-(**1**) with an optical purity of 99.4% (eq 3).[5]

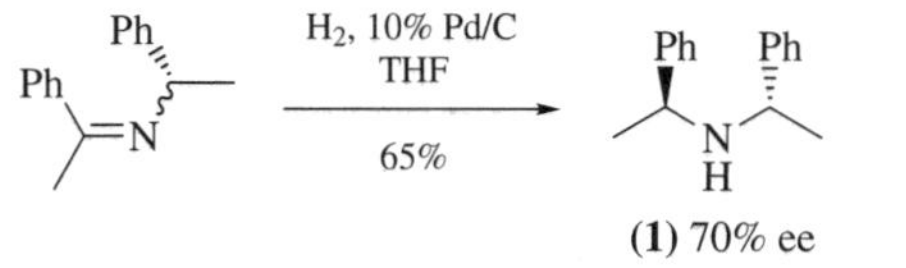

(1)

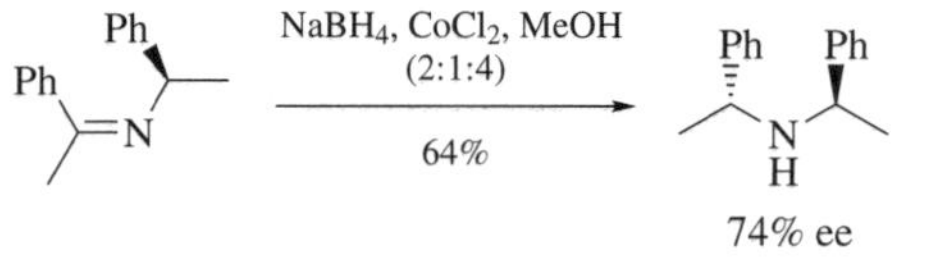

(2)

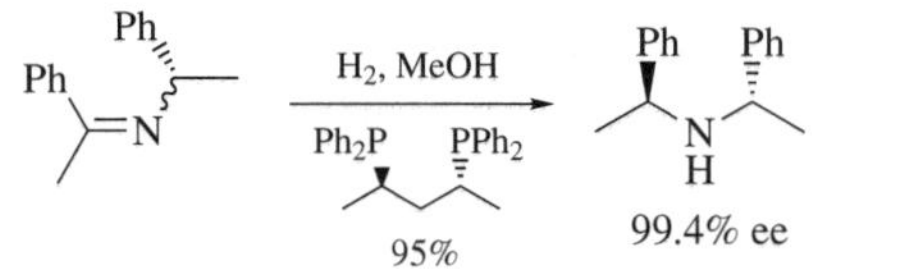

(3)

Purification: the free base can be distilled. The HCl salt can be recrystallized from water, which removes diastereomeric impurities. The benzoate salt can be recrystallized from isopropanol.

Handling, Storage, and Precautions: no special precautions have been noted in the literature. The free base is a clear distillable liquid that should be stored under an inert atmosphere to prevent air oxidation. Long term storage may lead to some coloration of the material.

Introduction. In most cases the (*R,R*) and (*S,S*) enantiomers of (**1**) possess similar synthetic applications. References to both enantiomers have been incorporated into this article, under the heading of (*S,S*). The equations depict the actual enantiomer used in each publication.

Asymmetric Deprotonation/Protonation of Ketones. Lithium amides of chiral amines have been used for performing asymmetric deprotonations of symmetrically substituted (prochiral) ketones.[7,8] The resulting optically active enols or enol derivatives (most frequently enol silanes) are highly versatile synthetic intermediates. Particularly useful for this purpose are chiral amines possessing C_2 symmetry, such as (**1**). For example, reaction of 4-*t*-butylcyclohexanone with the lithium amide of (*R,R*)-(**1**) (readily prepared in situ by treatment of (**1**) with *n-Butyllithium*) is highly stereoselective; the resulting enol silyl ether possesses an 88% ee (eq 4).[9]

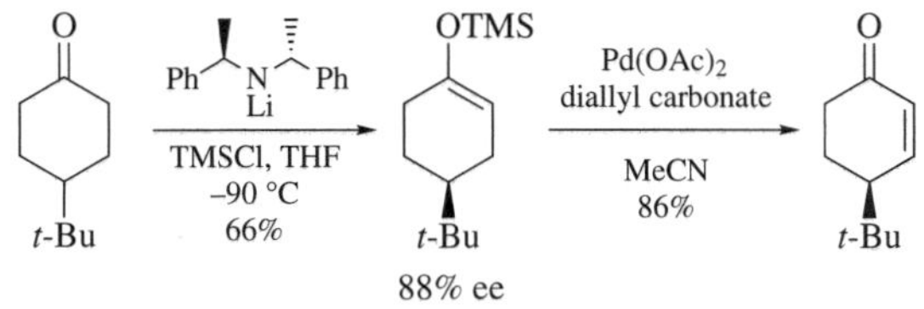

(4)

The most predictable results are obtained with conformationally rigid systems, such as those represented in eqs 5 and 6, which possess axially oriented α-protons.[1c,11] This minimizes complications resulting from the presence of diastereotopic α-protons, although unexpected modes of deprotonation have been described with related chiral amides, which may involve boat conformations.[12] To prevent enolate equilibration (with the resulting loss of stereoselectivity), Corey's internal quench method for enolate trapping with silyl chlorides is frequently used.[13] The stereospecificity of this deprotonation is highly dependent on solvent and temperature conditions. Best results are obtained at −100 °C or lower temperatures, with THF as the solvent.

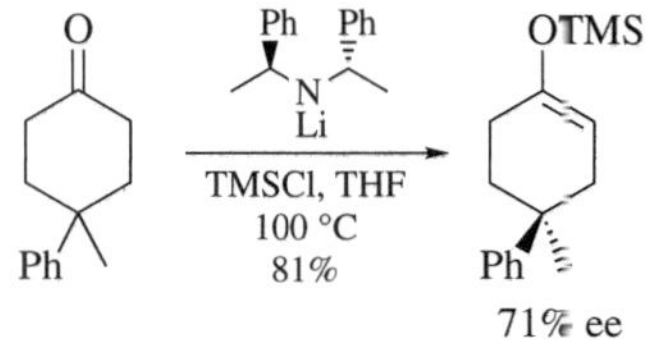

(5)

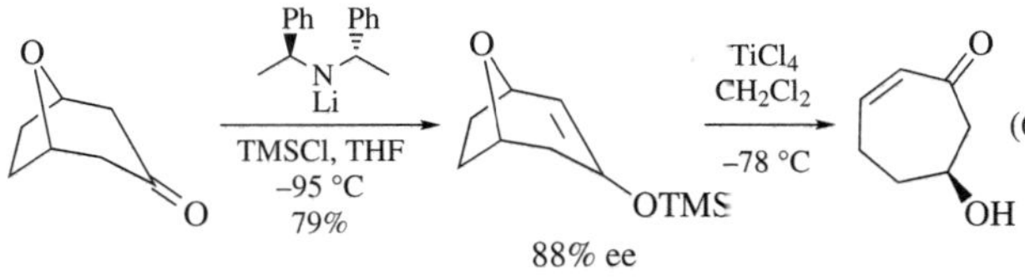

(6)

The lithium amide of (*S,S*)-(**1**) has been used to convert racemic α-substituted ketones into optically active ketones via sequential deprotonation/asymmetric protonation of rigid prochiral enolates. Enantiomeric enrichment may occur during the protonation step as a result of the tight coordination between the enolate and the lithium amide in the form of diastereomeric complexes (eq 7).[14]

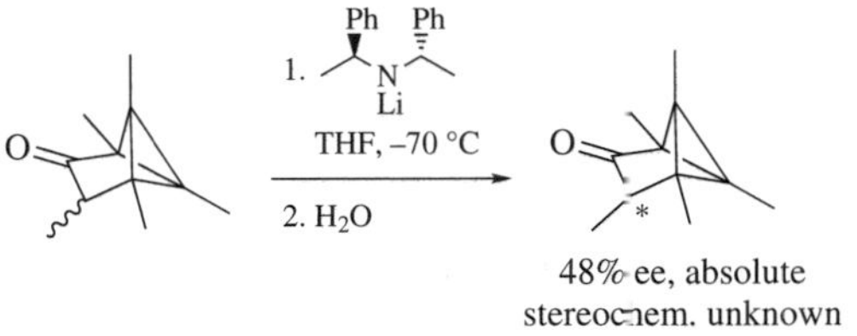

(7)

Alternatively, the enantiomeric enrichment derives from kinetic differences in the rate of deprotonation of the two ketone enantiomers (eq 8).[15] Ether is the best solvent for these reactions.

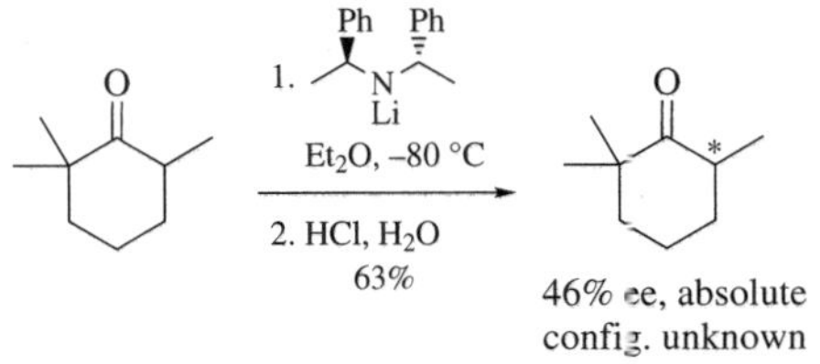

(8)

The lithium amide of (**1**) has also been used to perform the kinetic resolution of racemic lactams by selective kinetic deprotonation of one of the enantiomers, followed by reaction of the partially formed enolates with an electrophile.[16] These procedures have not proven to be particularly useful yet, since high enantiomeric purity is only achieved at low conversions of the starting materials (eq 9).[17]

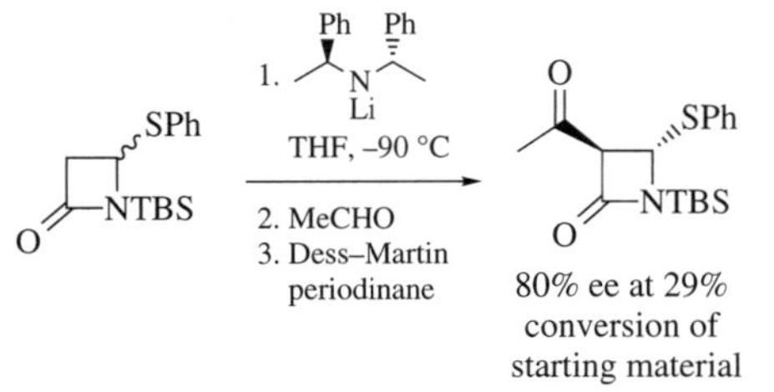

(9)

Stereoselective Alkylation of Prochiral Enolates. A limited amount of work has demonstrated the potential use of chiral amines in inducing stereoselectivity in the alkylation/carboxylation of prochiral enolates. The selectivity of these reactions, like those described above, is highly dependent on solvent and temperature conditions. The use of ether at −196 °C provides optimal results in a particular system (eq 10).[18]

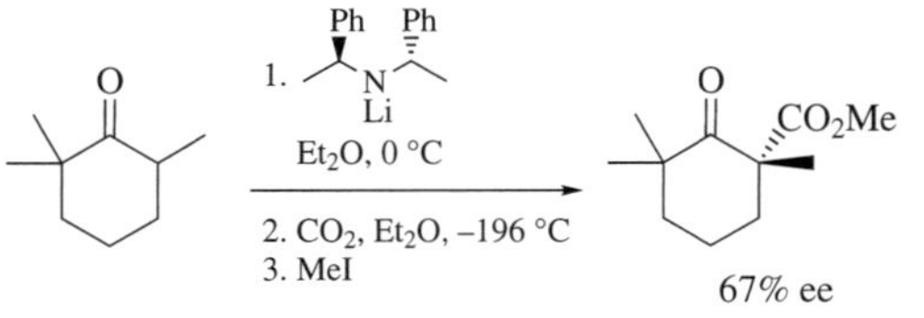

(10)

Asymmetric Induction in Organometallic Reactions. A number of chiral amines have been used as nontransferable ligands for the enantioselective conjugate addition of organocopper reagents, with optical yields as high as 95%.[19–22] (*R*,*R*)-(**1**) has also been used for this purpose, effecting the conjugate addition of organocopper reagents to enones with moderate to high enantioselectivity (eq 11).[23,24] The use of dimethyl sulfide as the solvent for this transformation is critical, since ether solvents produce products of low optical activity.

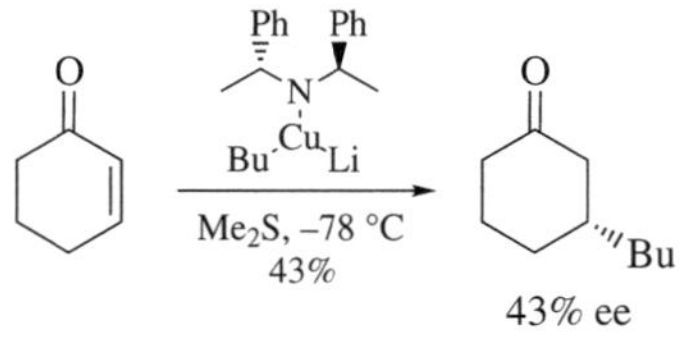

(11)

Although (**1**) itself has not been shown to be useful in the stereospecific 1,2-addition of organometallic reagents to carbonyl compounds, closely related amines, such as (*R*)-(α-methoxymethylbenzyl)-(*S*)-(α-methylbenzyl)amine, have been used to direct the addition of organolithium reagents to benzaldehyde with up to 95% stereoselectivity (eq 12).[25]

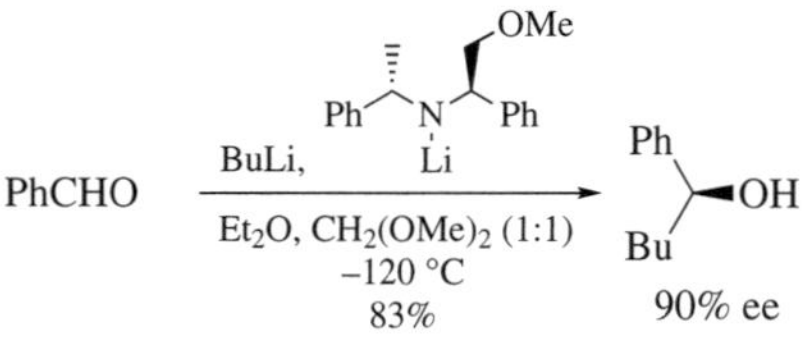

(12)

Enantioselective Conjugate Additions. (*R*,*R*)-(**1**) has been used in the synthesis of (*R*)-β-aminobutanoic acid. The conjugate addition of the lithium amide of (*R*,*R*)-(**1**) to (*E*)-methyl crotonate proceeds with complete diastereoselectivity. Catalytic reduction of the benzyl groups results in the formal stereospecific 1,4-addition of an amino group to an unsaturated ester (eq 13).[26] Although (**1**) has not been used extensively for this type of transformation, a variety of other chiral amines have been used for similar purposes.[8,27,28]

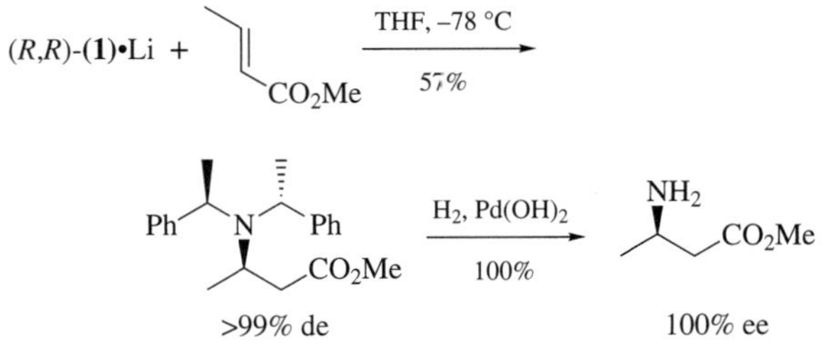

(13)

Chiral Auxiliary. (*R*,*R*)-(**1**) has been used as a chiral auxiliary to direct the stereochemistry of addition of a nucleophile to an acrylate moiety. Almost complete stereoselectivity is achieved in the addition of cyclopentanecarboxylic acid lithium dianion to the α-substituted acrylate substrate (eq 14).[29] This methodology allows stereochemical control at the α-position of a β-amino ester and thus complements the methodology described above[26] for the stereoselective formation of β-substituted β-amino esters.

Other Enantioselective Reactions. Enantioselective epoxide elimination by chiral bases has been demonstrated.[30] More recently, the enantioselective [2,3]-Wittig rearrangement of a 13-membered propargylic allylic ether has been performed using the lithium amide of (*R*,*R*)-(**1**) as the base for deprotonation (eq 15).[4] For this particular substrate, THF is a better solvent than ether, although pentane produces better results in a related transformation (eq 16).[4] In fact, a change in solvent in this type of reaction has been shown to lead to a reversal of the stereoselectivity of the transformation.[4]

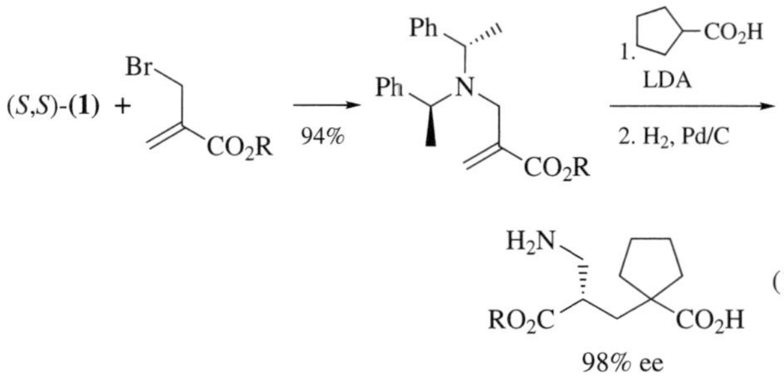

(14)

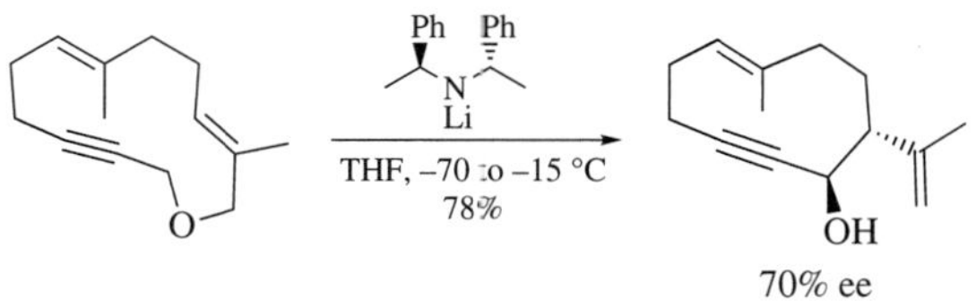

(15)

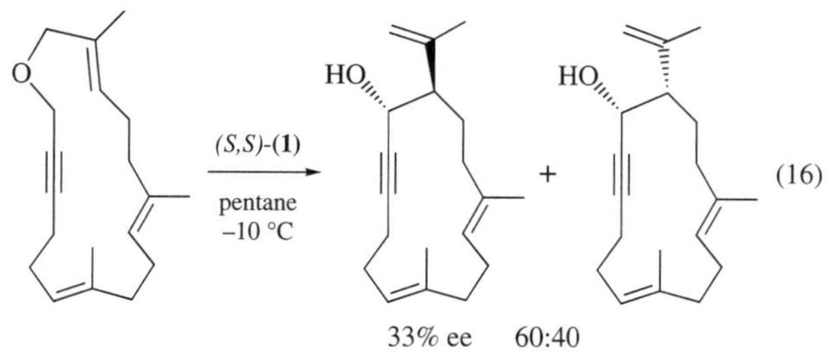

1. (a) Eleveld, M. B.; Hogeveen, H.; Schudde, E. P. *J. Org. Chem.* **1986**, *51*, 3635. (b) Yoshida, T.; Harada, K. *Bull. Chem. Soc. Jpn.* **1972**, *45*, 3706.
2. Periasamy, M.; Devasagayaraj, A.; Satyanarayana, N.; Narayana, C. *Synth. Commun.* **1989**, *19*, 565.
3. Raban, M.; Yamamoto, G. *J. Org. Chem.* **1975**, *40*, 3093.
4. Marshall, J. A.; Lebreton, J. *J. Am. Chem. Soc.* **1988**, *110*, 2925.
5. Lensink, C.; de Vries, J. G. *Tetrahedron: Asymmetry* **1993**, *4*, 215.
6. Overberger, C. G.; Marullo, N. P.; Hiskey, R. G. *J. Am. Chem. Soc.* **1961**, *83*, 1374.
7. Simpkins, N. S. *Chem. Ind. (London)* **1988**, 387.
8. Cox, P. J.; Simpkins, N. S. *Tetrahedron: Asymmetry* **1991**, *2*, 1.
9. Cain, C. M.; Cousins, R. P. C.; Coumbarides, G.; Simpkins, N. S. *Tetrahedron* **1990**, *46*, 523.
10. Honda, T.; Kimura, N.; Tsubuki, M. *Tetrahedron: Asymmetry* **1993**, *4*, 21.
11. Bunn, B. J.; Cox, P. J.; Simpkins, N. S. *Tetrahedron* **1993**, *49*, 207.
12. Sobukawa, M.; Nakajima, M.; Koga, K. *Tetrahedron: Asymmetry* **1990**, *1*, 295.
13. Corey, E. J.; Gross, A. W. *Tetrahedron Lett.* **1984**, *25*, 495.
14. Hogeveen, H.; Zwart, L. *Tetrahedron Lett.* **1982**, *23*, 105.
15. Eleveld, M. B.; Hogeveen, H. *Tetrahedron Lett.* **1986**, *27*, 631.
16. Coggins, P.; Simpkins, N. S. *Synlett* **1991**, 515.
17. Coggins, P.; Simpkins, N. S. *Synlett* **1992**, 313.
18. Hogeveen, H.; Menge, W. M. P. B. *Tetrahedron Lett.* **1986**, *27*, 2767.
19. Bertz, S. H.; Dabbagh, G.; Sundararajan, G. *J. Org. Chem.* **1986**, *51*, 4953.
20. Dieter, R. K.; Tokles, M. *J. Am. Chem. Soc.* **1987**, *109*, 2040.
21. Dieter, R. K.; Lagu, B.; Deo, N.; Dieter, J. W. *Tetrahedron Lett.* **1990**, *31*, 4105.
22. Ahn, K.-H.; Klassen, R. B.; Lippard, S. J. *Organometallics* **1991**, *9*, 3178.
23. Rossiter, B. E.; Eguchi, M. *Tetrahedron Lett.* **1990**, *31*, 965.
24. Rossiter, B. E.; Eguchi, M.; Miao, G.; Swingle, N. M.; Hernández, A. E.; Vickers, D.; Fluckiger, E.; Patterson, R. G.; Reddy, K. V. *Tetrahedron* **1993**, *49*, 965.
25. Eleveld, M. B.; Hogeveen, H. *Tetrahedron Lett.* **1984**, *25*, 5187.
26. Davies, S. G.; Ichihara, O. *Tetrahedron: Asymmetry* **1991**, *2*, 183.
27. Seebach, D.; Estermann, H. *Tetrahedron Lett.* **1987**, *28*, 3103.
28. Rudolf, K.; Hawkins, J. M.; Loncharich, R. J.; Houk, K. N. *J. Org. Chem.* **1988**, *53*, 3879.
29. Barnish, I. T.; Corless, M.; Dunn, P. J.; Ellis, D.; Finn, P. W.; Hardstone, J. D.; James, K. *Tetrahedron Lett.* **1993**, *34*, 1323.
30. Whitesell, J. K.; Felman, S. W. *J. Org. Chem.* **1980**, *45*, 755.

Juan C. Jaen
Parke-Davis Pharmaceutical Research, Ann Arbor, MI, USA

(4*S*)-2,2-Dimethyl-1,3-dioxolane-4-carboxaldehyde[1]

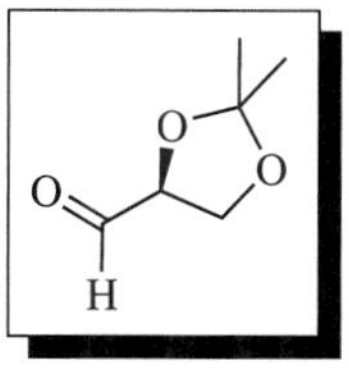

[22323-80-4] $C_6H_{10}O_3$ (MW 130.14)

(a fully oxygenated three-carbon chiral electrophile employed for a variety of uses: as a stereochemical probe in nucleophilic additions; as a chiral starting material in total synthesis of L-sugars and -nucleosides, β-lactams, and numerous complex natural products; as a source of other chiral building blocks)

Alternate Name: (*S*)-glyceraldehyde acetonide, L-glyceraldehyde acetonide, 2,3-*O*-isopropylidene-L-glyceraldehyde.
Physical Data: bp 64–66 °C/35 mmHg, $[\alpha]_D$ −75.4 ($c = 8$, benzene).[2]
Solubility: freely soluble in organic solvents; forms a readily soluble hydrate in water, readily soluble in alcohols as the corresponding hemiacetal.
Analysis of Reagent Purity: analytical methods for determination of enantiomeric purity have been reported.[5]
Preparative Methods: prepared in two steps from commercially available L-gulonolactone via ketalization and oxidative cleavage with **sodium periodate** at pH 5.5.[2] Also obtained from L-ascorbic acid via (i) ketalization, reduction (**lithium aluminum hydride**) and oxidative cleavage (**sodium periodate**),[3] or (ii) ketalization and oxidative fragmentation using **hydrogen peroxide** and **hypochlorous acid**.[4]
Purification: distilled under reduced pressure immediately prior to use. Partially polymerized material may be cracked by distillation under reduced pressure at 100 °C.[6]
Handling, Storage, and Precautions: to help prevent polymerization, anhydrous material is best stored dry at refrigerator or freezer temperatures and distilled immediately prior to use. Incompatible with acids, strong bases, oxidizing and reducing agents.

As a Stereochemical Probe in Nucleophilic Additions. Historically, the more synthetically available enantiomer, **(4*R*)-2,2-dimethyl-1,3-dioxolane-4-carboxaldehyde**, has been the compound of choice to probe stereochemistry in nucleophilic additions.[1] Nevertheless, several studies have employed the (4*S*)-aldehyde as a substrate. In analogy to its enantiomer, the reagent exhibits a moderate *si* enantiofacial preference for the addition of nucleophiles at the carbonyl, affording '*anti*' products. This preference for addition is predicted by Felkin-Ahn transition-state analysis,[7] and stands in contrast to that predicted by the Cram 'chelate' model.[8] Thus addition of the lithium (*Z*)-enolate shown (eq 1) to the reagent affords an 81:19 ratio of products with the 3,4-*anti* relationship predominating as a result of preferential *si*-face addition,[9] while the 2,3-*syn* relationship in each of the diastereomers is ascribed to a Zimmerman-Traxler-type chair transition state in the aldol reaction.[10]

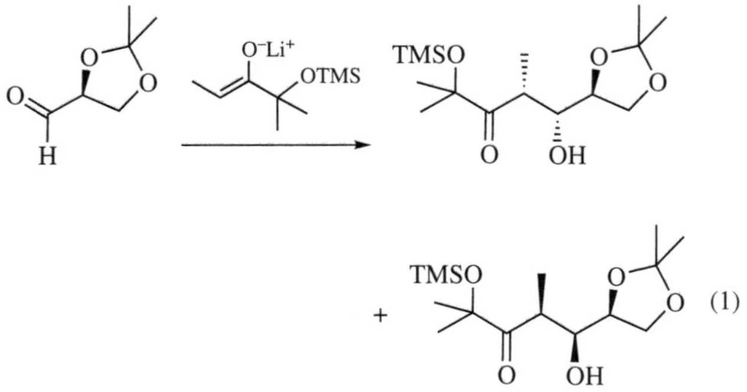

81:19

The *si* facial preference displayed by the reagent is enhanced in reactions proceeding through Lewis acid-catalyzed 'open' transition states.[11b] Thus, when reacted with the ketene silyl acetal shown (eq 2) under **zinc iodide** catalysis, a 96:4 ratio of products was obtained. The corresponding uncatalyzed reaction led to an 85:15 mixture of the same products in similar yield.[11]

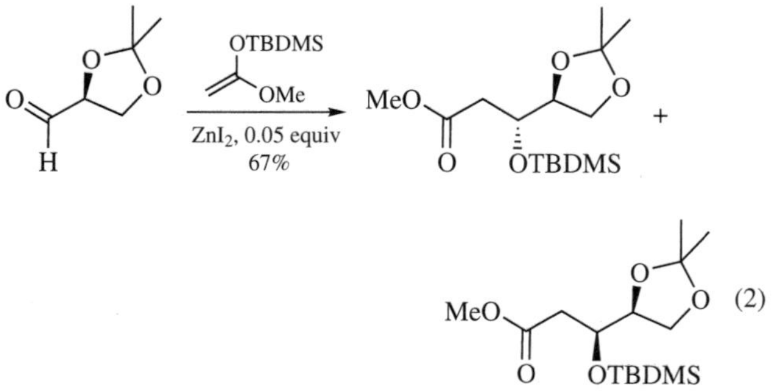

96:4

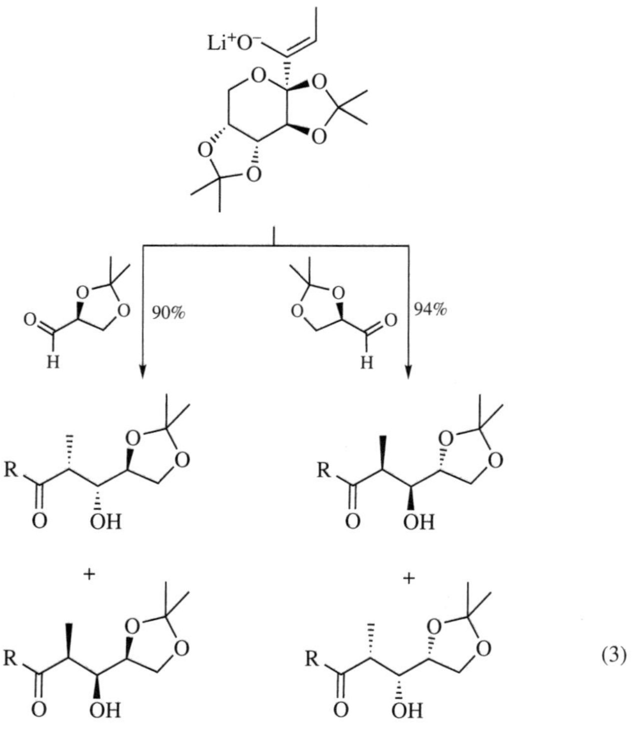

>97:3 61:28

Owing to its moderate facial preference, the reagent is an ideal choice for illustrating the concept of double asymmetric induction.[9] The chiral lithium (*Z*)-enolate, which also exhibits a moderate enantiofacial preference in reaction with achiral aldehydes, reacts with the reagent to afford a greater than 97:3 ratio of products (eq 3). This 'matched' case of amplified asymmetric induction occurs when the facial preferences of both compounds work in concert. The reaction with the enantiomer of the reagent afforded a 61:28 ratio of products, indicating 'mismatched' facial preferences working at cross purposes.[11]

As a Chiral Starting Material in L-Amino Sugar and L-Nucleoside Synthesis. The recent improved synthetic access[2] to the (4*S*)-aldehyde has facilitated non-natural sugar and nucleoside synthesis. Asymmetric synthesis of several L-amino sugars has been reported. Julia olefination of the (4*S*)-aldehyde with the sulfone afforded the key olefin intermediate as a 4:1 *E*/*Z* mixture, which was elaborated via Sharpless asymmetric dihydroxylation (SAD) and protecting group interchange to afford the protected 2-deoxy-2-amino-L-mannopyranose (eq 4).[12]

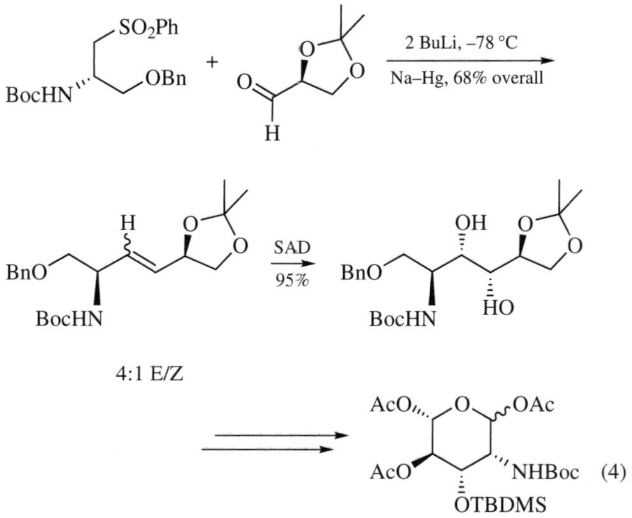

Reformatsky condensation of the reagent with ethyl bromodifluoroacetate afforded the 2,2-difluoro ester, which was further elaborated to the 2-deoxy-2-difluoro-L-ribofuranose. From there, various 2′-deoxy-2′,2′-difluoro-L-nucleosides were prepared (eq 5).[13]

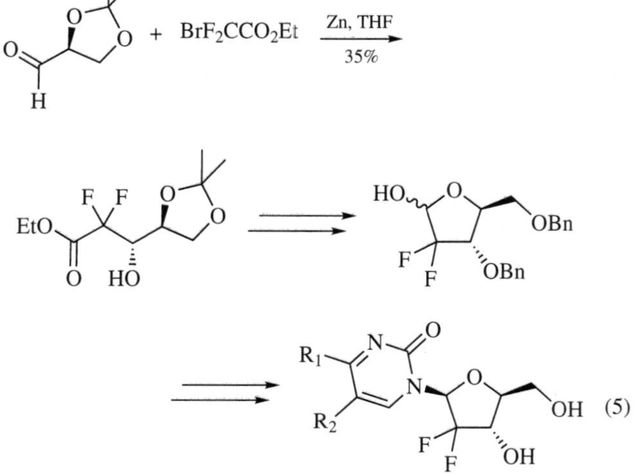

2′-Fluoro-2′,3′-unsaturated L-nucleosides have been prepared by condensing the reagent with a fluorophosphonate ester. The resulting vinyl fluoride was then transformed into the 2-fluorobutenolide, from which a variety of L-nucleosides could be prepared (eq 6).[14]

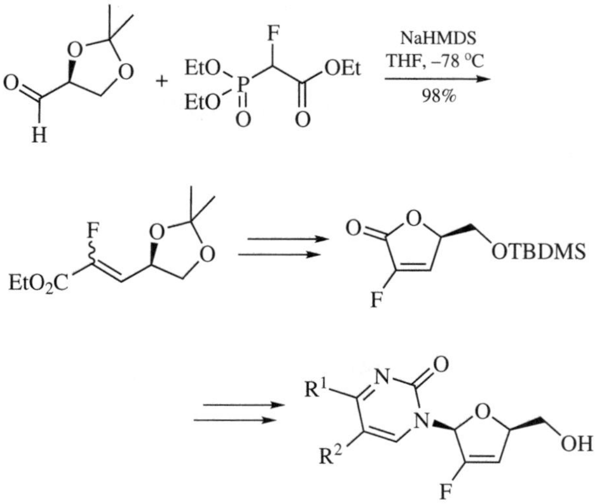

(6)

As a Chiral Starting Material in β-Lactam Synthesis. The reagent condenses with 2,4-dimethoxybenzylamine to form the corresponding imine, which undergoes a highly stereoselective [2+2] cyclization with the ketene of phthalimidoacetyl chloride (eq 7). The resulting β-lactam was elaborated into the antibiotic clinical candidate carumonan on a multikilogram scale.[15,2]

As a Chiral Starting Material in Total Synthesis. Two examples illustrate the reagent's use in total synthesis efforts. A Wittig olefination followed by an enolate Claisen rearrangement was employed to relay the reagent's chirality into key carbon-carbon bond stereochemistry in the total synthesis of (+)-ikarugamycin (eq 8).[16]

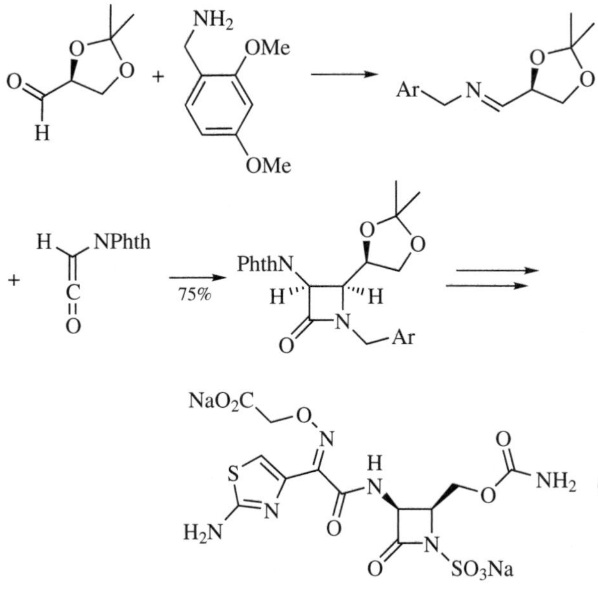

(7)

Carumonan

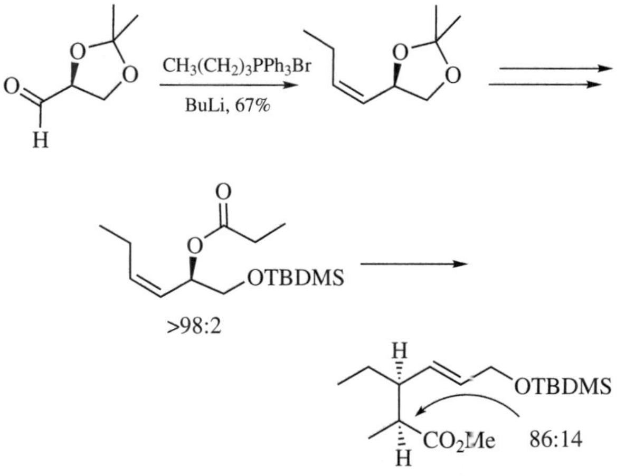

(8)

In studies directed toward the synthesis of phorboxazole A, the *si* facial preference of the reagent was evident, as a hetero-Diels-Alder reaction between the (4*S*)-aldehyde and the diene afforded the pyran shown as the major component of a 16:4:1 mixture of diastereomers (eq 9).[17]

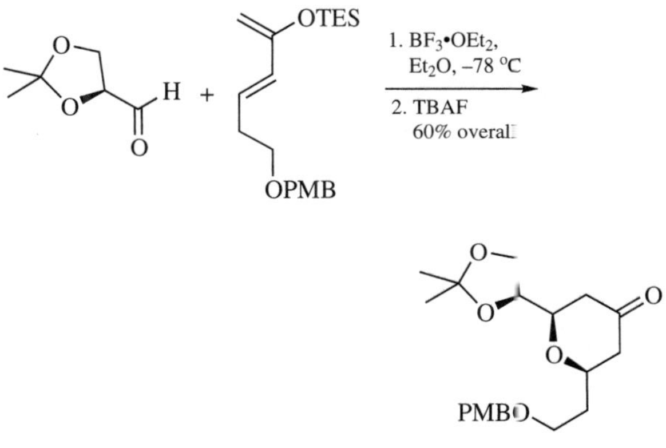

(9)

The (4*S*)-aldehyde has also been employed as a chiral starting material in total syntheses of several complex targets including levuglandin E2,[18] calyculin C,[19] (−)-rapamycin,[20] and tedanolide,[21] and in synthetic studies on spongistatin,[22] kijanolide, and tetronolide.[23]

As a Source of Other Chiral Building Blocks. The reagent is readily elaborated into several other key chirons, most notably the corresponding protected glycerol, (4*R*)-2,2-dimethyl-1,3-dioxolane-4-methanol (**1**) obtained by **sodium borohydride** reduction of aqueous solutions of the reagent.[2]

Reagent Alternatives—Variation of the Ketal Protecting Group. Analogous reagents have been prepared with different ketal protecting groups and which offer preparative and handling advantages over the isopropylidene ketal-derived reagent. Notable among them are (4*S*)-2,2-diethyl-1,3-dioxolane-4-carboxaldehyde (**2**),[24] and (2*S*)-dioxaspiro[4,5]decane-2-carboxaldehyde (**3**).[25] Both have found use in comparable synthetic situations since their introduction, though to a lesser extent than the reagent itself.

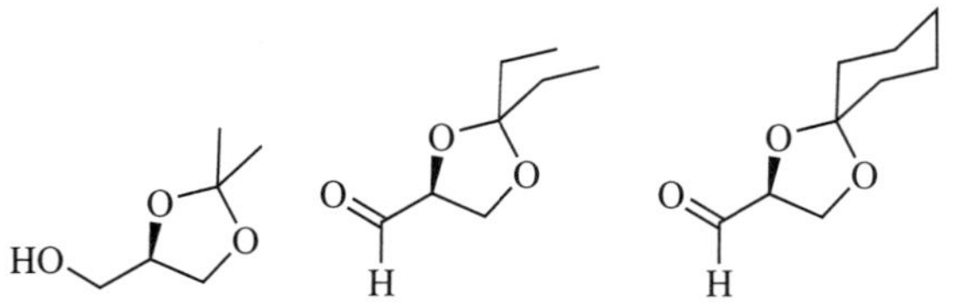

1 **2** **3**

1. (a) Jurczak, J.; Pikul, S.; Bauer, T. *Tetrahedron* **1986**, *42*, 447. (b) Mulzer, J. *Org. Synth. Highlights* **1991**, 243, CAN 116: 105726.
2. (a) Hubschwerlen, C. *Synthesis* **1986**, 962. (b) Hubschwerlen, C.; Specklin, J.-L.; Higelin, J. *Org. Synth.* **1995**, *72*, 1.
3. Jung, M. E.; Shaw, T. J. *J. Am. Chem. Soc.* **1980**, *102*, 6304.
4. Mizuno, I. Y.; Sugimoto, K. K., US Patent 4,567,282 (Jan. 28, 1986).
5. Geerlof, A.; Bert, J.; Van Tol, A.; Jongejan, J. A.; Duine, J. A. *J. Chrom.* **1993**, *648*, 119.
6. Schmid, C. R.; Bryant, J. D. *Org. Synth.* **1995**, *72*, 6.
7. (a) Cherest, M.; Felkin, H.; Prudent, N. *Tetrahedron Lett.* **1968**, *9*, 2199. (b) Anh, N. T.; Eisenstein, O. *Nouv. J. Chem.* **1977**, *1*, 61. (c) Ahn, N. T. *Top. Curr. Chem.* **1980**, *88*, 145.
8. Cram, D. J.; Kopecky, K. R. *J. Am. Chem. Soc.* **1959**, *81*, 2748.
9. (a) Heathcock, C. H.; White, C. T. *J. Am. Chem. Soc.* **1977**, *101*, 7076. (b) Heathcock, C. H.; White, C. T.; Morrison, J. J.; VanDerveer, D. *J. Org. Chem.* **1981**, *46*, 1296. (c) Masamune, S.; Choy, W.; Peterson, J. S.; Sita, L. R. *Angew. Chem., Int. Ed. Engl.* **1985**, *24*, 1.
10. (a) Zimmerman, H. E.; Traxler, M. D. *J. Am. Chem. Soc.* **1957**, *79*, 1920. (b) Dubois, M.-E.; Dubois, M. *Tetrahedron Lett.* **1967**, *8*, 4215.
11. (a) Kita, Y.; Tamura, O.; Itoh, F.; Yasuda, H.; Kishion, H.; Ke, Y. Y.; Tamura, Y. *J. Org. Chem.* **1988**, *53*, 554. (b) Kita, Y.; Yasuda, H.; Tamura, O.; Itoh, F.; Ke, Y. Y.; Tamura, Y. *Tetrahedron Lett.* **1985**, *26*, 5777.
12. (a) Ermolenko, L.; Sasaki, N. A.; Potier, P. *Tetrahedron Lett.* **1999**, *40*, 5187. (b) Ermolenko, L.; Sasaki, N. A.; Potier, P. *J. Chem. Soc., Perkin Trans. I* **2000**, 2465.
13. (a) Kotra, L. P.; Xiang, Y.; Newton, M. G.; Schinazi, R. F.; Cheng, Y-C.; Chu, C. K. *J. Med. Chem.* **1997**, *40*, 3635. (b) Xiang, Y.; Kotra, L. P.; Chu, C. K. *Bioorg. Med. Chem. Lett.* **1995**, *5*, 743.
14. (a) Lee, K.; Choi, Y.; Gullen, E.; Schlueter-Wirtz, S.; Schinazi, R. F.; Cheng, Y-C.; Chu, C. K. *J. Med. Chem.* **1999**, *42*, 1320. (b) Choi, Y.; Lee, K.; Hong, J. H.; Schinazi, R. F.; Chu, C. K. *Tetrahedron Lett.* **1998**, *39*, 4437.
15. (a) Hubschwerlen, C.; Schmid, G. *Helv. Chim. Acta* **1983**, *66*, 2206. (b) Hubschwerlin, C.; Specklin, J.-L. *Org. Synth.* **1995**, *72*, 14.
16. Boekman, R. K. Jr.; Napier, J. J.; Thomas, E. W.; Sato, R. I. *J. Org. Chem.* **1983**, *48*, 4152.
17. Cink, R. D.; Forsyth, C. J. *J. Org. Chem.* **1997**, *62*, 5672.
18. Salomon, R. G.; Miller, D. B.; Raychaudhuri, S. R.; Avasthi, K.; Lal, K.; Levison, B. S. *J. Am. Chem. Soc.* **1984**, *106*, 8296.
19. Scarlato, D. R.; DeMattei, J. A.; Chong, L. S.; Ogawa, A. K.; Lin, M. R.; Armstrong, R. W. *J. Org. Chem.* **1996**, *61*, 6139.
20. (a) Smith, A. B. III; Condon, S. M.; McCauley, J. A.; Leahy, J. W.; Leazer, J. L. Jr.; Maleczka, R. E. Jr. *Tetrahedron Lett.* **1994**, *35*, 4907. (b) Smith, A. B. III; Condon, S. M.; McCauley, J. A.; Leazer, J. L. Jr.; Leahy, J. W.; Maleczka, R. E. Jr. *J. Am. Chem. Soc.* **1997**, *119*, 947.
21. Smith, A. B. III; Lodise, S. A. *Org. Lett.* **1999**, *1*, 1249.
22. Smith, A. B. III; Lin, Q.; Nakayama, K.; Boldi, A. M.; Brook, C. S.; McBriar, M. D.; Moser, W. H.; Sobukawa, M.; Zhuang, L. *Tetrahedron Lett.* **1997**, *38*, 8675.
23. Roush, W. R.; Brown, B. B. *J. Am. Chem. Soc.* **1993**, *115*, 2268.
24. Schmid, C. R.; Bradley, D. A. *Synthesis* **1992**, 587.
25. Grauert, M.; Schollkopf, U. *Liebigs Ann. Chem.* **1985**, 1817.

Christopher R. Schmid
Eli Lilly and Company, Indianapolis, Indiana, USA

(4*R*)-2,2-Dimethyl-1,3-dioxolane-4-carboxaldehyde[1]

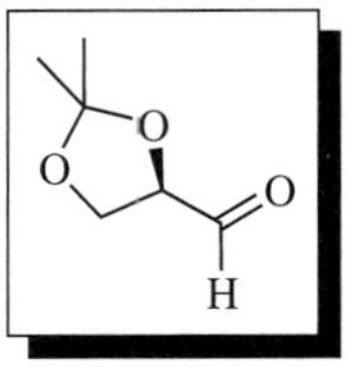

[15186-48-8] $C_6H_{10}O_3$ (MW 130.14)

(a fully oxygenated three-carbon chiral electrophile employed for a variety of uses: as a stereochemical probe in nucleophilic additions; as a chiral starting material in total synthesis of sugars and nucleosides, β-lactams, and numerous complex natural products; as a starting material for other chiral building blocks)

Alternate Name: (*R*)-glyceraldehyde acetonide, D-glyceraldehyde acetonide, 2,3-*O*-isopropylidene-D-glyceraldehyde.

Physical Data: bp 72–74 °C/30 mmHg; $[\alpha]_D$ + 80.1 (*c* 1.5, benzene)[2]

Solubility: freely soluble in organic solvents; forms a readily soluble hydrate in water, readily soluble in alcohols as the corresponding hemiacetal.

Analysis of Reagent Purity: analytical methods for determination of enantiomeric purity have been reported.[6]

Preparative Methods: prepared in two steps from D-mannitol via bis-ketalization to 1,2:5,6-bis-*O*-(1-methylethylidene)-D-mannitol, followed by oxidative cleavage with **sodium periodate** in dichloromethane.[2] Classically obtained from D-mannitol by bis-ketalization and oxidative cleavage with **lead tetraacetate**.[3] Bis-ketalization has been accomplished under a range of conditions;[4] a comparative study of the most commonly employed methods has appeared.[5]

Purification: distilled under reduced pressure immediately prior to use. Partially polymerized material may be cracked by distillation under reduced pressure at 100 °C.[2]

Handling, Storage, and Precautions: to help prevent polymerization, anhydrous material is best stored dry at refrigerator or freezer temperatures and distilled immediately prior to use. Incompatible with acids, strong bases, and oxidizing and reducing agents.

As a Stereochemical Probe in Nucleophilic Additions. The reagent has been the compound of choice to probe stereochemistry in nucleophilic additions.[1] It exhibits a moderate *re* enantiofacial preference for the addition of nucleophiles at the carbonyl, affording 'anti' products. This preference for addition is predicted by Felkin-Ahn transition-state analysis,[7] and stands in contrast to that predicted by the Cram 'chelate' model.[8] Thus on addition of alkyl-, allyl-, or phenylmagnesium, -lithium, or -zinc, *anti*/*syn* ratios ranging from 1:1 to 10:1 were observed (eq 1).[9] Curiously, PhTi(i-PrO)$_3$ gave a reversal of the ordinary trend, affording 1:3 and 1:10 *anti*/*syn* ratios depending on conditions (eq 2).[9]

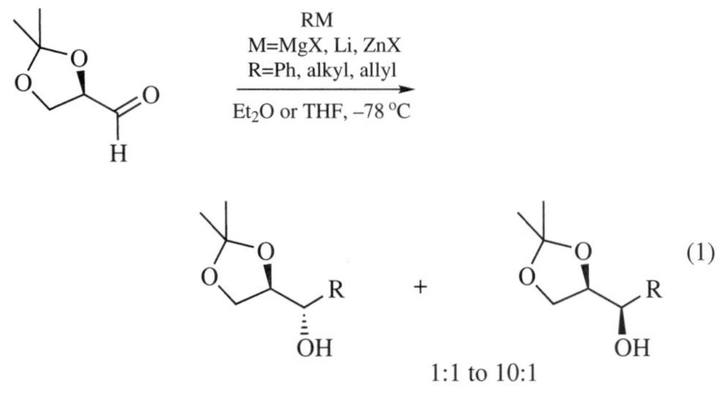

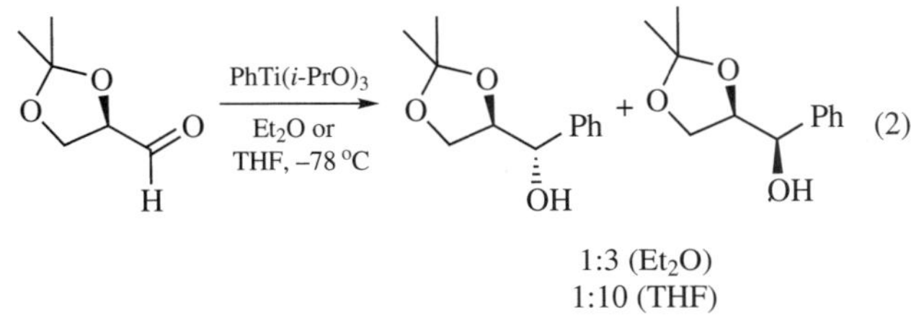

Addition of 2-furyllithium to the reagent afforded a 2:3 *anti*/*syn* ratio; on addition of various zinc halides, this very modest *si* facial preference was overturned, resulting in an almost exclusive *re*-face addition. The resulting *anti*-addition product was parlayed into D-ribulose in four steps (eq 3).[10]

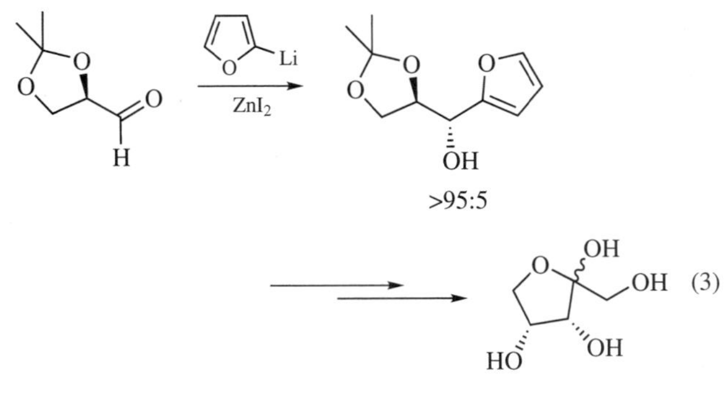

Aldol reactions employing the (4*R*)-aldehyde also proceed with *re* enantiofacial preference. In the case of the lithium (*Z*)-enolate shown, the 3,4-*anti*-relationship derives from the *re* face preference for nucleophilic attack, while the 2,3-*syn*-relationship is predicted by a Zimmerman-Traxler-type[11] chair transition state (eq 4).[12]

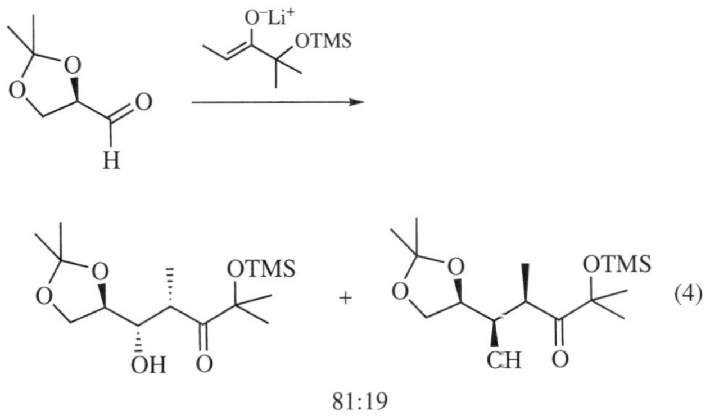

The *re* facial preference displayed by the reagent is enhanced in reactions proceeding through Lewis acid-catalyzed 'open' transition states.[13b] Thus, when reacted with the ketene silyl acetal (eq 5) under **zinc iodide** catalysis, a 96:4 ratio of products was obtained. The corresponding uncatalyzed reaction led to an 85:15 mixture of the same products in similar yield.[13]

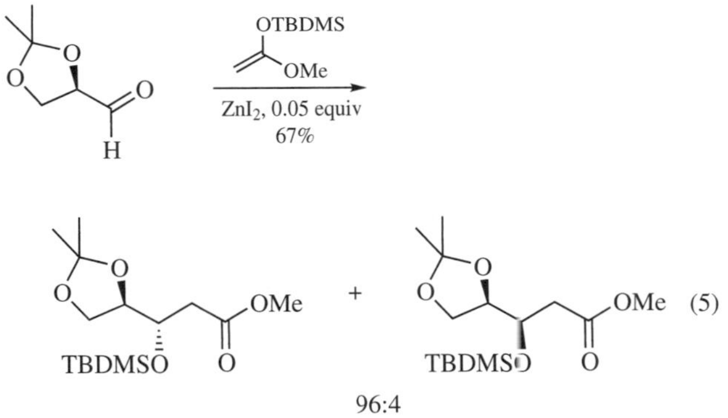

The reagent's moderate facial preference makes it an ideal choice for illustrating the concept of double asymmetric induction.[12] The chiral lithium (*Z*)-enolate, which also exhibits a moderate enantiofacial preference in reaction with achiral aldehydes, reacts with the reagent to afford a 61:28 ratio of products (eq 6). This 'mismatched' case of asymmetric induction indicates that the facial preferences of the two compounds are working at cross-purposes. With the reagent's enantiomer, **(4*S*)-2,2-dimethyl-1,3-dioxolane-4-carboxaldehyde**, a greater than 97:3 ratio of products is obtained, indicating 'matched' facial preferences.

An instance of the enantiofacial preference of the reacting partner overwhelming that of the reagent is shown in the case of the reagent's reaction with the tartrate-derived allyl boronates shown. Even in the 'mismatched' case, this example of 'reagent-based' stereocontrol affords a greater than 10:1 selectivity for the *syn* product (eq 7).[14]

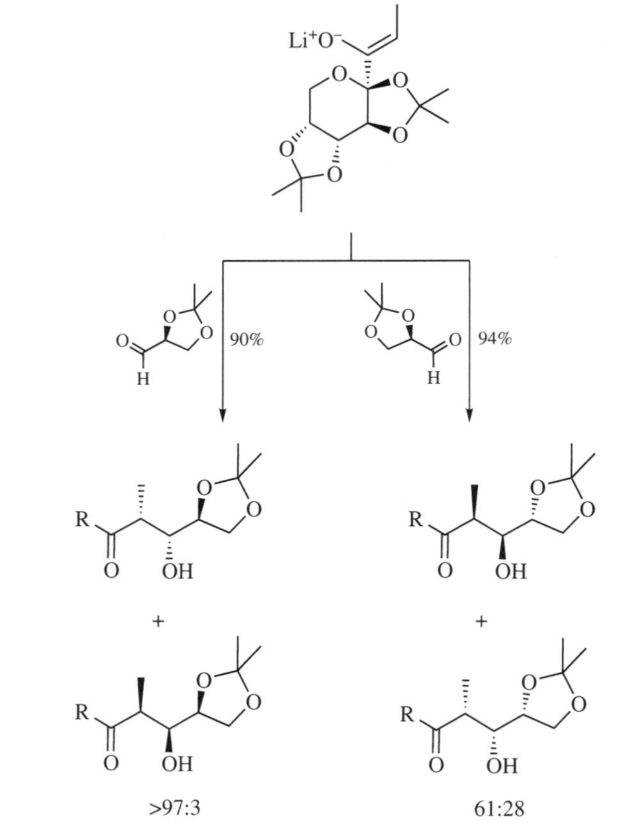

(6)

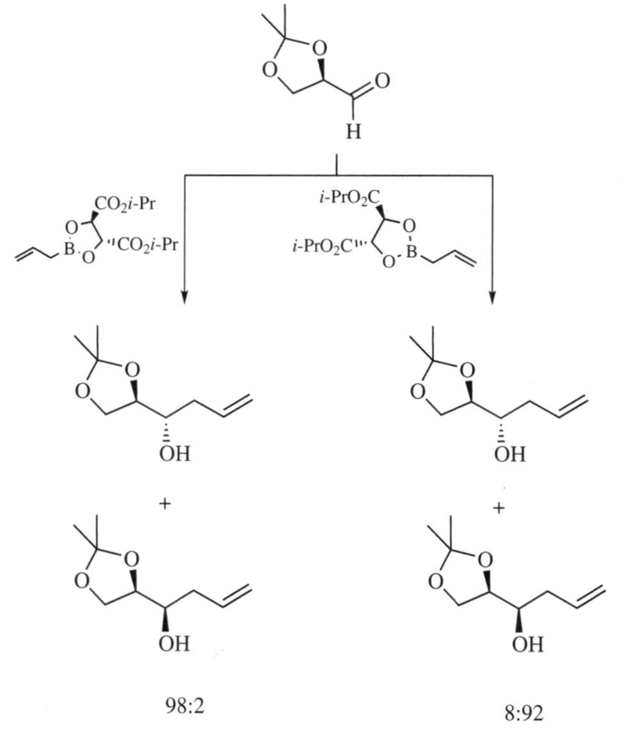

(7)

As a Chiral Starting Material in Sugar and Nucleoside Synthesis. Various D-sugars have been assembled using the reagent as the primary building block. Among the targets synthesized were 2-deoxyribose,[15] 2-deoxyribonolactone,[13] and 2-methyleneribose,[16] Erythrose, erythrulose, 2-deoxyribonolactone, ribonolactone, and lyxonolactone were prepared from addition of electrogenerated methyl dichloroacetate anion to the reagent, followed by subsequent divergent synthetic operations (eq 8).[17] In this instance, a greater than 95:5 *anti*/*syn* ratio of products was observed for the anion addition.

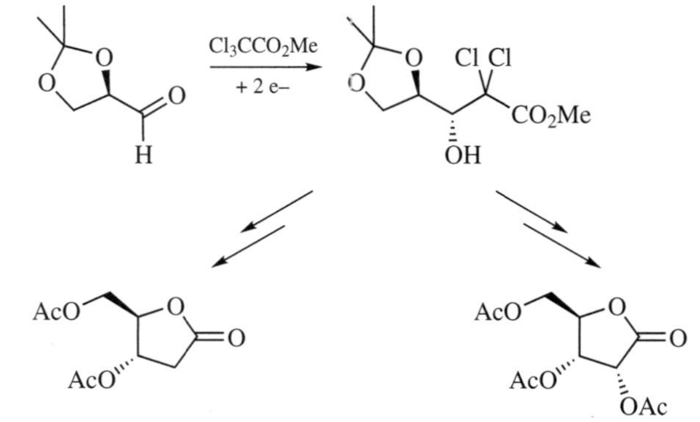

(8)

Hetero-Diels-Alder reactions have been employed with the reagent to afford pyrones that have been elaborated into D-sugars. Thus, **1-methoxy-3-(trimethylsilyloxy)buta-1,3-diene** reacted with the reagent under Lewis acid catalysis to afford the pyrone in 72% yield (eq 9). The pyrone was converted to 2-deoxy-D-ribonolactone to establish conclusively the stereochemistry of the newly formed center.[18]

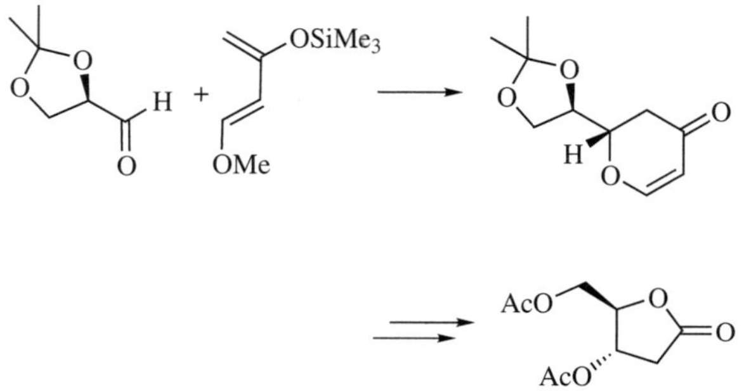

(9)

Various sugars have been constructed in a one-carbon iterative fashion starting from the reagent via condensation with thiazole anion functioning as a carbonyl anion synthon. Following protection, methylation, reduction, and hydrolysis, the resulting α-benzyloxy aldehyde erythrose was obtained (eq 10).[19] Products can be resubjected to the sequence, affording protected pentoses through octoses.[20] Using this methodology, a synthesis of the octulosonic acid KDO has been reported.[21]

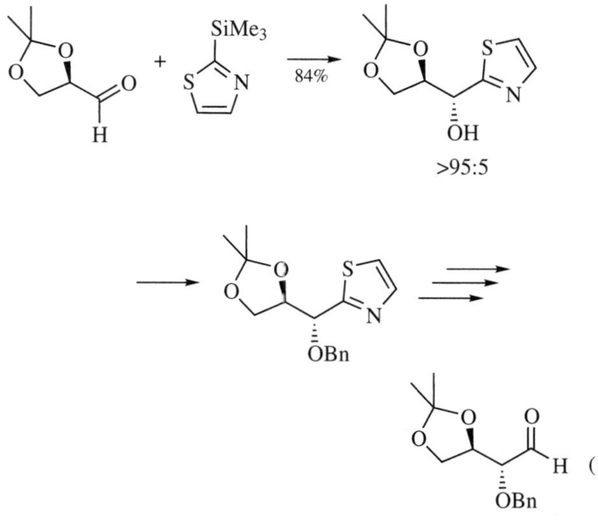

(10)

A strategy for the assembly of various carbasugars and aminocarbasugars employed condensation of the reagent with 2-silyloxyfurans and 2-silyloxy-*N*-protected pyrroles. The additions proceed in high yield and stereoselectivity to afford α,β-unsaturated lactones and lactams, respectively, which were parlayed into pseudo-D-gulopyranose, pseudo-D-xylofuranose, pseudo-D-gulopyranosylamine, and psuedo-D-xylofuranosylamine (eq 11).[22]

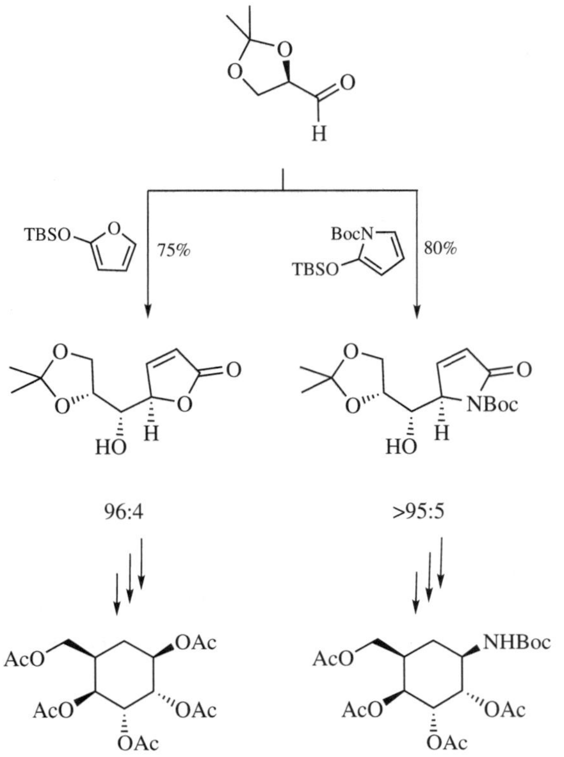

(11)

The HIV reverse-transcriptase inhibitor AZT was prepared via the (*Z*)-enone resulting from condensation of the reagent with **(ethoxycarbonylmethyl)triphenylphosphonium bromide**. Cyclization followed by Michael addition of **hydrazoic acid** afforded the azido lactone shown. Subsequent manipulations provided the target compound (eq 12).[23]

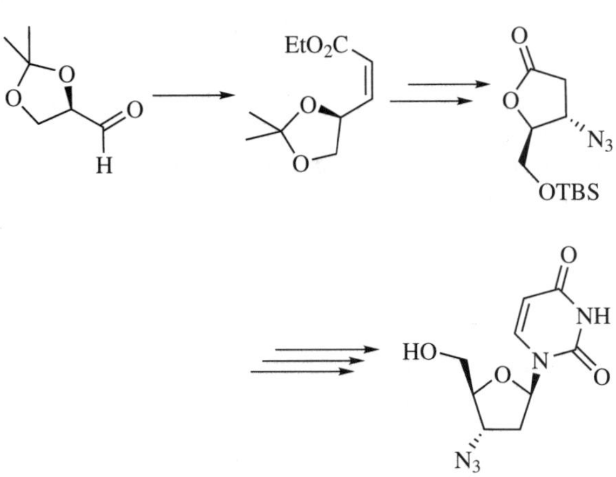

(12)

AZT

Access to 2-difluoro-2-deoxy-D-sugars and their derived nucleosides was realized by Reformatsky condensation of the reagent with ethyl bromodifluoroacetate. The resulting difluoro ester was obtained as a 2.5:1 *anti*/*syn* mixture. Following separation, the major isomer was transformed into 2-deoxy-2-difluoro-D-ribose. The sugar was then elaborated into various difluorodeoxynucleosides, including the oncolytic Gemzar(tm) (eq 13).[24]

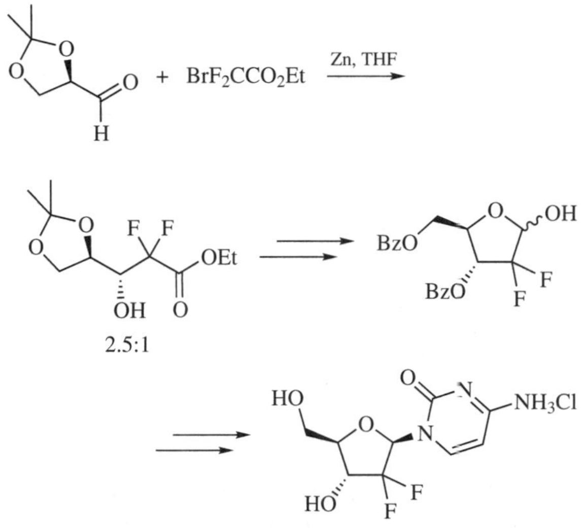

(13)

Gemzar™

As a Chiral Starting Material in β-Lactam Synthesis. Various β-lactams have been prepared via [2 + 2] cyclization of ketenes with aryl or benzyl imines derived from the reagent. Thus the *para*-methoxyphenyl imine derived from condensation of the corresponding aniline and the reagent underwent [2 + 2] cycloadditions with various ketenes to afford β-lactams in moderate yields but with very high stereoselectivity (eq 14).[25]

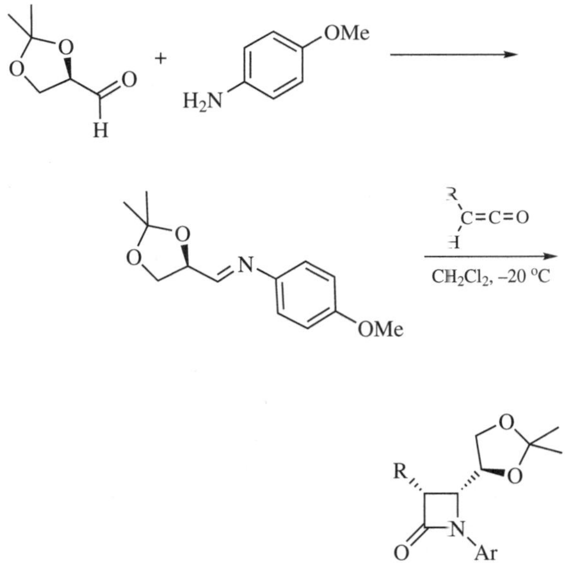

(14)

>95:5 ee
>95:5 de

R = N_3, 55%; R = OMe, 54%;
R = OPh, 50–55%; R = OAc, 50–55%

A model study for the direct synthesis of peptidyl nucleosides used the benzyl imine of the reagent and the requisite ketene in a [2 + 2] cycloaddition to prepare β-lactams, which were further elaborated through deprotection and oxidation. Again, the stereoselectivity of the [2 + 2] cyclization was very high (eq 15).[26]

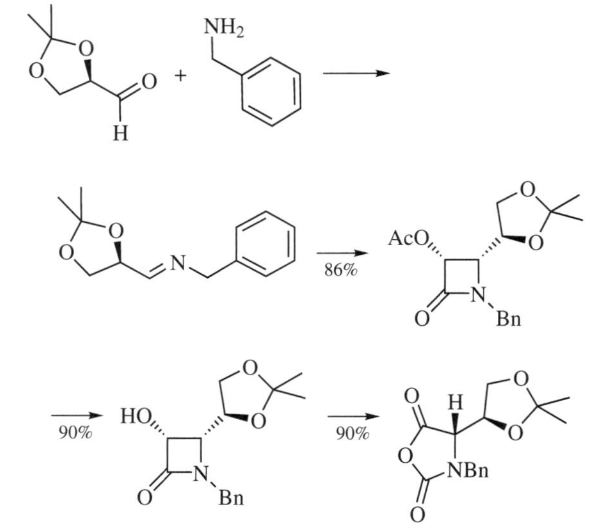

(15)

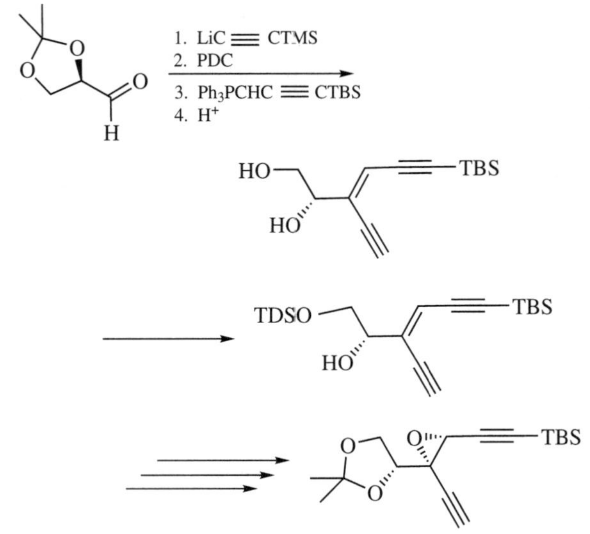

(17)

As a Chiral Starting Material in Total Synthesis. Three examples illustrate the widespread use of the reagent as a chiral starting material in total synthesis. In the total synthesis of (+)-CP-263,114, the reagent was treated with the organomagnesium shown and the resulting adduct oxidized to afford the ketone. This was then parlayed into a key vinyl bromide coupling partner in the synthesis via the epoxide (eq 16).[27]

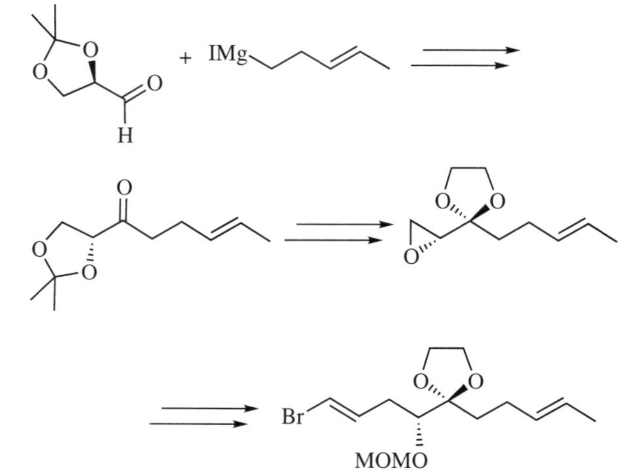

(16)

In the enantioselective synthesis of neocarzinostatin aglycone, the reagent served as the starting point for assembly of the crucial chiral epoxydiyne fragment shown, proceeding via sequential addition of lithium trimethylsilylacetylide, oxidation, and Wittig coupling. Following separation of olefin isomers, the acetonide was unmasked and then monoprotected to reveal the allylic alcohol, which underwent Sharpless asymmetric epoxidation. Reketalization delivered the chiral epoxydiyne (eq 17).[28]

In a total synthesis of (+)-brefeldin A, the reagent was elaborated into the α,β-enone shown, which participated in a palladium-mediated cyclopentene-forming reaction to afford the *exo*-olefin. Stereoselectivity in the ring-forming reaction was 3.5:1 in favor of the desired isomer over the alternative *trans*-cyclopentene. Ozonolysis and reduction of the resulting ketone, followed by protection afforded the MEM ether shown, where all the relevant stereocenters of the final target were established (eq 18).[29]

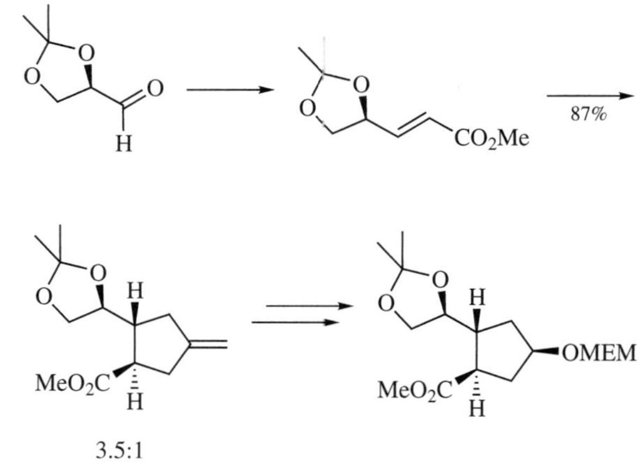

(18)

The (4*R*)-aldehyde has also been employed as a chiral starting material in total syntheses of numerous complex targets including prostaglandin PGE1,[30] PGF2α,[31] PGB1 methyl ester,[32] 11-(*R*)-HETE,[33] molybdenum cofactor,[34] calcimycin class polyether ionophores,[35] erythronolide B,[36] (+)-9,11-dehydroestrone methyl ester,[37] and 11,*O*(3)-dihydropseudopterolide.[38] Total synthesis studies on the nargenicins,[39] phorboxazole A,[40] macrolactin A,[41] neoliacinic acid,[42] tetronasin,[43] chlorothricolide,[44] and the annonacious acetogenins[45] have been undertaken using the reagent as a chiral building block or in key stereochemical studies.

As a Source of Other Chiral Building Blocks. The reagent is readily elaborated into several other key chiral building blocks, most notably the corresponding protected glycerol, (4*S*)-2,2-dimethyl-1,3-dioxolane-4-methanol (**1**), obtained by **sodium borohydride** reduction on aqueous solutions of the reagent.[2] The 3-[(4*S*)-2,2-dimethyl-1,3-dioxolan-4-yl]-2-propenoic acid esters **2**, mentioned previously, have also found significant use in synthesis.[46]

Alternative Reagents—Variation of the Ketal Protecting Group. Analogous reagents have been prepared that employ different ketal protecting groups and which offer preparative and handling advantages over the isopropylidene ketal-derived reagent. Notable amongst them are (4*R*)-2,2-diethyl-1,3-dioxolane-4-carboxaldehyde (**3**),[47] and (2*R*)-dioxaspiro[4,5]decane-2-carboxaldehyde (**4**).[48] Both have found use in comparable synthetic situations since their introduction, though to a lesser extent than the reagent itself.

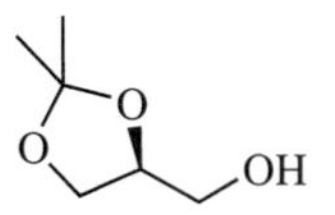

1

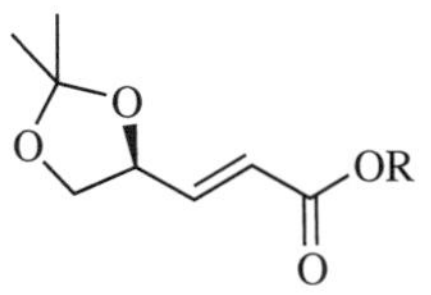

2a R=Me
2b R=Et

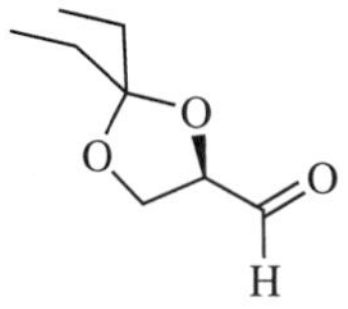

3

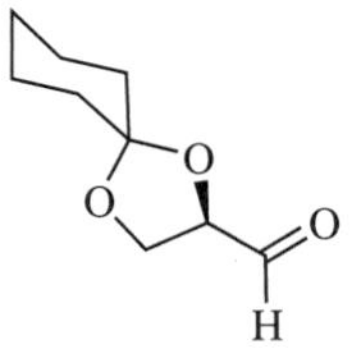

4

1. (a) Jurczak, J.; Pikul, S.; Bauer, T. *Tetrahedron* **1986**, *42*, 447. (b) Mulzer, J. *Org. Synth. Highlights* **1991**, 243, CAN 116:105726.
2. (a) Schmid, C. R.; Bryant, J. D.; Dowlatzadeh, M.; Phillips, J. L.; Prather, D. E.; Schantz, R. D.; Sear, N. L.; Vianco, C. S. *J. Org. Chem.* **1991**, *56*, 4056. (b) Schmid, C. R.; Bryant, J. D. *Org. Synth.* **1995**, *72*, 6. (c) Jackson, D. *Synth. Commun.* **1988**, *18*, 337. (c) Zanardi, F.; Battistini, L.; Rassu, G.; Auzzas, L.; Pinna, L.; Marzocchi, L.; Acquotti, D.; Casiraghi, G. *J. Org. Chem.* **2000**, *65*, 2048.
3. Baer, E.; Fisher, H. O. L. *Biol. Chem.* **1939**, *128*, 463.
4. (a) Chittenden, G. J. F. *Carbohydr. Res.* **1980**, *84*, 350. (b) DeBost, J.-L.; Gelas, J.; Horton, D. *J. Org. Chem.* **1983**, *48*, 1381. (c) Kohan, G.; Just, G. *Synthesis* **1974**, 192. (d) Morpain, C.; Nasser, B.; Laude, B.; Latruffe, N. *Org. Prep. Proc. Intl.* **1990**, *22*, 540. (e) Tipson, R. S.; Cohen, A. *Carbohydr. Res.* **1968**, *7*, 232.
5. Kuszmann, J.; Tomori, E.; Meerwald, I. *Carbohydr. Res.* **1984**, *128*, 87.
6. Geerlof, A.; Bert, J.; Van Tol, A.; Jongejan, J. A.; Duine, J. A. *J. Chrom.* **1993**, *648*, 119.
7. (a) Cherest, M.; Felkin, H.; Prudent, N. *Tetrahedron Lett.* **1968**, *9*, 2199. (b) Anh, N. T.; Eisenstein, O. *Nouv. J. Chem.* **1977**, *1*, 61. (c) Ahn, N. T. *Top. Curr. Chem.* **1980**, *88*, 145.
8. Cram, D. J.; Kopecky, K. R. *J. Am. Chem. Soc.* **1959**, *81*, 2748.
9. Mulzer, J.; Angermann, A. *Tetrahedron Lett.* **1983**, *24*, 2843.
10. Suzuki, Y.; Yuki, Y.; Mukaiyama, T. *Chem. Lett.* **1981**, 1529.
11. (a) Zimmerman, H. E.; Traxler, M. D. *J. Am. Chem. Soc.* **1957**, *79*, 1920. (b) Dubois, M.-E.; Dubois, M. *Tetrahedron Lett.* **1967**, *8*, 4215.
12. (a) Heathcock, C. H.; White, C. T. *J. Am. Chem. Soc.* **1977**, *101*, 7076. (b) Heathcock, C. H.; White, C. T.; Morrison, J. J.; VanDerveer, D. *J. Org. Chem.* **1981**, *46*, 1296. (c) Masamune, S.; Choy, W.; Peterson, J. S.; Sita, L. R. *Angew. Chem., Int. Ed. Engl.* **1985**, *24*, 1.
13. (a) Kita, Y.; Tamura, O.; Itoh, F.; Yasuda, H.; Kishion, H.; Ke, Y. Y.; Tamura, Y. *J. Org. Chem.* **1988**, *53*, 554. (b) Kita, Y.; Yasuda, H.; Tamura, O.; Itoh, F.; Ke, Y. Y.; Tamura, Y. *Tetrahedron Lett.* **1985**, *26*, 5777.
14. Roush, W. R.; Walts, A. E.; Hoong, L. K. *J. Am. Chem. Soc.* **1985**, *107*, 8186.
15. Harada, T.; Mukaiyama, T. *Chem. Lett.* **1981**, 1109.
16. (a) Depezay, J. C.; LeMerrer, Y. *Tetrahedron Lett.* **1978**, *19*, 2865. (b) Depezay, J. C.; LeMerrer, Y. *Carbohydr. Res.* **1980**, *83*, 51. (c) Depezay, J. C.; Sanier, M.; Mansuy, D. *Carbohydr. Res.* **1983**, *117*, 313.
17. (a) Shono, T.; Ohmizu, H.; Kise, N. *Tetrahedron Lett.* **1982**, *23*, 4801. (b) Shono, T.; Kise, N.; Suzumoto, T. *J. Am. Chem. Soc.* **1984**, *106*, 259.
18. Danishefsky, S.; Kobayashi, S.; Kerwin, J. F. Jr. *J. Org. Chem.* **1982**, *47*, 1981.
19. (a) Dondoni, A.; Merino, P. *Org. Synth.* **1995**, *72*, 21. (b) Dondoni, A.; Fantin, G.; Fogagnolo, M.; Medici, A.; Pedrini, P. *Synthesis* **1988**, 685.
20. (a) Dondoni, A.; Fantin, G.; Fogagnolo, M.; Medici, A. *Angew. Chem., Int. Ed. Engl.* **1986**, *25*, 835. (b) Dondoni, A.; Fantin, G.; Fogagnolo, M.; Medici, A.; Pedrini, P. *J. Org. Chem.* **1989**, *54* 693.
21. Dondoni, A.; Merino, P. *J. Org. Chem.* **1991**, *56*, 5294.
22. Rassu, G.; Auzzas, L.; Pinna, L.; Battistini, L.; Zanardi, F.; Marzocchi, L.; Acquotti, D.; Casiraghi, G. *J. Org. Chem.* **2000**, *65*, 6307.
23. Chu, C. K.; Beach, J. W.; Ullas, G. V.; Kosugi, Y. *Tetrahedron Lett.* **1988**, *29*, 5349.
24. (a) Hertel, L. W.; Kroin, J. S.; Misner, J. W.; Tustin, J. M. *J. Org. Chem.* **1988**, *53*, 2406. (b) Chou, T. S.; Heath, P. C.; Patterson, L. E.; Poteet, L. M.; Lakin, R. E.; Hunt, A. E. *Synthesis* **1992**, 565.
25. Bose, A. K.; Hegde, V. R.; Wagle, D. R.; Bari, S. S.; Manhas, M. S. *Chem. Commun.* **1986**, 161.
26. Palomo, C.; Oiarbide, M.; Esnal, A.; Landa, A.; Miranda, J. I.; Linden, A. *J. Org. Chem.* **1998**, *63*, 5838.
27. Chen, C.; Layton, M. E.; Sheehan, S. M.; Shair, M. D. *J. Am. Chem. Soc.* **2000**, *122*, 7424.
28. Myers, A. G.; Hammond, M.; Wu, Y.; Xiang, J.-N.; Harrington, P. M.; Kuo, E. Y. *J. Am. Chem. Soc.* **1996**, *118*, 10006.
29. Trost, B. M.; Lynch, J.; Renaut, P.; Steinman, D. H. *J. Am. Chem. Soc.* **1986**, *108*, 284.
30. Stork, G.; Takahasi, T. *J. Am. Chem. Soc.* **1977**, *99*, 1275.
31. Johnson, F.; Paul, K. G.; Favara, D.; Ciabatti, R.; Guzzi, U. *J. Am. Chem. Soc.* **1982**, *104*, 2190.
32. Mikolajczyk, M.; Mikina, M.; Jankowiak, A. *J. Org. Chem.* **2000**, *65*, 5127.
33. Corey, E. J.; Kang, J. *J. Am. Chem. Soc.* **1981**, *103*, 4618.
34. Taylor, E. C.; Ray, P. S.; Darwish, I. S.; Johnson, J. L.; Rajagopalan, D. V. *J. Am. Chem. Soc.* **1989**, *111*, 7664.
35. Boeckman, R. K.; Charette, A. B.; Asberom, T.; Johnston, B. J. *J. Am. Chem. Soc.* **1991**, *113*, 5337.
36. Mulzer, J.; Kirstein, H. M.; Buschmann, J.; Lehmann, C.; Luger, P. *J. Am. Chem. Soc.* **1991**, *113*, 910.
37. Mikami, K.; Takahashi, K.; Nakai, T.; Uchimaru, T. *J. Am. Chem. Soc.* **1994**, *116*, 10948.
38. Paquette, L. A.; Rayner, C. M.; Doherty, A. M. *J. Am. Chem. Soc.* **1990**, *112*, 4078.
39. Coe, J. W.; Roush, W. R. *J. Org. Chem.* **1989**, *54*, 915.
40. Williams, D. R.; Clark, M. P.; Emde, U.; Berliner, M. A. *Org. Lett.* **2000**, *2*, 3023.
41. Li, S.; Xu, R.; Bai, D. *Tetrahedron Lett.* **2000**, *41*, 3463.
42. Clark, S. J.; Dossetter, A. G.; Blake, A. J.; Li, W.-S.; Whittingham, W. G. *Chem. Commun.* **1999**, 749.
43. Okumura, K.; Okazaki, K.; Takeda, K.; Yoshii, E. *Tetrahedron Lett.* **1989**, *30*, 2233.
44. De Laszlo, S. E.; Ford, M. J.; Ley, S. V.; Maw, G N. *Tetrahedron Lett.* **1990**, *31*, 5525.
45. Zanardi, F.; Battistini, L.; Rassu, G.; Pinna, L.; Mor, M.; Culeddu, N.; Casiraghi, G. *J. Org. Chem.* **1998**, *63*, 1368.
46. (a) Matsunaga, H.; Sakamaki, T.; Nagaoka, H.; Yamada, Y. *Tetrahedron Lett.* **1983**, *24*, 3009. (b) Haefele, B.; Jaeger, V. *Liebigs Ann. Chem.* **1987**, 85. (c) Takano, S.; Kurotaki, A.; Takahashi, M.; Ogasawara, K. *Synthesis* **1986**, 403.
47. Schmid, C. R.; Bradley, D. A. *Synthesis* **1992**, 587.
48. Sugiyama, T.; Sugawara, H.; Watanabe, M.; Yamashita, K. *Agric. Biol. Chem.* **1984**, *48*, 1841.

Christopher R. Schmid
Eli Lilly and Company, Indianapolis, Indiana, USA

(*R*)-*N,N*-Dimethyl-1-[(*S*)-2-(diphenylphosphino)ferrocenyl]ethylamine

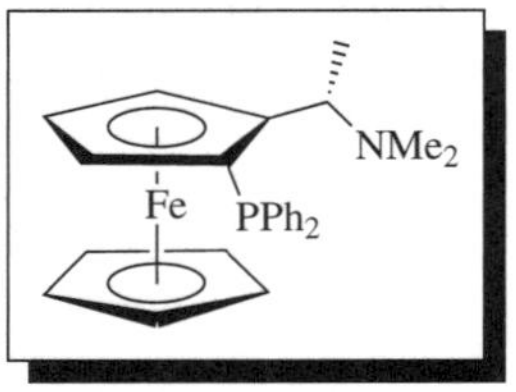

(*R*)-(*S*)-PPFA
[55700-44-2] $C_{26}H_{28}FeNP$ (MW 441.37) (*S*)-(*R*)-PPFA
[55650-58-3]

(effective chiral phosphine ligand[1] for nickel- or palladium-catalyzed asymmetric cross coupling of organomagnesium or -zinc reagents with alkenyl bromides,[2–4] and for palladium-catalyzed asymmetric hydrosilylation of 1,3-dienes[5])

Alternate Name: (*R*)-(*S*)-PPFA.
Physical Data: mp 139 °C; $[\alpha]_D^{25}$ −361° (*c* 0.6, ethanol).[1a]
Purification: recrystallization from ethanol.[1a]
Handling, Storage, and Precautions: stable in air for years, but best kept sealed in a refrigerator.

Asymmetric Cross Coupling. In the presence of a nickel catalyst, generated in situ from *Nickel(II) Chloride* and (*R*)-(*S*)-PPFA, secondary alkyl Grignard reagents, represented by 1-phenylethylmagnesium chloride, react with alkenyl halides to give optically active alkenes of up to 68% ee (eq 1). The isolated palladium complex, $PdCl_2$[(*R*)-(*S*)-PPFA], can be also used for the cross coupling.[2] Use of a zinc reagent in place of the Grignard reagent increases the enantioselectivity to 86% ee.[3]

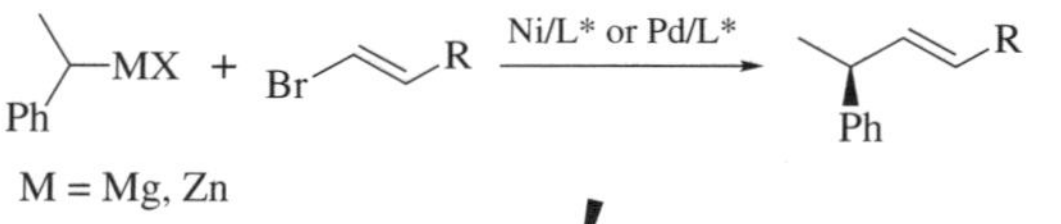
(1)

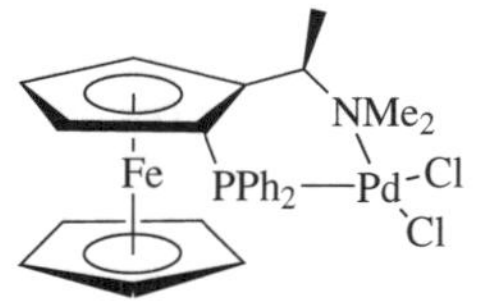

$PdCl_2$[(*R*)-(*S*)-PPFA]

The stereoselectivity obtained with the PPFA ligand is generally higher than that obtained with *(2,3-O-Isopropylidene)-2,3-dihydroxy-1,4-bis(diphenylphosphino)buta ne* (DIOP) (7–16% ee), 1,2-bis(diphenylphosphino)propane (prophos) (0% ee), and 2,2′-bis(diphenylphosphinomethyl)-1,1′-binaphthyl (NAPHOS) (11% ee).

The asymmetric cross coupling with the chiral ferrocenylphosphine–palladium catalyst has been successfully applied to the synthesis of optically active allylsilanes (eq 2).[4] The reaction of α-(trimethylsilyl)benzylmagnesium bromide with vinyl bromide and (*E*)-β-bromostyrene catalyzed by 0.5 mol % of $PdCl_2$[(*R*)-(*S*)-PPFA] gives a quantitative yield of the corresponding allylsilanes of 95% ee. High enantioselectivity is also observed in the cross coupling of 1-(triethylsilyl)ethylmagnesium chloride with (*E*)-β-bromostyrene.

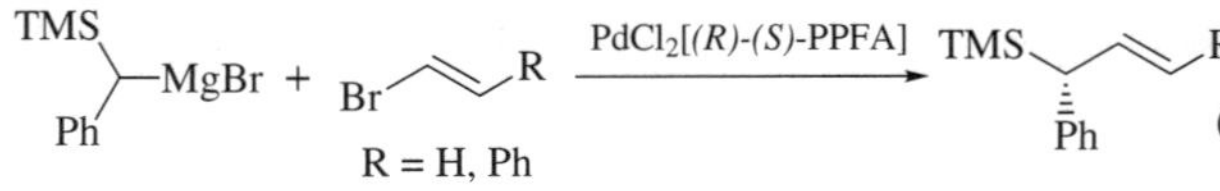
(2)

Asymmetric Hydrosilylation of Alkenes. The palladium complex $PdCl_2$[(*R*)-(*S*)-PPFA] catalyzes the asymmetric hydrosilylation of norbornene, styrene, and 1,3-dienes (eq 3).[5] The hydrosilylation of 1-phenyl-1,3-butadiene with *Trichlorosilane* proceeds regioselectively in a 1,4-fashion to give (*Z*)-1-phenyl-1-silyl-2-butene of 64% ee.

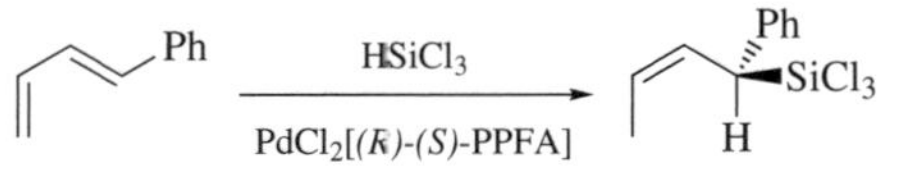
(3)

Optically active ferrocenylbisphosphines, (*R*)-*N,N*-dimethyl-1-[(*S*)-1′,2-bis(diphenylphosphino)ferrocenyl]ethylamine [(*R*)-(*S*)-BPPFA] and its derivatives, are efficient chiral bisphosphine ligands for rhodium-catalyzed asymmetric hydrogenation, palladium-catalyzed asymmetric allylic substitution reactions, and gold-catalyzed asymmetric aldol-type reactions of α-isocyano carboxylates.[1,6]

Synthesis of Chiral Phosphorane. (*S*)-(*R*)-PPFA has been converted to an enantiomerically pure ferrocenylphosphonium salt (**1**) in two steps in 54% yield (eq 4). The chiral phosphorane (**2**), generated in situ from (**1**) by *n-Butyllithium* in THF, reacts with aldehydes in the presence of a Lewis acid, *Diethylaluminum Ethoxide* to give vinylation products (**3**) (eq 5) with variable enantiomeric excess (up to 70%).[7]

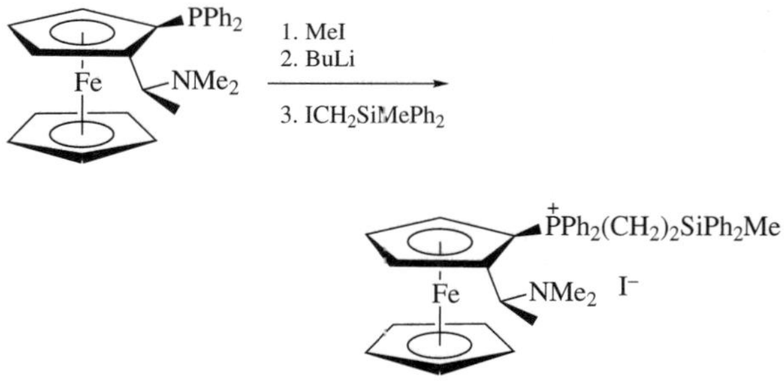
(4)

(**1**)

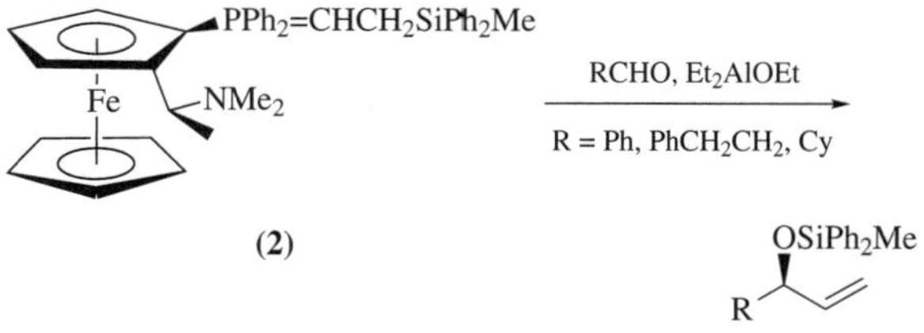

(**2**) (*S*)-(**3**) (5)

---

1. (a) Hayashi, T.; Mise, T.; Fukushima, M.; Kagotani, M.; Nagashima, N.; Hamada, Y.; Matsumoto, A.; Kawakami, S.; Konishi, M.; Yamamoto, K.; Kumada, M. *Bull. Chem. Soc. Jpn.* **1980**, *53*, 1138. (b) Hayashi, T. *Pure Appl. Chem.* **1988**, *60*, 7.
2. (a) Hayashi, T.; Tajika, M.; Tamao, K.; Kumada, M. *J. Am. Chem. Soc.* **1976**, *98*, 3718. (b) Hayashi, T.; Konishi, M.; Fukushima, M.; Mise, T.; Kagotani, M.; Tajika, M.; Kumada, M. *J. Am. Chem. Soc.* **1982**, *104*, 180.

3. Hayashi, T.; Hagihara, T.; Katsuro, Y.; Kumada, M. *Bull. Chem. Soc. Jpn.* **1983**, *56*, 363.
4. (a) Hayashi, T.; Konishi, M.; Ito, H.; Kumada, M. *J. Am. Chem. Soc.* **1982**, *104*, 3772. (b) Hayashi, T.; Konishi, M.; Okamoto, Y.; Kabeta, K.; Kumada, M. *J. Org. Chem.* **1986**, *51*, 3772.
5. (a) Hayashi, T.; Kabeta, K. *Tetrahedron Lett.* **1985**, *26*, 3023. (b) Hayashi, T.; Matsumoto, Y.; Morikawa, I.; Ito, Y. *Tetrahedron: Asymmetry* **1990**, *1*, 151.
6. (a) Ojima, I. *Catalytic Asymmetric Synthesis*; VCH: New York, in press. (b) Togni, A.; Hayashi, T. *Ferrocenes: From Catalysis to Materials Science*; VCH: New York, 1994.
7. Iio, H.; Fujii, A.; Ishii, M.; Tokoroyama, T. *Chem. Commun./J. Chem. Soc., Chem. Commun.* **1991**, 1390.

Tamio Hayashi
Hokkaido University, Sapporo, Japan
M. Mahmun Hossain & Anjan K. Saha
University of Wisconsin–Milwaukee, WI, USA

2-[(4*S*)-4-(1,1-Dimethylethyl)-4,5-dihydro-2-oxazolyl]-6-methylpyridine

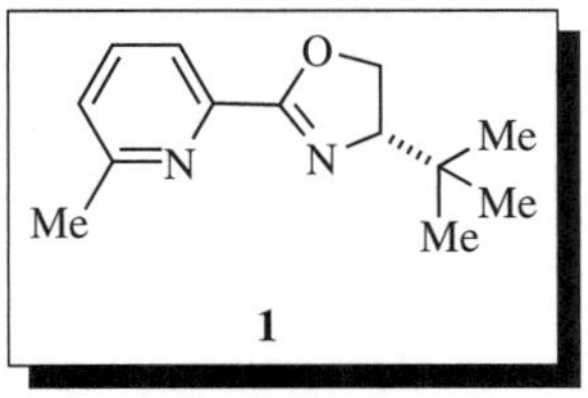

[199277-80-0] $C_{13}H_{18}N_2O$ (MW 218.30)

(chiral ligand for enantiocontrol of metal-catalyzed reactions)

Physical Data: mp 48 °C; $[\alpha]^{25}_D$ −75.8 (*c* 0.5, $CHCl_3$).
Solubility: soluble in most organic solvents.
Preparative Methods: 2-[(4*S*)-4-(1,1-Dimethylethyl)-4,5-dihydro-2-oxazolyl]-6-methylpyridine[1] was prepared in 58–75% yield by heating, under reflux for 24 h, a chlorobenzene solution of 2-cyano-6-methylpyridine with *tert*-leucinol in the presence of a catalytic amount of zinc chloride (eq 1).[1]

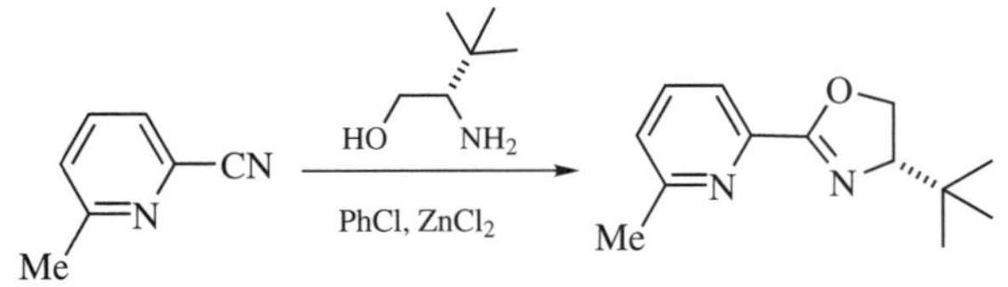

(1)

Handling, Storage, and Precautions: stable at ambient temperature.

Asymmetric Reactions

Allylic Substitution Use of 2-[(4*S*)-4-(1,1-dimethylethyl)-4,5-dihydro-2-oxazolyl]-6-methylpyridine(**1**) for enantioselective palladium-catalyzed allylic substitution has been reported.[1,2,3] The reactions were carried out using $[Pd(\eta^3\text{-}C_3H_5)Cl]_2$ as procatalyst in a mixture of dimethylmalonate, *N,O*-bis(trimethylsilyl)acetamide (BSA), and potassium acetate in methylene chloride (eq 2).[4] Under these conditions (*S*)-dimethyl 1,3-diphenylprop-2-enylmalonate was isolated in 92% yield and 91% enantioselectivity. The stereochemical outcome was rationalized by reaction through transition state **A**, in which the nucleophile attacks the allylic terminus trans to the oxazoline nitrogen.

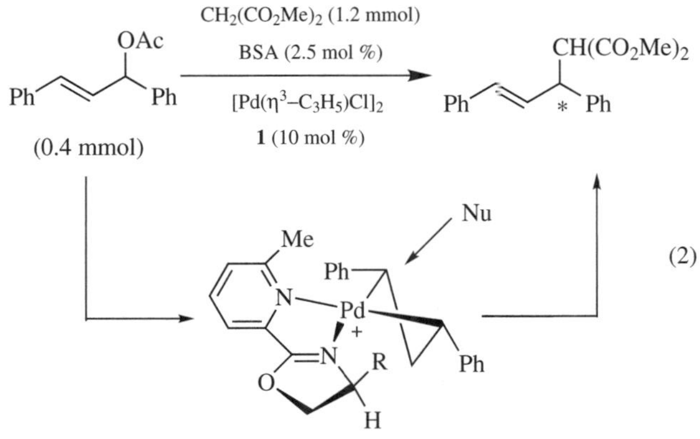

(2)

A

Cyclopropanation[5]

The copper(I) triflate complex of **1** has been evaluated in the asymmetric cyclopropanation of styrene with ethyl diazoacetate (eq 3). The *trans*- and *cis*-2-phenylcyclopropane carboxylates were isolated in 88% yield as a 70:30 ratio of diastereomers in 43% and 44% enantioselectivity. These enantioselectivities are not as high as observed with other bis(oxazoline) ligands.

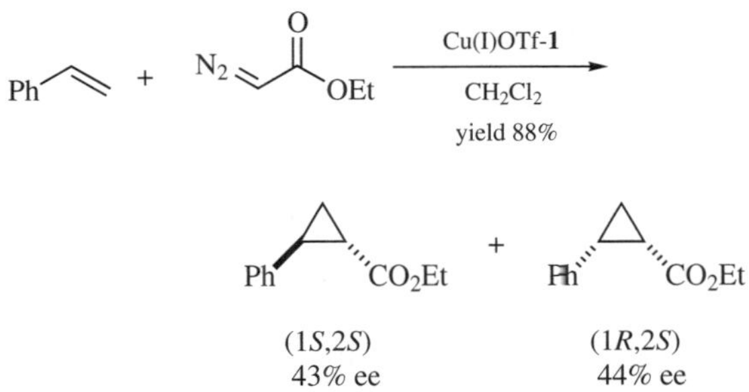

(3)

Enantioasymmetric Polymerizations[6]

The nickel complex of **1** has been employed as initiator to enantioasymmetrically polymerize γ-benzyl-glutamate-*N*-carboxyanhydride (NCA). This process allows polypeptides to be prepared with defined chain lengths and with narrow molecular weight distributions. The resultant products not only have pharmaceutical value but also possible relevance to the origins of handedness in biological macromolecules. The success of this ligand was determined by evaluating its ability to separately homopolymerize L- and D-Glu NCA, where it was assumed that k_D/k_L would be a good crude measure of initiator enantioselectivity (eqs 4 and 5). The nickel initiator prepared from **1** gave a $k_L/k_D = 2.5$ (0.1), indicating that it selectively polymerized L-Glu NCA faster than D-Glu NCA. Mechanistic data on the reason for this is unknown.

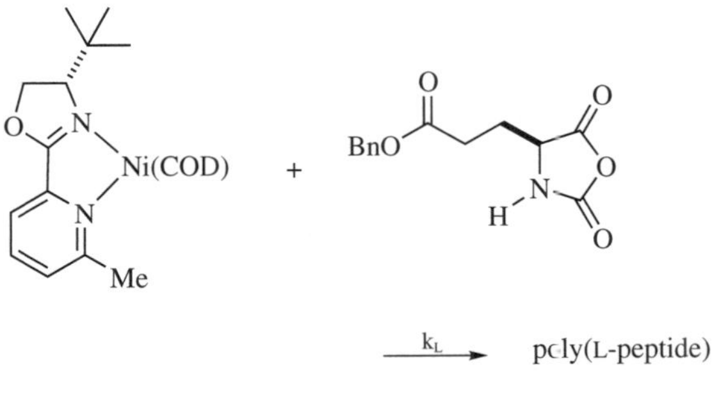

$\xrightarrow{k_L}$ poly(L-peptide) (4)

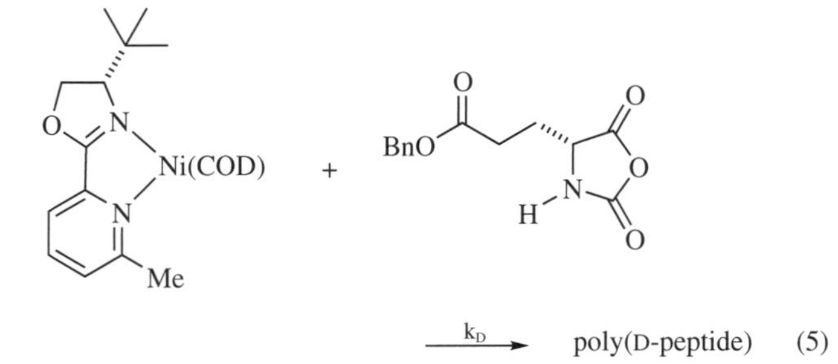

1. Chelucci, G.; Medici, S.; Saba, S. *Tetrahedron Asymm.* **1997**, *8*, 3183.
2. Chelucci, G.; Medici, S.; Saba, S. *Tetrahedron Asymm.* **1999**, *10*, 543.
3. Chelucci, G.; Deriu, S. P.; Saba, A.; Valenti, R. *Tetrahedron Asymm.* **1999**, *10*, 1457.
4. Trost, B. M.; Murphy, D. J. *Organometallics* **1985**, *4*, 1143.
5. Chelucci, G.; Sanna, M. G.; Gladiali, S. *Tetrahedron* **2000**, *56*, 2889.
6. Cheng, J.; Deming, T. J. *Macromolecules* **1999**, *32*, 4745.

Margaret M. Faul
Eli Lilly and Co., Indianapolis, IN, USA

(4*S*)-4-(1,1-Dimethylethyl)-2-{1-[(11b*S*)-dinaphtho[2,1-d:1′,2′-f][1,3,2]dioxaphosphepin-4-yloxy]-1-methylethyl}-4,5-dihydrooxazole

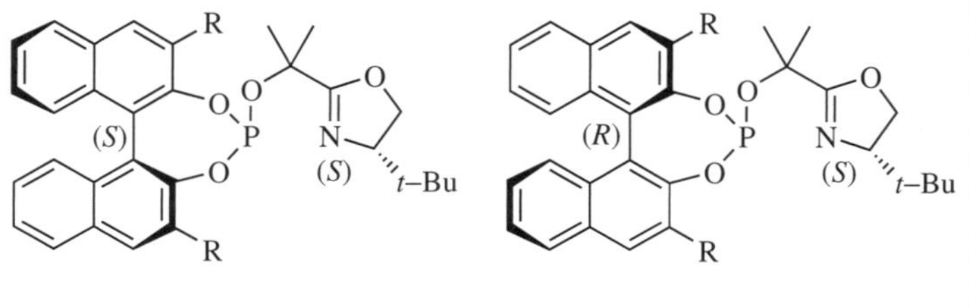

1–6a **1–5b**

(**1a**; (*S*,*S*), R = H);
[203399-79-5] $C_{31}H_{32}NO_3P$ (MW 497.56)
(**2a**; (*S*,*S*), R = Me);
[284019-78-9] $C_{33}H_{36}NO_3P$ (MW 525.62)
(**3a**; (*S*,*S*), R = Ph);
[284019-79-0] $C_{43}H_{40}NO_3P$ (MW 649.76)
(**4a**; (*S*,*S*), R = *p*-Ph-C_6H_4);
[284019-80-3] $C_{55}H_{48}NO_3P$ (MW 801.95)
(**5a**; (*S*,*S*), R = 2,4,6-$Me_3C_6H_2$);
[284019-81-4] $C_{49}H_{52}NO_3P$ (MW 733.92)
(**6a**; (*S*,*S*), R = 3,5-*t*$Bu_2C_6H_3$);
[284019-82-5] $C_{58}H_{70}NO_4P$ (MW 876.15)
(**1b**; (*R*,*S*), R = H);
[203312-03-2] $C_{31}H_{32}NO_3P$ (MW 497.56)
(**2b**; (*R*,*S*), R = Me);
[203312-04-3] $C_{33}H_{36}NO_3P$ (MW 525.62)
(**3b**; (*R*,*S*), R = Ph);
[284019-32-5] $C_{43}H_{40}NO_3P$ (MW 649.76)
(**4b**; (*R*,*S*), R = *p*-Ph-C_6H_4);
[284019-33-6] $C_{55}H_{48}NO_3P$ (MW 801.95)
(**5b**; (*R*,*S*), R = 2,4,6-$Me_3C_6H_2$);
[284019-35-8] $C_{49}H_{52}NO_3P$ (MW 733.92)

(modular chiral ligands for regio- and enantiocontrolled palladium-catalyzed allylic substitution reactions[1,2] and enantioselective copper-catalyzed 1,4-addition of organozinc reagents to enones[3,4])

Physical Data: (**1a**) colorless solid, mp 94 °C, $[\alpha]_D^{25}$+269 (*c* 3.10, $CHCl_3$); (**1b**) colorless solid, $[\alpha]_D^{25}$-360 (*c* 0.49, $CHCl_3$); (**2a**) colorless solid, $[\alpha]_D^{25}$+339 (*c* 0.45, $CHCl_3$); (**2b**) colorless solid, $[\alpha]_D^{25}$−379 (*c* 0.92, $CHCl_3$); (**3a**) colorless solid, mp 121 °C, $[\alpha]_D^{20}$+312 (*c* 0.46, $CHCl_3$); (**3b**) colorless solid, mp 106 °C, $[\alpha]_D^{25}$−366 (*c* 0.94, $CHCl_3$); (**4a**) colorless solid, mp 121 °C, $[\alpha]_D^{25}$ +253 (*c* 1.19, $CHCl_3$); (**4b**) colorless solid, mp 125 °C, $[\alpha]_D^{25}$−301 (*c* 0.89, $CHCl_3$); (**5a**) colorless solid, mp 130 °C, $[\alpha]_D^{23}$+74 (*c* 1.08, $CHCl_3$); (**5b**) colorless solid, mp 223 °C, $[\alpha]_D^{23}$−126 (*c* 0.52, $CHCl_3$); (**6a**) colorless solid, mp 113 °C, $[\alpha]_D^{23}$+164 (*c* 0.41, $CHCl_3$).

Purification: column chromatography on aluminum oxide (basic). Silica gel can also be used, however, with highly active silica gel, partial hydrolysis of the phosphite was observed.

Solubility: insoluble in H_2O; soluble in most organic solvents.

Handling, Storage, and Precautions: phosphite oxazolines of this type are sufficiently stable to be handled in air. For longer periods of time, they should be stored at −20 °C under nitrogen or argon.

Preparative Methods: Preparation of the phosphite-oxazoline ligands and metal complexes: the phosphite-oxazoline ligands are readily prepared in enantiomerically pure form from the BINOL derivative **7** and the oxazoline **8** (eq 1).[4] The BINOL derivative (**7**) is synthesized from the corresponding diol and phosphorus trichloride; oxazoline **8** is synthesized from commercially available (*S*)-*tert*-leucinol (eq 2).[5] By varying the R groups on **7**, a range of ligands can easily be synthesized. The modular design of the phosphite-oxazoline ligands allows a wide range of analogs to be readily prepared. Palladium and zinc complexes of the phosphite-oxazoline ligands are generally formed in situ. Palladium-allyl complexes have been prepared and characterized by NMR spectroscopy and X-ray diffraction.[2,6]

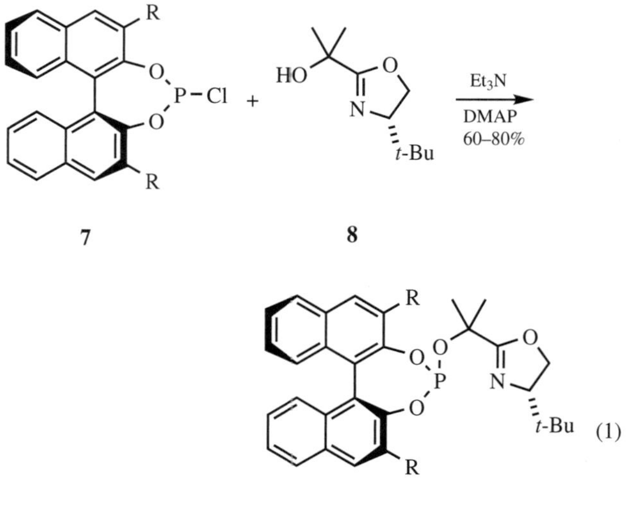

7 **8**

1–6a

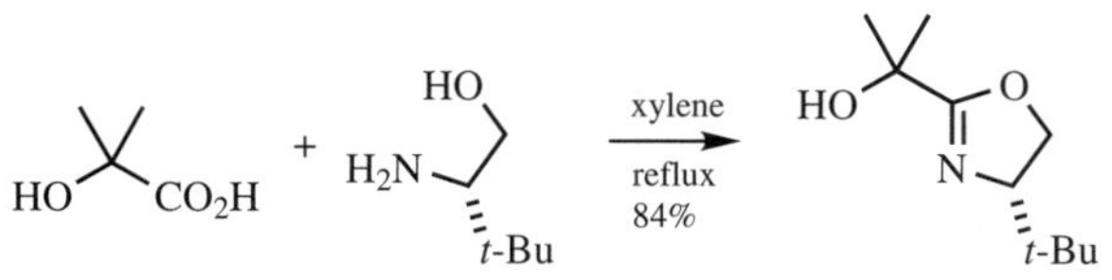

(2)

8

Regio- and Enantiocontrolled Palladium-Catalyzed Allylic Substitution Reactions.[7] Although a wide range of efficient catalysts are available for enantioselective allylic substitution reactions of substrates such as **9**, monosubstituted allylic substrates **10** and **11** generally react predominantly at the unsubstituted allyl terminus with these catalysts, producing achiral products (**13**) (eq 3). Palladium complexes of chiral phosphite-oxazoline ligands show improved regioselectivity favoring the chiral product with good enantioselectivity for monosubstituted aryl-allyl substrates.[1,2]

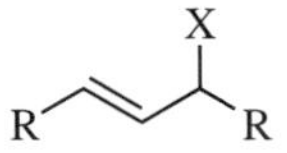

9 R = Me, Et, *i*-Pr, Ph

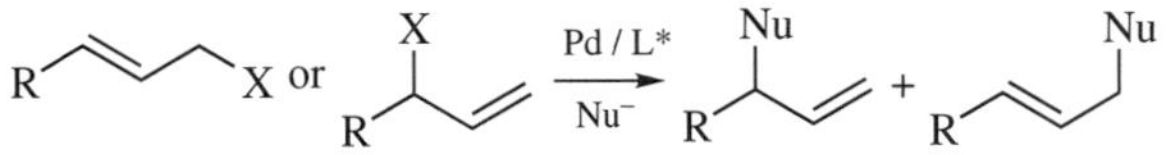

(3)

10 **11** **12** **13**

Reaction of the palladium complex of ligand **1a** with **10a** in the presence of *N*,*O*-bis(trimethylsilyl)acetamide (BSA), catalytic KOAc as the base, and dimethyl malonate results in good yield and high selectivities for **12a** (eq 4). The most efficient ligand in terms of regio- and enantioselectivity is **1a**. In benzene, the regio- and enantioselectivity are further improved (Table 1).

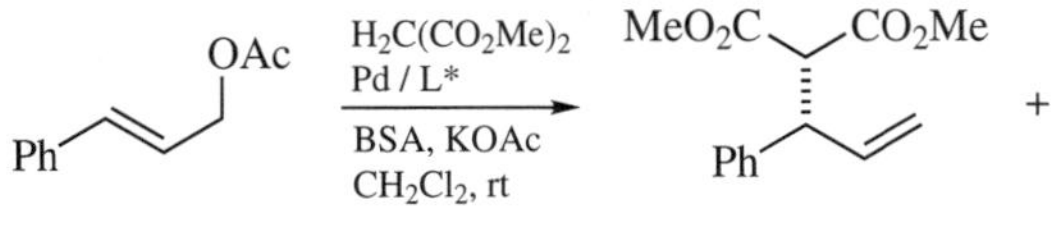

10a **12a**

L* = **1a**, 84%, **12a**/**13a** 69:31, **12a** 86% ee
L* = **1b**, 80%, **12a**/**13a** 46:54, **12a** 79% ee
L* = **2a**, 88%, **12a**/**13a** 55:45, **12a** 92% ee

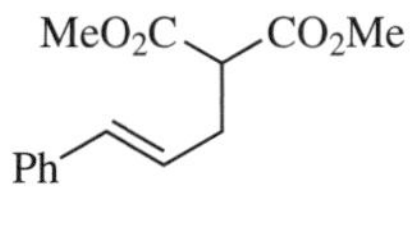

(4)

13a

Even better regio- and enantioselectivities were observed when 1-naphthyl-substituted allylic acetates (**10b** and **11b**) were used. The regio- and enantioselectivities were essentially the same using either the achiral substrates (**10**) or the racemic isomers (**11**) (eq 3, Table 1).

There have been several other reports of allylic substitution reactions that proceed with high selectivity for the chiral product **12**. Tungsten-phosphinooxazoline complexes give enantioselectivities of up to 96% ee and branched-to-linear ratios of up to 96:4 with aryl-allyl substrates.[2] Molybdenum-catalyzed allylic substitution reactions have been reported by Trost and by Pfaltz. Molybdenum complexes with a tetradentate nitrogen ligand (derived from *trans*-1,2-diaminocyclohexane) gave excellent branched to linear ratios (up to 99:1, generally >20:1) and high enantiomeric excesses (up to 99%) also for aryl-allyl substrates.[8] The related bisoxazolines with a *trans*-1,2-diaminocyclohexane backbone gave branched to linear ratios of 2:1 to 49:1 for a range of aryl- and alkyl-allyl substrates with enantiomeric excesses generally >90%.[9] Iridium complexes with phosphoramidite ligands developed by Helmchen are also efficient catalysts, giving branched to linear ratios of up to 99:1 with ees of up to 91%.[10]

Table 1 Allylic substitution of substrates **10** and *rac*-**11** using ligand **1a** [eq 3, X = OAc, Nu = $CH(CO_2Me)_2$][a]

	R	Yield (%)	**12**/**13**	ee of **12** (%)
10b	1-Naphthyl	87	95:5	94 (*S*)
11b	1-Naphthyl	91	96:4	96 (*S*)
10c	2-Naphthyl	72	77:23	88
11c	2-Naphthyl	71	74:26	89
10a	Ph	86	76:24	90[b]
11a	Ph	82	66:34	88 (*S*)

[a] 1 mol% $[Pd(C_3H_5)Cl]_2$, 2.4 mol% L , 50 C, CH_2Cl_2, 2 h; 2 equiv of $CH_2(CO_2Me)_2$ and *N*,*O*-bis(trimethylsilyl)-acetamide (BSA), 4 mol% KOAc, 23 C, 18 h.
[b] Reaction performed in benzene.

Table 2 Asymmetric conjugate addition to enones **14** (eq 5)

n	Ligand	GC yield (%)	ee (%) of **15**	Product con guration
3	**2b**	96	80	(−)
3	**4a**	99	83	(-)
3	**5a**	96	82	(−)
2	**2b**	96	90	(*R*)
2	**4a**	97	86	(*S*)
1	**3b**	49	91	(*R*)
1	**4b**	41	94	(*R*)
1	**5b**	7	25	(*S*)

Enantioselective 1,4-Addition of Organozinc Reagents to Enones.[11] Phosphite-oxazoline copper complexes are highly efficient catalysts for the 1,4-addition of organozinc reagents to 5-, 6- and 7-membered cyclic enones.[3,4] Both the chiral oxazoline and the chiral phosphite unit have a significant influence on the enantioselectivity.

The chiral ligands are used in a ligand to copper ratio of 1.2:1 along with 2-3 mol% of $Cu(OTf)_2$ and 1.3 equiv of diethylzinc in toluene, usually for 15 h. All ligands formed catalysts which were highly reactive in the reaction with cycloheptenone (**14**, $n = 3$) with enantiomeric excess reaching >80% (eq 5, Table 2). Surprisingly, the product configuration was reversed going from ligand **4a** to **5a**, whilst the enantiomeric excesses were almost identical. Excellent yields and enantioselectivities were obtained in the reaction with cyclohexenone (eq 5, $n = 2$, Table 2). In each of the above cases, there is no obvious correlation between steric bulk in the ligand and the observed enantioselectivity. Unsurprisingly, only moderate yields were obtained for the addition to cyclopentenone (eq 5, $n = 1$, Table 2). This is a general problem with this substrate; although the reaction goes to full conversion, a number of by-products are formed containing more than one cyclopentenone unit, because the enolate produced in the 1,4-addition has a high tendency to add to cyclopentenone. Bulky ligands resulted in reduced enantioselectivity and the (*R*,*S*) diastereoisomer (**2**–

5b) gave higher enantioselectivities than the corresponding (*S*,*S*) isomer (**2–5a**).

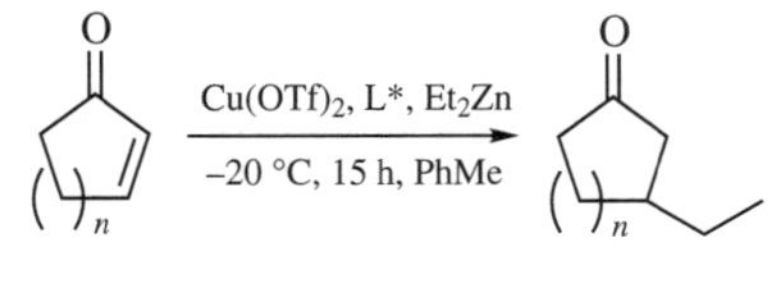

(5)

14 **15**

Acyclic substrates were also investigated and promising results were obtained with *trans*-4-phenylbut-3-en-2-one (eq 6).

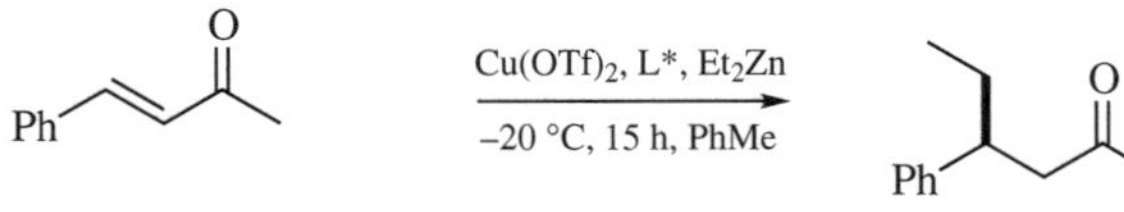

(6)

L* = **4a**, 90%, 58% ee
L* = **5a**, 70%, 59% ee
L* = **6a**, 99%, 87% ee

Several other phosphorus ligands produce high enantioselectivities in the 1,4-addition of organozinc reagents. A range of chiral phosphites has been investigated by Alexakis et al. with enantioselectivities of up to 96% for the addition of diethylzinc to cyclohenenone.[12] Yan and Chan have used chiral diphosphites and achieved enantiomeric excesses of 89-90% in the addition of diethylzinc to cyclohexenone and cyclopentenone.[13] Feringa has developed a range of phosphoramidites for the 1,4-diethylzinc additions.[14,15] Enantioselectivities of >98% have been reported for the addition to cyclohexenone and up to 82% for acyclic substrates. Hu et al. have used *P*,*N*-ligands derived from binaphthyl, recording enantiomeric excesses of 90% for the addition of diethylzinc to cyclohenanone and 98% for arylsubstituted acyclic enones.[16] The best reported method for the addition of a range of dialkylzincs to several different cyclopentenones has been reported by Degrado et al. using peptide-based *P*,*N*-ligands.[17] Isolated yields were 55-92% with enantioselectivities as high as >98%. The same ligands also gave excellent results (>95% ee) for the addition of dialkylzincs to cylohexenones and cycloheptenones.

Related Reagents. Phosphinoxazolines (PHOX ligands), BINAP, chiraphos, bisoxazolines.

1. Prétôt, R.; Pfaltz, A. *Angew. Chem.* **1998**, *110*, 337; *Angew. Chem., Int. Ed. Engl.* **1998**, *37*, 323.
2. Prétôt, R.; Lloyd-Jones, G. C.; Pfaltz, A. *Pure Appl. Chem.* **1998**, *70*, 1035.
3. Knöbel, A. K. H.; Escher, I.; Pfaltz, A. *Synlett* **1997**, 1429.
4. Escher, I. H.; Pfaltz, A. *Tetrahedron* **2000**, *56*, 2879.
5. (a) Allen, J. V.; Williams, J. M. J. *Tetrahedron: Asymm.* **1994**, *5*, 277. (b) Pridgen, L. N.; Miller, G. *J. Heterocyclic Chem.* **1983**, *20*, 1223.
6. Prétôt, R., PhD Thesis, University of Basel, 1997.
7. Pfaltz, A.; Lautens, M., In *Comprehensive Asymmetric Catalysis*; Jacobsen, E. N.; Pfaltz, A.; Yamamoto, H., Eds.; Springer: Berlin, 1999, Vol. 2, p 833.
8. Trost, B. M.; Hachiya, I. *J. Am. Chem. Soc.* **1998**, *120*, 1104.
9. Glorius, F.; Pfaltz, A. *Org. Lett.* **1999**, *1*, 141.
10. Bartels, B.; Helmchen, G. *Chem. Commum.* **1999**, 741.
11. Tomioka, K.; Nagaoka, Y., In *Comprehensive Asymmetric Catalysis*; Jacobsen, E. N.; Pfaltz, A.; Yamamoto, H., Eds.; Springer: Berlin, 1999, Vol. 3, p 1105.
12. Alexakis, A.; Burton, J.; Vastra, J.; Benhaim, C.; Fournioux, X.; van de Heuvel, A.; Levêque J.-M.; Mazé, F.; Rosset, S. *Eur. J. Org. Chem.* **2000**, 4011.
13. Yan, M.; Chan, A. S. C. *Tetrahedron Lett.* **1999**, *40*, 6645
14. Feringa, B. L. *Acc. Chem. Res.* **2000**, *33*, 346.
15. Arnold, L. A.; Imbos, R.; Mandoli, A.; de Vries, A. H. M.; Naasz, R.; Feringa, B. L. *Tetrahedron* **2000**, *56*, 2865.
16. Hu, X.; Chen, H.; Zhang, X. *Angew. Chem.* **1999**, *111*, 3720; *Angew. Chem., Int. Ed. Engl.* **1999**, *38*, 3518.
17. Degrado, S. J.; Mizutani, H.; Hoveyda, A. H. *J. Am. Chem. Soc.* **2001**, *123*, 755.

Jonathan A. Medlock & Andreas Pfaltz
University of Basel, Switzerland

Dimethyl L-Tartrate[1]

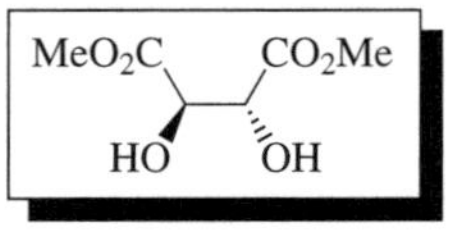

[608-68-4] $C_6H_{10}O_6$ (MW 178.14)

(synthon for chiral auxiliary/ligand preparation; chiral ligand for asymmetric catalysis)

Physical Data: mp 57–60 °C (dec); bp 163 °C/23 mmHg; *d* 1.238 g cm^{-3}; $[\alpha]^{23}$ +21° (c = 2.5, H_2O).
Solubility: sol most organic solvents, water.
Form Supplied in: moist white crystals.
Handling, Storage, and Precautions: may cause irritation; store in a cool, dry place.

Synthon for Chiral Auxiliary/Ligand Preparation. (*R*,*R*)-Dimethyl L-tartrate, a derivative of natural L-tartaric acid, is readily transformed into useful chiral auxiliaries and ligands for asymmetric synthesis.[1] Many of these transformations involve initial acetalization of the 2,3-vicinal hydroxy groups.[2–6] Thus $\alpha,\alpha,\alpha',\alpha'$-tetraaryl-1,3-dioxolane-4,5-dimethanols (TADDOLs) are prepared from the corresponding acetals of dimethyl L-tartrate (eq 1) and, as the Ti^{IV} complexes (see *2,2-Dimethyl-α,α,α′,α′-tetraphenyl-1,3-dioxolane-4,5-dimet hanolatotitanium Diisopropoxide*), are useful catalysts for enantioselective additions of dialkylzinc reagents to aldehydes,[5] Diels–Alder reactions,[6,7] asymmetric hydrocyanations of aldehydes,[8] and others.[9] The enolate of the acetonide of dimethyl L-tartrate alkylates with good selectivity to give, after acetonide hydrolysis, allyl- or benzylated derivatives (eq 2).[10] Orthoesters are also prepared from dimethyl tartrate, and acylate silyl enol ethers in good yield and selectivity to give monoprotected 1,3-diketones (eq 3).[11]

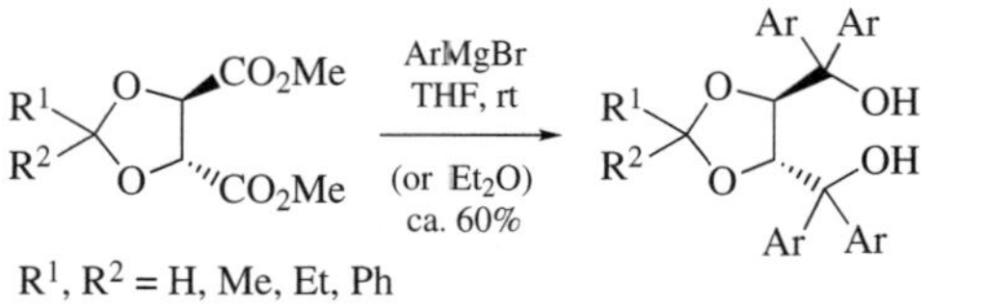

(1)

R^1, R^2 = H, Me, Et, Ph

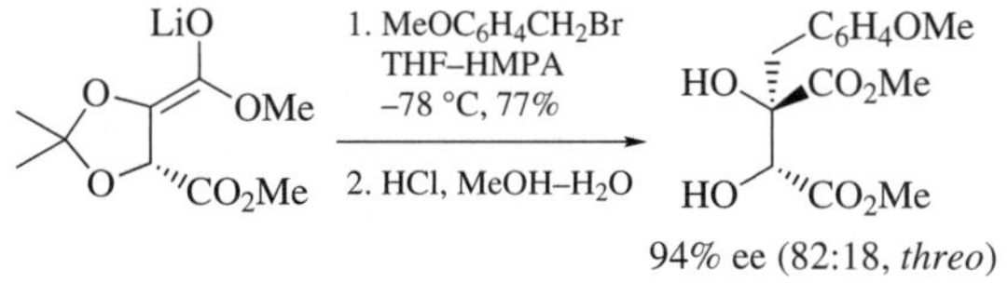

(2)

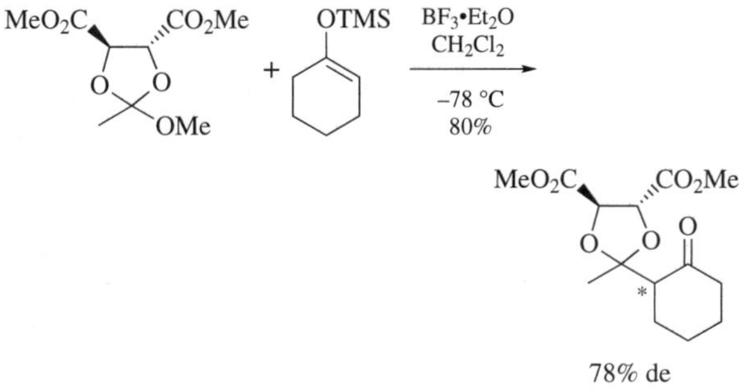

(3)

In an industrial asymmetric synthesis en route to the antiinflammatory agent naproxen, the dimethyl L-tartrate acetals of ethyl aryl ketones are brominated in high yield and selectivity to give the corresponding α-bromo derivatives.[12,13] Subsequent stereospecific Ag^{I}-promoted 1,2-aryl migration provides the 2-alkyl-2-arylacetic acid after hydrolysis of the tartrate auxiliary, which is recovered (e.g. eq 4).

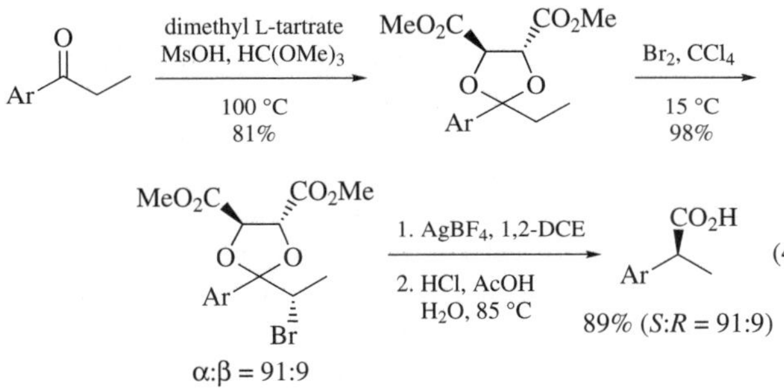

(4)

The vicinal diol cyclic sulfate from dimethyl tartrate undergoes nucleophilic opening to give substituted malate esters.[14,15] However, for this application diethyl and diisopropyl L-tartrates give superior yields and selectivities. The asymmetric cyclopropanation of the 1-alkenylboronic ester derived from dimethyl L-tartrate (eq 5) is another example where other tartaric acid derivatives surpass the performance of dimethyl tartrate.[16]

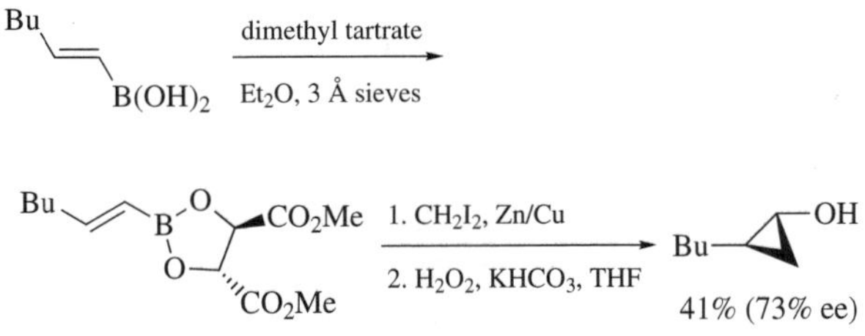

(5)

Chiral Ligand for Asymmetric Catalysis. Dimethyl L-tartrate is a demonstrated chiral ligand for the Ti^{IV}-catalyzed asymmetric epoxidation of allylic alcohols (Sharpless epoxidation),[17] and the Zn^{II}-mediated asymmetric cyclopropanation of allylic alcohols (Simmons–Smith reaction), see *Iodomethylzinc Iodide*.[18] Enantioselectivities in these reactions are, however, better with the use of diethyl L-tartrate as the chiral modifier.

1. Seebach, D.; Hungerbühler, E. *Modern Synthetic Methods*; Scheffold, R., Ed.; Salle-Sauerländer: Aarau, 1980; Vol. 2, pp 91–171.
2. Carmack, M.; Kelley, C. J. *J. Org. Chem.* **1968**, *33*, 2171.
3. Musich, J. A.; Rapoport, H. *J. Am. Chem. Soc.* **1978**, *100*, 4865.
4. Ott, J.; Ramos Tombo, G. M.; Schmid, B.; Venanzi, L. M.; Wang, G.; Ward, T. R. *Tetrahedron Lett.* **1989**, *30*, 6151
5. Seebach, D.; Plattner, D. A.; Beck, A. K.; Wang, Y. M.; Hunziker, D.; Petter, W. *Hel. Chim. Acta* **1992**, *75*, 2171, and references cited therein.
6. Narasaka, K.; Iwasawa, N.; Inoue, M.; Yamada, T.; Nakashima, M.; Sugimori, J. *J. Am. Chem. Soc.* **1989**, *111*, 5340.
7. Corey, E. J.; Matsumura, Y. *Tetrahedron Lett.* **1991**, *32*, 6289.
8. Minamikawa, H.; Hayakawa, S.; Yamada, T.; Iwasawa, N.; Narasaka, K. *Bull. Chem. Soc. Jpn.* **1988**, *61*, 4379.
9. Narasaka, K. *Synthesis* **1991**, 1.
10. Naef, R.; Seebach, D. *Angew. Chem., Int. Ed. Engl.* **1981**, *20*, 1030.
11. Longobardo, L.; Mobbili, G.; Tagliavini, E.; Trombini, C.; Umani-Ronchi, A. *Tetrahedron* **1992**, *48*, 1299.
12. Castaldi, G.; Cavicchioli, S.; Giordano, C.; Uggeri, F. *J. Org. Chem.* **1987**, *52*, 3018.
13. Giordano, C.; Coppi, L.; Restelli, A. *J. Org. Chem.* **1990**, *55*, 5400.
14. Gao, Y.; Sharpless, K. B. *J. Am. Chem. Soc.* **1988**, *110*, 7538.
15. Gao, Y.; Zepp, C. M. *Tetrahedron Lett.* **1991**, *32* 3155.
16. Imai, T.; Mineta, H.; Nishida, S. *J. Org. Chem.* **1990**, *55*, 4986.
17. Review: Finn, M. G.; Sharpless, K. B. In *Asymmetric Synthesis*; Morrison, J. D., Ed.; Academic: Orlando, 1985; Vol. 5, Chapter 8.
18. Ukaji, Y.; Nishimura, M.; Fujisawa, T. *Chem. Lett.* **1992**, 61.

Bruce A. Barner
Union Carbide Corporation, South Charleston, WV, USA

(*S,S*)-2,2′-(Dimethylmethylene)bis(4-*t*-butyl-2-oxazoline)[1]

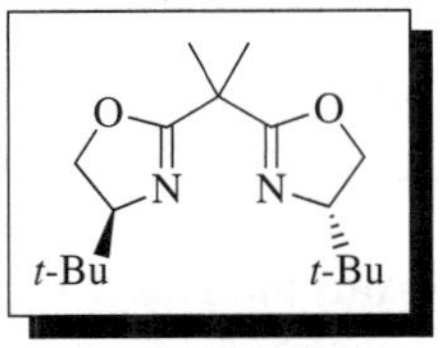

[131833-93-7] $C_{17}H_{30}N_2O_2$ (MW 294.49)

(versatile chiral ligands for enantiocontrol of metal-catalyzed reactions such as copper-catalyzed cyclopropanation[2–5] and aziridination of alkenes,[6] addition of cyanotrimethylsilane to aldehydes,[7] or Lewis acid-catalyzed Diels–Alder reactions[8,9])

Physical Data: mp 88–89 °C; $[\alpha]_D$ −108°; $[\alpha]_{365}$ −394° (*c* 0.97, CH_2Cl_2).

Solubility: insol H_2O; sol all common organic solvents.

Preparative Methods: ligand (**1**) and related C_2-symmetric bisoxazolines are readily prepared from chiral β-amino alcohols using standard methods for the synthesis of 2-oxazolines.[1] This is exemplified by the simple three-step procedure shown in eq 1, involving amide formation, conversion of the resulting bis(2-hydroxyalkyl)amide to the corresponding bis(2-chloroalkyl)amide, and subsequent base-induced cyclization.[3a,4,8a,10,11]

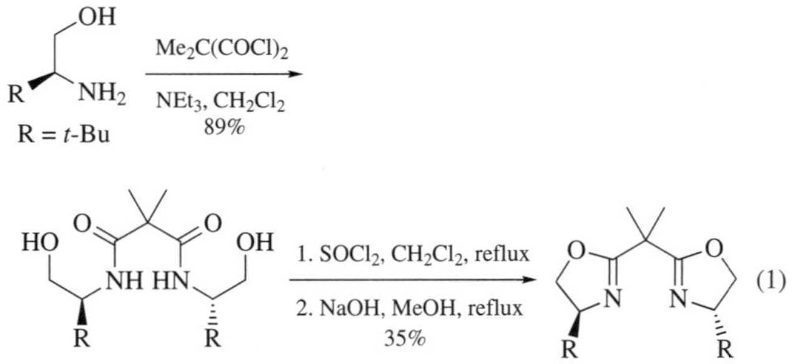

(1)

(1)

There are several other convenient one- and two-step syntheses leading to enantiomerically pure bisoxazolines, e.g. condensation of amino alcohols with dicarboxylic acids,[2,7] dinitriles,[10a,12] or diimino esters[4,10,13] (cf. eq 2),[4,10a] or acid-catalyzed cyclization of (2-hydroxyalkyl)amides.[8b] By these methods, various types of differently substituted bisoxazoline ligands are readily available in both enantiomeric forms, often in high overall yield.

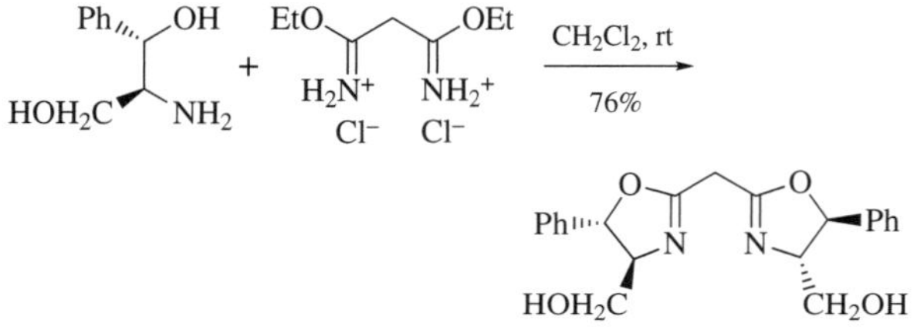

(2)

Purification: (**1**) can be purified by column chromatography (silica gel, hexane/EtOAc 7:3) and by recrystallization from pentane.

Handling, Storage, and Precautions: as a crystalline solid, (**1**) is stable at ambient temperature; for longer periods, storage at −20 °C is recommended.

C_2-Symmetric Bisoxazolines as Ligands in Asymmetric Catalysis. Methylenebis(oxazolines) such as (**1**), (**3**), and (**5**) are patterned after the semicorrins,[1] which have been successfully employed as ligands in enantioselective Cu-catalyzed cyclopropanations and other reactions (see *(1S,9S)-1,9-Bis{[(t-butyl)dimethylsilyloxy]methyl-}5-cyanosemico rrin*). The potential of bisoxazoline ligands of this type, which has been recognized independently by a number of research groups,[1–11,13–15] is demonstrated by a remarkable variety of different applications in asymmetric catalysis.

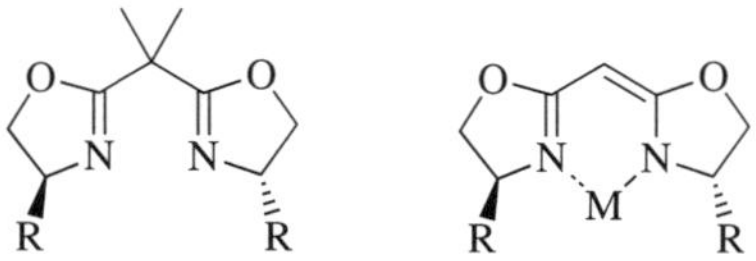

(**1**) R = t-Bu
(**2**) R = Ph

(**3**)

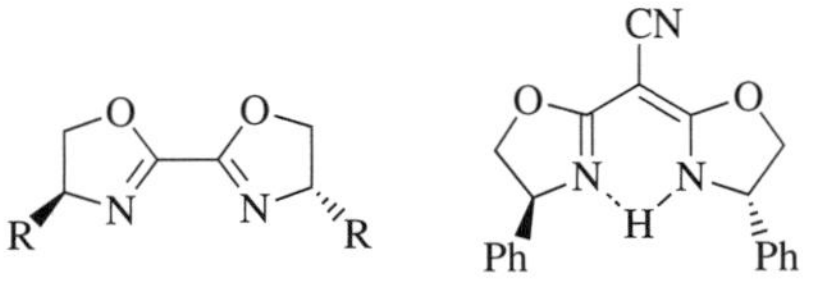

(**4**) (**5**)

The short and simple syntheses of these compounds and the ease of modifying their structures make them ideal ligands for the stereocontrol of metal-catalyzed reactions. Using different amino alcohols and dicarboxylic acid derivatives as precursors, the steric and electronic properties, as well as the coordination geometry, can be adjusted to the specific requirements of a particular application. The neutral methylenebis(oxazoline) ligands (**1**) and (**2**), which form six-membered chelate rings, the bioxazolines (**4**), a class of neutral ligands with π-acceptor properties forming five-membered chelate rings, and the anionic methylenebis(oxazolines) of type (**3**) and (**5**) are representative examples.

Enantioselective Cyclopropanation of Alkenes. Cationic Cu^{I} complexes of methylenebis(oxazolines) such as (**1**), which have been developed by Evans and co-workers,[3] are remarkably efficient catalysts for the cyclopropanation of terminal alkenes with diazoacetates. The reaction of styrene with ethyl diazoacetate in the presence of 1 mol % of catalyst, generated in situ from *Copper(I) Trifluoromethanesulfonate* and ligand (**1**), affords the *trans*-2-phenylcyclopropanecarboxylate in good yield and with 99% ee (eq 3). As with other catalysts, only moderate *trans/cis* selectivity is observed. Higher *trans/cis* selectivities can be obtained with more bulky esters such as 2,6-di-*t*-butyl-4-methylphenyl[3] or dicyclohexylmethyl diazoacetate[5] (94:6 and 95:5, respectively). The efficiency of this catalyst system is illustrated by the cyclopropanation of isobutene, which has been carried out on a 0.3 molar scale using 0.1 mol % of catalyst derived from the (*R,R*)-enantiomer of ligand (**1**) (eq 4).[3] The remarkable selectivity of >99% ee exceeds that of Aratani's catalyst[16] which is used in this reaction on an industrial scale.

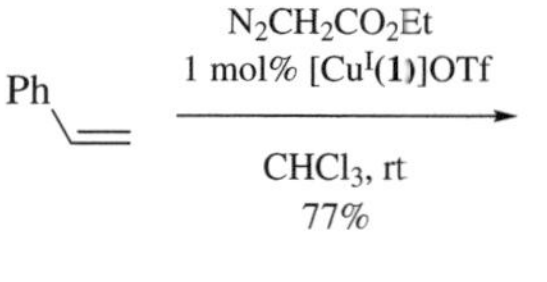

(3)

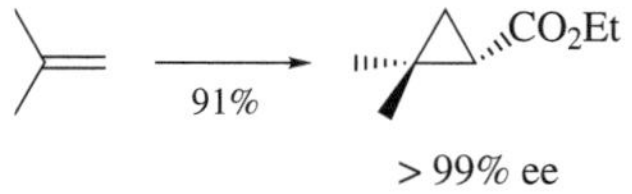

(4)

For the cyclopropanation of terminal mono- and disubstituted alkenes, the cationic Cu^{I} complex derived from ligand (**1**) is clearly the most efficient catalyst available today, giving consistently higher enantiomeric excesses than related neutral semicorrin[1,17] or bisoxazoline Cu^{I} complexes of type (**3**),[1,2,4] which can induce enantiomeric excesses of up to 92% ee in the cyclopropanation of styrene with ethyl diazoacetate. High enantioselectivities, ranging between the selectivities of the Evans catalyst (eq 3) and complex (**3**) (M = Cu^{I}, R = *t*-Bu), have also been observed with cationic Cu^{I} complexes of azasemicorrins.[1,10a,18]

For analogous cyclopropanation reactions of trisubstituted and 1,2-disubstituted (*Z*)-alkenes, ligand (**1**) is less well suited. In these cases, better results have been obtained with the bisoxazoline ligand (**6**).[5] This is illustrated by the enantioselective cyclopropanation of 1,5-dimethyl-2,4-hexadiene, leading to chrysanthemates (eq 5).[5] The enantioselectivity in this reaction is comparable to the best results reported for Aratani's catalyst.[16] Ligand (**6**) has also been reported to induce high enantiomeric excesses in the cyclopropanation of (*Z*)-4,4-dimethyl-2-pentene, (*Z*)-1-phenylpropene, and 1,1-dichloro-4-methyl-1,3-pentadiene with (−)-menthyl diazoacetate (92–95% ee).[5] A mechanistic model rationalizing the stereoselectivity of Cu catalysts of this type has been published;[17] a comparison of different cyclopropanation catalysts is also available.[19]

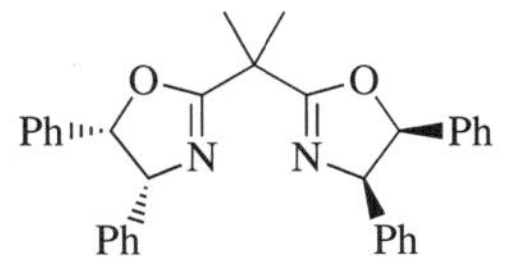

(**6**)

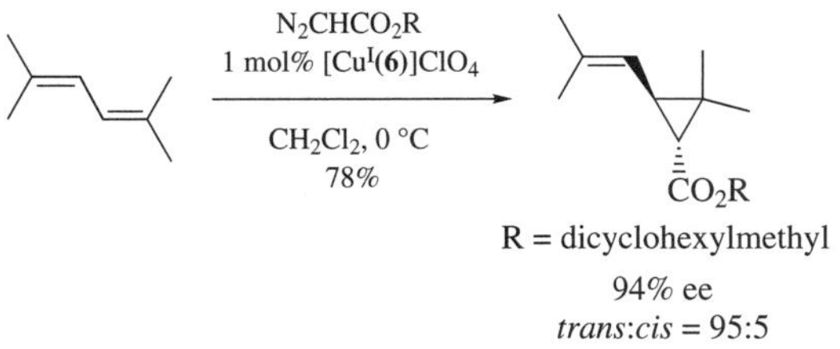

(5)

Enantioselective Aziridination of Alkenes. Copper complexes with neutral methylenebis(oxazoline) ligands (**1**) and (**2**) have also been employed as enantioselective catalysts for the reaction of alkenes with (*N*-tosylimino)phenyliodinane, leading to *N*-tosylaziridines.[6] The best results have been reported for cinnamate esters as substrates, using 5 mol % of catalyst prepared from CuOTf and the phenyl-substituted ligand (**2**) (eq 6). The highest enantiomeric excesses are obtained in benzene, whereas in more polar and Lewis basic solvents, such as acetonitrile, the selectivities are markedly lower. The chemical yield can be substantially improved by addition of 4Å molecular sieves. Both Cu^{I}– and Cu^{II}–bisoxazoline complexes, prepared from Cu^{I} or Cu^{II} triflate, respectively, are active catalysts, giving similar results. In contrast to the Cu-catalyzed cyclopropanation reactions discussed above, in which only Cu^{I} complexes are catalytically active, here Cu^{II} complexes are postulated as the actual catalysts.[6]

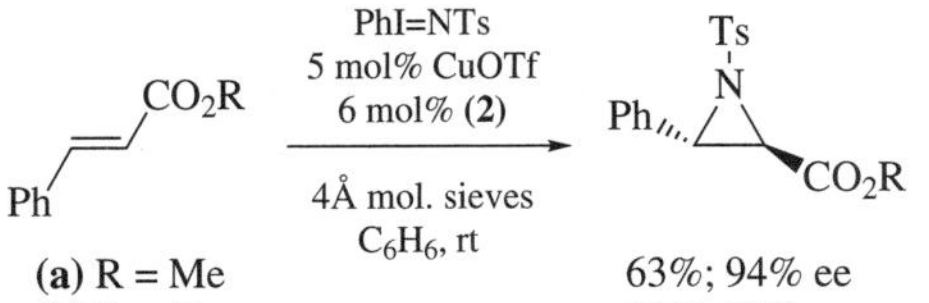

(6)

(**a**) R = Me — 63%; 94% ee
(**b**) R = Ph — 64%; 97% ee

Analogous naphthylacrylates also react with excellent enantioselectivity under these conditions. Styrene and (*E*)-β-methylstyrene afford the corresponding *N*-tosylaziridines with 63 and 70% ee, respectively. For these two substrates, the *t*-butyl-substituted bisoxazoline (**1**) rather than (**2**) proved to be the most effective ligand.

Similarly high enantioselectivities in aziridination reactions of this type have been reported for Cu catalysts with C_2-symmetric diimine ligands, derived from 1,2-diaminocyclohexane and aromatic aldehydes.[20] The best results in this case have been obtained with 7-cyano-2,2-dimethylchromene as substrate (>98% ee). At present, it is difficult to compare the diimine-based with the bisoxazoline-based catalysts because different substrates were examined in these studies, with the exception of styrene which gave very similar results with the two catalysts (66 and 63% ee, respectively).[20,6] Thus further work will be necessary to establish the full scope of these promising catalyst systems.

Enantioselective Diels–Alder Reactions. Methylenebis(oxazoline) complexes of Fe^{III}, Mg^{II}, and more recently also Cu^{II}, have been successfully employed as enantioselective Lewis acid catalysts in Diels–Alder reactions.[8,9] The most promising results have been obtained with Cu^{II} catalysts prepared from ligand (**1**) and *Copper(II) Trifluoromethanesulfonate* (eq 7).[9] In the presence of 10 mol % of catalyst in CH_2Cl_2 at −78 °C, acrylimide (**7a**) smoothly reacts with cyclopentadiene to afford the Diels–Alder product (**8a**) in 86% yield with excellent enantio and *endo*/*exo* selectivity. The crotonate derivative (**7b**) is less reactive, but at higher temperature also undergoes highly selective cycloaddition with cyclopentadiene. The fumarate (**7c**) gives similar results. In terms of selectivity and efficiency, this catalyst system can compete against the most effective chiral Lewis acid catalysts developed so far.[21]

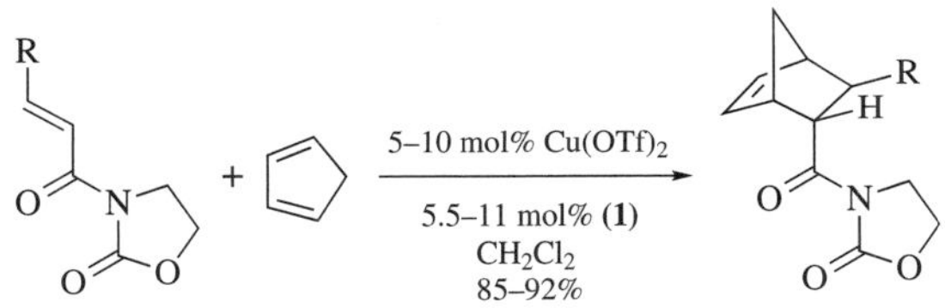

(7)

(**7**) (**8**)

(**a**) R = H	−78 °C	>98% ee	*endo*:*exo* = 98:2
(**b**) R = Me	−15 °C	97% ee	*endo*:*exo* = 96:4
(**c**) R = CO_2Et	−55 °C	95% ee	*endo*:*exo* = 94:6

The thiazolidine-2-thione analogs of (**7b**) and (**7c**) are more reactive dienophiles and, therefore, the cycloaddition can be carried out at lower temperature. However, the selectivities and yields are similar as with (**7b**) and (**7c**).[9] The corresponding cinnamate derivative (**7**) (R = Ph), on the other hand, reacts with substantially lower enantioselectivity than the corresponding thiazolidine-2-thione analog (90% vs. 97% ee).

The stereochemical course of these reactions has been rationalized assuming a chelate complex between the (bisoxazoline)Cu catalyst and the dienophile as the reactive intermediate, with square planar coordination geometry of the Cu^{II} ion.[9]

Enantioselective Cyanohydrin Formation. Magnesium complexes formed with the anionic semicorrin-type ligand (**5**) catalyze the addition of *Cyanotrimethylsilane* to aldehydes, leading to optically active trimethylsilyl-protected cyanohydrins.[7] In the presence of 20 mol % of the chloromagnesium complex (**9**), prepared from equimolar amounts of (**5**) and BuMgCl, cyclohexanecarbaldehyde is smoothly converted to the corresponding cyanohydrin derivative with 65% ee. Addition of 12 mol % of the bisoxazoline (**10**) results in a dramatic increase of enantioselectivity to 94% ee (eq 8). Replacement of (**10**) by its enantiomer reduces the selectivity to 38% ee. This remarkable effect has been proposed to arise from hydrogen-bond formation between the bisoxazoline (**10**) and HCN, which is generated in small amounts by hydrolysis of Me_3SiCN due to traces of water present in the reaction mixture. The chiral [(**10**)·sHCN] aggregate is postulated as the reactive species undergoing nucleophilic addition to the aldehyde which, at the same time, is activated by coordination with the chiral magnesium complex (**9**).

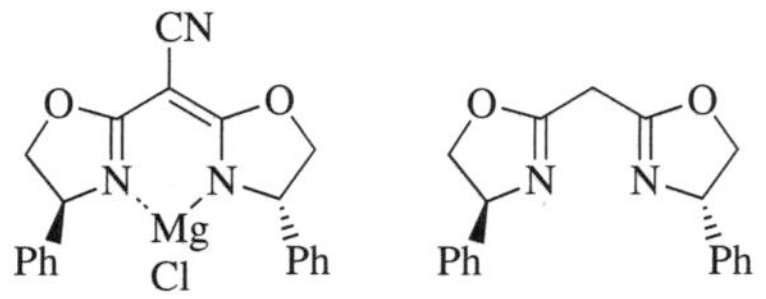

(9) **(10)**

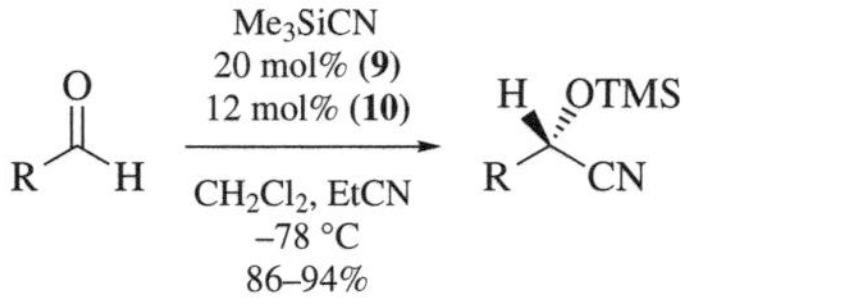

(8)

(**a**) R = C_6H_{13} 95% ee
(**b**) R = Et_2CH 91% ee
(**c**) R = Cy 94% ee

Heptanal, 2-ethylbutanal, and pivalaldehyde react with similarly high enantioselectivities, whereas benzaldehyde (52% ee) and certain α,β-unsaturated aldehydes such as geranial (63% ee) afford considerably lower enantiomeric excesses. Most other catalysts used for the addition of HCN or Me_3SiCN to aldehydes usually exhibit higher enantioselectivities with aromatic or α,β-unsaturated aldehydes than with alkyl carbaldehydes.[22]

Enantioselective Allylic Alkylation. Most ligands that have been employed in enantioselective Pd-catalyzed allylic substitutions are chiral diphosphines.[23] Recently, it has been found that chiral nitrogen ligands can also induce high enantioselectivities in such reactions.[1,18,24] The best results have been obtained with neutral azasemicorrin and methylenebis(oxazoline) ligands. In the presence of 1–2 mol % of catalyst, generated in situ from *Bis(allyl)di-μ-chlorodipalladium* and ligand (**11**), and a mixture of *N,O-Bis(trimethylsilyl)acetamide* (BSA) and catalytic amounts of KOAc in an apolar solvent like CH_2Cl_2 or toluene, racemic 1,3-diphenyl-2-propenyl acetate smoothly reacts with dimethyl malonate to afford the corresponding substitution product in high yield and with excellent enantioselectivity (eq 9).

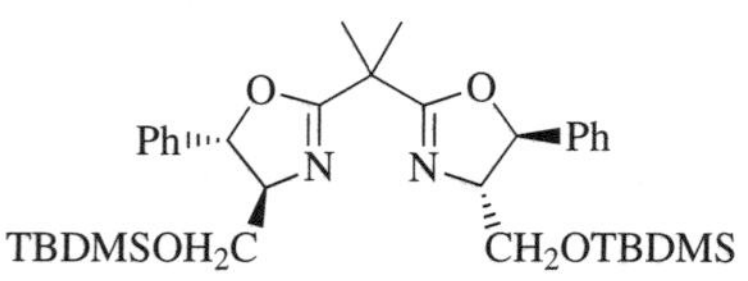

(11)

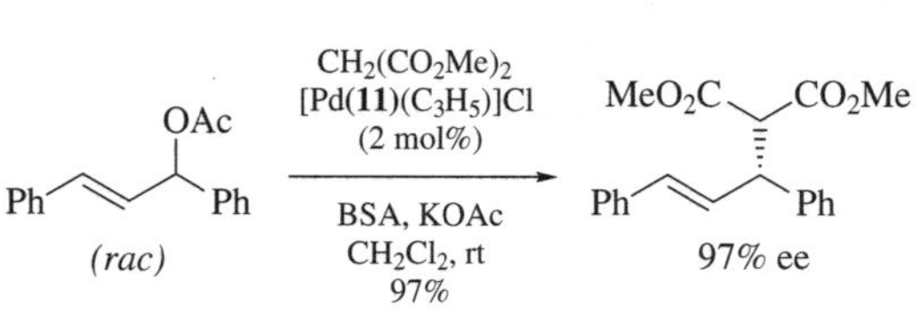

(9)

More recently, even higher selectivities of up to 99% ee have been achieved in this reaction with chiral phosphinooxazolines (see *(S)-2-[2-(Diphenylphosphino)phenyl]-4-phenyloxazoline*).[24-26] The application range of (bisoxazoline)Pd catalysts is limited to relatively reactive substrates such as aryl-substituted allylic acetates.[24] Analogous reactions of 1,3-dialkyl-2-propenyl acetates, for example, are impracticably slow and unselective. In this case, phosphinooxazolines have proved to be the ligands of choice.[24,25]

The crystal structures of some (allyl)Pd^{II}–bisoxazoline complexes have been determined by X-ray analysis.[1] The structural data of these complexes provide some clues about how the chiral ligand controls the stereochemical course of eq 9.

Other Applications. In the reactions discussed so far, methylenebis(oxazolines) were found to be superior to bioxazolines of type (**4**). However, there are some enantioselective metal-catalyzed processes for which the bioxazolines (**4**) are better suited than neutral or anionic methylenebis(oxazolines). Two examples, the Ir-catalyzed transfer hydrogenation of aryl alkyl ketones[4] and the Rh-catalyzed hydrosilylation of acetophenone,[11] are given in eq 10 and eq 11.

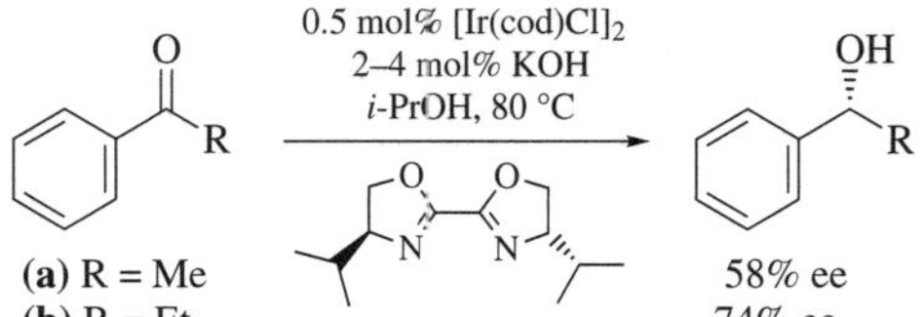

(10)

(**a**) R = Me 58% ee
(**b**) R = Et 74% ee
(**c**) R = *i*-Pr 91% ee

(12) (1.3 mol%)

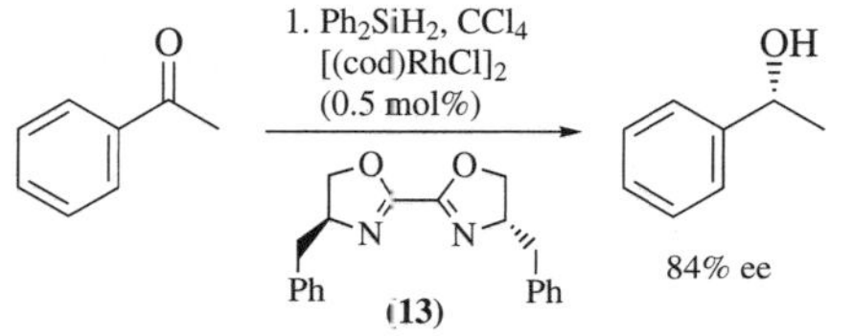

(11)

(13) (10 mol%)
2. H^+, H_2O
59%

84% ee

Using 1 mol % of catalyst generated in situ from *Di-μ-chlorobis(1,5-cyclooctadiene)diiridium(I)* and the bioxazoline (**12**) in refluxing isopropanol, various aryl alkyl ketones have been reduced in good yield with enantioselectivities ranging between 50–90% ee (eq 10).[10b] Dialkyl ketones are unreactive under these conditions. The highest enantiomeric excesses are obtained with phenyl isopropyl ketone (91% ee at 70% conversion, 88% ee at 93% conversion). Although these results compare favorably with the enantioselectivities reported for other Ir catalysts,[27] at present, (bioxazoline)Ir complexes cannot compete with the most efficient catalysts available for the enantioselective reduction of ketones[28] (see *Tetrahydro-1-methyl-3,3-diphenyl-1H,3H-pyrrolo[1,2-c][1,3,2]oxazaborole*). Recently, high enantioselectivities in the transfer hydrogenation of certain aryl alkyl ketones have been achieved with chiral samarium catalysts.[29]

The dibenzylbioxazoline derivative (**13**) has been found to induce up to 84% ee in the Rh-catalyzed hydrosilylation of acetophenone with diphenylsilane (eq 11).[11] A large excess of ligand relative to [Rh] is necessary for optimal selectivity. Analogous bithiazoline derivatives were also investigated, but gave lower selectivities. In this case too, there are more selective catalysts available which afford high enantiomeric excesses in the hydrolsilylation of a wide range of ketones.[30]

Bioxazolines have also been employed in the enantioselective dihydroxylation of alkenes with *Osmium Tetroxide*.[15] The best results have been obtained in the dihydroxylation of 1-phenylcyclohexene with a complex, formed between OsO_4 and the diisobutylbioxazoline (**4**) ($R = CH_2CHMe_2$), as a stoichiometric reagent (70% ee). Styrene and *trans*-stilbene afford enantioselectivities below 20% ee under these conditions (for highly enantioselective dihydroxylation catalysts,[31] see *Dihydroquinine Acetate* and *Osmium Tetroxide*).

Related Reagents. (*R*)-1,1′-Bi-2,2′-naphthotitanium Dichloride; (*R*)-1,1′-Bi-2,2′-naphthotitanium Diisopropoxide; Dihydroquinine Acetate; (1*S*,9*S*)-1,9-Bis{[(*t*-butyl)dimethylsilyloxy]-methyl}-5-cyanosemicorrin; 2,2-Dimethyl-α,α,α′,α′-tetraphenyl-1,3-dioxolane-4,5-dimethanolatotitanium Diisopropoxide.

1. Pfaltz, A. *Acc. Chem. Res.* **1993**, *26*, 339.
2. Lowenthal, R. E.; Abiko, A.; Masamune, S. *Tetrahedron Lett.* **1990**, *31*, 6005.
3. (a) Evans, D. A.; Woerpel, K. A.; Hinman, M. M.; Faul, M. M. *J. Am. Chem. Soc.* **1991**, *113*, 726. (b) Evans, D. A.; Woerpel, K. A.; Scott, M. J. *Angew. Chem.* **1992**, *104*, 439 *Angew. Chem., Int. Ed. Engl.* **1992**, *31*, 430.
4. Müller, D.; Umbricht, G.; Weber, B.; Pfaltz, A. *Helv. Chim. Acta* **1991**, *74*, 232.
5. Lowenthal, R. E.; Masamune, S. *Tetrahedron Lett.* **1991**, *32*, 7373.
6. Evans, D. A.; Faul, M. M.; Bilodeau, M. T.; Anderson, B. A.; Barnes, D. M. *J. Am. Chem. Soc.* **1993**, *115*, 5328.
7. Corey, E. J.; Wang, Z. *Tetrahedron Lett.* **1993**, *34*, 4001.
8. (a) Corey, E. J.; Imai, N.; Zhang, H.-Y. *J. Am. Chem. Soc.* **1991**, *113*, 728. (b) Corey, E. J.; Ishihara, K. *Tetrahedron Lett.* **1992**, *33*, 6807.
9. Evans, D. A.; Miller, S. J.; Lectka, T. *J. Am. Chem. Soc.* **1993**, *115*, 6460.
10. (a) Umbricht, G. Dissertation, University of Basel, 1993. (b) Müller, D. W. Dissertation, University of Basel, 1993.
11. Helmchen, G.; Krotz, A.; Ganz, K. T.; Hansen, D. *Synlett* **1991**, 257.
12. Witte, H.; Seeliger, W. *Justus Liebigs Ann. Chem./Liebigs Ann. Chem.* **1974**, 996.
13. Hall, J.; Lehn, J.-M.; DeCian, A.; Fischer, J. *Helv. Chim. Acta* **1991**, *74*, 1.
14. Onishi, M.; Isagawa, K. *Inorg. Chim. Acta* **1991**, *179*, 155.
15. Yang, R.; Chen, Y.; Dai, L. *Acta Chim. Sinica* **1991**, *49*, 1038 (*Chem. Abstr.* **1992**, *116*, 41 342v).
16. Aratani, T. *Pure Appl. Chem.* **1985**, *57*, 1839.
17. (a) Fritschi, H.; Leutenegger, U.; Pfaltz, A. *Helv. Chim. Acta* **1988**, *71*, 1553. (b) Piqué, C. Dissertation, University of Basel, 1993.
18. Leutenegger, U.; Umbricht, G.; Fahrni, C.; von Matt, P.; Pfaltz, A. *Tetrahedron* **1992**, *48*, 2143.
19. (a) Doyle, M. P. In *Catalytic Asymmetric Synthesis*; Ojima, I., Ed.; VCH: New York, 1993; pp 63–99. (b) Doyle, M. P. *Recl. Trav. Chim. Pays-Bas* **1991**, *110*, 305.
20. Li, Z.; Conser, K. R.; Jacobsen, E. N. *J. Am. Chem. Soc.* **1993**, *115*, 5326.
21. Kagan, H. B.; Riant, O. *Chem. Rev.* **1992**, *92*, 1007. Narasaka, K. *Synthesis* **1991**, 1.
22. (a) North, M. *Synlett* **1993**, 807. (b) Hayashi, M.; Miyamoto, Y.; Inoue, T.; Oguni, N. *J. Org. Chem.* **1993**, *58*, 1515.
23. (a) Consiglio, G.; Waymouth, R. M. *Chem. Rev.* **1989**, *89*, 257. (b) Howarth, J.; Frost, C. G.; Williams, J. M. J. *Tetrahedron: Asymmetry* **1992**, *3*, 1089.
24. von Matt, P. Dissertation, University of Basel, 1993.
25. von Matt, P.; Pfaltz, A. *Angew. Chem.* **1993**, *105*, 614 *Angew. Chem., Int. Ed. Engl.* **1993**, *32*, 566.
26. (a) Sprinz, J.; Helmchen, G. *Tetrahedron Lett.* **1993**, *34*, 1769. (b) Dawson, G. J.; Frost, C. G.; Williams, J. M. J.; Coote, S. J. *Tetrahedron Lett.* **1993**, *34*, 3149.
27. Zassinovich, G.; Mestroni, G.; Gladali, S. *Chem. Rev.* **1992**, *92*, 1051.
28. (a) Singh, V. K. *Synthesis* **1992**, 605. (b) Wallbaum, S.; Martens, J. *Tetrahedron: Asymmetry* **1992**, *3*, 1475.
29. Evans, D. A.; Nelson, S. G.; Gagné, M. R.; Muci, A. R. *J. Am. Chem. Soc.* **1993**, *115*, 9800.
30. Brunner, H.; Nishiyama, H.; Itoh, K. In *Catalytic Asymmetric Synthesis*; Ojima, I., Ed.; VCH: New York, 1993; pp 303–322.
31. (a) Johnson, R. A.; Sharpless, K. B. In *Catalytic Asymmetric Synthesis*; Ojima, I., Ed.; VCH: New York, 1993; pp 227–272. (b) Lohray, B. B. *Tetrahedron: Asymmetry* **1992**, *3*, 1317.

Andreas Pfaltz
University of Basel, Switzerland

[4*S*-(4α,5β)]-1-(1,3-Dimethyl-2-oxido-4,5-diphenyl-1,3,2-diazaphospholidine-2-yl)piperidine

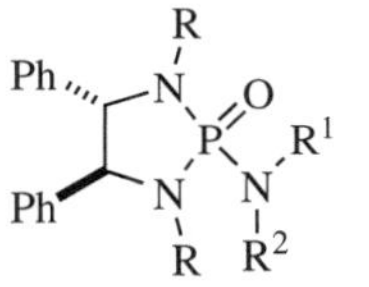

(*S,S*)-**1a**, R = Me, $R^1 = R^2 = -(CH_2)_5-$
(*S,S*)-**1b**, R = Ph, $R^1 = R^2 = -(CH_2)_5-$
(*S,S*)-**1c**, R = Me, $R^1 = R^2 = -(CH_2)_4-$
(*S,S*)-**1d**, $R = R^1 = R^2 = Me$
(*S,S*)-**1e**, R = Me, $R^1 = R^2 = i\text{-Pr}$
(*S,S*)-**1f**, R = Me, $R^1 = R^2 = n\text{-Pr}$
(*S,S*)-**1g**, R = Me, $R^1 = R^2 = Ph$
(*S,S*)-**1h**, R = Et, $R^1 = R^2 = -(CH_2)_5-$

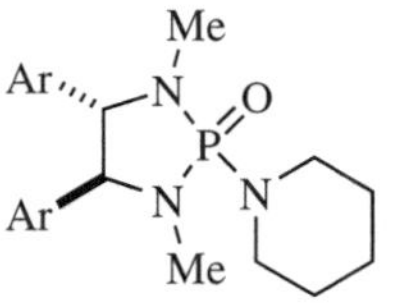

(*S,S*)-**2a**, Ar = $4\text{-}CF_3C_6H_4$
(*S,S*)-**2b**, Ar = $4\text{-}MeOC_6H_4$
(*S,S*)-**2c**, Ar = $3,5\text{-}Me_2C_6H_4$

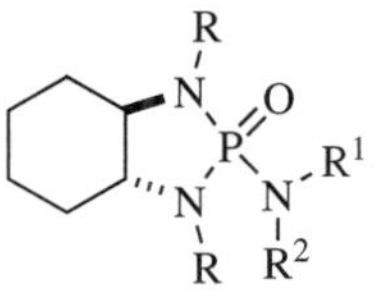

(*R,R*)-**3a**, R = Me, $R^1 = R^2 = -(CH_2)_5-$
(*R,R*)-**3b**, R = Me, $R^1 = -(CH_2)_4CH_3$, $R^2 = Me$
(*R,R*)-**3c**, R = Me, $R^1 = R^2 = i\text{-Pr}$
(*R,R*)-**3d**, R = Me, $R^1 = Me$, R^2 = (1′*S*)-PhCHMe
(*R,R*)-**3e**, R = CH_2Ph, $R^1 = R^2 = Me$

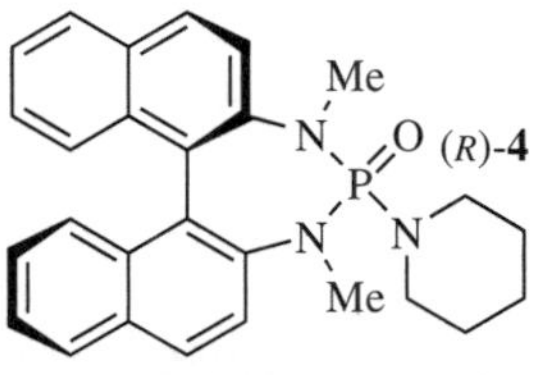

(*R*)-**4**

[180475-25-6] $C_{21}H_{28}N_3OP$ (MW 369.45)

(catalyst for asymmetric aldol additions of trichlorosilyl enolates to aldehydes,[1] catalyst for asymmetric allylations and crotylations of allylic trichlorosilanes to aldehydes,[2] catalyst for asymmetric ring opening of epoxides using silicon tetrachloride[3])

Physical Data: mp 110 °C; $[\alpha]_D^{22}$ +18.2 (*c* 1.3, $CHCl_3$).
Solubility: soluble in most organic solvents except hydrocarbons.
Form Supplied in: white needles.
Preparative Methods: The enantiopure phosphoramide (*S*,*S*)-**1a** is prepared from treatment of (*S*,*S*)-*N*,*N*′-dimethyl-1,2-diphenyl-1,2-ethanediamine (**5**) with piperidinylphosphoric dichloride (**6**) and triethylamine (eq 1).[4] Synthesis of related phosphoramides is easily accomplished by substituting the desired diamine backbone in the former procedure. Recrystallization of phosphoramide (*S*,*S*)-**1a** from hexane followed by drying over P_2O_5 allows for isolation of analytically pure material. Enantiopure 1,2-diphenyl-1,2-ethanediamine (**5**) can be prepared from benzil and cyclohexanone according to the method of Pikul & Corey.[5] Bis-formylation of the diamine with acetic formic anhydride followed by $LiAlH_4$ reduction affords *N*,*N*′-dimethyldiamine **5**. Alternatively, enantiopure **5** can be prepared directly from benzaldehyde and methylamine according to the method of Alexakis et al.[6] Piperidinylphosphoric dichloride (**6**) is prepared from piperidine, triethylamine and phosphorous oxychloride according to the modified procedure of Peyronel et al.[7]

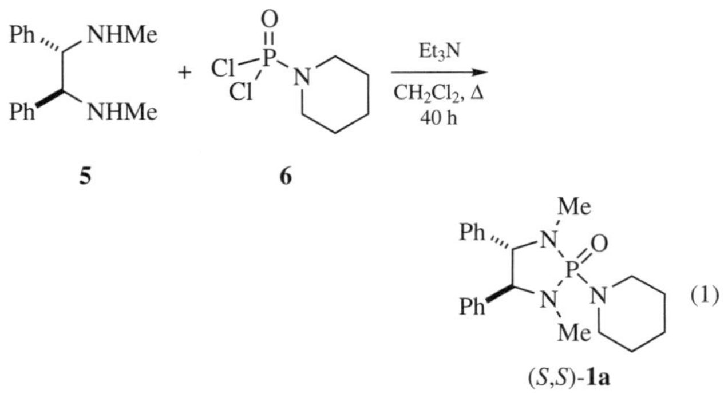

Handling, Storage, and Precautions: The phosphoramide is a stable, hygroscopic compound which is best stored in a dessicator or dry-box. Avoid prolonged exposure to air or moisture.

Introduction. [4*S*-(4α,5β)]-1-(1,3-Dimethyl-2-oxido-4,5-diphenyl-1,3,2-diazaphospholidine-2-yl)piperidine (**1a**) and its derivatives are an important class of phosphoramides which have seen much success as Lewis-basic catalysts for aldol additions,[8] allylations,[9] crotylations, and epoxide openings.[10] The three nitrogen subunits of the phosphoramide provide the opportunity for a large number of structurally diverse analogs, allowing a wide spectrum of properties and shapes to be customized.[11]

Ester Enolate Aldol Additions to Aldehydes. Among the first examples of aldol additions employing chiral Lewis bases as catalysts were the additions of trichlorosilyl ketene acetals to aldehydes.[12] Silyl ketene acetal **7** could be generated by metathesis of methyl tributylstannylacetate with $SiCl_4$. Treatment of **7** with benzaldehyde and 10 mol % of a phosphoramide in CH_2Cl_2 at −78 °C afforded aldol products in good to high yields with moderate enantioselectivities for all phosphoramides employed. Reaction of **7** with pivalaldehyde provided aldol products in similar yields and with slightly improved enantioselectivities. The increase in stereoselection is presumably attributed to a less competent background reaction inherent with additions to more sterically encumbered aldehydes (eq 2, Table 1).

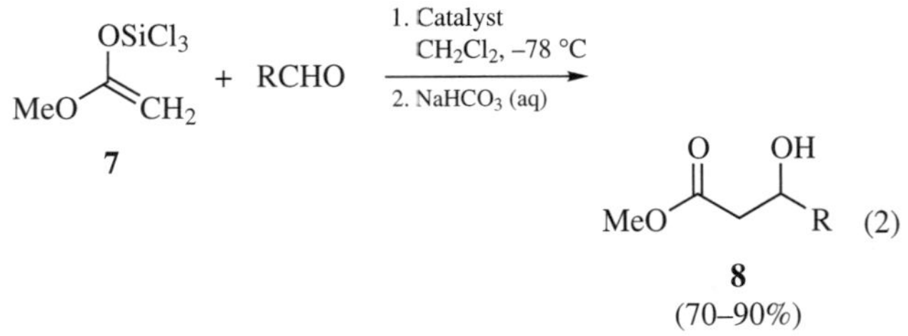

Table 1 Phosphoramide-catalyzed addition of ketene acetal **6** to aldehydes

Entry	Catalyst	Aldehyde	er (conÞguration)
1	(*S*,*S*)-**1**	PhCHO	2.0:1 (*S*)
2	(*R*,*R*)-**3**	PhCHO	1.5:1 (*R*)
3	(*R*)-**4**	PhCHO	1.6:1 (*S*)
4	(*S*,*S*)-**1**	*t*-BuCHO	2.2:1 (*S*)
5	(*R*,*R*)-**3**	*t*-BuCHO	1.4:1 (*R*)
6	(*R*)-**4**	*t*-BuCHO	2.9:1 (*S*)

er, enantiomeric ratio.

Ketone Enolate Aldol Additions to Aldehydes. Addition of methyl ketone trichlorosilyl enolate **9** to benzaldehyde in the presence of catalytic amounts of phosphoramide (*S*,*S*)-**1** affords aldol products in excellent yields.[13] The level of enantioselectivity was found to be dependent upon the amount of (*S*,*S*)-**1a** used, with higher loadings providing better selectivities. A typical procedure involves equilibration of a solution of trichlorosilyl enolate and (*S*,*S*)-**1a** in CH_2Cl_2 at −78 °C followed by addition of aldehyde. Reactions are quenched upon completion by quickly pouring the cold reaction mixture into a rapidly stirring aqueous $NaHCO_3$ solution placed in an ice bath. Alternatively, completed reactions may be quenched using a cold 1:1 mixture of a saturated aqueous KF solution and an aqueous 1 M KH_2PO_4 solution. In either case, rapid stirring of the cold quench mixture prevents acid-catalyzed β-elimination of the aldol products to give undesired α,β-unsaturated ketones (eq 3, Table 2).

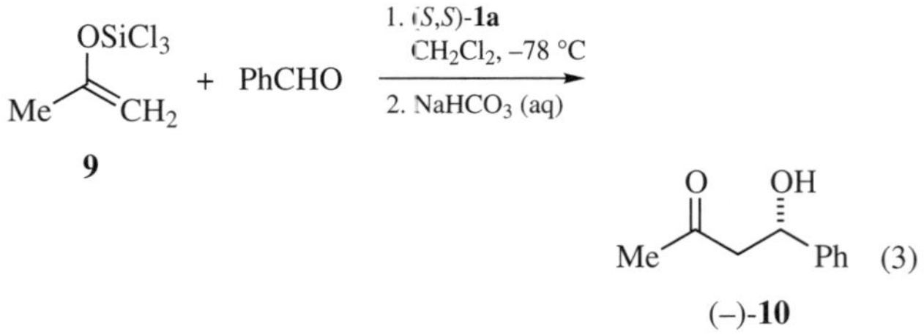

Table 2 Effect of catalyst loading on aldol additions with enolate **9**

Entry	Loading (mol %)	Conc. (M)	er	Yield (%)
1	10	0.1	11.3:1	90
2	20	0.1	12.3:1	89
3	10[a]	0.5	1:11.7	87
4	5	0.5	11.5:1	88
5	3	0.5	8.60:1	86
6	2	0.5	7.77:1	88
7	1	0.5	5.10:1	92

er, enantiomeric ratio. [a]Performed with 10 mol % (*R*,*R*)-**1a**.

The generality of methyl ketone enolate additions to benzaldehyde was demonstrated by varying the spectator portion of the nucleophile.[13] In all cases, high yields were achieved using as little as 5 mol % (*S*,*S*)-**1a**. The enantioselectivity of the process, however, was sensitive to the enolate structure with larger groups, such as *t*-butyl and phenyl (eq 4, Table 3, entries 5 and 6), providing lower enantioselectivities. The success of a functionalized enolate (eq 4, Table 3, entry 7) demonstrated the tolerance for oxygenated substituents in the phosphoramide-catalyzed aldol addition, whereby no deleterious effect was observed on either the yield or enantioselectivity as compared with unfunctionalized enolates.

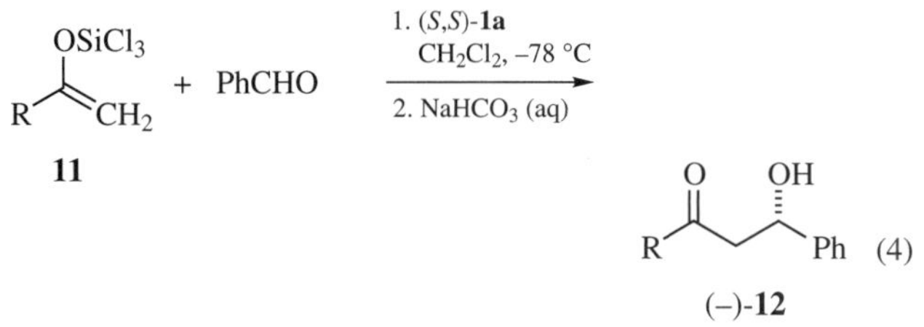

Table 3 Aldol additions of methyl ketone trichlorosilyl enolates to benzaldehyde catalyzed by (*S*,*S*)-**1a**

Entry	R	er	Yield (%)
1	Me	14.6:1	98
2	*n*-Bu	12.0:1	98
3	*i*-Bu	10.1:1	95
4	*i*-Pr	9.75:1	97
5	*t*-Bu	3.17:1	95
6	Ph	2.92:1	93
7	$TBSOCH_2$	13.5:1	94

er, enantiomeric ratio.

The phosphoramide-catalyzed aldol addition was less sensitive to variations on the electrophile. Using 2-hexanone trichlorosilyl enolate **13** as an example aldol additions to various aldehydes affords aldol products in good to high yields and with excellent enantioselectivities, using as little as 5 mol % catalyst.[13] Additions to branched aliphatic aldehydes, such as cyclohexanone and pivalaldehyde (eq 5, Table 4, entries 5 and 6), were sluggish when using 5 mol % catalyst. However, increasing the catalyst loading to 10 mol % provides good yields of the aldol products after 6 h with excellent enantioselectivities.

Table 4 Aldol additions of enolate **13** to various aldehydes catalyzed by (*S*,*S*)-**1a**

Entry	Aldehyde	Loading (mol %)	Time (h)	er	Yield (%)
1	(*E*)-PhCH CHCHO	5	2	14.6:1	94
2	(*E*)-PhCH C(Me)CHO	5	2	21.7:1	95
3	1-NaphthylCHO	5	2	13.1:1	92
4	4-PhC_6H_4CHO	5	2	12.7:1	95
5	$C_6H_{11}CHO$	10	6	17.5:1	79
6	*t*-BuCHO	10	6	24.0:1	81

er, enantiomeric ratio.

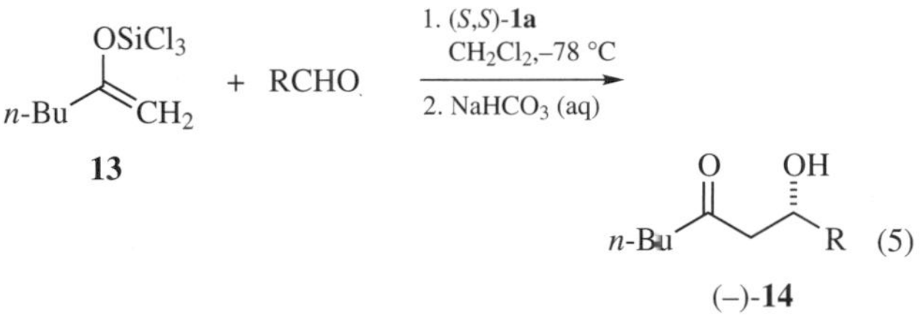

Recently, the synthetic utility of phosphoramide-catalyzed aldol additions of trichlorosilyl enolates to aldehydes has been enhanced by the ability to generate the delicate trichlorosilyl enolates in situ.[13] Treatment of the corresponding TMS enol ether with $Hg(OAc)_2$ followed by addition of a solution phosphoramide and aldehyde afford aldol products in high yields over two steps without affecting stereoselectivities. This is exemplified by the addition of chiral α-oxygenated trichlorosilyl enolates to benzaldehyde. Using the (*S*)-lactate-derived methyl ketone enol ether **15**, addition to benzaldehyde afforded aldol products in high diastereoselectivity for the matched-sense using (*R*,*R*)-**1a**.[13,14] Although the reaction was compatible with various oxygen functionalities, the stereoselectivity was highly dependent upon the nature of the protecting group, with the OTBS group superior to both the benzyloxy or pivaloyl protecting groups (eq 6, Table 5).

Table 5 Addition of TMS enol ether **15** catalyzed by 5 mol % **1a**

Entry	R	Catalyst	*syn*/*anti*	Yield (%)
1	TBS	(*S*,*S*)-**1a**	1.5:1	85
2	Piv	(*S*,*S*)-**1a**	3.4:1	78
3	Bn	(*S*,*S*)-**1a**	1:1.1	78
4	TBS	(*R*,*R*)-**1a**	73:1	85
5	Piv	(*R*,*R*)-**1a**	20:1	78
6	Bn	(*R*,*R*)-**1a**	11:1	77

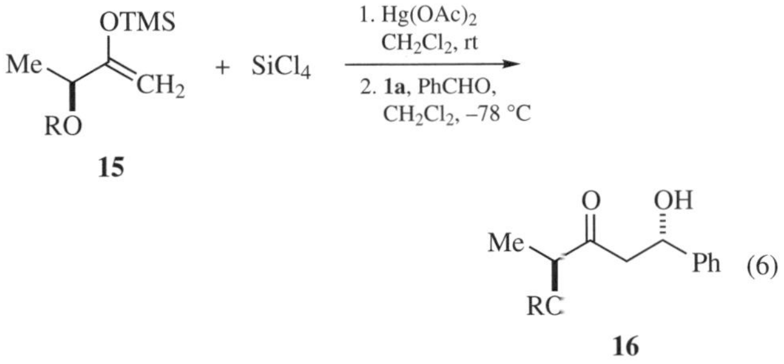

To further exemplify the dependence of the stereochemical course of chiral enolate additions catalyzed by phosphoramides, a detailed survey of additions to aldehydes using chiral β-hydroxyenolate **17** and 10 mol % of **1a** was performed.[15] Unlike additions of **15**, diastereoselectivities of the resultant aldol products using **17** could be switched depending upon the configuration of the phosphoramide employed (eq 7, Table 6).

Table 6 Addition of TMS enol ether **17** to aldehydes catalyzed by **1a**

Entry	Catalyst	Aldehyde	*syn*/*anti*	Yield (%)
1	(*R*,*R*)-**1a**	PhCHO	19.0:1	80
2	(*S*,*S*)-**1a**	PhCHO	1:7.33	75
3	(*R*,*R*)-**1a**	1-NaphthylCHO	15.7:1	85
4	(*S*,*S*)-**1a**	1-NaphthylCHO	1:8.09	83
5	(*R*,*R*)-**1a**	(*E*)-PhCH=CHCHO	8.00:1	81
6	(*S*,*S*)-**1a**	(*E*)-PhCH=CHCHO	1:4.26	82
7	(*R*,*R*)-**1a**	*t*-BuCHO	27.9:1	73
8	(*S*,*S*)-**1a**	*t*-BuCHO	1:6.51	78

Table 7 Aldol addition of enolate **19** catalyzed by 10 mol % (*S,S*)-**1** and (*S,S*)-**2**

Entry	Catalyst	Aldehyde	*anti/syn*	er (major)	Yield (%)
1	**1a**	C_6H_5CHO	61:1	27.6:1	95
2	**1b**	C_6H_5CHO	1:97	3.08:1	94
3	**1c**	C_6H_5CHO	8.1:1	7.33:1	89
4	**1d**	C_6H_5CHO	19:1	14.4:1	91
5	**1e**	C_6H_5CHO	1:32	1.00:1	90
6	**1f**	C_6H_5CHO	1:16	1.11:1	89
7	**1g**	C_6H_5CHO	1:1.9	2.39:1	86
8	**1h**	C_6H_5CHO	2.0:1	17.2:1	92
9	**2a**	C_6H_5CHO	2.5:1	12.0:1	80
10	**2b**	C_6H_5CHO	53:1	24.0:1	80
11	**2c**	C_6H_5CHO	10:1	9.50:1	98
12	**1a**	1-NaphthylCHO	>99:1	65.7:1	94
13	**1a**	(*E*)-PhCH=CHCHO	>99:1	15.7:1	94
14	**1a**	(*E*)-PhCH=C(Me)CHO	>99:1	24.0:1	98
15	**1a**	PhC CCHO	5.3:1	10.1:1	90

er, enantiomeric ratio.

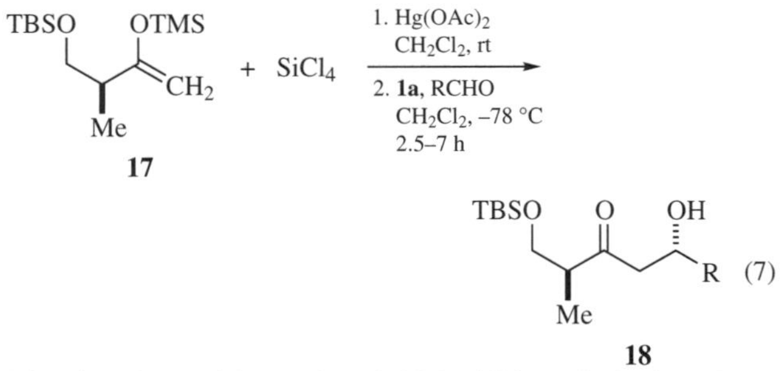

The phosphoramide-catalyzed aldol addition of substituted enolates to aldehydes provides the opportunity to generate disubstituted aldol products with high stereoselectivity. The (*S,S*)-**1a**-catalyzed addition of cyclohexanone trichlorosilyl enolate **19** to aldehydes has been well studied and demonstrates good substrate generality.[16] The aldol reactions of this *E*-enolate are rapid when 10 mol % of (*S,S*)-**1a** is used. Reactions are typically complete within 2 h affording *anti*-aldol products in high yields with excellent diastereo- and enantioselectivities. Although the typical procedure using methyl ketone enolates can be followed, higher diastereoselectivities may be achieved by slow addition of a solution of aldehyde to the mixture of enolate and catalyst in CH_2Cl_2 at −78 °C over several minutes.[17] Remarkably, a dramatic switch in diastereoselectivity is observed when phosphoramide **1b** (10 mol %) is employed as a catalyst (eq 8, Table 7, entry 2). Although the level of enantioselectivity was modest, the change in diastereoselectivity demonstrates that small structural changes in the phosphoramide can have profound effects on the stereochemical course of the reaction. This is further exemplified through a systematic substitution of the internal (phospholidino) nitrogens (eq 8, Table 7, entries 1 and 8), the external nitrogen (eq 8, Table 7, entries 1 and 3–7),[18,19] as well as changes to the electronic demand of the aryl substituent (eq 8, Table 7, entries 1 and 9–11).[20]

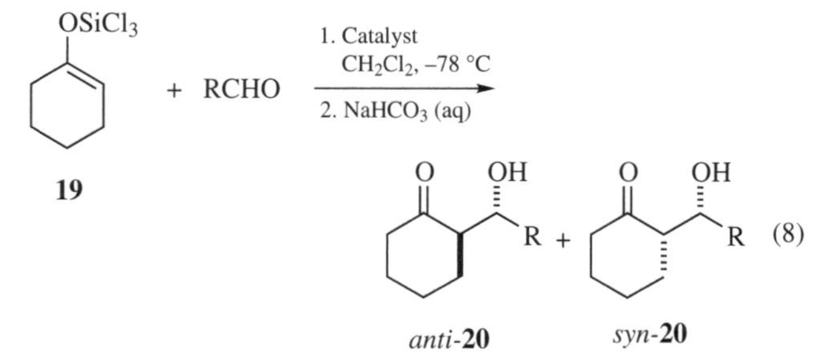

In contrast to the *E*-enolates derived from cyclic ketones, addition of propiophenone trichlorosilyl enolate (*Z*)-**21** to aldehydes requires longer reaction times and higher loadings of catalyst.[16] Although the yields of the aldol products remain high, both the diastereo- and enantioselectivities are attenuated as compared to their *E* counterparts (eq 9, Table 8).

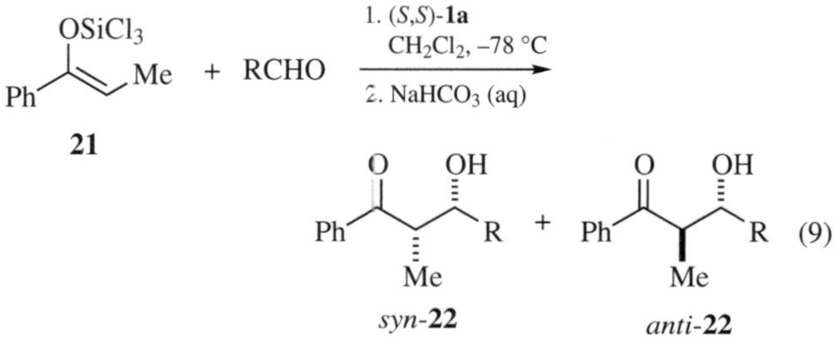

Table 8 Aldol addition of enolate **21** catalyzed by 15 mol % (*S,S*)-**1a**

Entry	Aldehyde	*syn/anti*	er (*syn*)	Yield (%)
1	C_6H_5CHO	18:1	39.0:1	95
2	1-NaphthylCHO	3.0:1	11.5:1	96
3	(*E*)-PhCH=CHCHO	9.4:1	24.0:1	97
4	PhC CCHO	1:3.5	3.76:1	92

er, enantiomeric ratio.

Allylic Additions to Aldehydes. Although phosphoramide **1** is capable of promoting the addition of allyltrichlorosilane to aldehydes, the enantioselectivities were modest at best compared with the cyclohexyldiamine analogs **3**.[21] Preliminary results showed the allylations to be high yielding but giving only moderate enantioselectivities when using one equivalent of phosphoramide as a promoter. As observed in the stereoselective aldol additions with phosphoramides, subtle changes to the nitrogen substituents also dramatically affect the stereochemical course of allylations using allylic trichlorosilanes (eq 10, Table 9, entries 7–9).[22] Under optimized conditions, as few as 0.25 equivalents of phosphoramide may be used with little erosion in yield or stereoselectivity (eq 10, Table 9, entries 3–5).

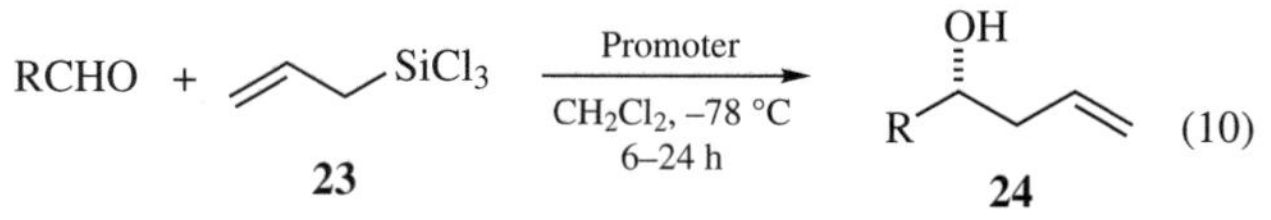

Table 9 Asymmetric allylation of aldehydes

Entry	Promoter	Equiv.	Conc. (M)	Aldehyde	er	Yield (%)
1	(*S,S*)-**1a**	1.0	1.0	C_6H_5CHO	2.0:1	NA
2	(*R,R*)-**3a**	1.0	1.0	C_6H_5CHO	4.0:1	81
3	(*R,R*)-**3a**	0.5	0.5	C_6H_5CHO	3.7:1	78
4	(*R,R*)-**3a**	0.25	0.5	C_6H_5CHO	3.9:1	74
5	(*R,R*)-**3a**	0.1	0.5	C_6H_5CHO	3.3:1	40
6	(*R,R*)-**3b**	1.0	1.0	C_6H_5CHO	3.1:1	73
7	(*R,R*)-**3c**	1.0	1.0	C_6H_5CHO	1.4:1	NA
8	(*R,R*)-**3d**	1.0	1.0	C_6H_5CHO	1.5:1	NA
9	(*R,R*)-**3e**	1.0	1.0	C_6H_5CHO	1.9:1	NA
10	(*R,R*)-**3a**	1.0	1.0	2-MeC_6H_4CHO	4.7:1	81
11	(*R,R*)-**3a**	1.0	1.0	4-$NO_2C_6H_4CHO$	1.5:1	76
12	(*R,R*)-**3a**	1.0	1.0	4-$MeOC_6H_4CHO$	3.0:1	80
13	(*R,R*)-**3a**	1.0	1.0	4-$NMe_2C_6H_4CHO$	2.0:1	69
14	(*R,R*)-**3a**	1.0	1.0	PhC CCHO	2.2:1	67

er, enantiomeric ratio; NA, not available.

The phosphoramide-catalyzed allylic additions to aldehydes have also been extended to include stereoselective crotylations. When using geometrically defined 2-butenyltrichlorosilanes, additions to benzaldehyde afford homoallylic alcohols in good yields and excellent diastereoselectivity (eq 11).[21a] Enantioselectivities, however, remain comparable to the results for allylic additions. Interestingly, the crotylation results suggest that addition proceeds through a cyclic transition structure.[23]

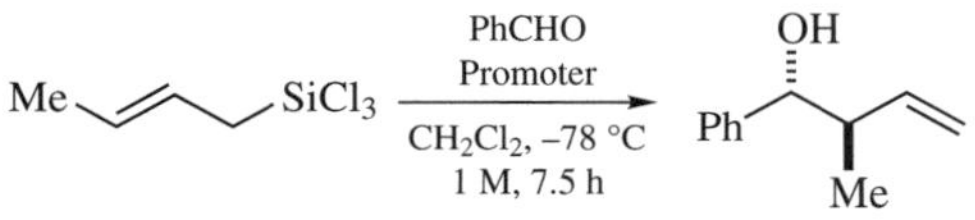

(*E*)-**25** >99:1 — 68% (98:2 *anti/syn*) 4.9:1 er (*anti*) — (1*R*,2*R*)-**26**

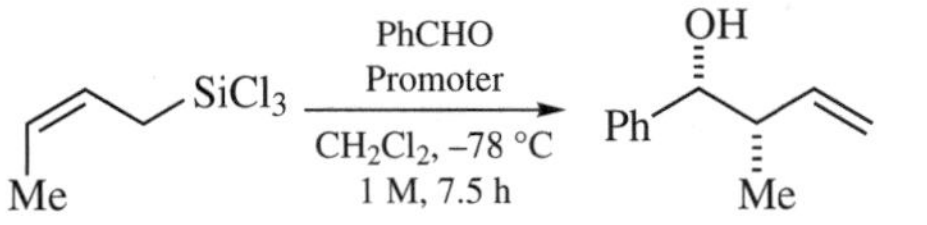

(11)

(*Z*)-**25** >99:1 — 72% (98:2 *syn/anti*) 4.0:1 er (*syn*) — (1*R*,2*S*)-**26**

Epoxide Openings Using $SiCl_4$. The binaphthyldiamine-derived phosphoramide (*R*)-**4** is an effective catalyst for the opening of various epoxides with $SiCl_4$ to afford chlorohydrins in high yields.[24] The rate and stereoselectivity for the ring opening of *meso*-epoxides are extremely dependent upon the structure of the epoxide. In the examples shown for epoxycycloalkanes, only cyclohexene oxide [10 mol % (*R*)-**4**] affords a chlorohydrin in high yield and good enantioselectivity (eq 12, Table 10, entry 2). In contrast, cyclopentene oxide and cyclooctene oxide give essentially racemic chlorohydrin (eq 12, Table 10, entries 1 and 3) with cyclooctene oxide requiring more than 5 days for complete reaction. In comparison to the cyclic epoxides, the acyclic substrates afford chlorohydrins with increased enantioselectivities (eq 12, Table 10, entries 4 and 5). In a typical procedure, a solution of epoxide in CH_2Cl_2 is added to a mixture of $SiCl_4$ and phosphoramide at $-78\,°C$ in CH_2Cl_2. To avoid opening of epoxides by adventitious HCl,[25] it is essential to maintain $SiCl_4$ acid free.[26]

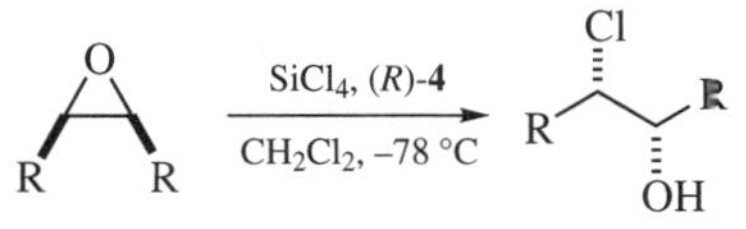

27 **28** (12)

Table 10 Stereoselective ring opening of *meso*-epoxides catalyzed by (*R*)-**4**

Entry	Epoxide (R)	Time (h)	er	Yield (%)
1	-$(CH_2)_3$-	0.3	1.16:1	87
2	-$(CH_2)_4$-	0.3	3.13:1	90
3	-$(CH_2)_6$-	132	1.04:1	95
4	Ph	3	14.4:1	94
5	$BnOCH_2$	4	5.94:1	95

er, enantiomeric ratio.

1. (a) Paterson, I.; Cowden C. J.; Wallace, D. J. In *Modern Carbonyl Chemistry*; Otera, J., Ed.: Wiley-VCH: Weinheim, 2000; Ch. 9. (b) Cowden, C. J.; Paterson, I. *Org. React.* **1997**, *51*, 1. (c) Heathcock, C. H.; Kim, B. M.; Williams, S. F.; Masamune, S.; Paterson, I.; Gennari, C. In *Comprehensive Organic Synthesis*; Trost, B. M.; Fleming, I., Eds.; Vol. 2, Pergamon Press: Oxford, 1991.
2. (a) Denmark, S. E.; Almstead, N. G. In *Modern Carbonyl Chemistry*; Otera, J., Ed.; Wiley-VCH: Weinheim, 2000, Ch.10. (b) Roush, W. R.; Chemler, C. R. In *Modern Carbonyl Chemistry*; Otera, J., Ed.; Wiley-VCH: Weinheim, 2000, Ch.11. (c) *Stereoselective Synthesis, Methods of Organic Chemistry (Houben-Weyl)*; Edition E21; Helmchen, G.; Hoffmann, R.; Mulzer, J.; Schaumann, E., Eds.; Thieme: Stuttgart, 1996, Vol. 3, pp 1357–1602. (d) Yamamoto, Y.; Asao, N., *Chem. Rev.* **1993**, *93*, 2207.
3. (a) Erden I. In *Comprehensive Heterocyclic Chemistry*, 2nd edn; Padwa, A., Ed.; Pergamon Press: Oxford, 1996, Vol. 1A, Ch. 1.03. (b) Bartók, M.; Lang, K. L. In *The Chemistry of Heterocyclic Compounds*; Weissberger, A.; Taylor, E. C., Eds.; Wiley: New York, 1985; Vol. 42, Part 3, p 1. (c) Rao, A. S.; Paknikar, S. K.; Kirtane, J. G., *Tetrahedron* **1983**, *39*, 2323.
4. Denmark, S. E.; Su, X.; Nishigaichi, Y.; Coe, D. M.; Wong, K.-T.; Winter, S. B. D.; Choi, J.-Y. *J. Org. Chem.* **1999**, *64*, 1958.
5. Pikul, S.; Corey, E. J. *Org. Synth.* **1991**, *71*, 22.
6. (a) Alexakis, A.; Aujard, I.; Mangeney, P. *Synlett* **1998**, 873. (b) Alexakis, A.; Aujard, I.; Mangeney, P. *Synlett* **1998**, 875.
7. Peyronel, J.-F.; Samuel, O.; Fiaud, J.-C. *J. Org. Chem.* **1987**, *52*, 5320.
8. Denmark, S. E.; Stavenger, R. A. *Acc. Chem. Res.* **2000**, *33*, 432.
9. (a) Iseki, K.; Kuroki, Y.; Takahashi, M.; Kobayashi, Y. *Tetrahedron Lett.* **1996**, *37*, 149. (b) Iseki, K.; Kuroki, Y.; Takahashi, M.; Kishimoto, S.; Kobayashi, Y. *Tetrahedron* **1997**, *53*, 13. (c) Iseki, K.; Mizuno, S.;

Kuroki, Y.; Kobayashi, Y. *Tetrahedron Lett.* **1998**, *39*, 67. (d) Hong, B.-C.; Hong, J.-H.; Tsai, Y.-C. *Angew. Chem., Int. Ed. Engl.* **1998**, *37*, 468.

10. For an additional example of Lewis-base catalyzed epoxide openings see: Tao, B.; Lo, M. M.-C.; Fu, G. C. *J. Am. Chem. Soc.* **2001**, *123*, 353.
11. (a) Ishihara, K.; Karumi, Y.; Kondo, S.; Yamamoto, H. *J. Org. Chem.* **1998**, *63*, 5692. (b) Verkade, J. G. *Acc. Chem. Res.* **1993**, *26*, 483.
12. Denmark, S. E.; Winter, S. B. D. *Synlett* **1997**, 1087.
13. Denmark, S. E.; Stavenger, R. A. *J. Am. Chem. Soc.* **2000**, *122*, 8837.
14. Denmark, S. E.; Stavenger, R. A. *J. Org. Chem.* **1998**, *63*, 6524.
15. Denmark, S. E.; Fujimori, S. *Synlett* **2001**, 1024.
16. Denmark, S. E.; Stavenger, R. A.; Wong, K.-T.; Su, X. *J. Am. Chem. Soc.* **1999**, *121*, 4982.
17. For a discussion on the molecularity and mechanism of phosphoramide catalyzed aldol additions, see: (a) Denmark, S. E.; Su, X.; Nishigaichi, Y. *J. Am. Chem. Soc.* **1998**, *120*, 12990. (b) Denmark, S. E.; Pham, S. M. *Helv. Chim. Acta* **2000**, *83*, 1846. For solid state and solution structural studies of phosphoramide-tin complexes, see: (c) Denmark, S. E.; Su, X. *Tetrahedron* **1999**, *55*, 8727.
18. Wong, K.-T., unpublished results from these laboratories.
19. Nishigaichi, Y., unpublished results from these laboratories.
20. Su, X., unpublished results from these laboratories.
21. (a) Denmark, S. E.; Coe, D. M.; Pratt, N. E.; Griedel, B. D. *J. Org. Chem.* **1994**, *59*, 6161. (b) Denmark, S. E.; Fu, J. *J. Am. Chem. Soc.* **2000**, *122*, 12021.
22. Coe, D. M., unpublished results from these laboratories.
23. Dimeric phosphoramides have recently been demonstrated to be superior to the monomeric versions for allylations and crotylations using allylic trichlorosilanes, see: reference 21b.
24. Denmark, S. E.; Barsanti, P. A.; Wong, K.-T.; Stavenger, R. A. *J. Org. Chem.* **1998**, *63*, 2428.
25. Cross, A. D. *Quart. Rev. Chem. Soc.* **1960**, *14*, 317.
26. A report by Brunel, Legrand, Reymond and Buono describes the use of phosphonamide-type catalysts to achieve excellent enantioselectivities in the stereoselective ring opening of *meso*-epoxides with $SiCl_4$. Recently, this group and others have been unable to verify the results by Buono et al., see: (a) Brunel, J. M.; Legrand, O.; Reymond, S.; Buono, G. *Angew. Chem., Int. Ed. Engl.* **2000**, *39*, 2554. (b) Denmark, S. E.; Wynn, T.; Jellerichs, B. G. *Angew. Chem., Int. Ed. Engl.* **2001**, *40*, 2255.

Scott E. Denmark & Son M. Pham
University of Illinois, Urbana, IL, USA

cis-3-[*N*-(3,5-Dimethylphenyl)benzene-sulfonamido]borneol[1]

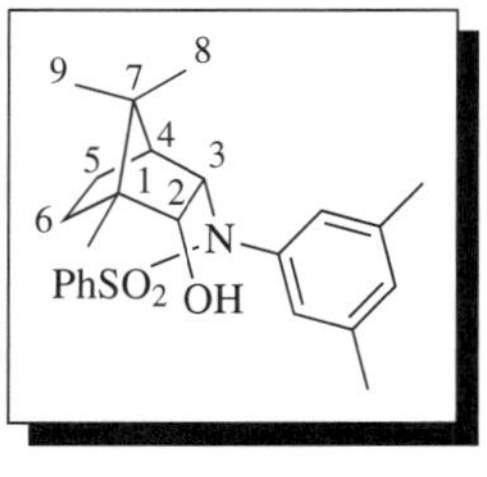

(1*R*) (*endo,endo*)
[87360-02-9] $C_{24}H_{31}NO_3S$ (MW 413.63)

(chiral auxiliary; ester enolate derivatives undergo stereoselective alkylations[2] and enantioselective *anti*-aldol reactions;[3] enoate derivatives undergo stereoselective 1,4-conjugate additions of organocopper reagents[4])

Alternate Name: *N*-(3,5-dimethylphenyl)-*N*-(3-hydroxy-4,7,7-trimethylbicyclo[2.2.1]hept-2-yl)benzenesulfonamide.
Physical Data: (1*R*) (*endo,endo*): mp 147–150 °C; $[\alpha]_D^{20}$ −26.0° (c = 4.0, $CHCl_3$).
Handling, Storage, and Precautions: the auxiliary is stable indefinitely at ambient temperatures in a sealed container.

Introduction. One of several auxiliaries that exploit the asymmetry of naturally occurring (+)-camphor, the 3-(*N*-(3,5-dimethylphenyl)benzenesulfonamido)borneol auxiliary has proven significant utility in the π-facial differentiation of ester enolates and enoate derivatives. The *endo* orientation of the C(2) and C(3) substituents places the reactive functionality within the concave pocket created by the bornane skeleton as well as the shielding ability of the *N*-arylbenzenesulfonamide.

Preparation of the Auxiliary. A synthesis of the (1*R*) auxiliary has been reported starting from (+)-camphor (eq 1). *Zinc* reduction of the intermediate imine (**2**) followed by sulfonylation and ketone reduction with $Ca(BH_4)_2$ afforded the *cis,endo* product in 70–75% overall yield from camphorquinone (**1**).

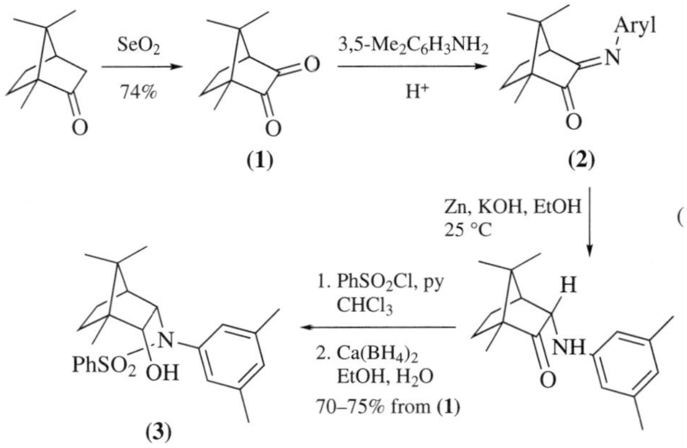

Preparation of Derivatives. Enoate derivatives were prepared by Horner–Wittig reactions between aldehydes and the ethyl phosphonate derived from the chloroacetyl ester of (**3**) in high (*E*) selectivity (97:3).[4b] Ester derivatives were obtained by treating alcohol (**3**) with the corresponding carboxylic acid chloride.[5]

α-Alkylation of Ester Derivatives[2]. Alkylation of ester enolate derivatives, prepared by metalation with lithium cyclohexylisopropylamide (LICA), proceeds with high stereoselectivity. The configuration of the product is dependent on the solvent employed (eqs 2 and 3). When performed in THF with the addition of HMPA the product with the (*S*) configuration was formed preferentially; however, without HMPA the (*R*) configuration predominated. Silyl chloride trapping studies suggest that in the presence of HMPA the (*E*)-ester enolate is stereoselectively formed, as opposed to the (*Z*)-ester enolate in THF alone.[2b] The stereochemical outcome has been explained by alkylation of the corresponding enolate π-face opposite to the shielding 3,5-dimethylphenyl moiety. *O*-Benzylglycolates have also been employed in stereoselective alkylations, affording diastereomeric excesses of 88–95%.[2c] In this case the solvent-dependent stereochemical reversal does not occur and the (*E*)-ester enolate is stereoselectively formed in both cases.

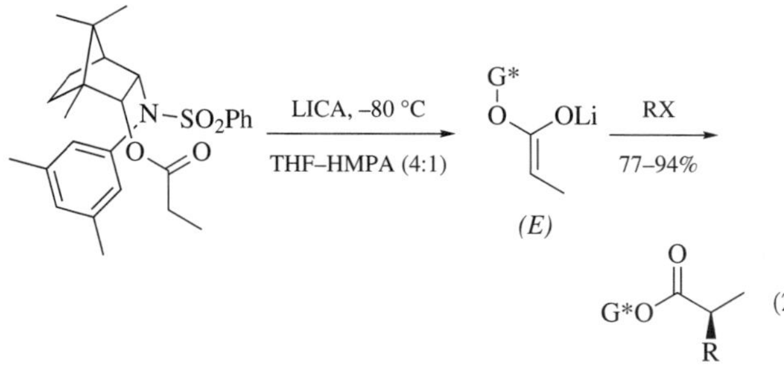

(2)

95:5–98:2 de

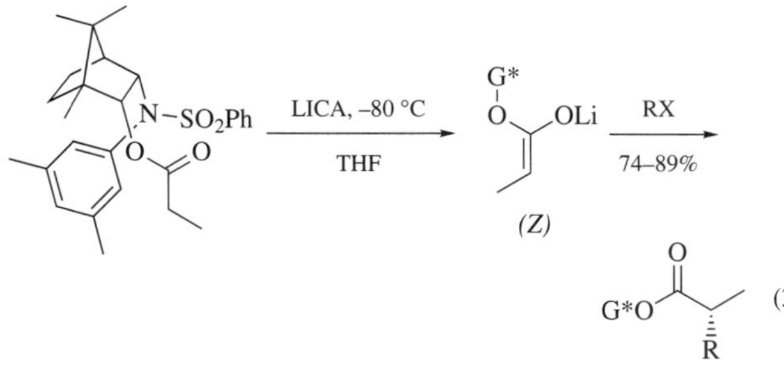

(3)

97:3 de

Aldol Reactions of Ester Derivatives[3]. The *Titanium(IV) Chloride*-catalyzed addition of aldehydes to *O*-silyl ketene acetals derived from acetate and propionate esters proceeds with high stereoselectivity. Formation of the silyl ketene acetal was found to be essential for high diastereoselectivity. Treatment of the silyl ketene acetal, derived from deprotonation of the acetate ester with LICA in THF and silyl trapping, with a corresponding aldehyde in the presence of $TiCl_4$ (1.1 equiv) afforded the addition products in 93:7 diastereoselectivity and moderate yield (51–67%). Similarly, the propionate ester provides the *anti*-aldol product in high *anti*/*syn* selectivity (14:1) and facial selectivity (eq 4).

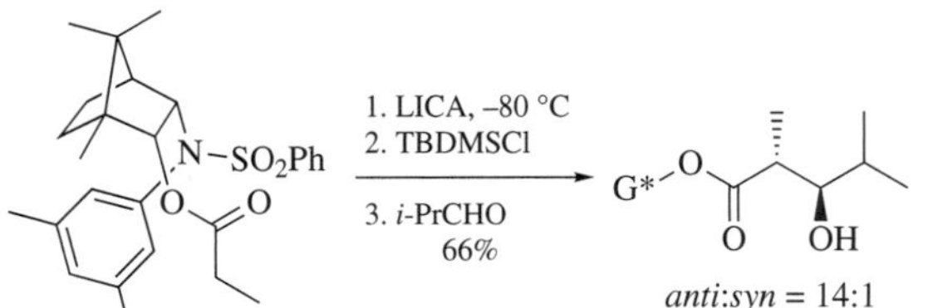

(4)

anti:*syn* = 14:1
97% de

1,4-Conjugate Additions to Enoate Derivatives[4]. High diastereoselectivity has been observed for *Boron Trifluoride*-promoted addition of alkyl and aryl organocopper reagents to enoate derivatives (eq 5).[4b] When the organocopper reagent was prepared from alkyl- or aryllithiums, diethyl ether was found to be the solvent of choice; however, with Grignard reagents, THF was superior. The addition of boron trifluoride exhibited little influence on reactivity of the copper reagent but did enhance the stereoselectivity of the addition. It is believed that the enoate adopts an *s-trans* conformation and the observed stereochemical preference results from approach of the organocopper reagent to the less sterically hindered face opposite the aryl moiety.

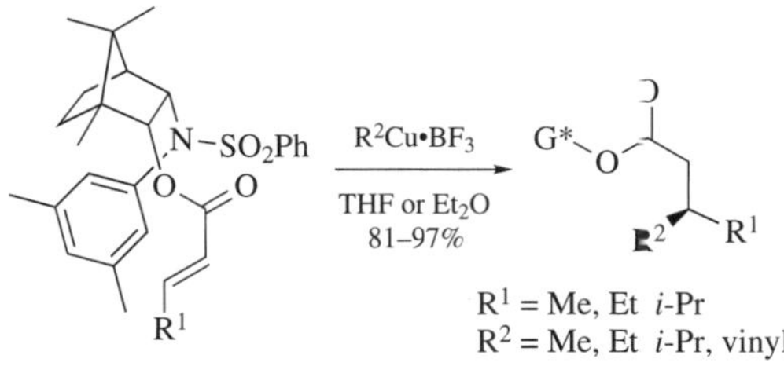

(5)

R^1 = Me, Et *i*-Pr
R^2 = Me, Et *i*-Pr, vinyl, Ph

Nondestructive Removal of the Auxiliary. Primary alcohols are obtained by *Lithium Aluminum Hydride* reduction of the corresponding chiral esters. Also, hydrolysis of the auxiliary under basic conditions, 2N KOH in methanol,[4b] provides the carboxylic acid and recovered alcohol (**3**).

Related Reagents. 3-Hydroxyisoborneol (1*R*,2*S*)-*N*-Methylephedrine

1. Oppolzer, W. *Tetrahedron* **1987**, *43*, 1969.
2. (a) Schmierer, R.; Grotemeier, G.; Helmchen, G.; Selim, A. *Angew. Chem., Int. Ed. Engl.* **1981**, *20*, 207. (b) Helmchen, G.; Selim, A.; Dorsch, D.; Taufer, I. *Tetrahedron Lett.* **1983**, *24*, 3213. (c) Helmchen, G.; Wierzchowski, R. *Angew. Chem., Int. Ed. Engl.* **1984**, *23*, 60.
3. Helmchen, G.; Leifauf, U.; Taufer-Knöpfel, I. *Angew. Chem., Int. Ed. Engl.* **1985**, *24*, 874.
4. (a) Rossiter, B.; Swingle, N. M. *Chem. Rev.* **1992**, *92* 771. (b) Helmchen, G.; Wegner, G. *Tetrahedron Lett.* **1985**, *26*, 6051.
5. Dorsch, D.; Kunz, E.; Helmchen, G. *Tetrahedron Lett.* **1985**, *26*, 3319.

Mark E. Schnute
Stanford University, CA, USA

(*S*)-(+)-5,5-Dimethyl-4-phenyl-2-oxazolidinone

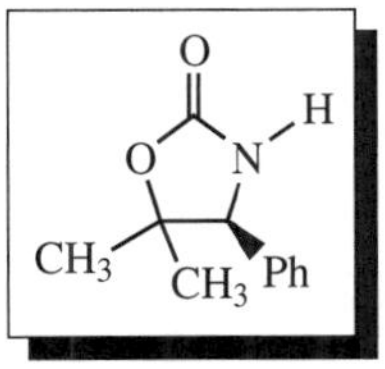

[168297-84-5] $C_{11}H_{13}NO_2$ (MW 191.23)

(versatile chiral auxiliary used for asymmetric synthesis[1–8] in diastereoselective enolate formation,[9–12] and Michael additions;[9,13] also used in the kinetic resolution of α-acetoxy carboxylic acids[14])

Alternate Name: (4*S*)-Phenyl SuperQuat.
Physical Data: mp 151–156 °C; $[\alpha]_D^{25}$ +71 (*c* 2 0, $CHCl_3$).
Solubility: THF, EtOAc, dichloromethane.
Form Supplied in: white crystalline solid; commercially available.
Analysis of Reagent Purity: ^1H NMR, ^{13}C NMR, IR, GCMS, chiral HPLC.
Preparative Methods: the original literature[9] reports that the desired 4-substituted-5,5-dimethyloxazolidin-2-one is readily accessible from the corresponding α-amino acid via esterification ($MeOH/SOCl_2$) followed by Grignard addition to afford

the 1,2-amino alcohol (eq 1). The formation of the oxazolidinone is then achieved either indirectly by treatment with tricholoracetyl chloride followed by base-catalyzed cyclization, or directly through reaction with carbonyldiimidazole.

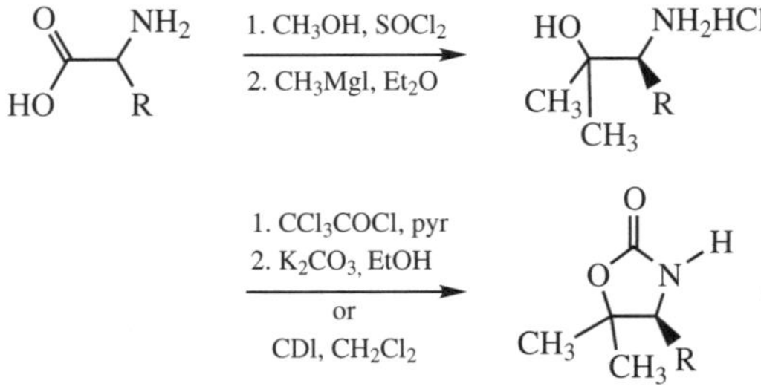

(1)

Table 1 Formation of 4-substituted-5,5-dimethyloxazolidin-2-ones

		% Overall yield	
R	% Yield (two steps)	Method A[a]	Method B[a]
Ph	41	25	34
Me	53	47	48
Bn	41	28	33

[a]Method A: CCl_3COCl, pyridine then K_2CO_3, EtOH.
Method B: CDI, CH_2Cl_2.

While this methodology is applicable to a variety of α-amino acids on a small scale, large-scale syntheses have proven problematic in that they are either low yielding or result in partial racemization of the desired auxiliary. In order to circumvent this difficulty, an alternative preparation has been developed (eq 2). Initially, an *N*-Boc-α-amino acid methyl ester is reacted with an excess of methylmagnesium iodide to generate the corresponding tertiary alcohol. Subsequent cyclization into the desired 4-substituted-5,5-dimethyloxazolidin-2-one upon treatment with *tert*-BuOK (eq 2) proceeds in good yield and with little or no racemization.[11,15]

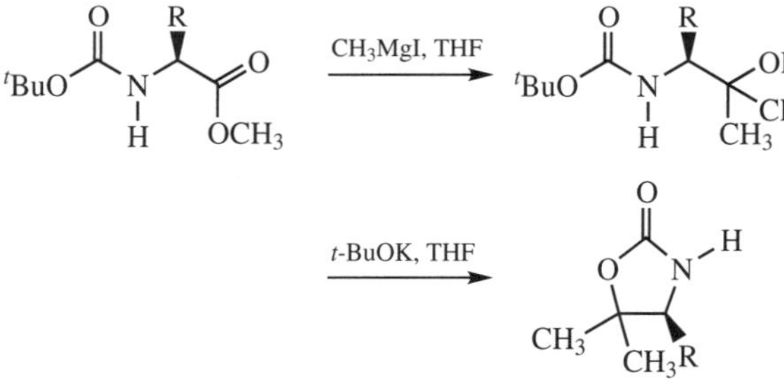

(2)

The *N*-Boc protecting group is critical in this synthetic strategy. Not only does it prevent racemization by disfavoring deprotonation at the α-center once the carbamate proton is removed, but it also serves as a carbonyl equivalent in the cyclization process. The major drawback to this methodology is that although many *N*-Boc protected α-amino acid methyl esters are commercially available, they tend to be significantly more expensive than the corresponding α-amino acids. They can, however, be synthesized easily from the parent α-amino acid, albeit in two steps.

It has also been reported[16] that 4-substituted-5,5-dimethyloxazolidin-2-ones can be prepared as illustrated in eq 3. Initially, stereoselective condensation of an *N*-acyloxazolidinone enolate with acetone affords a functionalized acyl fragment, which is then hydrolyzed to the carboxylic acid. Reaction of the hydroxy acid with DPPA at elevated temperatures yields the target via formation of the acyl azide, Curtius rearrangement and trapping of the isocyanate intermediate by the hydroxyl group (eq 3).

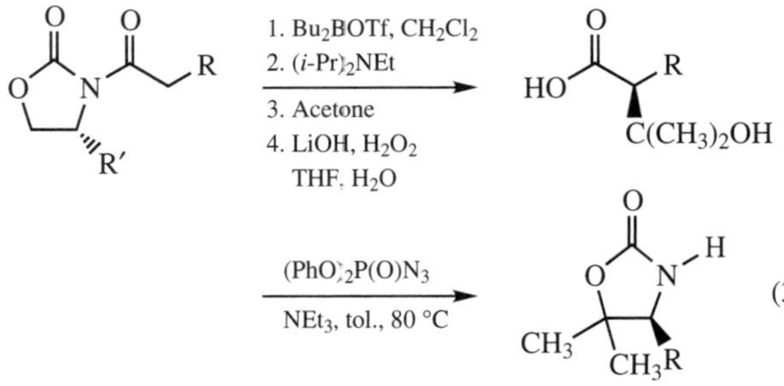

(3)

While this methodology is not intended for the preparation of oxazolidinones that can be generated in a more concise route from their parent α-amino acid (vide supra), it does allow for the preparation of 4-substituted-5,5-dimethyloxazolidin-2-ones in which the parent α-amino acid is either not commercially available or exceedingly expensive.

Purification: can be recrystallized from EtOAc with pentane.

Handling, Storage, and Precautions: stable for prolonged periods when stored in a cool, dry environment; easy to handle; solid; MSDS codes as irritant.

Introduction. The stoichiometric use of a chiral auxiliary has become one of the most prevalent and dependable methods to effect asymmetric transformations.[1–8] In this context, the use of homochiral 4-substituted oxazolidin-2-ones[2,17] has proven to be extremely effective in controlling facial diastereoselectivity in a wide variety of reactions of attached *N*-acyl fragments. While these 'Evans' auxiliaries' allow for facile attachment of the *N*-acyl fragment and impart a high degree of stereocontrol, their major drawback is the difficulty in removing of the chiral auxiliary from some products.[18] When the attached acyl fragment is either sterically demanding or branched at the α-position, there is a tendency for the auxiliary to undergo endocyclic hydrolysis. This affords the undesired ring-opened amide rather than the desired carboxylic acid and recovered auxiliary[19] resulting from exocyclic cleavage (eq 4).

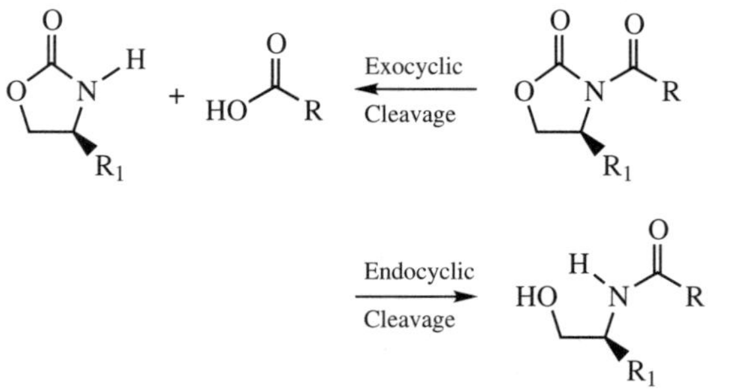

(4)

Although this problem can be overcome by using lithium hydroperoxide,[20] the use of this reagent on large scale can be hazardous. In order to completely circumvent this problem, Davies and Sanganee have developed 4-substituted-5,5-dimethyloxazolidin-2-ones, or 'super Quats'.[9] The key feature of this auxiliary is the *gem*-dimethyl groups at the C-5 position (eq 5). The rationale for the design of this auxiliary is three-fold: (i) the *gem*-dimethyl groups at C-5 prevent endocyclic ring opening by blocking the required Burgi–Dunitz (109°) approach of the incoming nucleophile to C-2 during hydrolysis; (ii) the presence of the *gem*-dimethyl groups serve to enhance the diasterofacial selectivity during enolate formation *via* a secondary interaction with the C-4 substituent; and (iii) the highly crystalline nature of these species makes them amenable to purification by recrystallization.

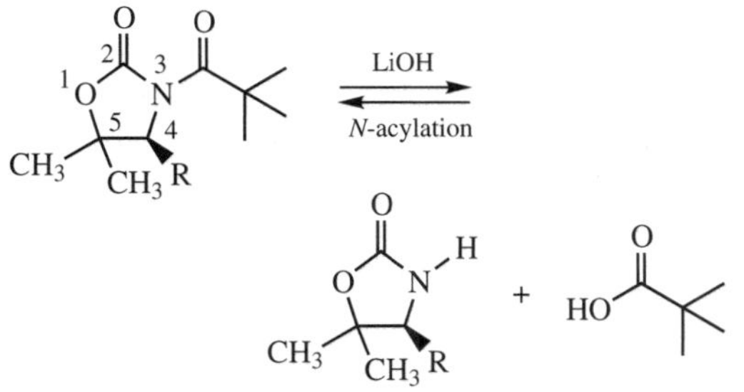

(5)

Diastereoselective Enolate Formation and Alkylation. These 'super Quat' auxiliaries are easily *N*-acylated via deprotonation with butyllithium followed by quenching with the desired acid chloride. Treatment of the *N*-acylated 'super Quat' with LDA followed by the addition of an alkylating agent results in the formation of the functionalized acyl fragment in good to excellent yield with a high de. Presumably, as with the Evans' auxiliaries, the high degree of asymmetric induction is a result of a carbonyl–metal-carbonyl transition state that results in the formation of a *Z*-enolate in which the C-4 substituent governs the diastereofacial bias of alkylation.[1,7] As illustrated in eqs 6 and 7, reaction of *N*-propionyl and *N*-hydrocinnamoyl 'super Quats' with LDA, followed by treatment with benzyl bromide or methyl iodide, respectively, affords the corresponding pairs of diastereomers in acceptable yields and % de. In all cases, the de was increased to >99% by a single recrystallization.[9]

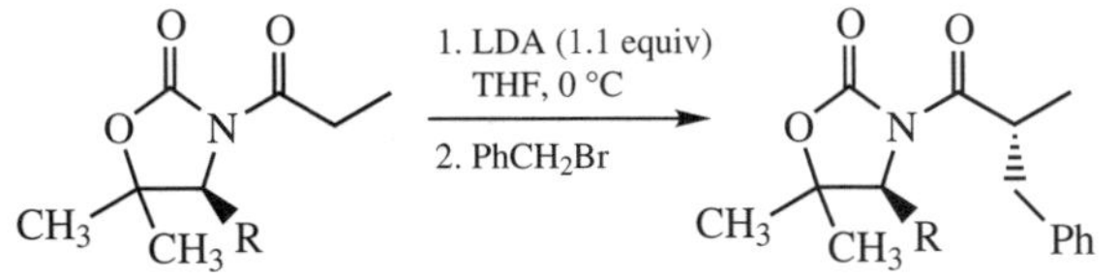

(6)

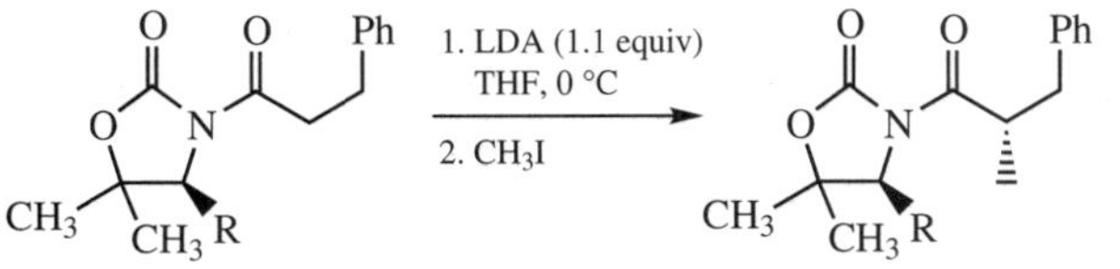

(7)

Unlike the Evans' auxiliaries, however, removal of the 'super Quat' auxiliary is easily accomplished upon treatment with either lithium hydroxide or lithium alkoxide. Thus, hydrolysis with LiOH affords the enantionmerically pure α-substituted carboxylic acid and near quantitative recovery of the chiral auxiliary (eq 8).

Diastereoselective Michael Additions. The 'super Quats' have also proven to be effective auxiliaries in diastereoselective conjugate additions to α,β-unsaturated carbonyl species.[9,13] The use of such auxiliaries for this type of 1,2-addition is best exemplified by the asymmetric synthesis of aplysillamide B, an antifungal, antibacterical alkaloid isolated from the marine sponge *Psammaplysilla purea*. Thus, (*S*)-(+)-5,5-dimethyl-4-phenyl-2-oxazolidinone was *N*-acylated via treatment with butyllithium followed by exposure to *trans*-crotonyl chloride to afford the desired *N*-substituted oxazolidinone. To this amide was added an organocuprate prepared from *n*-heptylmagnesium bromide according to the standard Hruby protocol.[21,22] The functionalized acyl fragment was next removed from the chiral auxiliary by treatment with 1,4-diaminobutane to afford the desired amino amide, which was converted to the target in two steps (eq 9).

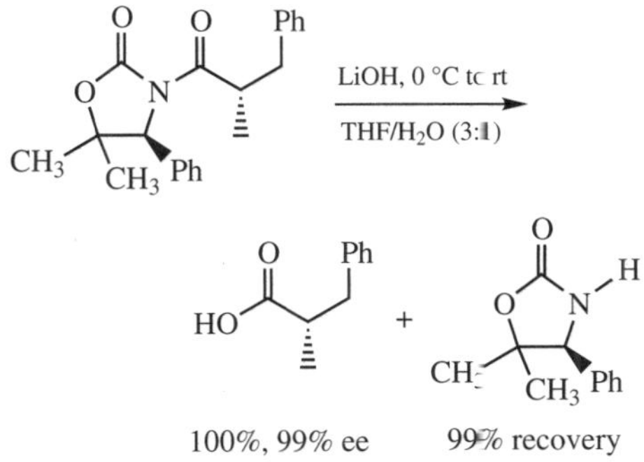

(8)

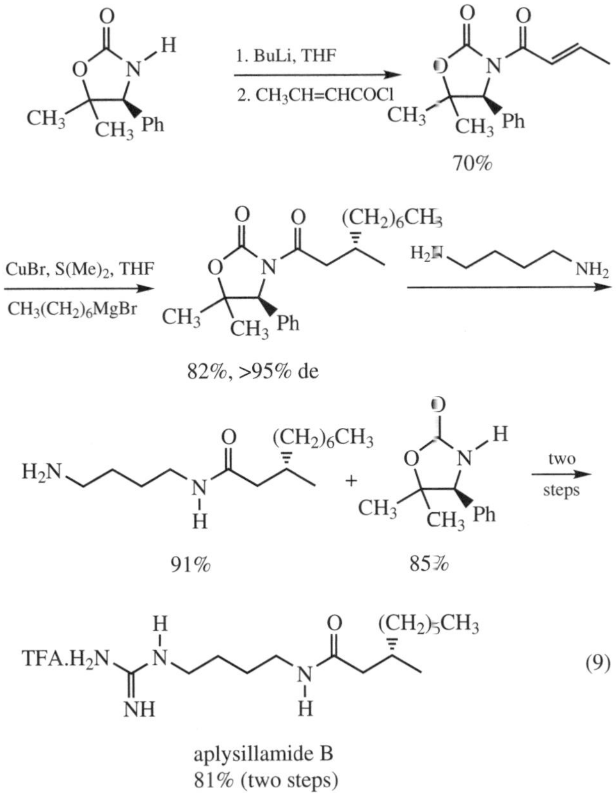

(9)

Kinetic Resolution of α-acetoxy Carboxylic Acids. One of the most recent applications for the 'super Quat' family of chiral auxiliaries is the kinetic resolution of α-substituted-α-acetoxy carboxylic acid chlorides.[14] Upon reaction of the lithium salt of the 'super Quat' auxiliary with 2 equiv of (±)-*O*-acetylmandelic chloride at −100 °C, the corresponding *N*-acylated 'super Quat' auxiliary was isolated in excellent yield with acceptable de (eq 10). The des that result from this type of resolution appear to be dependent on both solvent polarity and steric interactions at the α-position. The use of a less polar solvent causes a decrease in % de, while an increase in steric bulk tends to increase the % de. In all cases, however, a single recrystallization from hexane provide the *N*-acylated 'super Quat' auxiliary in >95% de.

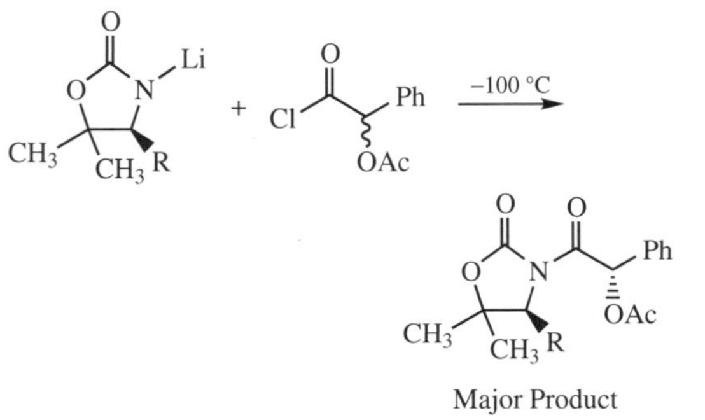

(10)

Related Reagents. (*S*)-4-Benzyl-5,5-dimethyl-2-oxazolidinone; (*S*)-5,5-dimethyl-4-*iso*-propyl-2-oxazolidinone; (*S*)-5,5-dimethyl-4-methyl-2-oxazolidinone; (*S*)-5,5-diphenyl-4-*iso*-propyl-2-oxazolidinone; (*R*)-4-benzyl-5,5-dimethyl-2-oxazolidinone.

1. (a) Evans, D. A. In *Asymmetric Synthesis*, Morrison, J. D., Ed.; Academic Press: New York, 1984; Vol. 3, Part B, pp 2–101. (b) Cowden, C. J. In *Organic Reactions*, Paquette, L. A., Ed.; John Wiley & Sons: New York, 1997; Vol. 51, pp 1–200.
2. Evans, D. A. *Aldrichimica Acta.* **1982**, *15*, 23.
3. Ager, D. J.; Prakash, I.; Schaad, D. R. *Chem. Rev.* **1996**, *96*, 835.
4. Ager, D. J.; Prakash, I.; Schaad, D. R. *Aldrichimica Acta.* **1997**, *30*, 3.
5. Seebach, D.; Hintermann, T. *Helv. Chim. Acta.* **1998**, *81*, 2093.
6. Regan, A. C. *J. Chem. Soc., Perkin Trans. 1* **1999**, *4*, 357.
7. Arya, P.; Qin, H. *Tetrahedron.* **2000**, *56*, 917.
8. Seebach, D.; Beck, A. K.; Heckel, A. *Angew. Chem., Int. Ed. Eng.* **2001**, *40*, 92.
9. Davies, S. G.; Sanganee, H. J. *Tetrahedron: Asymm.* **1995**, *6*, 671.
10. Bull, S. D.; Davies, S. G.; Key, M. S.; Nicholson, R. L.; Savory, E. D. *Chem. Comm.* **2000**, *18*, 1721.
11. Bull, S. D.; Davies, S. G.; Jones, S.; Sanganee, H. J. *J. Chem. Soc., Perkin Trans. 1* **1999**, *4*, 387.
12. Gibson, C. L.; Gillon, K.; Cook, S. *Tetrahedron Lett.* **1998**, *39*, 6733.
13. Davies, S. G.; Sanganee, H. J.; Szolcsanyi, P. *Tetrahedron.* **1999**, *55*, 3337.
14. Bew, S. P.; Davies, S. P.; Fukuzawa, S. I. *Chirality.* **2000**, *12*, 483.
15. Bull, S. D.; Davies, S. G.; Jones, S.; Polywka, M. C.; Prasad, R. S.; Sanganee, H. J. *Synlett.* **1998**, 519.
16. Takacs, J. M.; Jaber, M. R.; Vellekoop, A. S. *J. Org. Chem.* **1998**, *63*, 2742.
17. Evans, D. A.; Ennis, M. D.; Mathre, D. J. *J. Am. Chem. Soc.* **1982**, *104*, 1737.
18. Evans, D. A.; Britton, T. C.; Ellman, D. J. *Tetrahedron Lett.* **1987**, *28*, 6141.
19. Evans, D. A.; Bartroli, J. *Tetrahedron Lett.* **1982**, *23*, 807.
20. Evans, D. A.; Chapman, K. T.; Bisaha, J. *J. Am. Chem. Soc.* **1988**, *110*, 1238.
21. Hruby, V. J.; Russel, K. C.; Nicolas, E. *J. Org. Chem.* **1993**, *58*, 766.
22. Hruby, V. J.; Lou, B.; Lung, F. *J. Org. Chem.* **1995**, *60*, 5509.

Doug M. Krein & Todd L. Lowary
The Ohio State University, OH, USA

(1*R*,2*S*,4*R*,5*S*)-2,5-Dimethyl-7-phenyl-7-phosphabicyclo[2.2.1]heptane

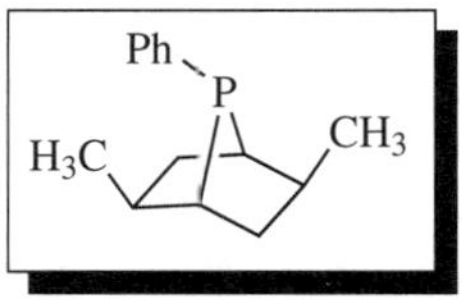

[189210-88-6] $C_{14}H_{19}P$ (MW 217.97)

(chiral, nonracemic phosphine ligand for asymmetric transition metal-catalyzed reactions)

Solubility: soluble in common organic solvents (i.e., benzene, toluene, CH_2Cl_2).
Analysis of Reagent Purity: ^{1}H-NMR.
Preparative Methods: prepared in four steps starting from *p*-xylene.[1] Birch reduction of *p*-xylene followed by asymmetic hydroboration-oxidation provides an optically pure diol. The diol is subsequently converted to the chiral phosphine by formation of the corresponding dimesylate and nucleophilic addition of Li_2PPh.
Purification: purification was accomplished by chromatography of the corresponding borane complex. Decomplexation using $HBF_4 \cdot O(C_2H_5)_2$ afforded the pure phosphine.
Handling, Storage, and Precautions: sensitive to atmospheric oxidation. Should be stored and handled under an inert atmosphere.

Introduction. Chiral phosphines have played a crucial role in the development of catalytic asymmetric reactions. In particular, the coordination of a resolved, chiral phosphine to a transition metal center has been exploited to produce highly enantioselective catalysts for a variety of catalytic processes.[2] Although there are many chiral monodentate and bidentate chiral phosphines available, (1*R*,2*S*,4*R*,5*S*)-2,5-dimethyl-7-phenyl-7-phosphabicyclo[2.2.1]heptane provides the advantage of a rigidified ring system that reduces the conformational flexibility present in many other phosphine ligands.

Transition Metal-Catalyzed Reactions. Application of this ligand to the Pd-catalyzed allylic alkylation of 1,3-diphenyl-2-propenyl acetate with dimethyl malonate provides an alkylated product in >99.5% enantiomeric excess (eq 1).[1] The enantioselectivity of the process is dependent on the ligand:Pd ratio, the palladium precursor, and the nature of the nucleophile. Optimal conditions employed $Pd(dba)_3$ as the Pd precursor and 2 equiv of phosphine ligand, suggesting that two phosphines coordinate to the active Pd catalyst. Replacement of 1,3-diphenyl-2-propenyl acetate with pent-3-en-2-yl acetate decreased the ee to 34% due to the reduced sterics of methyl relative to phenyl substituents. It is noteworthy that in contrast to this ligand, most monodentate ligands provide low enantioselectivity in this reaction.[3]

Phosphine-Catalyzed Reactions. This ligand has also been shown to be effective in the direct organocatalysis of asymmetric processes.[4] For example, the phosphine-catalyzed [3 + 2] annulation reaction of ethyl 2,3-butadienoate and isobutyl acrylate produces two cyclopentene regioisomers (**1** and **2**) (eq 2).[5] Isomer **1** generally predominates and enantiomeric excesses ranging from

86–93% are displayed. Similarly, the ligand induces enantiomeric excesses between 43–68% in the phoshine-catalyzed γ-addition reaction of 2-butynoates (eq 3).[6]

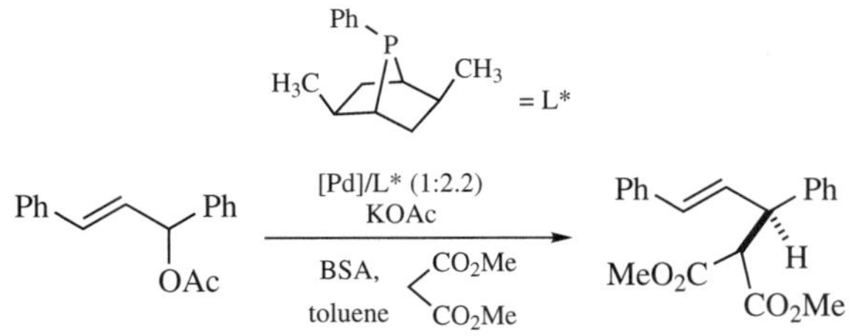

(1)

99% (97% ee)

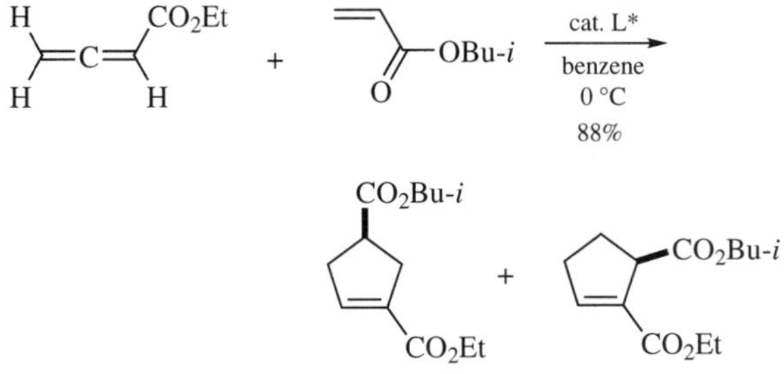

(2)

93% ee (100:0)

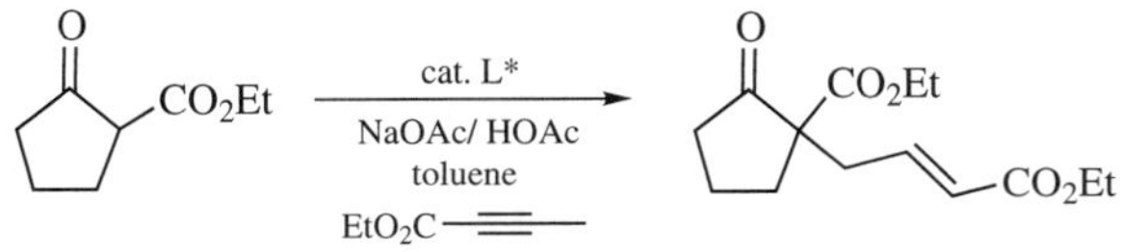

(3)

57% (68% ee)

1. Chen, Z.; Jiang, Q.; Zhu, G.; Xiao, D.; Cao, P.; Guo, C.; Zhang, X. *J. Org. Chem.* **1997**, *62*, 4521.
2. (a) Lee, S.; Hartwig, J. F. *J. Org. Chem.* **2001**, *66*, 3402. (b) Zhang, X. *Enantiomer* **1999**, *4*, 541. (c) Noyori, R. *Asymmetric Catalysis in Organic Synthesis*; Wiley: New York, 1994.
3. (a) Fiaud, J. C.; Legros, J. Y. *Tetrahedron Lett.* **1991**, *32*, 5089. (b) Fiaud, J. C.; Aribi-Zouioueche, L. *J. Organomet. Chem.* **1985**, *295*, 383.
4. Dalko, P. I.; Moisan, L. *Angew. Chem. Int. Ed.* **2001**, *40*, 3726.
5. Zhu, G.; Chen, Z.; Jiang, Q.; Xiao, D.; Cao, P.; Zhang, X. *J. Am. Chem. Soc.* **1997**, *119*, 3836.
6. Chen, Z.; Zhu, G.; Jiang, Q.; Xiao, D.; Cao, P.; Zhang, X. *J. Org. Chem.* **1998**, *63*, 5631.

Jon R. Parquette
The Ohio State University, Columbus, OH, USA

N,S-Dimethyl-*S*-phenylsulfoximine[1]

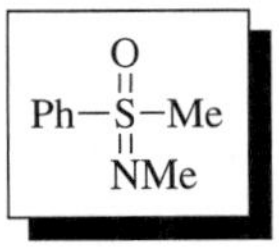

(±)
[30004-67-2] $C_8H_{11}NOS$ (MW 169.27)
(*R*)-(−)
[80482-67-3]
(*S*)-(+)
[33993-53-2]

(chiral sulfone analog useful in methylenation of carbonyl compounds[1,2] and resolution of ketones[1,3])

Physical Data: oil; $[\alpha]_D$ 184° (*c* 3, acetone); pK_a (DMSO) ~32.[1]
Solubility: sparingly sol water; sol dil acid, aq Cu^{II}; highly sol THF, alcohols, etc.
Preparative Methods: methyl phenyl sulfoxide is treated with *Hydrazoic Acid* (generated by addition of sulfuric acid to a slurry of sodium azide) in chloroform maintained at 45 °C to produce *S*-methyl-*S*-phenylsulfoximine.[4,5] The latter can be readily resolved using *10-Camphorsulfonic Acid*; from the (+)-acid the salt of (+)-(*S*)-*S*-methyl-*S*-phenylsulfoximine is obtained pure by recrystallization.[6,7] The Clarke–Eschweiler procedure using *Formaldehyde* and *Formic Acid* provides an effective method for conversion of the N–H sulfoximine to the title compound.[7]
Handling, Storage, and Precautions: due care should be taken in the synthetic step using hydrazoic acid. *N,S*-Dimethyl-*S*-phenylsulfoximine is apparently of low toxicity. The compound, which is stable to acids and bases, as well as to most oxidizing and reducing conditions, maintains chemical and enantiopurity on long term storage.

N,S-Dimethyl-*S*-phenylsulfoximine is rapidly deprotonated with *n-Butyllithium* in THF at 0 °C; the deprotonation can be conveniently monitored by use of a trace of Ph_3CH as an indicator. The lithio reagent is an excellent nucleophile, particularly with respect to addition to carbonyl compounds (eq 1). Addition occurs in high yields with a wide range of carbonyl compounds in the temperature range of −78 to 25 °C. The reaction is often reversible in the cases of hindered carbonyl compounds. The major side reaction is enolization. Both the enolization and reversibility problems can be circumvented by conducting the addition and subsequent quenching at low temperatures; in this manner, β-hydroxysulfoximines are obtained in high yield.[3] β-Hydroxysulfoximines are thermally unstable and revert to starting carbonyl compounds and sulfoximine in the temperature range 80–120 °C; the reversion is the basis for a ketone resolution method. The addition of optically pure sulfoximine (as its lithio derivative) to (±)-ketones which exhibit very high or complete diastereofacial selectivity results in the formation of two diastereomers which are generally responsive to separation by silica gel chromatography. Thermolysis of the separated diastereomers results in resolution of the ketone (eq 2).[3] Treatment of the separated diastereomers with *Raney Nickel* results in optically pure methyl carbinols (at the expense of destruction of the chiral *S* center).[8] The ketone resolution technique has reciprocity and a number of optically pure ketones, particularly (−)-menthone, have been found useful to resolve *N,S*-dimethyl-*S*-phenylsulfoximine.[3]

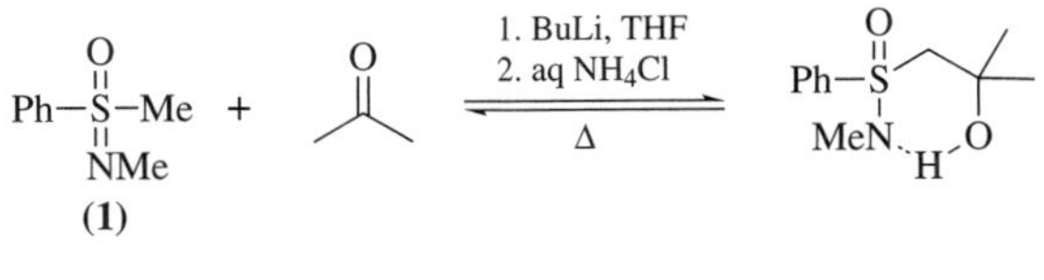

(1)

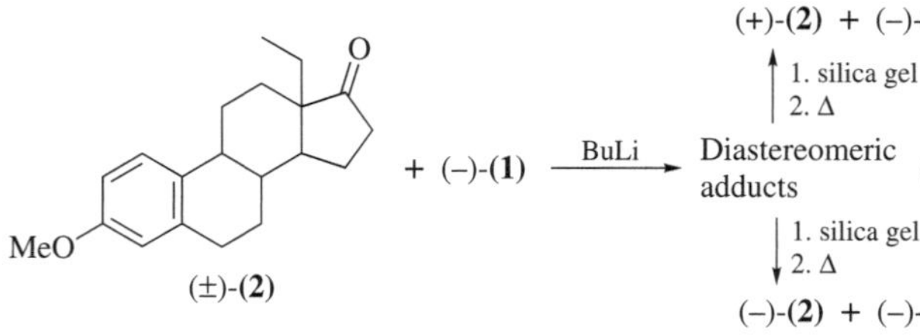

(2)

β-Hydroxysulfoximines undergo reductive elimination to yield alkenes upon treatment with *Aluminum Amalgam* in a mixture of THF, water, and acetic acid. In the case of conjugated enones and dienones the addition of the lithiosulfoximine occurs at the carbonyl; when the resulting adducts are reduced, dienes and trienes, respectively, are produced (eq 3).[9] The sulfoximine method often works where methylenation with triphenylphosphonium methylide fails[10] (eq 4).[11]

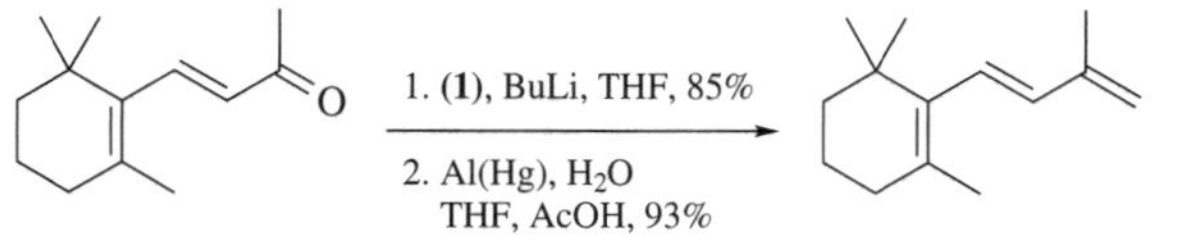

(3)

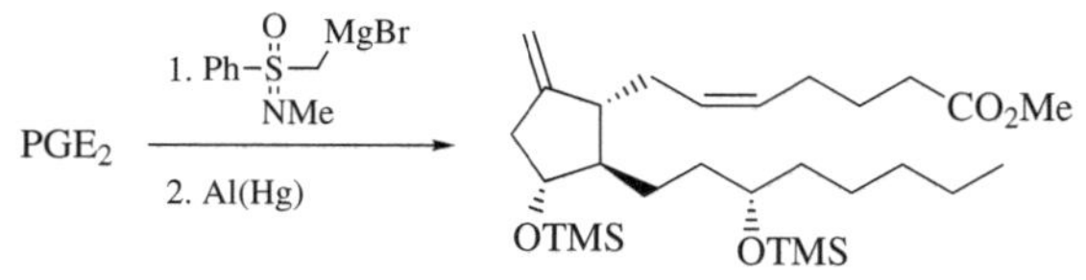

(4)

The combination of the chromatographic separation of enantiopure β-hydroxysulfoximine diastereomers and reductive elimination results in a method of ketone methylenation with optical resolution. The technique is illustrated in the synthesis of the ginseng sesquiterpene (−)-β-panasinsene and its enantiomer (eq 5).[12] The addition of the enantiopure lithiosulfoximine to prochiral enones or the diastereoface selective addition to racemic enones results in the formation of two diastereomeric adducts. The hydroxy group in these adducts can be used to direct the Simmons–Smith cyclopropanation (eq 6 and eq 7).[13] Catalytic osmylation of such adducts is directed by the *anti* effect of the hydroxy augmented by chelation by the methylimino group (eq 7).[14]

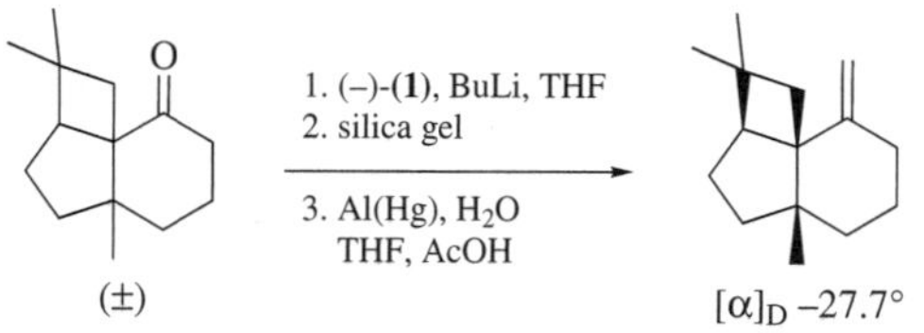

(5)

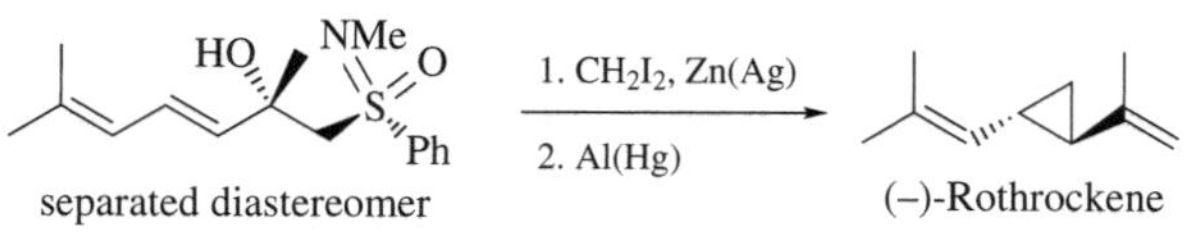

(6)

Ylides derived from the salts obtained by *N,N*-dialkylation of sulfoximines and anions derived from *N*-tosylsulfoximines are useful reagents for the synthesis of epoxides or cyclopropanes from aldehydes and ketones or enones.[1]

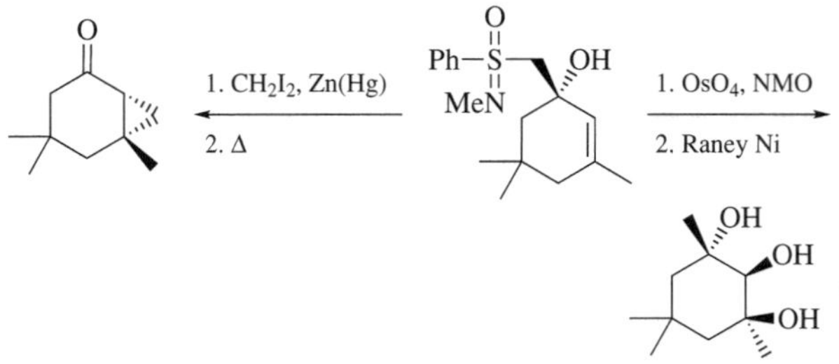

(7)

Related Reagents. Dibromomethane–Zinc–Titanium(IV) Chloride Methylenetriphenylphosphorane.

---

1. Johnson, C. R. *Aldrichim. Acta* **1985**, *18*, 2.
2. Johnson, C. R.; Kirchhoff, R. A. *J. Am. Chem. Soc.* **1979**, *101*, 3602.
3. Johnson, C. R.; Zeller, J. R. *Tetrahedron* **1984**, *40*, 1225.
4. Whitehead, J. K.; Bentley, H. R. *J. Chem. Soc.* **1952**, 1572.
5. Johnson, C. R.; Haake, M.; Schroeck, C. W. *J. Am. Chem. Soc.* **1970**, *92*, 6594.
6. Fusco, R.; Tenconi, F. *Chim. Ind. (Milan)* **1965**, *47*, 61 (*Chem. Abstr.* **1965**, *62*, 10 357h).
7. Johnson, C. R.; Schroeck, C. W.; Shanklin, J. R. *J. Am. Chem. Soc.* **1973**, *95*, 7424.
8. Johnson, C. R.; Stark, C. J. *J. Org. Chem.* **1982**, *47*, 1193.
9. Johnson, C. R.; Shanklin, J. R.; Kirchhoff, R. A. *J. Am. Chem. Soc.* **1973**, *95*, 6462.
10. Ansell, M. F.; Mason, J. S.; Caton, M. P. L. *J. Chem. Soc., Perkin Trans. 1* **1984**, 1061.
11. Bundy, G. L. *Tetrahedron Lett.* **1975**, 1957.
12. Johnson, C. R.; Meanwell, N. A. *J. Am. Chem. Soc.* **1981**, *103*, 7667.
13. Johnson, C. R.; Barbachyn, M. R. *J. Am. Chem. Soc.* **1982**, *104*, 4290. (b) Barbachyn, M. R.; Johnson, C. R.; Glick, M. D. *J. Org. Chem.* **1984**, *49*, 2726.
14. Johnson, C. R.; Barbachyn, M. R. *J. Am. Chem. Soc.* **1984**, *106*, 2459.

Carl R. Johnson
Wayne State University, Detroit, MI, USA

(S)-(−)-N-[(2,2′)-Dimethylpropionyl]-2-[(diphenylphosphino)methyl]pyrrolidine[1]

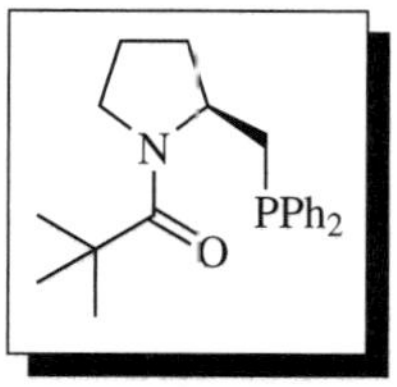

[145818-29-7] $C_{22}H_{28}NOP$ (MW 353.44)

(enantioselective conjugate addition,[1] *N*-tosylimine addition[2])

Physical Data: mp 97–97.5 °C; $[\alpha]_D$ −67.3 (*c* 1.45, $CHCl_3$).
Solubility: soluble in most organic solvents (CH_2Cl_2, Et_2O, THF, and toluene).
Form Supplied in: colorless prisms, not commercially available.
Analysis of Reagent Purity: NMR (^1H).

Preparative Methods: the ligand is prepared by the acylation of (*S*)-(−)-2-[(diphenylphosphino)methyl]pyrrolidine which is available from L-proline.[3]

Purification: the phosphine is oxygen-sensitive, however, the phosphine oxide by-product can be reduced back to the phosphine using Et_3SiH[4] or Cl_3SiH.[5]

Handling, Storage, and Precautions: the ligand is stable when stored under an inert atmosphere. The phosphine is air-sensitive and will oxidize to the phosphine oxide in the presence of oxygen or other oxidizing reagents.

Stoichiometric and Catalytic Chiral Ligand for the Asymmetric Conjugate Addition of Organocopper Reagents to Enones. Phosphine ligand (**1**) is a very effective stoichiometric chiral ligand for the asymmetric conjugate addition of organocuprates generated from organolithium reagents and CuI to α,β-unsaturated carbonyl derivatives. The addition reactions of simple cuprates to chalcones in the presence of ligand **1** proceed with some level of enantiocontrol.[6,7] In some cases, the introduction of a *gem*-dimethyl group at the 4-position of the pyrrolidine ring generates a ligand **2** that induces slightly higher enantioselectivities (eq 1).[8] Ligand **1** is also quite an effective chiral controller in the asymmetric conjugate addition of organocopper reagents to cycloalkenones, but at least 2 equiv of **1** are required (eq 2).[9]

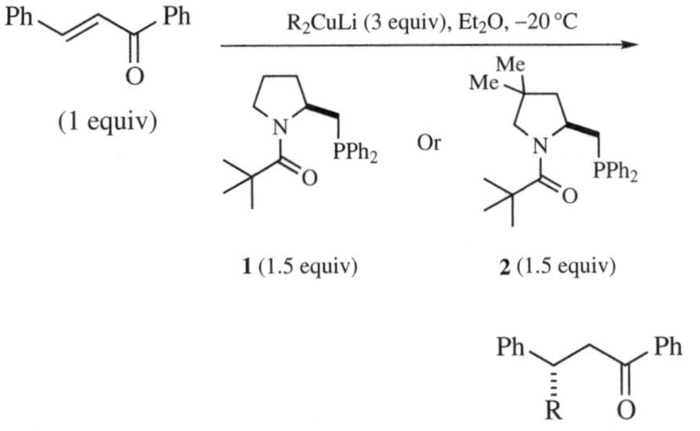

(1)

1; 84% ee (79%) (R = Me)
1; 24% ee (97%) (R = Bu)
2; 90% ee (99%) (R = Me)
2; 11% ee (95%) (R = Bu)

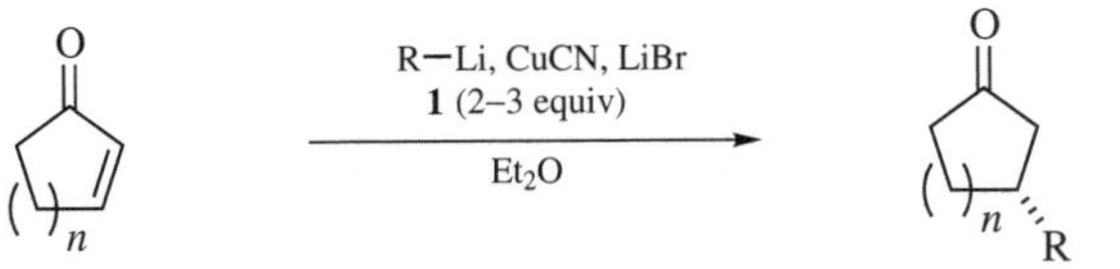

(2)

n = 2; R = Me	92% ee (66%)
n = 2; R = Et	91% ee (89%)
n = 2; R = Bu	90% ee (97%)
n = 1; R = Et	94% ee (90%)
n = 1; R = Bu	95% ee (99%)

It was later found that the amount of ligand **2** could be decreased to 0.32 equiv if the cuprate derived from a Grignard reagent and CuI was used (eq 3).[3,10]

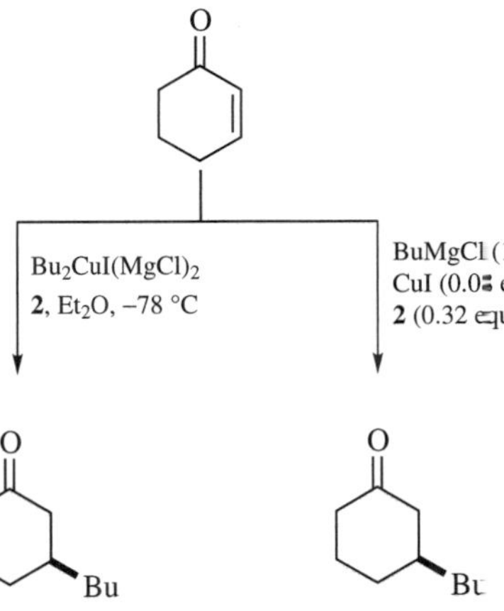

96% ee (96%) 92% ee (82%)

(3)

Catalytic, Enantioselective Addition of Arylboronic Acids to Cycloalkenones. A complex between ligand **1** and a rhodium(I) salt was found to catalyze the asymmetric 1,4-addition reaction of arylboronic acids to cyclohexenone and cycloheptenone. The reaction proceeds with high enantiocontrol and excellent yields (eq 4).[11] Lower enantiomeric excesses were observed with cyclopentenone (83% ee), but a variety of substituted phenylboronic acids could be used.

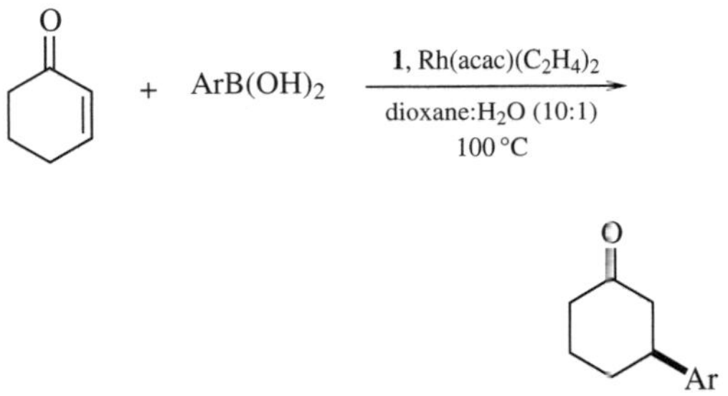

(4)

91–96% ee (90–99%)

Catalytic Asymmetric Addition of Organozincs to Imines. This class of ligands is also very effective in the copper-catalyzed addition of diethylzinc to *N*-sulfonylimine derivatives derived from aromatic aldehydes (eq 5).[12] It was found that the ligand bearing a *gem*-dibenzyl substituent at the 4-position of the pyrrolidine heterocycle produced the highest enantiomeric excesses. Cleavage of the *N*-sulfonyl group upon treatment with Red-Al in refluxing benzene for 12 h gave the secondary amine with slight racemization. The addition reaction proceeded almost equally well on the corresponding *N*-tosyl or *N*-trimethylsilylethylsulfonylimines. However, the advantage is significant in these latter two cases since the cleavage of the *N*-sulfonyl group to produce the secondary amine occurs without any racemization (SmI_2 in THF/HMPA or CsF in DMF, respectively). Replacement of the *gem*-dibenzyl substituent on the ligand by a *gem*-di(2,4,6-$Me_3C_6H_2CH_2$) group further increases the catalytic performance (eq 6).[13]

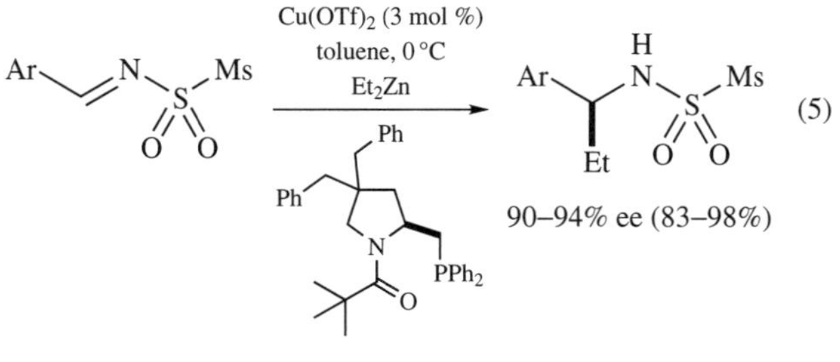

3 (3.6 mol %)

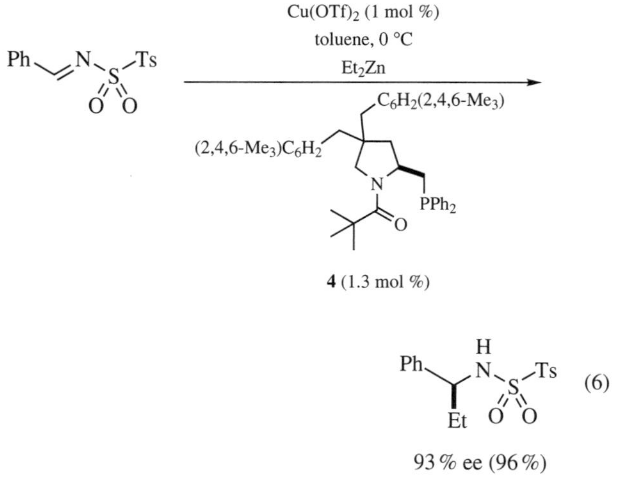

1. Krause, N.; Röder-Hoffmann, A. *Synthesis* **2001**, 171–196.
2. (a) Kobayashi, S.; Ishitani, H. *Chem. Rev.* **1999**, *99*, 1069–1094. (b) Bloch, R. *Chem. Rev.* **1998**, *98*, 1407–1438. (c) Denmark, S. E.; Nicaise, O. J. C. *Chem. Comm.* **1996**, 999–1004.
3. Kanai, M.; Nakagawa, Y.; Tomioka, K. *Tetrahedron* **1999**, *55*, 3843–3854.
4. Fritzsche, H.; Hasserodt, U.; Korte, F. *Chem. Ber.* **1965**, *98*, 1681–1687.
5. (a) Segall, Y.; Granoth, I.; Kalir, A. *Chem. Comm.* **1974**, 501–502. (b) Minami, T.; Okada, Y.; Nomura, R.; Hirota, S.; Nagahara, Y.; Fukuyama, K. *Chem. Lett.* **1986**, 613–616.
6. Kanai, M.; Koga, K.; Tomioka, K. *Tetrahedron Lett.* **1992**, *33*, 7193–7196.
7. Kanai, M.; Nakagawa, Y.; Tomioka, K. *Tetrahedron* **1999**, *55*, 3831–3842.
8. Nakagawa, Y.; Kanai, M.; Nagaoka, Y.; Tomioka, K. *Tetrahedron Lett.* **1996**, *37*, 7805–7808.
9. Kanai, M.; Tomioka, K. *Tetrahedron Lett.* **1994**, *35*, 895–898.
10. Nakagawa, Y.; Kanai, M.; Nagaoka, Y.; Tomioka, K. *Tetrahedron* **1998**, *54*, 10295–10307.
11. Kuriyama, M.; Tomioka, K. *Tetrahedron Lett.* **2001**, *42*, 921–923.
12. Fujihara, H.; Nagai, K.; Tomioka, K. *J. Am. Chem. Soc.* **2000**, *122*, 12055–12056.
13. Nagai, K.; Fujihara, H.; Kuriyama, M.; Yamada, K.-i.; Tomioka, K. *Chem. Lett.* **2002**, 8–9.

André B. Charette
Université de Montréal, QC, Canada

trans-2,5-Dimethylpyrrolidine

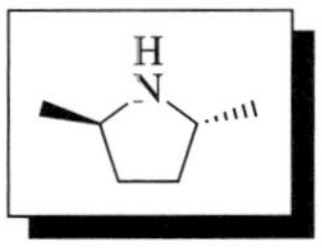

(racemate)
[62617-69-0; 39713-72-9] $C_6H_{13}N$ (MW 99.18)
(2*S*,5*S*)
[117968-50-0]
(2*R*,5*R*)
[62617-70-3]
(·HCl, racemate)
[114143-75-8; 4832-49-9] $C_6H_{14}ClN$ (MW 135.64)
(·HCl, 2*S*,5*S*)
[138133-34-3]
(·HCl, 2*R*,5*R*)
[70144-18-2]

(C_2 symmetric chiral pyrrolidine,[1] useful in optically active form as a chiral auxiliary in a variety of asymmetric reactions)

Physical Data: free amine: bp 102–103 °C; (2*S*,5*S*) $[\alpha]_D^{25}$ +10.6° (*c* 1.0, EtOH);[2] (2*R*,5*R*) $[\alpha]_D^{25}$ −11.5° (*c* 1.0, EtOH).[2] Hydrochloride: racemate mp 187–189 °C;[3] (2*S*,5*S*) mp 200–201 °C,[4] $[\alpha]_D^{25}$ −5.63° (*c* 0.67, CH_2Cl_2);[4] (2*R*,5*R*) mp 200–203 °C,[5] $[\alpha]_D^{25}$ +5.57° (*c* 1.18, CH_2Cl_2).[5]

Form Supplied in: colorless oil; commercially available as a mixture of (±)-*trans* and *cis* isomers (the mixture is not easily separated).[6]

Purification: the free amine can be purified by fractional distillation; the hydrochloride salt can be recrystallized from absolute ethanol and diethyl ether.

Handling, Storage, and Precautions: irritant; flammable. Use in a fume hood.

Synthesis. Several routes are available for the synthesis of *trans*-2,5-dimethylpyrrolidine.[2–9,22] Discussed below are preparative scale procedures for the synthesis of the pure *trans* compound in racemic and enantiomerically pure form.

The racemic hydrochloride salt can be prepared in four steps and 70% overall yield (eq 1).[3] The synthesis is carried out on 2 mmol scale and starts with commercially available 5-hexen-2-one. The key step involves a mercury-catalyzed intramolecular amidomercuration to form the pyrrolidine ring. If desired, the racemate can be resolved via the salts of *Mandelic Acid*.[2]

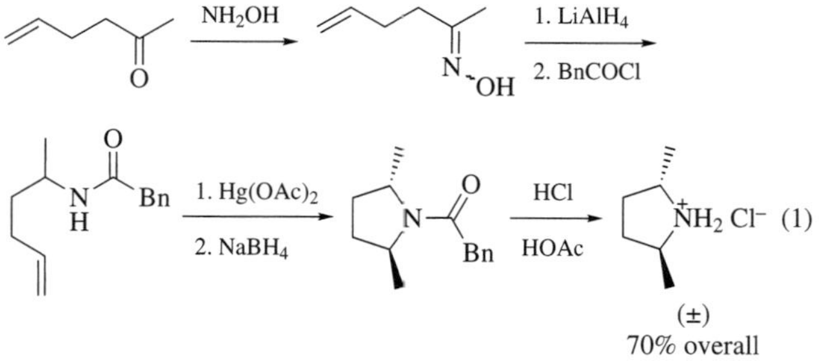

Alternatively, an efficient synthesis of either antipode starting from D- or L-alanine has been reported (eq 2).[9] The asymmetric synthesis conducted on 10 mmol scale involves a six-step sequence which incorporates the amidomercuration method.[3] The enantiomerically pure product is isolated as its hydrochloride salt in 44% overall yield. Furthermore, an optimization of the capricious cuprate reaction which improves both the yield and reproducibility has been described.[4]

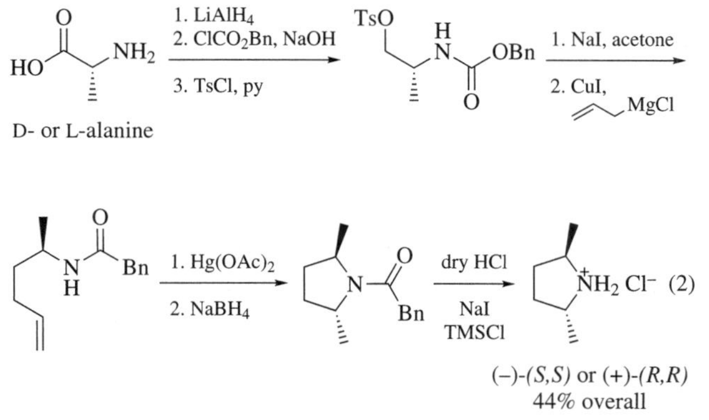

More recently, a four-step synthetic sequence which provides expedient access to the (−)-(*R*,*R*)-enantiomer in 42% overall yield has been reported.[5] This route is convenient for large-scale preparation (0.2 mol scale), and is highlighted by an asymmetric *Baker's Yeast* reduction of 2,5-hexanedione. Subsequent mesylation, *N*,*N*-dialkylation, and deprotection provides the enantiomerically pure free pyrrolidine (eq 3). Alternatively, either enantiomer of the chiral pyrrolidine can be obtained in 15% overall yield from an isomeric mixture of 2,5-hexanediol, via a similar sequence in which (*S*)-α-methylbenzylamine is used as a chiral auxiliary.[22] Also, an enantioselective route to either (2*S*,5*S*)- or (2*R*,5*R*)-hexanediol has been reported.[23]

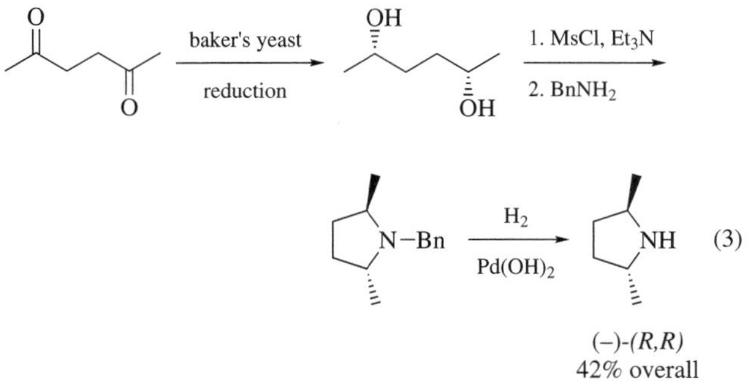

Asymmetric Alkylations and Michael Additions. Asymmetric alkylation of the cyclohexanone enamine derived from (+)-*trans*-2,5-dimethylpyrrolidine has been studied (eq 4).[2] Alkylation with *Iodomethane*, *n*-propyl bromide, and *Allyl Bromide* afforded the corresponding 2-*n*-alkylcyclohexanones in yields of 50–80% and with enantiomeric purities of 66, 86, and 64%, respectively.

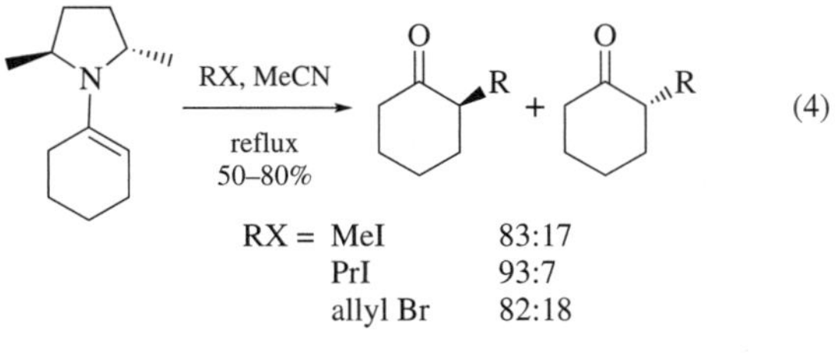

The lithium enolates from tetronic acid-derived vinylogous urethanes have been generated and their reactivity investigated with a variety of electrophiles (eq 5).[10,11] The reactions proceed with excellent regio- and diastereoselectivity and a variety of alkylating agents can be utilized.

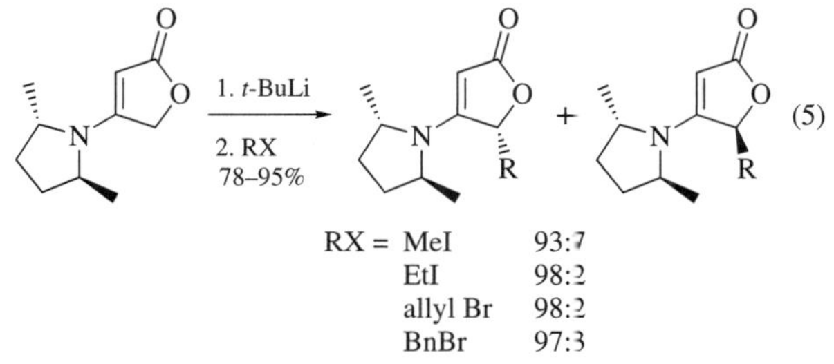

In the total synthesis of (−)-secodaphniphylline, an asymmetric [1,4]-conjugate addition was used to establish relative and absolute stereocontrol.[12] The lithium enolate of a *trans*-2,5-dimethylpyrrolidine-derived amide adds in a Michael fashion to a cyclic α,β-unsaturated ester, with subsequent enolate trapping, to afford the desired product in 64% yield and 92:8 diastereoselection (eq 6).

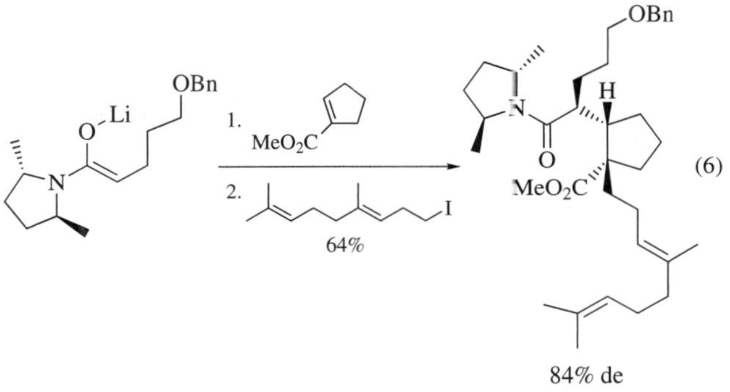

Asymmetric Radical Reactions. Several reports have documented the utility of nonracemic *trans*-2,5-dimethylpyrrolidine as a chiral auxiliary in asymmetric radical reactions.[13] For example, the addition of *n*-hexyl, cyclohexyl, and *t*-butyl radicals to the chiral acrylamide of 4-oxopentenoic acid provided four diastereomeric products resulting from α- and β-addition (eq 7).[14] The isomers resulting from β-addition were formed with no diastereoselectivity; however, the isomers resulting from α-addition were formed in ratios of 16:1, 24:1, and 49:1. Unfortunately, the application of this chemistry is limited due to the poor regioselectivity in the addition and difficulty in removal of the chiral auxiliary.

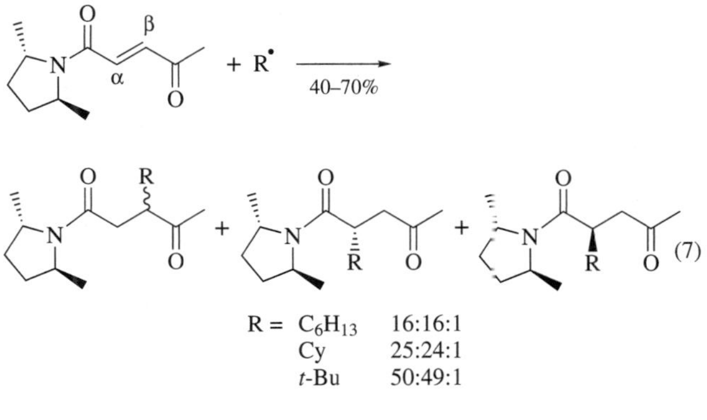

Similar results have been achieved in the addition of chiral amide radicals to activated alkenes.[13] For instance, a chiral amide radical, derived from (−)-*trans*-2,5-dimethylpyrrolidine, adds in a 1,4-fashion to ethyl acrylate in 35% yield and with 12:1 diastereoselectivity (eq 8).[15] Unfortunately, substantial amounts of higher oligomers are also formed. The radical telomerization of chiral acrylamides to afford nonracemic lower-order telomers ($n = 1$–5) has also been described.[16]

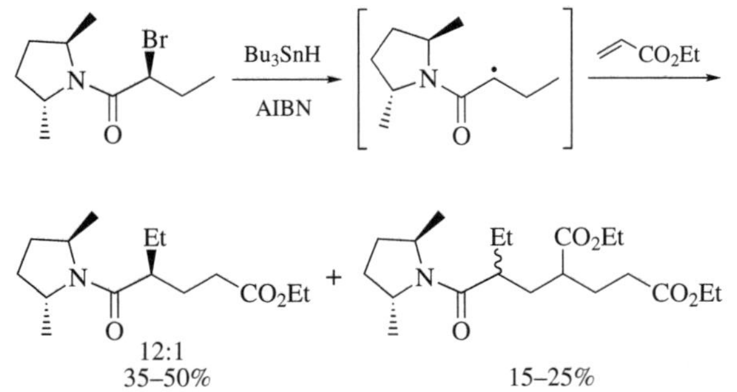

(8)

Asymmetric Pericyclic Reactions. Several reports illustrate the utility of *trans*-2,5-dimethylpyrrolidine as a chiral auxiliary in asymmetric Claisen-type rearrangements,[17] [4 + 2],[18,19] and [2 + 2] cycloaddition reactions.[20] The enantioselective Claisen-type rearrangement of *N,O*-ketene acetals derived from *trans*-2,5-dimethylpyrrolidine has been studied.[17] For example, the rearrangement of the *N,O*-ketene acetal, formed in situ by the reaction of *N*-propionyl-*trans*-(2*S*,5*S*)-dimethylpyrrolidine with (*E*)-crotyl alcohol, affords the [3,3]-rearrangement product in 50% yield and 10:1 diastereoselectivity (eq 9).

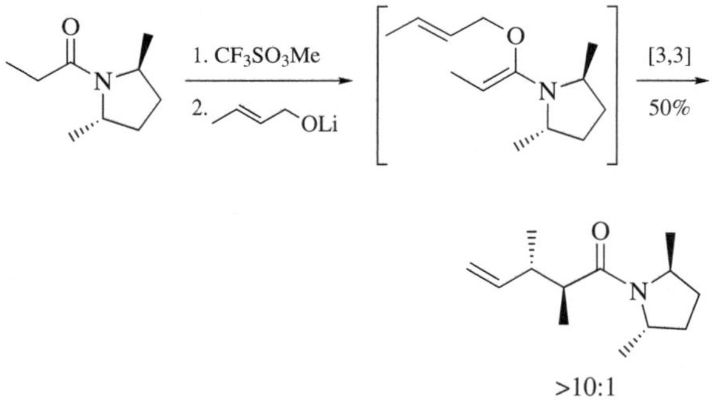

(9)

Carbamoyl nitroso dienophiles, derived from chiral pyrrolidines, have been generated and their reactivity with cyclohexadiene investigated.[18] Using (−)-*trans*-2,5-dimethylpyrrolidine as the auxiliary, the [4 + 2] cycloadduct is isolated in 82% yield and with 98% diastereomeric excess (eq 10). Similarly, chiral ynamine dienophiles have been utilized in asymmetric [4 + 2] cycloadditions with α,β-unsaturated nitroalkenes to afford cyclic nitronic esters.[19] The resulting esters subsequently undergo a rapid [1,3]-rearrangement to afford chiral cyclic nitrones in moderate yield and high diastereoselectivity (eq 11).

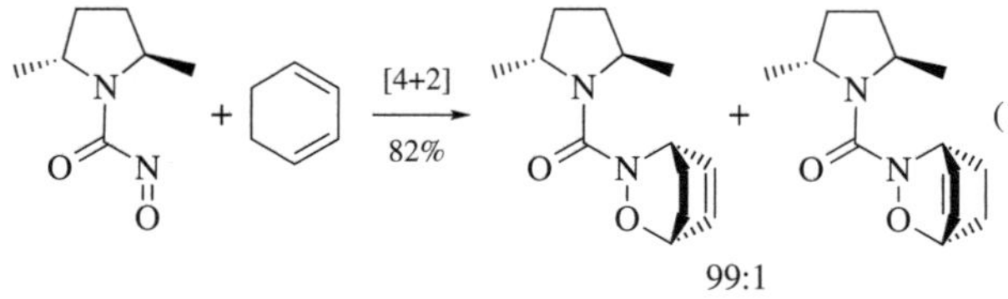

(10)

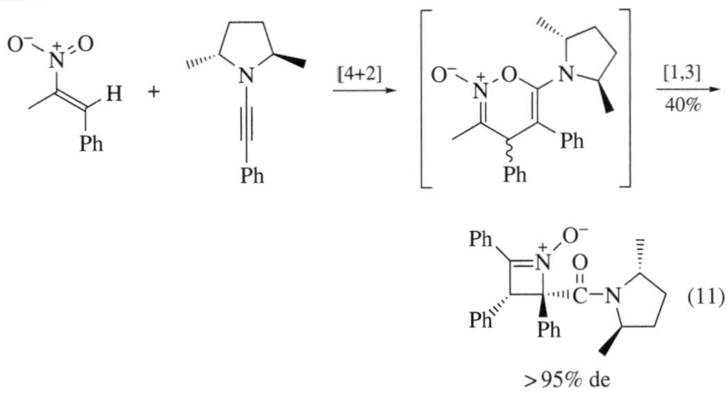

(11)

An asymmetric, thermal [2 + 2] cycloaddition of keteniminium salts derived from *trans*-2,5-dimethylpyrrolidine has been employed in the synthesis of prostaglandins.[20] An intramolecular [2 + 2] cycloaddition affords a *cis*-fused bicyclic system which is then further transformed into a common prostaglandin intermediate (eq 12).

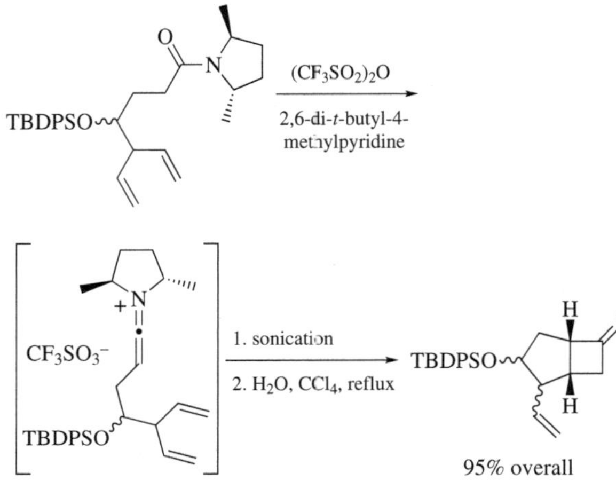

(12)

Miscellaneous. *trans*-2,5-Dimethylpyrrolidine has been utilized as a chiral auxiliary for an asymmetric iodolactonization in the total synthesis of (±)-pleurotin and (±)-dihydropleurotin.[21] The reaction affords the desired lactone in 47% yield and only 30% enantiomeric excess.

Related Reagents. *trans*-2,5-Bis(methoxymethyl)pyrrolidine.

---

1. Whitesell, J. K. *Chem. Rev.* **1989**, *89*, 1581.
2. Whitesell, J. K.; Felman *J. Org. Chem.* **1977**, *42*, 1663.
3. Harding, K. E.; Burks, S. R. *J. Org. Chem.* **1981**, *46*, 3920.
4. Yamazaki, T.; Gimi, R.; Welch, J. T. *Synlett* **1991**, 573.
5. Short, R. P.; Kennedy, R. M.; Masamune, S. *J. Org. Chem.* **1989**, *54*, 1755.
6. House, H. O.; Lee, L. F. *J. Org. Chem.* **1976**, *41*, 863.
7. Dervan, P. B.; Uyehara, T. *J. Am. Chem. Soc.* **1976**, *98*, 2003.
8. Gagné, M. R.; Stern, C. L.; Marks, T. J. *J. Am. Chem. Soc.* **1992**, *114*, 275.
9. Schlessinger, R. H.; Iwanowicz, E. J. *Tetrahedron Lett.* **1987**, *28*, 2083.
10. Schlessinger, R. H.; Iwanowicz, E. J.; Springer, J. P. *J. Org. Chem.* **1986**, *51*, 3070.
11. Schlessinger, R. H.; Iwanowicz, E. J.; Springer, J. P. *Tetrahedron Lett.* **1988**, *29*, 1489.

12. Heathcock, C. H.; Stafford, J. A. *J. Org. Chem.* **1992**, *57*, 2566.
13. Porter, N. A.; Giese, B.; Curran, D. P. *Acc. Chem. Res.* **1991**, *24*, 296.
14. Porter, N. A.; Scott, D. M.; Rosenstein, I. J.; Giese, B.; Veit, V.; Zeitz, H. G. *J. Am. Chem. Soc.* **1991**, *113*, 1791.
15. Porter, N. A.; Swann, E.; Nally, J.; McPhail, A. T. *J. Am. Chem. Soc.* **1990**, *112*, 6740.
16. Porter, N. A.; Breyer, R.; Swann, E.; Nally, J.; Pradhan, J.; Allen, T.; McPhail, A. T. *J. Am. Chem. Soc.* **1991**, *113*, 7002.
17. Yamazaki, T.; Welch, J. T.; Plummer, J. S.; Gimi, R. H. *Tetrahedron Lett.* **1991**, *32*, 4267.
18. Defoin, A.; Brouillard-Poichet, A.; Streith, J. *Helv. Chim. Acta* **1991**, *74*, 103.
19. Elburg, P. A.; Honig, G. W. N.; Reinhoudt, D. N. *Tetrahedron Lett.* **1987**, *28*, 6397.
20. Chen, L.-Y.; Ghosez, L. *Tetrahedron: Asymmetry* **1991**, *2*, 1181.
21. Hart, D. J.; Huang, H.-C.; Krishnamurthy, R.; Schwartz, T. *J. Am. Chem. Soc.* **1989**, *111*, 7507.
22. Mariël, E. Z.; Meetsma, A.; Feringa, B. L. *Tetrahedron: Asymmetry* **1993**, *4*, 2163.
23. Burk, M. J.; Feaster, J. E.; Harlow, R. L. *Tetrahedron: Asymmetry* **1991**, *2*, 569.

Lawrence R. Marcin
University of Illinois at Urbana-Champaign, IL, USA

2,2-Dimethyl-α,α,α′,α′-tetraphenyl-1,3-dioxolane-4,5-dimethanolatotitanium Diisopropoxide[1]

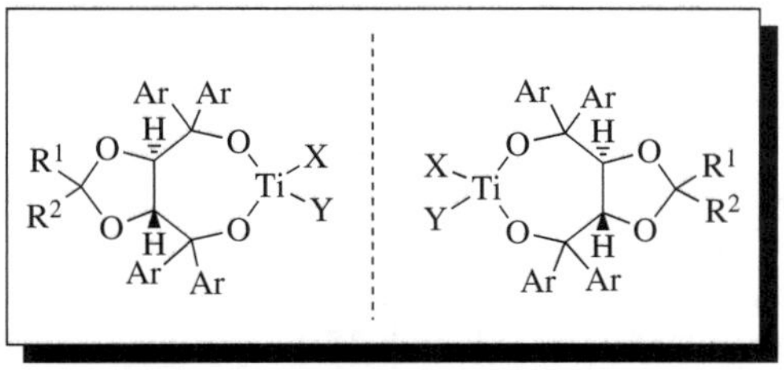

(**1a**; $R^1 = R^2 =$ Me; Ar = Ph; X = Y = *i*-PrO)
[144121-63-1] $C_{37}H_{42}O_6Ti$ (MW 630.66)
(**1b**; $R^1 = R^2 =$ Me; Ar = Ph; X = Y = Cl)
[109457-91-2] $C_{31}H_{28}Cl_2O_4Ti$ (MW 583.37)
(*ent*-**1b**; $R^1 = R^2 =$ Me; Ar = Ph; X = Y = Cl)
[139341-84-7] $C_{31}H_{28}Cl_2O_4Ti$ (MW 583.37)
(**1c**; $R^1 = R^2 =$ Me; Ar = Ph; X = *i*-PrO; Y = Cl)
[114031-33-3] $C_{34}H_{35}ClO_5Ti$ (MW 607.02)
(**1d**; $R^1 = R^2 =$ Me; Ar = 2-naphthyl; X = Y = *i*-PrO)
[144121-64-2] $C_{53}H_{50}O_6Ti$ (MW 830.90)
(**1e**; $R^1 =$ Me; $R^2 =$ Ar = Ph; X = Y = Cl)
[109414-72-4] $C_{36}H_{30}Cl_2O_4Ti$ (MW 645.44)
(**1f**; $R^1 = R^2 =$ Me; Ar = Ph; X = Cp; Y = Cl)
[132068-98-5] $C_{36}H_{33}ClO_4Ti$ (MW 613.02)

(chiral auxiliaries and Lewis acids for stoichiometric and catalytic enantioselective transformations)

Alternate Name: Ti-TADDOLates.

Introduction. Ti-TADDOLates are α,α,α′,α′-tetraaryl-2,2-disubstituted 1,3-dioxolane-4,5-dimethanolatotitanium derivatives. The most common substituents are R^1, R^2 = Me/Me and Ph/Me, Ar = Ph and 2-naphthyl, X, Y = Cl/Cl, *i*-PrO/Cl, Cp/Cl, and *i*-PrO/*i*-PrO. The corresponding TADDOLs (**2**) are available in both enantiomeric forms from tartrate esters which are acetalized (R^1R^2CO) and allowed to react with aryl Grignard reagents.[1a,g,2] The reactions performed in the presence of Ti-TADDOLates or with Ti-TADDOLate derivatives include: nucleophilic additions to aldehydes[1a,b,f,g,j,2a,3–5] and nitroalkenes[6] of alkyl,[1a,b,g,2a,3–5] aryl,[5] and allylic[1d,f] groups; aldol additions;[1d,f] hydrophosphonylations[7] and cyanohydrin reactions[1j] of aldehydes; inter- and intramolecular Diels–Alder additions;[1e,h,i,2a,8,9] [2 + 2] cycloadditions;[1e,h,i,10] intra-[1e,h,i] and intermolecular[11] ene reactions; iodolactonizations;[12] and transesterifications.[1h,i] Analogous compounds of other metals, and TADDOL derivatives containing one or two amino,[13] phosphinic, phosphonic, and/or phosphite groups,[13,14] have also been made and used for various reactions such as: $LiAlH_4$ reductions;[15] Grignard additions to ketones;[16] Li enolate additions to nitroalkenes;[17] hydrosilylations of ketones;[14a] Pd-catalyzed allylations;[14b] and metathesis reactions.[18] Finally, the TADDOLs themselves have been proved to be useful as NMR shift reagents;[2c,19] as components for enantioselective formation of host–guest complexes,[20,21] and for enantioselective solid-state reactions.[22] (*R,R*)-Ti-TADDOLates and (*P*)-Ti-BINOLates often give the same products in enantioselective reactions.[9]

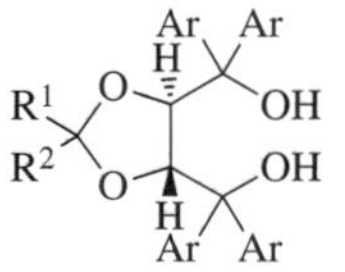

(**2**)

Preparation of Ti-TADDOLate Solutions. Five different procedures have been mostly used for the preparation of TADDOLates (**1**).

1. TADDOLate (**1f**) can be obtained from *Trichloro(cyclopentadienyl)titanium* and TADDOL (**2**), with removal of HCl (eq 1).[1d,f]

$$CpTiCl_3 + \mathbf{(2)} \xrightarrow{-HCl} \mathbf{(1f)} \quad (1)$$

2. TADDOLates (**1b**) and (**1e**) are prepared from (**2**) and *Dichlorotitanium Diisopropoxide*, without removal of the *i*-PrOH formed (eq 2). They are typically used in the presence of 4 Å molecular sieves.[1e,h,i,8b,c,9,10]

$$\mathbf{(2)} + (i\text{-PrO})_2TiCl_2 \longrightarrow \mathbf{(1b)} \text{ or } \mathbf{(1e)} + 2\ i\text{-PrOH} \quad (2)$$

3. Diisopropoxy Ti-TADDOLates (**1a**) and (**1d**) are conveniently made from (**2**) and *Titanium Tetraisopropoxide* with removal of *i*-PrOH by evaporation under reduced pressure or by azeotropic distillation (eq 3).[1a,b,g,2a,3e]

$$\mathbf{(2)} + (i\text{-PrO})_4Ti \xrightarrow{-i\text{-PrOH}} \mathbf{(1a)} \text{ or } \mathbf{(1d)} \quad (3)$$

4. TADDOLate (**1a**) can be synthesized alcohol-free from spirotitanate (**3**) and $(i\text{-PrO})_4Ti$ (eq 4).

$$\mathbf{(3)} + (i\text{-PrO})_4Ti \longrightarrow \mathbf{(1a)} \qquad (4)$$

5. Likewise, TADDOLate (**1b**) can be prepared from (**3**) and *Titanium(IV) Chloride* (eq 5).

$$\mathbf{(3)} + TiCl_4 \longrightarrow \mathbf{(1b)} \qquad (5)$$

An additional procedure leading from a titanate (**1**) (X = Y = *i*-PrO) to the corresponding dichloride (X = Y = Cl) is to treat the former with *Tetrachlorosilane* and pump off $(i\text{-PrO})_2SiCl_2$.[8a] A method in which the Ti-TADDOLate is present together with another Lewis acid (*Lithium Chloride*) is to treat a TADDOL (**2**) with 2 equiv *n-Butyllithium*, followed by $TiCl_4$.[9]

Spirotitanate (**3**) was obtained by reacting TADDOL and 0.5 equiv $(EtO)_4Ti$ or $(i\text{-PrO})_4Ti$, with azeotropic removal of the alcohol in refluxing toluene (eq 6). In the solid state it is rather stable to air (storage form);[3a,d,e] its crystal structure has been determined.[1g] Numerous TADDOLates of type (**1**) have been prepared and identified by NMR spectroscopy.[1g,3e,8a,23] Normally, they are used in situ in solvents such as toluene, petroleum ethers, CH_2Cl_2, Et_2O, or THF between −75 and +20 °C.

(**2**)
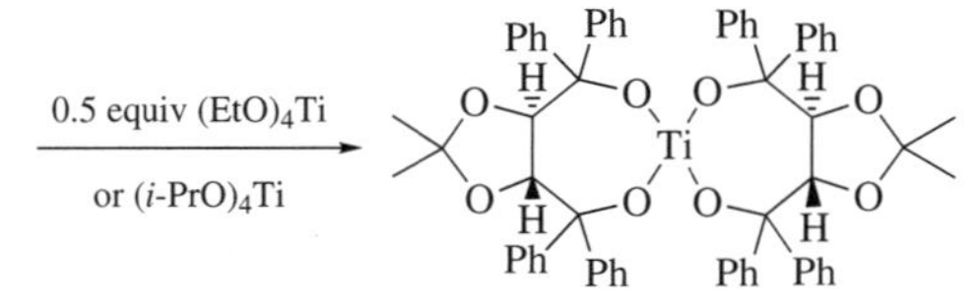

(**3**) (6)

Nucleophilic Additions of Polar Organometallic Compounds to Aldehydes. Ti-TADDOLates (**1**) have been used for both stoichiometric and catalytic nucleophilic additions to aldehydes. Thus the alcohols (**4**), (**9**), and (**10**) (eq 7) were obtained with the corresponding R_2Zn reagents[3d] in the presence of 0.2 equiv of (**1d**)[3c,e] or *ent*-(**1d**) and 1.2 equiv $(i\text{-PrO})_4Ti$. Alcohol (**5**) results from acetaldehyde and $(i\text{-PrO})_3TiPh$ mediated by 0.2 equiv of (**1a**).[5] (*R*)-Cyanohydrins such as (**6**) are formed from equimolar amounts of aldehydes, *Cyanotrimethylsilane*, and (**1e**).[1e,h–j,24] Alcohol (**7**) and threonine derivative (**8**) are the result of additions to aldehydes of the CpTi-TADDOLates prepared in situ from (**1f**) and crotyl Grignard reagent[1f,25] or a glycine Li enolate derivative.[1f] As can be seen, highly diastereoselective (dr = diastereomer ratio) and enantioselective (er = enantiomer ratio) conversions can be achieved. A number of examples have been reported in the literature.[1–7] (*R,R*)-TADDOL derivatives (**1**) always give rise to *Si* addition (rel. topicity *unlike*);[26] the mechanism of these reactions has been discussed.[1g,9]

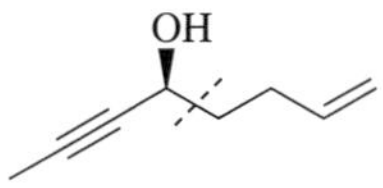

(**4**) er >98:2

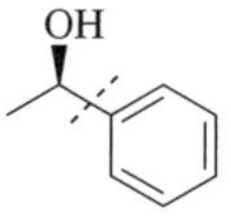

(**5**) er 98.5:1.5

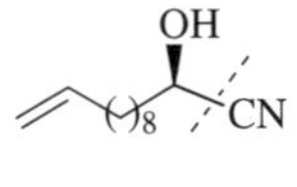

(**6**) er 97:3

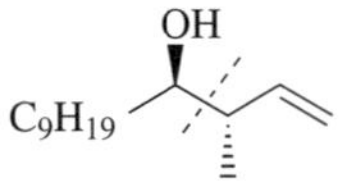

(**7**) dr >99:1; er >99:1

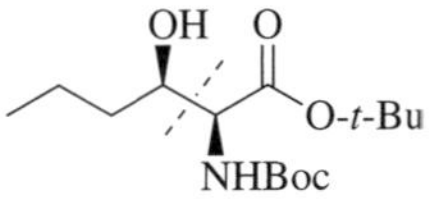

(**8**) dr 98:2; er 97:3

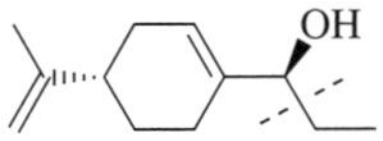

(**9**) dr 97:3

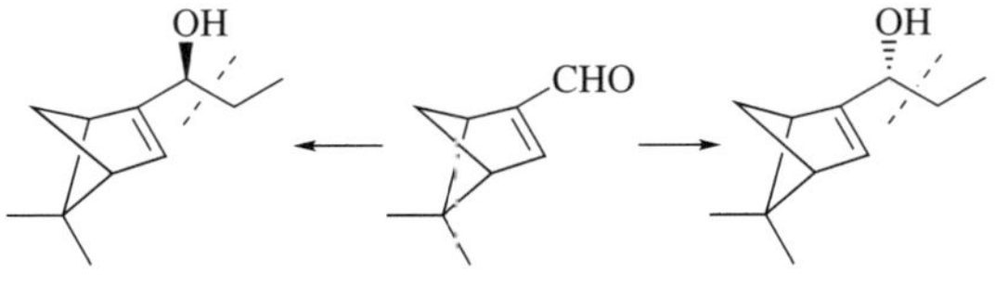

(7)

(**10a**) dr 95:5 (**10b**) dr 96:4

Cycloadditions and Ene Reactions. These reactions were mostly studied with the Cl_2Ti-TADDOLate (**1e**) as prepared by the procedure shown in eq 2. In all cases, conditions are critical. There are numerous examples in the literature,[1c,e,h,i] including the Diels–Alder addition of 3-crotonyl-1,2-oxazolidin-2-one to cyclopentadiene[8a,27] (leading to (**11**) (X = 1,3-oxazolidin-2-on-3-yl) in (**11**), (**12**), (**14**), and (**16**)) and to open-chain dienes,[27] as well as the intramolecular version of this reaction (see, for instance, (**12**)[1i]). It was shown that the Diels–Alder reaction leading to (**11**) can be done with (**1**), (Ar = 2-naphthyl, $R^1 = R^2 =$ Me, X = Y = Cl) with almost the same results as with (**1e**); with the analogous 1-naphthyl derivative the stereochemical course of the reaction reverses.[9] All these reactions are highly enantioselective and require ca. 10 mol % of the chiral catalyst. Some of these reactions have been carried out on a 45 mmolar scale.[9b,27] Other dienophiles such as methoxyquinones can also be employed. For the reactions leading to (**13**) and (**15**), somewhat different procedures were used.[8b,10b] Thus (**13**) was obtained with excellent enantioselectivity using stoichiometric amounts of a Ti-TADDOLate (**1e**).[8b] Likewise, [2 + 2] cycloadditions, again with the acyloxazolidinone (see the cyclobutene **14**)[10a] or with methoxyquinone (see **15**, a precursor to benzodihydrofurans),[10b] have been studied and occur with high regio-, diastereo-, and enantioselectivities; the nucleophilic components of these cycloadditions are electron-rich alkenes, allenes, or alkynes (cation-stabilizing substituents on the double or triple bond) such as ω-methylstyrenes,[10b] thioenol ethers,[1i] allenyl thioethers,[28] ketene dithioacetals,[28] or sulfenylated alkynes.[10a,28] The acyloxazolidinones derived from α,β-unsaturated carboxylic acids also lend themselves for intramolecular ene reactions. For instance, cyclopentane derivative (**16**) was formed from the corresponding open-chain 2,7-nonadienoic acid derivative.[29]

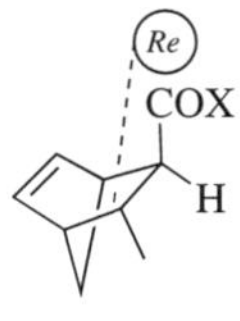

(**11**) dr 87:13; er 97:3

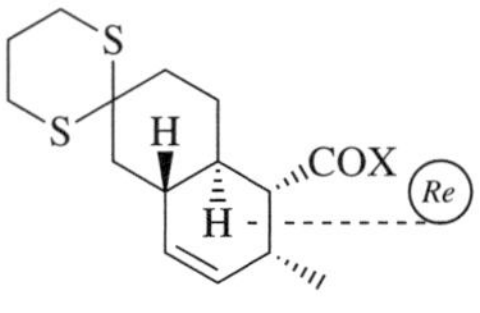

(**12**) er 99:1

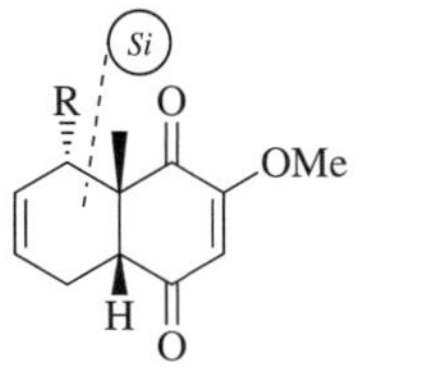

(13) R = *i*-Pr
dr >20:1; er 96:4

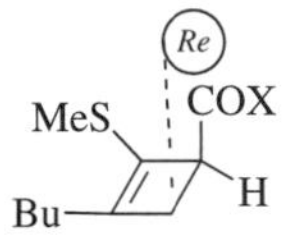

(14) er 99:1

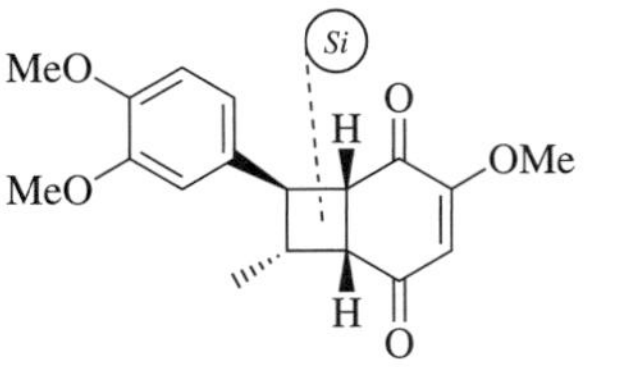

(15) er 96:4

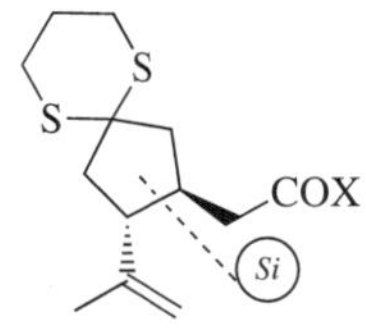

(16) er 98:2

As in other applications of *N*-acyl-1,3-oxazolidin-2-ones,[30] 2-thiones, and sthiazolidine-2-thiones,[31] chelation of the Lewis acid center for restricted rotation is considered decisive for the reactions occurring under the influence of the Ti-TADDOLates. Generally, the attack of the nucleophilic component (diene or ene) on the chelated electrophile occurs from the bottom face if the chelate ring is drawn as shown in structures (**17**) and (**18**),[8a,9,27] (**19**),[8b,f] and (**20**).[8d,e] For the oxazolidinones, this means that the trigonal α-carbonyl center is approached from the (*Re*)-face when an (*R,R*)-Ti-TADDOLate is used (rel. topicity *like*);[26] the mechanism of this reaction has been discussed.[1g,9]

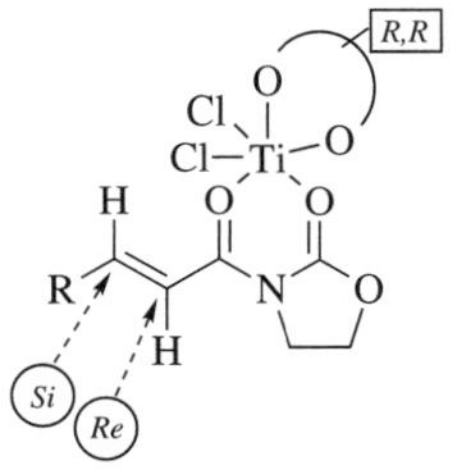

(17)

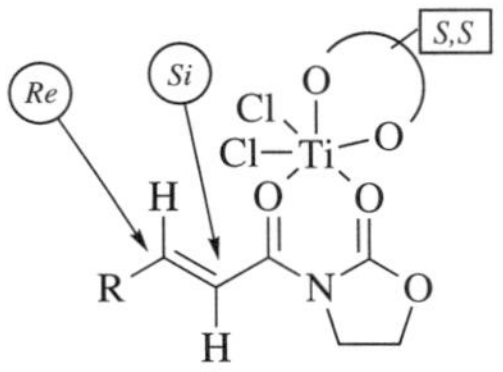

(18)

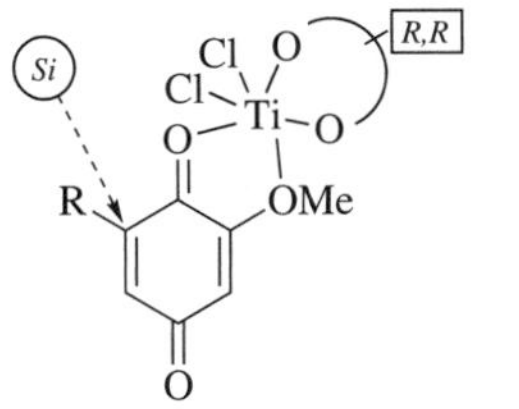

(19)

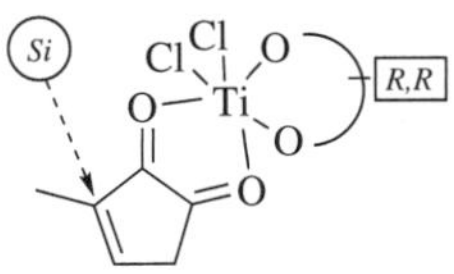

(20)

Other Enantioselective Transformations Mediated by Ti-TADDOLates. The iodolactonization of 2-allyl-2-hydroxy-4-pentenoic acid shown in eq 8 gives (**21**) in a 67% yield (after cyclization of some iodo isopropyl ester formed as a side product),[12] the iodolactone is a single (−)-diastereoisomer with a 5:1 (*S,S*)/(*R,R*) ratio. The TADDOLate generated in situ was employed in stoichiometric amount. The two enantiomers of 2-pyridyl 2-phenylthiobutyrate react with a rate difference of 39:1 with excess isopropanol in the presence of 0.1 equiv of a Ti-TADDOLate under the conditions specified in eq 9. This leads to the isopropyl ester (**22**) containing 96% of the (*R*)-enantiomer in a 69% yield.[32] Thus the complex (**23**) reacts much faster than the (*R,R*)/(*S*) isomer in the S/O transesterification.

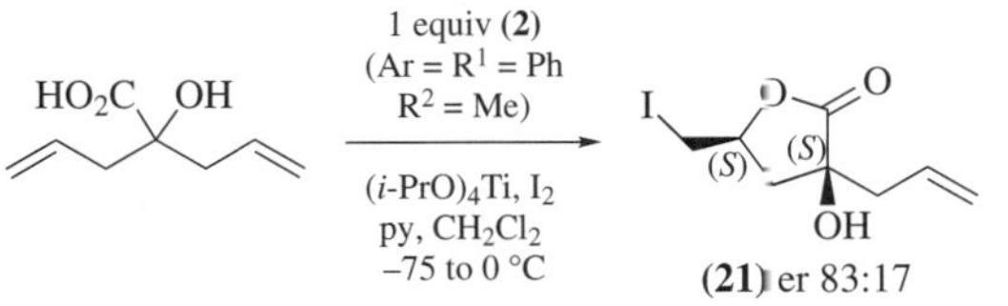

(8)

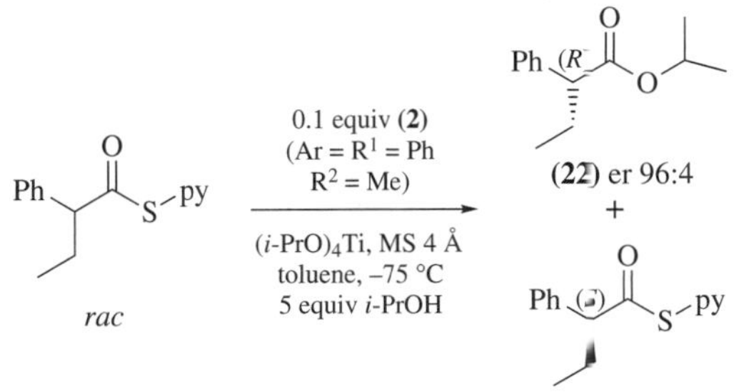

(9)

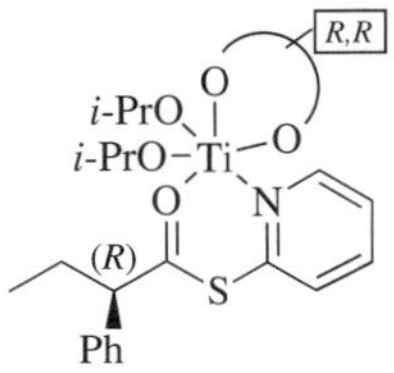

(23)

Use of TADDOLate Ligands on Other Metal Centers. Of the many possible and actually studied applications of the readily available TADDOLs, only two may be mentioned here: enantioselective reduction and Grignard addition to aryl ketones. A chiral *Lithium Aluminum Hydride* derivative prepared from (**2**) reduces aryl ketones in THF to the corresponding alcohols of (*S*)-configuration with an enantioselectivity of ca. 20:1 (eq 10).[15]

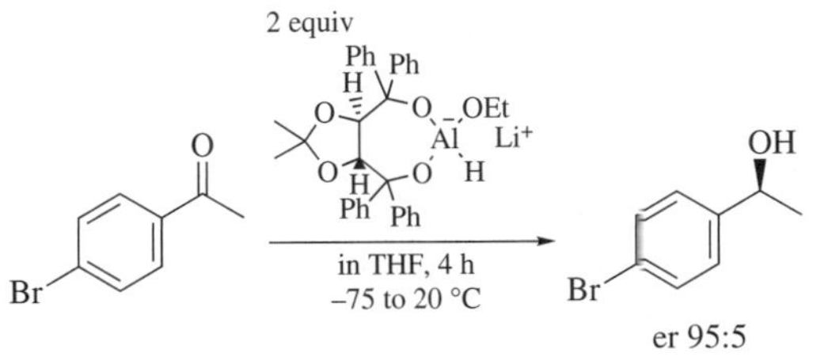

(10)

This procedure works with a smaller reagent excess and at higher temperatures than the analogous procedure using *(R)-1,1′-Bi-2,2′-naphthol* for similar results.[33] There is an added benefit in that the enantiomer excess of the alcohols formed in the reduction may sometimes be increased by a clathrating effect during the workup and isolation step (cf. the reviews).[20] Another useful application of the TADDOL (**2**) (Ar = Ph, $R^1 = R^2 =$ Me) is the highly enantioselective Grignard addition to aryl ketones. The procedure involves in situ reaction of 3 equiv RMgX with the TADDOL, followed by addition of the ketone at −100 °C in THF (eq 11).[16]

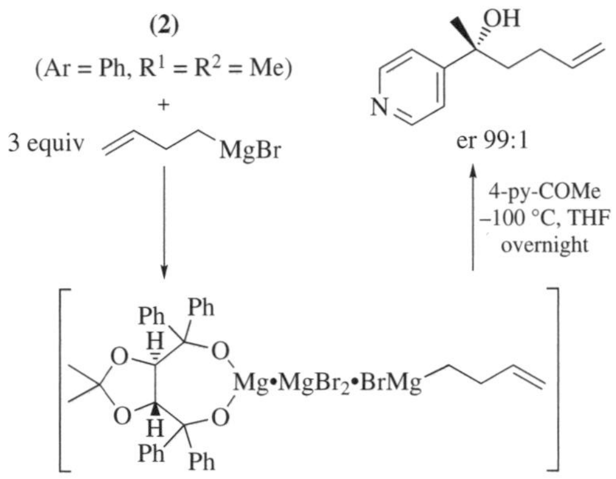

(11)

1. (a) Seebach, D.; Weidmann, B.; Widler, L. In *Modern Synthetic Methods*; Scheffold, R., Ed.; Salle & Sauerländer: Aarau (Switzerland) and Wiley: New York, 1983; Vol. 3, p 217. (b) Seebach, D.; Beck, A. K.; Schiess, M.; Widler, L.; Wonnacott, A. *Pure Appl. Chem.* **1983**, *55*, 1807. (c) Hayashi, Y.; Narasaka, K. *J. Synth. Org. Chem. Jpn.* **1990**, *48*, 988. (d) Duthaler, R. O.; Hafner, A.; Riediker, M. *Pure Appl. Chem.* **1990**, *62*, 631. (e) Narasaka, K. *Synthesis* **1991**, 1. (f) Duthaler, R. O.; Hafner, A. *Chem. Rev.* **1992**, *92*, 807. (g) Seebach, D.; Plattner, D. A.; Beck, A. K.; Wang, Y. M.; Hunziker, D.; Petter, W. *Helv. Chim. Acta* **1992**, *75*, 2171. (h) Narasaka, K. In *Organic Synthesis in Japan: Past, Present, and Future*; Noyori, R., Ed.; Kagaku Dozin: Tokyo, 1992; p 283. (i) Narasaka, K.; Iwasawa, N. In *Organic Synthesis: Theory and Applications*; Hudlicky, T., Ed.; JAI: London, 1993; Vol. 2, p 93. (j) North, M. *Synlett* **1993**, 807.
2. (a) Seebach, D.; Beck, A. K.; Imwinkelried, R.; Roggo, S.; Wonnacott, A. *Helv. Chim. Acta* **1987**, *70*, 954 (*Chem. Abstr.* **1988**, *108*, 203 984v). (b) Beck, A. K.; Bastani, B.; Plattner, D. A.; Petter, W.; Seebach, D.; Braunschweiger, H.; Gysi, P.; La Vecchia, L. *Chimia* **1991**, *45*, 238 (*Chem. Abstr.* **1991**, *115*, 279 866y). (c) Von dem Bussche-Hünnefeld, C.; Beck, A. K.; Lengweiler, U.; Seebach, D. *Helv. Chim. Acta* **1992**, *75*, 438.
3. (a) Schmidt, B.; Seebach, D. *Angew. Chem., Int. Ed. Engl.* **1991**, *30*, 99. (b) Seebach, D.; Behrendt, L.; Felix, D. *Angew. Chem., Int. Ed. Engl.* **1991**, *30*, 1008. (c) Schmidt, B.; Seebach, D. *Angew. Chem., Int. Ed. Engl.* **1991**, *30*, 1321. (d) Von dem Bussche-Hünnefeld, J. L.; Seebach, D. *Tetrahedron* **1992**, *48*, 5719. (e) Seebach, D.; Beck, A. K.; Schmidt, B.; Wang, Y. M. *Tetrahedron* **1994**, *50*, 4363.
4. (a) Takahashi, H.; Kawabata, A.; Niwa, H.; Higashiyama, K. *Chem. Pharm. Bull.* **1988**, *36*, 803. (b) Stanchev, S.; Hesse, M. *Helv. Chim. Acta* **1989**, *72*, 1052.
5. Weber, B. Dissertation No. 10663, ETH Zürich, 1994.
6. Schäfer, H. Dissertation No. 10822, ETH Zürich, 1994.
7. Yokomatsu, T.; Yamagishi, T.; Shibuya, S. *Tetrahedron: Asymmetry* **1993**, *4*, 1779.
8. (a) Corey, E. J.; Matsumura, Y. *Tetrahedron Lett.* **1991**, *32*, 6289. (b) Engler, T. A.; Letavic, M. A.; Takusagawa, F. *Tetrahedron Lett.* **1992**, *33*, 6731. (c) Narasaka, K.; Yamamoto, I. *Tetrahedron* **1992**, *48*, 5743. (d) Quinkert, G.; Del Grosso, M.; Bucher, A.; Bauch, M.; Döring, W.; Bats, J. W.; Dürner, G. *Tetrahedron Lett.* **1992**, *33*, 3617. (e) Quinkert, G.; Becker, H.; Del Grosso, M.; Dambacher, G.; Bats, J. W.; Dürner, G. *Tetrahedron Lett.* **1993**, *34*, 6885. (f) Tietze, L. F.; Ott, C.; Gerke, K.; Buback, M. *Angew. Chem., Int. Ed. Engl.* **1993**, *32*, 1485. (g) Engler, T. A.; Letavic, M. A.; Lynch, K. O., Jr.; Takusagawa, F. *J. Org. Chem.* **1994**, *59*, 1179.
9. Seebach, D.; Dahinden, R.; Marti, R. E.; Beck, A. K.; Plattner, D. A.; Kühnle, F. N. M. *J. Org. Chem.* **1995**, *60*, in press.
10. (a) Hayashi, Y.; Narasaka, K. *Chem. Lett.* **1990**, 1295. (b) Engler, T. A.; Letavic, M. A.; Reddy, J. P. *J. Am. Chem. Soc.* **1991**, *113*, 5068. (c) Hayashi, Y.; Otaka, K.; Saito, N.; Narasaka, K. *Bull. Chem. Soc. Jpn.* **1991**, *64*, 2122.
11. Mikami, K.; Terada, M.; Nakai, T. *J. Am. Chem. Soc.* **1990**, *112*, 3949.
12. Kitagawa, O.; Hanano, T.; Tanabe, K.; Shiro, M.; Taguchi, T. *Chem. Commun./J. Chem. Soc., Chem. Commun.* **1992**, 1005.
13. Seebach, D.; Hayakawa, M.; Sakaki, J.-i.; Schweizer, W. B. *Tetrahedron* **1993**, *49*, 1711.
14. (a) Sakaki, J.-i.; Schweizer, W. B.; Seebach, D. *Helv. Chim. Acta* **1993**, *76*, 2654. (b) Hayakawa, M. Dissertation No. 10352, ETH Zürich, 1993.
15. (a) Dahinden, R. Master's Thesis, ETH Zürich, 1991/92. (b) Hoffmann, M. Master's Thesis, ETH Zürich, 1992/93. (c) Beck, A. K., unpublished results, ETH Zürich, 1983 and 1991.
16. (a) Weber, B.; Seebach, D. *Angew. Chem., Int. Ed. Engl.* **1992**, *31*, 84. (b) Weber, B.; Seebach, D. *Tetrahedron* **1994**, *50*, 6117.
17. Juaristi, E.; Beck, A. K.; Hansen, J.; Matt, T.; Mukhopadhyay, T.; Simson, M.; Seebach, D. *Synthesis* **1993**, 1271.
18. McConville, D. H.; Wolf, J. R.; Schrock, R. R. *J. Am. Chem. Soc.* **1993**, *115*, 4413.
19. Tanaka, K.; Ootani, M.; Toda, F. *Tetrahedron: Asymmetry* **1992**, *3*, 709.
20. (a) Toda, F. *Top. Curr. Chem.* **1988**, *149*, 211. (b) Toda, F. *Bioorg. Chem.* **1991**, *19*, 157. (c) Toda, F.; Tohi, Y. *Chem. Commun./J. Chem. Soc., Chem. Commun.* **1993**, 1238. (d) Toda, F.; Tanaka, K.; Ootani, M.; Hayashi, A.; Miyahara, I.; Hirotsu, K. *Chem. Commun./J. Chem. Soc., Chem. Commun.* **1993**, 1413.
21. (a) Weber, E.; Dörpinghaus, N.; Goldberg, I. *Chem. Commun./J. Chem. Soc., Chem. Commun.* **1988**, 1566. (b) Goldberg, I.; Stein, Z.; Weber, E.; Dörpinghaus, N.; Franken, S. *J. Chem. Soc., Perkin Trans. 2* **1990**, 953. (c) Weber, E.; Dörpinghaus, N.; Wimmer, C.; Stein, Z.; Krupitsky, H.; Goldberg, I. *J. Org. Chem.* **1992**, *57*, 6825.
22. (a) Toda, F. In *Organic Synthesis in Japan: Past, Present, and Future*; Noyori, R., Ed.; Kagaku Dozin: Tokyo, 1992; p 473. (b) Toda, F. *Synlett* **1993**, 303.
23. Iwasawa, N.; Hayashi, Y.; Sakurai, H.; Narasaka, K. *Chem. Lett.* **1989**, 1581.
24. Minamikawa, H.; Hayakawa, S.; Yamada, T.; Iwasawa, N.; Narasaka, K. *Bull. Chem. Soc. Jpn.* **1988**, *61*, 4379.
25. (a) Hafner, A.; Duthaler, R. O.; Marti, R.; Rihs, G.; Rothe-Streit, P.; Schwarzenbach, F. *J. Am. Chem. Soc.* **1992**, *114*, 2321. (b) For an enantioselective hydroxylation of a (**1f**)-derived enolate by a dioxirane, see: Prechtl, F. PhD Thesis, University of Würzburg, 1993.
26. Seebach, D.; Prelog, V. *Angew. Chem., Int. Ed. Engl.* **1982**, *21*, 654.
27. Narasaka, K.; Iwasawa, N.; Inoue, M.; Yamada, T.; Nakashima, M.; Sugimori, J. *J. Am. Chem. Soc.* **1989**, *111*, 5340.
28. Narasaka, K.; Hayashi, Y.; Shimadzu, H.; Niihata, S. *J. Am. Chem. Soc.* **1992**, *114*, 8869.
29. Narasaka, K.; Hayashi, Y.; Shimada, S.; Yamada, J. *Isr. J. Chem.* **1991**, *31*, 261.
30. (a) Evans, D. A. In *Asymmetric Synthesis*; Morrison, J. D., Ed.; Academic: Orlando, 1984; Vol. 3, p 1. (b) Evans, D. A.; Rieger, D. L.; Bilodeau, M. T.; Urpi, F. *J. Am. Chem. Soc.* **1991**, *113*, 1047. (c) Nerz-Stormes, M.; Thornton, E. R. *J. Org. Chem.* **1991**, *56*, 2489. (d) Walker, M. A.; Heathcock, C. H. *J. Org. Chem.* **1991**, *56*, 5747.
31. Mukaiyama, T. *Challenges in Synthetic Organic Chemistry*; Clarendon: Oxford, 1990.
32. Narasaka, K.; Kanai, F.; Okudo, M.; Miyoshi, N. *Chem. Lett.* **1989**, 1187.
33. (a) Noyori, R.; Tomino, I.; Tanimoto, Y.; Nishizawa, M. *J. Am. Chem. Soc.* **1984**, *106*, 6709. (b) Noyori, R.; Tomino, I.; Yamada, M.; Nishizawa, M. *J. Am. Chem. Soc.* **1984**, *106*, 6717. (c) Singh, V. K. *Synthesis* **1992**, 605.

Robert Dahinden, Albert K. Beck & Dieter Seebach
Eidgenössische Technische Hochschule Zürich, Switzerland

S,S-Dimethyl-*N*-(*p*-toluenesulfonyl)-sulfilimine[1,2]

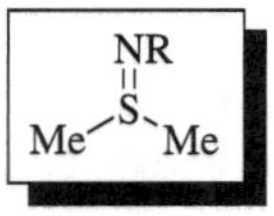

(**1**; R = *p*-$MeC_6H_4SO_2$)		
[13150-75-9]	$C_9H_{13}NO_2S_2$	(MW 231.37)
(**2**; R = 2,4-$(NO_2)_2C_6H_3$)		
[37873-98-6]	$C_8H_9N_3O_4S$	(MW 243.27)

(reagents for α-*S,S*-dimethylsulfuranylation of active methylene compounds,[2] and *ortho*-methylation of phenols;[3,4] (**1**) is a methylene transfer agent which converts carbonyl compounds into epoxides[5])

Alternate Name: (**1**) DMTS; (**2**) DMDNS.
Physical Data: (**1**) mp 154–155 °C; (**2**) mp 175–176 °C (dec).
Solubility: insol H_2O; sol ethyl and methyl alcohols, acetone, chloroform, and other common polar organic solvents.
Form Supplied in: (**1**) white solid; (**2**) orange solid.
Preparative Methods: (**1**) is conveniently prepared by adding a slight excess of *Dimethyl Sulfide* to an aqueous solution of *Chloramine-T*, and collecting the deposited crystals by filtration and recrystallization from ethanol.[1] The yield of white crystalline solid is >95% based on chloramine T.
For preparation of (**2**), *Phosphorus(V) Oxide* (60 mmol) is added with stirring to 25 mL of DMF at 0 °C. After 30 min, *Dimethyl Sulfoxide* (60 mmol) is added. After stirring for 1 h, 20 mmol of 2,4-dinitroaniline in 25 mL of DMF is added dropwise at 0 °C with continued stirring. After 3 h, 180 mmol of *Triethylamine* is added at 0–5 °C, and stirring is continued for 3 h. The deposited crystals are collected by filtration and recrystallized from THF. The yield of orange crystalline material is 96% based on 2,4-dinitroaniline.[2,3]
Handling, Storage, and Precautions: stable at rt in a sealed bottle, but storage at lower temperature is recommended for (**2**). Use in a fume hood.

Reagent (**2**) reacts with *p*-toluenesulfonamide (at 90 °C for 7 h in DMF) to give (**1**) (58%), in an ylide exchange reaction.

Reactions of (**1**) and (**2**) with active methylene compounds in DMF give the corresponding sulfuranes (eq 1). Reagent (**2**), which is more basic than (**1**), gives higher yields of sulfuranes. Furthermore, the yields of the ylide exchange reactions depend on the pK_a value of the active methylene compounds, as shown in Table 1. The lower the pK_a value, the higher the yield of sulfurane.[2,3]

$$Me_2S{=}NR + CH_2R^1R^2 \xrightarrow[90\ ^\circ C,\ 7\ h]{DMF} Me_2S{=}CR^1R^2 + H_2NR \quad (1)$$

R = Tos, 2,4-$(NO_2)_2C_6H_3$

Reactions of (**1**) and (**2**) with phenols give *o*-methylthiomethylated phenols (eq 2).[3,4] Mixtures of the phenols and 0.5 equiv of (**1**) or (**2**) are heated without solvent at 120–130 °C for 3–7 h. 2-Methylthiomethylphenols are obtained from 2- and 4-methyl, 2,5- and 3,5-dimethyl-, 2,3,5-trimethyl-, and 2-methoxyphenols in 55–95% yield (Table 2). In some case, significant amounts of bis(methylthiomethyl) products are also formed. The yields using (**2**) are higher than those using (**1**).

Table 1 Results of the Reactions of Sulfilimines (**1**) and (**2**) with Active Methylene Compounds in DMF at 90 °C for 9 h

Sulfilimine	Active methylene compound R^1	R^2	pK_a	Sulfurane yield (%)
(**2**)	COMe	COMe	8.94	94
(**2**)	COMe	CO_2Et	10.7	80
(**2**)	CN	CN	11.2	98
(**1**)	CN	CN	11.2	12.5
(**2**)	CO_2Me	CO_2Me	–	58
(**2**)	CO_2Et	CO_2Et	13.3	48
(**2**)	Ph	Ph	34.1	0

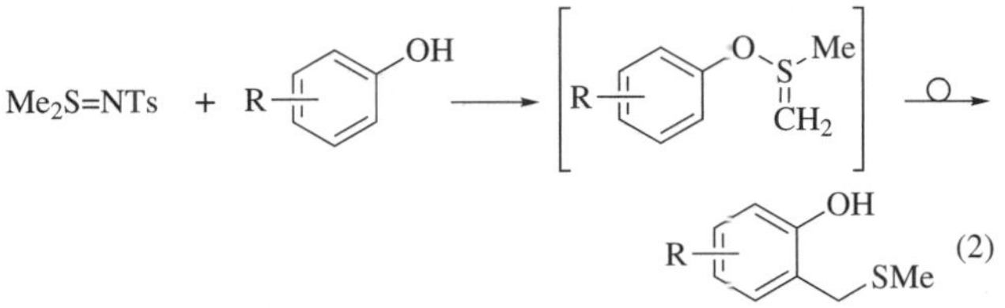

R = Tos, 2,4-$(NO_2)_2C_6H_3$

Table 2 Methylthiomethylation of Phenols with the Sulfilimines (**1**) and (**2**) at 120–130 °C for 3–7 h

Sulfilimine	Phenol	Yield (%)[a]	Sulfilimine	Phenol	Yield (%)[a]
(**2**)	2-MeO	82	(**1**)	2-MeO	78
(**2**)	2-Me	95	(**1**)	2-Me	73
(**2**)	4-Me	55 (24)[b]	(**1**)	2,5-Me_2	96
(**2**)	H	41	(**1**)	3,5-Me_2	58 (35)[b]
(**2**)	4-NO_2	0	(**1**)	2,3,5-Me_3	64

[a] Yields based on reacted sulfilimine. [b] Data in parentheses show the yields of bis(methylthiomethyl)phenols.

Like *S,S-Dimethyl-N-(p-toluenesulfonyl)sulfoximine*[6] and (dimethylamino)dimethyloxosulfonium tetrafluoroborate,[7] the *N*-tosylsulfilimine (**1**) reacts as a methylene transfer reagent, converting aldehydes and ketones to epoxides (eq 3). Thus (**1**) is heated at 80–90 °C for 0.5 h in DMSO in the presence of *Sodium Hydride*, and the resulting anion is allowed to react with carbonyl compounds to give 1-mono- and 1,1-disubstituted oxiranes in 46–56% yields.[5]

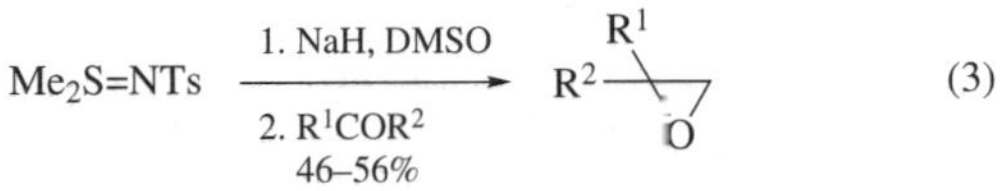

R^1 = Ph, R^2 = H, Me; $R^1R^2 = (CH_2)_6$

Related Reagents. *N,S*-Dimethyl-*S*-phenylsulfoximine; Dimethylsulfonium Methylide; Dimethylsulfoxonium Methylide; *S,S*-Dimethyl-*N*-(*p*-toluenesulfonyl)sulfoximine; *S,S*-Diphenylsulfilimine.

1. Mann, F. G.; Pope, W. J. *J. Chem. Soc.* **1922**, *121*, 1052.
2. Yamamoto, T.; Harigaya, Y.; Okawara, M. *Chem. Lett.* **1972**, 1009.
3. Yamamoto, T.; Harigaya, Y.; Okawara, M. *Tetrahedron* **1978**, *34*, 3097.
4. Yamamoto, T.; Okawara, M. *Bull. Chem. Soc. Jpn.* **1978**, *51*, 2443.
5. Tamura, Y.; Matsushima, H.; Ikeda, M.; Sumoto, K. *Synthesis* **1976**, 35.
6. Johnson, C. R.; Katekar, G. F. *J. Am. Chem. Soc.* **1970**, *92*, 5753.
7. Johnson, C. R.; Rogers, P. E. *J. Org. Chem.* **1973**, *38*, 1793.

Tamotsu Yamamoto
Kanto Gakuin University, Yokohoma, Japan

S,S-Dimethyl-*N*-(*p*-toluenesulfonyl)-sulfoximine[1]

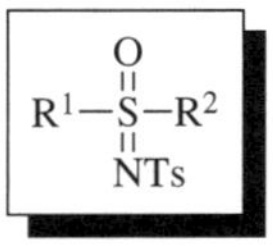

(**1**; R^1 = Me, R^2 = Me)		
[22236-45-9]	$C_9H_{13}NO_3S_2$	(MW 247.37)
(**2**; R^1 = Et, R^2 = Et)		
[42153-72-0]	$C_{11}H_{17}NO_3S_2$	(MW 275.43)
(**3**; R^1 = *i*-Pr, R^2 = *i*-Pr)		
[42153-73-1]	$C_{13}H_{21}NO_3S_2$	(MW 303.49)
(**4**; R^1 = Ph, R^2 = Me)		
[42153-74-2]	$C_{14}H_{15}NO_3S_2$	(MW 309.44)
(**5**; R^1 = Ph, R^2 = *c*-C_5H_9)		
[33332-99-9]	$C_{18}H_{21}NO_3S_2$	(MW 363.54)
(**6**; R^1 = Ph, R^2 = Cy)		
[33367-88-3]	$C_{19}H_{23}NO_3S_2$	(MW 377.57)
(**7**; R^1 = Ph, R^2 = Bn)		
[38764-59-9]	$C_{20}H_{19}NO_3S_2$	(MW 385.54)

(conversion of aldehydes and ketones to oxiranes,[1,2] ketones or oxiranes to oxetanes,[3] imines to aziridines,[1,2] and electrophilic alkenes to cyclopropanes[1,2])

Alternate Name: *S,S*-dimethyl-*N*-tosylsulfoximine.
Physical Data: (**1**) mp 167–169 °C, 170 °C (from ethanol);[2,4] (**2**) mp 89–91 °C; (**3**) mp 75–77 °C; (**4**) mp 107–109 °C; (**5**) mp 143–144 °C; (**6**) mp 145–146 °C; (**7**) mp 148–149 °C.
Solubility: moderately sol EtOH, THF, DMSO.
Form Supplied in: (**1**) white solid; commercially available.
Preparative Methods: excess DMSO containing *Copper(II) Chloride* (or another copper catalyst)[2,4] is treated with *Chloramine-T* trihydrate. (**1**) is obtained in 90% yield after aqueous EDTA workup and recrystallization from ethanol.The other *N*-tosylsulfoximines can be prepared by the tosylation of N–H sulfoximines with *p-Toluenesulfonyl Chloride* in the presence of base,[2] but the two most useful and general methods are the oxidation of *N*-tosylsulfilimines with basic *Hydrogen Peroxide*,[5] *m-Chloroperbenzoic Acid* anion,[6] *Sodium Hypochlorite*,[8] or *Ruthenium(IV) Oxide*/sodium metaperiodate[7] and the copper powder-promoted reaction of sulfoxides with *p-Toluenesulfonyl Azide*.[2,9]
Handling, Storage, and Precautions: (**1**) is a highly crystalline compound with no known toxicity and unlimited shelf life.

The generation of *N-p*-toluenesulfonyl-sulfonimidoyl-stabilized carbanions is best accomplished by stirring a slurry of *N*-tosylsulfoximine, e.g. (**1**), and *Sodium Hydride* in DMSO at rt until hydrogen evolution ceases (2–4 h). THF solutions of the lithium salts of *N*-tosylsulfoximines can be prepared by deprotonation with *n-Butyllithium*. These anions, which are quite stable at room or slightly elevated temperatures, form a class of nucleophilic alkylidene transfer reagents. The mechanism of these transfer reactions is similar to that of sulfonium ylide reactions but the leaving groups are water-soluble anions rather than neutral molecules (eq 1). The nucleophilic transfer chemistry of sodium *N*-tosylmethanesulfonimidoylmethide (**8**) is similar to that of dimethylsulfoxonium methylide[10] in regard to regio- and stereochemical selectivity in that the products reflect thermodynamic control.[11] Anion (**8**) has been reported to be superior to dimethylsulfonium and dimethylsulfoxonium methylides for reactions in which enolate formation is a serious problem.[12]

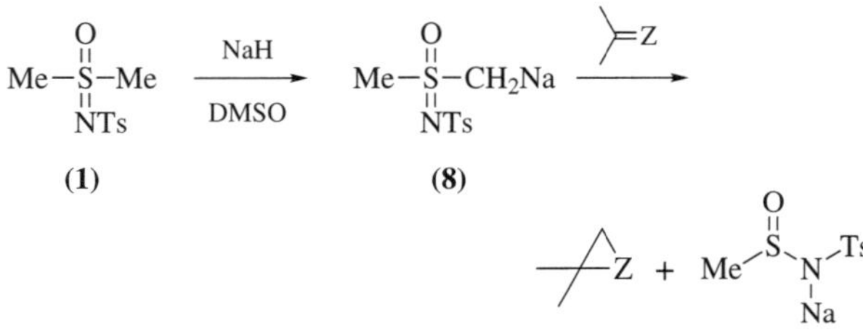

(1)

These salts have been used to prepare oxiranes from aldehydes and ketones (eqs 2 and 3),[2] cyclopropanes from enones (eq 4),[2] and aziridines from imines (eq 5).[2] Alkylidene groups which have been transferred using reagents in this series include methylene, ethylidene, isopropylidene, benzylidene, cyclopentylidene, and cyclohexylidene. Optically active versions of these reagents have been studied, but enantiomeric excesses of the resulting alkylidene transfer products have only been modest.[2] The reaction of carbanion (**8**) with epoxides results in the expansion of the ring by one carbon (eq 2). This unique oxetane synthesis, which can be carried out in one step by simply treating the ketone with 3 equiv of (**8**), is quite general and illustrates the use of (**1**) as a [$^-CH_2CH_2{}^+$] synthon.[3]

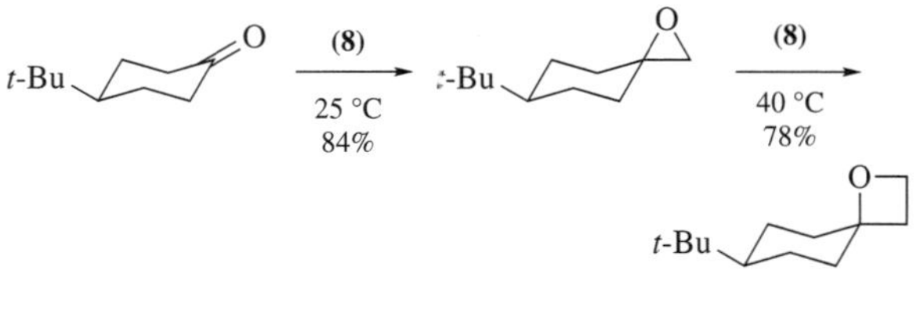

(2)

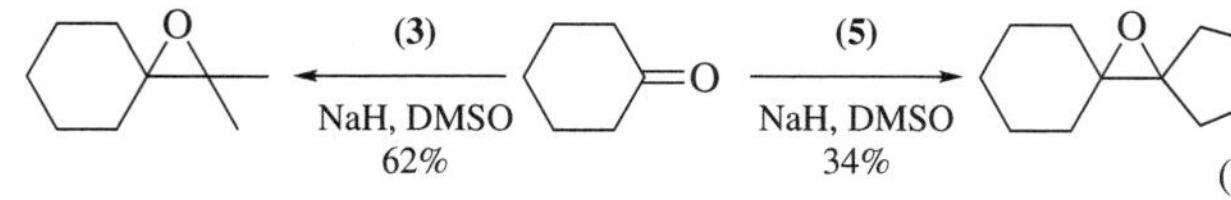

(3)

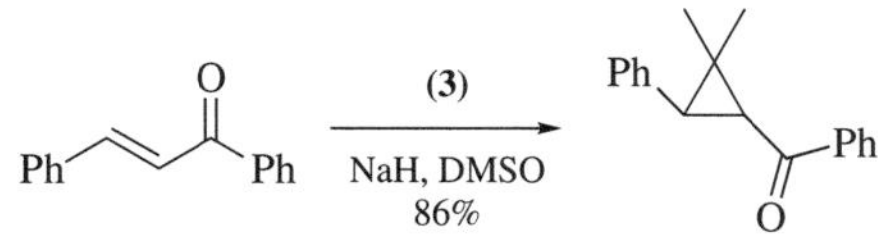

(4)

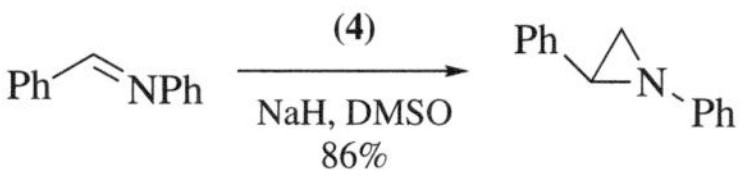

(5)

Reagent (**1**) can also be converted to an ethylene transfer reagent by condensation with benzonitrile, followed by reduction of the ketosulfoximine and dehydration (eq 6). The resulting *S*-vinyl-*N*-tosylsulfoximine reacts with stabilized anions to give cyclopropanes.[13] The *N*-tosyl group in *N*-tosylsulfoximines can be cleaved reductively using *Sodium Anthracenide*.[14]

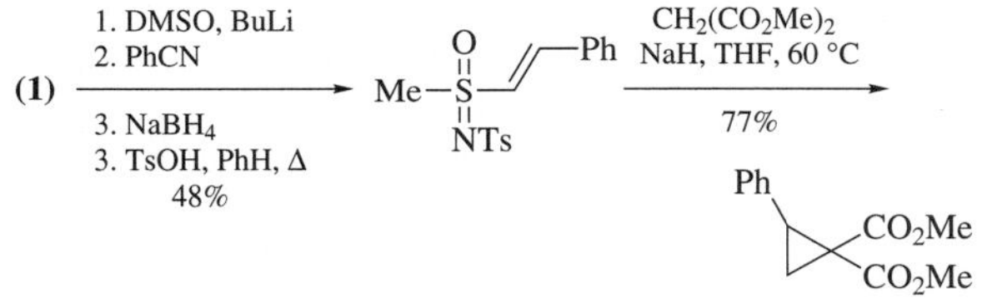

(6)

Related Reagents. *N,S*-Dimethyl-*S*-phenylsulfoximine; Dimethylsulfonium Methylide; Dimethylsulfoxonium Methylide; Diphenylsulfonium Methylide; Isopropyldiphenylsulfonium Tetrafluoroborate.

1. (a) Johnson, C. R. *Aldrichim. Acta* **1985**, *18*, 3. See also: (b) Kennewell, P. D.; Taylor, J. B. *Chem. Soc. Rev.* **1980**, *9*, 477; (c) Oae, S.; Furukawa, N. In *Sulfilimines and Related Derivatives*; American Chemical Society: Washington, 1983.
2. Johnson, C. R.; Krichhoff, R. A.; Reischer, R. J.; Katekar, G. F. *J. Am. Chem. Soc.* **1973**, *95*, 4287.
3. Welch, S. C.; Rao, A. S. C. P.; Lyon, J. T.; Assercq, J.-M. *J. Am. Chem. Soc.* **1983**, *105*, 252.
4. (a) Carr, D.; Seden, T. P.; Turner, R. W. *Tetrahedron Lett.* **1969**, 477; (b) Heintzelman, R. W.; Swern, D. *Synthesis* **1976**, 731.
5. Johnson, C. R.; Krichhoff, R. A. *J. Org. Chem.* **1979**, *44*, 2280.
6. Huang, S.-L.; Swern, D. *J. Org. Chem.* **1979**, *44*, 2510.
7. (a) Veale, H. S.; Levin, J.; Swern, D. *Tetrahedron Lett.* **1978**, 503. (b) Ketcha, D. M.; Swern, D. *Synth. Commun.* **1984**, *14*, 915.
8. Akutagawa, K.; Furukawa, N.; Oae, S. *J. Org. Chem.* **1984**, *49*, 2282.
9. Kwart, H.; Kahn, A. A. *J. Am. Chem. Soc.* **1967**, *89*, 1950.
10. Corey, E. J.; Chaykovsky, M. *J. Am. Chem. Soc.* **1965**, *87*, 1353.
11. Johnson, C. R.; Schroeck, C. W.; Shanklin, J. R. *J. Am. Chem. Soc.* **1973**, *95*, 7424.
12. Andersen, N. H.; Ladner, D. W.; Moore, A. L. *Synth. Commun.* **1978**, *8*, 437.
13. Johnson, C. R.; Lockard, J. P.; Kennedy, E. R. *J. Org. Chem.* **1980**, *45*, 264.
14. Johnson, C. R.; Lavergne, O. *J. Org. Chem.* **1989**, *54*, 986.

Carl R. Johnson
Wayne State University, Detroit, MI, USA

Claire Dufour & Viresh H. Rawal
The Ohio State University, Columbus, OH, USA

Di-(−)-(1*R*,2*S*)-2-phenyl-1-cyclohexyl Diazenedicarboxylate

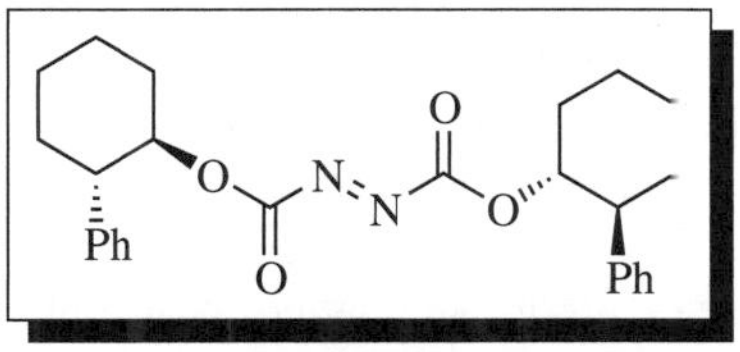

[206359-91-3] $C_{26}H_{30}N_2O_4$ (MW 434.53 (*E*))

(reagent used as a chiral azo-enophile in asymmetric azo-ene reactions)

Alternate Name: (1*R*-1α[*E*(1*R**,2*S**)],2β)-Bis(2-phenylcyclohexyl) diazenedicarboxylate.
Physical Data: $[\alpha]_D$ −56.9 (*c* 0.65, $CHCl_3$).
Solubility: soluble in CH_2Cl_2, diethyl ether, and most organic solvents.
Form Supplied in: yellow oil.
Analysis of Reagent Purity: ^{1}H NMR, IR, TLC, elemental analysis.
Preparative Methods: The title reagent is prepared[1] by reaction of (1*R*, 2*S*)-2-phenyl-1-cyclohexanol with excess phosgene in the presence of quinoline to afford a chloroformate which is treated directly with hydrazine monohydrate (0.5 equiv) to afford di-(−)-(1*R*, 2*S*)-2-phenyl-1-cyclohexyl diazanedicarboxylate. Oxidation of the diazanedicarboxylate to the diazenedicarboxylate is then readily effected using *N*-bromosuccinimide and pyridine (eq 1).

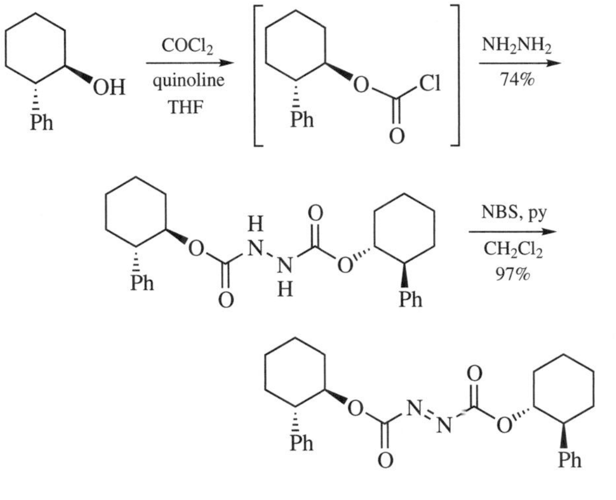

(1)

Purification: flash chromatography using hexane–ethyl acetate (9:1) as eluent.

Handling, Storage, and Precautions: store in closed vessels under an inert atmosphere in the refrigerator. Protect from light.

Azo-ene reactions. The ene reaction[1] provides a powerful method for C–C bond formation with concomitant activation of an allylic C–H bond. A variety of functionalized carbon skeletons can be constructed due to the range of enophiles which can be used. For example, carbonyl compounds give homoallylic alcohols[2] and imino derivatives of aldehydes afford homoallylic amines.[3] The azo-ene reaction offers a method for effecting allylic amination by treatment of an alkene with an azo-diester to afford a diacyl hydrazine which upon N–N cleavage furnishes a carbamate. Subsequent hydrolysis of the carbamate provides an allylic amine. Use of chiral diazenedicarboxylates provides a method for effecting stereoselective electrophilic amination.

Lewis acid-mediated ene reaction of di-(−)-(1*R*,2*S*)-2-phenyl-1-cyclohexyl diazenedicarboxylate with cyclohexene using tin(IV) chloride in dichloromethane at −60 °C for 5 min afforded the azo-ene adduct in 80% yield after purification by flash chromatography (eq 2).[4] The ^{1}H NMR spectrum of the azo-ene adduct recorded at 380 K in deuterated toluene established the presence of only one diastereomer. Further analysis of the ene adduct by HPLC on a Whatman Partisil 5 normal phase silica column using hexane–ethyl acetate (9:1) as eluent confirmed the presence of only one diastereomer.

Use of cyclopentene, *trans*-hex-3-ene and *trans*-oct-4-ene afforded the ene adducts in good yield with a diastereomeric excess of 86:14 in each case. The diastereoselectivity observed using di-(−)-(1*R*,2*S*)-2-phenyl-1-cyclohexyl diazenedicarboxylate as a chiral azo-enophile offered a significant improvement over the use of di-(−)-menthyl azodicarboxylate where the level of asymmetric induction achieved in Lewis acid-mediated ene reactions with simple alkenes was not impressive.[5] Moreover, it proved difficult to cleave the N–N bond in the menthyl ester azo-ene adducts whereas sodium/liquid ammonia was used to smoothly cleave the N–N bond in the diacylhydrazine adducts formed using di-(−)-(1*R*,2*S*)-2-phenyl-1-cyclohexyl diazenedicarboxylate as azo-enophile.

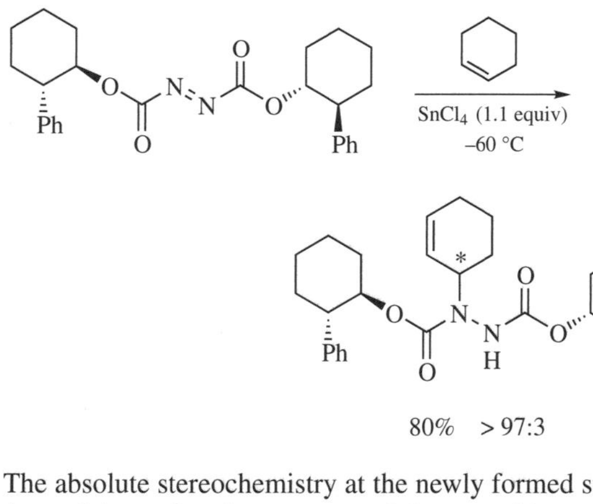

(2)

The absolute stereochemistry at the newly formed stereogenic carbon of the major diastereomer of the ene adduct can be predicted by analysis of the transition model for the ene reaction (eq 3). The (1*R*,2*S*)-2-phenyl-1-cyclohexyl chiral auxiliary adopts a chair conformation with equatorial placement of the bulky phenyl group. Complexation of the carbonyl group to the Lewis acid affords the more stable *s-trans* conformation about the C–N sigma bond. In this conformation, the phenyl group shields the N_β-*re*-face. Therefore the cyclic alkene preferentially attacks from the less hindered N_β-*si*-face. Ene reaction proceeds through a six-membered cyclic transition state affording the (1′*R*)-diastereomer of the ene adduct.

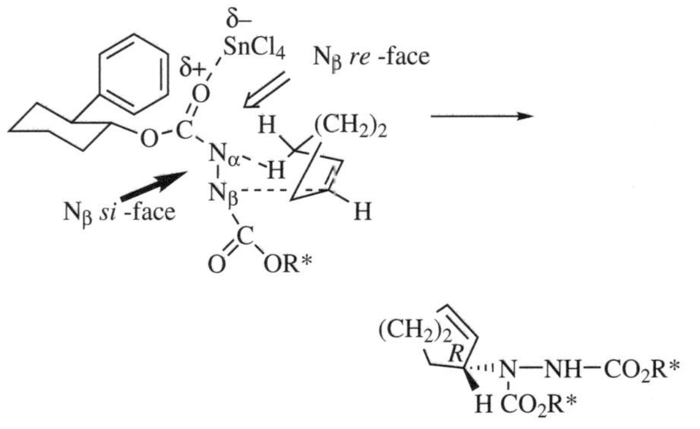

(3)

Related Reagents. The synthesis of chiral diazenedicarboxylates as potential chiral electrophilic aminating agents has received little attention. A series of chiral bornyl, isobornyl and menthyl diazenedicarboxylates has been reported[6] and their reaction with achiral enolates of esters and *N*,*N*-dimethyl amides afforded α-hydrazino acid derivatives with little or no selectivity. Incorporation of a chiral azodicarboxamide unit into a chiral bridging binaphthyl moiety afforded α-hydrazino acid derivatives with high stereoselectivity in reactions with achiral oxazolidinone anions.[7]

---

1. Snider, B. B. Ene Reactions with Alkenes As Electrophiles, in *Comprehensive Organic Synthesis*, Trost, B. M., Ed.; Pergamon: Oxford, 1991, Vol. 5, p1.
2. (a) Snider, B. B. The Prins and Carbonyl-Ene Reactions, in *Comprehensive Organic Synthesis*, Trost, B. M., Ed., Pergamon: Oxford, 1991, Vol. 2, p 527; (b) Mikami, K.; Shimizu, M. *Chem. Rev.* **1992**, *92*, 1021.
3. Borzilleri, R. M.; Weinreb, S. M. *Synthesis* **1995**, *4*, 347.
4. Brimble, M. A.; Lee, C. Y. K. *Tetrahedron: Asymmetry* **1998**, *9*, 873.
5. Brimble, M. A.; Heathcock, C. H.; Nobin, G. N. *Tetrahedron: Asymmetry* **1996**, *7*, 2007.
6. Harris, J. M.; Bolessa, E. A.; Mendonca, A. J.; Feng, S.-C.; Vederas, J. C. *J. Chem. Soc. Perkin Trans. 1* **1995**, 1945.
7. Harris, J. M.; McDonald, R.; Vederas, J. C. *J. Chem. Soc. Perkin Trans. 1* **1996**, 2669.

Margaret A. Brimble
The University of Auckland, Auckland, New Zealand

(*R*)-(–)-2,2-Diphenylcyclopentanol

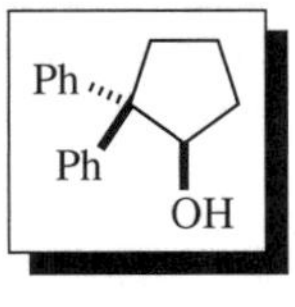

[126421-67-8] $C_{17}H_{18}O$ (MW 238.1358)

(chiral auxiliary in asymmetric synthesis)

Physical Data: a white solid,[1,2,3] mp 76–77 °C; $[\alpha]_D^{20}$ - 116 (*c* 0.97, EtOH).[1]

Solubility: soluble in most common organic solvents including acetone, DMSO, MeOH, EtOH, Et_2O, CH_2Cl_2, THF, and EtOAc.

Analysis of Reagent Purity: by ^1H NMR and X-ray analyses[1] of its (*R*)-α-methoxy-α-(trifluoromethyl)phenylacetic acid [(*R*)-MTPA] derivative;[2]chiral HPLC analysis; supercritical fluid chromatography (SFC).[4]

Preparative Methods: on a preparative scale (>97% ee) by borane reduction of 2,2-diphenylcyclopentanone in the presence of (*S*)-tetrahydro-1-methyl-3,3-diphenyl-1*H*,3*H*-pyrrolo[1,2-c][1,3,2] oxazaborole;[3,4] by asymmetric reduction of 2,2-diphenylcyclopentanone with (+)-β-chlorodiisopinocampheylborane;[1] by kinetic resolution of racemic acetate derived from the alcohol.[5]

Purification: recrystallization from hexane.[1,3,4]

General. The potential of (*R*)-(–)-2,2-diphenylcyclopentanol (DCP) (**1**) as a chiral auxiliary was first demonstrated by d'Angelo,[1] who designed and employed the alcohol in a highly diastereoselective synthesis of β-amido esters. Later, Zhang et al.[6] were able to access diastereomerically enriched cycloalkanones via Mn(III)-based oxidative free-radical cyclizations of β-keto DCP esters. Denmark and co-workers have extensively studied the use of DCP as a chiral auxiliary on vinyl ether dienophiles employed in the Lewis-acid-promoted tandem [4 + 2]/dipolar [3 + 2] cycloadditions with nitroalkenes. DCP has expanded the utility of the tandem nitroalkene cycloadditions, especially in the application of *Z*-propenyl ethers and exo [4 + 2] cycloadditions. The effectiveness of this auxiliary is attributed to the alcohol containing a single asymmetric center (that bears a hydroxyl group) and a quaternary carbon center (bearing two phenyl groups) α to the hydroxyl group. Because one of the two geminal aromatic nuclei is necessarily gauche (synclinal) to the adjacent hydroxyl function, the appropriate special relationship exists for masking one of the π-faces in the corresponding dienophile.[1]

Synthesis of Vinyl Ethers of (*R*)-(–)-2,2-Diphenylcyclopentanol. The preparation of DCP-derived vinyl ethers usually involves mercuric acetate-catalyzed transetherification reaction with DCP and a corresponding vinyl ether (eq 1).[4]

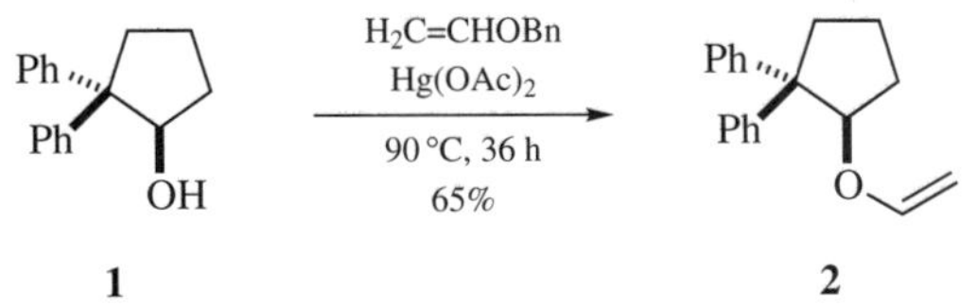

(1)

1 **2**

E- and *Z*-Propenyl ethers (**4**) of DCP have been prepared by the isomerization of the corresponding allyl ethers in the presence of Wilkinson's catalyst and DABCO (eq 2).[4]

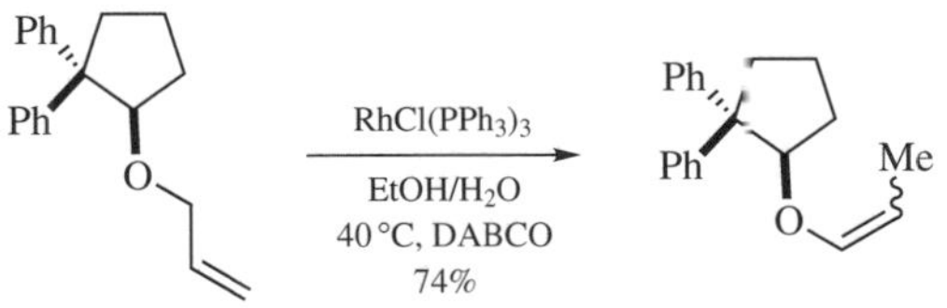

(2)

3 **4**

1.3:1 *Z*/*E*

Synthesis of Substituted Pyrrolidines. A cycloaddition/reduction sequence between nitroalkenes and vinyl ethers derived from DCP, i.e., **2** can effect the enantioselective synthesis of substituted pyrrolidines.[7,8] 2-Substituted 1-nitroalkenes undergo highly efficient and diastereoselective Lewis-acid-promoted [4 + 2] cycloaddition with DCP-derived vinyl ethers to afford cyclic nitronates **5** in high yields. Subsequent reduction with PtO_2 (7.5 mol %), under 160 psi of H_2 at room temperature for 24 h, affords the optically active 3-substituted pyrrolidines (**6**) (71–97%, both as the free base and *N*-protected derivatives), and the chiral auxiliary **1**[8] (eq 3).

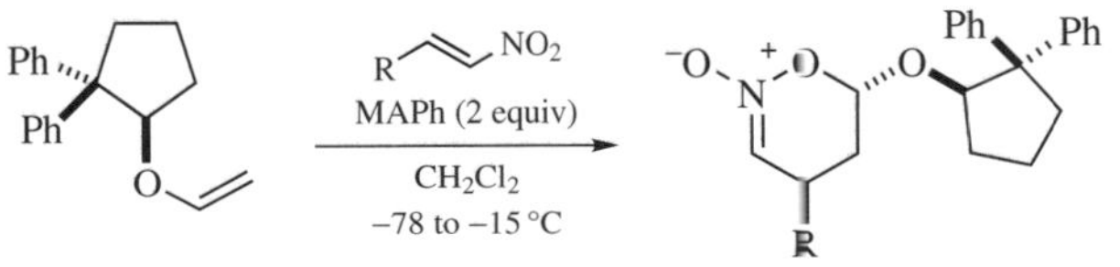

2 **5**

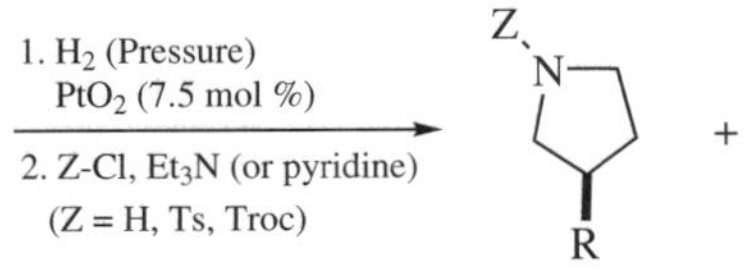

\+ **1** (3)

54–100% recovery

6

69–80% yield

71–97%ee

R = Ph, *t*-Bu, veratryl, *n*-pentyl, cyclohexyl, —$(CH_2)_4CO_2$-*t*-Bu

The choice of Lewis acid promoter for these reactions can change the sense of asymmetric induction.[4,8–12] For example, tandem [4 + 2]/ [3 + 2] cycloadditions (eq 4) mediated by Ti(O-*i*-Pr)$_2$Cl$_2$, followed by hydrogenolysis afforded tricyclic (–)-α-hydroxy lactam [(–)-**8**] in 98% ee. When mediated by methylaluminum-bis(2,6-diphenylphenoxide) (MAPh), the same reaction gave (+)-**8** in 93% ee. Importantly, the observed selectivity is not chiral auxiliary dependent.[4,8,9] Rather, it is attributed to a highly endo selective cycloaddition in the case of Ti compared to high exo selectivity in the case of MAPh.

The use of DCP-derived propenyl ethers in nitroalkene [4 + 2] cycloaddition allows for the installation of an additional stereogenic center in the tandem cycloadducts. The methyl substituent also provides a stereochemical marker to allow for the determination of endo/exo selectivity in the [4 + 2] cycloaddition.[4] DCP-derived *E*-propenylvinylether (*E*-**4**) has been employed in the asymmetric synthesis of 3,4-disubstituted pyrrolidines.[8] MAPh-promoted [4 + 2]-cycloaddition of the vinyl ether with *trans*-β-nitrostyrene provided a 20:1 mixture of diastereomeric nitronates **9** in 97% yield (eq 5). Subsequent room-temperature hydrogenolysis (160 psi H_2) with catalytic PtO_2 in EtOH provided a 20:1 mixture of trans- and cis-methyl-3-phenylpyrrolidine. Following this reduction, *N*-protection afforded the diastereomerically pure trans-4-methyl-3-phenylpyrrolidine (**10**) in 84% yield and 92% ee[8] along with **1** (94% recovery following SiO_2 chromatography).

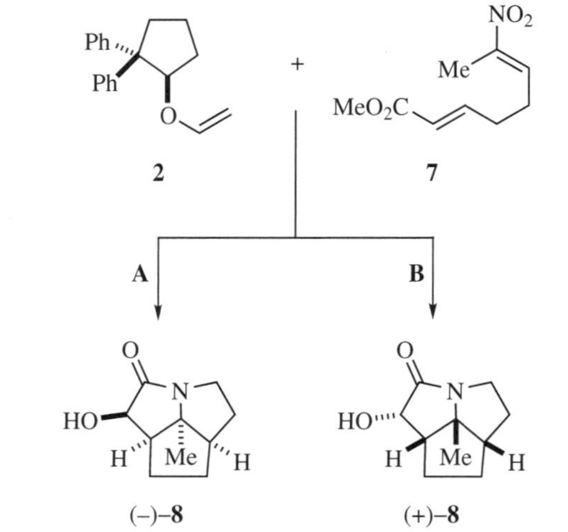

(4)

Conditions **A**: 1. Ti(O-*i*-Pr)$_2$Cl$_2$, CH$_2$Cl$_2$, −78 °C (89%)
2. H$_2$, Raney Ni, MeOH (70%)

Conditions **B**: 1. MAPh, CH$_2$Cl$_2$, −78 °C (86%)
2. H$_2$, Raney Ni, MeOH (74%)

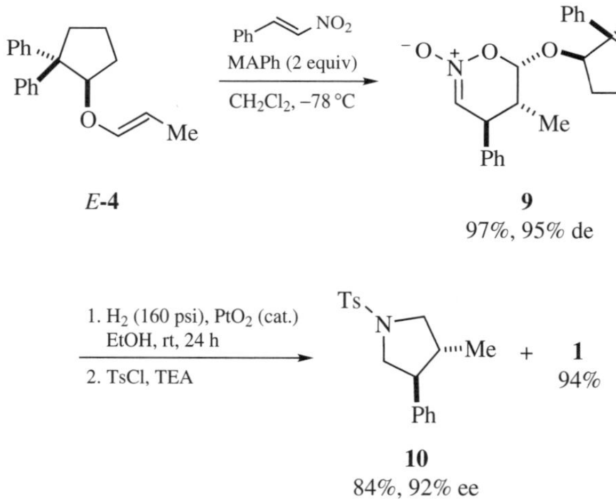

(5)

Synthesis of α-Hydroxy Lactams. Propenylethers of DCP have also been employed in the synthesis of α-hydroxy lactams.[4] The *Z*- and *E*-isomers show different levels of selectivity in the presence of MAPh or Ti(O-*i*-Pr)$_2$Cl$_2$(eq 6). When promoted by MAPh, the *Z*-propenyl ether undergoes exo selective [4 + 2] cycloadditions; in contrast, endo selective [4 + 2] cycloadditions are observed when the reactions are promoted by Ti(O-*i*-Pr)$_2$Cl$_2$.[4] MAPh-promoted cycloaddition of the *E*-propenyl ether afforded a single α-hydroxy lactam [(+)-**11**] derived from exclusive exo approach of the dienophile in the [4 + 2] cycloaddition.[4] Reactions of the *E*-propenylether is less selective with Ti(O-*i*-Pr)$_2$Cl$_2$, affording exo and endo products in the ratio of 2.3:1.0. Although the exo diastereomer [(−)-**12**] was found to be highly enantiomerically enriched (96% ee), this erosion of endo/exo selectivity can be viewed as a shortcoming of DCP (**1**) as a chiral auxiliary.

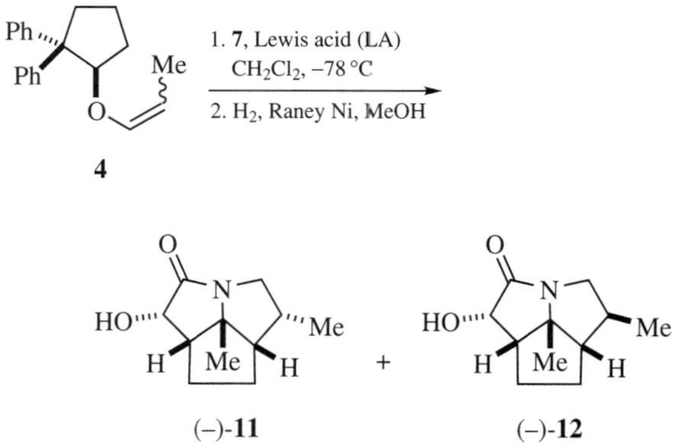

(6)

For *Z*-**4**:
w/ LA= Ti(O-*i*-Pr)$_2$Cl$_2$: (−)-**11** (92%ee)/(−)-**12** (65%ee) (endo:exo) ~8:1
w/ LA= MAPh: (−)-**11** (38%ee)/(+)-**12** (83%ee) (endo:exo) ~1:10.

For *E*-**4**:
w/ LA= Ti(O-*i*-Pr)$_2$Cl$_2$: (−)-**11** (66%ee)/(−)-**12** (96%ee) (endo:exo) 1:2.3
w/ LA= MAPh: (+)-**11** (74%ee) (exclusive product).

Synthesis and Reaction of 2-(Acyloxy) and 2-(Benzoyloxy) vinyl ethers of (*R*)-(−)-DCP. 2-(Acyloxy)vinyl ethers (**13**) of DCP[7] have been prepared (eq 7). Allylation of **1** followed by ozonolysis with a zinc/acetic acid reductive work-up affords the corresponding chiral aldehyde. Heating this aldehyde with the appropriate anhydride and sodium salt of the carboxylic acid gives the desired 2-(acyloxy)vinyl ethers.

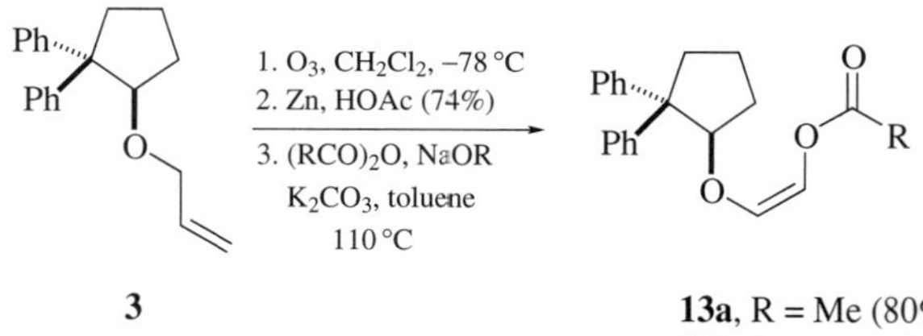

(7)

3

13a, R = Me (80%)
13b, R = *t*-Bu (39%)

A more efficient route to 2-(benzoyloxy)vinyl ether (**15**)[13] involves (0 °C, THF) conversion of the chiral alkoxy aldehyde **14** to its silyl enol followed by *O*-acylation with benzoyl fluoride and a catalytic amount of TBAF (2 mol %) to form a separable mixture of the *Z*-vinyl ether (81%) and *E*-vinyl ethers (6%) (eq 8).

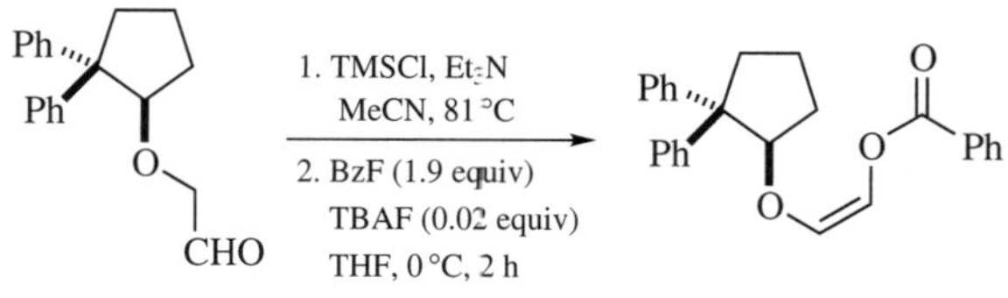

(8)

14

15
(81% *Z* + 6% *E*)

Compound **13a** exhibits high π-facial selectivity in the regioselective [4 + 2] cycloaddition (promoted by $SnCl_4$) with 2,2-

disubstituted aryl-1-nitroalkenes affording *N*-tosyl-4,4-disubstituted-3-hydroxypyrrolidines (**16**) in high enantiomeric excess (96%) (eq 9).[7]

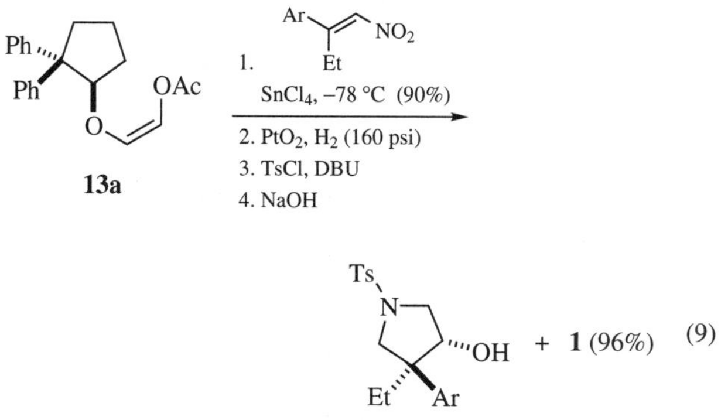

13a

16
75%, 96% ee

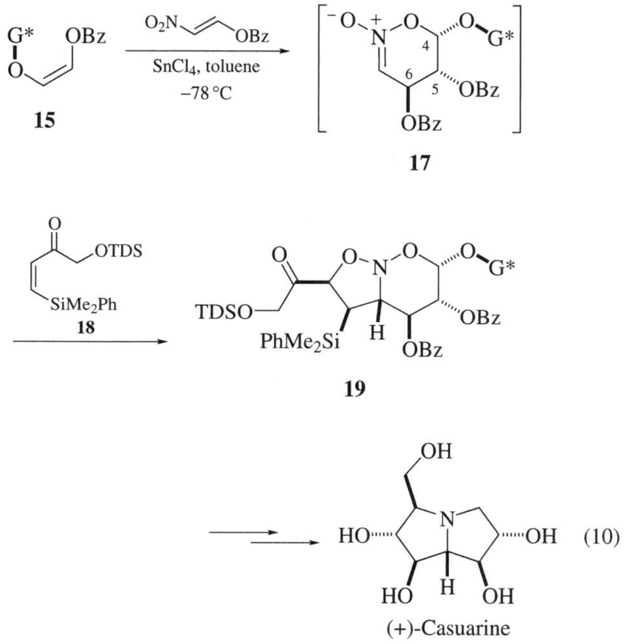

15 **17** **19**

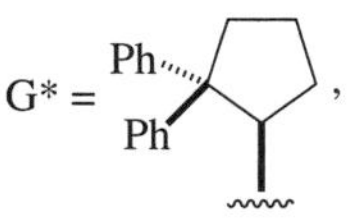

TDS = thexyl dimethylsilyl

DCP-based Chiral Auxiliaries in Total Synthesis. DCP-based chiral auxiliaries have proven amenable to asymmetric total synthesis, including Denmark's syntheses of of the pyrrolizidine alkaloid (–)-rosmarinecine[10] and the pentahydroxy pyrrolizidine alkaloid (+)-casuarine.[13,14] Denmark's synthesis of (+)-casuarine involves [4 + 2] cycloaddition of dienophile **15** with nitrobenzoate followed by [3 + 2] cycloaddition of the resulting nitronate **17** with a vinyl silane **18** (eq 10). During formation of the [4 + 2] cycloadduct, the relative configuration between C4 and C5 is a direct consequence of the vinyl ether geometry, while the stereochemistry at C6 is determined by the ability of the chiral auxiliary to differentiate the diastereotopic π faces (*Re* of Si) of the vinyl ether (termed internal diastereoselection). Thus, this tandem sequence establishes five of the six stereocenters present in the natural product. Moreover, the chiral auxiliary **1** is recovered in 99% yield after hydrogenolysis (260 psi H_2) with Raney nickel in MeOH followed by SiO_2 chromatography.

Synthesis of Chiral β-Amido Esters. The use of **1** as a chiral auxiliary in the asymmetric hydrogenation (H_2/PtO_2) of stereogenic β-acetamidocrotonates has also been reported.[1] Reaction of **1** with diketene in the presence of TEA and acetone as solvent, followed by saturation with NH_3, then Ac_2O-pyridine, and finally hydrogenation (PtO_2, 3–5 bars of H_2) afforded the β-amido esters (**22**) in high selectivity (96% de) (eq 11).

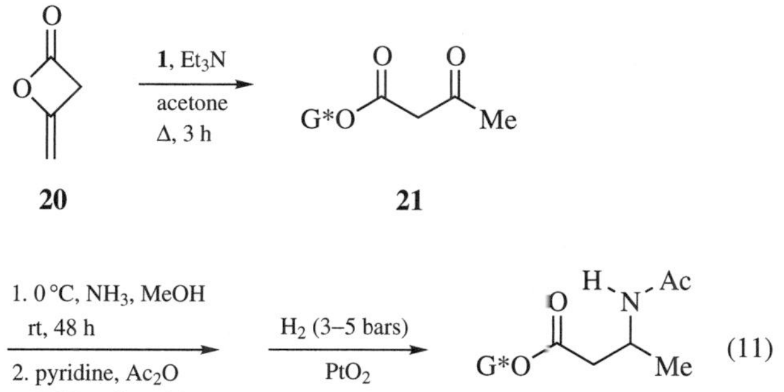

20 **21** **22**
96% de

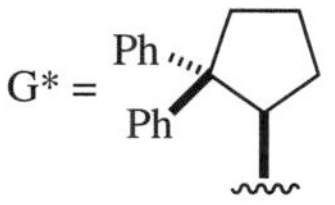

DCP as a Chiral Controller in Oxidative Free Radical Cyclizations. As a chiral auxiliary, DCP (**1**) is also reported to induce modest diastereoselection (60% de) in Mn(III)-based oxidative free-radical cyclizations[6] of β-keto esters (eq 12). Chiral β-keto ester **25** was prepared by transesterification reaction with methyl ester **23**, **1**, and 0.3 equiv of DMAP (catalyst) in anhydrous toluene at reflux for 3–5 d as described by Taber.[15] Oxidative cyclization of a 0.1 M solution of **24** in AcOH with 2 equiv of $Mn(OAc)_3{\cdot}2H_2O$ and 1 equiv of $Cu(OAc)_3{\cdot}H_2O$[6] provided bicyclo[3.2.1]octan-2-one (**25**).

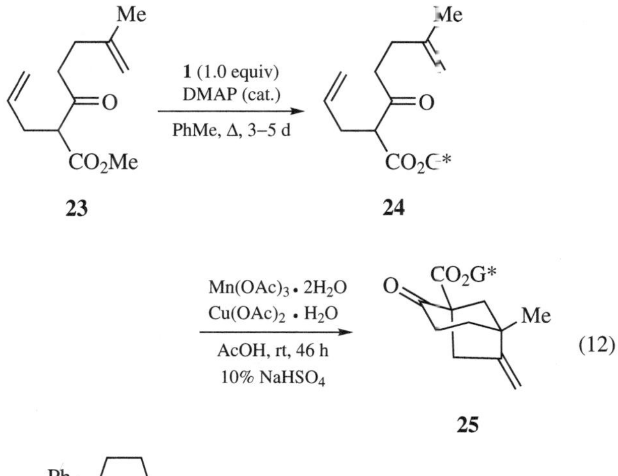

23 **24** **25**

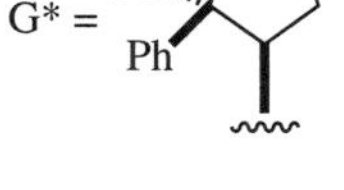

Related Reagents. Though not always as efficient as DCP (**1**), camphor derivatives (**26**),[4,7,11] (–)-8-phenylmenthol (8-PhM) (**27**);[1,5,6,10] (1*R*,2*S*)-2-phenylcyclohexanol (**28**);[1,4,5,7–9] and *trans*-2-(1-methyl-1-phenylethyl)cyclohexanol (**29**)[10] can also serve as chiral auxiliaries in asymmetric cycloadditions of vinyl and propenyl ethers with nitroalkenes (Figure 1). (*S*)-**1** can also be used, however, this enantiomer is relatively expensive to prepare by asymmetric borane reduction.

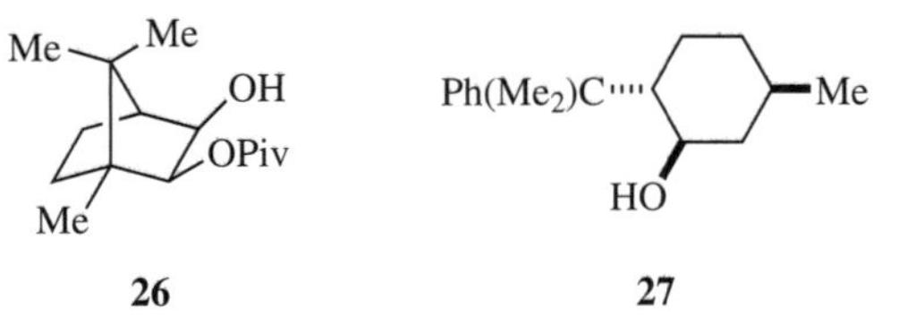

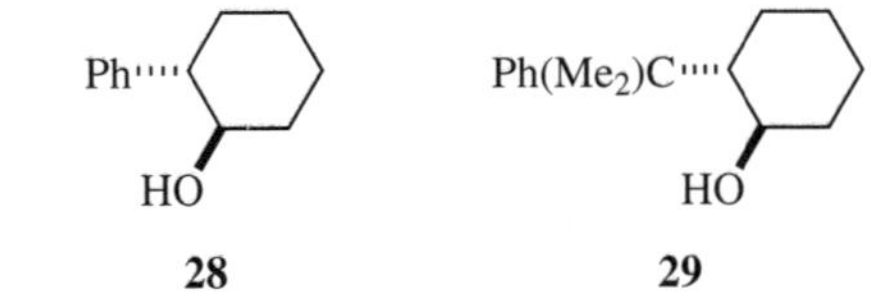

Figure 1

1. Pontin, D.; Dumas, F.; d'Angelo, J. N. *J. Am. Chem. Soc.* **1990**, *112*, 3483.
2. Dale, J. A.; Dull, D. L.; Mosher, H. S. *J. Org. Chem.* **1969**, *34*, 2543.
3. Denmark, S. E.; Marcin, L. R.; Schnute, M. E.; Thorarensen, A. *Org. Syn.* **1997**, *74*, 33.
4. Denmark, S. E.; Schnute, M. E.; Marcin, L. R.; Thorarensen, A. *J. Org. Chem.* **1995**, *60*, 3205.
5. Randrianasolo-Rakotozafy, L. R.; Azerad, R.; Dumas, F.; Potin, D.; d'Angelo, J. *Tetrahedron. Asym.* **1993**, *4*, 761.
6. Zhang, W.; Mohan, R. M.; Cook, L.; Kazanis, S.; Peisach, D.; Foxman, B. M.; Snider, B. B. *J. Org. Chem.* **1993**, *58*, 7640.
7. Denmark, S. E.; Schnute, M. E. *J. Org. Chem.* **1994**, *59*, 4576.
8. Denmark, S. E.; Marcin, L. R. *J. Org. Chem.* **1995**, *60*, 3221.
9. Denmark, S. E.; Schnute, M. E. *J. Org. Chem.* **1991**, *56*, 6738.
10. Denmark, S. E.; Thorarensen, A.; Middleton, D. S. *J. Am. Chem. Soc.* **1996**, *118*, 8266.
11. Denmark, S. E.; Thorarensen, A. *J. Org. Chem.* **1994**, *59*, 5672.
12. Denmark, S. E.; Schnute, M. E.; Senanayake, C. B. W. *J. Org. Chem.* **1993**, *58*, 1859.
13. Denmark, S. E.; Hurd, A. R. *Org. Lett.* **1999**, *1*, 1311.
14. Denmark, S. E.; Hurd, A. R. *J. Org. Chem.* **2000**, *65*, 2875.
15. Taber, D. F.; Amedio, J. C., Jr; Patel, Y. K. *J. Org. Chem.* **1985**, *50*, 3618.

Edith N. Onyeozili & Robert E. Maleczka, Jr
Michigan State University, East Lansing, MI 48824

(*R,R*)-1,2-Diphenyl-1,2-diaminoethane *N,N*′-Bis[3,5-bis(trifluoromethyl)-benzenesulfonamide]

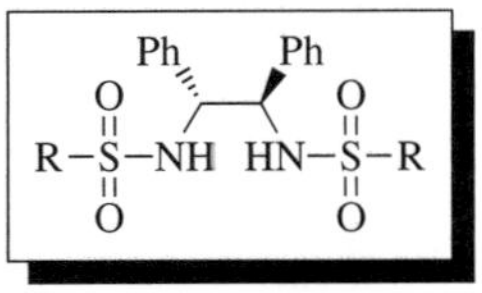

(**1**; R = 3,5-$(CF_3)_2C_6H_3$)		
[127445-51-6]	$C_{30}H_{20}F_{12}N_2O_4S_2$	(MW 764.66)
(**2**; R = CF_3)		
[121788-73-6]	$C_{16}H_{14}F_6N_2O_4S_2$	(MW 476.46)
(**3**; R = 4-MeC_6H_4)		
[121758-19-8]	$C_{28}H_{28}N_2O_4S_2$	(MW 520.72)
(**4**; R = 4-$NO_2C_6H_4$)		
[121809-00-5]	$C_{26}H_{22}N_4O_8S_2$	(MW 582.66)

(chiral controller group for enantioselective Diels–Alder reactions,[1] aldol additions,[2] Ireland–Claisen rearrangements,[3] ester-Mannich additions,[4] and carbonyl allylation[5] and propargylation[6])

Alternate Name: (*R,R*)-stilbenediamine *N,N*′-bis-3,5-bis(trifluoromethyl)benzenesulfonamide.

Physical Data: (**1**) mp 155–156 °C; α_D +83.7° (c = 1, $CHCl_3$). (**2**) mp 213–214 °C; α_D +6.6° (c = 1.4, $CHCl_3$). (**3**) mp 213–214 °C; α_D +43.9° (c = 1.74, $CHCl_3$). (**4**) mp 243 °C (dec); α_D 122° (c = 0.107, acetone).

Solubility: except for the nitro derivative, the sulfonamides are sol CH_2Cl_2.

Preparative Methods: the most convenient preparation of (*R,R*)-stilbenediamine is described in *Organic Syntheses*.[7] Condensation of benzil and cyclohexanone in the presence of ammonium acetate and acetic acid (eq 1) produces a spirocyclic 2*H*-imidazole (mp 105–106 °C). Reduction with *Lithium* in THF/NH_3 followed by an ethanol quench and hydrolysis with aqueous HCl (eq 2) affords the racemic diamine as a pale yellow solid (mp 81–82 °C). Resolution is achieved by multiple recrystallizations of the tartaric acid salts from water/ethanol. The sulfonamides are prepared by reaction of the enantiomerically pure diamine with the appropriate anhydride[1b] or sulfonyl chloride[2a] in CH_2Cl_2 in the presence of *Triethylamine* and a catalytic amount of *4-Dimethylaminopyridine* (eq 3).

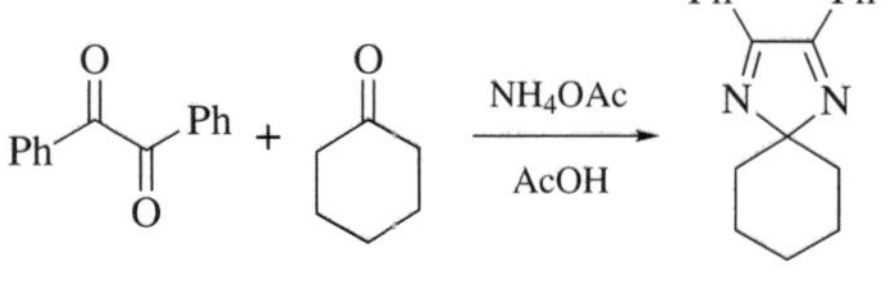

(1)

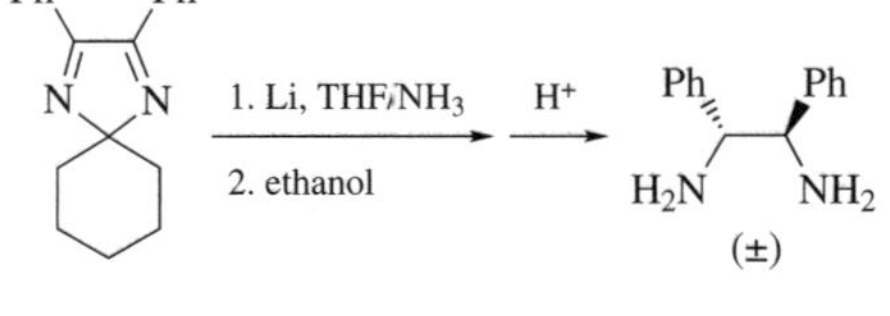

(2)

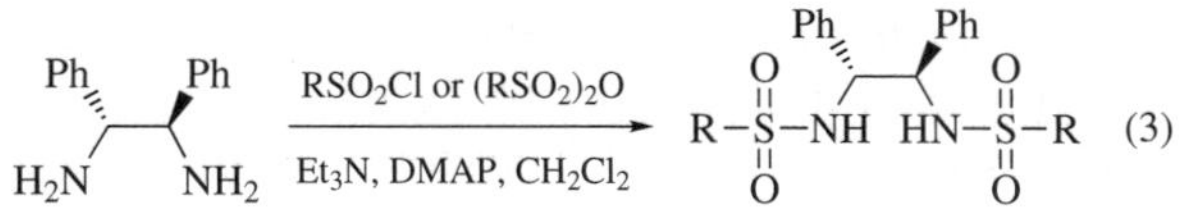

Handling, Storage, and Precautions: the sulfonamides are all stable, crystalline compounds that do not require any special precautions for storage or handling.

Diels–Alder Reactions. Reaction of the bis(triflamide) (**2**) with *Diisobutylaluminum Hydride* or *Trimethylaluminum* affords chiral Lewis acids that catalyze Diels–Alder reactions of acryloyl or crotonoyl derivatives with cyclopentadienes (eq 4).[1] The aluminum complex must be crystallized before use to remove traces of trimethylaluminum. High diastereo- and enantioselectivities are achieved with as little as 0.1 equiv of the Lewis acid, and the chiral sulfonamide is recoverable.

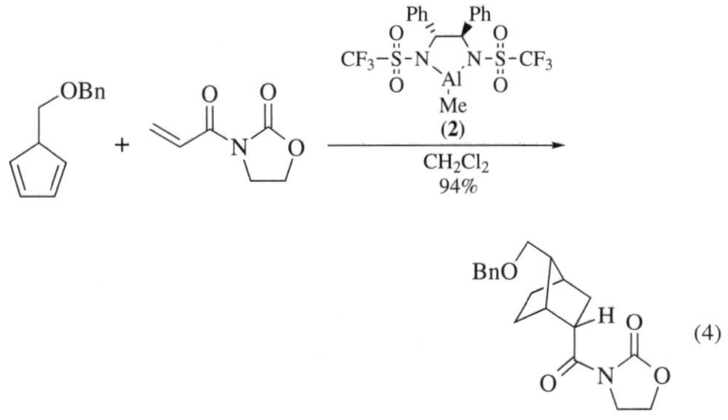

95% ee

Asymmetric Aldol Reactions. Reaction of (**1**) with *Boron Tribromide* in CH_2Cl_2 affords, after removal of solvent and HBr, a complex (**5**) useful for the preparation of chiral enolates (eq 5).[1a] Complex (**5**) is moisture sensitive and is generally prepared immediately before use. For propionate derivatives, either *syn* or, less selectively, *anti* aldol adducts may be obtained by selection of the appropriate ester derivative and conditions.[2a] Thus reaction of *t*-butyl propionate with (**5**) and triethylamine produces the corresponding *E*(O) enolate, leading to formation of *anti* aldol adducts upon addition to an aldehyde (eq 6). Selectivities may be enhanced by substitution of the *t*-butyl ester with the (+)-menthyl ester. Conversely, reaction of *S*-phenyl thiopropionate with (**5**) and *Diisopropylethylamine* affords the corresponding *Z*(O) enolates and *syn* aldol products (eq 7).[2a,c]

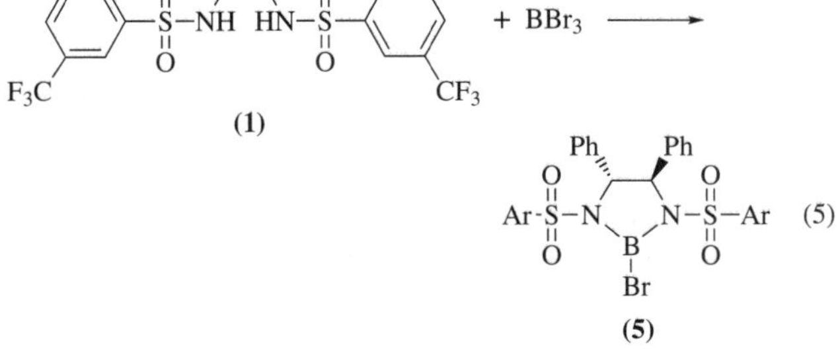

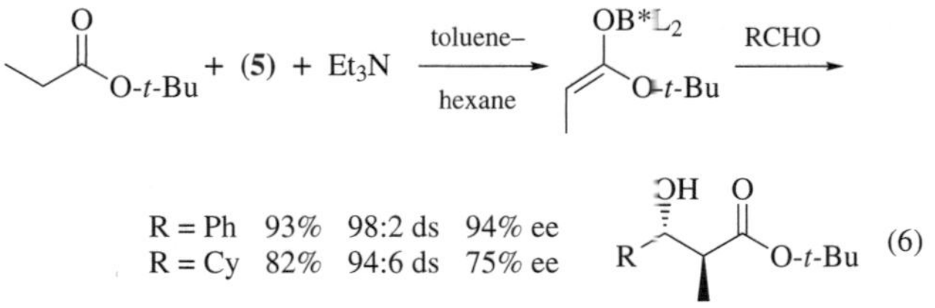

R = Ph 93% 98:2 ds 94% ee
R = Cy 82% 94:6 ds 75% ee

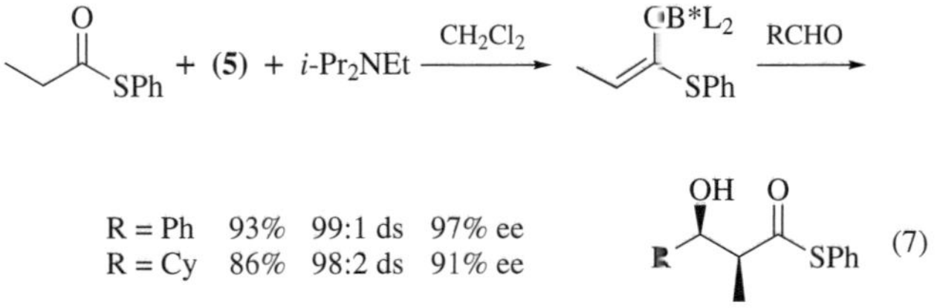

R = Ph 93% 99:1 ds 97% ee
R = Cy 86% 98:2 ds 91% ee

Products with low enantiomeric purity are obtained by direct application of this chemistry to unsubstituted acetate esters. However, aldol reactions of *t*-butyl bromoacetate mediated by (**5**) afford synthetically useful bromohydrins (**6**) with high selectivities (eq 8).[2b] These may be reductively dehalogenated or converted to a variety of compounds by way of the derived epoxides.

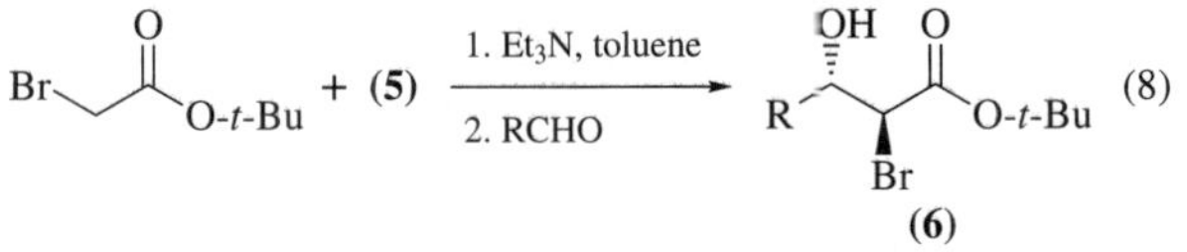

R = Ph 94% 99:1 ds 98% ee
R = Cy 65% 98:2 ds 91% ee

Asymmetric Ireland–Claisen Rearrangements. Chiral enolates derived from the boron complex (**5**) and allyl esters rearrange with excellent selectivity upon warming to −20 °C for a period of 1–2 weeks (eqs 9 and 10).[3] As discussed above, the geometry of the intermediate enolate can be controlled by appropriate choice of base and solvent, thus allowing access to either *syn* or *anti* configuration in the product. The reaction can be completed in 2–4 days with little erosion in selectivity when run at 4 °C.

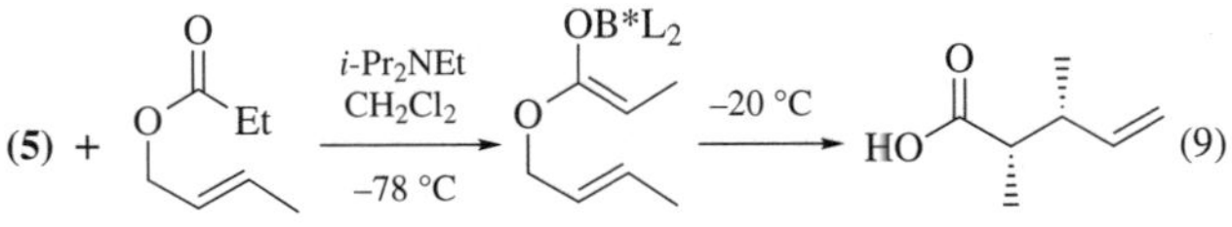

75%, 99:1 ds, >97% ee

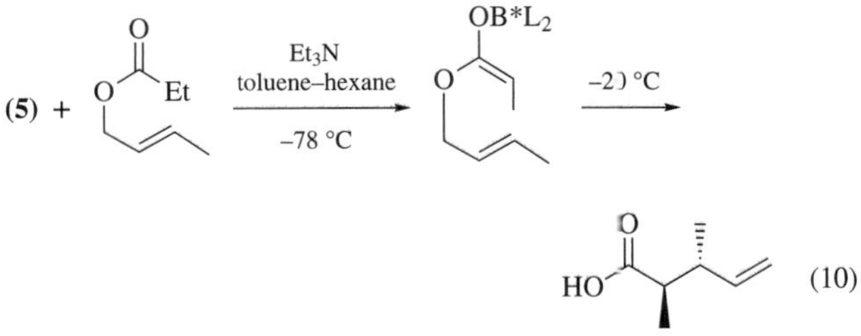

65%, 90:10 ds, 96% ee

Ester-Mannich Additions. The *E*(O) enolate (**7**) reacts with *N*-allyl or *N*-benzyl aldimines to afford chiral β-amino esters (eq 11).[4] As with the aldol reactions, best selectivities are achieved with imines derived from aromatic or unsaturated aldehydes. The method appears to have good potential for the synthesis of useful β-lactams if extended to other enolates.

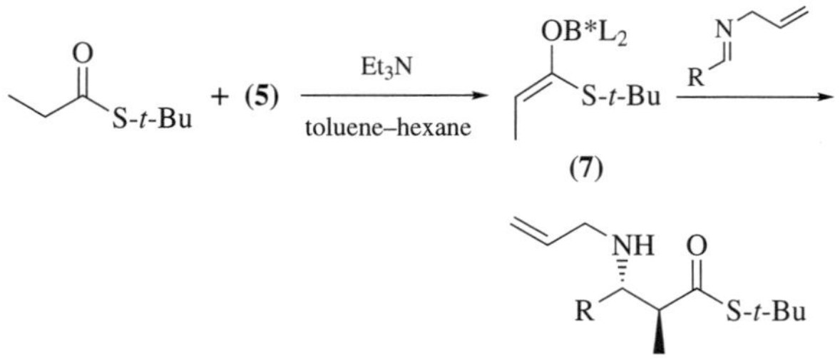
(11)

R = Ph	74%	>99:1 ds	90% ee
R = hydrocinnamyl	67%	97:3 ds	70% ee

Carbonyl Allylation and Propargylation. Boron complex (**8**), derived from the bis(tosylamide) compound (**3**), transmetalates allylstannanes to form allylboranes (eq 12). The allylboranes can be combined without isolation with aldehydes at −78 °C to afford homoallylic alcohols with high enantioselectivity (eq 13).[5] On the basis of a single reported example, reagent control might be expected to overcome substrate control in additions to aldehydes containing an adjacent asymmetric center. The sulfonamide can be recovered by precipitation with diethyl ether during aqueous workup. Ease of preparation and recovery of the chiral controller makes this method one of the more useful available for allylation reactions.

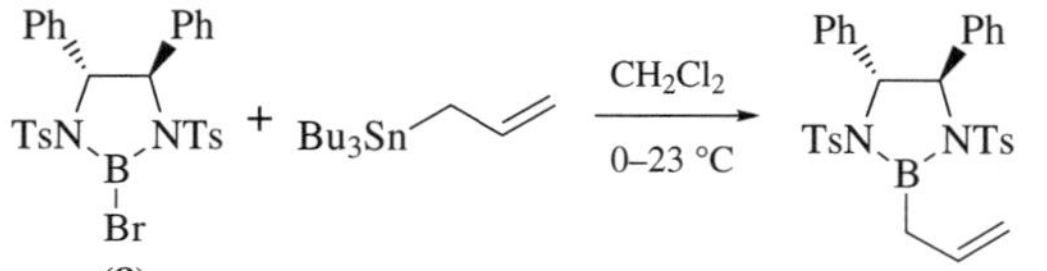
(12)

(**8**)

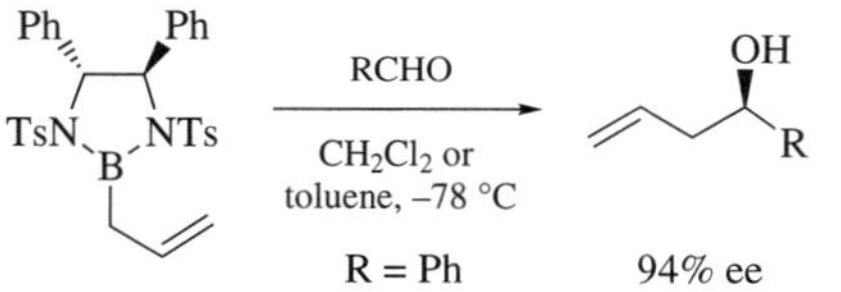
(13)

R = Ph	94% ee
R = cinnamyl	98% ee
R = cyclohexyl	93% ee
R = hexyl	90% ee

In the same way, reaction of (**8**) with allenyl- or propargylstannanes affords intermediate borane derivatives which, upon reaction with aldehydes, produce the expected adducts with high selectivities (eqs 14 and 15).[6]

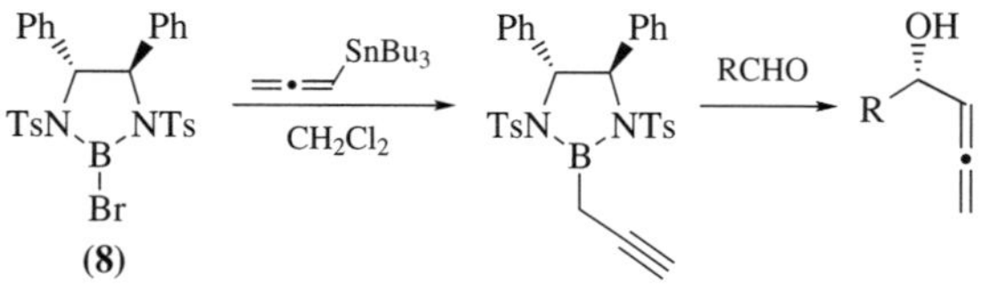
(14)

(**8**)

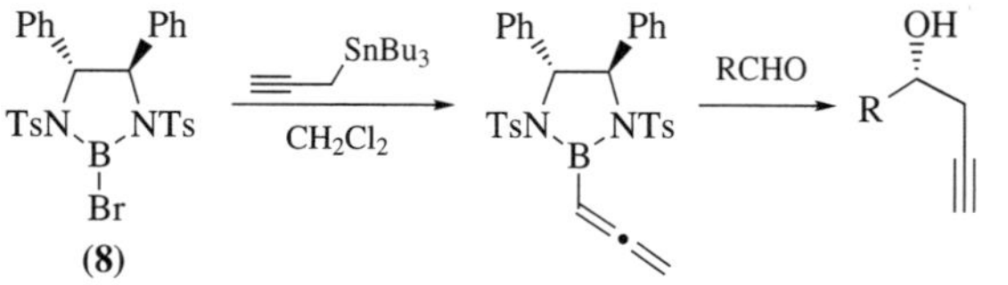
(15)

(**8**)

Other Applications. Other (R,R)-stilbenediamine derivatives have been used to direct the stereochemical course of alkene dihydroxylation[8] (with stoichiometric quantities of *Osmium Tetroxide* and epoxidation of simple alkenes with *Sodium Hypochlorite* and manganese(III) complexes.[9]

Related Reagents. *B*-Allyldiisopinocampheylborane; Chloro(η^5-cyclopentadienyl)[(4*R*,*trans*)-2, 2-dimethyl-α,α,α′,α′-tetraphenyl-1, 3-dioxolane-4,5-dimethanolato(2−)-O^{α}, $O^{\alpha'}$]titanium; Chloro(cyclopentadienyl)bis[3-*O*-(1, 2:5,6-di-*O*-isopropylidene-α- D-glucofuranosyl)]titanium; Diisopinocampheylboron Trifluoromethanesulfonate; Diisopropyl 2-Allyl-1,3,2-dioxaborolane-4,5-dicarboxylate; (4*R*,5*R*)-2,2-Dimethyl-4,5-bis-(hydroxydiphenylmethyl)-1,3-dioxolane; 2,2-Dimethyl-α,α,α′, α′- tetraphenyl-1,3-dioxolane-4, 5-dimethanolatotitanium Diisopropoxide.

1. (a) Corey, E. J.; Imwinkelried, R.; Pikul, S.; Xiang, Y. B. *J. Am. Chem. Soc.* **1989**, *111*, 5493. (b) Pikul, S.; Corey, E. J. *Org. Synth.* **1992**, *71*, 30.
2. (a) Corey, E. J.; Kim, S. S. *J. Am. Chem. Soc.* **1990**, *112*, 4976. (b) Corey, E. J.; Choi, S. *Tetrahedron Lett.* **1991**, *32*, 2857. (c) Corey, E. J.; Lee, D.-H. *Tetrahedron Lett.* **1993**, *34*, 1737.
3. Corey, E. J.; Lee, D.-H. *J. Am. Chem. Soc.* **1991**, *113*, 4026.
4. Corey, E. J.; Decicco, C. P.; Newbold, R. C. *Tetrahedron Lett.* **1991**, *32*, 5287.
5. Corey, E. J.; Yu, C.-M.; Kim, S. S. *J. Am. Chem. Soc.* **1989**, *111*, 5495.
6. Corey, E. J.; Yu, C.-M.; Lee, D.-H. *J. Am. Chem. Soc.* **1990**, *112*, 878.
7. Pikul, S.; Corey, E. J. *Org. Synth.* **1992**, *71*, 22.
8. Corey, E. J.; DaSilva Jardine, P.; Virgil, S.; Yuen, P.-W.; Connell, R. D. *J. Am. Chem. Soc.* **1989**, *111*, 9243.
9. Zhang, W.; Jacobsen, E. N. *J. Org. Chem.* **1991**, *56*, 2296.

James R. Gage

The Upjohn Company, Kalamazoo, MI, USA

(*R*,*R*)-1,2-Diphenyl-1,2-[di(pentafluorophenyl)phosphanoxy]ethane

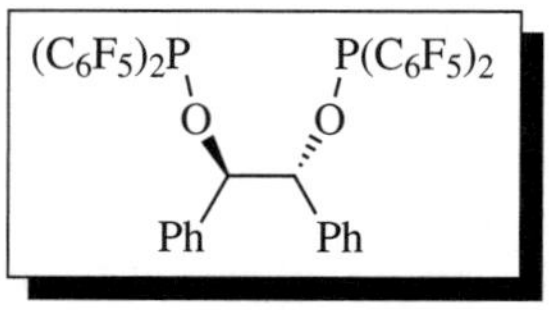

[220224-86-2] $C_{38}H_{12}F_{20}O_2P_2$ (MW 942.43)

(chiral C_2-symmetric bidentate phosphorus ligand with π-acceptor properties; used in the synthesis of transition metal based Lewis acids)

Physical Data: mp 118 °C; $[\alpha]_D$ +85 (*c* 2.26, CH_2Cl_2); ^{1}H NMR ($CDCl_3$, 400 MHz) : 6.98–7.20 (10 H), 5.18 (2 H); ^{31}P NMR ($CDCl_3$, 162 MHz) : 87.6 (quint, 40 Hz).

Solubility: high solubility in Et_2O, CH_2Cl_2, EtOAc, THF, acetone, $CHCl_3$; low solubility in pentane, hexane, cyclohexane and toluene; insoluble in methanol.

Form Supplied in: white solid.

Preparative Methods: [1] to a white suspension of (+)-(*R,R*)-hydrobenzoin[2] (901 mg, 4.21 mmol) and triethylamine (1.17 mL, 8.41 mmol) in diethyl ether (20 mL), a solution of bis(pentafluorophenyl)bromophosphine[3] (3.74 g, 8.41 mmol) in diethyl ether (20 mL) was added dropwise at −78 °C. A white precipitate formed immediately. The reaction mixture was allowed to warm to ambient temperature overnight. After filtration over Celite, the solvent was removed in vacuo to give (*R,R*)-BIPHOP-F (3.90 g, 98%). Further purification can be achieved by recrystallization. Both enantiomers are available.

Purification: recrystallization from pentane.

Handling, Storage, and Precautions: air stable for months at room temperature (no oxidation observed).

Electronic Properties. Ligands with the electron-poor pentafluorophenyl groups have good π-acceptor properties and electronically bridge the gap between phosphites and carbon monoxide. Other diols, with[4] or without[5] C_2-symmetry, have been used as ligand backbones. Pentafluorophenyl can also be replaced by other aromatic electron-withdrawing groups.[6]

Synthesis of Chiral Lewis Acids. BIPHOP-F is used in the synthesis of chiral transition metal Lewis acids. Because of its electronic properties, it enhances the acidity of the metal. Coordination of the bidentate ligand to the metal is accomplished by CO substitution (eq 1[1] and 2[7]). The cationic ruthenium or iron complexes are obtained after one or two additional steps (L is a labile ligand and X^- the counter anion).

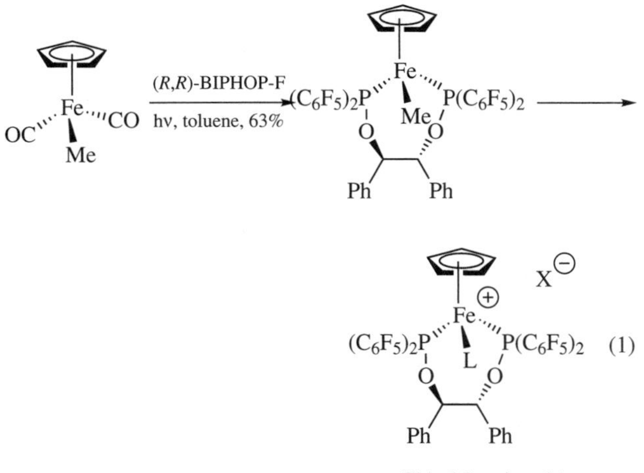

Chiral Lewis acid

The complexes are isolated, characterized and used as chiral Lewis acids. Dissociation of the labile ligand liberates a single coordination site at the metal center. These Lewis acids catalyze enantioselective Diels–Alder reactions.[1,7,8] For instance, reaction of methacrolein with cyclopentadiene in the presence of the cationic iron complex (L = acrolein) occurs with *exo* selectivity and an enantiomeric excess of the same order of magnitude as those obtained with the successful boron and copper catalysts (eq 3).[9]

The chiral environment around the coordination site of the catalyst is created by the perfluoroaryldiphosphinite ligand (crystallographic data).

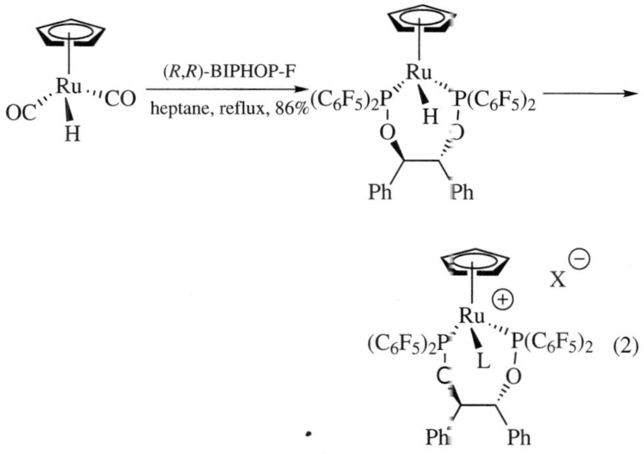

Chiral Lewis acid

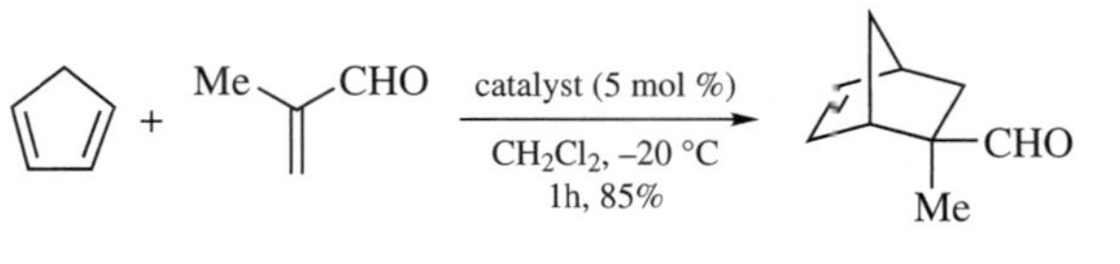

(2*R*)-*exo* ee = 97% (3)
exo/*endo* 98:2

Catalyst =

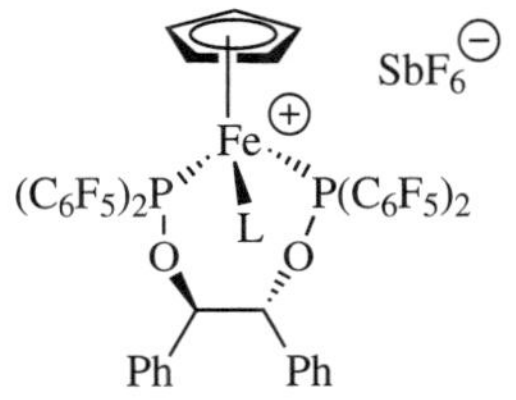

---

1. Bruin, M. E.; Kündig, E. P. *Chem. Commun.* 1998, 2635.
2. Sharpless, K. B.; Amberg, W.; Bennani, Y. L.; Crispino, G. A.; Hartung, J.; Jeong, K. S.; Kwong, H. L.; Morikawa, K.; Wang, Z. W.; Xu, D.; Zhang, X. L. *J. Org. Chem.* 1992, *57*, 2768.
3. (a) Fild, M.; Glemser, O.; Hollenberg, I. *Z., Naturforsch Teil B* 1966, *21*, 920. (b) Ali, R.; Dillon, K. B. *J. Chem. Soc., Dalton Trans.* 1990, 2593.
4. (a) Kündig, E. P.; Dupré, C.; Bourdin, B.; Cunningham Jr, A.; Pons, D. *Helv. Chim. Acta* 1994, *77*, 421. (b) Kündig, E. P.; Bourdin, B.; Bernardinelli, G. *Angew. Chem., Int. Ed. Engl.* 1994, *33*, 1856. (c) RajanBabu, T. V.; Radetich, B.; Kamfia, K. Y.; Timothy, A. A.; Casalnuovo, A. L.; Calabrese, J. C. *J. Org. Chem.* 1999, *64*, 3429.
5. Tolstikov, A. G.; Amosov, Y. I.; Tolstikova, O. V.; Khlebnikova, T. B.; Zakharova, I. V. *Russ. Chem. Bull.* 1997, *46*, 381.
6. (a) Clyne, D. S.; Mermet-Bouvier, Y. C.; Nomura N.; RajanBabu, T. V., *J. Org. Chem.* 1999, *64*, 7601. (b) Moloy, K. G.; Petersen, J. L., *J. Am. Chem. Soc.* 1995, *117*, 7696.
7. Kündig, E. P.; Saudan, C. M.; Bernardinelli, G. *Angew. Chem., Int. Ed. Engl.* 1999, *38*, 1220.
8. Kündig, E. P.; Saudan, C. M.; Viton, F. *Adv. Synth. Catal.* 2001, *343*, 51.
9. (a) Kündig, E. P.; Saudan, C. M., In *Handbook of Lewis Acids - Application in Organic Synthesis*; Yamamoto, H., Ed.; Wiley-VCH: Weinheim, 2000; pp 597-652. (b) Evans, D. A.; Johnson, J. S., In *Comprehensive Asymmetric Catalysis*; Jacobsen, E. N.; Pfaltz, A.; Yamamoto, H., Eds.; Springer: Berlin, 1999, Vol. III, pp 1177-1235.

Valérie Alezra
ENS-Lyon, Lyon, France
E. Peter Kündig
University of Geneva, Switzerland

(*S,S*)-1,2-Diphenylethylenediamine

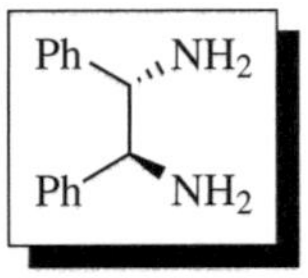

[35132-20-8]
[(R,R)-form]
[29841-69-8]
[(S,S)-form] $C_{14}H_{16}N_2$ (MW 212.29)

(chiral diamine ligand for transition[1,2] and main-group[3] metals; optical resolution agent;[4] chiral solvating agent in NMR analysis;[5,6] precursor of chiral auxiliaries[7])

Alternate Name: (*S,S*)-DPEN; (*S,S*)-1,2-diphenyl-1,2-ethanediamine; (*S,S*)-1,2-diamino-1,2-diphenylethane; (*S,S*)-stilbenediamine; (*S,S*)-α,β-diaminodihydrostilbene.

Physical Data: mp 85–86.5 °C;[8] $[\alpha]^{23}_D$ −106 ± 1 (*c* 1.1, MeOH).[9]

Solubility: soluble in benzene, chloroform, dichloromethane, ethanol, diethyl ether, methanol, THF; modestly soluble in hot hexane, hot water.

Form Supplied in: colorless solid.

Analysis of Reagent Purity: ^{1}H-NMR analysis of its salt with L-mandelic acid.[9,10]

Preparative Methods: (i) preparation of racemic DPEN and its optical resolution:[8,9] Reaction of benzil and cyclohexanone in the presence of ammonium acetate and acetic acid at reflux temperature gives a cyclic bis-imine (**1**) (eq 1).[9] Stereoselective reduction of the bis-imine with lithium in THF-liquid ammonia at −78 °C followed by addition of ethanol, then hydrolysis with hydrochloric acid and neutralization with sodium hydroxide produces the racemic diamine (**2**). Recrystallization of the L-tartaric acid salt from a 1:1 water–ethanol mixture followed by neutralization with sodium hydroxide, recrystallization from hexane results in (*S,S*)-DPEN (**3**) as colorless crystals.

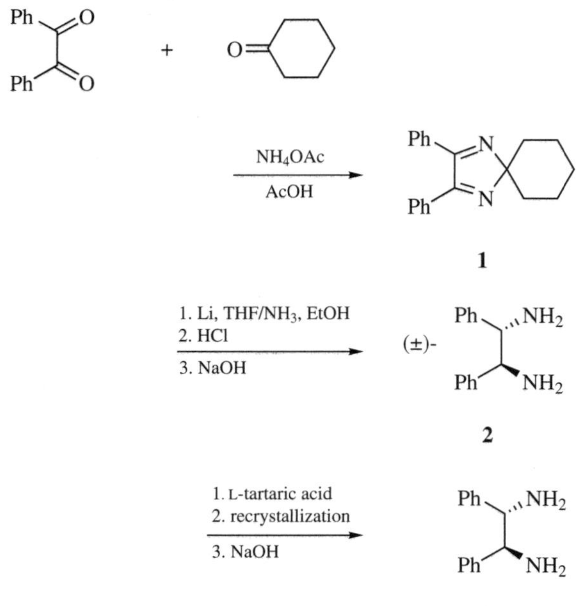

(1)

(ii) Transformation from (*R,R*)-diphenylethyleneglycol: (*R,R*)-Diphenylethyleneglycol prepared by Sharpless dihydroxylation[11] of *trans*-stilbene is treated with 2.4 equiv of *p*-tosyl chloride, and the resulting ditosylate is converted with 2.6 equiv of sodium azide to the *S,S*-configured diazide without racemization. The diazide is reduced by lithium aluminum hydride to afford (*S,S*)-DPEN (eq 2).[12] This transformation is also achieved via its cyclic sulfate[13] or sulfite.[14] Asymmetric borane reduction of a bisiminodiphenylethane derivative[15] and asymmetric imino pinacol-type coupling[16] are also reported.

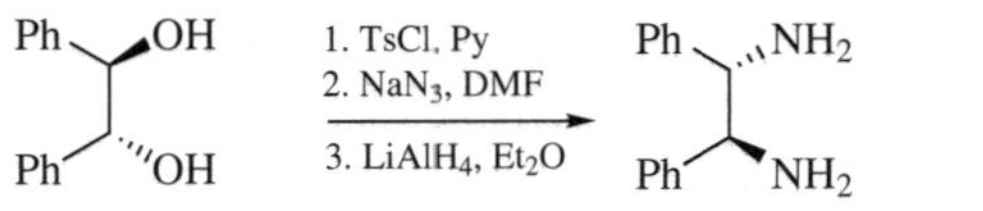

(2)

Purification: recrystallization from hexane.[8,9]

Handling, Storage, and Precautions: DPEN is substantially stable. No special handling care is required.

BINAP/DPEN/Ru(II)-catalyzed Asymmetric Hydrogenation of Simple Ketones.[1] Ru(II) complexes having a formula of *trans*-$RuCl_2$(diphosphine)(dpen) are most conveniently obtained by treatment of oligomeric $RuCl_2(\text{diphosphine})(\text{dmf})_n$ with 1.1 equiv of DPEN in DMF at room temperature[17] [diphosphine = 2,2′-bis(diphenylphosphino)-1,1′-binaphthyl (BINAP),[18] 2,2′-bis(di-4-tolylphosphino)-1,1′-binaphthyl (TolBINAP), 2,2′-bis(di-3,5-xylylphosphino)-1,1′-binaphthyl (XylBINAP),[19] 2,2′-bis(di-3,5-xylylphosphino)-1,1′-biphenyl (DM-BIPHEP)[20]]. The molecular structures of diastereomeric *trans*-$RuCl_2$[(*R*)-tolbinap][(*R,R*)- or (*S,S*)-dpen] have been elucidated by X-ray crystallographic analysis.[17] These chiral diphosphine/diamine/Ru(II) complexes act as excellent precatalysts for the asymmetric hydrogenation of simple ketones.[1] A range of aromatic, hetero-aromatic, amino, and α,β-unsaturated ketones are hydrogenated to the corresponding chiral alcohols quantitatively with excellent optical purity.

The diphosphine/diamine/Ru(II) complexes show exceptionally high catalytic activity for hydrogenation of simple ketones with an alkaline base in propan-2-ol. Acetophenone (601 g) is hydrogenated with *trans*-$RuCl_2$[(*S*)-tolbinap][(*S,S*)-dpen] (2.2 mg) and *t*-BuOK (5.6 g) in propan-2-ol (1.5 L) under 45 atm H_2 at 30 °C for 48 h to give (*R*)-1-phenylethanol in 80% ee quantitatively. The substrate concentration is as high as 30%. Under such conditions, the turnover number (TON) is at least 2 400 000, whereas the turnover frequency (TOF) at 30% conversion is 228 000 h^{-1} or 63 s^{-1} (eq 3).[17] The reaction can be conducted under 1 atm of H_2 with a substrate/catalyst (S/C) molar ratio of 800. When (*S*)-XylBINAP is used as a diphosphine ligand instead of (*S*)-TolBINAP, the ee value is increased to 99%.[21] Alkyl aryl ketones with various substituents are hydrogenated with a consistently high enantioselectivity.[1,22] The reaction is tolerant of aromatic halides and CF_3, OMe, COO-*i*-Pr, NO_2, and NH_2 groups. The high degree of enantioselectivity is a result of the synergistic effects of BINAPs and DPEN. In most cases, a combination of the (*S*)-BINAP derivative and (*S,S*)-DPEN or the *R*/*R,R* enantiomer results in the best enantioselectivity.

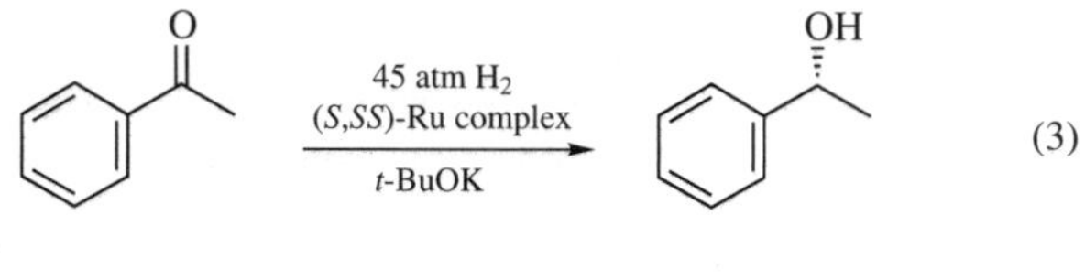

80% ee
TON = 2 400 000
TOF = 63 s^{-1}

(*S,SS*)-Ru complex:

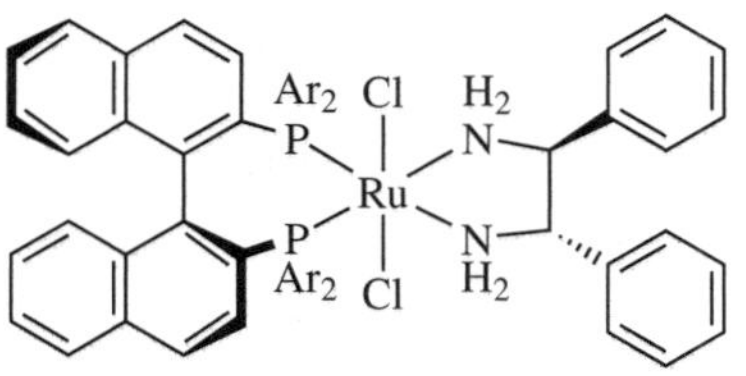

trans-$RuCl_2$[(*S*)-tolbinap][(*S,S*)-dpen]
Ar = 4-MeC_6H_4

Hydrogenation of open-chain α,β-unsaturated ketones proceeds selectively at the C=O linkage to afford chiral allylic alcohols with high ee.[22,23] β-Ionone, a dienone, is hydrogenated in the presence of *trans*-$RuCl_2$[(*S*)-xylbinap][(*S,S*)-dpen] and *t*-BuOK in propan-2-ol under 8 atm H_2 to give (*R*)-β-ionol in 93% ee (eq 4).[21] No saturation of the olefinic bond is observed. The related diamine-free BINAP/Ru complexes catalyze hydrogenation of the C=C unit of allylic alcohols,[24] and thus the presence of diamine dramatically reverses the chemoselectivity preference. In the hydrogenation of less hindered, base-sensitive enones, K_2CO_3, a relatively weak base, should be used to avoid production of polymeric compounds. In most cases, the degree of enantioselection with the XylBINAP/DPEN/Ru(II) catalyst is slightly lower than that with the XylBINAP/DAIPEN/Ru(II) catalyst (DAIPEN = 1,1-di-4-anisyl-2-isopropyl-1,2-ethylenediamine).[22,25]

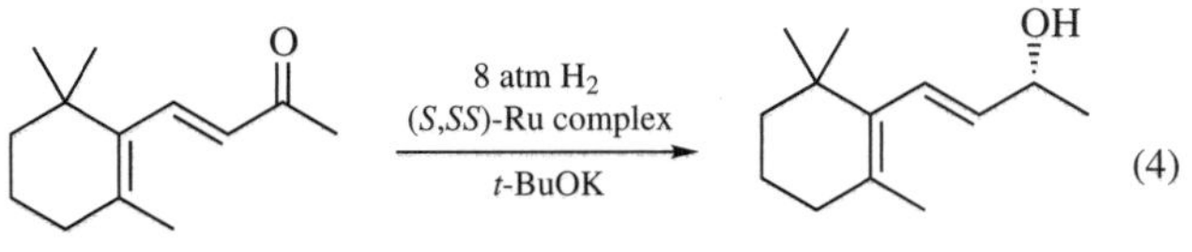

93% ee

(*S,SS*)-Ru complex = *trans*-$RuCl_2$[(*S*)-xylbinap][(*S,S*)-dpen]

A catalyst prepared in situ consisting of $RuCl_2$(diphosphine)-(dmf)$_n$, DPEN, and a base is also usable,[26] although it is less active than the diphosphine/DPEN/Ru(II) complex and base combination. Enantioselective hydrogenation of 2,4,4-trimethyl-2-cyclohexenone, a cyclic α,β-unsaturated ketone, with $RuCl_2$[(*R*)-tolbinap](dmf)$_n$, (*S,S*)-DPEN, and KOH under 8 atm H_2 at 0 °C results in (*S*)-2,4,4-trimethyl-2-cyclohexenol with 96% ee quantitatively (eq 5).[27] In this case, unlike the above mentioned example, the combined use of (*R*)-TolBINAP and (*S,S*)-DPEN (or *S* and *R,R*) is necessary for the high stereoselection. Reaction using (*R*)-TolBINAP and (*R,R*)-DPEN gives the *S* alcohol in only 26% ee.

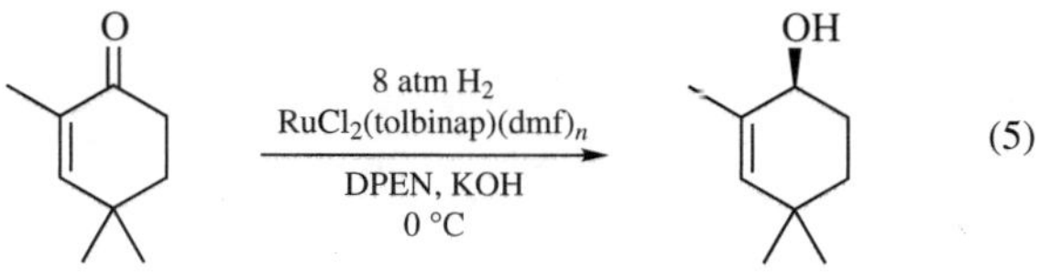

100% yield

TolBINAP	DPEN	% ee
R	*S,S*	96
R	*R,R*	26
±	*S,S*	95

When the cyclic enone is hydrogenated with the racemic TolBINAP/Ru(II) complex and (*S,S*)-DPEN under otherwise identical conditions, the *S* allylic alcohol is obtained in 95% ee and 100% yield.[28] The ee value is close to the 96% attained with the enantiomerically pure (*R*)-TolBINAP/(*S,S*)-DPEN system. Hydrogenation of *o*-methylacetophenone catalyzed by $RuCl_2$[(±)-tolbinap](dmf)$_n$ and (*S,S*)-DPEN results in the *R* alcohol in 90% ee and 100% yield (eq 6).[28] The *S*/*S,S* catalyst gives the *R* product in 97.5% ee.

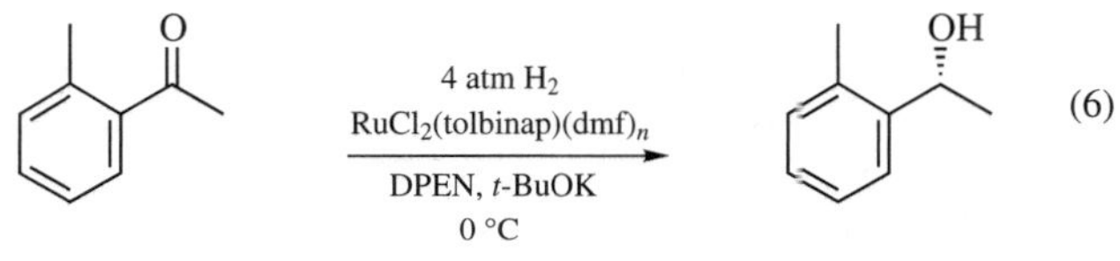

100% yield

TolBINAP	DPEN	% ee
S	*S,S*	97.5
S	*R,R*	8
±	*S,S*	90

DM-BIPHEP, a conformationally flexible diphosphine, is converted to racemic $RuCl_2$(dm-biphep)(dmf)$_n$.[20] Hydrogenation of 1′-acetonaphthone with a mixture of DM-BIPHEP/Ru(II) complex, (*S,S*)-DPEN, and KOH under 40 atm of H_2 at −35 °C results in the *R* alcohol and 92% ee in >99% yield (eq 7).

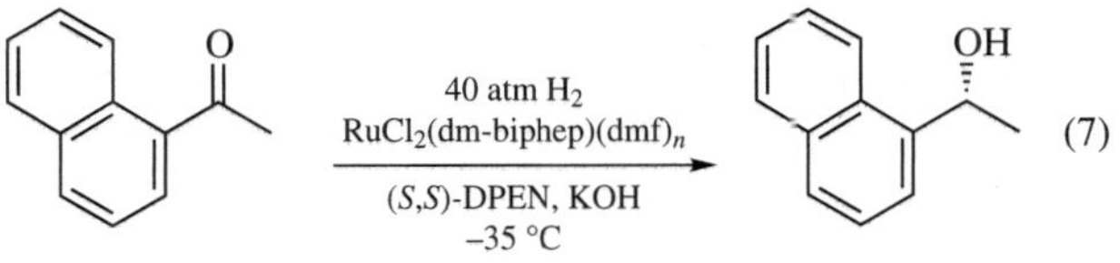

92% ee

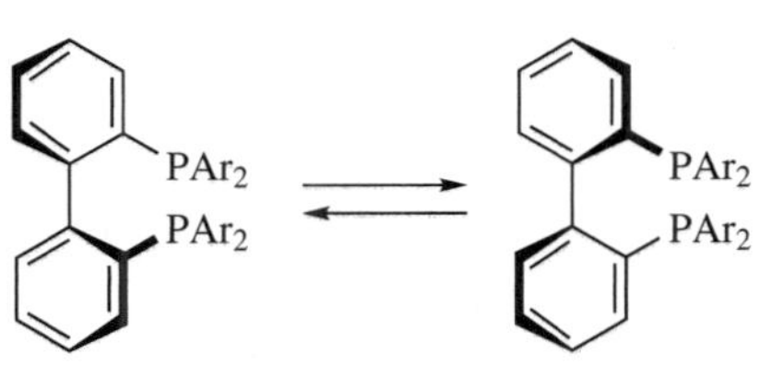

(*S*)-DM-BIPHEP (*R*)-DM-BIPHEP
Ar = 3,5-$Me_2C_6H_3$

Certain racemic ketones can be resolved kinetically through asymmetric hydrogenation. When racemic carvone is hydrogenated using an (*S*)-BINAP/Ru complex, (*R*,*R*)-DPEN, and KOH, it gives, at 54% conversion, the starting (*S*)-carvone in 94% ee (46%) together with (1*R*,5*R*)-carveol in 93% ee (50%) and some other minor alcohols (3.7%) (eq 8).[27] The extent of the enantiomer differentiation ability, k_{fast}/k_{slow}, is calculated to be 33.

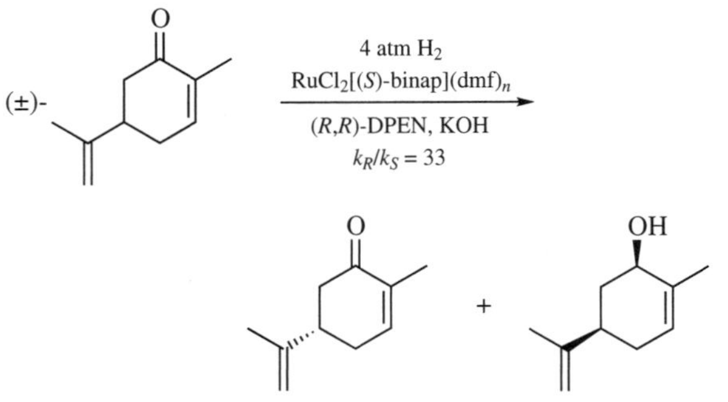

(8)

S, 94% ee
46%

R,*R*, 93% ee
50%

Dynamic kinetic resolution provides a stereoselective method to convert racemic ketones having an α-stereogenic center into a single stereoisomer quantitatively among the four possible stereoisomers.[29] Hydrogenation of 2-isopropylcyclohexanone with $RuCl_2$[(*S*)-binap](dmf)$_n$, (*R*,*R*)-DPEN, and KOH under 4 atm H_2 at room temperature results in a 99.8:0.2 mixture of the *cis* (1*R*,2*R*) alcohol in 93% ee and the *trans* (1*R*,2*S*) isomer in 28% ee (eq 9).[30] Under the conditions, the *R* ketone is hydrogenated 36 times faster than the *S* isomer and the slow-reacting *S* ketone undergoes in situ stereochemical inversion 47 times faster than it is hydrogenated.

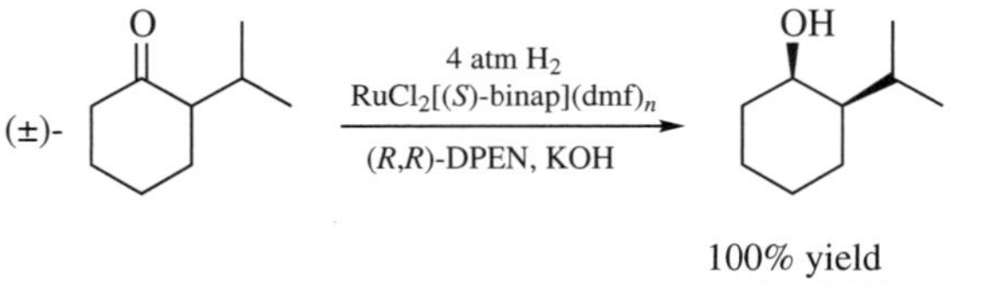

(9)

100% yield
1*R*,2*R*, 93% ee
cis/*trans* = 99.8:0.2

DPEN/metal-complex Catalyzed Asymmetric Reactions. A catalyst system generated in situ from Co(acac)$_2$ and (*R*,*R*)-DPEN accelerates enantioselective Michael addition of methyl 1-oxo-2-indanecarboxylate to methyl vinyl ketone to give the *R* adduct in up to 66% ee (eq 10).[2]

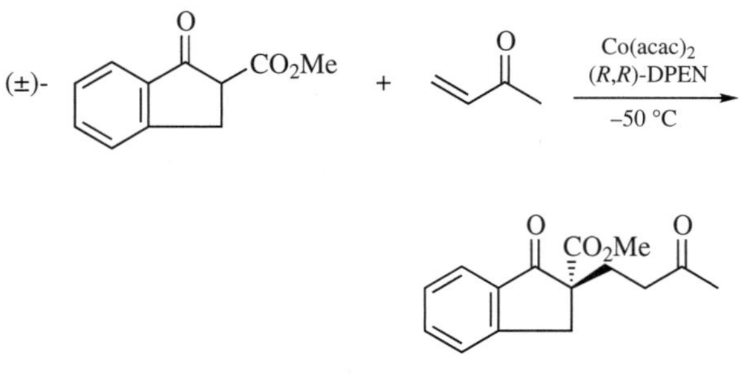

(10)

66% ee

A chiral Lewis acid prepared in situ from magnesium iodide and (*R*,*R*)-DPEN efficiently catalyzes asymmetric aza-Diels–Alder reaction of a methyl glyoxylate/*p*-anisidine derived imine with the Danishefsky diene to give the cyclic adduct in 97% ee (eq 11).[3]

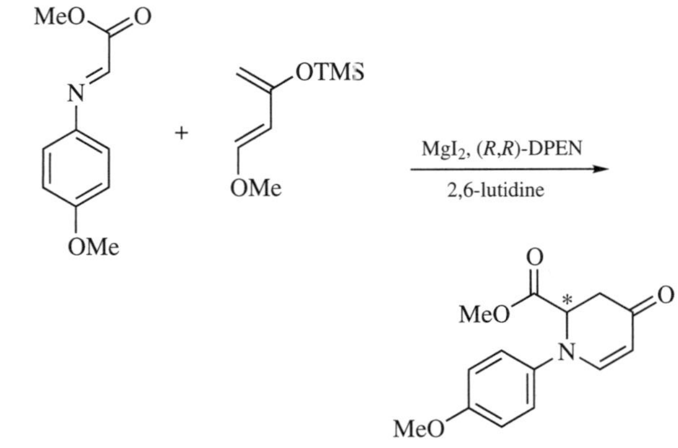

(11)

97% ee
absolute configuration
undetermined

Synthesis of Enantiomerically Pure C_2-symmetric Vicinal Diamines via Chirality Transfer from DPEN. Several C_2-symmetric vicinal diamines and their derivatives are prepared in optically pure form by chirality transfer from DPEN.[31] For example, condensation of (*S*,*S*)-DPEN with butane-2,3-dione in benzene at the reflux temperature is followed by stereoselective reduction with $NaBH_3CN$ and PPTS at −20 °C to afford the (2*S*,3*S*,5*R*,6*R*)-piperazine **4** and its diastereomer in a 15:1 ratio (eq 12). The crude product is purified by silica gel column chromatography. Formation of the biscarbamate followed by reductive cleavage of benzylic C–N bonds with lithium in liquid ammonia, and then removal of isobutyloxycarbonyl with HBr in acetic acid results in (*R*,*R*)-2,3-diaminobutane dihydrobromide **5** in 99% ee.

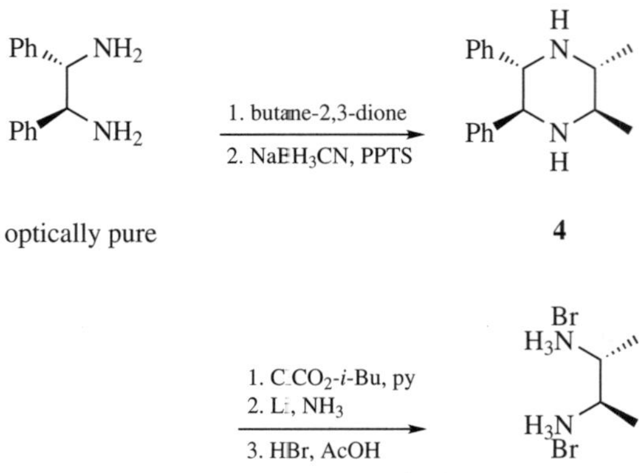

optically pure

4

(12)

5

99% ee

Optical Resolution Agent and NMR Chiral Solvating Agent. DPEN acts as an effective optical resolution agent of racemic 2,2′-

dihydroxy-1,1′-binaphthyl (BINAPHTHOL).[4] A 1:1 mixture of racemic BINAPHTHOL and (*R,R*)-DPEN in hot benzene produces colorless crystalline solid at room temperature. Recrystallization from benzene followed by the addition of HCl results in optically pure (*R*)-BINAPHTHOL (eq 13).

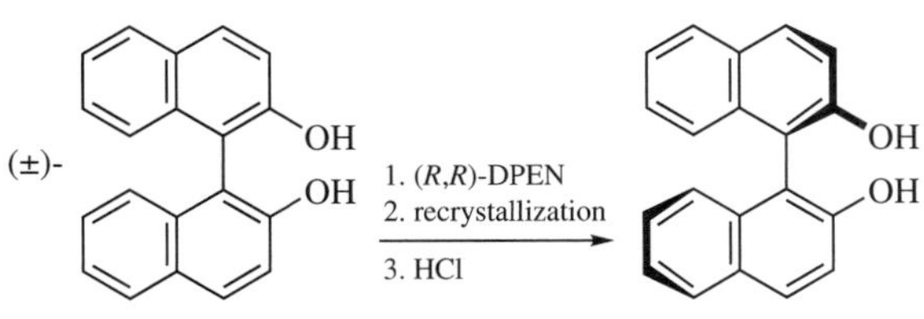

(13)

optically pure

In addition, DPEN is an efficient chiral solvating agent for determination of the enantiomeric excess in the ^{1}H NMR analysis of various chiral mono- and dicarboxylic acids including α-arylpropanoic and α-halo carboxylic acids.[5] The chemical-shift non-equivalence ($\delta\Delta$) in certain diastereomeric complexes is greater than 0.05 ppm. A DPEN/Pd(II) complex can be used for determination of enantiomeric excess of the non-protected chiral amino acids by ^{1}H and ^{13}C NMR analysis.[6] For example, $\{Pd[(S,S)\text{-dpen}](D_2O)_2\}^{2+}$ and racemic alanine with a base forms the square-planar complex (eq 14). The $\delta\Delta$ of ^{1}H-NMR resonance in the diastereomeric complexes in D_2O is 0.056 ppm, while this complex hardly dissolves in D_2O.

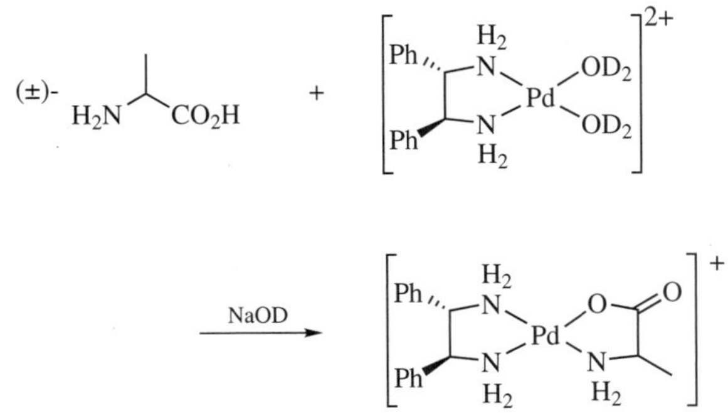

(14)

diastereomeric mixture

Precursor of Useful Chiral Ligands. DPEN is widely used for the preparation of chiral ligands.[7] Organometallic compounds with these ligands act as useful reagents or catalysts in asymmetric induction reactions such as dihydroxylation of olefins,[32] transfer hydrogenation of ketones and imines,[33] Diels–Alder and aldol reactions,[34] desymmetrization of *meso*-diols to produce chiral oxazolidinones,[35] epoxidation of simple olefins,[36,37] benzylic hydroxylation,[38] and borohydride reduction of ketones, imines, and α,β-unsaturated carboxylates.[39]

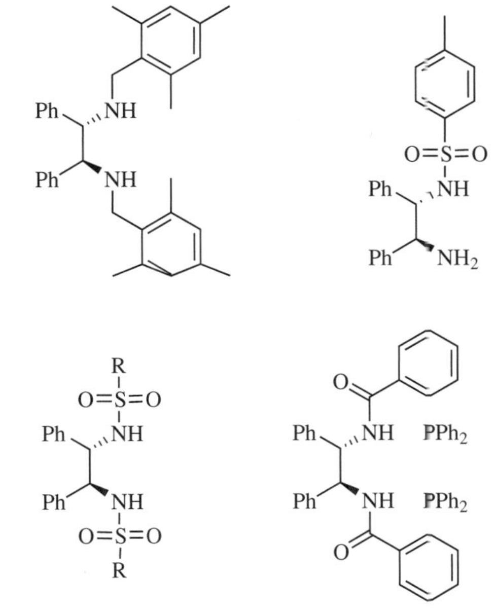

R = 4-MeC$_6$H$_4$, 4-NO$_2$C$_6$H$_4$, 3,5-(CF$_3$)$_2$C$_6$H$_3$, CF$_3$

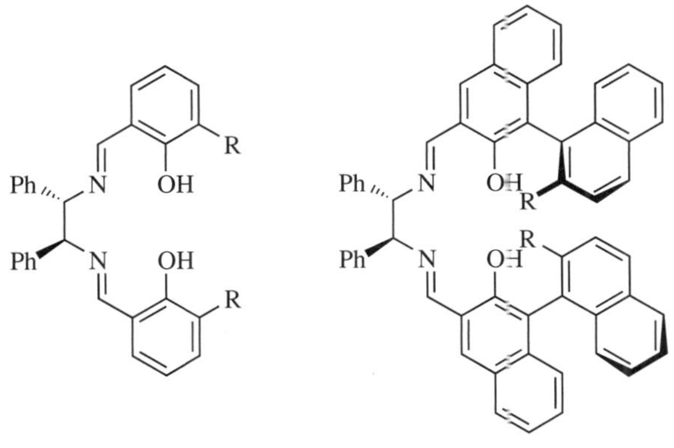

R = H, *t*-Bu, etc.

R = Me, Ph

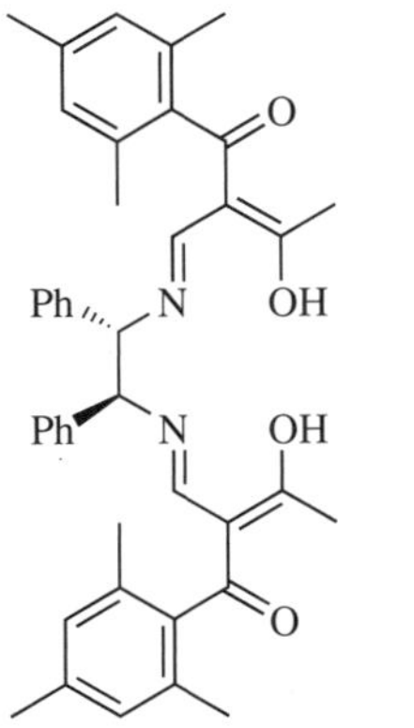

Related Reagents. (*R*)- and (*S*)-2,2′-Bis(diphenylphosphino)-1,1′-binaphthyl; 1,3,2-dioxathiolane 2,2-dioxide; (*R,R*)-1,2-diphenyl-1,2-diaminoethane *N,N′*-bis[3,5-bis(trifluoromethyl)benzenesulfonamide].

1. Noyori, R.; Ohkuma, T. *Angew. Chem., Int. Ed.* **2001**, *40*, 40.
2. Brunner, H.; Hammer, B. *Angew. Chem., Int. Ed. Engl.* **1984**, *23*, 312.
3. Bromidge, S.; Wilson, P. C.; Whiting, A. *Tetrahedron Lett.* **1998**, *39*, 8905.
4. Kawashima, M.; Hirata R. *Bull. Chem. Soc. Jpn.* **1993**, *66*, 2002.
5. Fulwood, R.; Parker, D. *J. Chem. Soc., Perkin Trans. 2*, **1994**, 57.
6. Staubach, B.; Buddrus, J. *Angew. Chem., Int. Ed. Engl.* **1996**, *35*, 1344.
7. (a) Alexakis, A.; Mangeney, P., In *Advanced Asymmetric Synthesis*; Stephenson, G. R., Ed.; Chapman & Hall: London, 1996, p 93. (b) Lucet, D.; Le Gall, T.; Mioskowski, C. *Angew. Chem., Int. Ed.* **1998**, *37*, 2580.
8. Saigo, K.; Kubota, N.; Takebayashi, S.; Hasegawa, M. *Bull. Chem. Soc. Jpn.* **1986**, *59*, 931.
9. Pikul, S.; Corey, E. J. *Org. Synth.* **1993**, *71*, 22.
10. Benson, S. C.; Cai, P.; Colon, M.; Haiza, M. A.; Tokles, M.; Snyder, J. K. *J. Org. Chem.* **1988**, *53*, 5335.
11. Kolb, H. C.; VanNieuwenhze, M. S.; Sharpless, K. B. *Chem. Rev.* **1994**, *94*, 2483.
12. Pini, D.; Iuliano, A.; Rosini, C.; Salvadori, P. *Synthesis* **1990**, 1023.
13. (a) Oi, R.; Sharpless, K. B. *Tetrahedron Lett.* **1991**, *32*, 999. (b) Lynch, N. J., In *Encyclopedia of Reagents for Organic Synthesis*; Paquette, L. A., Ed.; Wiley: Chichester, Vol. 3, 1995, p 2190
14. Lohray, B. B.; Ahuja, J. R. *J. Chem. Soc., Chem. Commun.* **1991**, 95.
15. Shimizu, M.; Kamei, M.; Fujisawa, T. *Tetrahedron Lett.* **1995**, *36*, 8607.
16. Shimizu, M.; Iida, T.; Fujisawa, T. *Chem. Lett.* **1995**, 609.
17. Doucet, H.; Ohkuma, T.; Murata, K.; Yokozawa, T.; Kozawa, M.; Katayama, E.; England, A. F.; Ikariya, T.; Noyori, R. *Angew. Chem., Int. Ed.* **1998**, *37*, 1703.
18. (a) Miyashita, A.; Takaya, H.; Souchi, T.; Noyori, R., *Tetrahedron* **1984**, *40*, 1245. (b) Kitamura, M.; Noyori, R., In *Encyclopedia of Reagents for Organic Synthesis*; Paquette, L. A., Ed.; Wiley: Chichester, Vol. 1, 1995, 509.
19. Mashima, K.; Kusano, K.; Sato, N.; Matsumura, Y.; Nozaki, K.; Kumobayashi, H.; Sayo, N.; Hori, Y.; Ishizaki, T.; Akutagawa, S.; Takaya, H., *J. Org. Chem.* **1994**, *59*, 3064.
20. Mikami, K.; Korenaga, T.; Terada, M.; Ohkuma, T.; Pham, T.; Noyori, R. *Angew. Chem., Int. Ed.* **1999**, *38*, 495.
21. Unpublished result.
22. Ohkuma, T.; Koizumi, M.; Doucet, H.; Pham, T.; Kozawa, M.; Murata, K.; Katayama, E.; Yokozawa, T.; Ikariya, T.; Noyori, R. *J. Am. Chem. Soc.* **1998**, *120*, 13529.
23. Ohkuma, T.; Ooka, H.; Ikariya, T.; Noyori, R. *J. Am. Chem. Soc.* **1995**, *117*, 10417.
24. Takaya, H.; Ohta, T.; Sayo, N.; Kumobayashi, H.; Akutagawa, S.; Inoue, S.; Kasahara, I.; Noyori, R. *J. Am. Chem. Soc.* **1987**, *109*, 1596 and 4129.
25. Wey, S.-J.; O'Connor, K. J.; Burrows, C. J. *Tetrahedron Lett.* **1993**, *34*, 1905.
26. Ohkuma, T.; Ooka, H.; Hashiguchi, S.; Ikariya, T.; Noyori, R., *J. Am. Chem. Soc.* **1995**, *117*, 2675.
27. Ohkuma, T.; Ikehira, H.; Ikariya, T.; Noyori, R. *Synlett* **1997**, 467.
28. Ohkuma, T.; Doucet, H.; Pham, T.; Mikami, K.; Korenaga, T.; Terada, M.; Noyori, R. *J. Am. Chem. Soc.* **1998**, *120*, 1086.
29. Noyori, R.; Tokunaga, M.; Kitamura, M. *Bull. Chem. Soc. Jpn.* **1995**, *68*, 36.
30. Ohkuma, T.; Ooka, H.; Yamakawa, M.; Ikariya, T.; Noyori, R., *J. Org. Chem.* **1996**, *61*, 4872.
31. Nantz, M. H.; Lee, D. A.; Bender, D.; Roohi, A. H. *J. Org. Chem.* **1992**, *57*, 6653.
32. Corey, E. J.; Jardine, P. D.; Virgil, S.; Yuen, P.-W.; Connell, R. D. *J. Am. Chem. Soc.* **1989**, *111*, 9243.
33. Noyori, R.; Hashiguchi, S. *Acc. Chem. Res.* **1997**, *30*, 97.
34. (a) Corey, E. J. *Pure Appl. Chem.* **1990**, *62*, 1209. (b) Gage, J. R., In *Encyclopedia of Reagents for Organic Synthesis*; Paquette, L. A., Ed.; Wiley: Chichester, Vol. 4, 1995, p 2207.
35. Trost, B. M.; Van Vranken, D. L. *Angew. Chem., Int. Ed. Engl.* **1992**, *31*, 228.
36. Zhang, W.; Loebach, J. L.; Wilson, S. R.; Jacobsen, E. N. *J. Am. Chem. Soc.* **1990**, *112*, 2801.
37. Hosoya, N.; Hatayama, A.; Irie, R.; Sasaki, H.; Katsuki, T., *Tetrahedron* **1994**, *50*, 4311.
38. Hamada, T.; Irie, R.; Mihara, J.; Hamachi, K.; Katsuki, T., *Tetrahedron* **1998**, *54*, 10017.
39. Yamada, T.; Nagata, T.; Ikeno, T.; Ohtsuka, Y.; Sagara, A.; Mukaiyama, T. *Inorg. Chim. Acta* **1999**, *296*, 86.

Takeshi Ohkuma & Ryoji Noyori
Nagoya University, Japan

(*S*)-Diphenyl(1-methylpyrrolidin-2-yl)-methanol[1]

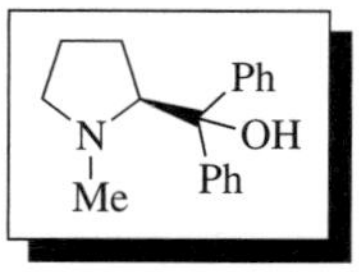

[110529-22-1] $C_{18}H_{21}NO$ (MW 267.40)

(chiral ligand for the enantioselective addition of dialkylzincs,[2] alkynylzinc,[3b] and cyanomethylzinc bromide[4] to aldehydes; chiral ligand for the enantioselective Reformatsky reaction;[5] chiral ligand for the enantioselective Diels–Alder reaction;[6] chiral auxiliary for asymmetric polymerization[7])

Alternate Name: DPMPM.
Physical Data: mp 68.5–68.9 °C; $[\alpha]_D^{23}$ + 57.0° (*c* 1.0, $CHCl_3$).
Solubility: sol hexane, benzene, toluene, Et_2O, cyclohexane, dichloromethane.
Form Supplied in: colorless crystals; available in either enantiomeric form.
Preparative Methods: reaction of *Phenylmagnesium Bromide* with (*S*)-*N*-[(benzyloxy)carbonyl]proline methyl ester and subsequent reduction with *Lithium Aluminum Hydride* affords the title compound in 83% overall yield.[2b]

Catalytic Enantioselective Addition of Dialkylzincs to Aldehydes. DPMPM (**1**) is a chiral amino alcohol which is a precursor to a chiral catalyst for the enantioselective addition of dialkylzincs to aldehydes.[2] In the presence of 2 mol % of (*S*)-(**1**), optically active alcohols of up to 100% ee are obtained from the enantioselective addition of dialkylzincs to aldehydes (eq 1, Table 1). When benzaldehyde is allowed to react with *Diethylzinc* using (*S*)-(**1**) (2 mol %), (*S*)-1-phenylpropan-1-ol with 97% ee is obtained in quantitative yield (entry 1). When the lithium alkoxide of (*S*)-(**1**) (5 mol %) is employed as a chiral ligand in the addition to aromatic aldehydes, ee's of the alcohols obtained increase to 99.5–100% ee (entries 4 and 5). Amino alcohol (**1**) is also effective in the enantioselective addition of Et_2Zn to the aliphatic aldehyde heptanal, and (*S*)-nonan-3-ol with 91% ee is obtained (entry 2). In the addition to aromatic aldehydes, enantioselectivities using DPMPM are comparable with those obtained with 3-*exo*-(dimethylamino)isoborneol (DAIB).[8] In the addition to heptanal, DPMPM (**1**) is more enantioselective than DAIB (61% ee). However, in the addition to aliphatic aldehydes of wider range, *N,N*-dibutylnorephedrine[9] is more enantioselective than (**1**).

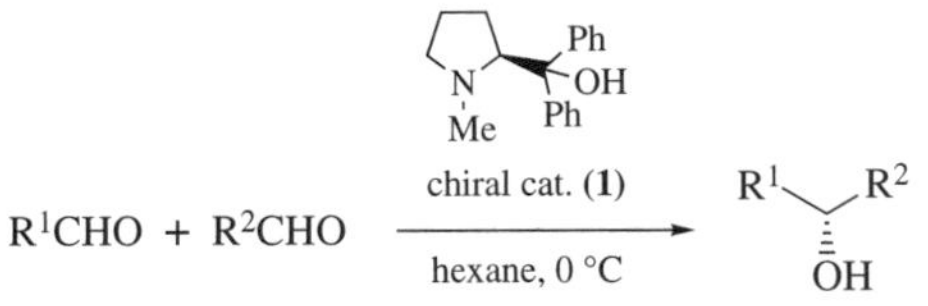

(1)

Table 1 Enantioselective Addition of Dialkylzincs to Aldehydes Using (**1**) or (**2**) as Chiral Catalysts

Entry	R^1	R^2	Catalyst	Yield (%)	ee (%)	Config.
1.[2b]	Ph	Et	(*S*)-(**1**)	100	97	(*S*)
2.[2b]	*n*-Hexyl	Et	(*S*)-(**1**)	96	91	(*S*)
3.[2b]	Ph	Et	(*S*)-(**2**)	100	100	(*R*)
4.[2b]	Ph	Et	Li-(*S*)-(**1**)	100	99.5	(*S*)
5.[2b]	4-$MeOC_6H_4$	Et	Li-(*S*)-(**1**)	96	100	(*S*)
6.[10]	4-$CF_3C_6H_4$	Et	(Li-*S*)-(**1**)	80	91	(*S*)
7.[11]	PhCDO	Et	(*S*)-(**1**)	86	91	(*S*)
8.[2a]	PhCH=CH	Et	(*S*)-(**1**)	91	97	(*S*)
9.[2b]	PhCH=CH	Me	(Li-*S*)-(**1**)	47	89	(*S*)
10.[3]	$Me_3SiC{\equiv}C$	Et	(*S*)-(**1**)	67	78	

The sense of the asymmetric induction is dependent on the structure of the catalyst. (1*R*,2′*S*)-Phenyl(1-neopentylpyrrolidin-2-yl)methanol (**2**) mediates the addition of Et_2Zn to aldehydes to afford (*R*)-alcohols in up to 100% ee (entry 3).[2b] By using (**1**) as a chiral ligand, optically active fluorine-containing alcohols[10] and deuterio alcohols[11] of high optical purities have been synthesized (entries 6 and 7). The enantioselective addition to α,β-unsaturated aldehydes (e.g. cinnamaldehyde) using (**1**) affords optically active allylic alcohols with 89–97% ee (entries 8 and 9).[2] Enantioselective addition of dialkylzincs to alkynyl aldehydes and furyl aldehydes using (**1**) as a chiral catalyst affords optically active alkynyl alcohols (78% ee, entry 10)[3] and furyl alcohols (88–94% ee).[12] When terephthalaldehyde is allowed to react with Et_2Zn using (**1**) as a chiral ligand, the corresponding optically pure diol is obtained.[13]

Unlike alkyllithium and Grignard reagents, dialkylzinc does not add to ketones even in the presence of (**1**). Thus the chemo- and enantioselective alkylation of a keto aldehyde (4-benzoylbenzaldehyde) with Et_2Zn using (*S*)-(**1**) affords the corresponding optically active hydroxy ketone with 93% ee in 99% yield.[14]

Enantioselective Addition of Cyanomethylzinc Bromide, Reformatsky Reagent, and Alkynylzinc Reagents to Aldehydes. The enantioselective additions of cyanomethylzinc bromide,[4] Reformatsky reagent (see *Ethyl Bromozincacetate*),[5] and alkynylzinc[3b] reagent to aldehydes using (**1**) as chiral catalyst or ligand afford optically active β-hydroxy nitrile (93% ee),[4] β-hydroxy ester (78% ee) (eq 2),[5] and alkynyl alcohol (43% ee).[3b]

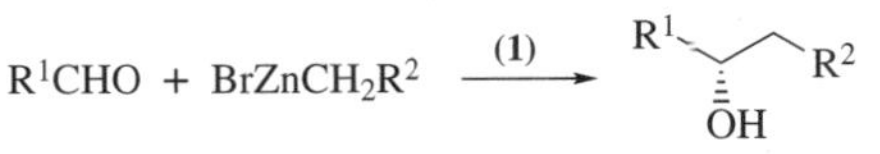

(2)

R^1 = Ph, R^2 = CN, 76%, 93% ee; R^1 = 2-naphthyl, R^2 = CO_2-*t*-Bu, 82%, 78% ee

Catalytic Asymmetric Diels–Alder Reaction. Amino alcohol (**1**) combined with *Boron Tribromide* generates a chiral catalyst for the asymmetric Diels–Alder reaction (97% ee) of unsaturated aldehydes and dienes.[6]

Asymmetric Polymerization. Polymerization of methacrylate derived from (**1**) affords optically active polymer of helical conformation of single screw sense.[7]

1. (a) Soai, K.; Niwa, S. *Chem. Rev.* **1992**, *92*, 833. (b) Noyori, R.; Kitamura, M. *Angew. Chem., Int. Ed. Engl.* **1991**, *30*, 49.
2. (a) Soai, K.; Ookawa, A.; Ogawa, K.; Kaba, T. *Chem. Commun./J. Chem. Soc., Chem. Commun.* **1987**, 467. (b) Soai, K.; Ookawa, A.; Kaba, T.; Ogawa, K. *J. Am. Chem. Soc.* **1987**, *109*, 7111
3. (a) Soai, K.; Niwa, S. *Chem. Lett.* **1989**, 481. (b) Niwa, S.; Soai, K. *J. Chem. Soc., Perkin Trans. 1* **1990**, 937.
4. Soai, K.; Hirose, Y.; Sakata, S. *Tetrahedron: Asymmetry* **1992**, *3*, 677.
5. Soai, K.; Kawase, Y. *Tetrahedron: Asymmetry* **1991**, *2*, 781.
6. Kobayashi, S.; Murakami, M.; Harada, T.; Mukaiyama, T. *Chem. Lett.* **1991**, 1341.
7. Okamoto, Y.; Nakano, T.; Ono, E.; Hatada, K. *Chem. Lett.* **1991**, 525.
8. Kitamura, M.; Suga, S.; Kawai, K.; Noyori, R. *J. Am. Chem. Soc.* **1986**, *108*, 6071.
9. (a) Soai, K.; Yokoyama, S.; Ebihara, K.; Hayasaka, T. *Chem. Commun./J. Chem. Soc., Chem. Commun.* **1987**, 1690. (b) Soai, K.; Yokoyama, S.; Hayasaka, T. *J. Org. Chem.* **1991**, *56*, 4264.
10. Soai, K.; Hirose, Y.; Niwa, S. *J. Fluorine Chem.* **1992**, *59*, 5.
11. Soai, K.; Hirose, Y.; Sakata, S. *Bull. Chem. Soc. Jpn.* **1992**, *65*, 1734.
12. (a) Soai, K.; Kawase, Y.; Niwa, S. *Heterocycles* **1989**, *29*, 2219. (b) Van Oeveren, A.; Menge, W.; Feringa, B. L. *Tetrahedron Lett.* **1989**, *30*, 6427.
13. Soai, K.; Hori, H.; Kawahara, M. *Chem. Commun./J. Chem. Soc., Chem. Commun.* **1992**, 106.
14. Soai, K.; Watanabe, M.; Koyano, M. *Chem. Commun./J. Chem. Soc., Chem. Commun.* **1989**, 534.

Kenso Soai
Science University of Tokyo, Japan

2′-(Diphenylphosphino)-*N,N*-dimethyl-[1,1′-binaphthalen]-2-amine[1]

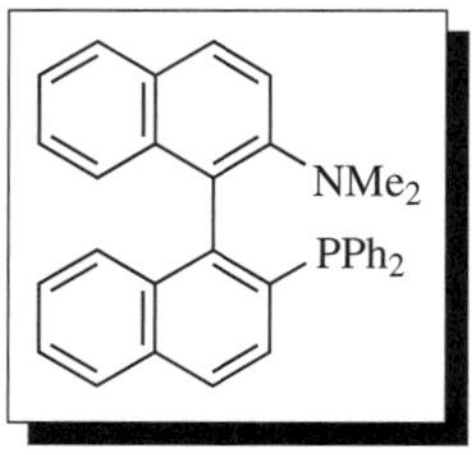

[216368-93-3], [233752-13-1] $C_{34}H_{28}NP$ (MW 481.57)

(chiral ligand[1])

Alternate Name: 2-(dimethylamino)-2′-(diphenylphosphino)-1,1′-binaphthyl; MAP.

Physical Data: (*R*)-(−) amorphous solid, $[\alpha]_D$ −19.0 (*c* 1.0, THF).[1] (*S*)-(+) amorphous solid +26.6 (*c* 1.0, THF).[2]

Solubility: (*R*)-(+) and (*S*)-(−) very well soluble in toluene, CH_2Cl_2, AcOEt, THF; well-soluble in ether; sparingly soluble in MeOH, EtOH, and hexane.

Preparative Methods: (*R*)-2′-(diphenylphosphino)-*N,N*-dimethyl[1,1′-binaphthalen]-2-amine (MAP) is conveniently prepared from the triflate of (*R*)-dimethyl-NOBIN by the Pd(0)-catalyzed coupling with $Ph_2P(O)H$ followed by reduction of the resulting phosphine oxide with Cl_3SiH (eq 1).[1] Practically identical procedure has been reported for the synthesis of (*S*)-MAP.[2] Direct coupling of the triflate with Ph_2PH was unsuccessful,[1] while the Ni(0)-catalyzed coupling with Ph_2PCl is capricious, giving 0–40% of MAP.[3]

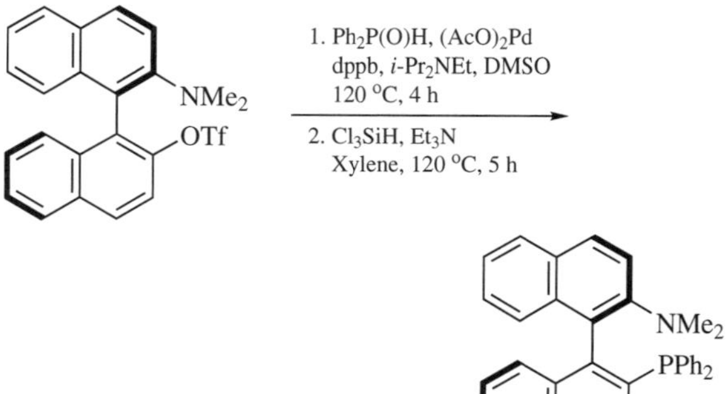

(1)

An analogous procedure, starting with NOBIN acetamide leads to desmethyl-MAP (eq 2).[1,4] A different approach to the same product relies on the Hofmann rearrangement of the corresponding amide (obtained by partial hydrolysis from the corresponding nitrile), followed by reduction of the P–O bond (eq 3).[5] Further analogues with various *N,N*-dialkyl and *P,P*-dialkyl/diaryl groups have also been described.[1,6] Their synthesis utilizes either the triflate coupling (as in eq 1) (ref 1) or the lithiation of the corresponding bromide with *t*-BuLi followed by quenching with R_2PCl.[6]

Drying: standard drying during the work up; not hygroscopic.

Handling, Storage, and Precautions: keep tightly closed, store in a cool dark place; deteriorates when exposed to direct sunshine and air.

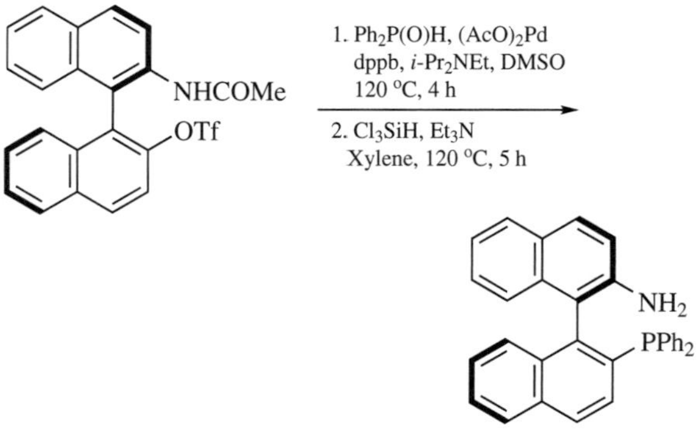

(2)

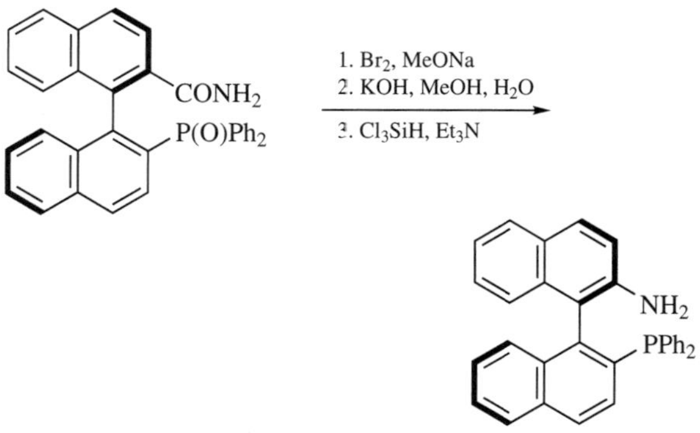

(3)

The Pd(0)-complexes of (*R*)-MAP and its *N,N*-dialkyl analogues[1] catalyze allylic substitution of allylic esters (acetates and carbonates; R′ = MeCO or MeOCO) with malonate nucleophiles (eq 4) in up to 73% ee (R = Ph).[1] Improved asymmetric induction (up to 91% ee) has been reported for H_8-MAP (5,5′,6,6′,7,7′,8,8′-octahydro-MAP), H_8-Xyl-MAP [with P(3,5-$Me_2C_6H_3$) group in place of PPh_2] (R = Ph, R″ = H),[7] and for MAP with chiral substituents on the nitrogen (86% ee).[8] MAP and H_8-Xyl-MAP are also efficient ligands when $NaN(CHO)_2$ is utilized as *N*-nucleophile (eq 5), giving up to 69% ee (note that 95% ee has been obtained in this case with BINAP as ligand).[9] Strong memory effects are observed in the case of cyclic substrates.[10]

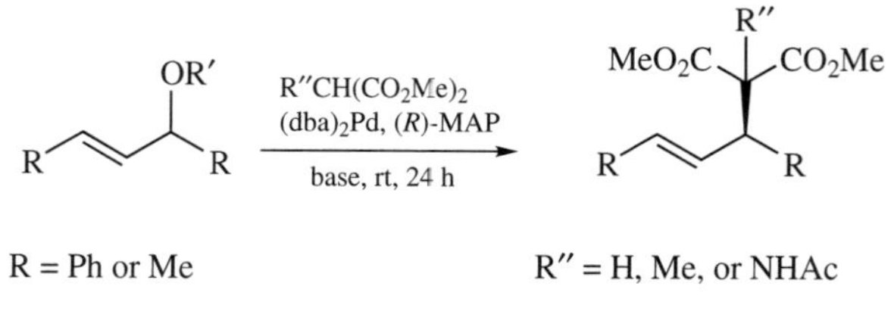

(4)

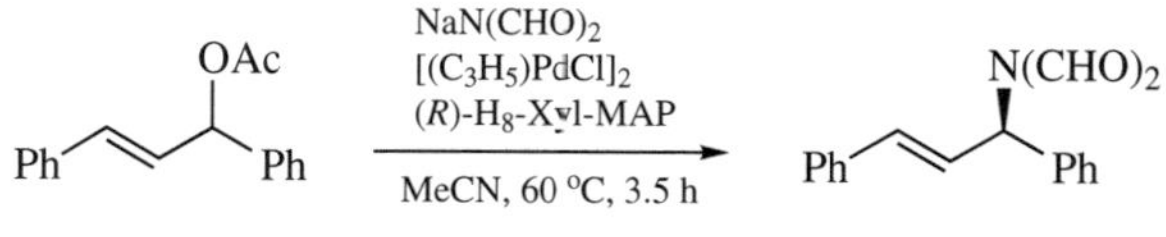

(5)

In allylic substitution and presumably in other reactions (vide supra), MAP acts as an *P,*C_{ipso}-ligand rather than *P,N*-ligand, as evidenced by NMR and X-ray crystallography (eq 6).[10,11] Strong memory effects, observed in the case of allylic substitution of cyclic substrates, are associated with this unusual coordination.[10]

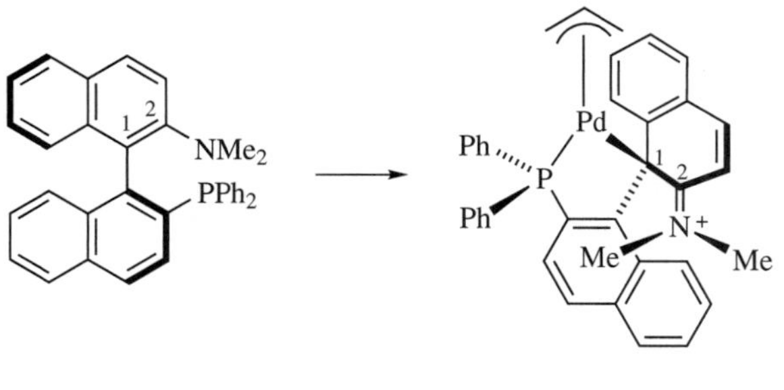

(6)

MAP and its analogues considerably accelerate the Hartwig–Buchwald amination of aromatic and heteroaromatic halides and triflates (eq 7).[1,6,11,12] Similar acceleration is observed for Suzuki–Miyaura coupling, which appears quite general, tolerating a number of functional groups (eq 8).[6,10] Further enhancement of the reaction rate is attained when the PPh_2 group in MAP is replaced by the more Lewis-basic PCy_2 group.[6]

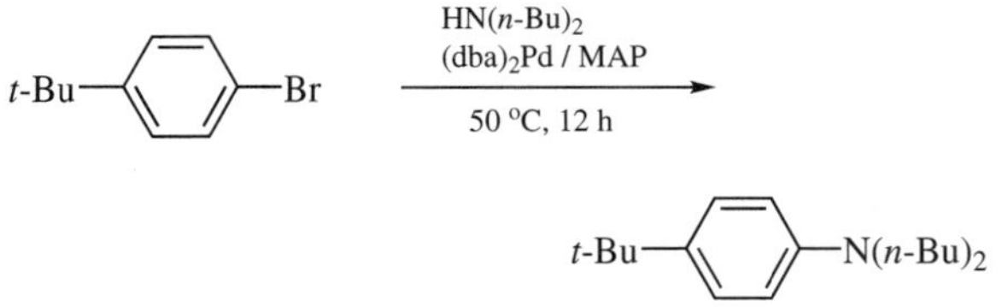

(7)

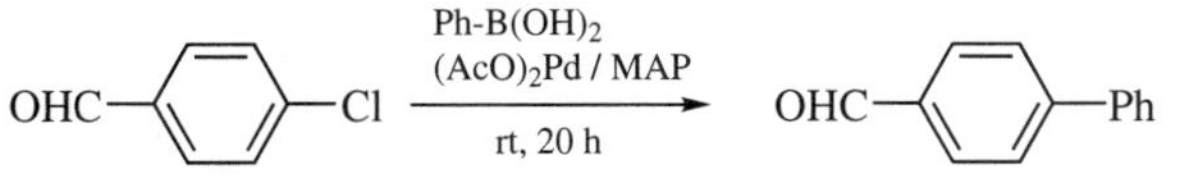

(8)

Asymmetric induction is attained for selected Suzuki–Miyaura aryl-aryl couplings (eq 9). In this case, more electron-rich MAP with PCy_2 group exhibits higher enantioselectivities (up to 87% ee) than its PPh_2 counterpart (75% ee).[13]

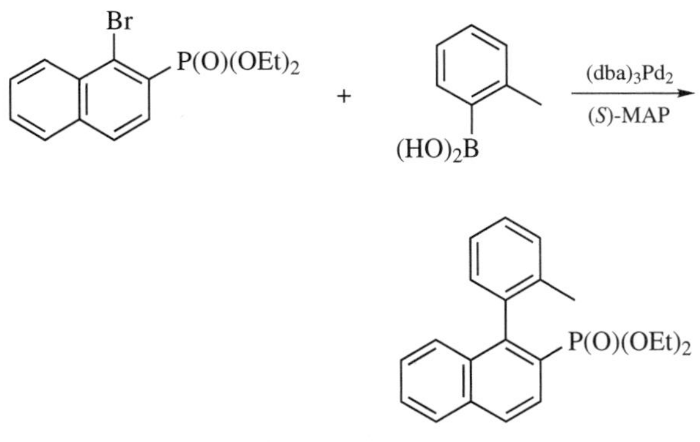

(9)

(+) < 87% ee

Pyridine amide, derived from (*S*)-desmethyl-MAP, induces high enantioselectivity in Cu-catalyzed conjugate addition of Et_2Zn to enones (eq 10).[4]

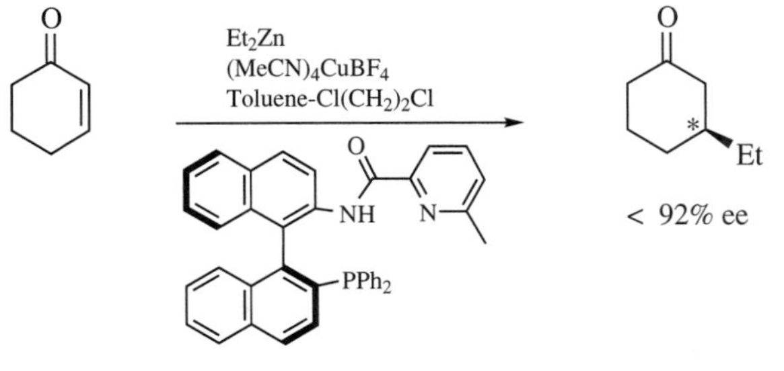

(10)

< 92% ee

MAP-type ligands also catalyze asymmetric vinylation of ketone enolates (eq 11) with 56% ee for MAP (PPh_2) and 90% ee for its PCy_2 analogue (96% ee at −20 °C).[6d]

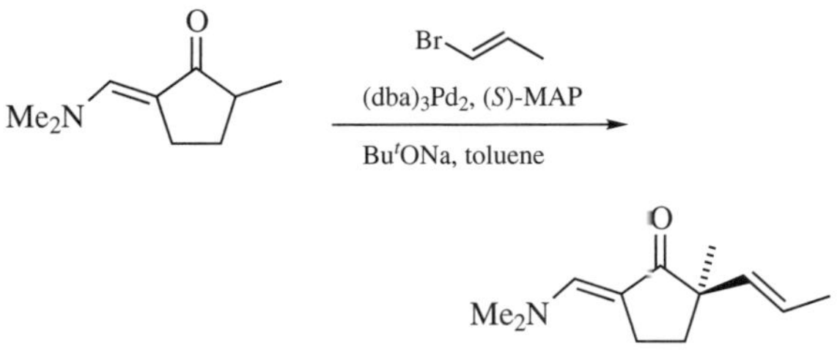

(11)

1. Vyskočil, Š.; Smrčina, M.; Hanuš, V.; Polášek, M.; Kočovský, P. *J. Org. Chem.* **1998**, *63*, 7738.
2. Ding, K.; Wang, Y.; Yun, H.; Liu, J.; Wu, Y.; Terada, M.; Okubo, Y.; Mikami, K. *Chemistry-Eur. J.* **1999**, *5*, 1734.
3. Vyskočil, Š.; Kočovský, P., unpublished results.
4. Hu, X.; Chen, H.; Zhang, X. *Angew. Chem., Int. Ed.* **1999**, *38*, 3518.
5. Sumi, K.; Ikariya, T.; Noyori, R. *Can. J. Chem.* **2000**, *78*, 697.
6. (a) Aranyos, A.; Old, D. W.; Kiyomori, A.; Wolfe, J. P.; Sadighi, J. P.; Buchwald, S. L. *J. Am. Chem. Soc.* **1999**, *121*, 4369. (b) Hu, X.; Chen, H.; Zhang, X. *Angew. Chem., Int. Ed.* **1999**, *38*, 3518. (c) Fox, J. M.; Huang, X.; Chieffi, A.; Buchwald, S. L. *J. Am. Chem. Soc.* **2000**, *122*, 1360. (d) Chieffi, A.; Kamikawa, K.; Åhman, J.; Fox, J. M.; Buchwald, S. L. *Org. Lett.* **2001**, *3*, 1897.
7. Wang, Y.; Guo, H.; Ding, K. *Tetrahedron: Asymmetry* **2000**, *11*, 4153.
8. Wang, Y.; Li, X., Ding, K. L. *Tetrahedron Lett.* **2002**, *43*, 159.
9. Wang, Y.; Ding, K. *J. Org. Chem.* **2001**, *66*, 3238.
10. Lloyd-Jones, G. C.; Stephen, S. C.; Murray, M.; Butts, C. P.; Vyskočil, Š.; Kočovský, P. *Chem. Eur. J.* **2000**, *6*, 4348.
11. Kočovský, P.; Vyskočil, Š.; Císařová, I.; Sejbal, J.; Tišlerová, I.; Smrčina, M.; Lloyd-Jones, G. C.; Stephen, S. C.; Butts, C. P.; Murray, M.; Langer, V. *J. Am. Chem. Soc.* **1999**, *121*, 7714.
12. Vyskočil, Š.; Smrčina, M.; Kočovský, P., *Tetrahedron Lett.* **1998**, *39*, 9289.
13. Yin, J. J.; Buchwald, S. L. *J. Am. Chem. Soc.* **2000**, *122*, 12051.

Pavel Kočovský
University of Glasgow, UK

(*S*)-2-[2-(Diphenylphosphino)phenyl]-4-phenyloxazoline

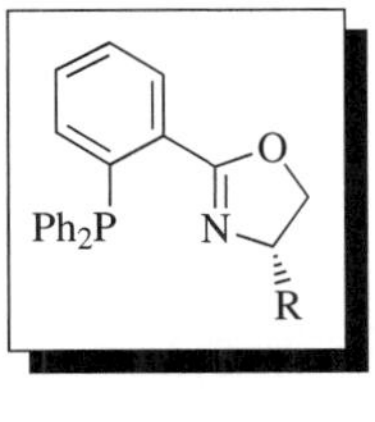

(**1**; R = Ph)		
[148461-15-8]	$C_{27}H_{22}NOP$	(MW 407.47)
(**2**; R = *i*-Pr)		
[148461-14-7]	$C_{24}H_{24}NOP$	(MW 373.46)
(**3**; R = *t*-Bu)		
[148461-16-9]	$C_{25}H_{26}NOP$	(MW 387.49)

(readily available chiral ligands for enantiocontrol of palladium-catalyzed allylic substitution reactions[1,2])

Physical Data: (**1**) amorphous solid; $[\alpha]_D$ +30.8° (c = 1.0, $CHCl_3$). (**2**) amorphous solid; $[\alpha]_D$ −44.9° (c = 1.4, $CHCl_3$). (**3**) mp 105 °C; $[\alpha]_D$ 58.2° (c = 1.2, $CHCl_3$).

Solubility: insol H_2O; sol most organic solvents.

Preparative Methods: chiral phosphinoaryloxazolines, a class of ligands developed independently in three different laboratories,[1,3,4] are readily prepared in enantiomerically pure form starting from chiral amino alcohols and aromatic carboxylic acids or nitriles. Several short, convenient syntheses from 2-bromobenzonitrile (eq 1),[1,5] 2-bromobenzoic acid (eq 2),[3] or 2-fluorobenzonitrile (eq 3)[4] have been described. Alternatively, derivatives such as (**2**), which do not contain any reactive groups in the oxazoline ring that are attacked by BuLi, can be prepared from aryloxazolines by orthometalation and subsequent reaction with Ph_2PCl (eq 4).[1] Phosphino-oxazolines with an additional stereogenic center at the phosphorus atom, such as (**4**), and phosphinomethyl oxazolines of type (**5**) have also been reported.[3] The different synthetic routes allow for a wide range of structural modifications at the oxazoline ring, the phosphine group, and the ligand backbone.

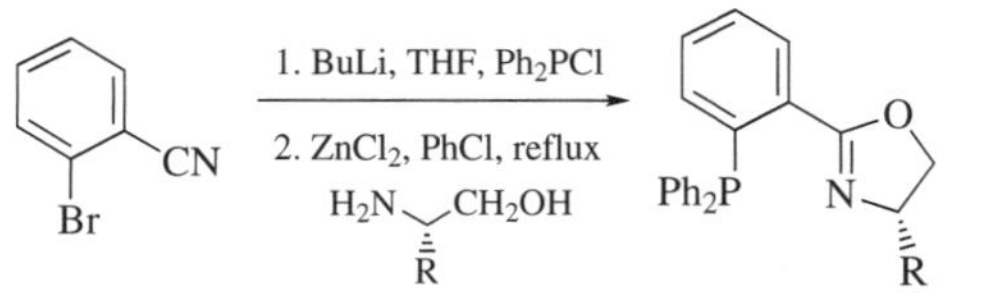

(1)

(**1**) R = Ph
(**2**) R = *i*-Pr
(**3**) R = *t*-Bu

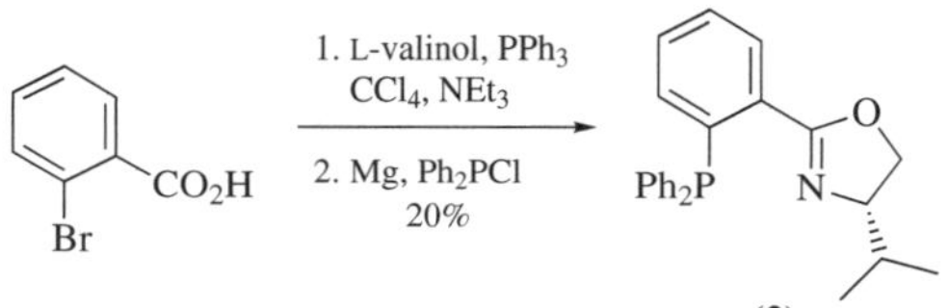

(2)

(**2**)

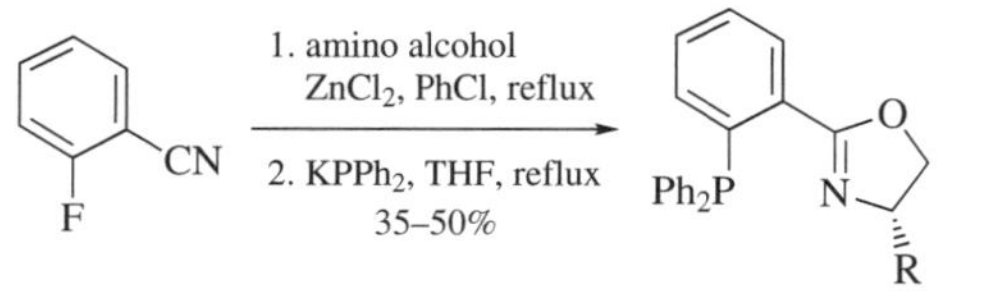

(3)

(**1**) R = Ph
(**2**) R = *i*-Pr
(**3**) R = *t*-Bu

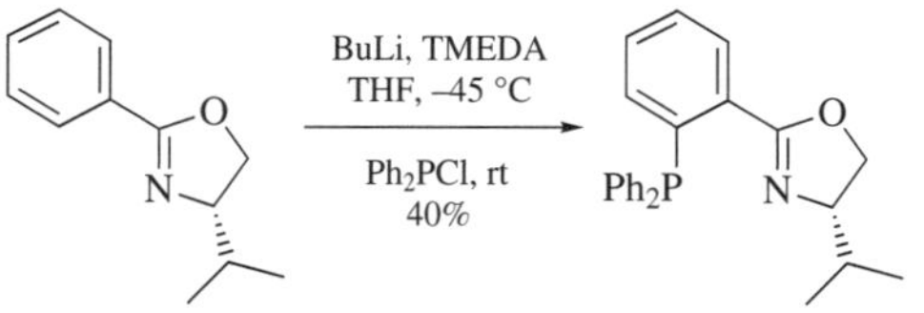

(4)

(**2**)

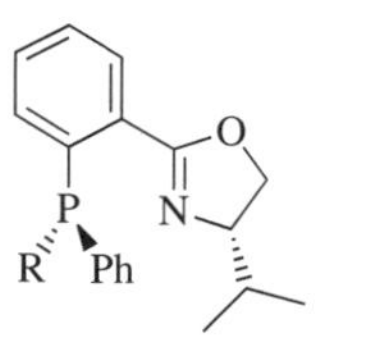

(**4**) R = 1-naphthyl

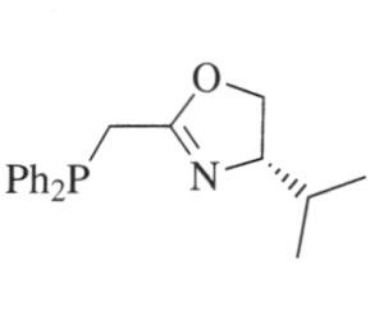

(**5**)

Purification: impurities such as phosphine oxides, formed by air oxidation, can be removed by column chromatography on silica gel or, for crystalline derivatives, by recrystallization.

Handling, Storage, and Precautions: Phosphino oxazolines of this type are sufficiently stable to be handled in air. For longer periods of time, they should be stored at −20 °C under argon.

Palladium-Catalyzed Allylic Substitution. Palladium complexes of chiral phosphino oxazolines are highly effective catalysts for enantioselective allylic substitution reactions.[1–4] The catalysts are usually prepared in situ from *Bis(allyl)di-μ-chlorodipalladium* and the corresponding ligands. In the presence of 1–2 mol % of catalyst and a mixture of *N,O-Bis(trimethylsilyl)acetamide* (BSA) and catalytic amounts of KOAc as a base, racemic 1,3-diphenyl-2-propenyl acetate reacts smoothly at rt with dimethyl malonate to afford the substitution product in essentially quantitative yield and with excellent enantioselectivity (eq 5). For this substrate, the phenyloxazoline (**1**) is the optimal ligand.[1] The observed ee of 99% exceeds the selectivities previously obtained with other ligands such as ferrocenyl phosphines,[6] chiraphos,[7] BINAP,[7] 5-azasemicorrins,[8] or bis(oxazolines).[8,9] The corresponding isopropyl- and *t*-butyl-oxazolines (**2**) and (**3**) afford slightly lower enantiomeric excesses (98 and 95% ee, respectively).[1,3] Acetylacetonate[1] and diethyl acetaminomalonate,[1] as well as *N*-nucleophiles such as benzylamine, *p*-toluenesulfonamide, benzoylhydrazine, and $(Boc)_2NNa$,[10] also react with excellent enantioselectivity (eq 6).

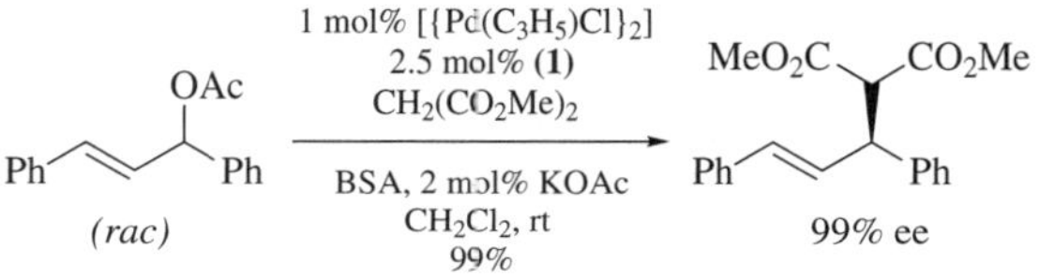

(5)

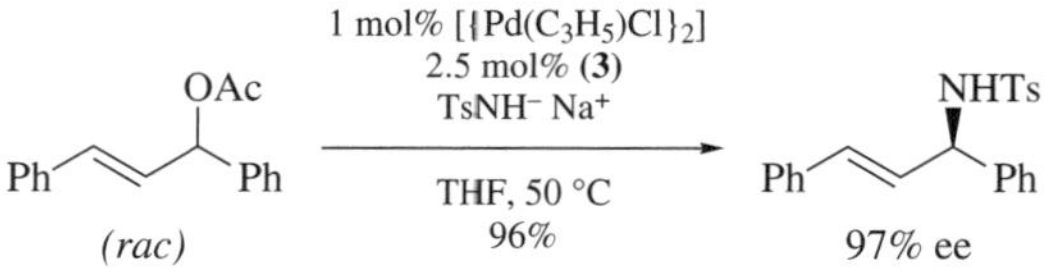

(6)

Moderate to high selectivities have been observed in allylic alkylations of 1,3-dialkyl-2-propenyl acetates.[1] Here, the *t*-butyloxazoline derivative (**3**) is the ligand of choice. Under standard conditions, *n*-alkyl-substituted allylic acetates smoothly react at rt with selectivities of 70–80% ee (eq 7). Similar results have been obtained with various *N*-nucleophiles.[10] The corresponding diisopropylallyl acetate is much less reactive, but under more vigorous conditions with $NaCH(CO_2Me)_2$ in DMF at 65 °C the reaction proceeds in good yield with high enantioselectivity (eq 8).[1] Analogous reactions of this substrate with *N*-nucleophiles are impracticably slow. In this case, the corresponding diethyl phosphate gives much better results (eq 9).[10]

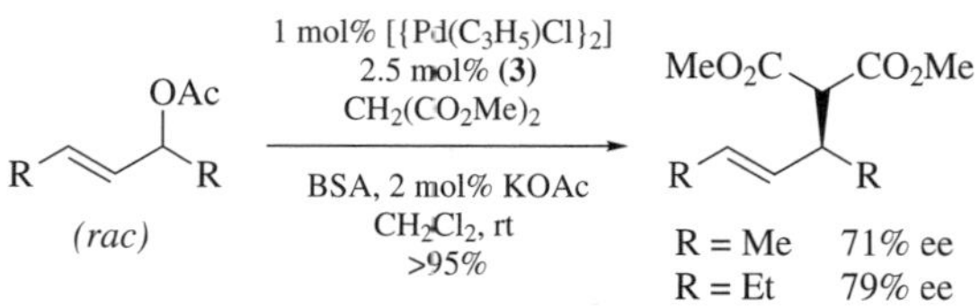

(7)

R = Me 71% ee
R = Et 79% ee
R = *n*-Pr 69% ee

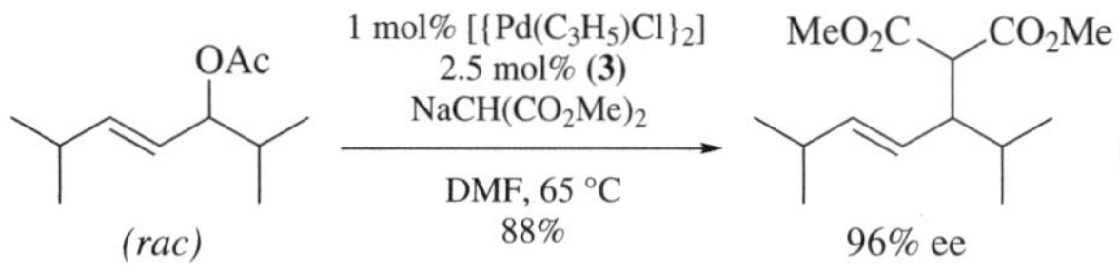

(8)

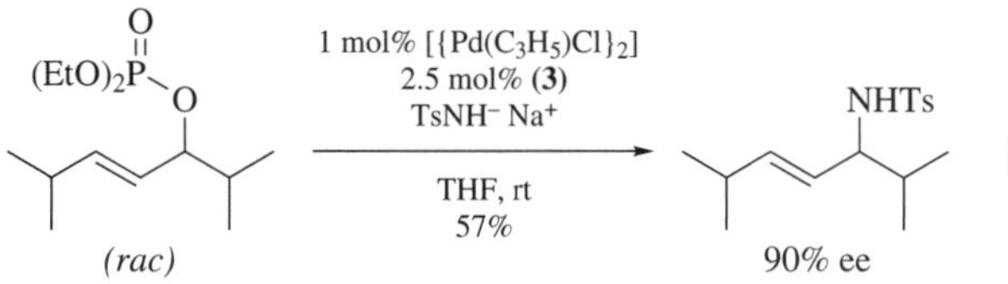

(9)

Related Reagents. 2,2′-Bis(diphenylphosphino)-1,1′-binaphthyl; 2,3-Bis(diphenylphosphino)butane; (*R*)-*N*-[2-(*N*,*N*-Dimethylamino)ethyl]-*N*-methyl-1-[(*S*)-1′,2-bis(diphenylphosphino)-ferrocenyl]ethylamine; (1S,9S)-1,9-{Bis[(*t*-butyl)dimethylsilyl-oxy]methyl}-5-cyanosemicorrin; (*R*)-*N*,*N*-Dimethyl-1-[(*S*)-2-(diphenylphosphino)ferrocenyl]ethylamine.

---

1. von Matt, P.; Pfaltz, A. *Angew. Chem.* **1993**, *105*, 614; *Angew. Chem., Int. Ed. Engl.* **1993**, *32*, 566.
2. Reiser, O. *Angew. Chem.* **1993**, *105*, 576; *Angew. Chem., Int. Ed. Engl.* **1993**, *32*, 547.
3. Sprinz, J.; Helmchen, G. *Tetrahedron Lett.* **1993**, *34*, 1769.
4. Dawson, G. J.; Frost, C. G.; Williams, J. M. J.; Coote, S. J. *Tetrahedron Lett.* **1993**, *34*, 3149.
5. von Matt, P. Dissertation, University of Basel, 1993.
6. Hayashi, T. *Pure Appl. Chem.* **1988**, *60*, 7.
7. Yamaguchi, M.; Shima, T.; Yamagishi, T.; Hida, M. *Tetrahedron Lett.* **1990**, *31*, 5049; *Tetrahedron: Asymmetry* **1991**, *2*, 663.
8. Leutenegger, U.; Umbricht, G.; Fahrni, Ch.; von Matt, P.; Pfaltz, A. *Tetrahedron* **1992**, *48*, 2143.
9. Pfaltz, A. *Acc. Chem. Res.* **1993**, *26*, 339.
10. von Matt, P.; Loiseleur, O.; Koch, G.; Pfaltz, A. Lefeber, C.; Feucht, T.; Helmchen, G. *Tetrahedron: Asymmetry* **1994**, *5*, 573.

Andreas Pfaltz
University of Basel, Switzerland

α,α-Diphenyl-2-pyrrolidinemethanol[1]

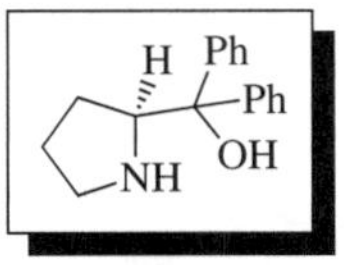

(*S*)		
[112068-01-6]	$C_{17}H_{19}NO$	(MW 253.37) (*R*)
[22348-32-9]		
((*S*) ·HCl)		
[16226-54-3]	$C_{17}H_{20}ClNO$	(MW 289.83)

(precursor to several chiral oxazaborolidine catalysts[1] used for the enantioselective reduction of prochiral ketones[2,3])

Alternate Name: diphenylprolinol.

Physical Data: mp 79–79.5 °C (hexane); 80–82 °C (EtOH). $[\alpha]_{589}$ −54.3° (*c* 0.261, MeOH); −68.1° (*c* 3.17, $CHCl_3$) for the (*S*)-enantiomer.

Solubility: very sol THF, CH_2Cl_2, MeOH, toluene.

Preparative Methods: addition of *Phenylmagnesium Bromide* to (*S*)-proline-*N*-carboxyanhydride (73% overall from proline).[4] Addition of *N*-benzyl-(*S*)-proline ethyl ester to phenylmagnesium chloride followed by catalytic hydrogenolysis (49% overall from proline).[5] Addition of *N*-(benzyloxycarbonyl)-(*S*)-proline methyl ester to phenylmagnesium chloride (50% overall yield from proline).[3a,c] Addition of phenylmagnesium bromide to *N*-(ethyloxycarbonyl)-(*S*)-proline methyl ester followed by alkaline hydrolysis (65% overall yield from proline).[6] Addition of (*S*)-proline ethyl ester hydrochloride to phenylmagnesium chloride (20–26% overall yield from proline, ca. 80% ee).[7] Enantioselective deprotonation of *N*-Boc-pyrrolidine with *s-Butyllithium/(−)-Sparteine* followed by reaction with benzophenone to give (*R*)-diphenylprolinol (63% yield from pyrrolidine).[8] Addition of phenylmagnesium chloride to methyl pyroglutamate followed by reduction with borane to give racemic diphenylprolinol (51% yield) which can be resolved as its *O*-acetylmandelate salt to give the (*R*)- and (*S*)-enantiomers (30% yield from the racemate).[3b] Addition of lithiated *N*-nitrosopyrrolidine to benzophenone to give racemic diphenylprolinol (58–60% yield based on benzophenone).[9]

Purification: recrystallization from hexane, ethanol, or methanol/water.

Handling, Storage, and Precautions: no special information available. In general, however, it is advisable that all reactions with this reagent be conducted in a well ventilated fume hood. Care should be exercised to avoid contact of this reagent and the derived oxazaborolidine catalyst with the eyes and skin.

Enantioselective Ketone Reduction. Following Itsuno's lead for enantioselective reductions using diphenylvalinol,[10] Kraatz was the first to describe the use of a 1:2 mixture of (*S*)-diphenylprolinol (**1**) and *Borane–Tetrahydrofuran* for the stoichiometric enantioselective reduction of ketone (**2**) to obtain the plant growth regulator triapenthenol (**3**) (eq 1).[2] Although not characterized at the time, the species responsible for the enantioselectivity observed was presumed to be an oxazaborolidine–borane complex.[10b]

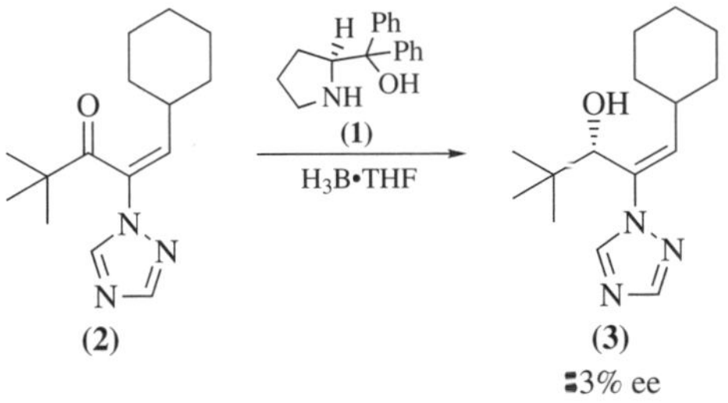

(1)

Diphenylprolinol (**1**) and borane–THF react in a multistep process to give the unsubstituted oxazaborolidine–borane complex (**7**) (eq 2). Formation of amine–borane complex (**4**) is exothermic, and this intermediate can be isolated as a stable crystalline solid.[4b] Subsequent conversion to oxazaborolidine (**6**) requires heating the THF solution under pressure (1.7 bar) at 70–75 °C for 48–72 h. Corey isolated and characterized free oxazaborolidine (**6**) as a solid (mp 107–124 °C), which was reported to be a mixture of monomer and dimer (NMR).[3a] Finally, addition of borane–THF to a solution of oxazaborolidine (**6**) affords oxazaborolidine–borane complex (**7**) which was not isolated and was identified based on ^{11}B NMR evidence.

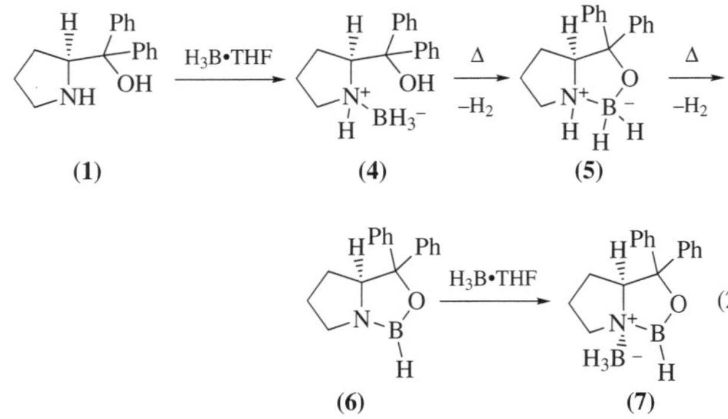

(2)

Corey demonstrated that oxazaborolidine (**6**) can be used catalytically (2.5–100 mol %) with excess borane (60–200 mol %) for the enantioselective reduction of prochiral ketones (eq 3 Table 1).[3a,c]

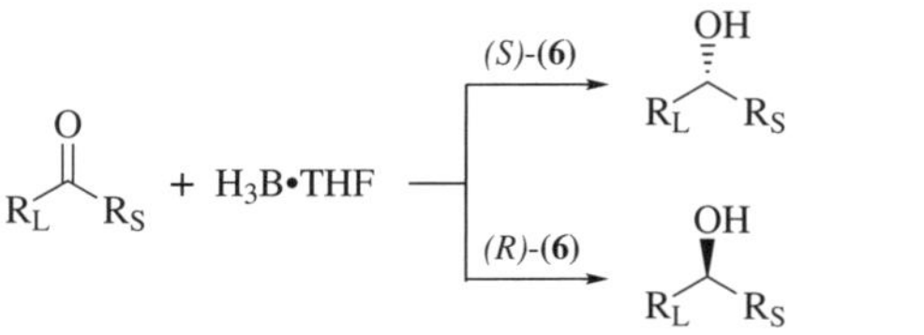

(3)

Table 1 Catalytic Enantioselective Reduction of Ketones using Oxazaborolidine (*S*)-(**6**)

Ketone	$H_3B{\cdot}THF$ (mol %)	Catalyst (*S*)-(**6**) (mol %)	Enantiomeric purity[a]	Absolute configuration
PhCOMe	100	10	97%	(*R*)
PhCOEt	60	5	90%	(*R*)
t-BuCOMe	60	10	92%	(*R*)
1-Tetralone	60	5	89%	(*R*)
$PhCOCH_2Cl$	60	5	97%	(*S*)

[a] Enantiomeric purities determined by capillary GC analysis of the (−)-menthyloxycarbonyl derivative, and have been corrected by adding 1% to the experimentally observed values to correspond to values for optically pure catalyst.[3a]

The following catalytic cycle was proposed to explain the behavior of the oxazaborolidine catalyst (eq 4).[3a,11] Oxazaborolidine (**6**) reacts with borane to give oxazaborolidine–borane complex (**7**). Coordination between the Lewis acidic ring boron and the carbonyl oxygen activates the ketone toward reduction. Intramolecular hydride transfer from the BH_3 coordinated to the ring nitrogen then occurs via a six-membered ring chair transition state.[12] Following hydride transfer, the alkoxy-BH_2 dissociates, and oxazaborolidine (**6**) is free to begin the cycle again. For a more detailed discussion, see the entry for *Tetrahydro-1-methyl-3,3-diphenyl-1H,3H-pyrrolo[1,2-c][1,3,2]oxazaborole*.

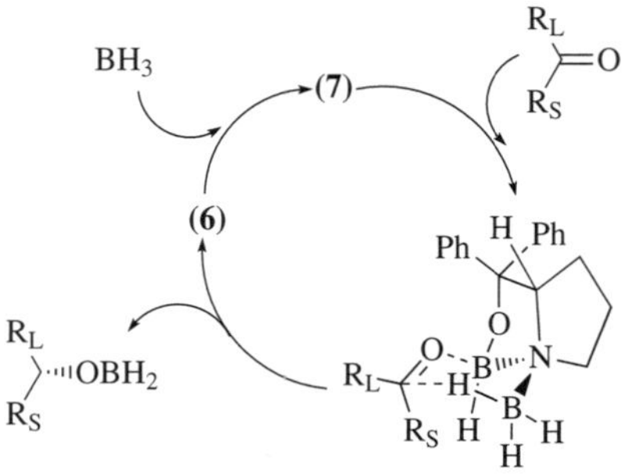

(4)

Problems with the preparation and stability of oxazaborolidine (**6**) led to the development of a series of *B*-substituted oxazaborolidines derived from diphenylprolinol. The *B*-methyl substituted oxazaborolidine (**9a**) was first prepared (eq 5) by reaction of diphenylprolinol (**1**) with methylboronic acid under dehydrating conditions (toluene at 23 °C in the presence of 4 Å molecular sieves or toluene at reflux using a Dean–Stark trap) followed by vacuum distillation (0.1 mmHg, 170 °C).[3a,c] Based on NMR evidence, the product (mp 74–87 °C) was reported to be a mixture of monomer and dimer.[3a] The corresponding *B*-butyloxazaborolidine (**9c**), prepared in a similar manner from *n*-butylboronic acid, was also reported to be a mixture of monomer and dimer.[13] Subsequent investigations demonstrated that the reported 'dimers' were in fact the intermediate (**8**) and the more stable disproportionation product (**10**) (eq 6).[4] Furthermore, the presence of (**8**) or (**10**) was demonstrated to be deleterious to the enantioselectivity of the catalyst.[14]

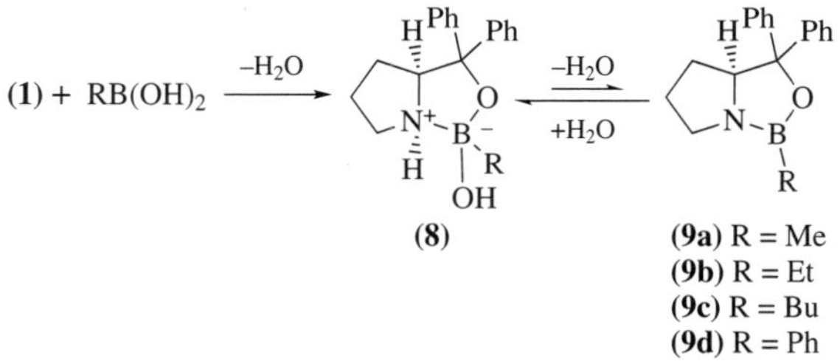

(5)

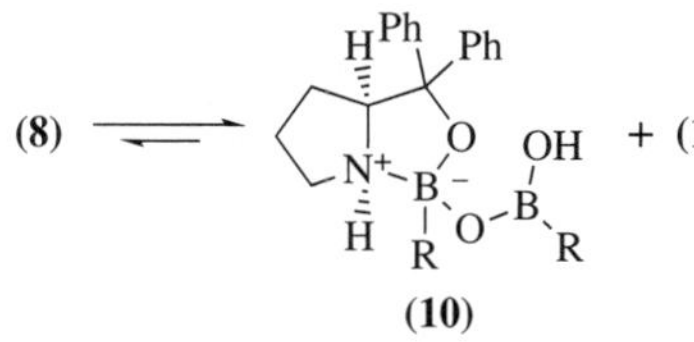

(6)

A small-scale procedure, based on the reaction of bis(trifluoroethyl) alkylboronates with diphenylprolinol (eq 7) was reported for the preparation of the *B*-ethyl- (**9b**) and *B*-butyl- (**9c**) oxazaborolidines.[15] The bis(trifluoroethyl) alkylboronates were prepared from the disproportionation of tris(trifluoroethyl) borate and the corresponding trialkylborane. Since trimethylborane is not commercially available, this procedure is not applicable for the preparation of *B*-methyloxazaborolidine (**9a**).

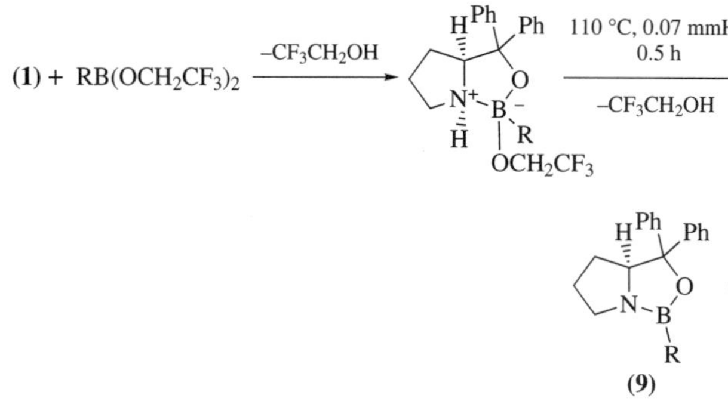

(7)

A practical large-scale process for the synthesis of *B*-methyloxazaborolidine (**9a**) was developed using commercially available trimethylboroxine (eq 8), which affords the product as an analytically pure, colorless crystalline solid (mp 79–81 °C).[4,16] The *B*-ethyl- (**9b**), *B*-butyl- (**9c**), and *B*-phenyloxazaborolidines (**9d**) were also prepared from the corresponding triethyl-, tributyl-, or triphenylboroxine. The free oxazaborolidines, thus prepared, are stable if rigorously protected from moisture.

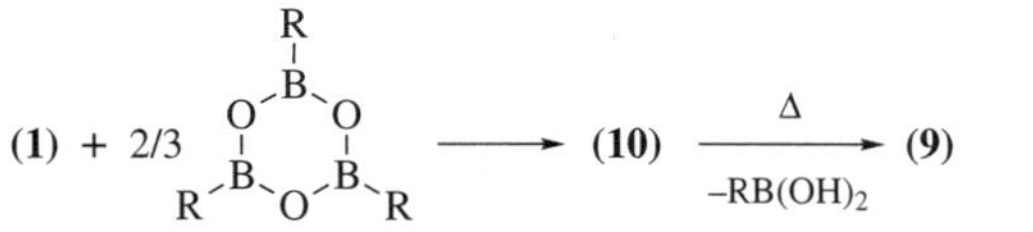

(8)

A significantly more stable form of the catalyst is the crystalline oxazaborolidine–borane complex (**11**).[4b,16,17] This borane complex is readily prepared from oxazaborolidine (**9a**) and *Borane–Dimethyl Sulfide* complex (BMS) (eq 9).

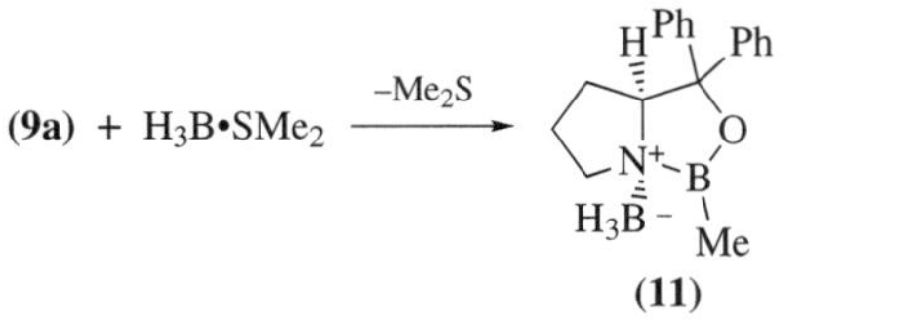

(9)

The enantioselectivities reported for the reduction of acetophenone and α-tetralone using the different catalysts (5–10 mol %) and borane–THF or BMS are summarized in Table 2. The best results are obtained by slowly adding the substrate (neat or as a solution in dichloromethane) to a solution of borane complex (**11**) (5 mol %) and BMS (0.6–1.0 mol equiv) in dichloromethane at −20 °C.[16]

Table 2 Enantioselective Reduction of Acetophenone and α-Tetralone

Catalyst	Prep.	mol %	Acetophenone	α-Tetralone
(**9a**) (R = Me)	eq 5	10	96.5% ee	83.3% ee
(**9a**) (R = Me)	eq 8	10	98% ee	94% ee
(**9b**) (R = Et)	eq 7	10	96% ee	–
(**9c**) (R = Bu)	eq 7	10	96% ee	–
(**9d**) (R = Ph)	eq 8	10	72% ee	94% ee
(**11**)	eqs 8 and 9	5	97.6% ee	99.2% ee
(**11**)	eqs 8 and 9	100	>99.8% ee	99.2% ee

In addition to these simple examples, oxazaborolidines derived from diphenylprolinol have been used as enantioselective catalysts for the preparation of prostaglandins,[3a] PAF antagonists,[3a] a key intermediate of ginkgolide B,[18] a key intermediate of forskolin,[19] (*R*)- and (*S*)-fluoxetine,[20] (*R*)- and (*S*)-isoproterenol,[21] vitamin D analogs,[22] the carbonic anhydrase inhibitor MK-0417,[14] the dopamine D1 agonist A-77636,[23] taxol,[24] the LTD_4 antagonists L-695,499 and L-699,392,[25] the β-adrenergic agonist CL 316,243,[26] and MK-0499.[27] Recently, Bringmann employed oxazaborolidines (**9a**) and (**9c**) to catalyze the atropo-enantioselective ring opening of achiral biaryl lactones (eq 10).[28]

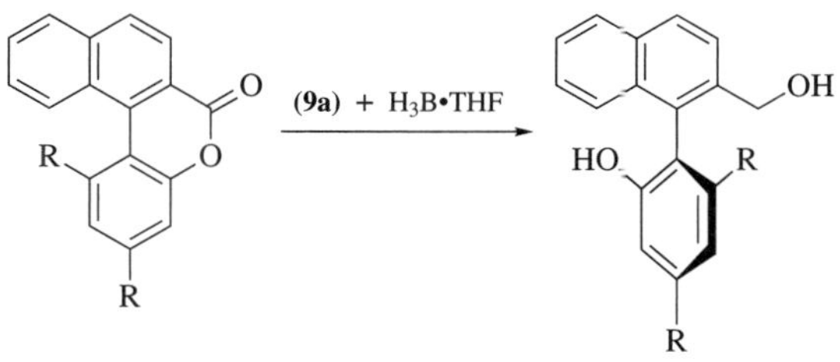

(10)

R = OMe, 94% ee
R = Me, 97% ee

Related Reagents. 2-Amino-3-methyl-1,1-diphenyl-1-butanol; Ephedrine-borane; Norephedrine–Borane; Tetrahydro-1-methyl-3,3-diphenyl-1*H*,3*H*-pyrrolo[1,2-*c*][1,3,2]oxazaborole.

1. (a) Wallbaum S.; Martens, J. *Tetrahedron: Asymmetry* **1992**, *3*, 1475. (b) Singh, V. K. *Synthesis* **1992**, 607. (c) Deloux L.; Srebnik M. *Chem. Rev.* **1993**, *93*, 763.
2. Kraatz, U. Ger. Patent 3 609 152, 1986 (*Chem. Abstr.* **1978**, *108*, 56 111c).
3. (a) Corey, E. J.; Bakshi, R. K.; Shibata, S. *J. Am. Chem. Soc.* **1987**, *109*, 5551. (b) Corey, E. J.; Bakshi, R. K.; Shibata, S.; Chen, C. P.; Singh, V. K. *J. Am. Chem. Soc.* **1987**, *109*, 7925. (c) Corey, E. J.; Shibata, S.; Bakshi, R. K. *J. Org. Chem.* **1988**, *53*, 2861.
4. (a) Mathre, D. J.; Jones, T. K.; Xavier, L. C.; Blacklock, T. J.; Reamer, R. A.; Mohan, J. J.; Jones, E. T. T.; Hoogsteen, K.; Baum, M. W.; Grabowski, E. J. J. *J. Org. Chem.* **1991**, *56*, 751. (b) Blacklock, T. J.; Jones, T. K.; Mathre, D. J.; Xavier, L. C. U.S. Patent 5 039 802, 1991. (c) Blacklock, T. J.; Jones, T. K.; Mathre, D. J.; Xavier, L. C. U.S. Patent 5 264 585, 1993.
5. Enders, D.; Kipphardt, H.; Gerdes, P.; Brena-Valle, L. J.; Bhushan, V. *Bull. Soc. Chim. Belg.* **1988**, *97*, 691.
6. Kanth, J. V. B.; Periasamy, M. *Tetrahedron* **1993**, *49*, 5127.
7. (a) Kapfhammer, J.; Matthes, A. *Hoppe-Seylers Z. Physiol. Chem.* **1933**, *223*, 43. (b) Roussel-Uclaf Fr. Patent 3638M (*Chem. Abstr.* **1969**, *70*, 106 375m).
8. Kerrick, S. T.; Beak, P. *J. Am. Chem. Soc.* **1991**, *113*, 9708.
9. (a) Seebach, D.; Enders, D.; Renger, B. *Ber. Dtsch. Chem. Ges./Chem. Ber.* **1977**, *110*, 1852. (b) Enders, D.; Pieter, R.; Renger, B. Seebach, D. *Org. Synth., Coll. Vol.* **1988**, *6*, 542.
10. (a) Itsuno, S.; Ito, K.; Hirao, A.; Nakahama, S. *Chem. Commun./J. Chem. Soc., Chem. Commun.* **1983**, 469. (b) Itsuno, S.; Hirao, A.; Nakahama, S.; Yamazaki, N. *J. Chem. Soc., Perkin Trans. 1* **1983**, 1673. (c) Itsuno, S.; Ito, K.; Hirao, A.; Nakahama, S. *J. Org. Chem.* **1984**, *49*, 555. (d) Itsuno, S.; Nakano, M.; Miyazaki, K.; Masuda, H.; Ito, K.; Hirao, A.; Nakahama, S. *J. Chem. Soc., Perkin Trans. 1* **1985**, 2039. (e) Itsuno, S.; Nakano, M.; Ito, K.; Hirao, A.; Owa, M.; Kanda, N.; Nakahama, S. *J. Chem. Soc., Perkin Trans. 1* **1985**, 2615.
11. Evans, D. A. *Science* **1988**, *240*, 420.
12. Jones, D. K.; Liotta, D. C.; Shinkai, I.; Mathre, D J. *J. Org. Chem.* **1993**, *58*, 799.
13. Corey, E. J.; Bakshi, R. K. *Tetrahedron Lett.* **1990**, *31*, 611.
14. (a) Jones, T. K.; Mohan, J. J.; Xavier, L. C.; Blacklock, T. J.; Mathre, D. J.; Sohar, P.; Jones, E. T. T.; Reamer, R. A.; Roberts, F. E.; Grabowski, E. J. J. *J. Org. Chem.* **1991**, *56*, 763. (b) Shinkai, I. *J. Heterocycl. Chem.* **1992**, *29*, 627.

15. Corey, E. J.; Link, J. O. *Tetrahedron Lett.* **1992**, *33*, 4141.
16. (a) Mathre, D. J.; Thompson, A. S.; Douglas, A. W.; Hoogsteen, K.; Carroll, J. D.; Corley, E. G.; Grabowski, E. J. J. *J. Org. Chem.* **1993**, *58*, 2880. (b) Blacklock, T. J.; Jones, T. K.; Mathre, D. J.; Xavier, L. C. U.S. Patent 5 189 177, 1993. (c) Carroll, J. D.; Mathre, D. J.; Corley, E. G.; Thompson, A. S. U.S. Patent 5 264 574, 1993.
17. Corey, E. J.; Azimioara, M.; Sarshar, S. *Tetrahedron Lett.* **1992**, *33*, 3429.
18. Corey, E. J.; Gavai, A. V. *Tetrahedron Lett.* **1988**, *29*, 3201.
19. Corey, E. J.; Jardine, P. D. S.; Mohri, T. *Tetrahedron Lett.* **1988**, *29*, 6409.
20. Corey, E. J.; Reichard, G. A. *Tetrahedron Lett.* **1989**, *30*, 5207.
21. Corey, E. J.; Link, J. O. *Tetrahedron Lett.* **1990**, *31*, 601.
22. (a) Kabat, M.; Kiegiel, J.; Cohen, N.; Toth, K.; Wovkulich, P. M.; Uskokovic, M. R. *Tetrahedron Lett.* **1991**, *32*, 2343. (b) Lee, A. S.; Norman, A. W.; Okamura, W. H. *J. Org. Chem.* **1992**, *57*, 3846.
23. DeNinno, M. P.; Perner, R. J.; Morton, H. E.; DiDomenico, S., Jr. *J. Org. Chem.* **1992**, *57*, 7115.
24. Nicolaou, K. C.; Hwang, C.-K.; Sorensen, E. J.; Clairborne, C. F. *Chem. Commun./J. Chem. Soc., Chem. Commun.* **1992**, 1117.
25. (a) Labelle, M.; Prasit, P.; Belley, M.; Blouin, M.; Champion, E.; Charette, L.; DeLuca, J. G.; Dufresne, C.; Frenette, R.; Gauthier, J. Y.; Grimm, E.; Grossman, S. J.; Guay, D.; Herold, E. G.; Jones, T. R.; Lau, Y.; Leblanc, Y.; Leger, S.; Lord, A.; McAuliffe, M.; McFarlane, C.; Masson, P.; Metters, K. M.; Ouimet, N.; Patrick, D. H.; Perrier, H.; Piechuta, H.; Roy, P.; Williams, H.; Wang, Z.; Xiang, Y. B.; Zamboni, R. J.; Ford-Hutchinson, A. W.; Young, R. N. *Bioorg. Med. Chem. Lett.* **1992**, *2*, 1141. (b) King, A. O.; Corley, E. G.; Anderson, R. K.; Larsen, R. D.; Verhoeven, T. R.; Reider, P. J.; Xiang, Y. B.; Belley, M.; Leblanc, Y.; Labelle, M.; Prasit, P.; Zamboni, R. J. *J. Org. Chem.* **1993**, *58*, 3731.
26. Bloom, J. D.; Dutia, M. D.; Johnson, B. D.; Wissner, A.; Burns, M. G.; Largis, E. E.; Dolan, J. A.; Claus, T. H. *J. Med. Chem.* **1992**, *35*, 3081.
27. Cai, D.; Tschaen, D.; Shi, Y.-J.; Verhoeven, T. R.; Reamer, R. A.; Douglas, A. W. *Tetrahedron Lett.* **1993**, *34*, 3243.
28. Bringmann, G.; Hartung, T. *Angew. Chem., Int. Ed. Engl.* **1992**, *31*, 761.

David J. Mathre & Ichiro Shinkai
Merck Research Laboratories, Rahway, NJ, USA

(*S*)-3,3-Diphenyl-1-[trimethylsilylmethyl] tetrahydro-1*H*,3*H*-pyrrolo[1,2-c][1,3,2] oxazaborole[1]

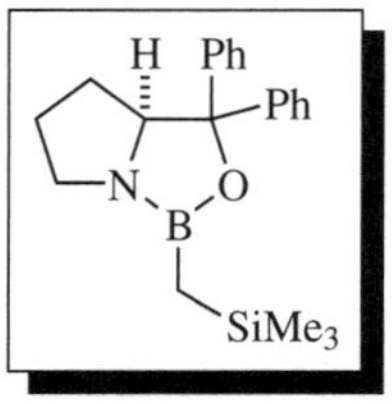

[174004-13-8] $C_{21}H_{28}BNOSi$ (MW 349.35)

(enantioselective carbonyl reduction[1])

Solubility: soluble in most organic solvents but the reactions are typically carried out in CH_2Cl_2.
Form Supplied in: colorless oil, not commercially available.
Analysis of Reagent Purity: NMR (^1H, ^{13}C, ^{11}B).
Preparative Methods: the reagent is easily prepared from trimethylsilylmethylboronic acid[2,3] and (*S*)-(−)-α,α-diphenyl-2-pyrrolidinemethano 1.[3,4] The other enantiomer is also readily available from (*R*)-(+)-α,α-diphenyl-2-pyrrolidinemethanol.
Purification: not easily purified since the reagent hydrolyzes slowly in the presence of moisture and slowly oxidizes in the presence of oxygen.
Handling, Storage, and Precautions: the reagent is stable when stored under an inert atmosphere. The catalyst is usually stored as a 0.2 M solution in toluene but the toluene is usually removed before the reaction solvent is added.

Enantioselective Reduction of α,β-Ynones. Oxazaborolidine ligands **1** are among the most effective catalysts for the enantioselective reduction of ketones to secondary alcohols.[3] Substitution of the methyl or butyl group on boron by a trimethylsilylmethyl group led to a much improved catalyst for the catecholborane mediated reduction of α,β-ynones. For example, the enantioselectivities for the reduction of an α,β-ynone was improved from 60% to 98.5% when the nature of the R group was modified (eq 1).[3]

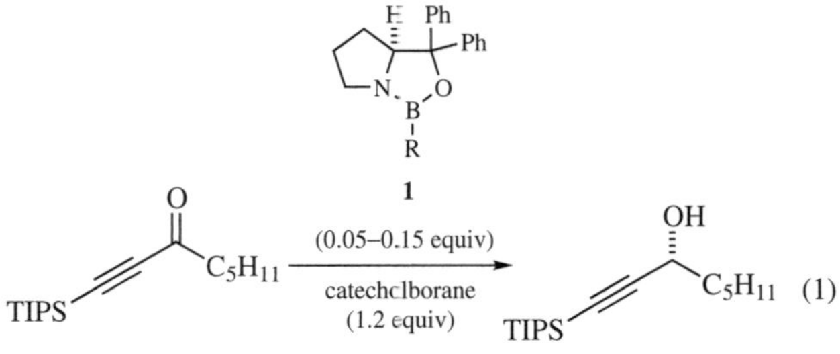

R = Me (**1a**)	60% ee
R = *n*-Bu (**1b**)	92% ee
R = CH_2SiMe_3 (**1c**)	97% ee

The level of enantioselectivity is quite dependant on the nature of the ketone and on the alkyne substituents. Sterically bulky substituents at the alkyne position usually give higher enantioselectivities (eq 2), whereas ketones bearing long alkyl chains are also reduced with higher enantiocontrol (eq 3).

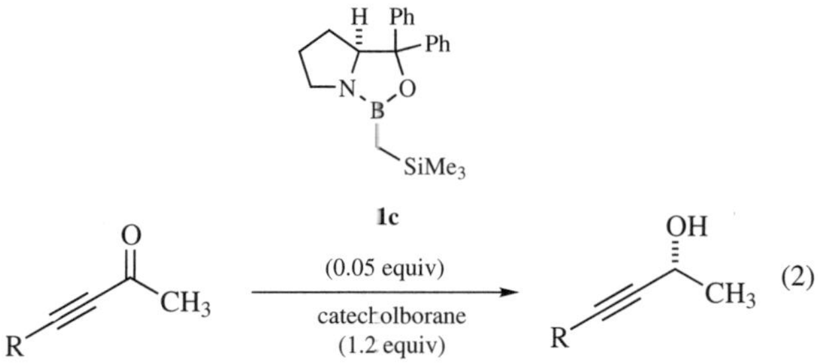

R = TMS	87% ee (92%)
R = TIPS	96% ee (95%)

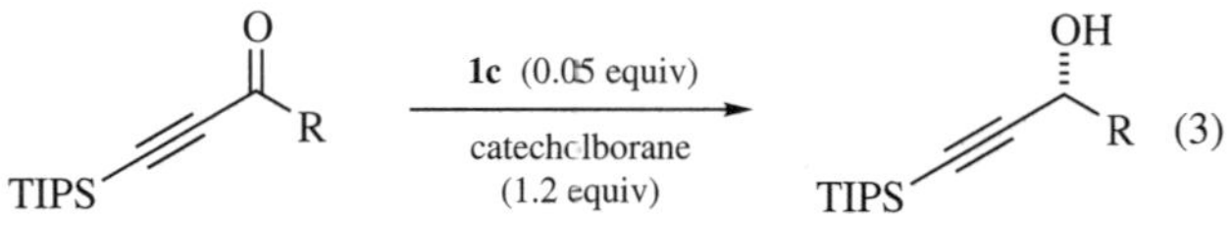

R = Me	95% ee
R = $CH_2(CH_2)_3CH_3$	97% ee
R = $CH_2CH_2CO_2Et$	90% ee

However, it would appear that further substitution at the α-position is detrimental for the level of enantioselection. In those cases, the use of the butyl-substituted catalyst resulted in a substantial improvement in the enantioselectivity (eq 4).[5]

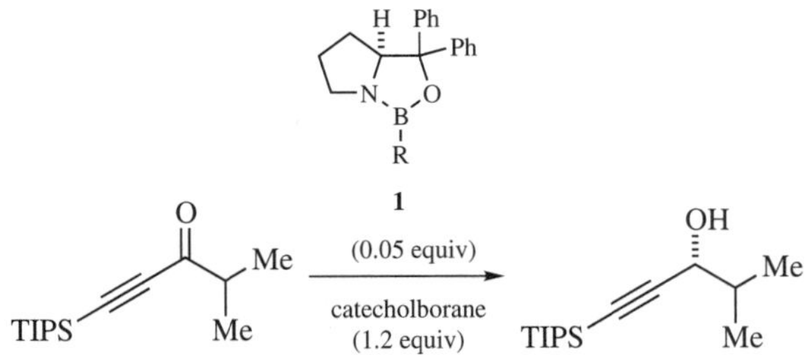

(4)

R = *n*-Bu (**1b**)	90% ee
R = CH_2SiMe_3 (**1c**)	72% ee

Enantioselective Reduction of α,β-Enones. Oxazaborolidine catalyst **1c** was also found to be a superior catalyst compared to **1a** and **1b** for the reduction of α,β-unsaturated ketone derivatives (eq 5).[6] It is interesting to note that the presence of a bulky substituent at the β-position is mandatory for high enantiocontrol. This reaction has been used as a key step in the synthesis of atractyligenin.[6] This catalyst has also been used to reduce α,β-unsaturated ketones bearing a trifluoromethyl group in 87% ee.[7]

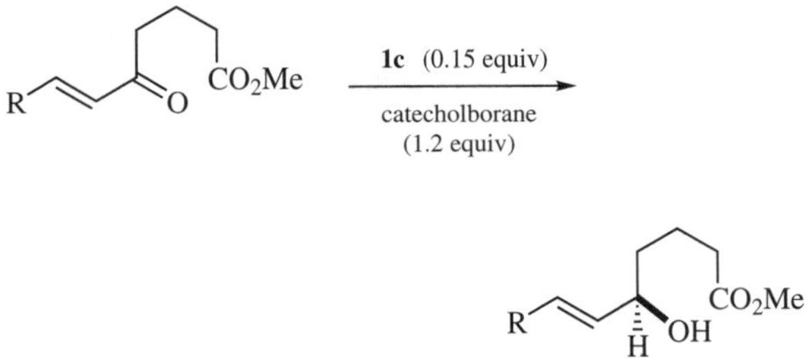

(5)

R = H	70% ee (30%)
R = $SiMe_3$	94% ee (90%)
R = $SnBu_3$	90% ee (94%)

1. (a) Corey, E. J.; Helal, C. J. *Angew. Chem. Int. Ed. Engl.* **1998**, *37*, 1987–2012. (b) Carboni, B.; Monnier, L. *Tetrahedron* **1999**, *55*, 1197–1248. (c) Fache, F.; Schulz, E.; Tommasino, M. L.; Lemaire, M. *Chem. Rev.* **2000**, *100*, 2159–2231.
2. (a) Matteson, D. S.; Majumdar, D. *Chem. Comm.* **1989**, 39–40. (b) Matteson, D. S. *Organometallics* **1983**, *2*, 236–241.
3. Helal, C. J.; Magriotis, P. A.; Corey, E. J. *J. Am. Chem. Soc.* **1996**, *118*, 10938–10939.
4. (a) Corey, E. J.; Shibata, S.; Bakshi, R. K. *J. Org. Chem.* **1988**, *53*, 2861–2863. (b) Mathre, D. J.; Jones, T. K.; Xavier, L. C.; Blacklock, T. J.; Reamer, R. A.; Mohan, J. J.; Jones, E. T. T.; Hoogsteen, K.; Baum, M. W.; Grabowski, E. J. J. *J. Org. Chem.* **1991**, *56*, 751–762. (c) Kanth, J. V. B.; Periasamy, M. *Tetrahedron* **1993**, *49*, 5127–2863. (d) Delaunay, D.; Corre, M. L. *J. Chem. Soc. Perkin Trans I* **1994**, 3041–3042. (e) Kaufman, T. S.; Ponzo, V. L.; Zinczuk, J. *Org. Prep. Proced. Int.* **1996**, *28*, 487–490.
5. Corey, E. J.; Helal, C. J. *Tetrahedron Lett.* **1997**, *38*, 7511–7514.
6. Corey, E. J.; Guzman-Perez, A.; Lazerwith, S. E. *J. Am. Chem. Soc.* **1997**, *119*, 11769–11776.
7. Nenajdenko, V. G.; Smolko, K. I.; Balenkova, E. S. *Tetrahedron: Asymmetry* **2001**, *12*, 1259–1266.

André B. Charette
Université de Montréal, QC, Canada

(2*R*,3*R*)-Dipivaloyltartaric Acid

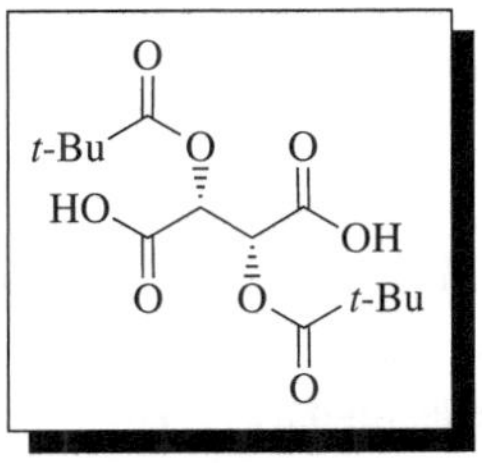

[65259-81-6] $C_{14}H_{22}O_8$ (MW 318.36)

(chiral auxiliary for enantioselective protonation (deracemization) and asymmetric transformation; starting material for synthesis of chiral succinimides and polyhydroxy compounds)

Alternate Name: (2*R*,3*R*)-DPTA.
Physical Data: mp 135 °C; $[\alpha]_D^{25}$ −24.2° (1.7 dioxane).[1]
Solubility: sol Et_2O, THF, aq $NaHCO_3$ solution; insol cold H_2O.
Preparative Methods: hydrolysis of the corresponding anhydride obtained by heating (2*R*,3*R*)-tartaric acid and pivaloyl chloride at 120–140 °C for 4 h.[1]
Drying: over P_2O_5 in vacuo, controlled by ^1H NMR and optical rotations.
Handling, Storage, and Precautions: the anhydrous solid can be stored at rt in the absence of moisture. It is retrievable from its basic solution (10 % aq $NaHCO_3$), after acidification (HCl), with a quantitative yield.[2] On heating it transforms into an anhydride with elimination of one molecule of water.

Chiral Auxiliary Used for Enantioselective Protonation (Kinetic Control). (2*R*,3*R*)-Dipivaloyltartaric acid has been essentially used for enantioselective protonations.[3] Surprisingly, for the creation of the chirality during the formation of the C–H linkage, this type of reaction has received attention only recently, unlike enantioselective hydrogenations or enantioselective reductions with hydride anion, widely used in asymmetric syntheses. (2*R*,3*R*)-DPTA was the first protonating reagent leading to appreciable enantioselections (up to 80 %) when applied to different classes of substrates such as enamines,[4] enolates of functionalized esters,[5] and carbonyl compounds.[4] At the present time, high enantioselectivities have also been reached with other protonating agents,[6] but often limited to one target molecule.

The first reported experiments concerned the protonation of enamines with (2*R*,3*R*)-DPTA leading, after hydrolysis of the iminium salts formed in situ, to optically active carbonyl compounds.[7] Starting from the (*Z*)- and (*E*)-morpholino enamines of 2-phenylpropanal, it was possible to establish, in spite of modest results (ee 13–18%), that the protonation step was kinetically controlled: the (*Z*)- and (*E*)-isomers led to 2-phenylpropanal with a reverse configuration, excluding an equilibrium of the intermediate diastereoisomeric salts (eq 1).[7]

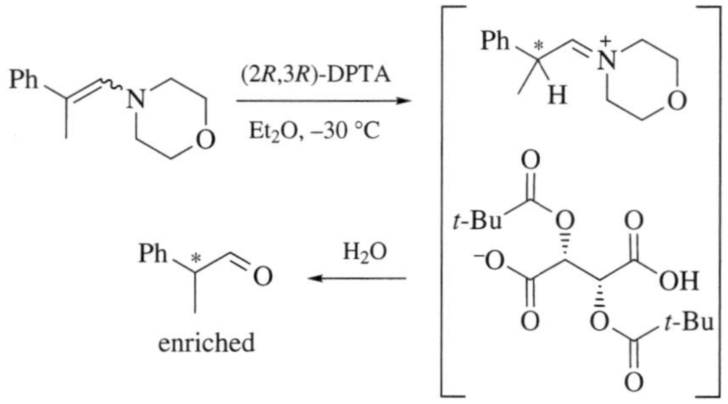

(1)

The protonation of lithium enolates of Schiff bases of racemic α-amino esters leads, after the workup, to α-amino acids of (*S*) configuration with ee as high as 70% (eq 2).[2,5,8,9]

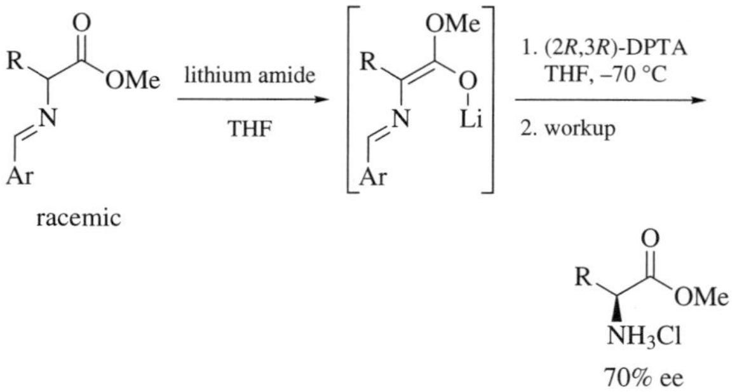

(2)

A first set of experiments, the study of the protonation of enolates obtained from benzaldehyde Schiff bases and *Lithium Diisopropylamide*, showed that the asymmetric induction was not significantly affected by the size of the R moiety of the amino acid (R = Me, Et, *i*-Pr, *n*-Bu, *t*-Bu, Ph; ee = 44–56%).[2,9] The two main factors improving the enantioselection were the Ar substituent of the Schiff base and the lithium amide used for the deprotonation.[5,8,10] The following results (Table 1) indicate clearly that the enantioselectivity increases with the electron-donating power of substituents *para* to the Schiff base (eq 3),[5] leading to 70% ee with the Schiff base of *p*-methoxybenzaldehyde derived from phenylglycine.[9]

Table 1 Influence of the *para*-Substituent

X	ee (%)	X	ee (%)
NC	12	Me	53
Cl	31	MeO	57 (70)[a]
H	50	Me_2N	61

[a] Protonation carried out at −105 °C.

The favorable effect of electron-donating substituents X, interpreted as the structure of the enolate becomes more rigid due to the increase of the coordination between the lithium and the nitrogen atoms, was confirmed on a series of α-amino acids.[9]

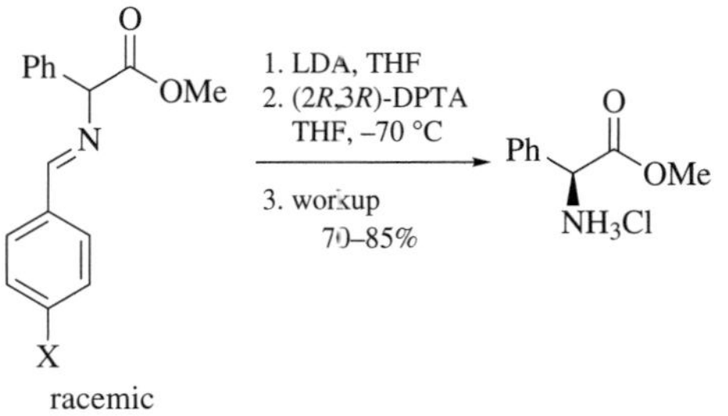

(3)

The enantioselectivity was dramatically affected by the structure of the lithium amide used for the deprotonation, indicating that the secondary amine liberated during the metalation step participates in the protonation step.[3,8,10] The utilization of chiral lithium amides allowed higher enantioselections than with classical LDA (eq 4) (Table 2).[8,10]

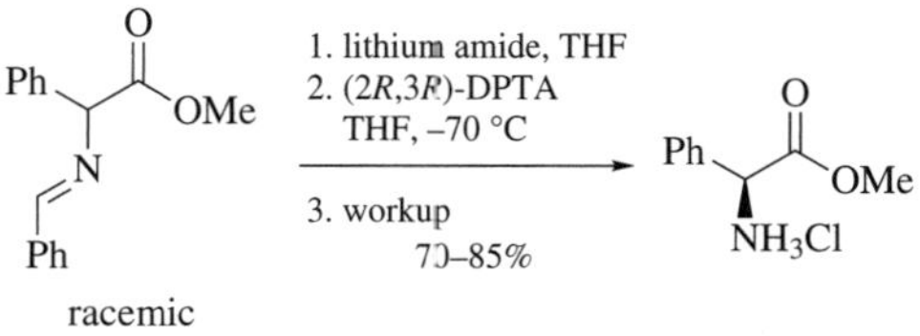

(4)

Table 2 Influence of the Lithium Amide used for Deprotonation

Lithium amide	ee (%)	Lithium amide	ee (%)
LiN(*i*-Pr)$_2$	50	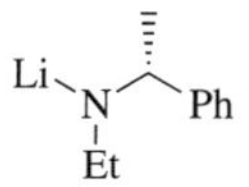	70
LiNEt$_2$	28	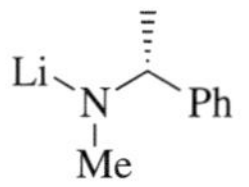	60
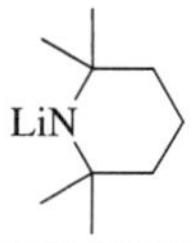	22		

The crucial role of the secondary liberated amine was also reported in experiments involving deprotonation with lithium (*R*)-*N*-ethyl(1-phenylethyl)amide and reprotonation at −70 °C with 2*R*,3*R*, racemic, and *meso*-DPTA, yielding, respectively, 70, 39, and 24% ee of the (*S*)-enantiomer. In the two last cases, significant inductions were obtained with the sole secondary chiral amine as chiral inductor in the medium.[8] Since these first results, chiral lithium amides have been widely used for asymmetric synthesis.

Owing to the importance of the amine, probably acting as a ligand of lithium or a proton carrier [ammonium salt of (2*R*,3*R*)-DPTA],[3,10] a process was proposed allowing the introduction of different amines and consequently a modification of the selectivity of the protonation: after deprotonation of a Schiff base of methyl valinate with *Lithium Hexamethyldisilazide* (LHMDS), the liberated HMDS was replaced by a more basic primary, secondary, or tertiary amine prior to the addition of (2*R*,3*R*)-DPTA (eq 5) (Table 3). In some cases, higher ee were observed compared to the classical procedure with LHMDS (34% ee) or LDA (47% ee).[10]

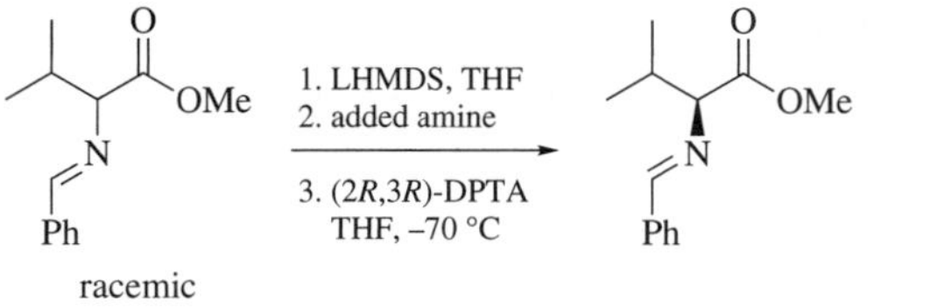
(5)

Table 3 Influence of the Added Amine

Added amine	ee (%)	Added amine	ee (%)
–	34		50
Et_2NH	44	$EtNH_2$	55
i-Pr_2NH	47	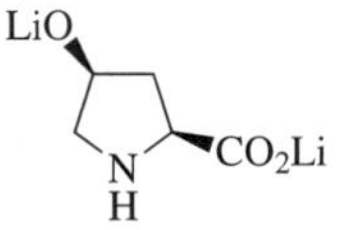	64

Enantioselective protonation of lithium dienolates obtained from Schiff bases of methyl α-aminobutenoates was carried out using (2*R*,3*R*)-DPTA, in order to synthesize vinylglycine by deconjugation (ee 36%).[9] Protonation of a cyclic lithium enolate derived from mandelic acid with (2*R*,3*R*)-DPTA was reported to occur with a low ee.[6b]

The potassium (*Z*)-enediolate obtained from racemic benzoin and *Potassium Hydride* when treated with (2*R*,3*R*)-DPTA affords (*S*)-benzoin with 80% ee (optically pure after one recrystallization in methanol).[11] In the reaction conditions, the enediolate is first *O*-protonated, then the resulting enediol slowly tautomerizes at low temperature into optically active benzoin with a high enantioselectivity. In the cases of incomplete tautomerization the residual enediol was immediately oxidized to benzil by the oxygen of the air during the workup. It was shown that the tautomerization of the enediol was hindered by the presence of an excess of (2*R*,3*R*)-DPTA (eq 6).[11]

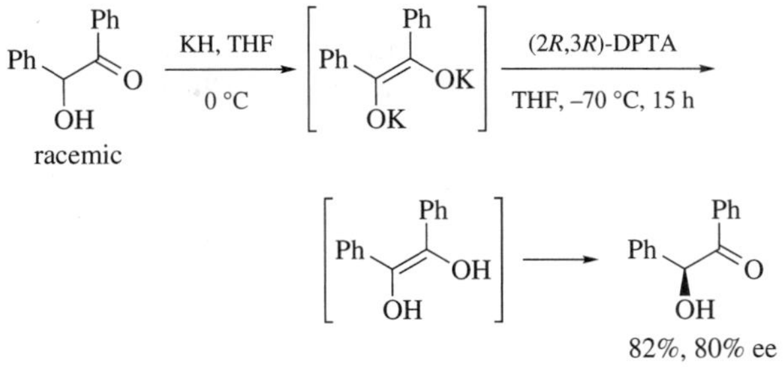
(6)

Finally, the configuration of the protonation product is predictable using the 'L Rule'. The prochiral substrate is represented according to the letter L, where the vertical line represents the C=C bond and the horizontal line the C–X bond (X = nucleophilic atom). In such a case, protonation with (2*R*,3*R*)-DPTA is favored on the L-face, i.e. on the side of the reader. As an example, the protonation of a lithium enolate of a Schiff base of an α-amino ester, leading to the (*S*)-enantiomer, is given in eq 7.[3]

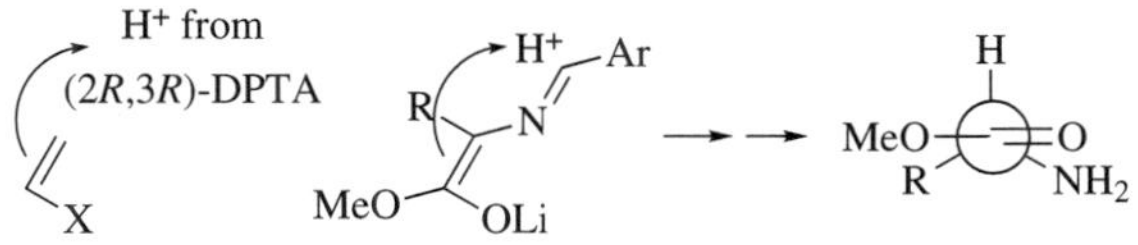
(7)

In enantioselective protonations, the final optically active product is generally chemically identical to the starting racemic material, precursor of the prochiral substrate on which the protonation is carried out. That is why the term deracemization was proposed for this type of process.[3,4,12] In a deracemization, the protonation step is kinetically controlled. Therefore a deracemization differs from an asymmetric transformation in which the reactions are thermodynamically controlled.[3,4]

Chiral Auxiliary Used for Asymmetric Transformations (Thermodynamic Control). Racemic *N*-benzyl-*N*-methyl-α-amino propiophenone mixed with (2*R*,3*R*)-DPTA in acetone or dichloromethane leads with 90% yield to the corresponding salt of the (*S*)-amino ketone, which was reduced over *Palladium on Carbon* to a mixture of ephedrine (80%) and pseudo-ephedrine (20%) (eq 8).[13]

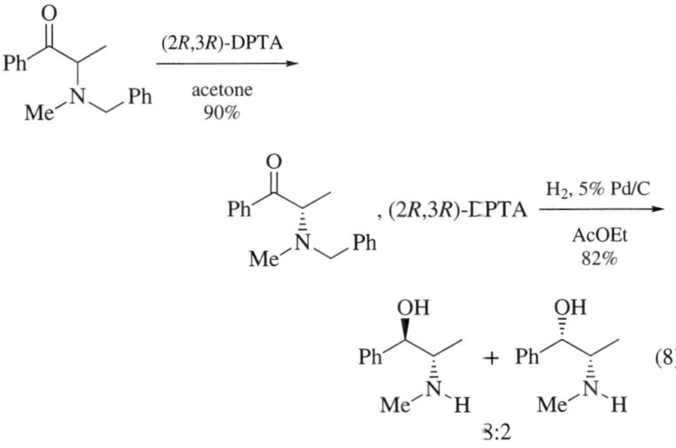
(8)

Starting Material for Asymmetric Syntheses. (3*R*,4*R*)-Dipivaloyltartaric anhydride, the direct precursor of (2*R*,3*R*)-DPTA, has been used as starting material for the synthesis of (3*R*,4*R*)-dipivaloyltartrimide,[14] its *N*-chloro and *N*-bromo derivatives (eq 9)[15], and epimeric (3*R*,4*R*)-dipivaloyl-5-alkyl lactones (eq 10), which are valuable intermediates for access to optically active polyhydroxy compounds.[16]

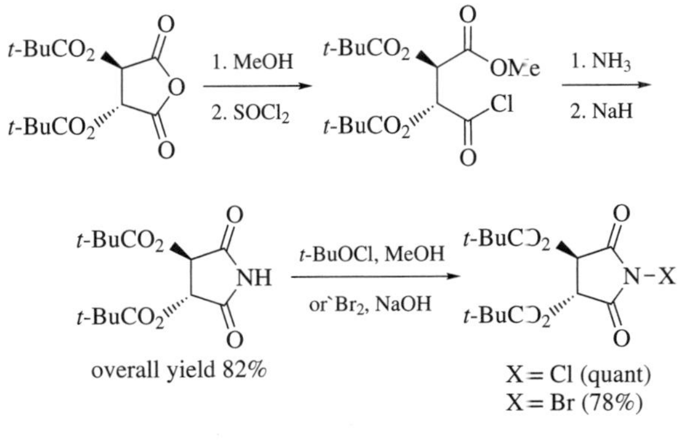
(9)

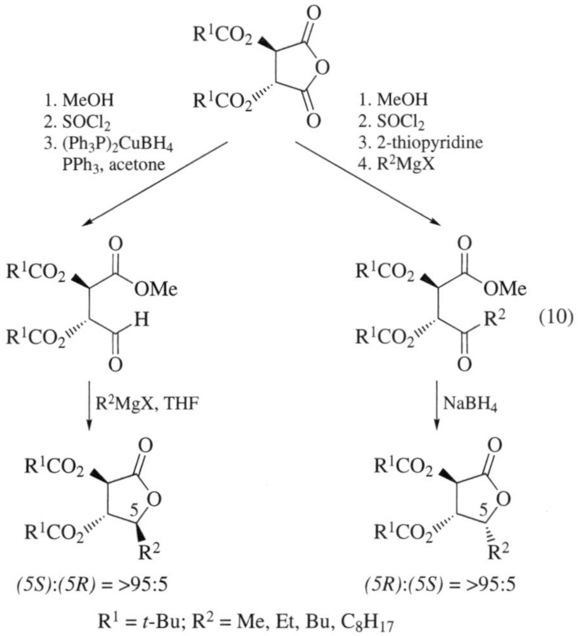

(10)

1. Duhamel, L.; Plaquevent, J. C. *Org. Prep. Proced. Int.* **1982**, *14*, 347.
2. Duhamel, L.; Plaquevent, J. C. *J. Am. Chem. Soc.* **1978**, *100*, 7415.
3. Duhamel, L.; Duhamel, P.; Launay, J. C.; Plaquevent, J. C. *Bull. Soc. Chem. Fr. Part 2* **1984**, 421.
4. Duhamel, L.; Plaquevent, J. C. *Bull. Soc. Chem. Fr. Part 2* **1982**, 69.
5. Duhamel, L.; Plaquevent, J. C. *Bull. Soc. Chem. Fr. Part 2* **1982**, 75.
6. (a) Fuji, K. *J. Am. Chem. Soc.* **1985**, *107*, 6404. (b) Gerlach, U.; Hunig, S. *Angew. Chem., Int. Ed. Engl.* **1987**, *26*, 1283. (c) Fehr, C.; Galindo, J. *J. Am. Chem. Soc.* **1988**, *110*, 6909. (d) Piva, O.; Pete, J. P. *Tetrahedron Lett.* **1990**, *31*, 5157. (e) Potin, D.; Williams, K.; Rebeck, J. *Angew. Chem., Int. Ed. Engl.* **1991**, *29*, 1420. (f) Matsumoto, K.; Otha, H. *Tetrahedron Lett.* **1991**, *32*, 4729. (g) Vedejs, E.; Lee, N. *J. Am. Chem. Soc.* **1991**, *113*, 5483. (h) Kumar, A.; Salumkhe, R. V.; Ramkrishna, A. R.; Suneel, Y. D. *Chem. Commun./J. Chem. Soc., Chem. Commun.* **1991**, 485. (i) Reymond, J. L.; Janda, K. D.; Lerner, R. A. *J. Am. Chem. Soc.* **1992**, *114*, 2257.
7. Duhamel, L.; Plaquevent, J. C. *Tetrahedron Lett.* **1977**, 2285.
8. Duhamel, L.; Plaquevent, J. C. *Tetrahedron Lett.* **1980**, *21*, 2521.
9. Duhamel, L.; Duhamel, P.; Fouquay, S.; Jamal Eddine, J.; Peschard, O.; Plaquevent, J. C.; Ravard, A.; Solliard, R.; Valnot, J. Y.; Vincens, H. *Tetrahedron* **1988**, *44*, 5495.
10. Duhamel, L.; Fouquay, S.; Plaquevent, J. C. *Tetrahedron Lett.* **1986**, *27*, 4975.
11. Duhamel, L.; Launay, J. C. *Tetrahedron Lett.* **1983**, *24*, 4209.
12. Duhamel, L. *C. R. Hebd. Seances Acad. Sci., Ser. C* **1976**, *282*, 125.
13. Noi, Y.; Ogura, S. Jap. Patent 63 91 352, 1988 (*Chem. Abstr.* **1989**, *110*, 7832w).
14. Duhamel, L.; Herman, T.; Angibaud, P. *Synth. Commun.* **1992**, *22*, 735.
15. Duhamel, L.; Plé, G.; Angibaud, P.; Desmurs, J. R. *Synth. Commun.* **1993**, *22*, 2473.
16. (a) Jacob, M. Diplôme d'Etudes Approfondies, Rouen, 1992. (b) Jacob, M.; Fernandez, A. M. to be published.

Lucette Duhamel
University of Rouen, Mont-Saint-Aignan, France

Dirhodium(II) Tetrakis(methyl 2-pyrrolidone-5(*S*)-carboxylate)

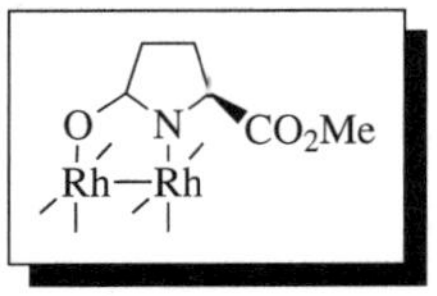

[131766-06-8] $C_{24}H_{36}N_4O_{12}Rh_2$ (MW 778.46)

(highly enantioselective catalyst for carbenoid reactions of diazo compounds)[1–3]

Physical Data: λ 615 nm, ϵ 211 ($ClCH_2CH_2Cl$). ^{1}H NMR ($CDCl_3$) of Rh_2(5*S*-MEPY)$_4$(MeCN)$_2$: δ 4.32 (dd, J = 8.8, 3.0 Hz, 2H), 3.95 (dd, J = 8.6, 2.1 Hz, 2H), 3.70 (s, 6H), 3.68 (s, 6H), 2.70–2.55 (m, 4H), 2.26 (s, 6H), 1.8–2.4 (m, 12H). $[\alpha]_D^{23} = -259.5°$ (MeCN, c = 0.098).
Solubility: sol MeOH, MeCN, acetone; slightly sol CH_2Cl_2, $ClCH_2CH_2Cl$, toluene.
Form Supplied in: red crystals as the bis-acetonitrile complex; blue solid after removal of the axial nitrile ligands.
Preparative Methods: from *Dirhodium(II) Tetraacetate* by ligand substitution with methyl 2-pyrrolidone-5(*S*)-carboxylate.[4,5]
Handling, Storage, and Precautions: air stable, weakly hygroscopic; stored in desiccator.

Introduction. The preparation of the title reagent, Rh_2(5*S*-MEPY)$_4$, is the same as that used for *Dirhodium(II) Tetraacetamide*[6] or *Dirhodium(II) Tetra(caprolactam)*.[7] Ligand exchange occurs in refluxing chlorobenzene, and the acetic acid that is liberated is trapped in a Soxhlet extraction apparatus by sodium carbonate. Purification occurs by chromatography on a CN-capped silica column; recrystallization from acetonitrile–2-propanol (1:1) provides Rh_2(5*S*-MEPY)$_4$(MeCN)$_2$(*i*-PrOH). Four 2-pyrrolidone-5(*S*)-carboxylate molecules ligate one dirhodium(II) nucleus; each rhodium is bound to two nitrogen and two oxygen donor atoms arranged in a *cis* configuration.[4] The methyl carboxylate substituents are positioned with a counterclockwise orientation on each rhodium face.

Metal Carbene Transformations. The effectiveness of Rh_2(5*S*-MEPY)$_4$ and its 5*R*-form, Rh_2(5*R*-MEPY)$_4$, is exceptional for highly enantioselective intramolecular cyclopropanation[8] and carbon–hydrogen insertion[9] reactions. Intermolecular cyclopropanation occurs with lower enantiomeric excesses[10] than with alternative chiral copper salicylaldimine[11] or C_2-symmetric semicorrin[12] or bis-oxazoline[13] copper catalysts, but intermolecular cyclopropenation exhibits higher enantiocontrol with Rh_2(MEPY)$_4$ catalysts.[14] The methyl carboxylate attachment of Rh_2(5*S*-MEPY)$_4$ is far more effective than sterically similar benzyl or isopropyl attachments for enantioselective metal carbene transformations.[4] The significant enhancement in enantiocontrol is believed to be due to carboxylate carbonyl stabilization of the intermediate metal carbene and/or to dipolar influences on substrate approach to the carbene center.

Enantioselective Intramolecular Cyclopropanation Reactions. The exceptional capabilities of the Rh_2(5*S*-MEPY)$_4$ and

$Rh_2(5R\text{-MEPY})_4$ catalysts for enantiocontrol are evident in results obtained with a series of allyl diazoacetates (eq 1).[5,8] Both high product yields and enantiomeric excess (ee's) are characteristic. Intramolecular cyclopropanation of (*Z*)-alkenes proceeds with a higher level of enantiocontrol than does intramolecular cyclopropanation of (*E*)-alkenes. In preparative scale reactions, less than 0.25 mol% of catalyst can be employed to achieve high yields of pure product.[5]

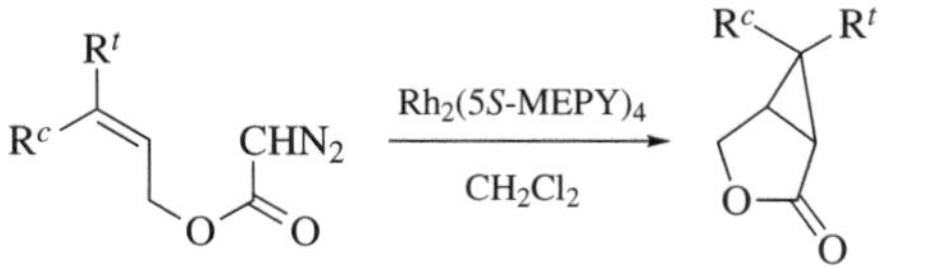

(1)

R^t	R^c	% ee
H	H	95
Me	Me	98
H	Ph, Et, Bn, Bu_3Sn	≥94
Pr	H	83
Ph	H	65

Similar success in enantiocontrol has been achieved for intramolecular cyclopropanation of homoallyl diazoacetates (eq 2).[15] With these substrates the enantiomeric excesses do not extend beyond 90%, but they are virtually independent of double bond substituents.

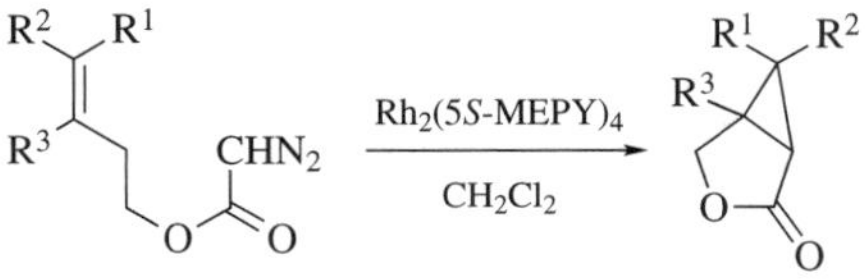

(2)

R^1	R^2	R^3	% ee
Me	Me	H	77
Et, Ph, Bn, TMS	H	H	80–90
H	Et, Ph	H	73–82
H	H	Me	79

Enantioselective Intermolecular Cyclopropenation Reactions. The use of $Rh_2(\text{MEPY})_4$ catalysts for intermolecular cyclopropenation of 1-alkynes results in moderate to high selectivity. With propargyl methyl ether (or acetate), for example, reactions with (−)-menthyl [(+)-(1*R*,2*S*,5*R*)-2-isopropyl-5-methyl-1-cyclohexyl] diazoacetate catalyzed by $Rh_2(5S\text{-MEPY})_4$ produces the corresponding cyclopropene product (eq 3) with 98% diastereomeric excess (de).[14,16]

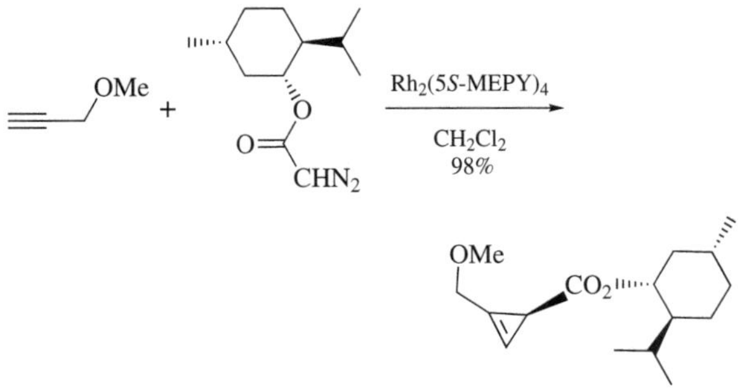

(3)

These reactions are subject to significant double diastereoselection with (+)- and (−)-menthyl diazoacetates. With ethyl diazoacetate, enantiomeric excesses are moderate (54–69% ee), but they increase up to 78% ee with *t*-butyl diazoacetate.[14] These are the first examples of enantioselective catalytic cyclopropenation reactions.

Enantioselective Intramolecular Carbon–Hydrogen Insertion Reactions. The suitability of $Rh_2(5S\text{-MEPY})_4$ and $Rh_2(5R\text{-MEPY})_4$ for enantioselective intramolecular C–H insertion reactions is evident in results with 2-alkoxyethyl diazoacetates (eq 4).[9] Both lactone enantiomers are available from a single diazo ester. Other examples have also been reported, especially those with highly branched diazo substrate structures.[9]

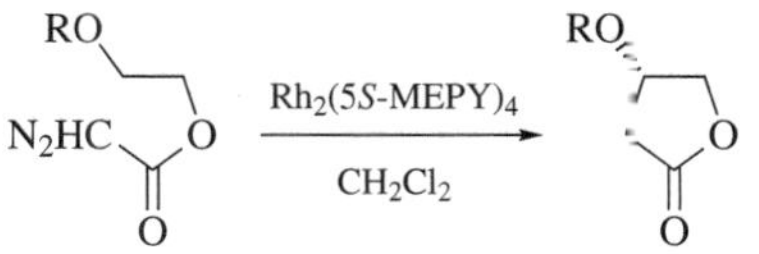

(4)

R = Me, 91% ee
R = Et, 89% ee
R = Bn, 87% ee

Diazoacetamides are robust diazo substrates, but they generally give lower enantioselection, and regioselectivity for γ-lactam formation is dependent on the substituents on carbon at which insertion occurs (e.g. eq 5).[17] With *N*-(*n*-butyl)-*N*-(*t*-butyl)diazoacetamide the ratio of γ:β-lactam is 88:12. A significant improvement in enantioselection (up to 78% ee) occurs with the use of the oxazolidinone analog of $Rh_2(5S\text{-MEPY})_4$.[17]

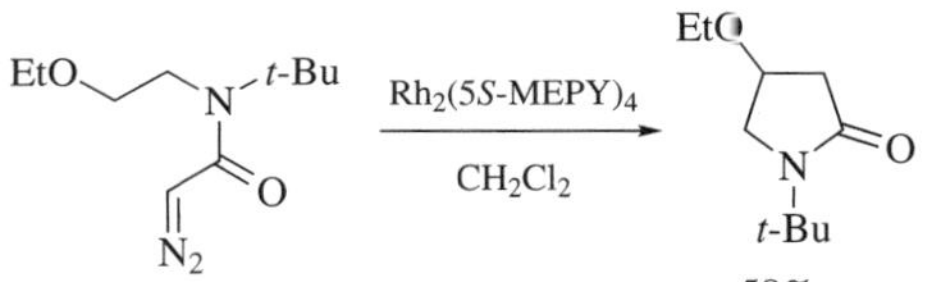

(5)

58% ee

Polyethylene-Bound, Soluble, Recoverable Dirhodium(II) 2-Pyrrolidone-5(*S*)-carboxylate. The homogeneous $Rh_2(5S\text{-MEPY})_4$ catalyst has been attached to a polyethylene chain that is soluble in organic solvents at about 70 °C.[18] Ligand displacement of 2-pyrrolidone-5(*S*)-carboxylate from $Rh_2(5S\text{-MEPY})_4$ by a soluble polyethylene-bound 2-pyrrolidone-5(*S*)-carboxylate produces a recoverable dirhodium(II) catalyst, PE–$Rh_2(5S\text{-PYCA})_4$, in high yield. The effectiveness of this catalyst has been demonstrated by high enantioselection for intramolecular cyclopropanation of 3-methyl-2-buten-1-yl diazoacetate (see eq 1) in refluxing benzene solution (98% ee) and for intramolecular C–H insertion of 2-methoxyethyl diazoacetate (see eq 4) under the same conditions (72% ee). For both transformations, reactions catalyzed by $Rh_2(5S\text{-MEPY})_4$ that occur at the same temperature give lower % ee values. Although diminished selectivity can occur with catalyst recovery and reuse under standard conditions, retention of catalyst effectiveness is achieved by using 2–3 mol % of the pyrrolidone ligand in up to seven subsequent runs with recovered, reused PE–$Rh_2(5S\text{-PYCA})_4$.

1. Doyle, M. P. In *Selectivity in Catalysis*; Davis, M. E.; Suib, S. L., Eds.; American Chemical Society: Washington, 1993.
2. Doyle, M. P. In *Catalytic Asymmetric Synthesis*; Ojima, I., Ed.; VCH: New York, 1993.
3. Doyle, M. P. *Recl. Trav. Chim. Pays-Bas* **1991**, *110*, 305.

4. Doyle, M. P.; Winchester, W. R.; Hoorn, J. A. A.; Lynch, V.; Simonsen, S. H.; Ghosh, R. *J. Am. Chem. Soc.* **1993**, *115*, 9968.

5. Doyle, M. P.; Winchester, W. R.; Protopopova, M. N.; Kazala, A. P.; Westrum, L. J. *Org. Synth.* **1994**, *73*, in press.

6. Doyle, M. P.; Bagheri, V.; Wandless, T. J.; Harn, N. K.; Brinker, D. A.; Eagle, C. T.; Loh, K.-L. *J. Am. Chem. Soc.* **1990**, *112*, 1906.

7. Doyle, M. P.; Westrum, L. J.; Wolthuis, W. N. E.; See, M. M.; Boone, W. P.; Bagheri, V.; Pearson, M. M. *J. Am. Chem. Soc.* **1993**, *115*, 958.

8. Doyle, M. P.; Pieters, R. J.; Martin, S. F.; Austin, R. E.; Oalmann, C. J.; Müller, P. *J. Am. Chem. Soc.* **1991**, *113*, 1423.

9. Doyle, M. P.; Van Oeveren, A.; Westrum, L. J.; Protopopova, M. N.; Clayton, T. W., Jr. *J. Am. Chem. Soc.* **1991**, *113*, 8982.

10. Doyle, M. P.; Brandes, B. D.; Kazala, A. P.; Pieters, R. J.; Jarstfer, M. B.; Watkins, L. M.; Eagle, C. T. *Tetrahedron Lett.* **1990**, *31*, 6613.

11. Aratani, T. *Pure Appl. Chem.* **1985**, *57*, 1839.

12. Pfaltz, A. *Acc. Chem. Res.* **1993**, *26*, 339.

13. (a) Evans, D. A.; Woerpel, K. A.; Hinman, M. M.; Faul, M. M. *J. Am. Chem. Soc.* **1991**, *113*, 726. (b) Lowenthal, R. E.; Masamune, S. *Tetrahedron Lett.* **1991**, *32*, 7373. (c) Müller, D.; Umbricht, G.; Weber, B.; Pfaltz, A. *Helv. Chim. Acta* **1991**, *74*, 232.

14. Protopopova, M. N.; Doyle, M. P.; Müller, P.; Ene, D. *J. Am. Chem. Soc.* **1992**, *114*, 2755.

15. Martin, S. F.; Oalmann, C. J.; Liras, S. *Tetrahedron Lett.* **1992**, *33*, 6727.

16. Doyle, M. P.; Protopopova, M. N.; Brandes, B. D.; Davies, H. M. L.; Huby, N. J. S.; Whitesell, J. K. *Synlett* **1993**, 151.

17. Doyle, M. P.; Protopopova, M. N.; Winchester, W. R.; Daniel, K. L. *Tetrahedron Lett.* **1992**, *33*, 7819.

18. Doyle, M. P.; Eismont, M. Y.; Bergbreiter, D. E.; Gray, H. N. *J. Org. Chem.* **1992**, *57*, 6103.

Michael P. Doyle
Trinity University, San Antonio, TX, USA

E

(1*R*,2*S*)-Ephedrine

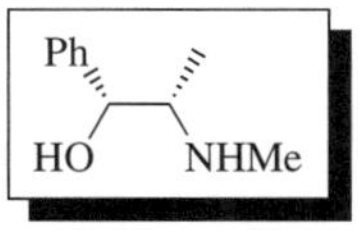

[299-42-3] $C_{10}H_{15}NO$ (MW 165.23)

(chiral auxiliary for the following: diastereoselective alkylation and reduction of chiral hydrazones; diastereoselective alkylation of chiral amides; diastereoselective conjugate addition of organometallic reagents to unsaturated amides and imidazolidinones; diastereoselective alkylation and cyclopropanation of oxazepinediones and oxazolidines; diastereoselective homoaldol addition of *N*-allylimidazolidinone, and asymmetric coupling reaction of Grignard reagents; chiral ligand for enantioselective conjugate addition of organometallic reagents to enones; chiral ligand for enantioselective addition of dialkylzincs to aldehydes)

Alternate Name: [*R*-(*R**,*S**)]-α-[1-(Methylamino)ethyl]benzenemethanol.
Physical Data: mp 37–39 °C; bp 255 °C; $[\alpha]_D^{21}$ −41° (*c* 5, 1 M HCl). Hydrochloride, mp 216–220 °C; $[\alpha]_D^{20}$ −34° (*c* 4, H_2O).
Solubility: sol alcohol, chloroform, ether, water.
Form Supplied in: waxy solid or crystals; also available as hydrochloride in either enantiomeric form.

General Features of Ephedrine. Ephedrine is a chiral β-amino alcohol which is available in either enantiomeric form. It is often utilized as a chiral auxiliary in asymmetric synthesis. Via bond formation with the amino group of ephedrine, ephedrine can be derived into chiral hydrazones[2,3] and amides.[5,6] Highly diastereoselective asymmetric reactions are known using these chiral compounds. In reactions using organometallic reagents, the hydroxy groups of hydrazones and amides become metal alkoxides. Metal atoms of the alkoxide may chelate with nitrogen or oxygen atoms of chiral hydrazones and amides. This chelation may reduce the number of possible conformations of reactive species, and this may increase the diastereoselectivities.

On the other hand, by bond formation with amino and hydroxy groups of ephedrine, ephedrine can be converted into chiral ring systems such as imidazolidinones,[12,13] oxazepinediones,[14–18] and oxazolidines.[20,21] Diastereoselective reactions of derivatives of these chiral ring systems afford compounds with high de. The relatively rigid conformation of these ring systems is one of the reasons for high diastereoselectivities.

Ephedrine becomes a chiral ligand of metal atoms by the deprotonation of the hydroxy group and by the presence of the nitrogen atom.[22–26] Highly enantioselective asymmetric reactions are known using chiral ephedrine-type ligands.

In addition, ephedrine is a chiral base catalyst because of the presence of the amine group.[29–32] A highly enantioselective base-catalyzed reaction is known.

Diastereoselective Alkylation and Reduction of Chiral Hydrazones Derived from Ephedrine.[2] *Methylmagnesium Bromide* adds to the chiral hydrazone derived from *N*-aminoephedrine and benzaldehyde to afford the optically active chiral hydrazine in almost 100% de. Hydrogenolysis of the chiral hydrazine gives (*R*)-α-phenylethylamine with more than 97% ee (eq 1). Ephedrine is recovered in good yield and without any loss of enantiomeric purity.

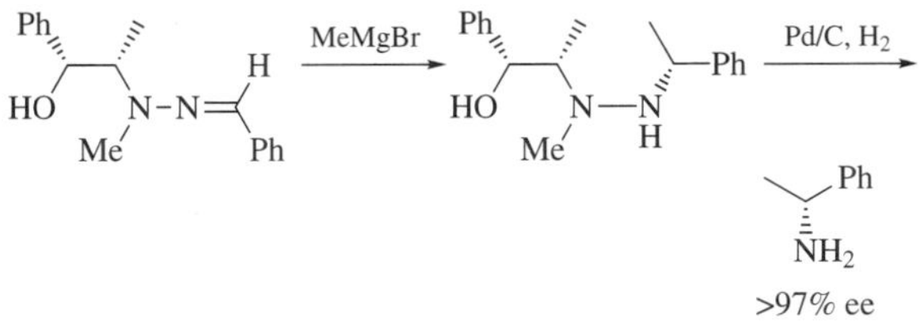

(1)

On the other hand, the diastereoselective reduction of the chiral hydrazone derived from *N*-aminoephedrine and acetophenone and subsequent hydrogenolysis affords (*S*)-α-phenylethylamine with 30% ee.[3] Optically active α-phenylethylamine with high ee is obtained from the diastereoselective alkylation of chiral hydrazones derived from (*R*)- or (*S*)-1-amino-2-(methoxymethyl)pyrrolidine.[4a]

Diastereoselective Alkylation of Chiral Amides Derived from Ephedrine. Chiral amides derived from ephedrine are converted to the corresponding dianion. The subsequent diastereoselective alkylation with alkyl iodides affords chiral α-substituted amides with >90% de.[5] Acid hydrolysis affords optically active α-substituted acids with 78% ee as a result of racemization in the cleavage step (eq 2).

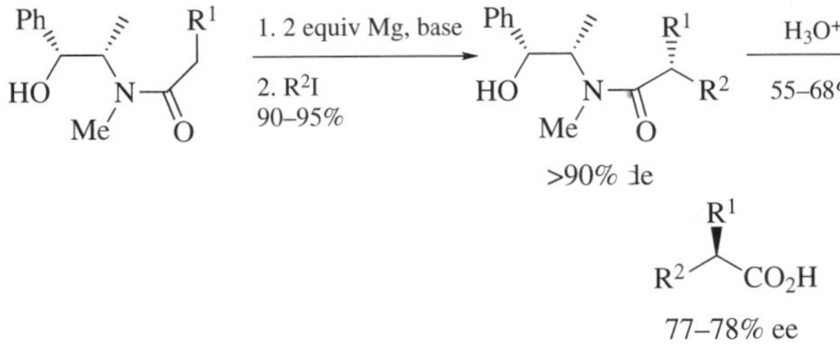

(2)

On the other hand, treatment with *Methyllithium* affords optically active methyl ketone in 44–74% ee, also as a result of racemization. α-Chiral ketones with higher ee (99% ee) are obtained from the diastereoselective alkylation of chiral hydrazones derived from (*R*)- or (*S*)-1-amino-2-methoxymethylproline.[4b]

Diastereoselective Conjugate Addition of Organometallic Reagents to Chiral α,β-Unsaturated Amides and Imidazolidinones Derived from Ephedrine. Grignard reagents (2 equiv) add to chiral α,β-unsaturated amides derived from ephedrine in a 1,4-addition manner with high diastereoselectivities. Subsequent acidic hydrolysis affords optically active β-substituted carboxylic acids with 85–99% ee (eq 3).[6]

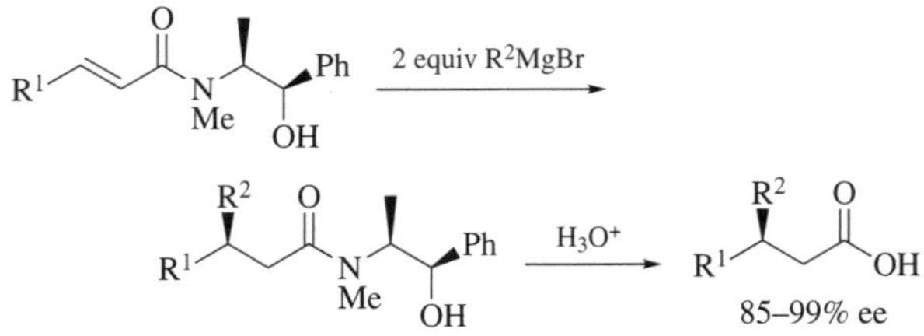

(3)

A seven-membered chelate intermediate is one of the reasons for the very high diastereoselectivities. The method is successfully applied to the asymmetric synthesis of malingolide.[7] Similar results are obtained in diastereoselective conjugate addition of Grignard reagents to unsaturated amides derived from (*S*)-2-(1-hydroxy-1-methylethyl)pyrrolidine. The presence of a tertiary amine (e.g., *1,8-Diazabicyclo[5.4.0]undec-7-ene*) increases the diastereoselectivity, and subsequent hydrolysis affords β-substituted carboxylic acids with up to 100% ee.[8] Conjugate additions of alkyllithium or Grignard reagents to chiral *N*-crotonoylproline,[9] imides,[10] and *N*-enoyl sultams[11] also afford β-substituted carboxylic acids with 60% ee, 96% ee, and 96% ee, respectively.

The chiral imidazolidinone[12] derived from urea and ephedrine hydrochloride is utilized in a diastereoselective conjugate methylation.[13] Subsequent hydrolysis affords optically pure (−)-citronellic acid (eq 4).

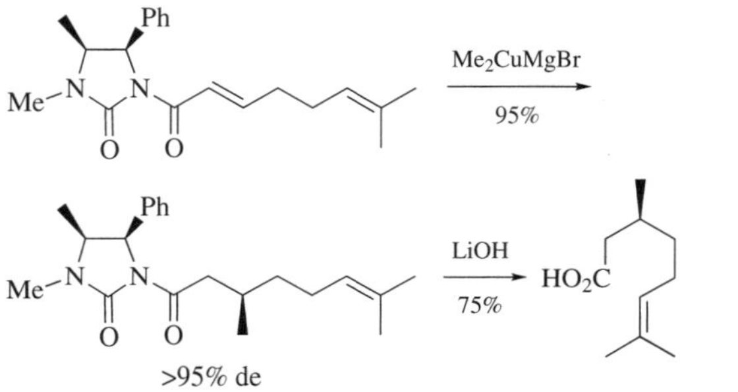

(4)

Diastereoselective Conjugate Additions to Chiral Oxazepinediones Derived from Ephedrine. Ephedrine can form a chiral seven-membered relatively rigid oxazepinedione ring by condensation with malonic acid monoester. Alkylidene derivatives of chiral oxazepinediones undergo highly diastereoselective additions with nucleophilic reagents. Grignard reagents in the presence of a catalytic amount of *Nickel(II) Chloride* add to chiral alkylideneoxazepinediones. Acid hydrolysis affords optically active β-substituted acids with up to >99% ee (eq 5).[14] The method is applied to the diastereoselective synthesis of (−)-indolmycin with 93% ee.[15]

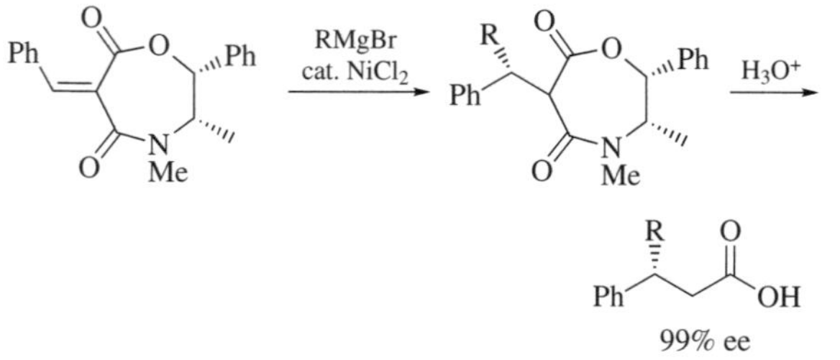

(5)

Diastereoselective addition of sulfoxonium ylides affords enantiomerically pure cyclopropanedicarboxylic acid diesters after removal of the chiral auxiliary (eq 6).[16]

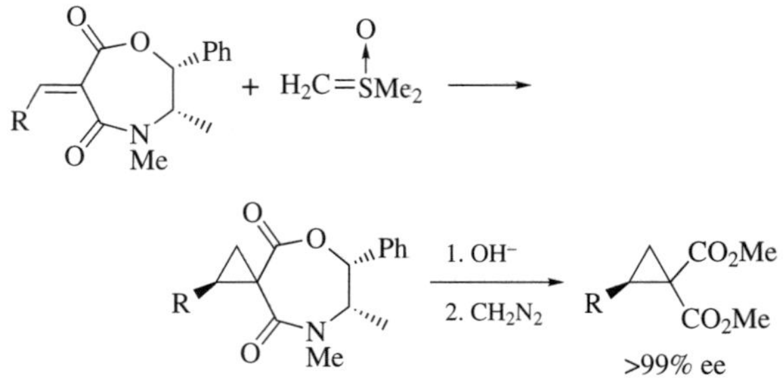

(6)

Diastereoselective addition of *Phenylthiomethyllithium* and subsequent treatment affords optically active lactones with >90% ee (eq 7).[17]

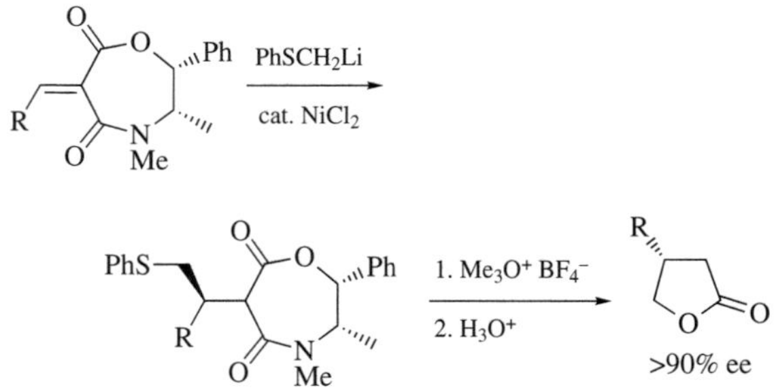

(7)

In addition, a chiral oxazepinedione plays the role of a nucleophile in the reaction with nitroalkenes in the presence of *Potassium t-Butoxide* and crown ether (eq 8).[18]

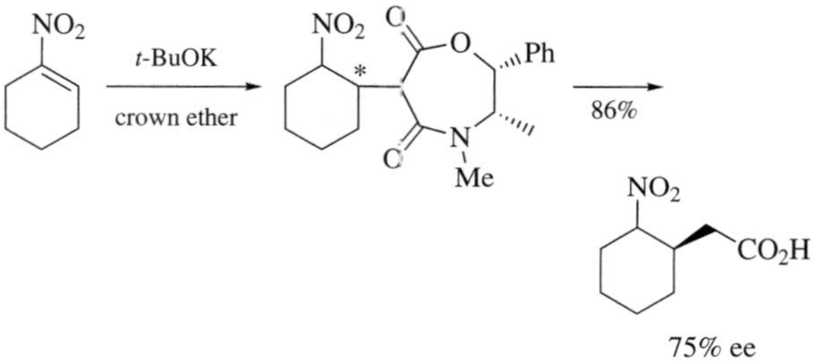

(8)

Homoaldol Addition with Chiral *N*-Allylimidazolidinone Derived from Ephedrine. The chiral allyltitanium compound derived from ephedrine reacts with carbonyl compounds with very high (>200:1) de. Subsequent hydrolysis and oxidation affords optically pure 4-substituted γ-lactones (eq 9).[12] 4-Substituted γ-lactones with 92% ee can also be synthesized by catalytic enantioselective alkylation of 3-formyl esters.[19]

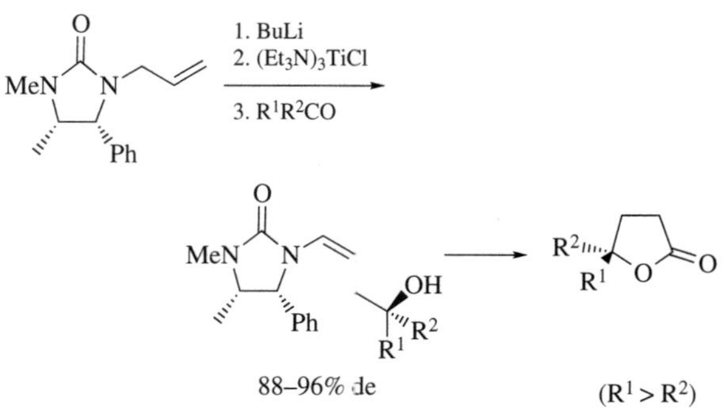

(9)

Diastereoselective Cyclopropanation and Alkylation of Chiral Oxazolidines Derived from Ephedrine. Ephedrine forms oxazolidines upon reaction with aldehydes. Chiral unsaturated oxazolidines derived from ephedrine and unsaturated aldehydes are treated with diazomethane in the presence of *Palladium(II) Acetate*. Hydrolysis of the oxazolidine ring affords optically active formylcyclopropanes with >90% ee (eq 10).[20]

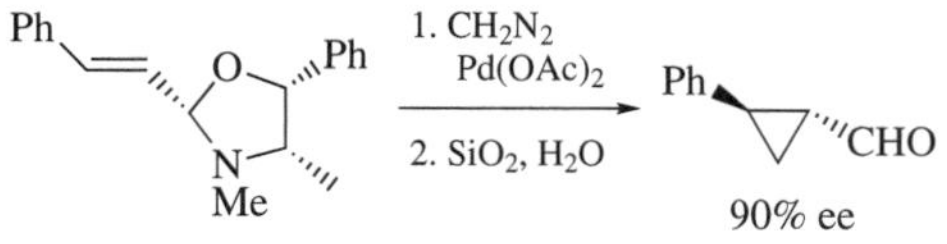
(10)

Diastereoselective addition of cuprate reagents to unsaturated oxazolidines and subsequent hydrolysis affords 3-substituted aldehydes with up to 81% ee (eq 11).[21]

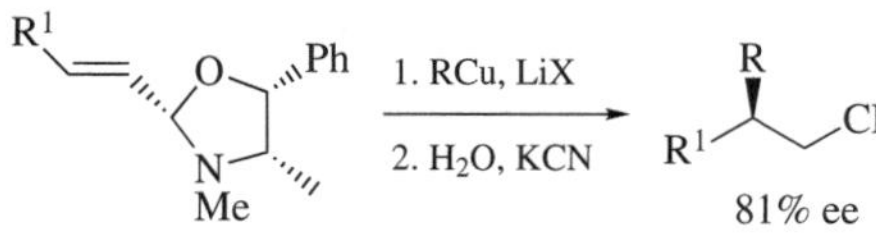
(11)

Asymmetric Coupling Reactions of Chiral Grignard Reagents Derived from Ephedrine Derivatives. Asymmetric coupling reactions of *Allyl Bromide* and chiral Grignard reagents derived from ephedrine methyl ether in the presence of *Copper(I) Iodide* (10 mol %) followed by oxidation affords optically active homoallyl alcohols with 60% ee (eq 12).[22]

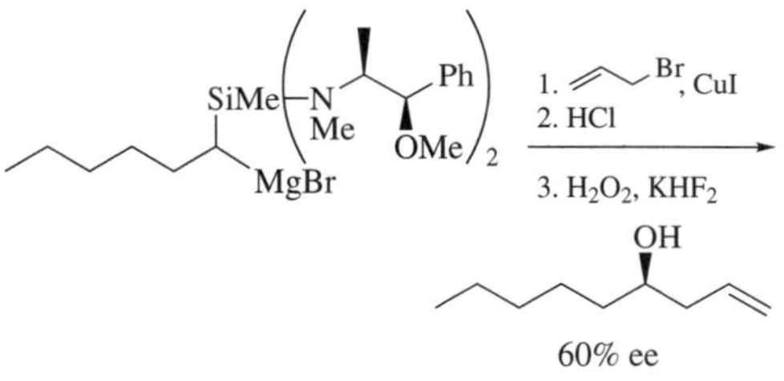
(12)

Enantioselective Conjugate Addition to Prochiral Enones of Organometallic Reagents Modified with Ephedrine. Enantioselective conjugate addition to 2-cyclohexenone with chiral organo(alkoxo)cuprates [MCu(OR*)R] has been studied.[1a] When the cuprate is prepared from the lithium alkoxide of ephedrine, *Phenyllithium*, and CuI, 3-phenylcyclohexanone with 50% ee is obtained.[23] The enantioselectivity reaches 92% ee in enantioselective ethylation when a chiral diamino alcohol derived from ephedrine is employed (eq 13).[24]

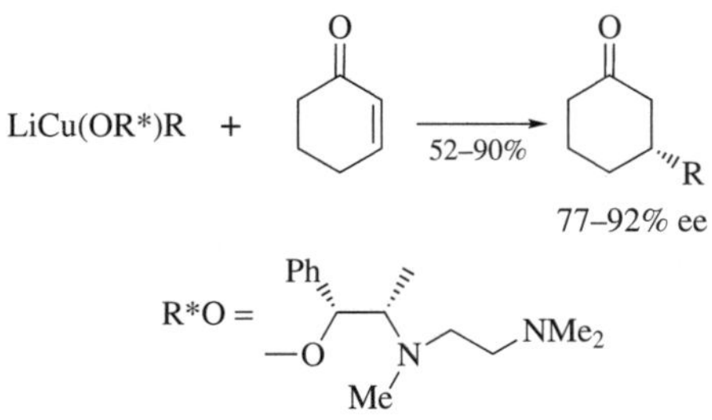
(13)

On the other hand, enantioselective conjugate addition to 2-cyclohexenone with lithium dibutylcuprates (having a noncovalently bound chiral phosphorus ligand derived from ephedrine) affords 3-butylcyclohexanone with up to 76% ee (eq 14).[25]

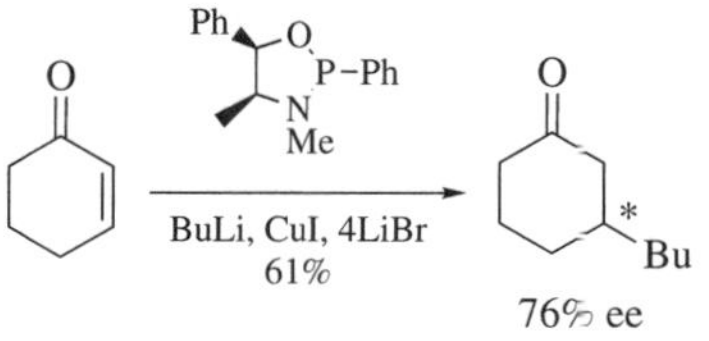
(14)

Isopropylmagnesium chloride adds to 2-cyclohexenone in 17% ee in the presence of a catalytic amount of chiral alkoxyzinc chloride derived from ephedrine and *Zinc Chloride* (eq 15).[26]

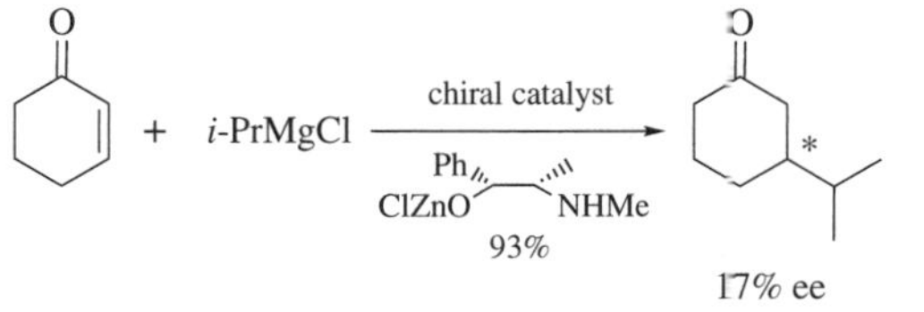
(15)

Concerning the catalytic enantioselective conjugate addition reaction, conjugate addition of dialkylzinc to chalcone in the presence of a catalytic amount of the chiral nickel complex derived from norephedrine affords β-substituted ketones with up to 90% ee (eq 16).[27]

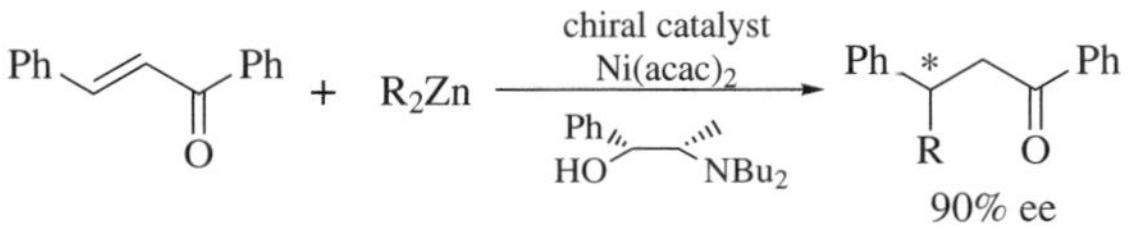
(16)

Enantioselective Addition of Dialkylzincs to Aldehydes Using Chiral Amino Alcohols Derived from Ephedrine. Nucleophilic addition of dialkylzinc to aldehydes is usually very slow. Amino alcohols facilitate the addition of *Diethylzinc* to benzaldehyde to afford 1-phenylpropanol.[1b,28] When chiral amino alcohols possessing the appropriate structure are used as a precatalyst, optically active secondary alcohols are obtained.[1b] Highly enantioselective chiral catalysts derived from ephedrine are known. (1*R*,2*S*)-*N*-Isopropylephedrine functions as a precatalyst for the enantioselective addition of diethylzinc to benzaldehyde to afford (*R*)-1-phenylpropanol with 80% ee in 72% yield.[29] The use of an excess amount of diethylzinc increases the enantioselectivity up to 97% ee (eq 17).[30]

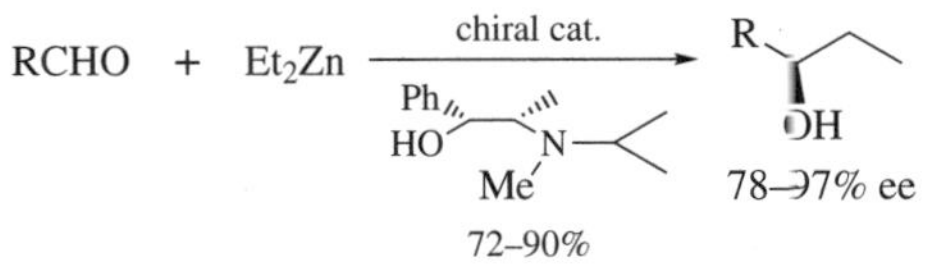
(17)

The lithium salt of (1*R*,2*S*)-*N*-[2-(dimethylamino)ethyl]-ephedrine acts as a precatalyst for the addition of diethylzinc to afford the alcohol with 90% ee (eq 18).[31]

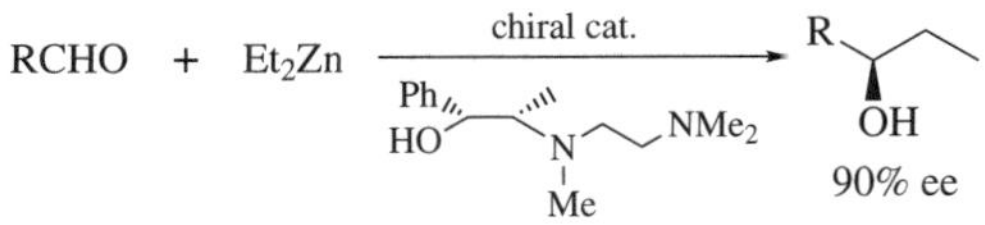
(18)

The dilithium salt of a chiral diaminodiol derived from ephedrine mediates the enantioselective addition of dialkylzinc to aldehydes to afford (*R*)-1-phenylethanol with 85% ee (eq 19).[32]

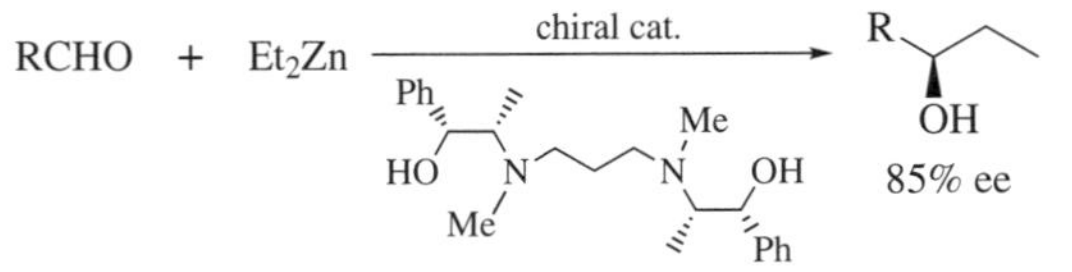

(19)

1. (a) Rossiter, B. E.; Swingle, N. M. *Chem. Rev.* **1992**, *92*, 771. (b) Soai, K.; Niwa, S. *Chem. Rev.* **1992**, *92*, 833.
2. Takahashi, H.; Tomita, K.; Noguchi, H. *Chem. Pharm. Bull.* **1981**, *29*, 3387.
3. Takahashi, H.; Tomita, K.; Otomasu, H. *Chem. Commun./J. Chem. Soc., Chem. Commun.* **1979**, 668.
4. (a) Enders, D.; Schubert, H.; Nübling, C. *Angew. Chem., Int. Ed. Engl.* **1986**, *25*, 1109. (b) Enders, D.; Eichenauer, H. *Angew. Chem., Int. Ed. Engl.* **1979**, *18*, 397.
5. Larcheveque, M.; Ignatova, E.; Cuvigny, T. *Tetrahedron Lett.* **1978**, 3961.
6. Mukaiyama, T.; Iwasawa, N. *Chem. Lett.* **1981**, 913.
7. Kogure, T.; Eliel, E. L. *J. Org. Chem.* **1984**, *49*, 576.
8. (a) Soai, K.; Machida, H.; Yokota, N. *J. Chem. Soc., Perkin Trans. 1* **1987**, 1909. (b) Soai, K.; Machida, H.; Ookawa, A. *Chem. Commun./J. Chem. Soc., Chem. Commun.* **1985**, 469.
9. Soai, K.; Ookawa, A. *J. Chem. Soc., Perkin Trans. 1* **1986**, 759.
10. Tomioka, K.; Suenaga, T.; Koga, K. *Tetrahedron Lett.* **1986**, *27*, 369.
11. (a) Oppolzer, W.; Mills, R. J.; Pachinger, W.; Stevenson, T. *Helv. Chim. Acta* **1986**, *69*, 1542. (b) Oppolzer, W.; Poli, G.; Kingma, A. J.; Starkemann, C.; Bernardinelli, G. *Helv. Chim. Acta* **1987**, *70*, 2201.
12. Roder, H.; Helmchen, G.; Peters, E. M.; Peters, K.; von Schnering, H. G. *Angew. Chem., Int. Ed. Engl.* **1984**, *23*, 898.
13. Stephan, E.; Pourcelot, G.; Cresson, P. *Chem. Ind. (London)* **1988**, 562.
14. (a) Mukaiyama, T.; Takeda, T.; Osaki, M. *Chem. Lett.* **1977**, 1165. (b) Mukaiyama, T.; Takeda, T.; Fujimoto, K. *Bull. Chem. Soc. Jpn.* **1978**, *51*, 3368.
15. Takeda, T.; Mukaiyama, T. *Chem. Lett.* **1980**, 163.
16. Mukaiyama, T.; Fujimoto, K.; Takeda, T. *Chem. Lett.* **1979**, 1207.
17. Mukaiyama, T.; Fujimoto, K.; Hirose, T.; Takeda, T. *Chem. Lett.* **1980**, 635.
18. (a) Mukaiyama, T.; Hirako, Y.; Takeda, T. *Chem. Lett.* **1978**, 461. (b) Takeda, T.; Hoshiko, T.; Mukaiyama, T. *Chem. Lett.* **1981**, 797.
19. Soai, K.; Yokoyama, S.; Hayasaka, T.; Ebihara, K. *Chem. Lett.* **1988**, 843.
20. Abdallah, H.; Gree, R.; Carrie, R. *Tetrahedron Lett.* **1982**, *23*, 503.
21. Berlan, J.; Besace, Y.; Pourcelot, G.; Cresson, P. *Tetrahedron* **1986**, *42*, 4757.
22. Tamao, K.; Kanatani, R.; Kumada, M. *Tetrahedron Lett.* **1984**, *25*, 1913.
23. Bertz, S. H.; Dabbagh, G.; Sundararajan, G. *J. Org. Chem.* **1986**, *51*, 4953.
24. Corey, E. J.; Naef, R.; Hannon, F. J. *J. Am. Chem. Soc.* **1986**, *108*, 7114.
25. Alexakis, A.; Mutti, S.; Normant, J. F. *J. Am. Chem. Soc.* **1991**, *113*, 6332.
26. Jansen, J. F. G. A.; Feringa, B. L. *J. Org. Chem.* **1990**, *55*, 4168.
27. Soai, K.; Hayasaka, T.; Ugajin, S. *Chem. Commun.* **1989**, 516.
28. (a) Sato, T.; Soai, K.; Suzuki, K.; Mukaiyama, T. *Chem. Lett.* **1978**, 601. (b) Mukaiyama, T.; Soai, K.; Sato, T.; Shimizu, H.; Suzuki, K. *J. Am. Chem. Soc.* **1979**, *101*, 1455.
29. Chaloner, P. A.; Perera, S. A. R. *Tetrahedron Lett.* **1987**, *28*, 3013.
30. (a) Chaloner, P. A.; Langadianou, E. *Tetrahedron Lett.* **1990**, *31*, 5185. (b) Chaloner, P. A.; Langadianou, E.; Perera, S. A. R. *J. Chem. Soc., Perkin Trans. 1* **1991**, 2731.
31. (a) Corey, E. J.; Hannon, F. J. *Tetrahedron Lett.* **1987**, *28*, 5233. (b) Corey, E. J.; Hannon, F. J. *Tetrahedron Lett.* **1987**, *28*, 5237.
32. Soai, K.; Nishi, M.; Ito, Y. *Chem. Lett.* **1987**, 2405.

Kenso Soai
Science University of Tokyo, Japan

Ephedrine-borane[1]

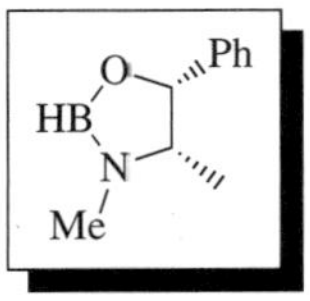

[126874-38-2] $C_{10}H_{14}BNO$ (MW 175.04)

(chiral Lewis acid catalyst for carbonyl additions, reductions, and hydroborations[1])

Alternate Name: (4*S*,5*R*)-3,4-dimethyl-5-phenyl-1,3,2-oxazaborolidine.
Physical Data: [2,4] colorless liquid, bp 110–112 °C/20 mmHg, 45 °C (bath)/0.1 mmHg; $[\alpha]_D^{20} = -108°$ (*c* 1.0 $CHCl_3$). For the corresponding reagent prepared from pseudoephedrine (1*S*,2*S*), $[\alpha]_D^{20} = +59°$ (*c* 1.0 $CHCl_3$). Spectral data:[2a,3,4] ^{1}H NMR selected data δ ($CDCl_3$) 3.68 (m), 5.58 (d, J = 8.5 Hz); ^{1}H NMR δ (C_6D_6) 0.43 (d, J = 6.6 Hz), 2.50 (s), 3.28 (m), 5.41 (d, J = 8.6 Hz); ^{11}B NMR δ +29 ppm (d, J = 147 Hz); ^{13}C NMR δ (C_6D_6) 15.42, 30.25, 59.62, 83.58, 126.59, 127.32, 128.13, 140.02; IR_{B-H} 2562 cm^{-1}. For the corresponding reagent from pseudoephedrine (1*S*,2*S*): ^{1}H NMR selected data δ ($CDCl_3$) 3.30 (m), 4.90 (d, J = 7.0 Hz).
Solubility: sol THF.
Form Supplied in: not commercially available.
Preparative Methods: [2] a solution of (1*R*,2*S*)-(−)-ephedrine (8.25 g, 50 mmol) in anhydrous THF (50 ml) was treated with *Borane–Dimethyl Sulfide* complex (50 mmol, 5 mL of 10 M solution). The reaction mixture was stirred at 25 °C for 1 h, at which time one equivalent of hydrogen had evolved. The volatiles were removed in vacuo to furnish a white solid, ^{11}B NMR (δ ~8 ppm). The solid was gradually heated to 100 °C and maintained at that temperature until the second equivalent of hydrogen had evolved. The product was distilled under reduced pressure to provide the pure oxazaborolidine (86%). An alternative procedure is available.[3]
Handling, Storage, and Precautions: the reagent is sensitive to moisture and should be handled under a dry, inert atmosphere.

Enantioselective Addition of Diethylzinc to Aldehydes. The reagent catalyzes the enantioselective addition of *Diethylzinc* to aldehydes (eq 1).[2]

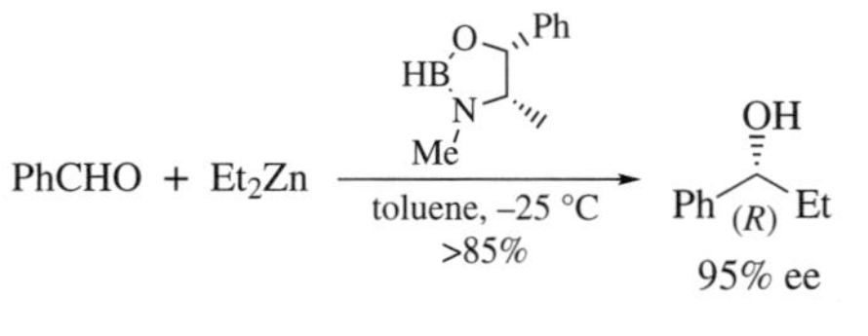

(1)

Enantioselective Alkene Hydroboration. Substituted styrene derivatives undergo rhodium-catalyzed hydroboration in the presence of a catalytic amount of the title reagent. However, optimal regio- as well as enantioselection is attained by using the corresponding reagent derived from pseudoephedrine (eq 2).[4]

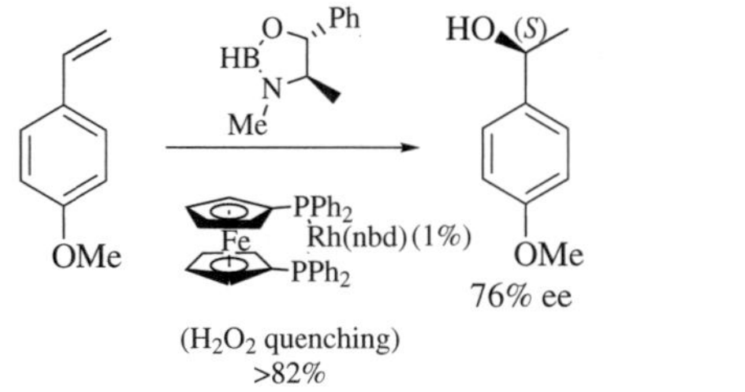

(2)

If the *N*-isopropyl analog of the borane reagent is used in conjunction with bis(4-methoxystyrene)rhodium chloride dimer as catalyst, the corresponding (*E*)-vinylborane and 4-methoxyethylbenzene are obtained in equal proportions (eq 3).[5]

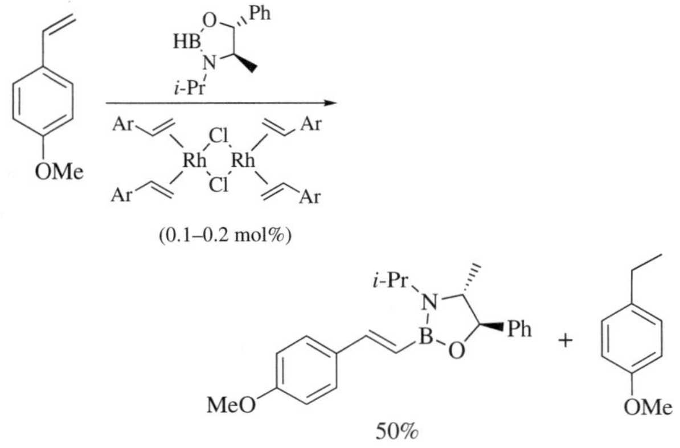

(3)

Enantioselective Carbonyl Reduction. The title reagent reacts with borane through N→B coordination. This complexation enhances the Lewis acidity at the ring boron atom, thereby triggering dimerization of the adduct via hydride bridging.[6] This complex is capable of reducing acetophenone with good enantioselectivity (eq 4). The proposed reactive complex features *anti* coordination of acetophenone to the ring boron atom and *syn* to the coordinated BH_3.[7]

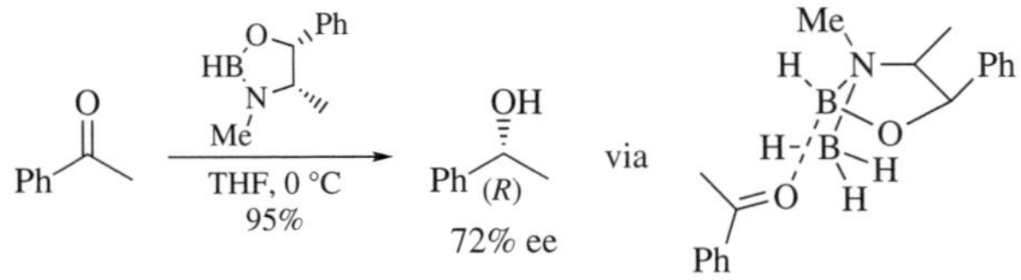

(4)

Several other 1,3,2-oxazaborolidines have been successfully used as chiral catalysts or reagents in borane-promoted reduction of ketones,[8] imines and oxime ethers,[9] and lactones[10] as well as in aldol condensations,[11] Diels–Alder cycloadditions,[12] and allylmetal additions to aldehydes.[13]

Related Reagents. α,α-Diphenyl-2-pyrrolidinemethanol; Norephedrine–Borane; Tetrahydro-1-methyl-3,3-diphenyl-1*H*,3*H*-pyrrolo[1,2-*c*][1,3,2]oxazaborole.

1. (a) Nishizawa, M.; Noyori, R. *Comprehensive Organic Synthesis* **1991**, *8*, Chapter 1.7. (b) Midland, M. *Asymmetric Synthesis*; Academic: New York, 1983; Vol. 2. (c) Midland, M. *Chem. Rev.* **1989**, *89*, 1553. (d) Seyden-Penne, J. *Reductions by the Alumino- and Borohydrides in Organic Synthesis*; VCH: New York, 1991. (e) Wallbaum, S.; Martens, J. *Tetrahedron: Asymmetry* **1992**, *3*, 1475.
2. (a) Joshi, N. N.; Srebnik, M.; Brown, H. C. *Tetrahedron Lett.* **1989**, *30*, 5551. (b) Soai, K.; Niwa, S. *Chem. Rev.* **1992**, *92* 833.
3. Tlahuext, H.; Contreras, R. *Tetrahedron: Asymmetry* **1992**, *3*, 727.
4. (a) Brown, J. M.; Lloyd-Jones, G. *Tetrahedron: Asymmetry* **1990**, *1*, 869. See also: (b) Burgess, K.; Van der Donk, W.; Ohlmeyer, M. J. *Tetrahedron: Asymmetry* **1991**, *2*, 613.
5. Brown, J. M.; Lloyd-Jones, G. C. *Chem. Commun./J. Chem. Soc., Chem. Commun.* **1992**, 710.
6. Tlahuext, H.; Contreras, R. *Tetrahedron: Asymmetry* **1992**, *3*, 1145.
7. Berenguer, R.; Garcia, J.; Gonzàlez, M.; Vilarrasa, J. *Tetrahedron: Asymmetry* **1993**, *4*, 13.
8. (a) Corey, E. J.; Link, J. O. *Tetrahedron Lett.* **1992**, *33*, 4141. (b) Martens, J; Dauelsberg, Ch.; Behnen, W.; Wallbaum, S. *Tetrahedron: Asymmetry* **1992**, *3*, 347. (c) Rao, A. V. R.; Gurjar, M. K.; Kaiwar, V. *Tetrahedron: Asymmetry* **1992**, *3*, 859. (d) Lohray, B. B.; Bhushan, V. *Angew. Chem.* **1992**, *104*, 740. (e) Jones, T. K.; Mohan, J. J.; Xavier, L. C.; Blacklock, T. J.; Mathre, D. J.; Sohar, P.; Turner Jones, E. T.; Reamer, R. A.; Roberts, R. A.; Roberts, F. E.; Grabowski, E. J. *J. Org. Chem.* **1991**, *56*, 763. (f) Mathre, D. J.; Jones, T. K.; Xavier, L. C.; Blacklock, T. J.; Reamer, R. A.; Mohan, J. J.; Turner Jones, E. T.; Hogsteen, K.; Baum, M. W.; Grabowski, E. J. *J. Org. Chem.* **1991**, *56*, 751. (g) De Ninno, M. P.; Perner, R. J.; Lijewski, L. *Tetrahedron Lett.* **1990**, *31*, 7415. (h) Rao, A. V. R.; Gurjar, M. K.; Sharma, P. A.; Kaiwar, V. *Tetrahedron Lett.* **1990**, *31*, 2341. (i) Corey, E. J.; Link, J. O. *Tetrahedron Lett.* **1989**, *30*, 6275. (j) Corey, E. J.; Chen, C.-P.; Reichard, G. A. *Tetrahedron Lett.* **1989**, *30*, 5547. (k) Youn, I. K.; Lee, S. W.; Pak, C. S. *Tetrahedron Lett.* **1988**, *29*, 4453. (l) Corey, E. J. *Pure Appl. Chem.* **1990**, *62*, 1209. (m) Corey, E. J.; Bakshi, R. K.; Shibata, S.; Chen, C.-P.; Singh, V. K. *J. Am. Chem. Soc.* **1987**, *109*, 7925. (n) Corey, E. J.; Link, J. O. *J. Am. Chem. Soc.* **1992**, *114*, 1906. (o) Stingl, K.; Martens, J.; Wallbaum, S. *Tetrahedron: Asymmetry* **1992**, *3*, 223. (p) Wallbaum, S.; Martens, J. *Synth. Commun.* **1991**, *2*, 1093. Behnen, W.; Dauelsberg, Ch.; Wallbaum, S.; Martens, J. *Synth. Commun.* **1992**, *22*, 2143. (q) Corey, E. J.; Yi, K. Y.; Matsuda, S. P. T. *Tetrahedron Lett.* **1992**, *33*, 2319. (r) Jones, D. K.; Liotta, D. C. *J. Org. Chem.* **1993**, *58*, 799. (s) Corey, E. J.; Shibata, S.; Bakshi, R. K. *J. Org. Chem.* **1988**, *53*, 2861. (t) Corey, E. J.; Bakshi, R. K. *Tetrahedron Lett.* **1990**, *31*, 611. (u) DeNinno, M. P.; Perner, R. J.; Morton, H. E.; DiDomenico, S., Jr. *J. Org. Chem.* **1992**, *57*, 7115. (v) Tanaka, K.; Matsui, J; Suzuki, H. *Chem. Commun./J. Chem. Soc., Chem. Commun.* **1991**, 1311. (w) Corey, E. J.; Bakshi, R. K.; Shibata, S. *J. Am. Chem. Soc.* **1987**, *109*, 5551.
9. (a) Cho, B. T.; Chun, Y. S. *Tetrahedron: Asymmetry* **1992**, *3*, 337. (b) Itsuno, S.; Sakurai, Y.; Ito, K.; Hirao, A.; Nakahama, S. *Bull. Chem. Soc. Jpn.* **1987**, *60*, 395.
10. Bringmann, G.; Hartung, T. *Angew. Chem., Int. Ed. Engl.* **1992**, *31*, 761.
11. (a) Parmee, E. R.; Tempkin, O.; Masamune, S. *J. Am. Chem. Soc.* **1991**, *113*, 9365. (b) Kiyooka, S.; Kaneko, Y.; Kume, K. *Tetrahedron Lett.* **1992**, *33*, 4927.
12. (a) Corey, E. J.; Loh, T.-P. *J. Am. Chem. Soc.* **1991**, *113*, 8966. (b) Takasu, M.; Yamamoto, H. *Synlett* **1990**, 194. (c) Sartor, D.; Saffrich, J.; Helmchen, G.; Richards, C. J.; Lambert, H. *Tetrahedron: Asymmetry* **1991**, *2*, 639. (d) Sartor, D.; Saffrich, J.; Helmchen, G. *Synlett* **1990**, 197. (e) Kagan, H. B.; Riant, O. *Chem. Rev.* **1992**, *92*, 1007.
13. Reetz, M. T.; Zierke, T. *Chem. Ind. (London)* **1988**, 663.

Giovanni Poli
Università di Firenze, Italy

Epichlorohydrin[1]

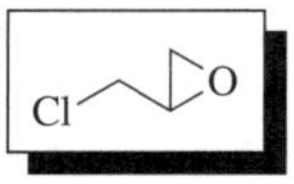

[106-89-8] C_3H_5ClO (MW 92.53)
(R)
[51594-55-9]
(S)
[67843-74-7]

(readily available three-carbon unit functionalized on every carbon; convenient HCl or HBr trap; linker for various polymers)

Alternate Name: chloromethyloxirane.
Physical Data: mp −57 °C; bp 115–117 °C; *d* 1.183 g cm^{-3}.
Solubility: 6.6 wt% in water; sol alcohol, acetone, THF, toluene, *n*-heptane.
Form Supplied in: neat liquid; both enantiomers available.
Handling, Storage, and Precautions: should only be handled in a well ventilated fume hood because of its low permissible exposure limit of 2 ppm and reports of allergic skin reactions and lung, liver, and kidney damage. MSDSs are available from the two principal manufacturers (Dow and Shell). The material is not moisture or air sensitive.

Introduction. Epichlorohydrin (**1**) is most widely used in polymer synthesis.[2] Other common uses include an in situ trapping agent for HCl, HBr,[3] or the alcohol generated during formation of Meerwein's reagent (eq 1).[4]

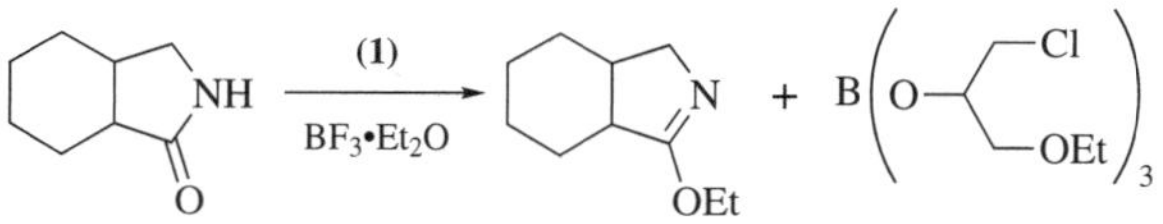
(1)

Reactions with Nucleophiles. The epoxide is, by far, the more reactive site and a wide variety of nucleophiles have been used (eq 2) to open the ring at C-3 such as HCl (96%),[5] HOAc (>50%),[6] H_2S (65% as cyclized product 3-thietanol),[7] HCN (66%),[8] ethanol (90%),[9] *t*-butanol (86%),[10] phenyl or benzyl thiol (99% or 93%, respectively),[11] and phenyl selenide (generated in situ from the diselenide and sodium hydroxymethyl sulfite) (>55%).[12] If desired, the epoxide is easily formed from the chlorohydrin by treatment with excess KOH or Et_3N.

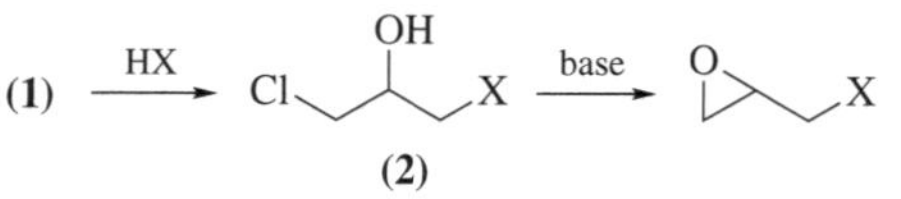
(2)

The epoxide is also opened at C-3 by various electrophilic reagents that fit into the generalized scheme in eq 3. Examples include *Chlorotrimethylsilane* (TMSCl) (85%),[13] TMSCl/NaBr (X = Br) (85%),[14] *Cyanotrimethylsilane* (91%),[15] *Azidotrimethylsilane* (83%),[16] *Thionyl Chloride* (70%),[17] H_2NCOCl (96%),[18] and MeCH=CHCOCl (80%).[19] The only report of unusual selectivity for opening the epoxide at C-2 was for *Sulfuryl Chloride* (eq 4).[20]

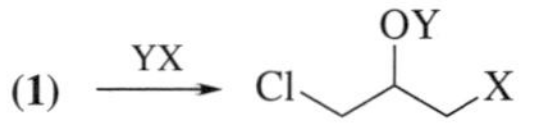
(3)

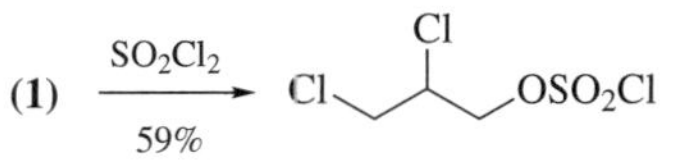
(4)

A number of special catalysts have been developed to facilitate ring opening and improve the regioselectivity for reaction at C-3. For example, Sn^{II} halides are useful in preparations of (**2**) (X = Cl, 70%; X = Br, 63%; X = I, 90%).[21] An equimolar mixture of *Lithium Bromide* and *Copper(II) Bromide* gave (**2**) (X = Br, 93%).[22] The ring can be opened selectively by anilines in the presence of other amines when *Cobalt(II) Chloride* is the catalyst.[23] $MgSO_4$ was found to catalyze the addition of 2 mol of CN^- to (**1**) to afford 3-hydroxyglutaronitrile.[24] CaF_2 supported on KF was used in the conversion of (**1**) to epifluorohydrin.[25] A catalyst composed of a 1:2 mole ratio of *Di-n-butyltin Oxide* and tributyl phosphate was developed for ring opening by alcohols.[26] Other catalysts shown to be of value for the examples given above include $FeCl_3$,[6] $LiClO_4$,[11] Et_3N,[11] CAN,[9] DDQ,[10] Ti(O-*i*-Pr)$_4$,[15b] $CoCl_2$,[13,19] $YbCl_3$,[15a] and Al(O-*i*-Pr)$_3$.[16]

A variety of carbon nucleophiles react at C-3 with high regioselectivity. Examples include Grignard reagents,[27] aryllithium,[28] alkynyllithium,[29] and others (eqs 5–8).[30–33]

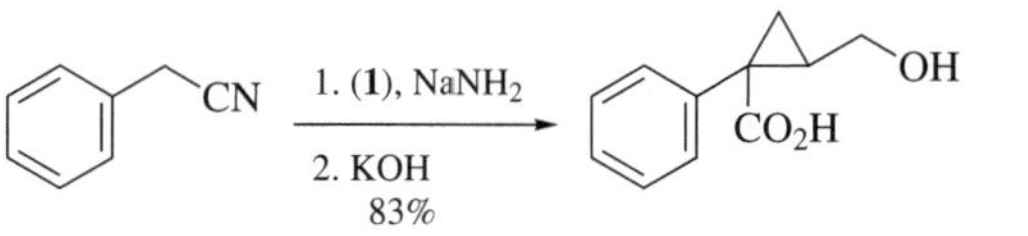
(5)

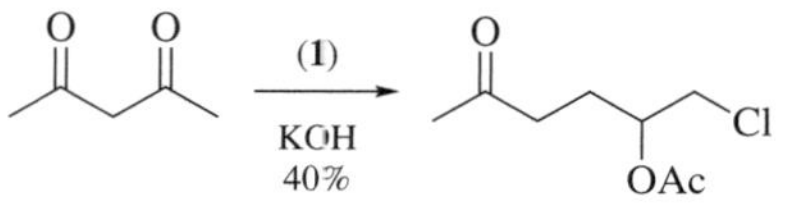
(6)

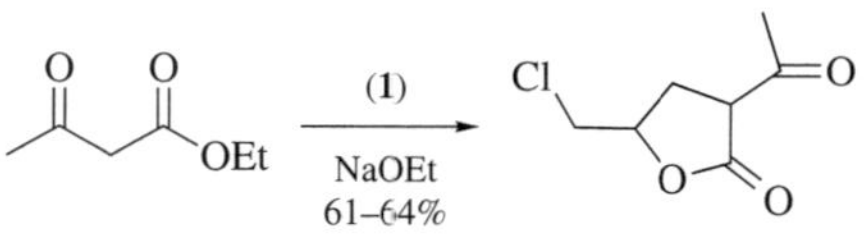
(7)

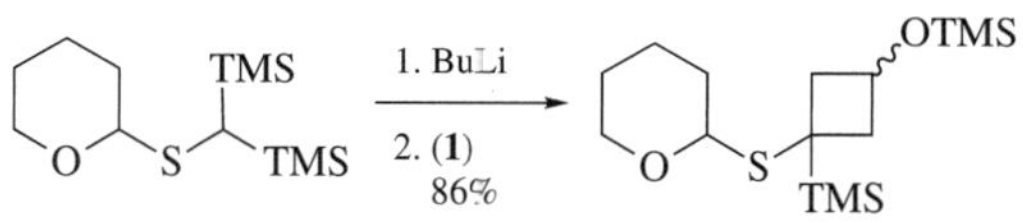
(8)

Compound (**1**) can be chain extended by one carbon by a Co-catalyzed CO insertion followed by reduction (eq 9).[34] A whole class of medicinally important compounds called β-blockers are prepared from (**1**) as illustrated in eq 10 for the synthesis of a propranolol analog.[35]

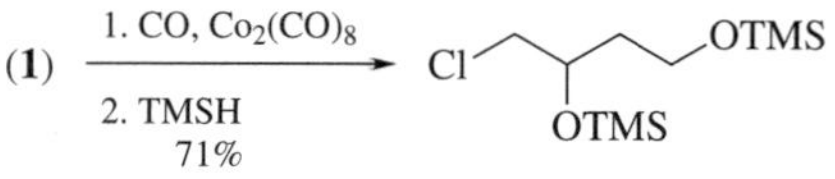
(9)

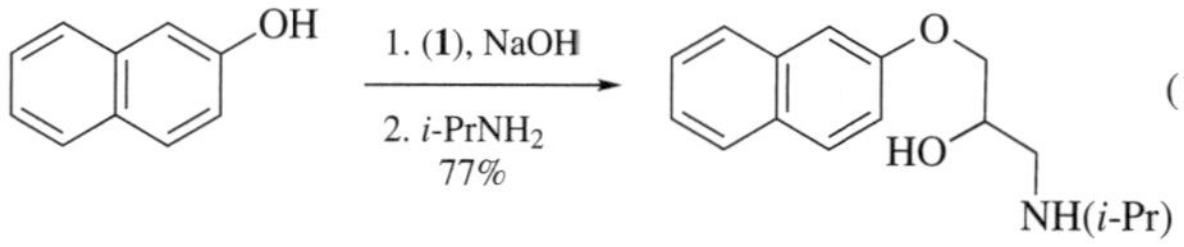
(10)

Preparation of Heterocycles. A wide array of heterocycles are available from (**1**). A few examples are shown in (eqs 11–16).[36–41]

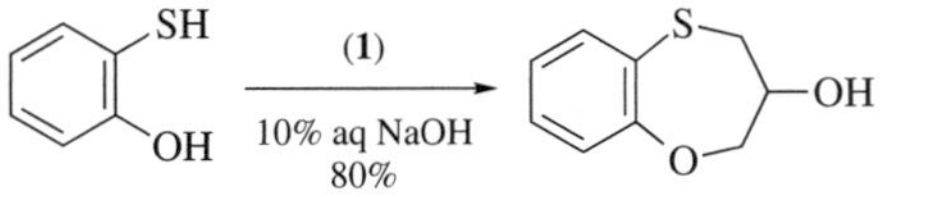
(11)

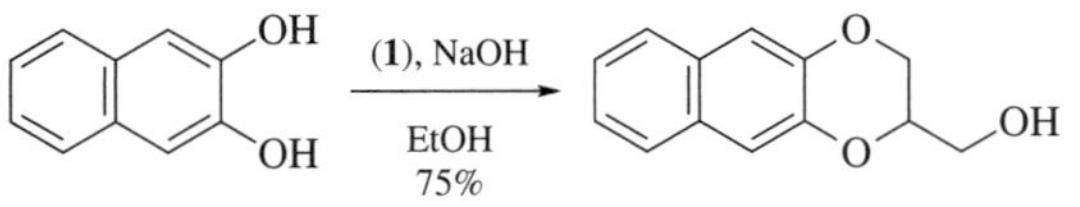
(12)

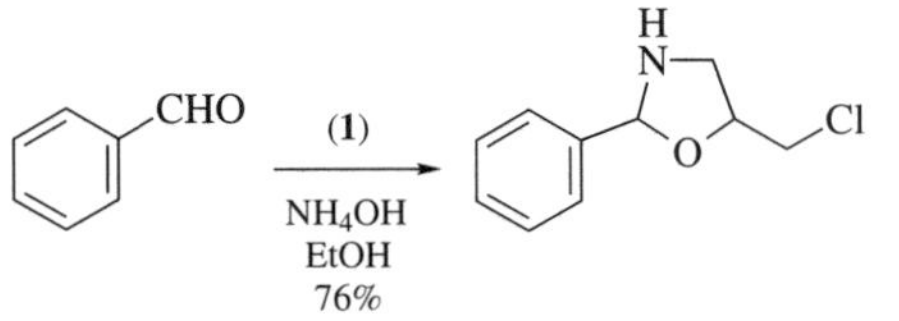
(13)

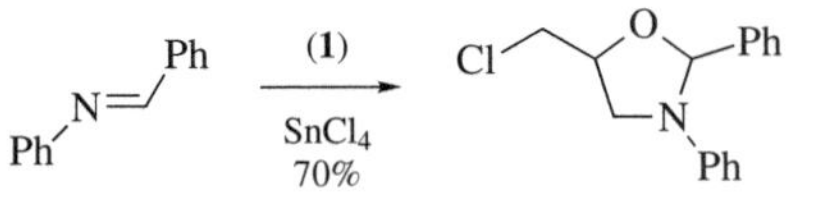
(14)

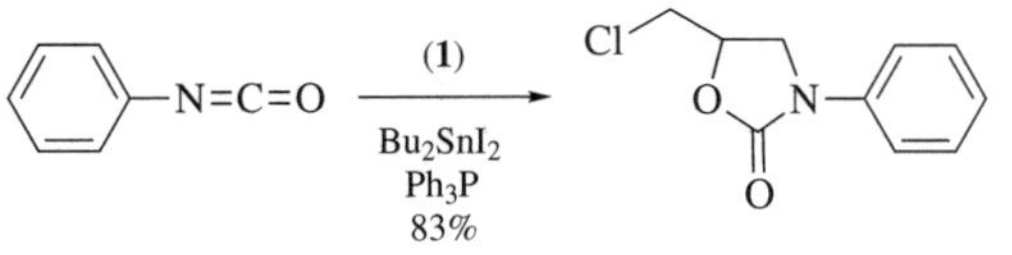
(15)

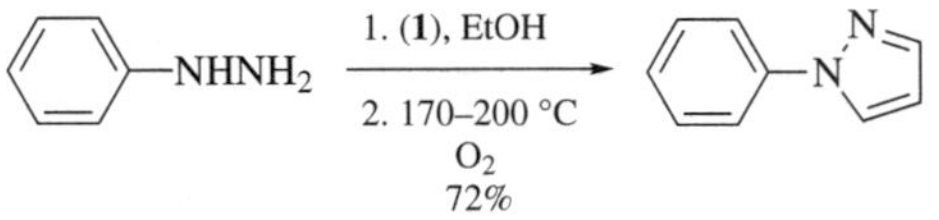
(16)

In some cases the reaction rate and yield are dramatically improved if the product can be trapped as the TMS ether (eq 17).[42] Also, (**1**) lends itself well to nucleophilic opening under phase transfer conditions (eq 18).[43]

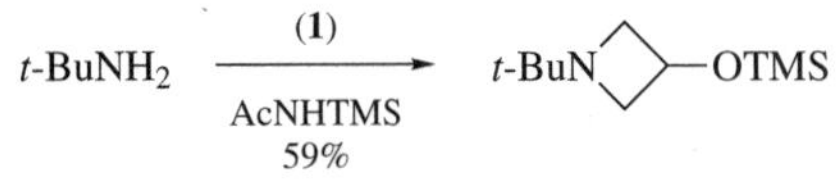
(17)

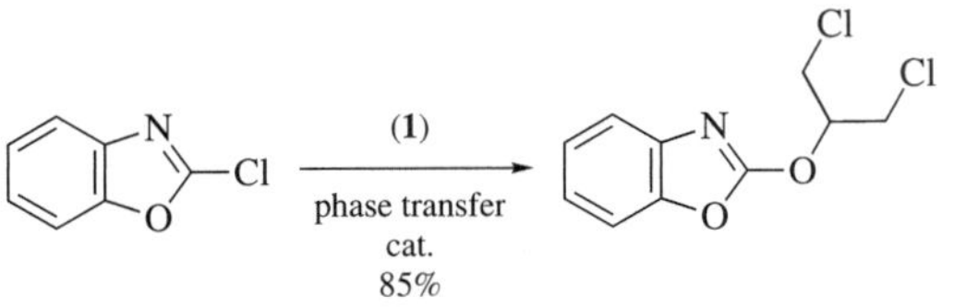
(18)

The ready availability[44] of both enantiomers of (**1**) has greatly enhanced its value as a synthetic intermediate. The pheromone (*S*)-(−)-ipsenol (**2**), prepared[45] in 16% overall yield in four steps from (*R*)-(**1**), is just one of many examples of this utility. In practice, either isomer can sometimes be used by adjusting the order of addition of the groups at C-1 and C-3. The synthesis of (−)-anisomycin (**3**) illustrates this point.[46]

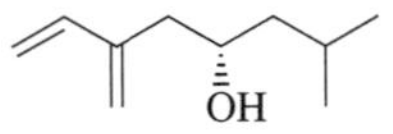
(**2**) (*S*)-(−)-Ipsenol

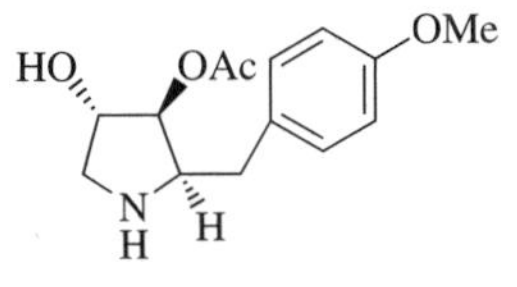
(**3**) (−)-Anisomycin

Related Reagents. Glycidol.

1. *Encyclopedia of Chemical Technology*, 3rd ed.; Wiley: New York, 1978; Vol. 5, pp 858–864; 4th ed., 1991; Vol. 2, pp 146 and 156; 1991; Vol. 6, pp 140–155.
2. For example, *Chem. Abstr.*, 12th Coll. Index lists 133 pages of references to polymers.
3. Sato, K.; Kojima, Y.; Sato, H. *J. Org. Chem.* **1970**, *35*, 2374.
4. (a) Petersen, S.; Tietze, E. *Liebigs Ann. Chem.* 1959, *623*, 166 (*Chem. Abstr.* **1960**, *54*, 14 257i). (b) Meerwein, H. *Org. Synth., Coll. Vol.* **1973**, *5*, 1080. (c) Curphey, T. J. *Org. Synth., Coll. Vol.* **1988**, *6*, 1019.
5. Spadlo, M.; *Przem. Chem.* **1990**, *69*, 164 (*Chem. Abstr.* **1990**, *113*, 190 697e).
6. Kozikowski, A. P.; Fauq, A. H. *Synlett* **1991**, 78[illegible].
7. Lamm, B.; Gustafsson, K. *Acta Chem. Scand.* **1974**, *B28*, 701.
8. Culvenor, C. C. J.; Davies, W.; Haley, F. G. *J. Chem. Soc.* **1950**, 3123.
9. Iranpoor, N.; Baltork, I. M. *Synth. Commun.* **1990**, *20*, 2789.
10. Iranpoor, N.; Baltork, I. M. *Tetrahedron Lett.* **1990**, *31*, 735.
11. Chini, M.; Crotti, P.; Giovani, E.; Macchia, F.; Pineschi, M. *Synlett* **1992**, 303.
12. Gasanov, F. G.; Aliev, A. Y.; Mamedov, E. G.; Akhmedov, I. M. *Azerb. Khim. Zh.* **1981**, *5*, 49 (*Chem. Abstr.* **1982**, *96*, 2[illegible]7 607v).
13. Iqbal, J.; Khan, M. A. *Chem. Lett.* **1988**, 1157.
14. Iqbal, J.; Khan, M. A.; Ahmad, S. *Synth. Commun.* **1989**, *19*, 641.
15. (a) Matsubara, S.; Onishi, H.; Utimoto, K. *Tetrahedron Lett.* **1990**, *31*, 6209. (b) Hayashi, M.; Tamura, M.; Oguni, N. *Synlett* **1992**, 663.
16. Emziane, M.; Lhoste, P.; Sinou, D. *Synthesis* **1988**, 541.
17. Etienne, A.; LeBerre, A.; Coquelin, J. *C. R. Hebd. Seances Acad. Sci., Ser. C* **1972**, *275*, 633 (*Chem. Abstr.* **1973**, *78*, 1[illegible]3 928b).
18. Boberg, F.; Schultze, G. R. *Ber. Dtsch. Chem. Ges.* **1955**, *88*, 275 (*Chem. Abstr.* **1956**, *50*, 1603e).
19. Iqbal, J.; Khan, M. A.; Srivastava, R. R. *Tetrahedron Lett.* **1988**, *29*, 4985.
20. Malinovskii, M. S. *J. Gen. Chem. USSR (Engl. Transl.)* **1947**, *17*, 1559 (*Chem. Abstr.* **1948**, *42*, 2229b).
21. Einhorn, C.; Luche, J. L. *Chem. Commun.* **1986**, 1368.
22. Ciaccio, J. A.; Heller, E.; Talbot, A. *Synlett* **1991**, 248.
23. Iqbal, J.; Pandey, A. *Tetrahedron Lett.* **1990**, *31*, 575.
24. Johnson, F.; Panella, J. P. *Org. Synth., Coll. Vol.* **1973**, *5*, 614.
25. Ichihara, J.; Matsuo, T.; Hanafusa, T.; Ando, T. *Chem. Commun.* **1986**, 793.
26. Otera, J.; Yoshinaga, Y.; Hirakawa, K. *Tetrahedron Lett.* **1985**, *26*, 3219.
27. DeCamp Schuda, A.; Mazzocchi, P. H.; Fritz, G.; Morgan, T. *Synthesis* **1986**, 309.
28. (a) Takano, S.; Yanase, M.; Sekiguchi, Y.; Ogasawara, K. *Tetrahedron Lett.* **1987**, *28*, 1783. (b) Takano, S.; Yanase, M.; Ogasawara, K. *Heterocycles* **1989**, *29*, 1825.
29. (a) South, M. S.; Liebeskind, L. S. *J. Am. Chem. Soc.* **1984**, *106*, 4181. (b) Hatakeyama, S.; Sugawara, K.; Kawamura, M.; Takano, S. *Synlett* **1990**, 691. (c) Russell, S. W.; Pabon, H. J. J. *J. Chem. Soc., Perkin Trans. 1* **1982**, 545.
30. Mouzin, G.; Cousse, H.; Bonnaud, B. *Synthesis* **1978**, 304.
31. Sangwan, N. K.; Dhindsa, K. S. *Org. Prep. Proced. Int.* **1989**, *21*, 241.
32. Zuidema, G. D.; vanTamelen, E.; VanZyl, G. *Org. Synth., Coll. Vol.* **1963**, *4*, 10.
33. Block, E.; Laffitte, J-A.; Eswarakrishnan, V. *J. Org. Chem.* **1986**, *51*, 3428.
34. Murai, T.; Kato, S.; Murai, S.; Toki, T.; Suzuki, S.; Sonoda, N. *J. Am. Chem. Soc.* **1984**, *106*, 6093.
35. Farina, J. S.; Jackson, S. A.; Cummings, C. L. *Org. Prep. Proced. Int.* **1989**, *21*, 173.
36. Cabiddu, S.; Melis, S.; Sotgiu, F. *Phosphorus Sulfur* **1983**, *14*, 151.
37. Parekh, K. B.; Shelver, W. H.; Tsai, A.-Y. S.; Rechelle, R. *J. Pharm. Sci.* **1975**, *64*, 875.

38. Mazzetti, F.; Lemmon, R. M. *J. Org. Chem.* **1957**, *22*, 228.
39. Oda, R.; Okano, M.; Tokiura, S.; Miyasu, A. *Bull. Chem. Soc. Jpn.* **1962**, *35*, 1216.
40. Baba, A.; Shibata, I.; Masuda, K.; Matsuda, H. *Synthesis* **1985**, 1144.
41. Finar, I. L.; Godfrey, K. E. *J. Chem. Soc.* **1954**, 2293.
42. Higgins, R. H.; Watson, M. R.; Faircloth, W. J.; Eaton, Q. L.; Jenkins, H. *J. Heterocycl. Chem.* **1988**, *25*, 383.
43. Jin, R.-H.; Nishikubo, T. *Synthesis* **1993**, 28.
44. Baldwin, J. J.; Raab, A. W.; Mensler, K.; Arison, B. H.; McClure, D. E. *J. Org. Chem.* **1978**, *43*, 4876.
45. Imai, T.; Nishida, S. *J. Org. Chem.* **1990**, *55*, 4849.
46. Takano, S.; Iwabuchi, Y.; Ogasawara, K. *Heterocycles* **1989**, *29*, 1861.

Joel E. Huber
The Upjohn Co., Kalamazoo, MI, USA

Esterases

(enzymes of the class of hydrolases, which catalyze the hydrolysis of carboxylic acid esters[1])

Solubility: insol cold and warm H_2O.
Form Supplied in: available from various sources (microorganisms and mammalian) as powders or water suspensions.
Handling, Storage, and Precautions: stable at a pH range 6–10; can be stored at 0–4 °C for months.

Esterase-Catalyzed Hydrolysis. Hydrolytic enzymes have been accepted in organic synthesis as valuable biocatalysts, since they are commercially available at relatively low price and possess a broad substrate specificity, without necessitating use of expensive cofactors.[2] Esterases are such useful enzymes and have been widely used for the preparation of enantiomerically pure chiral compounds, by hydrolytic resolution of racemic esters or asymmetrization of prochiral substrates.[3] Well defined experimental procedures for a pig liver esterase-catalyzed saponification have been documented.[4] Generally, the enzymatic hydrolysis is carried out in an aqueous buffer, sometimes containing cosolvents,[5] at pH 7–9 and keeping the temperature at 20–25 °C. Generally, the molar equivalent NaOH for the hydrolysis is added maintaining the pH constant with an automatic titrator and, after acidification, the product is extracted with organic solvents. Esterases are commercially available from various sources, either microbial or mammalian, and in some instances also crude acetone powders can be used for the same purpose.

Pig Liver Esterase (PLE). This is the more used carboxylesterase (carboxylic-ester hydrolase, EC 3.1.1.1, *CAS 9016-18-6*) which physiologically catalyzes the hydrolysis of carboxylic acid esters to the free acid anion and alcohol.[1] PLE is a serine hydrolase which has been widely used for the preparation of chiral synthons and these applications have been fully reviewed.[6] An active-site model for interpreting and predicting the specificity of the enzyme has been published.[7] In the pioneering studies of the enzyme applications field, PLE was used for the chiral synthesis of mevalonolactone.[8] Prochiral 3-substituted glutaric acid diesters are well suited for a PLE-catalyzed asymmetrization, which leads to optically active monoesters (eq 1).[9]

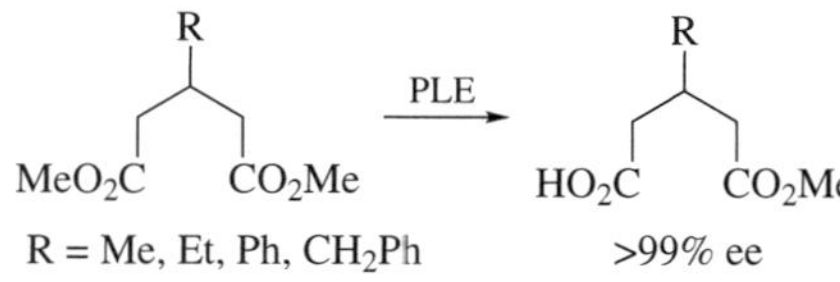

(1)

The asymmetrization of prochiral disubstituted malonates has been enantioselectively realized in the presence of PLE (eq 2).[10]

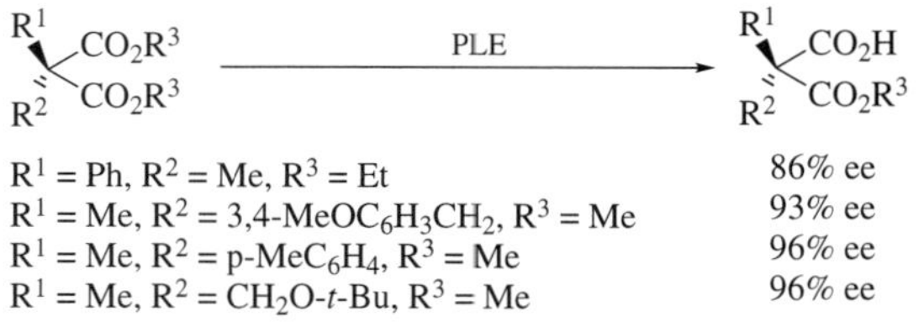

(2)

The asymmetric hydrolysis of several cyclic *meso*-diesters has been accomplished and optically pure monoesters have been obtained.[11] A classical example is the hydrolysis of dimethyl *cis*-4-cyclohexene-1,2-dicarboxylate, which affords the corresponding nearly optically pure half ester, a versatile synthon for various chiral cyclohexane derivatives (eq 3).[12]

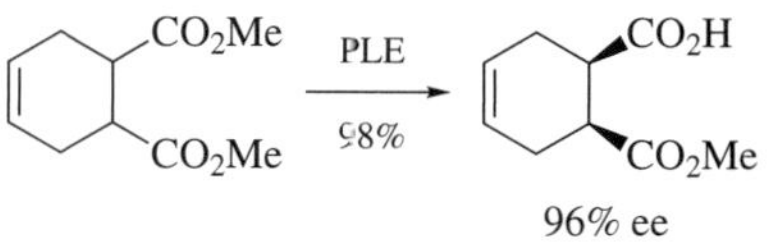

(3)

The PLE-catalyzed asymmetric hydrolysis of *meso*-1,3-*cis*-3,5-*cis*-1,3-diacetoxy-5-benzyloxycyclohexane afforded (1*S*,3*S*,5*R*)-1-acetoxy-5-benzyloxycyclohexan-3-ol, which could be used as chiral building block for the synthesis of the compactin lactone moiety and quinic acid (eq 4).[13]

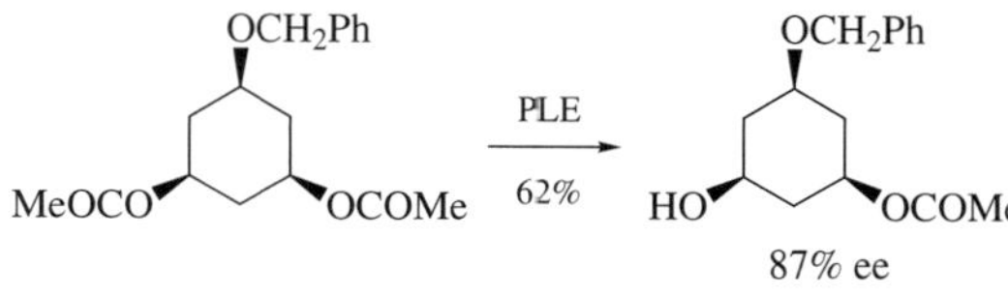

(4)

Organometallic *meso*-diesters can be asymmetrized as well to the corresponding half ester, as shown for an (arene)tricarbonylchromium diester (eq 5).[14]

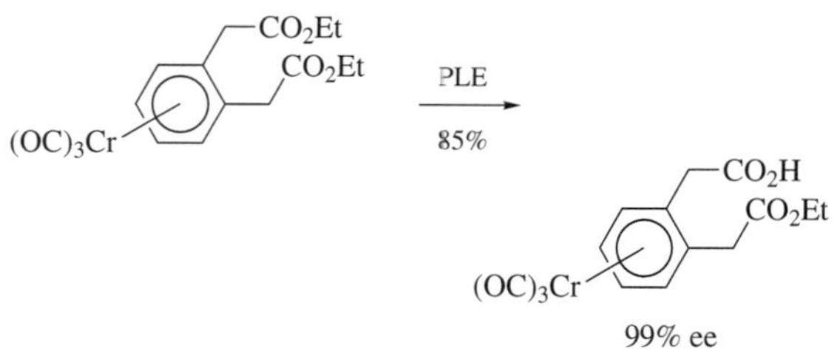

(5)

The resolution of racemic esters is catalyzed by PLE in a highly enantioselective fashion.[3,6] Several interesting applications of this method are available. The hydrolysis of *trans*-bicyclo[2.2.1]heptane diesters has been studied to ascertain the structural requirements for the PLE hydrolysis.[15] A bulky tricyclodecadienone ester can be resolved by an highly enantioselective reaction (eq 6).[16]

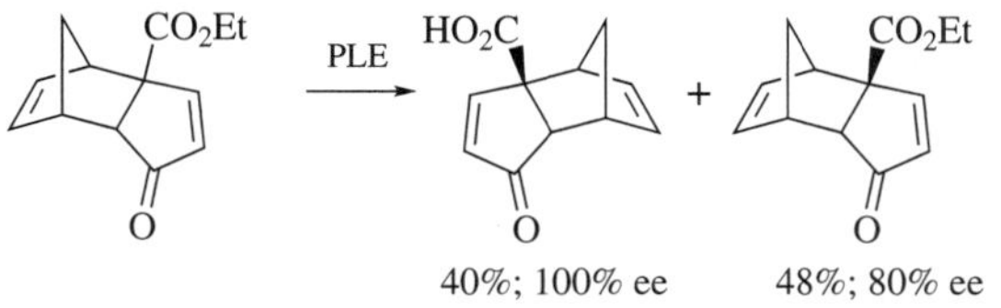

(6)

The resolution procedure applies to racemic organometallic esters[17] and to the esters of a thianucleoside, for the preparation of pure enantiomers of an antiviral agent (2′,3′-dideoxy-5-fluoro-3′-thiacytidine) (eq 7).[18]

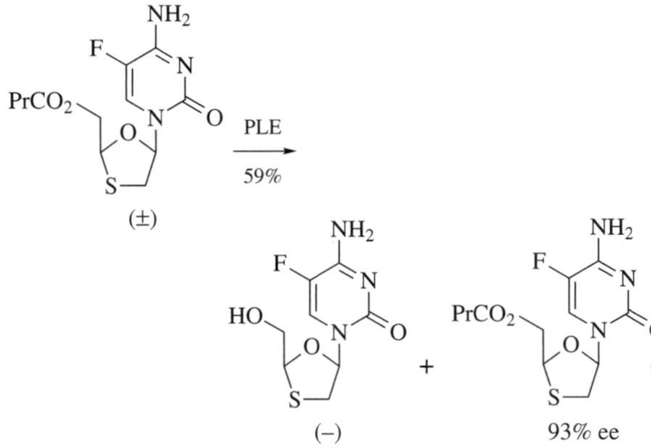

(7)

PLE has usually been applied to the enantioselective preparation of optically active compounds, but its use can be extended to chemo- or regioselective hydrolyses. A continuous process for the separation of a *cis/trans* unsaturated ester was realized using immobilized PLE (eq 8).[19]

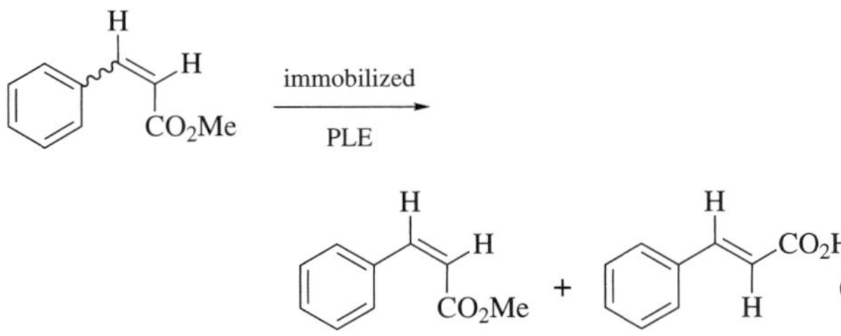

(8)

The chemoselective hydrolysis of an acetoxy group in the presence of a γ-lactone ring has been reported in the presence of PLE (eq 9).[20] In a benzylpenicillin, PLE catalyzes the chemoselective hydrolytic opening of the β-lactam ring, the methoxycarbonyl moiety remaining uneffected (eq 10).[21]

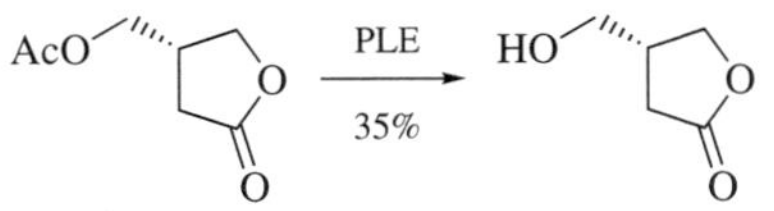

(9)

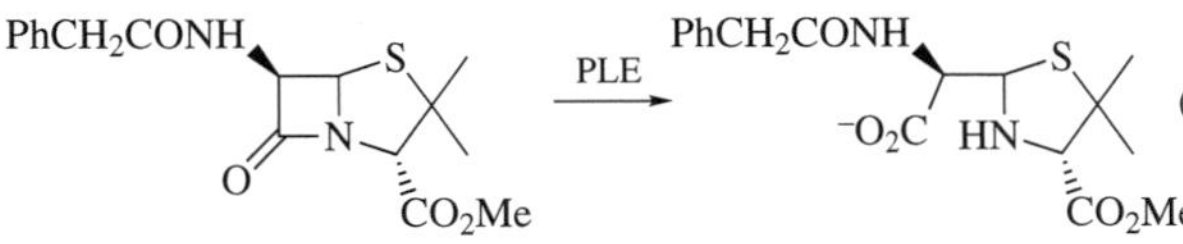

(10)

Dimethyl malate presents two ester functions α and β with respect to a hydroxy group, and PLE is able to regioselectively discriminate between these two moieties.[22]

Acetone Powder Containing Esterase Activity. The main advantage in using crude homogenates or acetone powders of organs such as liver is to have a cheap source of different enzymes. If one of these is desired for a specific substrate, the crude enzymatic mixture can be used with some advantage, compared to the purified enzymes. Pig liver acetone powder (PLAP), together with other extracts, is commercially available or can be prepared from fresh pig liver.[23] PLAP has been used for the enantioselective hydrolysis of the racemic acetate of *trans*-2-phenylcyclohexanol (eq 11).[24]

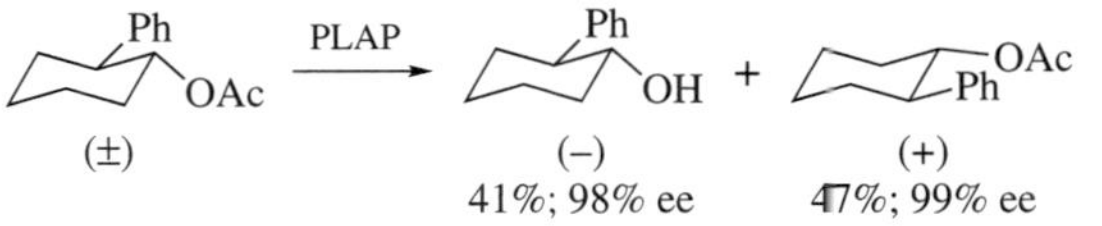

(11)

The acetates of 1-arylalkan-1-ols were successfully resolved by acetone powders (PLAP and goat liver acetone powder, GLAP) containing esterase activity (eq 12).[25]

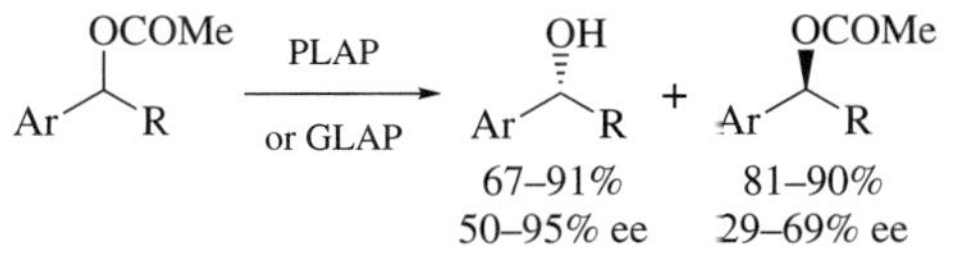

(12)

An interesting application of the esterase activity of horse liver acetone powder (HLE) has been the enantioselective hydrolysis of racemic lactones. The powder proved to be more effective than PLE in this hydrolysis, from which the unreacted lactone was recovered with high enantiomeric excess. The process seems more effective for δ and medium size lactones (eq 13).[26]

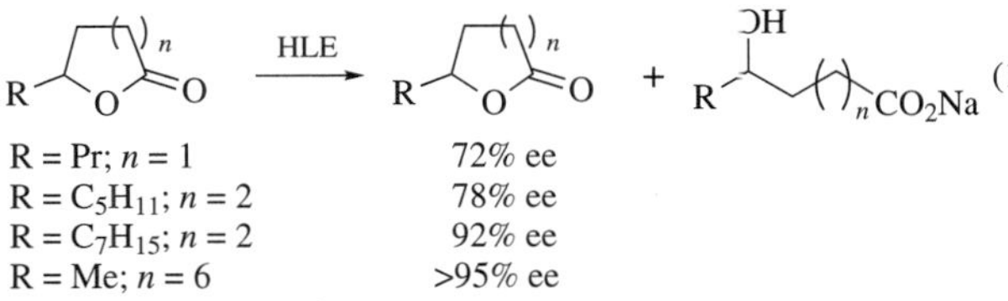

(13)

Cholesterol Esterase. This enzyme (EC 3.1.1.13; *CAS 9026-00-0*) physiologically catalyzes the hydrolysis of cholesterol esters, monoacylglycerols, and vitamin esters.[27] It has also been used for several cyclic and noncyclic substrates with variable enantioselectivity.[28] The resolution of racemic esters has been reported[29] and an interesting example is the application to the racemic acetate of an hemiacetal (eq 14).[30]

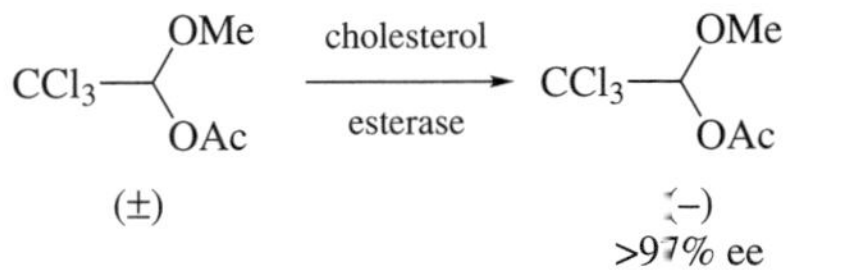

(14)

Acyl Cholinesterases. Acetylcholinesterase (AChE; EC 3.1.1.7; *CAS 9000-81-1*) is the serine esterase which catalyzes the hydrolysis of acetylcholine and possesses an esteratic site,[31] and which is responsible for unspecific hydrolyses of several substrates. Also, butyrylcholinesterase (EC 3.1.1.8; *CAS 9001-08-5*) has been sometimes used for asymmetric hydrolysis of esters.[32] Acetylcholinesterase has been used for the hydrolysis of noncyclic substrates and the results have shown satisfactory enantioselectivity.[32a,33] The enzyme from electric eel seems especially well suited to the hydrolysis of cyclic diols.[34] The asymmetrization of *cis*-3,5-diacetoxycyclopent-1-ene to (3*R*)-acetoxy-(5*S*)-hydroxycyclopent-1-ene (eq 15)[35] and the preparation of an optically active triol monoacetate starting from the triacetate of 1,3,6-trihydroxycyclohept-4-ene (eq 16)[36] are good examples of successful reactions catalyzed by acetylcholinesterase.

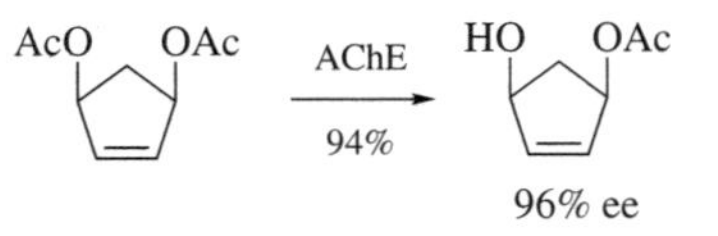

(15)

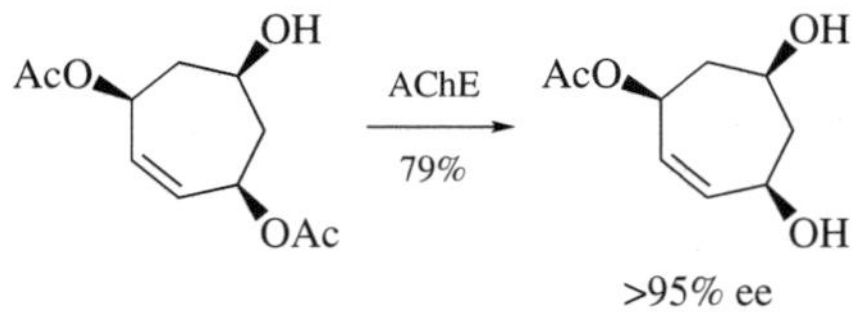

(16)

Other Esterases. Other less common esterases have been sometimes used for biocatalytic applications in organic synthesis.[37] The enzymatic approach can be the method of choice for the preparation of optically pure drugs, although sometimes special enzymes have to be prepared for this aim. By cloning a carboxylesterase into a microorganism, high level production of the esterase is made possible for the production of 2-(aryloxy)propionates and (*S*)-naproxen.[38] The esterase activity of rabbit plasma has been used for a chemoselective hydrolysis of a methylthiomethyl ester.[39] An esterase from *Candida lypolitica* has been used for the resolution of a tertiary α-substituted carboxylic acid ester.[40] Recently, a carboxyl esterase of molecular weight 30 000 ('Esterase 30 000') has been introduced for the asymmetric hydrolysis of diesters. A cyclopropyl malonate has been hydrolyzed by the esterase and the unreacted diester was recovered nearly optically pure (eq 17).[41] Diethyl 3-hydroxyglutarate, a substrate which is asymmetrized with modest enantioselectivity with PLE or other enzymes,[9e,f] has been enantioselectively hydrolyzed in the presence of Esterase 30 000 (eq 18).[42]

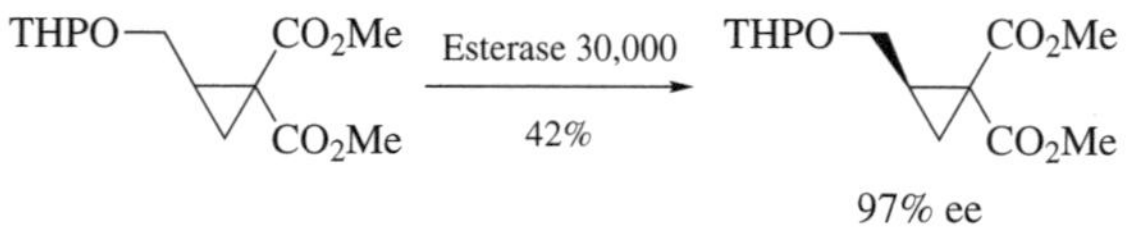

(17)

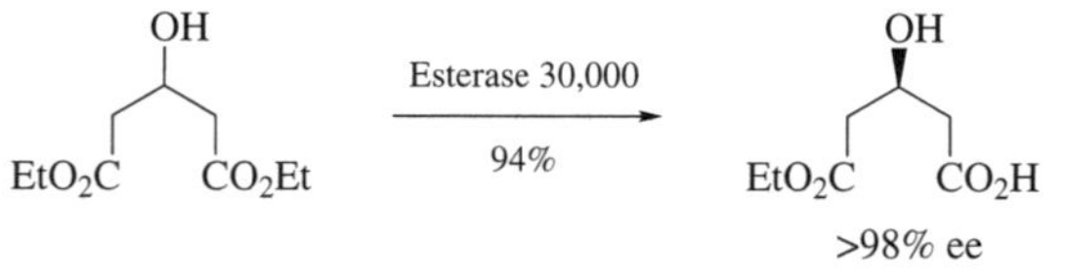

(18)

1. (a) Junge, W. In *Methods of Enzymatic Analysis*; Bergmeyer, H. U. Ed.; Verlag Chemie: Weinheim, 1984; Vol. IV, p 2. (b) *Enzyme Handbook*; Schomburg, D.; Salzmann, M., Eds.; Springer: Berlin, 1991; Vol. III.
2. Jones, J. B. *Tetrahedron* **1986**, *42*, 3351.
3. (a) Boland, W.; Frössl, C.; Lorenz, M. *Synthesis* **1991**, 1049. (b) Santaniello, E.; Ferraboschi, P.; Grisenti, P.; Manzocchi, A. *Chem. Rev.* **1992**, *92*, 1071.
4. Eberle, M.; Missbach, M.; Seebach, D. *Org. Synth.* **1990**, *69*, 19.
5. (a) Guanti, G.; Banfi, L.; Narisano, E.; Riva, R.; Thea, S. *Tetrahedron Lett.* **1986**, *27*, 4639. (b) Björkling, F.; Boutelje, J.; Hjalmarsson, M.; Hult, K.; Norin, T. *Chem. Commun./J. Chem. Soc., Chem. Commun.* **1987**, 1041. (c) Lam, L. K. P.; Hui, R. A. H. F.; Jones, J. B. *J. Org. Chem.* **1986**, *51*, 2047.
6. (a) Ohno, M.; Otsuka, M. *Org. React.* **1989**, *37*, 1. (b) Zhu, L.-M.; Tedford, M. C. *Tetrahedron* **1990**, *46*, 6587. (c) Jones, J. B. *Pure Appl. Chem.* **1990**, *62*, 1445.
7. Toone, E. J.; Werth, M. J.; Jones, J. B. *J. Am. Chem. Soc.* **1990**, *112*, 4946.
8. Huang, F.-C.; Hsu Lee, L. F.; Mittal, R. S. D.; Ravikumar, P. R.; Chan, J. A.; Sih, C. J.; Caspi, E.; Eck, C. R. *J. Am. Chem. Soc.* **1975**, *97*, 4144.
9. (a) Mohr, P.; Waespe-Šarčevič, N.; Tamm, C.; Gawronska, K.; Gawronski, J. K. *Helv. Chim. Acta* **1983**, *66*, 2501. (b) Brooks, D. W.; Palmer, J. T. *Tetrahedron Lett.* **1983**, *24*, 3059. (c) Francis, C. J.; Jones, J. B. *Chem. Commun./J. Chem. Soc., Chem. Commun.* **1984**, 579. (d) VanMiddlesworth, F.; Wang, Y. F.; Zhou, B.-N.; DiTullio, D.; Sih, C. J. *Tetrahedron Lett.* **1985**, *26*, 961. (e) Mohr, P.; Rösslein, L.; Tamm, C. *Helv. Chim. Acta* **1987**, *70*, 142. (f) Santaniello, E.; Chiari, M.; Ferraboschi, P.; Trave, S. *J. Org. Chem.* **1988**, *53*, 1567. (g) Andruszkiewicz, R.; Barrett, A. G. M.; Silverman, R. B. *Synth. Commun.* **1990**, *20*, 159. (h) Chênevert, R.; Desjardins, M. *Tetrahedron Lett.* **1991**, *32*, 4249.
10. (a) Schneider, M.; Engel, N.; Boensmann, H. *Angew. Chem., Int. Ed. Engl.* **1984**, *23*, 66. (b) Björkling, F.; Boutelje, J.; Gatenbeck, S.; Hult, K.; Norin, T. *Tetrahedron Lett.* **1985**, *26*, 4957. (c) Luyten, M.; Müller, S.; Herzog, B.; Keese, R. *Helv. Chim. Acta* **1987**, *70*, 1250. (d) De Jeso, B.; Belair, N.; Deleuze, H.; Rascle, M.-C.; Maillard, B. *Tetrahedron Lett.* **1990**, *31*, 653. (e) Fadel, A.; Canet, J.-L.; Salaün, J. *Synlett* **1991**, 60.
11. (a) Schneider, M.; Engel, N.; Hönicke, P.; Heinemann, G.; Görisch, H. *Angew. Chem., Int. Ed. Engl.* **1984**, *23*, 67. (b) Sabbioni, G.; Shea, M. L.; Jones, J. B. *Chem. Commun./J. Chem. Soc., Chem. Commun.* **1984**, 236. (c) Gais, H.-J.; Lukas, K. L.; Ball, W. A.; Braun, S.; Lindner, H. J. *Justus Liebigs Ann. Chem./Liebigs Ann. Chem.* **1986**, 687. (d) Naemura, K.; Takahashi, N.; Chikamatsu, H. *Chem. Lett.* **1988**, 1717. (e) Zemlicka, J.; Craine, L. E.; Heeg, M.-J.; Oliver, J. P. *J. Org. Chem.* **1988**, *53*, 937. (f) Brion, F.; Marie, C.; Mackiewicz, P.; Roul, J. M.; Buendia, J. *Tetrahedron Lett.* **1992**, *33*, 4889. (g) Hutchinson, E. J.; Roberts, S. M.; Thorpe, A. J. *J. Chem. Soc., Perkin Trans. 1* **1992**, 2245.
12. Kobayashi, S.; Kamiyama, K.; Iimori, T.; Ohno, M. *Tetrahedron Lett.* **1984**, *25*, 2557.
13. Suemune, H.; Matsuno, K.; Uchida, M.; Sakai, K. *Tetrahedron: Asymmetry* **1992**, *3*, 297.
14. Malézieux, B.; Jaouen, G.; Salaün, J.; Howell, J. A. S.; Palin, M. G.; McArdle, P.; O'Gara, M.; Cunningham, D. *Tetrahedron: Asymmetry* **1992**, *3*, 375.
15. Klunder, A. J. H.; van Gastel, F. J. C.; Zwanenburg, B. *Tetrahedron Lett.* **1988**, *29*, 2697.
16. Klunder, A. J. H.; Huizinga, W. B.; Hulshof, A. J. M.; Zwanenburg, B. *Tetrahedron Lett.* **1986**, *27*, 2543.
17. Alcock, N. W.; Crout, D. H. G.; Henderson, C. M.; Thomas, S. E. *Chem. Commun./J. Chem. Soc., Chem. Commun.* **1988**, 746.
18. Hoong, L. K.; Strange, L. E.; Liotta, D. C.; Koszalka, G. W.; Burns, C. L.; Schinazi, R. F. *J. Org. Chem.* **1992**, *57*, 5563.
19. Klibanov, A. M.; Siegel, E. H. *Enzyme Microb. Technol.* **1982**, *4*, 172.
20. Wang, Y.-F.; Sih, C. J. *Tetrahedron Lett.* **1984**, *25*, 4999.
21. Jones, M.; Page, M. I. *Chem. Commun./J. Chem. Soc., Chem. Commun.* **1991**, 316.
22. Papageorgiou, C.; Benezra, C. *J. Org. Chem.* **1985**, *50*, 1144.
23. (a) Adachi, K.; Kobayashi, S.; Ohno, M. *Chimia* **1986**, *40*, 311. (b) Seebach, D.; Eberle, M. *Chimia* **1986**, *40*, 315.
24. Whitesell, J. K.; Lawrence, R. M. *Chimia* **1986**, *40*, 318.
25. Basavaiah, D.; Raju, S. B. *Synth. Commun.* **1991**, *21*, 1859.
26. (a) Fouque, E.; Rousseau, G. *Synthesis* **1989**, 661. (b) Guibé-Jampel, E.; Rousseau, G.; Blanco, L. *Tetrahedron Lett.* **1989**, *30*, 67.

27. Rudd, E. A.; Brockman, H. L. In *Lipases*; Borgström, B.; Brockman, H. L., Eds.; Elsevier: Amsterdam, 1984; p 185.
28. Kazlauskas, R. J.; Weissfloch, A. N. E.; Rappaport, A. T.; Cuccia, L. A. *J. Org. Chem.* **1991**, *56*, 2656.
29. Chenault, H. K.; Kim, M.-J.; Akiyama, A.; Miyazawa, T.; Simon, E. S.; Whitesides, G. M. *J. Org. Chem.* **1987**, *52*, 2608.
30. Chênevert, R.; Desjardins, M.; Gagnon, R. *Chem. Lett.* **1990**, 33.
31. Sussman, J. L.; Harel, M.; Frolow, F.; Oefner, C.; Goldman, A.; Toker, L.; Silman, I. *Science* **1991**, *253*, 872.
32. (a) Dropsy, E. P.; Klibanov, A. M. *Biotechnol. Bioeng.* **1984**, *26*, 911. (b) Aragozzini, F.; Valenti, M.; Santaniello, E.; Ferraboschi, P; Grisenti, P. *Biocatalysis* **1992**, *5*, 325.
33. Santaniello, E.; Canevotti, R.; Casati, R.; Ceriani, L.; Ferraboschi, P; Grisenti, P. *Gazz. Chim. Ital.* **1989**, *119*, 55.
34. Danishefsky, S. J.; Cabal, M. P.; Chow, K. *J. Am. Chem. Soc.* **1989**, *111*, 3456.
35. Deardorff, D. R.; Matthews, A. J.; McMeekin, D. S.; Craney, C. L. *Tetrahedron Lett.* **1986**, *27*, 1255.
36. Johnson, C. R.; Senanayake, C. H. *J. Org. Chem.* **1989**, *54*, 735.
37. Senanayake, C. H.; Bill, T. J.; Larsen, R. D.; Leazer, J.; Reider, P. J. *Tetrahedron Lett.* **1992**, *33*, 5901.
38. Mutsaers, J. H. G. M.; Kooreman, H. J. *Recl. Trav. Chim. Pays-Bas* **1991**, *110*, 185.
39. Kamal, A. *Synth. Commun.* **1991**, *21*, 1293.
40. Yee, C.; Blythe, T. A.; McNabb, T. J.; Walts, A. E. *J. Org. Chem.* **1992**, *57*, 3525.
41. Fliche, C.; Braun, J.; Le Goffic, F. *Synth. Commun.* **1991**, *21*, 1429.
42. Monteiro, J.; Braun, J.; Le Goffic, F. *Synth. Commun.* **1990**, *20*, 315.

Enzo Santaniello, Patrizia Ferraboschi & Paride Grisenti
Università di Milano, Italy

(*R,R*)-[Ethylene-1,2-bis(η^5-4,5,6,7-tetrahydro-1-indenyl)]titanium (*R*)-1,1′-Bi-2,2′-naphtholate

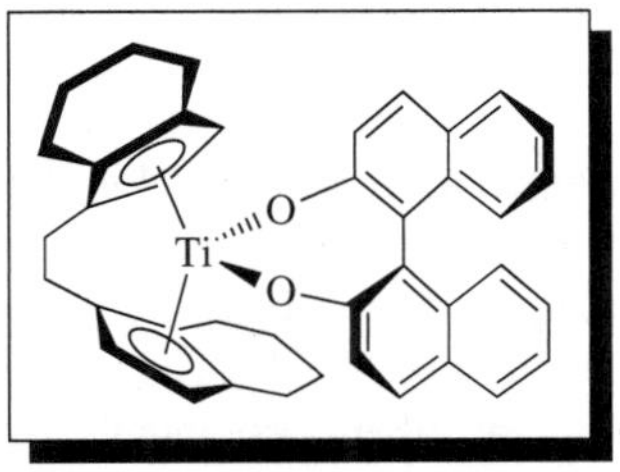

(R,R,R)
[143063-72-3] $C_{40}H_{36}O_2Ti$ (MW 596.64)
(S,S,S)
[83417-93-0]

(precursor to a catalyst for the asymmetric reduction of unsaturated organic molecules[1])

Physical Data: mp 274.5–277 °C (dec); $[\alpha]_{578}$, −3700° (c = 0.45 mg cm^{-3} in $CHCl_3$).

Solubility: sol THF, benzene, toluene; slightly sol ether; very slightly sol hexane.

Preparative Methods: originally prepared and characterized by Brintzinger et al. from the corresponding dichloride derivative and *(R)-1,1′-Bi-2,2′-naphthol*.[2] The dichloride is synthesized by reacting the dilithium salt of 1,2-bis(3-indenyl)ethane with *Titanium(IV) Chloride* followed by hydrogenation over *Platinum(IV) Oxide*.[2] Since the original report, two improved procedures for its preparation have appeared.[3,1b]

Handling, Storage, and Precautions: the complex is air and moisture stable and can be stored indefinitely.

Catalyst Generation and Handling. When complex (**1**) is allowed to react with 2 equiv of *n-Butyllithium* and 2.5–3 equiv of phenylsilane in THF under an inert atmosphere, an active reduction catalyst, complex (**2**), is formed (eq 1).

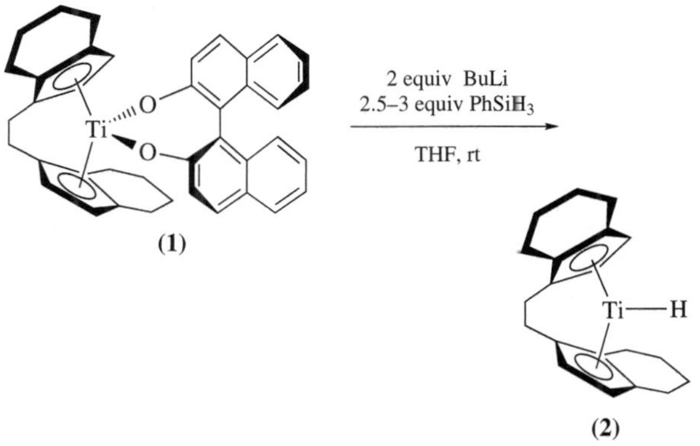

(1)

Complex (**2**) is believed to be a titanium(III) hydride and has not been isolated or characterized (no ^{1}H NMR signals for any titanium species are observable, probably as a result of the paramagnetic nature of the complex). This complex is extremely air sensitive and must be handled under rigorously oxygen-free conditions. Solutions of complex (**2**) are stable under inert atmosphere for at least 24 h and exhibit no sensitivity to light. For synthetic purposes it is most convenient to generate the active catalyst from complex (**1**) immediately prior to use.

Reduction of Unfunctionalized Alkenes. The asymmetric reduction of trisubstituted alkenes with this system has recently been investigated.[4] It was shown that high ee's could be achieved in this reaction, although high pressures and long reaction times were necessary. As an example, *trans*-methylstilbene can be reduced at 2000 psi of hydrogen and 65 °C with excellent chemical and optical yield (eq 2). This represents the first catalyst system for the reduction of unfunctionalized, trisubstituted alkenes with good to excellent enantioselectivity.

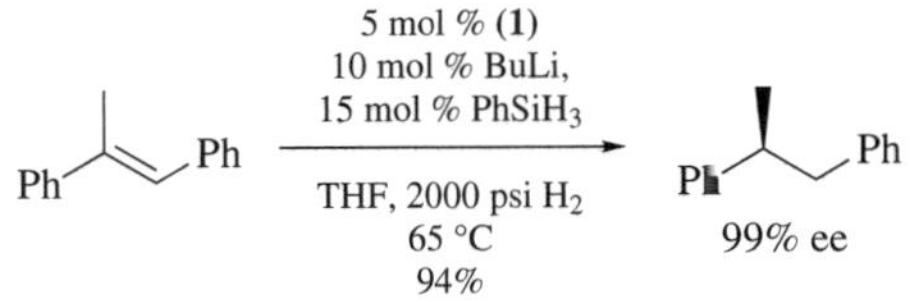

(2)

Reduction of Imines. This catalyst system is very effective for the asymmetric hydrogenation of imines.[1b] For example, *N*-(1-cyclohexyl)ethylidenebenzylamine (as a mixture of *anti* and *syn* isomers) can be reduced in excellent yield and good enantiomeric excess (eq 3). The reaction must be conducted at high pressures in order to achieve maximum enantioselectivity. This effect was found for several acyclic imines.

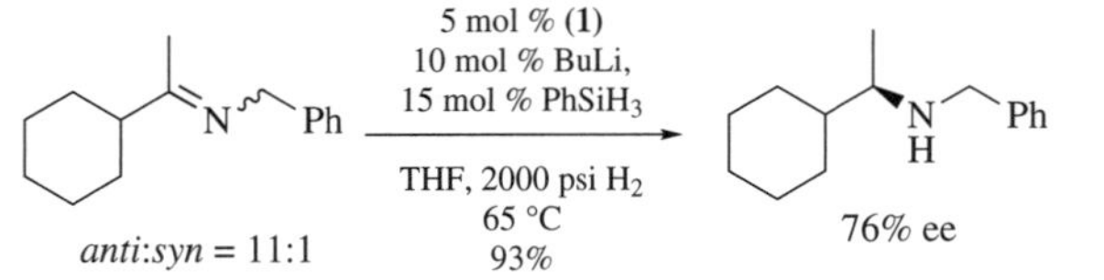

(3)

The reduction of cyclic imines with this system was found to proceed under much milder conditions.[5] For example, 2-phenylpyrroline was reduced at 80 psi of hydrogen to afford 2-phenylpyrrolidine in good yield and excellent enantiomeric excess (eq 4).

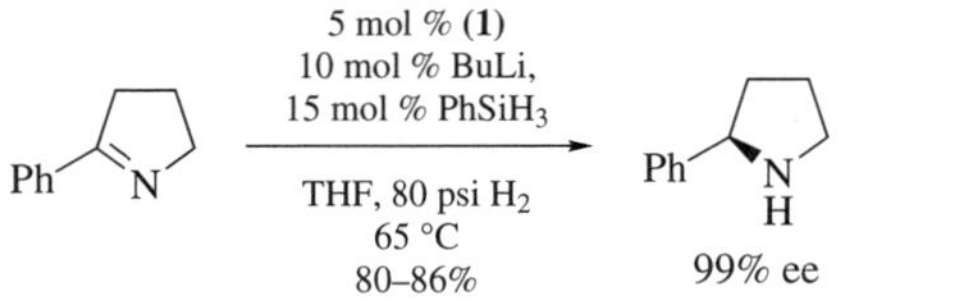

(4)

This reaction was found to be applicable to ring sizes of 5 to 7 (although more forcing conditions were required for six-membered rings) and was compatible with several functional groups. In all cases studied, ee's greater than 95% were observed. Among the functional groups investigated were acetals, silyl ethers, trisubstituted alkenes, and alcohols. Monosubstituted alkenes were completely reduced and disubstituted alkenes were partially reduced and isomerized under the standard conditions. The reaction has the practical advantage that the active catalyst can be generated in a Fisher–Porter bottle and the reaction can then be conducted in the same vessel. Thus no transfer of air-sensitive materials is necessary.

Related Reagents. (−)-[Ethylene-1,2-bis(η^5-4,5,6,7-tetrahydro-1-indenyl)]zirconium (*R*)-1,1′-Bi-2,2′-naphtholate; (±)-1,1′-Ethylenebis(4,5,6,7-tetrahydro-1-indenyl)zirconium Dichloride.

1. (a) Buchwald, S. L.; Kreutzer, K. A.; Willoughby, C. A.; Grossman, R. B.; Berk, S. C.; Spaltenstein, E.; Gutierrez, A. PCT Int. Appl. 92 09 545. (b) Willoughby, C. A.; Buchwald, S. L. *J. Am. Chem. Soc.* **1992**, *114*, 7562.
2. Wild, F. R. W. P.; Zsolnai, L.; Huttner, G.; Brintzinger, H. H. *J. Organomet. Chem.* **1982**, *232*, 233.
3. (a) Collins, S.; Kuntz, B. A.; Taylor, N. J.; Ward, D. G. *J. Organomet. Chem.* **1988**, *342*, 21. (b) Collins, S.; Kuntz, B. A.; Hong, Y. *J. Org. Chem.* **1989**, *54*, 4154.
4. Broene, R. D.; Buchwald, S. L. *J. Am. Chem. Soc.* **1993**, *115*, 12569.
5. Willoughby, C. A.; Buchwald, S. L. *J. Org. Chem.* **1993**, *58*, 7627.

Christopher A. Willoughby
Massachusetts Institute of Technology, Cambridge, MA, USA

(−)-[Ethylene-1,2-bis(η^5-4,5,6,7-tetrahydro-1-indenyl)]zirconium (*R*)-1,1′-Bi-2,2′-naphtholate[1]

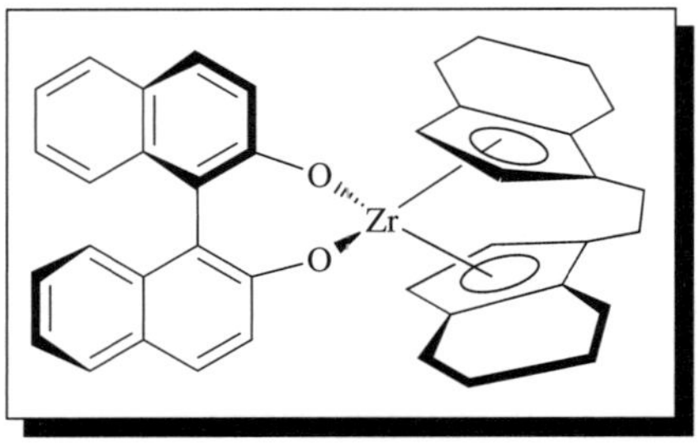

[133868-91-4] $C_{40}H_{36}O_2Zr$ (MW 639.98)

(in combination with methylaluminoxane it is an enantioselective hydrogenating[2] and cyclopolymerizing agent;[3] reagent for the resolution of racemic ethylenebis(tetrahydro-1-indenyl)]zirconium dichloride[4])

Physical Data: $[\alpha]^{25}_{436}$ −1761° (*c* 1.1, $CHCl_3$).

Preparative Methods: this reagent is an intermediate for the preparation of optically pure ethylenebis(tetrahydro-1-indenyl)zirconium dichloride.[4] That is, according to the procedure described for the kinetic resolution of ethylenebis(tetrahydro-1-indenyl)titanium dichloride with (*R,R*)- or (*S,S*)-binaphtholate,[5] 1 equiv of racemic ethylenebis(tetrahydro-1-indenyl)zirconium dichloride[6] can be resolved with 0.5 equiv of (*R,R*)-binaphthol in the presence of sodium metal in toluene to yield the optically active (*S,S*)-ethylenebis(tetrahydro-1-indenyl)zirconium dichloride and the (−)-[ethylenebis(tetrahydro-1(*R*)-indenyl]-zirconium (*R*)-binaphtholate. The separated optically pure (−)-[ethylenebis(tetrahydro-1(*R*)-indenyl]zirconium (*R*)-binaphtholate can be easily converted to the corresponding optically pure zirconocene dichloride upon treating with HCl gas.

Catalytic Hydrogenation and Cyclopolymerization. This zirconium reagent constitutes a homogeneous catalytic system (Ziegler–Natta catalyst) with methylaluminoxane (MAO) for the purpose of enantioselective hydrogenation[2] of alkenes or cyclopolymerization[3] of alkenic compounds to optically active polymers. In the hydrogenations, terminal alkenes substituted in the 2- or 3-positions and internal alkenes are hydrogenated with 23–65% ee in good yields. In the catalytic deuteration of styrene with this reagent, the *re*-face of styrene is deuterated in 65% optical purity (eq 1). This enantioselectivity in hydrogenation is opposite to propylene oligomerization with the same catalytic system. 1,5-Hexadiene is polymerized with this catalytic system in 91% enantioface selectivity to give poly(methylene-1,3-cyclopentane) (PMCP), which is inductive of a highly isotactic microstructure (eq 2). The absolute configuration of the polymer is tentatively assigned on the basis of the sign of the optical rotation of the model compound *trans*-(1*R*,3*R*) 1,3-dimethylcyclopentane.[3]

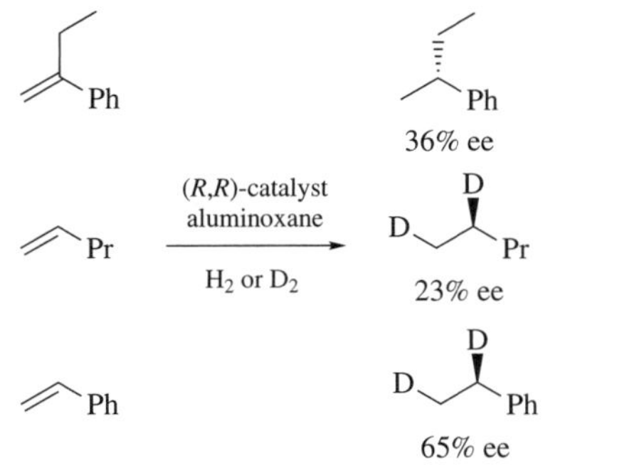

(1)

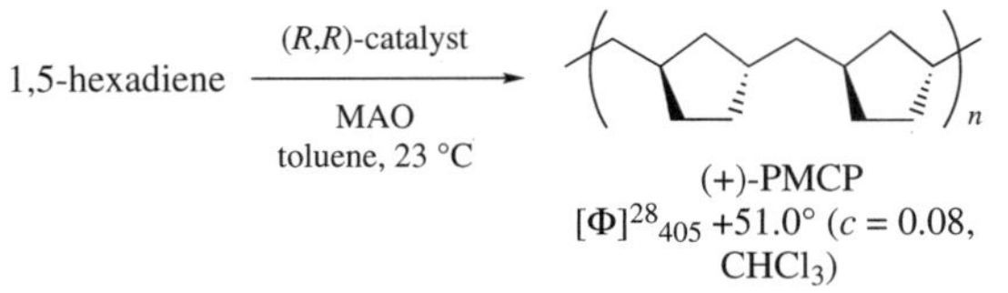

(2)

Related Reagents. (*R*,*R*)-[Ethylene-1,2-bis(η^5-4,5,6,7-tetrahydro-1-indenyl)]titanium (*R*)-1,1′-Bi-2,2′-naphtholate; (±)-1,1′-Ethylenebis(4,5,6,7-tetrahydro-1-indenyl)zirconium Dichloride.

1. Halterman, R. L. *Chem. Rev.* **1992**, *92*, 965.
2. Waymouth, R.; Pino, P. *J. Am. Chem. Soc.* **1990**, *112*, 4911.
3. Coates, G. W.; Waymouth, R. M. *J. Am. Chem. Soc.* **1993**, *115*, 91.
4. (a) Grossman, R. B.; Davis, W. M.; Buchwald, S. L. *J. Am. Chem. Soc.* **1991**, *113*, 2321. (b) For the improved procedure of resolution, see; Schäfer, A.; Karl, E.; Zsolnai, L.; Huttner, G.; Brintzinger, H. H. *J. Organomet. Chem.* **1987**, *328*, 87.
5. Wild, F. R. W. P.; Zsolnai, L.; Huttner, G.; Brintzinger, H. H. *J. Organomet. Chem.* **1982**, *232*, 233.
6. (a) Wild, F. R. W. P.; Wasiucionek, M.; Huttner, G.; Brintzinger, H. H. *J. Organomet. Chem.* **1985**, *288*, 63. (b) Collins, S.; Kuntz, B. A.; Taylor, N. J.; Ward, D. G. *J. Organomet. Chem.* **1987**, *342*, 21. (c) Ewen, J. A.; Haspeslagh, L.; Atwood, J. L.; Zhang, H. *J. Am. Chem. Soc.* **1987**, *109*, 6544. (d) Grossman, R. B.; Doyle, R. A.; Buchwald, S. L. *Organometallics* **1991**, *10*, 1501.

Takeo Taguchi & Yuji Hanzawa
Tokyo College of Pharmacy, Japan

(*S*)-Ethyl Lactate

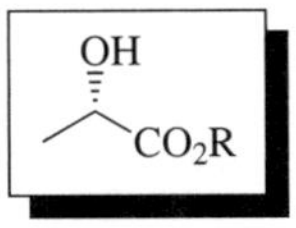

(*S*)-(R = Et) *[687-47-8]*	$C_5H_{10}O_3$	(MW 118.15)
(*R*)-(R = Et) *[97-64-3]*	$C_5H_{10}O_3$	(MW 118.15)
(*S*)-(R = Me) *[17392-83-5]*	$C_4H_8O_3$	(MW 104.12)
(*R*)-(R = Me) *[27871-49-4]*	$C_4H_8O_3$	(MW 104.12)
(*R*)-(R = Bu) *[34451-18-8]*	$C_7H_{14}O_3$	(MW 146.21)

(chiral pool reagent for synthesis; occasionally used as a chiral auxiliary)

Alternate Name: ethyl L-(−)-lactate.
Physical Data: (*S*)-(R = Et) bp 154 °C, 69 °C/36 mmHg, *d* 1.031 g cm^{-3};[20] (*R*)-(R = Et) bp 58 °C/20 mmHg, *d* 1.032 g cm^{-3};[20] (*S*)-(R = Me) bp 40 °C/11 mmHg, *d* 1.086 g cm^{-3};[25] (*R*)-(R = Me) bp 58 °C/19 mmHg, *d* 1.091 g cm^{-3};[20] (*R*)-(R = Bu) bp 77 °C/10 mmHg, *d* 0.974 g cm^{-3}.[27]
Solubility: sol water, alcohols, ethers, THF, and common organic solvents.
Form Supplied in: liquid; commercially available.

Use as a Chiral Pool Reagent. (*S*)-Ethyl lactate has been extensively used as a chiral pool reagent, often via transformation into a diverse array of simple, enantiomerically pure analogs. Principal among these are a variety of *O*-protected (*S*)-2-hydroxypropanals (eq 1).

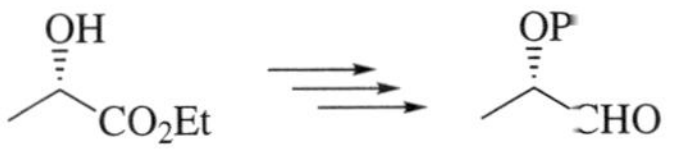

(1)

These have been prepared by various combinations of straightforward steps including ester to amide conversion, alcohol protection, direct reduction of the ester or amide to the aldehyde group, and reduction of the ester to the alcohol followed by reoxidation to the aldehyde. The sensitivity of the (*S*)-propanals to epimerization has been of paramount concern. One of the best procedures which avoids racemization and has been run on a preparative scale is noted (eq 2).[1]

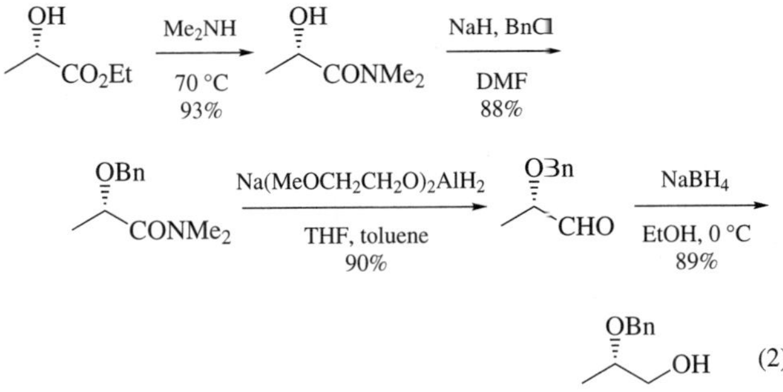

(2)

Subsequent reduction also affords (*S*)-2-benzyloxypropanol in 89% yield. NMR assay of the (*R*)- and (*S*)-Mosher esters indicated no racemization over the sequence. Synthesis via oxidation of (*S*)-benzyloxypropanol (eq 3) provides the benzyloxypropanal with <8% racemization.[2]

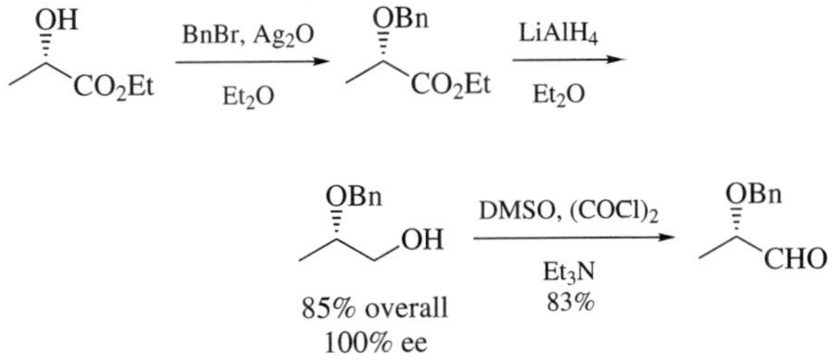

(3)

Attempted formation of the benzyl ether of (*S*)-ethyl lactate with NaH/BnBr results in considerable racemization (50–75% ee). This racemization is obviated by use of the amide analog noted in eq 2. *Diisobutylaluminum Hydride* has been used to convert the ester directly to the aldehyde employing the methoxymethyl,[3,4] benzyl,[2,5] 2,6-dichlorobenzyl,[6] *t*-butyldiphenylsilyl,[7] benzyloxymethyl,[8] THP,[9] trityl[10] and TBDMS[11] protecting groups. Protected (*S*)-2-hydroxypropanals have been used in synthetic studies relating to sugars,[12–15] amino sugars,[16] thiotetronic acids,[17] antimycin-A_3,[18] rhodinose,[19] aplysiatoxin via the (*R*)-lactate,[20] (−)-sarracenin,[21] and for preparation of enantiomerically pure 1-methyl-2-alkenyl-*N*,*N*-diisopropylcarbamates from the (*R*)- and (*S*)-lactates.[22]

(*S*)-Ethyl lactate has also been used as a ready source of (*S*)-propane-1,2-diol and (*S*)-methyloxirane (eq 4).[23,24] These compounds have been used for preparation of numerous natural products including nonactin,[25] sulcatol,[26] recifeiolide,[27] methyl-1,6-dioxaspiro[4,5]decanes, the pheromone components of *Paravespula vulgaris*,[28] and the rhynchosporosides.[29] The (*S*)-oxirane has also been used in the synthesis of chiral macrocyclic poly(ether diester)ligands.[30] A convenient procedure for preparation of the (*R*)-methyloxirane via mesylate activation, reduction, and internal inversion has been reported.[31]

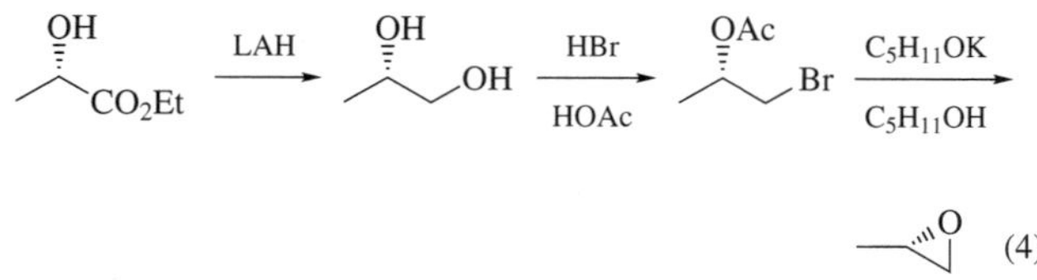
(4)

A variety of inverted analogs of (*S*)-ethyl lactate have been prepared by standard activation displacement procedures. Included are the (*R*)-propionyloxypropionate (mesylation/$EtCO_2$-Cs–DMF);[32] azide (Mitsunobu conditions);[33,34] aryloxy ethers (mesylation/aryl oxide);[35] chloride ($SOCl_2$–DMF);[36] bromide (sulfonation/$MgBr_2$);[37] mercapto analogs (Mitsonobu conditions);[38,39] amino analogs (triflate/amine);[40] hydroxylamines;[41,42] and selenides.[43]

Protected (*S*)-ethyl lactate cleanly acylates methyllithium to afford the 2-butanone with essentially complete enantiomeric fidelity and in nearly quantitative yield. Various diastereoselective constructions were achieved by nucleophilic addition to the ketone (eq 5).[44] For example, addition of vinyllithiums, followed by acetal formation and Lewis acid-mediated rearrangement, provided a ready entry into the indicated 3-acyltetrahydrofurans.

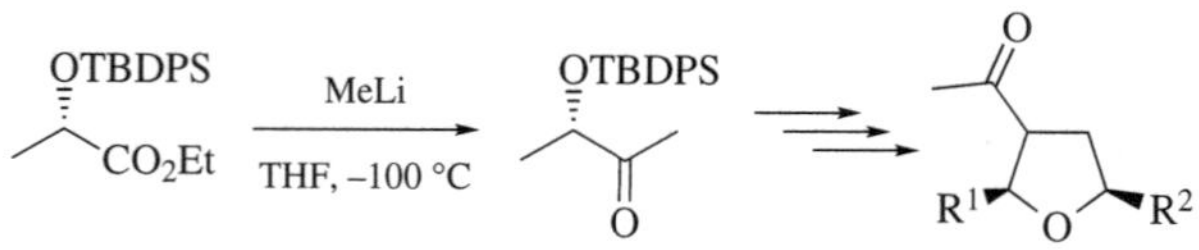
(5)

(*S*)-Ethyl lactate has been used to prepare (*S*)-2-methyloxetane in modest yield with <0.5% racemization by a series of standard transformations (eq 6).[45]

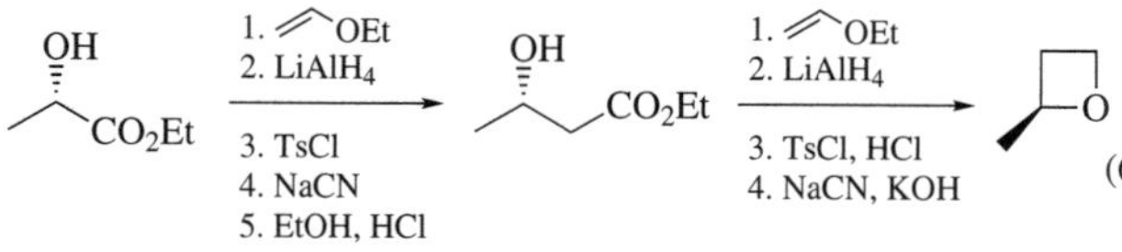
(6)

Other small chiral molecules have also been prepared by straightforward transformations (eq 7).[46]

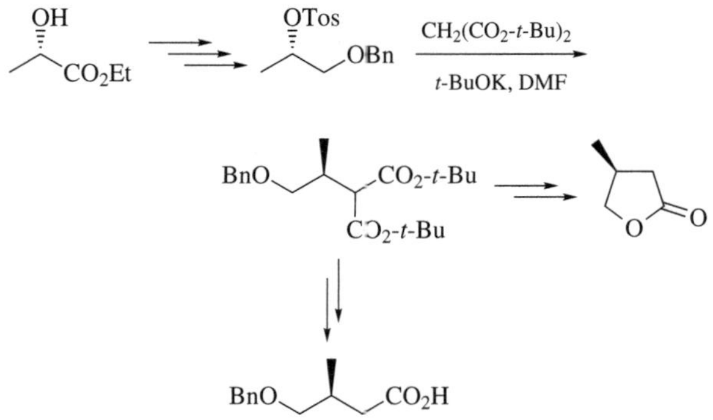
(7)

(*S*)-Ethyl lactate has also been used as a chiral fragment for numerous other studies. Included are synthetic efforts relating to salenomycin,[47] (−)-biopterin,[48] (+)-polyoxamic acid,[49] jaspamide,[50] the enantiomeric 2-pentanols,[51] pumilitoxin B,[52,53] D-ristosamine,[54] protomycinolide IV,[55] and tirandamycin.[56]

Use as a Chiral Auxiliary. (*S*)-Ethyl lactate has been used as a chiral auxiliary in a variety of simple Diels–Alder reactions.[57–60] As the fumaric acid diester, the de employing cyclopentadiene can almost be completely reversed by addition of *Titanium(IV) Chloride* (eq 8).[61] In general, superior de values are achieved using *(R)-Pantolactone* in this context, and also for base-mediated addition to ketenes.[62]

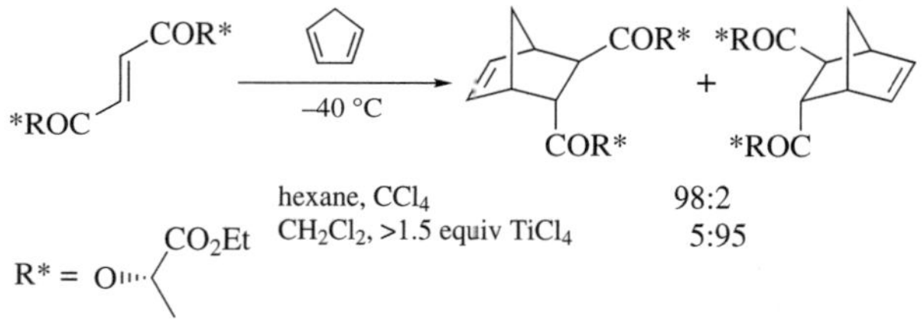
(8)

(*S*)-Ethyl lactate was used for diastereocontrol and asymmetric transmission in a sequential 2,3-Wittig–oxy-Cope rearrangement, affording product in 91% ee (eq 9).[63,64] Excellent asymmetric induction has also been noted in the Lewis acid-mediated ene reaction of (*S*)-ethyl lactate-derived intermediates (eq 10).[65]

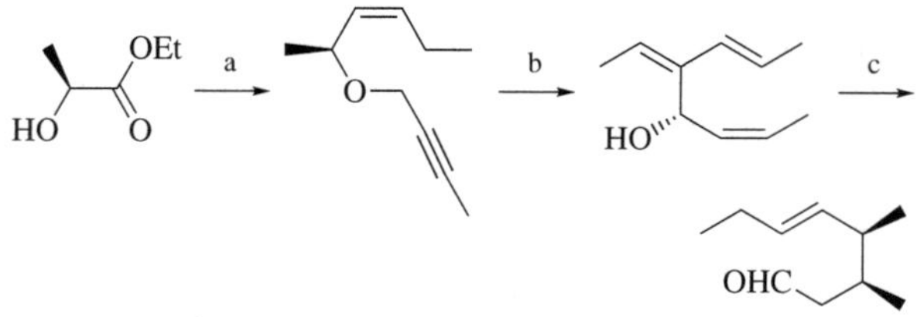
(9)

(a) MeC≡CCH_2OC(=NH)CCl_3, H^+, 85%; DIBAL, 82%; $PrPPh_3Br$, BuLi, −78 °C, 75%
(b) BuLi, −78 °C, 75%; H_2, P-2 Ni, 95%
(c) KH, 18-crown-6, 25 °C, 75%

(*S*)-Ethyl lactate has been used to enantioselectively protonate the indicated enolate at −100 °C to afford the (*R*)-ketone in 73% yield and 73% ee (eq 11).[66]

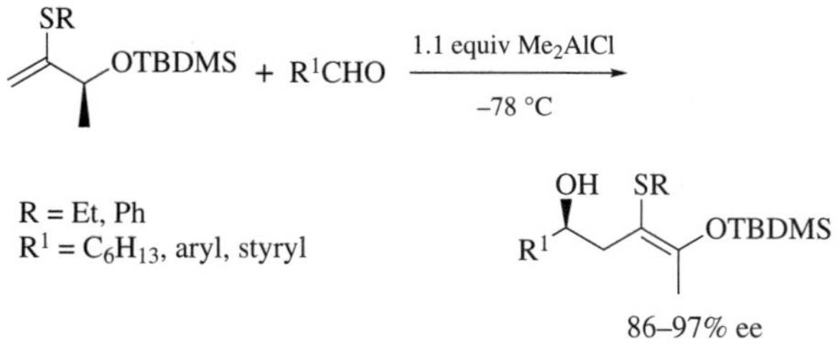

(10)

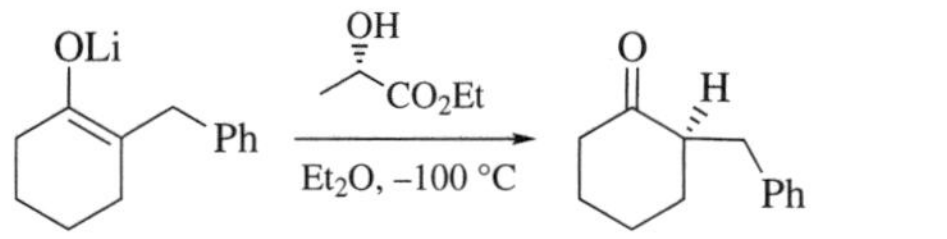

(11)

Applications to Products of Commercial Interest. (*S*)-Ethyl lactate has been incorporated in chiral syntheses of (*S*)-2-arylpropionic acids, an important class of nonsteroidal anti-inflammatory agents, including ibuprofen and naproxen (eq 12).[67–69] These syntheses, though elegant in concept, are unlikely to compete with existing industrial methods for production of the (*S*) enantiomers of these drugs.

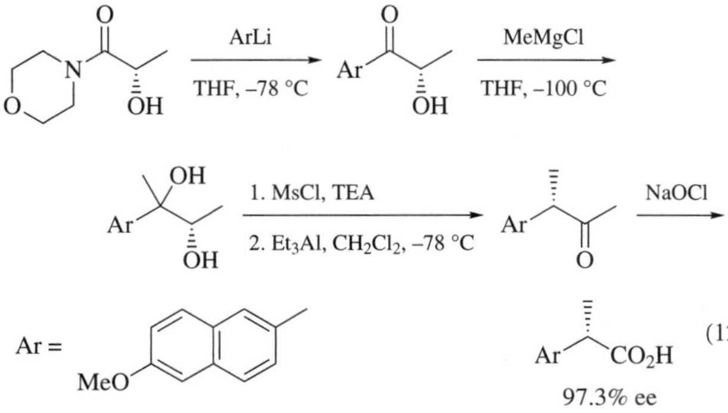

(12)

(*S*)-Ethyl lactate has also been used to synthesize the important 4-acetoxyazetidinone intermediate, crucial to numerous carbapenem syntheses. The key step in its use was the diketene addition to the (*S*)-lactaldehyde imine, which in the best case proceeded in 67% yield with a 10:1 ratio of diastereomers (eq 13).[70,71]

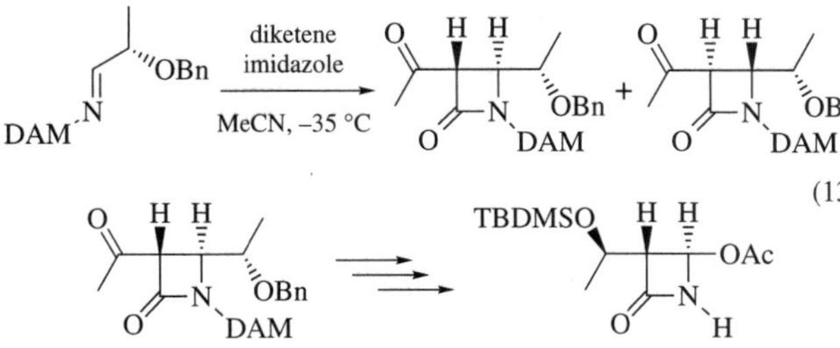

(13)

Other applications to β-lactam syntheses have been reported.[72–74]

Related Reagents. Ethyl Mandelate; (*R*)-Pantolactone.

1. Kobayashi, Y.; Takase, M.; Ito, Y.; Terashima, S. *Bull. Chem. Soc. Jpn.* **1989**, *62*, 3038.
2. Takai, K.; Heathcock, C. H. *J. Org. Chem.* **1985**, *50*, 3247.
3. Wasserman, H. H.; Gambale, R. J. *Tetrahedron* **1992**, *48*, 7059.
4. Iida, H.; Yamazaki, N.; Kibayashi, C. *Chem. Commun.* **1987**, 746.
5. De Amici, M.; Dallanoce, C. de M. C.; Grana, E.; Dondi, G.; Ladinsky, H.; Schiavi, G.; Zonta, F. *Chirality* **1992**, *4*, 230.
6. Chan, T. H.; Li, C. J. *Can. J. Chem.* **1992**, *70*, 2726.
7. Braun, M.; Moritz, J. *Synlett* **1991**, 750.
8. Brown, P. A.; Bonnert, R. V.; Jenkins, P. R.; Lawrence, N. J.; Selim, M. R. *J. Chem. Soc., Perkin Trans. 1* **1991**, 1893
9. Kang, S.-K.; Lee, D.-H. *Synlett* **1991**, 175.
10. Mori, K.; Kikuchi, H. *Justus Liebigs Ann. Chem./Liebigs Ann. Chem.* **1989**, 963.
11. Hirama, M.; Shigemoto, T.; Ito, S. *J. Org. Chem.* **1987**, *52*, 3342.
12. Hiyama, T.; Nishide, K.; Kobayashi, K. *Tetrahedron Lett.* **1984**, *25*, 569.
13. Guanti, G.; Banfi, L.; Narisano, E. *Gazz. chim. Ital.* **1987**, *117*, 681.
14. Yamamoto, Y.; Komatsu, T.; Maruyama, K. *J. Organomet. Chem.* **1985**, *285*, 31.
15. Guanti, G.; Banfi, L.; Guaragna, A.; Narisano, E *J. Chem. Soc., Perkin Trans. 1* **1988**, 2369.
16. Hiyama, T.; Kobayashi, K.; Nishide, K. *Bull. Chem. Soc. Jpn.* **1987**, *60*, 2127.
17. Chambers, M. S.; Thomas, E. J.; Williams, D. J. *Chem. Commun.* **1987**, 1228.
18. Wasserman, H. H.; Gambale, R. J. *J. Am. Chem. Soc.* **1985**, *107*, 1423.
19. Kelly, T. R.; Kaul, P. N. *J. Org. Chem.* **1983**, *48*, 2775.
20. Ireland, R. E.; Thaisrivongs, S.; Dussault, P. H. *J. Am. Chem. Soc.* **1988**, *110*, 5768.
21. Baldwin, S. W.; Crimmins, M. T. *J. Am. Chem. Soc.* **1982**, *104*, 1132.
22. Schwark, J-R.; Hoppe, D. *Synthesis* **1990**, 291.
23. Ellis, M. K.; Golding, B. T. *Org. Synth.* **1985**, *63*, 140.
24. Mori, K.; Senda, S. *Tetrahedron* **1985**, *41*, 541.
25. Schmidt, U.; Gombos, J.; Haslinger, E.; Zak, H. *Chem. Ber.* **1976**, *109*, 2628.
26. Johnston, B. D.; Slessor, K. N. *Can. J. Chem.* **1979**, *57*, 233.
27. Utimoto, K.; Uchida, K.; Yamaya, M.; Nozaki, H. *Tetrahedron Lett.* **1977**, 3641.
28. Hintzer, K.; Weber, R.; Schurig, V. *Tetrahedron Lett.* **1981**, *22*, 55.
29. Nicolaou, K. C.; Randall, J. L.; Furst, G. T. *J. Am. Chem. Soc.* **1985**, *107*, 5556.
30. Jones, B. A.; Bradshaw, J. S.; Izatt, R. M. *J. Heterocycl. Chem.* **1982**, *19*, 551.
31. Hillis, L. R.; Ronald, R. C. *J. Org. Chem.* **1981**, *46*, 3348.
32. Kruizinga, W. H.; Strijtveen, B.; Kellogg, R. M. *J. Org. Chem.* **1981**, *46*, 4321.
33. Viaud, M. C.; Rollin, P. *Synthesis* **1990**, 130.
34. Fabiano, E.; Golding, B. T.; Sadeghi, M. M. *Synthesis* **1987**, 190.
35. Burkard, U.; Effenberger, F. *Chem. Ber.* **1986**, *119*, 1594.
36. Biedermann, J.; Leon-Lomeli, A.; Borbe, H. O.; Prop, G. *J. Med. Chem.* **1986**, *29*, 1183.
37. Hanessian, S.; Kagotani, M.; Komaglou, K. *Heterocycles* **1989**, *28*, 1115.
38. Rollin, P. *Tetrahedron Lett.* **1986**, *27*, 4169.
39. Strijtveen, B.; Kellogg, R. M. *J. Org. Chem.* **1986**, *51*, 3664.
40. Effenberger, F.; Burkard, U.; Willfahrt, J. *Justus Liebigs Ann. Chem./Liebigs Ann. Chem.* **1986**, 314.
41. Feenstra, R. W.; Stokkingreef, E. H. M.; Nivard, R. J. F.; Ottenheijm, H. C. J. *Tetrahedron* **1988**, *44*, 5583.
42. Feenstra, R. W.; Stokkingreef, E. H. M.; Nivard, R. J. F.; Ottenheijm, H. C. J. *Tetrahedron Lett.* **1987**, *28*, 1215.
43. Fitzner, J. N.; Shea, R. G.; Fankhauser, J. E.; Hopkins, P. B. *J. Org. Chem.* **1985**, *50*, 417.
44. Hopkins, M. H.; Overman, L. E.; Rishton, G. M. *J. Am. Chem. Soc.* **1991**, *113*, 5354.
45. Hintzer, K.; Koppenhoefer, B.; Schurig, V. *J. Org. Chem.* **1982**, *47*, 3850.
46. Berens, U.; Scharf, H. D. *Synthesis* **1991**, 832.
47. Horita, K.; Nagato, S.; Oikawa, Y.; Yonemitsu, O. *Chem. Pharm. Bull.* **1989**, *37*, 1705.
48. Kikuchi, H.; Mori, K. *Agric. Biol. Chem.* **1989**, *53*, 2095.

49. Savage, I.; Thomas, E. *Chem. Commun.* **1989**, 717.
50. Chiarello, J.; Joullie, M. M. *Synth. Commun.* **1989**, *19*, 3379.
51. Cheskis, B.; Shpiro, N. A.; Moiseenkov, A. M. *Zh. Org. Khim.* **1990**, *26*, 1864.
52. Overman, L. E.; McCready, R. J. *Tetrahedron Lett.* **1982**, *23*, 2355.
53. Overman, L. E.; Bell, K. L.; Ito, F. *J. Am. Chem. Soc.* **1984**, *106*, 4192.
54. Hamada, Y.; Kawai, A.; Shiori, T. *Chem. Pharm. Bull.* **1985**, *33*, 5601.
55. Suzuki, K.; Tomooka, K.; Katayama, E.; Matsumoto, T.; Tsuchihashi, G. *J. Am. Chem. Soc.* **1986**, *108*, 5221.
56. Kelly, T. R.; Chandrakumar, N. S.; Cutting, J. D.; Goehring, R. R.; Weibel, F. R. *Tetrahedron Lett.* **1985**, *26*, 2173.
57. Avenoza, A.; Cativiela, C.; Mayoral, J. A.; Peregrina, J. M.; Sinou, D. *Tetrahedron: Asymmetry* **1990**, *1*, 765.
58. Rebiere, F.; Riant, O.; Kagan, H. B. *Tetrahedron: Asymmetry* **1990**, *1*, 199.
59. Avenoza, A.; Cativiela, C.; Mayoral, J. A.; Peregrina, J. M. *Tetrahedron: Asymmetry* **1992**, *3*, 913.
60. Cativiela, C.; Mayoral, J.; Avenoza, A.; Peregrina, J. M.; Lahoz, F. J.; Gimeno, S. *J. Org. Chem.* **1992**, *57*, 4664.
61. Helmchen, G.; Abdel Hady, A. F.; Hartmann, H.; Karge, R.; Krotz, A.; Sartor, K.; Urmann, M. *Pure Appl. Chem.* **1989**, *61*, 409.
62. Larsen, R. D.; Corley, E. G.; Davis, P.; Reider, P. J.; Grabowski, E. J. J. *J. Am. Chem. Soc.* **1989**, *111*, 7650.
63. Wei, S. Y.; Tomooka, K.; Nakai, T. *J. Org. Chem.* **1991**, *56*, 5973.
64. Wei, S. Y.; Tomooka, K.; Nakai, T. *Tetrahedron* **1993**, *49*, 1025.
65. Tanino, K.; Shoda, H.; Nakamura, T.; Kuwajima, I. *Tetrahedron Lett.* **1992**, *33*, 1337.
66. Matsumoto, K.; Ohta, H. *Tetrahedron Lett.* **1991**, *32*, 4729.
67. Yamauchi, T.; Hattori, K.; Nakao, K.; Tamaki, K. *Bull. Chem. Soc. Jpn.* **1987**, *60*, 4015.
68. Honda, Y.; Ori, A.; Tsuchihashi, G. *Bull. Chem. Soc. Jpn.* **1987**, *60*, 1027.
69. Brown, J. D. *Tetrahedron: Asymmetry* **1992**, *3*, 1551.
70. Ito, Y.; Kawabata, T.; Terashima, S. *Tetrahedron Lett.* **1986**, *27*, 5751.
71. Ito, Y.; Kobayashi, Y.; Kawabata, T.; Takase, M.; Terashima, S. *Tetrahedron* **1989**, *45*, 5767.
72. Okonogi, T.; Shibahara, S.; Murai, Y.; Inouye, S.; Kondo, S. *Heterocycles* **1990**, *31*, 791.
73. Okonogi, T.; Shibahara, S.; Murai, Y.; Yoshida, T.; Inouye, S.; Kondo, S.; Christensen, B. G. *J. Antibiot.* **1990**, *43*, 357.
74. Pfaendler, H. R., In *Recent Advances in the Chemistry of β-Lactam Antibiotics;* Gregory, G. I., Ed.; Royal Society of Chemistry: London, 1981, p 368.

Edward J. J. Grabowski
Merck Research Laboratories, Rahway, NJ, USA

(3a*R*,7a*R*)-2-Ethyloctahydro-1*H*-1,3-dineopentyl-1,3,2-benzodiazaphosphole Oxide

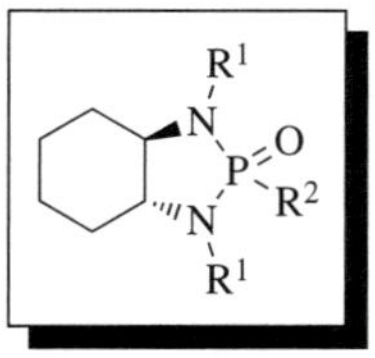

(**1**; R^1 = neopentyl, R^2 = Et)
[146397-34-4] $C_{18}H_{37}N_2OP$ (MW 328.54)
(**2**; R^1 = Me, R^2 = allyl)
[146098-95-5] $C_{11}H_{21}N_2OP$ (MW 228.31)
(**3**; R^1 = Me, R^2 = Et)
[91633-73-7] $C_{10}H_{21}N_2OP$ (MW 216.30)
(**4**; R^1 = Me, R^2 = Bn)
[146098-94-4] $C_{15}H_{23}N_2OP$ (MW 278.37)
(**5**; R^1 = Me, R^2 = $CH_2CH{=}CHMe$)
[–] $C_{12}H_{23}N_2OP$ (MW 242.34)
(**6**; R^1 = Bn, R^2 = Et)
[–] $C_{22}H_{29}N_2OP$ (MW 368.50)
(**7**; R^1 = Bn, R^2 = Pr)
[–] $C_{23}H_{31}N_2OP$ (MW 382.53)
(**8**; R^1 = Me, R^2 = CH_2Cl)
[146983-74-6] $C_9H_{18}ClN_2OP$ (MW 236.71)

(chiral α-alkyl bicyclophosphonamides useful for the asymmetric synthesis of alkenes,[1–3] of α,α′-substituted phosphonic acids,[4] and of α-amino-α-substituted phosphonic acids,[5,6] and for asymmetric conjugate additions of *C*-allyl and *C*-crotyl groups to α,β-unsaturated carbonyl compounds[7])

Physical Data: (**1**) mp 110–111 °C; $[\alpha]_D^{25}$ −98.8° (*c* 1.10, $CHCl_3$). (**2**) mp 45–46 °C; $[\alpha]_D^{25}$ −51.1° (*c* 1.85, $CHCl_3$). (**3**) mp 55–56 °C; $[\alpha]_D^{25}$ −98.9° (*c* 1.48, $CHCl_3$). (**4**) mp 104–105 °C; $[\alpha]_D^{25}$ −109.6° (*c* 1.17, $CHCl_3$). (**6**) mp 153–154 °C; $[\alpha]_D^{25}$ −68.2° (*c* 1.18, $CHCl_3$). (**7**) mp 117 °C; $[\alpha]_D^{25}$ −66.0° (*c* 1.02, $CHCl_3$). (**8**) mp 84 °C; $[\alpha]_D^{25}$ −109.8° (*c* 1.0, $CHCl_3$).

Solubility: sol chlorinated and dipolar aprotic solvents, and hydrocarbon solvents in some cases. Gradual hydrolysis in protic media.

Form Supplied in: colorless crystalline or waxy solids.

Preparative Methods:[3] to a solution of (1*R*,2*R*)-*N*,*N*′-dineopentyl-1,2-diaminocyclohexane (1.37 g, 5.40 mmol) and triethylamine (2.3 mL, 48.5 mmol) in 25 mL of benzene is added ethylphosphoryl dichloride (0.89 g, 6.06 mmol). The suspension is heated to reflux for 80 h, the salts are filtered, and the filtrate is washed successively with 10% aq HCl (2 × 10 mL), aq saturated $NaHCO_3$, and water. Drying, evaporation, and chromatographic separation (EtOAc) gives the title compound (1.1 g, 62%) as a crystalline solid, mp 110–111 °C (hexanes). Single crystal X-ray analysis confirmed the structure. For an alternative synthesis of related compounds, see Kueller and Spilling.[8] Other 2-alkyl derivatives can be similarly prepared.[1–4]

Handling, Storage, and Precautions: crystalline reagents are stable when stored under argon at 0 °C for several months.

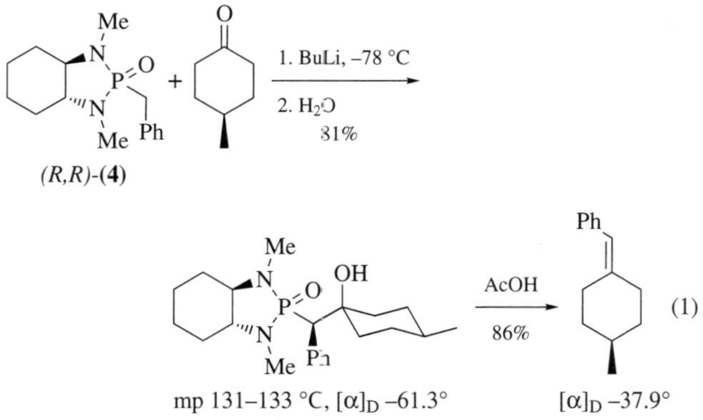

Asymmetric Alkenation of Alkylcyclohexanones.[1–3] Anions of the reagents (**1**), (**4**), (**6**), and (**7**) add to alkylcyclohexanones in THF solution at −78 °C to give intermediate β-hydroxyphosphonamide adducts, which can be isolated and purified by chromatography. Treatment of the adducts with aq acetic acid leads to the corresponding alkylidene alkylcyclohexanes in good to excellent enantiomeric or diastereomeric excesses. The alkenes can also be obtained directly from the original reaction mixtures (aq AcOH quench), without isolation of intermediates. Except in the case of reagent (**1**), using the reagents prepared from (*R*,*R*)-1,2-diaminocyclohexane gives the (a*R*)-alkenes with 4-substituted alkylcyclohexanones and the (*E*)-alkenes with other analogs, based on a transition state that favors equatorial attack of the least encumbered face of the anion on the cyclohexanone derivative. For steric reasons, the reverse is observed with reagent (**1**).[3] eq 1 illustrates a typical reaction and other examples of products are listed in Table 1 (enantiomeric excesses >99:1).[2,3]

Table 1 Products of Asymmetric Alkenation

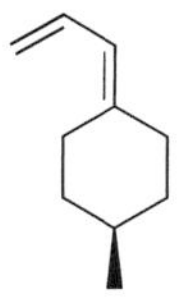

67%, $[\alpha]_{546}$ +42.3°

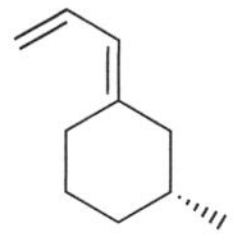

74%, $[\alpha]_{546}$ −41.9°

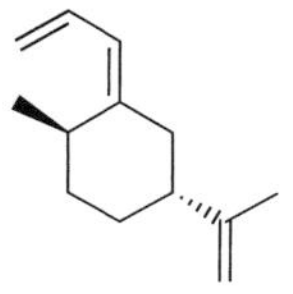

55%, 91:9

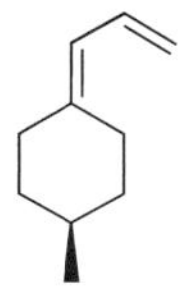

73%, $[\alpha]_{546}$ −42.7°

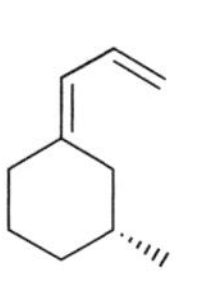

75%, $[\alpha]_{546}$ −141.6°

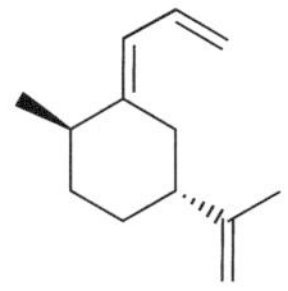

74%, $[\alpha]_{546}$ −155.9°

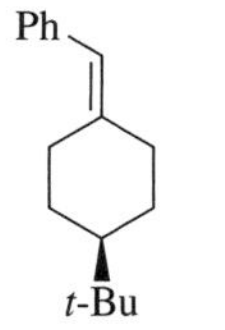

91%, $[\alpha]_D$ −33.3°

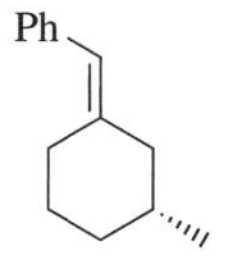

82%, $[\alpha]_D$ −72.0°

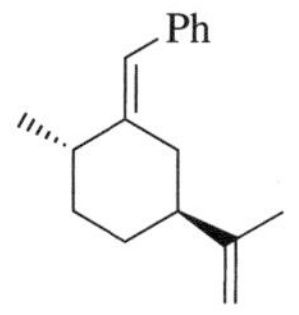

86%, $[\alpha]_D$ −86°

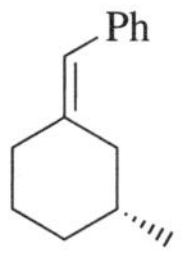

75%, $[\alpha]_D$ +30.3°

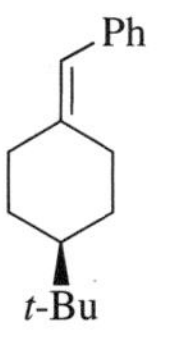

89%, $[\alpha]_D$ +32.3°

Kinetic Resolution.[2] When the reacting partners allow for a high degree of stereodifferentiation in the transition state, it is possible to achieve asymmetric alkenation by kinetic resolution (eq 2). This is best done with α-alkyl substituted cyclohexanones and 'bulky' anions such as that derived from (**4**). In a typical procedure, (±)-2-methylcyclohexanone (1 mmol) is treated with the anion of (**3**) (0.5 mmol, −78 °C, THF, 1 h; then AcOH, → 25 °C, and workup), to give (*E*,2*S*)-(2-methylcyclohexylidene)benzene; $[\alpha]_D^{25}$ −86.4° (*c* 1, $CHCl_3$), and the (*Z*)-isomer (>98:2 by capillary GC; 63% based on the reagent).

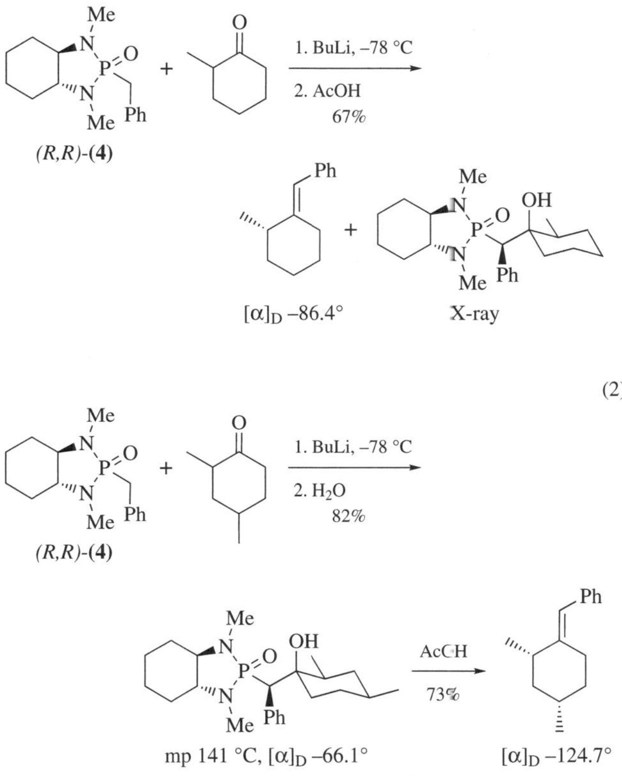

(2)

Sequential Asymmetric Alkenation and Ene Reactions.[3] Treatment of a number of (alkylcyclohexylidene)ethane derivatives with chiral nonracemic α-benzyloxy aldehydes results in the formation of branched alkylcyclohexene derivatives via a highly stereocontrolled ene reaction. Based on a chelated transition state, it is possible to predict the disposition of the double bond and the chirality of two new stereogenic centers, as illustrated in eq 3. The newly created stereogenic center bearing a *C*-methyl group from the ene reaction leads to interesting substitution patterns in relation to the existing *C*-methyl group. Oxidative cleavage of the double bond leads to acyclic counterparts with predictable disposition and chirality (e.g. 1,6-dimethyl, 1,5-dimethyl, etc.).[3]

α-Substituted α-Alkylphosphonic Acids[4] In general, α-substituted phosphonamides can be transformed into the corresponding α-alkyl derivatives of very high diastereomeric purity by alkylation of the anions at −78 °C or lower (eq 4, Table 2). In most cases the approach of the electrophile is favored from the least hindered side of the anion, leading to highly enriched diastereomers. These products can be subsequently hydrolyzed to the corresponding phosphonic acids. A typical procedure is as follows. To a solution of (**3**) (1 mmol) in THF, is added *n-Butyllithium* (1.40 mmol at −78 °C), the temperature is lowered to −100 °C, and *Allyl Bromide* (1.15 mmol) is added. After stirring for 15 min, the reaction mixture is quenched with MeOH, and the solution processed as usual. Chromatographic purification gives the expected product as a crystalline solid (X-ray), mp 87–88 °C, $[\alpha]_D^{25}$ −91.5° (*c* 1.0, $CHCl_3$).

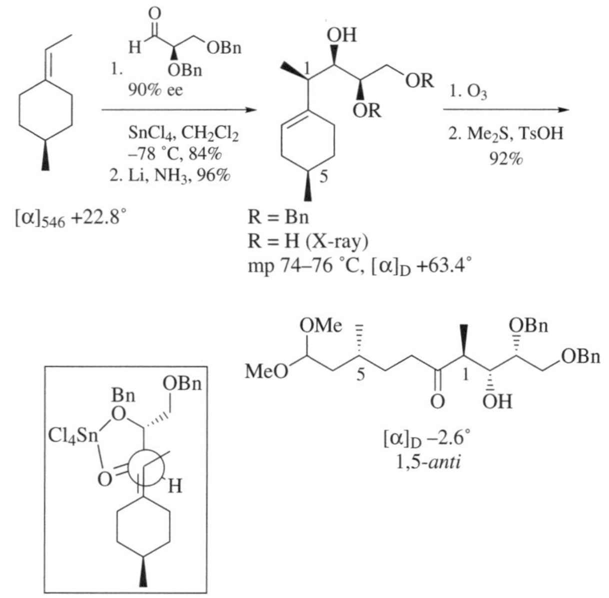

(3)

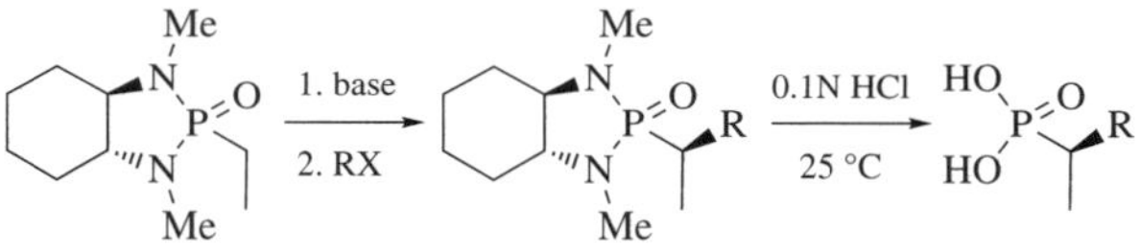

(4)

α-Chloro-α-alkylphosphonic Acids.[4] The reagent is prepared from α-chloromethylphosphonyl dichloride and (*R,R*)- or (*S,S*)-*N,N'*-dimethyl-1,2-diaminocyclohexane as described above. Alkylation is done as described above (BuLi, THF, −100 °C, followed by isolation, then hydrolysis, 0.1N HCl, 25 °C). Eq 5 and Table 3 illustrate some examples.[4]

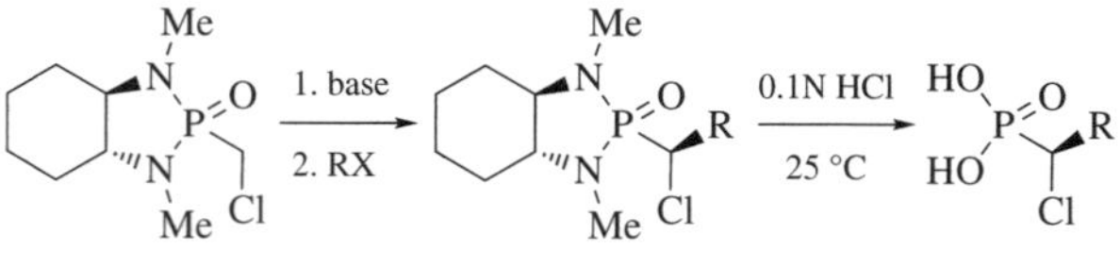

(5)

α-Amino-α-alkylphosphonic Acids.[5] Treatment of iminodithiolane derivatives (eq 6) with *Potassium Hexamethyldisilazide* in THF (−78 °C) generates the corresponding anions which, when treated with various alkyl halides, give the corresponding α-alkyl derivatives in high diastereomeric excess. Unlike other α-substituted phosphonamides discussed above, the alkylation of the (*R,R*)-α-iminodithiolane derivative gives products with the opposite orientation of the new alkyl chain, which are normally expected from the enantiomeric (*S,S*) series. This has been rationalized based on the intermediacy of a potassium chelate involving the phosphoryl oxygen and the imino nitrogen atoms, thus exposing the other face of the anion.[5] eq 6 illustrates a typical sequence and Table 4 lists some examples of α-amino-α-alkylphosphonic acids prepared using this sequence.

α-Amino-β-aryl phosphonic acids are accessible from the addition of the anion of chloromethylphosphonamide (**8**) to *N*-arylimines, followed by hydrogenolysis of the aziridine derivative.[6]

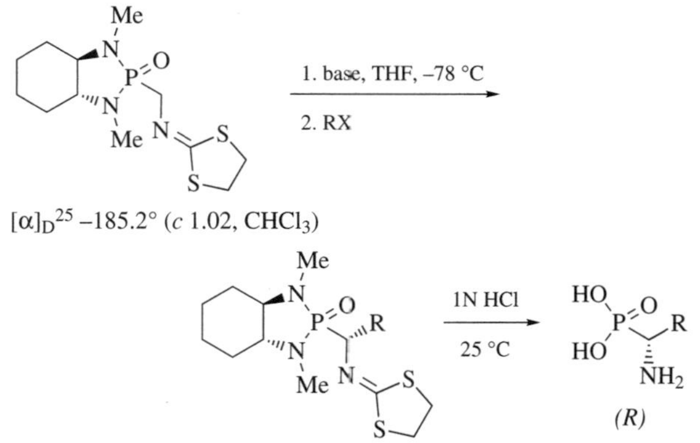

(6)

Asymmetric Conjugate Addition of Allyl- and Crotylphosphonamides.[7] The asymmetric *C*-allylation of α,β-unsaturated carbonyl compounds is a powerful tool for the functionalization of a carbonyl compound in the β-position. Since such a process normally leads to the corresponding enolate derivative when anionic reagents are used, there exists the possibility of trapping with an electrophile. Thus sequential addition and trapping can lead to vicinally substituted carbonyl compounds. Asymmetric allylation has been achieved previously with simple cycloalkenones using phosphorus[9] and sulfur[10] based reagents that must be prepared in diastereomerically pure form.

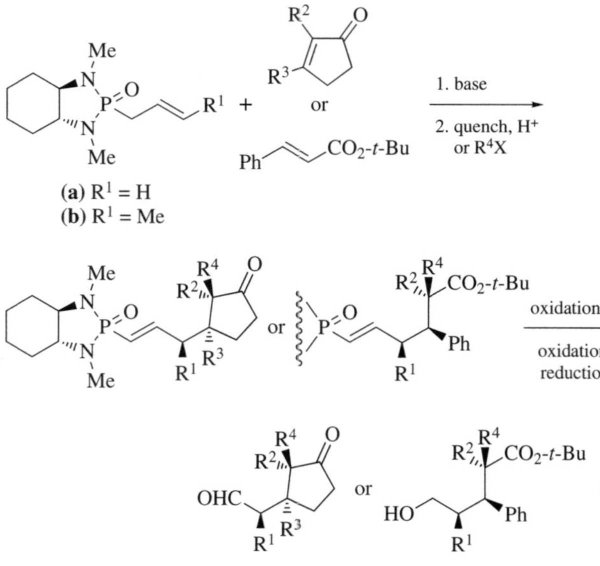

(7)

R^1, R^3 = H; R^2, R^4 = Me, alkyl, H
R^1 = Me, R^3 = H; R^2, R^4 = Me. alkyl, H (any combination)

The anions of allyl- and crotylphosphonamides, (**2**) and (**5**) respectively, show excellent selectivity toward a variety of α,β-unsaturated compounds, affording the diastereomerically pure or

Table 2 Synthesis of α-Substituted α-Alkylphosphonic Acids

RX	Alkylation (−100 °C)				Hydrolysis			
	Yield (%)	Ratio	Mp (°C)	$[\alpha]_D^{25}$	Yield (%)	Mp (°C)	$[\alpha]_D^{25}$	Config.
(*R*,*R*) Series								
EtI	76	95:5	95–96	−81.5° (*c* 1.0)	90	54–57	+6.2° (*c* 1.0)	(*R*)
CH_2=$CHCH_2Br$	82	94:6	87–88	−91.5° (*c* 1.0)	88	Oil	−1.1° (*c* 1.75)	(*R*)
BnBr	83	97:3	95–97	−35.2° (*c* 1.2)	86	122–125	+23.6° (*c* 2.2)	(*R*)
TBDPSO(CH_2)$_2$I	68	94:6	Oil	−32.5° (*c* 1.03)	94	35	+7.0° (*c* 1.25)	(*R*)
TBDPSO(CH_2)$_3$I	71	98:2	Oil	−35.5° (*c* 1.1)	92	Oil	+23.8° (*c* 1.1)	(*R*)
(*S*,*S*) Series								
EtI	70	5:95	92–95	+79.2° (*c* 1.0)	91	54–56	−6.1° (*c* 0.9)	(*S*)
CH_2=$CHCH_2Br$	83	4:96	86–88	+91.0° (*c* 1.0)	88	Oil	+1.4° (*c* 1.4)	(*S*)
BnBr	84	4:96	96–97	+33.6° (*c* 1.0)	82	122–124	−22.8° (*c* 1.0)	(*S*)

Table 3 Synthesis of α-Chloro-α-alkylphosphonic Acids

RX	Alkylation (−100 °C)				Hydrolysis			
	Yield (%)	Ratio	Mp (°C)	$[\alpha]_D^{25}$	Yield (%)	Mp (°C)	$[\alpha]_D^{25}$	Config.
(*R*,*R*) Series								
MeI	87	90:10	75–76	−99.6° (*c* 0.5)	95	35–38	+5.5° (*c* 1.0)	(*R*)
EtI	79	>99:1	123–125	−69.0° (*c* 0.5)	98	109	+39.6° (*c* 1.2)	(*R*)
PrI	83	>99:1	131–133	−62.0° (*c* 1.0)	97	87–88	+51.2° (*c* 1.0)	(*R*)
CH_2=$CHCH_2Br$	80	>99:1	136–137	−71.5° (*c* 0.6)	Quant.	Oil	+32.0° (*c* 1.0)	(*R*)
BnBr	86	91:9	103–105	−15.4° (*c* 1.0)	Quant.	104	+32.9° (*c* 0.9)	(*R*)
TBDPSO(CH_2)$_2$I	79	90:10	Oil	−21.0° (*c* 1.3)	95	35	+21.9° (*c* 1.65)	(*R*)
TBDPSO(CH_2)$_3$I	76	90:10	56–58	−34.0° (*c* 1.4)	88	Oil	+55.8° (*c* 1.1)	(*R*)

Table 4 Synthesis of α-Amino-α-alkylphosphonic Acids

RX	Alkylation (−78 °C)				Hydrolysis				
	Yield (%)	Ratio	Mp (°C)	$[\alpha]_D^{25}$	Yield (%)	Mp (°C)	$[\alpha]_{578}$	ee	Config.
(*R*,*R*) Series									
MeI	74	90:10	134	−183.6° (*c* 0.6)	82	278	−12.9°	81%	(*R*)
EtI	73	95:5	137	−189.0° (*c* 0.9)	84	277	−19.2°	91%	(*R*)
PrI	76	>99:1	129	−178.0° (*c* 0.7)	86	272	+8.5°	98%	(*R*)
CH_2=$CHCH_2Br$	82	>99:1	123	−163.2° (*c* 1.0)	88	272	−8.9°	–	(*R*)
i-BuOTf	77	95:5	136	−183.0° (*c* 1.0)	81	288	−21.6°	90%	(*R*)
BnBr	78	>99:1	140	−148.6° (*c* 0.5)	87	268	−47.6°	97%	(*R*)
HC≡CCH_2Br	78	>99:1	125	−137.2° (*c* 0.8)	–				
(*S*,*S*) Series									
MeI	75	8:92	139	+187.0° (*c* 1.0)	84	278	+14.3°	84%	(*S*)
BnBr	81	>1:99	142	+148.0° (*c* 0.75)	86	270	+50.9°	98%	(*S*)

enriched products. Quenching the enolates with various electrophiles gives vicinally substituted carbon centers. Oxidative cleavage of the phosphonamide moiety affords the equivalent of an acetaldehyde (α-methylacetaldehyde) anion 1,4-adduct to the original α,β-unsaturated carbonyl compound. Pertinent examples are shown in eq 7 and Table 5.

Table 5 Asymmetric Conjugate Addition of Allyl- and Crotylphosphonamides

Entry	Reagent	Substrate	Product	Ratio	Yield (%)
1	**(2)**	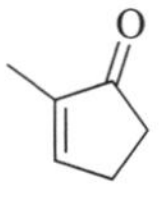	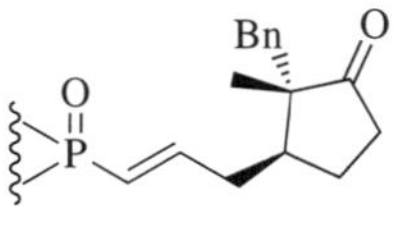 $[\alpha]_D$ +48.6° (*c* 1.20)	>99:1	80
2	**(2)**	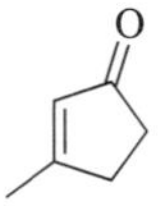	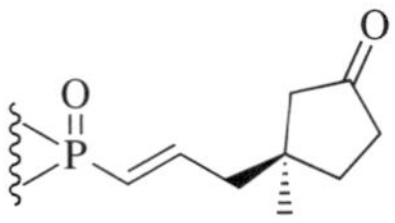 $[\alpha]_D$ +17.4° (*c* 1.22)	99:5	75
3	**(5)**	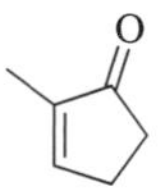	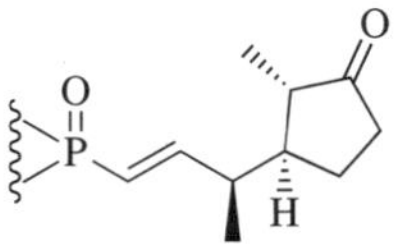 $[\alpha]_D$ +7.6° (*c* 2.00)	94:6	74
4	**(2)**		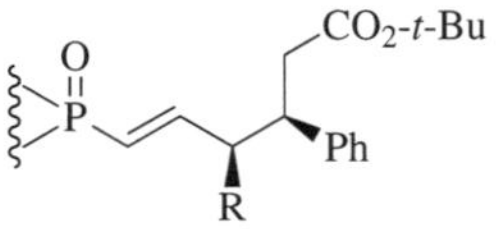 R = H, $[\alpha]_D$ −5.8° (*c* 1.90) R = Me, $[\alpha]_D$ −1.9° (*c* 1.10)	>99:1 >99:1	80 94
5	**(2)**	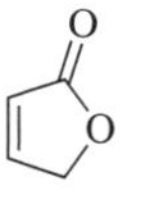	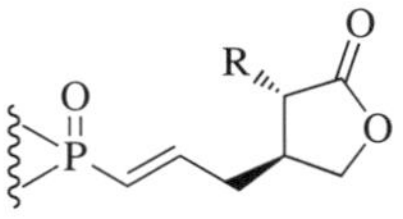 R = H, $[\alpha]_D$ −16.1° (*c* 1.00) R = Bn, $[\alpha]_D$ −8.7° (*c* 0.80)	>99:1 >99:1	93 67
6	**(5)**		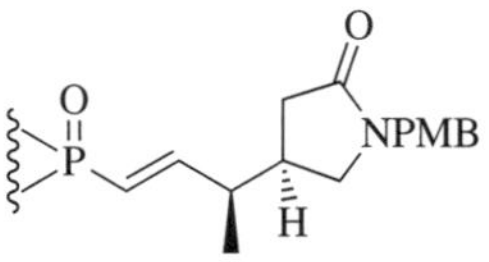 $[\alpha]_D$ −2.3° (*c* 2.20)	95:5	51

Conclusion. The C_2 symmetry of (*R,R*)- and (*S,S*)-1,2-diaminocyclohexane, readily available from the racemic compound by resolution,[11] has served as a versatile chiral motif in the design of topologically unique stereodifferentiating reagents such as the phosphonamide anions described here. Several other applications of these reagents via anion chemistry, or simply based on the exploitation of other effects offered by their structures and heteroatom functionality, can be explored (catalytic processes, chiral ligands, etc.). The *N,N′*-disubstituted 1,2-diaminocyclohexane motif has also been remarkably versatile in other asymmetric processes such as the dihydroxylation of alkenes,[12] and a variety of other C–C bond-forming reactions.[13]

1. Hanessian, S.; Delorme, D.; Beaudoin, S.; Leblanc, Y. *J. Am. Chem. Soc.* **1984**, *106*, 5754; *Chem. Ser.* **1985**, *88*, 1419.
2. Hanessian, S.; Beaudoin, S. *Tetrahedron Lett.* **1992**, *33*, 7655.
3. Hanessian, S.; Beaudoin, S. *Tetrahedron Lett.* **1992**, *33*, 7659.
4. Hanessian, S.; Bennani, Y.; Delorme, D. *Tetrahedron Lett.* **1990**, *31*, 6461.
5. Hanessian, S.; Bennani, Y. *Tetrahedron Lett.* **1990**, *31*, 6465.
6. Hanessian, S.; Bennani, Y.; Hervé, Y. *Synthesis* **1993**, 35.
7. Hanessian, S.; Gomtsyan, A.; Payne, A.; Hervé, Y.; Beaudoin, S. *J. Org. Chem.* **1993**, *58*, 5032.
8. Koeller, K.; Spilling, C. D. *Tetrahedron Lett.* **1991**, *32*, 6297.
9. Haynes, R. K.; Vonwiller, S. C.; Hambley, T. W. *J. Org. Chem.* **1989**, *54*, 5162.
10. Hua, D. H.; Chan-Yu-King, R.; McKie, J.-A.; Myer, L. *J. Am. Chem. Soc.* **1987**, *109*, 5026.
11. (a) Gasbol, F.; Seenbol, P.; Sorensen, B. S. *Acta Chem. Scand.* **1972**, *26*, 3605; (b) Asperger, R. G.; Liu, C. F. *Inorg. Chem.* **1965**, *4*, 1492.
12. Hanessian, S.; Meffre, P.; Girard, M.; Beaudoin, S.; Sancéau, J.-Y.; Bennani, Y. *J. Org. Chem.* **1993**, *58*, 1991.
13. (a) Rozema, M. J.; Eisenberg, C.; Lütjens, H.; Ostwald, R.; Belyk, K.; Knochel, P. *Tetrahedron Lett.* **1993**, *34*, 3115. (b) Jacobsen, E. N.; Zhang, W.; Muci, A. R.; Ecker, J. R.; Deng, L. *J. Am. Chem. Soc.* **1991**, *113*, 7063. (c) Denmark, S. E.; Stadler, H.; Dorw, R. L.; Kim, J.-H. *J. Org. Chem.* **1991**, *56*, 5063. (d) Alexakis, A.; Mutti, S.; Normant, J. F. *J. Am. Chem. Soc.* **1991**, *113*, 6332. (e) Bertz, S. H.; Dabbagh, G.; Sundararajan, G. *J. Org. Chem.* **1986**, *51*, 4953. (f) Takahashi, H.; Kawakita, T.; Yoshioka, M.; Kobayashi, S.; Ohno, M. *Tetrahedron Lett.* **1989**, *30*, 7095.

Stephen Hanessian
University of Montreal, Quebec, Canada

F

(+)-N-Fluoro-2, 10-(3,3-dichlorocamphorsultam)[1]

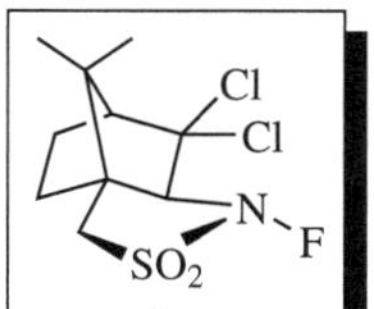

1

[151556-58-0] $C_{10}H_{14}Cl_2FNO_2S$ (MW 302.19)

(neutral, aprotic, electrophilic fluorinating agent for the asymmetric α-fluorination of enolates[1])

Physical Data: mp 161–162 °C, $[\alpha]_D^{20}$ +16.4 (*c* 1.3, $CHCl_3$).
Solubility: soluble in THF, CH_2Cl_2, $CHCl_3$; insoluble in hexane, pentane, H_2O.
Form Supplied in: white solid.
Preparative Methods: the title reagent **1** can be prepared by fluorination of the corresponding (+)-2,10-(3,3-dichlorocamphorsultam)[2] using 10% F_2/N_2 in dry chloroform in the presence of sodium fluoride.[3,4]
Purification: purified by silica gel chromatography using a mixture of CH_2Cl_2/*n*-pentane (70:30) as eluent.
Handling, Storage, and Precautions: can be stored in a bottle in the refrigerator for months without noticeable decomposition.

Fluorination of Ketone Enolates. (+)-*N*-Fluoro-2,10-(3,3-dichlorocamphorsultam) (**1**) reacts with ketone enolates to give α-fluoro ketones. For example, reaction of the sodium enolate of propiophenone **2** gives α-fluoropropiophenone **3** in 41% isolated yield (eq 1). No enantioselectivity, however, is observed due to racemization of the product under the reaction conditions.[4] When a tertiary substituted ketone such as α-methyltetralone (**4**) is employed, the desired α-fluorinated product (*S*)-(–)-**5** is obtained in 76% ee and 53% isolated yield (eq 2).[4] In this reaction, (+)-**1** was found to be more reactive, affording higher yields and better enantioselectivities than its parent (–)-*N*-fluoro-2,10-camphorsultam; i.e., 35% ee, < 5% yield.[5]

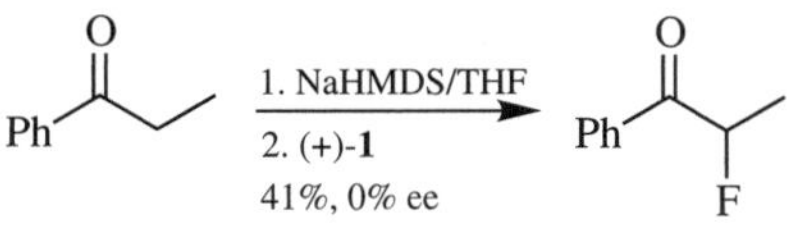

(1)

2 3

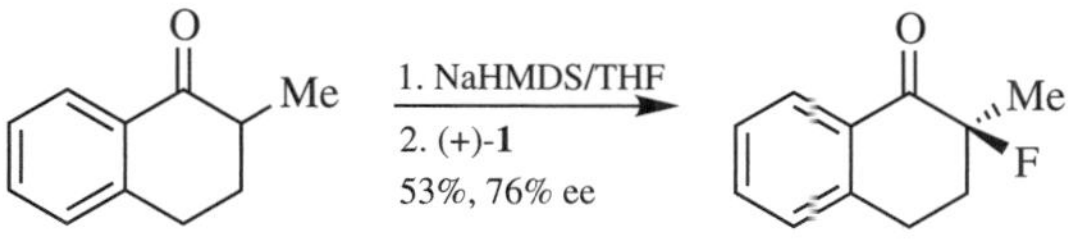

(2)

4 (*S*)-(–)-**5**

Fluorination of Ester Enolates. (+)-*N*-Fluoro-2,10-(3,3-dichlorocamphorsultam) (**1**) also reacts with the enolates generated from esters. For example, treatment of methyl 2-phenylpropionate **6** with NaHMDS in THF followed by addition of (+)-**1** affords the corresponding α-fluoro ester **7** in 54% yield and 33% ee (eq 3). The absolute configuration was not determined. The enantioselectivity observed with (+)-**1** is better than (–)-*N*-fluoro-2,10-camphorsultam; i.e., < 10% ee (30% yield).[5]

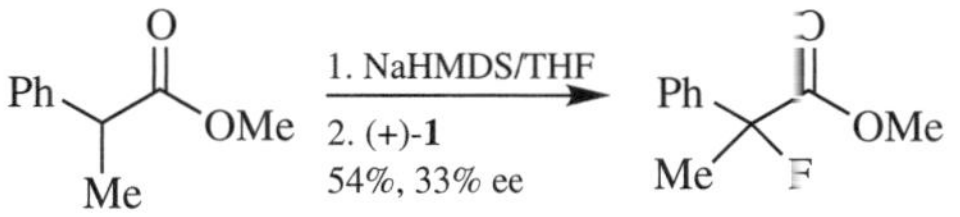

(3)

6 7

Fluorination of β-Ketoester Enolates. (+)-*N*-Fluoro-2,10-(3,3-dichlorocamphorsultam) (**1**) is reactive towards the enolates generated from β-ketoesters. Thus, treatment of the sodium enolate of **8** with (+)-**1** afforded the desired product **9** in 95% isolated yield and 46% ee with undetermined stereochemistry (eq 4). Reduced yields and enantioselectivities were noted under similar conditions for (-)-*N*-fluoro-2,10-camphorsultam (28% yield, 25% ee).[4]

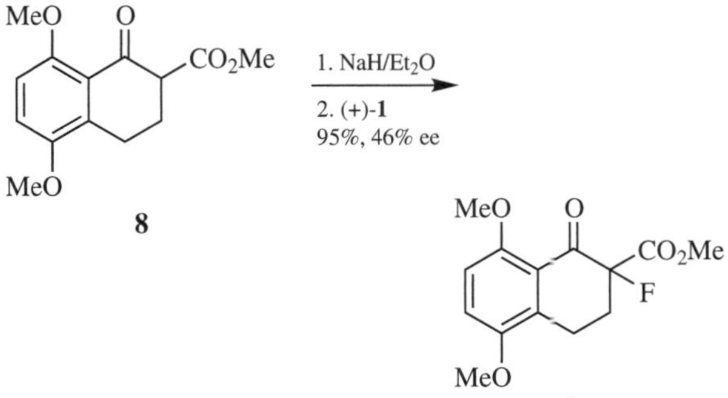

8 9 (4)

Related Reagents. (+)-*N*-fluoro-2,10-camphorsultam;[4,5] (–)-*N*-fluoro-2,10-(3,3-dimethoxycamphorsultam);[4] (3*S*)-(–)-*N*-fluoro-3-cyclohexyl-3-methyl-2,3-dihydrobenzo[1,2-*d*]isothiazole 1,1-dioxide;[6] *N*-fluoro-*o*-benzenesulfonimide;[7] *N*-fluorobenzenesulfonimide.[8]

---

1. (a) Davis, F. A.; Qi, H.; Sundarababu, G. In *Enantiocontrolled Synthesis of Fluoro-Organic Compounds: Stereochemical Challenges and biomedical Targets*; Soloshonok, V. A., Ed.; John Wiley & Sons Ltd: Chichester, 1999, pp 1-32. (b) Davis, F. A.; Kasu, P. V. N. *Org. Prep. Proc. Int.* **1999**, *31*, 125.

2. (a) Davis, F. A.; Zhou, P.; Chen, B.-C. *Phosphorus Sulfur Silicon* **1996**, *115*, 85. (b) Chen, B.-C.; Murphy, C. K.; Kumar, A.; ThimmaReddy, R.; Zhou, P.; Lewis, B. M.; Gala, D.; Mergelsberg, L.; Scherer, D.; Buckley, J.; Dibenedetto, D.; Davis, F. A. *Org. Synth.* **1995**, *73*, 159.
3. Davis, F. A.; Zhou, P.; Murphy, C. K. *Tetrahedron Lett.* **1993**, *34*, 3971.
4. Davis, F. A.; Zhou, P.; Murphy, C. K.; Sundarababu, G.; Qi, H.; Han, W.; Przeslawski, R. M.; Chen, B.-C.; Carroll, P. J. *J. Org. Chem.* **1998**, *63*, 2273.
5. Differding, E.; Lang, R. W. *Tetrahedron Lett.* **1988**, *29*, 6087.
6. Takeuchi, Y.; Suzuki, T.; Satoh, A.; Shiragami, T.; Shibata, N. *J. Org. Chem.* **1999**, *64*, 5708.
7. Davis, F. A.; Han, W. *Tetrahedron Lett.* **1991**, *32*, 1651.
8. Differding, E.; Ofner, H. *Synlett* **1991**, 187.

Franklin A. Davis
Temple University, Philadelphia, USA
Bang-Chi Chen
Discovery Chemistry, Bristol-Myers Squibb Pharmaceutical Research Institute, Princeton, NJ, USA

G

Glycidol[1]

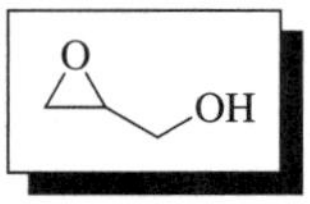

[556-52-5] $C_3H_6O_2$ (MW 74.08)
(*R*)
[57044-25-4]
(*S*)
[60456-23-7]
(±)
[61915-27-3]

(versatile bifunctional, three-carbon synthon)

Alternate Name: oxiranemethanol.
Physical Data: mp −53 °C; bp 161–163 °C (dec), 30 °C/1 mmHg, 54 °C/8 mmHg, 114 °C/114 mmHg; *d* 1.115 g cm^{-3}; $[\alpha]_D$ +15° (neat, L-(+)-glycidol).
Solubility: insol aliphatic hydrocarbons; sol H_2O, acetone, THF, toluene, most other organic solvents.
Form Supplied in: racemic, (*R*), and (*S*) forms; all as colorless, neat liquids. Solid derivatives: phenyl isocyanate, mp 60 °C; α-naphthyl isocyanate, mp 102 °C.
Analysis of Reagent Purity: ^{1}H NMR.
Handling, Storage, and Precautions: neat samples of glycidol should be stored in the freezer to slow the process of self-condensation; when stored neat, glycidol should be checked for purity before use and will usually require purification, which can be achieved by distillation under reduced pressure; self-condensation is greatly reduced by storage of glycidol in solutions, e.g. 50–70% in toluene or dichloromethane; distillation of glycidol should be done behind a safety shield; care should be taken when using glycidol under acidic conditions (e.g. acetic acid) since acid catalyzes self-condensation; use in a fume hood.

Glycidol and Glycidol Derivatives. Two excellent reviews of glycidol and glycidol derivatives are available. The first is a very thorough review of the properties and reactions of glycidol written by Kleemann and Wagner.[1a] In the second, the use of glycidol and glycidol derivatives as synthons, with a strong emphasis on nonracemic glycidol, is the subject of a superb review by sHanson.[1b]

Glycidol is a versatile three-carbon synthetic building block and its value is greatly expanded through derivatization of the hydroxyl group. The use in synthesis of derivatives such as *O*-aryl and *O*-arylmethyl (e.g. *O*-benzyl) ethers, sulfonates, carboxylates, and silyl ethers is integrated with those of glycidol for this review. In the following discussion, glycidol and derivatives are occasionally referred to collectively as glycidols. Also note that reactions of nonracemic glycidol are illustrated only with one enantiomer, but apply equally to use of both.

Glycidol, like all 2,3-epoxy alcohols, is susceptible to the Payne rearrangement when exposed to base. Payne rearrangement of (*R*)- or (*S*)-glycidol is degenerate; consequently racemization does not occur.

(*R*) and (*S*) are empirical designations of absolute configurations and in comparing glycidol and an *O*-substituted glycidol derivative having the same absolute configuration, the designation changes (see eq 1). For further discussion of this point, see Hanson's review.[1b]

Preparations. Racemic glycidol, (*R*) and (*S*)-glycidol, and a number of derivatives of each are commercially available. Preparations of these materials are described in the literature and a selected listing follows: (*S*)-glycidol via asymmetric epoxidation[2] and enzymatic kinetic resolution;[3] *O*-benzyl glycidol,[4] *O*-trityl glycidol;[5] (*R*)-(−)-*Glycidyl Tosylate*;[6] (*R*)-(−)-glycidyl 3-nitrobenzenesulfonate (a derivative whose optical purity is enhanced by recrystallization);[6] (*R*)-(−)-glycidyl *p*-nitrobenzoate (see eq 1).[2,7]

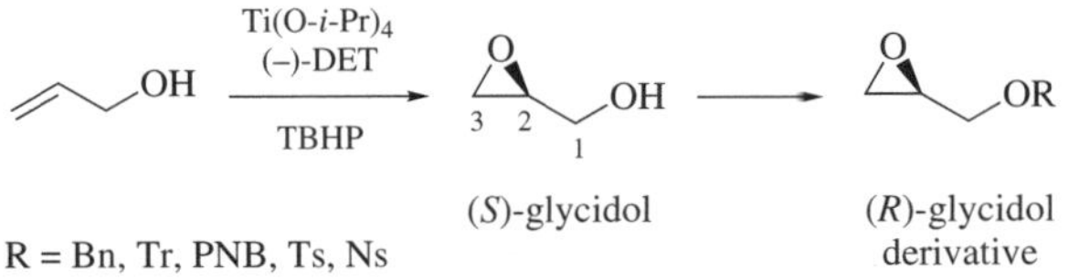

(1)

Reactions at C-1 of Glycidol. A number of *O*-derivatives of glycidol are described in the preceding section and may be prepared directly from racemic or (*R*)- or (*S*)-glycidol. Alternatively, if carrying out the laboratory preparation of (*R*)- or (*S*)-glycidol, convenient in situ methods for derivatization have been developed.[2,7,8] Derivatization as *O*-sulfonate esters (e.g. tosylates) activates the C-1 position and permits displacement by nucleophiles. An example is displacement by phenolates to generate *O*-aryl glycidol ethers (see eq 2),[6] which find extensive use as intermediates in the synthesis of a variety of pharmacologically active agents (see additions of nitrogen at C-3, below).

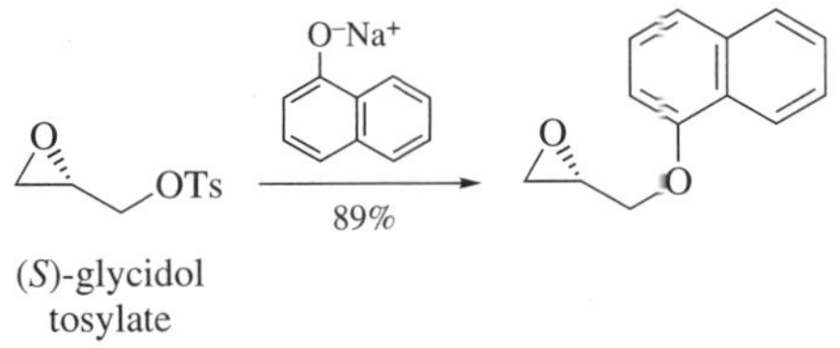

(2)

O-Aryl glycidol ethers can be prepared from glycidol by the Mitsunobu reaction with phenols (see eq 3)[9a] and are also made from direct displacement by glycidol on activated haloaryls.[9b]

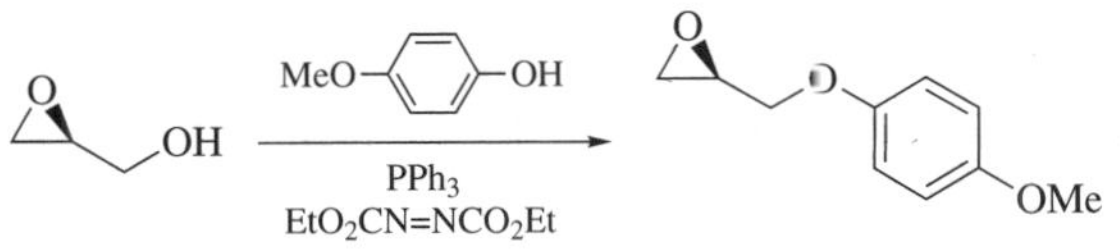

(3)

Addition of Hydrogen at C-3. Both catalytic reduction of glycidol over Pd/C[10] and reaction with MeLi/CuBr$(PBu_3)_2$[11]

give propane-1,2-diol as a consequence of addition of hydrogen at C-3. (*S*)-Glycidyl tosylate is reduced to (*S*)-propane-1,2-diol 1-monotosylate with *Borane–Tetrahydrofuran* and a catalytic amount of *Sodium Borohydride*, as shown in eq 4.[6]

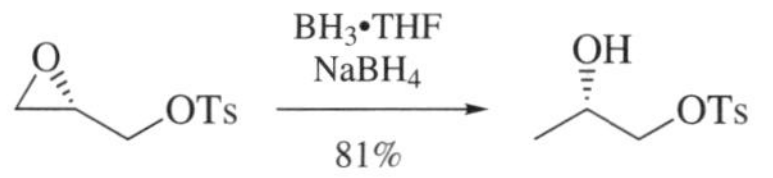

(4)

Nucleophilic Additions of Carbon at C-3. One of the few reported additions of a carbon nucleophile to underivatized glycidol is that of diethyl sodiomalonate. The initial addition at C-3 is followed by lactonization between the C-2 hydroxyl group and one of the malonate carboxylic esters (eq 5).[12] Far more numerous are the additions of carbon nucleophiles to glycidol derivatives such as *O*-benzyl, *O*-phenyl, or *O*-tosyl glycidol. In addition to the examples included below, many others may be found in Hanson's review.[1b]

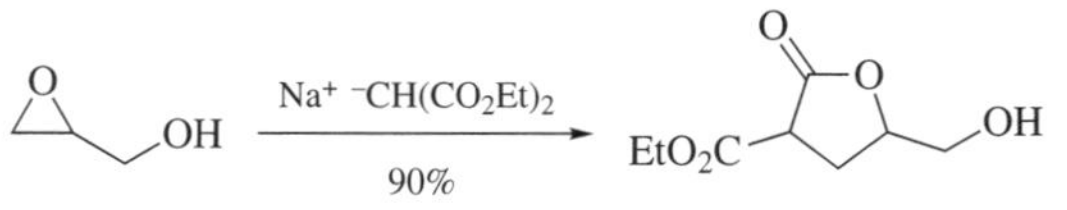

(5)

Single carbons can be added as cyanide using *Acetone Cyanohydrin*,[6,7] diethylaluminum cyanide (eq 6),[6,7] or *Lithium Cyanide*[13] or as methyl groups using an organocuprate (eq 7).[14] A single carbon may be added with dithiane salts and an example of addition of a substituted *1,3-Dithiane* to *O*-benzyl glycidol is shown in eq 8.[15]

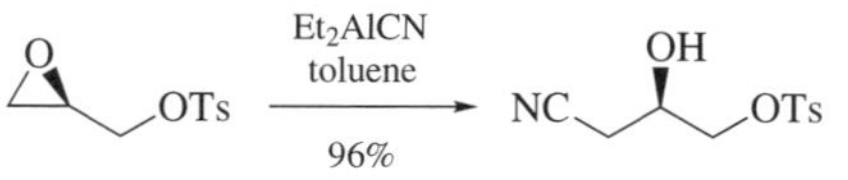

(6)

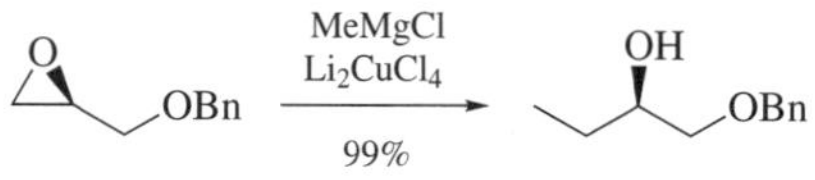

(7)

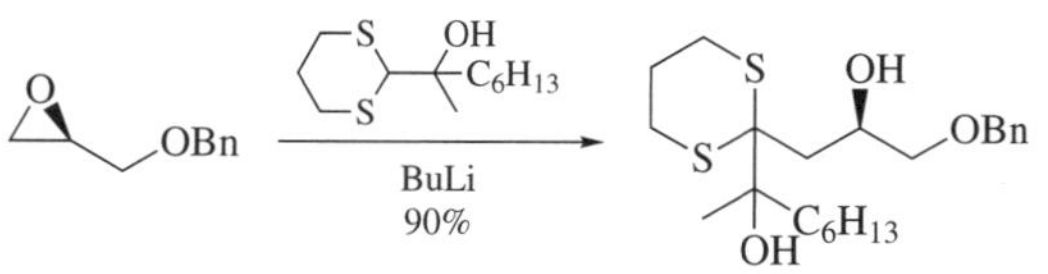

(8)

Other alkyl groups, alkenyl groups (eq 9), and aryl groups have been added to glycidol via organometallic reagents. The reactions with organometallic reagents often are sensitive to conditions and frequently are improved by the addition of Cu[I] or Cu[II] to the medium.[16] Alkynic salts add to glycidols, giving 3-alkynyl derivatives in yields which are generally good but which may be enhanced in some cases by the addition of a Lewis acid such as *Boron Trifluoride Etherate* to the reaction (eq 10).[17]

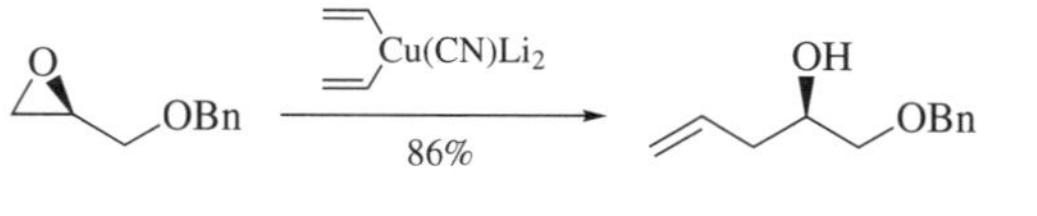

(9)

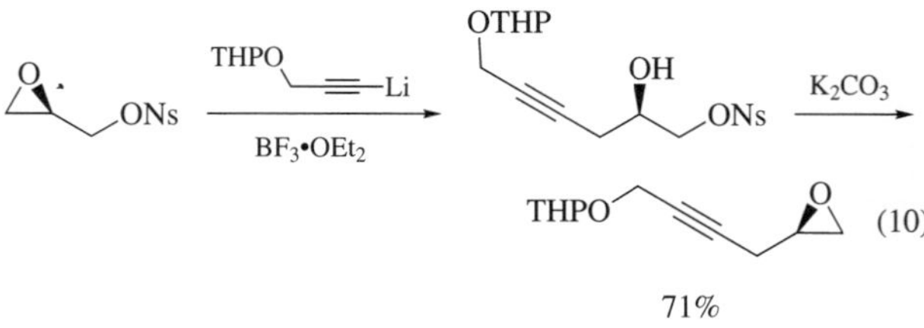

(10)

Opening at C-3 of glycidol sulfonates generates a 1,2-diol monosulfonate array which is ideally situated for closure under mildly alkaline conditions to a new epoxide group, as shown in eq 10 and also, below, in eq 18. Either the intermediate monosulfonate or the new epoxide present an activated electrophilic site for further synthetic transformations.

Carbon nucleophiles such as ester enolates and α-carboxylic acid anions add to glycidols by opening the oxirane ring and forming an intermediate C-2 alcohol. As shown above in eq 5, the intermediate can cyclize to a five-membered lactone via further reaction with the newly introduced carboxylic acid or ester.[18] Variations on the theme of intramolecular transformations following the initial addition to glycidol have been described. These include seven-membered lactone formation following addition of a sulfone-stabilized anion to *O*-benzylglycidol (eq 11),[19] and oxetane formation following addition of *Dimethylsulfoxonium Methylide* to glycidol (eq 12).[20]

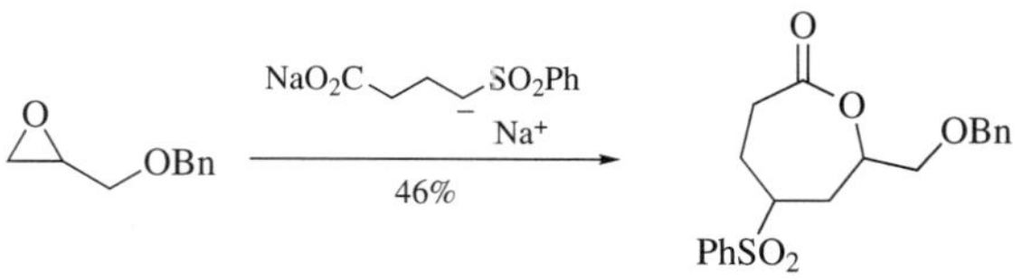

(11)

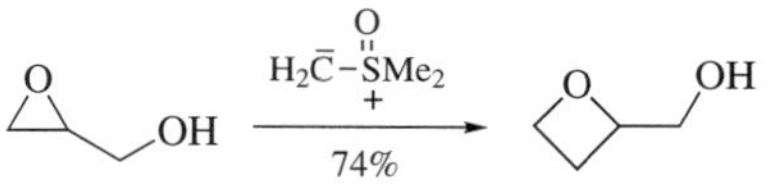

(12)

Other examples of carbon nucleophiles which have been added to a glycidol include the lithium salt of 1-trimethylsilyl-3-phenylthioprop-1-yne (eq 13),[21] the lithium salt of 1-phenylsulfonyl-2-trimethylsilylethane,[22] the lithium salt of pentacarbonyl(methoxymethylcarbene)chromium,[23] the dimsyl anion,[24] and the lithium salt of acetone dimethylhydrazone.[25]

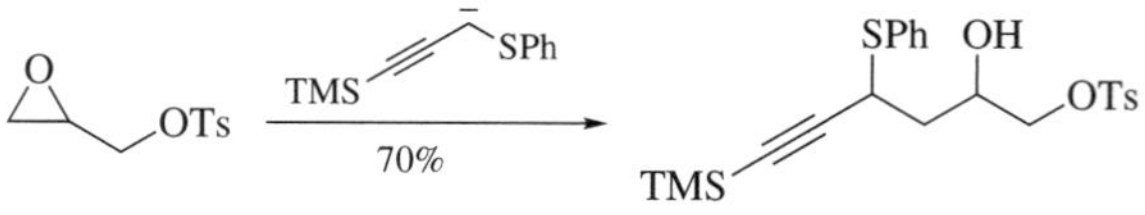

(13)

Nucleophilic Addition of Oxygen at C-3. Addition of water to glycidol or a glycidol derivative produces glycerol or a substituted glycerol, respectively. The oxygen nucleophiles used most frequently for addition to glycidols are alcohols, phenols, and carboxylic acids and their close relatives. For glycidol itself, Kleemann and Wagner summarize extensive studies of additions of these classes of compounds.[1a] Very good yields of products are achieved with all three classes when acid or, preferentially, basic catalysts are added to the reactions. Careful analyses of the reaction products reveal that in addition to the primary opening of the oxirane at C-3, most reactions include small (2–10%) amounts of product derived from opening at C-2. Other byproducts can result from self-reaction of glycidol with the reaction products. Opening of glycidol with primary alcohols with 0.5% NaOH as catalyst yields 70% of the 1-*O*-alkylglycerol together with 3% of the 2-*O*-alkylglycerol.[1a] Opening with phenols and 0.03% NaOH gives 70–80% yields of 1-*O*- and 2-*O*-arylglycerols in ratios of 90–95:5–10.[1a] Glycidol generated in situ from hydrolysis of the *p*-nitrobenzoate ester with MeOH/H_2SO_4 reacts further at C-3 with the MeOH to give 1-*O*-methoxyglycerol (eq 14).[7]

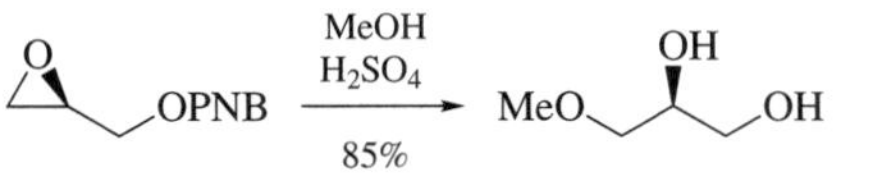

(14)

Lewis acid catalysis of additions to 2,3-epoxy alcohols often improves the regioselectivity of the ring-opening process.[26] $Ti(OR)_4$ catalyzed reaction of glycidol with primary alcohols gives 1-*O*-alkylglycerols in yields of 45–59%.[27] The addition of primary alcohols to (*R*)-glycidyl sulfonate esters give 1-*O*-alkylglycerol 3-sulfonates in yields of 73–89% when catalyzed with $BF_3{\cdot}OEt_2$ (eq 15).[28] Non-racemic glycidol, generated by catalytic asymmetric epoxidation of allyl alcohol with Ti(O-*i*-Pr)$_4$ and a (+)- or (−)-dialkyl tartrate, undergoes *Titanium Tetraisopropoxide* assisted reaction in situ with sodium phenolates to generate 1-*O*-arylglycerols (eq 16).[8,29] The $BF_3{\cdot}OEt_2$ addition of stearic anhydride to (*R*)-glycidyl tosylate gives (*R*)-1,2-distearoylglyceryl tosylate in 76% yield.[30]

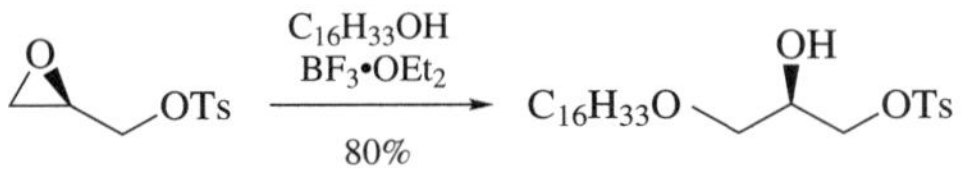

(15)

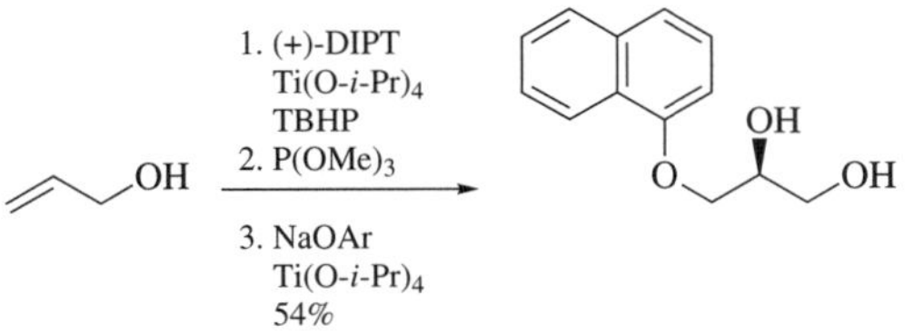

(16)

Examples of other oxygen nucleophiles that have been added at C-3 include phosphorylcholine (eq 17)[31] and ethyl *N*-hydroxyacetimidate (eq 18).[32]

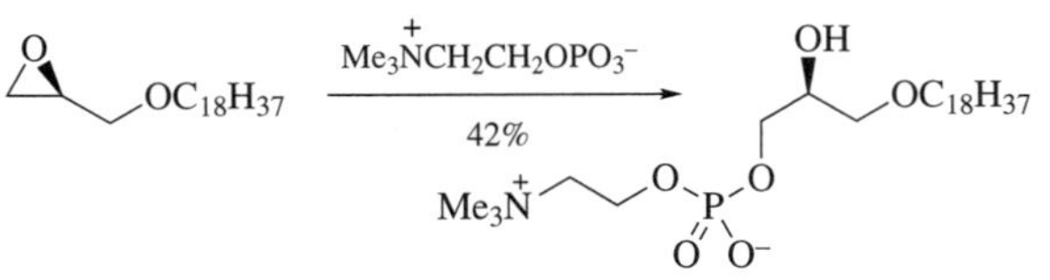

(17)

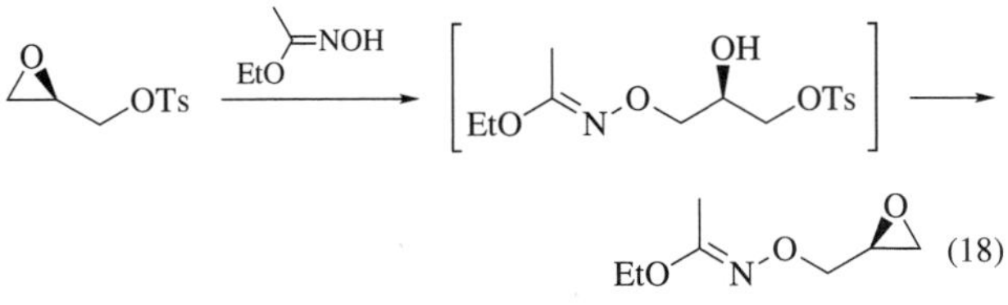

(18)

Nucleophilic Additions of Nitrogen at C-3. *Ammonia* and amines add readily to glycidol and glycidol derivatives, giving the 1-aminopropane-2,3-diols (eq 19).[33] With ammonia and primary amines, an excess of the amine often is used to reduce the amount of addition by a second glycidol to the 1-aminopropane-2,3-diol. Secondary amines are used with glycidols in an equimolar ratio. Azide ion also opens glycidols at C-3. Ti(O-*i*-Pr)$_4$ or *Aluminum Isopropoxide* assisted openings with *Azidotrimethylsilane* have been examined with glycidol and a variety of derivatives[34] and give excellent yields of 3-azido-2-hydroxypropane 1-*O*-derivatives (eq 20). *Sodium Azide* has also been used as a source of azide when combined with either *Pyridinium p-Toluenesulfonate*,[7] NH_4Cl,[9,35] or *Lithium Perchlorate*[36] to react with various glycidols.

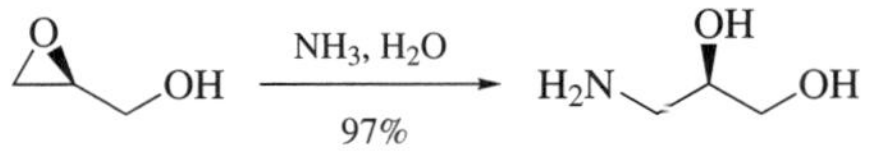

(19)

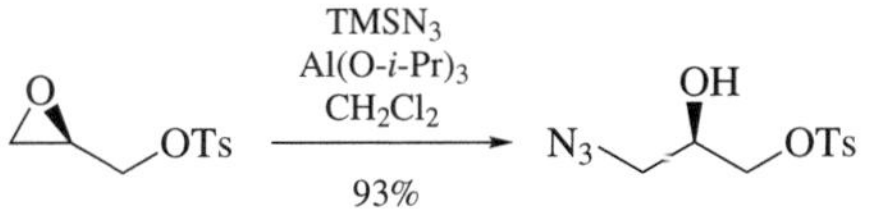

(20)

The opening of glycidols, especially of *O*-aryl glycidol ethers, at C-3 with amines has found extensive application in pharmaceutical research.[1b] A typical example is in the opening of *O*-(1-naphthyl)glycidol at C-3 with isopropylamine to generate the β-adrenergic blocking agent propranolol (eq 21).[29] A similar application is the addition of the 4-substituted piperazine to glycidol shown in eq 22.[37] With (*R*)- and (*S*)-glycidol now readily available, the synthesis of individual enantiomers or diastereoisomers of a pharmacological agent by methods such as those shown in eqs 21 and 22 becomes an attractive goal.

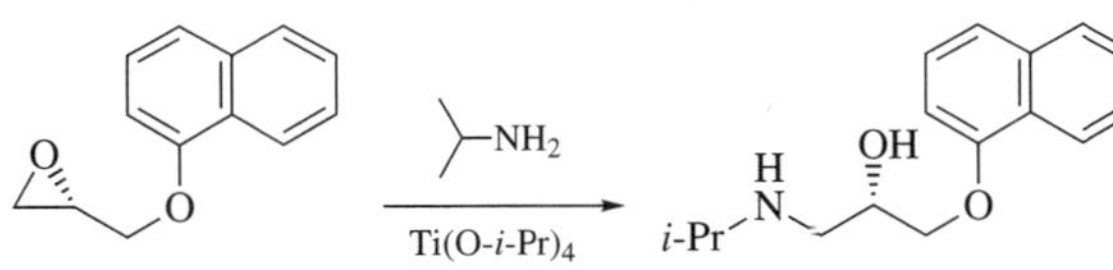

(21)

50% from
(*S*)-glycidol tosylate

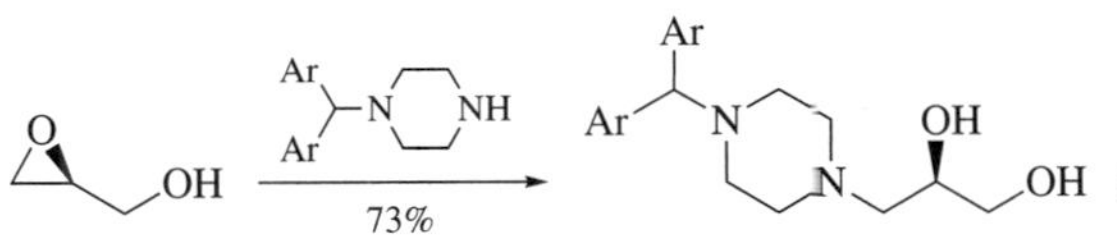

(22)

Other nitrogen nucleophiles added to glycidol include several heterocycles, an example of which is the addition of *Imidazole*.[38] The iminodioxolane shown in eq 23 adds to glycidol and then undergoes further cyclization to give a cyclic urethane.[39] *Acetonitrile* in a $BF_3{\cdot}OEt_2$-catalyzed reaction adds to glycidyl tosylate to form 2-methyl-4-(tosyloxy)methyloxazoline (eq 24).[40] Dibenzylamine adds via an amidocuprate at C-3 of *O*-phenylglycidol to give 3-dibenzylaminopropane-1,2-diol 1-*O*-phenyl ether in 94% yield.[41]

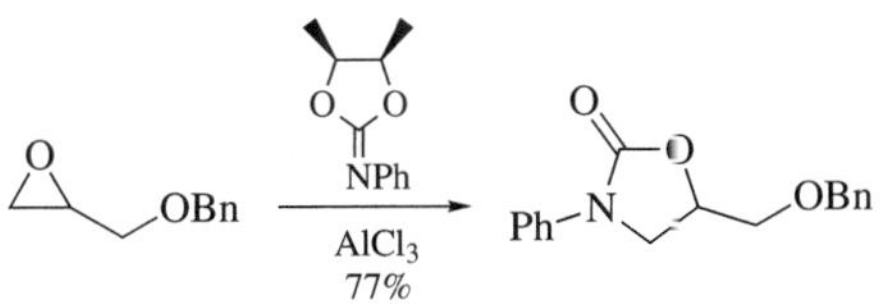

(23)

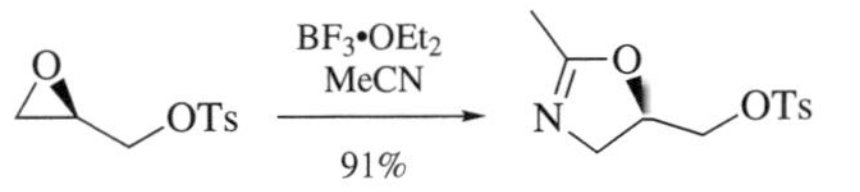

(24)

Additions of Other Nucleophiles at C-3. The halogens (F, Cl, Br, and I) and sulfur are the other elements most frequently found in C-3 additions to glycidols. Fluoride has been added to both glycidol and various glycidol (see eq 25) derivatives using tetrabutylammonium dihydrogentrifluoride.[42] Several methods have been used for the other three halogens, including reaction with the lithium salts in THF[43] or with the ammonium salts and $LiClO_4$ in AcCN.[44] Chlorine has also been added via HCl[45] or with *Benzoyl Chloride/Cobalt(II) Chloride*;[46] the latter reaction also adds the benzoyl group to give the 2-*O*-benzoate derivative (eq 26). Bromine has been added with dimethylboron bromide[47] and iodine has been added with *Sodium Iodide* in a NaOAc/HOAc/$EtCO_2H$ system.[7]

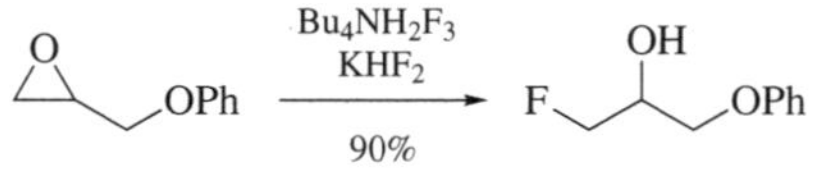

(25)

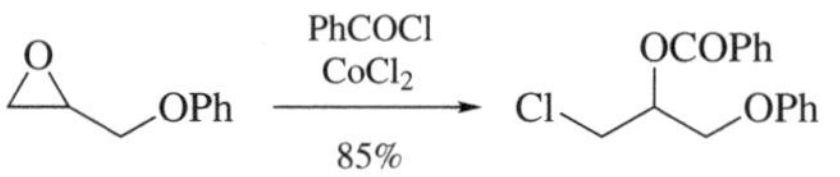

(26)

Most additions of sulfur to glycidols have been of arylthiolates and are performed under either acidic [$Ti(O\text{-}i\text{-}Pr)_4$ (eq 27),[8] $BF_3{\cdot}OEt_2$[48]] or alkaline conditions.[7,49] $LiClO_4$[50] or $CoCl_2$[51] have also been used as catalysts for addition of aryl thiols. The additions of lithium alkylthiolates and of thiobenzoic acid to *O*-trityl glycidol have been reported.[5a]

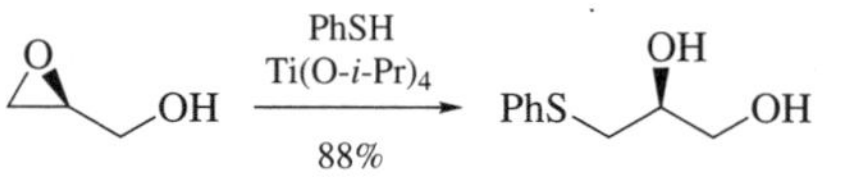

(27)

Oxidation of Glycidol. Glycidol is oxidized to glycidic acid with *Ruthenium(VIII) Oxide*.[52] Glycidaldehyde is a mutagenic compound that has been prepared in racemic form by epoxidation of *Acrolein*[53] and in nonracemic forms by the degradation of mannitol.[54] Alternately, (*R*)- and (*S*)-glycidaldehyde may be prepared and handled more conveniently via asymmetric dihydroxylation of acrolein benzene-1,2-dimethanol acetal followed by conversion of the diol to an epoxide (see eq 28).[55]

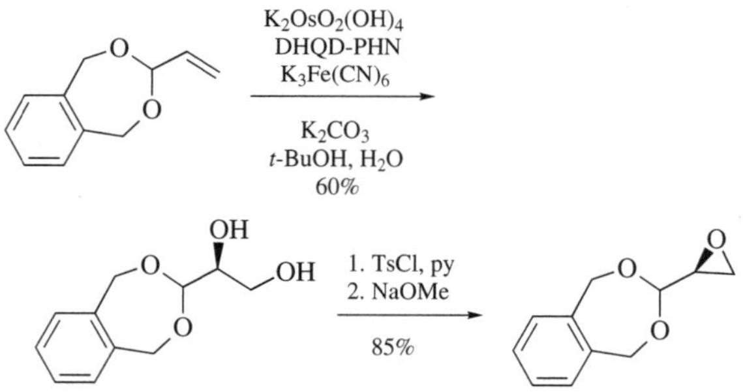

(28)

Miscellaneous. Glycidol reacts with dinitrogen pentoxide (N_2O_5) in CH_2Cl_2 in the presence of $AlCl_3$, giving trinitroglycerine (73%).[56] A useful review describing numerous synthetic transformations of 2,3-*O*-isopropylideneglyceraldehyde, a three-carbon synthon related to glycidol, has been published.[57]

1. (a) Kleemann, A.; Wagner, R. M. *Glycidol*; Hüthig: New York, 1981. (b) Hanson, R. M. *Chem. Rev.* **1991**, *91*, 437.
2. Gao, Y.; Hanson, R. M.; Klunder, J. M.; Ko. S. Y.; Masamune, H.; Sharpless, K. B. *J. Am. Chem. Soc.* **1987**, *109*, 5765.
3. (a) Ladner, W. E.; Whitesides, G. M. *J. Am. Chem. Soc.* **1984**, *106*, 7250. (b) Fu, H.; Newcomb, M.; Wong, C.-H. *J. Am. Chem. Soc.* **1991**, *113*, 5878.
4. Lipshutz, B. H.; Moretti, R.; Crow, R. *Org. Synth.* **1990**, *69*, 80.
5. (a) Hendrickson, H. S.; Hendrickson, E. K. *Chem. Phys. Lipids* **1990**, *53*, 115. (b) Kim, M.-J.; Choi, Y. K. *J. Org. Chem.* **1992**, *57*, 1605.
6. Klunder, J. M.; Onami, T.; Sharpless, K. B. *J. Org. Chem.* **1989**, *54*, 1295.
7. Ko, S. Y.; Masamune, H.; Sharpless, K. B. *J. Org. Chem.* **1987**, *52*, 667.
8. Ko, S. Y.; Sharpless, K. B. *J. Org. Chem.* **1986**, *51*, 5413.
9. (a) Swindell, C. S.; Krauss, N. E.; Horwitz, S. B.; Ringel, I. *J. Med. Chem.* **1991**, *34*, 1176. (b) McClure, D. E.; Engelhardt, E. L.; Mensler, K.; King, S.; Saari, W. S.; Huff, J. R.; Baldwin, J. J. *J. Org. Chem.* **1979**, *44*, 1826.
10. Kötz, A.; Richter, K. *JPR[2]* **1925**, *111*, 373.
11. Mitani, M.; Matsumoto, H.; Gouda, N.; Koyama, K. *J. Am. Chem. Soc.* **1990**, *112*, 1286.
12. Michael, A.; Weiner, N. *J. Am. Chem. Soc.* **1936**, *58*, 999.
13. Ciaccio, J. A.; Stanescu, C.; Bontemps, J. *Tetrahedron Lett.* **1992**, *33*, 1431.
14. Abushanab, E.; Sarma, M. S. P. *J. Med. Chem.* **1989**, *32*, 76.
15. Lipshutz, B. H.; Garcia, E. *Tetrahedron Lett.* **1990**, *31*, 7261.
16. Lipshutz, B. H.; Kozlowski, J. A. *J. Org. Chem.* **1984**, *49*, 1147.
17. Burgos, C. E.; Nidy, E. G.; Johnson, R. A. *Tetrahedron Lett.* **1989**, *30*, 5081.
18. Kraus, G. A.; Frazier, K. *J. Org. Chem.* **1980**, *45*, 4820.
19. Williams, K.; Thompson, C. M. *Synth. Commun.* **1992**, *22*, 239.
20. Fitton, A. O.; Hill, J.; Jane, D. E.; Millar, R. *Synthesis* **1987**, 1140.
21. Narjes, F.; Schaumann, E. *Synthesis* **1991**, 1168.
22. Lai, M.-t.; Oh, E.; Shih, Y.; Liu, H.-w. *J. Org. Chem.* **1992**, *57*, 2471.
23. Lattuada, L.; Licandro, E.; Maiorana, S.; Molinari, H.; Papagni, A. *Organometallics* **1991**, *10*, 807.
24. Takano, S.; Tomita, S.; Iwabuchi, Y.; Ogasawara, K. *Synthesis* **1988**, 610.
25. Takano, S.; Shimazaki, Y.; Takahashi, M.; Ogasawara, K. *Chem. Commun./J. Chem. Soc., Chem. Commun.* **1988**, 1004.
26. Caron, M.; Sharpless, K. B. *J. Org. Chem.* **1985**, *50*, 1557.
27. Johnson, R. A.; Burgos, C. E.; Nidy, E. G. *Chem. Phys. Lipids* **1989**, *50*, 119.
28. (a) Guivisdalsky, P. N.; Bittman, R. *J. Org. Chem.* **1989**, *54*, 4637. (b) Guivisdalsky, P. N.; Bittman, R. *J. Org. Chem.* **1989**, *54*, 4643. (c) Kazi, A. B.; Hajdu, J. *Tetrahedron Lett.* **1992**, *33*, 2291. (d) Liu, Y.-j.; Chu, T.-y.; Engel, R. *Synth. Commun.* **1992**, *22*, 2367.
29. Klunder, J. M.; Ko, S. Y.; Sharpless, K. B. *J. Org. Chem.* **1986**, *51*, 3710.
30. Ali, S.; Bittman, R. *J. Org. Chem.* **1988**, *53*, 5547.
31. Cimetiere, B.; Jacob, L.; Julia, M. *Tetrahedron Lett.* **1986**, *27*, 6329.
32. Stanek, J.; Frei, J.; Mett, H.; Schneider, P.; Regenass, U. *J. Med. Chem.* **1992**, *35*, 1339.
33. (a) Deveer, A. M. Th. J.; Dijkman, R.; Leuveling-Tjeenk, M.; van den Berg, L.; Ransac, S.; Batenburg, M.; Egmond, M.; Verheij, H. M.; de Haas, G. H. *Biochemistry* **1991**, *30*, 10034. (b) Sowden, J. C.; Fischer, O. L. *J. Am. Chem. Soc.* **1942**, *64*, 1291.
34. (a) Sutowardoyo, K. I.; Emziane, M.; Lhoste, P.; Sinou, D. *Tetrahedron* **1991**, *47*, 1435. (b) Sutowardoyo, K. I.; Sinou, D. *Tetrahedron: Asymmetry* **1991**, *2*, 437.
35. (a) Trinh, M.-C.; Florent, J.-C.; Grierson, D. S.; Monneret, C. *Tetrahedron Lett.* **1991**, *32*, 1447. (b) Konosu, T.; Oida, S. *Chem. Pharm. Bull.* **1992**, *40*, 609.
36. Chini, M.; Crotti, P.; Macchia, F. *Tetrahedron Lett.* **1990**, *31*, 5641.
37. Press, J. B.; Falotico, R.; Hajos, Z. G.; Sawyers, R. A.; Kanojia, R. M.; Williams, L.; Haertlein, B.; Kauffman, J. A.; Lakas–Weiss, C.; Salata, J. J. *J. Med. Chem.* **1992**, *35*, 4509.
38. Banfi, A.; Benedini, F.; Sala, A. *J. Heterocycl. Chem.* **1991**, *28*, 401.

39. Baba, A.; Seki, K.; Matsuda, H. *J. Org. Chem.* **1991**, *56*, 2684.
40. Delgado, A.; Leclerc, G.; Cinta Lobato, M.; Mauleon, D. *Tetrahedron Lett.* **1988**, *29*, 3671.
41. Yamamoto, Y.; Asao, N.; Meguro, M.; Tsukada, N.; Nemoto, H.; Sadayori, N.; Wilson, J. G.; Nakamura, H. *Chem. Commun./J. Chem. Soc., Chem. Commun.* **1993**, 1201.
42. Landini, D.; Albanese, D.; Penso, M. *Tetrahedron* **1992**, *48*, 4163.
43. Bajwa, J. S.; Anderson R. C. *Tetrahedron Lett.* **1991**, *32*, 3021.
44. Chini, M.; Crotti, P.; Gardelli, C.; Macchia, F. *Tetrahedron* **1992**, *48*, 3805.
45. Baldwin, J. J.; Raab, A. W.; Mensler, K.; Arison, B. H.; McClure, D. E. *J. Org. Chem.* **1978**, *43*, 4876.
46. Iqbal, J.; Srivastava, R. R. *Tetrahedron* **1991**, *47*, 3155.
47. Guindon, Y.; Therien, M.; Girard, Y.; Yoakim, C. *J. Org. Chem.* **1987**, *52*, 1680.
48. Guivisdalsky, P. N.; Bittman, R. *J. Am. Chem. Soc.* **1989**, *111*, 3077.
49. Takano, S.; Akiyama, M.; Ogasawara, K. *J. Chem. Soc., Perkin Trans. 1* **1985**, 2447.
50. Chini, M.; Crotti, P.; Giovani, E.; Macchia, F.; Pineschi, M. *Synlett* **1992**, 303.
51. Iqbal, J.; Pandey A.; Shukla, A.; Srivastava, R. R.; Tripathi, S. *Tetrahedron* **1990**, *46*, 6423.
52. Pons, D.; Savignac, M.; Genet, J. P. *Tetrahedron Lett.* **1990**, *31*, 5023.
53. Payne, G. B. *J. Am. Chem. Soc.* **1959**, *81*, 4901.
54. Schray, K. J.; O'Connell, E. L.; Rose, I. A. *J. Biol. Chem.* **1973**, *248*, 2214.
55. Oi, R.; Sharpless, K. B. *Tetrahedron Lett.* **1992**, *33*, 2095.
56. Golding, P.; Millar, R. W.; Paul, N. C.; Richards, D. H. *Tetrahedron* **1993**, *49*, 7037.
57. Jurczak, J.; Pikul, S.; Bauer, T. *Tetrahedron* **1986**, *42*, 447.

Roy A. Johnson
The Upjohn Company, Kalamazoo, MI, USA
Carmen E. Burgos-Lepley
Cortech, Denver, CO, USA

Glycidyl Tosylate[1]

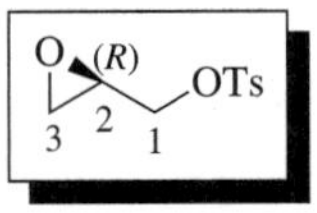

(*R*)-(−)		
[113826-16-5]	$C_{10}H_{12}O_4S$	(MW 228.29)
(*S*)-(+)		
[70987-78-9]		

(chiral C_3 synthon;[2] undergoes regio- and stereoselective ring opening at C-3 with alcohols in the presence of $BF_3{\cdot}OEt_2$,[3–8] Li_2CuCl_4-catalyzed Grignard reagents,[9,10] and carbanions (with[11] and without $BF_3{\cdot}OEt_2$);[12–15] undergoes direct attack at the C-1 position by nucleophiles, including aryl oxides[9,16–26])

Physical Data: mp 47.5–48.5 °C; racemate mp 37.5–39 °C.[9] $[\alpha]_D^{25}$ (*S*) enantiomer (≥97% enantiomeric excess) +18.1° (*c* 2.1, $CHCl_3$),[9] (*R*) enantiomer (~94% enantiomeric excess) −17.0° (*c* 2.75, $CHCl_3$).[4]

Solubility: v sol chloroform, methylene chloride, and THF; insol hexane.

Form Supplied in: white solid; commercially available.

Analysis of Reagent Purity: enantiomeric purity can be assessed by the following procedures. *Procedure A* (no derivatization): Chiral HPLC on a Diacel OD column (10 μm, 0.46 × 25 cm); flow, 0.7 mL min^{-1}; 99:1 hexane–2-propanol; t_R: (*S*) enantiomer, 51.2 min; (*R*) enantiomer, 53.9 min.[27] *Procedure B*: the reagent is opened to the iodohydrin; the crude iodohydrin is esterified with (*R*)-(+)-α-methoxy-α-(trifluoromethyl)phenylacetyl (MTPA) chloride; the resulting crude MTPA ester is analyzed by chiral HPLC and ^1H NMR.[28]

Preparative Methods: the title reagent (**1**) is obtained in 40% overall yield by the zeolite-modified Sharpless asymmetric epoxidation of allyl alcohol, using D-(−)-diisopropyl tartrate (DIPT) to obtain (*S*)-(**1**) and L-(+)-DIPT to obtain (*R*)-(**1**), followed by in situ low-temperature tosylation of glycidol.[28] Alternatively, (*R*)- and (*S*)-(**1**) can be prepared by direct sulfonylation of commercially available chiral *Glycidol*. Note that the relative configuration of (*R*)-glycidyl tosylate is the same as that of (*S*)-glycidol.

Purification: multiple recrystallizations from 5:1 ether–petroleum ether, seeding (before refrigeration) each time with pure material.[28]

Handling, Storage, and Precautions: mutagenicity via reactions with biological nucleophiles has been assessed by using the Ames test in *Salmonella typhimurium* and by analysis of in vivo chromosomal aberrations.[29] (*S*)-(**1**) was more mutagenic than (*R*)-(**1**).[29] Use in a fume hood.

Displacement Reactions. Aryl oxides attack (*R*)- or (*S*)-(**1**) at the C-1 position, displacing the tosylate group and affording aryloxymethyloxiranes (**2**) in good yield (see examples in Table 1). The latter react with amines in aqueous alcohol, yielding β-adrenergic blocking agents of high enantiomeric excess.[9] The regioselectivity for attack at the C-1 vs. C-3 position of (**1**) is very high, i.e. ≥97:3 in DMF at rt[9] and 85:15 in refluxing acetone.[18] The ratio of C-1:C-3 attack on the related substrates, (*S*)-glycidyl 3- or 4-nitrobenzenesulfonate, by 1-naphthol in DMF at rt is >99.8:0.02.[9] A comparison of the C-1:C-3 product ratio obtained from several glycidyl arenesulfonates suggested that attack by aryl oxide nucleophiles at C-1 is enhanced when the electron deficiency of the leaving group is increased (see Table 1).[9]

Reaction of *Catechol* with (*R*)- or (*S*)-(**1**) (*Potassium Carbonate*, DMF, 60 °C, 27 h;[25] or *Sodium Hydride*, DMF, rt, 10 h[26]) affords (*S*)- or (*R*)-2-hydroxymethyl-1,4-benzodioxane in yields of 72%[25] and 18–21%[26] (eq 1).

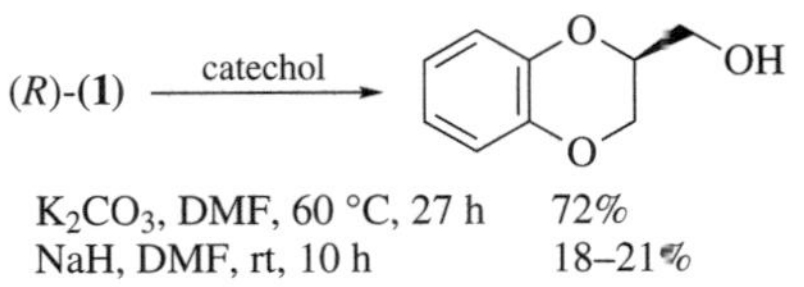

(1)

Ring-Opening Reactions. Epoxide opening takes place at the C-3 position of (**1**) with a wide variety of nucleophiles, as summarized below.

Table 1 Alkylation of Phenolic Hydroxy Groups by (**1**)

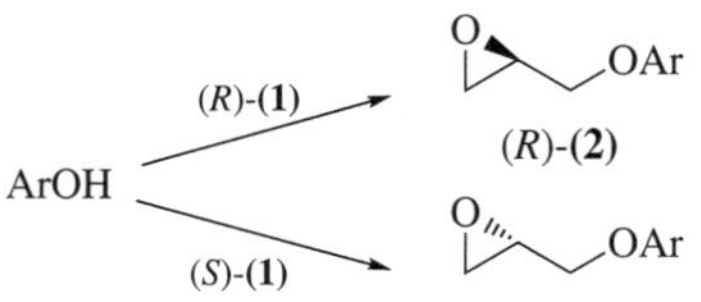

(*S*)-(**2**)

Reactant	Reagent	Conditions	Yield of glycidyl (**2**)	Ref
Phenol	(*R*)-(**1**) & (*S*)-(**1**)	NaH, DMF, 4.5–7 h	77–80	9, 16
$HOC_6H_4(CH_2)_2CO_2R$	(*S*)-(**1**)	K_2CO_3, DMF, 170 °C, 15 h	80	17
Methyl 4-hydroxy-3-methoxyphenylpropiolate	(*S*)-(**1**)	K_2CO_3, DMF, 70 °C, 1 h	82	17
3-Nitro-2-aminophenol	(*S*)-(**1**)	NaH, DMF, rt	70	18
2-Cyclopentylphenol	(*S*)-(**1**)	NaH, DMF, rt	86	9
1-Naphthol	(*R*)-(**1**) & (*S*)-(**1**)	NaH, DMF, rt	50–85	9, 19
8-Hydroxyisoquinoline	(*R*)-(**1**) & (*S*)-(**1**)	NaH, DMF, rt, 12 h	80	20
3-Acetyl-4-methoxy-1-naphthol	(*R*)-(**1**)	NaH, DMF, 50 °C, 4 h	75	21
4,8-Dibenzyloxy-6-methyl-1-naphthol	(*R*)-(**1**)	NaH, DMF, rt, 4 h	87	22
O^3-Benzyl-6β-naltrexol	(*R*)-(**1**)	KH, THF, 19 h	83	23
7-*O*-(3-*N*-Methyl-α-L-daunosaminyl)-β-rhodomycinone	(*R*)-(**1**)	heat	11	24

Alcohols and Benzenethiol. These react with (**1**) in methylene chloride or chloroform, generally at rt, in the presence of catalytic *Boron Trifluoride Etherate* (eq 2). The following alcohols have been used (yields in parentheses): benzyl alcohol (81–84%);[3b,4,8] 1-hexadecanol (80%);[3] $Cl(CH_2)_6OH$ (80%);[5] $Cl(CH_2)_{16}OH$ (85%);[5] methanol (100%);[7] $MeO_2C(CH_2)_5OH$ (89%).[6] Benzenethiol gives 3-phenylthio-2-hydroxy-1-tosyloxypropane in 81–83% yield.[3b] The 4-methylbenzene- and 3- nitrobenzenesulfonate derivatives of glycidol give exclusive formation of the C-3-opened product in the BF_3-mediated reaction,[3,4] whereas the *t*-butyldiphenylsilyl ether derivative of glycidol gives a regioselectivity (C-3 vs. C-2 isomer ratio) of 89:11 with benzyl alcohol, 93:7 with benzenethiol, and 90:10 with long-chain alcohols.[3b]

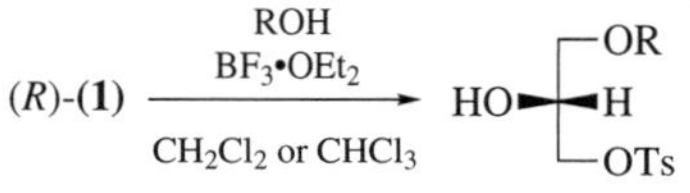

(2)

Halide and Azide Anions. These also open the epoxide regioselectively at the C-3 position. Addition of Li_2CuBr_4 results in bromide addition at the C-3 position of *rac*-(**1**), forming 3-bromo-1,2-propane 1-*O*-tosylate in 70–76% yield in THF or acetonitrile at rt, or 1,3-dibromo-2-propanol in 82% yield in refluxing acetonitrile.[30] Hydrofluorination takes place with KHF_2 under solid-liquid phase transfer conditions, but the yield of fluorohydrin is very low (eq 3).[31] *Azidotrimethylsilane* adds in the presence of a Lewis acid catalyst (eq 3).[32] Addition of cyanide ion is achieved by using *Diethylaluminum Cyanide* in toluene.[9]

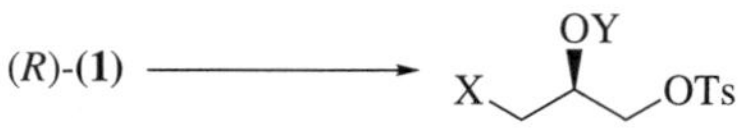

(3)

Hydrofluorination: KHF_2 (2 equiv), Bu_4N^+ $H_2F_3^-$ (1 equiv), 120 °C; X = F, Y = H, 6%

Azide opening: $TMSN_3$ (1.5 equiv), cat. Al(O-*i*-Pr)$_3$, CH_2Cl_2, 1 d, 93% or cat. Ti(O-*i*-Pr)$_4$, THF, 6 d, 76%

Addition of concentrated HCl to solid (*S*)-(**1**) at rt, followed by treatment of the ring-opened intermediate with sodium ethyleneglycolate, forms (*R*)-epichlorohydrin in 54% yield.[33a] The analogous reaction was carried out using (*S*)-glycidyl mesylate in 85%[33b] and 68% yield,[33c] and with *rac*-(**1**).[33d]

Carbanions. These add to the C-3 position of (*S*)-(**1**), affording epoxides (**3**) after intramolecular displacement of the tosylate group and in situ ring closure of the ring-opened intermediate (eq 4). Deprotonation of oxirane (**3**) leads to rearrangement to cyclopropane derivatives.[11a,12,13]

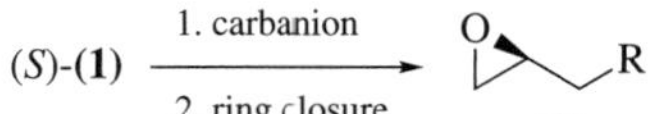

(**3**) (4)

Lithium carbanion	Overall yield of (**3**) (%)
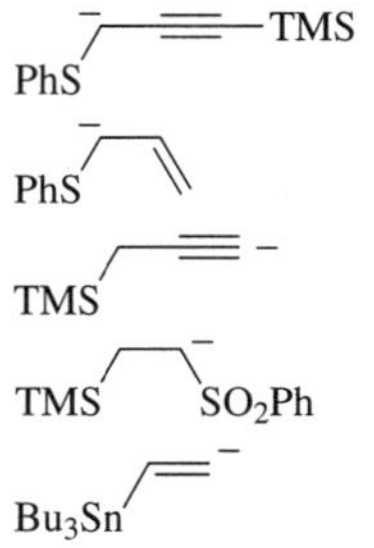	81
	58
	59
	56
	38

There are many examples of $BF_3{\cdot}OEt_2$ promoted openings of (**1**) by carbanions, including sulfone-stabilized anions,[11a] vinylic anions,[11b] allylic anions,[14] and phosphonate-stabilized anions.[15] For example, the lithium anion of *trans-1,2-Bis(tributylstannyl)ethylene* opens (*S*)-(**1**) in the presence of $BF_3{\cdot}OEt_2$ in THF at −78 °C, affording *trans*-1-(tributylstannyl)-5-tosyl-4-hydroxypent-1-ene in 50% yield; the latter is converted into oxirane (**3**) in 76% yield on treatment with powdered *Sodium Hydroxide* in monoglyme.[11b]

The carbanion derived from pentacarbonyl(methoxymethylcarbene)chromium(0) reacts with (*R*)-(**1**) in the presence of $BF_3{\cdot}OEt_2$ to give a lactone in low yield after oxidation of the ring-opened intermediate.[14]

Lithium diethylmethanephosphonate adds to *rac*-(**1**) in the presence of $BF_3{\cdot}OEt_2$ at −78 °C to afford a phosphonate ester in good yield (eq 5).[15]

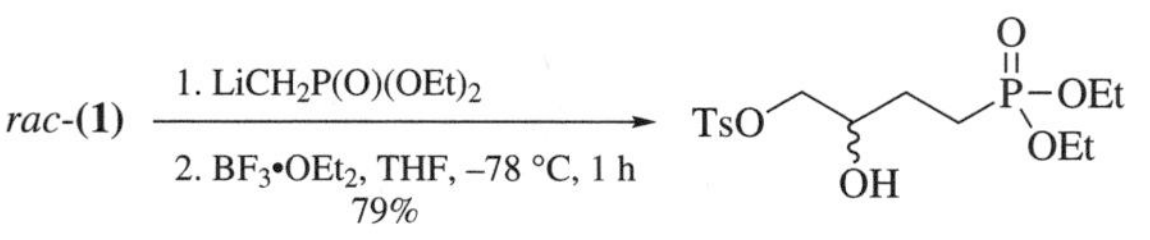
(5)

Organometallic Reagents. The major product of the *Dilithium Tetrachlorocuprate(II)*- or *Copper(I) Iodide*-catalyzed Grignard reaction in THF or ether at low temperature arises from epoxide opening rather than from direct tosylate displacement (Table 2).[9,10]

Table 2 Copper-Catalyzed Grignard Reactions of **(1)**[9]

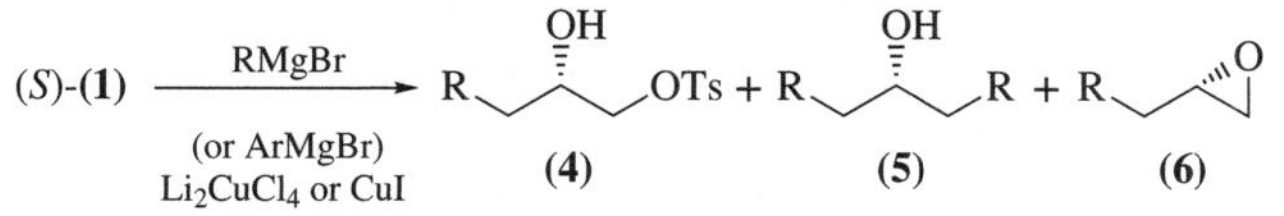

R	Conditions	Yields (%) (4)	(5)	(6)
Ph	Li_2CuCl_4, –30 °C, THF	84	9	5
Vinyl	Li_2CuCl_4, –35 °C, THF	49	–	–
Bn	Li_2CuCl_4, –15 °C, THF	56	–	–
Hexyl	Li_2CuCl_4, –50 °C, THF	74	–	–
5-Hexenyl	Li_2CuCl_4, 20 °C, THF	44[b]	–	–
Mesityl	CuI, –25 °C, THF–Et_2O	90	–	–
Cy	CuI, –40 °C, Et_2O	69[a]	–	–

[a] *rac*-**(1)** was used. [b] Ref. 10.

Organolithium reagents add to (*R*)- and (*S*)-**(1)** as shown in eq 6.[9]

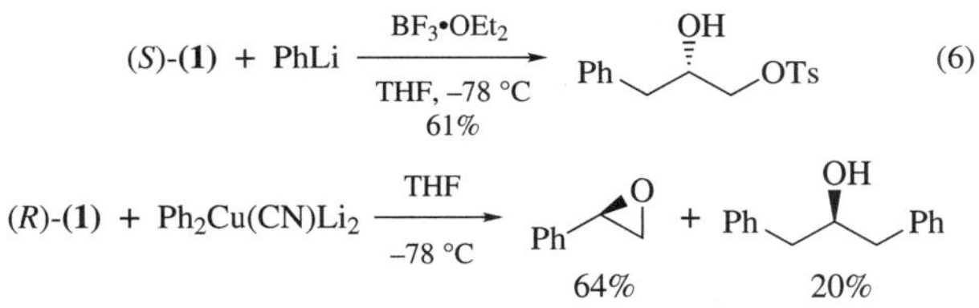
(6)

Hydride addition to (*S*)-**(1)** is achieved by use of BH_3 in THF with 0.05 equiv of *Sodium Borohydride*, forming (*S*)-1-*O*-tosyloxy-2-hydroxypropane in 81% yield.[9]

Miscellaneous Addition Reactions. Alkylation of MeC(OEt)=N–O$^-$ Na$^+$ with (*R*)-**(1)** gives (*S*)-*N*-(oxiranylmethoxy)ethanimidic acid ethyl ester in 34% yield.[34] Reaction of (*S*)-**(1)** with guanosine occurs at the N-7 position, giving (after deribosylation) (*S*)-7-(3-*O*-*p*-tolyl-2,3-dihydroxypropyl)guanosine in 56% yield.[35]

Fatty acid anhydrides react with (*R*)- and (*S*)-**(1)** in the presence of $BF_3{\cdot}OEt_2$, giving a glyceryl tosylate with two identical fatty acid ester linkages in 76% yield.[36]

Reaction of (*S*)-**(1)** with acetonitrile at low temperature in the presence of $BF_3{\cdot}OEt_2$ gives an oxazoline that is unstable at rt (eq 7).[9]

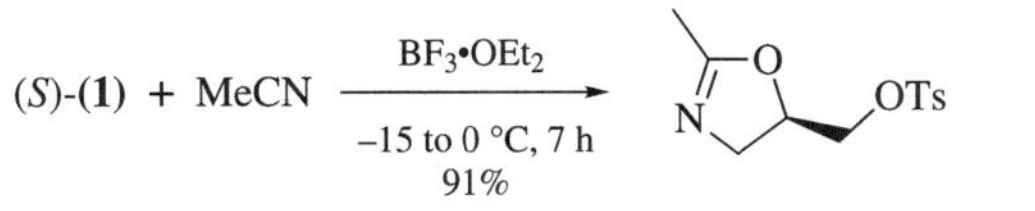
(7)

1. For a review of syntheses involving nonracemic glycidol and related 2,3-epoxy alcohols, see: Hanson, R. M. *Chem. Rev.* **1991**, *91*, 437.
2. For a review of recent syntheses of glycerolipids from **(1)** and other glycidyl derivatives and other precursors, see: Bittman, R. In *Phospholipids Handbook*; Cevc, G., Ed.; Dekker: New York, 1993; pp 141–232.
3. (a) Guivisdalsky, P. N.; Bittman, R. *Tetrahedron Lett.* **1988**, *30*, 4393. (b) Guivisdalsky, P. N.; Bittman, R. *J. Am. Chem. Soc.* **1989**, *111*, 3077. (c) Guivisdalsky, P. N.; Bittman, R. *J. Org. Chem.* **1989**, *54*, 4637. (d) Guivisdalsky, P. N.; Bittman, R. *J. Org. Chem.* **1989**, *54*, 4643.
4. Ali, S.; Bittman, R. *Biochem. Cell Biol.* **1990**, *68*, 360.
5. Berkowitz, W. F.; Pan, D.; Bittman, R. *Tetrahedron Lett.* **1993**, *34*, 4297.
6. Kazi, A. B.; Hajdu, J. *Tetrahedron Lett.* **1992**, *33*, 2291.
7. Deveer, A. M. Th. J.; Dijkman, R.; Leuveling-Tjeenk, M.; van den Berg, L.; Ransac, S.; Batenburg, M.; Egmond, M.; Verheij, H. M.; De Haas, G. H. *Biochemistry* **1991**, *30*, 10034.
8. Byun, H.-S.; Bittman, R. *Tetrahedron Lett.* **1989**, *30*, 2751.
9. Klunder, J. M.; Onami, T.; Sharpless, K. B. *J. Org. Chem.* **1989**, *54*, 1295.
10. Bertrand, P.; Gesson, J.-P. *Synlett* **1992**, 889.
11. (a) Baldwin, J. E.; Adlington, R. M.; Bebbington, D.; Russell, A. T. *Chem. Commun./J. Chem. Soc., Chem. Commun.* **1992**, 1249. (b) Biskupiak, J. E.; Grierson, J. R.; Rasey, J. S.; Martin, G. V.; Krohn, K. A. *J. Med. Chem.* **1991**, *34*, 2165.
12. Narjes, F.; Schaumann, E. *Synthesis* **1991**, 1168.
13. Narjes, F.; Bolte, O.; Icheln, D.; König, W. A.; Schaumann, E. *Organometallics* **1993**, *58*, 626.
14. Lattauda, L.; Licandro, E.; Maiorana, S.; Molinari, H.; Papagni, A. *Organomettalics* **1991**, *10*, 807.
15. Racha, S.; Li, Z.; El-Subbagh, H.; Abushanab, E. *Tetrahedron Lett.* **1992**, *33*, 5491.
16. Collington, E. W.; Finch, H.; Montana, J. G.; Taylor, R. J. K. *J. Chem. Soc., Perkin Trans. 1* **1990**, 1839.
17. Iguchi, S.; Iwamura, H.; Nishizaki, M.; Hayashi, A.; Senokuchi, K.; Kobayashi, K.; Sakaki, K.; Hachiya, K.; Ichioka, Y.; Kawamura, M.; *Chem. Pharm. Bull.* **1992**, *40*, 1462.
18. Hammadi, A.; Crouzel, C. *Tetrahedron: Asymmetry* **1990**, *1*, 579.
19. Krause, H. W.; Schmidt, U.; Foken, H. *Pharmazie* **1992**, *47*, 838.
20. Vo, D.; Wolowyk, M. W.; Knaus, E. E. *Drug Des. Discovery* **1992**, *9*, 69.
21. Gustavson, L. M.; Nelson, W. L. *Drug Metab. Dispos.* **1988**, *16*, 217.
22. Talaat, R. E.; Nelson, W. L. *Drug Metab. Dispos.* **1988**, *16*, 212.
23. Dasher, W. E.; Klein, P.; Nelson, W. L. *J. Med. Chem.* **1992**, *35*, 2374.
24. Gerken, M.; Grimm, M.; Raab, E.; Hoffmann, D.; Straub, R. Eur. Patent Appl. 485 894, 1992 (*Chem. Abstr.* **1993**, *117*, 131 503y).
25. Delgado, A.; Leclerc, G.; Lobato, C.; Mauleon, D. *Tetrahedron Lett.* **1988**, *29*, 3671.
26. Marciniak, G.; Delgado, A.; Leclerc, G.; Velly, J. Decker, N.; Schwartz, J. *J. Med. Chem.* **1989**, *32*, 1402.
27. Shaw, C. J.; Barton, D. L. *J. Pharm. Biomed. Anal.* **1991**, *9*, 793. For chiral HPLC on a Chiracel OB-H column, see Chen, J.; Shum, W. *Tetrahedron Lett.* **1993**, *34*, 7663.
28. Gao, Y.; Hanson, R. M.; Klunder, J. M.; Ko, S. Y.; Masamune, H.; Sharpless, K. B. *J. Am. Chem. Soc.* **1987**, *109*, 5765.
29. Sinsheimer, J. E.; Chen, R.; Das, S. K.; Hooberman, B. H.; Osorio, S.; You, Z. *Mutat. Res.* **1993**, *298*, 197.
30. Ciaccio, J. A.; Heller, E.; Talbot, A. *Synlett* **1991**, 248.
31. Landini, D.; Albanese, D.; Penso, M. *Tetrahedron* **1992**, *48*, 4163.
32. (a) Sutowardoyo, K. I.; Sinou, D. *Tetrahedron: Asymmetry* **1991**, *2*, 437. (b) Emziane, M.; Lhoste, P.; Sinou, D. *Synthesis* **1988**, 541. (c) Sinou, D.; Emziane, M. *Tetrahedron Lett.* **1986**, *27*, 4423.

33. (a) Takle, A.; Kocienski, P. *Tetrahedron* **1990**, *46*, 4503; an erroneous assignment of configuration of the starting material (**1**) was apparently made. (b) Baldwin, J. J.; Raab, A. W.; Menster, K.; Arison, B. H.; McClure, D. E. *J. Org. Chem.* **1978**, *43*, 4876. (c) Pirrung, M. C.; Dunlap, S. E.; Trinks, U. P. *Helv. Chim. Acta* **1989**, *72*, 1301. (d) Nakabayashi, N.; Masuhara, E.; Iwakura, Y. *Bull. Chem. Soc. Jpn.* **1966**, *39*, 413.
34. Stanek, J.; Caravatti, G.; Frei, J.; Capraro, H. G. *J. Med. Chem.* **1992**, *35*, 1339.
35. Sessler, J. L.; Magda, D. J.; Lynch, V.; Schiff, G. M.; Bernstein, D. I. *Nucleosides Nucleotides* **1989**, *8*, 431.
36. Ali, S.; Bittman, R. *J. Org. Chem.* **1988**, *53*, 5547.

Robert Bittman
Queens College of The City University of New York
Flushing, NY, USA

N-Glyoxyloyl-(2*R*)-bornane-10,2-sultam

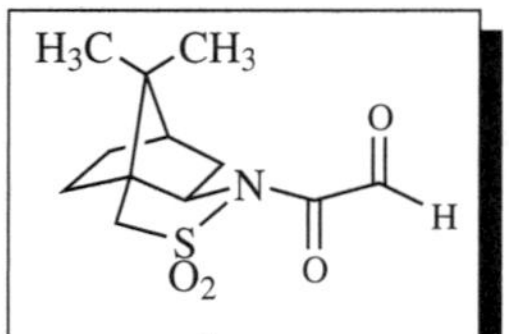

1

[123191-45-7] $C_{12}H_{17}NO_4S$ (MW 271.33)

(reagent used as a chiral glyoxylate derivative for various asymmetric organic reactions)

Physical Data: used in crude form from pyrolysis (110 °C, 0.1 mmHg) of its methanol hemiacetal; crystalline solid, mp 131–134 °C; $[\alpha]^{20}_D$ −103.6 (*c* 1.14, CH_2Cl_2).[1]
Solubility: soluble in most organic solvents.
Form Supplied in: available through synthesis.[1]
Handling, Storage, and Precautions: toxicity unknown; as with all organic chemicals, use only in a well-ventilated fume hood is recommended.

Introduction. Sultam **1** ($X_cC(O)CHO$) has been used as a chiral glyoxylate derivative for cycloadditions and similar carbon-carbon bond forming reactions. It is typically generated by pyrolysis of its methanol hemiacetal adduct just prior to use. In some instances, the methanol adduct is equally effective for in situ generation of **1** by the action of Lewis acid. Preference for *Si* face attack by reagents is commonly observed.

Cycloadditions. High levels of diastereoselection are observed in the hetero-Diels–Alder addition of **1** with electron-rich dienes. Lewis acid catalysis by europium(III) results in a 95:5 ratio of diastereomers (eq 1).[2,3] Under similar conditions, Danishefsky's diene also provides a single stereoisomer in good yield.[4]

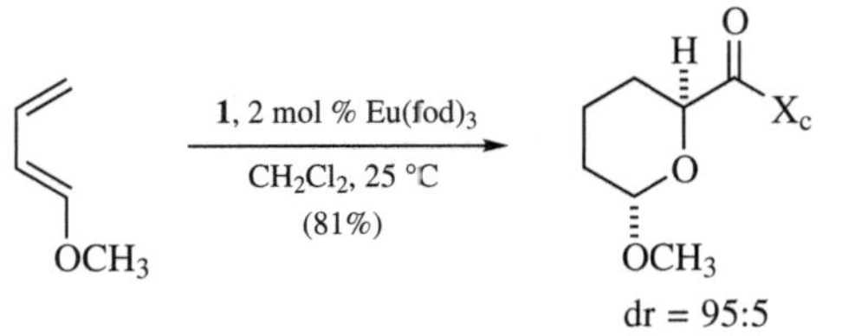

(1)

Carbon-Carbon Bond-Forming Reactions. Lewis acid catalysis by europium(III) also facilitates the highly diastereoselective Mukaiyama aldol reaction between trimethylsiloxyfuran and **1**. When methanol is used as cosolvent (1%), in situ desilylation of the aldol products results and a 97:3 ratio of diastereomers is produced (eq 2).[5] In the case of ene reactions with **1**, a stoichiometric amount of Lewis acid was required (eq 3). Only moderate levels of diastereoselection (up to 90:10) were observed using a range of Lewis acids.[6] Through use of a similar strategy, the allylation of **1** mediated by zinc(II) furnished the α-hydroxy ketone with good stereoinduction (eq 4).[7]

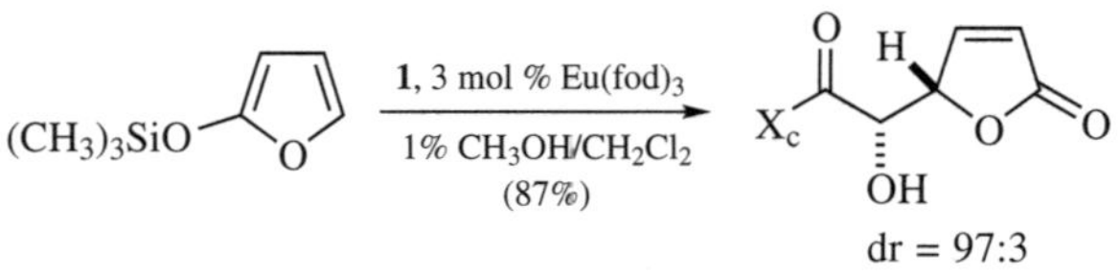

(2)

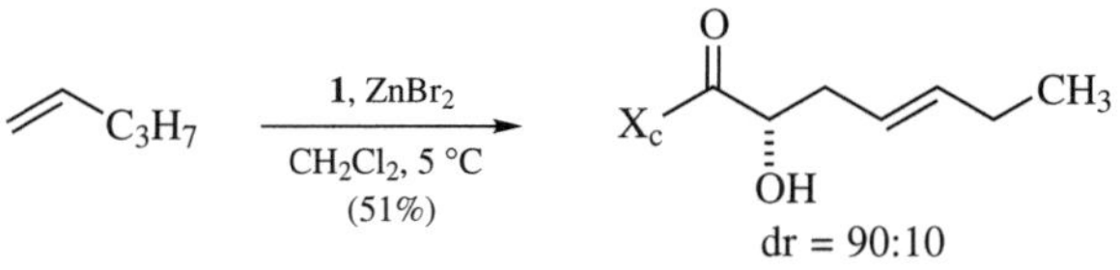

(3)

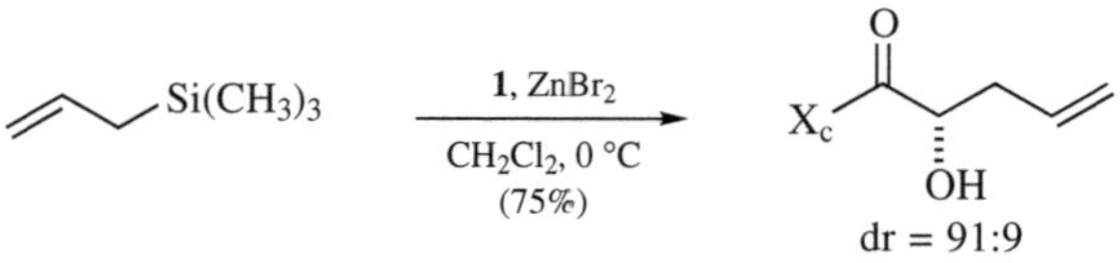

(4)

---

1. Bauer, T.; Jezewski, A.; Chapuis, C.; Jurczak, J. *Tetrahedron: Asymmetry* **1996**, *7*, 1385–1390.
2. Bauer, T.; Chapuis, C.; Kozak, J.; Jurczak, J. *Helv. Chim. Acta* **1989**, *72*, 482–486.
3. Bauer, T.; Chapuis, C.; Jezewski, A.; Kozak, J.; Jurczak, J. *Tetrahedron: Asymmetry* **1996**, *7*, 1391–1404.
4. Jurczak, J.; Jezewski, A. *Tetrahedron: Asymmetry* **1996**, *7*, 1413–1418.
5. Bauer, T. *Tetrahedron: Asymmetry* **1996**, *7*, 981–984.
6. Jezewski, A.; Chajewska, K.; Wielogorski, Z.; Jurczak, J. *Tetrahedron: Asymmetry* **1997**, *8*, 1741–1749.
7. Kiegiel, K.; Prokopowicz, P.; Jurczak, J. *Synth. Commun.* **1999**, *29*, 3999–4005.

Jeffrey N. Johnston
Indiana University, Bloomington, IN, USA

H

(2*S*)-(2α,3β,8aβ)-Hexahydro-3-(hydroxymethyl)-8a-methyl-2-phenyl-5*H*-oxazolo[3,2-*a*]-pyridin-5-one

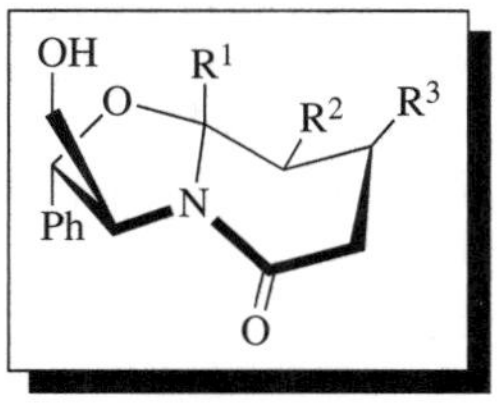

(2*S*)-(**1**; R^1 = Me, R^2 = H R^3 = H)		
[116950-01-7]	$C_{15}H_{19}NO_3$	(MW 261.35)
[101979-96-8]		
(2*S*)-(**2**; R^1 = Et, R^2 = H, R^3 = H)		
[101979-99-1]	$C_{16}H_{21}NO_3$	(MW 275.38)
(2*S*)-(**3**; R^1 = Me, R^2, R^3 = C_4H_4 (benzo))		
[127998-41-8]	$C_{19}H_{19}NO_3$	(MW 309.39)

(chiral templates for the synthesis of enantiopure 4,4-dialkyl-2-cyclohexenones, 6,6-dialkyl-2-cyclohexenones, and 4,4-dialkyl-1(4*H*)-naphthalenones[1])

Physical Data: (**1**) mp 98–99 °C; $[\alpha_D]$ +13.5°. (**2**) mp 83–85 °C; $[\alpha_D]$ +12.6°. (**3**) mp 91 °C; $[\alpha_D]$ +166.7°.

Preparative Methods: the chiral bicyclic lactams are easily procured via condensation of commercially available (1*S*,2*S*)-(+)-2-amino-1-phenyl-1,3-propanediol and the appropriate δ-keto acid (eq 1).[1,2] Similar bicyclic lactams have been prepared from other amino alcohols and have been extensively utilized in the stereocontrolled formation of quaternary carbon–carbon bonds.[1,3]

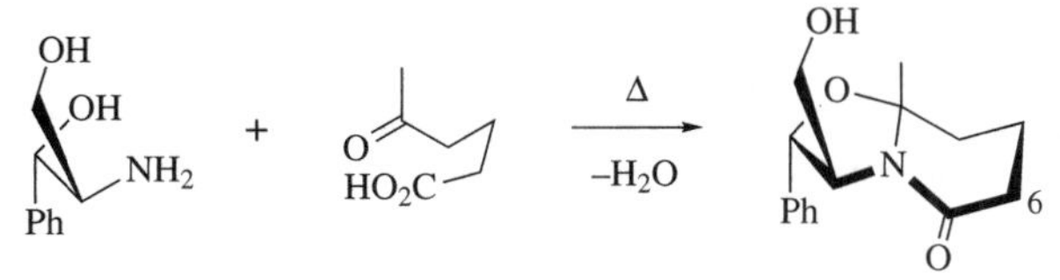

(1)

Handling, Storage, and Precautions: no special precautions are warranted.

These compounds can be alkylated twice at the 6-position in a stereocontrolled fashion (eq 2).[4,5] Treatment of the unsubstituted bicyclic lactam with *Lithium Diisopropylamide* and reaction of the enolate anion with an alkyl halide affords the monosubstituted product. The epimeric mixture is treated again with LDA and a second alkyl halide to give the dialkylated bicyclic lactam. The initial epimeric mixture is used directly in the second alkylation since this subsequent step proceeds via a planar enolate. It is the second alkylation that dictates the final diastereomeric ratio. The order of addition also affects the final ratio.[4] Inverting the order of addition provides the antipode at position 6, although the highest diastereoselectivity is observed when the larger electrophile is added last.[6] The monoalkylated bicyclic lactam can also be prepared by condensation of the amino alcohol and the appropriately substituted keto acid.[1]

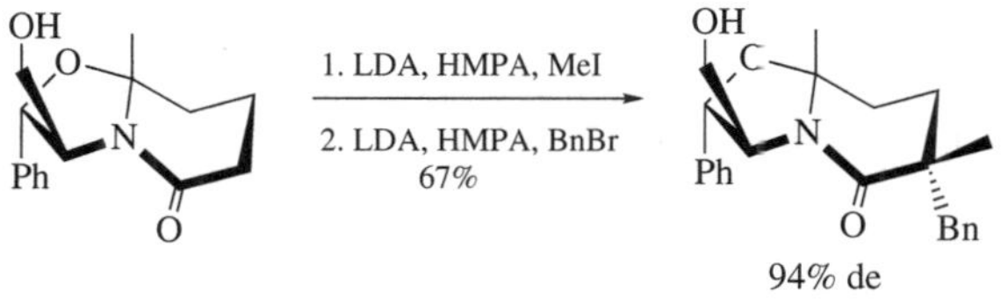

(2)

These compounds can be partially reduced to the carbinol amine, and then hydrolyzed and cyclized to afford the enantiomerically pure 4,4-dialkyl-2-cyclohexenones (eq 3).[4]

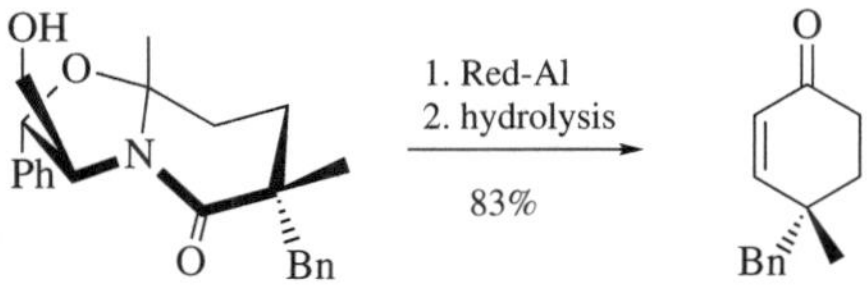

(3)

In addition, Grignard addition to the lactam followed by hydrolysis of the bicyclic lactam affords either 3,4,4- or 2,6,6-trisubstituted 2-cyclohexenones, depending upon the hydrolytic conditions employed (eq 4).[6]

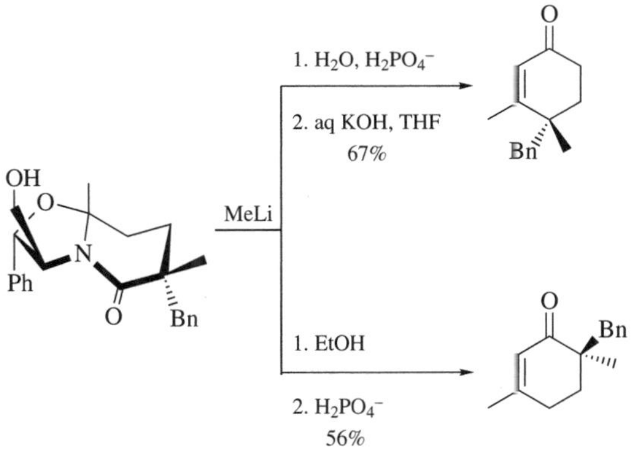

(4)

Similar manipulations with a dialkylated benzo tricyclic lactam lead to the corresponding 4,4-dialkylnaphthalenones (eq 5).[7]

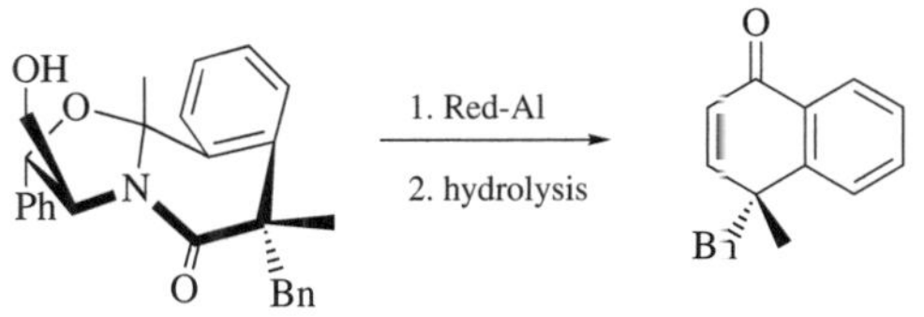

(5)

Related Reagents. (*S*)-1-Amino-2-methoxymethylpyrrolidine; *trans*-2,5-Bis(methoxymethyl)pyrrolidine; 10,2-Camphorsultam 10-Dicyclohexylsulfonamidoisoborneol; α-Methyltoluene-2,α-sultam; (3*S*,*cis*)-Tetrahydro-3-isopropyl-7a-methylpyrrolol[2,1-*b*]oxazol-5(6*H*)- one.

---

1. Romo, D.; Meyers, A. I. *Tetrahedron* **1991**, *47*, 9503.
2. Meyers, A. I.; Berney, D. *Org. Synth., Coll. Vol.* **1993**, *8*, 241.
3. (a) Other amino alcohols have been used to prepare bicyclic lactams such as the title reagent. These bicyclic lactams have served as precursors to a variety of enantiomerically pure compounds that possess a quaternary stereocenter, such as 2,2-dialkyl keto acids, cyclopropanes, cyclobutanes, cyclopentenones, cyclohexenones, indanones, naphthalenones, and 3,3-disubstituted dihydronaphthalenes. For an extensive review on the utility of chiral, nonracemic bicyclic lactams, see Ref. 1. (b) A valinol-derived bicyclic lactam has been used to prepare 4,4-disubstituted 2-cyclohexenones: Meyers, A. I.; Hanreich, R.; Wanner, K. T. *J. Am. Chem. Soc.* **1985**, *107*, 7776.
4. Meyers, A. I.; Lefker, B. A.; Wanner, K. T.; Aitken, R. A. *J. Org. Chem.* **1986**, *51*, 1936.
5. Meyers, A. I.; Berney, D. *J. Org. Chem.* **1989**, *54*, 4673.
6. Meyers, A. I.; Lefker, B. A. *Tetrahedron* **1987**, *43*, 5663.
7. Wünsch, T.; Meyers, A. I. *J. Org. Chem.* **1990**, *55*, 4233.

Todd D. Nelson & Albert I. Meyers
Colorado State University, Fort Collins, CO, USA

(4a*R*)-(4aα,7α, 8aβ)-Hexahydro-4,4,7-trimethyl-4*H*-1,3-benzoxathiin

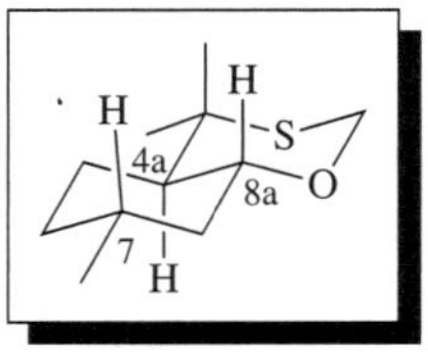

(4a*R*)-(4aα,7α,8aβ)
[79618-03-4] $C_{11}H_{20}OS$ (MW 200.38)
[59324-06-0]

(useful chiral auxiliary for asymmetric synthesis of tertiary[1] and secondary[2] alcohols[3])

Physical Data: mp 37–38 °C; bp 70–94 °C/0.1 mmHg (for diastereomeric mixture).
Solubility: very sol most organic solvents at 20 °C; slightly sol pentane at 0 °C; insol H_2O.
Preparative Methods: prepared in three steps from optically pure (+)-pulegone (eq 1).[4]

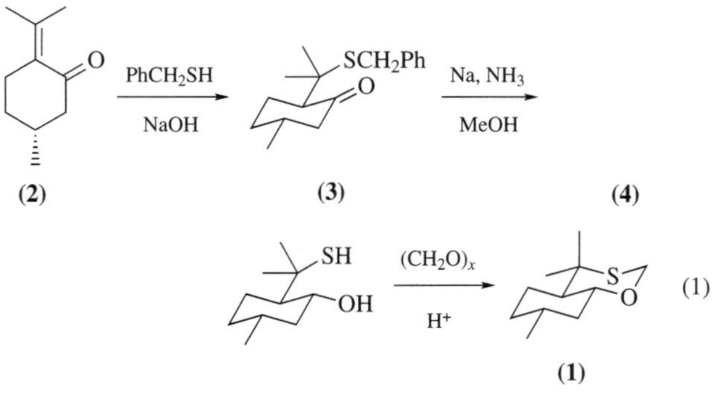

Handling, Storage, and Precautions: refrigerated storage is recommended. Use in a fume hood.

Asymmetric synthesis via the title reagent (**1**) is the result of two highly stereoselective reactions. The first involves the reaction of 2-lithio-1,3-oxathiane (**5**), with an aldehyde to give exclusively the equatorial addition product (**6**) (this reaction typically shows little aldehyde facial selectivity) (eq 2). The selectivity is due to the greatly enhanced stability of the equatorial lithium species (**5**) as compared to the axial isomer, a result of the stereoelectronics of the conformationally rigid oxathiane.[5] Swern oxidation of the resulting carbinols allows the preparation of 2-acyloxathianes as single stereoisomers in high yield.[6] Direct acylation of (**5**) has been achieved recently in selected cases by reaction with nitriles[7] or α-heteroatom-substituted esters.[8] Acyloxathianes (**7**) have also been prepared by acylation of the cuprate (**8**) (obtained from (**5**) and 0.5 equiv *Copper(I) Iodide*) with acid chlorides (eq 3).[9]

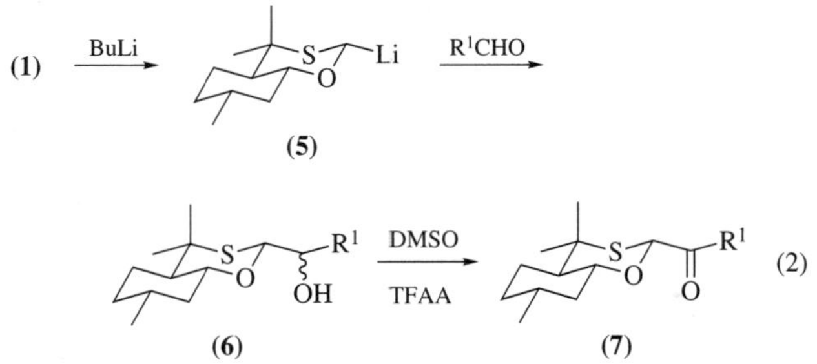

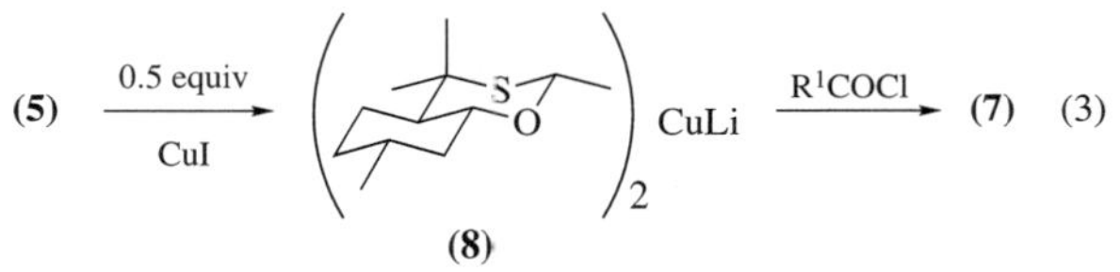

The ketones (**7**) undergo highly stereoselective additions with Grignard reagents to give tertiary alcohols or can be reduced stereoselectively to secondary alcohols. In the case of Grignard reactions, the addition generally shows very high selectivity, typically >9:1, often >95:5 (**9**):(**10**) (eq 4).[10] The major isomer is that predicted to be formed by Cram's chelate rule.[11,12] The kinetics of Grignard reactions with α-alkoxy ketones have been measured by Eliel et al.[13] and these studies strongly support the intermediacy of a chelate structure.

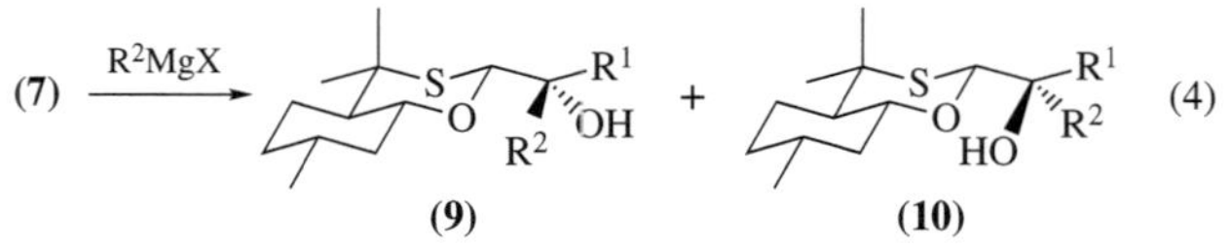

When the R group of ketone (**7**) contains a heteroatom capable of competing with the oxathianyl oxygen for magnesium chelation, a sharp decrease or reversal in the selectivity has been observed (eq 5).[14] The selectivity was restored when the heteroatom (in this case oxygen) was rendered incapable of chelation by protection with a bulky silyl group.

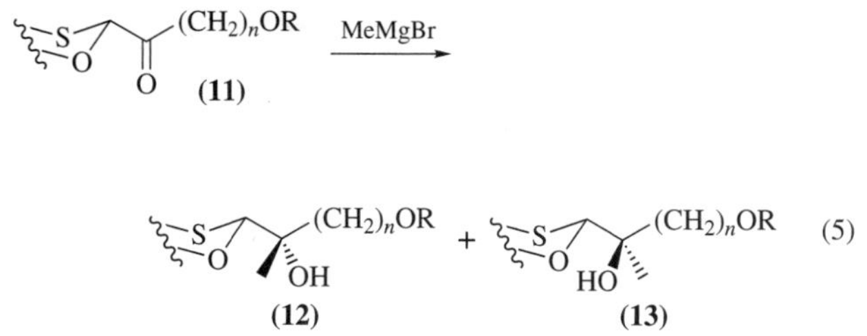

(11) (12) (13) (5)

n	R	(12) (%)
1	CH_2Ph	38
2	CH_2Ph	42
3	CH_2Ph	82
1, 2	$Si(i\text{-}Pr)_3$	98

Ketones having a heteroatom (O or S) at a second adjacent chiral center have also been studied.[15] High selectivity in the addition of a Grignard or lithium reagent was obtained in many cases, but the sterochemical outcome was found to depend on the configuration of the additional center, the organometallic reagent, and the heteroatom substituent (eq 6).[15]

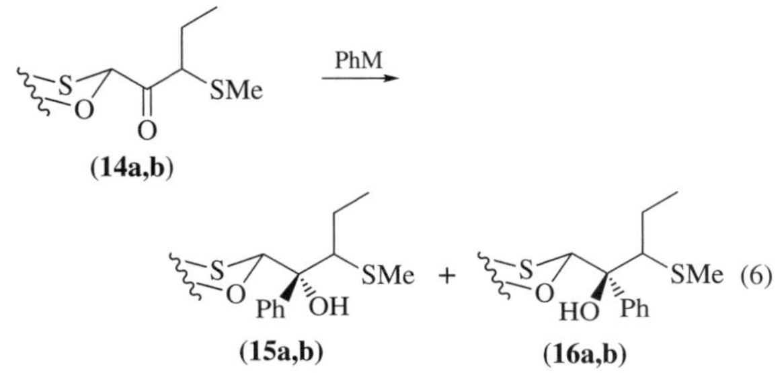

(14a,b) (15a,b) (16a,b) (6)

Side chain configuration	M	(15):(16)
(*R*)	Li	53:47
(*R*)	Mg	100:0
(*S*)	Li	0:100
(*S*)	Mg	100:0

Ytterbium-mediated additions of alkynyllithium or -magnesium reagents to (**7**) have also been reported to show high selectivity, but for the opposite diastereomer (**10**) (eq 4) from that obtained in Grignard reactions.[16]

Hydride reductions of (7) can be controlled to give either the (*R*) or (*S*) secondary hydroxy compound with good selectivity by choice of the reducing agent. *Lithium Tri-s-butylborohydride* (L-Selectride®) provided the (*S*)-alcohol (according to Cram's chelate rule) and *Diisobutylaluminum Hydride* (DIBAL) gave the (*R*)-carbinol in excess (eq 7).[2] The DIBAL results were rationalized in terms of the open-chain Cornforth dipole model.[17]

Hydrolysis of the 1,3-oxathiane moiety has been accomplished under mild conditions (0 °C, 5 min) by the use of *N-Chlorosuccinimide–Silver(I) Nitrate*.[18] This oxidative hydrolysis produces α-hydroxy aldehydes in good yields and, in addition, two diastereomeric sultines (**19**) (eq 8).[1] The use of *Iodine*–$AgNO_3$ for the oxidative hydrolysis of 1,3-oxathianes has also recently been reported.[19] The tertiary α-hydroxy aldehydes are easily oxidized directly to the acids (*Sodium Chlorite*)[20] or methyl esters (MeOH, I_2, KOH)[21] or are conveniently reduced to the diols by direct reduction of the hydrolysis mixture with *Sodium Borohydride*. The secondary α-hydroxy aldehydes could likewise be reduced to the glycols without racemization; however, oxidation required protection as the benzyl ether prior to hydrolysis. The sultines (after chromatographic separation) are reduced to the hydroxy thiol (**4**) by *Lithium Aluminum Hydride*.

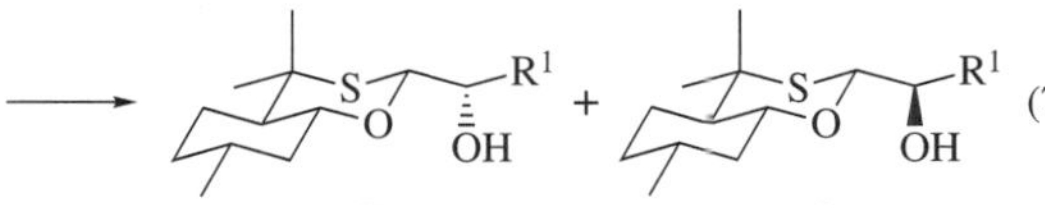

(7) (17) (18) (7)

R^1	Reducing agent	(17):(18)
Me	L-Selectride	21:79
Me	DIBAL	78:22
n-Hex	L-Selectride	11:89
n-Hex	DIBAL	87:13
i-Pr	L-Selectride	67:33
i-Pr	DIBAL	88:12
Cy	L-Selectride	52:48
Cy	DIBAL	89:11
t-Bu	L-Selectride	22:78
t-Bu	DIBAL	81:19

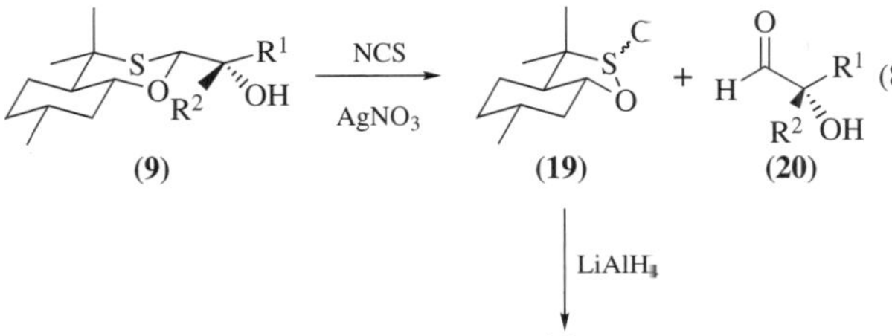

(9) (19) (20) (4) (8)

Advantage has been taken of the fact that the diastereomers (**9**) and (**10**) are often easily separated by silica gel chromatography, particularly when both enantiomers of a compound are desired in pure form. Nonselective addition of (**5**) to 2-hexanone followed by chromatographic separation of the diastereomeric carbinols and hydrolysis of each gave both (+)- and (−)-2-hydroxy-2-methylhexanal in optically pure form.[22]

Hydroxy thiol (4) likewise has been used to resolve dimethyl 4-oxocyclopentane-1,2-dicarboxylate by crystallization of the mixture of the derived oxathianes. This provided the (*R*,*R*) enantiomer in >99% purity (eq 9).[23]

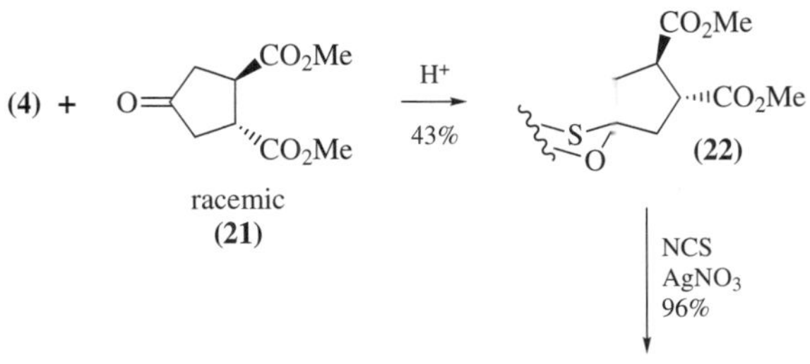

(4) + racemic (21) (22) (21) 99% ee (9)

An interesting 1,4-addition to the vinyl sulfone derivative of a related 1,3-oxathiane[24] has also been reported (eq 10).[25]

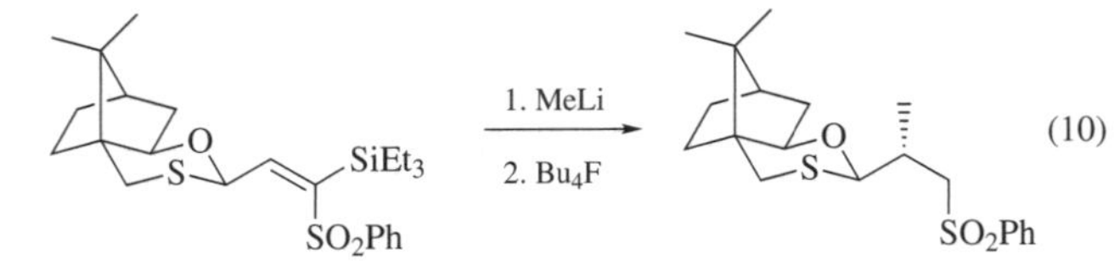

(23) (24)

The use of stoichiometric, covalently bound chiral auxiliaries as a method of asymmetric synthesis is generally impractical and cannot compete with catalytic methods on a commercial scale. However, at the laboratory scale, the oxathiane method provides a predictable method to obtain a desired enantiomer with high selectivity. Since the intermediate compounds prior to hydrolysis are diastereomeric, they are easily separated (often by crystallization) and thus enantiomerically pure compounds are readily obtained.

Related Reagents. Benzothiazole; Carbon Monoxide; *N,N*-Diethylaminoacetonitrile; *N,N*-Dimethyldithiocarbamoylacetonitrile; 2-Lithio-1,3-dithiane; Methylthiomethyl *p*-Tolyl Sulfone; 2-(Trimethylsilyl)thiazole.

---

1. Lynch, J. E.; Eliel, E. L. *J. Am. Chem. Soc.* **1984**, *106*, 2943.
2. Ko. K.-Y.; Frazee, W. J.; Eliel, E. L. *Tetrahedron* **1984**, *40*, 1333.
3. Reviews: Eliel, E. L. *Phosphorus Sulfur* **1985**, *24*, 73. Eliel, E. L.; Koskimies, J. K.; Frazee, W. J.; Morris-Natschke, S.; Lynch, J. E.; Soai, K. In *Asymmetric Reactions and Processes in Chemistry* Eliel, E. L.; Otsuka, S., Eds.; ACS: Washington, 1982; p 37. Eliel, E. L.; Frye, S. V.; Hortelano, E. R.; Chen, X.; Bai, X. *Pure Appl. Chem.* **1991**, *63*, 1591, and references therein.
4. Eliel, E. L.; Lynch, J. E.; Kume, F.; Frye, S. V. *Org. Synth.* **1987**, *65*, 215.
5. Abatjoglou, A. G.; Eliel, E. L.; Kuyper, L. F. *J. Am. Chem. Soc.* **1977**, *99*, 8262.
6. Eliel, E. L.; Koskimies, J. K.; Lohri, B. *J. Am. Chem. Soc.* **1978**, *100*, 1614.
7. Eliel, E. L.; Bai, X.; Abdel-Magid, A. F.; Hutchins, R. O.; Prol, J. *J. Org. Chem.* **1990**, *55*, 4951.
8. Bai, X.; Eliel, E. L. *J. Org. Chem.* **1992**, *57*, 5162.
9. Wei, J.; Hutchins, R. O.; Prol, J., Jr. *J. Org. Chem.* **1993**, *58*, 2920.
10. Morris-Natschke, S.; Eliel, E. L. *J. Am. Chem. Soc.* **1984**, *106*, 2937.
11. Cram, D. J.; Kopecky, K. R. *J. Am. Chem. Soc.* **1959**, *81*, 2748.
12. Eliel, E. L. In *Asymmetric Synthesis*, Morrison, J. D., Ed.; Academic: New York, 1983; Vol. 2, p 125.
13. Chen, X.; Hortelano, E. R.; Eliel, E. L.; Frye, S. V. *J. Am. Chem. Soc.* **1992**, *114*, 1778. Chen, X.; Hortelano, E. R.; Eliel, E. L.; Frye, S. V. *J. Am. Chem. Soc.* **1990**, *112*, 6130. Frye, S. V.; Eliel, E. L.; Cloux, R. *J. Am. Chem. Soc.* **1987**, *109*, 1862.
14. Frye, S. V.; Eliel, E. L. *J. Am. Chem. Soc.* **1988**, *110*, 484. Frye, S. V.; Eliel, E. L. *Tetrahedron Lett.* **1985**, *26*, 3907.
15. Bai, X.; Eliel, E. L. *J. Org. Chem.* **1992**, *57*, 5166.
16. Utimoto, K.; Nakamura, A.; Matsubara, S. *J. Am. Chem. Soc.* **1990**, *112*, 8189.
17. Cornforth, J. W.; Cornforth, R. H.; Mathew, K. K. *J. Chem. Soc.* **1959**, 112.
18. Corey, E. J.; Erickson, B. W. *J. Org. Chem.* **1971**, *36*, 3553.
19. Nishide, K.; Yokota, K.; Nakamura, D.; Sumiya, T.; Node, M.; Ueda, M.; Fuji, K. *Tetrahedron Lett.* **1993**, *34*, 3425.
20. Krause, G. A.; Roth, B. *J. Org. Chem.* **1980**, *45*, 4825.
21. Inch, T. D.; Ley, R. V.; Rich, P. *J. Chem. Soc. (C)* **1968**, *13*, 1693.
22. Cervantes-Cuevas, H.; Joseph-Nathan, P. *Tetrahedron Lett.* **1988**, *29*, 5535.
23. Solladie, G.; Lohse, O. *Tetrahedron: Asymmetry* **1993**, *4*, 1547.
24. Frazee, W. J.; Eliel, E. L. *J. Org. Chem.* **1979**, *44*, 3598.
25. Isobe, M.; Obeyama, J.; Funabashi, Y.; Goto, T. *Tetrahedron Lett.* **1988**, *29*, 4773.

Joseph E. Lynch
Merck Research Laboratories, Rahway, NJ, USA

(*S*)-(2-Hydroxy-*N,N*-dimethylpropanamide-*O,O′*)oxodiperoxymolybdenum(VI)

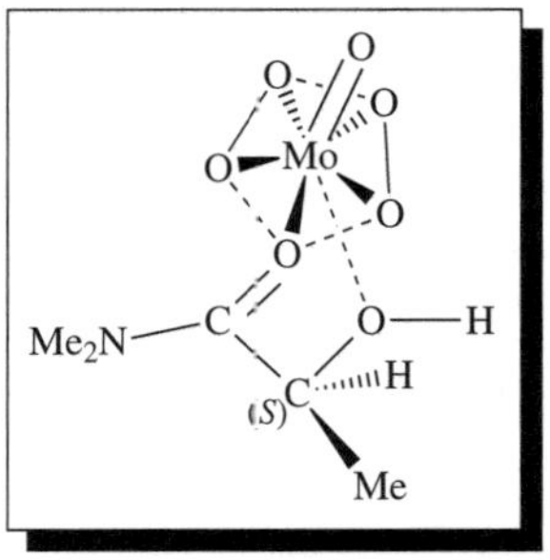

[70355-53-2] $C_5H_{11}MoNO_7$ (MW 293.11)

(enantioselective epoxidation of unfunctionalized simple alkenes[1])

Physical Data: mp 149 °C (dec). According to X-ray structural analysis, the molecule has the pentagonal bipyramidal geometry with sevenfold coordinated molybdenum.

Solubility: sol ethanol; insol ether.

Form Supplied in: yellow microcrystalline powder.

Drying: dried under vacuum.[1]

Preparative Methods: to 5mL aqueous hydrogen peroxide solution (30%) is added, in portions, 1 g molybdenum(VI) oxide at 20 °C. The mixture is stirred at 20 °C for 20 min and at 40 °C for 4 h. After most of the molybdenum oxide has dissolved the mixture is filtered. The yellow solution is treated at 10 °C with 1 equiv of ligand dissolved in approx. 4 mL methanol. The mixture is stirred for 30 min at 20 °C, concentrated and kept 12 h at 0 °C. When the complex does not crystallize after addition of dichloromethane or benzene, the mixture is diluted with ethanol and carefully (shield protection) concentrated on a rotary evaporator. The procedure is repeated five times. The mixture is then further concentrated. The complex is usually obtained as a yellow microcrystalline powder, but if it is obtained as an oil the addition of diethyl ether with stirring at 5 °C yields a yellow powder.

Handling, Storage, and Precautions: necessary care should be taken when preparing and handling peroxometal compounds. Fast and complete concentration of the mixture containing the compound may lead to deflagration of the complex. The complex decomposes at temperatures above 100 °C, changing color from yellow to blue and black. The complex is stored at 5 °C in <0.5 g portions.

Asymmetric Epoxidation[2]**.** Methods[3] are known for asymmetric epoxidation of prochiral alkenes having activated C=C double bonds (styrene, allyl alcohol, conjugated ketones, quinone). The title optically active metal–peroxo complex (**1**) is capable of asymmetric epoxidation of simple unfunctionalized alkenes. Simple prochiral alkenes such as propene or *trans*-2-butene are epoxidized stoichiometrically to optically active oxiranes by (**1**) in nitrobenzene at 20 °C/1 bar. The chemical yield is about 70%. The enantiomeric excess is around 30% and the configuration of the dominant oxirane enantiomer is (*R*). A marked increase in enantiomeric yield for trimethyloxirane from 2-methyl-2-butene is observed on reducing the reaction rate by lowering the temperature. The increasing steric hindrance of the alkyl group in 3-methyl-1-butene surprisingly leads to a decrease in the asymmetric induction. The continuously monitored enantiomeric composition of the oxiranes during the reaction remained constant within experimental accuracy. This shows that the alkene epoxidation is asymmetrically induced and that no enrichment of the enantiomers of the epoxide that is formed takes place by kinetic resolution. The asymmetric induction decreases in the order of propene > 1-butene > 3-methyl-1-butene with the preferential formation of (*R*)-alkyloxiranes.[1] The ee increases with inversion of prochiral recognition for 3,3-dimethyl-1-butene, resulting in the preferential formation of (*S*)-*t*-butyloxirane. *trans*-2-Butene undergoes higher asymmetric induction than does *trans*-2-pentene, whereby (2*S*,3*S*)-*trans*-2-methyl/ethyl-3-methyloxiranes are preferentially formed. The asymmetric epoxidation of *cis*-2-pentene leads to the preferential formation of (2*S*,3*R*)-2-ethyl-3-methyloxirane. Geminal ethyl/methyl disubstitution at the double bond in 2-methyl-1-butene shows no enantiofacial discrimination and, consequently, only racemic 2-ethyl-2-methyloxirane is formed. In connection with the synthesis of aranciamycinone,[4] complex (**1**) has been used for the asymmetric epoxidation of the nonfunctionalized alkene (**2**).[5] The alkene (**2**) (eq 1) is treated with diluted solutions (1 mmol in 300 mL of dichloromethane) of the molybdenum(VI)–oxodiperoxo complex. Decreasing the temperature to 0 °C raises the enantiomeric yield to 53% ee. The absolute configuration of the predominantly formed enantiomer (**3**) was determined by an unequivocal sequence of reactions to a known glycoside.

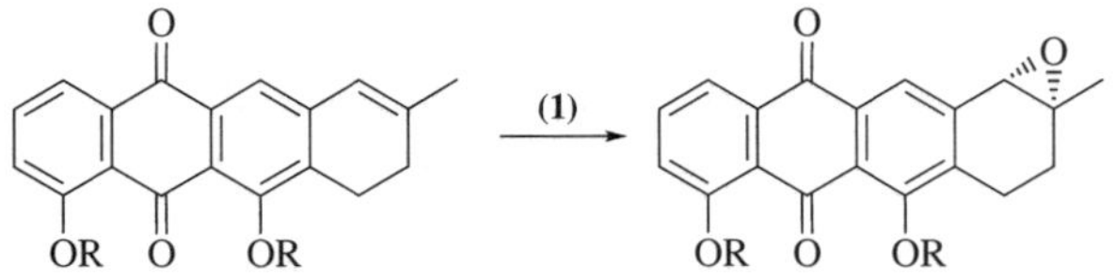

(1)

(**2**) R = CO-*t*-Bu (**3**)

1. Schurig, V.; Hintzer, K.; Leyrer, U.; Mark, C.; Pitchen, P.; Kagan, H. B. *J. Organomet. Chem.* **1989**, *370*, 81.
2. Kagan, H. B.; Mimoun, H.; Mark, C.; Schurig, V. *Angew. Chem., Int. Ed. Engl.* **1979**, *18*, 485.
3. (a) Ewins, R. C.; Henbest, H. B.; McKervey, M. A. *Chem. Commun./J. Chem. Soc., Chem. Commun.* **1967**, 1085. Montanari, F.; Moretti, I.; Torre, G. *Chem. Commun./J. Chem. Soc., Chem. Commun.* **1969**, 135. (b) Yamada, S.; Mashiko, T.; Terashima, S. *J. Am. Chem. Soc.* **1977**, *99*, 1988. Michaelson, R. C.; Palermo, R. E.; Sharpless, K. B. *J. Am. Chem. Soc.* **1977**, *99*, 1990. (c) Helder, R.; Hummelen, J. C.; Laane, R. W. P. M.; Wiering, J. S.; Wynberg, H. *Tetrahedron Lett* **1976**, 1831.
4. Krohn, K.; Broser, E. *Justus Liebigs Ann. Chem./Liebigs Ann. Chem.* **1982**, 1907.
5. Krohn, K.; Broser, E. *Tetrahedron Lett.* **1984**, *25*, 2463.

Tapan Ray
Sandoz Research Institute, East Hanover, NJ, USA

3-Hydroxyisoborneol[1]

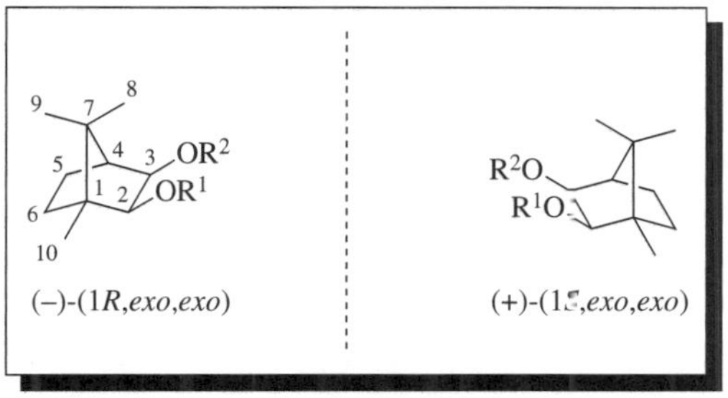

(**1**; R^1 = neopentyl, R^2 = H) (1*R*,*exo*,*exo*)
[85695-96-1] $C_{15}H_{28}O_2$ (MW 240.43)
(**2**; R^1 = neopentyl, R^2 = H) (1*S*,*exo*,*exo*)
[85718-76-9] $C_{15}H_{28}O_2$ (MW 240.43)
(**3**; R^1 = H, R^2 = neopentyl) (1*R*,*exo*,*exo*)
[85695-92-7] $C_{15}H_{28}O_2$ (MW 240.43)
(**4**; R^1 = benzyl, R^2 = H) (1*R*,*exo*,*exo*)
[104154-98-5] $C_{17}H_{24}O_2$ (MW 260.41)
(**5**; R^1 = H, R^2 = benzyl) (1*R*,*exo*,*exo*)
[73440-88-7] $C_{17}H_{24}O_2$ (MW 260.41)
(**6**; R^1 = Ph_2CH, R^2 = H) (1*R*,*exo*,*exo*)
[85695-93-8] $C_{23}H_{28}O_2$ (MW 336.51)
(**7**; R^1 = 1-naphthylmethyl, R^2 = H) (1*R*,*exo*,*exo*)
[85695-95-0] $C_{21}H_{26}O_2$ (MW 310.47)
(**8**; R^1 = 2-naphthylmethyl, R^2 = H) (1*R*,*exo*,*exo*)
[85695-94-9] $C_{21}H_{26}O_2$ (MW 310.47)

(chiral auxiliary; acrylate[2] and acyl nitroso[3] derivatives undergo stereoselective [4 + 2] cycloadditions; enoate derivatives undergo stereoselective 1,4-conjugate additions of organocopper reagents;[4] enol ether derivatives undergo stereoselective Pauson–Khand cyclizations,[5] [4 + 2][6] and [2 + 2][7] cycloadditions; alkynyl ether derivatives undergo stereoselective Pauson–Khand cyclizations[5])

Alternate Name: 1,7,7-trimethylbicyclo[2.2.1]heptane-2,3-diol.
Physical Data: (**1**) mp 4–5 °C; $[\alpha]_D^{25}$ (EtOH) −42.4° (c = 1.40). (**2**) mp 4–5 °C; $[\alpha]_D^{25}$ (EtOH) +42.6° (c = 2.52). (**3**) oil; $[\alpha]_D^{25}$ (EtOH) −18.8° (c = 1.08). (**4**) oil; bp 130 °C/0.05 mmHg; $[\alpha]_D^{25}$ ($CHCl_3$) −36.1° (c = 1.44). (**5**) mp 43 °C; $[\alpha]_D^{25}$ (EtOH) +0.4° (c = 4.99). (**6**) mp 57 °C; $[\alpha]_D^{25}$ (EtOH) −107.6° (c = 1.70). (**7**) mp 69–70 °C; $[\alpha]_D^{25}$ (EtOH) −79.2° (c = 0.90). (**8**) mp 70–71 °C; $[\alpha]_D^{25}$ (EtOH) −61.5° (c = 0.57).
Preparative Methods: the 3-hydroxyisoborneol derivatives are readily prepared from (+)- or (−)-camphor. The preparation of the 2-neopentyl ether derivative (**1**) is representative (eq 1).[2b]

Analogous alkylation with different electrophiles allows for easy variation of the shielding moiety.[2c,8] The corresponding 3-substituted derivatives have been prepared from either 3-hydroxyisoborneol after separation of the regioisomeric mixture or from 3-*exo*-hydroxycamphor regioselectively after reduction with *Lithium Tri-s-butylborohydride* (L-Selectride).[2c,9]

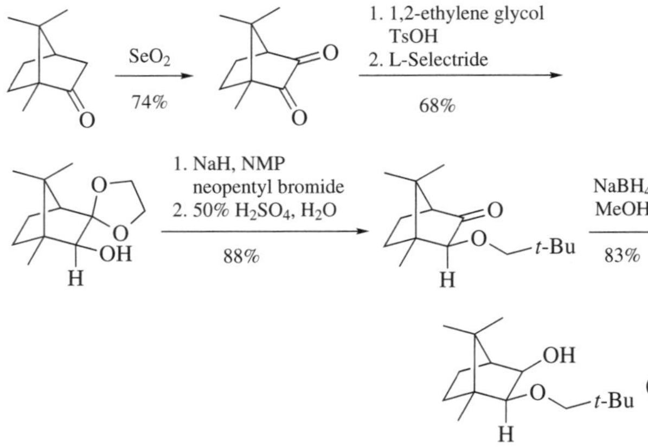

(1)

(1)

Handling, Storage, and Precautions: these auxiliaries vary from oils to white crystalline solids depending on the ether substituent and are stable indefinitely at ambient temperatures in sealed containers.

Introduction. The abundance of (+)-camphor in the chiral pool provided Oppolzer with an excellent framework to develop a chiral auxiliary which provides high levels of stereoselectivity in a wide range of reaction classes. The 3-hydroxyisoborneol skeleton provides two derivatizable positions at C-2 and C-3 of the molecule which are in close proximity to each other. By appending a reactive functionality to one and a sterically shielding appendage to the other, high stereodirecting ability can be envisioned. Likewise by reversing the roles of C-2 and C-3 it is possible to tune the auxiliary to fit the reaction parameters and desired product configuration. These characteristics have provided a means for π-facial differentiation to acrylates, enol ethers, and alkynyl ethers.

Preparation of Derivatives. Enoate derivatives are prepared from the corresponding chiral alcohol by treatment with acryloyl chloride in the presence of *Triethylamine* and catalytic *4-Dimethylaminopyridine* or the appropriate carboxylic acid chloride and *Silver(I) Cyanide*.[2b] Alkynyl ethers are readily available from the potassium alkoxide by treating with *Trichloroethylene*, in situ dechlorination with *n-Butyllithium*, and electrophilic trapping.[10] Trapping the intermediate anion with a proton source or *Iodomethane* followed by Lindlar reduction of the alkynyl ether affords the corresponding vinyl and 1-(*Z*)-propenyl ether, respectively, while reduction of the alkynyl ether with *Lithium Aluminum Hydride* affords the 1-(*E*)-propenyl ether.

[4 + 2] Cycloadditions of Acrylate Derivatives.[2] Acrylate derivatives undergo highly stereoselective Diels–Alder cycloadditions with 1,3-dienes when promoted by a Lewis acid, *Dichlorotitanium Diisopropoxide* or *Titanium(IV) Chloride* (eq 2). With the latter, care must be taken to avoid acid-mediated cleavage of the auxiliary ether linkage. Generally, 2-substituted auxiliaries (**10**) show higher facial and *endo* selectivity than the corresponding 3-substituted analogs. This has been rationalized by a buttressing effect caused by the C-10 methyl forcing the ether side chain into close proximity to the acrylate. Of the range of shielding moieties examined, the neopentyl ether was shown to provide the highest selectivity. The stereochemical outcome can be explained by assuming that the acrylate adopts an *s-trans* conformation on coordination of the Lewis acid[11] and that the diene approaches from the face opposite the neopentyl ether. It should be noted that the analogous cycloadditions with crotonate derivatives give very poor yields (<7%).[2b] Similar highly stereoselective Diels–Alder cycloadditions have also been reported for fumarate and allenic ester derivatives.[12]

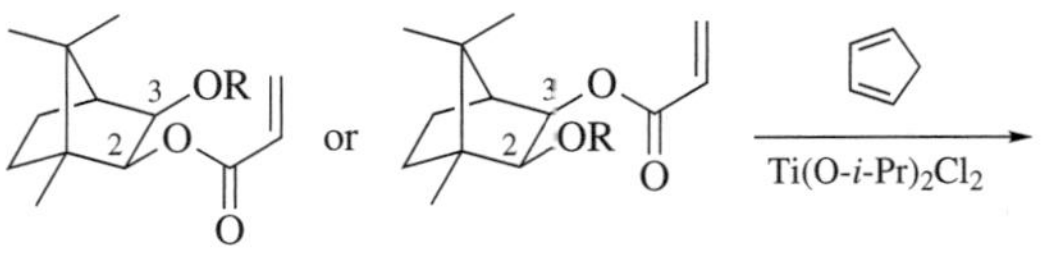

(9) (10)

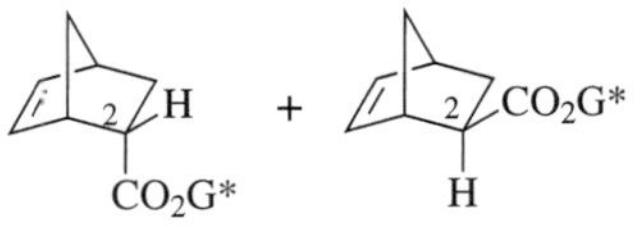

(2)

Acrylate	R	Temp (°C)	Yield (%)	*endo*:*exo*	*endo* adduct % de	*endo* adduct config.
(**9**)	Benzyl	0	91	86:14	46	(2*S*)
(**9**)	Diphenylmethyl	−20	94	86:14	64	(2*S*)
(**9**)	1-Naphthylmethyl	0	97	85:15	54	(2*S*)
(**9**)	2-Naphthylmethyl	−20	98	90:10	69	(2*S*)
(**9**)	Neopentyl	−20	95	96: 4	97	(2*S*)
(**10**)	Diphenylmethyl	−20	74	95: 5	91	(2*R*)
(**10**)	1-Naphthylmethyl	0	97	93: 7	88	(2*R*)
(**10**)	2-Naphthylmethyl	−20	98	95: 5	92	(2*R*)
(**10**)	Neopentyl	−20	96	96: 4	99	(2*R*)

[4 + 2] Cycloadditions of Enol Ether Derivatives.[6] Asymmetric, inverse electron demand Diels–Alder reactions between nitroalkenes and alcohol (**1**)-derived vinyl and 1-(*E*)- and 1-(*Z*)-propenyl ethers have been reported to proceed with high stereoselectivity (eq 3). The resulting cycloadducts undergo an intramolecular [3 + 2] cycloaddition at rt to afford nitroso acetals which, after hydrogenolytic cleavage, provide tricyclic α-hydroxy lactams in high enantiomeric excess. The auxiliary alcohol can be recovered in 86–92% yield. The overall sense of asymmetric induction is dependent on the Lewis acid promoter employed, either $Ti(O\text{-}i\text{-}Pr)_2Cl_2$ or *Methylaluminum Bis(2,6-diphenylphenoxide)* (MAPh).[6b] This has been rationalized by a switch from a highly *endo* selective cycloaddition with $Ti(O\text{-}i\text{-}Pr)_2Cl_2$ to a highly *exo* selective cyclization with MAPh. When promoted by $Ti(O\text{-}i\text{-}Pr)_2Cl_2$ the corresponding 1-(*E*)-propenyl ether shows exclusive *endo* selectivity and 99% facial selectivity; however, facial selectivity for the 1-(*Z*)-propenyl ether is only 50%.

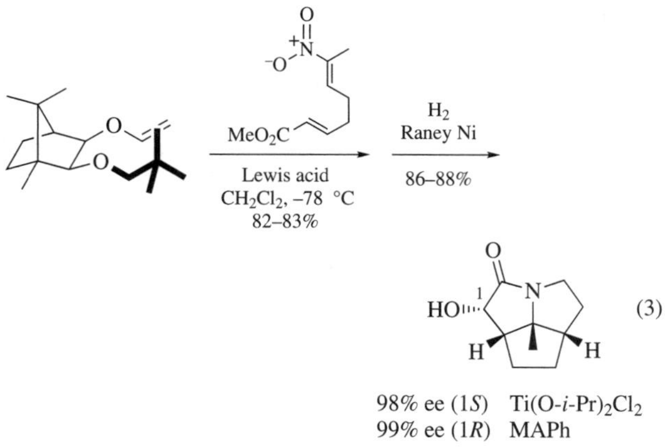

[4 + 2] Cycloadditions of Acyl Nitroso Derivatives.[3] In situ formation of the acyl nitroso derivative by oxidation of the hydroxy carbamic acid under Swern–Moffat conditions in the presence of a functionalized diene affords the corresponding cycloadduct in 94% yield and 96% diastereomeric excess (eq 4). The resulting cycloadduct can be further elaborated to prepare optically active functionalized amino alcohols.

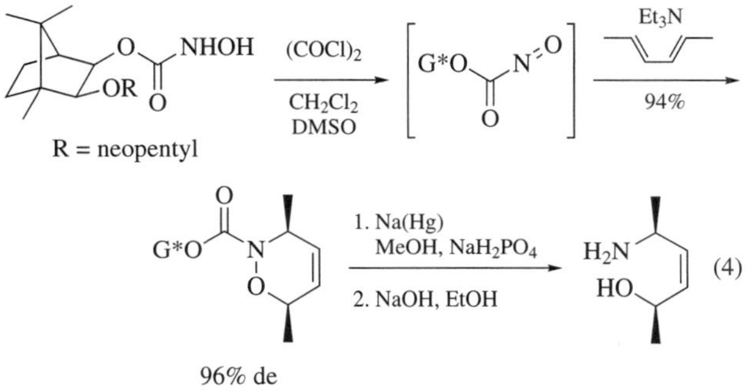

1,4-Conjugate Additions of Enoate Derivatives.[4] Boron trifluoride-mediated conjugate additions of organocopper reagents to (*E*)-enoates derived from auxiliary (**1**) proceed with high stereoselectivity, affording optically active carboxylic acids after saponification (eq 5). The organocopper reagent is formed by addition of an alkyllithium reagent to *Copper(I) Iodide*, *Tri-n-butylphosphine*, and *Boron Trifluoride Etherate* in equimolar amounts, where Bu_3P is believed to stabilize the reagent. The overall sense of asymmetric induction can be controlled by changing either the order of substituent introduction (R^1 and R^2) or the configuration of the auxiliary. The stereochemical course of the reaction has been rationalized by assuming that the enoate exists in an *s-trans* conformation and the organocopper reagent approaches from the face opposite to the neopentyl ether. Conjugate additions of this type have been applied to the total synthesis of several natural products.

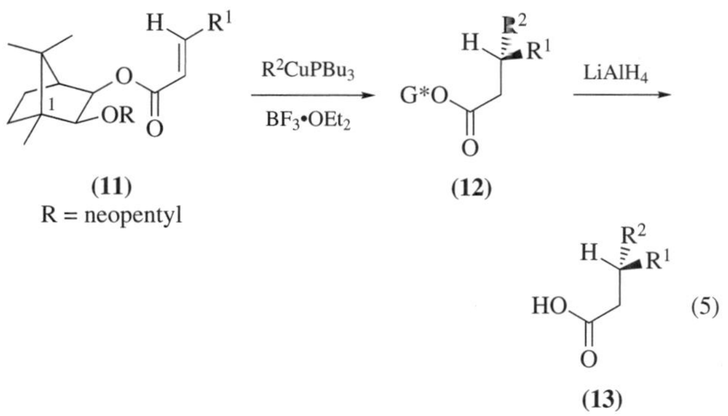

(11) config.	R^1	R^2	(12) yield (%)	(13) % ee	config.
(1*R*)	Bu	Me	82	94	(*R*)
(1*R*)	Et	Me	85	92	(*R*)
(1*R*)	$Me_2C{=}CH(CH_2)_2$	Me	90	92	(*S*)
(1*S*)	Me	$Me_2C{=}CH(CH_2)_2$	81	98	(*S*)
(1*R*)	Me	$H_2C{=}CH$	85	94	(*R*)

Pauson–Khand Bicyclization.[5] Alkynyl and enol ether derivatives have been studied in the cobalt-mediated intramolecular Pauson–Khand reaction and found to provide high diastereoselectivity, superior to previous work with the auxiliary 2-phenylcyclohexanol.[13] The 3-substituted auxiliary alcohol (**3**) provides higher selectivity than the 2-substituted analog. Also, the alkynyl ether derivatives exhibit higher reactivity and selectivity than the corresponding enol ether derivatives (eq 6).

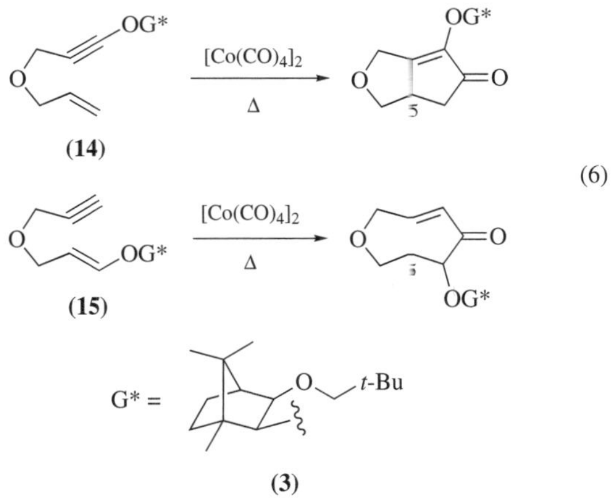

Enyne	Conditions	Yield (%)	Diastereomer ratio	Config.
(**14**)	18 °C, 2 h; 25 °C, 2 h; N_2	54	94: 6	(5*R*)
(**15**)	20 °C, 2 h; 50 °C, 12 h; CO	53	90:10	(5*S*)

Photochemical [2 + 2] Cycloadditions.[7] Photochemical [2 + 2] cycloadditions between alkenes and chiral phenylglyoxylate derivatives of 3-hydroxyisoborneol show minimal diastereoselectivity (16% de).[14] Better results are obtained in [2 + 2] cycloadditions between chiral enol ethers and *Dichloroketene* (eq 7). After ring expansion and expulsion of the auxiliary (*Diazomethane*, *Chromium(II) Perchlorate*), chiral α-chloro cyclopentenones are obtained in 60% yield. The observed diastereoselectivity is believed to arise from the enol ether *s-trans* conformation and approach of the ketene to the face opposite to the neopentyl ether.

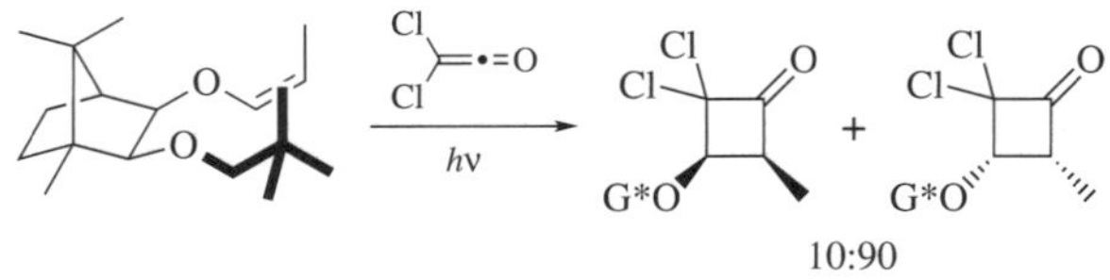

(7)

Non-destructive Auxiliary Cleavage. The high stability of the ether linkage to the shielding moiety generally allows for a very high recovery of the auxiliary alcohol. For acyl derivatives, primary alcohols can be obtained by $LiAlH_4$[2b] or AlH_3[15] reduction. Hydrolysis of the auxiliary under basic conditions providing the carboxylic acid has been accomplished with NaOH in aq. ethanol,[3] NaOH in methanol,[4b] or KOH in ethanol.[16] Intramolecular transesterification has been applied using KO-*t*-Bu in THF.[17] Enol ethers derived from Pauson–Khand cyclizations of alkynyl ether derivatives can be readily cleaved to the corresponding ketone and recovered auxiliary by catalytic HCl in methanol.[5]

Related Reagents. 10,2-Camphorsultam; 10-Dicyclohexylsulfonamidoisoborneol; (*S*)-Ethyl Lactate; α-Methyltoluene-2,α-sultam; (*R*)-Pantolactone.

1. Oppolzer, W. *Tetrahedron* **1987**, *43*, 1969.
2. (a) Oppolzer, W. *Angew. Chem., Int. Ed. Engl.* **1984**, *23*, 876. (b) Oppolzer, W.; Chapuis, C.; Dupuis, D.; Guo, M. *Helv. Chim. Acta* **1985**, *68*, 2100. (c) Oppolzer, W.; Chapuis, C.; Dao, G. M.; Reichlin, D.; Godel, T. *Tetrahedron Lett.* **1982**, *23*, 4781.
3. Martin, S. F.; Hartmann, M.; Josey, J. A. *Tetrahedron Lett.* **1992**, *33*, 3583.
4. (a) Rossiter, B. E.; Swingle, N. M. *Chem. Rev.* **1992**, *92*, 771. (b) Oppolzer, W.; Moretti, R.; Godel, T.; Meunier, A.; Löher, H. *Tetrahedron Lett.* **1983**, *24*, 4971.
5. Verdaguer, X.; Moyano, A.; Pericàs, M. A.; Riera, A.; Greene, A. E.; Piniella, J. F.; Alvarez-Larena, A. *J. Organomet. Chem.* **1992**, *433*, 305.
6. (a) Denmark, S. E.; Senanayake, C. B. W.; Ho G.-H. *Tetrahedron* **1990**, *46*, 4857. (b) Denmark, S. E.; Schnute, M. E.; Senanayake, C. B. W. *J. Org. Chem.* **1993**, *58*, 1859.
7. Greene, A. E.; Charbonnier, F. *Tetrahedron Lett.* **1985**, *26*, 5525.
8. Herzog, H.; Scharf, H.-D. *Synthesis* **1986**, 788.
9. (a) Oppolzer, W.; Kurth, M.; Reichlin, D.; Chapuis, C.; Mohnhaupt, M.; Moffatt, F. *Helv. Chim. Acta* **1981**, *64*, 2802. (b) Sasaki, S.; Kawasaki, M.; Koga, K. *Chem. Pharm. Bull.* **1985**, *33*, 4247.
10. Moyano, A.; Charbonnier, F.; Greene, A. E. *J. Org. Chem.* **1987**, *52*, 2919.
11. Loncharich, R. J.; Schwartz, T. R.; Houk, K. N. *J. Am. Chem. Soc.* **1987**, *109*, 14.
12. (a) Helmchen, G.; Schmieres, R. *Angew. Chem., Int. Ed. Engl.* **1981**, *20*, 205. (b) Oppolzer, W.; Chapuis, C. *Tetrahedron Lett.* **1985**, *24*, 4665.
13. Castro, J.; Sörensen, H.; Riera, A.; Morin, C.; Moyano, A.; Pericàs, M. A.; Greene, A. E. *J. Am. Chem. Soc.* **1990**, *112*, 9388.
14. Herzog, H.; Koch, H.; Scharf, H.-D.; Runsink, J. *Tetrahedron* **1986**, *42*, 3547.
15. Oppolzer, W.; Pitteloud, R.; Bernardinelli, G.; Baettig, K. *Tetrahedron Lett.* **1983**, *24*, 4975.
16. Cativiela, C.; López, P.; Mayoral, J. A. *Tetrahedron: Asymmetry* **1991**, *2*, 449.
17. Remiszewski, S. W.; Yang, J.; Weinreb, S. M. *Tetrahedron Lett.* **1986**, *27*, 1853.

Mark E. Schnute
Stanford University, CA, USA

(*S*)-3-Hydroxy-5-methyl-2,4-imidazolidinedione

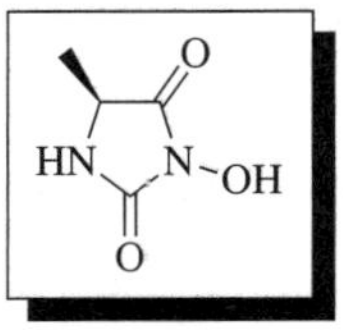

[30293-99-3] $C_4H_6N_2O_3$ (MW 130.12)

(acyl activating agent for asymmetrically selective peptide synthesis[1])

Alternate Name: (−)-3-hydroxy-5-methylhydantoin.
Physical Data: mp 163–164 °C; $[\alpha]_D$ −36.0°.
Preparative Methods: prepared from alanine (eq 1).[1]

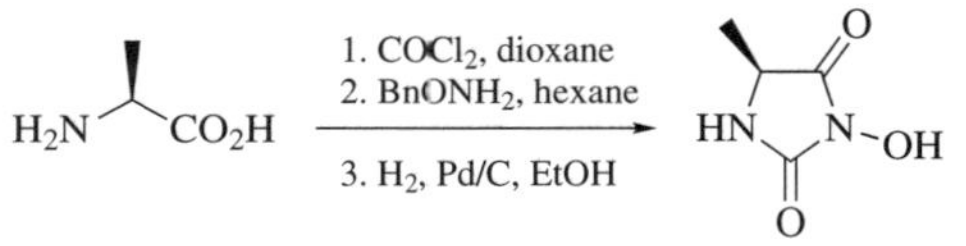

(1)

Asymmetric Peptide Synthesis.[1] The reagent activates amino acids through *1,3-Dicyclohexylcarbodiimide* (DCC) coupling to the *N*-hydroximide for subsequent coupling with chiral amino acids. The asymmetric center induces preferential reaction with L-amino acids and high optical purities of L-L-dipeptides can be achieved (eq 2). Enantioselectivity is improved if the 5-methyl group is replaced by isobutyl.[1]

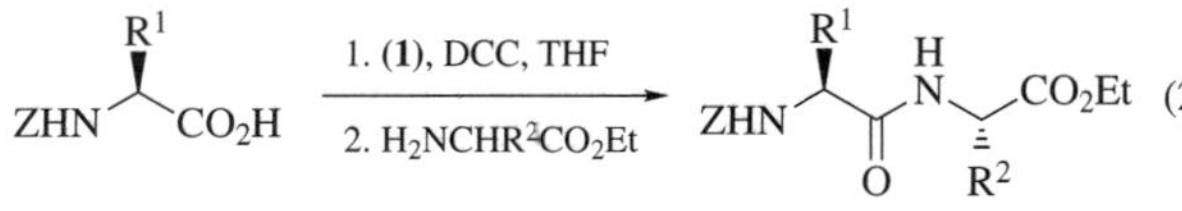

(2)

1. Teramoto, T.; Kurosaki, T. *Tetrahedron Lett.* **1977**, 1523.

Kenneth A. Murray
University of Cambridge, UK

(2*S*,2′*S*)-2-Hydroxymethyl-1-[(1-methylpyrrolidin-2-yl)methyl]pyrrolidine[1]

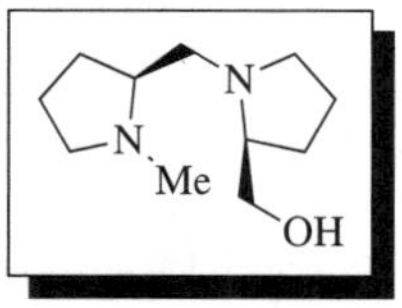

[66283-23-6] $C_{11}H_{22}N_2O$ (MW 198.35)

(chiral ligand for alkyllithium,[2] dialkylmagnesium,[3] alkynyllithium,[4] and functionalized organolithiums[5] in the enantioselective addition to aldehydes; accelerates the basicity of alkyllithiums;[5] catalyzes the addition of dialkylzinc to aldehyde[2])

Physical Data: bp 112 °C/4.5 mmHg; $[\alpha]_D^{28}$ −130° (*c* 0.36, EtOH).
Solubility: sol hexane, toluene, Me_2O, Et_2O, *n*-Pr_2O, THF, dimethoxymethane.
Preparative Methods: reduction of *N*-[*N*(-benzyloxycarbonyl)-prolyl]proline methyl ester with *Lithium Aluminum Hydride* affords the title reagent in 81% yield.

Enantioselective Addition of Alkyllithium Reagents to Aldehydes. (2*S*,2′*S*)-2-Hydroxymethyl-1-[(1-methylpyrrolidin-2-yl)methyl]pyrrolidine (**1**) is a chiral amino alcohol which binds well to alkyllithium reagents.[2] Enantioselective addition of *n-Butyllithium* to benzaldehyde in the presence of (**1**) in a mixed solvent of Me_2O and dimethoxymethane (DMM) (1:1) at −123 °C affords (*S*)-1-phenylpentan-1-ol with 95% ee in 77% yield.[2] In the addition to 3-methylbutanal, the corresponding (*S*)-alcohol is obtained in 80% ee (eq 1).

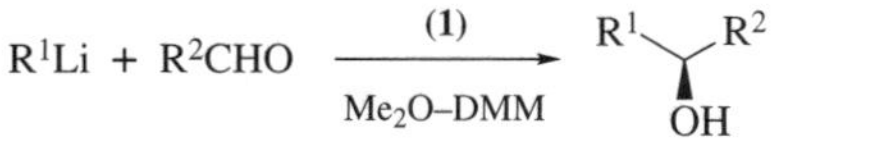

(1)

R^1 = Bu, R^2 = Ph, 77%, 95% ee; R^1 = Bu, R^2 = *i*-Bu, 57%, 80% ee

These ee's are higher than those of the preceding reports[1a] and are comparable with those of the reports appearing afterwards.[1b,c] It should be noted that the absolute configuration of the alcohol obtained in the addition of EtLi to PhCHO depends on the solvent employed [(*R*): EtOH; (*S*): DMM]. The sense of the enantioselectivity of the addition of *Methyllithium* is opposite to that of BuLi. When the derivative of (**1**) possessing a neopentyl group (**2**) is used in the addition of MeLi to PhCHO, (*R*)-1-phenylethanol with 86% ee is obtained in 82% yield.[2b]

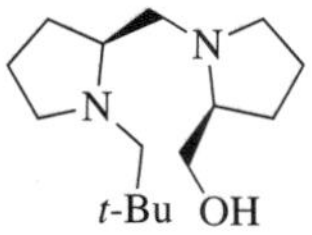

(**2**)

Enantioselective Addition of Dialkylmagnesium to Aldehydes. Amino alcohol (**1**) is a chiral ligand for dialkylmagnesium reagents in the enantioselective addition to aldehydes.[3] Reaction of Et_2Mg with PhCHO in the presence of (**1**) in toluene at −110 °C affords (*R*)-1-phenylpropan-1-ol with 92% ee in 74% yield. When *n*-Bu_2Mg is employed, (*R*)-1-phenylpentan-1-ol with 88% ee is obtained in 94% yield. It should be noted that the sense of the enantioselectivity of *n*-Bu_2Mg–(**1**) is opposite to that of *n*-BuLi–(**1**) (eq 2).

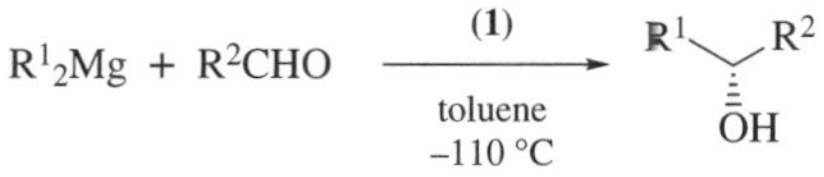

(2)

R^1 = Et, R^2 = Ph, 74%, 92% ee; R^1 = Bu, R^2 = Ph, 94%, 88% ee

Enantioselective Addition of Alkynyllithium to Aldehydes. The enantioselective addition of alkynyllithium to aldehydes in the presence of (**1**) provides optically active propargylic alcohols. (*S*)-1-Phenyl-2-propyn-1-ol with 92% ee is obtained in 87% yield from the enantioselective addition of *Lithium (Trimethylsilyl)acetylide* to PhCHO in the presence of (**1**) and the subsequent removal of the Me_3Si group (eq 3).[4a]

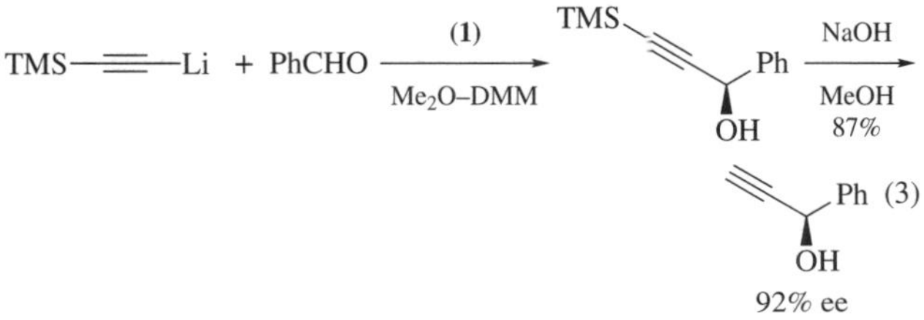

Optically active aliphatic propargylic alcohols are converted to corticoids (90% ee) via biomimetic polyene cyclization,[4b] and to 5-octyl-2(5*H*)-furanone.[4c] The ee's of propargylic alcohols obtained by this method are comparable with those of the enantioselective reduction of alkynyl ketones with metal hydrides,[6] catalytic enantioselective alkylation of alkynyl aldehydes with dialkylzincs using a chiral catalyst (*(S)-Diphenyl(1-methylpyrrolidin-2-yl)methanol*) (DPMPM),[7a] and the enantioselective alkynylation of aldehydes with alkynylzinc reagents using *N,N*-dialkylnorephedrines.[7b,c]

Enantioselective Addition of Functionalized Organolithiums to Aldehydes. Amino alcohol (**1**) is a chiral ligand for various functionalized organolithiums derived from acetonitrile and *N*-nitrosodimethylamine (eq 4).

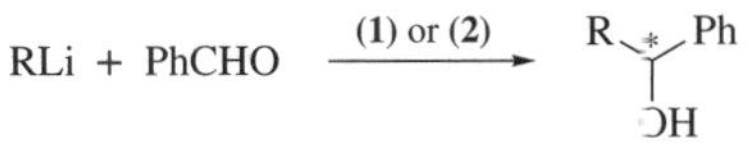

(4)

R = CH_2CN, 76%, 40% ee
R = $CH_2NMe(NO)$, 96%, 25% ee
R = CH_2SPh, 83%, 68% ee

An optically active β-hydroxy nitrile with 40% ee is obtained from the reaction of $LiCH_2CN$ with PhCHO in the presence of (**1**).[5] The same compound with higher ee (93% ee) is obtained using cyanomethylzinc bromide and the chiral ligand DPMPM.[8]

Amino alcohol (**1**) accelerates the basicity of BuLi. Thus methyl phenyl sulfide is deprotonated by BuLi in the presence of (**1**)

to afford $PhSCH_2Li$. Deprotonation does not occur without (**1**). Enantioselective addition of $PhSCH_2Li$ to PhCHO using (**1**) affords an optically active β-hydroxy sulfide, which is converted to (*R*)-2-phenyloxirane with 68% ee.[5]

Addition of Diethylzinc to Benzaldehyde. The addition of *Diethylzinc* to PhCHO without added catalysts is very sluggish. Amino alcohol (**1**) acts as a catalyst precursor for the addition of Et_2Zn to PhCHO under mild conditions to afford 1-phenylpropan-1-ol in 76% yield (eq 5).[2b,3] Although the obtained alcohol is racemic, the result of the addition of Et_2Zn to PhCHO in the presence of amino alcohol (**1**) led to the recently developed highly enantioselective addition of Et_2Zn to PhCHO using chiral amino alcohols.[1b,c]

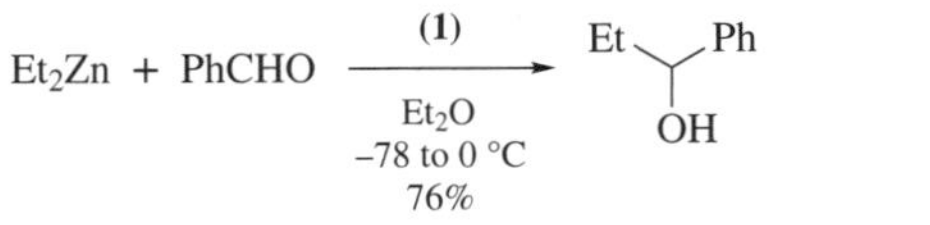

(5)

1. (a) Solladié, G. In *Asymmetric Synthesis*; Morrison, J. D., Ed.; Academic: New York, 1983; Vol. 2, pp 157–199. (b) Soai, K.; Niwa, S. *Chem. Rev.* **1992**, *92*, 833. (c) Noyori, R.; Kitamura, M. *Angew. Chem., Int. Ed. Engl.* **1991**, *30*, 49.
2. (a) Mukaiyama, T.; Soai, K.; Kobayashi, S. *Chem. Lett.* **1978**, 219. (b) Mukaiyama, T.; Soai, K.; Sato, T.; Shimizu, H.; Suzuki, K. *J. Am. Chem. Soc.* **1979**, *101*, 1455.
3. Sato, T.; Soai, K.; Suzuki, K.; Mukaiyama, T. *Chem. Lett.* **1978**, 601.
4. (a) Mukaiyama, T.; Suzuki, K.; Soai, K.; Sato, T. *Chem. Lett.* **1979**, 447. (b) Johnson, W. S.; Frei, B.; Gopalan, A. S. *J. Org. Chem.* **1981**, *46*, 1512. (c) Mukaiyama, T.; Suzuki, K. *Chem. Lett.* **1980**, 255.
5. Soai, K.; Mukaiyama, T. *Bull. Chem. Soc. Jpn.* **1979**, *52*, 3371.
6. Brinkmeyer, R. S.; Kapoor, V. M. *J. Am. Chem. Soc.* **1977**, *99*, 8339.
7. (a) Soai, K.; Niwa, S. *Chem. Lett.* **1989**, 481. (b) Niwa, S.; Soai, K. *J. Chem. Soc., Perkin Trans. 1* **1990**, 937. (c) Tombo, G. M. R.; Didier, E.; Loubinoux, B. *Synlett* **1990**, 547.
8. Soai, K.; Hirose, Y.; Sakata, S. *Tetrahedron: Asymmetry* **1992**, *3*, 677.

Kenso Soai
Science University of Tokyo, Japan

(1*S*,2*S*,5*S*)-2-Hydroxypinan-3-one

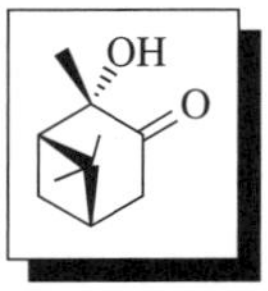

(1*S*,2*S*,5*S*)
[1845-25-6] $C_{10}H_{16}O_2$ (MW 168.26)
(1*R*,2*R*,5*R*)
[24047-72-1]

(chiral auxiliary for the asymmetric synthesis of α-substituted α-amino carboxylic acids,[1,2] phosphonic acids,[3] and phosphinic acids,[4] and of α-substituted benzylamines[5] and (2-pyridyl)-methylamines;[6] resolution of racemic α-amino acids[7])

Alternate Name: (1*S*,2*S*,5*S*)-2-hydroxy-2,6,6-trimethylbicyclo[3.1.1]heptan-3-one.
Physical Data: mp 36–38 °C; bp 245 °C; *d* 1.059 g mL^{-1}; $[\alpha]_D^{20}$ −37° (*c* = 0.5, $CHCl_3$).
Solubility: sol $CHCl_3$, CCl_4, ether, methanol, hot pentane.
Form Supplied in: available as the neat compound.
Analysis of Reagent Purity: NMR.
Preparative Methods: prepared by oxidation of α-pinene.[8]
Purification: distillation, or recrystallization from pentane.
Handling, Storage, and Precautions: no reported instability or toxicity.

Asymmetric Synthesis of α-Amino Acids. Chiral ketimines prepared from the title ketone and glycinates can be deprotonated and treated with electrophiles, such as alkyl halides (eq 1),[1] or Michael acceptors,[2] to give α-substituted α-amino acids with moderate to excellent levels of diastereoselectivity.

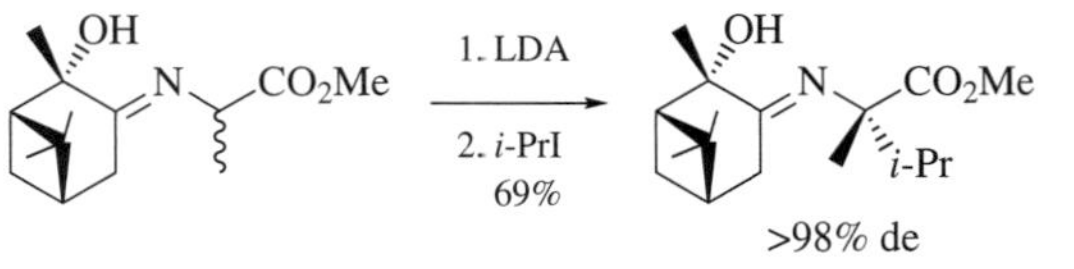

(1)

The product imine diastereomers can usually be separated by chromatography, which enables synthesis of enantiomerically pure α-amino acids even if the reaction is not completely diastereoselective, and provides an alternative to the resolution of racemic α-amino acids.[7] The imine is cleaved by mild hydrolysis with aqueous citric acid or by reaction with hydroxylammonium acetate.

This method is of special value for the synthesis of α,α-disubstituted α-amino acids.[1b,d] Analogous chiral ketimine glycinates prepared from camphor[1e,9a] or a protected D-galactodialdehyde[9b] are also synthetically useful, and in some cases give higher diastereoselectivities; with these reagents, however, separation of the imine diastereomers by chromatography does not seem to be possible. Several other chiral glycinate enolate equivalents have been reported, many of which give excellent levels of selectivity.[10] If the synthetic objective is to prepare β-hydroxy-α-amino acids by reaction of a chiral glycinate with a carbonyl compound, one of these alternative reagents should be chosen.

Asymmetric Synthesis of α-Substituted α-Amino Phosphonic and Phosphinic Acids. The title reagent can also be used to prepare chiral Schiff bases from α-amino phosphonic[3] and phosphinic[4] acid esters. Deprotonation and alkylation then gives α-substituted products with good to excellent diastereoselectivity (eq 2). Chromatographic separation of the imine diastereomers is often possible, giving access to enantiomerically pure products after hydrolysis. The corresponding Schiff bases prepared from camphor sometimes give higher diastereoselectivities in reactions with activated alkyl halides.[11] A useful alternative reagent based on a chiral phosphonamide has also been reported recently.[12]

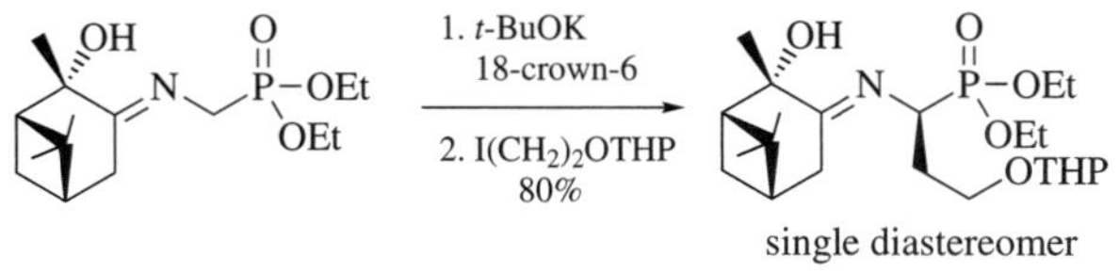

(2)

Asymmetric Synthesis of α-Substituted Benzylamines and (2-Pyridyl)methylamines. A strategy for the synthesis of chiral α-substituted benzylamines (eq 3)[5] and (2-pyridyl)methylamines[6] by alkylation of chiral ketimines has also been developed.

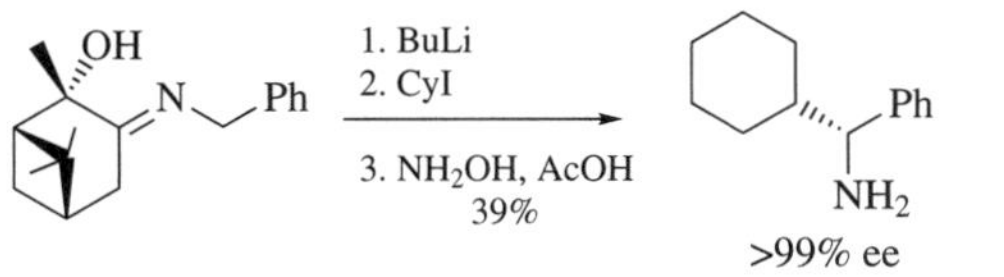

(3)

Ketimines derived from benzylamine and camphor can also be alkylated, but these reactions generally give lower diastereoselectivities.[13] Alternative approaches based on a chiral oxazoline or chiral oxazolidinones have been reported;[14] however, these reagents often give lower diastereoselectivities or problems with partial racemization during the reaction sequence required for cleaving off the chiral auxiliary.

---

1. (a) Oguri, T.; Kawai, N.; Shioiri, T.; Yamada, S. *Chem. Pharm. Bull.* **1978**, *26*, 803. (b) Bajgrowicz, J. A.; Cossec, B.; Pigiere, C.; Jacquier, R.; Viallefont, P. *Tetrahedron Lett.* **1983**, *24*, 3721. (c) Bajgrowicz, J.; El Achquar, A.; Roumestant, M.-L.; Pigiere, C.; Viallefont, P. *Heterocycles* **1986**, *24*, 2165. (d) Tabcheh, M.; El Achqar, A.; Pappalardo, L.; Roumestant, M.-L.; Viallefont, P. *Tetrahedron* **1991**, *47*, 4611. (e) Jiang, Y.; Zhou, C.; Piao, H. *Synth. Commun.* **1989**, *19*, 881.
2. Minowa, N.; Hirayama, M.; Fukatsu, S. *Bull. Chem. Soc. Jpn.* **1987**, *60*, 1761.
3. (a) Jacquier, R.; Ouazzani, F.; Roumestant, M.-L.; Viallefont, P. *Phosphorus Sulfur/Phosphorus Sulfur Sillicon* **1988**, *36*, 73. (b) Ouazzani, F.; Roumestant, M.-L.; Viallefont, P.; El Hallaoui, A. *Tetrahedron: Asymmetry* **1991**, *2*, 913.
4. McCleery, P. P.; Tuck, B. *J. Chem. Soc., Perkin Trans. 1* **1989**, 1319.
5. Chen, Y.; Mi, A.; Xiao, X.; Jiang, Y. *Synth. Commun.* **1989**, *19*, 1423.
6. Mi, A.; Xiao, X.; Wu, L.; Jiang, Y. *Synth. Commun.* **1991**, *21*, 2207.
7. Bajgrowicz, J. A.; Cossec, B.; Pigiere, C.; Jacquier, R.; Viallefont, P. *Tetrahedron Lett.* **1984**, *25*, 1789.
8. Carlson, R. G.; Pierce, J. K. *J. Org. Chem.* **1971**, *36*, 2319.
9. (a) McIntosh, J. M.; Leavitt, R. K.; Mishra, P.; Cassidy, K. C.; Drake, J. E.; Chadha, R. *J. Org. Chem.* **1988**, *53*, 1947. (b) Schöllkopf, U.; Tölle, R.; Egert, E.; Nieger, M. *Justus Liebigs Ann. Chem./Liebigs Ann. Chem.* **1987**, 399.
10. (a) Kim, B. M.; Williams, S. F.; Masamune, S. *Comprehensive Organic Synthesis* **1991**, *2*, Chapter 1.7.2. (b) Paterson, I. *Comprehensive Organic Synthesis* **1991**, *2*, Chapters 1.9.2 and 1.9.5. (c) Caine, D. *Comprehensive Organic Synthesis* **1991**, *3*, Chapter 1.1.6. (d) Fitzi, R.; Seebach, D. *Tetrahedron* **1988**, *44*, 5277. (e) Oppolzer, W.; Pedrosa, R.; Moretti, R. *Tetrahedron Lett.* **1986**, *27*, 831.
11. Schöllkopf, U.; Schütze, R. *Justus Liebigs Ann. Chem./Liebigs Ann. Chem.* **1987**, 45.
12. Hanessian, S.; Bennani, Y. L. *Tetrahedron Lett.* **1990**, *31*, 6465.
13. Jiang, Y.; Liu, G.; Liu, J.; Zhou, C. *Synth. Commun.* **1987**, *17*, 1545.
14. Gawley, R. E.; Rein, K.; Chemburkar, S. *J. Org. Chem.* **1989**, *54*, 3002.

Tobias Rein
The Royal Institute of Technology, Stockholm, Sweden

2-Hydroxy-1,2,2-triphenylethyl Acetate[1]

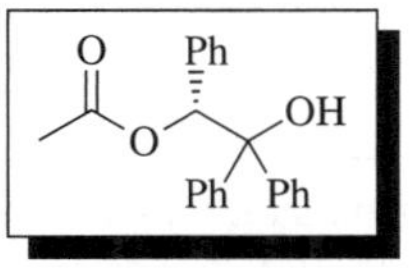

(*R*)		
[95061-47-5]	$C_{22}H_{20}O_3$	(MW 332.42)
(*S*)		
[95061-51-1]		

(enantiopure acetate for stereoselective aldol addition of the dilithioenolate to aldehydes to give chiral nonracemic β-hydroxy acids)

Alternate Name: HYTRA.
Physical Data: mp 239 °C (toluene); (*R*) $[\alpha]_D^{20}$ (c = 1, pyridine) +215° to +217°; (*S*) $[\alpha]_D^{20}$ (c = 1, pyridine) −214° to −216°.
Solubility: sol pyridine, boiling toluene; slightly sol THF, chloroform, cold toluene.
Form Supplied in: white solid; both isomers are commercially available.

Stereoselective Aldol Reactions. The (*R*)- and (*S*)-2-hydroxy-1,2,2-triphenylethyl acetates (HYTRA) offer a simple solution for a stereoselective aldol addition of α-unsubstituted enolates. When a suspension of HYTRA is treated in THF with 2 equiv of *Lithium Diisopropylamide*, a clear solution of the enolate forms (eq 1). Subsequent dilution with 2-methylbutane followed by the addition of 2-methylpropanal affords predominantly the (*R,R*)-diastereomeric adduct. Alkaline hydrolysis not only delivers (*R*)-3-hydroxy-4-methylpentanoic acid in 86–94% ee but also liberates the optically pure auxiliary reagent (*R*)-1,2,2-triphenylethane-1,2-diol, which can be removed and reused (eq 1).[2,3]

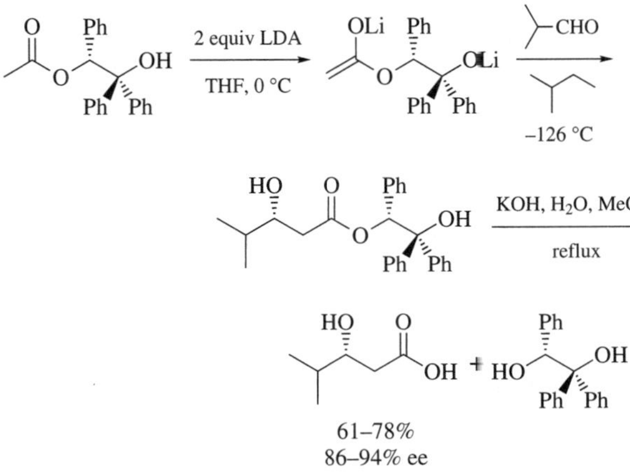

(1)

The diastereoselectivity has been enhanced by a transmetalation of the lithium enolate with *Magnesium Bromide* or *Magnesium Iodide* prior to the addition of benzaldehyde (eq 2).[3–5]

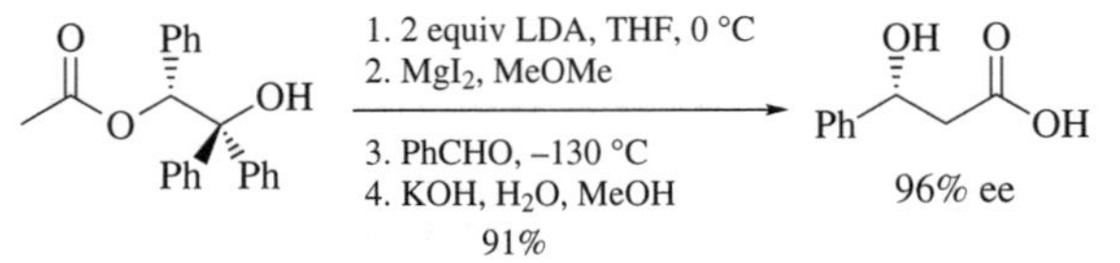

(2)

On the other hand, enantiomeric excesses up to 94% are reached even at −78 °C, provided that *Lithium Hexamethyldisilazide* (LHMDS) (3 equiv) is used for deprotonation instead of LDA (2 equiv) (eq 3).[6]

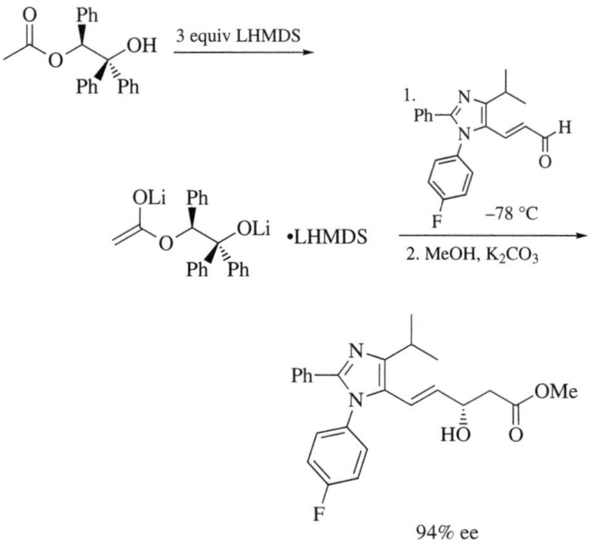

(3)

Predictable *lk*-topicity [i.e. (*R*)-enolate attacks predominantly the *re*-face of the carbonyl group whereas the (*S*)-reagent approaches predominantly from the *si*-face] has been observed in all additions of doubly deprotonated HYTRA to achiral as well as to enantiomerically pure aldehydes.[7,8] The aldol reaction of HYTRA has been used for the syntheses of natural products such as shikonin and alkannin,[9] D- and L-digitoxose (**1**),[10] FK506,[11] statin (**2**)[8] and analogs,[12] detoxinine (**3**)[13] and tetrahydrolipstatin (**4**).[14] Besides compactin and mevinolin (**5**),[15] a series of nonnatural HMG–CoA reductase inhibitors which serve as hypocholesterolemic agents have been synthesized by this method.[6,16] Peptide isosters,[17] deoxy- and aminodeoxyfuranosides,[18] 3-amino-2-hydroxybutanoic acid ('GABOB') (**6**)[19] and intermediates for the preparation of antidepressants[20] are available as well.

(**1**) L-Digitoxose

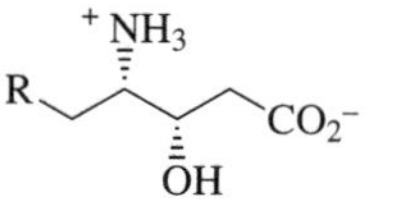

(**2**) R = $CHMe_2$: Statin
R = Ph, cyclohexyl: Statin analogs

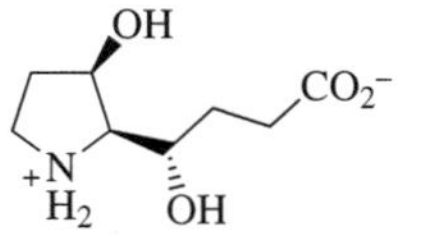

(**3**) Detoxinine

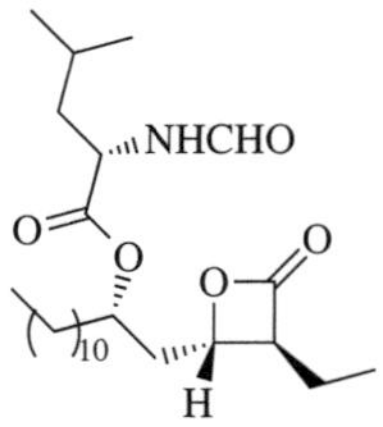

(**4**) Tetrahydrolipstatin

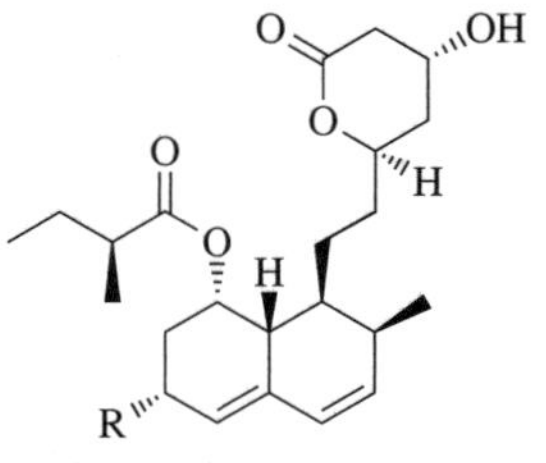

(**5**) R = H: Compactin
R = Me: Mevinolin

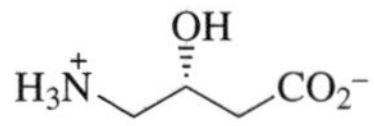

(**6**) (*R*)-GABOB

Related Reagents. Boron Triiodide; (*R*)-2-*t*-Butyl-6-methyl-4*H*-1,3-dioxin-4-one; (*S*)-4-Benzyl-2-oxazolidinone; (*R*,*R*)-1,2-Diphenyl-1,2-diaminoethane *N*,*N'*-Bis[3,5-bis-(trifluoromethyl)benzenesulfonamide]; trans-2,5-Bis(methoxymethyl)pyrrolidine; Chloro(cyclopentadienyl)bis[3-*O*-(1,2:5,6-di-*O*-isopropylidene-α-D-glucofuranosyl)]titanium; 10-Dicyclohexylsulfonamidoisoborneol; Diisopinocampheylboron Trifluoromethanesulfonate; α-Methyltoluene-2,α-sultam; 1,1,2-Triphenyl-1,2-ethanediol.

1. (a) Braun, M. *Angew. Chem.* **1987**, *99*, 24; *Angew. Chem., Int. Ed. Engl.* **1987**, *26*, 24. (b) Braun, M. In *Advances in Carbanion Chemistry*; Snieckus, V., Ed.; JAI: Greenwich, CT, 1992, Vol. 1, p 177–247.
2. (a) Braun, M.; Gräf, S.; Herzog, S. *Org. Synth.* **1993**, *72*, 32. (b) Braun, M.; Gräf, S. *Org. Syynth.* **1993**, *72*, 38.
3. Braun, M.; Devant, R. *Tetrahedron Lett.* **1984**, *25*, 5031.
4. Devant, R.; Mahler, U.; Braun, M. *Ber. Dtsch. Chem. Ges./Chem. Ber.* **1988**, *121*, 397.
5. Lynch, J. E.; Volante, R. P.; Wattley, R. V.; Shinkai, I. *Tetrahedron Lett.* **1987**, *28*, 1385.
6. Prasad, K.; Chen, K.-M.; Repic, O.; Hardtmann, G. E. *Tetrahedron: Asymmetry* **1990**, *1*, 703.
7. Mahler, U.; Devant, R. M.; Braun, M. *Ber. Dtsch. Chem. Ges./Chem. Ber.* **1988**, *121*, 2035.
8. Wuts, P. G. M.; Putt, S. R. *Synthesis* **1989**, 951.
9. Braun, M.; Bauer, C. *Justus Liebigs Ann. Chem./Liebigs Ann. Chem.* **1991**, 1157.
10. Braun, M; Moritz, J. *Synlett* **1991**, 750.
11. Mills, S.; Desmond, R.; Reamer, R. A.; Volante, R. P.; Shinkai, I. *Tetrahedron Lett.* **1988**, *29*, 281.
12. Devant, R.; Radunz, H. E. *Tetrahedron Lett.* **1988**, *29*, 2307.
13. Ewing, W. R.; Harris, B. D.; Bhat, K. L.; Joullié, M. M. *Tetrahedron* **1986**, *42*, 2421.
14. Barbier, P.; Schneider, F.; Widmer, U. *Helv. Chim. Acta* **1987**, *70*, 1412.
15. Lynch, J. E.; Shinkai, I.; Volante, R. P. (Merck and Co. Inc.) U.S. Patent 4 611 081, 1986 (*Chem. Abstr.* **1987**, *106*, 18 119n).
16. (a) Baader, E.; Bartmann, W.; Beck, G.; Below, P.; Bergmann, A.; Jendralla, H.; Kesseler, K.; Wess, G. *Tetrahedron Lett.* **1989**, *30*, 5115. (b) Jendralla, H.; Baader, E.; Bartmann, W.; Beck, G.; Bergmann, A.;

Granzer, E.; v. Kerekjarto, B.; Kesseler, K.; Krause, R.; Schubert, W.; Wess, G. *J. Med. Chem.* **1990**, *33*, 61. (c) Jendralla, H.; Granzer, E.; v. Kerekjarto, B.; Krause, R.; Schacht, U.; Baader, E.; Bartmann, W.; Beck, G.; Bergmann, A.; Kesseler, K.; Wess, G.; Chen, L.-J.; Granata, S.; Herchen, J.; Kleine, H.; Schüssler, H.; Wagner, K. *J. Med. Chem.* **1991**, *34*, 2962. (d) Baader, E.; Jendralla, H.; v. Kerekjarto, B.; Beck, G. (Hoechst A.-G.) Eur. Patent Appl. 324 347, 1989 (*Chem. Abstr.* **1990**, *112*, 21 003z). (e) Roth, B. D.; Blankley, C. J.; Chucholowski, A. W.; Ferguson, E.; Hoefle, M. L.; Ortwine, D. F.; Newton, R. S.; Sekerke, C. S.; Sliskovic, D. R.; Stratton, C. D.; Wilson, M. *J. Med. Chem.* **1991**, *34*, 357. (f) Roth, B. D. (Warner-Lambert Co.) Eur. Pat. Appl. 409 281, 1991 (*Chem. Abstr.* **1991**, *115*, 29 107u). (g) Patel, D. V.; Schmidt, R. J.; Gordon, E. M.; *J. Org. Chem.* **1992**, *57*, 7143. (h) Wright, J. J.; Sit, S. Y. (Bristol-Myers Co.) Ger. Offen. 3 805 801, 1988 (*Chem. Abstr.* **1989**, *110*, 114 836x). (i) Matsuo, M.; Manabe, T.; Okumura, H.; Matsuda, H.; Fujii, N. (Fujisawa Pharmacentical Co., Ltd.) PCT Int. Appl. 91 18 903, 1991 (*Chem. Abstr.* **1992**, *116*, 151 782w). (k) Natsugari, H.; Ikeda, H. (Takeda Chemical Industries, Ltd.) Eur. Pat. Appl. 424 929, 1991 (*Chem. Abstr.* **1991**, *115*, 114 373x).

17. (a) Allmendinger, T.; Felder, E.; Hungerbühler, E. *Tetrahedron Lett.* **1990**, *31*, 7301. (b) Allmendinger, T.; Hungerbühler, E.; Lattmann, R.; Ofner, S.; Schilling, W.; v. Sprecher, G.; Felder, E. (Ciba-Geigy A.-G.) Eur. Pat. Appl. 353 732, 1990 (*Chem. Abstr.* **1990**, *113*, 153 046w).
18. Gräf, S.; Braun, M. *Justus Liebigs Ann. Chem./Liebigs Ann. Chem.* **1993**, 1091.
19. Braun, M.; Waldmüller, D. *Synthesis* **1989**, 856.
20. Volante, R. P.; Corley, E.; Shinkai, I. (Merck and Co., Inc.) Eur. Pat. Appl. 251 714, 1988 (*Chem. Abstr.* **1988**, *108*, 150 455q).

Manfred Braun
Heinrich-Heine-Universität, Düsseldorf, Germany

(*R*)-2-Hydroxy-2′-methoxy-1,1′-binaphthyl

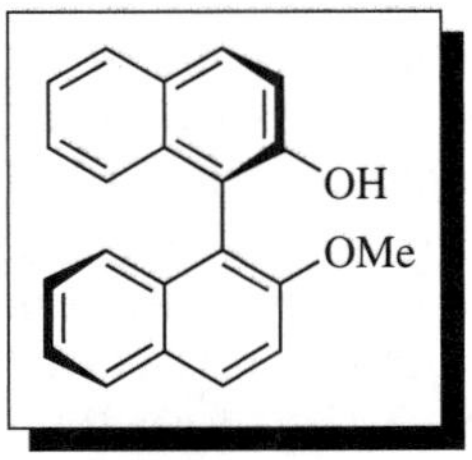

[79547-82-3] $C_{21}H_{16}O_2$ (MW 300.36)

(reagent used as chiral proton source or chiral ligand in several enantioselective reactions)

Physical Data: mp 89–91 °C; $[\alpha]_D^{27}$ +38.9 [*c* 0.68, THF, 99.3% ee (*R*)], $[\alpha]^{28}{}_D$ −44.8 (*c* 1.4, $CHCl_3$, 99.3% ee (*R*)]; HPLC [CHIRALCEL OD, 5% *i*-PrOH–hexane, retention time for the (*R*)-enantiomer, 25.12 min; for the (*S*)-enantiomer, 35.47 min]; IR (Nujol) 3478 cm^{-1}; ^1H NMR (300 MHz, $CDCl_3$) δ3.81 (s, 3H), 4.91 (s, 1H, exchangeable with D_2O), 7.04 (br d, 1H, $J=7.7$ Hz), 7.14–7.40 (m, 6H), 7.49 (d, 1H, $J=9.3$ Hz), 7.86 (br d, 1H, $J=8.0$ Hz), 7.90 (d, 2H, $J=8.5$ Hz), 8.06 (d, 1H, $J=9.1$ Hz); MS $m/z=300$ (M^+, 100%).

Solubility: soluble in alcohol, diethyl ether, toluene, and most organic solvents.

Form Supplied in: white solid.

Preparative Methods: (*R*)-2-hydroxy-2′-methoxy-1,1′-binaphthyl [(*R*)-BINOL-Me] can be prepared from commercially available (*R*)-1,1′-bi-2-naphthol [(*R*)-BINOL] by the use of 1 mol equiv of methyl iodide and sodium hydride in *N,N*-dimethylformamide (DMF)[1a] or by the use of Mitsunobu reaction.[1b]

Purification: recrystallization from toluene–hexane or purification by silica gel column chromatography from a hexane–AcOEt (8:1) eluent.

Handling, Storage, and Precautions: very stable in air.

Enantioselective Protonation using $SnCl_4$–BINOL Derivatives. Enantioselective protonation of silyl enol ethers is a very simple and attractive route for preparing optically active carbonyl compounds.[2] However, it is difficult to achieve high enantioselectivity using simple chiral Brønsted acids because of the conformational flexibility in the neighborhood of the proton. In 1994, the authors found that the *Lewis acid-assisted chiral Brønsted acid (LBA)* is a highly effective chiral proton donor for enantioselective protonation.[3a] The coordination of a Lewis acid to a Brønsted acid restricts the direction of the proton and increases its acidity. In the presence of a stoichiometric amount of (*R*)-BINOL–$SnCl_4$, the protonation of the trimethylsilyl enol ether derived from 2-phenylcyclohexanone proceeds in toluene at −78 °C to give the (*S*)-ketone with 97% ee. This reagent is applicable to various ketene bis(trialkylsilyl) acetals derived from α-arylcarboxylic acids. The observed absolute stereopreference can be understood in terms of the proposed transition state assembly. The trialkylsiloxy group is directed opposite to the binaphthyl moiety in order to avoid any steric interaction, and the aryl group stacks on this naphthyl group.

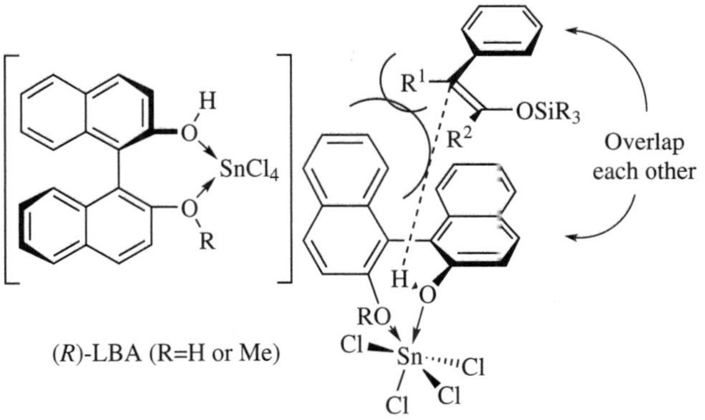

(*R*)-LBA (R=H or Me)

Taniguchi and Ogasawara have applied the enantioselective protonation using LBA to the asymmetrization of a *meso*-1,2-enediol bis(trimethylsilyl) ether having an *endo*-tricyclic [4.2.1.0^{2,5}]nonene framework (up to 90% ee) (eq 1).[3c] The enantioselectivity has been increased from 9% ee (R = H) to 90% ee (R = *i*-Pr) by screening the (*R*)-substituent of (*S*)-2-alkoxy-2′-hydroxy-1,1′-binaphthyl [(*S*)-BINOL-R]. The chiral acyloin thus obtained can be transformed into two versatile chiral building blocks, (−)-ketodicyclopentadiene and (−)-ketotricyclononene, in optically pure forms via a sequence involving concurrent enzymatic acetylation and optical purification.

The authors have succeeded in the enantioselective protonation using a stoichiometric amount of an achiral proton source and a catalytic amount of (*R*)-BINOL-Me in place of (*R*)-BINOL (eq 2).[4] In the presence of 8 mol % of $SnCl_4$, 10 mol % of (*R*)-

BINOL-Me, and stoichiometric amounts of 2,6-dimethylphenol as an achiral proton source, protonation of the ketene bis(trimethylsilyl)acetal derived from 2-phenylpropanoic acid proceeds at −80 °C to give the (*S*)-carboxylic acid with 94% ee. (*R*)-BINOL-Me is far superior to (*R*)-BINOL as a chiral proton source during the catalytic protonation, and 2,6-dimethylphenol is the most effective achiral proton source. In addition, it is very important that the molar quantity of $SnCl_4$ should be less than that of (*R*)-BINOL-Me to achieve a high enantioselectivity. For the reaction of 2-phenylcyclohexanone, however, the use of tin tetrachloride in molar quantities lower than BINOL-Me remarkably lowers the reactivity of the chiral LBA (eq 3). Excess $SnCl_4$ per chiral proton source, in contrast, promotes this protonation. In the protonation of silyl enol ethers less reactive than ketene bis(trialkylsilyl) acetals, chelation between excess tin tetrachloride and 2,6-dimethylphenol prevents the deactivation of the chiral LBA.

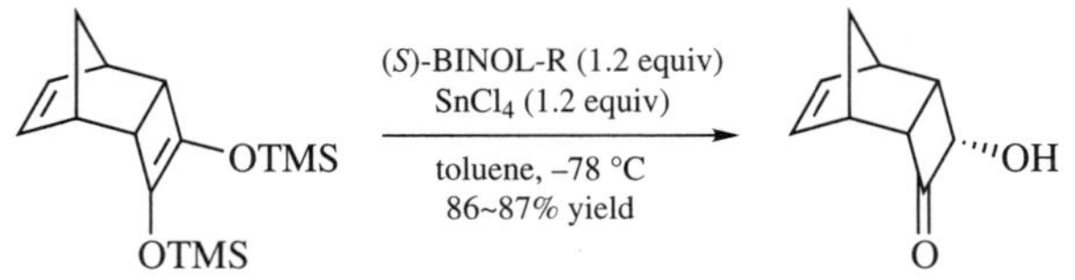

(1)

R=H: 9% ee; R=Me: 72% ee; R=*i*-Pr: 90% ee

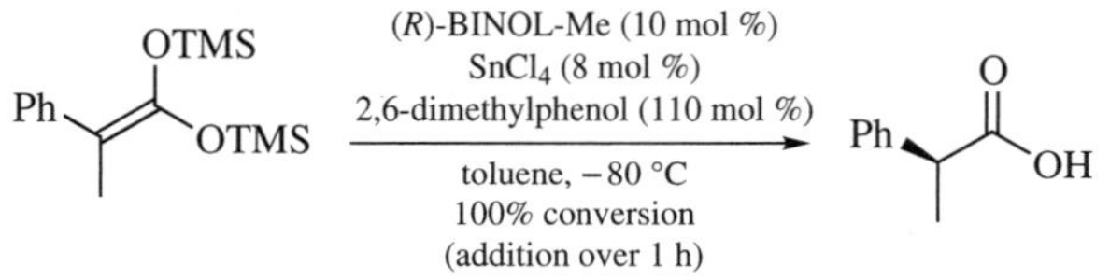

(2)

94% ee

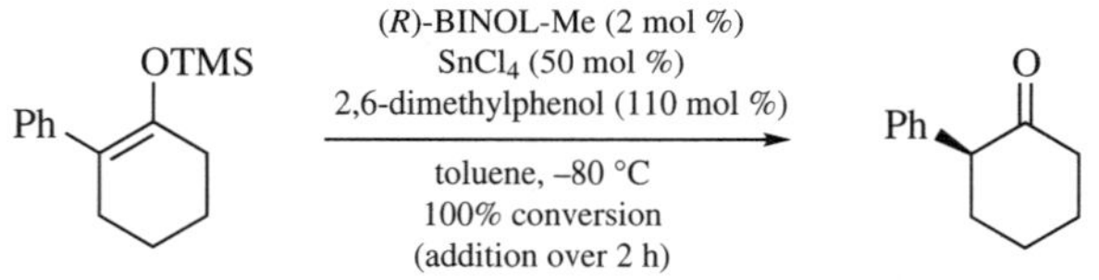

(3)

90% ee

The mechanism of the catalytic cycle has been investigated by ^{1}H NMR analysis of the 1 to 1 reaction mixtures of the silyl enol ether and chiral LBAs, (*R*)-BINOL–$SnCl_4$ and (*R*)-BINOL-Me–$SnCl_4$, at −78 °C. In the former case, two singlets for the TMS groups of Me_3SiCl and the mono trimethylsilyl ether of (*R*)-BINOL have been observed at a molar ratio of 15 to 85. In the latter case, only one singlet for TMSCl has been observed. The presence of Me_3SiCl suggests the generation of tin(IV) aryloxide intermediates. The catalytic cycle can be reasonably explained by assuming that the tin(IV) aryloxide intermediate is reconverted to the chiral LBA by receiving a proton and a chloride from 2,6-dimethylphenol and Me_3SiCl or $SnCl_4$, respectively (eq 4).

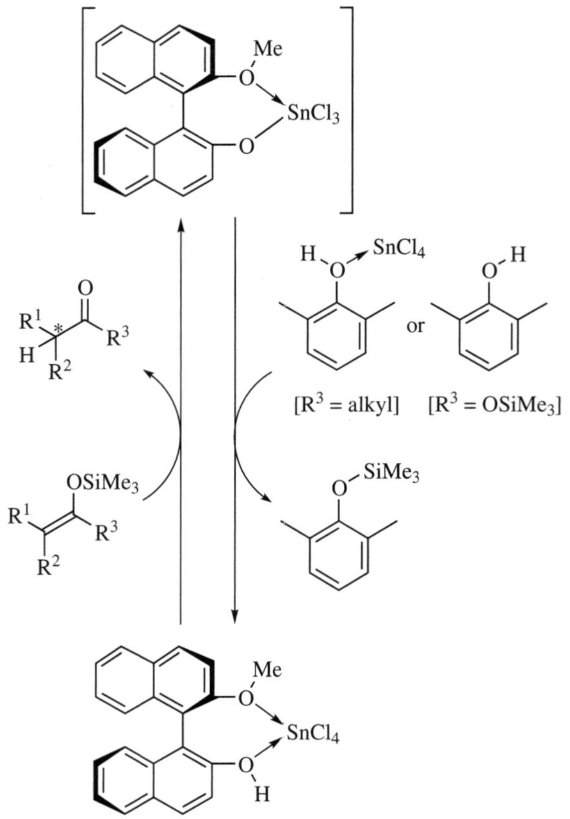

(4)

The LBAs, BINOL–$SnCl_4$ and BINOL-Me–$SnCl_4$, are highly effective proton donors for the enantioselective protonation of allyltrimethyltins to give optically active olefins.[3d] In the presence of 1.5 equiv of (*R*)-BINOL-Me–$SnCl_4$ in toluene, the protonation of (*E*)-3-phenyl-2-butenyltrimethyltin proceeds rapidly at −78 °C to form (*S*)-3-phenylbut-1-ene with good enantioselectivity and complete γ-regioselectivity (eq 5). The enantioselectivity is increased by lowering the reaction temperature to −90 °C in dichloromethane, and is dramatically decreased by using sterically bulky tin substituents. This latter tendency is interesting in that the enantioselectivity is independent of the steric features of the trialkylsilyl substituents in the protonation of silyl enol ethers with LBA. In the above protonation, a proton of (*R*)-LBA approaches the *si*-face of the γ-olefinic carbon of (*E*)-3-phenyl-2-butenyltrialkyltin, while it approaches the opposite enantioface in the protonation of the analogous ketene bis(trimethylsilyl) acetal derived from 2-phenylpropionic acid.[3a] In contrast, the enantioselectivity for the protonation of 1-(trimethylstannyl)methyl-2-phenylcyclohexene as a (*Z*)-allyltrimethyltin is moderate, and the absolute stereochemical selectivity is analogous to that in the protonation of silyl enol ethers derived from 2-phenylcyclohexanone (eq 6).

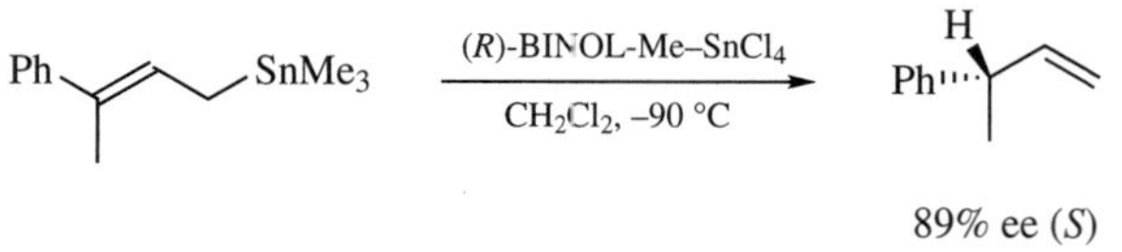

(5)

89% ee (*S*)

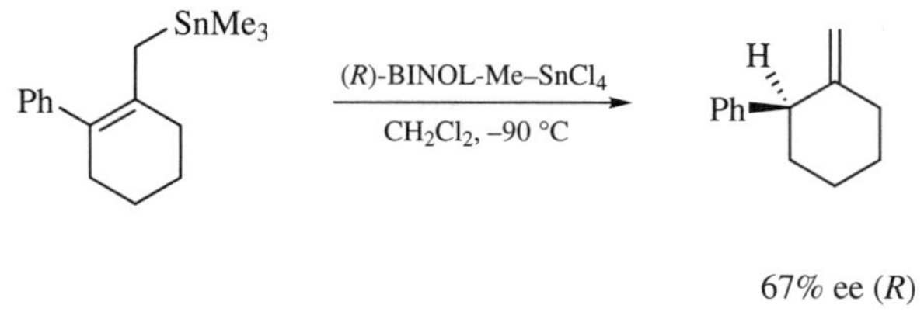
(6)

The *E*/*Z* substrate-dependent absolute stereochemistry and the steric influence of tin-substituents on the enantioselectivity observed in these reactions suggest that the mechanism is essentially different from that of silyl enol ethers. Although the detailed stereochemical course is not ascertained, it is possible that the protonation may occur via a two chlorine-bridged intermediate involving allyltrimethyltin and LBA.

Stereoselective Isomerization Catalyzed $SnCl_4$–biphenol Derivatives. Protodesilylation and isomerization are able to occur during the reaction of silyl enol ethers with a Brønsted acid. The thermodynamic equilibration of trimethylsilyl enol ethers catalyzed by a Brønsted acid was first reported by Stork and Hudrlik in 1968.[5] However, this equilibration was not established as a synthetically useful procedure, since the use of a Brønsted acid was seriously complicated by the concurrent formation of higher-molecular-weight materials and ketones. The greater stability of the Si–O bond in silyl enol ethers and the milder nucleophilicity of the conjugate base to the silicon atom favor the latter process. The authors have found that the regio and stereoselective isomerization of a kinetic silyl enol ether to a thermodynamic one is catalyzed by LBA.[6] Kinetic TBDMS enol ethers are isomerized to the thermodynamic ones in the presence of catalytic amounts of the coordinate complexes of tin tetrachloride and the monoalkyl ethers of BINOL or biphenol (BIPOL). On the other hand, use of the coordinate complexes with biphenol and other monoaryl alcohols affords predominantly the corresponding ketones. For the various structurally diverse substrates, the isomerization cleanly proceeds in the presence of 5 mol % of the achiral LBA, BIPOL-*i*-Pr–$SnCl_4$. The catalyst is effective not only for cyclic silyl enol ethers but also acyclic ones, and *Z*-isomers are stereoselectively produced (eq 7).

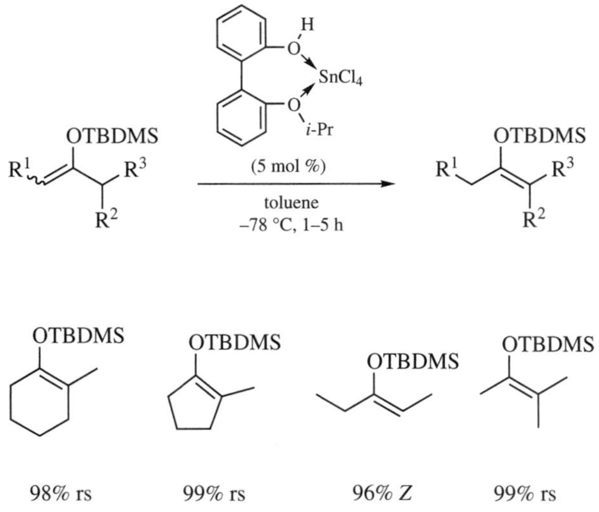
(7)

A one-pot procedure from the racemic silyl enol ether to (*S*)-2-phenylcyclohexanone has been realized by combination of the isomerization and subsequent enantioselective protonation catalyzed by (*R*)-BINOL-Me in the presence of 2,6-dimethylphenol, tin tetrachloride, and TMSCl (eq 8). Furthermore, the authors have succeeded in the enantiomer-selective isomerization of racemic silyl enol ethers. For example, during the isomerization of the same racemic silyl enol ether with 5 mol% of (*R*)-BINOL-Me–$SnCl_4$ at −78 °C for 2 min, the (*R*)-silyl enol ether is recovered in 42% yield with 97% ee. This absolute stereopreference is consistent with that in the above enantioselective protonation (eq 9).

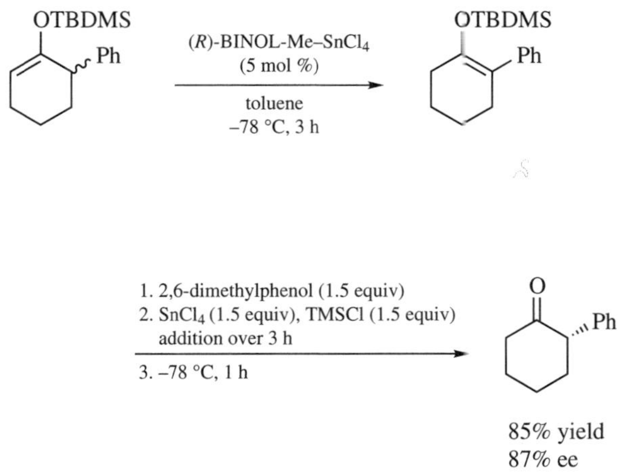
(8)

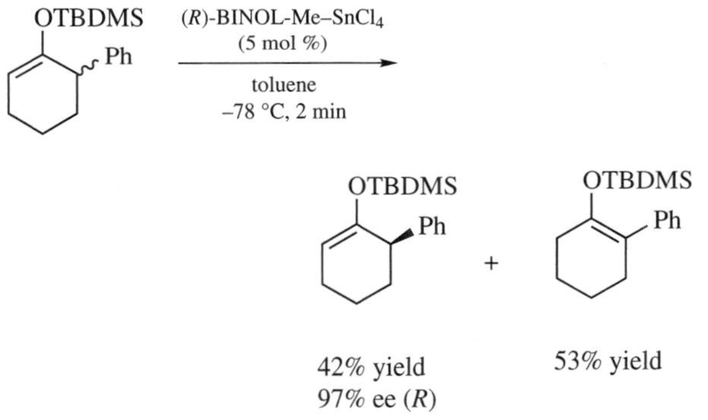
(9)

Enantioselective Polyene Cyclization Catalyzed $SnCl_4$–BINOL Derivatives. Non-enzymatic enantioselective polyene cyclizations are very attractive alternatives to the multistep synthesis from naturally occurring chiral synthons. The authors have succeeded in the first enantioselective biomimetic cyclization of polyprenoids catalyzed by LBA.[7] (−)-Ambrox® is the most important commercial substitute for ambergris, due to its unique olfactory and fixative properties. The successful preparation of (−)-ambrox® has been achieved by the enantioselective cyclization of homofarnesol promoted by (*R*)-BINOL-Me–$SnCl_4$, although the enantioselectivity and diastereoselectivity is moderate (eq 10).

Cyclization of the more reactive *o*-geranylphenol with (*R*)-BINOL–$SnCl_4$ gives the *trans*-fused tricyclic compound as a major diastereomer (36% ee, 84% ds) in good yield (eq 11). The enantioselectivity is improved to 50% ee by using (*R*)-BINOL-Me–$SnCl_4$. The monobenzoyl ester of (*R*)-BINOL [(*R*)-BINOL-Bz]-$SnCl_4$ complex is the most effective for controlling the absolute and relative stereochemistries (54% ee, 95% ds).

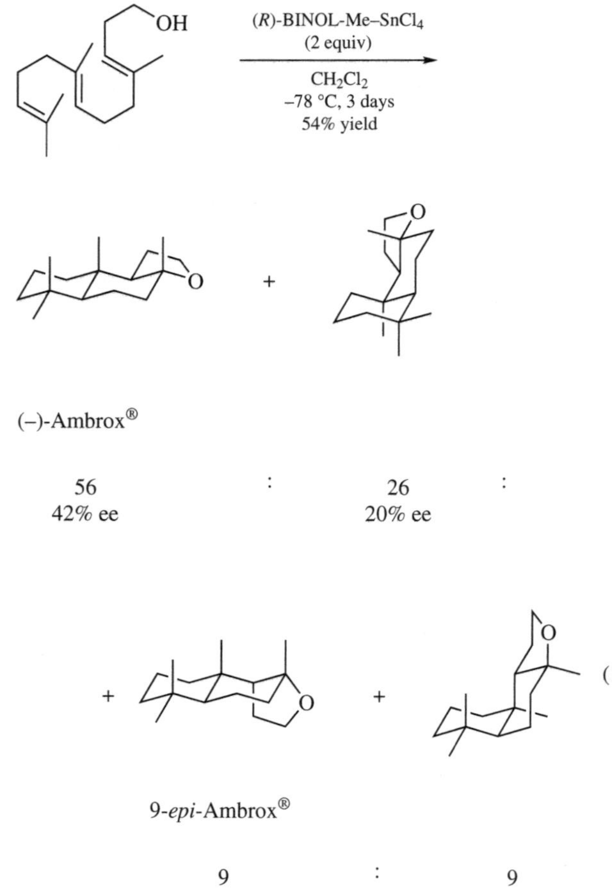

(10)

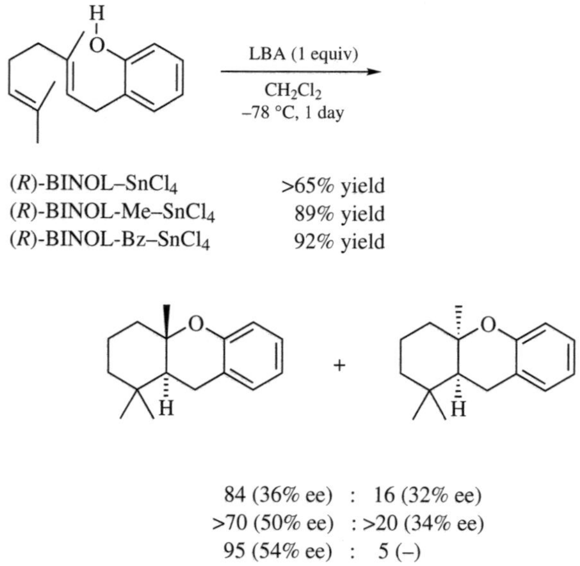

(11)

The authors have found that the same tricyclic ether is obtained with much better selectivity from geranyl phenyl ether (eq 12). Surprisingly, the reaction proceeded smoothly even in the presence of 20 mol % of this LBA to give the desired compound with 77% ee and 98% ds. This reaction is surmised to take place via a [1,3]-rearrangement and subsequent cyclization, although this has not yet been confirmed.

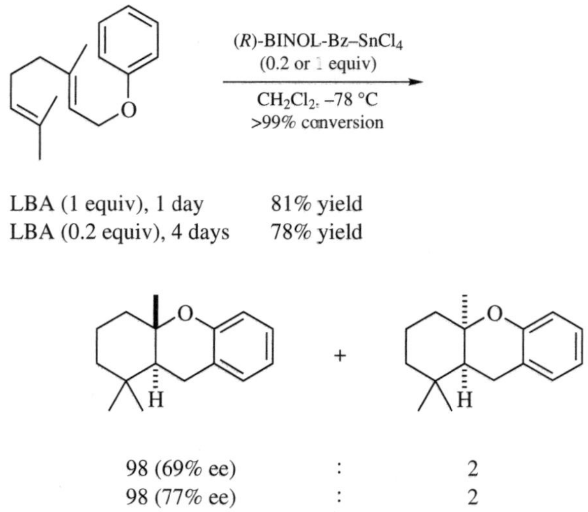

(12)

(−)-Chromazonarol, a minor constituent of the brown Pacific seaweed, has been synthesized using LBA-promoted enantioselective cyclization. The cyclization of 4-benzyloxyphenyl farnesyl ether with (*S*)-LBA gives the desired tetracyclic compound as the major diastereomer in 44% ee (eq 13).

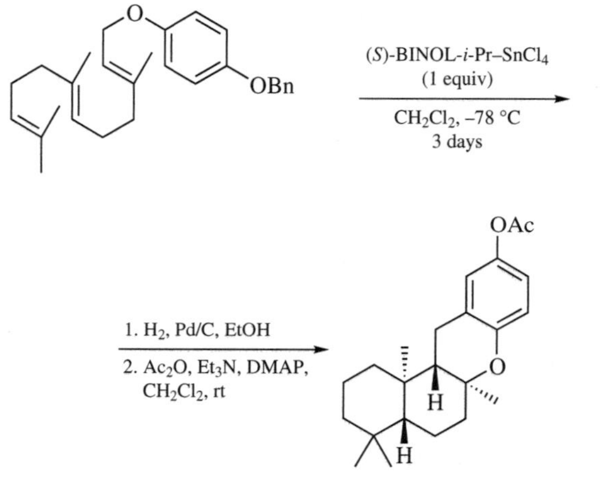

(13)

ca. 40% overall yield, 44% ee

This approach using LBA has been applied to the enantioselective cyclization of homo(polyprenyl)arenes possessing an aryl group that serves as a less-nucleophilic terminator than a hydroxy group.[8] The reaction of 1-homogeranyl-4-(*tert*-butyldiphenylsiloxy)benzene with (*R*)-BINOL-Me–$SnCl_4$ gives the desired tricyclic compound in 13% yield with 72% ee. The other products are monocyclization products. The enantioselectivity of the tricyclic compound is improved to 81% ee when mono(*o*-fluorobenzyl) ether of (*R*)-BINOL [(*R*)-BINOL-*o*-F-Bn] is used instead of BINOL-Me. The desilylation and subsequent diastereoselective cyclization of a crude mixture, which is obtained in the above enantioselective cyclization, gives the desired tricyclic compound in 78% ee and 94% yield in three steps. This compound can be converted to (+)-ferruginol (eq 14).

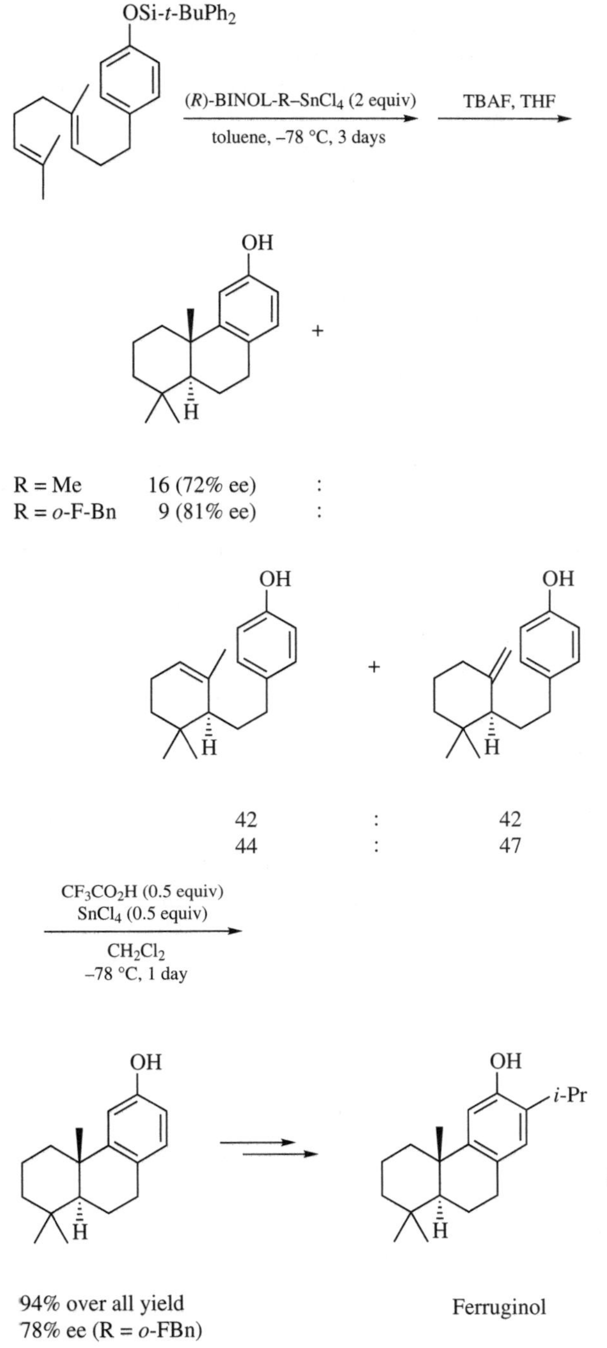

(14)

The enantioselective cyclization of 1-homogeranyl-3-(*tert*-butyldiphenylsiloxy)benzene with use of (*R*)-BINOL-*o*-F-Bn gives *trans*-fused tricyclic compound in 78% ee (*trans* only), together with the monocyclization products. The subsequent diastereoselective cyclization with $BF_3{\cdot}Et_2O$ gives (+)-13-acetoxy-podocarpa-8(14)-en-13-one, a versatile intermediate for synthesis of naturally occurring diterpenes.

Tetracyclic terpene from Eocene Messel shale (Germany) can be also synthesized by using the LBA-induced enantioselective cyclization of 3-homofarnesyltoluene as a key step (eq 16).

Direct Catalytic Enantioselective Aldol Reactions. Yamada and Shibasaki have found that a direct catalytic enantioselective aldol reaction of an aldehyde and unmodified ketone is promoted by a chiral barium complex (5 mol %) prepared from $Ba(O\text{-}i\text{-}Pr)_2$ and 2.5 mol equiv of (*R*)-BINOL-Me.[9] The possible structure of the barium catalyst which plays the role of a Lewis acid and a Brønsted base, has been characterized by LDI-TOFMS, ^{13}C-NMR spectroscopy and extensive studies of reaction conditions (eq 17).

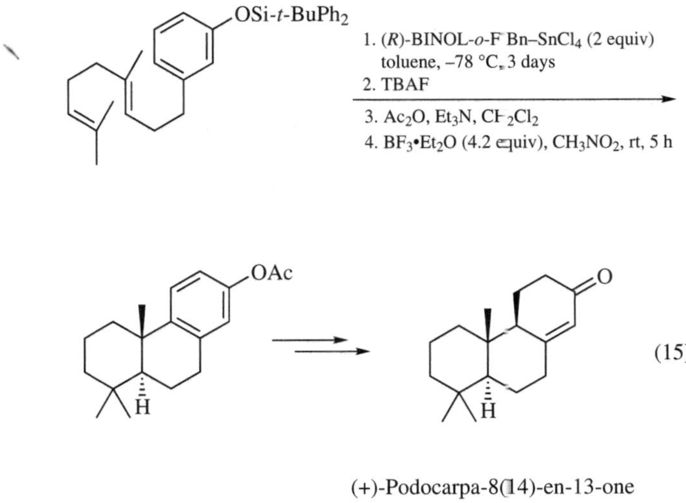

(15)

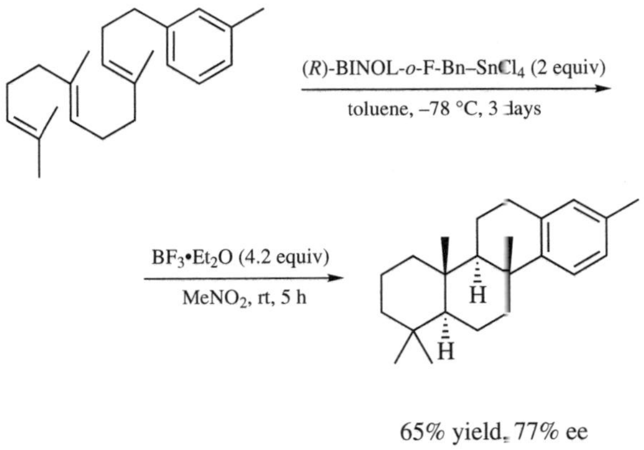

(16)

Enantioselective Intramolecular Cyclization (S_N2' reaction). The desymmetric transformation of *meso*-structures has been recognized as a versatile synthetic method for optically active compounds in organic enzymatic processes. The enantioselective intramolecular cyclization of the bis-phenyllithium species, which is generated by addition of butyllithium to a solution of *cis*-3,5-di(bromophenoxy)cyclopentene, has been attained by addition of lithium salt (1.2 equiv) of (*R*)-BINOL-Me to produce a cyclopenta[*b*]benzofuran with 87% ee (eq 18).[10]

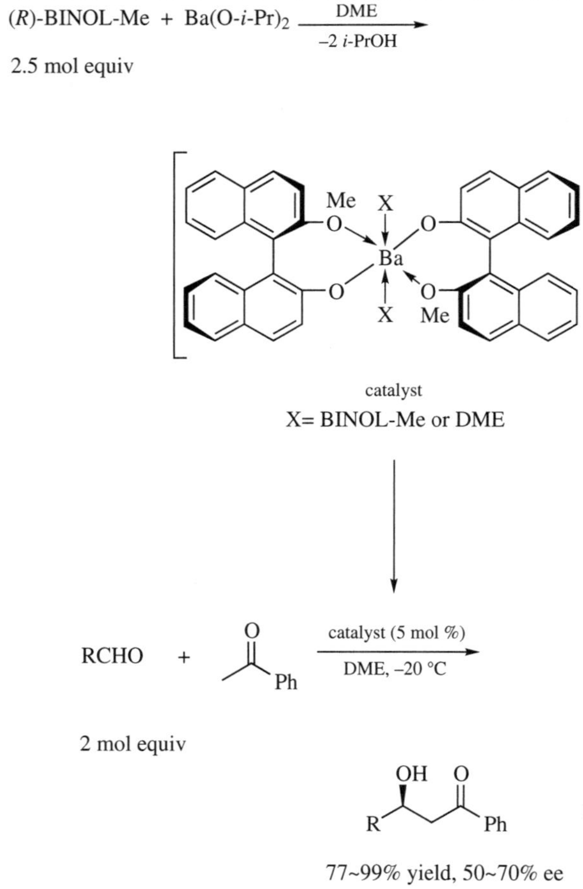

(17)

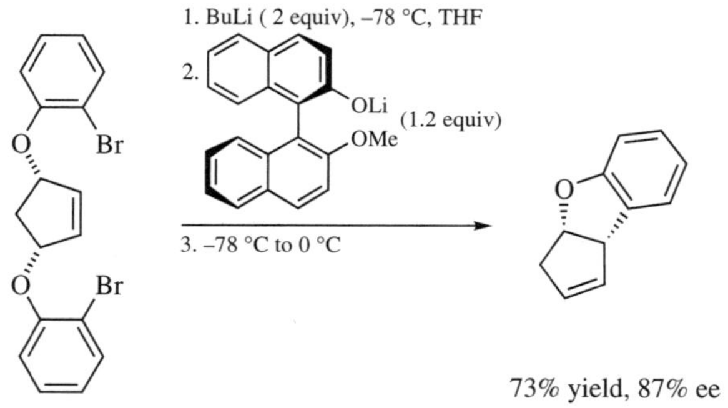

(18)

Related Reagents. (*R*)-2-Isoporpoxy-2′-hydroxy-1,1′-binaphthyl; 2-isopropoxy-2′-hydroxy-1,1′-biphenyl; (*R*)-2-*o*-fluorobenzyloxy-2′-hydroxy-1,1′-binaphthyl; (*R*)-2-benzoxy-2′-hydroxy-1,1′-binaphthyl.

1. (a) Tamai, Y.; Qian, P.; Matsunag, K.; Miyano, S. *Bull. Chem. Soc. Jpn.* **1992**, *65*, 817. (b) Takahashi, M.; Ogasawara, K. *Tetrahedron: Asymmetry* **1997**, *8*, 3125.
2. Fehr, C. *Angew. Chem. Int. Ed. Engl.* **1996**, *35*, 2566.
3. (a) Ishihara, K.; Kaneeda, M.; Yamamoto, H. *J. Am. Chem. Soc.* **1994**, *116*, 11179. (b) Ishihara, K.; Nakamura, S.; Yamamoto, H. *Croat. Chem. Acta* **1996**, *69*, 513. (c) Taniguchi, T.; Ogasawara, K. *Tetrahedron Lett.* **1997**, *38*, 6429. (d) Ishihara, K.; Ishida, Y.; Nakamura S.; Yamamoto, H. *Synlett* **1997**, 758.
4. (a) Ishihara, K.; Nakamura, S.; Kaneeda, M.; Yamamoto, H. *J. Am. Chem. Soc.* **1996**, *118*, 12854. (b) Yanagisawa, A.; Ishihara, K.; Yamamoto, H. *Synlett* **1997**, 411.
5. (a) Stork, G.; Hudrlik, P. F. *J. Am. Chem. Soc.* **1968**, *90*, 4462. (b) House, H. O.; Czuba, L. J.; Gall, M.; Olmstead, H. D. *J. Org. Chem.* **1969**, *34*, 2324.
6. Ishihara, K.; Nakamura, H.; Nakamura, S.; Yamamoto, H. *J. Org. Chem.* **1998**, *63*, 6444.
7. Ishihara, K.; Nakamura, S.; Yamamoto, H. *J. Am. Chem. Soc.* **1999**, *121*, 4907.
8. Ishihara, K.; Ishibashi, H.; Yamamoto, H. *J. Am. Chem. Soc.* **2001**, *123*, 1505.
9. Yamada, Y. M. A.; Shibasaki, M. *Tetrahedron Lett.* **1998**, *39*, 5561.
10. Nishiyama, H.; Sakata, N.; Motoyama, Y.; Wakita, H.; Nagase, H. *Synlett* **1997**, 1147.

Kazuaki Ishihara & Hisashi Yamamoto
Nagoya University, Japan

I

(2,3-*O*-Isopropylidene)-2,3-dihydroxy-1,4-bis(diphenylphosphino)butane[1]

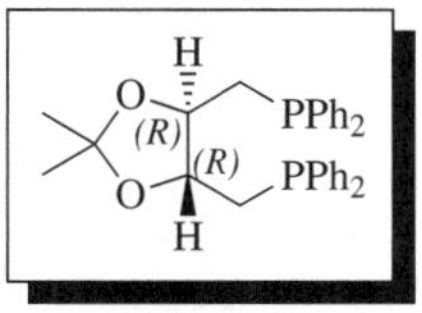

(*R*,*R*)
[*32305-98-9*] $C_{31}H_{32}O_2P_2$ (MW 498.57)
(*S*,*S*)
[*37002-48-5*]

(chiral bidentate phosphine, useful in asymmetric catalysis[1])

Alternate Name: DIOP.
Physical Data: mp 88–89 °C; $[\alpha]_D^{20}$ −12.5° (*c* 4.6, C_6H_6).
Solubility: sol most usual organic solvents.
Form Supplied in: white solid; both enantiomers available.
Preparative Methods: can be prepared in four steps from diethyl tartrate.[2] The two phosphorus groups are introduced in the last step of the reaction sequence using $LiPPh_2$,[2] $KPPh_2$,[3] or $LiP(BH_3)PPh_2$.[4] DIOP has also been prepared from 1,2:3,4-diepoxybutane.[5]
Handling, Storage, and Precautions: air stable.

Introduction. DIOP was the first example of a C_2 chelating diphosphine for transition metal complexes to be used in asymmetric catalysis. It was also one of the first examples of a useful C_2 chiral auxiliary.[6] DIOP can be considered as an example of the first generation of chelating diphosphine ligands with a chiral carbon skeleton, which were followed over the next 20 years by many examples of chelating diphosphines, one of the most efficient of which is BINAP [*2,2′-Bis(diphenylphosphino)-1,1′-binaphthyl*].[1d] The ready availability of DIOP has stimulated research in asymmetric catalysis beyond the area of asymmetric hydrogenation.

Asymmetric Hydrogenation. Conjugated acids (eq 1)[2,7] or various α-*N*-acyldehydroamino acids (eq 2)[2,8,9] are structural units which sometimes give quite high ee's in the presence of rhodium complexes formed in situ [such as Rh(Cl)(cod)(DIOP)] or isolated as cationic complexes, for example $[Rh(cod)(DIOP)]^+\ PF_6^-$. Hydrogenation of *N*-acetamidocinnamic acid using $[Rh(DIOP)_2]^+\ BF_4^-$ instead of $[Rh(cod)(DIOP)]^+\ BF_4^-$ as catalyst (80 °C, 1 bar H_2) gives a slower reaction but with a significant increase in the ee (94% ee instead of 82% ee).[10]

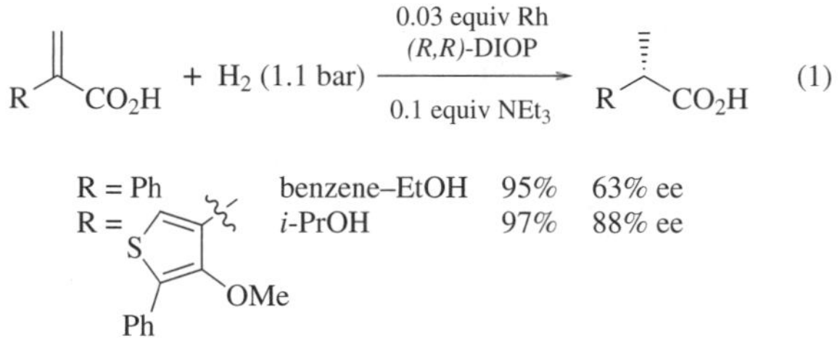

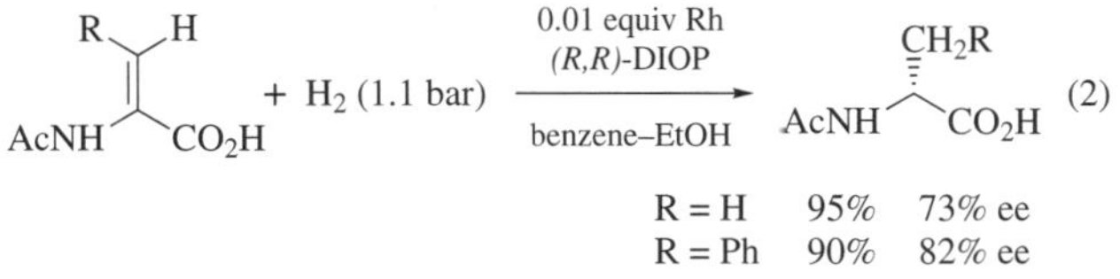

Enamides lacking a carboxy group on the double bond also act as excellent substrates in asymmetric hydrogenations, as exemplified in eq 3.[11]

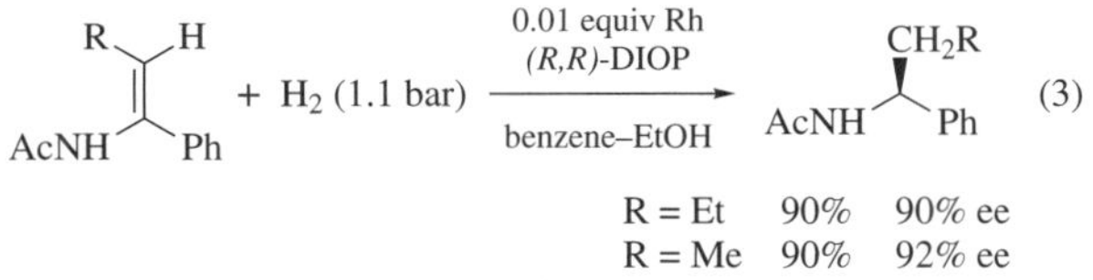

Ketones are known to be quite unreactive in homogeneous hydrogenations catalyzed by rhodium complexes. However, catalytic amounts of a base enhance the reactivity. In this way, acetophenone is hydrogenated to 2-phenylethanol in 80% ee in the presence of $[Rh(Cl)(cod)(DIOP)]/NEt_3$[12] or $N(CH_2OH)_3$.[13] Aromatic α-amino ketones are reduced to the alcohols with high ee's. For example, 2-naphthyl-*N*,*N*-diethylaminoethanol is produced in 95% ee by hydrogenation of the corresponding ketone.[13] Imines are very difficult to hydrogenate in the presence of rhodium catalysts, including [Rh(Cl)(cod)(DIOP)].[1] However, it was recently discovered that iridium complexes with a chiral chelating diphosphine are selective catalysts for the hydrogenation of imines. DIOP gives the best result (63% ee) for the reduction of $RN{=}C(Me)CH_2OMe$ ($R = 2,5\text{-}Me_2C_6H_3$).[14]

Asymmetric Hydrosilylation. Hydrosilylation of ketones catalyzed by chiral metal complexes, followed by hydrolysis, produces enantiomerically enriched alcohols. Rhodium complexes with chiral chelating diphosphines have been used successfully. In this context, DIOP was one of the ligands investigated. Aryl alkyl ketones provide the corresponding alcohols in low ee with Ph_2SiH_2, but α-$NpPhSiH_2$ gives ee's in the range of 50–60%.[12] This silane is also excellent for hydrosilylation of *i*-butyl levulinate (84% ee) and *n*-propyl pyruvate (85% ee).[15] Imines are transformed into amines (ee ≤ 65%) by Ph_2SiH_2 with a rhodium–DIOP catalyst.[16]

Asymmetric Hydroformylation. DIOP was very useful in the early stages of investigation of asymmetric hydroformylation of alkenes in the presence of rhodium or palladium catalysts. The combination of $PtCl_2$(diphosphine) and $SnCl_2$, where the diphosphine is a DIOP derivative (DIPHOL, eq 4), is an excellent system, although requiring high pressures.[17] In the case of the hydroformylation of styrene, the branched aldehyde is the major product. The various hydroformylations or hydroesterifications have been reviewed.[1e] Asymmetric hydroformylation of *N*-acylaminoacrylic acid esters is efficiently catalyzed by $[Rh(CO)(PPh_3)_3]$ + DIOP, giving the branched aldehyde in 60% ee.[18]

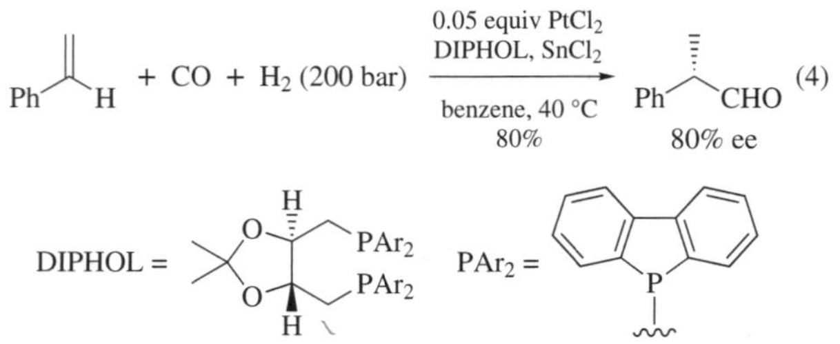

Asymmetric Hydroboration. Hydroborations of alkenes by catecholborane have been catalyzed by [Rh(Cl)(cod)-(diphosphine)].[19] For example, norbornene gives, after oxidation, *exo*-norborneol (82% ee) when a DIOP derivative (2-MeO-DIOP) was used (eq 5). Lower ee's were observed with DIOP, DIPAMP, and BINAP. The effectiveness of DIOP was also noticed in another report.[20]

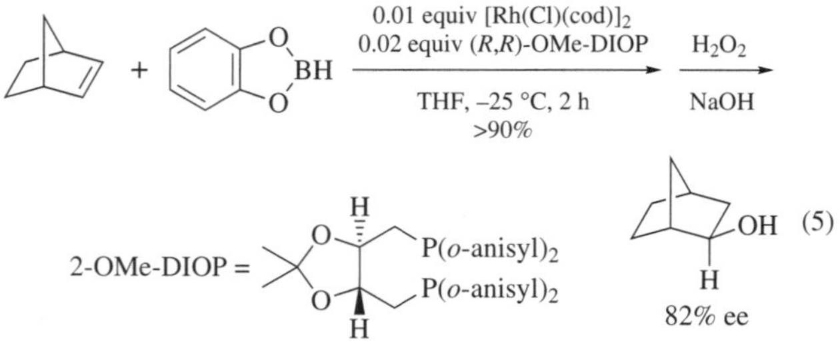

Miscellaneous Reactions. Hydrocyanation of norbornene is catalyzed by $[Pd(DIOP)_2]$, leading to *exo*-2-cyanonorbornane (16% ee), while $[Pd(BINAP)_2]$ gives 40% ee.[21] An asymmetric rearrangement was catalyzed by a nickel(0) complex bearing a diphosphine ligand (eq 6).[22] A DIOP derivative (MOD–DIOP) was more efficient than DIOP or BINAP. MOD–DIOP has previously been found to improve the enantioselectivity, with respect to DIOP, in rhodium-catalyzed hydrogenation of conjugated acids (ee's 90–95%).[23] Structural modifications of DIOP are very easy to perform by changing the nature of the aromatic rings or the acetal group, allowing tuning of the enantioselectivity. Many publications describe modified DIOP derivatives. DIOP has been utilized in several stoichiometric reactions, for example in an intramolecular Wittig reaction for the synthesis of the bis-nor-Wieland–Miescher ketone (52% ee).[24]

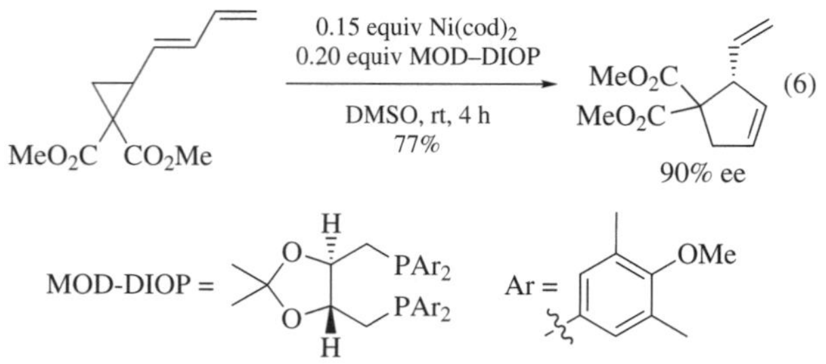

1. (a) Kagan, H. B., In *Comprehensive Organometallic Chemistry*; Wilkinson, G., Ed; Pergamon: Oxford, 1982, Vol. 8, pp 464–498. (b) Kagan, H. B., In *Asymmetric Synthesis*; Morrison, J. D., Ed; Academic: New York, 1985, Vol. 5, pp 1–39. (c) Brunner, H. *Top. Stereochem.* **1988**, *18*, 129. (d) Takaya, H.; Ohta, T.; Noyori, R., In *Catalytic Asymmetric Synthesis*; Ojima, I., Ed; VCH: New York, 1993, pp 1–39. (e) Ojima, I.; Hirai, K., In *Asymmetric Synthesis*; Morrison, J. D., Ed; Academic: New York, 1985, Vol. 5, pp 103–146.
2. Kagan, H. B.; Dang, T. P. *J. Am. Chem. Soc.* **1972**, *94*, 6429.
3. Murrer, B. A.; Brown, J. M.; Chaloner, P. A.; Nicholson, P. N.; Parker, D. *Synthesis* **1979**, 350.
4. Brisset, H.; Gourdel, Y.; Pellon, P.; Le Carre, M. *Tetrahedron Lett.* **1993**, *34*, 4523.
5. Zhang, S. Q.; Zhang, S. Y.; Feng, R. *Tetrahedron: Asymmetry* **1991**, *2*, 173.
6. Whitesell, J. K. *Chem. Rev.* **1989**, *89*, 1581.
7. Stoll, A. P.; Süess, R. *Helv. Chim. Acta* **1974**, *57*, 2487.
8. Gelbard, G.; Kagan, H. B.; Stern, R. *Tetrahedron* **1976**, *32*, 233.
9. Townsend, J. M.; Blount, J. F.; Sun, R. C.; Zawoiski, S.; Valentine, D., Jr. *J. Org. Chem.* **1980**, *45*, 2995.
10. James, B. R.; Mahajan, D. *J. Organomet. Chem.* **1985**, *279*, 31.
11. Sinou, D.; Kagan, H. B. *J. Organomet. Chem.* **1976**, *114*, 325.
12. Bakos, J.; Toth, I.; Heil, B.; Marko, L. *J. Organomet. Chem.* **1985**, *279*, 23.
13. Chan, A. S. C.; Landis, C. R. *J. Mol. Catal.* **1989**, *49*, 165.
14. Chan, Y. N. C.; Osborn, J. A. *J. Am. Chem. Soc.* **1990**, *112*, 9400.
15. Ojima, I.; Kogure, T.; Kumagai, M. *J. Org. Chem.* **1977**, *42*, 1671.
16. Kagan, H. B.; Langlois, N.; Dang, T. P. *J. Organomet. Chem.* **1975**, *90*, 353.
17. Consiglio, G.; Pino, P.; Flowers, L. I.; Pittman, C. U., Jr. *Chem. Commun.* **1983**, 612.
18. Gladiali, S.; Pinna, L. *Tetrahedron: Asymmetry* **1991**, *2*, 623.
19. Burgess, K.; van der Donk, W. A.; Ohlmeyer, M. J. *Tetrahedron: Asymmetry* **1991**, *2*, 613.
20. Sato, M.; Miyaura, N.; Suzuki, A. *Tetrahedron Lett.* **1990**, *31*, 231.
21. (a) Hodgson, M.; Parker, D.; Taylor, R. J.; Ferguson, G. *Organometallics* **1988**, *7*, 1761. (b) Elmes, P. S.; Jackson, W. R. *Aust. J. Chem.* **1982**, *35*, 2041.
22. Hiroi, K.; Arinaga, Y.; Ogino, T. *Chem. Lett.* **1992**, 2329.
23. Morimoto, T.; Chiba, M.; Achiwa, K. *Tetrahedron Lett.* **1989**, *30*, 735.
24. Trost, B. M.; Curran, D. P. *Tetrahedron Lett.* **1981**, *22*, 4929.

Henri Kagan
Université de Paris-Sud, Orsay, France

L

Lanthanum(III)-Lithium-BINOL Complex [(*R*)-LLB and (*S*)-LLB]

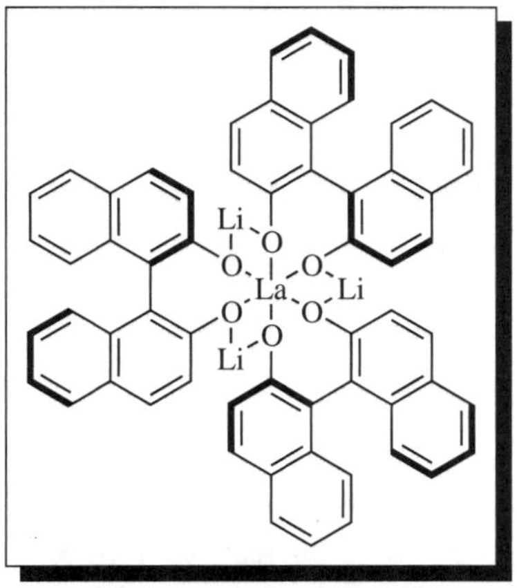

[161444-03-7]	$C_{60}H_{36}LaLi_3O_6$	(MW 1027.69)
[151736-98-0]	$C_{60}H_{36}LaLi_3O_6$	(MW 1027.69)

(heterobimetallic catalysts used for enantioselective organic transformations including carbon-carbon bond-forming reactions)[1–7]

Alternate Name: (*R*)-LLB [lithium tris{(1*R*)-[1,1′-binaphthalene]-2,2′-diolato(2−)-*O*, *O*′}-lanthanate(3−)], (*S*)-LLB[lithium tris {(1*S*)-[1,1′-binaphthalene]-2,2′-diolato(2−)-*O*, *O*′}-lanthanate(3−)], $LaLi_3$tris[(*R*)-binaphthoxide], Li_3(La[(*R*)-binol]$_3$).

Solubility: soluble in THF, toluene, and dichloromethane.

Form Supplied in: pale yellow solution in THF.

Analysis of Reagent Purity: ^{1}H NMR or ^{13}C NMR (not applicable to paramagnetic lanthanides).

Preparative Methods: several different procedures are available. The most common one employs lanthanum(III) triisopropoxide [La(O-*i*-Pr)$_3$] as a lanthanum source (eq 1).[8–10] Although this method (Procedure A) produces the complex in high yield, the availability of La(O-*i*-Pr)$_3$ might be limited. La(O-*i*-Pr)$_3$ is also sensitive to air and moisture, and its purity and activity vary depending on the provider (we purchase the ampules from Kojundo Chemical Laboratory Co., Ltd., Japan). Use of the alternative method (Procedure B) can avoid these problems (eq 2):[11,12] lanthanum(III) trichloride heptahydrate ($LaCl_3 \cdot 7H_2O$) is employed as a lanthanum source, which is much more accessible and much less expensive than La(O-*i*-Pr)$_3$. It is essential to use the hydrate of lanthanum chloride; otherwise the catalyst will not be formed efficiently in the absence of H_2O. Although Procedures A and B generate identical catalytic species, referral to the corresponding literature is recommended to determine which preparative method to adopt to obtain the best results.

Purification: used as prepared.

Handling, Storage, and Precautions: store under argon atmosphere at ambient temperature. Stable in the presence of small amounts of water. Although the color of the solution might darken, the catalytic activity is preserved for over 6 months.

Procedure A

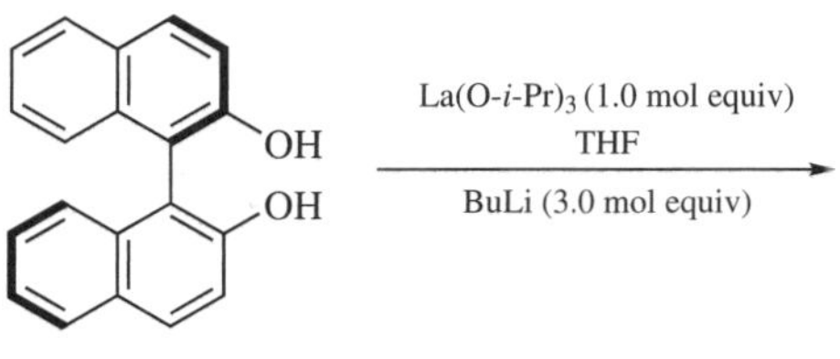

(3.0 mol equiv) (1)

Procedure B

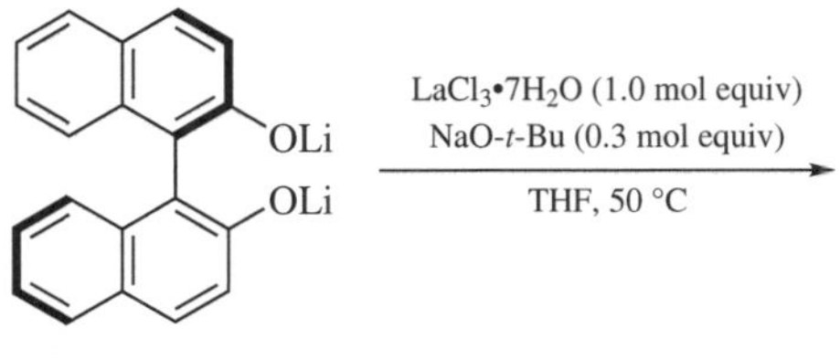

(2.7 mol equiv) (2)

Disposal: quench by the addition of diluted aqueous HCl. Dispose the aqueous layer after evaporation of THF and extraction of binaphthol with ethyl acetate.

Catalytic Enantioselective (Diastereoselective) Nitroaldol Reactions. LLB possesses both Lewis acidity and Brønsted basicity, and the synergistic effects of the two functions provide unique catalytic activity. The unreactive substrates that bear moderately acidic hydrogens undergo deprotonation to generate reactive nucleophiles. Nitromethane forms the corresponding nitronate in the presence of catalytic amounts of LLB, and reacts with a variety of aliphatic and aromatic aldehydes to afford nitroaldol adducts which are attractive building blocks for the synthesis of biologically important compounds.[1,8,11–21] For example, treatment of α-aryloxy aldehydes with nitromethane in the presence of LLB efficiently gives synthetic intermediates for β-blockers (eq 3).[14,16,17] The enantioselectivity of these processes can be optimized by choosing the proper lanthanide. Whereas LLB (Li_3[La(binol)$_3$]), a lanthanum (La)-based catalyst, produces the best results for aliphatic aldehydes, the use of samarium (Sm) or gadolinium (Gd) instead of La is more effective for aromatic aldehydes (eq 4).[21] Other nitroalkanes such as nitroethane and nitroethanol (eq 5) are also applicable, and derivatization of the BINOL (in LLB) at the 6- and 6′-positions improves the stereoselectivity.[18] Notably, the addition of catalytic amounts of BuLi and H_2O significantly accelerates the reactions (second-generation LLB), so that the catalyst loading can be decreased to 1 mol % (eq 5).[19] A tandem inter-intramolecular nitroaldol reaction forms a bicyclic compound with multiple newly formed stereogenic centers (eq 6).[20]

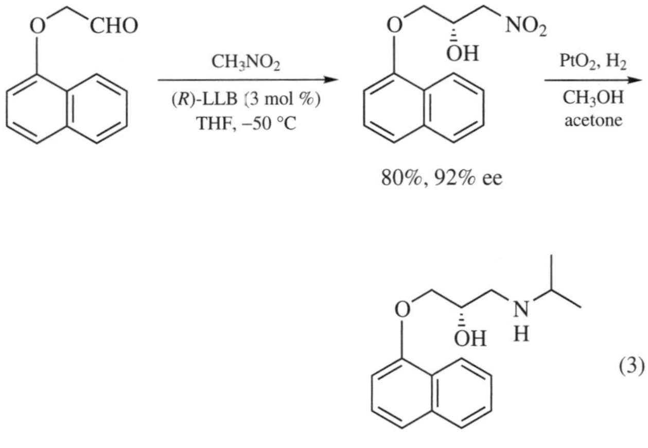

(3)

(*S*)-propranolol
90%, 92% ee

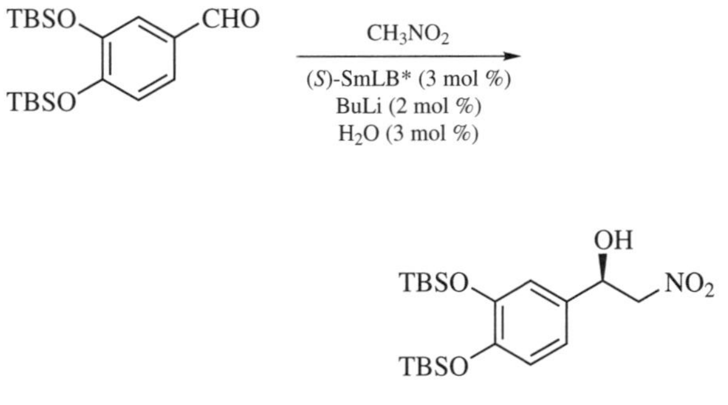

(4)

93%, 92% ee

SmLB* = $Li_3[Sm(binol^*)_3]$
[H_2binol* = 6,6'-bis(trimethylsilylethynyl)BINOL]

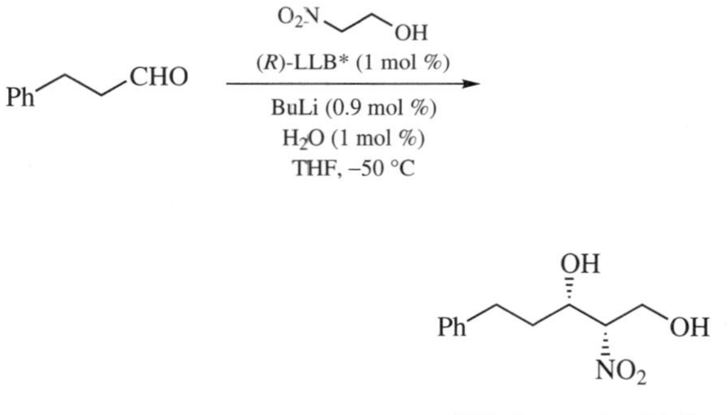

(5)

76% (*syn*:*anti* = 94:6)
syn = 96% ee

(*R*)-LLB* = $Li_3[La(R\text{-}binol^*)_3]$
[H_2binol* = 6,6'-bis(triethylsilylethynyl)BINOL]

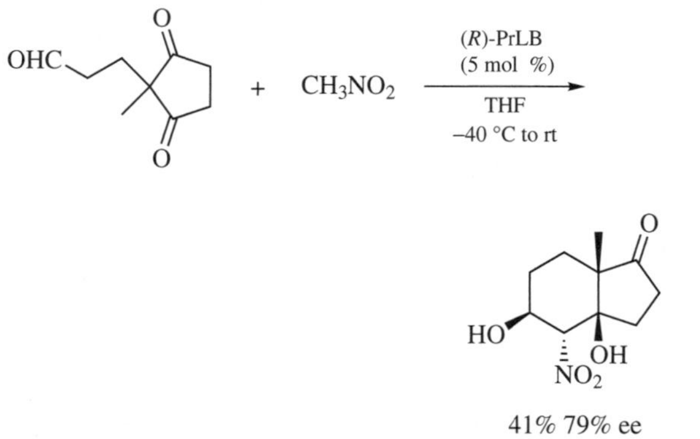

(6)

41% 79% ee
(isolated yield)

Catalytic Enantioselective Hydrophosphonylation of Aldehydes. LLB catalyzes the hydrophosphonylations of aldehydes with dimethyl phosphite to afford α-hydroxy phosphonates with high optical purity (eq 7).[22–24] In some cases, the aldehyde needs to be added slowly to the mixture of LLB and phosphite in THF. For some aromatic aldehydes, another catalyst, Li[Al(binol)$_2$] (ALB), gives better results.[25] Imines also react with dimethyl phosphite in a highly enantioselective manner when potassium-based complexes ($K_3[Ln(binol)_3]$, LnPB) are used as catalysts.[26–28]

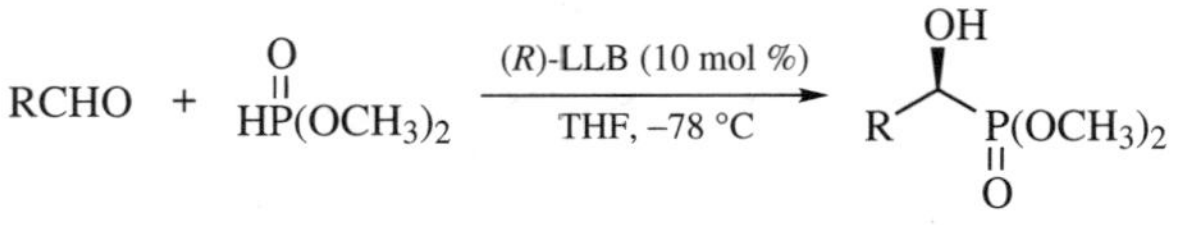

(7)

up to 94%,
up to 92% ee

Enantioselective Diels–Alder Reaction. LLB possesses moderate Lewis acidity sufficient to promote Diels–Alder reactions. Similar to the nitroaldol reactions, the selectivity can be optimized by introducing substituents at the 6- and 6′-positions of BINOL. The cycloaddition of cyclopentadiene and *N*-acryloyloxazolidinone is catalyzed by LLB (10 mol %) to afford the corresponding adduct in excellent yield with up to 86% ee (*endo*:*exo* = 36:1) (eq 8).[29]

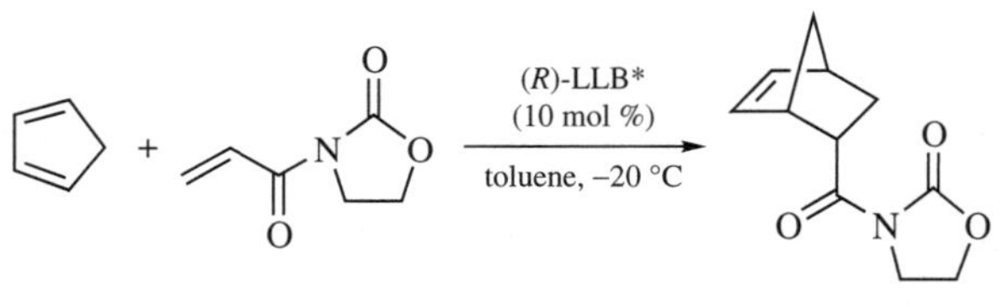

(8)

100%, *endo*:*exo* = 36:1
endo = 86% ee

LLB* = $Li_3[La(binol^*)_3]$
(H_2binol* = 6,6'-dibromo-BINOL)

Direct Catalytic Enantioselective Aldol Reactions. The unique catalytic activity of LLB enables aldol reactions to be performed without preformation of enolates (direct aldol reaction).[30–32] The reaction of methyl ketones such as acetone or acetophenone (eq 9) with aldehydes gives β-hydroxy ketones as the product,[9,31] and the reaction of α-hydroxy acetophenones produces α,β-dihydroxy ketones with excellent enantiomeric excess (eq 10).[10,33] In contrast to Mukaiyama-type aldol reactions,[34] ketones with a free hydroxyl group are suitable substrates in this direct reaction. The addition of KOH (generated in situ from KHMDS and H_2O) is required to enhance the catalytic activity (heteropolymetallic catalyst) in direct aldol reactions. β-Hydroxy-α-amino acid esters can also be synthesized using glycinate Schiff bases as donor substrates (eq 11).[35]

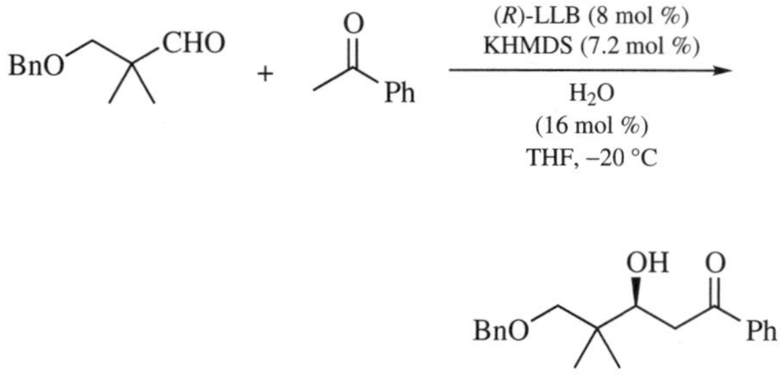

(9)

91%, 90% ee

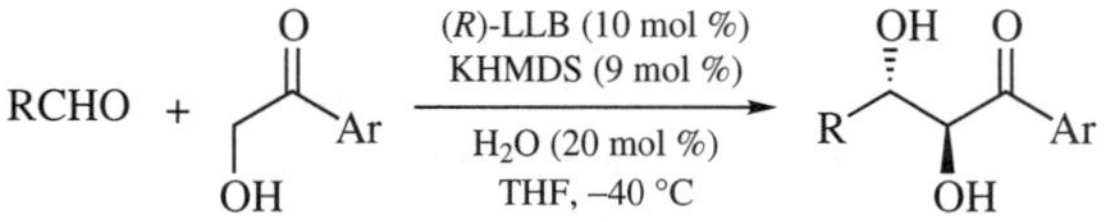

(10)

yield: up to 96%,
dr: up to 83:17
ee: up to 98% (*anti*)

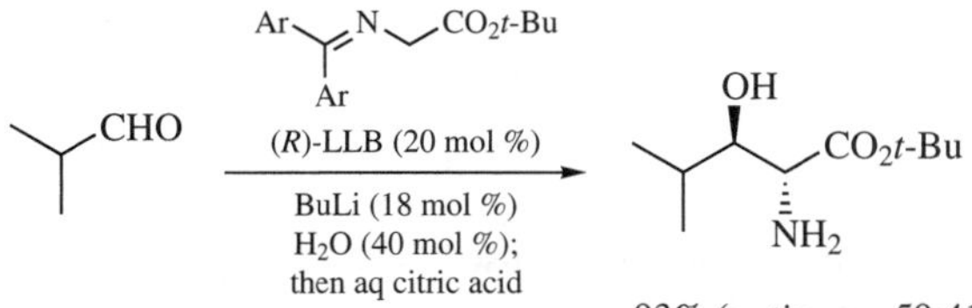

(11)

93% (*anti*:*syn* = 59:41)
anti = 74% ee, *syn* = 20% ee

Related Reagents. $Li_3\{La[6,6'\text{-bis(triethylsilylethynyl)binol}]_3\}$ (LLB*); $Li_3\{La[6,6'\text{-bis(trimethylsilylethynyl)binol}]_3\}$ (LLB*); $Li_3\{La[(6,6'\text{-dibromo})binol]_3\}$ (LLB*); $Li_3[Pr(binol)_3]$ (PrLB); $Li_3[Sm(binol)_3]$ (SmLB).

1. Sasai, H.; Suzuki, T.; Arai, S.; Arai, T.; Shibasaki, M. *J. Am. Chem. Soc.* **1992**, *114*, 4418–4420.
2. Shibasaki, M.; Yoshikawa, N. *Chem. Rev.* **2002**, *102*, 2187–2209.
3. Shibasaki, M. *Enantiomer* **1999**, *4*, 513–527.
4. Shibasaki, M. *Chemtracts-Org. Chem.* **1999**, *12*, 979–998.
5. Shibasaki, M.; Iida, T.; Yamada, Y. M. A. *J. Synth. Org. Chem. Jpn.* **1998**, *56*, 344–356.
6. Shibasaki, M.; Sasai, H.; Arai, T.; Iida, T. *Pure & Appl. Chem.* **1998**, *70*, 1027–1034.
7. Shibasaki, M.; Sasai, H.; Arai, T. *Angew. Chem., Int. Ed. Engl.* **1997**, *36*, 1236–1256.
8. Sasai, H.; Suzuki, T.; Itoh, N.; Tanaka, K.; Date, T.; Okamura, K.; Shibasaki, M. *J. Am. Chem. Soc.* **1993**, *115*, 10372–10373.
9. Yoshikawa, N.; Yamada, Y. M. A.; Das, J.; Sasai, H.; Shibasaki, M. *J. Am. Chem. Soc.* **1999**, *121*, 4168–4178.
10. Yoshikawa, N.; Suzuki, T.; Shibasaki, M. *J. Org. Chem.* **2002**, *67*, 2556–2565.
11. Sasai, H.; Suzuki, T.; Itoh, N.; Shibasaki, M. *Tetrahedron Lett.* **1993**, *34*, 851–854.
12. Sasai, H.; Watanabe, S.; Shibasaki, M. *Enantiomer* **1997**, *2*, 267–271.
13. Sasai, H.; Suzuki, T.; Itoh, N.; Arai, S.; Shibasaki, M. *Tetrahedron Lett.* **1993**, *34*, 2657–2660.
14. Sasai, H.; Itoh, N.; Suzuki, T.; Shibasaki, M. *Tetrahedron Lett.* **1993**, *34*, 855–858.
15. Sasai, H.; Kim, W. S.; Suzuki, T.; Shibasaki, M. *Tetrahedron Lett.* **1994**, *35*, 6123–6126.
16. Sasai, H.; Yamada, Y. M. A.; Suzuki, T.; Shibasaki, M. *Tetrahedron* **1994**, *50*, 12313–12318.
17. Sasai, H.; Suzuki, T.; Itoh, N.; Shibasaki, M. *Appl. Organomet. Chem.* **1995**, *9*, 421–426.
18. Sasai, H.; Tokunaga, T.; Watanabe, S.; Suzuki, T.; Itoh, N.; Shibasaki, M. *J. Org. Chem.* **1995**, *60*, 7388–7389.
19. Arai, T.; Yamada, Y. M. A.; Yamamoto, N.; Sasai, H.; Shibasaki, M. *Chem. Eur. J.* **1996**, *2*, 1368–1372.
20. Sasai, H.; Hiroi, M.; Yamada, Y. M. A.; Shibasaki, M. *Tetrahedron Lett.* **1997**, *38*, 6031–6034.
21. Takaoka, E.; Yoshikawa, N.; Yamada, Y. M. A.; Sasai, H.; Shibasaki, M. *Heterocycles* **1997**, *46*, 157–163.
22. Yokomatsu, T.; Yamagishi, T.; Shibuya, S. *Tetrahedron: Asymmetry* **1993**, *4*, 1783–1784.
23. Rath, N. P.; Spilling, C. D. *Tetrahedron Lett.* **1994**, *35*, 227–230.
24. Sasai, H.; Bougauchi, M.; Arai, T.; Shibasaki, M. *Tetrahedron Lett.* **1997**, *38*, 2717–2720.
25. Arai, T.; Bougauchi, M.; Sasai, H.; Shibasaki, M. *J. Org. Chem.* **1996**, *61*, 2926–2927.
26. Sasai, H.; Arai, S.; Tahara, Y.; Shibasaki, M. *J. Org. Chem.* **1995**, *60*, 6656–6657.
27. Gröger, H.; Saida, Y.; Arai, S.; Martens, J.; Sasai, H.; Shibasaki, M. *Tetrahedron Lett.* **1996**, *37*, 9291–9292.
28. Gröger, H.; Saida, Y.; Sasai, H.; Yamaguchi, K.; Martens, J.; Shibasaki, M. *J. Am. Chem. Soc.* **1998**, *120*, 3089–3103.
29. Morita, T.; Arai, T.; Sasai, H.; Shibasaki, M. *Tetrahedron: Asymmetry* **1998**, *9*, 1445–1450.
30. Alcaide, B.; Almendros, P. *Eur. J. Org. Chem.* **2002**, 1595–1601.
31. Shibasaki, M.; Yoshikawa, N.; Matsunaga, S. In *Comprehensive Asymmetric Catalysis*; Jacobsen, E.; Pfaltz, A.; Yamamoto, H., Eds.; Springer: New York; In press.
32. Yamada, Y. M. A.; Yoshikawa, N.; Sasai, H.; Shibasaki, M. *Angew. Chem., Int. Ed. Engl.* **1997**, *36*, 1871–1873.
33. Yoshikawa, N.; Kumagai, N.; Matsunaga, S.; Moll, G.; Ohshima, T.; Suzuki, T.; Shibasaki, M. *J. Am. Chem. Soc.* **2001**, *123*, 2466–2467.
34. Mukaiyama, T.; Uchiro, H.; Shiina, I.; Kobayashi, S. *Chem. Lett.* **1990**, 1019–1022.
35. Yoshikawa, N.; Shibasaki, M. *Tetrahedron*, In press.

Masakatsu Shibasaki & Naoki Yoshikawa
Graduate School of Pharmaceutical Sciences, The University of Tokyo, Japan

t-Leucine *t*-Butyl Ester[1]

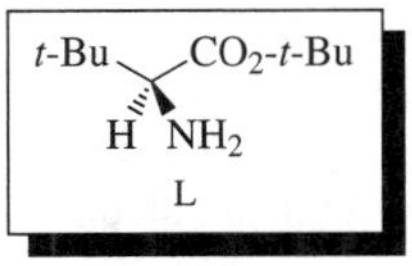

(D)
[61169-85-5] $C_{10}H_{21}NO_2$ (MW 187.32)
(L)
[31556-74-8]
(DL)
[99285-38-8]

(chiral auxiliary used in asymmetric alkylations,[1] 1,2-additions,[2] and 1,4-additions[3] of aldehyde and ketone derived Schiff bases)

Physical Data: D,[4a] $[\alpha]_D^{20}$ −56° bp 90–91 °C/21 mmHg. L,[4a] $[\alpha]_D^{20}$ +51.7° (92.3% ee); bp 87–89 °C/18 mmHg.

Solubility: sol hexanes, benzene.

Preparative Methods: prepared by esterification of *t*-leucine with *Isobutene* and conc *Sulfuric Acid* (under pressure).[4] Typical yields are 62–64% and 12–14% of recovered amino acid. *t*-Leucine itself is commercially available in racemic and optically pure forms. It can also be prepared by oxidation of pinacolone to trimethylpyruvic acid, followed by oxime formation and zinc reduction. Resolution of the *N*-formyl derivative of *t*-leucine has been carried out using brucine.[4a]

Handling, Storage, and Precautions: none.

Asymmetric Alkylations. The use of nitrogen derivatives of carbonyl compounds (imines, imides, amides, sultams, oxazolines) is often the most efficient procedure for achieving α-alkylations.[1] Chiral auxiliaries bearing heteroatoms in a 1,2-relationship appear to work best, as they have chelation sites for the metal cation. High levels of asymmetric induction can thus be achieved due to the system rigidity. Cyclic ketones have been alkylated via the lithiated enamine formed from L-*t*-leucine *t*-butyl ester (eq 1).[5] High enantiomeric excesses and predictability of absolute configuration make this method attractive.

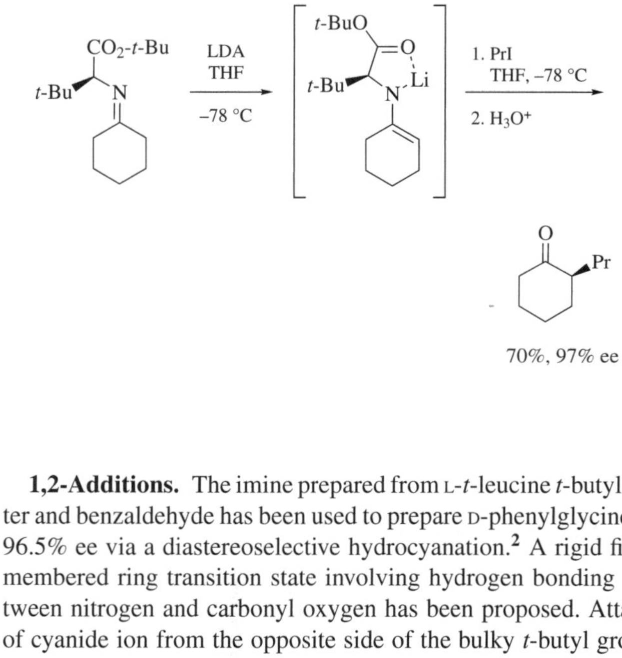

(1)

1,2-Additions. The imine prepared from L-*t*-leucine *t*-butyl ester and benzaldehyde has been used to prepare D-phenylglycine in 96.5% ee via a diastereoselective hydrocyanation.[2] A rigid five-membered ring transition state involving hydrogen bonding between nitrogen and carbonyl oxygen has been proposed. Attack of cyanide ion from the opposite side of the bulky *t*-butyl group accounts for the stereochemical outcome (eq 2).

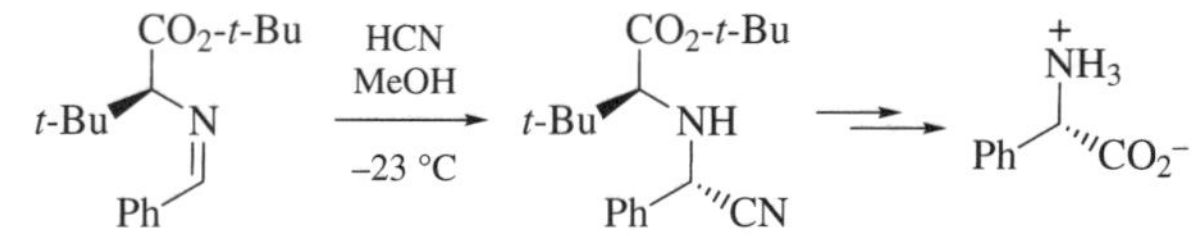

(2)

1,4-Additions. Asymmetric Michael additions[1] of Grignard reagents can be performed on α,β-unsaturated aldimines[3a] derived from either enantiomer of *t*-leucine *t*-butyl ester (eq 3).

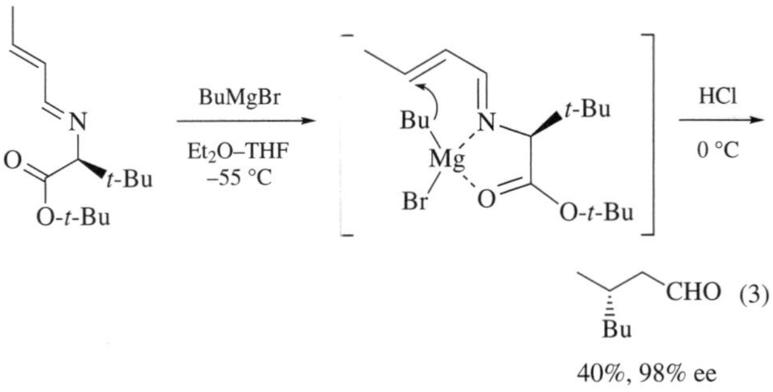

(3)

Similarly, malonate anions[3b,c] add to aldimines with reasonably high enantioselectivity. The new asymmetric center, however, has the opposite absolute configuration to that shown in eq 3. A chelated aldimine of (*E*) geometry is the proposed intermediate for this reversal of stereoselection (eq 4).

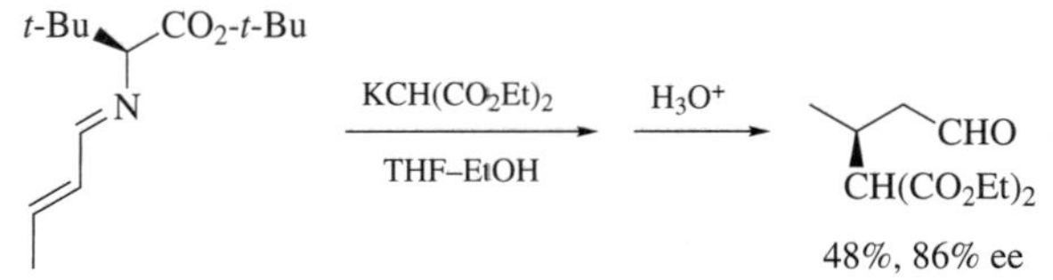

(4)

Cyclic aldimines can also be used, and subsequent alkylation of the magnesioenamine intermediate achieved with good to excellent diastereoselectivity.[3d,e,f] *Cis* or *trans* products can be obtained, depending on the procedure chosen (eq 5).

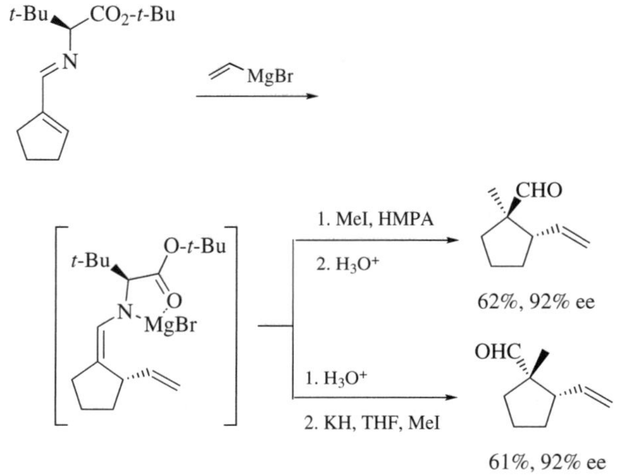

(5)

A few natural product syntheses feature the use of both acyclic[6] and cyclic[7,8] aldimines of either enantiomer of *t*-leucine *t*-butyl ester. Kinetic resolution of racemic aldehydes has also been achieved using L-*t*-leucine *t*-butyl ester.[8]

For the three types of reactions presented above, *t*-leucine *t*-butyl ester has been shown to be the most efficient amino acid derivative. It is often mentioned that valine *t*-butyl ester affords lower enantioselectivities. Work-up procedures allow recovery of reusable optically pure auxiliary.

1. (a) Ager, D. J.; East, M. B. *Tetrahedron* **1992**, *48*, 2803 and references cited therein. (b) ApSimon, J. W.; Lee Collier, T. *Tetrahedron* **1986**, *42*, 5157. (c) Tomioka, K.; Koga, K. In *Asymmetric Synthesis*; Academic: New York, 1983; Vol. 2. (d) Coppola, G. M.; Schuster, H. F. In *Asymmetric Synthesis*; Wiley: New York, 1987; Chapter 4.

2. Yamada, S.; Hashimoto, S. *Chem. Lett.* **1976**, 921.
3. (a) Hashimoto, S.; Yamada, S.; Koga, K. *J. Am. Chem. Soc.* **1976**, *98*, 7450. (b) Hashimoto, S.; Komeshima, N.; Yamada, S.; Koga, K. *Tetrahedron Lett.* **1977**, 2907. (c) Yamada, S.; Komeshima, N.; Yamada, S.; Koga, K. *Chem. Pharm. Bull.* **1979**, *27*, 2437. (d) Hashimoto, S.; Kogen, H.; Tomioka, K.; Koga, K. *Tetrahedron Lett.* **1979**, 3009. (e) Kogen, H.; Tomioka, K.; Hashimoto, S.; Koga, K. *Tetrahedron Lett.* **1980**, 4005. (f) Kogen, H.; Tomioka, K.; Hashimoto, S.; Koga, K. *Tetrahedron* **1981**, *37*, 3951.
4. (a) Hashimoto, S.; Yamada, S.; Koga, K. *Chem. Pharm. Bull.* **1979**, *27*, 771. (b) Roeske, R. W. *Chem. Ind. (London)* **1959**, 1121.
5. (a) Hashimoto, S.; Koga, K. *Tetrahedron Lett.* **1978**, 573. (b) Hashimoto, S.; Koga, K. *Chem. Pharm. Bull.* **1979**, *27*, 2760.
6. (a) Whittaker, M.; McArthur, C. R.; Leznoff, C. C. *Can. J. Chem.* **1985**, *63*, 2844. (b) Muraoka, O.; Fujiwara, N.; Tanabe, G.; Momose, T. *Tetrahedron: Asymmetry* **1991**, *2*, 357.
7. (a) Tomioka, K.; Masumi, F.; Yamashita, T.; Koga, K. *Tetrahedron Lett.* **1984**, *25*, 333. (b) Tomioka, K.; Masumi, F.; Yamashita, T.; Koga, K. *Tetrahedron* **1989**, *45*, 643.
8. (a) Paquette, L. A.; Macdonald, D.; Anderson, L. G.; Wright, J. *J. Am. Chem. Soc.* **1989**, *111*, 8037. (b) Paquette, L. A.; Macdonald, D.; Anderson, L. G. *J. Am. Chem. Soc.* **1990**, *112*, 9292. (c) Snider, B. B.; Yang, K. *Tetrahedron Lett.* **1989**, *30*, 2465. (d) Snider, B. B.; Yang, K. *J. Org. Chem.* **1990**, *55*, 4392.

Alyx-Caroline Guével
The Ohio State University, Columbus, OH, USA

Lipases[1]

[9001-62-1] (MW 30000–100000)

(catalyst for asymmetric transformations of chiral or prochiral alcohols or acids by hydrolysis or esterification reactions[1])

Solubility: powder sol aqueous solutions; suspension in organic media.
Form Supplied in: usually a white or brownish powder, but also immobilized on an appropriate support. Lipases from microbial sources are virtually homogeneous in terms of hydrolytic activity, while mammalian and plant lipase preparations contain several interfering enzymes including proteases and esterases.
Analysis of Reagent Purity: assay by titrimetry.[2]
Handling, Storage, and Precautions: must be stored in a refrigerator at 0–5 °C. Avoid breathing or inhaling dust. Avoid too vigorous stirring.

Lipase General Aspects. Enzymes are now widely recognized as practical catalysts for asymmetric synthesis.[1b,c] Lipases are among the most widely applied and versatile biocatalysts in organic synthesis as can be witnessed by a number of recent reviews.[1] There are several reasons for this. They are readily available, do not require cofactors, are inexpensive and highly stable, exhibit broad substrate specificity, do not require water-soluble substrates, mechanistically are relatively well understood and, finally, are splendidly suited to retain a high degree of activity in organic media.

More than 20 lipases are now commercially available, either free or immobilized, from animal, plant, and microbial sources.[2] Amongst the lipases, the pig pancreatic lipase (PPL), the yeast lipase from *Candida cylindracea* (*rugosa*) (CCL), and the bacteria lipases from *Pseudomonas fluorescens* (*cepecia*) (PFL) and other unclassified *Pseudomonas* species (PSL) have been most widely used. The experimental methods are very straightforward and little different in their execution from conventional chemical reactions. Hydrolysis reactions are conducted on the soluble lipase in buffered aqueous solutions, commonly in the presence of an organic cosolvent. In organic media the enzyme is added as a powder or in an immobilized form and the resulting suspension stirred or (better) shaken at approximately 40 °C. The enzyme is removed by filtration.

Since their action toward substrates in terms of chemo-, regio-, and enantioselectivity varies considerably, it is important to have a large selection of lipases to find the right enzyme for a specific reaction by traditional biocatalyst screening. Alternative strategies for improving enantioselectivity of the already existing commercial lipases have been developed,[1a,3] including product recycling,[4] solvent screening,[5] water content control,[6] and immobilization. In addition to this, several active site models have been proposed to predict the enantiopreference of certain lipases.[1a,7]

Lipases have been used in three main types of asymmetric transformations: kinetic resolution of racemic carboxylic acids or alcohols, enantioselective group differentiations of *meso* dicarboxylic acids or diols, and enantiotopic group differentiation of prochiral dicarboxylic acid and diol derivatives. Hydrolysis has been the most widely used technique, but complementary esterification or transesterification procedures are increasingly coming into use. Lipases are used most frequently in transformations involving chiral alcohols rather than acids, unlike the pig liver esterase (PLE),[1a,b,c] which is most frequently employed on esters of chiral carboxylic acids. Lipases are also gaining increasing importance in solving problems of regioselectivity of various polyol and carbohydrate compounds.[1c,e,f,8] They have found application in stereoselective transformations involving lactonization and oligomerization of hydroxy acids and esters.[1e,f] Finally, a minor but useful advantage of the lipases is their mildness, which is particularly important in transformations involving labile compounds.[1a]

The range of nucleophiles that lipases accept is not confined to water or alcohols. There are numerous examples of amines,[9] hydrazine,[10] phenols,[11] and hydrogen peroxide.[12] Proteases have frequently been used in biocatalytic transformations involving ester hydrolysis and esterification reactions and their different stereoselection often provides a useful complement to the lipases.[1a,c,f]

Kinetic resolution of racemic compounds is by far the most common transformation catalyzed by lipases, in which the enzyme discriminates between the two enantiomeric constituents of a racemic mixture. It is important to note that the maximum yield of a kinetic resolution is restricted to 50% for each enantiomer based on the starting material. The prochiral route and transformations involving *meso* compounds, 'the *meso*-trick', have the advantage of potentially obtaining a 100% yield of pure enantiomer. A theoretical quantitative analysis of the kinetics involved in the biocatalytic processes described above has been developed.[1a,d,e] The enantiomeric ratio (E), an index of enantioselectivity, can be calculated from the extent of conversion and the corresponding enantiomeric excess (ee) values of either the product or the remaining substrate. The results reveal that for an irreversible process, such as hydrolysis, the optimum in both chemical and optical yield for the faster hydrolyzed enantiomer is to be expected near 40% conversion, and for the remaining slower hydrolyzed enantiomer around 60% conversion. For a high enantiomeric ratio (>100), high enantioselectivity is expected for both enantiomers at 50% conversion.

Under almost anhydrous conditions in organic medium,[1e,f] lipases can be used in the reverse mode for direct ester synthesis from carboxylic acids and alcohols, as well as transesterifications (acyl transfer reactions) which can be divided into alcoholysis (ester and alcohol), acidolysis (ester and acid), and interesterification (ester–ester interchange). The direct esterification and alcoholysis in particular have been most frequently used in asymmetric transformations involving lipases. The parameters that influence enzymatic catalysis in organic solvents have been intensively studied and discussed.[1a,e,f]

Besides ester synthesis being favored over hydrolysis, there are several major advantages of undertaking biocatalytic reactions in anhydrous media: increased solubility of nonpolar substrates, ease of product and enzyme recovery, enhanced thermal stability of enzymes and substrate specificity, and enantioselectivity regulation by the solvent. The main disadvantages include lower catalytic activity in organic media and reversibility, which limits the yield and works against the kinetic resolution, lowering the enantioselectivity of such processes. There are several strategies available to overcome these problems.[1a,e,f] Enol esters, such as vinyl or isopropenyl esters, are by far the most commonly used acyl transfer agents to ensure irreversibility by tautomerization of the enol leaving group.[13,14] Anhydrides,[15] *S*-phenyl thioacetate,[16] acyloxypyridines,[17] and oximes[18] have also been applied in a similar manner as acyl donors. Active trifluoro-[19] and trichloroethyl[20] esters have similarly been used to suppress the reversibility by speeding the acyl–enzyme formation and generating the weakly nucleophilic trifluoro- or trichloroethanol. Primary alcohols have also been used as acyl acceptors in transesterifications (deacylations) involving esters of more bulky and less nucleophilic secondary alcohols.[21]

Kinetic Resolution by Hydrolysis. Until very recently, kinetic resolution of racemic alcohols as ester derivatives was by far the most common type of asymmetric transformations involving lipases.[1a] There are number of examples involving acyclic secondary alcohols, such as the glyceraldehyde derivative in eq 1[22] and various related alkyl- and aryloxy substituted chloride and tosylate glycerol derivatives.[22,23]

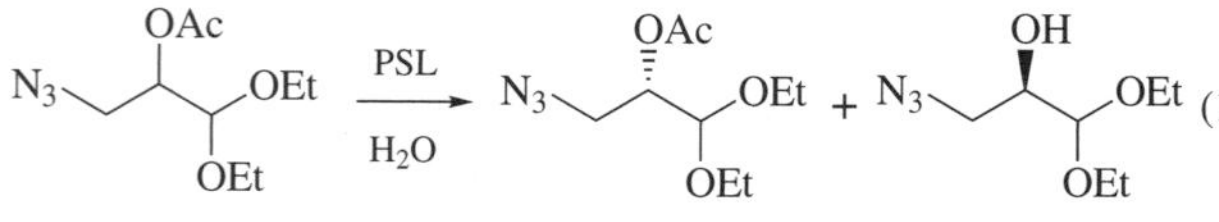

(1)

46% >98% ee   47% 97% ee

A wide variety of other alcohol substrates has been resolved,[1a] including aryl substituted secondary alcohols,[20,24] α-alkyl-β-hydroxy esters,[25] β-hydroxy nitriles,[26] and fluoroorganic compounds.[27] Active chloroacetate esters are commonly used to speed up the hydrolysis reactions, as exemplified in eq 2.[28] Primary acyclic alcohols possessing a stereogenic center that have been resolved include 2,3-epoxy alcohols,[29,30] 2-amino alcohols,[31] and crown ethers.[32]

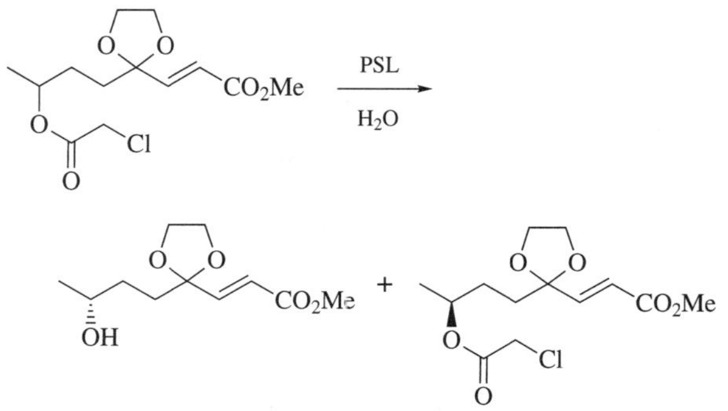

(2)

99% ee   99% ee

Lipase-catalyzed asymmetric hydrolysis has also been conducted on numerous monocyclic, variously substituted five-, six-, and seven-membered cycloalkane and cycloalkene secondary alcohols and diols.[1a] More recent reports include *cis*-4-acetoxyflavan,[33] substituted cyclopentenones,[34] and the 1,2-bis(hydroxymethyl)cyclobutanol derivative exemplified in eq 3.[35]

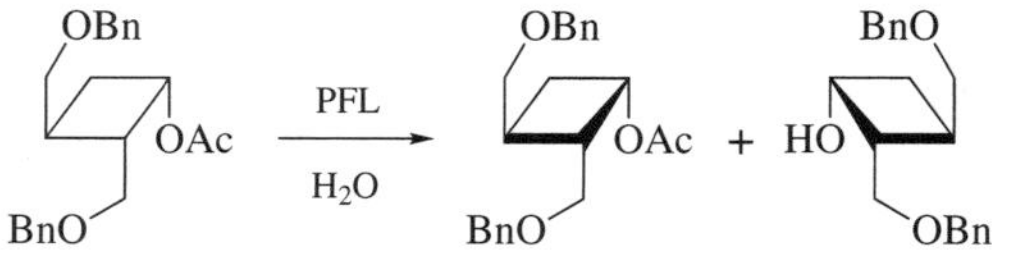

(3)

racemic   48% >99% ee   50% >99% ee

Various bicyclic racemic alcohols have been resolved by asymmetric hydrolysis of their corresponding esters. Generally, the *exo* isomers appear to be far inferior substrates compared with the *endo* substrates.[1a] eq 4 illustrates the resolution of a bicyclic derivative of the Corey lactone type.[36]

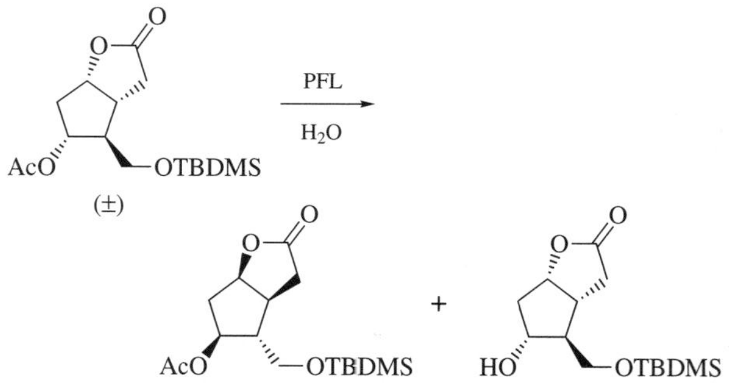

(4)

50% >99% ee   50% >99% ee

There are also several reports on the enantioselective hydrolysis of bicyclic secondary alcohols possessing the bicyclo[2.2.1]heptane and bicyclo[2.2.2] octane framework.[37] Again, with this type of substrate the lipases appear to exhibit strong preference for the *endo* isomers with the (*R*)-configured esters preferentially hydrolyzed.

Various chiral acids have also been resolved by lipase-catalyzed asymmetric hydrolysis.[1a] The reports include variously α-substituted acids[3,38] as well as the tertiary α-benzyloxy ester exemplified in eq 5.[39] Remethylation and repeated hydrolysis afforded the (*S*)-enantiomer in eq 5 optically pure. More recent examples include esters of glycidic acid,[40] β-aryl-β-hydroxy acid,[41] and sulfinyl alkanoates.[42]

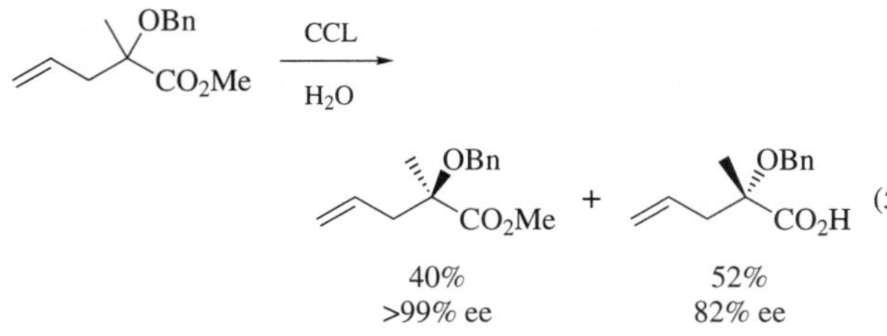

(5)

Kinetic Resolution by Transesterification. Asymmetric transformation involving acylation of chiral alcohols is by far the most common example of kinetic resolution by lipase-catalyzed transesterification, most commonly with irreversible vinyl esters.[1a,15] This field is now becoming the most widely applied technique involving lipases. Recent reports of the numerous secondary alcohol substrates include various monocyclic (eq 6)[43] and acyclic[44] compounds, cyanohydrins,[45] sulfones,[46] and glycals,[47] to name a few.

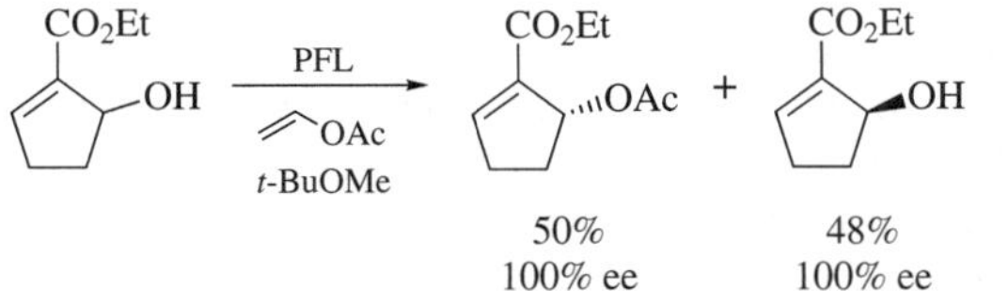

(6)

There are also several reports of enantioselective transesterification involving primary alcohols possessing stereogenic centers by similar acylation procedures, such as 2,3-epoxy alcohols (eq 7),[48] norbornene-derived iodolactones,[49] and 1,3-propanediols.[50]

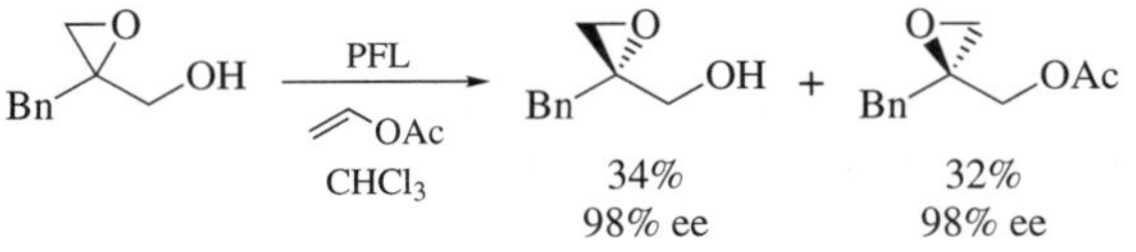

(7)

Enantioselective lipase-catalyzed transesterification involving deacylation of esters of racemic primary or secondary alcohols with primary alcohols, most frequently *n*-butanol, serving as an acyl acceptor, is fairly common.[1a] Recent examples include esters of amino alcohols,[51] isoserine,[52] chlorohydrins,[53] and various tosyloxybutanoate esters (eq 8).[54]

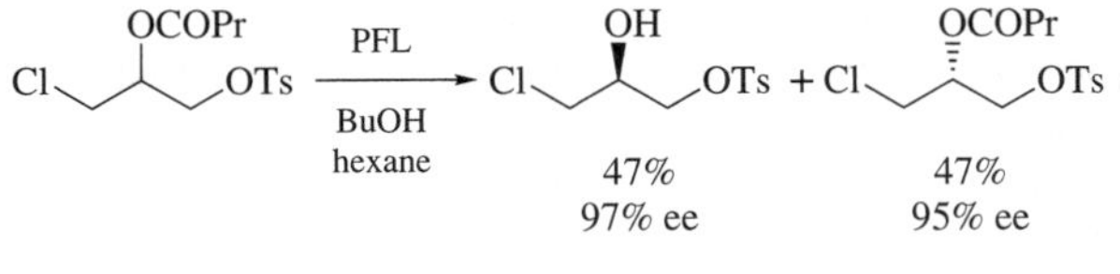

(8)

Kinetic resolution involving acidolysis of esters of racemic secondary alcohols and acids or transesterification of chiral acids does not have many examples in the literature.[1a]

Kinetic Resolution by Direct Esterification. This is the least common strategy for kinetic resolution and is most commonly executed on racemic alcohols with carboxylic acids in organic solvents.[1a] Reports include several alicyclic secondary alcohols such as menthol[55] and various aliphatic secondary alcohols.[56] Kinetic resolution of a variety of racemic saturated, unsaturated, and α-substituted carboxylic acids has also been effected by direct esterification with various alcohols.[20,57]

In addition to this, there are several reports of asymmetric esterification of racemic alcohols with anhydrides as acyl donors. Examples include various primary and secondary alcohols,[15] bicyclic secondary alcohols of the norbornane type,[58] amino alcohols,[59] and ferrocenes.[60] This is exemplified in eq 9 for 1-phenylethanol.[15]

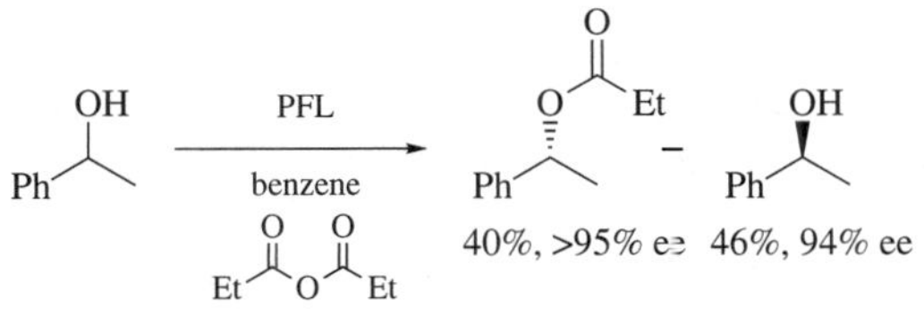

(9)

Prochiral Compounds. The enantiodifferentiation of prochiral compounds by lipase-catalyzed hydrolysis and transesterification reactions is fairly common, with prochiral 1,3-diols most frequently employed as substrates.[1a] Recent reports of asymmetric hydrolysis include diesters of 2-substituted 1,3-propanediols[61] and 2-*O*-protected glycerol derivatives.[8] The asymmetric transesterification of prochiral diols such as 2-*O*-benzylglycerol[8,13a] and various other 2-substituted 1,3-propanediol derivatives[13b,62] is also fairly common, most frequently with *Vinyl Acetate* as an irreversible acyl transfer agent.

There are also recent reports of the lipase-catalyzed enantioselective hydrolysis of prochiral diacid derivatives such as 2-substituted malonates,[63] barbiturates,[64] and highly substituted, sterically hindered 1,4-dihydropyridine derivatives using acyloxymethyl groups to enhance the reaction rate.[65] An example of a prochiral diester hydrolysis is illustrated in eq 10.[66]

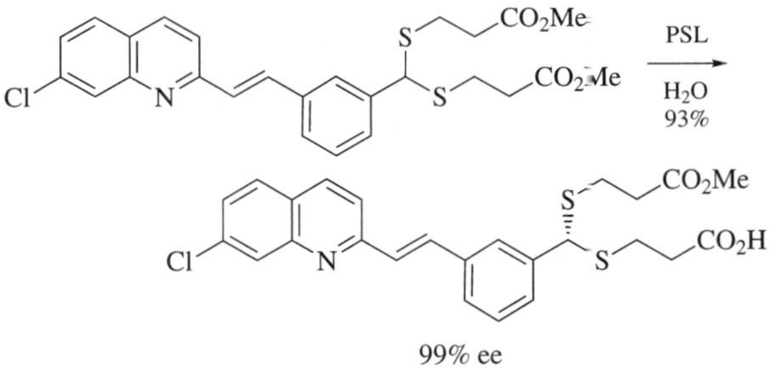

(10)

***Meso* Compounds.** Although pig liver esterase is by far the most suitable enzyme for asymmetric transformations involving *meso* compounds, especially diacids, there are several reports on the lipase-catalyzed hydrolysis and transesterification reactions of cyclic diol derivatives.[1a] The former includes variously substituted cycloalkene diacetates, cyclohexylidene protected erythritol diacetate,[67] piperidine derivatives,[68] and the *exo*-acetonide in eq 11.[69] Complementary results are clearly demonstrated in eq 11 and eq 12 for the hydrolysis and esterification processes.

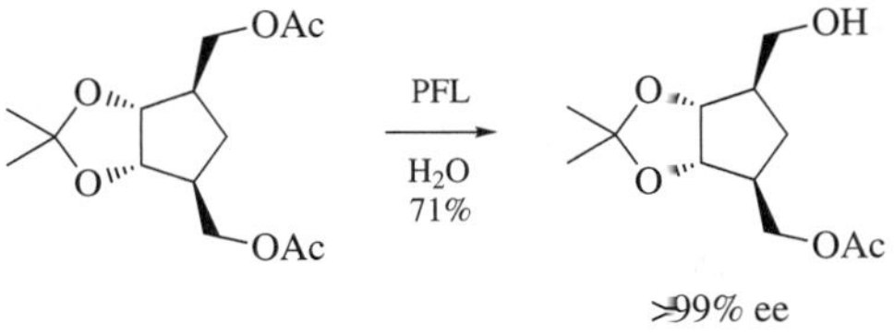

(11)

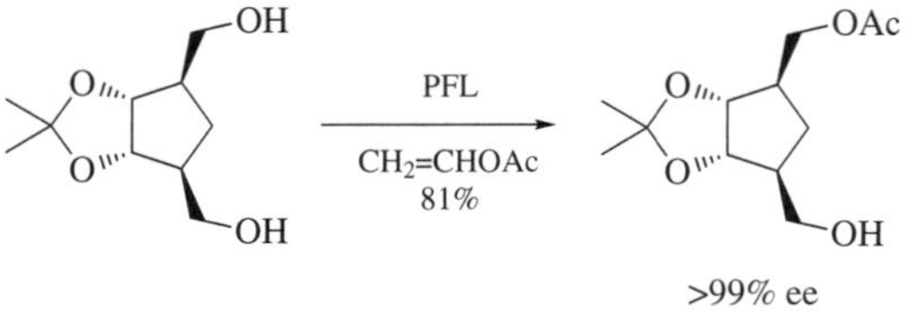

(12)

The asymmetric transesterification of cyclic *meso*-diols, usually with vinyl acetate as an irreversible acyl transfer agent, includes monocyclic cycloalkene diol derivatives,[70] bicyclic diols,[71] such as the *exo*-acetonide in eq 12,[69] bicyclic diols of the norbornyl type,[72] and organometallic 1,2-bis(hydroxymethyl)ferrocene possessing planar chirality.[73]

Regioselective Biotransformations with Lipases. Lipases are gaining increasing importance in solving problems of regioselectivity of various polyol and carbohydrate compounds.[1a,c,e,f,8] A variety of diols or the corresponding acetates as well as polyhydric phenol acetates[74] have been acylated or deacylated in a highly regioselective manner in high yields by lipase-catalyzed transesterification reactions. Regioselective direct esterification of aliphatic 1,2-diols[75] and inositol derivatives[76] using anhydrides as acylating agents has recently been reported. Primary hydroxyl groups are exclusively transformed, as would be anticipated on steric grounds. One example of a highly regioselective and at the same time highly enantioselective hydrolysis of a racemic diester is demonstrated in eq 13.[77]

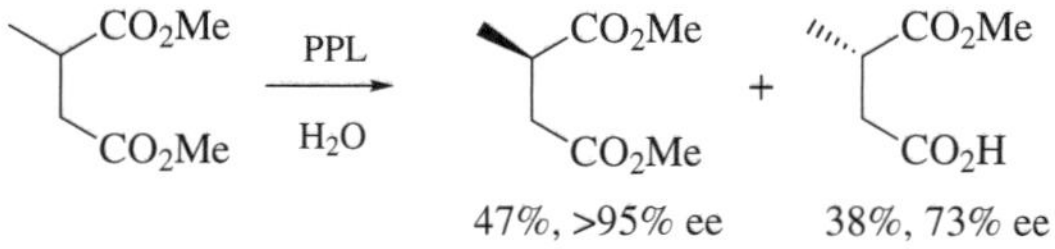

(13)

There are also several reports on highly regioselective transesterification of various steroid derivatives, one example being displayed in eq 14 in which butyration occurred exclusively at the 3β-hydroxyl group by *Chromobacterium viscosum* lipase (CVL).[78] Opposite regioselectivity toward the 17β-hydroxyl group was observed with subtilisin protease.[78]

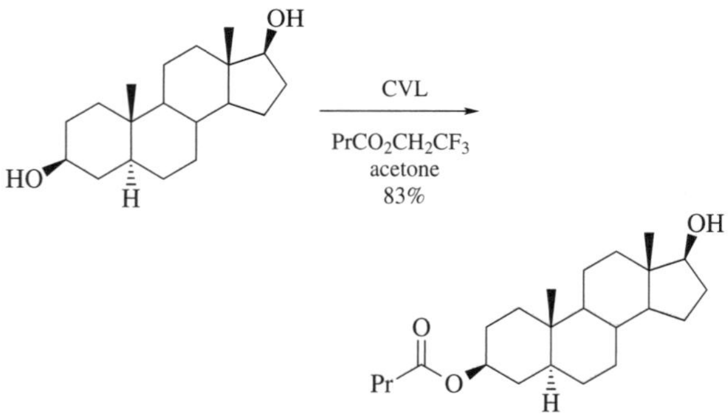

(14)

There are numerous examples of highly regioselective lipase-catalyzed hydrolysis and acylation/deacylation processes involving monosaccharide and carbohydrate derivatives.[1a,f,8] Usually, the biotransformation processes occur preferentially and in many cases exclusively on the primary hydroxyl group (eq 15),[79] but highly regioselective transformations have also been described on secondary alcoholic groups for various carbohydrate derivatives possessing an acyl or alkyl protection on the primary hydroxyl moiety. Recent reports include highly regioselective acetylation of pyranosidic and furanosidic monosaccharide derivatives[80] and alkoxycarbonylation of nucleosides with oxime carbonates.[18]

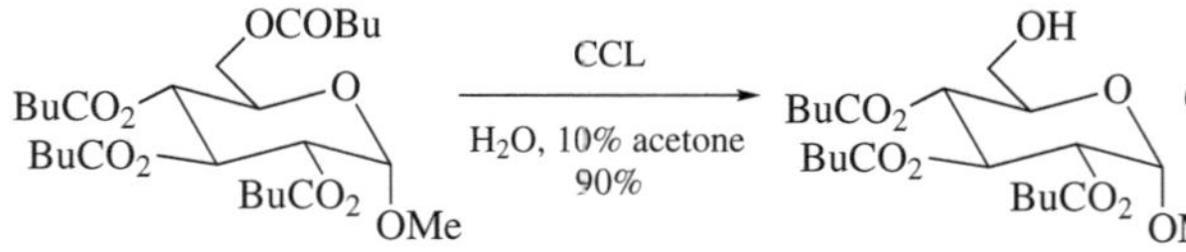

(15)

Lactonization and Polycondensation. The lipase-catalyzed intramolecular transesterification of a range of ω-hydroxy esters has been investigated extensively[1a,e,f] and was observed to be very dependent on the chain length of the substrate (eq 16).

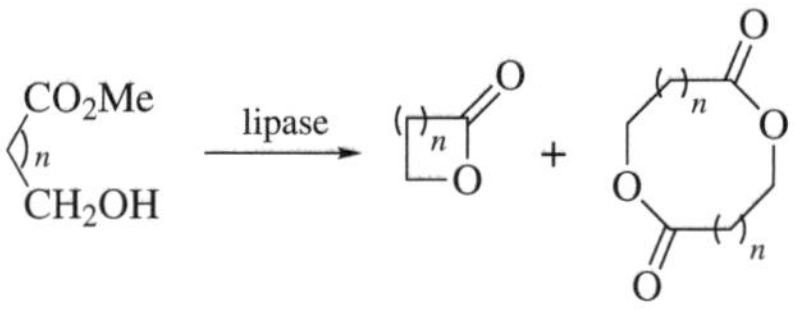

(16)

For longer-chain hydroxy esters ($n = 13$, 14) the corresponding macrolide was accomplished in high yield with very little diolide formed (diolide increased considerably with lower n). With medium-sized hydroxy esters the product profile became considerably more complex, consisting of a complex mixture of di-, tri-, tetra-, and pentalactones.[28,81] Shorter-chain unsubstituted β-, δ-, and ϵ-hydroxy esters almost exclusively underwent intermolecular transesterification to afford the corresponding oligomers. δ-Substituted δ-hydroxy esters[82] and γ-hydroxy esters[83] underwent lactonization with a high degree of enantioselectivity.

Prochiral γ-hydroxy diesters underwent enantioselective lactonization with PPL to afford the (*S*)-lactone in a highly enantioselective fashion (eq 17).[83a] Formation of macrocyclic lactones by the condensation of diacids or diesters with diols, leading to mono- and dilactones,[84] linear oligomeric esters, or high molecular weight optically active polymers,[85] depending upon type of substrates as well as reaction conditions, has also been described.

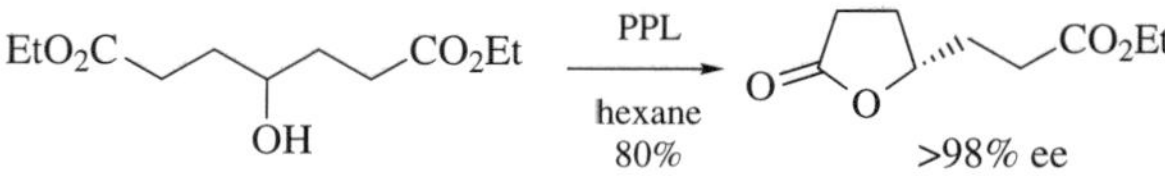

(17)

Mildness and Miscellaneous Reactions. The mildness of the lipases has been particularly well suited in transformations involving labile compounds that are likely to undergo decomposition when conventional chemical methods are applied,[1a] such as the long-chain polyunsaturated ω-3-type fatty acids[86] and highly labile prostaglandin precursor derivatives.[87] Under mild conditions, lipase was exploited to hydrolyze the peracetal protected hydroperoxy derivative in eq 18 to afford the corresponding acid without affecting the peracetal protection moiety.[88]

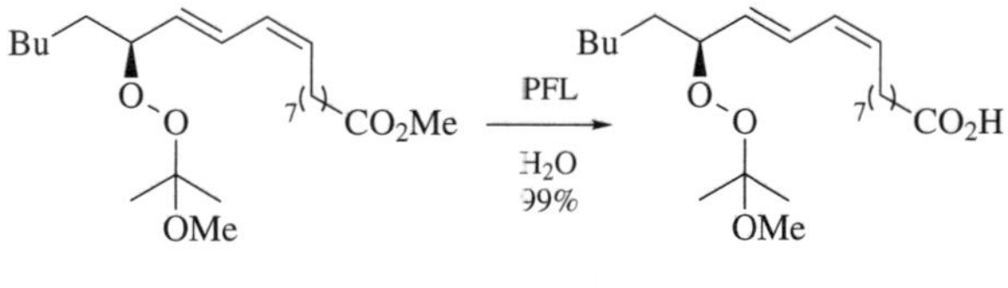

(18)

Various miscellaneous lipase-catalyzed reactions have been reported,[1a] including lipase-mediated epoxidation of alkenes,[12] transamidation,[89] thiotransesterification of thioesters for the preparation of optically active thiols,[90] regio- and chemoselective peptide acylation,[91] lactamization,[92] and highly enantioselective hydrolysis of racemic oxazolin-5-ones which undergo a rapid keto–enol tautomerism to afford optically pure amino acids, thus exceeding the 50% yield limit.[93]

Finally, lipases are able to differentiate enantiotopic faces of appropriately substituted enol esters to afford optically active ketones,[94] indicating that simultaneously upon hydrolysis of the acyl group, protonation occurs from one specified side of the double bond of the enol ester without formation of an enol intermediate (eq 19).[94a]

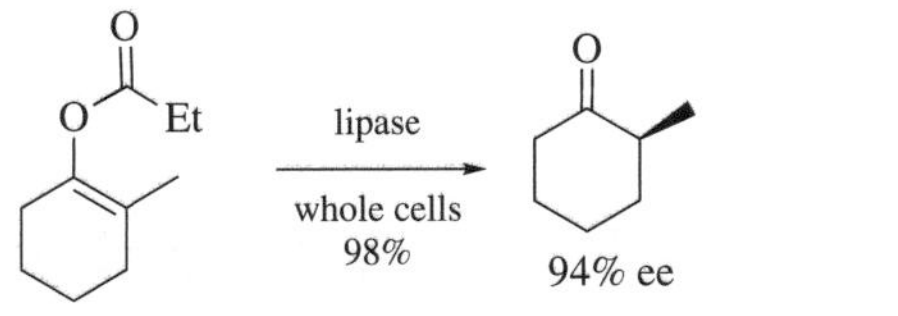

(19)

1. (a) Haraldsson, G. G. In *The Chemistry of the Functional Groups, Supplement B2: The Chemistry of Acid Derivatives*; Patai, S., Ed.; Wiley: Chichester, 1992; Vol. 2, Part 2, pp 1395–1473. (b) Jones, J. B. *Tetrahedron* **1986**, *42*, 3351. (c) Crout, D. H. G.; Christen, M. In *Modern Synthetic Methods*; Scheffold, R., Ed.; Springer: Berlin, 1989; Vol. 5, pp 1–114. (d) Sih, C. J.; Wu, S.-H. In *Topics in Stereochemistry*; Eliel, E. L.; Wilen, S. H., Eds.; Wiley: New York, 1989; Vol. 19, pp 63–125. (e) Chen, C.-S.; Sih, C. J. *Angew. Chem., Int. Ed. Engl.* **1989**, *28*, 695. (f) Klibanov, A. M. *Acc. Chem. Res.* **1990**, *23*, 114. (g) Xie, Z.-F. *Tetrahedron: Asymmetry* **1991**, *2*, 733. (h) Boland, W.; Frössl, C.; Lorenz, M. *Synthesis* **1991**, 1049.
2. Eigtved, P. In *Advances in Applied Lipid Research*; JAI: Greenwich, CT, 1992; Vol. 1, pp 1–64.
3. Wu, S.-H.; Guo, Z.-W.; Sih, C. J. *J. Am. Chem. Soc.* **1990**, *112*, 1990.
4. Chen, C. S.; Fujimoto, Y.; Girdaukas, G.; Sih, C. J. *J. Am. Chem. Soc.* **1982**, *104*, 7294.
5. Tawaki, S.; Klibanov, A. M. *J. Am. Chem. Soc.* **1992**, *114*, 1882.
6. Secundo, F.; Riva, S.; Carrea, G. *Tetrahedron: Asymmetry* **1992**, *3*, 267.
7. Kazlauskas, R. J.; Weissfloch, A. N. E.; Rappaport, A. T.; Cuccia, L. A. *J. Org. Chem.* **1991**, *56*, 2656.
8. Drueckhammer, D. G.; Hennen, W. J.; Pederson, R. L.; Barbas, III, C. F.; Gautheron, C. M.; Krach, T.; Wong, C.-H. *Synthesis* **1991**, 499.
9. Garcia, M. J.; Rebolledo, F.; Gotor, V. *Tetrahedron: Asymmetry* **1992**, *3*, 1519.
10. Astorga, C.; Rebolledo, F.; Gotor, V. *Synthesis* **1991**, 350.
11. Nicolosi, G.; Piattelli, M.; Sanfilippo, C. *Tetrahedron* **1992**, *48*, 2477.
12. Björkling, F.; Frykman, H.; Godtfredsen, S. E.; Kirk, O. *Tetrahedron* **1992**, *48*, 4587.
13. (a) Wang, Y.-F.; Wong, C.-H. *J. Org. Chem.* **1988**, *53*, 3127. (b) Wang, Y.-F.; Lalonde, J. J.; Momongan, M.; Bergbreiter, D. E.; Wong, C.-H. *J. Am. Chem. Soc.* **1988**, *110*, 7200.
14. Faber, K.; Riva, S. *Synthesis* **1992**, 895.
15. Bianchi, D.; Cesti, P.; Battistel, E. *J. Org. Chem.* **1988**, *53*, 5531.
16. Akita, H.; Umezawa, I.; Takano, M.; Matsukura, H.; Oishi, T. *Chem. Pharm. Bull.* **1991**, *39*, 3094.
17. Keumi, T.; Hiraoka, Y.; Ban, T.; Takahashi, I.; Kitajima, H. *Chem. Lett.* **1991**, 1989.
18. Morís, F.; Gotor, V. *Tetrahedron* **1992**, *48*, 9869.
19. Stokes, T. M.; Oehlschlager, A. C. *Tetrahedron Lett.* **1987**, *28*, 2091.
20. Kirchner, G.; Scollar, M. P.; Klibanov, A. M. *J. Am. Chem. Soc.* **1985**, *107*, 7072.
21. Bevinakatti, H. S.; Banerji, A. A.; Newadkar, R. V. *J. Org. Chem.* **1989**, *54*, 2453.
22. von der Osten, C. H.; Sinskey, A. J.; Barbas, III, C. F.; Pederson, R. L.; Wang, Y.-F.; Wong, C.-H. *J. Am. Chem. Soc.* **1989**, *111*, 3924.
23. (a) Pederson, R. L.; Liu, K. K.-C.; Rutan, J. F.; Chen, L.; Wong, C.-H. *J. Org. Chem.* **1990**, *55*, 4897. (b) Ader, U.; Schneider, M. P. *Tetrahedron: Asymmetry* **1992**, *3*, 201. (c) Ader, U.; Schneider, M. P. *Tetrahedron: Asymmetry* **1992**, *3*, 521.
24. Hiratake, J.; Inagaki, M.; Nishioka, T.; Oda, J. *J. Org. Chem.* **1988**, *53*, 6130.
25. Itoh, T.; Kuroda, K.; Tomasada, M.; Takagi, Y. *J. Org. Chem.* **1991**, *56*, 797.
26. Itoh, T.; Takagi, Y.; Nishiyama, S. *J. Org. Chem.* **1991**, *56*, 1521.
27. Bravo, P.; Resnati, G. *Tetrahedron: Asymmetry* **1990**, *1*, 661.
28. Ngooi, T. K.; Scilimati, A.; Guo, Z.-w.; Sih, C. J. *J. Org. Chem.* **1989**, *54*, 911.
29. Ladner, W. E.; Whitesides, G. M. *J. Am. Chem. Soc* **1984**, *106*, 7250.
30. Pawlak, J. L.; Berchtold, G. A. *J. Org. Chem.* **1987**, *52*, 1765.
31. Francalanci, F.; Cesti, P.; Cabri, W.; Bianchi, D.; Martinengo, T.; Foa, M. *J. Org. Chem.* **1987**, *52*, 5079.
32. Tsukube, H.; Betchaku, A.; Hiyama, Y.; Itoh, T. *Chem. Commun./J. Chem. Soc., Chem. Commun.* **1992**, 1751.
33. Izumi, T.; Hino, T.; Kasahara, A. *J. Chem. Soc., Perkin Trans. 1* **1992**, 1265.
34. Danda, H.; Nagatomi, T.; Maehara, A.; Umemura, T. *Tetrahedron* **1991**, *47*, 8701.
35. Chen, X.; Siddiqi, S. M.; Schneller, S. W. *Tetrahedron Lett.* **1992**, *33*, 2249.
36. Sugahara, T.; Satoh, I.; Yamada, O.; Takano, S. *Chem. Pharm. Bull.* **1991**, *39*, 2758.
37. Oberhauser, T.; Faber, K.; Griengl, H. *Tetrahedron* **1989**, *45*, 1679.
38. Kalaritis, P.; Regenye, R. W.; Partridge, J. J.; Coffen, D. L. *J. Org. Chem.* **1990**, *55*, 812.
39. Sugai, T.; Kakeya, H.; Ohta, H. *J. Org. Chem.* **1990**, *55*, 4643.
40. Gentile, A.; Giordano, C.; Fuganti, C.; Ghirotto, L.; Servi, S. *J. Org. Chem.* **1992**, *57*, 6635.
41. Boaz, N. W. *J. Org. Chem.* **1992**, *57*, 4289.
42. Burgess, K.; Henderson, I.; Ho, K.-K. *J. Org. Chem.* **1992**, *57*, 1290.
43. (a) Carrea, G.; Danieli, B.; Palmisano, G.; Riva, S.; Santagostino, M. *Tetrahedron: Asymmetry* **1992**, *3*, 775. (b) Takano, S.; Yamane, T.; Takahashi, M.; Ogasawara, K. *Tetrahedron: Asymmetry* **1992**, *3*, 837.
44. Morgan, B.; Oehlschlager, A. C.; Stokes, T. M. *J. Org. Chem.* **1992**, *57*, 3231.
45. Inagaki, M.; Hiratake, J.; Nishioka, T.; Oda, J. *J. Org. Chem.* **1992**, *57*, 5643.
46. Carretero, J. C.; Dominguez, E. *J. Org. Chem.* **1992**, *57*, 3867.
47. Berkowitz, D. B.; Danishefsky, S. J.; Schulte, G. K. *J. Am. Chem. Soc.* **1992**, *114*, 4518.
48. Ferraboschi, P.; Brembilla, D.; Grisenti, P.; Santaniello, E. *J. Org. Chem.* **1991**, *56*, 5478.
49. Janssen, A. J. M.; Klunder, A. J. H.; Zwanenburg, E. *Tetrahedron* **1991**, *47*, 5513.
50. Grisenti, P.; Ferraboschi, P.; Manzocchi, A.; Santaniello, E. *Tetrahedron* **1992**, *48*, 3827.
51. Kanerva, L. T.; Rahiala, K.; Vänttinen, E. *J. Chem. Soc., Perkin Trans. 1* **1992**, 1759.
52. Lu, Y.; Miet, C.; Kunesch, N.; Poisson, J. E. *Tetrahedron: Asymmetry* **1991**, *2*, 871.
53. Bevinakatti, H. S.; Banerji, A. A. *J. Org. Chem.* **1991**, *56*, 5372.
54. Chen, C.-S.; Liu Y.-C.; Marsella, M. *J. Chem. Soc., Perkin Trans. 1* **1990**, 2559.
55. Langrand, G.; Baratti, J.; Buono, G.; Triantaphylides, C. *Tetrahedron Lett.* **1986**, *27*, 29.

56. (a) Sonnet, P. E. *J. Org. Chem.* **1987**, *52*, 3477. (b) Lutz, D.; Guldner, A.; Thums, R.; Schreier, P. *Tetrahedron: Asymmetry* **1990**, *1*, 783.
57. Engel, K.-H. *Tetrahedron: Asymmetry* **1991**, *2*, 165.
58. Berger, B.; Rabiller, C. G.; Königsberger, K.; Faber, K.; Griengl, H. *Tetrahedron: Asymmetry* **1990**, *1*, 541.
59. Kamal, A.; Rao, M. V. *Tetrahedron: Asymmetry* **1991**, *2*, 751.
60. Izumi, T.; Tamura, F.; Sasaki, K. *Bull. Chem. Soc. Jpn.* **1992**, *65*, 2784.
61. Guanti, G.; Banfi, L.; Narisano, E. *J. Org. Chem.* **1992**, *57*, 1540.
62. Didier, E.; Loubinoux, B.; Ramos Tombo, G. M.; Rihs, G. *Tetrahedron* **1991**, *47*, 4941.
63. Gutman, A. L.; Shkolnik, E.; Shapira, M. *Tetrahedron* **1992**, *48*, 8775.
64. Murata, M.; Achiwa, K. *Tetrahedron Lett.* **1991**, *32*, 6763.
65. Holdgrün, X. K.; Sih, C. J. *Tetrahedron Lett.* **1991**, *32*, 3465.
66. Hughes, D. L.; Bergan, J. J.; Amato, J. S.; Bhupathy, M.; Leazer, J. L.; McNamara, J. M.; Sidler, D. R.; Reider, P. J.; Grabowski, E. J. J. *J. Org. Chem.* **1990**, *55*, 6252.
67. Gais, H.-J.; Hemmerle, H.; Kossek, S. *Synthesis* **1992**, 169.
68. (a) Chênevert, R.; Dickman, M. *Tetrahedron: Asymmetry* **1992**, *3*, 1021. (b) Momose, T.; Toyooka, N.; Jin, M. *Tetrahedron Lett.* **1992**, *33*, 5389.
69. Tanaka, M.; Yoshioka, M.; Sakai, K. *Chem. Commun./J. Chem. Soc., Chem. Commun.* **1992**, 1454.
70. (a) Mekrami, M.; Sicsic, S. *Tetrahedron: Asymmetry* **1992**, *3*, 431. (b) Harris, K. J.; Gu, Q.-M.; Shih, Y.-E.; Girdaukas, G.; Sih, C. J. *Tetrahedron Lett.* **1991**, *32*, 3941.
71. Theil, F.; Schick, H.; Winter, G.; Reck, G. *Tetrahedron* **1991**, *47*, 7569.
72. (a) Andreu, C.; Marco, J. A.; Asensio, G. *J. Chem. Soc., Perkin Trans. 1* **1990**, 3209. (b) Murata, M.; Uchida, H.; Achiwa, K. *Chem. Pharm. Bull.* **1992**, *40*, 2610.
73. Nicolosi, G.; Morrone, R.; Patti, A.; Piattelli, M. *Tetrahedron: Asymmetry* **1992**, *3*, 753.
74. Natoli, M.; Nicolosi, G.; Piattelli, M. *J. Org. Chem.* **1992**, *57*, 5776.
75. Bosetti, A.; Bianchi, D.; Cesti, P.; Golini, P.; Spezia, S. *J. Chem. Soc., Perkin Trans. 1* **1992**, 2395.
76. Ling, L.; Watanabe, Y.; Akiyama, T.; Ozaki, S. *Tetrahedron Lett.* **1992**, *33*, 1911.
77. Guibé-Jampel, E.; Rousseau, G.; Salaün, J. *Chem. Commun./J. Chem. Soc., Chem. Commun.* **1987**, 1080.
78. Riva, S.; Klibanov, A. M. *J. Am. Chem. Soc.* **1988**, *110*, 3291.
79. Sweers, H. M.; Wong, C.-H. *J. Am. Chem. Soc.* **1986**, *108*, 6421.
80. (a) Theil, F.; Schick, H. *Synthesis* **1991**, 533. (b) Chinn, M. J.; Iacazio, G.; Spackman, D. G.; Turner, N. J.; Roberts, S. M. *J. Chem. Soc., Perkin Trans. 1* **1992**, 661.
81. Guo, Z.-W.; Ngooi, T. K.; Scilimati, A.; Fülling, G.; Sih, C. J. *Tetrahedron Lett.* **1988**, *29*, 5583.
82. (a) Bonini, C.; Pucci, P.; Viggiani, L. *J. Org. Chem.* **1991**, *56*, 4050. (b) Henkel, B.; Kunath, A.; Schick, H. *Justus Liebigs Ann. Chem./Liebigs Ann. Chem.* **1992**, 809.
83. (a) Gutman, A. L.; Zuobi, K.; Bravdo, T. *J. Org. Chem.* **1990**, *55*, 3546. (b) Huffer, M.; Schreier, P. *Tetrahedron: Asymmetry* **1991**, *2*, 1157.
84. Guo, Z.-W.; Sih, C. J. *J. Am. Chem. Soc.* **1988**, *110*, 1999.
85. (a) Margolin, A. L.; Crenne, J.-Y.; Klibanov, A. M. *Tetrahedron Lett.* **1987**, *28*, 1607. (b) Margolin, A. L.; Fitzpatrick, P. A.; Dubin, P. L.; Klibanov, A. M. *J. Am. Chem. Soc.* **1991**, *113*, 4693.
86. (a) Haraldsson, G. G.; Höskuldsson, P. A.; Sigurdsson, S. Th.; Thorsteinsson, F.; Gudbjarnason, S. *Tetrahedron Lett.* **1989**, *30*, 1671. (b) Haraldsson, G. G.; Almarsson, Ö. *Synthesis* **1991**, *45*, 723.
87. (a) Porter, N. A.; Byers, J. D.; Holden, K. M.; Menzel, D. B. *J. Am. Chem. Soc.* **1979**, *101*, 4319. (b) Lin, C.-H.; Alexander, D. L.; Chidester, C. G.; Gorman, R. R.; Johnson, R. A. *J. Am. Chem. Soc.* **1982**, *104*, 1621.
88. Baba, N.; Yoneda, K.; Tahara, S.; Iwasa, J.; Kaneko, T.; Matsuo, M. *Chem. Commun./J. Chem. Soc., Chem. Commun.* **1990**, 1281.
89. Gotor, V.; Brieva, R.; González, C.; Rebolledo, F. *Tetrahedron* **1991**, *47*, 9207.
90. Bianchi, D.; Cesti, P. *J. Org. Chem.* **1990**, *55*, 5657.
91. Gardossi, L.; Bianchi, D.; Klibanov, A. M. *J. Am. Chem. Soc.* **1991**, *113*, 6328.
92. Gutman, A. L.; Meyer, E.; Yue, X.; Abell, C. *Tetrahedron Lett.* **1992**, *33*, 3943.
93. Gu, R.-L.; Lee, I.-S.; Sih, C. J. *Tetrahedron Lett.* **1992**, *33*, 1953.
94. (a) Ohta, H.; Matsumoto, K.; Tsutsumi, S.; Ihori, T. *Chem. Commun./J. Chem. Soc., Chem. Commun.* **1989**, 485. (b) Sugai, T.; Kakeya, H.; Ohta, H.; Morooka, M.; Ohba, S. *Tetrahedron* **1989**, *45*, 6135.

Gudmundur G. Haraldsson
University of Iceland, Reykjavik, Iceland

(1*R*,2*S*)-1-Lithio-1-phenylsulfonyl-2-{[(*tert*-butyldiphenyl)silyl]oxymethyl} Oxirane

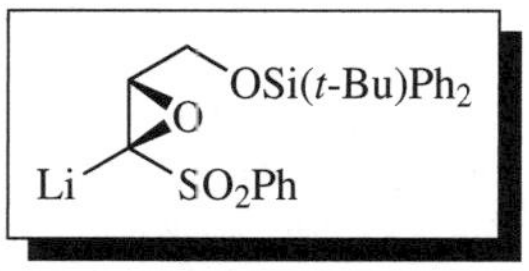

[181208-42-4] $C_{25}H_{27}LiO_4SSi$ (MW 458.08)

(oxiranyllithium; oxiranyl anion; nucleophilic epoxide; acyl anion equivalent; epoxy sulfone)

Solubility: soluble in THF, diethyl ether.

Preparative Methods: prepared by lithiation of (1*R*,2*S*)-1-phenylsulfonyl-2-{[(*tert*-butyldiphenyl)silyl]oxymethyl}oxirane (**1**, $[\alpha]_D$ +55.9°, *c* 1.0, $CHCl_3$) (1.0 equiv) in THF (0.15 M solution) with *n*-BuLi (1 equiv, 1.6 M solution in hexane) in the presence of DMPU or hexamethylphosphoramide (HMPA) (3.0 equiv) at −100 °C under argon. Deprotonation is completed within a few minutes (eq 1).[1]

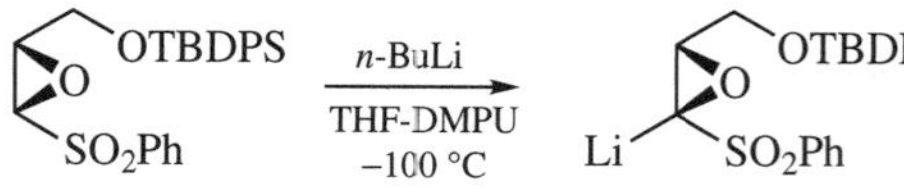

1 (1)

Handling, Storage, and Precautions: the oxiranyllithium is very unstable, even at −100 °C under argon, and should be reacted with electrophiles immediately. The reagent is also conformationally unstable and slowly isomerizes to the *trans*-isomer when addition of an electrophile is delayed (about 5% isomerization after 20 min at −100 °C). Elevated temperatures (>−78 °C) cause rapid decomposition.[1,2]

Introduction. Although epoxides are widely recognized as extremely versatile synthetic intermediates in view of their electrophilic nature, the reaction of an epoxide as a nucleophile, i.e. an oxiranyl anion, is less common. Recently, cumulative studies on the chemistry of oxiranyl anions have appeared and some aspects of the anions have been discussed.[3,4]

Preparation of (1*R*,2*S*)-1-Phenylsulfonyl-2-{[(*tert*-butyldiphenyl)silyl]oxymethyl}oxirane and Related Compounds. Epoxidation of (*Z*)-vinyl sulfone, which is available from the Peterson olefination of (*S*)-*O*-pentylideneglyceraldehyde[5] and phenyl trimethylsilylmethyl sulfone[6] in three steps (40% overall yield), with *t*-BuOOH/*t*-BuOK in THF gives epoxy sulfone (eq 2). Deprotection of the ketal group and recrystallization affords an optically pure epoxy diol, which is then treated with sodium periodate followed by sodium borohydride to give an alcohol. Protection of the resulting alcohol as its silyl ether yields (1*R*,2*S*)-1-phenylsulfonyl-2-{[(*tert*-butyldiphenyl)silyl]oxymethyl}oxirane (**1**).[7] Its enantiomer is available in the same manner starting from (*R*)-isopropylideneglyceraldehyde.[8]

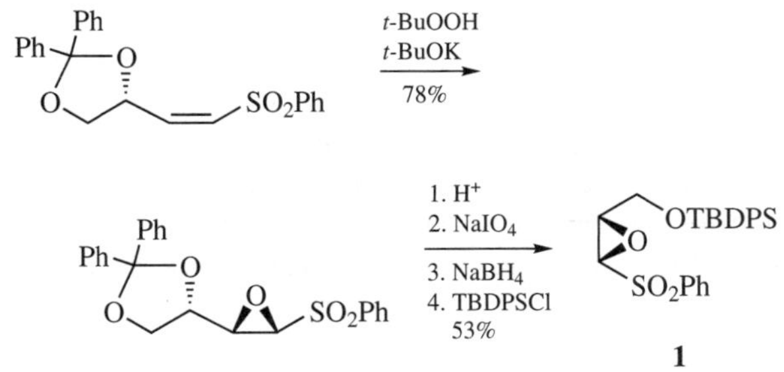

1 (2)

Racemic epoxy sulfone derivatives are easily prepared from allyl ethers by reaction with sodium *p*-toluenesulfinate in the presence of iodine followed by treatment with triethylamine, separation of *E*- and *Z*-isomers, and epoxidation with *t*-BuOOH and *n*-BuLi in THF (eq 3).[2]

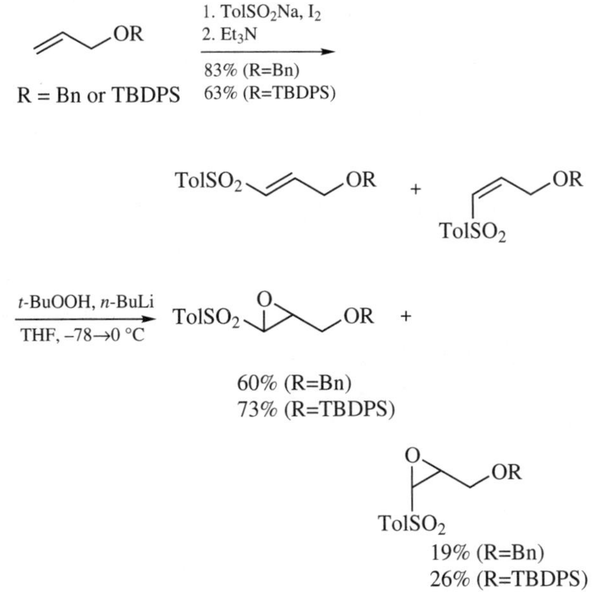

(3)

Reaction of Sulfonyl-Stabilized Oxiranyllithiums. Reaction of sulfonyl-stabilized oxiranyllithiums with primary alkyl halides gives acceptable yields of products.[1] More reactive alkyl triflates give generally better yields but, due to the instability of oxiranyllithiums, yields are often not reproducible when electrophiles are added to a solution of the preformed oxiranyllithiums. It is recommended that the alkylation reaction be carried out by an in situ trapping method.[2] Treatment of a solution of epoxy sulfone (1.0 equiv) and triflate (1.5 equiv) in THF-DMPU (or HMPA) at −100 °C under argon with *n*-BuLi (1.0 equiv) followed by stirring for 30 min affords the coupled product in high yield (eq 4).[9] The product can be converted to a tetrahydropyranone derivative by exposure to *p*-toluenesulfonic acid. The strong electron-withdrawing ability of the sulfonyl group works against the adjacent C-O bond-breaking in an acid-catalyzed epoxide ring-opening process and, consequently, favors the 6-*endo* mode pathway which yields the tetrahydropyranone after elimination of phenylsulfinic acid. Reaction with a halogenated metal Lewis acid yields a halo ketone instead of a cyclization product (eq 4).[10] These reactions demonstrate that the oxiranyllithium reagent serves as a functionalized acyl anion equivalent and a three-carbon building block.

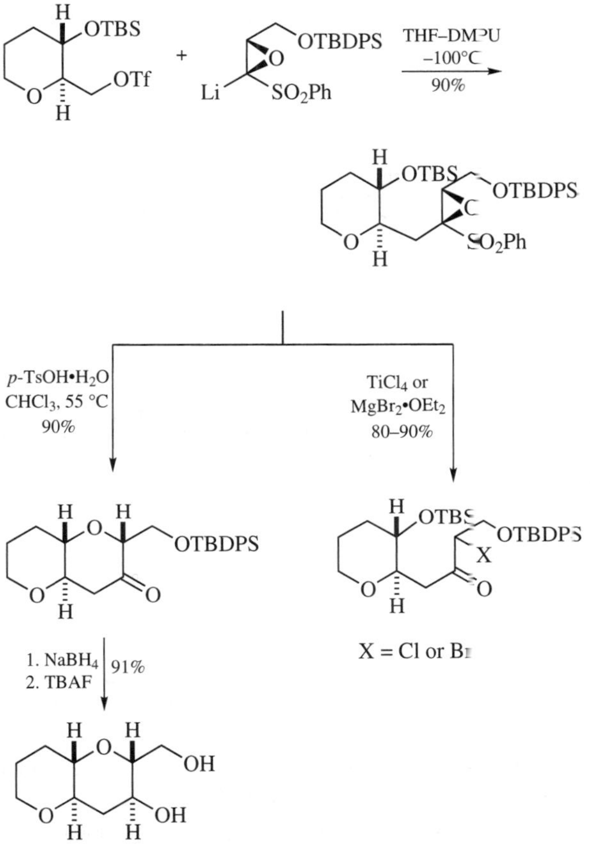

(4)

Reiterative application of this protocol has allowed the stereocontrolled construction of polytetrahydropyrans[9,10] and polycyclic ethers containing six- and seven-membered rings (eq 5).[7]

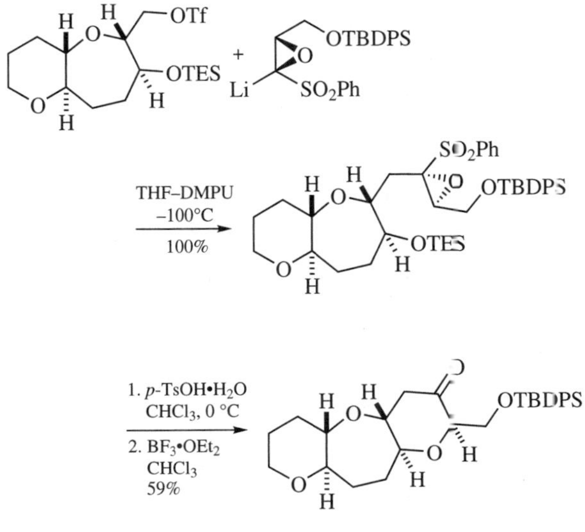

(5)

Reaction of the oxiranyllithium with aldehydes is also carried out by an in situ trapping method at very low temperatures in order to avoid decomposition of the reagent. Its applicability to a complex situation has been demonstrated in a synthesis of hemibrevetoxin B (eq 6).[11,12] It is noteworthy that deprotonation of **1** by *n*-BuLi is much faster than butyl addition to the aldehyde.

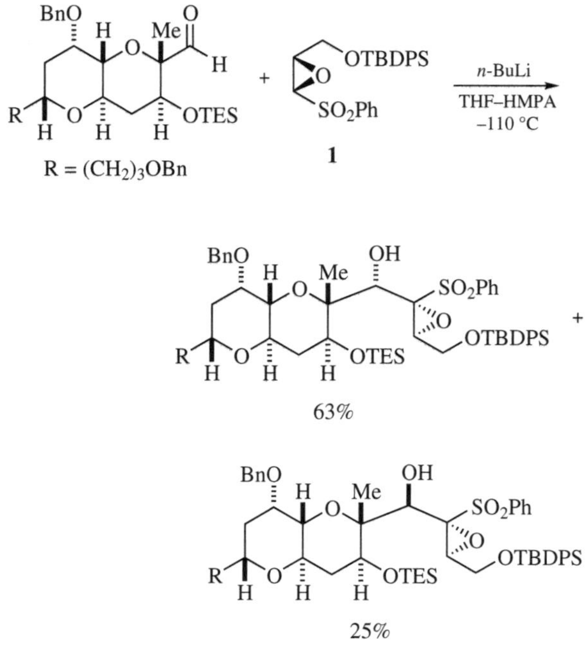

(6)

While alkylation of sulfonyl-stabilized oxiranyllithiums with primary alkyl triflates proceeds in high yield, the reaction towards epoxides is relatively slow (~2 h) and the decomposition of oxiranyllithium is marked, such that it decreases the yield, especially in the case of a *Z*-isomer (eq 7 and 8).[2] Addition of boron trifluoride diethyl etherate promotes this epoxide-epoxide coupling reaction. One of the diastereoisomers of eq 8 has been elaborated via 5-*endo* cyclization into a marine tetrahydrofuran isolated from a brown alga.[13]

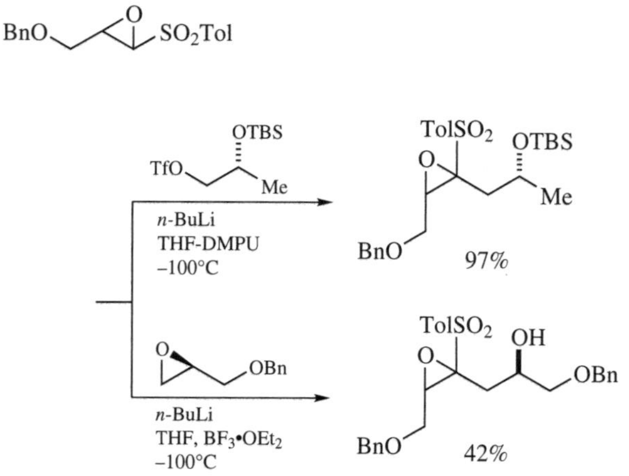

(7)

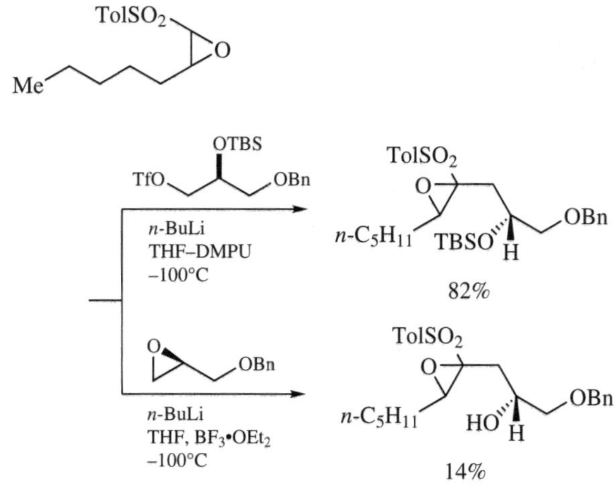

(8)

Related Reagents. Optically active trisubstituted sulfonyl-stabilized oxiranyllithiums can be generated by deprotonation of the corresponding epoxy sulfones[14] (eq 9). Due to the diminished reactivity of the reagents by steric hindrance, the reaction with triflates requires HMPA to obtain a high yield of product (eq 10).[12]

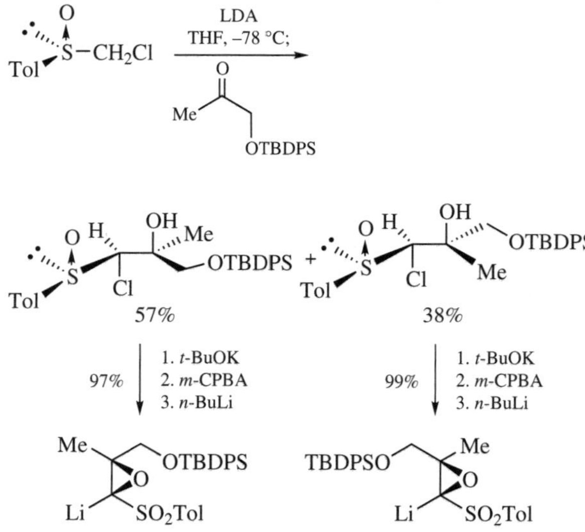

(9)

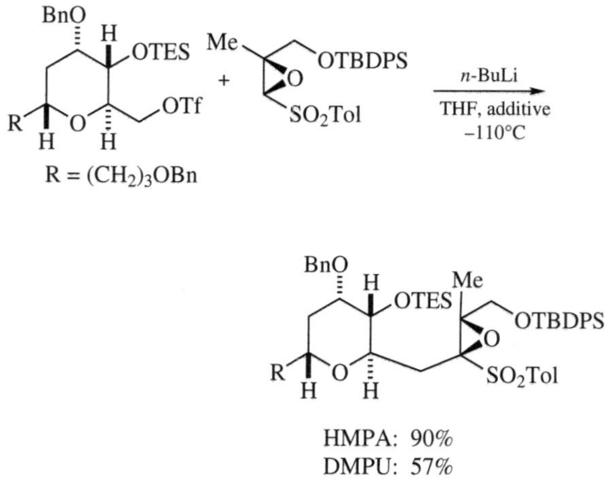

(10)

1. (a) Ashwell, M.; Clegg, W.; Jackson, R. F. W. *J. Chem. Soc., Perkin Trans. 1* **1991**, 897. (b) Dunn, S. F. C.; Jackson, R. F. W. *J. Chem. Soc., Perkin Trans. 1* **1992**, 2863.

2. Mori, Y.; Yaegashi, K.; Iwase, K.; Yamamori, Y.; Furukawa, H. *Tetrahedron Lett.* **1996**, *37*, 2605.
3. Satoh, T. *Chem. Rev.* **1996**, *96*, 3303.
4. Mori, Y. In *Reviews on Heteroatom Chemistry*; Oae, S., Ed.; MYU: Tokyo, 1997, Vol. 17, p 183.
5. Schmid, C. R.; Bradley, D. A. *Synthesis* **1992**, 587.
6. Craig, D.; Ley, S. V.; Simpkins, N. S.; Whitham, G. H.; Prior, M. J. *J. Chem. Soc., Perkin Trans. 1* **1985**, 1949.
7. Mori, Y.; Yaegashi, K. Furukawa, H. *Tetrahedron* **1997**, *53*, 12917.
8. Schmid, C. R.; Bryant, J. D.; Dowlatzadeh, M.; Philps, J. L.; Prather, D. E.; Schantz, R. E.; Sear, N. L.; Vianco, C. S. *J. Org. Chem.* **1991**, *56*, 4056.
9. Mori, Y.; Yaegashi, K.; Furukawa, H. *J. Am. Chem. Soc.* **1996**, *118*, 8158.
10. Mori, Y.; Yaegashi, K.; Furukawa, H. *Tetrahedron Lett.* **1999**, *40*, 7239.
11. Mori, Y.; Yaegashi, K.; Furukawa, H. *J. Am. Chem. Soc.* **1887**, *119*, 4557.
12. Mori, Y.; Yaegashi, K.; Furukawa, H. *J. Org. Chem.* **1998**, *63*, 6200.
13. Mori, Y.; Sawada, T.; Furukawa, H. *Tetrahedron Lett.* **1999**, *40*, 731.
14. Satoh, T.; Oohara, T.; Ueda, Y.; Yamakawa, K. *J. Org. Chem.* **1989**, *5* **4**, 3130.

Yuji Mori
Faculty of Pharmacy, Meijo University, Nagoya, Japan

Lithium Aluminum Hydride-2,2′-Dihydroxy-1,1′-binaphthyl

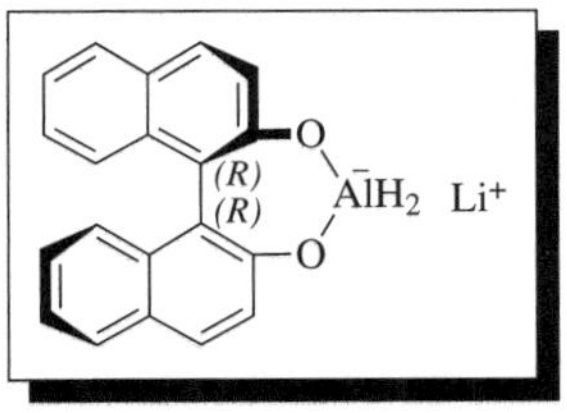

($LiAlH_4$)		
[16853-85-3]	AlH_4Li	(MW 37.96)
((*R*)-BINAL)		
[18531-94-7]	$C_{20}H_{14}O_2$	(MW 286.34)
((*S*)-BINAL)		
[18531-99-2]		

(used for enantioselective reduction of prochiral ketones to alcohols[1])

Alternate Name: BINAL-H.
Physical Data: BINAL: white solid, mp 208–210 °C. Also see *Lithium Aluminum Hydride*.
Solubility: sol THF.
Preparative Methods: prepared in situ from commercially available lithium aluminum hydride and BINAL.
Handling, Storage, and Precautions: sensitive to moisture (see *Lithium Aluminum Hydride*).

Overview and General Considerations. This article will cover the title reagent and other chiral reducing agents derived from lithium aluminum hydride and chiral additives, with initial emphasis on the title reagent. The enantioselective reduction of prochiral ketones is a reaction of considerable importance to the synthetic organic chemist and can now be accomplished by a variety of methods and reagents.[1,2] Particularly the use of chiral oxazaborolidines for the catalytic asymmetric reduction of ketones has received much recent interest. This method has been shown to be useful for the preparation of a variety of chiral alcohols with high optical purities. This transformation can also be realized using catalytic hydrogenation with a chiral catalyst or by use of chiral borane reducing agents such as *(R,R)-2,5-Dimethylborolane* and *B-3-Pinanyl-9-borabicyclo[3.3.1]nonane*. Enzyme-catalyzed transformations, for example *Baker's Yeast* reductions of carbonyl compounds, can also provide access to a range of chiral alcohols with high optical purities.

The use of complexes of lithium aluminum hydride (LAH) with various chiral ligands to achieve the enantioselective reduction of prochiral ketones has been extensively studied for over 40 years.[1] However, this method, with some exceptions, has not found widespread use due to a number of limiting factors. These factors vary from moderate to poor enantioselectivities, often observed in these reductions, to ready availability of only one antipode of a desired chiral ligand. The recovery of the often expensive chiral ligand that is used in stoichiometric quantities to form the LAH complex is obviously an important experimental concern. Also, in some cases the LAH complex with the chiral ligand may disproportionate to achiral reducing species under the reaction conditions, resulting in poor optical purities of the desired products. Further, no single complex appears to have a sufficiently broad substrate specificity. Aromatic and unsaturated ketones are in general the better substrates and they can be reduced with good enantioselectivities using this method. A useful article comparing the merits of some of the more promising asymmetric reducing agents known for ketones has been published.[3]

Chiral Alcohol Modifying Agents. Complexes of a variety of chiral alcohols (see Figure 1) with LAH have been prepared in situ and examined for their ability to effect enantioselective reduction of prochiral carbonyl compounds. However in most cases, the optical purities of the products obtained have not been satisfactory. This is in part due to the tendency of these chiral ligand–hydride complexes to disproportionate under reaction conditions yielding achiral reducing agents. An exception is the complex of LAH and (−)-menthol (**1**) which has been used to reduce α and β-aminoketones with good enantioselectivities.

The reduction of carbonyl compounds with LAH complexes of a number of chiral diols derived from carbohydrates and terpenes has been studied. In general, the enantioselectivities observed with such reagents have been low to moderate. Acetophenone, which is the model substrate in many of these reduction studies, is reduced by a complex of LAH and the glucose-derived diol (**2**) in about 71% ee under optimized conditions.

The reagent (*R*)- or (*S*)-BINAL-H (**7**), developed by Noyori, is undoubtedly the most useful LAH complex reported so far for the asymmetric reduction of a variety of carbonyl compounds.[4] The reagent is prepared from (*R*)- or (*S*)-2,2′-dihydroxy-1,1′-binaphthyl (**3**) (BINAL). Both enantiomers of BINAL are commercially available, although they are somewhat expensive. The chiral ligand, however, can be recovered after the reduction and reused. Equimolar quantities of BINAL and LAH are initially mixed together to form a LAH complex that has a C_2 axis of symmetry, which makes the two hydrogens on the aluminum homotopic. It is interesting to note that the 1:1 complex of BINAL and LAH is a reducing agent that exhibits extremely low enantioselectivity as seen in the case of acetophenone (2% ee). Replacement of one of the hydrogens with an alcohol, like methanol or ethanol, gives a single reducing agent (**7**), which exhibits much higher specificity in the reduction of prochiral ketones. Another useful observation is that reduction of carbonyls with the (*R*)-BINAL-H reagent tends to give the (*R*)-alcohol, while the (*S*)-reagent gives the (*S*)-alcohol. The use of lower reduction temperatures enhances optical purities of the product alcohols, but lowers the yields. Optimized conditions for reductions involve reaction of a ketone with 3 equiv of the reagent formed from LAH, BINAL, and ethanol (1:1:1) in THF for 1 h at −100 °C and then at −78 °C for 2 h.

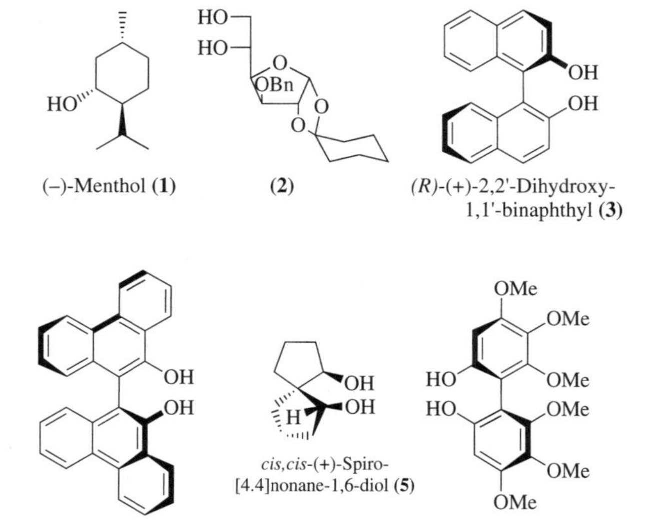

(−)-Menthol (**1**) (**2**) (*R*)-(+)-2,2′-Dihydroxy-1,1′-binaphthyl (**3**)

(*S*)-(−)-10,10′-Dihydroxy-9,9′-biphenanthryl (**4**) *cis,cis*-(+)-Spiro-[4.4]nonane-1,6-diol (**5**) (*S*)-2,2′-Dihydroxy-4,5,6,4′,5′,6′-hexamethoxybiphenyl (**6**)

Figure 1 Representative chiral alcohol modifying agents

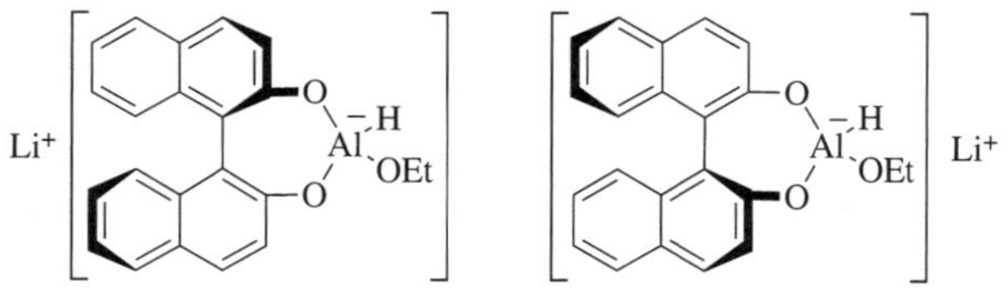

(*R*)-(**7**) (*S*)-(**7**)

A number of structurally diverse ketones have been reduced using BINAL-H. Some of the results are summarized in Table 1.[5] Aryl alkyl ketones, alkynic ketones, and α,β-unsaturated ketones are reduced to alcohols with good to excellent % ee, while aliphatic ketones give products with lower optical purities. The asymmetric reduction of a number of acylstannanes with (**7**) gives synthetically valuable α-alkoxystannanes with high optical purities after protection of the initially formed unstable alcohols as their MOM or BOM ethers.[6]

BINAL-H has been used to prepare deuterated primary alcohols with high optical purities. For example, benzaldehyde-*1-d* is reduced in 59% yield and 87% optical purity. β-Ionone is reduced with this reagent to the corresponding alcohol in 100% ee and 87% yield. Simple cyclic enones like 2-cyclohexenone are not reduced by the BINAL-H reagent under standard reduction conditions.[5]

Table 1 Reduction of Ketones with (**7**)

Ketone	(**7**)	Yield (%)	ee (%)	Product
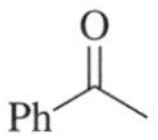	(*R*)	61	95	(*R*)
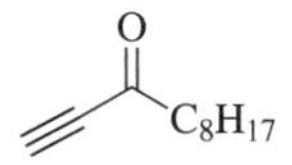	(*S*)	74	90	(*S*)
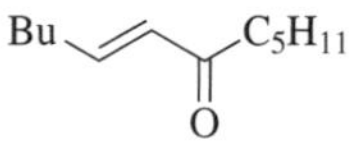	(*R*)	91	91	(*R*)
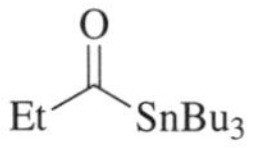	(*S*)	69	96	(*R*)
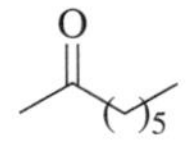	(*S*)	67	24	(*R*)

The chiral nonracemic enone (**8**) is reduced with (*S*)-(**7**) to give the (15*S*)-alcohol in 100% de and 88% yield. The product is a valuable intermediate in the synthesis of prostaglandins.[5]

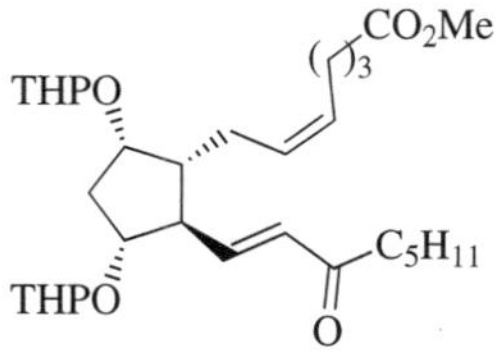

(**8**)

The asymmetric reduction of lactone (**9**) to give predominantly one atropoisomer can be achieved using 10 equiv of a complex prepared from LAH and BINAL (1:1) at −40 °C.[7] This reduction gives an 88:12 ratio of (**10a**):(**10b**) in good yield (80%). Reduction of the same substrate with 8 equiv of a complex of LAH with (*S*)-(+)-2-(anilinomethyl)pyrrolidine in ether at −40 °C leads to opposite stereochemical results (38:62 ratio of **10a**:**10b**).

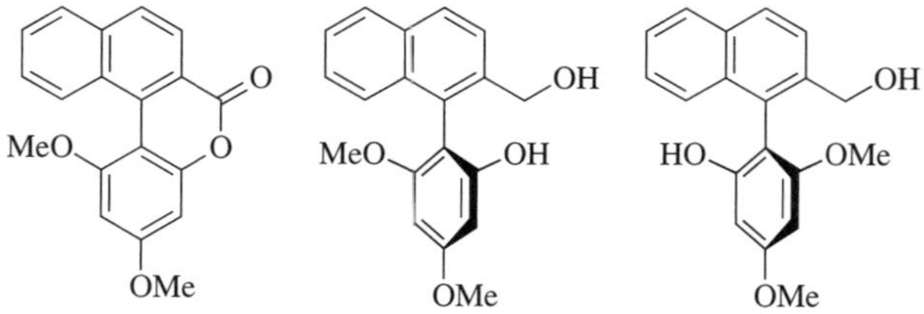

(**9**) (**10a**) (**10b**)

BINAL-H has also been used for the asymmetric reduction of methylaryl- and methylalkylphosphinylimines to the corresponding phosphinylamines in high % ee (Table 2).[8] Similar to the reduction of ketones, reduction of the imines with (*S*)-(**7**) produces the (*S*)-amine and reduction with (*R*)-(**7**) gives the (*R*)-amine.

Table 2 Reduction of Imines with **(7)**

R^1	R^2	**(7)**	Yield (%)	ee (%)	Product
Me	Ph	(*R*)	20	100	(*R*)
Me	Et	(*S*)	38	93	(*S*)
Me	C_5H_{11}	(*S*)	83	64	(*S*)

The complex of the biphenanthryl diol (**4**) with LAH has been prepared and its reduction properties have been examined.[9] This reagent gives excellent enantioselectivity in the reduction of aromatic ketones. For example, acetophenone is reduced in 75% yield with 97% ee. As with Noyori's reagent, reductions with the (*S*)-reagent give (*S*)-alcohols and aliphatic ketones are reduced with low enantioselectivity. Both enantiomers of this auxiliary can be readily prepared and can also be recovered for reuse at the end of the reduction.

The LAH complex of the chiral spirodiol (**5**) has recently been prepared. This complex exhibits excellent enantioselectivity in the reduction of some aromatic ketones.[10] Acetophenone is reduced at −80 °C in 98% ee and 80% yield. Reduction of other aryl alkyl ketones also gives excellent stereoselectivity, but the use of this reagent with a variety of ketones has not been studied. The chiral auxiliary can be recovered and reused.

Recently, the preparation of the chiral biphenyl (**6**) and its use as a modifying agent with LAH has been reported.[11] A complex of LAH–(**6**)–EtOH (1:1:1) at −78 °C gives the best enantioselectivities in the reduction of prochiral ketones. Similar to Noyori's reagent, use of the LAH complex with (*S*)-(**6**) leads to the (*S*)-alcohol. Enantioselectivity is usually high for aromatic ketones (acetophenone 97% ee, 93% yield). This reagent reduces 2-octanone in higher enantioselectivity (76% ee) than 3-heptanone (36% ee).

Chiral Amino Alcohol Modifying Agents. A number of chiral amino alcohols have been examined as ligands for the preparation of chiral LAH reducing agents (Figure 2). The complex of (−)-*N*-methylephedrine (**11**) with LAH has been widely studied and has shown promise for the asymmetric reduction of prochiral ketones. It has been found that addition of an achiral component such as 3,5-dimethylphenol (DMP), *N*-ethylaniline (NEA), or 2-ethylaminopyridine (EAP) to the complex of LAH with (**11**) can enhance the enantioselectivity observed in these reductions. Both enantiomers of (**11**) are commercially available and the ligand can be recovered subsequent to the reaction and reused.

Vigneron and co-workers have observed that a complex of LAH, (−)-(**11**), and DMP (1:1:2), in ether at −15 °C, appears to show the highest enantioselectivity in the reduction of a series of aromatic and alkynyl ketones to the corresponding (*R*)-alcohols (Figure 3).[12] Interestingly, the optical purities of the products obtained were lower both at higher and lower reaction temperatures.

The complex of LAH, (−)-(**11**), and DMP has also been used to reduce stereoselectively a steroidal alkynic ketone. Reduction of the alkynic ketone (**16**) with 3 equiv of the complex at −15 °C gave a 17:1 ratio of the two diastereomers (22*R*/22*S*) in 94% yield,

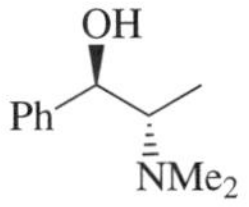

(−)-*N*-Methyl-ephedrine (**11**)

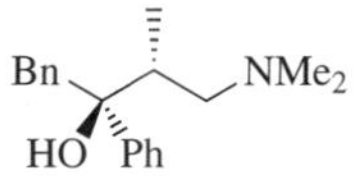

Chirald (**12**)

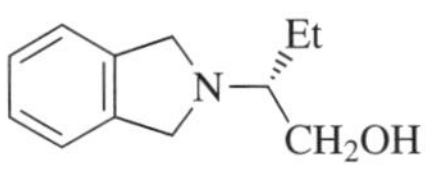

(**13**)

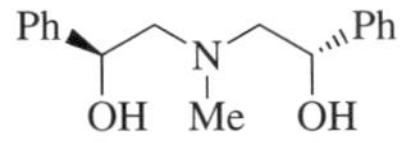

(**14**)

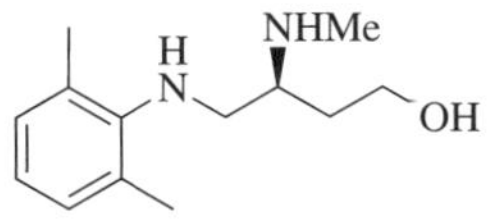

(**15**)

Figure 2 Representative chiral amino alcohol modify ng agents

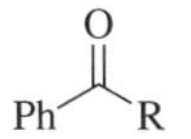

R = Me 84% ee
R = Pr 89% ee

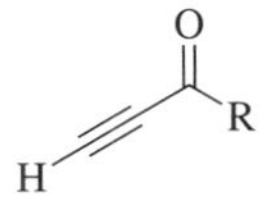

R = Me 79% ee
R = Et 86% ee
R = Bu 85% ee
R = *t*-Bu 90% ee
R = octyl 89% ee

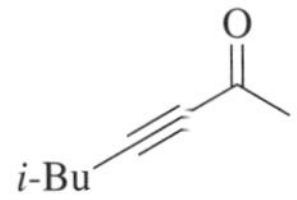

88% ee

Figure 3 Reduction of ketones with LAH/(−)-(**11**)/DMP (1:1:2) to give (*R*)-alcohols

to provide a key intermediate for the synthesis of a vitamin D_2 metabolite.[13]

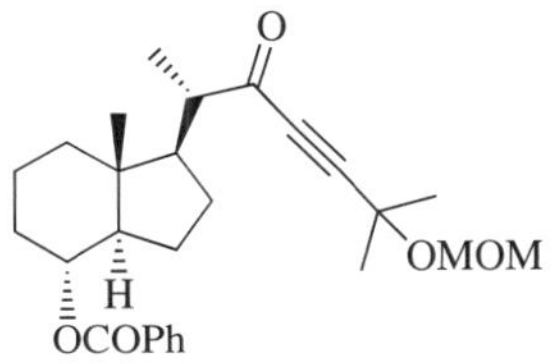

(**16**)

The enantioselective reduction of cyclic conjugated enones may be best accomplished using a complex of LAH with (**11**) to which EAP has been added.[14,15] Optimum conditions for these reductions involve treatment of the ketone with 3 equiv of a 1:1:2 complex of LAH–(−)-(**11**)–EAP in ether at −78 °C for 3 h (Table 3). However, under these conditions, acetophenone is reduced to the (*R*)-alcohol in only 54% ee.

Table 3 Reduction of Ketones with LAH/(−)-(**11**)/EAP to give (*R*)-Alcohols

Ketone	Yield (%)	ee (%)
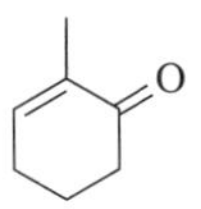	82	96
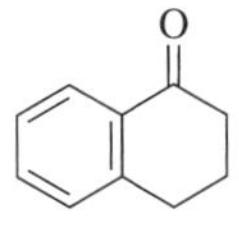	93	96
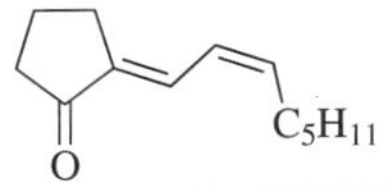	91	96

It has been found that the addition of 2 equiv of NEA to a 1:1 complex of LAH and (−)-(**11**) in ether produces a reagent capable of reducing some α,β-unsaturated ketones to the (*S*)-alcohols in good optical purities at −78 °C (Table 4).[16] It is interesting to note that, with this reagent, the (*S*)-alcohol is the product that is formed preferentially.

Table 4 Reduction of Ketones with LAH/(−)-(**11**)/NEA to give (*S*)-Alcohols

Ketone	Yield (%)	ee (%)
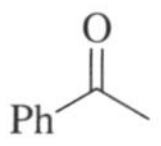	87	86
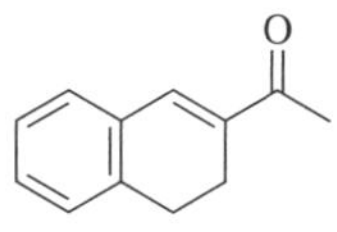	100	>90
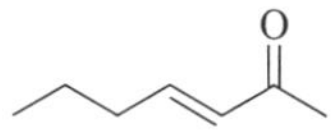	92	88
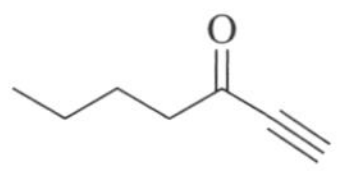	88	76
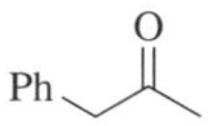	90	41

The preparation and use of a polymer supported LAH–ephedrine–DMP reducing reagent has been reported.[17] In preparing this reagent, ephedrine is attached to a 1% crosslinked polystyrene backbone prior to mixing with LAH and DMP. Careful control of the degree of functionalization of the polymer gives a reducing reagent comparable in efficacy to the analogous nonpolymeric complex.

The use of the complex formed between LAH and Chirald (often called Darvon alcohol in the literature) (**12**) for the reduction of conjugated enones and ynones was first reported by Yamaguchi and Mosher.[18] The mode of preparation of the complex, its age, and the precise experimental conditions of the reduction all appear to have significant impact on the enantioselectivities obtained using this reagent. Thus when 1.5 equiv of a freshly prepared complex of LAH and Chirald (1:2.3) is used to reduce acetophenone at 0 °C, the (*R*)-alcohol is obtained in 68% ee and nearly quantitative yield. If, however, the reagent is allowed to stir overnight, or is refluxed in ether prior to the addition of the ketone, the (*S*)-enantiomer is obtained in 66% ee and 43% yield. Unfortunately, this observed reversal in stereochemical outcome is not predictable. Hence, it may be preferable to use the complex of LAH with the enantiomer of Chirald to reverse the stereoselectivity of the reduction.[19]

A number of alkynic ketones have been reduced with the complex of LAH and (**12**) (1.1:2.5 equiv, ether, −78 °C, 30–60 min) to give the corresponding (*R*)-alcohols (Table 5).[20,21] Johnson and co-workers have reported[22] the reduction of ynone (**17**) to the (*R*)-alcohol in 84% ee and 95% yield with the LAH–Chirald complex. The resulting alcohol was an intermediate in an enantioselective synthesis of 11α-hydroxyprogesterone.[22] The thiophene ketone (**18**) is reduced by the same reagent in ether at −70 °C for 16 h to give the (*R*)-alcohol in 85–88% ee and 80–90% yield.[23] The resulting alcohol has been used in the synthesis of LY248686, an inhibitor of serotonin and norepinephrine uptake carriers.

Table 5 Reduction of Alkynic Ketones with LAH/(−)-(**12**) to give (*R*)-Alcohols

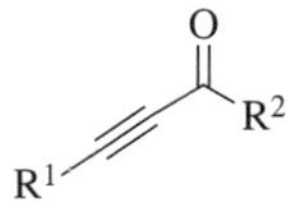

R^1	R^2	Yield (%)	ee (%)
H	C_5H_{11}	96	72
TMS	C_5H_{11}	96	66
C_5H_{11}	C_5H_{11}	97	62

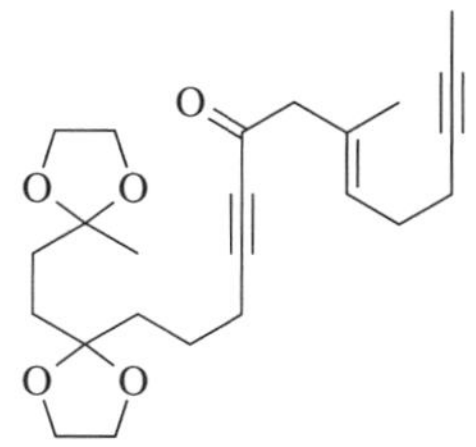

(**17**)

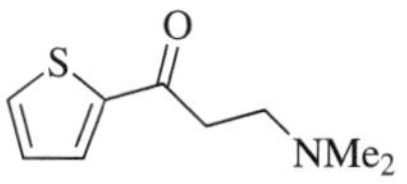

(**18**)

A macrocyclic alkynic ketone has been protected as the Co derivative and then reduced with the complex of LAH with (**12**) (eq 1). Deprotection gave the (*R*)-alcohol (71% ee) which was an important intermediate in a synthesis of (+)-α-2,7,11-cembratriene-4,6-diol.[24]

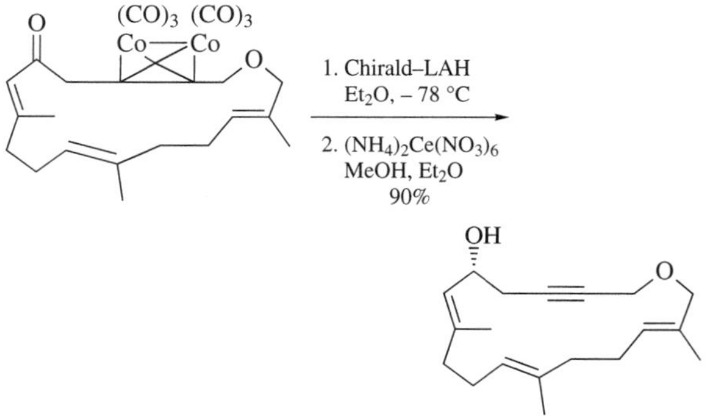

(1)

In general, structural variations to the backbone of the Chirald ligand have not led to the development of more selective or reliable LAH complexes for use in asymmetric reductions.[25] Other complexes of amino alcohols with LAH have been studied for their ability to achieve enantioselective reduction of prochiral ketones. However, in most cases the selectivities observed have been moderate.[26] The complex of LAH with the amino alcohol (**15**) reduces some enones, such as cyclohexenone and cyclopentenone, to the corresponding (*S*)-alcohols in high optical purities (100% and 82% ee, respectively).[27]

Chiral Amine Modifying Agents. Some chiral amine additives (Figure 4) have also been studied for their potential to give useful chiral LAH reagents, but the results so far have not been

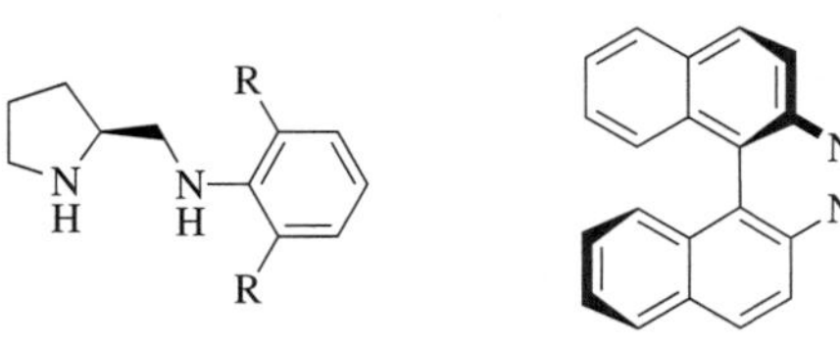

(19) (20)

Figure 4 Representative chiral amino modifying agents

very promising. An exception to this is the complex of LAH with the chiral aminopyrrolidine (**19**) (R = Me), which reduces aromatic ketones in good ee.[28] This reagent reduces acetophenone in 95% ee and 87% chemical yield. LAH complexes of diamine ligands (**20**), analogs of BINAL-H, have also been prepared and examined.[29] In general, the optical purities obtained with this reagent are significantly lower than those observed for BINAL-H in the reduction of aryl ketones.

1. (a) Nishizawa, M.; Noyori, R. *Comprehensive Organic Synthesis* **1991**, *8*, Chapter 1.7. (b) Grandbois, E. R.; Howard, S. I.; Morrison, J. D. *Asymmetric Synthesis*; Academic: New York, 1983; Vol. 2. (c) Nógrádi, M. *Stereoselective Synthesis*; VCH: Weinheim, 1986; Chapter 3. (d) ApSimon, J. W.; Collier, T. L. *Tetrahedron* **1986**, *42*, 5157. (e) Singh, V. K. *Synthesis* **1992**, 605. (f) Blaser, H.-U. *Chem. Rev.* **1992**, *92*, 935. (g) Haubenstock, H. *Top. Stereochem.* **1982**, *14*, 231. (h) Mukaiyama, T.; Asami, M. *Top. Curr. Chem.* **1985**, *127*, 133. (i) Rosini, C.; Franzini, L.; Raffaelli, A.; Salvadori, P. *Synthesis* **1992**, 503.
2. (a) Tomioka, K. *Synthesis* **1990**, 541. (b) Wallbaum, S.; Martens, J. *Tetrahedron: Asymmetry* **1992**, *3*, 1475. (c) Santaniello, E.; Ferraboschi, P.; Grisenti, P.; Manzocchi, A. *Chem. Rev.* **1992**, *92*, 1071.
3. Brown, H. C.; Park, W. S.; Cho, B. T.; Ramachandran, P. V. *J. Org. Chem.* **1987**, *52*, 5406.
4. Noyori, R.; Tomino, I.; Tanimoto, Y.; Nishizawa, M. *J. Am. Chem. Soc.* **1984**, *106*, 6709.
5. Noyori, R.; Tomino, I.; Yamada, M.; Nishizawa, M. *J. Am. Chem. Soc.* **1984**, *106*, 6717.
6. (a) Chan, P. C.-M.; Chong, J. M. *J. Org. Chem.* **1988**, *53*, 5584. (b) Chong, J. M.; Mar, E. K. *Tetrahedron* **1989**, *45*, 7709. (c) Chong, J. M.; Mar, E. K. *Tetrahedron Lett.* **1990**, *31*, 1981. (d) Marshall, J. A.; Gung, W. Y. *Tetrahedron Lett.* **1988**, *29*, 1657.
7. Bringmann, G.; Hartung, T. *Synthesis* **1992**, 433.
8. Hutchins, R. O.; Abdel-Magid, A.; Stercho, Y. P.; Wambsgans, A. *J. Org. Chem.* **1987**, *52*, 702.
9. Yamamoto, K.; Fukushima, H.; Nakazaki, M. *Chem. Commun./J. Chem. Soc., Chem. Commun.* **1984**, 1490.
10. Srivastava, N.; Mital, A.; Kumar, A. *Chem. Commun./J. Chem. Soc., Chem. Commun.* **1992**, 493.
11. Rawson, D.; Meyers, A. I. *Chem. Commun./J. Chem. Soc., Chem. Commun.* **1992**, 494.
12. (a) Vigneron, J. P.; Jacquet, I. *Tetrahedron* **1976**, *32*, 939. (b) Vigneron, J. P.; Blanchard, J. M. *Tetrahedron Lett.* **1980**, *21*, 1739. (c) Vigneron, J. P.; Bloy, V. *Tetrahedron Lett.* **1980**, *21*, 1735. (d) Vigneron, J.-P.; Bloy, V. *Tetrahedron Lett.* **1979**, 2683.
13. Sardina, F. J.; Mouriño, A. Castedo, L. *Tetrahedron Lett.* **1983**, *24*, 4477. (b) Sardina, F. J.; Mouriño, A.; Castedo, L. *J. Org. Chem.* **1986**, *51*, 1264.
14. Kawasaki, M.; Suzuki, Y.; Terashima, S. *Chem. Lett.* **1984**, 239.
15. Iwasaki, G.; Sano, M.; Sodeoka, M.; Yoshida, K.; Shibasaki, M. *J. Org. Chem.* **1988**, *53*, 4864.
16. (a) Terashima, S.; Tanno, N.; Koga, K. *Tetrahedron Lett.* **1980**, *21*, 2753. (b) Terashima, S.; Tanno, N.; Koga, K. *Chem. Commun./J. Chem. Soc., Chem. Commun.* **1980**, 1026. (c) Terashima, S.; Tanno, N.; Koga, K. *Chem. Lett.* **1980**, 981.
17. Fréchet, J. M.; Bald, E.; Lecavalier, P. *J. Org. Chem.* **1986**, *51*, 3462.
18. (a) Yamaguchi, S.; Mosher, H. S. *J. Org. Chem.* **1973**, *38*, 1870 (b) Yamaguchi, S.; Mosher, H. S.; Pohland, A. *J. Am. Chem. Soc.* **1972**, *94*, 9254.
19. Paquette, L. A.; Combrink, K. D.; Elmore. S. W.; Rogers, R. D. *J. Am. Chem. Soc.* **1991**, *113*, 1335.
20. Brinkmeyer, R. S.; Kapoor, V. M. *J. Am. Chem. Soc.* **1977**, *99*, 8339.
21. Marshall, J. A.; Salovich, J. M.; Shearer, B. G. *J. Org. Chem.* **1990**, *55*, 2398.
22. Johnson, W. S.; Brinkmeyer, R. S.; Kapoor, V. M.; Yarnell, T. M. *J. Am. Chem. Soc.* **1977**, *99*, 8341.
23. Deeter, J.; Frazier, J.; Staten, G.; Staszak, M.; Weigel, L. *Tetrahedron Lett.* **1990**, *31*, 7101.
24. Marshall, J. A.; Robinson, E. D. *Tetrahedron Lett.* **1989**, *30*, 1055.
25. Cohen, N.; Lopresti, R. J.; Neukom, C.; Saucy, G. *J. Org. Chem.* **1980**, *45*, 582.
26. (a) Brown, E.; Penfornis, A.; Bayma, J.; Touet, J. *Tetrahedron: Asymmetry* **1991**, *2*, 339. (b) Steels, I.; DeClercq, P. J.; Declercq, J. P. *Tetrahedron: Asymmetry* **1992**, *3*, 599. (c) Morrison, J. D.; Grandbois, E. R.; Howard, S. I.; Weisman, G. R. *Tetrahedron Lett.* **1981**, *22*, 2619.
27. (a) Sato, T.; Gotoh, Y.; Wakabayashi, Y.; Fujisawa, T. *Tetrahedron Lett.* **1983**, *24*, 4123. (b) Sato, T.; Goto, Y.; Fujisawa, T. *Tetrahedron Lett.* **1982**, *23*, 4111.
28. Asami, M.; Mukaiyama, T. *Heterocycles* **1979**, *12*, 499.
29. Kabuto, K.; Yoshida, T.; Yamaguchi, S.; Miyano, S.; Hashimoto, H. *J. Org. Chem.* **1985**, *50*, 3013.

Aravamudan S. Gopalan & Hollie K. Jacobs
New Mexico State University, Las Cruces, NM, USA

M

(−)-(1*R*,2*S*,5*R*)-Menthyl (*S*)-*p*-Toluenesulfinate

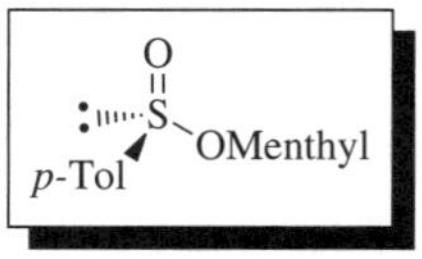

[1517-82-4] $C_{17}H_{26}O_2S$ (MW 294.50)

(agent used for the synthesis of chiral sulfoxides[1,3])

Physical Data: $[\alpha]_D = -202°$ (acetone, $c = 2.0$).

Preparative Methods: obtained by reaction of (−)-menthol with *p*-toluenesulfinyl chloride. This esterification showed no particular stereoselectivity, giving an equal amount of the two sulfinate diastereomers.[1] In order to avoid a chromatographic separation, it is possible to epimerize these sulfinate esters in acidic medium and displace the resulting equilibrium towards the less soluble isomer, (−)-menthyl (*S*)-*p*-toluenesulfinate, in 80% yield (eq 1).[2] This procedure was later extended to large scale preparation.[3]

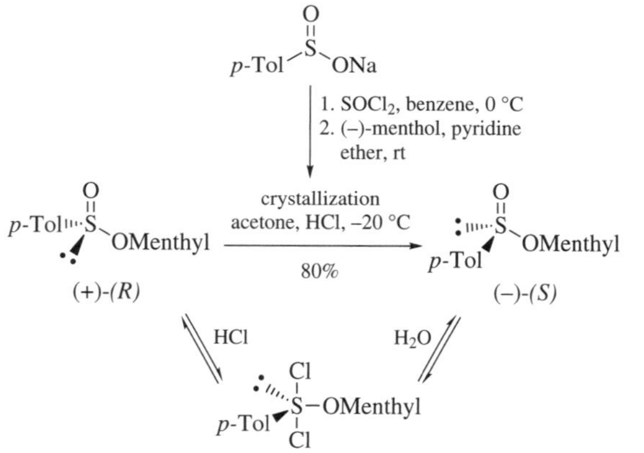

(1)

The absolute configuration of (−)-menthyl (*S*)-*p*-toluenesulfinate was established by correlation with (−)-menthyl *p*-iodobenzenesulfinate, known from X-ray diffraction analysis.[4]

Synthesis of Chiral Sulfoxides.

Alkyl Sulfoxides. Any Grignard reagent reacts with (−)-menthyl (*S*)-*p*-toluenesulfinate and displaces the menthoxy group with complete inversion of configuration at sulfur (eq 2; R = Me,[3,5] Et,[5,6] *n*-C_6H_{13}[7]).

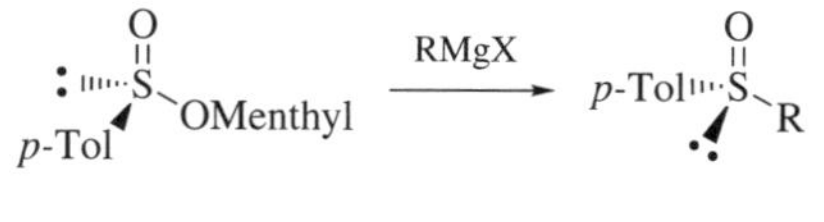

(2)

It was also reported that using methyllithium instead of the methyl Grignard could give some racemization of methyl *p*-tolyl sulfoxide as a result of methyl group exchange via a methylene sulfine intermediate.[8]

(*R*)-4-Substituted cyclohexylmethyl *p*-tolyl sulfoxide (**1**)[9] as well as (*R*)-4-hydroxybutyl *p*-tolyl sulfoxide (**2**)[10] and (*R*)-3-butenyl *p*-tolyl sulfoxide (**3**)[11] were also obtained by reaction of (−)-menthyl (*S*)-*p*-toluenesulfinate and the corresponding Grignard reagent.

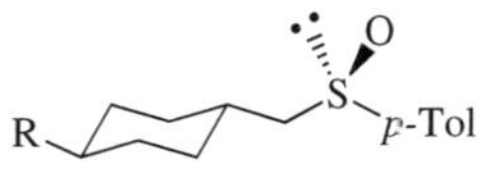

(1)

R = Me, $MeOCH_2$, $ClCH_2$

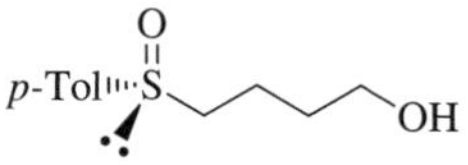

(2)

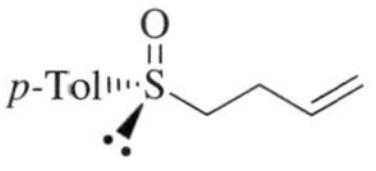

(3)

Vinyl Sulfoxides. A stereocontrolled preparation of (*E*)-1-alkenyl *p*-tolyl sulfoxide from (−)-menthyl-(*S*) *p*-toluenesulfinate was reported (eq 3).[12a]

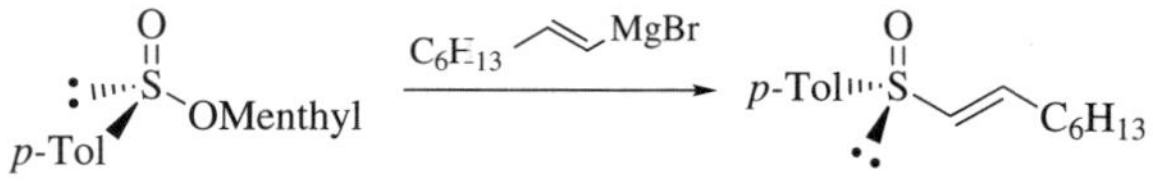

(3)

One example was also reported showing the formation of an (*E*)-alkenyl sulfoxide in the reaction of a vinylic lithium compound on menthyl sulfinate (eq 4).[12b]

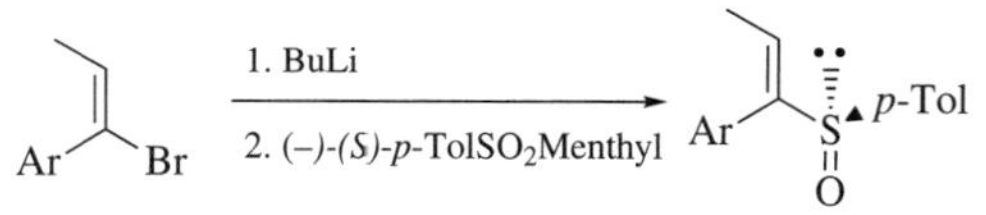

(4)

(+)-(*S*)-2-(*p*-Tolylsulfinyl)-2-cyclopentenone was also prepared by reaction of a vinyllithium derivative and menthyl sulfinate (eq 5).[13]

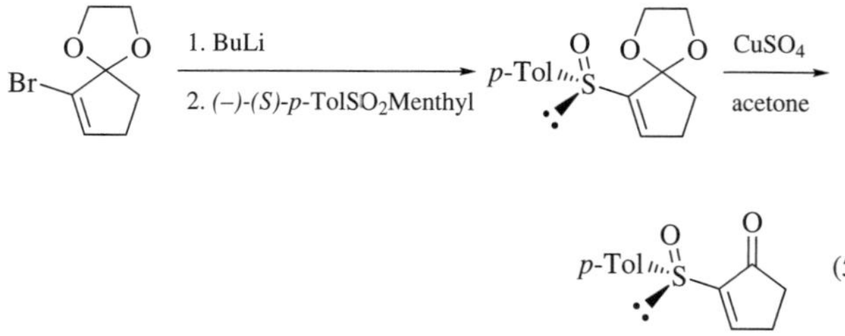

(5)

The preparation of optically pure (*E*)- and (*Z*)-1-alkenyl *p*-tolyl sulfoxides was described via stereoselective reduction of 1-alkynyl *p*-tolyl sulfoxides (eq 6).[14]

Alkynic sulfoxides have been made from trimethylsilylethynylmagnesium bromide and the resulting alkyne desilylated on silica gel (eq 7).[15]

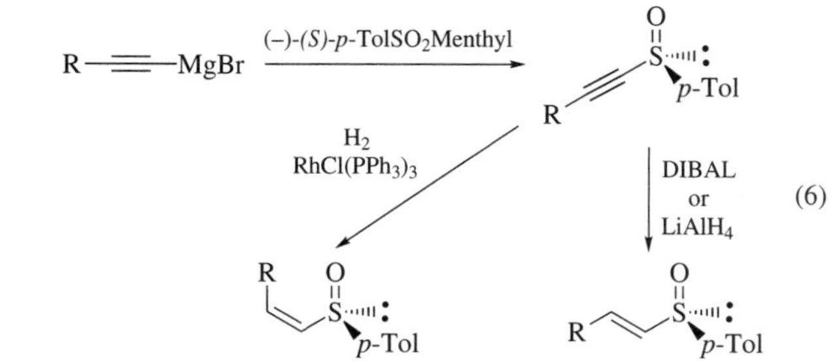

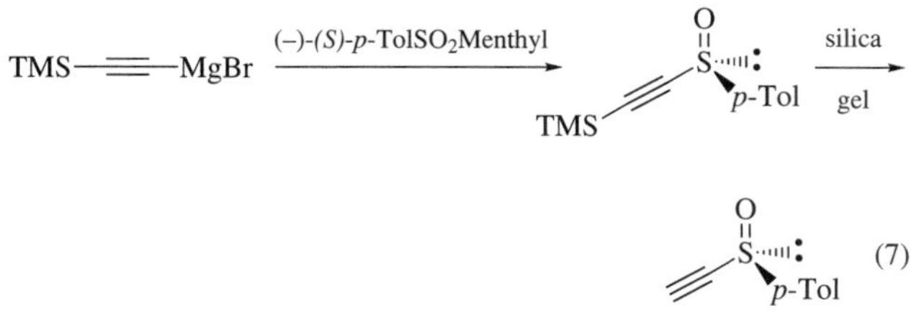

Chiral vinyl sulfoxides can also be prepared by Horner–Emmons reaction of carbonyl compounds with α-phosphoryl sulfoxides which are obtained from lithiated dimethyl methylphosphonate and (−)-menthyl (*S*)-*p*-toluenesulfinate (eq 8).[16] However, this reaction applied to carbonyl compounds often gives a mixture of the (*E*) and (*Z*) isomers of the vinylic sulfoxide.

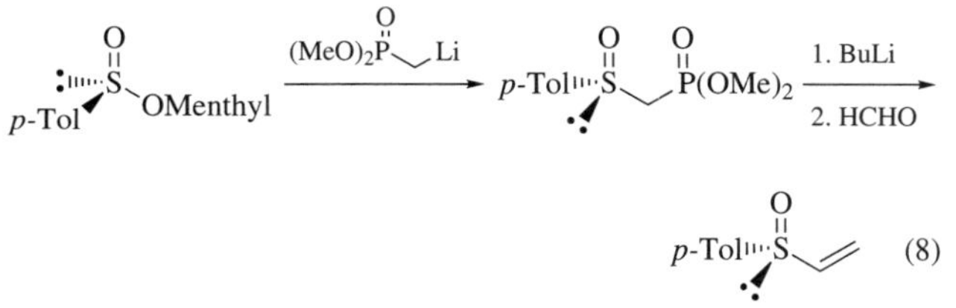

The reaction of α-phosphoryl sulfoxide with the dimethyl acetal of pyruvic aldehyde allowed the preparation of the corresponding vinylic sulfoxide as a 1:1 mixture of (*E*) and (*Z*) isomers which could be isomerized with *Lithium Diisopropylamide* to the lithiated (*E*) isomer, used for the asymmetric synthesis of α-tocopherol (eq 9).[17]

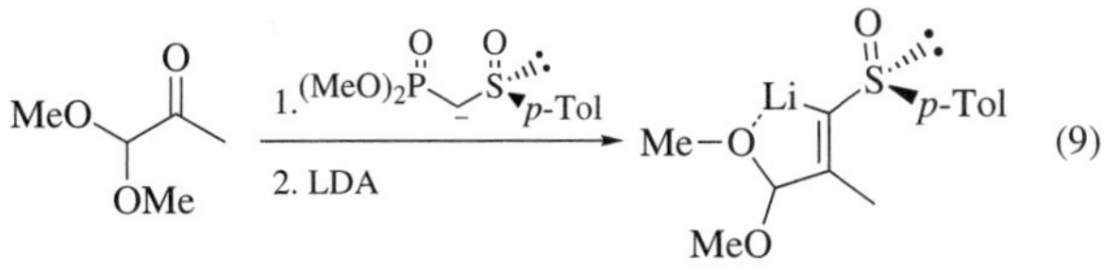

The Wittig reaction of an optically active sulfinylphosphonium ylide was reported to yield only the (*E*)-vinylic sulfoxides (eq 10).[18]

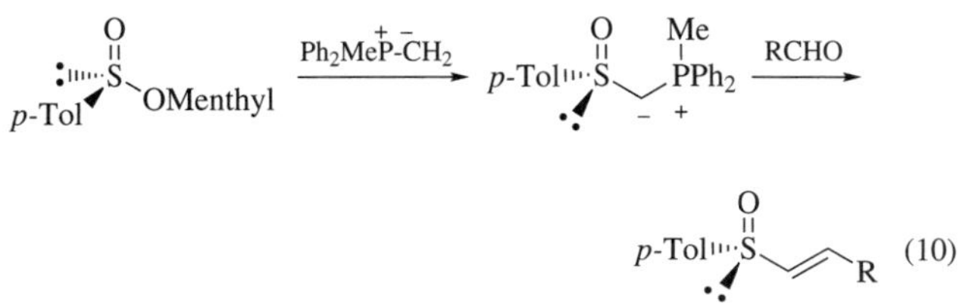

(*E*)

Diaryl Sulfoxides. Optically active diaryl sulfoxides are prepared by reaction of an aryl Grignard with (−)-menthyl (*S*)-*p*-toluenesulfinate: 2,5-dimethoxyphenyl *p*-tolyl sulfoxide (**4**), a precursor of sulfinyl quinones,[19] and 3-pyridyl *p*-tolyl sulfoxide (**5**), a precursor of sulfinyl dihydropyridines (studied as NADH model compounds)[20] are two typical examples.

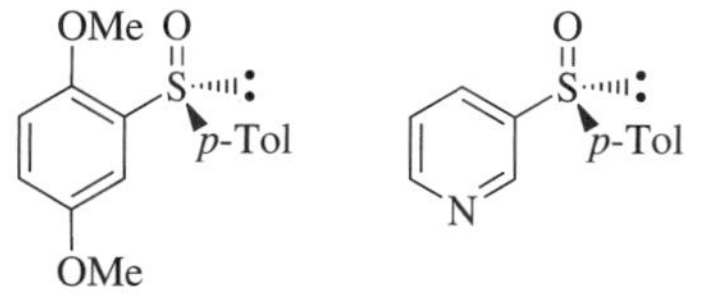

(**4**) (**5**)

Sulfinyl Esters and Derivatives. (*R*)-(+)-*t*-Butyl 2-(*p*-tolylsulfinyl)acetate is conveniently prepared by reaction of the magnesium enolate of *t*-butyl acetate (readily made with *Bromomagnesium Diisopropylamide*) with (−)-menthyl (*S*)-*p*-toluenesulfinate (eq 11).[21]

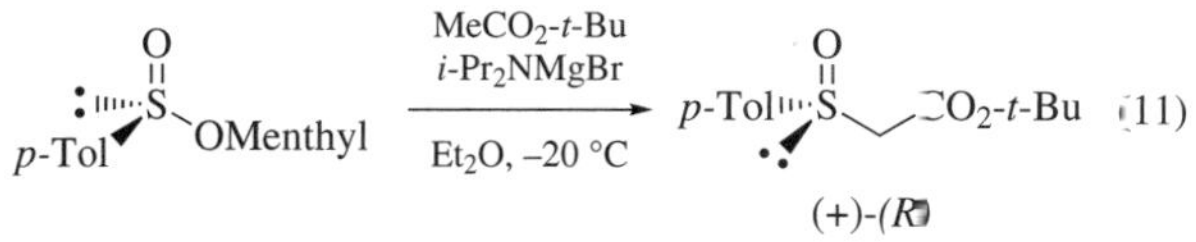

(+)-(*R*)

Substituted sulfinyl esters (**6**) have also been prepared by this reaction using the same base[22a] or lithium cyclohexyl-(isopropyl)amide,[22b] which gives higher yields.

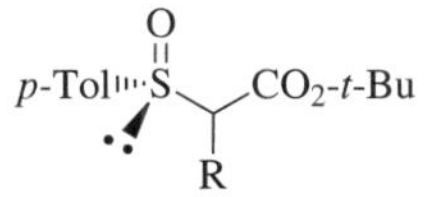

(**6**) R = Me, Et, $C_{14}H_{29}$

The anion of acetonitrile also reacts with (−)-menthyl (*S*)-*p*-toluenesulfinate to give the corresponding β-sulfinylacetonitrile (eq 12).[23]

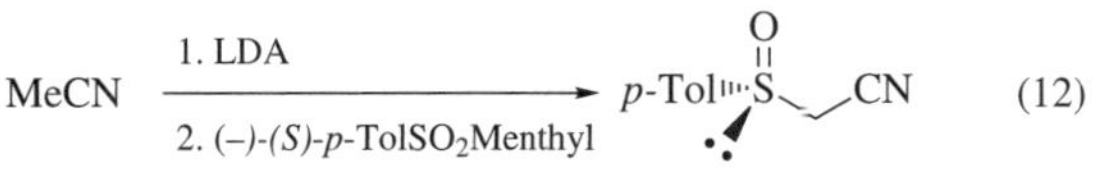

Similarly, *exo*-metalation with LDA of the racemic 3-methyl-4,5-dihydroisoxazole and reaction with (−)-menthyl (*S*)-*p*-toluenesulfinate afforded the sulfinyl-4,5-dihydroisoxazole as a diastereomeric mixture;[24] lithiated *N,N*-dimethylthioacetamide leads to the sulfinyl *N,N*-dimethylthioacetamide,[25] and lithiated ethyl *N*-methoxyacetimidate leads to *p*-tolylsulfinylethyl-*N*-methoxyacetimidate (eq 13).[26]

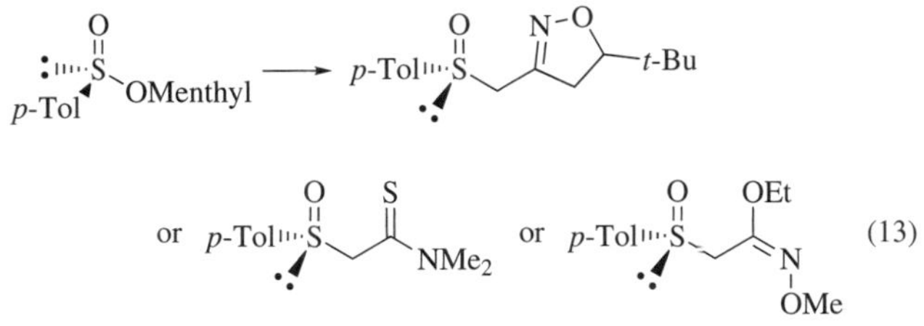

β-Keto Sulfoxides. Cyclic β-keto sulfoxides are readily obtained from the magnesium enolate of the ketone and (−)-menthyl (*S*)-*p*-toluenesulfinate[27] as a mixture of diastereomers in which the major epimer has the sulfoxide group in the equatorial orientation (eq 14).

By condensation of the dianion of *t*-butyl acetoacetate and (−)-menthyl (*S*)-*p*-toluenesulfinate, the corresponding β-keto sulfoxide was obtained in high yield (eq 15) and shown to be an efficient precursor of both enantiomers of β-hydroxybutyric acid via selective reduction of the ketone carbonyl group.[28] β,δ-Diketo sulfoxides were prepared in a similar way from diketone dianions (eq 16).[29]

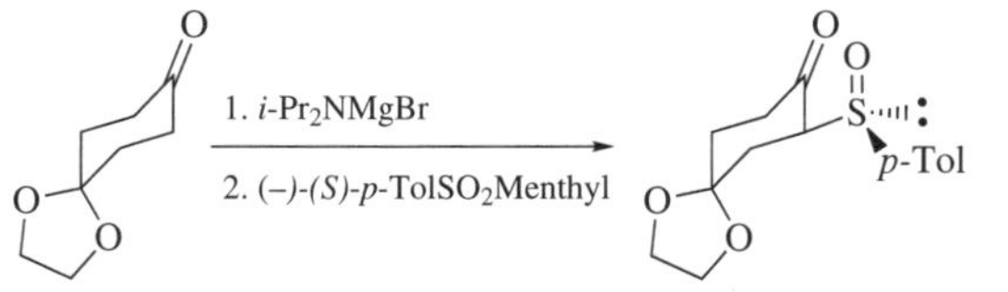

(14)

66% de

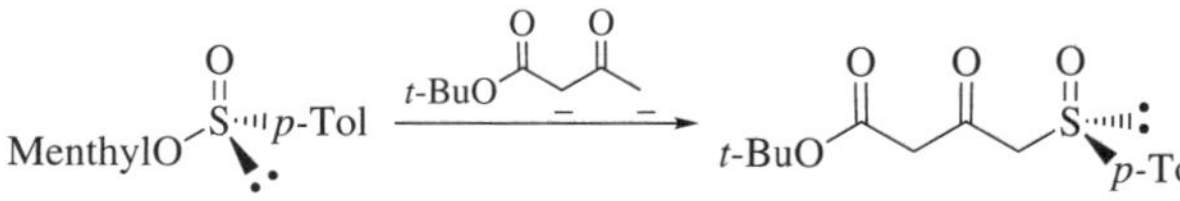

(15)

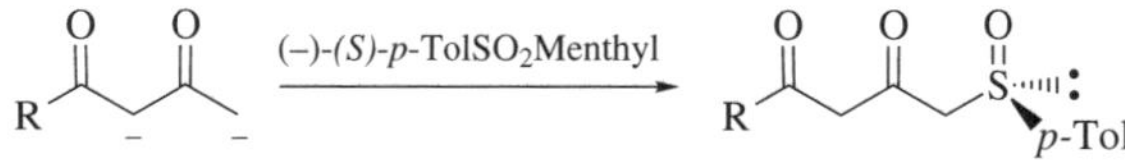

(16)

R = Me, Ph

Imino Sulfoxides. Metalated imines reacted with (−)-menthyl (*S*)-*p*-toluenesulfinate to yield the corresponding sulfinylimines as a diastereoisomeric mixture (eq 17).[30]

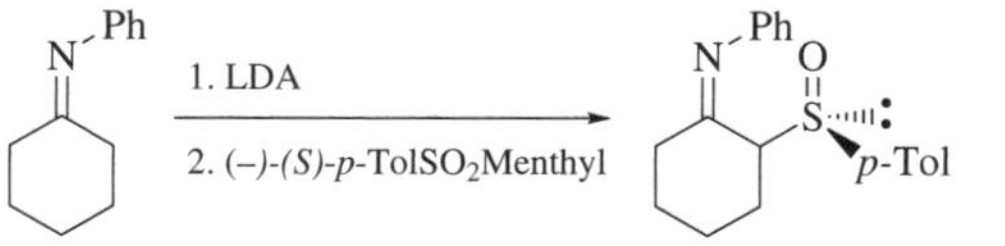

(17)

Similarly, *exo*-metalated cyclic imines afforded the sulfinylimines as an alkaloid precursor (eq 18).[31]

(18)

Miscellaneous. (*S,S*)-Bis(*p*-tolylsulfinyl)methane (**7**) is readily prepared from (−)-menthyl (*S*)-*p*-toluenesulfinate and (*R*)-methyl *p*-tolyl sulfoxide.[32] (+) (*S*)-*p*-Tolylsulfinylmethyl *t*-butyl sulfone (**8**) was made from the *t*-butyl methyl sulfone anion and (−)-menthyl (*S*)-*p*-toluenesulfinate.[33]

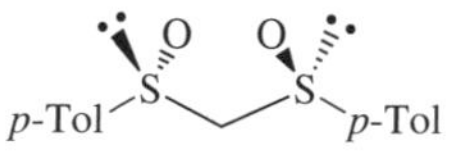

(**7**)

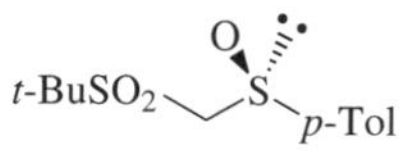

(**8**)

Chiral *N*-benzylidene *p*-toluenesulfinamides were prepared by reaction of benzonitrile with an alkyllithium followed by addition of (−)-menthyl (*S*)-*p*-toluenesulfinate and converted into optically active amines and amino acids (eq 19).[34]

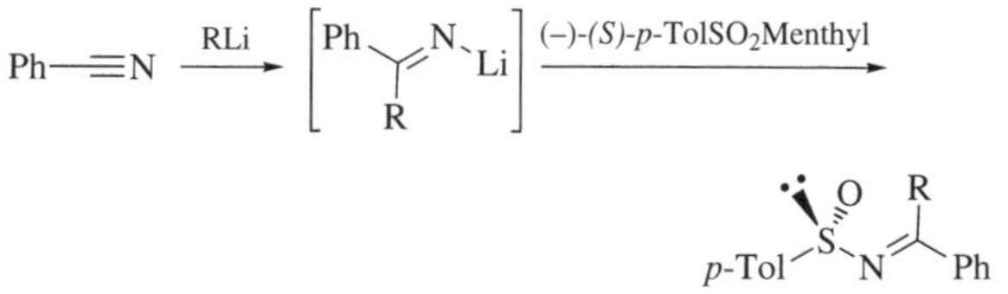

(19)

R = Me (50%), Bu (75%)

1. Solladié, G. *Synthesis* **1981**, 185.
2. Mioskowski, C.; Solladié, G. *Tetrahedron* **1980**, *36*, 227.
3. Solladié, G.; Hutt, J.; Girardin, A. *Synthesis* **1987**, 173.
4. Mislow, K.; Green, M. M.; Laur, P.; Melillo, J. T.; Simmons, T.; Ternay, A. L. *J. Am. Chem. Soc.* **1965**, *87*, 1958.
5. Andersen, K. K. *J. Org. Chem.* **1964**, *29*, 1953.
6. Arai, Y.; Kuwayama, S.; Takeuchi, Y.; Koizumi, T. *Tetrahedron Lett.* **1985**, *26*, 6205.
7. Bravo, P.; Resnati, G.; Viani, F.; Arnone, A. *Tetrahedron* **1987**, *43*, 4635.
8. Jacobus, J.; Mislow, K. *J. Am. Chem. Soc.* **1967**, *89*, 5228.
9. Solladié, G.; Zimmermann, R.; Bartsch, R. *Synthesis* **1985**, 662.
10. Iwata, C.; Fujita, M.; Hattori, K.; Uchida, S.; Imanishi, T. *Tetrahedron Lett.* **1985**, *26*, 2221.
11. Arnone, A.; Bravo, P.; Cavicchio, G.; Frigerio, M.; Viani, F. *Tetrahedron* **1992**, *48*, 8523.
12. (a) Posner, G. H.; Tang, P. W. *J. Org. Chem.* **1978**, *43*, 4131. (b) Marino, J. P.; Fernández de la Pradilla, R.; Laborde, E. *Synthesis* **1987**, 1088.
13. Hulce, M.; Mallamo, J. P.; Frye, L. L.; Kogan, T. P.; Posner, G. H. *Org. Synth.* **1986**, *64*, 196.
14. Kosugi, H.; Kitaoka, M.; Tagami, K.; Takahashi, A.; Uda, H. *J. Org. Chem.* **1987**, *52*, 1078.
15. Lee, A. W. M.; Chan, W. H.; Lee, Y. K. *Tetrahedron Lett.* **1991**, *32*, 6861.
16. (a) Mikolajczyk, M.; Grzejszczak, S.; Zatorski, A. *J. Org. Chem.* **1975**, *40*, 1979. (b) Mikolajczyk, M.; Midura, W.; Grzejszczak, S.; Zatorski, A.; Chefczyńska, A. *J. Org. Chem.* **1978**, *43*, 473.
17. Moine, G.; Solladié, G. *J. Am. Chem. Soc.* **1984**, *106*, 6097.
18. Mikolajczyk, M.; Perlikowska, W.; Omelańczuk, J.; Cristau, H. J.; Perraud-Darcy, A. *Synlett* **1991**, 913.
19. (a) Carreño, C. M.; García Ruano, J. L.; Urbano, A. *Tetrahedron Lett.* **1989**, *30*, 4003. (b) Carreño, C. M.; García Ruano, J. L.; Mata, J. M.; Urbano, A. *Tetrahedron* **1991**, *47*, 605.
20. Imanishi, T.; Hamano, Y.; Yoshikawa, H.; Iwata, C. *Chem. Commun.* **1988**, 473.
21. Mioskowski, C.; Solladié, G. *Tetrahedron* **1980**, *36*, 227.
22. (a) Solladié, G.; Matloubi-Moghadam, F.; Luttmann, C.; Mioskowski, C. *Helv. Chim. Acta* **1982**, *65*, 1602. (b) Nokami, J.; Ohtsuki, H.; Sokamoto, Y.; Mitsuoka, M.; Kunieda, N. *Chem. Lett.* **1992**, 1647.
23. Nokami, J.; Mandai, T.; Nishimura, A.; Takeda, T.; Wakabayashi, S.; Kunieda, N. *Tetrahedron Lett.* **1986**, *27*, 5109.
24. Annunziata, R.; Cinquini, M.; Cozzi, F.; Gilardi, A.; Restelli, A. *J. Chem. Soc., Perkin Trans. 1* **1985**, 2289.
25. Cinquini, M.; Manfredi, A.; Molinari, H.; Restelli, A. *Tetrahedron* **1985**, *41*, 4929.
26. Bernardi, A.; Colombo, L.; Gennari, C.; Prati, L. *Tetrahedron* **1984**, *40*, 3769.
27. (a) Carreño, M. C.; García Ruano, J. L.; Rubio, A. *Tetrahedron Lett.* **1987**, *28*, 4861. (b) Carreño, M. C.; García Ruano, J. L.; Pedregal, C.; Rubio, A. *J. Chem. Soc., Perkin Trans. 1* **1989**, 1335. (c) Carreño, M. C.; García Ruano, J. L.; Garrido, M.; Ruiz, M. P.; Solladié, G. *Tetrahedron Lett.* **1990**, *31*, 6653.
28. (a) Schneider, F.; Simon, R. *Synthesis* **1986**, 582. (b) Solladié, G.; Almario, A. *Tetrahedron Lett.* **1992**, *33*, 2477.
29. Solladié, G.; Ghiatou, N. *Tetrahedron: Asymmetry* **1992**, *3*, 33.
30. Carreño, M. C.; García Ruano, J. L.; Dominguez, E.; Pedregal, C.; Rodríguez, J. *Tetrahedron* **1991**, *47*, 10035.
31. (a) Hua, D. H.; Miao, S. W.; Bravo, A. A.; Takemoto, D. J. *Synthesis* **1991**, 970. (b) Hua, D. H.; Bharathi, S. N.; Panangadan, J. A. K.; Tsujimoto, A. *J. Org. Chem.* **1991**, *56*, 6998.
32. (a) Kunieda, N.; Nokami, J.; Kinoshita, M. *Bull. Chem. Jpn.* **1976**, *49*, 256. (b) Solladié, G.; Colobert, F.; Ruiz, P.; Hamdouchi, C.; Carreño, C. M.; García Ruano, J. L. *Tetrahedron Lett.* **1991**, *32*, 3695.

33. López, R.; Carretero, J. C. *Tetrahedron: Asymmetry* **1991**, *2*, 93.
34. Hua, D. H.; Miao, S. W.; Chen, J. S.; Iguchi, S. *J. Org. Chem.* **1991**, *56*, 4.

Guy Solladié & Françoise Colobert
University Louis Pasteur, Strasbourg, France

(*R,S,R,S*)-Me-PennPhos

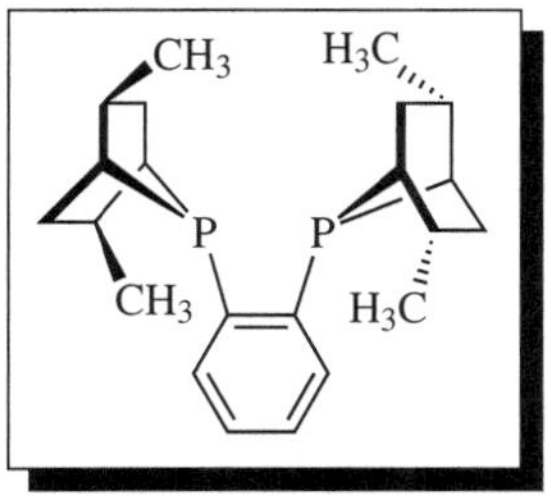

[201049-04-9] $C_{22}H_{32}P_2$ (MW 358.44)

(optically active reagent used as a ligand in homogeneous transition metal-chiral phosphine catalyst systems for asymmetric homogeneous reactions)

Alternate Name: (*R,S,R,S*)-*P,P*′-1,2-phenylenebis(*endo*-2,5-dimethyl-7-phospha-bicyclo[2.2.1]heptane).
Physical Data: viscous oil; $[\alpha]_D$ +221.8° (c = 1.01, $CHCl_3$).[1]
Solubility: soluble in $CHCl_3$ and other common organic solvents.
Form Supplied in: not commercially available.
Preparative Methods: it can be prepared from *p*-xylene[2] and **1,2-phenylenediphosphorane**[3] in four synthetic steps (eq 1).

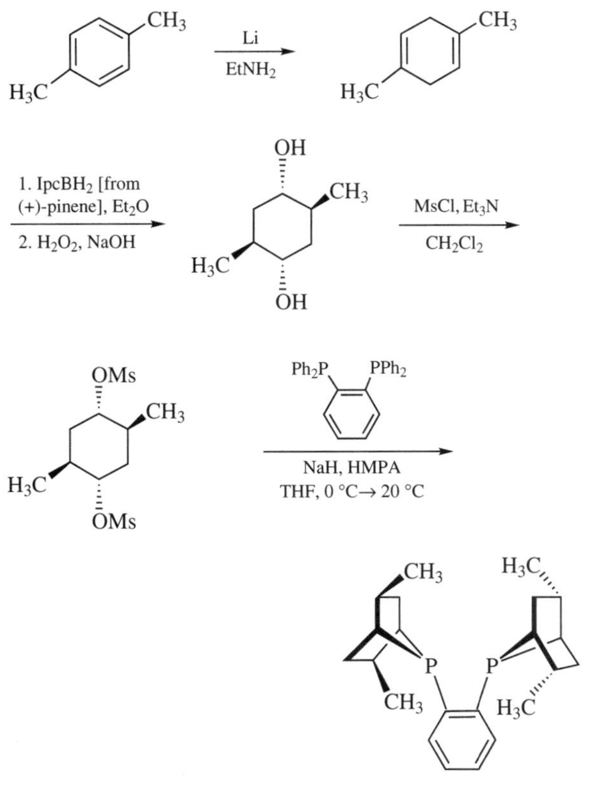

(1)

Purification: details regarding purification of PennPhos have not been published.

Handling, Storage, and Precautions: PennPhos is air-sensitive; storage of PennPhos and all operations that involve handling of PennPhos should be performed under an inert atmosphere. In general, aryldialkylphosphines are irritants; skin contact should be avoided, and care should be exercised to avoid vapor inhalation.

Enantioselective Hydrogenation of Alkenes. (*R,S,R,S*)-Me-PennPhos has been employed as catalyst in combination with Rh(I) for enantioselective hydrogenation of alkene carbon-carbon double bonds in a variety of substrates. A representative sampling of these asymmetric hydrogenations is shown in Table 1.[4,5] Changing solvents was found to have a small effect on the enantioselectivity of the hydrogenation reactions listed in Table 1. However, conversions (i.e., chemical yield of hydrogenation products) varied widely. Thus, when *N*-(3,4-dihydro-1-naphthyl)acetamide was employed as substrate (see Table 1, entry 3), conversion was highest (ca. 100%) in MeOH, CH_2Cl_2, and *i*-PrOH and lowest when toluene was employed as solvent.[5]

Among the various Rh-phosphine catalysts used to perform enantioselective hydrogenation of *N*-(3,4-dihydro-1-naphthyl)acetamide, Rh-(*R,S,R,S*)-Me-PennPhos afforded the desired hydrogenation product in highest optical yield. Thus, the use of Rh-L catalysts, where L = **(*S,S*)-(+)-DIOP (1)**,[6] **(*R*)-(+)-BINAP (2)**,[7] and **(*R,R*)-(−)-Me-DuPhos (3)**,[8] afforded *N*-(1,2,3,4-tetrahydro-1-naphthyl)acetamide in only 10% ee(*S*), 24% ee (*R*), and 1% ee (*R*, with 57% conversion).[5]

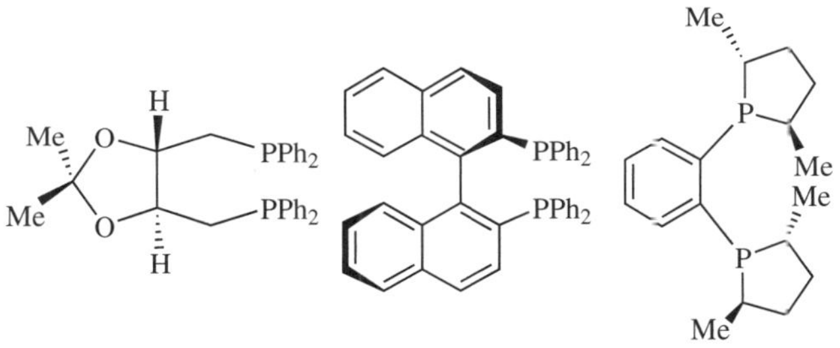

1 2 3

Enantioselective Hydrogenation of Ketones. (*R,S,R,S*)-Me-PennPhos has been employed as catalyst in combination with Rh(I) for highly enantioselective homogeneous hydrogenation of prochiral aralkyl ketones and purely aliphatic ketones.[1] This result is particularly significant in view of the fact that ketones behave as poor ligands toward Rh(I); accordingly, most Rh-phospholane complexes have proved to be ineffective as catalysts for hydrogenation of simple aliphatic ketones.[1,9] Optical yields have been optimized via addition of KBr or weak bases (e.g., 2,6-lutidine).[1] Optimal results obtained for Rh-(*R,S,R,S*)-Me-PennPhos–mediated hydrogenation of acetophenone are shown in eq 2.

Longer reaction times are required to achieve maximum conversion when aliphatic ketones are hydrogenated in the presence of Rh-(*R,S,R,S*)-Me-PennPhos catalyst. Thus, 2-hexanone is reduced by H_2 [30 atm, Rh-(*R,S,R,S*)-Me-PennPhos catalyst in the presence of KBr (1 equiv)] at room temperature. After 48 h, the chemical yield of (*S*)-2-hexanol is 96% (optical yield: 75% ee).[1]

In general, it was observed that introduction of bulky (i.e., branched) substituents into aliphatic methyl ketones of the type $RC(O)CH_3$ dramatically reduced the reactivity of the C=O group in these substrates toward H_2-Rh-(*R,S,R,S*)-Me-PennPhos

Table 1 Asymmetric hydrogenations of prochiral alkenes catalyzed by a (*R,S,R,S*)-Me-PennPhos-Rh(I) complex

Entry	Substrate	Rh(I) species	Conditions	Catalyst (substrate: [RH]:ligand ratio)	Product	ee (%) (configuration) [conversion, %]	Ref.
1	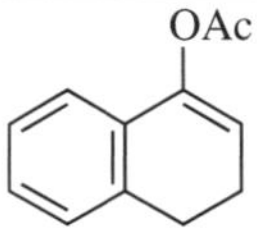	$[Rh(cod)_2]\ BF_4$	H_2 (1.7 bar) MeOH, 24 h, 25 °C	(*R,S,R,S*)-Me-PennPhos (100:1:1.1)	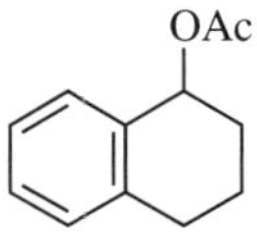	99.1 (*R*)[100]	4
2	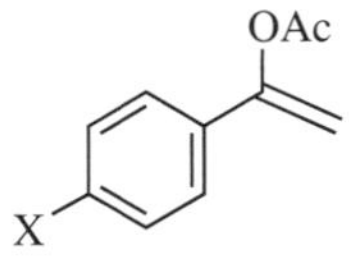 X = H, Me, OMe, Cl	$[Rh(cod)_2]\ BF_4$	H_2 (1.7 bar) THF, 24 h, 25 °C	(*R,S,R,S*)-Me-PennPhos (100:1:1.1)	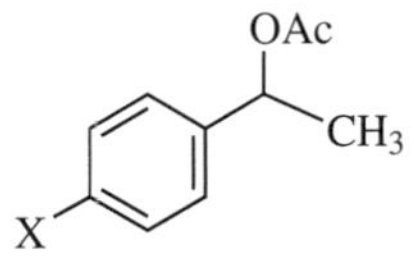	80.9Ð84.8 (*R*) [100]	4
3	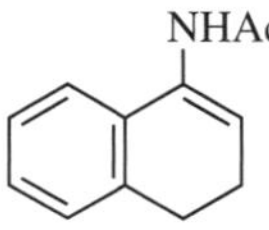	$[Rh(cod)_2]\ BF_4$	H_2 (40 psi(gauge)) MeOH, 24 h, 25 °C	(*R,S,R,S*)-Me-PennPhos (100:1:1.1)	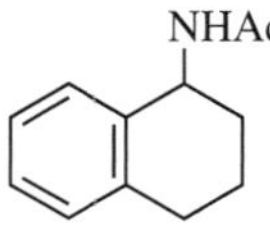	97 (*R*) [100]	5
4	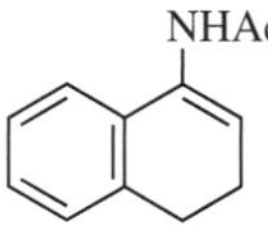	$[Rh(cod)\text{-}Cl]_2$	H_2 (40 psi(gauge)) MeOH, 24 h, 25 °C	(*R,S,R,S*)-Me-PennPhos (100:1:1.1)	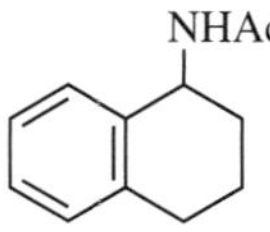	92 [100]	5
5	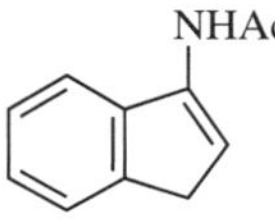	$[Rh(cod)_2]\ PF_6$	H_2 (40 psi(gauge)) MeOH, 20 h, 25 °C	(*R,S,R,S*)-Me-PennPhos (100:1:1.1)	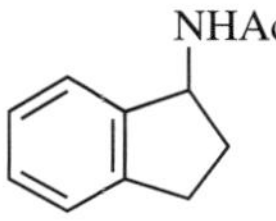	98 [100]	5
6	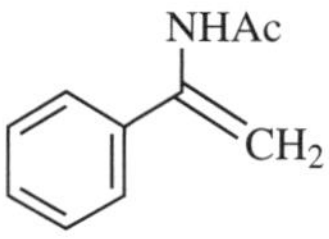	$[Rh(cod)_2]\ PF_6$	H_2 (40 psi(gauge)) MeOH, 20 h, 25 °C	(*R,S,R,S*)-Me-PennPhos (100:1:1.1)	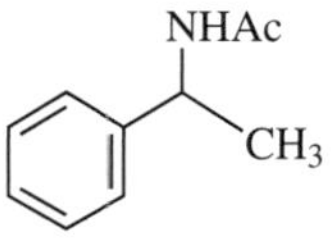	75 [100]	5

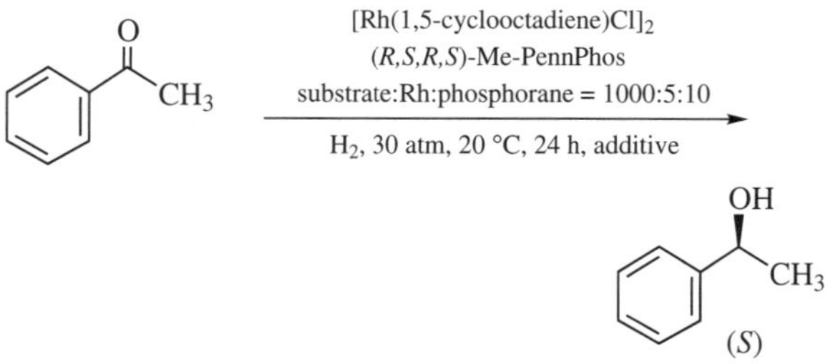

(2)

Additive = KBr Additive:Rh = 1:1 89% conversion; 92% ee	Additive = 2,6-lutidine Additive:Rh = 0.3:1 97% conversion; 95% ee

catalyst but resulted in an increased optical yield of the product alcohol. Thus, after a reaction time of 106 h, Rh-(*R,S,R,S*)-Me-PennPhos–mediated hydrogenation of cyclohexyl methyl ketone afforded the correponding (*S*)-carbinol in 90% chemical yield (92% ee).[1]

Related Reagents. (*S,S*)-(+)-DIOP;[6] (*R*)-(+)-BINAP;[7] (*R,R*)-(−)-Me-DuPhos.[8]

1. Jiang, Q.; Jiang, Y.; Ziao, D.; Cao, P.; Zhang, X. *Angew. Chem., Int. Ed. Engl.* **1998**, *37*, 1100–1103.
2. (a) Chen Z., Halterman, R. L. *Synlett* **1990**, 103–105. (b) Chen, Z.; Halterman, R. L. *J. Am. Chem. Soc.* **1992**, *114*, 2276–2277.
3. Kyba, E. P.; Liu, S.-T.; Harris, R. L. *Organometallics* **1983**, *2*, 1877–1879.

4. Jiang, Q.; Xiao, D.; Zhang, P. C.; Zhang, X. *Angew. Chem., Int. Ed. Engl.* **1999**, *38*, 516–518.
5. Zhang, Z.; Zhu, G.; Jiang, Q.; Xiao, D.; Zhang, X. *J. Org. Chem.* **1999**, *64*, 1774–1775.
6. Kagan, H. B.; Dang, T.-P. *J. Am. Chem. Soc.* **1972**, *94*, 6429–6433.
7. (a) Miyashita, A.; Yasuda, A.; Takaya, H.; Toriumi, K.; Ito, T.; Souchi, T.; Noyori, R. *J. Am. Chem. Soc.* **1980**, *102*, 7932–7934. (b) Miyashita, A.; Takaya, H.; Souchi, T.; Noyori, R. *Tetrahedron* **1984**, *40*, 1245–1253. (c) Takaya, H.; Masima, K.; Koyano, K.; Yagi, M.; Kumobayashi, H.; Taketomi, T.; Akutagawa, S.; Noyori, R. *J. Org. Chem.* **1986**, *51*, 629–635.
8. Burk, M. J. *J. Am. Chem. Soc.* **1991**, *113*, 8518–8519.
9. Fehring, V.; Selke, R. *Angew. Chem., Int. Ed. Engl.* **1998**, *37*, 1827–1830.

Alan P. Marchand
Jaroslaw Romanski
T. Pavan Kumar
University of North Texas, Denton, TX, USA
University of Lodz, Poland
University of North Texas, Denton, TX, USA

(*R, R*)-1,2-(Methanesulfonamido)-cyclohexane

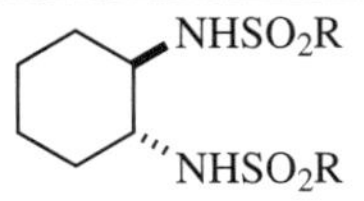

1a R = Me
1b R = CF_3 [CAS 122833-60-7]
1c R = *n*-Bu [CAS 290833-56-6]
1d R = 4-NO_2Ph [CAS 155237-72-2]
1e R = *i*-Pr [CAS 166109-85-9]
1f R = mesityl [CAS 263019-97-2]

[122833-58-3] $C_8H_{18}N_2O_4S_2$ (MW 270.36)

(catalyst for organozinc-mediated additions to aldehydes,[1] catalyst for Simmons–Smith type cyclopropanation of allylic alcohols[2])

Physical Data: mp 157 °C; $[\alpha]_D^{20}$ −20.1 (*c* 3.07, pyridine).
Solubility: soluble in most organic solvents except hydrocarbons.
Form Supplied in: white solid.
Preparative Methods: The enantiopure sulfonamide **1a** is prepared via sulfonylation of (*R*,*R*)-1,2-diaminocyclohexane **2** in the presence of an excess of triethylamine (eq 1).[3] Use of excess amine base is essential for obtaining a high yield of the bis-sulfonamide. Synthesis of related bis-sulfonamides is easily accomplished by substituting the desired sulfonyl chloride in the former procedure. Recrystallization of the bis-sulfonamide **1a** from hexane/ethyl acetate and drying over P_2O_5 allows for isolation of the analytically pure reagent. Methanesulfonyl chloride and (*R*,*R*)-1,2-diaminocyclohexane **2** are commercially available from a number of sources. However it should be noted that racemic 1,2-diaminocyclohexane **2** can be resolved via formation of the tartrate salt.[4] Typically, the diamine can be obtained in >99:1 enantiomeric ratio (er) after two crystallizations from water. Determination of the enantiopurity of the diamine is accomplished via formation of the bis-3-toluyl amide and analysis via chiral stationary phase HPLC (Chiralcel AD; hexane/*i*-PrOH; 95:5, 1.0 mL min^{-1}).

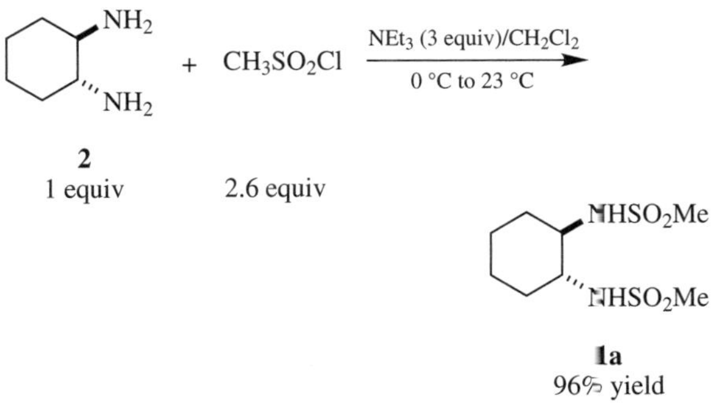

(1)

Handling, Storage, and Precautions: The sulfonamide is a shelf-stable, non-hygroscopic compound which does not require special precautions for storage or handling.

Introduction. The 1,2-bis-(methanesulfonamido)-cyclohexane **1a** is an important member of a larger class of C_2-symmetric bis-sulfonamide ligands which have had a powerful impact on the field of organozinc chemistry.[1,2] The success of these ligands is, in part, due to the straightforward installation of a variety of sulfonamide groups, providing access to a wide array of sterically and electronically diverse ligands.

Additions to Aldehydes. Alkylation of aromatic and aliphatic aldehydes with a combination of titanium tetraisopropoxide, Ti(O-*i*-Pr)$_4$, and diethylzinc, $ZnEt_2$, in the presence of a catalytic amount of the bis-sulfonamide **1a** leads to formation of (*S*)-1-phenyl-1-propanol **4** with high enantioselectivity (eq 2, Table 1).[5] Use of the (*R*,*R*)-1,2-(trifluoromethanesulfonamido)-cyclohexane **1b** [CAS 122833-60-7] allows for an equally selective reaction, but at exceptionally low catalyst loadings. In the case of aromatic aldehydes, these reactions are fairly rapid, requiring at most 2 hours to reach full conversion.

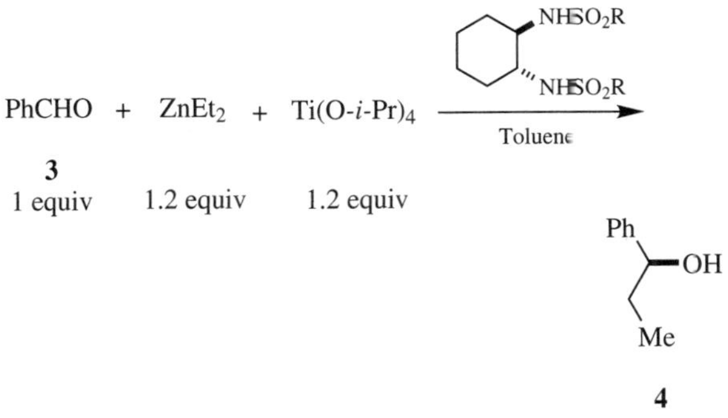

(2)

Table 1 Alkylation of benzaldehyde in the presence of sulfonamide catalysts **1a–b**

R	Cat. loading (%)	Temp. (°C)	Yield (%)	**4** (er)
Me (**1a**)	4	23	90	95:5
Me (**1a**)	4	0	97	83:17
CF_3 (**1b**)	4	0	99	99:1
CF_3 (**1b**)	0.05	−20	97	99:1

er, enantiomeric ratio.

Table 2 Substrate scope in the alkylation of aldehydes with sulfonamide **1b**

Entry	R^1	R^2	Loading (%)	Temp. (°C)	Yield (%)	Product	er
1	PhCH=CH (**5**)	CH_3CH_2	2	−50	85	**6**	>99:1
2	$PhCH_2CH_2$ (**7**)	CH_3CH_2	1	0	95	**8**	96:4
3	n-C_5H_{11} (**9**)	CH_3CH_2	4	−20	87	**10**	>99:1
4	Ph (**3**)	CH_3	4	0	99	**11**	86:14
5	Ph (**3**)	n-C_4H_9	2	−20	98	**12**	99:1
6	Ph (**3**)	n-C_5H_{11}	2	−50	99	**13**	>99:1

er, enantiomeric ratio.

Through the use of the bis-sulfonamide **1b**, the scope of the reaction has been expanded to include a larger number of aldehydes and organozinc reagents (Table 2). High yields and selectivities are obtained in the alkylations of conjugated aldehydes (**5**) as well as simple aliphatic aldehydes (**7**,**9**). The broad scope of this reaction with respect to the electrophile contrasts the slightly limited scope of the reaction when considering the structure of the nucleophile. The use of small alkylzinc reagents, such as dimethylzinc, leads to a depressed selectivity (entry 4). However, the use of larger alkylzinc reagents still provides the exceptional selectivity observed in the case of diethylzinc (entries 5 and 6).

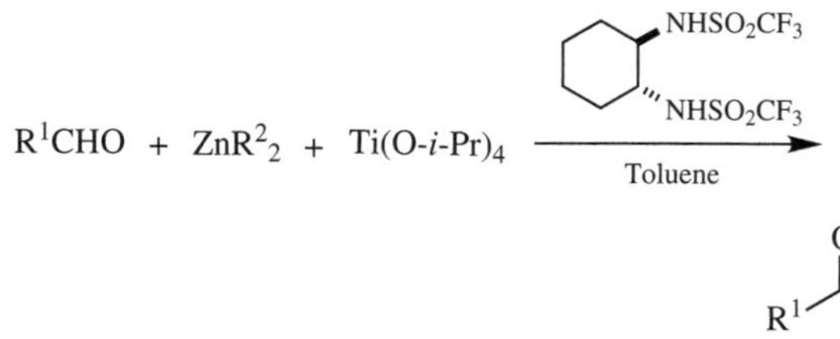

(3)

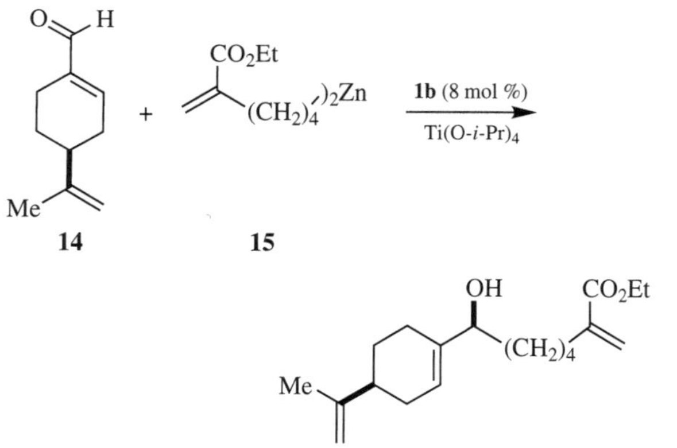

14 **15** (4)

16
71% yield (93:7 er)

The scope of the reactive partners has been fully explored and expanded to include a diversity of functionalized organozinc reagents. Preparation of the functionalized organozinc reagent proceeds via hydroboration and boron-zinc exchange of a simple terminal alkene. The resulting organozinc reagent can then be used in an identical manner to that shown above. In the presence of <10 mol % of catalyst **1b**, high yields and selectivities can be obtained (eq 4).[6] One drawback of this method is that 50% of the starting alkene must be sacrificed. However, recent reports have revealed that use of a mixed organozinc species, which is accessible by disproportionation of two symmetric organozinc reagents, obviates this wasteful complication (eq 5).[7]

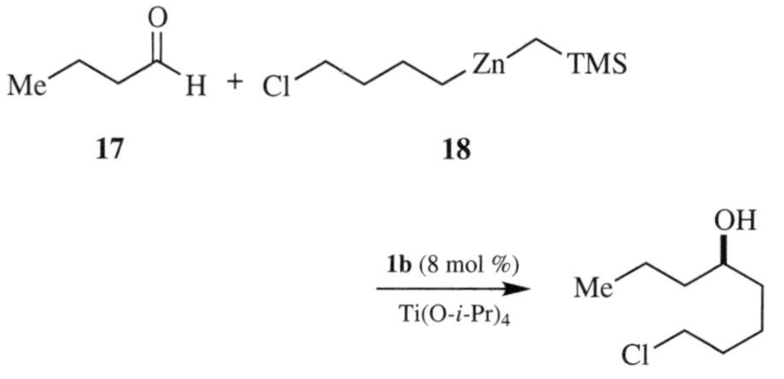

17 **18** (5)

19
86% yield (97:3 er)

Comparable selectivity can be obtained in the alkylation of benzaldehyde **3** with diethylzinc using the titanium TADDOL complex **20** (>99:1 er) or 3-*exo*-(dimethylamino)isoborneol, **21** (>99:1 er), although both methods employ higher catalyst loadings.[8,9] While benzaldehyde is illustrative, the substrate scope is equally broad in the case of these two catalysts.

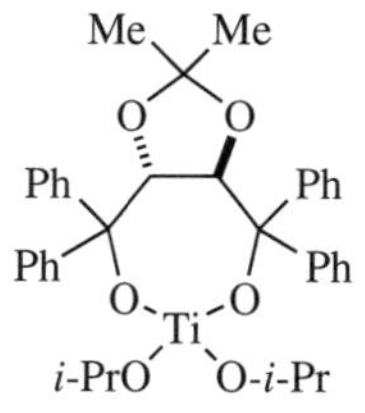

20 (95:5 er)

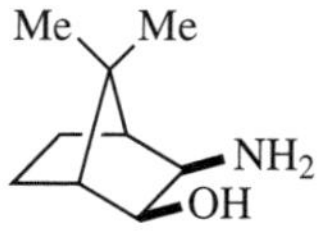

21 (>99:1 er)

Cyclopropanation of Allylic Alcohols. Simmons–Smith type cyclopropanation of the allylic alcohol **22** in the presence of a catalytic amount of the bis-sulfonamide **1a** leads to formation of the corresponding cyclopropane **23** in high yield and selectivity (eq 6, Table 3).[10] The reaction is rapid (<1 h) and can be performed at low temperature (either 0 °C or −20 °C). Substrate scope encompasses both di- and tri-substituted allylic alcohols (**24** and **26**). However, substitution at the 2 position, as in **28**, leads to a drastic decrease in selectivity. The presence of additional oxygenated functionality (**30**) in the proximity of the alkene also lessens selectivity.[11] The method is limited to the cyclopropanation of allylic alcohols. Other alkene-containing substrates, such as allylic ethers, homo-allylic alcohols and allylic carbamates, do not react with high selectivity.

Table 3 Substrate generality in the cyclopropanation using sulfonamide **1a**

Entry	Substrate	R^1	R^2	R^3	Yield (%)	Product	er
1	**22**	Ph	H	H	92	**23**	95:5
2	**24**	$PhCH_2CH_2$	H	H	88	**25**	95:5
3	**26**	Ph	Me	H	92	**27**	95:5
4	**28**	Ph	H	Me	91	**29**	51:49
5	**30**	$BnOCH_2$	H	H	70	**31**	68:32

er, enatiomeric ratio.

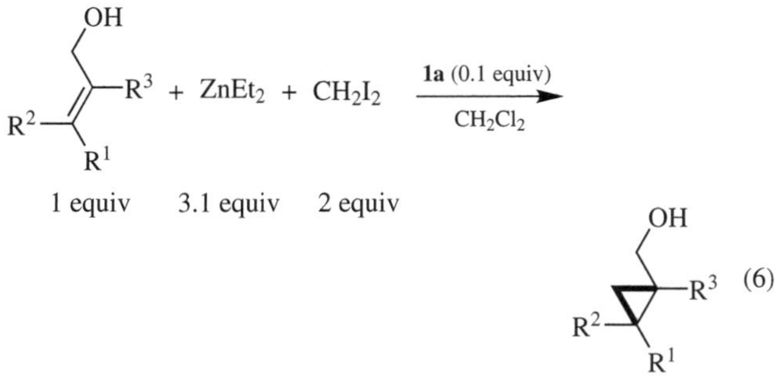

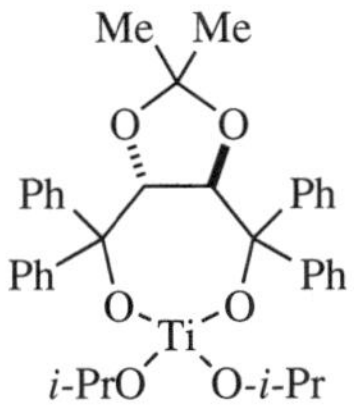

20 (95:5 er)

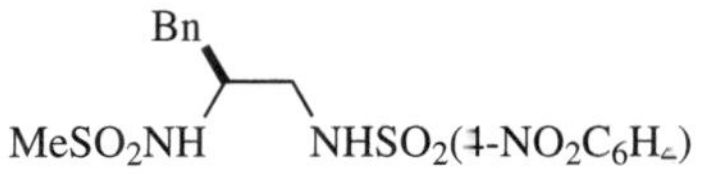

32 (93:7 er)

The optimal procedure calls for a three-flask protocol which segregates the individual reactive components. Pre-formation of the zinc alkoxide, zinc sulfonamide complex and the cyclopropanation reagent, $Zn(CH_2I)_2$, by combination of diethylzinc with the allylic alcohol, bis-sulfonamide and diiodomethane, respectively, is essential for high selectivity and reproducibility. While the individual reaction components are soluble in halogenated solvents such as dichloromethane, the zinc sulfonamide complex is a highly insoluble species which is prone to aggregation. Because of the nature of the zinc carbenoid, a heterogenous reaction is always observed. None of the related bis-sulfonamide catalysts shown in Table 4 are able to dissolve the precipitate. Still, a survey of catalyst structure reveals that large variations in sulfonamide structure can be tolerated without compromising selectivity (entries 1 and 2).[10,12] Bulky sulfonamide groups, however, clearly interfere with the selective cyclopropanation process (entries 3 and 4).

Table 4 Selectivity of various sulfonamides in the cyclopropanation of **22**

Entry	R	Compound	**23** (er)
1	*n*-Bu	**1c**	92:8
2	$4\text{-}NO_2C_6H_4$	**1d**	89:11
3	*i*-Pr	**1e**	86:14
4	$2,4,6\text{-}MeC_6H_2$	**1f**	62:38

This method is comparable to similar, catalytic Simmons–Smith-type methods employing the titanium TADDOL catalyst **20** (95:5 er) or the C_1-symmetric bis-sulfonamide catalyst **32** (93:7 er) for the cyclopropanation of the allylic alcohol **22** (eq 6).[13,14] However, due to the preliminary nature of these earlier investigations, substrate scope and generality have not been extensively documented. All of the aforementioned methods are limited by their dependence on the allylic alcohol functionality. Only one method for Simmons–Smith-type cyclopropanation of other substrate classes has been developed. Use of a stoichiometric, chiral dioxaborolane [CAS 161344-85-0] additive allows for selective cyclopropanation of allylic ethers, homo-allylic alcohols and allylic carbamates.[15]

1. (a) Pu, L.; Yu, H. B. *Chem. Rev.* **2001**, *101*, 757. (b) Soai, K.; Shibata, T. In *Comprehensive Asymmetric Catalysis II*; Jacobsen, E. N.; Pfaltz, A.; Yamamoto, H., Eds.; Springer Verlag: Heidelberg, 1999, Ch. 26.1, pp 911–922.
2. (a) Charette, A. B.; Beauchemin, A. *Org. React.* **2001**, *58*, 1–415. (b) Charette, A. B.; Lebel, H. In *Comprehensive Asymmetric Catalysis II*; Jacobsen, E. N.; Pfaltz, A.; Yamamoto, H., Eds.; Springer Verlag: Heidelberg, 1999, Ch. 16.3, p 581–603.
3. Denmark, S. E.; Christenson, B. L.; O'Connor, S. P. *Tetrahedron Lett.* **1995**, *36*, 2219.
4. (a) Glasbøl, F.; Steenbøl, P.; Søndergaard-Sørensen, B. *Acta Chem. Scand.* **1972**, *26*, 3605. (b) Whitney, T. A. *J. Org. Chem.* **1980**, *45*, 4214.
5. (a) Yoshioka, M.; Kawakita, T.; Ohno, M. *Tetrahedron Lett.* **1989**, *30*, 1657. (b) Takahashi, H.; Kawakita, T.; Ohno, M.; Yoshioka, M.; Kobayashi, S. *Tetrahedron* **1992**, *48*, 5691.
6. Langer, F.; Schwink, L.; Devasagayaraj, A.; Chavant, P.-Y.; Knochel, P. *J. Org. Chem.* **1996**, *61*, 8229.
7. (a) Lutz, C.; Knochel, P. *J. Org. Chem.* **1997**, *62*, 7895–7898. (b) Lutz, C.; Jones, P.; Knochel, P. *Synthesis* **1999**, 312.
8. (a) Schmidt, B.; Seebach, D. *Angew. Chem., Int. Ed. Engl.* **1991**, *30*, 99. (b) Seebach, D.; Beck, A. K.; Schmidt, B.; Wang, Y. M. *Tetrahedron* **1994**, *50*, 4363.
9. Kitamura, M.; Suga, S.; Kawai, K.; Noyori, R. *J. Am. Chem. Soc.* **1986**, *108*, 6071.
10. (a) Denmark, S. E.; O'Connor, S. P. *J. Org. Chem.* **1997**, *62*, 3390. (b) Denmark, S. E.; O'Connor, S. P. *J. Org. Chem.* **1997**, *62*, 584.
11. Takahashi, H.; Yoshioka, M.; Shibasaki, M.; Ohno, M.; Imai, N.; Kobayahsi, S. *Tetrahedron* **1995**, *51*, 12013.
12. Takahashi, H.; Yoshioka, M.; Ohno, M.; Kobayashi, S. *Tetrahedron Lett.* **1992**, *33*, 2575.
13. Charette, A. B.; Brochu, C. *J. Am. Chem. Soc.* **1995**, *117*, 11367.
14. Imai, N.; Sakamoto, K.; Maeda, M.; Kouge, K.; Yoshizane, K.; Nokami, J. *Tetrahedron Lett.* **1997**, *38*, 1423.
15. (a) Charette, A. B.; Juteau, H.; Lebel, H.; Molinaro, C. *J. Am. Chem. Soc.* **1998**, *120*, 11943. (b) Charette, A. B.; Lebel, H.; Gagnon, A. *Tetrahedron* **1999**, *55*, 8845.

Scott E. Denmark & Gregory Beutner
University of Illinois, Urbana, IL, USA

B-Methoxydiisopinocampheylborane[1]

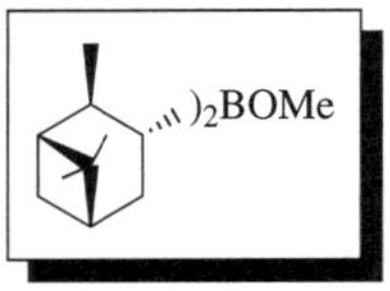

(+)
[85134-98-1] $C_{21}H_{37}BO$ (MW 316.39)
(−)
[99438-28-5]

(intermediate for the synthesis of *B*-allyl- and *B*-crotyldiisopinocampheylboranes;[2] reacts with potassium hydride to form an asymmetric reducing agent[3])

Physical Data: mp >110 °C.
Form Supplied in: white solid; commercially available; typical impurities include the disproportionation products $IpcB(OMe)_2$ and $B(OMe)_3$.
Analysis of Reagent Purity: ^{11}B NMR (δ +52).
Preparative Methods: prepared in two steps from either (+)- or (−)-α-pinene (eq 1).

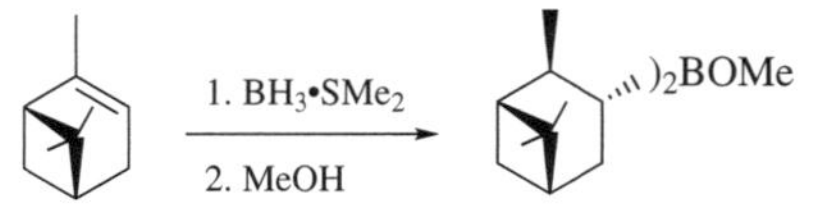

(+)-α-Pinene (−)-Ipc_2BOMe (1)

Preparation of *B*-Allyldiisopinocampheylborane and Related Reagents. Addition of allylic Grignard or potassium reagents to *B*-methoxydiisopinocampheylborane (Ipc_2BOMe) provides the corresponding allylic boranes (eqs 2 and 3).[2] Note that (−)-Ipc_2BOMe, derived from (+)-α-pinene, produces (+)-Ipc_2BAll and Ipc_2BCrt reagents, and (+)-Ipc_2BOMe gives the corresponding (−)-allylic boranes. These allyl and crotyl boranes condense with aldehydes to provide secondary homoallylic alcohols with high levels of enantioselection (see also *B-Allyldiisopinocampheylborane* and *B-Crotyldiisopinocampheylborane*).

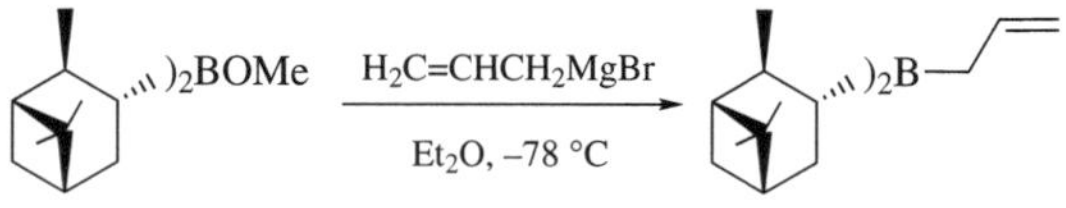

(2)

(+)-Ipc_2BAll

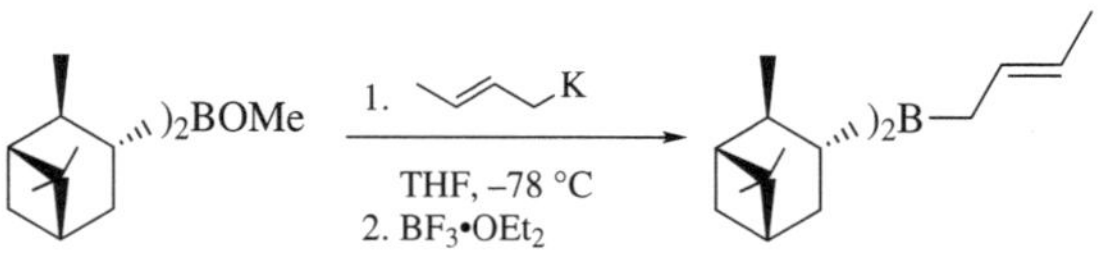

(3)

(*E*)-(+)-Ipc_2BCrt

Acetylide Coupling. *B*-Methoxydiisopinocampheylborane reacts with *Lithium (Trimethylsilyl)acetylide* to provide, after iodine-promoted rearrangement and desilylation, α-chiral monosubstituted alkynes in excellent yield (eq 4).[4]

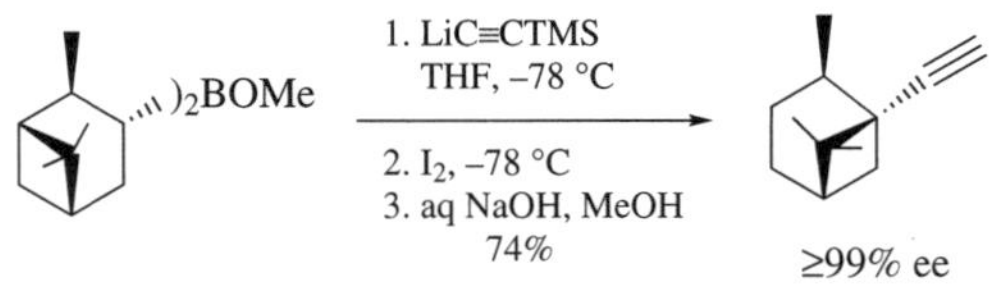

(4)

≥99% ee

***B*-Methoxydiisopinocampheylborohydrides.** Treatment of Ipc_2BOMe with an excess of *Potassium Hydride* produces the corresponding potassium *B*-methoxydiisopinocampheylborohydride.[3] The reduction of ketones with this reagent proceeds in high yield but with modest enantioselection (eq 5).

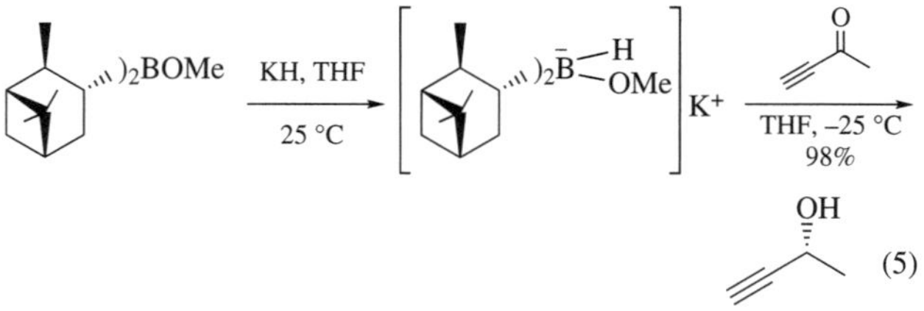

(5)

39% ee

1. Brown, H. C.; Ramachandran, P. V. *Pure Appl. Chem.* **1991**, *63*, 307.
2. (a) Jadhav, P. K.; Bhat, K. S.; Perumal, P. T.; Brown, H. C. *J. Org. Chem.* **1986**, *51*, 432. (b) Racherla, U. S.; Brown, H. C. *J. Org. Chem.* **1991**, *56*, 401. (c) Brown, H. C.; Bhat, K. S. *J. Am. Chem. Soc.* **1986**, *108*, 5919.
3. Cho, B. T. *Bull. Korean Chem. Soc.* **1991**, *12*, 662.
4. Brown, H. C.; Mahindroo, V. K.; Bhat, N. G.; Singaram, B. *J. Org. Chem.* **1991**, *56*, 1500.

Mark T. Goulet
Merck Research Laboratories, Rahway, NJ, USA

(*R*)-*N*-[2-(2-Methoxyethoxy)-ethyl]-α-phenyl-1-piperidineethanamine

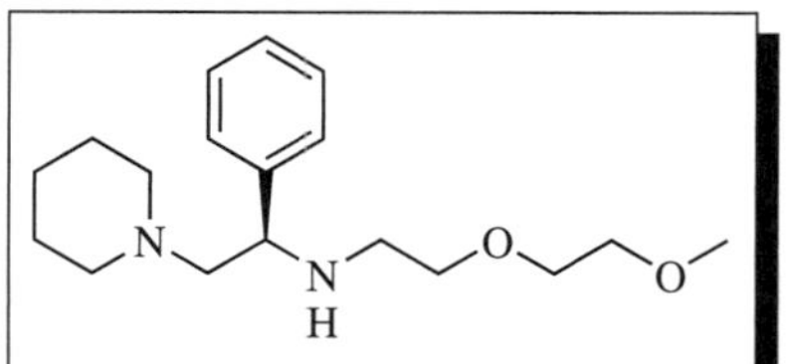

1

[132797-07-0] $C_{18}H_{30}N_2O_2$ (MW 306.44)

(chiral ligand for lithium enolate)

Physical Data: mp 140 °C (as dipicrate); $[\alpha]_D^{20} = -25.9$ (c 2.1, acetone, as dipicrate).
Solubility: soluble in organic solvents such as methanol, diethyl ether, and toluene.
Form Supplied in: not commercially available.

Preparative Methods: prepared from (*R*)-phenylglycine in six steps,[1] or from (*R*)-styrene oxide in a single step.[2]
Purification: recrystallized from methanol as dipicrate.

Asymmetric Alkylation of Ketone Enolates.[3] Alkylation of ketone enolates produced with the lithium amide of the amine **2** gives optically active α-substituted ketones (eq 1). The amine liberated during formation of a lithium enolate works as a chiral ligand of the enolate. When cyclohexanone reacts with benzyl bromide, the ee of the adduct depends on the reaction solvent; 62% ee (74% yield) is achieved in toluene at −20 °C. To attain higher ee, the presence of lithium bromide is indispensable and gives results up to 92% ee (89% yield). Almost comparable results are obtained by addition of the amine to the enolate solution (eq 2). Other than benzyl bromide, methyl iodide can also be added enantioselectively (71% yield, 88% ee). Although the selectivity is suggested to be insensitive to the conformation of enolate,[4] the ketones studied are limited to cyclohexanone analogs.

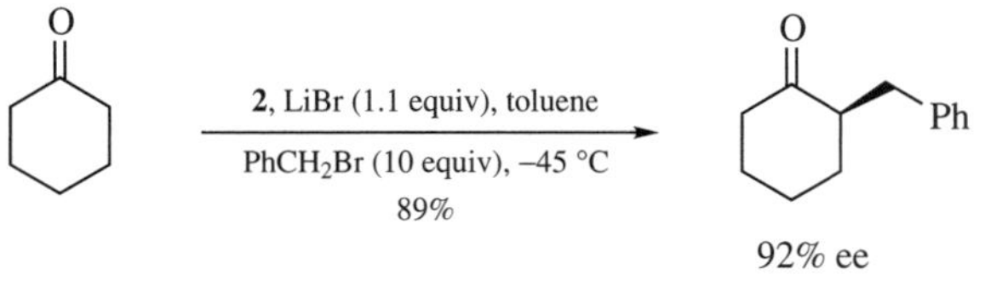

(1)

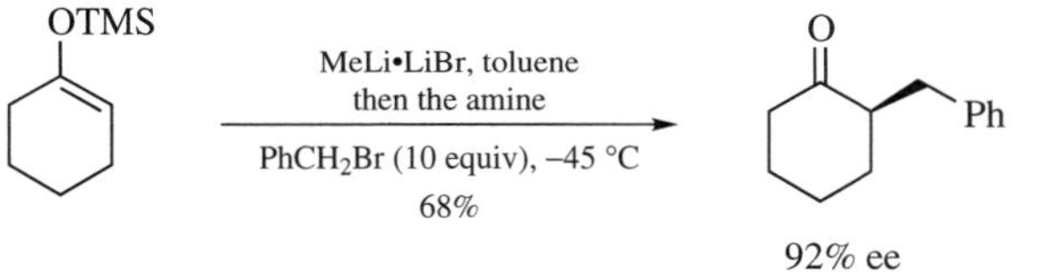

(2)

Asymmetric Aldol Reactions.[5] Lithium enolates, derived from an ester, and LDA react with aldehydes enantioselectively in the presence of the chiral amide **2** (eq 3). When benzaldehyde is employed, the major diastereomer is the *anti*-aldol with 94% ee, while the minor *syn*-aldol is only 43% ee. In this reaction, the lithium amide **2** coordinates to an additional lithium atom. There are four additional examples of aldehydes with the same ester enolate.

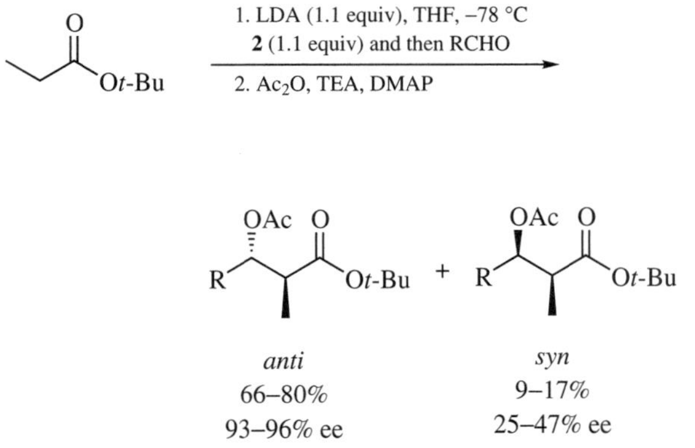

(3)

Related Reagents. (*R*)-*N*-[2-[*N*-(2-Dimethylaminoethyl)-*N*-methylamino]ethyl]-1-phenyl-2-pieridinoethylamine.

1. Shirai, R.; Aoki, K.; Sato, D.; Kim, H. D.; Murakata, M.; Yasukata, T.; Koga, K. *Chem. Pharm. Bull.* **1994**, *42*, 690.
2. Curthbertson, E.; O'Brien, P.; Towers, T. D. *Synthesis* **2001**, 693.
3. (a) Murakata, M.; Nakajima, M.; Koga, K. *J. Chem. Soc., Chem. Commun.* **1990**, 1657. (b) Murakata, M.; Yasukata, T.; Aoki, T.; Nakajima, M.; Koga, K. *Tetrahedron* **1998**, *54*, 2449.
4. Hasegawa, Y.; Kawasaki, H.; Koga, K. *Tetrahedron Lett.* **1993**, *34*, 1963.
5. Uragami, M.; Tomioka, K.; Koga, K. *Tetrahedron: Asymmetry* **1995**, *6*, 701.

Takashi Sugimura
Himeji Institute of Technology, Hyogo, Japan

(4*S*,5*S*)-4-Methoxymethyl-2-methyl-5-phenyl-2-oxazoline[1]

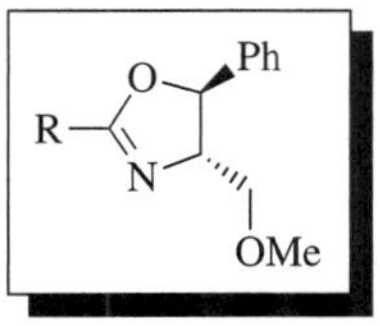

(**1**; R = Me)		
[52075-14-6]	$C_{12}H_{15}NO_2$	(MW 205.28)
(**2**; R = Et)		
[51594-37-7]	$C_{13}H_{17}NO_2$	(MW 219.31)
(**3**; R=ClCH$_2$)		
[54623-66-4]	$C_{12}H_{14}ClNO_2$	(MW 239.72)

(enantiopure carboxylic ester derivatives for synthesis of enantiomerically pure or enriched 2- and 3-substituted alkanoic acids, γ-butyrolactones, valerolactones, and benzovalerolactones by α-lithiation and asymmetric alkylations)

Physical Data: (**1**): bp 85–87 °C/0.20 mmHg; $[\alpha]_D^{CHCl_3}$ −118°. (**2**): bp 91–93 °C/0.25 mmHg; $[\alpha]_D^{CHCl_3}$−84.2°. (**3**): $[\alpha]_D^{CHCl_3}$−84.1°.
Preparative Methods: (**1**) and (**2**): cyclocondensation of commercially available (1*S*,2*S*)-2-amino-1-phenyl-1,3-propanediol with the appropriate orthoester, followed by methylation of the free hydroxy group (eq 1).[2a]

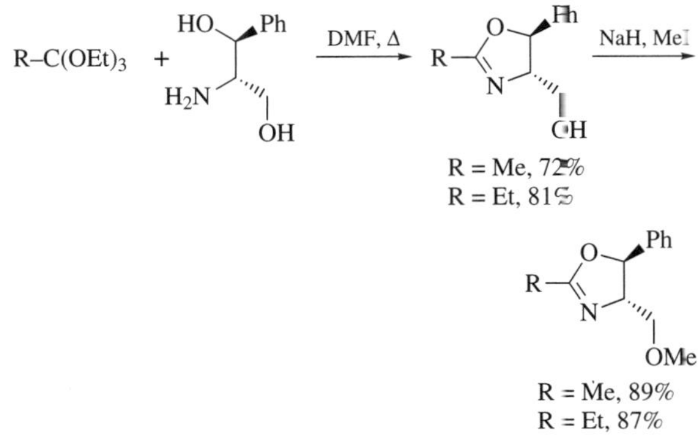

(1)

Handling, Storage, and Precautions: no special precautions.

Asymmetric Alkylations. The asymmetric synthesis of 2-alkanoic acids can be accomplished be treatment of the lithio salt of the 2-ethyl-2-oxazoline with alkyl halides followed by acidic hydrolysis of the oxazoline (eq 2).[2]

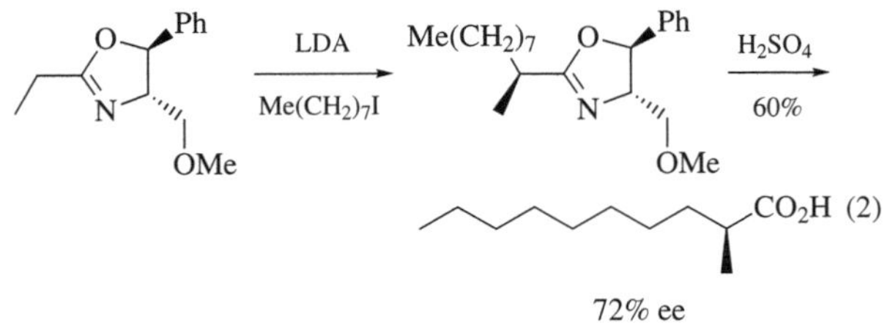

Alkylation of the 2-methyl-2-oxazoline (base/electrophile) results in homologated 2-oxazolines. A second alkylation sequence proceeds with asymmetric induction and results in the formation of highly substituted chiral 2-alkyl alkanoic acids. Use of ethylene oxide as the electrophile in this process allows for the formation of chiral α-substituted γ-butyrolactones and α-substituted γ-valerolactones with good stereoselectivity (60–80% ee; eq 3).[3]

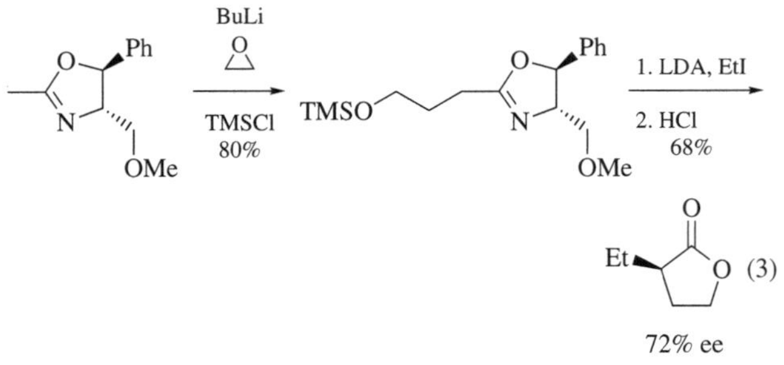

Conjugate Additions to 2-Vinyloxazolines. The 2-methyl-2-oxazoline can be converted into a phosphonate for Horner–Emmons alkenations. A variety of (*E*)-alkenes containing the chiral oxazoline auxiliary (α,β-unsaturated oxazolines) can be synthesized in high yields (80–93%) as the sole geometric isomer (eq 4).[4] Conjugate addition of alkyllithium reagents to these Michael acceptors affords, after oxazoline hydrolysis, the 3-substituted alkanoic acids (or corresponding alcohol) with a high level of stereoselectivity (91–99% ee; eq 4).[4]

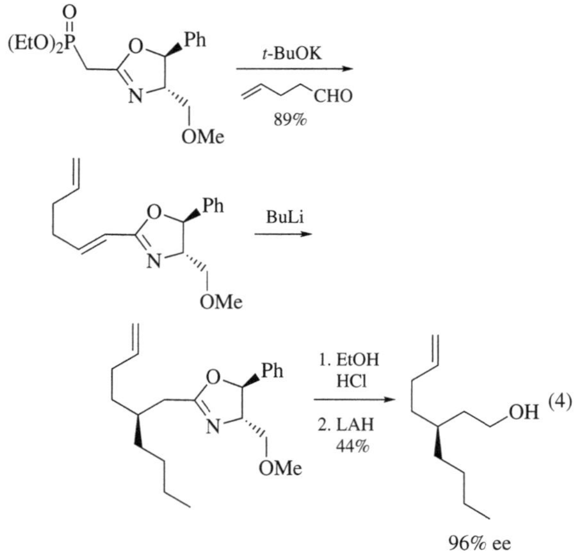

Use of appropriately substituted chiral α,β-unsaturated oxazolines allows access to 3-substituted δ-valerolactones and 4-substituted 2-chromanones with high stereoselectivity (95–98%; eq 5).[4a–c]

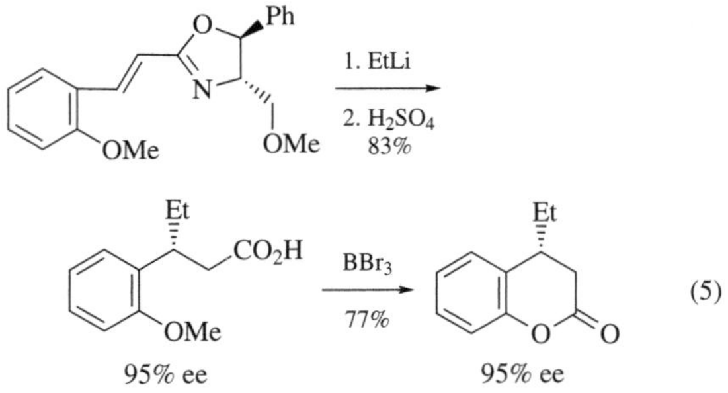

Asymmetric Aldol Additions. 2-Ethyl-2-oxazoline takes part in aldol condensations as its boron azaenolate. The *erythro* selectivity for this protocol is excellent (95:5 to 98:2) but the enantioselectivity is only moderate (29–71% ee; eq 6).[5]

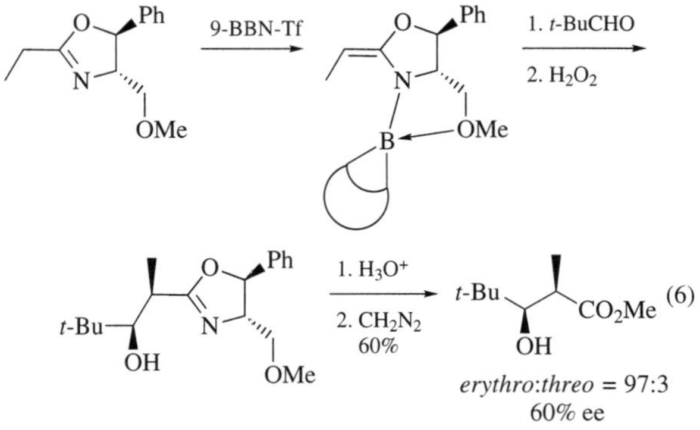

Enantioenriched α-Chloro Carboxylic Acids. Reaction of (1*S*,2*S*)-1-phenyl-2-amino-1,3-propanediol with the ethyl imidate of chloroacetonitrile gives (−)-2-chloromethyl-4-methoxymethyl-5-phenyl-2-oxazoline (eq 7).[6]

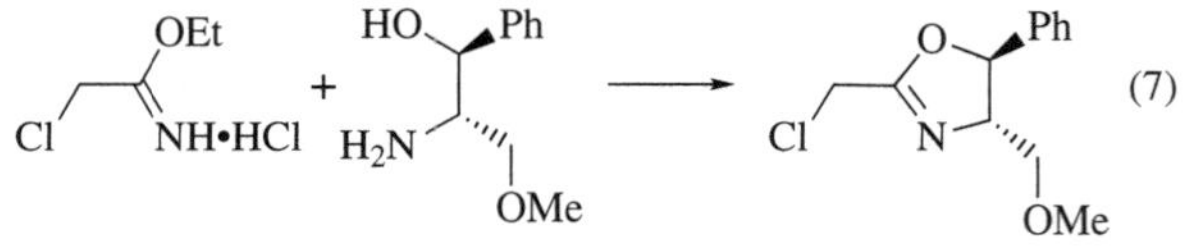

Alkylation of this oxazoline is accomplished by metalation with *Lithium Diisopropylamide* followed by adding a premixed solution of the electrophile and 2 equiv of HMPA (eq 8). Hydrolysis of the oxazoline moiety affords the enantioenriched 2-chloroalkanoic acids, albeit with low optical purity.[6]

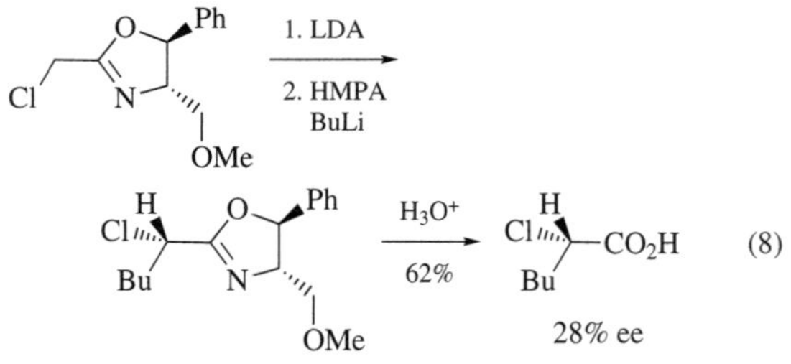

Related Reagents. (*R*)-(+)-*t*-Butyl 2-(*p*-Tolylsulfinyl)propionate; Chloro(cyclopentadienyl)bis[3-*O*-(1,2:5,6-di-*O*-isopropylidene-α-D-glucofuranosyl)]titanium; 10-Dicyclohexylsulfonamidoisoborneol; Diisopinocampheylboron Trifluoromethanesulfonate; (*R*,*R*)-2,5-Dimethylborolane; 2-(*o*-Methoxyphenyl)-4,4-dimethyl-2-oxazoline; (*S*)-4-Benzyl-2-oxazolidinone; 3-Propionylthiazolidine-2-thione; 2,4,4-Trimethyl-2-oxazoline.

1. Meyers, A. I. *Acc. Chem. Res.* **1978**, *11*, 375.
2. (a) Meyers, A. I.; Knaus, G.; Kamata, K.; Ford, M. E. *J. Am. Chem. Soc.* **1976**, *98*, 567. (b) Meyers, A. I.; Knaus, G. *J. Am. Chem. Soc.* **1974**, *96*, 6508. (c) Meyers, A. I.; Knaus, G.; Kamata, K. *J. Am. Chem. Soc.* **1974**, *96*, 268. (d) Meyers, A. I.; Mazzu, A.; Whitten, C. E. *Heterocycles* **1977**, *6*, 971. (e) Hoobler, M. A.; Bergbreiter, D. E.; Newcomb, M. *J. Am. Chem. Soc.* **1978**, *100*, 8182. (f) Meyers, A. I.; Snyder, E. S.; Ackerman, J. J. H. *J. Am. Chem. Soc.* **1978**, *100*, 8186. (g) Byström, S.; Högberg, H.-E.; Norin, T. *Tetrahedron* **1981**, *37*, 2249. (h) Liddell, R.; Whiteley, C. *Chem. Commun./J. Chem. Soc., Chem. Commun.* **1983**, 1535.
3. (a) Meyers, A. I.; Yamamoto, Y.; Mihelich, E. D.; Bell, R. A. *J. Org. Chem.* **1980**, *45*, 2792. (b) Meyers, A. I.; Mihelich, E. D. *J. Org. Chem.* **1975**, *40*, 1186.
4. (a) Meyers, A. I.; Smith, R. K.; Whitten, C. E. *J. Org. Chem.* **1979**, *44*, 2250. (b) Meyers, A. I.; Whitten, C. E. *Tetrahedron Lett.* **1976**, 1947. (c) Ziegler, F. E.; Gilligan, P. J. *J. Org. Chem.* **1981**, *46*, 3874. (d) Meyers, A. I.; Whitten, C. E. *J. Am. Chem. Soc.* **1975**, *97*, 6266. (e) Whittaker, M.; McArthur, C. R.; Leznoff, C. C. *Can. J. Chem.* **1985**, *63*, 2844. (f) Meyers, A. I.; Smith, R. K. *Tetrahedron Lett.* **1979**, 2749.
5. (a) Meyers, A. I.; Yamamoto, Y. *Tetrahedron* **1984**, *40*, 2309. (b) Meyers, A. I.; Yamamoto, Y. *J. Am. Chem. Soc.* **1981**, *103*, 4278. (c) via a lithium enolate: Meyers, A. I.; Reider, P. J. *J. Am. Chem. Soc.* **1979**, *101*, 2501.
6. Meyers, A. I.; Knaus, G.; Kendall, P. M. *Tetrahedron Lett.* **1974**, 3495.

Todd D. Nelson & Albert I. Meyers
Colorado State University, Fort Collins, CO, USA

(*S*)-2-Methoxymethylpyrrolidine[1]

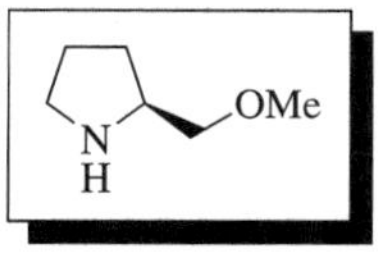

[63126-47-6] $C_6H_{13}NO$ (MW 115.20)

(chiral auxiliary; asymmetric syntheses with SMP enamines[2] and SMP amides;[3] asymmetric Birch reductions;[4] asymmetric Diels–Alder reactions[5])

Alternate Name: SMP.
Physical Data: bp 75 °C/40 mmHg; *d* 0.930 g cm^{-3}; n_D^{20} 1.4467; α_D^{20} −3 to −4° (neat).
Solubility: sol H_2O, ether, dichloromethane.
Form Supplied in: colorless liquid.
Handling, Storage, and Precautions: store at 0–4 °C under an argon atmosphere.

General Considerations. Since the pioneering times of the mid-1970s, (*S*)-2-methoxymethylpyrrolidine has been one of the most generally useful chiral auxiliaries in asymmetric synthesis, with a very broad range of applications. As a proline derivative, it generally shows high stereoselectivities due to the rigidity of the five-membered ring and the ability to coordinate metal fragments[6,7] [see also *(S)-1-Amino-2-methoxymethylpyrrolidine*, SAMP].

Lithiated SMP formamides and thioformamides have been used as acylanion equivalents (d^1 synthons) in the synthesis of enantiomerically pure α-hydroxy ketones and vicinal diols.[7] Metalated SMP aminonitriles have been used in nucleophilic acylation reactions to give α-hydroxy ketones.[8]

SMP enamines have a very broad range of applications as d^2 synthons. Cyclohexanone SMP enamine can be used for efficient Michael additions to nitroalkenes, Knoevenagel acceptors,[2a,b] and to a nitroallylic ester in a [3 + 3] carbocyclization[2c] with excellent stereoselectivities (eq 1). The synthesis of γ-oxo-α-amino acids using SMP enamines has been developed (eq 2).[2d]

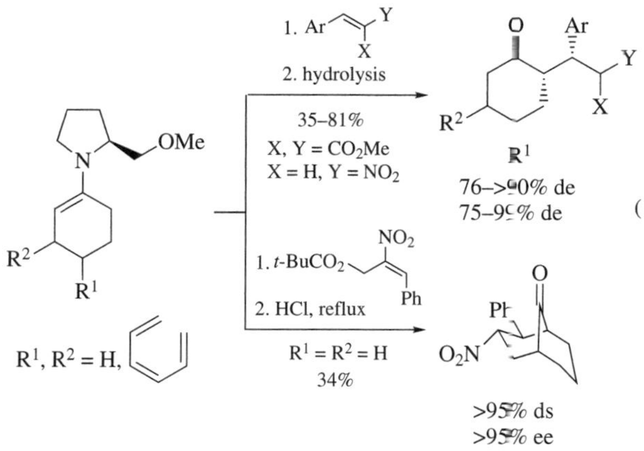

(1)

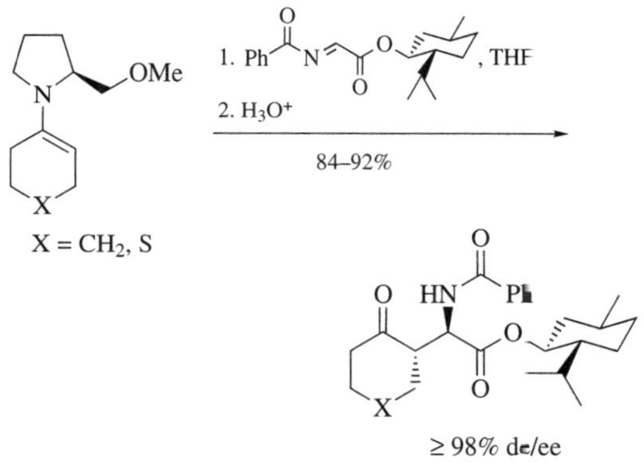

(2)

SMP amide enolates have been employed by several research groups. Alkylation of SMP amide enolates gives α-substituted acids (eq 3).[3a,b] Excellent yields and stereoselectivities are observed in the Birch reduction of aromatic SMP amides with subsequent alkylation (eq 4).[4]

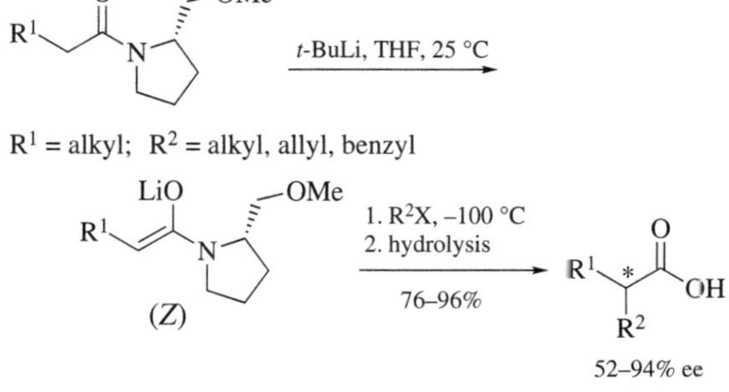

(3)

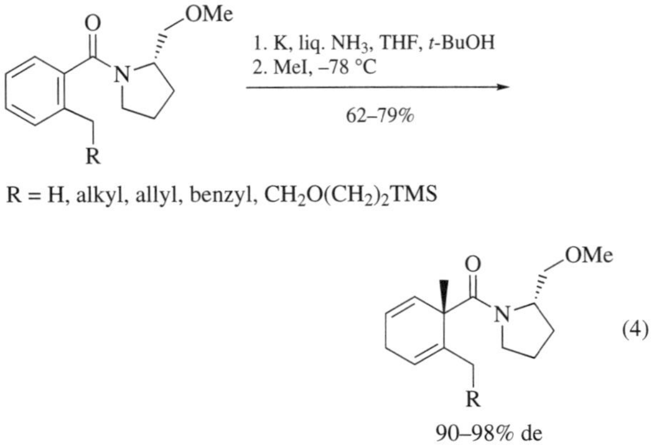

SMP amides have been used in vanadium(II)-promoted pinacol cross-coupling[3c,d] and in asymmetric oxidations with chiral oxaziridines.[3e] The diastereoselective addition of thiocarboxylic acids to 1-(2-methylacryloyl) SMP amides[3f] and the stereocontrolled addition of various organometallics to α-keto SMP amides[3g] have been studied.

Metalated SMP allylamines or enamines have been used as the first chiral homoenolate equivalents (d^3 synthons; eq 5).[9]

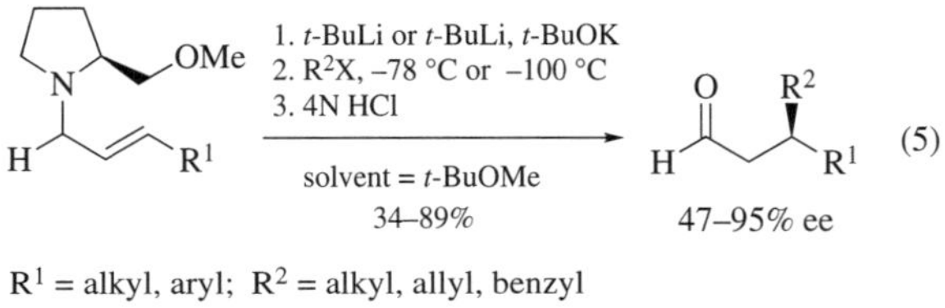

SMP is a useful chiral auxiliary in various cycloaddition reactions. Chiral 2-amino-1,3-dienes have been used in the Diels–Alder reaction with 2-aryl-1-nitroethylenes,[5a,b] and 5-aryl-2-methyl-substituted 4-nitrocyclohexanones were obtained in excellent enantiomeric purities (ee = 95–99%) and diastereoselectivities (ds = 75–95%; eq 6). The photo-Diels–Alder reaction of SMP acrylonitrile with 1-acetylnaphthalene has been carried out.[5b] After hydrolysis of the adduct, the 1,4-diketone was obtained in excellent enantiomeric purity (ee ≥ 97%; eq 7).

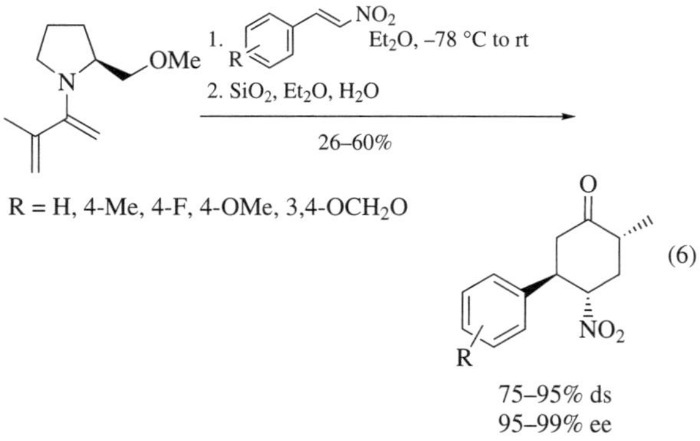

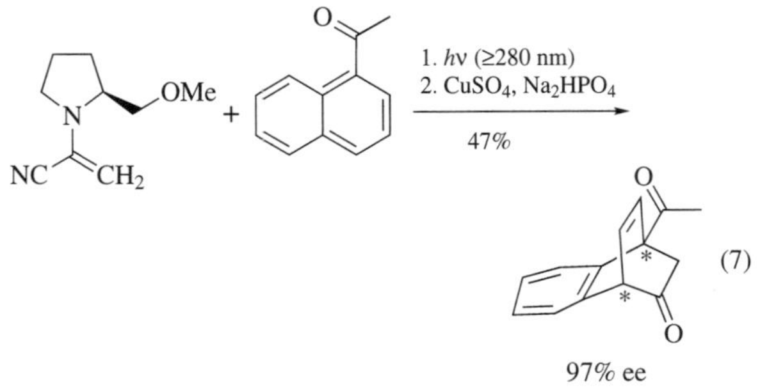

Stereoselective Diels–Alder reactions have been performed variously, using chirally modified sulfines as dienophiles,[5c] chiral ynamines,[5d] SMP enamines,[5e] SMP acrylamides,[5f] and the in situ preparation of SMP *N*-acylnitroso dienophiles.[9g,h,i] The [2 + 2] cycloaddition reactions of chiral keteniminium salts obtained from SMP amides with alkenes have been studied.[10]

Various metalated chiral organosilicon compounds bearing the SMP moiety have been alkylated to synthesize chiral alcohols.[11] Excellent regio- and stereoselectivities have been observed in the alkylation of chiral silylpropargyl anions (eq 8).[11f]

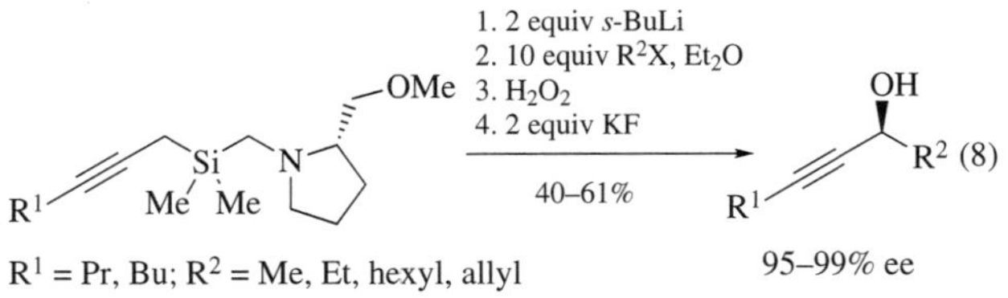

The elegant application of SMP as a chiral leaving group has been studied,[13] using chiral nitroalkenes in the reaction with zinc enolates.[12] The coupling products were obtained in very good yields and enantiomeric purities (eq 9). SMP methyl-2-cycloalken-1-ones undergo conjugate addition with lithium diorganocuprates followed by elimination of the chiral auxiliary to form optically active cycloalkanones (eq 10).

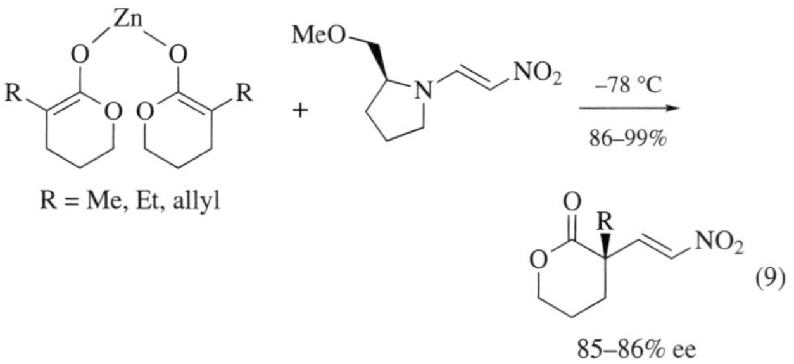

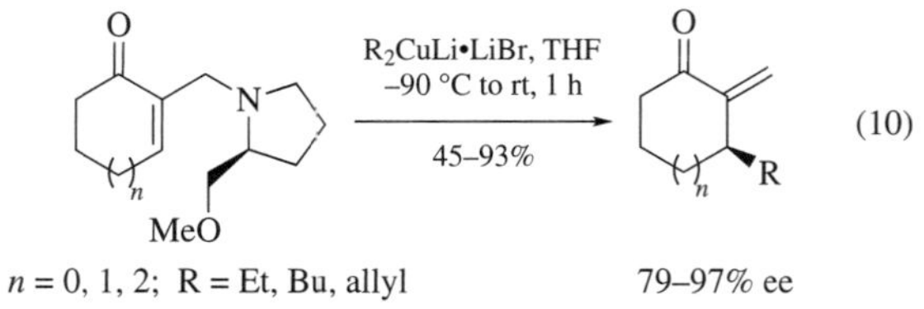

Various other applications are conjugate addition,[14] the ultrasound-promoted perfluoralkylation of SMP enamines,[15] the enantioselective fluorodehydroxylation of SMP 1-yl-sulfur trifluoride,[16] asymmetric telomerization of butadiene,[17] the chiral modification of ruthenium clusters,[18] and the application of SMP amide bases.[19]

Related Reagents. (*S*)-1-Amino-2-methoxymethylpyrrolidine.

1. Review (literature up to 1985): Enders, D.; Kipphardt, H. *Nachr. Chem. Tech. Lab.* **1985**, *33*, 882.
2. (a) Blarer, S. J.; Seebach, D. *Ber. Dtsch. Chem. Ges. Chem. Ber.* **1983**, *116*, 2250 and 3086. (b) Blarer, S. J.; Schweizer, W. B.; Seebach, D. *Helv. Chim. Acta* **1982**, *65*, 1693. (c) Seebach, D.; Missbach, M.; Calderari, G.; Eberle, M. *J. Am. Chem. Soc.* **1990**, *112*, 7625. (d) Kober, R.; Papadopoulos, K.; Miltz, W.; Enders, D.; Steglich, W.; Reuter, H.; Puff, H. *Tetrahedron* **1985**, *41*, 1637. (e) Risch, N.; Esser, A. *Justus Liegigs Am. Chem./Liebigs Ann. Chem.* **1992**, 233. (f) Hodgson, A.; Marshall, J.; Hallett, P.; Gallagher, T. *J. Chem. Soc., Perkin Trans. 1* **1992**, 2169. (g) Renaud, P.; Schubert, S. *Synlett* **1990**, 624.
3. (a) Sonnet, P. E.; Heath, R. R. *J. Org. Chem.* **1980**, *45*, 3137. (b) Evans, D. A.; Takacs, J. M. *Tetrahedron Lett.* **1980**, *21*, 4233. (c) Annunziata, R.; Cinquini, M.; Cozzi, F.; Giaroni, P. *Tetrahedron: Asymmetry* **1990**, *1*, 355. (d) Annunziata, R.; Cinquini, M.; Cozzi, F.; Giaroni, P.; Benaglia, M. *Tetrahedron* **1991**, *47*, 5737. (e) Davis, F. A.; Ulatowski, T. G.; Haque, M. S. *J. Org. Chem.* **1987**, *52*, 5288. (f) Effenberger, F.; Isak, H. *Ber. Dtsch. Chem. Ges./Chem. Ber.* **1989**, *122*, 553. (g) Fujisawa, T.; Ukaji, Y.; Funabora, M.; Yamashita, M.; Sato, T. *Bull. Chem. Soc. Jpn.* **1990**, *63*, 1894.
4. (a) Schultz, A. G.; Macielag, M.; Sundararaman, P.; Taveras, A. G.; Welch, M. *J. Am. Chem. Soc.* **1988**, *110*, 7828. (b) Schultz, A. G.; Green, N. J. *J. Am. Chem. Soc.* **1991**, *113*, 4931. (c) Schultz, A. G.; Taylor, R. E. *J. Am. Chem. Soc.* **1992**, *114*, 3937. (d) Schultz, A. G.; Hoglen, D. K.; Holoboski, M. A. *Tetrahedron Lett.* **1992**, *33*, 6611.
5. (a) Enders, D.; Meyer, O.; Raabe, G. *Synthesis* **1992**, 1242. (b) Barluenga, J.; Aznar, F.; Valdes, C.; Martin, A.; Garcia-Granda, S.; Martin, E. *J. Am. Chem. Soc.* **1993**, *115*, 4403. (c) Döpp, D.; Pies, M. *Chem. Commun./J. Chem. Soc., Chem. Commun.* **1987**, 1734. (d) Van den Broek, L. A. G. M.; Posskamp, P. A. T. W.; Haltiwanger, R. C.; Zwanenburg, B. *J. Org. Chem.* **1984**, *49*, 1691. (e) Van Elburg, P. A.; Honig, G. W. N.; Reinhoudt, D. N. *Tetrahedron Lett.* **1987**, *28*, 6397. (f) Bäckvall, J. E.; Rise, F. *Tetrahedron Lett.* **1989**, *30*, 5347. (g) Lamy-Schelkens, H.; Ghosez, L. *Tetrahedron Lett.* **1989**, *30*, 5891. (h) Brouillard-Poichet, A.; Defoin, A.; Streith, J. *Tetrahedron Lett.* **1989**, *30*, 7061. (i) Defoin, A.; Pires, J.; Tissot, I.; Tschamber, T.; Bur, D.; Zehnder, M.; Streith, J. *Tetrahedron: Asymmetry* **1991**, *2*, 1209. (k) Defoin, A.; Brouillard-Poichet, A.; Streith, J. *Helv. Chim. Acta* **1992**, *75*, 109.
6. (a) Enders, D.; Eichenauer, H. *Angew. Chem.* **1976**, *93*, 579. (b) Seebach, D.; Kalinowski, H.-O.; Bastani, B.; Crass, G.; Daum, H.; Dörr, H.; Du Preez, N. P.; Ehrig, V.; Langer, W.; Nüssler, C.; Oei, H. A.; Schmidt, M. *Helv. Chim. Acta* **1977**, *60*, 301. (c) Enders, D.; Fey, P.; Kipphardt, H. *Org. Prep. Proced. Int.* **1985**, *17*, 1.
7. (a) Enders, D.; Lotter, H. *Angew. Chem.* **1981**, *93*, 831. (b) Enders, D. In *Current Trends in Organic Synthesis*; Nozaki, H., Ed.; Pergamon: Oxford, 1983, p 151.
8. Enders, D.; Lotter, H.; Maigrot, N.; Mazaleyrat, J.-P.; Welvart, Z. *Nouv. J. Chim.* **1984**, *8*, 747. (b) Maigrot, N.; Mazaleyrat, J.-P.; Welvart, Z. *Chem. Commun.* **1984**, 40.
9. (a) Ahlbrecht, H.; Bonnet, G.; Enders, D.; Zimmermann, G. *Tetrahedron Lett.* **1980**, *21*, 3175. (b) Ahlbrecht, H.; Enders, D.; Santowski, L.; Zimmermann, G. *Ber. Dtsch. Chem. Ges./Chem. Ber.* **1989**, *122*, 1995. (c) Ahlbrecht, H.; Sommer, H. *Ber. Dtsch. Chem. Ges./Chem. Ber.* **1990**, *123*, 829.
10. (a) Saimoto, H.; Houge, C.; Hesbain-Frisque, A. M.; Mockel, A.; Ghosez, L. *Tetrahedron Lett.* **1983**, *24*, 2251. (b) Houge, C.; Frisque-Hesbain, A. M.; Ghosez, L. *J. Am. Chem. Soc.* **1984**, *104*, 2920.
11. (a) Chan, T. H.; Pellon, P. *J. Am. Chem. Soc.* **1989**, *111*, 8737. (b) Chan, T. H.; Wang, D. *Tetrahedron Lett.* **1989**, *30*, 3041. (c) Lamothe, S.; Chan, T. H. *Tetrahedron Lett.* **1991**, *32*, 1847. (d) Chan, T. H.; Nwe, K. T. *J. Org. Chem.* **1992**, *57*, 6107. (e) Lamothe, S.; Cook, K. L.; Chan, T. H. *Can. J. Chem.* **1992**, *70*, 1733. (f) Hartley, R. C.; Lamothe, S.; Chan, T. H. *Tetrahedron Lett.* **1993**, *34*, 1449.
12. (a) Fuji, K.; Node, M.; Nagasawa, H.; Naniwa, Y.; Terada, S. *J. Am. Chem. Soc.* **1986**, *108*, 3855. (b) Fuji, K.; Node, M.; Nagasawa, H.; Naniwa, Y.; Taga, T.; Machida, K.; Snatzke, G. *J. Am. Chem. Soc.* **1989**, *111*, 7921. (c) Fuji, K.; Node, M. *Synthesis* **1991**, 603.
13. (a) Tamura, R.; Katayama, H.; Watabe, K.; Suzuki, H. *Tetrahedron* **1990**, *46*, 7557. (b) Tamura, R.; Watabe, K.; Katayama, H.; Suzuki, H.; Yamamoto, Y. *J. Org. Chem.* **1990**, *55*, 408. (c) Tamura, R.; Watabe, K.; Ono, N.; Yamamoto, Y. *J. Org. Chem.* **1992**, *57*, 4895.
14. (a) Bertz, S. H.; Dabbagh, G.; Sundararajan, G. *J. Org. Chem.* **1986**, *51*, 4953. (b) Dieter, R. K.; Tokles, M. *J. Am. Chem. Soc.* **1987**, *109*, 2040. (c) Dieter, R. K.; Lagu, B.; Deo, N.; Dieter, J. *Tetrahedron Lett.* **1990**, *31*, 4105. (d) Quinkert, G.; Müller, T.; Königer, A.; Schultheis, O.; Sickenberger, B.; Dürner, G. *Tetrahedron Lett.* **1992**, *33*, 3469. (e) Schultz, A. G.; Harrington, R. E. *J. Am. Chem. Soc.* **1991**, *113*, 4926. (f) Schultz, A. G.; Lee, H. *Tetrahedron Lett.* **1992**, *33*, 4397. (g) Schultz, A. G.; Holoboski, M. A. *Tetrahedron Lett.* **1993**, *34*, 3021. (h) Schultz, A. G.; Lee, H. *Tetrahedron Lett.* **1993**, *34*, 4397.
15. Kitazume, T.; Ishikawa, N. *J. Am. Chem. Soc.* **1985**, *107*, 5186.
16. Hann, G. L.; Sampson, P. *Chem. Commun.* **1989**, 1650.
17. Keim, W.; Köhnes, A.; Roethel, T.; Enders, D. *J. Organomet. Chem.* **1990**, *382*, 295.
18. Süss-Fink, G.; Jenke, T.; Heitz, H.; Pellinghelli, M. A.; Tiripicchio, A. *J. Organomet. Chem.* **1989**, *379*, 311.
19. Hendrie, S. K.; Leonard, J. *Tetrahedron* **1987**, *43*, 3289.

Dieter Enders & Martin Klatt
RWTH Aachen, Germany

(*S*)-(−)-α-Methoxy-α-(trifluoromethyl)phenylacetic Acid

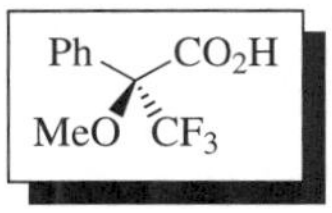

[17257-71-5] $C_{10}H_9F_3O_3$ (MW 234.19)

(determination of enantiomeric purity and absolute configuration of alcohols and amines[1])

Alternate Name: MTPA.

Physical Data: bp 115–117 °C/1.5 mmHg; $[\alpha]_D$ −71.8° (*c* 3.28, MeOH); *d* 1.344 g cm^{-1}.

Solubility: readily sol hexane, ether, THF, CH_2Cl_2, benzene.

Form Supplied in: both enantiomers are commercially available and generally possess similar applications.

Preparative Methods: both enantiomers are available by resolution of the racemic acid with α-phenylethylamine.[1] Other procedures have been described.[2]

Analysis of Reagent Purity: the enantiomeric purity of the reagent can be evaluated by capillary GC analysis of its methyl ester on a chiral stationary phase,[3] HPLC analysis of the corresponding 1-(α-naphthyl)ethylamide,[2] or by $LiAlH_4$ reduction to the corresponding alcohol, which is analyzed by chiral GC.[4]

Handling, Storage, and Precautions: very stable; commercial samples remain useful after extended periods of time.

Determination of the Enantiomeric Purity of Alcohols, Amines, and Other Compounds by Derivatization. The enantiomeric purity of a variety of chiral amines and alcohols can be assayed by reaction with chiral MTPA, followed by determination of the diastereomeric purity of the resulting amide or ester.[1] This is usually done by chromatographic (HPLC, GC, TLC, etc.) or spectroscopic methods (^{1}H NMR, ^{19}F NMR, etc.). Either type of experiment may provide the desired information and in fact they are often used in combination. MTPA is frequently converted to the corresponding acid chloride (MTPA-Cl) by refluxing in *Thionyl Chloride*. After distillation, MTPA-Cl is treated with the desired amine or alcohol.[1] An improved procedure for the microscale preparation of MTPA-Cl with *Oxalyl Chloride*, which does not require distillation of the MTPA-Cl prior to reaction with the alcohol/amine, has been described.[5] Alternatively, MTPA reacts directly with amines and alcohols in the presence of condensing agents such as *1,3-Dicyclohexylcarbodiimide*[6] or 2-chloro-1-methylpyridinium chloride.[7] Inherent in the successful use of these procedures is the need to ensure complete derivatization of the alcohol/amine, so that the diastereomeric purity of the derivative is truly reflective of the enantiomeric purity of the alcohol/amine under scrutiny.[8]

Alcohols. When using the ^{1}H NMR spectra of MTPA esters to determine the enantiomeric purity of alcohols, the MTPA methoxy peaks tend to be most useful. This technique can be sensitive enough to detect as little as 1% of the minor alcohol enantiomer. The enantiomeric purity of chiral alcohols (**1**)[9] and (**2**)[10] has been determined this way. The enantiomeric purity of primary alcohols (**3**)[11] and (**4**),[12,13] in which the asymmetric center is not the carbinol carbon, has also been determined by ^{1}H NMR analysis of their MTPA esters. A slight variation of this methodology is the use of shift reagents like Eu(fod)$_3$ to increase the chemical shift separation between diastereotopic MeO peaks; this procedure has been used in the analysis of alcohols (**5**)[14] and (**6**).[15]

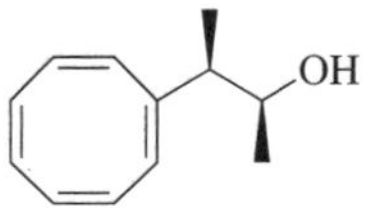
(**1**)

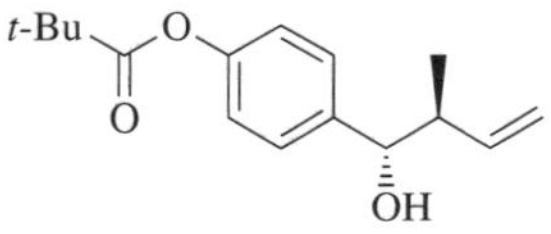
(**2**)

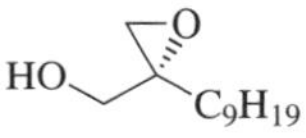
(**3**)

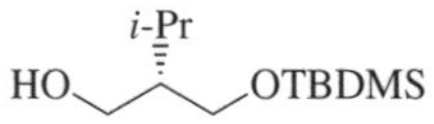
(**4**)

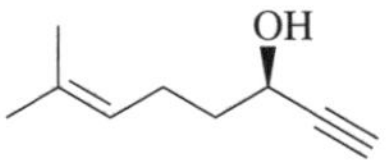
(**5**)

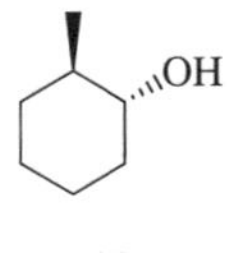
(**6**)

^{19}F NMR is also extremely useful in the analysis of MTPA esters, and is often used in combination with ^{1}H NMR.[1] The CF_3 peak(s) is easy to observe, being unencumbered by unrelated peaks. Enantiomeric analysis of primary alcohols, e.g. (**7**),[16,17] as well as secondary alcohols (**8**)[18] and (**9**)[19] has been performed utilizing this method. ^{19}F NMR analysis of MTPA esters in the presence of shift reagents has also been utilized to increase the separation between diastereomeric ^{19}F peaks.[15]

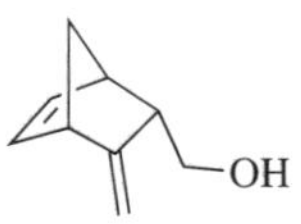
(**7**)

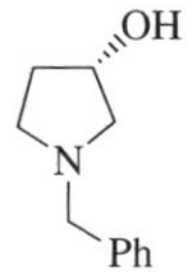
(**8**)

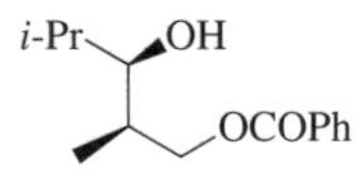
(**9**)

^{13}C NMR analysis of MTPA esters has not received much attention, but some examples have been described, e.g. the MTPA ester of (**10**).[20]

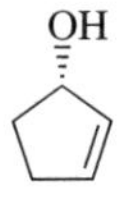
(**10**)

Almost every available chromatographic technique has been utilized in the analysis of MTPA esters. For example, capillary GLC was used to evaluate the diastereomeric composition of the MTPA esters of chiral alcohols (**5**),[14] (**11**),[21] and (**12**).[22] HPLC is routinely used, for analytical as well as preparative purposes (see below).

Amines. In a manner similar to alcohols, the enantiomeric purity of primary and secondary amines can be assayed by ^{1}H NMR analysis of their MTPA amides.[1] The technique has been particularly useful for amino acid derivatives,[23] e.g. (**13**),[24] (**14**),[25] and (**15**).[26]

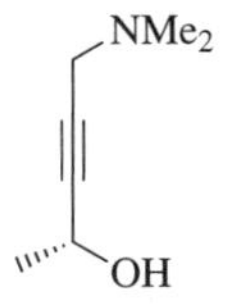
(**11**)

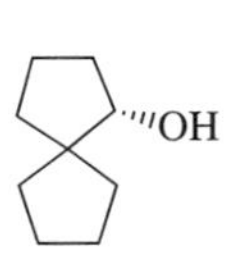
(**12**)

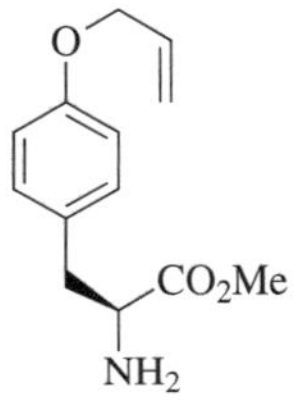
(**13**)

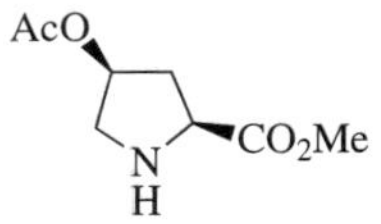
(**14**)

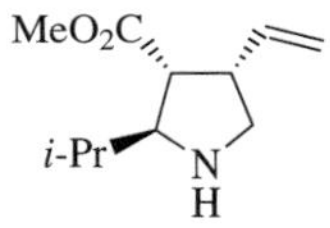
(**15**)

^{19}F NMR spectroscopy is extremely sensitive in the stereochemical evaluation of MTPA amides of a wide range of amino acids (as low as 0.05% of the minor isomer detectable).[27,28] HPLC analysis of diastereomeric MTPA amides may provide valuable analytical information on the enantiomeric composition of chiral primary and secondary amines.[29]

Other Compounds. In theory, any chiral compound with a reactive functional group can be derivatized with (*S*)- or (*R*)-MTPA in order to assess its enantiomeric purity. An example is the derivatization of cyclic carbamates, followed by ^{1}H NMR analysis (eq 1).[30] Similarly, axially chiral biaryls bearing amine or alcohol substituents, e.g. (**16**) and (**17**), have been analyzed via the corresponding MPTA derivatives.[31]

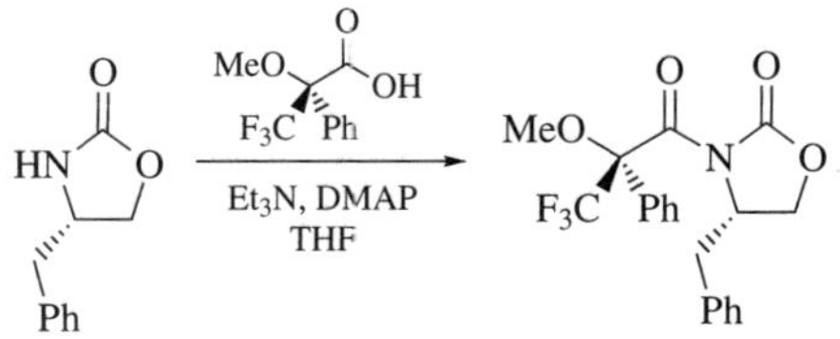

(1)

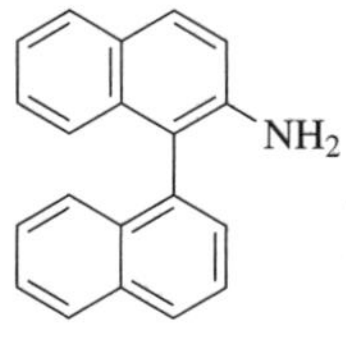

(16)

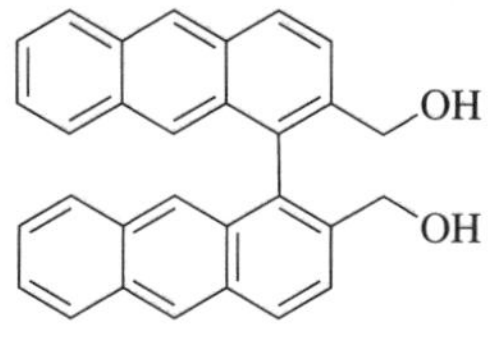

(17)

Noncovalent MTPA Derivatives. The enantiomeric purity of some chiral amines can be determined by ^{1}H NMR with (*S*)- or (*R*)-MTPA as a chiral solvating agent.[32,33] The method is particularly useful for chiral tertiary amines that are not amenable to conversion into MTPA amides, e.g. (**18**) and (**19**),[34] although it has been utilized for primary and secondary amines as well, e.g. (**20**).[35]

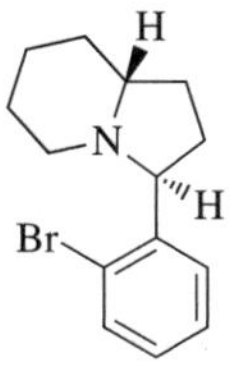

(18)

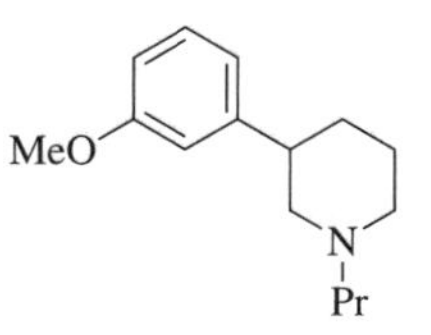

(19)

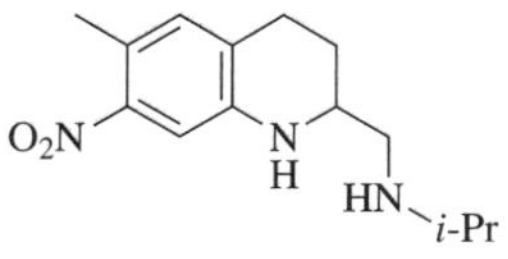

(20)

Determination of Absolute Configuration in Alcohols and Amines. The configuration of chiral alcohols and amines (where the heteroatom is attached to the stereocenter) has been correlated with the ^{19}F chemical shifts of their MTPA derivatives.[36] The model is not fully predictive, although the exceptions can often be rationalized. However, ^{1}H NMR spectroscopy of MTPA esters and amides has been more widely used in the assignment of configuration to chiral alcohols and amines.[37] Analysis of the chemical shifts of the MTPA methoxy peaks, in the presence of shift reagents, allows the correct stereochemical analysis of a series of bicyclic alcohols, such as (**6**).[15] The ^{1}H NMR shifts of the hydrogens directly attached to the carbinol carbons in MTPA esters have also been used to establish the configuration of chiral acyclic secondary alcohols.[38] Other peaks in the MTPA derivatives have been used in the stereochemical elucidation of a series of alcohol natural products.[39] Although less commonly used, ^{13}C NMR spectroscopy of MTPA derivatives has been used in the stereochemical study of a large group of chiral alcohols.[40] The stereochemistry of amino acids has also been amenable to study by ^{1}H NMR spectroscopy of their MTPA derivatives.[23,41]

Preparative Uses of MTPA Derivatives. Resolution of racemic compounds on a preparative scale is always a challenging endeavor. Conversion of the enantiomeric mixture into a mixture of diastereomers, each with unique physical properties, makes it possible to separate the components by a variety of physical methods, such as fractional recrystallization, distillation, or chromatography. One of the earliest uses of MTPA was the resolution of racemic alcohols via the separation of diastereomeric MTPA esters by preparative gas–liquid chromatography, followed by alcohol regeneration with *Lithium Aluminum Hydride* (eq 2).[1] More frequently, diastereomeric MTPA esters have been separated by high performance liquid chromatography (HPLC), followed by alcohol regeneration either by ester hydrolysis (eq 3)[42] or reduction (eq 4).[43]

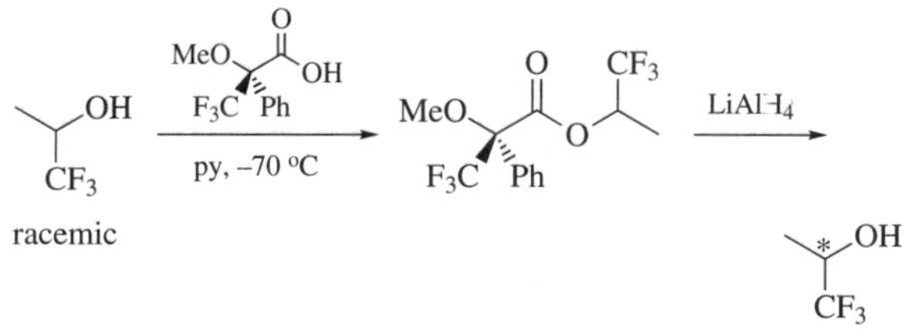

(2)

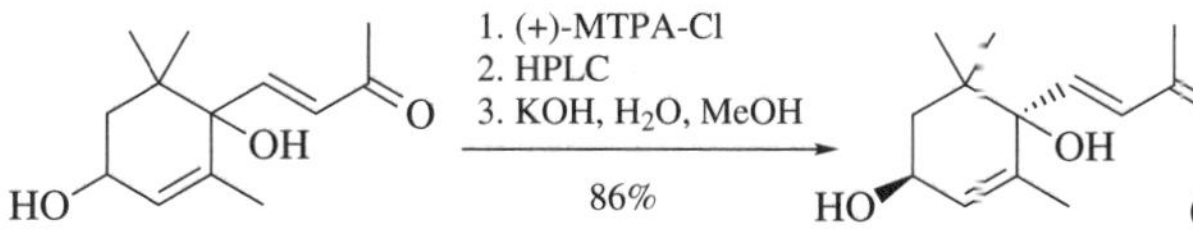

(3)

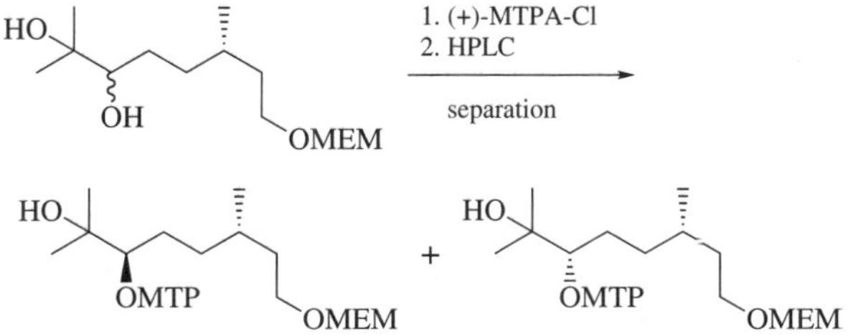

(4)

This procedure is useful even when the carbinol carbon is not the asymmetric center of the molecule (eq 5).[44]

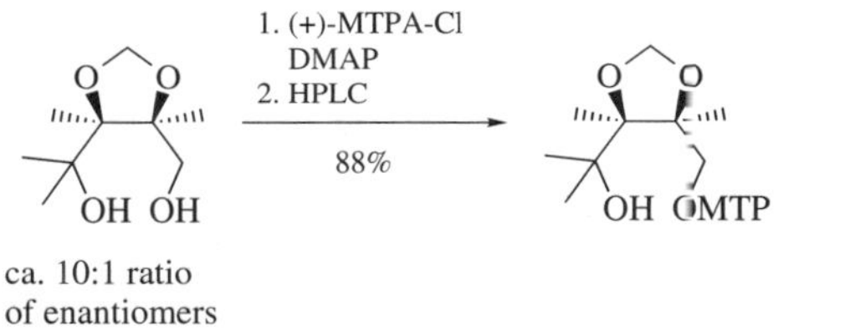

(5)

Far fewer examples of diastereomeric ester separations by fractional recrystallization have been described. However, this procedure is extremely practical for the resolution of a series of bromohydrins (eq 6).[45]

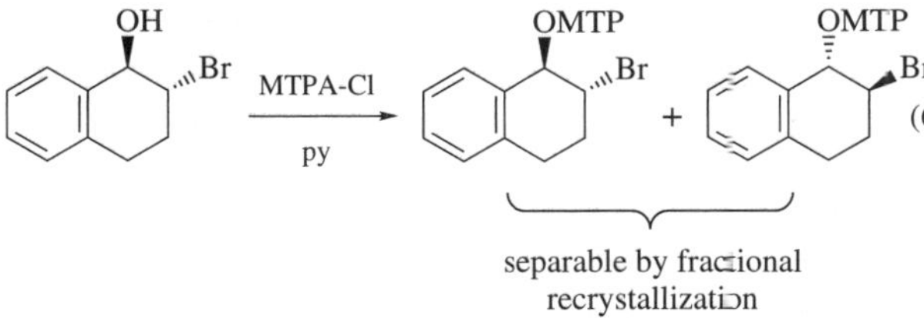

(6)

1. Dale, J. A.; Dull, D. L.; Mosher, H. S. *J. Org. Chem.* **1969**, *34*, 2543.
2. Ohta, H.; Miyamae, Y.; Kimura, Y. *Chem. Lett.* **1989**, 379.
3. König, W. A.; Nippe, K.-S.; Mischnick, P. *Tetrahedron Lett.* **1990**, *31*, 6867.
4. Jeanneret-Gris, G.; Pousaz, P. *Tetrahedron Lett.* **1990**, *31*, 75.
5. Ward, D. E.; Rhee, C. K. *Tetrahedron Lett.* **1991**, *32*, 7165.
6. Little, R. D.; Moeller, K. D. *J. Org. Chem.* **1983**, *48*, 4487.
7. Streinz, L.; Valterova, I.; Wimmer, Z.; Budesinsky, M. *Collect. Czech. Chem. Commun.* **1986**, *51*, 2207.
8. Svatos, A.; Valterova, I.; Saman, D.; Vrkoc, J. *Collect. Czech. Chem. Commun.* **1990**, *55*, 485.
9. Moore, J. S.; Gorman, C. B.; Grubbs, R. H. *J. Am. Chem. Soc.* **1991**, *113*, 1704.

10. Barrett, A. G. M.; Lebold, S. A. *J. Org. Chem.* **1991**, *56*, 4875.
11. Giese, B.; Rupaner, R. *Justus Liebigs Ann. Chem./Liebigs Ann. Chem.* **1987**, 231.
12. Ihara, M.; Takahashi, M.; Taniguchi, N.; Fukumoto, K.; Kametani, T. *Chem. Commun./J. Chem. Soc., Chem. Commun.* **1987**, 619.
13. Ihara, M.; Takahashi, M.; Taniguchi, N.; Yasui, K.; Fukumoto, K.; Kametani, T. *J. Chem. Soc., Perkin Trans. 1* **1989**, 897.
14. Mori, K.; Akao, H. *Tetrahedron Lett.* **1978**, 4127.
15. Kalyanam, N.; Lightner, D. A. *Tetrahedron Lett.* **1979**, 415.
16. Oppolzer, W.; Chapuis, C. *Tetrahedron Lett.* **1983**, *24*, 4665.
17. Chapuis, C.; Jurczak, J. *Helv. Chim. Acta* **1987**, *70*, 436.
18. Bhat, K. L.; Flanagan, D. M.; Joullié, M. M. *Synth. Commun.* **1985**, *15*, 587.
19. Wood, R. D.; Ganem, B. *Tetrahedron Lett.* **1982**, *23*, 707.
20. Wahhab, A.; Tavares, D. F.; Rauk, A. *Can. J. Chem.* **1990**, *68*, 1559.
21. Nilsson, B. M.; Vargas, H. M.; Hacksell, U. *J. Med. Chem.* **1992**, *35*, 2787.
22. Brown, H. C.; Chandrasekharan, J.; Ramachandran, P. V. *J. Am. Chem. Soc.* **1988**, *110*, 1539.
23. Yasuhara, F.; Kabuto, K.; Yamaguchi, S. *Tetrahedron Lett.* **1978**, *19*, 4289.
24. Erickson, S. D.; Simon, J. A.; Still, W. C. *J. Org. Chem.* **1993**, *58*, 1305.
25. Barrett, A. G. M.; Pilipauskas, D. *J. Org. Chem.* **1990**, *55*, 5170.
26. Cooper, J.; Knight, D. W.; Gallagher, P. T. *J. Chem. Soc., Perkin Trans. 1* **1991**, 705.
27. Hull, W. E.; Seeholzer, K.; Baumeister, M.; Ugi, I. *Tetrahedron* **1986**, *42*, 547.
28. Jackson, R. F. W.; Wishart, N.; Wood, A.; James, K.; Wythes, M. J. *J. Org. Chem.* **1992**, *57*, 3397.
29. Nordlander, J. E.; Njoroge, F. G.; Payne, M. J.; Warman, D. *J. Org. Chem.* **1985**, *50*, 3481.
30. Kano, S.; Yokomatsu, T.; Shibuya, S. *J. Org. Chem.* **1989**, *54*, 515.
31. Kabuto, K.; Yasuhara, F.; Yamaguchi, S. *Tetrahedron Lett.* **1980**, *21*, 307.
32. Maryanoff, B. E.; McComsey, D. F. *J. Heterocycl. Chem.* **1985**, *22*, 911.
33. Benson, S. C.; Cai, P.; Colon, M.; Haiza, M. A.; Tokles, M.; Snyder, J. K. *J. Org. Chem.* **1988**, *53*, 5335.
34. Villani, F. J.; Costanzo, M. J.; Inners, R. R.; Mutter, M. S.; McClure, D. E. *J. Org. Chem.* **1986**, *51*, 3715.
35. Baxter, C. A. R.; Richards, H. C. *Tetrahedron Lett.* **1972**, *13*, 3357.
36. Sullivan, G. R.; Dale, J. A.; Mosher, H. S. *J. Org. Chem.* **1975**, *38*, 2143.
37. Yamaguchi, S.; Yasuhara, F.; Kabuto, K. *Tetrahedron* **1976**, *32*, 1363.
38. Dale, J. A.; Mosher, H. S. *J. Am. Chem. Soc.* **1973**, *95*, 512.
39. Ohtani, I.; Kusumi, T.; Kashman, Y.; Kakisawa, H. *J. Am. Chem. Soc.* **1991**, *113*, 4092.
40. Doolittle, R. E.; Heath, R. R. *J. Org. Chem.* **1984**, *49*, 5041.
41. Kusumi, T.; Fukushima, T.; Ohtani, I.; Kakisawa, H. *Tetrahedron Lett.* **1991**, *32*, 2939.
42. Koreeda, M.; Weiss, G.; Nakanishi, K. *J. Am. Chem. Soc.* **1973**, *95*, 239.
43. Anderson, R. J.; Adams, K. G.; Chinn, H. R.; Henrick, C. A. *J. Org. Chem.* **1980**, *45*, 2229.
44. Niwa, H.; Ogawa, T.; Okamoto, O.; Yamada, K. *Tetrahedron* **1992**, *48*, 10531.
45. Balani, S. K.; Boyd, D. R.; Cassidy, E. S.; Greene, R. M. E.; McCombe, K. M.; Sharma, N. D. *Tetrahedron Lett.* **1981**, *22*, 3277.

Juan C. Jaen
Parke-Davis Pharmaceutical Research, Ann Arbor, MI, USA

(S)-α-Methylbenzylamine

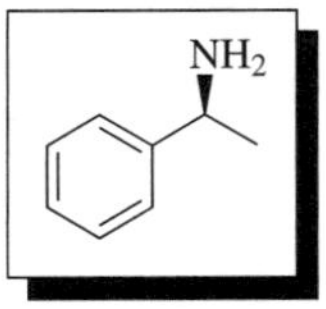

[2627-86-3] $C_8H_{11}N$ (MW 121.20)

(resolving agent for carboxylic acids;[7–11] determination of enantiomeric purity of carboxylic acids;[16,17] stereospecific reactions of carbonyl compounds;[18] reductive amination of carbonyl compounds[29,30])

Physical Data: bp 187 °C; *d* 0.940 g cm^{-3}; $[\alpha]_D$ −39° (neat).
Solubility: readily sol all organic solvents.
Form Supplied in: both enantiomers are commercially available.
Analysis of Reagent Purity: the enantiomeric purity of the reagent can be assessed by NMR analysis of the corresponding Mosher's amide.[4] Chiral complexing reagents (such as 1,1′-binaphthyl-2,2′-diylphosphoric acid) have also been used in the direct NMR analysis of the reagent.[5,6]
Preparative Methods: racemic α-methylbenzylamine has been resolved utilizing chiral acids such as tartaric acid[1] and (S)-(−)-carbamalactic acid,[2] among others. Several stereospecific syntheses have been reported.[3]
Handling, Storage, and Precautions: stable at rt for extended periods of time when stored under nitrogen.

Resolving Reagent for Carboxylic Acids and Other Types of Compounds. A large number of carboxylic acids have been resolved via their diastereomeric salts with (S)- or (R)-α-methylbenzylamine (**1**). The ready availability of both enantiomers of (**1**) guarantees access to both enantiomers of the desired acid. Compounds (**2**)–(**6**) are representative examples of acids obtained in high enantiomeric purity.[7–11] Alternatively, racemic carboxylic acids have been resolved by covalent derivatization with (**1**) and separation of the resulting diastereomeric amides by physical means such as chromatography (eq 1)[12] or fractional crystallization (eq 2).[13]

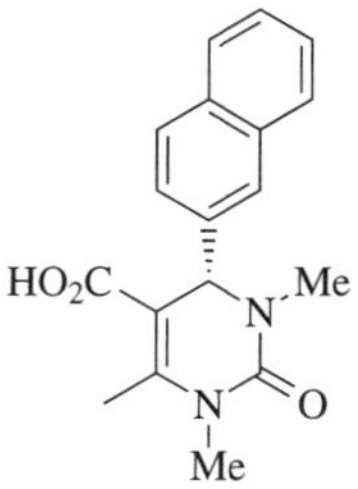

(**2**)

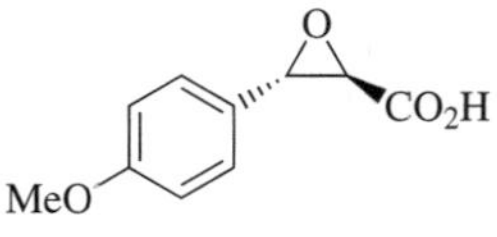

(**3**)

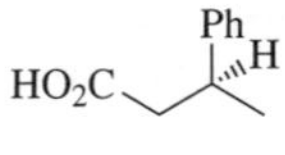

(**4**)

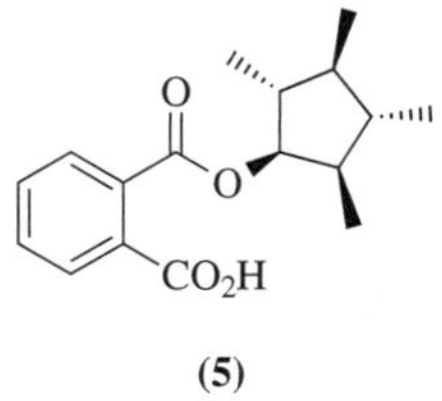

(**5**)

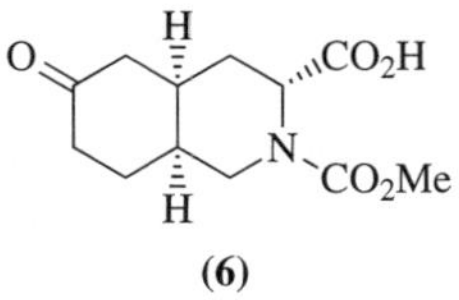

(**6**)

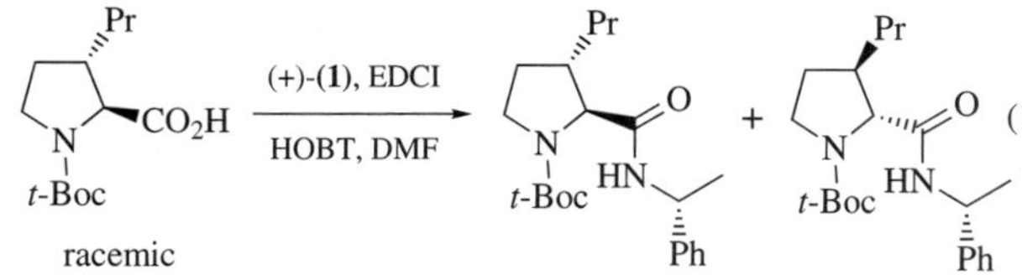
(1)

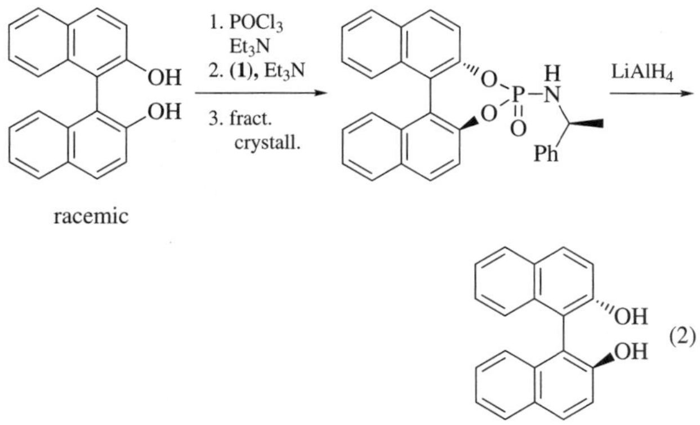
(2)

Racemic compounds other than carboxylic acids have also been resolved by reaction with enantiomerically pure (**1**) and separation of the corresponding diastereomeric mixtures by physical methods. For example, reaction of a racemic β-substituted γ-butyrolactone with (**1**) yields a mixture of hydroxy amides, which can be separated by fractional recrystallization and chromatography (eq 3).[14] Amide hydrolysis regenerates the chiral hydroxy acids, which spontaneously cyclize to produce the chiral lactones.

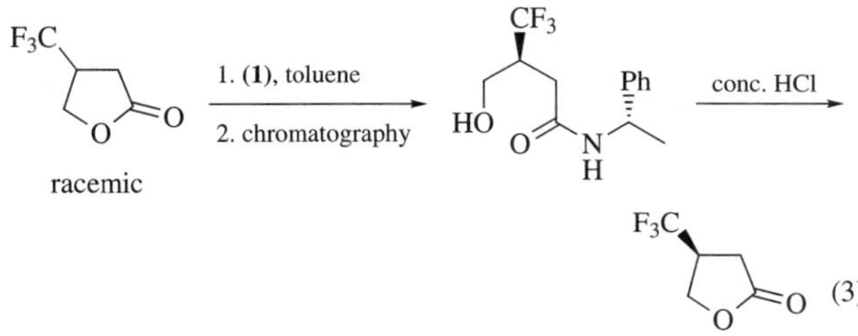
(3)

The displacement of a variety of leaving groups by (**1**) produces diastereomeric mixtures of amines, which can be separated into diastereomerically pure secondary amines and, following reductive removal of the α-methylbenzyl group, serve as a source of chiral primary amines (eq 4).[15]

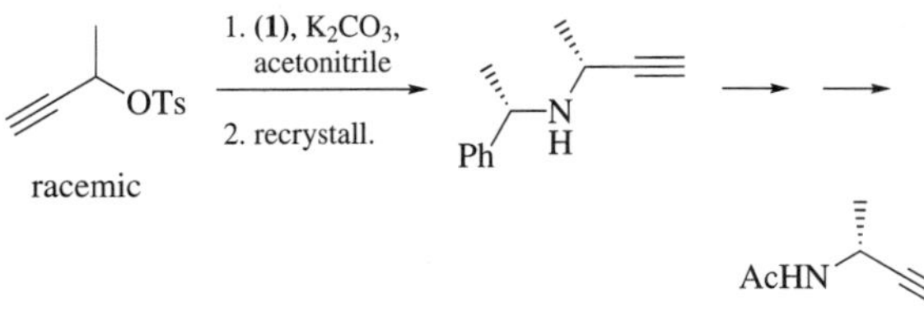
(4)

Reagent for the Determination of Enantiomeric Purity of Carboxylic Acids. Amine (**1**) is frequently used as a derivatizing reagent for determining the enantiomeric purity of carboxylic acids by HPLC, with limits of detection often as low as 1%. Most commonly used coupling methods include use of dehydrating agents such as *1,3-Dicyclohexylcarbodiimide* (eq 5)[16] and the mixed anhydride method (eq 6).[17]

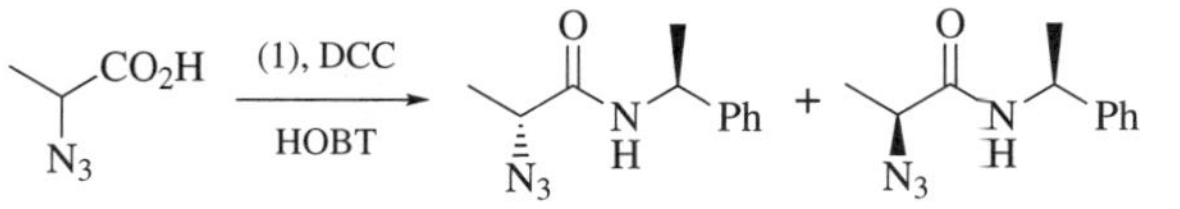
(5)

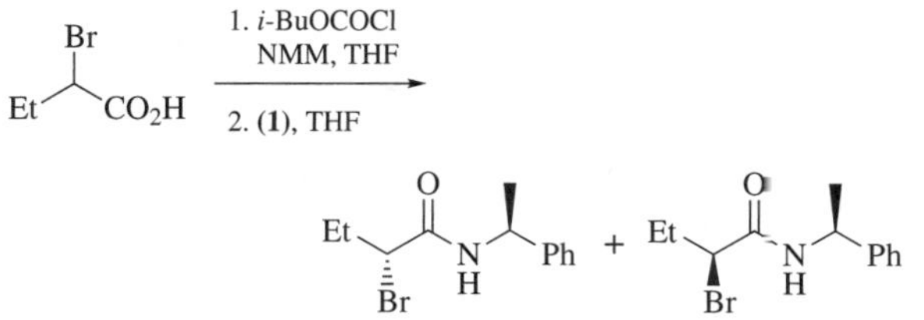
(6)

Stereospecific Reactions of Carbonyl Compounds. One of the most frequent uses of both enantiomers of reagent (**1**) is in promoting the stereospecific reaction of carbonyl compounds via the corresponding chiral imines. The transfer of chirality from (**1**) to the newly formed bonds is generally most effective in cyclization reactions. Some examples are the Lewis acid-catalyzed cyclization of ω-unsaturated aldehyde imines to produce amines of high enantiomeric purity (eq 7),[18] the enantioselective synthesis of γ,δ-unsaturated aldehydes via the aza-Claisen rearrangement of derivatives of (**1**) (eq 8),[19] and the asymmetric Lewis acid-catalyzed aza-Diels–Alder reaction of aldehyde imines with electron-rich dienes (eq 9).[20]

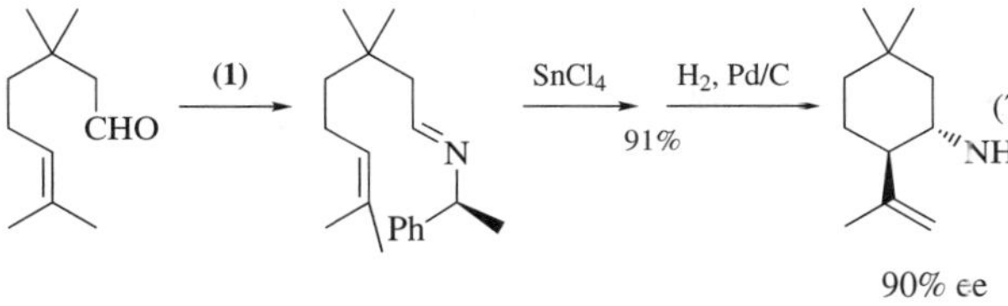
(7)

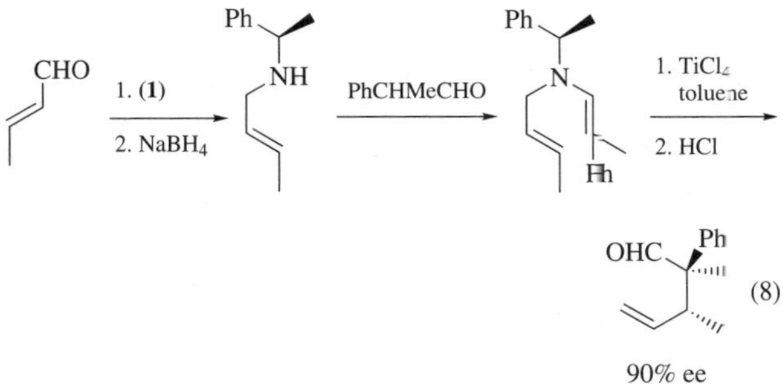
(8)

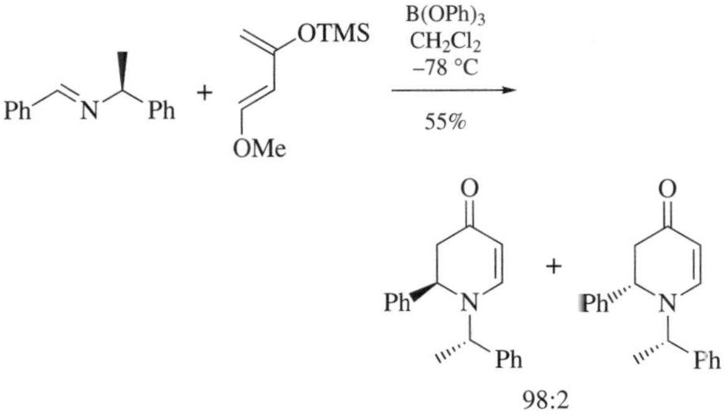
(9)

Enantiomerically pure disubstituted β-lactams are also available by cyclization of acyclic intermediates containing (**1**) as a chiral appendage, which is later removed by catalytic hydrogenation (eq 10).[21]

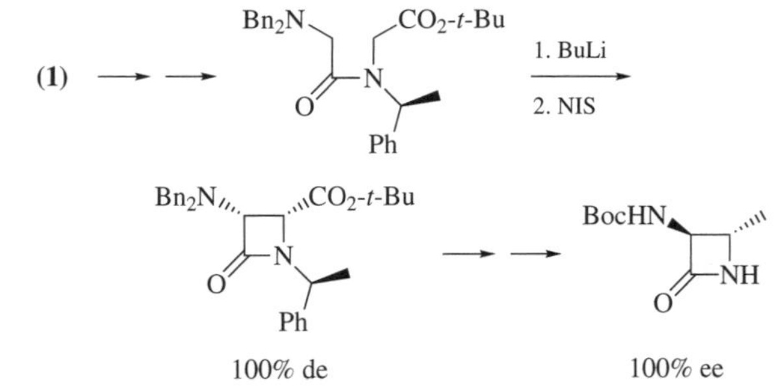

(10)

Examples of highly stereoselective acyclic reactions include the Zr-mediated coupling of aldehydes with imines of (**1**) to produce chiral amino alcohol derivatives (eq 11),[22] and the addition of cyanide to aldimines of (**1**) to yield intermediates that can be elaborated into enantiomerically pure α-amino acids (eq 12).[23]

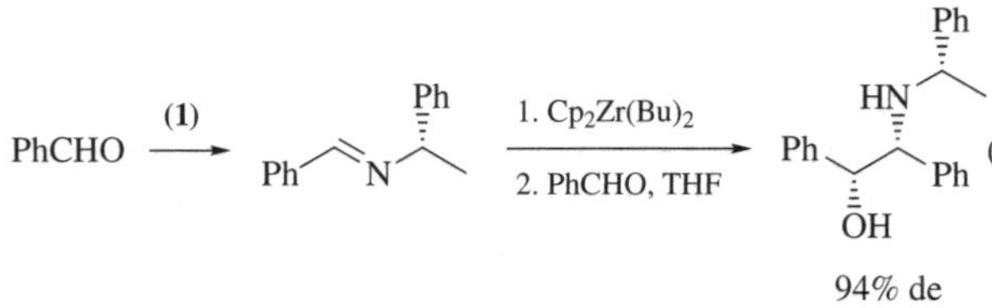

(11)

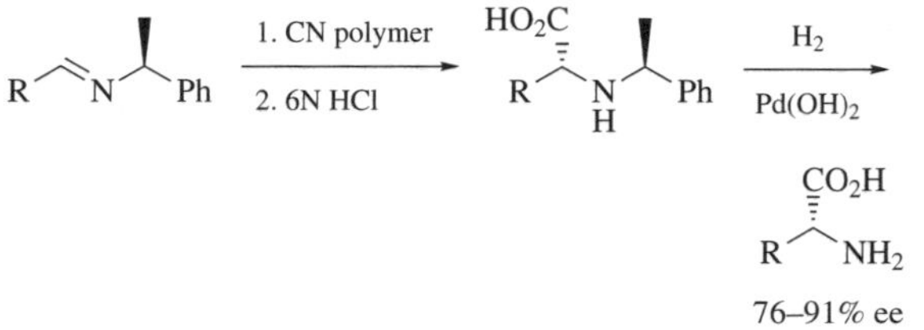

(12)

Another frequent use of (**1**) and its enantiomer is the stereospecific conjugate addition of carbonyl compounds to α,β-unsaturated systems. Most published examples contain chiral imine derivatives of cyclic ketones, which add to α,β-unsaturated esters and ketones in a highly stereoselective manner (eq 13 and eq 14).[24,25] When the ketone is not symmetrically substituted, reaction usually occurs at the most substituted α-position, including those cases where the ketone is α-substituted by oxygen (eq 15).[26] High stereoselectivity can also be achieved when the Michael acceptor is other than an unsaturated ketone or ester, such as a vinyl sulfone (eq 16).[27] Intramolecular variations of this transformation have also been described (eq 17).[28]

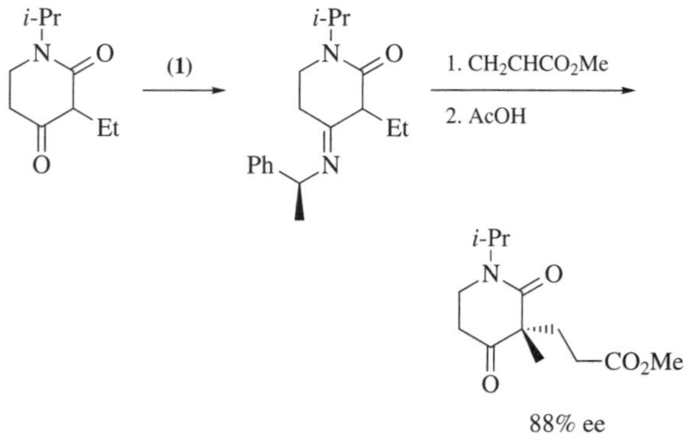

(13)

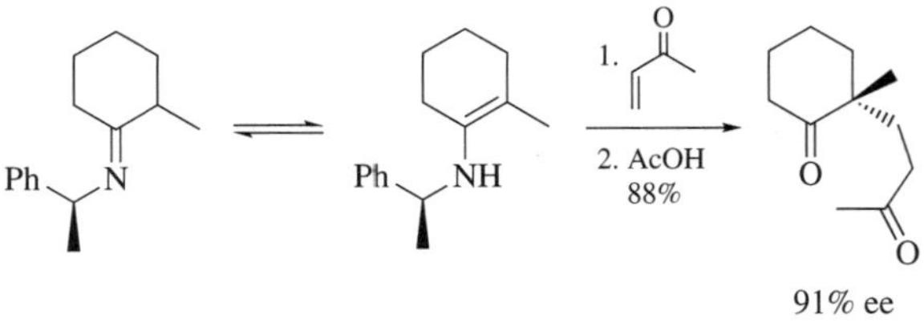

(14)

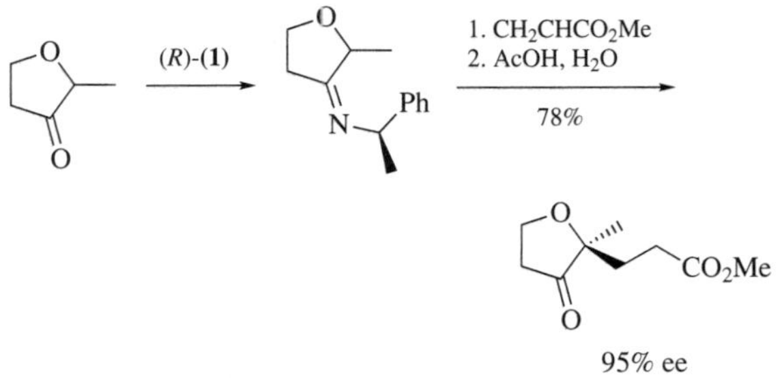

(15)

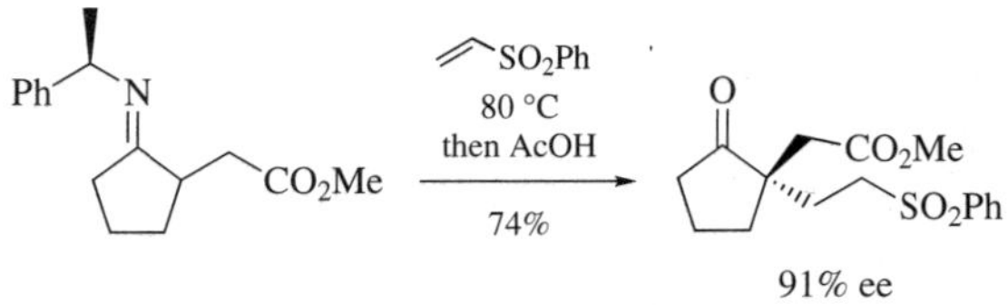

(16)

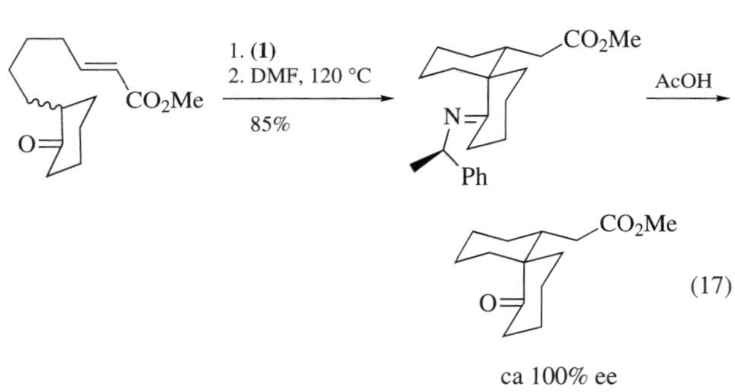

(17)

Stereospecific Reductive Amination of Carbonyl Compounds. Catalytic or chemical reduction of chiral imines derived from (**1**) often proceeds with high diastereoselectivity. Reductive removal of the α-methylbenzyl group yields chiral primary amines (eq 18 and 19).[29,30]

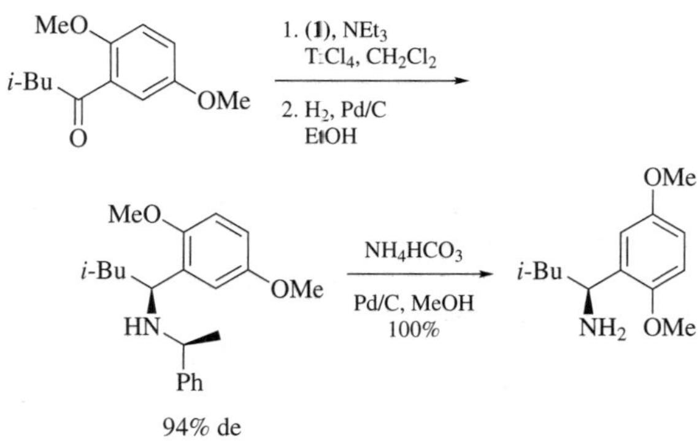

(18)

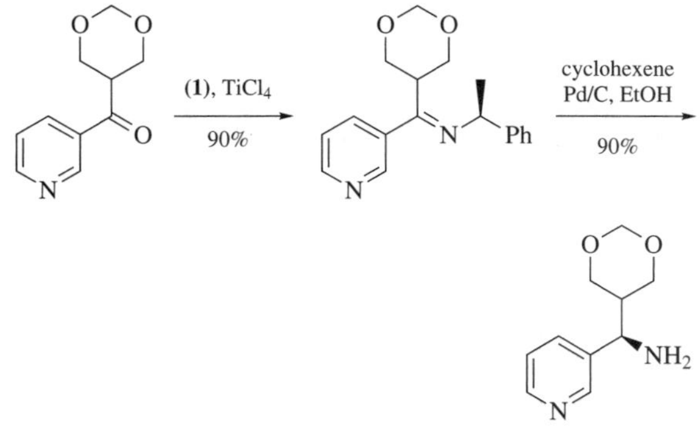

(19)

88% ee

Removable Chiral Appendage. Even in reactions that proceed with moderate stereoselectivity, incorporation of a chiral moiety such as (**1**) frequently provides an opportunity to easily separate diastereomeric products. For example, the introduction of (**1**) into an imidazolone structure allows the easy separation of diastereomers by chromatography. Reductive removal of the chiral appendage and imidazolone hydrolysis provides a synthesis of optically pure α-amino acids (eq 20).[31] In another example, even though the conjugate addition of (**1**) to methyl crotonate proceeds with low stereoselectivity, the diastereomeric conjugates are easily separated by chromatography and elaborated to provide optically active β-amino esters (eq 21).[32] Similarly, cycloaddition of the aldimine of (**1**) with a substituted ketene produces a mixture of β-lactams, which can be separated by chromatography as a source of optically active β-lactams (eq 22).[33]

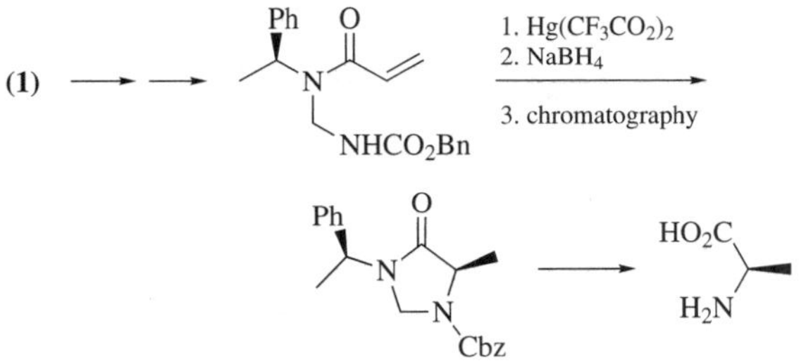

(20)

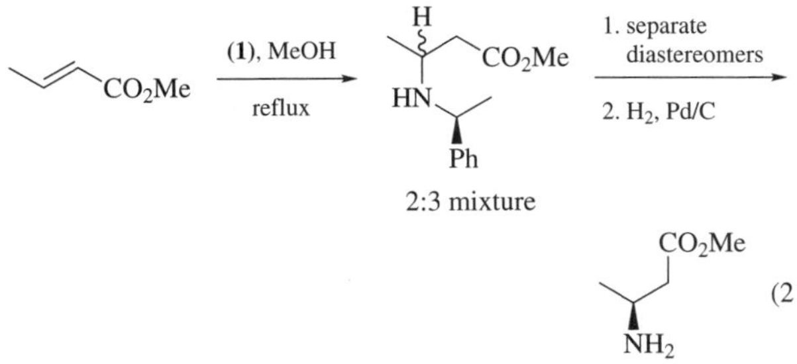

(21)

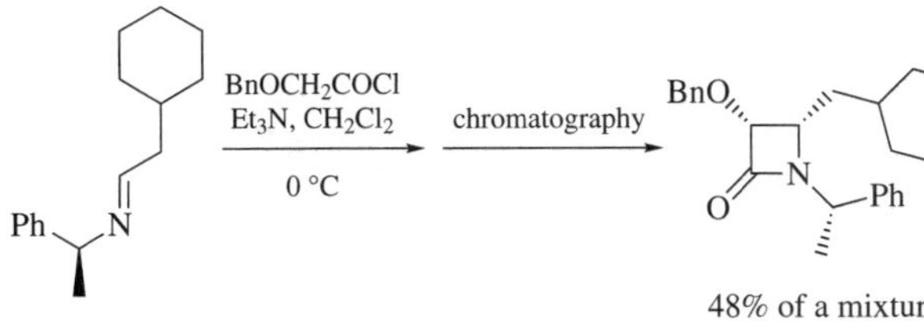

(22)

48% of a mixture of diastereomers

Miscellaneous Uses. Substituted derivatives of (**1**), e.g. (**7**), react with α,β-unsaturated carbonyl systems in a highly stereoselective manner to produce chiral β-aminocarbonyl compounds.[34] The lithium amides of a different type of substituted derivatives, e.g. (**8**), have been used to deprotonate symmetrical ketones, usually cyclic, in a highly stereoselective manner.[35]

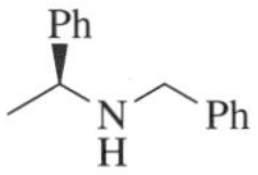

(**7**)

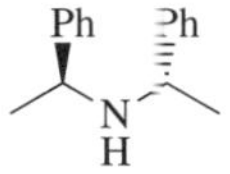

(**8**)

1. Newman, P. *Optical Resolution Procedures for Chemical Compounds*; O.R.I.C., Manhattan College: New York, 1978; Vol. 1 pp 79–82.
2. Brown, E.; Viot, F.; Le Floc'h, Y. *Tetrahedron Lett.* **1985**, *26*, 4451.
3. (a) Wu, M.-J.; Pridgen, L. N. *J. Org. Chem.* **1991**, *56*, 1340. (b) Hua, D. H.; Miao, S. W.; Chen, J. S.; Iguchi, S. *J. Org. Chem.* **1991**, *56*, 4.
4. Dale, J. A.; Dull, D. L.; Mosher, H. S. *J. Org. Chem.* **1969**, *34*, 2543.
5. Shapiro, M. J.; Archinal, A. E.; Jarema, M. A. *J. Org Chem.* **1989**, *54*, 5826.
6. Parker, D.; Taylor, R. J. *Tetrahedron* **1987**, *43*, 5451.
7. Kappe, C. O.; Uray, G.; Roschger, P.; Lindner, W.; Kratky, C.; Keller, W. *Tetrahedron* **1992**, *48*, 5473.
8. Yamamoto, M.; Hayashi, M.; Masaki, M.; Nohira, H. *Tetrahedron: Asymmetry* **1991**, *2*, 403.
9. Dharanipragada, R.; Nicolas, E.; Toth, G.; Hruby, V. *Tetrahedron Lett.* **1989**, *30*, 6841.
10. Hoffmann, N.; Scharf, H.-D. *Tetrahedron: Asymmetry* **1991**, *2*, 977.
11. Ornstein, P. L.; Arnold, M. B.; Augenstein, N. K.; Paschal, J. W. *J. Org. Chem.* **1991**, *56*, 4388.
12. Chung, J. Y. L.; Wasicak, J. T.; Arnold, W. A.; May, C. S.; Nadzan, A. M.; Holladay, M. W. *J. Org. Chem.* **1990**, *55*, 270.
13. Gong, B.; Chen, W.; Hu, B. *J. Org. Chem.* **1991**, *56*, 423.
14. (a) Taguchi, T.; Kawara, A.; Watanabe, S.; Oki, Y.; Fukushima, H.; Kobayashi, Y.; Okada, M.; Ohta, K.; Iitaka, Y. *Tetrahedron Lett.* **1986**, *27*, 5117. (b) Ishibashi, F.; Taniguchi, E. *Chem. Lett.* **1986**, 1771.
15. Nilsson, B. M.; De Boer, P.; Grol, C. J.; Hacksell, U. *Chirality* **1992**, *4*, 367.
16. Hoffman, R. V.; Kim, H.-O. *Tetrahedron* **1992**, *48*, 3007.
17. Compagnone, R. S.; Rapoport, H. *J. Org. Chem.* **1986**, *51*, 1713.
18. Sakane, S.; Maruoka, K.; Yamamoto, H. *Tetrahedron* **1986**, *42*, 2203.
19. Bailey, P. D.; Harrison, M. J. *Tetrahedron Lett.* **1989**, *30*, 5341.
20. Hattori, K.; Yamamoto, H. *Tetrahedron* **1993**, *49*, 1749.
21. (a) Kawabata, T.; Itoh, K.; Hiyama, T. *Tetrahedron Lett.* **1989**, *30*, 4837. (b) Kawabata, T.; Sumi, K.; Hiyama, T. *J. Am. Chem. Soc.* **1989**, *111*, 6843.
22. Ito, H.; Taguchi, T. Hanzawa, Y. *Tetrahedron Lett.* **1992**, *33*, 4469.
23. Saito, K.; Harada, K. *Tetrahedron Lett.* **1989**, *30*, 4535.
24. (a) Ambroise, L.; Chassagnard, C.; Revial, G.; d'Angelo, J. *Tetrahedron: Asymmetry* **1991**, *2*, 407. (b) d'Angelo, J.; Revial, G.; Volpe, T.; Pfau, M. *Tetrahedron Lett.* **1988**, *29*, 4427.
25. (a) Pfau, M.; Revial, G.; Guingant, A.; d'Angelo, J *J. Am. Chem. Soc.* **1985**, *107*, 273. (b) Revial, G. *Tetrahedron Lett.* **1989**, *30*, 4121
26. (a) Desmaële, D. *Tetrahedron* **1992**, *48*, 2925. (b) Desmaële, D.; d'Angelo, J. *Tetrahedron Lett.* **1989**, *30*, 345.
27. Pinheiro, S.; Guingant, A.; Desmaële, D.; d'Angelo, J. *Tetrahedron: Asymmetry* **1992**, *3*, 1003.
28. d'Angelo, J.; Ferroud, C. *Tetrahedron Lett.* **1989**, *30*, 6511.
29. (a) Bringmann, G.; Künkel, G.; Geuder, T. *Synlett* **1990**, 253. (b) Van Niel, J. C. G.; Pandit, U. K. *Tetrahedron* **1985**, *41*, 6005. (c) Bringmann, G.; Geisler, J.-P. *Tetrahedron Lett.* **1989**, *30*, 317.
30. Farkas, E.; Sunman, C. J. *J. Org. Chem.* **1985**, *50*, 1110.

31. Amoroso, R.; Cardillo, G.; Tomasini, C. *Tetrahedron Lett.* **1990**, *31*, 6413.
32. Estermann, H.; Seebach, D. *Helv. Chim. Acta* **1988**, *71*, 1824.
33. Kobayashi, Y.; Takemoto, Y.; Kamijo, T.; Harada, H.; Ito, Y.; Terashima, S. *Tetrahedron* **1992**, *48*, 1853.
34. Davies, S. G.; Ichihara, O. *Tetrahedron: Asymmetry* **1991**, *2*, 183.
35. Cain, C. M.; Cousins, R. P. C.; Coumbarides, G.; Simpkins, N. S. *Tetrahedron* **1990**, *46*, 523.

Juan C. Jaen
Parke-Davis Pharmaceutical Research, Ann Arbor, MI, USA

(*R*)-Methyl 2-*t*-Butyl-3(2*H*)-oxazole-carboxylate

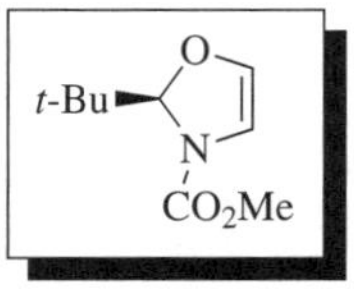

[104173-34-4] $C_9H_{15}NO_3$ (MW 185.25)

(chiral oxazoline with highly reactive double bond;[1–3] chiral derivative of α-hydroxy- or α-aminoacetaldehyde; reactions with acylating reagents,[1a,2] with carbenes,[1a,3] and with electron-poor alkenes;[1a,3] cycloadditions;[1a,3] building block for the synthesis of amino alcohols and aminohydroxy carboxylic acid derivatives[1–3])

Physical Data: bp 70 °C/0.02 mmHg; $[\alpha]_D^{rt} = +434°$ (*c* 1, $CHCl_3$).
Solubility: sol all common nonprotic organic solvents.
Preparative Methods: the acetal from serine methyl ester and pivaladehyde is *N*-methoxycarbonylated and saponified to give the *cis*-half ester (**2**) (61% yield). Oxidative decarboxylation by electrolysis in AcOH (1:3 by weight) in the presence of 5% Et_3N affords (**3**) (87%), which upon heating with NH_4Br in toluene yields the title oxazoline (**1**) (75% after distillation in vacuo) (eq 1).[4]

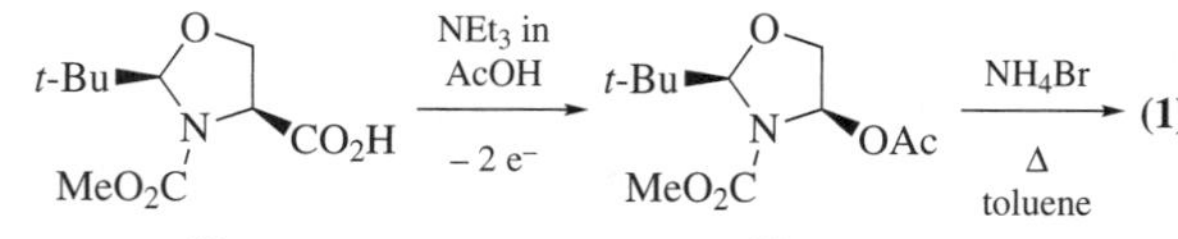
(1)

(**2**) (**3**)

Handling, Storage, and Precautions: stable over many months in a freezer under inert atmosphere (N_2 or Ar). Use in a fume hood.

Reactions with Electrophiles at C-4. The oxazoline (**1**) can be metalated on the double bond carbon atom next to the ester group, and the resulting Li derivative coupled with electrophiles (eq 2).[2]

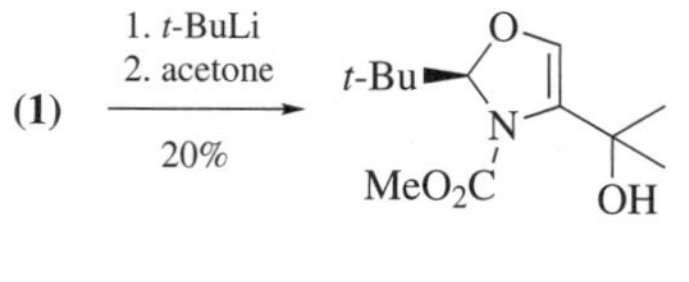
(2)

Direct Reactions with Electrophiles at C-5. Vilsmeier and Friedel–Crafts reactions lead to 5-acyl derivatives (eq 3),[2] which can be further elaborated[3] as indicated in eq 4; a single enantiopure diastereoisomer of the anhydride (**4**) is formed. Seebach and co-workers[3] erroneously assigned (**4**) as arising from an *exo* rather than the *endo* addition.

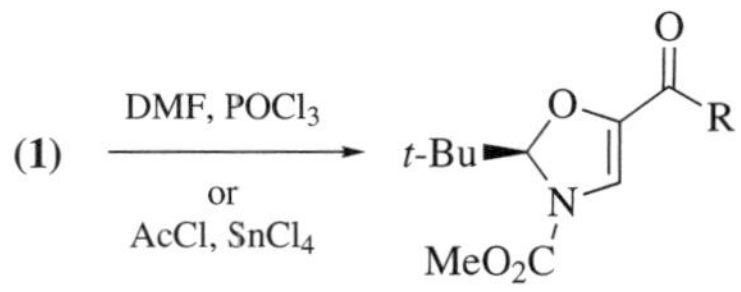
(3)

R = H, 75%
R = Me, 25%

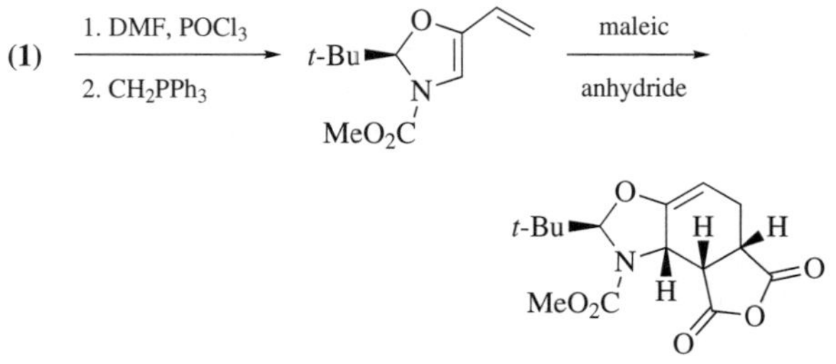
(4)

(**4**)

Cycloadditions.[3] [2 + 1], [2 + 2], and [4 + 2] cycloadditions with electrophilic reactants such as carbenes (eq 5), tetracyanoethylene (eq 6), and α-keto-β,γ-unsaturated nitriles[5] (eq 7) lead to interesting products which are all enantiopure; see, for instance, the 5-amino-6-hydroxy-2-keto acid (**5**).

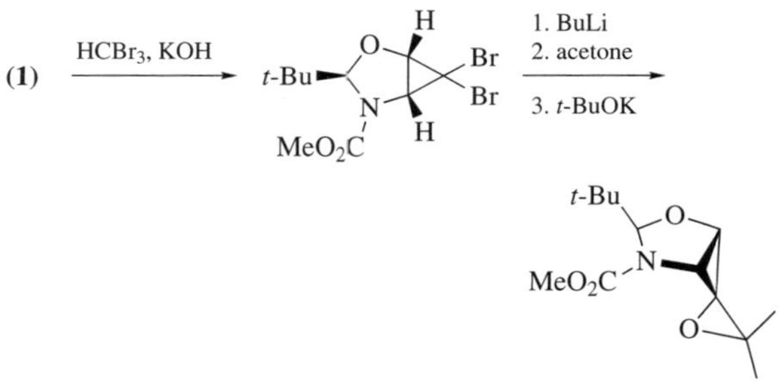
(5)

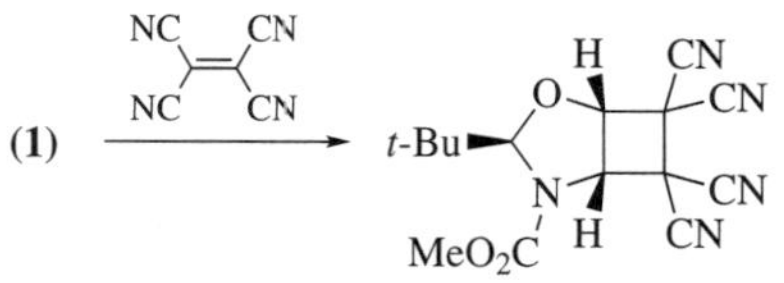
(6)

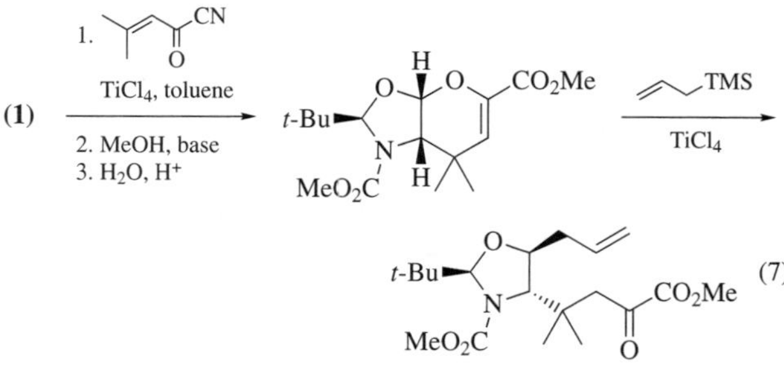
(7)

(**5**)

Other 2,3-Dihydrooxazoles and -thiazoles from α-Amino Acids. Table 1 shows oxazolines and thiazolines also prepared from amino acids (serine, threonine, or cysteine), with or without decarboxylation; common to all of them is the conversion of the original stereogenic center to a trigonal center, and chirality due to a *t*-butyl-substituted acetal carbon in the ring (see Table 1).[1–4,6,7] The structure and reactivity of such acetals has been discussed.[8,9]

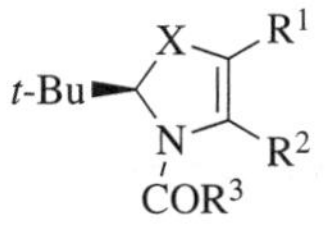

(6)

Table 1 2,3-Dihydrooxazoles and -thiazoles (**6**) from α-Amino Acids

X	R^1	R^2	R^3	Amino acid
O	H	H	OMe	Serine[1–4]
O	Me	H	OMe	Threonine[1–4]
O	H	CO_2Me	OMe	Serine[1a,2]
O	Me	CO_2Me	OMe	Threonine[1a,2]
O	Me	Me	OMe	Threonine[2]
S	H	H	OCH_2Ph	Cysteine[6]
S	H	CO_2Me	OCH_2Ph	Cysteine[6,7]
S	H	CO_2Me	OMe	Cysteine[7]
SO	H	H	OCH_2Ph	Cysteine[6,7]
SO_2	H	H	OCH_2Ph	Cysteine[7]

Related Reactions. The photochemistry (Paterno–Büchi reactions) of the achiral oxazoline (**7**) has been studied.[10] The analogous urethane (**8**), which is chiral by attachment of an apocamphanoyl group, shows an intriguing stereoselectivity pattern in its reaction with electrophiles.[11] For another case of an oxidative decarboxylation as a key step in the application of the SRSC (self-regeneration of stereogenic centers) principle, see the preparation of the dihydropyrimidone (**9**) from aspartic acid.[12]

(**7**)

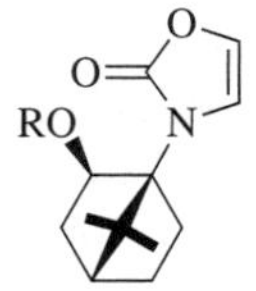

(**8**)

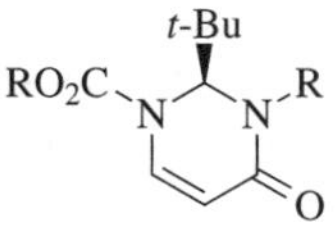

(**9**)

Related Reagents. 1-Benzoyl-2-*t*-butyl-3,5-dimethyl-4-imidazolidinone; (*S*2,4*S*)-3-Benzoyl-2-*t*-butyl-4-methyl-1,3-oxazolidin-5-one; *N*-*t*-Butoxycarbonyl-*N*-methylaminomethyllithium; (*R*)-2-*t*-Butyl-6-methyl-4*H*-1,3-dioxin-4-one; (*R*,*R*)-2-*t*-Butyl-5-methyl-1,3-dioxolan-4-one.

1. (a) Seebach, D.; Stucky, G. *Angew. Chem., Int. Ed. Engl.* **1988**, *27*, 1351. (b) Renaud, P.; Seebach, D. *Angew. Chem., Int. Ed. Engl.* **1986**, *25*, 843.
2. Stucky, G.; Seebach, D. *Ber. Dtsch. Chem. Ges./Chem. Ber.* **1989**, *122*, 2365 (*Chem. Abstr.* **1990**, *112*, 55 674j).
3. Seebach, D.; Stucky, G.; Pfammatter, E. *Ber. Dtsch. Chem. Ges./Chem. Ber.* **1989**, *122*, 2377 (*Chem. Abstr.* **1990**, *112*, 77 008x).
4. Seebach, D.; Stucky, G.; Renaud, P. *Chimia* **1988**, *42*, 176 (*Chem. Abstr.* **1989**, *110*, 173 705b).
5. (a) John, R. A.; Schmid, V.; Wyler, H. *Helv. Chim. Acta* **1987**, *70*, 600. (b) Zhuo, J.-C.; Wyler, H. *Helv. Chim. Acta* **1993**, *76*, 1916.
6. Seebach, D.; Jeanguenat, A.; Schmidt, J.; Maetzke, T. *Chimia* **1989**, *43*, 314 (*Chem. Abstr.* **1990**, *112*, 217 495f).
7. Jeanguenat, A.; Seebach, D. *J. Chem. Soc., Perkin Trans. 1* **1991**, 2291.
8. Seebach, D.; Lamatsch, B.; Amstutz, R.; Beck, A. K.; Dobler, M.; Egli, M. et al. *Helv. Chim. Acta* **1992**, *75*, 913.
9. Lamatsch, B.; Seebach, D.; Ha, T.-K. *Helv. Chim. Acta* **1992**, *75*, 1095 (*Chem. Abstr.* **1992**, *117*, 150 330f).
10. Weuthen, M.; Scharf, H.-D.; Runsink, J.; Vassen, R. *Ber. Dtsch. Chem. Ges./Chem. Ber.* **1988**, *121*, 971 (*Chem. Abstr.* **1988**, *108*, 221 625h).
11. (a) Ishizuka, T.; Ishibuchi, S.; Kunieda, T. *Tetrahedron Lett.* **1989**, *30*, 3449. (b) Ishizuka, T.; Ishibuchi, S.; Kunieda, T. *Tetrahedron* **1993**, *49*, 1841.
12. (a) Negrete, G. R.; Konopelski, J. P. *Tetrahedron: Asymmetry* **1991**, *2*, 105. (b) Chu, K. S.; Negrete, G. R.; Konopelski, J. P.; Lakner, F. J.; Woo, N.-T; Olmstead, M. M. *J. Am. Chem. Soc.* **1992**, *114*, 1800.

Joachim Podlech & Dieter Seebach
Eidgenössische Technische Hochschule, Zürich, Switzerland

(*R*)-4-Methylcyclohexylidenemethyl-copper

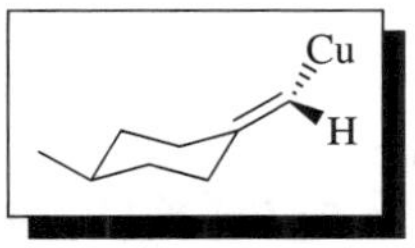

[60164-96-7] $C_8H_{13}Cu$ (MW 172.76)

(optically active vinylcopper reagent;[1] useful in stereospecific synthesis of axially dissymmetric dienes and alkenes[1,2])

Solubility: dark black suspensions of the reagent (0.26 M) have been prepared in THF–pentane.[1,2]

Preparative Methods: (−)-(*R*)-(4-methylcyclohexylidene)methyl bromide[3,4] in THF is treated with *s-Butyllithium* in pentane at −75 °C to give a solution of (*R*)-4-methylcyclohexylidenemethyllithium. Addition of 1 equiv of *Copper(I) Iodide* to this mixture, followed by stirring at −35 °C for 30 min, gives a dark black suspension of (*R*)-4-methylcyclohexylidenemethylcopper (eq 1).[2] (*S*)-4-Methylcyclohexylidenemethylcopper may be prepared from (+)-(*S*)-(4-methylcyclohexylidene)methyl bromide[3,4] using a similar procedure.[1,2]

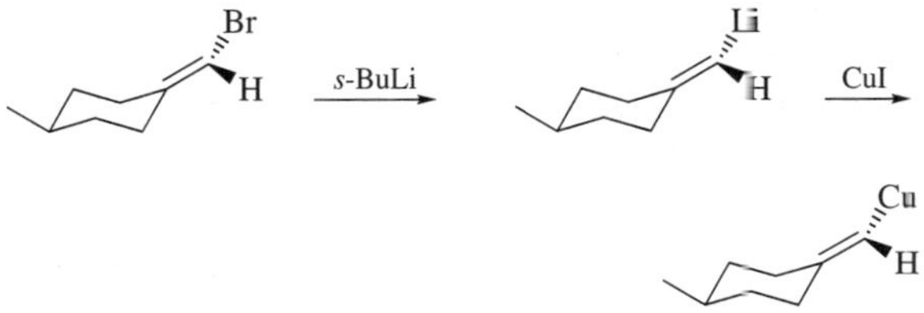

(1)

Handling, Storage, and Precautions: the reagent is prepared under inert gas atmosphere at temperatures of −35 °C or below. Upon heating to rt, or exposure to oxygen, the reagent decomposes with formation of bis(4-methylcyclohexylidene)ethane. Use in a fume hood.

Oxidative Coupling: Synthesis of Optically Active Dienes of Biaxial Dissymmetry. Alkenylcopper reagents undergo thermal or oxidative coupling to give 1,3-dienes with retention of configuration at the alkenic bond.[5] Passage of a stream of oxygen through a solution of (*R*)-4-methylcyclohexylidenemethylcopper affords (*aR,aR*)-bis(4-methylcyclohexylidene)ethane in 30% yield (eq 2).[1,2] The dissymmetry of the latter compound results from the combination of two chiral alkenic axes which are oriented in a planar transoid conformation. The chiroptical properties of this and closely related planar acyclic 1,3-dienes have been reported.[2,6] Additional members of this novel class of optically active dienes could be synthesized by oxidative coupling of analogous chiral vinylcopper reagents.

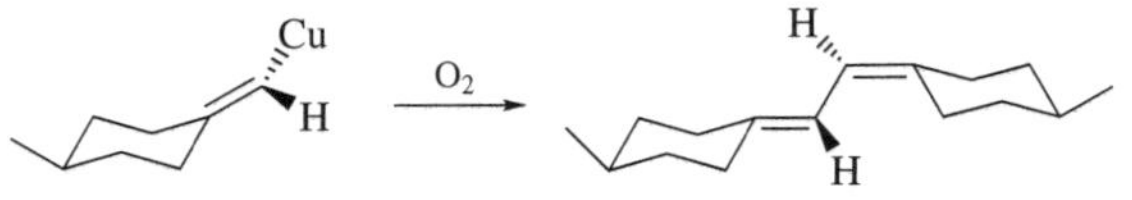

(2)

Alkylation: Formation of Optically Active 4-Methylcyclohexylidenealkanes. Alkenylcopper reagents can be alkylated by treatment with alkyl halides and other electrophiles.[7] Reaction of (*S*)-4-methylcyclohexylidenemethylcopper (enantiomer of the title reagent) with *Iodomethane* gives (+)-(*S*)-4-(methylcyclohexylidene)ethane with 98% retention of configuration (eq 3).[1,2] Alkylation of (*R*)-4-methylcyclohexylidenemethylcopper with methyl iodide and other alkyl halides is expected to show similar stereospecificity.

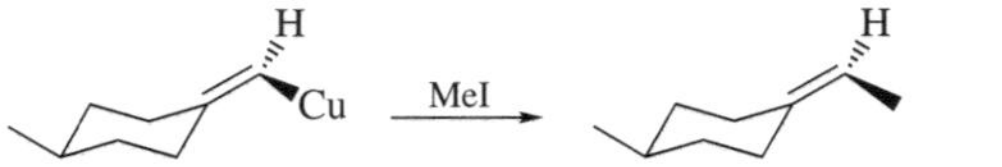

(3)

---

1. Banks, R. B.; Walborsky, H. M. *J. Am. Chem. Soc.* **1976**, *98*, 3732.
2. Walborsky, H. M.; Banks, R. B.; Banks, M. L. A.; Duraisamy, M. *Organometallics* **1982**, *1*, 667.
3. Perkin, W. H.; Pope, W. J. *J. Chem. Soc.* **1911**, *99*, 1510.
4. Solladie, G.; Zimmermann, R. G. *Tetrahedron Lett.* **1984**, *25*, 5769.
5. (a) Normant, J. F. *Synthesis* **1972**, 63. (b) Kaufmann, T. *Angew. Chem., Int. Ed. Engl.* **1974**, *13*, 291.
6. (a) Duraisamy, M.; Walborsky, H. M. *J. Am. Chem. Soc.* **1983**, *105*, 3252. (b) Duraisamy, M.; Walborsky, H. M. *J. Am. Chem. Soc.* **1983**, *105*, 3264. (c) Duraisamy, M.; Walborsky, H. M. *J. Am. Chem. Soc.* **1983**, *105*, 3270. (d) Reddy, S. M.; Goedken, V. L.; Walborsky, H. M. *J. Am. Chem. Soc.* **1986**, *108*, 2691. (e) Gawronski, J. K.; Walborsky, H. M. *J. Org. Chem.* **1986**, *51*, 2863.
7. (a) Posner, G. H. *Org. React.* **1974**, *22*, 253. (b) Lipshutz, B. H.; Sengupta, S. *Org. React.* **1992**, *41*, 135.

R. Bruce Banks
University of North Carolina at Greensboro, NC, USA

(*S*)-1-Methyl-2-[(dihydroisoindol-2-yl)methyl]pyrrolidine

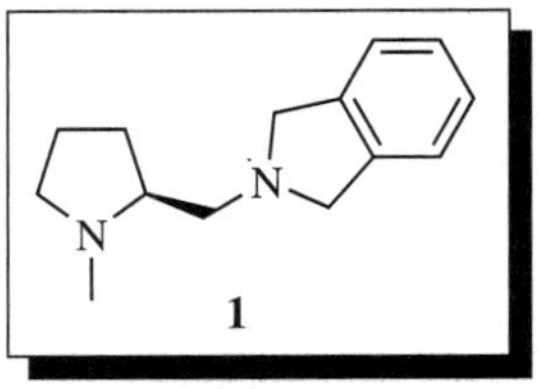

[159497-37-7] $C_{14}H_{20}N_2$ (MW 216.32)

(asymmetric catalyst for the Mukaiyama aldol reaction,[1,2] kinetic resolution of secondary alcohols,[3] acylation of *meso*-diols,[4] and palladium-catalyzed rearrangement of allylic imidates to allylic amides[5])

Alternate Name: (*S*)-2-(isoindolinylmethyl)-*N*-methylpyrrolidine.
Physical Data: bp 98–99 °C/ 0.08 mm Hg, $[\alpha]_D^{28}$ -76.5 (*c* 1.3, EtOH).
Solubility: soluble in CH_2Cl_2, CH_3CN.
Form Supplied in: not commercially available; colorless oil.
Analysis of Reagent Purity: by NMR and specific rotation determination.
Preparative Methods: can be prepared by DCC coupling of *N*-Boc-(*S*)-proline and isoindoline followed by $LiAlH_4$ reduction.[5]
Purification: can be purified by distillation.

Asymmetric Aldol Reaction.[1,2] A number of proline-derived chiral diamines have been prepared and utilized in stereoselective syntheses since they can effectively create chiral environments by forming structurally rigid chelate complexes. Optically active *syn*-2,3-dihydroxy thioesters are prepared from (*Z*)-1-(ethylthio)-1-(trimethylsiloxy)-2-(*tert*-butyldimethylsiloxy)-ethene and various aldehydes in the presence of $Sn(OTf)_2$, chiral diamine (**1**), and $Bu_2Sn(OAc)_2$ (eq 1). This reaction is general with respect to the aldehyde component and both diastereo and enantioselectivities are excellent. The absolute configuration of the product diol is 2*R*,3*S*. The use of (*S*)-1-methyl-2-[(indolinyl)methyl]pyrrolidine as a chiral ligand gives the *syn* aldol adducts with the reversed absolute configuration (2*S*,3*R*).

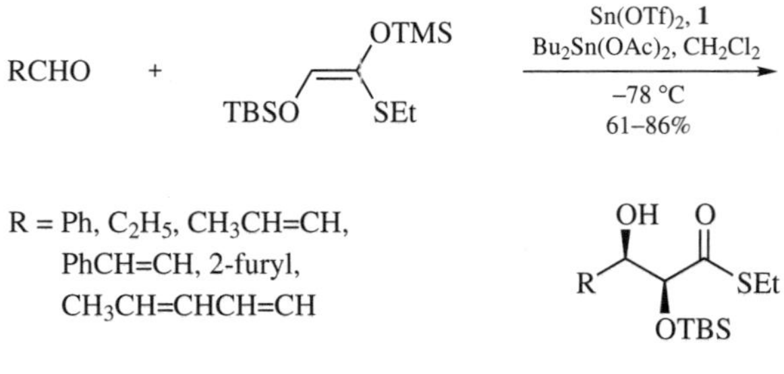

(1)

>96% de, >96% ee

Kinetic Resolution of Racemic Secondary Alcohols. Racemic cyclic[3] and acyclic[3] secondary alcohols and β-halohydrins[6] are kinetically resolved in good chemical yields with modest-to-excellent enantioselectivity (eqs 2 and 3).

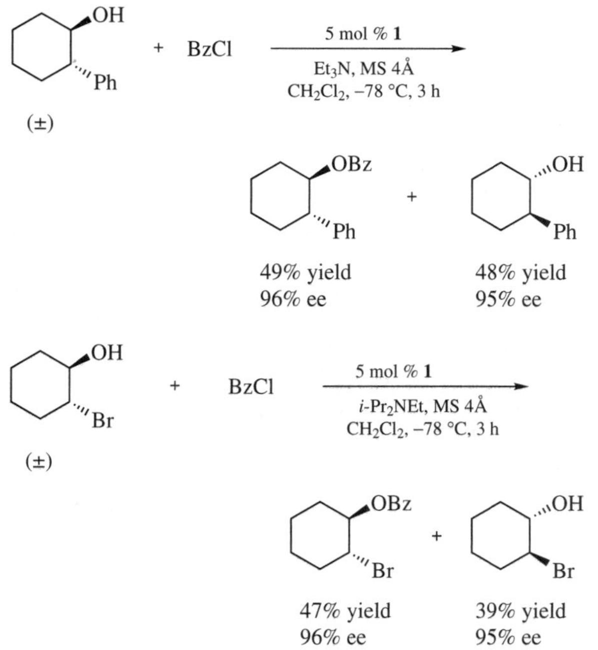

(2)

(3)

Asymmetric Acylation of *meso*-Diols. Both cyclic[4] and acyclic[4] *meso*-1,2-diols are desymmetrized by acylation in the presence of a stoichiometric amount of this ligand with modest-to-excellent enantioselectivity (eq 4). In a special case, *cis*-5,5-dimethyl-2-cyclopentene-1,4-diol[7] was monobenzoylated in the presence of a catalytic amount of this ligand in good yield and with perfect enantioselection (87%, > 99.5% ee).

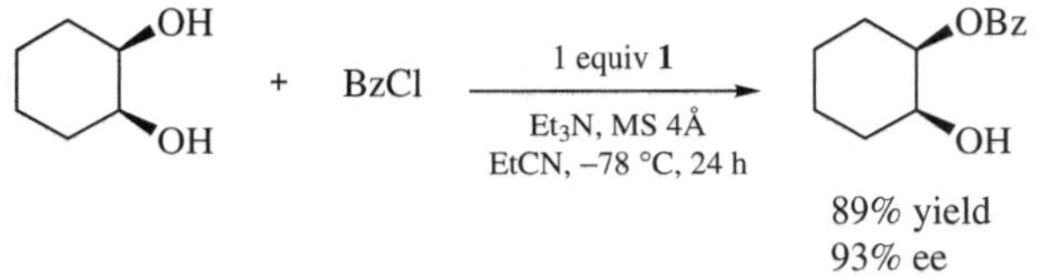

(4)

Rearrangement of Allylic Imidates.[5] Overman and co-workers reported an interesting application of this chiral diamine to induce enantioselectivity in Pd(II)-catalyzed rearrangement of allylic imidates to allylic amides (eq 5).

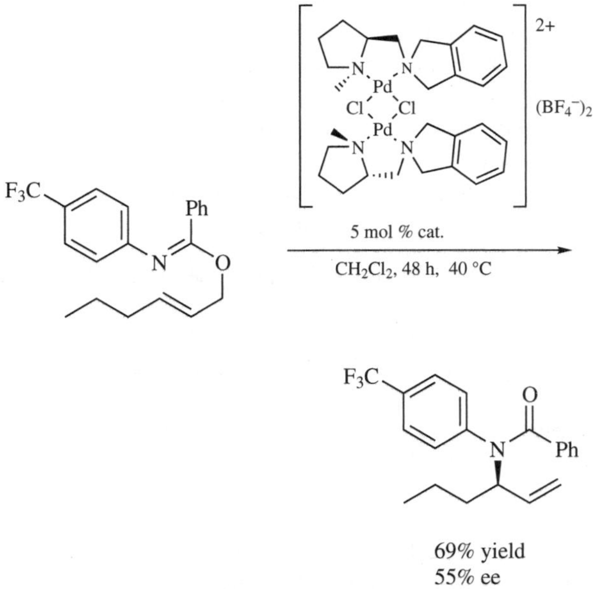

(5)

Related Reagents. (*S*)-1-Methyl-2-[(indolinyl)methyl]pyrrolidine; (*S*)-1-methyl-2-[(1-benz[*cd*]indolinyl)methyl]pyrrolidine; (*S*)-1-methyl-2-[(1-naphthylamino)methyl]pyrrolidine; (*S*)-1-methyl-2-[(1-piperidin-1-yl)methyl]pyrrolidine; (*S*)-1-ethyl-2-[(1-piperidin-1-yl)methyl]pyrrolidine; (*S*)-1-propyl-2-[(1-piperidin-1-yl)methyl]pyrrolidine.

1. Kobayashi, S.; Horibe. M. *J. Am. Chem. Soc.* **1994**, *116*, 9805.
2. Kobayashi, S.; Horibe. M. *Chem. Eur. J.* **1997**, *3*, 1472.
3. Sano, T.; Imai, K.; Ohashi, K.; Oriyama. T. *Chem. Lett.* **1999**, 265.
4. Oriyama, T.; Imai, K.; Hosoya, T.; Sano, T. *Tetrahedron Lett.* **1998**, *39*, 397.
5. Calter, M.; Hollis, T. K.; Overman, L. E.; Ziller, J.; Zipp, G. G. *J. Org. Chem.* **1997**, *62*, 1449.
6. Sano, T.; Miyata, H.; Oriyama, T. *Enantiomer* **2000**, *5*, 119.
7. Oriyama, T.; Hosoya, T.; Sano, T. *Heterocycles* **2000**, *52*, 1065.

Erik J. Sorensen, HirofumiSeike & Jason M. Rohde
The Scripps Research Institute, La Jolla, CA, USA

Methyl (4*R*,5*R*)-(*E*)-3-(1,3-Dimethyl-4,5-diphenyl-2-imidazolidinyl)propenoate

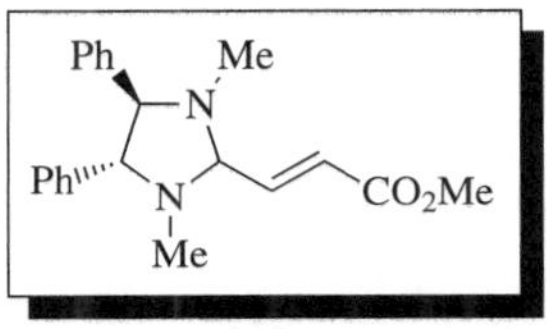

[135212-29-2] $C_{21}H_{24}N_2O_2$ (MW 336.47)

(a chiral synthetic equivalent of methyl (*E*)-4-oxo-2-butenoate)

Physical Data: colorless needles (MeOH); mp 109–110 °C; $[\alpha]_D^{23}$ +45.0° (*c* 0.6, $CHCl_3$);[1] IR, ^1H NMR, ^{13}C NMR, MS spectral data, and X-ray structural data are also available.[1]
Solubility: very sol dichloromethane, chloroform, ethyl acetate, acetone; sol methanol, ethanol; hardly sol hexane.
Analysis of Reagent Purity: chiral HPLC.[1]
Preparative Methods: both the (4*R*,5*R*) and (4*S*,5*S*) compounds are readily available by condensation of optically active 1,2-bis(methylamino)-1,2-diphenylethanes[2] with methyl (*E*)-4-oxo-2-butenoate.[3]

Pyrrolidine Synthesis. The lithium (*Z*)-enolate of methyl *N*-benzylideneglycinate undergoes a cycloaddition rapidly at −78 °C with methyl (4*R*,5*R*)-(*E*)-3-(1,3-dimethyl-4,5-diphenyl-2-imidazolidinyl)propenoate (**1**) to give pyrrolidine-2,4-dicarboxylate in quantitative yield and as a single stereoisomer (eq 1).[1] Removal of the chiral auxiliary from the cycloadduct can be readily performed by treatment with conc *Sulfuric Acid* in MeOH at rt to produce the optically pure pyrrolidine-2,4-dicarboxylate bearing an acetal substituent at the 3-position.

Asymmetric Nitrile Oxide Cycloadditions. Although the dipolar cycloadditions of nitrile oxides with (**1**) are poor both in reactivity and regioselectivity, regioisomeric isoxazolines are obtained as single diastereomers (eq 2).[4] Removal of the chiral auxiliary can again be performed by an acetal exchange reaction under reflux in MeOH in the presence of H_2SO_4.

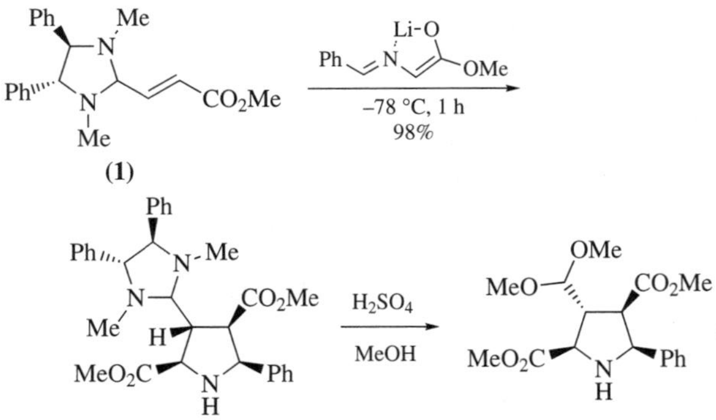

(1)

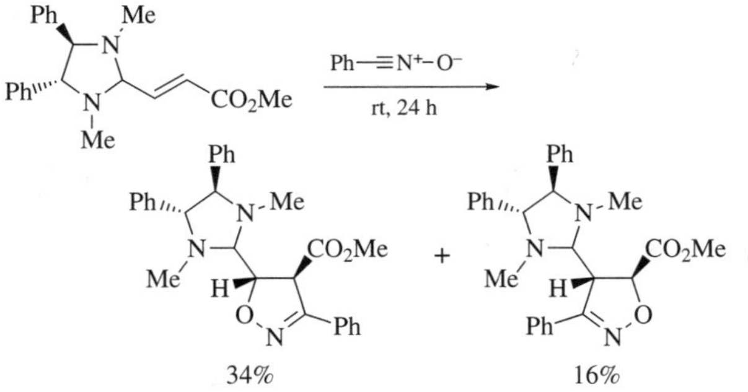

(2)

Related α,β-Unsaturated Esters. Similar α,β-unsaturated esters bearing a heterocyclic chiral auxiliary of α-amino acid origin at the β-position are known and have been utilized in asymmetric synthesis. Effective asymmetric conjugate additions of cuprates to (**2**),[5] (**3**),[6] and (**5**),[7] epoxidations of (**3**),[8] and dipolar cycloadditions of (**2**)[9] have been reported. Although oxazolidine (**4**)[10] is only obtained as an 86:14 equilibrating mixture of stereoisomers, reactions with the lithium (*Z*)-enolate of methyl *N*-benzylideneglycinate (see *Ethyl N-Benzylideneglycinate*) are exclusively diastereoselective.

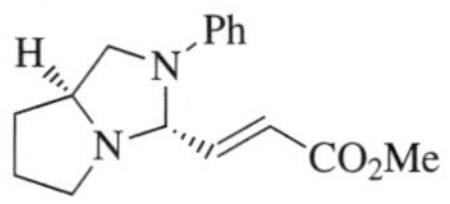

(**2**)

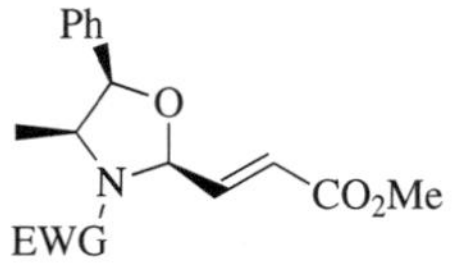

(**3**) EWG = BnOCO, *p*-Ts

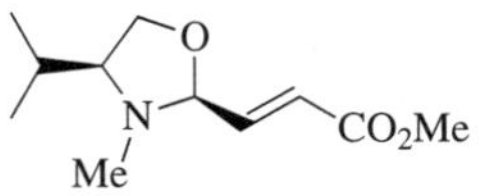

(**4**)

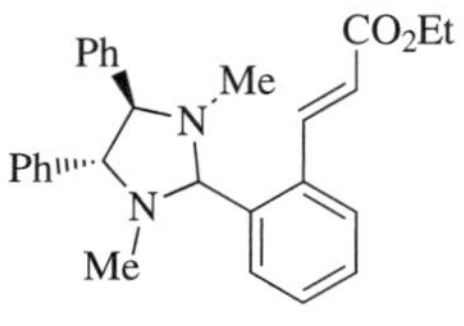

(**5**)

1. Kanemasa, S.; Hayashi, T.; Tanaka, J.; Yamamoto, H.; Sakurai, T. *J. Org. Chem.* **1991**, *56*, 4473.
2. Mangeney, P.; Grojean, F.; Alexakis, A.; Normant, J. F. *Tetrahedron Lett.* **1988**, *29*, 2675.
3. Bohlmann, F.; Inhoffen, E. *Ber. Dtsch. Chem. Ges./Chem. Ber.* **1956**, *89*, 1276.
4. Kanemasa, S.; Hayashi, T.; Yamamoto, H.; Wada, E.; Sakurai, T. *Bull. Chem. Soc. Jpn.* **1991**, *64*, 3274.
5. (a) Asami, M.; Mukaiyama, T. *Chem. Lett.* **1979**, 569. (b) Mukaiyama, T. *Tetrahedron* **1981**, *37*, 4111.
6. (a) Bernardi, A.; Cardani, S.; Poli, G.; Scolastico, C. *J. Org. Chem.* **1986**, *51*, 5041. (b) Bernardi, A.; Cardani, S.; Pilati, T.; Poli, G.; Scolastico, C. *J. Org. Chem.* **1988**, *53*, 1600.
7. Alexakis, A.; Sedrani, R.; Mangeney, P. *Tetrahedron Lett.* **1990**, *31*, 345.
8. Cardani, S.; Bernardi, A.; Colombo, L.; Gennari, C.; Scolastico, C.; Venturini, I. *Tetrahedron* **1988**, *44*, 5563.
9. Kanemasa, S.; Yamamoto, H. *Tetrahedron Lett.* **1990**, *31*, 3633.
10. Kanemasa, S.; Yamamoto, H.; Wada, E.; Sakurai, T.; Urushido, K. *Bull. Chem. Soc. Jpn.* **1990**, *63*, 2857.

Shuji Kanemasa

Kyushu University, Kasuga, Japan

(1*R*,2*S*)-*N*-Methylephedrine

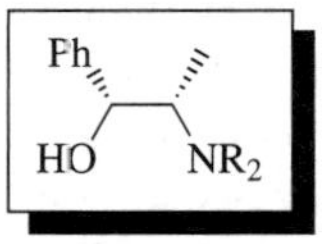

(**1**; R = Me)		
[552-79-4]	$C_{11}H_{17}NO$	(MW 179.29)
(**2**; R = *n*-Bu)		
[115651-77-9]	$C_{17}H_{29}NO$	(MW 263.47)
(**3**; R = CH_2=$CHCH_2$–)		
[150296-38-1]	$C_{15}H_{21}NO$	(MW 231.37)
(**4**; R = $Ph(CH_2)_4$)		
[132284-82-3]	$C_{29}H_{37}NO$	(MW 415.67)
(**5**; R = $-(CH_2)_5-$)		
[133576-76-8]	$C_{14}H_{21}NO$	(MW 219.36)

(chiral ligand for the enantioselective reduction of ketones with lithium aluminum hydride; chiral auxiliary for the diastereoselective aldol condensation; chiral catalyst for the enantioselective Darzens reaction; chiral catalyst for the enantioselective alkylation of aldehydes with dialkylzincs; chiral catalyst for the enantioselective conjugate addition of dialkylzincs to enones; chiral catalyst for the enantioselective alkylation of imines with dialkylzincs; chiral catalyst for the enantioselective Michael addition of nitromethane to α,β-unsaturated ketones)

Alternate Name: [*R*-(*R**,*S**)]-α-[1-(dimethylamino)ethyl]benzenemethanol.

Physical Data: (**1**) mp 86–88 °C; $[\alpha]_D^{21}$ −29.2° (*c* 5, MeOH). (**2**) bp 170 °C/2 mmHg; $[\alpha]_D^{25}$ + 24.4° (*c* 2, hexane).

Solubility: sol many organic solvents.

Form Supplied in: (**1**) colorless crystals; (**2**) colorless oil; (**1**) and (**2**) are commercially available.

Enantioselective Reduction of Ketones with Lithium Aluminum Hydride–*N*-Methylephedrine. Aryl alkyl ketones and α-alkynic ketones are reduced enantioselectively by a chiral complex of *Lithium Aluminum Hydride*, *N*-methylephedrine (**1**), and 3,5-dimethylphenol (molar ratio, 1:1:2) to afford optically active alcohols with 75–90% ee (eq 1).[2]

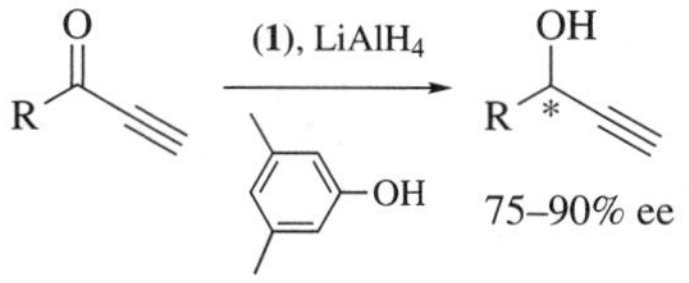
(1)

The optically active alkynyl alcohols are converted into the corresponding optically active 4-alkyl-γ-butyrolactones[3] and 4-alkylbutenolides.[4]

A chiral complex of (**1**), $LiAlH_4$, and *N*-ethylaniline (molar ratio, 1:1:2) reduces aryl alkyl ketones to optically active alcohols in high ee.[5] α,β-Unsaturated ketones are reduced enantioselectively to afford optically active (*S*)-allylic alcohols with 80–98% ee. An intermediate in an anthracyclinone synthesis is prepared in 92% ee by the enantioselective reduction of a cyclic α,β-unsaturated ketone (eq 2).[6]

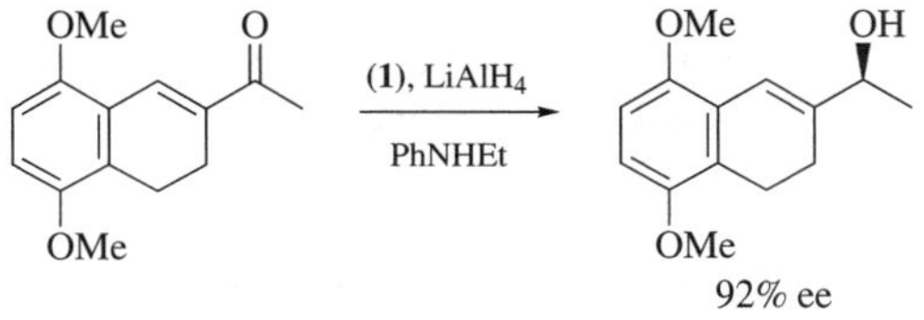
(2)

A chiral complex of (**1**), $LiAlH_4$, and 2-ethylaminopyridine (molar ratio, 1:1:2), prepared in refluxing ether for 3 h, reduces cyclic ketones to (*R*)-alcohols in 75–96% ee.[7] Advantages of the enantioselective reduction of ketones with $LiAlH_4$ modified with (**1**) and additives are the ready availability of (**1**) in either enantiomeric form and easy removal of (**1**) from the reaction mixture by washing with dilute acid.

***anti*-Selective Aldol Condensation and Related Reactions.** Silyl ketene acetals react with aldehydes in the presence of *Titanium(IV) Chloride* to give β-hydroxy esters.[8] The silyl ketene acetal derived from (1*R*,2*S*)-(**1**)-*O*-propionate reacts with benzaldehyde in the presence of $TiCl_4$ and *Triphenylphosphine* to afford the *anti*-α-methyl-β-hydroxy ester in 94% de (eq 3).[9]

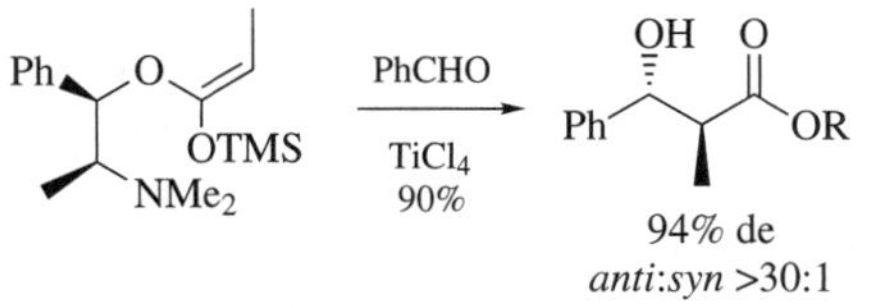
(3)

When the same silyl ketene acetal is reacted with benzylideneaniline in the presence of $TiCl_4$, the *anti*-β-amino ester is obtained (*anti*/*syn* > 10/1). Cyclization of the β-amino ester affords the *trans*-β-lactam in 95% ee (eq 4).[10]

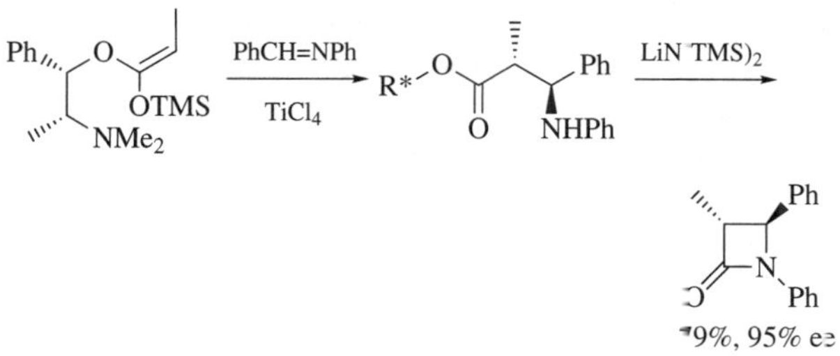
(4)

The reaction of this silyl ketene acetal with *Di-t-butyl Azodicarboxylate* in the presence of $TiCl_4$ affords the adduct in 45–70% yield with ca. 90% de. The subsequent treatment of the adduct with *Trifluoroacetic Acid* and *Lithium Hydroxide* affords (*R*)-α-hydrazo acids (eq 5).[11]

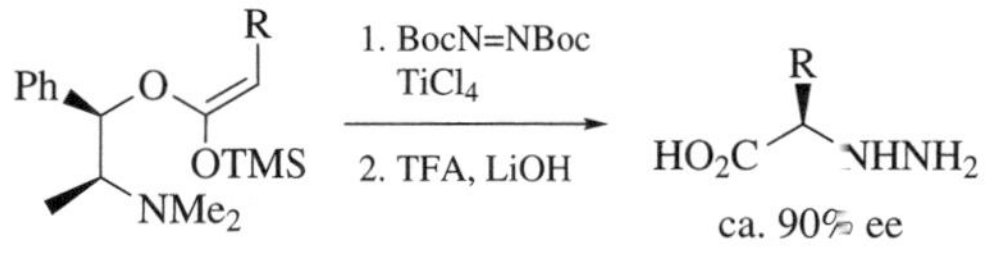
(5)

Enantioselective Butylation of Carbonyl Compounds with Lithium Tetra-*n*-butylaluminate Modified with (1). The reaction between lithium tetra-*n*-butylaluminate and (**1**) forms the chiral lithium alkoxytri-*n*-butylaluminate. This chiral ate complex reduces carbonyl compounds to form secondary and tertiary alcohols in 8–31% ee (eq 6).[12]

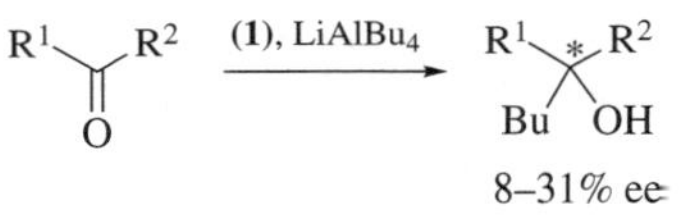
(6)

Enantioselective Darzens Reaction. An enantioselective Darzens reaction between ethyl methyl ketone and chloromethyl *p*-tolyl sulfone in the presence of a chiral ammonium salt derived from (**1**) and chloromethylpolystyrene affords an optically active α,β-epoxy sulfone in 23% ee.[13]

Catalytic Enantioselective Alkylation of Aldehydes with Dialkylzincs[1]. The chiral *N*,*N*-dialkylnorephedrines, analogs of (**1**), are highly efficient catalysts for the enantioselective addition of dialkylzincs to aliphatic and aromatic aldehydes.[14,15] Optically active aliphatic and aromatic secondary alcohols with high ee are obtained using *N*,*N*-dialkylnorephedrines (4–6 mol%) as chiral catalyst precursors. When (1*S*,2*R*)-*N*,*N*-dialkylnorephedrine is used as a chiral catalyst precursor, prochiral aldehydes are attacked at the *si* face to afford (*S*)-alcohols (when the priority order is $R^1 > R^2$) (eq 7).

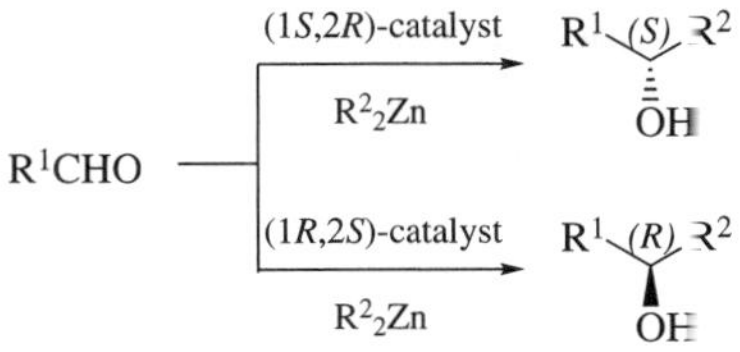
(7)

N-Alkyl substituents on the (1*S*,2*R*)-*N*,*N*-di-*n*-alkylnorephedrines have a significant effect on the enantioselectivity of the addition of diethylzinc to aldehydes (3-methylbutanal).

As shown in Table 1, the optical purity of the product [(*S*)-5-methylhexane-3-ol] increases as the chain length of the *N*-*n*-alkyl substituent increases and reaches a peak of 93% ee at a chain length of four carbons (Table 1, entry 4). Thus, among *N*,*N*-di-*n*-alkylnorephedrines examined, (1*S*,2*R*)-*N*,*N*-di-*n*-butylnorephedrine (DBNE) (**2**) is the best chiral catalyst precursor.[14,15]

Table 1 Effect of *N*-Alkyl Substituents of (1*S*,2*R*)-*N*,*N*-Dialkylnorephedrine as Chiral Ligand for the Addition of Diethylzinc to 3-Methylbutanal to Yield (*S*)-5-Methylhexan-3-ol

Entry	*N*-Alkyl substituent	Yield (%)	ee (%)
1	Me (**1**)	53	53
2	Et	95	83
3	*n*-Pr	90	87
4	*n*-Bu (DBNE) (**2**)	92	93
5	n-C_5H_{11}	91	85
6	n-C_6H_{13}	85	83
7	n-C_7H_{15}	80	79
8	n-C_8H_{17}	53	76
9	$-(CH_2)_5-$ (**5**)	81	70

As shown in Table 2 (eq 7), the advantages of *N*,*N*-di-*n*-alkylnorephedrines (most typically DBNE) over other chiral catalysts for the enantioselective addition of dialkylzincs to aldehydes are as follows:

1. DBNE is highly enantioselective for the alkylation of *aliphatic* aldehydes (Table 2, entries 5–11) as well as for the alkylation of *aromatic* aldehydes (Table 2, entries 1–4). Most of the other types of chiral catalysts are effective only for the alkylation of *aromatic* aldehydes. Thus, various types of optically active *aliphatic* alcohols are first synthesized using DBNE (Table 2, entries 5–11). (It should be noted that the structures of aliphatic alcohols synthesized by asymmetric reduction of ketones or by asymmetric hydroboration of alkenes have been somewhat limited.)

2. The dialkylzinc additions catalyzed by *N*,*N*-di-*n*-alkylnorephedrines (most typically DBNE) are not limited to *primary* organometallic reagents. Diisopropylzinc (with a *secondary* alkyl substituent) adds to benzaldehyde in the presence of a catalytic amount of DBNE to afford the corresponding alcohol with high ee (entry 4).[15] The reaction of diisopropylzinc in the presence of other types of catalysts may result in the reduction of aldehydes.

3. *N*,*N*-Di-*n*-alkylnorephedrines are readily synthesized in a one-pot reaction between norephedrine and alkyl iodide in the presence of potassium carbonate.[14,15] (DBNE is commercially available.)

4. Either enantiomer of the *N*,*N*-di-*n*-alkylnorephedrines are available. Therefore by using the appropriate enantiomer of *N*,*N*-di-*n*-alkylnorephedrine as a chiral catalyst precursor, the optically active alcohol of the desired configuration with the same ee can be synthesized (entries 8 and 9).

Table 2 Enantioselective Addition of Dialkylzincs to Aldehydes using Norephedrine-Derived Chiral Catalyst (eq 7)

Entry	R^1	R^2	Catalyst (6 mol %)	Yield (%)	ee (%)	Config.
1	Ph	Et	(1*S*,2*R*)-DBNE	100	90	(*S*)
2	2-$MeOC_6H_4$	Et	(1*S*,2*R*)-DBNE	100	94	(*S*)
3	Ph	Et	(1*S*,2*R*)-DBNE	94	95	(*S*)
4	Ph	*i*-Pr	(1*S*,2*R*)-DBNE	73	91	(*S*)
5	Me_2CHCH_2	Et	(1*S*,2*R*)-DBNE	92	93	–
6	n-C_6H_{13}	Me	(1*S*,2*R*)-DBNE	70	90	(*S*)
7	n-C_6H_{13}	*n*-Pr	(1*S*,2*R*)-DBNE	100	90	–
8	n-C_8H_{17}	Et	(1*S*,2*R*)-DBNE	95	87	(*S*)
9	n-C_8H_{17}	Et	(1*R*,2*S*)-DBNE	99	87	(*R*)
10	n-C_8H_{17}	Et	(1*S*,2*R*)-(**6**)	87	>95	(*S*)
11	n-C_8H_7	Et	(1*S*,2*R*)-(**3**)	61	88	(*S*)
12	4-$CF_3C_6H_4$	Et	(1*S*,2*R*)-DBNE	92	91	(*S*)
13	4-FC_6H_4	Et	(1*S*,2*R*)-DBNE	83	93	–
14	PhCDO	Et	(1*S*,2*R*)-DBNE	92	94	(*S*)
15	$Me(CH_2)_5CDO$	Et	(1*S*,2*R*)-DBNE	79	84	(*S*)

Optically active fluorine-containing alcohols (91–93% ee) (entries 12 and 13)[16] and deuterio alcohols (84–94% ee) (entries 14 and 15)[17] are synthesized, respectively, by the enantioselective alkylation of fluorine-containing aldehyde and deuterio aldehyde using DBNE.

(1*S*,2*R*)-1-Phenyl-2-(1-pyrrolidinyl)propan-1-ol (**6**) (entry 10) and (1*S*,2*R*)-*N*,*N*-diallylnorephedrine (**3**) (entry 11) are also highly enantioselective catalyst precursors.[15]

Enantioselective Addition of Various Organozinc Reagents to Aldehydes. Catalytic enantioselective addition of dialkynylzinc reagents to aldehydes using (1*S*,2*R*)-DBNE (20 mol %) affords optically active (*R*)-alkynyl alcohols with 43% ee in 99% yield.[18] When an alkylalkynylzinc is used with (1*S*,2*R*)-DBNE (5 mol %), an alkynyl alcohol with 40% ee is obtained.[18] When 2-phenylzinc bromide is reacted with an aldehyde in the presence of 1 equiv of the lithium salt of (1*R*,2*S*)-(**1**), the corresponding alkynyl alcohol is obtained in 88% ee (eq 8).[19]

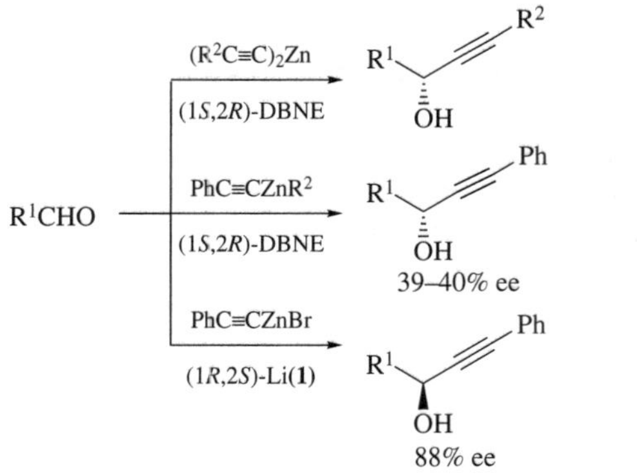

(8)

Alkenylzinc bromides add to aldehydes to afford optically active allyl alcohols with 88% ee in 80% yield using a stoichiometric amount of the lithium salt of (1*S*,2*R*)-(**1**) (eq 9).[20]

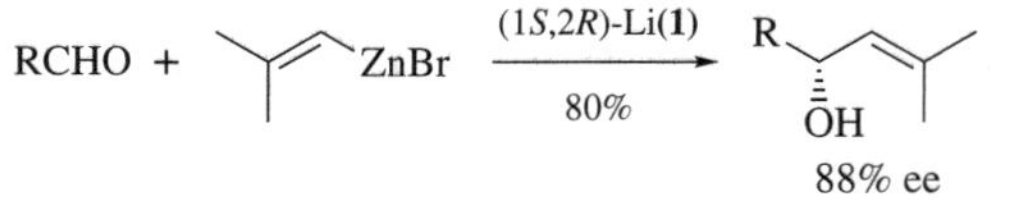

(9)

A mixture of phenyl Grignard and zinc halide adds to aldehydes in the presence of a stoichiometric amount of (1*R*,2*S*)-DBNE to afford optically active phenyl alcohols with 82% ee in 90% yield (eq 10).[21]

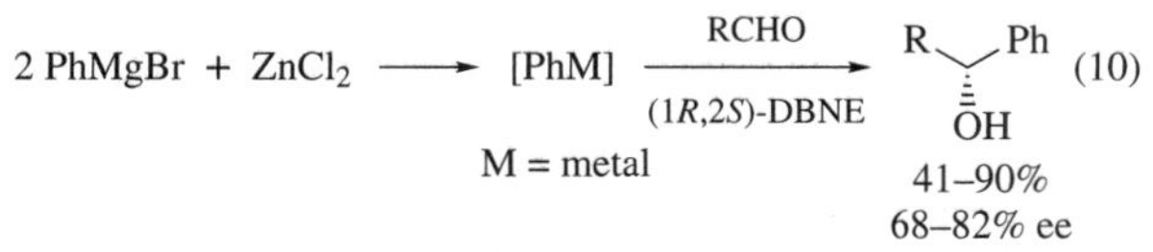

Difurylzinc adds to aldehydes in the presence of a stoichiometric amount of the lithium salt of (1*S*,2*R*)-*N*,*N*-bis(4-phenylbutyl)norephedrine (**4**) to afford optically active furylalcohols with 73% ee in 58% yield (eq 11).[22a]

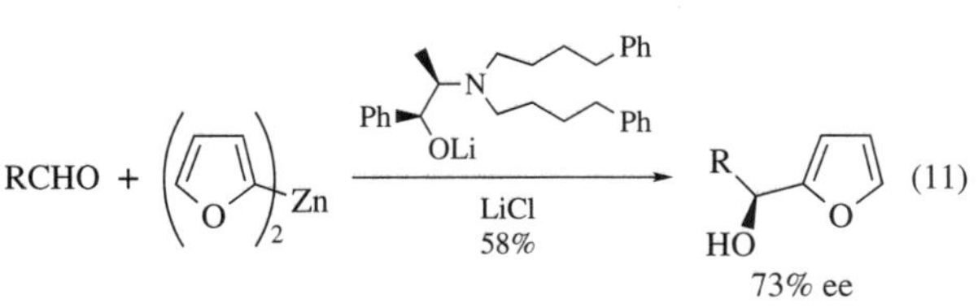

Enantioselective addition of a Reformatsky reagent to aldehydes[22b] and ketones[22c] in the presence of DBNE or *N*,*N*-diallylnorephedrine (**3**) affords the corresponding β-hydroxy esters in up to 75% ee (eq 12).

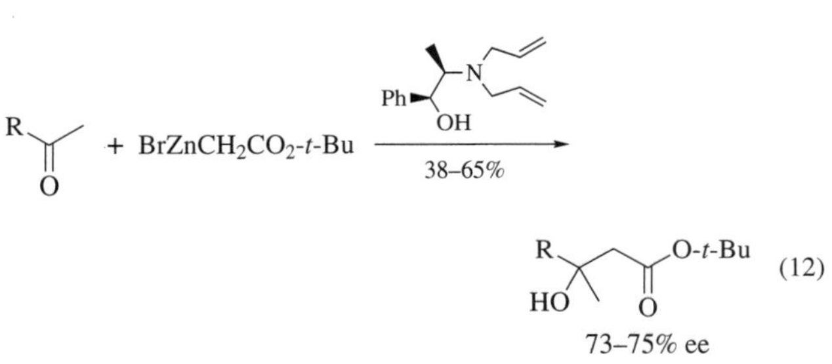

Enantioselective Addition of Dialkylzincs to Aldehydes with Functional Groups. Enantioselective and chemoselective addition of dialkylzincs to formyl esters using (1*S*,2*R*)-DBNE as a catalyst affords optically active hydroxy esters. The subsequent hydrolysis of the esters affords the corresponding optically active alkyl substituted lactones with up to 95% ee (eq 13).[23]

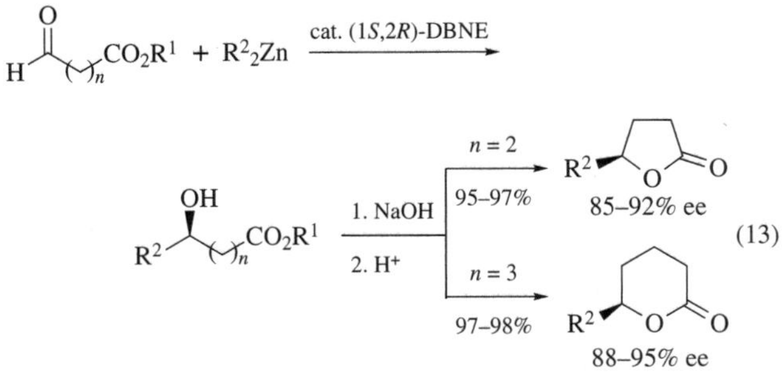

Enantio- and chemoselective addition of diethylzinc to keto aldehydes using DBNE as a chiral ligand affords optically active hydroxy ketones with 91% ee in 84% yield (eq 14).[24] This reaction cannot be realized by Grignard reagents or alkyllithium reagents because of the strong reactivity towards both aldehydes and ketones.

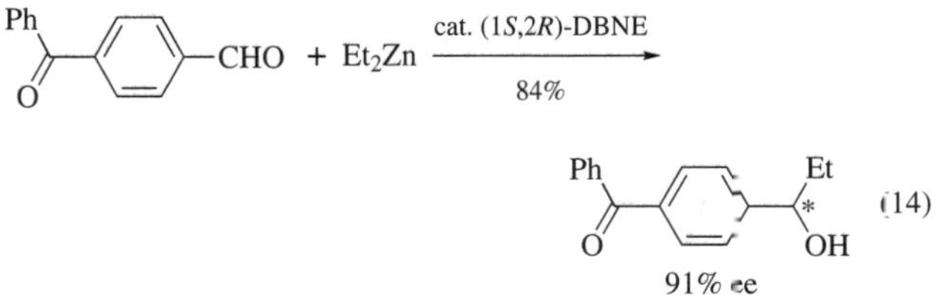

Enantioselective addition of dialkylzinc to furyl aldehydes using DBNE as a chiral catalyst affords optically active furyl alcohols in up to 94% ee (eqs 15 and 16).[25]

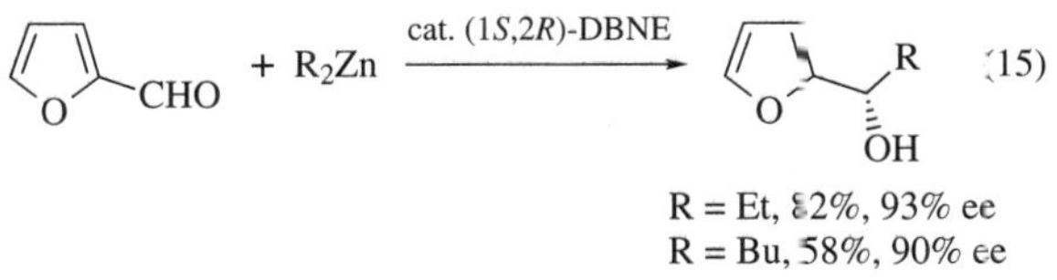

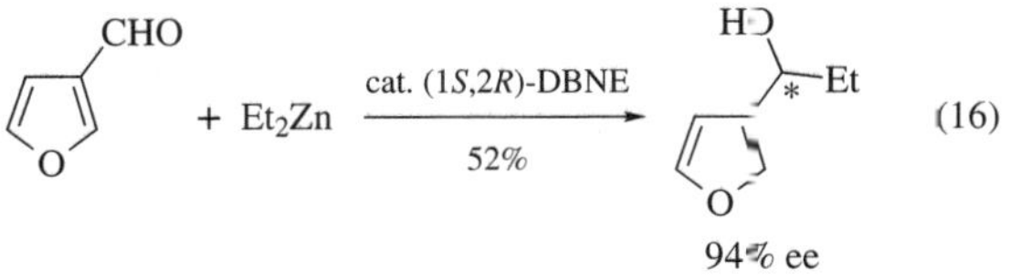

Enantioselective additions of dialkylzincs to 4-(diethoxymethyl)benzaldehyde,[26] 3-pyridinecarbaldehyde,[27] terephthalyl aldehyde,[28] and 2-bromobenzaldehyde[29] using DBNE as a chiral catalyst afford, after appropriate treatment, optically active hydroxy aldehydes,[26] pyridyl alcohols (eq 17),[27] diols (eq 18),[28] and 3-alkylphthalides (eq 19),[29] respectively, with high ee.

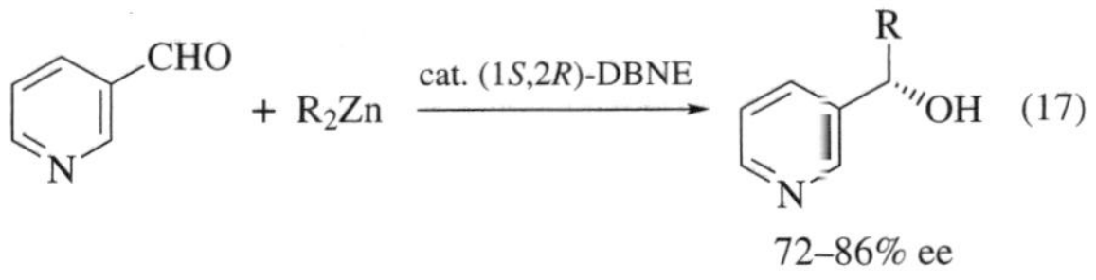

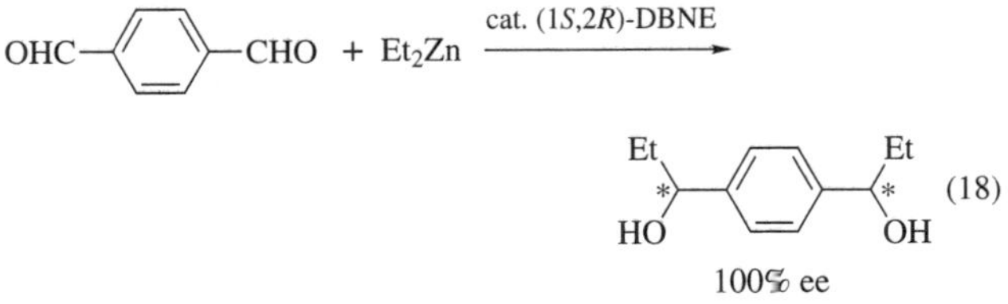

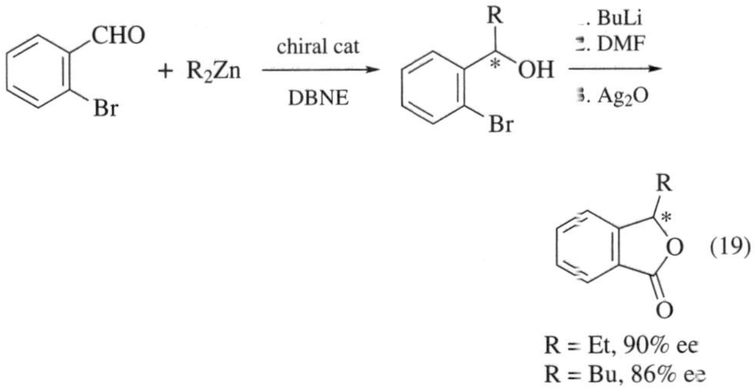

A highly functionalized chiral aldehyde when treated with Et_2Zn using (1*R*,2*S*)-DBNE as a chiral catalyst affords the optically active alcohol with 82% de in 98% yield (eq 20).[30] The alcohol has been further elaborated into (+)-lepicidin.

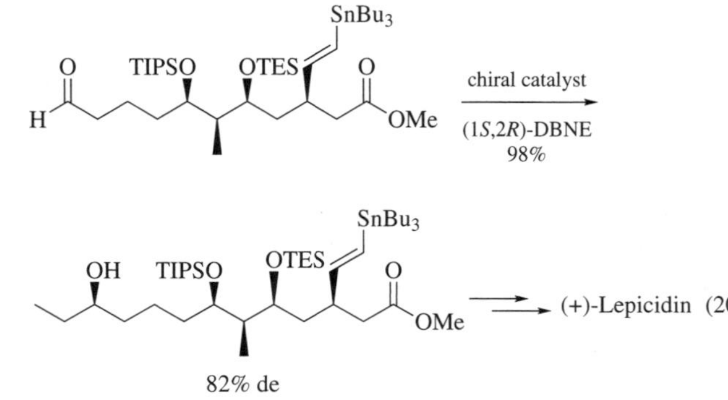

(20)

Stereoselective Addition of Dialkylzincs to Chiral Aldehydes. Stereoselective addition of dibutylzinc to racemic 2-phenylpropanal using (1*S*,2*R*)-DBNE as a chiral catalyst affords optically active alcohols (84% ee, 92% ee) as a result of the *si* face attack of the aldehyde regardless of its configuration (eq 21).[31]

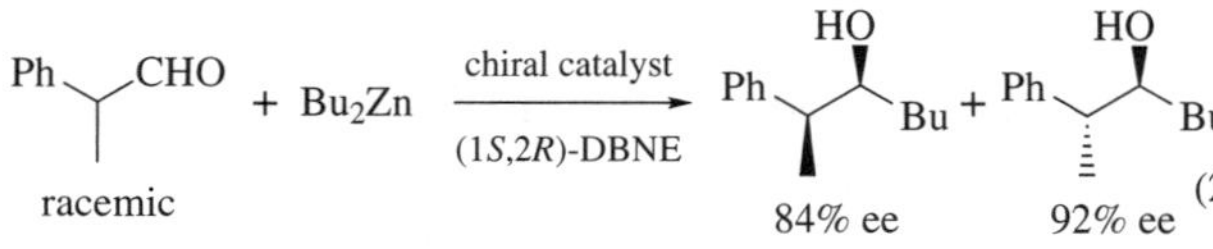

(21)

By changing the configuration of the chiral catalyst precursor (DBNE), stereoselective synthesis of optically active *syn* (78% de) and *anti* (91% de) 1,3-diols has been reported in the addition of diethylzinc to optically active β-alkoxyaldehyde (eq 22).[32] The method has an advantage over the R_2Zn–$TiCl_4$ method,[33] which is only *anti* selective.

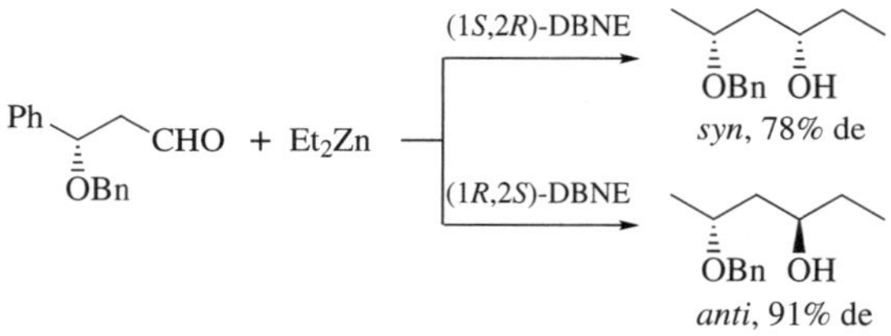

(22)

Catalytic Enantioselective Conjugate Addition of Dialkylzincs to Enones. A chiral nickel complex modified with DBNE and an achiral ligand such as 2,2′-bipyridyl in acetonitrile/toluene is an highly enantioselective catalyst for the addition of dialkylzincs to enones.[34] β-Substituted ketones with up to 90% ee are obtained (eq 23).[34c] The method is the first highly enantioselective catalytic conjugate addition of an organometallic reagent to an enone.

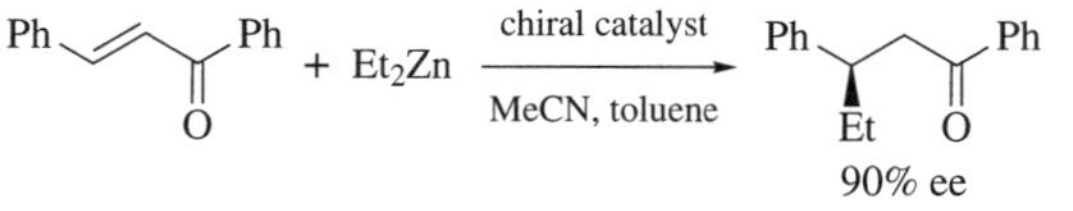

(23)

chiral catalyst (1*S*,2*R*)-DBNE–Ni(acac)$_2$–2,2'-bipyridyl

In addition, a chiral amino alcohol [1-phenyl-2-(1-piperidinyl)propan-1-ol] mediates the reaction without using any nickel compound to afford the adduct in 94% ee (eq 24).[34d]

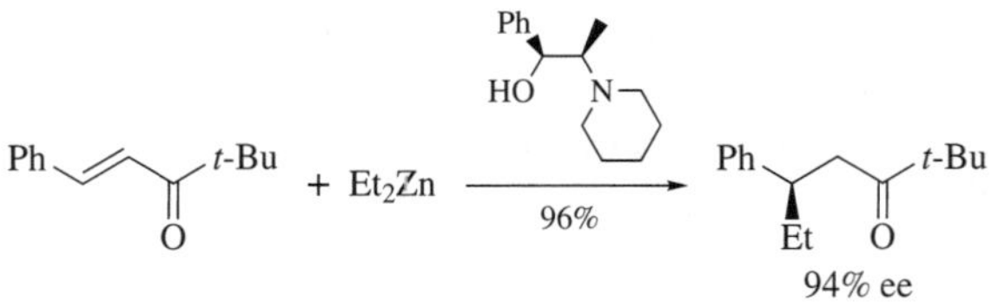

(24)

Enantioselective Addition of Dialkylzincs to Imines. Enantioselective addition of dialkylzincs to *N*-diphenylphosphinoylimines in the presence of DBNE or its analog affords optically active phosphoramides. Subsequent hydrolysis affords optically active amines in up to 91% ee (eq 25).[35] When the amount of DBNE is catalytic (10 mol %), the enantioselectivity is 75% ee. One of the advantages of this method over the alkyllithium method[36] is the use of a lesser amount of chiral ligand.

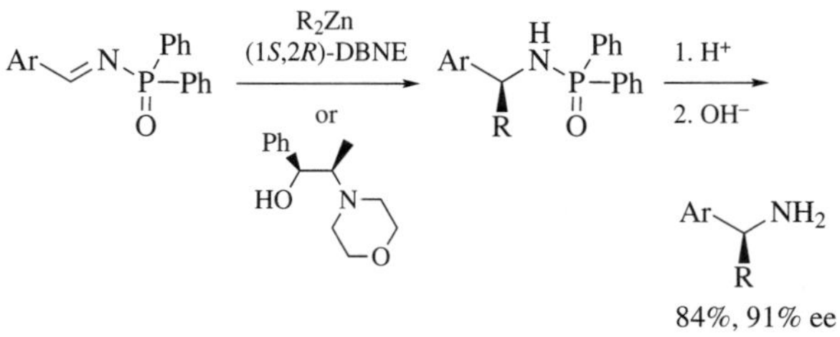

(25)

Diethylzinc also adds to *N*-(amidobenzyl)benzotriazoles (masked *N*-acylimines) in the presence of DBNE to afford an optically active amide with 76% ee (eq 26).[37]

(26)

Asymmetric Michael Addition of Nitromethane to Enone. *N*-Methylephedrinium fluoride catalyzes the Michael addition of nitromethane to chalcone to afford the adduct with 23% ee in 50% yield (eq 27).[38]

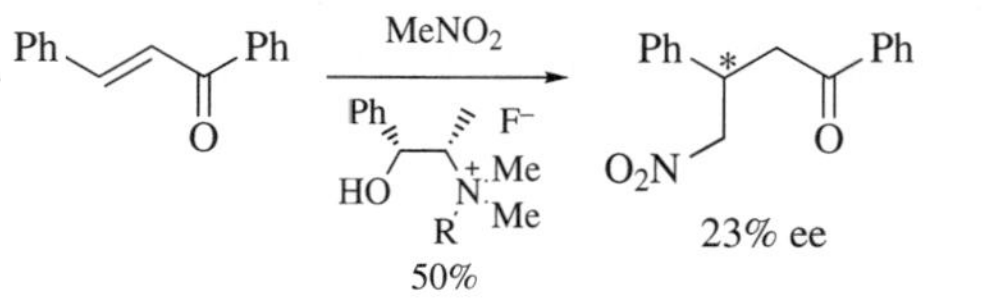

(27)

1. Soai, K.; Niwa, S. *Chem. Rev.* **1992**, *92*, 833.
2. (a) Jacquet, I.; Vigneron, J. P. *Tetrahedron Lett.* **1974**, 2065. (b) Vigneron, J. P.; Bloy, V. *Tetrahedron Lett.* **1979**, 2683.
3. Vigneron, J. P.; Bloy, V. *Tetrahedron Lett.* **1980**, *21*, 1735.
4. Vigneron, J. P.; Méric, R.; Dhaenens, M. *Tetrahedron Lett.* **1980**, *21*, 2057.
5. Terashima, S.; Tanno, N.; Koga, K. *Chem. Commun./J. Chem. Soc., Chem. Commun.* **1980**, 1025.
6. (a) Terashima, S.; Tanno, N.; Koga, K. *Chem. Lett.* **1980**, 981. (b) Terashima, S.; Hayashi, M.; Koga, K. *Tetrahedron Lett.* **1980**, *21*, 2749, 2753.
7. Kawasaki, M.; Suzuki, Y.; Terashima, S. *Chem. Lett.* **1984**, 239.

8. Mukaiyama, T. *Org. React.* **1982**, *28*, 203.
9. Gennari, C.; Bernardi, A.; Colombo, L.; Scolastico, C. *J. Am. Chem. Soc.* **1985**, *107*, 5812.
10. Gennari, C.; Venturini, I.; Gislon, G.; Schimperna, G. *Tetrahedron Lett.* **1987**, *28*, 227.
11. Gennari, C.; Colombo, L.; Bertolini, G. *J. Am. Chem. Soc.* **1986**, *108*, 6394.
12. Boireau, G.; Abenhaïm, D.; Bourdais, J.; Henry-Basch, E. *Tetrahedron Lett.* **1976**, 4781.
13. Colonna, S.; Fornasier, R.; Pfeiffer, U. *J. Chem. Soc., Perkin Trans. 1* **1978**, 8.
14. Soai, K.; Yokoyama, S.; Ebihara, K.; Hayasaka, T. *Chem. Commun./J. Chem. Soc., Chem. Commun.* **1987**, 1690.
15. (a) Soai, K.; Yokoyama, S.; Hayasaka, T. *J. Org. Chem.* **1991**, *56*, 4264. (b) Soai, K.; Hayase, T.; Takai, K.; Sugiyama, T. *J. Org. Chem.* **1994**, *59*, 7908.
16. Soai, K.; Hirose, Y.; Niwa, S. *J. Fluorine Chem.* **1992**, *59*, 5.
17. Soai, K.; Hirose, Y.; Sakata, S. *Bull. Chem. Soc. Jpn.* **1992**, *65*, 1734.
18. Niwa, S.; Soai, K. *J. Chem. Soc., Perkin Trans. 1* **1990**, 937.
19. Tombo, G. M. R.; Didier, E.; Loubinoux, B. *Synlett* **1990**, 547.
20. Oppolzer, W.; Radinov, R. N. *Tetrahedron Lett.* **1991**, *32*, 5777.
21. Soai, K.; Kawase, Y.; Oshio, A. *J. Chem. Soc., Perkin Trans. 1* **1991**, 1613.
22. (a) Soai, K.; Kawase, Y. *J. Chem. Soc., Perkin Trans. 1* **1990**, 3214. (b) Soai, K.; Kawase, Y. *Tetrahedron: Asymmetry* **1991**, *2*, 781. (c) Soai, K.; Oshio, A.; Saito, T. *Chem. Commun./J. Chem. Soc., Chem. Commun.* **1993**, 811.
23. Soai, K.; Yokoyama, S.; Hayasaka, T.; Ebihara, K. *Chem. Lett.* **1988**, 843.
24. Soai, K.; Watanabe, M.; Koyano, M. *Chem. Commun./J. Chem. Soc., Chem. Commun.* **1989**, 534.
25. (a) Soai, K.; Kawase, Y.; Niwa, S. *Heterocycles* **1989**, *29*, 2219. (b) Van Oeveren, A.; Menge, W.; Feringa, B. L. *Tetrahedron Lett.* **1989**, *30*, 6427.
26. Soai, K.; Hori, H.; Kawahara, M. *Tetrahedron: Asymmetry* **1990**, *1*, 769.
27. Soai, K. Hori, H.; Niwa, S. *Heterocycles* **1989**, *29*, 2065.
28. Soai, K.; Hori, H.; Kawahara, M. *Chem. Commun./J. Chem. Soc., Chem. Commun.* **1992**, 106.
29. Soai, K.; Hori, H.; Kawahara, M. *Tetrahedron: Asymmetry* **1991**, *2*, 253.
30. Evans, D. A.; Black, W. C. *J. Am. Chem. Soc.* **1992**, *114*, 2260.
31. (a) Soai, K.; Niwa, S.; Hatanaka, T. *Chem. Commun./J. Chem. Soc., Chem. Commun.* **1990**, 709. (b) Niwa, S.; Hatanaka, T.; Soai, K. *J. Chem. Soc., Perkin Trans. 1* **1991**, 2025.
32. Soai, K.; Hatanaka, T.; Yamashita, T. *Chem. Commun./J. Chem. Soc., Chem. Commun.* **1992**, 927.
33. Reetz, M. T.; Jung, A. *J. Am. Chem. Soc.* **1983**, *105*, 4833.
34. (a) Soai, K.; Yokoyama, S.; Hayasaka, T.; Ebihara, K. *J. Org. Chem.* **1988**, *53*, 4148. (b) Soai, K.; Hayasaka, T.; Ugajin, S.; Yokoyama, S. *Chem. Lett.* **1988**, 1571. (c) Soai, K.; Hayasaka, T.; Ugajin, S. *Chem. Commun./J. Chem. Soc., Chem. Commun.* **1989**, 516. (d) Soai, K.; Okudo, M.; Okamoto, M. *Tetrahedron Lett.* **1991**, *32*, 95.
35. Soai, K.; Hatanaka, T.; Miyazawa, T. *Chem. Commun./J. Chem. Soc., Chem. Commun.* **1992**, 1097.
36. (a) Tomioka, K.; Inoue, I.; Shindo, M.; Koga, K. *Tetrahedron Lett.* **1991**, *32*, 3095. (b) Itsuno, S.; Yanaka, H.; Hachisuka, C.; Ito, K. *J. Chem. Soc., Perkin Trans. 1*, **1991**, 1341.
37. Katritzky, A. R.; Harris, P. A. *Tetrahedron: Asymmetry* **1992**, *3*, 437.
38. Colonna, S.; Hiemstra, H.; Wynberg, H. *Chem. Commun./J. Chem. Soc., Chem. Commun.* **1978**, 238.

Kenso Soai
Science University of Tokyo, Japan

[2,2′-(1-Methylethylidene)[(4*S*)-4-(1,1-dimethylethyl)-4,5-dihydrooxazole-*N*³] copper(2+)bis[hexafluorophosphate], [2,2′-(1-Methylethylidene)[(4*S*)-4-(1,1-dimethylethyl)-4,5-dihydrooxazole-*N*³] copper(2+) bis(triflate)[1–4]

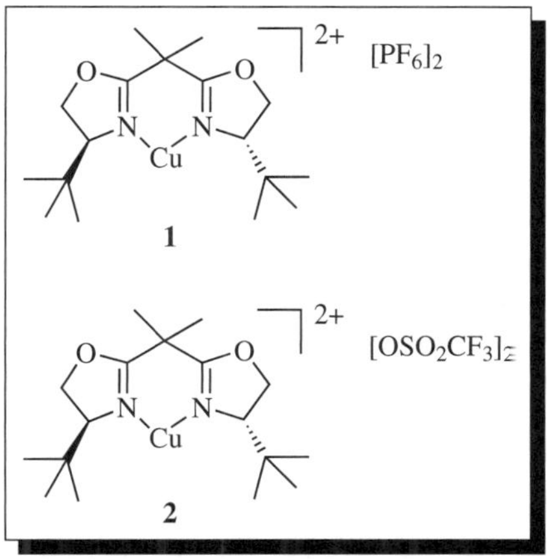

1

2

[165275-72-9]	$C_{17}H_{30}CuF_{12}N_2O_2P_2$	(MW 647.91)
[172323-63-6]	$C_{19}H_{30}CuF_6N_2O_8S_2$	(MW 656.11)

(versatile chiral C_2-symmetric Lewis acid catalyst for numerous asymmetric reactions)

Preparative Methods: can be prepared immediately before use from the commercially available bis(oxazoline) ligand and $Cu(PF_6)_2$ or $Cu(OTf)_2$ by simply combining the two reagents in an appropriate solvent and stirring until complexation is complete (>2 h).[5,6] Addition of 2 equiv of water to the bis(triflate) complex affords a di-aquo derivative that is bench-stable for several months.

Handling, Storage, and Precautions: the reagent is most effective under anhydrous conditions. The hydrated bis(triflate) complex can be activated in the presence of molecular sieves.

Aziridination. The bis(triflate) complex (**2**) catalyzes the aziridination of styrene derivatives in the presence of [*N*-(*p*-toluenesulfonyl)imino]phenyliodinane (eq 1).[7] Spectroscopic studies revealed that catalysts prepared from both $Cu^{I}OTf$ and $Cu^{II}(OTf)_2$ were identical under the reaction conditions, thus leading to the conclusion that the Cu(II) complex is the catalytically active species.

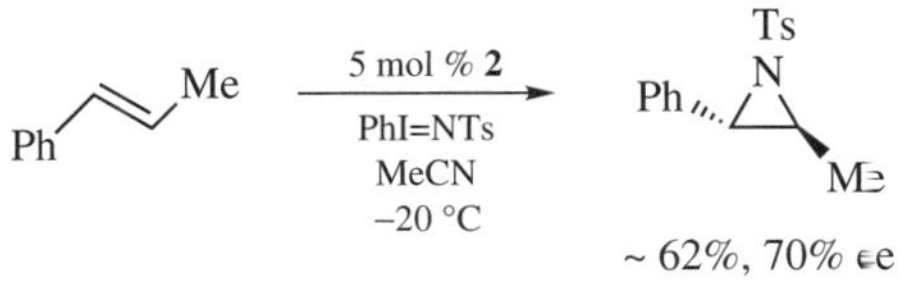

(1)

Diels–Alder Cycloaddition Reactions. Catalysts **1** and **2** facilitate asymmetric Diels–Alder cycloaddition reactions between propenyl oxazolidinones (and thio-oxazolidinones) and dienes such as cyclopentadiene (eq 2).[5,6,8] Both **1** and **2** display similar catalytic abilities and high *endo*:*exo* selectivities (~95:5). Isolated products are obtained in good chemical yields and in very high enantiomeric excesses. It is noteworthy, however, that both catalysts are less efficient than the related copper bis(oxazoline) complex possessing SbF_6 counterions.[9] In addition to cyclopentadiene, less reactive cyclic dienes (e.g., 1,3-cyclohexadiene) and acyclic dienes [e.g., 1,3-pentadiene and 2,4-hexadiene (eq 3)] give the reaction as well.[6] A significant solvent effect was observed in the reaction of oxazolidinone (**3**) and 1,3-cyclohexadiene catalyzed by complex **2**. Switching from CH_2Cl_2 to more polar CH_3NO_2 resulted in a decrease in reaction time and an increase in enantioselectivity (eq 4).[10] The formation of a square-planar catalyst–substrate complex (**4**) has been proposed to account for the high diastereoselectivities and enantioselectivities encountered in Diels–Alder reactions catalyzed by **2**. This model for stereoinduction is supported by results from X-ray crystallographic studies as well as double-stereodifferentiating experiments.[9] Versions of catalyst **2** attached to a poly(ethylene glycol) support also show catalytic efficacy in the context of Diels–Alder (and other) reactions, although with diminished levels of enantioselectivity.[11]

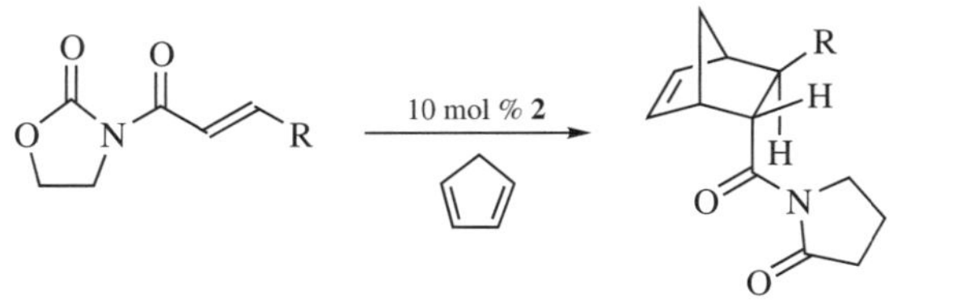

(2)

R = H, 86%, >98% ee
R = Me, 85%, 97% ee
R = CO_2Et, 92%, 95% ee

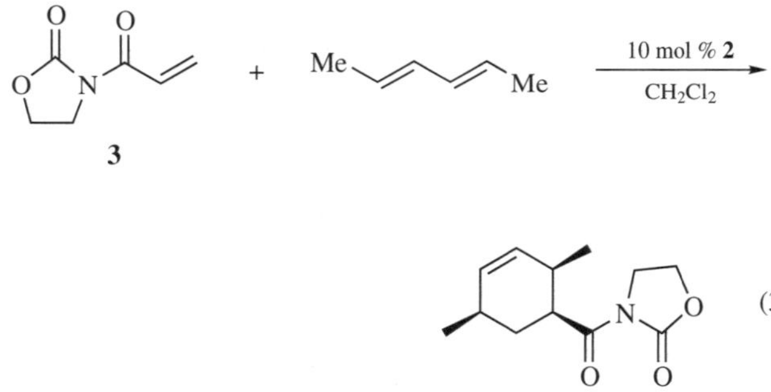

(3)

51%, 84% ee

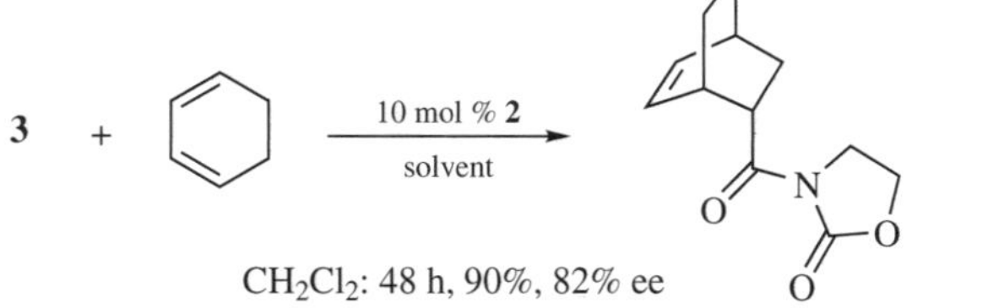

(4)

CH_2Cl_2: 48 h, 90%, 82% ee
CH_3NO_2: 4 h, 90%, 92% ee

Hetero-Diels–Alder Cycloaddition Reactions. Copper-bis-(oxazoline) complex (**2**) facilitates a variety of asymmetric hetero-Diels–Alder reactions between a diverse range of substrates. A crucial requirement for successful transformations is the ability of one reactant susceptible to Lewis acid activation to coordinate in a bidentate fashion to the Cu(II) center. Thus, pyruvate esters undergo hetero-Diels–Alder reactions with electron-rich dienes (e.g., Danishefsky's diene) in the presence of **2** in high yield and in excellent enantiomeric excess (eq 5).[12,13] In contrast to carbon-based Diels–Alder reactions in which SbF_6 is the counterion of choice for the copper catalyst, the bis(triflate) complexes are superior in these instances. Glyoxylate esters also participate in hetero-Diels–Alder reactions with dienes in the presence of **2** in nitroalkane solvents (eq 6).[14] Products of a competitive ene reaction are obtained along with dihydropyran derivatives depending on the nature of the diene reaction partner.[15] Dienes such as unsaturated α-ketophosphonates (eq 7)[16–18] and α-ketoesters (eq 8)[19,20] also participate in inverse electron demand hetero-Diels–Alder reactions in the presence of **2**. Electron-rich dienophiles (e.g., vinyl ethers) are suitable 2π reaction partners, and endo diastereomers are obtained virtually exclusively. Catalyst **2** is effective at promoting the hetero-Diels–Alder reaction between vinyl ethers fixed to a solid support and has successfully been applied in the syntheses of dihydropyran libraries.[21] Finally, enantiomerically pure piperidinones have been prepared from propenyl oxazolidinones and aza-dienes (eq 9).[22] In this instance cycloaddition proved to be highly exo selective.

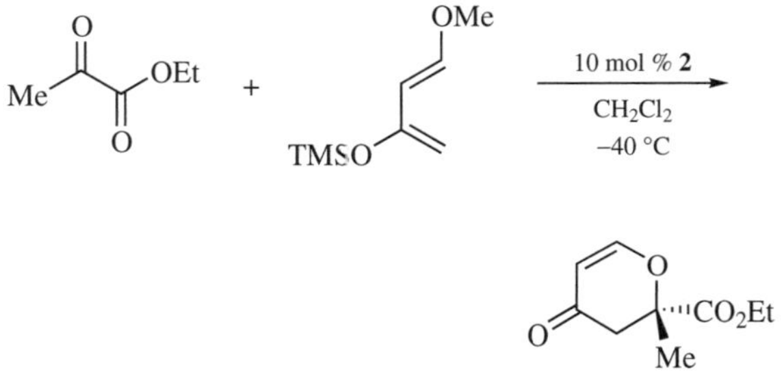

(5)

78%, 99% ee

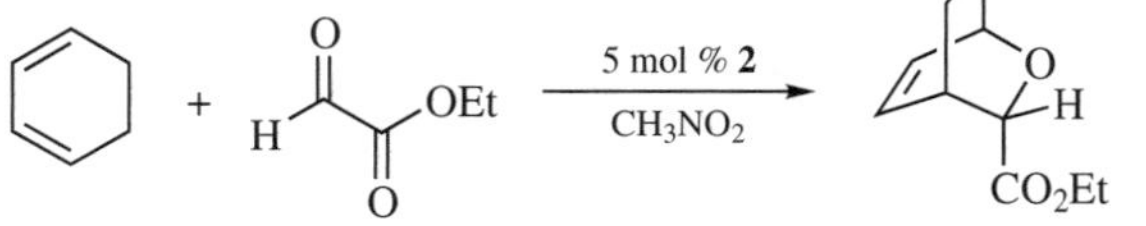

(6)

66%, 97% ee

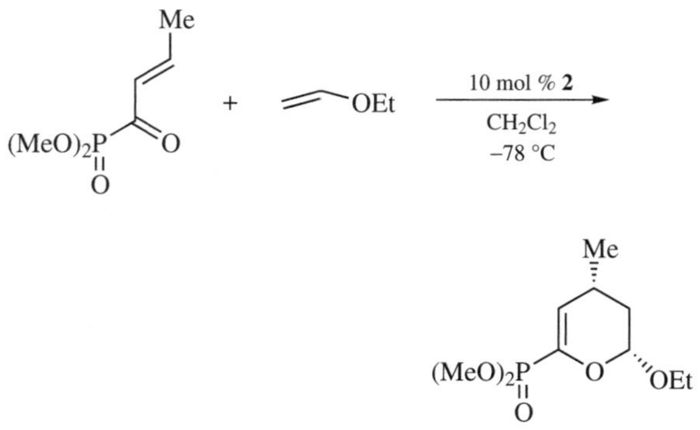

(7)

89%, 99% ee

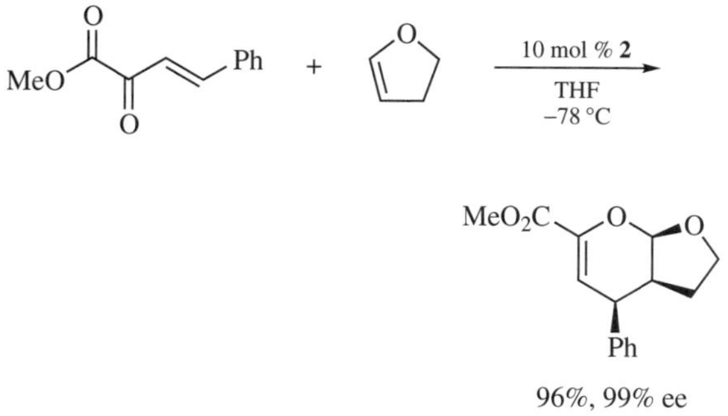
(8)

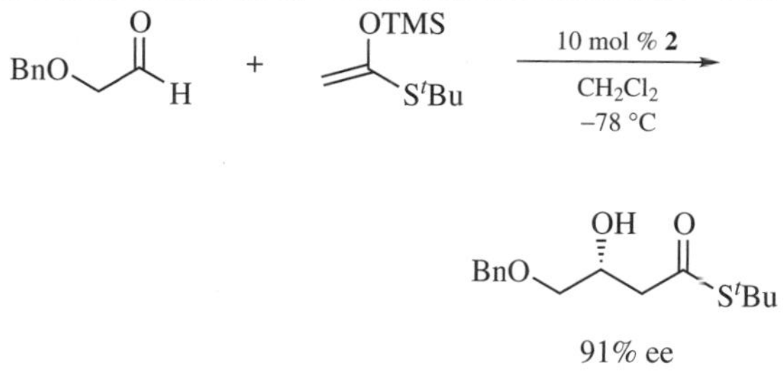
(10)

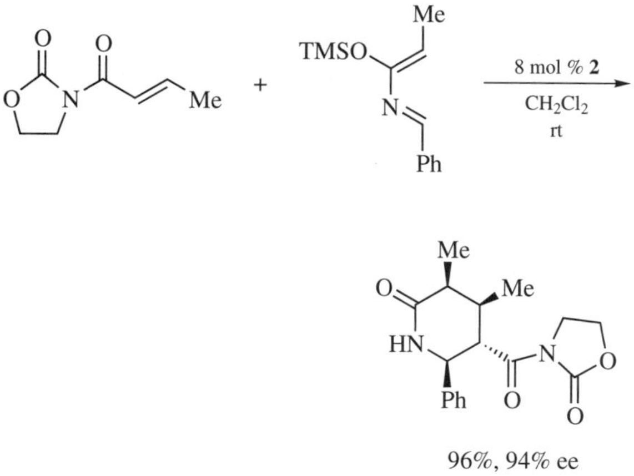
(9)

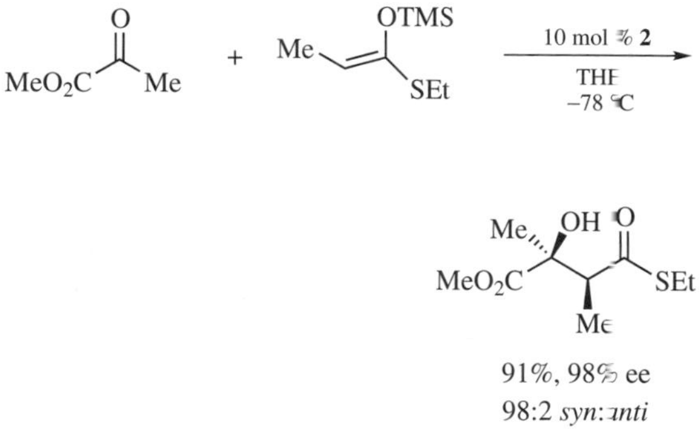
(11)

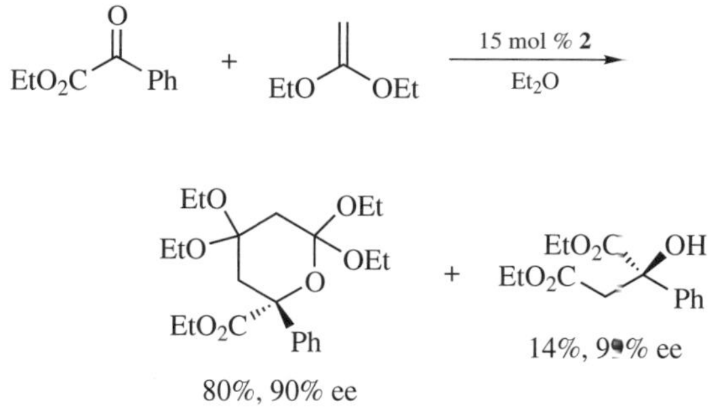
(12)

Aldol Reactions. Bis(oxazoline) complex (**2**) is a viable catalyst for the Mukaiyama aldol reaction between enol silanes and aldehydes. As shown in eq 10, treatment of benzyloxyacetaldehyde with the trimethylsilylketene acetal of *tert*-butyl thioacetate in the presence of 10 mol % of **2** affords the aldol adduct in 91% ee favoring the (*R*)-enantiomer.[23,24] The presence of the α-benzyloxy moiety is important for successful stereoinduction mediated via the formation of a bidentate catalyst–substrate complex. Pyruvate esters are also excellent substrates for catalyzed Mukaiyama aldol reactions. A great deal of structural variation in both the pyruvate ester and enol silane reaction components is tolerated. Tetrahydrofuran is the solvent of choice for these reactions and catalyst **2** is selective for formation of *syn*-aldol products with substituted enol silanes favoring the (*S*)-configuration at the quaternary center (eq 11).[25,26] Vicinal diketones participate in the asymmetric aldol reaction as well. In the case of 2,3-pentanedione, enol silane addition occurs with high regioselectivity and enantioselectivity at the methyl-substituted ketone.[26] Complex **2** also catalyzes a sequential aldol addition/cyclization reaction, leading to δ-lactone derivatives. The best results were obtained with 15 mol % **2** in Et_2O (eq 12).[27] Bis(triflate) (**2**) is reported to be inferior to the analogous hexafluoroantimonate complex in the catalysis of Mukaiyama–Michael reactions and Michael reactions of alkylidine malonates.[28,29] An exception is encountered in the reaction of propenyl oxazolidinone Michael acceptors with trimethylsilyloxyfuran (eq 13).[30] Mechanistic investigations performed by the Evans group, however, indicate that the product shown in eq 13 is most likely formed via a sequential 4 + 2 cycloaddition/retro-aldol process.[28] It is notable that poor chemical yields were reported in the absence of hexafluoroisopropanol.

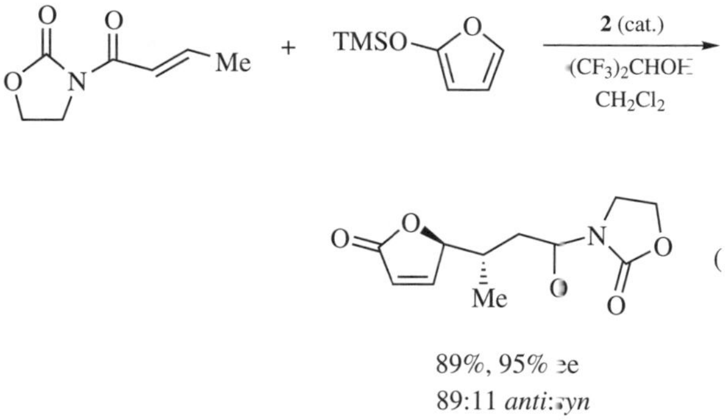
(13)

Amination of Enol Silanes. Enol silanes can be aminated by azidodicarboxylate derivatives in the presence of **2** (eq 14).[31] Similar to the reaction illustrated in eq 13, this process is thought to involve initial formation of a Diels–Alder adduct followed by ring-opening. Breakdown of the cycloadduct is facilitated by the inclusion of a relatively acidic alcohol additive (trifluoroethanol) in the reaction.

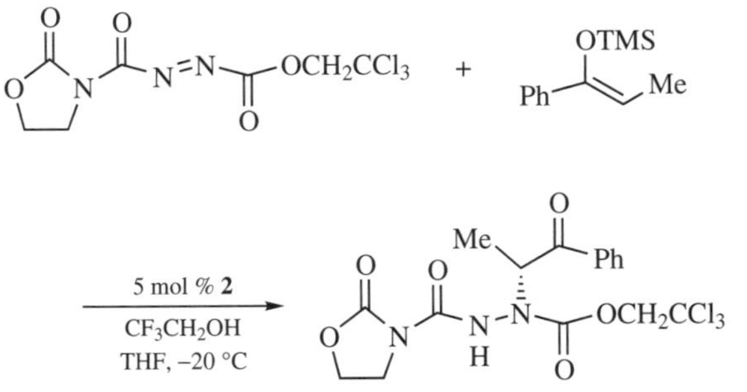
(14)

95%, 99% ee

Ene Reactions. Bis(triflate) complex **2** catalyzes the ene reaction of glyoxylate esters (eq 15).[32] Catalyst turnover was not observed at low temperature. As is the case with Diels–Alder reactions, the related bis(hexafluoroantimonate) complex is a more efficient catalyst for this transformation.

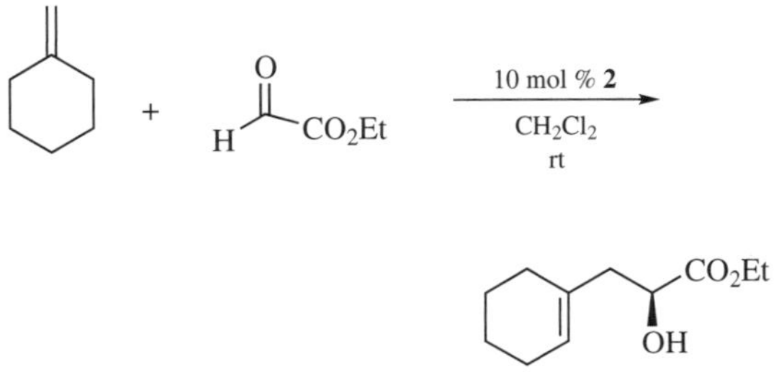
(15)

94%, 86% ee

Friedel–Crafts Alkylation Reactions. The activation of glyoxylate esters,[33] trifluoromethyl pyruvate esters,[34] and unsaturated α-ketoesters[35] by catalyst **2** converts these materials into effective electrophiles for asymmetric Friedel–Crafts alkylation reactions with activated arenes (eqs 16 and 17). In fact, bis(triflate) (**2**) is far superior to the bis(hexafluoroantimonate) complex at catalyzing the enantioselective alkylation of benzene derivatives.[33] Aniline and anisole derivatives both give the reaction, as do heterocyclic aromatic compounds such as indole and furan.

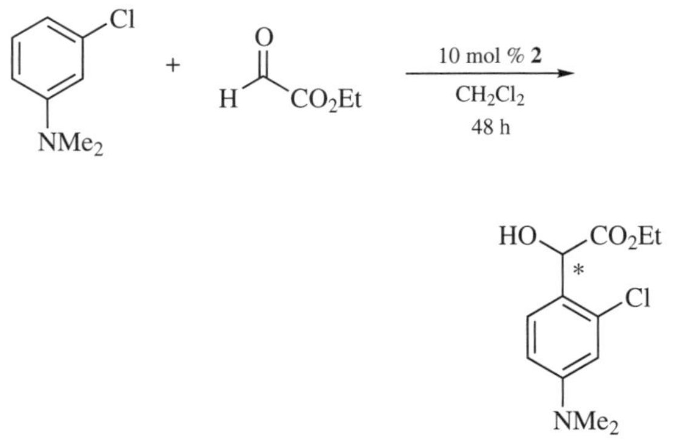
(16)

84%, 93% ee

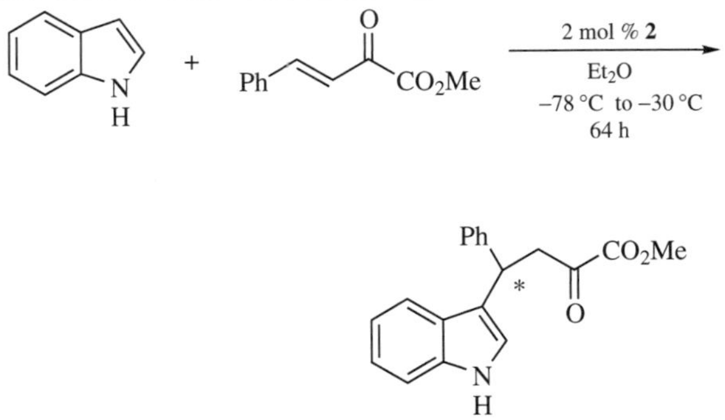
(17)

77%, >99% ee

[2 + 2] Cycloaddition Reactions. Bis(oxazoline) copper complexes such as **2** (and its hydrated congener) facilitate the [2 + 2] cycloaddition between silylketenes and glyoxylate/pyruvate esters (eq 18).[36] The reaction is tolerant to various silyl substituents and structural variation on the dicarbonyl reactant.

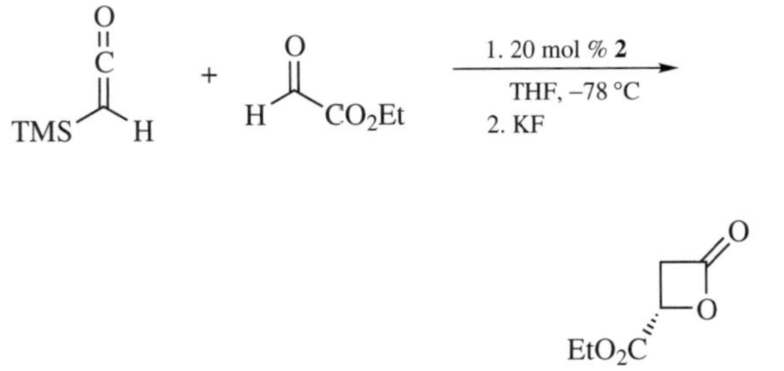
(18)

99%, 95% ee

[3 + 2] Cycloaddition Reactions. Bis(oxazoline) copper complex **2** catalyzes the dipolar cycloaddition reaction between electron deficient nitrones and electron rich alkenes. While *exo*:*endo* selectivities are marginal, products can be obtained in as high as 94% enantiomeric excess (eq 19).[37] Based on the stereochemical outcome of the reaction, a five-coordinate intermediate has been postulated in which both the nitrone (as a bidentate ligand) and alkene are coordinated to the Cu^{II} center.

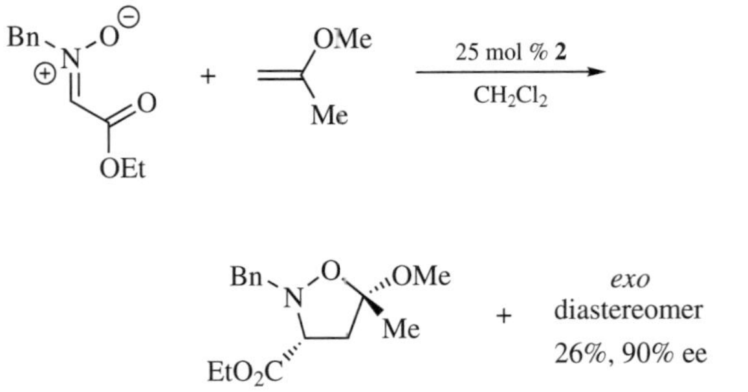
(19)

57%, 94% ee

Related Reagents. [2,2′-(1-Methylethylidene)[(4*S*)-4-(1,1-dimethylethyl)-4,5-dihydrooxazole-N^3]copper (2+) bis-[hexafluoroantimonate].

---

1. Rovis, T.; Evans, D. A. *Prog. Inorg. Chem.* **2001**, *50*, 1.
2. Johnson, J. S.; Evans, D. A. *Acc. Chem. Res.* **2000**, *33*, 325.
3. Evans, D. A.; Rovis, T.; Johnson, J. S. *Pure Appl. Chem.* **1999**, *71*, 1407.

4. Jørgensen, K. A.; Johannsen, M.; Yao, S.; Audrain, H.; Thorhauge, J. *Acc. Chem. Res.* **1999**, *32*, 605.
5. Evans, D. A.; Miller, S. J.; Lectka, T. *J. Am. Chem. Soc.* **1993**, *115*, 6460.
6. Evans, D. A.; Barnes, D. M.; Johnson, J. S.; Lectka, T.; von Matt, P.; Miller, S. J.; Murray, J. A.; Norcross, R. D.; Shaughnessy, E. A.; Campos, K. R. *J. Am. Chem. Soc.* **1999**, *121*, 7582.
7. Evans, D. A.; Faul, M. M.; Bilodeau, M. T.; Anderson, B. A.; Barnes, D. M. *J. Am. Chem. Soc.* **1993**, *115*, 5328.
8. Evans, D. A.; Murray, J. A.; von Matt, P.; Norcross, R. D.; Miller, S. J. *Angew. Chem. Int. Ed. Engl.* **1995**, *34*, 798.
9. Evans, D. A.; Miller, S. J.; Lectka, T.; von Matt, P. *J. Am. Chem. Soc.* **1999**, *121*, 7559.
10. Johannsen, M.; Jørgensen, K. A. *J. Chem. Soc., Perkin Trans. 2* **1997**, 1183.
11. Annunziata, R.; Benaglia, M.; Cinquini, M.; Cozzi, F.; Pitillo, M. *J. Org. Chem.* **2001**, *66*, 3160.
12. Johannsen, M.; Yao, S.; Jørgensen, K. A. *Chem. Commun.* **1997**, 2169.
13. Yao, S.; Johannsen, M.; Audrain, H.; Hazell, R. G.; Jørgensen, K. A. *J. Am. Chem. Soc.* **1998**, *120*, 8599.
14. Johannsen, M.; Jørgensen, K. A. *Tetrahedron* **1996**, *52*, 7321.
15. Johannsen, M.; Jørgensen, K. A. *J. Org. Chem.* **1995**, *60*, 5757.
16. Thorhauge, J.; Johannsen, M.; Jørgensen, K. A. *Angew. Chem. Int. Ed. Engl.* **1998**, *37*, 2404.
17. Audrain, H.; Thorhauge, J.; Hazell, R. G.; Jørgensen, K. A. *J. Org. Chem.* **2000**, *65*, 4487.
18. Evans, D. A.; Johnson, J. S.; Olhava, E. J. *J. Am. Chem. Soc.* **2000**, *122*, 1635.
19. Evans, D. A.; Olhava, E. J.; Johnson, J. S.; Janey, J. M. *Angew. Chem. Int. Ed. Engl.* **1998**, *37*, 3372.
20. Evans, D. A., Johnson, J. S. *J. Am. Chem. Soc.* **1998**, *120*, 4895.
21. Stavenger, R. A.; Schreiber, S. L. *Angew. Chem. Int. Ed. Engl.* **2001**, *40*, 3417.
22. Jnoff, E.; Ghosez, L. *J. Am. Chem. Soc.* **1999**, *121*, 2617.
23. Evans, D. A.; Murray, J. A.; Kozlowski, M. C. *J. Am. Chem. Soc.* **1996**, *118*, 5814.
24. Evans, D. A.; Kozlowski, M. C.; Murray, J. A.; Burgey, C. S.; Campos, K. R.; Connell, B. T.; Staples, R. J. *J. Am. Chem. Soc.* **1999**, *121*, 669.
25. Evans, D. A.; Kozlowski, M. C.; Burgey, C. S.; MacMillan, D. W. C. *J. Am. Chem. Soc.* **1997**, *119*, 7893.
26. Evans, D. A.; Burgey, C. S.; Kozlowski, M. C.; Tregay, S. W. *J. Am. Chem. Soc.* **1999**, *121*, 686.
27. Audrain, H.; Jørgensen, K. A. *J. Am. Chem. Soc.* **2000**, *122*, 11543.
28. Evans, D. A.; Scheidt, K. A.; Johnston, J. N.; Willis, M. C. *J. Am. Chem. Soc.* **2001**, *123*, 4480.
29. Evans, D. A.; Rovis, T.; Kozlowski, M. C.; Downey, C. W.; Tedrow, J. S. *J. Am. Chem. Soc.* **2000**, *122*, 9134.
30. Kitajima, H.; Katsuki, T. *Synlett* **1997**, 568.
31. Evans, D. A.; Johnson, D. S. *Org. Lett.* **1999**, *1*, 595.
32. Evans, D. A.; Tregay, S. W.; Burgey, C. S.; Paras, N. A.; Vojkovsky, T. *J. Am. Chem. Soc.* **2000**, *122*, 7936.
33. Gathergood, N.; Zhuang, W.; Jørgensen, K. A. *J. Am. Chem. Soc.* **2000**, *122*, 12517.
34. Zhuang, W.; Gathergood, N.; Hazell, R. G.; Jørgensen, K. A. *J. Org. Chem.* **2001**, *66*, 1009.
35. Jensen, K. B.; Thorhauge, J.; Hazell, R. G.; Jørgensen, K. A. *Angew. Chem. Int. Ed. Engl.* **2001**, *40*, 160.
36. Evans, D. A.; Janey, J. M. *Org. Lett.* **2001**, *3*, 2125.
37. Jensen, K. B.; Hazell, R. G.; Jørgensen, K. A. *J. Org. Chem.* **1999**, *64*, 2353.

F. Christopher Pigge
University of Missouri-St. Louis, St. Louis, MO, USA

(*R*)-(−)-2-(-1-Methylhydrazino)-butan-1-ol

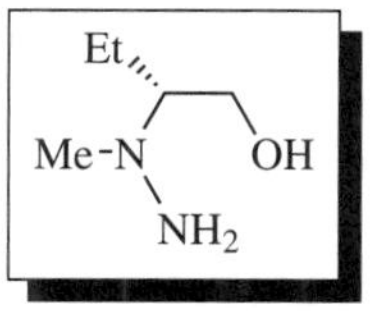

[(2R)-[211987-91-6]] $C_5H_{14}N_2O$ (MW 118.14)

(chiral reagent for the synthesis of enantiomerically enriched α-arylalkanamines)

Physical Data: colorless oil, $[\alpha]_D^{20}$ −21.6(*c* 1.1, MeOH).
Solubility: soluble in CH_2Cl_2, THF, alcohols.
Purification: none; immediate use is recommended following its preparation because of the inherent instability of the reagent.
Handling, Storage, and Precautions: unstable compound, must be used without purification. Probably toxic.

Introduction. The enantioselective addition of organometallic reagents to chiral hydrazones, followed by hydrogenolytic cleavage of the N–N bond of the resulting hydrazine, constitutes an attractive method for the preparation of optically active amines. The general synthetic strategy disclosed by Takahashi and his coworkers[1] as early as 1979 is still in use: the chiral hydrazones are most generally derived from an enantiopure secondary amine by *N*-nitrosation followed by reduction of the NO group to an NH_2 group and reaction with an appropriate aldehyde.[1–5]

Racemic 2-aminobutan-1-ol (**1**) is a cheap chemical which can be easily resolved into both its enantiomers on an industrial scale. The asymmetric synthesis of chiral amines from hydrazines derived from (*R*)-(−)-2-aminobutan-1-ol [(*R*)-(−)-**1**], using the general strategy disclosed in early works,[1] is summarized here. The title hydrazine (**4**) is prepared as follows (eq 1). Treatment of the amino alcohol [(*R*)-(−)-**1**] with excess ethyl formate followed by LAH reduction of the intermediate formamide gives the *N*-methylamine [(*R*)-(−)-**2**].[6] *N*-Nitrosation of the latter afforded (*R*)-(+)-**3** which is next reduced to the hydrazine [(*R*)-(−)-**4**] by means of LAH.[7] Being unstable, the hydrazine (**4**) must be used immediately without purification.

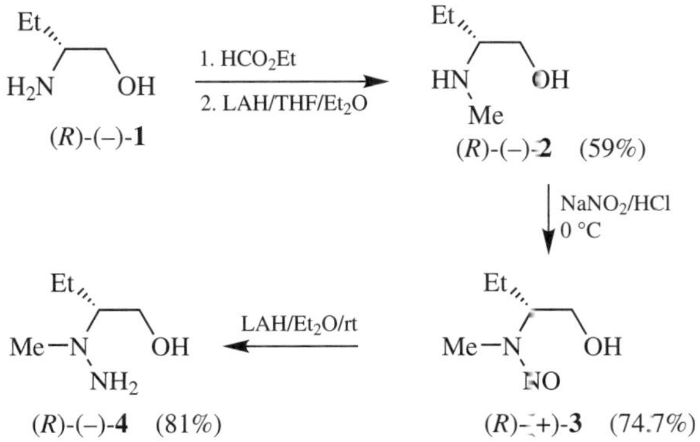

(1)

Chiral α-phenylalkanamines. The hydrazine [(*R*)-(−)-**4**] was transformed into the hydrazone [(*R*)-(−)-**5**] upon reaction with benzaldehyde in the presence of anhydrous $MgSO_4$ in dichloromethane (eq 2). The hydrazone (**5**) was next treated with a tenfold molar excess of various *n*-alkyl Grignard reagents in refluxing ether for 15 h. This led to the corresponding seven trisubstituted liquid hydrazines [(*R*,*R*)-**6a–g**] in yields ranging between 70 and 89% in all cases but one. The use of smaller quantities of Grignard reagents (i.e. fivefold molar excess) gave mixtures of starting hydrazone (**5**) and trisubstituted hydrazine (**6**), the latter having rather average diastereomeric excesses. Examination of the high resolution ^{1}H and ^{13}C NMR spectra of the hydrazines (**6a–g**) (prepared with a tenfold excess of Grignard reagents) revealed that they were diastereomerically pure (de=100% in all cases). The absolute *R*,*R* configuration of the hydrazines (**6a–g**) was assigned on the basis of the tentative mechanistic proposal depicted in eq 2.

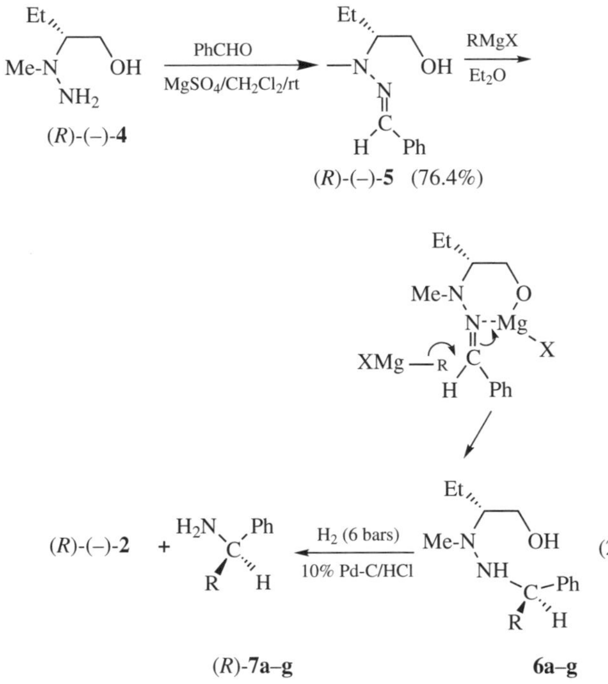

(2)

	a	b	c	d	e	f	g
R	Et	*n*-Pr	*n*-Bu	*n*-C_5H_{11}	*n*-C_6H_{13}	*n*-C_7H_{15}	*n*-C_8H_{17}

Being rather unstable, the hydrazines (**6a–g**) were used directly in the following step without purification. Thus, hydrogenolysis of the crude colorless hydrazines (**6a–g**) was carried out in the presence of concentrated HCl and 10% Pd-C catalyst under hydrogen (6 bars) at ca. 60 °C for 16 h. This afforded the crude amines [(*R*)-**7a–g**)] which were purified by chromatography over silica gel in the presence of triethylamine in order to avoid racemization.[7] The amines (*R*)-(+)-**7a**,[8] (*S*)-(−)-**7b**,[9] and (*S*)-(−)-**7c**[10] are known compounds, which made it possible to confirm the *R*,*R* absolute configuration allotted to the starting hydrazines (**6**). It is assumed that the other amines (**7d–g**) also have the *R* configuration. The latter amines have been described in racemic form only.[11] Gas chromatography using a chiral column revealed that the ees of the amines (**7a, c–g**) were in the range 90–92%, which implies that some racemization must have occurred during the final hydrogenolysis step.[1a]

Chiral Ring-substituted α-arylalkanamines. Following the reaction scheme (eq 3), the hydrazones [(*R*)-(−)-**8–15**] (pure *anti*-isomers) were prepared in 63–86% yields from the hydrazine [(*R*)-(−)-**4**] and the corresponding substituted aromatic aldehydes, using the previously described experimental conditions (anhydrous $MgSO_4$/TsOH/CH_2Cl_2/20 °C/17 h).The addition of Grignard reagents to the hydrazones (**8–15**) was carried out as above (10 equiv RMgX/Et_2O/reflux/17 h). The eight trisubsituted hydrazines [(*R*,*R*)-**16a–h**] (eq 4) were thus obtained in 51–83% yields and with a de=100% in all cases (as evidenced by ^{1}H and ^{13}C NMR). The addition of EtMgBr to the hydrazones (**14** and **15**) could not be carried out to completion and gave inseparable mixtures of trisubstituted hydrazine and starting hydrazone.

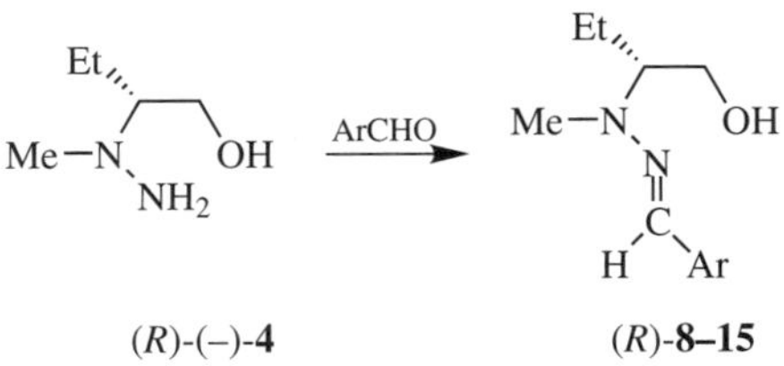

(3)

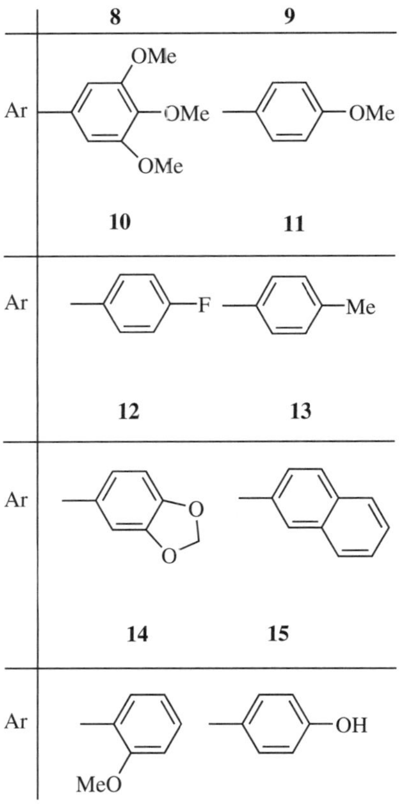

None of the ring-substituted hydrazines (**16a–h**) could be hydrogenolyzed under the conditions which were previously developed for the hydrazines (**6a–g**) (H_2, 6 bar/HCl/EtOH/60 °C/17 h). The temperature proved to be the determining factor: indeed, hydrogenolysis of the hydrazines (**16a–f**) at 110–120 °C, in the presence of a 10% Pd-C catalyst and concentrated HCl in EtOH under 6 bar for 5 h, afforded the corresponding (*R*)-α-arylalkanamines [(*R*)-(+)-**17a–f**] in 35–47% yield after purification by chromatography. Under the same conditions, hydrogenolysis of the hydrazines (**16g** and **16h**) gave inseparable mixtures. The enantiomeric excesses of the three amines (**17a,d,f**) were found to be within the range 90–93% by means of chiral GPC using a Restek β dex column. The other three amines (**17b,c,e**) could not be resolved using this or other chiral columns, or by running the ^{1}H NMR spectra in the presence of the chiral shift reagent Eu(hfc)$_3$. It can be assumed that the enantiomeric excesses of the amines (**17b,c,e**) are also in the range 90–93%, and that the six amines (**17a–f**) all belong to the *R*-series, analogous with the α-phenylalkanamines (**7a–g**), and in agreement with the addition mechanism which was previously put forth. The α-arylalkanamines [(*R*)-**17a–d**] were known in racemic form only. The amines [(*R*)-**17e,f**] are new compounds.[12]

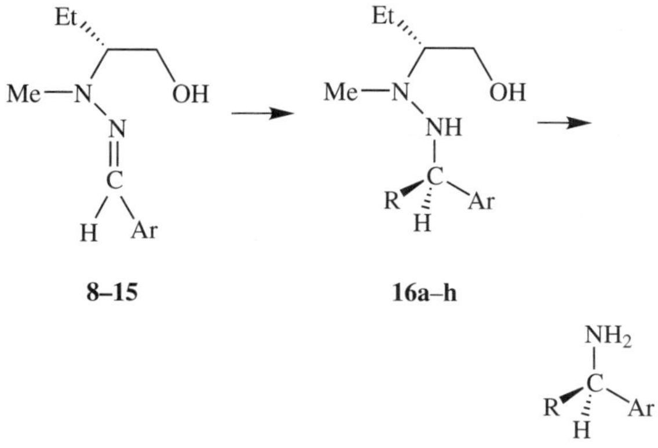

8–15 **16a–h** **17a–f** (4)

	a	b	c
Ar	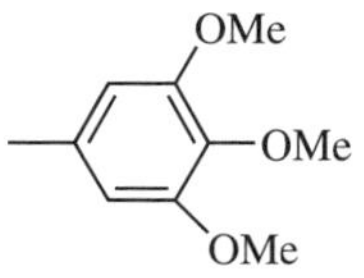	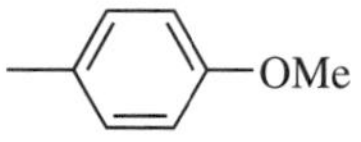	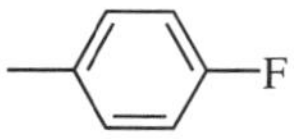
R	Et	Et	Et

	d	e	f
Ar	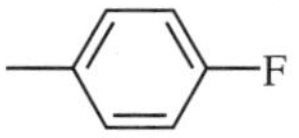	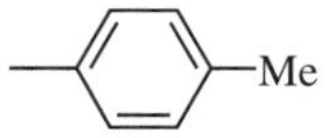	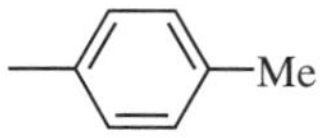
R	*n*-Bu	Et	*n*-Bu

	g	h
Ar	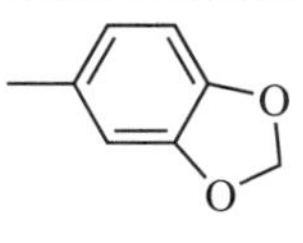	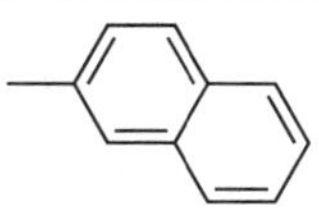
R	Et	Et

Since 2-aminobutan-1-ol (**1**) is readily available in both enantiomeric forms on an industrial scale, the above strategy can be applied to the synthesis of α-arylalkanamines belonging to both the *R*- and *S*-series.

The final hydrogenolysis step leading to the required α-arylalkanamine also yields *N*-methyl-2-aminobutan-1-ol (**2**) which can be recovered and distilled in view of recycling via its transformation into the hydrazine (**4**).

Related Reagents. RAMP; SAMP; (−)-*N*-aminoephedrine.

1. (a) Takahashi, H.; Tomita, K.; Otomasu, H. *J. Chem. Soc., Chem. Commun.* **1979**, 668. (b) Takahashi, H.; Tomita, K.; Noguchi, H. *Chem Pharm. Bull.* **1981**, *29*, 3387. (c) Takahashi, H.; Inagaki, H. *Chem. Pharm. Bull.* **1982**, *30*, 922.
2. Takahashi, H.; Suzuki, Y. *Chem. Pharm. Bull.* **1983**, *31*, 4295.
3. (a) Enders, D.; Schubert, H.; Nübling, C. *Angew. Chem.* **1986**, *98*, 1118. (b) Enders, D.; Reinhold, U. *Tetrahedron: Asymmetry* **1997**, *8*, 1895. (c) Enders, D.; Nübling, C.; Schubert, H. *Liebigs Ann. Recueil* **1997**, 1089.
4. (a) Denmark, S. E.; Weber, T.; Piotrowski, D. W. *J. Am. Chem. Soc.* **1987**, *109*, 2224. (b) Denmark, S. E.; Nicaise, O.; Edwards, J. P. *J. Org. Chem.* **1990**, *55*, 6219.
5. Kim, Y. H.; Choi, J. Y. *Tetrahedron Lett.* **1996**, *37*, 5543.
6. Touet, J.; Baudouin, S.; Brown, E. *Tetrahedron: Asymmetry* **1992**, *3*, 587.
7. Bataille, P.; Paterne, M.; Brown, E. *Tetrahedron: Asymmetry* **1998**, *9*, 2181.
8. Nohira, H.; Nohira, M.; Yoshida, S.; Osaka, A.; Terunuma, D *Bull. Chem. Soc. Jpn* **1988**, *61*, 1395.
9. Tanaka, H.; Inoue, K.; Pokorski, U.; Taniguchi, M.; Torii, S. *Tetrahedron Lett.* **1990**, *31*, 3023.
10. Yang, T. K.; Chen, R. Y.; Lee, D. S.; Peng, W. S.; Jiang, Y. Z. *J. Org. Chem.* **1994**, *59*, 914.
11. de Roocker, A.; de Radzitzky, P. *Bull. Soc. Chim. Belg.* **1963**, *72*, 202.
12. Bataille, P.; Paterne, M.; Brown, E. *Tetrahedron: Asymmetry* **1999**, *10*, 1579.

Eric Brown, Patricia Bataille & Michel Paterne
Laboratoire de Synthése Organique (UMR-CNRS 6011), Faculté des Sciences, Avenue Olivier Messiaen, Francé

2-(*S*)-[(4-Methylphenyl)sulfinyl]-2-cyclopenten-1-one (Ar = Tol), 2-(*S*)-[(4-Methoxyphenyl)sulfinyl]-2-cyclopenten-1-one (Ar = *p*-anisyl), 2-(*S*)-(1-Naphthylsulfinyl)-2-cyclopenten-1-one (Ar = 1-naphthyl), 2-(*S*)-[(2,4,6-Trimethylphenyl)sulfinyl]-2-cyclopenten-1-one (Ar = 2,4,6-trimethylphenyl), 2-(*S*)-[(2,4,6-Triisopropylphenyl)sulfinyl]-2-cyclopenten-1-one (Ar = 2,4,6-triisopropylphenyl)

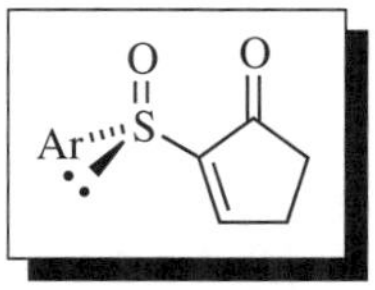

[79681-26-8],	$C_{12}H_{12}O_2S$,	(MW 220.29),
[93366-59-7],	$C_{12}H_{12}O_3S$,	(MW 236.29),
[82136-10-5],	$C_{15}H_{12}O_2S$,	(MW 256.32),
[178670-85-4],	$C_{14}H_{16}O_2S$,	(MW 248.34),
[151951-76-7]	$C_{20}H_{28}O_2S$	(MW 332.49)

(useful intermediates in the synthesis of chiral cyclic compounds)

Physical Data: Ar = Tol,[1] mp 125–126 °C; $[\alpha]_D^{25}$ +148° (*c* 0.11, $CHCl_3$). Ar = *p*-anisyl,[2] mp 120.5–121.5 °C; $[\alpha]_D^{25}$ +141° (*c* 1.45, acetone). Ar = 1-naphthyl,[3] mp 96.5–97.0 °C; $[\alpha]_D^{25}$ + 291.5° (*c* 1.30, acetone). Ar = 2,4,6-trimethylphenyl,[4] mp 131.6–132.0 °C; $[\alpha]_D^{20}$ +354.5° (*c* 0.416, $CHCl_3$). Ar = 2,4,6-triisopropylphenyl,[4] mp 139.5–140.4 °C; $[\alpha]_D^{17}$ +229.2° (*c* 0.216, $CHCl_3$).

Analysis of Reagent Purity: Ar = Tol, IR (CCl_4) cm^{-1}: 1715, ^{1}H NMR ($CDCl_3$) δ 2.2–2.5 (m, 2H), 2.30 (s, 3H), 2.6–2.8 (m, 2H) 7.19 and 7.58 (2d, 4H), 8.03 (t, 1H).

Preparative Methods: prepared from 2-bromo-2-cyclopenten-1-one in three steps (eq 1),[1] involving a key reaction of the vinyllithium and (*S*)-menthyl *p*-toluenesulfinate which proceeds with complete inversion at sulfur.[5]

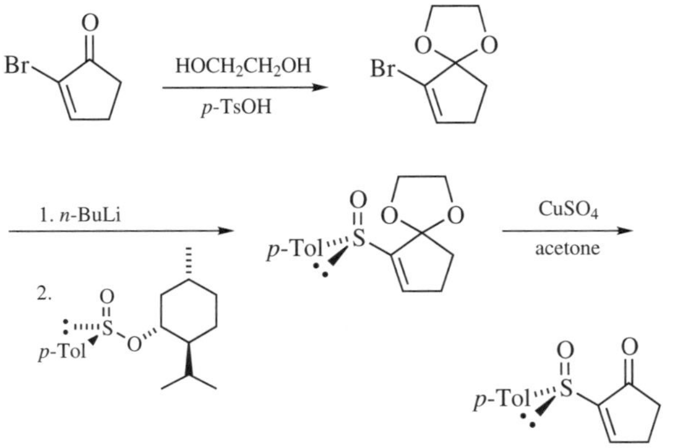

(1)

1

~50% overall yield

Handling, Storage, and Precautions: 2-(*S*)-[(4-methylphenyl)sulfinyl]-2-cyclopenten-1-one can be stored in vials in a desiccator at 0 °C for more than 1 year without decomposition. It became discolored after several weeks when exposed to the atmosphere at room temperature.

Michael Additions. The title compound 2-(*S*)-[(4-methylphenyl)sulfinyl]-2-cyclopenten-1-one (**1**) gives the conjugate addition products with excellent stereoselectivity (eq 2).[6] **1** is either treated directly with Grignard reagents or first treated with $ZnBr_2$ and subsequently with Grignard reagents to give, after reductive removal of the sulfinyl group, (*R*)-3-substituted cyclopentanones in excellent enantiomeric purity (Table 1).[3,7] The (*R*)-3-substituted cyclopentanones are formed through a chelated intermediate **2**. On the other hand, the conjugate addition with MeMgI or R_2Mg occurs from the diastereotopic face opposite to the bulky *p*-tolyl group in conformation **3** having the sulfoxide and carbonyl dipoles oriented in opposite directions, to give (*S*)-3-substituted cyclopentanones (Table 1).[8] Replacing the *p*-tolyl to the *p*-anisyl group, which would stabilize the chelate more effectively, causes a noticeable increase in diastereoselectivity (Table 1).[2,9] The reaction of the cyclopentenone having the bulky 1-naphthylsulfinyl group with Me_2CuLi gives the (*S*)-product in 57% ee (Table 1).[7a]

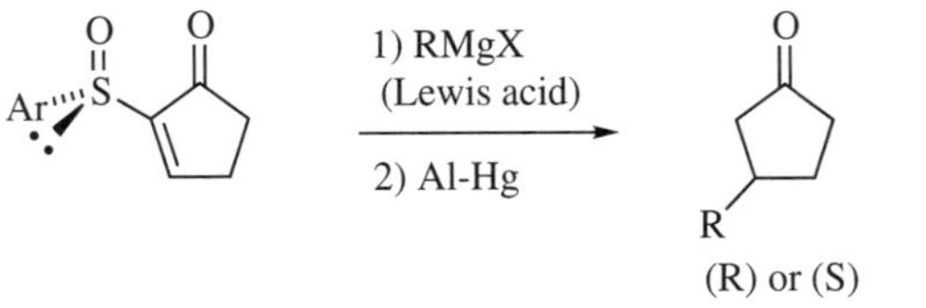

(2)

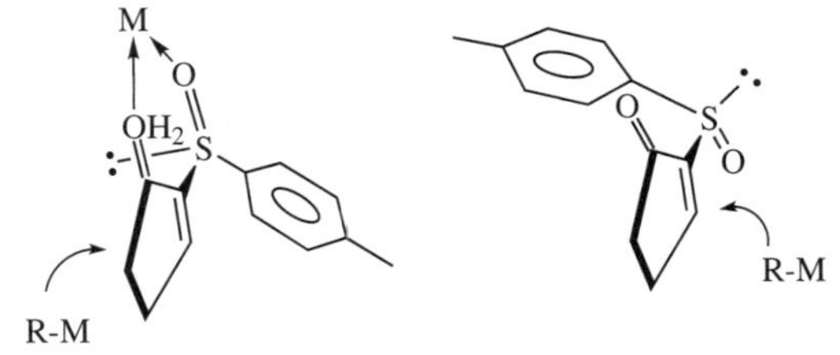

2 **3**

Table 1 Stereoselective Michael addition to 2-(*S*)-sul nylcyclopentenones

Ar	Reagent	Lewis acid	Optical yield (% ee)
Tol	Me_2Mg	–	97 (*S*)
Tol	Ph_2Mg	–	>98 (*S*)
Tol	MeMgCl	–	>98 (*R*)
Tol	MeMgI	–	72 (*S*)
Tol	MeMgI	$ZnBr_2$	87 (*R*)
Tol	PhMgBr	$ZnBr_2$	92 (*R*)
Tol	TolMgBr	$ZnBr_2$	58 (*R*)
4-Anisyl	TolMgBr	$ZnBr_2$	69 (*R*)
1-Naphthyl	Me_2CuLi	–	57 (*S*)

The conjugate addition with a naphthyl group affords the addition product. Methylation, reductive cleavage of the sulfinyl group, and alkylation give the optically pure steroid intermediate (eq 3).[3]

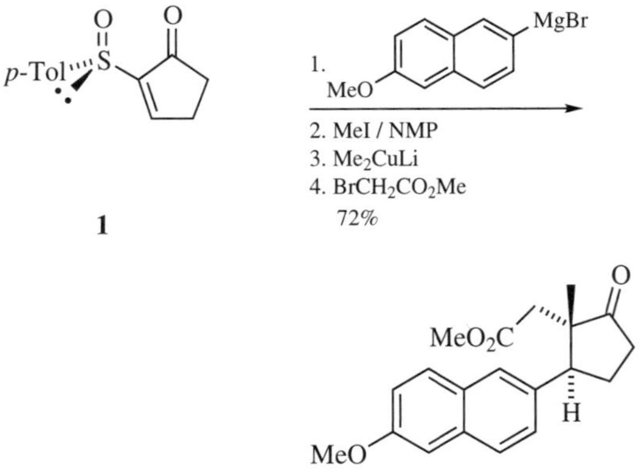

(3)

Optically pure

Reagent **1** also undergoes asymmetric Michael additions with enolate ions.[10] Michael additions with disubstituted lithium enolates proceed with almost complete π-facial diastereoselectivity. Starting with these Michael additions, (−)-methyl jasmonate[11] (eq 4) and (−)-estrone methyl ether[12] (eq 5) can be obtained in high enantiomeric purities.

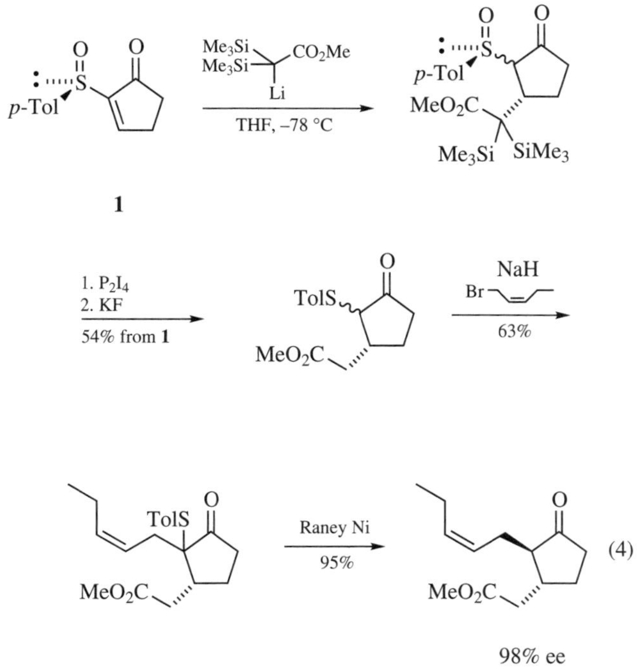

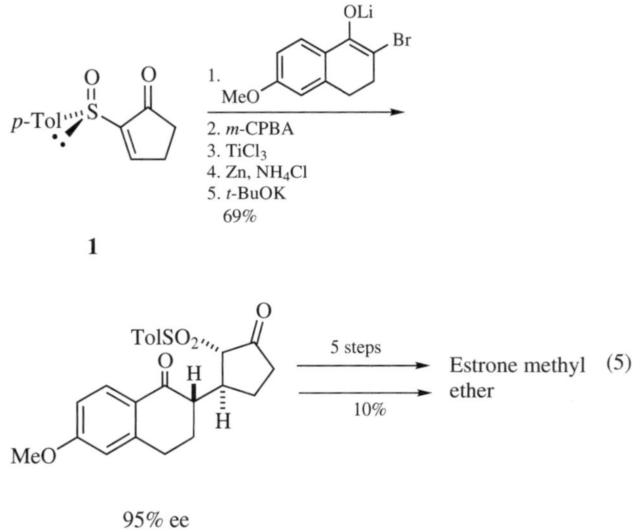

The $TiCl_4$-catalyzed reaction of **1** with crotylsilanes proceeds with high diastereoselectivity to give the (3*S*)-products (eq 6).[13] Reaction of 2-(phenylsulfinyl)-2-cyclopenten-1-one with LiOO-*t*-Bu gives the epoxide with low diastereoselectivity.[14]

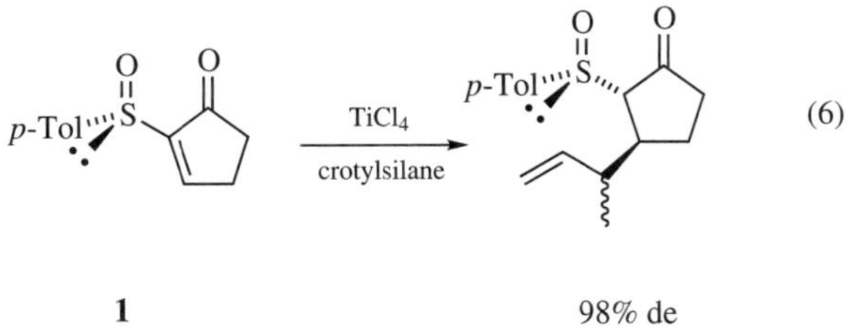

Diels–Alder Reactions. Reagent **1** is useful as an efficient chiral dienophile in asymmetric Diels–Alder reactions. Reaction of **1** with cyclopentadiene in the presence of a Lewis acid occurs with high stereoselectivity.[15] Reaction with 6-methoxy-1-vinyl-3,4-dihydronaphthalene in the presence of $EtAlCl_2$ proceeds with complete regioselectivity and *endo* selectivity (eq 7).[16] This stereochemical result can be explained in terms of a chelated conformer which directs the stereochemistry of approach of the dienophile.

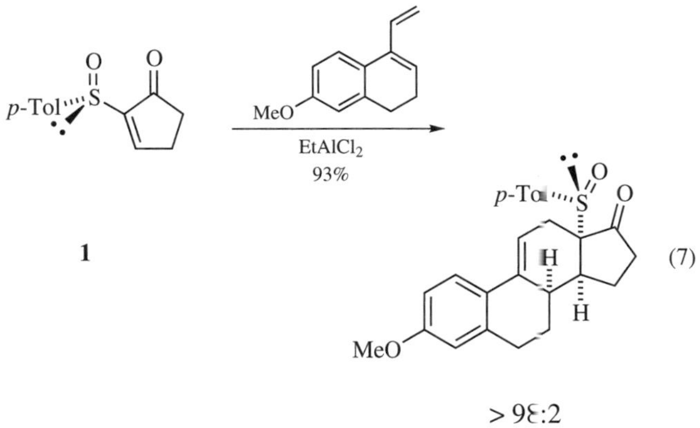

Treatment of 2-(phenylsulfinyl)-2-cyclopenten-1-one with phthalic anhydride in the presence of NaH gives the [4 + 2] cycloaddition product (eq 8).[17]

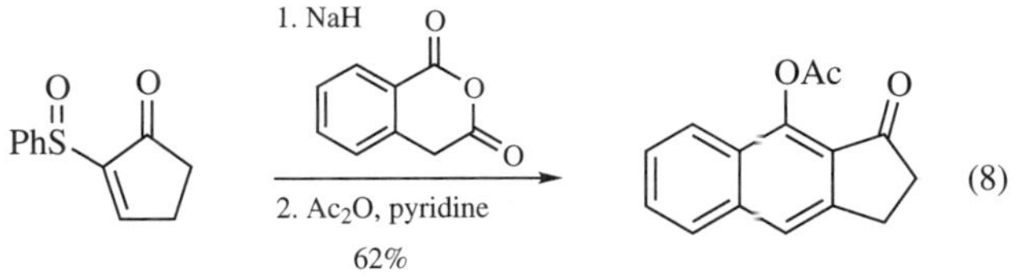

Reaction with Alkyl Radicals. The addition of alkyl radicals to **1** affords the products in good yields but with low stereoselectivity (eq 9). In order to shield one diastereotopic face effectively toward the attack of alkyl radicals, the cyclopentenone should have a sterically bulkier aryl group on the sulfur, i.e. the sulfoxides having an *ortho*-substituted aryl group show high diastereoselection. Thus, 2-(*S*)-[(2,4,6-triisopropylphenyl)sulfinyl]-2-cyclopenten-1-one (**4**) and 2-(*S*)-[(2,4,6-trimethylphenyl)sulfinyl]-2-cyclopenten-1-one (**5**) give (*R*)-3-alkylated cyclopentanones with extremely high diastereoselectivities in reactions with alkyl radicals such as a *t*-butyl, isopropyl, or cyclohexyl radical, and even with the less bulky ethyl radical (Table 2).[4,18] The reaction of **5** with a *tert*-butyl radical in the presence of $EtAlCl_2$ or $TiCl_2(O\text{-}i\text{-}Pr)_2$ completely reverses the face selection, giving only the (3*S*)-3-*tert*-butylcyclopentanone (Table 2). The change in the product distribution is apparently due to the conformation fixed by chelation with a Lewis acid between the carbonyl and sulfinyl oxygens. However, $ZnBr_2$, which is an efficient chelating Lewis acid in the Michael additions, causes only a small change or none in the product ratio in these radical reactions.

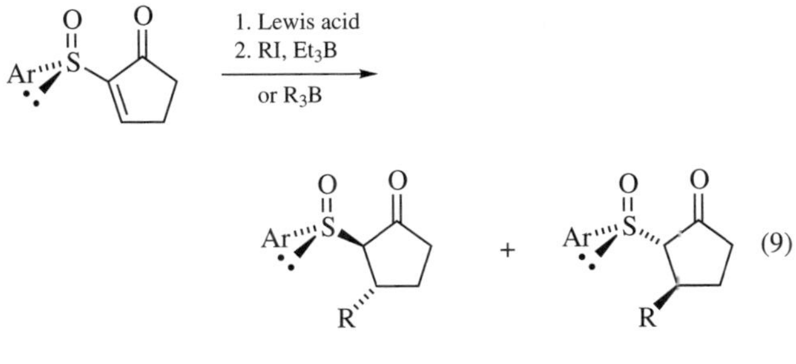

Table 2 Stereoselective radical addition to 2-(s)-sul nylcyclopentenones

Ar	R	Lewis acid	Yield (%)	Ratio (*R*/*S*)
Tol	*t*-Bu	–	93	67:33
Tip	*t*-Bu	–	95	>98:2
Tip	*i*-Pr	–	95	>98:2
Tip	Et	–	94	>98:2
Mes	*t*-Bu	–	99	>98:2
Mes	*t*-Bu	$EtAlCl_2$	99	2:>98

Tip = 2,4,6-triisopropylphenyl.

The photo-induced reaction of 2-sulfinylcyclopentenones in alcohols in the presence of Ph_2CO gives addition products (eq 10). Reagent **1** shows low stereoselectivity, whereas complete diastereoselection can be achieved in the reaction of **4** and **5** (Table 3).[19]

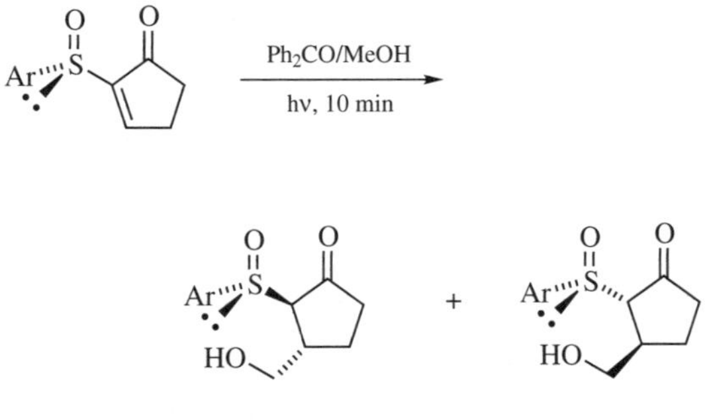

(10)

(3*R*) (3*S*)

Table 3 Photo-induced radical addition to 2-(*S*)-sul nylcyclopentenones

Ar	Yield (%)	(3*R*)/(3*S*)
Tol	93	53:47
Tip	96	>98:2
Mes	75	>98:2

Tip = 2,4,6-triisopropylphenyl.

1. Hulce, M.; Mallomo, J. P.; Frye, L. L.; Kogan, T. P.; Posner, G. H. *Org. Synth., Coll. Vol.* **1990**, *7*, 495.
2. Posner, G. H.; Frye, L. L.; Hulce, M. *Tetrahedron* **1984**, *40*, 1401.
3. Posner, G. H.; Mallamo, J. P.; Hulce, M.; Frye, L. L. *J. Am. Chem. Soc.* **1982**, *104*, 4180.
4. Mase, N.; Watanabe, Y.; Ueno, Y.; Toru, T. *J. Org. Chem.* **1997**, *62*, 7794.
5. Andersen, K. K.; Gaffield, W.; Papanikolau, N. E.; Foley, J. W.; Perkins, R. I. *J. Am. Chem. Soc.* **1964**, *86*, 5637.
6. Posner, G. H. *Acc. Chem. Res.* **1987**, *20*, 72.
7. (a) Posner, G. H.; Mallamo, J. P.; Miura, K. *J. Am. Chem. Soc.* **1981**, *103*, 2886. (b) Posner, G. H.; Hulce, M.; Mallamo, J. P.; Drexler, S. A.; Clardy, J. *J. Org. Chem.* **1981**, *46*, 5244. (c) Posner, G. H. In *Asymmetric Synthesis*; Morrison, J. D., Ed.: Academic: New York, 1983, Vol. 2, p 239.
8. Posner, G. H.; Hulce, M. *Tetrahedron Lett.* **1984**, *25*, 379.
9. Posner, G. H.; Frye, L. L. *Isr. J. Chem.* **1984**, *24*, 88.
10. (a) Posner, G. H.; Wayne, H. *J. Chem. Soc., Chem. Commun.* **1985**, 1786. (b) Posner, G. H.; Weitzberg, M.; Hamill, T. G.; Asirvatham, E. *Tetrahedron* **1986**, *42*, 2919. (c) Posner, G. H.; Weitzberg, M.; Jew, S. S. *Synth. Commun.* **1987**, *17*, 611.
11. (a) Posner, G. H.; Asirvatham, E. *J. Org. Chem.* **1985**, *50*, 2589. (b) Posner, G. H.; Asirvatham, E.; Ali, S. F. *J. Chem. Soc., Chem. Commun.* **1985**, 542.
12. Posner, G. H.; Switzer, C. *J. Am. Chem. Soc.* **1986**, *108*, 1239.
13. Pan, L. R.; Tokoroyama, T. *Tetrahedron Lett.* **1992**, *33*, 1469.
14. Fernandez de la Pradilla, R.; Castro, S.; Manzano, P.; Martin-Ortega, M.; Priego, J.; Viso, A.; Rodriguez, A.; Fonseca, I. *J. Org. Chem.* **1998**, *63*, 4954.
15. Alonso, I.; Carretero, J. C.; Garcia Ruano, J. L. *Tetrahedron Lett.* **1989**, *30*, 3853.
16. Alonso, I.; Carretero, J. C.; Garcia Ruano, J. L.; Martin Cabrejas, L. M. *Tetrahedron Lett.* **1994**, *35*, 9461.
17. Fujioka, H.; Akai, S.; Kita, Y. *J. Org. Chem.* **2000**, *65*, 89.
18. (a) Toru, T.; Watanabe, Y.; Tsusaka, M.; Ueno, Y. *J. Am. Chem. Soc.* **1993**, *113*, 19464. (b) Toru, T.; Watanabe, Y.; Mase, N.; Tsusaka, M.; Hayakawa, T.; Ueno, Y. *Pure Appl. Chem.* **1996**, *68*, 711.
19. Mase, N.; Watanabe, Y.; Toru, T. *Bull. Chem. Soc. Jpn.* **1998**, *71*, 2957.

Takeshi Toru

Nagoya Institute of Technology, Nagoya, Japan

(*S*)-1-Methyl-2-(piperidinomethyl)-pyrrolidine

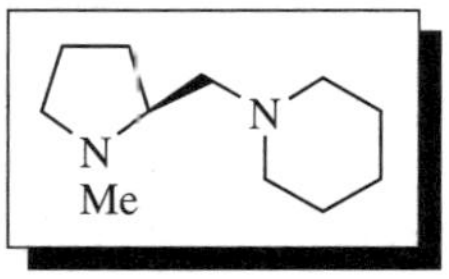

[84466-85-3] $C_{11}H_{22}N_2$ (MW 182.31)

(reagent used as a chiral ligand in asymmetric synthesis of organic compounds)

Physical Data: bp 121–123 °C/22 mmHg; d 0.909 g cm^{-3}; $[\alpha]_D^{20}$ −69 (*c* 0.5, EtOH).

Solubility: soluble in alcohol, diethyl ether, and most organic solvents.

Form Supplied in: colorless liquid; commercially available form Tokyo Kasei Kogyo Co., Ltd. (TCI) and Aldrich.

Purification: usable without further purification.

Handling, Storage, and Precautions: (*S*)-1-methyl-2-(piperidinomethyl)pyrrolidine and its derivatives can be synthesized from L-proline. Methods for the preparation are described in refs 1b and 2b. Properties of chiral diamines and synthetic intermediates are shown in refs 1b, 2d, and 3b.

The Tin(II) Enolate Mediated Asymmetric Aldol Reaction. Chiral auxiliaries derived from (*S*)-proline are particularly attractive since they possess conformationally rigid pyrrolidine rings. In particular, chiral diamines derived from (*S*)-proline have been successfully employed for the generation of an efficient chiral environment since almost all the main and transition metals having vacant d orbitals are capable of accepting a bidentate ligand. An intermediate derived from the chiral ligand and organometallic reagent would have a conformationally restricted *cis*-fused five-membered ring chelate and would afford the optically active organic compounds by reaction with appropriate substrates. Here, highly stereoselective asymmetric reactions employing a chiral diamine, (*S*)-1-methyl-2-(piperidinomethyl)pyrrolidine (**1**), and their application to the syntheses of optically active natural products are described.

The asymmetric aldol reaction is one of the most powerful tools for the construction of new carbon-carbon bonds with control of absolute configurations at new chiral centers, and the utility of this reaction has been demonstrated by a number of applications to the synthesis of natural products such as macrolides and polyether antibiotics, carbohydrates, etc. In the asymmetric aldol reactions reported, chiral auxiliary groups are usually attached to the reacting ketone equivalent molecules. Until early 1980s, there had not been an example of aldol-type reaction where two achiral carbonyl compounds were used to form a chiral molecule with the aid of a chiral ligand. The enantioselective aldol reaction *via* tin(II) enolates coordinated with chiral diamines was explored in 1982.[1] In the presence of chiral diamine **1**, various optically active aldol adducts were produced by the reaction between aromatic ketones and aldehydes (eqs 1 and 2). This is the first example of the formation of crossed aldol products in high optical purity which started from two achiral carbonyl compounds and employed chiral diamines as chelating agents.

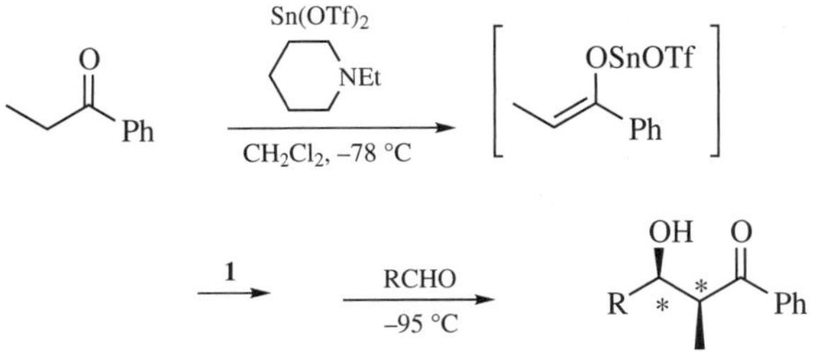

(1)

57~78%
syn/*anti* 80:20~100:0
75~90% ee (*syn*)

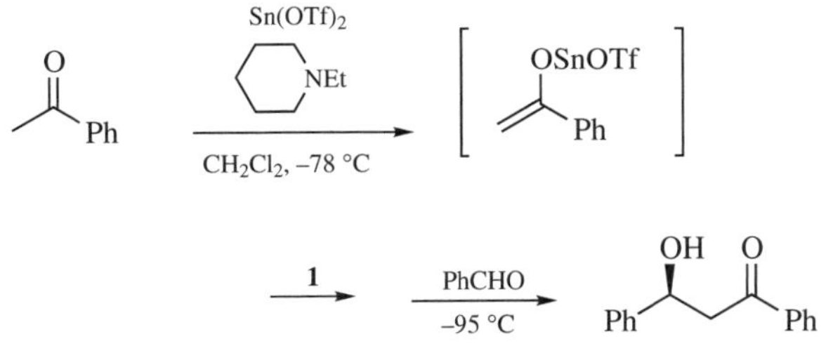

(2)

35%, 75% ee

This protocol has been successfully applied to the reactions of carboxylic acid derivatives such as thioamides and thione esters (eqs 3 and 4).[4] 3-Acetylthiazolidine-2-thiones are quite suitable substrates for the tin(II) enolate mediated asymmetric aldol reaction and various optically active β-hydroxy 3-acetylthiazolidine-2-thiones are obtained by using chiral diamine **1** (eq 5).[5]

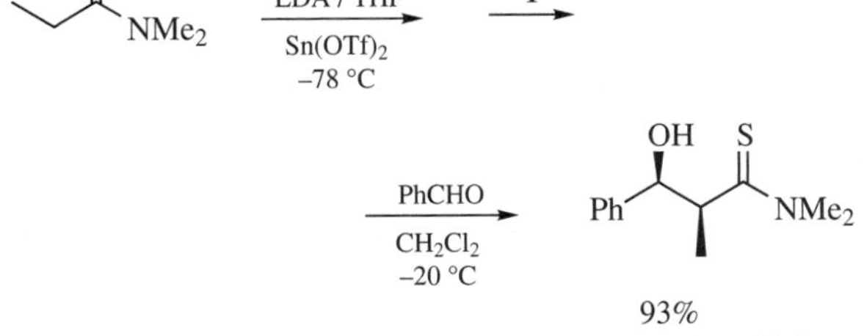

(3)

93%
syn/*anti* 92:8
85% ee (*syn*)

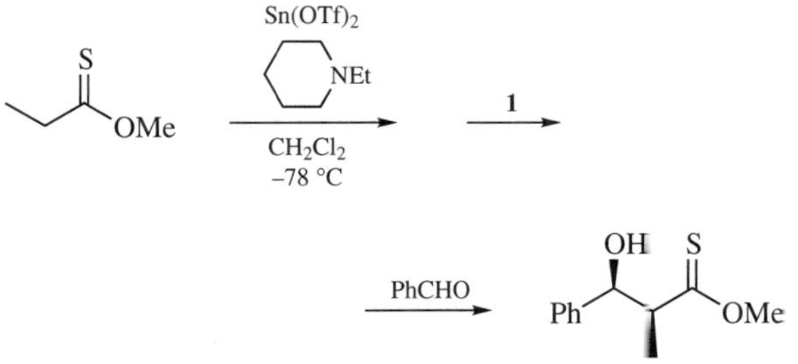

(4)

73%
syn/*anti* 78:22
90% ee (*syn*)

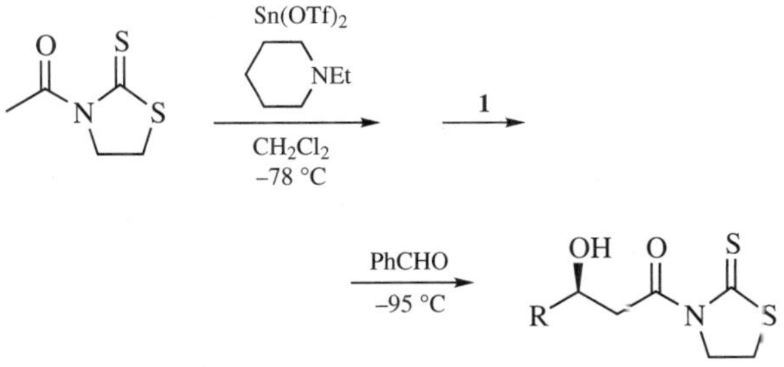

(5)

63~81%
65~>90% ee

When 3-(2-benzyloxyacetyl)thiazolidine-2-thione is treated under the above reaction conditions, the corresponding *anti*-diol units are produced with good diastereo- and high enantioselectivities upon the addition of chiral diamine **1** (eq 6).[6]

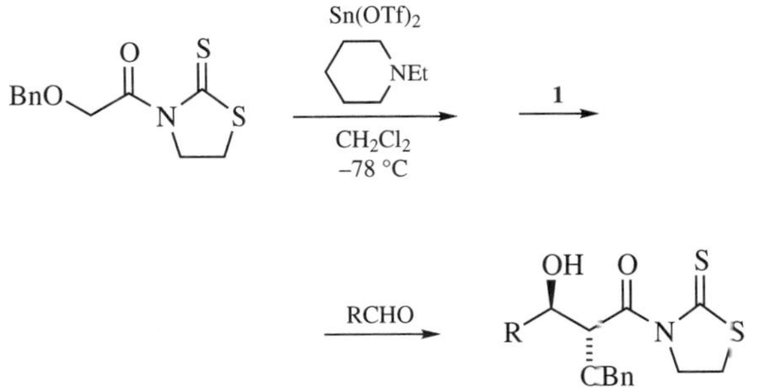

(6)

68~93%
syn/*anti* 19:81~7:93
87~94% ee (*anti*)

Because tin(II) enolates of thioesters are generated upon reaction of tin(II) thiolates with ketenes, the optically active β-hydroxy thioesters are also easily synthesized by way of the aldol reaction with aldehydes in the presence of tin(II) trifluoromethanesulfonate and chiral diamine **1** (eq 7).[7]

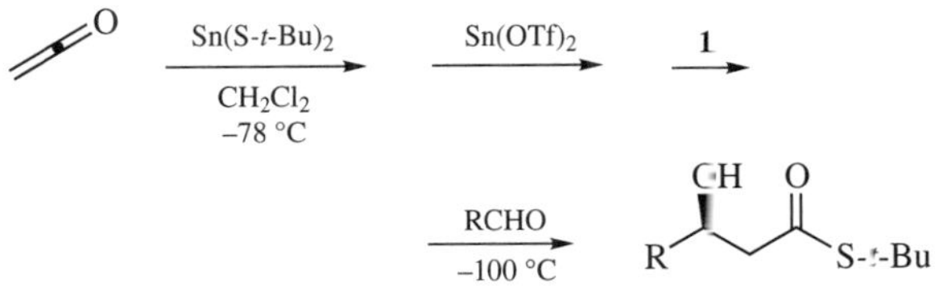

(7)

55~70%
58~80% ee

In the presence of chiral diamine **1**, the sulfurization of tin(II) enolates of ketones or 3-acetylthiazolidine-2-thiones by use of thiosulfonates proceeds smoothly to give the corresponding β-keto sulfides with high enantioselectivities (eq 8 and eq 9).[8]

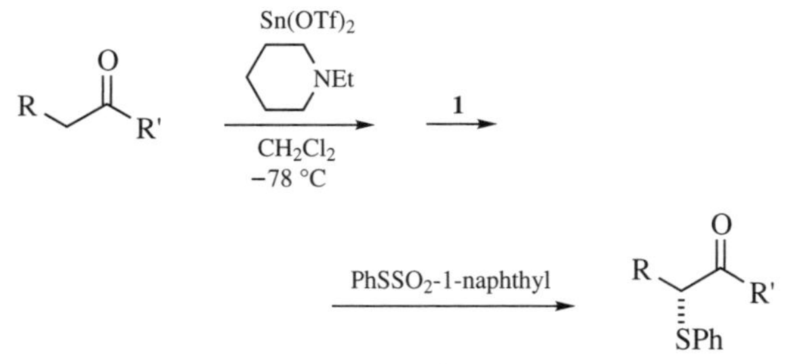

(8)

52~80%
50~85% ee

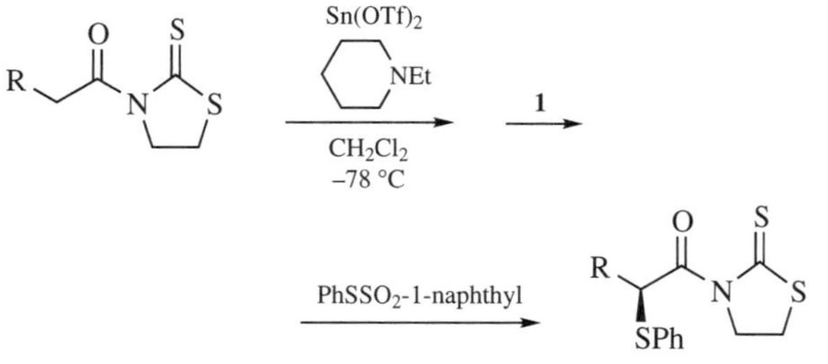

(9)

91~93%
81~82% ee

The enantioselective Michael addition reaction of tin(II) enolates to α,β-unsaturated ketones is also successfully achieved by employing the chiral diamine **1** to yield the desirable optically active adduct (eq 10).[9]

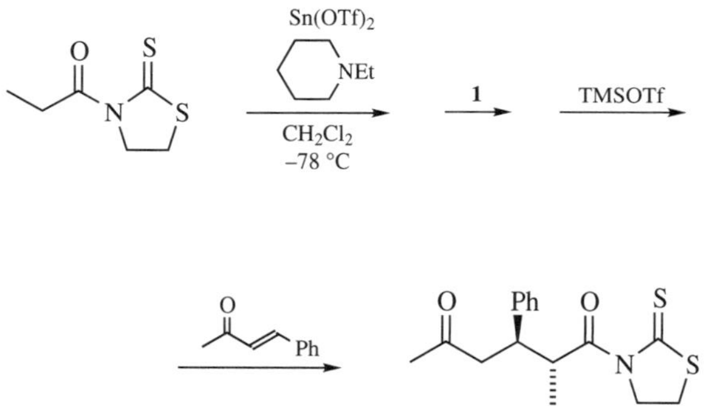

(10)

70%
syn/*anti* 5:>95
80% ee (*anti*)

Asymmetric Reduction of Prochiral Ketones. The asymmetric reduction of prochiral ketones with chiral hydride reagents has been widely examined for producing the optically active alcohols and a number of methods have therefore been reported. In general, the chiral hydride reagent is generated in situ by the reaction of a suitable metal hydride with chiral ligands such as alkaloids, sugar derivatives, amino alcohols, chiral oxazolines, tartaric acid derivatives, chiral amines, and chiral diols. A novel chiral reducing agent prepared from tin(II) chloride, the chiral diamine (**1**) and diisobutylaluminum hydride was developed in 1984 and various optically active secondary alcohols were obtained effectively by the asymmetric reduction of prochiral ketones (eq 11).[10]

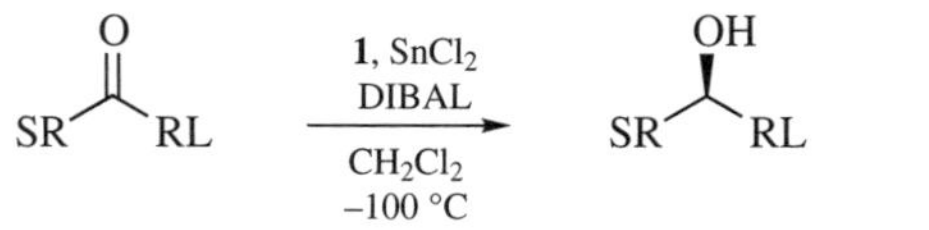

(11)

67~91%
61~78% ee

Asymmetric reduction of prochiral α-ketoesters as well as α- and β-alkoxy ketones using the above reagent including chiral diamine **1** affords the corresponding functionalized secondary alcohols in good yields with high enantioselectivities (eqs 12–14).[11]

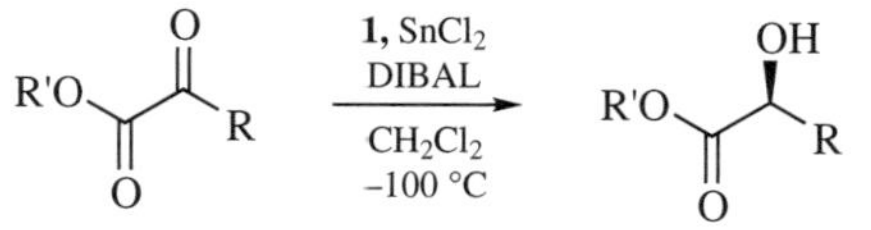

(12)

60~90%
40~89% ee

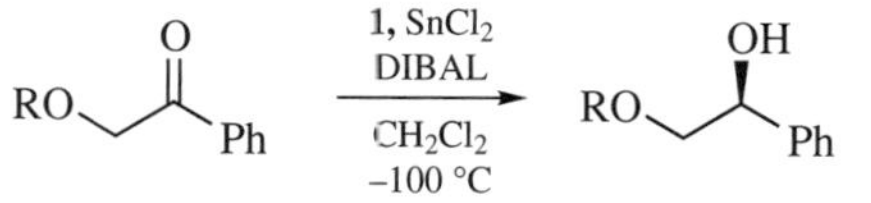

(13)

63~85%
32~93% ee

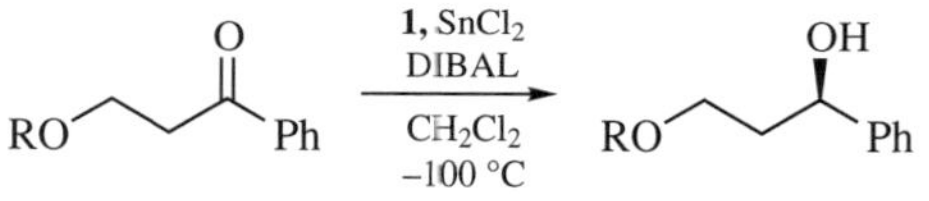

(14)

65~92%
50~81% ee

Successive work of the asymmetric reduction of prochiral ketones shows that (*S*)-enantiomers are produced by using the combination of chiral diamine **1** with zinc(II) chloride whereas (*R*)-enantiomers are obtained by using the combination of chiral diamine **1** with tin(II) chloride under the same reaction conditions (eq 15).[12]

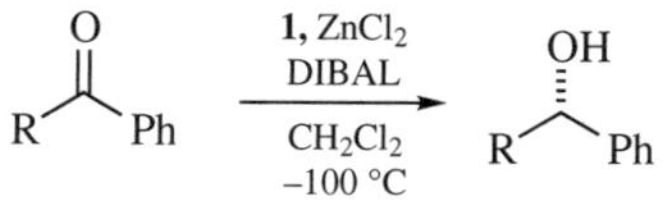

(15)

59~98% (conversion yield)
79~83% ee

Asymmetric Allylation of Prochiral Aldehydes. The asymmetric allylation reaction is useful for stereoselective carbon-carbon bond formation, and therefore development of an effective method for the synthesis of optically active homoallyl alcohols was sought. A chiral allylating agent, readily generated from tin(II) trifluoromethanesulfonate, chiral diamine **1**, and allyldiisobutylaluminum, was efficiently employed in the asymmetric allylation of aldehydes in 1996 (eq 16).[13]

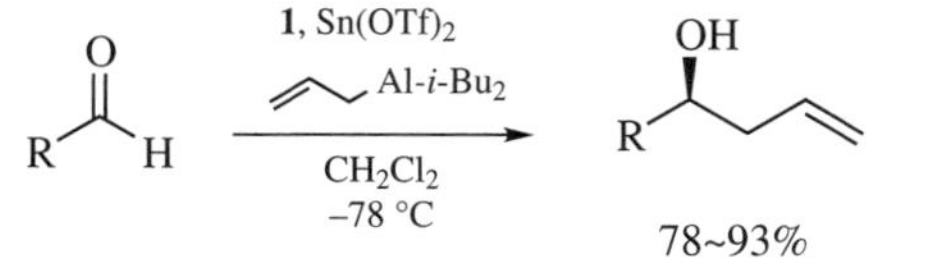

(16)

78~93%
53~84% ee

Reaction of a chiral methylating agent with aldehydes also proceeds smoothly to afford the corresponding homoallyl alcohols in good yields and with high enantioselectivities (eq 17). According to the similar procedure, asymmetric propargylation of aldehydes gives homopropargyl alcohols in good optical purities (eq 18).[14]

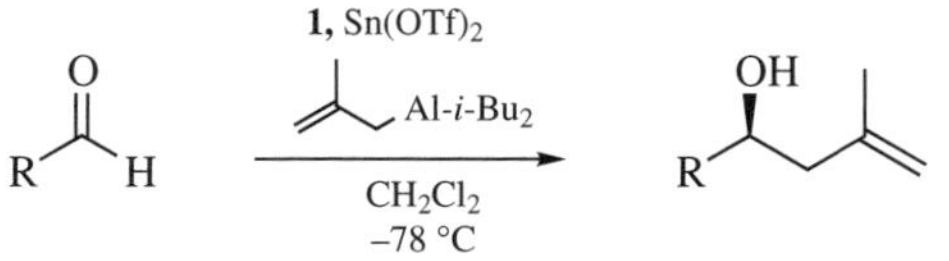

(17)

70~85%
67~88% ee

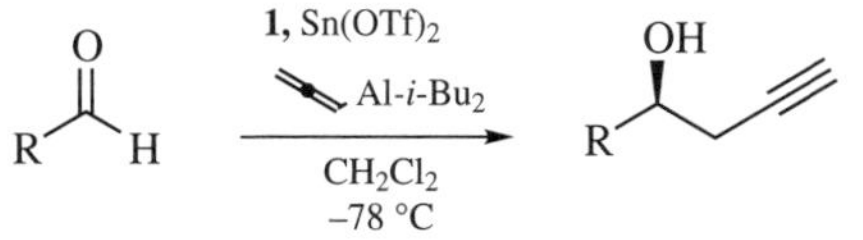

(18)

45~77%
48~60% ee

The asymmetric allylation reaction of achiral aldehydes with diallyltin dibromide has been achieved by using the chiral diamine **1** as a promoter (eq 19). The enantiomeric excess increases up to 79% by using a chiral diamine similar to **1** which possesses an *n*-butyl group on the nitrogen of the pyrrolidine ring.[15]

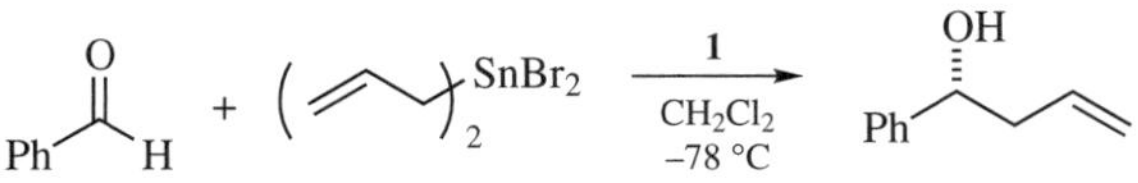

(19)

79%, 68% ee

Asymmetric Aldol Reaction using Tin(II) Complex as a Lewis Acid. Chiral tin(II) Lewis acids prepared in situ by the coordination of chiral pyrrolidine derivatives to tin(II) trifluoromethanesulfonate was developed for the promotion of asymmetric aldol reactions in 1989. The quite important key for the asymmetric aldol reaction is the choice of the chiral Lewis acid. Some chiral Lewis acids were already reported and fruitful results were observed particularly in the field of the Diels-Alder and related reactions in late 1980s. The chiral Lewis acids employed consisted of rather strong and hard acidic metals such as aluminum and titanium. Since these metals were strongly coordinated with oxygen, smooth metal exchange from hard metals to silicon would hardly take place. On the other hand, a chiral tin(II) Lewis acid, which is prepared in situ by the chelation of chiral diamine **1** to tin(II) trifluoromethanesulfonate, might be quite effective because tin(II) is a soft metal and this complex has one vacant d orbital to be coordinated with oxygen in carbonyl group of aldehyde without losing the favorable asymmetric environment. Based on this consideration, various efficient asymmetric aldol reactions between achiral enol silyl ethers and achiral carbonyl compounds have been developed as follows.

The asymmetric aldol reaction of enol silyl ethers of thioesters with aldehydes is performed in high enantiomeric excess by employing a chiral promoter, tin(II) trifluoromethanesulfonate coordinated with chiral diamine **1** and tri-*n*-butyltin fluoride (eqs 20 and 21).[3] Highly enantioselective aldol reactions of achiral ketene silyl acetals with achiral aldehydes are carried out by means of the same chiral promoter (eq 22).[2]

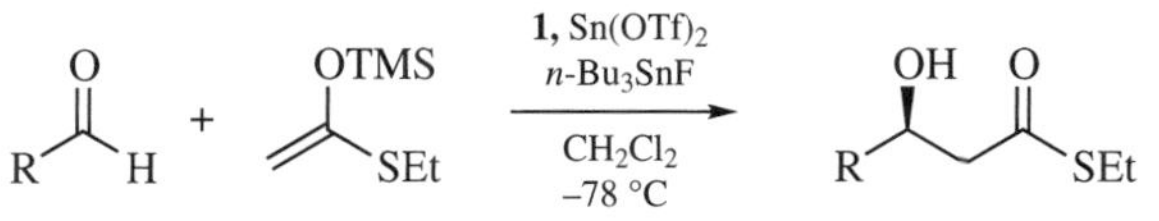

(20)

70~90%
78~>95% ee

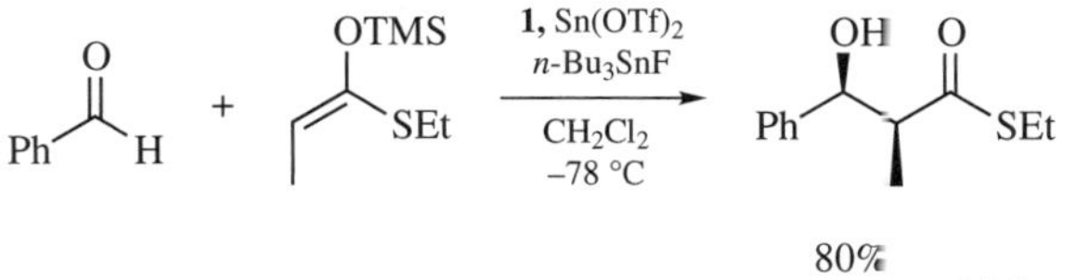

(21)

80%
syn/*anti* 93:7
80% ee (*syn*)

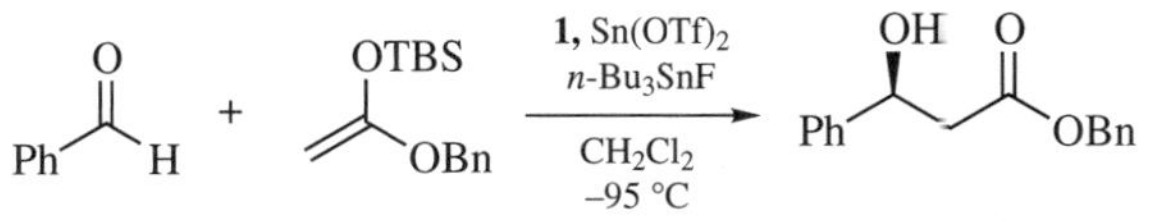

(22)

81%
82% ee

In the presence of promoter including the chiral diamine **1**, the enol silyl ether of thioesters reacts with α-ketoesters to afford the corresponding aldol-type adducts, 2-substituted malates, in good yields with excellent enantiomeric excess (eq 23).[16]

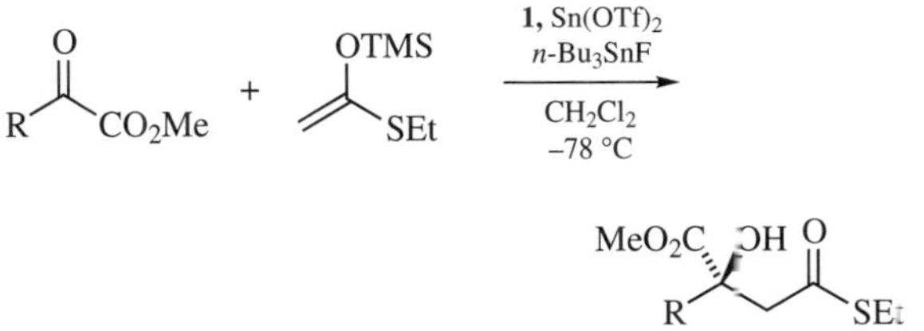

(23)

73~76%
83~>98% ee

Asymmetric Synthesis of *syn*- and *anti*-1,2-diol Derivatives. Optically active 1,2-diol units are often observed in nature as carbohydrates, macrolides or polyethers, etc. Several excellent asymmetric dihydroxylation reactions of olefins using osmium tetroxide with chiral ligands have been developed to give the optically active 1,2-diol units with high enantioselectivities. However, there still remain some problems, for example, preparation of the optically active *anti*-1,2-diols and so on. The asymmetric aldol reaction of an enol silyl ether derived from α-benzyloxy thioester with aldehydes was developed in order to introduce two hydroxyl groups simultaneously with stereoselective carbon-carbon bond formation by using the chiral tin(II) Lewis acid. For example, various optically active *anti*-α,β-dihydroxy thioester derivatives are obtained in good yields with excellent diastereo-

and enantioselectivities when the chiral diamine **1**, tin(II) trifluoromethanesulfonate, and di-*n*-butyltin diacetate are employed together (eq 24).[17] According to the present aldol methodology, two hydroxyl groups can be stereoselectively introduced in 1,2-position during new carbon-carbon bond formation.

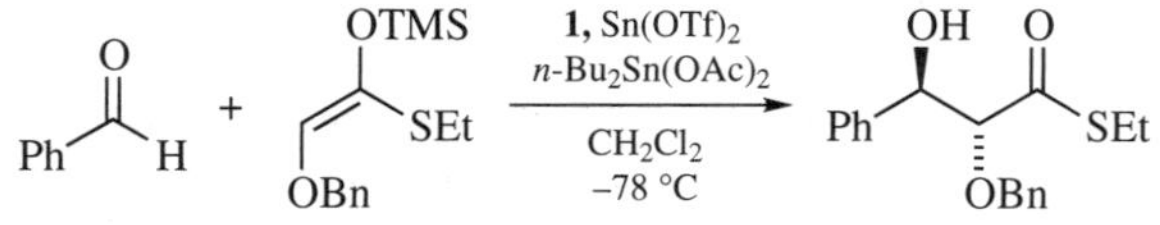
(24)

74%
syn/*anti* 1:99
96% ee (*anti*)

On the other hand, several *syn*-aldol adducts are obtained under the same reaction conditions: namely, in the presence of chiral diamine **1**, tin(II) trifluoromethanesulfonate, and di-*n*-butyltin diacetate, the reaction of an enol silyl ether possessing a *tert*-butyldimethylsiloxy group at the 2-position with achiral aldehydes proceeds smoothly to give the corresponding *syn*-α,β-dihydroxy thioester derivatives in high yields with good stereoselectivities (eq 25). When a chiral diamine that is similar to **1** possessing a *n*-propyl group on the nitrogen of the pyrrolidine ring is used, the enantiomeric excess increases up to 90%.[18] Now it becomes possible to control the enantiofacial selectivity of the enol silyl ethers derived from α-alkoxy thioesters just by choosing the appropriate protective groups of alkoxy parts of the enol silyl ethers, and the two diastereomers of the optically active α,β-dihydroxy thioesters can be synthesized.[19]

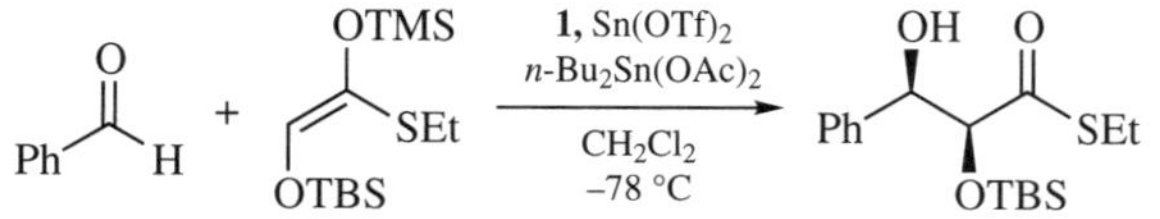
(25)

83%
syn/*anti* 91:9
49% ee (*syn*)

Furthermore, it is found that enol silyl ethers derived from phenyl alkoxyacetates react with aldehydes to afford the corresponding *anti*-1,2-diol derivatives with high diastereo- and enantioselectivities through use of a tin(II) Lewis acid in the presence of chiral diamine **1** (eqs 26 and 27).[20]

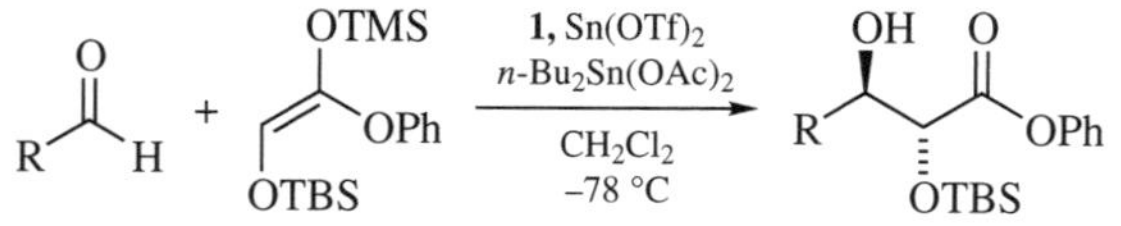
(26)

61~87%
syn/*anti* 7:93~6:94
90~96% ee (*anti*)

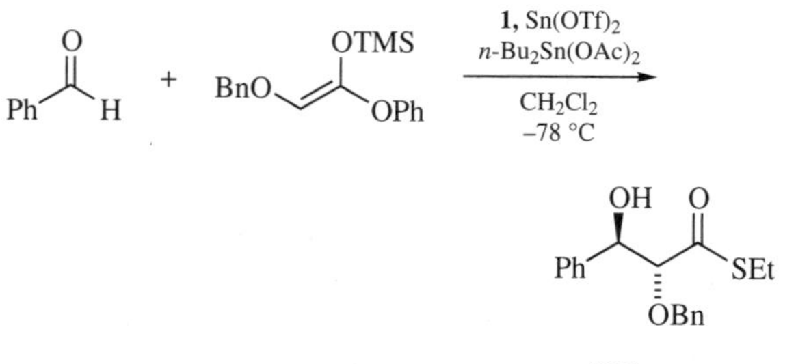
(27)

90%
syn/*anti* 29:71
78% ee (*anti*)

The method for producing chiral 1,2-diol units is also applicable to the construction of asymmetric quaternary carbons contained in aldol units. In the presence of a chiral promoter consisting of the chiral diamine **1**, tin(II) trifluoromethanesulfonate, and di-*n*-butyltin diacetate, various optically active α-alkoxy-α-methyl-β-hydroxy thioesters and esters are synthesized in good yields with high stereoselectivities (eqs 28 and 29).[21]

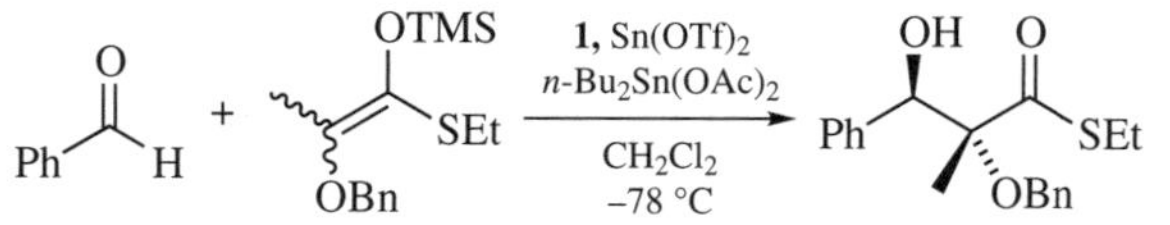
(28)

58%
syn/*anti* 2:98
97% ee (*anti*)

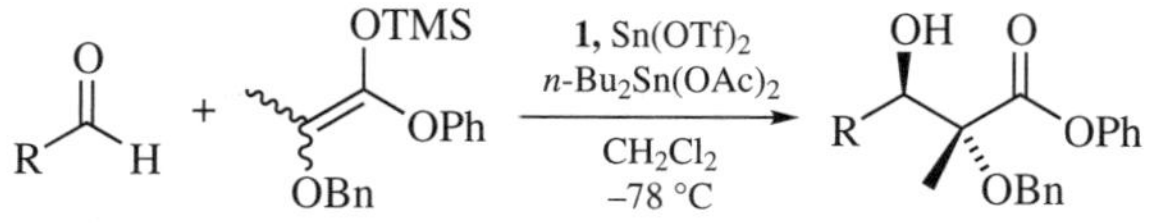
(29)

44~72%
syn/*anti* 26:74~8:92
73~95% ee (*anti*)

The diastereo- and enantioselective synthesis of both stereoisomers of α-alkoxy-β-hydroxy-β-methyl thioesters is also attained by reaction of enol silyl esters possessing alkoxy groups at the 2-position with a tin(II) Lewis acid and the chiral diamine **1** as promoters (eqs 30 and 31).[22]

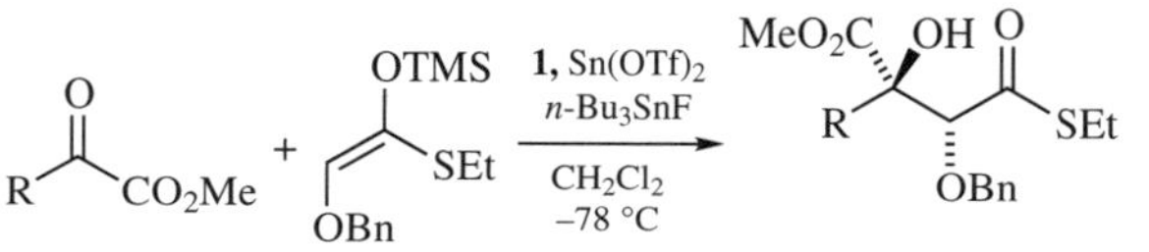
(30)

59~64%
syn/*anti* 14:86~8:92
83~88% ee (*anti*)

By these reactions, optically active 1,2-diol units can be positioned efficiently on the desired carbon skeletons. Recently, the above methodologies have been successfully utilized for the stereoselective syntheses of natural and unnatural poly-oxy compounds such as monosaccharides,[23] leinamycin,[24] paclitaxel (Taxol®),[25] and a part of rapamycin.[26]

Related Reagents. (*S*)-1-Methyl-2-(naphthylaminomethyl) pyrrolidine (commercially available form TCI);[3d] (*S*)-1-ethyl-2-(piperidinomethyl)pyrrolidine;[2b,3d] (*S*)-1-propyl-2-(piperidinomethyl)pyrrolidine;[3d] (*S*)-1-butyl-2-(piperidinomethyl)pyrrolidine.[3d]

1. (a) Iwasawa, N.; Mukaiyama, T. *Chem. Lett.* **1982**, 1441. (b) Mukaiyama, T.; Iwasawa, N.; Stevens, R. W.; Haga, T. *Tetrahedron* **1984**, *40*, 1381.
2. (a) Kobayashi, S.; Sano, T.; Mukaiyama, T. *Chem. Lett.* **1989**, 1319. (b) Mukaiyama, T.; Kobayashi, S.; Sano, T. *Tetrahedron* **1990**, *46*, 4653.
3. (a) Kobayashi, S.; Mukaiyama, T. *Chem. Lett.* **1989**, 297. (b) Mukaiyama, T.; Uchiro, H.; Kobayashi, S. *Chem. Lett.* **1989**, 1001. (c) Mukaiyama, T.; Kobayashi, S. *J. Organomet. Chem.* **1990**, *382*, 39. (d) Kobayashi, S.; Uchiro, H.; Fujishita, Y.; Shiina, I.; Mukaiyama, T. *J. Am. Chem. Soc.* **1991**, *113*, 4247.
4. Iwasawa, N.; Yura, T.; Mukaiyama, T. *Tetrahedron* **1989**, *45*, 1197.
5. Iwasawa, N.; Mukaiyama, T. *Chem. Lett.* **1983**, 297.

6. Mukaiyama, T.; Iwasawa, N. *Chem. Lett.* **1984**, 753.
7. Mukaiyama, T.; Yamasaki, N.; Stevens, R. W.; Murakami, M. *Chem. Lett.* **1986**, 213.
8. Yura, T.; Iwasawa, N.; Clark, R.; Mukaiyama, T. *Chem. Lett.* **1986**, 1809.
9. Yura, T.; Iwasawa, N.; Mukaiyama, T. *Chem. Lett.* **1988**, 1021.
10. (a) Oriyama, T.; Mukaiyama, T. *Chem. Lett.* **1984**, 2071. (b) Falorni, M.; Lardicci, L.; Piroddi, A. M.; Giacomelli, G. *Gazz. Chim. Ital.* **1989**, *119*, 511. (c) Falorni, M.; Giacomelli, G.; Marchetti, M.; Culeddu, N.; Lardicci, L. *Tetrahedron: Asymmetry* **1991**, *2*, 287.
11. (a) Mukaiyama, T.; Tomimori, K.; Oriyama, T. *Chem. Lett.* **1985**, 813. (b) Mukaiyama, T.; Tomimori, K.; Oriyama, T. *Chem. Lett.* **1985**, 1359.
12. Falorni, M.; Giacomelli, G.; Lardicci, L. *Gazz. Chim. Ital.* **1990**, *120*, 765.
13. Mukaiyama, T.; Minowa, N.; Oriyama, T.; Narasaka, K. *Chem. Lett.* **1986**, 97.
14. Minowa, N.; Mukaiyama, T. *Bull. Chem. Soc. Jpn.* **1987**, *60*, 3697.
15. Kobayashi, S.; Nishio, K. *Tetrahedron Lett.* **1995**, *36*, 6729.
16. Kobayashi, S.; Fujishita, Y.; Mukaiyama, T. *Chem. Lett.* **1989**, 2069.
17. Mukaiyama, T.; Uchiro, H.; Shiina, I.; Kobayashi, S. *Chem. Lett.* **1990**, 1019.
18. Mukaiyama, T.; Shiina, I.; Kobayashi, S. *Chem. Lett.* **1991**, 1901.
19. Mukaiyama, T.; Shiina, I.; Uchiro, H.; Kobayashi, S. *Bull. Chem. Soc. Jpn.* **1994**, *67*, 1708.
20. (a) Kobayashi, S.; Kawasuji, T. *Tetrahedron Lett.* **1994**, *35*, 3329. (b) Kobayashi, S.; Hayashi, T. *J. Org. Chem.* **1995**, *60*, 1098. (c) Kobayashi, S.; Horibe, M. *Tetrahedron: Asymm.* **1995**, *6*, 2565.
21. (a) Kobayashi, S.; Shiina, I.; Izumi, J.; Mukaiyama, T. *Chem. Lett.* **1992**, 373. (b) Mukaiyama, T.; Shiina, I.; Izumi, J.; Kobayashi, S. *Heterocycles* **1993**, *35*, 719.
22. (a) Kobayashi, S.; Horibe, M. *Synlett* **1994**, 147. (b) Kobayashi, S.; Horibe, M.; Saito, Y. *Tetrahedron* **1994**, *50*, 9629.
23. (a) Mukaiyama, T.; Shiina, I.; Kobayashi, S. *Chem. Lett.* **1990**, 2201. (b) Mukaiyama, T.; Anan, H.; Shiina, I.; Kobayashi, S. *Bull. Soc. Chim. Fr.* **1993**, *130*, 388. (c) Kobayashi, S.; Onozawa, S.; Mukaiyama, T. *Chem. Lett.* **1992**, 2419. (d) Kobayashi, S.; Kawasuji, T. *Synlett* **1993**, 911.
24. (a) Kanda, Y.; Fukuyama, T. *J. Am. Chem. Soc.* **1993**, *115*, 8451. (b) Fukuyama, T.; Kanda, Y. *Yuki Gosei Kagaku Kyokaishi* (English edition) **1994**, *52*, 888.
25. (a) Mukaiyama, T.; Shiina, I.; Sakata, K.; Emura, T.; Seto, K.; Saitoh, M. *Chem. Lett.* **1995**, 179. (b) Mukaiyama, T.; Shiina, I.; Iwadare, H.; Saitoh, M.; Nishimura, T.; Ohkawa, N.; Sakoh, H.; Nishimura, K.; Tani, Y.; Hasegawa, M.; Yamada, K.; Saitoh, K. *Chem. Eur. J.* **1999**, *5*, 121.
26. White, J. D.; Deerberg, J. *Chem. Commun.* **1997**, 1919.

Teruaki Mukaiyama & Isamu Shiina
Science University of Tokyo, Kagurazaka, Shinjuku-ku Tokyo, Japan

β-Methyl-β-propiolactone[1]

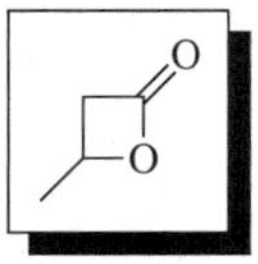

(±)
[3068-88-0] $C_4H_6O_2$ (MW 86.10)
(*R*)
[32082-74-9]
(*S*)
[65058-82-4]

(three-carbon homologating reagent for synthesis of chiral β-methyl carboxylic acids[1])

Physical Data: mp −43.5 °C; bp 71–73 °C/29 mmHg; *d* 1.056 g mL^{-1}. (*R*)-isomer: $[\alpha]_D^{20}$ −27.8° (*c* 4.14, $CHCl_3$); (*S*)-isomer: $[\alpha]_D^{22}$ +28.8° (*c* 4.30, $CHCl_3$).
Solubility: misc alcohol, acetone, ether, chloroform.
Form Supplied in: colorless oil; the racemic form is widely available.
Preparative Methods: by ring-closure of 3-bromobutyric acid with *Sodium Carbonate*,[2a] or by hydrogenation of *Diketene*.[2b] The optically active forms are obtained in the same manner starting from (*R*)- or (*S*)-3-bromobutyric acid, which may be resolved with the (*S*) form of *1-(1-Naphthyl)ethylamine*.[3] Asymmetric aldol condensation using an enantiopure iron acetyl complex followed by cyclization,[3c] or asymmetric hydrogenation of diketene catalyzed by a chiral ruthenium complex,[3d] also gives the optically active β-lactone.
Handling, Storage, and Precautions: can be stored in the refrigerator for several months without noticeable changes; cancer suspect reagent; should be handled with due care.

General Discussion. β-Methyl-β-propiolactone is particularly useful as a reactive four-carbon building block. Like *β-Propiolactone*, β-methyl-β-propiolactone undergoes a variety of ring-opening reactions in which the regiochemistry is dependent on the nature of the nucleophile (eq 1). Addition to the carbonyl carbon predominates in reactions with organolithium[4] or Grignard reagents,[5] giving α,β-unsaturated ketones (**2**) or 1,3-diols (**3**). Organocadmium reagents[4] effect C–O bond fission to give β-methyl carboxylic acids (**4**).

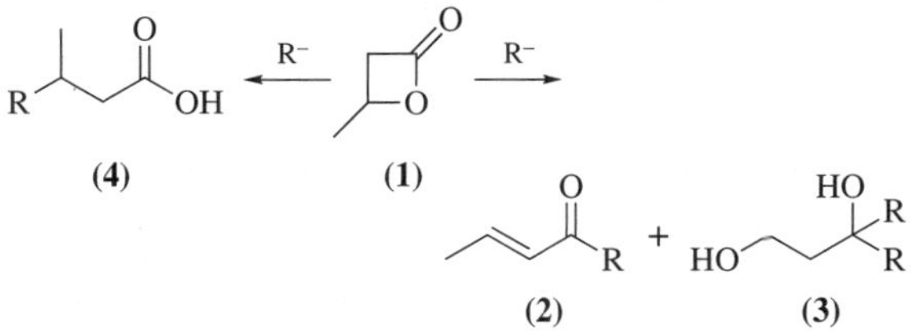

(1)

Selective β-attack is best accomplished by the use of organocopper reagents (eq 2).[6] Organocuprates prepared from 2 equiv of Grignard reagents and 1 equiv of *Copper(I) Iodide* give β-methyl carboxylic acids (**4**) in better yields than when the corresponding organolithium reagents are used. In the presence of a catalytic amount of a copper(I) salt, Grignard reagents also attack at the β-carbon to give the same products in good yields.[7]

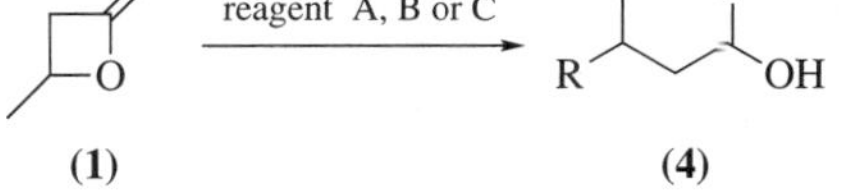

(2)

A: R_2CuMgX, THF•Me_2S
B: RMgX, cat CuX, THF
C: RCu•PBu_3, Et_2O

The *Potassium* complex of *18-Crown-6* or *Potassium Naphthalenide* effects ring-opening to give acetates or their alkylated derivatives in good yield (eq 3). Treatment of the reaction mixture obtained from β-methyl-β-propiolactone and potassium–18-crown-6 with hydrochloric acid or alkyl halides gives the acetate (**5**) or its alkylated derivative (**6**), respectively.[8] The α,β-unsaturated carboxylic acid (**7**) or its ester (**8**) is formed by the action of the potassium naphthalenide–18-crown-6 complex.[9]

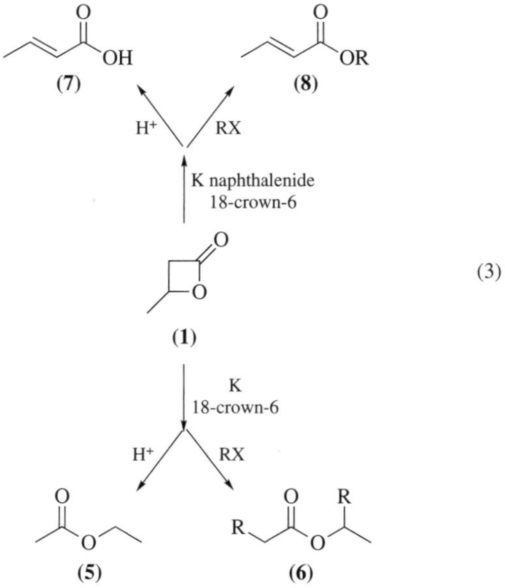

(3)

β-Methyl-β-propiolactone is useful as a four-carbon building block for terpenoid synthesis (eq 4). Citronellic acid (**9**) is prepared by reaction with the homoprenyl Grignard reagent; pulegone (**10**), citronellol (**11**), geraniol, and nerol (**12**) can be obtained by further functional group manipulations.[10]

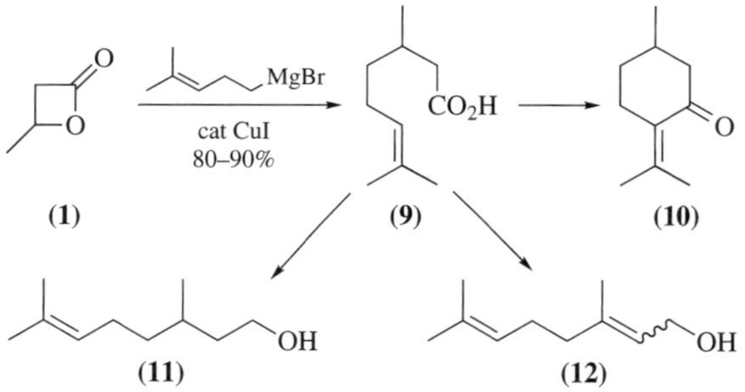

(4)

Optically active (*R*)- and (*S*)-β-methyl-β-propiolactone serve as versatile reagents for the synthesis of 3-sulfinylbutyric acid[11] and for various natural products in optically active form. (*S*)-*ar*-Turmerone (**13**) is obtained by reaction of the (*R*)-enantiomer with a di-*p*-tolylcopper reagent followed by functional group manipulation (eq 5).[3a] (*R*,*R*)-Phytol (**15**) is prepared by ring-opening with the Grignard reagent (**14**) followed by the chain-elongation reaction (eq 6).[12] (*R*,*Z*)-Trogodermal (**18**), an insect pheromone of *Trogoderma inclusum*, is synthesized by using the *cis*-alkenylcopper reagent (**17**) (eq 7).[13] The enantiomeric purity of the final natural products is 83–84% ee. Comparison of this value with that of the starting (*S*)-(+)-3-bromobutyric acid (90% ee) indicates that these magnesiocuprate coupling reactions occur with ≥92% inversion of configuration at the β-position of the (*R*)-(+)-β-methyl-β-propiolactone.[3a]

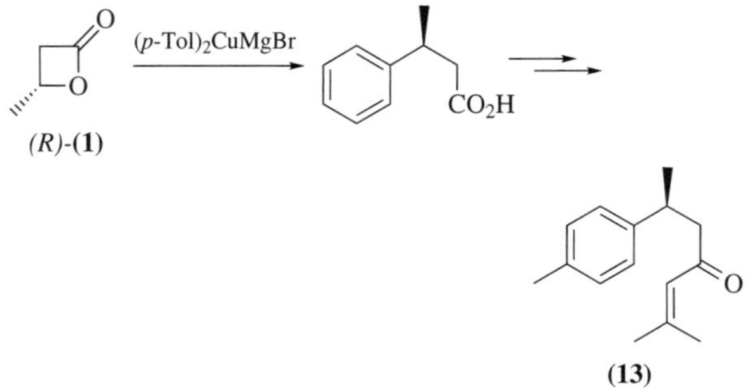

(5)

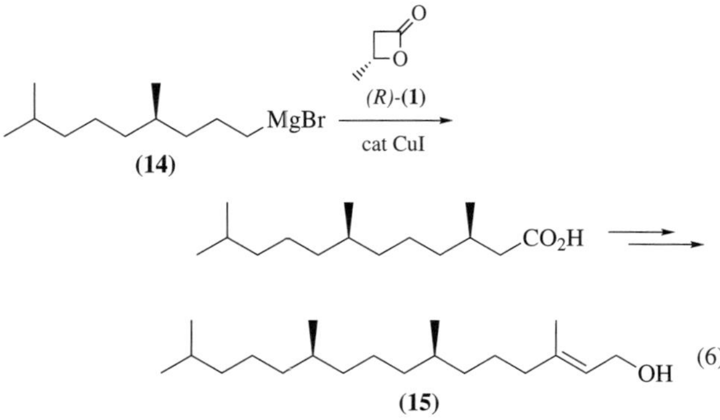

(6)

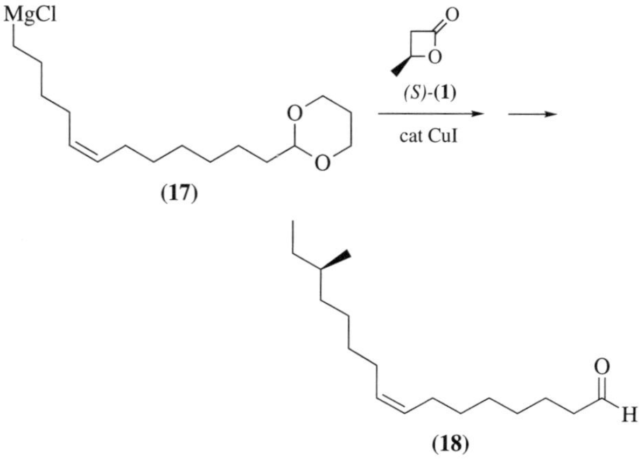

(7)

1,3-Dimethyl-2,9-dioxabicyclo[3.3.1]nonane (**19**), an insect pheromone, has been synthesized starting from (*R*)-(**1**) via ring-opening at the acyl carbon followed by conjugate addition to the resulting α,β-unsaturated ketone (eq 8).[14]

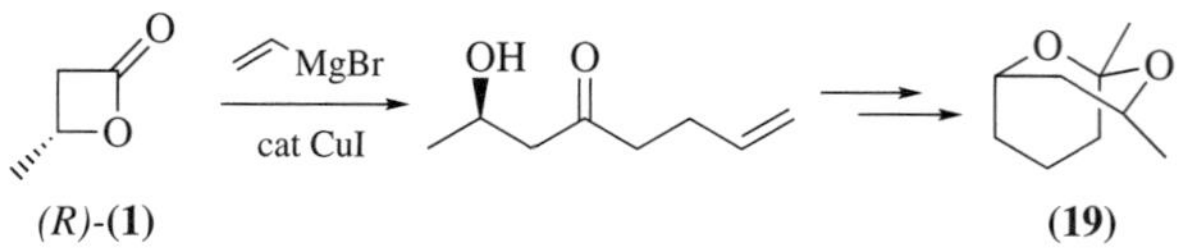

(8)

Related Reagents. *N*-Benzyloxycarbonyl-L-serine β-Lactone; α,β-Butenolide; γ-Butyrolactone; Dihydro-5-(hydroxymethyl)-2(3*H*)-furanone; β-Ethynyl-β-propiolactone; β-Propiolactone.

1. (a) Zaugg, H. E. *Org. React.* **1954**, *8*, 305. (b) Fujisawa, T.; Sato, T. *J. Synth. Org. Chem. Jpn.* **1982**, *40*, 618 (*Chem. Abstr.* **1982**, *97*, 198 011y). (c) Pommier, A.; Pons, J. M. *Synthesis* **1993**, 441.
2. (a) Agostini, D. E.; Lando, J. B.; Shelton, J. R. *J. Polym. Sci. A-1* **1971**, *9*, 2775. (b) Sixt, J. U.S. Patent 2 763 664, **1956**, (*Chem. Abstr.* 1957, *51*, 5115b).

3. (a) Sato, T.; Kawara, T.; Nishizawa, A.; Fujisawa, T. *Tetrahedron Lett.* **1980**, *21*, 3377. (b) Sato, T.; Naruse, K.; Fujisawa, T. *Tetrahedron Lett.* **1982**, *23*, 3587. (c) Davies, S. G. *Aldrichim. Acta* **1990**, *23*, 31. (d) Ohta, T.; Miyake, T.; Takaya, H. *Chem. Commun./J. Chem. Soc., Chem. Commun.* **1992**, 1725.
4. Stuckwisch, C. G.; Bailey, J. V. *J. Org. Chem.* **1963**, *28*, 2362.
5. (a) Gresham, T. L.; Jansen, J. E.; Shaver, F. W.; Bankert, R. A. *J. Am. Chem. Soc.* **1949**, *71*, 2807. (b) Kutner, A.; Perlman, K. L.; Lago, A.; Sicinski, R. R.; Schnoes, H. K.; DeLuca, H. F. *J. Org. Chem.* **1988**, *53*, 3450.
6. Fujisawa, T.; Sato, T.; Kawara, T.; Kawashima, M.; Shimizu, H.; Ito, Y. *Tetrahedron Lett.* **1980**, *21*, 2181. Kawashima, M.; Sato, T.; Fujisawa, T. *Tetrahedron* **1989**, *45*, 403.
7. Sato, T.; Kawara, T.; Kawashima, M.; Fujisawa, T. *Chem. Lett.* **1980**, 571. Normant, J. F.; Alexakis, A.; Cahiez, G. *Tetrahedron Lett.* **1980**, *21*, 935.
8. Jedlinski, Z.; Kowalczuk, M.; Misiolek, A. *Chem. Commun./J. Chem. Soc., Chem. Commun.* **1988**, 1261.
9. Kowalczuk, M.; Kurcok, P.; Glowkowski, W.; Jedlinski, Z. *J. Org. Chem.* **1992**, *57*, 389.
10. Fujisawa, T.; Sato, T.; Kawara, T.; Noda, A.; Obinata, T. *Tetrahedron Lett.* **1980**, *21*, 2553.
11. Breitschuh, R.; Seebach, D. *Synthesis* **1992**, *25*, 1170.
12. Fujisawa, T.; Sato, T.; Kawara, T.; Ohashi, K. *Tetrahedron Lett.* **1981**, *22*, 4823.
13. Sato, T.; Naruse, K.; Fujisawa, T. *Tetrahedron Lett.* **1982**, *23*, 3587.
14. Sato, T.; Itoh, T.; Hattori, C.; Fujisawa, T. *Chem. Lett.* **1983**, 1391.

Tamotsu Fujisawa & Makoto Shimizu
Mie University, Japan

(S)-(−)-4-(2-Methylpropyl)-2-(2-pyridyl)-2-oxazoline

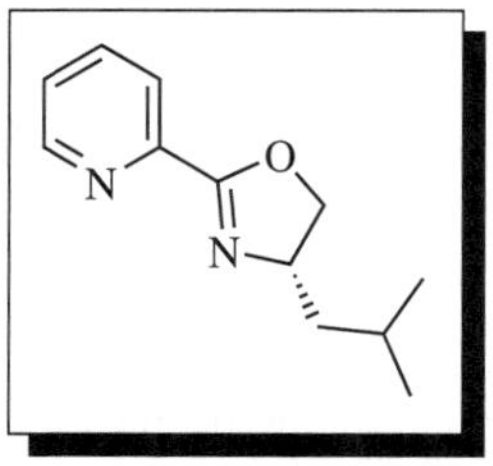

[108915-07-7] $C_{12}H_{16}N_2O$ (MW 204.27)

(chiral ligand for enantiocontrol of metal-catalyzed reactions)

Physical Data: $[\alpha]^{20}_D$ −93.4 (*c* 20.0, PhMe), bp 160–170 °C/0.1 Torr.
Solubility: soluble in aromatic (PhH, PhMe) and chlorinated hydrocarbon solvents.
Form Supplied in: oil.
Preparative Methods: reaction of pyridine carboximidate with (*S*)-leucinol in MeOH at 60 °C generates (*S*)-(−)-4-(2-methylpropyl)-2-(2-pyridyl)-2-oxazoline in 68% yield (eq 1).[1]

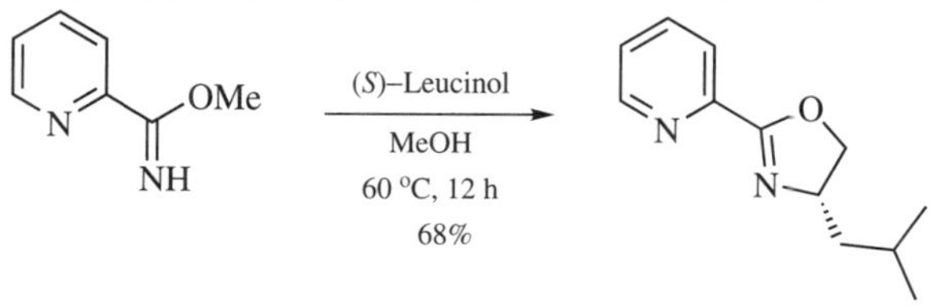

(1)

Purification: can be purified by Kugelrhor distillation.
Handling, Storage, and Precautions: stable at ambient temperature.

Asymmetric Reactions

Monophenylation of cis-Diols. Monophenylation of *cis*-diols derived from cyclopentane and cyclohexane using triphenylbismuthdiacetate, (*S*)-(−)-4-(2-methylpropyl)-2-(2-pyridyl)-2-oxazoline as the ligand (*L**), and $Cu(OAc)_2$ as a co-catalyst affords the products in moderate yields (38–54%) and enantioselectivities (13–44% ee) (eq 2).[2]

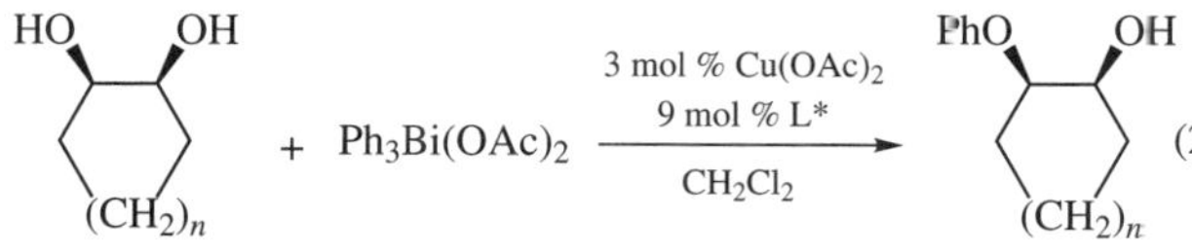

(2)

n = 1 54% yield
44% ee
n = 2 38% yield
13% ee

Epoxidation. Epoxidation of olefins with sodium periodate, using catalytic amounts of ruthenium(III) and employing (*S*)-(−)-4-(2-methylpropyl)-2-(2-pyridyl)-2-oxazoline as the chiral ligand (*L**) for the metal, afforded products in 44–50% yield and enantioselectivities of 11–15% (eqs 3 and 4).[3]

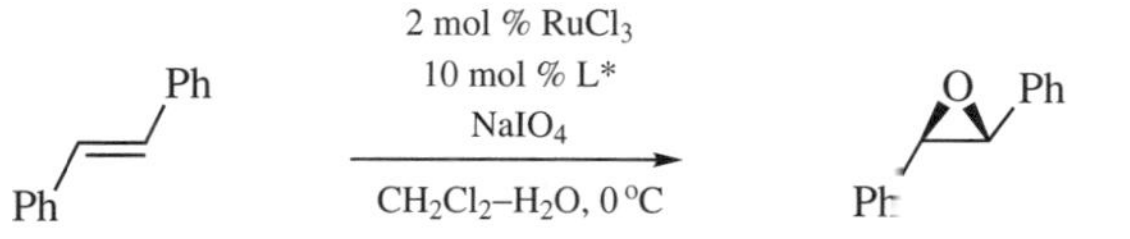

(3)

50% yield
15% ee

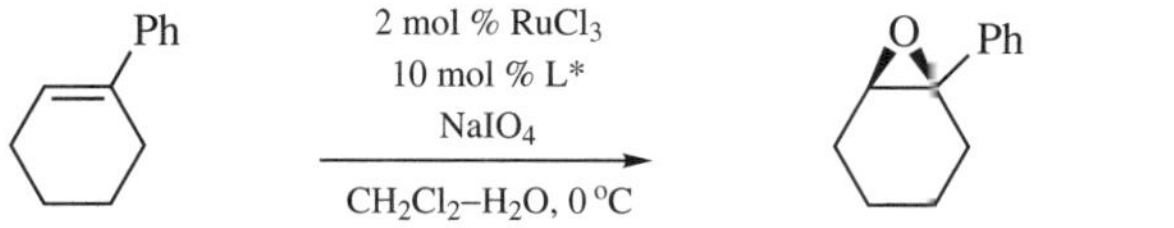

(4)

44% yield
11% ee

Crotylsilane Addition. Stoichiometric amounts of (*S*)-(−)-4-(2-methylpropyl)-2-(2-pyridyl)-2-oxazoline serve as the asymmetric Lewis base directing the addition of crotyltrichlorosilane to aryl aldehydes.[4] Of the various pyridyl oxazolines screened, this ligand yielded the highest optical purity and conversion of benzaldehyde to product (eq 5). Other homoallylic aryl alcohols were prepared in 61–91% yield using this ligand in ee's ranging from 36–74% (Table 1; with permission of the Royal Society of Chemistry, this table was reproduced from ref. 4).

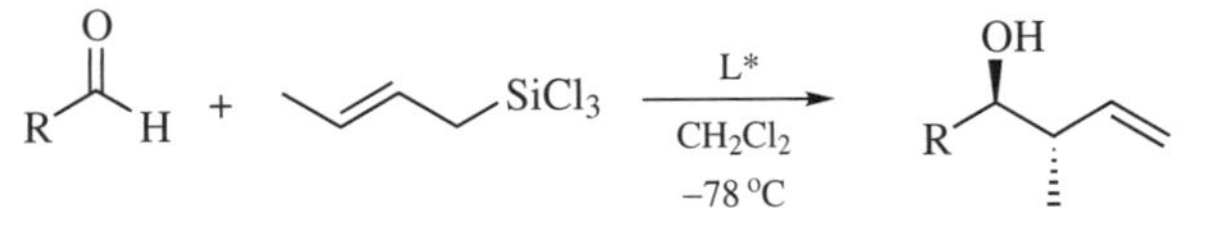

(5)

61–91% yield
36–74% ee

Table 1 Homoallylic alcohols from aryl aldehydes

R	Yield (%)	ee (%)
Ph	72	74
4-MePh	70	72
4-MeOPh	79	46
4-O_2NPh	66	36
4-FPh	61	74
PhCH=CH	91	60

Allylic Oxidation. The Kharasch-Sosnovsky reaction[5] involves oxidation of the allylic position while the olefin remains intact. In the presence of catalytic copper (II) salts, treatment of olefins with peresters affords acylated allylic alcohols. When (*S*)-(−)-4-(2-methylpropyl)-2-(2-pyridyl)-2-oxazoline was involved, (*R*)-cyclohexenyl benzoate was isolated in 57% yield and 28% ee (eq 6).[6]

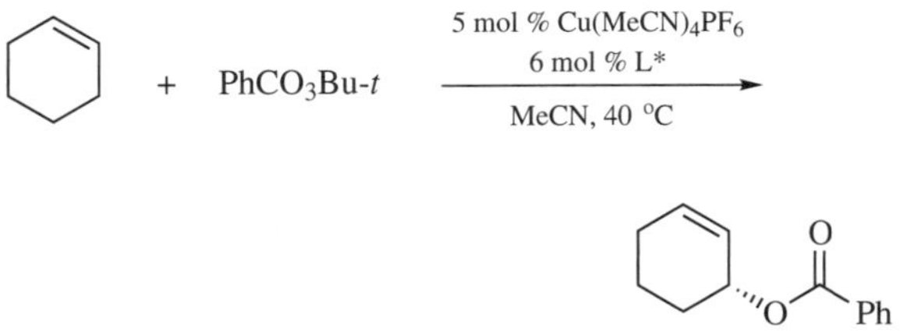

(6)

57% yield
28% ee

Hydrosilation of Ketones. (*S*)-(−)-4-(2-Methylpropyl)-2-(2-pyridyl)-2-oxazoline provides a high degree of stereocontrol in the hydrosilation of acetophenone catalyzed by rhodium (eq 7).[1] Analysis of 1-phenylethanol, isolated in 89% yield after hydrolysis, revealed 71% ee. Cationic cobalt catalysts have been used to facilitate the same transformation with this ligand generating product in 76% yield and 14% ee.[7]

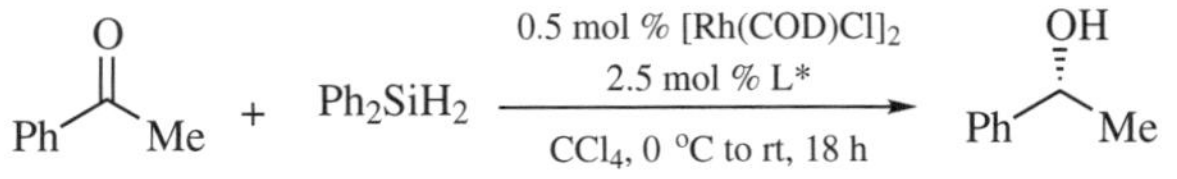

(7)

89% yield
71% ee

Cyclization/Hydrosilation. Recently, (*S*)-(−)-4-(2-methylpropyl)-2-(2-pyridyl)-2-oxazoline has been applied to the cyclization/hydrosilation process catalyzed by palladium (eq 8).[8] The precatalyst (L*-Pd(Me)Cl), purified and characterized after preparation by ligand exchange with (COD)Pd(Me)Cl, afforded product in >95% de and 77% ee.

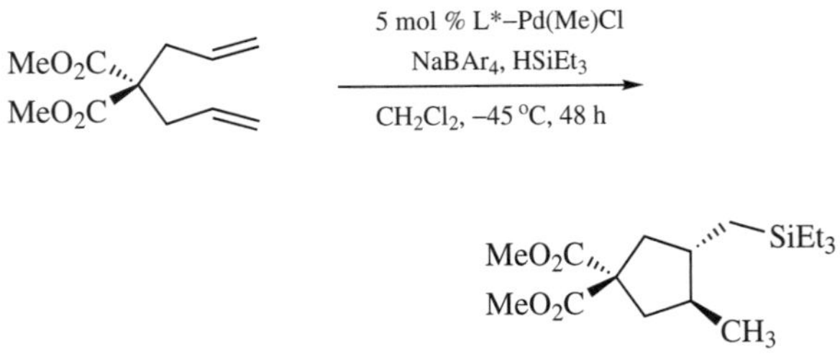

(8)

84% yield
>95% de
77% ee

1. Brunner, H.; Obermann, U. *Chem. Ber.* **1989**, *122*, 499.
2. Brunner, H.; Obermann, U.; Wimmer, P. *Organometallics* **1989**, *8*, 821.
3. Yang, R. Y.; Dai, L. X. *J. Mol. Catal.* **1994**, *87*, L1.
4. Angell, R. M.; Barrett, A. G. M.; Braddock, D. C.; Swallow, S.; Vickery, B. D. *Chem. Comm.* **1997**, *10*, 919.
5. Rawlinson, D. J.; Sosnovsky, G. *Synthesis* **1972**, 1.
6. Clark, J. S.; Tolhurst, K. F.; Taylor, M.; Swallow, S. *J. Chem. Soc., Perkin Trans. 1* **1998**, *7*, 1167.
7. Brunner, H.; Amberger, K. *J. Organomet. Chem.* **1991**, *417*, C63.
8. Perch, N. S.; Pei, T.; Widenhowfer, R. A. *J. Org. Chem.* **2000**, *65*, 3836.

Jeremy T. Cooper
Eli Lilly and Co., Indianapolis, IN, USA

(+)-(*S*)-*N*-Methylsulfonylphenylalanyl Chloride

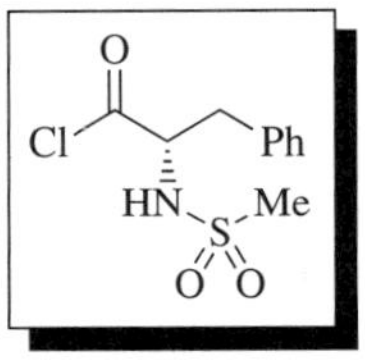

[63640-54-0] $C_{10}H_{12}ClNO_3S$ (MW 261.75)

(chiral reagent for the resolution of racemic alcohols via separation of the corresponding diastereomeric esters[1])

Physical Data: mp 84–85 °C; $[\alpha]_D^{20}$ +4.3° (*c* 1.6, THF).
Solubility: readily sol THF, benzene, ether.
Form Supplied in: pale yellow needles; not available commercially.
Preparative Methods: prepared from (*S*)-phenylalanine by reaction with *Methanesulfonyl Chloride*, followed by *Phosphorus(V) Chloride*.[1b]
Purification: the crude compound can be recrystallized from hexane/ether.
Handling, Storage, and Precautions: best if prepared immediately prior to use. Can be stored at 0 °C under nitrogen for several days without appreciable decomposition.

Introduction. Enantiomerically pure alcohols can frequently be obtained by physical separation of the diastereomeric esters prepared from the racemic alcohols and chiral acids.[2] Chiral acids that have been successfully employed for this purpose and are available in either enantiomeric form include *ω*-camphanic acid[3] and the monomethyl ester of diacetyltartaric acid.[4] No reagent can be considered generally applicable to all alcohols, since the ease of separation of the diastereomeric esters frequently depends on their crystallinity and/or chromatographic properties. A successful resolution is frequently the result of multiple trials and errors with a variety of acids.

***N*-Sulfonylated α-Amino Acids.** *N*-Protected derivatives of the natural α-amino acids offer a wide range of potential derivatizing agents.[5] Particularly useful are *N*-arylsulfonyl-α-amino acids,[6] many of which are commercially available and produce crystalline ester mixtures from which pure diastereomers can often be isolated by recrystallization. *N*-Methylsulfonyl-α-amino acids or the corresponding acid chlorides are generally not commercially available, but in some cases have been shown to be superior to the corresponding *N*-tosyl derivatives (eq 1).[1b,7]

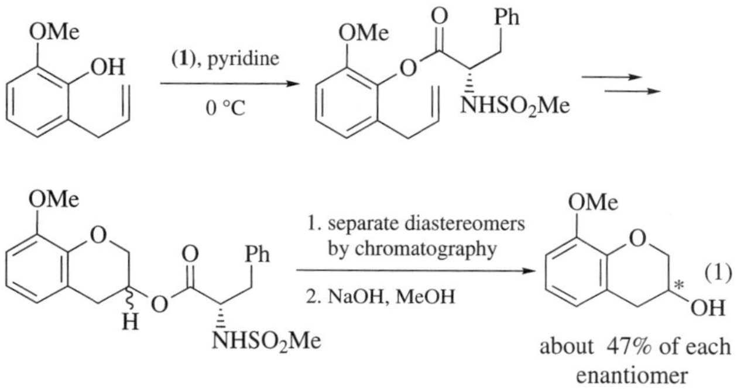

N-Methylsulfonylphenylalanyl chloride (**1**) is particularly useful in the derivatization of *meso*-diols. Mixtures of diastereomeric monoesters can be obtained, from which pure diastereomers are usually isolated by fractional recrystallization and/or chromatography. Chemical transformation of the free hydroxy group, followed by removal of the chiral auxiliary, allows the selective transformation of each prochiral hydroxy group. Isolation of the other diastereomeric ester from the mother liquors, followed by a series of protection–deprotection steps, provides the flexibility of converting 100% of the *meso*-material into one single enantiomer of the product. Alternatively, by rearranging the order of the chemical transformations, both enantiomers of the product can be obtained (eqs 2 and 3).[1,8]

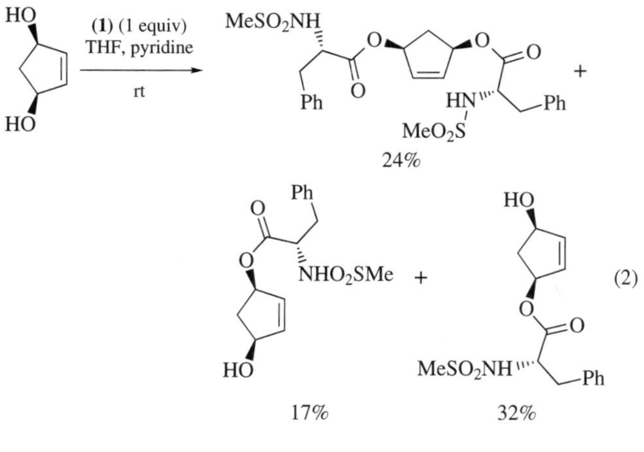

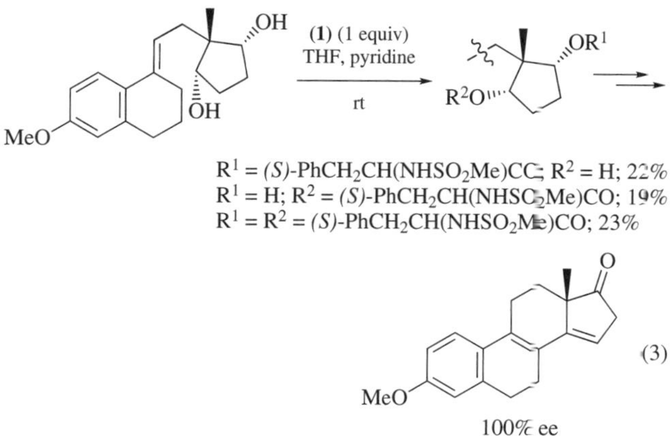

1. (a) Terashima, S.; Yamada, S. *Tetrahedron Lett.* **1977**, 1001. (b) Nara, M.; Terashima, S.; Yamada, S. *Tetrahedron* **1980**, *36*, 3161.
2. *Enantiomers, Racemates and Resolutions*; Jacques, J.; Collet, A.; Wilen, S. H., Eds.; Wiley: New York, 1981; pp 332–335.
3. Wilen, S. H.; Collet, A.; Jacques, J. *Tetrahedron* **1977**, *33*, 2725.
4. Hübner, M.; Ponsold, K.; Siemann, H. J.; Schwartz, S. *Z. Chem.* **1968**, *8*, 380.
5. Hashimoto, S.; Kase, S.; Shinoda, T.; Ikegami, S. *Chem. Lett.* **1989**, 1063.
6. (a) Jermyn, M. A. *Aust. J. Chem.* **1967**, *20*, 2283. (b) Halpern, B.; Westley, J. W. *Aust. J. Chem.* **1966**, *19*, 1533.
7. Kawamura, K.; Ohta, T.; Otani, G. *Chem. Pharm. Bull.* **1990**, *38*, 2088.
8. Nara, M.; Terashima, S.; Yamada, S. *Tetrahedron* **1980**, *36*, 3171.

Juan C. Jaen
Parke-Davis Pharmaceutical Research, Ann Arbor, MI, USA

α-Methyltoluene-2,α-sultam[1]

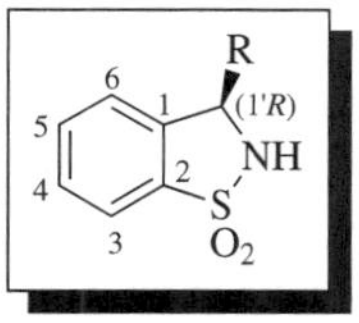

(1′*S*)-(**1a**; R = Me)		
[130973-57-8]	$C_8H_9NO_2S$	(MW 183.25)
(1′*R*)-(**1a**)		
[130973-53-4]		
(1′*S*)-(**1b**; R = *t*-Bu)		
[137694-01-0]	$C_{11}H_{15}NO_2S$	(MW 225.34)
(1′*R*)-(**1b**)		
[137694-00-9]		

(chiral auxiliary: *N*-enoyl derivatives undergo highly stereoselective Diels–Alder reactions with cyclopentadiene[2] and 1,3-dipolar cycloadditions with nitrile oxides;[3] enolates of *N*-acyl derivatives participate in highly stereoselective alkylations, acylations, and aldolizations[4])

Physical Data: (**1a**) mp 92 °C. (1′*S*)-(**1a**) $[\alpha]_D^{20}$ −30.0° (*c* 1.21, $CHCl_3$). (1′*R*)-(**1a**) $[\alpha]_D^{20}$ +31.0° (*c* 0.6, EtOH). (**1b**) mp

129–130 °C. (1′*S*)-(**1b**) $[\alpha]_D^{20}$ −53.9° (*c* 1.00, $CHCl_3$). (**1b**) has been incorrectly assigned.[3]

Preparative Methods: both enantiomers of the α-methyl sultam may be prepared on a multigram scale in optically pure form by asymmetric hydrogenation of imine (**2a**) followed by simple crystallization (eq 1).[5] The (*R*)-enantiomer of the α-*t*-butyl sultam may also be prepared in enantiomerically pure form by asymmetric reduction of imine (**2b**) followed by fractional crystallization.[3] However, multigram quantities of either enantiomer of the α-*t*-butyl sultam may be prepared by derivatization of the racemic auxiliary (obtained in 98% yield from reaction of (**2b**) with *Sodium Borohydride* in MeOH) with *10-Camphorsulfonyl Chloride*, separation of the resulting diastereomers by fractional crystallization, and acidolysis.[3] Prochiral imines (**2a**) and (**2b**) are readily prepared from inexpensive Saccharine by treatment with *Methyllithium* (73%) and *t-Butyllithium* (66%), respectively.

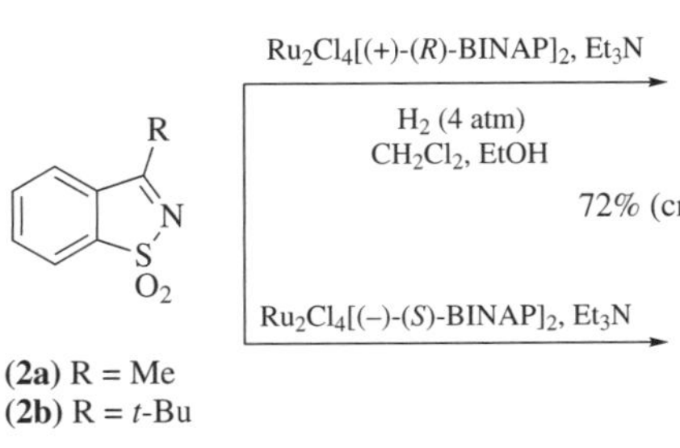

(**2a**) R = Me
(**2b**) R = *t*-Bu

72% (cryst) [>99% ee (crude)]

(1)

71% (cryst) [>99% ee (crude)]

Handling, Storage, and Precautions: these auxiliaries are white crystalline solids which are stable indefinitely at ambient temperature in sealed containers.

Introduction. The toluene-2,α-sultams are recently introduced relatives of the well established *10,2-Camphorsultam* chiral auxiliary and have been designed to provide similar high levels of face discrimination in reactions of pendent prochiral functionality. Features that distinguish them include high crystallinity and facile NMR and HPLC analysis of derivatives, favorable acylation and aldolization characteristics of derived *N*-acyl enolates, and improved cleavage characteristics.

Preparation of Derivatives. *N*-Enoyl[2,3] and *N*-acyl[4] sultam derivatives are readily prepared using either *Sodium Hydride*–acid chloride or *Triethylamine*–acid chloride single-step protocols. Various alternative derivatization procedures that work for the 10,2-camphorsultam auxiliary would also be expected to be effective.

Reactions of *N*-Enoyl and *N*-Acyl Derivatives.

[4 + 2] Diels–Alder Cycloadditions (Alkene → Six-Membered Cycloadduct).[2] *N*-Acryloyl-α-methyltoluene-2,α-sultam (**3a**) participates in highly *endo* and C(α)-*re* π-face selective Lewis acid promoted Diels–Alder reactions with *Cyclopentadiene*, *1,3-Butadiene*, and *Isoprene* (eq 2 and Table 1). These levels of induction compare favorably with most alternative auxiliaries, including the 10,2-camphorsultam. However, *N*-crotonyl-α-methyltoluene-2,α-sultam (*ent*-**3b**) reacts with cyclopentadiene with only moderate π-face selectivity (cf. 93% de with 10,2-camphorsultam). Unusually high *endo* selectivity is observed for the non-Lewis acid-catalyzed reaction of sultam (**3a**) with cyclopentadiene, but again the π-face selectivity is only moderate. The corresponding reactions of both α-*t*-butyl- and α-benzyltoluene-2,α-sultams are less selective.

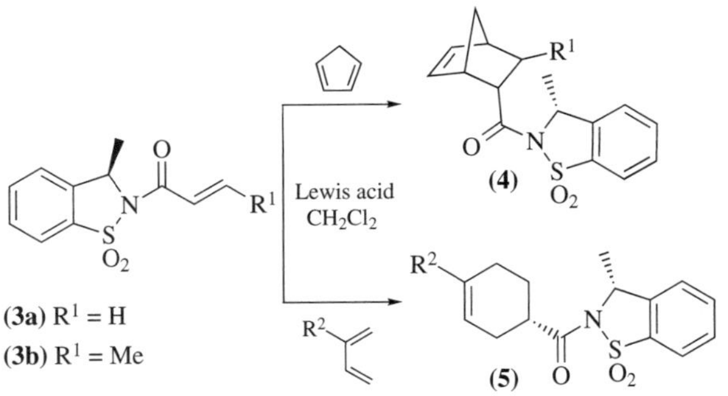

(**3a**) R^1 = H
(**3b**) R^1 = Me

(**4**)

(**5**)

(2)

1,3-Dipolar Cycloadditions with Nitrile Oxides (Alkene → Isoxazoline).[3] 1,3-Dipolar cycloaddition reactions of *N*-acryloyl-α-*t*-butyltoluene-2,α-sultam (**6**) with various nitrile oxides give isoxazolines with extremely high C(α)-*re* π-facial control (eq 3). The levels of selectivity exceed those obtainable with the 10,2-camphorsultam auxiliary and are comparable to the highest levels reported for such cycloadditions.[6] The corresponding reactions of α-methyltoluene-2,α-sultams are less selective.

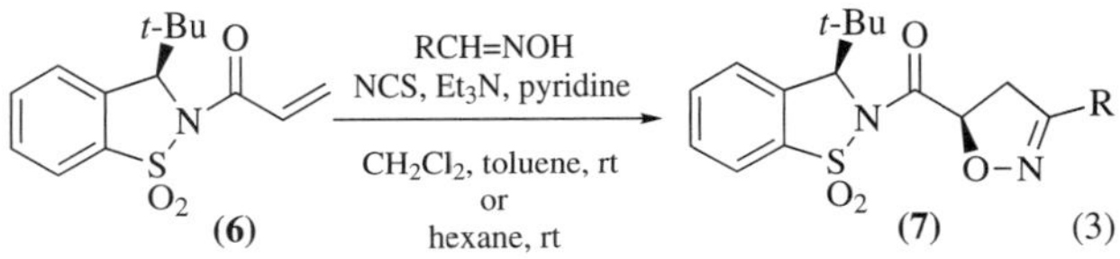

(**6**)

(**7**)

(3)

R = Me*, Et*, *t*-Bu, Ph, CH_2O-*t*-Bu
77–88% (cryst) [90–96% de (crude)]
*using *ent*-(**6**) → *ent*-(**7**)

Acylation, Alkylation, and Aldolization (Acyl Species → α-, β-, or α/β-Functionalized Acyl Product).[3] Alkylation reactions of sodium enolates of various *N*-acyl-α-methyltoluene-2,α-sultams with selected (both "activated" and "nonactivated") alkyl iodides and bromides proceed with good C(α)-*re* stereocontrol (90–99% de). Analogous acylations with various acid chlorides can also be performed, giving β-keto products (97–99% de). Selective reduction of these latter products with *Zinc Borohydride* (chelate controlled, 82.6–98.2% de) or N-Selectride (nonchelate controlled, 95.8–99.6% de) can provide *syn*- and *anti*-aldol derivatives, respectively.[3]

Syn-aldol derivatives may also be obtained directly from boryl enolates of the same *N*-acyl-α-methyltoluene-2,α-sultams by condensation with aliphatic and aromatic aldehydes (eq 4).[3,7] The high C(α)-*si* topicity of these reactions parallels but exceeds that when using the 10,2-camphorsultam auxiliary and is the result of an analogous transition state.[3] It is noteworthy, however, that aldolizations of α-methyltoluene-2,α-sultam derivatives generally proceed to completion with just a small excess of aldehyde (1–1.2 equiv, cf. 2–3 equiv when 10,2-camphorsultam mediated). This may be ascribed to the lack of acidic protons α to the SO_2 group in the Saccharine-derived auxiliary.

Table 1 Intermolecular Diels–Alder Reactions of *N*-Enoyl Sultams (**3a**) or (**3b**) → (**4**) and (**3a**) → (**5**) (eq 2)

Dienophile	Diene	Lewis acid[a]	Temp (°C)	Time (h)	Adduct	Yield crude (cryst.) (%)	de crude (cryst.) (%)
(**3a**)	Cyclopentadiene	None	25		(**4**) (R^1 = H)	95[b]	62
(**3a**)	Cyclopentadiene	Me_2AlCl	−98	0.2	(**4**) (R^1 = H)	97[c] (83)	93 (>99)
(**3a**)	1,3-Butadiene	$EtAlCl_2$	−78	18	(**5**) (R^2 = H)	79	90
(**3a**)	Isoprene	Me_2AlCl	−78	7	(**5**) (R^2 = Me)	87	92
ent-(**3b**)	Cyclopentadiene	Me_2AlCl	−78	24	*ent*-(**4**) (R^1 = Me)	74[d] (58)	59 (>99)

[a] 1.6–2.0 equiv. [b] 96% *endo*. [c] >99% *endo*. [d] 97% *endo*.

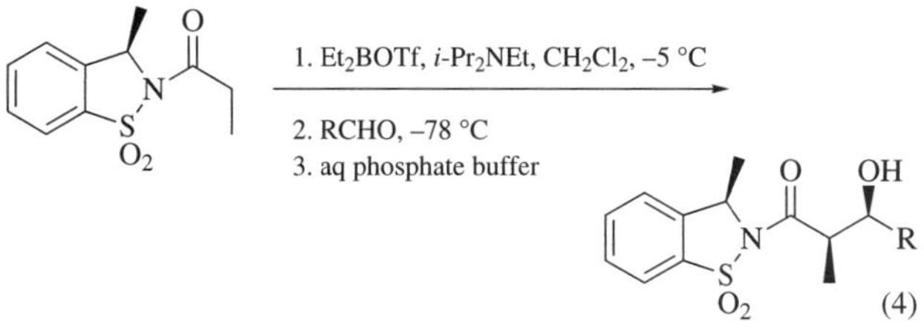

R = Me, *i*-Pr, *i*-Bu, Ph
71–84% (cryst); [R = *i*-Pr, 95% (FC)]
(>99% of crude)

Nondestructive Auxiliary Cleavage. The toluene-2,α-sultam auxiliaries are even more readily cleaved from derivatives than the 10,2-camphorsultam auxiliary. Following *N*-acyl bond cleavage, simple extraction and crystallization usually effect almost quantitative recovery of enantiomerically pure auxiliary which may be re-used if desired.

Enantiomerically pure carboxylic acids are routinely obtained from *N*-acylsultams by *Hydrogen Peroxide* assisted saponification with *Lithium Hydroxide* in aqueous THF.[2,4] Alternatively, transesterification can be effected under 'neutral' conditions in allyl alcohol containing *Titanium Tetraisopropoxide*, giving the corresponding allyl esters which can be isomerized/hydrolyzed with Wilkinson's catalyst (*Chlorotris(triphenylphosphine)rhodium(I)*) in $EtOH–H_2O$. This provides a convenient route to carboxylic acids containing base-sensitive functionality.[8] Primary alcohols are obtained by treatment with L-Selectride (*Lithium Tri-s-butylborohydride*) in THF at ambient temperature.[3]

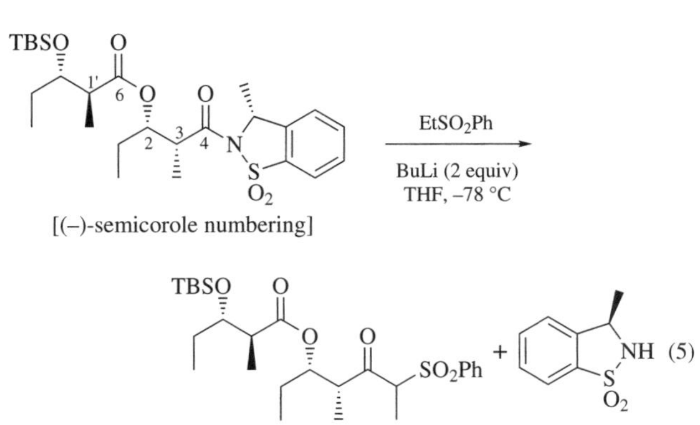

(**8**) 72% (FC) 89% (FC)

The α-methyltoluene-2,α-sultam auxiliary is also displaced by a variety of dilithiated alkyl phenyl sulfones.[7,9] This unique procedure provides direct access to synthetically useful β-oxo sulfones which may be further functionalized or simply subjected to reductive desulfonation to give alkyl ketones. A particularly striking use of this method is the preparation of β-oxo sulfone (**8**), a key intermediate in a concise synthesis of (−)-probably should be semicorrole (eq 5).[7] Remarkably, the $MeCLi_2SO_2Ph$ reagent attacks selectively the C(4)-imide C=O group in preference to the C(6)-ester C=O group and no epimerization occurs at C(3) or C(1′).

Related Reagents. 10,2-Camphorsultam; 10-Dicyclohexylsulfonamidoisoborneol; 2-Hydroxy-1,2,2-triphenylethyl Acetate; (4*S*,5*S*)-4-Methoxymethyl-2-methyl-5-phenyl-2-oxazoline; (*S*)-4-Benzyl-2-oxazolidinone.

1. Ganem, B. *Chemtracts–Org. Chem.* **1990**, 435.
2. (a) Oppolzer, W.; Wills, M.; Kelly, M. J.; Signer, M.; Blagg, J. *Tetrahedron Lett.* **1990**, *31*, 5015. (b) Oppolzer, W.; Seletsky, B. M.; Bernardinelli, G. *Tetrahedron Lett.* **1994**, *35*, 3509.
3. Oppolzer, W.; Kingma, A. J.; Pillai, S. K. *Tetrahedron Lett.* **1991**, *32*, 4893.
4. Oppolzer, W.; Rodriguez, I.; Starkemann, C.; Walther, E. *Tetrahedron Lett.* **1990**, *31*, 5019.
5. Oppolzer, W.; Wills, M.; Starkemann, C.; Bernardinelli, G. *Tetrahedron Lett.* **1990**, *31*, 4117.
6. Curran, D. P.; Jeong, K. S.; Heffner, T. A.; Rebek, J., Jr. *J. Am. Chem. Soc.* **1989**, *111*, 9238.
7. Oppolzer, W.; Rodriguez, I. *Helv. Chim. Acta* **1993**, *76*, 1275.
8. Oppolzer, W. Lienard, P. *Helv. Chim. Acta* **1992**, *75*, 2572.
9. Oppolzer, W.; Rodriguez, I. *Helv. Chim. Acta* **1993**, *76*, 1282.

Alan C. Spivey
University of Cambridge, UK

(*R*)-(+)-Methyl *p*-Tolyl Sulfoxide

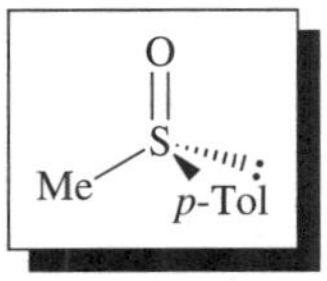

[1519-39-7] $C_8H_{10}OS$ (MW 154.25)

(agent used in the synthesis of chiral β-keto sulfoxides[1b,10])

Physical Data: $[\alpha]_D$ +146° (acetone, c = 2), +192° ($CHCl_3$, c = 1.2).

Preparative Methods: the most popular means of preparing this reagent is the nucleophilic displacement of *(−)-(1R,2S,5R)-Menthyl (S)-p-Toluenesulfinate* with methyl Grignardwith complete inversion of configuration at sulfur (eq 1).[1]

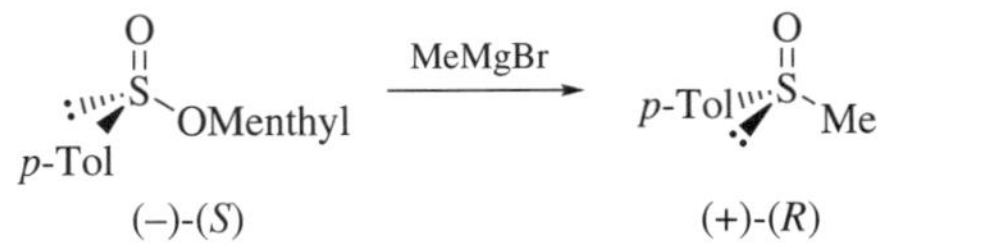

(1)

This reagent is also prepared by the reaction of *Methylmagnesium Bromide* with optically active (*S*)-*N*-sulfinyloxazolidinone, which is obtained by asymmetric synthesis[2] from the oxazolidinone derived from (4*R*,5*S*)-norephedrine with a low diastereoselectivity (70:30) (eq 2).

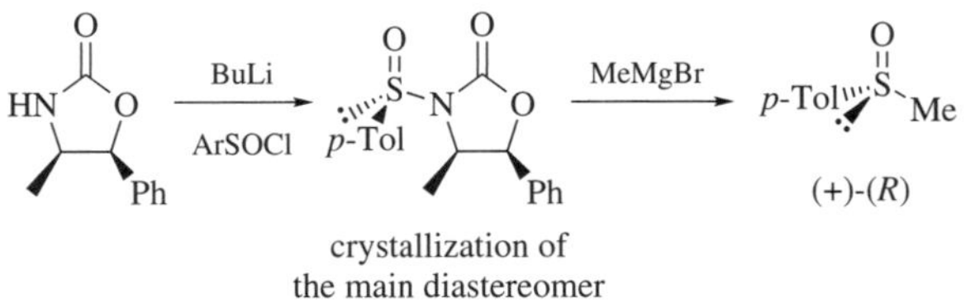

(2)

Both enantiomers of methyl *p*-tolyl sulfoxide are also prepared from diacetyl D-glucose giving, with mesyl chloride, and according to the base used, the (*S*)-methyl sulfinate with diisopropylethylamine or the (*R*)-methyl sulfinate with pyridine, which are then transformed with *p*-tolylmagnesium bromide into the corresponding (*S*)- or (*R*)-methyl *p*-tolyl sulfoxide (eq 3).[3]

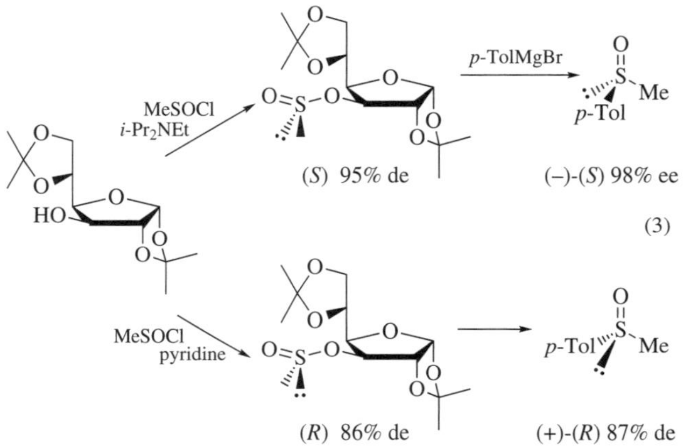

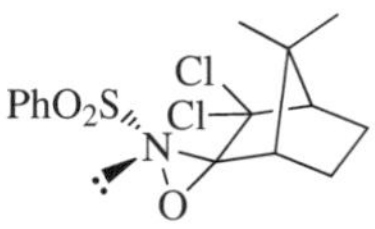

(1)

(*R*)-(+)-methyl *p*-tolyl sulfoxide is obtained by asymmetric oxidation of the corresponding sulfide with *t-Butyl Hydroperoxide* in the presence of a stoichiometric amount of a modified Sharpless reagent (*Titanium Tetraisopropoxide*)–(+)-(*R*,*R*)-diethyl tartrate–H_2O in a ratio of 1:2:1) in 96% ee.[4](−)-α,α-Dichlorocamphorsulfonylo xaziridine (**1**) was shown to be a highly efficient reagent for the asymmetric oxidation of methyl *p*-tolyl sulfide, giving the corresponding (+)-(*R*)-sulfoxide in 95% ee.[5]

It was shown recently that chloroperoxidase-catalyzed oxidation of methyl *p*-tolyl sulfide, using *Hydrogen Peroxide* or *t*-BuOOH as the stoichiometric oxidant, afforded the corresponding (+)-(*R*)-sulfoxide in 99% ee.[6]

Synthesis of β-Keto Sulfoxides. Optically active β-keto sulfoxides are very useful building blocks (eq 4) because they can be stereoselectively reduced to afford either diastereomer of the corresponding β-hydroxy sulfoxide under appropriate conditions (*Diisobutylaluminum Hydride* or *Zinc Chloride*/DIBAL)[8] and thus give access to a wide variety of compounds: chiral carbinols[7] by desulfurization with *Raney Nickel* or *Lithium*/ethylamine in the case of allylic alcohols;[8b] epoxides[8a] via cyclization of the derived sulfonium salt; butenolides[7b] by alkylation of the hydroxy sulfoxide; 1,2-diols via a Pummerer rearrangement followed by reduction of the intermediate.[9]

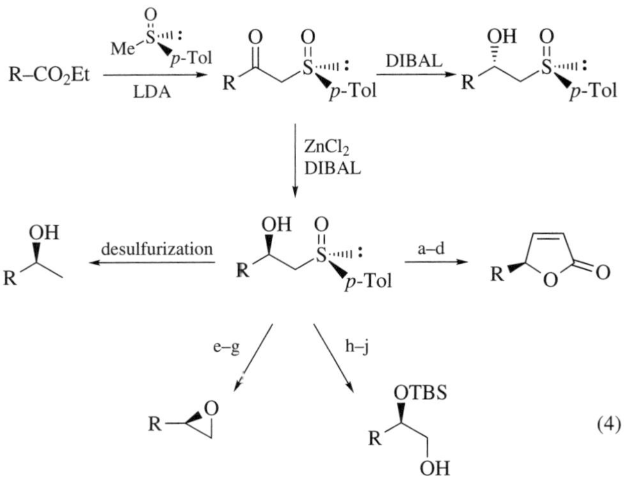

(4)

a) oxidation to sulfone; b) alkylation with sodium iodoacetate;
c) lactonization; d) sulfone elimination;
e) sulfoxide reduction to sulfide with $LiAlH_4$;
f) sulfur methylation with Me_3OBF_4; g) cyclization with a base;
h) OH protection; i) Pummerer rearrangement in Ac_2O;
j) reduction of the Pummerer intermediate.

Numerous applications to total synthesis of natural products have been reported. In the case of the macrolide (*R*)-lasiodiplodin, the achiral ester (eq 5) was reacted with the (+)-(*R*)-methyl *p*-tolyl sulfoxide derived anion to give the corresponding β-keto sulfoxide, which was then reduced with DIBAL to give, after desulfurization, the seco-ester of (*R*)-lasiodiplodin (eq 5).[10] This is an example showing that the chirality can be introduced at the end of the synthesis in the desired configuration.

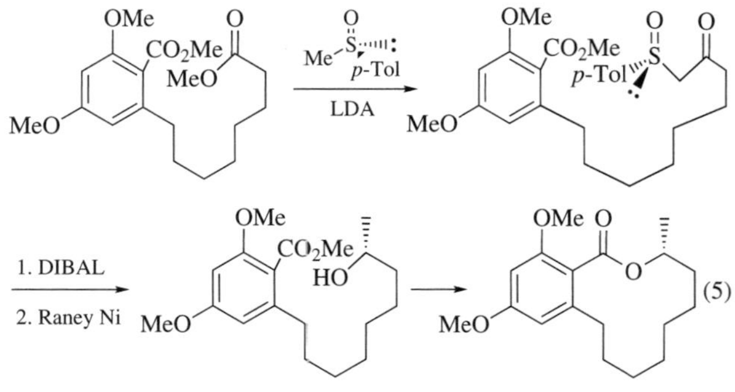

(5)

(*R*)-Lasiodiplodin

In the synthesis of (*S*)-zearalenone[11] and of a chiral spiroacetal, (2*S*,6*R*)-2-methyl-1,7-dioxaspiro[5.6]dodecane,[12] the starting product was a functionalized β-keto sulfoxide resulting from the reaction of glutaric anhydride with lithiated (+)-(*R*)-methyl *p*-tolyl sulfoxide (eq 6).

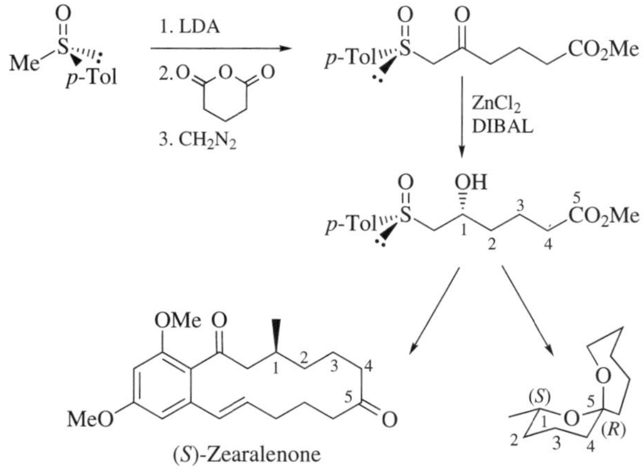
(6)

It was also shown in the enantioselective synthesis of the macrolide patulolide A[13] that the anion of methyl *p*-tolyl sulfoxide was more reactive towards the imidazolide, prepared from the hemi ethyl sebacate, than the ester group (eq 7).

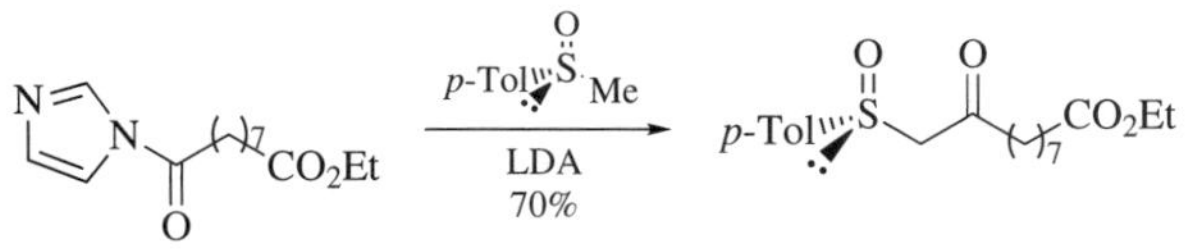
(7)

In a similar way,[9] lithiated (+)-(*R*)-methyl *p*-tolyl sulfoxide was able to react only with the methyl ester group in presence of a *t*-butyl ester, as shown in the case of *t*-butyl methyl octadioate (eq 8).

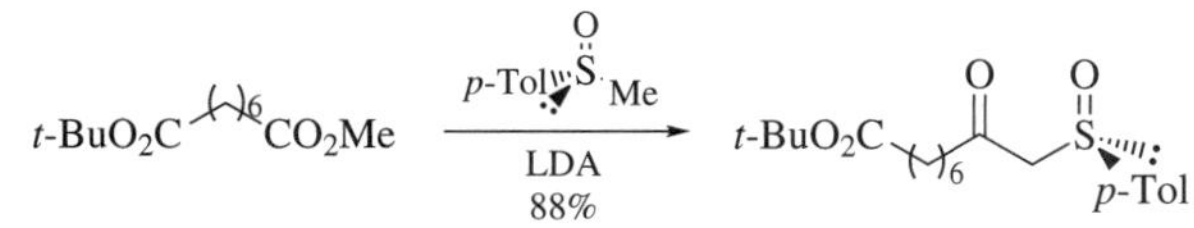
(8)

The enantioselective syntheses of yashabushiketol[14] and gingerols[15] showed the synthetic utility of chiral epoxides obtained from (*R*)-methyl *p*-tolyl sulfoxide (eq 9).

Methyl chloroacetate reacts with the anion of (*R*)-methyl *p*-tolyl sulfoxide to give the corresponding δ-chloro-β-keto sulfoxide (eq 10), which can be easily transformed into the corresponding β-hydroxy sulfoxide which gives, in presence of a base, the optically active α-sulfinyl epoxides.[16] As illustrated here, α-sulfinyl epoxides can be opened by cuprates, leading to chiral homoallylic alcohols.[17]

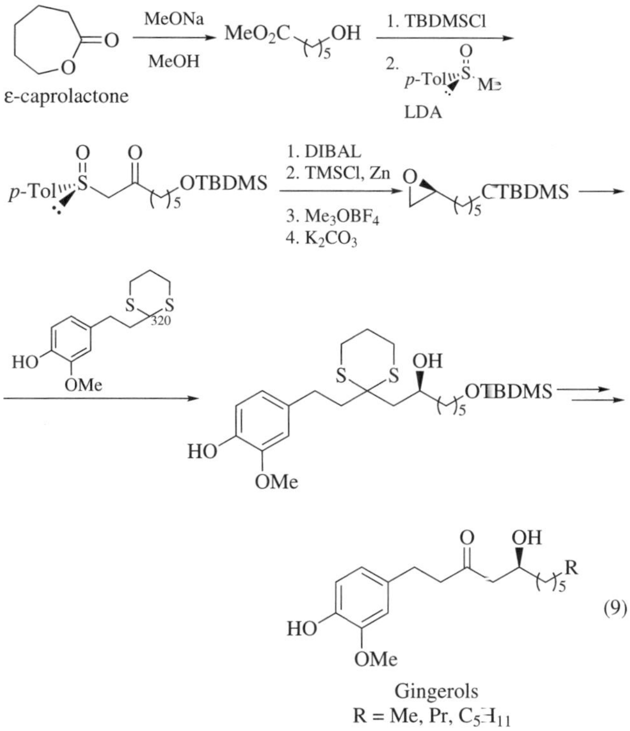
(9)

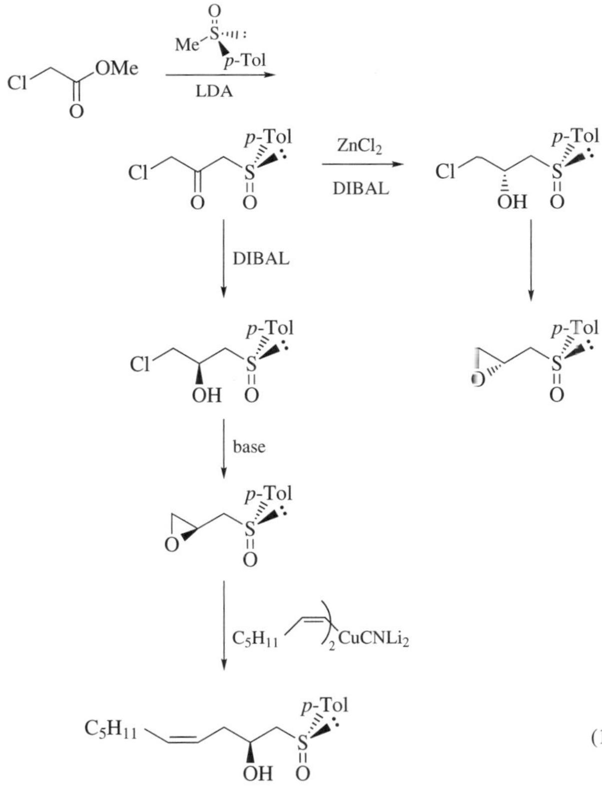
(10)

(*R*)-Methyl *p*-tolyl sulfoxide anion also reacts with β-keto esters to give the corresponding β,δ-diketo sulfoxides,[18] which are useful in the preparation of optically active 1,3-diols (eq 11).[19]

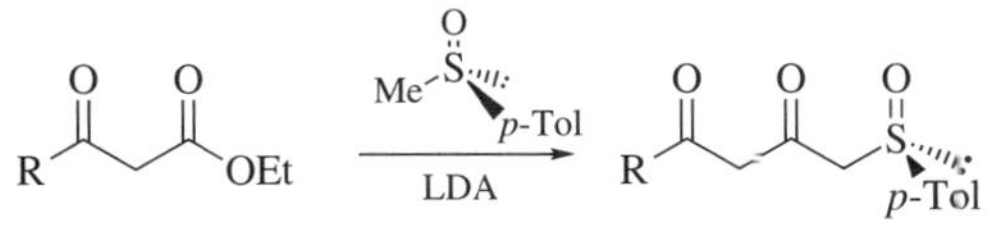
(11)

Difluoroalkyl sulfinylmethyl ketones have been prepared in enantiomerically pure form from (+)-(*R*)-methyl *p*-tolyl sulfoxide in high yield (eq 12).[20] The ketone function was then reduced with complete diastereoselectivity.

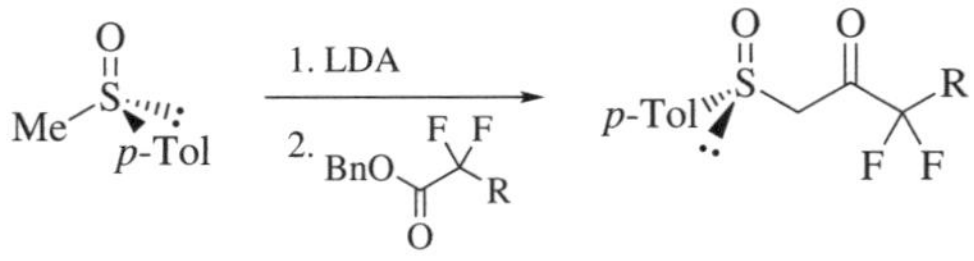
(12)

The absolute configuration at C-32 in recently isolated triterpenoids was assigned by reduction of a β-keto sulfoxide (eq 13).[21]

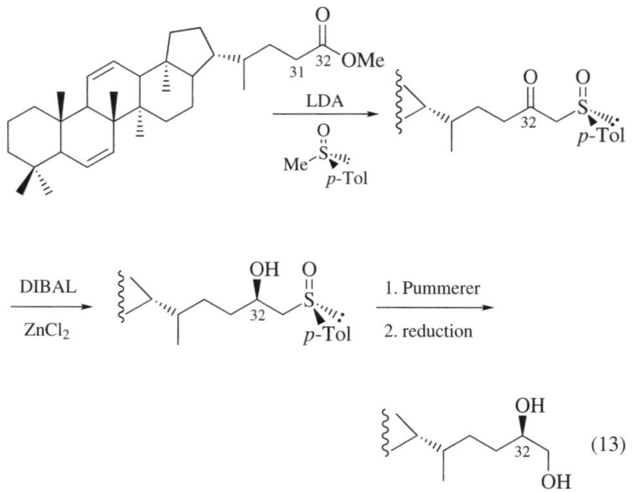
(13)

Miscellaneous. A vinylic sulfoxide was prepared from (*R*)-methyl *p*-tolyl sulfoxide and benzophenone. Cyclopropanation of the double bond with *Dimethylsulfoxonium Methylide* gave a good diastereoselectivity (eq 14).[22]

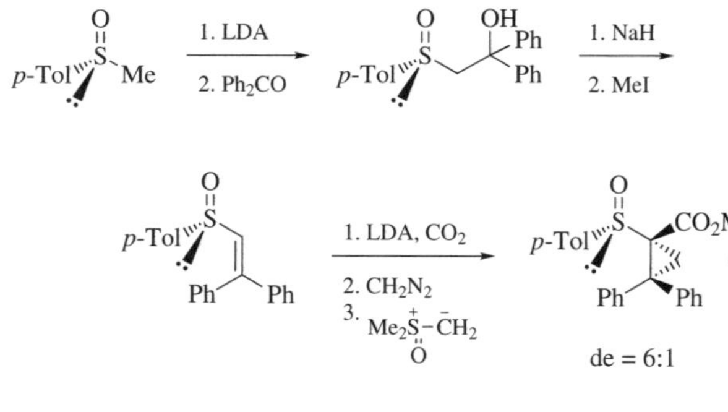
(14)

3-Sulfinylpropionic acid was made from lithium 2-bromoacetate and (*R*)-methyl *p*-tolyl sulfoxide (eq 15).[23]

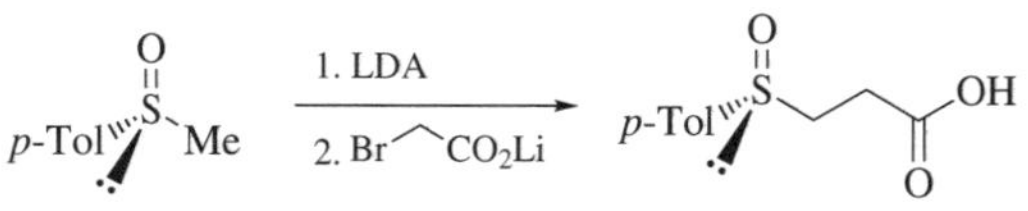
(15)

Sulfinyl dienes were made from α,β-unsaturated esters and methyl *p*-tolyl sulfoxide followed by enolization of the ketone group (eq 16).[24]

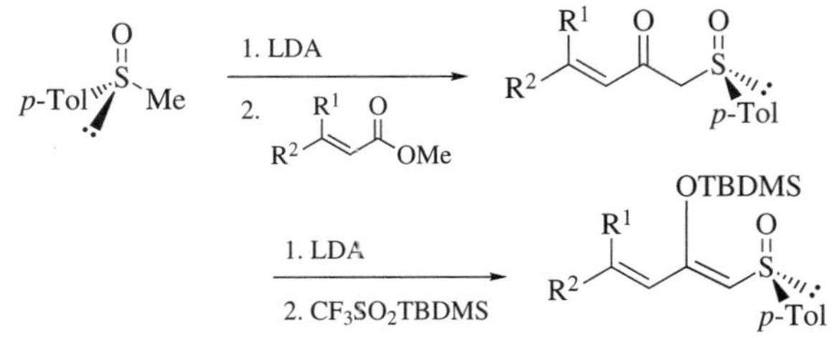
(16)

Although the carbanion of (*R*)-methyl *p*-tolyl sulfoxide reacted with aldehydes and ketones with a poor diastereoselectivity,[25] it reacts with imines with a much higher stereoselectivity[25] as long as the imine substituent is an aromatic ring.[26] (*R*)-(+)-Tetrahydropalmatine was synthesized by addition of (*R*)-methyl *p*-tolyl sulfoxide carbanion to 3,4-dihydro-6,7-dimethoxyisoquinoline (eq 17).[27]

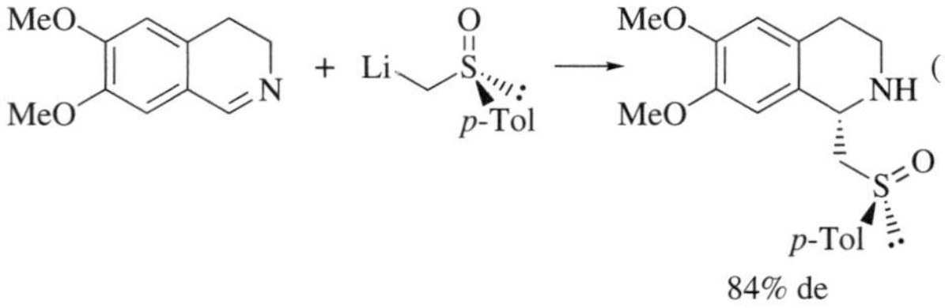
(17)

1. (a) Andersen, K. K.; Gaffield, W.; Papanikolaou, N. E.; Foley, J. W.; Perkins, R. I. *J. Am. Chem. Soc.* **1964**, *86*, 5637. (b) Solladié, G.; Hutt, J.; Girardin, A. *Synthesis* **1987**, 173.
2. Evans, D. A.; Faul, M. M.; Colombo, L.; Bisaha, J. J.; Clardy, J.; Cherry, D. *J. Am. Chem. Soc.* **1992**, *114*, 5977.
3. (a) Llera, J. M.; Fernández, I.; Alcudia, F. *Tetrahedron Lett.* **1991**, *32*, 7299. (b) Fernández, I.; Khiar, N.; Llera, J. M.; Alcudia, F. *J. Org. Chem.* **1992**, *57*, 6789.
4. Zhao, S. H.; Samuel, O. K.; Kagan, H. B. *Tetrahedron* **1987**, *43*, 5135.
5. Davis, F. A.; Reddy, R. T.; Han, W.; Carroll, P. J. *J. Am. Chem. Soc.* **1992**, *114*, 1428.
6. (a) Fu, H.; Kondo, H.; Ichikawa, Y.; Look, G. C.; Wong, C. H. *J. Org. Chem.* **1992**, *57*, 7265. (b) Colonna, S.; Gaggero, N.; Casella, L.; Carrea, G.; Pasta, P. *Tetrahedron: Asymmetry* **1992**, *3*, 95.
7. (a) Solladié, G.; Greck, C.; Demailly, G.; Solladié-Cavallo, A. *Tetrahedron Lett.* **1982**, *23*, 5047. (b) Solladié, G.; Fréchou, C.; Demailly, G.; Greck, C. *J. Org. Chem.* **1986**, *51*, 1912, 1914.
8. (a) Solladié, G.; Demailly, G.; Greck, C. *Tetrahedron Lett.* **1985**, *26*, 435. (b) Solladié, G.; Demailly, G.; Greck, C. *J. Org. Chem.* **1985**, *50*, 1552. (c) Kosugi, H.; Konta, H.; Uda, H. *Chem. Commun./J. Chem. Soc., Chem. Commun.* **1985**, 211. (d) Solladié-Cavallo, A.; Suffert, J.; Adib, A.; Solladié, G. *Tetrahedron Lett.* **1990**, *31*, 6649.
9. Solladié, G.; Fernandez, I.; Maestro, C. *Tetrahedron: Asymmetry* **1991**, *2*, 801.
10. Solladié, G.; Rubio, A.; Carreño, M. C.; García Ruano, J. L. *Tetrahedron: Asymmetry* **1990**, *1*, 187.
11. Solladié, G.; Maestro, M. C.; Rubio, A.; Pedregal, C.; Carreño, M. C.; García Ruano, J. L. *J. Org. Chem.* **1991**, *56*, 2317.
12. Solladié, G.; Almario, A.; Colobert F. *Synlett* **1992**, 167.
13. Solladié, G.; Gerber, C. *Synlett* **1992**, 449.
14. Solladié, G.; Ziani-Chérif, C.; Jesser, F. *Tetrahedron Lett.* **1992**, *33*, 931.
15. Solladié, G.; Ziani-Chérif, C. *J. Org. Chem.* **1993**, *58*, 2181.
16. Solladié, G.; Hamdouchi C.; Vicente, M. *Tetrahedron Lett.* **1988**, *29*, 5929.
17. Solladié, G.; Hamdouchi, C.; Ziani-Chérif, C. *Tetrahedron: Asymmetry* **1991**, *2*, 457.

18. Solladié, G.; Ghiatou, N. *Tetrahedron: Asymmetry* **1992**, *3*, 33.
19. Solladié, G.; Ghiatou, N. *Tetrahedron Lett.* **1992**, *33*, 1605.
20. (a) Bravo, P.; Pregnolato, M.; Resnati, G. *J. Org. Chem.* **1992**, *57*, 2726. (b) Bravo, P.; Pregnolato, M.; Resnati, G. *Tetrahedron: Asymmetry* **1991**, *2*, 1105.
21. Peiseler, B.; Rohmer, M. *J. Chem. Soc., Perkin Trans. 1* **1991**, 2449.
22. Hamdouchi, C. *Tetrahedron Lett.* **1992**, *33*, 1701.
23. Albinati, A.; Bravo, P.; Ganazzoli, F.; Resnati, G.; Viani, F. *J. Chem. Soc., Perkin Trans. 1* **1986**, 1405.
24. Solladié, G.; Maugein, N.; Morreno, I.; Almario, A.; Carreño, C.; García Ruano, J. L. *Tetrahedron Lett.* **1992**, *33*, 4561.
25. Solladié, G. *Synthesis* **1981**, 185.
26. Ronan, B.; Marchalin, S.; Samuel, O.; Kagan, H. B. *Tetrahedron Lett.* **1988**, *29*, 6101.
27. Pyne, S. G.; Dikic, B. *J. Org. Chem.* **1990**, *55*, 1932.

Guy Solladié & Françoise Colobert
University Louis Pasteur, Strasbourg, France

(*R*)-*B*-Methyl-4,5,5-triphenyl-1,3,2-oxazaborolidine

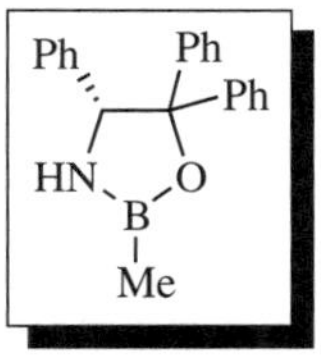

[155268-88-5] $C_{21}H_{20}BNO$ (MW 313.21)

(catalyst used for the borane-mediated stereoselective reduction of ketones)

Solubility: soluble in most organic solvents, e.g. THF, diethyl ether, $CHCl_3$, toluene.

Analysis of Reagent Purity: ^{1}H NMR (C_6D_6).

Preparative Methods: (*R*)-1,1,2-Triphenyl-2-aminoethanol, the precursor of the oxazaborolidine, is prepared in 60–73% yield by portionwise addition of solid methyl (*R*)-phenylglycinate hydrochloride to an excess of **phenylmagnesium bromide** (3 M in diethyl ether) at 0 °C. The amino alcohol (>99% ee, after recrystallization from ethanol) is treated with trimethylboroxine in refluxing toluene in a flask provided with a Dean-Stark trap and under argon. Removal of all the volatiles under vacuum gives (*R*)-*B*-methyl-4,5,5-triphenyl-1,3,2-oxazaborolidine [(*R*)-**1**]as a colorless oil, which is then diluted with toluene up to a known concentration.[1]For reductions, a sample of that solution is transferred via cannula to another flask and the solvent is removed under vacuum and replaced by THF under argon. (*S*)-*B*-Methyl-4,5,5-triphenyl-1,3,2-oxazaborolidine [(*S*)-**1**] can be obtained from methyl (*S*)-phenylglycinate hydrochloride in a similar way. Both enantiomers of methyl phenylglycinate hydrochloride are commercially available at moderate prices.

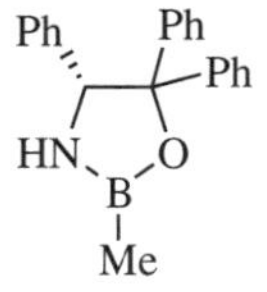

(*R*)-**1**

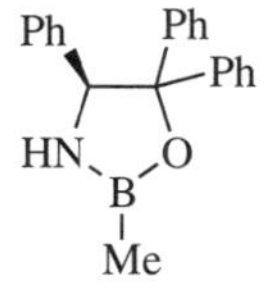

(*S*)-**1**

Purification: Occasionally in the ^{1}H NMR spectrum of the oxazaborolidine (C_6D_6), besides the expected signal at δ 5.38 ppm, a singlet at δ 4.93 ppm can be observed which is due to hydrolyzed, ring-cleaved material. The product may be purified by further treatment with a small amount of trimethylboroxine in refluxing toluene.

Handling, Storage, and Precautions: This oxazaborolidine is sensitive to water and to oxygen. However, its toluene solution can be stored at room temperature under argon for months with negligible loss of its catalytic activity. Care should be exercised to avoid contact of this compound with eyes and skin, and it should be manipulated in a well-ventilated fume hood.

Enantioselective Ketone Reduction. After the pioneering work of Itsuno et al.,[2] Corey's group isolated the 1,3,2-oxazaborolidine derived from chiral **α,α-diphenyl-2-pyrrolidinemethanol** (**2**) and applied it (and also other related *B*-alkyl compounds) to the stereoselective reduction of ketones with **borane-tetrahydrofuran**, **borane-dimethyl sulfide** (BMS) or **catecholborane.**It was named the CBS method (after Corey, Bakshi, and Shibata).[3] Since then, the CBS method has become a standard and has been extensively used, specially for aromatic and α,β-unsaturated ketones, not only in academic laboratories but also in industrial processes.[4]

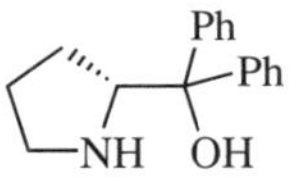

(*R*)-**2**

Among the diverse 1,2-amino alcohols described as precursors of oxazaborolidines,[5] the use of both enantiomers of highly enantioenriched 1,1,2-triphenyl-2-aminoethanol, which lead to oxazaborolidines (*R*)-**1** and (*S*)-**1**, is especially attractive as they arise from inexpensive (*R*)- or (*S*)-phenylglycine, respectively. Oxazaborolidines (*R*)-**1** and (*S*)-**1**are efficient catalysts in the borane-mediated stereoselective reduction of some types of prochiral ketones.

Typically, reductions are performed by slow addition (~15-30 min) of the ketone (1.0 mmol) to a solution of BMS (1.0 mmol) and 0.1-1.0 mmol of **1** (~1 M in THF) under argon at 0 °C (eq 1). Yields are excellent in general after stirring for a further few minutes. The slow addition of ketone appears to enhance the stereoselectivity and in many cases causes the ee noted with 0.1 mmol of **1** to be similar or only slightly lower than that in the stoichiometric case.

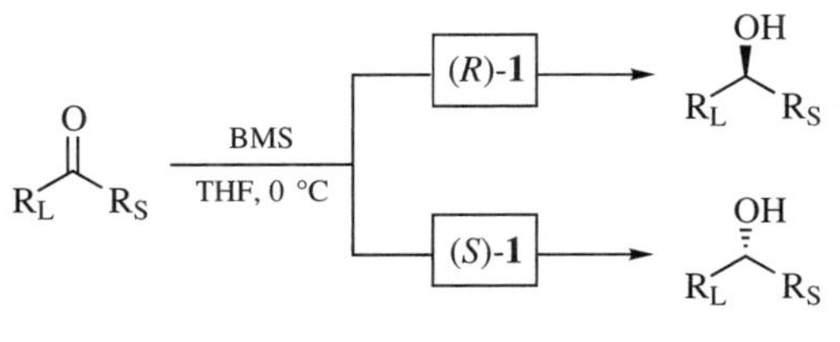

(1)

Alcohols **3–8**, obtained by the reduction of the corresponding ketones with equimolar amounts of BMS and (*R*)-**1**, are obtained with high ees (ee values given are obtained using 0.1 equiv of (*R*)-**1**). Enantioselectivity is excellent (often similar or only slightly lower than those reported in the CBS reduction) for aromatic and hindered methyl ketones,[1a,6] (e.g. **3–5**) and is also good for linear and α-monobranched enones[7] (e.g. **7** and **8**), but lower for linear methyl ketones like 2-octanone (**6**). In should be noted that in the reduction of unsaturated ketones, the time of addition is critical (the optimum being around 15-20 min) in order to avoid concomitant olefin hydroboration. In sharp contrast to the CBS process, the use of **catecholborane** (instead of BMS) or alternative solvents proved to be detrimental.

3 R = Me, 96% ee (96% ee) **5** R = *t*-Bu, 93% ee (92% ee)
4 R = Et, 96% ee (94% ee) **6** R = Hexyl, 72% ee (70% ee)

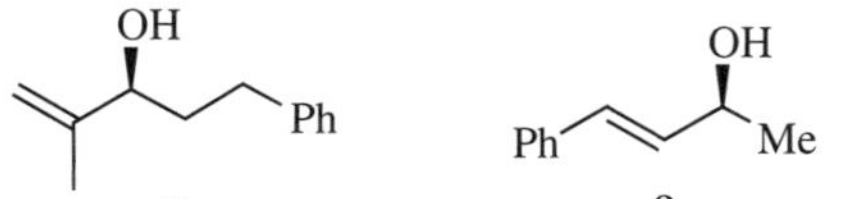

7 91% ee (91% ee) **8** 88% ee (82% ee)

As far as the stereochemical course of the reaction is concerned, the configuration of the emergent stereocenter may be explained in terms of the mechanism proposed by Corey et al. for similar oxazaborolidine mediated reactions.[4] Thus, the transition state operates such that the bigger group (R_L) is located remotely from the methyl group on the boron atom (eq 2).

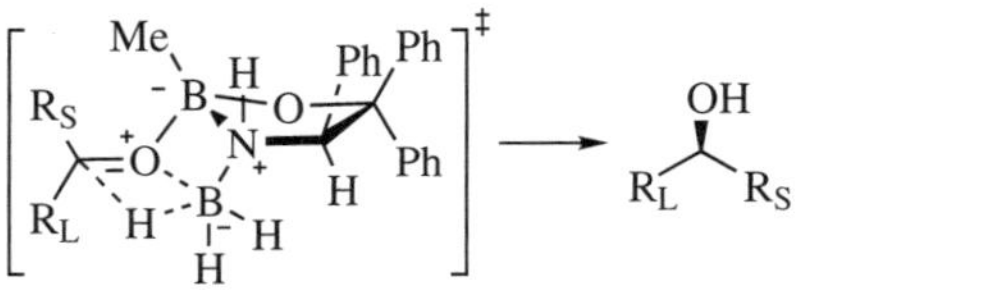

(2)

Accordingly, the experience gained with oxazaborolidine **1** suggests an order of 'empirical' size of R groups.[8] Obviously, better enantioselectivities in the reduction of the ketone carbonyl group are achieved when the substituents R_L and R_S are dissimilar.

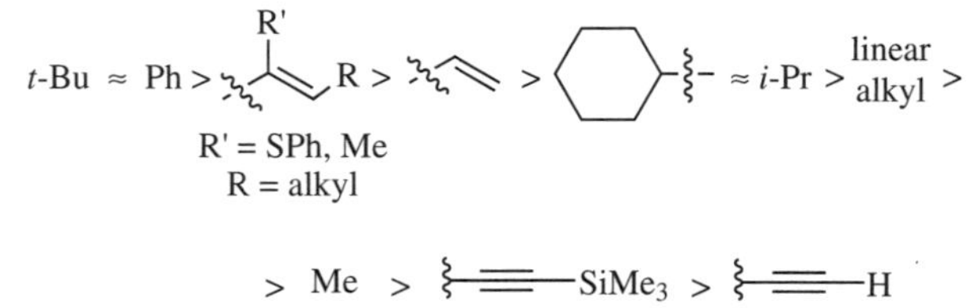

The reduction of α-phenylthio enones constitutes a recent application of these findings to the preparation of chiral α-hydroxy thioesters (eq 3).[9]

Reduction of Acetylenic Ketones. The enantioselective reduction of α,β-acetylenic ketones (R–CO–CC–R′, R′ = H or TMS) with BMS and **1**affords the corresponding propargylic alcoholsin good to excellent yields and >90% ee (see eq eq 4 as and example).[1b] In some cases, the reductions of more sterically crowded hexacarbonyldicobalt complexes (e.g.**9**) derived from the acetylenic ketones also lead, after decomplexation with Cerium (IV) Ammonium Nitrate (CAN), to the same alcohols. However, the use of an oxazaborolidine with an α-face more available for complexation, such as those derived from commercially available (1*S*,2*R*)-2-amino-1,2-diphenylethanol, is required (eq 5).[10] Remarkably, the temporary transformation of the acetylenic moiety into its $Co_2(CO)_6$ complex, not only reverses the stereoselectivity in the reduction step, but also enhances it.

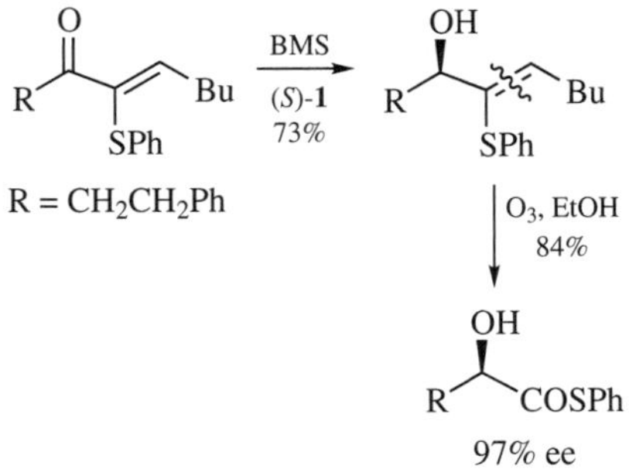

(3)

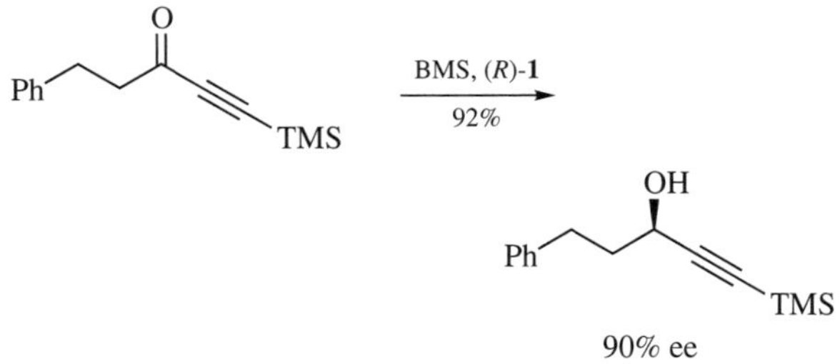

(4)

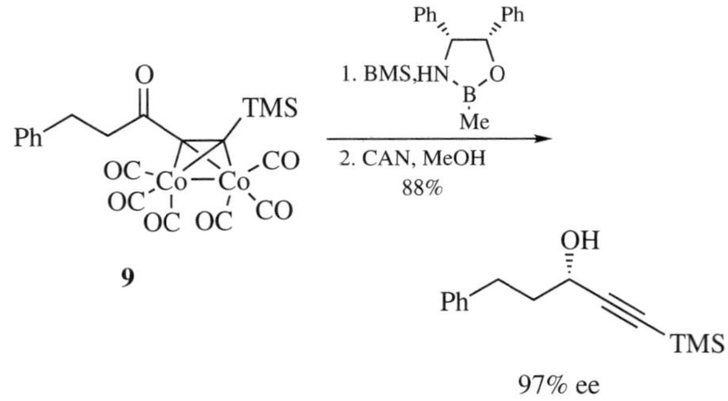

(5)

In addition, the highly enantioenriched propargylic alcohols obtained in such a way are versatile building blocks. Theyhave been applied to the syntheses of the alkyl side chains of zaragozic acids A and C,[11]several metabolites isolated from marine sponges,[12] and the octalactin A ring.[13]

On the other hand, the BMS/(*R*)- or (*S*)-**1** mixture is capable of displaying high asymmetric induction in the reduction of acetylenic ketones, thereby overriding the normally small diastereofacial selectivity of a chiral α-substituted 1-trialkylsilyl-1-alkyn-3-one (e.g. **10**) in a predictable and controlled manner (reagent control) (eq 6). Remarkably, the stereoselectivity noted in such reductions has shown strong dependence upon the steric requirement of the C(1) substituent. Thus, an increasing stereoselectivity has been noted in the reduction of ketones **10**as R changes from Me to Et to *i*-Pr. An explanation for such an unexpected remote effect has been suggested based on abinitio calculations.[14]

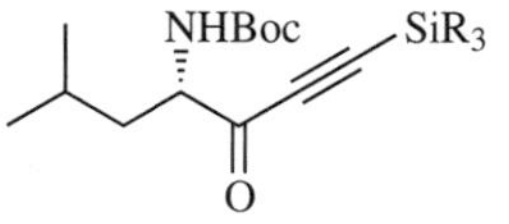

10

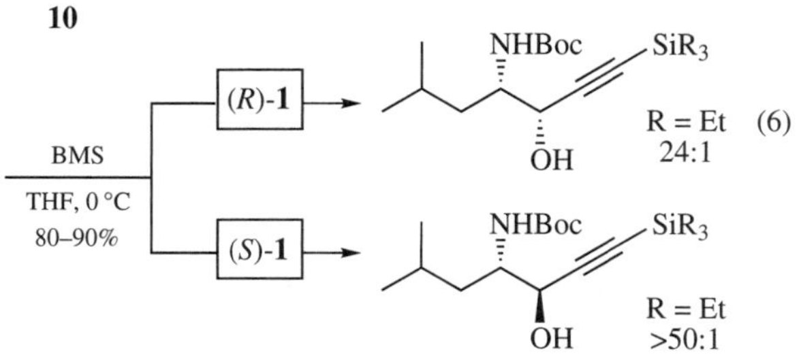

Further work according to this double asymmetric strategy has led to the establishment of a stereodivergent route to β-hydroxy γ-substituted carboxylic acids and α-hydroxy β-substituted carboxylic acids (including *N*-Boc-statine and *N*-Boc-norstatine).[15]

Reduction of 1,4-Diketones. Synthetic access to C_2-symmetric 1,4-diols, useful building blocks for the preparation of chiral 2,5-disubstituted pyrrolidines and phospholanes, involves reduction of the parent 2-alkane-1,4-diones, or, even better, reduction of the related (*E*)-alk-2-ene-1,4-diones (**11**) (eq 7) or 2-alkyne-1,4-diones (**12**) (eq 8), followed by catalytic hydrogenation.[16]

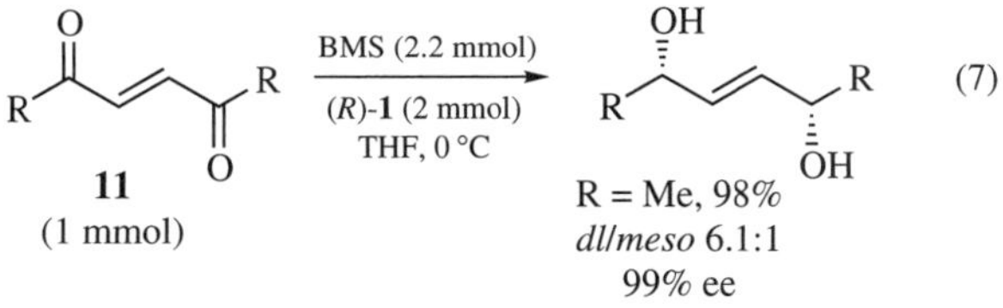

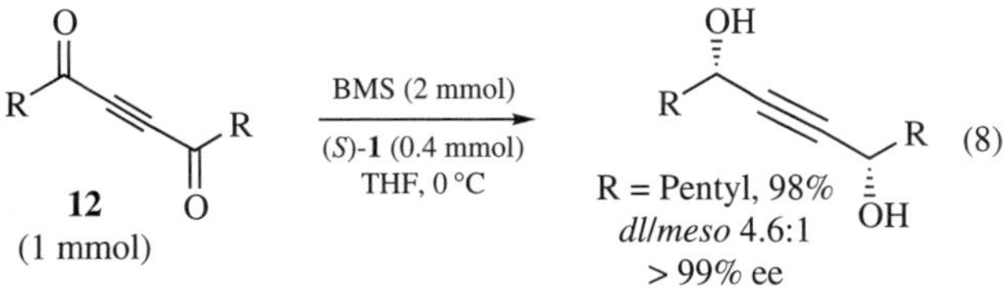

Generally, reduction of diketones **12** (or in some cases their hexacarbonyldicobalt complexes) yields better stereoselectivities than the related ethylenic diketones **11**, especially when R is a sterically demanding group. In addition, the propargylic diols obtained can be easily transformed not only into the saturated 1,4-diols, but also into (*Z*)- or (*E*)-alk-2-ene-1,4-diols. The C_2-symmetric allylic 1,4-diols have been very recently used as building blocks in a formal synthesis of (–)-methylenolactocin and (–)-phaseolinic acid.[17] Related to this, it should be noted that reduction of the unstable (*Z*)-alk-2-ene-1,4-diones (**13**) is an unsuitable route to (*Z*)-alk-2-ene-1,4-diols since a considerable amount of 1,4-reduction (eq 9) is also observed.[17,18]

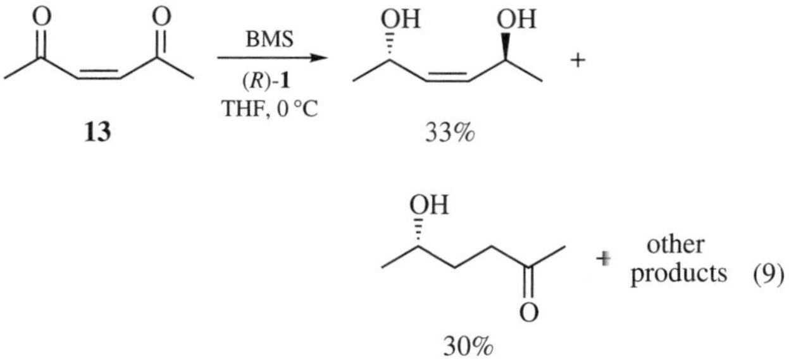

1. (a) Berenguer, R.; Garcia, J.; Vilarrasa, J. *Tetrahedron: Asymmetry* **1994**, *5*, 165. (b) Bach, J.; Berenguer, R.; Garcia, J.; Loscertales, T.; Vilarrasa, J. *J. Org. Chem.* **1996**, *61*, 9021.
2. Itsuno, S.; Sakurai, Y.; Ito, A.; Hirao, S.; Nakahama, S. *Bull. Chem. Soc. Jpn* **1987**, *60*, 395 and references therein.
3. Corey, E. J.; Bakshi, R. K.; Shibata, S. *J. Am. Chem. Soc.* **1987**, *109*, 5551.
4. For a review, see: Corey E. J.; Helal, C. J. *Angew. Chem., Int. Ed. Engl.* **1998**, *37*, 1986.
5. For reviews, see: (a) Deloux, L.; Srebnik M. *Chem. Rev.* **1993**, *93*, 763. (b) Wallbaum, J.; Martens, J. *Tetrahedron: Asymmetry* **1992**, *3*, 1475.
6. Berenguer, R.; Garcia, J.; Gonzàlez, M.; Vilarrasa, J. *Tetrahedron: Asymmetry* **1993**, *4*, 13.
7. (a) Bach, J.; Berenguer, R.; Garcia, J.; Vilarrasa, J. *Tetrahedron Lett.* **1995**, *36*, 3425. (b) Bach, J.; Berenguer, R.; Farràs, J.; Garcia, J.; Meseguer, J.; Vilarrasa, J. *Tetrahedron: Asymmetry* **1995**, *6*, 2683.
8. Bach, J.; Berenguer, R.; Garcia, J.; López, M.; Vilarrasa, J., unpublished results.
9. Berenguer, R.; Cavero, M.; Garcia, J.; Muñoz, M. *Tetrahedron Lett.* **1998**, *39*, 2183.
10. Quallich, G. J.; Woodall, T. M. *Tetrahedron Lett.* **1993**, *34*, 4145.
11. Bach, J.; Galobardes, M.; Garcia, J.; Romea, P.; Tey, C.; Urpí, F.; Vilarrasa, J. *Tetrahedron Lett.* **1998**, *39*, 6765.
12. (a) Garcia, J.; López, M.; Romeu, J. *Tetrahedron: Asymmetry* **1999**, *10*, 2617. (b) Garcia, J.; López, M.; Romeu, J. *Synlett* **1999**, *4*, 429.
13. Bach, J.; Garcia, J. *Tetrahedron Lett.* **1998**, *39*, 6761.
14. Alemany, C.; Bach, J.; Farràs, J.; Garcia, J. *Org. Lett.* **1999**, *1*, 1831.
15. Alemany, C.; Bach, J.; Garcia, J.; López, M.; Rodriguez, A. B. *Tetrahedron* **2000**, *56*, 9305.
16. (a) Bach, J.; Berenguer, R.; Garcia, J.; Loscertales, T.; Manzanal, J.; Vilarrasa, J. *Tetrahedron Lett.* **1997**, *38*, 1091. (b) Bach, J.; Berenguer, R.; Garcia, J.; López, M.; Manzanal, J.; Vilarrasa, J *Tetrahedron* **1998**, *54*, 14947.
17. Ariza, X.; Garcia, J.; López, M. *Synlett* **2001**, *6*, 9305.
18. Berenguer, R.; Garcia, J., unpublished results.

Jordi Garcia
University of Barcelona, Barcelona, Spain

Monoisopinocampheylborane[1]

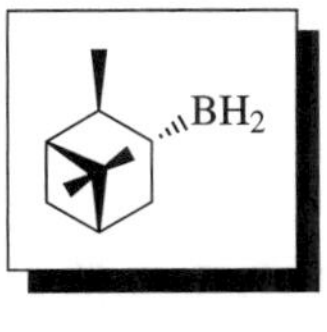

[64234-27-1] $C_{10}H_{19}B$ (MW 150.10)

(asymmetric hydroboration of *trans* and trisubstituted alkenes;[2] asymmetric reduction of ketones[3])

Physical Data: a crystalline adduct with *N,N,N′,N′-Tetramethylethylenediamine* ($2IpcBH_2 \cdot$ TMEDA) has mp 140.5–141.5 °C; $[\alpha]_D^{23}$ +69.03° (*c* 9.33, THF).

Solubility: sol THF, Et_2O.

Analysis of Reagent Purity: the optical purity of the reagent is assayed by oxidation (NaOH, H_2O_2) to give isopinocampheol, $[\alpha]_D^{27}$ −35.8° (*c* 0.9, benzene).

Handling, Storage, and Precautions: best used as prepared. Guidelines for the handling of air- and moisture-sensitive materials should be followed.[1c]

Preparation. Hydroboration of alkenes with *Borane–Tetrahydrofuran* or *Borane–Dimethyl Sulfide* proceeds rapidly past the monoalkylborane stage with all but the most sterically demanding alkenes. Thus, attempts to prepare a solution of monoisopinocampheylborane ($IpcBH_2$) by simple admixture of 1:1 (+)-α-pinene:borane result in an equilibrium mixture of $IpcBH_2$, diisopinocampheylborane (Ipc_2BH), and borane.[4] Although several indirect methods for the preparation of $IpcBH_2$ have been devised,[5a–f] the preparation of $IpcBH_2$ by displacement of α-pinene from Ipc_2BH with TMEDA is recommended (eq 1).[5e,6] It is unique to this procedure that the crystalline adduct[7] incorporates two $IpcBH_2$ units which are of higher optical purity than the starting (+)-α-pinene used. Analysis of the mother liquor reveals that the minor enantiomer accumulates in the more soluble diastereomeric adduct.

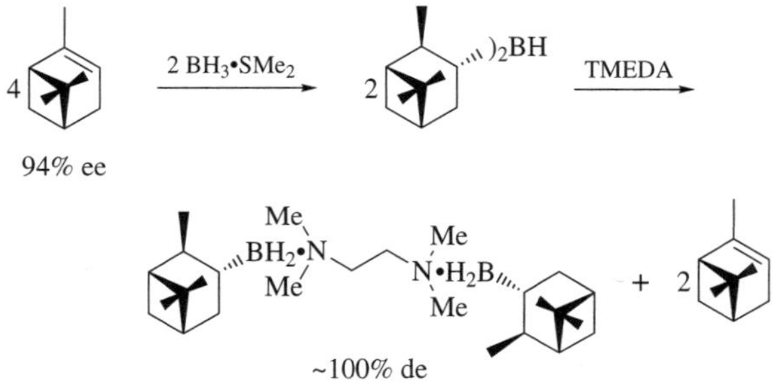

(1)

$IpcBH_2$ of essentially 100% ee is liberated from the TMEDA adduct by addition of *Boron Trifluoride Etherate* in THF. Filtration of the TMEDA · $2BF_3$ adduct provides a solution of $IpcBH_2$ in THF ready for subsequent hydroboration reactions (eq 2).

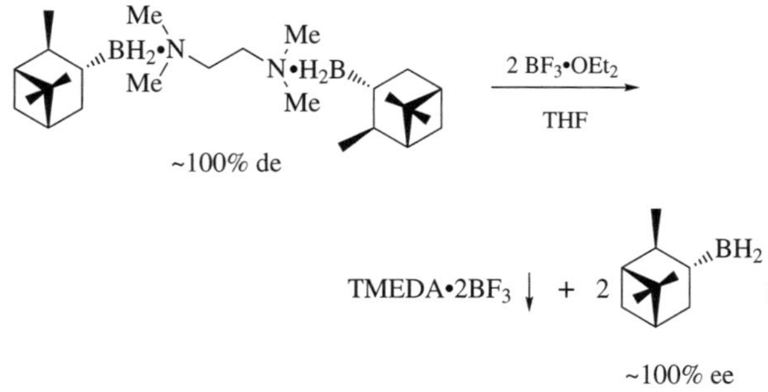

(2)

Asymmetric Hydroboration. The steric requirements of $IpcBH_2$ are such that hydroboration of *trans* and trisubstituted alkenes proceeds with little or no displacement of α-pinene from the reagent, a phenomenon which is observed with the more hindered *Diisopinocampheylborane* (Ipc_2BH). Ipc_2BH is most effective for the hydroboration of relatively unhindered *cis* alkenes, which are hydroborated with low asymmetric induction by $IpcBH_2$ (eq 3). The two reagents are therefore complementary in this respect. (For a reagent which hydroborates *cis, trans*, and trisubstituted alkenes with excellent asymmetric induction, see also *(R,R)-2,5-Dimethylborolane*)

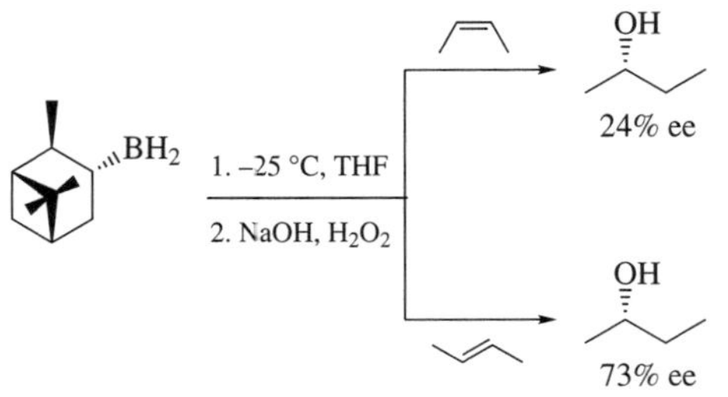

(3)

Representative *trans* alkenes are asymmetrically hydroborated by $IpcBH_2$ derived from (+)-α-pinene to give (*S*)-alcohols in the range of 65–76% ee. With highly hindered *trans* alkenes, the enantioselectivity can be somewhat higher (eq 4).[2]

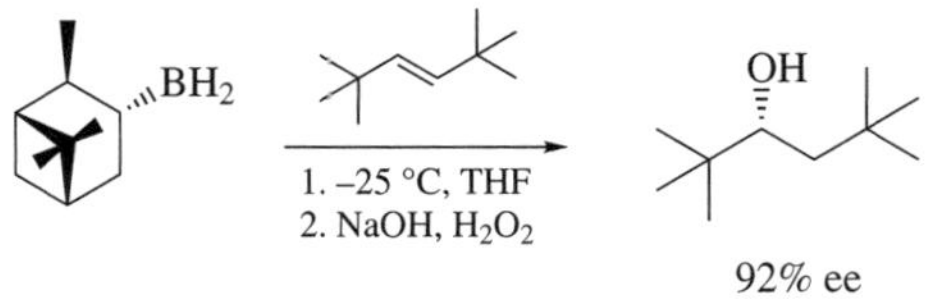

(4)

Aliphatic trisubstituted alkenes are likewise hydroborated to give (*S*)-alcohols in the range of 53–72% ee. When the trisubstituted alkene bears a phenyl substituent, a significant increase in enantioselectivity is observed (eq 5).[2,8]

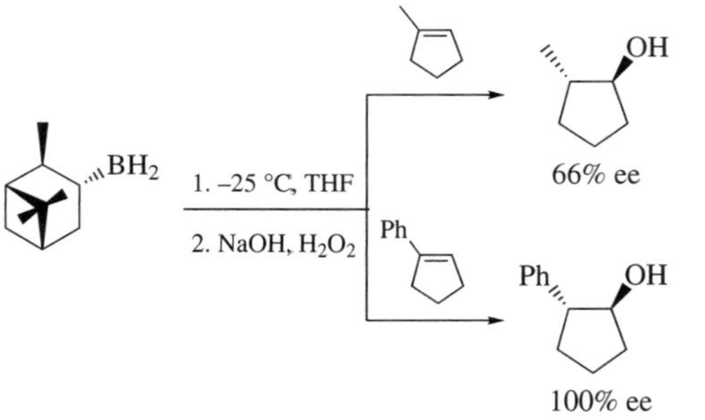

(5)

In the asymmetric hydroboration of 1-heteroarylcycloalkenes,[9] $IpcBH_2$ exhibits enantioselectivities of 83–90% ee, comparable to the phenyl-substituted alkenes examined (eq 6).

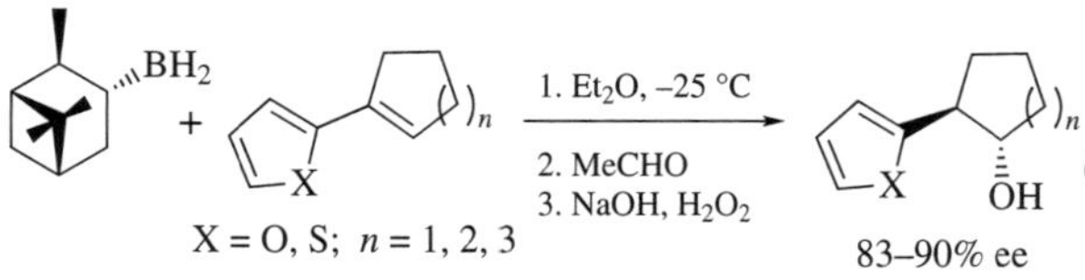

(6)

Many intermediate dialkylboranes derived from hydroboration with $IpcBH_2$ can be recrystallized to enantiomeric purities approaching 100%, thus giving alcohols of 98–99% ee upon oxidation.[10] If, instead of being oxidized in situ, the dialkylborane intermediate is treated with *Acetaldehyde*,[11] α-pinene is displaced for recovery and a chiral boronate bearing the R group of the alkene is obtained (eq 7).

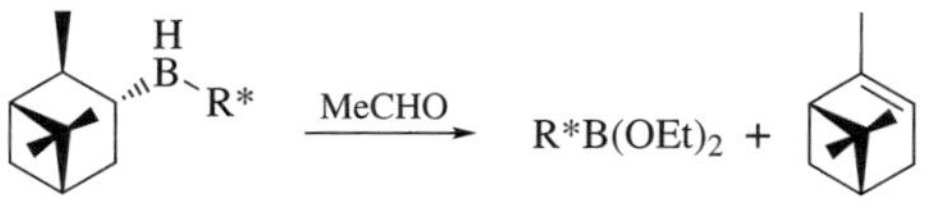

(7)

This reaction is general, and these boronic esters are versatile synthetic intermediates in their own right.[12]

Asymmetric Reduction of Ketones. The reduction of prochiral ketones with $IpcBH_2$ is mechanistically complex. Although the secondary alcohols obtained are consistantly enriched in the (*S*) enantiomer when the reagent is prepared from (+)-α-pinene, the degree of asymmetric induction observed, 11–46% ee, varies with the reaction stoichiometry.[3] This has been attributed to the ability of the 1:1 $IpcBH_2$:ketone addition product to serve as a reducing agent for an additional equivalent of ketone (eq 8).

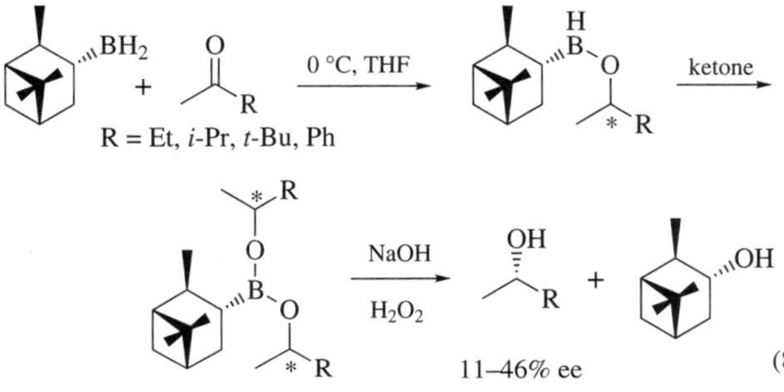

(8)

1. (a) Brown, H. C.; Jadhav, P. K.; Mandal, A. K. *Tetrahedron* **1981**, *37*, 3547. (b) Brown, H. C.; Jadhav, P. K. In *Asymmetric Synthesis*; Morrison, J. D., Ed.; Academic: New York, 1983; Vol. 2, Chapter 1. (c) Brown, H. C.; Kramer, G. W.; Levy, A. B.; Midland, M. M. In *Organic Synthesis via Boranes*; Wiley: New York, 1975. (d) Smith, K.; Pelter, A. *Comprehensive Organic Synthesis* **1991**, *8*, Chapter 3.10.
2. Brown, H. C.; Jadhav, P. K.; Mandal, A. K. *J. Org. Chem.* **1982**, *47*, 5074.
3. Brown, H. C.; Mandal, A. K. *J. Org. Chem.* **1984**, *49*, 2558.
4. Mandal, A. K.; Yoon, N. M. *J. Organomet. Chem.* **1978**, *156*, 183.
5. (a) Brown, H. C.; Yoon, N. M. *J. Am. Chem. Soc.* **1977**, *99*, 5514. (b) Singaram, B.; Schweir, J. R. *J. Organomet. Chem.* **1978**, *156*, C1. (c) Brown, H. C.; Mandal, A. K. *Synthesis* **1978**, 146. (d) Pelter, A.; Ryder, D. J.; Sheppard, J. H.; Subrahmanyam, C.; Brown, H. C.; Mandal, A. K. *Tetrahedron Lett.* **1979**, *49*, 4777. (e) Brown, H. C.; Mandal, A. K.; Yoon, N. M.; Singaram, B.; Schweir, J. R.; Jadhav, P. K. *J. Org. Chem.* **1982**, *47*, 5069. (f) Jadhav, P. K.; Desai, M. C. *Heterocycles* **1982**, *18*, 233.
6. Brown, H. C.; Schweir, J. R.; Singaram, B. *J. Org. Chem.* **1978**, *43*, 4395.
7. X-ray crystal structure: Soderquist, J. A.; Hwang-Lee, S.-J.; Barnes, C. L. *Tetrahedron Lett.* **1988**, *29*, 3385.
8. Mandal, A. K.; Jadhav, P. K., Brown, H. C. *J. Org. Chem.* **1980**, *45*, 3543.
9. Brown, H. C.; Gupta, A. K.; Vara Prasad, J. V. N. *Bull. Chem. Soc. Jpn.* **1988**, *61*, 93.
10. Brown, H. C.; Singaram, B. *J. Am. Chem. Soc.* **1984**, *106*, 1797.
11. Brown, H. C.; Jadhav, P. K.; Desai, M. C. *J. Am. Chem. Soc.* **1982**, *104*, 4303.
12. Brown, H. C.; Jadhav, P. K.; Singaram, B. In *Modern Synthetic Methods*; Scheffold, R., Ed.; Springer: Berlin, 1986; Vol 4, pp 307–356.

Robert P. Short
Polaroid Corporation, Cambridge, MA, USA

N

[(*R*)-α-(2-Naphthyl)aminomethyl] ferrocene

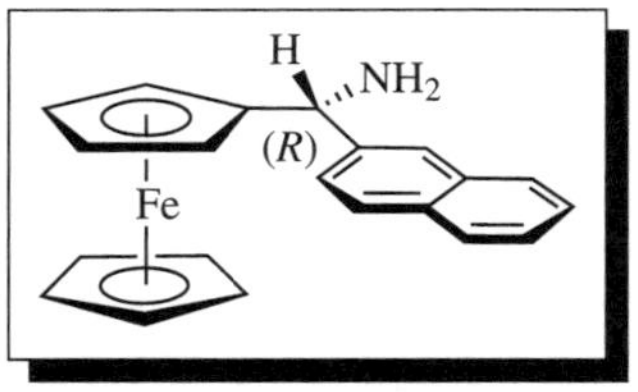

[221528-09-2] $C_{21}H_{19}NFe$ (MW 341.24)

(the reagent is used as a chiral ligand for asymmetric S_N2' addition of zinc organometallic reagents to allyl chlorides in the presence of catalytic $CuBr{\cdot}SMe_2$ in allylic substitution reactions)

Physical Data: limited, mp 97 °C.
Solubility: soluble in common organic solvents such as THF and Et_2O.
Form Supplied in: orange solid, see patent (reference 1).
Purification: it can be purified by free basing the initial hydrochloride salt of the above title compound via aq NaOH followed by biphasic extraction using Et_2O and aqueous medium.

Preparation. [(*R*)-α-(2-Naphthyl)aminomethyl]ferrocene was prepared in three steps from ferrocenyl 2-naphthyl ketone featuring an asymmetric CBS reduction[2] with > 99% ee (eq 1).[3,4] After protection of the secondary hydroxyl group with an acetyl group, a nucleophilic displacement of the acetoxy group with an amino group proceeded with retention of stereochemistry. A range of different variations of [(*R*)-α-(2-naphthyl)aminomethyl]ferrocene could be prepared using this sequence with similar efficiency.

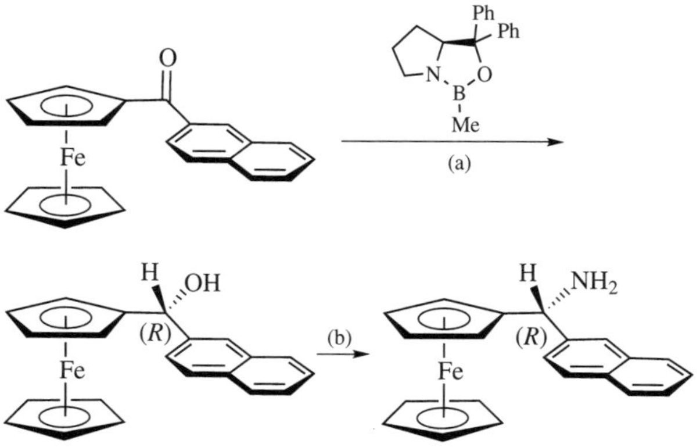

(1)

89% (>99% ee) 66%

Conditions:
(a) $BH_3{\bullet}SMe_2$, THF, 0 °C, 30 min.
(b) (i) Ac_2O, pyridine; (ii) NH_3, CH_3CN, rt, 24 h.

Other Related Substrates:

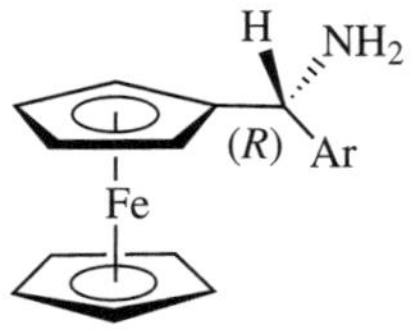

Ar = Ph, *o*-tolyl, 1-naphthyl, *p*-binphenyl, *o*-bromophenyl, *p*-*tert*-butylphenyl, 3,5-di-*tert*-butylphenyl, 3,5-dimethylphenyl.

Applications in Asymmetric Allylic Substitutions. [(*R*)-α-(2-Naphthyl)aminomethyl]ferrocene has been used in various enantioselective S_N2' allylic displacements of allyl chlorides using organozinc reagents. As shown in eq 2, the addition of various organozinc reagents (1.2 equiv) to allyl chlorides could be rendered enantioselective by using 10 mol % of [(*R*)-α-(2-naphthyl) aminomethyl]ferrocene along with 1 mol % of $CuBr{\cdot}SMe_2$.

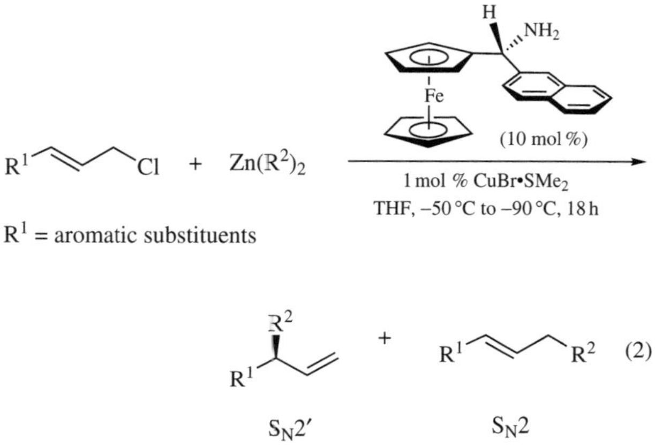

(2)

S_N2':S_N2 (90:10 to >99:1)
yields: 45–72%
ee for S_N2' products: 37–87%

The Best Result:

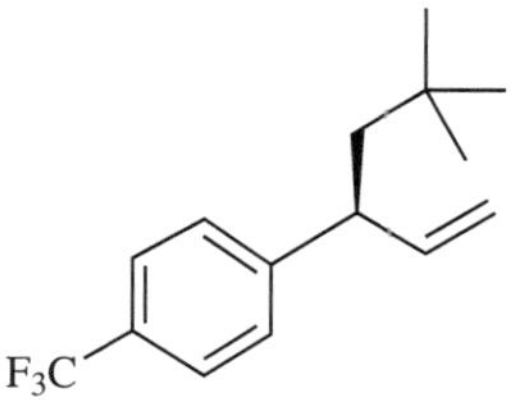

S_N2':S_N2 (97:3) yields: 72%
ee for S_N2' products: 87%
The Ideal Temperature: −90 °C
The Best Nucleophile: $Zn(R^2)_2$ = $Zn(neopentyl)_2$

[(*R*)-α-(2-Naphthyl)aminomethyl]ferrocene was the best chiral ligand, for it provided better ee's than when the 2-naphthyl group was replaced with a Ph, *o*-tolyl, or 1-naphthyl group. Although a range of different allyl chlorides could be used, the best ee's were achieved at 87% using *p*-trifluoromethyl cinnamyl chloride. In general, the R^1 groups in eq 2 are aromatic substituents, and the

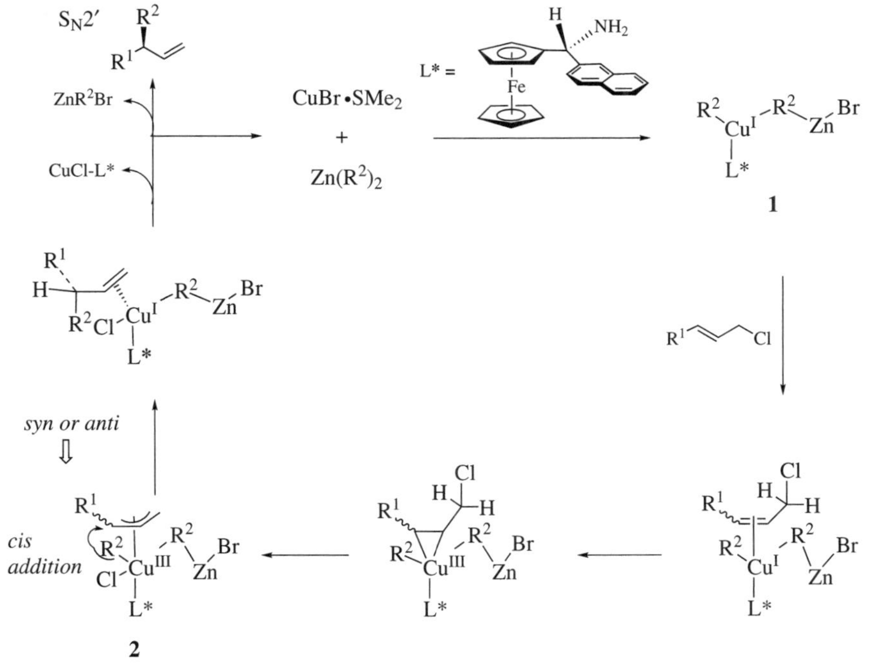

(3)

best leaving group proved to be chloride over bromide, carbonate, phosphate, and xanthate.

Although details are not known or given in the literature, these reactions likely involve additions of organocopper reagents after an initial transmetallation with the zinc reagent, or a putative Cu/Zn couple reagent (**1**) as shown in eq 3. Thus, regioselectivity is very high in favor of the S_N2' displacement over S_N2. Based on proposals made by Nakamura regarding Cu^{III} species as the active intermediate,[5] the key intermediate could be the Cu allyl complex **2** after oxidative addition to allyl chlorides.

The stereochemistry of intermediate **2** with regard to the R^1 group (*syn* or *anti* with respect to the π-allyl complex), as well as the π facial selectivity of the Cu complex (it is shown here as complexing to the α-face), should be affected by the chiral ligand (L*). These factors coupled with cis addition of the R group should lead to the final observed enantioselectivity.

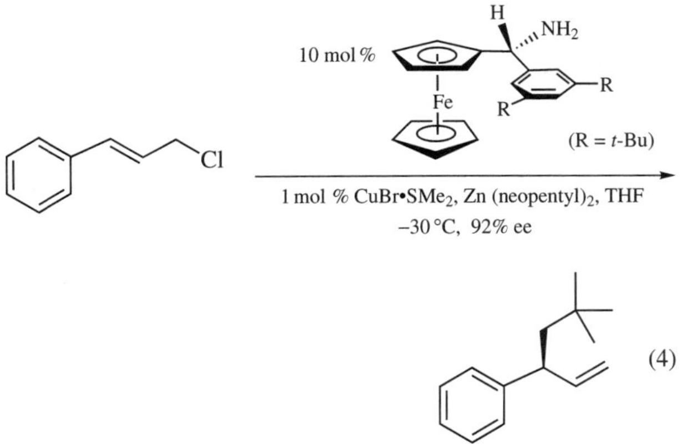

(4)

S_N2'

S_N2':S_N2 (98:2)
yield: 82%

Limitations. The initial limitations were (1) a very hindered nucleophile such as Zn(neopentyl)$_2$ had to be used to achieve reasonable ee's, and (2) the temperature had to be −90 °C or the ee drops to as low as 25% at rt (eq 2).

Modifications. A diverse array of chiral ligands related to [(*R*)-α-(2-naphthyl)aminomethyl]ferrocene was screened (see those mentioned in eq 1).[4] It was found that when the 2-naphthyl group was replaced with the 3,5-di-*tert*-butyl phenyl group, the ee's improved from 82% to 92% in the case of cinnamyl chloride when Zn(neo-pentyl)$_2$ was used. It is equally significant to note that the temperature could be elevated to −30 °C (eq 4).[4]

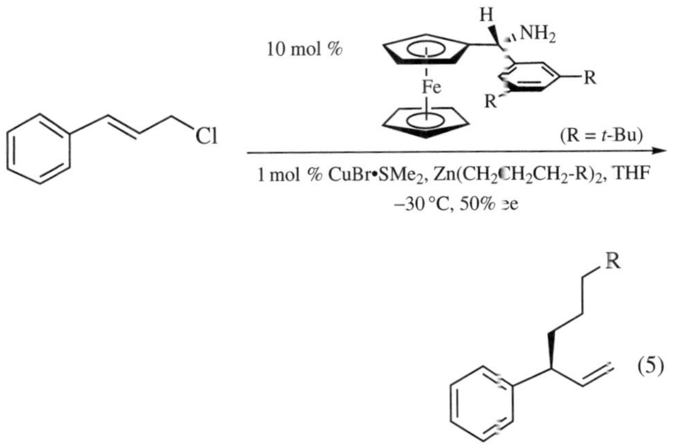

(5)

R = —CH_2-OAc (yield: 63%)
R = —CO_2Et (yield: 75%)

Using the optimum chiral ligand, ee's improved from 26% to 65%, 10% to 44%, and 45% to 72% when Zn(neopentyl)$_2$, ZnEt$_2$, and Zn(*i*-Pr)$_2$ were used, respectively, thereby diversifying the range of nucleophiles that could be used. This asymmertric S_N2' displacement reaction is no longer limited to the hindered nucleophiles. Further examples in illustrating a wider scope of nucleophiles that can used are shown in eq 5, although ee's were 50% for these two new zinc reagents.

General Procedures. To a solution of [(*R*)-α-(2-naphthyl) aminomethyl]ferrocene (0.2 mmol) and CuBr·SMe_2 (0.02 mmol) in 5 mL of THF at −90 °C was added the organozinc reagent (2.4 mmol) and the allyl chloride (2.0 mmol) in succession. After being stirred at −90 °C for 18 h, the reaction mixture was worked up via standard biphasic work-up of aqueous/organic solvent. The crude residue was purified by silica gel column chromatography [eluent used is Et_2O:pentane (1:50)].

In the modified procedure for reactions carried out at −30 °C, a solution of the organozinc reagent (2.4 mmol) in THF and the allyl chloride (2.0 mmol) were added simultaneously over 3 h.

1. Knochel, P.; Duebner, F. *Copper-Catalyzed Enantioselective Allylic Substitution Reactions,* PCT Int. Appl. **2000**.
2. (a) Corey, E. J.; Bakshi, R. K.; Shibata, S. *J. Am. Chem. Soc.* **1987**, *109*, 5551. (b) Corey, E. J.; Bakshi, R. K.; Shibata, S. *J. Am. Chem. Soc.* **1987**, *109*, 7925. (c) Corey, E. J.; Shibata, S.; Bakshi, R. K. *J. Org. Chem.* **1988**, *53*, 2861.
3. Duebner, F.; Knochel, P. *Angew. Chem, Int. Ed.* **1999**, *38*, 379.
4. Duebner, F.; Knochel, P. *Tetrahedron Lett.* **2000**, *41*, 9233.
5. Nakamura, E.; Mori, S. *Angew. Cehm. Int. Ed.* **2000**, *39*, 3750.

Richard P. Hsung & Jiashi Wang
University of Minnesota, Minneapolis, MN, USA

1-(1-Naphthyl)ethylamine

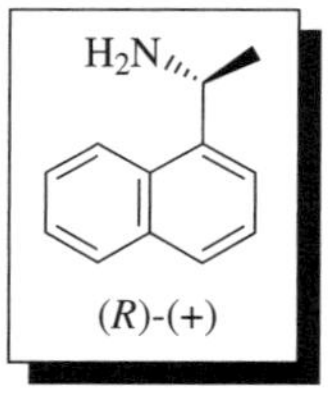

(±)
[42882-31-5] $C_{12}H_{13}N$ (MW 171.26)
(+)
[3886-70-2]
(−)
[10420-89-0]

(chiral reagent used for resolution of carboxylic acids,[1] alcohols,[2] and lactones;[3] a chiral derivatization agent used for chromatographic resolution of carboxylic acids[4] and alcohols;[5] can serve as a chiral solvating agent[6])

Alternate Name: NEA.
Physical Data: (±) form: bp 156 °C/15 mmHg; *d* 1.063 g cm^{-3}. Oxalate: mp 221 °C (dec). (*S*)-(−) isomer: bp 153 °C/11 mmHg; *d* 1.060 g cm^{-3}; $[\alpha]^{25}$ −80.8° (neat); $[\alpha]^{20}$ −59° (c = 5, MeOH). Oxalate: mp 232 °C (dec). (*R*)-(+) isomer: bp 153 °C/11 mmHg; *d* 1.060 g cm^{-3}; $[\alpha]^{25}$ +82.8° (neat); $[\alpha]^{20}$ +60° (c = 5, MeOH). Oxalate: mp 240 °C.
Solubility: sol alcohol, ether; insol H_2O.
Form Supplied in: clear liquid; both enantiomers and the racemate are all widely available.
Analysis of Reagent Purity: the enantiomeric purity of the reagent can be determined by either NMR or HPLC analysis of derivatives produced from α-methoxy-α-(trifluoromethyl)benzyl isocyanate,[7] α-methoxy-α-(trifluoromethyl)benzyl acid chloride,[8] or 2′-methoxy-1,1′-binaphthyl-2-carboxylic acid chloride.[9]
Preparative Methods: racemic 1-(1-naphthyl)ethylamine can be resolved with camphoric acid,[10] tartaric acid,[11] L-menthyl hydrogen phthalate,[11] di-*O*-isopropylidene-2-ketogulonic acid,[11,12] and (*S*)-(−)-(2-phenylcarbamoyloxy)propionic acid.[13] These procedures can also be used to enhance the enantiomeric purity of the reagent.
Handling, Storage, and Precautions: use in a well-ventilated fume hood.

Resolution of Carboxylic Acids. The enantiomers of 1-(1-naphthyl)ethylamine are used to resolve racemic carboxylic acids by selective crystallization of diastereomeric salts. For example, crystallization of racemic 3-bromobutyric acid with (*R*)-(+)-NEA followed by acidification of the diastereomeric salt afforded (*S*)-(+)-3-bromobutyric acid (eq 1).[1] In the same manner, resolution with (*S*)-(−)-NEA yielded (*R*)-(−)-3-bromobutyric acid after liberation of the amine (eq 1).[1]

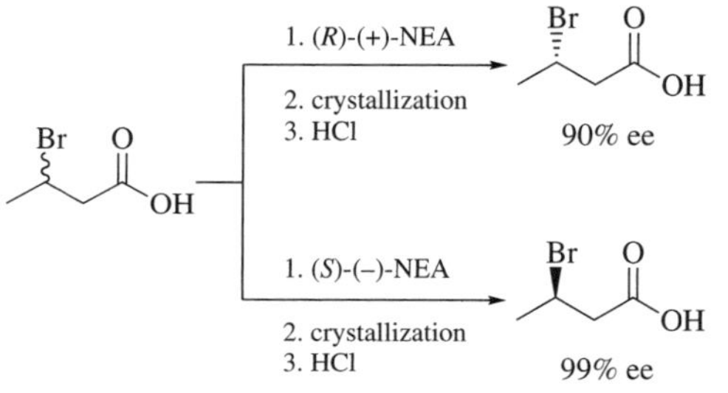

(1)

Resolution of alcohols can be achieved following derivatization with phthalic anhydride.[2] Racemic 1-undecyn-3-ol was converted into a phthalic monoester derivative and resolved with (*R*)-(+)-NEA (eq 2). Liberation of the resolved phthalic ester and saponification yielded (*R*)-(+)-alcohol in 92% optical purity (eq 2). Similarly, the (*S*)-(−) alcohol is obtained upon resolution with (*S*)-(−)-NEA (eq 2).

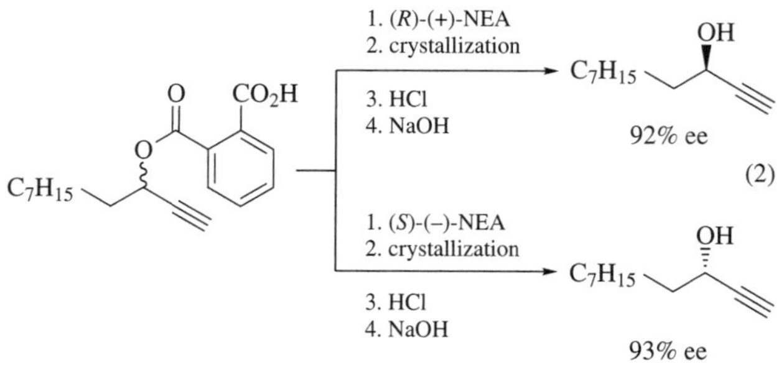

(2)

Optically active lactones are also readily available through this classical resolution technique. The racemic lactone is hydrolyzed to the hydroxy acid and resolved with (*S*)-(−)-NEA (eq 3).[3] After crystallization, the *dextro* (+)-lactone is regenerated upon acidification of the chiral salt. Resolution of the lactone with (R)-(+)-NEA leads to the *levo* (−)-lactone (eq 3).[3]

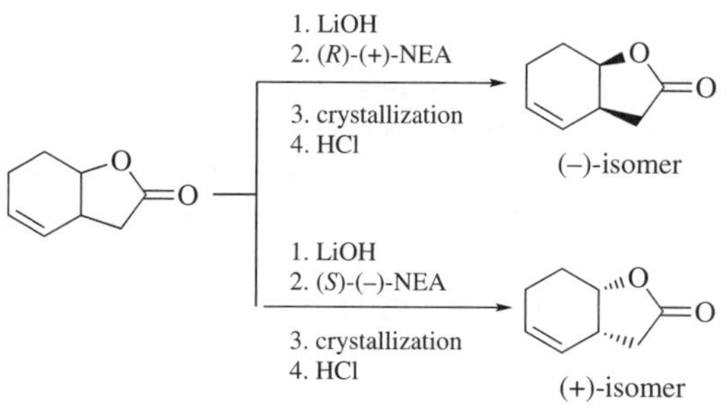

(3)

Chromatographic Resolutions. 1-(1-Naphthyl)ethylamine serves as a chiral derivatization agent useful in preparing diastereomeric amides from racemic acids for chromatographic resolution.[4] For example, various terpenoid acids, after conversion to the diastereomeric amides using (*R*)-(+)-NEA, were analyzed by HPLC to define the enantiomeric composition (eq 4).[4] Application of the procedure has been used to analyze the enantiomeric purity of several carboxylic acid derivatives.[14]

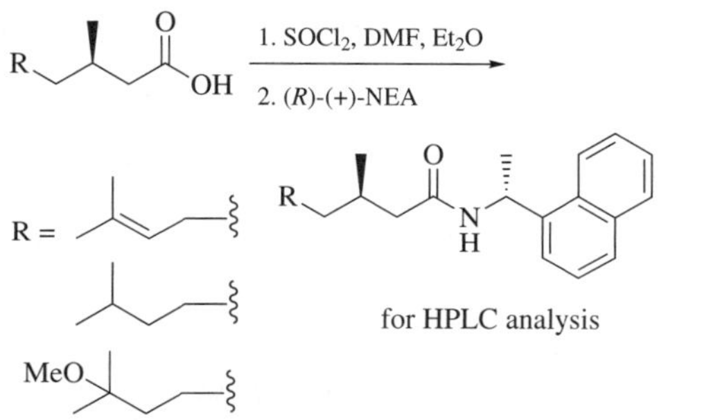

(4)

In some cases the resolution of the diastereomeric amides on silica gel is sufficiently large to achieve preparative separation. The preparative separation of diastereomeric hydroxy amides has proved useful in supplying quantities of enantiomerically pure lactones (eq 5).[15]

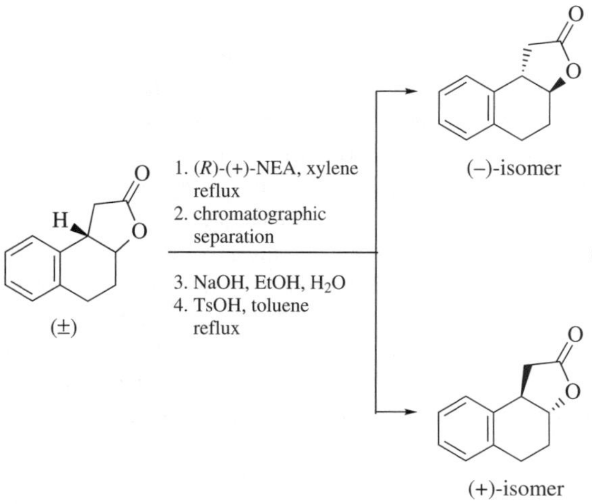

(5)

Nonracemic NEA is equally useful in preparing diastereomeric carbamates. Typically, the carbamates are derived from alcohols and (*R*)-(+)-naphthylethyl isocyanate,[16] which is conveniently prepared from (*R*)-(+)-NEA (see *(R)-1-(1-Naphthyl)ethyl Isocyanate*). However, the diastereomeric carbamates may also be produced by treating the chloroformate derivative of a racemic alcohol with (*R*)-(+)-NEA.[5] Preparative separation of racemic 1-heptyn-3-ol was achieved through chromatographic separation of the diastereomeric carbamates prepared via the chloroformate derivative (eq 6).[17] The carbamates are conveniently cleaved by the action of trichlorosilane (see *Trichlorosilane*).[18]

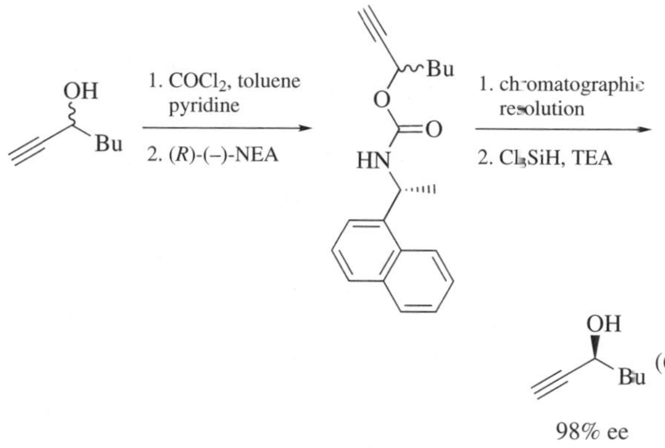

(6)

Chiral Solvating Agent. NEA is an effective chiral solvating agent for NMR determination of enantiomeric purity.[6,19] The combination of enantiomerically pure NEA (3–5 mol excess) and racemic solute causes the NMR spectra of the diastereomerically solvated enantiomers to differ. Since NEA is an efficient hydrogen-bond acceptor, it solvates better if the solute is a hydrogen-bond donor. (*R*)-(+)-NEA has been used to determine the enantiomeric purity of a variety of substrates.[20]

Chiral Stationary Phases for GC and HPLC. Enantiomerically pure NEA has been used to prepare a variety of chiral stationary phases for liquid,[21] gas,[22] and supercritical fluid[23] chromatography. These stationary phases are used to separate enantiomers without derivatization of the substrate with a chiral agent.

Related Reagents. (*R*)-1-(1-Naphthyl)ethyl Isocyanate.

1. (a) Sato, T.; Kawara, T.; Nishizawa, A.; Fujisawa, T. *Tetrahedron Lett.* **1980**, *21*, 3377. (b) Sato, T.; Naruse, K.; Fujisawa, T. *Tetrahedron Lett.* **1982**, *23*, 3587.
2. Mori, K.; Nukada, T.; Ebata, T. *Tetrahedron* **1981**, *37*, 1343.
3. Corey, E. J.; Snider, B. B. *J. Org. Chem.* **1974**, *39*, 256.
4. Bergot, B. J.; Anderson, R. J.; Schooley, D. A.; Henrick, C. A. *J. Chromatogr.* **1978**, *155*, 97.
5. Pirkle, W. H.; Adams, P. E. *J. Org. Chem.* **1979**, *44*, 2169.
6. Burlingame, T. G.; Pirkle, W. H. *J. Am. Chem. Soc.* **1966**, *88*, 4294.
7. Nabeya, A.; Endo, T. *J. Org. Chem.* **1988**, *53*, 3358.
8. Dale, J. A.; Dull, D. L.; Mosher, H. S. *J. Org. Chem.* **1969**, *34*, 2543.
9. Miyano, S.; Okada, S.; Hotta, H.; Takeda, M.; Suzuki, T.; Kabuto, C.; Yasuhara, F. *Bull. Chem. Soc. Jpn.* **1989**, *62*, 3886.
10. Samuelson, E. *Chem. Abstr.* **1924**, *18*, 1833.
11. Newman, P. *Optical Resolution Procedures for Chemical Compounds*; Manhattan College: New York, 1978; Vol. 1, p 230.
12. Mohacsi, E.; Leimgruber, W. *Org. Synth.* **1976**, *55*, 80.
13. Brown, E.; Viot, F.; Le Floc'h, Y. *Tetrahedron Lett.* **1985**, *26*, 4451.
14. (a) Eberhardt, R.; Glotzmann, C.; Lehner, H.; Schlögl, K. *Tetrahedron Lett.* **1974**, 4365. (b) Bergot, B. J.; Baker, F. C.; Lee E.; Schooley, D. A. *J. Am. Chem. Soc.* **1979**, *101*, 7432. (c) Mori, K.; Masuda, S.; Suguro, T. *Tetrahedron* **1981**, *37*, 1329. (d) Vandewalle, M.; Van der Eycken, J.;

Oppolzer, W.; Vullioud, C. *Tetrahedron* **1986**, *42*, 4035. (e) Hiratake, J.; Inagaki, M.; Yamamoto, Y.; Oda, J. *J. Chem. Soc., Perkin Trans. 1* **1987**, 1053.

15. Pirkle, W. H.; Adams, P. E. *J. Org. Chem.* **1980**, *45*, 4111.
16. Pirkle, W. H.; Hoekstra, M. S. *J. Org. Chem.* **1974**, *39*, 3904.
17. Overman, L. E.; Bell, K. L.; Ito, F. *J. Am. Chem. Soc.* **1984**, *106*, 4192.
18. Pirkle, W. H.; Hauske, J. R. *J. Org. Chem.* **1977**, *42*, 2781.
19. (a) Pirkle, W. H.; Beare, S. D. *J. Am. Chem. Soc.* **1967**, *89*, 5485. (b) Pirkle, W. H.; Burlingame, T. G. *Tetrahedron Lett.* **1967**, 4039. (c) Pirkle, W. H.; Beare, S. D. *Tetrahedron Lett.* **1968**, 2579.
20. (a) Weisman, G. R. In *Asymetric Synthesis*; Morrison, J. D., Ed.; Academic: New York, 1983; Vol. 1, Chapter 8. (b) Pirkle, W. H.; Hoover, D. J. In *Topics in Stereochemistry*; Wiley: New York, 1982; Vol. 13, p 263.
21. (a) Oi, N.; Nagase, M.; Doi, T. *J. Chromatogr.* **1983**, *257*, 111. (b) Pirkle, W. H.; Hyun, M. H. *J. Org. Chem.* **1984**, *49*, 3043. (c) Pirkle, W. H.; Hyun, M. H. *J. Chromatogr.* **1985**, *322*, 295. (d) Lloyd, M. J. B. *J. Chromatogr.* **1986**, *351*, 219. (e) Dappen, R.; Meyer, V. R.; Arm, H. *J. Chromatogr.* **1986**, *361*, 93.
22. (a) Weinstein, S.; Feibush, B.; Gil-Av, E. *J. Chromatogr.* **1976**, *126*, 97. (b) Oi, N.; Kitahara, H.; Inda, Y.; Doi, T. *J. Chromatogr.* **1981**, *213*, 137. (c) Oi, N.; Kitahara, H.; Inda, Y.; Doi, T. *J. Chromatogr.* **1982**, *237*, 297.
23. Bradshaw, J. S.; Aggarwal, S. K.; Rouse, C. A.; Tarbet, B. J.; Markides, K. E.; Lee, M. L. *J. Chromatogr.* **1987**, *405*, 169.

John M. McGill
Eli Lilly and Company, Lafayette, IN, USA

(*R*)-1-(1-Naphthyl)ethyl Isocyanate

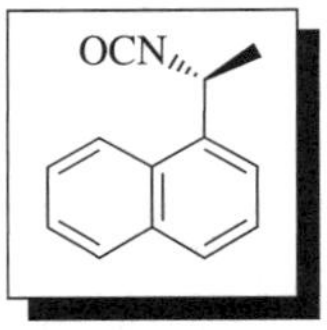

(*R*)-(−) [*42340-98-7*]	$C_{13}H_{11}NO$	(MW 197.25)
(*S*)-(+) [*73671-79-1*]		

(enantiomeric resolving agent for alcohols,[1,2] thiols,[1,3] sulfonamides,[4] and amines[5] (forms diastereomers separable by chromatography or crystallization); used in creating enantiomeric stationary phases for liquid chromatography;[6,7] reagent in the synthesis of chiral nonracemic allenes[8])

Alternate Name: NEI; (*R*)-1-isocyano-1-naphthylethane.

Physical Data: bp 106–108 °C; fp 93 °C; *d* 1.13 g cm^{-3}; optical rotation $[\alpha]_D^{20} = -47°$ (*c* 3.5, toluene), $[\alpha]_D^{24.1} = -50.5°$ (*c* 27.9, benzene).

Solubility: freely sol benzene, toluene, and a variety of dry organic solvents; reacts with water.

Form Supplied in: colorless liquid; available commercially >99% pure.

Preparative Methods: from (*R*)-(+)-*1-(1-Naphthyl)ethylamine* by reaction with *Phosgene* or by a method using *Trichlorosilane*.[1]

Handling, Storage, and Precautions: stable when stored between 0 and 4 °C; reacts with water; highly toxic; should be handled with proper skin and eye protection in a well-ventilated fume hood.

Resolution of Alcohols. The reagent (*R*)-(−)-1-(1-naphthyl)ethyl isocyanate, (*R*)-NEI, as well as its enantiomer, (*S*)-NEI, forms diastereomeric carbamates with racemic secondary alcohols (eq 1);[1,2] the reaction may be facilitated by base or Lewis acid catalysts.[4] The diastereomers can usually be separated by liquid chromatography on silica or alumina, providing a convenient means for analysis or preparative purification of the enantiomers: the resolved alcohols are recovered in high yield under mild, or nonracemizing, conditions by treatment with trichlorosilane/triethanolamine (TEA) (eq 1).[9]

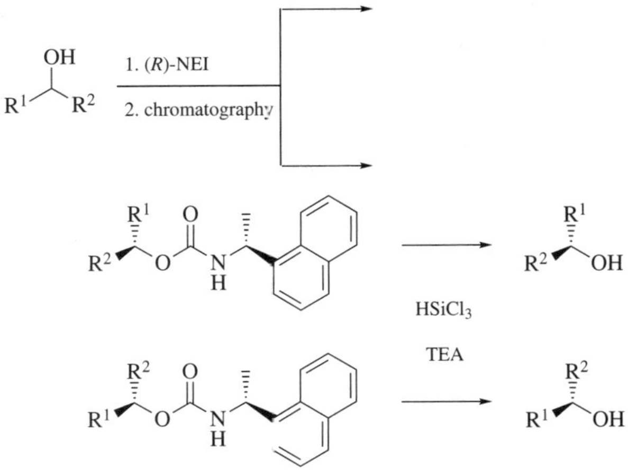

(1)

This general strategy has found wide application in synthesis. Enantiomerically pure cyano alcohols obtained in this way are starting materials in the synthesis of uni- and multicyclic lactones.[10–13] High purity enantiomeric epoxides can be prepared by resolution of the appropriate alcohol precursor followed by ring closure.[14] Propargylic alcohol enantiomers resolved by this technique are intermediates in the synthesis of the four stereoisomers of 1,2,3-decanetriol,[15] and of the sesquiterpene fungal metabolite (+)-sterpurene.[16,17] The purification of stereoisomers of secondary alkan- and alkenols by this method is an essential step in the synthesis of biologically active enantiomers of 8-methyl-2-decanol propanoate,[18] and of the germination inhibitor (−)-gloeosporone.[19] The general method given in eq 1 is used to isolate enantiomerically pure intermediates in the synthesis of the fungal metabolites ascofuranone and ascofuranol,[20] of the fungitoxic hydroquinone zonarol,[21] and of naturally occurring *C*-nucleosides and their analogs.[22] A hindered *endo* alcohol is also resolved after derivatization with (*R*)-NEI.[23] In the synthesis of vitamin D_3 metabolites, an enynyl alcohol intermediate is resolved with the help of both (*R*)- and (*S*)-NEI using crystallization rather than chromatography to separate the diastereomers.[24]

Indirect analysis of the enantiomeric ratio of an alcohol can be accomplished via separation and quantitation of the diastereomers of eq 1 by high-performance liquid chromatography (HPLC), gas–liquid chromatography (GLC), or supercritical fluid chromatography (SFC). For example, diastereomeric diacylglycerol 1-(1-naphthyl)ethyl carbamates are separated by HPLC on a silica gel column,[25] as are the diastereomeric derivatives of the tertiary monoterpene alcohol linalool,[26] diastereomers of (unsymmetrically substituted) 1,3-dialkylglycerol ethers are separable by GLC,[27] and secondary alcohol enantiomers are derivatized with (*R*)-NEI and separated by SFC using several different stationary phases.[28] The use of a bonded amine HPLC column is reported for the analysis of enantiomers of several 2,3-epoxy-1-propanols following derivatization with (*R*)- and (*S*)-NEI.[29] Enantiomeric purity of a thiol, a derivative of the enkephalinase inhibitor thiorphan, is also determined after derivatization with (*R*)-NEI.[3]

Resolution of Amines. Amines react with (*R*)- or (*S*)-NEI to form the corresponding urea diastereomers which can be separated in a manner analogous to the alcohols in eq 1. Secondary amines thus resolved can be recovered from the diastereomers by hydrolysis, as in the synthesis of chiral nonracemic lactams,[5] or by decomposition in refluxing alcohols, as demonstrated by the resolution of several amine drugs.[30]

Analysis of the enantiomeric ratios of several β-blocking drugs (1-aryloxy-3-isopropylamino-2-propanol derivatives) is carried out by HPLC with UV or fluorescence detection after derivatization with (*R*)-NEI or (*R*)-(+)-1-(1-phenyl)ethyl isocyanate (in a reversed-phase system),[31,32] or (*S*)-NEI (on silica gel);[33] only the amine function of the drugs reacts with the NEI; the hydroxy group does not.[31] Similar schemes for HPLC determination of enantiomeric purity of tetrahydrofolate derivatives[34] and of fluoxetine[35] are also reported.

In analysis, the original compound need not be recovered after separation, and therefore primary amines such as amino acids can be derivatized and analyzed in the same manner.[36] (*S*)-NEI is employed as a derivatizing reagent in an Edman-like sequencing scheme to assess the extent of racemization of amino acid residues in synthetic peptides.[37]

Enantiomeric Stationary Phases. Chiral nonracemic chromatographic stationary phases prepared from β-cyclodextrin, derivatized with (*R*)- and (*S*)-NEI, and covalently bonded to a silica support are useful for the direct separation of enantiomers of a wide variety of compounds in both normal-phase and reversed-phase HPLC.[6,7,38]

Determination of Absolute Configuration of Enantiomeric Compounds. Empirical rules for the elution order in normal-phase chromatography of diastereomeric carbamates are used to assign absolute configurations of chiral nonracemic compounds resolved by the general method of eq 1.[2] Owing to the inflexibility of the carbamate linkage, the relative positions of the most hydrophobic (or repulsive) group on the original compound and the naphthyl group of NEI are fixed in either a *syn* or an *anti* configuration in the two diastereomers. The *syn* conformer is likely to elute last, as its less repulsive moieties are more directly available for interaction with the silica surface.[2] The rules fail occasionally.[39] ^{1}H NMR[39,40] and single crystal X-ray diffraction[41,42] of purified diastereomers are also used in determining absolute conformation.

Synthesis of Chiral Nonracemic Allenes. Enantiomerically enriched chiral allenes can be prepared by derivatization of a racemic propargyl alcohol with (*R*)-NEI, followed by chromatographic separation of the resulting diastereomers and reaction of the purified diastereomers with lithium dialkylcuprates at −78 °C, as shown in eq 2.[8,43]

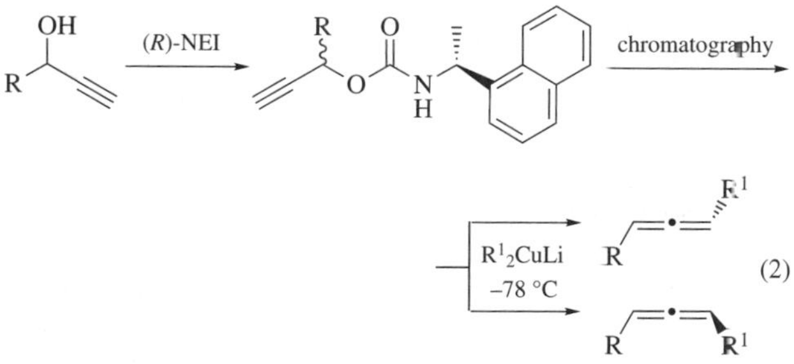

(2)

Alternative Reagents. 1-(1-Phenyl)ethyl isocyanate (PEI) is a cheaper alternative to NEI; however, diastereomers formed from NEI are usually easier to separate by liquid chromatography.[2,5,9] Other alternatives are also used, e.g. in determining the enantiomeric purity of 3-aminoquinuclidine, PEI, NEI, 2,3,4,6-tetraacetyl-β-D-glucopyranosyl isothiocyanate, and (*R,R*)- and (*S,S*)-*O,O*-dibenzoyltartaric acid anhydride are all employed successfully.[44] *Mandelic Acid* and Mosher's reagent, α-methoxytrifluoromethylphenylacetyl chloride, may not be quite as effective as NEI when resolutions are carried out by liquid chromatography.[2]

1. Pirkle, W. H.; Hoekstra, M. S. *J. Org. Chem.* **1974**, *39*, 3904.
2. Pirkle, W. H.; Finn, J. In *Asymmetric Synthesis*; Morrison, J. D., Ed.; Academic: New York, 1983; Vol. 1, pp 87–124.
3. Evans, D. A.; Mathre, D. J.; Scott, W. L. *J. Org. Chem.* **1985**, *50*, 1830.
4. Irie, H.; Nishimura, M.; Yoshida, M.; Ibuka, T. *J. Chem. Soc., Perkin Trans. 1* **1989**, 1209.
5. Pirkle, W. H.; Robertson, M. R.; Hyun, M. H. *J. Org. Chem.* **1984**, *49*, 2433.
6. Armstrong, D. A.; Stalcup, A. M.; Hilton, M. L.; Duncan, J. D.; Faulkner Jr., J. R.; Chang, S.-C. *Anal. Chem.* **1990**, *62*, 1610.
7. Armstrong, D. W.; Chang, C.-D.; Lee, S. H. *J. Chromatogr.* **1991**, *539*, 83.
8. Pirkle, W. H.; Boeder, C. W. *J. Org. Chem.* **1978**, *43*, 1950.
9. Pirkle, W. H.; Hauske, J. R. *J. Org. Chem.* **1977**, *42*, 2781.
10. Pirkle, W. H.; Adams, P. E. *J. Org. Chem.* **1978**, *43*, 378.
11. Pirkle, W. H.; Adams, P. E. *J. Org. Chem.* **1979**, *44*, 2169.
12. Pirkle, W. H.; Adams, P. E. *J. Org. Chem.* **1980**, *45*, 4111.
13. Mori, K.; Sasaki, M. *Tetrahedron* **1980**, *36*, 2197.
14. Pirkle, W. H.; Rinaldi, P. L. *J. Org. Chem.* **1978**, *43*, 3803.
15. Rinaldi, P. L.; Levy, G. C. *J. Org. Chem.* **1980**, *45*, 4348.
16. Gibbs, R. A.; Okamura, W. H. *J. Am. Chem. Soc.* **1988**, *110*, 4062.
17. Gibbs, R. A.; Bartels, K.; Lee, R. W. K.; Okamura, W. H. *J. Am. Chem. Soc.* **1989**, *111*, 3717.
18. Sonnet, P. E.; Carney, R. L.; Henrick, C. *J. Chem. Ecol.* **1985**, *11*, 1371.
19. Matsushita, M.; Yoshida, M.; Zhang, Y.; Miyashita, M.; Irie, H.; Ueno, T.; Tsurushima, T. *Chem. Pharm. Bull.* **1992**, *40*, 524.
20. Mori, K.; Takechi, S. *Tetrahedron* **1985**, *41*, 3049.
21. Mori, K.; Komatsu, M. *Bull. Soc. Chim. Belg.* **1986**, *95*, 771
22. Sato, T.; Hayakawa, Y.; Noyori, R. *Bull. Chem. Soc. Jpn.* **1984**, *57*, 2515.
23. Kluge, A. F.; Kertesz, D. J.; O-Yang, C.; Wu, H. Y. *J. Org. Chem.* **1987**, *52*, 2860.
24. Lee, A. S.; Norman, A. W.; Okamura, W. H. *J. Org. Chem.* **1992**, *57*, 3846.
25. Laakso, P.; Christie, W. W. *Lipids* **1990**, *25*, 349.
26. Rudmann, A. A.; Aldrich, J. R. *J. Chromatogr.* **1987**, *407*, 324.

27. Sonnet, P. E.; Piotrowski, E. G.; Boswell, R. T. *J. Chromatogr.* **1988**, *436*, 205.
28. Sakaki, K.; Hirata, H. *J. Chromatogr.* **1991**, *585*, 117.
29. Kennedy, J. H.; Weigel, L. O. *Chirality* **1992**, *4*, 132.
30. Schönenberger, B.; Brossi, A. *Helv. Chim. Acta* **1986**, *69*, 1486.
31. Gübitz, G.; Mihellyes, S. *J. Chromatogr.* **1984**, *314*, 462.
32. Jira, T.; Toll, C.; Vogt, C.; Beyrich, T. *Pharmazie* **1991**, *46*, 432.
33. Piquette-Miller, M.; Foster, R. T. *J. Chromatogr.* **1990**, *533*, 300.
34. Rees, L.; Suckling, C. J.; Valente, E.; Wood, H. C. S. In *Chemistry and Biology of Pteridines: Pteridines and Folic Acid Derivatives: Proceedings of the 7th International Symposium on Pteridines and Folic Acid Derivatives, Chemical, Biological, and Clinical Aspects*; Blair, J. A., Ed.; de Gruyter: Berlin, 1983.
35. Robertson, D. W.; Krushinski, J. H.; Fuller, R. W.; Leander, J. D. *J. Med. Chem.* **1988**, *31*, 1412.
36. Dunlop, D. S.; Neidle, A. *Anal. Biochem.* **1987**, *165*, 38.
37. Davies, J. S.; Enjalbal, C.; Llewellyn, G. *J. Chem. Soc., Perkin Trans. 2* **1992**, 1225.
38. Berthod, A.; Chang, S. C.; Armstrong, D. W. *Anal. Chem.* **1992**, *64*, 395.
39. Pirkle, W. H.; Simmons, K. A.; Boeder, C. W. *J. Org. Chem.* **1979**, *44*, 4891.
40. Sonnet, P. E.; Dudley, R. L.; Osman, S.; Pfeffer, P. E.; Schwartz, D. *J. Chromatogr.* **1991**, *586*, 255.
41. Brooks, D. W.; Bevinakatti, H. S.; Powell, D. R. *J. Org. Chem.* **1985**, *50*, 3779.
42. Tacke, R.; Wuttke, F.; Henke, H. *J. Organomet. Chem.* **1992**, *424*, 273.
43. Pirkle, W. H.; Boeder, C. W. *J. Org. Chem.* **1978**, *43*, 2091.
44. Demian, I.; Gripshover, D. F. *J. Chromatogr.* **1989**, *466*, 415.

Firoz D. Antia
The R. W. Johnson Pharmaceutical Research Institute, Spring House, PA, USA

Norephedrine–Borane[1]

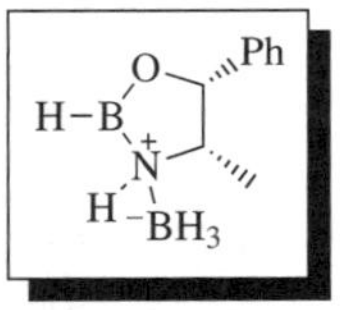

(+)
[154145-14-9] $C_9H_{15}B_2NO$ (MW 174.87)
(−)
[161512-50-1]

(chirally modified hydride reagent[1])

Alternate Name: (4*S*,5*R*)-(4-methyl-5-phenyl-1,3,2-oxazaborolidine)·borane

Physical Data: the title reagent has never been isolated; the structure shown above has been assigned on the basis of related studies[2] and by analogy to other well-characterized 1,3,2-oxazaborolidines.[3]

Solubility: THF.

Preparative Methods: a 1M THF solution of BH_3 (3.0 mmol) was added to a THF solution of (−)-norephedrine (1.5 mmol) at −30 °C and the resulting mixture was warmed to 20 °C; the thus-formed chiral hydride reagent was used in situ for enantioselective reductions.

Handling, Storage, and Precautions: moisture sensitive; handle under a dry, inert atmosphere.

Enantioselective Reductions. This chiral hydride reagent reduces aromatic ketones to the corresponding alcohols with high enantioselectivity (eq 1).[4]

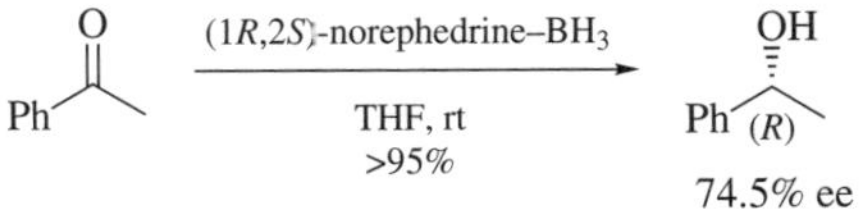

(1)

Anti and *syn* ketoxime ethers are also reduced by this reagent to the corresponding amines in up to 92% ee. The absolute configuration of the resulting amine is dependent on the geometry of the starting oxime ethers: *anti* oximes give (*S*) configurated amines whereas *syn* oximes afford the (*R*) antipodes (eqs 2 and 3); see Table 1.[4,5]

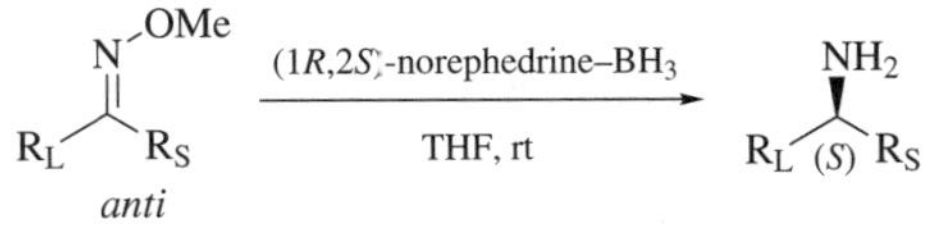

(2)

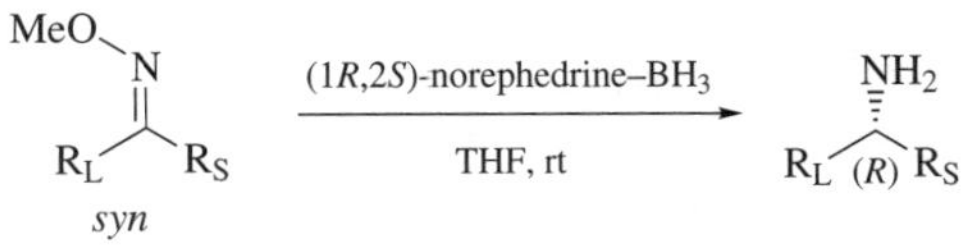

(3)

Table 1 Enantioselective Oxime Reductions[2,3]

Oxime	R_L	R_S	Config	ee (%)	Yield (%)
anti	Ph	*p*-TolCH$_2$	*S*	92	64
syn	*p*-TolCH$_2$	Ph	*R*	92	58
anti	2-Naphthyl	Me	*S*	92	73
syn	Me	2-Naphthyl	*R*	92	–
anti	Ph(CH$_2$)$_2$	Me	*S*	86	40
syn	Me	Ph(CH$_2$)$_2$	*R*	81	46
anti	Ph	Me	*S*	93	99

Similar results can be obtained if the borane is substituted for the system *Sodium Borohydride/Aluminum Chloride*.[6]

On the basis of experimental[3b] as well as theoretical[7] studies on similar systems, the mechanism of carbonyl reduction is expected to involve:

1. In situ formation of the 1,3,2-oxazaborolidine from borane and the 1,2-amino alcohol with 2 equiv of H_2 uptake;
2. coordination between the N atom of the oxazaborolidine and a second equivalent of borane;
3. coordination at the heterocyclic B atom by the oxygen of the solvent and then by that of the carbonyl compound;
4. hydride transfer from the NBH_3^- unit to the substrate via a six-membered transition state.

Although these reactions are expected to be catalytic in oxazaborolidine, the reported examples use the chiral ligand in stoichiometric amounts.

Related Reagents. Tetrahydro-1-methyl-3,3-diphenyl-1*H*,3*H*-pyrrolo[1,2-*c*][1,3,2]oxazaborole.

1. (a) Nishizawa, M.; Noyori, R. *Comprehensive Organic Synthesis* **1991**, *8*, Chapter 1.7. (b) Midland, M. *Asymmetric Synthesis*; Academic: New York, 1983; Vol. 2. (c) Midland, M. *Chem. Rev.* **1989**, *89*, 1553. (d) Seyden-Penne, J. *Reductions by the Alumino- and Borohydrides in Organic Synthesis*; VCH: New York, 1991. (e) Wallbaum, S.; Martens, J. *Tetrahedron: Asymmetry* **1992**, *3*, 1475.
2. Itsuno, S.; Hirao, A.; Nakahama, S.; Yamazaki, N. *J. Chem. Soc., Perkin Trans. 1* **1983**, 1673.
3. (a) Joshi, N. N.; Srebnik, M.; Brown, H. C. *Tetrahedron Lett.* **1989**, *30*, 5551. (b) Corey, E. J.; Bakshi, R. K.; Shibata, S. *J. Am. Chem. Soc.* **1987**, *109*, 5551. (b) Corey, E. J.; Azimioara, M.; Sarshar, S. *Tetrahedron Lett.* **1992**, *33*, 3429.
4. Didier, E.; Loubinoux, B.; Ramos Tombo, G. M.; Rihs, G. *Tetrahedron* **1991**, *47*, 4941. (b) Komeyoshi, Y.; Suzukamo, T.; Hamada, K.; Nishioka, T. Jpn. Patent 62 10 024 [87 10 024] (*Chem. Abstr.* **1987**, *106*, 175 410t).
5. (a) Sakito, Y.; Yoneyoshi, Y.; Suzukamo, G. *Tetrahedron Lett.* **1988**, *29*, 223. (b) Sakito, Y; Suzukamo, G.; Yoneyoshi, Y. Eur. Pat. Appl. 237 305 (*Chem. Abstr.* **1988**, *108*, 150 040a).
6. Konya, N.; Suzukamo, K.; Komeyoshi, Y. Jpn. Patent 02 311 446 [90 311 446] (*Chem. Abstr.* **1991**, *114*, 228 361b).
7. Nevalainen, V. *Tetrahedron: Asymmetry* **1992**, *3*, 1441.

Giovanni Poli
Università di Firenze, Italy

(*R,R*)-(−)-NORPHOS, (*S,S*)-(+)-NORPHOS

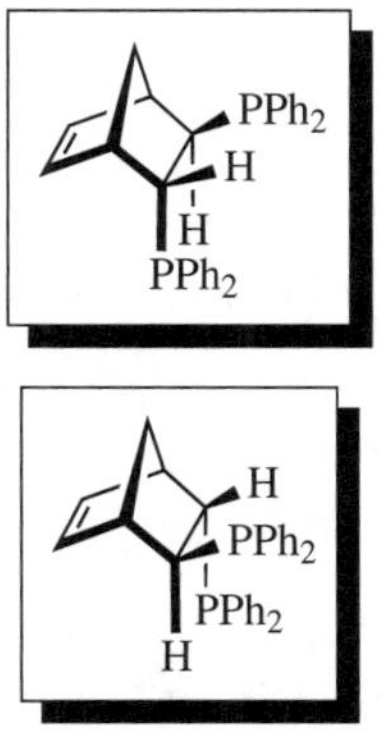

[71042-54-1]	$C_{31}H_{28}P_2$	(MW 462.51)
[71042-55-2]	$C_{31}H_{28}P_2$	(MW 462.51)

(optically active reagent used as a ligand in homogeneous transition metal-chiral phosphine catalyst systems for asymmetric reactions)

Alternate Name: (2*R*,3*R*)-(−)-2,3- Bis(diphenylphosphino)bicyclo[2.2.1]hept-5-ene, (2*S*,3*S*)-(+)-2,3- Bis(diphenylphosphino)bicyclo[2.2.1]hept-5-ene.

Physical Data: (*R,R*)-(−)-NORPHOS: $[\alpha]^{20}{}_{578}$ −43.5 °(*c* 1, $CHCl_3$), mp 129–130 °C;[1] (*S,S*)-(+)-NORPHOS: $[\alpha]^{20}{}_{578}$ +45 °(*c* 1, $CHCl_3$), mp 129–130 °C.[1]

Solubility: soluble in chloroform and other common organic solvents.

Form Supplied in: both (*S,S*)-(+)- and (*R,R*)-(−)-NORPHOS are available from Strem Chemicals, 7 Mulliken Way, Dexter Industrial Park, Newburyport, MA 01950-9899. (*S,S*)-(+)-NORPHOS is supplied as a colorless microcrystalline solid: mp 112–115 °C; (*R,R*)-(−)-NORPHOS is supplied as a colorless microcrystalline solid: mp 116–119 °C.

Preparative Methods: from the Diels–Alder adduct of cyclopentadiene and *trans*-vinylenebis(diphenylphosphorane oxide)[2] followed by optical resolution of the resulting [4 + 2] cycloadduct and subsequent reduction with trichlorosilane (eq 1).[3]

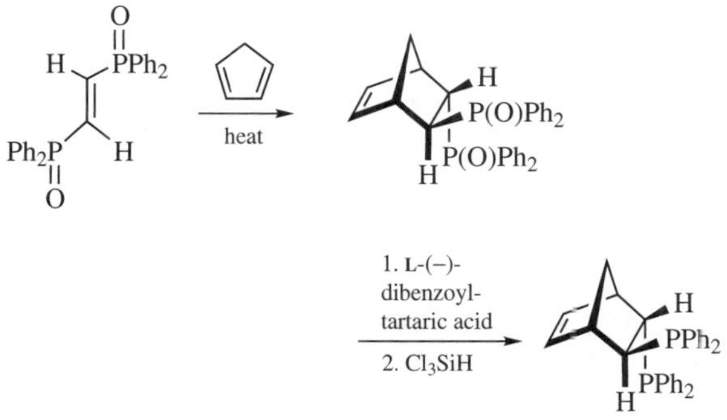

(1)

Purification: can be purified by recrystallization from acetone.[1]

Handling, Storage, and Precautions: NORPHOS is air-sensitive; storage, handling, and other operations that involve NORPHOS should be performed under an inert atmosphere. In general, alkyldiarylphosphines are irritants; skin contact should be avoided, and care should be exercised to avoid vapor inhalation.

Enantioselective Hydrogenation of Alkene C=C Double Bonds. (*S,S*)-NORPHOS has been employed as a catalyst in combination with Rh(I) for enantioselective hydrogenation of alkene carbon-carbon double bonds in a variety of substrates. A representative sampling of these asymmmetric hydrogenations is shown in Table 1.[4–8]

In addition, a Rh-(*R,R*)-NORPHOS catalyst has been used to promote enantioselective transfer hydrogenation of the C=C double bond in (*Z*)-α-(acetylamino)cinnamic acid and in (*Z*)-α- and (*E*)-α-(benzoylamino)-2-butenoate by using 80% aqueous formic acid as the source of H_2.[9] Optical yields were improved by the addition of sodium formate; representative results are presented in Table 2.[9] Comparable, but generally somewhat lower, optical yields were obtained by using other Rh-(biphosphine ligand) catalysts, e.g., biphosphine ligand = (*R,S*)-(+)-**BPPFA** (**2**),[10] (*R*)-(+)-**PROPHOS**(**3**),[11] or (*R,R*)-(−)-**DIOP** (**1**).[12]

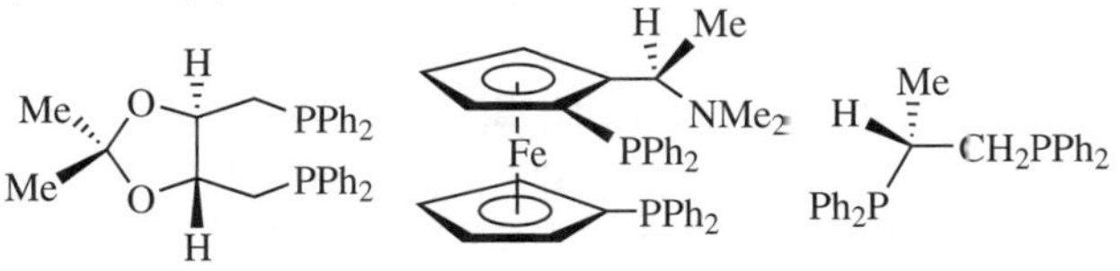

1 **2** **3**

Insoluble, immobilized hydrogenation catalysts have been prepared by impregnating a variety of solid supports (e.g.,

Table 1 Enantioselective hydrogenations of prochiral alkenes catalyzed by a transition-metal–NORPHOS complex

Substrate	Catalyst	Conditions	Reactant ratios	Product	ee (%) (configuration) [conversion, %]	Ref.
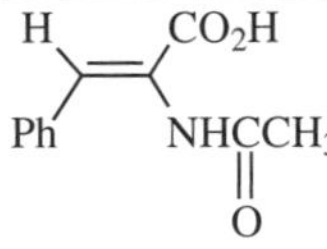	$[Rh(cod)Cl]_2$-(-)(*R,R*)-NORPHOS	H_2 (1.1 bar) EtOH, 10–20 h, 25 °C	substrate: Rh: ligand = (80:1:2.2)	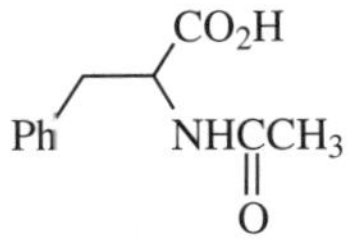	95 (*R*) [100]	4
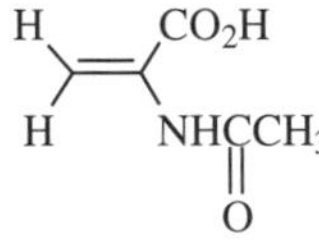	$[Rh(cod)Cl]_2$-(-)(*R,R*)-NORPHOS	H_2 (1.1 bar) EtOH, 10–20 h, 25 °C	substrate: Rh: ligand = (130:1:2.2)	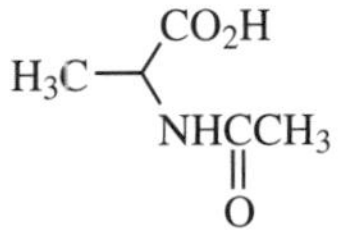	90 (*R*) [100]	4
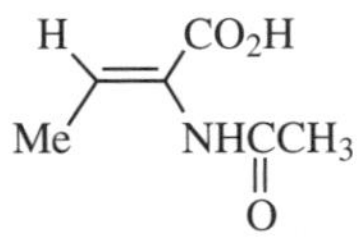	$[Rh(cod)Cl]_2$-(-)(*R,R*)-NORPHOS	80% aq HCO_2H, HCO_2Na, 120 °C, 16 h (transfer hydrogenation)	1.8 mmol substrate; Rh:ligand = 1:1.09-1.17; Rh:substrate = 1:36–45	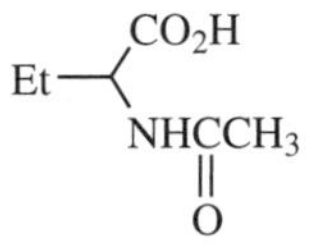	47 (*S*) [- -]	5
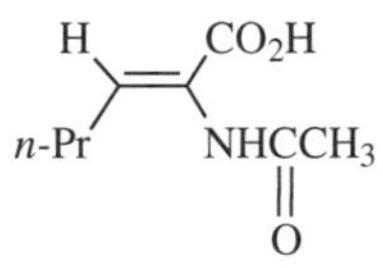	$[Rh(norbornadiene)Cl]_2$-(-)(*R,R*)-NORPHOS	H_2 (49 psi(gauge)), MeOH, 9 min, 25 °C	substrate: catalyst = 100:1	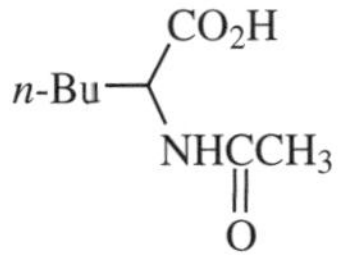	79 (*R*) [- -]	6
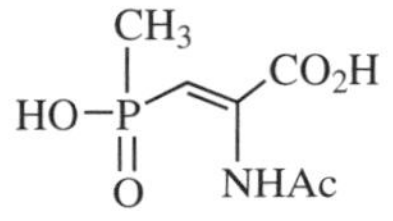	$[Rh(cod)_2]$ BF_4-(-)(*R,R*)-NORPHOS	H_2 (0.25–0.30 MPa), MeOH, 22 h, 25–50 °C	0.125–0.25 M substrate; substrate: catalyst = 300:1	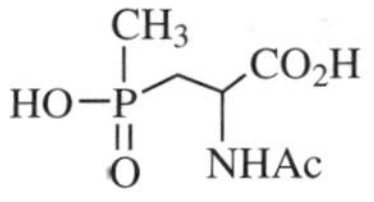	87.2 (*R*) [100]	7
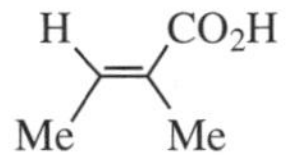	Ru(2-methylallyl)$_2$-(-)(*R,R*)-NORPHOS	H_2 (3 atm), MeOH, 24–60 h, 20 °C	1 mmol substrate; quantity of catalyst not specified	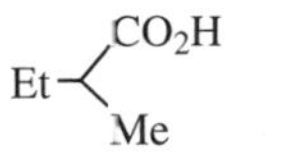	4 (*S*) [100]	8

$BaSO_4$, cellulose, silica gel, alumina, AgCl, charcoal) with {[Rh(COD)[(*R,R*)-NORPHOS]}PF_6 and with {[Rh(COD)[(*S,S*)-NORPHOS]}PF_6 catalysts.[13] Interestingly, while the degree of optical induction into the reduction product was observed to increase with the first few repeated uses, catalytic hydrogenation activity decreased with repeated use.[13]

Diastereoselective Hydrogenation of C=N Double Bonds. Immobilized Rh-NORPHOS catalysts have been employed for diastereoselective heterogeneous hydrogenation of the C=N double bonds in the pyrazine ring of folic acid (eq 2).[14] With (*S,S*)-NORPHOS, optically active tetrahydrofolic acid that possesses the (6*S*)- configuration at the newly formed asymmetric center was obtained in 96–98% chemical yield (18–21% de). When (*R,R*)-NORPHOS was used for this purpose, tetrahydrofolic acid with the (6*R*)- configuration at the new asymmetric center was obtained in 98–99% chemical yield (11–13% de). Best results were obtained with (*S*)-(−)-1,4-bis(diphenylphosphoanyl)pentane(**4**) [i.e., (*S*)-(−)-**1,4-BDPP**],[15] which afforded (6*S,S*)-tetrahydrofolic acid in 96–98% chemical yield (20–24% ee).[14]

Enantioselective Hydrosilylation of C=N Double Bonds in Ketoximes and Ketimines. Homogeneous enantioselective hydrosilylation of prochiral alkyl aryl ketoximes has been carried out by using a Rh-(*R,R*)-NORPHOS catalyst.[16] Thus, hydrosilylation of *t*-butyl phenyl ketoxime in the presence of $[Rh(COD)Cl]_2$-(*R,R*)-NORPHOS followed by aqueous acidic work-up afforded the corresponding amine (eq 3) [16.5% ee, (*S*)], which became inverted to 15.0% ee (*R*) when this reaction was performed in the presence of added ammonium hexafluorophosphate (Rh:NH_4PF_6 = 1.1, CH_2Cl_2 solvent).[16]

Table 2 Enantioselective transfer-hydrogenations of prochiral alkenes catalyzed by a transition-metal–NORPHOS complex

w Substrate	Catalyst	Hydrogen transfer agent (conditions)	Reactant ratios	Product	ee (%)	Ref.
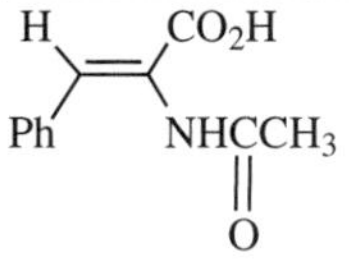	$[Rh(cod)Cl]_2$-(-)(*R,R*)-NORPHOS	80% aq HCO_2H, HCO_2Na, 120 °C, 6–16 h	Rh:ligand = 1:1.02-1.12; Rh:substrate = 1:33-41	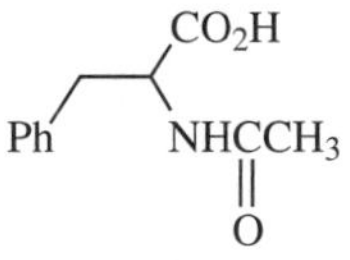	67 ± 5 (*S*)	9
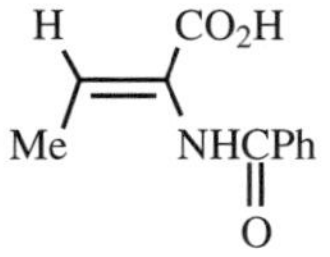	$[Rh(cod)Cl]_2$-(-)(*R,R*)-NORPHOS	80% aq HCO_2H, HCO_2Na, 120 °C, 14.5–25 h	Rh:ligand = 1:1.09-1.17; Rh:substrate = 1:36-45	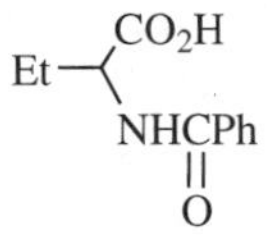	47 ± 9 (*S*)	9
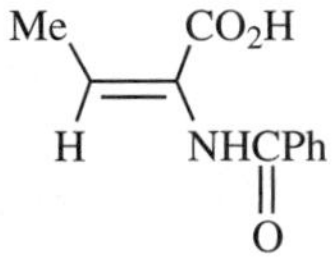	$[Rh(cod)Cl]_2$-(-)(*R,R*)-NORPHOS	80% aq HCO_2H, HCO_2Na, 120 °C, 6.5 h	Rh:ligand = 1:1.02-1.14; Rh:substrate = 1:41-42	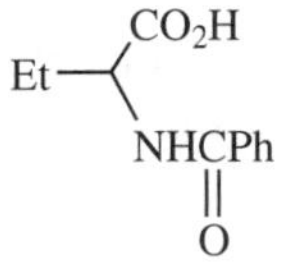	9 ± 2 (*S*)	9

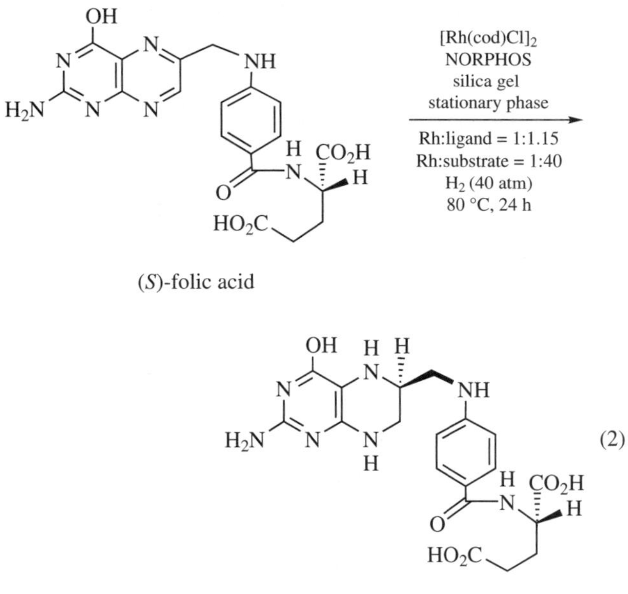

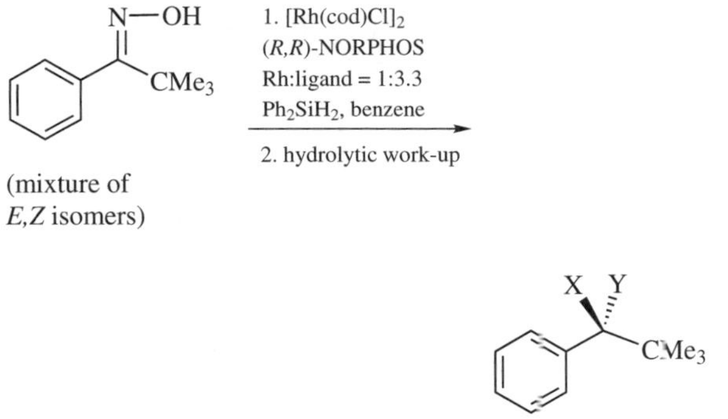

(*S*) [X = NH_2, Y = H], 68%, 16.5±0.8% ee, obtained by using Rh:substrate = 1:100, 0 °C → 20 °C, 96 h
(*R*) [X = H, Y = NH_2], 35%, 15.0±4.0% ee, obtained by using Rh:substrate = 1:200, −10 °C → 20 °C, 96 h

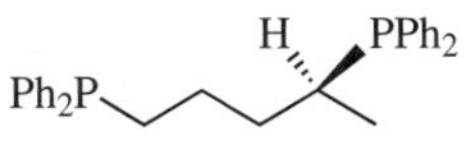

4

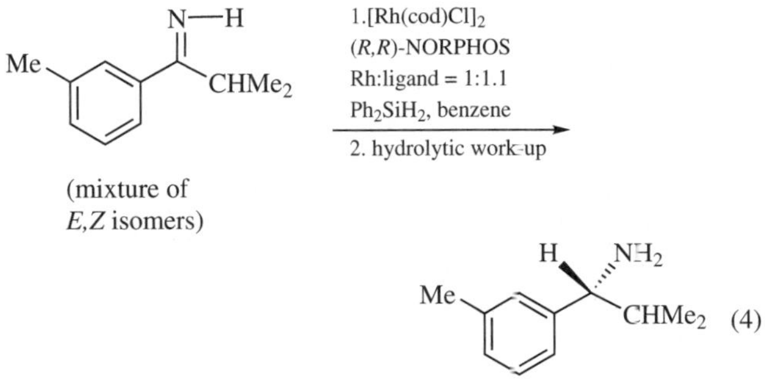

(*R*), 50%, 1.5±0.5% ee, 0 °C → 20 °C, 72 h

Similar enantioselective hydrosilylation of isopropyl *m*-tolyl ketimine afforded the corresponding (*R*) amine in 50% chemical yield but with only 1.5% ee (eq 4). By way of contrast, the corresponding hydrosilylation reaction, when performed in the presence of $[Rh(COD)Cl]_2$-**1**, produced the same (*R*)- amine in 60% chemical yield and with 13.8±1.1% ee.[16]

Enantioselective Hydrosilylation of C=O Double Bonds in Ketones. The use of Rh-phosphorane catalyst systems to promote asymmetric hydrosilylation of prochiral ketones with silanes of the type R_3SiH has met with only limited success. Thus, hydrosilylation of acetophenone with Ph_2SiH_2 promoted by $[Rh(COD)Cl]_2$-(*S,S*)-DIOP[12] catalyst afforded the (*S*)-(−)-phenylmethylcarbonyl with an optical yield of 32% ee.[17] Similarly, the use of a Rh-NORPHOS catalyst in this reaction proceeded with an optical induction of only 16% ee.[17,18]

Enantioselective Baeyer–Villiger Oxidation. A cationic platinum-(*R,R*)-NORPHOS catalyst has been reported to promote enantioselective Baeyer–Villiger oxidation of cyclic ketones to lactones. Thus, $\{[(R,R)\text{-NORPHOS}]Pt(\mu\text{-OH})\}_2(BF_4)_2$ catalyzes enantioselective oxidation of 4-methyl- and 4-phenylcyclohexanones by 35% aqueous H_2O_2 to produce the corresponding substituted ε-caprolactone in low optical yield (eq 5).[19] Replacement of (*R,R*)-NORPHOS in the Pt complex by other optically active diphosphines, e.g., (*R*)-**BINAP** (**5**) and (*S,S*)**-BPPM** (**6**) raised the optical yield of product lactone to ca. 50–70%.[19,20,21]

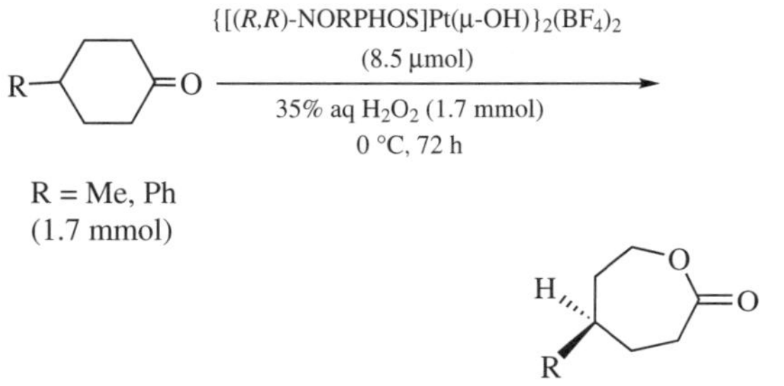

(5)

R = Me: 0.13 mmol;
optical yield: 7% ee (+)
R = Ph: 0.030 mmol;
optical yield: 13% ee (+)

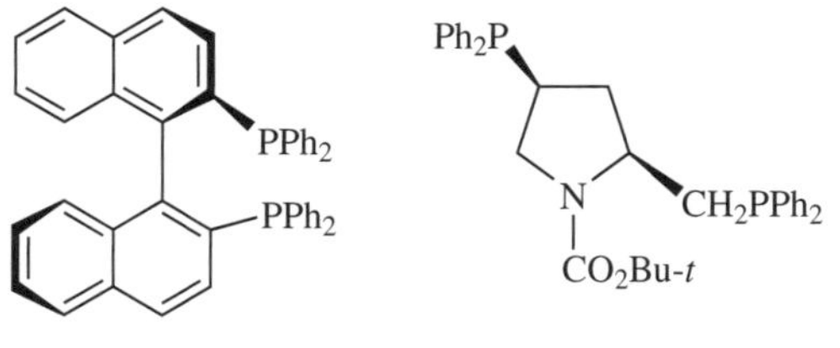

5 **6**

Enantioselective Conjugate Addition. A Cu-(*S,S*)-NORPHOS catalyst has been used to promote conjugate addition of diethylzinc to α,β-unsaturated ketones (eq 6), e.g., cyclohexen-2-one, chalcone, and benzalacetone.[22] The use of (*S,S*)-**CHIRAPHOS** (**7**) and (*R*)-**PROPHOS** (**8**) afforded (*S*)-3-ethylcyclohexanone in somewhat improved chemical and optical yields relative to those obtained with Cu-(*S,S*)-NORPHOS catalyst.[22,23,24]

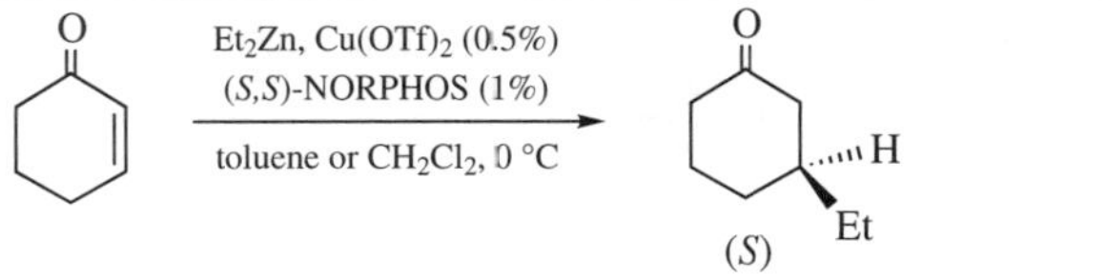

(6)

toluene, 0 °C, 3 h:
chemical yield, 75% (28% ee)
CH_2Cl_2, 0 °C, 1.5 h:
chemical yield, 81% (44% ee)

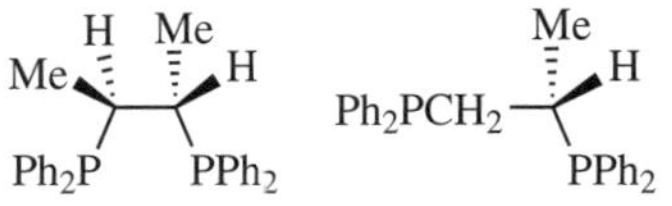

7 **8**

Enantioselective Allylic Alkylation (Allylation). Asymmetric allylation of the benzophenone imine of glycine methyl ester has been performed by using Pd-NORPHOS catalysts. When the reaction was performed by using $Pd(dba)_2$ with (*R,R*-NORPHOS), the corresponding allylated product was obtained (*S*-configuration, optical yield 32% ee) (eq 7).[25] When **1** was employed instead as the chiral ligand in this reaction, the same allylation product was obtained in slightly lower chemical yield (50%) but in higher optical yield (55% ee).[25]

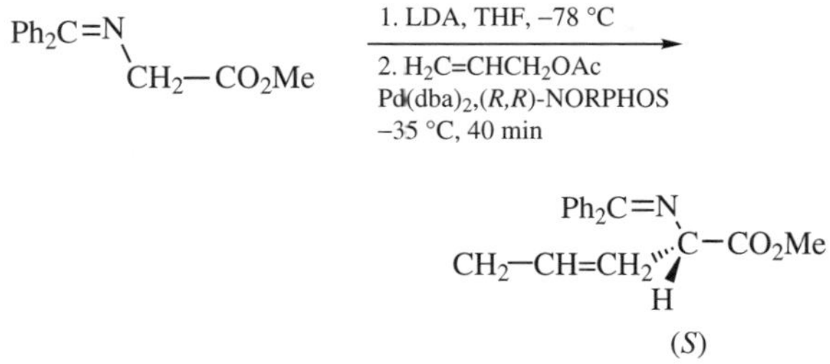

(7)

chemical yield: 55%
optical yield: 32% ee

Similarly, a $\{Pd[(S,S)\text{-NORPHOS}](\eta^3\text{-}C_3H_5)\}ClO_4$ precursor prepared from 2-acetoxy-4,4-diphenylbut-3-ene has been shown to react with sodium dimethylmalonate, a soft nucleophile. The corresponding '[Pd^0(chiral phosphine)]' species is thereby generated in situ, which serves to initiate the catalytic cycle that results in allylic alkylation of the nucleophile.[26] The resulting allylation product is formed in 76% ee (eq 8).[26]

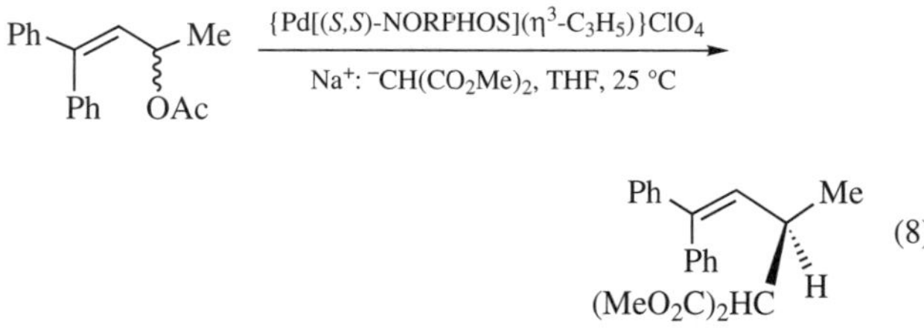

(8)

(*R*) (76% ee)

Enantioselective Hydroarylation/Hydroalkenylation of Alkene C=C Bonds (Heck Reaction). Enantioselective intramolecular hydroalkenylation of a prochiral vinyl iodide by using $Pd_2(dba)_3 \cdot CHCl_3$ in the presence of a chiral diphosphine ligand and silver phosphate in various solvents has been used to prepare optically active indolizidine derivatives (eq 9).[27] However, both the optical and chemical yields are low when the reaction is performed by using (*R,R*)-NORPHOS as catalyst in this reaction. Indeed, the highest optical yield (up to 86% ee) was obtained when (*R,S*)-**BPPFOH**[10] was used as chiral ligand in this reaction. Optically active decalin derivatives also have been prepared in low chemical and optical yields in this fashion by using (*R,R*)-NORPHOS as catalyst.[28]

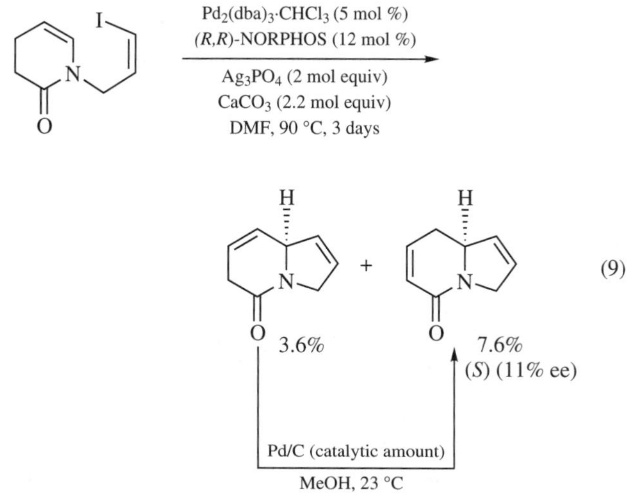

Enantioselective hydrophenylation of the alkene C=C double bond in 7-oxabenzonorborndiene has been carried out by using a variety of Ph(chiral diphosphine)Pd^+OTf^- catalysts (eq 10).[29] Moderate chemical and optical yields (68% ee and 43% ee, respectively) are obtained when the reaction is performed by using (*R,R*)-NORPHOS as catalyst. Highest optical yields were obtained when **5**[20] (Pd:ligand = 1:2.1) was used as chiral ligand in this reaction.[29]

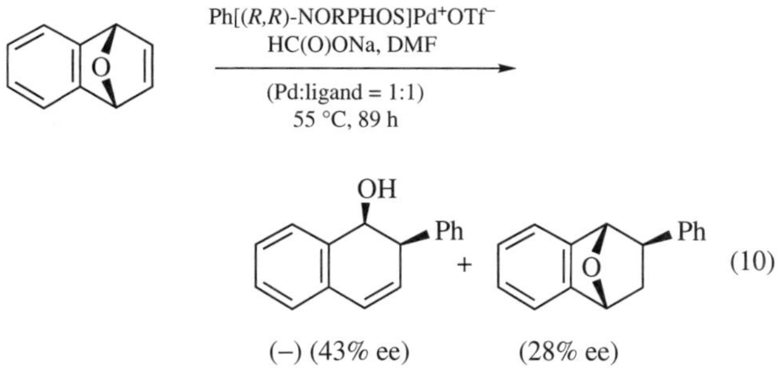

Similarly, enantioselective hydrophenylation of the alkene C=C double bond in norbornene and in norbornadiene has been performed by using a Pd-NORPHOS catalyst (eq 11).[30] The use of other optically active phosphine ligands [eg., (*R,S*)-**BPPFOH** (**9**)] generally afforded slightly higher chemical yields of hydrophenylated products with somewhat lower optical yields.[30]

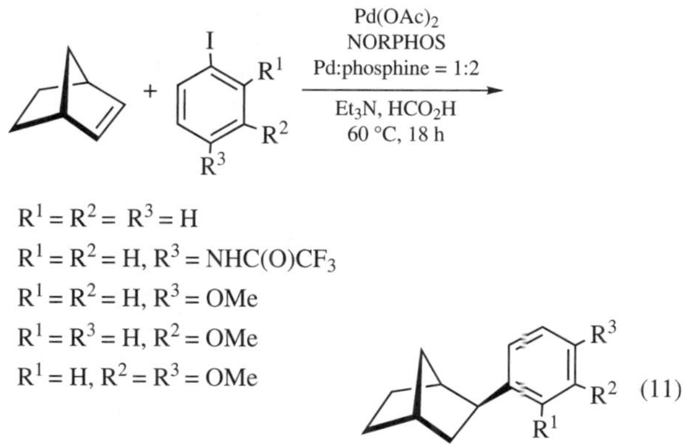

(*R,R*)-NORPHOS: chemical yield 50–70%; 36.8–37.7% ee (+)
(*S,S*)-NORPHOS: chemical yield 52%; 34.6–34.7% ee (−)

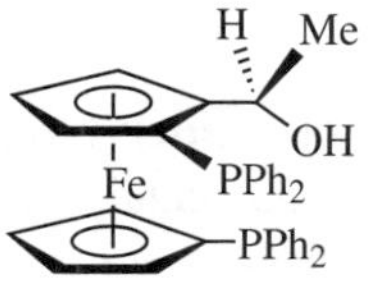

9

Enantioselective Cross-Coupling. A Ni-(*R,R*)-NORPHOS catalyst has been used to promote cross-coupling of Grignard reagents, RMgX (X = Br or Cl) with aryl halides (PhX, X = Br or Cl).[31] Reaction of PhX with EtCHMe-MgX afforded (*R*)-PhCHEtMe (X = Br: 50.7% ee; X = Cl: 26.7% ee) with concomitant formation of 10–12% of an isomerized product, $Ph(CH_2)_3CH_3$ (eq 12).[31] Similarly, Ni[(*R,R*)-NORPHOS]Cl_2 promotes coupling of racemic PhCH(Me)MgCl with vinyl bromide, thereby affording (*S*)-3-phenyl-1-butene in 95% chemical yield with 67% ee.[32]

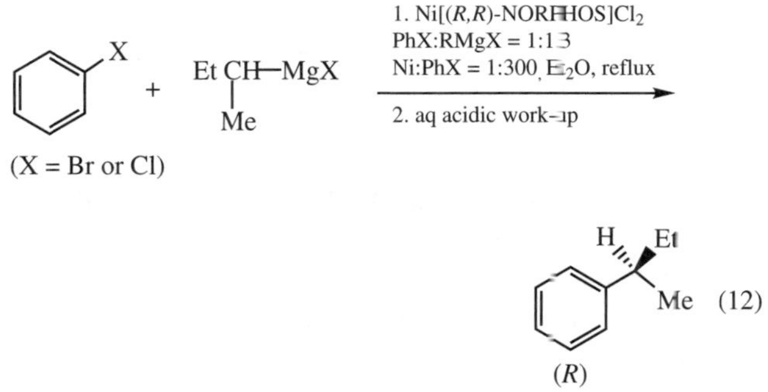

Enantioselective Telomerization. A Pd-(*R,R*)-NORPHOS catalyst has been reported to promote enantioselective telomerization of 2 equiv of both butadiene and formaldehyde (eq 13).[33] Both *trans*- and *cis*-2,5-divinyltetrahydropyrans are obtained (total 55% chemical yield). The trans isomer is formed preferentially (85% de), albeit in low optical yield (3% ee), whereas the minor telomer (cis isomer) is formed in 36% ee.[33]

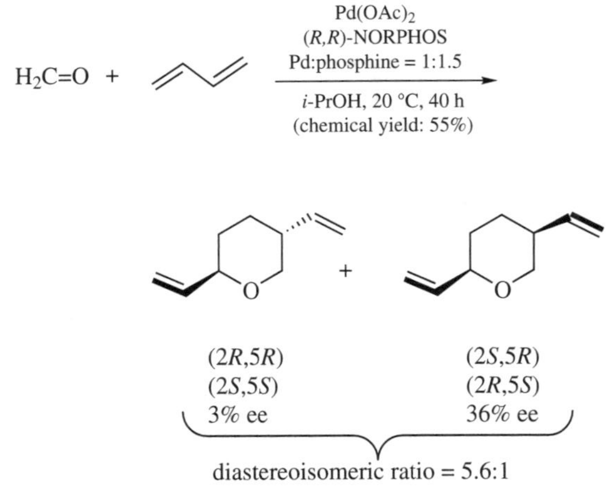

(13)

The results obtained from subsequent studies indicate that this same approach can be used to promote telomerization of butadiene with β-dicarbonyl compounds, nitroalkanes, and enamines.[34]

Enantioselective Diels–Alder Cycloaddition. A cationic palladium-(*R,R*)-NORPHOS catalyst has been reported to promote enantioselective Diels–Alder cycloaddition of cyclopentadiene to acrolein (eq 14). Both endo and exo [4+2] cycloadducts are produced (88% conversion, *endo*:*exo* = 7.3:1), albeit in low optical yield (*endo*: 2% ee; exo: 8% ee).[35]

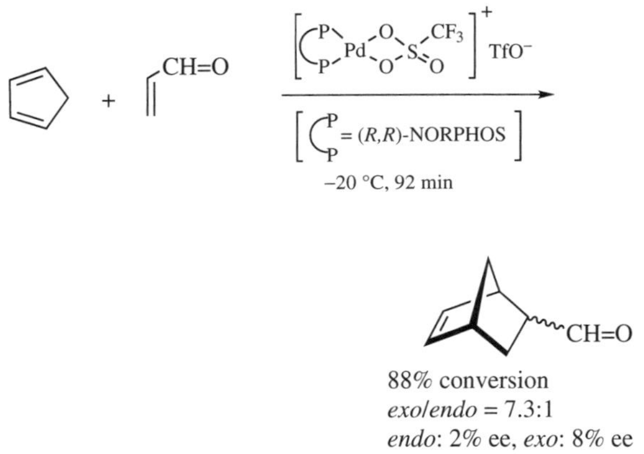

(14)

Enantioselective Homo-Diels–Alder Cycloaddition. Reaction of norbornadiene with a Co-(*S,S*)-NORPHOS catalyst (0.2–0.3 mol %, norbornadiene:phenylacetylene:(*S,S*)-NORPHOS:cobalt = 500:500:1.5:1), when performed in the presence of diethylaluminum chloride, produces the corresponding subtituted deltacyclene ([2+2+2] cycloadduct) in quantitative chemical yield and excellent optical yield (98.4% ee) (eq 15).[36] Similar results were obtained when either **7** or **(−)-BDDP** (**10**) was used in place of (*S,S*)-NORPHOS.[23,37,38] However, the use of other optically active diphosphines, e.g., **3**,[11] **(−)-MENO (11)**,[39] or **6**,[21] generally afforded 4-phenyldeltacyclene in lower optical yield.[38] Interestingly, the use of [Rh(norbornadiene) (*R,R*)-NORPHOS]PF_6 as catalyst did not result in deltacyclene formation; instead, 3-phenylethynylnortricyclene was produced in 45% chemical yield and 60% optical yield (eq 15).[40]

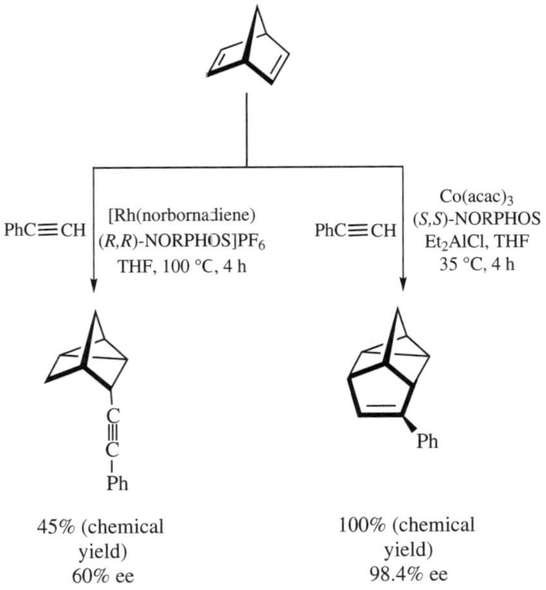

(15)

10 **11**

Enantioselective Aldol Addition. A Rh(I)-NORPHOS catalyst has been used to promote catalytic enantioselective aldol addition of enolsilanes to benzaldehyde (eq 16).[41] Although the aldol addition product is obtained in good chemical yield (>75%), the enantioselectivity of this reaction is modest, at best.

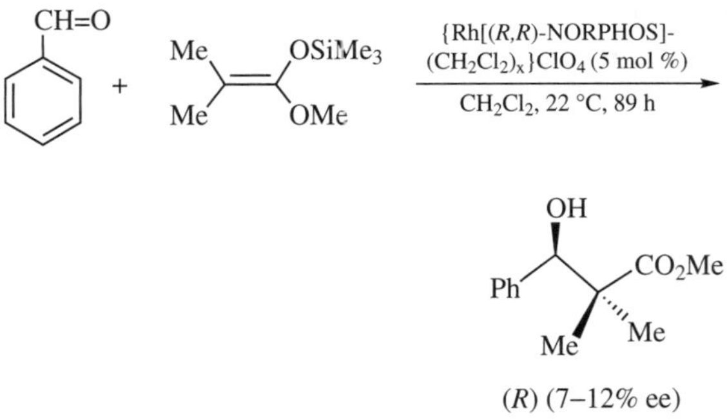

(16)

Enantioselective Hydroformylation. Enantioselective hydroformylation of a 4-vinyl-β-lactam, i.e., (3*S*,4*R*)-3-[(*R*)-1-(*tert*-butyl-dimethylsilyloxy)ethyl]-4-vinyl-2-azetidinone, has been achieved by using an Rh(I)-NORPHOS catalyst system (eq 17).[42] The optically active hydroformylation products thereby obtained are of interest as intermediates in the synthesis of 1-methylcarbapenem antibiotics.

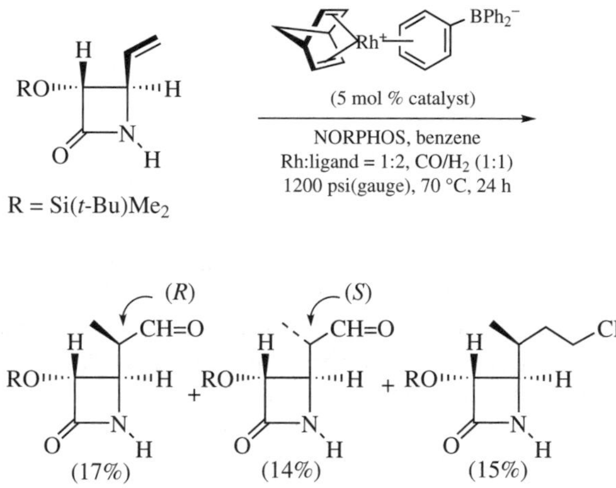

(17)

Related Reagents. (−)-DIOP; (+)-NMDPP; (*R,S*)-BPPFA; (*R,S*)-BPPFOH; (*R*)-BINAP; (+)-PROPHOS; (−)-MENO; (−)-BPPM; (−)-CHIRAPHOS; (−)-BDDP; (*S*)-(−)-1,4-BDPP.

1. Brunner, H.; Pieronczyk, W. *Angew. Chem., Int. Ed. Engl.* **1979**, *18*, 620–621.
2. (a) Nesterova, N. P.; Medved, T. Y.; Polikarpov, Y. M.; Kabachnik, M. I. *Izv. Akad. Nauk SSSR, Ser. Khim.* **1974**, *10*, 2295–2300; *Chem. Abstr.* **1975**, *82*, 43521f. (b) Arkhipova, L. I.; Berkman, Z. A.; Bertina, L. E.; Kabachnik, M. I.; Kossykh, V. G.; Medved, T. Y.; Nesterova, N. P.; Polikarpov, Y. M.; Rozen, A. M.; Yudina, K. S. *Dokl. Akad. Nauk SSSR* **1973**, *209*, 1093–1096; *Chem. Abstr.* **1973**, *79*, 35572w.
3. Naumann, K.; Zon, G.; Mislow, K. *J. Am. Chem. Soc.* **1969**, *91*, 7012–7022.
4. Brunner, H.; Pieronczyk, W.; Schönhammer, B.; Streng, K.; Bernal, I.; Korp, J. *Chem. Ber.* **1981**, *114,* 1137–1149.
5. Brunner, H.; Kunz, M. *Chem. Ber.* **1986**, *119*, 2868–2873.
6. Scott, J. W.; Keith, D. D.; Nix, G., Jr; Parrish, D. R.; Remington, S.; Roth, G. P.; Townsend, J. M.; Valentine, D., Jr; Yang, R. *J. Org. Chem.* **1981**, *46*, 5086–5093.
7. Zeiss, H.-J.; *J. Org. Chem.* **1991**, *56*, 1783–1788.
8. Genet, J. P.; Mallart, S.; Pinel, C.; Juge, S.; Laffitte, J. A. *Tetrahedron. Asymm.* **1991**, *2*, 43–46.
9. Brunner, H.; Kunz, M. *Chem. Ber.* **1986**, *119*, 2868–2873.
10. (a) Hayashi, T.; Mise, T.; Fukushima, M.; Kagotani, M.; Nagashima, N.; Hamada, Y.; Matsumoto, A.; Kawakami, S.; Konishi, M.; Yamamoto, K.; Kumada, M. *Bull Chem. Soc. Jpn.* **1980**, *53*, 1138–1151. (b) Hayashi, T.; Kumada, M. *Acc. Chem. Res.* **1982**, *15*, 395–401. (c) Hayashi, T.; Konishi, M.; Fukushima, M.; Mist, T.; Kagotani, M.; Tajika, M.; Kumada, M. *J. Am. Chem. Soc.* **1982**, *104*, 180–186.
11. Fryzuk, M. D.; Bosnich, B. *J. Am. Chem. Soc.* **1978**, *100*, 5491–5494.
12. Kagan, H. B.; Dang, T.-P. *J. Am. Chem. Soc.* **1972**, *94*, 6429–6433.
13. Brunner, H.; Bielmeier, E.; Wiehl, J. *J. Organometal. Chem.* **1990**, *384*, 223–241.
14. Brunner, H.; Huber, C. *Chem. Ber.* **1992**, *125*, 2085–2093.
15. Brunner, H.; Lautenschlager, H.-J. *Synthesis* **1989**, 706–709.
16. Brunner, H.; Becker, R.; Gauder, S. *Organometallics* **1986**, *5*, 739–746.
17. Brunner, H. *Angew. Chem., Int. Ed. Engl.* **1983**, *22*, 897–907.
18. Dumont, W.; Poulin, J. C.; Kagan, H. B. *J. Am. Chem. Soc.* **1973**, *95*, 8295–8299.
19. Paneghetti, C.; Gavagnin, R.; Pinna, F.; Strukul, G. *Organometallics* **1999**, *18*, 5057–5065.
20. (a) Miyashita, A.; Yasuda, A.; Takaya, H.; Toriumi, K. Ito, T.; Souchi, T.; Noyori, R. *J. Am. Chem. Soc.* **1980**, *102*, 7932–7934. (b) Miyashita, A.; Takaya, H.; Souchi, T.; Noyori, R. *Tetrahedron* **1984** *40*, 1245–1253. (c) Takaya, H.; Masima, K.; Koyano, K.; Yagi, M.; Kumobayashi, H.; Taketomi, T.; Akutagawa, S.; Noyori, R. *J. Org. Chem.* **1986**, *51*, 629–635.
21. Achiwa, K. *J. Am. Chem. Soc.* **1976**, *98*, 8265–8266.
22. Alexakis, A.; Burton, J.; Vastra, J.; Mangeney, P. *Tetrahedron: Asymm.* **1997**; *8*, 3987-3990.
23. Fryzuk, M. D.; Bosnich, B. *J. Am. Chem. Soc.* **1977**, *99*, 6262–6267.
24. Fryzuk, M. D.; Bosnich, B. *J. Am. Chem. Soc.* **1978**, *100*, 5491–5494.
25. Genet, J. P.; Ferroud, D.; Juge, S.; Montes, J. R. *Tetrahedron Lett.* **1986**, *27*, 4573–4576.
26. Auburn, P. A.; Mackenzie, P. B.; Bosnich, B. *J. Am. Chem. Soc.* **1985**, *107*, 2033–2046.
27. Nukui, S.; Sodeoka, M.; Shibasaki, M. *Tetrahedron Lett.* **1993**, *34*, 4965–4968.
28. Sato, Y.; Nukui, S.; Sodeoka, M.; Shibasaki, M. *Tetrahedron* **1994**, *50*, 371–382.
29. Moinet, C., Fiaud, J.-C. *Tetrahedron Lett.* **1995**, *36*, 2051–2052.
30. Brunner, H.; Kramler, K. *Synthesis* **1991**, 1121–1124.
31. Consiglio, G.; Morandini, F.; Piccolo, O. *Tetrahedron* **1983**, *39*, 2699–2707.
32. Brunner, H.; Pröbster, M. *J. Organometal. Chem.* **1981**, *209*, C1–C3.
33. Keim, W.; Meltzow, W.; Koehnes, A.; Roethel, T. *J. Chem. Soc., Chem. Commun.* **1989**, 1151–1152.
34. Keim, W.; Koehnes, A.; Roethel, T.; Enders, D. *J. Organometal. Chem.* **1990**, *382*, 295–301.
35. Pignat, K.; Vallotto, J.; Pinna, F.; Strukul, G. *Organometallics* **2000**, *19*, 5160–5167.
36. Brunner, H.; Mushiol, M.; Prester, F. *Angew Chem., Int. Ed. Engl.* **1990**, *29*, 652–653.
37. Bakos, J.; Toth, I.; Heil, B.; Marko, L. *J. Organomet. Chem.* **1985**, *279*, 23–29.
38. Brunner, H.; Prester, F. *J. Organometal. Chem.* **1991**, *414*, 401–409.
39. Aviron-Violet, P.; Golleuille, Y.; Varagnet, J. *J. Mol. Catal.* **1979**, *5*, 41–50.
40. Brunner, H.; Prester, F. *Tetrahedron: Asymm.* **1990**, *9*, 589–592.
41. Reetz, M.; Vougioukas, A. E. *Tetrahedron Lett.* **1987**, *28*, 793–796.
42. Park, H. S.; Alberico, E.; Alper, H. *J. Am. Chem. Soc.* **1999**, *121*, 11697–11703.

Alan P. Marchand
Jaroslaw Romanski
T. Pavan Kumar
University of North Texas, Denton, TX, USA
University of Lodz, Poland
University of North Texas, Denton, TX, USA

O

[(2*S*)-(2α,3aα,4α,7α,7aα)]-2,3,3a,4,5,6,7,7a-Octahydro-7,8,8-trimethyl-4,7-methanobenzofuran-2-ol

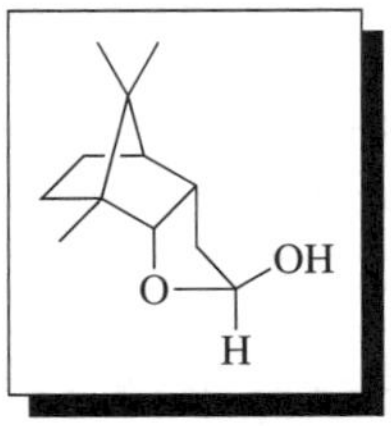

([(2*S*)-(2α,3aα,4α,7α,7aα)]-isomer)
[81925-09-9] $C_{12}H_{20}O_2$ (MW 196.32)
(enantiomer)
[108031-75-0]
(2*S*) acetal dimer (**1**)
[87248-50-8] $C_{24}H_{38}O_3$ (MW 374.62)
(enantiomeric acetal dimer)
[108031-79-4]

(useful for resolving racemic alcohols via formation of diastereomeric acetals,[1] and also for determining the absolute configuration of certain types of chiral alcohols[2])

Alternate Name: MBF-OH.
Physical Data: monomeric lactol: bp 120 °C/0.005 mmHg, $[\alpha]_D^{20}$ + 100° (*c*, 11.24, THF); acetal dimer: mp 150–151 °C, $[\alpha]^{21}$ + 199.1° (*c*, 2.25, THF).
Solubility: monomeric lactol: sol ether, $CHCl_3$, THF; acetal dimer: sol $CHCl_3$, THF, hot petroleum ether.
Form Supplied in: the acetal dimer, and its enantiomer, are available as the neat solids.
Analysis of Reagent Purity: NMR,[3] mp.
Preparative Methods: the [(2*S*)-(2α,3aα,4α,7α,7aα)]-isomer is prepared from (+)-camphor,[3a] and the (2*R*)-enantiomer from (−)-borneol via oxidation to (−)-camphor.[3b]
Purification: the monomeric lactols can be purified by distillation, and the acetal dimers by recrystallization from petroleum ether.
Handling, Storage, and Precautions: no reported instability or toxicity.

Resolving Agent for Chiral Alcohols. The title lactol reagent (known as the Noe reagent) has been used mainly for resolving racemic alcohols through formation of diastereomeric, separable acetals.[1] From each alcohol enantiomer, a single diastereomer of the product acetal is formed with high selectivity. Separation of the diastereomeric derivatives by chromatography or crystallization, followed by mild methanolysis, gives the resolved alcohol in good yield and high enantiomeric purity. If an excess of the racemic alcohol is employed, 'enantiomer-selective' acetal formation results in one of the product diastereomers being produced in excess, which increases the yield of pure diastereomer obtained (eq 1).[4]

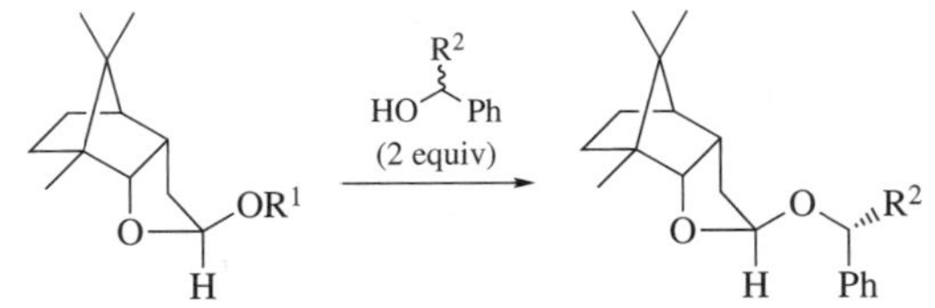

(1)

R^1 = H or MBF

major diastereomer

R^2 = Me, 49%; R^2 = *t*-Bu, 64%

The commercially available acetal dimer (**1**), and the enantiomeric dimer, can also be used as reagents instead of the lactols.[1] As an alternative to the (2*R*)-enantiomer of the *endo*-lactol, one can use the *exo*-lactol (**2**), or the corresponding acetal dimer.[5] Compound (**2**) is prepared from (+)-camphor, as is the (2*S*)-enantiomer of the *endo*-lactol, but the two reagents show opposite sense of enantiomer selectivity in acetal formation.

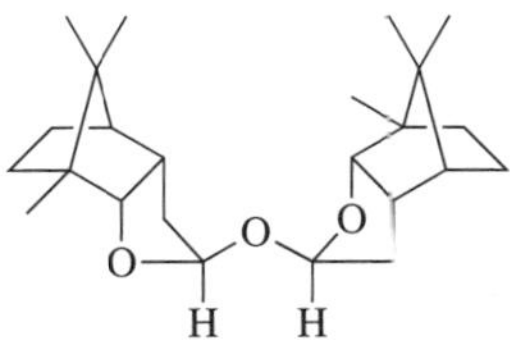

(**1**)

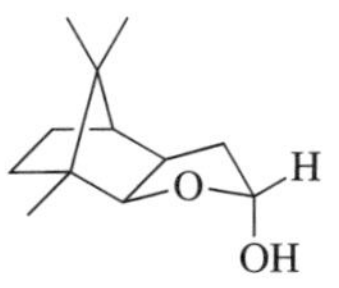

(**2**)

Using these reagents, resolutions of various types of racemic alcohols, including alkylarylcarbinols,[1] alkylthienylcarbinols,[6] cyanohydrins,[1,7] α-hydroxyalkynes,[2b] and α-hydroxyphosphonates,[8] and also of a thiol[1] and an amine,[1] have been performed.

An alternative procedure has been reported for the resolution of alcohols through separation of diastereomeric *O*-methylmandelates;[9] a drawback of this approach is that partial racemization of the mandelic acid derivative sometimes occurs during the esterification. In contrast, the Noe reagents are configurationally stable; also, complete separation of the derived diastereomeric acetals can usually be achieved, which gives access to enantiomerically pure alcohols when the selectivity of the acetal formation is low. An additional advantage of the title reagents is that the derived acetal can be used as an ordinary hydroxyl-protecting group in subsequent synthetic operations, with a reactivity similar to a THP group.[7,10] Of other alternative methods for resolution of racemic alcohols, kinetic resolutions by enzyme-catalyzed transformations often give very high selectivities.[11]

Determination of Absolute Configuration for Chiral Alcohols. Using NMR data for the derived acetals, in combination with the known sense of enantiomer selectivity for the Noe reagent used, one can also determine the absolute configuration of the starting alcohol.[2] Gas chromatography data can be used as well.[2c] This method for determining absolute configuration should be a useful complement to the more commonly used methods based on NMR analysis of Mosher esters[12] or *O*-methylmandelates.[9]

---

1. Noe, C. R. *Ber. Dtsch. Chem. Ges./Chem. Ber.* **1982**, *115*, 1591.
2. (a) Noe, C. R.; Knollmüller, M.; Wagner, E.; Völlenkle, H. *Ber. Dtsch. Chem. Ges./Chem. Ber.* **1985**, *118*, 1733. (b) Noe, C. R.; Knollmüller, M.; Oberhauser, B.; Steinbauer, G.; Wagner, E. *Ber. Dtsch. Chem.*

Ges./Chem. Ber. **1986**, *119*, 729. (c) Schönauer, K. J.; Walter, P.; Noe, C. R. *Monatsh. Chem.* **1986**, *117*, 127.

3. (a) Noe, C. R. *Ber. Dtsch. Chem. Ges./Chem. Ber.* **1982**, *115*, 1576. (b) Noe, C. R.; Knollmüller, M.; Göstl, G.; Oberhauser, B.; Völlenkle, H. *Angew. Chem., Int. Ed. Engl.* **1987**, *26*, 442.
4. Knollmüller, M.; Noe, C. R.; Steinbauer, G.; Dungler, K. *Synthesis* **1986**, 501.
5. Noe, C. R.; Knollmüller, M.; Steinbauer, G.; Jangg, E.; Völlenkle, H. *Ber. Dtsch. Chem. Ges./Chem. Ber.* **1988**, *121*, 1231.
6. Noe, C. R.; Knollmüller, M.; Dungler, K.; Miculka, C.; Gärtner, P. *Monatsh. Chem.* **1991**, *122*, 705.
7. Noe, C. R.; Knollmüller, M.; Göstl, G.; Gärtner, P. *Monatsh. Chem.* **1991**, *122*, 283.
8. Hammerschmidt, F.; Völlenkle, H. *Justus Liebigs Ann. Chem./Liebigs Ann. Chem.* **1989**, 577.
9. Trost, B. M.; Belletire, J. L.; Godleski, S.; McDougal, P. G.; Balkovec, J. M. *J. Org. Chem.* **1986**, *51*, 2370.
10. Greene, T. W.; Wuts, P. G. M. *Protective Groups in Organic Synthesis* 2nd ed.; Wiley: New York, 1991; pp 37–38.
11. (a) Santaniello, E.; Ferraboschi, P.; Grisenti, P.; Manzocchi, A. *Chem. Rev.* **1992**, *92*, 1071. (b) Frykman, H.; Öhrner, N.; Norin, T.; Hult, K. *Tetrahedron Lett.* **1993**, *34*, 1367.
12. (a) Dale, J. A.; Mosher, H. S. *J. Am. Chem. Soc.* **1973**, *95*, 512. (b) Ohtani, I.; Kusumi, T.; Kashman, Y.; Kakisawa, H. *J. Am. Chem. Soc.* **1991**, *113*, 4092.

Tobias Rein
The Royal Institute of Technology, Stockholm, Sweden

S-(1-Oxido-2-pyridinyl)-1,1,3,3-tetramethylthiouronium Hexafluorophosphate (HOTT)

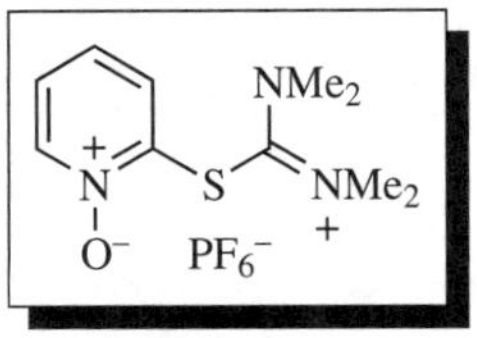

[212333-72-7] $C_{10}H_{16}N_3OSPF_6$ (MW 371.28)

(reagent used to convert carboxylic acids into Barton esters and for peptide coupling/amidation)

Physical Data: mp 115–116 °C.
Solubility: acetonitrile, CH_2Cl_2, THF (moderate solubility).
Form Supplied in: white solid.
Purification: recrystallized from CH_2Cl_2.
Handling, Storage, and Precautions: it is advised to protect the solid from prolonged exposure to light since 2-mercaptopyridine *N*-oxide and related compounds can be light-sensitive.
Preparative Methods: the title reagent can be prepared by slowly adding Et_3N to a dry CH_2Cl_2 solution of 2-mercaptopyridine-*N*-oxide and tetramethylchloroformamidinium hexafluorophosphate.[1] After removal of CH_2Cl_2 from the reaction mixture, the resulting solid mass is pulverized, washed with $CHCl_3$, and then filtered to give a white solid of sufficient purity to be used in subsequent reactions.[2]

Barton Esterification: Reductive Decarboxylation. *O*-Acyl thiohydroxamates or Barton esters are useful precursors of carbon-centered radicals via thermolysis or photolysis.[3] Several different methods are available for converting carboxylic acids into Barton esters (eq 1).[4] These reactions generally proceed via the attack of a 2-mercaptopyridine-*N*-oxide salt on an activated carboxylic acid that has either been preformed (acid chloride, mixed anhydride) or generated in situ (with 1,3-dicyclohexylcarbodiimide or tri-*n*-butylphosphine + 2,2′-dithiodipyridine-1,1′-dioxide). However, HOTT has the distinct advantages of (1) being easy to prepare and handle without the need for any special precautions, (2) facilitates efficient Barton esterification of carboxylic acids, and (3) simplifies subsequent work-up and purifications by avoiding the need to remove by-products like 1,3-dicyclohexylurea.

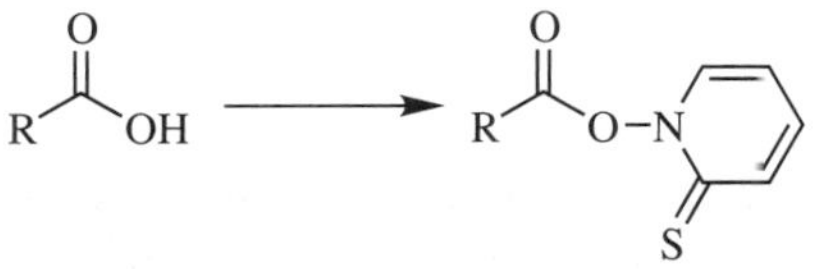

(1)

The HOTT reagent has been shown to significantly improve the yields of reductive decarboxylations of 2,3:4,6-di-*O*-isopropylidene-2-keto-L-gulonic acid (eq 2) and peracetylated *N*-acetylneuraminic acid (eq 3).[2] In both cases, the yield of reduced product nearly doubled when HOTT was used to esterify these hindered carboxylic acids.

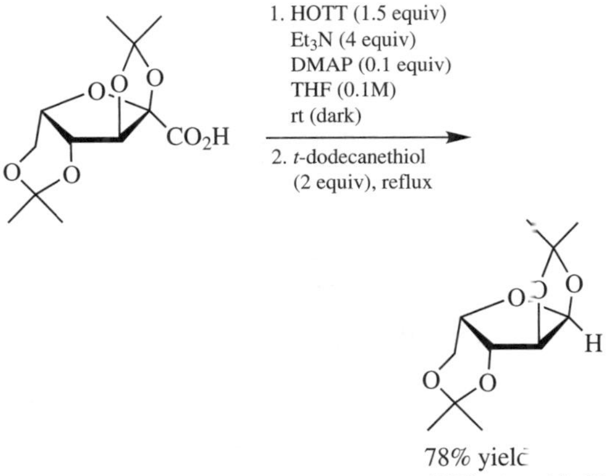

78% yield
(42% via acid chloride) (2)

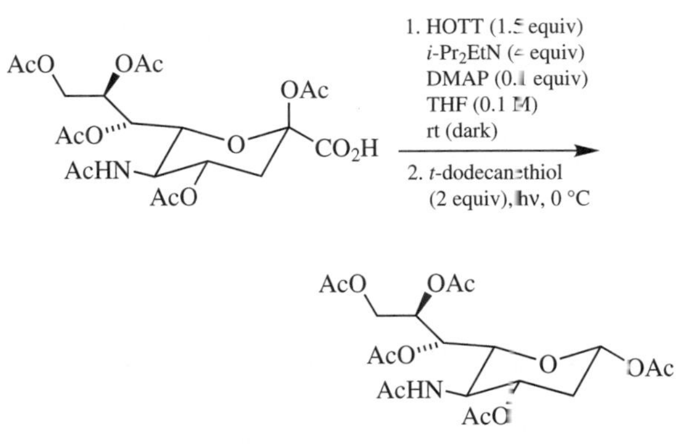

(3)

50% yield
(27% via carbodiimide)

Barton Esterification: Oxidative Decarboxylation. HOTT-mediated Barton esterification was coupled to oxidative decarboxylation in a synthesis of the sesquiterpene (+)-culmorin (eq 4).[5] Use of the HOTT reagent was clearly superior with this hindered substrate when compared with the acid chloride method.

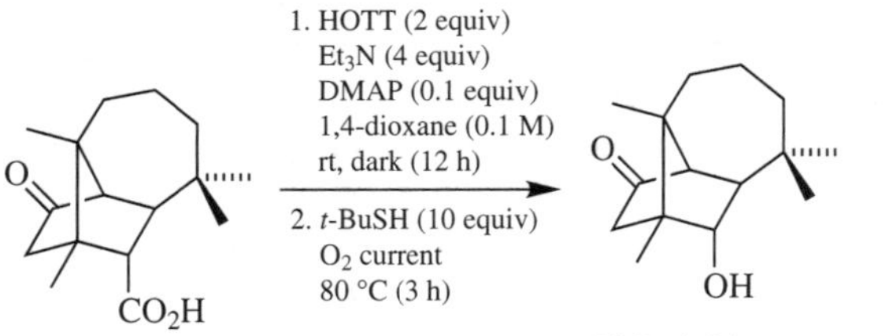

82% yield
(48% via acid chloride) (4)

Barton Esterification: Radical Addition. One of the best examples illustrating the benefits of HOTT for this transformation is shown in eq 5. The Barton esterification of this very hindered acid was followed by IR spectroscopy by monitoring the disappearance of the carbonyl stretch of the acid (1740 cm^{-1}) and the appearance of the carbonyl stretch of the Barton ester (1810 cm^{-1}). Barton esterification using HOTT was complete within 20 min, whereas over 4 h was required when using the combination of DCC and 2-mercaptopyridine-*N*-oxide.

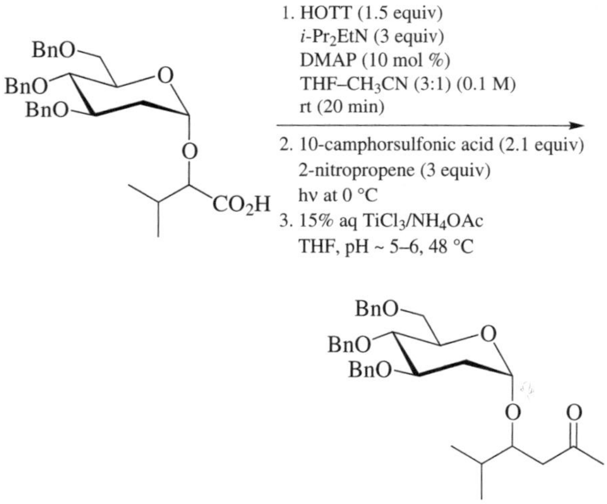

(5)

73% yield
(47% via carbodiimide)

HOTT was used to effect Barton esterification in a novel approach to 1, 3, 5, ... (2*n*+1) polyols based on iterative stereocontrolled homologation of chiral hydroxyalkyl radicals (eq 6).[6]

General notes: As with most Barton esterifications, the reaction should be performed in the dark and under anhydrous conditions.

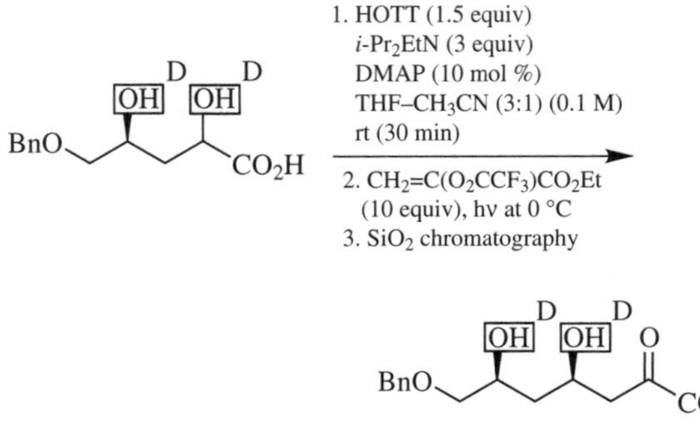

(6)

80% yield, ds = 9:1

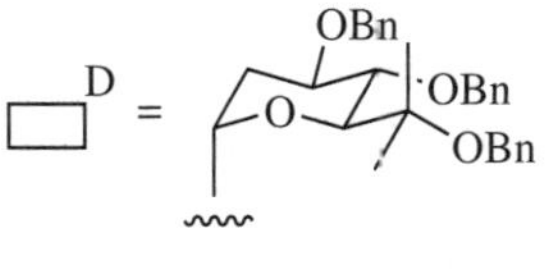

Peptide Coupling. The HOTT reagent, as well as the corresponding tetrafluoroborate salt (TOTT), have also been reported to be inexpensive alternatives to uronium- and phosphonium-based peptide coupling reagents (eq 7).[7] Yields were generally on the same order as those observed with standard peptide coupling reagents. An advantage of these reagents—at least in some instances—may be a reduced propensity of the *N*-protected amino acid component to racemize during the coupling reaction.

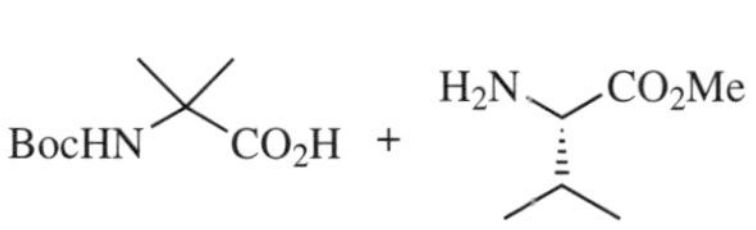

(1 equiv of each)

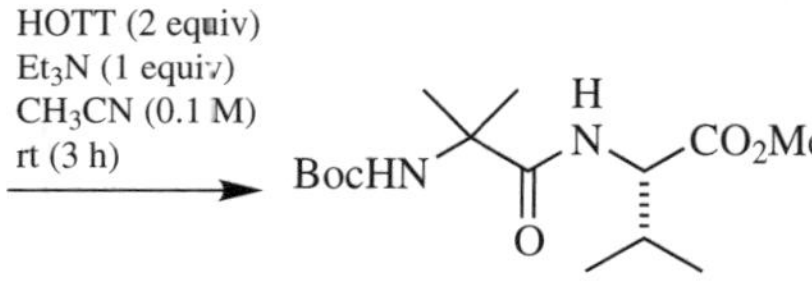

(7)

58% yield

Synthesis of Primary Amides. Carboxylic acids can be converted to their primary amides using HOTT (or TOTT) as the coupling agent (eq 8 and 9).[8] The reaction conditions are very mild and do not adversely affect other functionality prone to nucleophilic attack by ammonia. A simple extractive work-up is sufficient to obtain the primary amides in pure form.

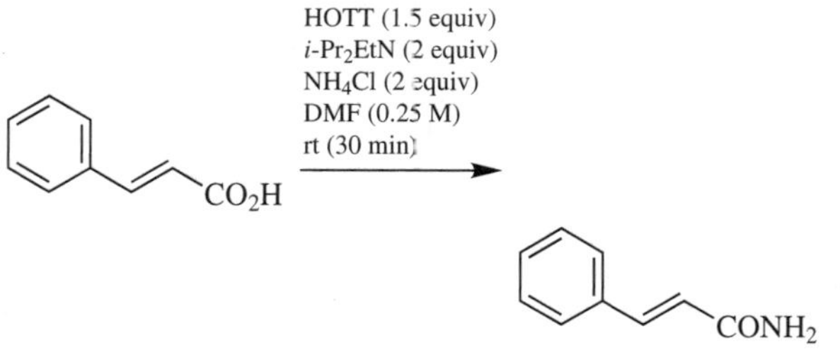

(8)

93% yield

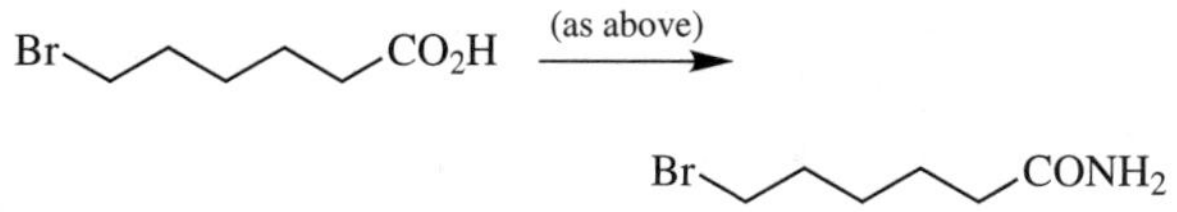

(9)

93% yield

1. Dourtoglou, V.; Gross, B.; Lambropoulou, V.; Zioudrou, C. *Synthesis* **1984**, 572.
2. Garner, P.; Anderson, J. T.; Dey, S.; Youngs, W. J.; Galat, K. *J. Org. Chem.* **1998**, *63*, 5732.
3. Crich, D. *Aldrichimica Acta* **1987**, *20*, 35.
4. Barton, D. H. R.; Samadi, M. *Tetrahedron* **1992**, *48*, 7083.
5. Takasu, K.; Mizutani, S.; Noguchi, M.; Ihara, M. *J. Org. Chem.* **2000**, *65*, 4112.
6. Garner, P.; Anderson, J. A. *Org. Lett.* **1999**, *1*, 1057.
7. Bailén, M. A.; Chinchilla, R.; Dodsworth, D. J.; Nájera, C. *J. Org. Chem.* **1999**, *64*, 8936.
8. Bailén, M. A.; Chinchilla, R.; Dodsworth, D. J.; Nájera, C. *Tetrahedron Lett.* **2000**, *41*, 9809.

James T. Anderson, Subhakar Dey & Philip Garner
Case Western Reserve University, Cleveland, OH, USA

P

(*R*)-Pantolactone[1]

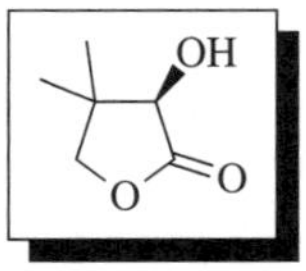

[599-04-2] $C_6H_{10}O$ (MW 130.16)

(effective chiral auxiliary in diastereoselective Diels–Alder reactions,[1] and for diastereoselective addition to ketenes;[2] used as a chiral pool reagent; also used as a covalently bound resolving agent[3])

Alternate Name: (*R*)-dihydro-3-hydroxy-4,4-dimethyl-2(3*H*)-furanone.
Physical Data: mp 92 °C; bp 120–122 °C/15 mmHg; $[\alpha]_D^{25}$ −50.7° (*c* 2.05, H_2O).
Solubility: sol water, alcohols, benzene, ether, chlorocarbons, THF.
Form Supplied in: crystalline white solid; commercially available.
Handling, Storage, and Precautions: hygroscopic.

Availability. Although commercially available via the degradation of pantothenic acid, (*R*)-pantolactone is also conveniently prepared by enantioselective reduction of its corresponding keto lactone employing homogeneous catalysis,[4a–g] or by microbial methods.[5] The (*S*)-enantiomer has been prepared by inversion of the natural product in 90% yield and 97% ee via triflate activation, acetate displacement, and *Lithium Hydroxide* hydrolysis.[6] The enantiomers were also prepared by resolution of the racemate with (*R*)- and (*S*)-phenethylamine.[7] A gas chromatographic method exists for ee determination.[8]

Diels–Alder Reactions. (*R*)-Pantolactone is one of the most effective chiral auxiliaries for preparative scale Diels–Alder additions of simple enoate esters in the presence of Lewis acids (eq 1).[9]

Endo–exo selectivity typically ranges from 20:1 to 45:1 with a maximum of 97.5:2.5 diastereoselection. Preparatively convenient reaction conditions are employed (CH_2Cl_2, CH_2Cl_2/cyclohexane; temp. approx. 0 °C; ca. 0.3 M concentration; and 0.1–1.0 molar equiv of Lewis acid). Products are typically crystalline and brought to high optical purity by recrystallization. Epimerization-free hydrolysis is effected with LiOH in THF/water. This procedure has been successfully applied in a nine-step synthesis of cyclosarkomycin in 17% overall yield (eq 2),[10] and to syntheses of the sandalwood fragrances.[11]

The cyclohexane unit of the C(30) stereocenter of the C(18)–C(35) segment of FK-506 was established in excellent yield and de employing the same concept (eq 3).[6]

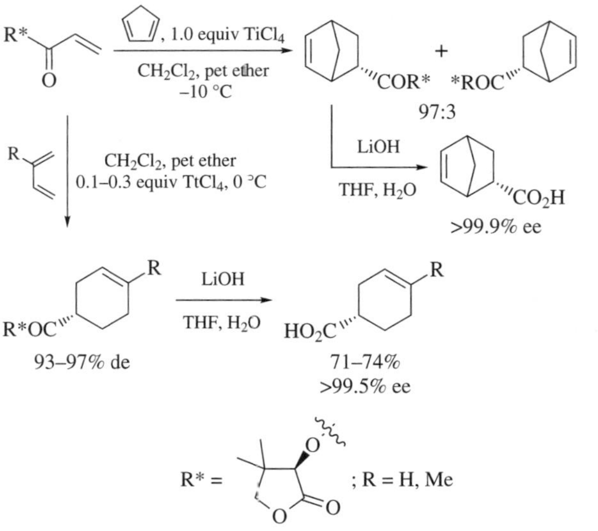
(1)

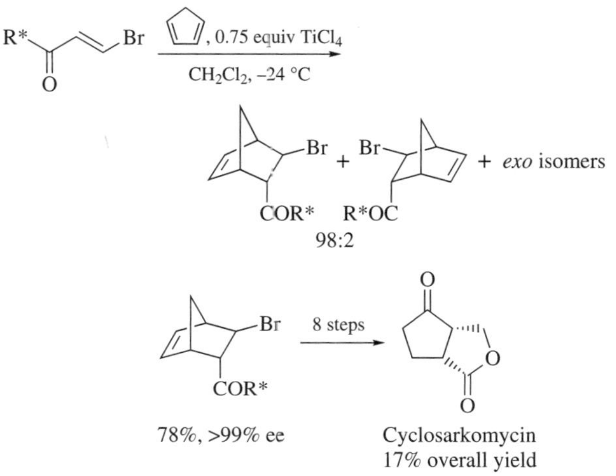
(2)

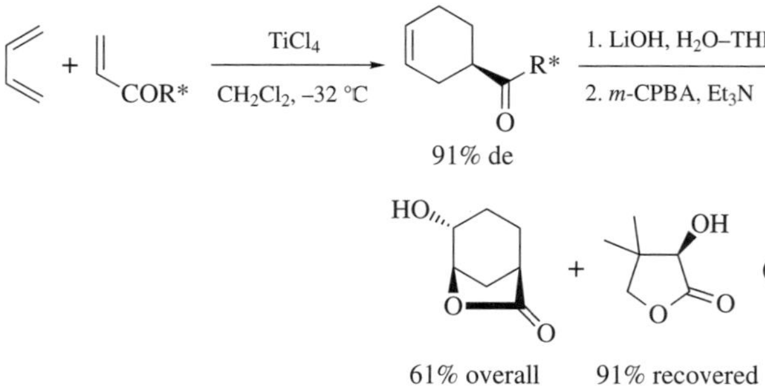
(3)

(*E*)-2-Cyanocinnamates have been similarly used as dienophiles. An *endo–exo* selectivity of 85:15 at a diastereoselectivity of 99:1 was obtained (eqs 4 and 5).[12,13]

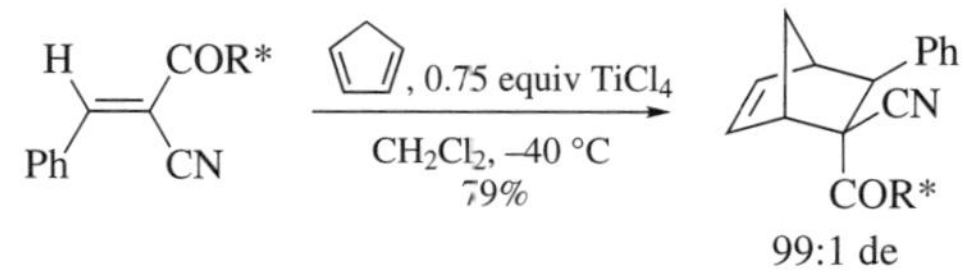
(4)

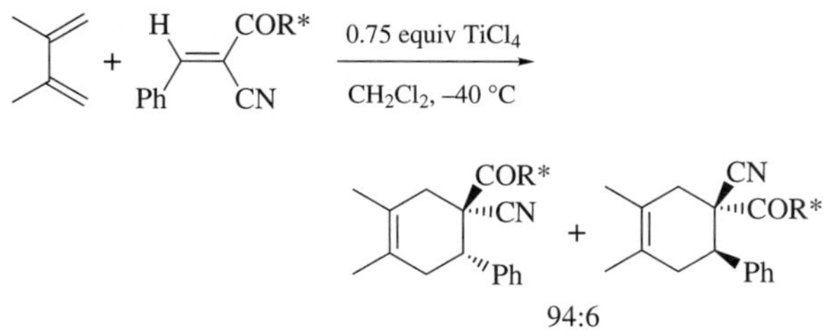
(5)

Further variations in dienophile have been equally successful (eqs 6 and 7),[14] including applications to the Michael reaction (eq 7)[15] and in the synthesis of a prostaglandin intermediate (eq 8).[16]

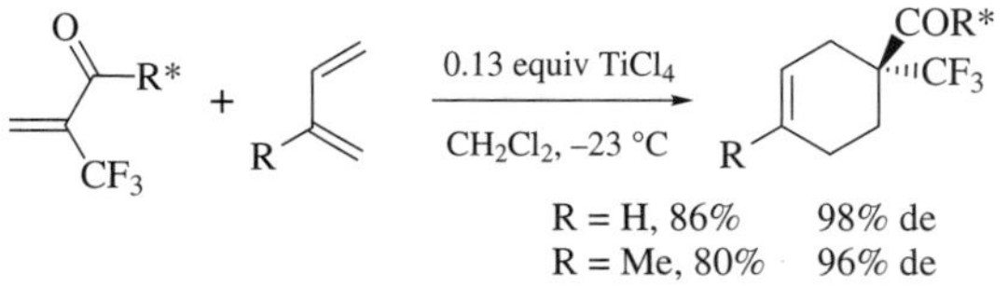
(6)

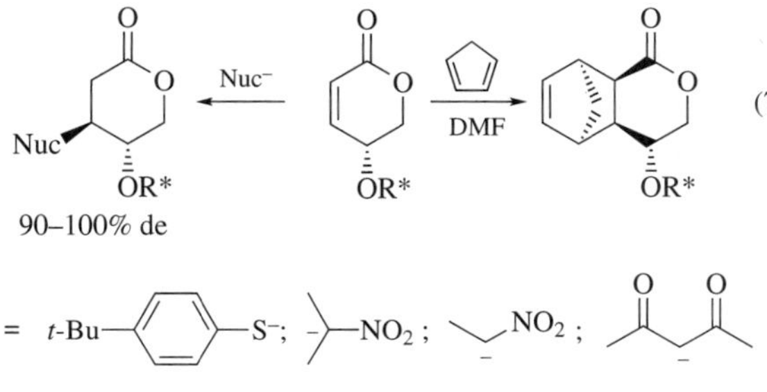
(7)

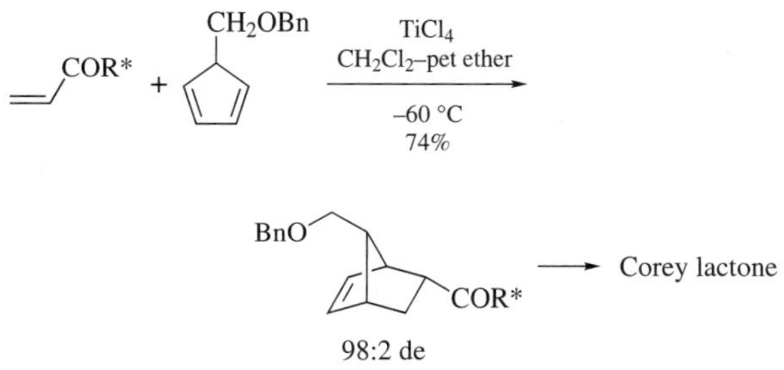
(8)

Ketene Additions. Reaction of the ketene derived from ibuprofen (Ar = *p*-isobutylphenyl) with (*R*)-pantolactone in the presence of simple tertiary amine bases in apolar solvents yielded >99% de favoring the (*R,R*)-ester (eq 9).[3] The reaction is first order in each component and possesses a pronounced deuterium isotope effect ($k_H/k_D \approx 4$). The ketene from naproxen (Ar = 2-(6-methoxynaphthyl)) affords a de of 80% under similar conditions.

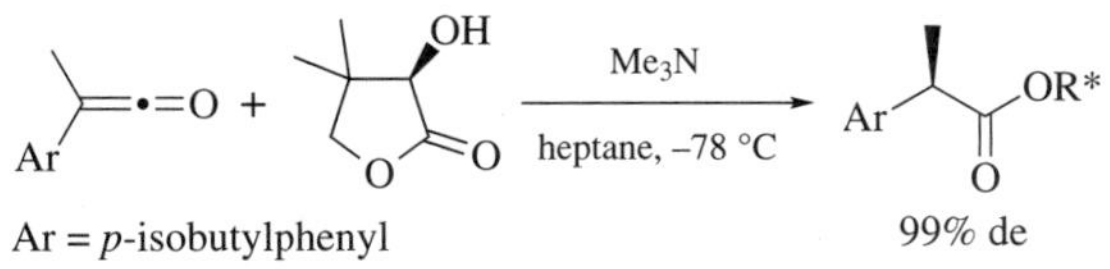
(9)

Extension of this work to a series of bromo- and iodoketenes proceeds with good to excellent de (eq 10).[17] Reaction of the products with azide ion affords a ready entry into amino acid synthesis (eq 11). However, with R = aryl, no selectivity was noted, possibly due to base-mediated epimerization under the reaction conditions.

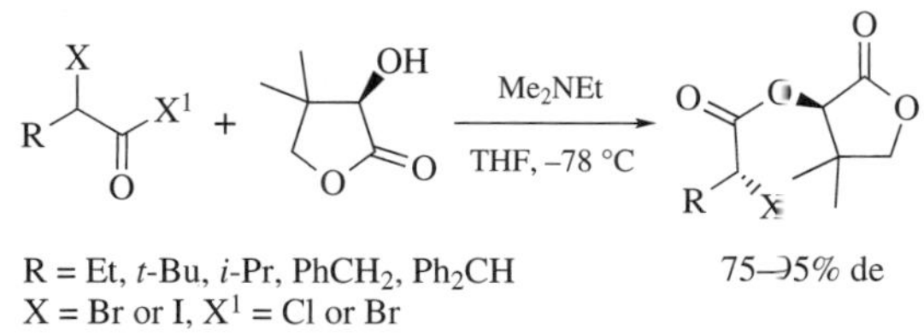
(10)

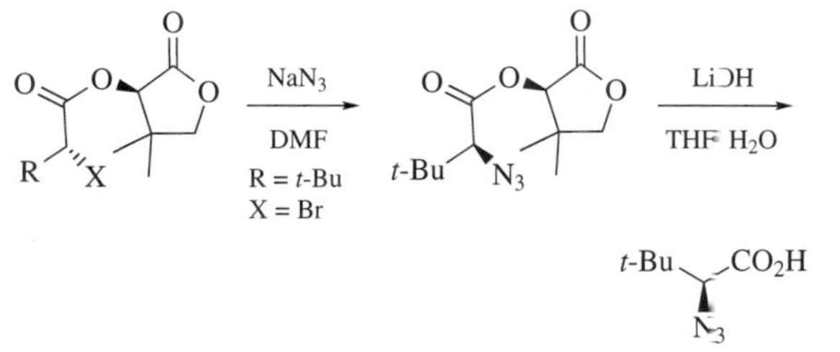
(11)

Chiral Pool Reagent. (*R*)-Pantolactone has been used as a source of chiral fragments for synthesis. Applications include use in the syntheses of the elfamycins (eq 12)[18] and the bryostatins (eq 13).[19a,b] It has also been used to prepare potentially useful chiral epoxide synthons possessing a quaternary *gem*-dimethyl carbon.[20]

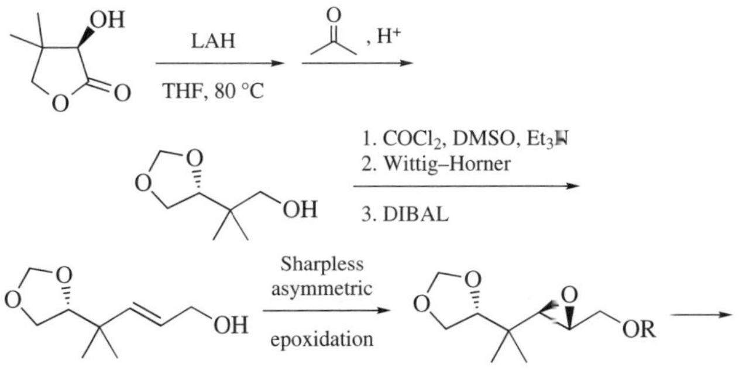
Elfamycins (12)

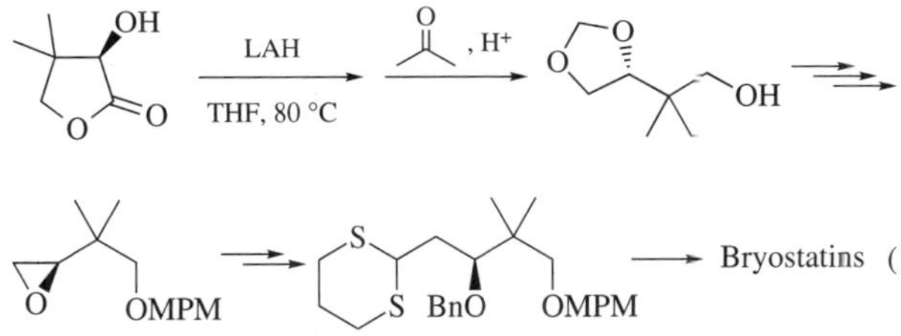
(13)

Miscellaneous Applications. Only one attempt to use (*R*)-pantolactone as an enantioselective protonating agent for enolates has been reported.[21] A series of structurally diverse chiral alcohols afforded modest ee's with (*R*)-pantolactone affording the largest ee noted for the series. The complexities of attempting a protonation of this sort in the presence of base and under exchanging conditions are discussed. Finally, the lactone has been used to resolve chiral acids by crystallization and chromatographic techniques applied to the (*R*)-pantolactone-derived esters.[3,22,23]

Related Reagents. (*S*)-Ethyl Lactate; Ethyl Mandelate; 3-Hydroxyisoborneol.

---

1. Helmchen, G.; Hady, A. F. A.; Hartmann, H.; Karge, R.; Krotz, A.; Sartor, K.; Urmann, M. *Pure Appl. Chem.* **1989**, *61*, 409.
2. Larsen, R. D.; Corley, E. G.; Davis, P.; Reider, P. J.; Grabowski, E. J. J. *J. Am. Chem. Soc.* **1989**, *111*, 7650.

3. Duke, C. C.; Wells, R. J. *Aust. J. Chem.* **1987**, *40*, 1641.
4. (a) Ojima, I.; Kogure, T.; Yoda, Y. *Org. Synth.* **1985**, *63*, 18. (b) Ojima, I.; Kogure, T.; Terasaki, T.; Achiwa, K. *J. Org. Chem.* **1978**, *43*, 3444. (c) Morimoto, T.; Takahashi, H.; Fujii, K.; Chiba, M.; Achiwa, K. *Chem. Lett.* **1986**, 2061. (d) Hatat, C.; Karim, A.; Kokel, N.; Mortreux, A.; Petit, F. *Tetrahedron Lett.* **1988**, *29*, 3675. (e) Genet, J. P.; Pinel, C.; Mallart, S.; Juge, S.; Cailhol, N.; Laffitte, J. A. *Tetrahedron Lett.* **1992**, *33*, 5343. (f) Takahashi, H.; Hattori, M.; Chiba, M.; Morimoto, T.; Achiwa, K. *Tetrahedron Lett.* **1986**, *27*, 4477. (g) Hatat, C.; Karim, A.; Kokel, N.; Mortreux, A.; Petit, F. *Nouv. J. Chim.* **1990**, *14*, 141.
5. Shimizu, S.; Yamada, H.; Hata, H.; Morishita, T.; Akutsu, S.; Kawamura, M. *Agric. Biol. Chem.* **1987**, *51*, 289.
6. Corey, E. J.; Huang, H. C. *Tetrahedron Lett.* **1989**, *30*, 5235.
7. Nohira, H.; Nohira, M.; Yoshida, S.; Osada, A.; Terunuma, D. *Bull. Chem. Soc. Jpn.* **1988**, *61*, 1395.
8. Brunner, H.; Forster, St. *Monatsh. Chem.* **1992**, *123*, 659.
9. Poll, T.; Sobczak, A.; Hartmann, H.; Helmchen, G. *Tetrahedron Lett.* **1985**, *26*, 3095.
10. Linz, G.; Weetman, J.; Hadey, A. A. F.; Helmchen, G. *Tetrahedron Lett.* **1989**, *30*, 5599.
11. Krotz, A.; Helmchen, G. *Tetrahedron: Asymmetry* **1990**, *1*, 537.
12. Cativiela, C.; Mayoral, J. A.; Avenoza, A.; Peregrina, J. M.; Lahoz, F. J.; Gimeno, S. *J. Org. Chem.* **1992**, *57*, 4664.
13. Avenoza, A.; Cativiela, C.; Mayoral, J. A.; Peregrina, J. M. *Tetrahedron: Asymmetry* **1992**, *3*, 913.
14. Hanzawa, Y.; Suzuki, M.; Kobayashi, Y.; Taguchi, T.; Iitaka, Y. *J. Org. Chem.* **1991**, *56*, 1718.
15. Knol, J.; Jansen, J. F. G. A.; Van Bolhuis, F.; Feringa, B. L. *Tetrahedron Lett.* **1991**, *32*, 7465.
16. Miyaji, K.; Arai, K.; Ohara, Y.; Takahashi, Y. U.S. Patent 4 837 344, 1989.
17. Durst, T.; Koh, K. *Tetrahedron Lett.* **1992**, *33*, 6799.
18. Dolle, R. E.; Nicolaou, K. C. *J. Am. Chem. Soc.* **1985**, *107*, 1691.
19. (a) DeBrabander, J.; Vanhessche, K.; Vandewalle, M. *Tetrahedron Lett.* **1991**, *32*, 2821. (b) Roy, R.; Rey, A. W.; Charon, M.; Molino, R. *Chem. Commun./J. Chem. Soc., Chem. Commun.* **1989**, 1308.
20. Lavallée, P.; Ruel, R.; Grenier, L.; Bissonnette, M. *Tetrahedron Lett.* **1986**, *27*, 679.
21. Gerlach, U.; Hünig, S. *Angew. Chem., Int. Ed. Engl.* **1987**, *26*, 1283.
22. Allan, R. D.; Bates, M. C.; Drew, C. A.; Duke, R. K.; Hambley, T. W.; Johnston, G. A. R.; Mewett, K. N.; Spence, I. *Tetrahedron* **1990**, *46*, 2511.
23. Mash, E. A.; Arterburn, J. B.; Fryling, J. A.; Mitchell, S. H. *J. Org. Chem.* **1991**, *56*, 1088.

Edward J. J. Grabowski
Merck Research Laboratories, Rahway, NJ, USA

(2*R*,4*R*)-2,4-Pentanediol[1]

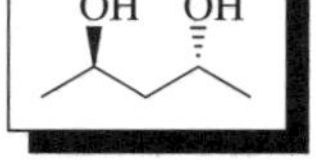

[42075-32-1] $C_5H_{12}O_2$ (MW 104.17)

(diol used for the preparation of chiral acetals[1])

Physical Data: mp 48–50 °C; bp 111–113 °C/19 mmHg.
Form Supplied in: white solid; widely available.
Preparative Methods: via asymmetric hydrogenation of *2,4-Pentanedione*.[2] Its enantiomer *[72345-23-4]* is also available by the same method.
Purification: recrystallization from ether.
Handling, Storage, and Precautions: should be stored in a tightly closed container since it is hygroscopic.

Cleavage of Acetals. Acetals of 2,4-pentanediol are easily prepared from aldehydes via standard procedures (e.g. cat. PPTS, PhH, Dean–Stark removal of H_2O). These acetals have been cleaved with a variety of nucleophiles in the presence of Lewis acids to yield hydroxy ethers with high (typically 90–95% de) diastereoselectivities. Oxidation and β-elimination then provides enantiomerically enriched alcohols (eq 1). Nucleophiles have included allylsilanes to produce homoallylic alcohols,[3] alkynylsilanes to give propargylic alcohols,[4] Me_3SiCN to provide (after hydrolysis) α-hydroxy acids,[5] and enol silyl ethers, α-silyl ketones, or silyl ketene acetals to yield aldol-type products.[6] The same strategy using organometallic reagents/Lewis acid combinations (e.g. $RCu/BF_3{\cdot}OEt_2$,[7] $RMgX/TiCl_4$,[8] $RLi/TiCl_4$,[8,9] $R_2Zn/TiCl_4$[9]) is a general route to secondary alcohols. Other nucleophile/Lewis acid combinations that have been used include alkynylstannanes/$TiCl_4$[10] and zinc enolates/$TiCl_4$.[11] These acetals have also been used in polyene cyclizations.[3a,12]

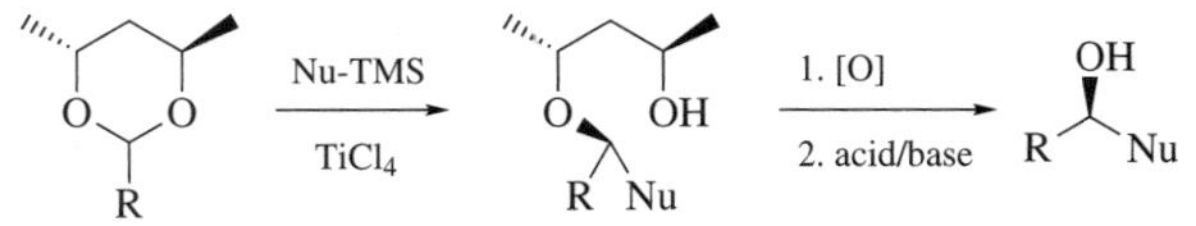

(1)

2,4-Pentanediol is often superior to other diols such as 2,3-butanediol for these reactions because of higher distereoselectivities in reactions with nucleophiles and the more facile cleavage of the resulting hydroxy ether by oxidation–β-elimination.[3] Removal of the chiral auxiliary is usually carried out with *Pyridinium Chlorochromate* oxidation followed by β-elimination using KOH,[3] K_2CO_3,[13] piperidinium acetate,[6] dibenzylammonium trifluoroacetate,[14] or DBU.[4c] In some cases, 1,3-butanediol is preferred because the final β-elimination may be effected under milder conditions.[14]

A detailed study of the mechanism and origin of stereoselectivity in reactions of allyltrimethylsilane with dioxane acetals has been published.[15]

Reduction of Acetals. Reductions of acetals of 2,4-pentanediol can provide (after removal of the chiral auxiliary by oxidation and β elimination) secondary alcohols with good enantioselectivity. The choice of reagents dictates the configuration of the final product. Use of *Dibromoalane* gives products from selective *syn* cleavage of the acetal while *Triethylsilane/Titanium(IV) Chloride* gives the more usual *anti* cleavage products (eq 2).[13]

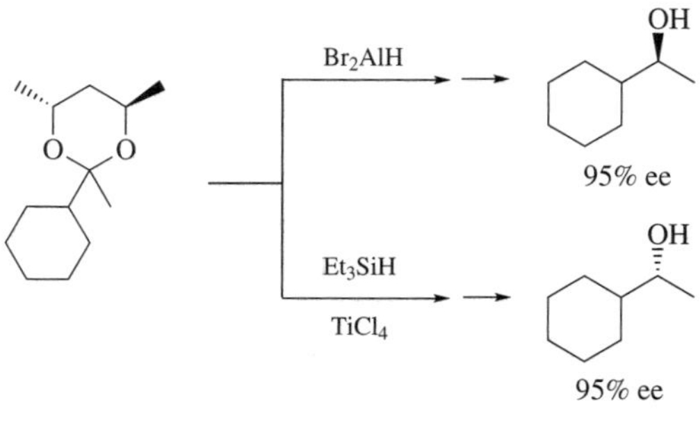

(2)

Elimination of Acetals. Treatment of 2,4-pentanediol acetals of *meso* ketones with *Triisobutylaluminum* gives enol ethers with high diastereoselectivities (eq 3).[16]

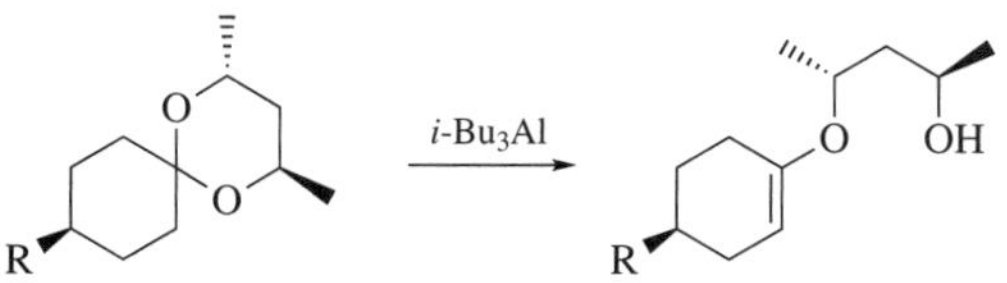

(3)

>80% de

Acetals as Chiral Auxiliaries. There have been many applications of acetals of 2,4-pentanediol as chiral auxiliaries to control the diastereoselectivity of reactions on another functional group.[1] Examples include cyclopropanation of alkenyl dioxanes,[17] lithium amide-mediated isomerization of epoxides to allylic alcohols,[18] and addition of dioxane-substituted Grignard reagents[19] or organolithiums[20] to aldehydes.

Other Uses. Acetals of 2,4-pentanediol have also been prepared in order to determine the enantiomeric purity of aldehydes and ketones by analysis of diastereomers by GC or NMR.[21] *2,3-Butanediol*[22] is more commonly used for this purpose but has been shown to be less effective in some cases.

1. Alexakis, A.; Mangeney, P. *Tetrahedron: Asymmetry* **1990**, *1*, 477.
2. (a) Ito, K.; Harada, T.; Tai, A. *Bull. Chem. Soc. Jpn.* **1980**, *53*, 3367. (b) Tai, A.; Kikukawa, T.; Sugimura, T.; Inoue, Y.; Osawa, T.; Fujii, S. *Chem. Commun./J. Chem. Soc., Chem. Commun.* **1991**, 795, 1324.
3. (a) Bartlett, P. A.; Johnson, W. S.; Elliott, J. D. *J. Am. Chem. Soc.* **1983**, *105*, 2088. (b) Johnson, W. S.; Crackett, P. H.; Elliott, J. D.; Jagodzinski, J. J.; Lindell, S. D.; Natarajan, S. *Tetrahedron Lett.* **1984**, *25*, 3951.
4. (a) Johnson, W. S.; Elliott, R.; Elliott, J. D. *J. Am. Chem. Soc.* **1983**, *105*, 2904. (b) Tabor, A. B.; Holmes, A. B.; Baker, R. *Chem. Commun.* **1989**, 1025. (c) Holmes, A. B.; Tabor, A. B.; Baker, R. *J. Chem. Soc., Perkin Trans. 1* **1991**, 3301, 3307.
5. (a) Elliott, J. D.; Choi, V. M. F.; Johnson, W. S. *J. Org. Chem.* **1983**, *48*, 2295. (b) Choi, V. M. F.; Elliott, J. D.; Johnson, W. S. *Tetrahedron Lett.* **1984**, *25*, 591. (c) Solladié-Cavallo, A.; Suffert, J.; Gordon, M. *Tetrahedron Lett.* **1988**, *29*, 2955.
6. (a) Johnson, W. S.; Edington, C.; Elliott, J. D.; Silverman, I. R. *J. Am. Chem. Soc.* **1984**, *106*, 7588. (b) Elliott, J. D.; Steele, J.; Johnson, W. S. *Tetrahedron Lett.* **1985**, *26*, 2535.
7. (a) Alexakis, A.; Mangeney, P.; Ghribi, A.; Marek, I.; Sedrani, R.; Guir, C.; Normant, J. F. *Pure Appl. Chem.* **1988**, *60*, 49. (b) Normant, J. F.; Alexakis, A.; Ghribi, A.; Mangeney, P. *Tetrahedron* **1989**, *45*, 507.
8. Lindell, S. D.; Elliott, J. D.; Johnson, W. S. *Tetrahedron Lett.* **1984**, *25*, 3947.
9. Mori, A.; Marvoka, K.; Yamamoto, H. *Tetrahedron Lett.* **1984**, *25*, 4421.
10. Yamamoto, Y.; Abe, H.; Nishii, S.; Yamada, J. *J. Chem. Soc., Perkin Trans. 1* **1991**, 3253.
11. (a) Basile, T.; Tagliavini, E.; Trombini, C.; Umani-Ronchi, A. *Chem. Commun./J. Chem. Soc., Chem. Commun.* **1989**, 596. (b) Basile, T.; Tagiavini, E.; Trombini, C.; Umani-Ronchi, A. *Synthesis* **1990**, 305.
12. Johnson, W. S.; Elliott, J. D.; Hanson, G. J. *J. Am. Chem. Soc.* **1984**, *106*, 1138.
13. (a) Ishihara, K.; Mori, A.; Arai, I.; Yamamoto, H. *Tetrahedron Lett.* **1986**, *27*, 983. (b) Ishihara, K.; Mori, A.; Yamamoto, H. *Tetrahedron Lett.* **1986**, *27*, 987. (c) Ishihara, K.; Mori, A.; Arai, I.; Yamamoto, H. *Tetrahedron* **1987**, *43*, 755.
14. Silverman, I. R.; Edington, C.; Elliott, J. D.; Johnson, W. S. *J. Org. Chem.* **1987**, *52*, 180.
15. Denmark, S. E.; Almstead, N. G. *J. Am. Chem. Soc.* **1991**, *113*, 8089.
16. (a) Naruse, Y.; Yamamoto, H. *Tetrahedron Lett.* **1986**, *27*, 1363. (b) Mori, A.; Yamamoto, H. *J. Org. Chem.* **1985**, *50*, 5446. (c) Naruse, Y.; Yamamoto, H. *Tetrahedron* **1988**, *44*, 6021. (d) Kaino, M.; Naruse, Y.; Ishihara, K.; Yamamoto, H. *J. Org. Chem.* **1990**, *55*, 5814. (e) Underiner, T. L.; Paquette, L. A. *J. Org. Chem.* **1992**, *57*, 5438.
17. (a) Arai, I.; Mori, A.; Yamamoto, H. *J. Am. Chem. Soc.* **1985**, *107*, 8254. (b) Mori, A.; Arai, I.; Yamamoto, H. *Tetrahedron* **1986**, *42*, 6447.
18. Yoshikawa, M.; Sugimura, T.; Tai, A. *Chem. Lett.* **1990**, 1003.
19. Kaino, M.; Ishihara, K.; Yamamoto, H. *Bull. Chem. Soc. Jpn.* **1989**, *62*, 3736.
20. Chikashita, H.; Yuasa, T.; Itoh, K. *Chem. Lett.* **1992**, 1457.
21. (a) Fukutani, Y.; Maruoka, K.; Yamamoto, H. *Tetrahedron Lett.* **1984**, *25*, 5911. (b) Furuta, K.; Kanematsu, A.; Yamamoto, H.; Takaoka, S. *Tetrahedron Lett.* **1989**, *30*, 7231. (c) Nonoshita, K.; Banno, H.; Maruoka, K.; Yamamoto, H. *J. Am. Chem. Soc.* **1990**, *112*, 316.
22. Lemière, G. L.; Dommisse, R. A.; Lepoivre, J. A.; Alderweireldt, F. C.; Hiemstra, H.; Wynberg, H.; Jones, J. B.; Toone, E. J. *J. Am. Chem. Soc.* **1987**, *109*, 1363.

J. Michael Chong
University of Waterloo, Ontario, Canada

N-Phenylcampholylhydroxamic Acid

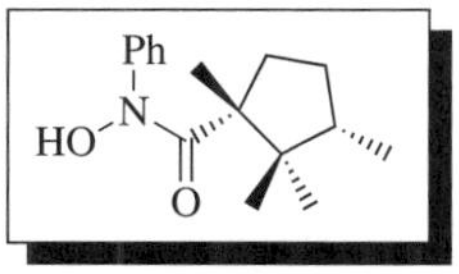

[62668-00-2] $C_{16}H_{23}NO_2$ (MW 261.40)

(chiral ligand in asymmetric epoxidation of allylic alcohols[1])

Solubility: sol toluene, acetonitrile.
Preparative Methods: phenylhydroxylamine (2 equiv) is added to a solution of campholyl chloride (1 equiv) in acetonitrile.[1]

Asymmetric Epoxidation. *N*-Phenylcampholylhydroxamic acid has been used as a chiral ligand in the transition metal-catalyzed asymmetric epoxidation of allylic alcohols.[1] Three allylic alcohols [geraniol (**1**), (*E*)-α-phenylcinnamyl alcohol (**2**), and 1-hydroxymethyl-2-methylcyclohexene (**3**)] have been oxidized to the corresponding epoxides in the presence of the chiral hydroxamate complex of vanadium. The best asymmetric induction (50% ee) is observed for alcohol (**2**) at −78 °C. Generally, low temperature reactions give higher asymmetric inductions but lower yields. The optimum inductions are attained when a 5:1 ratio of hydroxamic acid and *Vanadyl Bis(acetylacetonate)* is used. *t-Butyl Hydroperoxide* gives substantially better inductions than *Cumyl Hydroperoxide*. The asymmetric oxidation is proposed to go through the intermediate (**4**),[1,2] in which the alcohol is coordinated to the metal during the oxygen atom transfer step.

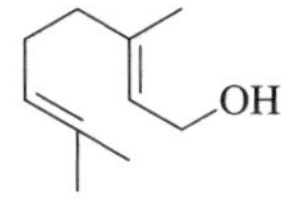

(**1**)

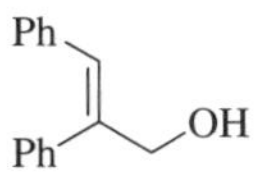

(**2**)

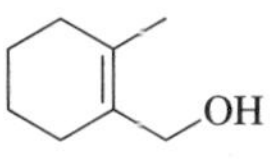

(**3**)

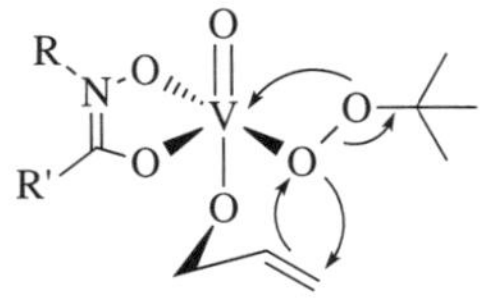

(4)

1. Michaelson, R. C.; Palermo, R. E.; Sharpless, K. B. *J. Am. Chem. Soc.* **1977**, *99*, 1990.
2. Chong, A. O.; Sharpless, K. B. *J. Org. Chem.* **1977**, *42*, 1587.

Sangho Koo
Myong Ji University, Seoul, Korea

(*S*)-(−)-5-(α-Phenylethyl)semioxamazide

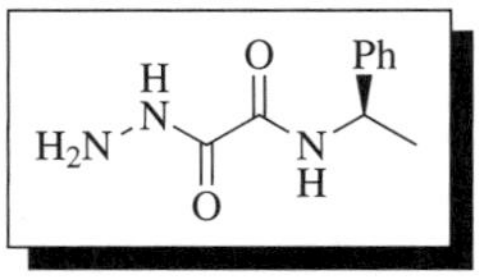

[6152-25-6] $C_{10}H_{13}N_3O_2$ (MW 207.26)

(optical resolution of carbonyl compounds[2,3])

Physical Data: $[\alpha]_D$ −103° (*c* 1.0, $CHCl_3$).

Solubility: sol polar aprotic solvents; limited sol hexane, toluene, etc.

Preparative Methods: conveniently prepared by sequential addition of (*S*)-(−)-α-phenylethylamine and *Hydrazine* to *Diethyl Oxalate* (eq 1).[1]

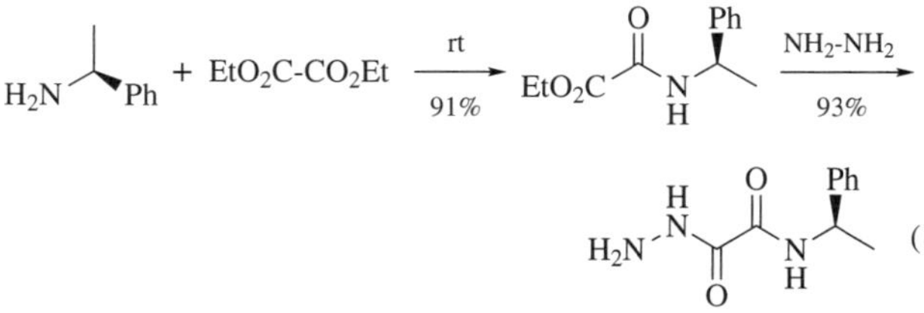

(1)

Optical Resolution. This reagent is commonly used for the optical resolution of various compounds. Racemic spiro[4.4]nonane-1,6-dione was the first compound to be resolved (55% overall yield).[2] The corresponding semioxamazone was obtained in optically active pure form in two or three recrystallizations and hydrolyzed to (−)-(*S*)-spiro[4.4]nonane-1,6-dione in a refluxing methanol–water mixture in the presence of *Iodine* (eq 2).

In the field of arene tricarbonyl chromium complexes, racemic aldehydes can be resolved quantitatively by chromatographic separation of the corresponding semioxamazones. The first optical resolution was carried out on the semioxamazone made from the chromium tricarbonyl complex of *o*-anisaldehyde.[3,4] The separation of the diastereomers was done by silica gel chromatography (eq 3). Chromium tricarbonyl complexes of *m*-anisaldehyde,[4] of the trifluoromethyl analog of *o*-anisaldehyde,[5] and of *o*-trimethylsilylbenzaldehyde[6] have also been resolved by the same procedure.

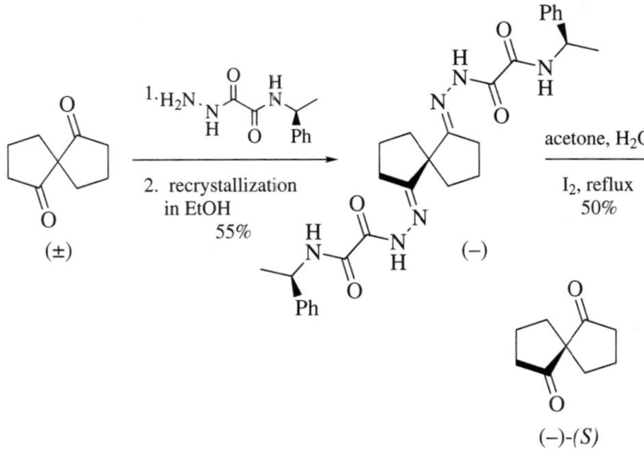

(2)

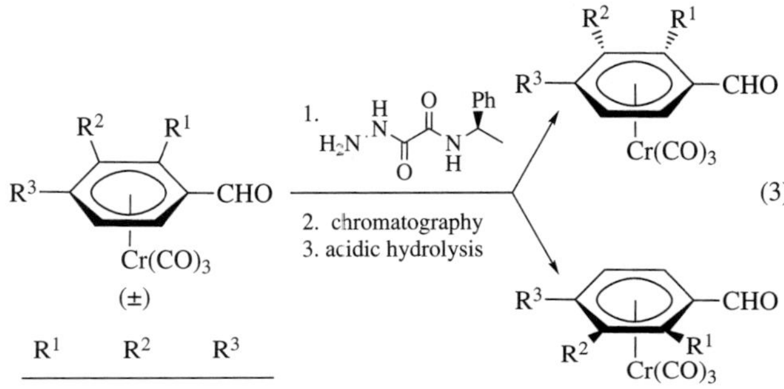

(3)

R^1	R^2	R^3
OMe	H	H
H	OMe	H
CF_3	H	H
TMS	H	H
OMe	OMe	H
H	OMe	OMe

The iron tricarbonyl complex of 2-formylbutadiene has been resolved by chromatographic separation of its chiral semioxamazones (eq 4).[7]

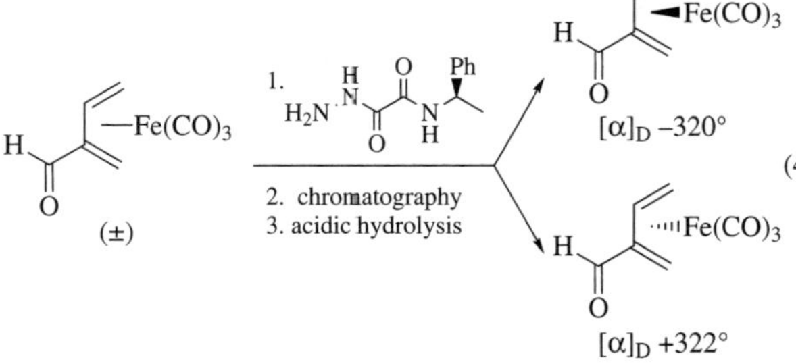

(4)

2,3-Dihydro-2-phenyl-4(1*H*)-quinolone has also been resolved by fractional crystallization of the corresponding chiral semioxamazones (eq 5).[8]

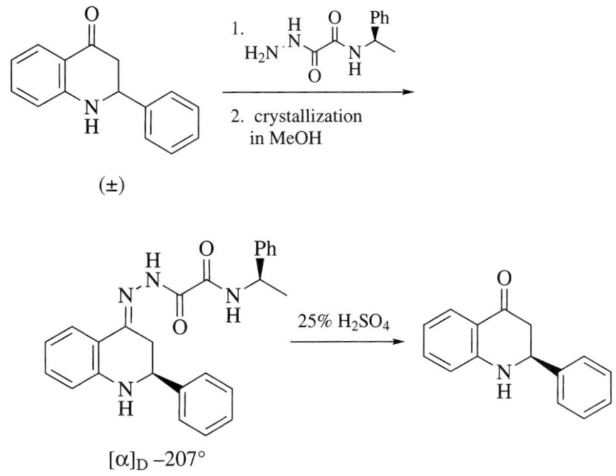

(5)

1. Leonard, N. J.; Boyer, J. H. *J. Org. Chem.* **1950**, *15*, 42.
2. Harada, N.; Ochiai, N.; Takada, K.; Uda, H. *Chem. Commun./J. Chem. Soc., Chem. Commun.* **1977**, 495.
3. Solladié-Cavallo, A.; Solladié, G.; Tsamo, E. *J. Org. Chem.* **1979**, *44*, 4189.
4. Solladié-Cavallo, A.; Solladié, G.; Tsamo, E. *Inorg. Synth.* **1985**, *23*, 85.
5. Solladié-Cavallo, A.; Farkhani, D.; Dreyfuss, A. C.; Sanch, F. *Bull. Soc. Chem. Fr. Part 2* **1986**, 906.
6. Mukai, C.; Cho, W. J.; Kim, I. J.; Kido, M.; Hanakoa, M. *Tetrahedron* **1991**, *47*, 3007.
7. Franck-Neumann, M.; Martina, D.; Heitz, M. P. *J. Organomet. Chem.* **1986**, *301*, 61.
8. Tokés, A. L.; Szilágyi, L. *Synth. Commun.* **1987**, *17*, 1235.

Guy Solladié & Françoise Colobert
University Louis Pasteur, Strasbourg, France

(–)-8-Phenylmenthol

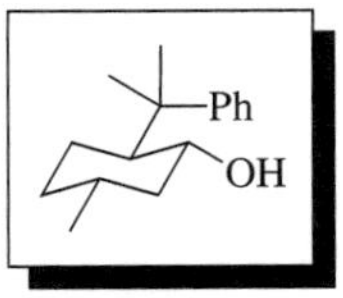

(1*R*,2*S*,5*R*) [65253-04-5] (1*S*,2*R*,5*R*) [100101-42-6] (1*S*,2*R*,5*S*) [57707-91-2]	$C_{16}H_{24}O$	(MW 232.37)

(chiral auxiliary for asymmetric induction)

Alternate Name: (1*R*,2*S*,5*R*)-5-methyl-2-(1-methyl-1-phenylethyl)cyclohexanol.
Physical Data: $[\alpha]_D$ −26° (*c* 2, EtOH); *d* 0.999 g cm^{-3}.
Solubility: sol organic solvents.
Form Supplied in: commercially available as a colorless oil (98%) and as the chloroacetate ester.
Analysis of Reagent Purity: NMR and $[\alpha]_D$.
Preparative Methods: (1*R*,2*S*,5*R*)-(–)-8-phenylmenthol is prepared by the reaction of *Phenylmagnesium Bromide* (cat. *Copper(I) Iodide*) with (+)-pulegone, equilibration of the resulting conjugate addition product, and reduction of the ketone (*Sodium*, isopropanol).[1] The (1*R*,2*S*,5*R*)-isomer is accompanied by the (1*S*,2*R*,5*R*)-isomer and is conveniently separated by recrystallization of the chloroacetate ester.[2] This separation is essential to obtaining high optical yields since it has been shown that the two diastereomers have opposite chiral directing ability.[3] The preparation of the enantiomeric (1*S*,2*R*,5*S*)-(–)-8-phenylmenthol from (+)-pulegone has been reported.[4]
Handling, Storage, and Precautions: no special precautions are necessary other than those used for combustible organic compounds.

Chiral Auxiliary for Asymmetric Induction. Numerous derivatives of (–)-8-phenylmenthol have been utilized for asymmetric induction studies. These include inter-[5] and intramolecular[6] Diels–Alder reactions, dihydroxylations,[7] and intramolecular ene reactions[8] of α,β-unsaturated 8-phenylmenthol esters. These reactions usually proceed in moderate to good yield with high diastereofacial selectivity. α-Keto esters of 8-phenylmenthol (see *8-Phenylmenthyl Pyruvate*) have been used for asymmetric addition to the keto group,[9] as well as for asymmetric [2 + 2] photoadditions[10] and nucleophilic alkylation.[11] Ene reactions of α-imino esters of 8-phenylmenthol with alkenes provide a direct route to α-amino acids of high optical purity.[12]

Vinyl and butadienyl ethers of 8-phenylmenthol have been prepared and the diastereofacial selectivity of nitrone[13] and Diels–Alder[14] cycloadditions, respectively, have been evaluated. α-Anions of 8-phenylmenthol esters also show significant diastereofacial selectivity in aldol condensations[15] and enantioselective alkene formation by reaction of achiral ketones with 8-phenylmenthyl phosphonoacetate gives de up to 90%.[16]

Related Reagents. 8-Phenylmenthyl Crotonate; 8-Phenylmenthyl Glyoxylate; 8-Phenylmenthyl Pyruvate.

1. Corey, E. J.; Ensley, H. E. *J. Am. Chem. Soc.* **1975** *97*, 6908.
2. (a) Herzog, H.; Scharf, H. D. *Synthesis* **1986**, 420. (b) Ort, O. *Org. Synth.* **1987**, *65*, 203. (c) See also Cervinka, O.; Svatos, A.; Masojidkova, M. *Collect. Czech. Chem. Commun.* **1990**, *55*, 491.
3. Whitesell, J. K.; Liu, C. L.; Buchanan, C. M.; Chen, H. H.; Minton, M. A. *J. Org. Chem.* **1986**, *51*, 551. Whitesell, J. K. *Chem. Rev.* **1992**, *92*, 953.
4. Ensley, H. E.; Parnell, C. A.; Corey, E. J. *J. Org. Chem.* **1978**, *43*, 1610.
5. Oppolzer, W.; Kurth, M.; Reichlin, D.; Chapuis, C.; Mohnhaupt, M.; Moffat, F. *Helv. Chim. Acta* **1981**, *64*, 2802.
6. Roush, W. R.; Gillis, H. R.; Ko, A. I. *J. Am. Chem. Soc.* **1982**, *104*, 2269.
7. Hatakeyama, S.; Matsui, Y.; Suzuki, M.; Sakurai, K.; Takano, S. *Tetrahedron Lett.* **1985**, *26*, 6485.
8. (a) Oppolzer, W.; Robbiani, C.; Bättig, K. *Helv. Chim. Acta* **1980**, *63*, 2015. (b) Oppolzer, W.; Robbiani, C.; Bättig, K. *Tetrahedron* **1984**, *40*, 1391.
9. (a) Grossen, P.; Herold, P.; Mohr, P.; Tamm, C. *Helv. Chim. Acta* **1984**, *67*, 1625. (b) Sugimura, H.; Yoshida, K. *J. Org. Chem.* **1993**, *58*, 4484. (c) Solladie-Cavallo, A.; Bencheqroun, M. *Tetrahedron: Asymmetry* **1991**, *2*, 1165. (d) Comins, D. L.; Baevsky, M. F.; Hong, H. *J. Am. Chem. Soc.* **1992**, *114*, 10971.

10. (a) Koch, H.; Scharf, H. D.; Runsink, J.; Leismann, H. *Ber. Dtsch. Chem. Ges./Chem. Ber.* **1985**, *118*, 1485. (b) Nehrings, A.; Scharf, H.-D.; Runsink, J. *Angew. Chem., Int. Ed. Engl.* **1985**, *24*, 877.
11. Hamon, D. P. G.; Massy-Westropp, R. A.; Razzino, P. *Tetrahedron* **1992**, *48*, 5163.
12. Mikami, K.; Kaneko, M.; Yajima, T. *Tetrahedron Lett.* **1993**, *34*, 4841.
13. Carruthers, W.; Coggins, P.; Weston, J. B. *Chem. Commun./J. Chem. Soc., Chem. Commun.* **1991**, 117.
14. (a) Thiem, R.; Rotscheidt, K.; Breitmaier, E. *Synthesis* **1989**, 836. (b) Danishefsky, S.; Bednarski, M.; Izawa, T.; Maring, C. *J. Org. Chem.* **1984**, *49*, 2290.
15. Corey, E. J.; Peterson, R. T. *Tetrahedron Lett.* **1985**, *26*, 5025.
16. (a) Gais, H. J.; Schmiedl, G.; Ball, W. A. *Tetrahedron Lett.* **1988**, *29*, 1773. (b) Rehwinkel, H.; Skupsch, J.; Vorbrüggen, H. *Tetrahedron Lett.* **1988**, *29*, 1775.

Harry E. Ensley, Matthew Beggs & Yinghong Gao
Tulane University, New Orleans, LA, USA

8-Phenylmenthyl Acrylate

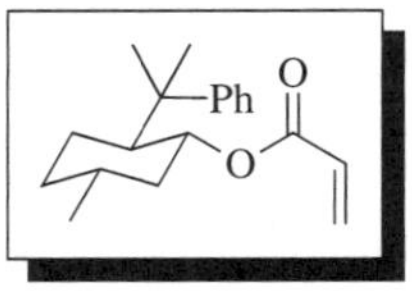

[72526-00-2] $C_{19}H_{26}O_2$ (MW 286.45)

(reagent used in asymmetric synthesis for cycloadditions and conjugate additions)

Alternate Name: (1*R*,2*S*,5*R*)-2-(1-methyl-1-phenylethyl)-5-methylcyclohexyl acrylate.

Solubility: sol CH_2Cl_2, toluene, most organic solvents.

Analysis of Reagent Purity: ^{1}H and ^{13}C NMR.[1] $[\alpha]_D$ +16.21° (*c* 1.68, CH_2Cl_2).[2]

Preparative Methods: prepared by the reaction of *(−)-8-Phenylmenthol*, acryloyl chloride, *Triethylamine*, and *4-Dimethylaminopyridine* in CH_2Cl_2 at 0 °C. Following an aqueous workup, the compound is purified by chromatography on silica gel.[1]

Handling, Storage, and Precautions: acrylates are prone to polymerization and are best stored below room temperature under N_2.

[4 + 2] Cycloadditions. The asymmetric Diels–Alder reaction[3] of phenylmenthyl acrylate with 5-benzyloxymethylcyclopentadiene in the presence of *Aluminum Chloride* produces an 89% yield of the *endo* cycloadduct (eq 1), accompanied by 7% of the *exo* adduct. This provides a useful intermediate for the preparation of various prostaglandins.[2] The *Tin(IV) Chloride* and *Titanium(IV) Chloride* catalyzed reactions with *Cyclopentadiene* deliver a mixture of *endo* and *exo* adducts in 89% de, and 90% de, respectively (eq 2). The $TiCl_4$ reaction gives an 89:11 *endo*:*exo* ratio, while the $SnCl_4$ reaction gives an 84:16 *endo*:*exo* ratio. From a practical point of view, the titanium and tin catalysts are the best of the various Lewis acids surveyed.[4] The use of $TiCl_4$ is also the most effective for the reaction of the acrylate with *1,3-Butadiene* (eq 3).[5] The increased asymmetric induction over the simpler menthyl acrylate is attributed to the shielding of the C(α)-*re* face of the dienophile by the phenyl ring.[6]

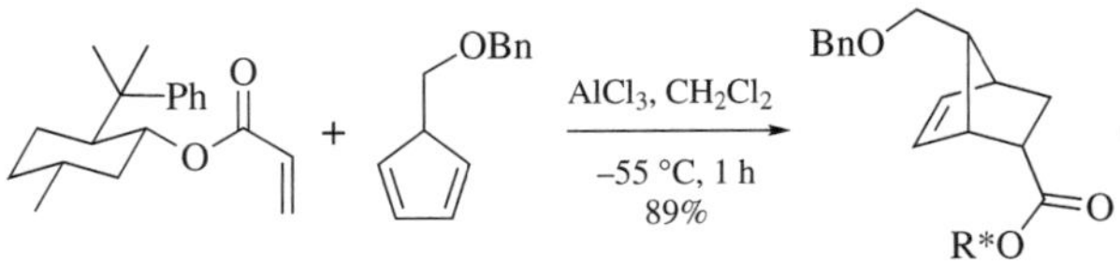
(1)

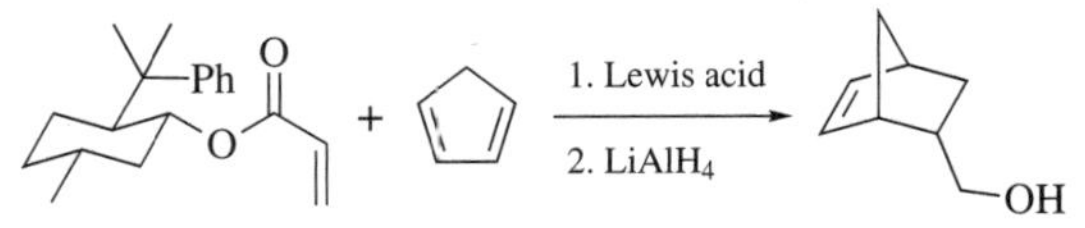
(2)

1.5 equiv $SnCl_4$, 95%; 89% ee
1.5 equiv $TiCl_4$, 83%; 90% ee

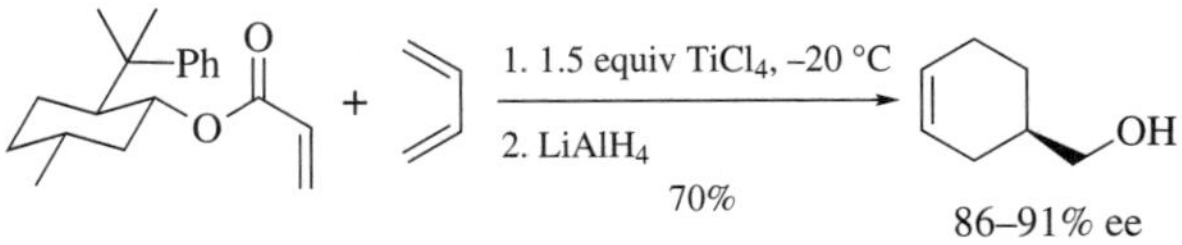
(3)

86–91% ee

1,3-Dipolar Cycloadditions. The asymmetric induction for a 1,3-dipolar cycloaddition of phenylmenthyl acrylate is not as good as in the [4 + 2] cycloadditions. The thermal decomposition of diazofluorene in the presence of the acrylate produces the spirocyclopropane in 96% yield, but with only a 20% de (eq 4).[7]

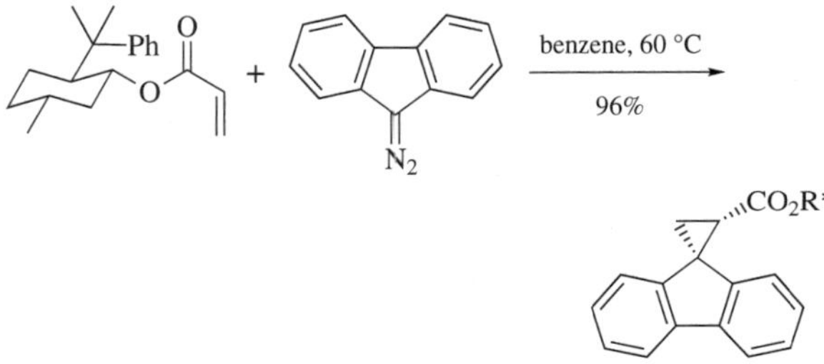
(4)

20% de

Conjugate Additions. The reaction of this acrylate derivative with lithium *t*-butyl hydroperoxide (generated from anhydrous *t-Butyl Hydroperoxide* and *n-Butyllithium*) in THF leads to the corresponding epoxide (eq 5) in 95% yield with a de of 40%.[8]

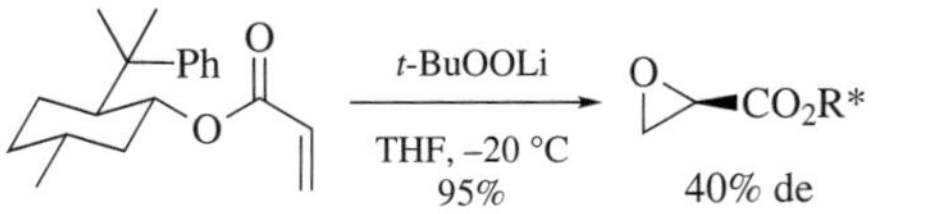
(5)

40% de

In an asymmetric approach to the bicyclo[2.2.2]octane ring system, a double Michael addition has been employed using phenylmenthyl acrylate as the initial Michael acceptor. The condensation of the dienolate, generated with *Lithium Diisopropylamide*, reacts with the acrylate to afford the bicyclo[2.2.2]octane derivative (eq 6). The de for the reaction is only 50%; however, it is highly *endo* selective (>95%).[9]

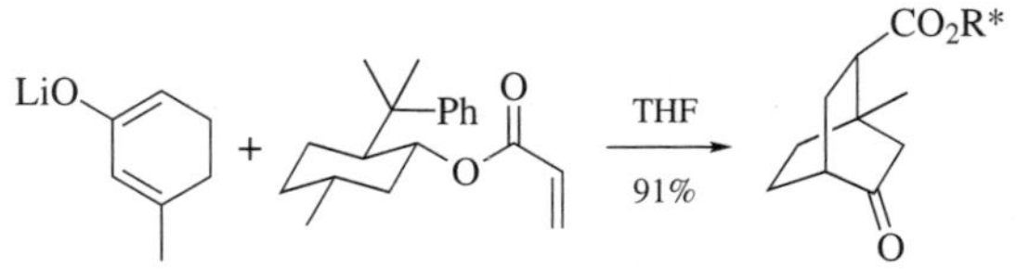
(6)

50% de

A Lewis acid-mediated two-fold asymmetric Michael addition allows access to *cis*-decalin derivatives. The reaction of the trimethylsilylenol ether of acetylcyclohexene with phenylmenthyl acrylate in the presence of *Diethylaluminum Chloride* (eq 7) yields the decalone in 64% yield (70% de). This has been shown not to be a Diels–Alder reaction. If the reaction is worked-up early, the initial Michael adduct can be isolated.[10]

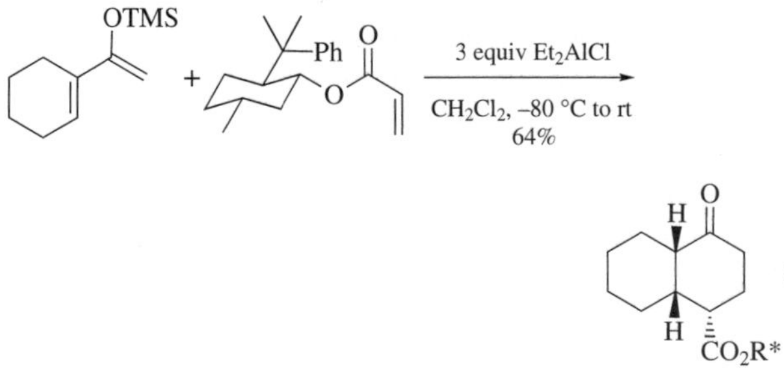

(7)

70% de

Phenylmenthyl acrylate has been used as a component in an asymmetric Baylis–Hillman reaction. Treatment of the acrylate with *1,4-Diazabicyclo[2.2.2]octane* and benzaldehyde at 8 kbar of pressure delivers the α-(hydroxyalkyl)acrylate (eq 8). The product obtained has an 86% de. Menthyl acrylate is superior to the phenylmenthyl acrylate in this particular application.[11] In a radical-mediated addition, phenylmenthyl acrylate gives rise to the α-pyridyl sulfide in 68% yield (eq 9).[12] The final product is isolated with a 56% de.

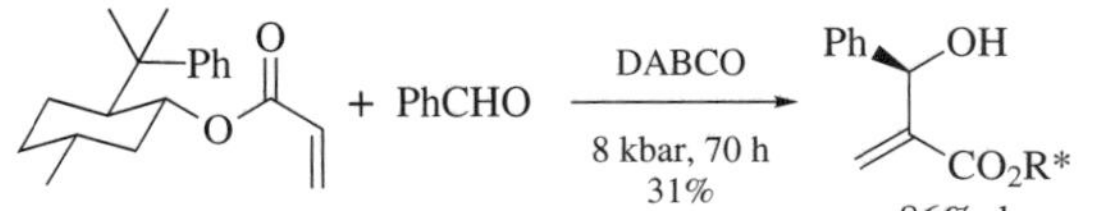

(8)

86% de

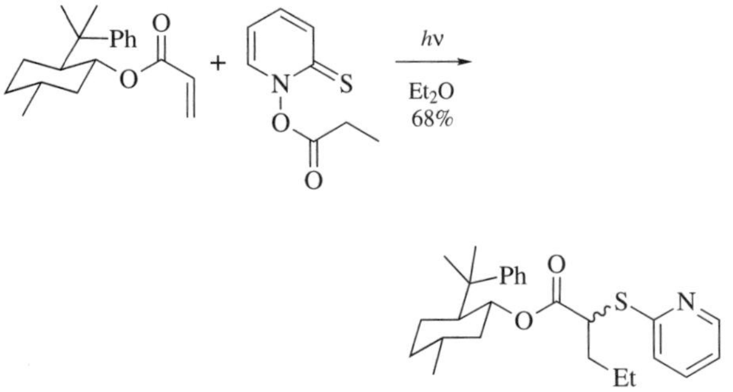

(9)

56% de

Miscellaneous. The acrylate provides a synthon for the preparation of *8-Phenylmenthyl Glyoxylate*, which is useful for asymmetric ene reactions.[1] Thus ozonolysis and remoeval of the water of hydration produces the glyoxylate in 89% yield (eq 10).

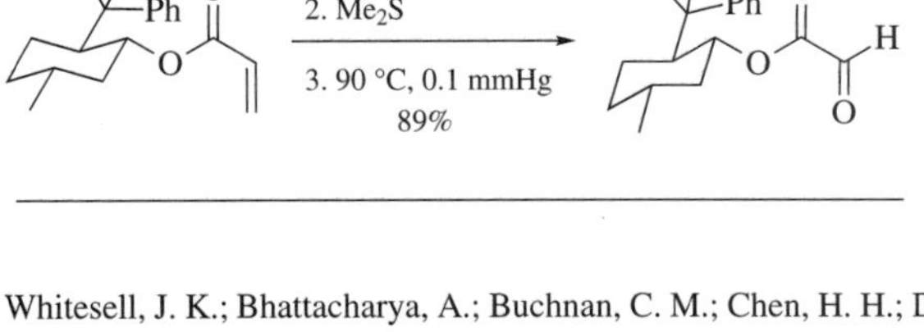

(10)

1. Whitesell, J. K.; Bhattacharya, A.; Buchnan, C. M.; Chen, H. H.; Deyo, D.; James, D.; Liu, C.-L.; Minton, M. A. *Tetrahedron* **1986**, *42*, 2993.
2. Corey, E. J.; Ensley, H. E. *J. Am. Chem. Soc.* **1975**, *97*, 6908.
3. (a) Oppolzer, W. *AGE* **1984**, *23*, 876. (b) Paquette, L. A. *Asymmetric Synthesis*; Morrison, J. D., Ed.; Academic: Orlando, 1984; Vol. 3B, pp 455–501. (c) Helmchen, G.; Karge, R.; Weetman, J. *Modern Synthetic Methods*; Scheffold, R. Ed.; Springer: Berlin, 1986; Vol. 4, pp 262–306. (d) Taschner, M. J. *Organic Synthesis: Theory and Applications*; Hudlicky, T., Ed.; JAI: Greenwich, CT, 1989; Vol. 1, pp 1–101.
4. Oppolzer, W.; Kurth, M.; Reichlin, D.; Moffatt, F. *Tetrahedron Lett.* **1981**, *22*, 2545.
5. (a) Boeckman, R. K.; Naegely, P. C.; Arthur, S. D. *J. Org. Chem.* **1980**, *45*, 752. (b) Kocienski, P.; Stocks, M.; Donald, D.; Perry, M. *Synlett* **1990**, 38.
6. Oppolzer, W.; Kurth, M.; Reichlin, D.; Chapius, C.; Mohnhaupt, M.; Moffatt, F. *Helv. Chim. Acta* **1981**, *64*, 2802.
7. Okada, K.; Samizo, F.; Oda, M. *Chem. Lett.* **1987**, 93.
8. (a) Clark, C.; Hermans, P.; Meth-Cohn, O.; Moore, C.; Taljaard, H. C.; van Vuuren, G. *Chem. Commun./J. Chem. Soc., Chem. Commun.* **1986**, 1378. (b) Meth-Cohn, O.; Moore, C.; Taljaard, H. C. *J. Chem. Soc., Perkin Trans. 1* **1988**, 2663.
9. Spitzner, D.; Wagner, P.; Simon, A.; Peters, K. *Tetrahedron Lett.* **1989**, *30*, 547.
10. Hagiwara, H.; Akama, T.; Okano, A.; Uda, H. *Chem. Lett.* **1989**, 2149.
11. Gilbert, A.; Heritage, T. W.; Isaacs, N. S. *Tetrahedron: Asymmetry* **1991**, *2*, 969.
12. Crich, D.; Davies, J. W. *Tetrahedron Lett.* **1987**, *28*, 4205.

Michael J. Taschner
The University of Akron, OH, USA

8-Phenylmenthyl Crotonate

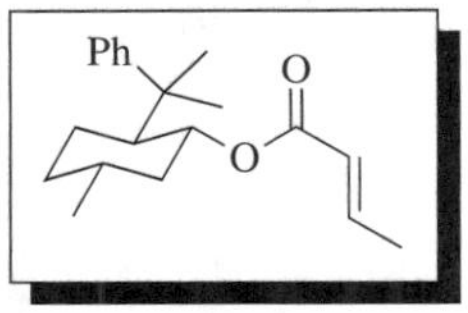

[81002-19-9] $C_{20}H_{26}O_2$ (MW 300.46)

(chiral ester for asymmetric induction)

Alternate Name: (1*R*,2*S*,5*R*)-5-methyl-2-(1-methyl-1-phenylethyl)cyclohexyl crotonate.
Solubility: sol organic solvents.
Form Supplied in: not commercially available.
Preparative Methods: isolated as an oil from the esterification of *(−)-8-Phenylmenthol* with 2-butenoic acid (DCC, DMAP, CH_2Cl_2).[1]
Analysis of Reagent Purity: NMR[2] and $[\alpha]_D$.
Handling, Storage, and Precautions: no special precautions are necessary other than those used for combustible organic compounds.

General Discussion. Reaction of the crotyl ester of (−)-8-phenylmenthol with organocopper reagents in the presence of *Boron Trifluoride* gives good chemical yields and >99% de of the 1,4-addition product where addition has taken place from the *re* face (eq 1).[1] Extension of this methodology to the 8-phenylmenthyl *cis*-butenoate gives significantly lower de.[1] Epoxidation of 8-phenylmenthyl crotonate with *t-Butyl*

Hydroperoxide–n-Butyllithium gives 50% de.[2] Chiral β-amino esters of 8-phenylmenthol have been prepared in 50–60% de by the addition of amines to the *re* face of 8-phenylmenthyl crotonate under 14–15 kbar pressures (eq 2). Much higher (75 to >99%) de is obtained using 8-(β-naphthyl)menthol crotonate. The β-amino esters obtained are of the proper configuration for conversion to biologically active β-lactams.[3]

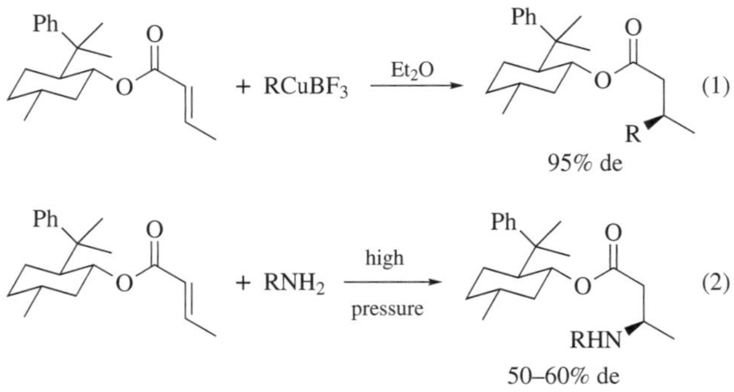

1. (a) Oppolzer, W.; Löher, H. J. *Helv. Chim. Acta* **1981**, *64*, 2808. (b) Carpita, A.; De Magistris, E.; Rossi, R. *Gazz. Chim. Ital.* **1989**, *119*, 99.
2. Meth-Cohn, O.; Moore, C.; Taljaard, H. C. *J. Chem. Soc., Perkin Trans. 1* **1988**, 2663.
3. d'Angelo, J.; Maddaluno, J. *J. Am. Chem. Soc.* **1986**, *108*, 8112.

Harry E. Ensley & Yunjie Dang
Tulane University, New Orleans, LA, USA

8-Phenylmenthyl Glyoxylate[1]

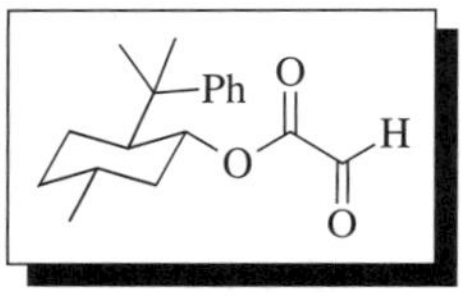

[129444-92-4] $C_{18}H_{24}O_3$ (MW 288.42)

(two-carbon unit utilized as a versatile chiral reagent in ene reactions,[2] Diels–Alder reactions,[3] various nucleophilic additions,[4] and aromatic substitutions.[5])

Physical Data: bp 135–140 °C/0.3 mmHg; $[\alpha]^{20}_{589}$ −169° (*c* 0.51, benzene).[6]
Solubility: sol most common organic solvents.
Preparative Methods: ozonolysis of *8-Phenylmenthyl Acrylate*, NaOAc-catalyzed elimination of nitrite ion from nitrate esters, and direct esterification with glyoxylic acid.[7]
Handling, Storage, and Precautions: heating the glyoxylate monohydrate to 90 °C/0.1 mmHg for 4.5 h or distillation provides material which is sufficiently anhydrous for most practical applications. Due to its tendency to form a monohydrate and/or polymerize, 8-phenylmenthyl glyoxylate should be stored under nitrogen in the refrigerator.

Ene Reactions. Lewis acid (*Tin(IV) Chloride* or *Titanium(IV) Chloride*) catalyzed ene reactions of the glyoxylate esters of 8-phenylmenthol and 2-*epi,ent*-8-phenylmenthol, as well as *trans*-2-phenylcyclohexanol, with terminal, monosubstituted alkenes proceed with excellent levels of stereocontrol (eq 1).[8] The phenyl group presumably forms a π-complex with the aldehyde–Lewis acid complex thereby blocking one face of the dicarbonyl, rendering the opposite face accessible to incoming reagents. This has been demonstrated through the systematic study of the effect of auxiliary structure on asymmetric induction in the glyoxylate–ene reaction, as well as by photophysical studies involving a family of α-carbonyl esters of 8-phenylmenthol.[7,9] The ene reactions of glyoxylates with nonterminal alkenes proceed with a high level of relative asymmetric induction to give the (*S*)-configuration at the newly created stereocenter with 8-phenylmenthol and the (*R*)-configuration with *trans*-2-phenylcyclohexanol. The reaction of 8-phenylmenthyl glyoxylate with *trans*-2-butene produces a 93:7 mixture of diastereomers at C-3 in 85% yield; introduction of a TMS group into *trans*-2-butene increases the *anti*-selectivity up to ~100% (eq 2).[8f,g] Double bond isomerization in the substrate under the reaction conditions, as well as contamination of halogen adducts in the products, point to a cationic mechanism for these ene reactions.[8e]

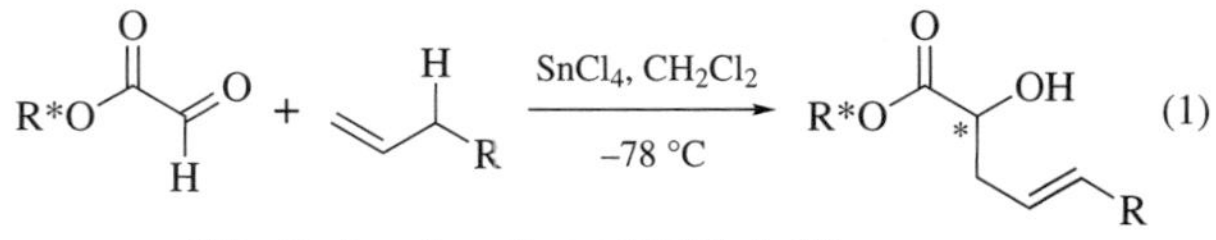

R* = 8-phenylmenthyl, >99.8% de *(S)*
2-*epi,ent*-8-phenylmenthyl, >99.8% de *(R)*
trans-2-phenylcyclohexyl, >97% de *(R)*

The asymmetric glyoxylate–ene reactions have been exploited in the total synthesis of (−)-specionin, which involves asymmetric desymmetrization of a prochiral diene (eq 3), and (−)-xylomollin, which involves an efficient kinetic resolution of a racemic diene (eq 4).[10a–c]

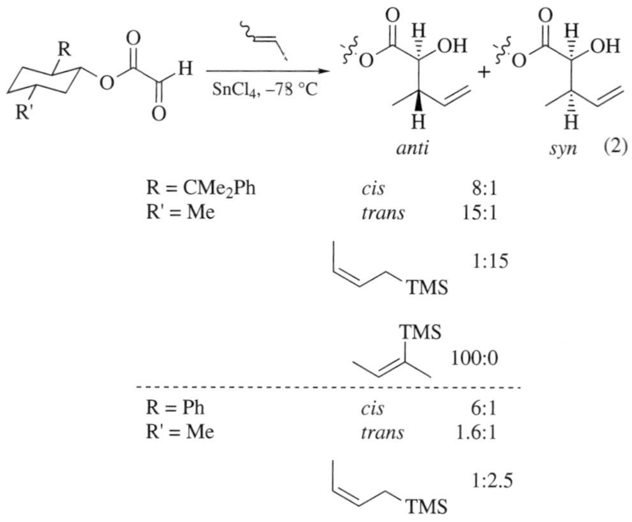

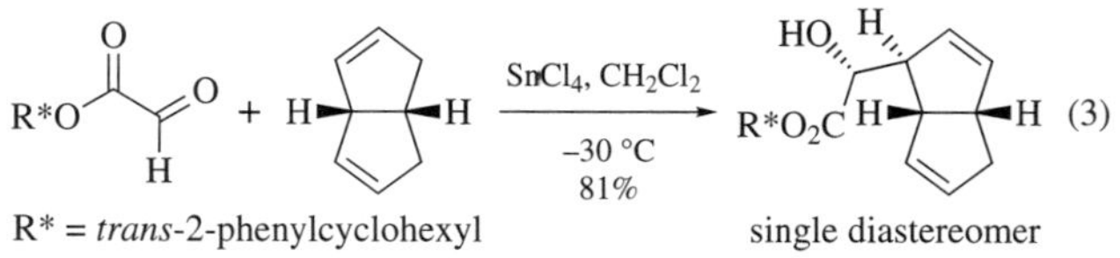

R* = *trans*-2-phenylcyclohexyl

single diastereomer

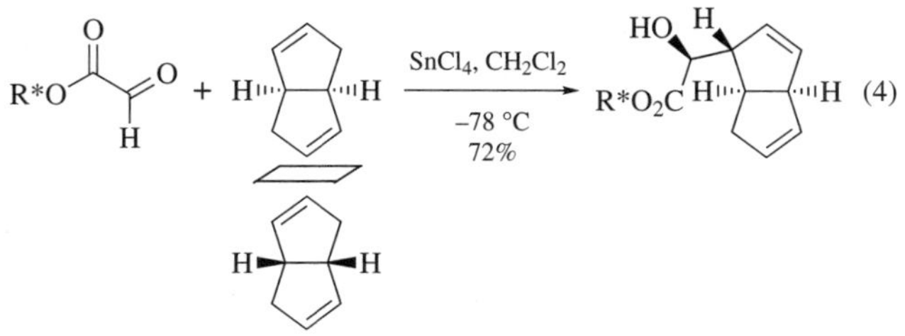

R* = 8-phenylmenthyl

Nucleophilic Additions. A variety of nucleophiles have been added to 8-phenylmenthyl (and *trans*-2-phenylcyclohexyl) glyoxylates with high levels of asymmetric induction. These include organomagnesium[4] and organotin reagents,[11] as well as nitroalkane anions (eq 5).[12] Other applications of 8-phenylmenthyl glyoxylate include asymmetric hetero-Diels–Alder reactions to produce chiral dihydropyran derivatives[3] and *o*-hydroxylation of phenols producing the corresponding chiral 2-hydroxymandelic acid derivatives.[5] α-Imino esters derived from 8-phenylmenthyl glyoxylate undergo nucleophilic additions[13] as well as ene reactions[14] to produce nonproteinogenic α-amino acid derivatives.

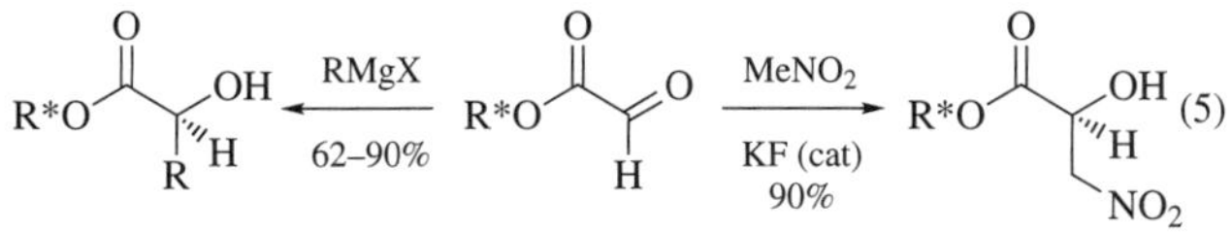

>95% de R* = 8-phenylmenthyl >95% de

1. Whitesell, J. K. *Chem. Rev.* **1992**, *92*, 953.
2. Mikami, K.; Shimizu, M. *Chem. Rev.* **1992**, *92*, 1021.
3. Cervinka, O.; Svatos, A.; Trska, P.; Pech, P. *Collect. Czech. Chem. Commun.* **1990**, *55*, 230.
4. (a) Whitesell, J. K.; Bhattacharya, A.; Henke, K. *Chem. Commun./J. Chem. Soc., Chem. Commun.* **1982**, 988. (b) Whitesell, J. K.; Deyo, D.; Bhattacharya, A. *Chem. Commun./J. Chem. Soc., Chem. Commun.* **1983**, 802. (c) Whitesell, J. K.; Buchanan, C. M. *J. Org. Chem.* **1986**, *51*, 5443.
5. Bigi, F.; Casnati, G.; Sartori, G.; Dalprato, C.; Bortolini, R. *Tetrahedron: Asymmetry* **1990**, *2*, 861.
6. Cervinka, O.; Svatos, A.; Masojidkova, M. *Collect. Czech. Chem. Commun.* **1990**, *55*, 491.
7. Whitesell, J. K.; Lawrence, R. M.; Huang-Hsing, C. *J. Org. Chem.* **1986**, *51*, 4779.
8. (a) Whitesell, J. K.; Bhattacharya, A.; Aguilar, D. A. A.; Henke, K. *Chem. Commun./J. Chem. Soc., Chem. Commun.* **1982**, 989. (b) Whitesell, J. K.; Bhattacharya, A.; Buchanan, C. M.; Chen, H. H.; Deyo, D. *Tetrahedron* **1986**, *42*, 2993. (c) Whitesell, J. K.; Chen, H. H.; Lawrence, R. M. *J. Org. Chem.* **1985**, *50*, 4663. (d) Whitesell, J. K.; Liu, C.; Buchanan, C.; Chen, H-H.; Minton, M. M. *J. Org. Chem.* **1986**, *51*, 551. (e) Whitesell, J. K. *Acc. Chem. Res.* **1985**, *18*, 280. (f) Mikami, K.; Wakabayashi, H.; Nakai, T. *J. Org. Chem.* **1991**, *56*, 4337. (g) Grossen, P.; Herold, P.; Mohr, P.; Tamm, C. *Helv. Chim. Acta* **1984**, *67*, 1625.
9. Whitesell, J. K.; Younathan, J. N.; Hurst, J. R.; Fox, M. A. *J. Org. Chem.* **1985**, *50*, 5499.
10. (a) Whitesell, J. K.; Allen, D. E. *J. Org. Chem.* **1985**, *50*, 3025. (b) Whitesell, J. K.; Allen, D. E. *J. Am. Chem. Soc.* **1988**, *110*, 3558. (c) Whitesell, J. K.; Minton, M. A. *J. Am. Chem. Soc.* **1986**, *108*, 6802.
11. (a) Yamamoto, Y.; Maeda, N.; Maryuma, K. *Chem. Commun./J. Chem. Soc., Chem. Commun.* **1983**, 774. (b) Yamamoto, Y.; Yatagai, H.; Ishihara, Y.; Maeda, N.; Maryuma, K. *Tetrahedron* **1984**, *40*, 2239.
12. (a) Solladie-Cavallo, A.; Khiar, N. *Tetrahedron Lett.* **1988**, 2189. (b) Solladie-Cavallo, A.; Khiar, N.; Fischer, J.; DeCian, A. *Tetrahedron* **1991**, *47*, 249. (c) Solladie-Cavallo, A.; Khiar, N. *J. Org. Chem.* **1990**, *55*, 4750.
13. Yamamoto, Y.; Ito, W. *Tetrahedron* **1988**, *44*, 5415.
14. Mikami, K.; Kaneko, M.; Yajima, T. *Tetrahedron Lett.* **1993**, *34*, 4841.

Apurba Bhattacharya
Hoechst-Celanese Corporation, Corpus Christi, TX, USA

8-Phenylmenthyl Pyruvate

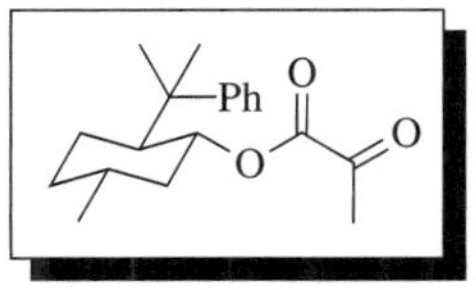

(1*R*,2*S*,5*R*)
[88292-41-5] $C_{19}H_{26}O_3$ (MW 302.45)
(1*S*,2*R*,5*R*)
[100101-44-8]

(chiral ester for asymmetric induction)

Alternate Name: (1*R*,2*S*,5*R*)-5-methyl-2-(1-methyl-1-phenylethyl)cyclohexyl pyruvate.
Solubility: sol organic solvents.
Form Supplied in: not commercially available.
Preparative Methods: isolated as an oil from the acid-catalyzed esterification of *(−)-8-Phenylmenthol* with pyruvic acid.[1]
Handling, Storage, and Precautions: susceptible to hydrate formation on exposure to moisture. No other special precautions other than those for combustible organic compounds are necessary.

Chiral α-Hydroxy Acids. Grignard reagents add to the pyruvate ester of (−)-8-phenylmenthol from the *si*-face to afford the *S*-α-hydroxy esters (eq 1).[2] Chelation of the pyruvate with magnesium ensures the *s-cis* conformation and is required for high diastereoselectivity.[1] The high degree of diastereoselectivity is attributed to a favorable FMO interaction of the pyruvate with the phenyl moiety.[3] The corresponding *R*-α-hydroxy ester is available by reversing the order of bond formation starting with 8-phenylmenthyl glyoxylate.[4]

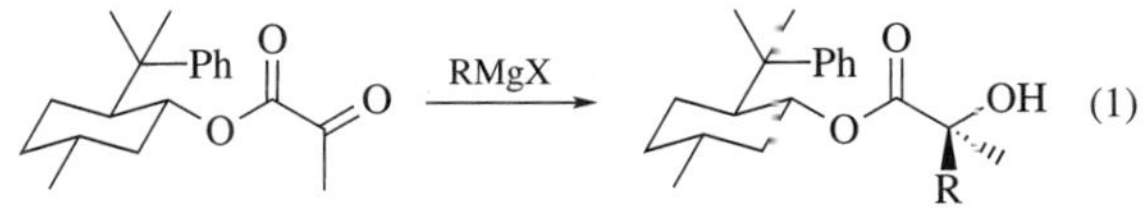

Reduction of 8-phenylmenthyl pyruvate with *Potassium Triisopropoxyborohydride* in THF gives 90% de of the (*R*)-lactate ester (eq 2), but other reducing reagents show little selectivity.[2] The stereochemical outcome of this reduction is explained as occurring in the *s-trans* conformation of the pyruvate.

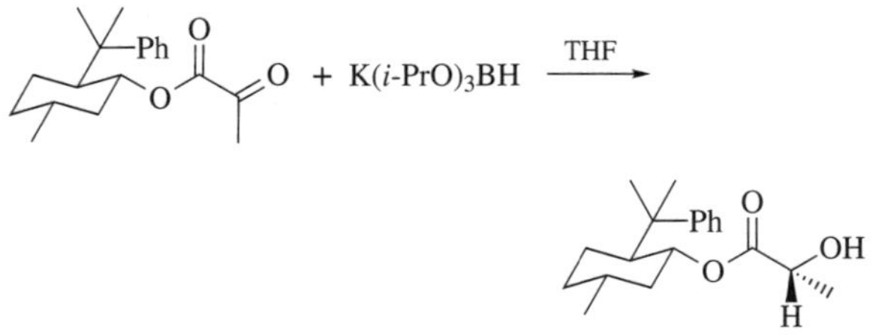

(2)

90% de

Tin(IV) Chloride-catalyzed ene reaction of the pyruvate ester of (−)-8-phenylmenthol with 1-hexene gives a single diastereomer of the unsaturated (*S*)-α-hydroxy ester (de > 90%) (eq 3).[2] Similarly, tin(IV) chloride-catalyzed ene reactions of 8-phenylmenthyl glyoxylates afford unsaturated secondary alcohols with 93–98% de of the (*S*) configuration.[5]

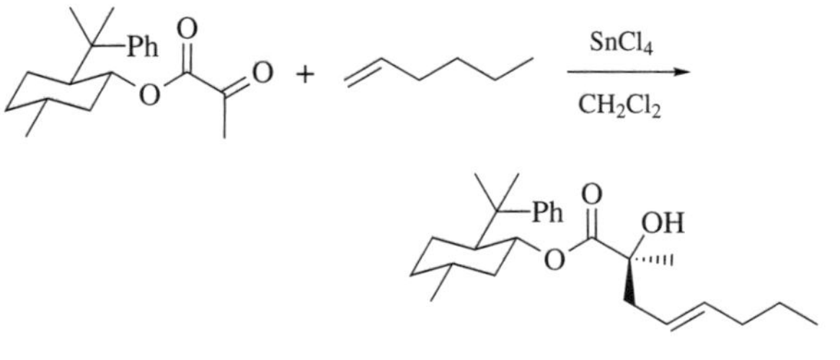

(3)

90% de

1. (a) Whitesell, J. K.; Bhattacharya, A.; Henke, K. *Chem. Commun./J. Chem. Soc., Chem. Commun.* **1982**, 988. (b) Chen, M.-Y.; Fang, J.-M. *J. Org. Chem.* **1992**, *57*, 2937.
2. Whitesell, J. K.; Deyo, D.; Bhattacharya, A. *Chem. Commun./J. Chem. Soc., Chem. Commun.* **1983**, 802.
3. (a) Runsink, J.; Koch, H.; Nehrings, A.; Scharf, H.-D.; Nowack, E.; Hahn, T. *J. Chem. Soc., Perkin Trans. 2* **1988**, 49. (b) Whitesell, J. K.; Younathan, J. N.; Hurst, J. R.; Fox, M. A. *J. Org. Chem.* **1985**, *50*, 5499.
4. Whitesell, J. K.; Buchanan, C. M. *J. Org. Chem.* **1986**, *51*, 5443.
5. Whitesell, J. K.; Bhattacharya, A.; Aguilar, D. A.; Henke, K. *Chem. Commun./J. Chem. Soc., Chem. Commun.* **1982**, 989.

Harry E. Ensley & Shivkumar Mahadevan
Tulane University, New Orleans, LA, USA

(*S*)-(+)-1-Phenyl-2-propylamine[1,2]

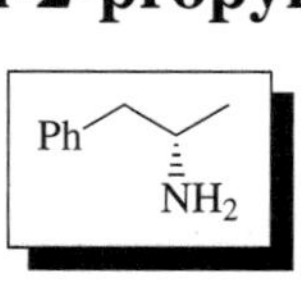

[51-64-9] $C_9H_{13}N$ (MW 135.23)

(chiral reagent for the resolution of racemic acids and aldehydes; chiral directing group for the enantioselective conjugate addition of ketones)

Alternate Name: (+)-amphetamine; *d*-amphetamine; dexamphetamine; dextroamphetamine.

Physical Data: bp 200–203 °C; *d* 0.913 g cm^{-3}; $[\alpha]_D^{19}$ +38° (benzene); mp (HCl salt) 154–155 °C.[3]

Solubility: readily sol all common organic solvents and acids; slightly sol water.

Form Supplied in: the free base (**1**) is a colorless liquid. A number of salts of (**1**) and its enantiomer (including the hydrochloride and sulfate) are commercially available.

Preparative Methods: enantiomerically pure (**1**) can be obtained by resolution of racemic (**1**) with (+)-tartaric acid.[4] Several highly stereospecific syntheses have been described.[5]

Analysis of Reagent Purity: enantiomeric and chemical purity of the reagent can be assayed by GC or HPLC analysis of its *(S)-(−)-α-Methoxy-α-(trifluoromethyl)phenylacetic Acid* amide.[3,6]

Handling, Storage, and Precautions: the reagent is a central stimulant and should be handled with gloves in a well-ventilated hood. Both the free base and its salts are stable at room temperature for extended periods of time.

Reagent for the Resolution of Carboxylic Acids. Reagent (**1**) and its enantiomer have been used, although not as extensively as the more common *(S)-α-Methylbenzylamine*, as resolving agents for carboxylic acids via fractional crystallization of the corresponding diastereomeric salts.[7] Examples of acids resolved this way include (**2**)–(**6**).[8–10] Additional examples, such as mandelic, hydratopic, and α-aryloxypropionic acids, can be found in the literature.[11,12]

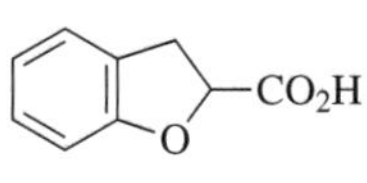

(**2**)

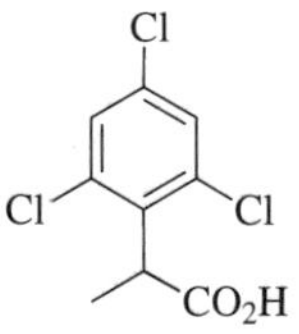

(**3**)

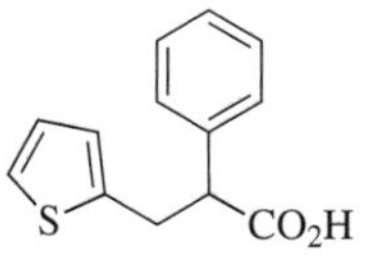

(**4**)

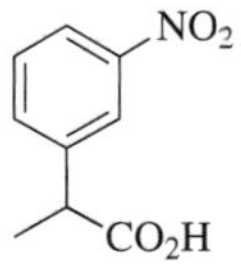

(**5**)

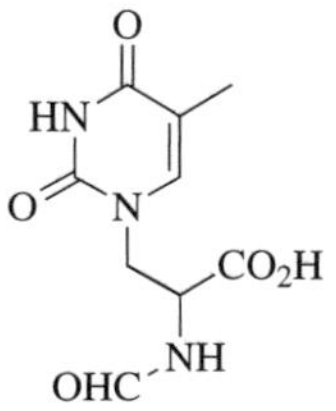

(**6**)

Resolution of Asymmetric Aldehydes. The resolution of the aldehyde-containing natural product (±)-gossypol has been accomplished by chromatographic separation of (**7**), the diastereomeric condensation product between (**1**) and gossypol hexaacetate.[13] Other chiral primary amines commonly used for the resolution of aldehydes and ketones by physical separation of diastereomeric imines include 2-amino-1-butanol, α-methylbenzylamine, and Betti's base (**8**).[14,15]

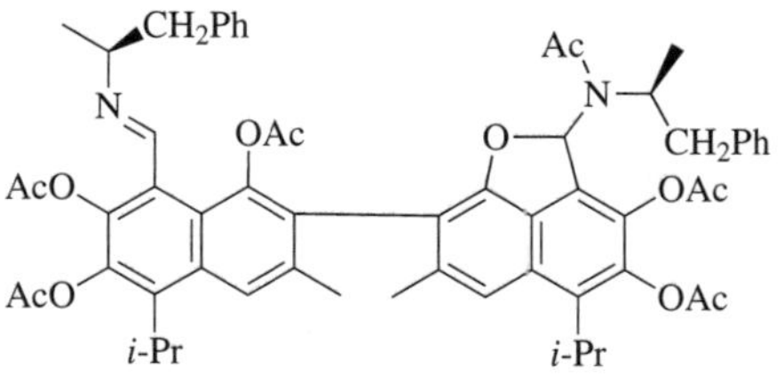

(**7**)

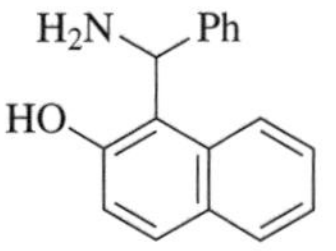

(**8**)

Stereoselective Synthesis of Biaryl Compounds. The best known application of (**1**) to the asymmetric oxidative coupling of phenolic compounds is the copper(II)-catalyzed synthesis of 1,1′-binaphthyl-2,2′-diol in greater than 95% ee (eq 1).[16,17]

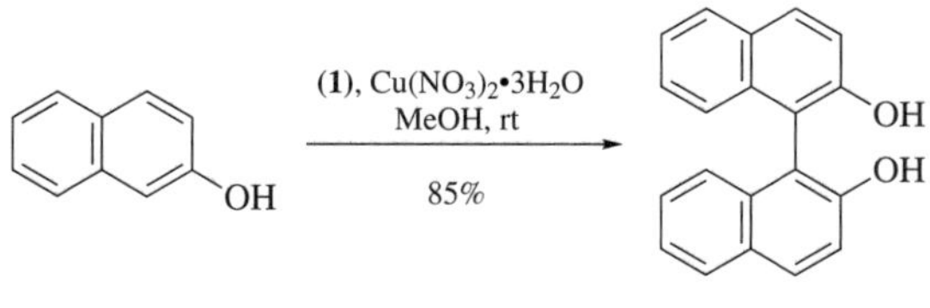

(1)

>95% ee

Enantioselective Conjugate Additions. The use of chiral imines for the enantioselective conjugate addition of carbonyl compounds to α,β-unsaturated systems is well established, mostly with imines derived from α-methylbenzylamine.[18–20] Recently, (**1**) has been used to effect the Michael addition of a 4-piperidone to acrylonitrile and methyl acrylate (eq 2).[21]

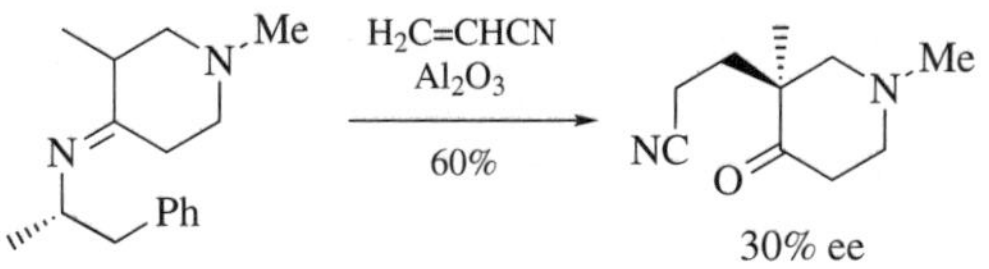

(2)

30% ee

1. Leithe, W. *Ber. Dtsch. Chem. Ges./Chem. Ber.* **1932**, *65*, 660.
2. (a) Karrer, P.; Ehrhardt, K. *Helv. Chim. Acta* **1951**, *34*, 2202. (b) Smith, H. E.; Cook, S. L.; Warren, M. E., Jr. *J. Org. Chem.* **1964**, *29*, 2265.
3. Buckley, T. F., III; Rapoport, H. *J. Am. Chem. Soc.* **1981**, *103*, 6157.
4. (a) Ernst, R. E.; O'Connor, M. J.; Holm, R. H. *J. Am. Chem. Soc.* **1968**, *90*, 5735. (b) Červinka, O.; Kroupová, E.; Bělovský, O. *Collect. Czech. Chem. Commun.* **1968**, *33*, 3551.
5. (a) Denmark, S. E.; Weber, T.; Piotrowski, D. W. *J. Am. Chem. Soc.* **1987**, *109*, 2224. (b) Rangaishenvi, M. V.; Singaram, B.; Brown, H. C. *J. Org. Chem.* **1991**, *56*, 3286.
6. Dale, J. A.; Dull, D. L.; Mosher, H. S. *J. Org. Chem.* **1969**, *34*, 2543.
7. Jacques, J.; Collet, A.; Wilen, S. H. *Enantiomers, Racemates and Resolutions*; Wiley: New York, 1981.
8. Wilen, S. H. *Tables of Resolving Agents and Optical Resolutions*; Univ. of Notre Dame Press: Notre Dame, IN, 1972.
9. Fredga, A. *Acta Chem. Scand.* **1969**, *23*, 2216.
10. Buttrey, J. D.; Jones, A. S.; Walker, R. T. *Tetrahedron* **1975**, *31*, 73.
11. Beckett, A. H.; Choulis, N. H. *J. Pharm. Sci.* **1966**, *55*, 1155.
12. Leclercq, M.; Jacques, J. *Bull. Soc. Chem. Fr. Part 2* **1975**, 2052.
13. Kai, Z. D.; Kang, S. Y.; Ke, M. J.; Jin, Z.; Liang, H. *Chem. Commun./J. Chem. Soc., Chem. Commun.* **1985**, 168.
14. Wilen, S. H. *Top. Stereochem.* **1971**, *6*, 107.
15. Betti, M. *Org. Synth., Coll. Vol.* **1941**, *1*, 381.
16. Brussee, J.; Jansen, A. C. A. *Tetrahedron Lett.* **1983**, *24*, 3261.
17. Trost, B. M.; Murphy, D. J. *Organometallics* **1985**, *4*, 1143.
18. (a) Ambroise, L.; Chassagnard, C.; Revial, G.; d'Angelo, J. *Tetrahedron: Asymmetry* **1991**, *2*, 407. (b) d'Angelo, J.; Revial, G.; Volpe, T.; Pfau, M. *Tetrahedron Lett.* **1988**, *29*, 4427.
19. (a) Pfau, M.; Revial, G.; Guingant, A.; d'Angelo, J. *J. Am. Chem. Soc.* **1985**, *107*, 273. (b) Revial, G. *Tetrahedron Lett.* **1989**, *30*, 4121.
20. (a) Desmaële, D. *Tetrahedron* **1992**, *48*, 2925. (b) Desmaële, D.; d'Angelo, J. *Tetrahedron Lett.* **1989**, *30*, 345.
21. Gaidarova, E. L.; Grishina, G. V. *Synlett* **1992**, 89.

Juan C. Jaen
Parke-Davis Pharmaceutical Research, Ann Arbor, MI, USA

(*R*)-(+)-Phenyl (*p*-Toluenesulfinyl)acetate

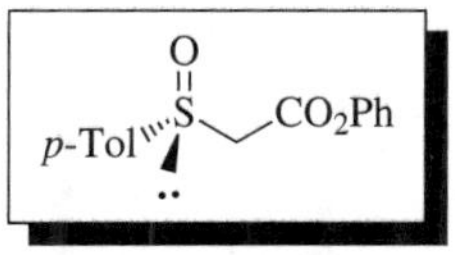

[75340-59-9] $C_{15}H_{14}O_3S$ (MW 274.36)

(asymmetric aldol-type condensations[1])

Physical Data: $[\alpha]_D$ +87° ($CHCl_3$, *c* 0.95).

Preparative Methods: conveniently prepared by deprotonation of (*R*)-(+)-*p*-tolyl methyl sulfoxide and treatment of the resulting sulfinyl carbanion with phenyl chloroformate (eq 1).[1]

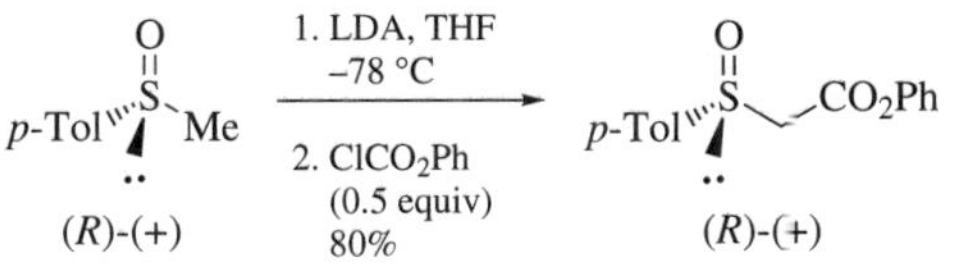

(1)

Asymmetric oxidation of phenyl (*p*-toluenesulfinyl)acetate with bovine serum albumin (BSA) gave a poor ee (eq 2).[2]

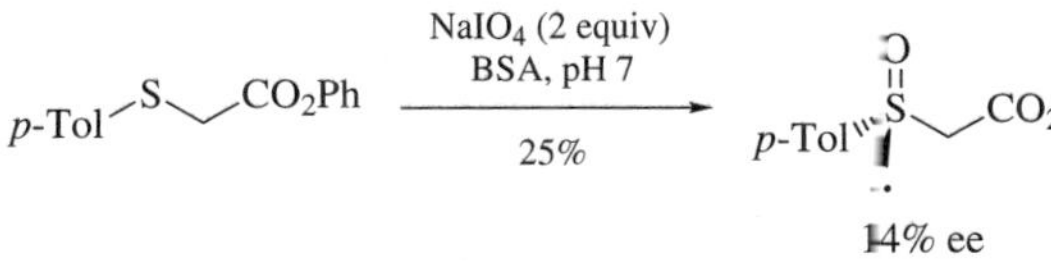

(2)

14% ee

General Discussion. Aldol-type condensation of the magnesium enolate of (*R*)-(+)-phenyl (*p*-toluenesulfinyl)acetate, prepared with *t*-butylmagnesium bromide, with the aldehyde precursor of maytansine afforded, after desulfurization with *Aluminum Amalgam*, the desired 4,5-unsaturated 3-(*S*)-hydroxy ester in high yield and high diastereoselectivity (eq 3).[1]

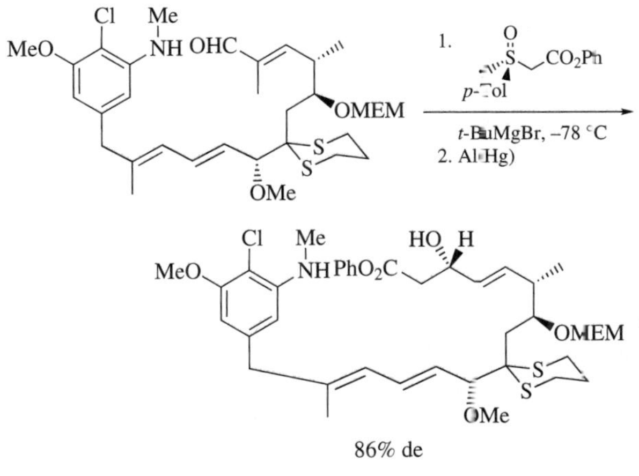

(3)

86% de

1. Corey, E. J.; Weigel, L. O.; Chamberlin, R.; Cho, H.; Hua, D. H. *J. Am. Chem. Soc.* **1980**, *102*, 6615.
2. Colonna, S.; Banfi, S.; Fontana, F.; Sommaruga, M. *J. Org. Chem.* **1985**, 769.

Guy Solladié & Françoise Colobert
University Louis Pasteur, Strasbourg, France

B-3-Pinanyl-9-borabicyclo[3.3.1]nonane

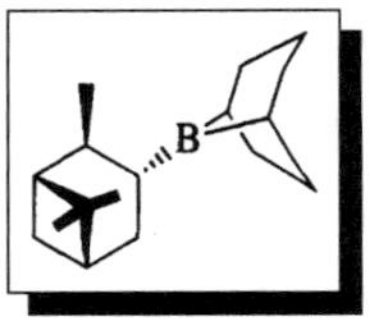

(*R*)
[173624-47-2] $C_{18}H_{31}B$ (MW 258.30)
(*S*)
[42371-63-1]

(asymmetric reducing agent which is particularly effective for aldehydes[1] and alkynic ketones[2])

Alternate Name: Alpine-Borane®.
Solubility: sol most organic solvents.
Form Supplied in: 0.5 M solution in THF. (*R*)-Alpine-Borane is prepared from (+)-α-pinene, and (*S*)-Alpine-Borane from (−)-α-pinene. High purity α-pinene is available commercially.
Preparative Methods: readily prepared by hydroboration of either (+)- or (−)-α-pinene[4] with *9-Borabicyclo[3.3.1]nonane* (9-BBN) (eq 1).

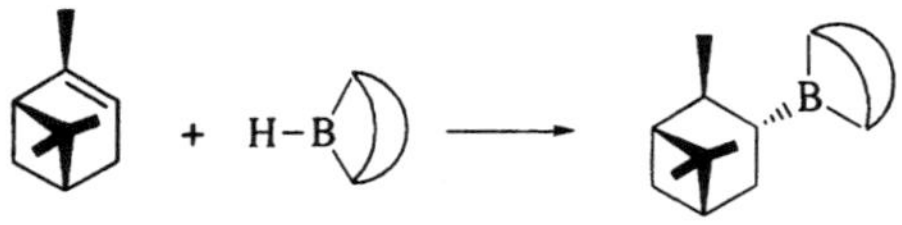

(1)

(+)-α-Pinene 9-BBN (*R*)-Alpine-Borane

Since the neat reagent is most effective, the solvent is usually removed before reduction of the ketone. Brown has reported a synthesis using neat α-pinene and solid 9-BBN.[5] The deuterium- or tritium-labeled compound may be prepared by hydroboration with labeled 9-BBN.[6] Alternatively *B*-methoxy-9-BBN may be reduced with $LiAlD_4$ (see *Lithium Aluminum Hydride*) in the presence of α-pinene.[7,8]
Handling, Storage, and Precautions: organoboranes are spontaneously flammable in air and must be handled under an inert atmosphere. They are generally stable to moisture. Alpine-Borane can slowly undergo dehydroboration.[3] Use in a fume hood.

Reduction of Aldehydes. Stereospecificially labeled primary alcohols are useful in biochemical and physical organic studies. Such compounds may be prepared by enzymatic reduction of a labeled aldehyde using yeast.[9] However, isolation of the product is often tedious. Alpine-Borane greatly simplifies the process and provides compounds of high enantiomeric purity. It is the most efficient reagent available for reduction of aldehydes. The limiting factor is often the enantiomeric purity of the starting α-pinene. Either enantiomer of the labeled primary alcohol may be obtained by using either (+)- or (−)-α-pinene or by placing the label either on the aldehyde or on the reducing agent (eq 2).

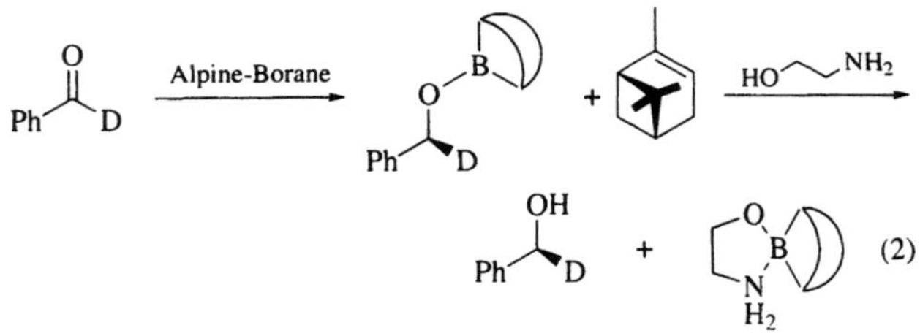

(2)

The reduction is bimolecular and thus the rate is dependent on concentration. Running the reaction neat provides the fastest rates. Usually an excess of Alpine-Borane is used to insure that the reaction does not become excessively slow at the end of the reduction. The excess organoborane may be destroyed by addition of an aldehyde such as *Acetaldehyde*. The resulting alkoxy-9-BBN may be treated with *Ethanolamine* to liberate the alcohol and precipitate the majority of the 9-BBN. Any remaining borane impurities may be removed by oxidation with basic *Hydrogen Peroxide*.

The absolute configuration of the product may be predicted by using a simple six-membered ring transition state model (structures **1** and **2**). In this model the predicted transition state resembles a boat cyclohexane with the small group occupying an axial-like position.

(**1**) favored (**2**) disfavored

Alkynic Ketones. The reagent is very sensitive to the steric requirements of the carbonyl group. Ketones are reduced at considerably slower rates than aldehydes.[10] Alkynic ketones are reduced at somewhat slower rates than aldehydes, but generally proceed at 25 °C. An alkynic ketone may be reduced in the presence of a methyl ketone (eq 3).

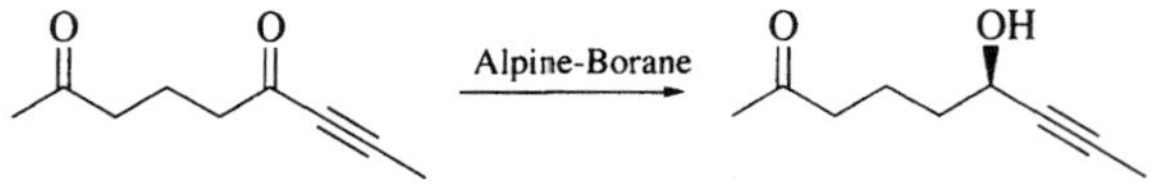

(3)

Aromatic, *n*-alkyl, and branched alkyl alkynic ketones are effectively reduced (eq 4).

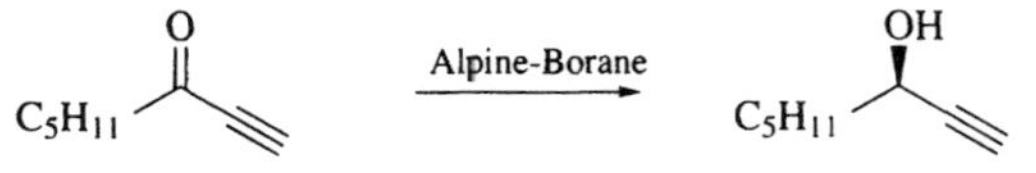

(4)

90–100% ee

Methyl alkynic ketones are reduced with slightly lower efficiency and *t*-butyl alkynic ketones are reduced very slowly. In the latter case, dehydroboration of Alpine-Borane to give 9-BBN competes with the rate of reduction and the liberated 9-BBN reduces the ketone to give products of lower enantiomeric purity. This problem may be overcome by using high pressure[11] or by using *B*-10-*cis*-myrtanyl-9-BBN (eqs 5 and 6).[12]

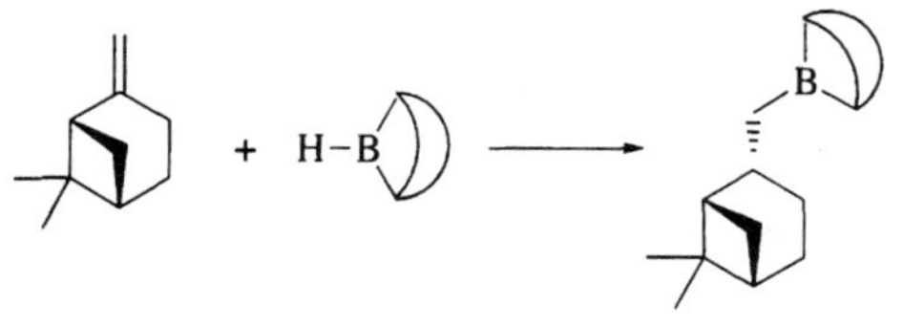

(5)

(–)-β-Pinene   9-BBN   Myrtanyl Borane

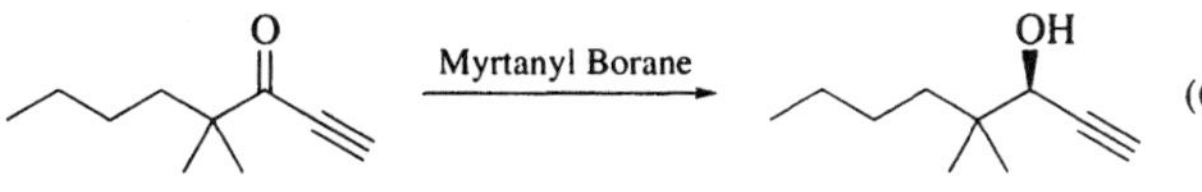

(6)

88% ee

α-Keto Esters. In general, electron-withdrawing groups enhance the rate of reduction of ketones with Alpine-Borane. Thus α-keto esters are generally good substrates for reduction. Methyl pyruvate is reduced within 4 h at rt with neat Alpine-Borane.[13] The use of *t*-butyl pyruvate increases the efficiency (eq 7).

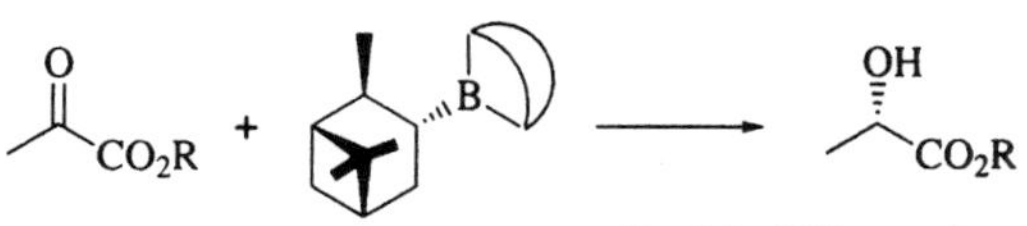

(7)

R = Me, 79% ee; *t*-Bu, 86% ee

Methyl, *n*-alkyl, and isobutyl behave as small groups in the transition state model for reduction, while isopropyl or aromatic groups behave as large groups (eq 8).

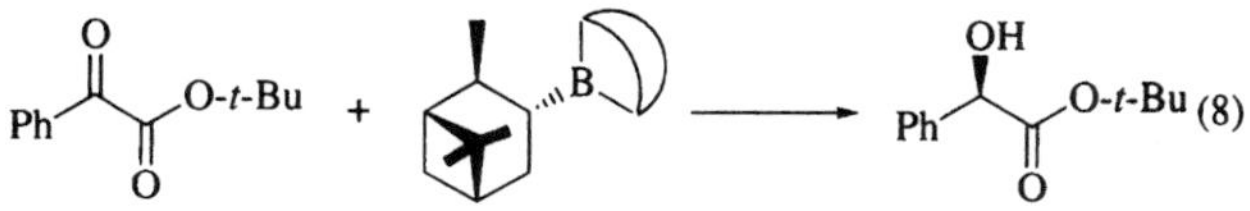

(8)

92% ee

Other Ketones. Ketones such as acetophenone are reduced rather slowly by THF solutions of Alpine-Borane (eq 9). A competing dehydroboration process leads to reduction via 9-BBN (eq 10).

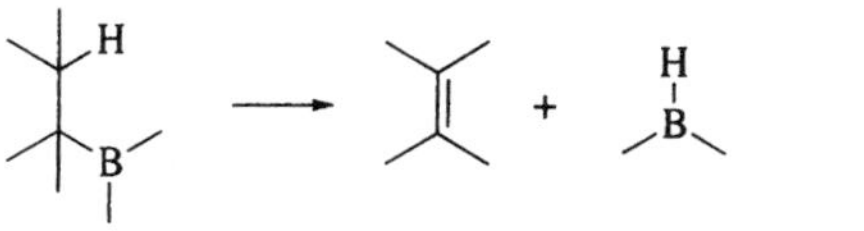

(9)

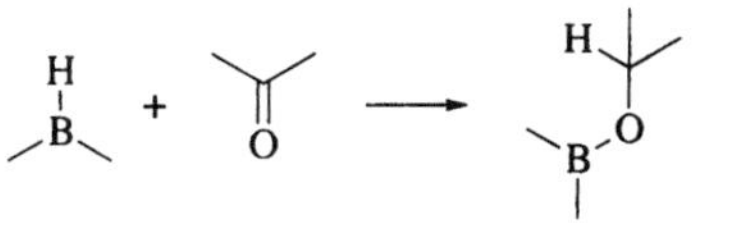

(10)

At 65 °C, Alpine-Borane undergoes 50% dehydroboration in 500 min.[3] At rt there is approximately 1–2% dehydroboration per day. Running the reaction neat increases the rate of the favorable bimolecular reduction.[5] Alternatively, high pressure may be used to increase the rate of the bimolecular process and retard the rate of the dehydroboration reaction (eq 11).[11]

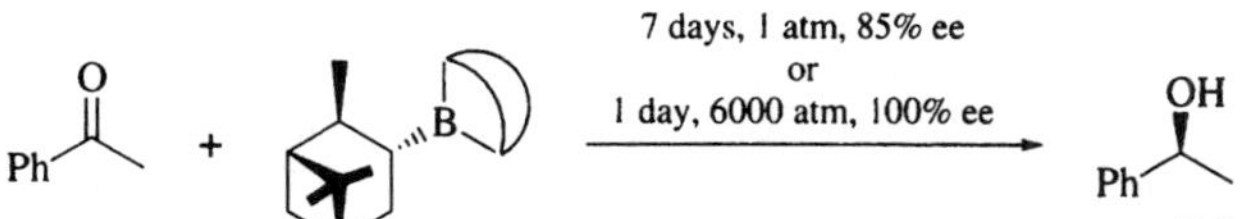

(11)

The simple steric model for the transition state may be used to predict the absolute configuration of the product. The related reagent *(+)-B-Chlorodiisopinocampheylborane* reduces ketones with greater ease and efficiency (eq 12).[14]

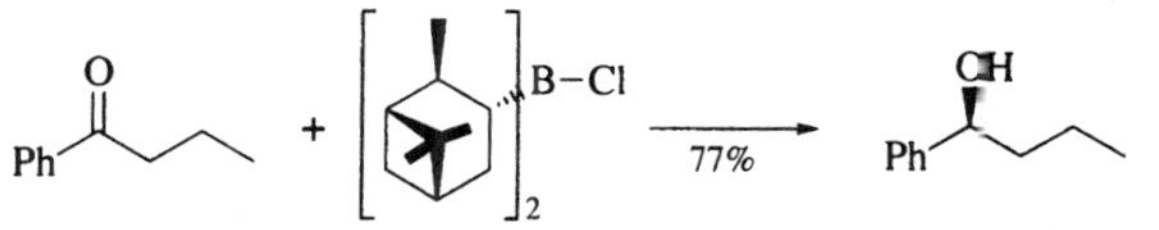

(12)

Chlorodiisopinocampheylborane   98.2% ee

Related Reagents. 2-[2-[(Benzyloxy)ethyl]-6,6-dimethylbicyclo[3.3.1]-3-nonyl]-9-borabicyclo[3.3.1]nonane.

---

1. Midland, M. M.; Greer, S.; Tramontano, A.; Zderic, S. A. *J. Am. Chem. Soc.* **1979**, *101*, 2352.
2. (a) Midland, M. M.; McDowell, D. C.; Hatch, R. L.; Tramontano, A. *J. Am. Chem. Soc.* **1980**, *102*, 867. (b) Midland, M. M.; Tramontano, A.; Kazubski, A.; Graham, R.; Tsai, D. J-S.; Cardin, D. B. *Tetrahedron* **1984**, *40*, 1371. (c) Midland, M. M; Graham, R. S. *Org. Synth.* **1984**, *63*, 57. (d) Midland, M. M.; Graham, R. S. *Org. Synth., Coll. Vol.* **1990**, *7*, 402. (e) Midland, M. M. *Chem. Rev.* **1989**, *89*, 1553.
3. Midland, M. M.; Petre, J. E.; Zderic, S. A.; Kazubski, A. *J. Am. Chem. Soc.* **1982**, *104*, 528.
4. α-Pinene of high optical purity may be obtained from Aldrich or by enrichment: Brown, H. C.; Yoon, N. M. *Isr. J. Chem.* **1976/1977**, *15*, 12.
5. Brown, H. C.; Pai, G. G. *J. Org. Chem.* **1985**, *50*, 1334.
6. (a) Midland, M. M.; Tramontano, A.; Zderic, S. A. *J. Organomet. Chem.* **1978**, *156*, 203. (b) Midland, M. M.; Greer, S. *Synthesis* **1978**, 845.
7. Althouse, V. E.; Feigl, D. M.; Sanderson, W. A.; Mosher, H. S. *J. Am. Chem. Soc.* **1966**, *88*, 3595.
8. Midland, M. M.; Asirwatham, G; Cheng, J. C.; Miller, J. A.; Morell, L. A. *J. Org. Chem.* **1994**, *59*, 4438.
9. Singaram, B.; Cole, T. E.; Brown, H. C. *J. Am. Chem. Soc.* **1985**, *107*, 460.
10. Midland, M. M.; Tramontano, A. *J. Org. Chem.* **1978**, *43*, 1470.
11. Midland, M. M.; McLoughlin, J. I.; Gabriel, J. *J. Org. Chem.* **1989**, *54*, 159.
12. Midland, M. M.; McLoughlin, J. I. *J. Org. Chem.* **1984**, *49*, 4101.
13. Brown, H. C.; Pai, G. G.; Jadhav, P. K. *J. Am. Chem. Soc.* **1984**, *106*, 1531.
14. Brown, H. C., Chandrasekharan, J.; Ramachandran, P. V. *J. Am. Chem. Soc.* **1988**, *110*, 1539.

M. Mark Midland
University of California, Riverside, CA, USA

(*S*)-Proline[1]

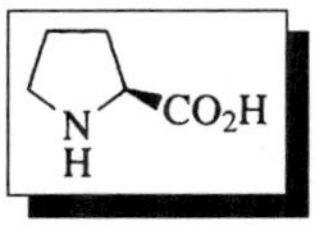

[147-85-3] $C_5H_9NO_2$ (MW 115.15)

(chiral auxiliary[1] in asymmetric synthesis)

Avoid Skin Contact with All Reagents

Physical Data: mp 228–233 °C (dec.); $[\alpha]_D^{20} = -84°$ ($c = 4$, H_2O); ninhydrin yellow in color.
Solubility: sol H_2O, alcohol; insol ether.
Form Supplied in: white solid; widely available; inexpensive.
Analysis of Reagent Purity: measurement of optical rotation; mp.
Handling, Storage, and Precautions: cold and dry storage.

General Considerations. In addition to its use in peptide chemistry, (*S*)-proline is often applied as a chiral precursor in the total syntheses of natural products, e.g. odorin,[2] pumiliotoxin,[3] petasinecine,[4] or threonine.[5] Some highly effective pharmaceuticals, such as optically pure ACE inhibitors, are prepared from L-proline.[6] In the last two decades, (*S*)-proline has attracted much attention as an optically active auxiliary in asymmetric synthesis.

Asymmetric Aldolization. Proline mediates the asymmetric aldol cyclization of the prochiral triketone (**1a**) to the optically active bicyclic enedione (*S*)-(**3a**) (eq 1).[7,8] Optical yields up to 94% are realized, depending on the solvent used. The enedione (**3**) is prepared directly in the presence of an acid such as $HClO_4$. In the absence of acid, the aldol (**2**) is frequently isolated. Among the amino acids tested, (*S*)-proline gives the best results in almost every case. With (*S*)-proline the (*S*) configured products (**3**) are usually obtained. Enediones such as (**3**) are important building blocks for the synthesis of steroids or alkaloids[9] because natural steroids have the same configuration at C-13. Analogously, the Wieland–Miescher ketone (*S*)-(**3e**) is prepared with 70% ee from the cyclic prochiral triketone (**1e**) in the presence of (*S*)-proline.[10] Optically pure enedione (*S*)-(**3e**) is obtained by a single crystallization of the product mixture having an enantiomeric excess over 50%.

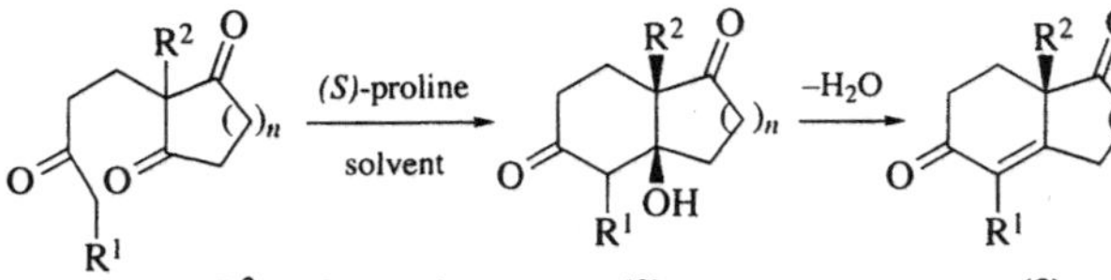

(1)

(1)   (2)   (3)

(**1a**) R = H, R^2 = Me, n = 1
(**1b**) R = H, R^2 = Et, n = 1
(**1c**) R = $(CH_2)_2CH{=}CH_2$, R^2 = Me, n = 1
(**1d**) R = $(CH_2)_2$-2-(6-Me-pyridyl), R^2 = Me, n = 1
(**1e**) R = H, R^2 = Me, n = 2
(**1f**) R = Me, R^2 = Me, n = 2

An asymmetric aldolization was successfully applied to the preparation of gibbane.[11] The total synthesis of the macrolide antibiotic erythromycin was developed involving an asymmetric aldolization step catalyzed by proline.[12] Since the mid-1970s, a flood of papers has appeared dealing with the asymmetric aldolization of various triketones. Some results are listed for comparison in Table 1.

The intramolecular asymmetric cyclization of open chain symmetrical triketone (**4**) leading to (*R*)-(**5**) proceeds with 16% ee (eq 2).[18]

Even acyclic 1,5-diketones (**6**) are cyclized enantioselectively in the presence of (*S*)-proline.[19] Depending on the structure of the cyclic α,β-unsaturated ketone, (R)-(**7**) is obtained in up to 43% (R = Me) optical yield (eq 3).

Table 1 Asymmetric Aldolization of Prochiral Triketones (**1**)

Educt	(*S*)-Proline (mol %)	Solvent	Product	Chemical yield (%)	Optical yield (%)
(**1a**)	4	DMF	(**2a**)[13]	97	94
(**1a**)	47/$HClO_4$	MeCN	(**3a**)[13]	90	82
(**1a**)	100	MeCN	(**2a**)[8]	97	97
(**1a**)	50/$HClO_4$	MeCN	(**3a**)[7]	87	84
(**1a**)	3	DMF	(**2a**)[8]	100	93
(**1b**)	30	DMF	(**2b**)[8]	71	100
(**1b**)	50/HCl	DMF	(**3b**)[7]	76	80
(**1c**)	100/$HClO_4$	MeCN	(**3c**)[14]	22	20
(**1d**)	120/$HClO_4$	MeCN	(**3d**)[15]	67	26
(**1e**)	5	DMSO	(**3e**)[10]	71	70
(**1e**)	50/$HClO_4$	MeCN	(**3e**)[7]	83	71
(**1f**)	13.7	DMSO	(**3f**)[16]	43	38
(**1f**)	50/$HClO_4$	MeCN	(**3f**)[17]	74	15

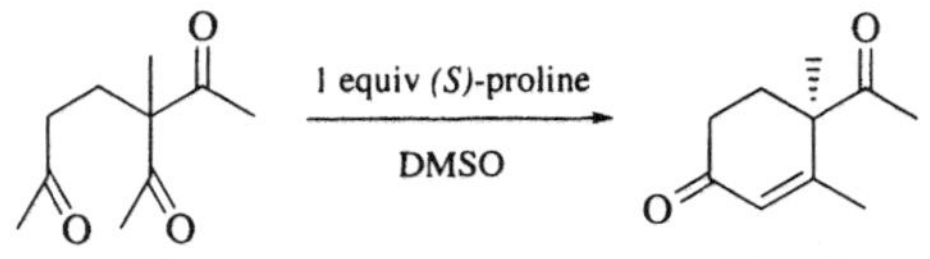

(2)

(**4**)   (*R*)-(**5**)

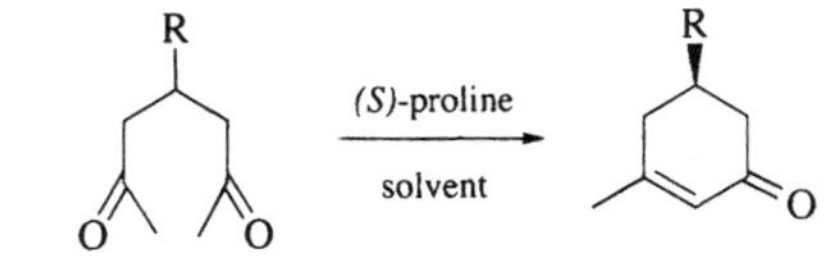

(3)

(**6**)   (*R*)-(**7**)

The mechanism of the proline-catalyzed enantioselective aldol reaction has been studied.[20] An extension of the asymmetric aldolization deals with the cyclization of diketones.[21] Also investigated was the dehydration of racemic β-ketols in the presence of (*S*)-proline and a kinetic resolution was observed.[22]

Asymmetric Michael Addition. An intramolecular Michael reaction catalyzed by (*S*)-proline leads to the chiral thiadecalin (**9**) and thiahydrindan (**11**) and (**12**).[23] Enone (**8**) undergoes cyclization in the presence of (*S*)-proline to give exclusively the *trans* isomer (**9**) (eq 4). The thiahydrindandions (**11**) and (**12**) are obtained from (**10**) as a 1:1 mixture of the *cis* and *trans* isomers (eq 5).

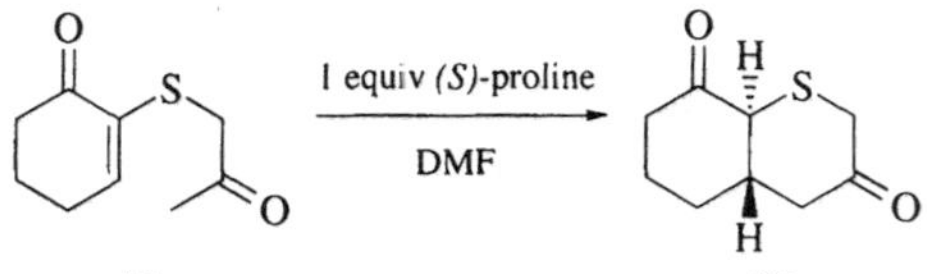

(4)

(**8**)   (**9**)

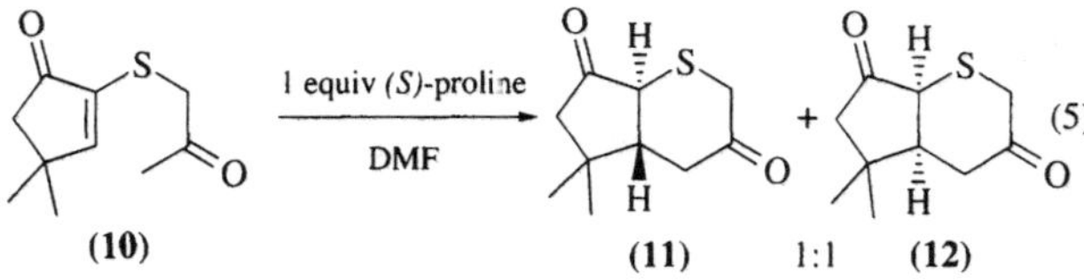

(5)

(**10**)   (**11**)   1:1   (**12**)

The intramolecular asymmetric Michael reaction of acyclic compounds obtained from chiral alkaloid building blocks using amines and (*S*)-proline has been investigated.[24] The Michael addition of dimethyl malonate to α,β-unsaturated aldehydes proceeds smoothly with a catalytic amount of (*S*)-proline lithium salt.[25] However, no asymmetric induction was observed.

Asymmetric Halolactonization. An asymmetric halolactonization reaction using proline as a chiral auxiliary has been reported.[26] Optically active α-hydroxy acids (**16**) are prepared from α,β-unsaturated acids via the corresponding (*S*)-proline amide (**13**) involving an asymmetric bromolactonization step (eq 6).[26a]

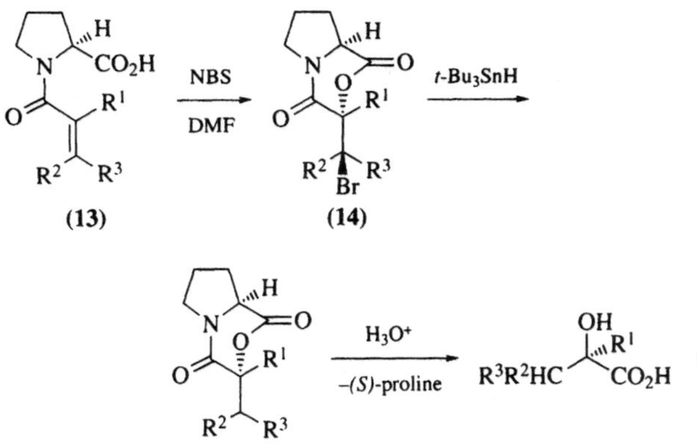

(13) (14) (15) (16) (6)

The unsaturated carboxylic acid (**13**) undergoes an asymmetric bromolactonization when treated with *N-Bromosuccinimide* in DMF. The bromolactone (**14**) and its diastereomer are obtained in a 94.5:5.5 ratio. Reduction and hydrolysis yields the α-hydroxy acid (**16**) in an overall optical yield of 90%. The same procedure gives chiral α-hydroxy ketones.[26c]

A modification of the asymmetric bromolactonization leads to optically active α,β-epoxy aldehydes (**18**).[26d,e] Treatment of the bromolactone (**14**) with *Sodium Methoxide* results in the formation of the epimeric epoxy ester (**17**) in a ratio of 2:1 (eq 7).

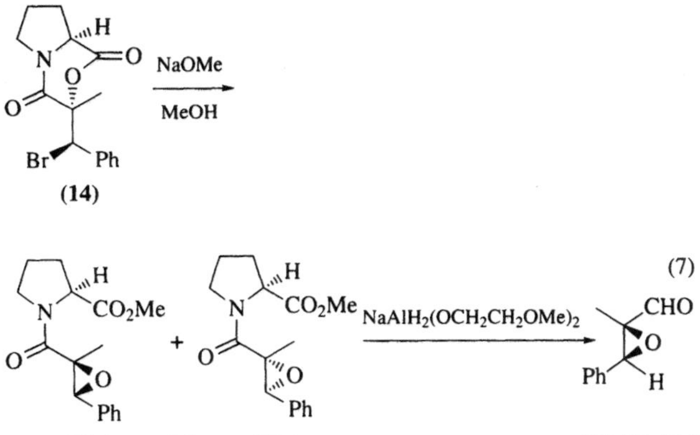

(14) (17a) 2:1 (17b) (2*R*,3*S*)-(18) (7)

The reductive cleavage of the proline derivative yields the chiral α,β-epoxy aldehyde (2*R*,3*S*)-(**18**) in 98% ee. Even natural product syntheses can be realized utilizing the bromolactonization procedure.[26h,i]

Reduction of C=O and C=N Bonds. Asymmetric reductions of prochiral ketones (**19**) to the corresponding chiral alcohols (**20**) using (*S*)-proline-modified borohydride reagents as the reductant have been published. The borane reductions of ketones (**19**) employing (*S*)-proline as chiral mediator proceeds with enantiomeric excesses up to >95%. It is proposed that the in situ produced (*S*)-prolinol reacts with borane to form the oxazaborolidine (*S*)-(**21**) as the reducing catalyst (eq 8).[27]

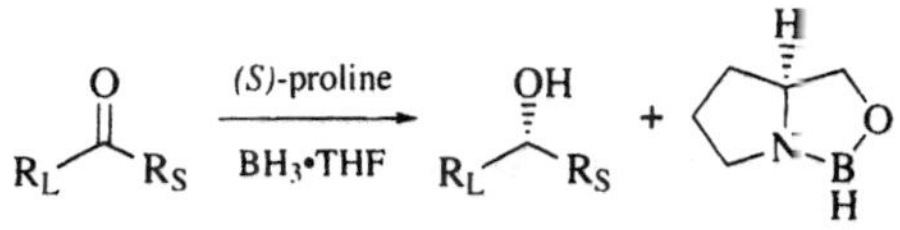

(19) (*R*)-(20) (*S*)-(21) (8)

The (*S*)-prolinate–borane complex (*S*)-(**22**) reduces ketones to the corresponding alcohols with optical yields up to 50%.[28] The asymmetric reduction of cyclic imines (**24**) with chiral sodium triacyloxyborohydride (*S*)-(**23**) was utilized to prepare optically active alkaloids (**25**) with optical yields up to 86% (eq 9).[29]

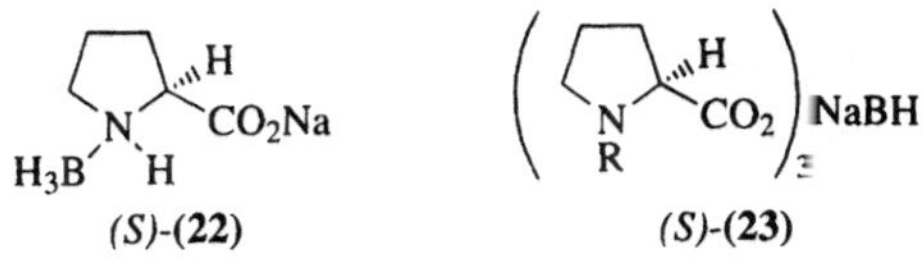

(*S*)-(22) (*S*)-(23)

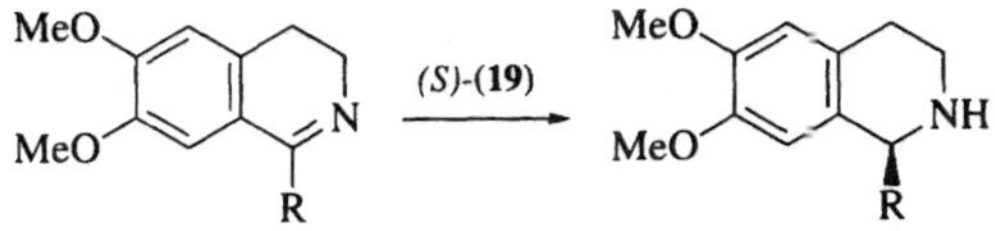

(24) (25) (9)

The hydrogenation of various ketones with heterogeneous *Palladium on Carbon* or *Raney Nickel* catalysts in the presence of (*S*)-proline proceeds to produce the corresponding optically active alcohols with low optical yields (up to 23%).[30]

Reduction of C=C Bonds. The reduction of the C=C double bond of isophorone (**26**) with Pd/C in the presence of (*S*)-proline yields the saturated ketone (**27**) with 60% optical purity (eq 10).[30a,31] With (*S*)-proline ester/Pd (or Pt) systems the hydrogenation of ethyl pyruvate, an α-keto ester, was investigated, but only insignificant enantioselectivities were reached.[32]

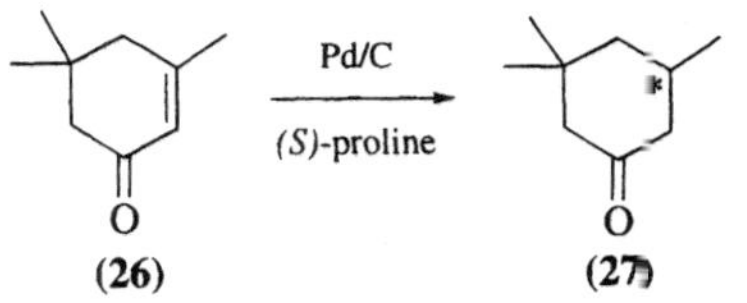

(26) (27) (10)

An efficient synthesis of (*S*)-amino acids from α-keto acids via a diastereoselective hydrogenation step with (*S*)-proline as the chiral inducer was reported (eq 11).[33] Optical yields up to 90% were reached.

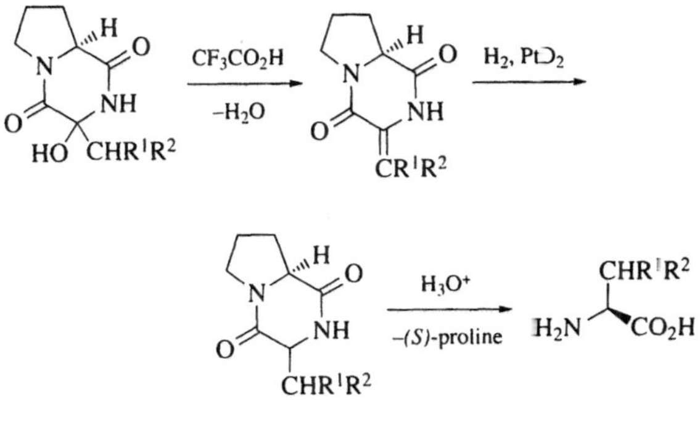

(11)

Racemization of Amino Acids. The synthesis of (*R*)-alanine was achieved starting from (*S*)-alanine via formation of the imidazoline with (*S*)-proline.[34] This result can be explained in terms of epimerization and stereoselective protonation with asymmetric induction by the chiral center originating from (*S*)-proline.

Resolution of Amino Acids. For the optical resolution of racemic threonine via replacing resolution, (*S*)-proline was utilized as an optically active cosolute although the structure of the imino acid is different from that of threonine.[35] The same procedure was applied less sucessfully to the resolution of (*R,S*)-thiazolidine carboxylic acid.[36]

Synthesis of Unnatural (*S*)-Proline Derivatives. The condensation of pivaladehyde with (*S*)-proline yields stereoselectively, after lithiation and reaction with an electrophile, the bicyclic compound (**28**), which is a versatile educt for the synthesis of many α-substituted proline analogs (**29**) (eq 12).[37] The reactions proceed via the formation of a chiral lithium enolate without the use of a chiral auxiliary (self-reproduction of chirality). The reaction with a variety of electrophiles *cis* to the *t*-Bu group yields a plethora of α-substituted (*S*)-proline derivatives (**29**). A limitation of this strategy is the acetal cleavage of some substituted products (**28**).[38]

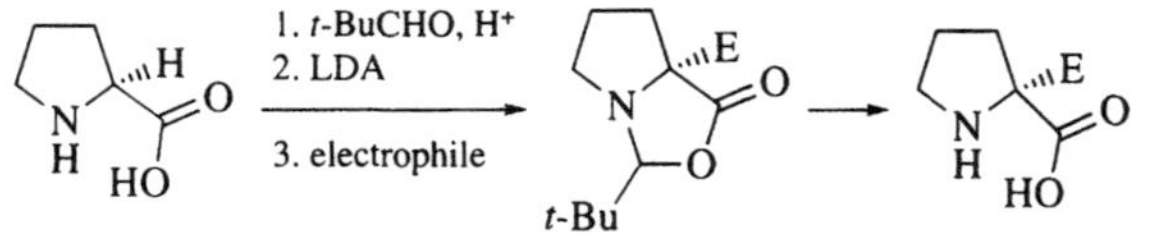

(**28**) (**29**) (12)

The β-amino acid homoproline can be synthesized via an Arndt–Eistert reaction from (*S*)-proline.[39]

Synthesis of Optically Active Phophorus Compounds. A series of chiral organophosphorus compounds (**33**) have been prepared in which the phophorus atom is the stereogenic center (eq 13).[40] The best stereoselectivity is reached with (*S*)-proline esters (**30**) as the chiral auxiliary. The reaction of phosphonic acid chloride (**31**) with (*S*)-proline ethyl ester affords a mixture of diastereomeric amides (**32**) in high stereoselectivity. The diastereomers can easily be purified by chromatography. The chiral organophosphorus compounds (**33**) are obtained from hydrolysis of (**32**).

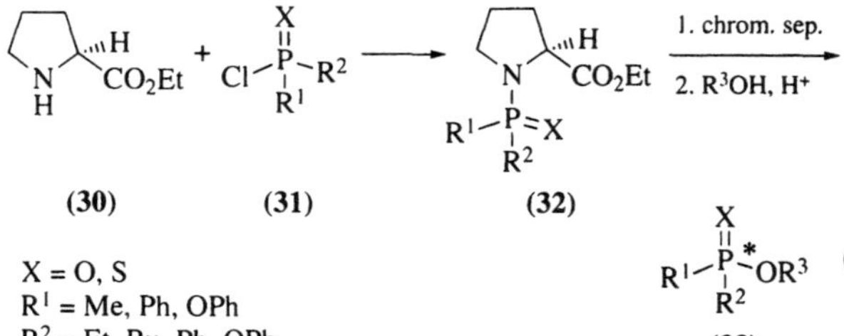

(**30**) (**31**) (**32**) (**33**) (13)

X = O, S
R^1 = Me, Ph, OPh
R^2 = Et, Bu, Ph, OPh
R^3 = Me, Et, Pr, *i*-Pr

Alkylations and Allylations. The asymmetric alkylation of chiral enamines derived from (*S*)-proline esters has been disclosed.[41] The α-alkylation of cyclohexanone proceeds with an optical purity of 59%. (*S*)-Proline catalyzes the alkylation of xanthopurpurin (**34**) by 2-hydroxytetrahydropyran[42] yielding (**35**), which was later used in the synthesis of the racemate of the pigment averufanin (eq 14).[43] With phenylacetaldehyde, two molecules react to build up an anomeric mixture of lactols with a new pyran ring.[44] In each case, no enantioselectivity is detected.

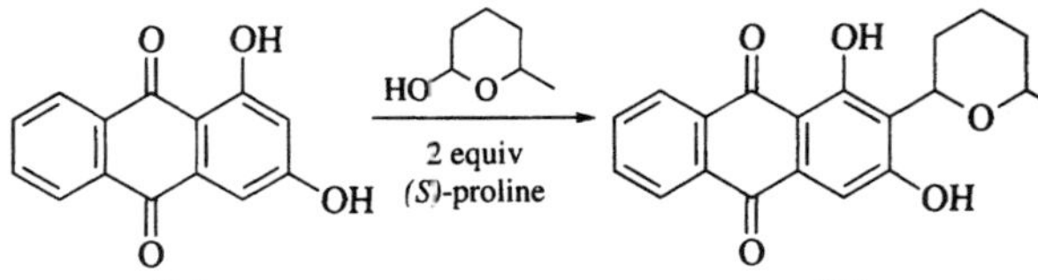

(**34**) (**35**) (14)

The stereoselective allylation of aldehydes was reported to proceed with allyltrifluorosilanes in the presence of (*S*)-proline.[45] The reaction involves pentacoordinate silicate intermediates. Optical yields up to 30% are achieved in the copper-catalyzed allylic acetoxylation of cyclohexene with (*S*)-proline as a chiral ligand.[46]

The intramolecular asymmetric palladium-catalyzed allylation of aldehydes, including allylating functionality in the molecules, via chiral enamines prepared from (*S*)-proline esters has been reported (eq 15).[47] The most promising result was reached with the (*S*)-proline allyl ester derivative (**36**). Upon treatment with *Tetrakis(triphenylphosphine)palladium(0)* and PPh_3 in THF, the chiral enamine (**36**) undergoes an intramolecular allylation to afford an α-allyl hemiacetal (**37**). After an oxidation step the optically active lactones (**38**) with up to 84% ee were isolated in high chemical yields. The same authors have also reported sucessful palladium-catalyzed asymmetric allylations of chiral allylic (*S*)-proline ester enamines[48] and amides[49] with enantiomeric excesses up to 100%.

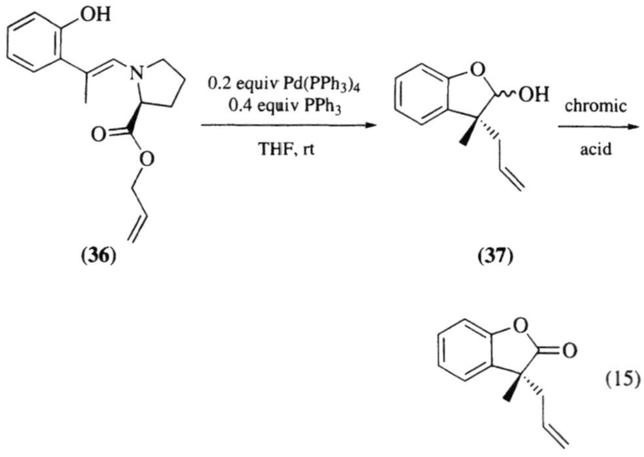

(**36**) (**37**) (**38**) (15)

---

1. (a) Coppola, G. M.; Schuster, H. F. *Asymmetric Synthesis: Construction of Chiral Molecules Using Amino Acids*; Wiley: New York, 1987; pp 267–345. (b) Drauz, K.; Kleemann, A.; Martens, J. *Angew. Chem., Int. Ed. Engl.* **1982**, *21*, 584. (c) Martens, J. *Top. Curr. Chem.* **1984**, *125*, 165.
2. Babidge, P. J.; Massy-Westropp, R. A.; Pyne, S. G.; Shiengthong, D.; Ungphakorn, A.; Veerachat, G. *Aust. J. Chem.* **1980**, *33*, 1841.
3. (a) Ito, A.; Takahashi, R.; Baba, Y. *Chem. Pharm. Bull.* **1975**, *23*, 3081. (b) Overman, L. E.; Bell, K. L. *J. Am. Chem. Soc.* **1981**, *103*, 1851. (c) Overman, L. E.; Bell, K. L.; Ito, F. *J. Am. Chem. Soc.* **1984**, *106*, 4192.
4. Mulzer, J.; Shanyoor, M. *Tetrahedron Lett.* **1993**, *34*, 6545.
5. (a) Berlokon, Y. N.; Zeltzer, I. E.; Ryzhov, M. G.; Saporovskaya, M. B.; Bakhmutov, V. I.; Belikov, V. M. *Chem. Commun./J. Chem. Soc., Chem.*

Commun. **1982**, 180. (b) Belokon, Y. N.; Bulychev, A. G.; Vitt, S. V.; Struchkov, Y. T.; Batsanov, A. S.; Timofeeva, T. V.; Tsyryapkin, V. A.; Ryzhov, M. G.; Lysova, L. A.; Bakhmutov, V. I.; Belikov, V. M. *J. Am. Chem. Soc.* **1985**, *107*, 4252.

6. (a) Kim, D. H. *J. Heterocycl. Chem.* **1980**, *17*, 1647. (b) Cushman, D. W.; Cheung, H. S.; Sabo, E. F.; Ondetti, M. A. *Biochemistry* **1977**, *16*, 5484. (c) Suh, J. T.; Skiles, J. W.; Williams, B. E.; Youssefyeh, R. D.; Jones, H.; Loev, B.; Neiss, E. S.; Schwab, A.; Mann, W. S.; Khandwala, A.; Wolf, P. S.; Weinryb, I. *J. Med. Chem.* **1985**, *28*, 57.
7. Eder, U.; Sauer, G.; Wiechert, R. *Angew. Chem., Int. Ed. Engl.* **1971**, *10*, 496.
8. Hajos, Z. G.; Parrish, D. R. *J. Org. Chem.* **1974**, *39*, 1615.
9. (a) Nagasawa, K.; Hiroi, K.; Yamada, S. *Yakugaku Zasshi* **1975**, *95*, 46. (b) Micheli, R. A.; Hajos, Z. G.; Cohen, N.; Parrish, D. R.; Portland, L. A.; Sciamanna, W.; Scott, M. A.; Wehrli, P. A. *J. Org. Chem.* **1975**, *40*, 675. (c) Cohen, N.; Banner, B. L.; Eichel, W. F.; Parrish, D. R.; Saucy, G.; Cassal, J.-M.; Meier, W.; Fürst, A. *J. Org. Chem.* **1975**, *40*, 681. (d) Eder, U.; Sauer, G.; Haffer, G.; Ruppert, J.; Wiechert, R.; Fürst, A.; Meier, W. *Helv. Chim. Acta* **1976**, *59*, 999. (e) Eder, U.; Gibian, H.; Haffer, G.; Neef, G.; Sauer, G.; Wiechert, R. *Ber. Dtsch. Chem. Ges./Chem. Ber.* **1976**, *109*, 2948.
10. Gutzwiller, J.; Buchschacher, P.; Fürst, A. *Synthesis* **1977**, 167.
11. Takano, S.; Kasahara, C.; Ogasawara *Chem. Commun./J. Chem. Soc., Chem. Commun.* **1981**, 635.
12. Woodward, R. B.; Logusch, E.; Nambiar, K. P.; Sakan, K.; Ward, D. E.; Au-Yeung, B.-W.; Balaram, P.; Browne, L. J.; Card, P. J.; Chen, C. H.; Chênevert, R. B.; Fliri, A.; Frobel, K.; Gais, H.-J.; Garratt, D. G.; Hayakawa, K.; Heggie, W.; Hesson, D. P.; Hoppe, D.; Hoppe, I.; Hyatt, J. A.; Ikeda, D.; Jacobi, P. A.; Kim, K. S.; Kokube, Y.; Kojima, K.; Krowicki, K.; Lee, V. J.; Leutert, T.; Malchenko, S.; Martens, J.; Matthews, R. S.; Ong, B. S.; Press, J. B.; Rajan Babu, T. V.; Rousseau, G.; Sauter, H. M.; Suzuki, M.; Tatsuta, K.; Tolbert, L. M.; Truesdale, E. A.; Uchida, I.; Ueda, Y.; Uyehara, T.; Vasella, A. T.; Vladuchick, W. C.; Wade, P. A.; Williams, R. M.; Wong, H. N.-C. *J. Am. Chem. Soc.* **1981**, *103*, 3210, 3213, 3215.
13. Buchschacher, P.; Cassal, J.-M.; Fürst, A.; Meier, W. *Helv. Chim. Acta* **1977**, *60*, 2747.
14. Shimizu, I.; Naito, Y.; Tsuji, J. *Tetrahedron Lett.* **1980**, *21*, 487.
15. Danishefsky, S.; Cain, P. *J. Am. Chem. Soc.* **1976**, *98*, 4975.
16. Coisne, J.-M.; Pecher, J.; Declercq, J.-P.; Germain, G.; Van Meerssche, M. *Bull. Soc. Chim. Belg.* **1981**, *90*, 481.
17. Uma, R.; Rajagopalan, K.; Swaminathan, S. *Tetrahedron* **1986**, *42*, 2757.
18. Terashima, S.; Sato, S.; Koga, K. *Tetrahedron Lett.* **1979**, 3469.
19. Agami, C.; Platzer, N.; Sevestre, H. *Bull. Soc. Chem. Fr.* **1987**, 358.
20. (a) Agami, C. *Bull. Soc. Chem. Fr.* **1988** 499. (b) Agami, C.; Puchot, C. *J. Mol. Catal.* **1986**, *38*, 341. (c) Puchot, C.; Samuel, O.; Duñach, E.; Zhao, S.; Agami, C.; Kagan, H. B. *J. Am. Chem. Soc.* **1986**, *108*, 2353. (d) Agami, C.; Levisalles, J.; Puchot, C. *Chem. Commun./J. Chem. Soc., Chem. Commun.* **1985**, 441. (e) Agami, C.; Sevestre, H. *Chem. Commun./J. Chem. Soc., Chem. Commun.* **1984**, 1385. (f) Agami, C.; Puchot, C.; Sevestre, H. *Tetrahedron Lett.* **1986**, *27*, 1501.
21. Agami, C.; Levisalles, J.; Sevestre, H. *Chem. Commun./J. Chem. Soc., Chem. Commun.* **1984**, 418.
22. Agami, C.; Puchot, C. *Tetrahedron* **1986**, *42*, 2037.
23. Kozikowski, A. P.; Mugrage, B. B. *J. Org. Chem.* **1989**, *54*, 2274.
24. Hirai, Y.; Terada, T.; Yamazaki, T.; Momose, T. *J. Chem. Soc., Perkin Trans. 1* **1992**, 509.
25. Yamaguchi, M.; Yokota, N.; Minami, T. *Chem. Commun./J. Chem. Soc., Chem. Commun.* **1991**, 1088.
26. (a) Terashima, S.; Jew, S. *Tetrahedron Lett.* **1977**, 1005. (b) Jew, S.-S.; Terashima, S.; Koga, K. *Tetrahedron* **1979**, *35*, 2337. (c) Jew, S.-S.; Terashima, S.; Koga, K. *Tetrahedron* **1979**, *35*, 2345. (d) Terashima, S.; Hayashi, M.; Koga, K. *Tetrahedron Lett.* **1980**, 2733. (e) Hayashi, M.; Terashima, S.; Koga, K. *Tetrahedron* **1981**, *37*, 2797. (f) Terashima, S.; Jew, S.; Koga, K. *Chem. Lett.* **1977**, 1109. (g) Terashima, S.; Jew, S.; Koga, K. *Tetrahedron Lett.* **1977**, 4507. (h) Jew, S.; Terashima, S.; Koga, K. *Chem. Pharm. Bull.* **1979**, *27*, 2351. (i) Terashima, S.; Jew, S.; Koga, K. *Tetrahedron Lett.* **1978**, 4937. (j) Rüeger, H.; Benn, M. *Heterocycles* **1982**, *19*, 23.
27. (a) Wallbaum, S.; Martens, J. *Tetrahedron: Asymmetry* **1992**, *3*, 1475. (b) Mehler, T.; Behnen, W.; Wilken, J.; Martens, J. *Tetrahedron: Asymmetry* **1994**, *5*, 185.
28. Umino, N.; Iwakuma, T.; Itoh, N. *Chem. Pharm. Bull.* **1979**, *27*, 1479.
29. (a) Yamada, K.; Takeda, M.; Iwakuma, T. *Tetrahedron Lett.* **1981**, *22*, 3869. (b) Yamada, K.; Takeda, M.; Ohtsuka, H.; Iwakuma, T. *Chem. Pharm. Bull.* **1983**, *31*, 70. (c) Yamada, K.; Takeda, M.; Iwakuma, T. *J. Chem. Soc., Perkin Trans. 1* **1983**, 265.
30. (a) Tungler, A.; Kajtar, M.; Mathé, T.; Toth, G.; Fogassy, E.; Petró, J. *Catal. Today* **1989**, *5*, 159. (b) Tungler, A.; Tarnai, T.; Máthé, T.; Petró, J. *J. Mol. Catal.* **1991**, *67*, 277.
31. Tungler, A.; Máthé, T.; Petró, J.; Tarnai, T. *J. Mol. Catal.* **1990**, *61*, 259.
32. Tungler, A.; Tarnai, T.; Máthé, T.; Petró, J. *J. Mol. Catal.* **1991**, *70*, L5.
33. (a) Bycroft, B. W.; Lee, G. R. *Chem. Commun./J. Chem. Soc., Chem. Commun.* **1975**, 988. (b) Poisel, H.; Schmidt, U. *Ber. Dtsch. Chem. Ges./Chem. Ber.* **1973**, *106*, 3408.
34. (a) Shibata, S.; Matsushita, H.; Noguchi, M.; Saburi M.; Yoshikawa, S. *Chem. Lett.* **1978**, 1305. (b) Shibata, S.; Matsushita, H.; Kato, K.; Noguchi, M.; Saburi, M.; Yoshikawa, S. *Bull. Chem. Soc. Jpn.* **1979**, *52*, 2938.
35. Shiraiwa, T.; Yamauchi, M.; Yamamoto, Y.; Kurokawa, H. *Bull. Chem. Soc. Jpn.* **1990**, *63*, 3296.
36. Shiraiwa, T.; Yamauchi, M.; Tatsumi, T.; Kurokawa, H *Bull. Chem Soc. Jpn.* **1992**, *65*, 267.
37. Seebach, D.; Naef, R. *Helv. Chim. Acta* **1981**, *64*, 2704.
38. Seebach, D.; Boes, M.; Naef, R.; Schweizer, W. B. *J Am. Chem. Soc.* **1983**, *105*, 5390.
39. Rüeger, H.; Benn, M. *Heterocycles* **1982**, *19*, 1677.
40. (a) Koizumi, T.; Kobayashi, Y.; Amitani, H.; Yosshii, E. *J. Org. Chem.* **1977**, *42*, 3459. (b) Koizumi, T.; Amitani, H.; Yoshii, E. *Tetrahedron Lett.* **1978**, 3741. (c) Koizumi, T.; Amitani, H.; Yoshii, E. *Synthesis* **1979**, 110. (d) Koizumi, T.; Takagi, H.; Yoshii, E. *Chem. Lett.* **1980**, 1403. (e) Koizumi, T.; Yanada, R.; Takagi, H.; Hirai, H.; Yoshii, E. *Tetrahedron Lett.* **1981**, 477 and 571.
41. (a) Yamada, S.; Hiroi, K.; Achiwa, K. *Tetrahedron Lett.* **1969**, 4233. (b) Hiroi, K.; Achiwa, K.; Yamada, S.-I. *Chem. Pharm. Bull.* **1972**, *20*, 246. (c) Hiroi, K.; Yamada, S.-I. *Chem. Pharm. Bull.* **1973**, *21*, 47. (d) Hiroi, K.; Yamada, S.-I. *Chem. Pharm. Bull.* **1973**, *21*, 54.
42. Castonguay, A.; Berger Y. *Chem. Commun./J. Chem. Soc., Chem. Commun.* **1978**, 951.
43. Castonguay, A.; Berger Y. *Tetrahedron* **1979**, *35*, 1557.
44. Castonguay, A.; Berger Y. *Aust. J. Chem.* **1979**, *32*, 2681.
45. Kira, M.; Sato, K.; Sakurai H. *J. Am. Chem. Soc.* **1990**, *112*, 257.
46. Muzart, J. *J. Mol. Catal.* **1991**, *64*, 381.
47. Hiroi, K. Abe J. *Heterocycles* **1990**, *30*, 283.
48. (a) Hiroi, K.; Suya, K.; Sato, S. *Chem. Commun./J. Chem. Soc., Chem. Commun.* **1986**, 469. (b) Hiroi, K.; Abe, J.; Suya, K.; Sato, S. *Tetrahedron Lett.* **1989**, *30*, 1543.
49. Hiroi, K.; Maezuru, K.; Kimura, M.; Ito, N. *Chem. Lett.* **1989**, 1751.

Sabine Wallbaum & Jürgen Martens
Universität Oldenburg, Germany

N-Propenoyl Camphor-10,2-sultam[1]

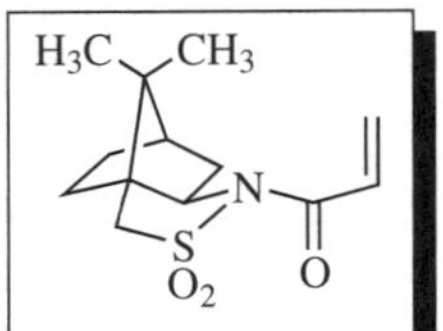

1

[94594-91-9] $C_{13}H_{19}NO_3S$ (MW 269.36)

(reagent used as a chiral acrylate derivative for various asymmetric organic reactions)

Physical Data: crystalline solid, mp 196–197 °C; $[\alpha]^{26}_D$ −100.9 (*c* 0.98, $CHCl_3$).
Solubility: soluble in most organic solvents.
Form Supplied in: available through synthesis.[2]
Handling, Storage, and Precautions: toxicity unknown; as with all organic chemicals, use of a well-ventilated fume hood is recommended.

Introduction. Sultam derivative **1** ($X_cC(O)CH{=}CH_2$) has been exploited as a chiral auxiliary for a variety of reactions. Both antipodes are available in optically pure form from the chiral pool. Stereoselection of non-catalyzed reactions is usually consistent with the model advanced by Curran and co-workers[3] in which bond formation occurs at the *Re* face.

Cycloadditions. Oppolzer first used this chiral acrylate derivative as an auxiliary in the Diels–Alder reaction with cyclopentadiene. Promotion by Lewis acids such as $TiCl_4$ $SnCl_4$, and Et_2AlCl provides the adduct in greater than 90% de (eq 1).[2] Lithium perchlorate-promoted [4+2] reaction between **1** and 1-acetoxybutadiene was similarly effective.[4] More recently, an exo-selective Diels–Alder addition of **1** with 2-acylamino dienes provided a single diastereomer in 80% yield.[5] Cyclopentane formation is possible through exposure of **1** to methylenecyclopropane and Ni(0) (eq 2).[6] An example of a higher-order cycloaddition with **1** gave only low diastereoselection (78:22)[7] for the endo product.

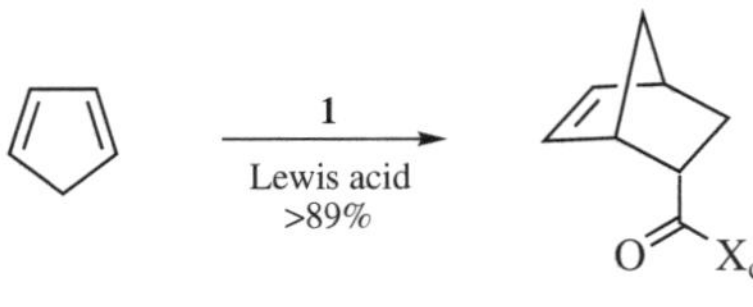

>90% de (1)

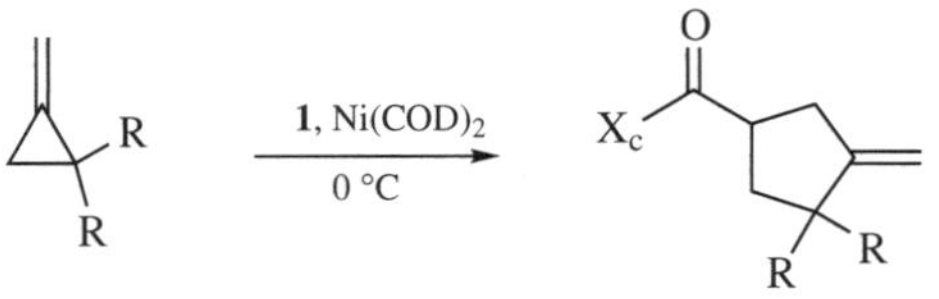

R = H 86% de (76%)
R = CH_3 98% de (69%) (2)

Both nitrile oxide (eq 3)[3] and silyl nitronate (eq 4)[8] cycloadditions are highly diastereoselective processes. The use of either approach enables access to Δ^2-isoxazolines in good yield. Unfortunately, the corresponding nitrone cycloadditions are only slightly selective (dr = 78:22)[9,10] The enantiomeric sultam was implemented effectively in azomethine ylide cycloadditions to gain access to bridged pyrrolidines with high levels of diastereoselection (eq 5).[11]

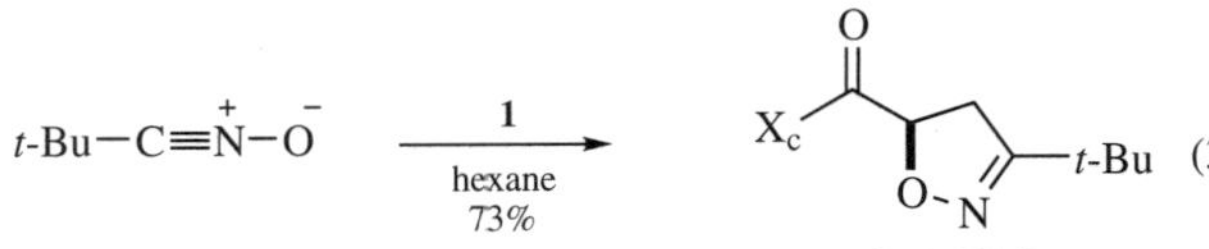

dr = 95:5 (3)

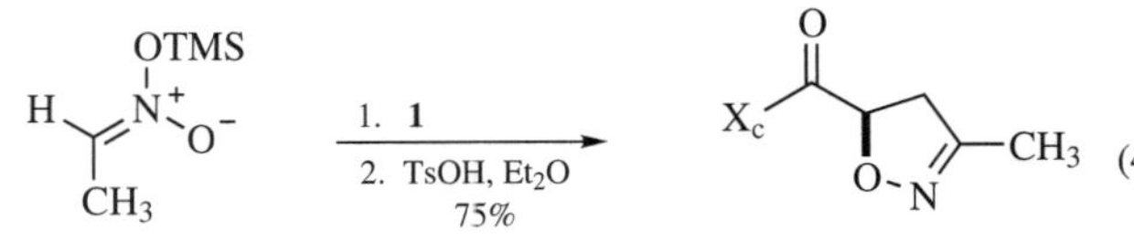

dr = 93:7 (4)

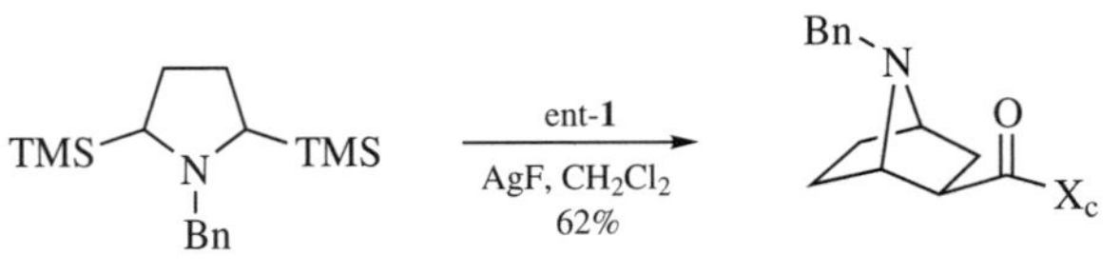

dr = 98:2 (5)

Carbon-Carbon Bond-Forming Reactions. Several asymmetric ring-forming reactions using **1** have been developed to give products with high stereohomogeneity. In the Morita-Baylis-Hillman reaction of **1** with acetaldehyde, the auxiliary is cleaved in situ to give the hydroxy acid acetal in high ee (eq 6).[12] Michael addition to **1** followed by alkylation with an α-imino ester again results in cleavage of the auxiliary and formation of β-lactams stereoselectively (eq 7).[13,14] Cyclohexenes are conveniently accessed via palladium-mediated annulation (eq 8).[15]

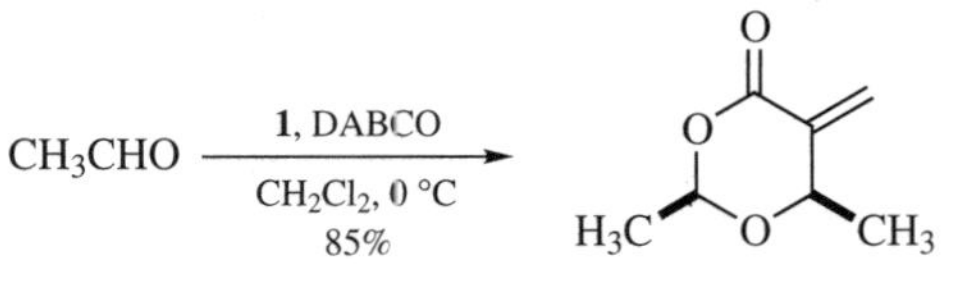

>99% ee (6)

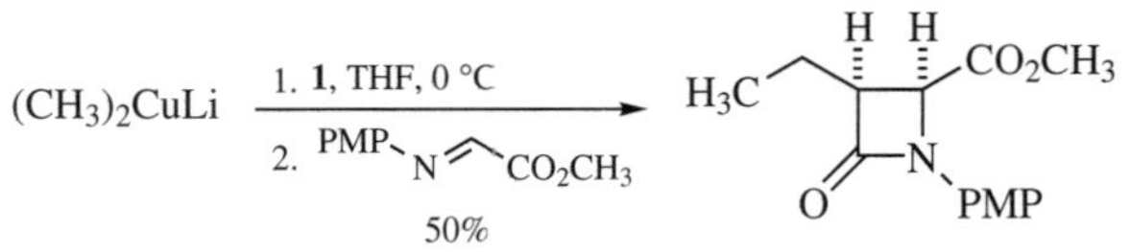

dr = 88:12
97% ee (7)

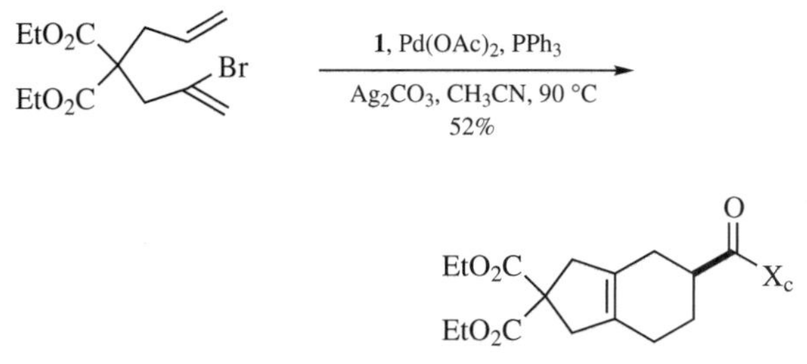

(8)

>95% de

Oxidation. A single example involving aziridination of **1** has been reported, using *N*-aminophthalimide and Pb(IV) (eq 9). Good diastereoselection was observed (89:11) for production of the hydrazine derivative.[16]

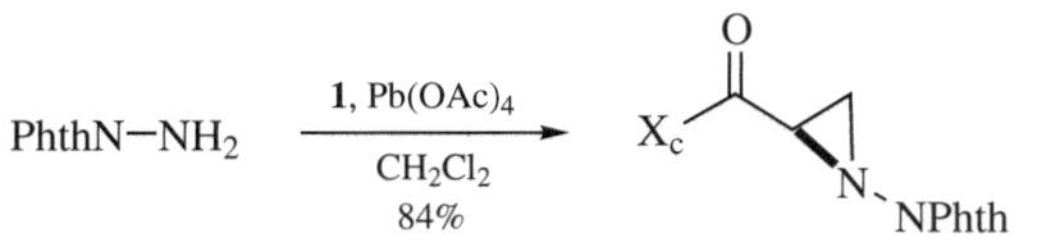

(9)

1. (a) Oppolzer, W. *Pure Appl. Chem.* **1990**, *62*, 1241. (b) Kim, B. H.; Curran, D. P. *Tetrahedron* **1993**, *49*, 293.
2. Oppolzer, W.; Chapuis, C.; Bernardinelli, G. *Helv. Chim. Acta* **1984**, *67*, 1397–1401.
3. Curran, D. P.; Kim, B. H.; Daugherty, J.; Heffner, T. A. *Tetrahedron Lett.* **1988**, *29*, 3555–3558.
4. Mayer, S. C.; Pfizenmayer, A. J.; Cordova, R.; Li, W. R.; Joullie, M. M. *Tetrahedron: Asymmetry* **1994**, *5*, 519–522.
5. Ha, J. D.; Kang, C. H.; Belmore, K. A.; Cha, J. K. *J. Org. Chem.* **1998**, *63*, 3810–3811.
6. Binger, P.; Schaefer, B. *Tetrahedron Lett.* **1988**, *29*, 529–530.
7. Rigby, J. H.; Ateeq, H. S.; Charles, N. R.; Henshilwood, J. A.; Short, K. M.; Sugathapala, P. M. *Tetrahedron* **1993**, *49*, 5495–5506.
8. Kim, B. H.; Lee, J. Y. *Tetrahedron: Asymmetry* **1991**, *2*, 1359–1370.
9. Tejero, T.; Dondoni, A.; Rojo, I.; Merchan, F. L.; Merino, P. *Tetrahedron* **1997**, *53*, 3301–3318.
10. Merino, P.; Anoro, S.; Franco, S.; Merchan, F. L.; Tejero, T.; Tunon, V. *J. Org. Chem.* **2000**, *65*, 1590–1596.
11. Pandey, G.; Laha, J. K.; Mohanakrishnan, A. K. *Tetrahedron Lett.* **1999**, *40*, 6065–6068.
12. Brzezinski, L. J.; Rafel, S.; Leahy, J. W. *J. Am. Chem. Soc.* **1997**, *119*, 4317–4318.
13. Palomo, C.; Aizpurua, J. M.; Gracenea, J. J.; Garcia-Granda, S.; Pertierra, P. *Eur. J. Org. Chem.* **1998**, 2201–2207.
14. Palomo, C.; Aizpurua, J. M.; Gracenea, J. J. *J. Org. Chem.* **1999**, *64*, 1693–1698.
15. (a) Ang, K. H.; Braese, S.; Steinig, A. G.; Meyer, F. E.; Llebaria, A.; Voigt, K.; de Meijere, A. *Tetrahedron* **1996**, *52*, 11503–11528. (b) Voigt, K.; Lansky, A.; Noltemeyer, M.; de Meijere, A. *Liebigs Ann.* **1996**, 899–911.
16. Kapron, J. T.; Santarsiero, B. D.; Vederas, J. C. *J. Chem. Soc., Chem. Commun.* **1993**, 1074–1076.

Jeffrey N. Johnston
Indiana University, Bloomington, IN, USA

Pseudoephedrine

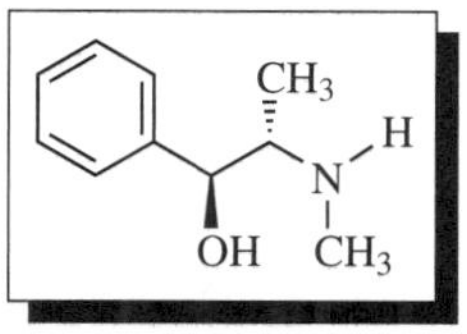

[1S,2S]-(+) [90-82-4],
[1R, 2R]-(−) [321-97-1] $C_{10}H_{15}ON$ (MW 165.24)

(reagent used as a practical chiral auxiliary for asymmetric synthesis)

Alternate Name: α-[1-(methylamino)ethyl] benzenemethanol; Ψ-ephedrine; isoephedrine.
Physical Data: mp 118–120 °C.
Solubility: sparingly soluble in water, soluble in ether, alcohol, and many other organic solvents.
Form Supplied in: white crystalline solid; widely available.
Purification: recrystallization from water.
Handling, Storage, and Precautions: stable; combustible; incompatible with strong oxidizing agents; eye, skin, and respiratory irritant; toxicity (oral) rat LD_{50}: 660 mg kg^{-1}.

Asymmetric Alkylation. *d*-Pseudoephedrine ([1*S*, 2*S*]-(+)) is a commodity chemical employed in over-the-counter medications with annual worldwide production in excess of 300 metric tons. The enantiomer, *l*-pseudoephedrine, is also readily available in bulk and is inexpensive. Pseudoephedrine has been shown to be highly effective as a chiral auxiliary in asymmetric alkylation reactions.[1,2] Treatment of either enantiomer of pseudoephedrine with carboxylic acid chlorides and anhydrides leads to efficient and selective *N*-acylation to form the corresponding tertiary amide derivatives (Table 1).[2] Typically, the only by-product in the acylation reactions is a small amount (<5%) of the *N*,*O*-diacylated product, which is easily removed by crystallization or flash column chromatography. Because intramolecular *O*→*N* acyl transfer within pseudoephedrine β-amino esters occurs rapidly, and because the *N*-acyl form is strongly favored under neutral or basic conditions,[3] products arising from (mono)acylation on oxygen rather than nitrogen are not observed.

Pseudoephedrine amides undergo efficient and highly diastereoselective alkylation reactions with a wide range of alkyl halides as substrates (Table 2).[2] Alkylation of pseudoephedrine amides is accomplished by dianion formation with lithium diisopropylamide (LDA) in tetrahydrofuran (THF) in the presence of lithium chloride (6 equiv), followed by the addition of an alkylating agent.[4] The use of lithium chloride leads to a substantial acceleration in the rate of alkylation and is essential for complete reaction. In addition, *O*-alkylation of the secondary hydroxyl group of the pseudoephedrine auxiliary is suppressed in the presence of lithium chloride. Although the specific role of lithium chloride in the reaction is not known, there is ample precedent in the literature, notably in the work of Seebach and co-workers, documenting the beneficial influence of lithium chloride in enolate alkylation reactions. These studies suggest that lithium chloride modifies the aggregation state, and thereby the reactivity of an enolate in solution.[5–8]

Table 1 Preparation of pseudoephedrine amides

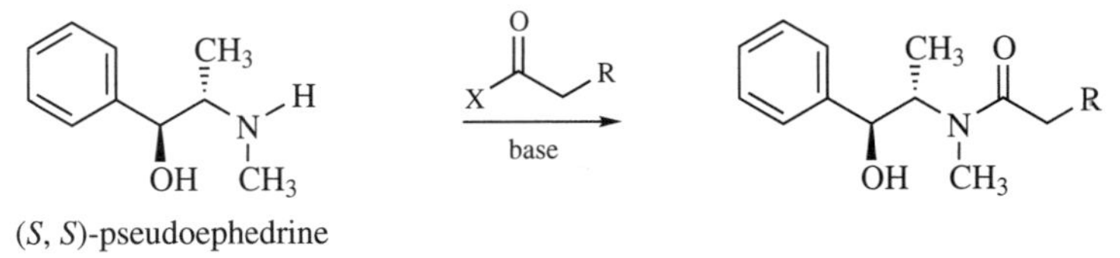

R	X	Isolated yield (%)
CH_3	RCH_2CO_2	95
CH_3	CH_3O	89
i-Pr	Cl	92
3-Pyridyl	*t*-$BuCO_2$	72

Table 2 Diastereoselective alkylation of pseudoephedrine amides with alkyl halides

R	R'X	Isolated de (%)	Isolated yield (%)
CH_3	*n*-BuI	≥99	80
CH_3	BOMBr	98	80
t-Bu	BnBr	≥99	84
2-Thiophene	CH_3I	95	88

A useful mnemonic for deriving the preferred diastereomer formed in the alkylation reaction of pseudoephedrine amide enolates with alkyl halides is as follows: the alkyl halide enters from the same face as the methyl group of the pseudoephedrine auxiliary when the (putative) (*Z*)-enolate is drawn in a planar, extended conformation (eq 1).[1]

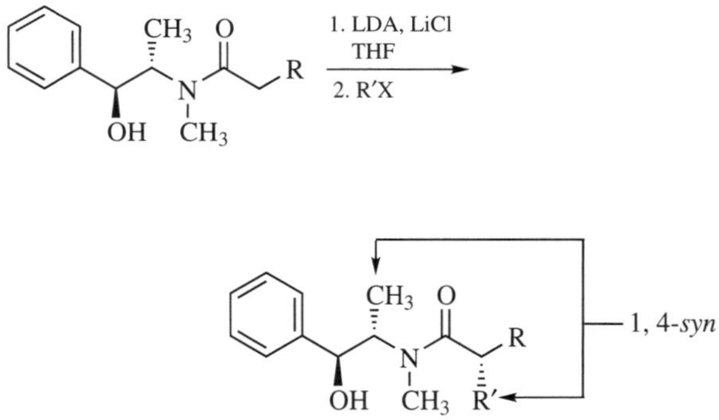

(1)

The superior nucleophilicity and excellent thermal stability of pseudoephedrine amide enolates make possible alkylation reactions with substrates that are ordinarily unreactive with the corresponding ester and imide-derived enolates, such as β-branched primary alkyl iodides.[2] Also, alkylation reactions of pseudoephedrine amide enolates with chiral β-branched primary alkyl iodides proceed with high diastereoselectivity for both the matched and mismatched cases (Table 3).[9]

Epoxides can also be used as substrates in pseudoephedrine amide enolate alkylation reactions, but react with opposite diastereofacial selectivity (suggesting a change in mechanism, proposed to involve delivery of the epoxide electrophile by coordination to a side-chain associated lithium ion), and are more limited in scope (Tables 4 and 5).[10]

A pictorial representation of the opposing diastereoselectivities of alkyl halides and epoxides is shown in Figure 1.[10] A similar electrophile dependence upon diastereoselectivity was first noted in the alkylation of prolinol amide enolates.[11]

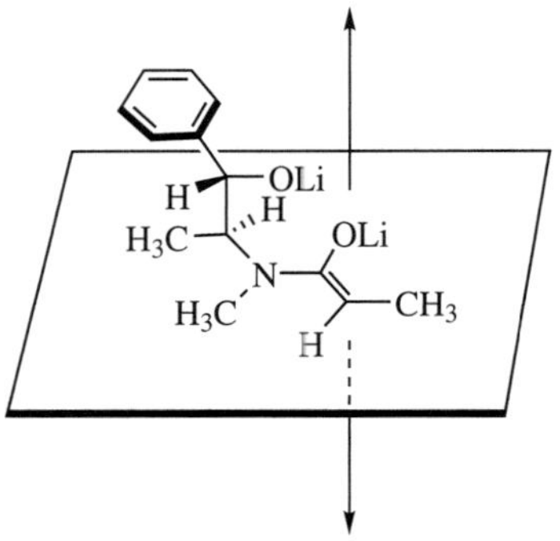

Figure 1

Although alkylation reactions of pseudoephedrine amide enolates are successful with a broad range of electrophiles, a few problematic substrates have been identified. Among these are secondary alkyl halides, such as cyclohexyl bromide, and alkyl halides that are both β-branched and β-alkoxy substituted.[2] However, there is evidence that the thermal stability of pseudoephedrine amide enolates may be such that extended reaction times at ambient temperature, or even heating, may be tolerated;

Table 3 Diastereoselcetive alkylation of pseudoephedrine amides β -branced electrophiles

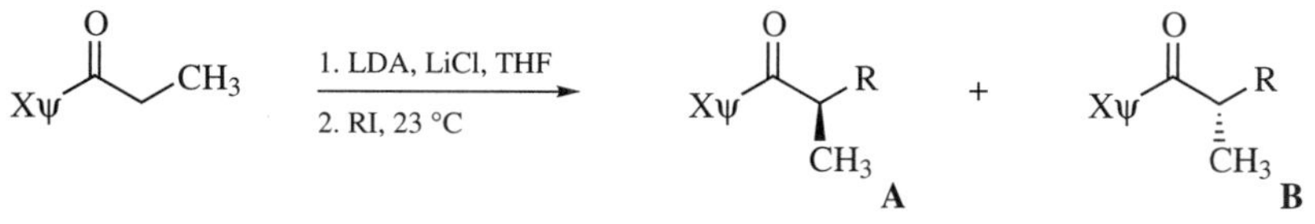

Xψ = pseudoephedrine auxiliary

RI	Product	Ratio of **A**:**B**	Isolated yield (%)
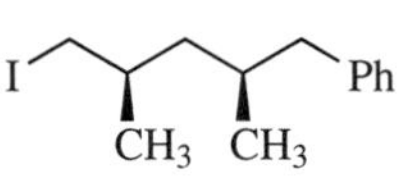	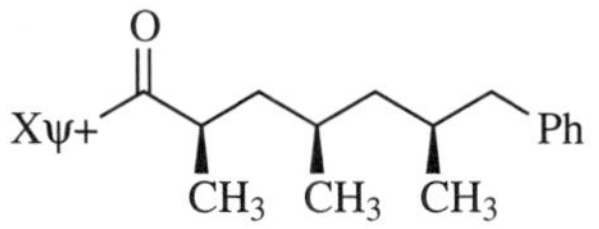	142:1 (matched)	93
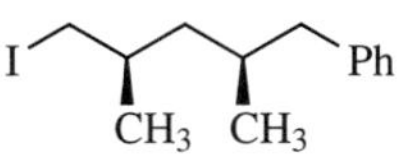	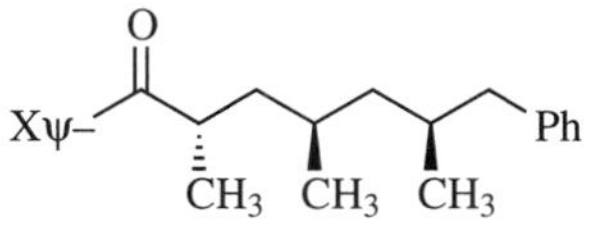	1:70 (mismatched)	96
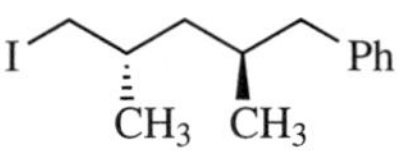	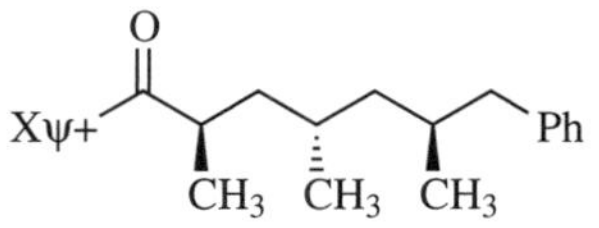	66:1 (mismatched)	93
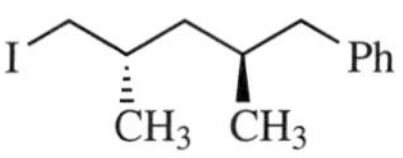	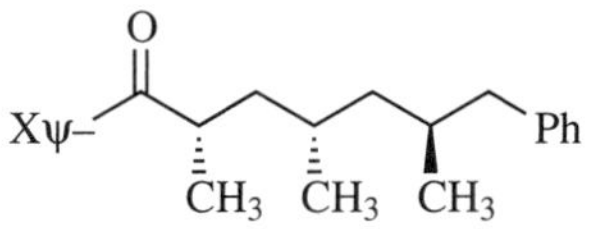	1:199 (matched)	94

Table 4 Diastereoselective alkylation of pseudoephedrine amides with matched epoxides

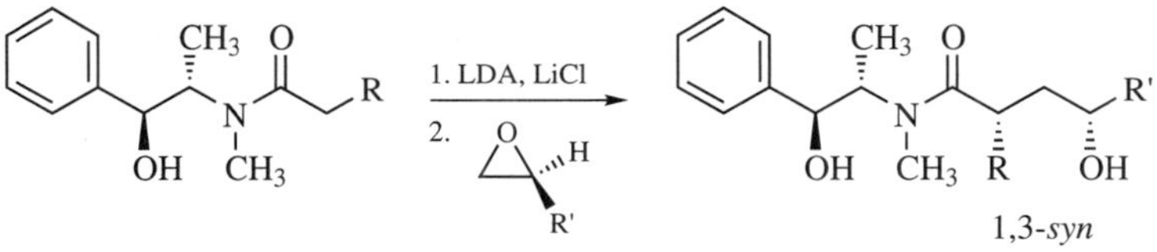

R	R'	Isolated de (%)	Isolated yield (%)
CH_3	CH_3	93	88
CH_3	CH_2OTBS	96	84
Bn	C_6H_6	≥95	86
Bn	CH_2OBn	≥95	87

both approaches have led to successful alkylation reactions with problematic electrophiles (eqs 2, 3, and 4).[12,2,13]

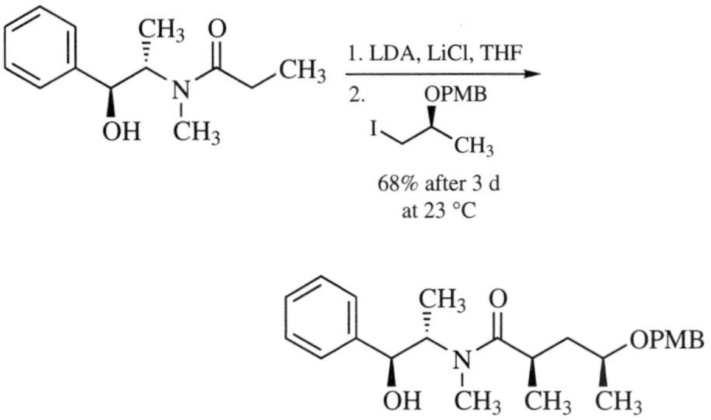

(2)

~13:1

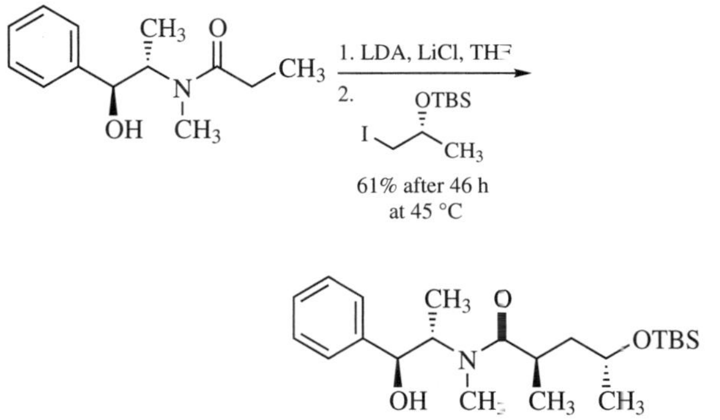

(3)

99% de

Pseudoephedrine amides with a wide variety of α-substituents, including aryl,[1] branched alkyl,[14] chloro,[12] fluoro (described in

Table 5 Diastereoselective alkylation of pseudephedrine amides with mismatched epoxides

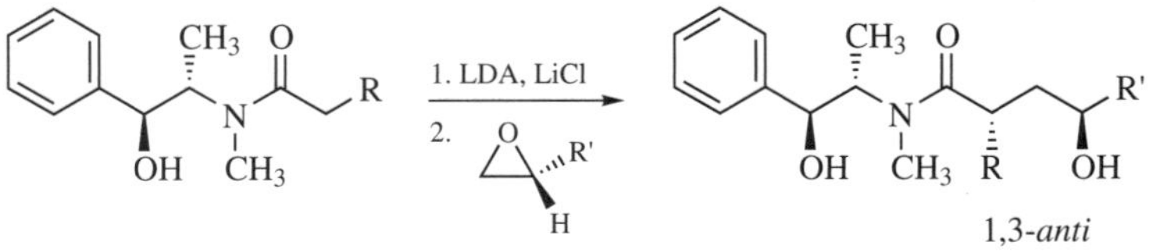

R	R'	Isolated de (%)	Isolated yield (%)
CH_3	CH_3	73	86
CH_3	CH_2OTBS	12	78
Bn	C_6H_6	46	72
Bn	CH_2OBn	36	80

detail in the section Asymmetric Synthesis of Organofluorine Compounds), amino (described in detail in the section Synthesis of α-Amino Acids), and 2-pyridyl groups,[2] undergo highly diastereoselective alkylation reactions. However, to date, no general solution has emerged for the diastereoselective alkylation of pseudoephedrine amides with an α-oxygenated substituent. Enolization of pseudoephedrine α-hydroxyacetamide with 3.2 equiv of LDA furnishes a presumed trianion, with partial decomposition of the starting material. Alkylation of the resulting enolate (1.65 equiv) with benzyl bromide (limiting reagent) then produces the corresponding *C*-benzylated product with 82% de (eq 5).[2]

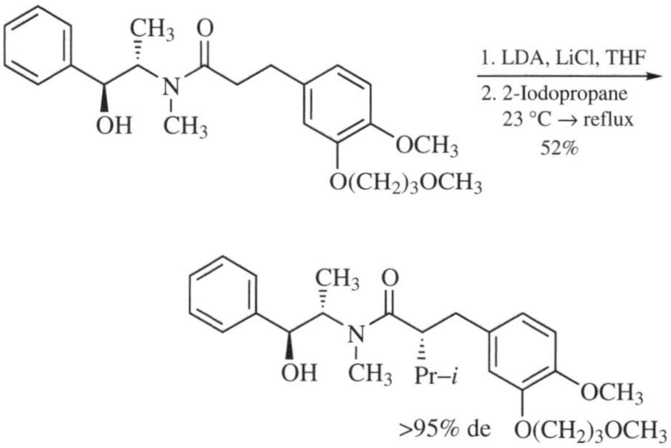

(4)

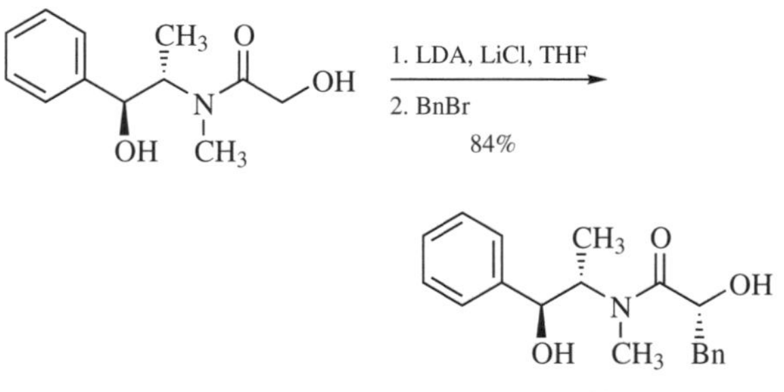

(5)

82% de

The diastereoselectivity of the reaction is lower than that obtained in benzylations of pseudoephedrine amide enolates lacking the α-hydroxyl group. Although an extensive series of *O*-protected derivatives of α-hydroxyacetamide has been examined in a search for an alternative alkylation substrate [TBS, TBDPS, THP, Bn, BOM, Piv, and methyl(1-methoxyethyl)], none has provided satisfactory results nor offered any improvement over pseudoephedrine α-hydroxyacetamide itself.[2]

α,β-Unsaturated pseudoephedrine amides undergo γ-deprotonation when subjected to standard conditions for pseudoephedrine amide enolate formation. The resulting enolate can be α-alkylated with high diastereoselectivity to provide β,γ-unsaturated alkylated products (eq 6).[15]

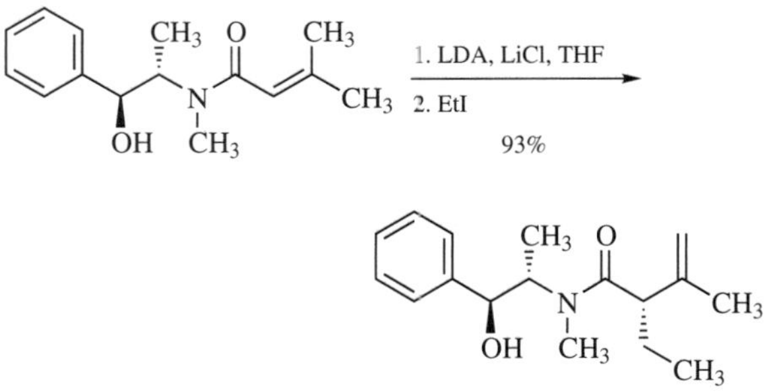

(6)

Transformations of Alkylated Pseudoephedrine Amides. Alkylation products of pseudoephedrine amides are readily transformed in a single operation into highly enantiomerically enriched carboxylic acids, aldehydes, ketones, lactones or primary alcohols.[1,2] Alkylated pseudoephedrine amides can be hydrolyzed under acidic or basic conditions to form carboxylic acids. Simply heating a pseudoephedrine amide at reflux in a 1:1 mixture of sulfuric acid (9–18 N) and dioxane affords the corresponding carboxylic acid in excellent chemical yield with little or no epimerization (eq 7).[16] Under these conditions, the substrate initially undergoes a rapid *N*→*O* acyl transfer reaction followed by rate-limiting hydrolysis of the resulting β-ammonium ester intermediate to form the carboxylic acid.[3,17]

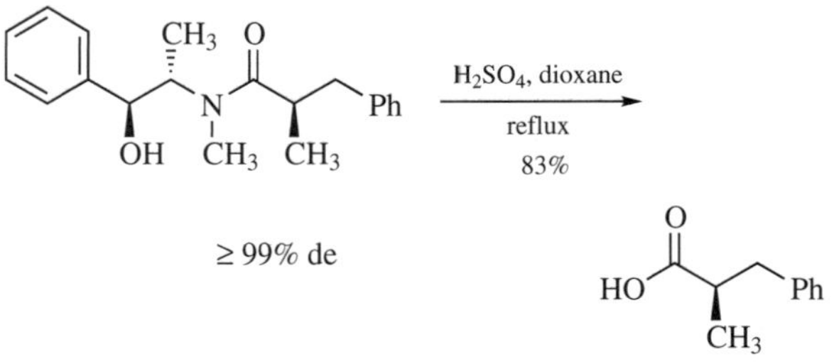

(7)

97% ee

Basic conditions for the hydrolysis of pseudoephedrine amides typically involve heating the substrate with tetra-*n*-butylammonium hydroxide in a mixture of *tert*-butyl alcohol and water

(Table 6).[1,2] Where the expense of tetra-*n*-butylammonium hydroxide is a consideration, or in cases where the product carboxylic acid is poorly soluble in ether (making extractive removal of tetra-*n*-butylammonium salts difficult), an alternative procedure employing sodium hydroxide in a mixture of water, methanol, and *tert*-butyl alcohol can be used. The mechanism of the base-induced hydrolysis reaction is believed to involve initial rate-limiting intramolecular $N \rightarrow O$ acyl transfer, followed by rapid saponification of the resulting β-amino ester.[3]

Pseudoephedrine amides can be converted directly into highly enantiomerically enriched aldehydes[1,2] using Brown and Tsukamoto's lithium triethoxyaluminum hydride reagent[18] (eq 8).[19,20]

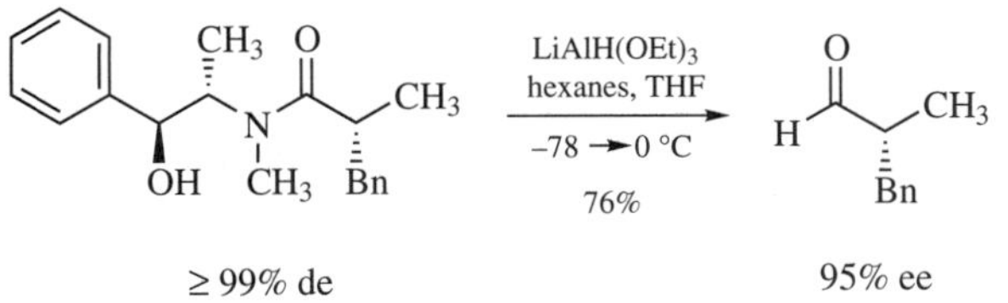

(8)

Addition of alkyllithium reagents to pseudoephedrine amides leads to the formation of enantiomerically enriched ketones[1,2,21] (eqs 9 and 10).[19,20] The protocol developed to transform alkylated pseudoephedrine amides into ketones was optimized to avoid premature breakdown of the tetrahedral intermediate generated following addition of the organolithium species to the amide.[23]

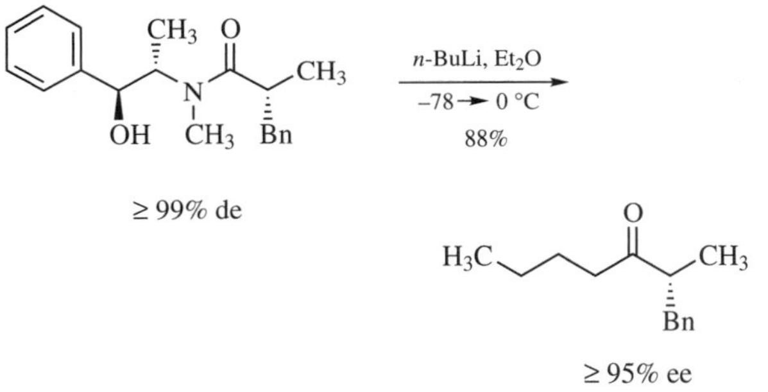

(9)

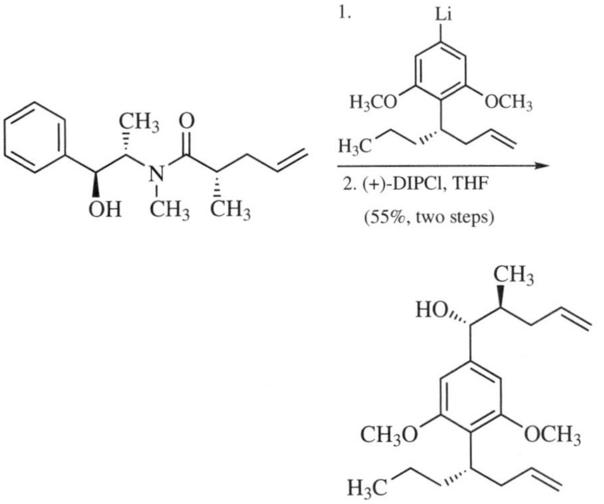

(10)

Reduction of pseudoephedrine amides with metal amide-borane complexes,[1] and lithium amidotrihydroborate (LAB) in particular,[2,24] furnishes the corresponding primary alcohols in high yield. In the initial report, LAB was prepared by deprotonation of the commercial, solid reagent borane–ammonia complex,[25] using slightly less than 1 equiv of butyllithium as base (eq 11).[24] In more recent work,[9] an improved preparation of the reagent has been developed that uses 1 equiv of LDA as base in the reaction (eq 12).[26] The greater efficiency of reductions using LDA as base is attributed to the propensity of *n*-butyllithium to form butylboron side-products in the reaction and, ultimately, butylboron alkoxide products that are difficult to hydrolyze.

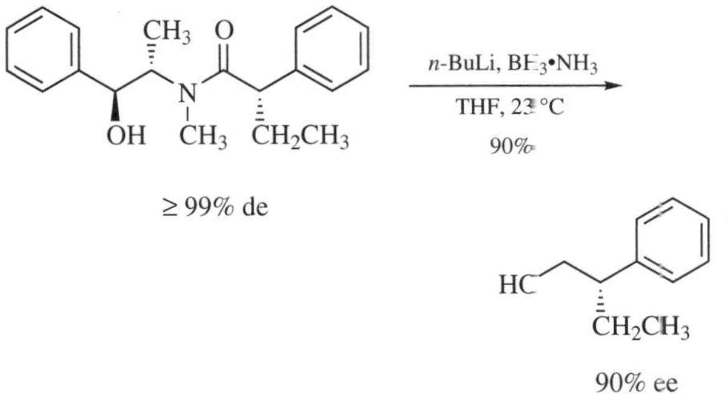

(11)

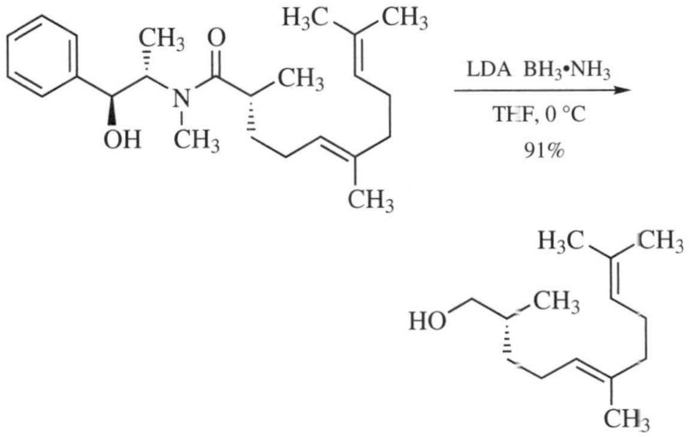

(12)

γ,δ-Unsaturated pseudoephedrine amides are efficiently converted into γ-lactones by cleavage of the auxiliary through halolactonization reactions (eqs 13 and 14).[27,28]

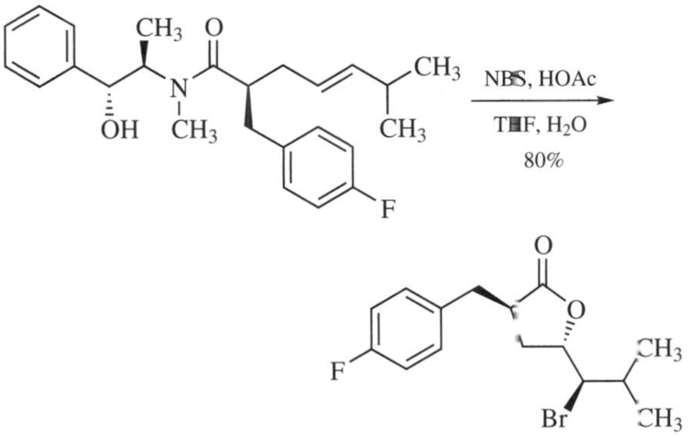

(13)

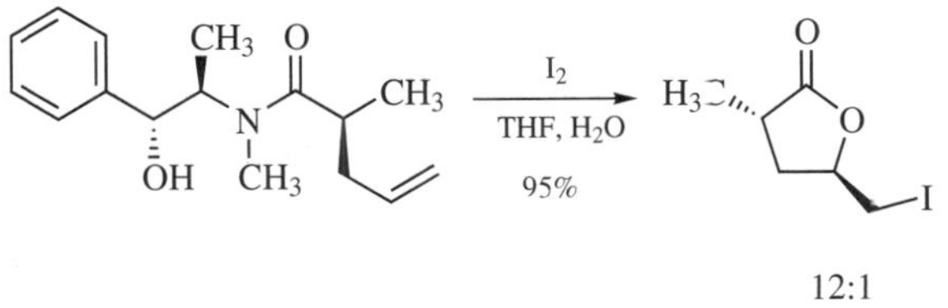

(14)

The efficiency and practicality of pseudoephedrine-based asymmetric alkylation reactions has been exploited in syntheses of several complex natural products, including cylindrocyclophane A,[22,29] fumonisin B_2,[30] pironetin,[15] epothilones A and B,[31] salicylihalamide A,[32] 6,7-dideoxysqualestatin H5,[33] saframycin A,[34,35] and terpestacin.[28]

Table 6 Basic hydrolysis of pseudoephedrine amides

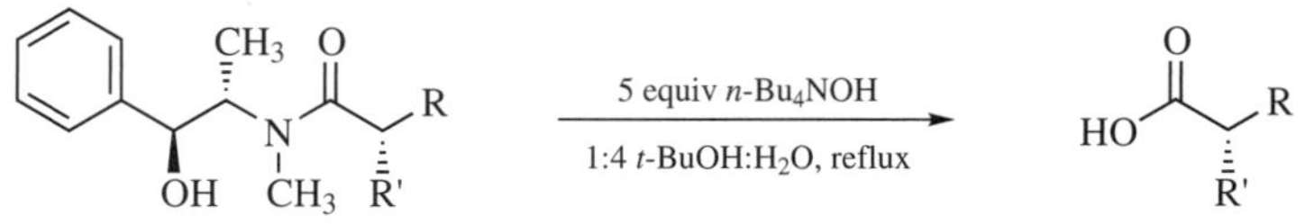

R	R'	Isolated ee (%)	Isolated yield (%)
CH_3	Bn	94	93
CH_3	BOM	69	92
n-Bu	CH_3	93	88
Ph	Et	64	82

Synthesis of α-Amino Acids. The diastereoselective alkylation of enolates derived from pseudoephedrine glycinamide has been shown to be an effective method for the preparation of α-amino acids of high enantiomeric purity.[36,37] Pseudoephedrine glycinamide hydrate can be easily prepared in a single step by the condensation of pseudoephedrine with the free-base form of glycine methyl ester in the presence of lithium chloride and base (*n*-butyllithium[36] or lithium methoxide[37]). The primary by-product in the reaction is the dipeptide pseudoephedrine glycylglycinamide, formed to the extent of < 10%. The crude acylation reaction mixture can be directly purified by selective crystallization of pseudoephedrine glycinamide hydrate from hot aqueous tetrahydrofuran. An improved preparation of pseudoephedrine glycinamide hydrate entails the direct treatment of glycine methyl ester hydrochloride with lithium *tert*-butoxide.[38] This procedure is advantageous because it obviates the need to use the hygroscopic reagent lithium chloride and it eliminates difficulties associated with the handling of the free-base form of glycine methyl ester, which is prone to polymerization.

Enolization of pseudoephedrine glycinamide is complicated by the presence of two other acidic sites in the molecule: the secondary hydroxyl group and the primary amino group. The enolization protocol originally reported requires the addition of a carefully measured amount of LDA to a thoroughly dried solution of pseudoephedrine glycinamide and lithium chloride.[37] The strict use of less than 2 equiv of base avoided partial cleavage of the auxiliary from pseudoephedrine glycinamide. Several practical laboratory-scale preparations of enantiomerically enriched α-amino acids, including *l*-azatyrosine[39] and *l*-allylglycine (eq 15),[40] have been executed based on this methodology.

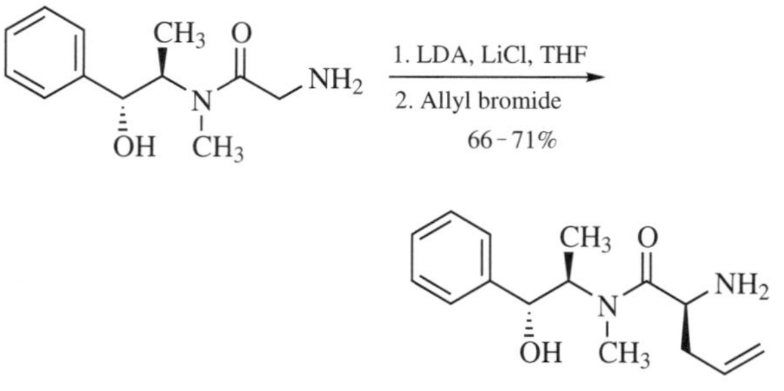

(15)

≥ 99% de

A modified procedure has since been developed that involves the direct alkylation of pseudoephedrine glycinamide hydrate.[38] In this operationally simpler procedure, excess lithium hexamethyldisilazide (LHMDS) is added to a solution of anhydrous lithium chloride and pseudoephedrine glycinamide hydrate. In situ generation of LHMDS•LiCl from lithium metal, hexamethyldisilazane (HMDS), and hexyl chloride can also been used for the enolization and subsequent alkylation of pseudoephedrine glycinamide hydrate.[38] These procedures for the alkylation of pseudoephedrine glycinamide reliably afford good yields of alkylated products (Table 7). The procedure employing commercial LHMDS has been used in the total synthesis of saframycin A (eq 16).[34,35]

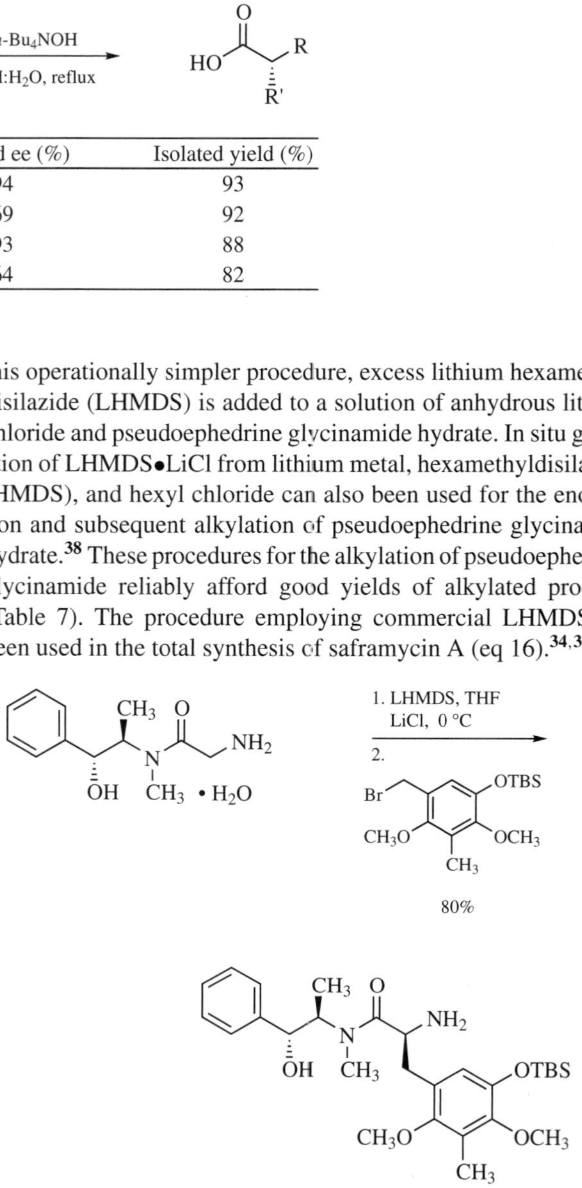

(16)

97% de

Alkylation of pseudoephedrine sarcosinamide can be used to prepare enantiomerically enriched *N*-methyl-α-amino acids.[36,37] Anhydrous pseudoephedrine sarcosinamide has been prepared by the addition of sarcosine methyl ester to a mixture of pseudoephedrine, lithium chloride, and lithium methoxide. In contrast to the preparation of pseudoephedrine glycinamide, the amount of dipeptide by-product produced in the reaction is minimal, perhaps due to the increased steric hindrance of the *N*-methyl group of sarcosine. Thus, pure anhydrous pseudoephedrine sarcosinamide can be obtained from the crude acylation reaction mixture by precipitation from toluene and subsequent drying. Like anhydrous pseudoephedrine glycinamide, anhydrous pseudoephedrine sarcosinamide can be handled in the atmosphere for brief periods without consequence, but should be stored with scrupulous avoidance of moisture to prevent hydration.

Table 7 Alkylation of pseudoephedrine glycinamide hydrate

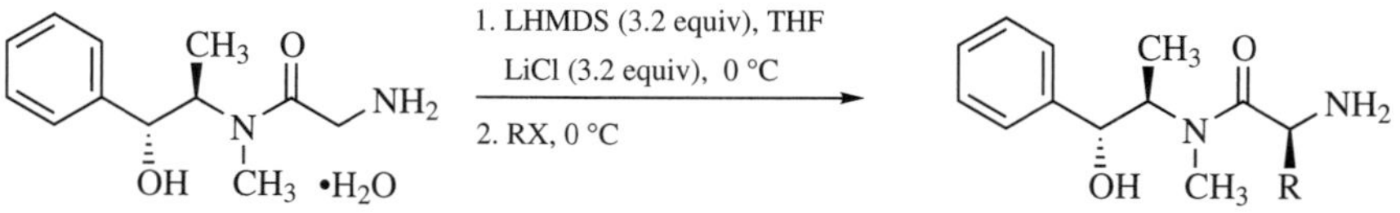

RX	LHMDS	Isolated de (%)	Isolated yield (%)
CH_2=$CHCH_2Br$	Commercial solution (1.0 M in THF)	93	86
CH_2=$CHCH_2Br$	Generated in situ (Li, HMDS, *n*-HexCl)	93	82
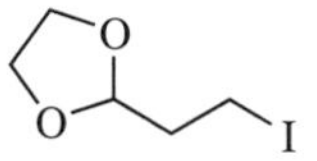	Commercial solution (1.0 M in THF)	97	65
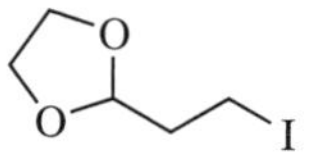	Generated in situ (Li, HMDS, *n*-HexCl)	96	62

The alkylation of anhydrous pseudoephedrine sarcosinamide is similar to the alkylation of anhydrous pseudoephedrine glycinamide, with one important experimental modification, wherein the reaction is conducted in the presence of 1 equiv of *N*-methylethanolamine. The optimum conditions for alkylation of anhydrous pseudoephedrine sarcosinamide involve the addition of *n*-butyllithium or LDA (2.95 equiv) to a suspension of anhydrous pseudoephedrine sarcosinamide (1 equiv), anhydrous lithium chloride (6.00 equiv), and *N*-methylethanolamine (1.00 equiv) in THF at −78 °C, followed by warming the resulting slurry to 0 °C and the addition of an alkylating agent (1.1–1.5 equiv) (eq 17).[37] The presence of *N*-methylethanolamine in the alkylation reaction is necessary to achieve reproducible diastereoselectivity and may function by facilitating anionic equilibration.

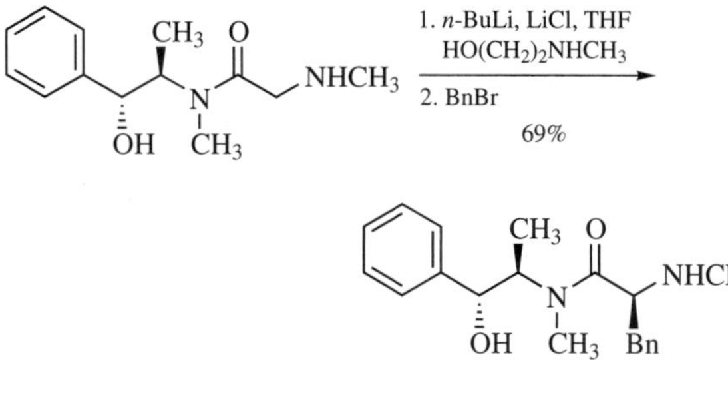

(17)

99% de

Many functional groups are stable under conditions for the alkylation of pseudoephedrine glycinamide enolates, including aryl benzenesulfonate esters (eq 18),[39] *tert*-butyl carbamate and *tert*-butyl carbonate groups (eq 19),[41] *tert*-butyldimethylsilyl ethers,[42] benzyl ethers,[37] *tert*-butyl ethers,[37] methoxymethyl ethers,[36] and alkyl chlorides.[36] The stereochemistry of the alkylation reactions of pseudoephedrine glycinamide and pseudoephedrine sarcosinamide is the same as that observed in alkylations of simple *N*-acyl derivatives of pseudoephedrine.

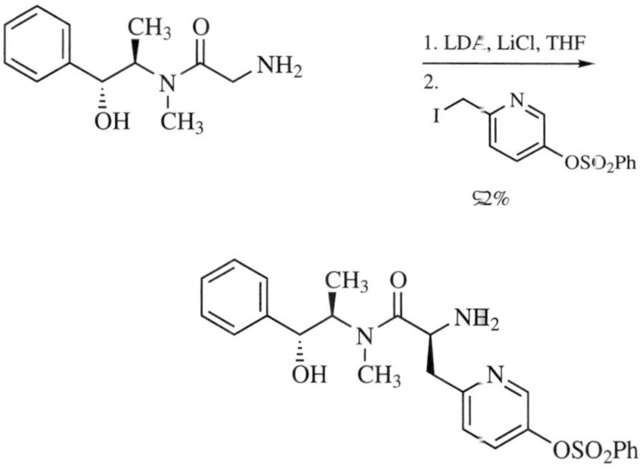

(18)

90–91% de

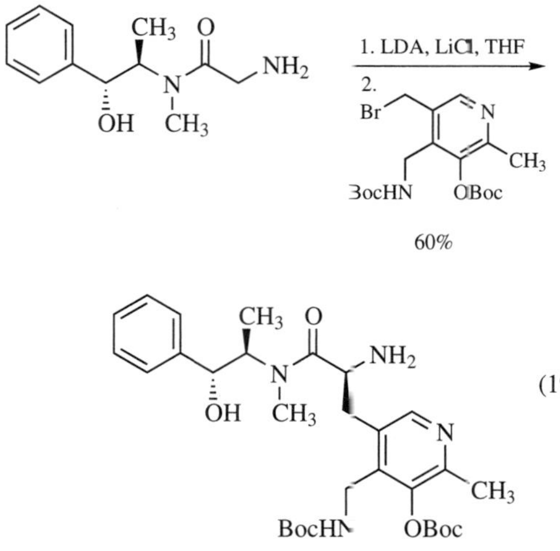

(19)

>10:1

Hydrolysis reactions of alkylated pseudoephedrine glycinamides are more rapid than the hydrolysis of pseudoephedrine amides without α-amino groups. It is believed that this reflects the inductive influence of the amino group, enhancing the electrophilicity of the amide group.[37] It is significant that this rate enhancement is not accompanied by an increased rate of racemization. Typically, alkaline hydrolysis of the alkylation products occurs upon heating at reflux in aqueous sodium hydroxide solution (0.5 M, 2 equiv).[36] Upon cooling, the pseudoephedrine auxiliary is easily recovered by extraction of the aqueous product slurry with dichloromethane (typically, 96% of the pseudoephedrine auxiliary is recovered, and 83–86% after one recrystallization from water). After extraction of the auxiliary, the alkaline aqueous product solution can be treated with an acylating agent to furnish the corresponding *N*-protected α-amino acid derivative directly. *N-tert*-Butoxycarbonyl (*N*-Boc) and *N*-(9-fluorenylmethoxy)-carbonyl (*N*-Fmoc) protected α-amino acids are prepared efficiently by this method (Table 8).[37] Free α-amino acids can be obtained simply by refluxing the alkylation products in pure water. Extraction of the aqueous reaction mixture with dichloromethane, lyophilization of the aqueous layer, and trituration of the solid residue with ethanol (to remove any remaining pseudoephedrine) then provides the pure α-amino acids (Table 9).[36]

The asymmetric amination of pseudoephedrine amide enolates has been introduced as an alternative method for the synthesis of α-amino acids.[43] Lithium enolates, generated by the addition of LDA to pseudoephedrine amides, can be efficiently aminated with di-*tert*-butyl azodicarboxylate (DTBAD). The amination reaction is complete within a few minutes at low temperature and does not require the use of lithium chloride. Cleavage of the Boc groups within the adducts using trifluoroacetic acid (TFA) and hydrogenolysis of the resulting α-hydrazino derivatives then provides α-amino acids in good yield following acidic hydrolysis and ion exchange chromatography (eq 20).[43]

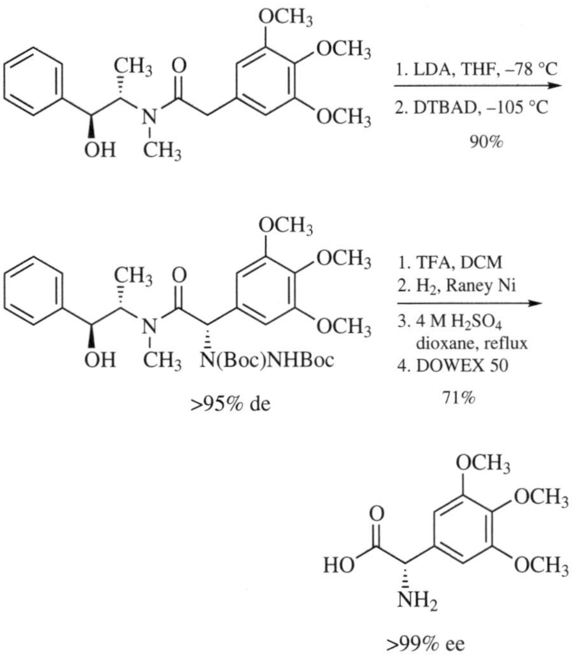

(20)

Recently, the bis(methylthio)methylene imine of pseudoephedrine glycinamide was shown to undergo diastereoselective alkylation at 23 °C with lithium *tert*-butoxide or sodium ethoxide as base and various alkyl halides as electrophiles (eq 21).[44] This procedure was used to prepare enantiomerically enriched α-amino acids.

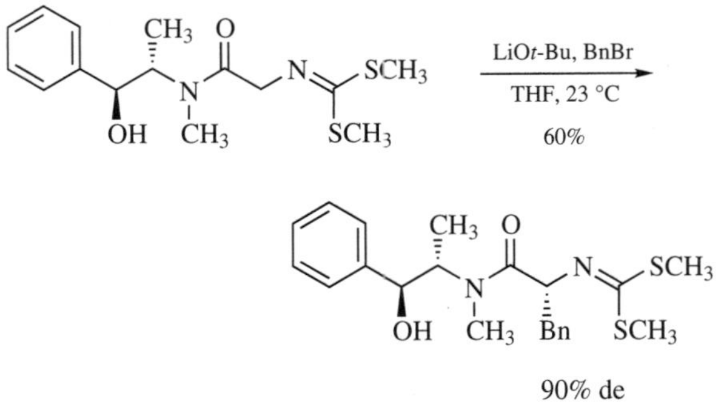

(21)

Alkylation reactions of pseudoephedrine amides offer many practical advantages over existing procedures for the asymmetric construction of α-amino acids. These include the high crystallinity of many pseudoephedrine amides, the low cost of pseudoephedrine, the high diastereoselectivity of the alkylation reactions, a simple protocol for recovering the auxiliary, and the ease of hydrolytic, racemization-free removal of the chiral auxiliary. The methodology is also advantageous because it requires no protecting group for the α-amine. Thus, in many instances, alkylation of pseudoephedrine glycinamide has been deemed the method of choice for the preparation of enantiomerically enriched α-amino acids in quantity (eq 22).[45]

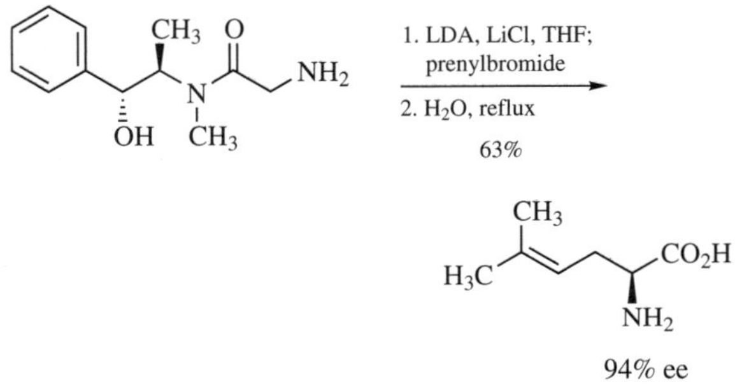

(22)

β-Amino Acids. Pseudoephedrine has been used as a chiral auxiliary for the preparation of both α-substituted and α,β-disubstituted β-amino acids. Alkylation of β-alanine was shown to furnish an efficient, inexpensive, and enantioselective route to α-alkyl β-amino acids (eq 23).[46]

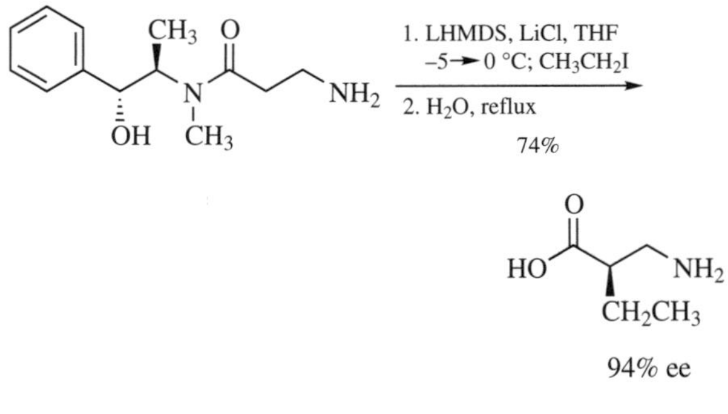

(23)

In addition, the lithium enolate derived from pseudoephedrine propionamide has been shown to undergo highly diastereoselective Mannich reactions with *p*-(methoxy)phenyl aldimines to form enantiomerically enriched α,β-disubstituted β-amino acids (Table 10).[47] As observed in alkylation reactions using alkyl halides as electrophiles, lithium chloride is necessary for the reaction of aldimines. With respect to the enolate, the stereochemistry of the alkylation reactions is the same as that observed with

Table 8 Basic hydrolysis of pseudoephedrine amides followed by *N*-protection

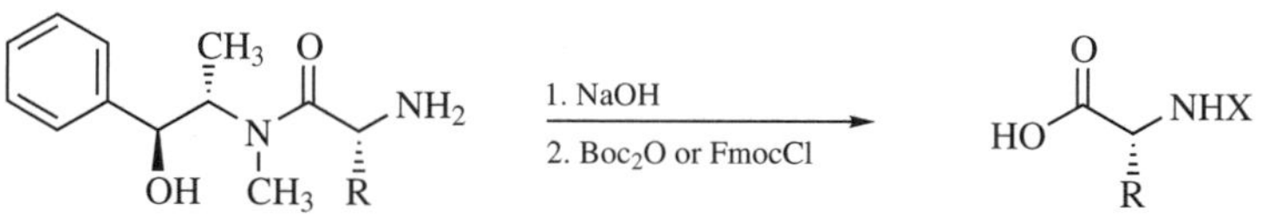

R	X	Isolated ee (%)	Isolated yield (%)
CH_3CH_2	Boc	≥99	97
CH_3CH_2	Fmoc	≥99	99
$(CH_3)_2CHCH_2$	Boc	≥99	97
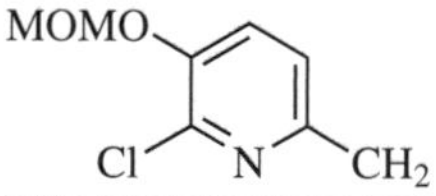	Fmoc	96	73

Table 9 Hydrolysis of pseudophedrine amides in water

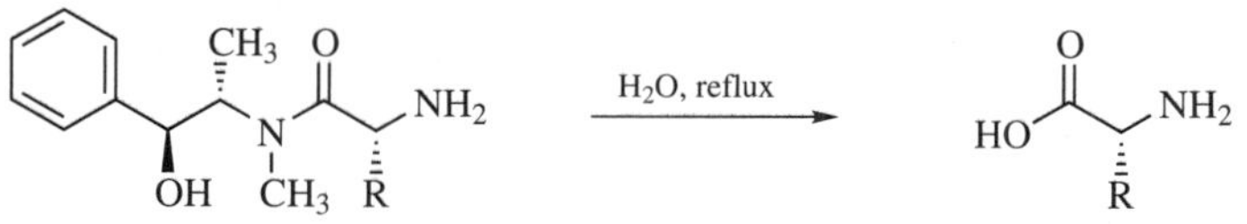

R	Isolated ee (%)	Isolated yield (%)
CH_2=$CHCH_2$	≥99	87
c-$C_3H_5CH_2$	≥98	79
$(CH_3)_3SiCH_2$	≥99	77
2-$CH_3OC_6H_4CH_2$	≥99	71

alkyl halides; reactions of *p*-(methoxy)phenyl aldimines are further characterized by a preference for the formation of 2,3-anti products, a unique and highly useful feature of these reactions.

Aldol Reactions. Pseudoephedrine amide enolates have been shown to undergo highly diastereoselective aldol addition reactions, providing enantiomerically enriched β-hydroxy acids, esters, ketones, and their derivatives (Table 11).[48,49] The optimized procedure for the reaction requires enolization of the pseudoephedrine amide substrate with LDA followed by transmetalation with 2 equiv of $ZrCp_2Cl_2$ at −78 °C and addition of the aldehyde electrophile at −105 °C. It is noteworthy that the reaction did not require the addition of lithium chloride to favor product formation as is necessary in many other pseudoephedrine amide enolate alkylation reactions. The stereochemistry of the alkylation is the same as that observed with alkyl halides and the formation of the 2,3-*syn* aldol adduct is favored. The tendency of zirconium enolates to form *syn* aldol products has been previously reported.[50,51,52] The β-hydroxy amide products obtained can be readily transformed into the corresponding acids, esters, and ketones as reported with other alkylated pseudoephedrine amides. An asymmetric aldol reaction between an (*S,S*)-(+)-pseudoephedrine-based arylacetamide and paraformaldehyde has been used to prepare enantiomerically pure isoflavanones.[53]

Asymmetric Synthesis of Organofluorine Compounds. Asymmetric alkylation of fluorinated pseudoephedrine amides has been employed to synthesize a variety of enantiomerically enriched α-fluoro carboxylic acid derivatives. Pseudoephedrine α-fluoroacetamide, a nonvolatile, crystalline compound, can be readily prepared by the acylation of pseudoephedrine with ethyl fluoroacetate. (CAUTION: Fluoroacetic acid and derivatives of fluoroacetic acid are exceedingly toxic, causing convulsions and ventricular fibrillation upon inhalation and should be used only under adequate supervision and in an appropriate fume hood. Although the specific toxicities of pseudoephedrine α-fluoroacetamide and other fluorinated pseudoephedrine derivatives are unknown, extreme caution in their preparation and handling is urged.) Pseudoephedrine α-fluoroacetamide can be enolized with LHMDS in the presence of anhydrous lithium chloride and the resulting enolate can be efficiently trapped with reactive electrophiles, such as benzyl bromide, to form the corresponding alkylated products with high diastereoselectivity (eq 24).[54] Interestingly, enolization of pseudoephedrine α-fluoroacetamide with LDA in the presence of anhydrous lithium chloride and subsequent trapping of the resulting enolate with reactive electrophiles resulted in the formation of alkylated products with diminished diastereoselectivity. The basis for the improved selectivity in alkylations conducted with LHMDS versus LDA is not known; however, the stereochemistry of enolate formation is proposed to be the selectivity-determining step in these reactions. Presumably, the enolization of pseudoephedrine α-fluoroacetamide with LHMDS, be it kinetically or thermodynamically controlled, exhibits a strong preference for the *Z*-configuration. The stereochemistry of the subsequent alkylation reaction is then consistent with the model proposed for the alkylation of simple *N*-acyl derivatives of pseudoephedrine. Unlike other pseudoephedrine amide enolates, the enolate derived from pseudoephedrine α-fluoroacetamide exhibits limited thermal sta-

Table 10 Mannich reaction of pseudoephedrine propionamide enolate with *p*-(methoxy)phenyl aldimides

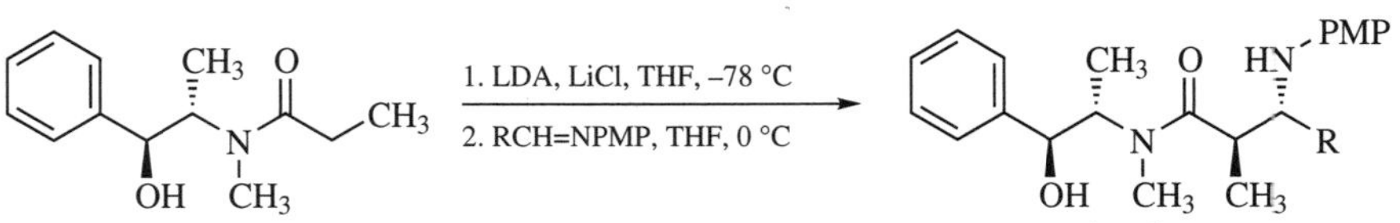

Major diastereomer

R	*anti/syn*	*anti/anti**[a]	Yield (%)
Ph	>99/1	>99/1	86
2-furyl	>99/1	>99/1	80
t-Bu	>99/1	>99/1	69

[a]Ratio of major anti diastereomer (shown) to doubly epimeric minor anti diastereomer.

Table 11 Pseudophedrine-based asymmetric aldol reactions

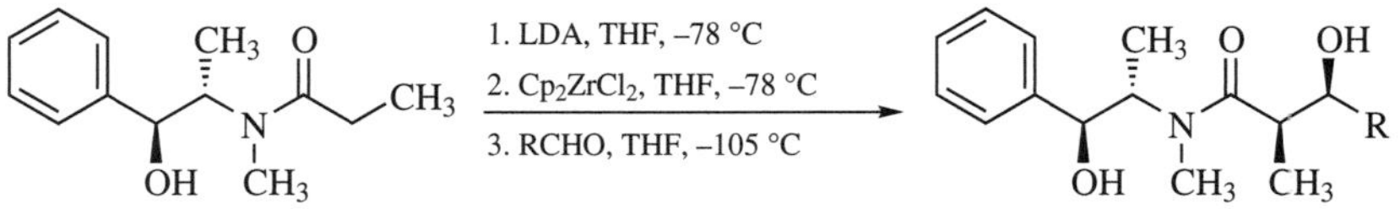

Major diastereomer

R	*syn/anti*	*syn/syn**[a]	Yield (%)
Ph	94/6	>99/1	90
Et	96/4	>99/1	90
i-Pr	>99/1	>99/1	94

[a]Ratio of major syn diastereomer (shown) to doubly epimeric minor syn diastereomer.

bility above ∼−40 °C and, as a consequence, alkylation reactions with relatively unreactive electrophiles, such as ethyl iodide, proceed poorly. However, Michael addition with 1-nitro-3-phenyl-1-propene does occur, even at −78 °C, forming two of the four possible diastereomeric conjugate addition products (eq 25).[55] These products were demonstrated to be stereoisomeric at the β-carbon, and had the same configuration at the α-carbon, that expected based upon addition of simple alkyl halides to the *Z*-enolate derived from pseudoephedrine α-fluoroacetamide.

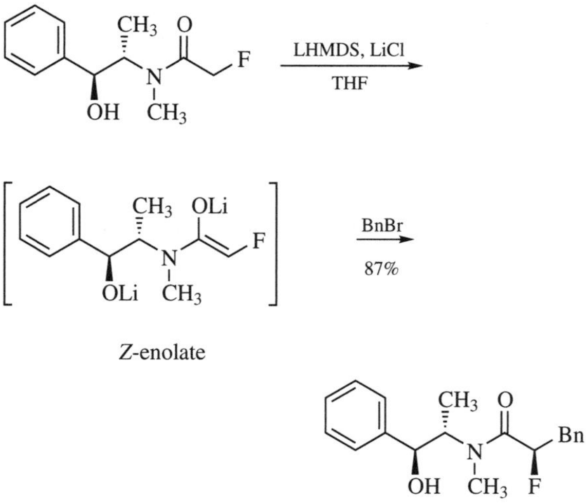

(24)

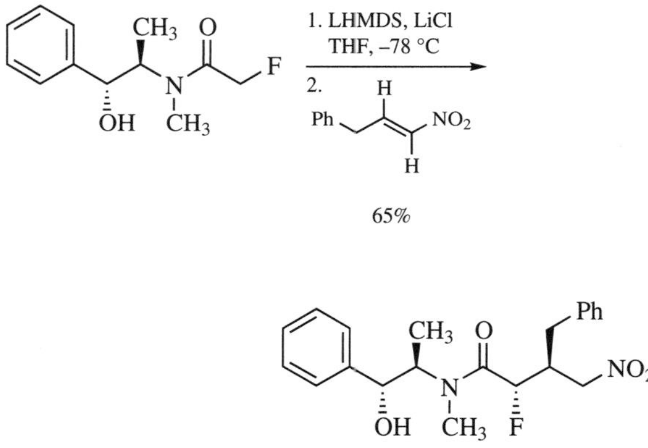

(25)

1.7:1 *anti:syn*

Inductive activation of the amide by the adjacent fluorine atom allows for the basic hydrolysis of the amide bond under relatively mild conditions (warming to ∼75 °C in a biphasic solution of 2 N sodium hydroxide in a 2:2:1 mixture of water, *tert*-butyl alcohol, and methanol) to form carboxylic acids with high enantiomeric excess (eq 26).[54]

Alkylation of pseudoephedrine α-fluoropropionamide can be used to prepare enantiomerically enriched tertiary alkyl fluoride centers (eq 27).[56] In contrast to the alkylation of pseudoephedrine α-fluoroacetamide, alkylation of pseudoephedrine α-fluoropropionamide proceeds with high diastereoselectivity when LDA in used as the base in the reaction and low diastereoselectivity when LHMDS is used. In these reactions, deprotonation of pseudoephedrine α-fluoropropionamide with LDA, proposed to occur under kinetic control, is believed to form the corresponding

E-enolate. Electrophilic attack by alkyl halides then occurs opposite the enolate π-face occupied by the side-chain alkoxide group, as observed with other pseudoephedrine amide enolates.

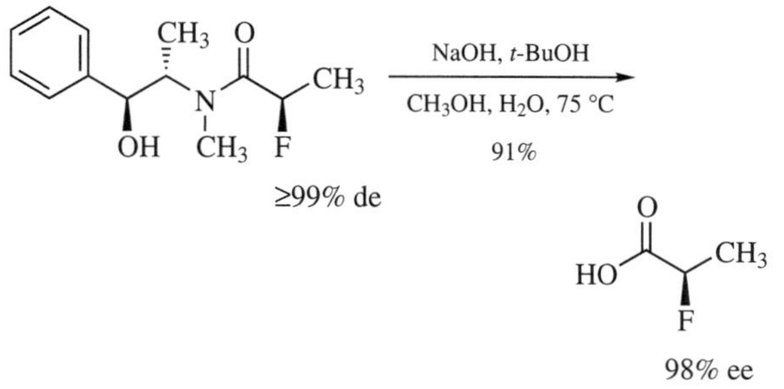

(26)

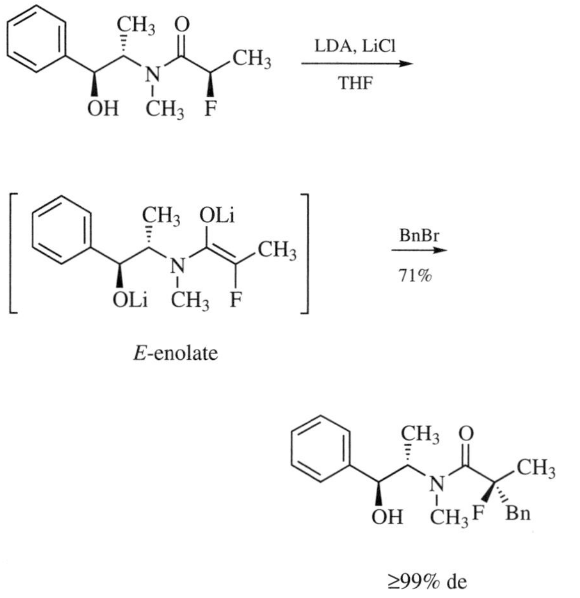

(27)

Related Reagents. Prolinol; ephedrine; oxazolidinones; camphorsultams; camphor-derived auxiliaries; and oxazolines.

1. Myers, A. G.; Yang, B. H.; Chen, H.; Gleason, J. L. *J. Am. Chem. Soc.* **1994**, *116*, 9361.
2. Myers, A. G.; Yang, B. H.; Chen, H.; McKinstry, L.; Kopecky, D. J.; Gleason, J. L. *J. Am. Chem. Soc.* **1997**, *119*, 6496.
3. Welsh, L. H. *J. Am. Chem. Soc.* **1947**, *69*, 128.
4. Myers, A. G.; Yang, B. H. *Org. Synth.* **1999**, *77*, 22.
5. Seebach, D.; Bossler, H.; Gründler, H.; Shoda, S.-I. *Helv. Chim. Acta.* **1991**, *74*, 197.
6. Miller, S. A.; Griffiths, S. L.; Seebach, D. *Helv. Chim. Acta.* **1993**, *76*, 563.
7. Bossler, H.; Seebach, D. *Helv. Chim. Acta.* **1994**, *77*, 1124.
8. Rück, K. *Angew. Chem., Int. Ed. Engl.* **1995**, *34*, 433.
9. Myers, A. G.; Yang, B. H.; Chen, H.; Kopecky, D. J. *Synlett* **1997**, 457.
10. Myers, A. G.; McKinstry, L. *J. Org. Chem.* **1996**, *61*, 2428.
11. Askin, D.; Volante, R. P.; Ryan, K. M.; Reamer, R. A.; Shinkai, I. *Tetrahedron Lett.* **1988**, *29*, 4245.
12. Lee, D.-H.; Rho, M.-D. *Tetrahedron Lett.* **2000**, *41*, 2573.
13. Sandham, D. A.; Taylor, R. J.; Carey, J. S.; Fässler, A. *Tetrahedron Lett.* **2000**, *41*, 10091.
14. Ravn, M. M.; Coates, R. M.; Jetter, R.; Croteau, R. B. *Chem. Commun.* **1998**, 21.
15. Keck, G. E.; Knutson, C. E.; Wiles, S. A. *Org. Lett.* **2001**, *3*, 707.
16. Bach, J.; Galobardes, M.; Garcia, J.; Romea, P.; Tey, C.; Urpí, F.; Vilarrasa, J. *Tetrahedron Lett.* **1998**, *39*, 6765.
17. Mitchell, W. *J. Chem. Soc.* **1940**, 1153.
18. Brown, H. C.; Tsukamoto, A. *J. Am. Chem. Soc.* **1964**, *86*, 1089.
19. Myers, A. G.; Yang, B. H.; Chen, H. *Org. Synth.* **1999**, *77*, 29.
20. Paterson, I.; Febner, K.; Finlay, M. R. V. *Tetrahedron Lett* **1997**, *38*, 4301.
21. Martín, M.; Mas, G.; Urpí, F.; Vilarrasa, J. *Angew. Chem., Int. Ed. Engl.* **1999**, *38*, 3086.
22. Smith, A. B., Jr; Kozmin, S. A.; Adams, C. M.; Paone, D V. *J. Am. Chem. Soc.* **2000**, *122*, 4984.
23. Myers, A. G.; Yoon, T. *Tetrahedron Lett.* **1995**, *36*, 9429.
24. Myers, A. G.; Yang, B. H.; Kopecky, D. J. *Tetrahedron Lett* **1996**, *37*, 3623.
25. Andrews, G. C.; Crawford, T. C. *Tetrahedron Lett.* **1980**, *21*, 693.
26. Whitlock, G. A.; Carreira, E. M. *Helv. Chim. Acta.* **2000**, *83*, 2007.
27. Dragovich, P. S.; Prins, T. J.; Zhou, R.; Fuhrman, S. A.; Patick, A. K.; Matthews, D. A.; Ford, C. E.; Meador, J. W., Jr; Ferre, R. A.; Worland, S. T. *J. Med. Chem.* **1999**, *42*, 1203.
28. Myers, A. G.; Siu, M.; Ren, F. *J. Am. Chem. Soc* **2002**, *124*, 4230.
29. Smith, A. B., III; Kozmin, S. A.; Paone, D. V. *J. Am. Chem. Soc.* **1999**, *121*, 7423.
30. Shi, Y.; Peng, L. F.; Kishi, Y. *J. Org. Chem.* **1997**, *62*, 5666.
31. Bode, J. W.; Carreira, E. M. *J. Am. Chem. Soc.* **2001**, *123*, 3611.
32. Snider, B. B.; Song, F. *Org. Lett.* **2001**, *3*, 1817.
33. Martin, S. F.; Naito, S. *J. Org. Chem.* **1998**, *63*, 7592.
34. Myers, A. G.; Kung, D. W.; Zhong, B.; Movassaghi, M.; Kwon, S. *J. Am. Chem. Soc.* **1999**, *121*, 8401.
35. Myers, A. G.; Kung, D. W. *J. Am. Chem. Soc.* **1999**, *121*, 10828.
36. Myers, A. G.; Gleason, J. L.; Yoon, T. *J. Am. Chem. Soc.* **1995**, *117*, 8488.
37. Myers, A. G.; Gleason, J. L.; Yoon, T.; Kung, D. W. *J. Am. Chem. Soc.* **1997**, *119*, 656.
38. Myers, A. G.; Schnider, P.; Kwon, S.; Kung, D. W. *J. Org. Chem.* **1999**, *64*, 3322.
39. Myers, A. G.; Gleason, J. L. *J. Org. Chem.* **1996**, *61*, 813.
40. Myers, A. G.; Gleason, J. L. *Org. Synth.* **1999**, *76*, 57.
41. Sinha Roy, R.; Imperiali, B. *Tetrahedron Lett.* **1996**, *37*, 2129.
42. Kearney, P. C.; Nowak, M. W.; Zhong, W.; Silverman, S. K.; Lester, H. A.; Dougherty, D. A. *Mol. Pharmacol.* **1996**, *50*, 1401.
43. Vicario, J. L.; Badía, D.; Domínguez, E.; Crespo, A.; Carrillo, L.; Anakabe, E. *Tetrahedron Lett.* **1999**, *40*, 7123.
44. Guillena, G.; Nájera, C. *Tetrahedron: Asymmetry* **2001**, *12*, 181.
45. Smith, A. B., III; Benowitz, A. B.; Favor, D. A.; Sprengeler, P. A.; Hirschmann, R. *Tetrahedron Lett.* **1997**, *38*, 3809.
46. Nagula, G.; Huber, V. J.; Lum, C.; Goodman, B. A. *Org. Lett.* **2000**, *2*, 3527.
47. Vicario, J. L.; Badía, D.; Carrillo, L. *Org. Lett.* **2001**, *3*, 773.
48. Vicario, J. L.; Badía, D.; Domínguez, E.; Carrillo, L. *Tetrahedron Lett.* **1998**, *39*, 9267.
49. Vicario, J. L.; Badía, D.; Domínguez, E.; Rodríguez, M.; Carrillo, L. *J. Org. Chem.* **2000**, *65*, 3754.
50. Evans, D. A.; McGee, L. R. *Tetrahedron Lett.* **1980**, *21*, 3975.
51. Katsuki, T.; Yamaguchi, M. *Tetrahedron Lett.* **1985**, *26*, 5807.
52. Murphy, P. J.; Procter, G.; Russell, A. T. *Tetrahedron Lett.* **1987**, *28*, 2037.
53. Vicario, J. L.; Badía, D.; Domínguez, E.; Rodríguez, M.; Carrillo, L. *Tetrahedron Lett.* **2000**, *41*, 8297.
54. Myers, A. G.; McKinstry, L.; Barbay, J. K.; Gleason, J. L. *Tetrahedron Lett.* **1998**, *39*, 1335.

55. Myers, A. G.; Barbay, J. K.; Zhong, B. *J. Am. Chem. Soc.* **2001**, *123*, 7207.
56. Myers, A. G.; McKinstry, L.; Gleason, J. L. *Tetrahedron Lett.* **1997**, *38*, 7037.

Andrew G. Myers & Mark G. Charest
Harvard University, Cambridge, MA, USA

(1*R*,2*S*)-*N*-Pyrrolidinylnorephedrine

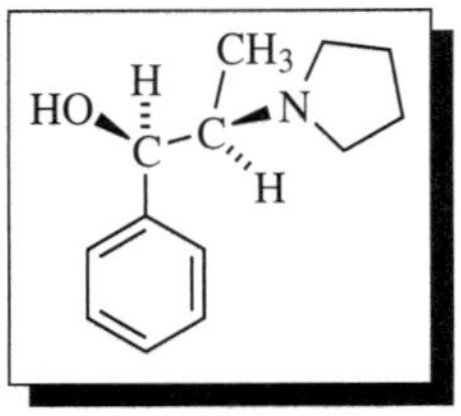

[127641-25-2] $C_{13}H_{19}NO$ (MW 205.30)

(chiral ligand for the enantioselective addition of dialkylzinc reagents to aromatic and aliphatic aldehydes, alkynes to aromatic aldehydes, and lithium acetylides to an aromatic ketone)

Alternate Name: (1*R*,2*S*)-1-phenyl-2-(1-pyrrolidinyl)-1-propanol, [*R*-(*R**,*S**)]-β-methyl-α-phenyl-1-pyrrolidine ethanol.
Physical Data: mp 45–46 °C, $[\alpha]_D^{25}$ + 15 (c 2, $CHCl_3$).
Solubility: soluble in heptane and toluene.
Form Supplied in: off-white crystalline solid.
Analysis of Reagent Purity: H-NMR, C-NMR, elemental analysis.
Preparative Methods: the title reagent is prepared[1] by the treatment of (1*R*, 2*S*)-(−)-norephedrine with 1,4-dibromobutane and $NaHCO_3$ in toluene. This approach can be modified to form a wide variety of cyclic and acyclic tertiary amine derivatives. The enantiomeric reagent has also been formed from (+)-norephedrine.
Purification: the hydrochloride salt can be precipitated from toluene solution, and the free base is dissolved in heptane upon treatment with NaOH. Concentration and cooling of the cold heptane solution affords the crystalline solid.
Handling, Storage, and Precautions: no particular precautions are recommended for this relatively safe compound.

Additions of Dialkylzinc Reagents. (1*R*,2*S*)-*N*-Pyrrolidinylnorephedrine (**1**) is an effective catalyst for the enantioselective addition of dialkylzinc reagents to aromatic aldehydes (eq 1).[2,3] Optimized conditions involve reaction in toluene at 0 °C with 10 mol % of the ligand and 2.2 equiv of the dialkylzinc reagent. Normal work-up after 20 h affords the product from addition to the *Si* face of the aldehyde. Product yields for a variety of alkylzinc reagents (1° and 2°) and an array of aromatic aldehydes are normally 80–100% with ee being nearly 90%. While similar results can be obtained for pyrazole-4-carbaldehydes,[4] aliphatic aldehydes,[2] and 1,2-phthalic dicarbaldehydes,[5] the optimal ligand structure may involve variation of the amine substitution pattern (aliphatic tertiary amine rather than pyrrolidine structure).

Alkynylation of Aromatic Aldehydes. The title ligand **1** can also be used as a catalyst for the enantioselective addition of terminal alkynes to aromatic aldehydes (eq 2).[6] Using a mixture of toluene and THF as the solvent practically eliminates the competitive addition of the alkylzinc. The zinc acetylide does not need to be preformed.

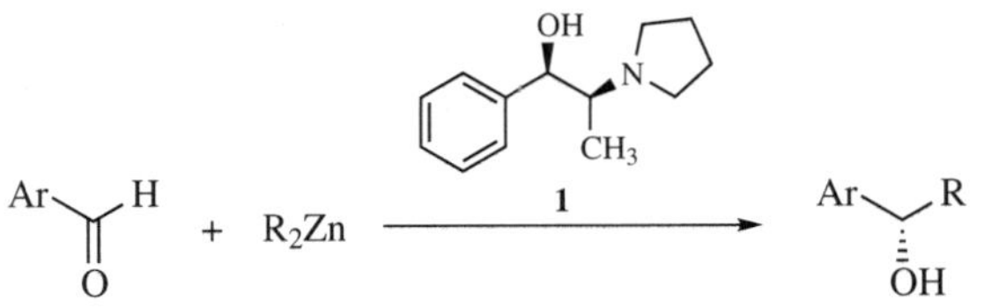

(1)

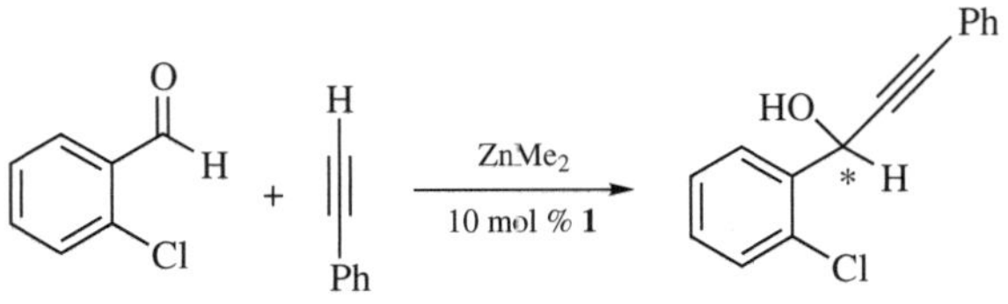

(2)

In a typical reaction, a solution of alkyne in THF is cooled to −20 °C for 5 min. An equimolar amount of the dialkylzinc is added in toluene (ratio of THF:toluene = 1:3). After 15 min, 10 mol % of the ligand is added, followed by the aldehyde. HPLC analysis shows complete reaction usually within 18 h. Both electron-rich and electron-poor aldehydes have been used along with aromatic and aliphatic alkynes. Yields are normally 70–90% with ee being 65–85%. Once again the optimal ligand structure may involve variation of the amine substitution pattern.

Addition of Acetylide to an Aromatic Ketone. The synthesis of efavirenz, a potent HIV transcriptase inhibitor, required the enantioselective addition of lithium cyclopropylacetylide to the carbonyl carbon of a trifluoroacetophenone (eq 3). Careful control of reaction conditions and the use of the lithium salt of the title ligand affords the desired alcohol in 91% yield and >99.5% ee.[7,8]

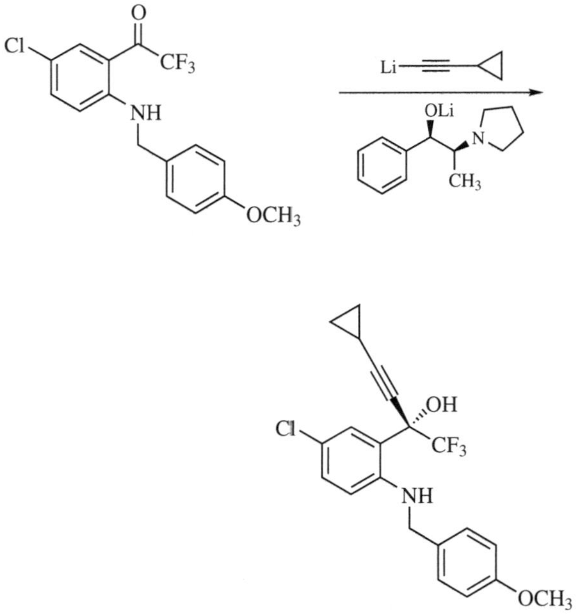

(3)

A solution of the lithium alkoxide of the ligand and the acetylide (2 equiv each compared to ketone) was aged at 0 °C in THF for 30 min. The solution was cooled to −55 °C and the ketone added, with the temperature reaching no higher than −50 °C. Further reaction at −55 °C for 60 min and citric acid work up affords the optimized results.

Substantial structural and mechanistic work has been carried out on the acetylide–ligand aggregation complex.[9–11] The reactive species responsible for this remarkable degree of selectivity is sensitive to the conditions of formation, and both yield and selectivity depend on strict adherence to the reaction protocol.

Enantioselective Ring Opening of Epoxides. In two steps, (1*R*,2*S*)-*N*-pyrrolidinylnorephedrine can be converted into the analogous diamine **2**. Treatment with butyllithium affords an amide salt that can be used as a chiral base. In a limited study, the opening of achiral epoxides to chiral allylic alcohols proceeds in high yield and with good enantioselectivity (eq 4).[12]

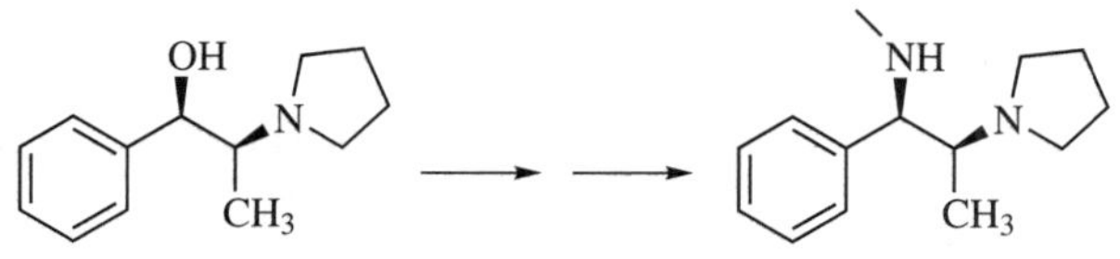

1 **2**

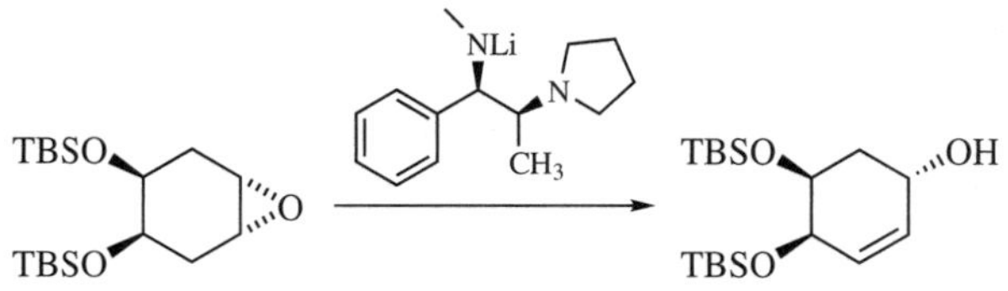

(4)

94% yield
94% ee

Related Reagents. As mentioned above, other alkylating agents can form open chained and other ringed derivatives of the title compound. Under some circumstances, these reagents may give better enantioselectivity.

1. Zhao, D.; Chen, C.; Xu, F.; Tan, L.; Tillyer, R.; Pierce, M. E.; Moore, J. R. *Org. Synth.* **1999**, *77*, 12.
2. Soai, K.; Yokoyama, S.; Hayasaka, T. *J. Org. Chem.* **1991**, *56*, 4264.
3. Soai, K.; Konishi, T.; Shibata, T. *Heterocycles* **1999**, *51*, 1421.
4. Tanji, S.; Aoyagi, H.; Tabira, H.; Sato, I.; Soai, K. *Heterocycles* **2000**, *53*, 381.
5. Kleijn, H.; Jastrzebski, J. T. B. H.; Boersma, J.; Koten, G. v. *Tetrahedron Lett.* **2001**, *42*, 3933.
6. Li, Z.; Upadhyay, V.; DeCamp, A. E.; DiMichele, L.; Reider, P. J. *Synthesis* **1999**, 1453.
7. Tan, L.; Chen, C.; Tillyer, R. D.; Grabowski, E. J. J.; Reider, P. J. *Angew. Chem., Int. Ed.* **1999**, *38*, 711.
8. Pierce, M. E.; Parsons, R. L., Jr.; Radesca, L. A.; Lo, Y. S.; Silverman, S.; Moore, J. R.; Islam, Q.; Choudhury, A.; Fortunak, J. M. D.; Nguyen, D.; Luo, C.; Morgan, S. J.; Davis, W. P.; Confalone, P. N.; Chen, C.; Tillyer, R. D.; Frey, L.; Tan, L.; Xu, F.; Zhao, D.; Thompson, A. S.; Corley, E. G.; Grabowski, E. J. J.; Reamer, R.; Reider, P. J. *J. Org. Chem.* **1998**, *63*, 8536.
9. Thompson, A.; Corley, E. G.; Huntington, M. F.; Grabowski, E. J. J.; Remenar, J. F.; Collum, D. B. *J. Am. Chem. Soc.* **1998**, *120*, 2028.
10. Xu, F.; Reamer, R. A.; Tillyer, R.; Cummins, J. M.; Grabowski, E. J. J.; Reider, P. J.; Collum, D. B.; Huffman, J. C. *J. Am. Chem. Soc.* **2000**, *122*, 11212.
11. Sun, X.; Winemiller, M. D.; Xiang, B.; Collum, D. B. *J. Am. Chem. Soc.* **2001**, *123*, 8039.
12. De Dousa, S. E.; O'Brien, P.; Steffens, H. C. *Tetrahedron Lett.* **1999**, *40*, 8423.

Richard T. Taylor
Miami University, Oxford, OH, USA

Q

Quinine

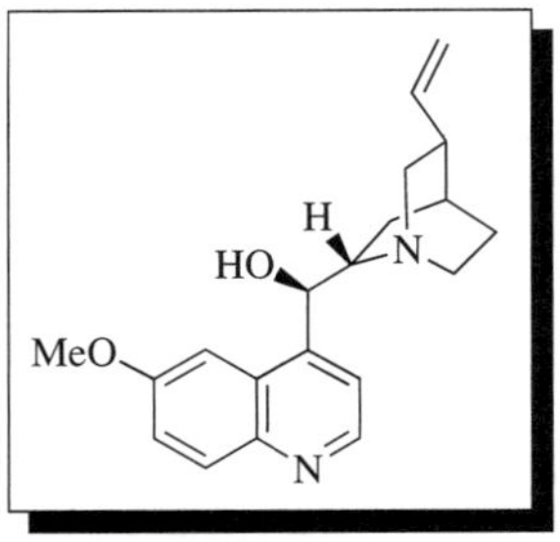

[130-95-0]	$C_{20}H_{23}N_2O_2$	(MW 324.45)

(chiral catalyst[1–14])

Physical Data: mp 173–175 °C.
Solubility: sol hot water, methanol, benzene, chloroform, ether, glycerol; insol pet ether.
Form Supplied in: crystalline solid; 90% purity.
Analysis of Reagent Purity: NMR, mp.
Preparative Methods: commercially available from several sources.
Purification: recrystallize from absolute ethanol.
Handling, Storage, and Precautions: toxic; irritant.

Asymmetric Diels–Alder Reactions. Chiral bases, including quinine, have been used as catalysts in Diels–Alder reactions (eq 1).[1] The reactions take place at room temperature or below and require 1–10% equiv of the alkaloid. The asymmetric induction that is observed can be attributed to complex formation between the achiral dienolate and the chiral amine.[1]

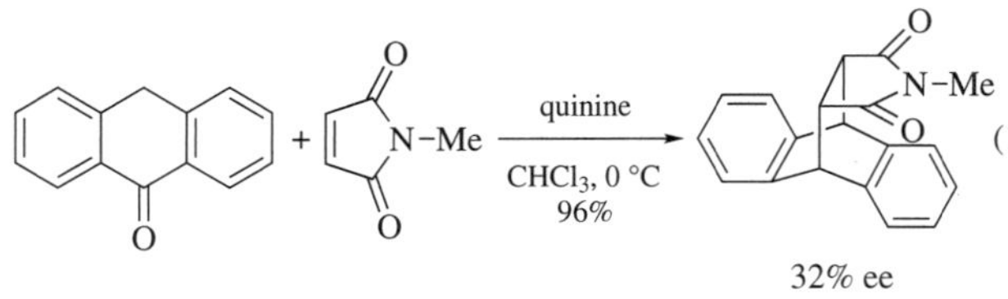

(1)

Preparation of Chiral Sulfinates. Optically active sulfinates can be prepared by reaction of a symmetrical sulfite with *t-Butylmagnesium Chloride* in the presence of an optically active amino alcohol. The best enantioselectivity has been observed using quinine as the optically active amine (eq 2).[2] An alternative approach to this new enantioselective asymmetric synthesis of alkyl *t*-butylsulfinates would be reaction of a racemic sulfinate with *t*-butylmagnesium chloride complexed by optically active alkaloids (eq 3).[2] In this case, kinetic resolution of the racemic sulfinate leads to an optically active sulfinate and an optically active sulfoxide.

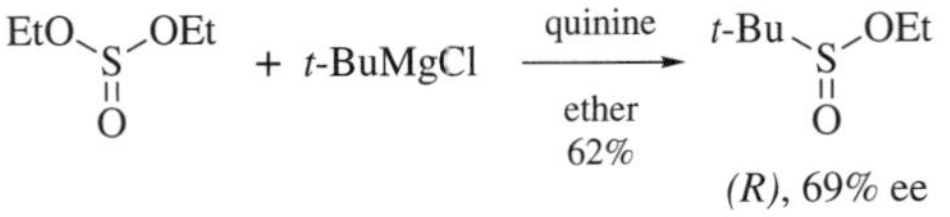

(2)

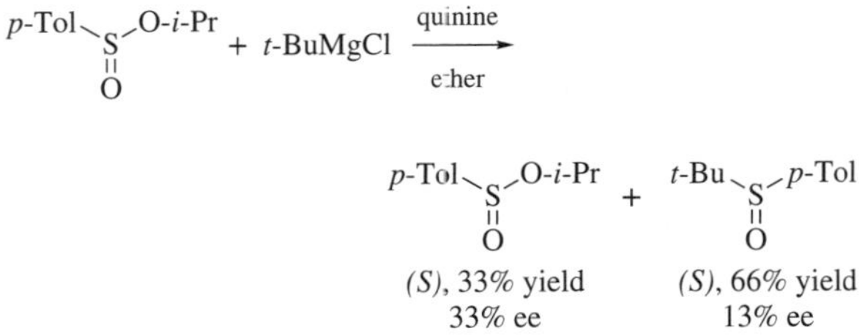

(3)

Stereoselective Addition of Diethylzinc to Aldehydes. Wynberg has found that the cinchona alkaloids catalyze the reaction of *Diethylzinc* and aldehydes to form optically active alcohols (eq 4).[3] The highest enantiomeric excess obtained was from reactions which used quinine as the catalyst. Results show that the hydroxyl group of the catalyst hydrogen bonds with the aldehyde and that the diethylzinc interacts with the vinyl group of the catalyst as well, but it has not been determined if one or two catalyst molecules are involved in the transition state. Similar results have been obtained using a furan aldehyde.[4]

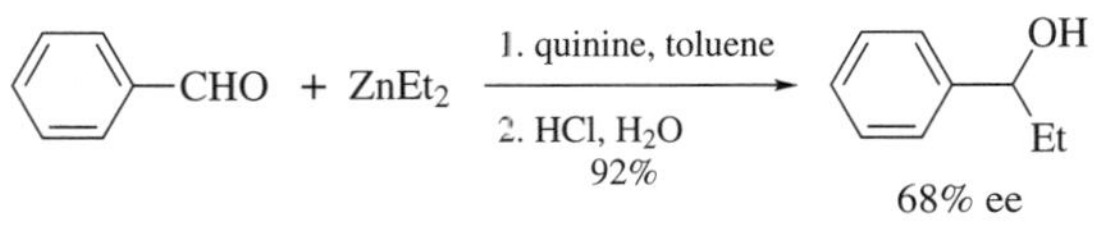

(4)

Synthesis of Optically Active Epoxides. Alkaloids and alkaloid salts have been successfully used as catalysts for the asymmetric synthesis of epoxides. The use of chiral catalysts such as quinine or quinium benzylchloride (QUIBEC) have allowed access to optically active epoxides through a variety of reaction conditions, including oxidation using *Hydrogen Peroxide* (eq 5),[5] Darzens condensations (eq 6),[6] epoxidation of ketones by *Sodium Hypochlorite* (eq 7),[6] halohydrin ring closure (eq 8),[6] and cyanide addition to α-halo ketones (eq 9).[6] Although the relative stereochemistry of most of the products has not been determined, enantiomerically enriched materials have been isolated. A more recent example has been published in which optically active 2,3-epoxycyclohexanone has been synthesized by oxidation with *t-Butyl Hydroperoxide* in the presence of QUIBEC and the absolute stereochemistry of the product established (eq 10).[7]

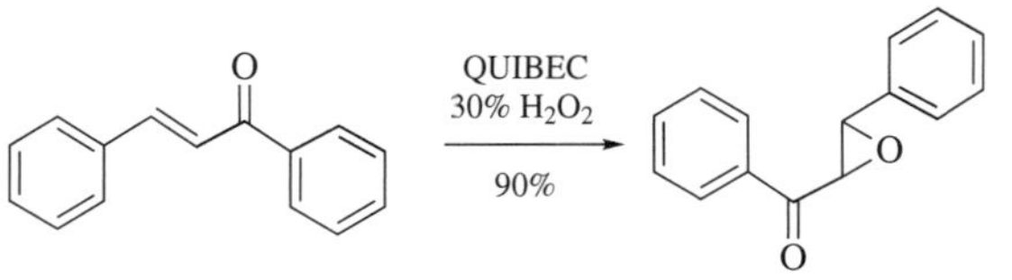

(5)

(optical purity not determined)

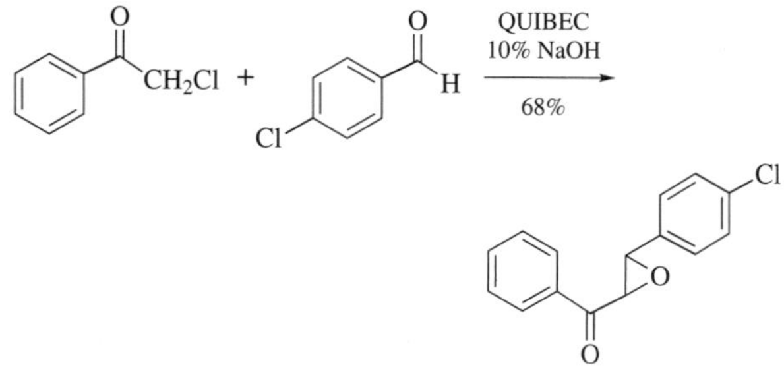

(6)

8% ee

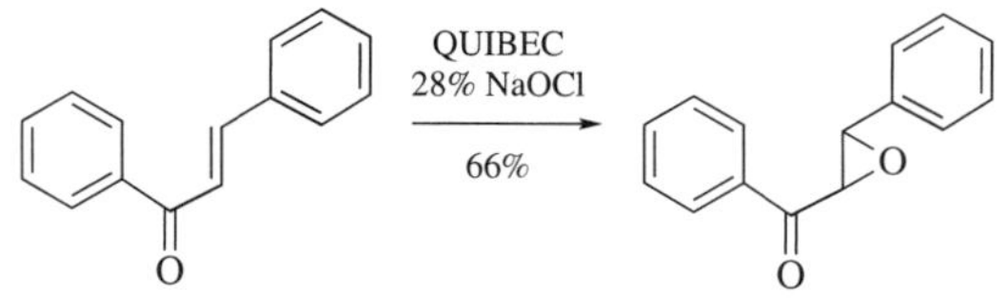

(7)

25–30% ee

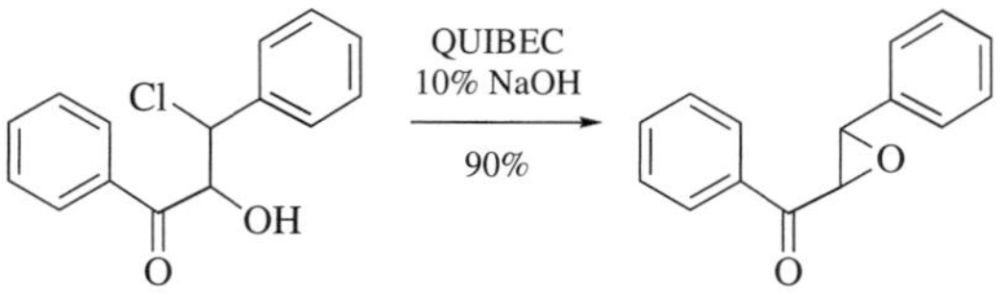

(8)

6% ee

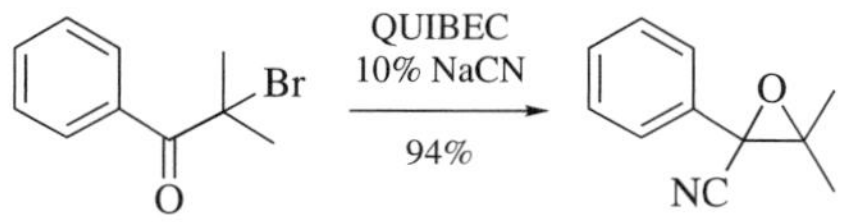

(9)

(optical purity not determined)

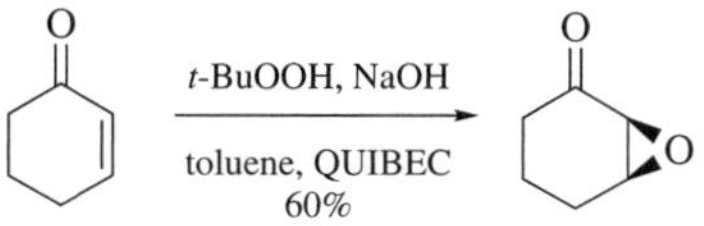

(10)

20–23% ee

Asymmetric Michael Reactions. Asymmetric induction has been observed in Michael-type addition reactions that are catalyzed by chiral amines.[8] The *N*-benzyl fluoride salt of quinine has been particularly successful since the fluoride ion serves as a base and the aminium ion as a source of chirality.[9] Drastic improvements in optical purity (1–23%) have resulted by changing from quinine to the *N*-benzyl fluoride salt (eq 11).[9]

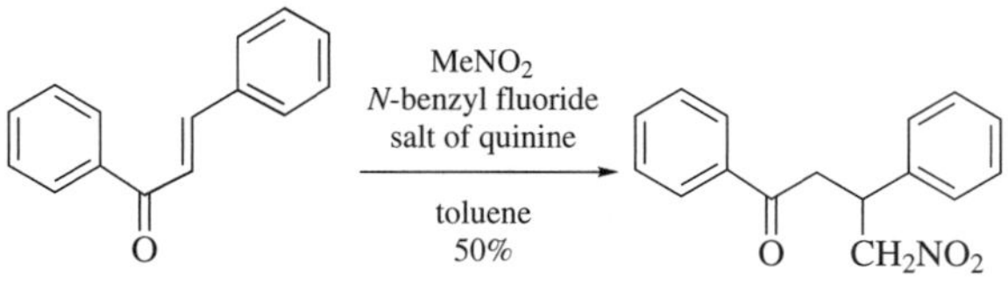

(11)

23% ee

Asymmetric Synthesis of β-Keto Sulfides. Quinine can be used to catalyze asymmetrically the addition of thiols to cyclohexenone, thus forming β-keto sulfides (eq 12).[10] The absolute stereochemistry of the products has not been determined.

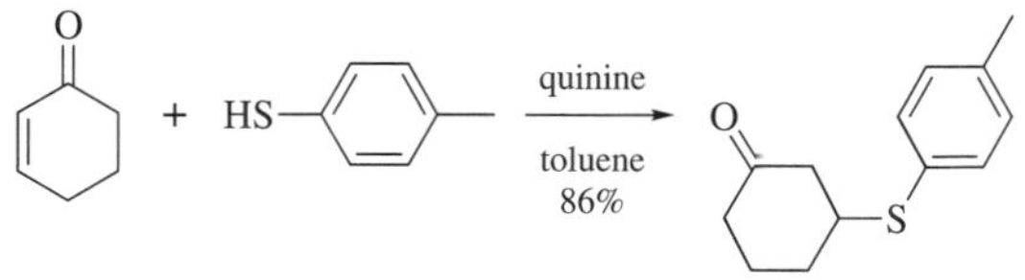

(12)

46% ee

Asymmetric Reduction of Ketones. Alkyl phenyl ketones can be asymmetrically reduced to the corresponding alcohol using *Sodium Borohydride* under phase-transfer conditions in the presence of a catalytic amount of QUIBEC (eq 13).[11] The results indicate that the asymmetric reduction is due to the rigidity of the catalyst as well as the β-position of the hydroxyl group on the quinine molecule. The asymmetric induction is much lower with a γ-hydroxyl group.[11]

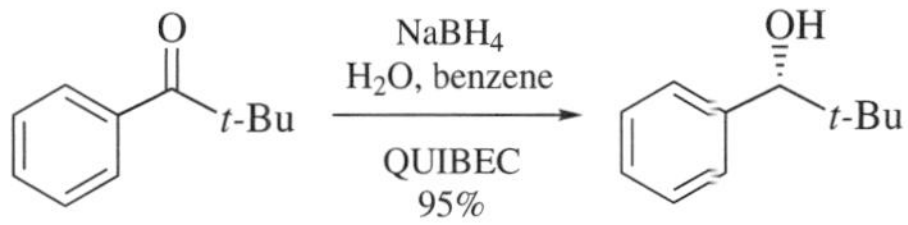

(13)

32% ee

Synthesis of Optically Active β-Hydroxy Esters. Chiral amino alcohols such as quinine have been used in the enantioselective synthesis of β-hydroxy esters via an indium-induced Reformatsky reaction (eq 14).[12] Although the enantioselectivities are not particularly high, aromatic aldehydes have produced the best results to date. The absolute stereochemistry of the products has not yet been assigned.

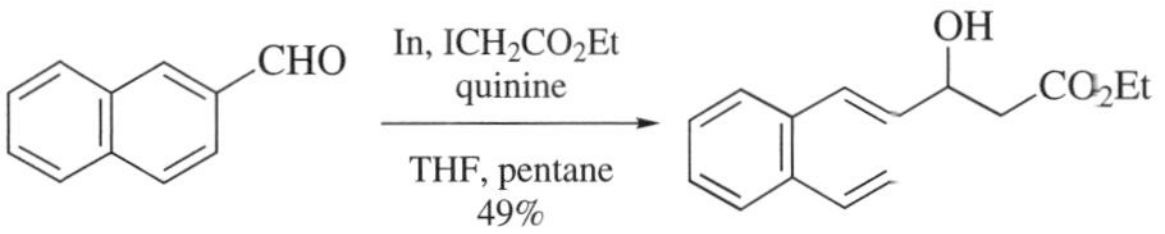

(14)

49% ee

Preparation of Polymeric Catalyst. A quinine/*Acrylonitrile* copolymer has been successfully synthesized via radical polymerization using *Azobisisobutyronitrile* (AIBN) as initiator (eq 15).[13] The polymer can be prepared such that the vinyl group is the connecting site and the amino alcohol portion can either be free or protected. These copolymers are thermally stable and are soluble in polar aprotic solvents such as DMF and DMSO, but insoluble in common organic solvents. Preliminary experiments have shown that these copolymers can be used as asymmetric catalysts.[13]

Asymmetric Addition of Thioglycolic Acid to Nitro Alkenes. Quinine has been used to catalyze the addition of thioglycolic acid to nitro alkenes (eq 16).[14] Enantiomerically enriched materials have been isolated, although the absolute stereochemistry of the products has not been assigned. The direction and extent of asymmetric induction seems to be dependent on the catalyst/acid ratio, thereby pointing to interaction between the carbonyl of the acid and the alkaloid nitrogen as being responsible for the asymmetric induction.[14]

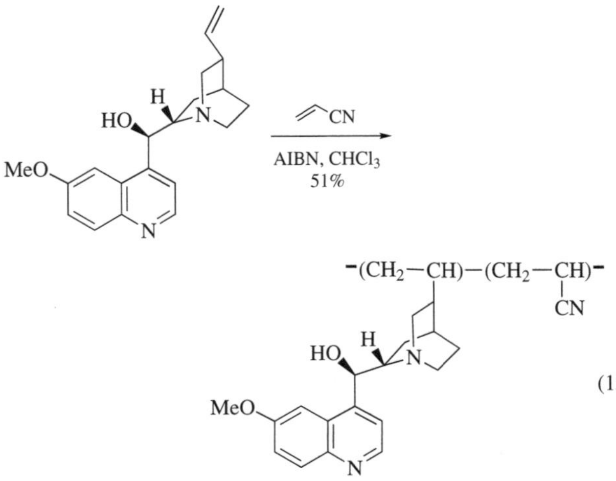

(15)

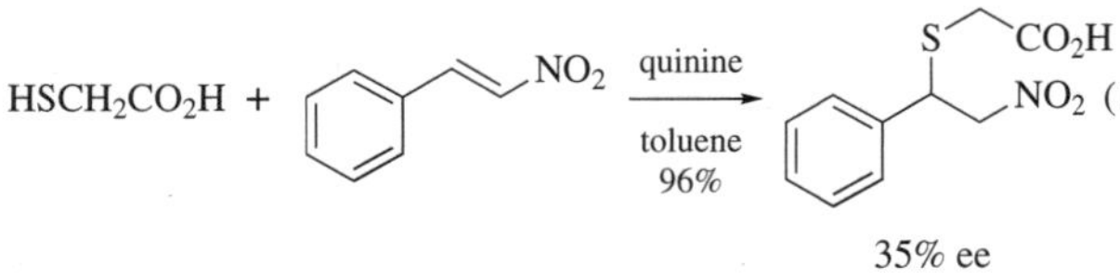

(16)

1. Riant, O.; Kagan, H. B. *Tetrahedron Lett.* **1989**, *30*, 7403.
2. Drabowicz, J.; Legedź, S.; Mikołajczyk, M. *Tetrahedron* **1988**, *44*, 5243.
3. Smaardijk, A. A.; Wynberg, H. *J. Org. Chem.* **1987**, *52*, 135.
4. van Oeveren, A.; Menge, W.; Feringa, B. L. *Tetrahedron Lett.* **1989**, *30*, 6427.
5. Helder, R.; Hummelen, J. C.; Laane, R. W. P. M.; Wiering, J. S.; Wynberg, H. *Tetrahedron Lett.* **1976**, 1831.
6. Hummelen, J. C.; Wynberg, H. *Tetrahedron Lett.* **1978**, 1089.
7. Wynberg, H.; Marsman, B. *J. Org. Chem.* **1980**, *45*, 158.
8. Wynberg, H.; Helder, R. *Tetrahedron Lett.* **1975**, 4057.
9. Colonna, S.; Hiemstra, H.; Wynberg, H. *Chem. Commun./J. Chem. Soc., Chem. Commun.* **1978**, 238.
10. Helder, R.; Arends, R.; Bolt, W.; Hiemstra, H.; Wynberg, H. *Tetrahedron Lett.* **1977**, 2181.
11. Colonna, S.; Fornasier, R. *J. Chem. Soc., Perkin Trans. 1* **1978**, 371.
12. Johar, P. S.; Araki, S.; Butsugan, Y. *J. Chem. Soc., Perkin Trans. 1* **1992**, 711.
13. Kobayashi, N.; Iwai, K. *J. Am. Chem. Soc.* **1978**, *100*, 7071.
14. Kobayashi, N.; Iwai, K. *J. Org. Chem.* **1981**, *46*, 1823.

Ellen M. Leahy
Affymax Research Institute, Palo Alto, CA, USA

S

Sodium Hypochlorite–*N,N′*-Bis(3,5-di-*t*-butylsalicylidene)-1,2-cyclohexanediaminomanganese(III) Chloride[1]

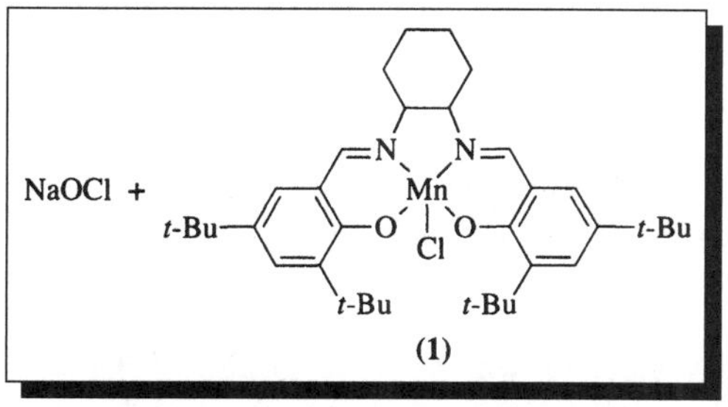

(NaOCl)		
[7681-52-9]	ClNaO	(MW 74.44)
(*R,R*)-(**1**)		
[138124-32-0]	$C_{36}H_{52}ClMnN_2O_2$	(MW 635.29)
(*S,S*)-(**1**)		
[135620-04-1]		

(catalytic system for enantioselective epoxidation of unfunctionalized alkenes[2])

Physical Data: NaOCl: see *Sodium Hypochlorite*. (**1**): mp 330–332 °C.

Solubility: (**1**) freely sol CH_2Cl_2, *t*-butyl methyl ether, acetonitrile, ethyl acetate.

Form Supplied in: both enantiomers of (**1**) are commercially available as brown powders, 98% synthetic preparations may contain up to one solvated molecule of ethanol or DMF for each manganese.

Analysis of Reagent Purity: (salen)Mn^{III} complexes are paramagnetic and do not provide readily interpretable NMR data. (**1**): $R_f = 0.63$ (SiO_2, ethanol); purity may be established by elemental analysis.

Preparative Methods: over 100 chiral manganese(III) salen complexes have been reported;[1] the general procedure for their preparation involves condensation of a 1,2-diamine with 2 equiv of a salicylaldehyde derivative, followed by addition of $Mn(OAc)_2$ in the presence of air.[3] Yields of (salen)Mn^{III} complexes usually exceed 90%.

Purification: (**1**) can be recrystallized from toluene or CH_2Cl_2/heptane. The solvent content of (**1**) and related complexes does not influence their effectiveness as epoxidation catalysts,[4] but heating to >80 °C for 3 h under vacuum results in liberation of solvated molecules.

Handling, Storage, and Precautions: (**1**) is sensitive to acid, but indefinitely stable to air, moisture, and light.

Epoxidation Method. Epoxidation of a variety of conjugated and nonconjugated alkenes may be effected in a biphasic reaction system consisting of aqueous bleach at pH > 9.5 and an organic phase bearing catalytic levels of a soluble manganese(III) complex.[4,5] The ideal pH range appears to be 10.5–11.5 for most applications, with nonwater-miscible solvents such as CH_2Cl_2, *t*-butyl methyl ether, or ethyl acetate as the organic solvent. At pH ≤ 11.5, no phase transfer catalysts are necessary for epoxidation to occur, due to the presence of significant equilibrium concentrations of HOCl.[5] At low pH, equilibrium levels of Cl_2 can produce chlorinated byproducts. Reactions with alkenes are carried out in air, without the need for precautions to exclude moisture or trace impurities. The substrate and catalyst are dissolved in the organic solvent and combined with the bleach solution at 0 °C or room temperature. Catalyst turnover numbers and product yields may be improved in the epoxidation of certain substrates by the addition of substoichiometric levels of a pyridine *N*-oxide derivative (Table 1).[3a,6] Isolation of the epoxide is accomplished by separation of the organic phase and purification by distillation, crystallization, or chromatography.

Table 1 Asymmetric Epoxidation of Representative Alkenes by Catalyst (**1**)

Alkene	*N*-Oxide[a] (mol %)	(**1**) (mol %)	Yield (%)	ee of major epoxide (%)
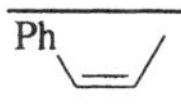	0	4	84[b]	92
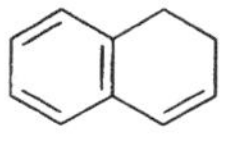	20	4	67	86
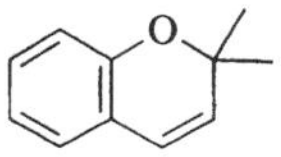	0	2	87	98
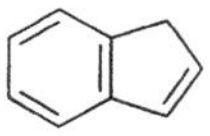	0	1	80	88
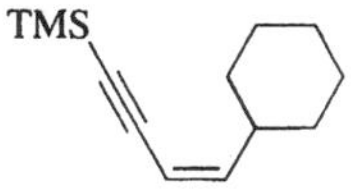	0	4	65[c]	98
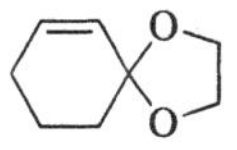	20	15	63	94
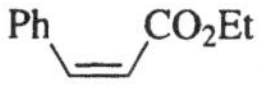	20	8	67[d]	97

[a]*N*-Oxide employed is 4-phenylpyridine *N*-oxide. [b]Isolated yield of epoxide mixture (*cis*:*trans* = 11.5:1). [c]Isolated yield of epoxide mixture (*cis*:*trans* = 1:5.2). [d]Isolated yield of epoxide mixture (*cis*:*trans* = 5:1).

Substrate Scope. Best results in the (salen)Mn^{III}-catalyzed epoxidation reaction have been obtained with *cis*-disubstituted, conjugated alkenes (Table 1). Epoxidation of 2,2-dimethylchromene derivatives occurs with especially high selectivity (>97% ee).[7] *trans*-Disubstituted alkenes are epoxidized with low selectivity (20–50% ee), as are simple alkyl-substituted alkenes.

Mechanistic Considerations. A stepwise mechanism involving a nonpolar intermediate has been proposed for the oxygen atom transfer event in (salen)Mn^{III}-catalyzed epoxidations (eq 1).[6a,8] Consistent with this proposal, acyclic *cis*-alkenes afford mixtures of *cis*- and *trans*-epoxides, and conjugated alkenes are 1–2 orders of magnitude times more reactive than isolated alkenes.[1] In the case of dienes and enynes, the *trans*-epoxide can in fact constitute the major product.[7]

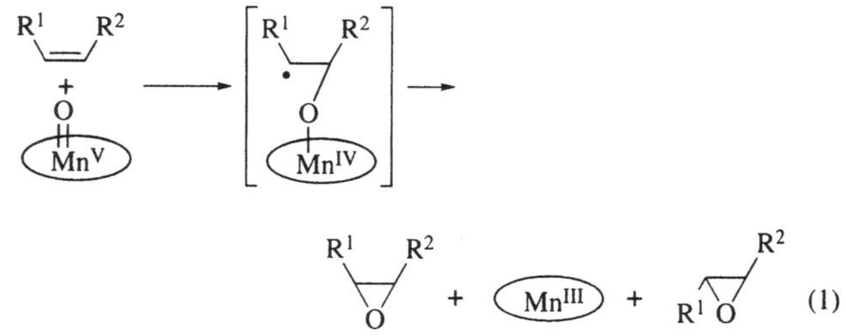

In the case of 1,2-disubstituted alkenes, the nonstereospecificity of the epoxidation reaction results in formation of diastereomeric epoxides. In contrast, for terminal alkenes the *trans* pathway results in partitioning to enantiomers. Thus, diminished enantioselectivity observed in the epoxidation of terminal alkenes such as styrene (50–70% ee) relative to sterically similar *cis*-disubstituted alkenes can be attributed to enantiomeric leakage due to the *trans* pathway. Suppression of this pathway has not been accomplished successfully, and synthetically useful enantioselectivities with terminal alkenes have not yet been achieved using the chiral (salen)Mn^{III} systems.

1. Jacobsen, E. N. in *Catalytic Asymmetric Synthesis*; Ojima, I., Ed.; VCH: New York, 1993; in press.
2. (a) Zhang, W.; Loebach, J. L.; Wilson, S. R.; Jacobsen, E. N. *J. Am. Chem. Soc.* **1990**, *112*, 2801. (b) Jacobsen, E. N.; Zhang, W.; Muci, A. R.; Ecker, J. R.; Deng, L. *J. Am. Chem. Soc.* **1991**, *113*, 7063.
3. (a) Boucher, L. *J. Inorg. Nucl. Chem.* **1974**, *36*, 531. (b) Deng, L.; Jacobsen, E. N. *J. Org. Chem.* **1992**, *57*, 4320.
4. Zhang, W.; Jacobsen, E. N. *J. Org. Chem.* **1991**, *56*, 2296.
5. Banfi, S.; Montanari, F.; Quici, S. *J. Org. Chem.* **1989**, *54*, 1850.
6. (a) Samsel, E. G.; Srinivasan, K.; Kochi, J. K. *J. Am. Chem. Soc.* **1985**, *107*, 7606. (b) Irie, R.; Ito, Y.; Katsuki, T. *Synlett* **1991**, 265.
7. Lee, N. H.; Muci, A. R.; Jacobsen, E. N. *Tetrahedron Lett.* **1991**, *32*, 5055.
8. Lee, N. H.; Jacobsen, E. N. *Tetrahedron Lett.* **1991**, *32*, 6533.

Eric N. Jacobsen
Harvard University, Cambridge, MA, USA

(−)-Sparteine[1]

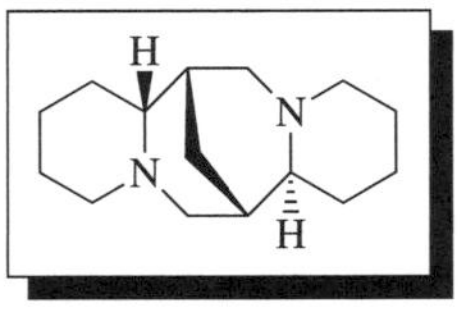

[90-39-1]	$C_{15}H_{26}N_2$	(MW 234.43)
(sulfate pentahydrate) *[6160-12-9]*	$C_{15}H_{38}N_2O_9S$	(MW 422.62)

(reagent for chiral modification of organo-lithium, -magnesium, and -zinc reagents[1,2])

Alternate Name: [(7*S*)-(7α,7aα,14α,14aβ)]-dodecahydro-7,14-methano-2*H*,6*H*-dipyrido[1,2-*a*:1′,2′-*e*][1,5]diazocine.
Physical Data: bp 137–138 °C/1 mm Hg; *d* 1.02 g cm^{-3}; $[\alpha]_D^{20}$ −17.5° (*c* = 2, EtOH). X-Ray structures of several complexes of metal salts,[3] alkyllithium derivatives,[4] and of allylpalladium[5] and studies on the conformation in solution[6] and a NMR study on the structure of the 2-propyllithium–ether–(−)-sparteine complex[7] have been reported.
Solubility: 0.3 g/100 ml H_2O at 20 °C; sol ether, hexane.
Form Supplied in: free base: colorless viscous fluid.
Handling, Storage, and Precautions: highly toxic in the digestive tract. Keep in refrigerator at 0 °C. Moderately hygroscopic; dehydration by drying an ethereal solution over *Calcium Hydride*. Is easily recovered by extraction of alkaline aqueous solutions. Use in a fume hood.

Chiral Modification of Achiral Organometallic Reagents. The addition of *n-Butyllithium* or *Ethylmagnesium Bromide* to aldehydes or ketones in the presence of (−)-sparteine resulted in the formation of optically active secondary or tertiary alcohols with 20% ee or lower.[8] Optically active acyl sulfoxides (≤15% ee) were obtained by acylation of *p-Tolylsulfinylmethyllithium*.[9] The asymmetric Reformatsky reaction of ethyl bromoacetate with benzaldehyde proceeds with 95% ee,[10] in an exceptional case (eq 1).[11]

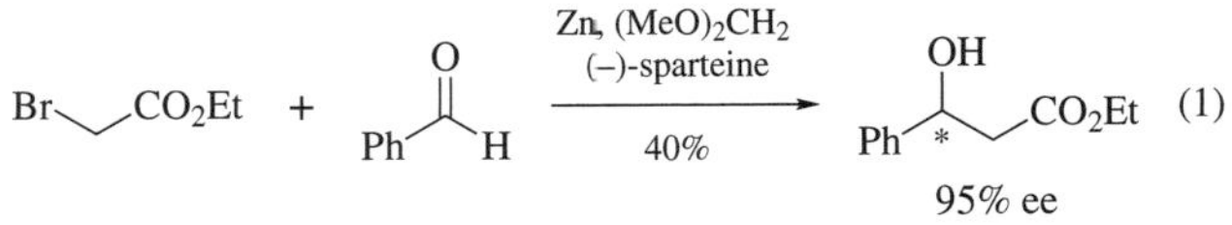

Equilibration of Configurationally Labile Organolithium Reagents. The equilibration of diastereomeric pairs of alkyllithium–(−)-sparteine complexes and trapping by achiral electrophiles gives enantioenriched products. Examples are α-(*N,N*-diisopropylcarbamoyloxy)benzyllithium in ether,[12] not in THF,[13] 1-phenylethyllithium,[8a] and the dilithium salt of *N*-methyl-3-phenylpropanoic acid amide (eq 2).[14]

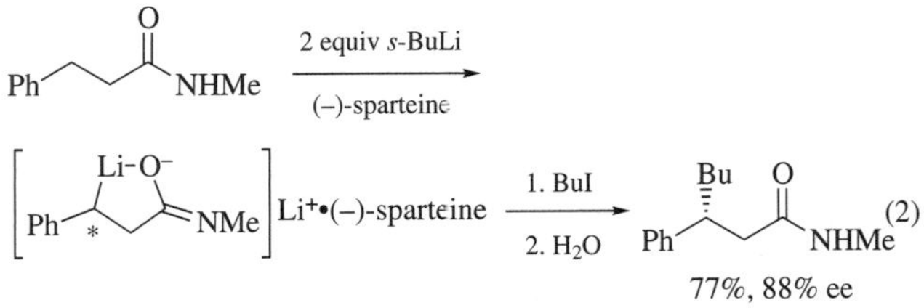

The deprotonation[15] of (*E*)-2-butenyl *N,N*-diisopropylcarbamate leads to (1*S*,2*E*)-1-(*N,N*-diisopropylcarbamoyloxy)-2-butenyllithium–(−)-sparteine[16] with ≥90% de after crystallization, combined with a second-order asymmetric transformation (eq 3).[4d] It has been applied in the enantioselective synthesis of γ-lactones,[16] such as (+)-eldanolide (eq 3),[17] dihydroavermectin B_{1b},[18] and doubly branched sugar analogs.[19]

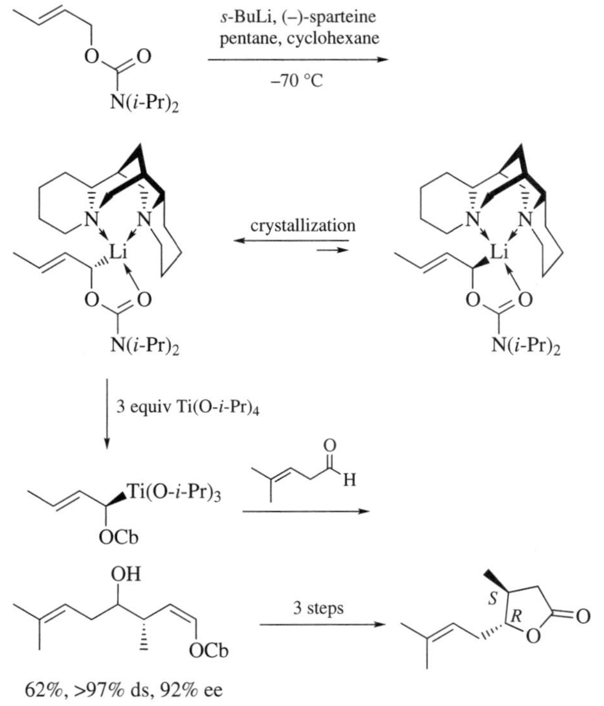

(3)

Generation of Enantioenriched, Configurationally Stable Organolithium Reagents.[15,20] (1*S*,2*E*)-1-(*N*,*N*-Diisopropylcarbamoyloxy)-1-methyl-2-butenyllithium–(−)-sparteine is configurationally stable in solution and is obtained by kinetic resolution of the racemic 2-alkenyl carbamate by *n*-butyllithium–(−)-sparteine with ≥80% de (eq 4).[21] The enantioenriched allylstannane, obtained on γ-stannylation, was used as chiral homoenolate reagent.[21a] The methoxycarbonylation (α, inversion) yields enantioenriched 3-alkenoates.[21b]

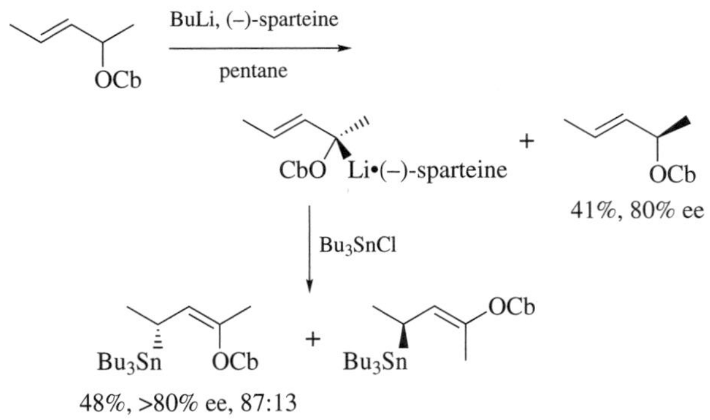

(4)

Alkyl carbamates, derived from 2,2,4,4-tetramethyl-1,3-oxazolidine (R–CH_2–OCby), are deprotonated by *s-Butyllithium*–(−)-sparteine with differentiation between the enantiotopic protons (eq 5).[22,20] The pro-*S* proton is removed with high stereoselectivity and reliability, and, subsequently, stereospecifically substituted by electrophiles with stereoretention to give enantiomerically enriched secondary alcohols (≥95% ee) after deprotection.[22b]

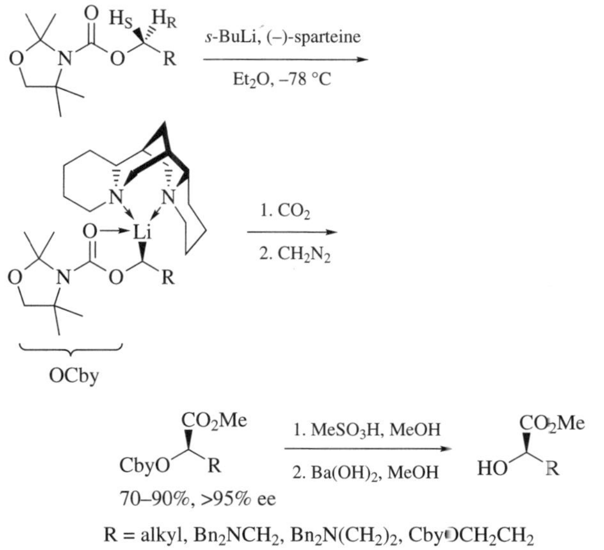

(5)

The ee values in the enantioselective deprotonation are independent of the size of the attached alkyl residue. The method tolerates several substituents, e.g. 2-[23] or 3-dibenzylamino,[24] 3- or 4-(*N*,*N*-dialkylcarbamoyloxy),[25] or 4-TBDMSO.[25a] Essentially enantiopure 2-hydroxy acids,[22a] β-amino alkanols,[24] γ-amino alkanols,[23] cyclopropyl carbamates,[25a] and 2-hydroxy-4-butanolides[25a] were obtained. Extraordinary high (>70) kinetic H/D isotope effects were observed in the deprotonation of chiral 1-deuteroalkyl carbamates.[26] Kinetic resolution of racemic alkyl carbamates was achieved.[27]

N-Boc-pyrrolidines are similarly deprotonated and furnish enantioenriched 2-substituted pyrrolidines (eq 6).[28]

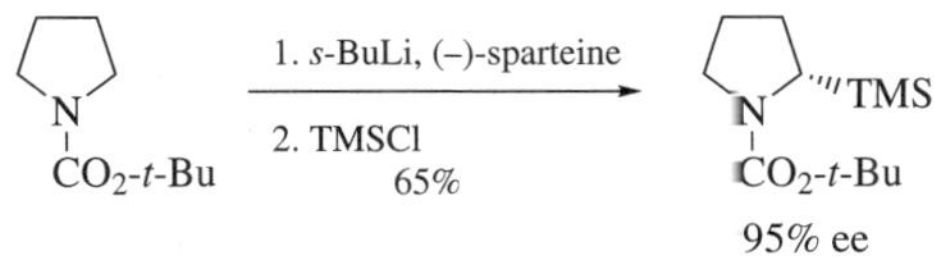

(6)

Further Applications. Chiral 1,1-diaryl-2-propynols are resolved by mutual crystallization with (−)-sparteine.[29] Low ee values were achieved in Pd-mediated alkylations.[30] Numerous attempts at enantioselective, alkyllithium-catalyzed polymerizations of alkenes in the presence of (−)-sparteine have been reported.[31]

Related Reagents. (+)-Sparteine (pachycarpine[32]) is best prepared by resolution of (±)-sparteine, obtained from *rac*-lupanine[33] or by total synthesis[34] with (−)-10-camphorsulfonic acid.[35].

1. Boczon, W. *Heterocycles* **1992**, *33*, 1101.
2. Review: Tomioka, K. *Synthesis* **1990**, 541.
3. For leading references see: Review: Kuroda, R.; Mason, S. F. *J. Chem. Soc., Dalton Trans.* **1977**, 371.
4. (a) Engelhardt, L. M.; Leung, W.-P.; Raston, C. L.; Salem, G.; Twiss, P.; White, A. H. *J. Chem. Soc., Dalton Trans.* **1988**, 2403. (b) Byrne, L. T.; Engelhardt, L. M.; Jacobsen, G. E.; Leung, W.-P.; Papasergio, R. I.; Raston, C. L.; Skelton, B. W.; Twiss, P.; White, A. H. *J. Chem. Soc., Dalton Trans.* **1989**, 105. (c) Marsch, M.; Harms, K.; Zschage, O.; Hoppe, D.; Boche, G. *Angew. Chem.* **1991**, *103*, 338; *Angew. Chem., Int. Ed. Engl.* **1991**, *30*, 321. (d) Ledig, B.; Marsch, M.; Harms, K.; Boche, G. *Angew. Chem.* **1992**, *104*, 80; *Angew. Chem., Int. Ed. Engl.* **1992**, *31*, 79.

5. Togni, A.; Rihs, G.; Pregosin, P. S.; Ammann, C. *Helv. Chim. Acta* **1990**, *73*, 723.
6. (a) Bohlmann, F.; Schumann, D.; Arndt, C. *Tetrahedron Lett.* **1965**, 2705. (b) Wiewiorowski, M.; Edwards, O. E.; Bratek-Wiewiorowska, M. D. *Can. J. Chem.* **1967**, *45*, 1447.
7. Gallagher, D. J.; Kerrick, S. T.; Beak, P. *J. Am. Chem. Soc.* **1992**, *114*, 5872.
8. (a) Nozaki, H.; Aratani, T.; Noyori, R. *Tetrahedron* **1971**, *27*, 905. (b) Nozaki, H.; Aratani, T.; Toraya, T. *Tetrahedron Lett.* **1968**, 4097. (c) Aratani, T.; Gonda, T.; Nozaki, H. *Tetrahedron* **1970**, *26*, 5453.
9. Kunieda, N.; Kinoshita, M. *Phosphorous Sulfur/Phosphorous Sulfur Silicon* **1981**, *10*, 383.
10. Guetté, M.; Capillon, J.; Guetté, J.-P. *Tetrahedron* **1973**, *29*, 3659.
11. Hansen, M. M.; Bartlett, P. A.; Heathcock, C. H. *Organometallics* **1987**, *6*, 2069.
12. Hoppe, D.; Retzow, S., unpublished.
13. Zhang, P.; Gawley, R. E. *J. Org. Chem.* **1993**, *58*, 3223.
14. Beak, P.; Du, H. *J. Am. Chem. Soc.* **1993**, *115*, 2516.
15. Reviews: Hoppe, D.; Krämer, T.; Schwark, J.-R.; Zschage, O. *Pure Appl. Chem.* **1990**, *62*, 1999. (b) Kunz, H.; Waldmann, H. *Chemtracts Org. Chem.* **1990**, *3*, 421.
16. (a) Zschage, O.; Hoppe, D. *Tetrahedron* **1992**, *48*, 5657. (b) Hoppe, D.; Zschage, O. *Angew. Chem.* **1989**, *101*, 67; *Angew. Chem., Int. Ed. Engl.* **1989**, *28*, 69.
17. Paulsen, H.; Hoppe, D. *Tetrahedron* **1992**, *48*, 5667.
18. Férézou, J. P.; Julia, M.; Khourzom, R.; Pancrazi, A.; Robert, P. *Synlett* **1991**, 611.
19. Peschke, B.; Lüssmann, J.; Dyrbusch, M.; Hoppe, D. *Ber. Dtsch. Chem. Ges./Chem. Ber.* **1992**, *125*, 1421.
20. Review: Knochel, P. *Angew. Chem.* **1992**, *104*, 1486; *Angew. Chem., Int. Ed. Engl.* **1992**, *31*, 1459.
21. (a) Zschage, O.; Schwark, J.-R.; Krämer, T.; Hoppe, D. *Tetrahedron* **1992**, *48*, 8377. (b) Zschage, O.; Hoppe, D. *Tetrahedron* **1992**, *48*, 8389. (c) Zschage, O.; Schwark, J.-R.; Hoppe, D. *Angew. Chem.* **1990**, *102*, 336; *Angew. Chem., Int. Ed. Engl.* **1990**, 29, 296.
22. (a) Hoppe, D.; Hintze, F.; Tebben, P. *Angew. Chem.* **1990**, *102*, 1457; *Angew. Chem., Int. Ed. Engl.* **1990**, *29*, 1422. (b) Hintze, F.; Hoppe, D. *Synthesis* **1992**, 1216.
23. Schwerdtfeger, J.; Hoppe, D. *Angew. Chem.* **1992**, *104*, 1547; *Angew. Chem., Int. Ed. Engl.* **1992**, *31*, 1505.
24. Sommerfeld, P.; Hoppe, D. *Synlett* **1992**, 764.
25. (a) Paetow, M.; Ahrens, H.; Hoppe, D. *Tetrahedron Lett.* **1992**, *33*, 5323. (b) Ahrens, H.; Paetow, M.; Hoppe, D. *Tetrahedron Lett.* **1992**, *33*, 5327.
26. Hoppe, D.; Paetow, M.; Hintze, F. *Angew. Chem.* **1993**, *105*, 430; *Angew. Chem., Int. Ed. Engl.* **1993**, *32*, 394.
27. Haller, J.; Hense, T.; Hoppe, D. *Synlett* **1993**, 726.
28. Kerrick, S. T.; Beak, P. *J. Am. Chem. Soc.* **1991**, *113*, 9708.
29. (a) Toda, F.; Tanaka, K.; Ueda, H.; Oshima, T. *Chem. Commun./J. Chem. Soc., Chem. Commun.* **1983**, 743. (b) Toda, F.; Tanaka, K.; Ueda, H.; Oshima, T. *Isr. J. Chem.* **1985**, *25*, 338.
30. Trost, B. M.; Dietsche, T. J. *J. Am. Chem. Soc.* **1973**, *95*, 8200.
31. For leading references see: Nakano, T.; Okamoto, Y.; Hatada, K. *J. Am. Chem. Soc.* **1992**, *114*, 1318.
32. Orechoff, A.; Rabinowitch, M.; Konowalowa, R. *Ber. Dtsch. Chem. Ges. Chem. Ber.* **1933**, *66*, 621.
33. Clemo, G. R.; Raper, R.; Short, W. S. *J. Chem. Soc.* **1949**, 663.
34. van Tamelen, E. E.; Foltz, R. L. *J. Am. Chem. Soc.* **1960**, *82*, 1960.
35. Ebner, T.; Eichelbaum, M.; Fischer, P.; Meese, C. O. *Arch. Pharm. (Weinheim, Ger.)* **1989**, *322*, 399.

Dieter Hoppe
University of Münster, Germany

(1*R*,5*R*,6*R*)-Spiro[4.4]nonane-1,6-diyl Diphenylphosphinous acid Ester (*R*-SpirOP)

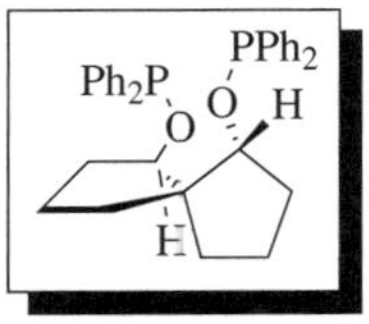

[197159-86-7] $C_{33}H_{34}O_2P_2$ (MW 524.578)

(spirocyclic phosphinite ligand used as a rhodium catalyst in the asymmetric hydrogenation of prochiral olefins)

Physical Data: mp 96–96.5 °C; $[\alpha]_D$ −43.2 (*c* 0.104, $CHCl_3$).
Solubility: soluble in alcohol, ether, and most organic solvents.
Analysis of Reagent Purity: ^{1}H NMR (400 MHz, $CDCl_3$) δ 1.33 (m, 2H), 1.65–1.82 (m, 10H), 4.53 (d, $J = 5.3$ Hz, 2H), 7.10–7.52 (m, 20H). ^{13}C NMR (100 MHz, $CDCl_3$) δ 20.9, 32.5 (d, $^3J_{P-C} =$ 7.9 Hz), 33.1, 63.2, 87.3 (d, $^2J_{P-C} = 18.9$ Hz), 128.5 (d, $^3J_{P-C} =$ 6.9 Hz), 128.7, 128.8 (d, $^3J_{P-C} = 8.8$ Hz), 129.6, 130.0 (d, $^2J_{P-C} = 21.8$ Hz), 130.9 (d, $^2J_{P-C} = 23.4$ Hz), 143.4 (d, $^1J_{P-C} =$ 12.0 Hz), 145.2 (d, $^1J_{P-C} = 21.8$ Hz). ^{31}P NMR (160 MHz, $CDCl_3$) δ 102.8. IR (KBr), 3060, 2968, 2907, 2868, 1486, 1440, 1348, 1104, 1006, 940, 742, 703 cm^{-1}. Analytically calculated for $C_{33}H_{34}O_2P_2$: C, 75.57; H, 6.48; P, 11.83. Found: C, 75.23; H, 6.41; P, 11.71. MS: *m/z* 524 (M^+).
Preparative Methods: (−)-(1*R*,5*R*,6*R*)-(*cis,cis*)-spiro[4.4]nonane-1,6-diol[1] (78 mg, 0.50 mmol), 4-*N,N*-dimethylaminopyridine (12.4 mg, 0.10 mmol), and triethylamine (101.9 mg, 1.00 mmol) in THF (3 mL) were charged to a 10 mL Schlenk flask under a nitrogen atmosphere. This flask was cooled in an ice-cooled water bath. A solution of chlorodiphenylphosphine (0.18 mL, 1.0 mmol) in THF (1 mL) was added dropwise to the above solution, the ice-water bath was removed and the mixture was stirred at room temperature for 8 h. The solution was filtered to remove the solid triethylamine hydrochloride. THF was removed in vacuo and the residue was dissolved in approximately 10 mL of anhydrous diethyl ether with heating. After cooling in a refrigerator, white needle shaped crystals were obtained (218 mg, 83%).
Handling, Storage, and Precautions: *R*-SpirOP is obtained as a white solid that is stable in air but decomposes gradually in alcoholic solution. To help prevent oxidation, storage under a nitrogen atmosphere is recommended. The stability of *R*-SpirOP has been tested in methanol solution through a ^{31}P NMR study. Observable decomposition occurred in 2 h and complete decomposition in 24 h.

Hydrogenation of Amidoacrylic Acids.[2] When a cationic rhodium catalyst containing *R*-SpirOP is used in the asymmetric hydrogenation of 2-acetamidoacrylic acid at ambient temperature and under 1 atm of H_2 in methanol, the desired 2-acetamidopropionic acid is obtained in >99.9% ee. Under similar conditions, the asymmetric hydrogenation of the methyl ester of 2-acetamidoacrylic acid gave 100% conversion to the corresponding hydrogenation product in 99.0% ee (eq 1).

Table 1 The Rh(*R*-SpirOP)$^+$-catalyzed asymmetric hydrogenation of (*Z*)-2-acetamido-3-arylacrylic acids[a]

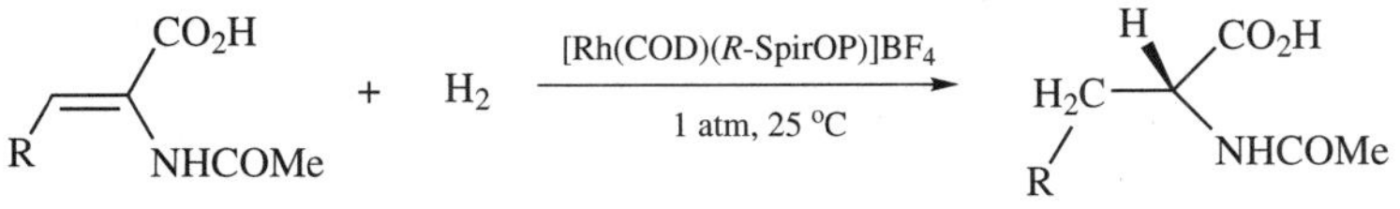

Entry	Substrate (R)	ee (%)[b]
1	Ph	97.9
2	4-Cl-Ph	97.3
3	2-Cl-Ph	97.3
4	3-Cl-Ph	97.4
5	4-NO_2-Ph	97.0

[a]Reaction conditions: 1 atm of H_2, ambient temperature, 10 min reaction time, substrate/catalyst = 100 (M/M), solvent used is methanol, and 100% conversion was observed in all cases.
[b]The ee values were determined by chiral GLC with a Chrompack Chirasil-L-Val column after converting the products to the corresponding methyl esters. The *R* configuration was obtained for all products.

Table 2 The Rh(*R*-SpirOP)$^+$-catalyzed asymmetric hydrogenation of (*Z*)-2-acetamido-3-arylacrylic acid methyl esters[a]

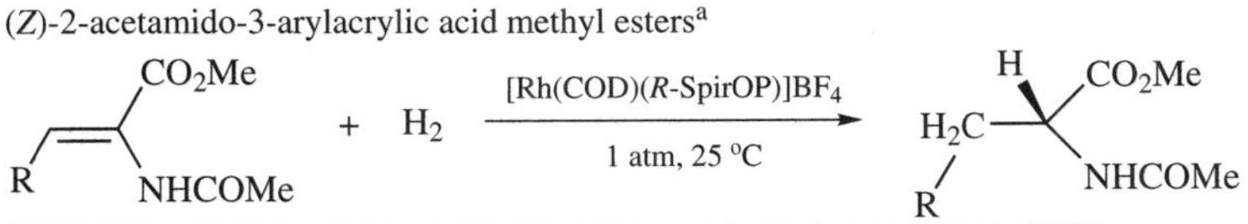

Entry	Substrate (R)	ee (%)[b]
1	Ph	95.7
2	4-Cl-Ph	94.2
3	4-F-Ph	95.5
4	4-Br-Ph	96.3
5	4-MeO-Ph	96.2
6	4-Me-Ph	95.6
7		97.2
8	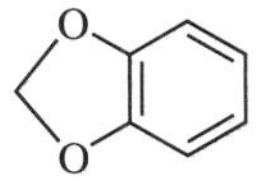	94.9

[a]Reaction conditions: 1 atm of H_2, ambient temperature, 10 min reaction time, substrate/catalyst = 100 (M/M), solvent used is methanol, and 100% conversion was obtained in all cases.
[b]The ee values were determined by GLC with a Chrompack Chirasil-L-Val column. The *R* configuration was obtained for all products.

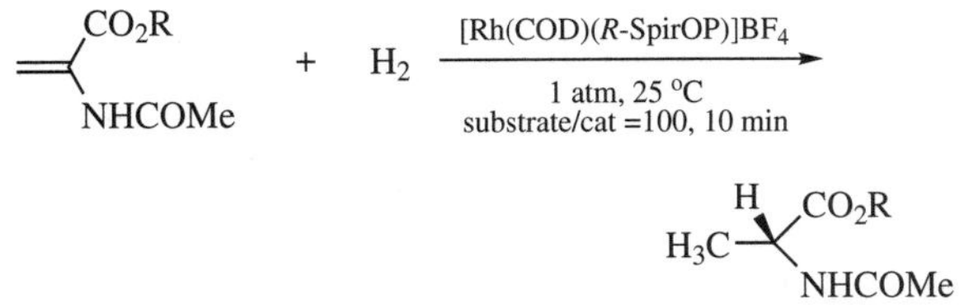

(1)

> 99.9% conversion
R = H, > 99.9% ee
R = Me, 99.0% ee

Further studies for the hydrogenation of other prochiral amidoacrylic acids confirmed that the high enantioselectivity of the catalyst is quite general. Several (*Z*)-2-acetamido-3-arylacrylic acids were hydrogenated with this catalyst and in all cases the desired products were found to have ee values of over 97%. More detailed data are summarized in Table 1.

The enantioselectivities of Rh(*R*-SpirOP)$^+$ in the asymmetric hydrogenation of the methyl esters of (*Z*)-2-acetamido-3-arylacrylic acids were also found to be very high (Table 2).

In addition to the high enantioselectivity, the rate of the hydrogenation using Rh(*R*-SpirOP)$^+$ catalyst is also very fast. When a substrate/catalyst ratio of 10 000 was used and when the reaction was carried out at ambient temperature under 200 psi H_2, >99.9% conversion of 2-acetamidoacrylic acid to 2-acetamidopropionic acid (96.8% ee) was observed in 1 h. The detailed results are shown in Table 3.

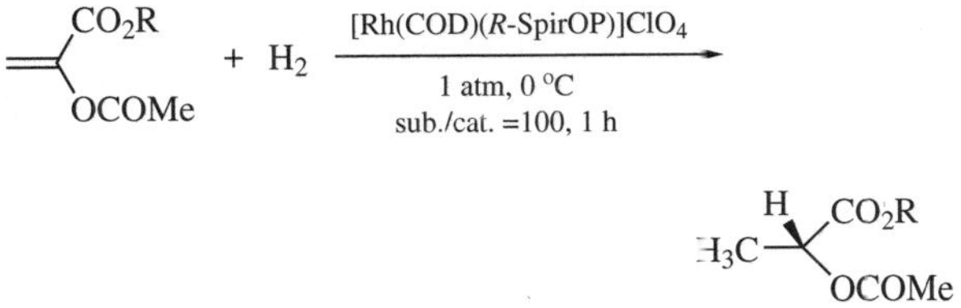

(2)

R = CH_3, 100% conversion, 96.9% ee (*R*)
R = C_2H_5, 68.0% conversion, 97.9% ee (*R*)

Hydrogenation of Enol Esters. In addition to the high activity and excellent enantioselectivity being obtained in the asymmetric

Table 3 The Rh(*R*-SpirOP)$^+$-catalyzed asymmetric hydrogenation of 2-acetamidoacrylic acid with different S/C ratio[a]

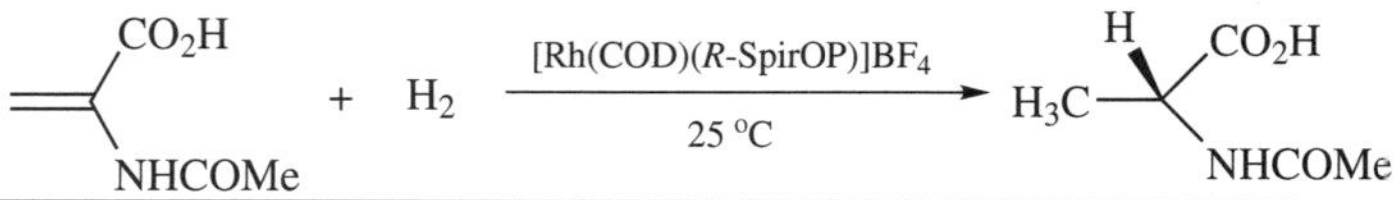

Entry	S/C	Time	H_2 (psi)	ee (%)[b]
1	100	10 min	14	99.9
2	5,000	30 min	100	97.0
3	10,000	1 h	200	96.8
4	25,000	10 h	200	94.0
5	50,000	48 h	200	93.0

[a]Reactions conditions: ambient temperature, solvent used is methanol, and 100% conversion was obtained in all cases.
[b] The ee values were determined by GLC with a Chrompack Chirasil-L-Val column. The *R* configuration was obtained for all products.

Table 4 Asymmetric hydrogenation of α-arylenamides by [Rh(COD)(*R*-SpirOP)]ClO_4[a]

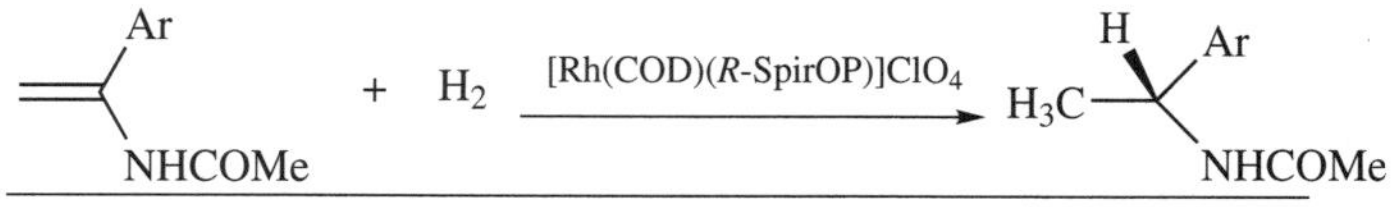

Entry	Substrate (R)	ee (%)[b]
1	Ph	89.0
2	4-Me-Ph	86.5
3	3-Me-Ph	85.6
4	4-Cl-Ph	86.1
5	4-F-Ph	87.9
6	4-CF_3-Ph	90.0
7		97.4

[a]Reaction conditions: 1 atm of H_2, 0 °C, 10 min reaction time, substrate/catalyst = 100 (M/M), solvent = isopropanol, and 100% conversion was observed in all cases.
[b] The ee values were determined by chiral GLC with a Chrompack Chirasil-L-Val column. The *R* configuration was obtained for all products.

hydrogenation of amidoacrylic acid substrates, the cationic complex Rh(*R*-SpirOP)$^+$ afforded excellent ee in the hydrogenation of enol esters (eq 2). The best result was derived from the hydrogenation of enol esters in acetone at 0 °C and under 1 atm of H_2 for 1 h.

Hydrogenation of Enamides.[3] Except for the asymmetric hydrogenation of specific cyclic enamides with Ru(BINAP) catalyst which shows high enantioselectivity,[4] the successful enantioselective hydrogenation of simple α-substituted enamides has been relatively rare.[5] The Ru(R-SpirOP)$^+$ catalyst was found to be effective in the asymmetric hydrogenation of a series of α-substituted enamides (Table 4). The best result (97.4% ee) was achieved in the hydrogenation of N-acetyl-α-(2-furanyl)ethenamine (entry 7 of Table 4). The high enantioselectivity (97.4% ee) and the fast hydrogenation rate (10 min, 100% conversion) compares favorably with the Rh-Me-DuPHOS system (96.1% ee, 15 h, 100% conversion).[6]

Hydrogenation of Itaconic Acid. Compared to the hydrogenation of amidoacylic acids, enols, and enamides, the Rh(*R*-SpirOP)$^+$ catalyzed hydrogenation of itaconic acid was less successful. After optimizing the hydrogenation conditions, 76.8% ee of the corresponding product was obtained in isopropanol at ambient temperature under 100 psi H_2 for 2 h (eq 3).

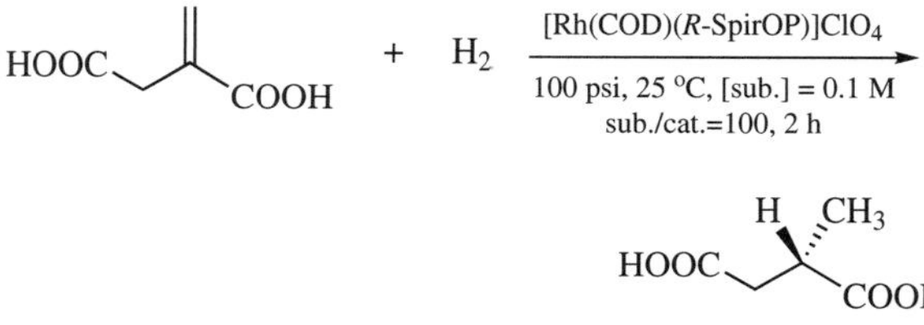

(3)

76.8% ee (*S*)

1. Chan, A. S. C.; Lin, C. C.; Sun, J.; Hu, W.; Li, Z.; Pan, W.; Mi, A.; Jiang, Y.; Huang, T.-M.; Yang, T.-K.; Chen, J.-H.; Wang, Y.; Lee, G.-H. *Tetrahedron: Asymmetry* **1995**, *6*, 2953, and references therein.
2. Chan, A. S. C.; Hu, W.; Pai, C.-C.; Lau, C.-P.; Jiang, Y.; Mi, A.; Yan, M.; Sun, J.; Lou, R.; Deng, J. *J. Am. Chem. Soc.* **1997**, *119*, 9570.
3. Hu, W.; Yan, M.; Lau, C.-P.; Yang, S. M.; Chan, A. S. C. *Tetrahedron Lett.* **1999**, *40*, 973.
4. (a) Noyori, R.; Ohta, M.; Hsiao, Y.; Kitamura, M.; Ohta, T.; Takaya, H. *J. Am. Chem. Soc.* **1986**, *108*, 7117. (b) Kitamura, M.; Hsiao, T.; Noyori, R.; Takaya, H. *Tetrahedron Lett.* **1987**, *28*, 4829. (c) Heiser, B.; Broger, E. A.; Crameri, Y. *Tetrahedron: Asymmetry* **1991**, *2*, 51. (d) Tschaen, D. M.; Abramson, L.; Cai, D.; Desmond, R.; Dolling, U.-H.; Frey, L.; Karady, S.; Shi, Y.-J.; Verhoeven, T. R. *J. Org. Chem.* **1995**, *60*, 4324.
5. Zhang, F.-Y.; Pai, C.-C.; Chan, A. S. C. *J. Am. Chem. Soc.* **1998**, *120*, 5808.
6. Burk, M. J.; Wang, Y. M.; Lee, J. R. *J. Am. Chem. Soc.* **1996**, *118*, 5142.

Wai Him Kwok, Wen Hao Hu & Albert S. C. Chan
The Hong Kong Polytechnic University, Hong Kong

T

(3*S*,*cis*)-Tetrahydro-3-isopropyl-7a-methylpyrrolol[2,1-*b*]oxazol-5(6*H*)-one[1]

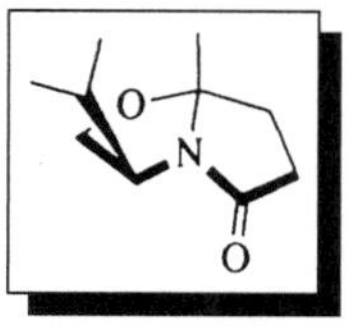

(3*S*,*cis*)
[98203-44-2] $C_{10}H_{17}NO_2$ (MW 183.28)
(3*R*,*cis*)
[123808-97-9]

(chiral template for synthesis of enantiomerically pure cyclopropanes, cyclobutanes, cyclopentenones, pyrrolidines, pyrrolidinones, and α,α-disubstituted γ-keto acids[1])

Physical Data: bp 76–80 °C/0.05 mmHg; $[\alpha]_D$ 95.5°.
Preparative Methods: the richly functionalized chiral bicyclic lactam is easily procured by condensation of commercially available (*S*)-valinol and levulinic acid in 86% yield (eq 1).[2] Similar bicyclic lactams have been prepared from other amino alcohols.[3] These bicyclic lactams have served as precursors to a variety of enantiomerically pure compounds that possess quaternary stereocenters. An extensive review on the utility of chiral, nonracemic bicyclic lactams is available.[1]

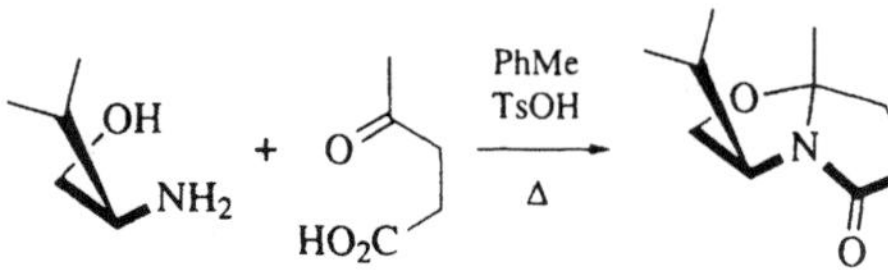

(1)

General Considerations. The title reagent can be sequentially alkylated α to the carbonyl group in a stereocontrolled fashion (eq 2).[1] Lithiation of the parent bicyclic lactam with *s-Butyllithium* and reaction with an alkyl halide affords the monoalkylated product. The epimeric mixture is treated again with *s*-BuLi and a second alkyl halide to give the dialkylated bicyclic lactam. The initial epimeric mixture is used directly in the second alkylation since this step proceeds via a planar enolate. It is the second alkylation that dictates the final diastereomeric ratio. The opposite stereochemistry at C-6 can be obtained by inverting the order of electrophile addition.

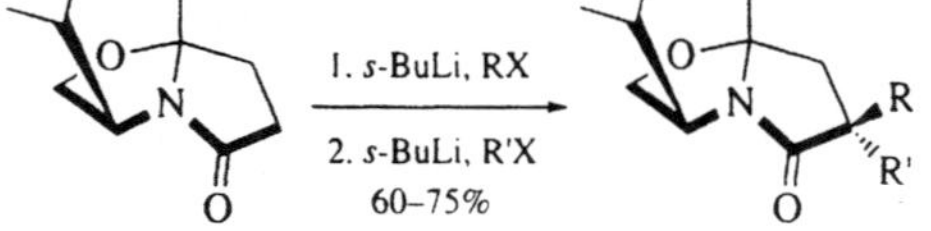

(2)

Reduction and hydrolysis of the bicyclic lactam followed by aldol cyclization affords enantiomerically pure 4,4-dialkyl-2-cyclopentenones (eq 3).[2,4]

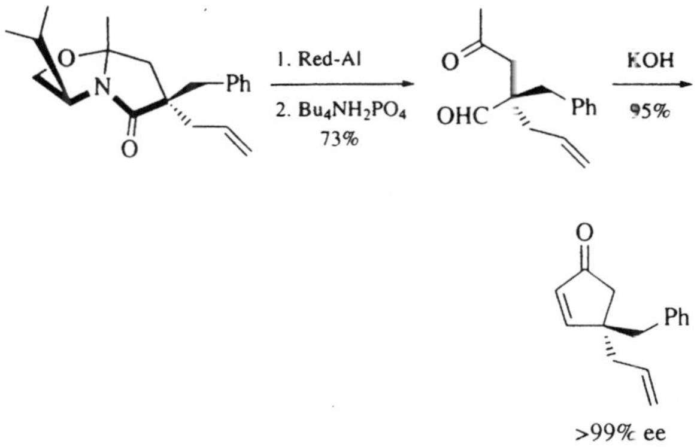

(3)

More highly functionalized cyclopentenones can be accessed by organolithium addition to the carbonyl group instead of hydride reduction with *Sodium Bis(2-methoxyethoxy)aluminum Hydride* (Red-Al).[5]

α-Substituted γ-keto acids, upon condensation with β-amino alcohols, afford bicyclic lactams containing α-substitutents such as aryl groups. In this case, only one metalation–alkylation sequence is required to form the chiral, nonracemic α,α-disubstituted bicyclic lactam (eq 4).[4b]

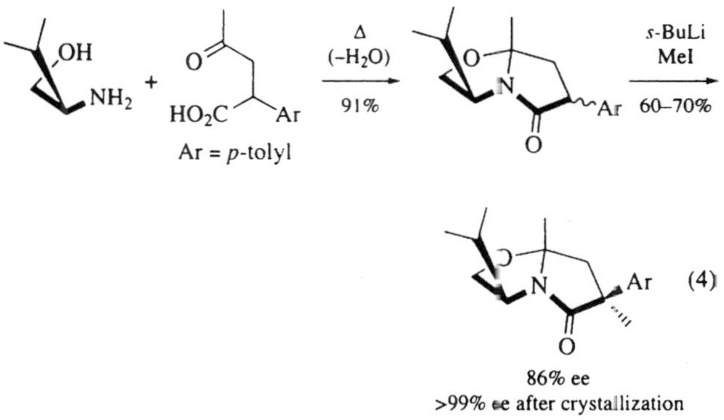

(4)

The angular 7a-phenyl bicyclic lactam can be prepared by the cyclocondensation of 3-benzoylpropionic acid and (*S*)-valinol in 85% yield.[6] Dialkylation of this lactam also affords cleanly the α,α-disubstituted compound. Lactam hydrolysis releases chiral, nonracemic α,α-disubstituted γ-keto carboxylic esters (or acids) (eq 5) and 3,3-disubstituted dihydronaphthalenes may be obtained via cyclization.[6]

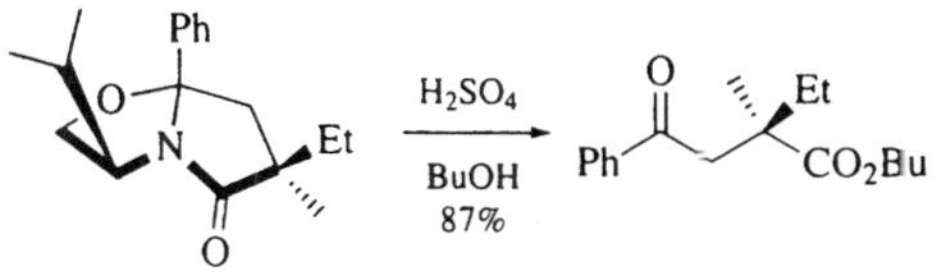

(5)

When the bicyclic lactam is substituted with a 3-hydroxypropyl group in the 6-position, acidic hydrolysis gives a bridged bicyclic acetal lactone (eq 6).[7]

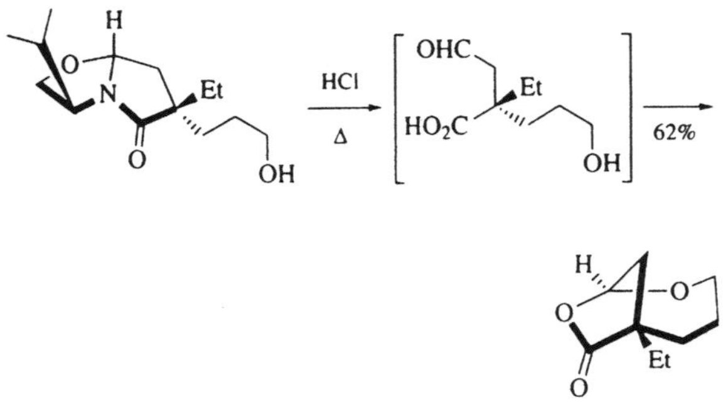
(6)

An α-(4-bromobutyl) group can be used as a latent organolithium species by means of bromine–lithium exchange. Intramolecular addition of the organometallic tether to the carbonyl group, followed by lactam hydrolysis and aldol cyclization, affords enantiomerically pure hydrinden-2-ones (eq 7).[8]

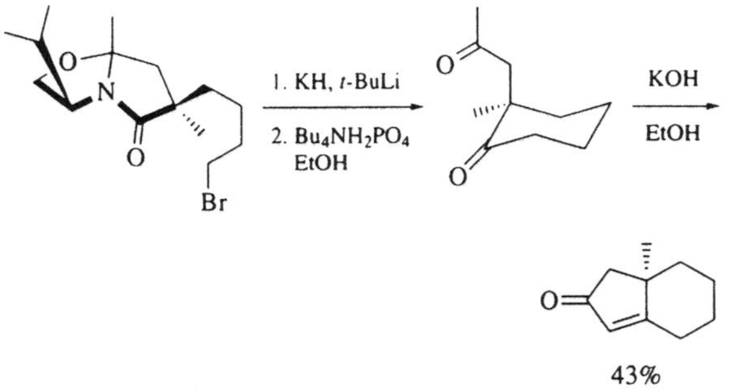
(7)

α,β-Unsaturation may be introduced into the bicyclic lactams by standard α-selenation–oxidation methodology. The lactam can now behave as a chiral enone in photochemical [2 + 2] cycloadditions. The lactam moiety can be easily detached owing to its amide and aminal features; thus chiral, nonracemic cyclobutanes are obtained upon hydrolysis (eq 8).[9]

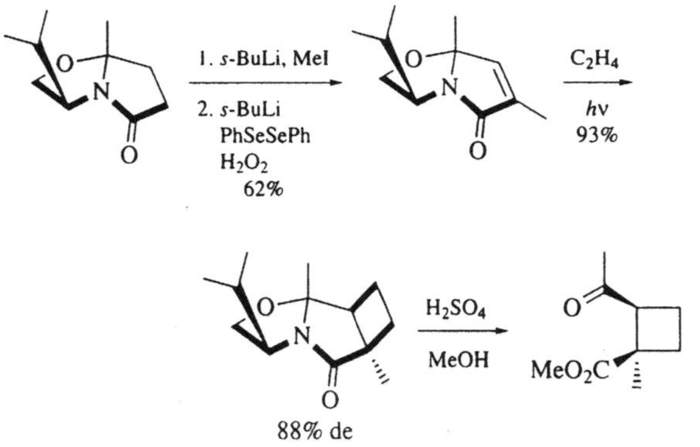
(8)

These unsaturated, bicyclic lactams are also precursors to a variety of chiral nonracemic cyclopropanes. Treatment of the parent α,β-unsaturated lactam with *Dimethylsulfoxonium Methylide* generates the *endo* cyclopropanated adduct (eq 9).[3b,10]

Diels–Alder cycloadditions occur on the *endo* face when the unsaturated bicyclic lactam is treated with 1,3-dienes such as *Isoprene* (eq 10).[11] Ester reduction followed by organolithium addition to the lactam carbonyl group and subsequent hydrolysis affords a variety of enantiopure functionalized cyclohexenes.[11]

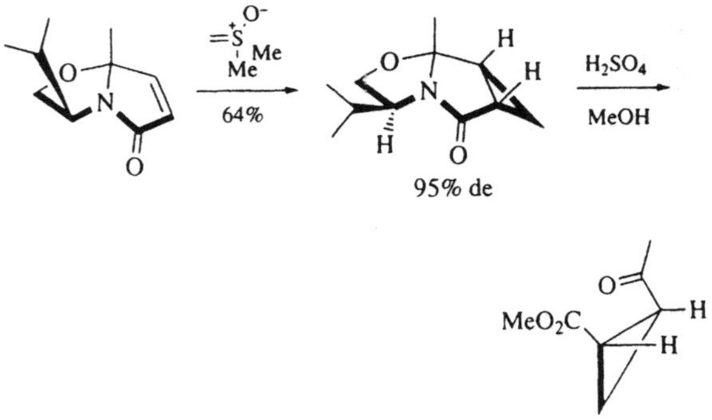
(9)

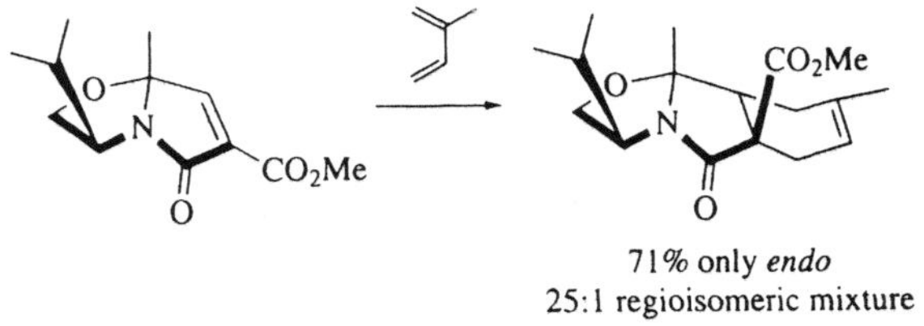
(10)

Cyclopropyl-containing carbocycles can be prepared from the initial [4 + 2] cycloadducts by an *N*-acyliminium ion–enamide rearrangement. The unsaturated bicyclic lactam also undergoes 1,3-dipolar cycloadditions with azomethine ylides.[12] Reduction of the bicyclic lactam with alane followed by hydrogenation affords enantiomerically pure 2-substituted pyrrolidines (eq 11).[13]

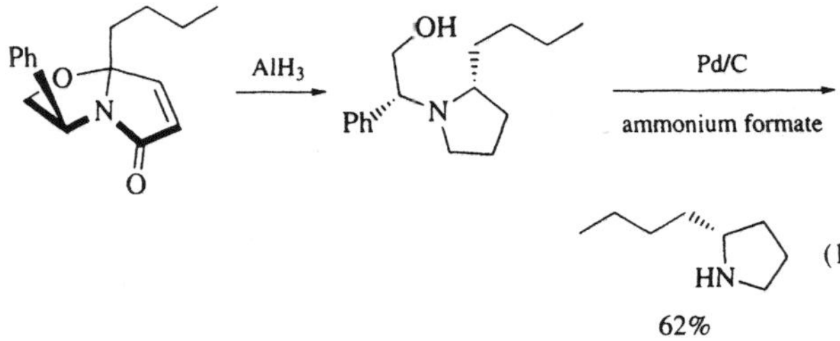
(11)

5,5-Disubstituted pyrrolidinones are formed when the bicyclic lactam is treated with *Allyltrimethylsilane/Titanium(IV) Chloride*. The remaining phenylglycinol moiety is cleaved with Li/NH_3 (see *Lithium Amide*) (eq 12).[14] Further reduction with *Lithium Aluminum Hydride* affords 2,2-disubstituted pyrrolidines.

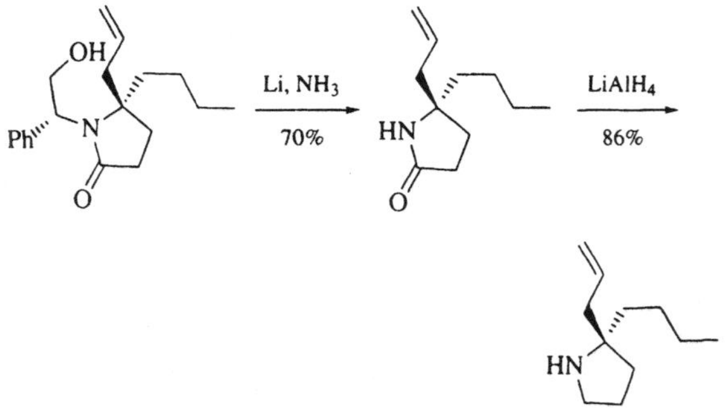
(12)

Reduction with *Triethylsilane* allows for the formation of enantiomerically pure 5-substituted pyrrolidinones and 2-substituted pyrrolidines in the same manner.[15]

Conjugate addition of organocuprates to the unsaturated bicyclic lactams (see above) affords rapid access to chiral, nonracemic 3- and 4-substituted pyrrolidines[16] and *trans*-2,3-disubstituted pyrrolidines.[17]

Related Reagents. (*S*)-1-Amino-2-methoxymethylpyrrolidine; *trans*-2,5-Bis(methoxymethyl)pyrrolidine; 10,2-Camphorsultam; 10-Dicyclohexylsulfonamidoisoborneol; (2*S*)-(2α,3β,8aβ)-Hexahydro-3-(hydroxymethyl)-8a-methyl-2-phenyl-5*H*-oxazolo[3,2-*a*]pyridin-5-one; (4*S*,5*S*)-4-Methoxymethyl-2-methyl-5-phenyl-2-oxazoline; α-Methyltoluene-2,α-sultam; (*R*,*R*)-1,2-Diphenyl-1,2-diaminoethane *N*,*N*′-Bis[3,5-bis(trifluoromethyl)benzenesulfonamide].

1. (a) Romo, D.; Meyers, A. I. *Tetrahedron* **1991**, *47*, 9503. (b) Meyers, A. I.; Berney, D. *Org. Synth., Coll. Vol.* **1993**, *8*, 241.
2. Meyers, A. I.; Wanner, K. T. *Tetrahedron Lett.* **1985**, *26*, 2047.
3. (a) Meyers, A. I.; Lefker, B. A.; Sowin, T. J.; Westrum, L. J. *J. Org. Chem.* **1989**, *54*, 4243. (b) Meyers, A. I.; Romo, D. *Tetrahedron Lett.* **1989**, *30*, 1745.
4. (a) Meyers, A. I.; Lefker, B. A. *J. Org. Chem.* **1986**, *51*, 1541. (b) Meyers, A. I.; Bienz, S. *J. Org. Chem.* **1990**, *55*, 791.
5. (a) Meyers, A. I.; Lefker, B. A. *Tetrahedron* **1987**, *43*, 5663. (b) Meyers, A. I.; Lefker, B. A. *Tetrahedron Lett.* **1987**, *28*, 1745.
6. (a) Meyers, A. I.; Wallace, R. H.; Harre, M.; Garland, R. *J. Org. Chem.* **1990**, *55*, 3137. (b) Meyers, A. I.; Harre, M.; Garland, R. *J. Am. Chem. Soc.* **1984**, *106*, 1146.
7. Meyers, A. I.; Romine, J.; Robichaud, A. J. *Heterocycles* **1990**, *30*, 339.
8. Meyers, A. I.; Snyder, L. B. *Synlett* **1991**, 863.
9. Meyers, A. I.; Fleming, S. A. *J. Am. Chem. Soc.* **1986**, *108*, 306.
10. (a) Meyers, A. I.; Romine, J. L.; Fleming, S. A. *J. Am. Chem. Soc.* **1988**, *110*, 7245. (b) Meyers, A. I.; Wallace, R. H. *J. Org. Chem.* **1989**, *54*, 2509. (c) Romo, D.; Romine, J. L.; Midura, W.; Meyers, A. I. *Tetrahedron* **1990**, *46*, 4951. (d) Romo, D.; Meyers, A. I. *J. Org. Chem.* **1992**, *57*, 6265.
11. (a) Meyers, A. I.; Busacca, C. A. *Tetrahedron Lett.* **1989**, *30*, 6973. (b) Meyers, A. I.; Busacca, C. A. *Tetrahedron Lett.* **1989**, *30*, 6977. (c) Busacca, C. A.; Meyers, A. I. *J. Chem. Soc., Perkin Trans. 1* **1991**, 2299.
12. Fray, A. H.; Meyers, A. I. *Tetrahedron Lett.* **1992**, *33*, 3575.
13. Meyers, A. I.; Burgess, L. E. *J. Org. Chem.* **1991**, *56*, 2294.
14. Burgess, L. E.; Meyers, A. I. *J. Am. Chem. Soc.* **1991**, *113*, 9858.
15. Burgess, L. E.; Meyers, A. I. *J. Org. Chem.* **1992**, *57*, 1656.
16. Meyers, A. I.; Snyder, L. *J. Org. Chem.* **1993**, *58*, 36.
17. Meyers, A. I.; Snyder, L. *J. Org. Chem.* **1992**, *57*, 3814.

Todd D. Nelson & Albert. I. Meyers
Colorado State University, Fort Collins, CO, USA

Tetrahydro-1-methyl-3,3-diphenyl-1*H*,3*H*-pyrrolo[1,2-*c*][1,3,2]oxazaborole[1]

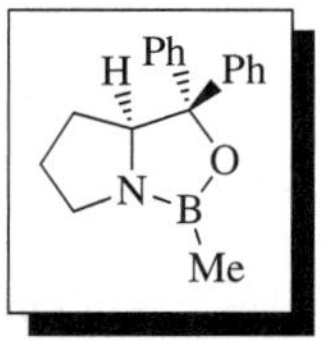

(*S*)
[112022-81-8] $C_{18}H_{20}BNO$ (MW 277.20)
($\cdot BH_3$)
[112022-90-9]
(*R*)
[112022-83-0]

(one of many chiral oxazaborolidines/chiral Lewis acids useful as enantioselective catalysts for the reduction of prochiral ketones,[1–3] imines,[4] and oximes,[2e,f,5] and the reduction of 2-pyranones to afford chiral biaryls;[6] other chiral oxazaborolidines have been used for the addition of diethylzinc to aldehydes,[7] asymmetric hydroboration,[8a,b] the Diels–Alder reaction,[9–11] and the aldol reaction[12,13])

Physical Data: mp 79–81 °C.
Solubility: very sol THF, CH_2Cl_2, toluene.
Preparative Methods: see text.
Purification: Kugelrohr distillation (50 °C/0.001 mbar)
Handling, Storage, and Precautions: the free oxazaborolidine must be rigorously protected from exposure to moisture. The crystalline borane complex is more stable, and is the preferred form to handle and store this catalyst.

Enantioselective Ketone Reduction. The major application of chiral oxazaborolidines has been the stoichiometric (as the oxazaborolidine–borane complex) (eq 1) and catalytic (in the presence of a stoichiometric borane source) (eq 2) enantioselective reduction of prochiral ketones.[1] These asymmetric catalysts work best for the reduction of aryl alkyl ketones, often providing very high (>95% ee) levels of enantioselectivity.

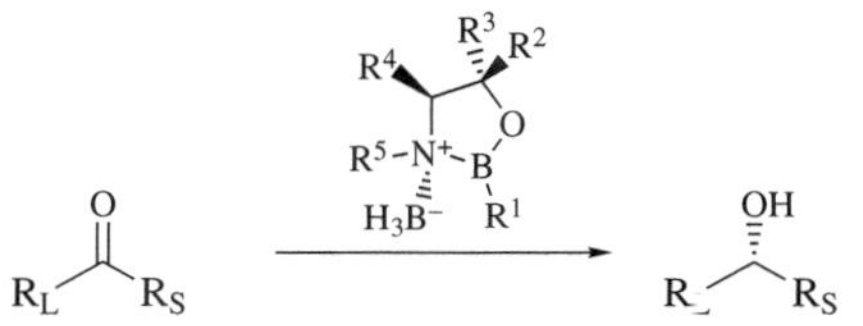

(1)

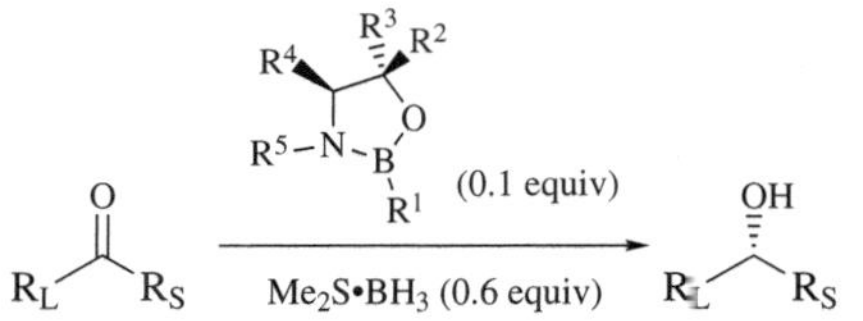

(2)

Following from the work of Itsuno[2] and Corey,[3] over 75 chiral oxazaborolidine catalysts have been reported for the reduction of prochiral ketones [(**1**),[2,3a,14,15a,e,f,16d–f,17b] (**2**),[16d,18b] (**3**),[3,6,19b–e,20,21,26c] (**4**),[16a] (**5**),[1b,16c,22] (**6**),[22b] (**7**),[3d,18a] (**8**),[16b] (**9**),[23] (**10**),[24] (**11**),[24] (**12**),[19a]]. Oxazaborolidines derived from proline (**3**) (see *α,α-Diphenyl-2-pyrrolidinemethanol*) and valine (**1**; $R^4 = i$-Pr) (see *2-Amino-3-methyl-1,1-diphenyl-1-butanol*) have received the most attention.

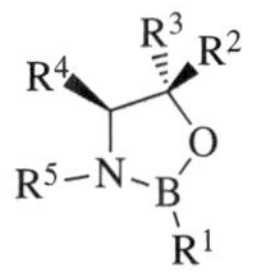

(**1**)

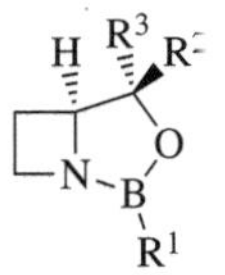

(**2**)

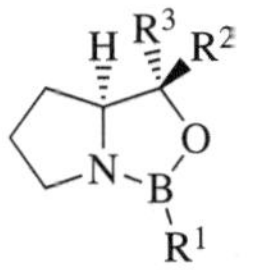

(**3**)

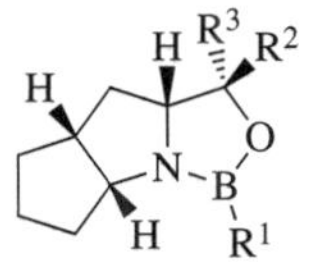

(**4**)

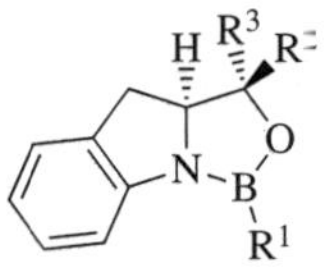

(**5**)

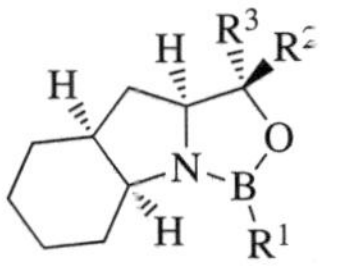

(**6**)

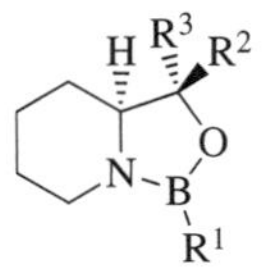
(7)

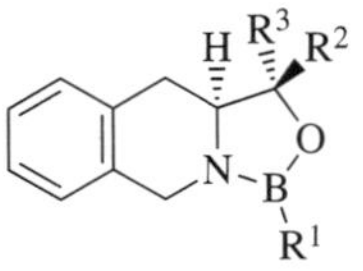
(8)

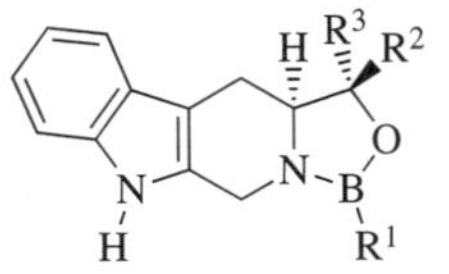
(9)

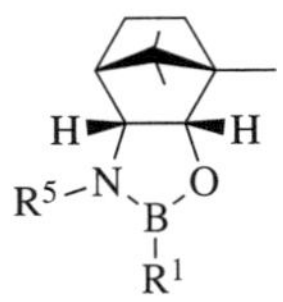
(10)

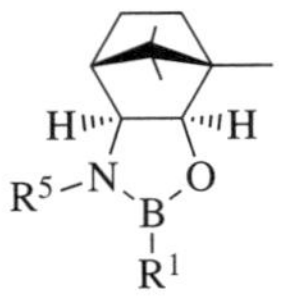
(11)

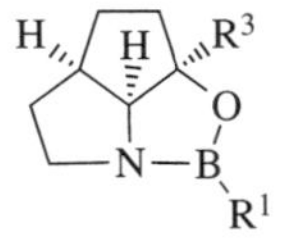
(12)

Unsubstituted (B–H) oxazaborolidines (**16**) are prepared from a chiral β-amino alcohol (**13**) and a source of borane (*Diborane*, *Borane– Tetrahydrofuran*, *Borane–Dimethyl Sulfide*, or $H_3B{\cdot}NMe_3$) via a multistep process (eq 3). Formation of the initial amine–borane complex (**14**) is generally exothermic, and this intermediate can often be isolated. Gentle heating with the loss of one mole of hydrogen results in the formation of (**15**). Continued heating with the loss of a second mole of hydrogen then affords oxazaborolidine (**16**). When R^4 and R^5 are connected, forming a four- or five-membered ring, more forcing conditions (70–75 °C, 1.7 bar, 48–72 h) are required to effect this conversion due to the additional ring strain. [*Caution*: under these conditions, borane or diborane in the vapor phase can begin to decompose.[25]] Finally, additional borane is added to afford the oxazaborolidine–borane complex (**17**).

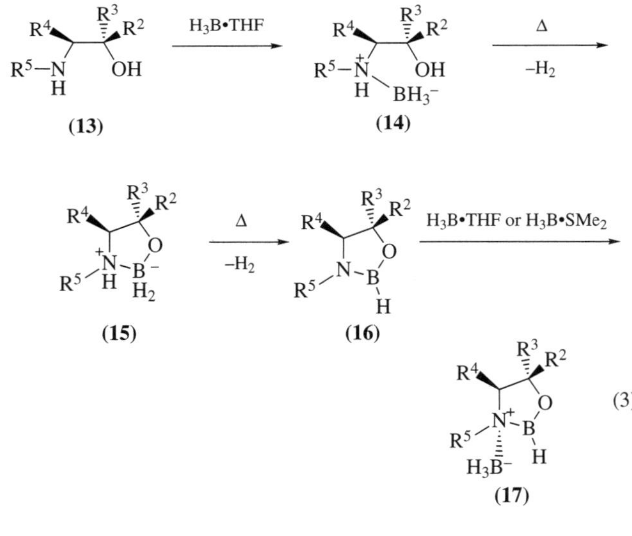
(3)

Free oxazaborolidine (**16**), by itself, will not reduce ketones. Furthermore, (**16**) is not particularly stable, reacting with moisture (H_2O), air (O_2), unreacted amino alcohol, other alcohols,[8c] or, depending on the substituents, with itself to form various dimers.[3a,8c,d,15d,26,27a] This instability is due to the strain of a partial double bond between nitrogen and boron (eq 4). Formation of the oxazaborolidine–borane complex (**17**) tends to release some of this strain. As such, (**16**) and (**17**) are generally prepared and used in situ without isolation; in many cases, they have not been fully characterized.[17c]

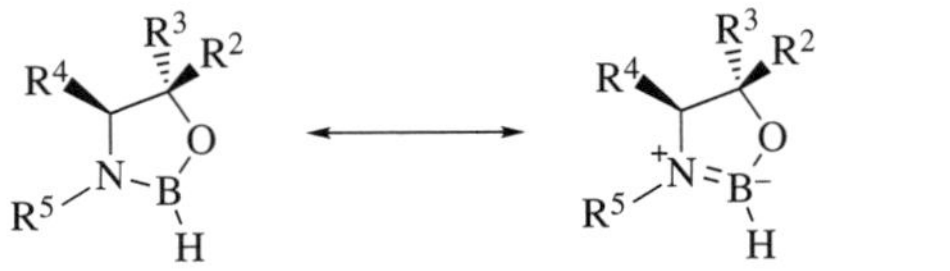
(4)

Oxazaborolidines substituted at boron (**1**; R^1 = alkyl, aryl) are prepared from a chiral β-amino alcohol and the corresponding boronic acid in a two-step process (eq 5).[3b,9] Heat and an efficient method of water removal (i.e. azeotropic distillation, molecular sieves) are required to drive the second step. When R^4 and R^5 are connected, more forcing conditions are necessary, both to complete the second step and to prevent the intermediate from proceeding to an alternate disproportionation product.[21] Alternative procedures using bis(diethylamino)-phenylborane (eq 6),[26a,b] trisubstituted boroxines (eq 7),[21,27] and ethyl or butyl bis(trifluoroethyl)boronate esters (eq 8)[19e] have been developed to circumvent these problems. The substituted oxazaborolidines are more stable than unsubstituted (B–H) oxazaborolidines (i.e. they can be handled in the presence of air, and do not form dimers), but are still prone to decomposition by moisture (H_2O).[21] In many cases the substituted oxazaborolidines have been isolated, purified, and characterized.

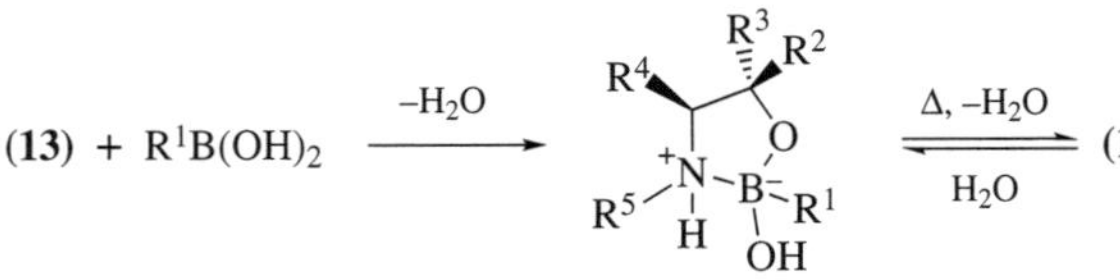
(5)

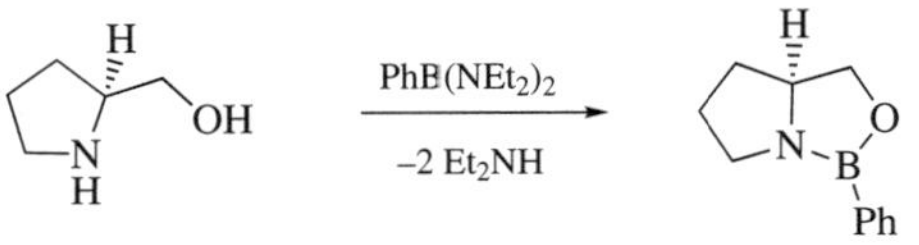
(6)

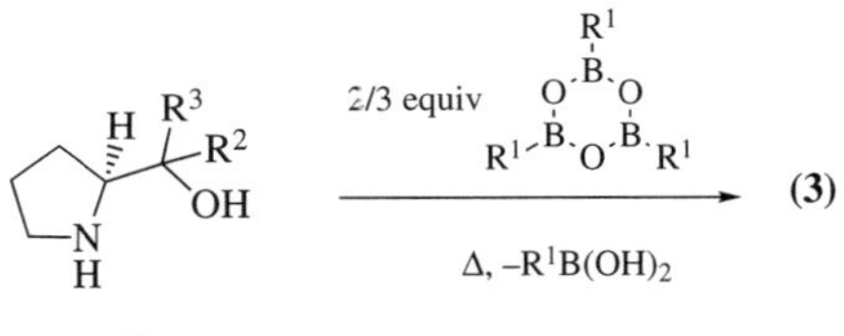
(7)

(18)

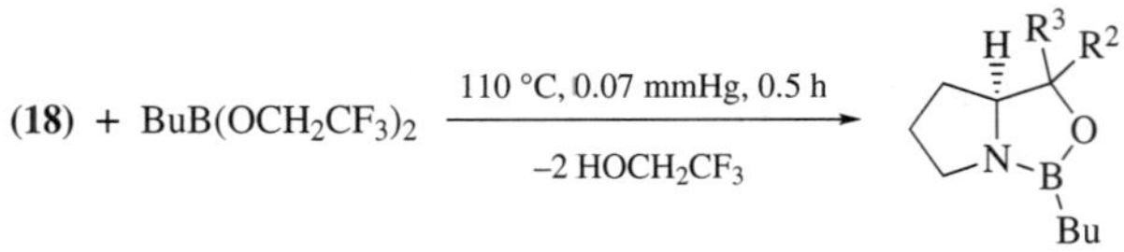
(8)

Substituted oxazaborolidines also react with borane (B_2H_6, $H_3B{\cdot}THF$, or $H_3B{\cdot}SMe_2$) to form an oxazaborolidine–borane complex (**19**) (eq 9).[3b,27] The oxazaborolidine–borane complex, by releasing the strain of the partial double bond between the ring boron and nitrogen, is more stable than the free oxazaborolidine, and in many cases exists as a stable crystalline solid.[21c,27,28]

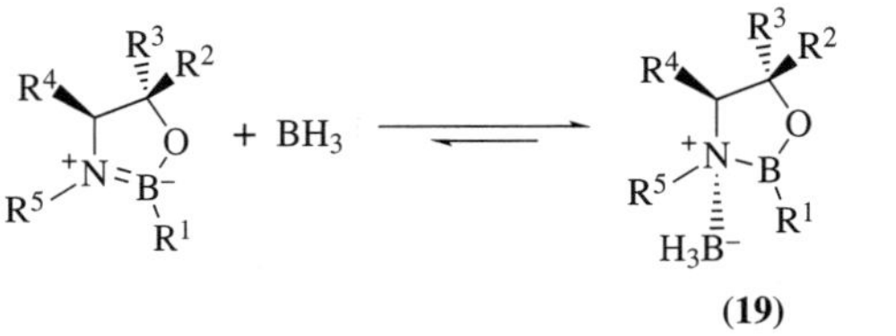
(9)

(19)

The oxazaborolidine–borane complex (**19**) can be used stoichiometrically (eq 1) or catalytically (eq 10) for the enantioselective reduction of prochiral ketones.[27a] When used catalytically, the oxazaborolidine–borane complex (**19**) is the second intermediate in the catalytic cycle (eq 10) proposed to explain the behavior of the oxazaborolidine catalyst.[3a,29] Subsequent coordination between the Lewis acidic ring boron and the carbonyl oxygen activates the ketone toward reduction. Intramolecular hydride transfer from the BH_3 coordinated to the ring nitrogen then occurs via a six-membered ring chair transition state.[17b,27a,30] Following hydride transfer, the alkoxy–BH_2 dissociates, and oxazaborolidine (**1**) is free to begin the cycle again. The diastereomeric transition state model (**20**), leading to the enantiomeric carbinol product, is disfavored due to unfavorable 1,3-diaxial steric interactions between R_L and R^1. Additional work will be required to better understand the catalytic cycle and the intermediates involved to further improve the oxazaborolidine catalysts. The behavior of the catalysts has been the subject of molecular orbital calculations in a series of 12 papers.[31] It should be noted, however, that not all of the results and conclusions are supported by experimental observations.

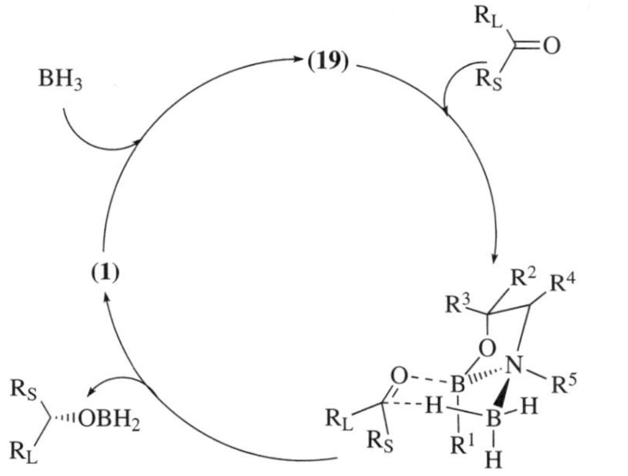

(10)

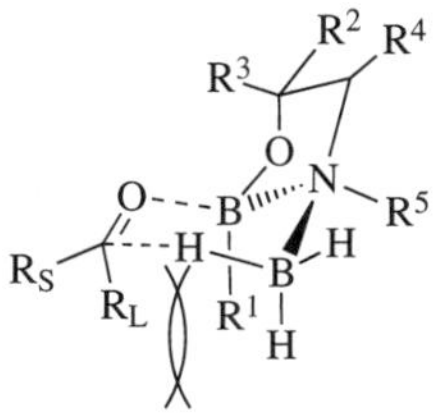

(20)

The enantioselectivities reported for the reduction of acetophenone and 1-tetralone using several representative chiral (4*S*)-oxazaborolidine catalysts are summarized in Table 1. The oxazaborolidines derived from (*S*)-azetidinecarboxylic acid and (*S*)-proline provide the best results. It is interesting to note the reversal in enantioselectivity going from catalyst (**5a**) to (**6a**).

Oxazaborolidine catalyzed reductions are generally performed in an aprotic solvent, such as dichloromethane, THF, or toluene. When the reactions are run in a Lewis basic solvent, such as THF, the solvent competes with the oxazaborolidine to complex with the borane, which can have an effect on the enantioselectivity and/or rate of the reaction.[27a] The solubility of the oxazaborolidine–borane complex can be the limiting factor for reactions run in toluene, although this problem has been circumvented by using oxazaborolidines with more lipophilic substituents (R^1 = *n*-Bu; R^2, R^3 = 2-naphthyl).[19b–d] We have found dichloromethane to be the best overall solvent for these reactions.[27a]

The reactions are typically performed using $H_3B{\cdot}THF$, $H_3B{\cdot}SMe_2$, or *Catecholborane*[19d] as the hydride source. When using $H_3B{\cdot}THF$ or $H_3B{\cdot}SMe_2$, two of the three hydrides are effectively utilized.[27a] This is only true for reactions run at temperatures greater than −40 °C. At lower temperatures, only one hydride is transferred at a reasonable rate. When two hydrides are used, there is some evidence that the enantioselectivity for transfer of the second hydride is different, and may in fact be lower.[27a] Whether this implies that an alternative catalytic cycle operates, whereby the alkoxy–BH_2 intermediate generated during the first hydride transfer remains coordinated to the oxazaborolidine, and then transfers the second hydride (with a different degree of enantioselectivity), or that some other intermediate present is active, but not as an enantioselective reducing agent, will require further investigation. In any event, the amount of BH_3 used should be at least 0.5 mole per mole of ketone plus an amount equal to the oxazaborolidine catalyst, with the possibility that 1 mole per mole provides slightly higher enantioselectivity. When catecholborane is used as the hydride source, a 50–100% excess of this reagent is used.

The mode of addition and the reaction temperature both affect the enantioselectivity of the reaction. The best results are obtained when the ketone is added slowly to a solution of the oxazaborolidine (or oxazaborolidine–borane complex) and the borane source, at as low a temperature that provides a reasonable reaction rate.[27a] This is in contrast to a previous report that indicated that oxazaborolidine-catalyzed reductions 'lose stereoselectivity at lower temperatures'.[19d] With unsubstituted (R^1 = H) oxazaborolidines, higher temperatures may be required due to incomplete formation of the catalyst, the presence of dimers, and/or other intermediates.[26c]

In their role as enantioselective catalysts for the reduction of prochiral ketones, chiral oxazaborolidines have been used for the preparation of prostaglandins,[3a] PAF antagonists,[3a] a key intermediate of ginkgolide B,[32a] bilobalide,[32b] a key intermediate of forskolin,[32c] (*R*)- and (*S*)-fluoxetine,[32d] (*R*)- and (*S*)-isopreterenol,[19c] vitamin D analogs,[33] the carbonic anhydrase inhibitor MK-0417,[21b] the dopamine D1 agonist A-77636,[20b] taxol,[34] the LTD_4 antagonists L-695,499 and L-699,392,[35] the β-adrenergic agonist CL 316,243,[36] and the antiarrhythmic MK-0499.[37] They have also been used for the synthesis of chiral amines,[38,39] α-hydroxy acids,[19d,40a] benzylic thiols,[40c] the enantioselective reduction of trihalomethyl ketones,[40a,b,d] and ketones containing various heteroatoms.[17a,21b,27a,35,37]

Enantioselective Reduction of Imines and Ketoxime *O*-Ethers. In addition to the reduction of prochiral ketones, chiral oxazaborolidines have been employed as enantioselective reagents and catalysts for the reduction of imines (eq 11)[4,23] and ketoxime *O*-ethers (eq 12)[2e,f,5] to give chiral amines. It is interesting to note that the enantioselectivity for the reduction of ketoxime *O*-ethers is opposite that of ketones and imines. For more information, see *2-Amino-3-methyl-1,1-diphenyl-1-butanol*.

Enantioselective Addition of Diethylzinc to Aldehydes. Oxazaborolidines derived from ephedrine have been used to catalyze the addition of *Diethylzinc* to aldehydes (eq 13).[7] Both

Table 1 Chiral Oxazaborolidine Catalyzed Reduction of Acetophenone and 1-Tetralone

Catalyst	R^1	R^2,R^3	R^4 (mol %)	Catalyst (ee %)	Acetophenone (ee %)	1-Tetralone
(**1a**)[3a]	H	Ph	*i*-Pr	10	94.7 (*R*)	–
(**2a**)[16d]	H	Ph	–	10	98 (*R*)	–
(**3a**)[3a]	H	Ph	–	10	97 (*R*)	89 (*R*)
(**3b**)[21b]	Me	Ph	–	10	98 (*R*)	94 (*R*)
(**3b**)·BH_3[27a]	Me	Ph	–	5	97.6 (*R*)	99.0 (*R*)
(**3b**)·BH_3[27a]	Me	Ph	–	100	99.8 (*R*)	99.2 (*R*)
(**5a**)[22b]	H	Ph	–	10	96 (*R*)	79 (*R*)
(**6a**)[22b]	H	Ph	–	10	90 (*S*)	79 (*S*)
(**7a**)[18a]	H	Ph	–	10	87 (*R*)	–
(**8a**)[16b]	H	Ph	–	10	71 (*R*)	44 (*R*)
(**9a**)[23]	H	H	–	110	88 (*R*)	–
(**12a**)[19a]	Me	Ph	–	10	97.5 (*R*)	95.3 (*R*)

the rate and enantioselectivity are optimized when $R^1 = H$. Aromatic aldehydes generally react faster than aliphatic aldehydes, and the enantioselectivity for aromatic aldehydes is good to excellent (86–96% ee).

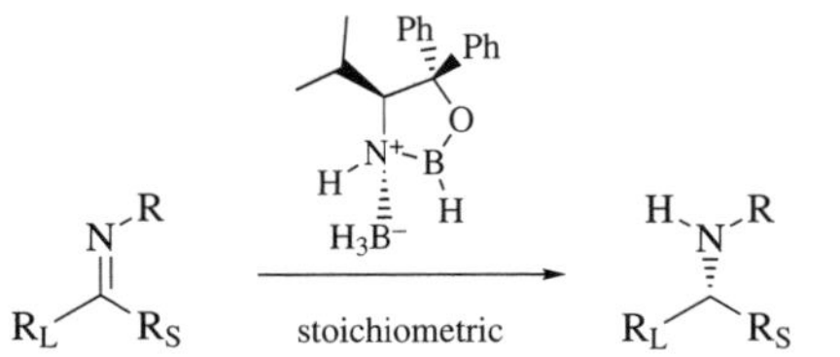

(11)

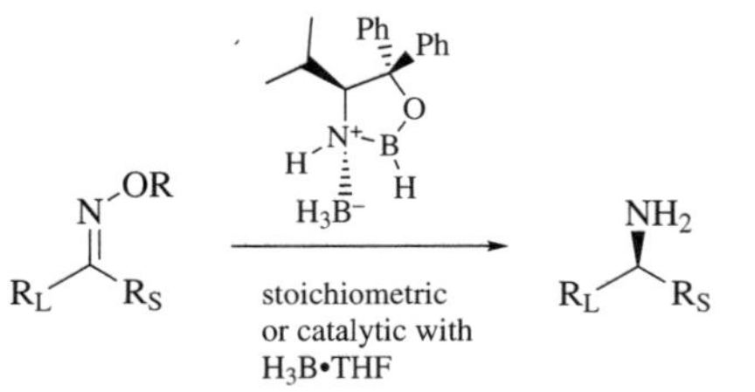

(12)

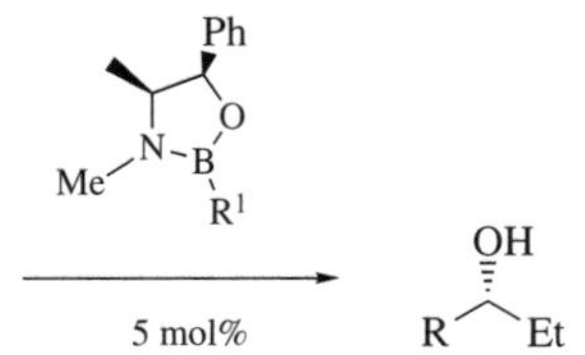

(13)

Other Applications. Chiral oxazaborolidines derived from ephedrine have also been used in asymmetric hydroborations,[8a,b] and as reagents to determine the enantiomeric purity of secondary alcohols.[8c] Chiral 1,3,2-oxazaborolidin-5-ones derived from amino acids have been used as asymmetric catalysts for the Diels–Alder reaction,[9–11] and the aldol reaction.[12,13]

Related Reagents. 2-Amino-3-methyl-1,1-diphenyl-1-butanol; α,α-Diphenyl-2-pyrrolidinemethanol; Ephedrine-borane; Norephedrine–Borane.

---

1. (a) Wallbaum, S.; Martens, J. *Tetrahedron: Asymmetry* **1992**, *3*, 1475. (b) Singh, V. K. *Synthesis* **1992**, 605. (c) Deloux, L.; Srebnik M. *Chem. Rev.* **1993**, *93*, 763.
2. (a) Hirao, A.; Itsuno, S.; Nakahama, S.; Yamazaki, N. *Chem. Commun./J. Chem. Soc., Chem. Commun.* **1981**, 315. (b) Itsuno, S.; Hirao, A.; Nakahama, S.; Yamazaki, N. *J. Chem. Soc. Perkin Trans. 1* **1983**, 1673. (c) Itsuno, S.; Ito, K.; Hirao, A.; Nakahama, S. *Chem. Commun./J. Chem. Soc., Chem. Commun.* **1983**, 469. (d) Itsuno, S.; Ito, K.; Hirao, A.; Nakahama, S. *J. Org. Chem.* **1984**, *49*, 555. (e) Itsuno, S.; Nakano, M.; Miyazaki, K.; Masuda, H.; Ito, K.; Hirao, A.; Nakahama, S. *J. Chem. Soc. Perkin Trans. 1* **1985**, 2039. (f) Itsuno, S.; Nakano, M.; Ito, K.; Hirao, A.; Owa, M.; Kanda, N.; Nakahama, S. *J. Chem. Soc. Perkin Trans. 1* **1985**, 2615.
3. (a) Corey, E. J.; Bakshi, R. K.; Shibata, S. *J. Am. Chem. Soc.* **1987**, *109*, 5551. (b) Corey, E. J.; Bakshi, R. K.; Shibata, S.; Chen, C. P.; Singh, V. K. *J. Am. Chem. Soc.* **1987**, *109*, 7925. (c) Corey, E. J.; Shibata, S.; Bakshi, R. K. *J. Org. Chem.* **1988**, *53*, 2861. (d) Corey, E. J. U.S. Patent 4 943 635, 1990.
4. (a) Cho, B. T.; Chun, Y. S. *J. Chem. Soc. Perkin Trans. 1* **1990**, 3200. (b) Cho, B. T.; Chun, Y. S. *Tetrahedron: Asymmetry* **1992**, *3*, 337.
5. (a) Itsuno, S.; Sakurai, Y.; Ito, K.; Hirao, A.; Nakahama, S. *Bull. Chem. Soc. Jpn.* **1987**, *60*, 395. (b) Itsuno, S.; Sakurai, Y.; Shimizu, K.; Ito, K. *J. Chem. Soc. Perkin Trans. 1* **1989**, 1548. (c) Itsuno, S.; Sakurai, Y.; Shimizu, K.; Ito, K. *J. Chem. Soc. Perkin Trans. 1* **1990**, 1859.
6. Bringmann, G.; Hartung, T. *Angew. Chem., Int. Ed. Engl.* **1992**, *31*, 761.
7. Joshi, N. N.; Srebnik, M.; Brown, H. C. *Tetrahedron Lett.* **1989**, *30*, 5551.
8. (a) Brown, J. M.; Lloyd-Jones, G. C. *Tetrahedron: Asymmetry* **1990**, *1*, 869. (b) Brown, J. M.; Lloyd-Jones, G. C. *Chem. Commun.* **1992**, 710. (c) Brown, J. M.; Leppard, S. W.; Lloyd-Jones, G. C. *Tetrahedron: Asymmetry* **1992**, *3*, 261. (d) Brown, J. M.; Lloyd-Jones, G. C.; Layzell, T. P. *Tetrahedron: Asymmetry* **1993**, *4*, 2151.
9. Takasu, M.; Yamamoto, H. *Synlett* **1990**, 194.
10. (a) Sartor, D.; Saffrich, J.; Helmchen, G. *Synlett* **1990**, 197. (b) Sartor, D.; Saffrich, J.; Helmchen, G.; Richards, C. J.; Lambert, H. *Tetrahedron: Asymmetry* **1991**, *2*, 639.
11. (a) Corey, E. J.; Loh, T.-P. *J. Am. Chem. Soc.* **1991**, *113*, 8966. (b) Corey, E. J.; Loh, T.-P.; Roper, T. D.; Azimioara, M. D.; Noe, M. C. *J. Am. Chem. Soc.* **1992**, *114*, 8290.
12. Kiyooka, S.; Kaneko, Y.; Komura, M.; Matsuo, H.; Nakano, M. *J. Org. Chem.* **1991**, *56*, 2276.
13. Parmee, E. R.; Tempkin, O.; Masamune, S.; Abiko, A. *J. Am. Chem. Soc.* **1991**, *113*, 9365.
14. Mandal, A. K.; Kasar, T. G.; Mahajan, S. W.; Jawalkar, D. G. *Synth. Commun.* **1987**, *17*, 563.
15. (a) Grundon, M. F.; McCleery, D. G.; Wilson, J. W. *J. Chem. Soc. Perkin Trans. 1* **1981**, 231. (b) Mancilla, T.; Santiesteban, F.; Contreras, R.; Klaebe, A. *Tetrahedron Lett.* **1982**, *23*, 1561. (c) Tlahuext, H.; Contreras,

R. *Tetrahedron: Asymmetry* **1992**, *3*, 727. (d) Tlahuext, H. Contreras, R. *Tetrahedron: Asymmetry* **1992**, *3*, 1145 (e) Cho, B. T.; Chun, Y. S. *Tetrahedron: Asymmetry* **1992**, *3*, 1539 (f) Berenguer, R.; Garcia, J.; Gonzalez, M.; Vilarrasa, J. *Tetrahedron: Asymmetry* **1993**, *4*, 13.

16. (a) Wallbaum, S.; Martens, J. *Tetrahedron: Asymmetry* **1991**, *2*, 1093. (b) Stingl, K.; Martens, J.; Wallbaum, S. *Tetrahedron: Asymmetry* **1992**, *3*, 223. (c) Martens, J.; Dauelsberg, C.; Behnen, W.; Wallbaum, S. *Tetrahedron: Asymmetry* **1992**, *3*, 347. (d) Behnen, W.; Dauelsberg, C.; Wallbaum, S.; Martens, J. *Synth. Commun.* **1992**, *22*, 2143. (e) Mehler, T.; Martens, J. *Tetrahedron: Asymmetry* **1993**, *4*, 1983. (f) Mehler, T.; Martens, J. *Tetrahedron: Asymmetry* **1993**, *4*, 2299.
17. (a) Quallich, G. J.; Woodall, T. M. *Tetrahedron Lett.* **1993**, *34*, 785. (b) Quallich, G. J.; Woodall, T. M. *Tetrahedron Lett.* **1993**, *34*, 4145. (c) Quallich, G. J.; Woodall, T. M. *Synlett* **1993**, 929.
18. (a) Rao, A. V. R.; Gurjar, M. K.; Sharma, P. A.; Kaiwar, V. *Tetrahedron Lett.* **1990**, *31*, 2341. (b) Rao, A. V. R.; Gurjar, M. K.; Kaiwar, V. *Tetrahedron: Asymmetry* **1992**, *3*, 859.
19. (a) Corey, E. J.; Chen, C. P.; Reichard, G. A. *Tetrahedron Lett.* **1989**, *30*, 5547. (b) Corey, E. J.; Link, J. O. *Tetrahedron Lett.* **1989**, *30*, 6275. (c) Corey, E. J.; Link, J. O. *Tetrahedron Lett.* **1990**, *31*, 601. (d) Corey, E. J.; Bakshi, R. K. *Tetrahedron Lett.* **1990**, *31*, 611. (e) Corey, E. J.; Link, J. O. *Tetrahedron Lett.* **1992**, *33*, 4141.
20. (a) DeNinno, M. P.; Perner, R. J.; Lijewski, L. *Tetrahedron Lett.* **1990**, *31*, 7415. (b) DeNinno, M. P.; Perner, R. J.; Morton, H. E.; DiDomenico, Jr., S. *J. Org. Chem.* **1992**, *57*, 7115.
21. (a) Mathre, D. J.; Jones, T. K.; Xavier, L. C.; Blacklock, T. J.; Reamer, R. A.; Mohan, J. J.; Jones, E. T. T.; Hoogsteen, K.; Baum, M. W.; Grabowski, E. J. J. *J. Org. Chem.* **1991**, *56*, 751. (b) Jones, T. K.; Mohan, J. J.; Xavier, L. C.; Blacklock, T. J.; Mathre, D. J.; Sohar, P.; Jones, E. T. T.; Reamer, R. A.; Roberts, F. E.; Grabowski, E. J. J. *J. Org. Chem.* **1991**, *56*, 763. (c) Blacklock, T. J.; Jones, T. K.; Mathre, D. J.; Xavier, L. C. U.S. Patent 5 039 802, 1991. (d) Blacklock, T. J.; Jones, T. K.; Mathre, D. J.; Xavier, L. C. U.S. Patent 5 264 585, 1993. (e) Shinkai, I. *J. Heterocycl. Chem.* **1992**, *29*, 627.
22. (a) Youn, I. K.; Lee, S. W.; Pak, C. S. *Tetrahedron Lett.* **1988**, *29*, 4453. (b) Kim, Y. H.; Park, D. H.; Byun, I. S.; Yoon, I. K.; Park, C. S. *J. Org. Chem.* **1993**, *58*, 4511.
23. Nakagawa, M.; Kawate, T.; Kikikawa, T.; Yamada, H.; Matsui, T.; Hino, T. *Tetrahedron* **1993**, *49*, 1739.
24. Tanaka, K.; Matsui, J.; Suzuki, H. *Chem. Commun./J. Chem. Soc., Chem. Commun.* **1991**, 1311.
25. (a) Long, L. H. *J. Inorg. Nucl. Chem.* **1970**, *32*, 1097. (b) Fernandez, H.; Grotewold, J.; Previtali, C. M. *J. Chem. Soc., Dalton Trans.* **1973**, 2090. (c) Gibb, T. C.; Greenwood, N. N.; Spalding, T. R.; Taylorson, D. *J. Chem. Soc., Dalton Trans.* **1979**, 1398.
26. (a) Bielawski, J.; Niedenzu, K. *Synth. React. Inorg. Met.-Org. Chem.* **1980**, *10*, 479. (b) Cragg, R. H.; Miller, T. J. *J. Organomet. Chem.* **1985**, *294*, 1. (c) Brunel, J. M.; Maffei, M.; Buono, G. *Tetrahedron: Asymmetry* **1993**, *4*, 2255.
27. (a) Mathre, D. J.; Thompson, A. S.; Douglas, A. W.; Hoogsteen, K.; Carroll, J. D.; Corley, E. G.; Grabowski, E. J. J. *J. Org. Chem.* **1993**, *58*, 2880. (b) Blacklock, T. J.; Jones, T. K.; Mathre, D. J.; Xavier, L. C. U.S. Patent 5 189 177, 1993. (c) Carroll, J. D.; Mathre, D. J.; Corley, E. G.; Thompson, A. S. U.S. Patent 5 264 574, 1993.
28. Corey, E. J.; Azimioara, M.; Sarshar, S. *Tetrahedron Lett.* **1992**, *24*, 3429.
29. Evans, D. A. *Science* **1988**, *240*, 420.
30. Jones, D. K.; Liotta, D. C.; Shinkai, I.; Mathre, D. J. *J. Org. Chem.* **1993**, *58*, 799.
31. Nevalainen, V. *Tetrahedron: Asymmetry* **1993**, *4*, 2001; and references contained therein.
32. (a) Corey, E. J.; Gavai, A. V. *Tetrahedron Lett.* **1988**, *29*, 3201. (b) Corey, E. J.; Su, W.-G. *Tetrahedron Lett.* **1988**, *29*, 3423. (c) Corey, E. J.; Jardine, P. D. S.; Mohri, T. *Tetrahedron Lett.* **1988**, *29*, 6409. (d) Corey, E. J.; Reichard, G. A. *Tetrahedron Lett.* **1989**, *30*, 5207.
33. (a) Kabat, M.; Kiegiel, J.; Cohen, N.; Toth, K.; Wovkulich, P. M.; Uskokovic, M. R. *Tetrahedron Lett.* **1991**, *32*, 2343. (b) Lee, A. S.; Norman, A. W.; Okamura, W. H. *J. Org. Chem.* **1992**, *57*, 3846.
34. Nicolaou, K. C.; Hwang, C.-K.; Sorensen, E. J.; Clairborne, C. F. *Chem. Commun./J. Chem. Soc., Chem. Commun.* **1992**, 1117.
35. (a) Labelle, M.; Prasit, P.; Belley, M.; Blouin, M.; Champion, E.; Charette, L.; DeLuca, J. G.; Dufresne, C.; Frenette, R.; Gauthier, J. Y.; Grimm, E.; Grossman, S. J.; Guay, D.; Herold, E. G.; Jones, T. R.; Lau, Y.; Leblanc, Y.; Leger, S.; Lord, A.; McAuliffe, M.; McFarlane, C.; Masson, P.; Metters, K. M.; Ouimet, N.; Patrick, D. H.; Perrier, H.; Pechuta, H.; Roy, P.; Williams, H.; Wang, Z.; Xiang, Y. B.; Zamboni, R. J.; Ford-Hutchinson, A. W.; Young, R. N. *Bioorg. Med. Chem. Lett.* **1992**, *2*, 1141. (b) King, A. O.; Corley, E. G.; Anderson, R. K.; Larsen, R. D.; Verhoeven, T. R.; Reider, P. J.; Xiang, Y. B.; Belley, M.; Leblanc, Y.; Labelle, M.; Prasit, P.; Zamboni, R. J. *J. Org. Chem.* **1993**, *58*, 3731.
36. Bloom, J. D.; Dutia, M. D.; Johnson, B. D.; Wissner, A.; Burns, M. G.; Largis, E. E.; Dolan, J. A.; Claus, T. H. *J. Med. Chem.* **1992**, *35*, 3081.
37. Cai, D.; Tschaen, D.; Shi, Y.-J.; Verhoeven, T. R.; Reamer, R. A.; Douglas, A. W. *Tetrahedron Lett.* **1993**, *34*, 3243.
38. Chen, C.-P.; Prasad, K.; Repic, O. *Tetrahedron Lett.* **1991**, *32*, 7175.
39. Thompson, A. S.; Humphrey, G. R.; DeMarco, A. M.; Mathre, D. J.; Grabowski, E. J. J. *J. Org. Chem.* **1993**, *58*, 5886.
40. (a) Corey, E. J.; Cheng, X. M.; Cimprich, K. A.; Sarshar, S. *Tetrahedron Lett.* **1991**, *32*, 6835. (b) Corey, E. J.; Link, J. O. *Tetrahedron Lett.* **1992**, *33*, 3431. (c) Corey, E. J.; Cimprich, K. A. *Tetrahedron Lett.* **1992**, *33*, 4099. (d) Corey, E. J.; Link, J. O.; Bakshi, R. K. *Tetrahedron Lett.* **1992**, *33*, 7107.

David J. Mathre & Ichiro Shinkai
Merck Research Laboratories, Rahway, NJ, USA

(*S*)-(+)-2-(2,4,5,7-Tetranitro-9-fluorenylideneaminooxy)propionic Acid

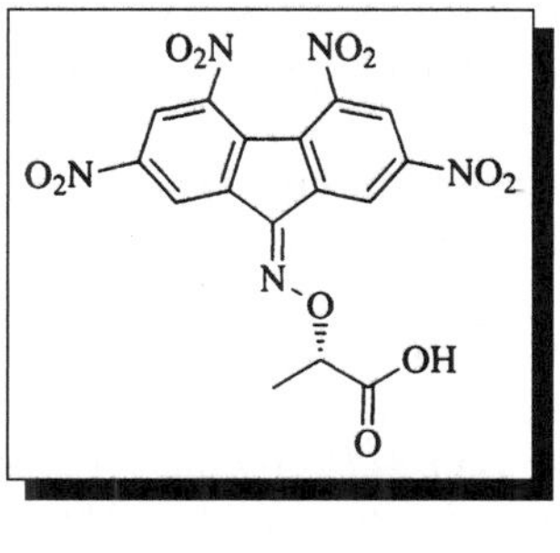

(*S*)-(+)
[50996-73-1] $C_{16}H_9N_5O_{11}$ (MW 447.30)
(*R*)-(−)
[50874-31-2]

(used as a π-acidic resolving agent for racemic π-bases, especially carbohelicenes[1] and heterohelicenes,[1a,2] and also for resolving certain types of amines[3])

Alternate Name: TAPA.
Physical Data: mp 195 °C (dec.); $[\alpha]^{25}$ +92° (*c* = 1.6, dioxane).
Solubility: sol $CHCl_3$, dioxane, hot acetic acid; slightly sol CH_2Cl_2, toluene.
Form Supplied in: available as the neat solid.
Analysis of Reagent Purity: NMR,[4] mp, [α].
Preparative Methods: prepared from acetone oxime, ethyl 2-bromopropionate, and 2,4,5,7-tetranitrofluorenone.[4] The inter-

mediate racemic 2-(isopropylideneaminooxy)propionic acid is resolved with *(1R,2S)-Ephedrine*.

Handling, Storage, and Precautions: should be stored under protection from light. No reported toxicity.

Resolving Agent for π-Bases. The title reagent was originally designed as an agent for resolving hexahelicene, through formation of diastereomeric charge-transfer complexes.[5] It has subsequently been used for resolution of several different types of chiral, racemic π-bases not containing any of the functional groups usually needed to effect resolution by conventional reagents. Carbohelicenes[1] and heterohelicenes (e.g. **1**),[1a,2] [2,2]paracyclophanes,[6] naphthalenophanes,[7] highly substituted alkenes,[8] derivatives of polycyclic aromatic hydrocarbons (e.g. **2**),[9] porphyrin derivatives,[10] an alkyl aryl ether,[11] and an α-aryl ester[11] have been successfully resolved using this reagent, either by chromatographic separation on columns coated with TAPA or by fractional crystallization. Analytical determination of the enantiomeric purity of hexahelicene and several of its derivatives is also possible using HPLC.[12]

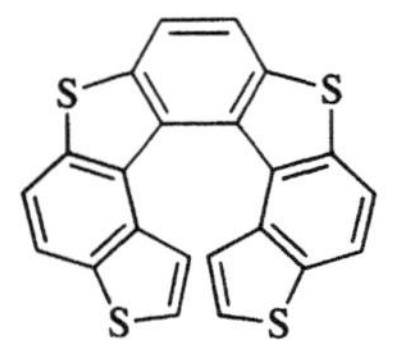

(1)

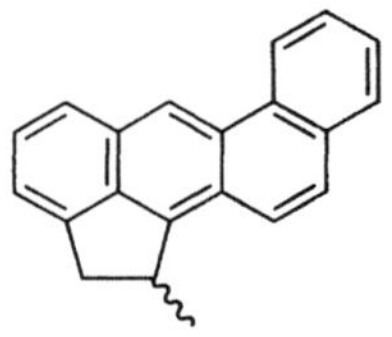

(2)

Resolution of helicenes has been effected using chromatography on other types of chiral stationary phases, including columns derivatized with analogs of TAPA containing larger alkyl groups at the stereocenter, but TAPA is generally the most effective agent.[1]

Resolving Agent for Certain Types of Amines. TAPA has also been used as resolving agent for some amines that formed either unstable, insoluble, or noncrystalline salts with common resolving acids. Compounds (**3**) and (**4**) were among those resolved with TAPA, whereas camphor-10-sulfonic acid, 3-bromo-8-camphorsulfonic acid, *O,O*-di-*p*-toluoyl-(+)-tartaric acid, (+)-tartaric acid, and (+)-camphoric acid could not be used.[3]

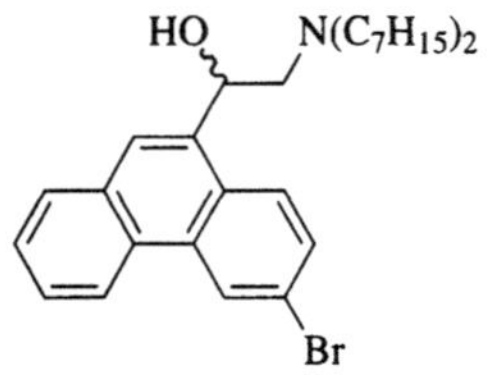

(3)

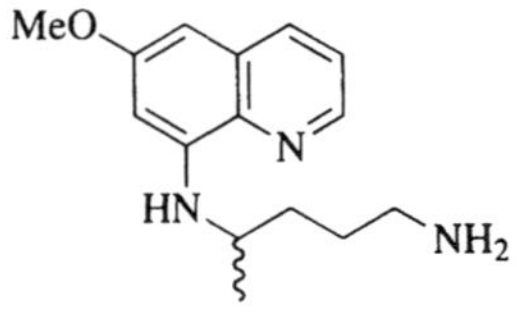

(4)

---

1. (a) Laarhoven, W. H.; Prinsen, W. J. C. *Top. Curr. Chem.* **1984**, *125*, 86. (b) Newman, M. S.; Mentzer, R. G.; Slomp, G. *J. Am. Chem. Soc.* **1963**, *85*, 4018. (c) Goedicke, C.; Stegemeyer, H. *Tetrahedron Lett.* **1970**, 937. (d) Laarhoven, W. H.; Cuppen, T. J. H. M.; Nivard, R. J. F. *Tetrahedron* **1974**, *30*, 3343. (e) Mikes, F.; Boshart, G.; Gil-Av, E. *Chem. Commun./J. Chem. Soc., Chem. Commun.* **1976**, 99. (f) Mikes, F.; Boshart, G.; Gil-Av, E. *J. Chromatogr.* **1976**, *122*, 205.
2. (a) Numan, H.; Helder, R.; Wynberg, H. *Recl. Trav. Chim. Pays-Bas* **1976**, *95*, 211. (b) Nakagawa, H.; Ogashiwa, S.; Tanaka, H.; Yamada, K.; Kawazura, H. *Bull. Chem. Soc. Jpn.* **1981**, *54*, 1903. (c) Nakagawa, H.; Yamada, K.; Kawazura, H. *Chem. Commun./J. Chem. Soc., Chem. Commun.* **1989**, 1378.
3. Carroll, F. I.; Berrang, B.; Linn, C. P. *J. Med. Chem.* **1978**, *21*, 326.
4. Block, P., Jr.; Newman, M. S. *Org. Synth., Coll. Vol.* **1973**, *5*, 1031.
5. (a) Newman, M. S.; Lutz, W. B.; Lednicer, D. *J. Am. Chem. Soc.* **1955**, *77*, 3420. (b) Newman, M. S.; Lednicer, D. *J. Am. Chem. Soc.* **1956**, *78*, 4765.
6. (a) Longone, D. T.; Reetz, M. T. *Chem. Commun./J. Chem. Soc., Chem. Commun.* **1967**, 46. (b) Rebafka, W.; Staab, H. A. *Angew. Chem.* **1973**, *85*, 831; *Angew. Chem., Int. Ed. Engl.* **1973**, *12*, 776 (*Chem. Abstr.* **1974**, *80*, 47 529w).
7. Meurer, K.; Luppertz, F.; Vögtle, F. *Ber. Dtsch. Chem. Ges./Chem. Ber.* **1985**, *118*, 4433.
8. Feringa, B.; Wynberg, H. *J. Am. Chem. Soc.* **1977**, *99*, 602.
9. (a) Kim, Y. H.; Tishbee, A.; Gil-Av, E. *Chem. Commun./J. Chem. Soc., Chem. Commun.* **1981**, 75. (b) Newman, M. S.; Wotring, Jr., R. W.; Pandit, A.; Chakrabarti, P. M. *J. Org. Chem.* **1966**, *31*, 4293.
10. Risch, N.; Reich, H. *Tetrahedron Lett.* **1979**, 4257.
11. Newman, M. S.; Lutz, W. B. *J. Am. Chem. Soc.* **1956**, *78*, 2469.
12. Prinsen, W. J. C.; Laarhoven, W. H. *J. Chromatogr.* **1987**, *393*, 377.

Tobias Rein
The Royal Institute of Technology, Stockholm, Sweden

(*R*)-(+)-*p*-Tolylsulfinylacetic Acid

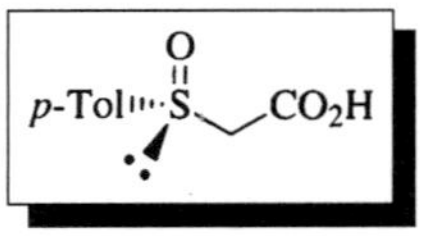

[88981-65-1] $C_9H_{10}O_3S$ (MW 198.26)

(optical resolution of amines[1,3,4])

Physical Data: $[\alpha]_D = +143.5°$ (acetone, $c = 33.0$). The absolute configuration of this reagent (and other arylsulfinyl acetic acids) is characterized by two CD Cotton effects which are observed in the presence of the metal cluster $[Mo_2(OAc)_4]$ in DMSO solution above 300 nm.[2]

Preparative Method: prepared by carboxylation of *(R)-(+)-Methyl p-Tolyl Sulfoxide* carbanion generated with *Lithium Diisopropylamide* (eq 1).[1]

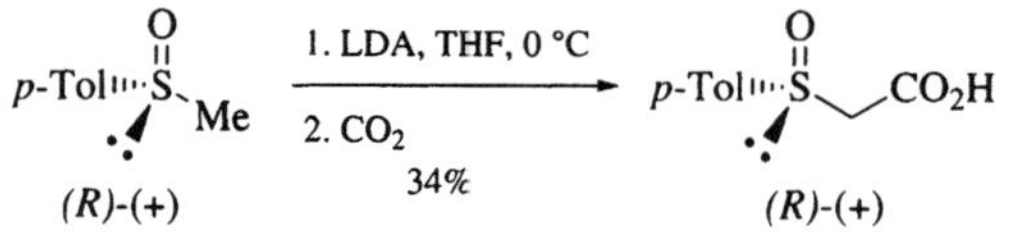

(1)

Optical Resolution of Indole Alkaloid Precursors. This reagent has been used for optical resolutions of tetracyclic alkaloids such as 10,22-dioxokopsane (**1**)[1] and kopsinine (**2**).[3]

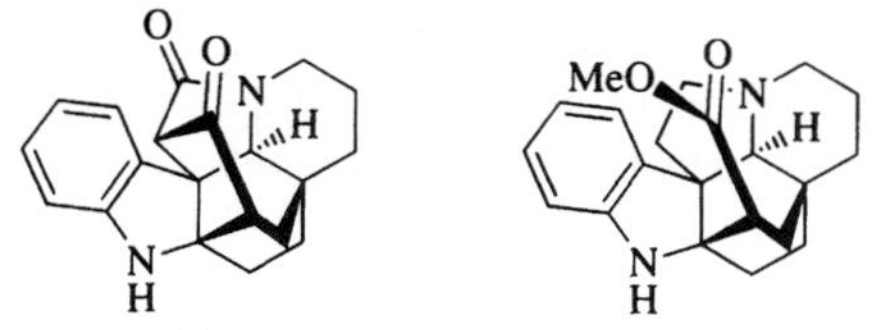

(1) (2)

The racemic tetracyclic amine (**3**) can be coupled to (*R*)-(+)-*p*-tolylsulfinylacetic acid using the modified carbodiimide reagent *1-Cyclohexyl-3-(2-morpholinoethyl)carbodiimide Metho-p-toluenesulfonate* to give the diastereomeric sulfinyl amides (**4**) and (**5**) (eq 2), which are readily separated by HPLC.[1,3]

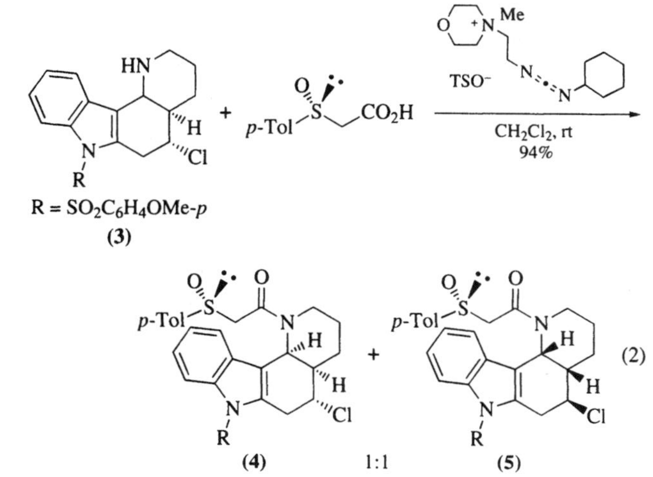

Similarly, in a total synthesis of strychnine, the optical resolution was carried out by separation of the sulfinyl lactam diastereomers (**7**) and (**8**), which were obtained from the heptacyclic indole alkaloid precursor (**6**); this was first transformed with (*R*)-(+)-*p*-tolylsulfinylacetic acid into the corresponding sulfinyl amide and then converted to the diastereomeric lactams (**7**) and (**8**) by an intramolecular conjugate addition (eq 3).[4]

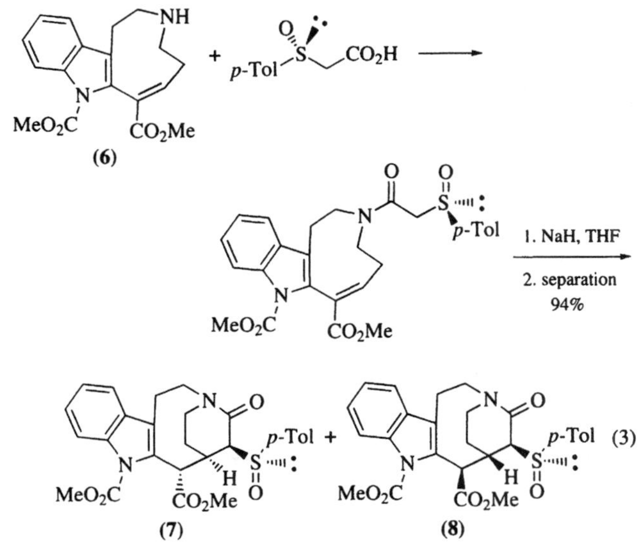

1. Magnus, P.; Gallagher, T.; Brown, P.; Huffman, J. C. *J. Am. Chem. Soc.* **1984**, *106*, 2105.
2. Drabowicz, J.; Mikolajczyk, M. *Croat. Chem. Acta* **1989**, *62*, 423.
3. Magnus, P.; Brown, P. *Chem. Commun./J. Chem. Soc., Chem. Commun.* **1985**, 184.
4. Magnus, P.; Melvyn, G.; Bonnert, R.; Kim, C. S.; McQuire, L.; Merritt, A.; Vicker, N. *J. Am. Chem. Soc.* **1992**, *114*, 4403.

Guy Solladié & Françoise Colobert
University Louis Pasteur, Strasbourg, France

(*R*)-(+)-α-(*p*-Tolylsulfinyl)-*N*,*N*-dimethylacetamide

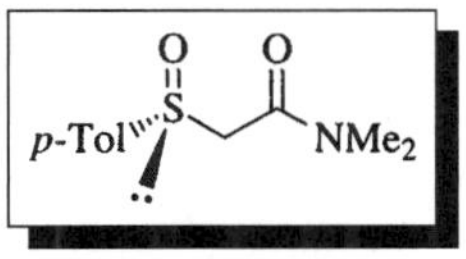

[72298-22-7] $C_{11}H_{15}NO_2S$ (MW 225.34)

(asymmetric aldol-type condensation[1,2])

Physical Data: $[\alpha]_D = +194.7°$ ($CHCl_3$, $c = 1$).
Preparative Methods: prepared[1,2] by reaction of *Lithio-N,N-dimethylacetamide* with (*S*)-(−)-menthyl *p*-toluenesulfinate (see *(−)-(1R,2S,5R)-Menthyl (S)-p-Toluenesulfinate*) (eq 1).

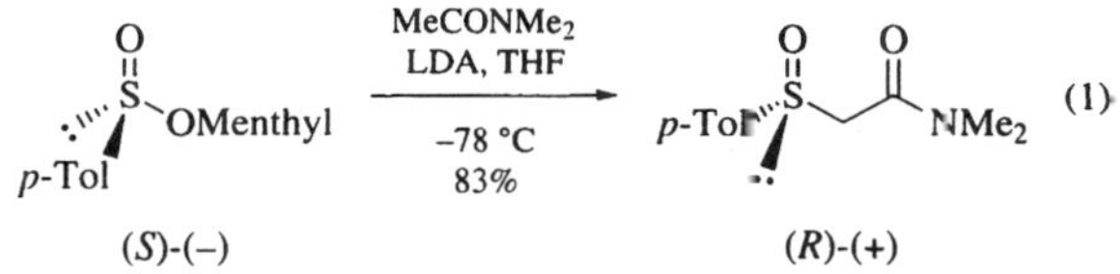

The Pummerer reaction of optically active (*R*)-(+)-α-(*p*-tolylsulfinyl)-*N*,*N*-dimethylacetamide with *Acetic Anhydride* in the presence of *1,3-Dicyclohexylcarbodiimide* is highly stereoselective, affording the corresponding α-acetoxy sulfide in moderate yield but with nearly 70% ee (eq 2).[3,4] The recovered starting sulfoxide is obtained in 63% yield.

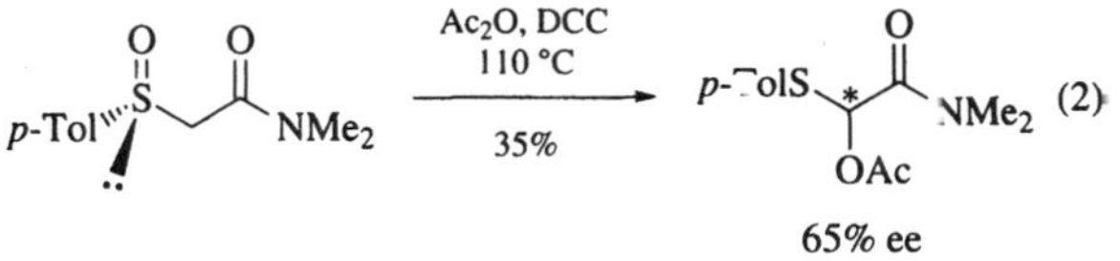

Good asymmetric induction is also observed during the aldol-type condensation of the magnesium enolate of (*R*)-(+)-α-(*p*-tolylsulfinyl)-*N*,*N*-dimethylacetamide with aldehydes (eq 3).[1,2]

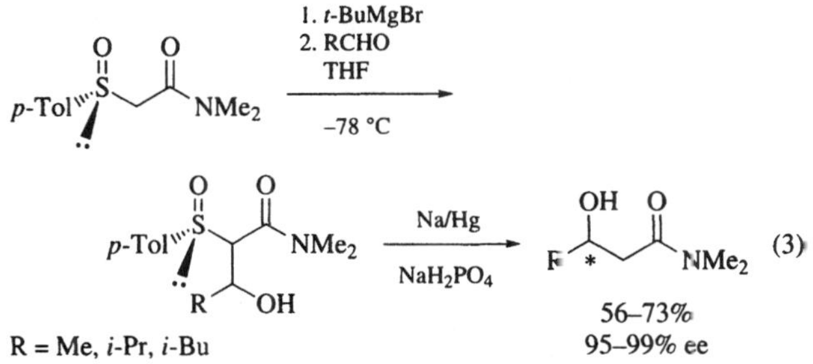

A model (**1**), similar to that proposed for aldol-type condensation of α-sulfinyl esters,[5] has been proposed to predict the chirality of the resulting β-hydroxy amides.

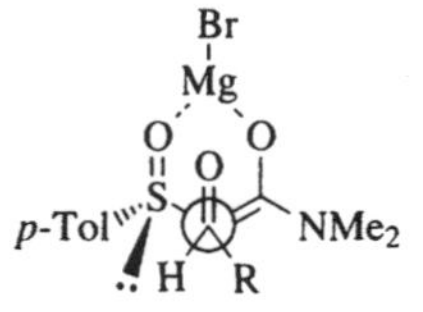

(**1**)

1. Annunziata, R.; Cinquini, M.; Cozzi, F.; Montanari, F.; Restelli, A. *Chem. Commun./J. Chem. Soc., Chem. Commun.* **1983**, 1138.
2. Annunziata, R.; Cinquini, M.; Cozzi, F.; Montanari, F.; Restelli, A. *Tetrahedron* **1984**, *40*, 3815.
3. Numata, T.; Itoh, O.; Oae, S. *Tetrahedron Lett.* **1979**, *21*, 1869.
4. Numata, T.; Itoh, O.; Yoshimura, T.; Oae, S. *Bull. Chem. Soc. Jpn.* **1983**, *56*, 257.
5. Mioskowski, C.; Solladié, G. *Tetrahedron* **1980**, *36*, 227.

Guy Solladié & Françoise Colobert
University Louis Pasteur, Strasbourg, France

(3*R*)-(*p*-Tolylsulfinyl)-*N*-methoxyacetimidic Acid Ethyl Ester

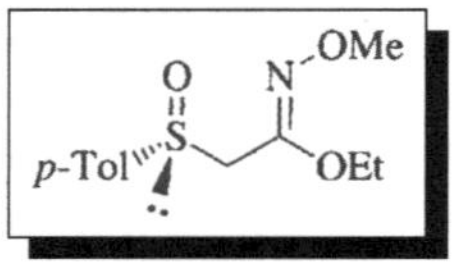

[95614-76-9] $C_{12}H_{17}NO_3S$ (MW 255.37)

(asymmetric aldol-type condensation[1])

Alternate Name: ethyl (*p*-tolylsulfinyl)-*N*-methoxyacetimidate.
Physical Data: $[\alpha]_D = +28°$ ($c = 1$, $CHCl_3$).
Preparative Methods: prepared in 89% yield by reacting the lithium enolate of ethyl *N*-methoxyacetimidate with (−)-(*S*)-menthyl *p*-toluenesulfinate (see *(−)-(1R,2S,5R)-Menthyl (S)-p-Toluenesulfinate*) (eq 1).[1]

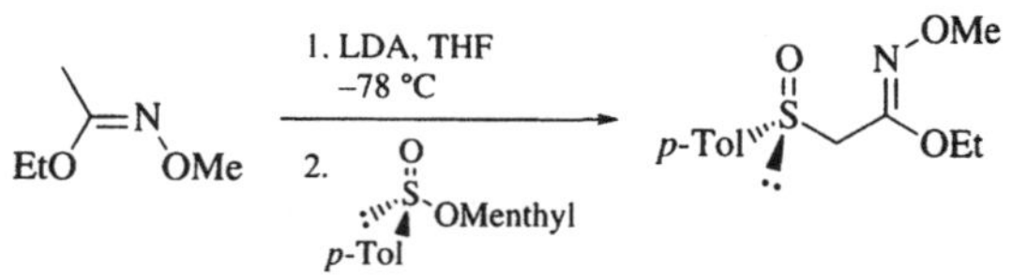

(1)

Aldol-Type Condensation. Aldol-type condensation of the lithium enolate of ethyl (*R*)-(*p*-tolylsulfinyl)-*N*-methoxyacetimidate (prepared with *n-Butyllithium*) with aldehydes affords, after desulfurization with *Raney Nickel*, β-hydroxy esters with high enantioselectivity (eq 2).[1]

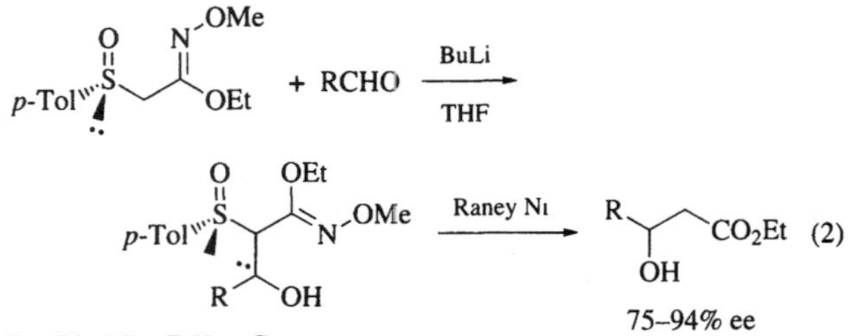

R = Ph, *i*-Pr, C_5H_{11}, Cy

Anti diastereoselectivity gives the optically active (*S*)-β-hydroxy ester while *syn* diastereoselectivity leads to the (*R*)-β-hydroxy ester, via a chelated six-membered transition state (eq 3). Since the *anti* intermediate is more stable, the (*S*)-β-hydroxy ester predominates under thermodynamic conditions (Table 1, entry 1). Higher diastereoselectivity is achieved by changing the counterion from lithium to a more chelating one such as zinc (Table 1, entry 2). On the other hand, in order to obtain diastereoselection under kinetic control, zirconium enolates (prepared by treating the lithium enolate with *Dichlorobis(cyclopentadienyl)zirconium*) are used, leading to the (*R*)-β-hydroxy ester (Table 1, entry 3) in high yield.

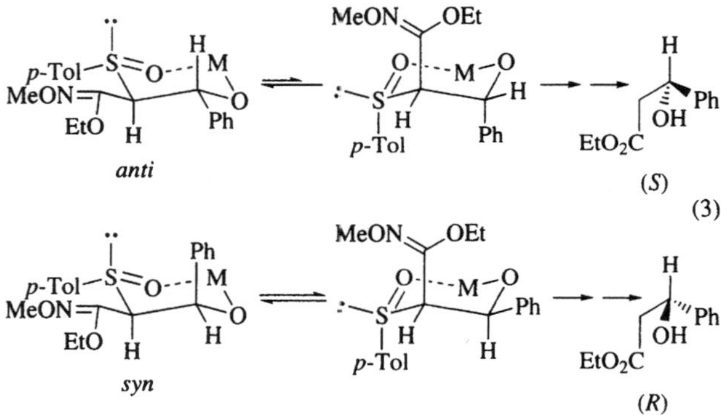

1. Bernardi, A.; Colombo, L.; Gennari, C.; Prati, L. *Tetrahedron* **1984**, *40*, 3769.

Guy Solladié & Françoise Colobert
University Louis Pasteur, Strasbourg, France

Table 1 Diastereoselection in the Aldol-Type Reaction

Entry	Aldehyde	Metalation conditions	Condensation conditions	Abs. conf.	ee
1	R = Ph	BuLi, −78 °C, 15 min	−78 °C to 0 °C, 4 h	(*S*)	75%
2	R = Ph	BuLi, −78 °C, 15 min 1 equiv. $ZnCl_2$, −78 °C, 30 min	−78 °C to 0 °C, 4 h 20 °C, 30 min	(*S*)	86%
3	R = Ph	BuLi, −78 °C, 15 min 2 equiv. Cp_2ZrCl_2, −78 °C, 30 min	−78 °C, 2.5 h	(*R*)	88%

(*R*)-(+)-3-(*p*-Tolylsulfinyl)propionic Acid

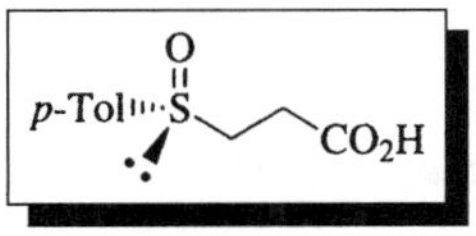

[90334-31-9] $C_{10}H_{12}O_3S$ (MW 212.29)

(asymmetric aldol-type condensation[2])

Physical Data: $[\alpha]_D = +180°$ ($CHCl_3$, $c = 0.7$), $[\alpha]_D = +188°$ (MeOH, $c = 0.7$).

Preparative Methods: conveniently prepared in 76% yield by addition of a suspension of lithium bromoacetate to a solution of the anion of *(R)-(+)-Methyl p-Tolyl Sulfoxide* (eq 1).[1,2]

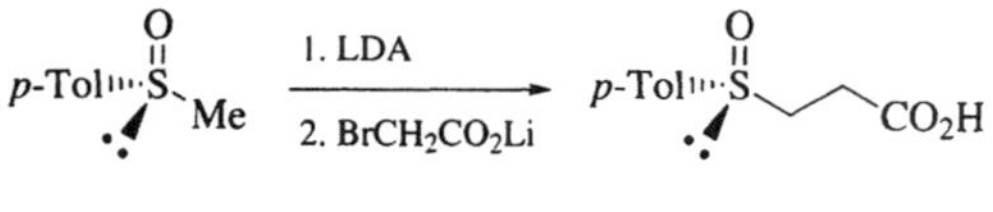

(1)

Aldol-Type Condensation. Dimetalation of (*R*)-(+)-3-(*p*-tolylsulfinyl)propionic acid with *Lithium Diisopropylamide* produces a chiral homoenolate dianion equivalent which reacts with carbonyl compounds to afford β-sulfinyl-γ-hydroxy acids; these spontaneously cyclize to give the corresponding β-sulfinyl γ-lactones (eq 2).[1,2]

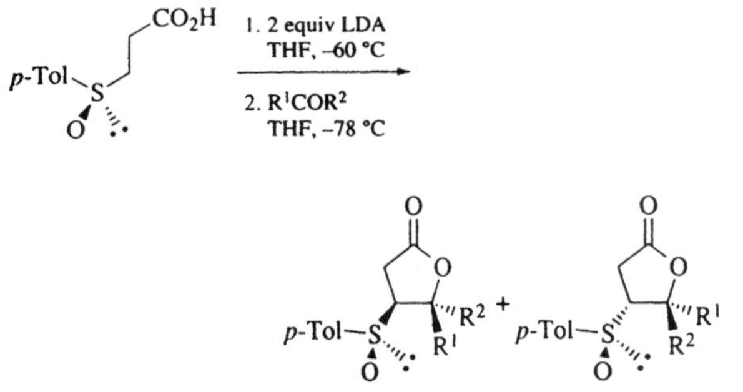

(2)

(3*S*,4*R*,R_S) (3*R*,4*R*,R_S)

R^1 = H, R^2 = *t*-Bu; R^1 = H, R^2 = Ph; R^1 = *t*-Bu, R^2 = Me

Two new chiral carbon atoms are formed in the condensation and four diastereoisomeric β-sulfinyl γ-lactones can therefore in principle be obtained. However, only two diastereoisomers, (3*S*,4*R*,R_S) and (3*R*,4*S*,R_S), are isolated when the carbanion is condensed with pivalic aldehyde, benzaldehyde, or pinacolone (yield 65–70% for aldehydes, ratio 53:47; yield 47% for pinacolone, ratio 81:19). The diastereoselectivity decreases when the two substituents of the carbonyl group are sterically similar. However, single diastereoisomers can easily be separated through chromatography and transformed in high yield into both enantiomers of optically pure saturated (by desulfurization) and α,β-unsaturated γ-lactones (by pyrolytic sulfoxide elimination) (eq 3). The relative and absolute stereochemistry of all the products have been determined by circular dichroism, nuclear Overhauser effects, and X-ray analyses.

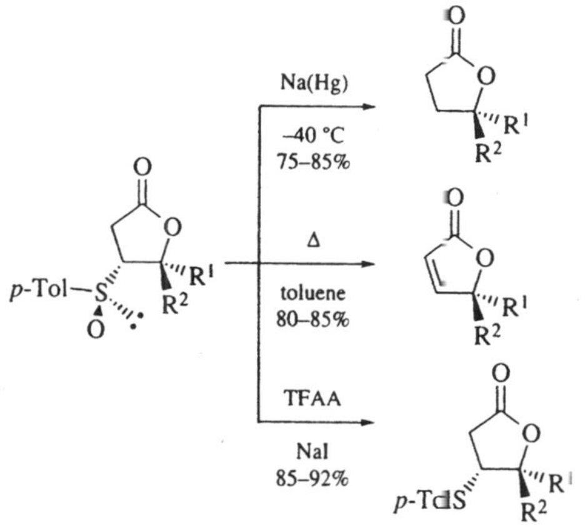

(3)

The condensation of the dilithio derivative of (*R*)-(+)-3-(*p*-tolylsulfinyl)propionic acid with protected glycoaldehydes (*O*-*t*-butyl and *O*-benzyl) gives 5-alkoxy-4-hydroxy-3-(*p*-tolylsulfinyl)pentanoic acids, which spontaneously cyclize to the corresponding 3-sulfinyl-4-alkoxymethyl butanolides (eq 4).[3] Pure diastereomers can be separated by flash chromatography and are obtained in comparable amounts. The corresponding optically pure butenolides are obtained by pyrolytic elimination of the sulfoxides and then transformed into natural (–)-(*R*)-umbelactone (eq 5).

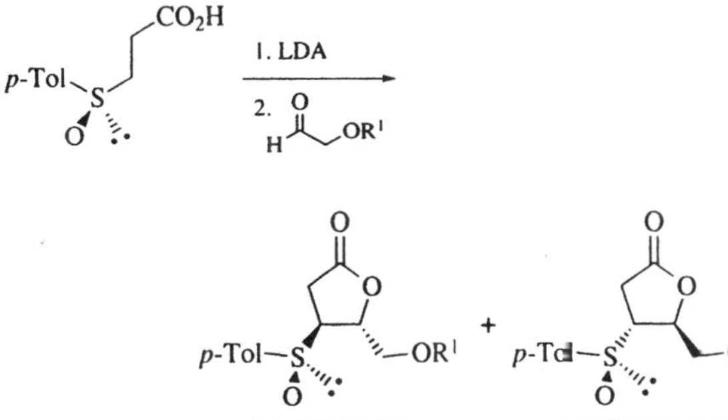

(4)

(+)-(3*S*,4*R*,R_S) (+)-(3*R*,4*S*,R_S)

R^1 = *t*-Bu, 36%; R^2 = CH_2Ph, 52%

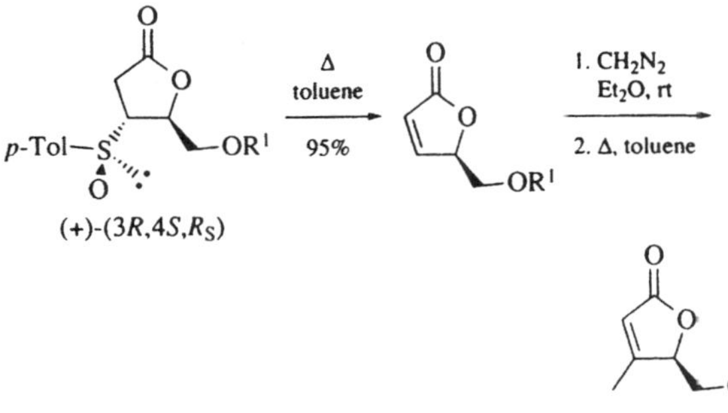

(5)

(+)-(*R*), 89%

1. Bravo, P.; Carrera, P.; Resnati, G.; Ticozzi, C. *Chem. Commun./J. Chem. Soc., Chem. Commun.* **1984**, 19.
2. Albinati, A.; Bravo, P.; Ganazolli, F.; Resnati, G. Viani, F. *J. Chem. Soc., Perkin Trans. 1* **1986**, 1405.
3. Bravo, P.; Resnati, G.; Viani, F. *Gazz. Chim. Ital.* **1987**, *117*, 747.

Guy Solladié & Françoise Colobert
University Louis Pasteur, Strasbourg, France

N-[4-(Trifluoromethyl)benzyl]-cinchoninium Bromide[1]

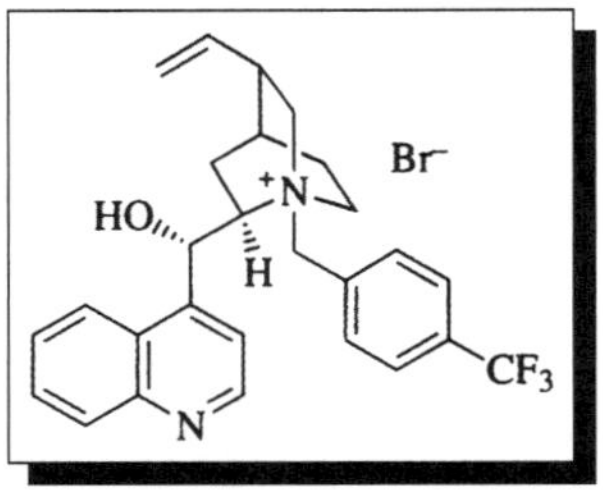

[*95088-20-3*] $C_{27}H_{28}BrF_3N_2O$ (MW 533.47)

(chiral phase-transfer catalyst for asymmetric alkylations,[2a] amino acid synthesis,[7] hydroxylations,[6] Michael additions,[3] and Robinson annulations[2b])

Physical Data: mp 245 °C (dec).
Solubility: $<10^{-5}$ M in toluene; 20 mM in toluene as dimer.[2a]
Form Supplied in: crystalline salt; commercially available. Can contain anywhere from 0 to 25 mol % of the dihydro analog, usually 15 mol %.
Handling, Storage, and Precautions: do not breath dust; avoid contact with skin and eyes.

Asymmetric Alkylation.[1,2] *N*-[4-(Trifluoromethyl)benzyl]-cinchoninium bromide (**1**) has been used as chiral phase-transfer catalyst[1,2] in the alkylation of indanones (eq 1).[2a] For the alkylation of α-aryl-substituted carbonyl compounds the diastereomeric *N*-[4-(trifluoromethyl)benzyl]cinchonidinium bromide (**2**) was used to obtain the opposite stereochemistry (eqs 2 and 3).[5] The asymmetric alkylation of oxindoles was used as the key step in an asymmetric synthesis of (−)-physostigmine (eq 4).[4]

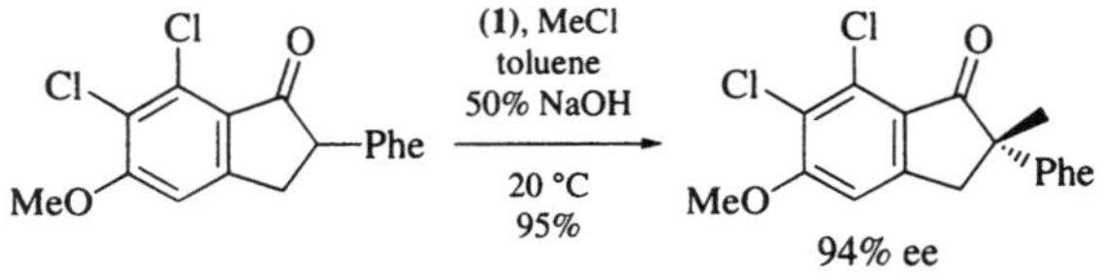

(1)

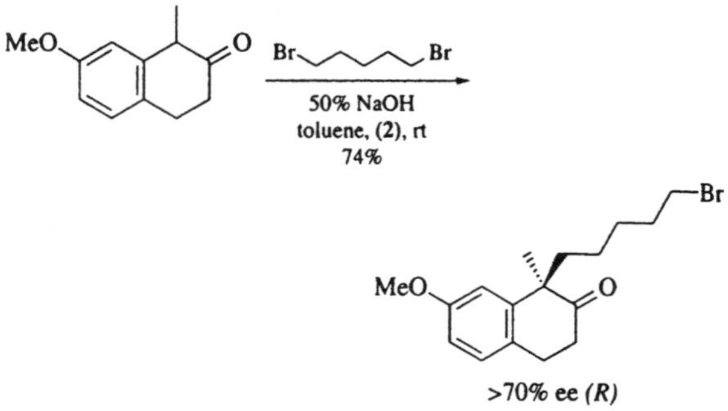

(2)

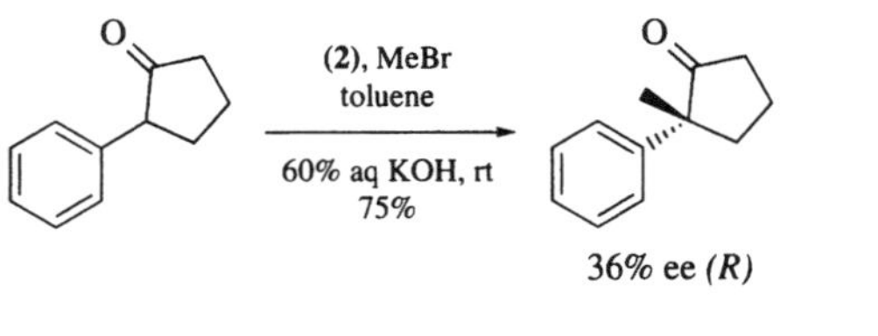

(3)

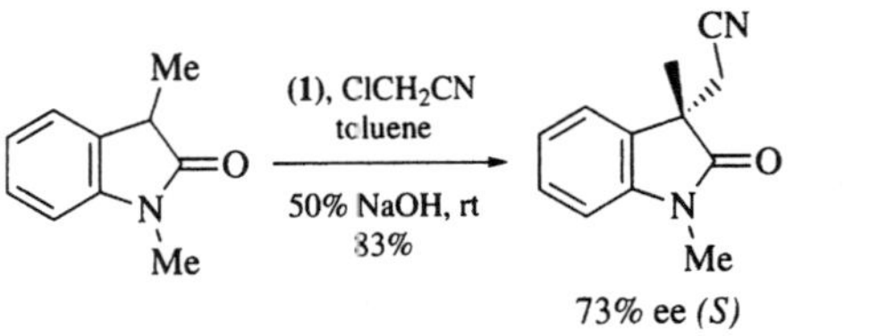

(4)

In all cases it was reported that the trifluoromethyl group enhances the interaction in the prochiral ion pair, resulting in higher ee. The exception appears to be the asymmetric synthesis of α-amino acids via alkylation of the benzophenone Schiff base of glycine alkyl esters with allyl bromide, which produced a 56% ee with the trifluoromethyl-substituted catalyst compared to 66% with the unsubstituted catalysts *N*-benzylcinchoninium chloride (**3**) or *N*-benzylcinchonidinium chloride (**4**) (eq 5).[7]

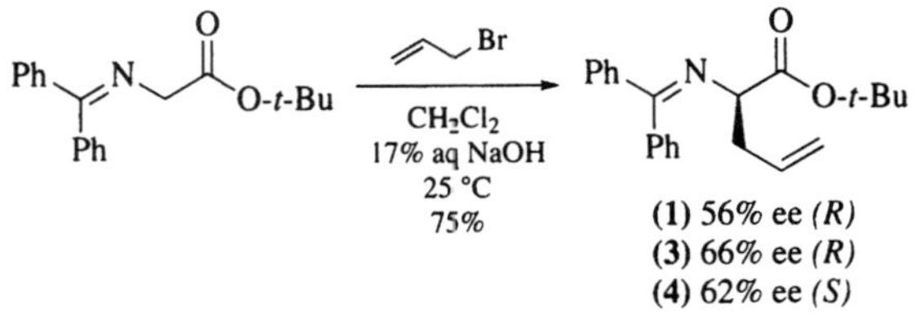

(5)

The unsubstituted catalyst (**3**) was also used in an asymmetric Gabriel synthesis of α-amino acids via solid–liquid chiral phase-transfer alkylation of potassium phthalimide with 2-bromocarboxylates.[10]

In general, nonpolar solvents and less reactive alkyl halides (Cl > Br > I) give higher ee values. Aqueous 50% NaOH is the preferred base, as it acts as a dehydrating agent and keeps the organic solvent dry, thus promoting a tight ion pair. During the reaction the catalyst is extracted into the organic layer as a dimer of ammonium bromide and its zwitterionic oxide,[2a] but it reacts as a monomer resulting in an order of 0.5 for the catalyst in the alkylation reaction. The catalyst degrades during the reaction via Hofmann elimination to tertiary amines, which are readily removed by acid extractions during the workup.[2a] Therefore the catalyst usually cannot be recovered. An economical alternative catalyst is *N*-(3,4-dichlorobenzyl)cinchoninium chloride, which may give equivalent results.[2a,3,4a,9] Reduction in catalyst concentration and reaction times may be achieved by addition of a very small amount of a PEG cocatalyst (PEG 400 or PEG *p*-isooctylphenyl ether).[2e,9] The ee will depend on the interaction of the catalyst with the substrate in the ion pair and in certain cases an electron-donating substituent on the benzyl group may provide optimal interaction, as was found for an asymmetric Michael reaction using *N*-methyl-*N*-benzyl ephedrinium bromide as the phase-transfer catalyst.[8]

Asymmetric Michael Addition. A chiral catalytic addition of methyl vinyl ketone to 2-propylindanone in 93% yield and 80% ee (*S* enantiomer) has been reported (eq 6).[3]

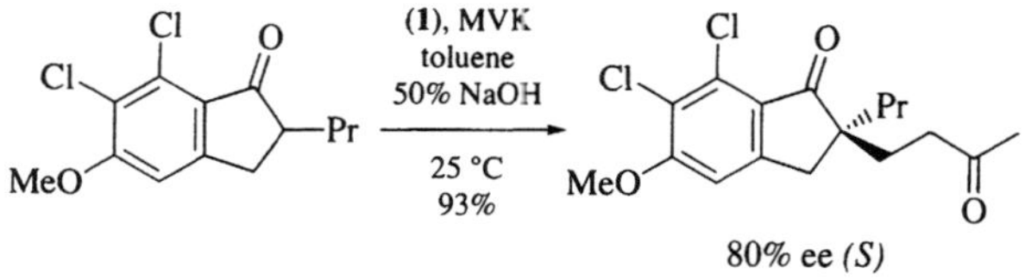

(6)

The (*R*) enantiomer was prepared with 40% ee using the cinchonidine catalyst (**2**). When the vinyl group of the catalyst was hydrogenated to the ethyl group (*N*-[4-(trifluoromethyl)benzyl]dihydrocinchonidinium bromide, **5**), the ee improved to 52%. Equally good results were obtained using the basic catalyst dimer[2a] in a homogeneous system which allowed the use of Michael acceptors not compatible with hydroxide bases.

Asymmetric Hydroxylation. The catalyst has been used for asymmetric α-hydroxylations of indanones and α-tetralones using the standard conditions in combination with oxygen and *Triethyl Phosphite* (eq 7).[6] Substituents on the aromatic ring of the substrates will influence the π–π interaction in the ion pair and affect the ee. Similarly, (*E*)-2-ethylidene-1-tetralone was oxidized to the α-hydroxy ketone (eq 8).

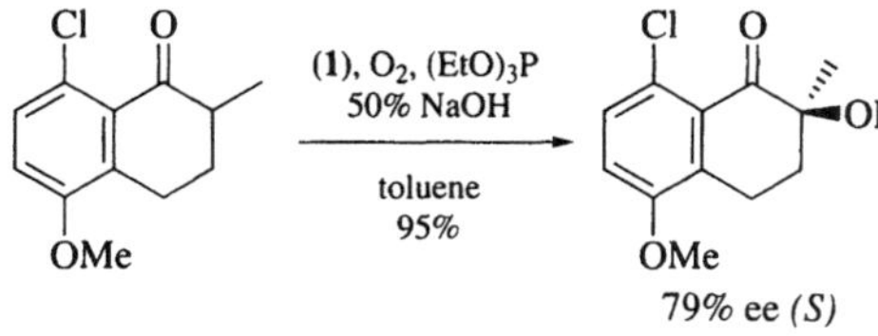

(7)

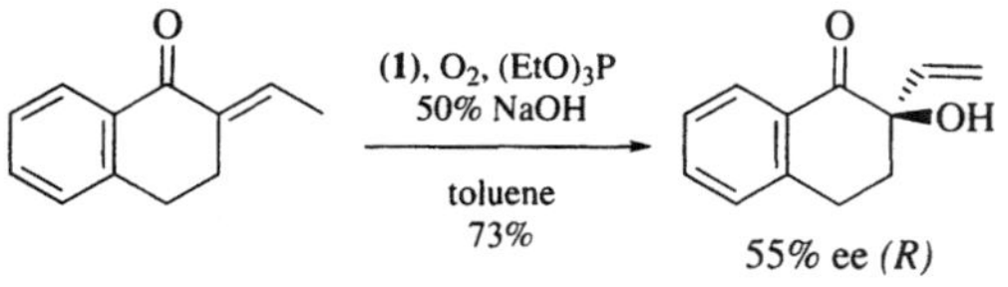

(8)

Asymmetric Robinson Annulation. 2-Propyl-1-indanone undergoes Robinson annulation with the catalyst and methyl vinyl ketone (eq 6).[3] Higher ee values were achieved using 1,3-dichloro-2-butene (Wichterle Reagent) as an MVK surrogate for the Michael addition and overall Robinson annulation (eq 9).[2b,d,e]

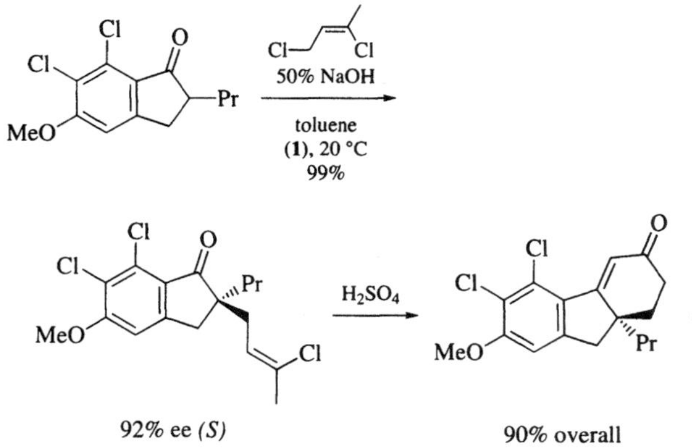

(9)

Using (**2**) as catalyst provided the (*R*) enantiomer in 99% yield, 78% ee. The key introduction of asymmetry during the synthesis of (+)-podocarp-8(14)-en-13-one was the phase-transfer-catalyzed Robinson annulation of 6-methoxy-1-methyl-2-tetralone with ethyl vinyl ketone. The authors carried out a comparative study of the *N*-(4-trifluoromethyl)benzyl derivatives of cinchonine, cinchonidine, dihydrocinchonine, and dihydrocinchonidine and found that (**5**) produced the highest ee of the desired (*S*) enantiomer at −45 °C using toluene and 60% aq KOH (eq 10).[5]

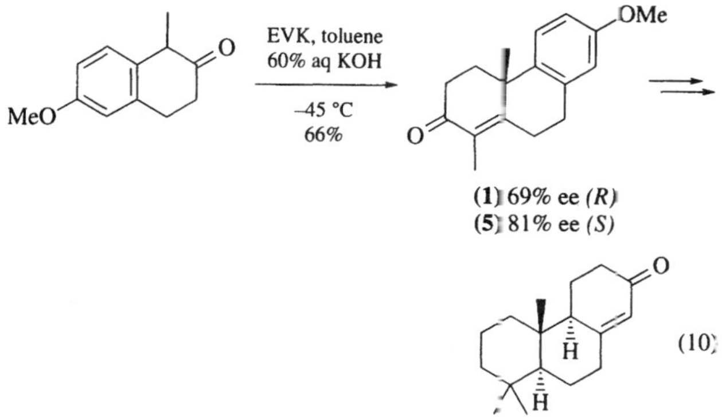

(10)

1. Dehmlow, E. V.; Dehmlow, S. S. *Phase Transfer Catalysis*; VCH: Weinheim, 1993; pp 80–91.
2. (a) Hughes, D. L.; Dolling, U.-H.; Ryan, K. M.; Schoenewaldt, E. F.; Grabowski E. J. J. *J. Org. Chem.* **1987**, *52*, 4745. (b) Bhattacharya, A.; Dolling, U.-H.; Grabowski, E. J. J.; Karady, S.; Ryan, K. M.; Weinstock, L. M. *Angew. Chem., Int. Ed. Engl.* **1986**, *25*, 476. (c) Dolling, U.-H.; Davis, P.; Grabowski, E. J. J. *J. Am. Chem. Soc.* **1984**, *106*, 446. (d) Dolling, U.-H.; Hughes, D. L.; Bhattacharya, A.; Ryan, K. M.; Karady, S.; Weinstock, L. M.; Grabowski, E. J. J. In *Phase-Transfer Catalysis*; Starks, C. M., Ed.; American Chemical Society: Washington, 1987; pp 67–81. (e) Dolling, U.-H.; Hughes, D. L.; Bhattacharya, A.; Ryan, K. M.; Karady, S.; Weinstock, L. M.; Grenda, V. J.; Grabowski, E. J. J. In *Catalysis of Organic Reactions*; Rylander, P. N.; Greenfield, H.; Augustine, R. L.; Eds.; Dekker: New York, 1988; pp 65–86.
3. Conn, R. S. E.; Lovell, A. V.; Karady, S.; Weinstock, L. M. *J. Org. Chem.* **1986**, *51*, 4710.
4. (a) Lee, T. B. K.; Wong, G. S. K. *J. Org. Chem.* **1991**, *56*, 872. (b) Chen, B. H.; Ji, Q. E. *Acta Chim. Sinica* **1989**, *47*, 350 (*Chem. Abstr.* **1989**, *111*, 194 508).
5. Nerinckx, W.; Vandewalle, M. *Tetrahedron: Asymmetry* **1990**, *1*, 265.
6. Masui, M.; Ando, A.; Shioiri, T. *Tetrahedron Lett.* **1988**, *29*, 2835.
7. (a) O'Donnell, M. J.; Bennett, W. D.; Wu, S. *J. Am. Chem. Soc.* **1989**, *111*, 2353. (b) O'Donnell, M. J.; Wu, S. *Tetrahedron: Asymmetry* **1992**, *3*, 591. (c) Imperiali, B.; Prins, T. J.; Fisher, S. L. *J. Org. Chem.* **1993**, *58*, 1613.
8. Loupy, A.; Zaparucha, A. *Tetrahedron Lett.* **1993**, *34*, 473.
9. Dolling, U.-H. U. S. Patent 4 605 761, 1986 (*Chem. Abstr.* **1987**, *106*, 4697).
10. Guifa, S.; Lingchong, Y. *Synth. Commun.* **1993** *23*, 1229.

Ulf-H. Dolling
Merck Research Laboratories, Rahway, NJ, USA

N,N,N′-Trimethyl-*N′*-(2-{[(1*R*)-1-phenyl-2-(1-piperidinyl)ethyl]amino}ethyl)-1,2-ethanediamine

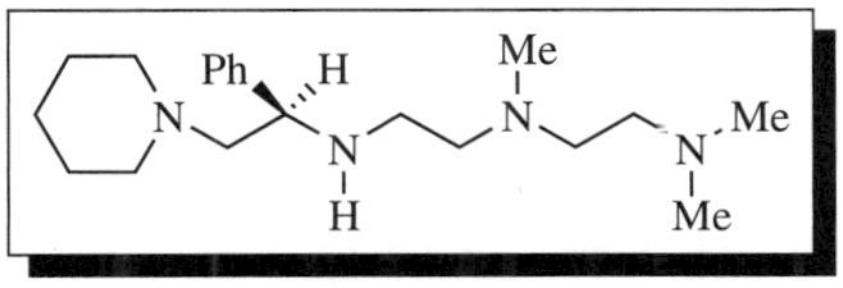

[157303-88-3] $C_{20}H_{36}N_4$ (MW 332.53)

(chiral tetradentate ligand that has been shown to be an effective auxiliary for enantioselective alkylation,[1–7] Michael additions,[7–9] and aldolization[10] in stoichiometric and in some cases catalytic amounts)

Alternate Name: (*R*)-(−)-*N*-{2-[*N*-(2-dimethylaminoethyl)-*N*-methylamino]ethyl}-1-phenyl-2-piperidinoethylamine; (*R*)-*N*′-[2-(dimethylamino)ethyl]-*N*-methyl-*N*-[1-phenyl-2-(1-piperidinyl)ethyl]-1,2-ethanediamine; *N*-[(2*R*)-6,9-dimethyl-2-phenyl-3, 6, 9-triazadecyl]piperidine.

Physical Data: $[\alpha]_D^{25}$ −57.1 (*c* 2.0, benzene).

Solubility: most organic solvents.

Form Supplied in: clear, colorless oil; not commercially available.

Analysis of Reagent Purity: ^{1}H NMR; Elemental Analysis.

Purification: column chromatography (silica, $CHCl_3$/MeOH 9:1 then $CHCl_3$/*i*-$PrNH_2$ 20:1) followed by bulb-to-bulb distillation (290 °C bath temperature, 0.8 mmHg).

Preparative Methods: the original literature[11] reports that (*R*)-(−)-*N*-{2-[*N*-(2-dimethylaminoethyl)-*N*-methylamino] ethyl}-1-phenyl-2-piperidinoethylamine (**1**) can be prepared from (*R*)-phenylglycine in six steps. Thus, (*R*)-phenylglycine is first protected as the *N*-Cbz-amino acid, and is then condensed with piperidine in the presence of diethylphosphorocyanidate (DEPC) and triethylamine to provide the corresponding amide. Removal of the Cbz protecting group under acidic conditions gives amino amide, which is subsequently reduced with $LiAlH_4$. The primary amine is amidated upon treatment with *N*-[2-(dimethylamino)ethyl]-*N*-methylglycine and DEPC, and the resulting product is reduced with BH_3·THF to afford **1** (eq 1).

Handling, Storage, and Precautions: presumably, as with all amines, air-oxidation will occur over time; store in a cool, dry place away from light; avoid oxidizing agents.

Introduction. There have been numerous studies focused on asymmetric methods in synthetic organic chemistry.[12] These investigations can be classified into two main categories: either diastereoselective or enantioselective. In the diastereoselective strategy, an appropriate substrate is covalently attached to a chiral auxiliary and the incipient stereogenic center is introduced via an intramolecular bias established by the chiral appendage. In the enantioselective approach, an achiral substrate is directly transformed into a chiral product via an intermolecular interaction it establishes with the chiral auxiliary. Koga *et al.* have shown that chiral tetradentate amines such as **1** can be used in enantioselective synthesis.[1–10] Treatment of an achiral lithium enolate with **1** and lithium bromide generates a ternary complex, which reacts in an enantioselective manner with electrophiles.

The structure of **1** is similar to lithium diisopropyl amide (LDA) in that there are two bulky alkyl groups attached to the amide nitrogen.[6,11] In **1**, however, one of the alkyl groups has been modified to contain a chiral center at the α-position and a piperidinyl substituent at the β-position. Upon deprotonation, the tertiary amino group of the piperidine acts as an internal ligation site for lithium (eq 2). The *N*-lithio derivative of **1** has a number of useful characteristics: (i) in solution it will form a stable, five-membered chelated structure; (ii) because the α-phenyl substituent will, for steric reasons, orient itself exclusively *trans* to the other alkyl appendage, the lone pair electrons residing on the amide nitrogen must reside *cis* to the phenyl ring in the chelate, thus making this nitrogen chiral; (iii) in solution, aggregates of the complex will form to satisfy lithium's valency; and (iv) the use of an external additive could be used to control the degree of aggregation in solution.[6,11]

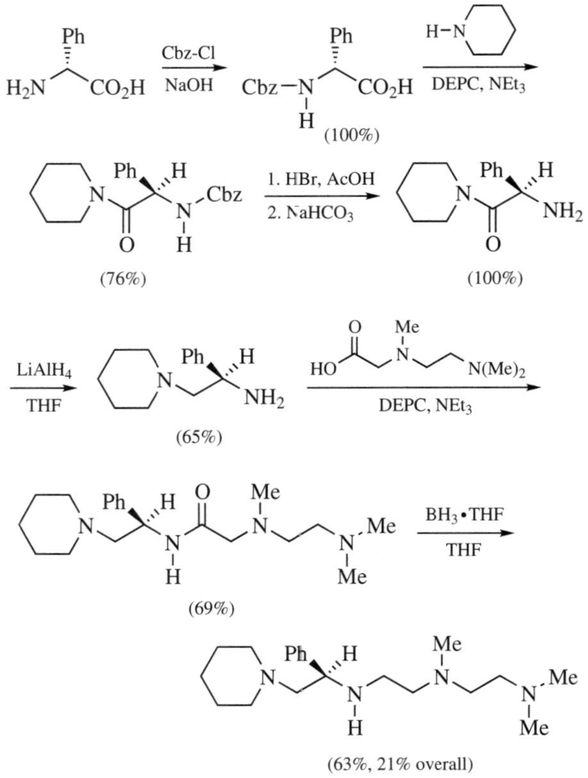

(1)

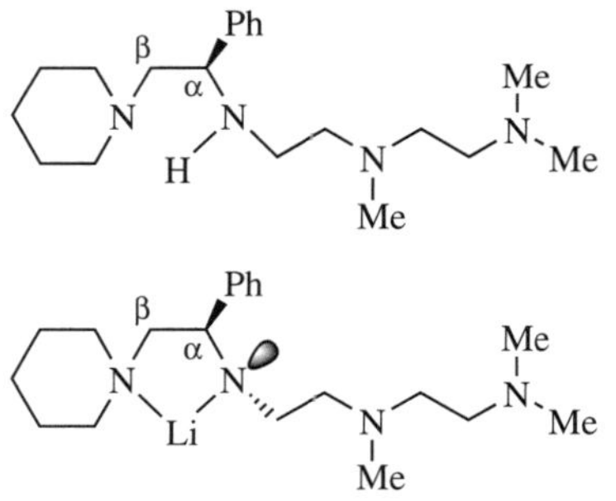

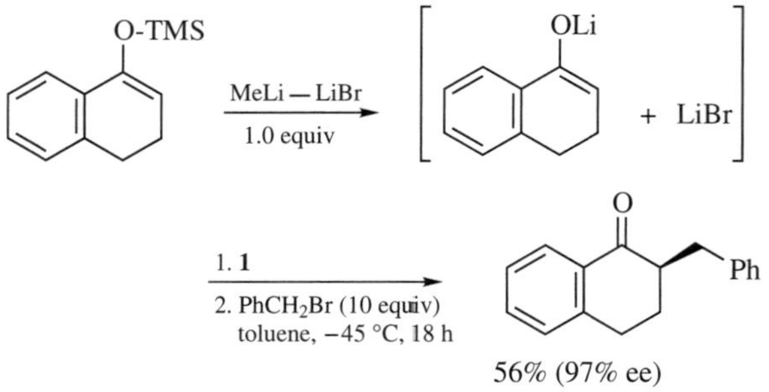

(2)

Enantioselective Alkylations and Catalytic Asymmetric Alkylations. α-Substitution of a carbonyl-containing substrate via generation of an enolate ion followed by subsequent reaction with an electrophile remains one of the most fundamental transformations in synthetic organic chemistry. A more recent advance to this type of transformation is the ability to perform this conversion in an enantioselective manner. This type of alkylation is illustrated by the reaction of **1** with the lithium enolate derived from 1-tetralone (eq 2).

The lithium enolate of cyclohexanone[1,4,6] has been used as an efficient substrate for this same reaction; 53% (90% ee). Alternatively, the parent carbonyl compound can be employed if the lithium amide of **1**, prepared by treating **1** with 1.0 equiv of *n*-BuLi, is used instead of the amine.[6] Both the chemical yield and the degree of asymmetric induction are dependent on reaction conditions, e.g. solvent and reaction time. It has been observed[4,6] that in strongly ligating solvents (e.g. DME, THF, or diethyl ether), the yield is higher, however, the degree of asymmetric induction tends to be lower. Opposite trends are observed in non-ligating solvents (e.g. toluene). Additionally, as the reaction time is lengthened the degree of asymmetric induction increases. This observation has been correlated to the concentration of lithium bromide present in the reaction mixture. Initially, there is no lithium bromide in solution, however, as the alkylation proceeds lithium bromide is generated in situ. Accordingly, if lithium bromide is added at the beginning of the reaction, the % ee is dramatically increased. It is, therefore, most convenient to perform this reaction using the silyl enol ether substrate and to treat it with methyllithium-lithium bromide to generate the lithium enolate-lithium bromide complex (eq 3, Table 1).

Table 1 Reaction of lithium enolates with R-X

Solvent	R–X	Yield (%)	% ee
Toluene[a]	$PhCH_2Br$	86	96
DME	$PhCH_2Br$	95	87
Toluene	MeI	64	98
DME	MeI	75	56

[a]DME (8.0 equiv) was added to complete desilylation.

This methodology has also been applied to the alkylation of five- and six-membered *N*-alkylated lactams and lactones[3] (eq 4 and 5). In both cases, **1** is first converted to the corresponding lithium amide and pre-complexed with lithium bromide. Furthermore, in the case of the lactams, it was observed that the use of 2,2,5,5-tetramethyltetrahydrofuran (TMTHF) as the solvent resulted in higher yields and greater enantiomeric excess.

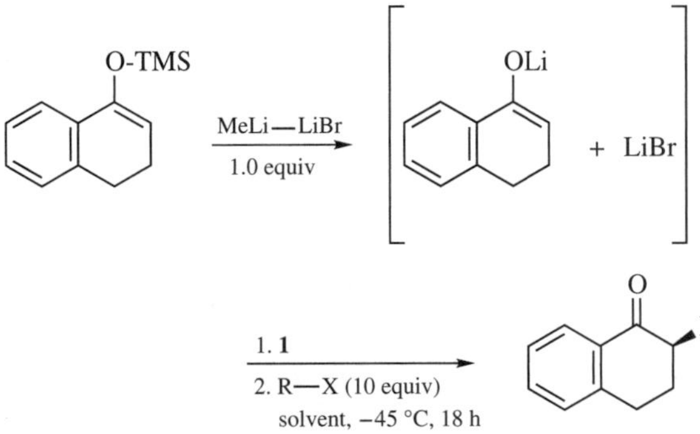

(3)

Solvent	R—X	Yield(%)	% ee
toluene*	$PhCH_2Br$	86	96
DME	$PhCH_2Br$	95	87
toluene	MeI	64	98
DME	MeI	75	56

*DME (8.0 equiv) was added to complete desilylation

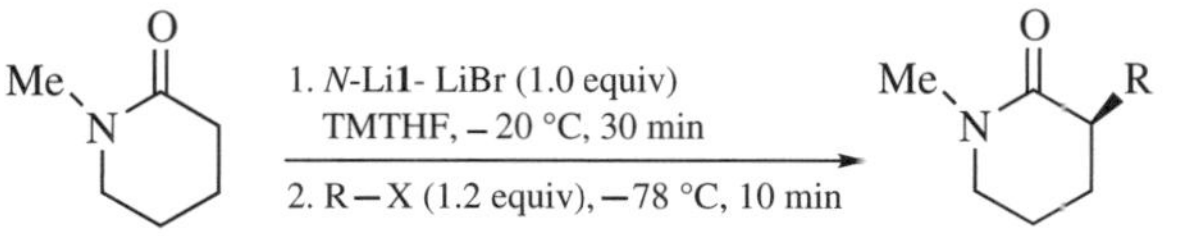

(4)

R—X	Yield(%)	% ee
$PhCH_2Br$	74	89
$PhCH{=}CHCH_2Br$	55	96
(2- Naphthyl)CH_2Br	55	97

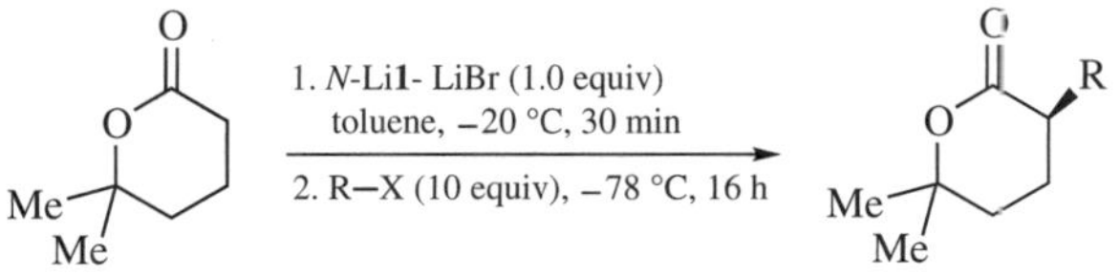

(5)

R—X	Yield(%)	% ee
$PhCH_2Br$	74	90
$PhCH{=}CHCH_2Br$	64	85
(2- Naphthyl)CH_2Br	63	90

The use of **1** for the preparation of a chiral quaternary center via asymmetric alkylation has also been investigated.[2] Although **1** has proven to be an effective reagent for the enantioselective generation of tertiary centers, its use for generating quaternary centers has been of only marginal use (eq 6). However, other chiral tetradentate amines can be used for this purpose.[2]

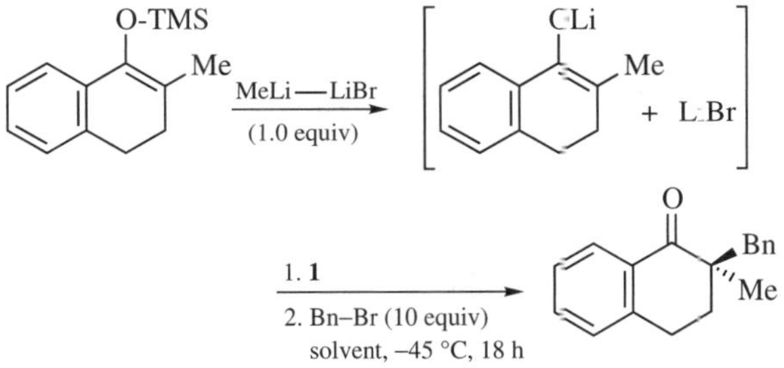

(6)

In an extension of this methodology, it has been demonstrated that in some cases the enantioselective alkylation of lithium enolates can be achieved by means of a catalytic amount of **1**.[1,5–7] As in the stoichiometric version (*vide supra*), the reaction conditions play a crucial role in determining the yield and % ee. One fundamental modification in the catalytic version is the addition of two equiv of an achiral bidentate amine [e.g *N,N,N',N'*-tetramethylethylenediamine (TMEDA) or *N,N,N',N'*-tetramethylpropylene diamine (TMPDA)] to trap the large excess of lithium bromide present at the beginning of the reaction. This catalytic asymmetric variant is illustrated by the reaction of the lithium enolate of 1-tetralone with a variety of electrophiles (eq 7). In this example, the optimal reaction conditions were determined to be 0.05 equiv of **1**, 2.0 equiv of TMPDA, and 10.0 equiv of the alkyl halide.

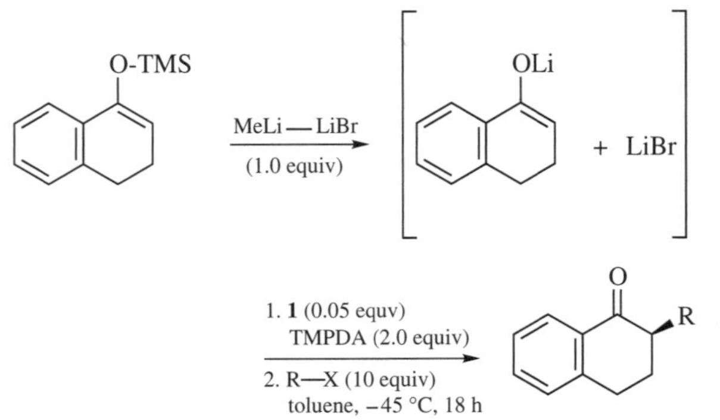

(7)

R—X	Yield(%)	% ee
$PhCH_2Br$	82	95
$CH_2{=}CH{-}CH_2Br$	69	96
$Me_2C{=}CH{-}CH_2Br$	62	97
$PhCH{=}CHCH_2Br$	84	81
$MeO_2C{-}CH_2Br$	63	90

Enantioselective Aldol Reactions. The use of **1** for generating two contiguous stereocenters via an asymmetric aldol condensation has also been investigated,[10] but only with marginal success. For example, reaction of the lithium enolate derived from *tert*-butyl propionate with the *N*-lithio derivative of **1**, followed by condensation with benzaldehyde, provided a mixture of *anti* and *syn* aldol products in poor-to-modest % ee (eq 8).

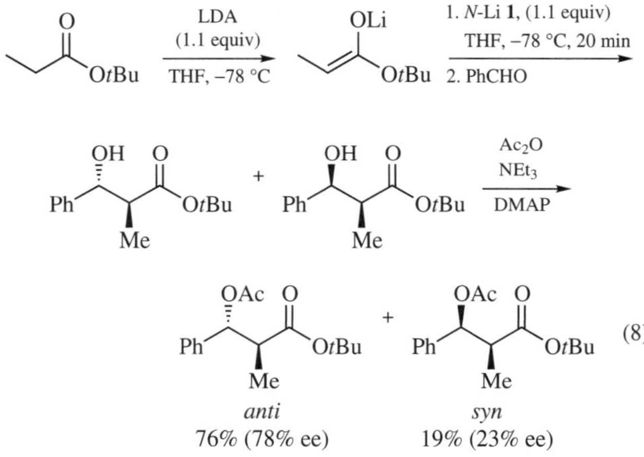

(8)

Although **1** is of only limited utility, further studies have shown that other chiral tetradentate amines can perform this type of transformation with yields for the *anti* product greater than 80% and in greater than 95% ee.

Enantioselective Michael Additions. Amine **1** has also been used as an effective ligand for enantioselective Michael reactions of ketone lithium enolate donors with various benzylidene acceptors.[9] As representative examples, the lithium enolates of aryl methyl ketones were reacted with dimethyl benzylidenemalonate in the presence of **1** (eq 9). The lithium enolate was generated from the corresponding ketone by treatment with hexamethyldisilazide in the presence of lithium bromide in toluene. The resulting enolate was then exposed to **1** and allowed to stir for 30 min to form the desired ternary complex. After addition of the benzylidene acceptor, the desired products were isolated in acceptable yields and with high % ee.

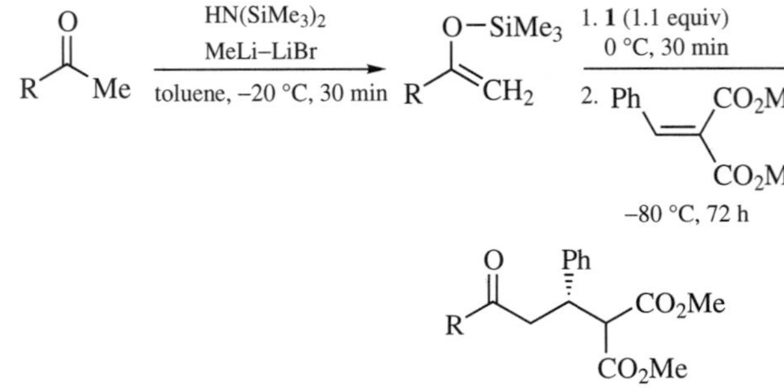

(9)

R	Yield (%)	% ee
Ph	93	92
4-Me-Ph	80	94
4-MeO-Ph	52	93
2-Naphthyl	76	90

In an analogous manner,[8] α -substituted phenyl ketones have been used to afford Michael adducts containing two vicinal chiral tertiary centers in both high diastereo- and enantioselectivity (eq 10, Table 2).

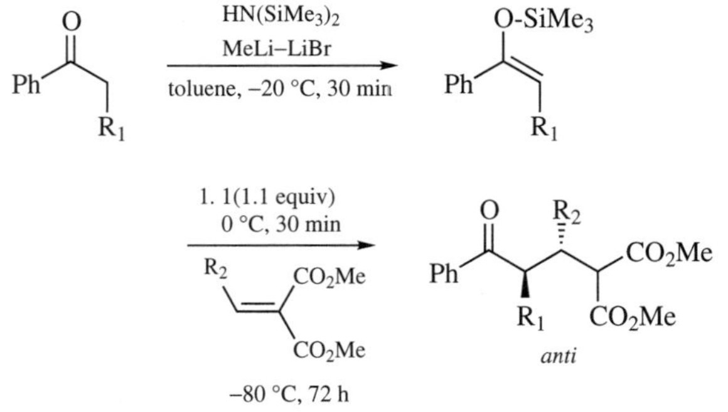

(10)

R_1	R_2	Yield (%)	anti/syn	% ee
Me	Ph	99	99/1	99
Me	Me	98	96/4	96
Me	Et	97	97/3	98
Me	^{i}Pr	96	84/16	81
Pr	Ph	99	98/2	99
Bn	Ph	95	96/4	96

Table 2 Michael addition of -substituted phenyl ketones

R^1	R^2	Yield (%)	*anti/syn*	% ee
Me	Ph	99	99/1	99
Me	Me	98	96/4	96
Me	Et	97	97/3	98
Me	*i*-Pr	96	84/16	81
Pr	Ph	99	98/2	99
Bn	Ph	95	96/4	96

Related Reagents. (*R*)-*N*-[2-(2-methoxyethyloxy)ethyl]-1-phenyl-2-piperidinoethylamine; (*R*)-*N*-[2-(2-dimethylaminoethyloxy)ethyl]-1-phenyl-2-piperidinoethylamine.

1. Imai, M.; Hagihara, A.; Kawasaki, H.; Manabe, K.; Koga, K. *Tetrahedron* **2000**, *56*, 179.
2. Yamashita, Y.; Odashima, K.; Koga, K. *Tetrahedron Lett.* **1999**, *40*, 2803.
3. Matsuo, J.; Kobayashi, S.; Koga, K. *Tetrahedron Lett.* **1998**, *39*, 9723.

4. Murakata, M.; Yasukata, T.; Aoki, T.; Nakajima, M.; Koga, K. *Tetrahedron* **1998**, *54*, 2449.

5. Imai, M.; Hagihara, A.; Kawasaki, H.; Manabe, K.; Koga, K. *J. Am. Chem. Soc.* **1994**, *116*, 8829.

6. Shindo, M.; Koga, K. *J. Synth. Org. Chem., Jpn.* **1995**, *53*, 1021.

7. Odashima, K.; Koga, K. *Yakugaku Zasshi* **1997**, *117*, 800.

8. Yasuda, K.; Shindo, M.; Koga, K. *Tetrahedron Lett.* **1997**, *38*, 3531.

9. Yasuda, K.; Shindo, M.; Koga, K. *Tetrahedron Lett.* **1996**, *37*, 6343.

10. Uragami, M.; Tomioka, K.; Koga, K. *Tetrahedron Asym.* **1995**, *6*, 701.

11. Shirai, R.; Aoki, K.; Sato, D.; Kim, H.; Murakata, M.; Yasukata, T.; Koga, K. *Chem. Pharm. Bull.* **1994**, *42*, 690.

12. (a) O'Brian, P. *J. Chem. Soc., Perkin Trans. 1* **2001**, 95. (b) Seebach, D.; Beck, A. K.; Heckel, A. *Angew. Chem., Int. Ed. Engl.* **2001**, *40*, 92. (c) Arya, P.; Qin, H. *Tetrahedron* **2000**, *56*, 917. (d) Regan, A. C. *J. Chem. Soc., Perkin Trans. 1* **1999**, 357. (e) Seebach, D.; Hintermann, T. *Helv. Chim. Acta.* **1998**, *81*, 2093. (f) Wills, M. *J. Chem. Soc., Perkin Trans. 1* **1998**, 3101. (g) O'Brian, P. *J. Chem. Soc., Perkin Trans. 1* **1998**, 1439. (h) Regan, A. C. *Contemp. Org. Synth.* **1997**, *4*, 1. (i) Ager, D. J.; Prakash, I.; Schaad, D. R. *Aldrichimica Acta* **1997**, *30*, 3. (j) Ager, D. J.; Prakash, I.; Schaad, D. R. *Chem. Rev.* **1996**, *96*, 835. (k) Evans, D. A. *Aldrichimica Acta* **1982**, *15*, 23. (l) Evans, D. A., In *Asymmetric Synthesis*; Morrison, J. D., Ed.; Academic Press: New York, 1984; Vol. 3, Part B, 2-101. (m) Cowden, C. J., In *Organic Reactions*; Paquette, L. A., Ed.; John Wiley & Sons: New York, 1997; Vol. 51, 1-200.

13. Ireland, R. E.; Mueller, R. H.; Williard, A. K. *J. Am. Chem. Soc.* **1976**, *98*, 2868.

14. Seebach, D.; Beck, A. K.; Heckel, A. *Angew. Chem., Int. Ed. Engl.* **1988**, *27*, 1624.

Douglas M. Krein & Todd L. Lowary
The Ohio State University, Columbus, Ohio, USA

1,1,2-Triphenyl-1,2-ethanediol[1,2]

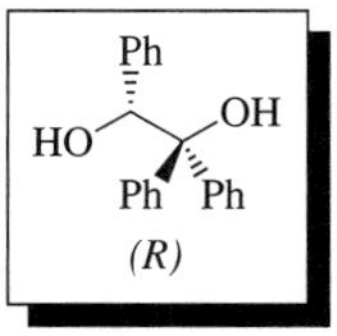

[95061-46-4] $C_{20}H_{18}O_2$ (MW 290.38)

(derived chiral monoesters undergo stereoselective aldol additions; formation of *O*-silyl orthoesters and cyclic phosphonates)

Physical Data: mp 126 °C. (*R*): $[\alpha]_D^{25}$ +214° (*c* = 1RM, ethanol), +220° (*c* = 1, 95% ethanol); (*S*): $[\alpha]_D^{25}$ −217° (*c* = 1, ethanol).
Solubility: sol dichloromethane, chloroform, THF, ethanol; insol hexane.
Form Supplied in: white solid; the (*R*)-form is commercially available.
Preparative Methods: (*R*)-1,1,2-triphenylethane-1,2-diol [(*R*)-(**1**)] is easily available from commercial (*R*)-*Mandelic Acid*, which is first esterified to give methyl mandelate and then treated with *Phenylmagnesium Bromide* (3.5 equiv). In an analogous way, (*S*)-(**1**) is accessible from (*S*)-mandelic acid, which is also commercially available (eq 1).[2]

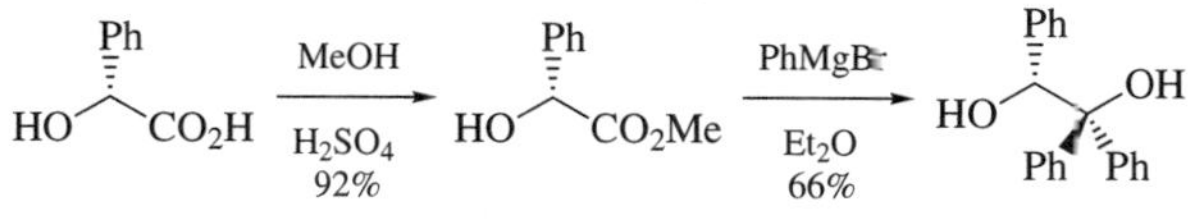
(1)

***anti*-Selective and Diastereofacially Selective Aldol Additions.** 2-Trimethylsilyloxy-1,2,2-triphenylethyl propionate, which is prepared from (*R*)-(**1**) by esterification with propionyl chloride and subsequent silylation of the tertiary hydroxy group, reacts in a highly stereoselective manner upon deprotonation, transmetalation with *Dichlorobis(cyclopentadienyl)zirconium*, and addition to 2-methylpropanal. The diastereoselectivity is 96:4, which is the ratio of the major product to the sum of all other diastereomers. Subsequent reduction with *Lithium Aluminum Hydride* affords (2*S*,3*R*)-2,4-dimethyl-1,3-pentanediol in 95% ee (eq 2).[3] *anti*-Selective aldol additions which deliver chiral nonracemic products have been a longstanding problem of asymmetric synthesis.[4] Doubly deprotonated 2-hydroxy-1,2,2-triphenylethyl propionate has been applied in a total synthesis of dolastatin.[5]

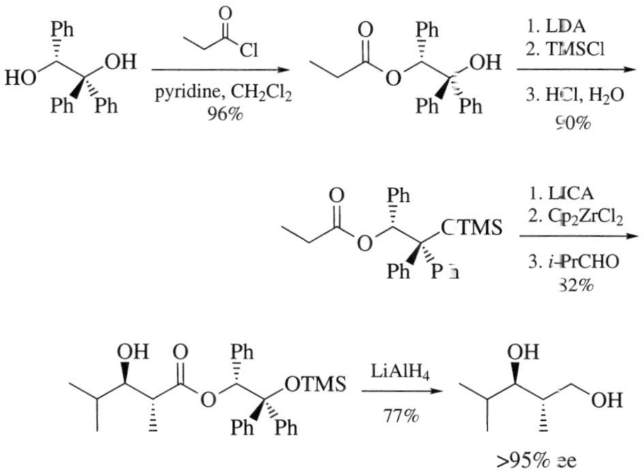
(2)

When 1,1,2-triphenylethane-1,2-diol-derived esters are submitted to a monodeprotonation and subsequently treated with *Chlorotrimethylsilane*, the formation of 2-trimethylsilyloxy-1,3-dioxolanes results. The orthoester moiety thus obtained serves as a protecting group for carboxylic acids (eq 3); it is stable towards alkyllithium reagents and can be cleaved under nonacidic conditions by alkaline hydrolysis.[6]

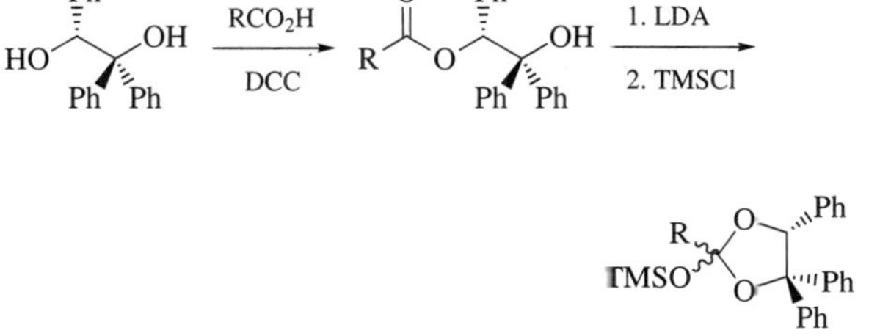
(3)

Methanephosphonyl dichloride reacts with (*R*)-(**1**) to give 2-methyl-4,4,5-triphenyl-2-oxo-1,3,2-dioxaphospholane (eq 4); the (R_P,R_C) diastereomer forms predominantly (9:1).[7]

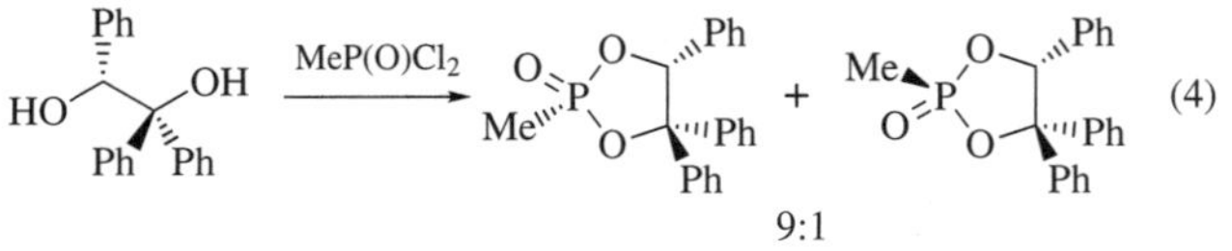

A series of enantiomerically pure 1,1-diaryl-2-phenylethane-1,2-diols is available from methyl mandelate by addition of the corresponding substituted arylmagnesium bromides or aryllithium reagents.[2b,8]

Related Reagents. 10,2-Camphorsultam; (*R*)-2-*t*-Butyl-6-methyl-4*H*-1,3-dioxin-4-one; (*R*)-(+)-(*t*)-Butyl 2-(*p*-Tolysulfinyl)propionate; Chloro(cyclopentadienyl)bis[3-*O*-(1,2:5,6-do-*O*-isopropylidene-α-D-glucofuranosyl)]titanium; 10-Dicyclohexylsulfonamidoisoborneol; Diisopinocampheylboron Trifluoromethanesulfonate; (*R,R*)-2,5-Dimethylborolane; 2-Hydroxy-1,2,2-triphenylethyl Acetate; α-Methyltoluene-2-α-sultam; (*S*)-4-Benzyl-2-oxazolidinone; 3-Propionylthiazolidine-2-thione; (*R,R*)-1,2-Diphenyl-1,2-diaminoethane *N,N′*-Bis[3,5-bis-(trifluoromethyl)benzenesulfonamide]; *trans*-2,5-Bis(methoxymethyl)pyrrolidine.

1. (a) McKenzie, A.; Wren, H. *J. Chem. Soc.* **1910**, *97*, 473. (b) Roger, R.; McKay, W. B. *J. Chem. Soc.* **1931**, 2229.
2. (a) Braun, M.; Devant, R. *Tetrahedron Lett.* **1984**, *25*, 5031. (b) Devant, R.; Mahler, U.; Braun, M. *Ber. Dtsch. Chem. Ges./Chem. Ber.* **1988**, *121*, 397. (c) Braun, M.; Gräf, S.; Herzog, S. *Org. Synth.* **1993**, *72*, 32.
3. (a) Braun, M.; Sacha, H. *Angew. Chem.* **1991**, *103*, 1369; *Angew. Chem., Int. Ed. Engl.* **1991**, *30*, 1318. (b) Sacha, H.; Waldmüller, D.; Braun, M. *Ber. Dtsch. Chem. Ges.* **1994**, *127*, 1959.
4. For reviews, see: (a) Braun, M., In *Advances in Carbanion Chemistry*; Snieckus, V., Ed.; JAI: Greenwich, CT, 1992, Vol. 1, pp 177–247; (b) Braun, M.; Sacha, H. *J. Prakt. Chem.* **1993**, 653.
5. (a) Pettit, G. R.; Singh, S. B. U.S. Patent 4 978 744, 1990 (*Chem. Abstr.* **1991**, *114*, 164 824v). (b) Pettit, G. R.; Singh, S. B.; Herald, D. L.; Lloyd-Williams, P.; Kantoci, D.; Burkett, D. D.; Barkóczy, J.; Hogan, F.; Wardlaw, T. R. *J. Org. Chem.* **1994**, *59*, 6287.
6. Waldmüller, D.; Braun, M.; Steigel, A. *Synlett* **1991**, 160.
7. Brodesser, B.; Braun, M. *Phosphorus Sulfur/Phosphorus Sulfur Silicon* **1989**, *44*, 217.
8. Prasad, K.; Chen, K. M.; Repic, O.; Hardtmann, G. E. *Tetrahedron: Asymmetry* **1990**, *1*, 703.

Manfred Braun
Heinrich-Heine-Universität, Düsseldorf, Germany

Tris(acetylacetonato)cobalt–Diethylaluminum Chloride–NORPHOS

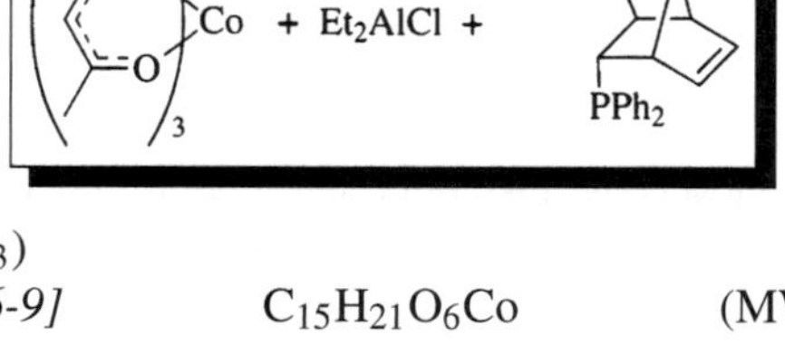

(Co(acac)$_3$)
[21679-46-9] $C_{15}H_{21}O_6Co$ (MW 356.29)
(Et$_2$AlCl)
[96-10-6] $C_4H_{10}AlCl$ (MW 120.57)
((+)-NORPHOS)
[71042-54-1] $C_{31}H_{28}P_2$ (MW 462.53)

(catalyst for the formation of optically active deltacyclene derivatives by homo Diels–Alder reaction of norbornadiene with substituted acetylenes[1,2])

Physical Data: see *Diethylaluminum Chloride* and *(+)-trans-(2S,3S)-Bis(diphenylphosphino)bicyclo[2.2.1]hept-5-ene* (NORPHOS).
Solubility: sol THF, benzene, toluene.
Preparative Methods: synthesized in situ from the three components, using a 1.5 excess of the chelate phosphine; the components are commercially available or can be easily prepared.
Handling, Storage, and Precautions: the in situ catalyst should be used in an atmosphere of dry nitrogen or argon, in a fume hood. The catalyst components can be stored indefinitely. They are air stable except Et$_2$AlCl, which should be kept under exclusion of air.

The Homo Diels–Alder Reaction of Norbornadiene with Acetylenes. [2+2+2] Cycloadditions of dienes such as norbornadiene with the double bonds in 1,4-position are called homo Diels–Alder reactions. Using an in situ catalyst (consisting of Co(acac)$_3$–Et$_2$AlCl–bis(diphenylphosphino)ethane) the products obtained with monosubstituted acetylenes, such as phenyl, *i*-propyl-, *n*-butyl-, *t*-butyl-, and trimethylsilylacetylene, are 4-substituted deltacyclenes.[1,2] In the formation of the polycyclic deltacyclene skeleton, six new stereo centers are generated in one step. Thus enantiocontrol by using optically active phosphine ligands as cocatalysts allows the synthesis of optically active cycloadducts,[3–5] as shown for the reaction of norbornadiene with phenylacetylene to give 4-phenyldeltacyclene (eq 1).

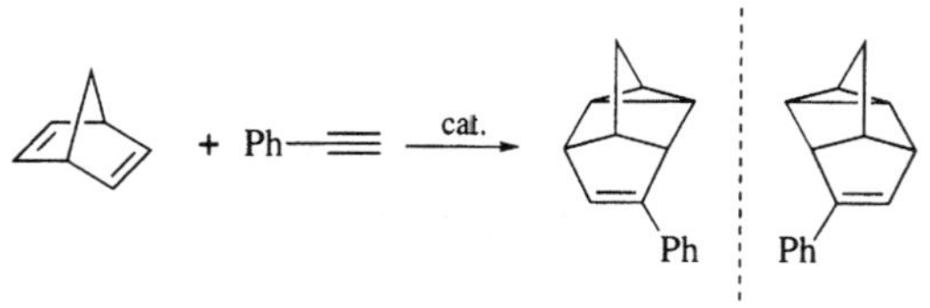

(1)

Preparation of the in situ Catalyst and Catalytic Reaction. (This is a typical procedure for norbornadiene and phenylacetylene or 1-hexyne). Tris(acetylacetonato)cobalt (7.1 mg, 2.0×10^{-2} mmol) and (+)-NORPHOS (13.8 mg, 3.0×10^{-2} mmol) are dissolved in 1 ml of THF under dry nitrogen, using standard Schlenk techniques. Norbornadiene (1.0 mL, 10.0 mmol) and 10 mmol of phenylacetylene or 1-hexyne are added. The reaction is started by addition of 5 mL of a 1M solution of diethylaluminum chloride in hexane. The reaction mixture is kept at 35 °C for 4 h. Then 5 mL of isopropanol are added and the volatile components are removed in vacuum. The oily residue is distilled at 80 °C in high vacuum in a Kugelrohr apparatus. Chemical yield >99% enantiomeric excess 98.4–99.6% for (+)-4-phenyldeltacyclene and 97.6–98.0% for (+)-4-*n*-butyldeltacyclene.

Product Analysis. The distilled product is dissolved in 2 mL of dry methylene chloride and an internal standard is added (naphthalene for 4-phenyldeltacyclene, biphenyl for 4-*n*-butyldeltacyclene). The chemical yield and the enantiomeric excess of the product can be determined by GLC using a 40 m column of perpentylated β-cyclodextrin. 4-Phenyldeltacyclene: column temperature 104 °C, retention time (−)-isomer 124.7 min, (+)-isomer 128.2 min; 4-*n*-butyldeltacyclene: column temperature 65 °C, retention time (−)-isomer 78.3 min, (+)-isomer 80.8 min.

Variation of the Optically Active Phosphine Ligand (Cocatalyst) and the Solvent.[5] In the synthesis of 4-phenyldeltacyclene, (+)-NORPHOS as the optically active ligand and THF as the solvent gave the best results, as indicated in the typical procedure. Used in benzene, the NORPHOS-containing catalyst also gives extremely high enantioselectivities but low chemical yields. With PROPHOS as the ligand, quantitative conversion in THF is only achieved after long reaction times. The enantiomeric excess in this case was ca. 80%. CHIRAPHOS and BDPP as cocatalysts result in slow conversions of the starting materials, BDPP giving high optical yields. In the case of DIOP in benzene, only low chemical and optical yields are obtained. In THF, DIOP-containing catalysts are inactive, as are BINAP-containing catalysts in benzene/THF.

For 4-*n*-butyldeltacyclene, no other cocatalyst gives the quantitative yield and 98–99% enantiomeric excess observed in the NORPHOS-containing system. The PROPHOS system gives 80% enantiomeric excess.

Variation of the Procatalyst (Metal Component) and the Acetylenic Substrate. The in situ catalysts $Co(acac)_3$–Et_2AlCl–phosphine have proven to be well-suited for the synthesis of 4-aryl- and 4-alkyl-substituted deltacyclenes. The catalysts tolerate remote oxygen functionalities in the acetylenic substrate.[4] However, they could not be used with functionalized acetylenes such as propargylic acid derivatives.

Recently, new procatalysts have been reported. They contain different cobalt sources and reducing agents. The procatalyst CoI_2–Zn has proven valuable in the preparative homo Diels–Alder reaction.[6] It has been shown that monodentate and bidentate ligands of the aminophosphine and phosphite type, e.g. ValNOP and ProliNOP, give high optical yields in the synthesis of 4-phenyldeltacyclene and 4-*n*-butyldeltacyclene.[7] With these new procatalysts, an extension of the homo Diels–Alder reaction to functionalized acetylenes is possible. High chemical and optical yields are obtained in the reaction of norbornadiene with substrates such as propargylic and homopropargylic ethers and esters.[8]

1. Lyons, J. E.; Myers, H. K.; Schneider, A. *Chem. Commun./J. Chem. Soc., Chem. Commun.* **1978**, 636.
2. Lautens, M.; Crudden, C. M. *Organometallics* **1989**, *8*, 2733.
3. Brunner, H.; Muschiol, M.; Prester, F. *Angew. Chem.* **1990**, *102*, 680; *Angew. Chem., Int. Ed. Engl.* **1990**, *29*, 652.
4. Lautens, M.; Lautens, J. C.; Smith, A. C. *J. Am. Chem. Soc.* **1990**, *112*, 5627.
5. Brunner, H.; Prester, F. *J. Organomet. Chem.* **1991**, *414*, 401.
6. Duan, I.-F.; Cheng, C.-H.; Shaw, J.-S.; Cheng, S.-S.; Liou, K. F. *Chem. Commun./J. Chem. Soc., Chem. Commun.* **1991**, 1347.
7. Pardigon, O.; Buono, G. *Tetrahedron: Asymmetry* **1993**, *4*, 1977.
8. Buono, G.; personal communication.

Henri Brunner
Universität Regensburg, Germany

L-Tyrosine Hydrazide

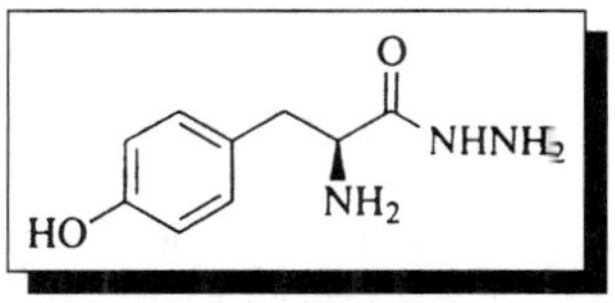

[7662-51-3] $C_8H_{13}N_3O_2$ (MW 195.24)

(resolution of carboxylic acids and amino acids[1])

Physical Data: crystalline solid; mp 196–198 °C; $[\alpha]_D^{25}$ +76° (*c* 4.2, 1N HCl)

Solubility: slightly sol cold methanol or ethanol; readily sol hot methanol or ethanol.

Form Supplied in: the free base (98% purity) is available from several commercial sources. No additional purification before use is required.

Preparative Methods: the synthesis of L-tyrosine hydrazide from L-tyrosine has been described.[2] D-Tyrosine hydrazide has been obtained by resolution of DL-tyrosine hydrazide with Cbz-L-proline.[3]

Handling, Storage, and Precautions: the free base is stable in amber bottles, at room temperature, for indefinite periods of time. No special handling precautions have been described.

Introduction. L-Tyrosine hydrazide (L-Tyr-$NHNH_2$) (**1**) is useful in the resolution of simple carboxylic acids and amino acid derivatives. It often forms highly crystalline salts with these compounds, which yield diastereomerically pure salts in just one or two recrystallizations. The yields of resolved acids tend to be high, and in many cases both enantiomers can be obtained from the same operation (the more soluble diasteromeric salt that remains in the mother liquors is often quite pure). The tyrosine hydrazide can be recovered without appreciable loss of optical activity.

Resolution of Carboxylic Acids. A variety of racemic monofunctional carboxylic acids have been resolved with chiral α-amino acid hydrazides, including L-Tyr-$NHNH_2$ and L-leucine hydrazide, which produce mandelic acid analogs with greater than 99% ee.[4] Other examples of resolutions of simple carboxylic acids have appeared in the patent literature (eq 1).[5]

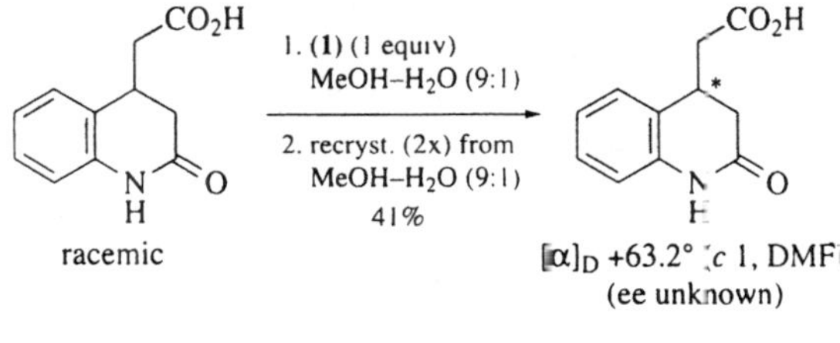

(1)

Resolution of Cyclic Amino Acid Derivatives. L-Tyr-NHNH$_2$ has been used many times in the resolution of all types of *N*-functionalized amino acids. The high crystallinity of the salts formed has been found to be a great advantage in situations where many other common resolving agents have failed. As indicated above, multiple recrystallizations of the diastereomeric salts formed by L-Tyr-NHNH$_2$ are rarely necessary to obtain amino acids of high optical purity. In most cases, D-α-amino acid derivatives form less soluble salts with L-Tyr-NHNH$_2$ than the corresponding L-α-amino acid derivatives. One of the earliest works in this area was the resolution of (±)-Cbz-proline (eq 2).[3] (±)-Cbz-Alanine and (±)-Cbz-isoleucine (eq 3) have also been resolved.[3] In all cases, the unnatural D-α-amino acids are obtained. This procedure also allows the isolation of D-tyrosine hydrazide by resolution of (±)-Tyr-NHNH$_2$ with Cbz-L-proline.[3] Most other amino acids resolved with L-tyrosine hydrazide also have their amino group protected as the Cbz derivative. For example, both enantiomers of azetidine-2-carboxylic acid are readily available, in high yield and about 100% ee, after one single crystallization of the racemate with L-Tyr-NHNH$_2$ (eq 4).[6] Similarly, both enantiomers of pipecolic acid are obtained by resolution of (±)-*N*-Cbz-pipecolic acid with L-Tyr-NHNH$_2$ (eq 5).[7]

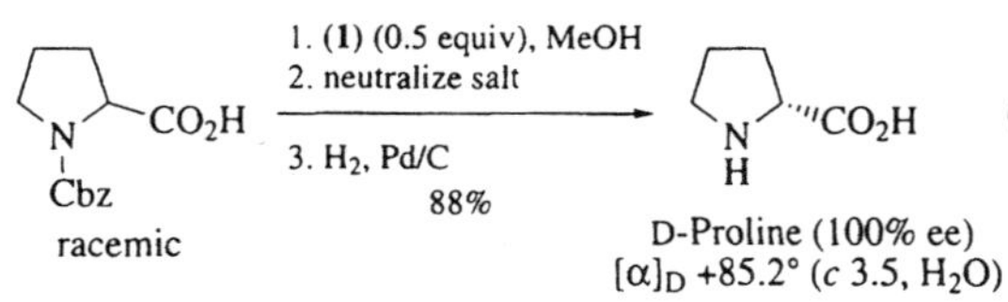

(2)

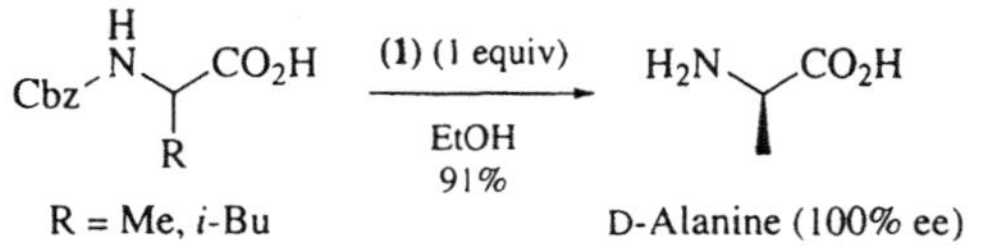

(3)

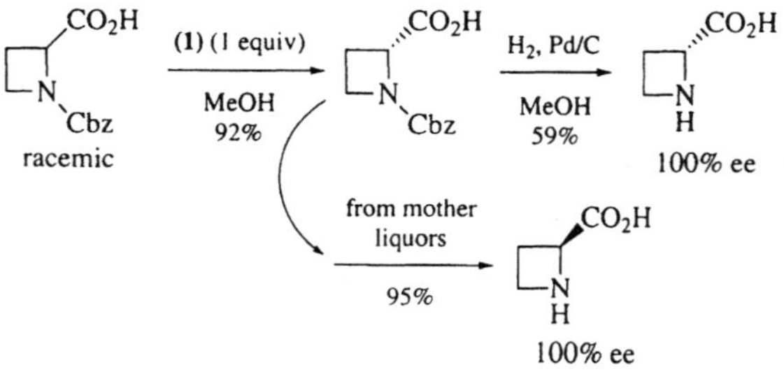

(4)

Resolution of Acyclic Amino Acid Derivatives. D-Homoserine is readily available by resolution of (±)-*N*-Cbz-homoserine (eq 6).[8] Both enantiomers of *threo*-2-amino-3,4-dihydroxybutyric acids are available via a similar resolution with L-Tyr-NHNH$_2$.[9] *N*-Cbz-derivatives of amino dicarboxylic acids, such as α-aminosuberic acid, have been resolved with D-Tyr-NHNH$_2$. In this case, the L-amino acid derivatives crystallize preferentially with the resolving agent.[10] Finally, tetrazole analogs of α-amino acids have also been resolved with L-Tyr-NHNH$_2$. For example, racemic tetrazole analogs of *N*-protected alanine, leucine, phenylalanine, and valine are resolved with L-Tyr-NHNH$_2$ to yield, except for the phenylalanine analog, the D-enantiomers (eq 7).[11]

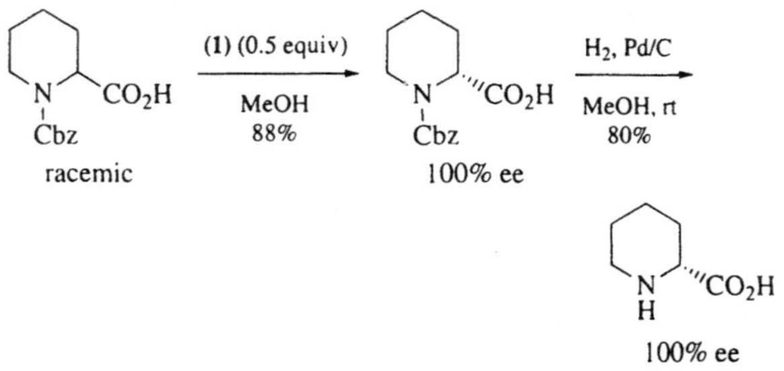

(5)

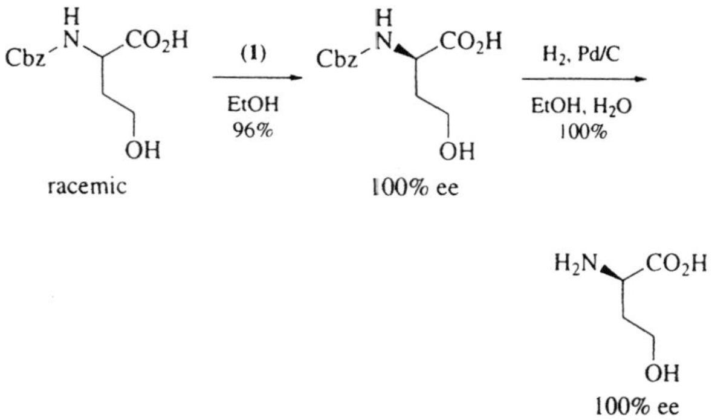

(6)

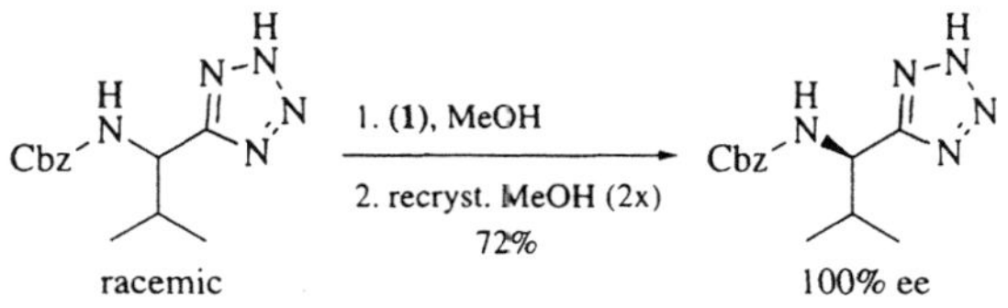

(7)

1. All chemical yields indicated for resolution steps represent the % of the theoretical amount of pure enantiomer. For a review of resolving agents used for acids and amino acids, see Wilen, S. H. In *Tables of Resolving Agents and Optical Resolutions*; Eliel, E., Ed.; Univ. of Notre Dame Press Notre Dame, 1972.
2. Curtius, T.; Donselt, W. *J. Prakt. Chem.* **1917**, *95*, 349.
3. Vogler, K.; Lanz, P. *Helv. Chim. Acta* **1966**, *49*, 1348.
4. Jap. Patent 01 221 345 (*Chem. Abstr.* **1990**, *112*, 118 459r).
5. Manoury, P.; Obitz, D.; Peynot, M.; Frost, J. Eur. Pat. Appl. 364 327, 1990 (*Chem. Abstr.* **1990**, *113*, 191 392p).
6. Rodebaugh, R. M.; Cromwell, N. H. *J. Heterocycl. Chem.* **1969**, *6*, 993.
7. Balaspiri, L.; Penke, B.; Petres, J.; Kovacs, K. *Montash. Chem.* **1970**, *101*, 1177.
8. Curran, W. V. *Prep. Biochem.* **1981**, *11*, 269.
9. Okawa, K.; Hori, K.; Hirose, K.; Nakagawa, Y. *Bull. Chem. Soc. Jpn.* **1969**, *42*, 2720.
10. Hase, S.; Kiyoi, R.; Sakakibara, S. *Bull. Chem. Soc. Jpn.* **1968**, *41*, 1266.
11. Grzonka, Z.; Liberek, B. *Tetrahedron* **1971**, *27*, 1783.

Juan C. Jaen
Parke-Davis Pharmaceutical Research Division, Ann Arbor MI, USA

V

Vitamin B_{12}[1-3]

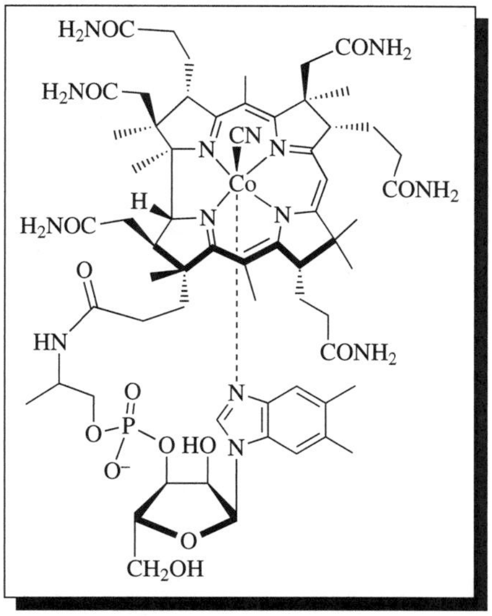

[68-19-9] $C_{63}H_{88}CoN_{14}O_{14}P$ (MW 1355.40)

(radical source via carbon–cobalt bond homolysis; stoichiometric and catalytic radical C–C bond formation; enantioselective catalyst for molecular rearrangements)

Physical Data: odorless and tasteless, hygroscopic, dark red solid; does not have a defined melting point; darkens at 210–220 °C but is not melted at 300 °C.

Solubility: sol H_2O (1 g/80 mL), alcohol; insol acetone, chloroform, ether.

Form Supplied in: powder or crystalline solid; available from biologically oriented chemical suppliers.

Handling, Storage, and Precautions: hygroscopic; absorbs moisture from air. Hydrated crystals are air stable. Aqueous solutions slowly decompose. May be harmful by inhalation, ingestion, or skin absorption and can cause eye and skin irritation. Keep containers tightly closed and store in a cool dry place. Use in a fume hood with safety goggles and chemically resistant gloves and clothing.

Stoichiometric Processes. In 1964, Schrauzer published the first of many papers on the synthesis and properties of alkyl cobaloximes.[1] This work led to the development of cobaloximes and related compounds as vitamin B_{12} model compounds, e.g. (**1**)–(**4**), summarized in a review in 1976.[2] By the mid 1970's, many of the fundamental reactions of vitamin B_{12} and model complexes were well established. This work is summarized in several reviews.[3]

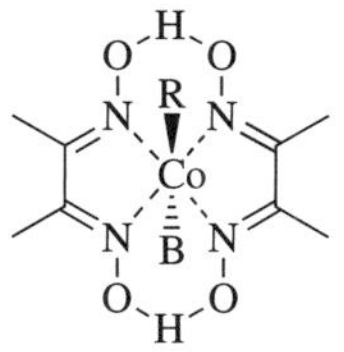

$RCo^{III}(dmgH)_2B$
(cobaloxime complex)
(**1**)

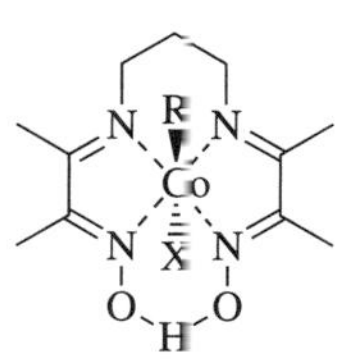

$RCo^{III}\{(DO)(DOH)PN\}X$
('DODOH' or Costa's complex)
(**2**)

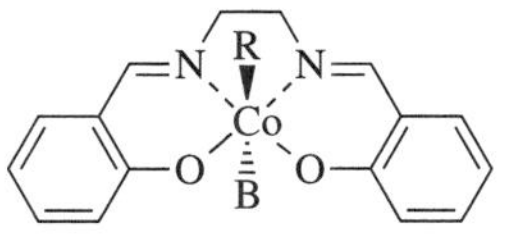

$RCo^{III}(salen)B$
(salen complex)
(**3**)

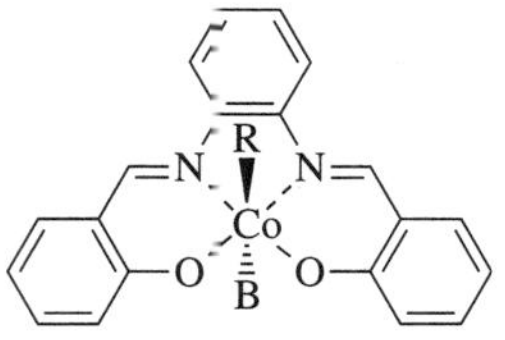

$RCo^{III}(salophen)B$
(salophen complex)
(**4**)

Alkylcobalt complexes provide an easy bridge between two-electron ionic chemistry and one-electron radical chemistry. This has made them popular and useful radical precursors; ionic reactions provide alkylcobalt complexes which then provide alkyl radicals via C–Co bond homolysis. In the mid 1970s and early 1980s, radical chain S_H2' reactions of allylcobaloximes were studied (eq 1).[4] These reactions were not applied to specific synthetic problems.[5]

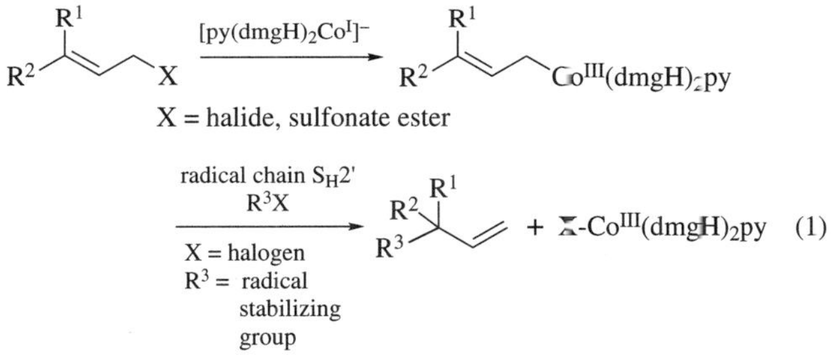

In the late 1980s, and early 1990s, stoichiometric nonchain organocobalt reactions have been developed.[6] Cobalt-based radicals, formed by carbon–cobalt bond homolysis, continue to participate in multistep radical processes to guide reaction pathways into particular directions.[7] The main practical benefits are: (1) radical–alkene additions are possible; (2) polymerization is inhibited; and (3) the alkene is regenerated in the final product by cobalt-mediated β-H elimination. A tandem radical cyclization (eq 2)[8] illustrates the main features. Oxidative addition of the cobalt anion to the alkyl bromide generates the alkyl radical which undergoes cyclization followed by trapping by the Co^{II} radical. This type of reaction was first observed in earlier studies of the mechanism of oxidative addition of Co^{I} to hindered alkyl halides.[9] Photolytic homolysis of the C–Co bond produces the alkyl radical which undergoes cyclization followed by Co^{II}-mediated β-H elimination.

A similar strategy has been applied to the synthesis of kainoids and related compounds (eq 3).[10] Both stoichiometric and catalytic amounts of cobalt reagents have been used in these and other cyclization studies. Several examples of these types of cyclizations have been published,[11] including cyclizations using aryl halides as precursors to aryl radicals.[12]

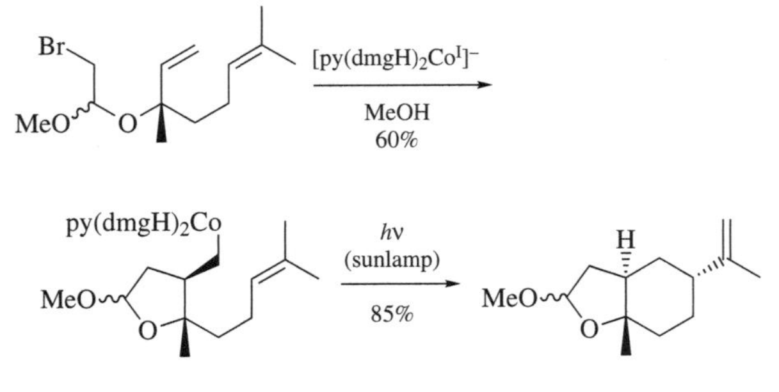

(2)

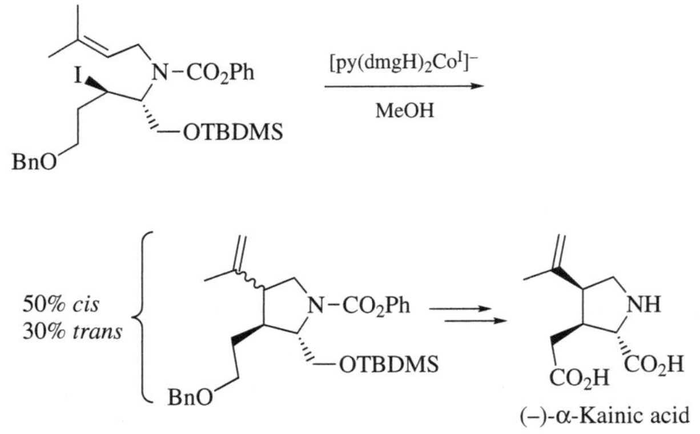

(3)

Alkylcobalt reagents are often prepared from the reaction of anionic Co^{I} complexes with alkyl halides or sulfonate esters. They can also be prepared by conjugate addition of anionic Co^{I} complexes to α,β-unsaturated carbonyl compounds and nitriles, placing the cobalt β to the activating group, or by addition of neutral Co^{III} hydrides to activated alkenes (carbonyl, nitrile, and aryl activating groups), placing the cobalt α to the activating group.[13] Anionic Co^{I} anions open epoxides regioselectively and the resulting cobaloximes show different patterns of reactivity under thermolysis versus photolysis (eq 4).[14] A mechanistic study indicates that cobalt-mediated cyclizations proceed via radicals,[15] but the reaction mechanism in the reactions in eq 4 has not been studied in detail. Acyl radicals can be generated from readily prepared acyl cobalt complexes.[16] The key step in a formal synthesis of racemic thienamycin is illustrative (eq 5).[17]

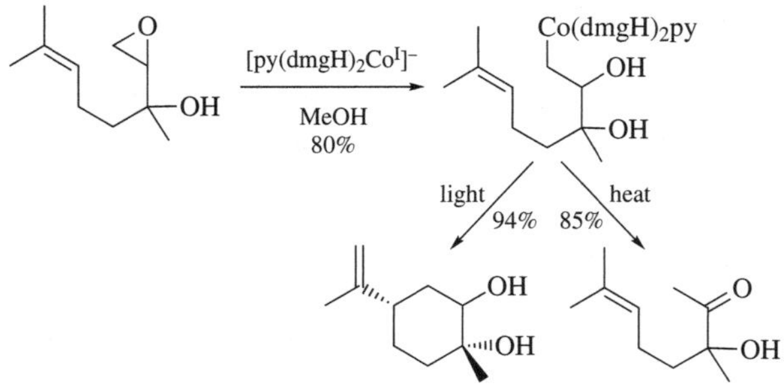

(4)

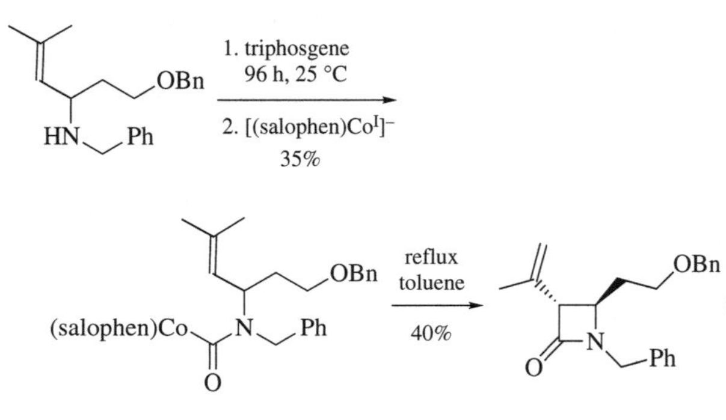

(5)

Intermolecular cross-coupling reactions have been developed.[18] One application to the synthesis of KDO takes advantage of alkene regeneration to allow further synthetic elaboration (eq 6).[19] Similar reactions have been developed for C–C bond constructions at the anomeric center of hexopyranoses,[20] leading to the production of *C*-glycosides, at C-1 of an open-chain pentose, leading to a synthesis of KDO,[21] and at C-3 of ribofuranoses.[22] Nonalkene cross-coupling partners have been used, specifically protonated heteroaromatics[23] and nitroalkyl anions.[24] Nitroalkyl anion cross couplings have been used to prepare *C*-disaccharides (eq 7).[25]

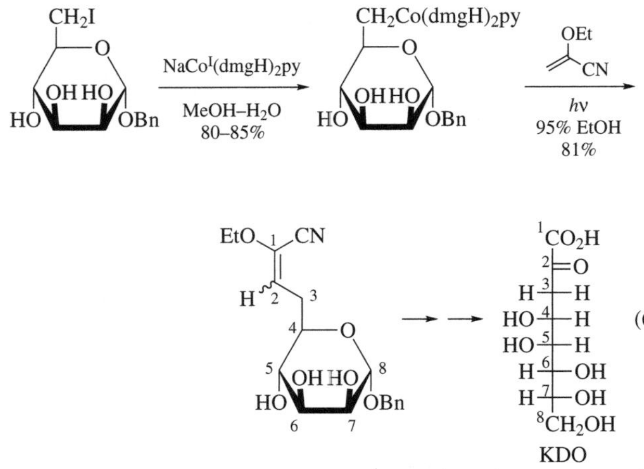

(6)

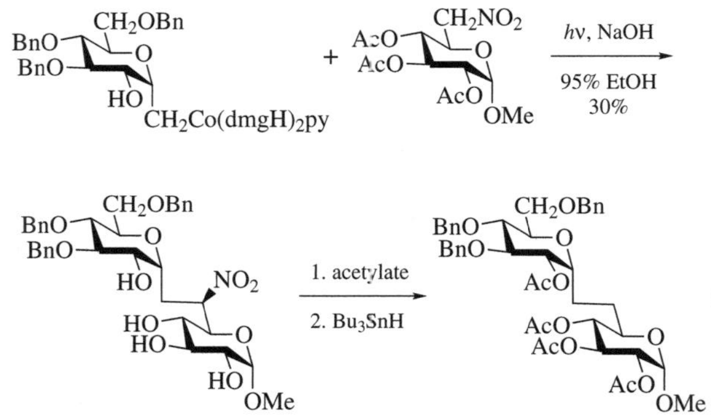

(7)

Catalytic Processes. Catalytic processes lead to intramolecular and intermolecular C–C bond constructions which are usually directly analogous to the stoichiometric reactions. This topic was reviewed in 1983.[26] Catalytic processes often lead to reduction rather than alkene regeneration; this is more likely to happen with B_{12} as a catalyst than it is with a cobaloxime. Scheffold pioneered the use of vitamin B_{12} as a catalyst for C–C bond formation,[27] and Tada pioneered the use of model complexes such as cobaloximes.[28] Several of the reactions described in the section on stoichiometric reactions have also been performed catalytically, as mentioned in that section. Commonly used chemical reductants include *Sodium Borohydride* and *Zinc* metal. Electrochemical reduction has also been used.[29] A novel catalytic system with a Ru^{II} trisbipyridine unit covalently tethered to a B_{12} derivative has been used for photochemically driven catalytic reactions using triethanolamine as the reductant.[30] A catalytic system using DODOH complexes can lead to reduction products or alkene regeneration depending upon the reaction conditions.[31] These catalytic B_{12} and model complex systems all utilize a

Co^{I}–Co^{II}–Co^{III} redox shuttle, shown in eq 8. Several other publications have described catalytic systems such as these,[32] including B_{12}-catalyzed alkene acylations via addition of acyl radicals to activated alkenes[33] and the use of hydrophobic B_{12} derivatives which are designed to provide a binding pocket for enzyme-like catalytic reactions, usually skeletal rearrangement reactions designed to mimic reactions catalyzed by B_{12}-containing enzymes.[34] A catalytic system utilizing a Co^{II}–Co^{III} redox shuttle has been described.[35] A catalytic system for alkene oligomerization has been developed.[36] Cobalt complexes are known to catalyze radical alkene polymerization.[37]

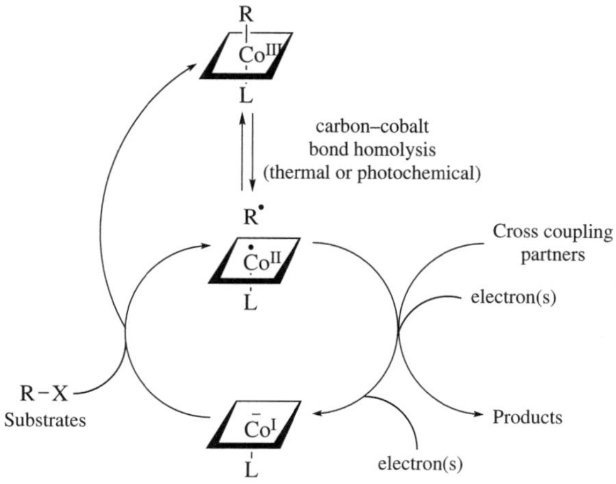

(8)

Examples of the use of vitamin B_{12} as a catalyst for enantioselective processes have been reported. The rearrangement of aziridines can proceed catalytically with ee's of up to 95% (eq 9).[38] Analogous rearrangements on achiral epoxides (typically 60% ee)[39] and achiral peroxides (low ee's)[40] have been reported.

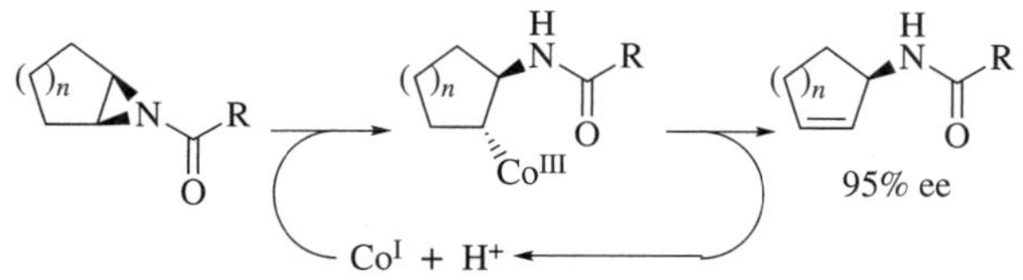

(9)

1. Schrauzer, G. N.; Kohlne, J. *Ber. Dtsch. Chem. Ges./Chem. Ber.* **1964**, *97*, 3056.
2. Schrauzer, G. N. *Angew. Chem., Int. Ed. Engl.* **1976**, *15*, 417.
3. (a) Dodd, D.; Johnson, M. D. *J. Organomet. Chem.* **1973**, *52*, 1. (b) Kemmitt, R. D. W.; Russell, D. R. In *Comprehensive Organometallic Chemistry*; Wilkinson, G.; Stone, F. G. A.; Abel, E. W., Eds.; Pergamon: New York, 1982; Vol. 5, pp 80–131. (c) Toscano, P.; Marzilli, L. G. *Prog. Inorg. Chem.* **1984**, *31*, 105. (d) Gupta, B. D.; Roy, S. *Inorg. Chim. Acta* **1988**, *146*, 209.
4. (a) Johnson, M. D. *Acc. Chem. Res.* **1983**, *16*, 343, and references cited therein. (b) Veber, M.; Duong, K. N. V.; Gaudemer, F.; Gaudemer, A. *J. Organomet. Chem.* **1979**, *177*, 231.
5. Work continues in this area: Gupta, B. D.; Roy, S. *J. Chem. Soc., Perkin Trans. 2* **1988**, *2*, 1377.
6. A review of Pattenden's early contributions: Pattenden, G. *Chem. Soc. Rev.* **1988**, *17*, 361.
7. This phenomenon has been termed the 'persistent radical effect'. See Branchaud, B. P.; Yu, G.-X. *Organometallics* **1993**, *12*, 4262, and references cited therein.
8. Ali, A.; Harrowven, D. C.; Pattenden, G. *Tetrahedron Lett.* **1992**, *33*, 2851.
9. (a) Tada, M.; Okabe, M. *Chem. Lett.* **1980**, 201. (b) Okabe, M.; Tada, M. *Bull. Chem. Soc. Jpn.* **1982**, *55*, 1498.
10. (a) Baldwin, J. E.; Li, C. S. *Chem. Commun./J. Chem. Soc., Chem. Commun.* **1987**, 166. See also: (b) Baldwin, J. E.; Moloney, M. G.; Parsons, A. F. *Tetrahedron* **1991**, *47*, 155. (c) Baldwin, J. E.; Moloney, M. G.; Parsons, A. F. *Tetrahedron* **1990**, *46*, 7263. (d) Baldwin, J. E.; Li, C. S. *Chem. Commun./J. Chem. Soc., Chem. Commun.* **1988**, 261.
11. (a) Bhandal, H.; Patel, V. F.; Pattenden, G.; Russell, J. J. *J. Chem. Soc., Perkin Trans. 1* **1990**, 2691. (b) Patel, V. F.; Pattenden, G. *J. Chem. Soc., Perkin Trans. 1* **1990**, 2703. (c) Begley, M. J.; Bhandal, H.; Hutchinson, J. H.; Pattenden, G. *Tetrahedron Lett.* **1987**, *28*, 1317. (d) Patel, V. F.; Pattenden, G. *Tetrahedron Lett.* **1987**, *28*, 1451. (e) Branchaud, B. P.; Meier, M. S.; Malekzadeh, M. N. *J. Org. Chem.* **1987**, *52*, 212. (f) Bhandal, H.; Pattenden, G.; Russell, J. J. *Tetrahedron Lett.* **1986**, *27*, 2299.
12. (a) Clark, A. J.; Jones, K. *Tetrahedron* **1992**, *48*, 6875, and references cited therein. The first publication on cobalt-mediated aryl radical cyclizations: (b) Patel, V. F.; Pattenden, G.; Russell, J. J. *Tetrahedron Lett.* **1986**, *27*, 2303.
13. (a) Howell, A. R.; Pattenden, G. *J. Chem. Soc., Perkin Trans. 1* **1990**, 2715. (b) Bhandal, H.; Pattenden, G. *Chem. Commun./J. Chem. Soc., Chem. Commun.* **1988**, 1110.
14. Harrowven, D. C.; Pattenden, G. *Tetrahedron Lett.* **1991**, *32*, 243.
15. Giese, B.; Hartung, J.; He, J.; Hueter, O.; Koch, A. *Angew. Chem., Int. Ed. Engl.* **1989**, *28*, 325.
16. (a) Patel, V. F.; Pattenden, G.; Thompson, D. M. *J. Chem. Soc., Perkin Trans. 1* **1990**, 2729. (b) Coveney, D. J.; Patel, V. F.; Pattenden, G.; Thompson, D. M. *J. Chem. Soc., Perkin Trans. 1* **1990**, 2721. (c) Gill, G. B.; Pattenden, G.; Reynolds, S. J. *Tetrahedron Lett.* **1989**, *30*, 3229. (d) Patel, V. F.; Pattenden, G. *Tetrahedron Lett.* **1988**, *29*, 707. (e) Coveney, D. J.; Patel, V. F.; Pattenden, G. *Tetrahedron Lett.* **1987**, *28*, 5949.
17. Pattenden, G.; Reynolds, S. J. *Tetrahedron Lett.* **1991**, *32*, 259.
18. (a) Patel, V. F.; Pattenden, G. *Chem. Commun./J. Chem. Soc., Chem. Commun.* **1987**, 871. (b) Branchaud, B. P.; Meier, M. S.; Choi, Y. L. *Tetrahedron Lett.* **1988**, *29*, 167. (c) Bhandal, H.; Howell, A. R.; Patel, V. F.; Pattenden, G. *J. Chem. Soc., Perkin Trans. 1* **1990**, 2709.
19. (a) Branchaud, B. P.; Meier, M. S. *Tetrahedron Lett.* **1988**, *29*, 3191. (b) Branchaud, B. P.; Meier, M. S. *J. Org. Chem.* **1989**, *54*, 1320.
20. Ghosez, A.; Göbel, T.; Giese, B. *Ber. Dtsch. Chem. Ges./Chem. Ber.* **1988**, *121*, 1807.
21. (a) Giese, B.; Carboni, B.; Göbel, T.; Muhn, R.; Wetterich, F. *Tetrahedron Lett.* **1992**, *33*, 2673. (b) Veit, A.; Giese, B. *Synlett* **1990**, 166.
22. Branchaud, B. P.; Yu, G.-X. *Tetrahedron Lett.* **1991**, *32*, 3639.
23. Branchaud, B. P.; Choi, Y. L. *J. Org. Chem.* **1988**, *53*, 4638.
24. (a) Branchaud, B. P.; Yu, G.-X. *Tetrahedron Lett.* **1988**, *29*, 6545. (b) Ref. 22.
25. Martin, O. R.; Xie, F.; Kakarla, R.; Benhamza, R. *Synlett* **1993**, 165.
26. Scheffold, R.; Rytz, G.; Walder, L. In *Modern Synthetic Methods*; Scheffold, R., Ed.; Wiley: New York, 1983; Vol. 3, pp 355–440.
27. (a) Auer, L.; Weymuth, C.; Scheffold, R. *Helv. Chim. Acta* **1993**, *76*, 810. (b) Yamamoto, K.; Abrecht, S.; Scheffold, R. *Chimia* **1991**, *45*, 86. (c) Scheffold, R. *Nachr. Chem., Tech. Lab.* **1988**, *36*, 261. (e) Scheffold, R.; Abrecht, S.; Orlinski, R.; Ruf, H. R.; Stamouli, P.; Tinembart, O.; Walder, L.; Weymuth, C. *Pure Appl. Chem.* **1987**, *59*, 363. (d) Scheffold, R. *Chimia* **1985**, *39*, 203.
28. (a) Okabe, M.; Abe, M.; Tada, M. *J. Org. Chem.* **1982**, *47*, 1775. (b) Okabe, M.; Tada, M. *J. Org. Chem.* **1982**, *47*, 5382.
29. In addition to the work of Scheffold, see: (a) Fry, A. J.; Sirisoma, U. N.; Lee, A. S. *Tetrahedron Lett.* **1993**, *34*, 809. (b) Fry, A. J.; Sirisoma, U. N. *J. Org. Chem.* **1993**, *58*, 4919.
30. Steiger, B.; Eichenberger, E.; Walder, L. *Chimia* **1991**, *45*, 32.
31. Giese, B.; Erdmann, P.; Göbel, T.; Springer, R. *Tetrahedron Lett.* **1992**, *33*, 4545.

32. (a) Hu, C.-M.; Qui, Y.-L. *J. Org. Chem.* **1992**, *57*, 3339. (b) Erdmann, P.; Schäfer, J.; Springer, R.; Zeitz, H.-G.; Giese, B. *Helv. Chim. Acta* **1992**, *75*, 638. (c) Inokuchi, T.; Tsuji, M.; Kawafuchi, H.; Torii, S. *J. Org. Chem.* **1991**, *56*, 5945.
33. Walder, L.; Orlinski, R. *Organometallics* **1987**, *6*, 1606.
34. Murakami, Y.; Hisaeda, Y.; Song, X.-M.; Takasaki, K.; Ohon, T. *Chem. Lett.* **1991**, 977. (b) Murakami, Y.; Hisaeda, Y. *Pure Appl. Chem.* **1988**, *60*, 1363.
35. Branchaud, B. P.; Detlefsen, W. D. *Tetrahedron Lett.* **1991**, *32*, 6273.
36. Bandaranayake, W. M.; Pattenden, G. *Chem. Commun./J. Chem. Soc., Chem. Commun.* **1988**, 1179.
37. (a) Suddaby, K. G.; O'Driscoll, K. F.; Rudin, A. *J. Polym. Sci., Part A: Polym. Chem.* **1992**, *30*, 643. (b) Suddaby, K. G.; Sanayei, R. Amin; Rudin, A.; O'Driscoll, K. F. *J. Appl. Polym. Sci.* **1991**, *43*, 1565. (c) Sanayei, R. A.; O'Driscoll, K. F. *J. Macromol. Sci., Chem.* **1989**, *A26*, 1137.
38. Zhang, Z.; Scheffold, R. *Helv. Chim. Acta* **1993**, *76*, 2602.
39. Bonhôte, P.; Scheffold, R. *Helv. Chim. Acta* **1991**, *74*, 1425.
40. Essig, S.; Scheffold, R. *Chimia* **1991**, *45*, 30.

Bruce P. Branchaud & Gregory K. Friestad
University of Oregon, Eugene, OR, USA

List of Contributors

Valérie Alezra — *ENS-Lyon, Lyon, France*
- (*R*,*R*)-1,2-Diphenyl-1,2-[di(pentafluorophenyl)phosphanoxy]ethane 302

James T. Anderson — *Case Western Reserve University, Cleveland, OH, USA*
- S-(1-Oxido-2-pyridinyl)-1,1,3,3-tetramethylthiouronium Hexafluorophosphate (HOTT) 463

Pher G. Andersson — *Uppsala University, Sweden*
- 2,2-Bis{[2-[4(*S*)-*tert*-butyl-1,3-oxazolinyl]}propane 108

Firoz D. Antia — *The R. W. Johnson Pharmaceutical Research Institute, Spring House, PA, USA*
- (*R*)-1-(1-Naphthyl)ethyl Isocyanate 452

Luca Banfi — *Università di Genova, Italy*
- 9-*O*-(1,2;5,6-Di-*O*-isopropylidene-α-D-glucofuranosyl)-9-boratabicyclo[3.3.1 nonane, Potassium Salt 236

R. Bruce Banks — *University of North Carolina at Greensboro, NC, USA*
- (*R*)-4-Methylcyclohexylidenemethylcopper 411

Bruce A. Barner — *Union Carbide Corporation, South Charleston, WV, USA*
- Dimethyl L-Tartrate 268

Patricia Bataille — *Laboratoire de Synthése Organique (UMR-CNRS 6011) Faculté des Sciences, Avenue Olivier Messiaen, France*
- (*R*)-(−)-2-(-1-Methylhydrazino)butan-1-ol 423

Albert K. Beck — *Eidgenössische Technische Hochschule, Zürich, Switzerland*
- (*R*)-2-*t*-Butyl-6-methyl-4*H*-1,3-dioxin-4-one 164
- 2,2-Dimethyl-α,α,α′,α′-tetraphenyl-1,3-dioxolane-4,5-dimethanolatotitanium Diisopropoxide 289

Matthew Beggs — *Tulane University, New Orleans, LA, USA*
- (−)-8-Phenylmenthol 471

Gregory Beutner — *University of Illinois, Urbana, IL, USA*
- (*R*,*R*)-1,2-(Methanesulfonamido)cyclohexane 395

Apurba Bhattacharya — *Hoechst-Celanese Corporation, Corpus Christi, TX, USA*
- 8-Phenylmenthyl Glyoxylate 474

Robert Bittman — *Queens College of The City University of New York, Flushing, NY, USA*
- Glycidyl Tosylate 349

Armin Börner — *Institut für Organische Katalyseforschung, Rostock, Germany*
- (*S*,*S*)-1,2-Bis(2,5-diethylphospholano)benzene 119

Mary Ellen Bos — *R. W. Johnson Pharmaceutical Research Institute, Raritan, NJ, USA*
- *N*-Benzylquininium Chloride 72

John Boukouvalas — *Université Laval, Québec, Canada*
- Dihydro-5-(hydroxymethyl)-2(3*H*)-furanone 216

J. A. Gladysz — *University of Utah, Salt Lake City, UT, USA*
- Cyclopentadienyl(3,5-dimethoxybenzyl)(nitrosyl)(triphenylphosphine)rhenium 198

Aravamudan S. Gopalan — *New Mexico State University, Las Cruces, NM, USA*
- Lithium Aluminum Hydride-2,2′-Dihydroxy-1,1′-binaphthyl 385

Mark T. Goulet — *Merck Research Laboratories, Rahway, NJ, USA*
- *B*-Allyldiisocaranylborane 26
- *B*-Methoxydiisopinocampheylborane 398

Edward J. J. Grabowski — *Merck Research Laboratories, Rahway, NJ, USA*
- (*S*)-Ethyl Lactate 335
- (*R*)-Pantolactone 466

Gareth J. Griffiths — *Lonza, Visp, Switzerland*
- 2-Azabicyclo[2.2.1]hept-5-en-3-one 44

Paride Grisenti — *Università di Milano, Italy*
- Baker's Yeast 45
- Esterases 330

Alyx-Caroline Guével — *The Ohio State University, Columbus, OH, USA*
- L-Aspartic Acid 42
- *t*-Leucine *t*-Butyl Ester 375

Ronan Guével — *The Ohio State University, Columbus, OH, USA*
- Dichloro[2,3-*O*-isopropylidene-2,3-dihydroxy-1,4-bis(diphenylphosphino)-butane]palladium(II) 213

Srinivas Reddy Gurrala — *The Ohio State University, Columbus, OH, USA*
- (*S*)-2,2′Binaphthoyl(*R,R*)-di(1-phenylethyl)aminoylphosphine 95

Andreas Hafner — *Ciba-Geigy, Marly, Switzerland*
- Chloro(cyclopentadienyl)bis[3-*O*-(1,2:5,6-di-*O*-isopropylidene-α-D-glucofuranosyl)] titanium 189
- Chloro (η^5-cyclopentadienyl) [(4*R,trans*)-2,2-dimethyl-α,α,α′,α′-tetraphenyl-1,3-dioxolane-4,5-dimethanolato(2−)-O^{α},$O^{\alpha'}$]titanium 191

Stephen Hanessian — *University of Montreal, Quebec, Canada*
- (3a*R*,7a*R*)-2-Ethyloctahydro-1*H*-1,3-dineopentyl-1,3,2-benzodiazaphosphole Oxide 338

Yuji Hanzawa — *Tokyo College of Pharmacy, Japan*
- (−)-[Ethylene-1,2-bis(η^5-4,5,6,7-tetrahydro-1-indenyl)]zirconium (*R*)-1,1′-Bi-2,2′-naphtholate 333

Gudmundur G. Haraldsson — *University of Iceland, Reykjavik, Iceland*
- Lipases 377

W. Hartwig — *Bayer, Wuppertal, Germany*
- (2*S*)-(+)-2,5-Dihydro-2-isopropyl-3,6-dimethoxypyrazine 219

Tamio Hayashi — *Hokkaido University, Sapporo, Japan*
- (*R*)-*N,N*-Dimethyl-1-[(*S*)-2-(diphenylphosphino)ferrocenyl]ethylamine 264

Kwok-Kan Ho — *Texas A & M University, College Station, TX, USA*
- (2*R*,3*R*)-(*Z*)-*cyclo*-Phenylalanine 200

Jens Holz — *Institut für Organische Katalyseforschung, Rostock, Germany*
- (*S,S*)-1,2-Bis(2,5-diethylphospholano)benzene 119

Dieter Hoppe — *University of Münster, Germany*
- (−)-Sparteine 502

Eusebio Juaristi
Centro de Investigación y de Estudios Avanzados del Instituto Politécnico Nacional, México, D.F., México
- 1-Benzoyl-2(*S*)-*tert*-butyl-3-methylperhydropyrimidin-4-one 53

Henri Kagan
Université de Paris-Sud, Orsay, France
- (2,3-*O*-Isopropylidene)-2,3-dihydroxy-1,4-bis(diphenylphosphino)butane 371

David C. Kammler
Indiana University, Bloomington, Indiana, USA
- (4*R*,5*R*)-2-Bromo-1,3-bis[(4-methylphenyl)sulfonyl]-4,5-diphenyl-1,3,2-diazaborolidine 147

Shuji Kanemasa
Kyushu University, Kasuga, Japan
- (*S*)-4-Benzyl-2,2,5,5-tetramethyloxazolidine 73
- Methyl (4*R*,5*R*)-(*E*)-3-(1,3-Dimethyl-4,5-diphenyl-2-imidazolidinyl)propenoate 413

Annette S. Kim
Harvard University, Cambridge, MA, USA
- (*S*)-4-Benzyl-2-oxazolidinone 57

Masato Kitamura
Nagoya University, Japan
- (*R*)- & (*S*)-2,2′-Bis(diphenylphosphino)-1,1′-binaphthyl 128

Martin Klatt
RWTH Aachen, Germany
- (*S*)-1-Amino-2-methoxymethylpyrrolidine 32
- (*S*)-2-Methoxymethylpyrrolidine 401

Michael Klinge
University of Alberta, Edmonton, AB, Canada
- *N*-Benzyloxycarbonyl-L-serine β-Lactone 68

Pavel Kočovský
University of Glasgow, UK
- 2′-(Diphenylphosphino)-*N*,*N*-dimethyl[1,1′-binaphthalen]-2-amine 310

Joseph P. Konopelski
University of California, Santa Cruz, CA, USA
- Benzyl(methoxymethyl)methylamine 56

Sangho Koo
Myong Ji University, Seoul, Korea
- *N*-Phenylcampholylhydroxamic Acid 469

Gerard van Koten
Debye Institute, Utrecht University, The Netherlands
- (*R*)-2-[1-(Dimethylamino)ethyl]benzenethiol 238

Cyrille Kouklovsky
Université de Paris-Sud, Orsay, France
- (1*S*,2*S*)-1,2-Diaminocyclohexane 202

Douglas M. Krein
The Ohio State University, Columbus, Ohio, USA
- (*S*)-(+)-5,5-Dimethyl-4-phenyl-2-oxazolidinone 279
- *N*,*N*,*N*′-Trimethyl-*N*′-(2-{[(1*R*)-1-phenyl-2-(1-piperidinyl)ehtyl]amino}ethyl}-1,2-ethanediamine 519

Grant R. Krow
Temple University, Philadelphia, Pennsylvania, USA
- (1*R*,2*S*,3*R*,4*S*)-3-Dimethylamino-1,7,7-trimethylbicyclo[2.2.1]heptan-2-ol [(−)DAIB] 243

T. Pavan Kumar
University of North Texas, Denton, TX, USA
- (*R*,*S*)-CAMPHOS 188
- (*R*,*S*,*R*,*S*)-Me-PennPhos 393
- (*R*,*R*)-(−)-NORPHOS, (*S*,*S*)-(+)-NORPHOS 455

E. Peter Kündig
University of Geneva, Switzerland
- (*R*,*R*)-1,2-Diphenyl-1,2-[di(pentafluorophenyl)phosphanoxy]ethane 302

Wai Him Kwok
The Hong Kong Polytechnic University, Hong Kong
- (1*R*,5*R*,6*R*)-Spiro[4.4]nonane-1,6-diyl Diphenylphosphinous acid Ester (*R*-SpirOP) 504

Moriyasu Masui
Aburahi Laboratories, Shionogi & Co., Ltd., Japan
- 3-Amino-2,6,6-trimethylbicyclo[3.1.1]heptan-2-ol 38

David J. Mathre
Merck Research Laboratories, Rahway, NJ, USA
- 2-Amino-3-methyl-1,1-diphenyl-1-butanol 36
- α,α-Diphenyl-2-pyrrolidinemethanol 313
- Tetrahydro-1-methyl-3,3-diphenyl-1*H*,3*H*-pyrrolo[1,2-*c*][1,3,2]oxazaborole 509

Patrick G. McDougal
Reed College, Portland, OR, USA
- *trans*-2,5-Bis(methoxymethyl)pyrrolidine 138

John M. McGill
Eli Lilly and Company, Lafayette, IN, USA
- 1-(1-Naphthyl)ethylamine 450

Jonathan A. Medlock
University of Basel, Basel, Switzerland
- (4*S*)-4-(1,1-Dimethylethyl)-2-{1-[(11b*S*)-dinaphtho[2,1-d:1′,2′-f][1,3,2]dioxaphosphepin-4-yloxy]-1-methylethyl}-4,5-dihydrooxazole 266

Albert I. Meyers
Colorado State University, Fort Collins, CO, USA
- (*S*)-*N,N*-Dimethyl-*N*′-(1-*t*-butoxy-3-methyl-2-butyl)formamidine 251
- (2*S*)-(2α,3β,8aβ)-Hexahydro-3-(hydroxymethyl)-8a-methyl-2-phenyl-5*H*-oxazolo-[3,2-*a*]pyridin-5-one 353
- (4*S*,5*S*)-4-Methoxymethyl-2-methyl-5-phenyl-2-oxazoline 399
- (3*S,cis*)-Tetrahydro-3-isopropyl-7a-methylpyrrolol[2,1-*b*]oxazol-5(6*H*)-one 507

M. Mark Midland
University of California, Riverside, CA, USA
- 2-[2-[(Benzyloxy)ethyl]-6,6-dimethylbicyclo[3.3.1]-3-nonyl]-9-borabicyclo[3.3.1]nonane 70
- *B*-3-Pinanyl-9-borabicyclo[3.3.1]nonane 477

Koichi Mikami
Tokyo Institute of Technology, Japan
- (*R*)-1,1′-Bi-2,2′-naphthol 86
- (*R*)-1,1′-Bi-2,2′-naphthotitanium Dichloride 91
- (*R*)-1,1′-Bi-2,2′-naphthotitanium Diisopropoxide 94

Scott J. Miller
Harvard University, Cambridge, MA, USA
- (Bicyclo[2.2.1]hepta-2,5-diene)[1,4-bis(diphenylphosphino)butane]rhodium(I) Tetrafluoroborate 76

J. Mittendorf
Bayer, Wuppertal, Germany
- (2*S*)-(+)-2,5-Dihydro-2-isopropyl-3,6-dimethoxypyrazine 219

Yuju Mori
Faculty of Pharmacy, Meijo University, Nagoya, Japan
- (1*R*,2*S*)-1-Lithio-1-phenylsulfonyl-2-{[(*tert*-butyldiphenyl)silyl]oxymethyl} Oxirane 382

James P. Morken
UNC Chapel Hill, North Carolina, USA
- 1,2-Bis((2*S*,5*S*)-2,5-dimethylphospholano)benzene (*S,S*)-Me-DuPhos, 1,2-Bis((2*R*,5*R*)-2,5-dimethylphospholano)benzene (*R,R*)-Me-DuPhos 123

Yukihiro Motoyama
Tokyo Institute of Technology, Japan
- (*R*)-1,1′-Bi-2,2′-naphthol 86

Teruaki Mukaiyama
Science University of Tokyo, Kagurazaka, Shinjuku-ku, Tokyo, Japan
- (*S*)-1-Methyl-2-(piperidinomethyl)pyrrolidine 428

Kenneth A. Murray
University of Cambridge, UK
- (*S*)-3-Hydroxy-5-methyl-2,4-imidazolidinedione 360

Andrew G. Myers
Harvard University, Cambridge, MA, USA
- Pseudoephedrine 485

Akira Nakamura
Osaka University, Japan
- Bis(α-camphorquinone dioximato)cobalt 98

K. Narasaka
The University of Tokyo, Japan
- (4*R*,5*R*)-2,2-Dimethyl-4,5-bis(hydroxydiphenylmethyl)-1,3-dioxolane–Titanium(IV) Chloride 245

Enrica Narisano
Università di Genova, Italy
- 9-*O*-(1,2;5,6-Di-*O*-isopropylidene-α-D-glucofuranosyl)-9-boratabicyclo[3.3.1]nonane, Potassium Salt 236

Antonio Navarro
The Ohio State University, Columbus, OH, USA
- *N,N′*-(1*R*,2*R*)-1,2-Cyclohexanediylbis-2-pyridinecarboxamide 194

Todd D. Nelson
Colorado State University, Fort Collins, CO, USA
- (*S*)-*N,N*-Dimethyl-*N′*-(1-*t*-butoxy-3-methyl-2-butyl)formamidine 251
- (2*S*)-(2α,3β,8aβ)-Hexahydro-3-(hydroxymethyl)-8a-methyl-2-phenyl-5*H*-oxazolo-[3,2-*a*]pyridin-5-one 353
- (4*S*,5*S*)-4-Methoxymethyl-2-methyl-5-phenyl-2-oxazoline 399
- (3*S*,*cis*)-Tetrahydro-3-isopropyl-7a-methylpyrrolol[2,1-*b*]oxazol-5(6*H*)-one 507

Hisao Nishiyama
Toyohashi University of Technology, Japan
- 2,6-Bis[(*S*)-4′-isopropyloxazolin-2′-yl](pyridine)rhodium Trichloride 136

Ryoji Noyori
Nagoya University, Japan
- 1,1′-Binaphthyl-2,2′-diyl Hydrogen Phosphate 97
- (*R*)- & (*S*)-2,2′-Bis(diphenylphosphino)-1,1′-binaphthyl 128
- (*S*,*S*)-1,2-Diphenylethylenediamine 304

Takeshi Ohkuma
Nagoya University, Aichi, Japan
- (*S*,*S*)-1,2-Diphenylethylenediamine 304

Edith N. Onyeozili
Michigan State University, East Lansing, MI 48824
- (*R*)-(−)-2,2-Diphenylcyclopentanol 297

Steven D. Paget
The Ohio State University, Columbus, OH, USA
- (−)-Dichloro(ethylene)(α-methylbenzylamine)platinum(II) 212

Jon R. Parquette
The Ohio State University, OH, USA
- (1*R*,1′*R*,2*R*,2′*R*)-[1,1′-Bicyclopentyl-2,2′-diylbisdiphenylphosphine 81
- (1*R*,2*S*,4*R*,5*S*)-2,5-Dimethyl-7-phenyl-7-phosphabicyclo[2.2.1]heptane 282

Stephen D. Pastor
Ciba-Geigy Corporation, Ardsley, NY, USA
- Bis(cyclohexyl isocyanide)gold(I) Tetrafluoroborate–(*R*)-*N*-[2-(*N,N*-Dimethylamino)-ethyl]-*N*-methyl-1-[(*S*)-1′,2-bis(diphenylphosphino)ferrocenyl]ethylamine 115
- (*R*)-*N*-[2-(*N,N*-Dimethylamino)ethyl]-*N*-methyl-1-[(*S*)-1′,2-bis(diphenylphosphino)ferrocenyl]ethylamine 240

Michel Paterne
Laboratoire de Synthése Organique (UMR-CNRS 6011), Faculté des Sciences, Avenue Olivier Messiaen, Francé
- (*R*)-(−)-2-(-1-Methylhydrazino)butan-1-ol 423

Eduardo Peña-Cabrera
Emory University, Atlanta, GA, USA
- Dibornacyclopentadienyltrichlorozirconium 209

Tang-Sheng Peng
University of Utah, Salt Lake City, UT, USA
- Cyclopentadienyl(3,5-dimethoxybenzyl)(nitrosyl)(triphenylphosphine)rhenium 198

Peter J. Sinclair
Merck Research Laboratories, Rahway, NJ, USA
- 3-Bromo-5,6-diphenyl-2,3,5,6-tetrahydro-4*H*-oxazin-2-one 152
- 4-*t*-Butoxycarbonyl-5,6-diphenyl-2,3,5,6-tetrahydro-4*H*-oxazin-2-one 158

Kenso Soai
Science University of Tokyo, Japan
- (*S*)-4-Anilino-3-methylamino-1-butanol 40
- (*S*)-2-(Anilinomethyl)pyrrolidine 41
- (*S*)-Diphenyl(1-methylpyrrolidin-2-yl)methanol 308
- (1*R*,2*S*)-Ephedrine 323
- (2*S*,2′*S*)-2-Hydroxymethyl-1-[(1-methylpyrrolidin-2-yl)methyl]pyrrolidine 361
- (1*R*,2*S*)-*N*-Methylephedrine 414

Guy Solladié
University Louis Pasteur, Strasbourg, France
- (*R*)-(+)-*t*-Butyl 2-(*p*-Tolylsulfinyl)acetate 168
- (*R*)-(+)-*t*-Butyl 2-(*p*-Tolylsulfinyl)propionate 169
- (−)-(1*R*,2*S*,5*R*)-Menthyl (*S*)-*p*-Toluenesulfinate 390
- (*R*)-(+)-Methyl *p*-Tolyl Sulfoxide 431
- (*S*)-(−)-5-(α-Phenylethyl)semioxamazide 470
- (*R*)-(+)-Phenyl (*p*-Toluenesulfinyl)acetate 477
- (*R*)-(+)-*p*-Tolylsulfinylacetic Acid 514
- (*R*)-(+)-α-(*p*-Tolylsulfinyl)-*N*,*N*-dimethylacetamide 515
- (3*R*)-(*p*-Tolylsulfinyl)-*N*-methoxyacetimidic Acid Ethyl Ester 516
- (*R*)-(+)-3-(*p*-Tolylsulfinyl)propionic Acid 517

Erik J. Sorensen
The Scripps Research Institute, La Jolla, CA, USA
- [Bis(4*R*,5*S*)-4,5-diphenyl-1,3-oxazolin-2-yl]methane [Bis(4*S*,5*R*)-4,5-diphenyl-1,3-oxazolin-2-yl]methane 126
- (*S*)-1-Methyl-2-[(dihydroisoindol-2-yl)methyl]pyrrolidine 412

Alan C. Spivey
University of Cambridge, UK
- 10,2-Camphorsultam 178
- 10-Dicyclohexylsulfonamidoisoborneol 214
- α-Methyltoluene-2,α-sultam 437

Andrea Rolf Sting
Eidgenössische Technische Hochschule, Zürich, Switzerland
- (2*S*,4*S*)-3-Benzoyl-2-*t*-butyl-4-methyl-1,3-oxazolidin-5-one 51
- (*R*,*R*)-2-*t*-Butyl-5-methyl-1,3-dioxolan-4-one 166

Armido Studer
Eidgenössische Technische Hochschule, Zürich, Switzerland
- 1-Benzoyl-2-*t*-butyl-3,5-dimethyl-4-imidazolidinone 50
- *t*-Butyl 2-*t*-Butyl-3-methyl-4-oxo-1-imidazolidinecarboxylate 162

Takashi Sugimura
Himeji Institute of Technology, Hyogo, Japan
- (*R*)-7,7′-Bis(diphenylphosphinomethyl)-2,2′-dimethoxy-1,1′-binaphthyl 133
- (*R*)-*N*-[2-(2-Methoxyethoxy)-ethyl]-α-phenyl-1-piperidineethanamine 398

Michinori Suginome
Kyoto University, Japan
- Bis(bicyclo[2.2.1]hepta-2,5-diene)rhodium Perchlorate–(*R*)-1-(*S*)-1′,2-Bis(diphenylphosphino)ferrocenylethanol 104
- Bis(1,5-cyclooctadiene)rhodium Tetrafluoroborate–(*R*)-2,2′-Bis(diphenylphosphino)-1,1′-binaphthyl 118
- (1,5-Cyclooctadiene)[(3*R*,4*R*)-3,4-bis(diphenylphosphino)-1-methylpyrrolidine]rhodium Tetrafluoroborate 197

Takeo Taguchi
Tokyo College of Pharmacy, Japan
- (−)-[Ethylene-1,2-bis(η^5-4,5,6,7-tetrahydro-1-indenyl)]zirconium (*R*)-1,1′-Bi-2,2′-naphtholate 334

Tetsuaki Tanaka
Osaka University, Suita, Japan
- (1*R*,5*R*)-2*H*-1,5-Benzodithiepin-3(4*H*)-one 1,5-Dioxide 48

Dan Yang
The University of Hong Kong, Hong Kong

Naoki Yoshikawa
Graduate School of Pharmaceutical Science, The University of Tokyo, Japan

Reagent Formula Index

Subject Index